ABSTRACTS OF PAPERS
Part II

**Fourth Chemical Congress
of
North America**
ISBN 8412-2114-6

New York, NY
August 25-30, 1991

NUCLEAR CHEMISTRY & TECHNOLOGY

Fourth Chemical Congress of North America

New York, NY
August 25 - 30, 1991

T. T. Sugihara, S. W. Yates,
Program Chairman

MONDAY MORNING AND AFTERNOON
● **Symposium on Neutrino Science: Recent Advances Neutrino Masses from Beta Decay. I and II**
J. B. Wilhelmy, R. L. Hahn, Presiding   Papers 1-3, 10-13

● **Symposium on Nucleus-Nucleus Collision Mechanisms Overview**
J. M. Alexander, Presiding   Papers 4-9

MONDAY AFTERNOON
● **Symposium Nucleus-Nucleus Collision Mechanisms Synthesis of Heavy Elements**
G. T. Seaborg, Presiding   Papers 14-18

TUESDAY MORNING AND AFTERNOON
● **Symposium on Neutrino Science: Recent Advances Solar Neutrinos. I and II.**
R. Davis, Jr., Presiding   Papers 19-23, 30-33

● **Symposium on Nucleus-Nucleus Collision Mechanisms Intermediate Energies**
R. Vandenbosch, Presiding   Papers 24-29

● **Symposium on Nucleus-Nucleus Collision Mechanisms**
V. E. Viola, Presiding   Papers 34-40

WEDNESDAY MORNING AND AFTERNOON
● **Symposium on the Use of Ion Beams in the Analysis and Modification of Materials Laser Ionization Mass Spectrometry**
N. Winograd, Presiding   Papers 41-45

WEDNESDAY MORNING
● **Symposium on Nucleus-Nucleus Collision Mechanisms Relativistic Phenomena**
W. Loveland, Presiding   Papers 46-51

WEDNESDAY AFTERNOON
● **Symposium on the Use of Ion Beams in the Analysis and Modification of Materials**
Cluster Ion Impact Phenomena Used for Surface Modification and Analysis
E. A. Schweikert, R. Remillieux, Presiding   Papers 52-57

● **Symposium in Memory of Nathan Sugarman**
N. T. Porile, Presiding   Papers 58-63

THURSDAY MORNING AND AFTERNOON
● **Symposium on the Use of Ion Beams in the Analysis and Modification of Materials**
Elemental/Isotopic Analysis by Nuclear Reactions/Scattering
M. R. Weller, Presiding   Papers 64-67

● **Symposium on Additional Aspects of Nuclear Science**
C. Chung, L. R. Morss, Presiding   Papers 68-76, 82-90

● **Elemental Analysis by Proton-Induced X-ray Emission (PIXE)**
T. A. Cahill, D. S. Burnett, Presiding   Papers 77-81

FRIDAY MORNING
● **Symposium on the Use of Ion Beams in the Analysis and Modification of Materials**
Composition Produced by Ion Bombardment ("Ion Mixing")
Y-T. Cheng, F. Saris, Presiding   Papers 91-94

NUCLEAR CHEMISTRY AND TECHNOLOGY

1. NEUTRINO SCIENCE.
   Sheldon L. Glashow, Mellon Professor of the Sciences, Harvard University Cambridge, MA 02138.

   A most remarkable development in physics is the convergence between the studies of the largest and smallest structures: particle physics and cosmology. Neutrino physics reveals the confluence most clearly. We observe these particles (or their effects) in natural radioactivity, at reactors, accelerator and cosmic rays, and coming from the Sun and a supernova. Cosmological theory constrains both the number of neutrino species and their masses. The 'solar neutrino problem' --- the principal subject of this meeting --- is tightly linked to many aspects of particle physics and cosmology. Its resolution may lead to a successor to the standard model of particle physics. However, we also face the challenge of understanding the role and properties of an alleged 17-keV neutrino, if indeed it exists.

2. THE UNBEARABLE HEAVINESS OF NEUTRINOS. J. J. Simpson, Dept. of Physics, University of Guelph, Guelph, Ontario N1G 2W1 Canada

   This talk will describe the evidence, obtained by studying β spectra of ^{3}H and ^{35}S, for the existence of a 17-keV/c^2 component in the electron neutrino, and will discuss some of the implications of this finding.

3. EVIDENCE FOR A MASSIVE NEUTRINO IN NUCLEAR BETA DECAY.*
   Eric B. Norman, Nuclear Science Division, Lawrence Berkeley Laboratory, Berkeley, CA 94720
   We[1] have studied the β-spectrum of ^{14}C using a germanium detector containing a crystal with ^{14}C dissolved in it. We find a feature in the β-spectrum 17 keV below the endpoint which can be explained by the hypothesis that there is a heavy neutrino emitted in the β-decay of ^{14}C with a mass of 17±1 keV and an emission probability of 1.3±0.3%. In addition, we[2] have studied the inner bremsstrahlung spectrum of ^{55}Fe and also find indications of the emission of a ~17-keV neutrino. These results are consistent with observations of similar anomalies in the decays of ^{3}H (Refs. 3,4), ^{35}S (Refs. 5,6), and ^{71}Ge (Ref. 7).
   *Supported by the United States Department of Energy under Contract No. DE-AC03-76SF00098.
   [1] B. Sur et al., Phys. Rev. Letters (in press).
   [2] E. B. Norman et al., Proc. XIV Europhys. Conf. on Nuclear Physics, Bratislava, Czech., 1990, (in press).
   [3] J. J. Simpson, Phys. Rev. Lett. 54, 1891 (1985).
   [4] A. Hime and J.J. Simpson, Phys. Rev. D 39, 1837 (1989).
   [5] J. J. Simpson and A. Hime, Phys. Rev. D 39, 1825 (1989).
   [6] A. Hime and N. A. Jelley, Phys. Lett. B (in press).
   [7] I. Zlimen et al., Proc. XIV Europhys. Conf. on Nuclear Physics, Bratislava, Czech., 1990, (in press).

4. TO FUSE OR NOT TO FUSE.* R. Vandenbosch, Department of Chemistry, University of Washington, Seattle, Washington, 98195

   The factors governing the probability for two complex nuclei to fuse will be considered over the whole bombarding energy range from sub-barrier energies to the Fermi energy domain. At sub-barrier energies the collective degrees of freedom play the dominant role in determining the probability to fuse. At energies above 10 MeV/A the single-particle degrees of freedom begin to play the dominant role in determining both the probability for fusion-like processes to occur and in determining what portion of the projectile plus target mass ends up in the composite system. Pre-equilibrium particle emission eventually limits the excitation energy which can be deposited in the composite system.

* Work supported in part by the U.S. Department of Energy.

5. STATISTICS AND DYNAMICS IN MULTIFRAGMENTATION. L. G. Moretto and G. J. Wozniak, Nuclear Science Division, Lawrence Berkeley Laboratory, Berkeley, CA 94720.

Multifragment decay may be the manifestation of a variety of instabilities. Due to Raileigh instabilities, dynamically generated cigar-like or pancake-like shapes may breakup into many fragments. Also, density fluctuations in an expanding system may be amplified and lead to the coalescence of droplets. Alternatively, very hot systems could decay statistically giving rise to the simultaneous or sequential formation of several fragments. It is likely that dynamics and statistics play a complementary role in these reactions, dynamics dominating the early stage and statistics controlling the latter stages. A combination of a Landau-Vlasov calculation with sequential statistical decay seems to give an adequate reproduction of rather asymmetric collisions. Strong hints of dynamic instabilities are visible in more symmetric collisions.

6. HOW AND WHY DO NUCLEI FRAGMENT? A. Bohnet, J. Jaenicke, W. Trautmann and J. Aichelin, Inst. f. Theor. Physik, Philosophenweg 19, 6900 Heidelberg, Germany, GSI Darmstadt, Postfach 110552, 6100 Darmstadt 11, Germany

The Quantum Molecular Dynamics approach (QMD) is presently the only microscopical theory which is able to describe the time evolution of a heavy ion reaction from the initial separation of target and projectile to the final distribution of fragments, protons, pions and other reaction products. Comparison with experiments at high energies 1 GeV > E_{kin} > 200 MeV/N have yielded a remarkably good agreement (J. Aichelin et al., Phys. Lett. B224 (1989) 34 and references therein).

We have improved the model to allow for the simulation of heavy ion collisions of asymmetric systems at lower and higher energies. At higher energies we find a beam energy independence of the target spectator fragmentation above E_{kin} = 2 GeV/N. In the target rapidity regime the calculations at 2 GeV are in good agreement with the data at 200 GeV/N of the WA80 collaboration. Even at energies as low as 84 MeV/N the system does not reach thermal equilibrium. There are strong correlations between the fragments and between protons and fragments, which we have also seen experimentally.

The fragmentation mechanism at 84 MeV/N is quite different from that at high energies despite a great similarity of the mass yield curves and other observables. Combining these investigations we are now able to understand fragmentation of asymmetric collisions from threshold until the highest available energies. We can explain how different experimental observables are connected and how they give a consistent picture of fragmentation processes.

7. ASPECTS OF DISSIPATIVE REACTIONS FROM LOW ENERGY TO THE FERMI DOMAIN.
L. TASSAN-GOT

The problem of mass and charge drifts in heavy ion collisions around 10 MeV/u will be discussed. The experimental results will be interpreted in the framework of the nucleon exchange theory and the role of the potential energy surface will be emphasized. At higher bombarding energy, in the Fermi domain, reactions exhibit several aspects of deep inelastic reactions. A description of these reactions by a model based on the nucleon exchange theory will be presented.

8. NUCLEAR EQUATION OF STATE AND FINITE RANGE INTERACTION. D. Sperber, Rensselaer Polytechnic Institute, Troy, NY 12180-3590.

We determine the nuclear equation of state using a finite range momentum dependent interaction and a method which does not depend on the mean field approximation. This interaction reproduces the bulk properties of cold nuclei. We obtain the equation of state of nuclear matter by minimizing the Helmholtz free energy. For nuclear matter we determine: a) the equilibrium density, b) the effective mass and c) the compressibility, all as a function of temperature. We also determine the critical temperature as about 9 MeV. Turning to finite charged nuclei we minimize the free energy for a fixed number of nucleons. We develop in iterative procedure starting with the information gained for nuclear matter. We consider a different depth for the potential between like and unlike nucleons and include the Coulomb interaction. For a fixed number of nucleons we determine, as a function of temperature, the proton and neutron density profiles and the limiting temperature.

9. FLUCTUATIONS OF THE SINGLE-PARTICLE DENSITY IN NUCLEAR DYNAMICS. G. F. Burgio, Ph. Chomaz and J. Randrup, Nuclear Science Division, Lawrence Berkeley Laboratory, University of California, Berkeley, CA 94720.

In recent years semiclassical methods have been developed to study heavy-ion collisions in the framework of the Boltzmann-Uehling-Uhlenbeck theory, in which the collisionless mean field evolution has been augmented by a Pauli-blocked Nordheim collision term. Since these models describe the average dynamical trajectory, they cannot be applied to describe fluctuations of one-body observables, correlations in the emission of light particles and catastrophic processes like multifragmentation. We have developed a new method in order to include the stochastic part of the collision integral into BUU-type simulations of the nuclear dynamics. We apply this method to a two-dimensional gas of fermions on a torus, for which the time evolution of the mean trajectory and the associated correlation function are calculated; the variance of the phase-space occupancy follows closely the predictions of the corresponding Fokker-Planck equation and relaxes towards the appropriate quantum-statistical limit. The breaking of the translational and spherical symmetry in our model permits the study of unstable situations in phase-space. The introduction of the non-linear one-body field allows us to explore dynamical instabilities and bifurcations. Therefore the model can be appropriate for studying nuclear multifragmentation.

10. LIMIT ON THE ELECTRON ANTINEUTRINO MASS FROM MEASUREMENT OF THE BETA DECAY OF MOLECULAR TRITIUM - D. A. Knapp, Lawrence Livermore National Laboratory, Livermore, CA 94550, J. F. Wilkerson, T. J. Bowles, J. L. Friar, R.G.H. Robertson, G. J. Stephenson, Jr., and D. L. Wark, Los Alamos National Laboratory, Los Alamos, NM 87545

We have obtained an upper limit on the mass of the electron antineutrino of 9.4 eV (95% confidence level). The mass is measured by a study of the shape of the beta decay sptrum of free molecular tritium. Achieving this sensitivity requires precise determinations of all processes that modify the shape of the observed spectrum. Our results in in clear disagreement with a reported neutrino mass of 26(5) eV.

11. THE TRITIUM BETA DECAY SPECTRUM FROM END TO END AND THE NEUTRINO MASS. W. STOEFFL and D. J. DECMAN, Lawrence Livermore National Laboratory, Livermore, CA 94550

We set a new upper limit for the mass of the electron antineutrino by measuring the beta decay of gaseous molecular tritium. Our system uses five superconducting magnets to guide the beta decay electrons from a five meter long source tube into a large toroidal field magnetic spectrometer. The spectrometer has a resolution of ~12 eV for 23 keV electrons. We have measured the entire beta decay spectrum of molecular tritium from the endpoint to zero energy. The results near zero electron energy, the first data of this kind for any beta-decay study, are a measurement of the final state distribution, a major source of systematic error for neutrino mass experiments. We will present these data and the results from our endpoint measurement.
This work was performed by LLNL under the auspices of the U.S. Department of Energy under Contract No. W-7405-Eng-48.

12. DOUBLE BETA DECAY OF ^{100}Mo AND ^{150}Nd. S. R. Elliott, M. K. Moe, M. A. Nelson, and M. A. Vient, Physics Department, University of California, Irvine, CA 92717.

The central role of the neutrino in particle physics, astrophysics and cosmology, makes its behavior a topic of wide general interest. The shape of the double beta decay energy spectrum is a sensitive probe of neutrino properties such as mass, particle-antiparticle identity, and possible coupling to new particles. ^{100}Mo and ^{150}Nd are two $\beta\beta$-unstable isotopes with high transition energies and relatively fast decay rates with good discrimination against background. Their spectra as measured in an underground time projection chamber will be presented.

13. A REPORT OF SOME RECENT DOUBLE-BETA DECAY EXPERIMENTS. F. T. Avignone, III, J. I. Collar, C. K. Guerard, Department of Physics and Astronomy, University of South Carolina, Columbia, SC 29208, R. L. Brodzinski, H. S. Miley, J. H. Reeves, Pacific Northwest Laboratory, Richland, WA 99352.

An update of experiments searching for 2ν and 0ν $\beta\beta$-beta decay will be given. Observations of 2ν $\beta\beta$-decay of ^{82}Se, ^{76}Ge, and ^{100}Mo have thus far been reported. New efforts involving kilogram quantities of Ge, isotopically enriched to 86% in ^{76}Ge will also be discussed. New results from the search of 2ν $\beta\beta$-decay of ^{100}Mo, to the first excited 0^+ state in ^{100}Ru, will also be reported. A ~3σ positive effect has been observed in this experiment. The meaning of these results in the context of nuclear structure theory will also be reviewed. The methods of radioactive background reduction used by the PNL/USC group will also be presented.

14. THE DARMSTADT PROGRAM TO INVESTIGATE HEAVY ELEMENTS. S. Hofmann, GSI, Planckstrasse 1, D-6100 Darmstadt, Fed. Rep. Germany.

During the past ten years isotopes of the elements 107, 108, and 109 have been identified unambiguously by means of alpha-decay chains. Velocity separation of the reaction products after fusion and their consecutive implantation into position-sensitive silicon detectors increased the sensitivity to a level, that is required to identify a single atom produced with a cross-section of a few picobarn. In this talk the stability of the observed nuclei and the trends in the measured reaction cross-sections will be discussed. Based on extrapolated data the production and identification of the new element 110 is feasible with the experimental techniques currently prepared at the UNILAC in Darmstadt.

15. THE FUSION OF Co AND Bi TO MAKE ELEMENT 110, A. Ghiorso, Lawrence Berkeley Laboratory, Berkeley, CA 94720

A gas-filled double-focussing magnetic spectrometer has been constructed and tested that will be used to make the new nuclide, ^{267}110, via the fusion of ^{59}Co with ^{209}Bi and the emission of one neutron. The cross section is expected to be less than one picobarn so that it is mandatory that the overall efficiency of the instrument be close to 100% to be able to complete the experiment in a reasonable time (a few weeks). It is also necessary to use a large beam current through the low-melting target. Both of these objectives have been achieved. We are confident that we should be able to detect a single atom of element 110 in the focal plane every few days for a cross section of one picobarn using a beam current of 3×10^{12} ^{59}Co ions/sec. The fusion recoil is imbedded into one of the fifty focal plane crystal detectors such that the geometry is 45% for a high energy alpha particle to escape from the detector without depositing all of its energy. This means that a full-energy alpha particle from ^{267}110 would be observed about once a week under the above conditions.

16. PRODUCTION OF ACTINIDE ISOTOPES VIA TRANSFER REACTIONS*, Darleane C. Hoffman, Marvin M. Hoffman, and John D. Leyba, Nuclear Science Division, 70A-3307, Lawrence Berkeley Laboratory, Berkeley, CA 94720

Production cross sections for actinide isotopes have now been measured radiochemically for reactions of several actinide targets with heavy ion projectiles at energies near and somewhat above the Coloumb barriers. Calculations based on a binary transfer mechanism have shown that many of the target-like products have excitation energies which are small enough to greatly reduce losses due to prompt fission and particle emission compared to production by compound nucleus formation. Maxima of the isotopic distributions generally occur at the reaction channel which involves exchange of the fewest number of nucleons and has a positive excitation energy. A simple binary transfer model has been used to calculate the geometric cross sections for Rutherford trajectories expected to result in nucleon exchange. The values of the calculated cross sections are in good agreement with the existing experimental data for a variety of projectiles with ^{248}Cm targets.

*This work was supported in part by the Director, Office of Energy Research, Division of Nuclear Physics of the US Department of Energy under Contract DE-AC03-76SF00098.

17. REACTIONS LEADING TO FORMATION OF HEAVY ELEMENTS. Y.T. Oganessian, Joint Institute for Nuclear Research, Dubna, USSR.

The current status of research on the properties and synthesis of the heaviest elements will be discussed.

18. A model for transfer reactions leading to heavy actinides. M.T.Magda Chemistry Department, SUNY, Stony Brook, NY 11794

A mechanism is proposed to describe the multinucleon transfer reactions leading to heavy actinides, based on the assumption that the clusters separated from the projectile are captured as a whole by the target nucleus. The primary distributions of isotopes are corrected for neutron emission. The predictions of the model are compared with and succeed in describing the cross sections as well as the distributions of the Z= 96 - 103 isotopes produced in various reactions.

19. NEUTRINOS AND CHEMISTRY: A BRIEF TUTORIAL. Richard L. Hahn, Chemistry Department, Brookhaven National Laboratory, Upton, NY 11973.

There are several fascinating, basic ideas in solar neutrino research, some of which involve chemistry. The purpose of this talk is to present a tutorial on these concepts, by way of introduction to the subsequent proceedings. Key topics include: 1. The nuclear fusion reactions that power the Sun and the stars produce neutrinos and "build" the chemical elements. 2. The extremely difficult task of detecting neutrinos on Earth is done either by radiochemical methods ("black box" approach) or by instrumental methods ("real time" approach); the principles of these methods, and their respective advantages and disadvantages, will be contrasted. 3. Solar neutrino experiments have very low event rates; controlling backgrounds and impurities by chemical and nuclear methods of analysis is imperative.

This research was carried out at Brookhaven National Laboratory under contract DE-AC02-76CH00016 with the U.S. Department of Energy and supported by its Office of High Energy and Nuclear Physics.

20. LOW-ENERGY TESTS OF NEUTRINO MASSES. W. C. Haxton, Department of Physics, University of Washington, Seattle, Washington 98195

Recent solar neutrino and beta decay studies may be revealing the effects of massive neutrinos. Some of the theoretical issues that arise in interpreting solar neutrino, beta decay, and double beta decay experiments will be discussed. For instance, what are the relations between the various "masses" probed in β-decay, neutrino oscillation, and double beta decay experiments? The constraints imposed by the ^{37}Cl, Kamioka II, and ^{71}Ga experiments on solar neutrino oscillations will be explored. I will also describe some exotic scenarios that could lead to time variations in the solar neutrino flux.

21. RESULTS FROM THE HOMESTAKE SOLAR NEUTRINO OBSERVATORY. B. Cleveland, T. Daily, R. Davis, J. Distel, K. Lande, C. K. Lee, P. Wildenhain, University of Pennsylvania, Philadelphia, PA 19104, and J. Ullman, Lehman College, CUNY, Bronx, NY 10468.

The Homestake chlorine detector measures the neutrino flux emitted by fusion reactions in the solar core, via the reaction $\nu_e + {}^{37}Cl \rightarrow e^- + {}^{37}Ar$. The continuous twenty year record of these measurements indicates an average ^{37}Ar production rate of 0.49 ± 0.03 per day corresponding to 2.2 ± 0.3 SNU. The ν_e flux appears to vary with the 11-year solar activity cycle, with higher ν_e fluxes during solar quiet periods and lower ν_e fluxes during solar active periods. When the Homestake data is combined with the Kamiokande results, the region of overlap between the two experiments is for an observed to predicted ^{8}B neutrino flux ratio of about 0.4 and very little low-energy neutrino flux. If the Kamiokande results are corrected for MSW effects, neutral current scatterings by non-electron neutrinos, then both the ^{8}B and the ^{7}Be fluxes are about 1/3 of the Standard Solar Model predictions.

22. THE MOLYBDENUM-TECHNETIUM SOLAR NEUTRINO EXPERIMENT. Norman C. Schroeder, Kurt Wolfsberg and Donald J. Rokop, Los Alamos National Laboratory, Los Alamos, New Mexico, 87545.

We are attempting to measure the time-averaged ^{8}B solar-neutrino flux over 10 Myr by measuring ^{98}Tc produced through the ^{98}Mo(ν,e$^-$) reaction in a deeply buried molybdenum deposit. This will test the prediction of periodic mixing of the Sun's core over long time intervals. To separate technetium from 10,000-ton quantities of Henderson ore, we have taken advantage of the commercial processing of molybdenite. Technetium, volatilized during roasting of molybdenite to MoO_3, was scrubbed from the gas stream and collected on anion exchange columns. After sample reduction and chemical separation and purification we measured technetium, as TcO_4^-, using negative thermal ionization mass spectrometry. Measurement of ^{99}Tc in spiked and ^{98}Tc in unspiked fractions from one sample gives an apparent solar neutrino production rate of 95.8 SNU. However, roaster memory probably invalidates this result.

23. SOLAR NEUTRINO RESULTS FROM THE KAMIOKANDE-II DETECTOR.
W. Frati,* University of Pennsylvania, Philadelphia, PA 19104

The Kamiokande II detector is briefly described and the results of 1040 live-days of data are presented. Among the results are: 1) Observation of the directional correlation of the solar neutrino induced electron events with respect to the Sun. 2) Measurement of the differential electron energy-spectrum of the solar neutrino induced signal. 3) Search for variations of the solar neutrino signal during the 1040 day observation period. 4) Search for day/night and semi-annual variations of the solar neutrino signal. 5) Constraints on neutrino oscillation parameters in the Mikheyev-Smirnov-Wolfenstein scenario.

* For the Kamiokande Collaboration.

24. HEAVY RESIDUE PROPERTIES IN INTERMEDIATE ENERGY NUCLEAR COLLISIONS WITH GOLD.
K. Aleklett[a], J.O. Liljenzin[b], W. Loveland[c], L. Sihver[a], G.T. Seaborg[d] and R. Yanez[a], Uppsala University[a], Chalmers University of Technology[b], Oregon State University[c], Lawrence Berkeley Laboratory[d].

We have measured the target fragment production cross section, angular distributions and heavy residue energies for the interactions of 95 MeV/N argon and 45 MeV/N xenon with gold. The fragment isobaric yield distribution, moving frame angular distributions and velocities have been deduced from these data. For the argon - gold interaction we find, when compared with reactions at lower energies, that the fission cross sections decrease with increasing projectile energy and the heavy residue cross section increases. While the fragment mass yield distribution for the 95 MeV/A Ar + Au reaction is similar to that seen in relativistic Ar + Au collisions, the larger fragment energies, momentum widths, etc. do not support a participant-spectator mechanism. For all reactions we have found that the ratio $v_{||}/v_{c.n.}$ increases approximately linearly with mass removed from the target.

25. TARGET FRAGMENTATION OF COPPER BY INTERMEDIATE ENERGY HEAVY IONS. J. Whitfield and N. T. Porile, Department of Chemistry, Purdue University, W. Lafayette, IN 47907*

Recoil ranges and cross sections of products from the interaction of copper with intermediate-energy heavy ions have been measured by off-line γ-ray spectrometry. The results were used to obtain the linear momentum transfer and the mass-yield distribution. The dependence of these quantities on projectile mass for a total projectile center-of-mass kinetic energy of ~430 MeV is studied using ^{6}Li, ^{12}C, ^{14}N, ^{15}O, and ^{22}Ne beams. The dependence on projectile energy is studied with ^{12}C ions up to an energy of 90 MeV/nucleon.

*Supported by the U.S. Department of Energy

26. MULTIFRAGMENT EMISSION IN LIGHT-ION-INDUCED REACTIONS.* V.E. Viola, Department of Chemistry and IUCF, Indiana University, Bloomington, IN 47405.

Multifragment events for IMFs ($3 \leq Z \leq 12$) with multiplicity up to four have been observed in the reaction of 0.90 and 3.6 GeV ^{3}He ions with ^{nat}Ag nuclei. Events are detected in which IMFs account for up to 75% of the total charge of the system. Total IMF kinetic energies for multiplicities equal to or greater than two extend up to 400 MeV and cannot be accounted for in terms of simple Coulomb arguments.

Fragment energy spectra, angular distributions and charge distributions are found to be dependent on event multiplicity.

* work supported by the U.S. Department of Energy

27. PROJECTILE BREAKUP IN 25 MEV/A REACTIONS, PROMPT OR SEQUENTIAL?* J. Barreto[+], R. Charity, L.G. Sobotka, D.G. Sarantites, D.W. Stracener, A. Chbihi[++], and N.G. Nicolis, Department of Chemistry *Washington University*, St. Louis, Missouri 63130, R. Auble, C. Baktash, J.R. Beene, F. Bertrand, M. Halbert, D.C. Hensley, D. Horen, C. Ludermann, M. Thoennessen and R. Varner *Oak Ridge National Laboratory*, Oak Ridge, Tennessee 37831.

The breakup of the projectiles in the 25 MeV/A reactions $^{16}O + ^{159}Tb$ and $^{28}Si + ^{100}Mo$ is studied using the *Dwarf Ball/Wall* 4π plastic-CsI array. The breakup of ^{16}O into 4 α-particles is studied in detail and is shown to be a sequential process. The dynamics of the more complicated channels present in the reaction using the ^{28}Si projectile will be discussed. The division of excitation energy between the projectile and target is both deduced with the aid of two body kinematics and examined directly via the measured exclusive charged particle multiplicity. These exclusive charged particle data are also used to verify the projectile fragmentation process and discriminate against more complex transfer processes.

* Work supported by the U.S. Department of Energy
+ On leave from Instituto de Física da UFRJ - RJ- Brazil
++ Present Address: Laboratoire Ganil - BP 5027 - Caen 14021 - France

28. MULTIFRAGMENTATION OF NUCLEI WITH A = 70. K. Hagel, M. Gonin, Y. Lou, J. B. Natowitz, R. Wada, D. Utley, M. Gui, Cyclotron Institute, Texas A&M University, College Station, TX 77843-3366; G. Nebbia, D. Fabris, G. Prete, J. Ruiz, INFN-Legnaro, I-35020, Legnaro, Italy; D. Drain, B. Chambon, B. Cheynis, D. Guinet, X. C. Hu, A. Demeyer, C. Pastor, INP-Lyon, 69622 Villeurbanne Cedex, France; A. Giorni, A. Lleres, P. Stassi, B. Viano, ISN-Grenoble, 38026 Grenoble-Cedex, France; P. Gonthier, Hope College, Holland, MI 49423.

The 4π Multidetector AMPHORA at the SARA facility in Grenoble has been used to detect multifragmentation following the reactions of 35 MeV/u ^{40}Ca with ^{40}Ca. Approximately 10^5 events with high multiplicity and essentially complete event characterization have been obtained. Comparisons of the data with model calculations assuming either sequential binary deexcitation or simultaneous multifragmentation of the hot intermediate composite nucleus suggest that both mechanisms may be required to explain the observed fragmentation patterns.

29. INTERMEDIATE ENERGY HEAVY ION REACTIONS AND RI-BEAM EXPERIMENTS. M. Ishihara, RIKEN, Wako, Saitama 351-01 Japan.

Intense radioactive beams of projectile fragments have become available at RIKEN for studies on exotic nuclei. Experiments have been made along three major directions: 1) Study of the unique properties of neutron-halo nuclei using RI-beam secondary reactions. 2) Measurement of unstable nucleus reaction rates of astrophysical interest. 3) Production and application of spin-polarized RI-beams. Several reaction mechanisms characteristic of intermediate energy heavy ion collisions have been essential to facilitate these studies. Such useful mechanisms and, in particular, the spin-polarization mechanism in projectile fragmentation reactions will be discussed.

30. THE SAGE SOLAR NEUTRINO EXPERIMENT. Richard T. Kouzes, Department of Physics, Princeton University, Princeton, NJ 08544.

The Solar Neutrino Problem has been with us for almost 20 years, and will not be resolved for several more. The Soviet American Gallium Experiment (SAGE), a radiochemical experiment based on gallium sited at the Baksan Neutrino Observatory INR AS USSR, is sensitive to the primary pp neutrino flux from the sun and has been producing data since 1990. The results indicate that the solar neutrino puzzle is deepening. The next few years will tell whether the solar neutrino problem is new physics or poor astrophysics.

31. DESIGN AND STATUS OF GALLEX. E. Henrich, K. Ebert, Kernforschungszentrum Karlsruhe GmbH, Germany, (for the GALLEX Collaboration)

Low-energy solar neutrinos are measured with a radiochemical Ga detector via the Ga-71 (ν_e, e-) Ge-71 reaction. In the Gran Sasso underground laboratory in Italy, 30 tons of gallium are used by the "GALLEX" Collaboration in the form of 100 tons of 8.1M $GaCl_3$/1.9M HCl solution. About one radioactive Ge-71 atom per day is expected to be produced by solar neutrinos. The technical facilities for the periodic separation, concentration and conversion of germanium to GeH_4 counting gas, as well as the low-level counters, have been operating without technical difficulties since last year. Prior to a reliable determination of the Ge-71 production rate, all potential background contributions must be characterized. Efficient removal of cosmic-ray produced Ge-68 (long-lived) is indispensable, to achieve a sufficiently low counting background. Besides the astro- and particle-physics aspects of the experiment, the chemistry at such ultra-low concentrations is of considerable significance.

GALLEX collaboration: Germany, Italy, France, U.S.A. and Israel.

32. SUDBURY NEUTRINO OBSERVATORY, The SNO Collaboration, presenter: J. B. Wilhelmy, (Los Alamos National Laboratory, Los Alamos, New Mexico 87545).

The Sudbury Neutrino Observatory (SNO) is a large scale experiment designed to study neutrino production in the sun. It will consist of an active detector volume containing a 1000 tons of D_2O that will be placed 6800 feet underground in an operating Ni mine located near Sudbury, Ontario. The heavy water will be contained in a 12 m diameter acrylic vessel which is viewed by nearly 10,000 8" phototubes located in a 22 m barrel shaped vessel filled with H_2O. This will be the first solar neutrino experiment capable of measuring the three basic neutrino interactions:

I) Neutrino Scattering: $\nu_e + e^- \longrightarrow \nu_e + e^-$
II) The Charged Current: $\nu_e + d \longrightarrow p + p + e^-$ - 1.44 MeV
III) Neutral Current: $\nu_x + d \longrightarrow \nu_x + p + n$ - 2.22 MeV

The first two reactions are observed by recording the Čerenkov radiations emitted by the relativistic recoiling electrons. The neutral current reaction is observed by measuring radiation emitted following capture of the liberated neutron. Funding for the project has been received from Canadian and U.S. agencies. Excavation for the detector site was begun in February 1990 and the experiment is scheduled to begin at the end of 1994. The current status of the project will be reviewed.

33. BOREXINO: LOW ENERGY SOLAR NEUTRINO SPECTROSCOPY. R. S. Raghavan[*], AT&T Bell Laboratories, Murray Hill, NJ 07974.

Borexino is a liquid scintillation detector with about 100 t fiducial volume, being developed for operation at Gran Sasso, Italy. Its mission is solar neutrino spectroscopy with signal and event thresholds above 0.25 MeV. The strongest signal in the detector will arise from ^7Be neutrinos at the rate of about 50/day (in the standard solar model) via neutrino-electron scattering. High energy neutrinos such as ^8B will also yield useful information. In addition, antineutrinos can be detected with high sensitivity. Borexino will be uniquely placed to tackle physics questions needing real-time low-energy sensitivity and high signal rates (inaccessible to present or planned experiments) such as: magnitudes and short-term

variations (day/night, seasonal) of the ^{7}Be neutrino flux, neutrino-electron scattering cross-sections at sub-MeV energies, and strong limits on antineutrinos from the sun. Answers from Borexino on these topics will have specific and critical implications for neutrino mass/mixing and magnetic moments.
*For the Borexino Collaboration: (AT&T Bell Labs, Argonne Natl. Lab., Drexel U., JINR Dubna, INFN/U. Genoa, LNGS Gran Sasso, U. Hawaii, MIT, INFN/U. Milan, Tech. U. Munich, INFN/U. Pavia and Charles U. Prague).

34. USE OF A TIME-PROJECTION CHAMBER IN MULTIFRAGMENTATION EXPERIMENTS AT THE BEVALAC.
N. T. Porile, Department of Chemistry, Purdue University, W. Lafayette, IN 47907, and the EOS Collaboration.*

An exclusive study of multifragmentation is described. The moments of the fragment charge distribution are used to extract the critical exponents associated with the phase transition to which the breakup is ascribed. The fragmentation of 1 GeV/nucleon La and Au is studied by reverse kinematics using a carbon target. Fragments with $Z \leq 6$ will be identified with the EOS time projection chamber (TPC) while heavier fragments will be identified with a multiple sampling ionization chamber (MUSIC). The experimental setup using these detectors will be described.

*Supported by the U.S. Department of Energy

35. THE MSU MINIBALL: A TOOL FOR THE CHARACTERIZATION OF NUCLEAR MULTIFRAGMENTATION;
Romualdo T. de Souza, Michigan State University, East Lansing, MI 48824

In the past year, a new generation 4 pi detector array has been completed. This detector, the MSU Miniball, is a granular, low threshold phoswich device. Such a device is essential in the understanding of multifragment decays observed in intermediate energy heavy-ion reactions. Recent experiments with the Miniball have yielded interesting preliminary results. Complete disintegration of the nuclear system has been observed in events associated with large multiplicity. The detected charge in these large multiplicity events can be as much as the total charge of the system. The average number of intermediate mass fragments (Z=3) in central collisions is five. The dependence of first and second moments of the intermediate mass fragment (IMF) distribution on total multiplicity will be presented. Systematics are explored for the system ^{36}Ar + ^{197}Au at E/A = 35,50,80, and 110 MeV, as well as ^{129}Xe + C, Al, Cu, and Au at E/A = 50 MeV. Analyses based on full event characterization will be discussed. Timescales for the reactions will be extracted from IMF-IMF correlations. Comparisions to both equilibrium and dynamical theoretical calculations will be presented.

36. THE INDIANA SILICON SPHERE (ISiS) 4π DETECTOR.*K. Kwiatkowski, K. Komisarcik, J. Brzychczyk, K. Morley, E. Renshaw, D. Bracken and V.E. Viola, Department of Physics and Chemistry and IUCF, Indiana University, Bloomington, IN 47405.

A 4π detector array for the study of multifragment emission processes in light-ion-induced reactions is described. The detector array is based on ion-implanted passivated silicon detector technology, combined with gas-ionization and CsI(Tl) scintillator devices to measure light-charged particles up to 100 MeV/A and all heavier fragments with discrete Z identification up to $Z \leq 14$. The design criteria include good granularity (162 elements), very low energy thresholds (0.1-0.2 MeV/A), excellent energy resolution and stability, broad dynamic range and ease of calibration.

* work supported by the U.S. Department of Energy

37. STOPPING POWER AND RANGES OF IONS IN CONDENSED MATTER. J. Biersack, Hahn-Meitner Institute, D-1000 Berlin 39, Germany.

No abstract available at this time.

38. THE EVOLUTION OF HEAVY ION RECOIL AND FRAGMENT SEPARATORS. J. Nolen, National Superconducting Cyclotron Laboratory, Michigan State University, East Lansing, MI 48824

Research in heavy ion nuclear science is evolving towards more extensive use of reaction recoil and fragment separators. The emphasis has been on the development of devices which promptly separate reaction products according to A or A/q at or very near zero degrees. Probably the most important performance criterion for such devices is the degree to which weak reaction products are separated from the beam and other abundant species. Another important design consideration is the optimization for the beam and reaction product energy or rigidity and the related dominant reaction mechanisms. Separator configuration and size requirements vary rapidly as the research emphasis changes from fission fragments, to slow fusion products (light beam on heavy target), to fast fusion products (heavy on light), to deep inelastic, and finally to fragmentation products at 50 to 500 MeV per nucleon. Examples of on-going and new research programs with such devices around the world will be discussed along with plans for nearly-completed and recently-designed facilities at several heavy ion labs.
Research supported by the U.S. National Science Foundation.

39. ENTRANCE CHANNEL EXCITATIONS IN THE ^{28}Si + ^{28}Si REACTION, P. Decowski, Smith College, Northampton, MA 01063, USA, E. Gierlik, University of Warsaw, Poland, P.F. Box, R. Kamermans, G.J. van Nieuwenhuizen, University of Utrecht, the Netherlands, R.J. Meijer, GSI, Darmstadt, Germany, K.A. Griffioen, University of Pennsylvania, H.W. Wilschut, KVI, Groningen, the Netherlands, A. Giorni, C. Morand, ISN, Grenoble, France, A. Demeyer, D. Guinet, IPN, Lyon, France.

Velocity spectra of heavy ions produced in the ^{28}Si + ^{28}Si reaction at bombarding energies of 19.7 and 30 MeV/nucleon were measured and interpreted within the Q-optimum model extended by the inclusion of particle evaporation from excited fragments. Regions of forward angle spectra corresponding to the mutual excitation of the reaction partners with net mass transfer zero projected onto the Q-value variable show an enhancement at Q-values of -60 - -80 MeV (excitation energies of the reaction partners equal to 30 - 40 MeV). This energy range coincides with the region of $2\hbar\omega$ - $3\hbar\omega$ excitations characteristic for giant oscillations. This selective excitation, which occurs at a very early stage of the reaction (the cross section is the largest at very forward angles), provides an important doorway to other dissipative processes.

40. DISPERSION RELATION AND THE LOW-ENERGY BEHAVIOR OF HEAVY-ION OPTICAL MODEL POTENTIAL
Brian Robson, The Australian National University, Canberra, ACT 2601, Australia and Malgorzata Zielińska-Pfabé, Smith College, Northampton, MA 01063, USA.

The probability of exciting the first 3^- state in ^{208}Pb has been studied using the optical model potential. In the neighborhood of the Coulomb barrier the strength of the imaginary part was assumed to vary linearly with bombarding energy[1]. The strength of the real part has been calculated using the dispersion relation. An excellent fit to experimental excitation functions has been obtained for ^{12}C + ^{208}Pb and ^{16}O + ^{208}Pb reactions. The calculated angular distributions for various bombarding energies also agree very well with experimental data.
1. M.A. Nagarajan, C.C. Mahaux, G.R. Satchler, Phys. Rev. Lett., 54(1985)1136.

41. STUDIES OF THE FUNDAMENTAL ASPECTS OF ION/SOLID COLLISIONS USING LASER SPECTROSCOPY. N. Winograd, Department of Chemistry, Penn State University, 152 Davey Laboratory, University Park, PA 16802

The ion/solid interaction is a complex event. Our group has been involved in elucidating the details of this process by experimentally determining the energy and angular distributions of atoms desorbed from single crystal surfaces. Detection of these atoms is made possible by state-selective multiphoton resonance ionization techniques. These detailed measurements are then directly compared to molecular dynamics computer simulations of the ion impact event. This talk will focus on recent data obtained for ground state and excited state distributions of Rh atoms ejected from Rh{100}. From the model calculations, we show that the excitations arise from collision events both inside the solid and in the near surface region. The significance of these results will be discussed with respect to the ejection of molecules from ion-bombarded surfaces and the formation of secondary ions observed in SIMS.

42. SPUTTER-INITIATED RESONANCE IONIZATION SPECTROSCOPY AS AN ANALYTICAL TOOL: INSTRUMENTAL REQUIREMENTS AND SALIENT RESULTS. Norbert Thonnard, Atom Sciences, Inc., Oak Ridge, TN 37830

Sputter-Initiated Resonance Ionization Spectroscopy (SIRIS), an analytical techniqe with extremely high element specificity and sensitivity, is becoming recognized as an emerging tool with wide applications. In the SIRIS technique, the abundant neutral atoms released by the sputtering process are ionized by precisely-tuned laser beams and counted in a mass spectrometer. The exceptionally high ionization efficiency and element specificity enables measurements that frequently are very difficult or impossible with other techniques. SIRIS has almost all the advantages of secondary ion mass spectrometry (SIMS) and other surface analytical techniques, while making significant improvements in the following SIMS shortcomings: efficiency, matrix dependance, isobaric and molecular interferences, and sensitivity. With SIRIS, it is possible to localize with high spatial resolution ultra-trace concentrations of a selected element to the sub-parts per billion level. We will describe the implementation of SIRIS to solve a number of analysis problems, and illustrate its salient characteristics with data from a wide range of applications.

43.
SPUTTERING PROCESSES OF ION-BOMBARDED ELECTRONIC MATERIALS. S.W. Downey and A.B. Emerson, AT&T Bell Laboratories, Murray Hill, NJ 07974

Resonance ionization mass spectrometry (RIMS) of neutral atoms sputtered from a solid is used in conjunction with secondary ion mass spectrometry (SIMS) to probe atoms and ions ejected from solids during ion sputtering. The detection of atoms, molecules and ions provides information about a wide variety of sputtering phenomena. We have studied speciation and quantum state distribution of sputtered products from a wide variety of electronic materials such as Si, GaAs and InP. The primary ion mass (e.g., Xe^+ vs. Ar^+) has an effect on several processes including the sputtered atom energy distribution and atomization efficiency for many important matrices.

44. CONCRETE RESULTS THROUGH VIRTUAL STATES: SURFACE ANALYSIS WHEN INTENSE LASERS MISS THE SURFACE. C. H. Becker, Molecular Physics Laboratory, SRI International, Menlo Park, California 94025

We have developed a sensitive method of surface analysis that uses nonselective photoionization of sputtered atoms and molecules above the surface, followed by time-of-flight mass spectrometry. We call this method surface analysis by laser ionization or SALI. In particular, nonresonant multiphoton ionization (NRMPI) is employed with focused pulsed laser intensities of typically 10^{10} W/cm^2 for ns systems, and more recently 10^{13} to 10^{14} W/cm^2 with a powerful amplified ps system. A range of examples from different types of materials will be presented, including chemical imaging of surfaces with submicrometer lateral resolution, for all elements simultaneously.

45. MOLECULAR IDENTIFICATION IN SURFACE ANALYSIS.* M.J. Pellin, K.R. Lykke, P. Wurz, D.H. Parker, J.C. Hemminger,+ D.M. Gruen, Materials Science/Chemistry Divisions, Argonne National Laboratory, Argonne, IL 60439; +University of California-Irvine.

Elemental surface analysis, using directed energy atomization sources (sputtering or laser vaporization) followed by resonant laser ionization, has proven to be uniquely sensitive and discriminative. Recently, we have begun to apply these techniques to molecules at surfaces. In comparison to atoms, broad molecular absorption features lead to a loss of specificity in the laser ionization process. We can now demonstrate sensitive and discriminative detection using a high field (7 Tesla) Fourier Transform Mass Spectrometer (FTMS). This FTMS system can have mass resolutions approaching $m/\Delta m = 5 \times 10^5$ even for volume ionization processes such as laser or electron ionization. This technique will be demonstrated for molecules ejected from both metal and polymer surfaces.

*Work supported by the U.S. Department of Energy, BES-Materials Sciences, under Contract W-31-109-ENG-38.

46. SYNTHESIS OF HIGH-ENERGY PROJECTILE AND TARGET FRAGMENTATION. D.J. Morrissey, NSCL and Dept. of Chemistry, Michigan State University, East Lansing, MI, USA, 48824-1321.

For historical and technical reasons, the processes of projectile and target fragmentation have been studied and thought of in different lights. None the less, the physical processes that give rise to these products must be the same although the difference in the appropriate rest-frame velocities masks this fact. The results of recent summaries [1,2] of the production cross sections and the linear momenta of all of these reaction products show that they are consistent with one another. Moreover, all the results are consistent with the mechanism of sudden creation of the products in highly excited states and subsequent statistical (evaporative) decay. The challenge for the future is to understand the transition and connection to reactions below the Fermi energy. This work was supported by the National Science Foundation under grant PHY 89-13815.

1] K. Sümmerer, et al., Phys. Rev. C42, 2546 (1990).
2] D.J. Morrissey, Phys. Rev. C39, 460 (1989).

47. FRAGMENTATION OF GOLD PROJECTILES AT 600 MEV/NUCLEON. W. Trautmann, GSI Darmstadt, D-6100 Darmstadt, Germany.

In a first experiment with the ALADIN spectrometer at SIS the fragmentation of Au projectiles interacting with targets of C, Al, and Cu at an incident energy of 600 MeV/nucleon was studied. The employed inverse kinematics allowed a nearly complete detection of projectile fragments with atomic number $Z \geq 2$.

A maximum mean multiplicity of 3 to 4 intermediate mass fragments is observed for all three targets. The corresponding impact parameters range from central for the C target to more peripheral for the heavier targets. The correlations of the fragment multiplicity with other observables such as the associated multiplicity of light charged particles, the largest atomic number Z_{max} within an event, or the shape of the fragment spectra are independent of the target. This universal behaviour indicates a large degree of equilibration of the projectile fragment prior to its decay.

48. **TRANSPARENCY AND ABRASION IN HIGH-ENERGY NUCLEUS-NUCLEUS COLLISIONS.**
Paul J. Karol, Department of Chemistry, Carnegie Mellon University, Pittsburgh, Pennsylvania 15213.

The soft spheres geometric model of nucleus-nucleus collisions has been used to incorporate a tapered nuclear density distribution into considerations of transparency and of the abrasion step of the (hard spheres) abrasion-ablation model. The useful concept of an average target transparency $<T_{tr}>$ required a careful definition to accommodate the fact the transparency becomes unity as the impact parameter approaches infinity. The result is a simple numerical calculation that gives $<T_{tr}>$ as a function of the soft spheres parameter χ. The latter depends in a known way on the target and projectile dimensions and on the energy-dependent nucleon-nucleon cross section. In the specific case of projectile = target, the result is an analytical expression reminiscent of the Fernbach, Serber and Taylor result for hard spheres.

$$<T_{tr}> = \frac{1 - e^{-\chi}(\chi + 1)}{\chi^2}$$

Applications are expected to be valid for any projectile-target combination at intermediate (several hundred MeV/A) through ultrarelativistic (several hundred EeV/A) energies.

49. ELECTROMAGNETIC DISSOCIATION WITH RELATIVISTIC HEAVY IONS. J. C. Hill and F. K. Wohn, Ames Laboratory, Iowa State University, Ames, IA 50011, and A. R. Smith, Lawrence Berkeley Laboratory, Berkeley, CA 94720.

Electromagnetic dissociation (ED) occurs following absorption by the nucleus of a virtual photon when a relativistic heavy ion passes close to, but outside the range, of the strong force. The usual result is the excitation of the giant E1 resonance. Results from a series of experiments at the Bevalac, AGS, and CERN-SPS to determine the dependence of the ED cross section on projectile charge and energy will be given for one- and two-neutron removal reactions on Au and Co targets using projectiles ranging from ^{12}C to ^{238}U. The measured results will be compared with calculations using the Weizsäcker-Williams virtual photon procedure. Results of calculations for ED at RHIC energies where ED is expected to be a major factor in the lifetime of stored beams will be presented. Also, a discussion will be given of ED production of e^+e^- pairs which also effects lifetimes for RHIC beams since the e^- can be recaptured by the ion in the beam.

50. TARGET FRAGMENTATION OF SILVER BY 14.6 GEV/NUCLEON ^{16}O IONS M. Bronikowski and N. T. Porile, Department of Chemistry, Purdue University, W. Lafayette, IN 47907[*]

Cross sections and recoil properties of 90 products of the interaction of silver with 14.6 GeV/nucleon ^{16}O ions were determined by off-line γ-ray spectrometry. The data were used to obtain the isobaric-yield and mass-yield distributions as well as the two-step model values of the mean remnant velocities and the mean kinetic energies associated with the deexcitation step. The results are compared with lower energy data.

[*]Supported by the U.S. Department of Energy

51. FORWARD AND TRANSVERSE ENERGIES IN RELATIVISTIC HEAVY ION COLLISIONS AT 14.6 GeV/c PER NUCLEON
James B. Cumming, Brookhaven National Laboratory, Upton, NY 11973.

Transfer of beam energy into produced particles, primarily pions, during the interactions of 14.6 A GeV/c ^{28}Si ions with targets from ^{27}Al to ^{197}Au was studied with the downstream calorimeter, ZCAL, and the midrapidity lead glass array, PbGl, of experiment E802 at the Brookhaven AGS. Forward and transverse energy spectra and the correlation between these two global variables will be discussed in terms of the idea of projectile stopping in heavy nuclei and compared with calculations based on the Fritiof model.

Supported by U.S. Department of Energy under contract DE-AC02-76CH00016.

52. INTRODUCTION TO CLUSTER PHENOMENA. J. Remillieux, Institut de Physique Nucleaire de Lyon, IN2P3-CNRS et Universite Claude Bernard, 43 bd du 11 Novembre 1918, F-69622 Villeurbanne Cedex, FRANCE.

The interest of atomic cluster beams for basic and applied researches in surface physics and chemistry will be presented. Then a short review of the methods currently used to produce and accelerate various cluster species will be given. Finally, the collective effects expected under cluster impact will be described over a large range of projectile velocities: primary effects for the projectile (explosion and slowing down of the fragments) and secondary effects in the target (shock wave, electron emission, desorption, ...).

53. CLUSTER ENERGY LOSS IN SOLIDS. N. R. Arista, Division Colisiones Atomicas, Instituto Balseiro and Centro Atomico Bariloche, RA-8400, Bariloche, Argentina.

The energy loss of ion clusters in dense media shows important differences (vicinage effects) as compared with the stopping power of isotachic single ions. These effects have been studied for small ion clusters at energies above and below the stopping power maximum. More recently, experiments with large molecular clusters, and possible applications in nuclear fusion studies, give rise to new interest in studying the beam-energy deposition in solid targets. The energy loss of ion clusters in solids will be discussed, starting from earlier experimental and theoretical work. The theory will be extended to clusters of larger size in cold or heated media.

54. INTERACTION PROCESSES OF FAST (V# V_0) HYDROGEN CLUSTERS WITH SOLID MATERIALS. Jean-Paul Thomas, Institut de Physique Nucleaire de Lyon, IN2P3-CNRS, Universite Claude Bernard Lyon I, 43 Boulevard du 11 Novembre 1918, 69622 Villeurbanne Cedex, FRANCE.

Hydrogen clusters with masses as high as 99 at a maximum energy of 600 keV have been used to investigate various consequences of interactions with solids, namely charge exchange, energy loss and secondary ion emission. We will particularly emphasize the role of collective effects in the energy loss process from transmission experiments through thin self-supported films and from secondary ion emission induced at the surface of dielectric materials by the deposited energy.

55. CLUSTER IMPACT ON SOLID SURFACES, S. Della-Negra, Y. Le Beyec, Institut de Physique Nucleaire, Orsay, B.P. No. 1, F-91406, Orsay-Cedex, FRANCE.

Cluster ion bombardment of a solid surface has proved to be very efficient to emit secondary ions. Nonlinear effects occur in the collision processes which result in a large enhancement of secondary yield experiment using different polyatomic ions as projectiles will be described. It is intended in the near future to increase the mass and the energy of these projectiles to explore the phenomena created by cluster impacts. A particle accelerator is being transformed for this purpose at the Institute.

56. MOLECULAR ANIONS AS SURFACE PROBES. J. E. Delmore, A. D. Appelhans, EG&G Idaho, Inc., P.O. Box 1625, M/S 2208, Idaho Falls, ID 83415.

Our group at the INEL has developed ion sources producing 5- to 7-atom molecular anion beams for use in secondary ion mass spectrometry (SIMS) of insulating materials. Anion sources were chosen because the buildup of electrical charge on insulating samples is much more readily controlled with an anion probe than with a cation probe. Both sulfur hexafluoride and perrhenate molecular anions have proven well adapted to SIMS analysis of insulators, particularly when combined with pulsing of the secondary ion extraction voltage. Pulsed extraction with an anion beam permits the sample surface to be easily held constant to within a few tenths of a volt. We have successfully used these beams for analysis of a wide variety of bulk plastics, biological specimens, organics, pharmaceuticals, and polymer films. In particular, our studies have shown sulfur hexafluoride anions are a factor of 30 times more efficient than cesium ions for desorbing tertiary amines.

57. THE CLUSTER-SOLID INTERACTION FROM THE PERSPECTIVE OF CHEMICAL ANALYSIS. E. A. Schweikert, Center for Chemical Characterization & Analysis, Department of Chemistry, Texas A&M University, College Station, TX 77843-3144.

Bombardment of solids with polyatomic projectiles (e.g., Cs_2I^+) results in high secondary ion (SI) yields. For projectiles of 5 to 30 keV, SI yields as high as 10% have been observed; they are up to 50 times higher than those obtained with equal velocity monoatomic ions. Moreover, SIs emitted from a projectile impact are spatially correlated. Thus, when the SIs are identified (by TOF-MS) in an event-by-event coincidence counting mode, spatial and chemical relationships can be examined. The application of this approach for examining chemical microhomogeneity of surfaces has been demonstrated at the level of a few hundred nanometers. Perspectives opened by these features for surface characterization will be discussed.

58. NUCLEAR CHEMISTRY, THE MET LAB, AND NATHAN SUGARMAN - A RETROSPECTIVE. Ellis P. Steinberg, Argonne National Laboratory, 9700 So. Cass Avenue, Argonne, IL 60439

The evolution of nuclear chemistry will be traced briefly, with special emphasis on the exciting and highly productive period of the war-time Metallurgical Laboratory from 1942 to 1946. In particular, the Fission Product Radiochemistry section at The University of Chicago, which underwent sequential fissions of its own to Oak Ridge and Los Alamos, will provide a major focus. The post-war spread of nuclear chemistry throughout the country and the establishment of the National Laboratories provided the setting for the Golden Age of the field. Throughout this period, the personality and character of Nathan Sugarman was clearly evident. Whether as teacher, researcher, colleague, critic, counselor, friend, or acquaintance, "Sug's" intelligence, warmth, humor, high standards, and quiet leadership made a lasting impression on a generation of nuclear chemists.

59. RECOIL STUDIES AND HIGH-ENERGY REACTION MODELS*.
James B. Cumming, Brookhaven National Laboratory, Upton, NY 11973.

The thick-target thick-catcher technique provides a convenient method for the determination of mean kinetic properties of nuclear reaction products. Its development and limitations will be reviewed and its application illustrated by studies of the fragmentation of copper targets by high-energy protons, alpha particles, and heavy ions. Mean kinetics energies and momentum transfers derived for these systems will be analyzed in terms of a simple model for the fragmentation process and discussed in the context of the hypotheses of limiting fragmentation and factorization.

*Supported by the U.S. Department of Energy, Office of Basic Energy Sciences under contract no. DE-AC02-76CH00016.

60. RECOIL STUDIES REVISITED: HEAVY ION INDUCED REACTIONS. James J. Hogan, Department of Chemistry, McGill University, Montreal, Quebec, Canada H3A 2K6.

Sug, along with John Alexander, Jim Cumming, and Lester Winsberg, may be considered to be the founding fathers, and leading proponents, of recoil studies as a probe of nuclear reaction mechanisms. After a number of years in which "electronics chemistry" seemed to prevail, the recoil approach has been revived, making use of large germanium detectors and computer analysis to study large numbers of individual products from given reactions. Both spallation and fission have been studied in heavy ion induced reactions, where the recoil velocity of the intermediate excited nucleus may be correlated with the mass transfer in the initial projectile-target interaction. Taken together with the yields, the angular distributions and kinetic energies of individual reaction products measured by recoil techniques may be used to produce kinematic models of these reactions.

A variety of recent heavy ion induced spallation and fission studies will be reviewed with emphasis on the contribution of recoil studies to an understanding of the predominant mechanisms.

61. THE DOUBLE BETA DECAY OF U-238. Anthony L. Turkevich[1,2,3], Thanasis Economou[1], and George A. Cowan[3,4]. [1]Enrico Fermi Institute and [2]Department of Chemistry, University of Chicago, Chicago, IL 60637; [3]Los Alamos National Laboratory, Los Alamos, NM 87545, and [4]Santa Fe Institute, Santa Fe, NM 87501.

Plutonium has been isolated from 8.47 kg of 33 yr old uranyl nitrate. The alpha particles of added ^{239}Pu measured the recovery and efficiency of the operations. A small (~0.1 c/day) activity of 5.5 MeV alpha particles in the isolated plutonium corresponds to a half-life of 2×10^{21} yr for the double beta decay of ^{238}U. This is 40 times longer than our previous limit but shorter than the most recent theoretical estimates. No source of extraneous production of ^{238}Pu has been identified.

62. EVOLUTION OF THE REACTION ^{40}Ar + Ag FROM E/A=7 TO 34 MeV. J. M. Alexander, Nuclear Chemistry Group at State University of New York, Stony Brook, NY 11794-3400, AMPHORA and EMRIC Groups at Grenoble and Lyon, France.

The 4π charged-particle multidetector AMPHORA has been used to study the reaction ^{40}Ar + ^{nat}Ag from 270-1356 MeV. Charged-particle multiplicity distributions show a low-multiplicity group associated with peripheral collisions and a high multiplicity group associated with central collisions. Average multiplicities for central collisions increase with increasing projectile energy, indicating ever-increasing collision violence. Angular distributions of emitted protons are essentially isotropic for $\theta \geq 80°$ in a reference frame characterized by the empirical systematics of linear momentum transfer (i.e. $\approx 100\%$ to $\approx 70\%$ from 7-34 MeV/nucleon). Spectra of these protons at side angles are evaporation-like in shape and indicate relative effective temperatures of 3,6,8, and 12 MeV for beam energies of 7, 17, 27 and 34 A MeV respectively.

Azimuthal angular correlations between various particle pairs are consistent with spin-driven emission from emitter sources of reasonable spin values. In short, these results support a classical picture of extensively thermalized emitter nuclei even for initial excitation energies of ≈ 5 MeV per system nucleon and spins of $\geq 100\hbar$.

63. SEARCH FOR QUARK-GLUON PLASMA IN $\bar{P}$-P COLLISIONS AT FERMILAB. <u>N. T. Porile</u>, Department of Chemistry, Purdue University, W. Lafayette, IN 47907, and the E-735 Collaboration.*

A survey of the search for quark-gluon plasma (QGP) signals in recent results of Fermilab E-735 is presented. The basic data are the inclusive transverse momentum (p_t) spectra of centrally produced π, K, $\bar{p}$, $\bar{K}$, and γ as a function of total charged particle multiplicity in the collision. From these, the dependence of $<p_t>$ and particle ratios on multiplicity are derived. Results for π-π correlations, multiplicity distribution, and intermittency will be presented.

*Supported by the U.S. Department of Energy

64. **PARTICLE SCATTERING FOR MEASUREMENT OF SURFACE COMPOSITION AND STRUCTURE,** <u>Marcus H. Mendenhall</u> and Robert A. Weller, Vanderbilt University, Nashville, TN 37235

This paper will present a brief review of the status of ion backscattering as an analytical tool, discussing the techniques of Rutherford Backscattering Spectrometry (RBS), Heavy-Ion RBS, and similar techniques. This discussion will include a survey of the various methods by which information can be extracted from the data acquired from such measurements. It will then discuss recently made advances in the field of Time-of-Flight Medium Energy Backscattering Spectrometry (ToF MEBS) as applied to analysis of surfaces. This technique provides higher surface sensitivity, higher depth resolution, lower beam-induced material damage and lower cost than traditional RBS. It is especially well adapted to the analysis of films with thicknesses less than about 100 nm, such as optical coating and electronic thin films. It is also well suited to the measure of diffusion into surfaces, since it provides depth profile information with a typical resolution 1 nm near the surface of a sample.

65. HIGH-ENERGY ION SCATTERING FOR MATERIALS ANALYSIS. <u>J. C. Barbour</u>, J. A. Knapp, and B. L. Doyle, Division 1111, Sandia National Laboratories, Albuquerque, NM 87185.

Conventional ion-beam techniques for studying the chemistry of thin-film reactions and near-surface regions are Rutherford Backscattering Spectrometry (RBS) and Nuclear Reaction Analysis (NRA). RBS is useful for high-Z elements particularly in or on lighter substrates, whereas NRA is useful for higher sensitivity of low-Z elements. Although NRA has been used for analyzing elements from H to Al, NRA is usually more difficult and time intensive than RBS. Recently, simpler high-energy ion-beam techniques have been developed for low-Z materials analysis. High-Energy Non-Rutherford Backscattering Spectrometry (HEN-RBS) has been applied to accurately measure the oxygen content in high temperature superconductors, the nitrogen content in plasma-deposited silicon nitride, and the carbon and nitrogen content in the surface of corrosion/wear-resistant stainless steels. In addition, HEN-RBS can be used in conjunction with the micron size analysis spot afforded a nuclear microprobe to gain both depth and lateral resolution in the analysis of

superconductors. This talk will examine the use and applications of HEN-RBS in comparison with NRA. Also, the use of heavy-ion beams for light element detection with Elastic Recoil Detection (ERD) and Time-of-Flight ERD will be discussed.
This work was supported by the U.S. Department of Energy under contract DE-AC04-76DP00789

66.
CRYSTALLOGRAPHIC INFORMATION FROM ION CHANNELING
Carl J. Maggiore, *Los Alamos National Laboratory, Los Alamos, NM*

When a collimated beam of ions is incident on an oriented single crystal, the scattering yield is modified by the non-uniform flux of ions within the crystal. The change in backscattering or nuclear reaction yield as a function of crystal orientation provides detailed structural information about the near surface region of a solid. Lattice site as well as lattice disorder can be analysed by the use of channeled ions. The technique has found widespread and varied application in the areas of thin film growth, ion implantation, interfaces, epitaxial structures, and surfaces.

67. **NUCLEAR REACTION ANALYSIS OF HYDROGEN IN MATERIALS: PRINCIPALS AND APPLICATIONS.** W. A. Lanford, Department of Physics, SUNY Albany, Albany, New York 12222.

Analysis for hydrogen in materials is difficult by most traditional analytic methods. Because hydrogen has no Auger transitions, no X-ray transitions, does not neutron activate, and does not backscatter ions, it is invisible in analytical methods based on these effects. In addition, since hydrogen is a universal contaminant in vacuum systems, techniques based on mass spectrometry are difficult unless extreme measures are taken to reduce hydrogen backgrounds. Because of this situation, methods have been developed for analyzing for hydrogen in solid materials based on nuclear reactions between bombarding ions and hydrogen atoms (protons) in the samples. The nuclear reaction methods are now practiced at laboratories around the world. The basic principals of nuclear reaction analysis will be briefly presented. This method will be illustrated by applications to problems ranging from basic physics, to geology, to materials science, and to art history and archeology.

68. **STRUCTURAL STUDIES OF Li_4NpO_5 AND Li_5NpO_6 BY NEUTRON AND X-RAY POWDER DIFFRACTION.** L. R. Morss, E. H. Appelman, R. R. Gerz, and D. Martin-Rovet, Chemistry Division, Argonne National Laboratory, 9700 S. Cass Ave., Argonne, Illinois 60439.

As part of a program to synthesize and characterize complex oxides with transuranium elements in high oxidation states, we have prepared Li_4NpO_5 and Li_5NpO_6 by reacting low-fired NpO_2 with $^7Li_2O_2$ in flowing O_2 at ~700 and 400 °C respectively, and characterized the compounds by both X-ray and neutron powder diffraction. The structure of Li_4NpO_5 is consistent with that reported earlier by X-ray powder diffraction (Keller, Koch, and Walter, *J. Inorg. Nucl. Chem.* **27**, 1205, 1965) and with the structure of Li_4UO_6 refined by neutron powder diffraction (Hoekstra and Siegel, *J. Inorg. Nucl. Chem.* **26**, 693, 1964): space group I4/m, a = 6.698(3) Å, c = 4.415(2) Å, Np-O bond distances: 2 axial 2.207(2) Å; 4 equatorial 1.998(3) Å [uncertainties 1σ]. The comparable bond distances in Li_4UO_5 are 2.23 and 1.99 Å. The structure of Li_5NpO_6 is not consistent with that reported earlier by X-ray powder diffraction (Keller and Seiffert, *Inorg. Nucl. Chem. Lett.* **5**, 51, 1969); other structural models have been evaluated. This work was sponsored by the U.S. Department of Energy, Office of Basic Energy Sciences, Divisions of Chemical and Materials Sciences, under Contract W-31-109-ENG-38.

69. CHARACTERIZATION OF FERRISILICATES BY MOSSBAUER SPECTROSCOPY
ARRIOLA S. HUMBERTO, NAVA E. NOEL; GUZMAN C. LOURDES; LORENZO JOAQUIN* FACULTAD DE QUIMICA, DEPg., UNAM. C.U. MEXICO 04510 D.F.,*DIVISION DE GEOFISCA, IMP. APDO. POSTAL 14-805, MEXICO, D.F.

A ferricilicate of the $Nax[(MO_4)x (SiO_4)y]$ type was sintetized, where the sum x+y fits into the structure ZSM-5 formed by iron and Si atoms, in the centre of a tetrahedron, surrounded by 4 oxigens, producing chains which in place form tridimensional structures interconected by a channel system. This catalyst was studied by Mossbauer spectroscopy in order to determine if and how the iron atoms are bond to the structure. Several samples were analyzed in which the atomic ratio x/y was changed, characterizing in this way the structure in each case for the different Mossbauer parameters and the position of the iron atoms.

70. THE GAMMA RADIOLYSIS OF α-KETOGLUTARIC ACID IN AQUEOUS SOLUTIONS.
A. Negrón-Mendoza, G. Albarrán and S. Castillo-Rojas.
Instituto de Ciencias Nucleares, UNAM.
A. Postal 70-543, México, D.F. 04510 México

The purpose of this work is to study the radiolytic behavior of aqueous solutions of ^{14}C-α-ketoglutaric acid (oxygen-free). Several parameters were varied such as radiation dose, concentration and pH.
Base on the identified of the major non volatile products we proposed three main pathways for the radiolysis of the alpha-ketoglutaric acid: 1) reduction of the keto group yielding hydroxyglutaric acid; 2) loss of the CO group yielding a saturated carboxylic acid (succinic acid); 3) addition of hydrated electron to the alpha-carbonyl group yielding a diol. The yields of these products were influenced by the irradiation dose, concentration of the target compound and pH of the solution.

71. HIGH DOSE DOSIMETRY. COMPARATIVE TECHNIQUES
MARTINEZ TRINIDAD., CANIZAL GERARDO., GALVAN G. ADRIANA
FACULTAD DE QUIMICA, UNAM., CD. UNIVERSITARIA, D.F. MEXICO 04510

High dose dosimetry (5×10^4 Gy >) is a matter of present interest due to its applications in the search of new materials for advanced reactor as well as for nuclear accident dosimetry. likewise, future high intensity irradiation techniques (gamma rays and electrons) will re - quire precise methods for dose determination. Therefore, some research becomes important for the high dose region. This work describes the experiments performed to design solid dosimetric systems using pure salts of heptahydrated ferrous sulphate and tetrahydrated cerium (IV) sulphate. The evaluation of induced gamma radiolytic changes was made spectrophotometrically. Some conclusion are obtained about the cha - racteristic of dosimetric systems, linearity, dose rate influence. - etc. In the heptahydrated ferrous sulphate system a comparisson was made between the sensibility of this measurement method of Fe (III) and that by means of Mössbauer spectroscopy.

72. RADIOLYSIS OF ADSORBED ACETIC ACID ON Na-MONTMORILLONITE.
A. Negrón-Mendoza[1], S. Ramos[2] and G. Albarrán[1]
[1]Instituto de Ciencias Nucleares,UNAM. A. Postal 70-543 México,D.F. México. [2]Facultad de Ciencias, UNAM. Circuito Exterior, C.U. México, D.F. México.

In the present work we investigate the influence of doses and water content on the radiolysis of acetic acid adsorbed in a clay surface (Na-montmorillonite). The reaction was followed by the formation of CO_2, also for other non-volatile radiolytic products from ^{14}C-acetic acid. The main reaction observed was a decarboxylation reaction. The mechanism of this heterogeneous catalysis is complex. This reaction can be promoted by a loss of electrons from the carbonyl group of the acid and could be accepted by acids sites on clay. Other mechanism that are currently under study involve energy transfer process and free radical initiator by the water radiolytic products trapped in the clay lattice.

73. SEPARATION STUDIES OF SELECTED ACTINIDE IONS WITH 4-BENZOYL-5-METHYL-2-PHENYL-PYRAZOL-3-THIONE AND TRI-n-OCTYLPHOSPHINE OXIDE.* Nancy J. Hannink and Darleane C. Hoffman, Nuclear Science Div., 70A-3307, Lawrence Berkeley Laboratory, Berkeley CA 94720 and Barbara F. Smith, Analytical Chemistry Group, Los Alamos National Laboratory, Los Alamos NM 87545.

The first measurements of distribution coefficients (K_d) for Bk(III), Cf(III), Es(III), and Fm(III) between aqueous perchlorate solutions and solutions of 4-benzoyl-5-methyl-2-phenyl-pyrazol-3-thione (BMPPT) and the synergist tri-n-octyl-phosphine oxide (TOPO) in toluene are reported. ^{250}Bk, ^{249}Cf, ^{254}Es, and ^{253}Fm were used in these studies. The K_d for ^{241}Am was also measured and is in agreement with previously published results. Our new results show that the K_d's decrease somewhat with increasing atomic number for the actinides with a dip at Cf. In general, the K_d's for these actinides are about a factor of 5 to 10 greater than the K_d's for the homologous lanthanides. The larger K_d's for the actinides are consistent with greater covalent bonding between the actinide metal ion and the sulfur bonding site in the ligand.

*Work supported in part by a Department of Education Fellowship and in part by the U. S. Department of Energy under Contract No. DE-AC03-76SF00098.

74. EXTRACTION OF Rf (ELEMENT 104) AND ITS HOMOLOGS WITH TBP*, K. R. Czerwinski, K. Gregorich, T. Hamilton, N. J. Hannink, C. D. Kacher, B. A. Kadkhodayan, S. A. Kreek, D. Lee, M. Nurmia, A. Türler, and D. C. Hoffman, Nuclear Science Division, 70A-3307, Lawrence Berkeley Laboratory, Berkeley, CA 94720

The extraction of tetravalent Rf, Zr, Hf, Th and Pu into tributylphosphate (TBP) in benzene from aqueous acidic chloride solutions has been studied. The HCl, chloride and hydrogen ion concentrations were varied between 8 and 12 M. Extraction of all the elements studied increased as a function of HCl concentration. Studies of the extraction vs. chloride concentration at 8 M hydrogen ion concentration showed Rf to behave differently than Zr, Hf, and Th, and most like Pu. The Rf extraction with hydrogen ion concentration held constant at 12 M, peaked at 10 M chloride concentration, and decreased at 12 M chloride concentration. An increase in the hydrogen ion concentration increased the extraction of Rf, but had no noticeable effect on the extraction of Zr and Hf.

*This work was supported in part by the Director, Office of Energy Sciences, Chemical Sciences Division of the US Department of Energy under Contract DE-AC03-76SF00098.

75. DEPOSITO ELECTROLITICO DE SAMARIO Y PLUTONIO.Arturo Becerril(1), Yunny Meas(2), Jean Truffy(3). (1) Instituto Nacional de Investigaciones Nucleares, CMRI-LPR. A.P.116-006, 01141, México, D.F.(2) Universidad Autónoma Metropolitana-Iztapalapa. Area de Electroquí_mica. Apdo. Postal 55-534, 09340, México, D. F.(3) Centre d'Etudes Nucléaires de Saclay. DTA - DAMRI.B.P. 6,09430 Gif Sur Yvette Cedex, France.

La calidad espectroscópica y la estabilidad de las fuentes radiactivas de Plutonio depende del método empleado para su elaboración. El método más empleado es la obtención de películas delgadas electrodepositadas sobre discos metálicos. Los estudios previos y la optimización misma del proceso, se lleva_ron a cabo sustituyendo al Plutonio por el Samario. Este elemento tiene propiedades físicoquímicas

similares al Plutonio presentando la ventaja de no ser radiotóxico y estratégico. En este trabajo se presentan resultados experimentales que muestran la similitud del comportamiento de ambos elementos dentro del sistema electroquímico utilizado. Se estudio el deposito de Samario en función del pH, de la densidad de corriente, del tiempo de depósito y de la velocidad de rotación de cátodo.
Aplicando al Plutonio los parámetros experimentales optimizados obtenidos para el depósito de Samario, se obtuvieron fuentes con una calidad espectroscópica superior a la obtenida en los procesos reportados y utilizados hasta ahora.

76. 3-IODOBENZYLAMINE CONJUGATE OF 2-NITROIMIDAZOLE, A NEW POTENTIAL HYPOXIA IMAGING AGENT. A. Najafi, A. Sosa, M.M. Alauddin, M.E. Siegel. USC-LAC Division of Nuclear Medicine, Los Angeles, CA 90033.

Different analogs of 2-nitroimidazole, including misonidazole, have been reported to have a radiosensitizing effect and when radiolabeled, useful for imaging hypoxic tissue. Recently (J.Nucl.Med. 31:815, 1990), 4-fluorobenzylamine conjugate of 2-nitroimidazole was proposed suitable PET hypoxia imaging agent. We, therefore, synthesized the 3-iodo counterpart of this compound for use in planar imaging. 1-(2-nitroimidazol)-acetyl-3-iodobenzylamide (NAIB), was synthesized by first reacting 3-iodobenzylamine and bromoacetylbromide. The product was then reacted with 2-nitroimidazole. The final product was purified and identified by Mass Spectroscopy (MS), NMR, and Microanalysis which showed C37.26, H2.92, and N14.26%; calculated for $C_{12}H_{11}IN_4O_3$: C37.30, H2.85, and N14.51%. Radioiodine exchange failed using all reported methods. However, a modification of the solid state exchange method was highly successful, yielding more than 95% radioiodine exchange, even when a very small amount of cold precursor was used to obtain a high specific-activity product.

77. THE STATUS OF PIXE IN GEOCHEMISTRY. D. S. Burnett[1] and D. S. Woolum[2]. 1. Geology, Caltech, Pasadena, CA 91125.
2. Physics, Cal State Fullerton, CA 92634.

PIXE is only one of several in-situ microanalytical techniques available to geochemists, but it occupies a unique niche in terms of a combination of accuracy and sensitivity. PIXE is 1-2 orders of magnitude more sensitive than electron probe analysis and has comparable accuracy at present. In the future, the simpler PIXE X-ray production physics could allow an order of magnitude greater accuracy for major element analysis. Secondary ion mass spectrometry has greater sensitivity, but, although progress has recently been made, SIMS has no general scheme for correcting for matrix effects on ion yields. In contrast, the conversion of observed X-ray counts to concentrations for PIXE is based on well understood physics. Synchrotron X-ray fluorescence has comparable accuracy and perhaps slightly better sensitivity than PIXE, but has much worse depth resolution. The excellent signal to noise inherent in PIXE could be fully realized with the use of efficient crystal spectrometers instead of the Si(Li) detectors utilized in essentially all studies to date.

78. PIXE AS AN ANALYTICAL/EDUCATIONAL TOOL, Evan T. Williams, Department of Chemistry, Brooklyn College of CUNY, Brooklyn, N.Y. 11210.

The advantages and disadvantages of PIXE as an analytical method will be summarized. We will also discuss the advantages of PIXE as a means of providing interesting and feasible research projects for undergraduate science students.

79. AIR SAMPLERS AND COMPLEMENTARY ANALYSES FOR PIXE. Bruno Jensen, Bob Leonard and J. William Nelson ; Physics Department, Florida State University, Tallahassee FL 32306 and PIXE International Corporation, 507 S. Woodward Ave., Tallahassee, FL 32316

Analyses of micrograms of matter in minutes of time make PIXE an ideal method for air particulate elemental determinations. Particle size selective and time sequential air samplers especially suited for ion beam analyses are described as developed at the Florida State University and the PIXE International Corporation. For elements not readily accessible by PIXE, complementary analysis methods such as proton scattering and optical densitometry are also reviewed.

80. STATISTICAL SUMMARY OF PIXE PROGRAMS 1970 - 1980 - 1990. Thomas A. Cahill, Roberto Morales, Javier Miranda, Crocker Nuclear Laboratory, University of California, Davis, CA 95616-8569.

A survey has been made of all programs using particle-induced x-ray emission (PIXE) for the period 1981-1990. This survey extends an earlier survey that covered all PIXE programs through 1980.[3] In that survey, approximately 80 PIXE programs were identified using standard millimeter-sized ion beams, while 20 programs were using PIXE with micrometer-sized ion beams. Since that survey, there has been a sizeable increase in the use of PIXE with proton microprobes. Other types of PIXE programs, however, have become larger in size but, at this time, fewer in number. Results will be presented on the nature of the programs and current applications.

[1] Facultad de Ciencias, Universidad de Chile
[2] UNAM, Mexico City, D.F.
[3] T.A. Cahill, "Proton microprobes and particle-induced x-ray analytical systems", Ann. Reviews Nuclear and Particle Science 30:211-252 (1980).

81. APPLICATIONS OF PIXE IN DENDROCHRONOLOGY. P. Revesz, J.W. Mayer, G. Vizkelethy, Bard Hall, Cornell; P.I. Kuniholm, C.B. Griggs, H.E. Kuniholm, M.W. Newton, S.L. Tarter, Goldwin Smith Hall, Cornell University, Ithaca, NY 14853.

The Bard Hall Ion Beam FAcility has developed external proton beams at 1-3 MeV for ion-induced x-ray emission studies of artists' pigments and metallic printers inks. We have extended these PIXE measurements to analysis of Athenian Tetradrachms (5th and 6th centuries BC). In both pigment and coin studies we have compared PIXE analyses with those from XRF and electron micro probe for calibration purposes. The Aegean Dendrochronology Project has developed over 5500 years of tree-ring chronologies for the Aegean, Balkans, and Eastern Mediterranean. Included are forest sites adjacent to mining and smelting operations, also sites at the upper timberline where the annual growth-rings might be expected to have picked up fallout from volcanic activity around the world. A preliminary examination of selected rings by PIXE has shown remarkable differences from year to year in the deposition of various elements, notable Zn, Fe, and Cu. We propose to look at a number of selected annual ring sequences to see a) whether we can chart the progress of known metallurgical activity; b) whether we can find evidence for volcanic effects, both for absolutely-dated volcanoes such as Krakatoa, Tambora, and Sumbawa, and for problematic volcanoes such as Santorini/Thera.

82. INITIAL ABUNDANCES OF URANIUM-235 IN METEORITES AND LUNAR SAMPLES. P. K. Kuroda and William A. Myers, University of Arkansas, Fayetteville, Arkansas 72701

In 1956, Patterson reported that, within experimental error, meteorites had one age as determined by three independent methods and the most accurate method ($^{207}Pb/^{206}Pb$) gave an age of $(4.55 \pm 0.07) \times 10^9$ years. Re-examination of a vast amount of lead and xenon isotope data which have been accumulated since the Apollo 11 landing on the moon in July 1969 reveals, however, that some of the lunar fines and breccias may have started to retain their radiogenic lead and fissiogenic xenon isotopes 4,900 to 5,000 million years ago, when the ratios of ^{235}U and ^{244}Pu to ^{238}U in the early solar system were about 4 and 2 atoms per 10 atoms of ^{238}U, respectively.

83. INITIAL ABUNDANCES OF PLUTONIUM-244 IN METEORITES AND LUNAR SAMPLES. William A. Myers and P. K. Kuroda, University of Arkansas, Fayetteville, AR 72701

Over 3,500 mass-spectrometric analyses of xenon released from bulk samples, temperature fractions and acid residues of meteorites and lunar samples, which have been carried out in various laboratories since the 1960's, are being re-examined by the use of a computer method instead of conventional graphical methods. The results obtained so far show that carbonaceous chondrites contain 22 to 34 x 10^{-12} ccSTP ^{136}Xe per gram of ^{244}Pu fission xenon and they appear to have started to retain their xenon more than 4,800 million years ago when the ratio of ^{244}Pu to ^{238}U in the solar system was as high as 0.170 ± 0.010 (atom/atom). Some of the lunar samples contain as much as 1000 x 10^{-12}ccSTP ^{136}Xe per gram and the ^{244}Pu ages of lunar samples vary from 3,965 to 4,909 million years.

84. THORIUM-230 CHRONOLOGY OF NATURAL WATERS AT THE NEVADA TEST SITE. S. N. Bakhtiar, Health Physics Department, Reynolds Electrical and Engineering Company, P. O. Box 98521, Las Vegas, Nevada 89193-8521.

Carbon-14 dating was used in the past to estimate the ages of ground water from the Paleozoic aquifer underlying the Nevada Test Site (NTS) and adjacent areas. Grove et al. reported that a generally good correlation existed between the carbon-14 dates and the hydraulic interpretations. The so-called "adjusted" carbon-14 ages of ten water samples taken from wells and springs at the NTS and adjacent areas vary from -1600 to 23,400 years. It seems, however, the aquifer is heterogeneous and is insufficiently defined to permit detailed predictions of ground water velocities or flow patterns. An alternative dating method measured the concentrations of Th-230, Th-232, U-234, and U-238 in several water samples from the wells and springs, and calculated the Th-230 ages. (Grove, D. B., M. Rubin, B. B. Hanshaw, W. A. Beetem, U. S. Geol. Survey Prof. Paper 650-C, C215-C218, 1969).

85. LOW-LEVEL DETERMINATION OF STRONTIUM-90 IN A HIGH-SALT ENVIRONMENT AT THE NEVADA TEST SITE. R. K. Guimon, Reynolds Electrical and Engineering Co. Inc., P. O. Box 98521, Las Vegas, Nevada 89193-8521.

Concentration of Sr-90 in water samples taken from wells, reservoirs, and natural springs are generally far below the detection limit of 300 fCi per liter when one-liter samples and routine Nevada Test Site (NTS) analytical procedures are used. By taking 10 and 100 liter samples respectively, the detection limit can be lowered to 30 fCi and 3 fCi per liter. Concentrations of Sr-89 and Sr-90 in rainwater can be as low as 10 fCi per liter when taking a 10-liter sample of rain. (Guimon et al. 1985) In the case of NTS natural water with a high-salt concentration, it is difficult isolating extremely low activities of Sr-90 from large sample volumes. We have recently modified the procedures used by Guimon et al. 1985, and Burchfield et al. 1983, for the determination of Sr-90. A carbonate metathesis was discontinued and a carbonate fusion was substituted at this point in the procedure. Samples spiked with .333 dpm/20 liters were analyzed and their activities determined. These spiked samples had a 16 to 22 percent absolute error. (Guimon R. K., Z. Z. Sheng, L. A. Burchfield, and P. K. Kuroda. 1985. Geochem. Journal 19:229-235.) (Burchfield L. A. and P. K. Kuroda. 1983. Geochem. Journal 17:63-671.)

86. A PLASMA REACTOR DESIGN. Juan G. Lartigue, Guadalupe R. Cisneros. Facultad de Química, UNAM, Ciudad Universitaria, DF, 04510, México.

This paper presents a plasma reactor design developed for the Faculty of Chemistry of the National University with the collaboration of the Plasma Physics Institute of Checoslovaquia. Materials and dimensions of the reactor are specified as well as the associated equipment required and the double probe technique used for plasma diagnosis. Some considerations about high temperature reaction kinetics are also included.

87. CASCADAS RETICULARES.
García C. Miguel Angel, Turnbull A. Roberto y Palma G. Federico. Instituto Nacional de Investigaciones Nucleares, Sierra Mojada 447, C.P.11010,México,D.F. México.

Al experimentar recientemente la concentración electrolítica intermitente de agua pesada, procesando repetidas veces volúmenes de agua con una relación constante de concentración, tanto para efectuar concentraciones sucesivas como para acumular soluciones intermedias, se visualizó la necesidad de desarrollar un procedimiento de cálculo para una cascada cuyo diagrama de flujo tiene una apariencia reticular. Ello ocurre porque par-- tiendo de una o varias cascadas, los productos pasan a otras etapas para reconcentrarse y las colas a cascadas adyacentes para recuperar el deuterio contenido en los gases -- oxhídricos producidos. En estas condiciones las consideramos mono o poli alimentadas. Idealmente la concentración de etapa a etapa aumenta según un factor constante, de modo que el cálculo de las concentraciones se facilita, no así el de las cantidades a procesar. Para esto último se obtuvieron ecuaciones y matrices numéricas para calcular pro-- ducto, alimentación y residuo de cualquier etapa y cascada de la red.

88. FRACTAL THEORY OF RADON EMANATION FROM SOLIDS. Thomas M. Semkow, Wadsworth Center for Laboratories and Research, New York State Department of Health, Albany, NY 12201-0509, and State University of New York at Albany.

We developed a fractal theory of Rn emanation from solids, based on α recoil from the α decay of Ra. Range straggling of the recoiling Rn atoms in the solid state is included and the fractal geometry is used to describe the roughness of the emanating surface. A fractal dimension D of the surface and the median projected range become important parameters in calculating the radon emanating power E_R from solids. A relation between E_R and the specific surface area measured by the gas adsorption is derived for the first time, assuming a uniform distribution of the precursor Ra throughout the samples. It is suggested that the E_R measurements can be used to determine D of the surfaces on the scale from tens to hundreds of nm. One obtains, for instance, $D = 2.17 \pm 0.06$ for Lipari volcanic glass and $D = 2.83 \pm 0.03$ for pitchblende. In addition, we suggest a new process of penetrating recoil and modify the role of indirect recoil. The penetrating recoil may be important for rough surfaces, in which case Rn loses its kinetic energy by penetrating a large number of small surface irregularities. The indirect recoil may be important at the very last stage of energy-loss process, for kinetic energies below ~ 5 keV.

89. A SIMPLE COMPUTER PROGRAM TO FIT EXPERIMENTAL RESULTS.
ARRIOLA S. HUMBERTO. VARGAS CH. VICTOR
FACULTAD DE QUIMICA, DEPg., UNAM, C.U. MEXICO 04510, D.F.

In this work we describe a program that fits typical experimental results found in nuclear sciences, which can be also used in other areas. It fits given funtions such as "linear" correlations, (linear, semi-log, log-log, etc). Gaussian and Lorentzian functions with the possibility of deconvolving up to three of those curves. It uses an iterative, interactive least-squares method. Examples are given in each case including X-Ray Fluorescence, Mossbauer Spectroscopy and finally a general method to fit any type of function is discussed.

90. MIGRATION OF ACTIVATION PRODUCTS IN DISCHARGE POOL IN T.H.O.R. NUCLEAR RESEARCH REACTOR SITE BOUNDARY C. Chung & C. C. Chan, Nuclear Science and Technology Development Center, National Tsing Hua University, Hsinchu 30043, Taiwan R.O.C.

Due to material degradation in tubing, the heat exchanger for the primary coolant of the 1 MW Tsing Hua Open-pool Reactor(THOR) has minor leakage to the secondary cooling system since 1970. In the past twenty years, trace amount of Co-58,60, Cr-51, Cs-137, Mn-54, Sc-46, and Zn-65 have been leaked through the cooling system, accumulated and trapped eventually in the discharge pool right in front of the THOR facility. The distribution of these activation products in mud at different depths and various locations in the pool was measured with standard procedures of radioactive soil sampling and counting techniques. Concentration of activation products, with no more than 40 kBq/kg.dry at the hottest spot, was contour-mapped to reveal the migration of these trace level radioactive products in a period of 20 years.

91. MOLECULAR DYNAMICS COMPUTER SIMULATION STUDIES OF COLLISION CASCADES IN SOLIDS.*
T. Diaz de la Rubia and M. W. Guinan, University of California, Lawrence Livermore National Laboratory, L-644, Livermore, CA 94550.

We review recent molecular dynamics simulation studies of energetic displacement cascades in metals. These simulation studies have revealed details of the atomic structure and dynamics of cascades. Notably, we have shown that melting takes place at the core of displacement cascades in metals. We demonstrate how these ideas provide a simple model with which to explain many radiation effects in metals. We show that recent studies at subcascade formation energies provide evidence for a new fundamental mechanism of defect formation in metals and present results of the formation of interstitial and vacancy dislocation loops as a direct result of 25 keV cascades in Cu. We relate the simulation results to experimental and phenomenological models of ion beam mixing and defect production in metals and discuss the dependence of these processes on energy density.

*Work performed by LLNL under U.S. DoE Contract No. W-7405-Eng-48.

92. THERMODYNAMIC PARALLELS BETWEEN SOLID-STATE AMORPHIZATION AND MELTING.
P.R. Okamoto, Argonne National Laboratory, Argonne, IL 60439

A thermodynamic approach which views melting and amorphization as an isobaric and an isothermal expansion process, respectively, is discussed. The approach is based on experimental observations and molecular dynamics simulations of melting and amorphization which show that volume expansion is a generic response of most crystalline materials to heating at constant pressure and to structural disordering at fixed temperatures. A systematic application of this point of view to melting and amorphization leads to an extended temperature-volume phase diagram which takes into account the volume dependence of the elastic constants and the effects of atomic mobility on the kinetics of crystal-to-glass transitions. The extended phase diagram brings out clearly the parallels between the two phenomena and suggests that, in principle, crystalline materials can undergo two different types of crystal-to-glass transitions which are the isothermal analogues of thermodynamic and mechanical melting. Experimental observations and molecular dynamic simulations which support the thermodynamic approached adopted here will be presented.
Work supported by the U.S. DOE, BES-Materials Science, Contract W-31-109-Eng-38.

93. TRIBOLOGICAL PROPERTIES OF STRUCTURES FORMED BY ION IRRADIATION. Michael Nastasi, Materials Science and Technology Division, Los Alamos National Laboratory, Los Alamos, NM 87545.

The tribological response of two surfaces in sliding contact will be dependent on the chemical and mechanical interactions occurring at the sliding interface. Ion irradiation techniques provide a mechanism for altering the surface sensitive tribological properties of a material without affecting bulk properties. Techniques such as ion implantation, ion beam mixing and ion beam assisted deposition can taylor surface composition while producing a variety of microstructures including amorphous, nanocrystalline and equilibrium phases.

This talk will focus on recent experiments which have utilized ion beam surface modification techniques to improve tribological properties.

94. ION BEAM INDUCED INTERFACIAL CHEMICAL COMPLEXES AND ADHESION. A. A. Galuska, Division 1823, Sandia National Laboratory, POB 5800, Albuquerque, NM 87185

In recent years, ion bombardment during and after deposition has been shown to produce large increases in thin film adhesion. However, the critical mechanisms for this adhesion enhancement have remained unclear. In this paper, evidence will be presented which indicates that interfacial chemistry (complex formation) is the most critical factor in ion-induced adhesion enhancements. Various methods of forming adhesive interfacial complexes will be discussed. Particular emphasis will be placed on the use of reactive ion beams to form adhesive interfacial complexes in nonreactive film/substrate systems. Furthermore, thermodynamic criteria for choosing a reactive ion for a given application will be discussed.

*This work was performed at Sandia National Laboratories, supported by the U. S. Department of Energy under contract # DE-AC04-76-DP00789.

ORGANIC CHEMISTRY

Fourth Chemical Congress of North America

New York, NY
August 25 - 30, 1991

F. A. Davis,
Program Chairman

SUNDAY EVENING AND MONDAY MORNING AND AFTERNOON
- **Poster Session**

F. A. Davis, Presiding   Papers 1-119

MONDAY MORNING AND AFTERNOON
- **Plenary Session**

W. P. Dailey III, Presiding   Papers 120-132

TUESDAY MORNING AND AFTERNOON
- **Cope Award Symposium**

J. A. Marshall, Presiding   Papers 133-143

WEDNESDAY MORNING AND AFTERNOON
- **Poster Session**

F. A. Davis, Presiding   Papers 144-264

WEDNESDAY MORNING AND AFTERNOON
- **Plenary Session**

T. S. Widlanski, J. H. Rigby, Presiding   Papers 265-278

THURSDAY MORNING AND AFTERNOON
- **Plenary Session**

K. Steliou, G. Delgado, Presiding   Papers 279-292

FRIDAY MORNING
- **Plenary Session**

William F. Berkowitz, Presiding   Papers 293-298

ORGANIC CHEMISTRY

1. EFFICIENT TOTAL SYNTHESIS OF THE CYTOTOXIC HALOGENATED MONOTERPENE APLYSIAPYRANOID D Michael E. Jung and Willard Lew, Department of Chemistry, University of California, Los Angeles, CA 90024.

The efficient total synthesis of the cytotoxic halogenated monoterpene Aplysiapyranoid D **1d** will be described. In particular, the rationale for the cyclization of the chloroolefinic alcohol **2** to give predominantly the desired tetrahydropyran **3** rather than the corresponding tetrahydrofuran will be discussed. Also the preparation of various analogues of **1d** will be described.

2. STUDIES TOWARD THE TOTAL SYNTHESIS OF GINKGOLIDE B.
Michael T. Crimmins*, Philippe G. Nantermet, James B. Thomas, Department of Chemistry, The University of North Carolina, Chapel Hill, North Carolina 27599-3290.

Progress toward the total synthesis of Ginkgolide B will be presented. The cyclopentenone **3** is formed using a zinc homoenolate conjugate addition-cyclization on the suitable acetylenic ester. Conversion of **3** to the intermediate **2** is accomplished by the intramolecular photocycloaddition of the enone-furan system. Elaboration of **2** to Ginkgolide B is in progress.

3. SYNTHESIS AND ABSOLUTE CONFIGURATION OF POLYPROPIONATE METABOLITES OF SIPHONARIA AUSTRALIS
Uma N. Sundram and Kim F. Albizati,* Dept. of Chemistry, Wayne State University, Detroit, MI

The synthesis of **1** via thermodynamic equilibration has been studied under a variety of conditions. The absolute configuration of **1** and **2** have been determined by an unambiguous synthesis of **2** coupled with the use of the exciton chirality method for the determination of the absolute configuration of acyclic allylic alcohols.

4. **SYNTHETIC STUDIES ON ACTINOBOLIN.** Dale E. Ward and Brian F. Kaller, University of Saskatchewan, Saskatoon, Saskatchewan, CANADA, S7N 0W0.

Actinobolin and Bactobolin are broad spectrum antibiotics which also display significant antitumor activity. The synthesis of these compounds has attracted considerable interest recently. Synthetic studies based on the following approach will be presented.

actinobolin: R_1=Me, R_2=H
bactobolin: R_1=CHCl$_2$, R_2=Me

5. TOWARDS THE SYNTHESIS OF GERMINE. Keith Jones, Roger Newton, and Christopher Yarnold, Department of Chemistry, King's College London, Strand, London WC2R 2LS, U.K.

A synthetic route to a model of the DEF rings of the alkaloid germine will be described which uses an intramolecular [3+2] nitrile oxide cycloaddition as the key step via the oxidation of oxime [1]. We have used this cycloaddition reaction to prepare other nitrogen containing cyclic systems.

6. **A TOTAL SYNTHESIS OF (±)-β-CHAMIGRENE.** Julian Adams, Carole Lepine-Frenette and Denice M. Spero. Boehringer Ingelheim Pharmaceuticals Inc., 90 East Ridge, Ridgefield, CT 06877 and BioMega Inc., Laval, Quebec, Canada H7S2G7

The transannular cyclopropanation of a ketocarbene, generated by Rh$_2$(OAc)$_4$ catalysis on a bicyclic dihydropyran (3) nucleus provided a key oxa-tetracyclic ketone (2) intermediate for the synthesis of the [6,6]spirocyclic ring construction. Selective fragmentation of the cyclopropane followed by hydrolytic cleavage of the C-O bond provided the spirocyclic skeleton. Functional group manipulations led to a total synthesis of (±)-β-chamigrene(1).

7. **THE FIRST TOTAL SYNTHESIS OF AMARYLLIDACEAE ALKALOIDS OF THE 5,11-METHANOMORPHANTHRIDINE TYPE. AN EFFICIENT TOTAL SYNTHESIS OF (±)-PANCRACINE.** Larry E. Overman and Jaechul Shim, Department of Chemistry, University of California, Irvine, CA 92717.

Tandem aza-Cope rearrangement - Mannich cyclization (**1** → **2**) is the central step in a concise total synthesis of (±)-pancracine.

8. **STUDIES DIRECTED TOWARD THE TOTAL SYNTHESIS OF ISOLAUREPINNACIN.** Larry E. Overman and Dan M. Berger, Department of Chemistry, University of California, Irvine, California 92717

The Lewis acid catalyzed cyclization of mixed acetals **1** produces 2,3,6,7-tetrahydrooxepins **2** containing halogen substituents on both alkyl side chains. The scope of this chemistry, and its application to the synthesis of isolaurepinnacin, a marine natural product, will be described.

9. **Total Asymmetric Synthesis of 15,20-Dihydrosecodine, Tetrahydropresecamine and Tetrahydrosecamine.**
William G. Bornmann and Martin E. Kuehne
Dept. of Chemistry, University of Vermont, Burlington, VT. 05405
The total asymmetric synthesis of 15,20 Dihydrosecodine,Tetrahydropresecamine (**1**) and Tetrahydrosecamine (**2**) has been achieved in 13 steps with an overall yield for the **S** stereoisomers of 14.85% while that of the **R** stereoisomers was 15.33%.

10. **REGIOSELECTIVE ADDITION OF NITRILE-STABILIZED CARBANIONS TO 4-AMINO-SUBSTITUTED INDOLE CHROMIUM COMPLEXES AND THE TOTAL SYNTHESIS OF (-)-INDOLACTAM V.** M. F. Semmelhack and Hakjune Rhee, Department of Chemistry, Princeton University, Princeton, NJ 08544

Nucleophilic addition of stabilized carbanions occurs regioselectively at the C-7 position of 4-amino-substituted indole chromium tricarbonyl complexes. The synthesis of (-)-indolactam V was accomplished via enantiospecific S_N2 displacement of a chiral triflate with substrate **1** and asymmetric coupling with a chiral bislactim ether **2**. The total synthesis of teleocidins A-1, A-2 should be facilitated by this methodology.

11. **STEREOSELECTIVE S_N2' ADDITION OF ORGANOCUPRATE REAGENTS TO ALKYNYL OXIRANES: SYNTHESIS OF A PRECURSOR OF (+)-VERRUCOSIDIN.** James A. Marshall and Kevin G. Pinney, Dept. of Chemistry, University of South Carolina, Columbia, S.C. 29208.

The addition of organocuprate reagents to alkynyl oxiranes **1** proceeds in high yield to afford allenols **2** with good diastereoselectivity (>9:1). The reaction is general. A variety of functionalized alkynyl oxiranes have been converted to their corresponding allenols which readily cyclize to dihydrofurans such as **3** upon treatment with AgNO3. This methodology has been successfully employed in an efficient synthesis of the highly substituted tetrahydrofuran core unit of (+)-Verrucosidin.

12. **A PHOTOCHEMICAL APPROACH TO TAXOL.** W.F. Berkowitz, P. Hamilton, Y.Y. Jiang, P.J. Wilson, Department of Chemistry, Queens College (CUNY), Flushing, N.Y. 11367.

Our model studies have demonstrated a successful photochemical (de Mayo) approach to the synthesis of the Taxane skeleton. Application to the synthesis of Taxol will require synthon <u>1</u>. We have prepared ketoester <u>2</u>, by cyclization of geraniol derivatives, to use as a convenient starting point.

13. THE CHEMISTRY OF SOME UNUSUAL RING-OPENED RIFAMYCIN DERIVATIVES. Benjamin B. Mugrage, Clive G. Diefenbacher, Michael D. Grim, and Xiaobei Yang, Research Department, Ciba-Geigy Pharmaceuticals Division, Summit, NJ, 07901.

The chemistry associated with an unusual ring-opened rifamycin derivative will be described. The known open-chain compound **1**, available in four steps from the antibiotic rifamycin S, was used as a common intermediate in the synthesis of ester and amide analogs. Variations of the oxazole substituent and of the piperazine substituent will be detailed. Additionally, the acid degradation products of **1** and its analogs will be discussed.

14. CONVERSION OF LACTONES INTO SUBSTITUTED CYCLIC ETHERS. K.Tsushima, K.Araki, and A.Murai*, Department of Chemistry, Faculty of Science, Hokkaido University, Sapporo, 060 JAPAN

We describe the results of a new synthetic strategy aimed to medium-sized cyclic ethers involving conversion of lactones into substituted cyclic ethers via enol triflates.

Six to eight-membered lactones could efficiently be converted to the corresponding enol triflates, which were smoothly coupled with various lithium dialkylcuprates to give the corresponding alkylated cyclic enol ethers. This strategy was effectively applied to the synthesis of (+)-lauthisan.

15. THE CONSTRUCTION OF FK-506 ANALOGS FROM RAPAMYCIN-DERIVED MATERIALS
M.T. Goulet and D.W. Hodkey, Department of Basic Medicinal Chemistry
Merck Sharp and Dohme Research Laboratories, P.O. Box 2000, Rahway, NJ 07065.

The degradation of rapamycin, a tricarbonyl containing macrolide similar in structure to the potent immunosuppressant FK-506 is described. Modification of the fragments derived from this process is discussed, including reassembly to form a macrocyclic analog of FK-506.

16. AVERMECTIN ANALOGS WITH A SPACER BETWEEN THE AGLYCONE AND THE DISACCHARIDE
T. Blizzard, G. Margiatto, B. Linn, H. Mrozik, and M. Fisher
Merck Sharp & Dohme Research Laboratories R50G-231 P.O.Box 2000 Rahway NJ 07065

Conversion of ivermectin (1) to spacer containing analogs (2) will be described.
Biological activity of the novel derivatives will be discussed briefly.

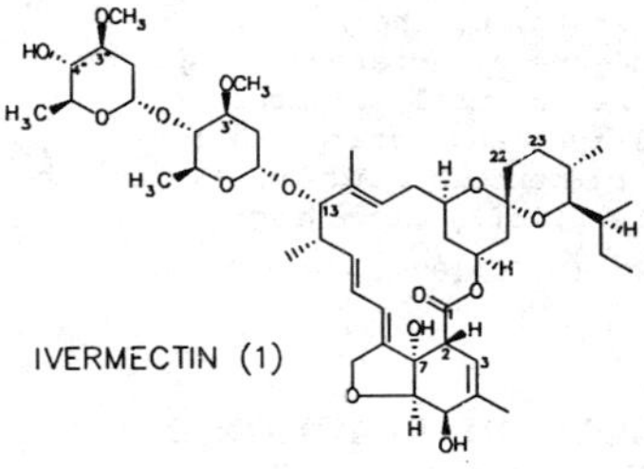

IVERMECTIN (1)

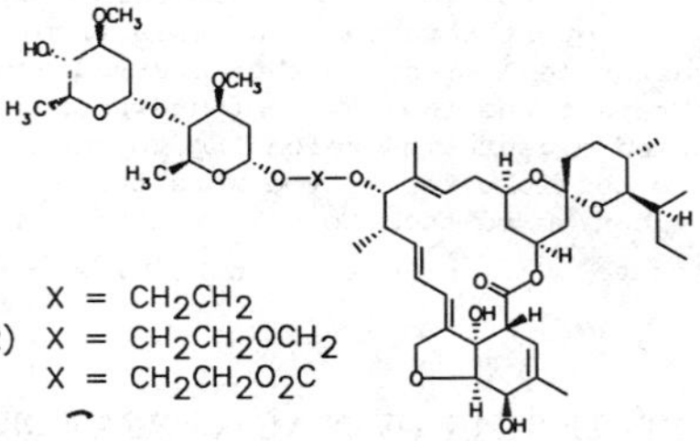

(2) X = CH_2CH_2
X = $CH_2CH_2OCH_2$
X = $CH_2CH_2O_2C$

17. APPROACHES TO THE SYNTHESIS OF (±) PEPEROMIN C. Raymundo Cruz-Almanza and
Fernando Padilla-Higareda. Instituto de Química, Universidad Nacional Autónoma
de México, Circuito Exterior, Ciudad Universitaria, Coyoacán 04510, México,
D.F., México.

Peperomin C 1, a novel lignan having unusual seco structure was isolated along with
peperomins A and B from Peperomia Japonica Makino (Het., 29, 411 (1989). The aqueous
and alcoholic decoctions of the hole herbs are used as a folk medicine for the treatment of malignant tumors. In the present work we describe our efforts to the synthesis
of peperomin C. Our approach is mainly based on the stereoselective Michael addition
of carbanion A to lactone 2. A new procedure for the preparation of lactone 2 is also
reported.

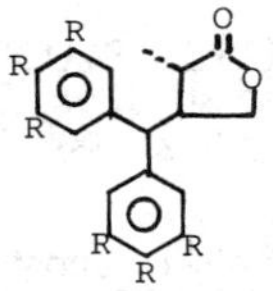

R = OMe

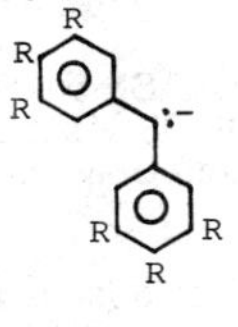

18. ACID INDUCED REARRANGEMENTS OF THE NATURAL MELAMPOLIDE SCHKUHRIOLIDE. AN
ALTERNATIVE APPROACH TO THE OPLOPANE SKELETON. Salvador Guzmán, Rubén A. Toscano
and Guillermo Delgado.* Instituto de Química de la Universidad Nacional Autónoma
de México, Cd. Universitaria, Circuito Exterior, Coyoacán 04510, México, D. F.

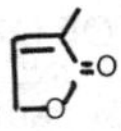

Acid catalyzed reaction of the natural melampolide 1 provided compounds 2-5. Additional
experimental data indicate the sequence 1 → 2 + 3; 3 → 4 → 5 for the transformation. The
preparation of the oplopanolide 5 from the germacradiene 4 suggests this possible
alternative biogenetic route for oplopanoid natural products.

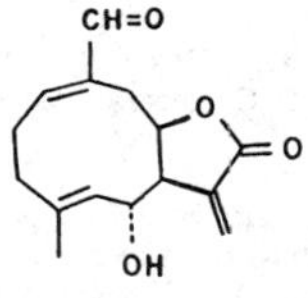

1 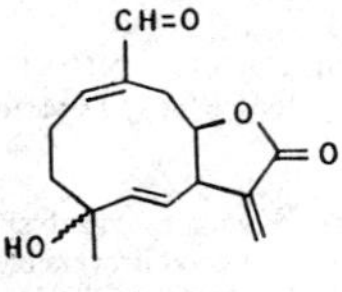   2 β-OH   3 α-OH 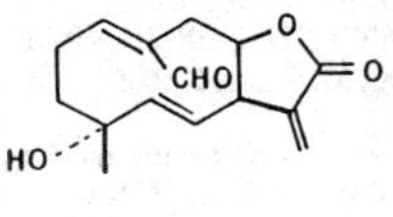   4 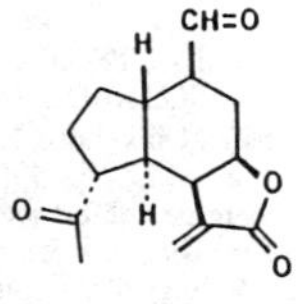   5

19. CHEMICAL TRANSFORMATIONS OF TRITERPENES FROM GUAYULE. Martínez R., Arellano G. Rosario, Espinosa P. Georgina*and Martínez V. Mariano. Instituto de Química, Universidad Nacional Autónoma de México. Circuito Exterior, Ciudad Universitaria Coyoacán, 04510 México, D. F. *FES-Cuautitlán, UNAM. Sec.Quím.Orgánica. Campo 1.

In continuation of our synthetic work dealing with chemical transformations of the triterpenes from guayule to obtain heterocyclic compounds with potential biological activities. We wish to report here the synthesis from its enol acetate, of the 2-1, 16,24-diacetate argentatine A which is the intermediate to obtain several heterocyclic products. Also we report the isolation of unexpected products when natural mixtures of triterpenes named argentatins belonging to the cycloartenol-type and isoargentatins which belong to the lanosterol-type were treated with m-Cl perbenzoic acid in alkaline medium. The mechanism related to this transformation will be briefly discussed.

Acknowledgments to DGAPA,UNAM for financial suport (Proyect IN-202189).

20. **SOLID STATE NMR OF N-ACYLUREAS DERIVED FROM THE REACTION OF HYALURONIC ACID WITH ISOTOPICALLY-LABELED CARBODIIMIDES.**
Tara Pouyani, Gerard S. Harbison, and Glenn D. Prestwich, Department of Chemistry, State University of New York, Stony Brook, New York 11794-3400

Hyaluronic acid (HA) is a naturally-occurring linear polysaccharide consisting of alternating D-glucuronic acid and N-acetyl-D-glucosamine residues. Reaction of the carboxyl group of the glucuronate residues with 1-ethyl-3-(3-dimethyl-aminopropyl)carbodiimide (EDC) in the presence of primary amines yielded only the N-acylurea adducts rather than the expected amide coupling products. To determine the nature of this linkage unambiguously, and to deduce the primary structure of the N-acylurea products, ^{13}C-and ^{15}N-labeled EDC were synthesized. The isotopically-labeled carbodiimides coupled to the carboxyl group of HA (molecular weight ca. 2,000,000 daltons) in water at pH = 4.75, and the modified polysaccharides were isolated, purified, and examined by CP-MAS solid state ^{13}C and ^{15}N NMR. The bond orders and chemical shifts confirmed the presence of two isomeric N-acylureas in unequal amounts, and ruled out the presence of any unrearranged O-acylurea product.

21. **INHIBITION OF OXIDOSQUALENE CYCLASE BY PSEUDOSUBSTRATES: STUDIES OF 26- AND 29-FUNCTIONALIZED 2,3-OXIDOSQUALENES.** Xiao-yi Xiao and Glenn D. Prestwich*, Department of Chemistry, State University of New York at Stony Brook, Stony Brook, New York 11794-3400

Abstract: Conversion of squalene to (3S)-2,3-oxidosqualene and subsequently to lanosterol by squalene epoxidase (SE, EC 1.14.99.7) and oxidosqualene cyclase (OSC, EC 5.4.99.7.) are two key steps in the biosynthesis pathway of cholesterol. As potential mechanism-based inhibitors to OSC, 26- and 29-hydroxy-, oxo-, methylidene- and 1',2'-ethanediyl-2,3-oxidosqualene (and the respective 2,3,22,23-bisoxidosqualenes) were synthesized by an efficient route starting from squalene. The potency of these analogs acting as substrates and inhibitors for OSC from pig-liver microsomes or baker's yeast suspension was investigated.

22. ACTIVE-SITE-DIRECTED ENZYME-ACTIVATED-SUBSTRATE INHIBITION OF TRYPSIN BY N-NITROSOAMIDE DERIVATIVES. Emil H. White and Yulong Chen, Department of Chemistry, The Johns Hopkins University, Baltimore, Maryland 21218

N-Nitrosoamide inhibitors of trypsin based on the mechanism of enzymatic hydrolysis of substrates and the primary specificity of trypsin were designed and synthesized. N-Nitrosoamide I was a good irreversible inhibitor of trypsin; it had little effect on the activity of α-chymotrypsin. A molar ratio of the inhibitor I to trypsin(5.2 : 1) resulted in 50% of inhibition of trypsin with reference to the tryptic hydrolysis of p-toluenesulfonyl-L-arginine methyl ester. The extent of inhibition was decreased in the presence of the competitive inhibitor, benzamidine, indicating that the nitrosoamide inhibitor was binding to the active site of trypsin. Preliminary studies showed that the D-enantiomer produced higher inhibition than the racemate. These inhibitors generate highly reactive species in the active sites of enzymes, which then alkylate a variety of residues. The inhibition mechanism is probably similar to that proposed for the inhibition of α-chymotrypsin by deaminatively produced carbonium ions (White, E.H. et al. *J. Am. Chem. Soc.* 1981, *103*, 4231-4239).

23. SYNTHESIS OF SERINE-AMC-CARBAMATE, A NOVEL FLUOROGENIC TRYPTOPHANASE SUBSTRATE. C. Preston Linn, Patrick D. Mize, Randal A. Hoke, J. Michael Quante, and J. Bruce Pitner, Becton Dickinson Research Center, Research Triangle Park, NC 27709.

The title compound, **1**, releases the fluorescent moiety 7-amino-4-methylcoumarin (AMC) on reaction with tryptophanase in the presence of the cofactor pyridoxal-5'-phosphate. Coupling of the lithium salt of N-BPOC-serine (**2**) with 7-isocyanato-4-methylcoumarin (**3**) was the key step in the preparation of **1**. This synthesis and the use of **1** in a tryptophanase assay will be presented.

24. **SYNTHESIS OF MACROCYCLIC BIFUNCTIONAL CHELATING AGENTS**
Thomas J. McMurry, Martin Brechbiel, Krishan Kumar, and Otto Gansow
National Institutes of Health, Radiation Oncology Branch, Bethesda, Maryland 20892

A convenient synthesis of 4-nitrobenzyl substituted macrocyclic tetra- amines and their conversion to bifunctional polyaminocarboxylate chelates is described. Cyclization of 4-nitrobenzylethylene diamine with appropriate BOC protected amino di-N-hydroxysuccinimide esters in dioxane at 90°C resulted in the formation of 12 and 14 membered ring diamides in 41, and 44% yield, respectively. A 12 membered macrocyclic triamide was also prepared in 44% yield by cyclization of N-2-aminoethyl-4-nitrophenylalanine amide with disuccinimidyl-N-tert-butoxycarbonyl iminodiacetate. Deprotection (HCl/dioxane) and reduction with borane gave the substituted macrocyclic amines which were then alkylated with either bromoacetic acid or t-butyl bromo acetate. The preparation of the isothiocyanate derivatives and C-14 labeled chelating agents are described. Attempts to prepare of a 9 membered macrocyclic diamide using this cyclization technique instead resulted in a 20% yield of a 10:1 mixture of isomeric fused 5,6 ring acylamidines. Deprotection (HCl/dioxane) and reduction with borane gave a substituted piperazine derivative in 55% yield.

25. STERIC COURSE OF THE METHYL COENZYME M REDUCTASE REACTION IN *METHANOBACTERIUM THERMOAUTOTROPHICUM*. Y. Ahn, J. Krzycki and H.G. Floss, Department of Chemistry, BG-10, Univesity of Washington, Seattle, WA 98195, and Department of Microbiology, The Ohio State University, Columbus, OH 43210

The enzyme, methyl coenzyme M reductase ("methyl reductase"), catalyzes the conversion of methyl coenzyme M into methane and coenzyme M. This reaction is the last step in the reduction of various one-carbon substrates, e.g., CO_2, methanol, methylamine or acetate, to methane by methanogenic bacteria. Methyl reductase represents a very complex enzyme system consisting of several proteins and involving a number of unique cofactors; its detailed mode of operation is still not fully understood. To help shed light on this intriguing process we determined the steric course of this reduction, using as substrate R- and S-[1-^{2}H$_1$,^{3}H]ethyl-CoM. Anaerobic incubation with a cell-free extract of *Methanobacterium thermoautotrophicum* gave [1-^{2}H$_1$,^{3}H]ethane at a rate of about 20% of that of methyl-CoM reduction; this product was degraded chemically to acetic acid, which was subjected to configurational analysis by the method of

Cornforth and Arigoni. The results indicated enzymatic replacement of the -SCoM by hydrogen in an inversion mode, consistent with a mechanism in which the methyl group is displaced nucleophilically from the sulfur of methyl-CoM by the nickel of factor F_{430} (inversion) followed by protonolytic cleavage of the CH_3-Ni bond (retention). (Work supported by NIH grant GM 32333).

26. SYNTHESIS AND PROPERTIES OF A NOVEL SPIN-LABELED NUCLEOSIDE. D.A. Alessi[2], J.E.T. Corrie[1], J. Feeney[1], I.P. Trayer[2] and D.R. Trentham[1], [1]National Institute for Medical Research, Mill Hill, London, London NW7 1AA, U.K. and [2]Department of Biochemistry, University of Birmingham, Birmingham, B15 2TT, U.K.

The 2',3'-O-spiroketal of uridine and 4-oxo-2,2,6,6-tetramethyl-1-piperidinyloxy was prepared via acid-catalysed addition of 5'-O-benzoyluridine to 1-acetoxy-4-methoxy-2,2,6,6-tetramethyl-3-piperideine. The latter compound was prepared by an unusually facile loss of methanol from the corresponding 4,4-dimethoxy-ketal. The N-acetoxy-spiroketal product of the addition reaction exists as two conformers which interconvert slowly on the NMR time scale at 25°C, and the origins of this conformational barrier will be discussed. Prolonged alkaline hydrolysis and concomitant aerial oxidation gave the required nitroxyl spiroketal. Nucleotide analogues of this nitroxyl derivative are expected to be useful both as EPR and NMR probes on account of the restricted conformational mobility of the nitroxyl.

27. PROCESS DEVELOPMENT OF BI-RG-587: A POTENT AND SELECTIVE REVERSE TRANSCRIPTASE INHIBITOR OF HIV-1. S. C. Mauldin, K. G. Grozinger, V. U. Fuchs, J. Vitous, J. Adams, K. D. Hargrave, Boehringer Ingelheim Pharmaceuticals, Inc., 90 East Ridge, P. O. Box 368, Ridgefield, CT 06877.

Reverse transcriptase (RT) of retroviruses is a necessary component for early proviral synthesis. In most cases, the molecules that are effective as RT inhibitors are nucleoside analogues (converted to triphosphates by cellular enzymes) which act as chain terminators and not as direct enzyme inhibitors. We have synthesized a non-nucleoside which is a potent inhibitor of HIV-1 reverse transcriptase. The compound, BI-RG-587, has a IC50 of 84 nM against HIV-1 RT and is currently in clinical trials. In an effort to improve the synthetic route for bulk scale production, we have developed other synthetic routes in addition to the original procedure published in Science, 250, 1411 (1990). Details of this research will be presented.

28.
INHIBITOR DESIGN I: INVESTIGATION OF VARIOUS MOIETIES AS STABLE 3-PHOSPHATE MIMICS IN AROMATIC ANALOGS OF EPSP. Ajit S. Shah, Michael J. Miller, Darryl G. Cleary, Diane S. Braccolino, Jose L. Font, Joel E. Ream and James A. Sikorski, New Products Division, Monsanto Agricultural Company, A Unit of Monsanto Company, 800 N. Lindbergh Boulevard, St. Louis, Missouri 63167.

As part of a multidisciplinary biorational approach to herbicide discovery based on the inhibition of the enzyme 5-enolpyruvoylshikimate-3-phosphate (EPSP) synthase, we have designed and synthesized aromatic structural analogs of EPSP. The necessity of a 3-phosphate group in EPSP synthase inhibitors for effective inhibition together with the hydrolytic instability of these phosphates led to the synthesis of compounds containing stable groups with the potential to serve as phosphate mimics. Based on molecular modeling, we have incorporated both a malonate ether and a hydroxymalonate group as stable phosphate mimics into aromatic analogs of EPSP. These moieties were demonstrated to be unique phosphate mimics particularly useful for enzyme recognition.

29. EPSP SYNTHASE INHIBITOR DESIGN II: SYNTHESIS AND EVALUATION OF AROMATIC MIMICS OF THE TETRAHEDRAL INTERMEDIATE AND EVALUATION OF A 3-PHOSPHATE MIMIC. Paul D. Pansegrau, Michael J. Miller, Jose L. Font, Joel E. Ream and James A. Sikorski, New Products Division, Monsanto Agricultural Company, A Unit of Monsanto Company, 800 N. Lindbergh Blvd. St. Louis, Mo. 63167.

Eight new inhibitors of EPSP synthase were prepared which resemble the cataltyic tetrahedral intermediate formed by this enzyme. Important structural substitutions which were studied versus the intermediate were: 1.) Substitution of an aromatic ring for a shikimate ring, and 2.) Incorporation of a stable 3-phosphate mimic. A study of these issues has resulted in an understanding of the steric limitations at the active site of EPSP synthase.

30.
Toward the Synthesis of A Potential Inhibitor of Chorismate Synthase.

Craig B. Fryhle, Derin C. D'Amico, and Kathleen M. Brandt, Department of Chemistry, Pacific Lutheran University, Tacoma, Washington 98447

The shikimate biosynthetic pathway in plant and microbial metabolism is responsible for the production of aromatic amino acids, isoprenoid quinones and a host of other secondary metabolites. Chorismate synthase catalyzes the penultimate step in a linear sequence of transformations in the shikimate pathway beginning with glucose and leading to chorismate at a branch point in the pathway. As such, chorismate synthase occupies a key position in the metabolism of plants and microbes. The detailed mechanism of chorismate synthase is not well understood at present. This work in progress involves the synthesis of a proposed inhibitor of chorismate synthase. The proposed inhibitor is expected to shed further light on the mechanism of chorismate synthase.

31. **AN EFFECTIVE, ONE-POT SYNTHESIS OF CIS N-ALKYLPERHYDROQUINOLINES FROM CYCLOHEXANONES.** R. A. Atkins, J. W. Frazier, L. O. Weigel* *Chemical Process Research and Development, Lilly Research Laboratories*, Indianapolis, IN 46285
We shall describe methodology for the annulation of cyclohexanones into cis perhydroquinolines. Conversion of cyclohexanones into N-alkylketimines followed by: (a) LDA, I(CH$_2$)$_3$Cl, -78 °C; (b) NaBH$_4$, IPA, -25 °C; (c) 25 °C affords predominantly the cis fused perhydroquinoline (admixed with the trans isomer). **EXAMPLE SEQUENCE:**

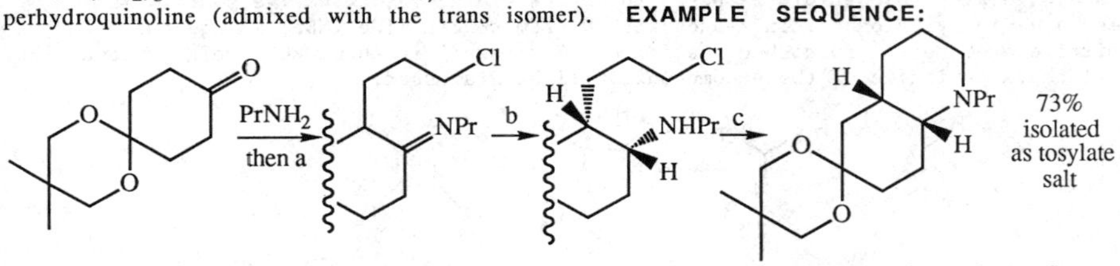

We will present five examples (yields 63-80% using benzylamine and S-α-methylbenzylamine with other cyclohexanones), proof of cis stereochemistry, and experimental details.

32.   δ-Diketones: Potential Synthons for Heterocycles.
R.Umarani, Department of Chemistry, The American College Madurai, India.

δ-Diketones are of interest as chemical intermediates in the synthesis of heterocycles. The cyclisation of δ-diketones to γ-pyrans has been long overlooked and tried. They tend to undergo intramolecular aldol cyclisation to cyclohexenone derivatives. Suitably substituted δ-diketones i.e.. benzylidene bis(acetophenones) serve as potential precursor for γ-pyran synthesis. Complementary method for the synthesis of γ-pryarns with reactive formyl groups will be detailed. The electrophilic reaction of δ-diketones with well known reagents phosphorusoxychloride in dimethylformamide give access to these compounds in a facile and convenient and rapid manner. Mechanism, Substituent effect and various methods of preparation of pyran heterocycles form δ-diketones of varying utility will be described.

33. THE [2,3] SIGMATROPIC REARRANGEMENT OF SULFILIMINES FROM 3-EXOMETHYLENE CEPHAMS. Nancy J. Snyder, Jonathan W. Paschal, Thomas K. Elzey, and Douglas O. Spry, Lilly Research Laboratories, Lilly Corporate Center, Eli Lilly and Company, Indianapolis IN 46285-1523.

The reaction of 3-exomethylenecephams with nitrene derivatives was studied. For example, the sulfide 1 was reacted with N-aminophthalimide in the presence of lead tetraacetate to give the 1,2,6-thiadiazepine azetidinone 3 which is derived from the sulfilimine 2, presumably via a [2,3] sigmatropic rearrangement. Further reaction of 3 with N-aminophthalimide in the presence of lead tetraacetate gave the ring-opened azetidinone 4. The synthesis, structure determination, and the mechanism of formation will be discussed.

34. A Facile Synthesis of Azetidine-2,4-diones. S.S. Bari, I.R. Trehan and A.K. Sharma, Department of Chemistry, Panjab University, Chandigarh, India; M.S. Manhas, Department of Chemistry and Chemical Engineering, Stevens Institute of Technology, Hoboken, NJ 07030.

Azetidine-2,4-diones which possess anti-inflammatory and sedative properties have been previously synthesised by photochemical and thermolytic methods. We have developed a facile synthesis of this category of compounds. The annelation of dithiocarbonimidates 2 with an acid chloride 1 in the presence of triethylamine affords 4,4,-dialkylthiozetidin-2-ones 3 in 45-70% yield. The oxidative hydrolysis of these 4,4-dialkylthioazetidin-2-ones with N-bromosuccinimide (NBS) in aqueous acetinitrile at 0 to -5° furnished azetidine-2,4-diones 4 in about 90% yield. Unique spectral characteristics of these compounds will be discussed.

35. THE SYNTHESIS OF SUBSTITUTED SACCHARINS THROUGH THE o-LITHIATIVE SULFONAMIDATION OF N,N-DIETHYLBENZAMIDES. Dennis J. Hlasta, John J. Court, and Ranjit C. Desai, Department of Medicinal Chemistry, Sterling Research Group, Rensselaer, NY 12144.

To prepare analogues of our novel mechanism-based inhibitors of human leukocyte elastase we needed general and efficient synthetic methods to prepare saccharins. Many syntheses of substituted saccharins are reported in the literature, although none were sufficiently general. We have developed a method for the synthesis of a variety of substituted saccharins in good overall yields in a two step process from readily available N,N-diethylbenzamides. The scope of this reaction will be described.

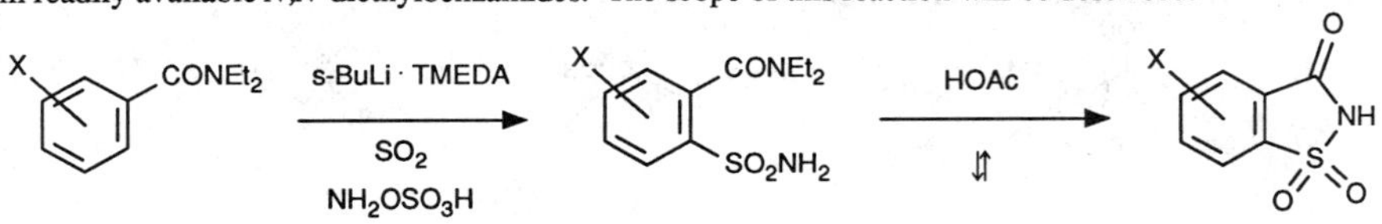

36. **A GENERAL SYNTHESIS OF 4-ALKYL/ARYL SUBSTITUTED SACCHARINS:** Chakrapani Subramanyam* and Malcolm R. Bell., Department of Medicinal Chemistry, Sterling Research Group, 81 Columbia Turnpike, Rensselaer, N.Y. 12144.

Human leukocyte elastase (HLE) is a serine proteinase that has been invoked in the etiology of a number of pulmonary disorders. In our efforts to synthesise novel mechanism based inhibitors of HLE, a number of 4-alkyl/aryl substituted saccharins were needed. Since no general methods are reported in the literature, we have developed a short and efficient synthesis of the title compounds.

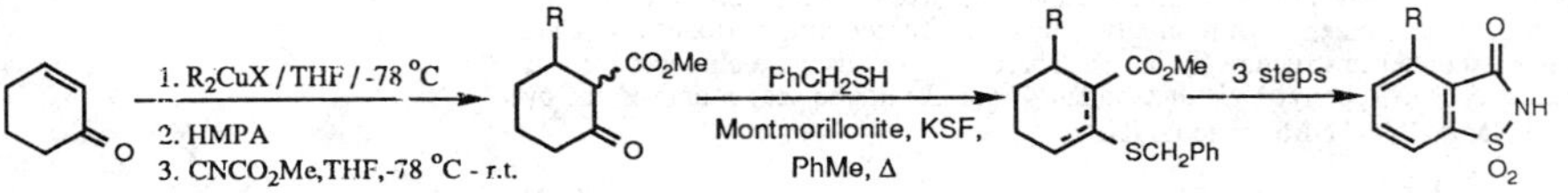

37. SYNTHESIS AND REACTIONS OF Δ^3-THIAZOLINES SUBSTITUTED IN THE POSITION 2. Leticia Contreras S., Elvira Santos S., Fernando León C. y Patricia Elizalde G. Depto. Química Orgánica, División de Estudios de Posgrado, Facultad de Química, UNAM, Ciudad Universitaria, 04510 México, D.F.

In this work we describe a modified route (based on the route of Martens) to obtain Δ^3-thiazolines (I), since the normal route described by Asinger was not totally succesfull. This type of compounds mono- or di- substituted in the position 2 were synthetized using as starting material vinyl acetate, this is transformed to thioglycolic aldehyde and the later compound was condensed with ammonium hydroxide and aldehydes or ketones as shown below.
Also in the same work we effectuated nucleophilic addition of HCN on the imino group to obtain thiazolidines (II), which are useful intermediates to synthesize aminoacids:

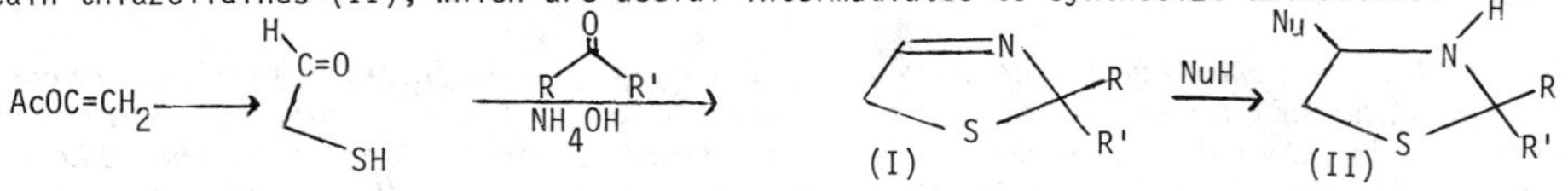

38. HIGHLY SELECTIVE ACYLATION OF DI- AND POLYHYDROXYL COMPOUNDS BY 3-ACYLTHIAZOLIDINE-2-THIONES, Shinji YAMADA, Department of Materials Science, Faculty of Science, Kanagawa University, Hiratsuka, Kanagawa, 259-12, Japan

Acylation of hydroxyl groups is one of the most fundamental and important reactions in organic synthesis, therefor a number of methods have been developed. In general, acid anhydrides and acid chlorides have been widely used because of their highly reactivity for hydroxyl groups. However, these reagents are not always effective for selective acylation of di- and polyhydroxyl compounds.
In this presentation, we wish to report that 3-acylthiazolidine-2-thiones serve well as highly primary-selective acylating reagents for di- and polyhydroxyl compounds. The selective acylation of them by 3-acylthiazolidine-2-thiones were performed in the presence of NaH in THF at room temperature to give

primary acylated compounds in high yields. Extremly highly selectivity was obtained in the pivaloylation with 3-pivaloylthiazolidine-2-thione.

39. SYNTHESIS OF STEROIDAL THIAZOLIDONES

A.H.Siddiqui , K. vijaysena Reddy
Department of Chemistry, Nizam College,
Osmania university, Hyderabad-500001, INDIA

As thiazolidones have been found to be associated with a variety of biological activies such as antitubercular, hypnotic, anaesthetic, amoebicidal, antiviral and cardiovascular activities, we synthesised extranuclear steroidal thiazolidones. The steroidal 17-ketones, 17-oxo-5-androsten-3β-yl acetate (1a), 3β-chloro-17-oxo-5-androstene (1b) and 17-oxo-5-andosten-3β-methyl ether (1c) on reaction with hydrazine hydrate gave their respective 17-hydrazone derivatives 1d, 1e &1f). Subsequent base catalysed condensation's of these 17-hydrazones with benzaldehyde gave their respective schiff bases (1g, 1h & 1i) whose reaction with thioglycollic acid afforded thiazolidones (2a, 2b & 2c) which may exist either in structure (i) or (ii). These compounds were characterised as (i) i.e. 3-substituted 17 (2'-phenyl-1',3' thiazolidan-4'-one-3-ylidene aza)-5-androstene by study of IR, ^{1}H NMR & ^{13}C NMR spectral data.

40. SYNTHESIS OF 5-PERFLUOROALKYL-3-ALKYL-2-ARYL-1,3-THIAZOLI- DINES. R. H Hesse, M. M. Pechet and G. Subba Reddy , Research Institute for Medicine & Chemistry, 49 Amherst Street, Cambridge, MA 02142.

We are reporting a general route to a variety of 5-perfluoroalkyl thiazolidines (a new class of compounds) having substituents in 2, 3, 4 and 5 positions. 5-Perfluoro alkyl-3-alkyl-2-aryl-1,3-thiazolidines are successfully synthesised in 7 steps. The key step involves the β-addition of RSH to the α -nitro, β-perfluoroalkyl alkene derived from the condensation of nitroalkane and a perfluorocarbonyl compound followed by dehydration.

41. APPLICATION OF ENANTIOPURE TRÖGER'S BASE METHOSULFATE AS AN INCLUSION HOST. RESOLUTION OF ALCOHOLS BY INCLUSION COMPOUND FORMATION.
Samuel H. Wilen and Jian Zong Qi, Department of Chemistry, The City College, City University of New York, New York, N.Y. 10031

Racemic alcohols enantioselectively form inclusion compounds with the dimethyl sulfate salt (2) of enantiopure Tröger's Base (1) on heating. The enantiomeric excesses (10-85% e.e) were measured by NMR on the alcohols in presence of 2 and on Mosher esters of the alcohols recovered from the inclusion compounds by reaction with aq. NaOH.

Anisochronic signals due to the alcohol were observed in the ^{1}H NMR spectra of most of the inclusion compounds formed. Chemical shift differences between signals due to the included alcohol enantiomers were sufficient to determine the e.e. of the inclusion compounds. Chemical shift differences were larger than those observed with diastereomeric association complexes formed by the identical alcohols and **1** itself.

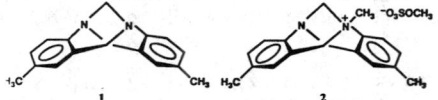

42. **ASYMMETRIC SYNTHESIS OF LACTONES WITH HIGH ENANTIOSELECTIVITY BY INTRAMOLECULAR CARBON-HYDROGEN INSERTION REACTIONS OF ALKYL DIAZOACETATES CATALYZED BY CHIRAL RHODIUM(II) CARBOXAMIDES** Michael P. Doyle*, Arjan van Oeveren, Larry J. Westrum, Marina N. Protopopova, and Thomas W. Clayton, Jr., Department of Chemistry, Trinity University, San Antonio, TX 78212

Chiral dirhodium(II) carboxamides possessing 2-pyrrolidone-5-carboxylate ligands are exceptionally effective catalysts for enantioselective carbon–hydrogen insertion reactions that form γ-lactones. Enantiomeric excesses of 50 to 85% were achieved in reactions of 2-alkoxyethyl diazoacetates catalyzed by dirhodium(II) tetrakis[methyl 2-pyrrolidone-5-(S)-carboxylate], $Rh_2(5S\text{-MEPY})_4$, which enriched the S configuration of the resulting β-alkoxy-γ-lactone, and its enantiomer, $Rh_2(5R\text{-MEPY})_4$, which enriched the γ-lactone's R-configuration. Diazo esters, including 2,3,4-trimethyl-3-pentyl diazoacetate and 2-phenyl-2-propyl diazoacetate, that undergo insertion into a C–H bond vicinal to the incipient chiral center formed γ-lactones with enantioselectivities of 60-76%. In catalytic insertion reactions of branched diazoacetates with the use of dirhodium(II) tetrakis[neopentyl 2-pyrrolidone-5-(S)-carboxylate], $Rh_2(5S\text{-NEPY})_4$, enantioselectivities for lactone formation are up to 12% above those obtained with $Rh_2(5S\text{-MEPY})_4$.

43. **THE FIRST KINETIC RESOLUTION OF CHIRAL AZIDES.** Alex Pasternak, Julio Perez, Stephen R. Wilson. Dept. of Chemistry, New York University, New York, NY 10003.

Novel chiral phosphines are being examined with regard to asymmetric induction in phosphine mediated reactions. For example, the Staudinger reaction involves the treatment of an alkylazide **1** with tertiary phosphines to form phosphazenes **2**, which are normally hydrolyzed to give amines **3**. We have found that chiral phosphines effect kinetic resolution of racemic azides in the Staudinger reaction. Enantiomeric excesses as high as 60% have been achieved. The design, synthesis and implementation of these phosphines and the rationalization for chiral recognition will be discussed.

$$R^1\text{—}N_3 \xrightarrow{(R^2)_3P, -N_2} \underset{R^2}{\overset{R^2}{\underset{|}{R^2\text{—}P}}}=N\text{—}R^1 \xrightarrow{H_2O} R^1\text{—}NH_2 + (R^2)_3P=O$$

      1   2   3

44. **SYNTHESIS AND CHARACTERIZATION OF CATIONIC TRINUCLEAR BINAP—RUTHENIUM(II) COMPLEXES AND THEIR ROLES IN CATALYTIC ASYMMETRIC HYDROGENATION OF β-KETO ESTERS.** Kazushi Mashima, Takahiro Hino, and Hidemasa Takaya, Department of Industrial Chemistry, Faculty of Engineering, Kyoto University, Yoshida, Kyoto 606, Japan.

BINAP—Ru(II) complexes bearing halogen bound to ruthenium have been used as excellent catalysts for asymmetric hydrogenation of olefinic and ketonic substrates. Catalytically active species derived from these complexes are dependent mainly on reaction conditions such as the kind of substrates, solvents, and additives. Cationic trinuclear complexes [Ru₃X₅((S)-binap)₃]Y (X = Cl, Br; Y = Cl, Br, BF₄), prepared by heating monomeric BINAP—Ru complexes in methanol, were characterized by spectral data, conductivity, and X-ray analysis. Catalytic activity of these BINAP—Ru complexes for asymmetric hydrogenation of β-keto esters has been investigated and the relationship between the structure of BINAP—Ru complexes in solution and catalytic activity will be discussed.

45. ASYMMETRIC OXIDATIONS CATALYZED BY A D_4-SYMMETRICAL MANGANESE-TETRAPHENYLPORPHYRIN R. L. Halterman and H. L. Nimmons Dept. of Chemistry, Boston University 590 Commonwealth Ave. Boston, MA. 02215.

The D_4-Symmetrical Manganese-Tetraphenylporphyrin **1**, was used in the catalytic asymmetric oxidation of aryl alkyl sulfides to chiral sulfoxides. Using iodosylbenzene as the stoichiometric oxidant yields above 82% and enantioselectivities ranging from 40 to 68% ee were found. The sulfoxide/sulfone ratio was ≥34:1 for all but one substrate which gave 13:1. The catalytic asymmetric oxidation of benzylic positions was also carried out using **1** and iodosylbenzene as the stoichiometric oxidant. Yields ranging from 8-50% and enantioselectivities of 9 to 53% ee were observed.

46. ASYMMETRIC ALKYLATION, ACYLATION AND ALDOL REACTION OF CHIRAL β-LACTAM ESTER ENOLATES
Iwao Ojima*, Elke Schoffers, Lamont Scott, and Thierry Brigaud,
Department of Chemistry, State University of New York at Stony Brook, NY 11794-3400

Asymmetric alkylation, acylation and aldol condensation of the enolate of the β-lactam esters **1** afforded the corresponding β-lactam esters **2** and **3** in good yields with good to excellent stereoselectivity, which are precursors of dipeptides bearing α-alkyl-, α-acyl-, and β-hydroxy-α-amino acid residues.

47. ASYMMETRIC SYNTHESIS OF 3-HYDROXY-4-SUBSTITUTED-β-LACTAMS AS POTENTIAL PRECURSORS OF NORSTATINE, STATINE AND THEIR ANALOGS
Iwao Ojima*, Mangzhu Zhao, and Young-Hoon Park
Department of Chemistry, State University of New York at Stony Brook, Stony Brook, NY 11794

Chiral ester enolate-imine cyclocondensation protocol is successfully applied to the asymmetric synthesis of a variety of 3-hydroxy-4-substituted-β-lactams with high enantiomeric purity. This method is found to be also effective for enolizable imines, which has opened a new and efficient routes to the precursors of norstatine, statine and their analogs.

48. **NOVEL APPROACH TO THE ASYMMETRIC SYNTHESIS OF PEPTIDE FRAGMENTS BEARING α-HYDORXY-β-AMINO ACID RESIDUES**
Iwao Ojima* and Mangzhu Zhao
Department of Chemistry, State University of New York at Stony Brook, Stony Brook, NY 11794

The reactions of N,O-di-t-BOC-3-hydroxy-β-lactam or N-t-BOC-O-TIPS-3-hydroxy-β-lactam with amino esters give the corresponding protected dipeptides bearing α-hydroxy-β-amino acid residues in high yields. This reaction provides a novel method for peptide synthesis based on ring-opening coupling of N-acylated β-lactams.

49. **SYNTHESIS OF BICYCLIC NITROGEN HETEROCYCLES BY MEANS OF HOMOGENEOUS CATALYTIC CARBONYLATION.** Iwao Ojima, Anna Korda, Masakatsu Eguchi, and Zhaoda Zhang, *Department of Chemistry, State University of New York at Stony Brook, Stony Brook, NY 11794-3400*

Hydrocarbonylation of α-ethenyl and α-ethynyllactams catalyzed by $HRh(CO)(PPh_3)_3$ gave the corresponding bicyclic O-ethyl hemiamidals which are the intermediates for the synthesis of izidine alkaloids. The synthesis of basic skeleton of pyrrolizidine alkaloid bearing hydroxymethyl group was carried out through the combination of silyformylation and hydrocarbonylation.

50. **THE SYNTHESIS OF 2-OXO-2-PROPIONYL-1,3,2-DIOXAPHOSPHORINANES AND THE SYNTHESIS AND ALDOL REACTIONS OF A CHIRAL 2-OXO-2-PROPIONYL - 1,3,2-OXAZAPHOSPHORINANE.** Neil J. Gordon and Slayton A. Evans, Jr.*, Department of Chemistry, The University of North Carolina, Chapel Hill, North Carolina 27599-3290, USA.

The synthesis of the title compounds will be described and some insight into the stereospecificity of the Arbusov reaction presented. The results of Aldol reactions with the chiral propionyl-oxazaphosphorinane (1) will also be shown.

51. **THE REGIOSELECTIVE SUBSTITUTION OF 1,2-PROPANEDIOL WITH TRIMETHYLSILYL REAGENTS USING THE $1,3,2\lambda^5$-DIOXAPHOSPHOLANE METHODOLOGY.** I. Mathieu-Pelta and S. A. Evans, Jr. William Rand Kenan, Jr., Laboratories of Chemistry, University of North Carolina, Chapel Hill, North Carolina 27599-3290.

The development of highly regioselective procedures of substitution of unsymmetrical 1,2-diols in a "single synthetic event" offers the potential for effecting a host of valuable synthetic transformations. We describe the kinetically-controlled opening of a 4-substituted 1,3,2λ^5-dioxaphospholane **1**, formed from the transoxyphosphoranylation of 1,2-propanediol with DTPP, with a wide range of trimethylsilyl reagents: Me$_3$SiX (X = PhS, I, Cl, Br, CN and N$_3$). The substitutions are highly regio-selective and afford the thermodynamically less stable C-2 substituted derivatives of the 1,2-diol.

52. **ENZYMES IN ORGANIC SYNTHESIS: SYNTHESIS OF ENANTIOMERICALLY PURE 1,2-EPOXY ALDEHYDES, THIIRANES, AND AZIRIDINES.** Richard L. Pederson Bioprocess Division, Bend Research Inc., Bend, OR 97701, and Chi-Huey Wong, Department of Chemistry, The Research Institute of Scripps Clinic, La Jolla, CA 92037.

This poster describes the chemoenzymatic procedure for the synthesis of (R)- and (S)-2-(diethoxymethyl)oxirane (glycidaldehyde diethyl acetal) in >98% enantiomeric excess. LP-80 lipase enantioselectively hydrolyzes 2-acetoxy-3-chloropropanal diethyl acetal (1) to (S)-3-chloro-2-hydroxypropanal diethyl acetal (2). Products (2) and (3) were separated, and converted by a simple chemical step to (R)-glycidaldehyde diethyl acetal (4) and (S)-glycidaldehyde diethyl acetal (5); respectively. These chiral epoxides were converted to chiral thiiranes and chiral aziridines.

53. **CATALYSIS OF ENOLATE REACTIONS BY AN ARTIFICIAL ENZYME**
Anne M. Kelly-Rowley, Larry A. Cabell, Eric V. Anslyn
Department of Chemistry, The University of Texas, Austin TX 78712

Polypyrrole cleft **1** binds enolate anions in acetonitrile with binding constants on the order of 10^3 to 10^4. The cleft possesses convergent NH groups which act as hydrogen bond donors to the anionic oxygens of 1,3-diketo enolate functionalities. The binding of the enolates is significantly stronger than that of the neutral diketo compounds. The increase in binding of the reactive enolate form results in catalysis of several reactions such as alkylations and conjugate additions.

54. **POLYAZA CLEFTS FOR THE RECOGNITION AND CLEAVAGE OF PHOSPHODIESTERS**
Lisa S. Flatt, Katsuhiko Ariga, Eric V. Anslyn
The University of Texas, Austin TX 78712

Polyazaclefts possess convergent 2-aminopyridines or guanidines were prepared in six to seven synthetic steps. The clefts bind phosphodiesters with binding constants on the order of 10^3 to 10^4 in chloroform. The binding constants were determined by following the ^{31}P NMR chemical shifts of the phosphodiesters. The phosphodiesters are bound via hydrogen bonding and salt pairing in a variety of solvents including CHCl$_3$, CH$_3$CN and H$_2$O. Complexes of greater than 1:1 stoichiometry can often be formed. The protonated clefts are designed to act as templates for promoting phosphodiester cleavage via catalyzing multiple proton transfers. The syntheses, binding studies and catalysis will all be presented.

55. **SYNTHESIS OF SILICON-CONTAINING CAGES.** Ruert Damrauer and Joseph Hankin, Chemistry Department, University of Colorado at Denver, Denver, CO 80217-3364

We have prepared 1 and 2. These are the first examples of such a four-atom bridged ring system having their substituent in the thermodynamically less stable outside position. We have studied metal catalyzed cyclotrimerizations of RSi(OCH$_2$CH$_2$CH$_2$C≡CH)$_3$ to prepare 1 and 2. Synthetic studies

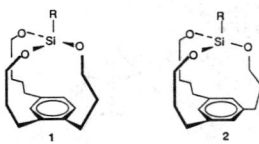

have examined catalyst systems, the effects of different R groups on cyclotrimerization, and isomer distribution patterns for cyclotrimerization. X-ray structures and full characterization information will be reported as will information on the preparation of precursor molecules for the cyclotrimerization and on efforts to prepare other silicon-containing caged compounds.

56. OLEFIN AZIRIDINATION AND EPOXIDATION CATALYZED BY CHIRAL METAL SALEN COMPLEXES BEARING BULKY SILYL GROUPS. Kenneth J. O'Connor, Shiow-Jyi Wey, and Cynthia J. Burrows, Department of Chemistry, SUNY at Stony Brook, Stony Brook, NY, 11794-3400.

While chiral manganese salen complexes have been reported to catalyze asymmetric epoxidation, there have been no examples of metal salen-catalyzed olefin aziridination. Accordingly, a series of chiral salen ligands derived from chiral diamines and bearing bulky silyl groups on the arene ring have been synthesized and studied in Mn(III) and Ni(II)-catalyzed aziridination and epoxidation. Good chemical yield of both aziridines and epoxides were obtained in Mn-catalyzed reactions while Ni salens catalyzed only epoxidation. Despite moderate ee's observed in olefin epoxidation, no enantioselectivity was observed for aziridination. A full account of this investigation and results obtained with other metal salen complexes in which the chiral diamine was varied will be presented.

57. SYNTHESIS OF CHOLIC ACID DERIVATIVES FOR FORMATION OF MOLECULAR RECOGNITION COMPOUNDS. Ruey-fen Liao, Hsing-Pang Hsieh, Cynthia J. Burrows, Department of Chemistry, SUNY at Stony Brook, Stony Brook, NY, 11794-3400, and Suzanne Evans and Carol A. Venanzi, Department of Chemical Engineering, Chemistry and Environmental Science, New Jersey Institute of Technology, Newark, NJ 07012.

Cholic acid is an attractive architectural unit for the construction of synthetic molecular receptors because of its rigid steroid framework and the positioning of three hydroxyl groups on its α face. In order to form hydrophilic cavities, cholic acid units have been linked with either m- or p-xylylene diamines to yield dimers. Both the parent compound, cholic acid, and a side chain degradation product in which two methylene groups were removed have been studied. Molecular mechanics calculations suggest that these more rigid structures present fewer low energy conformations appropriate for binding substrates such as glucosides; in addition, quite different molecular shapes are predicted for meta vs. para-linked cholamide dimers. Progress in NMR studies of binding of glucosides to the synthetic receptors will be presented. Recent work aimed at converting cholic acid derivatives to steroid diamines and triamines will also be discussed.

58. **SYNTHESIS AND CHARACTERIZATION OF A NEW HEXAPYRROLIC EXPANDED PORPHYRIN: RUBYRIN.** Jonathan L. Sessler, Takashi Morishima, and Vincent Lynch. Department of Chemistry and Biochemistry, The University of Texas, Austin, Texas 78712.

The hexapyrrolic expanded porphyrin, **1**, has been synthesized by an acid-catalyzed [4]+[2] MacDonald-type oxidative condensation and characterized structurally as its dihydrochloride complex. The compound, which has been assigned the trivial name rubyrin in light of its bright red color, displays a near planar structure as would be expected for a large porphyrin-like aromatic macrocycle.

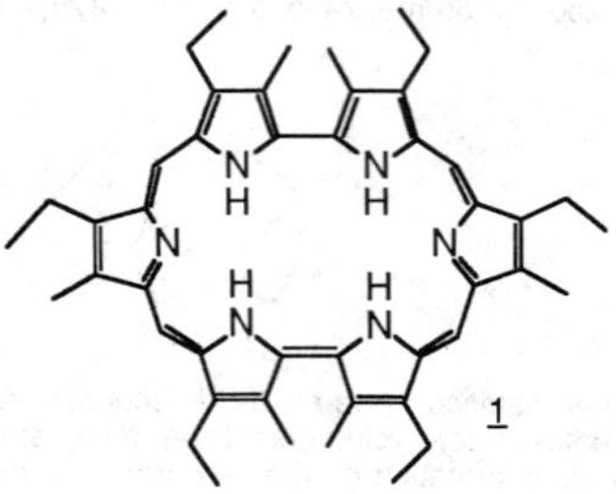

59. **MOLECULAR RECOGNITION VIA BASE-PAIRING: NONCOVALENT APPROACHES TO PHOTOSYNTHETIC MODEL SYSTEMS.** Yuji Kubo, Darren J. Magda, Anthony Harriman, and Jonathan L. Sessler, Department of Chemistry and Biochemistry and Center for Fast Kinetic Research, University of Texas, Austin, Texas 78712.

The syntheses and photophysical properties of several guanine - porphyrin and cytosine containing porphyrin, pyrene, and quinone systems are described. Upon mixing in aprotic solvents, self-association of nucleic acid base occurs to produce various noncovalent ensembles, which undergo light induced donor to acceptor energy and/or electron transfer as shown below.

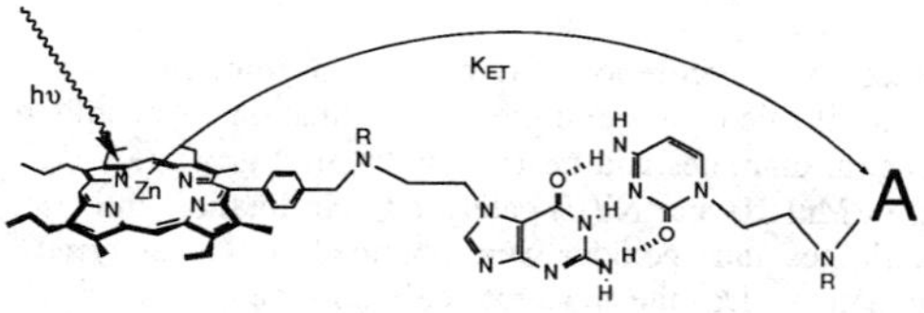

R = $CH_2C_6H_4OCH_2CH_2OCH_2CH_2OCH_3$
or $CH_2CH_2OCH_2CH_2OCH_3$

A = Substituted Porphyrin, Quinone, or Pyrene

60. **SYNTHESIS OF A TETRAIMIDAZOLE CONTAINING MACROCYCLE: A NEW LIGAND CLASS.** J. Paul Street, University of Houston, Dept. of Chemistry, Houston, TX 77204

The first member of a new coronand class containing imidazole donor groups has been prepared and spectrally characterized. Macrocyclization is achieved concurrently with imidazole ring construction, utilizing the parent reaction of benzylamines with oximino-β-dicarbonyl compounds. Regioselective N-alkylation of the macrocycle can be directed to the periphery or the interior, by varying the metal cation of the base which is employed. This effect correlates directly to the chelate hole size of the macrocycle. Further derivatives, and metal ion binding studies will be presented.

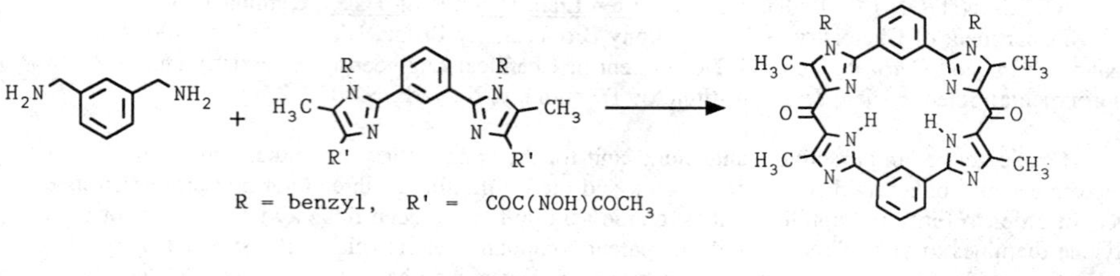

R = benzyl, R' = $COC(NOH)COCH_3$

61. **COMPLEXATION OF BENZAMIDINIUM SALTS BY A SYNTHETIC RECEPTOR.** Vincent J. Santora, Thomas W. Bell, Department of Chemistry, State University of New York, Stony Brook, NY 11794-3400.

A new molecular receptor (**1**) that combines hydrogen bonding and π-stacking interactions to bind benzamidines (**2**) has been synthesized by condensation of bis-α,α'-oxyamino-*meta*-xylene with diketone **3**. The molecular framework of this receptor orients naphthyridine and oxime nitrogen lone pairs to form four hydrogen bonds with benzamidinium ions. The spectroscopic changes accompanying complexation and potential application of analogs as optical sensors are also discussed.

62. DESIGNED SPIROCYCLIC IONOPHORES FOR LITHIUM. Gary Hiel, Heung-Jin Choi and Thomas W. Bell, Department of Chemistry, State University of New York, Stony Brook, NY 11794-3400.

Bisspirodiamide/-spirodiimide macrocycles 3 and 4 represent a new class of hosts designed for lithium selectivity. These new macrocycles were synthesized by [2+2] condensation of spirodiamide 1 or spirodiimide 2 with dichloride 5. Intermediates 1 and 2 are prepared in two steps from acrylonitrile and diethylmalonate or malononitrile, respectively. Separation, characterization and results of preliminary binding studies will also be described.

63. SYNTHESIS AND COMPLEXATION OF PYRROLE-CONTAINING RECEPTORS AND PORPHYRIN ANALOGUES. Andrew Papoulis, Thomas W. Bell, Department of Chemistry, State University of New York, Stony Brook, NY 11794-3400.

A synthetic approach to pyrrole-containing receptors involving Piloty rearrangement of azines is described. For example, urea receptor 2 was prepared in three steps from intermediate 1 in 92 % yield overall. Complexation of 2 towards urea derivatives, and its use as an intermediate in the synthesis of porphyrin analogue 3 are described.

64. SYNTHESIS OF TRI-ARYL SUBSTITUTED TORANDS. Richard T. Ludwig, Thomas W. Bell. Department of Chemistry, State University of New York, Stony Brook, N.Y. 11794-3400.

The tri-n-buty-torand 1 has proven to be an extremely efficient metal cation sequestering host (Bell, T.W.; Firestone, A., J. Am. Chem. Soc., 1986, *108*, 8109). A new series of tri-aryl substituted torands (2 X= H, OMe, Cl, NO_2) have been synthesized from readily available octahydroacridine derivatives. These new torands allow for additional functionalization of the three aromatic substituents. The primary goal is the synthesis of discotic liquid crystaline, water soluble and chromogenic torands. The synthesis of these macrocycles, as well as the complexation properties of some key intermediates will be illustrated.

65. **EXPANDED HETEROHELICENES: MOLECULAR COILS THAT FORM CHIRAL COMPLEXES**
Hélène Jousselin, Thomas W. Bell, Department of Chemistry, State University of New York, Stony Brook, NY 11794-3400.

The first examples of a new class of preorganized, helical polypyridines are reported. These new ligands are composed of fused six-membered rings and are analogous to the helicenes in that they cannot adopt planar conformations. The combination of angular with linear ring fusion forms cavities lined with pyridine nitrogen atoms. Two helical compounds, an "expanded" heterohelicene containing seven pyridine rings (6) and a 5-pyridine homologue (12), were synthesized in several steps from 9-n-butyl-1,2,3,4,5,6,7,8-octahydroacridin-4-one and 5-benzylidene-9-n-butyl-1,2,3,4,5,6,7,8-octahydroacridin-4-one.

66. **SYNTHESIS OF AMPHIPATHIC SURFACES FROM AMINO ACIDS AND STEROIDS FOR CONSTRUCTION OF HYDROPHOBIC BINDING SITES AND SELF-ORGANIZING STRUCTURES**

Alan W. Schwabacher, Haiyan Lei, Jinho Lee, Department of Chemistry, Iowa State University, Ames, IA 50011

We have synthesized bisphenylalanyl phosphinic acid, a new bis amino acid, in a suitable form for incorporation into peptides to make binding sites. The V shaped diarylphosphinic acid moiety provides a hydrophobic internal surface, and polar external functionality. The differentially protected bis amino acid is obtained in five steps from phenylalanine. The key synthetic transformation involves sequential coupling of two different aryl iodides with methyl phosphinate under palladium catalysis to generate methyl diarylphosphinate. This new coupling reaction is quite general. Self-assembled metal complexes of bisphenylalanyl phosphinate, and self-organized structures containing this subunit, demonstrate interesting hydrophobic binding properties. Cholic acid-derived appendages provide additional hydrocarbon surface for association in H$_2$O.

67. **NUCLEOSIDE SYNTHESIS FROM THIOGLYCOSIDES** Spencer Knapp* and Wen-Chung Shieh, Department of Chemistry, Rutgers University, New Brunswick, New Jersey 08903.

Acetylated phenyl and methyl 1-thiofurano- and pyranosides react with silylated uracil (1), acetylcytosine, and benzoyladenine under the influence of NIS / triflic acid to generate the β-nucleoside derivatives in good yield. An application (2 → 3) to a projected synthesis of capuramycin is described.

68. **SYNTHESIS OF MANNOSTATIN A.** Spencer Knapp* and T. G. Murali Dhar, Department of Chemistry, Rutgers University, New Brunswick, New Jersey 08903.

The unique mannosidase inhibitor mannostatin A (**4**) was efficiently synthesized from D-ribonolactone via allylic alcohol **1**. An iodocyclization sequence gave **2** (Ar = *p*-methoxyphenyl), which was converted to **3** by NaSMe displacement and further to **4** by deprotection.

69. **FACE SELECTIVITY IN THE PROTONATION OF GLYCALS.** Neelu Kaila, Richard W. Franck, Department of Chemistry, Hunter College/CUNY, 695 Park Ave. New York, NY 10021-5024

Glucals are known to react with a variety of nucleophiles in the presence of catalytic amount of triphenylphosphine hydrobromide to give 2-deoxyglucosides (Falck, J. R.; Lee, S. -G.; Bolitt, V.; Mioskowski, C. *J. Org. Chem.* **1990**, *55*, 5812). When deuterated methanol (CH_3OD) was reacted with tribenzylglucal in these conditions, NMR analysis of the products showed that protonation occurs from the bottom face of the glucal. The reaction has been studied with other deuterated nucleophiles and different glycal substrates. Details will be presented.

70. **THE MECHANISM OF β-GLYCOSIDASES: A REASSESSMENT OF SOME SEMINAL PAPERS.** Richard W. Franck, Department of Chemistry, Hunter College / CUNY, 695 Park Ave., New York, NY 10021-5024

There is a generally accepted mechanism for β-glycosidases acting upon their natural substrates. As shown, the enzyme protonates the exocyclic oxygen to make it a good leaving group; and then the enzyme supplies the force necessary to distort the normal chair form so that nucleophilic attack with inversion can occur to a glycosyl-enzyme adduct. In the mid-1980's, Karplus and Post using dynamics calculations, and Fleet using evidence from the behavior of natural enzyme inhibitors , proposed an alternate endocyclic protonation and cleavage pathway, eventually leading to the same glycosyl-enzyme adduct. Our paper will reanalyze four key papers of the β-glycosidase literature to demonstrate that the data is more consistent with the endocyclic proposal than with the scheme that is more widely accepted by the bio-organic community.

71. SYNTHESES OF AMINODEOXY, CYANODEOXY AND BRANCHED-CHAIN AMINO SUGARS BY EPOXY AND TRIFLATE SUGAR INTERMEDIATES. S.N.H. KAZMI, Z. AHMED and A. MALIK, H.E.J. Research Institute of Chemistry, University of Karachi, Karachi-75270, Pakistan.

The chemistry of epoxy sugars and sugar triflates have progressively become more apparent to the synthetic carbohydrate chemists as a number of examples of these have so far been synthesized. The easy accessibility to the epoxy sugars, coupled with stereo rigidity influenced by the oxirane ring, made them ideal starting materials for the synthesis of a variety of aminodeoxy sugars, the essential constituents of a large variety of highly effective antibiotics and cyanodeoxy sugars which assume a key role in the syntheses of C-glycosyl compounds and branched-chain sugars besides serving as precursors for most of the naturally occurring C-nucleoside antibiotics and their analogs. On the other hand the displacement of trifluoromethanesulfonyloxy (triflyl) group in suitably protected sugar triflates has been affected by a variety of nucleophiles. The difficulties which, somtimes, are encountered during displacement reactions in carbohydrate systems were absent.

72. SYNTHESIS OF ALTROSE AND ALLOSE. T. Ebata, K. Matsumoto, K. Koseki, H. Kawakami and H. Matsushita, Life Science Research Laborarory, Japan Tobacco Inc., 6-2, Umegaoka, Midori-ku, Yokohama, Kanagawa 227, Japan

Levoglucosenone (1,6-anhydro-3,4-dideoxy-β-D-glycero-hex-3-enopyranos-2-ulose) is known widely as a pyrolytic product of cellulose. This compound is a very useful chiral source for synthesizing natural products because of its highlyfunctionalized structure, which contains one chiral center. We will present a synthesis of altrose and allose from levoglucosenone in high overall yield.

73. CONVENIENT DEGRADATION TO LOWER FURANOSE SUGARS VIA OZONIZATION OF PROTECTED GLYCALS. Zbigniew Kaluza, Khaled Barakat, Maghar S. Manhas and Ajay K. Bose, Department of Chemistry, Stevens Institute of Technology, Hoboken, NJ 07030.

Variously protected 1,2-glycals (**1**) which are readily available have been ozonized under reductive conditions to obtain lower sugars (**2**) with a formate ester substituent. Treatment with methanol and triethylamine led to selective hydrolysis of the formate and formation of furanose sugars (**3**). Thus, the pentoses and tetroses (**2**) were prepared in 70-80% yield from hexoses and pentoses (**1**). The new sugars in their aldehyde form (**2**) are versatile, fully protected, chiral synthons. Thus, β-lactams (e.g., **4**) can be obtained via Schiff bases from **2a** and higher sugars, via chain elongation of **2** using the Wittig reaction.

1a $R_1=CH_2OTr; R_2=R_5=H; R_3=R_4=OBn$ **2a**
1b $R_1=H; R_2=R_4=H; R_3=R_5=OBn$ **2b**

74. STEREOSELECTIVE CARBOHYDRATE SYNTHESIS VIA PALLADIUM HYDROXIDE CATALYZED EPOXIDE HYDROGENOLYSIS.
J. Gabriel Garcia, Ronald J. Voll and Ezzat S. Younathan, Department of Biochemistry, Louisiana State University, Baton Rouge, LA 70803-1806.

2,5:3,4-Dianhydro-D-altritol (**1**) reacts with LAH to produce a 50:50 mixture of the corresponding epimeric alcohols (**2**) and (**3**) in yields up to 74%. However, when **1** is catalytically hydrogenated in the presence of Pearlman's catalyst, the ratio of the epimers formed was 99:1 respectively (**2:3** as determined by

GC of the corresponding acetylated compounds) in a yield of 92%. The stereochemistry of the epimers is verified by NMR spectroscopy. Attempts to explain this result mechanistically will be reported.

	LAH	50	50
	H$_2$/Cat.	99	1

75. **STEREOCONTROLLED SYNTHESIS OF C-LINKED DEOXYRIBOSIDES OF 2-HYDROXYPYRIDINE AND 2-HYDROXYQUINOLINE.** Paul B. Hopkins and Marjorie S. Solomon, Department of Chemistry, University of Washington, Seattle, WA 98195

C-Nucleosides 1 and 2 were prepared by a general and preparatively useful route. Addition of 2-fluoro-3-lithio-pyridine or -quinoline to 4,5-isopropylidene-3-*tert*-butyldimethylsilyl-protected 2-deoxyribose afforded an alcohol, which was coverted to the mesylate and deprotected to yield the fluorinated C-nucleoside. Displacement of fluoride with methoxy- or benzyloxy-anion and subsequent treatment with trimethylsilyl iodide afforded the C-nucleosides 1 and 2 in good yields. The relative configuration at the anomeric carbon was determined by ^{1}H NMR. Attempts to incorporate 1 and 2 into DNA are underway, and once accomplished, the resulting DNAs will be tested for their ability to form stable duplexes.

76. **N-FLUOROPYRIDINIUM PYRIDINE HEPTAFLUORODIBORATE: A USEFUL FLUORINATING AGENT.** A. J. Poss, W. J. Wagner and R. L. Frenette, Allied-Signal Inc., Buffalo Research Laboratory, 20 Peabody Street, Buffalo, New York, 14210.

N-Fluoropyridinium pyridine heptafluorodiborate (NFPy) has recently been developed as a safe and inexpensive source of electrophilic fluorine. This novel reagent reacts under mild conditions to selectively transfer fluorine to the reactive sites of enol derivatives.

77. **N-FLUORO-O-BENZENEDISULFONIMIDE (NFOBS): A USEFUL NEW FLUORINATING REAGENT.** Franklin A. Davis and Wei Han, Department of Chemistry, Drexel University, Philadelphia, PA 19104

The titled compound (NFOBS), readily prepared from *o*-benzenedisulfonimide and fluorine in greater than 90% yield, is a stable, highly efficient source of "electrophilic" fluorine which fluorinates enolates, azaenolates and carbanions in good to excellent yields. Its synthesis, properties and applications will be presented.

78. **SYNTHETIC STUDIES OF ATIPAMEZOLE ANALOGS: FLUORINATION OF 5-SUBSTITUTED INDANONES AND INDANDIONES.** Joel D. Enas, John M. Gerdes* and Chester A. Mathis, Research Medicine and Radiation Biophysics Division, Lawrence Berkeley Laboratory, University of California, Berkeley, CA 94720.

Atipamezole (**1a**) is a potent α_2-adrenoreceptor antagonist and preliminary studies have targeted the related 5-fluoroatipamezole **1b** as a biologically active analog. The ^{18}F-radiolabeled form of **1b** (^{18}F-**1b**) is considered a potential α_2-adrenoreceptor imaging agent. In order to devise a synthetic route to **1b** which would also facilitate the preparation of ^{18}F-**1b** with ^{18}F-fluoride ion ($t_{1/2}$ = 110 min), we undertook a model study of the syntheses of the indanones **2a-b** and **3a-b**, transformation of these substrates to the 5-fluoroketones **2c** and **3c** by nucleophilic aromatic fluoride substitution, and formation of the reduced form **4** from the fluoroketones. These synthetic transformations and the potential for ^{18}F-**1b** formation employing a similar sequence will be discussed.

1a X = H
1b X = F

a. X = NO$_2$ b. X = N(CH$_3$)$_3$OTf c. X = F

79. **APPLICATION OF THE PETERSON OLEFINATION REACTION TO THE SYNTHESIS OF FLUOROOLEFIN DIPEPTIDE ISOSTERES.** John T. Welch, Bart De Corte, Rayomand H. Gimi and Agnes Gyenes. Department of Chemistry, The University at Albany, State University of New York, Albany, NY 12222.

The Peterson olefination reaction of 2,4,6-trimethylphenyl α-fluoro-α-trimethylsilylacetate **1** with 2-(2-trimethylsilylethoxymethyl)-cyclopentanone **2** stereoselectively forms the corresponding α-fluoro-α,β-unsaturated ester **3**. This reaction has been applied to the construction of the fluoroolefin dipeptide isostere **4** of the prolyl-glycyl dipeptide **5**.

80. **STEREOSELECTIVE SYNTHESIS OF 3-FLUORO-3-HYDROXYALKYLAZETIDINONES** Koichi Araki, Robert Kawecki, John T. Welch, Department of Chemistry, The University at Albany, State University of New York, Albany, New York 12222.

The reaction of the enolate of 3-fluoroazetidinone **1** with aldehydes and ketones gives a mixture of two aldol diastereoisomers differing in configuration at the side chain. The very high stereoselectivity of this reaction at C-3 of the ring was remarkable. Investigation of the nature of the carbanion derived from **1** sheds light on the origin of this selectivity. The conditions for stereoselective reduction of the 3-acetyl derivative of **1** with various reducing agents will be described.

81. ASYMMETRIC SYNTHESIS WITH 3-FLUOROAZETIDINONE. Koichi Araki, Ferenc Gyenes, John C. O'Toole, John T. Welch,* Department of Chemistry, State University of New York at Albany, Albany, New York, 12222.

Optically active 3-fluoroazetidinone **1** prepared by the ketene-imine cycloaddition method is a convenient building block for asymmetric synthesis. This azetidinone has been alkylated in a highly stereoselective manner at the 3-position. The preparation of fluorinated synthons based on these building blocks will be described. Syntheses employing **1** include the transformation of **1** into 1-O-acetyl-5-O-(*tert*-butyldimethylsilyl)-2,3-dideoxy-2-fluoro-3-(4-methoxy-phenyl)amino-D-*xylo*-furanose, **3**, and the gentosamine derivative 1,4-di-O-acetyl-2,3-dideoxy-2-fluoro-3-(4-methoxy-phenyl)amino-D-*xylo*-pyranose, **4**.

82. REGIOSPECIFIC AROMATIC SUBSTITUTION VIA ARYL 1,2-DIPOLAR SYNTHONS. NICKEL (O) CATALYZED CROSS COUPLING OF ARYL TRIFLATES WITH GRIGNARD REAGENTS.

Saumitra Sengupta, Claude A. Quesnelle and Victor Snieckus.
Guelph-Waterloo Centre for Graduate Work in Chemistry, University of Waterloo, Waterloo, Ontario, CANADA N2L 3G1.

In pursuit of regioselective functionalization of aromatic nucleii, we are investigating a general theme of sequential electrophilic-nucleophilic substitution utilizing aryl 1,2-dipolar synthons (3). As part of the nucleophilic component we disclose here the methodological developments of Ni-catalyzed cross coupling of aryl triflates with Grignard reagents which provides a facile entry into contiguously substituted aromatics.

83. DIRECTED ortho METALATION. THE O-THIOCARBAMATE AS A VERSATILE NEW METALATION DIRECTOR FOR THE SYNTHESIS OF POLYSUBSTITUTED AROMATICS. F. Beaulieu and V. Snieckus, Guelph-Waterloo Centre for Graduate Work in Chemistry, University of Waterloo, Waterloo, Ontario, Canada N2L 3G1.

Previous work has established the O-carbamate as a useful Directed Metalation Group (Sibi, M.P.; Snieckus, V. *J. Org. Chem.* 1983, *48*, 1935-37).

Recently we have demonstrated that aryl O-thiocarbamate **1** also undergo facile *ortho*-lithiation and the resulting *ortho*-lithiated species can be trapped with a variety of electrophiles to give products **2**. The scope and limitations of this new metalation director as well as the conversion of **2** into S-thiocarbamate **3** will be discussed.

E^+ = TMSCl, MeI, PhCHO, PhSSPh, I_2, Et_2NCOCl

84. DEHYDRATIVE CYCLIZATION OF DIPHENYL SULFOXIDE BY LITHIUM DIISOPROPYLAMIDE: A SIMPLE ROUTE TO DIBENZOTHIOPHENE. Songman Hsu and L. M. Tolbert*, School of Chemistry and Biochemistry, Georgia Institute of Technology, Atlanta, GA 30332-0400

Dibenzothiophene (**DBT**) is prepared in quantitative yield from diphenyl sulfoxide (Ph_2SO) and lithium diisopropylamide in tetrahydrofuran. Similarly, diphenyl sulfone (Ph_2SO_2) also reacts with lithium diisopropyl amide to form **DBT** in high yield. In contrast, diphenyl sulfide is inert to these reaction conditions. A possible mechanism involves initial deprotonation at the α-carbon, followed by cyclization and elimination of hydroxide. A similar mechanism for Ph_2SO_2 produces dibenzothiophene sulfoxide (**DBTSO**), which undergoes reduction by diisopropylamide anion to **DBT**. Separate studies with **DBTSO** and **DBTSO**$_2$ confirmed that reduction to **DBT** takes place readily under the reaction conditions, supporting the proposed mechanism.

85. REDIRECTED DIRECTED LITHIATION, III. D. W. Slocum, R. Moon and M. G. Slocum, Department of Chemistry, Western Kentucky University, Bowling Green, KY 42101 and A. Siegel and S. Payne, Department of Chemistry, Indiana State University, Terre Haute, IN 47809.

Relatively few investigations have been published describing physical and theoretical studies of the directed metalation reactions. Of the factors that can influence an aromatic compound's ability to undergo directed metalation, five have been varied in this investigation: (1) coordinating ability of the directing group, (2) ortho H acidity-enhancing ability of the directing group, (3) the size of the coordinated ring system formed upon metalation, (4) structure of the metalating system, (5) reaction conditions. Interplay of these factors for the substituents -F, -OMe, -CH_2NMe_2, -C(O)NMe_2, -NMe_2 and -CN has revealed the following: (1) dramatic increase of the initial rate of directed metalation in one system (PhOMe) by use of TMEDA, (2) effectiveness of TMEDA in catalytic amounts, (3) acceleration of the rate of directed metalation provided by a p-substituent (-F), (4) examples of the reversal of the site of metalation when two substituents are competing (-OMe, -CH_2NMe_2). Other reactions that compete with the directed metalation reaction have also been studied.

86. FACE SELECTION IN ENOLATE CHEMISTRY. Vani R. Bodepudi and William J. le Noble Department of Chemistry, State University of New York at Stony Brook, Stony Brook, New York, 11794-3400

The enolates of E & Z-2-benzoyl-5-phenyladamantanes react stereoselectively with D_2O to give Z & E-2-benzoyl-2-deuterio-5-phenyladamantanes in the ratio 59:41. 2-Methylene (methoxy, phenyl)-5-phenyladamantane on ketonization under acidic conditions yields 2-benzoyl-5-phenyladamantanes with a Z : E ratio of 57:43. In both instances, the major isomer results from the attack of proton by the **en** face of the enol intermediate. These results, which are in contrast to other examples which we have previously ascribed to transition state hyperconjugation, will be discussed.

87. **SILICON MEDIATED ALKYLATIONS: A CONVENIENT SYNTHESIS OF 9,9-DIALKYL-9,10-DIHYDROANTHRACENES.** Peter W. Rabideau, R. K. Dhar, Department of Chemistry, Louisana State University, Baton Rouge, LA 70803.

A trimethylsilyl (TMS) substituent is used to control the regiochemistry of alkylation in 9, 10-dihydroanthracenes (9,10-DHA) to produce 9,9-dialkyl-10-TMS-9,10-DHAs. The TMS group is subsequently removed resulting in the first convenient synthesis of a variety of 9,9-dialkyl-9,10-DHAs in good to excellent yields. Studies on the migration of TMS in various 9-alkyl-9-TMS-DHAs have also been carried out to better understand this rearrangement process.

88. NEW STRATEGY FOR CYCLIZATION—O-FUNCTIONALIZATION OF 5-, 6-, and 7-MEMBERED CARBOCYCLIC RINGS. David MaGee and Sidima Kabanyane, Department of Chemistry, University of New Brunswick, Fredericton, New Brunswick, Canada E3B 6E2.

The generality of the sequence 1 - 2 - 3 has been studied for both the intra- and intermolecular reactions. It has been established how to control the regiochemical problems associated with the functionalization of dianion 2. The utility of this sequence was demonstrated in its application towards the synthesis of 5-, 6-, and 7-membered rings. The possibility of utilizing this sequence for the production of diosphenols was also investigated.

R,R',R''= H or alkyl

89. TRANSMETALATION REACTIONS OF ALKYL ZIRCONOCENES: COPPER-CATALYZED CONJUGATE ADDITION TO ENONES. Peter Wipf,* and Jacqueline H. Smitrovich, Department of Chemistry, University of Pittsburgh, Pittsburgh, PA 15260

Organocuprates are among the most useful organometallic derivatives applied in organic synthesis. However, most of the ligands that are transferred via both higher and lower order cuprates originate from organolithium or organomagnesium species. This report demonstrates a novel *in situ* transmetalation/conjugate addition protocol that converts alkenes to cuprates via readily obtained alkyl zirconocene intermediates. Differences in the properties of Schwartz's reagent from various sources are discussed.

90. **METHYL AS A BLOCKING GROUP IN THE REACTION OF MIXED ALKYLMETHYLZINCS WITH ALDEHYDES.** Morris Srebnik and Alan D. J. Griffiths, Department of Chemistry, University of Toledo, Toledo, Oh, 43606.

We have determined the relative migratory tendency of a various alkyl groups attached to zinc in the reaction of mixed alkylmethylzincs, RZnMe, with aldehydes. The transfer of the methyl group from zinc to an aldehyde is relatively slow compared to the other groups investigated. Thus when R=Et, i-Pr, n-Bu, Bnz, cyclohexyl or cyclopentyl, essentially exclusive transfer of the R groups occurs in preference to Me, enabling the use of the latter group as a inexpensive blocking group. The mixed alkylmethylzincs are readily available in-situ by the reaction of the corresponding Grignards or organolithium reagents. Salt-free solutions of RZnMe, necessary for the asymmetric version of this reaction, have been obtained by treating the in-situ prepared zinc reagents with the stoichiometric amounts of 1,4-dioxane.

$$RZnMe + R'CHO \longrightarrow R\underset{}{\overset{OH}{-}}R'$$

91. **A PRACTICAL REVERSED POLARITY ALTERNATIVE TO ORGANOCUPRATE CONJUGATE ADDITION**
J. R. Johnson, P. S. Tully, Peter B. Mackenzie, Dept of Chemistry, Northwestern U., Evanston, IL.

Conjugated enals and enones react with chlorotrialkylsilanes and bis(1,5-cyclooctadiene)nickel(0) to afford 1-(trialkylsilyloxy)allyl(chloro)(pyridine)nickel(II) complexes, which in turn react with halocarbons to yield the products of net homoenolate anion equivalent coupling. The scope, chemoselectivity, mechanism and applications of this chemistry will be discussed.

$$NiCl_2(H_2O)_6 \xrightarrow[-6\ H_2O]{C_5H_5N} NiCl_2(C_5H_5N)_4 \xrightarrow{\text{one-pot}}$$

1) 2 Na, xs COD
2) $R^1CHR^2CHR^4CO$, Me_2R^3SiCl
3) RX, hv

Product: alkene with R^1, R^2, $OSiMe_2R^3$, R, R^4 substituents, 63-84%

92. **ALDOL CHEMISTRY VIA INITIAL DOUBLE BOND MIGRATION**

Y. D. Ward, J. A. Faunce, Peter B. Mackenzie, Dept of Chemistry, Northwestern U., Evanston, IL.

Optically active allyl alkyl ethers and achiral allyloxysilanes and allylborates are isomerized to the corresponding (E)-enol ether or enolate derivatives by cationic bis(phosphine)iridium catalysts (eq 1). The scope and mechanism of this chemistry will be discussed, as will its utility in subsequent stereoselective aldol/oligoaldol/polyaldol chemistry.

$$\text{CH}_2\text{=CHCH}_2\text{OM}^1 \xrightarrow{ML_n} \left[\begin{array}{c} \text{OM}^1 \\ | \\ ML_n \\ | \\ H \end{array} \right] \xrightarrow{-ML_n} \text{CH}_3\text{CH=CHOM}^1 \qquad (1)$$

M^1 = 8-phenylmenthyl, $SiMe_n(OCH_2CH=CH_2)_{4-n}$, $B(OR)_2$; $ML_n = [(MePh_2P)_2Ir(THF)_n]PF_6$

93. RECENT DEVELOPMENTS IN THE SYNTHESIS OF STRAINED CYCLIC CUMLENES. Mary M. Kirchhoff, Richard P. Johnson, William C. Shakespeare, Tina M. Andro, Stephen G. Swartz, Jr., Department of Chemistry, University of New Hampshire, Durham, NH 03824

Progress toward the synthesis of several new strained, cyclic cumulenes (**1-4**) will be reported. Further examples of the reactions of 1,2,3-cyclohexatriene (**5**) will also be presented.

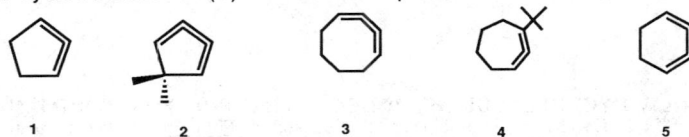

94. HETEROCYCLIC HOMOCONJUGATED POLYACETYLENES. Mark J. Cooney and Lawrence T. Scott, *Department of Chemistry, University of Nevada, Reno, Nevada 89557.*

The synthesis and properties of heterocycles **1-4**, other related macrocycles, and acyclic reference compounds will be described. Special attention will be given to the cyclic delocalization of electrons in such compounds.

95. SYNTHESIS AND REACTIVITY OF SUBSTITUTED ENEDIYNE SYSTEMS. Kamal N. Bharucha, Robert G. Bergman, Department of Chemistry, University of California, Berkeley, California 94720.

Dibromomaleic anhydride has been coupled with tin acetylides in high yields to give a series of functionalized enediyne compounds. Anhydride **1** is a versatile starting material for a variety of enediyne systems. Extended enediyne systems such as **2** have also been prepared in high yield using palladium-catalyzed couplings of terminal acetylenes with vinyl halides. 1,4-Dehydrobenzene intermediates derived from these functionalized enediynes are being studied.

96. CYCLIC CONJUGATED ENEDIYNES VIA THE ELIMINATION OF A THIONOCARBONATE IN A LATENT Z-HEX-3-ENE-1,5-DIYNE UNIT. M. F. Semmelhack and J. J. Gallagher, Department of Chemistry, Princeton University, Princeton, NJ 08544.

New methodology for the synthesis of the ene-diyne toxins is based on introduction of the ene unit at a late stage in synthesis procedures. The ketal-diyne dianion **1** was prepared and used in alkylations and additions to carbonyl units in the construction of ene-diyne precursors such as the simple version, **2**. Conversion to the thionocarbonate and elimination induced by Ni(0) or phosphines (Corey-Winter reaction) produced the ene-diynes under mild conditions.

97. SYNTHESIS OF NEW CYCLIC ENEDIYNE MODEL COMPOUNDS FOR BERGMAN CYCLIZATION STUDIES. M. F. Semmelhack, T. Neu and F. Foubelo. Frick Chemical Laboratory, Princeton University, Princeton, NJ 08544.

Aromatic and quinone analogs (e.g., **1** and **2**) of the ene-diyne toxins were prepared and examined with regard to rates and efficiency of Bergman cyclization to the diradicals (e.g., **3**). While the benzene, naphthalene (**1**) and anthracene analogs show high barriers to the cyclization, the corresponding naphthoquinone (**2**) and anthraquinone derivatives rearrange under mild conditions : $t_{1/2}$ = 20h at 55°C for **2**.

98. AN EFFICIENT SYNTHESIS OF ortho-ISOTOLUENES AND DIENE-ALLENES. Yemane W. Andemichael and Kung K. Wang*, Department of Chemistry, West Virginia University, Morgantown, WV 26506.

Treatment of **3**, prepared from **1** and **2**, with KH afforded o-isotoluene **5**. The electrocyclic rearrangement of **4** was very facile at room temperature. Diene-allene **6** was produced from **3** by using H_2SO_4 to induce the Peterson olefination reaction.

99. STEREOSELECTIVE SYNTHESIS OF ENYNES AND ENYNE-ALLENES. Yu Gui Gu, Yemane W. Andemichael, and Kung K. Wang*, Department of Chemistry, West Virginia University, Morgantown, WV 26506.

Condensation of **1** with **2**, followed by treatment with 2-aminoethanol, afforded **3** with high diastereoselectivity. The subsequent Peterson olefination reaction, promoted by KH or $NaHSO_3$, provided enyne-allene **4** or **5**, respectively. Enyne-allenes, such as **5**, have been found to have interesting DNA-cleaving properties. Enynes were similarly synthesized by condensation with simple aldehydes.

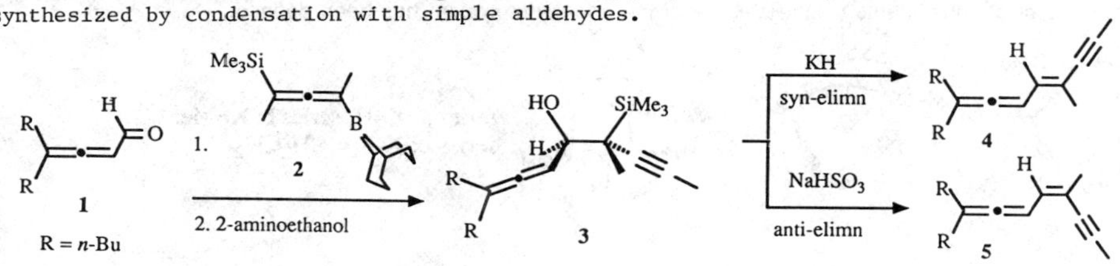

R = n-Bu

1. Me₃Si-B (2)
2. 2-aminoethanol

KH, syn-elimn
NaHSO₃, anti-elimn

100. A MULTIPLE EQUILIBRATION STRATEGY FOR POLYPROPIONATE SYNTHESIS
<u>Norman E. Pratt</u>, Yun-bo Zhao, and Kim F. Albizati,* Dept. of Chemistry, Wayne State University, Detroit, Michigan, 48202.

A new method for the stereocontrolled formation of polypropionate natural products, based on a multiple equilibration has been developed. A new method for the rapid formation of 1,3,5 - triones and their subsequent ring closure and multiple equilibration will be presented.

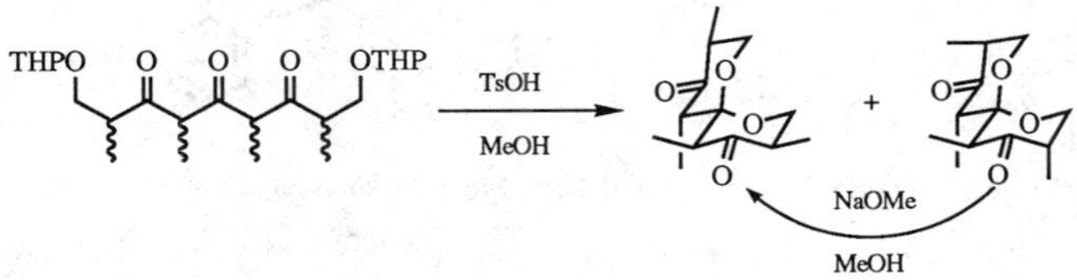

TsOH / MeOH

NaOMe / MeOH

101. THE CLAISEN REARRANGEMENT OF 4-ALLYLOXYISOQUINOLINES.
Richard W. Franck and <u>Tony E. Nicolas</u>. Department of Chemistry, Hunter College/CUNY, New York, N.Y., 10021.

In the context of our efforts towards the synthesis of Sakyomicin A, an antiviral agent with demonstrated inhibitory properties towards the HIV virus, intermediate 4-alkoxy -3-alkenyl substituted isoquinolines were required. We have found that the Claisen rearrangement of 4-allyloxyisoquinolines provides a convenient access to the required intermediates. The synthesis of a variety of 4-allyloxyisoquinolines and the results of our study on their Claisen rearrangement will be presented.

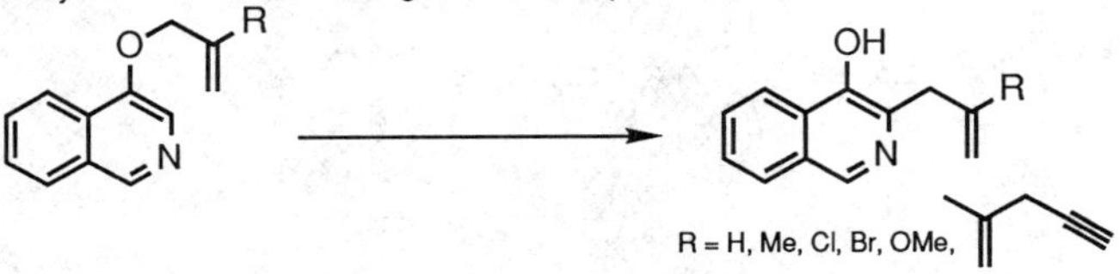

R = H, Me, Cl, Br, OMe,

102. MODELS FOR THE ENE-DIYNE TOXINS WITH AN ENOL-ETHER TRIGGER.
M. F. Semmelhack, Tatsuya Minami, and <u>Takashi Date</u>, Department of Chemistry, Princeton University, Princeton, NJ 08544

Functional models have been designed for the core ene-diyne unit of the natural toxins, calicheamicin and esperimicin. The trigger is a enol ether (1) to ketone (2) hydrolysis, activating Bergman rearrangement to the diradical (3). The synthesis of three examples (1a-c) has been pursued, and a surprising effect of substituents on the stability of the ene-diyne system has been defined.

series a: X=H; series b: X=Me;
series c: X-X = CH_2CH_2

103. MECHANISTIC STUDY ON THE FORMATION OF ALKYL ESTERS AND AMIDOESTERS OF (ANILINOMETHYLENE)PROPANEDIOIC ACIDS BY DIRECT CONDENSATION. Carmen Estrella, Héber Muñoz and Joaquín Tamariz*. Departamento de Química Orgánica, Escuela Nacional de Ciencias Biológicas, I.P.N. Carpio y Plan de Ayala, 11340 México, D.F.

Ethyl esters of (anilinomethylene)propandioic acids (1) have been prepared by direct condensation of diethyl malonate, ethyl ortoformate, and acetic anhydride (2) with the corresponding aniline, under Lewis acid catalysis ($ZnCl_2$). En the absence of 2 and $ZnCl_2$, amidoesters 3 were obtained. Analysis of these processes by proton NMR and tlc, isolation of some intermediates and crossed experiments, have allowed to establish the concurrence of several equilibria between intermediates like formamidines and theirs malonate addition derivatives. A mechanism on the formation of 1 and 3 is proposed.

E = CO_2Et; R = H, p-Me, m-Cl, m-NO_2, m-Me, p-MeO, p-NO_2

104. SYNTHETIC STRATEGIES FOR THE CONSTRUCTION OF NOVEL AZATRICYCLIC SYSTEMS: NOVEL 5-HT3 AND 5-HT4 SEROTONERGIC AGENTS. Daniel P. Becker, Daniel L. Flynn, Roger Nosal, Dan Zabrowski. Searle Research & Development, Skokie, IL 60077.

Our drug discovery efforts have led us to pursue conformationally-restricted analogs of serotonin (5-HT) which has culminated in the discovery of novel azatricyclic compounds such as SC-50410 which are potent and selective ligands at serotonin 5-HT3 and 5-HT4 receptors. Synthesis of SC-50410 and the aminoazatricycle 1 has been accomplished as outlined in the scheme. Azabicyclo[3.3.0]octane 3 was prepared by three complementary routes employing 1) a palladium-catalyzed [3+2] annulation; 2) an atom-transfer radical cyclization; and 3) a cobalt-mediated Pauson-Khand double annulation. Stereoselective elaboration of 3 into the azabicycle 2 was accomplished by reaction with bis(tosyl)sulfodiimide and hydroboration. Compound 2, which contains four contiguous centers in the proper orientation for conversion into the azatricycle, was converted in quantitative yield to the tosyl-protected azatricycle.

105. PHENOXYDIAZIRINES AS USEFUL CARBENE PRECURSORS FOR PHOTOAFFINITY LABELING. Yasumaru Hatanaka, Makoto Hashimoto and Yuichi Kanaoka Faculty of Pharmaceutical Sciences, Hokkaido University, Sapporo, 060 JAPAN

Although diazirines are already used for photoaffinity labeling, only limited varieties of useful diazirines are known. A series of new diazirines 1 has been synthesized. The thallation of 1 and subsequent nitration, iodination, or palladium-catalyzed carbonylation gave several useful diazirines 2,3 for photoaffinity labeling. The relative ease of derivatization may contribute to the widespread use of diazirines for photoaffinity labeling. Photoreactivities of these diazirines will also be discussed.

$R = CH_2CO_2CH_3, CH_3$
$R' = I, NO_2$

106. **SYNTHESIS OF UNSYMMETRIC UREAS OR CARBAMATES, SULFAMIDES USING TRIPHOSGENE, CBMIT, AND SBMIT.** Xiu C. Wang[1] and Dale J. Kempf[2], [1]Chemical And Agricultural Products Division, [2]Pharmaceutical Products Division, Abbott Laboratories, Abbott Park, IL 60064

Unsymmetric ureas or carbamates are synthesized in a sequential one-pot reaction of triphosgene or CBMIT (Rapoport, J Am Chem Soc, 111, 4856, 1989) with amines or alcohols. At low temperature, the reaction is chemoselective, and for aminoalcohols, no protection of the hydroxy group is required. A new reagent, 1,1'-sulfonylbis(3-methylimidazolium) triflate (**SBMIT**) has been prepared and used for the synthesis of unsymmetric sulfamides.

$R^1NH_2 \xrightarrow[THF, -78°C]{COX_2 /NMM} R^1NHCOX \xrightarrow[RT]{R^2NH_2 \text{ or } R^2OH} R^1NHCOYR^2$ (COX_2 = triphosgene; CBMIT; Y=NH, O)

Imidazole $\xrightarrow[\text{2. } CH_3OTf]{\text{1. } SO_2Cl_2;}$ **SBMIT** $\xrightarrow[\text{2. } R^2NH_2, RT]{\text{1. } R^1NH_2, NMM, THF, -78°C;} R^1NHSO_2NHR^2$

107. **NON-AQUEOUS PREPARATION OF β-HYDROXYNITRILES FROM EPOXIDES WITH LITHIUM CYANIDE IN THF.** James A. Ciaccio* and Catherine Stanescu, Dept. of Chemistry, Fordham University, Bronx, NY, 10458.

A variety of epoxides are regioselectively converted to β-hydroxynitriles in good yield upon treatment with LiCN in refluxing tetrahydrofuran (THF). The anhydrous conditions permit a one-pot conversion of epoxides to γ-hydroxyamines via hydride reduction of the corresponding β-hydroxynitriles. The scope and limitations of this epoxide opening reaction in comparison with current methods will be presented and discussed.

108. **A CONVENIENT HYDRODEAMINATION AND CHLORODEAMINATION OF PRIMARY AMINES.** D. Wei and F.S. Guziec, Jr., Department of Chemistry, New Mexico State University, Las Cruces, New Mexico 88003.

Treatment of p-toluenesulfonamides of primary amines with chloroamine at room temperature in the presence of base leads to reductive deamination. If excess chloroamine is present, the corresponding alkyl or aryl halides are obtained instead in good yields. Both reactions proceed via tosylhydrazine intermediates and the course of the reaction is often governed by steric hindrance. Optimal conditions for obtaining selective halogenation or reduction will be discussed.

109. A NEW ORGANOBORON REAGENT FOR THE PREPARATION OF PROPARGYLIC ALCOHOLS. Jonathan C. Evans[a], Christian T. Goralski[a] and Dennis L. Hasha[b], [a]Pharmaceuticals Process Research, [b]Analytical Sciences, The Dow Chemical Co., Midland, Michigan 48674

Propargylic alcohols are important intermediates in the synthesis of a number of natural products. A new organoboron reagent, B-(2-(trimethylsilyl)ethynyl)-9-borabicylco-[3.3.1]nonane (2), has been prepared from the lithium anion of trimethylsilylacetylene (1) and B-methoxy-9-borabicylco[3.3.1]nonane. This new reagent has been completely characterized as a 1:1 complex with tetrahydrofuran by ^1H, ^{11}B, ^{13}C and ^{29}Si NMR. This reagent has been shown to react with aldehydes and ketones to give excellent yields of the corresponding trimethylsilyl substituted propargylic alcohols. The reaction of reagent 1 with chiral aldehydes was also examined.

110. THERMOLYSIS OF O-ETHYL ESTERS OF PHOSPHORAMIDIC ACIDS; GENERATION OF ETHYL META-PHOSPHATE. Stefan Jankowski and Louis D. Quin, Department of Chemistry, University of Massachusetts, Amherst, MA 01003.

Two phosphoramidates of structure 1 have been found to undergo decomposition on heating in toluene in the range 55-100° C. From the products formed, and from kinetics measurements on the decomposition, it appears that the primary process involves loss of the amino group with the formation of ethyl metaphosphate (2).

1a, R=R'=Et, 1b, R=Mes, R'=H

This is supported by trapping experiments for this highly reactive species. Thus, when ethanol is present, the ethyl metaphosphate is trapped as $(EtO)_2P(O)OH$. Silica gel, through the numerous OH groups on its surface, also is an effective trapping agent, as we have recently shown in two other processes where a metaphosphate is generated. Phosphoramidates such as 1 therefore represent a new, potentially useful source of alkyl metaphosphates for service as phosphorylating agents.

111. **Solid and Liquid State Photochemistry of an N-Acyl-1-azadiene.** Min Teng, Joseph W. Lauher and Frank W. Fowler *Department of Chemistry, State University of New York at Stony Brook, Stony Brook, New York 11713.*

In contrast to most compounds containing carbon-nitrogen double bonds, compound 1 is remarkably photochemically active. Direct irradiation in solution results in a rapid and nearly quantitative photochemical conversion to the novel bicyclo[3.2.0]hexene derivative 2 whereas irradiation of the 1 in the crystalline state gave mainly the 2 + 2 cycloadduct 3. This unusual behavior allows for an investigation of topochemical control of chemical reactivity.

112. **SYNTHESIS OF SUBSTITUTED 2-CARBOMETHOXY-4'-METHYLBIPHENYLS, 2-CYANO-4'-METHYLBIPHENYLS AND 2-(TETRAZOL-5-YL)-4'-METHYLBIPHENYLS.** Nancy J. Kevin, Ralph A. Rivero, and William J. Greenlee, Department of Exploratory Chemistry, Merck Sharp and Dohme Research Laboratories, P.O. Box 2000, Rahway, NJ 07065

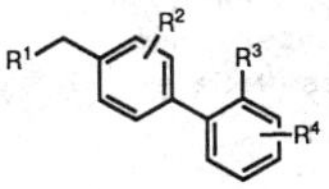

R^1 = H, OTBS
R^2 = H, Me
R^3 = CO_2Me, CN, CN_4CPh_3
R^4 = NO_2, NH_2, Cl, Br, F, Allyl

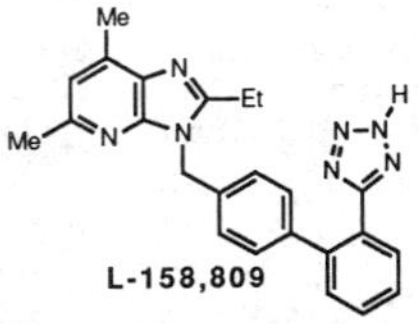

L-158,809

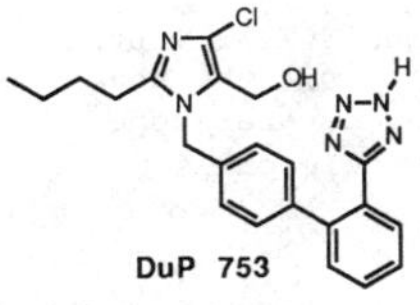

DuP 753

A series of substituted 2-carbomethoxy-4'-methylbiphenyls, 2-cyano-4'-methylbiphenyls and 2-(tetrazol-5-yl)-4'-methylbiphenyls were prepared to investigate the effect of substitution on the biphenyl element of the potent angiotensin II receptor antagonists L-158,809 and DuP 753. The key step in the construction of most of these substituted biphenyls is the biaryl coupling of an arylhalide with an arylmetal. The synthesis of these pharmaceutically important intermediates will be described.

113. **A NOVEL SYNTHETIC APPROACH TO PYRROLO[2,3-d]PYRIMIDINE ANTIFOLATES.** Tetsuo Miwa, Takenori Hitaka, and Hiroshi Akimoto, Chemistry Research Laboratories, Research and Development Division, Takeda Chemical Industries, Ltd., Jusohonmachi, Yodogawa-ku, Osaka 532, Japan

A novel and efficient synthetic method for pyrrolo[2,3-d]pyrimidine antifolates will be described. The key reaction of this method is the photo-initiated free radical addition of bromomalononitrile or ethyl bromocyanoacetate to an enol ether 3 to afford the back-bone skeleton of the targeted antifolate molecule. The key intermediates 2 are smoothly converted to the pyrrolo[2,3-d]pyrimidine antifolates 1 in 4 steps and in high overall yield.

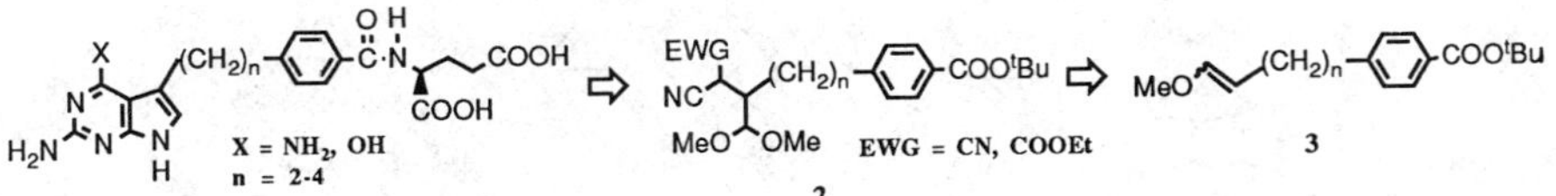

X = NH_2, OH
n = 2-4

EWG = CN, COOEt

114. **THE SYNTHESIS OF A CONFORMATIONALLY RIGID, CALCIUM ENTRY BLOCKER.** Joel C. Barrish, Steven H. Spergel, Suzanne Moreland, and Andrew Pudzianowski, Bristol-Myers Squibb Pharmaceutical Reseach Institute, Princeton, NJ 08543.

Recently, a series of 1-benzazepin-2-one calcium entry blockers (2), related to diltiazem (1), were prepared. In order to test hypotheses for the receptor bound conformation of the benzazepinones, compounds that replace the benzazepinone seven membered ring with a rigid bicyclic framework of general formula 3 were synthesized. The synthesis involved elaboration of the known ketoester 5 to aminoketone 4, followed by the introduction of the 4-methoxyphenyl group and subsequent stereoselective reduction.

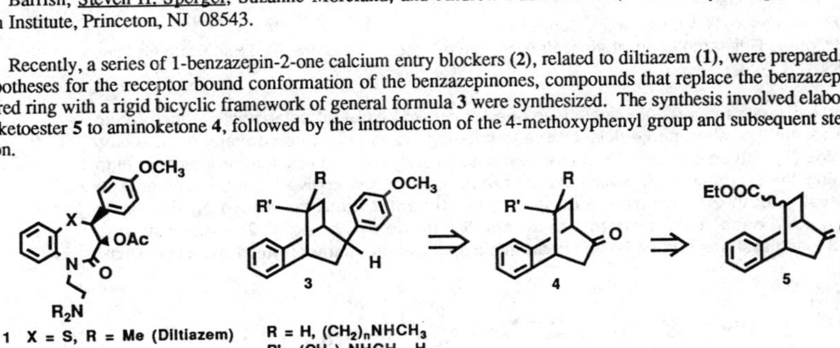

1 X = S, R = Me (Diltiazem)
2 X = CH_2

R = H, $(CH_2)_nNHCH_3$
R'= $(CH_2)_nNHCH_3$, H

115. **ENANTIOSELECTIVE ROUTES TO *syn* AND *anti*-ß-PHENYLCYSTEINES**
M. Amparo Lago, James Samanen[#] and John D. Elliott[*], Departments of
[*]Medicinal Chemistry and [#]Peptidomimetic Research, SmithKline Beecham
Pharmaceuticals, 709 Swedeland Road, Swedeland, PA 19479, USA

Stereocontrolled routes to both *syn*- and *anti*-ß-phenylcysteines **1** and **2** have been developed via the enantioselective synthesis of the corresponding ß-phenylserines, followed by stereospecific S_N2 displacement of their respective mesylates with a sulfur nucleophile.

116. SYNTHETIC STUDIES TOWARDS THE IMMUNOSUPPRESANT FK-506
Zhaoyin Wang, Merck Frosst Centre For Therapeutic Research
P. O. Box 1005, Pointe Claire/Dorval, Québec, Canada H9R 4P8

Several subunits of FK-506 were synthesized from readily available starting materials. The cyclohexyl portion **1** and the allylic iodide **2** were prepared from chiral oxazolidone **4**. The δ-lactone **3** was synthesized from *L*-arabinose. Coupling of these subunits to form the advanced intermediate corresponding to the C_{10}-C_{34} fragment of FK-506 will be presented.

117.
Synthesis of 4 - Amido 2 - carboxytetrahydroquinolines

G. I. Stevenson, P. D. Leeson, R. W. Carling, I. Sanderson.
Merck Sharp and Dohme Research Laboratories, Neuroscience Research Centre, Terlings Park, Harlow, Essex, CM20 2QR, U.K.

4 - Amido 2 - carboxytetrahydroquinolines have been shown to be potent antagonists at the glycine site of the NMDA receptor. A short and efficient route to this ring system has been developed involving reaction of the imine (1) with an enamide (2) under Lewis acid catalysis. The reaction proceeds in high yield to afford mainly the *cis* product (3), which can be readily epimerised to give the *trans* isomer. The reaction is general to both electron rich and electron deficient enamides, affording the same regiochemistry in each case. This route has been applied to the synthesis of 3, 4 disubstituted analogues, C2-C3 fused systems and variations in aromatic substitution, many of which were previously unavailable.

118. **SYNTHESIS OF THE PUTATIVE METABOLITES OF THE CYCLOPENTA[a]-PHENANTHRENES. SYNTHESIS OF THE trans-3,4-DIHYDRODIOL OF THE MUTAGEN 15,16-DIHYDROCYCLOPENTA[a]PHENANTHREN-17-ONE.** Masato Koreeda and Stephen A. Woski, Department of Chemistry, The University of Michigan, Ann Arbor, Michigan 48109

The first synthesis of the *trans*-dihydrodiol derivative (**1**) of a biologically active cyclopenta[a]-phenanthrene has been achieved. The cyclopenta[a]phenanthrene skeleton is rapidly and efficiently assembled utilizing the Lewis acid-catalyzed Diels-Alder reaction of **2** with an α-heterosubstituted cyclopentenone **3**, a "cyclopentynone" equivalent.

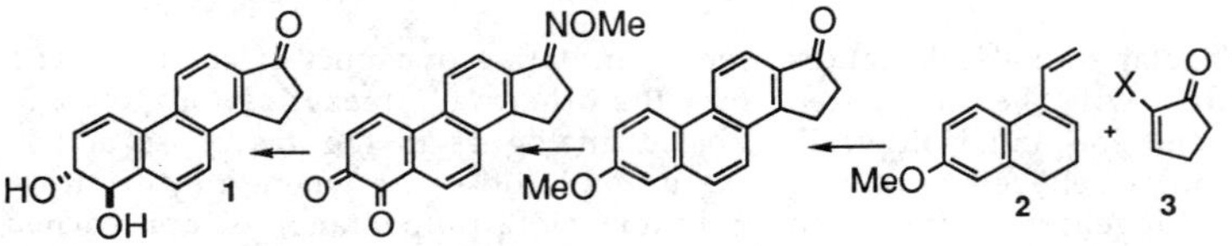

119. **SYNTHETIC ROUTES TO IMIDAZOLE-5-ACRYLIC ACIDS: POTENT SMALL MOLECULE ANGIOTENSIN II RECEPTOR ANTAGONISTS** Joseph A. Finkelstein, Dimitri Gaitanopoulos and Richard M. Keenan*, Department of Medicinal Chemistry, SmithKline Beecham Pharmaceuticals, P.O. Box 1539, King of Prussia, PA 19406-0939

A novel series of 1-benzylimidazole-5-acrylic acid angiotensin II receptor antagonists has recently been reported. This series, exemplified by SKF 108566 (**2**, X = 4-CO_2H, R_1 = H, R_2 = 2-thienylmethyl), exhibit high affinity and specificity for the AII-1 receptor. A variety of synthetic routes have been utilized to attach the acrylate moiety to the benzylimidazole nucleus, including functionalization of the imidazole-5-carboxaldehyde as well as a palladium catalyzed coupling of a stannylimidazole with β-substituted acrylate derivatives.

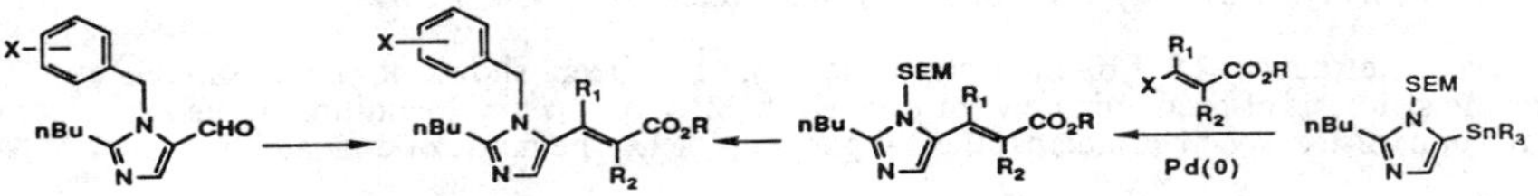

120. **MECHANISMS OF REACTIONS AT HETEROATOMS; DETERMINATION OF THE STERIC COURSE OF FORMAL NUCLEOPHILIC SUBSTITUTIONS AT NONSTEREOGENIC ATOMS BY THE ENDOCYCLIC RESTRICTION TEST.** Peter Beak, Department of Chemistry, University of Illinois at Urbana-Champaign, Urbana, IL 61801

In this lecture determinations of the geometries of transition structures for formal substitution reactions at heteroatoms will be discussed. Experimental results using the endocyclic restriction test, for the conversion illustrated below in which N is a formal nucleophile, L is a leaving group, and Y is a nonstereogenic heteroatom which is transferred from L to N, will be reported. Use of the information obtained by this approach to distinguish between concerted, associative and dissociative mechanisms for these reactions will be presented.

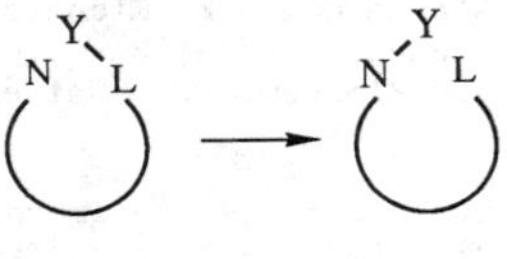

121. **IONIC ORGANIC CHEMISTRY IN THE GAS PHASE. WHAT CAN WE LEARN FROM REACTIONS WITHOUT SOLVENTS,** John I. Brauman, Department of Chemistry, Stanford University, Stanford, California 94305-5080.

Some of the most common ionic organic reactions—including proton transfers, nucleophilic displacement, and carbonyl addition—take place readily in the gas phase. These reactions show substantial similarities as well as differences when compared with their solution analogs. Study of these reactions can thus reveal intrinsic behavior and also solvation effects on ionic reactions.

122. **UNSYMMETRICAL ACCESS TO A SYMMETRICAL REACTION COORDINATE; A POSSIBLE DOMAIN FOR DYNAMIC EFFECTS.**
Barbara A. Lyons, Jörg Pfeifer, Thomas H. Peterson, and Barry K. Carpenter, Department of Chemistry, Baker Laboratory, Cornell University, Ithaca, New York 14853-1301.

Ab initio molecular orbital calculations, semiquantitative dynamics calculations, and experimental results will be presented for the thermal deazetization of 2,3-diazabicyclo[2.2.1]hept-2-ene and isotopically labeled analogues in the gas phase and in solution. The reaction is believed to represent an example of a case where a symmetrical reaction coordinate (degenerate ring flipping in bicyclo[2.1.0]pentane) is approached unsymmetrically from a diazenyl biradical. The asymmetry in the kinetic product ratio (revealed by isotopic labeling) is thought to arise from dynamic effects. This reaction raises several general questions of terminology such as the meaning of an intermediate and the distinction between stepwise and concerted processes. Perhaps more importantly, it also raises questions of whether the current methods for analyzing the kinetics of thermal reactions are sufficient to describe their mechanisms accurately.

123. **INTRAMOLECULAR 2π + 4π CYCLOADDITION REACTIONS OF ALLYL CATIONS AND CATION RADICALS GENERATED FROM 1,3-DIENES.** Paul G. Gassman, Department of Chemistry, University of Minnesota, Minneapolis, MN 55455.

The acyclic tetraene, 1,3,8,10-undecatetraene (**1**), has been shown to be a suitable precursor for the synthesis of predominately (or exclusively) **2**, **3**, or **4** via cycloaddition of the a-b, b-c, or c-d portion of **1**, respectively, across the 1-4 positions of **1**. The mode of cycloaddition has

been shown to be a function of substitution pattern, catalyst, and temperature. The intermediacy of allyl cations and of 1,3-diene cation radicals in these processes will be discussed.

124. TOWARDS A MOLECULAR-SIZE 'TINKERTOY' CONSTRUCTION SET: SYNTHESIS AND CHARACTERIZATION OF THE RODS. Josef Michl, Piotr Kaszynski, Andrienne C. Friedli, Prabha Ibrahim, Tomasz Janecki, Steven Hamrock, Richard W. Morrison, Donald E. David, Allan J. McKinley, Joey D. Pierce, Karel Base, Gudipati S. Murthy, and V. Balaji, Department of Chemistry and Biochemistry, University of Colorado, Boulder CO 80309-0215

Progress to date will be described on the synthesis and characterization of doubly terminally functionalized linear molecules for use as rods in the assembly of molecular size mechanical objects analogous to those children build from the 'Tinkertoy' construction set elements. The molecular rods are based on rigid linear strings of bicyclic or polycyclic cages. Simple examples of assembled structures will be given as well.

Acknowledgement. The work has been supported by the National Science Foundation, the Welch Foundation, and the Texas Advanced Research Program. Most of it was performed while the authors were at The University of Texas at Austin.

125 DESTABILIZED KETENES. Thomas T. Tidwell, Department of Chemistry, University of Toronto, Toronto, Ontario, Canada M5S 1A1

Substituents affect both the thermodynamic stability and the rate of reaction of ketenes. In order to separate these two effects investigations have carried out on ketenes **1** that are either totally unknown or are formed only as transient intermediates in solution to determine which of these ketenes are truly "destabilized".

$$\begin{array}{c} R \\ \diagdown \\ C=C=O \\ \diagup \\ H \end{array} \quad \textbf{1} \quad R = CF_3, CN, R^1C \equiv C, CH_2=CH, NO_2$$

The thermodynamic stability of ketenes has been measured by 6-31G*//6-31G* MO calculations for the isodesmic reaction $RCH=C=O + CH_3CH=CH_2 \rightarrow CH_3CH=C=O + RCH=CH_2$. A linear relationship has been found between ΔE and the Boyd-Edgecombe (*JACS* **1988** 4182) group electronegativities of the substituents, such that ketene stabilization increases with the electropositive character of the substituent. Ketenes are also stabilized by π acceptors, and destabilized by π donors. These substituent effects are compared to those for β ethyl carbocations and for diazomethanes.

126. ORGANIC PHOTOCHEMISTRY IN RESTRICTED REACTION SPACES. CHEMISTRY AT THE LIQUID-LIQUID, LIQUID-SOLID, AND SOLID-GAS INTERFACES. N.J. Turro, Department of Chemistry, Columbia University, New York, NY 10027.

The photochemistry of organic molecules adsorbed in the restricted reaction spaces provided by micelles and porous solids has been investigated. For the systems investigated, the primary photochemical events are velieved not to be generally significantly altered compared to homogeneous solution, i.e., geminate radical pairs are produced, but the course of the reaction of the intermediates produced depends strongly on the absorption site, the geometry of the space and the diffusional and rotational dynamics of the pair in the restricted space. These points will be discussed with examples given.

127. TRANSFORMATIONS OF ORGANIC COMPOUNDS MEDIATED BY TRANSITION METAL COMPLEXES. Robert G. Bergman, Department of Chemistry, University of California, Berkeley, California 94720

Many stoichiometric and catalytic reactions are now known in which organotransition metal complexes mediate the interconversion of organic compounds. These processes have been used in both industrial and academic chemistry to convert simple organic compounds into more complex functionalized materials, in some cases with high degrees of stereoselection. A small number of fundamental bond-breaking and bond-cleavage steps form the mechanistic "building blocks" for the more complicated transformations that result in the conversion of one organic molecule to another. This talk will focus on the search for new examples of such fundamental primary organometallic reactions and on studies of the mechanisms of these processes. Current research is directed toward the insertion of metals into carbon-hydrogen bonds in simple alkanes, the chemistry of heterodinuclear organometallic complexes, and the chemistry of complexes containing bonds between metals and heteroatoms such as oxygen, nitrogen, and sulfur.

128. TUNING THE MULTIPLET ENERGY GAP IN NON-KEKULE MOLECULES. SYNTHESIS, STRUCTURE, REACTIVITY, AND SPIN. Jerome A. Berson, Department of Chemistry, Yale University, New Haven, Conn., 06511

The disjoint biradical tetramethyleneethane (TME), first synthesized by Dowd, forms a conceptual template for the construction of a series of non-Kekulé heterocyclic molecules 3,4-dimethylenefuran, 3,4-dimethylenethiophene, and 3,4-dimethylene-N-substituted pyrroles. Ab initio and semi-empirical theory suggest that the singlet-triplet gap in TME should be small and that the variable interaction of the heteroatom lone-pair electrons with the TME NBMOs in the heterocyclic compounds therefore should permit control of the singlet-triplet energy separation and ordering. Recent experimental tests of these ideas will be reported.

129. OLEFIN DIMERIZATIONS, OLIGOMERIZATIONS AND POLYMERIZATIONS CATALYZED BY Co(III) AND Rh(III) COMPLEXES. M. Brookhart, Department of Chemistry, University of North Carolina at Chapel Hill, Chapel Hill, NC 27599-3290

Agostic cobalt(III) complexes of the type $C_5Me_5(L)CoCH_2CH_2-H^+$ (L = PR_3, $P(OR)_3$) function as ethylene polymerization catalysts and α-olefin oligomerization catalysts while rhodium(III) hydride complexes of the type $C_5Me_5(L)Rh(CH_2CH_2)H^+$ (L = $P(OR)_3$, PR_3, C_2H_4) serve as olefin dimerization catalysts. The basic catalytic mechanisms functioning for each of these systems will be presented. Results of recent studies which will be discussed include (a) the stabilization of the cobalt catalyst by highly stable counterions, (b) the living polymerization of ethylene, (c) the mechnism of oligomerization of 1-hexene, (d) copolymerization of ethylene and 1-hexene, (e) synthesis of terminally functionalized ethylene oligomers and polymers and, (f) a highly efficient rhodium-catalyzed tail-to-tail dimerization of methyl acrylate.

130. NEW REACTIONS OF ORGANORHENIUM COMPOUNDS. Charles P. Casey, Hiroyuki Sakaba, Chae S. Yi, and Todd L. Underiner, Department of Chemistry, University of Wisconsin, Madison, Wisconsin 53706.

The reactions of a number of new organorhenium compounds will be discussed. The synthesis and reactions of π-propargyl rhenium cations will be described. Several reactions that involve the formation of agostic alkyl rhenium intermediates will be discussed. The synthesis and reactions of new compounds with rhenium-rhenium double bonds will be reported.

131. PHOTOREACTIVITY IN SUPERCRITICAL FLUIDS. Kevin O'Shea, Jimmie Combs, Keith Johnston, Marye Anne Fox, Departments of Chemical Engineering and Chemistry, University of Texas, Austin, Texas 78712.

Supercritical fluids constitute a unique environment for exploring solvation effects since small pressure changes near the critical point cause large changes in density-dependent properties (solubility, viscosity, refractive index, and dielectric). In characterizing photoreactions of molecules in supercritical fluids, we take advantage of the pronounced effect of pressure on density for the purposeful alteration of chemical potential, partial molar volume, and diffusional kinetics. Examples which will illustrate these effects include solvation effects on [2+2] photodimerizations, on cage effects in a Norrish Type 1 photocleavage, and on observable rates of very fast intermolecular photoinduced electron transfers.

132. DEVELOPMENT AND MECHANISTIC STUDY OF NEW ASYMMETRIC OXIDATION CATALYSTS. Eric N. Jacobsen, Dept. of Chemistry, University of Illinois, Urbana, IL 61801.

Highly enantioselective reactions of practical value have been developed using chiral salen complexes as catalysts. In particular, epoxidation of a variety of alkenes is now possible in good yield and with >90% ee. Subtle steric and electronic modifications to the salen ligand template are readily accomplished, and have permitted catalyst optimization through systematic ligand variation. The epoxidations take place via oxo transfer, and the high stereoselectivities that are observed provide a previously unavailable mechanistic tool for studying this important reaction class. In particular, energetically favored trajectories for substrate approach have been mapped out from the available data, and the intermediacy of radical and/or zwiterionic species has been evaluated. The mechanistic insights gleaned from the epoxidation process have in turn been applied to the rational design of other potentially useful asymmetric oxidation reactions.

133. EFFECTS OF STRUCTURAL CHANGES ON RADICAL STABILITIES. F. G. Bordwell, Department of Chemistry, Northwestern University, Evanston, Illinois 60028-3113

The ephemeral nature of radicals has made it difficult to obtain quantitative estimates of relative radical stabilities. The best estimates have been obtained from the determination of relative homolytic bond dissociation energies (BDEs) of small molecules in the gas phase by a variety of methods. In our laboratory we have developed a method of estimating BDEs in DMSO solution for weak acids, H-A, that are comparable to gas phase values, by use of equation (1)

$$BDE = 1.37\ pK_{HA} + 23.06\ E_{ox}(A^-) + 56 \qquad (1)$$

This method is potentially applicable for estimates of thousands of weakly acidic H-A bonds in simple or complex molecules. The BDEs thus obtained have been found to give excellent guides to relative radical stabilities. Examples will be given of the application of these BDE estimates to assess the importance of donor and acceptor effects on carbon- and nitrogen-centered radicals.

134. ORGANIC SYNTHESIS USING GROUP 4 TRANSITION METAL COMPLEXES. Stephen L. Buchwald, Department of Chemistry, Massachusetts Institute of Technology, Cambridge, MA 02139.

This talk will focus on our recent work which utilizes group 4 transition metal complexes as reagents for the preparation of a variety of polyfunctionalized molecules.

135. WAVELENGTH DEPENDENT ORGANIC PHOTOCHEMISTRY
William G. Dauben, Department of Chemistry, Berkeley, CA 94720

The study of the wavelength dependence of the photochemistry of polyenes and conjugated unsaturated ketones has led to a more detailed understanding of the mechanism of these photochemical processes. Examples of the results will be presented.

136. SELECTIVE OXYGENATIONS WITH METALLOPORPHYRINS. FROM MODEL ENZYMES TO MODEL MEMBRANES   John T. Groves, Department of Chemistry, Princeton University, Princeton, NJ 08544, USA.

Considerations of the heme iron active site of cytochrome P-450 have led to the development of a family of metalloporphyrin based systems for the oxygenation of hydrocarbons. Reactive oxometalloporphyrin complexes have been characterized and do show the requisite reactivity for olefin epoxidation and alkane hydroxylation. A *trans*-dioxo ruthenium(VI) porphyrin complex has been prepared and has been shown to be an efficient catalyst for the aerobic epoxidation of olefins. We have also shown that this complex will catalyze the hydroxylation of alkanes under appropriate conditions. When oxometalloporphyrins were placed in a chiral environment, both catalytic asymmetric epoxidation and catalytic asymmetric hydroxylation have been demonstrated. Amphiphilic metalloporphyrin complexes have been designed which bind with predetermined orientations in phospholipid bilayer membranes. Selective oxygenations of ordered, amphiphilic substrates such as steroids and fatty acids have been observed in the resulting catalytic assemblies. The strategy of biocompatible catalysis in which a redox enzyme such as pyruvate oxidase was used to activate a synthetic, membrane bound catalyst has been demonstrated in these lipid assemblies. Recent advances in understanding the mechanisms of oxygen transfer with metalloporphyrins and examples of selective oxygenations will be discussed.

137. RECOGNITION AND REPLICATIONS IN MODEL SYSTEMS. Julius Rebek, Jr., Department of Chemistry, Massachusetts Institute of Technology, Cambridge, MA 02139

For the past several years we have been involved in projects designed to bring forces and chemical reactions together in space and time. For this purpose we have generated a number of cleft like molecules which feature convergent functionalities at fixed distances. These act as model receptors for a large number of a biorelevnat target structures. More recently we have used these molecules to demonstrate autocatalysis in their own replication. The lecture will trace recent developments along these lines and emphasize molecular shape and its implications for bioorganic and prebioorganic chemistry

138. THE FRUITFUL INTERPLAY BETWEEN COMPUTATIONAL AND EXPERIMENTAL CHEMISTRY, Paul von Ragué Schleyer, Institut für Organische Chemie der Friedrich-Alexander-Universität Erlangen-Nürnberg, Henkestrasse 42, D-8520 Erlangen, Germany.

Chemistry no longer is an exclusively experimental science. The combination of highly sophisticated quantum mechanics programs and ever more powerful and cost effective computers now enable the chemist not only to help interpret the results of experimental investigations, but also to discover entirely new formulations of matter, and to do so systematically and reliably. Since the vast majority of possible chemical compounds in unknown, calculational investigations can lead to structures which are fundamentally different from those already described. Our predictions of molecules with, e.g., "anti van't Hoff" structures, bridging lithiums, octet-rule violating stoichiometries, and unanticipated geometries, are now being verified experimentally with increasing frequency. This underscores the ability of quantitative theory to lead rather than merely to follow experiment. Nevertheless, when used in conjunction, theory and experiment are particularly effective. New examples will be presented in this lecture.

139. SYNTHESIS AND PROPERTIES OF EXPANDED PORPHYRINS. J. L. Sessler, A. K. Burrell, M. J. Cyr, D. Ford, G. Hemmi, B. G. Maiya, T. Murai, T. D. Mody, T. Morishima, and K. Schreder, Department of Chemistry and Biochemistry, University of Texas, Austin, TX 78712.

The syntheses and properties of a variety of new and previously reported expanded porphyrins, including texaphyrin (1), pentaphyrin (2), and rubyrin (3) will be presented. Applications of these systems to a variety of problems in biotechnology and medicine will also be discussed.

140. **SYNTHESIS OF ARCHITECTURALLY COMPLEX NATURAL AND UNNATURAL PRODUCTS**
Amos B. Smith, III

Department of Chemistry, The Monell Chemical Senses Center and The Laboratory for Research on the Structure of Matter, University of Pennsylvania, Philadelphia, Pennsylvania 19104

Recent progress in the synthesis of structurally complex natural and unnatural products possessing significant bioregulatory properties will be presented. Topics not yet selected from the areas of breynolide, FK506, calyculin and/or non-peptide peptidomimetics will be discussed.

141. ORGANIC TRANSFORMATIONS INVOLVING ORGANOCOBALT: MECHANISMS AND SYNTHESIS. K. Peter C. Vollhardt, Department of Chemistry, University of California, Berkeley, California 94720.

New developments in the application of cobalt-mediated [2+2+2]cycloadditions to synthesis include the extremely facile assembly of molecules of theoretical and medicinal interest, the discovery of novel fused ring migrations, the observation of dramatic enantioselectivity caused by proximal enantiomerically pure auxiliaries in the conversion of prochiral alkenes, and the isolation of the pivotal organometallic mechanistic link between alkyne substrates and catalytically generated arene products. Some of the structures illustrating these advances are shown below.

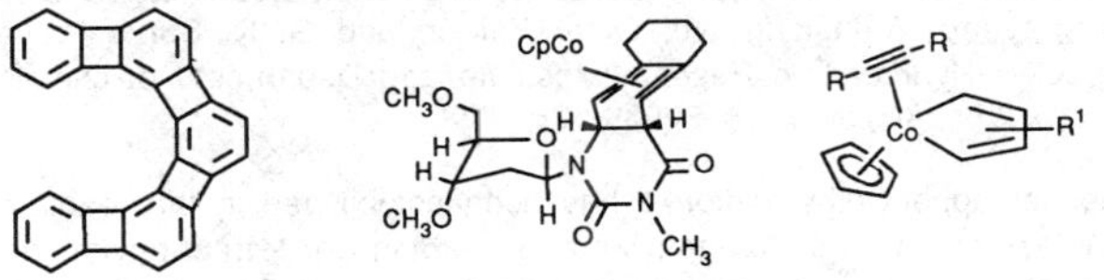

142. ADVANCES IN ORGANIC PHOTOCHEMISTRY

Howard E. Zimmerman, Department of Chemistry, University of Wisconsin, Madison, Wisconsin 53706.

Our recent research in the area of organic photochemistry has been in several directions which include: (1) Unusual rearrangements affording heterocycles, (2) The effect of the reaction environment on the course of photochemical rearrangements; we term this "Photochemistry in a Box" and "Spiderweb" photochemistry, (3) Reactions proceeding via diradicals, (4) $\perp$-π* and π-$\perp$* photochemistry, (5) Radical ion photochemistry, and (6) Theoretical studies of photochemical mechanisms. A survey of this work will be presented.

143.
HOW ARE WE DOING IN PREDICTING ABSOLUTE RATES OF INTRAMOLECULAR ELECTRON AND ENERGY TRANSFER? G.L. Closs, Department of Chemistry, The University of Chicago, Chicago, Illinois 60637.

Much work in our laboratory over the last decade has been aimed at being able to predict absolute rates of intramolecular electron and energy transfer from structural parameters and reaction conditions. The problem is basically one of being able to predict the nuclear overlap functions (Franck Condon factors) and the electronic coupling matrix elements. Starting with classical Marcus theory and expanding it to a semiclassical model, including high frequency modes, has gone a long way to solve the nuclear overlap problem. Modern MO theory promises to provide an equally important step in achieving the final goal.

144. **THERMAL ISOMERIZATION OF A VINYLCYCLOBUTENE TO A CYCLOHEXADIENE.**
Gitendra C. Paul and Joseph J. Gajewski,* Department of Chemistry Indiana University, Bloomington, Indiana 47405.

Above 100°C, 5-methylenespiro[3.5]nona-1-ene(**1**) undergoes cyclobutene ring opening to E-triene **2** and an apparent 1,3-shift to a hexahydronaphthalene (**3**) in a 3 to 1 ratio. E-triene **2** gives **3** in a competitive reaction. The activation energies for all three processes are obtained from a kinetic simulation at three temperatures. It is likely that the Z-triene **4** also derived by cyclobutene ring opening, is an intermediate in the **1** to **3** conversion. Speculation on the **2** to **3** conversion focuses on the possible intermediacy of a 3-vinylcyclohexenocyclobutene (**5**). The apparent non-involvement of a concerted 1,3-shift in the **1** to **3** reaction is discussed.

145. **EVIDENCE FOR THE INTERMEDIACY OF THE PENTAPHENYLETHYL CATION IN THE REACTION OF PERDEUTERATED TRIPHENYLMETHYL CATION WITH DIPHENYLDIAZOMETHANE OR DIPHENYLKETENE,**
George A. Olah, Mesfin Alemayehu, An-hsiang Wu, Omar Farooq and G. K. Surya Prakash Donald P. and Katherine B. Loker Hydrocarbon Research Institute and Department of chemistry, University of Southern California Los Angeles, CA 90089-1661

Abstract: Intermediacy of the pentaphenylethyl cation 1 has been established in the reaction of triphenyl cation with diphenyldiazomethane or diphenylketene leading to tetraphenylethylene. Reacting perdeuterated triphenylmethyl cation with diphenyldiazomethane gave tetraphenylethylene containing about 10% labeled tetraphenylethylene with two pentadeuterophenyl groups. These results confirm the intermediacy of 1 undergoing slow 1,2-phenyl exchange through the corresponding phenonium ion.

146. **DIPOLE VS DIRADICAL CHARACTER OF TRIMETHYLENEMETANES. NEW REACTION PATHWAYS INVOLVING THE DIPOLAR FORM.** Lynn M. Brown, M. R. Masjedizadeh, and R. D. Little, Department of Chemistry, UCSB, Santa Barbara, CA 93106

During the past decade, Little and coworkers have extensively studied inter- and intramolecular diyl trapping reactions. These reactions lead to the construction of one or more rings, and they do so in a predictable manner. We have learned many details concerning the nature and chemistry of the diyl and have been able to use them in the total synthesis of several antitumor-antibiotics. Existing data clearly indicates the *diradical* nature of the intermediates. Particularly compelling evidence comes from the detection of a triplet diradical by ESR, even in cases where the diyl could have been intercepted intramolecularly at a rate too rapid to allow its detection. Attempts to generate the diyl in water rather than an organic solvent, however, leads to the formation of products which are best be explained through the intervention of a *dipolarar diradical*. In those cases, we have determined: (a) the preferred site of capture of both the "H" and the "OH", (b) that water attacks the five-membered ring of the intermediate with equal probability from the top and bottom faces, and (c) the conditions necessary to switch between the two types of intermediates: diradical behavior is observed when the dielectric is approximately 30 or less, while dipolar behavior is observed above 40.

* = •/• or +/-, diradical, dipolar diradical

147. **A NEW MECHANISM FOR REACTIONS OF CARBENES AND BICYCLO[1.1.0]BUTANES**

Linxiao Xu, <u>Thomas Miebach</u>, Udo H. Brinker*, *Department of Chemistry, State University of New York at Binghamton, Binghamton, New York 13902-6000*;
William B. Smith, *Department of Chemistry, Texas Christian University, Fort Worth, Texas 76129*.

Reaction of tricyclo[4.1.0.0^{2,7}]heptane with dihalocarbenes and bromine are reported. A new *side bond opening* mechanism for reactions of carbenes and bicyclo[1.1.0]butanes is proposed.

148. **EVALUATION OF ACIDITY OF STRONG ACID CATALYSTS**; Dan Fărcaşiu and <u>Anca Ghenciu</u>, Dept. of Chemical and Petroleum Engineering, University of Pittsburgh, Pittsburgh, PA 15261.

The Hammett indicator method has limitations when applied to liquid acid catalysts, and it is inapplicable <u>in principle</u> to solid acids. Acidities can be evaluated from the C-13 NMR measurement of the hydronation equilibrium of α,β-unsaturated ketones, like 4-hexen-3-one (Ia) and 4-methyl-3-penten-2-one (Ib). Determination of the chemical shift difference δ(C-β)-δ(C-α), Δδ, in two acids allows a comparison of their acidities corrected for medium effects on chemical shifts. The variation of Δδ with concentration of I is linear; extrapolation to [I] = O (Δδ°) allows us to establish a thermodynamically meaningful acidity scale. We also found that the slope of the Δδ <u>vs.</u> [I] plot depends on acidity, being the steepest in the acid where I is half-protonated. For comparison of liquid and solid acids Δδ is determined at stoichiometric ratio of I to acid molecules or sites.

R-CH$_2$-CO-CH=CR'Me   I   <u>a</u>: R=Me, R'=H;   <u>b</u>: R=H, R'=Me

149. DEVELOPMENT OF NMR ACTIVE INDICATORS FOR MEASUREMENT OF CALCIUM ION CONCENTRATION. <u>L.A. LEVY</u>, B. Raju, E. Murphy, and R.E. London, Laboratory of Molecular Biophysics, NIEHS, Research Triangle Park, NC 27709.

The use of fluoro substituted 1,2 *bis*(o-aminophenoxy)ethane N,N,N',N'tetraacetic acid (F-BAPTA) has made possible the *in vivo* measurement of [Ca^{+2}] by ^{19}F NMR spectroscopy. In order to increase the sensitivity of these indicators by means other than increasing the number of magnetically equivalent nuclei we have synthesized several new fluorinated derivatives or modifications of the BAPTA structure. These new structures were suggested by consideration of the limiting effect of the relaxation time (T$_1$) of the observed nucleus on the rate of acquisition of spectra and the effect of the dynamics involved in the formation and dissociation of the calcium-chelator complex on the lineshape of the signals of these species involved in this equilibrium.

150. Search for the Conformational Effect Responsible for the Eclipsing in <u>cis-2-tert-</u>Butyl-5-(<u>tert</u>-Butylsulfonyl)-1,3-dioxane. <u>Bàrbara Gordillo</u>,[1] Eusebio Juaristi,[1] Roberto Martínez,[1] Rubén A. Toscano,[2] Peter White[3] and Ernest L. Eliel.[3] [1]Depto. Química, CINVESTAV-IPN, Apdo. Postal 14-740, 07000 México, D.F. [2]Instituto de Química, UNAM, Cd. Universitaria. 04510 México, D.F. [3]Department of Chemistry, UNC-CH, North Carolina 27599-3290, USA.

A crystal structure study of sulfones (<u>1</u>-<u>4</u>) was performed in order to determine their preferred rotational conformation. These structures were compared with the most stable conformations predicted by MMP2 calculations. From the results of this study, steric and stereoelectronic interactions appear to be responsible of the S-O/C-C bond eclipsing in sulfones 1 and 5 (structure A).

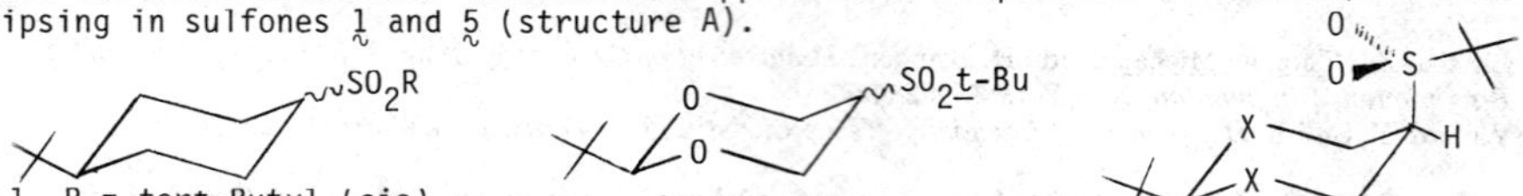

1, R = tert-Butyl (cis)
2, R = tert-Butyl (trans)
3, R = Methyl (cis)

4, (trans)
5, (cis)

A, X = CH_2, O

151. **PHTHALIDE CATION IS NOT A STABLE SPECIES DURING SOLUTION PROTONATION OF 3-HYDROXPHTHALIDE.** Treliant Fang, AT&T Bell Labs, Engineering Research Center, P.O. Box 900, Princeton, New Jersey 08540.

Protonation of 3-hydroxphthalide by triflic acid generates a short-lived phthalide cation (I) intermediate followed by nucleophlilic attack from the parent alcohol to form a stable diphthalyl ether (IV) under anhydrous condition. Traces of water in the acid facilitates the hydrolysis reaction toward open the lactone ring of the hydroxyphthalide and the ether IV to form protonated 2-carboxybenzaldehyde, which is stable under ambient conditions. No cyclic phthalide cation has been detected by COLOC experiments, contrary to the results reported by Scaiano et al. (Boate, D.R.; Johnston, L.J.; Kwong, P.C.; Lee-Ruff, E.; and Scaiano, J.C.J. Am. Chem. Soc. 1990, 112, 8858.)

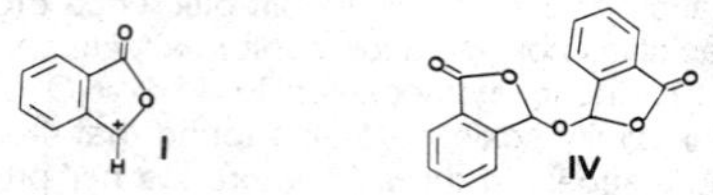

152. **KINETICS OF THE HYDROLYSIS OF SUBSTITUTED BENZYLIDENEMALONO-DIALDEHYDES.** Claude F. Bernasconi and Francis X. Flores, Department of Chemistry and Biochemistry, University of California, Santa Cruz, CA 95064

The kinetics of the reversible reaction of p- and m-substituted benzylidenemalonodialdehydes with water and OH^- have been determined in aqueous solution at 25°C. Brønsted-type plots of log k_1^{OH}

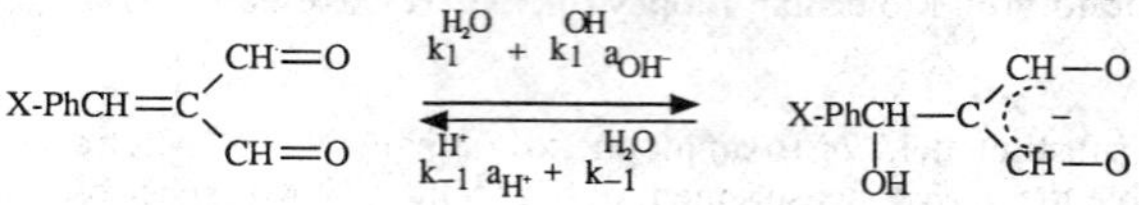

and log $k_1^{H_2O}$ vs. log $K_1^{H_2O}$ reveal negative deviations for X = p-NMe_2 and p-MeO. The negative deviations indicate a reduction in the intrinsic rate constant of the reaction, k_o ($k_o = k_1 = k_{-1}$ when $K_1 = 1$). This reduction may be understood to arise from a transition state imbalance whereby the loss of resonance in the olefinic substrate is ahead of bond formation.

153. **CONCERTED PERICYCLIC ELIMINATION WITH A CARBOCATION-LIKE TRANSITION STATE FOR THE REACTION OF TERTIARY CUMYL DERIVATIVES.** John P. Richard and Tina L. Amyes, *University of Kentucky, Department of Chemistry, Lexington, KY 40506-0055.*

We have determined the effect of changing the aromatic ring substituent on the amount of elimination to give the α-methylstyrene for the reaction of XArC(Me)$_2$Y (Y$^-$ = substituted benzoates and chloride) in 50:50 (v:v) trifluoroethanol/water. For $\sigma_x^+ < -0.08$ ionization of the substrate gives diffusionally-equilibrated carbocation intermediates, XArC(Me)$_2^+$, but no detectable yield of the α-methylstyrene (≤ 1%). For $-0.08 \leq \sigma_x^+ \leq 0.12$, there are 4 - 5% yields of the α-methylstyrenes, which arise from deprotonation of XArC(Me)$_2^+$ by the leaving group within an ion pair. An increase to $\sigma_x^+ > 0.12$ leads to the appearance of another pathway for the formation of the α-methylstyrenes, and their yields increase to a maximum of 30% for 3,5-di-CF$_3$ArC(Me)$_2$Cl. The rate constants for this elimination reaction are correlated by ρ^+_{elim} = -4.6, for XArC(Me)$_2$Cl, and m_{elim} = 0.70 for the reaction of 4-NO$_2$ArC(Me)$_2$Cl in MeOH/H$_2$O solvents. There is a large kinetic deuterium isotope effect [$(k_H/k_D)_{elim}$ = 4.4] on the elimination reaction of 3-FArC(CD$_3$)$_2$pentafluorobenzoate. The transition state for the elimination reaction is less polar than that for solvolysis of cumyl chlorides, for which ρ^+_{solv} = -6.1 and m_{solv} = 1.0. The structure-reactivity data and and the isotope effect excludes mechanisms that involve highly polar, nonselective, ion pair intermediates. The data are consistent with a *concerted pericyclic* elimination reaction through a carbocation-like transition state, which provides a pathway that avoids the formation of very highly destabilized cumyl carbocation intermediates.

154. **ABSENCE OF NUCLEOPHILIC ASSISTANCE BY SOLVENT AND AZIDE ION TO THE REACTION OF CUMYL DERIVATIVES: MECHANISM OF NUCLEOPHILIC SUBSTITUTION AT TERTIARY CARBON.** John P. Richard, <u>Tina L. Amyes</u> and Tomas Vontor, *University of Kentucky, Department of Chemistry, Lexington, KY 40506-0055*

k_{obsd} for reaction of tertiary cumyl derivatives, XArC(Me)$_2$Y, in 50:50 (v:v) trifluoroethanol/water is independent of added N$_3^-$ in the range 0 - 0.50 M. Cumyl derivatives with $\sigma_x^+ \leq$ -0.08 react by an S$_N$1 (D$_N$ + A$_N$) mechanism through the liberated cumyl carbocations. The product selectivity decreases sharply for $\sigma_x^+ \leq$ -0.08 and then levels off to a constant limiting value of $k_{az}/k_s \approx$ 0.75 M^{-1} for $\sigma_x^+ \geq$ 0.12. The selectivity of 4-NO$_2$ArC(Me)$_2$Cl toward 1-propanethiol is $k_{PrSH}/k_s \approx$ 0.3 M^{-1}. The product selectivity k_{MeOH}/k_{TFE}, determined in 5:45:50 (v:v:v) MeOH/TFE/H$_2$O, decreases with decreasing carbocation stability for $\sigma_x^+ \leq$ 0.34 and then levels off to a constant limiting selectivity of $k_{MeOH}/k_{TFE} \approx$ 2 for $\sigma_x^+ \geq$ 0.78. There is no evidence for onset of a nucleophilically assisted reaction with azide ion, 1-propanethiol or methanol when the carbocation becomes very short-lived. It is concluded that there is also no assistance by the more weakly nucleophilic solvent TFE/H$_2$O.

155. **NUCLEOPHILIC ADDITION OF ALKYL THIOLATE IONS TO β-NITROSTYRENES: RADICALOID TRANSITION STATE OR PRINCIPLE OF NONPERFECT SYNCHRONIZATION?** Claude F. Bernasconi and <u>David F. Schuck</u>, Department of Chemistry and Biochemistry, University of California, Santa Cruz, CA 95064

Rate (k_1) and equilibrium constants (K_1) for the nucleophilic addition of alkyl thiolate ions to phenyl-substituted β-nitrostyrenes have been determined in water. Hammett and Brønsted plots give reasonable correlations except that the point for 4-Me$_2$N shows strong positive deviations. Within the framework of the Principle of Nonperfect Synchronization (PNS) one may interpret these deviations in terms of π-donations by the 4-Me$_2$N group which "preorganizes" the substrate and facilitates charge delocalization into the nitro group at the transition state, thereby lowering the typically high intrinsic barrier of nitronate ion forming reactions.[1] Another interpretation, suggested by Hoz[2] in a similar situation, is in terms of a radicaloid transition state that would be stabilized by the π-donor substituent. The absence of a substantial positive deviation from our plots for the 4-MeS and 4-CN substituents, both known as radical stabilizers, does not support the notion of a radicaloid transition state, and hence we favor the PNS interpretation.
(1)   Bernasconi, C. F.; Renfrow, R. A.; Tia, P. R. *J. Am. Chem. Soc.* **1986**, *108*, 4541.
(2)   Hoz, S.; Gross, Z. *J. Am. Chem. Soc.* **1988**, *110*, 7489.

156. **RADICAL REACTIVITY OF CYCLIC ACETALS**, [1]J.M. Antonucci, [1]J.W. Stansbury, [2]B.B. Reed, [1]NIST, [2]ADAHF, Gaithersburg, MD, 20899
Two types of cyclic acetals (CA) were synthesized: I, nonvinyl CAs and II, exomethylene CAs. Radical reactivities of type I CAs were assessed by their reaction with N-bromosuccimide (NBS) using a catalytic amount of azobis-isobutyronitrile (AIBN) and by their radical copolymerization with methyl methacrylate (MMA) and styrene (ST). Radical reactivities of type II CAs were evaluated by their ability to both homo- and copolymerize. Type I CAs with NBS/AIBN undergo hydrogen abstraction (HA) and ring cleavage to generate an ester radical intermediate which is quenched by bromine. Type I did not copolymerize with MMA or ST indicating that intermolecular HA did not occur. Polymerization of type II CAs occurred by several competing pathways that varied according to reaction conditions: (a) 1,2-addition, (b) 1,2-addition

with ring opening to keto-ether structures and (c) ring opening with molecular elimination. Monomer and polymer structures were characterized using IR and ^{1}H NMR spectroscopy. A difunctional type II CA yielded a crosslinked polymer. Type II CAs of varying structure can be easily prepared to yield polymers with minimal polymerization shrinkage and having a wide range of properties. Supported by NIDR Grant DE09322, ADAHF, and NIST.

157. RADICAL RING OPENING CHARACTERISTICS OF CYCLIC VINYL MONOMERS, [1]B.B. Reed, [2]J.W. Stansbury, [2]J.M. Antonucci, [1]ADAHF, [2]NIST, Gaithersburg, MD 20899
The goal of this study was to synthesize exomethylene substituted cyclic monomers capable of efficient free radical ring-opening polymerization. To this end, 2-methylene-4-phenyl-1,3-dioxolane (I) and 5,6-benzo-2-methylene-1,3-dioxepane (II) were synthesized and characterized by ^{1}H NMR and IR spectroscopy. These monomers offer access to stabilized benzylic propagation radicals upon ring opening and their polymerizations have been studied previously. Polymerizations of these monomers with t-butylperoxide at 120°C were reported to yield 100% ring opened polyesters. In this study polymerizations were conducted with azobisisobutyronitrile (AIBN) at 60-65°C. The bulk homopolymer of monomer I yielded only about 30% ring opening. The ring opening occurred primarily through the stabilized benzylic radical. The bulk homopolymer of monomer II exhibited 72% ring opening demonstrating that the ring size of a cyclic monomer can exert an influence over the mode of polymerization. Thus ring size is a parameter of primary importance in the design of efficient ring-opening monomers. The differences in ring opening efficiency observed for these two compounds can be attributed to ring strain and steric effects. Supported by NIDR Grant DE09322, ADAHF, and NIST.

158. MECHANISM OF THE ACID CATALYZED CHLORINATION OF 1-METHYLPYRROLE WITH N-CHLOROBENZAMIDES. Michael De Rosa* and Manuel Marquez, Department of Chemistry, The Pennsylvania State University, Delaware County Campus, 25 Yearsley Mill Road, Media, PA 19063.
The acid catalyzed (7 aliphatic carboxylic acids) chlorination of 1-methylpyrrole with 11 *para*- and *meta*-substituted N-chlorobenzamides was carried out in chloroform. A mixture of mono- and dichloropyrroles was formed. Reaction was first order with respect to pyrrole, N-Chlorobenzamide and acid. In a preliminary study a primary kinetic isotope effect was observed. This indicated that the rate determining step was the deprotonation of the σ-complex. Hammett and Bronsted correlations were carried out. The values of ρ(-0.85±.16) and α(0.49±.05) indicated that there was no change in the mechanism or rate determining step as the substituent or acid was varied.

159. EFECTO DE LA PRESION EN LAS PROPIEDADES DE CONDUCCION EN POLIMEROS ZWITTERIONICOS.
Cardoso, J., Blas, J., González, L., Huanosta, A.
Instituto de Investigaciones en Materiales, UNAM.
Apdo. Postal 70-360, Coyoacán 04510, México, D. F.

En el estudio de relajamiento dieléctrico de los polímeros zwitteriónicos se ha encontrado que las interacciones dipolo-dipolo y la presencia de sitios preferenciales para la conducción son los responsables de las propiedades dieléctricas que muestran estos materiales en estado sólido.
En este trabajo se analiza el efecto de la presión, P, utilizada para formar una pastilla de polímero, en sus propiedades de conducción. Existe un valor de P que optimiza el valor de la conductividad en polímeros derivados del ácido metacrílico conteniendo grupos del tipo N-óxidos neutralizados con NaCl. El estudio se hizo para $8.85 \leq P(ton/cm^2) \leq 2.65$, $-5 < T(°C) < 100$ y $0.005 \leq f(kHz) \leq 13000$. Este efecto podría interpretarse en términos de la existencia de una distancia crítica entre sitios preferenciales.

160. **REACTION OF 2,3-DIBROMOCARBONYL COMPOUNDS WITH NUCLEOPHILES**
ABDEL-HAMID A. YOUSSEF, SABER M. SHARAF, SAMIR K. EL-SADANY AND EZZAT A. HAMED. DEPARTMENT OF CHEMISTRY, FACULTY OF SCIENCE, ALEXANDRIA UNIVERSITY, P.O. BOX 426 IBRAHIMIA, ALEXANDRIA 21321, EGYPT.

The reactions of <u>erythro</u>-methyl-2,3-dibromo-3-arylproponates **1**a-d, and erythro-2,3-dibromo-1,3-diarylpropan-1-ones **2** a-g with different nucleophiles in DMF were studied. The nucleophiles used range from strong ones as thiophenoxide (n=9.9) to weak ones as phenoxide ions(n=5.8), together with piperidine (n=7.5) as a moderate one. The thiophenoxide ion gave initially <u>trans</u>-dehalogenation products followed by Michael addition to give the corresponding saturated ones. The phenoxide ion, on the other hand, gave the corresponding 3-phenoxy-2-propenoates and 3-phenoxy-2-propen-1-ones, respectively. Using piperidine in a 1:2 molar ratio cpd. **1** gave exclusively (Z)-methyl-2-bromo-3-aryl-2-propenoates, whereas a 1:3- molar ratio gave a mixture of <u>erythro</u>- and <u>threo</u>-methyl-2,3-<u>bis</u>-piperidino 3-arylproponates. Cpd. **2** gave either 2-piperidino-3-aryl-1-phenyl-2-propen 1-one or a mixture of this and <u>bis</u>-2,3-piperidino-3-aryl-1-phenylpropan-1-one depending on the substituents on the phenyl and benzoyl groups.
The mechanism of the reactions will be discussed.

161.
TOPOLOGICAL REQUIREMENTS OF PREMICELLAR PORPHYRIN SOLUBILIZATION AS DETECTED BY J-AGGREGATION. <u>David C. Barber</u>, Ruth A. Freitag-Beeston, Thomas A. Owens and David G. Whitten, University of Rochester, Department of Chemistry, Rochester, NY 14627.

The 4,0 atropisomer of intermediate (C_6-C_7) to long (C_{16}) chain picket fence porphyrins exhibits red shifted absorption, and reduced singlet and triplet lifetimes in dilute aqueous surfactant solutions (i.e. below the critical micelle concentration). The spectral and photophysical behavior of these species is consistent with formation of premicellar aggregates containing two or more porphyrin chromophores lying in an offset, face-to-face arrangement (λ_{max} ~ 436 nm, v. sharp, for C_6-C_7 compounds) which differ from normal aqueous porphyrin aggregates (λ_{max} ~ 426 nm, broad) or monomeric porphyrin in homogeneous or micellar solution (λ_{max} = 418-420 nm, sharp). The most characteristic behavior is noted at intermediate chain lengths for the free base or Pd(II) complexes of 4,0 meso-tetrakis(2-hexylamidophenyl)porphyrin (4,0 THex) and for the free base of 4,0 THept, suggesting an optimal molecular hydrophobicity, side chain length and molecular topology for complex formation. These complexes, formed selectively from the 4,0 atropisomer, exhibit an SDS/porphyrin ratio of 3.2 for 4,0 THex. Acid-base titration behavior and Cu^{2+} incorporation rate trends suggest a structure in which the porphyrin core is isolated from the anionic surfactant head groups in a relatively hydrophobic microenvironment.

162. **NAPHTHOLS WITH PROTON-MEDIATING SIDE-CHAINS.** <u>Lilia Cuesta-Harvey</u> and Laren M. Tolbert*, School of Chemistry and Biochemistry, Georgia Institute of Technology, Atlanta, GA 30332-0400

The role of solvent structure in the dynamics of photoexcited proton transfer is a subject of intense and continuing interest. Naphthol derivatives containing side chains which constrain the available geometries for solvent cluster formation provide one approach for examining the effect of solvent. In particular, 1-(3-hydroxypropyl)-2-naphthol and 1-(2,3-dihydroxypropyl)-2-naphthol have been prepared and studied in mixed water/ethanol solvents. These "internally solvating" photoacids exhibit significantly different photodynamic behavior relative to the parent 2-naphthol.

163. SPECTROSCOPIC AND CHEMICAL STUDIES OF THE PHOTODECOMPOSITION PATHWAYS OF ADAMANTYLDIAZIRINE. M.S. Platz and Scott C. Morgan, Department of Chemistry, The Ohio State University, 120 W. 18th Avenue, Columbus, Ohio 43210.

The participation of the excited states of diazirines has been suggested previously as a possible explanation for different product distributions in reactions of photo versus thermal decompositions of diazirines. During the course of a laser flash photolysis study of the absolute kinetics of adamantylidene (2-adamantylcarbene), it was observed that added alcohols seemed to be intercepting a photogenerated precursor to the carbene. This observation has led to an examination of the photodecomposition pathway of the diazirine. Spectroscopic and chemical support for the formation of transient 2-diazoadamantane has been found and will be presented. Additionally, fluorescence emission believed to be attributable to the diazirine has been observed at 77 K, thus supporting the idea that an excited state of the diazirine could be participating (i.e. reacting with added alcohols) in the laser flash photolysis studies of the carbene.

164. SOLVENT EFFECTS ON THE SIDE SELECTIVITY OF SINGLET OXYGEN WITH α, β UNSATURATED ESTERS: EVIDENCE FOR A PEREPOXIDE INTERMEDIATE. M. Stratakis and M. Orfanopoulos, Department of Chemistry, University of Crete, 71110 Iraklion, Crete, Greece.

The ene reaction of singlet oxygen with α, β unsaturated esters, ketones, carboxylic acids, and sulfones has recently received considerable attention. In all these substrates it was found that the major product is formed regioselectively by preferential hydrogen abstraction from the alkyl group which is geminal to the electron - withdrawing functionality. In this work we have discovered a significant and mechanistic useful solvent effect on the side selectivity of the ene reaction of singlet oxygen with α, β unsaturated esters. Variation of solvent polarity causes an increase in the percentage of the minor ene adduct from 5% up to 20% in DMSO. When only one side of the double bond is reactive, the ene product ratio is insensitive to solvent polarity. These results are consonant with the formation of a perepoxide intermediate in a rather limiting step, followed by rearrangement to the ene products.

165. PHOTOCHEMISTRY OF BENZYL PHENYL SULFIDE
Steven A. Fleming, Anton W. Jensen
Department of Chemistry, Brigham Young University, Provo, UT 84602

The nucleoside transport inhibitor 6-[(4-nitrobenyzyl)thio]-9-(ß-D-ribofuranosyl)-purine, NBMPR, has been used successfully in photoaffinity labeling. Synthesis and photolysis of para- and meta-substituted derivatives of benzyl phenyl sulfide as analogs of NBMPR will be reported. Hammett plots based on nitro, cyano, trifluoromethyl, methyl, and methoxy substituents provide a mechanistic understanding that should assist in the application of the benzyl sulfide moiety for photoaffinity labeling.

166. A PHOTOCHEMICAL STUDY OF N-(12-DODECANOIC ACID) BENZOYLFORMAMIDE IN HOMOGENEOUS AND MICROHETEROGENEOUS MEDIA. W.G. Richard, C.A. Chesta, D.G. Whitten, Department of Chemistry, University of Rochester, Rochester, N.Y. 14627-0216,

An investigation of the photoreactivity of the amphiphilic α-oxoamide, N-(12-dodecanoic acid) benzoylformamide, (1), in aqueous solution and its inclusion with various hosts is presented. In all media photolysis of 1 yields predominantly the fragmentation products, mandelamide and the corresponding aldehyde. The most reasonable mechanism for the reaction is an initial electron transfer quenching of the "ketone" excited state with the amide serving as the electron donor followed by a proton transfer and a second electron transfer, to form products indicated. Compound 1 forms two types of complexes with β-cyclodextrin and shows enhanced quantum efficiencies for reaction. 1 also forms a complex with carboxymethylamylose. Here again reactivity is enhanced for the photoreaction of 1 upon complexation indicating that neither the electron transfer quenching of the excited states nor reaction of subsequent intermediates is prohibited in the complexes. Mechanistic implications of these results are discussed.

167. QUANTUM YIELDS OF SINGLET OXYGEN PRODUCTION BY TRIPLET CYCLOHEXENONES. Christopher A. Rhodes, David I. Schuster and Christopher S. Foote, Department of Chemistry and Biochemistry, University of California, Los Angeles, L. A., Ca. 90024 and Department of Chemistry, New York University, N. Y., N. Y. 10003.

Absolute singlet oxygen quantum yields have been determined for several triplet cyclohexenones by detection of the singlet oxygen emission at 1268 nm. In acetonitrile, singlet oxygen yields vary from 0.58 to 1.0 and oxygen quenching rates of the triplets (K_{ox}) are between 1/9 and 4/9 diffusion control (K_{dif}). The high rates of quenching suggest involvement of the triplet encounter complex between triplet enone and oxygen in addition to the singlet complex. This is expected to reduce singlet oxygen yields since only singlet encounter complexes can dissociate to produce singlet oxygen. However, high singlet oxygen yields are obtained for the cyclohexenones which exhibit high rates of quenching by oxygen. A mechanism which can account for this behavior involves interconversion of the singlet and triplet encounter complexes, possibly via a charge-transfer interaction.

168. BENZOFUROXAN PHOTOCHEMISTRY: DIRECT OBSERVATION OF 1,2-DINITROSOBENZENE BY STEADY STATE SPECTROSCOPY - A NEW PHOTOCHROMIC REACTION.
Nigel P. Hacker, IBM Research Division, Almaden Research Center, 650 Harry Road, San Jose, CA 95120-6099

Photolysis of benzofuroxan (I) isolated in inert matrices (Ar or Xe), or as a film, at $\lambda = 365$ nm gives 1,2-dinitrosobenzene (II), which was characterized by IR and UV absorption spectroscopy. Warming 1,2-dinitrosobenzene from 12 - 80 K in Xe or as a film, regenerates benzofuroxan. Also irradiation of 1,2-dinitrosobenzene in Ar at 14 K at $\lambda = 254$ nm or 313 nm regenerates benzofuroxan.

$$ \text{I} \xrightleftharpoons[h\nu', \Delta]{h\nu} \text{II} \xrightarrow{h\nu''} CO_2 + C_3O_2 + CO $$

This is an example of a new photochromic system which only operates at cryogenic temperatures. Prolonged photolysis of benzofuroxan-1,2-dinitrosobenzene mixtures does not eject nitric oxide and yield benzyne (c.f. nitrosobenzene), but gives carbon dioxide, carbon suboxide and carbon monoxide.

169. SIMULATED ANNEALING OF RINGS Stephen R. Wilson and Frank Guarnieri, Department of Chemistry, New York University, Washington Square, New York, NY 10003

The starting geometry dependence of almost all energy minimizations requires the selection of "good" structures on which to base calculations. Conformational searching of very flexible molecules usually requires that large numbers of starting structures be examined. Similar analysis of ring compounds has the additional complication of non-independent degrees of freedom due to closure constraints. We have recently interfaced our simulated annealing program [S. R. Wilson and W. Cui, J. Comp. Chem, 12, 342 (1991).] with an algorithm for exact calculation of ring closure. We will discuss the use of simulated annealing for conformation searching and the location of the global minimum of rings up to seventeen-membered. In addition we will describe some applications to protein chain and loop distortions.

170. **Modelling transition states with force fields: development of the best possible parameters.**
Jonathan M. Goodman. University Chemical Laboratory, University of Cambridge, Lensfield Road, Cambridge, CB2 1EW, UK

A method is presented that aids the development of force field parameters to model transition states from *ab initio* calculations. It is based on fitting the parameters to reproduce the energies and the derivatives of the potential function using a least squares method. This allows a quantitative estimate to be made of the goodness of fit of these parameters. This study of torsion parameter fitting demonstrates the utility of the method for the study of the boron-mediated aldol condensation.

171. HARTREE FOCK DESCRIPTON OF SUBSTITUED AZOMETHINE YLIDES.
L.D. Burke, S.D. Kahn, W.J. Hehre,
Department of Chemistry, University of California Irvine, Irvine, CA 92717.

Knowing the electronic structure of substituted azomethine ylides is important for understanding and modeling the stereo and regio chemistries of their cycloaddition reactions. To understand the details of azomethine cycloaddition chemistry, we have computed the electronic structure of monosubstituted azomethine ylides which have a pi electron acceptor or a donor substituent. The structures were computed at the restricted and unrestricted Hartree Fock levels of theory as well as at a level which includes electron correlation i.e.; restricted second order Moller Plesset Perturbation theory(RMP2). The zwitterionic and biradicaloid contributions to their electronic structure will be discussed. In addition, RHF//3-21G calculations have been preformed on the four azomethine ylides for which crystal structure data is known.

172. RELATIVE CONFORMATIONAL STABILITIES OF CHIRAL (Z)-ALKENE
Mark A. Wolf, Andrew J. Peat, and Benjamin W. Gung, Department of Chemistry, Miami University, Oxford, OH 45056

Chiral (Z) alkenes often exhibit greater diastereofacial selectivity than the corresponding (E) isomers. An ab initio MO study of the ground state conformations of (Z)-3-penten-2-ol, **A** to **F**, has been completed. The hydrogen eclipsed conformer, **A**, is the most stable form according to basis sets STO-3G and 6-31G*. However, the 3-21G basis set predicts that the C-O eclipsed form, **B**, is the most stable isomer. All the calculations agree that the conformation C (with C-C-C=C constrained at 0°) has the highest energy.

173. STRAIN-COMPENSATED HEATS OF HYDROGENATION, D. W. Rogers, Chemistry Department, Long Island University, Brooklyn NY 11201

Heats (enthalpies) of hydrogenation Hh have long been used as a quantitative measure of stabilization effects due to conjugation, resonance, and other non-classical corrections to simple additive bonding schemes. The experimental value of Hh, however, is always accompanied by a heat component brought about by differences in strain energy between reactant and product, exothermic if strain is relaxed or endothermic if strain is increased. With the advent of molecular mechanics, it is possible to calculate the strain change upon hydrogenation, to subtract the strain component from the observed Hh and so arrive at a strain-compensated Hh. Strain-compensated Hh values often yield a rather different picture of hyperconjugation, homoaromaticity, and antiaromaticity from the one we currently accept.

174. A COMPARISON OF DISTANCE GEOMETRY AND MOLECULAR DYNAMICS SIMULATION TECHNIQUES FOR CONFORMATIONAL ANALYSIS OF BETA-CYCLODEXTRIN
David A. Wertz[+], Chen-Xi Shi, and Carol A. Venanzi Chemistry Dept., New Jersey Institute of Technology, Newark, N.J. 07102 +FMC Corp., Box 8, Princeton, N.J. 08543.

Distance geometry and molecular dynamics simulation techniques were compared in their ability to search the conformational potential energy surface of beta-cyclodextrin. Structures generated by the DISGEO program were minimized using three different atomic point charge sets. Some of these structures were used as starting points for molecular dynamics simulation in vacuo at 298K. The distance geometry results showed that the global features of the conformational potential energy surface were generally independent of the point charge set. The distance geometry technique was able to find structures of lower energy than direct minimization of the X-ray or neutron diffraction structures. However, the molecular dynamics simulation technique was consistently able to find structures of lower energy than those generated by distance geometry.

175. MOLECULAR ORBITAL ESTIMATION OF THE ACTIVATION ENTHALPIES FOR INTRAMOLECULAR HYDROGEN TRANSFER AS FUNCTIONS OF SIZE OF THE CYCLIC TRANSITION STATE AND C-H-C ANGLE, Xiao.Ling Huang and J. J. Dannenberg, Department of Chemistry City University of New York -Hunter College and The graduate School, 695 Park Avenue New York, NY 10021

The AM1 molecular orbital method was used to compt activation enthalpies for H-transfer between carbons as a function of C-H-C angle and size of the cyclic transition state for intramolecular H-transfer. In the case of intramolecular H-transfer, reaction of a primary radical site with primary, secondary and tertiary C-H's were considered for H-shifts to C_1 from C_3, C_4, C_5, C_6, C_7, and C_8. The activation enthalpies are insensitive to C-H-C angle in the range 145-180 degrees. Activation enthalpies are lowest for intramolecular H-transfers involving 1-5 and 1-6 H-shifts. The higher activation enthalpies for the other internal H-transfers are attributed to C-H-C strain for 1-3 and 1-4 H-transfers only, and conformational effects other than C-H-C angle in the transition states.

176. THE TOTAL SYNTHESIS OF RACEMIC PARVIFOLIN. Adrián Covarrubias and Luis A. Maldonado, División de Estudios de Posgrado, Facultad de Química, UNAM, 04510 México D.F., MEXICO.

The benzocyclooctene parvifolin $\underline{1}$, a sesquiterpene phenol isolated from Coreopsis parvifolia, Pereziae carpholepsis, Pereziae alamani var. oolepis and Pereziae longifolia Blake, has been synthesized as the racemate in ten steps, starting from the known symmetrical dimethyl-dimethoxy naphthalene $\underline{2}$. The key step in the synthesis involves a modified Stork-Landesman $6 \rightarrow 8$ ring expansion sequence.

177. A TOTAL SYNTHESIS OF (-) α-KAINIC ACID AND (-) α-ALLOKAINIC ACID. N. Jeong, S. -e. Yoo, S. H. Lee, K. Kim†, Korea Research Institute of Chemical Technology, Daedeog danji, Daejeon, 305-606, Korea and †Department of Chemistry, Pohang Institute of Science and Technology, Pohang, Kyungbuk, 790-784, Korea

(-) α-Kainic acid **1**, a potent neuroexcitatory transmitter in the central nervous system, and its epimer (-) α-allokainic acid were synthesized in optically pure form employing a 3-azabicyclo [3,3,0]-oct-5-en-7-one **4** as a versatile common intermediate. The key intermediate **4** was obtained efficiently from the N-propargyl-N-allyl-N-tosylamide **3** by the aid of cobalt carbonyl (Pauson-Khand reaction).

178. **SYNTHESIS AND REACTIONS OF STEROIDAL OXAZOLINES.**
N.I. Carruthers*, D. R. Andrews*, S. Garshasb and R. A. Giusto.
Schering-Plough Research, Union, NJ 07083.

The development of efficient fermentation processes that use soy sterols has made 9α-hydroxyandrost-4-en-3,17-dione (**1**) readily available for the synthesis of corticoids. The conversion of (**1**) to the steroidal oxazolines (**2**) and (**3**) will be described. Further reactions of the oxazolines, including conversion to pregnanes and corticoids will be presented.

179. EFFICIENT SYNTHESES OF (+/-)- AND (-)-FASTIGILIN-C. Steven P. Tanis [a], Edward D. Robinson [a], and Mark C. McMills [b], a) Medicinal Chemistry Research, The Upjohn Co., Kalamazoo, MI 49001, b) Department of Chemistry, Ohio University, Athens OH 45701

For the past several years we have been investigating furan terminated cationic cyclizations as the key step in the construction of natural products. Fastigilin-C **1**, a complex helenanolide, has been reported to exhibit extremely potent cytotoxic and antineoplastic activity, thus making it an attractive target for total synthesis. We wish to report the first total syntheses of (+/-)- and (-)-Fastigilin-C. In the forward direction a Mukaiyama Michael-Aldol protocol affords **3** with complete control of relative stereochemistry. Fastigilin-C **1** is ultimately produced in a 17 step (14.9% overall yield) sequence. In a similar fashion (-)-**1** is constructed from (S)-(+)-4-methoxy-2-methyl-2-cyclopentenone.

180. **A SHORT SYNTHESIS OF THE SYNTHETIC PROGESTIN ST-1435.** Gustavo A. García de la Mora, Yvonne Grillasca R., Jose C. Ramirez R., Alejandrina Acosta H.; Departmento de Química Orgánica, Division de Estudios de Posgrado, Facultad de Química, UNAM, 04510 Mexico, D. F., México.

The synthesis of ST-1435 starting from the cheap commercially available progestin NET (I) by a short route will be presented. ST-1435, a potent synthetic progestin that has not effect by oral administration, can be used when lactating women are under contraceptive treatment. Previous synthesis are at least twelve steps long from commercial steroids. One of the key steps is a selective reaction on the △-16-20-keto in the presence of the △-4-3-keto.

ST-1435

181. **SYNTHESIS OF (±)-MUSCONE FROM [12]METACYCLOPHANE.**
Yu Li, Thomas C.W. Mak, Henry N.C. Wong, Tze-Lock Chan, Department of Chemistry, The Chinese University of Hong Kong, Shatin, N.T., Hong Kong

The usefulness of cyclophanes as synthetic building blocks for natural products has been demonstrated in a short efficient synthesis of muscone from [12]metacyclophane. The requisite cyclophane can be conveniently obtained in multi-gram quantities from 2,13-dithia[14]metacyclophane bissulfone, without recourse to FVP, by a newly uncovered variant (CBr_2F_2, KOH/Al_2O_3, t-BuOH) of the Meyers modification of the Ramberg-Bäcklund reaction followed by hydrogenation of the resulting [12]metacyclophane-1,11-diene. The five-step sequence from [12]metacyclphane to (±)-muscone, in 43% overall yield, involves: (1) Birch reduction to 15,18-dihydro[12]metacyclophane, (2) ozonolysis to cyclopentadecane-1,3-dione, (3) O-methylation to the corresponding keto methyl enol ether, (4) treatment with methyllithium followed by aqueous acid demethylation and dehydration to 3-methyl-2-cyclopentadecenone, and (5) hydrogenation to muscone.

182. **NEW DIALKYLATED SCALARINS FROM THE SPONGE PHYLLOSPONGIA FOLIASCENS**
Longmei Zeng, Xiong Fu, Jingyu Su and Francis J. Schmitz*
Dept. of Chemistry, Zhongshan University, Guangzhou, China
*Dept. of Chemistry, University of Oklahoma, Norman, OK 73019 U.S.A.

Sesterterpenes possessing a scalarane skeleton with methyl groups at C-19 or 20 and C-24 occour in marine sponge of the order Dictyoceratida. From the sponge Phyllospongia foliascens, collected off the Xisa Islands of South China Sea, we have isolated and reported a new dialkylated scalarin sesterterpene, phyllofenone A. In this paper, we report the structures of four novel dialkylated scalarin sesterterpenes, namely, phyllofenone B, phyllofolactones -A and -B, and phylloketal and a known compound, scalardysin-B. The structures were determined by IR, MS and extensive NMR, including one and two-dimensional NMR techniques.

183. **TOTAL SYNTHESIS AND STEREOCHEMICAL ASSIGNMENT of (±)-EPIDERSTATIN.** Robert L. Dow, Marcella A. Hain and John A. Lowe III, Central Research Division, Pfizer Inc., Groton, CT 06340.

A recent paper has detailed the isolation of a novel member of the glutarimide antibiotic family which potently inhibits the mitogenic response in cells stimulated by epidermal growth factor (*J. Antibiotics* **1989**, *42*, 1599). Spectroscopic studies on this natural product, termed epiderstatin, led to its structural assignment as **I**; however, the investigators were unable to establish the C_3/C_5 relative and absolute stereochemistries (*J. Antibiotics* **1989**, *42*, 1607). We report here the total synthesis and C_3/C_5-relative stereochemical assignment of ($\pm$)-epiderstatin.

184. **STRUCTURE AND ABSOLUTE STEREOCHEMISTRY OF VARICULANOL: A NOVEL SESTERTERPENOID FROM *ASPERGILLUS VARIECOLOR*.** Sheo B Singh*, Robert A. Reamer, Deborah L. Zink, Dennis M. Schmatz, Anne W. Dombrowski, and Michael A. Goetz. *Merck Sharp & Dohme Research Laboratories, P. O. Box 2000, Rahway NJ 07065*

The structure, absolute stereochemistry and conformation of variculanol (**1**), a sesterterpenoid with novel skeleton isolated from *Aspergillus variecolor* has been described. The molecular structure was deduced from extensive application of 2D NMR methods, in particular HMBC, which established the unambiguous assignment of the novel [5/12/5] ring system. NOEDS measurements were very useful in establishing the relative stereochemistry. The NMR-mandelate method and CD spectral analysis of 1,17-bis (4-bromo)benzoate were used to determine the absolute stereochemistry. A unique solution phase conformation was determined both from extensive NOE measurements and MM2 calculations and optimization.

185. COMPOSICION QUIMICA DE LAS PLANTAS PERTENECIENTES A LA FAMILIA Scrophulariaceae DE MEXICO. Jiménez E. Manuel, Navarro O. Arturo y Lira-Rocha Alfonso. Instituto de Química, Universidad Nacional Autónoma de México, Circuito Exterior, Ciudad Universitaria, Coyoacán, 04510 México, D. F.

La familia Scrophulariaceae taxonómicamente es muy compleja y está bien representada en México ya que de los 200 géneros que se mencionan a nivel mundial aproximadamente 65 (33%) se encuentran en nuestro país. Muchas de estas se usan como remedio en la medicina tradicional de muchos países. En los últimos años ha habido un creciente interés por estudiarlas. En esta ocasión informamos de las plantas que hemos trabajado y de los constituyentes que se han identificado después de su laboriosa separación y purificación. Los iridoides se han caracterizado por sus datos de RMN, IR, UV, EM y rayos X. Se describe la composición química de los Penstemon rosseus, Penstemon gentianoides, Penstemon apateticus, Penstemos barbatus, Penstemon centrathifolius, Benth, Lamourouxia Multifida H.B.K., Lamourouxia rhinanthifolia, H.B.K. y Castilleja tenuiflora, Benth; caracterizándose 15 iridoides y 2 alcaloides, algunos se encuentran en cantidades apreciables y los empleamos en síntesis.

186. CHEMICAL COMPOUNDS FROM ANTHERS OF SPATHODEA CAMPANULATA FLOWERS. Rivera M. Carmen, Sanjurjo B. Marisol, Giral G. Francisco. Departamento de Farmacia, División de Estudios de Posgrado, Facultad de Química, Universidad Nacional Autónoma de México, 04510 México, D.F.

This report, is part of a systematic phytochemical study of tropical flowers. The Spathodea campanulata flowers, were set apart in their constituents: chalice, corolla, pistil and stamens. The last ones, were separated in filaments and anthers. In the Third Congress of North America (Toronto, Canada), was recorded the study of the pistil of this Bignoneaceae flower.
The dried anthers, were extracted with several solvents. The dehydro acetic acid present in others flowers, was not found (Experientia vol 32 pag 2490 (1976)). From the chloroformic extract were isolated and purified several compounds. Due to the very small amounts of the mixtures, the identification was made by Gas Liquid Chromatography-Mass Spectrometry.

187. SECOIRIDOIDS AND OTHER CONSTITUENTS OF *Fraxinus uhdei*.
Romo de Vivar R. Alfonso, Pérez C. Ana Lidia, Escalona Sergio and Núñez Odilón. Instituto de Química, Circuito Exterior, Ciudad Universitaria, Coyoacán, 04510 México, D. F., México.

Fraxinus uhdei (oliaceae) is a beautiful tree which grows wild and cultivated in the high lands of Mexico. Their leaves are used as medicine for rheumatism and other diseases. Chemical studies of other *Fraxinus* species have shown to contain glucosides of secoiridoids and phenyl propanoids as common constituents. Some secoiridoid glucosides have interesting biological properties such as inmunosuppressors. With these antecedents in mind we undertook chemical studies of seed and leaves of *Fraxinus uhdei* collected in the University campus. The seeds afforded manitol, 2(p-hydroxyphenyl)ethanol, the secoiridoid 10-hydroxyligustroside and the phenylpropanoid verbacoside together with other secoiridoids which are under study. The leaves have shown to contain 2(p-acetoxyphenyl) ethanol, 2(p-hydroxyphenyl)ethylacetate and several secoiridoids which are under study in our laboratory.

188. NEW DIALKYLATED SCALARINS FROM THE SPONGE PHYLLOSPONGIA FOLIASCENS
Longmei Zeng, Xiong Fu, Jingyu Su and Francis J. Schmitz*
Dept. of Chemistry, Zhongshan University, Guangzhou, China
*Dept. of Chemistry, University of Oklahoma, Norman, OK 73019 U.S.A.

Sesterterpenes possessing a scalarane skeleton with methyl groups at C-19 or 20 and C-24 occour in marine sponge of the order Dictyoceratida. From the sponge Phyllospongia foliascens, collected off the Xisa Islands of South China Sea, we have isolated and reported a new dialkylated scalarin sesterterpene, phyllofenone A. In this paper, we report the structures of four novel dialkylated scalarin sesterterpenes, namely, phyllofenone B, phyllofolactones -A and -B, and phylloketal and a known compound, scalardysin-B. The structures were determined by IR, MS and extensive NMR, including one and two-dimensional NMR techniques.

189. THE FIRST WATER-SOLUBLE PORPHYRINYL-MONONUCLEOSIDE AND OTHER NUCLEOSIDE-SUBSTITUTED PORPHYRINS. Leszek Czuchajowski, Jan Habdas, Halina Niedbala and Vinay Wandrekar, Department of Chemistry, University of Idaho, Moscow, ID 83843.

Continuing the interest of our group in porphyrinyl-nucleosides, we now report the synthesis of the first water-soluble representative of this new class of compounds, namely, the 5'-O-[5-p-phenylene-10,15,20-tri-(N-methyl-4-pyridinium)porphyrin]-uridine. Also the first representatives of porphyrins di- and tetra-substituted with nucleosides were obtained as water-insoluble compounds. These were the meso-di-p-tolyl-di-(p-phenylene-5'-O-2',3'-isopropylidene-uridine)porphyrin, its adenosine and thymidine analogs, the meso-tetra-(p-phenylene-5'-O-2',3'-isopropylidene-uridine)porphyrin and the respective tetra-adenosine porphyrin. In addition, the porphyrinyl-monothymidine was obtained which as the 3'-phosphoramidite derivative can be 5'-terminally linked to a polynucleoside/nucleotide strand, while the respective porphyrinyl-dithymidine bears the potential of joining two strands.

190. NAPHTHOQUINONE DERIVATIVES AS SELECTIVE AGENTS FOR DNA ALKYLATION. Moneesh Chatterjee and Steven Rokita. Department of Chemistry, SUNY, Stony Brook, NY 11794.

Site specific alkylation of DNA can be readily achieved by simply conjugating together a recognition and reactive component. This technique has been adopted in our efforts to develop such an activity that will be compatible with a cellular environment and will require biological activation. By using an oligonucleotide-based model system, a series of naphthoquinone derivatives has been found to express the desired characteristics. Modification was controlled by a reductive signal and non-targeted sequences did not react. Now, chemical and enzymatic methods of triggering DNA alkylation have been examined, mechanistic questions have been investigated and reaction products have been identified.

191. SOLVOLYSIS TRIGGERED ALKYLATION OF DNA USING A SILYL PHENOL. Tianhu Li and Steven Rokita. Department of Chemistry, SUNY, Stony Brook, NY 11794.

A silyl protected phenol has been developed as a stable precursor for site specific generation of a highly electrophilic quinone methide. Model studies were first performed to identify the most suitable species. The best candidate was then conjugated to an oligonucleotide in order to effect selective alkylation of a chosen DNA sequence. Simple hybridization of the probe and target strands did not result in detectable reaction. Instead, derivatization was controlled as desired by manipulating the ionic nature of the medium. Addition of KF, NaF, CsF, NaCl, KBr, or $LiClO_4$ initiated solvolysis and conversion of the latent electrophile. A non-complementary sequence of DNA did not serve as a target for the reactive intermediate under any conditions examined.

192. SYNTHETIC STUDIES FOR CARBON-11-LABELED ALFENTANIL. Anthony L. Feliu, Institute for Nuclear Chemistry (ICH-1), Jülich Research Center GmbH, D-5170 Jülich, FRG

Alfentanil (4) is a fast-acting second-generation mu-opioid suggested as an alternative to fentanyl for neuroleptanaesthesia. In order to study its cerebral distribution and in vivo receptor binding kinetics with positron emission tomography, rapid synthetic methods were investigated for labeling with no-carrier-added (NCA) carbon-11 (t½ = 20.4 min). The remarkable suseptibility of 1 to undergo N-to-O propionyl migration with base (NaH, DMF, room temp, << 5 min) thwarted a direct C-11 O-methylation strategy. And attempted [^{11}C]-propionylation of 3 failed, a result consistent with the known low-reactivity of this 2°-amine. Thus, new conditions for one-pot methylation and propionylation were devised (NaH, 15-crown-5, THF, 70°, 10 min/step). Starting from 2 and NCA [^{11}C]MeI radiochemical yield of [O-methyl-^{11}C]4: 28-50% (isolation by prep-HPLC, ≤10 nmol carrier detected, GC-MS product identification).

193. FLUORINATED PREGNANE DERIVATIVES WITH ANTIANDROGENIC ACTIVITY. Eugene A. Bratoeff, Pablo Martinez and Elisa Carrillo, Universidad Nacional Autonoma de Mexico, Departamento de farmacia, Facultad de Química, 04510 Mexico, D.F., Mexico.
The antiandrogens are synthetic substances used for the treatment of prostate cancer and acne. These compounds tend to reduce the concentration of the cancer producing androgens by competing for the same protein receptor. In our laboratory we have determined the activity profile of several antiandrogens on the basis of experimental data obtained from a variety of steroidal derivatives.
The compounds which show high antiandrogenic activity are derivatives of pregnane with electronegative substituents at C-6, having the 1,4,6-trien-3-one moiety or a cyclomethylene group at C_1-C_2 and a bulky substituent at C-17.
In view of the fact that the 16-dehydropregnenolone acetate is an inexpensive commercially available raw material. This compound was used for the synthesis of 17α-acetoxy-1,2-cyclomethylene-20,20-difluoro-6-halo-16β-methyl-4,6-pregnadien-3-one.

194. **NUCLEOPHILIC SUBSTITUTION BY AROMATIC AMINES ON THE ULTIMATE HEPATACARCINOGEN,N-(SULFONATOOXY)-2-(ACETYLAMINO)FLUORENE, K. S. Rangappa and M. Novak**, Department of Chemistry, Miami University, Oxford, OH 45056

Carcinogenic metabolites of polycyclic aromatic amines and amides, including the title compound, are apparently responsible for the generation of the characteristic "C-8 Adducts" with guanosine that are obtained in low yield from in vivo and in vitro studies. We have previously shown that monocyclic models of these metabolites generate similar adducts in high yields from reaction with aromatic amines in methanol by an Sn2 mechanism (J. Am. Chem. Soc. **1991**, 113, 3459.). The title compound, an apparent carcinogenic metabolite of 2-(acetylaminofluorene, also reacts with aniline and N,N-dimethylaniline in high yield in methanol. In this paper the evidence for the structures of these adducts will be described and the mechanism of their formation will be discussed.

195 BISCHLER-NAPIERALSKI REACTION UNDER MICROWAVE OVEN. Maghar S. Manhas, Shamsher S. Bari, Vegesna S. Raju, Malay Ghosh, Ajay K. Bose, Department of Chemistry and Chemical Engineering, Stevens Institute of Technology, Hoboken, NJ 07030.

Many organic reactions usually conducted in refluxing hydrocarbon solvents for several hours can be completed in DMF solutions in minutes at ordinary pressure in an open vessel in a microwave oven (Bose et. al., Heterocycles, 1990). Since hydrocarbons are poorly heated in a microwave oven, we have introduced the dipolar solvents chloro-, 1,2-dichloro- and 1,2.4-trichlorobenzene at 20-30°C below their b.p. as efficient energy transfer agents under microwave irradiation. The Bischler-Napieralski reaction ($\underline{1}$ + P_2O_5 → $\underline{2}$) can be completed in 70-75% yield in 2-5 min in dichlorobenzene. Mixtures of $\underline{2}$ and the intermediate amidine $\underline{3}$ in varying amounts are obtained when chlorobenzene or dichloroethane is the reaction medium.

196. A new and facile method for the synthesis of 3 - substituted and 1,3 - disubstituted pyrroles. Alfredo Vazquez M.,Blas Flores P.,Fernando Leon C.,Arturo Gonzalez H., G.A. Garcia de la Mora., Jose M. Mendez S.* Departamento de Quimica Organica,Division de Estudios de Posgrado,Facultad de Quimica,UNAM., 04510 Mexico,D.F. MEXICO.
A short and facile new versatile synthesis of 3 - substituted and 1,3 - disubstituted pyrroles is described. Alkylation of nitriles (1) with bromoacetaldehyde diethyl acetal followed by reduction and hydrolysis afforded 1,4 - dialdehydes (2),treatment of these compounds with ammonia or primary amines gave the corresponding 3 - substituted or 1,3 - disubstituted pyrroles (3).

R = Alkyl,Aryl.   R' = H,Alkyl,Aryl.

197. **A NEW AND SHORT METHOD FOR THE SYNTHESIS OF 3-SUBSTITUTED PYRROLES.** Gustavo A. Garcia de la Mora, Yvonne Grillasca R., Alejandrina Acosta H., Jose M. Mendez S., Departamento de Quimica Organica, Division de Estudios de Posgrado, Facultad de Quimica, UNAM, 04510 Mexico, D. F., Mexico.

The importance of 3-substituted pyrroles without substitution at 2 or 5 position has had an increased value due its potential use as conducting organic polymers. A short route to substituted 1,4-dialdehydes which are the intermediates for the synthesis and easily converted to pyrroles will be presented. This method allows almost any kind of substitution on 1 and/or 3 position in a three step process.

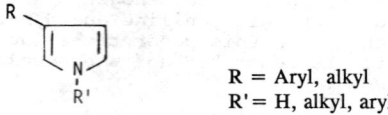

R = Aryl, alkyl
R' = H, alkyl, aryl

198. SYNTHESIS OF 7,7-DIMETHYL-QUINO[2,3,4-kl]-5a,6,7,8-TETRAHYDROBENZ[c]ACRIDINE 5N-OXIDE. V. Martínez Roberto[1], Sánchez E. Olivia[2] and Linzaga E. Irma[2].
[1]Instituto de Química, Universidad Nacional Autónoma de México, Circuito Exterior, Ciudad Universitaria, Coyoacán, 04510 México, D. F.
[2]División de Estudios Superiores, FCQI, UAEM, Cuernavaca, Morelos, México.

Aminoacridines are useful as dye stuff, antibacterial and the treatment of Alzheimer's disease. As we have described an one-step method for the synthesis of benzo[c]acridin-8-ones from dimedone, 2-naphthylamine and substituted benzaldehydes, it was deemed interest to prepare an aminoacridine derivative from the appropriate benzo[c]acridin-8-one. However, hydrogenation of the 9,9-dimethyl-7-(o-nitrophenyl)-8,9,10,11-tetrahydrobenz[c]acridin-8-one in the presence of Pd/C and ethanol, unexpectedly, produce the hitherto unknown 7,7-dimethyl-quino[2,3,4-kl]-5a,6,7,8 tetrahydrobenz[c]acridine 5N-oxide instead of aminoalcohol derivative. The structure was confirmed using ir, nmr and ms spectroscopy.

199. **A TANDEM S$_N$2-MICHAEL APPROACH TO THE SYNTHESIS OF 5- AND 6-RING NITROGEN AND SULFUR HETEROCYCLES**, Richard A. Bunce, Christopher J. Peeples and Paul B. Jones, Department of Chemistry, Oklahoma State University, Stillwater, OK 74078-0447.

A tandem S$_N$2-Michael addition sequence has been developed for the preparation of 5- and 6-ring nitrogen and sulfur heterocycles (2 and 3) from ethyl ω-halo-2-alkenoates (1, X = Br, I; n = 1, 2). The process is useful for the preparation of fused and spiro as well as monocyclic rings having an acetate residue at C-2 relative to the heteroatom. The reactions proceed in 60-80% yields and can be carried out in a single reaction flask. The mechanism, stereochemistry and scope of the reaction will be presented.

A: 2 Y = N-Bn, R = Et
B: 3 Y = S, R = H

A: BnNH$_2$, Et$_3$N, EtOH
B: (H$_2$N)$_2$C=S, EtOH; (2) 40% KOH; (3) H$_3$O$^+$

200. *A CONVENIENT SYNTHESIS OF 2-ACETYL-1-METHYL-5-NITROIMIDAZOLE. A. Shafiee, B. Pirouzzadeh, F. Ghasemian, K. Parang. Department of Chemistry, College of Pharmacy, Tehran Medical Sciences University, Tehran, I.R. Iran.*

Manganese dioxide oxidation of 2-hydroxymethyl-1-methyl-5-nitroimidazole gave 1-methyl-5-nitroimidazole-2-carboxaldehyde (1) in high yield. Reaction of diazomethane with **1** afforded in addition to compound **3** the title compound **2** in low yield. Treatment of ethyl acid malonate with two equivalents of isopropylmagnesium bromid in THF and subsequent addition to 1-methyl-5-nitroimidazole-2-carbonyl imidazolide **(4)** yielded ethyl (1-methyl-5-nitroimidazole-2-carbonyl)acetate **(5)** in 70% yield. Hydrolysis and decarboxylation of compound **5** gave the desired compound **2** in a 97% yield.

201. **FACILE SYNTHESIS OF A CARBOXYLIC LINKER ON HALOFUGINONE.**
Loyd D. Rowe, Ross C. Beier, A. Barry Astroff, Elizabeth A. Weaver-Delamater, and Marcel H. Elissalde, *USDA, ARS, Food Animal Protection Research Laboratory, Route 5, Box 810, College Station, TX 77845.*

The quinazolinone halofuginone (HF) **(1)** is a FDA approved feed additive used for the prevention of coccidiosis in the poultry industry. The availability of an immunoassay for HF would increase monitoring efficiency and decrease monitoring costs. A succinic acid derivative of HF was produced which could be covalently bound to carrier proteins. We recently prepared the carboxylic acid **4** from **1** in three steps. The protected alcohol **2** is prepared from **1** by the addition of TMSI. The alcohol **2** is converted to the acid **3** by addition of succinyl anhydride. Removal of the alcohol protection by H^+ affords the carboxylic acid derivative **4** of HF in high yield.

202. **SYNTHESIS OF THE FIRST BRANCHED QUATERTHIENYLS.** Daniel M. Perrine, David M. Bush, Eugene P. Kornak, *Department of Chemistry, Loyola College, Baltimore MD 21210;* Ming Zhang, Young H. Cho, Jacques Kagan *University of Illinois at Chicago, Chicago IL 60680.*

The first branched quaterthienyls: 5´-(2-thienyl)-2,2´:3´,2˝-, 5´-(2-thienyl)-2,2´:3´3˝-, 5´-(2-thienyl)-2,3´:2´,3˝-, and 5´-(2-thienyl)-3,2´:3´,3˝-terthiophene, **1-4**, were synthesized via the Stetter reaction and their structures confirmed by 2D COSY spectra.

203. 1-ACETAMIDO-2,x-DIMETHYLPYRIDINIUM N-BETAINES: PREPARATION AND SPECTRAL CHARACTERISTICS. A. K. Saha and J. Bhattacharyya*, *Department of Chemistry, Southeastern Massachusetts University, N. Dartmouth, MA and *Laboratorio de Tecnologia Farmaceutica, Universidade Federal da Paraiba, 58.059 Joao Pessoa, PB, Brazil.*

1-Amino-2-alkylpyridinium salts react with acid chlorides in pyridine to give pyrazolo-[1,5-a]pyridine ring system with acylation at C-3. It was of interest to us to study the reactivity of the free bases liberated from various 1-amono-2-alkylpyridinium salts towards this cyclization reaction as well as the general applicability of this approach in the synthesis of various pyrazolo[1,5-a]azines. Thus, the free bases obtained from 1-amino-2-methyl(or ethyl)pyridinium salts react readily with acetic anhydride to give 3-acetyl(or methyl)pyrazolo[1,5-a]pyridine in good yield. Surprisingly, however, the bases from 1-amino- 2,3-, 2,4-, 2,5-, and 2,6-dimethyl pyridinium salts failed to cyclize with acetic anhydride under identical condition. Instead, the corresponding 1-amino-2,x-dimethylpyridinium N-betaines were obtained. Cyclization of these betaines could not be affected even after prolonged heating. The structures of these betaines have been established by analytical data and spectroscopy (uv, ir, ^{1}H and ^{13}C NMR). The spectroscopic data and probable mechanism of formation of the betaines will be presented. (CNPq)

204. **ONE-POT SYNTHESIS OF CYCLIC AMIDINIUM TETRAFLUOROBORATES AND HEXAFLUOROPHOSPHATES; THE SIMPLEST MODELS OF N^5,N^{10}-METHENYLTETRAHYDROFOLATE COENZYME.** Shahrokh Saba,* AnneMarie Brescia, and Moses K. Kaloustian.* Department of Chemistry, Fordham University, Bronx, New York 10458.

Reaction of triethyl orthoesters with various N,N'-dialkyl-α,ω-alkanediamines in the presence of ammonium tetrafluoroborate or hexafluorophosphate, all in the same molar ratio, affords cyclic amidinium salts in excellent yields.

$$R^1\text{-}C(\text{OEt})_3 + R^2NH(CH_2)_nNHR^2 \xrightarrow{^+NH_4 X^-} \begin{array}{c}\text{cyclic amidinium salt}\end{array}$$

n = 2, 3, 4

205. TRAPPING TETRAZOLINYLIDENES WITH NITROGEN ELECTROPHILES
Rainer H. Lowack and Robert Weiss*, *Institut für Organische Chemie der Universität Erlangen-Nürnberg, Henkestr. 42, W-8520 Erlangen, Federal Republic of Germany.*

The redox system formazanide ion / tetrazolium ion is conceived as a general device to manipulate electronically both electron-deficient and -rich centers attached to the carbon atom of this system. In order to introduce this redox substituent as a nucleophilic entity into organic substrates, we generated 2,3-diaryltetrazolinylidenes as novel heterocyclic carbenes in solution (a) by dephosphonionation of a (triphenylphosphonio)tetrazolium salt, (b) via oxidation of a tetrazolium-5-thiolate, and (c) by deprotonation of 2,3-diaryltetrazolium salts. The Carbenes were trapped by protonation, deuteronation, halogenation, oxygen transfer and coupling with diazonium ions or tosylazide. Like isomeric tetrazolinylidenes, the carbenes open up in the absence of electrophilic trapping agents, yielding the hitherto unknown 1-cyanoazimines. This ring opening and that of isomeric tetrazolinylidenes is discussed on the basis of semiempirical calculations.

206.
DEVELOPMENT OF AN EFFICIENT PROCESS FOR THE PRODUCTION OF CHIRAL AZETIDINONES Thomas M. Eckrich, J. Brennan, Erin E Strouse, Brett D. Boyer, Chemical Process Research and Development Division, Lilly Research Laboratories, Lilly Corporate Center, Indianapolis, Indiana 46285

The preparation and resolution of chiral azetidinones through a Staudinger cycloaddition reaction and chemical isolation of the ββ enantiomer is described.

The optimal yield for the 2 + 2 cycloaddition from statistical methods is also presented.

207.
HOMOGENEOUS NUCLEOPHILE EXCHANGE: A NOVEL, FACILE ROUTE TO HIGHLY OPTICALLY PURE GLYCIDOL DERIVATIVES. Roger C. Hahn, Department of Chemistry, Syracuse University, Syracuse, New York 13244-1200

The homogeneous nucleophile exchange (HNE) methodology by which we previously have made simple primary heterodihalides and haloalkyl arenesulfonates is applicable to direct synthesis of glycidyl camphorsulfonate (**2**) from epibromohydrin (**1**). The melting points of the diastereomers of **2** differ by 30°; the higher-melting isomer is readily recrystallized from a 1:1 mixture to >98% optical purity. Some uses of this chiral building block will be presented. Mother liquors from crystallization of **2** can be randomized to a 1:1 mixture of diastereomers by a special bromide-camphorsulfonate exchange technique, thereby establishing a cycle through which the mixture can be converted totally to a single isomer.

OCas is one camphorsulfonate enantiomer

208. **SYNTHESIS OF ENANTIOMERICALLY PURE BAY-REGION 10,11-DIOL-8,9-EPOXIDE DIASTEREO-MERS OF CARCINOGENIC DIBENZ[a,h]ACRIDINE (DB[a,h]ACR).** S. Kumar and P. L. Kole, Division of Environ Toxicology and Chemistry, SUNY College, Buffalo, N. Y. 14222; S. K. Balani and D. M. Jerina, NIH, NIDDKD/LBC, Bethesda, MD 20892.

The key intermediate, ($\pm$)-trans-10,11-dihydroxy-10,11-dihydroDB[a,h]ACR (**1**), was converted to its diastereomeric mixture of bis(-)menthyloxy esters (**2**), and separated b short bed/continuous development preparative TLC. The purity of each diastereomer was checked by NMR and HPLC. The enantiomeric dihydrodiols obtained by hydrolysis of diastereomers were converted to (+)- and (-)-enantiomers of diol epoxide-1 (**3**) and diol epoxide-2 (**4**). Assignment of the absolute configuration of the optically active dihydr diols and their epoxides will be presented.

1. R = H
2. R = (-)menthyloxyacetyl

209.
Asymmetric Michael Addition Catalyzed by Chiral Azacrown Ethers

Ernesto Brunet*, Dolores Rabasco
Departamento de Quimica, C-I. Facultad de Clencias
Universidad Autonoma. 28049-Madrid (Spain) Fax 341-397-3966

We report our recent results about the use of chiral azacrown ethers as catalysts in enantioselective synthesis. The Michael addition of phenylacetate to acrylate was catalyzed with azacrowns of different sizes, derived from commercial (+)-camphor, obtaining ee's up to 63%. The R,S. configuration of the resulting about depended on the endo-exo stereochemistry of the catalyst and on the alkaline metal of the base as well. Results of molecular mechanics calculations (MM2) will be presented to give an insight of the possible transition states involved.

210. SYNTHESIS OF (*R*)- AND (*S*)-3-ALKYLBUTYROLACTONES. K. R. Buszek, Department of Chemistry, Kansas State University, Manhattan, KS 66506.

The synthesis of (*R*)- and (*S*)-3-alkylbutyrolactones from a common precursor is described. Treatment of lactone **1** with lithium diisopropyl amide at -78°C and quenching with an appropriate alkyl halide gave **2** in good diastereomeric ratios. Desilylation, reduction, and oxidative cleavage gave the lactols which were subsequently oxidized under Swern conditions to afford the desired lactones (*R*)-**3** in good yields. The antipodes were similarly obtained by first quenching the enolates derived from (*R*)-**2** with saturated sodium sulfate to yield (*S*)-**2**. Reaction as above afforded lactones with the S configuration.

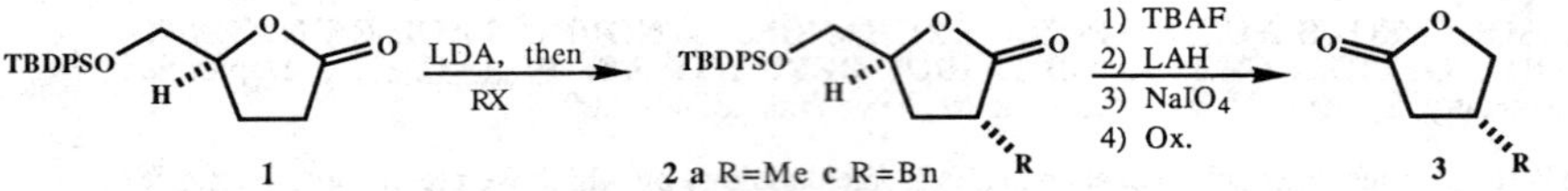

211. CHELATION-INDUCED REARRANGEMENT OF ALLYLIC METHOXY-ACETATES: A NOVEL TYPE OF ALLYLIC TRANSPOSITION CATALYZED BY Eu(fod)$_3$

Masato Koreeda and Brian K. Shull, Department of Chemistry, The University of Michigan, Ann Arbor, Michigan 48109

Eu(fod)$_3$ has been shown to be a highly efficient catalyst for the stereoselective rearrangement of allylic methoxyacetate. These esters undergo [3,3]-sigmatropic rearrangements in the presence of 0.1 equiv of Eu(fod)$_3$ at room temperature in CHCl$_3$ as shown below. The results of the mechanisitc studies as well as a comparison with the similar rearrangement reaction catalyzed by Pd (II) will be presented.

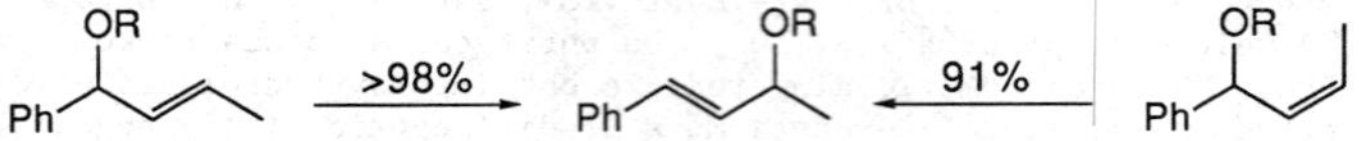

R = MeOCH$_2$C(=O)-

212.
1-SUBSTITUTED BOREPINS AND THEIR Mo(CO)$_3$ COMPLEXES. Arthur J. Ashe, III.

Jeff W. Kampf, Jennifer M. Pace and Yasuhiro Nakadaira, Departments of Chemistry of The University of Michigan, Ann Arbor, MI 48109 and The University of Electro-Communications, 1-5-1 Chofugooka, Chofu, Tokyo 182, Japan.

The exchange reaction of 1,1-dibutylstannepin (**1**) with phenylboron dibromide give 1-phenylborepin (**2**), which can be converted to the Mo(CO)$_3$ complex **3**. The X-ray structure of **3** shows that the Mo is η^7-coordinated to the borepin ring. The C–C bond distances range from 1.39-1.42 Å indicating that the borepin serves as an aromatic 6π-ligand.

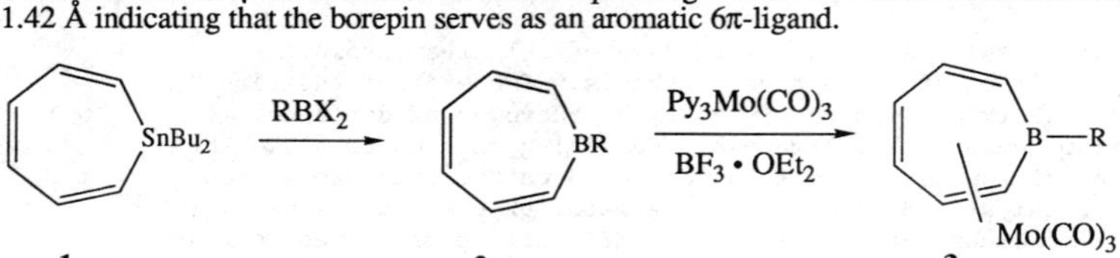

213. **LEWIS ACID-PROMOTED ADDITIONS OF ALLYL(CYCLOPENTADIENYL)IRON(II) DICARBONYL TO ALDEHYDES AND KETONES.**
Edward Turos, Gregory E. Agoston, Songchun Jiang, and Maria P. Cabal
Department of Chemistry, State University of New York at Buffalo, Buffalo, NY 14214

In the presence of $BF_3\text{-}Et_2O$, allyliron(II) reagent **I** reacts with *unactivated* aldehydes and ketones to afford zwitterionic iron-olefin complexes **II** as isolable yellow salts. Treatment of complexes **II** with NaI in acetone provides homoallylic alcohols in good to excellent overall yields. By virtue of the metal moiety being retained within the addition product **II**, new opportunities are available for directly incorporating additional functionality onto the double bond. We are examining the application of this methodology for the construction of novel polypropionates.

214. **REGIOSELECTIVE SYNTHESIS OF (η^6-ARENE)TRICARBONYLCHROMIUM COMPOUNDS.** Zang Y. Own[1], Su M. Wang[1], Je F. Chung[1], Lih J. Huang[1], Dwight W. Miller[2], and Peter P. Fu[1,2]. [1]Institute of Applied Chemistry, Providence University, Taiwan, and [2]National Center for Toxicological Research, Jefferson, AR 72079.

Reaction of chromium hexacarbonyl with seventeen polycyclic aromatic hydrocarbons (PAHs) was studied. The PAHs employed for study contain two to four aromatic benzo-rings and include parent PAHs, mono- and di-methylated PAHs, PAHs with one or two rings saturated, γ-keto-hexacyclic and phenyl substituted derivatives. In general, a cyclohexyl substituent, presumably due to its electron-donating character, can direct the reaction regioselectively at the substituted aromatic ring and that steric hindrance may not be an important factor. A methyl substituent, also with the electron-donating character, exhibits a similar effect. Due to the electron-withdrawing character of the keto functional group, a γ-ketocyclohexyl ring inhibits the reaction rate and directs the reaction away from the substituted aromatic ring. Our results, thus, have demonstrated that reaction of chromium hexacarbonyl with PAHs generates (η^6-arene)tricarbonylchromium compounds in a highly regioselective manner, preferentially at the aromatic ring with the lowest localization energy.

215. **CHARACTERIZATION OF (η^6-ARENE)TRICARBONYLCHROMIUM COMPOUNDS BY PROTON NMR SPECTROSCOPY.** Dwight W. Miller[1], Peter P. Fu[1,2], Zang Y. Own[2], Su M. Wang[2], Je F. Chung[2], and Lih J. Huang[2]. [1]National Center for Toxicological Research, Jefferson, AR 72079, and [2]Institute of Applied Chemistry, Providence University, Taiwan.

The high resolution 500 MHz proton NMR of (η^6-arene)tricarbonylchromium compounds derived from a series of modified polycyclic aromatic hydrocarbons (PAHs) was studied. The PAHs employed for study contain two to four aromatic benzo-rings. In general, due to the electron-withdrawing effect of the chromiumtricarbonyl group, the protons of the aromatic ring coordinated with the chromium are shielded approximately 2.0 ppm compared with the parent PAHs, while the other aromatic protons receive little or no change. The NMR assignments have allowed us to demonstrate that reaction of chromium hexacarbonyl with PAHs generates (η^6-arene)tricarbonylchromium compounds in a highly regioselective manner, preferentially at the aromatic ring with the lowest localization energy.

216. **CYCLOPROPYLTHIOCARBENE-METAL COMPLEXES: SUPER-HOMO-MICHAEL ACCEPTORS**; James W. Herndon, Margaret D. Reid, Gautam Chatterjee, Department of Chemistry & Biochemistry, University of Maryland, College Park, Maryland 20742

The reaction of *in situ* generated cyclopropylthiocarbene-chromium and -tungsten complexes with nucleophiles has been investigated. Reaction of cyclopropylacetoxycarbene complexes with hard nucleophiles such as alcohols or amines leads to high yields of the expected carbene complexes, while reaction with thiols leads to ring opened products accompanied by trace amounts of the desired thiocarbene complexes. Treatment of thiocarbene complexes with iodide or chloride ion provides the ring opened products, suggesting that these complexes readily undergo the homo-Michael addition reaction. The scope, limitations, and mechanistic studies of this reaction will be discussed.

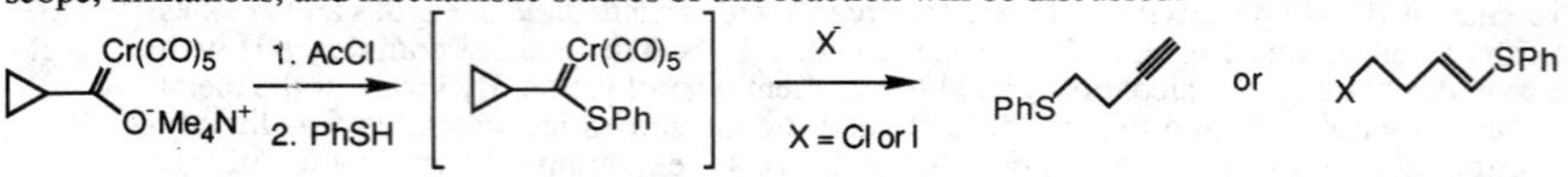

217.
SYNTHESIS OF 3,4-DISUBSTITUTED BENZOFURANS VIA THE INTRAMOLECULAR COUPLING OF A ZIRCONOCENE BENZYNE COMPLEX WITH ALKENES AND ALKYNES

Delton Cox and Stephen L. Buchwald*, Department of Chemistry, Massachusetts Institute of Technology, Cambridge, MA 02139

The in situ generated zirconocene-benzyne complex has been shown to be a useful intermediate in carbon-carbon bond forming reactions. We now describe the intramolecuar insertion of olefins into zirconocene-benzyne complexes to form diastereomerically pure metallacycles, **1**. We also show that **1** can be converted to 3,4-disubstituted benzofuran derivatives. The intramolecular insertion of alkynes gives metallacycle **2**.

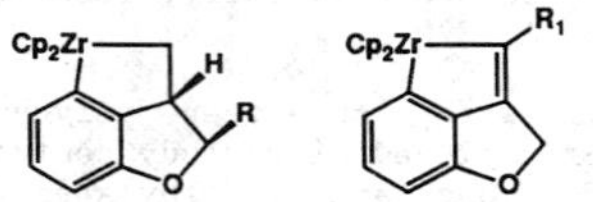

1 R = Me, C_5H_{11}

2 R_1 = Me, Et, TMS

218. Novel Silyl-Carbocyclization (SiCAC) of Alkynes and Alkenynes Catalyzed by Rh and Co-Rh Mixed Metal Complexes
Iwao Ojima*, Robert J. Donovan, William R. Shay, Zhaoda Zhang, *Department of Chemistry, State University of New York at Stony Brook, Stony Brook, NY 11794-3400*

In the course of our study on the new synthetic methods based on the reactions of hydrosilanes, CO, and alkynes or alkenynes promoted by $Rh_4(CO)_{12}$, $Co_2Rh_2(CO)_{12}$, and $Co(CO)_4$-$Rh(CNCMe_3)_4$, we have found that novel silyl-carbocyclizations (**SiCAC**) proceed smoothly under mild conditions to give the corresponding carbocycles or heterocycles. It is found that $Co(CO)_4Rh(CNCMe_3)_4$ is an excellent catalyst for **SiCAC**. Possible mechanisms of the novel SiCAC reactions will be discussed.

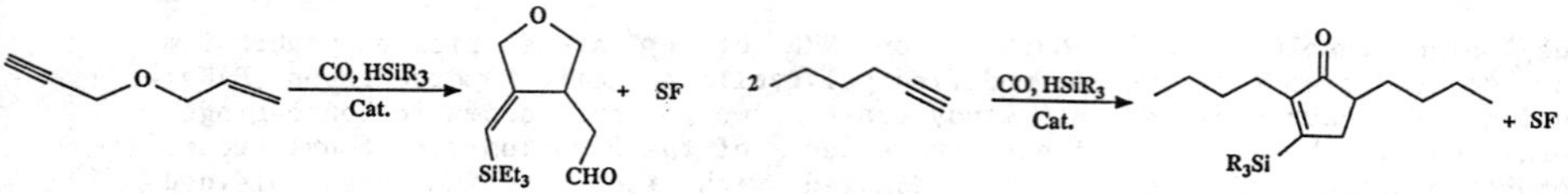

219. REACTIVITY STUDIES OF NEW NITRIDO AND IMIDO COMPLEXES OF OSMIUM(VI).
Jeanine M. Shusta, Robert W. Marshman and Patricia A. Shapley, Department of Chemistry, University of Illinois, Urbana, Illinois 61801

The first cyclopentadienyl-nitrido complexes of a transition metal, $(\eta^5\text{-}C_5H_5)Os(N)R_2$ and $(\eta^5\text{-}C_5Me_5)Os(N)R_2$ (R= Me, CH_2SiMe_3), were synthesized by the reactions between NaC_5H_5 and LiC_5Me_5 and $Os(N)R_2Cl$. The latter compound was prepared by the addition of $AgBF_4$ to $[NBu^n_4][Os(N)R_2Cl_2]$. The nitrogen atom in $CpOs(N)(CH_2SiMe_3)_2$ is a "soft" Lewis base which binds reversibly to BF_3 forming a 1:1 adduct, while with silver(I) salts it binds to give a 2:1 adduct. Alkylation of the nitrogen atom in $CpOs(N)R_2$ and $Cp^*Os(N)R_2$ with $CH_3OSO_2CF_3$ produced novel alkylimido complexes, $[CpOs(NMe)R_2][OSO_2CF_3]$ and $[Cp^*Os(NMe)R_2][OSO_2CF_3]$, respectively. These imido complexes react with trialkylphosphines, triarylphosphines, and olefins. Increasing steric bulk in the cyclopentadienyl ring or in the alkyl groups decreases the rate of reactions. Products arising from transfer of the methylimido group to the olefin have been identified in reactions with norbornene and styrene.

220. A NEW CATALYST FOR THE EPIMERIZATION OF SECONDARY ALCOHOLS; CARBON-HYDROGEN BOND ACTIVATION IN RHENIUM ALKOXIDE COMPLEXES (η^5-C_5R_5)Re(NO)(PPh$_3$)(OCHRR'). Isabel Saura-Llamas, Charles M. Garner, and J. A. Gladysz, Department of Chemistry, University of Utah, Salt Lake City, Utah 84112

Exo- and endo-borneol, endo-norborneol, cis-2-methylcyclohexanol, and cis-4-methylcyclohexanol are epimerized to mixtures of diastereomers in C_6H_5R at 65-90 °C in the presence of 10 mol% (η^5-C_5R_5)Re(NO)(PPh$_3$)(OCH$_3$) (1; R = H, Me). The methoxide ligand of 1 first exchanges with the alcohol substrate to give alkoxide complexes (η^5-C_5R_5)Re(NO)(PPh$_3$)(OCHRR') (2). Authentic samples of diastereomerically and enantiomerically pure 2 are prepared, where OCHRR' is derived from (+)- and (-)-exo-borneol, and (+)- and (-)-endo-borneol. NMR data show that epimerization occurs first at rhenium (35 °C) and then carbon (65 °C). Rate experiments show that PPh$_3$ initially dissociates from 2. An intermediate with a trigonal planar rhenium, which can either undergo β-hydride elimination to a ketone hydride complex, or return to 2, is proposed.

221. SEQUENTIAL BENZANNULATION / NUCLEOPHILIC AROMATIC SUBSTITUTION MEDIATED BY CHROMIUM (0)
William D. Wulff and Steven A. Chamberlin. Department of Chemistry, The University of Chicago, Chicago, Illinois, 60637.

Benzannulation of Fischer carbene complexes with alkynes and nucleophilic aromatic substitution by stabilized carbanions on Cr(CO)$_3$-complexed arenes form carbon-carbon bonds under mild conditions. We report the first use of the chromium-complexed arenes resulting from benzannulation as substrates for aromatic substitution. Though the metal fragment of the phenolic complexes initially generated is easily removed, *in situ* protection of the arene complexes resulting from alkenyl carbene complexes, 1, by a variety of electrophiles leads to stable complexes, 2, in yields approaching those of the generation of the free arenes via benzannulation. When R is TBDMS or TIPS, the complexes undergo nucleophilic aromatic substitution efficiently ; the product partition between 3 and 4 can be controlled due to the sensitivity of the substitution reaction to electronic and steric factors. Progress in the use of this sequence in the synthesis of anthracyclinones and spirocycles will be presented.

222. A "ONE POT" GEM-DIMETHYLATION OF THE CARBONYL GROUP. Stanley H. Pine, Gregory S. Shen, Huiling L. Zhang, Gia Kim, Huan Hoang, Department of Chemistry and Biochemistry, California State University, Los Angeles, CA 90032

The titanium - aluminum metallacycle known as the Tebbe reagent is shown to effectively convert a carbonyl group to gem-dimethyl in a one pot synthesis sequence.

$$>C=O + Cp_2TiClCH_2Al(CH_3)_2 \longrightarrow >C<^{CH_3}_{CH_3}$$

Aldehydes and ketones undergo the transformation. The process is believed to involve initial conversion of the carbonyl to a methylidene which then forms a metallacycle with a second equivalent of the reagent. Protonolysis of the metallacycle provides the gem-dimethyl product. Formation of the metallacycle is hindered by steric crowding so that hindered carbonyls lead to a lower yield of product.

223. **PREPARATION AND REACTION OF DILITHIO-2,4-OXAZOLIDINEDIONE WITH α-HALOKETONES. A VERSATILE SYNTHESIS OF 3-HYDROXY-5(2H)-FURANONES** Arie Zask, Medicinal Chemistry Department, Wyeth-Ayerst Research, CN 8000, Princeton, New Jersey 08543-8000

The dianion of 2,4-oxazolidinedione **1** was prepared by deprotonation with t-butyllithium/lithium chloride in tetrahydrofuran. Reaction of **1** with α-haloketones **2** (X = Cl, Br) gave allylic alcohols **3** which were hydrolyzed to the corresponding 3-hydroxy-5(2H)-furanones **4**. The natural product WF-3681 (**4**; R_1 = Ph, R_2 = -$CH_2CH_2CO_2H$), an aldose reductase inhibitor, was synthesized by this method.

224. **N- AND 4-ARYLATION OF N-(4-METHOXYPHENYL)-2-BUTENAMIDES BY VARIOUS ARYNES.** A. Rakeeb Deshmukh, Hongming Zhang, and Ed Biehl*, Department of Chemistry, Southern Methodist University, Dallas, TX 75275.

We report that arynes, generated at low temperatures (ca. -40 °C) from LDA and bromobenzene and certain of its methoxy and dimethoxy derivatives as well as 4-bromo-2,6-dimethoxy-3-methylpyridine, are readily trapped by dilithiated N-(4-methoxyphenyl)-2-butenamide (**1**) yielding mixtures of the corresponding N-(4-methoxyphenyl)-4-aryl-2- and 3-butenamides. In most cases, the individual amides were isolated by flash chromatography; all mixtures could be readily reduced to the respective N-(4-methoxyphenyl)butanamide. 2-Bromo-1,4-dimethylbenzene treated similarly gives N-(4-methoxyphenyl)-4-(2,5-dimethylphenyl)-2-butenamide and 2(1-[2,5-dimethylphenyl]-4,7-dimethylphenyl)indanylamide, most likely via a tandem addition-rearrangement aryne pathway. 3,4,5,6-Tetramethylbenzyne, which requires a temperature of 0°C for its generation, reacts with **1** yielding mainly N-(4-methoxyphenyl)-N-(2,3,4,5-tetramethylphenyl)-2-butenamide and small amounts of the corresponding 4-arylated 2-butenamide. A discussion on the factors which influence the extent of N- and 4-arylations will be presented.

225. MIXED METAL SYNTHESIS OF 2-ETHYLHEXYLLITHIUM. C. W. Kamienski, R. C. Morrison, B. T. Dover, Technology Department, FMC Lithium Division, Bessemer City, North Carolina 28016.

It was discovered that by incorporating a significant amount of sodium dispersion (5 to 50 mole % with lithium dispersion - 95 to 50 mole %) the reaction between 2-ethylhexyl chloride and the metals proceeded efficiently to produce hydrocarbon soluble 2-ethylhexyllithium. Sodium was converted to insoluble sodium chloride via metal/metal exchange reactions and was removed along with lithium chloride and residual metals by filtration. The reaction can be carried out efficiently in refluxing or non-refluxing solvent with optimum yields being 90 to 95% when employing lithium/sodium dispersion (83/17 mole %) with little or no residual alkyl halide remaining in solution. Thermal stability testing indicated 2-ethylhexyllithium to be comparable to n-butyllithium. Neat 2-ethylhexyllithium (95% pure) was found to be a liquid and also to be less pyrophoric than concentrated n-butyllithium. Thus, the neat product could be sold and dissolved by the customers in the solvent of their choice.

226. A HIGH PURITY, NON-PYROPHORIC PHENYLLITHIUM SOLUTION. R. C. Morrison and D. E. Sutton, Technology Department, FMC Lithium Division, Bessemer City, North Carolina 28016.

A stable, non-pyrophoric essentially halide-free form of phenyllithium was synthesized by the reaction of chlorobenzene and lithium metal dispersion in dibutyl ether (DBE) as solvent. By maintaining a butyl ether to chlorobenzene ratio of 1.7 to 1.0 or greater, yields of >90% phenyllithium were achieved. Industrial applications of phenyllithium have been limited by the hazards associated with previous commercially available formulations in the highly flammable solvent, diethyl ether, and by the presence of significant amounts of lithiated aromatic impurities. Unexpectedly, these new phenyllithium solutions were found to be indefinitely stable in a totally ether solvent, even at 40°C, and contained no lithiated aromatic impurities. Phenyllithium was determined to be highly soluble in DBE (greater than 30 wt. %). Concentrations of phenyllithium in DBE of 15 wt. % and below were found to be non-pyrophoric.

227.
ADDITION OF ORGANOLITHIUM COMPOUNDS TO N-THP PROTECTED NITRONE
Anwer Basha*, James D. Ratajczyk and Dee W. Brooks. Immunosciences Research Area, D-47K, Abbott Laboratories, Abbott Park, IL 60064

A novel synthesis of α-branched primary hydroxylamines by addition of organolithium reagents to N-THP protected nitrone is described.

$$R\text{-}Li + \overset{\bar{O}}{\underset{}{N}}\text{-}THP \xrightarrow[2) H^+]{1) THF / -40°C} \underset{CH_3}{\overset{R}{\diagup}}\text{-}NHOH$$

Application of this methodology is illustrated by the synthesis of zileuton (A-64077). Zileuton is a 5-lipoxygenase inhibitor, currently in clinical trials for treating human diseases involving leukotrienes as mediators.

228. TANDEM MICHAEL ADDITION-ETHOXYCARBONYLATION-CARBONYL DEBLOCKING OF ANIONS OF CARBETHOXY PROTECTED CYANOHYDRINS WITH ENONES. SYNTHETIC APPLICATIONS. Héctor M. Torres[2], José M. Méndez[1] and Luis A. Maldonado[1] 1)División de Estudios de Posgrado, Facultad de Química, UNAM, 04510, México,D.F.; 2)Instituto Mexicano del Petróleo, Eje Lázaro Cárdenas 152, México,D.F. 07730.

Conjugate addition of selected anions of carbethoxy protected cyanohydrins to cycloalkenones gives regiospecifically generated 2-carbethoxy-3-acyl-Δ^2- cycloalkeno enolates which can be O -or C - trapped with appropriate electrophilic reagents. The procedure involves one-pot consecutive 1,4-addition of the acyl carbanion equivalent, carbethoxy migration to the 2 - position with concomitant HCN loss to free the acyl group, and the initially formed β - keto ester. This reaction sequence has been applied to the synthesis of linear polynuclear compounds.

229.
GENERATION OF *ORTHO*-QUINONE METHIDES FROM CYCLOBUTENONES
Meng Taing, Haiji Xia, Simon L. Xu and Harold W. Moore*. Department of Chemistry, University of California, Irvine, CA 92717

Three new routes to the *ortho*-quinone methides from appropriately substituted cyclobutenones are described. The first one involves the electrocyclic ring opening of allene substituted cyclobutenone 1, followed by electrocyclic ring closure. The second and third arise from the electrocyclic ring opening of cyclobutenones 3 and 5, followed by a series of rearrangements through radical intermediates. The quinone methide intermediates 2, 4 and 6 undergo self-condensation (dimerization and trimerization) and can be trapped with dienophiles through hetero Diels-Alder reaction to form chromans.

230. 1,3-DIPOLAR CYCLOADDITION REACTIONS WITH CAPTODATIVE OLEFINS.

Cadena J.L., Jiménez R., Tamariz J., and Salgado-Zamora*. Departamento de Química Orgánica, Escuela Nacional de Ciencias Biológicas, I.P.N. Carpio y Plan de Ayala, 11340 México, D. F.

The 1,3-dipolar reactions of benzonitrile oxide $\underline{I}$, with captodative olefins $H_2C=C(COCH_3)OCOR$, $\underline{2}$, R=pnitrophenyl and p-chlorophenyl were investigated. From these reactions, the fully unsaturated 5-acetyl-3-phenylisoxazole $\underline{3}$ was the only isolated product. The observed regiochemistry of the reaction is analysed in terms of the FMO theory and an explanation involving the polarisability factor is also offered.

$Ph-C=\overset{+}{N}-O^{-}$
$\underline{1}$

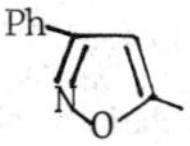

[structure of 5: 3-phenyl-5-acetylisoxazole with $COCH_3$] $\underline{5}$

231. INTRAMOLECULAR DIELS-ALDER REACTIONS OF VINYLSULFONAMIDES. Yuqi GU, Richard A. Hudson, Center for Drug Design and Development, Department of Medicinal and Biological Chemistry, The University of Toledo, College of Pharmacy, Toledo, OH 43606

The intramolecular Diels-Alder reaction of several N-substituted furfurylamine vinylsulfonamides were investigated. The nitrogen substituents play an important role in reaction rate. When R_2 is benzyl or phenyl, the reaction rate increased dramatically. Compound 1a gave 25% yield of 2a while compound 1b gave 70% and 1c 100% yield of 2b and 2c respectively after 24 h in CDCl$_3$ at 55 °C. Vinylsulfonamides were better dienophiles than acrylamides. Furfurylamine vinylsulfonamide undergoes intramolecular Diels-Alder reaction in CDCl$_3$ at 60 °C while furfurylamine acrylamide gave no products even in refluxing benzene. When R_1 is phenyl, no cycloadducts were obtained. The synthetic utility of these cycloadducts will be discussed.

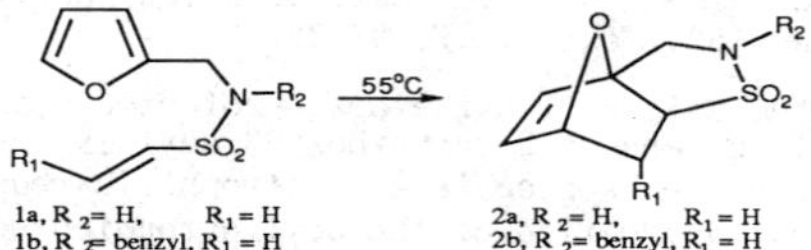

1a, R_2 = H, R_1 = H
1b, R_2 = benzyl, R_1 = H
1c, R_2 = phenyl, R_1 = H

2a, R_2 = H, R_1 = H
2b, R_2 = benzyl, R_1 = H
2c, R_2 = phenyl, R_1 = H

232. **SYNTHESIS AND CHEMISTRY OF SOME TRICYCLIC CYCLOPROPENES.** Philip J. Chenier, Shawn D. Balding, Michael J. Bauer, Gregory M. Canard, Christina L. Hodge, and Dale A. Southard, Jr., Department of Chemistry, University of Wisconsin-Eau Claire, Eau Claire, WI 54702-4004.

In the last several years a number of workers have synthesized highly strained olefins in an attempt to determine the limits of ring strain which can be overcome before a double bond can no longer be possible. Some of these olefins cannot be isolated but have been proven to exist by their subsequent reactivity and chemistry. Recently we have synthesized or are synthesizing olefins **1**, generated from the corresponding dibromides **2**. Their chemistry has proven to be complex and illusive. Diels-Alder reactivity with diphenylisobenzofuran, the "ene" reaction of the olefins with themselves, and other attempted chemistry will be discussed as well as the syntheses of the dibromide precursors.

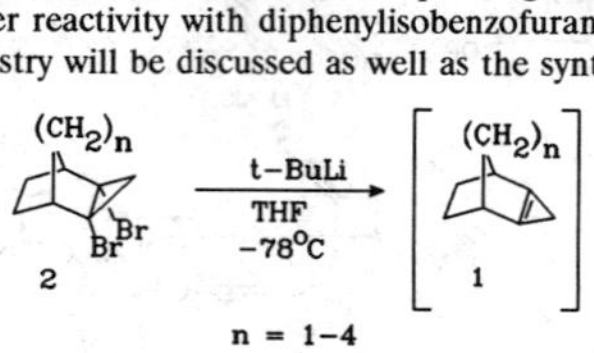

$$2 \xrightarrow[\text{THF}\,-78°C]{t\text{-BuLi}} [\,1\,]$$

n = 1-4

233. **Diastereofacial Selectivity in the Diels-Alder reaction of 2,3-Dialkoxy-1,3-butadienes.** John E. Kerrigan, Patrick G. McDougal and Don VanDerveer, *School of Chemistry and Biochemistry, Georgia Institute of Technology, Atlanta, Georgia 30332.*

Facial selectivity in the Diels-Alder reaction of dialkoxy dienes (1) will be presented. The alcohol (R=H) exhibits a preference for the *like* diastereomer (2) with PTAD, which is solvent and temperature dependent (de's > 90%). Conversely, when R=TMS selectivity reverses to favor the *unlike* diastereomer (3). Possible reasons for this selectivity reversal as well as reactivity with other dienophiles will be discussed.

234. **The Diels-Alder Reaction of α-Cyano-1-azadienes** Nicholas J. Sisti, Frank W. Fowler and David Grierson *Institut de Chimie des Substances Naturelles, CNRS, Gif-sur-Yvette 91190, FRANCE and the Department of Chemistry, State University of New York, Stony Brook, New York 11794*

An α-cyano substituent has been demonstrated to provide sufficient activation to induce 1-azadienes to participate in the Diels-Alder reaction. These azadienes display a remarkable reactivity in that they react with electron rich as well as electron poor dienophiles. The intramolecular version of the Diels-Alder reaction has also been demonstrated with an unactivated dienophile. The simplicity of preparation of the reactants, the generally high yields of the reactions and the synthetic utility of the α-cyano substituent for further structural elaboration support this Diels-Alder reaction as being a general solution to problems involving the preparation of piperidine rings.

235. **Intramolecular [4+4] Photocycloaddition of Tethered 2-Pyridones.**
Scott McN. Sieburth, Jian-long Chen and Dora P. Kuan, Department of Chemistry, State University of New York at Stony Brook, Stony Brook, New York 11794-3400

Photolysis of two 2-pyridones tethered at the 3 and 6' positions results in an efficient [4+4] cycloaddition. The reaction product is a 1,5-cyclooctadiene *trans* fused to a five or six-membered ring. Studies concerning the control of stereogenesis and the effects of tether length will be presented.

236. UNEXPECTEDLY HIGH DIASTEREOSELECTIVE OXIDATION OF LOCKED-CONFORMATION STYRENES. STEREOSELECTIVE SYNTHESES OF EXO- OR ENDO POLYCYCLIC EPOXIDES. Michael J. Martinelli, Barry C. Peterson, Vien V. Khau, and Darrell R. Hutchison, Chemical Process R&D, Lilly Research Laboratories, Indianapolis, IN 46285

Selectivity in organic synthesis is the focal point in many laboratories. Investigations have been aimed at both an understanding, as well as an exploitation of the principles which afford selectivity in an organic reaction. More recently, the literature has witnessed a growing body of evidence suggesting that very subtle stereoelectronic effects have resulted in substantially high facial selectivities in reactions of sterically unbiased substrates. We recently encountered extraordinarily high diastereoselectivities with several oxidation protocols (epoxidation, osmylation) in a series of partial ergot olefin substrates. For example, epoxidation (mcpba) of **1** was exo selective to provide the epoxide **5** (exo) in 97% yield with 99:1 diastereofacial selectivity (eqn 1). Likewise, the complementary endo epoxide **6** was formed by a two step procedure (NBS/H_2O/DMSO; NaOH, toluene) with equally high diastereoselectivity (1:99). The epoxidation was insensitive to solvent effects or reagent. Furthermore, osmylation followed the same exo selective trend, as determined by x-ray analysis. The selectivity was not affected by conjugation to an electron-withdrawing or donating group. We have now examined a number of "locked-conformation styrene" substrates to determine the generality of this effect, and summarize our initial results in this poster.

237. ELECTRONIC CONTROL OF FACE SELECTION IN THE REDUCTION OF 5-AZAADAMANTAN-2-ONE. Juliet M. Hahn, William J. le Noble, Department of Chemistry, State University of New York at Stony Brook, Stony Brook, N.Y. 11794-3400

5-Azaadamantan-2-one has been used to probe the electronic effect to stereoselectivity in nucleophilic addition. The reduction of this compound (NaBH$_4$ E/Z 58%:42%, MeLi E/Z 55%:45%) and the reduction of its derivatives (N-methyl-5-azaadamantan-2-one iodide (E/Z 92%:8%), 5-azaadamantan-2-one amine oxide (E/Z >90%:<10%), glycine derivative of 5-azaadamantan-2-one(E/Z >90%:<10%)} show a dramatic change from E/Z product ratio of 58%:42% to >90%:<10% as the free nitrogen lone pair is bound. The effect is proven not to be due to ion pairing by the internal salts, the amine oxide and the glycine derivative. The salt ketones also all show hydration at the carbonyl site in water as demonstrated by ^{13}C NMR spectra which suggests a hydration-dehydration preequilibrium before reduction. These results can be well accommodated by the application of transition state hyperconjugation.

238. FACE SELECTIVITY IN 5-,7-DISUBSTITUTED ADAMANTANONES. In H. Song and William J. le Noble. Department of Chemistry, State University of New York at Stony Brook, Stony Brook, NY 11794

With substituents of opposite polarity, we hope to demonstrate further the importance of electronic effects in determining the stereochemistry of the reactions.

r_1 r_2 r_t

239. FACE SELECTION IN THE THIO-CLAISEN REARRANGEMENT. Ashis Mukherjee and William J. le Noble. Department of Chemistry, State University of New York at Stony Brook, Stony Brook, NY 11794-3400

Allyl 5-fluoroadamantylidenemethyl sulfoxide has been prepared as a 50-50 mixture of two diastereomers. Upon heating the mixture, each of the components produces a mixture of two sulfines. The four sulfines obtained are achiral stereoisomers differing in the configuration at C_2 and in the $\underline{E}$ or $\underline{Z}$ nature of the C=S=O group. Analysis of these mixtures allows one to assess the magnitude of the preference for approach to the zu face at C_2, and for the pseudoequatorial position of oxygen in the transition state.

240. STEREOELECTRONIC CONTROL OF DIASTEREOSELECTIVITY IN ELECTROPHILIC AND DIELS-ALDER REACTIONS R. L. Halterman, M. A. McEvoy, and B. A. McCarthy, Department of Chemistry, Boston University, Boston, MA 02215

In an attempt to separate steric from electronic factors for the induction of asymmetry in diastereoselective reactions, we have studied the reactivity of a series of substrates based on the 2,2'-diphenylcyclopentane framework. The electrophilic osmylation of cyclopentene **1** has resulted in diastereoselectivities up to 70:30. For the peracetic acid epoxidation of cyclopentene **1**, we have observed diastereomeric ratios up to 73:27. Hydroboration reactions have also been examined. Diels-Alder cycloadditions have been performed on the cyclopentadiene **2**, resulting in diastereomeric ratios as high as 69:31. With the exception of the epoxidation reaction which demonstrates anti-Cieplak selectivity, addition occurs opposite the more electron rich aromatic ring.

241. CONFORMATION DEPENDENT DIELS-ALDER DIENOPHILE FACE SELECTIVITY.

J. Gabriel Garcia and Mark L. McLaughlin, Department of Chemistry, Louisiana State University, Baton Rouge, LA 70803-1804.

Diels-Alder reactions of a series of flexible remotely substituted cyclooctenes with hexachlorocyclopentadiene (HCCP) and 5,5-dimethoxy-1,2,3,4-tetrachlorocyclopenta-2,4-diene (DMTCCP) indicate that dienophile π-face selectivity results from conformation dependent transmission of π–σ–π electronic interactions. Ratios (syn:anti) from: **1** (X=Y=Cl) and **2** (X=Cl,Y=OCH$_3$), 1:4; **3** (X=OCH$_3$, Y=H), 1:12; **4** (X=O, Y=H) and **5** (X=Y=H), 1:≥20; and **6** (X=Y=H, sat'd norbornane), 1:2.5.

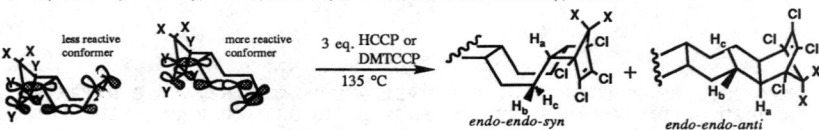

242. A REDUCTIVE AMINATION-LACTAMIZATION PROCEDURE USING BOROHYDRIDE REAGENTS
Ahmed F. Abdel-Magid, Bruce D. Harris and Cynthia A. Maryanoff, Department of Chemical Development The R. W. Johnson Pharmaceutical Research Institute, Spring House, PA 19477

Although there are several reports describing the synthesis of lactams from aminoacids and aminoesters, few have dealt with tandem reactions in which an aminoacid or aminoester produced from a reductive amination reaction is cyclized directly under the reaction conditions to the corresponding lactam. This paper presents our preliminary results of a systematic study to develop this methodology.

243. THE REACTION OF SULFIDES WITH PERMANGANATE ION.
Donald G. Lee and Tao Chen, Department of Chemistry, University of Regina, Regina Saskatchewan Canada S4S 0A2

The current literature suggests that the reaction of sulfides with permanganate involves an "electrophilic oxygen transfer" in an S_N2 like mechanism. However, a study of the oxidation of substituted phenylmercaptoacetic acids by permanganate under basic conditions provides experimental evidence that can more reasonably be accommodated by a mechanism in which the reaction is initiated by the formation of a coordinate covalent bond utilizing an unshared pair of sulfur electrons and the empty d-orbitals of manganese. Consistent with this conclusion is the observation of a negative Hammett rho value (-1.25) and a large entropy of activation, $-\Delta S = 176$ $J\ mol^{-1}\ K^{-1}$.

244. A MILD, OSMIUM TETRAOXIDE-CATALYZED PROCESS FOR THE OXIDATION OF SULFIDES TO SULFONES, Stephen W. Kaldor and Marlys Hammond, Lilly Research Laboratories, Indianapolis, IN 46285

Although several researchers have observed that sulfides are generally inert to oxidation by osmium tetraoxide under stoichiometric conditions, we have recently discovered that treatment of a variety of sulfides with one mole percent of OsO_4 in the presence of the co-oxidant N-methylmorpholine-N-oxide results in rapid, room temperature oxidation to the corresponding sulfones in nearly quantitative yields. This process is tolerant of a variety of other functional groups; chemoselective oxidation of a sulfide in the presence of an olefin is even possible in some instances.

245. SYNTHESIS OF PYRIDINES BY BENTONITE-SUPPORTED METALLIC NITRATES : INDUCED OXIDATION OF HANTZSCH ESTERS (DIALKOXYCARBONYLDIHIDROPYRI-DINES) Miranda, R. ; Gutiérrez,C. ; Martínez,R. ; Angeles,E. ; Cervantes,R. ; Facultad de Estudios Superiores Cuautitlán - Universidad Nacional Autónoma de México, Campo 1 , Cuautitlán Izcalli , Estado de Mé - xico , México .

In this work we wish to report the results of several induced oxidations on Hantzsch Esters with four metallic nitrates supported on "Tonsil" a mex-xican bentonitic earth , which has been succesfully applied by our group either support or catalyst (R.Miranda , H.Cervantes , P.Joshep-Nathan , Synth.Commun.,21 , (1990),153 ; M.Salmón , E.Angeles,R.Miranda , J.Chem . Soc.Chem.Commun.,(1990) , 1188 . The products were characterized by ^{1}H-NMR

EIMS and IR spectral data . Some of these Hantzsch derivatives may possess interesting biological activities . This research was generously supported by the GOBIERNO DEL ESTADO DE MEXICO , MEXICO .

246. THE INFLUENCE OF HETEROATOM ON THE STEREOCHEMISTRY OF NICKEL BORIDE REDUCTION OF 2,6-DIHALO-9-OXABICYCLONONANES AND RELATED OXAADAMANTANES.
S. Kugabalasooriar and Raymond C. Fort, Jr., Department of Chemistry University of Maine, Orono, Maine 04469

Nickel boride reduction of anti, anti; anti, syn; syn, syn-2,6-Diiodo-9-oxabicyclononanes (1-3) and anti, anti-2,6-Dibromo-9-oxabicyclononane (4) in THF/MeOD solution gives the same product mixture (5) stereoselectively (65-72% of syn and 35-82% of anti). The reduction of 4,8-Dibromo-2-oxaadamantane under the similar conditions gives the product with opposite stereoselectivity (35% of syn and 65% of anti). These results imply reduction on the surface of the nickel boride, participation of the heteroatom, and isomerization of 3 to 2 and to 1.

```
1 : X=I  ; anti, anti
2 : X=I  ; anti, syn
3 : X=I  ; syn, syn
4 : X=Br ; anti, anti
```

1-4

6

247. PHOTOREDUCTION OF ALDEHYDES WITH SEMICONDUCTOR AS CATALYST. Jeff Pross, Cheryl Joyce-Pruden, and Yuzhuo Li, Department of Chemistry, Clarkson University, Potsdam, New York 13699-5810.

A group of aldehydes was quantitatively reduced to alcohols using colloidal and submicron sized titanium dioxide as photocatalyst. The photoreduction is initiated by a single electron transfer from excited state of titanium dioxide to aldehyde and followed by protonation to yield an a-hydroxy radical. The radical is then further reduced by a second electron transfer and protonation. The second electron reduction sluggishly competes with deprotonation and back electron transfer reactions. In the presence of a strong hydrogen donor, further reduction of the radical can be accomplished by a simple hydrogen abstraction and quantum efficiency are substantially enhanced.

248. FREE RADICAL HALOGENATION OF HYDRIDOPHOSPHORANES. Massoud Garrossian, Institute of Chemistry, The University of Mazandaran, Babolsar 311, Iran.

Within the past several years a great deal of works have been done on phosphoranyl radicals ($Z_4P.$) which can be trapped by nonscission process, these are typically spiro or bicylclic ones. This work is dealing with the initial results on the reaction of hydridophosphoranes (Z_4PH) with carbontetrachloride, bromine and trichlorobromomethane, by UV-light or AIBN. These reactions could be inhibited by ⍺-methylstyrene and in certain cases will give the corresponding halophosphorane (Z_4PX) with high yields.

249. ELECTROCHEMICAL ALKYLATION AND METHOXYLATION OF ACTIVE METHYLENE COMPOUNDS. S.Nakajima, S.Suzuki, and M.Kato. Department of Organic Chemistry, Hoshi University, Ebara 2-4-41, Shinagawa-ku, Tokyo, Japan 142

Two new electrochemical substitution methods of the active methylene compounds were developed. (1)Triphenylmethane was found to afford a new electrogenerated base upon cathodic reduction. And, with the aid of this base, methyl p-methoxyphenylacetate, 3,4-dimethoxyphenylacetate, 3,4-dimethoxyphenylacetonitrile were either methylated or ethylated. (2)Methyl phenylacetate, methyl p-chlorophenylacetate, methyl P-methoxyphenyl acetate , and methyl 1-naphthaleneacetate were electrochemically sucessfully derived, in one-step, to each methoxylated products by anodic oxidation, in the anolyte containing potassium iodide as an electron carrier, and sodium methoxide as an enolizer and a methoxylating agent.

250. ELECTRODUCTION OF o. NITROBENZONITRILE
ESTRADA M. EVA, RIEKER ANTON
FACULTAD DE QUIMICA, UNAM LAB. TECNOLOGICO, CIUDAD UNIVERSITARIA MEXICO 04510 D.F. MEXICO
INSTITUT FUR ORGANISCHE CHEMIE.- UNIVERSITAT TUBINGEN,-GERMANY, AUF DER MORGENSTELLE 18, D-4700 TUBINGEN, GERMANY

o, Nitrobenzonitrile can be reduced electrochemically forming different products. Their formation depends on the reaction conditions. The reduction at controlled potential (-700mV) in a mercury pool, in an acid medium forms 2.2 methyl 3.4-dihydro-4-oxoquinazoline. At -600 mV under slow potential increase the main product is 2.2'- dicyanoazoxibenzene. This compound derives from the hydroxylamine and it proves its formation as an intermediate.

251. SINGLE ELECTRON TRANSFER CATALYZED CLEAVAGE OF THE C-S AND S-S BONDS.
S. Munavalli[1], D. I. Rossman[2], D. K. Rohrbaugh[2], C. P. Ferguson[2] and H. D. Banks[2] (1) Geo-Centers, Inc. and (2) US Army , CRDEC, Attn: SMCCR-RSC-O, Aberdeen Proving Ground, MD 21010

The S-S bond of the polysulfides, which are ubiquitous in nature, plays a vital role in life's processes. As such, over the years, the chemistry of the S-S bond has received considerable attention. Examination of the molecular models of the trisulfides suggests that they can exist in "cis and trans" forms (a and b). X-ray analysis favors the trans form. The reaction of the trisulfides with organolithiums at -78º has now been found to cause both C-S and S-S bond cleavages to furnish mixed mono-, di-, and tri-sulfides(c). These reactions appear to be exothermic and can be satisfactorily rationalized on the basis of the single electron transfer process. This represents the first study which demonstrates that both C-S and S-S bond scission can occur simultaneously at -78º. The mechanism of cleavage and the mass spectral fragmentation of the mixed sulfides are described.

252. SULFUR-SULFUR BOND SCISSION BY A COPPER REAGENT.
S.Munavalli[1], D.I.Rossman[2], D.K.Rohrbaugh[2], C.P.Ferguson[2], and F-L.Hsu[2].
(1) Geo-Centers, Inc, Ft Washington, MD 20744 and (2) US Army , CRDEC, Attn: SMCCR-RSC-O, Aberdeen Proving Ground, MD 21010-5423.

Since sulfur-sulfur bridged compounds occur widely in nature and play a vital role in the biological processes, the chemistry of the S-S bond has attracted considerable interest. It is well known that organometallic reagents cleave the S-S bond of the disulfides and yield substituted sulfides.   We wish to report that the thiyl radicals generated from the trifluoromethylthiocopper precursor cause displacement on the sulfur atom of the disulfides to furnish unsymmetrical disulfides. The yields of the mixed disulfides, in general, depend on the steric bulk of the substituent. However, the attack of the thiyl radicals on diaryl disulfides appears to take place more easily than on dialkyl disulfides. Two mechanisms -(1) a step-wise free radical process and(2) a synchronous process(Scheme 1)- can be envisaged for the S-S bond scission. This paper discusses the mechanism of the free radical induced cleavage of the S-S bond and the mass spectra of the mixed disulfides.

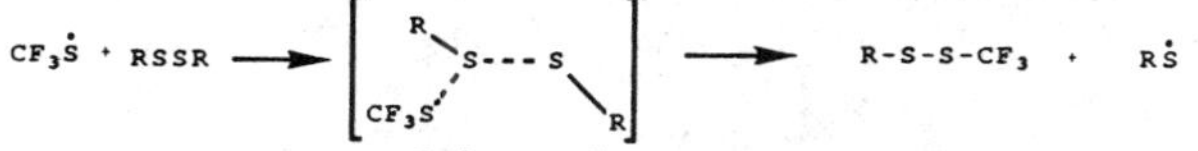

$$CF_3\dot{S} + RSSR \longrightarrow \begin{bmatrix} R \\ CF_3S' \end{bmatrix} S\text{---}S \begin{bmatrix} \\ R \end{bmatrix} \longrightarrow R\text{-}S\text{-}S\text{-}CF_3 + R\dot{S}$$

Scheme 1

253. **GRIGNARD REAGENT PROMOTED SCISSION OF THE C-S AND S-S BONDS.**
S. Munavalli[1], D. I. Rossman[2], D. K. Rohrbaugh[2], C. P. Ferguson[2] and L. J. Szafraniec[2]. (1) Geo-Centers, Inc. and (2) US Army, CRDEC, Attn: SMCCR-RSC-O, Aberdeen Proving Ground, MD 21010

Since the sulfur-sulfur bridge determines the tertiary structure of proteins and enzymes and brings about life-sustaining bio-transformations through redox reactions of the S-S bond, its chemistry has attracted considerable attention. In continuation of our interest in the reactions of the S-S bond, we have investigated the behavior of the trisulfides with Grignard reagents and for first time have observed the scission of both the C-S and S-S bonds(1). Contrary to an earlier claim that all free radical displacement reactions of the trisulfides involve the central sulfur atom(2), the mixed sulfides have been identified. The formation of the mixed mono-, di-, and tri-sulfides can be rationalized and explained on the basis of a single electron transfer process. The mechanism of the formation of the above products and their mass spectral fragmentation are discussed.

$$RS_3R + R'MgX \longrightarrow RSR' + RSSR' + RSSSR'$$
      1

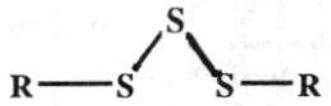

$$R-S\overset{S}{\underset{}{\diagdown}}S-R$$
      2

254. **ORGANOMETALLIC APPROACHES TO SEROTONIN ANALOGS.** Youhua Yang, Arnold R. Martin, David L. Nelson* Hong B. Li* and John Regan*. Departments of Pharmaceutical Sciences and *Pharmacology & Toxicology, College of Pharmacy, University of Arizona, Tucson, Arizona 85721.

A number of 5-aryl indoles were prepared by cross coupling of 5-indole boronic acid with various aryl halides in the presence of palladium(0) and base. 5-Bromo-3-iodoindole was tandem coupled with various trimethylarylstannanes and aryl boronic acids to give 3- and 5-diarylated indoles catalyzed by palladium(0). After selective reduction of the pyridyl ring, 5-arylated serotonin analogs were obtained.

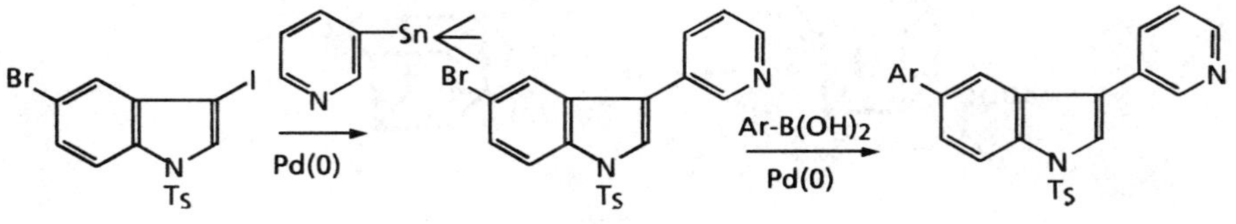

255. A NEW SYNTHESIS OF α,β-UNSATURATED AMIDES. A.G. Chaudhary, M.S. Manhas and E.W. Robb, Department of Chemistry and Chemical Engineering, Stevens Institute of Technology, Hoboken NJ 07030

In continuation of our studies on the synthesis of β-lactams we have observed a hitherto unreported reaction. When a Schiff base (1) was treated with ethyl phenylacetate and excess of LDA, the product obtained was not the expected β-lactam (3), but an unsaturated amide (2). This method is of general applicability and provides a means of introducing an alkylidene group at the α-position of substituted acetic acids. The use of 1 molar equivalent of LDA results in the formation of the β-lactam, which is stable to further treatment with LDA. Thus the formation of β-lactam and of the unsaturated amide appear to proceed through independent pathways.

(1) (2) (3)

256. REGIOSELECTIVE CARBON-CARBON BOND CLEAVAGE OF CUBANES
A. Bashir-Hashemi, P. R. Dave, GEO-CENTERS, INC., 762 Route 15 South, Lake Hopatcong, NJ 07849, T. Axenrod, City College of the City University of New York, New York, NY 10031.

A series of strained 'Cage' molecules employing cubanes as starting materials was prepared. Selective carbon-carbon bond cleavage of substituted cubanes leading to the synthesis of difficultly accessible polycyclic compounds will be presented.

257. SAFE AND PRACTICAL SYNTHESIS OF NITROCUBANES
R. Damavarapu, S. Iyer, Army Research & Development Center, Picatinny Arsenal, NJ 07806-5000, A. Bashir-Hashemi, GEO-CENTERS, INC., 762 Route 15 South, Lake Hopatcong, NJ 07849.

The chemical transformation of commercially-available dicarbomethoxycubane to the corresponding dinitrocubane requires the formation of a deadly explosive acyl azide intermediate. A safe and large-scale synthesis of nitrocubanes which avoids the formation of acyl azides will be presented.

258. THE EFFECT OF HYDROXYL PROTECTION ON STEREOCHEMISTRY AND STEREOCONTROL DURING AMIDE IODOLACTONIZATION. David P. Rotella, Department of Pharmacognosy, School of Pharmacy, University of Mississippi, University, MS 38677.

Previous work in this laboratory indicated that esterification of 2-alkyl-3-hydroxy-4-pentenamides with long chain (C_6 or C_{18}) alkyl groups was a useful approach for improving stereocontrol during kinetic iodolactonization of the anti diastereomer of such substrates. The present study investigates additional C2 substituents (e.g. iPr, benzyl, phenyl) and TBDMS-protected derivatives to probe the effect of such moieties on the stereochemistry and level of stereocontrol. As reported by others, it was observed that, in general, the anti diastereomer of the unprotected hydroxyl compound cyclized with less stereoselectivity than the syn isomer. However, derivatization of the alcohol as a TBDMS ether not only improved stereocontrol in the anti isomer, but also strengthened the trans-directing effect of the C2 substituent. Thus, this approach represents a generally useful strategy for improved stereocontrolled during amide iodolactonization.

259. **PHASE-MANAGED ORGANIC SYNTHESIS: ACTIVATION OF α-CHLORO, α-HYDROXY AND N-BLOCKED α-AMINO ACIDS.** W.K. Fife and Zheng-Yun Zhan, Department of Chemistry, Indiana University-Purdue University at Indianapolis, 1125 East 38th Street, Indianapolis, IN 46205.

The activation of α-amino- and α-hydroxy-acids for conversion to derivatives such as amides, esters and peptides is an important problem in organic synthesis. We report new results from our continuing development of convenient and efficient multiple-phase methodology for synthesis of these acid derivatives. The general reaction scheme is illustrated below.

$$\text{RCH(X)COOH} + \text{ARCOCl} \xrightarrow[\text{2) } R^1R^2NH]{\text{1) P4-VP}} \text{RCH(X)CONR}^1R^2 \qquad X = \text{Cl, OH, NHB}\ell$$

We have found that amines attack the mixed anhydrides of aryl and α-substituted alkyl carboxylic acids regioselectively to give the amides of the latter acids in high yields (≥ 80%). This presentation will highlight four aspects of these reactions: (1) reaction conditions, (2) effect of structure of mixed anhydrides on regioselectivity, (3) effect of amine structure on regioselectivity, and (4) the protection/deprotection requirements for α-hydroxyacids.

260. A GENERAL SYNTHESIS OF HYDROPEROXIDES. S. Prasad Peri, Kyeong-Eun Jung, and Dee Ann Casteel,* Division of Medicinal and Natural Products Chemistry, College of Pharmacy, University of Iowa, Iowa City, Iowa 52242

A general method for the preparation of hydroperoxides from alcohols has been developed. The approach involves initial conversion of the alcohol to its mesylate and subsequent reaction with hydrazine followed by hydrogen peroxide. The intermediate hydrazino products are not purified but are carried forward. Examples of the reaction at primary, secondary, tertiary, and allylic positions will be described. The usefulness of leaving groups other than mesylate has also been explored.

261. 18-CROWN-6 AS A CATALYST IN THE DIALKYLATION OF *o*- NITROPHENACYL DERIVATVES Girija Prasad and Patrick E. Hanna, Dept. of Medicinal Chemistry, and Wayland E. Noland and Shankar Venkatraman, Dept. of Chemistry, Univ. of Minnesota, Minneapolis, MN 55455.

A new method of synthesizing α,α-dialkyl *o* -nitrophenacyl derivatives (**3a-h**) from commercially available *o* -nitrophenacyl derivatives has been developed. Dialkylation of *o* -nitrophenacyl ketone (**1a**), *o* -nitrophenylacetate ester (**1b**), and *o* -nitrophenylacetonitrile (**1c**) with alkyl halides and potassium *tert* -butoxide proceeds in good yields with 18-crown-6 as a catalyst. The dialkylation can be carried out in one step on one equivalent of **1a-c** using 2.2 equivalents of potassium *tert* -butoxide and 2.2 equivalents of alkyl halide in 60-97% yields in the presence of 0.25 equivalent of18-crown-6. The dialkylation can also be achieved by alkylation of the monoalkyl derivatives (**2a-g**) with the same or different alkyl halide in 85-97% yields in the presence of 18-crown-6. The steric limits of the method are reached in the dialkylation of **1a** with ethyl iodide or allyl bromide, which gave the *C,O* -dialkylated products **4a** (60%) and **4b** (55%), respectively.

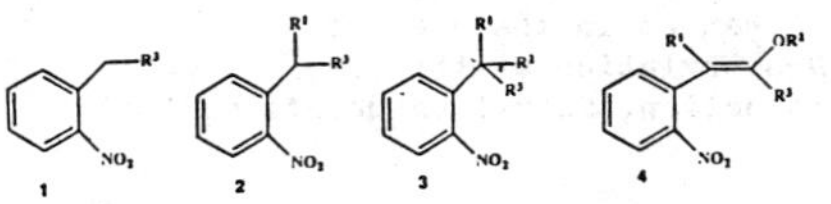

1 2 3 4

	R^3
1a	COPh
b	COOMe
c	CN

2a-g: R^3=COPh,COOMe,CN; R^1=Me,Et,allyl,Bn

3a-h: R^3=COPh,COOMe,CN; R^1,R^2=Me,Et,allyl,Bn

	R^1=R^2	R^3
4a	Et	Ph
b	allyl	Ph

262. SYNTHESES OF SOME ORGANOBORON COMPOUNDS AS FLAME-RETARDANTS [illegible] ZHENG, Z.P.MAO, DEPARTMENT OF TEXTILE CHEMICAL ENGINEERING, CHINA TEXTILE UNIVERSITY, SHANGHAI, 200051, China

Several inorganic boron compounds are well-known for their flame-retarding properties. The reported work on organoboron compunds in flame-retarding applications is limited. Our efforts in the use of organoboron compounds as flame-retardants involved the syntheses of them. According to our design some of hydrolytically stable and structurally unsophisticated organoboron compounds, e.g. cyclic organic boronate esters, have been prepared. These organoboron compounds involved, 2-(2,3-dibromopropoxy)-1, 2-oxaborolane, tris-(2, 3-dibromopropoxy) borate, and etc. Starting material 2-allyloxy-1, 2-oxaborolane (BO-1) was prepared on the basis of previously published papers in reference. Their structures were determined by elemental analysis and IR, ^{1}H-NMR spectra. The evaluation of flame retardancy was obtained by means of LOI. The resultant organoboron compounds impart good flame retardancy. The total seven organoboron compounds are of no environmental pollution.

263. REACTION OF PHOSPHORUS PENTAHALIDES WITH CYCLIC KETONES: A ONE-STEP PREPARATION OF 1,1,2-TRICHLOROCYCLOALKANES. Alan B. Brown, Chris W. Chronister, Diana M. Watkins, Richard J. Mazzaccaro, Scott R. Rajski, and Martha G. Fountain, Department of Chemistry, Florida Institute of Technology, Melbourne, FL 32901.

Small-ring cyclic ketones react with excess phosphorus pentachloride in carbon tetrachloride at reflux over 1-3 d to give the corresponding 1,1,2-trichlorocycloalkanes, but react with phosphorus penta<u>brom</u>ide to give α,α'-dibrominated and $\alpha,\alpha,\alpha',\alpha'$-tetrabrominated ketones.

[Scheme showing: Cl,Cl-substituted cycloalkane with Cl, (CH$_2$)$_{n-3}$, n = 4-8 ← PCl$_5$ — cyclic ketone O=C with (CH$_2$)$_{n-3}$, n = 4-8 — PBr$_5$ → α,α'-dibromoketone with Br, (CH$_2$)$_{n-3}$, n = 5-8 + α,α,α',α'-tetrabromoketone with Br, Br, Br, Br, (CH$_2$)$_{n-3}$, n = 5]

264. ASYMMETRIC REDUCTIONS WITH CHIRAL NADH MODELS. G. Gelbard, S. Zehani and J. Lin, Laboratoire des Matériaux Organiques - C.N.R.S., B.P. 24, 69390 Vernaison (France).

A new series of NADH models have been devised for the asymmetric reduction of prochiral ketones such as methyl benzoylformates and 2-acetyl pyridine. These coenzymes models are 1,4-dihydropyridines related to Hantzsch esters and amides where the chiral substituent at 3- and 5- positions on the dihydropyridine ring derived from amines or simple sugars. Chirality at the 4-position was introduced in using monosaccharide derived aldehydes as chiral synthons.
These NADH models were obtained by a direct cyclisation reaction with chiral acetoacetates, acetoacetamides or monocaccharide aldehydes as building blocks.
The place of the chiral substituents made important changes in the e.e. of the and the nature of the prepondrant enantiomer; N-methylation at the ring nitrogen increased both reactivity and asymmetric induction. Methyl mandelate with enantiomeric excess of 88% could be obtained.

265. MULTIFUNCTIONAL ENZYME MIMICS AND ACTIVATORS.
R. Breslow, Department of Chelistry, Columbia University, New York, NY 10027.

The combination of coenzyme units with binding groups and/or other catalytic functionalities affords interesting enzyme mimics that can show good selectivity with respect to substrate and product. The combination of appropriate polar groups with hydrophobic spacers produces enzyme activators that show interesting potential as chemotherapeutic agents. Recent progress in these two areas will be described.

266. THE ASSEMBLY OF CARBON ATOMS INTO RODS, RINGS, NETS, AND SPHERES: THE NEW GENERATIONS OF CARBON ALLOTROPES. François Diederich, Department of Chemistry and Biochemistry, University of California, Los Angeles, CA 90024-1569.

The progress in the synthesis and study of monomeric and polymeric all-carbon materials will be discussed.

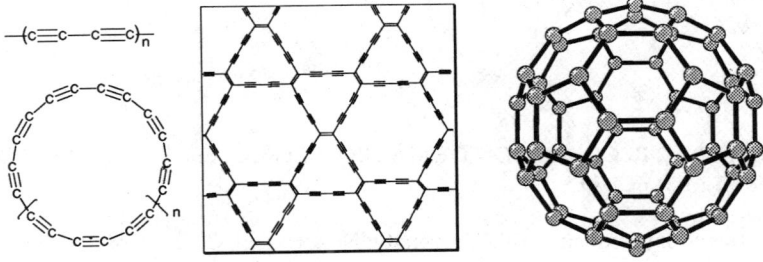

267. PROBING ENZYMATIC GLYCOSIDE HYDROLYSIS. Bruce Ganem, Department of Chemistry, Baker Laboratory, Cornell University, Ithaca, NY 14853.

Exo and endoglycosidases play important roles in the breakdown of carbohydrate foodstuffs, the biosynthesis of glycoproteins, the metabolism of glycolipids and glycosaminoglycans, as well as in the alteration of bacterial cell wall during growth and division. Current developments in the design and synthesis of new glycosidase inhibitors suggest that such compounds may well find important therapeutic applications.

268. RECENT ADVANCES IN THE SYNTHESIS OF SMALL BIOLOGICALLY ACTIVE INDOLE DERIVATIVES. Martha S. Morales-Ríos, M. Alvina Bucio-Vázquez and Pedro Joseph-Nathan, Departamento de Química, Centro de Investigación y de Estudios Avanzados, Instituto Politécnico Nacional, Apartado 14-740, Mexico City, 07000 México.

With the experience gained during the syntheses of the human antihypertensive drug candidate 3-amino-2-(5-methoxy-1H-indol-3-yl)propionic acid monohydrate [U.S. Patent 4,803,284 (1989)] we became interested in using an analogous synthetic intermediate (1) for the development of a new route that allows to complete the construction of the physostigmine alkaloid skeleton. Two successive key reaction steps for this purpose are the selective oxidation of methyl 2-cyano-2-(1-carbomethoxy-1H-indol-3-yl)acetate (1) to give methyl 2--cyano-2-(1-carbomethoxy-2,3-dihydro-2-hydroxy-1H-indol-3-ylidene)acetate (2) followed by the introduction of a methyl group to afford methyl 2-cyano-2-(1-carbomethoxy-2,3-dihydro-2-hydroxy-3-methyl-1H-indol-3-yl)acetate (3). The structure of 2 and the stereochemistry of 3 were verified by single crystal X-ray diffraction studies, while 2 and several related molecules were studied in detail by nmr measurements, which show the selective broadening of some aromatic hydrogen signals when measured at 300 MHz in the presence of minor amounts of base.

269. CHEMICAL PRINCIPLES AS A DRIVING FORCE IN THE DESIGN OF PROTEINASE INHIBITORS. A. Krantz, Syntex Inc., 2100 Syntex Court, Mississauga, Ontario L5N 3X4.

Proteinases have been classified into four major subtypes: serine, cysteine, metallo, and aspartyl proteinases. Each of these types has their own unique chemistry related to the active-site residues participating directly in bond-breaking and bond-making events during the course of catalysis. A challenge to chemists is to devise tactics which will exploit the chemical proclivities of specific proteinases taking into account their three-dimensional structure in order to design potent and specific inhibitors. Selected examples from Syntex Research Canada will be discussed in order to illustrate design principles with potential utility.

270. **MOLECULAR RECOGNITION OF MACROMOLECULES: THE UTILITY OF HYBRID MOLECULES.**
Alanna Schepartz
Department of Chemistry, Yale University, New Haven, Connecticut 06511.

Molecules which combine recognition domains with functional or reporter domains have widespread application as probes in research at the chemistry/biology interface. Our group uses these multifaceted molecules to investigate several topics including the identification of ligand binding domains on protein receptors, the detection and analysis of conformational changes resulting from protein denaturation, the recognition and analysis of structured RNAs, and the design of semi-synthetic, sequence-specific DNA-binding peptides. Experiments illustrating the utility of hybrid molecules in these areas will be discussed.

271. ENZYMES IN ORGANIC SYNTHESIS, C.-H. Wong, Department of Chemistry, Scripps Research Institute, La Jolla, California 92037.

This lecture will cover recent progress on the development of enzymatic catalysts for organic synthesis. Specifically, new approaches to the synthesis of azasugars and novel oligosaccharides based on aldolases and glycosyltransferases, asymmetric oxidoreductions based on monooxygenases and alcohol dehydrogenases, and irreversible strategies for peptide segment condensation and asymmetric transformations based on engineered enzymes will be covered.

272. FROM CYCLOPENTADIENE, BENZENE AND CYCLOHEPTATRIENE TO CARBOHYDRATES: A CHEMOENZYMATIC APPROACH. Carl R. Johnson, Joseph P. Adams, Adam Golebiowski, Mark A. Scialdone, Darryl H. Steensma and Michael C. Van Zandt, Department of Chemistry, Wayne State University, Detroit, MI 48202 USA

The title cyclic polyenes are transformed to functionalized meso-diols which are subjected to enzymatic dissymmetrization. The resulting chiral products are further oxidized and ring cleaved to unveil optically pure sugars and/or sugar derivatives. For example,

273. BISUBSTRATE REACTION TEMPLATES. T. Ross Kelly, Dept. of Chemistry, Boston College, Chestnut Hill, MA 02167

The formation of bonds between organic substrates is a central operation in organic synthesis. The ability to construct templates that foster such bond formations would make possible numerous reactions that are beyond the scope of current methodology. The underlying design rationale and the results of initial studies aimed at the development of such templates will be presented; **1**, a molecule that promotes the reaction between **2** and **3**, is one template that will be described.

274. **SYNTHESIS OF SUBSTITUTED AROMATICS VIA UNSATURATED KETENE INTERMEDIATES USING TRANSITION METAL MEDIATED PROCESSES.** L. S. Liebeskind, M. A. Huffman, and D. Krysan, Department of Chemistry, Emory University, Atlanta, Georgia 30322

Highly substituted phenols can be prepared via various unsaturated ketene intermediates. In a reaction that apparently proceeds via metal vinylketene complexes, alkynes and cyclobutenones react to give substituted phenols in the presence of catalytic $Ni(COD)_2$. In a different metal mediated process, cyclobutenones bearing 4-unsaturated substituents are easily prepared via transition metal catalyzed cross-coupling reactions of 4-chlorocyclobutenones and organostannanes. When the cross-coupling is conducted near 100 °C, the intermediate 4-substituted cyclobutenones transform directly into highly substitued phenols.

275. **NUCLEOPHILE-PROMOTED CYCLIZATION REACTIONS OF ALKYNES. ASYMMETRIC SYNTHESIS AND PHARMACOLOGICAL STUDIES OF PUMILIOTOXIN A ALKALOIDS AND CONGENERS.** Larry E. Overman, Department of Chemistry, University of California, Irvine, CA 92717.

The reactivity of alkynes towards intramolecular electrophiles can be selectively "tuned" by external nucleophiles.[1] The application of nucleophile-promoted alkyne-iminium ion cyclizations (**1** → **2**) for the asymmetric synthesis of cardiotonic pumiliotoxin A alkaloids (e.g. **3**) and congeners as well as the development of a molecular model for cardiotonic activity[2] in this series will be described.

1. Overman, L.E.; Sharp, M.J. *J. Am. Chem. Soc.* **1988**, *110*, 612.
2. Daly, J.W.; Overman, L.E.; Rossignol, D.P. *Biochem. Pharmacol.* **1990**, *40*, 315.

276. **HETEROCYCLIC SYNTHESIS USING RHODIUM STABILIZED CARBENOIDS** Albert Padwa, Department of Chemistry, Emory University, Atlanta GA 30322.

The reaction of keto carbenoids with heteroatoms which possess a lone pair of electrons is rapidly gaining prominence as an efficient method for heterocyclic synthesis. The high efficiency of the tandem cyclization-cycloaddition sequence coupled with the simplicity of the procedure provides a variety of heterocyclic rings in excellent yield. The scope and limitations of the method and its application to the total synthesis of a number of alkaloids will be described.

277. **COMBINED DIRECTED ortho METALATION - CROSS COUPLING TACTICS. APPLICATION IN NATURAL PRODUCT SYNTHESIS.** J.-m. Fu, B. Zhao, X. Wang, W. Wang, and V. Snieckus, Guelph-Waterloo Centre for Graduate Work in Chemistry, University of Waterloo, Waterloo, Ontario, Canada N2L 3G1.

The efficient and regiospecific construction of a variety of biaryl 1 and heterobiaryls may be achieved by combination of directed *ortho* metalation and transition metal catalyzed cross coupling methodologies. Subsequent remote (C-2′) metalation leads to cyclization (DMG = $CONR_2$) to fluorenones or ring-to-ring carbamoyl transfer (DMG = $OCONEt_2$). The scope and limitation of these new processes and their application to the synthesis of diverse natural products, e.g. eupolauramine (**2**), gymnopusin (**3**), and dengibsin (**4**) will be described.

DMG = Directed Metalation Group
= $CONR_2$, $OCONR_2$

278. **THE WITTIG REACTION: HOW DOES IT REALLY WORK?** Edwin Vedejs, Department of Chemistry, University of Wisconsin, 1101 University Avenue, Madison, WI 53706

The connection between Wittig stereoselectivity and the kinetic selectivity at the stage of intermediate oxaphosphetanes will be explored. New reagents for controlling alkene geometry in the reactions of ketones and aldehydes will be described. The possible intervention of various transient intermediates will be evaluated using experimental probes.

279. STUDIES IN ASYMMETRIC OLEFINATION AND ENE REACTIONS. S. Hanessian, Department of Chemistry, Université de Montréal, P.O. Box 6128, Montréal, Québec, CANADA

In spite of the remarkable achievements in the area of asymmetric C-C bond formation over the past 15 years, little if any attention has been devoted to the development of asymmetric olefination reagents. Following up on our original disclosure in 1984 (JACS, **1984**, *106*, 5754), we now describe recent studies on the formation of enantiomerically and diastereomerically enriched (or pure) olefins derived from substituted cyclohexanones and topologically unique chiral alkyl phosphonamides. These compounds are then subjected to an ene reaction with chiral non-racemic aldehydes which provides functionally interesting and stereochemically distinct cyclohexenes with appropriate appendages. Oxidative cleavage then leads to acyclic carbon chains with a predictable pattern of substitution such as can be found in several of nature's biosynthetic pathways. (Collaborators, S. Beaudoin, Y. Bennani, Y. Hervé.)

280. DITERPENOIDS WITH NEW SKELETONS RELATED TO A CLERODANIC PRECURSOR, FROM MEXICAN SALVIA SPECIES. Rodríguez-Hahn Lydia, Esquivel R. Baldomero, Cárdenas P. Jorge. Instituto de Química de la Universidad Nacional Autónoma de México. Circuito Exterior, Ciudad Universitaria, Coyoacán 04510 México, D. F.

The Salvia species found in Mexico, have been classified in the subgenus Calosphace (90 sections). Our systematic phytochemical study of Mexican Salvia species, revealed an interesting chemotaxonomic relationship between the diterpenoid content of the species studied and the section to which it belongs. Most of the diterpenoids isolated so far, from Mexican Salvia spp. have a clerodanic skeleton or can be biogenetically related to a clerodanic precursor. The isolation and structural determination of diterpenoids with unusual skeleton arrangements, isolated from species of the Section Angulatae, will be presented. A possible biogenetic relationship with a clerodanic precursor will be proposed.

281. MECHANISM OF INACTIVATION OF RIBONUCLEOTIDE REDUCTASE BY 2'-AZIDO 2'-DEOXY-NUCLEOTIDES. J. Stubbe, H. Baker, J.M. Bollinger, Dept. of Chemistry, M.I.T, Cambridge, MA 02139; M. J. Robins, V. Samano, Dept. of Chemistry, Brigham Young University, Provo, UT 84602.

2'-N_3UDP is a stoichiometric inactivator of ribonucleotide reductase (RNR) resulting in its conversion to uracil, PPi, 3-methylene-3(2H) furanone and N_2. This process is initiated by 3' carbon hydrogen bond cleavage of the nucleotide and is accompanied by loss of the tyrosyl radical on the B_2 subunit of RNR and formation of a new nitrogen centered radical derived from the azide moiety of N_3UDP. Studies using [2'-^{13}C,^{15}N]N_3 UDP and [1',2',3',4'-^{2}H]N_3UDPs and EPR and ESEEM spectroscopies revealed that during this transformation the 2' C-N bond is cleaved and that the nitrogen centered radical is within 6A° of the 1' and 4' Hs originally present in N_3UDP. Studies using C225S-B_1·B_2 complex reveal an altered interaction with N_3UDP in which N_3^- or $N_3^·$ is released. C225 of B_1 appears to play a crucial role in the conversion of $N_3^{-(·)}$ to N_2 and ·N. The kinetics of these transformations has been investigated and a mechanism is proposed to account for all of the available information.

282. STRATEGIES IN THE SYNTHESIS OF MEDIUM-SIZED CARBOCYCLES. APPLICATIONS TOWARD THE TOTAL SYNTHESIS OF NOVEL TERPENES

David R. Williams, C. Richard Nevill and Leslie A. Robinson
Department of Chemistry, Indiana University, Bloomington, IN 47405

Our interest in strategies for the stereocontrolled synthesis of medium-ring carbocycles has led to explorations toward anadensin (1), a terpene related to the ophiobolin class. Many of these fused 5-8-5 tricycles were uncovered as naturally occurring phytotoxins. However recent citations report stimulation of seed germination, chlorophyll biosynthesis, sugar and ion transport and ethylene emission. A general pathway for preparation of 5-11 bicyclic carbocycles will be discussed followed by elaboration for synthesis of 1.

283. **UNUSUAL STEREOFACIAL SELECTIVITY IN THE BIOSYNTHESIS AND SYNTHESIS OF THE BREVIANAMIDE/PARAHERQUAMIDE CLASS OF MYCOTOXINS: IN SEARCH OF THE BIOSYNTHETIC DIELS-ALDER CONSTRUCTION.**
Robert M. Williams, Tomasz Glinka, Timothy Cushing and Ewa Kwast, Department of Chemistry, Colorado State University, Fort Collins, Colorado 80523

The brevianamide/paraherquamide class of mycotoxins are produced by *Penicillium brevicompactum* and *Penicillium paraherquei*, respectively. The bicyclo[2.2.2] nucleus of these substances has been proposed to arise via oxidative [4+2] cycloadditon. The subject of our investigations is to establish the exact biosynthetic pathway to these substances and to unravel the unusual enantiodivergent occurrence of the bicyclo[2.2.2] ring system. Facial selectivities in the S_N2' reaction has been exploited for the synthesis of paraherquamide and several shunt metabolites to be used in biosynthetic work.

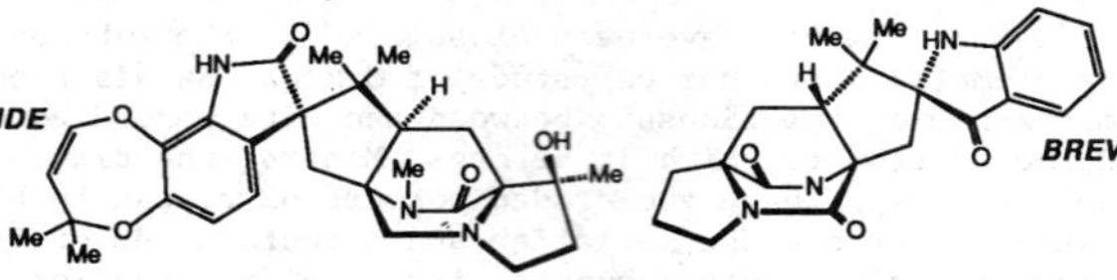

PARAHERQUAMIDE

BREVIANAMIDE B

284. **Control of Stereoselectivity in Samarium Metal Induced Cyclopropanations. Synthesis of 1,25 Dihydroxycholecalciferol.**
M. Kabat, J. Kiegiel, N. Cohen, K. Toth, P.M. Wovkulich and M. R. Uskoković
Roche Research Center, Hoffmann-La Roche Inc., Nutley, New Jersey 07110, USA

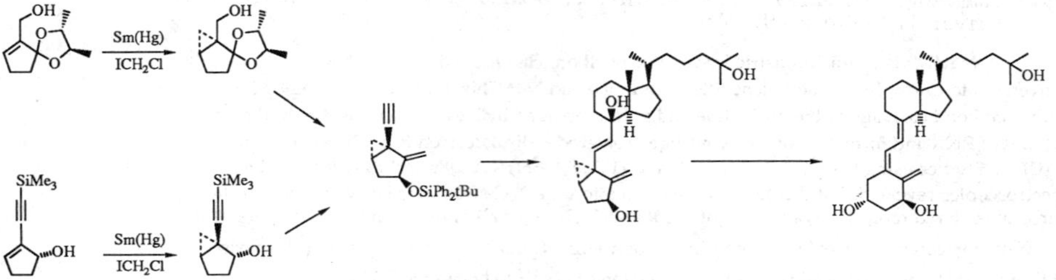

285. **SYNTHESIS OF MACROLACTONE PYRROLIZIDINE ALKALOIDS.** James D. White, Department of Chemistry, Oregon State University, Corvallis, Oregon 97331-4003

Macrolactone pyrrolizidine alkaloids (MPAs) are widely distributed in the plant kingdom and their pronounced hepatotoxicity is known to present a serious animal health problem. Previous work from this laboratory has focused on routes to the necic acid components of MPAs which led to syntheses of (-)-integerrimine and (+)-usaramine. Our current research is directed toward synthesis of the highly functionalized MPA swazine. A route to swazinecic acid from (S)-(-)-citronellol will be discussed along with related chemistry that addresses (a) methodology for coupling necic acids to necine bases and (b) the role of dehydro MPAs in crosslinking DNA.

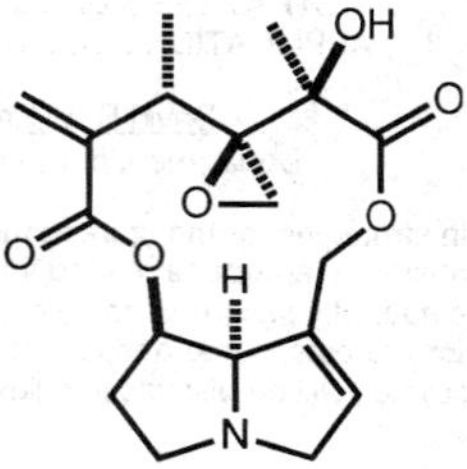

Swazine

286. **ANNULATION STRATEGIES FOR THE SYNTHESIS OF HIGHLY SUBSTITUTED CARBOAROMATIC AND HETEROAROMATIC COMPOUNDS.** Rick L. Danheiser, Department of Chemistry, Massachusetts Institute of Technology, Cambridge, Massachusetts 02139.

Highly substituted aromatic rings are key structural features in many biologically significant and commercially important compounds. Although classical synthetic approaches to such compounds have generally relied on linear substitution strategies, recently highly convergent annulation methods have emerged as a powerful alternative strategy for the assembly of highly substituted aromatic and heteroaromatic compounds. The intrinsic convergent nature of annulation strategies facilitates the efficient assembly of highly substituted aromatics that would have required long, multistep routes using classical substitution methodology. This lecture will describe the development of several new aromatic annulation methods, as well as their application in the total synthesis of biologically active natural products.

287. **SYNTHESIS OF BIOLOGICALLY INTERESTING GLYCOSIDES.** S.J. Danishefsky, Department of Chemistry, Yale University, New Haven, Connecticut 06511-8118.

The products of nature continue to constitute a rich storehouse of fascinating structures. As one contemplates issues associated with the synthesis of such systems in the laboratory, deficiencies in the capabilities of synthetic methodology are often uncovered.
In some instances the mechanisms by with structurally novel natural products react with their receptors, provides inspiration to the chemist to design comparable effectors which might be operationally viable in living systems.
As the sophistication of theory increases, the de novo (unprompted) design of receptor-effector ensembles will become more and more powerful. However, even in recent times it is the products of nature (cf taxol, anthracyclines, FK-506, rapamycin, calicheamicin, allosamidin etc.) which have provided the most inspiring instances of novel biological action.
This lecture will attempt to demonstrate that natural products continue to provide exciting opportunities for invention.

288. TRANSANNULAR DIELS-ALDER REACTION, A POWERFUL SYNTHETIC STRATEGY.
Pierre Deslongchamps. Département de chimie, Université de Sherbrooke, Sherbrooke, Qué., Canada J1K 2R1

The synthetic potential of transannular Diels-Alder reaction on macrocyclic trienes having various geometries and substituents will be presented. Experimental evidence on the conformation of the macrocyclic triene required for the Diels-Alder transition state will be presented.

289. **THE ASYMMETRIC SYNTHESIS OF CALYCULIN A**
D. A. Evans, J. R. Gage, J. L. Leighton
Department of Chemistry, Harvard University, Cambridge, MA 02138

This lecture will describe the asymmetric synthesis of the illustrated natural product which has been isolated from the marine sponge *Discoderma calix*.

290. TOTAL SYNTHESIS OF (±)-FR-900482. Tohru Fukuyama, Lianhong Xu and Shunsuke Goto, Department of Chemistry, Rice University, P.O. Box 1892, Houston, Texas 77251.

A stereocontrolled total synthesis of (±)-FR-900482 (**1**), a potent antitumor antibiotic related to mitomycin C (**2**), will be discussed. Readily available aminophenol **3** was used as our starting material.

1   **2**   **3**

291. ENANTIOCONTROLLED TAXANE CONSTRUCTION. A COMPLEMENTARY BIFURCATE APPROACH TO TAXUSIN AND TAXOL. Leo A. Paquette, Keith D. Combrink, Steven W. Elmore, and Richard C. Thompson, Department of Chemistry, The Ohio State University, Columbus, Ohio 43210.

The structurally complex, biologically active taxanes fall into two broadly definable classes. The first, exemplified by taxusin (1), features a C-H bond at the non-olefinic bridgehead position. Characteristic of the second group of which taxol (2) is a member is a hydroxyl substituent at this site. Synthetic methodology that is capable of accomplishing these twin objectives in a complementary and enantioselective manner will be described. The importance of substituent control of atropisomerism, as well as computational and crystallographic analysis, to progress toward these synthetic objectives will be described.

292. ABSOLUTE STEREOCHEMICAL CONTROL IN ALLYLIC FUNCTIONALIZATION James K. Whitesell, Joel Carpenter, H. Kenan Yaser, Edmund Hudson, and Mark A. Minton, Department of Chemistry, The University of Texas at Austin, Austin, TX 78712

In 1986 we reported on a sequence that effects allylic oxidation of *cis*-alkenes with the introduction of a hydroxyl group (A). The functionalization proceeds with high levels of control of absolute stereochemistry, regiochemistry, and double bond geometry using our chiral auxiliary *trans*-2-phenylcyclohexanol. Recently, we have been ably to modify this sequence to afford allylic amines both with (B), and without (C) migration of the double bond with good to excellent control of stereochemistry.

293. ASYMMETRIC SYNTHESIS OF α-HYDROXY CARBONYL COMPOUNDS USING ENANTIOPURE N-SULFONYLOXAZIRIDINES Franklin A. Davis, Bang-Chi Chen, Anil Kumar, Christopher K. Murphy, R. Thimma Reddy and Michael C. Weismiller, Department of Chemistry, Drexel University, Philadelphia, PA 19104

The asymmetric enolate oxidation protocol employing an enantiopure (camphorylsulfonyl)oxaziridine derivative 1 is highly successful in the synthesis of enantiomerically enrich α-hydroxy carbonyl compounds. The application of this protocol to the synthesis of 2-hydroxy-2-substituted-1-tetralones of biological interest and the factors responsible for the molecular recognition will be presented.

294. **FURFURYL ALCOHOL-TO-PYRANONE OXIDATION. AN APPROACH TO THE STEREOSELECTIVE SYNTHESIS OF HETEROCYCLIC NATURAL PRODUCTS.**

Philip DeShong, David A. Simpson, Chuljin Ahn, Steven J. Shimshock, Robert E. Waltermire, Yvonne Class. Department of Chemistry and Biochemistry, University of Maryland, College Park, Maryland 20742

Peracid oxidation of furfuryl alcohol derivatives yields pyranones which are versatile intermediates in the synthesis of oxygen heterocycles. Application of the furfuryl alcohol-to-pyranone methodology for the stereoselective synthesis of heterocyclic natural products will be presented.

295. **RECENT CONFORMATIONAL STUDIES OF SIX-MEMBERED CARBO- AND HETERO-CYCLES.** Eusebio Juaristi, Depto. Química, Centro de Investigación y de Estudios Avanzados del Instituto Politécnico Nacional, Apdo. Postal 14-740, 07000 México, D.F.

This presentation will consist of two parts:
(1) Reexamination of the conformational preference of the benzyl group in cyclohexane, which describes the spectroscopic and force-field work that confirms a smaller $-\Delta H°(CH_2Ph)$ relative to $-\Delta H°(CH_3)$, as well as a significant $\Delta S°(CH_2Ph)$ contribution.

(2) Effect of the coordination at phosphorus in the conformational equilibria of 2-P-substituted-1,3-dithianes, which shows an excellent linear correlation between the estimated S-C-P(X) anomeric effects and the corresponding $\sigma_I(X)$ values (X = O, S, BH$_3$, lone pair). This result appears to support the importance of 2-electron stereoelectronic interactions in these systems.

296. STEREOSPECIFIC ENAMMONIUM-IMINIUM REARRANGEMENTS. A MECHANISTIC INVESTIGATION OF STEREOCONTROLLED PROTON MIGRATION FROM NITROGEN TO CARBON. Bruce E. Maryanoff, David F. McComsey, Kirk L. Sorgi, Cynthia A. Maryanoff, Gregory C. Leo, Martin S. Mutter, and Harold R. Almond, Jr., Medicinal Chemistry and Chemical Development Departments, R. W. Johnson Pharmaceutical Research Institute, Spring House, PA 19477

Although reduction of 1a·HBr or 1b·HBr with borane-THF in trifluoroacetic acid gave a 68:32 mixture of amines 2a and 2b, the reduction of 3a·HBr or 3b·HBr gave an 8:92 ratio of 4a:4b. The improved trans-selectivity with 3 results from an increased bias for a cis ring-fusion in the N-protonated 5,6 system (5b) relative to the 6,6 system (6b). By proton NMR, we observed dehydration of 1a·HBr to a 76:24 mixture of enammonium salts 6a:6b, each of which rearranged stereospecifically to give a 75:25 mixture of iminium salts 7a:7b. Similar NMR work was applied to 3a and 3b. Rate studies and mechanistic details will be discussed.

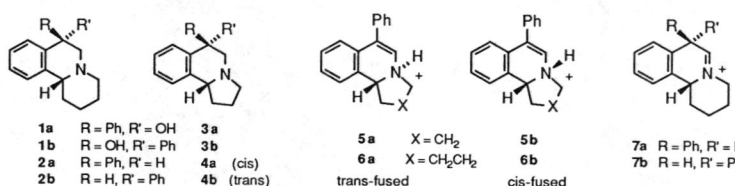

297. **DESIGN AND DEVELOPMENT OF PRACTICAL ASYMMETRIC SYNTHESES**
Ichiro Shinkai, *Merck Sharp & Dohme Res. Labs., Rahway, New Jersey 07065, U.S.A.*

During recent years, there has been a growing interest in the replacement of peptide bonds by isosteres which impart greater activity, selectivity, and stability to peptides with interesting pharmacological properties. Of particular interest were compounds incorporating the "hydroxyethylene dipeptide isostere". In the course of our work on protease inhibitors of human immunodeficiency viruses (HIV), we required a stereocontrolled synthesis of hydroxyethylene dipeptide isostere. A one-step, chelation-controlled, three-carbon homologation to the desired hydroxyester oxidation state has not been reported. Typically, control of the 4S stereocenter has been problematic. Our work focused on chelation-controlled homoenolate anion addition to the aldehyde.

Our studies show that stereoselectivity is strongly dependent on the titanium ligands. For example, the (4S)/(4R) reaction diastereoselectivity increased as the number of chlorines on the titanium homoenolate anion increased.

298. CALICHEAMICIN: EXPERIMENTAL PROGRESS TOWARD UNDERSTANDING ITS SEQUENCE-SELECTIVE BINDING AND CLEAVAGE OF DNA. C. A. Townsend, Department of Chemistry, The Johns Hopkins University, Baltimore, Maryland 21218.

Much has been written and said about the origins of the sequence-selective binding and cleavage of duplex DNA by calicheamicin, an important member of the diynene class of antitumor antibiotics. How do these proposals fare against the test of actual experiments? A variety of experimental methods are being applied in attempts to answer several of the questions posed by this intriguing chemical system. Progress made to date in this work will be discussed.

PETROLEUM CHEMISTRY

Fourth Chemical Congress of North America

New York, NY
August 25 - 30, 1991

G. J. Antos,
Program Chairman

MONDAY MORNING

● **Symposium on Advances in Zeolites and Pillared Clay Synthesis**
M. L. Occelli, Presiding
Papers 1-6

● **Symposium on Alkylation, Aromatization, Oligomerization and Isomerization of Short Chain Hydrocarbons over Heterogeneous Catalysts**
H. J. Lovink, Presiding
Papers 7-12

● **Symposium on Catalyst Supports: Chemistry, Forming and Characterization**
S. A. Bradley, Presiding
Papers 13-18

MONDAY AFTERNOON

● **Symposium on Advances in Zeolites and Pillared Clay Synthesis**
H. E. Robson, Presiding
Papers 19-24

● **Symposium on Alkylation, Aromatization, Oligomerization and Isomerization of Short Chain Hydrocarbons over Heterogeneous Catalysts**
C. P. Nicolaides, Presiding
Papers 25-30

● **Symposium on Catalyst Supports: Chemistry, Forming and Characterization**
E. Gulari, Presiding
Papers 31-35

TUESDAY MORNING

● **Symposium on Advances in Zeolites and Pillared Clay Synthesis**
Z. Gabelica, Presiding
Papers 36-42

● **Symposium on Alkylation, Aromatization, Oligomerization and Isomerization of Short Chain Hydrocarbons over Heterogeneous Catalysts**
G. J. Antos, Presiding
Papers 43-48

● **Symposium on Catalyst Supports: Chemistry, Forming and Characterization**
R. G. Anthony, Presiding
Papers 49-53

TUESDAY AFTERNOON

● **Symposium on Advances in Zeolites and Pillared Clay Synthesis**
D. E. W. Vaughan, Presiding   Papers 54-60

● **Symposium on Alkylation, Aromatization, Oligomerization and Isomerization of Short Chain Hydrocarbons over Heterogeneous Catalysts**
J. P. Van den Berg, Presiding   Papers 61-67

● **Symposium on Catalyst Supports: Chemistry, Forming and Characterization**
J. Schwartz, Presiding   Papers 68-73

WEDNESDAY MORNING

● **Symposium on Advances in Zeolites and Pillared Clay Synthesis**
G. W. Skeels, Presiding   Papers 74-81

● **Symposium on Akylation, Aromatization, Oligomerization and Isomerization of Short Chain Hydrocarbons over Heterogeneous Catalysts**
F. Nierlich, Presiding   Papers 82-87

● **Symposium on Catalyst Supports: Chemistry, Forming and Characterization**
D. E. W. Vaughan, Presiding   Papers 88-92

WEDNESDAY AFTERNOON

● **Symposium on Advances in Zeolites and Pillared Clay Synthesis**
H. Kessler, Presiding   Papers 93-99

● **Symposium on Alkylation, Aromatization, Oligomerization and Isomerization of Short Chain Hydrocarbons over Heterogeneous Catalysts**
R. Galiasso, Presiding   Papers 100-105

THURSDAY MORNING

● **Symposium on Advances in Zeolites and Pillared Clay Synthesis**
T. J. Pinnavaia, Presiding   Papers 106-111

● **Symposium on Alkylation, Aromatization, Oligomerization and Isomerization of Short Chain Hydrocarbons over Heterogeneous Catalysts**
E. F. Sousa-Aguiar, Presiding   Papers 112-117

● **Symposium on Modern Analytical Techniques for Analysis of Petroleum**
F. P. Di Sanzo, Presiding   Papers 118-124

THURSDAY AFTERNOON
● **Symposium on Advances in Zeolites and Pillared Clay Synthesis**
A. Clearfield, Presiding   Papers 125-129

● **Symposium on Alkylation, Aromatization, Oligomerization and Isomerization of Short Chain Hydrocarbons over Heterogeneous Catalysts**
H. J. Lovink, Presiding   Papers 130-135

● **Symposium on Modern Analytical Techniques for Analysis of Petroleum**
P. Grey, Presiding   Papers 136-141

FRIDAY MORNING
● **Symposium on Advances in Zeolites and Pillared Clay Synthesis**
S. Yamanaka, Presiding   Papers 142-145

● **Symposium on Modern Analytical Techniques for Analysis of Petroleum**
L. Wolfram, Presiding   Papers 146-151

PETROLEUM CHEMISTRY, INC.

1. SYNTHESIS AND CHARACTERIZATION OF VPI-6 ANOTHER INTERGROWTH OF HEXAGONAL AND CUBIC FAUJASITE, M. E. Davis, Department of Chemical Engineering, Virginia Polytechnic Institute and State University, Blacksburg, VA 24061

   We report here the synthesis procedures used to crystallize the zeolite VPI-6. VPI-6 is synthesized in the absence of organic materials at low temperatures (70-100°C) with short heating times (0.5-3 hours). This new zeolite is another intergrowth of cubic and hexagonal faujasite. VPI-6 can be indexed on a hexagonal unit cell of dimensions a = 17.655(2) Å and c = 27.483(2) Å. Solid-state NMR and physical adsorption data from VPI-6 (SiO_2/Al_2O_3 = 2.4) are reported and compared to faujasite. A summary of all zeolites related to either the hexagonal and/or cubic faujasite is presented.

2. SYNTHESIS AND CHARACTERIZATION OF BRECK STRUCTURE SIX, (BSS). Gary W. Skeels, UOP, Tarrytown Technical Center, Tarrytown, NY 10591 and Kathleen B. Reuter and Nancy K. McGuire, Union Carbide, Tarrytown Technical Center, Tarrytown, NY 10591.

   Synthesis of pure Breck Structure Six (BSS) was recently reported by Delprato et al. We have followed their basic synthesis method and have prepared BSS. Basic characterization data are reported for this material including x-ray powder diffraction and unit cell, thermal analysis, hydroxyl region and framework region infrared spectra and wet chemical analyses. In addition, electron diffraction and high resolution electron microscopy were used to confirm whether the Delprato synthesis procedure produces pure hexagonal packing of sodalite cages (BSS), or whether some cubic packing (faujasite) is retained. The relationship between this material and other members of the "faujasite family" is discussed.

3. Synthesis of zeolite Y in the gelatin solution. Y.Goto, H.Saezusa*, M.Koizumi, Department of materials chemistry, Ryukoku Universtiy, Otsu-shi 520-21 Japan, *Department of chemistry, Gunma University, Kiryu-shi 376, Japan.

   The Synthesis of zeolite Y was attempted at 100 °C in the gelatin solution with a concentration of 0.1 to 10wt%. Hydrothermal reaction for 1 to 10h. formed zeolite Y as a single phase. The crystallization rate decreased at 7.5 and 10wt% gelatin. The crystal of zeolite Y was smaller in size and narrower in size distribution than that produced by the ordinary method. The increment of gelatin concentration caused the increment of Si/Al ratio of zeolite Y. It appears that the role of gelatin are not the template but the supply of many crystallization sites and the contraint of diffusion of species.

4. THE CHEMISTRY OF NaY CRYSTALLIZATION FROM SODIUM SILICATE SOLUTIONS. D. M. Ginter and A. T. Bell, Center for Advanced Materials, Lawrence Berkeley Laboratory, and Department of Chemical Engineering, University of California, Berkeley, CA 94720

   A study has been conducted of NaY zeolite synthesis from seeded gels. The objective of this work was to understand the physical and chemical transformations occurring in the seed and synthesis gels. The composition and structure of the solid component of the gels were determined by elemental analysis, MAS-NMR spectroscopy, FTIR spectroscopy, and X-ray diffraction. Precursors to NaY nuclei could be produced in a highly alkaline (pH = 14) gel. ^{29}Si MAS-NMR spectroscopy revealed evidence of structures similar to those found in NaY, after 1 d of aging at room temperature. The solid component of the seed gel contains a high concentration of Na^+ ions and water. Addition of a small amount of the seed gel to a low pH (pH = 11.3) feedstock resulted in the nucleation of NaY crystallization at 100°C. The physical and chemical transformations occurring in the synthesis gel during crystallization, and the influence of pH on crystallization kinetics, will be discussed.

5. LOW TEMPERATURE HYDROTHERMAL CRYSTALLISATION OF MFI ZEOLITES IN AN ALKALINE FLUORIDE MEDIUM. M.J. Eapen, P.N. Joshi, S.V. Awate, A.N.Kotasthane and V.P. Shiralkar*, Catalysis Group, National Chemical Laboratory, Pune 411008, India.

Low temperature (369 K) hydrothermal crystallisation of MFI framework zeolite in an alkaline fluoride medium has been carried out under unstirred condition using the gel composition $30Na_2O:Al_2O_3: 100 SiO_2: x K_2O: y R_2O: 5700 - 7300 H_2O$. From the view of commercial application, the optimization of the quantities of (X), different organic templates (Y) and seed crystals was carried out. As a result fully crystalline product exhibiting -ir and XRD of typical ZSM-5 was obtained with solid yield upto 83% with 18 moles of KF within 4-12 days depending on the nature and the combination of templates as well as seed crystals. The dilution of the gel exerted marked influence on the crystallisation rate and the product yield. The crystalline ZSM-5 zeolites were characterized by SEM, TG/DTA and sorption measurements.

6. ^1H - ^{27}Al Cross-Polarization and Very Fast ^{27}Al MAS NMR Studies of Dealuminated Zeolite Y , João Rocha,[1] Stuart W. Carr[2] and Jacek Klinowski[1]
(1) Department of Chemistry, University of Cambridge, Lensfield Road, Cambridge CB2 1EW, U.K.
(2) Unilever Research, Port Sunlight Laboratory, Quarry Road East, Bebington,

The use of ^{27}Al solid-state NMR with very fast (> 12 kHz) magic-angle spinning (MAS) is a prerequisite for the quantification of aluminium in dealuminated zeolite Y and in tandem with ^1H - ^{27}Al cross-polarization (CP/MAS) allows distinction between framework and extra-framework Al species to be made. The ^{27}Al signal at ca. 30 ppm from steamed (ultrastable) zeolite Y is an independent NMR resonance, probably due to 5-coordinated non-framework Al, and not part of a second-order quadrupole lineshape as previously reported. We conclude that the extra-framework species contain 4-, 5- and 6-coordinated aluminium.

7. THE OLIGOMERIZATION AND ISOMERIZATION OF OLEFINS BY HOMOGENEOUS NICKEL-BASED CATALYSTS, S.J. Brown[1], G.G. Humphreys[1], C. Hinton[1], A.F. Masters[1], P.A. Tregloan[2], J.I. Sachinides[2], Department of Inorganic Chemistry, University of Sydney, N.S.W., 2006, Australia, [2]School of Chemistry, University of Melbourne, Parkville, Vic., 3502, Australia.

The oligomerization of low molecular weight olefins by "so-called" homogeneous catalysts is one of the most widespread applications of homogeneous catalysts. Several processes are currently in commercial practice. An additional potential application of olefin oligomerization is as one step in the conversion of natural gas to gasoline and distillate fuels.

This paper discusses the oligomerization and isomerization of low molecular weight olefins by homogeneous catalysts, with particular emphasis on catalytic systems derived from [Ni(sacsac)(PBu$_3$)Cl] and related species (sacsac = $C_8H_7S_2^-$ = pentane-2,4-dithionate anion = dithioacetylacetonate anion). In the presence of an appropriate cocatalyst, these compounds generate extremely active catalysts, with activities of 1000 moles of ethylene converted per mole of nickel per second at 25°C and 1 atmosphere. The catalyst performance can be controlled by the selection of catalyst, such that catalysts with near quantitative selectivity for dimerization, isomerization or oligomerization can be chosen. The performance of these systems as oligomerization and isomerization catalysts in terms of activity, selectivity, conversions, catalysts lifetime and poisons resistance will be discussed.

8. CATALYSTS AND CONDITIONS FOR THE HIGHLY EFFICIENT AND STABLE HETEROGENEOUS OLIGOMERIZATION OF ETHYLENE. J. Heveling, C.P. Nicolaides and M.S. Scurrell, Catalysis Programme, Division of Energy Technology, CSIR, P.O. Box 395, Pretoria 0001, Republic of South Africa.

The oligomerization of ethylene into petrol and diesel range products over nickel catalysts at high pressure (35 bar) and low temperature (120°C) will be described. In these studies, the catalysts used were obtained by nickel (II)-exchange or impregnation of two differently prepared silica-alumina supports. Under the above reaction conditions and at MHSV=2, a conversion level of 99% is obtained with the products formed having almost exclusively an even number of carbon atoms, C_4-C_{22}. The percentage of products in the diesel range, C_{10+}, varied from 23-41% by mass, depending on the type of catalyst used. Most significantly, the catalysts are extremely stable in use showing no detectable drop in conversion after 40+ days on stream. The work has been extended to include propylene as feed for the oligomerization reaction over our catalysts.

9. THE OLIGOMERIZATION OF LOWER OLEFINS BY HETEROGENEOUS CATALYSIS, F. Nierlich, J. Neumeister, T. Wildt, HULS AG, W-4370 Marl, Germany

The oligomerization of lower olefins, especially n-butenes, is investigated and compared on heterogeneous acidic and nickel modified inorganic catalysts. The structures of the products are in line with a carbenium ion and Ziegler type reaction mechanism respectively. In the investigated 1-butene/2-butene concentration range their ratio has hardly any influence on the product distribution. The rate of reaction is inverse proportional to the carbon chain length and is of no technical relevance anymore for olefins exceeding C_8 on the catalysts employed. Causes of catalyst deactivation are discussed. Successful commercialization of this development work offers a refiner an additional possibility to produce gasoline components. Petrochemical applications include superior PVC plasticizers and detergent intermediates.

10. ETHYLENE DIMERIZATION OVER HIGHLY DISPERSED METAL CATALYSTS. A. Guerrero-Ruiz, I. Rodríguez-Ramos and J.L.G. Fierro, Instituto de Catálisis yPetroleoquímica, CSIC, Serrano 119, 28006 Madrid, Spain.

Selective conversion of ethylene into butenes is an attractive process since C_4 olefins can then be used in petrochemistry for many purposes. The reaction seems, in principle, simple but the problem to stop it just in the butene still remains. This communication will discuss recent results obtained in our laboratory on the catalytic behavior of Ni, Rh and Pd metals supported on NaY, HY, SiO_2 and Al_2O_3 carriers. From the catalytic experiments there is evidence that (i) acidity of the carrier plays a key role in hydrogen transfer from strongly held species to intermediates, yielding not only isomerization products but also butane, (ii) c/t 2-butene ratio can be modified by adequate selection of metal/carrier couple, (iii) catalyst deactivation seems to be due to poisoning by strongly held species.

11. HETEROGENEOUS OLIGOMERIZATION OF PROPENE OVER HETEROPOLY ACIDS, J.C.Q. Fletcher, J.S. Vaughan, C.T. O'Connor, Department of Chemical Engineering, University of Cape Town, Private Bag, Rondebosch, 7700, South Africa.

The large diversity of heteropoly acid formulations possible has made these compounds interesting candidates for application as both homogeneous and heterogeneous catalysts. Heteropoly compounds of molybdenum and tungsten peripheral atoms have been most widely studied as heterogeneous catalysts.

Early studies by Verstappen and Waterman showed dodecatungstosilicic acid (HSiW) to be active for propene oligomerization and the formation of significant quantities of C_{12+} product. Recently, Ratnasamy and Subramanian have reported silicotungsten-supported HSiW to be an efficient catalyst for the production of distillate fuels by light olefin oligomerization. This paper reports the activity of HPW, HSiW, HPMo and various salts of these acids for propene oligomerization. Data for the selectivity to distillate fractions and the quality (cetane rating) of these fuels is presented.

12. DEHYDROGENATIVE COUPLING OF CH_4 IN A HOT WIRE DIFFUSION COLUMN
H. D. Gesser and L. A. Morton, Department of Chemistry, University of Manitoba, Winnipeg, MB, R3T 2N2.

Methane is thermally decomposed by a hot tungsten wire in a vertical thermal diffusion column. The products identified in an upward flow were primarily aromatic and unsaturated hydrocarbons. With the wire at about 850°C and the column walls at -40°C it was possible to increase the proportion of linear hydrocarbons. A GC/MS analysis of the oil has been made when the system is run in a partial recycle mode by circulating the products through a trap at -78°C.

13. PILLARED CLAYS AS SUPPORTS IN HYDROCRACKING CATALYST SYSTEMS.
Takeshi Hashiguchi and Joseph Shabtai, Department of Fuels Engineering, University of Utah, Salt Lake City, Utah 84112-1183.

Factors affecting the thermal stability and acidity of pillared clay supports, e.g., structural type of the starting clay, type of exchangeable cation, type and size of the pillaring cation, degree of pillaring, mode and conditions of pillared clay preparation, etc., were systematically investigated. Starting clays included several Wyoming and Cheto type U.S. montmorillonites, as well as Japanese montmorillonites and beidellites. Conditions for preparation of ultra-stable pillared clays, showing no major change in basal spacing at 700°C, were developed. Selected pillared products were functionalized as hydrocracking catalysts by incorporation of hydroxy-NiCrMo oligocations, using methods developed in this Department [J. Shabtai and J. Fijal, U.S. Patent 4,579,832 (1986)]. The hydrocracking activities of the resulting catalysts were determined using n-hexadecane as model feed. Some of the new catalysts showed higher activity than that of commercially available samples.

14. PROPENE ALKYLATION OF BIPHENYL CATALYZED BY PILLARED CLAYS. Jean-Rémi Butruille and Thomas J. Pinnavaia, Department of Chemistry and Center for Fundamental Materials Research, Michigan State University, East Lansing, MI 48824.

Traditional liquid-phase Friedel-Crafts alkylation catalysts such as $ALCl_3$ and $FeCl_3$ suffer the practical disadvantages of corrosivity and the formation of undesired effluent by-products. Consequently, it is desireable to substitute these catalysts by regenerable solid acids. Owing to their facile synthesis and adaptability, pillared clays are a particularly attractive class of microporous acid catalysts for alkylation reactions. We report here the properties of a series of alumina pillared smectite clays for the alkylation of biphenyl. This substrate is well suited as a probe of shape selectivity, in part, because the kinetic diameter of the molecule is comparable to the pore size typical of most pillared clays. The selectivity toward ortho-alkylated products was correlated with pore structure of the pillared clay.

15. **AL MODIFIED SEPIOLITE AS CATALYST OR CATALYST SUPPORT**, J.-B.d'Espinose and J. J. Fripiat, Department of Chemistry and Laboratory of Surface Studies, University of Wisconsin-Milwaukee, P.O. Box 413, Milwaukee, Wisconsin 53201

Fibrous mineral sepiolite is formed of staggered talc sheets. The discontinuous octahedral layers provide for infinite channels along the fiber axis with cross-section $\sim (10 \times 4)$ Å^2. During crystal growth, a second set of irregularly shaped pores, with a diameter around 100 Å, is created within the fiber. Zeolitic water fills the narrow and regular channels, and is eliminated by outgassing at 100°C. At that stage, the specific surface area is ~ 300 m^2/g. Upon activation at increasing temperature, the structure folds, and ultimately collapses (~ 500°C), leaving a surface area ~ 150 m^2/g. The natural mineral lattice is almost neutral, having a cation-exchange capacity (CEC) in the order of 0.07 meq/g.

It is shown that mild treatment with NaAlO$_2$ results in partial Si-by-Al substitution in the tetrahedral layer. The CEC is almost tripled and the residual surface area after calcination is about the same as in the original material. The alumination process generated Lewis sites as evidenced by pyridine adsorption study.

The mechanism of alumination, as well as the structural modifications upon calcination, are discussed on the basis of ^{27}Al and ^{29}Si MAS-NMR spectroscopy. Possible uses of the Al modified sepiolite as catalyst and catalyst support are put forward.

16. **Electrically Conductive Catalyst Supports From Fibrous Precursors.** S. Ahn, A. Krishnagopalan and B.J. Tatarchuk, Department of Chemical Engineering, Auburn University, AL 36849.

A process has been explored wherein a mixture of small diameter (ca. 2um) metal fibers, support fibers or particulates, and cellulose fibers are cast into a free-standing preform sheet using paper-making procedures. High temperature firing is subsequently used to remove the cellulose binder/pore former causing a sinter-bonding of metal fibers and an associated entrapment of secondary support materials. Supports of either e.g., activated carbon or nonconductive metal oxides are used depending on the application and desired electrical properties. Electrochemical means such as cyclic voltammetry, A.C. impedance and potential step response have been used in conjunction with various probe reactions and both liquid and solid electrolytes. Equivalent circuit models of catalyst/electrode response have been developed and compared to experiment. Fiber-based supports possess low resistances to mass transport, low pressure drop and high overall reaction rates. Due to enhanced accessibility of the catalyst, high reactivity persists even when loadings of impregnated precious metals are reduced. Regardless of the support composition, the conductive metal fiber network provides a direct means for heating and controlling the rate of the catalytic reaction.

17. **SYNTHESIS AND CATALYTIC ACTIVITY WHEN USING A TYPE 2 TITANATE**, R.G. Anthony, E.F. Gonzalez, R.G. Dosch, Texas A&M University, College Station, TX, Sandia national Laboratories, Albuquerque, NM.

Anthony and Dosch reported the synthesis of five new crystalline titanates at the International Meeting on Catalysts Preparation, September, 1990. The synthesis of Type 2 titanate, and its use as a catalyst support will be discussed. The Type 2 titanate is prepared from a synthesis mixture containing aluminum and titanium compounds in the molar ratio of 0.05:1. The procedure for synthesis was modified in scaling from 1 to ^{2}g preparations to 100 g preparations. The rate of addition of the alkoxides to the remaining portion of the synthesis mixture and the methods used in washing and drying the resulting material were found to be important for obtaining high surface area material. TGA and DSC studies indicate an initial weight loss of approximately 40%, but after the initial weight loss there was essentially no weight loss and no major phase changes on heating to 600°C. However, there was significant loss in surface area and the XRD pattern matched that of a sodium titanium oxide. Elemental analysis for aluminum indicated only a small amount of aluminum was retained in the sample. However, this material when loaded by ion exchange with Mo and Ni, was found to be of comparable activity for hydrogenation of pyrene as the hydrous metal oxides.

18. **THE PREPARATION AND CHARACTERIZATION OF NOVEL CATALYST SUPPORT MATERIALS: THE HYDROUS METAL OXIDE ION-EXCHANGERS**[*] Charles H. F. Peden, Sandia National Laboratories Albuquerque, NM 87185.

Hydrous metal oxides, that are prepared from solution by hydrolysis and condensation of metal alkoxides, exhibit some unique properties that make them especially attractive as catalyst supports. In particular, they have high surface area, good chemical stability, and the ability to bind catalytic metals by ion exchange. These ion-exchange reactions involve the replacement of alkali or alkaline earth metals, that charge compensate the hydrous titanate (HTO) precipitate, with transition metal ions. As such, it is important to characterize the acid/base behavior of these materials in solution. In addition, we are concerned with the determination of the molecular structure of the amorphous hydrous titanate because this will determine the type and relative abundance of ion-exchange sites in the material. For this purpose, we are using Raman, EXAFS and ^{17}O NMR spectroscopies. With these tools, we can determine, and thus control, the nature of the sites by varying the synthesis conditions used in making the HTO materials. Studies of the material and chemical properties of HTO-supported catalysts will also be described in this presentation. These have led to the preparation of improved catalytic materials based in part on insights from the spectroscopic studies.

19. NOVEL APPROACH TO THE SYNTHESIS OF MOLECULAR SIEVES: RECRYSTALLIZATION OF THE INTERMEDIATES ISOLATED FROM THE GEL PHASE. Z.Gabelica, L.Maistriau, T.Florent and E.G.Derouane, Dept. of Chemistry, University of Namur, 61, Rue de Bruxelles, B-5000 Namur, Belgium.

A series of gel phases of various compositions, precursors to aluminosilicate (NaY, ZSM-5) or (silico)aluminophosphate (SAPO-5, SAPO-37, VPI-5) molecular sieves, have been partly dehydrated at low temperature by the freeze-drying technique and further stored at ambient temperature as new potential starting materials. Either as-dehydrated or rehydrated at 20°C to various extents, the gels readily yield crystalline porous materials upon appropriate heating. The nature of the latter can be independant on the degree of rehydration (NaY, ZSM-5) but also strongly affected by the final amount of water left or further added to the hydrogel (AlPO's and SAPO's). When excess of water is added to the freeze dried gels that normally yield SAPO-5 or VPI-5, the same layered SAPO type phase H3 is obtained. The importance of pre- and post freeze drying ageing for VPI-5, NaY and SAPO-37 is questioned and the corresponding crystallization mechanisms rediscussed accordingly.

20. CLEAR SOLUTION SYNTHESIS OF AlPO$_4$ MOLECULAR SIEVES: IDENTIFICATION OF TRANSIENT PHASES. R. Szostak, B. Duncan, Zeolite Research Program, Georgia Tech Research Institute, Atlanta, Georgia 30332 USA, R. Aiello, A. Nastro, Dept. of Chemistry, University of Calabria, Rende, Italy, K. Vinje, K. Lillirud, Dept. of Chemistry, University of Oslo, Oslo, Norway.

Synthesis from clear solution is used to identify transient phases formed during the crystallization of AlPO$_4$ molecular sieves. AlPO$_4$-11 forms from clear solutions with a batch composition: Al$_2$O$_3$:1.26 P$_2$O$_5$:1.5 HF:2.4 DPA:70 H$_2$O. Crystallization occurred readily at the highest temperature producing low yields of very thin 1 mm size needles after 2 hours. After 18 hours the yield increases 50 to 75% at the expense of the crystal size with homogeneous 50 micron crystals resulting. Such result suggests that a secondary nucleation occurs with redistribution of nutrients between the larger initially formed crystals and the smaller growing crystals. The solution phase plays a critical role in AlPO$_4$ crystallization while transient amorphous and crystalline phases appear to act as a source of Al and P nutrients. When the intermediate solid is removed from the liquid phase or when other solid crystalline AlPO$_4$ phases (s.a. AlPO$_4$-H$_3$) are added to the clear solution, microporous crystalline AlPO$_4$-11 still results; in low yield in the former and with a yield equivalent to the addition of the extra nutrients in the latter. In the synthesis of CoAPO-11, the initially precipitated amorphous phase is depleted of the Al and P components with Co depletion to a lesser extent as the crystalline phase grows. The CoAPO-11/amorphous Co/Al/P product results. In the case of AlPO$_4$-H1 (VPI-5) synthesis (0.8 P$_2$O$_5$:1 AL$_2$O$_3$:1.5 HCl:50 H$_2$0), no transient amorphous phase is observed. Rapid formation of the crystalline phase from the clear solution after 3-4 hours occurs.

21. ZEOLITE NU-87: ASPECTS OF ITS SYNTHESIS, CHARACTERISATION, STRUCTURE AND PROPERTIES. JL Casci, MD Shannon, PA Cox and SJ Andrews, ICI Chemicals and Polymers Ltd, PO Box 90, Wilton, Middlesbrough, TS6 8JE U.K.

In the last decade there have been a number of significant advances in the science of zeolite molecular sieves. Arguably two of the most important of these have been the preparation of the family of aluminophosphate zeotypes and the work on structure solution, which has resulted in a dramatic increase in the number of known molecular sieve topologies.

What is of particular interest is that what were the first three high-silica zeolites to be prepared: zeolites Beta, ZSM-5 and ZSM-11 are the only materials of this type to have multi-dimensional channel systems based on 10, or 12, T-atom rings.

This paper will describe aspects of the preparation, characterisation and properties of a new multi-dimensional structure: zeolite NU-87 (JL Casci and A Stewart, EPA 377 291, 1990). The framework topology, of NU-87, has been determined and details of its unique pore system will be presented together with aspects of its synthesis and characterisation; with information being presented from powder x-ray diffraction, chemical and thermal analysis and electron microscopy.

22. ZEOLITE SYNTHESIS IN THE Na-TMA SYSTEM AT LOW TEMPERATURE. P. Donald Hopkins, Amoco Oil Company, MS H-9, P.O. Box 3011, Naperville, IL 60566-7011

Previous studies, at Amoco and elsewhere, have shown that the tetramethylammonium cation (TMA) is a structure directing agent for zeolites containing sodalite and gmelinite cage 14-hedra. Although TMA is not required for 14-hedra formation at low Si/Al ratios in the reactant mixture, it becomes necessary as the Si/Al ratio increases. In zeolite X synthesis we find that less substitution of TMA for sodium is required to change the structure from FAU to LTA than was found previously for zeolites Y and ZK-4 type recipes. Direct evidence for sodium ion participation in FAU synthesis has been found. We have also observed that the FAU structure apparently is a more ubiquitous predecessor to other zeolites than has been reported in the literature. Some discussion of the implications of these and related observations on the mechanisms of zeolite synthesis may be made.

23. ALUMINUM- AND BORON-CONTAINING SSZ-24, THE ALL-SILICA ANALOG CF AlPO-5. Robert A. Van Nordstrand and Stacey I. Zones; Chevron Research and Technology Co., Richmond, CA 94802-0627.

Recently we reported (1) synthesis of SSZ-24, an all-silica material, isostructural with the aluminum phosphate mol sieve, AlPO-5. Direct introduction of Al in the synthesis was not successful; other, competing, zeolites resulted. However, boron-containing beta zeolite may be converted using the adamantammonium template to B-SSZ-24 (2). This is then treated with an aluminum nitrate solution, whereby the B is replaced by Al (3). The B and Al forms of SSZ-24 have catalytic utility. They are distinguished from the all-Si version by XRD measurement of lattice constants, by NMR and by catalytic studies. These techniques show the Al or B is present in tetrahedral sites.
References:
1. Van Nordstrand,R.A.;Santilli,D.S.;Zones,S.I.;ACS Symp Ser 368, 1988.
2. Zones,S.I.,et al; Europ.Pat.Appl.,Intl.Pub.WO 91/00844 Jan.24, 199_.
3. Pochen Chu, U.S.Patent 4,912,073; 27 March, 1990.

24. ON THE CATALYTIC SITES FORMED BY THE REACTION OF METAL ALKYLS WITH DEBORONATED ZEOLITES. R. de Ruiter and H. van Bekkum, Laboratory of Organic Chemistry, Delft University of Technology, Julianalaan 136, 2628 BL Delft, The Netherlands.

Stable homogeneously distributed hydroxyl nests are created by mild extraction of framework boron out of high silica (Si/B ≥ 23) [B]-MFI single crystals. These silanol nests are thermally stable till at least 400 °C in flowing nitrogen and therefore appropriate for post-synthesis gasphase addition of catalytically active elements.

In this study post-synthesis "vacancy-addition" with metalchlorides, metalalkyls and metal compounds containing mixed chloride and alkyl ligands is compared.

For a series of the model compounds $SiCl_xMe_{4-x}$ the reaction with nest-hydroxyl groups occurs if $x \leq 2$ which can be understood on the basis of physical and chemical properties of the guest-molecules and zeolite channels/pores.

Work on e.g. Al and Zn compounds is in progress. Reaction of the silanol nests with the guest-molecules have been characterized with single crystal FT-IR microscopy as well as MAS NMR spectroscopy.

Preliminary results of the same procedure using zeolite β will be presented.

25. THE SHELL POLYGASOLINE AND -KERO PROCESS, J.P. van den Berg, A.H. Klazinga, P.M.M. Blauwhoff, Koninklijke/Shell Laboratorium, Amsterdam, Shell Research B.V.

The newly developed Shell Polygasoline and -Kero (SPGK) Process is a process in which light olefins ($C_2^=$-$C_5^=$ e.g. from FCC units) are oligomerised into liquid transportation fuels using a novel zeolite catalyst system. The process is very flexible with respect to product selectivity. Yields can be varied between gasoline and middle distillate products. In this paper the basic mechanisms of the process chemistry will be discussed as well as effects of feedstock composition, poisons and process conditions. Furthermore, the product properties will be discussed.

26. COMPARATIVE STUDY OF THE OLIGOMERIZATION OF C_2-C_6 OLEFINS ON HZSM-5, N.S. Gnep, F. Bouchet, M. Guisnet, URA CNRS DO 350, Catalyse en chimie organique, Universite de Poitiers, 40 Avenue du recteur Pineau, 86022 Poitiers Cedex, France

The oligomerization of various olefins: ethene, propene, isobutene, and 1-hexene was carried out at 250°C on a HZSM-5 zeolite (Si/Al=30) in a flow reactor under atmospheric pressure. A wide range of contact times was used so as to obtain the product distribution for conversions between 5 and 80-95%. The distribution as a function of the number of carbon atoms was obtained by gas chromatography, the branching of the products estimated by HNMR. Whatever the reactant, C_3-C_{14} olefins were formed. With C_2-C_4 olefins, dimers and trimers appeared as primary products. The other olefinic products resulted from secondary transformation of oligomers (cracking, alkylation). With 1-hexene, isomerization was the main reaction. The rate of oligomerization depended on the olefinic reactant: propene was about 1.5 times more reactive than isobutene and 5 times more than ethylene and hexene. Reaction mechanisms are proposed to explain the reactivities of the olefins and the reaction schemes.

27. DEACTIVATION BY COKE FORMATION OF ZSM-5 TYPE CATALYSTS DURING OLIGOMERIZATION OF ETHENE, T. Dypvik*, A. Holmen*, Y.B. Taarit*, Dept. of Industrial Chemistry, Norwegian Institute of Technology, N-7034 Trondheim, Norway, *Institut de Recherches sur la Catalyse, 2, av. A.Einstein, F-69626 Villeurbanne, France.

Coke formation on HZSM-5 and modified ZSM-5 catalysts was studied during oligomerization of ethene at 473 - 773 K using a microbalance reactor. The reactant mixture contained about 7% ethene and N_2 was used as the diluent gas. The experimental setup permitted to study the coke formation in situ measured as weight increase of the catalyst sample, as a function of time on stream. At low temperatures, less than about 625K, a rapid weight increase was observed in the first few minutes of reaction. Most of the adsorbed species was easily removed in N_2. However, the last part of the deposit could only be removed with O_2 or air at high temperatures. At higher reaction temperatures only this resistant carbonaceous material was formed. On catalysts modified by ion-exchange with Ni, the amount of resistant coke increased with increasing temperature. The same effect was observed for Cu-exchanged catalysts, but to a lesser degree. The results are compared with rates and selectivities obtained in a fixed bed reactor.

28. SULFATED ZIRCONIA SUPERACID CATALYSTS: THERMAL ANALYSIS AND CHARACTERIZATION STUDIES, R. Srinivasan, B.H. Davis, University of Kentucky, Center for Applied Energy Research, 3572 Iron Works Pike, Lexington, KY 40511-8433.

Zirconium oxide is claimed to be the only oxide catalyst that possesses four chemical properties: acidic, basic, oxidizing and reducing properties. The activity and selectivity of zirconia catalysts are greatly influenced by the preparation methods and the activating treatments. Zirconia catalysts impregnated with sulfate anions have been found to exhibit superacid behavior as such, and to exhibit a high activity for isomerization of hydrocarbons, and even for conversion of methanol into hydrocarbons. Tanabe suggested that the superacid sites are generated by the interaction between the oxide and sulfate ions. It was reported that the presence of sulfate ions is, in some way, responsible for inhibiting the recrystallization of the ZrO_2, and the sulfate-treated zirconia samples retained a much greater fraction of the original specific surface area than the untreated ones. In the present investigation, zirconia was impregnated with different amounts of sulfate ions, and each sample was characterized by thermal analysis, (TGA/DTA, M.S.), and reaction studies.

29. DIENE SYNTHESIS BY SOLID ACID CATALYSTS. T.Yamaguchi and C.Nishimichi, Department of Chemistry, Faculty of Science, Hokkaido University, Sapporo 060, JAPAN.

A condensation reaction of aldehyde and olefin, which is known as a Prins reaction, is one of the useful methods to obtain conjugated dienes. A homogeneous reaction system indicates this reaction occurs in acidic media. In fact, the reaction of propene with formaldehyde to yield isoprene is known to be catalyzed by solid acid catalysts. We examined the catalytic activity and selectivity of various catalysts which include solid (super) acids and solid bases as well as some metallic catalysts for the synthesis of octadiene from isobutyraldehyde and isobutene, which was taken as a model reaction of the Prins reaction. As a result, though solid bases were totally inactive, solid acids exhibited high activity to yield dienes. Acid strength required for the diene synthesis was found to be in the range of intermediate strength, $-8.2 < Ho < -3.0$. Niobic acid and supported tungsten oxide were found to be excellent catalysts, specifically the latter yielded octadienes with 98-99% selectivity. Reactions from furan derivatives and octanol were also investigated.

30. CATALYTIC PERFORMANCE OF PILLARED INTERLAYERED CLAY MOLECULAR SIEVE WITH HIGH THERMAL STABILITY, J. Jiao, Z. Gao, Department of Modern Chemistry, University of Science and Tech of China, Hefei, Anhui, 230026, P.R. China.

The surface acidity of a kind of pillared interlayered clay molecular sieve with high thermal and hydrothermal stabilities (AlR) has been studied using titration and IR methods, and the experimental data obtained have been compared with that of HY and USY zeolites. The acid site density of AlR amounts to about 0.45 mmole g^{-1}, which is 4-5 times less than that of HY and USY zeolites. There are two types of acid sites with medium and weak acid strengths. IR pyridine adsorption spectra of AlR show that the acid sites on AlR are mainly Lewis acid sites from the alumina pillars.

The catalytic behavior of pillared interlayered clay molecular sieve was compared with those of HY and USY zeolites for acid-catalyzed reactions such as dehydration of isopropanol, conversion of cyclohexene, cracking of cumene and decalin, and dealkylation of bulky molecules.

31. MICROPOROUS OXIDES BY THE SOL-GEL PROCESS: SYNTHESIS AND APPLICATIONS. L. C. Klein, C. Yu, R. Woodman, R. Pavlik, Rutgers-The State University of New Jersey, Ceramics Department, P.O. Box 909, Piscataway, NJ 08855-0909.

Several oxides, notably silica, alumina and zirconia, can be processed by the sol-gel process to produce microporous materials. The sol-gel process involves hydrolysis and polymerization of metal alkoxides. These reactions lead to the gelling of an oxide network in the presence of a solvent phase. The solvent is removed by evaporation or hyper-critical evacuation. The rigid oxide material remaining is 50 to 90% porous. The way in which the process is carried out, from start to finish, controls the generation of the microstructure such that a set of conditions can be chosen for a specific application. The applications for these materials include porous coatings for optical effects, gas separation media, absorbents, chemical sensors and catalyst supports. The porosity can be infiltrated with polymers or indicators to make devices. One important commercial application is supported membranes.

32. **SOL-GEL PROCESSING OF CONTROLLED PORE SIZE OXIDES,** C. J. Brinker, G. C. Frye, C. S. Ashley, R. A. Assink, Sandia National Laboratories, Albuquerque, NM 87185-5800, D. M. Smith, P. M. Davis, R. Deshpande, and S. Hietala, Center for Micro-Engineered Ceramics, University of New Mexico, Albuquerque, NM, 87131.

Sol-gel processing involves the hydrolysis and condensation of inorganic or metal organic precursors. Often polymer growth occurs by kinetically limited growth processes such as cluster-cluster aggregation leading to highly ramified structures described by a mass fractal dimension. We have used ^{1}H spin relaxation techniques and ^{29}Si MAS NMR to characterize the porosity of wet silicate gels **in situ** during aging and drying. We find that aging in alcohol results in depolymerization accompanied by a significant reduction in the pore size distribution and increase in surface area. Aging in water has the reverse effect. Within a narrow compositional range near Al:Si = 1:1, Al_2O_3-SiO_2 gels form closed-pore solids when dried from water-rich fluids. By varying the fluid surface tension the porosity may be changed from closed, to microporous, to mesoporous. Prior to gelation the inorganic sols may be deposited as thin films by spinning or dipping. N_2 adsorption measurements employing SAW techniques show that the pore size of the deposited films depends on such factors as the size and mass fractal dimension of the inorganic polymers, the relative rates of condensation and evaporation during deposition, the magnitude of the capillary pressure, and the shear rate imposed by the deposition process or surface tension gradient driven flows. For a system of mutually opaque fractal clusters, we show that the pore size may be systematically varied from <0.4 nm to >3.0 nm.

33. **ACTIVATION OF WIDE PORE ALUMINA SUPPORTS,** M.S. Goldstein and J.D. Carruthers American Cyanamid Company Chemical Research Division 1937 W. Main Street Stamford CT 06904

Wide Pore supports are frequently used to make catalysts for applications where intra-particle mass transfer tends to limit the rate of the chemical reactions that occur over them. Many of these catalysts are supported on alumina. One way to increase their pore diameters is to activate the alumina supports at high temperatures and in the presence of high concentrations of water vapor.

Recently, new procedures have been developed to manufacture alumina powders with inherently larger pore diameters. Results reported here show that catalyst supports prepared from both the newer, inherently wider-pored, alumina powders and the earlier, inherently smaller-pored, alumina powders behave similarly to activation at high temperatures and in the presence of high concentrations of water vapor.

34. **THE ALKALI EARTH SALTS AS CATALYST SUPPORTS AND STABILIZERS.** Alvin B. Stiles, Center for Catalytic Science and Technology, Department of Chemical Engineering, University of Delaware, Newark, DE 19716.

The alkali earths are only infrequently used as supports despite the fact that they have many properties favoring their use. This paper's objective is to enumerate these attributes while at the same time warning of the adverse factors which may be encountered. Among the attributes is the high melting point of the oxides with magnesia having a melting point well above alumina (2045°C) and silica or silica-aluminas and almost as high as thoria (2800°C vs. 3050°C). On the other hand the alkali earths are basic and may react with other oxides at rather low temperatures to form spinels; for example, $Mg\cdot Al_2O_4$ or other ionic crystals. These may have desirable properties as well. The hydroxides are even more basic and may, as a consequence, temporarily adsorb or chemisorb an acidic gas for at-site catalytic action. This will be recognized as support-catalyst-interaction and may beneficially be designed into a catalyst. The oxides can form a labile peroxide which in fact can also be exploited. Be and Ra are members of the family but well-known problems preclude their use so they will not be a part of the discussion.

35. EFFECT ON HIGH TEMPERATURE LEAN AGING ON THE PERFORMANCE OF Pt, Rh/CeO$_2$ AND RARE EARTH/ALKALINE EARTH DOPED Pt, Rh/CeO$_2$ CATALYSTS. J.G. Nunan, Allied-Signal Research and Technology, 50 East Algonquin Road, P.O. Box 5016, Des Plaines, IL 60017-5016

Current state-of-the-art automotive emission control catalysts contain noble metals and a variety of promoters such as rare earths and alkaline earths. Ce is the most important promoter used in automotive emmission control catalysts. Recently La has also been added to automotive catalysts; however, it is known that La can interact negatively with the Ce component after high temperature lean aging. We have doped the CeO$_2$ lattice of Pt, Rh/CeO$_2$ catalysts with La and other rare earth and alkaline earth metals and have found that the presence of these ions can have a large detrimental effect on the catalyst performance. Extensive characterization of the doped catalysts using XPS, STEM and TPR has shown that the presence of the dopant ions results in loss of noble metal availability and to a decrease in the reducibility of both the metal and surface Ce.

36. FROM CLATHRASILS TO LARGE-PORE ZEOLITES: THE EFFECTS OF SYSTEMATICALLY ALTERING A TEMPLATE STRUCTURE. Y. Nakagawa and S.I. Zones, Chevron Research and Technology Co., 100 Chevron Way, Richmond, California 94802-0627.

Organic templates are believed to play an important role in the process of high-silica zeolite crystallization. We have prepared a series of related compounds using a versatile synthetic method in an effort to study how the template structure affects the final product obtained in hydrothermal zeolite preparations. By making small, systematic modifications on the parent structure A, we have been able to suppress formation of a clathrasil (Nonasil) structure and favor the crystallization of large-pore zeolitic materials. The effects of varying substituting framework atoms will also be presented.

37. 12-RING CHANNEL TEMPLATED MAZZITE: A STABILIZED CALCINED STRUCTURE. D.E.W.Vaughan and K.G.Strohmaier, Exxon Research and Engineering Company, Annandale, N.J., 08801.

Synthetic MAZ, made with tetramethylammonium cations trapped in gmelinite cages (ZSM-4, Omega), undergoes partial structural degradation when the template is burned out at high temperature (600°C), damage which is significantly increased in the gallium substituted form. If, however, dimethyl-diethyl ammonium cations or other similarly sized templates are used,the larger templates locate in the 12-ring channel and are removed at a temperature (400-480°C) which is too low to induce framework hydrolysis and the associated structural breakdown. Structural degradation is shown by XRD peak broadening and lower hydrocarbon sorption capacities, the latter directly related to template burn-off temperatures. Isotherm shape is highly variable between syntheses, showing "tailing" associated with high aspect ratio crystals and "partial blocking" of the channels.

38. COMMON FACTORS IN THE SYNTHESIS OF Na-CHABAZITE AND K-MAZZITE. F. Di Renzo, F. Fajula, F. Figueras, M.A. Nicolle and T. Des Courieres. URA 418 du CNRS School of Chemistry. 8, rue de l'Ecole Normale. 34053 Montpellier. France

The selectivity of crystallisation of the silicoaluminate zeolites can be directed by governing the supply of aluminium. Two examples are discussed in detail. Ageing of the silica source before the addition of aluminium allows to reverse the selectivity of a system yielding zeolite Y towards the formation of Na-chabazite. In another example Na,K,TMA-mazzite is formed instead of offretite when aluminium is provided through the controlled dissolution of a crystalline source or by an external supply. In both cases the degree of polymerization of the silicoaluminate species in solution and the surface reactivity of the amorphous precursors are affected. These factors are expected to determine the orientation of the nucleation.

39. INVESTIGATION OF SEED PROPERTIES INFLUENCE ON THE RATE OF ZSM-5 CRYSTALLIZATION. K.G. Ione, S.V. Dudarev, A.V. Toktarev. Institute of Catalysis, Ak.Lavrentiev Av.,5, Novosibirsk, 630090, USSR.

Study of the role of physico-chemical properties of the seeds on the kinetics of ZSM-5 zeolite synthesis has been done. The main parts of this study are: preparation of the zeolite seeds with different properties ; measuring of full kinetic curves of crystallization at several temperatures ; use of the complex of chemical and physico- chemical methods (chemical dissolution, adsorption technique, TEM, SEM, XRD, IR, multinuclear NMR, TG) to characterize the crystallization.
It has been shown the influence of the seed properties on the properties of the final synthesized products, the parameters of the crystallization process, the nucleation energies and the energies of crystal growth.
The structure of prepared seeds and the model of nucleation and crystal growth mechanism of ZSM-5 crystallization in presence of seeds have been proposed.

40. PRECURSOR PHASES AND NUCLEATION OF ZEOLITE TON. F. Di Renzo, F. Remoué, P. Massiani, F. Fajula, F. Figuéras and T. Des Courières. URA 418 du CNRS. School of Chemistry. 8, rue de l'Ecole Normale. 34053 Montpellier. France.

The kinetics of crystallisation of zeolite TON have been studied under various experimental conditions. The complete characterization of the synthesis media allowed to establish some patterns of parallel evolution of the morphology and the composition of the gel. An intermediate amorphous phase containing the organic agent is always observed before the crystallisation of the zeolite. All data indicate that the induction time of the zeolite nucleation is essentially controlled by the kinetics of formation of the intermediate amorphous phase. Not only the crystallisation steps but also all previous gel transformations are of paramount importance in the synthesis of zeolite TON. Could similar considerations apply to the crystallisation of other zeolites?

41. KINETIC STUDIES OF THE CRYSTALLIZATION OF THE ZEOLITES MORDENITE, L, OMEGA AND OFFRETITE, H.Lechert[*] and H.Weyda[**]
[*]Institute of Physical Chemistry, University of Hamburg, Germany
[**]SÜD-CHEMIE AG, Katalyse Labor, 8206 Heufeld Bruckmühl, Germany

Systematic studies of the kinetics and the fields of crystallization have been carried out for the zeolites Mordenite, L, Omega and Offretite. For the Mordenite thorough experiments for the relations of the batch

compositions and the Si/Al-ratios in the products have been done. The kinetic data have been analyzed with the Avrami theory. The crystallization of zeolite L has been optimized. For the crystallization of zeolite Omega detailed studies have been carried out in dependence on the content of TMA-Cl in the batch.
The crystallization of all zeolites is discussed together with the formation of Offretite in a common crystallization diagram with the content of the cations as parameter.

42. ^{13}C and ^1H Magic-angle-spinning NMR Studies of the Conversion of Methanol to Olefins over Offretite/Erionite Intergrowths ., Michael W. Anderson,[1,2] Mario L. Occelli[3] and Jacek Klinowski[1]

(1) Department of Chemistry, University of Cambridge, Lensfield Road, Cambridge CB2 1EW, U.K.
(2) .Department of Chemistry, UMIST, P.O. Box 88, Manchester M60 1QD, U.K.
(3) ·Unocal Corporation, Science and Technology Division, Brea, CA 92621, U.S.A.

^{13}C and ^1H nuclear magnetic resonance with magic-angle spinning (MAS NMR) has been used to monitor the conversion of methanol to low molecular weight hydrocarbons over intergrowths of the zeolites offretite and erionite. Four samples with varying content of the two structures have been studied. The sample richest in offretite shows the lowest selectivity to light olefins; this is correlated to the ease of formation of long-chain polymers within the intracrystalline space of offretite. We confirm the presence of such polymers is confirmed by ^{13}C MAS NMR and attribute it to the uninterrupted one-dimensional pore system.

43. FACTORS AFFECTING THE DEACTIVATION OF GALLIUM CONTAINING HZSM-5 ZEOLITES USED FOR THE AROMATIZATION OF SMALL ALKANES, C.R. Bayense,
J.H.C. van Hooff, Eindhoven University of Technology, Lab. of Inorganic Chemistry and Catalysis, PO Box 513, 5600 MB, Eindenhoven, The Netherlands

Application of gallium containing HZSM-5 zeolites for alkane aromatization results in high aromatics selectivities. Introduction of gallium, however, also accelerates catalyst deactivation by coke formation, because of the dehydrogenating properties of the gallium. This undesirable behavior can be influenced by: Increase of the hydrogen partial pressure in the catalyst bed by addition of hydrogen to the feed, which leads to an enhanced catalyst stability but also a reduction of activity and aromatics selectivity; Introduction of gallium in the zeolite by isomorphous substitution, which gives gallium located in the zeolite pores, where shape selective restriction prevent fast coke formation. Other methods yield gallium deposited on the outer surface of the zeolite particles, where rapid coke formation can take place.

^{71}Ga magic Angle Spinning NMR and XPS measurements show that subsequent reaction/regeneration cycles or steam treatments cause a loss of framework gallium. The so-formed extra-framework gallium, however, remains in the zeolite pores, and is thus still protected against coke formation.

44. NAPHTENES TRANSFORMATION OVER H-[Al]-ZSM-5 CATALYSTS, G. Giannetto, J.A. Perez[2], R. Scalamanna, L. Garcia, R. Galiasso[2], R. Monque[2], Universidad Central de Venezuela, Escuela de Ingenieria Quimica, P.O. Box 47100, Caracas 1040, Venezuela, [2]INTEVEP, S.A. Seccion de Catalisis Aplicada, Los Teques, Edo, Mirando, PO Box 76343, Caracas 1070A, Venezuela.

The transformations of propane, n-hexane, cyclohexane, n-heptane and methylcyclohexane were studied over acidic H-[Al]-ZSM-5 catalysts at 530°C, 1 bar and with a N_2/hydrocarbon molar ratio of 4. Results show that the aromatic molar product distribution (AMPD) is affected by the type of catalyst. On H-[Al]-ZSM-5 catalyst, propane, n-hexane and cyclohexane, for one side and n-heptane and methylcyclohexane, for the other side, show very similar AMPD. Over Ga/[Al]ZSM-5 while, the AMPD for propane is almost the same as on H-[Al]ZSM-5, it changes for liquid hydrocarbons. In these cases, the fraction of the aromatic isomer of the transformed molecule is higher than with acidic catalysts. This change is more pronounced for cycloparaffins and it is related to the dehydrogenating activity of gallium species which are catalyzing the direct aromatization of hydrocarbons rather than its previous cracking and subsequent aromatization.

45. AROMATIZATION OF ETHANE AND OF PROPANE ON HZSM-5. EFFECT OF GALLIUM, M.R.Guisnet, D.A. Taleb, J.Y. Doyemet, N.S. Gnep, URA CNRS DO 350, Catalyse en chime organique, Universite de Poitiers, 40 Avenue du recteur Pineau, 86022 Poitiers cedex, France

The transformations of ethane and of propane were carried out on pure and on gallium-doped HZSM-5 zeolites (Si/Al=40) at 530°C under a hydrocarbon pressure of 1 bar. On a pure HZSM-5, propane was about 30 times more reactive than ethane. With a 4 wt% Ga HZSM-5, the rate of propane transformation was about 3 times greater than on pure HZSM-5 and the rate of ethane transformation about 8 times. For both reactants, Ga increases greatly the rate of formation of olefins (ethene and propene) and improves the selectivity for benzenic hydrocarbons. We show that the aromatization of ethane and of propane can be considered as bifunctional processes. Gallium species catalyse the dehydrogenation steps (dehydrogenation of ethane and of propane into olefins and of naphthenes into aromatics) while the acid sites of HZSM-5 catalyse the oligomerization of ethene or of propene as well as the cyclization of oligomers.

46. EVOLUTION OF THE GALLIUM PHASE OF GaHMFI AROMATIZATION CATALYSTS IN THE COURSE OF THE PREPARATION PROCEDURE AND REACTION-REGENERATION CYCLES. H. Ajot, F. Alario, J.F. Joly, E. Merlen and F. Raatz, Institut Français du Pétrole, BP311, 92506 Rueil Malmaison, France.

The transformation of lower alkanes on GaHMFI catalysts is a new important route to aromatics. A large number of studies have been reported on the determination of aromatization reaction pathways using GaHMFIcatalysts. It is now well established that such catalysts act as bifunctional systems : the gallium phase (namely Ga_2O_3) is responsible for alkanes and napthenes dehydrogenation while the acidic groups of the zeolite are responsible of olefins oligomerization and cyclization (and methane production). Very little is known about the effect of the gallium phase dispersion on the aromatic selectivities and on the evolution of this phase when hydrogen is present in the feed as it is the case in aromatization reactions. The present work aims to determine the main parameters controlling the dispersion of the gallium phase of GaHMFI aromatization catalysts and the following items have been examined:
1. the way of gallium introduction in HMFI zeolites: ionic exchanges at various pH, incipient wetness technique,
2. the effect of hydrogen activation treatments,
3. the effect of reaction-regeneration cycles in propane aromatization.
The dispersion and chemical state of the galium phase have been studied by X ray diffraction, electron microscopy (STEM mode) and temperature programmed reduction.

47. ZEOLITE BETA AND ISOMORPHOUSLY SUBSTITUTED GA-BETA AS A CATALYST FOR PROPANE AROMATIZATION - INFRARED SPECTROSCOPIC INVESTIGATIONS. F. Schüth, R. Spichtinger, Institut für Anorganische und Analytische Chemie, Universität Mainz, Becherweg 24, 65oo Mainz, Germany.

Zeolite beta is known now for more than 2o years, but has not been very widely used in catalytic reactions. There are only few reports in the open literature, and about ten examples in the patent literature describing its catalytic properties. In our studies we employed zeolite beta and isomorphously substituted Ga-beta synthesized by different procedures for propane aromatization. These catalysts are quite active compared to ZSM-5 catalysts also synthesized in our group, and show a very promising coking behavior. The activity and the coking properties, however, are strongly dependent on the mode of preparation. The catalysts are characterized by ex situ IR-spectroscopy to obtain information about the concentration, type,and strength of acidic sites, and in situ IR-spectroscopy under HV conditions as well as under atmospheric pressure. The differences of the catalysts will be explained in terms of the results of these measurements. The results will also be compared to studies of ZSM-5 type catalysts investigated under similar conditions.

48. INFLUENCE OF THE ZINC CONTENT ON AROMATICS FORMATION OVER ZINC-CONTAINING ULTRA-STABLE Y ZEOLITES, P.A. Arroyo[1], E.F. Sousa-Aguiar[2], J.L.Fontes Monteiro[1], [1]Federal University of Rio de Janeiro, UFRJ, COPPE., [2]Petrobras Research Center, CENPES-DICAT., Ilha do Fundao, Quadra 7, Cidade universitaria-CEP 21910, Rio de Janeiro, Brazil

The preparation and evaluation of zinc-containing USY zeolites was studied. NH_4-Y was zinc-exchanged before calcination aiming at different metal contents. After calcination, at constant Na_2O levels, it was observed that the presence of zinc prevents zeolite dealumination. Also, both the collapse temperature (measured by DTA) and the Bronsted/Lewis ratio decreased with the increasing zinc content. N-hexane cracking test was used to evaluate the zeolites. One observes that the higher the zinc content, the higher the aromatics selectivity at constant conversion. Furthermore, higher olefinicity and methane formation were observed. The hydrogen production ratio was plotted against the aromatics selectivity. A straight line with slope 3.5 was obtained, indicating that naphthenes might undergo dehydrogenation over zinc sites.

49. STRONG SOLID ACID CATALYSTS. Russell S. Drago, and Steven C. Petrosius, Department of Chemistry, University of Florida, Gainesville, Florida 32611.

Studies of a strong solid acid, consisting of tetrahedral chloro-aluminum sites on silica gel are described. Measurements indicate that the material is amongst the strongest solid acids known. A series of other supported strong Lewis acids have been prepared and their reactivity will be described.

50. AMORPHOUS ALUMINUM PHOSPHATE GELS. Rimantas Glemza, Yves O. Parent, William A. Welsh, W. R. Grace & Co.-Conn., Davison Chemical Division, 7379 Route 32, Columbia, MD 21044.

Amorphous aluminum phosphate gels have been synthesized which have structures analogous to amorphous silica gels. Crystalline aluminum phosphates are known to have structures isomorphous with most SiO_2 structures and therefore the similarity of amorphous material was anticipated. In the present work, amorphous aluminum phosphate gels have been prepared with a P/Al atomic ratio of ~1 with higher pore volume and pore diameter than previously reported, making them more suitable for certain catalyst support and adsorbent applications. Maintenance of a P/Al atomic ratio of ~1 was done with the intention of synthesizing a uniform $AlPO_4$ phase rather than a mixture of discrete $AlPO_4$, Al_2O_3 and P_2O_5 regions. Surface areas ranging from 250-350 m^2/g were routinely obtained in conjunction with pore volumes from 0.4-1.8 c/g. Surface chemistry was studied by NMR and ESCA techniques. Use of the amorphous aluminum phosphate gels as ethylene polymerization catalysts and adsorbents are described.

51. OPEN MORPHOLOGY ZIRCONIUM PHOSPHATES, ALUMINAS, AND MIXED OXIDES. W. H. Quayle, J. A. Lowe, Aluminum Company of America, Alcoa Technical Center CSD-C-31, Alcoa Center, PA 15069.

Particles derived from the intergrowth of high aspect ratio crystals can exhibit open pore structure morphologies. The wide access channels in such particles can allow facile intraparticle transport. Rapid access to active sites is needed for diffusion limited reactions. Wide channels also allow intraparticle reactions of large molecules, and can help produce low pressure drops in packed columns. We will discuss nominally 8 μ diameter spherulitic particles of intergrown, lamellar crystals of zirconium hydrogen phosphate, and the pseudomorphic, thermal transformation of these particles to zirconium pyrophosphate. We will also describe nominally 100 μ diameter spherulites of transition alumina, and mixed oxide precursors. Our characterization methods include nitrogen and mercury porosimetries, x-ray diffraction, and solid state nuclear magnetic resonance.

52. ACIDITY OF SUBSTITUTED ZIRCONIAS. S. L. Soled and G. B. McVicker, Exxon Research and Engineering Co., Corporate Research, Rt. 22 East, Annandale, NJ 08801.

In exploring the use of zirconia-based catalyst supports, the ability to produce catalysts with high surface area and a controllable range of acidity is important. We have studied the substitution of silicon into zirconia as well as the deposition of silica on the surface of zirconia and measured differences in surface area, acidic properties and surface charge. The isomer distribution during the conversion of 2-methylpentene-2 provides a probe reaction to monitor acidity changes. Comparison of these acidity parameters with those of alumina- and titania- based supports yields a basis for comparative ranking of these solid acids. The enhancement of acidity with silica substituion is compared for bulk versus surface impregnation. Zeta potential measurements monitors changes in zirconia surface properties as a function of silica substitution.

53. SUPPORTED FLUOROCARBONSULFONIC ACID CATALYSTS, J. D. Weaver, E. L. Tasset, and W. E. Fry, Organic Product Research, Dow U.S.A., B-1215 Building, Freeport, TX 77541.

For nearly 25 years, fluorocarbonsulfonic acid polymers have been used in such applications as the chlor-alkali process and the hydrogen-oxygen fuel cell process. Their remarkable ability to catalyze a wide range of chemical reactions has also been demonstrated. These materials possess several physical characteristics that are very desirable in heterogeneous catalysts, such properties as extraordinarily high acid strength, stability at high temperatures, and resistance to solvent and oxidative attack. However, their relatively high cost and low efficiency have prevented their widespread use in industrial applications. This paper will discuss acid catalysts prepared by supporting fluorocarbonsulfonic acid polymers on inert carriers, and the resulting improvements in efficiency and reactivity.

54. MOLECULAR CHEMICAL ASPECTS OF SILICA GEL FORMATION
Peter W.J.G. Wijnen, Theo P.M. Beelen, Rutger A. van Santen, Schuit Institute of Catalysis, Eindhoven University of Technology, P.O. Box 513, 5600 MB Eindhoven, The Netherlands

Formation of silica gels implies polycondensation reactions of dissolved silicate species. Using in-situ ^{29}Si-MAS-NMR, a mechanism for the oligomerisation of monomeric silicic acid and formation of cubic octameric silicate species is proposed. Aggregation of oligomerised silicate species was investigated by small angle scattering of X-rays (SAXS), a technique used to investigate length-scales of 1-100 nm (colloidal). The influence of (polyvalent) cations was studied, showing differences in aggregation kinetics in the case that magnesia and alumina cations are present in the reacting solution. Aluminum cations cause an inhibition in aggregation kinetics.
Through the combination of SAXS, NMR and N_2-sorption measurements it is possible to follow reactions involved in the aging process of aqueous silica gels. Aging implies re-organisation of particles constructing the aggregates. Here, a new mechanism for aging of silica gels is presented showing that aggregates become more dense in the core and less dense at the periphery. The silica gel structure is strengthened and as such (meso)pores are preserved upon drying.
Results of the application of SAXS to the study of the synthesis of zeolite A will be presented as well.

55. ZINCOPHOSPHATES AND BERYLLOARSENATES...ZEOLITES AND OTHER MICROPOROUS STRUCTURES. T. Gier, W.T.A. Harrison, T.M. Nenoff, and G.D. Stucky, Department of Chemistry, University of California, Santa Barbara, CA 93106.

Zinc and beryllium phosphates and arsenates have been found to form micro-porous structures analogous to the commercially important aluminosilicates (zeolites). Compounds isotypic with sodalite, gismondine, edingtonite, and zeolites X, RHO, and Li-A(BW) are readily accessible, and preliminary results indicate that interesting chemistry can be carried out within the pores. Many of these materials, although thermally less stable than the aluminosilicates, can be synthesized as reasonable sized crystals at moderate temperature (-20°C to 100°C). As with the aluminophosphate materials, several of these structures appear to have no zeolite counterparts.

56. **HYDROTHERMAL SYNTHESIS OF MANGANESE OXIDES WITH TUNNEL STRUCTURES.** Chi-Lin O'Young, Texaco Inc., PO Box 509, Beacon, NY, 12508

Manganese oxides with tunnel structures, such as hollandites, nsutite (γ-MnO_2), and pyrolusite (β-MnO_2) were synthesized by the hydrothermal method under mild conditions, 60 to 180 °C and autogenous pressures. The materials were prepared by the redox reaction between permanganate ion (MnO_4^-) and manganous ion (Mn^{2+}). The type and concentration of counter cation, pH, and temperature were identified as important parameters. Template effects of the counter cations, mainly alkali and alkali earth ions, were observed. By adjusting the pH and temperature, hollandites could be synthesized with a counter cation which has an ionic diameter between 2.3 and 4.6 Å. Otherwise, nsutite and pyrolusite would be formed.

57. **SYNTHESIS AND CHARACTERIZATION OF NEW METAL OXIDES.** Edward W. Corcoran, Jr., Corporate Research Science Labs, Exxon Research & Engineering Company, Clinton Township, Route 22 East, Annandale, NJ 08801

The focus of our program over the past several years has been the synthesis and identification of new, microporous materials. Since the properties these types of materials exhibit are intimately tied to their framework geometries, we have been particularly interested in developing new structural arrangements since such compositions are likely to have desirable application characteristics. The approach we have taken to accomplish this goal has been to introduce novel framework components into "traditional" zeolite synthesis schemes thereby forcing the formation of previously unsynthesized structures. The first system of this type that we examined was $M_2O/SnO_2/SiO_2$ (M = alkali metal or quaternary ammonium cation) where octahedral SnO_6 was incorporated into a lattice with tetrahedral silicon. Several new phases were produced which were found to have properties related to those of small pore zeolites in that they reversibly sorbed small molecules and were capable of ion-exchange. We next concentrated our efforts on phosphate based materials, beginning with a series of both two- and three-dimensional molybdenum phosphates containing octahedral Mo(V) corner-sharing with phosphate tetrahedra. We have recently extended this work to a variety of other metal phosphate systems and have conducted single-crystal structural determinations of several new phases. This talk will detail these studies as well as describing related results from earlier work.

58. SMALL-ANGLE NEUTRON SCATTERING STUDIES OF THE TEMPLATE-MEDIATED CRYSTALLIZATION OF ZSM-5 TYPE ZEOLITE[†]

L. E. Iton[1]*, F. Trouw[2], T. O. Brun[1], and J. E. Epperson[1], J. W. White[3]* and S. J. Henderson

Materials Science Division and Pulsed Neutron Source Division

Argonne National Laboratory, Argonne, Illinois 60439, U. S. A., The Australian National University, Canberra, Australia.

Small-angle neutron scattering is a useful new approach to the study of zeolite crystallization from aluminosilicate gels and the action of template molecules. It has been applied to gels for synthesis of zeolite ZSM-5 using tetrapropylammonium ions as templates where the scattering length densities of the gel particles and their texture were determined using contrast variation methods. Gels formulated from soluble silicate incorporate template molecules promptly into an amorphous "embryonic" structure and crystallization ensues via a solid hydrogel transformation mechanism. Gels formulated from colloidal silica show different scattering behavior, and a liquid phase transport mechanism is inferred.

59. DESIGN AND SYNTHESIS OF CARBON MOLECULAR SIEVES FOR SEPARATION AND CATALYSIS. H. C. Foley, D. S. Lafyatis, R. K. Mariwala, Center for Catalytic Science and Technology, Department of Chemical Engineering, University of Delaware, Newark, DE 19716.

Carbon molecular sieves are ultramicroporous materials that do not have the extended structural order of zeolites, but which do separate small molecules on the basis of size, shape, and molecular properties. Most recently the commercial success of the separation of nitrogen from air over carbon molecular sieve has focused new attention on the CMS materials.

This paper will review the commercial methods of CMS preparation, and their application in catalysis. In particular, the aim will be to highlight shape selective catalysis with little or no acidity. The results of our own work with polymer-derived materials will be summarized. This will include the synthesis methods, and the effects of preparation variables on the shape-selective properties of the materials. The interplay of shape-selective effects based on restricted diffusional transport through the structure and the reactivity at the catalytic sites will be placed in context with earlier work and our model of the pore structure in CMS catalysts.

60. Characterization of Molecular Sieve Frameworks and Products of Catalytic Reactions Using Two-dimensional NMR Techniques
Waclaw Kolodziejski and Jacek Klinowski
Department of Chemistry, University of Cambridge, Lensfield Road, Cambridge CB2, UK.

Several two-dimensional *solid-state* NMR techniques, such as COSY, NOESY and spin diffusion are shown to provide important information about molecular sieves. The details of these methods are given and their performance discussed. Thus the 2D solid-state J-scaled COSY NMR spectrum of highly siliceous mordenite reveals the tetrahedral site connectivities, permitting the unambiguous assignment of the ^{29}Si spectrum. 2D solid-state spin-diffusion ^{13}C NMR experiment *in situ* on the products of methanol conversion into gasoline over zeolite H-ZSM-5 substantially aids the spectral assignment and provides new information concerning the distribution of organic species in the intracrystalline space of the catalyst.

61. USE OF A KINETIC MODEL FOR INVESTIGATION OF LIGHT OLEFIN AROMATIZATION REACTIONS OVER HZSM-5 ZEOLITES, D.B. Luk'yanov, V.I. Shtral, Karpov Institute of Physical Chemistry, Ulitsa Obukha, 10, 103064, Moscow, K-64 USSR.

Conversion of light olefins over HZSM-5 at high temperatures leads to the formation of aromatics and light paraffins. The variety and complicated character of the chemical transformations involved in the aromatization process make it very difficult to use the experimental data directly for the characterization of hydrocarbon reactivities and catalyst activity in the different reaction steps. The possibilities of the characterization of the catalyst activity and hydrocarbon reactivities can be broadened significantly by means of comparison of the experimental data with the results of a kinetic model which describes the reaction pathway. In this paper we discuss the results of experimental study and the kinetic model for light olefin aromatization reaction over HZSM-5 zeolites. Experimental data are used in conjunction with mathematical modeling results for quantitative characterization of 1) the reactivities of $C_2^=$-$C_{10}^=$ olefins, and 2) the catalytic activity of various HZSM-5 in the chemical transformations of different types (olefin oligomerization, cracking and cyclization, hydrogen transfer).

62. SELECTIVE AROMATIZATION OF LIGHT HYDROCARBONS ON POLYFUNCTIONAL METALLOSILICATE CATALYSTS, T. Inui, A. Matsuoka, Dept. of Hydrocarbon Chemistry, Faculty of Engineering, Kyoto University, Sakyo-ku, Kyoto 606, Japan

Effective aromatization from $n-C_4H_{10}$ and $n-C_5H_{12}$ was investigated on Pt-incorporated H-Pt·Ga- or H-Pt·Zn-bimetallosilicate comparing with that on Pt-ion-exchanged H-Ga- or H-Zn-silicate (Pt/H-Ga-silicate, Pt/H-Zn-silicate). The conversion of C_2-C_5 hydrocarbons on these H-Pt·Ga- or H-Pt·Zn-bimetallosilicate was lower than that on Pt/H-Ga- or Pt/H-Zn-silicate; however, still larger than that on Pt-non-loaded Ga- or Zn-silicate catalyst. On the bimetallosilicate catalysts, the selectivity to light aromatics and olefinic hydrocarbons increased, indicating that activities of cracking, alkylation of benzene ring, and dehydrogenation caused by Pt could be moderated. On the other hand, activities of dehydrogenation and aromatization caused by Ga or Zn exerted more explicitly. This was supported by the difference of the Pt state between Pt-incorporated and Pt-ion-exchanged ones which were compared through measurements of CO adsorption, redox response, and TEM observation.

63. THE CONCEPT OF HYDROGEN BACK SPILLOVER (HBS): TOWARDS THE DEVELOPMENT OF HIGH PERFORMANCE CATALYSTS FOR THE AROMATIZATION OF LIGHT ALKANES AND OLEFINS, R. Le Van Mao, L. Dufresne, J. Yao, D. Ly, Department of Chemistry & Biochemistry, Catalysis Research Laboratory, Concordia University, 1455 De Maisonneuve Blvd. W., Montreal (Quebec), Canada H3Q 1M8

Hybrid catalysts, made up of two components (acidic ZSM-5 zeolite and supported Zn or Ga oxide) mechanically mixed and bound within a clay matrix, exhibit enhanced activity and selectivity towards BTX aromatics in the conversion of light olefins and alkanes. Such an exceptionally high aromatic production has been attributed to a long distance hydrogen transfer from the zeolite acid sites to the metal oxide surface. It is assumed that the role of the metal oxide component is to lower the energetic barrier at the zeolite particle surface. The feeds used are: propane or n-butane, ethylene (pure or produced by the Bioethanol-to Ethylene process). Kinetic and mechanistic studies show that the rate determining steps are the hybride abstraction and dehydrocyclization for the paraffinic and olefinic feeds, respectively.

64. NEW MODIFICATION METHOD OF PT/KL ZEOLITE CATALYST FOR LIGHT NAPHTHA REFORMING. T. Fukunaga H. Katsuno and M. Sugimoto, Central Research Laboratories, Idemitsu Kosan Co., Ltd, 1280 Kamiizumi, Sodegaura-machi, Kimitsu-gun, Chiba, 299-02 Japan

Platinum supported on KL zeolite (Pt/KL) is known as a light naphtha reforming catalyst. We now wish to report that treatment of Pt/KL with halocarbons improves the activity and selectivity for the hexane aromatization. The halocarbon treatment is also effective for the stability of the catalyst.
Transmission electron microscope observations have revealed that the halocarbon treatment results in high dispersion of platinum on the freshly reduced catalyst. This may lead to the high activity and selectivity of the treated catalyst. Elemental analysis indicates much less carbon deposit on the halocarbon treated catalyst during the reaction than on the untreated catalyst. Polymerization of the cracking products is considered to be responsible for the coke formation and hence the stability increase may be accounted for by the low hydrogenolysis activity as a result of the high selectivity for aromatization.

65. SHAPE EFFECTS IN THE ETHANOL CONVERSION TO HYDROCARBONS OVER ALKALINE-FREE SYNTHESISED ZSM-5 AND ALUMINATED SILICALITE-1 CRYSTALS. P. Voogd, J.C. Sollie, A. Doelwijt and H. van Bekkum, Laboratory of Organic Chemistry, Delft University of Technology, Julianalaan 136, 2628 BL Delft, The Netherlands.

The diffusional effect on conversion rates and product distributions in the ethanol to gasoline reaction on zeolite H-ZSM-5 was investigated. A sample of large ZSM-5 crystals having a narrow particle size distribution was hydrothermally synthesised in an alkaline-free medium, applying ammonium cations. X-ray data were indicative of a considerable heterogeneous aluminium distribution through the large ZSM-5 crystal.

In the ethanol conversion, comparison with conventional small H-ZSM-5 crystallites revealed that compounds having small critical diameters predominated in the product mixture. This is caused by their high relative mobility in the ZSM-5 channels. High *para*-selectivities have been found, especially at low contact times. Large ZSM-5 crystals deactivated rather quickly, mostly within 24 hours, because of coke deposition on the outer crystal surface.

Large silicalite-1 crystals, having a near-homogeneous Al distribution were obtained by applying a post-synthesis treatment with $AlCl_3$. It was concluded that Bronsted acidity was obtained inside of these silicalite-1 crystals.

66. INVESTIGATION OF REACTIONS AND CATALYSTS OF C_3-C_4 PARAFFIN DEHYDROGENATION AND AROMATIZATION, N.R. Bursian, S.B. Kogan, V.K. Daragan, O.M. Oranskaya, N.M. Podkletnova, NPO Lenneftekhim, 193148 Leningrad, U.S.S.R

DEHYDROGENATION The nature of the coke tolerance effect of the "second" metals in bimetallic platinum-alumina catalysts for paraffin dehydrogenation has been investigated. A method for separate estimation of geometrical and electronic effects on the basis of IR spectroscopy of adsorbed CO has been proposed.
AROMATIZATION The influence of Zn and Ga on H-ZSM-5 for C_3-C_4 paraffins aromatization selectivity has been studied. Peculiarities of the influences of each metal upon conversion chemistries have been established.

67. ZEOLITE-CONTAINING CATALYSTS FOR C_1-C_4 AROMATIZATION: CHARACTERIZATION AND GENERAL CATALYTIC FEATURES, Kh.M. Minachev, E.S. Shpiro, Zelinsky Institute of Organic Chemistry, USSR Academy of Sciences, Leninsky prospect, 47, Moscow V-334 USSR

The data on both catalytic behavior and active sites nature of multicomponent zeolite catalysts for C_1-C_4 aromatization have been reviewed. The main attention was focused on the following points: a) factors affecting activity and aromatics selectivity of modified zeolite catalysts; b) the state, distribution and formation mechanism of dehydrogenation sites (Pt, Ga, Zn, Cr, Cd, PtCr, PtGa) in pentasils and metallosilicates; c) mutual influence of metal and acid sites and synergy effects in dependence on metal dispersion, location and metal-support interaction; d) hydrogen spillover and its role to maintain high selectivity and long catalyst life; e) the role of the reaction mixture in formation of the active sites, redispersion and migration of active components over the zeolite crystal. The possible routes for preparation of highly selective catalysts preparation for aromatization of lower alkene have been considered.

68. IN-SITU PORE STRUCTURE CHARACTERIZATION DURING SOL-GEL SYNTHESIS OF CONTROLLED POROSITY MATERIALS, D.M. Smith, R. Deshpande, P.D. Majors, UNM/NSF Center for Micro-Engineered Ceramics, University of New Mexico, Albuquerque, NM, 87131, C.J. Brinker, Division 1846, Sandia National Laboratories, Albuquerque, NM, 87185, W.L. Earl, P.J. Davis, CLS-1, Los Alamos National Laboratory, Los Alamos, NM, 87545.

Using low-field NMR relaxation measurements of pore fluid, the pore volume, surface area, and pore size distribution of wet materials may be measured in-situ. In this manner, changes in pore structure may be directly observed during various aging steps (i.e., temporal, thermal, solvent exchange, pH) of sol-gel derived porous solids. By combining these relaxation measurements with Nuclear Magnetic Resonance Imaging (NMRI), the spatial variation of pore structure may be observed. This allows the study of various catalyst-related problems such as "skin" effects, and catalyst supports which have spatially-varied pore structure in order to minimize mass transfer limitations.

69. SURFACE AREA AND POROSITY CHARACTERIZATION OF CATALYSTS SUPPORTS
A.F, Venero, VTI Corporation, 15211 N.W. 60th Ave., Miami Lakes, FL. 33014

Since the publication of the BET model in 1938 and the Washburn equation for mercury porosimetry in 1921, these two methods have been used almost exclusively for the determination of surface area and pore size distribution of porous catalysts supports. Except for some minor modifications of the BET equation and a few new models designed specifically for the characterization of microporous supports, especially zeolites, these two classic methods are still in use virtually unchanged. However, several innovative techniques have been developed recently to circumvent the slowness of the sorption process used for determining the isotherms for BET analysis and the cumbersome handling procedures needed when working with mercury, two of the important defficiencies noted for these methods. Descriptions of these new techniques, such as NMR, thermoporometry, small angle X-ray scattering (SAXS), and positron annihilation are given together with their relative merits and range of validity.

70. ANALYSIS OF POROUS CARBON WITH ^{129}XE AND N_2 POROSIMETRY. H. C. Foley, Navin Bansal, David Lafyatis, Cecil Dybowski, Center for Catalytic Science and Technology, Department of Chemical Engineering, Department of Chemistry and Biochemistry, University of Delaware, Newark, DE 19716.

Porous carbons are in fact useful as adsorbent materials but also as solid supports for catalytically active metals. Transport, along with accessible surface area, into and through these materials is a crucial factor in their performance. Although the microstructures of these materials are complex, including a chaotic mixture of graphite microcrystalline domains and amorphous carbon, one usually can approximate them with models. The parameters, that derive from these models, are evaluated experimentally and can be used as quantitative, comparative measures of the important physical properties. In this case we have examined several types of activated carbon with a combination of ^{129}Xe nuclear magnetic resonance spectroscopy and nitrogen porosimetry. Using the two techniques in tandem allows us to provide more quantitative information about microstructure than either technique alone.

71. SURFACE CHARACTERIZATION OF TITANIA-SILICA-BASED DENO$_x$ CATALYSTS, J.P. Solar, P. Basu, and M.P. Shatlock, W.R. Grace and Co.-Conn., Research Division,7379 Route 32, Columbia, MD 21044.

SYNOX™ titania-silica-based catalysts have been developed for the selective catalytic reduction (SCR) of nitrogen oxides. These new catalysts have been demonstrated to provide a 50% increase in overall catalytic activity. The improved performance of SYNOX™ is related to a reduction in diffusional resistance within the catalyst walls, while surface chemistry plays a critical role in intrinsic activity. This paper describes our work on the surface characterization of the SYNOX™ titania-silica supports as well as the catalysts formed by impregnation of V_2O_5. XRD, Laser Raman and ^{51}V NMR spectroscopies have been utilized to study the structure of the titanium and vanadium phases on the silica support. The surface chemistry has been studied using XPS and FTIR. The presence of Lewis and Brønsted acid sites was determined using pyridine adsorption. The properties of these catalysts will be compared to the widely studied vanadia/titania catalysts.

72. THE CHARACTERIZATION OF MODEL ALUMINA AND ITS COMPARISON WITH OTHER ALUMINAS. S. Srinivasan, A. K. Datye, Department of Chemical & Nuclear Engineering, University of New Mexico, Albuquerque, NM 87131

Although alumina polymorphs are widely used as catalysts the understanding of their structures is still limited. They are generally assumed to have a close packed oxygen sublattice with Al^{3+} located in the octahedral and tetrahedral interstices. Defects in the lattice and the small size of the crystalline domains lead to broad XRD peaks that complicate structure analysis. In order to improve the characterization of transitional alumina, we have studied model alumina powders grown by oxidation of aluminum metal. The resulting nonporous powders have been studied by high resolution TEM, XRD, solid state NMR and their surface chemistry probed by IR spectroscopy and reactivity measurements. Preliminary results from this study will presented including a comparison with high surface area γ alumina.

73. Electron Energy Loss Spectrometry of Alumina and Silica Catalyst Supports.
R.E. Lakis, C.E. Lyman, and H.G. Stenger, Dept of Materials Science & Engr and Chemical Engineering, Lehigh University, Bethlehem, PA 18015

Analytical electron microscopy (AEM) is a versatile technique well-suited for the characterization of catalyst systems. AEM may be thought of as a microchemical analysis laboratory where the structure of the specimen can be determined by electron diffraction and the elements present can be determined by either x-ray emission spectrometry (XES) or electron energy loss spectrometry (EELS). The demand for high lateral spatial resolution in both XES and EELS has led to the use of very high electron current densities on the specimen. This raises questions concerning the physical and chemical stability of the various catalyst components, including the support, during observations in AEM.

Electron beam damage to the specimen during analysis may take the form of crystallinity loss by radiolysis, knock-on damage to light elements, or mass loss into the microscope vacuum by electron stimulated desorption and sputtering. Time-resolved parallel EELS provides the ability to monitor mass loss and phase transitions with sub-second time resolution. Results indicate that in a conventional transmission electron microscope (TEM) with a thermionic electron source little damage is apparent at 300kV.

74. COMPETITIVE ROLE OF CATIONS IN ZEOLITE CRYSTALLIZATION. Prabir K. Dutta and Reza Asiaie, The Ohio State University, Chemistry Deptartment, 120 W. 18th Ave. Columbus, OH 43210.

Cations play a significant role in nucleation and crystallization of zeolites. Their specificity for a particular framework is well recognized e.g. Na^+ for faujasite type, K^+ for chabazite type and TPA^+ for pentasil zeolites. In this study, we have focused on the competitive role of cations in zeolite synthesis by focusing on multi-cation systems. Zeolite Na-Y has been the system of choice as far as reactant composition is concerned. We find that with Li^+ as the sole cation in this composition system, a zeolite resembling Li-ABW is formed. Presence of both Na^+ and Li^+ lead to effective destruction of nuclei of both Na-Y and Li-ABW, resulting in an amorphous solid. Perturbations of this system at low levels of K^+, Cs^+, TMA^+ lead to stabilization of Li-ABW, whereas the presence of crown ether leads to formation of zeolite Y. At high levels of K^+ and TMA^+, zeolite L and mazzite are formed. Polar organic solvents also lead to stabilization of Li-ABW. The competitive role of these cations and their influence on mechanism of zeolite growth will be discussed.

75. ORIENTING EFFECT OF FLUORIDE ANIONS IN THE SYNTHESIS OF A NEW GALLOPHOSPHATE OF THE LTA TYPE. A. Merrouche, J. Patarin, M. Soulard, H.Kessler and D. Anglerot*, Ecole Nationale Supérieure de Chimie, 68093 Mulhouse Cedex, France, *Groupe Elf-Aquitaine, GRL, Lacq, BP 34, 64170 Artix, France.

Several microporous gallophosphates have been reported in the litterature. The addition of fluoride anions to the synthesis mixture enabled us to prepare selectively new structure types, especially the LTA type. The latter was obtained in the presence of di-n-propylamine (DPA) as a template and F^- anions, whereas in the absence of fluoride UCC $GaPO_4$-a was formed. The substitution of Ga and/or P in the LTA material by Al or Si was possible. A variety of techniques such as thermal analysis, electron microprobe analysis, multinuclear solid state NMR... were used to characterize the new materials. The crystal structure of the as-synthesized LTA $GaPO_4$ was determined by the Rietveld procedure. DPA and F^- were localized ; the latter found in the D4R of the LTA structure play presumably a co-structuring role beside their mineralizing effect.

76.
GALLOSILICATE : A NOVEL DERIVATIVE OF EUO FRAMEWORK ZEOLITES.
G.N. Rao, V.P. Shiralkar, A.N.Kotasthane and P.Ratnasamy,
Catalysis Group, National Chemical Laboratory, Pune 411008, India.

A novel crystalline gallium(III) analog of EUO framework zeolites have been synthesised from Al-free hydrogel systems in the presence of dibenzyldimethylammonium or hexamethonium cations. Evidence for the presence of Ga(III) in the lattice framework has been obtained utilising XRD, framework-ir, 29-Silicon and 71-Gallium MASNMR techniques. The insertion of the larger gallium atoms in the EUO framework increases the unit cell volume. The material has a significant ion exchange capacity. With increasing Ga-content the framework (T-O) asymmetric stretching frequencies shift to lower wavenumbers. The MASNMR spectrum of 71-Gallium also suggests the presence of tetrahedrally coordinated Ga(III) in the EUO framework zeolites.

77. ON THE VARIOUS STRUCTURAL STATES OF GALLIUM IN AS-SYNTHESIZED AND MODIFIED MFI-GALLOSILICATES. Z.Gabelica, C.Mayenez, R.Monque, R.Galiasso and G.Giannetto, Dept. of Chemistry, University of Namur, 61, Rue de Bruxelles, B-5000 Namur, Belgium, and INTEVEP S.A., Los Teques, Venezuela.

Various synthesis methods leading to pure MFI-type gallosilicates or aluminogallosilicates have been tested in terms of the effiency of Ga or Al framework incorporation. In each case, the amount of tetrahedrally coordinated Ga could be quantitatively measured by ^{71}Ga-MAS NMR. Extra framework Ga species that contribute to bulk catalytic properties of gallosilicate zeolites can be generated by exchanging the corresponding Al-MFI with Ga^{3+} salts, by various impregnation-calcination procedures or by a controlled heating of the as-synthesized gallosilicates. In this latter case, the progressive partial or total removal of framework Ga can be monitored by carefully controlling the heating conditions : temperature, time, nature of the flowing atmosphere and presence of steam, and its extent measured by combining ^{71}Ga NMR and chemical analysis. Such extra framework Ga species, probably located within the zeolite channels or macropores, could not be detected by ^{71}Ga NMR, even after various regenerating treatments.

78. SYNTHESIS OF ETS-10, A NEW WIDE-PORE TITANIUM SILICATE MOLECULAR SIEVE, S.M. Kuznicki[1], D.T. Hayhurst[2], [1]Specialty Chemicals Div., Engelhard Corp. Iselin, NJ 08830, [2]Dept. of Chemical Engineering, Cleveland State University, Cleveland OH 44115

ETS-10 is a new microporous crystalline solid with uniform 8A pores similar in dimension to classical large-pore zeolites. Like alumino-silicate zeolites, this new titanium-silicate is synthesized under hydrothermal conditions in an alkaline environment. Based on compositional analyses, it is concluded that the titanium is octahedrally coordinated in the framework. This requires that two counter-balancing cations are found in the channels per titanium. These cations are readily exchanged by an alkaline, alkaline-earth and multivalent metal cations to a level exceeding 4 meq/g. ETS-10 reversibly adsorbs a wide variety of species, taking in molecules up to 8A in diameter. Adsorption capacities are found to exceed 20cc adsorbate per 100g of dry sieve. Neither adsorption nor ion-exchange appreciably distort the crystalline lattice. High resolution transmission electron microscopy indicated uniform three-dimensional pore structure. This synthesis represents the discovery of a new family of molecular sieves.

79. ADSORPTIVE PROPERTIES OF ETS-10, A NEW LARGE-PORE TITANIUM SILICATE, D.T. Hayhurst[1], M. Mansour[1], S.M. Kuznicki[2], [1]Department of Chemical Engineering, Cleveland State University, Cleveland, OH 44115, [2]Specialty Chemicals Div., Engelhard Corp., Menlo Park, CN28, Edison, NJ 08818

Adsorptive properties are reported for several ion-exchange forms of ETS-10, a unique new large-pore molecular sieve titanium silicate constructed from non-traditional primary building units. ETS-10 is formed from tetrahedral silica linked to octahedral titanium forming a porous large-pore three-dimensional structure. This sieve contains two charge-balancing exchangeable cations per octahedral titanium. ETS-10 was exchanged with $Na^+, K^+, Mg^{+2}, Ca^{+2}$, Sr^{+2}, and Ba^{+2}. Adsorptions capacities were measured at 295K for water, benzene, o-xylene and triethylamine. With the exception of the K-form, all ETS-10 samples adsorbed large amounts of these probe molecules, in many cases capacities exceed 20cc/100g. Pore volumes were determined at 77K using oxygen as the adsorbate. Adsorption isotherms were also measured for N_2, O_2 and Ar at 295K and pressure up to 6 MPa. Nitrogen adsorption for Ca-ETS-10 was found to be similar to CaA, however, the adsorption isotherms for water and hydrocarbons show very different trends than those observed for classical zeolites. Although structurally quite different, ETS-10 demonstrates molecular sieving properties that are similar to traditional large-pore zeolites. Adsorbate-adsorbent interactions however show substantial differences when compared to alumino-silicate zeolites.

80. **MICROWAVE HEATING OF ZEOLITE SYNTHESIS MIXTURES.** J.C. Jansen, A. Arafat and H. van Bekkum, Laboratory of Organic Chemistry, Delft University of Technology, Julianalaan 136, 2628 BL Delft, The Netherlands.

Zeolite syntheses have been performed using i) an ordinary household microwave oven (1000 W) containing a 500 ml Teflon autoclave including a temperature controlling unit and a safety relief valve to prevent over-pressurization; (ii) a special purpose microwave oven (2500 W) with a glassfiber reenforced polyetheretherketon (PEEK) autoclave of 5 ml for high temperatures and pressures. Details on both set-ups will be presented.

Low Si/Al-zeolites like Na-A and hydroxysodalite were prepared within ten minutes using high heating rates. High Si/Al-zeolite syntheses mixtures were subjected to microwave radiation.

The influence of very high heating rates on the unexpected nucleation of other zeolite phases in syntheses mixtures will be discussed.

81. EXAFS Studies of Isomorphously Substituted Zeolites with the ZSM-5 Structure

Sean A. Axon,[1] Katharine K. Fox,[2] Stuart W. Carr[2] and Jacek Klinowski[1]

(1) Department of Chemistry, University of Cambridge, Lensfield Road, Cambridge CB2 1EW, U.K. (2) Unilever Research, Port Sunlight Laboratory, Quarry Road East, Bebington, Wirral, Merseyside, L63 3JW, U.K.

Extended X-ray absorption fine structure (EXAFS) used in combination with Mössbauer spectroscopy and X-ray diffraction confirms the substitution of Si by Ga and Fe in the frameworks of zeolites ZSM-5 prepared via the "fluoride route". On the other hand, Co, Mn and several other elements occupy extra-framework positions. In [Si,Fe]-ZSM-5 each framework Fe atom is coordinated to four oxygens with an Fe-O distance of 1.83Å, and to four Si atoms in the second coordination shell: two at 3.2Å and two at 3.38Å from Fe, compared with an average Si-Si distance in silicalite of 3.1Å. For the first time, the distances between the framework Fe atom and some of the atoms in the template molecule could be determined. Thus the distance from Fe to the nitrogen of TPA+ is 3.24Å. Of the carbons in the propyl chains two are 2.87Å and one 2.69Å, two at 3.71Å and one at 3.94Å from Fe.

82. **ALKYLATION OF ISOBUTANE WITH PENTENES USING SULFURIC ACID AS A CATALYST: CHEMISTRY AND REACTION MECHANISMS**, <u>L.F. Albright</u>, Purdue University, West Lafayette, IN 47907, K.E. Kranz, STRATCO, Inc., 4601 College Blvd., Leawood, KA 66211

Isobutane was alkylated at conventional conditions with sulfuric acid as the catalyst using pure pentenes, mixtures of pentenes and butenes, and pure butenes. Alkylates produced from normal pentenes and isopentenes had research octane numbers of about 90.5 to 91.7. These values are significantly lower than those of alkylates produced from isobutylene and especially normal butenes. Compositions of the alkylates produced from 1-pentene, 2-pentenes, and 2-methyl-2-butene (2MB2) were greatly different. Alkylates from 2-pentenes contained 50-60% C_9 isoparaffins, but alkylates from 2MB2 contained only 11-13%. The results of this investigation clarify the alkylation mechanisms of C_5 olefins. These mechanisms show some similarities but also major differences as compared with C_4 olefins. These investigations also show that cyclopentene is a highly undesired olefin for alkylation.

83. FREE RADICAL ALKYLATION IN THE PRESENCE OF H_2S AS A HYDROGEN TRANSFER CATALYST, D.G. Peferoen, K.P. de Jong, Koninklijke/Shell-Laboratorium, Amsterdam (Shell Research B.V.), Amsterdam, The Netherlands

Increasingly stringent legislative, environmental and performance constraints will more and more dictate the gasoline composition of the future. Of all hydrocarbons, light, highly branched alkanes are among the most desirable motor gasoline blending components in view of their high octane, low density, low volatility and cleanliness. In addition to the present prime numbers routes to light iso-alkanes, i.e. acid-catalyzed alkylation and isomerization, highly branched alkanes can also be produced by free radical alkylation. The thermal alkylation of isobutane with ethene, propene and isobutene has been studied in the presence of H_2S acting as a homogeneous gas phase catalyst. It was found that the addition of small amounts of H_2S, catalyzing the transfer of hydrogen between hydrocarbon reactant and free radical, improved not only the overall reaction rate but also and especially the selectivity towards the primary alkylation products. The positive effect of H_2S was most pronounced when ethene was used as the alkylating agent, improving the selectivity towards 2,2-dimethylbutane by a factor of more than 5 in certain cases. With propene and isobutene the side reactions to oligomers in the case of propene and 1,1,3-trimethylcyclopentane in the case of isobutene, could only be suppressed to a limited extent by the addition of H_2S.

84. SULFURIC ACID CATALYZED ALKYLATION OF PENTENES AND OTHER LIGHT OLEFINS, K.E. Kranz, K.R. Masters, STRATCO, Inc., 4601 College Blvd. Suite 300, Leawood, KA 66211

Recent requirements on gasoline vapor pressure and interest in reformulated gasoline have prompted refineries to consider including pentenes (amylenes) in their alkylation feedstock. Published data for estimating alkylate quality and sulfuric acid consumption is limited. Extensive laboratory investigations have been conducted with mixtures of pentenes and butenes as well as pure pentenes only. The effect of operating conditions and feedstock compositions on alkylate quality and sulfuric acid consumption was determined from this investigation. This paper also describes the chemical mechanisms of the different pentene isomers as well as integrating pentene alkylation into a commercial alkylation process.

85. THE EFFECT OF SULFURIC ACID COMPOSITION ON THE KINETICS AND MECHANISM OF SULFURIC ACID-CATALYZED ALKYLATIONS OF ISOBUTANE WITH ALKENES. David J. amEnde and Lyle F. Albright, School of Chemical Engineering, Purdue University, West Lafayette, IN 47907-1283.

For the alkylation of isobutane with olefins, a different optimum sulfuric acid composition exists for the production of maximum quality alkylate or for highest rates of alkylate formation. Furthermore, different optimum acid compositions were found for various C_3–C_5 olefins. Information obtained in the present investigation clarifies the mechanism for the alkylation of these olefins. The effects of water content, conjunct polymer content, acidity, and additives will be discussed relative to the chemistry of the reaction and to the interfacial area where most of the reactions occur. The additives used exhibited both surfactant and hydride transfer characteristics.

86. PRIMARY FORMATION AND SECONDARY TRANSFORMATION OF P-ETHYLTOLUENE DURING THE TOLUENE ETHYLATION ON MODIFIED ZSM-5, F. Lonyi, J. Engelhardt, D. Kallo, Central Research Institute for Chemistry. Hungarian Academy of Sciences, Budapest, Hungary.

Ethylation of toluene was studied on ZSM-5 in an excess of toluene at atmospheric pressure using hydrogen as diluent in order to avoid coke formation. Space time, reaction temperature, catalyst pretreatment temperature, catalyst aging and poisoning with organic bases, introduction of inorganic additives containing phosphorus and/or magnesium into the catalyst were varied. The change of yields and selectivities points to the primary formation of p-ethyltoluene. Secondary reactions are the isomerization to m-ethyltoluene and dealkylation of ethyltoluenes. Disproportionation of toluene is a simultaneous transformation. The para-selectivity increased when the side reactions were suppressed by the reduction of Bronsted acidity.

87. REACTION KINETICS OF PROPANE DEHYDROGENATION OVER Zn OR Mo SUPPORTED ON SILICALITE, B-Z. Wan, C.C. Wang and H-M. Chu, Department of Chemical Engineering, National Taiwan Univ., Taipei, Taiwan, R.O.C. 10764

By using differential reactor, the reaction kinetics of propane dehydrogenation over Zn or Mo on silicalite were studied. it was found that the rate order with respect to propane was 1.0 on Zn. However, the rate dependence was divided into two regions on Mo catalyst: when propane partial pressure was lower than 0.25 atm, the rate order was 1.0; when propane partial pressure was larger than 0.25 atm, the dehydrogenation rate was increased dramatically with the increase of propane concentration then converged to a first order rate order relationship. Furthermore, it was found that propane dehydrogenation rate decreased with the increase of hydrogen concentration on Zn. Nevertheless, it increased with the increase of hydrogen concentration on Mo. Based on these observations, the reaction kinetics models of propane dehydrogenation over Zn or Mo were proposed in this study. The attempt was made to explain how the reaction rates were changed with the concentration of propane and hydrogen.

88. ROLE OF SPRAY DRYING IN PRODUCTION OF CATALYSTS AND CATALYST SUPPORTS. F.V. Shaw, Niro Atomizer Inc. 9165 Rumsey Road, Columbia, Maryland 21045.

Spray drying is the most direct route for conversion of a slurry or solution to a dry flowable powder. In the production of catalysts or catalyst support materials, the particle size and shape in the resulting powder is critical to the performance of the final catalyst. Manipulation of spray drying process parameters gives the producer control of the powder particle characteristics.

89. CARRIER AND CATALYST FORMING PROCESSES -- OR, HOW WE GO FROM THE BENCH TOP TO COMMERCIAL REALITY, T. Szymanski, Norton Company, Chemical Process Products, P. O. Box 350, Akron, Ohio   44309

Many products which enhance the quality of our modern life directly or indirectly owe their existence to heterogeneous catalysis. Transferring this technology from the researcher's bench to commercial reality can follow a long and torturous path. This paper will introduce the various processes used to prepare catalyst carriers and supports. We will compare and contrast these processes highlighting the advantages and disadvantages of each technology with special emphasis on scale up concerns. We will conclude by discussing an imagineered scale up program.

90.   EXTRUDED MONOLITIC CATALYST SUPPORTS
              Irwin M. Lachman and Jimmie L. Williams
        Corning, Inc., Sullivan Park, DV-01-9, Corning, N.Y.   14831

Extruded monolithic catalyst substrates are widely used in automotive and stationary emissions control reactors. They are increasingly being used, developed, and evaluated in many other reactor applications such as for the chemical processing industries, for catalytic combustion, for ozone abatement, and others.
This paper will compare the variety of extruded monoliths used for these applications and will review the forming and characterization of them. Emphasis is on the materials used for automotive substrates, catalysts for selective catalytic reduction of NO_x and for new high surface area substrates.

91. FORMING OF HIGH SURFACE AREA TiO_2 TO CATALYST SUPPORTS. M. Bankmann, R. Brand, B.H. Engler and J. Ohmer, Degussa AG, R & D Chemical Catalyst Dept., Zweigniederlassung Wolfgang, Postfach 1345, D-6450 Hanau 1, Germany

The attention which has been payed to the application of TiO_2 as a catalyst support has been increasing over the past several years. High surface area titanium dioxides are synthesized by flame hydrolysis or precipitation of $TiCl_4$ or other precursers. The high purity is characteristic of the flame hydrolytic produced TiO_2. Formed catalyst supports can be manufactured by extrusion processes using raw materials of both high purity and conventional qualities. Four main steps are normally part of an extrusion process: Kneading, extrusion, drying and calcination. Typical process parameters of each production step are varied. Intermediates and final products are characterized. The influence of different process conditions and additives on catalytically important parameters such as internal surface area, pore volume and pore size distribution is discussed. It is shown how crushing strength and attrition resistance can be varied. Special attention is given to the TiO_2 phase modification in the formed product. It can be varied from mainly anatase via different anatase -rutile-ratios to pure rutile. Using formed TiO_2 catalyst supports, precious metal catalysts show new catalytic properties in selective hydrogenation.

92. PHYSICAL CHARACTERIZATION OF TiO_2 AND Al_2O_3 PREPARED BY PRECIPITATION AND SOL-GEL METHODS.
Rodriguez Oscar(1), Gonzalez Felipe(1), Bosch Pedro(2), Portilla Margarita(3), Viveros Tomas (1)
(1)Area de Ing. Quimica, UAM-Iztapalapa, Apdo. Postal 55-534, Mexico 09340 D.F., (2)Depto. Quimica, UAM-Iztapalapa, Apdo. Postal 55-534, Mexico 09340, D.F., (3)Fac. Quimica UNAM, Mexico 04510 D.F.

Alumina and titania supports were prepared by precipitation of the corresponding chlorides and by the sol-gel method using aluminum tri-secbutoxide and titanium isopropoxide. In the sol-gel method samples were prepared varying H_2O/M ratio H^+/M ratio (M= Al or Ti) and calcination temperature. We report the results on the characterization by differential thermal analysis, thermogravimetric analysis, X-ray diffraction, BET surface areas and pore size distribution. The results obtained using both preparation methods are compared and discussed.

93. INVESTIGATION OF THE CRYSTALLIZATION OF VERY LARGE PORE ALUMINOPHOSPHATE PHASES AND THEIR PHYSICO-CHEMICAL CHARACTERIZATION: E. Jahn, D. Müller, J. Richter-Mendau, Zentralinstitut für physikalische Chemie Berlin, Rudower Chaussee 5, Berlin, 1199

On the basis of the investigation of various synthesis procedures it is concluded that the preparation of very large pore aluminophosphate molecular sieves involves techniques known for pillaring of layered solids. In the case of VPI-5 this means a) the partial transformation of the starting pseudoboehmite into a layered aluminophosphate phase, b) the expansion of the latter material by substances containing amine and phosphate molecules and c) the crystallization of the expanded product. The use of compounds which contain P-O-P linkages instead of monophosphoric acid as P source improves the conditions for occuring of steps a) and b). In the present work the sequence of the different synthesis parts is characterized by n.m.r. spectroscopy and X-ray diffraction. The suggestion to apply the model of pillaring layered substances to the preparation of very large pore aluminophosphate phases is supported by a plate-like morphology of VPI-5 crystals.

94. A SOLID STATE NMR STUDY OF 18-RING LARGE PORE ALUMINOPHOSPHATE MOLECULAR SIEVES, J.O. Perez, P.-J Chu and A. Clearfield, Department of Chemistry, Texas A&M University, College Station, Texas 77843

A magic angle spinning solid state NMR study of 18-ring aluminum phosphates has been carried out as a function of the hydration level of the solid phases. Two samples of aluminum phosphate were considered. One has a one to one ratio of P to Al and is thermally stable while the other has a deficiency of P and readily converts to $AlPO_4$-8. The two phases are designated as VPI-5 and H1, respectively. Both solids yield a ^{31}P spectrum, when completely dry, which consists of three resonances. These spectra cannot arise from the 18-ring VPI-5 framework of alternating 4-coordinate Al and P. It is suggested that in the dry state the P-O-Al bond in every other 4-ring is broken to produce a P=O site and 3-coordinate aluminum. At low water contents six resonances are observed and these eventually transform to the familiar three peak 1:1:1 pattern. The changes observed are described on the basis of a healing of the broken bond and hydration of Al in the 4-ring which transforms to 6-coordination of hydration of the aluminum atoms in alternate 6-membered rings at high levels of water content.

95. IN SITU MONITORING OF THE DEGREE OF TRANSFORMATION OF VPI-5 TO AlPO$_4$-8 AT VARIOUS TEMPERATURES BY MULTINUCLEAR SOLID STATE NMR AND DTA Z.Gabelica, L.Maistriau and E.G.Derouane, Dept. of Chemistry, University of Namur, 61, Rue de Bruxelles, B-5000 Namur, Belgium.

Combined DTA with ^{31}P- and ^{27}Al-MAS NMR proved choice techniques to follow the in situ solid state transformation of VPI-5 to AlPO$_4$-8 either under progressive heating or isothermally, at different temperatures. The transformation itself is always proceeded by a slight structural perturbation of the VPI-5 stucture due to the destruction of the water triple helix configuration in the channels. The resulting unit cell symmetry increase is detected by a slight DTA exotherm and by characteristic changes of the ^{27}Al- and ^{31}P-NMR lines. This first transition is completely reversible and essentially depends on the actual amount of water left in the structure rather than on temperature. Variable temperature ^{31}P-NMR experiments confirm that VPI-5 further readily and irreversibly transforms to AlPO$_4$-8 when most of the structural water is rapidly removed at or below 100°C. By contrast, VPI-5 can remain stable at temperatures higher than 120°C for long periods of time provided the structural water is not completely removed.

96
Solid-State ^{27}Al NMR Studies of Hydrated AlPO$_4$-8

João Rocha and Jacek Klinowski

Department of Chemistry, University of Cambridge, Lensfield Road, Cambridge CB2 1EW, U.K.

Hydrated AlPO$_4$-8 obtained by calcination of DPA-VPI-5 has been studied by fast ^{27}Al magic-angle spinning (MAS) NMR at 4.7 and 9.4 T, by ^{1}H-^{27}Al CP/MAS NMR and quadrupole nutation MAS NMR. A careful inspection of the individual rows of the nutation spectrum allows the different ^{27}Al signals to be assigned. The results suggest the presence in AlPO$_4$-8 of four or possibly five crystallographically inequivalent Al sites, and are compared with those of double-rotation ^{27}Al NMR of VPI-5 and AlPO$_4$-8.

97. PAPER WITHDRAWN

98. SYNTHESIS OF VPI-7: A NOVEL 3-MEMBERED RING CONTAINING MOLECULAR SIEVE, <u>Michael J. Annen</u> and Mark E. Davis, Department of Chemical Engineering, Virginia Polytechnic Institute and State University, Blacksburg, VA 24061

VPI-7 is the first microporous zincosilicate molecular sieve to contain 3-membered rings (3MR). The framework of VPI-7 contains 3, 4 and 5 T-atom rings which form unidimensional 8- and intersecting 9MR channels. It possesses tetragonal symmetry with unit-cell dimensions of $a=7.179(1)$Å and $c=40.62(1)$Å. VPI-7 can be synthesized using a range of reaction mixtures including a gel of composition: 0.44 Na_2O: 0.15 K_2O: 0.04 TEA_2O: 0.039 ZnO: SiO_2: 22 H_2O. The addition of TEAOH (tetraethylammonium hydroxide) serves only to enhance the rate of crystallization. Thus, unlike the other 3MR-containing molecular sieves, lovdarite and ZSM-18, VPI-7 can be synthesized in the absence of potentially toxic materials or highly specific organic structure directing agents. We report here the various synthetic procedures used to crystallize VPI-7 and physicochemical characterization data from the resulting materials.

99. ISOMORPHOUS SUBSTITUTIONS OF SILICON AND COBALT IN MOLECULAR SIEVE $AlPO_4$-5. <u>K.J. Chao</u>, S.P. Hsu, S.H. Chen, J.C. Lin, J. Lievens and Y. Wang, Department of Chemistry, Tsinghua University, Hsinchu, Taiwan.

The substitutions of Si(IV) for P(V) and Co(II) for Al(III) in $AlPO_4$ molecular sieves result in anionic tetrahedral frameworks coupled with exchangeable cations and Brønsted acid sites. The syntheses and characterizations of silicoaluminophosphate SAPO-5 and Co-aluminophosphate CoAPO-5 are reported. The CoAPO-5 crystals were accompanied by side products (depending upon the reactant composition used; CoAPO-34, $AlPO_4$-5 and amorphous oxides). The incorporation of Si in the framework of SAPO-5 was enriched using the monomeric silicon reactant instead of colloidal silica and confirmed by ^{29}Si NMR and Brønsted acidity measurements. The cation sites were found to be close to the oxygen (O3) of the four-membered ring of the tetrahedral framework in the dehydrated Cs form of SAPO-5 by using Rietveld analysis of powder XRD pattern. The nature of 12 membered ring channels of SAPO-5 and CoAPO-5 were compared with that of $AlPO_4$-5 using ^{129}Xe NMR and xenon adsorption isotherm.

100. PARA SELECTIVE ALKYLATIION OF MONO ALKYLBENZENES OVER SILYLATED GALLOSILICATE ZEOLITES. <u>A.B. Halgeri</u>, Y.S. Bhat, S. Unnikrishnan and T.S.R. Prasada Rao, Research Centre, Indian Petrochemicals Corporation Ltd., Baroda - 391 346, INDIA

The shape selective alkylation of mono alkylbenzene is of industrial importance owing to the formation of very useful paradialkylbenzenes. One of the recently emerged techniques to achieve a high selectivity (95-99%) of zeolite is by Chemical Vapour Deposition of silica, which involves blocking of nonselective external surface and pore mouth sites, and fine controlling of pore opening. The present work deals with a comprehensive study on enhancement of paraselectivity features of gallosilicate by silylation. The study also encompasses the silylation period required to achieve a desired para selectivity and zeolite crystal size, the relation between reaction conditions and para selectivity for various alkylation reactions such as; methylation of toluene, ethylation of toluene and ethylation of ethylbenzene, and the involved reaction mechanisms.

101. ALKYLATION OF ISOPARAFFINS WITH OLEFINS OVER ZEOLITE CATALYSTS, S.N. Khadzhiev, <u>I.M. Gerzeliev</u>, Groz NII, Grozny, 364913, USSR.

The most desirable catalyst for alkylation of isobutane with olefins is exchanged Y zeolite. There were developed effective ways of introduction of Ca^{+2} and RE^{+3} cations into S1 of Y zeolite crystalline structure by ion-exchange at high temperature without intermediate calcinations. With ethylene not more than 20%

of direct alkylation products is formed. Propylene produces substantial quantities of desired fraction (isoheptane), but it is characterized by yielding abnormal reaction products (isodecane fractions). Butene-1 gives substantial yields of desired isooctanes, as for abnormal reaction products (i-C_5-, i-C_7 and i-C_9+) they are distributed uniformly. On the basis of detailed studies of reactions of alkylation of isoparaffins and aromatic hydrocarbons with olefins, considerations on mechanism of the processes are suggested.

102. SULFATE-CONTAINING SOLID SUPERACIDS FOR ACID-CATALIZED HYDROCARBON CONVERSIONS. I. Rodríguez-Ramos, A. Guerrero-Ruiz and J.L.G. Fierro, Instituto de Catálisis y Petroleoquímica, CSIC, Serrano 119, 28006 Madrid, Spain.

Sulfate- promoted metal oxides of the type Fe_2O_3, ZrO_2, etc. have been reported to display unusual superacid properties in several isomerization, cycling and dehydration reactions. As the rate of these reactions is expected to be proportional to the number of surface acid sites, supporting the active ingredient seems to be an attractive way to explore catalyst performances. Therefore, active carbon and alumina-supported Fe_2O_3 and ZrO_2 oxides, to which ammonium sulfate was added, have been prepared and tested in isomerization of 1-butene, isobutene alkylation with methanol to yield MTBE and n-butane isomerization. From this study we can conclude: (i) S-ZrO_2 systems show larger acidity than its S-Fe_2O_3 counterpart, (ii) MTBE and isomerization of 1-butene reactions occur on S-ZrO_2 catalysts, but n-butane isomerization is highly impeded, (iii) as revealed by XPS, sulfide poisoning, through reduction of sulfate by H_2 in the fed, and coking are the principal catalyst deactivation sources.

103. TOLUENE DISPROPORTIONATION REACTION OVER ZEOLITES: EFFECT OF Si/Al RATIO AND COKE FORMATION, K. Ej-Jennane, C. Marcilly, C. Travers, F. Raatz, Institut Francais du Petrole, 1-4 Avenue du Bois-Preau, BP 311, 92506 Rueil-Malmaison

The toluene disproportionation reaction has been studied over a series of H-Mordenites and H-ZSM-5 with different framework Al content. The reaction was carried out under operating temperature and pressure close to industrial conditions. It appeared that the reaction rate and the apparent turnover frequency increase over H-Mordenites with dealumination. On H-ZSM-5, the reaction rate decreased with Si/Al ratio and the apparent turnover frequency is low and independent of Al content. Two explanations involving diffusion phenomena and catalyst acidity are considered.

104. A NEW CYCLOHEXANOL PROCESS VIA CHYLOHEXENE FROM BENZENE, Y. Fukuoka, H. Nagahara, Chemical Development Laboratory, Asahi Chemical Industry Co., Ltd., Kurashiki-shi 711, Japan.

Cyclohexanol is an important raw material of Nylon. The route of cyclohexanol via cyclohexene from benzene is simpler, so this route has been investigated for many years. However, the elementary processes of hydrogenation and hydration have proven more difficult than expected, all of proposed patents are far from commercially usable because of the STY, single-pass conversion rate, and yield. We have developed highly selective cyclohexanol process via cyclohexene from benzene, (cyclohexene selectivity is 80% at benzene conversion of more than 60% in a batch reaction, cyclohexanol selectivity is >99% at cyclohexene conversion of 10-15%). We will talk about 1) new catalysts of both reactions, 2) some factors that affect selectivity, 3) pilot studies carried out by using continuous flow reactors with recycling systems, 4) some advantages of this process. Asahi Chemical built a 60,000 tons/year plant using this method in August of 1990.

105. PROPERTIES OF ALKANE ISOMERIZATION PRODUCTS, C.L. Li, Z.L. Zhu, L. Shi, J.G. Cao, Research Institute of Petroleum Processing, East China University of Chem. Tech., Shanghai 200237, P.R. China.

There are five isomers for hexane. Dimethyl butanes are preferred owning to their much higher octane number. Kinetic results could be used to compare selectivities for those isomers. Data on alkane hydrosomerization over metal loaded zeolite catalysts were obtained in a pressurized microreactor. Rate constants were calculated by an improved eigenvector method. Experimental results under different temperatures could thus be well predicted. This method were also used to estimate selectivities of gaseous products in pentane isomerization reactions. Product prediction for n-pentane isomerization was also given. This information was useful in catalyst preparation and process optimation for alkane isomerization.

106. PILLARING OF LAYERED INORGANIC COMPOUNDS. A. Clearfield, R.-C. Wang, J.D. Wang, M. Kuchenmeister, and K. Wade, Department of Chemistry, Texas A&M University, College Station, Texas 77843.

Several types of layered non-clay classes of compounds have been pillared. These include layered group 4 and 14 phosphates, titanates, layered perovskites and layered double hydroxides. The first three classes of com,pounds are cation exchangers and do not swell in water. Therefore, the interlayer distances are enlarged by amine intercalation followed by exchange of the amine by the pillaring cation. The pore sizes and temperature stabilities will be discussed as well as ion uptakes.

Layered double hydroxides have been pillared in both aqueous and non-aqueous media. In some cases lauryl sulfate is first intercalated and then exchanged out by Keggin ions. In other cases direct pillaring was obtained. These differences will be discussed in relationship to the layered double hydroxide structures.

107. INTERCALATING PROPERTIES OF LAYERED METAL(IV) PHOSPHATES
TOWARDS ORGANIC BASES, METAL COMPLEXES AND METAL OXIDES.
C. Ferragina, A. Frezza, A. La Ginestra, M.A. Massucci, G. Mattogno and P. Patrono.

Metal(IV) (Zr, Ti, Sn, etc.) hydrogenphosphates are able to intercalate between their non-rigid layers organic bases (amines, aminoacids, aminoalcohols, alcohols, etc.). Transition metal ions can be coordinated to these ligands forming "in situ" complex-pillars.

The properties of the various prepared materials depend on the crystal structure of the guest, on the characteristics of the host (basicity, steric hindrance, hydrphobicity, etc) and the complexes formed depend on their stability constant. The insertion of inorganic oxides (Zr,Al,Ti) has been made possible in different ways and inorganic pillared layered compounds can be prepared.

108. DIRECT SYNTHESIS OF NOVEL INTERCALATED LAYER SILICATES.
W. Schwieger and D. Freude, Universität Halle, Schloßberg 2, O-4020 Halle, and Universität Leipzig, Linnéstraße 5, O-7010 Leipzig, Germany

A family of novel layered silicates of metal silicate hydrate type has been synthesized from the quarternary system $Na_2O - SiO_2 - H_2O$ - polymeric organic cation. The polymeric organic cation was intercalated into the silicate during crystallization. The structure as given by X-ray diffraction pattern and the composition of the new silicate do not fit in with any known layered silicates. The novel silicate is distinguished by a high stability against a wide range of structure breaking effects. The multi layer structure of the silicate was studied by electron microscopy (HERM). A model of the sheet structure is given. The crystallization is studied by means of physical and chemical methods including NMR.

109. TEM ANALYSIS OF PILLARED RECTORITES. J. M. Dominguez[1] and M. L. Occelli[2]. [1]IMP, Eje Central Lazaro Cardenas N152, 07730 Mexico, D.F., [2]Unocal Corporation, P. O. Box 76, Brea, CA 92621.

Transmission electron Microscopy (TEM) together with x-ray diffraction (XRD) and BET surface area measurements have been used to characterize a sample of natural rectorite pillared with Al_2O_3 clusters. After drying and calcination at 400°C/10h, the pillared clay had BET surface area between 150 and 200 m_2/g and a d001 value of about 28:41Å. As reported in the literature, pillared rectorites are stable to 800°C in air as well as in the presence of steam. After steam aging, the d001 value decreased only to 27Å and a 93% retention of surface area was observed.

TEM analysis has shown that the pillared rectorite is an interstratified layered silicate mineral consisting of a regular stacking of mica-like and montmorillonite-like layers. The nature of the layers as well as their stacking sequence is difficult to predict because of the variation found within and between samples. In addition to variation in the 1:1 stacking sequence of the two elementary layers, stacking disorders such as bending and folding of the silicate layers have been observed.

110. LAYERED SILICATES: THE PROTONATION BEHAVIOR OF $KHSi_2O_5$
Dean M. Millar, Juan M. Garces, David W. Susnitzky, 1776 Building CR-Catalysis Laboratory, The Dow Chemical Company, Midland, MI 48674

As part of a larger effort exploring the structure and chemistry of layered silicates, we have investigated the synthesis and protonation behavior of potassium hydrogen disilicate, $KHSi_2O_5$ (1). Crystals of 1 were prepared hydrothermally and obtained in high yield. Protonation of this material with excess acid rapidly produces crystalline $H_2Si_2O_5$. Although the bulk crystal morphology is retained during protonation, significant reorganization of the silicate layers has occurred, including the formation of a previously unreported $H_2Si_2O_5$ phase. These transformations have been studied by carefully controlling the acid concentration during protonation and characterizing the resulting structural changes using electron diffraction and solid state NMR techniques.

111. PREPARATION AND CHARACTERIZATION OF BORATE PILLARING ANIONIC CLAY. S. Cheng and J.-T. Lin, Department of Chemistry, National Taiwan University, Taipei, Taiwan, R.O.C. 107

Recent developments in the intercalation of anionic clays by robust polyoxometal oligomers stem from their potential use as catalytic materials. Many of the anionic clays are strong bases. In the literatures, a polyoxometalate of transition metals, such as molybdenum, tungsten or vanadium was usually the subject to introduce into the interlayer space of anionic clays. However, the composition of these polyoxometalates is pH dependent, and these oligomers become unstable in basic environment. In the present work, the oligomers of tetranuclear borate were introduced into the layers of a magnesium aluminum hydroxide hydrotalcite-like compound. Because the tetraborate ions were formed in a basic solution, pillared hydrotalcite of good crystallinity was obtained. The metal hydroxide layer structure was found to remain intact until 723 K. A systematic study on the preparation and structural characterization of borate pillaring hydrotalcite was carried out. The catalytic behaviors of these materials on 2-butanol decomposition were also examined.

112. NEW DEVELOPMENTS IN C_5-C_7 PARAFFIN ISOMERIZATION, J.P. van den Berg, G.J. den Otter, P.M.M. Blauwhoff, Koninklijke/Shell Laboratorium Amsterdam, (Shell Research B.V.).

Hydroisomerization of the C_5/C_6 refinery tops stream is an attractive, well known route to boost its octane level. This type of process is currently of considerable interest due to the pronounced world-wide trend towards abandonment of lead anti-knock additives in the gasoline pool as a result of new environmental legislation.

Due to legislation coming up to also reduce the maximum gasoline benzene and total aromatics content, there is a strong drive to increase the catalytic reformer feed initial boiling point. This results in substantial amounts of C_7 paraffins to be routed to the isomerization units. However, different temperatures are required for C_5 isomerization. The fundamentals of the effects of process conditions in the single HYSOMER reactor option will be discussed and, dependent on feedstock composition, the optimum determined. Recently developed options for C_5-C_7 paraffin isomerization, comprising multi HYSOMER reactor process schemes, will be discussed.

113. DESIGN OF Pt-MORDENITE CATALYSTS FOR THE ISOMERIZATION OF C_5 AND C_6 N-ALKANES IN LIGHT STRAIGHT RUN, A. Corma[1], J. Frontela[2], J. Lazaro[2], M. Perez[2], [1]Instituto de tecnologia Quimica, UPV-CSIC, Universidad, Politecnica de Valencia, Camino de vera, s/n, 46071 Valencia, Spain, [2]CEPSA S.A., Picos de Europa 7, Poligono Industrial S. Fernando de Hanares II, Madrid, Spain.

In the present work a refinery Light Straight Run (LSR) has been isomerized in a laboratory fixed bed reactor, and in a pilot plant scale, using Pt-Mordenite zeolites as catalysts. The influence of the final framework Si/Al ratio of the zeolite and the procedures of dealumination, i.e., steam, steam-acid, acid-steam, on activity and selectivity have been studied.

114. HYDROISOMERIZATION OF N-HEXANE IN THE PRESENCE OF AROMATICS OVER Pt/ZEOLITES, A.M. Martin[1], J.K. Chen[2], V.T. John[3], [1]J.M. Montgomery Consulting Engineers, 3501 N. Causeway, Metairie, LA 70002, [2]Industrial Technology Research Institute, Hsinchu, Taiwan, [3]Tulane University, New Orleans, LA 70118

In complex feeds processing over zeolite catalysts, competitive reaction between feed components results in observed reaction rates and selectivities that cannot be predicted from pure component data alone. Our research has focused on competitive reactions between paraffins and aromatics over Pt/zeolites. The specific reaction considered here is the hydroisomerization of n-hexane over Pt/zeolites in the presence of aromatics. The zeolites used in this work include ZSM-5, Zeolite Y, and Mordenite. Earlier we found that the presence of benzene can significantly inhibit the n-hexane reaction over Pt/mordenite, and reported the effect of a series of aromatics on n-hexane reaction rates and selectivities.

We now present results of n-hexane hydroisomerization reactions in the presence of a series of aromatics over Pt/ZSM-5 and Pt/zeolite Y. The aromatics used in these experiments consist of benzene, toluene, p-xylene, and mesitylene. These results are compared to the earlier results obtained for the case of Pt/mordenite. The results indicate a very interesting and clear trend between n-hexane reaction inhibition and aromatic size. Additionally, relative differences in results between the different catalysts indicate inhibition as a function of both pore size and zeolite/aromatic steric considerations.

115. IMPROVEMENTS ON BOTH AMORPHOUS AND ZEOLITIC ISOMERIZATION CATALYST, G. Szabo, F.X. Cormerais, Total-France, Centre de Recherches, B.P. 27, 76700 Harfleur, France.

Lead suppression in gasolines is a prime concern of the refining industry. Refiners are obliged to evaluate gasoline pool octane upgrading alternatives. Among the many alternatives available for gasoline pool octane improvement, light straight run naphtha isomerization is one of the most effective. The catalysts at the present time used in C_5-C_6 paraffin isomerization operate in the presence of hydrogen and are of two types: very active, Friedel-Crafts acidic type amorphous catalysts; and molecular sieve based bifunctional catalyst which is intrinsically active at moderate temperature. The present paper describes the preparation and the activity of the zirconia promoted platinum-alumina-aluminium chloride catalyst, as well as of the zirconium modified, platinum loaded H-mordenite type dual-function catalyst.

116. KINETICS OF HYDROISOMERIZATION OF N-HEXANE ON NICKEL CONTAINING ZEOLITES. H. Exner, E. Nagy,* F. Fetting, Inst. f. Chem. Techn., Darmstadt D-6100, *Res. Inst. Techn. Chem. Hung. Acad. Sci. Veszprem, H-8201 P.O.Box 125.

The isomerization of n-paraffin is given in the literature as a consecutive reversible reaction: n-hexane ⇄ methylpentane ⇄ dimethylbutane → crack product. For the kinetic scheme of the reaction via carbenium ions on a bifunctional catalyst the following three steps are assumed : dehydrogenation on metallic sites, isomerization on acid sites and hydrogenation on metallic sites. Froment et al. (see e.g. Catalysis Today, 1, 455, 1987) have intensively investigated the kinetics of reaction for n-paraffins up to n-decane. They have concluded, that the isomerization is the so called rate limiting step. However, the kinetic model suggested by them does not fit well the experimental data if the reaction takes place in presence of inert gas, e.g. nitrogen. We have developed a new kinetic model assuming that the dehydrogenation is the rate limiting step. The values of kinetic constants were calculated by parameter estimation using nonlinear regression. The kinetic equations were solved by modified Runge-Kutta method. The experimental data - the experiments were carried out in packed bed reactor using nickel containing mordenite and Y-zeolite - obtained under different conditions agree well with data calculated by the modell.

117. STUDIES ON THE ACTIVITY AND SURFACE ACIDITY OF HM ZEOLITE AND Pd/HM CATALYSTS, Z.M. Chai, C.L. Li, G.X. Huang, Research Institute of Petroleum Processing, East China University of Chem. Tech., Shanghai 200237, P.R. China.

Palladium exchanged H-form mordenite (Pd/HM) alkane isomerization catalysts were prepared. Temperature programmed desorption (TPD) technique and FTIR have been used in the measurement of concentrations and strength of acid sites. Formations and modification of their acid properties in various steps for the manufacture of these catalysts were investigated. Preparation parameters have a marked influence on their activities and selectivities for C_5/C_6 alkane isomerization. This is mainly attributed to difference in acidity. It was proved that only OH group which was present in the main channels of mordenite and gave a peak at 3607 cm^{-1} in FTIR spectra, was relevant to Bronsted acid sites, although three different OH groups have been formed. Moderately strong acid sites have been considered to be active sites for hexane isomerization catalysts. Loaded palladium could not only delay deactivation by coking but also increase strength of acid sites and B-acids to L-acids ratio by the interaction between palladium and HM.

118. SIMULTANEOUS DETECTION OF AROMATICS, SULFUR AND HYDROCARBONS IN DIESEL FUELS BY GAS CHROMATOGRAPHY. R. J. BOHLER, A. J. McCORMACK, J. M. McCANN. Texaco Inc., P. O. Box 509, Beacon, New York, 12508.

A unique application in the use of multidetector gas chromatography has been employed for the analysis of diesel fuels. Three detectors, installed in series, are used to analyze diesel fuels for their aromatic, sulfur and hydrocarbon constituents. The effluent from a chromatographic column enters a photoionization detector. Because of the non-destructive nature of this detector, column flow can subsequently be directed to a flame ionization detector and a chemiluminescence detector. Three different chromatographic fingerprints can be obtained simultaneously from all three detectors with a single injection. A description of the system parameters will be given along with a comparison with results obtained using both FIA and ASTM D2549.

119. APPLICATIONS OF GC-AES/BOILING POINT DISTRIBUTION ANALYSIS FOR CHARACTERIZATION OF REFINERY STREAMS AND HEAVY OILS. Joseph J. Kosman, BP Research, 4440 Warrensville Center Road, Cleveland, OH 44128.

The Hewlett Packard atomic emission detector provides analysts with a convenient means to characterize the elemental distribution of petroleum related samples. Since gas chromatography provides the separation technique for the detector, the elemental output can be related to boiling point using simulated distillation methodology. This facilitates characterization of samples in terms that are important to refining processes. The multielement SIMDIS technique used in this laboratory for over two years will be described along with comparison of the elemental accuracy of the technique for characterization of samples with final boiling points up to 1000°F. Results obtained from extension of the technique to higher boiling samples including crude oils using high temperature gas chromatography will also be described. It will also be shown that by the conversion of GC elemental data to boiling point, correlational analyses (such as cluster analysis) can be used to increase the amount of information which can be extracted from the multielement data. Results will be shown that demonstrate the power of this technique for environmental analysis as well as the methodology used to implement it.

120.
SIMULTANEOUS DETERMINATION OF HYDROCARBON DISTRIBUTION AND SULFUR CONTENT IN PETROLEUM AND PETROLEUM RESIDUES. N.G. Johansen, M.F. Novelli, R.S. Hutte, Sievers Research, Inc., 1930 Central Ave., Ste. C, Boulder, CO 80301.

Determination of both sulfur content and hydrocarbon distribution in crude oils and petroleum residues provides valuable information for optimization of production processes. Petroleum feedstocks and residues are not only increasingly valuable, but also highly variable in their composition. High temperature gas chromatography and simulated distillation techniques (GC-SIMDIS) are widely utilized for the determination of boiling point distributions of these materials. To date however, there has been no technique which allows the simultaneous measurement of the boiling point distribution and the sulfur content of a complex petroleum sample in a single chromatographic analysis. Use of the Sulfur Chemiluminescence Detector (SCD) with the Varian Model 3410 high temperature gas chromatograph permits simultaneous detection of the hydrocarbon components (FID) and low level sulfur compounds (SCD) without post-column splitting. Applications of these techniques for the analysis of petroleum feedstocks and residues will be discussed.

121. High Temperature Gas Chromatography of Crudes and Residua Using Atomic Emission Detection, D.W. Hausler and D.H. Renfro, Phillips Research Center, Phillips Petroleum Company, Bartlesville, OK 74004.

Aluminum-clad fused silica capillary columns (0.54 and 0.25 mm id) were used to separate crudes and residua by gas chromatography with atomic emission detection. Carbon, hydrogen, sulfur, vanadium and nickel were detected. High temperature (425 C) separations of a Polywax 1000 alkane showed elution of a C98 hydrocarbon (BP 1321 F, 716 C). Detection limits for nickel were shown to be approximately 2 ppb. A California crude with 30 percent 'polars' was shown to have nearly all of the volatile nickel components to be in the 'polars' fraction. Both on-column and splitless injection techniques were examined. Hydrotreated product and feed residua fractions were examined for the distribution of carbon, hydrogen and sulfur. Residua from a North Sea crude and a West Texas crude were separated into 8 fractions from 850 to 1300 F. The nickel, vanadium, sulfur, carbon and hydrogen simulated distillation distributions were obtained and compared.

122. DETERMINATION OF MOTOR OIL VOLATILITY USING HIGH-TEMPERATURE GAS CHROMATOGRAPHY. A. J. McCORMACK, J. M. McCANN, E. K. CHIANESE, E. N. CHEN, JR., AND R. J. BOHLER, Texaco Inc., P. O. Box 509, Beacon, New York, 12508.

A high temperature, gas chromatography method was developed to determine the volatility of crankcase motor oils. Although the method was developed to obtain volatility data at 371C, it can also be used to determine an oil's volatility at other temperatures after modification. The volatility data obtained on several grades of motor oil show a significant improvement in precision over ASTM D2887, the test method specified by the automotive industry to measure a crankcase motor oil's volatility. A comparison of data obtained using this method with both capillary and packed columns will be presented.

123. **CHARACTERIZATION OF WAXY CRUDES AND WAXES.** Randy F. Alex, Bryan J. Fuhr & Marvin Rawluk, Alberta Research Council, Box 8330, Stn F, Edmonton, Canada, T6H 5X2. Harish Kalra, D.B. Robinson & Assoc. Ltd., 9419-20 Ave., Edmonton, Canada T6N 1E5. Ali I. Majeed, Norsk Hydro, Box 200, N-1321, Stabekk, Oslo, Norway.

An analytical protocol has been developed for the detailed molecular characterization of waxy crudes and waxes. It involves initial separation of the crude by spinning band distillation, followed by fractionation of some of the cuts into saturates, aromatics and polars using silica-alumina column chromatography. Asphaltenes are first removed from the highest boiling cut using n-pentane. The paraffin-naphthene ratio of the saturates, and carbon number distributions of saturates and aromatics fractions are determined using gas chromatography-mass spectrometry with either electron impact or nitrous oxide chemical ionization. For the wax samples the higher boiling n-paraffins are initially separated from the remaining oil, and their carbon number distribution determined by supercritical fluid chromatography with flame ionization detection. The remainder of the saturates and aromatics are analyzed by the previously described techniques. The compositional data obtained by this protocol have been successfully employed in a model for predicting wax formation and deposition from North Sea crudes.

124. DETERMINATION OF SULFUR COMPONENTS IN LIGHT PETROLEUM STREAMS
BY HIGH RESOLUTION GC WITH CHEMILUMINESCENCE DETECTION

B. Chawla and F. P. Di Sanzo, Mobil Research and Development Corporation
P. O. Box 480, Paulsboro, NJ 08066-0480

A versatile high resolution gas chromatographic system utilizing a universal sulfur chemiluminescence detector (USCD) coupled with a flame ionization detector (FID) has been optimized for the determination of sulfur components in light petroleum process streams.

The normalized area precision for individual species was found to be approximately 1.5-6.0% relative standard deviation for the major components at the total 1.7 wt% S level. The system is capable of speciating sulfur components in the petroleum process streams containing 0.01 to 3.13 wt% sulfur. The linear relationship (r^2 = 0.998) between the USCD absolute sulfur response and the total sulfur content of several samples of the petroleum process streams clearly indicated that, in addition to speciation, total sulfur can simultaneously be determined with reasonable accuracy.

125. SUPERGALLERY PILLARED CLAYS. Thomas J. Pinnavaia, Department of Chemistry and Center for Fundamental Materials Research, Michigan State University, East Lansing, MI 48824.

Metal oxide pillared clays typically are prepared by the intercalation of a polycation into a smectite clay host and subsequent thermal conversion of the cation to an oxide-like aggregate. This process normally affords microporous derivatives in which the gallery height is comparable to the thickness of the clay layers. Supergallery clays have been defined as derivatives in which the gallery heights are about two or more times as large as the thickness of the host layers. This paper will examine a number of promising synthetic routes to this new generation of pillared clay materials. Products derived from the direct intercalation of metal oxide sol particles, molecularly regular oxides, and exceptionally large polycations in smectite will be presented.

126. SYNTHETIC CONTROL AND ANALYTICAL DETERMINATION OF PILLAR DENSITIES IN ALUMINA PILLARED MONTMORILLONITE. J.R.Jones, R.S.Mason and J.H.Purnell, Department of Chemistry, University College, Swansea, U.K. SA2 8PP.

Pillaring of layered structures such as the montmorillonite clays commonly involves ion-exchange of an ionic pillar precursor into the interlamellar region followed by some procedure, usually calcination, that converts the precursor into the pillar. The number of pillars that may be created clearly depends on both the c.e.c. of the host and the charge on the precursor. There are several ways to modify the c.e.c. of a given host, the simplest being controlled (time and temperature) calcination. Texas montmorillonite has been so treated to provide a family of materials with c.e.c. ranging over a factor of ten which have then been alumina pillared. Chemical analysis and the novel technique of Glow Discharge Mass Spectrometry (GDMS) have been used to assess the resulting pillar densities. GDMS is shown to have considerable potential for rapid analysis of alumino-silicates and related solids.

127. ANIONIC CLAYS : TRENDS IN PILLARING CHEMISTRY. A. de Roy, C. Forano, J.P. Besse, Laboratoire de Physico-Chimie des Matériaux (U.R.A.444) Université Blaise Pascal, 63177 AUBIERE CEDEX FRANCE.

Compounds of the anionic clay family are also designated as lamellar double hydroxides (LDH). Such compounds exist as minerals as well as synthetic phases and present wide anionic exchange properties. During the last few years, research works have been developed, mainly for catalytic applications and electrochemical devices. More recently promising results were obtained in the field of magnetic properties.

This talk will give first a general presentation of the synthetic anionic clays insisting on the structural background. Then we propose to speak about the various preparative methods (metal oxide - metal salt reaction, coprecipitation, anionic exchange, hydrothermal synthesis) to obtain these materials. The specificity of each synthesis process will be pointed out in relation to the chemical composition of the LDH. Finally, trends in pillaring anionic clays will be developed from both chemical and structural points of view. The speach will be enlightened by results from our proper research works completed with data from the literature.

128. SYNTHESIS OF DIOCTAHEDRAL 2:1 LAYER SILICATES IN ACID AND FLUORINATED MEDIUM. L. Huve, R. Le Dred, D. Saehr and J. Baron, Laboratoire de Matériaux Minéraux, Ecole Nationale Supérieure de Chimie, 68093 Mulhouse Cedex, France.

The most common method for preparing layer silicates is the hydrothermal synthesis, from a neutral to basic aluminated hydrogel. Hydroxide ions are involved as mineralizing agents of the structural elements. Fluoride ions are also mineralising agents, not only in basic but also in acid medium. The composition of the aluminosilicated hydrogel is similar to that of the layer silicate desired and includes fluoride ions. Its pH is between 0 and 7. After a maturing stage at room temperature, the hydrothermal treatment was carried out at a temperature around 220 °C. Well crystallized and pure dioctahedral 2:1 layer silicates were, thus, prepared in acid and fluorinated medium. Their general chemical formula is : $M_{x/m}(Al_2\square_1)(Si_{(4-x)}Al_x)O_{10}(OH_{(2-y)}F_y)$, where x = 0-0.9 and y = 0-2. The substitution of fluorine atoms for hydroxyl groups increases the thermal stability. These layer silicates are the first stage towards the synthesis of pillared clays.

129. MICROPOROUS PILLARED MICA WITH CATION-INCORPORATED SILICATE SURFACES. Kazuo Urabe, Hiroaki Sakurai, Kazuki Kawabe and Yusuke Izumi, Department of Applied Chemistry, School of Engineering, Nagoya University, Furo-cho, Chikusa-ku, Nagoya 464, Japan.

Pillared clay is the most attractive and interesting group of new microporous crystals due to its wide applications from catalysts to molecular devices [1]. The number of materials have still been expanded with the use of specific clays, such as rectorite and synthetic swellable mica. Recently we found that the interlayer cations such as La^{3+} are peculiarly and irreversibly fixed by heating onto the exchange sites in tetrasilicic mica (TSM) and the fixed cations confer pillared TSM strong acidity. In contrast, the fixation of cations did not occur in a well-known pillared montmorillonite. Such resulted microporous pillared mica possesses, in principle, all of its acid sites only on the silicate surfaces but not on the pillars. Here we report the chemical and structural descriptions of this novel interlayer cation-fixation method and its significance in a 'pillared clays by design' approach.

[1] I.V. Mitchell (Ed.), "PILLARED LAYERED STRUCTURES; Current Trends and Applications", Elsevier (1990).

130. DEACTIVATION BY COKING AND REGENERATION OF Pd/HM CATALYSTS FOR ALKANE ISOMERIZATION, C.L. Li, L. Shi, R.Y. Wang, L.D. Li, East China University of Chem. Tech., 130 Meilon Road, Shanghai 200237, P.R. China.

C_5/C_6 alkane hydroisomerization processes have been performed in microreactors, laboratory reactors and a pilot plant. The latter unit was designed to treat 1000t/y feed. The catalyst is a novel 0.5% palladium loaded H-form mordenite (Pd/HM). It is possible to give almost the equilibrium product composition under 533-553K for any C_5-C_6 alkane feed. Stability test indicated that no catalyst deterioration occurred for 1000 hrs and five months on stream in the laboratory reactor and pilot plant respectively. Accelerated deactivations were performed under severe operating conditions. Deactivated catalysts were taken from different sections in the reactor. It is concluded by catalyst characterization that only carbon laydown is the cause of deactivation. Proper conditions have been presented to burn off deposited coke and refresh the activities of Pd/HM catalysts. Deactivation kinetic results were given.

131. SURFACE ORGANOMETALLIC CHEMISTRY ON ZEOLITES: SYNTHESIS AND CHARACTERIZATION OF MORDENITES MODIFIED BU TETRABUTYLTIN. APPLICATION TO THE C8 CUT ISOMERIZATION. M. Liddell[1], A. Théolier[1], C. Nedez[1], F. Lefebvre[1], A. Choplin[1], J.M. Basset[1], F. Raatz[2], C. Travers[2] and J.F. Joly[2]. 1: Institut de Recherche sur la Catalyse, 2 av. A. Einstein, 69626 Villeurbanne Cedex, France. 2: Institut Français du Pétrole, BP311, 92506 Rueil Malmaison Cedex, France.

Modification of a H-mordenite (HM) has been achieved by the controlled reaction of bulky organometallic complexes such as dineopentylmagnesium and tetrabutyltin with the external surface. For both complexes, the first observed reaction is a reaction of solvolysis leading to the formation of a grafted alkyl complexe, i.e. →Si-O-MgNp and →Si-O-SnBu$_3$ respectively, and to the liberation of one mole of alkane per atom M (M=Sn,Mg). Full characterization of these surface alkyl complexes has been achieved by IR, ^{13}C MAS NMR, analysis of the evolved gases, elemental analysis. Thermal decomposition at 350°C under vacuum leads to surface oxidized tin (Sn(II) or Sn(IV)) well dispersed on the external surface of the mordenite. These grafted mordenites mixed with a 0.5% Pt/Al$_2$O$_3$, exhibit higher performances in the C_8 (orthoxylene+ethylbenzene) cut isomerization compared with non modified mordenites. The activity is found to be higher and the dismutation yields have been strongly reduced, regeneration of grafted tetrabutyltin mordenite leads to additional reduction of dismutation.

132. INFLUENCE OF THE Si/Al RATIO ON THE CATALYTIC PROPERTIES OF SMALL AND LARGE PORT MORDENITES IN C_8 AROMATICS ISOMERIZATION, M.F.Ribeiro, F.Lemos, F.Ramoa Ribeiro, Ch.Travers, F.Raatz, Ch.Marcilly, GRECAT, Instituto Superior Técnico, Lisboa

Starting from HM zeolites with various Si/Al ratios, bifunctionnal Pt/HM catalysts have been evaluated in C8 aromatics isomerization under industrial-like conditions. Different reactions occur simultaneously under such conditions: isomerization, disproportionation, dealkylation of xylenes and ethylbenzene and cracking side reactions of partly saturated cycles. Global conversion as well as disproportionation and dealkylation of the C8 cut steadily increases with the Si/Al ratio of the zeolitic component of the catalysts, while the selectivity for cracking side reactions shows an optimum with this Si/Al ratio. The same optimum is obtained when n–C7 is used as a feed. These results indicate that the same type of acidic sites are likely to be responsible for the cracking of naphtenic intermediates in the course of the transformation of C8 aromatics under pressure and the cracking of n–C7 under atmospheric pressure. Moreover our results would indicate, that for isomerization, disproportionation, and dealkylation of C8 aromatics, the T.O.N. per acidic site increases up to Si/Al ratios higher than about 100.

133. SUPERACID AND CATALYTIC PROPERTIES OF SULFATED ZIRCONIA, F.R. Chen[1], G. Coudurier[1], J.F. Joly[2], J.C. Vedrine[1], [1]Institut de Recherches sur la Catalyse, CNRS, 2 avenue A. Einstein, F-69626 Villeurbanne, Cedex, [2]Institut Francais du Petrole, Laboratorie de Catalyse, 1-4 avenue de Bois Preau F-92506 Rueil-Malmaison, Cedex.

Zirconium hydroxide was sulfated by H_2SO_4 aqueous solution of different normality. The presence of sulfur was observed to slow down surface area decrease upon calcination and the tetragonal to monoclinic phase transformation of zirconia with calcination temperature.

Acidic properties were studied by Hammett's indicator technique which was observed to fail, presumably because of the presence of Lewis sites. Charge transfer complex formation was studied by ESR technique using benzene as a probe since it has a high ionizing potential value (9.24 eV). The ESR signal was assigned to biphenyl cation with a_H=6.74; 3.37 and 0.52G for 2,4 and 4 protons respectively. Its intensity was observed to be maximum for a calcination temperature of 600°C and a sulfur content equal to 1.5 to 3 wt%. Catalytic properties were studied for n-butane to i-butane isomerization reaction in the 150 to 300°C range. The highest catalytic activity in flow reactor was observed for the 600°C calcined samples as for benzene ionization properties, which indicates a close correlation. Hydrogen was observed to sharply decrease the deactivation rate suggesting that the active sites do not correspond to reduced species and that Bronsted sites rather than Lewis sites seem to be involved in the reaction.

134. CATALYTIC REACTIONS OF HYDROCARBONS UNDER TRANSIENT CONDITIONS, J.L. Margitfalvi, M. Hegedus, Central Research Institute for Chemistry of the Hungarian Academy of Sciences, 1525 Budapest, POB 17, Hungary

Transient response method was used to investigate the conversion of n-hexane and n-butane on an industrial Pt/Al_2O_3 catalyst. It was demonstrated that the type of transient response curves for a given reaction product strongly depended on the reaction temperature and the presence or absence of hydrogen on the catalyst surface prior to the introduction of the hydrogen-hydrocarbon mixture. The differences above indicate that the rate determining steps of the formation of the main reaction products in n-hexane or n-butane conversion are different and strongly depend on the reaction temperature.

135. SCIENTIFIC FOUNDATIONS FOR SYNTHESIS OF HYDROCARBON ISOMERIZATION CATALYSTS, N.R. Bursian, P.N. Borutsky, V.Yu Georgievsky, V. Sh. Gruver, B.V. Krasii, G.A. Lastovkin, G.B. Martynova, D.S. Orlov[a], V.N. Brovko[b], [a]NPO Lenneftekhim, Zheleznodorozhny pr., 40, 193148 Leningrad, USSR, [b]PO Kirishinefteorgsintez, 187110 Kirishi, Leningrad Oblast, USSR

Platinum-alumina catalysts promoted with halogens and in certain cases containing mordenite have been investigated. An increase in isomerizing activity of alumina and mordenite when being promoted with fluorine is connected with an increase in acidity and with platinum state alteration (IR spectroscopy data). It is explained by an increase in acceptor properties of ligands which interact with platinum. An increase in electron deficiency of platinum when promoting platinum-alumina catalyst with fluorine and sulphur is manifested in Pt-C bond weakening, suppression of hydrogenolysis reactions and in an increase in selectivity and stability as observed for industrial isomerization catalysts which operate from 5 to 8 years without regeneration. The same effect is achieved by other means, such as the method of introduction of halogens and sulphur, as well as by the use of crystalline ferrisilicate instead of aluminosilicate, as observed for C_8 aromatics isomerization.

136. MULTINUCLEAR SOLID-STATE NMR INVESTIGATION OF BORON-SUBSTITUTED ZEOLITES AND SILICATES. J.C. EDWARDS, C.L. O'YOUNG, P.J. GIAMMATTEO, Texaco Inc., P. O. Box 509, Beacon, New York, 12508.

We have recently been investigating the structural chemistry of boron-substituted zeolites, and amorphous borosilicates due to their importance as isomerization catalysts. ^{11}B MAS NMR, obtained by cross polarization and Bloch decay methods, revealed the presence of three types of boron in the zeolites, while the ^{23}Na, ^{1}H and ^{29}Si MAS NMR, obtained by Bloch decay and cross polarization, revealed distinct differences between the zeolites and the amorphous borosilicates. These differences and the effect of heat treatment on the boron environments of the zeolites will be discussed.

137. DERIVATIVE SPECTROSCOPY AS A ANALYTICAL TOOL FOR HYDROCARBON IDENTIFICATION. M. V. Reyes, Texaco EPTD, Houston, Texas 77401

Analysis using Total Scanning Fluorescence (TSF) has been useful in fingerprinting oils and characterizing other polyaromatic hydrocarbon mixtures. Recent new developments in fluorescence technology, however, have enabled even greater differentiation of hydrocarbon mixtures than techniques previously available. Now, the normal fluorescence spectra (fingerprint) can be further analyzed to yield sets of higher order spectra using DERIVATIVE SPECTROSCOPY. These resultant higher order spectra can be used to differentiate very similar hydrocarbon mixtures such as light oil from diesel, different brands and grades of gasolines, and weathered oil from fresh crude. This paper addresses the development of derivative fluorescence spectroscopy for environmental applications, including its usage in identifying hydrocarbon contaminant type and source in incidents such as underground storage tank leaks and oil spills.

138. ADSORPTION KINETICS OF ASPHALTENES - A LIQUID/WATER INTERFACIAL TENSION MEASUREMENT. M.M. De Tar E.Y. Sheu, and D.A. Storm, Texaco Research Center, Beacon, N.Y. 12508.

Asphaltenes derived from vacuum residue behave similarly to surfactant molecules. However, the self-association and the physical properties of the resulting colloidal-like particles are not well understood. We have carefully measured the surface tension of a series of asphaltene/apolar solvent, to experimentally prove the self-association propensity. In this work, we examined the surface activity of asphaltenes by measuring the dynamic interfacial tension of an asphaltene/toluene solution against aqueous solutions. We have measured the interfacial tension as a function of asphaltene concentration and pH of the aqueous solutions respectively. By using various theories for adsorption kinetics, we were able to characterize the adsorption kinetics of asphaltenes using a kinetic parameter. The results suggest two distinctive kinetic regimes, namely, the diffusion limited and the reaction limited regime. The correlation between the molecular structure of asphaltene and the obsevered kinetics will be discussed.

139. DETERMINATION OF TRACE METALS
IN LIQUEFIED PETROLEUM GAS
by
E.H. Homeier, P.O. Hennes, and P.V. Tota

UOP, Des Plaines, Illinois

The development of a reliable method for determination of parts per billion (1 in 10^9) Hg, As, Sb, Cu, and Pb in liquefied petroleum gas (LPG) is described. The method, UOP 906, covers the adsorption of metals on a resin followed by desorption and the analysis of the resulting aqueous-acid solutions. Both graphite furnace-atomic absorption spectrometry (GF-AAS) and inductively coupled plasma-mass spectrometry (ICP-MS) are satisfactory finishes, although ICP-MS is preferred because it accelerates the multielement determinations. Tests with synthetic LPG and metal blends verified adsorption, and repeatability tests for Sb, As, and Hg confirmed that parts per billion concentrations were determined quantitatively. Preliminary tests with a second adsorbent are also discussed. The elimination of this proprietary adsorbent significantly accelerated the procedure without affecting the lower limit of detection (~0.1 ppb).

140. SURFACE LAYERS FROM LUBRICANT ADDITIVES IN FRICTION/WEAR EXPERIMENTS. R. O. Carter III, N. E. Lindsay and P. A. Willermet Ford Motor Company, P.O. Box 2053, SRL S-3061, Dearborn, MI 48121

Progress on the ongoing effort to identify friction/wear induced surface films and modifications are reported. Initial studies using reference oils in a controlled cam/tappet identified certain experimental requirements for artifact free infrared reflection-absorption analysis of such experiments. Experiments were repeated with polished tappet surfaces with reasonable mechanical results and clear infrared spectra. These samples were subsequently analyzed by UHV surface techniques which added detail to the infrared results and highlighted the need for surface mapping studies. The original infrared mapping was performed on a commercial grazing incidence micro-sampler and lead us to design and have built a different type of reflection-absorption accessory.

The help and assistance of the following are gratefully acknowledged: J. deVries, L. Haak, P. Schmitz, Ford Research Surface Analysis, W. Weber, B. Poindexter, Ford Research Physics Department, J. Katon, A. Summer, Miami U. MML, and R. Messerschmidt, Connecticut Instruments.

141. TPOP-IRMS STABLE CARBON ISOTOPE ANALYSIS OF HEAVY PETROLEUM RESIDUES BIGOIS M., CASABIANCA H., EIMOUCHNINO J., FIXARI B., LE PERCHEC P. CNRS Laboratory of Organic Material and Central Analytical Service 69390 Vernaison (FRANCE)

Stable C^{13}-C^{12} ratios ($\delta\ C^{13}$) were determined for Hydroconverted Heavy Oil Vacuum Residues under various catalytic and non-catalytic conditions. Temperature Programmed Oxidative Pyroanalysis (TPOP)-IRMS afford quantitative CO_2 emission of, respectively, $500^-/500^+C°$ and Residual Carbon fractions. A clear isotopic carbon distribution is associated with the deepness of cracking of the alkyl chain and with conversion. The increase for $\delta\ C^{13}$ depletion of the gas fraction is thus associated with an increase in C^{13} concentration of the $500^-C°$ fraction while a stand-by value of the Residual Carbon content of $\delta\ C^{13}$ = -27 is observed over a large range of thermal severity. Also interesting was the change in correspondence observed between thermal conversion and $\delta\ C^{13}$ in the presence of various amounts of catalyst, probably due to a mechanism change in comparison with free-radical pure thermal process. TPOP-IRMS stable carbon isotope ratios provide a new way to appreciate the conversion and quality factor during Heavy Petroleum transformations and underlines the mechanism of evolution which might be involved in the presence of catalyst.

142. OPTIMIZATION OF THE PILLARING OF A SAPONITE, S. Chevalier, R. Franck, H. Suquet, C. Marcilly, D. Barthomeuf, Laboratoire de Reactivite de Surface et Structure, Universite Paris 6, 4 place Jussieu 75252, Paris Cedex 05, France.

Al-pillared clays thermally stable up to 750°C have been prepared from easily available starting materials: a natural clay, namely the Ballarat saponite, and a commercial aqueous solution of Chlorhydrol.

The Al content fixed by the pillared clay increases with the pH and with the Al/clay ratio. The best thermal stability of Al-pillared saponites is obtained for the following experimental conditions: clay concentration ≤ 5 g.l^{-1}, Al/clay ≤ 5 mmol.g^{-1} and pH values between 4.8 and 6.0. A very careful thermal treatment is required up to 500°C (36°/h heating rate) to transform the intercalated species into stable oxide pillars. At 750°C, the surface areas are about 150-250m^2.g^{-1} and basal spacings about 17.3Å. According to these values, Al-pillared saponites may be used as catalysts of 7-8Å pore space.

143. **Ga_{13}, $GaAl_{12}$ AND Al_{13} POLYOXOCATIONS AND PILLARED CLAYS** Susan M. Bradley, Ronald A. Kydd* and Raghav Yamdagni, Department of Chemistry, University of Calgary, Calgary, Alberta T2N 1N4 CANADA, and Colin A. Fyfe, Department of Chemistry, University of British Columbia, Vancouver, British Columbia V6T 1Y6 CANADA

Studies of the hydrolysis of aqueous gallium and mixed gallium/aluminum solutions have resulted in the synthesis of Ga_{13} and $GaAl_{12}$ cations which are isostructural with the Al_{13} species which has been the principal ion utilized in clay mineral pillaring studies. These ions have been characterized through the use of solution NMR studies, as well as MAS NMR, powder X-ray diffraction and infrared investigations of their sulfate salts. Their relative stabilities in solution have been found to follow the order $Ga_{13} << Al_{13} << GaAl_{12}$, which also appears to correlate with the order of increasing symmetry of their overall structures. The thermal stabilities of their corresponding PILC's were characterized through the use of powder X-ray diffraction, differential thermal analyses and surface area measurements. These relative stabilities were also found to increase in the order Ga_{13}-PILC < Al_{13}-PILC < $GaAl_{12}$-PILC. Their relative acidities have been examined through infrared pyridine sorption studies. An overview of our studies of the new pillaring agents themselves, and of their corresponding pillared clays will be presented.

144. OXYGEN ADSORPTION PROPERTIES OF MICROPOROUS SILICA DERIVED FROM LAYERED SILOXENE BY OXIDATION. S. Yamanaka, H. Itoh and M. Hattori, Department of Applied Chemistry, Faculty of Engineering, Hiroshima University, Higashi-Hiroshima 724, Japan.

Layer structured siloxene prepared by the reaction of $CaSi_2$ with hydrochloric acid was further oxidized by a prolonged treatment with water. During the treatment, the Si-Si bonds of the siloxene were oxidized to Si-O-Si bonds and the interlayers were cross-linked by the condensation of the Si-OH groups. The resulting compounds have high surface areas with pores in the micro- to mesoporous ranges. The compounds calcined in a vacuum at high temperatures keep the high surface area and show peculiar ESR centers on adsorption of oxygen, which are attributed to the formation of silicon dangling bonds. The ESR centers disappear reversibly on desorption of oxygen.

145. SURFACE AND CRACKING PROPERTIES OF EXPANDED CLAYS DRIED USING A SUPERCRITICAL FLUID. M. L. Occelli[1], K. Takahama[2], M. Yokoyama[2] and S. Hirao[2]. [1]Unocal Corporation, 376 S. Valencia, Brea, CA 92621, [2] Matsushita Ltd., Osaka, Japan.

By expanding montmorillonite and saponite with large $SiO_2 \cdot TiO_2$ clusters, it is possible to generate expanded clays with BET surface area in the 450-650 m^2/g range that are thermally and hydrothermally stable. When conventional air drying is replaced by removal of the solvent at supercritical condition (with CO_2), a macrospace is generated with Hg surface area in the 350-400 m^2/g range. At MAT conditions these materials have high cracking activity. However, catalyst density as well as coke selectivity will have to be increased significantly before these types of clays can compete with commercially available FCC.

146. APPLICATIONS OF NMR SPECTROSCOPY IN SECONDARY PROCESSING OF PETROLEUM FRACTIONS. Indra D. Singh and Prem V Dogra, Indian Institute of Petroleum, Dehradun-248005, India

The application of NMR Spectroscopy to provide information on the structural properties and chemical composition of crude oils & their fractions is well documented. In recent years, enormous work is being reported applying this technique to solve various industrial problems related to petroleum industry. The structural investigations of petroleum fractions employing NMR technique before and after processing can make an important contribution in selection of proper operating parameters including catalyst to achieve improved product quality and yield. Such investigations also provide necessary information to understand the reaction pathways during the secondary processes. An upto date review of applications of NMR in various petroleum secondary processes will be presented. These applications will be demonstrated using typical examples from visbreaking of short residues, hydrogenation of kerosene and mild hydrotreating of lube base stocks.

147. ELUCIDATION OF THE NATURE OF NAPHTHENO-AROMATIC GROUPS IN HEAVY PETROLEUM FRACTIONS BY ^{13}C NMR AND CATALYTIC DEHYDROGENATION. J. Gallacher[1], C.E. Snape[1], P.R. Dennison[1], J. Boyle[2] and K.A. Holder[2]. [1]University of Strathclyde, Department of Pure & Applied Chemistry, Glasgow G1 1XL, UK. [2]BP International plc, Sunbury Research Centre, Sunbury on Thames, Middlesex TW16 7LN, UK.

^{13}C NMR spectral editing techniques have been used to determine the concentrations of aliphatic CH, CH_2 and CH_3 groups directly in aromatic fractions from a light and heavy cut of a North Sea waxy heavy distillate (vacuum gas oil) as well as hydrotreated samples of these distillates. To help resolve effective and poor hydrogen donor naphthenic groups, these fractions were catalytically dehydrogenated and re-examined by NMR. The results indicate that the naphthenic groups in the initial aromatic fraction account for a maximum of 40% of the total carbon but only 10% of the carbon is present in effective H-donors and that, in contrast to earlier structural models for heavy petroleum fractions, the naphthenic groups are relatively small comprising of no more than two rings on average.

148. A COMPREHENSIVE ANALYSIS OF DIESEL FRACTIONS, V.P. NERO*, G.L. CHAPMAN**, S.H. CHAO**, C.S. MacMAHON*, AND E.J. POPIEL*, *Texaco Inc. P. O. Box 509, Beacon, New York, 12508; **Texaco Inc., Savannah Avenue at Highway 73, Port Arthur, Texas, 77641

A hydrotreated diesel fuel was distilled into ten equal volume fractions, which were comprehensively analyzed by a variety of techniques to establish aromatic and saturate hydrocarbon types, sulfur and nitrogen content, and cetane number. The data were required to determine whether distillation controls and hydrotreating would be helpful in producing a diesel product which would meet new California air control standards. The naphthalenes, biphenyls, fluorenes, and phenanthrenes, which are the major polynuclear aromatic components, have been shown to correlate with diesel particulate emissions; therefore, they were of prime interest. Mass spectral analysis showed the mononuclear aromatics were concentrated in the lower boiling fractions and the polynuclear aromatics were primarily in the higher fractions. Paraffins content, which directly correlates with cetane number, increased as the boiling temperature increased.

149. 1- AND 2-DIMENSIONAL NMR ANALYSES OF COAL OIL HYDROPROCESSING PRODUCTS. Heather D. Dettman and Michael F. Wilson, EMR Canada, CANMET Energy Research Laboratories, Ottawa, ON., Canada, K1A 0G1.

The depletion of conventional crude oil reserves necessitates development of nonconventional resources for the generation of fuels. Due to the low quality of synthetic crude oil (SCO) derived from sources such as coal, extensive hydroprocessing is needed to produce quality transportation fuels. Current hydroprocessing technology is costly. Significant research effort is directed towards the development of new upgrading processes. Nuclear magnetic resonance (NMR) spectroscopy is an ideal technique for studying the changes in the composition of the molecular species of the SCO before and after processing. In particular, 1- and 2-dimensional ^{13}C NMR techniques provide detailed fingerprints of the molecular components present. The analyses of a coal-derived oil will be used to demonstrate the application of DEPT and HETCOR NMR techniques to aid the development of new hydroprocessing technologies.

150. MEASUREMENT OF THE MOLECULAR WEIGHT DISTRIBUTION OF VACUUM RESIDUE USING MASS SPECTROMETRY - A HETEROGENEOUS ANALYTICAL MODEL. E.Y. Sheu D.A. Storm, and M.M. De Tar, Texaco Research Center, Beacon, N.Y. 12508.

The molecular weight distribution of vacuum residue is an inportant for petroleum research. Mass spectrometry (MS) appears to be an appropriate technique for this purpose. However, the MS spectrum usually does not represent the true molecular weight distribution. This is due to error arising from incomplete introduction of the sample into the inlet system and fragmentation during the ionization process. In a previous work, we developed an analytical scheme to correct these errors for a homogeneous system. In this study, we extend the model for the treatment of a heterogeneous system, which is more often the case. In addition to the two experimental axes we used in the previous work, we made use of the electron current as another axis. By proper combination of these axes, we were able to obtain the true molecular weight distribution by decomposing the observed MS spectrum. Moreover, the ionization cross section and its dependence on electron energy were included, this makes the model applicable not only for petroleum products, but also for any heterogeneous system.

151. STRUCTURE/PROPERTY RELATIONSHIPS IN BASE OIL PROCESSING: MONITORING DEWAXING PROCEDURES VIA SPECTROSCOPY AND CHROMATOGRAPHY. G. MARSH, P.J. GIAMMATTEO, V.P. NERO, A.J. McCORMACK, R.J. TAYLOR[1], Texaco Inc., P. O. Box 509, Beacon, New York, 12508; [1]Texaco Inc., Savannah Avenue at Highway 73, Port Arthur, Texas, 77641.

Differences in wax removal comparing solvent dewaxing (SD) and catalytic dewaxing (CD) processes were studied using quantitative ^{1}H/^{13}C NMR, mass spectral type analyses (ASTM D2786, D3239) and gas chromatography.

```
              ---> SD ---> *SD Oil + Wax
              |            |-------> SD ----> *SDSD Oil + *Wax
*Starting Base Oil
              |---> CD ---> *CD Oil-------> SD ----> *CDSD Oil + *Wax
(*samples analyzed)
```

Samples recovered from the processing scheme outlined above were analyzed. Discussion will focus on characterization of base oil and wax components and their relationships to processing considerations (pour points, Viscosity Index, carbon numbers, etc.).

PHYSICAL CHEMISTRY

Fourth Chemical Congress of North America

New York, NY
August 25 - 30, 1991

H. F. Schaefer,
Program Chairman

SUNDAY EVENING
● **Symposium on Superconductivity**
R. L. Martin, Presiding
Papers 1-3

● **Symposium on Comparison of Ab Initio Quantum Chemistry With Experiment**
Session I. Clusters
J. L. Gole, Presiding
Papers 4-8

MONDAY MORNING
● **Symposium on Superconductivity**
R. L. Martin, Presiding
Papers 9-12

● **Symposium on Comparison of Ab Initio Quantum Chemistry With Experiment**
Session II. Symmetry Breaking in Molecules
R. J. Bartlett, Presiding
Papers 13-17

● **Symposium on Energy Transfer and Relaxation in Condensed Phases**
J. M. Drake, Presiding
Papers 18-23

MONDAY AFTERNOON
● **Symposium on Superconductivity**
R. L. Martin, Presiding
Papers 24-28

● **Symposium on Comparison of Ab Initio Quantum Chemistry With Experiment**
Session III. Vibrational Spectra
V. Smith, Presiding
Papers 29-33

● **Symposium on Energy Transfer and Relaxation in Condensed Phases**
R. Kopelman, Presiding
Papers 34-39

● **Symposium on New Developments and Applications of Magnetic Resonance and Optical Spectroscopies**
R. Tycko, Presiding
Papers 40-44

MONDAY EVENING
● **Symposium on Energy Transfer and Relaxation in Condensed Phases**
R. Kopelman, Presiding

● **Symposium on Superconductivity**
R. L. Martin, Presiding   Papers 45-49

TUESDAY MORNING
● **Symposium on Superconductivity**
M. L. Martin, Presiding   Papers 50-54

● **Symposium on Comparison of Ab Initio Quantum Chemistry With Experiment**
Session IV. Cations, Anions and Spectra
H. b. Schlegel, Presiding   Papers 55-59

● **Symposium on Energy Transfer and Relaxation in Condensed Phases**
G. Zumofen, Presiding   Papers 60-65

● **Symposium on New Developments and Applications of Magnetic Resonance and Optical Spectroscopies**
W. S. Warren, Presiding   Papers 66-70

● **Symposium on Gas-Phase Metal Reactions I.**
A. Fontijn, Presiding   Papers 71-74

TUESDAY AFTERNOON
● **Symposium on Comparison of Ab Initio Quantum Chemistry With Experiment**
Session V. Excited States and Electronic Spectra
R. Boyd, Presiding   Papers 75-79

● **Symposium on Energy Transfer and Relaxation in Condensed Phases**
P. Levitz, Presiding   Papers 80-85

● **Symposium on Gas-Phase Metal Reactions II.**
R. S. Miller, Presiding   Papers 86-90

● **Chemical Vapor Deposition of Diamond (and c-BN, SiC) Growth Techniques**
J. J. Cuomo, Presiding   Papers 91-95

● **Symposium on Atomic Imaging: STM and Related Techniques**
P. Avouris, Presiding   Papers 96-101

WEDNESDAY MORNING
● **Symposium on Comparison of Ab Initio Quantum Chemistry With Experiment**
Session VI. Weakly Bonded Species
W. Ermler, Presiding   Papers 102-106

● **Symposium on Energy Transfer and Relaxation in Condensed Phases**
K. Lindenberg, Presiding   Papers 107-112

- **Symposium on New Developments and Applications of Magnetic Resonance and Optical Spectroscopies**
K. W. Zilm, Presiding   Papers 113-117

- **Chemical Vapor Deposition of Diamond (and c-BN, SiC) Material and Defect Characterization**
J. T. Glass, Presiding   Papers 118-122

- **Symposium on Atomic Imaging: STM and Related Technniques**
T. Sakurai, Presiding   Papers 123-128

WEDNESDAY AFTERNOON

- **Symposium on Energy Transfer and Relaxation in Condensed Phases**
A. Genack, Presiding   Papers 129-135

- **Symposium on New Developments and Applications of Magnetic Resonance and Optical Spectroscopies**
P. A. Mirau, Presiding   Papers 136-140

- **Symposium on Gas-Phase Metal Reactions III.**
P. Davidovits, Presiding   Papers 141-144

- **Chemical Vapor Deposition of Diamond (and c-BN, SiC) Diagnostics of Growth Environment**
S. J. Harris, Presiding   Papers 145-151

- **Symposium on Atomic Imaging: STM and Related Techniques**
J. E. Demuth, Presiding   Papers 152-154
                                                                     156-159

WEDNESDAY EVENING

- **SCI-MIX**
                                                              Papers 160

THURSDAY MORNING

- **Symposium on Comparison of Ab Initio Quantum Chemistry With Experiment**
Session VII. Reactive Scattering
T. Lee, Presiding   Papers 161-165

- **Symposium on New Developments and Applications of Magnetic Resonance and Optical Spectroscopies**
G. P. Drobny, Presiding   Papers 166-170

- **Symposium on Gas-Phase Metal Reactions IV.**
M. A. Duncan, Presiding   Papers 171-175

- **Chemical Vapor Deposition of Diamond (and c-BN, SiC) Growth Mechanisms**
S. J. Harris, Presiding   Papers 176-181

- **Symposium on Atomic Imaging: STM and Related Techniques**
R. M. Feenstra, Presiding   Papers 182-187

THURSDAY AFTERNOON

• Symposium on Comparison of Ab Initio Quantum Chemistry With Experiment
Session VIII. Potential Energy Surfaces and Dynamics
J. D. Goddard, Presiding — Papers 188-192

• Symposium on New Developments and Applications of Magnetic Resonance and Optical Spectroscopies
Jl. Baum, Presiding — Papers 193-197

• Symposium on Gas-Phase Metal Reactions 4 V.
J. V. Michael, Presiding — Papers 198-201

• Chemical Vapor Deposition of Diamond (and c-BN, SiC) Surface Studies
J. T. Yates, Jr., Presiding — Papers 202-205

• Symposium on Atomic Imaging: STM and Related Techniques
R. J. Colton, Presiding — Papers 206-212

THURSDAY EVENING

• Poster Session/Social Hour
H. G. Schaefer, Presiding — Papers 213-222, 224-338

FRIDAY MORNING

• Symposium on Comparison of Ab Initio Quantum Chemistry With Experiment
Session IX. Hyperpolarizabilities and Non-Linear Optics of Molecules
R. J. Bartlett, Presiding — Papers 339-343

• Symposium on New Developments and Applications of Magnetic Resonance and Optical Spectroscopies
D. B. Zax, Presiding — Papers 344-348

• Symposium on Gas-Phase Metal Reactions VI
M. C. Heaven, Presiding — Papers 349-353

• Chemical Vapor Deposition of Diamond (and c-BN, SiC) Theory
M. R. Pederson, Presiding — Papers 354-359

• Symposium on Atomic Imaging: STM and Related Techniques
J. S. Murday, Presiding — Papers 360-366

• Symposium on Gas-Phase Metal Reactions VII
A. Fontijn, Presiding — Papers 367-369

• Symposium on New Developments and Applications of Magnetic Resonance and Optical Spectroscopies
T. M. Duncan, Presiding — Papers 370-375

PHYSICAL CHEMISTRY

1. THEORY OF PAIRING IN HTC MATERIALS. Alan R. Bishop, MS B262, Los Alamos National Laboratory, Los Alamos, NM 87545, USA.

Much of the theoretical discussion of superconductivity in the cuprate HTC materials has focused on a simple 2-dimensional, 1-band Hubbard model. We review the current understanding of this model and argue that it probably does *not* contain all the ingredients to produce high-temperature superconductivity. We suggest that additional physics is needed for this, and review the evidence for strong, *nonlinear* electron-phonon coupling. This represents charge-transfer-lattice relaxation coupling, and high-T superconductivity driven by the proximity of a ferroelectric phase rather than an antiferromagnetic one. In particular, we point to the importance of *biphonon*-electron coupling mechanisms, and the influence of neighboring structural phase transitions.

2. Physical Properties of the Organic Superconductors. Paul Chaikin.

Abstract not available.

3. TERNARY NITRIDES F. J. DiSalvo, **Department of Chemistry, Cornell University, Ithaca, NY 14853**

The nitride ion is expected to be the ion most like oxide in solids, since the size, polarizability, and electronegativity of nitrogen are the closest to oxygen. Since their valences are different and since the N_2 bond energy is twice that of O_2, the chemistry and structures of nitrides might be expected to be unusual. Since ternary solid state nitrides are virtually unexplored, certainly compared to oxides, we are examining their chemistry and properties as potential analogues to high temperature superconductors and to look for unusual properties in general. We have found that many ternary nitrides can be prepared if an electropositive metal is combined with a transition metal or post-transition metal and nitrogen. I will discuss the synthesis, structure and properties of many novel nitrides discovered by Cornell students. These include compounds such as Ca_3BiN, Ca_2ZnN, $CaNiN$, Ca_3CrN_3, Na_3MoN_3, $LiMoN_2$, and $NaNbN_2$.

Support of this work by NSF grant DMR-8920583 and by the Office of Naval Research is greatly appreciated.

4. GOLD6. R. E. Smalley, Rice Quantum Institute and Departments of Chemistry and Physics, Rice University, Houston, Texas 77251

For an electron traveling above a flat metal surface there are a number of bound stages due to the potential caused by interaction of the electron with its image charge in the metal. For spherical metal particles these states should also exist even as the sphere approaches 1 nm in size. We believe we have found an example in the near-threshold photodetachment spectrum of the negative ion of a six-atom cluster of gold.

5. THE STUDY OF THE SPECTROSCOPY OF SMALL MOLECULES BY
   AB INITIO METHODS.
   S.R.Langhoff, C. W. Bauschlicher, Jr. and H. Partridge NASA Ames Research Center, Moffett Field, CA 94035.

   Ab initio calculations are reported for the dimers of the simple metals B_2 and Al_2 as well as the transition-metal systems Ti_2 and Zr_2. For the simple metal systems we are able to definitively determine the ground states, determine accurate spectroscopic constants for the bound states, and characterize the dipole-allowed transitions to aid experimental assignment of the bands. In contrast, calculations on the transition-metal dimers, especially Ti_2, are much more challenging. A definitive determination of the identity of the ground state of Ti_2 is complicated by the significant number of low-lying states of different multiplicity and the large differential contributions from inner-shell correlation to the stability of the states.

6. VIBRATIONAL STRUCTURE OF FULLERENES. M. Moskovits, K. Akers, L. Cousins, X. Gu, Dept. of Chemistry, Univ. of Toronto, Toronto, Ontario M5S 1A1

   Surface enhanced Raman was used to obtain good quality spectra of C_{60}, C_{70}, and C_{84}. Molecules were vaporized under ultra-high vacuum conditions. Samples enriched in the different molecules studied were prepared by vaporization at different source temperatures. Because of the weakening of selection rules in SERS, more lines were observed in the SERS spectrum of C_{60} than in the ordinary Raman or infrared spectra.

   Scanning Tunnelling Microscopy was also used to image C_{60}. These will be shown.

7. INFRARED LASER SPECTROSCOPY OF SUPERSONICALLY COOLED CARBON CLUSTERS. J.R. Heath and R.J. Saykally, Department of Chemistry, Univeristy of California, Berkeley, California 94720

   A new experiment for measuring rovibrational spectra of covalent clusters has been developed. Clusters are generated and cooled to 10–50 K by laser vaporization into a supersonic expansion. Infrared diode lasers are used to measure direct absorption spectra using pulse-averaging techniques. Extensive spectra have been measured for the linear C_3, C_4, C_5, C_7, and C_9 clusters.[1-6] Analysis reveals structures, bonding, and dynamical information.

   [1] J.R. Heath, A.L. Cooksy, M.H.W. Gruebele, C.A. Schmuttenmaer, and R.J. Saykally, Science 244, 564 (1989).
   [2] J.R. Heath, R.A. Sheeks, A.L. Cooksy, and R.J. Saykally, Science 249, 895 (1990).
   [3] J.R. Heath and R.J. Saykally, J. Chem. Phys. (in press).
   [4] J.R. Heath and R.J. Saykally, J. Chem. Phys. 93, 8392 (1990).
   [5] J.R. Heath and R.J. Saykally, J. Chem. Phys. (in press).
   [6] C.A. Schmuttenmaer, R.C. Cohen, N. Pugliano, J.R. Heath, A.L. Cooksy, K.L. Busarow, and R.J. Saykally, Science, 249, 897 (1990).

8. HIGH LEVEL THEORETICAL STUDIES OF CARBON CLUSTERS WITH COUPLED-CLUSTER METHODS. John D. Watts, Jurgen Gauss, John F. Stanton and Rodney J. Bartlett, University of Florida, Quantum Theory Project, Departments of Chemistry and Physics, 362 Williamson Hall, Gainesville, FL 32611-2085.

   We report results of coupled-cluster calculations including triple excitation effects on several neutral and charged small carbon clusters. Specific problems addressed include: (1) Properties of low-lying states of C_2; (2) The structure and harmonic force field of C_3; (3) The C_4 linear-rhombic energy difference and properties of the two isomers; (4) Structures, electronic, and vibrational spectra of C_4 and some larger carbon clusters; (5) Photodetachment spectra of some carbon cluster anions.

This work has been supported by the United States Office of Naval Research Contract No. ONR-N00014-91-J-1282.

9. **SYNTHESIS AND T_c IN SIMPLE SUPERCONDUCTING OXIDE SYSTEMS.** <u>D.G. Hinks</u>, J.D. Jorgensen, B. Dabrowski, Shiyou Pei*, P. Radaelli* and D. Richards*, Materials Science Division and Science and Technology Center for Superconductivity*, Argonne National Laboratory, Argonne IL 60439.

The synthesis of the oxide superconductors involves the incorporation of electrically active defects into the normally insulating lattice. In simple oxide systems, such an $La_{2-x}Sr_xCuO_{4\pm\delta}$, as the dopant concentration increases oxidizing the CuO_2 planes, T_c increases. Above a critical concentration of dopant, further oxidization depresses T_c. In this overdoped region it is felt that the depression of T_c is due to the increased carrier concentration. Our results indicate this is not so: the behavior of T_c in this region depends on the synthesis procedure. In this metastable, overoxidized region, different subtle effects can occur (e.g. phase separation, crystallographic phase changes, etc.) which determine T_c.

10. **OXYGEN ISOTOPE EFFECT AND STRUCTURAL PHASE TRANSITIONS IN LA_2CuO_4 BASED SUPERCONDUCTORS.** <u>M. K. Crawford</u>, W. E. Farneth, E. M. McCarron, R. L. Harlow, Du Pont Experimental Station, PO Box 80356, Wilmington, DE 19880-0356; A. H. Moudden, Brookhaven National Laboratory, Upton, Long Island, NY 11794 and Laboratoire Leon Brillouin, CEA-CNRS Saclay 91191, Gif sur Yvette, France; M. Kunchur and S. J. Poon, University of Virginia, Department of Physics, Charlottesville, VA 22901

We have measured the oxygen isotope effect exponent, α_O, on the superconducting transition temperature, $T_c(T_c \sim M^{-\alpha_O}$, where M is the average oxygen mass), in a series of materials of composition $La_{2-x}M_xCuO_4$ (M=Sr,Ba,Ca). We find that α_O exhibits a narrow maximum at x=0.12 in each system [M. K. Crawford et al., Phys. Rev. B<u>41</u>, 282 (1990); M. K. Crawford, et al., Science <u>250</u>, 1390 (1990)]. The maximum in α_O versus x($\alpha_O > 0.5$) is well correlated with the region in which the Bmab→P4$_2$/ncm lattice instability occurs in $La_{2-x}Ba_xCuO_4$ [J. D. Axe et al., Phys. Rev. Lett. <u>62</u>, 2751 (1989)]. Synchrotron x-ray diffraction studies demonstrate that ^{18}O substitution produces no change in crystal symmetry and little change in lattice parameters, ruling out such changes as the origin for the large isotope shifts of T_c. The unusual isotope effects presumably arise either from lattice anharmonicity or electronic density of states variations near the Fermi surface associated with the LTO→LTT instability. We will also discuss recent studies of newly discovered structural phase transitions in closely related systems which yield additional insight into the relationship between lattice instability and high T_c superconductivity in these materials.

11. **TRANSIENT PHOTOINDUCED CONDUCTIVITY IN SINGLE CRYSTALS OF $YBa_2Cu_3O_{6.3}$: "PHOTO-DOPING" TO THE METALLIC STATE.** <u>A. J. Heeger</u>, G. Yu and C. H. Lee, Institute for Polymers and Organic Solids, University of California, Santa Barbara 93106 and N. Herron and E. M. McCarron, E. I. du Pont de Nemours and Co., Inc. Central Research, Wilmington, Delaware 19898.

The temperature dependences of the transient *photo-induced conductivity in $YBa_2Cu_3O_{6.3}$ at different light intensities* (I_L) and of the *doping-induced conductivity in $YBa_2Cu_3O_{7-\delta}$ at different* δ are similar, indicative of "photo-doping". Signatures of the photo-induced transition to metallic behavior are observed at $I_L > 5 \times 10^{15}$ photons/cm^2; the deep resistivity minimum below 100K, reminiscent of the onset of superconductivity in inhomogeneously doped samples, is interpreted in terms of phase separation and metallic droplet formation. A longitudinal magnetic field (≤ 0.5 Tesla) reduces both the resistivity minimum and the superlinear contribution to the transient photoconductance.

12.
NONLINEAR ELECTRON-PHONON INTERACTIONS IN HIGH-TEMPERATURE SUPERCONDUCTORS

Annette Bussmann-Holder, Max-Planck-Institut für Festkörperforschung, Heisenbergstr. 1, D-7000 Stuttgart 80, FRG

The nonlinear polarizability of the oxygen ion induces an onsite nonlinear potential in the electron-ion interaction. This potential can be treated within a selfconsistent phonon approximation (SPA) which quantitatively describes soft-phonon induced structural phase transitions. The use of Eliashberg equations admits to solve for the electron-phonon coupling λ resulting from soft phonon modes and to calculate the superconducting transition temperature T_c. It is found that definite limitations on T_c are imposed using SPA while high T_c's can be obtained by considering electron multiple-phonon interactions.

13. MOLECULAR REARRANGEMENTS AND PSEUDOROTATION. E. R. Davidson, Department of Chemistry, Indiana University, Bloomington, Indiana 47405.

CASSCF calculations with a 6-31G* basis for the Cope rearrangement shows both concerted and Dewar reaction paths. Higher level results favor a symmetrical transition state. Excitation exchange in molecules will be discussed in the context of the model system $Zn^* + C_2H_4 \rightarrow Zn + C_2H_4^*$. Finally, the vibronic levels of the Jahn-Teller distorted molecule CH_4^+ will be discussed using globally and locally diabatic wavefunctions.

14. CONFIGURATION INTERACTION STUDIES OF SYMMETRY BREAKING IN SMALL MOLECULES.* Byron H. Lengsfield, Theoretical Atomic and Molecular Physics Group, Lawrence Livermore National Laboratory, Livermore, California 94550.

The problems associated with the accurate determination of the equilibrium geometry and asymmetric vibrational frequencies of small molecules which exhibit symmetry breaking will be examined with both single reference and multireference CI wavefunctions. This talk will focus on the determination of the molecular orbitals that can be used to study both the equilibrium properties and the reactions of molecules that exhibit symmetry breaking. In particular, the ability of unstable SCF and MCSCF solutions to fulfill this goal will be discussed.
*This work was performed under the auspices of the U.S. Department of Energy at Lawrence Livermore National Laboratory under contract number W-7405-Eng-48.

15 COUPLED CLUSTER TREATMENT OF SYMMETRY BREAKING. J.F. Stanton, J.D. Watts J. Gauss, R.J. Bartlett

A common problem in the theoretical description of open-shell molecules is the phenomenon of orbital symmetry breaking, in which the zeroth-order wavefunction [either restricted open-shell or unrestricted Hartree-Fock (UHF and ROHF, respectively)] localizes unpaired spins on one of a set of symmetrically equivalent centers. As a result, the total HF wavefunction does not transform as a pure irreducible representation of the molecular point group, a property which must be rigorously obeyed by the exact wavefunction. As a result of the poor zeroth-order orbitals, convergence towards the exact answer in a many-body perturbation theory or configuration interaction treatment can be excruciatingly slow. However, methods based on the coupled-cluster approximation offer significant advantages in the treatment of orbital relaxation effects and are capable of nearly full recovery from the qualitatively incorrect features of the zeroth-order wavefunction. Even better results can be obtained in CC calculations based on reference functions which are not symmetry broken, such as those composed of quasi-restricted Hartree-Fock (QRHF) orbitals, symmetrized natural orbitals, and other possibilities. In this talk, the classic symmetry breaking cases found in certain electronic states of nitrogen dioxide and nitrogen trioxide radicals are used to discuss the suitability of CC-based approaches to these problems. Special emphasis is placed on the choice of reference function and the convergence of results with increasing sophistication of the CC treatment.

16. SYMMETRY BREAKING IN RADICALS STUDIED BY THE FOCK-SPACE MULTIREFERENCE COUPLED CLUSTER METHOD. <u>Uzi Kaldor</u>, School of Chemistry, Tel Aviv University, 69978 Tel Aviv, Israel

The geometry and harmonic frequencies of several radicals in their ground and excited states were studied by the Fock-space coupled cluster method with single and double excitations. Hartree-Fock orbitals for a closed shell are used in this formalism, with one or two electrons added or ionized to yield the state of interest. This feature serves to avoid the bias toward symmetry breaking observed in UHF functions. The orbital relaxation is included by summing the T_1 operator to all orders. Results will be reported and compared with experimental and theoretical work.

17. SHORT-LIVED MOLECULES OF RELEVANCE IN THE EARTH'S ATMOSPHERE: EXPERIMENTAL STUDIES OF THE SPECTROSCOPY OF NO_3, Cl_2O_2, AND Cl_2O_3. <u>R. R. Friedl</u>, Jet Propulsion Laboratory, Pasadena, California 91109.

The physical properties of the Earth's atmosphere are profoundly influenced by a variety of theoretically interesting molecules which are produced from photochemically initiated processes. The nitrate radical, NO_3, which serves as the primary chemical intermediate in the NO_x-catalyzed destruction of ozone, is one important atmospheric constituent for which an adequate description of its physical properties has remained elusive. We have studied the rovibronic structure of this species using high-resolution Fourier transform spectroscopy. Data has been obtained between 700-6000 cm^{-1} and includes temperature dependent band strengths. The significance of this data with respect to the ground state geometry of NO_3 will be discussed.
The dramatic destruction of polar ozone has focussed attention, for the first time, on the possible importance of higher oxides of chlorine in the atmosphere. Submillimeter wave spectroscopic data between 380 and 435 GHz will be presented for Cl_2O_2 and Cl_2O_3 along with a challenge to theoreticians.

18. EXCITED STATE STRUCTURE AND TRANSPORT DYNAMICS OF PHOTOSYNTHETIC UNITS: SPECTRAL HOLE BURNING STUDIES. <u>G. J. Small</u>, Ames Laboratory USDOE and Department of Chemistry, Iowa State University, Ames, Iowa 50011.

In recent years, spectral hole burning spectroscopies (photochemical, non-photochemical) have been used to improve the resolution in the electronic absorption spectra of chlorophylls (Chls) and other photocatalytic cofactors in photosynthetic units by up to 4-orders of magnitude [1]. It is now routinely possible to get "under the skin" of broad (~ 200-500 cm^{-1}) absorption bands to determine the site inhomogeneous broadening (Γ_I) and homogeneous broadening (Γ_H) contributions to the bandwidth. It is also possible to determine the contributions from linear electron-phonon coupling and exciton-level structure to Γ_H. The interplay between Γ_H and Γ_I is important for determining the role of heterogeneity in the temperature dependence of energy (optical excitation) and electron-transfer dynamics in the antenna and reaction center complexes. This talk will discuss the principles and theory of spectral hole burning and recent data for photosynthetic units from green plants and photosynthetic bacteria which provide new insights on their excited electronic state structure and transport dynamics (occurring on a 0.1-1 ps time scale).

1. S. G. Johnson, I.-J. Lee and G. J. Small, in *Chlorophylls*; Ed., H. Scheer; CRC Publishers, 1991.

19. ON THE CONVERSION OF SOLAR TO ELECTRIC ENERGY BY THE OTHER PHOTOSYNTHETIC SYSTEM IN NATURE, BACTERIORHODOPSIN (bR). <u>M. A. El-Sayed</u>, Department of Chemistry and Biochemistry, University of California, Los Angeles, California 90024

bR undergoes a photochemical cycle upon light absorption involving changes in the retinal (with its protonated Schiff base (PSB)) and protein conformations. This leads to: 1) rapid isomerization of retinal, 2) charge separation, 3) proton dissociation of the PSB, and 4) proton pumping from inside to the outside surface of the membrane. Proton gradients are thus created and used in converting ADP into ATP, the fuel of life. It is shown that, contrary to the generally accepted belief for the past 15 years, the CD spectrum cannot be explained by the exciton theory in bR as the prediction from such model are found to contradict observations on the retinal-retinal distance, linear dichroism, the isomerization time scale, and above all, the CD spectrum itself if the uncertainty principle is considered. Thus, unlike chlorophyll, bR does not have an antenna system.

Energy transfer studies between tryptophan (TRP) and retinal are carried out to follow the protein conformation changes during the photocycle. Furthermore, in order to determine the location of the metal cations, which are important to the dissociation of the PSB, time resolved studies involving Eu^{3+}-regenerated bR are carried out.

20. ELECTRONIC EXCITATION TRANSPORT IN PHOTOSYNTHETIC ANTENNAE. S. Lin, P. A. Lyle, H. van Amerongen, and W. S. Struve, Iowa State University, Ames, IA 50011.

Our recent ultrafast pump-probe experiments on photosynthetic antennae have been focused on the issues of exciton coherence and phonon involvement in electronic excitation transport (EET) processes. The first of these issues was addressed in a study of the 740 nm BChl c light-harvesting antenna in chlorosomes from the green bacterium Chloroflexus aurantiacus. The absorption difference spectrum (which is dominated by excited state absorption and ground state photobleaching respectively at wavelengths less than and greater than ~ 730 nm) shifts dynamically to the blue during the first 7 ps. This blue shift is consistent with relaxation between exciton components in the BChl c aggregate, and not with downhill EET between different BChl c pigment pools. Phonon participation in antenna EET was investigated in a temperature-dependent study of the photosystem I Chl a antenna between 38 and 300 K.

21. FLUORESCENCE MECHANISMS AND PHOTOINDUCED ELECTRON TRANSFER IN COVALENTLY LINKED POLYNUCLEAR AROMATIC - NUCLEOSIDE AND OLIGONUCLEOTIDE COMPLEXES. N.E. Geacintov, L. Margulies, M. Cosman, B. Mao, N.-Q. Ya, R. Zhao and P.F. Pluzhnikov, Chemistry Department, New York University, New York, NY 10003.

The diol epoxide metabolites of polycyclic aromatic hydrocarbons are potent mutagens and carcinogens. We have studied the physico-chemical characteristics of adducts derived from the binding of the diol epoxide (+)-trans-7,8-diol-anti-9,10-epoxy-benzo[a]pyrene (BPDE), the ultimate carcinogenic metabolite of benzo[a]pyrene, to deoxyguanosine (dG). (BPDE and dG are covalently linked via the C-10 position of BPDE and N^2-dG). Picosecond laser flash photolysis studies indicate that when the pyrene-like residue of the BPDE-dG adduct is excited at 355 nm, an electron is transferred from dG to the pyrenyl moiety, decreasing the fluorescence yield by a factor of ≈ 30); this exciplex decays within 3-4 ns. The fluorescence quantum yields, as well as the characteristics of the fluorescence emission spectra, depend strongly on (1) the stereochemical characteristics of the adducts (the relative conformations of the guanosyl and pyrenyl moieties), and (2) on the nature of the bases flanking the BPDE-dG adducts on either side in deoxyoligonucleotides three or more bases long.

22. TOWARDS SPATIALLY RESOLVED ENERGY TRANSFER AND SCANNING MOLECULAR EXCITON MICROSCOPY. R. Kopelman, Department of Chemistry, University of Michigan, Ann Arbor, MI 48109.

Subwavelength light sources have been constructed with the aid of luminescent and exciton transporting materials. These can be used as scanning, light emitting tips of nanometer dimensions or as scanning exciton donor tips. The theoretical resolution limit for near-field exciton microscopy is on the molecular scale. The detection limit is a single molecule where this molecule could be identified spatially as well as spectrally. The probes consist of aluminum coated glass micropipettes with active crystal tips (anthracene, tetracene, perylene, etc.). As it is scanned over a sample, it senses a variety of perturbations such as quenching or external heavy atom effects. Biological and chemical applications will be discussed, with emphasis on single molecule energy transfer, single site reactions and in-situ DNA sequencing.

23. ENERGY TRANSFER IN MOLECULAR AGGREGATES: INFLUENCE OF COLORED NOISE
P. Reineker, V. Kraus and Ch. Warns, Abteilung Theoretische Physik, Universität Ulm, Albert-Einstein-Allee 11, 7900 Ulm, Germany

The transport of electronic excitation energy is of importance in various fields of condensed matter physics. To investigate the influence of a vibrating environment on the transport properties in a dimer, we consider a model, in which the electronic interaction between its molecules is described by a transfer matrix element and the vibrations give rise to fluctuations of the molecular excitation energy. These fluctuations are modeled by a dichotomic stochastic process with colored noise. The equation for the density operator of the model is solved numerically and the behaviour of the solution in time is discussed in dependence of the parameters of the stochastic process. Recently we have obtained exact analytical expressions for the eigenvalues of the density operator which allow a detailed discussion of the crossover between coherent and incoherent motion.

24. Overview of Organic Superconductors: Syntheses, Structures and Physical Properties
H. Hau Wang, K. Douglas Carlson, U. Geiser, Arvinda M. Kini, Jack M. Williams, and James E. Schirber,[+] Chemistry/Materials Science Divisions, Argonne National Laboratory, Argonne, IL, U.S.A. 60439 [+]Sandia National Laboratories, Albuquerque, NM 87185, U.S.A.

The critical temperature of organic superconductors has been steadily increasing at a rate of ~1 K per year since the discovery of $(TMTSF)_2PF_6$ ($T_c = 0.9$ K, 12 kbar) at 1979. Near forty organic superconductors have been prepared and the highest T_c known at present is 13 K (0.3 kbar) in the κ-$(d_8$-$ET)_2Cu[N(CN)_2]Cl$. These materials have two-dimensional layer type crystal structures similar to that of the high T_c oxide superconductors. Physical property measurements also reveal their 2D nature. The synthetic strategy, typical packing motifs, conductivity, specific heat, and ESR measurements of these materials will be discussed.

*Work performed under the auspices of the Office of Basic Energy Sciences, Division of Materials Sciences, of the U. S. Department of Energy under Contract W-31-109-ENG-38.

25. THE IMPORTANCE OF PRESSURE IN THE CREATION AND STUDY OF SUPERCONDUCTIVITY IN ORGANIC MATERIALS. James E. Schirber, Org. 1090, Sandia National Laboratories, Albuquerque, NM 87185.

Pressure has played a key role in the study of organic superconductivity since its initial discovery at 0.9 K and 11 kbar a decade ago. The sensitivity of the electronic structure and phase diagram to modest pressures has been demonstrated in several families of organic materials. The historical evolution of the field with emphasis on the role of pressure will be reviewed. Recent studies where regions of the phase diagram displaying record high superconducting temperatures for organic materials can be accessed with modest hydrostatic pressure will be discussed.

Work was supported by the U.S. Department of Energy under contract #DE-AC04-76DP00789.

26. **GROWTH AND CHARACTERIZATION OF HIGH T_c CUPRATE CRYSTALS** Lynn F. Schneemeyer and Joseph V. Waszczak, AT&T Bell Laboratories, Murray Hill, NJ 07974

Much of the progress in our understanding of the chemistry and properties of high T_c superconductors has come from studies of single crystals. The growth of crystals of these cuprates is complicated, however, by the extensive substitutional chemistry and defect chemistry of these complex materials. We discuss recent progress in the growth of crystals of high T_c oxides with emphasis on the 92K superconductor, $Ba_2YCu_3O_7$, and its isomorphs. Approaches to improving the size and quality of crystals include the use of melt additives such as alkali halides, variations in the oxygen partial pressure during growth, and control of growth parameters such as soak time and temperature. Results of the characterization of these crystals including correlations between oxygen stoichiometry and critical currents will also be presented.

27. CATION-DOPANT SITE SELECTIVITY, OXYGEN STOICHIOMETRY, AND Cu VALENCE IN THE HIGH T_c SUPERCONDUCTORS. George H. Kwei, Los Alamos National Laboratory, Los Alamos, NM 87545.

From the outset, it was clear that changes in the hole concentration by cation or oxygen doping in the high T_c cuprate superconductors were important in determining the superconducting properties. In the last few years, many x-ray and neutron diffraction experiments have been done to determine the site selectivities of cation dopants and the occupancies of the individual oxygen sites. In many cases, the determination of cation site occupancy has been difficult because of the lack of sufficient contrast between the dopant ions and copper. In these cases, the use of isotopically labelled dopants or Cu in neutron diffraction experiments, or of joint refinement of neutron and anomalous x-ray diffraction data, provides the greater contrast necessary to establish the site preference. The oxygen stoichiometry has not been as much of a problem and is generally available from neutron diffraction experiments. The large body of information that has been obtained allows us to examine and discuss the relationship between cation content and ordering, the oxygen stoichiometry, the estimated valences and hole concentrations, and T_c. We will also present direct Cu site valence determinations for $YBa_2Cu_3O_{6+\delta}$ as a function of oxygen stoichiometry using a new joint neutron/anomalous x-ray technique we have introduced. In these experiments, the neutron data provides the structural information while the valence shift of the absorption edge provides information about the site-specific valence.

28. STRUCTURE-PROPERTY CORRELATIONS IN CUPRATE SUPERCONDUCTORS. M.-H. Whangbo, Department of Chemistry, N. C. State University, Raleigh, NC 27695-8204, and C. C. Torardi, Central Research, E. I. du Pont de Nemours and Co., Experimental Station, DE 19880.

We examine structural and electronic factors governing the properties of cuprate superconductors, and point out several structure-property relations that a satisfactory theory of cuprate superconductivity must accommodate.

29. HIGH OVERTONE SPECTRA: THE SIGNIFICANCE OF PERTURBATIONS. Ian M. Mills, Department of Chemistry, University of Reading, RG6 2AD, England

Observing high overtone vibrational spectra by direct absorption from the ground state requires long path lengths or special detection techniques such as laser photoacoustic spectroscopy in order to observe weak bands. However persistent application of these methods yeilds valuable results: for the HC≡CF molecule we have now observed more than 200 bands in the region up to 16 000 cm^{-1}, of which about 75% have been rotationally analysed, providing information on more than 100 vibrational states.

Resonances play a key role in the analysis of such spectra, and they are believed to control the molecular dynamics of energy flow within the molecule. In HC≡CF we see both weak localized rotational resonances, some of which we have been able to completely analyse, and a complex pattern of strong anharmonic (Fermi type) resonances, which we are also studying. I shall describe these results. The strong resonances control the short time (ps to fs) vibrational relaxation, and the weak resonances control the long time (ns) relaxation.

30. VARIATIONAL ROVIBRATIONAL STUDIES OF 3- AND 4- ATOM MOLECULES. Nicholas C. Handy Department of Chemistry, University of Cambridge, Lensfield Road, Cambridge CB2 1EW

We report studies using the variational method for the rovibrational energy levels of 3- and 4- atom molecules. A representation of the kinetic energy operator in valence coordinates is used. The purpose is to obtain accurate representation of the potential energy surfaces by adjusting parameters to fit known experimental data, and then to help assign other unknown features of the spectra.

We shall discuss results from studies on HCN, HCP, NH_2, CH_2, H_2S_2, HCCH, HCCF on which we have collaborated with other scientists.

31. MATRIX INFRARED SPECTRA AND AB INITIO FREQUENCIES OF SULFUR-CONTAINING REACTIVE SPECIES, L. Andrews, R. Bohn, P. Hassanzadeh, Department of Chemistry, University of Virginia, Charlottesville, VA 22901.

A variety of new sulfur-containing reactive species have been prepared by combining discharge and matrix isolation techniques. Ab initio calculations were performed using the HONDO 7.0 programs with STO-3G and DZP basis sets to find optimized structures and fundamental frequencies. Comparisons will be made for S_3, SF, SF_3, HSCS, C_2S_2, SNO, NS, NS_2 and N_2S. Properly scaled calculated DZP frequencies agree quite well with matrix infrared spectra. Using S_3 as an example, DZP frequencies are 772, 670 and 292 cm^{-1}. A scale factor of 0.87 derived from the Raman observation of ν_1 at 583 cm^{-1} predicts ν_3 at 671 cm^{-1}, as compared to the 674.5 cm^{-1} observed value. The S_3 valence angle of 116 ± 2° is calculated from four sets of sulfur isotopic ν_3 fundamentals, which is in agreement with the DZP value of 117.3°.

32. Cyanomethylene and Benzyne: Is It Possible to Escape From the Tar Baby? Henry F. Schaefer, Center for Computational Quantum Chemistry, University of Georgia, Athens, Georgia 30602.

Abstract not available

33. AB INITIO CALCULATIONS AND THE INTERPRETATION OF VIBRATIONAL SPECTRA. Willis B. Person, Krystyna Szczepaniak, M. Szczesniak, Janet E. Del Bene[a], and J. Leszczynski[b], Dept. of Chemistry, University of Florida, Gainesville, FL 32611-2046, [a]Youngstown State University, Youngstown, OH, and [b]Jackson State University, Jackson, MS.

Calculations of the vibrational spectra of pyrimidine and purine molecules made using 3-21G or better 6-31G(d) basis sets have proven to be accurate enough to be of considerable help in the assignment and interpretation of the vibrational spectra of these large organic molecules. Such calculations have now been made for enough molecules that it is possible to examine the calculated force constants and intensity parameters (atomic polar tensors) for similar molecules to determine the extent of transferability of these parameters, and to try to relate the changes to the chemical properties of the molecules involved. Here we shall try to review results for pyrimidines uracil, thymine, and their derivatives and water complexes, and for the purine guanine and its derivatives.

We are grateful for support from NIH Research Grant No. GM-32988.

34. **TIME-DOMAIN ANALYSIS OF THE DYNAMICS OF FRENKEL EXCITONS IN DISORDERED SYSTEMS: T = 0 K.** D. L. Huber, Department of Physics, University of Wisconsin-Madison, Madison, WI 53706.

The dynamics of Frenkel excitons in disordered systems is studied in the time domain at zero temperature. Particular emphasis is placed on the evolution of the integrated fluorescence in the forward direction following pulsed, broadband optical excitation. Methods for calculating the fluorescence by integrating the coupled equations of motion for the dipole-moment correlation functions are outlined. Two models are studied in detail: a binary system where the two components have different transition frequencies and a single-component system with a random distribution of trapping centers. In the binary model, the conditions for oscillatory decay of the fluorescence are established for a 50-50 mixture. In the trapping problem, the asymptotic decay rates are compared with the results from lowest-order perturbation theory and theories based on the coherent potential and average T-matrix approximations. The effect of a Gaussian distribution of transition frequencies on the trapping is also investigated.

35. STATISTICAL MECHANICAL STUDIES OF ELECTRONIC LINE BROADENING IN LIQUIDS. R.F. Loring, N. Shemetulskis, and A.M. Walsh, Department of Chemistry, Baker Laboratory, Cornell University, Ithaca, NY 14853.

A relation is presented between the inhomogeneously broadened electronic absorption spectrum of a solute at infinite dilution in a solvent and an equilibrium free energy of solvation of a fictitious solute. This formally exact result permits the techniques of equilibrium fluid theory to be applied to the calculation of solvation effects in electronic spectroscopy. Monte Carlo calculations of the inhomogeneous line shape of a polar solute in a polar solvent are used to test the accuracy of analytical approaches. The simulations demonstrate that a linear response calculation of the solvation free energy works semiquantitatively for moderately strong solute-solvent coupling. This work suggests that electronic spectra of solvated molecules may be calculated with the same level of rigor as equilibrium thermodynamic properties of fluids. Our analysis is based on the assumption that electronic spectra in room temperature fluids are predominantly inhomogeneously broadened. This assumption is investigated with calculations of electronic dephasing times in nonpolar fluids.

36. MOLECULAR THEORY OF INHOMOGENEOUS SPECTRAL BROADENING IN LIQUIDS AND GLASSES. J. L. Skinner, Department of Chemistry, University of Wisconsin, Madison, WI 53706.

We review phenomenological theories of inhomogeneous broadening of the absorption lineshape for a solute in a condensed phase. We then discuss a microscopic statistical mechanical theory, appropriate for liquids or glasses, that focuses on the molecular origin of the correlation between the inhomogeneous energy distributions of different solute quantum states. We then generalize these ideas to consider the correlation between different transitions, and discuss their relevance to nonresonant fluorescence- and phosphorescence-line-narrowing and hole-burning studies. Within a similar theoretical framework we also discuss some interesting pressure effects on hole-burning spectra of solutes in polymeric glasses. Finally, we consider the correlation between transitions on different solutes, and discuss its relevance to problems of coherent and incoherent energy transfer.

37. FAST VIBRATIONAL RELAXATION FOR POLAR SOLUTES IN POLAR SOLVENTS. James T. Hynes[1], Margaret Bruehl[1], Robert M. Whitnell[2] and Kent R. Wilson[2], [1]Department of Chemistry, University of Colorado, Boulder, CO 80309-0215; [2]Department of Chemistry, University of California, San Diego, La Jolla, CA 92093-0339.

The vibrational relaxation of dipolar and charged solutes in water and other polar solvents is an important but relatively neglected area. We report on theoretical studies of very fast vibrational relaxation for methyl chloride in water (RMW, KRW, JTH), and for a model hydrogen-bonded complex in a model dipolar aprotic solvent (MB, JTH). The microscopic origin for this rapid relaxation is discussed, as is the applicability of Landau-Teller theory and Isolated Binary Collision theory. This work was supported by NSF grant CHE88-07852 to JTH. MB is a NSF Postdoctoral Fellow.

38. EXCITON COHERENCE-SIZE AND OPTICAL NONLINEARITIES IN RESTRICTED GEOMETRIES. S. Mukamel, Department of Chemistry, University of Rochester, Rochester, NY 14627.

The magnitude of optical nonlinearities of molecular microstructures is determined by a characteristic *coherence-length* which controls the cooperativity of the optical response. Equations of motion describing the evolution of the optical polarization coupled to two-exciton variables are derived, and used to calculate the third order response ($\chi^{(3)}$) in a one dimensional molecular crystal. We calculate the coherence length, and explore its dependence on the dephasing mechanism and the limitations of the local-field approximation.

39. PAIRED DISORDER AND THE METALLIC STATE IN POLYMERS LACKING A DEGENERATE GROUND STATE. P. Phillips, Department of Chemistry, M.I.T., Cambridge, Massachusetts 02139.

We consider here a site-disordered 1-dimensional tight-binding model in which neighboring pairs of sites are assigned one of two energies. We show that this random dimer model posesses a mobility edge as a function of the energy difference between the sites.[1] Transport is shown to arise because $\sqrt{N}$ of the electronic states at a particular energy remain unscattered by the disorder. We then show explicitly that the disorder inherent in a randomly protonated strand of the conducting polymer, polyaniline, is described by the random dimer model with additional off-diagonal disorder. We demonstrate exactly that the location in the energy band of the unscattered or conducting states of the random dimer model of polyaniline coincides with recent calculations of the location of the Fermi level in the protonated form of the polymer. We argue then that the random dimer model is capable of explaining the insulator-metal transition in polyaniline. We use the results of this model to obtain the temperature dependence of the experimentally-observed dc conductivity.

[1] D. H. Dunlap, H. L. Wu and P. Phillips, Phys. Rev. Lett. 65, 88 (1990).

40. NMR, COHERENT LASER SPECTROSCOPY, AND LASER NMR SPECTROSCOPY. Warren S. Warren, Chemistry Department, Princeton University, Princeton, N.J. 08544

This talk will emphasize comparisons between coherent optics and modern NMR. For example, pulse shape effects in two-level systems are well understood in both fields. Our recent work on NMR multilevel systems (quadrupolar nuclei, dipolar or scalar coupled spins) has led to a variety of shaped pulse sequences which outperform rectangular pulse excitation. We have also shown experimentally that picosecond frequency swept laser pulses generate larger inversions than unmodulated pulses, but deviations from two-level approximations make it important to match the modulation to the energy level spectrum. We will also compare macroscopic cooperative effects (radiation damping in NMR, super-radiance in optics). Finally, we will show the effects of off resonant c.w. laser irradiation on complex NMR spectra, to determine the feasibility of selective broadening or frequency shifts in crowded spectra without degradation.

41. MAGNETIC RESONANCE EXPERIMENTS WITH LASER POLARIZED RARE GASES G.D.Cates, A. Barton, P. Bogorad, B. Driehuys, M. Gatzke, W. Happer, N.R. Newbury, and B. Saam, Department of Physics, Princeton University, Princeton, New Jersey 08544

The nuclear spins of rare gases, *e.g.* ^{3}He, ^{21}Ne, ^{83}Kr, ^{129}Xe, ^{131}Xe, can be conveniently polarized by spin exchange with electrons of alkali-metal vapors which are maintained in a state of high polarization by optical pumping with circularly polarized laser light. Magnetic resonance studies benefit greatly from nuclear polarizations that can approach 100%, and gaseous spin-relaxation times that range from seconds to days. Once polarized, gaseous ^{129}Xe can be *frozen* with negligible loss of polarization. In the condensed phase, at temperatures below about 120 K, spin-relaxation of frozen ^{129}Xe is dominated by a nuclear spin-rotation interaction $\gamma_I I \cdot N$. At temperatures approaching 4 K, very long relaxation times are expected opening the possibility that existing lasers can be used to produce as much as 20 g of frozen ^{129}Xe per day. Laser-polarized noble gases can be a powerful tool in many diverse disciplines. Examples include the study of surfaces, solids, neutron scattering, and van der Waals molecules, to name just a few. — This work was supported by the U.S. AFOSR.

42. NEW APPROACHES TO ULTRASENSITIVE MAGNETIC RESONANCE. Daniel P. Weitekamp, A.A. Noyes Laboratory of Chemical Physics, Caltech, Pasadena, CA 91125.

Surface species, transient intermediates, molecular beam clusters, trapped ions, and semiconductor defects are all examples where structural insight from NMR is needed, but is unavailable for lack of sensitivity. We are developing several methods for overcoming this problem by coupling spin to other more readily detected degrees of freedom. These include the coupling to rotational motion through nuclear spin statistics, to photon polarization through hyperfine couplings, and to translational motion through spin–dependent forces. General principles will be described along with applications. Parahydrogen spin labelling on the surface of ZnO provides enhanced proton NMR spectra of a molecular chemisorption site, while the ordinary NMR spectrum is dominated by physisorbed molecules. A new scheme for optical NMR in GaAs is demonstrated, which simultaneously optimizes sensitivity and resolution and appears to be applicable to dilute defects. A novel ion spectroscopy will be described which promises to extend the single-molecule sensitivity of mass spectroscopies to magnetic resonance.

43. OPTICAL SPECTROSCOPY AT 1 MHz RESOLUTION. D. W. Pratt, Department of Chemistry, University of Pittsburgh, Pittsburgh, PA 15260.

Resolution is the key to obtaining information from both optical and magnetic resonance spectroscopy. Recent developments in technology have made possible great improvements in the resolving power of both techniques. This talk will describe one example, a new molecular beam laser spectrometer that has a resolving power of 3×10^8 in the ultraviolet, and its application to several problems in molecular structure and dynamics.

44. THE RELATION BETWEEN T_1 AND T_2 RELAXATION TIMES. J. L. Skinner, Department of Chemistry, University of Wisconsin, Madison, WI 53706.

In several different time-domain spectroscopies (e.g. NMR, vibrational, electronic) one can measure both the population and phase relaxation times for a particular transition, which are known as T_1 and T_2 respectively. The usual relation between these two times implies that T_2 is less than or equal to $2T_1$. In order to understand at a fundamental level the origin of this inequality, we have studied theoretically two quantum states coupled either to a stochastic or quantum-mechanical bath. We show that in some instances $T_2 > 2T_1$. By implication, then, the linewidth of a transition can be narrower than its "lifetime-limited" value.

45. Buckyball. R. E. Smalley, Department of Chemistry, Rice University, Houston, Texas 77251.

Abstract not available

46. SUPERCONDUCTIVITY IN ALKALI-METAL-DOPED FULLERENES, A.F. Hebard, AT&T Bell Laboratories, Murray Hill, NJ 07974.

The recent unexpected discovery of superconductivity in the alkali-metal-doped fullerenes, K_xC_{60} and Rb_xC_{60}, has generated keen interest in the research community not only because a new class of superconducting compounds, the fullerides, has been identified but also because the broad potential of fullerene research in general has been promoted. These novel superconductors with relatively high transition temperatures ($T_c \geq 28K$ in Rb_xC_{60}) provide unique materials challenges for the preparation of both film and bulk samples. This talk will describe preparative techniques and measurement results on these highly air-sensitive materials. Physical attributes such as structure, stoichiometry, electron transport, magnetic response, optical absorbance, uniformity of composition, and sensitivity of conductivity to alkali-metal dopant will be discussed and reviewed in the context of possible superconducting mechanisms.

47. Synthesis and Characterization of M_xC_{60}. R. Whetten, Department of Chemistry, University of California, Los Angeles, CA 90024.

Abstract not available

48. Structural Studies of Binary Compounds of C_{60} and Alkali Metals. J. E. Fischer, University of Pennsylvania.

Abstract not available

49. ELECTRONIC AND STRUCTURAL STUDIES OF FULLERENE-BASED SOLIDS: C_{60}, C_{70}, AND THE ALKALI-METAL FULLERIDES. J.H. Weaver, Department of Materials Science and Chemical Engineering, University of Minnesota, Minneapolis, MN 55455

Photoemission and inverse photoemission studies of condensed C_{60} and C_{70} reveal a rich distribution of occupied and empty electronic states derived from the molecular orbitals of the molecule. Those features will be discussed and their origin will be identified through comparison to band structure calculations. Equivalent studies conducted as a function of doping of the fullerenes with the alkali metals Na, K, and Cs show the non-rigid-band-like filling of the LUMO-derived band, an effect we attribute to disorder in the site occupation of the alkali ion. Parallel electrical measurements make it possible to correlate changes in the electronic structure with the electrical resistance. Analogous investigations for C_{70} reveal differences in energy band filling. Scanning tunneling microscopy results show growth and equilibrium structures for thin films of these high temperature van der Waals solids. STM images obtained as a function of K incorporation for K_xC_{60} show a highly ordered surface for pure C_{60}, an equally ordered surface for the metallic phase, but a disordered configuration for the K_6C_{60} phase, consistent with the conversion from the fcc to the bcc lattice structure.

50. MEASUREMENT OF THE MAGNETIC FIELD PENETRATION DEPTH IN $YBa_2Cu_3O_x$ AND $Bi_2Sr_2CaCu_2O_x$ SUPERCONDUCTORS BY ELECTRON SPIN RESONANCE LINE BROADENING OF SURFACE PARAMAGNETIC PROBES, Jerzy T. Masiakowski, Micky Puri and Larry Kevan, Department of Chemistry and the Texas Center for Superconductivity, University of Houston, Houston, Texas 77204-5641, USA.

The electron spin resonance line broadening of a paramagnetic probe on an oxide superconductor below the superconducting critical transition temperature is shown to depend on the parallel or perpendicular orientation of the probe to the applied magnetic field. The line broadening is associated with the magnetic flux lattice formed in the superconductor and leads to a measure of the magnetic field penetration depth. It is shown that the temperature below the superconducting critical transition temperature at which significant line broadening from the flux lattice occurs is best identified as the magnetic flux lattice melting temperature. This is particularly clear in the BiSrCaCuO superconductor in which the flux lattice melting temperature lies significantly below the superconducting critical transition temperature.

51. EXPERIMENTAL CONSTRAINTS ON THE THEORY OF HIGH TEMPERATURE SUPERCONDUCTIVITY. Z. Schlesinger, R. T. Collins, L. D. Rotter, F. Holtzberg, C. Feild, IBM Research, Yorktown Heights, New York 10598; U. Welp, G. W. Crabtree, J. Z. Liu, and Y. Fang, STCS & MSD, Argonne National Lab, Argonne, Illinois 60439

Over the past several years significant progress has been made in measuring the fundamental properties of high T_c superconductors. These data place substantial constraints on the theory of both the normal and superconducting states. We will discuss and relate results from infrared measurements, NMR, penetration depth, tunnelling, resistivity, and Raman scattering. Within these data one finds evidence for an anomalously large superconducting energy gap ($2\Delta/kT_c \simeq 8$), unusual temperature dependence near the superconducting transition, a highly unconventional normal state, and very weak electron-phonon coupling, (see e.g., Z. Schlesinger et al., Phys. Rev. Lett. **65**, 801 (1990)). The data strongly suggest that conventional approaches will not be adequate to describe either the normal state or the mechanism of superconductivity in superconductors with $T_c \sim 90$ K (e.g., $Y_1Ba_2Cu_3O_7$ and $Bi_2Sr_2CaCu_2O_{8-y}$).

52. INVESTIGATION OF SUPERCONDUCTING YBa$_2$Cu$_3$O$_7$ USING RAMAN SPECTROSCOPY[‡] K. F. McCarty, Sandia National Labs., Livermore, CA 94551; R. N. Shelton and J. Z. Liu, Dept. of Physics, U.C. Davis, Davis, CA. 95616; and H. B. Radousky, Lawrence Livermore National Lab., Livermore, CA. 94550

The rich chemistry and physics of the cuprate superconductors can be profitably explored by Raman spectroscopy. In addition to identifying the Raman-active phonons and characterizing the phase purity of samples, the technique can measure fundamental properties such as the degree of electron-phonon coupling and the superconducting energy gap. In this study, 14 of the 15 Raman-allowed phonons have been unambiguously assigned using single crystals of YBa$_2$Cu$_3$O$_7$ that are free of twins. Electronic Raman scattering was investigated using polarization along and perpendicular to the c axis of the crystals. A broad continuum was found for both polarizations but only the continuum from the a-b plane dramatically redistributed due to formation of the energy gap below T_c = 93 K. The magnitude of the energy gap (2Δ) was bounded between 440 and 500 cm^{-1} using the temperature dependence of phonon linewidths. This energy gap ($2\Delta = 6.8 - 7.7 k_B T_c$) is extremely large compared to the weak-coupling BCS value of $2\Delta = 3.5 k_B T_c$, highlighting the unconventional nature of YBa$_2$Cu$_3$O$_7$.
[‡]Work supported by the USDOE (U.C. Davis and LLNL) and USDOE, OBES, Division of Materials Sciences (SNL).

53. MODEL FOR THE ELECTRONIC STATES OF HIGH Tc CUPRATES AS PROBED BY X-RAY ABSORPTION. E.B. Stechel, Div. 1151, Sandia Nat'l Labs., Alb., NM, 87185 and M.S. Hybertsen, M.C. Foulkes and M. Schlutter, AT&T Bell Labs, Murray Hill, NJ, 07974.

The evolution of the electronic states in the cuprates as carrier concentration increases from the parent insulators has been quite controversial. Recent X-ray absorption measurements of C.T. Chen, et al [Phys. Rev. Lett. **66** 104 (1991)] follow the systematic evolution of the O p-like states as Sr is substituted for La in La2CuO4. We present a detailed theory of the X-ray absorption results starting from the three-band Hubbard model, mapping this with core-valence interactions and the absorption process onto a single band Hubbard model with material specific parameters. Within this model and without any adjustable parameters the X-ray absorption spectrum is calculated as a function of carrier concentration. The experimental pre-edge peak intensities and positions are semi-quantitatively reproduced as a function of Sr concentration. The combination of theory and experiment gives new and strong evidence that a single band Hubbard model adequately describes the electronic states of the high Tc cuprates well into the doping range of superconductivity. An interpretation of the single band model will be given. Work at Sandia Nat'l Labs supported by US DOE contract DE-AC04-76DP00789.

54. QUANTUM SIMULATIONS OF HUBBARD AND HOLSTEIN MODELS. R.T. Scalettar, Department of Physics, University of California, Davis, CA 95616.

We review recent quantum monte carlo simulations of models of interacting electrons and phonons in two dimensions, focusing on the competition between spin density wave and Peierls/charge density wave order, on the one hand, and superconductivity on the other. An overview of the current status of numerical work, where the progress has been and what bottlenecks remain, will be given. We will describe quantitative results for the magnetic and charge correlations, the gap to excitations, and dynamic/transport properties for these systems.

55. NEGATIVE ION PHOTOELECTRON SPECTROSCOPIC STUDIES OF THE STRUCTURE OF FREE RADICALS

W. C. Lineberger, Mark L. Polak, Mary K. Gilles and Joe Ho, Department of Chemistry and Biochemistry, University of Colorado, and Joint Institute for Laboratory Astrophysics, Boulder, Colorado 80309-0440

Negative ion photoelectron spectroscopy can provide direct information on the low-lying electronic states of transient neutral species. When coupled with measurements of gas-phase acidities, these data also provide precise measurements of homolytic bond strengths. In this lecture, we report experimental bond strengths and comparison with *ab initio* theory for a variety of C_2 and C_3 hydrocarbon radicals. We also discuss the low-lying electronic states of a number of main group metal clusters, including Bi_n, Pb_n, Sn_n and some mixed metal compounds.

56. ANION-MOLECULE AND ELECTRON-MOLECULE COMPLEXES, Jack Simons, Jeff Nichols, and Maciej Gutowski, Department of Chemistry, University of Utah, Salt Lake City, Utah 84112

Several examples of interesting anion-molecule and electron-molecule complexes will be discussed. These species include:

1. The $H^- + HF \rightarrow H^-\cdot\cdot HF \rightarrow HH\cdot\cdot\cdot F^- \rightarrow HH + F^-$ energy surface as it relates to "transition region spectroscopy" probes of the $H + HF \rightarrow HH + F$ transition state.
2. The $H^-\cdot\cdot H_3N \rightarrow (NH_4^+)^=$, $H^-\cdot\cdot H_2O \rightarrow (H_3O^+)^=$, etc. Double-Rydberg energy surfaces.
3. The role of "frustrated channels" in determining near-threshold rotational auto detachment rates : $H_2CCN^- (J,M,K) \rightarrow H_2CCN (J^1,M^1,K^1) + e^-$.

57. LASER AND FOURIER TRANSFORM SPECTROSCOPY OF TRANSIENT MOLECULES. Peter F. Bernath, Department of Chemistry, University of Waterloo, Waterloo, Ontario Canada, N2L 3G1.

The detection of a variety of free radicals and high temperature molecules will be discussed. The first part of the talk will focus on the gas phase inorganic chemistry of new alkaline-earth containing free radicals such as CaN_3, $CaBH_4$ and CaSH. These molecules were synthesised by laser-driven reactions of metal atoms in a Broida oven and detected by laser induced fluorescence. The second part of the talk will be about the Fourier transform spectroscopy of transient molecules including C_3, C_5, C_{60}, SiC and BeF_2.

58. ONE-ELECTRON PICTURES OF MOLECULAR ELECTRONIC STRUCTURE THROUGH ELECTRON PROPAGATOR THEORY; J. V. Ortiz, Department of Chemistry, University of New Mexico, Albuquerque, New Mexico 87131, U.S.A.

Electron propagator theory provides for direct calculation of correlated electron binding energies, Feynman-Dyson amplitudes, one-electron properties and total energies without evaluation of wavefunctions. A perturbative approach based on Hartree-Fock reference states leads to several accurate and efficient approximations which have been applied to simple organometallics, polysilane oligomers, double Rydberg anions and anion-molecule complexes. The use of coupled cluster reference states enables applications to systems where electron correlation plays a qualitatively important role, such as open shell molecules. In addition to its success as a computational framework for calculating observables, electron propagator theory offers interpretational advantages through association of electron binding energies to Feynman-Dyson amplitudes. The one-electron picture emergent from molecular orbital theory is thereby superseded in a quantitative sense without abandoning some of its desirable qualitative advantages.

59. NEGATIVE ION PHOTOELECTRON SPECTROSCOPY: COMPARISON OF NEW RESULTS WITH THEORY. D. M. Neumark, Department of Chemistry, University of California, Berkeley, CA 94720

We have obtained photoelectron spectra of several anions and anion clusters. The spectra yield new vibrational and electronic energies for both the anion and neutral species formed by photodetachment which permit a detailed comparison with recent *ab initio* results. Results will be presented for NO_3 as well as several carbon and silicon cluster anions.

60. Chord distribution analysis of a biphasic random medium : Relation with small angle scattering and direct dipolar energy transfer in porous solids.
P. Levitz, D. Tchoubar, C.R.M.D., C.N.R.S., 1B, rue de la Férollerie, 45071 Orléans France.

Biphasic random media such as some porous solids can be though as oriented surfaces filling the 3D space in a complex way. The "void" and the "solid" part of the porous matrix lay on each site of this oriented interface. The space filling morphology of an isotropic and bimodal porous medium can then be investigated using the bulk and the interface autocorrelation functions. In one hand, the bulk autocorrelation function is related to the small angle scattering (SAS) of the porous matrix. In another hand the surface autocorrelation function is probed by direct dipolar energy transfer between excited optical donors and grounded state acceptors randomly distributed on the interface. We present a chord distribution model of a biphasic random medium, allowing the direct computation of the SAS spectrum and the surface autocorrelation function. The present study follows and extends an early attempt to develop a "lineal" analysis of multiphase medium which can be related with small angle scattering data [1]. Application to different types of randomness (pseudo periodicity, Debye randomness, self similarity) are critically discussed and predictions compared with available experimental data.
[1] J.Mering, D.Tchoubar-Vallat, C.R.Acad.Sci.Paris, 262, 1703(1966);

61. END-TO-END DISTANCE DISTRIBUTIONS AND DIFFUSION WITHIN MACROMOLECULES OBSERVED BY INTRAMOLECULAR ENERGY TRANSFER USING FREQUENCY-DOMAIN FLUOROMETRY.

J. R. Lakowicz, J. Kusba, I. Gryczynski, H. Szmacinski, and W. Wiczk. University of Maryland School of Medicine, Department of Biological Chemistry and Center for Fluorescence Spectroscopy, Baltimore, MD 21201 and M. L. Johnson, University of Virginia, Department of Pharmacology, Charlottesville, VA 22908.

The remarkable progress in time-resolved fluorescence spectroscopy during the last few years provides a unique opportunity to investigate molecular processes characterized by non-exponential decays, like transient effects, lifetime distributions, or energy transfer. In particular, frequency-domain fluorometry operating with extended modulation frequencies to several GHz (Rev. Sci. Instruments 57:2499, 1986; 61:2331, 1990) has allowed the study of distance distributions between labeled sites on proteins and flexible bichromphoric molecules. We now describe use of the phenomena of fluorescence resonance energy transfer (FRET) to recover both the end-to-end distance distribution and diffusion coefficient of flexible molecules, model peptides (T-Gly$_6$-D,T-Pro$_6$·D) and the proteins mellitin (random coil and α-helix states) and TnI (native and denatured), using only the donor decay kinetics. The differential equation describing the donor population was solved numerically. The α-helix state of mellittin was induced by the presence of methanol in the solution. For melittin, the apparent donor (Trp, residue 19) to acceptor (N-terminal α-amino group labeled with a dansyl residue) diffusion coefficients are 0.16 and 99.9 x 10^{-6}cm^2/s in the random coil and α-helical states, respectively. The larger donor to acceptor diffusion in the α-helical state may be a consequence of the limited range of motion, $\leq$9Å in the α-helix as compared with $\leq$27Å in the random coil state. Denatured TnI displays faster donor-acceptor diffusion than native TnI.

62. **ELECTRONIC ENERGY RELAXATION AT INTERFACES** T. W. Scott, Y. J. Chang and J. Martorell, Department of Chemistry, New York Univ., New York, NY 10003.

Large perturbations in electronic relaxation rates have been observed when molecular adsorbates are subjected to the physical adsorption forces at a liquid-dielectric solid interface. These perturbations have been studied using picosecond multiple light scattering from high surface area metal oxides as well as by time resolved surface second harmonic generation from macroscopic interfaces. The adsorption forces are found to hinder relaxation of stilbene, but accelerate relaxation of the substituted stilbene containing 4-nitro and 4'-dimethylamino groups. The extended S_1 lifetime for unsubstituted cis and trans isomers is attributed to surface binding constraints which favor a planar molecular geometry and impede non-radiative relaxation by torsional pathways. The reduced S_1 lifetime for the disubstituted stilbene is attributed to a dramatic energy stabilization of the dipolar excited singlet state, accompanied by an increase in the Franck-Condon overlap integrals for non-radiative electronic relaxation.

63. INFLUENCE OF MULTIPOLAR CORRELATIONS AND SURFACE IMPERFECTIONS ON THE EFFICIENCY OF DIFFUSION-CONTROLLED REACTIVE PROCESSES ON MOLECULAR ORGANIZATES AND COLLOIDAL CATALYSTS. John J. Kozak, Department of Chemistry, University of Georgia, Athens 30602.

Theoretical studies of encounter-controlled reactive processes mediated by a colloidally-dispersed catalyst or molecular organizate have usually been based on the assumption that the diffusing coreactant in its motion is strictly confined to the surface of the host particle. This assumption is relaxed in the present study and the kinetic consequences of coreactant excursions into the double layer are considered explicitly. The role of intermolecular forces, operative between a diffusing coreactant and a stationary target molecule, the influence of temperature, the dielectric properties of the medium and perturbations induced by surface defects are studied systematically and (relative) changes in the rate constants quantified. From these calculations, spatial bounds on the reaction space for which short-range attractive potentials (the angle-averaged ion-dipole and dipole-dipole potentials) can compete in effectiveness with the long-range coulombic potential in influencing the efficiency of the underlying reactive process are determined.

64. PHOTON CORRELATION AND LOCALIZATION AND ITS IMPLICATIONS FOR ENERGY TRANSFER A.Z. Genack and N. Garcia, Department of Physics, Queens College of CUNY, Flushing, NY 11367, G. Kurizki, Department of Chemical Physics, Weizmann Institute of Science, Rehovot, 76100 Israel and J.M. Drake, Exxon Research and Engineering Company, Annandale, NJ 08801

Coherent backscattering dramatically modifies propagation and energy transfer in random and periodic structures. Measurements of pulsed optical transmission indicate that in random titania samples the transit time through the sample is lengthened by interference in the medium. We find that the character of propagation is determined by the ratio, κ, of the typical level spacing to the width of the EM modes of the medium. Measurements of spectral and spatial intensity correlation are described as a perturbation expansion in the parameter κ. κ is also a measure of the proximity to the localization threshold. We find that microwave localization is achieved in random mixtures of metal and dielectric spheres when κ = 2. We expect that suppressed transport of localized light may enhance energy transfer between molecules within a localization length. On the other hand resonant dipole-dipole interactions are suppressed in periodic dielectric structures in which spontaneous emission is inhibited at the resonant optical transitions.

PHYS

65. ELECTRONIC ENERGY TRANSFER BETWEEN DONOR AND ACCEPTOR ADLAYERS ON SINGLE-CRYSTAL SURFACES. D.R. Haynes, A. Tokmakoff and S.M. George, Department of Chemistry, Stanford University, Stanford California 94305

Laser induced fluorescence techniques and relative quantum yield measurements were used to examine the distance dependence of electronic energy transfer between two-dimensional donor and acceptor adlayers on $Al_2O_3(0001)$ in ultra high vacuum. The donor adlayer was p-terphenyl (TP), the acceptor adlayer was 9,10-diphenyl anthracene (DPA) and n-butane was the variable spacer layer. The electronic energy transfer rate versus spacer thickness was determined at both 30 K and 85 K. The butane spacer experiments showed that the TP donor energy transfer rate decreased with $1/d^3$, where d is the thickness of the spacer layer. Given a Forster energy transfer mechanism, a $1/d^3$ distance dependence is consistent with resonance electronic energy transfer from a two-dimensional TP donor adlayer to a three-dimensional array of DPA acceptors. The spacer measurements yielded a critical transfer distance of $d_0 = 38$ Å ± 2 Å at 30 K and $d_0 = 27$ Å ± 5 Å at 85 K. Critical transfer distances were also calculated theoretically and were in excellent agreement with the experimental measurements.

66. MULTIPLE-PULSE FEMTOSECOND SPECTROSCOPY. Keith A. Nelson, Department of Chemistry, Massachusetts Institute of Technology, Cambridge, MA 02158

Multiple-pulse techniques are commonplace and in magnetic resonance but still rare in optical spectroscopy. The technology now exists for automated (i.e. computer controlled) generation of complex optical waveforms with femtosecond or picosecond time resolution. By the end of the 1990's, pulse-shaping should be a standard option for commercial laser systems.
Femtosecond spectroscopy involving multiple-pulse techniques will be discussed. Phase-coherent vibrational and electronic excitation have been demonstrated recently. Current efforts are aimed at generation of ultrahigh-frequency acoustic waves, excitation of large-amplitude coherent vibrations, mediation of photochemical reactions to control reaction yields and branching ratios, and generation of complex waveforms in the far-IR spectral region.

67. MOLECULAR GEOMETRIC PHASE DEVELOPMENT AND PHASE-CONTROLLED OPTICAL PULSES. Jeffrey A. Cina, The James Franck Institute, The University of Chicago, 5640 S. Ellis Avenue, Chicago, IL 60637

Calculations will be presented which show that internal Berry phase development in Jahn-Teller active molecules is accessible to measurement with sequences of vibrationally abrupt phase-controlled optical pulses. Strategies for preparing specific candidate systems in the requisite time dependent states will be discussed. Comparison will be made with previous wave packet interferometry experiments on iodine vapor and with magnetic resonance interferometry experiments verifying the sign change under spin rotation by 2π. Prospects for observing inter-conformation tunneling and the nonadiabatic transition amplitude will be detailed.

68. NMR STUDIES OF NONCLASSICAL TRANSITION METAL POLYHYDRIDES: ROTATIONAL TUNNELLING AND QUANTUM EXCHANGE Kurt W. Zilm, Linda Wisniewski, Souheil Inati and D. Michael Heinekey, Dept. of Chemistry, Yale Univ., New Haven, CT. 06511

There has been a resurgence of interest in the chemistry of transition metal hydrides since the discovery by Kubas in 1980 that molecular hydrogen can bind to a transition metal center as an intact ligand. Due to the fact that the dihydrogen ligand is so light, its rotation is dominated by tunnelling effects. In a series of related transition metal trihydrides the effects of yet another tunnelling phenomenon, quantum mechanical exchange, have also been observed in solution phase NMR spectra. Using a combination of solid state and solution NMR methods we have been able to accurately characterize the potential energy surfaces that these quantum mechanical motions occur on. From these potential energy curves one can obtain much insight into the nature of the bonding in these nonclassical polyhydrides, i.e. the importance of backbonding or cis-hydride effects. In addition these systems display a striking structural diversity, for instance at least three distinct modes of dihydrogen coordination have been identified. The impact of the observation of quantum mechanical motions in these systems on concepts of structure and fluxionality will be discussed.

69. NMR MEASUREMENT OF COOPERATIVE MOTIONS IN SOLIDS

Patricia Lani Lee, Asher Schmidt, and Jacob Schaefer
Department of Chemistry; Washington University; St. Louis, MO 63130

The dynamic NMR properties of a spin in a polymer chain depend on the motions of that chain, which, in the solid state, are strongly influenced by the packing and motions of neighboring chains. Because most chemical-shift and relaxation NMR parameters are short range, measurement of cooperative events by NMR is generally difficult. Conventional analysis of these motions involves observing *isolated* ^{13}C's or ^{2}D and interpreting the results of NMR relaxation experiments by modeling the motion of a single chain interacting with its neighbors. The cooperativity between chains enters into the analysis only by inference. However, direct measurement of the cooperativity between chains is possible if *pairs* of spins are observed, one spin of the pair on one chain, and the second on a neighboring chain. Schemes to observe pairs of spins include detection of homonuclear dipolar dephasing of residual protons in perdeuterated polymers, and detection of heteronuclear dipolar dephasing of a rare-spin label on one chain by a second rare-spin of a different type on another chain.

70. SOLUTION AND SOLID STATE ^{13}C NMR STUDIES OF C_{60} AND C_{70}

D. S. Bethune, †P. P. Bernier, C. A. Brown, ‡H. C. Dorn, R. D. Johnson, §G. Meijer, J. R. Salem and C. S. Yannoni IBM Almaden Research Center, San Jose, CA 95021

One- and two-dimensional ^{13}C NMR experiments have been used to study the structure and dynamics of C_{60} and C_{70}, both in solution and solid state. The results have included confirmation of the icosahedral geometry of C_{60} and strong support for the "rugby ball" structure of C_{70}. The bonding connectivities in C_{70} have been mapped and the C-C coupling constants obtained along with an unambiguous assignment of the five resonances. Spectra of solid C_{60} show rapid "isotropic" reorientation at ambient, but motion slow enough at 77K to permit a measurement of the chemical shielding tensor. Spectra at intermediate temperatures indicate a distribution of reorientational correlation times. Dipolar spectroscopy experiments permitted the only measurement to date of the bond lengths in C_{60}.

Permanent address: †Univ. at Montpelier; ‡Virginia Polytechnic Institute; §Univ. at Nijmegen

71. CHEMICAL KINETICS AND DYNAMICS OF THE MESOSPHERIC SODIUM NIGHTGLOW. Charles E. Kolb, Douglas R. Worsnop, Mark S. Zahniser, Garry N. Robinson, Center for Chemical and Environmental Physics, Aerodyne Research, Inc., Billerica, MA 01821, Xiangian Shi and Dudley R. Herschbach, Department of Chemistry, Harvard University, Cambridge, MA 02138.

The mesospheric sodium nightglow is believed to be produced by the chemiluminescent reaction of sodium monoxide with atomic oxygen. Sodium monoxide is formed by the reaction of atomic sodium with ozone. Using fast flow reactor techniques we have measured the reaction rate constant for atomic sodium with ozone between 293 and 216 K and examined the reaction rate constants and excited state sodium branching ratios for NaO with O and CO. We have also used molecular beam magnetic deflection analysis to examine the product electronic states for the reaction of atomic Na, K and Rb with ozone. The impact of these chemical kinetic and dynamic studies on our understanding of the sodium nightglow will be discussed.

72. A COMPARISON BETWEEN THE OXIDATION REACTIONS OF ALKALI AND ALKALINE EARTH ATOMS. J.M.C. Plane, The School of Environmental Sciences, University of East Anglia, Norwich NR4 7TJ, United Kingdom.

This paper will review the progress that has been made during the last ten years in studying the reactions of *ground-state* Group 1 and 2 metal atoms. Particular attention will be paid to the reactions of these metals with N_2O and O_2, which have now been studied using a variety of techniques over considerable ranges of temperature and pressure. There are significant differences in kinetic behavior between the two Groups, which can be explained by the open- and closed-shell nature of the Group 1 and 2 atoms, respectively. Thus, the Mg reactions are extremely slow, and in the case of Mg and Ca recombining with O_2 to form the metal superoxides, the rate coefficients show complex temperature dependences. The reactions of both Groups with N_2O sometimes exhibit positive non-Arrhenius curvature at high temperatures, which has been attributed to vibrationally-excited N_2O enhancing the reactions. However, it remains a problem to explain why this occurs for Li, K and Ca, but not for Na, Mg, Cs and Ba. Where appropriate, comparison will be made with the analogous reactions of the metals in excited states, and with theoretical treatments of these types of reactions.

73. THE FLAME CHEMISTRY OF ALKALI AND ALKALINE EARTH METALS, Keith Schofield, Department of Chemistry, University of California, Santa Barbara, CA 93106.

The alkali and alkaline earth metals display oxide and hydroxide formation in clean flames. Their behavior, although having some similarities, is quite different. Fuel type is not a factor other than its control over temperature and radical concentration levels. For the alkalis there is a sensitivity to stoichiometry and under fuel lean conditions displays a complex chemistry involving MOH, MO_2, MO and M. However, if fuel rich, the alkali is mainly atomic with only small amounts of hydroxide. The kinetics are sufficiently fast at atmospheric pressure to attain steady state distributions that track the radical decay. These distributions can be perturbed by other elements such as halides that have equally stable molecular forms. The stabilities of the alkaline earth oxides and hydroxides are such that these are dominant even under fuel rich flame conditions and display distributions involving MO, MOH, $M(OH)_2$ and M. Kinetic rate data generally are not available for these but at atmospheric pressure the system is well described by partial equilibrium expressions. The ionization mechanisms of the two systems are particularly interesting. The alkalis represent a classic case of a thermal mechanism where as the alkaline earths exemplify chemi-ionization mechanisms. Important consequences of these differences will be outlined.

74. INFRARED SPECTRA OF METAL-ETHYLENE COMPLEXES ISOLATED IN SOLID ARGON, L. Manceron and L. Andrews, Department of Chemistry, University of Virginia, Charlottesville, VA 22901.

The reactivity of Group I, II and III metal atoms toward ethylene has been investigated using the matrix isolation technique by cocondensation of ground state metal atoms and ethylene molecules in solid argon. Infrared spectroscopy leads to the observation of two kinds of complexes: i) with group II metals and the heavy alkali metals (from Na to Cs) only very weak Van der Waals complexes were detected where the ethylene moiety is virtually unperturbed, and ii) with group III metal atoms and lithium strong pi-complexes with symmetrical structures were observed. Using isotopic substitution, it was possible to estimate quantitatively the perturbations of the ethylene geometry within the complex. The C=C bond is most affected and seems intermediate between a single and double bond. The perturbations change very little from one metal to the other for this class of complexes.

An analysis of the infrared absorption intensities reveals differences in the electrooptical parameters linked to the metal-to-pi-system charge transfer. This is indicative of differences in the metal-ethylene bonding, which are not characterized using the position of the vibrational energy levels alone.

75. DIENE EXCITED STATE ENERGETICS, STRUCTURES, AND DYNAMICS. R. McDiarmid, National Institutes of Health, Bethesda, MD 20892.

Because only limited experimental data can be obtained relating to the electronic excited states of conjugated dienes and theoretical calculations on these states have been found to be exceedingly difficult, it has not been possible to either evaluate the results of these calculations or confidently describe the energetics, structures, and dynamics of these states. Recently the spectrum of cyclopentadiene has been shown to contain significantly more information that does that of butadiene. Results derived from these spectra will be used to provide structural and dynamical information on the excited states of the molecule and to evaluate corresponding theoretical calculations. The relations of these results to those obtained for butadiene will also be discussed.

76. LONG-RANGE INTRAMOLECULAR INTERACTIONS; K. D. Jordan[a] and M. N. Paddon-Row;[b] [a]Department of Chemistry, University of Pittsburgh, Pittsburgh, PA 15260; [b]School of Chemistry, University of New South Wales, Kensington, NSW, Australia 2033

Much progress in understanding the role of electronic factors in electron transfer processes has come from studies of series of donor-bridge-acceptor (D-B-A) compounds, with the same donor and acceptance groups, but with bridges of different lengths. It is generally assumed that the electronic coupling depends exponentially on the bridge length (or the number of bonds) separating the D and A groups. In this talk we describe the results of theoretical calculations of the splittings between the π cation states and between the π^* anion states of a series of dienes with bridges containing 4 to 12 C-C bonds separating the ethylenic groups. Although the splittings for the 8, 10, and 12-bond bridges are consistent with an exponential dependence on the bridge length that for the shorter bridges is not. The non-exponential behavior is more pronounced when using non-minimal basis sets, indicating that long-range interactions are important. The splittings between the two lowest excited singlet states and between the two lowest triplet states are also examined.

77. INTRAMOLECULAR ELECTRON TRANSFER AT FIXED DISTANCES* John R. Miller, Chemistry Division, Argonne National Laboratory, Argonne, IL 60439

The study of electron transfer (ET) is an appropriate topic for this symposium because of the importance of an interplay between theory and experiment. I will review briefly how theorists led experimentalists to an understanding of how the motions of nucleii control electron transfer rates. But now the situation is reversed: Several fascinating observations have been made concerning the control of electron transfer rates by electronic factors, but theory lags behind in providing an understanding of those observations. Dependence of ET rates on distance is sometimes puzzlingly insensitive to some factors, but sensitive to others. A similar situation occurs for orientation effects on ET rates. A close interplay between experiment and theory is essential to develop a conceptual basis for progress in the field.

*Work at Argonne performed under the auspices of the Office of Basic Energy Sciences, Division of Chemical Science, US-DOE under contract number W-31-109-ENG-38.

78. STRUCTURE AND SPECTROSCOPY OF SMALL CARBON CLUSTERS IN MATRICES. Martin Vala, Department of Chemistry, University of Florida, Gainesville, FL 32611-2046.

Small carbon clusters (C_3-C_9) have been prepared by pulsed Nd/YAG laser vaporization of graphite and subsequently isolated in argon matrices. Two different but interrelated methods of investigation of these species have been pursued. 1) Mixed isotope (^{12}C:^{13}C) studies together with normal coordinate calculations have led to the geometry determination and infrared band assignments for linear C_5, C_6, and C_9 clusters, and an ionic, cyclic C_5 species. 2) Using FTIR and UV-visible spectrometers in a crossed beam configuration, IR and UV-visible bands of the clusters have been followed as a function of matrix annealing and initial preparation conditions. Correlations between known IR bands and unassigned UV-visible spectral features have been found for linear C_3, C_5, and C_9.

79. THEORETICAL STUDY OF SILICON CLUSTER ELECTRON AFFINITIES.
C. M. Rohlfing, Sandia National Laboratories, Livermore, CA 94551-0969;
K. Raghavachari, AT&T Bell Laboratories, Murray Hill, NJ 07974.

Ab initio calculations have been performed to investigate the structures and energies of silicon cluster anions. There is very good agreement between calculated and experimental adiabatic electron affinities. Variations in electron affinities with cluster size are interpreted in terms of low-lying states of the neutral species.

80. ENERGY TRANSFER AND RELAXATION IN MOLECULAR GLASSES AND POLYMER SOLIDS,
D.M. Hanson, S.K. Kook, and Y. Lin, Department of Chemistry, State University of New York, Stony Brook, NY 11794-3400.

Spectral diffusion in 4-bromo-4'-chlorobenzophenone aggregates in polystyrene films at 4.2K is characterized using steady-state and time-resolved phosphorescence spectroscopy. The spectral features and energy-transfer dynamics provide evidence for delocalized aggregate states, localized aggregate states, and discrete molecule states. Conclusions are made regarding the coupling of these states and their energy and spatial distributions.

Similar studies were conducted on glassy methylbenzophenone. These experiments reveal that the spectral diffusion efficiency increases abruptly at a critical mole fraction of energy acceptors. Although low-dimensional energy transfer dynamics has been found in other studies of disordered systems, these results for glassy methylbenzophenone are consistent with the rate of spectral diffusion being controlled by a three-dimensional exchange interaction and the emission of phonons and/or vibrons

81. TRIPLET ENERGY TRANSFER IN ISOTOPICALLY MIXED NAPHTHALENE: HIGH PRESSURE EFFECTS AND DELAYED FLUORESCENCE ODMR. H.C. Brenner, Department of Chemistry, New York University, New York, NY 10003.

Naphthalene-h₈ (N-h) in naphthalene-d₈ crystals show a well-known critical concentration effect, in which the phosphorescence yield of a dilute sensor trap, betamethylnaphthalene, monitored as a function of N-h concentration, rises abruptly near 10% N-h. High pressure (4 kbar) has been previously shown in our laboratory to significantly lower the critical concentration, and has been used to probe the distance dependence of the interactions giving rise to the critical threshold, thus testing theoretical models such as percolation and Anderson delocalization. Recent results concerning decay kinetics of phosphorescence and delayed fluorescence (arising from the annihilation of N-h trap triplet excitons) will be presented and discussed in terms of fractal models for the transfer. We have also shown that the polarity of delayed fluorescence ODMR is sensitive to the onset of efficient N-h intertrap transfer, and we analyze our results in terms of simple kinetic models which enable some of the rate parameters describing the annihilation to be determined. Work supported by ACS-PRF.

82. DISPERSIVE ENERGY TRANSPORT - RELATION TO SPIN GLASS RELAXATION, Christian von Borczyskowski, Department of Physics, Freie Universität Berlin, Arnimallee 14, D-1000 Berlin 33

Triplet excitation energy transport on the substitutionally disordered lattice of p-dichlorobenzene in p-dibromobenzene follows a Kohlrausch

-type relaxation. Temperature variation reveals β=1/3 at low temperatures. We describe this behavior either by random-walk processes affected by Campbell et al. (1) for spin glasses. Excitation energy transport in heavily doped molecular crystals can be described at the critical concentration by dynamic percolation. At this concentration the percolating cluster is a fractal object with a related spectral dimension (2). Comparison with recent Monte Carlo simulation (3) reveals that the spectral dimension depends critically on the range of the interaction responsible for the transport.
(1)I.A.Campbell,J.M.Flesselles,R.Julien,R.Botet,Phys.Rev.B 37,3825 (1988); (2)C.v.Borczyskowski,T.Kirski,Phys.Rev.Lett.60,1578 (1988);(3)R.Brown,J.L. Garitey,F.Dupuy,P.Pee,J.Phys.C.21,1191 (1988);(4)C.v.Borczyskowski,T.Kirski,Phys.Rev.B40,11335 (1989)

83. THE VIBRATIONAL POPULATION RELAXATION TIME OF IONS IN PROTIC SOLVENTS MEASURED BY TRANSIENT INFRARED SPECTROSCOPY. J. Owrutsky and R. M. Hochstrasser, Department of Chemistry, University of Pennsylvania, Philadelphia, PA 19104-6323

Two-color transient infrared spectroscopy with very short pulses permits the study of T_1 relaxation in ions. The 5 μm infrared pump pulse having a width as narrow as 500 fs is generated by mixing an amplified dye laser pulse with a regeneratively amplified YLF pulse. The infrared probe is accomplished by non linear gating of a cw diode laser as in our previous transient IR studies (Hochstrasser, et al., in Ultrafast Phenomena VII, 1990). This approach permits two color transient infrared spectroscopy with a single dye laser. The vibrational population relaxation times (T_1) of the ν_3 (antisymmetric stretch) fundamental band of the azide ion (N_3^-) in solutions of water (D_2O and H_2O), methanol and ethanol have been measured. The fast T_1 times, which are observed to be 3 ps under all the conditions investigated, are indicative of relaxation through strong solvent interactions. A discussion of the relationships between solvent structure, intramolecular perturbations and the T_1 and T_2' relaxation will be presented.

84. FEMTOSECOND SOLVATION DYNAMICS OF CH_3Cl AND Cl^- AT MULTILAYER SURFACES. M. Kwini, J. P. Cowin, M. Iedema, Pacific Northwest Labs, Richland, WA 99352, T. L. Gilton, Micron Technologies, Boise, ID 83706.

The photoelectron induced dissociation of CH_3Cl adsorbed on top of a multilayer deposit ejects methyl radicals to the gas phase. The kinetic energies of these methyls vary with the identity of the underlying multilayer (H_2O, hexane, CH_3Cl, xenon), from .44 to .7 eV at the peak, and are much higher than seen for the gas phase dissociative electron attachment to this molecule. The additional energy is understood in terms of the effects of the "prompt" solvation (<20 fs) on the anionic repulsive curve. The attachment cross section for condensed CH_3Cl is also orders of magnitude greater for condensed CH_3Cl than for the gaseous molecule, which is also understood in terms of the solvation modified repulsive curve altering the relative probability of dissociation versus electron autodetachment for the transient anion.

85. ENERGY DISSIPATION IN LIQUID STATE CHEMICAL REACTIONS ON THE FEMTOSECOND TIMESCALE. C. B. Harris, Department of Chemistry, University of California, Berkeley, CA 94720.

A series of picosecond and femtosecond experiments will be presented which test Isolated Binary Collision (IBC) theories and the Generalized Langevin Equations (GLE) for vibrational relaxation and redistribution in highly excited molecules in simple solvents such as liquid Xe. In addition, a connection between collision theories and binary part of continuum theories will be demonstrated. Finally, some surprising new results on energy redistribution and relaxation in complex molecules in liquids will be presented.

86. KINETIC STUDIES OF BORON AND ALUMINUM SPECIES. Nancy L. Garland, Chemistry Division, Code 6111, Naval Research Laboratory, Washington, D.C. 20375

Both boron and aluminum make ideal advanced fuels and fuel additives because of the tremendous energy released on combustion of these species. Knowledge of the rates and mechanisms of the key chemical reactions in the combustion processes will aid in the exploitation of these fuels. Our recent experimental and theoretical kinetic studies of the reactions of the oxides and hydrides of B and Al will be described. The implication of these results on the modelling of the pertinent combustion systems will be discussed.

87. KINETICS OF METAL ATOM AND RADICAL REACTIONS OVER WIDE TEMPERATURE RANGES: MEASUREMENTS, CORRELATIONS, AND PREDICTIONS. Arthur Fontijn and Peter M. Futerko, High-Temperature Reaction Kinetics Laboratory, The Isermann Department of Chemical Engineering, Rensselaer Polytechnic Institute, Troy, NY 12180-3590.

Measurements have been made by the HTFFR (high-temperature fast-flow reactor) and metals HTP (high-temperature photochemistry) techniques. Results will be shown for a considerable number of oxidation reactions of Al, AlO, AlCl and BCl. Various correlations between them will be discussed. For example, if we express k(T) for MX = AlCl, BCl reactions with oxygen oxidants OY in the $AT^n \exp(-E/RT)$ form, and fix n, we find that $E = a + bD(M-X) + c[IP(MX) - EA(OY)]$. Here a,b, and c are constants, D is bond energy, IP is ionization potential, and EA is electron affinity. Literature values for BF and BH reactions with O_2 follow this trend as well. Resonance theory can be used to explain this relationship. We will show that the resonance theory approach allows correlations and predictions for activation energies in several further series of reactions, such as those between metal atoms and N_2O. Work supported by AFOSR and NSF.

88. POTENTIAL ENERGY SURFACES FOR BCl REACTIONS. Wei Chen, W. L. Hase and H. B. Schlegel, Department of Chemistry, Wayne State University, Detroit, MI 48202.

Ab initio molecular orbital theory has been used to investigate BCl and AlCl reactions with $^3O_2$ and CO_2. Potential energy surfaces for the reactions have been calculated at MP4/6-31G*, QCISD/6-31G* and MP4/6-31+G(2df) levels of theory; the geometries of the reactants, transition states and products were optimized at HF/6-31G* and MP2/6-31G* levels. A revised heat of formation of OBCl has been determined from the calculations. The mechanism and kinetics of these reactions will be discussed.

89. REACTIVE COLLISIONS OF Al, Mg, Si AND C ATOMS. C. Naulin, M. Costes and G. Dorthe, URA 348 Photophysique et Photochimie Moléculaire, Université Bordeaux I, 33405 Talence cedex, France.

Reactive collisions of ground-state Al, Mg, Si and C atoms with oxidant molecules were studied over a wide range of collision energies in pulsed, crossed, supersonic molecular beams. Threshold energies could be determined for endoergic reactions (Al + CO_2 → AlO + CO, Al + SO_2 → AlO + SO, Si + NO → SiN + O) or exoergic reactions with significant potential energy barriers such as Mg + N_2O → MgO + N_2. The dependence of product energy distribution upon collision energy was also analysed. The dynamics of two exoergic reactions C + OCS → CO + CS and Mg + N_2O → MgO + N_2 were found quite unusual. Atomic carbon reaction dynamics can be explained in terms of covalent potential energy surfaces and experimental results check theoretical calculations when available (C + NO → CN + O). However, the unexpected behaviour of the Mg + N_2O reaction can be understood through multiple crossings of covalent and ionic potential energy surfaces.

90. REACTIONS OF BORON AND ALUMINUM ATOMS WITH SMALL MOLECULES. Paul Marshall and Peter B. O'Connor, Department of Chemistry, University of North Texas, Denton, Texas 76203, Wai-to Chan, Peter V. Kristof and John D. Goddard, Department of Chemistry and Biochemistry, University of Guelph, Guelph, Ontario, CANADA N1G 2W1

The reactions of B and Al with O, O_2 and CO_2 have been investigated at Hartree-Fock and many body perturbation (MP2 and MP4) levels of theory, with the 3-21G*, 6-31G* and larger basis sets. Some configuration interaction and multiconfiguration SCF approaches to electron correlation were also employed. The status of ab initio studies on these radicals will be discussed. Reactions with O_2 and CO_2 are calculated to proceed via formation of excited adducts which may either be stabilized collisionally or may further dissociate to products: intermediates and transition states have been characterized. The results are combined with transition state/RRKM theories to predict the kinetics. Comparison with experimental data suggests directions for further study.

91. NUCLEATION AND GROWTH OF DIAMOND FROM HYDROCARBON GASES. J.C. Angus, J.A. Mann, M. Sunkara, Z. Li and R. Gat, Chemical Engineering Department, Case Western Reserve University, Cleveland, OH 44106.

Factors influencing the nucleation and growth of diamond from hydrocarbon gases have been studied experimentally and through computer modelling. The influence of different substrates (diamond, silicon, graphite and diamond-like carbon) and substrate pretreatment with organic nucleating agents will be described. The inter-relationship between observed crystalline morphologies, sequences of stacking errors and growth rates are examined and a comparison between the structure of diamond and the so-called "diamond-like" phases is made. The overall reaction environment in a hot-filament reactor, including convective and diffusive transport of energy and chemical species and chemical reactions is modelled and comparisons made with experimental reactors.

92. MICROWAVE CVD DEPOSITION OF DIAMOND THIN FILMS WITH APPLICATIONS IN X-RAY LITHOGRAPHY. J.E. Yehoda, C.R. Guarnieri, S.J. Whitehair, J.J. Cuomo, IBM Research Division, P.O. Box 218, Yorktown Heights, NY 10598.

The advent of microwave CVD processing has opened up a new regime for thin film deposition. It provides an efficient source for radical generation and atomic hydrogen production. The various microwave CVD diamond deposition methods will be discussed and compared. The process parameters such as CH_4/H_2 ratio and the introduction of other gas species are important in activating the plasma and process. How these effect both the intrinsic properties of diamond as well as the physical properties, such as morphology and texturing, will be elucidated. A technology where diamond has potentially a great impact is x-ray lithography. In addition to being flat and smooth, the membranes for x-ray lithography masks should possess high thermal conductivity, mechanical stiffness, radiation hardness, and optical transparency. These are diamond material properties which make it an attractive contender to replace boron doped Si membranes, the current state-of-the-art.

93. SYNTHESIS OF THICK, LARGE AREA DIAMOND FILM. R.L. Woodin, Norton Co., Goddard Rd., Northboro, MA 01532.

Diamond's rare combination of physical toughness, high thermal conductivity and low resistivity make it an intriguing material for many applications ranging from cutting tools, electronic substrates, domes and window coatings to semiconductor microelectronic devices. Recent progress in chemical vapor deposition (cvd) of diamond films now brings these potential applications into the realm of commercial practicality.

The most common techniques for cvd diamond are hot filament and microwave plasma assisted deposition. For thin coatings (1-20 um hr^{-1}). A broader range of applications opens when thick (500 um), large area (100 cm^2) films can be produced at reasonable rates. In order to have acceptable rates to produce thick diamond film, high flux methods such as dc arc jets, oxy-acetylene torches and rf plasma torches can be used. In this talk we will compare the methods currently used to make thick, large area diamond film and the current accomplishments and limitations. Recent measurements and modelling of diamond deposition in dc arc jets will be presented.

94. DEPOSITIONS AND CHARACTERIZATIONS OF BORON NITRIDE FILMS. G.L. Doll, Physics Dept., General Motors Research Laboratories, Warren, MI 48090.

Boron nitride crystallizes in at least four structures analogous to carbon; hexagonal (hBN), wurtzite (wBN), thombohedral (rBN), and cubic (cBN). Whereas hBN is an insulating analog to graphite, cBN exhibits physical characteristics similar to diamond including large indirect bang gaps, extreme hardnesses, high thermal conductivities, high melting points, and chemical inertness. Due to these properties, cBN films are espoused for such applications as coatings for high temperature and corrosive environments, grinding or machining applications, device packaging, and as components in high-temperature electronic devices. Many different growth techniques have been used to deposit BN films, but typically, only those growth techniques that employ (1) a reactive nitrogen (or nitrogen-containing) plasma and/or (2) ion beam interaction with BN films have succeeded in producing the cBN phase. A survey of the deposition techniques that have succeeded in growing cubic boron nitride films will be presented, and several characterization experiments particularly well suited to the boron nitride system will be discussed.

95. ON CVD GROWTH OF SiC. R.I. Fuentes, IBM Research Division, T.J. Watson Research Center, P.O. Box 218, Yorktown Heights, NY 10598.

Over the past few years the interest in SiC has steadily increased. It is now being considered for a host of new applications, including optoelectronics, high temperature devices, and microwave devices, to x-ray masks and mirrors. Such new applications are made possible mainly by advances in the techniques to grow the material.

In this paper key aspects that relate to the growth or deposition of SiC materials will be discussed. Processes to grow SiC, from Acheson's to today's modern sublimation, PECVD, sputtering, and CVD will be reviewed. Significant progress and problems encountered will be addressed. Issues such as precursors, the lattice mismatch in heteroepitaxial growth, inversion domain boundaries, polytypism and stacking disorder will be considered. Emphasis will be placed on the growth of SiC on Si. Growth of SiC on other substrate materials will also be discussed. Finally, a brief summary of current and prospective applications will be given.

96. **THEORETICAL STUDIES OF SINGLE ADSORBED ATOMS IN THE STM.** N. D. Lang, IBM Research Division, Thomas J. Watson Research Center, Yorktown Heights, New York 10598

We discuss the way in which STM images differ for chemically different atoms, including rare gases. We also discuss in particular the recently studied case of Xe, and conclude that the 6s resonance, although lying close to the vacuum level, is the origin of the Fermi-level local state density which renders this atom visible in the STM. We consider in addition the effects on adsorbed atoms of relatively large biases applied between tip and sample. The large-bias calculations also permit study of the tip as a point electron source.

97. TUNNELING MICROSCOPIES: FEM, FEES, FIM, A-P, STM, STS. Osamu Nishikawa, M. Tomitori and F. Iwawaki, Graduate School at Nagatsuta, Tokyo Institute of Technology, 4259 Nagatsuta, Midori-ku, Yokohama 227 Japan

A field emission microscope(FEM), a field ion microscope(FIM) and a scanning tunneling microscope(STM) utilize electron tunneling and have realized the atomically high resolution which has been attained by confining the tunneling area to the size of an atoms. Unique features are:
1. Direct projection of the arrangement of apex atoms by FIM and topographic depiction of individual atoms on a flat specimen surface by STM.
2. Electron spectroscopy of a small apex area with FEM(field emission electron energy spectroscopy: FEES) and individual surface atoms with STM(scanning tunneling spectroscopy: STS).
3. Mass analysis of individual apex atoms by an atom-probe(A-P), the combined instrument of an FIM and a mass spectrometer.

On the other hands, these microscopies have many restrictions such as:
1. Small specimen area of FEM and FIM: the apex of a sharp tip.
2. High sensitivity of STM images and STS spectra to the tip apex.
3. Incomprehensible tunneling characteristics of polymers and biomolecules.

The unique features and imposed restrictions of the microscopies will be discussed based on the experimental results.

98. STUDIES OF KINETIC PROCESSES AT SURFACES USING ATOMIC-LEVEL IMAGING. Max G. Lagally, Department of Materials Science and Engineering, University of Wisconsin-Madison, Madison, WI 53706.

The ability of scanning tunneling microscopy (STM) to resolve individual atoms while also allowing fields of view over many micrometers makes possible investigations of processes involving atomic transport and atomic interactions at the most microscopic level. Such processes are of particular importance in crystal growth and in surface chemistry. Quantitative measurements of surface diffusion, transport of atoms over monatomic or multiatomic steps, interactions of atoms with steps, and the formation of two-dimensional layers and very small three-dimensional crystals are demonstrated using Si, Ge, and GaAs surfaces. The talk will illustrate how simple statistical analysis of STM images can provide information on surface kinetics. Because these kinetic mechanisms are universal, the analytical methods presented in this talk can be applied to any material that can be imaged in the STM.

*Research Supported by ONR, Chemistry Program.

99. SURFACE DIFFUSION AND PHASE TRANSITION ON THE Ge(111) SURFACE STUDIED BY SCANNING TUNNELING MICROSCOPY,* R. M. Feenstra, IBM T.J. Watson Research Center, P.O. Box 218, Yorktown Heights, NY 10598

The scanning tunneling microscope has been used to image the Ge(111) surface at elevated temperatures up to 400° C. The surface is initially prepared in the c2×8 structure, consisting of Ge adatoms on top of an ideal (111) plane. Discrete motion of the adatoms is observed at temperatures in the range 100 – 200° C, with the motion occuring along the <011> surface directions. The 2× spacing of the adatoms is preserved in this diffusion, and the 8× spacing is broken. Near 300° C, we observe the well-known phase transition to the disordered 1×1 state. We find that the transition nucleates at domain boundaries of the c2×8 structure, and "pre-melting" of the adatom layer near the domain boundaries is observed.

*in collaboration with A. J. Slavin, G. A. Held, and M. A. Lutz

100. GROWTH OF SILICIDE LAYERS ON SI SURFACES, Y. Kuk, Seoul National University, Seoul, Korea

Nickel and Cobalt silicide layers were grown on various silicon surfaces. The layers grown are epitaxial on Si(111) surface and change structures with increasing coverage. Grown structures on Si(100) are rather complicated and show many phase boundaries. Ballistic Electron Emission Microscopy (BEEM) of the silicides-silicone interfaces has been studied to understand interface structures.

PHYS

101. ATOMIC SCALE SURFACE ANALYSIS WITH THE ATOM-PROBE FIM: <u>T. T. Tsong</u>*, Physics Institute, Academia Sinica, Nankang, Taipei, Taiwan, ROC

The time-of-flight atom-probe field ion microscope is a combination of a field ion microscope, an siming probe-hole and a single ion detection sensitivity time-of-flight spectrometer. Using this instrument surface atoms and surface layers selected from an atomically resolved field ion image can be chemically analyzed one at a time. When one uses laser pulses for the pulsed field evaporation needed for the ToF atom-probe operation, one can also use the atom-probe for measuring the binding energy of surface atoms. We have been using it to analyze the chemical compositional variation of the near surface layers in surface segregation of alloys, in the formation of compound layers on surfaces, as well as in studies of simple surface chemical reactions etc.

* Distinguished Chair of The National Science Council of ROC: On leave from The Pennsylvania State University.

102. SPECTROSCOPY OF NON-RIGID MOLECULAR COMPLEXES, David Yaron and <u>William Klemperer</u>, Department of Chemistry, Harvard University, Cambridge, Massachusetts 02138.

In this talk we discuss molecular complexes where effects of non-rigidity are observed spectrally. A common occurrence to be discussed are proton interchange motions. In addition we discuss systems in which appreciable nuclear density occurs in classically forbidden regions. In these situations, bizarre isotope effects are observed.

103. AB INITIO CALCULATIONS ON HYDROGEN-BONDED COMPLEXES.
<u>P. Botschwina</u>, Institute of Physical Chemistry, University of Göttingen, D-3400 Göttingen, Germany

Potential energy and electric dipole moment surfaces have been computed for a number of hydrogen-bonded complexes such as OC---HX, HCN---HX (X=F, Cl), and HCN---HCN by means of the Coupled Electron Pair Approximation (CEPA). Therefrom, various spectroscopic properties (vibrational frequencies, rotational and centrifugal distortion constants, vibration-rotation coupling constants, l-type doubling constants, electric dipole moments in different vibrational states and IR intensities) have been calculated either variationally or by means of perturbation theory. The results are compared with available experimental data.

104. INTERMOLECULAR FORCES: DO WE REALLY NEED TO COMPARE "AB INITIO" CALCULATIONS WITH EXPERIMENT? <u>G. Scoles</u>, Department of Chemistry, Princeton University, Princeton, New Jersey 08544-1009

The current experimental and theoretical sources of information on intermolecular forces will be briefly reviewed, discussing in particular the precision and the completeness of the information. The complexity of a full a priori analysis will be pointed out together with the usefulness of the exploration of approximate models provided these allow us to identify the approximations. The usefulness of models is greatly enhanced when these permit us to transfer some of our knowledge from the area where it was tested to another area where the available information is more limited (as in the case of gas-surface interactions, for instance).

105. WEAK MOLECULAR INTERACTIONS. <u>B. Liu</u> and A.D. McLean, IBM Almaden Research Center, 650 Harry Road, San Jose, CA 95120-6099

The Interacting Correlated Fragments method for systematic and accurate determination of potentials for van der Waal's and hydrogen bonded interactions will be illustrated by applications to a variety of molecular systems.

106. CHEMICAL, CONJUGATIVE AND COOPERATIVE EFFECTS IN HYDROGEN BONDING: THE NATURAL BOND ORBITAL DONOR-ACCEPTOR PERSPECTIVE. F. Weinhold, Theoretical Chemistry Institute and Department of Chemistry, University of Wisconsin, Madison, WI 53706.

The chemical, valence-like nature of H-bonding was stressed by Lewis, Pimentel, Coulson, and others, yet many current workers favor the alternative "electrostatic" picture in which H-bonds are interpreted (and modelled) using classical Coulombic expressions for interacting monomer point charges or distributed multipoles. The inadequacy of the latter picture is suggested by Natural Bond Orbital (NBO) analysis of *ab initio* wavefunctions, which shows the important role of orbital "donor-acceptor" interactions, closely related to Pimentel's 3-center MO model. The chemical picture is also supported by experimental evidence of significant monomer rearrangements and non-pairwise-additive cooperativity effects, particularly in H-bonds of biological significance. We describe NBO and Natural Resonance-Theoretic (NRT) analysis of model amide clusters to illustrate the remarkable coupling between H-bonds and monomer electronic structure (even complete changes of chemical identity!) characteristic of peptides and related groups. Such conjugative coupling effects, lying outside a classical electrostatic perspective, are probably essential to the biological functioning of proteins.

107. CONFORMATION AND DYNAMICS OF PARTIALLY FOLDED AND UNFOLDED PROTEIN MOLECULES. Elisha Haas, Department of Life Sciences, Bar-Ilan University, Ramat-Gan 52900, Israel

The folding of polypeptide chains of globular proteins into compact, stable native conformations is assumed to be a programed transition. It is assumed that this transition follows a pathway directed by the sequence of each polypeptide. The pathway is composed of sparsely populated, partially folded intermediate conformers. The intermediates are marginally stabilized by specific short and long range interactions. In order to decipher the instructions for the folding pathway, an attempt is made to characterize the conformations of the intermediate states, in terms of the intramolecular distances and rates of their changes at the subnanosecond time scale. Time resolved energy transfer measurements between donor - acceptor probes, chemically attached to specific sites on the polypeptide chain are used. Global analysis of series of measurements yields the series of intramolecular distances distributions and rates of segmental intramolecular Brownian motions. Synopsis of data collected for a series of labeled derivatives of the bovine pancreatic trypsin inhibitor (BPTI), revealed very early folding intermediates in conformational states assumed to be purely unfolded. The unfolded BPTI molecule shows very stable interactions between the chain ends and between sites of secondary structures. The N terminal segment of the unfolded protein (15 residues) has a random coil type conformations, while a hydrophobic core in the center of the chain seems to be involved in the formation of a folding initiation site (FIS). Both chain termini are strongly attracted by the newly formed FIS's and thus a major step in the folding pathway is completed.

108. **ENERGY TRANSFER IN LATEX MICRODOMAINS.** M. A. Winnik, Department of Chemistry, University of Toronto, Toronto, Canada M5S 1A1

A common morphology for two component latex particles is the interpenetrating network-type [IPN] structure. In these kinetically-controlled structures, the component present in smaller amounts can form an interconnected discrete phase penetrating throughout the core of the particle composed of the major component. In two systems we have examined, 1 μm diameter poly(methyl methacrylate) [PMMA] latex containing small amounts (2 wt%) of polyisobutylene [PIB], and 300 nm diameter poly(vinyacetate) [PVAc] containing small amounts (8 wt%) of poly(2-ethylhexyl methacrylate) [PEHMA], the minor component forms an IPN-like structure in the particle interior. These domains are of such small size that donor-acceptor dye pairs confined to this domain exhibit energy transfer in restricted dimensions. This talk will examine the scope of this phenomenon and show how various solvents, selective for the minor component, can perturb the interface between the two polymers.

109. COMPUTER SIMULATION OF ELECTRONIC ENERGY TRANSFER IN POLYMERS: COMPARISON OF THEORY AND EXPERIMENT. S.E. Webber, Dept. of Chemistry and Biochemistry and Center for Polymer Research, University of Texas at Austin, Austin, Texas, 78712

Computer simulation of electronic energy transfer in polymers involves two steps: 1) modelling of a polymer coil, 2) calculation of the time-dependent survival probability of an excited state that can migrate between chromophores before encountering a trap. The efficiency of energy trapping depends on the distance over which energy can be transferred and the probability of cross-chain transfers. Calculations will be described in which the polymer is treated as a self-avoiding random walk using the "pivot algorithm" with an inter-segmental interaction energy. The time-dependence is computed using the Lanczos algorithm to solve the resulting Master equation. These calculations demonstrate the sensitivity of down-chain energy transfer to the number of non-bonding segmental contacts, which in turn depend on the solvent-polymer interaction parameter. Comparisons between these simulations and experimental data will be presented.

110. FRACTALS AND EXCITATION TRANSFER IN LANGMUIR-BLODGETT FILMS.
Iwao Yamazaki, Naoto Tamai and Tomoko Yamazaki,
Faculty of Engineering, Hokkaido University, Sapporo 060, Japan

Electronic excitation transfer in Langmuir-Blodgett films has been studied with the picosecond time-resolved fluorescence spectra and the decay curves. The fractal behaviors are essential in spatial distribution of chromophores or in site-energy distribution. The fluorescence decay curves can be analyzed by using equations including a stretched exponential function. The mechanisms of singlet excitation-energy relaxation were interpreted in terms of (1) the energy migration among energetically disordered monomer sites represented by ultrametric space (hierarchical energy distribution) or the Gaussian distribution of the density of excited states and (2) the energy trapping by two-dimensional aggregates. The diffusion length in the excitation energy transfer was discussed.

111. DOMAINS AND ORDER IN LANGMUIR-BLODGETT-TYPE CHROMOPHORE SYSTEMS.
D. Weiß, R. Kietzmann, W. Storck, J. Lehnert and F. Willig,
Fritz-Haber-Institut der MPG, Faradayweg 4-6, W-1000 Berlin 33, FRG

A great number of promising applications are currently under study for molecular films assembled with the well-known Langmuir-Blodgett (LB) technique. We report spectral and picosecond-time resolved fluorescence measurements on LB systems of simple well-matched amphiphilic molecules. The latter are anthracene chromophores attached at the 2-position to fatty acid tails. The combined information from fluorescence measurements, from the X-ray structure of LB-type sublimation crystals of these molecules and from polarisation micrographs gives insight into domain structure and packing of these molecules within the domains of such LB-type layers. Consequences for transport of excitons and charge carriers in such LB layers are discussed.

112. COMPUTATIONAL METHODS FOR POLYMER SIMULATIONS. R.A. Friesner, Department of Chemistry, Columbia University, New York, New York 10027.

We describe methods for efficiently generating long polymer chains in a lattice model of a polymer with attractive or repulsive nearest neighbor interactions. Polymer conformations are generated via the pivot algorithm, and a hierarchical blocking of the polymer sites into groups allows a linear (as opposed to quadratic) dependence of the cost in evaluating the interaction energy as a function of chain length. The polymer conformations are then used to simulate the effects of excitation transfer and trapping on fluoresence decay curves, utilizing a novel approach to solution of the master equation for the excitation dynamics.

113. **MULTI-DIMENSIONAL NMR OF ISOTOPICALLY ENRICHED PROTEINS**
Ad Bax, Frank Delaglio, Mitsuhiko Ikura, Lewis E. Kay, Marius Clore, Angela Gronenborn and Dennis Torchia*, Laboratory of Chemical Physics, NIDDK, and *Bone Research Branch, NIDR, National Institutes of Health, Bethesda, Maryland 20892

In principle, protein structures can be determined from two-dimensional ^{1}H NMR spectra. In practice, spectral overlap limits the size of proteins that can be studied in this manner to less than ~10 kDa. Advances in molecular biology now make it possible to obtain most proteins of interest from bacterial growth media, which offer a straighforward avenue for incorporation of the stable isotopes ^{13}C and ^{15}N. With these isotopically enriched proteins it becomes feasible to record spectra of higher dimensionality (3D and 4D) which have greatly reduced spectral overlap compared to their conventional 2D counterparts. In addition, the presence of ^{15}N and ^{13}C in the protein provides new information on the structure and dynamics of the protein.

114. TWO DIMENSIONAL ELECTRON SPIN RESONANCE, J.H. Freed, Baker Laboratory of Chemistry, Cornell University, Ithaca, NY 14853

Recent years have witnessed an explosive growth in new ESR techniques based on modern technologies. We address the development of Fourier Transform (FT) and Two-Dimensional ESR which may be expected to revolutionize chemical applications of ESR much as they did NMR. This development presented major instrumental challenges compared to NMR: time-scales of ESR relaxation are of the order of ns. compared to ms. for nuclei; spectral extents are of the order of 100 MHz requiring ns. pulses; and microwave technology is more complex than rf technology. The new 2D-ESR techniques will be described and their applications to molecular dynamics and structure will be discussed. These include molecular dynamics in liquid crystals, model membranes, and on surfaces. The greater resolution of 2D-ESR for structural studies in disordered solids will be illustrated. The possibilities of spatially-resolved 2D-FT-ESR for studies of molecular dynamics in condensed media will be discussed.

115. TWO–DIMENSIONAL MICROWAVE SPECTROSCOPY. A. Bauder, Laboratorium für Physikalische Chemie, Eidgenössische Technische Hochschule, CH–8092 Zürich, Switzerland.

True 2D rotational spectra in the gas phase have been recorded with pulsed Fourier transform techniques in the microwave range. Different pulse sequences have been applied to probe connections between rotational transitions with widely separated or closely spaced frequencies. Introduction of two– and four–step phase cycles removed artifacts from the 2D spectra efficiently. Experiments have included 2D single–quantum correlation, 2D n–quantum filtered correlation, 2D double quantum correlation, and 2D autocorrelation spectroscopy. The method has been applied to study relaxation processes and to hole burning in inhomogeneously broadened lines. A theoretical description of the experiments was given in terms of a density matrix formalism.

116. NEW APPLICATIONS OF COHERENT PULSED ELECTRON PARAMAGNETIC RESONANCE. Michael K. Bowman, Chemistry Division, Argonne National Laboratory, 9700 S. Cass Avenue, Argonne, Illinois 60439

All of the parameters in the Hamiltonian for a system influence the conventional absorption spectrum in terms of spectral peak positions, intensities and lineshapes. Frequently, only a few of those parameters can be successfully extracted from the absorption spectrum because of limited spectral resolution and sensitivity. This problem can be circumvented by generating spectra using several coherent pulses in a sequence crafted so as to emphasize the Hamiltonian parameters of interest. This strategy has formed the basis of modern pulsed nuclear magnetic resonance spectroscopy and is now being applied in pulsed electron paramagnetic resonance (EPR). Examples will be shown of EPR nutation spectroscopy to measure electron spin multiplicity and

light-induced spin alignment, coherent Raman beat spectroscopy to measure strictly forbidden transitions, and the '2+1' and '3+1' experiments to measure the intensity and dynamics of very weak electron spin-spin interactions. All of these techniques are generalizable to coherent spectroscopies at other frequencies.
This work was supported by the U.S. Department of Energy, Office of Basic Energy Sciences, Division of Chemical Sciences under contract W-31-109-Eng-38.

117. TWO-DIMENSIONAL INFRARED (2D IR) SPECTROSCOPY. Isao Noda, A.E. Dowrey, and C. Marcott, The Procter & Gamble Company, PO Box 398707, Cincinnati, OH 45239-8707.

Two-dimensional infrared (2D IR) spectroscopy is a novel analytical technique developed to study submolecular-level interactions. In 2D IR, a spectrum defined by two independent wavenumbers is generated by applying a correlation analysis to the dynamic fluctuation of infrared signals induced by an external perturbation. Typically, a small dynamic strain with an acoustic-range frequency is employed to create such fluctuation; many other types of stimuli (e.g., electrical, thermal) could be used as well. Because of the specificity of IR absorption to chemical functional groups, fluctuations of IR signals contain detailed information about highly localized submolecular-level responses which are specific to individual functional groups. Notable features of 2D IR spectra are: simplification of complex spectra consisting of many overlapped peaks; enhancement of spectral resolution by spreading peaks over the second dimension; and establishment of unambiguous assignments through correlation of bands selectively coupled by various interaction mechanisms. The formalism developed here is a very general one, which could be extended to many other multidimensional spectroscopy applications where the conventional multiple pulse approach so successfully used in NMR is not readily available.

118. CHARACTERIZATION OF DIAMONDS, SYNTHETIC DIAMOND FILMS AND RELATED CARBON MATERIALS BY RAMAN SPECTROSCOPY. Diane S. Knight, Materials Research Laboratory, Pennsylvania State University, University Park, PA 16802

Raman spectroscopy has evolved into the principle method of characterizing diamond films made by plasma assisted CVD and other processes. A single sharp vibrational mode at 1332cm^{-1} allows cubic diamond to be easily identified. Raman spectra of natural diamond, crystalline graphite, and other carbon materials are given to compare with that of diamond films. Shifts in peak position may be caused by stress, isotopic composition, and temperature. The Raman line width varies with the sample and may be related to the degree of structural order. If a small amount of silver is sputtered onto a sample, surface enhanced Raman spectroscopy can be used to identify very thin diamond and amorphous carbon coatings which are not discernable by normal Raman scattering experiments. The amount of enhancement depends on the sample and its thickness.

119. STRUCTURAL CHARACTERIZATION OF CVD DIAMOND. R. E. Clausing, L. Heatherly, L. L. Horton, K. L. More, E. D. Specht, and Z. L. Wang, Oak Ridge National Laboratory, P. O.Box 2008, Oak Ridge, TN. 37831-6093, (615) 574-5084

The structure of diamond films grown by activated chemical vapor deposition vary greatly depending on the growth conditions. The development of two commonly observed polycrystalline morphologies, which give rise to <110> textures, will be described as well as special films grown to produce <100>, <111> and "near <100>" textures with various combinations of growth facets. Twining appears to have an important role in the nucleation and growth of most common films. Crystallites grown from {111} facets in these films contain numerous stacking faults and twins. Growth on flat {100} facets does not produce twins or stacking faults. The presentation will illustrate the use of X-ray diffraction,TEM, SEM, and optical characterization techniques.

The research was jointly sponsored by the Exploratory Studies Program of Oak Ridge National Laboratory; and by the Division of Materials Science and the Assistant Secretary for Conservation and Renewable Energy, Office of Transportation Technology, as part of the High Temperature Materials Laboratory Program,U.S. Department of Energy under contract DE-AC05-84OR21400 with Martin Marietta Energy Systems, Inc.

120. APPLICATIONS OF TRANSMISSION ELECTRON MICROSCOPY TO THE STUDY OF DIAMOND.
P. Pirouz, Department of Materials Science and Engineering, Case Western Reserve University, Cleveland, OH 44106.

Transmission electron Microscopy (TEM) has been applied to investigate the microstructure of diamond since the early sixties. In this talk, we show examples of recent studies of defects in natural diamond. These include line defects such as dislocations, planar defects such as stacking faults and impurity (nitrogen?) platelets, and three-dimensional defects such as voidites. In addition, examples are presented of the applications of TEM to synthetic diamond films grown on various substrates to study the microstructure of the films and, in cases, the interfacial layers between the diamond film and the substrate surface. The examples presented include various techniques of electron microscopy such as strain contrast, high resolution, and analytical TEM.

121. SURFACE AND MICROSTRUCTURAL CHARACTERIZATION OF THE NUCLEATION AND GROWTH OF DIAMOND THIN FILMS, B. E. Williams, T. Tachibana, B. R. Stoner, and J. T. Glass, Department of Materials Science and Engineering, North Carolina State University, Raleigh, NC 27695-7907.

A study of both immersed and downstream microwave plasma CVD growth of diamond was accomplished using in-vacuo x-ray photoelectron spectroscopy (XPS) and Auger electron spectroscopy (AES) to examine the surface of the sample at selected intervals of the nucleation and growth process. Ex-situ analysis consisted of scanning electron microscopy (SEM), Raman spectroscopy, transmission electron microscopy (TEM), and high resolution TEM. The effects of different substrate pretreatments on nucleation have also been investigated using XPS, AES, and SEM. The results of this investigation indicate the enhanced nucleation of diamond may be associated with the presence of a carbon precursor on the surface. Analysis of the structure of the diamond/silicon interface has been accomplished using cross-sectional TEM. The density of defects in these films is very sensitive to the growth conditions, with higher growth rates typically producing higher defect densities. The quality of CVD diamond films can be assessed in terms of the amount of non-diamond, sp^2-bonded carbon in the films. Raman spectroscopy has been utilized extensively in this regard, and in this research AES and XPS-EELS have been employed as well. Semi-quantitative analysis of the sp^2/sp^3 ratio in a series of diamond films grown under different growth conditions has been accomplished using curve-fitting routines combined with analysis of standards. This research was primarily supported by SDIO/IST through ONR.

122. WHAT'S ON THE SURFACE DURING DIAMOND GROWTH? David N. Belton, Physical Chemistry Department, General Motors Research Laboratories, Warren, MI 48090.

In order to construct a detailed mechanism for CVD diamond growth we must know what types of species are present on the substrate and the diamond film during the growth process. It is relatively straightforward to use surface analysis techniques and in vacuo sample transfer to characterize the time evolution of stable species during the growth process. In fact, we have successfully employed a combination of X-ray photoelectron spectroscopy (XPS), electron energy loss spectroscopy (EELS), and low energy electron diffraction (LEED) to accomplish that goal. However, it is much more difficult to identify the adsorbed growth intermediates, such as methyl or vinyl groups, which are responsible for the formation of diamond and/or diamond-substrate interlayers. In order to understand these intermediates, we have examined the thermochemistry of some adsorbed species, including radicals. We use molecular mechanics to obtain both enthalpies and entropies which allows us to calculate ΔG for a number of gas-surface reactions. By combining these data with estimates of rate constants we can calculate surface coverages of the different adsorbed species during growth. This combination of experimental surface analysis and thermochemical calculations give a fairly complete description of both the stable deposits and adsorbed intermediates present during diamond growth.

123. IMAGING OF INDIVIDUAL ATOMS AND CLUSTERS WITH THE FIM.* Gert Ehrlich, Dep't of Materials Science and Engineering and Materials Research Lab., University of Illinois at Urbana-Champaign, Urbana, IL 61801

The field ion microscope is unequalled in its ability to reveal the location of individual metal atoms on a surface. By drawing upon this capability, it has for the first time become possible to obtain quantitative information on the atomic level about a variety of important surface phenomena. This will be illustrated by recent studies concerned with the location of atomic binding sites on close-packed planes, the structure and properties of atomic clusters, as well as with the elementary steps in atomic migration.

* Supported by the Department of Energy under Grant DEFG02-91ER45439

124. TWO-DIMENSIONAL PHASES OF SULFUR FORMED BY COALESCENCE OF ATOMS ON Re(0001) SURFACES. M. Salmeron. Center for Advanced Materials. Materials Science Division. Lawrence Berkeley Laboratory. Berkeley. CA 94720.

The structures formed by Sulfur Chemisorbed on Re(0001) as a function of coverage have been studied by LEED and STM. At coverages of 0.25 ML (monolayers) and below, a (2X2) structure consisting of sulfur monomers is formed. This observation, together with the observation of islands at lower coverages indicates the existence of repulsive first and second nearest neighbor interactions and attractive third nearest neighbor interactions. Depending on the bias voltage the S atoms appear as round balls or as three armed stars. As the coverage becomes larger than 0.25 ML S is forced to occupy nearest neighbor positions or to rearrange in a (V3XV3) structure as observed on other compact surfaces. Instead new structures formed by aggregates of three and more S atoms form. The absence of dimer formation indicates that substrate mediated three and more many body interactions are dominant at coverages above 0.25 and up to 0.5 where saturation occurs.

125. REAL-SPACE IMAGING OF MOLECULAR ORGANIZATION: NAPHTHALENE vs. AZULENE ON PT(111). Vickie M. Hallmark, IBM Research Division, Almaden Research Center, San Jose, CA 95120-6099.

Scanning tunneling microscopy of naphthalene adsorbed onto the Pt(111) surface has revealed a wealth of information regarding molecular orientation, binding site, and two-dimensional organization. STM allows direct comparison of ordered and disordered adsorbate structures and yields a qualitative understanding of both rotational and translational diffusion in this system. Adsorption onto Pt(111) of the related molecule azulene provides an interesting comparison system. Azulene, with an aromatic structure consisting of a seven-carbon ring attached to a five-carbon ring, is an isomer of naphthalene (two fused six-carbon rings). Azulene forms a sequence of ordered structures with increasing surface coverage, quite different from its naphthalene analog. Comparison of data from the two systems provides an experimental test of STM sensitivity to small changes in adsorbate spacial and electronic structure.

126. CHEMISORPTION INDUCED PHASE TRANSFORMATIONS OF METAL SURFACES STUDIED BY STM, J. Wintterlin, J.V. Barth, H. Brune, R. Schuster, G. Ertl, Fritz-Haber-Institut der Max-Planck-Gesellschaft, Berlin, Germany, R. J. Behm, Universität München, Germany

A variety of metal surfaces are known for their structural instability, with often only small energy differences between the "truncated bulk" structures and reconstructions. In many of these surfaces when the adsorption energies are larger than these energy differences, chemisorption is known to induce structural transformations. Phase transitions of this type were studied by STM for chemisorption on the (110) surfaces of Pt, Cu, and Au. The data display substantial variations in the transformation mechanisms for the different systems. It will be discussed how these are determined through studies of surface mobility, nucleation and growth phenomena, and adsorbate-adsorbate interactions. Surface self-diffusion plays a major role, as the two-dimensional density of the topmost surface layers changes in these reactions. The STM results also allow for clear distinctions between long disputed structure models for these surface phases.

127. THE SURFACE DYNAMICS OF DIFFUSION AND CORROSION AT ATOMIC LAYER RESOLUTION STUDIED BY *IN-SITU* ELECTROCHEMICAL SCANNING TUNNELING MICROSCOPY. Dennis J. Trevor, Christopher E. D. Chidsey AT&T Bell Laboratories, Murray Hill NJ 07974, and I. Oppenheim and K. Sieradski, The Johns Hopkins University, Materials Science and Engineering, Baltimore, MD 21218

Crystal growth, annealing and dissolution are intrinsically dynamic process that occur at liquid- or gas-surface interfaces. We have used electrochemical scanning tunneling microscopy (ESTM) to study the kinetics of these processes at atomic layer resolution at liquid-solid interfaces of 0.1 M $HClO_4$ with Au(111) or dilute Ag/Au alloys. In these systems we have observed fast diffusion on the metal electrodes. Our ESTM studies suggest mechanisms for these surface reactions which are driven by the motion of step adatoms along the edge of terrace steps and not adatoms across terraces. We will also present a model that predicts a size dependence for this mechanism of pit or island diffusion.

128. ATOMIC FORCE MICROSCOPE STUDIES OF UNDERPOTENTIAL DEPOSITION PROCESSES. Chun-hsien Chen, Scott M. Vesecky, and Andrew A. Gewirth, Department of Chemistry, University of Illinois, 505 S. Mathews Avenue, Urbana, IL 61801.

We report atomic resolution Atomic Force Microscope (AFM) images of the underpotential deposition of Ag and Cu adatoms on Au(111) surfaces obtained in solution and under potential control. Cu underpotentially deposited on Au(111) in sulfuric acid electrolyte exhibits a ($\sqrt{3}$ x $\sqrt{3}$)–R30° adlattice. Alternatively, upd of Cu in perchlorate electrolyte yields an adlattice with the same 0.29 +/- 0.02 nm spacing as the underlying Au(111) surface but which is rotated by 30 +/- 10°. This implies that the Cu is in an incommensurate close packed layer atop the Au. Ag deposited on Au(111) in sulfate containing electrolyte exhibits a 3x3 structure, while deposition in carbonate and nitrate yields a 4x4 lattice. These different structures can be related to differences in specific adsorption of electrolyte onto the electrode surface, changes in anion-anion repulsion, and differing degrees of charge transfer between electrolyte and adatom.

129. TORSIONAL SOLITONS IN ORGANIC CRYSTALS. Katja Lindenberg, Department of Chemistry and Institute for Nonlinear Science, University of California at San Diego, La Jolla, CA 92093; David W. Brown, Institute for Nonlinear Science, University of California at San Diego, La Jolla, CA 92093; Alexander V. Zolotaryuk, Institute for Theoretical Physics of the Ukrainian SSR Academy of Sciences, 252130 Kiev, USSR

Experiments on *l*-alanine [A. Migliori et al., Phys. Rev. **38**, 13464 (1988)] indicate that energy in the low-frequency vibrations might spontaneously localize. The near perfection of the material indicates that this is an intrinsic property unrelated to impurity modes. This observation leads us to consider the role of nonlinearities. Experimental facts suggest the importance of nonlinear effects: 1) Two modes of different frequencies share a population of oscillator quanta, a transition from one to another occurring with increasing temperature, an effect consistent with a softening potential. 2) The relevant modes are identified as librational, frequently known to red-shift with increasing amplitude due to anharmonicities. 3) *l*-alanine displays an asymmetry in its thermal expansion coefficients, with the usual thermal expansion along two crystal axes but a *contraction* along the third. This is consistent with an anharmonic interaction between librations and acoustic phonons. We discuss models for the observed effects that may lead to spontaneous spatial energy localization.

130. SCALING PROPERTIES IN ANOMALOUS DIFFUSION. G. Zumofen, Physical Chemistry Laboratory, ETH, CH-8092 Zurich, Switzerland, A. Blumen, Physics Institute and BIMF, University of Bayreuth, D-8580 Bayreuth, F.R.G., and J. Klafter, School of Chemistry, Tel-Aviv University, Ramat-Aviv, 69978 Israel.

For anomalous, non-Brownian diffusion the time evolution of the mean-squared displacement shows a non-linear behavior: $<r^2(t)> \sim t^\alpha$, $\alpha \neq 1$. Furthermore, the propagator

$P(r,t)$, the probability to be at **r** at time t, deviates from a Gaussian. In the sublinear, dispersive regime, such as found for fractals, two different scaling forms have been proposed for $P(r,t)$. Our detailed numerical analysis demonstrates that the two approaches hold in different domains of the scaling variable. We also analyze the enhanced, superlinear diffusion regime. Examples here are slowly decaying velocity-velocity correlation functions or broad distributions of stepping lengths. These effects are included in two representative models: Motion in random velocity fields and Lévy walks.

131. **ENERGY TRANSPORT IN A ONE-DIMENSIONAL INORGANIC SOLID: EXCITON TRAPPING AND ANNIHILATION IN CRYSTALS OF $(CH_3)_4NMnCl_3$ (TMMC),** Gary L. McPherson, Ali Gharavi, Abofazl Amini and Michael F. Herman, Department of Chemistry, Tulane University, New Orleans, LA 70118.

The salt, $(CH_3)_4NMnCl_3$(TMMC), adopts a linear chain structure where the interchain contact between Mn^{2+} ions is almost three times longer than the intrachain contact (9.15Å vs 3.25Å). In photoexcited crystals the exciton derived from the lowest ligand field excited state of the Mn^{2+} ion is thermally mobile. The dynamics of exciton migration are probed by examining the time-resolved luminescence from TMMC doped with Cu^{2+} (exciton trapping) and undoped TMMC under high power excitation (exciton annihilation). A theoretical treatment for successfully fitting the luminescence decay curves has been developed. The results indicate that intrachain migration has a room temperature rate of roughly $10^{12} sec^{-1}$ with an activation energy of 1000 to 1200 cm^{-1}. Interchain migration is considerably slower, on the order of 10^3 to 10^4 sec^{-1} at room temperature. The trapping rate at the Cu^{2+} impurities falls in the 10^8 to 10^9 sec^{-1} range with an activation energy of 500 to 600 cm^{-1}. The details and conclusions of this study will be discussed.

132. RELAXATION AND REACTIONS IN LOW-DIMENSIONAL SYSTEMS: Igor M. Sokolov*, Horst Schnörer and Alexander Blumen, Physikalisches Institut und BIMF, Universität Bayreuth, Postfach 101251, D-8580 Bayreuth, FRG
*Permanent address: P.N. Lebedev Physical Insitute of the Academy of Sciences of the USSR, Leninsky Prosp., 53, Moscow 117924, USSR

We consider relaxation mechanisms having the form of the A+B→0 irreversible reaction (e.g. electron-hole annihilation or relaxation of radiative defects in solids). For equal concentrations of A and B reactants as is well-known the kinetics of relaxation follows the power-law $t^{-d/4}$, where d is spatial dimension. Here we investigate the case when one of the reactants is in excess; we find the decay law to be stretched-exponential and the upper critical dimension for this behavior is d=2. For different concentrations of A and B particles the reaction displays two main types of behavior, depending on wheter the majority species moves or is in rest. The talk will discuss the results of Monte-Carlo simulations of the reaction in different basic situations and present the analytical approach to the problem via (in general nonlinear) diffusion equations.

133. COMPUTATIONAL AND ULTRAFAST STUDIES OF ELECTRONIC ENERGY TRANSPORT IN SPATIALLY DISORDERED SYSTEMS. P.T. Rieger and R.J.D. Miller, Department of Chemistry, University of Rochester, Rochester, New York 14627.

We have used the resources of the NSF's supercomputer centers to obtain numerical solutions to the Pauli master equation for systems consisting of up to 5000 molecular sites at various number densities. The results serve as a benchmark for several analytical solutions, which agree only in the short-time regime. We show how sensitive the results are to the volume of the molecules, within what limits the energy migration is diffusive, and that the first-order cumulant approximation is the most consistent in predicting the transport. In addition, we are currently using polarization grating methods to study coherent and incoherent energy transport on the femtosecond timescale.

134. RECOVERY OF DISTRIBUTIONS OF FLUORESCENCE LIFETIMES IN HOMOGENEOUS AND HETEROGENEOUS SYSTEMS. W.R. Ware and B.D. Wagner, Photochemistry Unit, Department of Chemistry, The University of Western Ontario, London, Ontario, Canada, N6A 5B7. A. Siemiarczuk, Photon Technology International, 347 Consortium Crt., London, Ontario, Canada, N6E 2S

The recovery of distributions of fluorescence lifetimes from time correlated single photon decay data will be discussed. Applications of the Maximum Entropy Method and the Exponential Series Method will be described for Forster energy transfer systems in rigid media and in solvents where diffusion is important. Also, the use of lifetime distribution analysis in micellar photophysics will be discussed in connection with the determination of the distribution function for quenchers, the kinetics of quenching, and micelle size distributions.

135. ENERGY TRANSFER FROM ORGANIC ADSORBATES TO DEFECT SITES IN SILICA SUPPORTS. A. R. Leheny*, N. J. Turro*, J. M. Drake**, *Department of Chemistry, Columbia University, New York, NY 10027; **Exxon Research and Engineering Company, Annandale, NJ 08801.

We have studied the fluorescence from excited aromatic molecules adsorbed to porous silica substrates. We have found that the excited-state relaxation is non-exponential and have demonstrated that this is due to dipole-dipole coupling between the excited adsorbate and self-trapped exciton defect sites in the silica lattice. We have made an independent study of these defects and suggest that they are interstitial oxygen atoms forming peroxide radicals and bridges between the silicon atoms. By using the Forster equation to model the coupling, we can fit the relaxation of the fluoresence of the adsorbates to determine the strength of the interaction as well as the population and distribution of the defects.

136. EXCITATION METHODS USEFUL FOR INVESTIGATING MOLECULAR MOTION, G. L. Hoatson* and R. L. Vold, Department of Physics, College of William and Mary, Williamsburg, Virginia 23186

Wideline nuclear magnetic resonance spectroscopy is frequently used to study static and dynamic properties of physically, chemically and biologically important and relevant systems. Conventional studies of lineshapes, relaxation rates and 2D spectra provide motional correlation times over an extraordinarily large dynamic range, 10^{-12}s-1s. Extending this range requires the development of new techniques and it appears that tailored excitation is particularly promising.

Two illustrative ^{2}H-NMR examples will be considered which demonstrate the ideas and concepts involved. (1) Use of selective inversion (π-pulse) followed by an evolution period, τ, and hard quadrupolar echo detection. (2) Use of composite pulse broadband Jeener Broekaert (JB) sequences with high power pulses and echo detection. This new method provides a powerful way to obtain the anisotropy in the spin lattice relaxation time, T_{1Q}. For a 3-level system (I=1) JB sequences create zero and double quantum coherences and, while this violates quantum mechanical selection rules, they provide unique and interesting information.

Experimental results will be compared to calculated excitation profiles. The scope and limitations of each technique will be discussed with special attention paid to dynamic range, molecular motion occurring during excitation pulses, the effects of coherence mixing, and resonance offsets. Phase coherent inversion will be contrasted with the incoherent saturation (hole burning) used in optical spectroscopy. Successful generation of optical analogues of multiple quantum coherences would allow investigation of previously inaccessible transitions.

137. MULTIDIMENSIONAL NMR STUDIES OF STRUCTURE AND DYNAMICS OF SOLID POLYMERS. H.W. Spiess, Max–Planck–Institut für Polymerforschung, P.O.Box 3148, D–6500 Mainz, FRG

Multidimensional NMR spectroscopy, which has proven its value already in structural elucidation of biopolymers also offers many advantages for the study of solids. First of all the introduction of new frequency dimensions increases the spectral resolution. This is particularly important in solid state NMR, where the resolution is especially low. However, multidimensional spectroscopy also provides new information, unavailable from 1D spectra even in the limit of high resolution. Experiments have been designed which correlate different spin interactions or, through

exchange, relate the various states taken up by the molecule during different time periods. These two types of experiment provide structural and dynamic information, respectively.
Two–, three–, and reduced four–dimensional techniques will be introduced and their use demonstrated on partially ordered, semi–crystalline, amorphous as well as liquid crystalline polymers.
Reference
H.W.Spiess, Chem. Rev. (1991), submitted.

138. **Radiative Recombination and Vibronic Relaxation in *Sigma*-Delocalized Silicon Backbone Network Polymers.** William L. Wilson and T. W. Weidman, AT&T Bell Laboratories, 600 Mountain Avenue, Murray Hill, NJ 07974

We report a detailed study of ultrafast energy relaxation in (n-hexyl)polysilyne, a prototypical *sigma*-delocalized alkylsilicon network polymer. This disordered, silicon backbone material exhibits strong near UV to visible bandedge absorption and a high quantum yield of visible emission. The time evolution of the emission band is studied over four decades of time using a gated multichannel detection technique. The data indicate that while there is an "intrinsic" Stokes shift immediately after photoexcitation, (<10ps), intramolecular phonon assisted hopping on a nanosecond timescale is the dominant mechanism for excited state relaxation. The data can be understood in terms of theoretical predictions for energy relaxation in disordered materials. The large temporal range of our experiment permits us to explore both the short and long time limits of the theory.

139. OPTICAL AND NMR STUDIES OF LOCAL POLYMER DYNAMICS IN DILUTE SOLUTION. M.D. Ediger, Department of Chemistry, University of Wisconsin--Madison, Madison, WI 53706.

Rates of conformational transitions are related to the macroscopic properties of polymers, e.g., viscosity and the glass transition temperature. Time-resolved optical spectroscopy on the ps and ns time scales have been used to observe the local dynamics of polyisoprene and polystyrene labeled with anthracene. C-13 NMR relaxation measurements have been used to study the local dynamics of unlabeled chains. Solvents spanning a range of 100 in viscosity have been used in each study, and a wide range of temperatures were investigated. For polystyrene, the Kramers' equation in the high friction limit adequately describes the experiments. In polyisoprene, the Kramers' approach is inappropriate and frequency dependent friction can be used to understand the results. A recently developed technique for the superposition of T_1 values at different Larmor frequencies is used to interpret the NMR results in a model independent manner.

140. CHARACTERIZATION OF POLYMER SEGMENTAL MOTION VIA DEUTERIUM NMR
A. D. ENGLISH, Du Pont Central Research and Development, Experimental Station, Wilmington, DE 19880-0356

^{2}H NMR lineshapes, when coupled with relaxation experiments, can uniquely characterize both the rate and amplitude of molecular motion. The anisotropy of T_{1Q} relaxation can differentiate jump and diffusive models of motion even in the fast exchange limit. In semi-crystalline nylon 66 polymers, ^{2}H magnetization from crystalline and amorphous domains is isolated via their different T_1s. N-D and C-D bonds in nylon 66 crystals are found to undergo spatially heterogeneous librational motion (not discreet jumps) at all temperatures. Near the melting point the C-D bonds execute very fast ($<\tau_c> \sim 40$ psec), very large amplitude librational motions (~60°). N-D bonds in crystals, which also rapidly librate, are still quite restricted (~10°) in amplitude. In the amorphous domains discreet conformational isomerization of individual methylene groups is observed and is identified as the origin of the γ relaxation.

141. KINETICS OF THE MOLECULAR OXYGEN REACTION WITH SODIUM, MAGNESIUM AND COPPER ATOMS. C. Vinckier, P. Christiaens and M. Hendrickx, Department of Chemistry, K.U. Leuven, 3001 Heverlee, Belgium

The gas phase combination reaction of ground state sodium, magnesium and copper atoms with molecular oxygen is investigated in a fast flow reactor. Metal atoms are generated by thermal evaporation and plasma afterglow atomization in the temperature range from 300 to 900 K. Atomic absorption spectroscopy is used as detection technique. The influence of various experimental parameters such as the initial absorbance, plasma composition, carrier gas, reactor pressure and temperature on the derived rate constants will be described. The reactivity of these metals will be compared to others of the same group and will be discussed in view of current theoretical models.

142. KINETICS OF HIGH TEMPERATURE, HYDROCARBON ASSISTED BORON COMBUSTION. Robert C. Brown, Charles E. Kolb, Center for Chemical and Environmental Physics, Aerodyne Research, Inc., Billerica, MA 01821, Seog-Yeon Cho, Richard A. Yetter, Herschel Rabitz, and Frederick L. Dryer, Department of Mechanical and Aerospace Engineering or Department of Chemistry, Princeton University, Princeton, NJ 08544.

Boron/hydrocarbon slurries possess very high volumetric energy densities and thus have attracted a great deal of interest as advanced propellants for aerospace systems. Hydrocarbon assisted boron combustion is a complicated process involving both heterogeneous (gas/surface) and homogeneous (gas phase) kinetics. We have evaluated available measurements of elemental boron, boron oxide and boron oxyhydride reaction kinetics, estimated rate parameters for unmeasured processes and built a comprehensive model of single particle boron combustion. Kinetic sensitivity analysis techniques have been used to identify critical heterogeneous and homogeneous kinetic processes requiring further study.

143. REACTIONS OF METAL ATOMS IN PLASMA ETCHING AND CHEMICAL VAPOR DEPOSITION: Judith A. Halstead, Department of Chemistry and Physics, Skidmore College, Saratoga Springs, N.Y. 12866

Many of the processes currently used in the semiconductor industry for the fabrication of microdevices involve chemical reactions at a solid gas interface. The role of gas phase metal atom reactions in the heterogeneous processes, chemical vapor deposition and dry etching, will be examined. Kinetic information available on the dry etching of Al, Cu, W and other metals will be reviewed and compared with the more extensively studied etching of Si. Selective and blanket chemical vapor deposition (CVD) of metals such as Cu and W will be described and compared to CVD of silicon.

144. COMPETITION BETWEEN HOMOGENEOUS AND HETEROGENEOUS REACTIONS IN METAL VAPOR OXIDATION. Oleg E. Kashireninov, Institute of Structural Macrokinetics, USSR Academic of Sciences, 142432 Chernogolovka, Moscow Region, USSR.

Prolonged controversy regarding the macroscopic mechanism of metal vapor chemical reactions can be solved through the macrokinetic approach. The combined consideration of the kinetics of the elementary stages, diffusion and heat transfer for gas-gas and gas-surface processes, together with the dynamics of condensed product formation demonstrates that the structure of the reaction zone and size distribution of the condensed product particles are predominantly determined by the kinetics of product vapor formation. In reasonable ranges of pressures and temperatures, as well as metal vapor concentrations and size of product particles, the gas-phase reaction (heterogeneously, as a rule), is dominating. The process peculiarities caused by the chemical nature of metal and oxidizer are analyzed.

145. MOLE FRACTION MEASUREMENTS OF H, CH_3 AND OTHER SPECIES BY MOLECULAR-BEAM-MASS-SPECTROMETRY DURING HOT-FILAMENT ASSISTED DIAMOND FILM GROWTH.* Wen L. Hsu, Sandia National Laboratories, Livermore, CA, USA.

The molecular-beam-mass-spectrometry technique has been used successfully in the past to study gas-phase kinetics in a variety of reaction systems. We have applied this technique to study hot-filament assisted diamond film growth, and in this paper we will present our studies on the formation of stable, as well as free radical, gas-phase species during the growth process. The system has been designed to operate at realistic diamond growing conditions, i.e. sampling up to process pressures of 60 Torr and temperature of the substrate can be controlled up to 1000 °C. Relying on reported electron impact ionization cross-sections for each species, mole fractions have been determined for H, CH_3, CH_4, H_2 and C_2H_2. For the case of H, mole fractions as low as 10^{-5} can be measured. We have measured the species' concentrations at a variety of operating conditions, and we will compare these with experimentally and numerically calculated values reported by other authors.
*This work was supported by US DOE under contract #DE-AC0476DP00789 and DARPA/DSO-Material Science Program.

146. TEMPERATURE AND DENSITY DISTRIBUTION OF H_2 AND H IN A HOT FILAMENT CVD REACTOR OF DIAMOND. M.-C. Chuang, K.-H. Chen, C. M. Penney, and W. F. Banholzer, GE CR&D, P.O. Box 8, Schenectady, NY 12301.

The temperature and density distribution of hydrogen in a hot-filament CVD reactor of diamond have been measured using coherent anti-Stokes Raman scattering (CARS). By using the folded BOX CARS configuration and gating on the detector as well as spatial filtering, the bright background from the filament was rejected. The CARS spectra provided direct and accurate measurements of the hydrogen temperature and density distributions. Assuming the ideal gas behavior and constant pressure in the system, the density distribution of atomic hydrogen was also determined. These temperature and density distributions are essential for the understanding and modeling of the diamond growth processes. The effects of varying added methane concentration and the filament power of the system were also studied. It was found that the assumption of thermodynamic equilibrium failed to describe the observed atomic hydrogen density distribution. A diffusion controlled model appeared to explain the experimental measurements very well.

147. LASER-INDUCED FLUORESCENCE MEASUREMENTS FOR THE CHEMICAL MODELING OF DIAMOND DEPOSITION.* Jay B. Jeffries, Molecular Physics Laboratory, SRI International, 333 Ravenswood Ave, Menlo Park, California 94025

Laser-induced fluorescence (LIF) measurements provide *in situ*, nonintrusive, spatially resolved measurements of species concentrations and gas temperature. Diamond chemical vapor deposition environments all have large temperature gradients between the regions of gas activation and temperature controlled substrate. Since reaction rate constants are proportional to $\exp(-E_a/RT)$, spatially resolved temperature measurements are crucial input to models of the gas phase chemistry of diamond deposition. LIF measurements of reactive species concentrations provide sensitive tests of these models. LIF diagnostic measurements on CH, C_2, and OH in plasma arc-jet and hot filament diamond reactors are the examples discussed.

*Supported by the US Army Research Office.

148. GAS PHASE ANALYSIS OF DIAMOND-PRODUCING PLASMAS. Linda S. Plano, Crystallume, 125 Constitution Drive, Menlo Park, California 94025.

Diamond film can be deposited by a wide range of techniques, including thermally and plasma-driven systems. The plasma based systems can be particularly difficult to analyze because of the presence of charged species and because they are highly reactive chemically. Any diagnostic instrument which interfaces physically (as opposed to optically) with the plasma can be expected to perturb the plasma, at least logically.

To compensate for such perturbations, three different diagnostic instruments were used to study diamond-producing DC plasmas. DC plasmas were selected because they are better understood than other plasma techniques such as microwave. The diagnostic techniques used include optical emission spectroscopy, Langmuir probe, and mass spectrometry. These techniques have been applied to diamond producing DC plasmas to establish correlations between plasma processes and resulting film characteristics for process control purposes. Relationships between various plasma species, plasma electrical structure, and film properties will be discussed.

149. **IN-SITU DIAGNOSTICS OF THE FLAME GROWTH OF DIAMOND FILAMENTS.** P.W.Morrison, Jr., J.E. Cosgrove, J.R. Markham, and P.R. Solomon, Advanced Fuel Research, Inc., 87 Church Street, East Hartford, CT.

A variety of different in-situ diagnostics help elucidate the growth of diamond filaments. The growth reactor consists of an oxyacetylene torch impinging on a cooled surface. Throughout the growth, a video camera focused through a microscope records the nucleation and growth of the diamond. The growing diamond glows red hot, and the brighter it glows, the faster it grows. By watching the growth process in real time, one can produce a filament of diameter 200-250 μm and length of 1 mm in a 6-8 hours; in principle, there is no limit to the length of the filament. XRD shows that the filament is cubic diamond with a substantial degree of single crystallinity. In another set of experiments, in-situ Fourier transform infrared (FT-IR) spectroscopy helps elucidate the flame chemistry. A combination of emission/transmission FT-IR reveals the concentrations of CO, CO_2, H_2O, and OH and the gas temperature. Tomographic reconstruction yields local data and shows that the cooled surface significantly perturbs the species distributions and the gas temperature. Emission FT-IR of the growing surface shows that the emissivity of the deposit slowly increases during the growth of poor material but is relatively constant for high quality diamond; in both cases, the temperature remains constant.

150. **THE EFFECTS OF AMMONIA ON CVD DIAMOND FILM GROWTH.** Ching-Hsong Wu, M.A. Tamor, T.J. Potter, Research Staff, Ford Motor Company, P.O. Box 2053, MD 3028, Dearborn, MI 48121-2053

Understanding of physical and chemical processes occurring in chemical vapor deposition (CVD) of diamond films is essential for advancement in CVD diamond technology. As a part of this effort, we have been investigating the gas-phase chemical transformations and effects of deposition parameters on film properties and growth rate. One of our approaches involves the "tailoring" of gas composition by selective addition of certain chemical species in the gas-phase during film deposition. This method allows preferential enhancement or suppression of the film growth or modification of the film characteristics. This paper will describe results of such a study, namely, addition of ammonia (NH_3) into hydrocarbon/hydrogen feed gases. Strong effects were observed on gas compositions and the resultant film morphology even for small amounts of ammonia: reduction in hydrocarbon concentrations, several orders of magnitude increase in electrical resistivity of the deposited film, and substantial enhancement of secondary nucleation and film smoothness. The relation between gas-phase chemistry and film properties will be discussed.

151. LASER SPECTROSCOPIC DIAGNOSTICS FOR CHEMICAL VAPOR DEPOSITION OF DIAMOND FILMS. Robert W. Shaw, W. B. Whitten, and J. M. Ramsey, Analytical Chemistry Division, Oak Ridge National Laboratory, Oak Ridge, TN 37831-6142.

A hot filament chemical vapor deposition (CVD) reactor has been assembled for growth of diamond films via decomposition of methane in hydrogen. Optical access is provided for process diagnostic techniques based on laser spectroscopy. Speciation of the bulk reactor gases and intermediates immediately above the substrate as well as identification of molecular fragments on the growing film surface will permit a fundamental understanding of the CVD growth process. Atomic hydrogen has been detected using 3 + 1 resonantly enhanced multiphoton ionization (REMPI) in the vicinity of the filament. Through-the-substrate orifice sampling into a REMPI/time of flight mass spectrometer for identification of species in the active layer immediately above the substrate is under development.

* Research supported by U.S. Department of Energy, Office of Basic Energy Sciences, under contract DE-AC05-84OR21400 with Martin Marietta Energy Systems, Inc.

152. PRINCIPLES AND EXAMPLES OF SCANNING FORCE MICROSCOPY. D. Sarid, T. Chen, L. Yi, S. Howells, M. Gallagher, and D. Lichtenberger*, Optical Sciences Center, *Department of Chemistry, University of Arizona, Tucson, AZ 85721.

The talk presents the principles of operation of scanning force microscopy where a raster scanning sharp tip mounted on a cantilever interacts with forces across a surface. The systems developed use electron tunneling and capacitance, optical homodyne, heterodyne, laser-feedback, deflection, and polarization to sense the minute deflections of the force-sensing tip. Results obtained with several of these implementations will be shown, and future trends will be discussed. In particular, images of semiconductor microstructures, buckeyballs, magnetic domains, and biological samples will be presented.

153. ATOMIC FORCE MICROSCOPY OF MOLECULES ON SURFACES. C.M. Mate, V.J. Novotny, M.R. Philpott, IBM Almaden Research Center, San Jose, California 95120; G.S. Blackman, DuPont Company, Wilmington, Delaware 19880; G. Rudd, Department of Ceramic Engineering, Rutgers University, Piscataway, New Jersey, 08855.

We have been using the atomic force microscope (AFM) to study molecular films on surfaces. Experiments have been carried out in both air and under ultra high vacuum conditions. We find that the forces acting on the tip are dramatically influenced by the type of molecular film covering a surface. The films examined so far fall into three categories: liquid for unbound films of perfluoropolyether, intermediate for bonded films of perfluoropolyether, and soft solid for multilayers of cadmium arachidate. From AFM experiments, many important physical properties of the films can be determined such as surface tension, liquid film thickness, disjoining pressure, molecular conformation, microhardness, and surface stiffness. We will also discuss recent work on measuring forces on an STM tip during STM images of molecules on surfaces under UHV conditions.

154 PROBING THE SURFACE FORCES OF MATERIALS USING ATOMIC FORCE MICROSCOPY, Richard J. Colton, Nancy A. Burnham, Barbara I. Gans, Hubert M. Pollock, Judith A. Harrison, and Donald W. Brenner, Chemistry Division, Code 6177, Naval Research Laboratory, Washington, DC 20375.

New tip-based proximal probes or force microscopes are capable of measuring or mapping electromagnetic forces between surfaces on the length scale of 10^{-11} to 10^{-7} meters. The commonly measured forces include van der Waals, Born repulsion, electrostatic and magnetic forces, friction and adhesion. This paper describes the attractive, repulsive and adhesive forces between a tip and flat surface as a function of separation. These forces are measured with nanometer-scale resolution using instrumentation derived from scanning tunneling microscopy and compared to calculated forces obtained by computer simulation. The mechanical properties of a variety of materials -- metals, organic thin films and diamond -- have been studied. The data consist of force curves from which the mechanical properties of the materials including monolayer films can be determined. In addition, we will discuss possible contact deformation mechanisms which are relevant to 'atomic resolution' imaging in force microscopy.

156. ATOMIC FORCE MICROSCOPY STUDIES OF FRICTION AND WEAR AT THE NANOMETER SCALE. Yun Kim and Charles M. Lieber, Department of Chemistry, Columbia University, New York, NY 10027

Atomic force microscopy (AFM) has been used to study the tribological properties of two dimensional metal dichalcogenide materials, MX_2. We have observed layer-by-layer nanometer scale wear at defects or dopant sites on the surface of these materials while rastering AFM tip. Large area images show clearly step heights corresponding to the layer repeat length where the material has been removed. In addtion, the wear phenomena occurs with a rate characteristic for each MX_2 material. The implication of these new data to the mechanism of friction and wear will be discussed.

157. IMAGING DNA MOLECULES UNDER POTENTIAL CONTROL BY STM AND AFM. S.M. Lindsay, Department of Physics, Arizona State University, Tempe, Arizona 85287-1504.

Coulomb interactions can be strong enough to maintain a macromolecule in place on a surface under the enormous forces of an STM or AFM probe. When the surface charge is controlled potentiostatically, we can deposit DNA molecules onto a gold surface reversibly. The high ionic strength at the interface leads to conformational transitions that are a function of the ions used in the supporting electrolyte, the DNA concentration and the electrode charge. Reversible transitions can be visualized directly and repeatably.

158. SCANNING TUNNELING AND ATOMIC FORCE MICROSCOPY OF BIOLOGICAL SURFACES. Joseph A. N. Zasadzinski, Department of Chemical and Nuclear Engineering, University of California, Santa Barbara, California, 93106.

We prepared biomembranes by freeze-fracture replication techniques in which the fluid, non-conductive membrane is replaced with a metal replica. We have used this technique to image the three-dimensional structure of the $P_{\beta'}$ or ripple phase of dimyristoylphosphatidylcholine in water with superior resolution. We have also used the AFM to image bilayers and monolayers of lipids prepared by Langmuir-Blodgett (L-B) deposition on rigid substrates. With the L-B technique, the distribution, anchoring, and the rigidity of lipids and proteins can be controlled, thereby overcoming the usual limitations encountered in AFM imaging of biomolecules. We present a detailed study of the organization of dimyristoylphosphatidylethanolamine bilayers at better than .4 nm resolution. We have also demonstrated a new, non-destructive method of analyzing defects in cadmium arachidate monolayers with resolution 5 times better than that of electron microscopy techniques.

Acknowledgements: This work was done in collaboration with P. K. Hansma, H. G. Hansma, D. Schwartz, J. Garnaes, B. Dixon-Northern and M. Longo. Funding was provided by a Whitaker Foundation Biomedical Engineering Grant, NSF CBT86-57444, and ONR N00014-90-J-1551.

159. ATOMIC SCALE IMAGING OF MOLECULES ON SURFACES BY SCANNING TUNNELING MICROSCOPY. John D. Baldeschwieler, Department of Chemistry, California Institute of Technology, Pasadena, California 91125

The rapid development of the scanning tunneling microscope and its derivatives in the past five years has created new techniques not only for imaging directly atoms and molecules on surfaces, but also for surface modification, lithography, and performing surface chemistry and spectroscopy at microscopic levels never before attainable. Recent studies at Caltech including imaging and spectroscopy of MoS_2 and GaAs surfaces and atomic scale imaging of DNA will be described. Understanding the mechanisms involved in these processes will be key to the successful development of the capabilities potentially available with this technology.

160. COMPUTERIZATION OF MATERIAL AND CHEMICAL DATA. Malcolm W. Chase, Standard Reference Data, A323 Physics Building, NIST, Gaithersburg, Maryland 20899, and Nancy P. Korendyke, Database Development, Maxwell Online, Inc., 8000 Westpark Drive, McLean, Virginia 22102

Numeric data lies at the core of much of the chemical literature and is used extensively by the scientific community in solving problems. The utility of these data is dependent on the accompanying information, such as sample purity, temperature scales, and experimental procedures; in these days of electronic networking, the format of these data is also a significant consideration. Within ASTM Committee E49 on the Computerization of Materials Data, Subcommittee E49.53 is working on writing standards for presenting and exchanging numeric and ancillary data for physical and chemical properties. A sample exchange format for heat capacity is proposed.

161. QUANTUM THEORY OF CHEMICAL REACTIONS: REACTIVE SCATTERING AND TRANSITION STATE THEORY. W. H. Miller, Department of Chemistry, University of California, Berkeley California 94720

Quantum mechanical reactive scattering provides the most complete and rigorous description of a chemical reaction. Recent developments in its theory and applications are presented. Transition state-type theories avoid the complete state-to-state description of a reaction and focus for the average (microcanonical or canonical) rate constant directly. Recent discoveries in this area are also presented, e.g., an exact representation for the average reaction rate in terms of Siegert eigenstates related to the transition state.

162. STATE-TO-STATE REACTION DYNAMICS BEYOND A + BC. James J. Valentini, Department of Chemistry, Columbia University, New York, New York 10027.

We present the results of experimental studies of the state-to-state reaction dynamics in systems involving polyatomic reactants and polyatomic products. The systems that will be discussed include the H + alkane (methane, ethane, propane) hydrogen abstraction reactions, where the focus will be on the unusual correlation between rotational and vibrational state distributions of the H_2 product. We will also present measurements of the photofragment state distributions in the photodissociation of methyl iodide and butadiene, with an emphasis on the dissociation of vibrationally excited parent species that are produced by stimulated Raman excitation. The emphasis in the talk will be on comparing and contrasting the behavior observed for these many-atom systems with the better understood behavior of ABC systems.

163. **DYNAMICS OF H_2 ELIMINATION IN THE PHOTODISSOCIATION OF ETHYLENE.**
B.A. Balko, E.F. Cromwell, A. Stolow, J. Zhang, M.J.J. Vrakking, and Y.T. Lee, Department of Chemistry, University of California, and Chemical Sciences Division, Lawrence Berkeley Laboratory, Berkeley, California 94720 USA.

The dynamics of H_2 elimination in the photodissociation of ethylene at 193 nm was investigated through measurements of the translational energy distribution and rovibrational state distribution of H_2 products. Using 1,1 $D_2C=CH_2$ and 1,2 $HDC=CDH$, it was shown that both 4-centered and 3-centered elimination of H_2 could take place with an acetylene/vinylidene formation ratio of approximately two-thirds. Limited 1,2 H-migration takes place during the fragmentation, presumably through an ethylidene type structure. The relatively high rotational excitation of the H_2 fragment suggests that the transition state is not symmetric, as the two ethylene H atoms approach each other and reach the transition state one C-H bond should be significantly longer than the other. The vibrational energy distributions can be roughly characterized by a vibrational temperature of 4800K and an average translational energy, which is dependent on the rovibrational state of H_2, exceeding 20 kcal/mole.

164. UNIMOLECULAR REACTION DYNAMICS. Young S. Choi, Sang-kyu Kim, Ned Lovejoy and C. Bradley Moore, Chemical Sciences Division of the Lawrence Berkeley Laboratory and Department of Chemistry, University of California, Berkeley, CA 94720

Many aspects of simple unimolecular reactions may now be studied experimentally in complete quantum-state-resolved detail and theoretically with accurate ab initio surfaces and dynamical methods. The opportunities for comparison of experiment and theory will be illustrated with recent experimental results on HFCO and H_2CCO. Quasi-stable, extreme-motion out-of-plane bending states in HFCO, carbon-atom exchange in ketene, and fine-structures in dissociation rate vs energy for triplet ketene will be discussed.

165. **AB INITIO QUANTUM CHEMISTRY AND DYNAMICS.** *Donald G. Truhlar*, Department of Chemistry and Supercomputer Institute, University of Minnesota, Minneapolis, MN 55455-0431.

We are working on several systems where dynamics calculations based on *ab initio* potential energy surfaces may be compared to experiment. These include the chemical reactions $D + H_2 \rightarrow HD + H$, $F + H_2 \rightarrow HF + H$, $H + DCl \rightarrow HCl + D$ and $HD + Cl$, $H + DBr \rightarrow HBr + D$ and $HD + Br$, $OH + CH_4 \rightarrow H_2O + CH_3$, and $Cl^-(H_2O) + D_2O \rightarrow Cl^-(D_2O) + H_2O$ and the energy transfer collisions of HF with HF and of He with I_2. In addition we are making *ab initio* studies of the properties of hydrogen bonded complexes that may be compared to experiment; this includes the dimers of HF and NH_3 and the mixed dimer of formamidine and water. Recent progress on one or more of these projects involving the comparison of *ab initio* quantum chemistry with experiment will be presented at the Symposium.

I am grateful to my current collaborators Jan Almlöf, Frank Brown, Bruce Garrett, Angels Gonzalez-Lafont, Mark Gordon, Gillian Lynch, Steven Mielke, A. V. Nemukhin, K. A Nguyen, David Schwenke, Rozeanne Stedkler, Thanh Truong, Susan Tucker, Michael Unekis, Antonio Varandas, J.-g. Zhao, and Xin Gui Zhao for their work on these projects.

This work was supported in part by the National Science Foundation and the U.S. Department of Energy, Office of Basic Energy Sciences.

166. NMR RELAXATION, FLUORESCENCE DEPOLARIZATION AND THE DYNAMICS OF PROTEINS AND MEMBRANES: THEORY AND EXPERIMENT. Attila Szabo, Lab. of Chem. Phys., National Institutes of Health, Bethesda, MD 20892

Time-resolved fluorescence depolarization reflects the reorientational dynamics of electronic transition dipoles in the time domain. NMR dipolar relaxation parameters reflect reorientational dynamics of bond vectors in the frequency domain. The problem of analyzing such experiments to yield information about the time scale and amplitude of internal motions in proteins and lipid bilayers will be discussed. Applications of the model free approach and its extensions will be presented and the results will be compared with the predictions of molecular dynamics simulations.

167. DYNAMICS STUDIES OF THE PROTEIN ECHISTATIN. J. Baum[+], X.Y. Chen[+], D. Kominos[+], R. Levy[+], S. Pitzenberger[*], [+]Rutgers University, New Brunswick, NJ 08855 and [*]Merck, Sharp & Dohme, West Point, PA 19486.

The dynamics of a small (49 amino acid residues) blood clotting protein, Echistatin, have been studied by ^{1}H NMR. It is found that the overall motion of the protein is anisotropic and that large portions of the protein contain flexible regions. The dynamics measurements and the effects of motion on structure determination will be described.

168. MODEL FREE ANALYSIS OF THE HIGH FREQUENCY DYNAMICS IN PROTEINS. Diane M. Schneider, Martin J. Dellwo and A. Joshua Wand, Department of Biochemistry, University of Illinois at Urbana-Champaign, Urbana, IL 61801.

In order to understand macromolecular function it is necessary to obtain detailed dynamic as well as structural information. NMR relaxation parameters such as longitudinal relaxation time (T1) and the nuclear Overhauser enhancement (NOE) depend on molecular events occurring with a frequency of 10^9 per second or greater. Model-free analysis (Lipari & Szabo) of this type of NMR relaxation data allows one to describe subnanosecond internal motion in terms of spatial restriction and rate without making any assumptions about the nature of the motion. We are using this method to analyze 2D-heteronuclear T1 and NOE data in order to develop a detailed picture of the fast internal dynamics of several proteins. Proteins being studied include human ubiquitin and reduced cytochrome c. We are interested not only in quantifying internal dynamics on the timescale of NMR relaxation but are also examining the relationship between motions occurring on this timescale and timescales represented by hydrogen exchange and crystallographic temperature factors. Results of ^{15}N and ^{13}C relaxation studies will be presented.

169. ULTRAFAST OPTICAL STUDIES OF VIBRATIONAL RELAXATION AND STRUCTURAL DYNAMICS OF HEME PROTEINS. R. J. Dwayne Miller, Department of Chemistry, The University of Rochester, Rochester, NY 14627-0216

Protein systems provide an interesting test for chemical principles developed for condensed phase processes. For example, large amplitude motions of hundreds to thousands of atoms occur in a correlated fashion as part of the function of many important biological functions. Examples include DNA unwinding, immunoglobin binding to antigents, and cooperative effects such as binding of oxygen to hemoglobin. How is energy channeled to accomplish these motions for these large molecules? How is the potential energy surface of proteins crafted to perform specific functions? In order to answer these questions, the mechanism of vibrational energy propagation in proteins and biopolymers needs to be experimentally addressed as well as the dynamics for the various protein motions. Picosecond phase grating spectroscopy provides a new approach to following both the energetics and structural changes in biological systems. This method provides a direct measurement of the time evolution of the protein strain which is the main parameter used to describe protein motion. Results will be presented for vibrational energy relaxation processes and optically triggered tertiary and quaternary structure changes in heme proteins which serve as model systems for understanding protein dynamics.

170. OPTICAL SPECTROSCOPY OF THE BACTERIAL PHOTOSYNTHETIC REACTION CENTER. R.A. Friesner, Department of Chemistry, Columbia University, New York, NY 10027

We describe simulations of the optical properties of the bacterial photosynthetic reaction center. The simulations are carried out using a model Hamiltonian containing both electronic and vibrational terms which is treated via an efficient algorithm for obtaining spectral lineshapes of coupled vibronic systems. Simulations will be presented for absorption, circular dichroism, Stark spectroscopy, and polarized absorption, with a special focus on the properties of the special pair of bacteriochlorophyll molecules. The anomalous temperature dependence of the absorption spectrum of the special pair is explained in terms of thermal expansion of the protein. Finally, we investigate the spectroscopy of the reaction center of the heterodimer mutant, in order to assess the validity of our model involving an internally charge separated species of the special pair. The implications of these results for the mechanism of charge separation in the reaction center will then be discussed.

171. STUDY OF LIGAND-LIGAND INTERACTIONS IN ML_n^+ SYSTEMS.
C.W. Bauschlicher, Jr., S. R. Langhoff and H. Partridge NASA Ames Research Center, Moffett Field, CA 94035.

Ab initio calculations are used to study the bonding in, $Mg(H_2O)_n^+$, $Al(H_2O)_n^+$, $Cu(H_2O)_n^+$ and $Na(H_2O)_n^+$. The calculations show that the differences between Cu^+ and Na^+ arise from $sd\sigma$ hybridization on the Cu and changes in the ligand-ligand repulsion. The $Mg(H_2O)_n^+$ and $Al(H_2O)_n^+$ ions have lower symmetry structures than the analogous Cu or Na compounds due to metal $3s$ polarization. Systems with other multiple ligands, such as benzene, acetone, formaldehyde and methanol, are also discussed.

172. THE ROLE OF COOPERATIVE EFFECTS IN THE BINDING OF SMALL MOLECULES TO METAL ATOMS. K. K. Sunil[a] and K. D. Jordan[b]; [a]Department of Medicinal Chemistry, Berlex Laboratories, Inc., 110 East Hanover Ave., Cedar Knolls, NJ 07927; [b]Department of Chemistry, University of Pittsburgh, Pittsburgh, PA 15260.

A detailed theoretical study of the role of cooperative effects in the binding of CO, C_2H_2, and C_2H_4 to non-transition metal atoms, in particular, beryllium, has been carried out. Although the interaction of a single CO, C_2H_2 or C_2H_4 molecule with Be is van der Waals in nature and, hence, very weak, the $Be(CO)_3$, $Be(C_2H_2)(CO)_2$ and $Be(C_2H_4)(CO)_2$ complexes are found to be strongly bound. Moreover, it is also found that the $(CO)_2Be-Be(CO)_2$ molecule has a strong ($D_e > 1$ eV) Be-Be bond. The role of rehybridization in the bonding in these systems is discussed.

173. THE UNIQUE COMPLEXATION AND OXIDATION OF METAL BASED CLUSTERS. J. L. Gole, Kangkang Shen, C. B. Winstead, D. Grantier, T. C. Devore, S. H. Cobb, and J. R. Woodward, School of Physics, Georgia Institute of Technology, Atlanta, GA 30332.

Oven and laser vaporization based techniques are used to form small metal clusters and to study their unusual oxidation behavior. Metal and metalloid clusters are found to oxidize to form not only a distinct class of metal atom grouped cluster oxides and halides (M_nO, M_nX, M=Na,Cu,Ag,B,Si,Mn,...) under kinetically as opposed to thermodynamically controlled conditions but unusual excited electronic state product distributions. An effort is underway to characterize and compare the internal mode structure demonstrated by the first vibrationally resolved optical signatures for several "asymmetric" metal cluster oxides with the corresponding structure of their thermodynamically more stable symmetric counterparts. A subset of the oxidation studies touches on the sodium trimer - halogen atom reactions, [Na_3-X(Cl,Br,I)], which create a continuous electronic population inversion based on the chemical pumping of Na_2 with laser amplifiers throughout the visible and ultraviolet. These studies graphically demonstrate the dramatic and unexpected oxidation behavior characteristic of small metal cluster reactions and point to the potential for developing new insights on the nature of chemical reactivity.

174. OXIDATION REACTIONS OF METAL AND SEMI-METAL CLUSTER IONS. P. A. Hintz, M. B. Sowa, J. Christian, Z. Wan, and S. L. Anderson, Department of Chemistry, State University of New York at Stony Brook, Stony Brook, NY 11794

Over the past few years we have undertaken a concerted effort to understand the physical and chemical properties of cluster ions of boron, carbon, and aluminum. For physical characterization, we measure cluster stabilities and ionization potentials, and have calculated structures and some electronic properties. Ion beam studies of the chemistry with 14 small molecules have been used to get a handle on reaction mechanisms and energetics, and how these relate to cluster physical properties. This talk will concentrate on oxidation reactions together with some new spectroscopic probes of cluster structure. The results will be compared with new *ab initio* calculations of structures and energetics.

175.
KINETIC PROCESSES WITH METAL CLUSTER PARTICIPATION IN SHOCK WAVES.
I.S. Zaslonko and V.N. Smirnov, Semenov Institute of Chemical Physics, The USSR Academy of Sciences, Moscow 117334.

Abstract not received.

176. HALOGENATION OF DIAMOND (100) USING ATOMIC BEAMS, Andrew Freedman, Center for Chemical and Environmental Physics, Aerodyne Research, Inc., 45 Manning Road, Billerica, MA 01821

We have halogenated the annealed diamond (100) surface using atomic beams of fluorine and chlorine under UHV conditions. XPS measurements indicate that fluorine atoms reach a saturation level of ~3/4 of a monolayer at room temperature. The carbon fluoride adlayer is stable to 700 K, decomposing at temperatures above this. The chlorine coverage, on the other hand, is highly temperature sensitive; saturation coverage monotonically decreases above 200 K. There is no evidence of the formation of a distinct C-Cl moiety. Molecular fluorine and chlorine are unreactive under these low flux conditions. Initial studies on the basal plane (0001) of graphite indicate that an ionic carbon-fluorine species which is stable up to 550 K is formed. No evidence, though, for subsurface penetration is found. These results will be linked to recent advances in the formation of diamond thin films.

177.
KINETICS OF DIAMOND FILM GROWTH. L. R. Martin, The Aerospace Corporation, P.O. Box 92957, Los Angeles, CA 90009.

A high temperature passivated quartz flowtube is a powerful tool for learning about the mechanism of diamond film growth at low pressures. We have shown previously that diamonds may be grown by adding methane to a flow of atomic hydrogen at 800 °C. In recent experimental work, we have measured the rate of diamond growth and the rate of hydrogen atom decay simultaneously at a series of different flowtube temperatures. The results were modeled with a highly simplified chemical model. Satisfactory agreement with the data is seen only when a non-productive surface catalyzed loss of methyl radicals is included in the model.

178.
FILAMENT-ASSISTED DIAMOND GROWTH KINETICS. Stephen J. Harris, Physical Chemistry Dept., General Motors Research Labs, Warren, MI 48090

We have measured the growth rate of CVD diamond films under filament-assisted conditions using a microbalance. The pressure was varied from 20 to 200 torr, and the ratio R of CH_4 to H_2 was varied from 0.2% to 1%. Raman spectra showed only diamond features for our films. We found that for $R > 0.2\%$, where filament carburization was not an issue, the growth rate scaled as R^α, where α is an empirical constant that varies from about 1 at 20 torr to about 0.5 at 200 torr. A comparison of these results to predictions of our gas-phase/gas-surface model for diamond growth shows that the model accurately predicts the value of α and how α varies with pressure. Reasons for the success of our very simple model are discussed.

179. MODELING DIAMOND GROWTH IN PEDESTAL CVD REACTORS. D. S. Dandy, M.E. Coltrin, R. J. Kee, Sandia National Laboratories, Livermore, California 94551-0969.

A stagnation flow model has been developed to describe the deposition of diamond film in hot-filament and DC arc-jet pedestal reactors. Through a similarity transformation, the full three-dimensional equations governing conservation of momentum, mass and energy in the gas phase collapse to one-dimensional form. Included in these gas-phase equations are detailed pyrolysis chemistry, multicomponent transport properties, and a mass- and energy-transfer balance at the deposition surface. An elementary surface reaction mechanism describing the formation of diamond through adsorption of C, CH_3 and C_2H_2 is implemented, as well as a competing mechanism for the formation of graphitic (sp^2) carbon. The calculations for the hot filament reactor are compared to other results (Goodwin and Gavillet, *J. Appl. Phys.*, 1991) obtained using different growth mechanisms (Harris, *Appl. Phys. Lett.*, 1990; Frenklach and Wang, *Phys. Rev. B*, 1991). The results will also be quantitatively compared to data obtained by molecular beam mass spectroscopy in a concurrent experimental project.

180. SURFACE CHEMISTRY OF DIAMOND GROWTH. Michael Frenklach, Department of Materials Science and Engineering, The Pennsylvania State University, University Park, PA 16802.

Chemical vapor deposition of diamond thin films will be discussed in terms of elementary chemical reactions taking place on the growing surface. The principal postulate on similarity between gas-phase and surface reactions of carbonaceous materials, introduced by us several years ago, will be reviewed in light of the present findings. The mathematical formalism for detailed description of gas-surface and surface reaction kinetics, based on the concept of active site reactivity and the algorithm of chemical lumping, will be presented. Specific reaction pathways responsible for diamond growth and nucleation will be discussed. The work was supported in part by SDIO/IST via the Office of Naval Research, Contract No. N00014-86-K-0443.

181. SIMULATION OF CVD DIAMOND GROWTH. D. G. Goodwin, Division of Engineering and Applied Science, California Institute of Technology, Pasadena, CA 91125.

Numerical modeling can serve as a useful complement to experimental diagnostics in gaining insight into the complex, nonequilibrium environment in which diamond growth occurs. This talk will discuss recently-developed computational models of the diamond growth environment, which simultaneously solve for gas-phase chemistry, surface chemistry, and diffusive and convective transport through the boundary layer above the substrate. The models have been applied to investigate diamond growth in hot-filament, arcjet, oxyacetylene flame, and inductively-coupled plasma environments. The simulation results indicate that a methyl growth mechanism can account reasonably well for the growth rates measured in each of these environments. The simulations also predict that transport of H to the substrate is diffusion-limited in high pressure environments, which may have significant implications for film growth rates, morphology, and reactor scale-up.

182. FIELD ION-SCANNING TUNNELING MICROSCOPY(FI-STM). Toshi Sakurai, T. Hashizume & Y. Hasegawa, Institute for Materials Research, Tohoku Univ., Sendai 980 Japan.

The scanning tunneling microscope (STM) invented in 1982 by Binnig and Rohrer, at IBM Zurich, has revolutionized the research approach of surface science and its related fields. The enormous potential and power of the STM, however, come with a discouraging and frustrating weakness: unpredictably ill-defined resolution and poor reliability in atomic-resolution performance. The problem was due to the ill-prepared tip sharpness. Realizing that an field ion microscope(FIM) could be the best candidate to overcome this problem, we have constructed the FI-STM, which is the high-performance STM combined with a room-temperature(RT) FIM for the in-situ characterization of a scanning tip on an atomic scale. The FI-STMs perform superbly, guaranteeing a long tip lifetime and atomic resolution (T. Sakurai et al., Prog. Surf. Sci. __33__, 3-89 (1990)). The FI-STMs have been applied for various chemisorption systems on Si and GaAs surfaces. In this talk, alkali metal (Li/K/Cs) adsorption on the both Si(111)7x7 and Si(100)2x1 surfaces will be discussed and the similarities and differences are presented.

183. SCANNING TUNNELING MICROSCOPY OF ALKALI METALS ON III-V(110) SURFACS: FROM 1- AND 2-D INSULATORS TO 3-D METALS*. Joseph A. Stroscio, National Institute of Standards and Technology, Gaithersburg, MD 20899.

The interaction of Cs with GaAs and InSb(110) is shown to yield a rich variety of structures with increasing alkali density ranging from: 1) one-atom wide linear structures with lengths greater than 100 nm, 2) 2-dimensional monolayer films composed of ordered arrays of polygon-like alkali clusters, and 3) a disordered 3-dimensional bilayer with saturation exposure. The 1-D and 2-D alkali structures are shown to be insulating with a tunneling gap that narrows in going from 1- to 2-D; metallic characteristics are not observed until 3-D structures are formed. In addition, we demonstrate that the alkali atoms can be induced to diffuse under the action of a pulsed electric field generated by the STM probe tip. This field induced directional diffusion occurs towards the tip during the voltage pulse due to the local potential energy gradient arising from the interaction of the adsorbate dipole moment (both static and induced) with the electric field gradient at the surface.
* In collaboration with L. J. Whitman, R. A. Dragoset, and R. J. Celotta. This work is supported in part by the Office of Naval Research.

184. PROBING AND INDUCING SILICON SURFACE CHEMISTRY WITH THE STM. In-Whan Lyo and Ph. Avouris, IBM Watson Research Center, Yorktown Heights, NY 10598.

We have used scanning tunneling microscope (STM) images and spectra to probe the surface chemistry of silicon with atomic resolution. We also have found that tip-induced perturbations of the electronic structure of surface sites can be used to induce local chemical changes. As examples we used the reactions of Si(111)-7x7 with O_2 and H_2O. We will discuss the mechanisms and site selectivities of these reactions as revealed by the STM studies and demonstrate that the dissociation products of the above thermal reactions can be further modified or desorbed individually by electronic tip-induced interactions. We have found that even substrate Si atoms can be manipulated in this way. We will show how clusters of Si atoms ranging in size from tens of atoms down to single atoms can be removed by the tip and then be redeposited at predetermined surface sites. The size of the clusters can be controlled by varying the tip-sample distance and the magnitude of applied voltage pulses while the direction of atom transfer is determined by the direction of the electric field. The mechanisms by which the above sample manipulations take place will be discussed.

185. ORDERED BUCKLED DIMER CONFIGURATIONS ON SI(100) OBSERVED WITH A NEW VARIABLE TEMPERATURE STM. Robert A. Wolkow, AT&T Bell Laboratories, 600 Mountain Avenue, Murray Hill, NJ 07974

Although it is now generally accepted that reconstruction of the Si(100) surface involves dimer formation, the preferred orientation of the dimers has remained unclear despite considerable experimental and theoretical effort. Using a new variable temperature STM, the issue has been resolved in favour of buckled dimers. At 100K the majority of dimers are buckled. The sense of buckling alternates along rows in a highly correlated manner. Regions of (1x2), p(2x2), and c(2x4) are observed.

186. ADSORPTION AND GROWTH ON SI AND GE SURFACES. T.-C. Chiang, Department of Physics, University of Illinois, Urbana, Illinois, 61801.

The adsorption of atoms and subsequent growth of overlayers on semiconductor surfaces can be observed directly with the use of scanning tunneling microscopy (STM). From such studies, a detailed understanding of the atomic processes on surfaces and the resulting structure can be obtained. There are limitations, however. For instance, it is sometimes difficult to identify the various kinds of atoms present in the system, and this can be a problem for reactive and/or disordered systems involving several atomic species. Examples, including the deposition of Ag and Sb and the chemisorption of oxygen on Si and Ge surfaces, will be discussed to illustrate the successes and difficulties. Complementary information gathered from other studies, such as photoemission, is very useful for the construction of structural models.

187. ROLE OF BOND STRAIN IN THE CHEMISTRY OF HYDROGEN ON SI(111) AND SI(100) SURFACES. J. J. Boland, IBM Research Division, P.O. Box 704, Yorktown Heights, New York 10598

In this paper we present an STM study of the nature and origin of the different silicon hydrides species produced by the interaction of atomic hydrogen with the Si(111)-7x7 and Si(100)-2x1 surfaces. We show how the weakening of bonds induced by surface reconstruction controls the chemistry of these surfaces. In the case of the Si(111) surface all SiH$_x$ species are observed and are directly associated with the rearrangement of the reconstructed surface following H adsorption. The Si(100) surface, on the other hand, exhibits a 2x1 monohydride phase, a 1x1 dihydride phase, and a mixed 3x1 phase. The relationship between these phases will be discussed including the role of bond strain in determining the stability and intercovertability of these phases.

188. THE DYNAMICS OF THE DECOMPOSITION AND ISOMERIZATION OF AZIDES AND ISO-CYANATES. Millard H. Alexander and Terrence Hemmer, University of Maryland, and Hans-Joachim Werner, Fakultät für Chemie, Universität Bielefeld, Germany. We shall report the results of continuing *ab initio* MCSCF+CI calculations with large atomic orbital bases of the topology of the potential surfaces relevant to both spin-allowed and spin-forbidden decomposition of HN$_3$, N$_3$, HNCO, and NCO. The lowest energy decomposition process involves crossing in the exit channel with a state of higher spin multiplicity (triplet in the case of HN$_3$ and HNCO; quartet in the case of N$_3$ and NCO) which correlates with the ground electronic state of either NH($X^3\Sigma^-$) or N(^{4}S). Barriers can also be present in the spin-allowed decomposition channel. The calculated barrier heights will be compared with those inferred from concurrent infrared overtone pumping experiments and from temperature dependent quenching studies. Modified RRKM and Franck-Condon-like models will be used to predict product energy disposal, spin-allowed/spin forbidden branching ratios, and decomposition and quenching rates. Of additional interest is the isomerization of the NCO radical (NCO ↔ CNO ↔ NOC) on both ground and excited potential energy surfaces.
Supported by AFOSR under contract F49620-88-C-0056 and ARO under grant DAAG29-85-K-0102.

189. THE DYNAMICS OF THE O(^{3}P) + NH$_2$ (ND$_2$) REACTION. Paul J. Dagdigian, Dipti Patel-Misra, and Deborah G. Sauder, Department of Chemistry, The Johns Hopkins University, Baltimore MD 21218.

The O(^{3}P) + NH$_2$ reaction can proceed by formation of HNO + H or OH + NH. The internal state distribution of the OD product from O + ND$_2$ and of HNO from O + NH$_2$ was determined in a crossed-beam experiment. The imidogen reagent was prepared by 193 nm photolysis of ammonia (seeded in a diluent) at the nozzle of a pulsed supersonic beam, while oxygen atoms were produced in a microwave discharge source. The internal state distributions of the products were determined from laser fluorescence excitation spectra. The state distributions for both products were found to be quite cold and to exhibit less excitation than that predicted by phase-space theory. These reactions are expected to proceed through H$_2$NO and HNOH intermediates. The implication of these experimental results for the dynamics of this reaction will be discussed.

190. QUANTUM DYNAMICS OF THE HYDRATED ELECTRON, Peter J. Rossky, Department of Chemistry, The University of Texas at Austin, Austin, Texas 78712.

The molecular and electronic description of the hydrated electron, and of excess electrons in polar fluids in general, has been the subject of intense experimental and theoretical study for decades. Recent advances in theory and in computer simulation now permit the investigation of the equilibrium and dynamical behavior of such species via a fully molecular description based only on a plausible set of interactions among the component species, *and* allow the direct evaluation of experimental observables. In this talk, results obtained from such quantum mechanical simulation methods for the hydrated electron system will be described. In particular, the nature of the electronic state manifold, and the resulting steady state and transient spectroscopy will be described with emphasis on comparison to recent results from experimental femtosecond spectroscopy.

191. AB INITIO STUDIES OF THE ^{1}A$_2$ AND ^{1}B$_1$ STATES OF OZONE: A CONICAL INTERSECTION AND ITS EFFECT ON PHOTOABSORPTION AND DYNAMICS. M. Braunstein, P. J. Hay, R. T Pack and R. L. Martin, Theoretical Division, Los Alamos National Laboratory, Los Alamos, NM 87545

Ab initio configuration interaction calculations are reported on the ^{1}A$_2$ and ^{1}B$_1$ states of ozone near their C$_{2v}$ minima and along C$_s$ geometries leading to dissociation. These states have a seam intersection near the ^{1}B$_2$ minimum which becomes a conical intersection when C$_s$ geometries are included. The influence of this intersection on the Chappuis and Wulf bands of ozone is investigated and a new interpretation of these bands is presented.

192. CALCULATIONS OF THE GEOMETRIES OF TRANSITION-METAL COMPLEXES. Michael B. Hall, Zhenyang Lin, Arthur A. Low, Andrew L. Sargent, Jun Song, and Rodney L. Williamson, Department of Chemistry, Texas A&M University, College Station, Texas 77843-3255.

Ab Initio calculations with modest basis sets and modest amounts of electron correlation have proven useful in elucidating the puzzling structure of a number of transition-metal complexes. A wide variety of first and second row complexes have been surveyed to establish relative errors of a modest basis set suitable for initial calculations. This basis set and extension of it have been used to explore a variety of experimental structural problems: the origin of the pair-wise inequality of the Cr-CO bond lengths in trans-Cr(CO)$_4$Cl(CCH$_3$), the apparently flattened methyl in the electron-diffraction structure of Cl$_3$TiCH$_3$, the unusual structure reported for the blue green bond-stretch "isomers" of M(XR$_3$)$_3$Cl$_2$O^n (M=Mo, X=P, n=0; M=W, X=N, n=1+) complexes, structures of transition-metal polyhydrides, and the apparently planar structure of the dimer (μ-NO)$_2$(CoCp)$_2$.

193. LIGAND DYNAMICS IN MYOGLOBIN. L. Rothberg, M. W. Roberson and T. M. Jedju, AT&T Bell Laboratories, Murray Hill, New Jersey 07974.

Picosecond transient infrared spectroscopy is used to observe the behavior of the CO stretching mode after photodissociation of carboxymyoglobin. The experiment is performed at both low temperature (150 K) and at physiological temperatures. In both cases, rebinding inside the protein is observed and lifetimes of these metastable configurations are measured. The angle of the CO with respect to the normal to the porphyrin plane in the metastable state is also measured using polarization spectroscopy. We use these experimental parameters in conjunction with theoretical molecular energetics calculations which incorporate the x-ray structure of the protein to speculate on where the CO is in the metastable state and the pathway by which it enters and leaves the protein.

194. ULTRAFAST VIBRATIONAL DYNAMICS USING TRANSIENT INFRARED SPECTROSCOPY. T. Lian, B. Locke, M. Iannone, J. Owrutsky, M. Li and R. M. Hochstrasser, Department of Chemistry, U. of Pennsylvania, Philadelphia, PA 19104-6323

An account will be given of various approaches for infrared spectroscopy with ultrafast time resolution. Optically pumped samples can be probed in the IR with time resolution of a few 100 fs either by gating CW carbon monoxide lasers or by generating IR pulses at high rep rate using a parametric and regenerative amplifier. We have also accomplished tunable pump: probe IR spectroscopy with low energy pulses at high rep rate. These methods are applicable to a wide range of questions in biological dynamics. In particular, recent results on the photodissociation of hemoglobin (Hb) bound to CO and NO show interesting new effects relating to the geometrical structure and heme pocket fluctuations. Intrinsic coherent coupling effects show up in many spectra that provide information about excited state (e.g. IR + VIS) two-photon absorption. A basic question regarding dynamics in proteins concerns the structural relaxations occurring after photolysis: These are directly addressed by our experiments. The goal of current work is to study the various substates of "free" NO and CO in hemoproteins and relate them to the associated dynamics.

195. NMR INVESTIGATIONS OF MACROMOLECULAR DYNAMICS. T. Alam, J. Dumais, T. Jarvie, J. Orban, J. Shiels, D. Wilson, G. Drobny, Department of Chemistry, University of Washington, Seattle, WA 98195.

The description of DNA has evolved from the uniform, static double helix originally presented by Watson and Crick nearly forty years ago. The determination of X-ray crystal structures of several synthetic DNA oligomers reveals that while these molecules have overall geometries which are similar to the structures obtained by fiber diffraction, there is considerable sequence-dependent variability of the local structure. An intriguing prospect is that this phenomenon provides the molecular basis for the specificity of protein-DNA interactions. Therefore, it is important to determine how well-defined are these structures, and the degree to which deviations occur from average structures. In other words, what are the internal dynamics of DNA?

This talk will review the use of solid state NMR, and particularly deuterium NMR, to probe the internal dynamics of synthetic oligonucleotides and high molecular weight DNA, and will acquaint the audience with the potential of solid state NMR to contribute to the ever growing picture of the internal dynamics of polynucleotides.

196. ROTATIONAL RESONANCE: DISTANCE MEASUREMENTS IN LARGE PROTEINS. A. McDermott, F. Creuzet, L.K. Thompson, R.G.S. Spencer, K. Halverson, M. Auger, M.H. Levitt, P. Lansbury, R.G. Griffin, Francis Bitter National Magnet Laboratory and Dept. of Chemistry, MIT, Cambridge, MA 02139, USA; R. Gebhard, I. van der Hoef, J. Lugtenburg, Dept. of Chemistry, Leiden University, Leiden, The Netherlands; J. Herzfeld, Dept. of Chemistry, Brandeis University, Waltham, MA 02254, USA

We describe an approach, termed rotational resonance, to determine internuclear distances in the range 3-6 Å such as are found in proteins, nucleic acids, membranes, etc. Specifically, when homonuclear dipolar-coupled spin pairs are introduced into solids, R2 occurs when the spinning speed is adjusted so that the condition $\delta = n\omega_r$ is satisfied. Here δ is the isotropic shift difference, ω_r is the spinning speed, and n is an integer which determines the order of the resonance. Under these conditions a rapid oscillatory exchange of Zeeman order between the dipolar coupled spins is observed. The time dependence of the exchange can be simulated numerically to obtain information on internuclear distances. Experimental data will be presented which demonstrate that the technique will be useful for measuring $^{13}C-^{13}C$ distances up to 5 Å in the membrane protein bacteriorhodopsin, and in an insoluble peptide related to Alzheimer's disease.

197.   DRUG-DNA INTERACTIONS:PROBING BINDING MODES AND DYNAMICS BY NMR
        David E Wemmer, Patricia Fagan and Bernhard Geierstanger, Dept of Chemistry University of California, Berkeley, California 94720

It has been known for many years that minor groove binding drugs, such as distamycin-A have a preference for binding to A-T rich DNA sequences. Recent NMR work has shown that there are significant variations in binding modes and kinetics for different numbers and combinations of A-T base pairs in the binding site. For understanding the basis for these differences it is crucial to have a method, such as NMR, which allows identification and characterization of multiple binding sites which are occupied simultaneously. Use of NOESY (EXCSY) and ROFSY for analysis of the complexes formed with a variety of different DNA sequences will be discussed. The implications for understanding the molecular basis for specificity will be discussed.

198.   ASSOCIATION REACTIONS OF TRANSITION METAL ATOMS WITH SIMPLE MOLECULES NEAR ROOM TEMPERATURE. S.A. Mitchell, Steacie Institute for Molecular Sciences, National Research Council of Canada, 100 Sussex Dr., Ottawa, Ont., CANADA, K1A 0R6.

Association complexes formed between isolated transition metal atoms and simple molecules are intriguing chemical species whose properties are relevant to fundamental studies of metal-ligand binding in organometallic and surface chemistry. Such complexes may be investigated using a combination of approaches, including gas phase kinetics, matrix-isolation spectroscopy and quantum chemistry calculations. Our work on gas phase kinetic studies uses a laser photolysis - laser fluorescence technique in which metal atoms are produced by multiphoton dissociation of a volatile organometallic precursor in a static pressure reaction cell. Association reactions of metal atoms including Cr, Fe, Co, Ni and Cu with ligand molecules including O_2, NO, CO, NH_3, C_2H_4 and C_2H_2 have been investigated in the temperature range 285-335 K, and at total pressures in the range 1-700 Torr. Information on binding energies has been obtained by observation of equilibrium in the dissociation reaction, or by modeling limiting low pressure association rate constants using unimolecular reaction theory.

199.   COMPARISON OF REACTIVITY WITH HYDROCARBONS OF TRANSITION METAL ATOMS IN DIFFERENT CHARGE STATES: M, M^+, and M^{2+}. J. C. Weisshaar, Department of Chemistry, University of Wisconsin–Madison, Madison, WI 53706.

The chemistry of monopositive transition metal cations with hydrocarbons is relatively well studied, both in terms of qualitative reactivity and the effects of specific kinds of electronic excitation. Recent work in our group and others has begun to explore the chemistry of neutral and dipositive transition metal atoms as well. Neutral metal atoms are relatively unreactive because of barriers to close approach arising from the typical $d^{n-2}s^2$ ground state configuration, which appears inert-gas-like at long range. Dipositive cations may fail to achieve close approach because of long-range surface intersections leading to exothermic electron or hydride transfer products. However, comparisons are beginning to suggest that when transition metal atoms gain close proximity to hydrocarbons, the propensity to cleave C-H and C-C bonds is common to *all* charge states, M, M^+, and M^{2+}.

200. COMPARISON OF MAIN GROUP AND TRANSITION METAL ION CHEMISTRY, P. B. Armentrout, Chemistry Department, University of Utah, Salt Lake City, UT 84112

Reactions of various atomic ions with H_2, D_2 and HD have been studied as a function of kinetic energy by using guided ion beam mass spectrometry. Both main group (e.g. B^+, Al^+, Si^+, Sn^+) and transition metal atomic ions have been examined. The observed behavior falls into several distinct groups (statistical, direct and impulsive) that can be used to characterize the potential energy surfaces for the reactions. These categories facilitate the comparison of the reactivity of the main group metals to that for the transition metals.

201. METAL ION CHEMISTRY IN FLAMES. John M. Goodings, Department of Chemistry, York University, 4700 Keele Street, North York, Ontario, M3J 1P3, Canada.

Small quantities of metals introduced into premixed and diffusion flames near 2000 K produce neutral oxides and hydroxides as well as atoms. Metallic flame ions, both positive and negative, have been observed by mass-spectrometric sampling. Thermal ionization of alkali metals and chemi-ionization of alkaline earths have been studied in hydrogen flames where the degree of natural flame ionization is low. Additional channels exist for the chemical ionization of metallic neutrals in hydrocarbon flames where the natural ionization level is much higher. Mechanisms include proton and electron transfer, addition reactions and electron attachment. Accordingly, metallic ions have also been observed for elements of group 13 (Al, Tl and non-metallic B), group 14 (Ge, Sn, Pb and non-metallic Si), and first-row transition metals (Sc − Cu, not Zn) as well as oxygenated anions of some higher members (Mo, W). The flame environment favours metal-oxygen, but not metal-carbon or metal-hydrogen, bond formation. The ion chemistry of metals in flames will be reviewed and discussed.

202. STRUCTURE AND CHEMISTRY OF NATURAL DIAMOND SURFACES*, Glenn D. Kubiak, Michelle T. Schulberg, and Richard H. Stulen, Sandia National Laboratories, Livermore, CA 94551-0969. The C (diamond) (100) and (111) surfaces chemisorb hydrogen, which stabilizes their truncated-bulk surface structures. Near 1200 K hydrogen desorbs, causing both surfaces to reconstruct to new phases of (2x1) symmetry. The electronic and geometric structures of both the (100) and (111) (2x1) reconstructions have been investigated using a variety of techniques and the results will be discussed. The kinetics of H_2, HD, and D_2 desorption from the (111) and (100) surfaces has also been examined using temperature programmed desorption (TPD) following exposure of the surface to atomic beams of H and D. We find that *atomic* hydrogen not only saturates the surface dangling bonds but also enters the bulk crystal easily. On the C(100) surface, one desorption peak is observed, shifting smoothly from 1200 K to 925 K as coverage increases from 0.06 monolayers (ML) to 8 ML(1 $ML_{C(100)}$ = 2H/C = 3.14 x 10^{15} H/cm^2). The integrated desorption flux from the polished C(100) surface yields the equivalent of 10 ML. TPD of hydrogen from the polished C(111) surface yields an integrated desorption flux of > 20 ML (1 $ML_{C(111)}$ = 1 H/C = 1.85 x 10^{15} H/cm2) appearing as one broad peak at ~1200 K. TPD spectra after exposure of the C(111) surface to H or D show two peaks: one at 1200 K and one at 800 K. Total desorption yield is again > 20 ML, with neither peak achieving saturation. Kinetic parameters have been extracted from these spectra and their relevance to CVD diamond growth will be discussed.
*Work supported by DARPA and the U. S. Department of Energy Office of Basic Energy Science, Division of Materials Science.

203. **MOLECULAR BEAM STUDIES ON DIAMOND FILM SYNTHESIS.** Deon Anex, Matt Lewin, David Gosalvez, Laura Smoliar, and Y.T. Lee, Department of Chemistry, University of California, and Chemical Sciences Division, Lawrence Berkeley Laboratory, Berkeley, California 94720 USA.

Using intense beams of H atoms and CH_3 radicals of adjustable kinetic energies produced by supersonic expansion from high temperature ovens, the kinetic processes of diamond film synthesis are investigated as functions of the kinetic energies of H atoms and CH_3 radicals and surface temperature. Two steps which are of particular interest in this investigation are the attachment of CH_3 radicals to "diamond" surface and the removal of H atoms from the CH_3 group on the surface by the reaction with energetic H atoms to form H_2.

204. **SILICON CARBIDE THIN FILM FORMATION ON Si(100)-THE BEHAVIOR OF DANGLING BONDS.** John T. Yates, Jr., Department of Chemistry, University of Pittsburgh, Surface Science Center, Pittsburgh, PA 15260.

The role of surface dangling bonds in controlling the surface chemistry of Si(100) has been investigated using surface science methods. The adsorption and decomposition of acetylene and ethylene on Si(100) has been studied and it is shown that adsorption occurs by means of a mobile precursor mechanism, leading to a di-σ chemisorbed species occupying Si dimer sites on Si(100). This acetylene-derived species decompose at elevated temperatures to produce adsorbed carbon and hydrogen (800 K) and then silicon carbide (1000 K). Adsorbed ethylene mainly desorbs without decomposition. Studies of the elementary steps in the decomposition of acetylene are coupled to the growth of SiC thin films.

205. HOT FILAMENT ACTIVATED CVD OF BORON NITRIDE AND CARBON. R. R. Rye, Division 1114, Sandia National Laboratories, Albuquerque, NM 87185.

Diamond film growth occurs for a hydrogen rich hydrocarbon ambient in the presence of a hot tungsten filament. Similar film growth also occurs as the result of a hot tungsten filament using a single gaseous species. For hot filament (~1400 C) activation of borazine ($B_3N_3H_6$) in direct line-of-sight with a silicon substrate, boron nitride films as thick as 25,000 Å have been grown for substrate temperatures as low as 100 C. XPS spectra give a B:N ratio of 1:1 with no other impurities above the 0.5% level. Raman spectra of films produced for substrate temperatures of 450 C are characteristic of h-BN with a particle size of ~17 Å. For these small particle films both Raman and infrared spectra show the presence of small amounts of hydrogen detected as B-H and N-H bands. Comparable BN results have been obtained for direct low temperature pyrolysis of borazinyl amine polymers, and comparable hot filament activated film growth of carbon is obtained for pure hydrocarbons. The effect of this low crystallinity and hydrogen content on reactivity and subsequent film growth will be discussed.

206. **LOCAL STRUCTURE and ELECTRONIC STATES OF HIGH-T_c SUPERCONDUCTORS BY TUNNELING MICROSCOPY and SPECTROSCOPY.** Charles M. Lieber, Department of Chemistry, Columbia University, New York, New York 10027.

Scanning tunneling microscopy (STM) and scanning tunneling spectroscopy (STS) have been used to study single crystal $Bi_2Sr_2CaCu_2O_8$, and metal-substituted and oxygen-doped $Bi_2Sr_2CaCu_2O_8$ superconductors at the atomic level. The evolution of the structural and electronic properties with metal doping has been determined by STM and STS for Pb(Bi), La(Sr), and Fe(Cu) substituted materials. These techniques have also been used to characterize for the first time spatial variations in the structure and electronic states due to oxygen doping. The relationship of these microscopic data obtained by STM and STS to the macroscopic superconducting properties will be discussed.

207. **LOCAL ATOMIC STRUCTURE OF DEFECTS ON OXIDE SURFACES.** D. A. Bonnell, G. Rohrer, Q. Zhong, Department of Materials Science, University of Pennsylvania, Philadelphia, PA 19104

Scanning tunneling microscopy (STM) can be used to characterize the geometric and electronic structures of defects on oxides with high spatial resolution. This is demonstrated with results of analysis of ZnO and TiO_2 surfaces structures. Factors affecting the spatial resolution of STM on wide bandgap semiconductors will be illustrated. Images and local bonding information from tunneling spectroscopy of several types of defects will be compared; including steps on ZnO (0001), crystallographic shear planes intersecting TiO_2 (110), facets on TiO_2 (001) and adsorbed impurities on polycrystalline ZnO and reduced rutile.

208. **OPTICALLY-EXCITED SCANNING TUNNELING MICROSCOPY: IMAGING SUB-SURFACE AND NON-EQUILIBRIUM ELECTRONIC PROPERTIES OF SEMICONDUCTORS.** Robert J. Hamers, Department of Chemistry, University of Wisconsin-Madison, Madison, WI 53706.

Scanning tunneling microscopy (STM) has become a widely-used tool for imaging the topography of surfaces at the atomic level. When applied to semiconductor surfaces, however, STM has several limitations; among these are the inability to determine semiconductor doping, the inability to probe sub-surface electronic properties, and the inability to probe the dynamic processes which are so important to real semiconductor devices. In recent studies, we have coupled STM with optical excitation techniques in order to overcome many of these limitations. By illuminating the semiconductor surface and using the STM as a local detector, it is possible to obtain several new kinds of information about semiconductor doping, the nature of the sub-surface "space-charge" layer, and about charge transport between surface states and bulk states. By combining STM with time-resolved optical excitation and using optical correlation techniques, it is even possible to directly measure the lifetime of the surface charge carriers, thereby pushing the timescale of STM into the nanosecond and picosecond regime. This talk will focus on the combination of STM with optical excitation as a means of obtaining new information about semiconductor surfaces on the atomic- and nanometer-distance scales.

209. **STM AND AFM EXTENSIONS,** H.K. Wickramasinghe, Research Division, IBM, T.J. Watson Research Center, Yorktown Heights, New YorK 10598

As device sizes get smaller and smaller,conventional optical techniques for characterisation are reaching their limit.There is an ever increasing need to perform such measurements on the nanometer scale. The Scanning Tunneling Microscope and the Atomic Force Microscope has taught us how to position a tip with angstrom accuracy over a sample surface and scan it in X,Y and Z in order to form an image. It is possible to use this principle to image other physical properties of a surface than the density of electronic states provided by the Tunneling Microscope. In this talk, we shall review several extensions of the STM and AFM that we have demonstrated recently within our group. These include high resolution magnetic force imaging, c/v imaging, thermal imaging , photo-voltage microscopy, chemical potential microscopy, optical and phonon absorption microscopy. The principles of operation of these microscopies will be described together with representative results obtained on various samples.

210. **PHOTON SCANNING TUNNELING MICROSCOPY.** T. L. Ferrell, Health and Safety Research Division, Oak Ridge National Laboratory, Oak Ridge, Tennessee 37831-6123

Photon scanning tunneling microscopy (PSTM) is shown to be the photon analogue of the electron STM (ESTM). Photons are made to tunnel from a dielectric surface on which an evanescent field is created by externally-supplied photon fields to a fiber-optic tip. The transmitted light signal is strongly dependent upon the tip to interface distance and can be regulated by feedback control. Scanning this tip can therefore provide a three-dimensional topographic map of optical surfaces at subwavelength resolution (Reddick, et al., Phys. Rev. B39, 1989, p. 767). The signal received is also influenced by local absorption, fluorescence, scattering, and index-of-refraction changes, and can be separated by proper scanning and excitation techniques. The ability to directly and nondestructively probe these fields provides uniquely powerful capabilities for integrated optics. Finally, spectroscopic information can yield important chemical and physical data at a higher resolution than is otherwise available (Paisler, et al., Phys. Rev. B42, 1990, p. 6750). Research sponsored by the Office of Health and Environmental Research, U.S. Department of Energy, under contract DE-AC05-84OR21400.

211. SCANNING TUNNELING MICROSCOPY/SPECTROSCOPY STUDIES OF SINGLE CRYSTAL $Bi_2Sr_2CaCu_2O_8$ HIGH - T_c SUPERCONDUCTORS. <u>Zhe Zhang</u> and C.M. Lieber.

ABSTRACT NOT AVAILABLE.

212. ON THE INTERPRETATION OF SCANNING TUNNELING MICROSCOPE IMAGES SHOWING ANOMALOUS PERIODIC STRUCTURES. <u>John F. Womelsdorf</u>, Makoto Sawamura, Walter C. Ermler. Department of Chemistry and Chemical Engineering, Stevens Institute of Technology, Hoboken, New Jersey 07030.

Scanning tunneling microscope images of graphite surfaces containing anomalous long-range structures are shown to originate from two distinct mechanisms. A twisted top layer graphite configuration produces super structures with divergent periods from 101 to 760 and decaying amplitudes. A graphite-flake-contaminated tip yields long range ordered arrays with constant periodicity, constant amplitudes, and abrupt domain termination. These arrays are consistently reproduced using a theoretical crystalline tip model. Hexagonal closed-packed patterns with periods ranging from 48 to 220 are experimentally observed and theoretically simulated using this model. The origin of such images is discussed in detail and a class of the anomalous long-range periodicities observed is attributed to a defect-mediated tip-substrate convolution phenomenon.

213. DETERMINACIÓN DEL MECANISMO DE ELECTROREDUCCION DE LA PEREZONA. <u>Aceves</u> J. M., Miranda R. R., Gonzalez B., F. J., Fac. de Est. Sup. Cuautitlan, Campo 1, U. N. A. M., Calle Quetzaltcoatl S/N, Cuautitlan Izcalli, Edo. de México.

Se presentan los resultados experimentales obtenidos usando diferentes técnicas electroquímicas, mismos que permitieron dilucidar el mecanismo de electroreducción de la perezona. La perezona es una quinona que no habia sido estudiada por metodos electroquímicos de manera sistemática. El problema fundamental en este tipo de compuestos quinónicos es la determinación de la manera en que se intercambian dos electrones y dos protones, ya sea mediante el mecanismo ECE o el mecanismo de dismutación. Estos mecanismos fueron comparados experimentalmente y al final se comprobo que solo uno de ellos ocurria en las condiciones en que se efectuo el trabajo.

214. ELECTRONIC STRUCTURES AND STABILITIES OF MIXED CLUSTERS. <u>M. A. Al-Laham</u> and K. Raghavachari, AT&T Bell Laboratories, Murray Hill, NJ 07974.

Electronic structures and stabilities of small Mg_nS_n, Al_nP_n and Ga_nAs_n (n=1,4) clusters are explored by means of accurate quantum chemical calculations. The effects of polarization functions and electron correlation are included in these calculations. The resulting structures are compared to those of valence isoelectronic silicon clusters Si_2, Si_4, Si_6 and Si_8. The differences between the different types of clusters are discussed in terms of their ionic characters.

215. **ESR Spectroscopy at Sub-9 GHz Microwave Frequencies. Advantages and Prospects.** M.G. Alonso-Amigo, Janusz Bednarek and <u>**Shulamith Schlick**</u>, **Department of Chemistry, University of Detroit, Detroit, Michigan 48221.**

ESR spectra at frequencies **lower** that 9 GHz have been measured to improve the spectral resolution in disordered systems, where local structural inhomogeneities ("strain") lead to a distribution in the g- and hyperfine tensors, and to line broadening. The advantage of multifrequency ESR (MESR) spectroscopy was explored in three polymeric systems. (a) In crosslinked polyacrylamide gels an analysis of ESR spectra of the Cu^{2+} probe at X-band (9.4 GHz) and S-band (2.4 GHz) has been used to determine the distribution parameters of the g- and ^{63}Cu hyperfine tensors, and to quantify the gradual change in the water properties as a function of the distance from the network. (b) In perfluorinated ionomers ESR spectra of Cu^{2+} at X, S and L-band (1.2 GHz) have been measured and simulated, in order to divide out the contribution of the g- and hyperfine tensor distributions to the line width, and to isolate the broadening associated with paramagnetic neighbors. (c) The gradual replacement of oxygen ligands from water by ^{14}N ligands from CH_3CN and CD_3CN in Nafion neutralized by Cu^{2+}, and the number of ^{14}N ligands, has been deduced from ESR spectra at X-band and at **three lower** frequencies. MESR can also be used to control the temperature range of a dynamic effect, in systems where the spectral range of the averaging process depends on the g-anisotropy.

Subject Area: Magnetic Resonance

216. H_2O DIFFUSION IN THE H_2O-D_2O-GELATIN SYSTEM
Brian Antalek, Andreas Langner, and Joseph P. Hornak
Rochester Institute of Technology, Rochester, NY 14623

The behavior of water in a gelatin matrix was studied by magnetic resonance imaging. Spin relaxation times in a ternary H_2O-D_2O-gelatin system were determined to enable the interpretation of signal profiles from 1) the counter diffusion of H_2O and D_2O in gelatin, and 2) the drying of a water-gelatin matrix. A relation for the concentration dependence of T_1 and T_2 on H_2O, D_2O, and gelatin content was established. Both T_1 and T_2 displayed an increase with decreasing gel concentration and increased dilution of H_2O by D_2O. However, at D_2O concentrations greater than 70% both T_1 and T_2 decreased for high gel concentrations. The counter diffusion of H_2O into a D_2O-gelatin matrix was studied using a 1.5 T GE Signa imager. It was determined that for the concentration range of gelatin, 1.5% to 10% by weight, the diffusion coefficient varied little between the bulk liquid and gelatin phases. This indicates that in this concentration range the gelatin matrix, although a solid, is largely composed of water which behaves like a liquid. The drying of a cylindrical gelatin rod was imaged using a 2.0 T GE CSI imager. Concentration profiles were modeled by way of a nonlinear diffusion equation with a concentration dependant diffusivity. The diffusivity appears to decrease as a function of time in the initial stages of drying. Diffusion at the outer surface is considerably more rapid than in the bulk of the gelatin matrix.

217. ELECTROCHEMICAL BEHAVIOR OF PbS AT THE FLOTATION CONDITIONS. Huerta C. Antonio y Genescá Ll. Juan. Facultad de Química, Departamento de Metalurgia, Edificio "D". Cd. Universitaria, 04510, México, D.F.

Kinetic and mechanism of the anodic oxidation of ethyl xanthate on galena have been studied in a boric acid-borax buffer solution (pH= 9.1) giving as a postulate a two step mechanism which propose the PbSX ads formation as an intermediate product and the PbX_2 formation as a final product. At the same time it is proposed a 4 step mechanism for a catodic reduction of O_2 on PbS with a (O_2H^+) ads as intermediate. It was verified the valid application of the mixed potential theory for galena-ethylxanthate-oxygen system.

218. SURFACE MODIFICATION BY METALLIC ION IMPLANTATION ON-CdTe. Elsa Arce. Div. Ing. Metalúrgica. ESIQIE-IPN. A.P. 75-876, 07300, México, D.F. Jorge Ibáñez. Dpto. Ing. Química. Universidad Iberoamericana. Reforma 880, México D.F. Yunny Meas. Dpto. química. Universidad Autónoma Metropolitana. I. A.P. 55-534, 09340 México D.F.

Surface modification by metallic ion implantation has been applied to the reduction of photocorrosion of semiconducting materials by promoting alternative reactions that compete with the semiconductor's decomposition reaction. Here, we have modified n-CdTe surfaces with Pt, Bi, Hg and Au by an immersion procedure and with Pt, Hg and Au by cathodic electrodeposition. The effect that these metals have upon the oxidation of hydroquinone and upon the semiconductor's decomposition reaction is discussed. The semiconductor's stability and stability and stability coefficient were evaluated as a function of time, surface composition and the Cd content of the solution for each case with the aid of the following techniques; rotating ring-disk electrode, variation of photocurrent with time surface analysis and atomic absorption. The cathodic displacements of the potentials for the oxidation of hydroquinone favor the semiconductor's stability; this is the case for the surface modifications with Pt, Bi, Hg and Au that increase the generated photocurrent.

219. THE EFFECT OF CRYSTALLIZATION INDUCED ON CORROSION BEHAVIOR OF METGLASS 2826MB AND MBF 15. Elsa Arce, David Jaramillo and Sebastían Díaz de la Torre. División de Ingeniería Metalúrgica. ESIQIE-IPN, A.P. 75-876, 07300, México, D.F.

Two commercially obtained vitreous alloys were studied by electrochemical techniques to analize their corrosion behavior and their pit formation tendency. The alloys involved were Metglass 2826 MB (Fe-40; Ni-35; Mo-4; B-15) and MBF 15 (Ni-75, Cr-13; Fe-4.5; Si-4.5; B-2.8) The effect of the crystallization induced by heating to 370, 500, 600, 700 and 1040 °C for the Fe-Ni-based alloy was also analysed. The experiments were carried out in H_2SO_4 1M solution with and without Cl^-. When Cl^- was absent an active-passive-transpasive behaviour was observed. However, when Cl^- was present the passive region was drastically reduced, this fact, is probably due to the rupture of the passive film by the Cl^-. It was also observed that the localized corrosion in the vitreous material took place preferentially in the surface irregularities (channels) present in the as-produced material. In the case of the crystallized samples, however, a larger tendency to the pit formation was detected; possible due, to the presence of the grain boundaries formed. In general an increment on the corrosion tendency was introduced by cristallinity. The Fe-based samples were in general less resistant to corrosion.

220. PICOSECOND TRANSIENT IR SPECTROSCOPY OF UV PHOTOLYZED $Cr(CO)_6$ IN SOLUTION Steven M. Arrivo,
J.R. Sprague, X. Zhu, C. Wu, and K.G. Spears Dept of Chemistry, Northwestern Univ., Evanston, IL 60208-3113

Picosecond transient infrared absorption has been used to study the solution phase dynamics of UV generated $Cr(CO)_5$. The effects of solvent structure and composition on the dynamics of the coordinatively unsaturated $Cr(CO)_5$ photofragment were investigated in cyclohexane, perfluoroheptane, CCl_4 and THF. Transient decay times were found to differ greatly, 100-400ps, and are assigned as vibrational relaxation of $Cr(CO)_5 \cdot S$ vibrational modes. Kinetics of bound solvent reorientation were studied by UV photodissociation of $Cr(CO)_6$ in THF. Visible photodissociation of the stable $Cr(CO)_5 \cdot S$ complex, allowed study of the $Cr(CO)_5$ species without interference from vibrationally excited $Cr(CO)_6$. This involves a 532nm pulse 1ns after the initial UV photolysis pulse. These results will be compared to visible photodissociation of the stable $Cr(CO)_5 \cdot THF$ complex. A full spectral assignment based upon these experiments will be presented.

221. **MATERIALES SEMICONDUCTORES POLICRISTALINOS DENTRO DE CELDAS FOTOELECTROQUIMICAS. I. CdS PURO E IMPURIFICADO.** Ramos M. Aurora y Castro-Acuña Mauricio. DEPg, Depto. de Fisicoquímica. Facultad de Química. UNAM. Ciudad Universitaria, D.F. México, CP 04510.

Se trabaja CdS puro e impurificado con óxido férrico o sulfuro ferroso. El CdS es un semiconductor tipo n que funciona como fotoánodo cuando se ilumina dentro de un sistema electroquímico. Los electrodos semiconductores policristalinos se obtienen por prensado y sinterizado de polvos. Si bien la eficiencia es menor que la de los electrodos monocristalinos, el costo es también mucho menor, lo que justifica su utilización. Las impurezas, cuya naturaleza y composición están perfectamente determinadas, causan un defecto en el material, lo que modifica el número de portadores de carga. Con los electrodos, se efectúan voltametrías cíclicas en condiciones de oscuridad e iluminación. Al iluminar se genera una fotocorriente que representa una conversión de energía luminosa en eléctrica y química. Finalmente se establece si estos electrodos se pueden utilizar eficientemente en celdas solares.

222. **DYNAMICAL CHARACTER OF PHYSICOCHEMICAL PROCESSES UNDER HIGH PRESSURE: (6) ELECTRODE PROCESSES IN SOLUTION AND MATERIAL CHARACTERIZATION.** *Sardar Khan Bahador*

B&A research scheme for advanced energy materials and systems, Bahador research room at Kumonoryo room B-311, Dept. Indus. Chem. Kyoto Univ., P.O. Box 51, Sakyo Kyoto 606 Japan

Studies with respect to the effects of pressure on energy and/or electron transfer reaction rate and equilibrium in ionic solutions and CTC formation processes, as well as in many liquid-solid and solid-solid state phase transitions which also means crystal growth, materials preparation and characterization via optical, magnetic, electrical, and x-ray techniques, have gained significance. Of particular interest are synthesis of semiconducting and high temperature superconductors, investigating the mechanism of electrical conduction in orgnic and inorganic solids, as well as the influence of pressure on photo- and solide state chemical processes. In all studies performed, standard and some home built instrumentations were employed. Computer simulations are carried out. Entities such as $\Delta V^{\neq}$, k_s, $D_{o,r}$ found for electron transfer processes in compressed solution at pressures up to 100 MPa, by using Iron complexes on Pt microelectrodes. The electrical conductivity σ of polyacetylene thin film measured at pressures up to 12 Kbar in relation with temperature. The results of both case studies will be reported. The nature and fates of the critical parameters are to be discussed.

References: (1) S.K. Bahador, in: Book Abst. Spring Meet. Electrochem. Soc. Japan, Apr. 1991, Chiba Japan.
(2) S.K. Bahador, in: Book Abst., Spring Meet. Chemical Soc. Japan, Apr. 1991, Yokohama, Japan.(Vol. I).
(3) S.K. Bahador, in: Proc. 42nd I.S.E. Ann. Meet., Aug, 1991, Montreux Switzerland.
(4) For a short biography, please refer to, IUPAC, Chem.Int., 10(3,5), 85, 170 (1988).
* *The author himself is fully responsible for all financio - scientific aspects of this work.*

223. **COMPUTERIZATION OF MATERIAL AND CHEMICAL DATA.** <u>Malcolm W. Chase</u>, Standard Reference Data, A323 Physics Building, NIST, Gaithersburg, Maryland 20899, and Nancy P. Korendyke, Database Development, Maxwell Online, Inc., 8000 Westpark Drive, McLean, Virginia 22102

Numeric data lies ~~~~~~~~~ re of much of the chemical literature and is use~ WITHDRAWN by the scientific community in solving problems. Th~ ~ of these data is dependent on the accompanying informat~ such as sample purity, temperature scales, and experimental procedures; in these days of electronic networking, the format of these data is also a significant consideration. Within ASTM Committee E49 on the Computerization of Materials Data, Subcommittee E49.53 is working on writing standards for presenting and exchanging numeric and ancillary data for physical and chemical properties. A sample exchange format for heat capacity is proposed.

224. **ANION BINDING TO POLAR MOLECULES**, <u>Harold Basch</u>, Department of Chemistry, Bar-Ilan University, Ramat-Gan, Israel. Morris Krauss and Walter J. Stevens, National Institute of Standards and Technology, Gaithersburg, MD 20899

The interaction of phosphoric acid and chlorine substituted phosphoric acid with hydroxyl and chlorine anions yields a number of hydrogen-bonded and five-coordinate adducts. Stable hydrogen-bonded structures are found both for adducts with OH^- and Cl^- and for proton transferred structures involving H_2O in the complexes, $H_4PO_5^-$, $H_3PO_4Cl^-$ and $H_2PO_3Cl_2^-$. The hydrogen bond energy of the OH^- and Cl^- adducts is calculated to be large due to the polar character of the P-O and O-H bonds. Ligand field analysis of the hydrogen bonding determines the electrostatic interaction is dominant for these adducts. Five-coordinate phosphorus structures corresponding to reaction intermediates are also found at a higher energy than the hydrogen-bonded structures. The phosphorane bond structure is analyzed and related to the leaving group behavior of the dissociating transition state.

225. SYNTHESIS, CHARACTERIZATION AND CATALYTIC ACTIVITY OF $LaBO_3$ (B = Ni, Mn, Co, Fe) PEROVSKITES. M. Beltrán*, G. Tavizón*, R. Trinidad**, T. Viveros**, L. Vicente*.
* Facultad de Química, División de Posgrado, UNAM, 04510 México, D.F.
**Universidad Autónoma Metropolitana, Ing. Química, Apdo. Postal 55-534, 09340 México, D.F.

$LaBO_3$ perovskites (B = Ni, Mn, Co, Fe) were prepared by pyrolisis of the corresponding nitrates at several pH values. The oxide structures were studied by powder X-ray diffraction and surfaces areas were measured by Nitrogen adsorption. The results show that the perovskite structure was obtained, the surface areas are affected by the pH of the precipitating solution and are in the range of $2 - 4 \, m^2/g$. The catalytic activity of the compounds were tested in the propylene oxidation reaction varying the propylene/oxygen ratio. The samples were active, giving basically complete oxidation to CO_2, and with the exception of $LaFeO_3$, all the others have an ignition temperature around 510 K.

226. THERMODYNAMIC PROPERTIES OF Li_3. Louis Biolsi, Department of Chemistry, University of Missouri-Rolla, Rolla, MO 65401 and Paul M. Holland, General Research Corporation, P.O. Box 6770, Santa Barbara, CA 93160-6770.

The thermodynamic properties of Li_3 are calculated by using a statistical thermodynamic method that relates the thermodynamic properties to the second and third virial coefficients. These virial coefficients depend on the potential energy of Li_3 and the interaction potential between two Li atoms. We have accurately fit either spectroscopically determined or SCF-CI potential energy curves for the Li-Li interaction, using the Hulburt-Hirschfelder potential, and the LEPS potential function for Li_3. The thermodynamic properties as a function of temperature are obtained from the equilibrium constant which is related to the virial coefficients.

227. NEW CUBIC COMPOUND IN THE Bi-Sr-Cu-O SYSTEM. Bockimi, L. Salinas and R. Pérez, Instituto de Física, U.N.A.M., A.P. 20-364, 1000 México, D.F., México, M. Portilla, DEPg., Facultad de Química, U.N.A.M. Ciudad Universitaria, 04510, México, D.F., México.

We report the existence and characterization of a new compound with cubic structure in the strontium rich region of the Bi-Sr-Cu-O system. The compound has an ordered structure based in the perovskite. We find ordering both in the A and B type cations (ABO_3 is the formula for the perovskite). Under different heat treatments it is possible to change the oxygen content in the sample and with it, to transform the cubic structure into one laminar. We study the oxygen behavior in the samples with thermogravimetric and differential thermal analysis. The structure is studied by X-ray diffraction and electron microscopy. We present also simulations of X-ray diffraction and electron microscopy high resolution images. The electrical and the magnetic behavior of the samples are measured at temperatures between 10 and 300 K. We find solid solutions by substitution of strontium by copper.

228.
HYDROCARBON RADICAL REACTIONS WITH OXYGEN: COMPARISON OF ALLYL, FORMYL AND VINYL TO ALKYL. Joseph W. Bozzelli and Anthony M. Dean, Exxon Research and Engineering, Annandale, NJ 08801

Recent work has strongly suggested that alkyl radical reactions with O_2 occurs via addition -- H shift isomerization -- elimination reaction; rather than the direct hydrogen transfer reaction, as previously thought. In this study we extend the analysis of these reactions to other types of organic radicals. It is shown that the variations in R--OO bond strength ranging from ca. 18 Kcal/mole for allyl to ca. 40 Kcal/mole for vinyl and formyl strongly influence the overall rate constants and product distributions. The vinyl and allyl cases also present the added possibility of cyclization as a new pathway to compete with the H-shift reactions. A chemical activation analysis is performed and the predictions compared to available literature data. We find that the weaker R--OO bond for the allyl case significantly decreases the energy available for reactions induced by the chemical activation process and the reaction rate constants decrease substantially relative to alkyl reactions. The increased well depths in the vinyl and formyl cases lead to faster rate constants and opens possibilities for additional important channels.

229.
QRRK CHEMICAL ACTIVATION ANALYSIS ON REACTIONS OF CH WITH N_2, O_2 AND NO.
Joseph W Bozzelli, Md Hasan Karim and Anthony M. Dean[*], Dept. of Chem. Eng. and Chemistry, New Jersey Inst. of Tech., Newark, NJ 07102 and [*]Exxon Res. and Eng. Co, Annandale, NJ 08801

The reactions of CH, methylidyne radical, are reported to be of importance to prompt NO formation in flames, as well as in the atmospheric chemistry of Titan and other planetary atmospheres. In this study we investigate the reactions of CH with N_2, O_2 and NO and we compare the results of our analysis with experimental data where available. These reactions produce chemically activated intermediates and the competition between reaction and stabilization is treated using a bimolecular Quantum Rice Ramsperger Kassel - QRRK approach. Our objective is to understand the chemistry of CH reactions over the temperature and pressure ranges where kinetic measurements exist and then to use this understanding as a basis for extrapolating to other temperature and pressure regimes. Results for CH + N_2 indicate that the literature data can be explained by formation of an HCN_2 adduct, which can be stabilized, dissociate back to reactants or proceed over a barrier to HCN + N.

230. STM STUDIES OF CHARGE TRANSFER SALTS, George W. Flynn and J. J. Breen, Department of Chemistry, Columbia University, New York, New York 10027.

The scanning tunneling microscope has been used to investigate the topography and electronic structure of both single crystals and vacuum deposited thin films of a number of charge transfer salts. Donor molecules are either tetrathiafulvalene (TTF) or tetramethyltetraselenafulvalene (TMTSF) and acceptor molecules include tetracyanoquiodimethane (TCNQ), tetracyanoethylene (TCNE), tetrachloro-o-benzoquinone and iodine. In these studies, performed under ambient conditions, molecular resolution is often attainable. Comparisons with previous structural characterizations and simple molecular orbital calculations will be made.

231. UTILIZATION OF ELECTRON SPIN ECHO MODULATION SPECTROSCOPY AS A NOVEL METHOD TO DETECT FRAMEWORK SUBSTITUTION OF PARAMAGNETIC IONS IN MICROPOROUS ALUMINOPHOSPHATE MOLECULAR SIEVES. Guillaume Brouet, Xinhua Chen and Larry Kevan, Department of Chemistry, University of Houston, Houston, Texas 77204-5641, USA.

Electron spin echo modulation analysis of 0.1 mole % Mn (relative to P) in the aluminophosphate molecular sieve MnAPO-11 with adsorbed D_2O shows two deuteriums at 0.24 nm and two at 0.36 nm from Mn. This suggests that two waters are coordinated to Mn in a D-O bond configuration which is that expected for a negatively charged site. When Mn substitutes for Al in the AlPO-11 framework it forms a negatively charged site, whereas Mn would form a positively charged site in an extra-framework position. Thus the electron spin echo modulation analysis indicates that Mn at low concentration substitutes in a framework position of MnAPO-11 during synthesis.

232. MAGNETIC RELAXATION COUPLING IN DYNAMICALLY HETEROGENEOUS SYSTEMS. Robert G. Bryant, Departments of Biophysics and Chemistry, University of Rochester, Rochester, NY 14642

Nuclear magnetic spin-lattice relaxation in heterogeneous systems like tissues is complicated by the effects of magnetic coupling between liquid components and the solid components of the system. Previous work has shown that this coupling causes generally nonexponential decay of longitudinal magnetization so that the usual descriptions of T_1 are inaccurate. The magnetic resonance response of the rotationally immobilized components of the system is characteristic of a solid in that the proton-proton dipole-dipole couplings are unaveraged making the proton spectrum a pseudo-homogeneous line that may be dramatically affected by irradiation anywhere in the line shape. This coupling permits a form of indirectly detected spectroscopy that reads the lineshape of the solid using the response of the liquid. The coupling also accounts for the magnetic field dependence of the relaxation in heterogeneous systems such as tissues and provides a noninvasive way to alter the information content of a magnetic image.

233. THE NONPLANARITY OF DOPED PI SYSTEMS. L.A.Burke, K.Krogh-Jespersen Dept of Chemistry, Rutgers University, Camden, NJ 08102

Optimized geometries and calculated vibrational frequencies using ab initio MO methods are presented for a series of conjugated polyenes, C_nH_{2n+2}, n=3-9 (even n:neutral; odd n:anion) with Li and Na atoms. Also presented are results for furan and difuran with Li or Cl. All structures show marked departure from planarity within the pi system. This has major implications for the structural aspects of conducting polymers.

234. TEMPERATURE DEPENDENT RATE CONSTANTS FOR ELECTRON ATTACHMENT REACTIONS OF METAL-CENTERED COMPOUNDS. Steven J. Burns and David L. McFadden, Department of Chemistry, Boston College, Chestnut Hill, MA 02167

Thermal energy electron attachment reactions are studied over the temperature range 300-1200 K. The experiments are performed in a discharge-flow system coupled to an electron paramagnetic resonance spectrometer. Electrons are generated by Penning ionization and detected by means of electron cyclotron resonance absorption. Rate constants are determined for several compounds including halides of both main group and transition metal elements. Electronic structures of both neutral reactants and negative ion products are also calculated.

235. SPECTROSCOPIC STUDIES OF FLOPPY MOLECULES: C_3 AND SiC_2.[*] Thomas J. Butenhoff and Eric A. Rohlfing, Combustion Research Facility, Sandia National Laboratories, Livermore, CA 94551.

We shall present results from studies in which several laser spectroscopies (LIF, dispersed fluorescence, SEP) are used to characterize the internal dynamics of the jet-cooled, nonrigid triatomics C_3 and SiC_2. These systems have highly anharmonic ground-state potential energy surfaces and exhibit large-amplitude motion. They provide ideal tests for two forms of *ab initio* theory: accurate quantum chemical calculations of the potential surface and dynamical or variational theories for calculation of the rovibrational eigenstates. C_3 is an extremely floppy, nominally linear molecule for which we have measured vibrational term energies up to 17 000 cm^{-1} and rovibrational structure as high as 8930 cm^{-1}. Analysis of our data demonstrates the dramatic coupling of the low-frequency bend to rotation and to the high-frequency stretches. SiC_2 is nominally cyclic (C_{2v}) but the linear isomer is nearly isoenergetic. We shall present spectroscopic evidence for large-amplitude pseudorotation that takes place at low energies on the ground-state surface.

*This work supported by the U. S. DOE, Office of Basic Energy Sciences, Chem. Sciences Division.

236. ENERGIA LIBRE DE TRANSFERENCIA DE CuSO$_4$ EN AGUA / AGUA-ACETONA
      IRMA CAHERO, ISABEL CARRANCO, ANTONIO REYES CHUMACERO
      DEPTO. FISICOQUIMICA; FAC. DE QUIMICA ; U.N.A.M.
      CD. UNIVERSITARIA   MEXICO D.F.   04510   MEXICO

Se ha determinado la FEM del sistema formado por sulfato cúprico en solución acuosa
y acuoso-acetónica. Las concentraciones del CuSO$_4$: 0.01 - 0.2 ; % en masa de acetona
0.0 - 0.12 ; para diferentes temperaturas : 20, 25 y 300 C .
Se utilizó una celda sin transferencia : Cu / CuSO$_4$(S1/S2) / HgSO$_4$ / Hg
Los datos de la FEM se obtuvieron en un baño de temperatura constante, mediante un --
potenciómetro K-4 modelo 7554, un detector de punto nulo modelo 2437, una fuente de
voltaje constante de precisión modelo 9878 y otros accesorios, todos de Leeds & Nortrup.

De los datos obtenidos de la FEM para las dos celdas, se calcularon las propiedades --
Termodinámicas de Transferencia del CuSO$_4$. La obtención de los coeficientes de activi-
dad, permitió comparar los calculados por otros modelos semiteóricos.

237. A THEORETICAL INVESTIGATION OF THE LOW-LYING STATES OF BUTADIENE RADICAL
      CATION. Robert J. Cave and Michael G. Perrott, Department of Chemistry,
      Harvey Mudd College, Claremont, CA 91711-5990.

   Polyene radical cations are of interest as models for conducting polymers as
well as testing grounds for the prediction of electronic excitation energies. Results
are presented here from multi-configuration SCF, multi-reference singles and doubles
CI (MRSDCI), quasi-degenerate variational perturbation theory (QDVPT), and multi-
reference averaged coupled pair functional theory (MRACPF) calculations on the ground
and low-lying excited states of $C_4H_6^+$. Comparisons are made with experimental assignments, and predictions are made for the positions of previously unassigned states. It
is shown that QDVPT, MRACPF and Davidson-corrected MRSDCI give nearly identical excitation energies in all cases considered. In addition, the differences between the
low-lying π states of the cation and those of the neutral parent, s-trans 1,3-butadiene
will be considered.

238. GEOMETRIES AND MOLECULE PROPERTIES OF XCN (X = H, Cl, Br) IN VARIOUS IONIC
      STATES. F.T. Chau, Department of Applied Biology and Chemical Technology,
      Hong Kong Polytechnic, Hung Hom, Kowloon, Hong Kong.

   The ionic geometries and molecular properties of XCN$^+$ with X = H, Cl and Br in
the first two lowest energy electronic states were deduced through Franck-Condon analyses
on photoelectron spectra of the corresponding ionization processes as well as vibrational
analyses on ionic frequency data from laser and emission spectroscopic studies. The
calculated structural parameters and rotational constants are compared with observed
values if available. Furthermore, the change in molecular properties of XCN in various
ionic states are explained in light of the respective bonding properties. In this work,
the harmonic oscillator approximation was found to be valid for vibrational intensity
calculations on ionization processes with appreciable changes in geometry.

239. Studies of Halogen Substitution in High-T_c Superconductors
Chia-Chun Chen and Charles M. Lieber,
Department of Chemistry, Columbia University, New York, N.Y. 10027

Substitution of halogens such as fluorine for oxygen allow the role of charge transfer and defects to be studied systematically in the high-T_c copper-oxide-based materials. Sintered sample of high T_c superconductors $YBa_2Cu_3O_x$, $Bi_2Sr_2CaCu_2O_x$ and $La_{2-x}Sr_xCuO_y$ have been treated by Br_2 and NF_3 gas. These materials have been characterized by X-Ray diffraction, thermal gravimetric analysis, resistivity, and magnetic susceptibility measurements. The dependence of Tc and the critical current density on halogen doping will be discussed.

240. INTERNAL ENERGY DISTRIBUTIONS OF SO($X^3\Sigma^-$) FROM THE 193 nm PHOTODISSOCIATION OF $SOCl_2$, Xirong Chen, Hongxin Wang and Brad R. Weiner, Department of Chemistry, University of Puerto Rico, Río Piedras, PR 00931.

Photodissociation of thionyl chloride ($SOCl_2$) at 193 nm has been shown to produce SO($X^3\Sigma^-$) in the absence of collisions. We have measured the nascent vibrational and rotational energy distributions of SO($X^3\Sigma^-$) following 193 nm photofragmentation of $SOCl_2$, both in a bulb and in a pulsed supersonic free jet expansion to ascertain the effect of parent molecule temperature on the photofragment energy distributions. These results will be discussed in light of detailed photofragmentation mechanisms for thionyl chloride.

Work supported by NSF-EPSCOR under Grant No. R11-8610677 and AFOSR under Grant No. F49620-89-C0070.

241. VIBRATIONAL MODE, ELECTRONIC STATE, AND COLLISION ENERGY EFFECTS ON ION-MOLECULE REACTIONS. Yu-hui Chiu, Baorui Yang, Hunghsin Fu, and Scott L. Anderson. Department of Chemistry, State University of New York at Stony Brook, Stony Brook, NY 11794-3400.

We have developed a method for studying the effects of reactant vibrational mode and collision energy on polyatomic ion-molecule reactions. Both integral cross sections and product recoil energy distributions can be measured. The technique involves a combination of multiphoton ionization, photoelectron spectroscopy, and guided-ion beam mass spectrometry. Results will be presented for reactions of mode-selectively excited OCS^+ and $C_2H_2^+$ ions with H_2, CH_4, Xe, Ar, C_2H_2, and OCS. The information extracted provides information on the dynamics and energy disposal in these reactions.

242. INTERNAL AND TRANSLATIONAL ENERGY EFFECTS ON CLUSTER ION CHEMISTRY. James Christian, Zhimin Wan, and Scott L. Anderson. Department of Chemistry, State University of New York at Stony Brook, Stony Brook, NY 11794-3400.

A laser vaporization/variable temperature radio frequency storage source has been developed, which allows production of beams of cluster ions with independently variable internal and translational energies. The source is coupled to a triple sector tandem guided-beam mass spectrometer that allows ion beam chemistry studies with high mass and energy resolution. The design and performance of the instrument, and characterization of the variable temperature source will be presented. Results on the effects of internal and translational energy on the chemistry of boron and carbon cluster ions will also be presented.

243. ROLE OF EXCITED STATES OF MAXIMAL d-SHELL OCCUPANCY IN THE Ru + + H2 REACTION. F.Colmenares-Landín, J.M.Martínez-Magadán* and O. Novaro,* FES-Cuautitlán UNAM. Instituto Mexicano del Petróleo, Lázaro Cárdenas # 152 Apdo. Postal 14-805. 07730 México, D.F.

Pseudopotential-CI calculations including relativistic effects have been performed to study the interaction of a hydrogen molecule with one ruthenium atom. The present results are for all the channels of C_{2v} symmetry which correlate with the two lowest excited states of the reactants: $Ru(^3F;d^7s^1)$ + H2 and $Ru(^3F;d^8)$ + H2. For all these symmetry channels, potential energy curves obtained for the approach of H2 molecule to the metal, keeping the H-H distance fixed at the isolated molecule equilibrium value, show the existence of avoided crossings between these two states. The main features of the equilibrium geometries for the bent electronic states of the RuH2 system, such as the leading configurations, are then explained in terms of the avoided crossings mentioned above, showing that the metallic $Ru(^3F;d^8)$ state is responsible for the reactivity of these bent channels for this reaction.

244. ACTIVACION DEL ENLACE C-H POR Ni EN LA MOLECULA DE METANO. Cuan H. M.A., Castillo A. S., Poulain G. E. y Ramírez-Solis A. Instituto Mexicano del Petróleo, Eje Central Lazaro Cardenas 152, CP 07730.

Para la interacción Ni-CH4, se presentan las curvas de energía potencial (MRCI-MP2) de los estados singuletes A' correspondientes a los sistemas $Ni\ ^1D(d^9s^1)$ + CH4 y $Ni\ ^1S(d^{10})$ + CH4 de los fragmentos libres. En la estructura más estable para H-Ni-CH3 predomina la configuración $Ni(d^{10})$, que proviene de la existencia de un cruce evitado por simetría, entre las dos curvas de potencial correspondientes al estado A' para el sistema NiCH4 de los fragmentos separados. Con lo cual se puede ver la importancia de los estados exitados en la descripción de los mecanismos de reacción del sistema.

245. ESR INVESTIGATION OF RADICAL PROCESSES IN ORGANIC POLYCRYSTALS UNDER SHEAR DEFORMATION AT HIGH PRESSURE. A.A. Dadali and H.C. Brenner, Department of Chemistry, New York University, New York, NY 10003.

Mechanistic studies of organic reactions initiated by mechanical action are becoming an increasingly important area of solid state organic chemistry. In this work, EPR techniques have been used to study the kinetics of radical generation and decay under shear deformation at high pressure (up to 1.2 GPa). The spectra were recorded simultaneously with development of the deformation directly in the cavity of a modified EPR spectrometer, which included a dielectric solid resonator in the form of a Bridgman anvil and a mechanical press. Radicals were generated in the process of mechano-induced dissociation of arylindandione dimer molecules. The partitioning of the energy consumed during shear deformation of the sample into heat, elastic deformation and rupture of chemical bonds was quantitatively estimated. The shapes of the kinetic curves for radical decay and recombination are dependent strongly on pressure, the angle and rate of shear deformation. Studies were also made of radical processes under shear deformation developed in the shock regime. A semi-phenomenological theoretical kinetic model was suggested, which explained the main features of the experiments.

246. THE PHYSICAL CHEMISTRY INSTRUMENTATION LABORATORY AT JOHNS HOPKINS. Paul J. Dagdigian, and Betty J. Gaffney, Department of Chemistry, The Johns Hopkins University, Baltimore MD 21218.

The co-authors bring expertise in the areas of biophysical chemistry and chemical physics to this undergraduate laboratory. Several experiments in these areas recently developed or adapted for this laboratory will be described. These include oxygen binding to hemoglobin, solution calorimetry, an NMR study of inhibitor binding to an enzyme, magnetic resonance imaging using unpaired electrons, stopped flow kinetics of metal complexation, excited state photophysics, the sodium atom absorption spectrum, and multiphoton infrared dissociation. The recent acquisition of major instrumentation was enabled, in part, through a grant from the Howard Hughes Medical Institute to improve undergraduate education in the biological sciences.

247. **AM1 STUDIES OF THE NUCLEATION OF CRYSTALS OF 1,3-DIONES.** J.J Dannenberg and Laszlo Turi, Department of Chemistry, City University of New York, Hunter College and The Graduate School, 695 Park Avenue, New York, NY 10021

The AM1 molecular orbital method is used to calculate the intermolecular interactions (particularly hydrogen-bonding) of dimers and higher aggregates of various 1,3-diones, which often crystallize in the enol form. Several different relative orientations of the individual molecules have been considered. The results indicate that cooperative effects increase the stabilization of adding an additional unit to a growing aggregate, as expected. However the extent of the cooperativity varies for the several different relative orientations of the precrystalline aggregates studies. Remarkably, the preliminary results suggest that the most stable orientations of the smaller aggregates may not dictate the eventual crystal structure as the greater cooperativity of the initially less stable orientations may eventually be overcome by the larger cooperative effects in others.

The experimental observation that hexane-1,3-dione crystallizes from benzene as a cyclic hydrogen-bonded hexamer with one molecule of benzene in the center is reproduced by the molecular orbital calculations. Comparison with other experimental data will be made.

The effects of strategic substitutions upon the cooperativities of the various interactions will also be reported.

248. **AN *AB INITIO* STUDY OF THE REACTION PATHWAYS FOR Si^+ + SiX_4 (X=F,Cl).** Cynthia L. Darling and H. Bernhard Schlegel, Wayne State University, Department of Chemistry, Detroit, MI 48202.

The reaction pathways of Si^+ + SiF_4 and Si^+ + $SiCl_4$ have been studied by *ab initio* molecular orbital methods. Equilibrium geometries and transition structures were fully optimized at the Hartree-Fock level with the 3-21G and 6-31G* basis sets. Barrier heights have been computed with fourth order Møller-Plesset perturbation theory, with and without annihilation of spin contamination. Direct attack of Si^+ on F or Cl forms a cluster $X_3Si\text{-}X\text{-}Si^+$ which rearranges to $X_3Si\text{-}SiX^+$. This second intermediate can dissociate into SiX_3 and SiX^+ or it can rearrange further to yield $X_2Si\text{-}SiX_2^+$ which can dissociate to SiX_2 and SiX_2^+.

249. **TIME-RESOLVED LASER SPECTROSCOPY IN THE UNDERGRADUATE LABORATORY.** Julio C. de Paula,* Jeffrey Lind, and Matthew Gardner, Department of Chemistry, Haverford College, Haverford, PA 19041.

We will describe an affordable, computer-controlled luminescence instrument, which consists of a pulsed N_2/dye laser, a 0.25-m single monochromator, a photomultiplier detector, and a boxcar integrator. The response time of the entire instrument is approximately 40 ns. Under computer control are: (i) the stepper motors driving gratings in the dye laser and in the monochromator; and (ii) the time delay between a trigger signal to the boxcar and the variable-width gate. These features allow the experimenter to obtain luminescence lifetimes, emission spectra, and excitation spectra. Our luminescence instrument was tested in the classroom with a study of the photo-induced electron transfer reactions between RuL_3^{2+} complexes [L= 2,2'-bipyridine (bpy) or 1,10-phenanthroline (phen)] and Fe(III) in aqueous acidic solutions. By using ligands bearing different substituents, the effect of thermodynamic driving force on the rate of electron transfer was assessed. *This work was supported by grants from the Pew Foundation and the Howard Hughes Foundation.*

250. **ENERGY TRANSFER IN ULTRAHOT SOLIDS: VIBRATIONAL COOLING AND MULTIPHONON UP PUMPING,** Dana D. Dlott, University of Illinois at Urbana Champaign, Box 37-1 Noyes Lab, 505 S. Mathews Ave., Urbana, IL 61801.

A novel method of producing an enormous ultrafast temperature jump in condensed matter has been developed at Illinois. The method uses a highly stable organic dye molecule which absorbs at 1.064 microns, the Nd:YAG laser fundamental. Owing to a sub picosecond nonradiative relaxation process, each molecule can absorb $>10^2$ infrared photons. A rapid redistribution of this excess energy among the vibrations of molecule produces an initial temperature of 5,000 degrees. The picosecond time scale vibrational cooling produces a bulk temperature jump in a surrounding polymer or liquid of hundreds of degrees. Picosecond time scale optical absorption and Raman experiments, molecular dynamics and anharmonic lattice calculations are used to investigate the cooling of these ultrahot molecules, and heating of the surrounding material. The temperature dependent absorption spectra of molecular thermometers embedded in a polymer matrix is used to investigate multiphonon up pumping and thermal conduction over nanometer length scales.

251. TWO-PHOTON IONIZATION OF $(NO)_n Ar_m$ CLUSTERS. Sunil R. Desai, C.S. Feigerle, Department of Chemistry, University of Tennessee, Knoxville, TN 37996 and J.C. Miller, Chemical Physics Section, Oak Ridge National Laboratory, Oak Ridge, TN 37831.

Binary Clusters of the type $(NO)_n Ar_m$ have been formed in a neutral supersonic expansion and ionized using the 266 nm beam of a picosecond Nd:YAG laser. In order to ensure against further clustering around the photoion products, photoionization was performed far downstream of the skimmed beam in a differentially pumped region. Nascent cluster photoions with combinations of n = 1-4 and m = 0-21 were subsequently detected by time-of-flight mass spectrometry. The observed cluster distributions are contrasted with those obtained in other laboratories by condensation around bare NO^+ ions as well as with distributions of other $X^+ Ar_m$ species. In addition, differences have been observed in the relative intensities of the $NO^+ Ar_m$ and $(NO)_2^+ Ar_m$ ions as a function of m and are discussed in terms of solvation shells and possible multiphoton effects.

252. **ELECTRONIC STRUCTURE AND CHEMICAL BONDING IN TERNARY TRANSITION-METAL COMPOUNDS.** Fernando Estrada, Sergio Pérez and Alejandro Pisanty. Departamento de Física y Química Teórica, Facultad de Química, U.N.A.M., México D.F. 04510, México.

In the present study, we calculate the band structures and derived magnetic and bonding properties of ordered compounds of formulae $Co_x Ni_y M_z$, M= Pd or Pt, x,y,z integers 0-4. We describe them as cubic, formed by substitution on the simple cubic lattice with four atoms per unit cell which equivalent, for the pure elements, to the fcc lattice. We use the LMTO method the "combined correction" to the ASA for the calculation of band structures and related quantities. The calculations are selfconsistent and spin-polarized. They are repeated for several values of the lattice constant for each composition, in order to determine the equilibrium volume. Atomic sphere ratios equal to those of the pure elements (fcc for Co) are used all over, i.e., no variation from the Vegard law is assumed. The approximation of von Barth and Hedin is used for the exchange-correlation potential; scalar relativistic corrections are included.

253. **ELECTRONIC STRUCTURE AND MAGNETISM IN TERNARY COMPOUNDS OF Ni-Fe-Pt, Ni-Fe-Pd, Fe-Co-Pt, AND Fe-Co-Pd.** Fernando Estrada and Alejandro Pisanty. Departamento de Física y Química Teórica, Facultad de Química, U.N.A.M., México D.F. 04510, México.

We present band structure calculations, within the LMTO formalism, to study the electronic structure and magnetismo of ordered ternary compounds with formula $Fe_{2m} M1_n M2_{4-n-m}$ (M1=Ni or Co and M2=Pt or Pd). The calculations are selfconsistent and spin-polarized. They are repeated for several values of the lattice constant for each composition, in order to determine the equilibrium volume. We present the local density of states an the values of magnetic moments at equilibrium. The magnetic moment of each atom depends strongly on the environment, increasing with the number of nonmagnetic first neighbors; this trend is particularly important for Fe atoms. We find Stoner and covalent magnetism on different atoms. The alignment of the magnetic moments in a unit cell is parallel. The magnetic moment on Pd and Pt are approximate the same.

254. "EQUATION FOR THE SQUARE ROOT OF ELECTRONIC DENSITY" José-Antonio Flores and Jaime Keller. Departamento de Física y Química Teórica, Facultad de Química, Universidad Nacional Autónoma de México, Ciudad Universitaria, Coyoacán (04510), D. F. Mxico.

In the context of density functional theory, Levy Perdew and Sahni proposed an equation for the square root of electronic density. In this context we suggest a potential that contains constrains over the total particles number and system symetry. We demonstrate the validity of the equation and show results for the atomic case.

PHYS

255. **THE ORIGIN OF YO CHEMILUMINESCENCE FROM THE LASER ABLATION OF YTTRIUM OXIDE AND CHLORIDE IN OXYGEN.**

Daniel Fried and Gene P. **Reck**, Department of Chemistry, **Gyungsoo Kim** and Erhard W. **Rothe**, Department of Chemical Engineering, **Toshimoto Kushida**, Institute of Manufacturing Research, Wayne State University 48202

Excimer laser deposition is one of the most successful techniques of producing high quality superconducting thin films of $YBa_2Cu_3O_7$. Using this method, films may be deposited in oxidant environments such as ozone, nitrous oxide, or oxygen, directly producing films of sufficient oxygen content without a post annealing procedure in oxygen. In the standard method of depositing these films, an oxygen pressure of 100 to 200 mTorr is used.

In vacuum, before the introduction of oxygen, the plume generated from the ablation is bluish - white in color. After the introduction of oxygen the plume changes to reddish-white and takes on a flame-like appearance. The color change is due to the intense yttrium oxide chemiluminescence generated from the reaction of ablated yttrium atoms with oxygen. In our study we used high ablation fluences, simulating the conditions at which the $YBa_2Cu_3O_7$ superconducting thin films are deposited, to photoablate a pellet of pure Y_2O_3. The fluence used in this study, $1.5\ J/cm^2$, is sufficient to generate excited yttrium atoms and ions. We used a gated diode array spectrometer and a gated intensified CCD imaging camera to determine the origin of this emission.

256. COBALT SPECTROSCOPY AT THE ACTIVE SITE OF CARBONIC ANHYDRASE, David R. Garmer, Center for Advanced Research in Biotechnology, 9600 Gudelsky Drive, Rockville, MD 20850. Morris Krauss, National Institute of Standards and Technology, Gaithersburg, MD 20899.

Theoretical models of divalent cobalt bound to the active site of the enzyme, carbonic anhydrase, are constructed to study the effect of the first shell geometry and coordination number on the experimentally observed visible spectra. Favorable comparisons of the ab initio calculated spectra of the hexa-aquo complex of Co(II) with experimental results provides a test of the method. The d-d spectra variation with pH in the enzyme is found to be determined by the first-shell complex geometry as obtained from energy optimization. The low pH complex is predicted to be predominately five-coordinate with two waters in the first-shell while the high pH complex is predominately four-coordinate. Equilibria between four- and five-coordinate structures at low pH is not indicated by the calculation.

257. **EVALUACION DE LA CORROSION DE LAS DIFERENTES ETAPAS DEL LAVADO QUIMICO DE UNA CALDERA POR TECNICAS ELECTROQUIMICAS.** *Esteban M. Garcia O., Jose Ma. Malo A. Martinez Villafañe. INSTITUTO DE INVESTIGACIONES ELECTRICAS. DEPTO. FISICOQUIMICA APLICADA. Interior Internado Palmira, Cuernavaca Mor. México*

En este trabajo se muestra que durante el lavado químico de una caldera, el fenómeno de corrosión cambia continuamente dependiendo de la etapa de limpieza, y expone la manera de detectar dichos cambios de forma instantánea, por medio de técnicas electroquímicas.

258. FAST KINETICS INVESTIGATION OF IRIDIUM BROMIDE COMPLEXES, Mohammad R. Gholami, Charles L. Crawford and Robert J. Hanrahan, Chemistry Department, University of Florida, Gainesville, FL 32611

Transient processes initiated by attack of OH• radicals on $IrBr_6^{3-}$ or e^-_{aq} on $IrBr_6^{2-}$ and $IrBr_6^{3-}$ were studied in aqueous solutions using the technique of pulse radiolysis. Pulse irradiation of $IrBr_6^{3-}$ under oxidizing conditions (N_2O saturated) produces a species stable for 100 μsec or more, with an absorption spectrum identical to that of $IrBr_6^{2-}$ indicating that the Ir(III) is oxidized to Ir(IV) and that the transfer of the electron does not change the structure of the complex ion or the identity of the ligands. Rate constants of $1.60 \times 10^{10}\ M^{-1}s^{-1}$ and $2.61 \times 10^9\ M^{-1}s^{-1}$ for the reduction of $IrBr_6^{2-}$ and $IrBr_6^{3-}$, respectively, were obtained by following the decay of the hydrated electron transient, e^-_{aq}, using a 632.8 nm He-Ne laser source. Methods of preparation and purification of the iridium bromide complexes will be discussed.

259. NEAR ROOM TEMPERATURE DEPOSITION OF COPPER FILMS BY H-ATOM REACTION WITH $Cu(FOD)_2$ AND $Cu(HFA)_2$: Judith Ann Halstead, Department of Chemistry and Physics, Skidmore College, Saratoga Springs, N.Y. 12866 and Robert R. Reeves, Hongwen Li and Peter Locke, Rensselaer Polytechnic Institute, Troy, N.Y. 12180.

The mechanism of nonselective chemical vapor deposition of copper films by H-atom reaction with $Cu(FOD)_2$ and $Cu(HFA)_2$ at near room temperature has been studied. The concentration of copper atoms in the gas phase above the substrate was determined during deposition by atomic absorption spectroscopy. The gas phase formation of copper atoms is proposed to be important in the formation of films highly adherent to a wide variety of surfaces including pyrex and teflon. Elemental analysis of copper films was carried out by ESCA and film morphology characterization was carried out by SEM.

260. **High Pressure NMR Studies of the Glass Transition in the Diamond Anvil Cell.** K.E. Halvorson*, D.P. Raffaelle**, R.F. Marzke**, and G.H. Wolf*, * Department of Chemistry, ** Department of Physics, Arizona State University, Tempe, AZ. 85287-1604

NMR studies of condensed materials at very high pressures have been extremely limited. It has recently been demonstrated, however, that it is possible to measure T_1 and T_2 proton relaxation times of low viscosity fluids in a high-pressure diamond anvil cell, capable of reaching pressures in excess of 10 GPa [Lee et al., *Rev. Sci. Instr.*, 58, 415 (1987)]. We report here NMR measurements of the T_1 and T_2 proton relaxation times of liquid glycerol as a function of pressure in the diamond anvil cell at 51 and 400 MHz. Data were obtained at room temperature up to and through the pressure-induced glass transition. We also report the first measurement of high-resolution chemical shift data obtained in a diamond anvil cell. We find a marked increase with pressure in the isotropic chemical shift difference between the hydrogen bonded and non hydrogen bonded protons. High-pressure NMR studies of other simple liquids will also be presented.

261. SURFACE PHOTOCHEMICAL DYNAMICS OF CH_3Br ADSORBED ON Pt(111). Ian Harrison and Thomas J. Long, Department of Chemistry, University of Virginia, Charlottesville, VA 22901.

A XeCl exciper laser was used to photolyse submonolayers of CH_3Br adsorbed on Pt(111) at 25 - 120 K. The translational and angular distributions of CH_3 photofragments from the 308 nm photolysis were obtained from mass-selected time-of-flight spectra. The intra-adsorbate bond dissociation energy, $D_0(H_3C-Br)$, could be educed and the orientation and photochemical dynamics of the CH_3Br/Pt(111) system will be discussed.

262. SIZE DEPENDENT PROPERTIES OF BORON CLUSTER IONS: REACTIVITY, STRUCTURE AND PHOTODISSOCIATION STUDIES. Paul A. Hintz, Marianne B. Sowa, Scott L. Anderson. Department of Chemistry, State University of New York at Stony Brook, Stony Brook, NY 11794-3400.

Employing guided ion beam techniques, reactions of boron cluster ions (B_n^+, n=1-14) with various neutral molecules have been examined as a function of collision energy from 0.1-10 eV. In recent experiments, the reactant molecules used included CO_2, N_2O, and CF_4. In the experiments with N_2O, we have recently probed the reactivity of clusters up to 24 atoms in size. Dramatic variations in the product branching ratios are observed as the cluster size changes. The effects of pressure on the relative product distribution has also been explored. Through studies of this nature, it has been possible to gain information concerning thermochemistry, dynamics and structure of boron cluster ions. Initial results on photodissociation of boron cluster ions will be presented. The effects of laser wavelength and power on the fragmentation patterns are explored and compared to our previous studies on collision induced dissociation of boron cluster ions.

263. **QUASICLASSICAL TRAJECTORY STUDIES OF THE H+CH$_4$→H$_2$+CH$_3$ REACTION**
Jingrong Huang and James J. Valentini, Department of Chemistry, Columbia University, New York, NY 10027; James T. Muckerman, Chemistry Department, Brookhaven National Laboratory, Upton, NY 11973

Quasiclassical trajectory calculations are being carried out for the H+CH$_4$->H$_2$+CH$_3$ reaction, using a potential energy surface developed by D.G.Truhlar and coworkers, which is based on quantum electronic structure calculations at the saddle point and has been extensively calibrated by comparison with experimental kinetic rate constants. All four H atoms are treated as possible reaction sites and all possible product channels, not just abstraction reaction itself, are considered in the calculation. The results include state-to-state reaction cross sections and the internal energy distributions in H$_2$ product. The calculations are motivated by recent experimental measurements in our group of the state-to-state dynamics of the H+CH$_4$->H$_2$+CH$_3$ reaction as well as similar reactions with C$_2$H$_6$ and C$_3$H$_8$ These perimental results show an anomalous positive correlation of product rotational and vibrational excitation. Explanations for the observed energy partitioning, product branching ratios and the abnormal experimental quantum state distributions of the H$_2$ product will be presented.

264. **AN EPR LINE WIDTH STUDY OF THE ANISOTROPIC REORIENTATION OF VANADYL ACETYLACETONATE IN TOLUENE.** Jimmy S. Hwang and M. Habibur Rahman, King Fahd University of Petroleum & Minerals, Department of Chemistry, Dhahran 31261, Saudi Arabia.

A detailed EPR line widths study of vanadyl acetylacetonate (VOAA) in toluene was carried out as a function of temperature. An analysis employing anisotropic rotational diffusion in the fast-motion limit (when nonsecular terms are unimportant) suggests that VOAA exhibits axially symmetric rotational diffusion with $N = 2.8 \pm 1$ at an axis $z'=Y$, where N is the ratio of $R_{\parallel}/R_{\perp}$ and $R_{\parallel}$ is the rotational diffusion constant along the molecular Z axis and $R_{\perp}$ is the rotational diffusion constant perpendicular to the molecular Z axis. The N value obtained is consistent, within experimental error, with the Stokes-Einstein model and the allowed-values equation (AVE) $\rho_x = 0.634\rho_y - 0.910$ with $\rho_x = 1$ and $\rho_y = 3$.

265. **LIFETIME AND CONCENTRATION QUENCHING OF TRIPLET METHYLBENZENES IN SOLUTION, USING A FLUORESCENCE PROBE.** Guilherme L. Indig and Thérèse Wilson The Biological Laboratories, Harvard University, Cambridge, MA 02138.

9,10-dibromoanthracene (DBA) permits the easy measurement of the singlet and triplet lifetimes of methylbenzenes in solution at 20 °C, by the time-resolved single photon counting method. Excitation at 285 nm of benzene derivatives and DBA (0.5-5x10^{-4}M) in cyclohexane generates ^{1}DBA*, resulting in DBA fluorescence; its decay follows triple exponential functions, with well-separated constants and positive amplitudes. The prompt component (ca. 2 ns) corresponds to direct excitation of DBA. The slower components (ca. 15-40 ns and .1-3 μs) result from sensitization of DBA fluorescence by either singlet or triplet excited methylbenzene, via SS or ST energy transfer (by way of a DBA T$_n$). The decay constants of the sensitized emissions depend on [DBA]. Extrapolation to [DBA]=0 yield values of τ_s of the methylbenzenes which agree with the literature. The decay rates of triplet o-xylene and mesitylene depend linearly on their concentration, with k_q=2x10^6 and 7x10^6 M^{-1}s^{-1}, respectively. At infinite dilution, $\tau_T \geq 10$ μs. In contrast, the ≤10 μs lifetime of triplet durene is concentration independent, up to ca. 0.4 M (solubility limit). There is no evidence either for "self quenching" of triplet durene in acetonitrile or n-hexane.

266. **QUENCHING OF NH A$^3\Pi_i$ IN HIGH ROTATIONAL LEVELS.** Ellen L. Chappell, Jay B. Jeffries, and David R. Crosley, Molecular Physics Laboratory, SRI International, 333 Ravenswood Avenue, Menlo Park, California 94025

Previous experiments have shown that collisional quenching of the excited A$^3\Pi_i$ state of the NH molecule decreases with increasing rotational quantum number N'. This behavior has been attributed to effects of an anisotropic attractive surface. These measurements have been extended to N' = 24, much higher than before. NH ($X^3\Sigma^-$) is produced up to at least N" = 30 by 193 nm photolysis of NH_3, and after a short delay excited to the A state by a tunable dye laser. Quenching cross sections σ_Q are determined from the pressure dependence of the time decay of the fluorescence. For both NH_3 and CH_4 colliders, σ_Q continues to decrease with increasing N', although the rate of change is much slower for high than for low N'. This suggests that rapid rotation averages the potential surface anisotropies.

This research was supported by the Basic Energy Sciences Division of the Department of Energy.

267. **ADSORBANT INTERACTIONS AND SPIN-LATTICE RELAXATION OF SODIUM CATIONS IN ZEOLITE-Y PROBED BY DOUBLE-ROTATION NMR.** Raz Jelinek, Geoffrey A. Ozin[+] and Alexander Pines. Materials Sciences Division, Lawrence Berkeley Laboratory and Department of Chemistry, University of California, Berkeley, CA 94720 and [+]Lash Miller Chemistry Laboratories, University of Toronto, 80 St. George Street, Toronto, Ontario, Canada M5S 1A1.

High resolution Na-23 double-rotation (DOR) NMR experiments were performed on a series of NaY-zeolite structures containing adsorbed "guest" compounds. Extra-framework sodium cations yield distinct DOR resonances that enable the detection of fine changes in shifts and spin-lattice relaxation rates at specific sites. The information helps us achieving a better insight onto anchoring sites of the adsorbant molecules and mobility of the sodium atoms inside the zeolite framework. Spectra obtained upon insertion of $Mo(CO)_6$ and the semiconductor precursor $W(CO)_6$ into the zeolite cages indicate specific sodium binding sites within the alpha-cage. Significantly different shifts and spin--lattice relaxation rates of certain sodium sites occur as the $Mo(CO)_6$ is ligand exchanged with trimethylphosphine, and upon photooxidation and thermal reduction of the $W(CO)_6$ clusters. These changes, observed upon progressive chemical treatment, yield valuable information on the chemical environment, structure and ordering of this adsorbant-adsorbate system.

268. LASER GENERATED HYDROCARBON PLASMAS: DETERMINATION OF TEMPERATURES BY C_2 EMISSION SPECTROSCOPY, A. C. Jones, W. Conaway, C. Stevens, Lawrence Livermore National Laboratory, P.O. Box 808, Livermore, CA 94550

Thin, free-standing hydrocarbon foils are heated uniformly by two oppositely directed laser beams resulting in a uniform, Lambertian plasma expansion. We find that a Boltzman equilibrium distribution is established among the energy levels of these freely expanding plasmas. The radiant emission is viewed through an optically thin path and spectrally dispersed through a gated array spectrograph. The observed ro-vibrational radiance spectra is associated with the $A^3\Pi g$ to $A^3\Pi u$ Swan bands of C_2. These bands are compared to spectral calculations which allow us to determine plasma temperatures in the 3000K to 10,000K range. The temperatures are compared with thermodynamic predictions based upon the equation-of-state TIGER code.

269. COMPLEX KINETICS OF OXIDATION OF THYMOL BLUE WITH BROMATE ION IN ACIDIC SOLUTIONS.

S.B. Jonnalagadda and R.C. Chinake
Department of Chemistry, University of Zimbabwe
P O Box MP 167, Harare, Zimbabwe

There is a growing appreciation for the possibility of using kinetic analytical methods, particularly the catalytic reaction rate methods in determination of trace concentrations of various species. This presentation will process the results of kinetics of the uncatalysed, ruthenium(III) and vanadium(V) catalysed reactions between thymol blue, a sulphonephthalein class of dye (λ_{max} 544 nm) and bromate ion in aqueous sulfuric acid solution. Reactions mechanisms and rate laws will be described based on the experimental data, explaining the complex kinetic behaviour exhibited system and the dual role played by bromide ion as an inhibitor and as an autocatalyst. Procedures for kinetic analysis of low concentrations of Ru(III) [10^{-8}M] and V(V) [10^{-7}M] will be described. The scope and limitations of proposed methods will be discussed.

270. STUDY OF GALVANIZED STEEL/SODIUM CHLORIDE INTERFACE BY ELECTROCHEMICAL IMPEDANCE SPECTROSCOPY. Genescá Ll. Juan, Rodríguez G. Francisco J., and Gómez A. Xicotencatl*. Facultad Química. U.N.A.M. Ciudad Universitaria. 04510 México D.F. * Dpto. Investigación y Desarrollo Nacional de Conductores Eléctricos. México D.F.

The behaviour of galvanized steel in NaCl solution has been studied by Electrochemical Impedance Spectroscopy (E.I.S.). The impedance results are analyzed in terms of the parameters of the equivalent circuit which best describe the spectra obtained during the time of experimentation (60 days). Several equivalent circuit models are considered for analysis. Three electrical circuits were chosen to model the different steps of corrosion process. The model that seems more consistent and which best describes the different measured impedance spectra take into account the presence of a coating porous layer, being Θ the fraction of surface electrode covered by this porous layer

271. X-RAY CRYSTALLOGRAPHY: INSTRUCTIONAL MATERIALS. M. E. Kastner, Department of Chemistry, Bucknell University, Lewisburg, PA 17837

As part of the Chemistry Program of the Mid-Atlantic Cluster sponsored by the Pew Charitable Trust, a series of instructional materials in X-ray Crystallography are being produced. These include a videotape describing the process used to determine the structure from a single crystal and a series of Hypercard Stacks (for Macintosh computers) describing the interpretation of precession photographs and the symbols of the International Tables of Crystallography. A grant from the NSF-ILI Program (Grant #USE-8951058) is also acknowledged.

272. EVIDENCE THAT "HYDROGEN ION" IS $(H_3O)^+$. A.S. Keston, Department of Physiological Chemistry, Mt. Sinai School of Medicine, New York, N.Y. 10029.

An electrical transference cell consisting of two chambers is connected by a very small hole. One chamber contains an electrolyte in heavy water e.g. DCl, the other an electrolyte e.g. HCl in ordinary water. When the cation in the D_2O chamber (deuterium D^+ ion is electrically transferred to the other chamber, for each deuteron transported there is also a molecule of D_2O. This finding is consistent with the formula for hydrogen ion being $(H_3O)^+$. Chloride ion (Cl^-), (I^-) and sulfate ($SO_4^=$) were found to be unhydrated by this method.

273. DIFFUSION CONTROLLED REACTION KINETICS IN A CAPILLARY: REACTANT SEGREGATION AND ANOMALOUS TIME EXPONENTS. Yong-Eun Koo and Raoul Kopelman, Department of Chemistry, University of Michigan, Ann Arbor, Michigan 48109-1055.

An experimental investigation of one-dimensional, diffusion-limited A + B → C chemical reactions is reported. The persistence of reaction segregation and the formation of a depletion zone is observed and expressed in terms of the universal time exponents: α (motion of the boundary zone), β (width of instantaneous product formation zone), γ (rate of instantaneous local formation of product), δ (rate of instantaneous global formation of product), ε (reactant front propagation), etc. There is good agreement with the recently predicted anomalous time exponents. Furthermore, classically the segregation would not be preserved and there would be no formation of a depletion zone and no motion (just dissipation) of the reaction zone.

274. **FLUORESCENCE AND EXCITATION SPECTRA PATTERN AND PHASE TRANSITION IN TETRACENE CRYSTALS AND TETRACENE DOPED POLYMERIC MATRICES.** Seong-Keun Kook and Raoul Kopelman, Department of Chemistry, University of Michigan, Ann Arbor, Michigan 48109-1055.

A study on tetracene single crystals and polycrystals, and tetracene doped into polymethylmethacrylate (PMMA) matrices reveals unusual spectroscopic and spatial patterns with an interesting temperature and phase dependence. These involve both radiative and non-radiative energy transfer and provide information on the nature of the tetracene crystals and aggregates in the different phases.

275. MOLECULAR STRUCTURE OF ASPARAGINE BY PSEUDOSPECTRAL GRADIENT METHOD. Jung-Goo Lee and Richard A. Friesner, Department of Chemistry, Columbia University, New York, N.Y. 10027.

It is now well known that the Pseudospectral(PS) Hartree-Fock method can predict electronic energies and structures for fairly large molecules much more rapidly than conventional ab initio methods. Recently, we have determined an equilibrium geometry of an asparagine molecule($NH_2COCH_2CHNH_2COOH$) with 6-31G** basis function by the PS gradient method. The structure on the asparagine molecule obtained by the PS gradient method turned out to be in good agreement with that by the GAUSSIAN90 ab initio program, and the cpu time by the PS gradient calculation was faster than that by the GAUSSIAN90 calculation (by more than 2 times).

276.

AN AB INITIO STUDY OF THE MOLECULAR STRUCTURE, VIBRATIONAL SPECTRUM AND BINDING ENERGY OF THE LOWEST SINGLET AND TRIPLET STATES OF MgC_2. Timothy J. Lee and Peter R. Taylor NASA Ames Research Center, Moffett Field, California 94035

The minimum energy structures, binding energies and vibrational spectra, including infrared intensities, have been determined for the lowest singlet and triplet states of MgC_2. High levels of *ab initio* quantum mechanical theory have been used in these evaluations. Specifically, a triple zeta plus double polarized basis set has been used in conjunction with complete active space self-consistant field (CASSCF) and single and double excitation coupled-cluster (CCSD) wave functions. In addition, the importance of connected triple excitations has been investigated with the CCSD(T) correlation method. Accurate energy differences have been computed using large atomic natural orbital (ANO) basis sets.

277. **TWO-DIMENSIONAL PURE QUADRUPOLE CORRELATION SPECTROSCOPY (NQR-COSY)** Ming-Yuan Liao and Gerard S. Harbison Dept. of Chemistry, SUNY At Stony Brook, Stony Brook, NY 11794

The pure quadrupole spectra of nuclei of spin higher than $\frac{3}{2}$ display distinct resonances for each of the possible $\Delta m = 1$ transitions at zero field. The ratios of these frequencies are a function of the asymmetry parameter (η) of the electric field gradient tensor at the nucleus. In materials where there are several distinct quadrupolar species, assignment becomes a problem, there being no ab initio way to determine which resonances belong to the same spin species. We have therefore developed a two-dimensional double resonance NQR experiment which allows us to correlate connected NQR transitions.

The pulse sequence is superficially similar to the well known heteronuclear correlation experiment in NMR, although the spin dynamics are quite different. Briefly, a $\pi/2_x - t_1 - \pi/2_x$ sequence is applied to one of the two quadrupolar transitions, resulting in a z polarization for that transition which is proportional to $\cos(\omega_1 t_1)$, where ω_1 is the off resonance frequency. If two transitions are connected, the intensity of the second transition (which is observed immediately thereafter via a simple $\pi/2 - \tau - \pi$ echo sequence) will also have a dependence on $\cos(\omega_1 t_1)$. Two dimensional Fourier transformation gives a spectrum with the two transition frequencies along orthogonal axes. The experiment can be phase-cycled so that only the transferred polarization is observed; in addition, full quadrature can be implemented in the ω_1 dimension.

The experiment has been successfully applied to spin $\frac{5}{2}$, $\frac{7}{2}$ and $\frac{9}{2}$ species, including samples of biological interest. We have also used it to indirectly observe $\Delta m = 2$ transitions, which are weakly allowed at zero field in species with non-zero η. The method has been applied to materials with lines that are inhomogeneously broadened by stress or secondary isotope effects, allowing us to quantify the effect of such perturbations on the full quadrupolar tensor. Finally, such multiple dimensional experiments allow the precise *simultaneous* measurement of several transition frequencies, which may allow more unambiguous determination of the nuclear hexadecapole coupling constant.

PHYS

278.
DIODE LASER PROBING OF N₂O FOLLOWING COLLISIONS WITH O(^{1}D).
<u>Yongsik Lee</u>, Lei Zhu, and George W. Flynn,
Department of Chemistry, Columbia University, New York, NY 10027

Time domain diode laser absorption spectroscopy has been used to measure state-specific energy deposition in the vibrational, rotational, and translational degrees of freedom of N$_2$O following the 248 nm photolysis of ozone in a low pressure mixture. Nascent rotational population distributions have been measured in a number of low-lying N$_2$O vibrational levels. In addition, measurements of translational excitation have been obtained for the rovibrational states which were probed. The prompt absorption signals observed so far arise from inelastic scattering between translationally hot O(^{1}D) and N$_2$O.

279. **HIGH RESOLUTION ELECTRON ENERGY LOSS SPECTROSCOPY STUDIES OF HIGH TEMPERATURE SUPERCONDUCTORS**
<u>Yonghong Li</u> and Charles M. Lieber,
Department of Chemistry, Columbia University, New York, N.Y. 10027

The value of the energy gap in the Cu-O-based high Tc superconductors is widely disputed but of fundamental importance to the origin and nature of superconductivity. Recently, High Resolution Electron Energy Loss Spectroscopy (HREELS) has emerged as an important experimental method to study fundamental electronic excitations at surfaces. We have used HREELS to determine the properties of the surfaces of Bi$_2$Sr$_2$CaCu$_2$O$_8$ superconductors. Our preliminary studies show that the energy gap has a magnitude and temperature dependence that differ significantly from BCS theory.

280. **ESTUDIO DE REDUCCION ELECTROQUIMICA DEL 1-NITRO-NAFTALENO CON ELECTRODOS METALICOS Y SEMICONDUCTORES.** Godínez M. Luis, <u>Ramos M. Aurora</u> y Castro-Acuña Mauricio. DEPg, Depto. de Fisicoquímica, Facultad de Química UNAM. Cd. Universitaria, D.F. México, CP 04510.

Los mecanismos de reacción por los que proceden las reducciones electroquímicas de algunos grupos funcionales de compuestos orgánicos son, en general, fuertemente dependientes de factores tales como el material del cátodo y el electrolito. En éste trabajo se estudia la influencia que ejercen sobre la reducción electroquímica del 1-nitro-naftaleno seis materiales de electrodo: oro, platino, carbón vítreo, cobre y dos semiconductores, TiO$_2$ y Fe$_2$O$_3$. El estudio realizado comprende análisis por voltametría cíclica en un medio de acetonitrilo, agua y (CH$_3$CH$_2$)$_4$NBr como electrolito soporte. La experimentación con TiO$_2$ y Fe$_2$O$_3$ contempló un estudio bajo iluminación. De los resultados obtenidos se propone un mecanismo de reducción del grupo nitro en dos etapas. En la primera se forma el nitroso-naftaleno, y en la segunda, éste último se reduce a la hidroxilamina junto con el azoxi-naftaleno formado por condensación.

281. ESTUDIO ESTADISTICO DE TECNICAS ELECTROQUIMICAS DE ACEROS MICROALEADOS EN ACIDO SULFURICO. *I.Q. LIB: 'RIA MARIACA R..DR. JORGE URUCHURTU CH.*INSTITUTO DE INVESTIGA-CIONES ELECTRICAS.. 'ISION FUENTES DE ENERGIA, DEPTO. FI 'COQUIMICA APLICADA.*Interior Internado Palmira. Cuernacaca Morelos.* **México**.

En el presente trabajo se muestra el an. ,is estadístico realizado a datos de velocidades de corrosión obtenidos por técnicas electroquímicas de resistencia de polarización y ruido electroquímico compara ıs con pérdida de peso, en aceros microaleados (corten, s-ten, 1018) inmersos en ácido sulfúrico a diferente temperaturas y concentraciones. Este análisis se realizó con el fin de determinar la confiabilidad de la información y correlación existente entre cada material y técnica. El análisis muestra la influencia relativa de los parámetros de inter para las diversas condiciones estudiadas. Las tres técnicas presentar. in comportamiento muy similar para cada ma. rial a las mismas condiciones ambientales, existiendo una correlación directa hasta concentraciones de 50%. A may es concentraciones el ruido electroquímico es inversamente proporcional a la resistencia de polarización y pérdida de peso. Estos resultados confirman el carácter complementario de l. s técnicas utilizadas.

282. **AB-INITIO CI STUDY OF THE INTERACTION BETWEEN THE Zn, Zn^+ AND Zn^{2+} + H_2 MOETIES.** J.M.Martínez-Magadán, A.Ramírez Solís, O.Novaro, Instituto Mexicano del Petróleo, Lázaro Cárdenas # 152 Apdo. Postal 14-805. 07730 México, D.F. Also at FES-Cuautitlán UNAM.

The potential energy curves for the interaction of Zn, Zn^+ and Zn^{2+} with the H_2 molecule were calculated at the MRCI + MP2 level using the RECP method of Durand et al., and double-ζ basis sets. The spectroscopic states of Zn, Zn^+ and Zn^{2+} were well reproduced. The $^1A_1$, $^3B_2$ and $^1B_2$ curves for ZnH_2, the $^2A_1$, $^2B_2(^2P)$ and $^2B_2(^2D)$ curves for ZnH_2^+ and the $^1A_1$, $^3B_2$, and $^1B_2$ curves for ZnH_2^{2+} were studied keeping the hydrogen molecule at its equilibrium distance, 1.41 a_0. The $^1A_1$ ground state of ZnH_2 and the $^2A_1$ ground state of ZnH_2^+ are repulsive while the $^1A_1$ ground state of ZnH_2^{2+} is found to have an attractive well of 26.8 Kcal/mol with respect to the dissociation limit. The $^3B_2$ and the $^1B_2$ states of ZnH_2 show wells of 2.7 and 8.3 Kcal/mol, while the $^2B_2(^2P)$ curve of ZnH_2^+ and the $^3B_2$ curve of ZnH_2^{2+} have wells of 45.8 and 2.8 Kcal/mol respectively. The $^2B_2(^2D)$ state of ZnH_2^+ is repulsive.

283. **PICOSECOND NONLINEAR OPTICAL STUDIES OF SEMICONDUCTOR-LIQUID INTERFACES** J. Martorell, Y. J. Chang and T. W. Scott, Department of Chemistry, New York University, New York, New York 10003.

Two different nonlinear optical techniques have been used to probe the surface and bulk relaxation kinetics of photogenerated carriers in single crystal CdS. Picosecond near infrared absorption measurements of the excess carriers provides a novel measure of bulk transport and surface recombination velocities. Surface recombination measurements are favored by one-photon excitation while bulk relaxation is favored by simultaneous two-photon excitation which creates excess carriers uniformly throughout the crystal. Picosecond _surface_ second harmonic generation has been used to characterize the adsorption of malachite green. Selective light polarization and oriented crystals are used to suppress the bulk second harmonic signal and selectively detect the adsorbates. Time resolved surface optical harmonic experiments are in progress to directly measure charge separation kinetics at the semiconductor-liquid junction.

284. MEASUREMENT OF THE MAGNETIC FIELD PENETRATION DEPTH IN $YBa_2Cu_3O_x$ AND $Bi_2Sr_2CaCu_2O_x$ SUPERCONDUCTORS BY ELECTRON SPIN RESONANCE LINE BROADENING OF SURFACE PARAMAGNETIC PROBES, Jerzy T. Masiakowski, Micky Puri and <u>Larry Kevan</u>, Department of Chemistry and the Texas Center for Superconductivity, University of Houston, Houston, Texas 77204-5641, USA.

The electron spin resonance line broadening of a paramagnetic probe on an oxide superconductor below the superconducting critical transition temperature is shown to depend on the parallel or perpendicular orientation of the probe to the applied magnetic field. The line broadening is associated with the magnetic flux lattice formed in the superconductor and leads to a measure of the magnetic field penetration depth. It is shown that the temperature below the superconducting critical transition temperature at which significant line broadening from the flux lattice occurs is best identified as the magnetic flux lattice melting temperature. This is particularly clear in the BiSrCaCuO superconductor in which the flux lattice melting temperature lies significantly below the superconducting critical transition temperature.

285. ELECTROCHEMICAL REMOVAL OF A TIN LAYER FROM COPPER SUBSTRATE. Yunny Meas, Jaime R. Esquivel and Ulises Morales. Area de Electroquímica, Depto. de Química, Universidad Autónoma Metropolitana-Iztapalapa, A.P. 55-534, 09340 México, D.F.

Copper substrate is frequently tinned to improve its solderability, low contact resistivity and corrosion resistance. In order to recover the copper, several processes were developed. These methods where based on acid or basic chemical attack. Nevertheless, they are not totally selective and copper dissolution can occur. An electrochemical detinning process in acidic media was also proposed, but the same inconvenient could be present.
In this work, we propose an electrochemical detinning method in basic media. This method is based on the fact that copper can be passivated in alkalin media. Although massive tin may be passivated in these media, we show that the passive behavior of thin tin layers deposited on copper is different and tin dissolution can occur. Studies by voltametric and chronopotentiometric techniques show that different stages are involved in the process. These stages depend on the media concentration, operation temperature and applied current density. The changes are due to the formation of a stable cupric oxide or an unstable copper hydroxide in the copper passivation.

286. EXCESS HEAT PRODUCTION BY THE ELECTROLYSIS OF AN AQUEOUS POTASSIUM CARBONATE ELECTROLYTE AND THE IMPLICATIONS FOR COLD FUSION. Randell L. Mills, Mills Technologies, Mills Technologies, The Griest Building, Suite 700 I, N. Queen Street, Lancaster, PA 17603.

According to a novel atomic model, the predominant source of heat of the phenomenon denoted Cold Fusion is the electrocatalytically induced reaction whereby hydrogen atoms undergo transitions to quantized energy levels of lower energy than the conventional "ground state". These lower energy states correspond to fractional quantum numbers. The hydrogen electronic transition requires the presence of an energy hole of approximately 27.21 eV provided by electrocatalytic reactant(s) (such as Pd^{2+}/Li^+, Ti^{2+}, or K^+/K^+), and results in "shrunken atoms" analogous to muonic atoms. In the case of deuterium, fusion reactions of shrunken atoms yielding predominantly tritium are possible. Calorimetry of pulsed current and continuous electrolysis of aqueous potassium carbonate (K^+/K^+ electrocatalytic couple) at a nickel cathode was performed in single cell dewar calorimetry cells. Excess power out exceeded input power by a factor greater than 37.

287. *AB INITIO* STUDY OF THE VINYLIDENE-ACETYLENE ISOMERIZATION PROCESS. John A. Montgomery, Jr., United Technologies Research Center, East Hartford, CT 06108 and G. A. Petersson, Hall-Atwater Laboratories of Chemistry, Wesleyan University, Middletown, CT 06457.

Ab initio calculations, including extrapolation to the complete basis set limit, are used to study the energetics of the vinylidene-acetylene isomerization process. Similar calculations gave RMS errors of 0.5 kcal/mol for the dissociation energies, ionization potentials and electron affinities of the first-row diatomics and hydrides. These calculations give an acetylene-vinylidene isomerization energy of 43.9 kcal/mol and a classical barrier to isomerization of 2.2 kcal.mol, in excellent agreement with the experimental results of Lineberger and coworkers. Calculations of the triplet vinylidene-acetylene isomerization process will also be presented.

Supported in part by AFOSR Contract F49620-89-C-0019.

288. ELECTRONIC STRUCTURE OF $LaBO_3$ (B = Mn, Fe, Co, Ni) PEROVSKITES. P. de la Mora*, J.A. Flores**, L. Vicente**, C. de Teresa**.
*Dpto. de Física, Fac. de Ciencias, UNAM, Cd. Universitaria, 04510 México, D.F.
**División de Posgrado, Fac. de Química, UNAM. Cd. Universitaria, 04510 México D.F.

The electronic structure of $LaBO_3$ (B = Mn, Fe, Co, Ni) perovskite has been studied inside a tight binding scheme using the extended Hückel Method. The band structure and electronic density of states show the crystal field splitting due to the octahedral symmetry of the metal site. An analysis of the meta-oxygen bond is also presented as well as the electronic structure of the surface in order to establish a correlation between electronic properties and catalytic activity of these compounds.

289. **A theoretical foundation of the *Al* sitting and pairing avoidance principles in zeolites.**, J.Morales, M.Bonilla-Marín and P. Pimentel, INSTITUTO MEXICANO DEL PETROLEO, Subdirección General de Investigación Aplicada, Apartado Postal 14-805, 07730, MEXICO, D. F.

Substitutional energies for multiple *Al* -> *Si* replacements were obtained by **MNDO** on cluster models of the Secondary Building Unit **(SBU)** of the ZSM5, ZSM11 and ZSM23 framework. The choice of the semiempirical method is justified from the study of single aluminium substitution on monomer, and hexamer aggregates whose replacement energies are qualitatively the same as those from *ab-initio* calculations. According to the results, a site occupancy order in the unit cell is derived from sundry *Al* replacements in the corresponding **SBU**. Likewise, a Si/Al limit ratio with the related *Al* distribution pattern in the different aluminosilicate networks considered are also obtained. Both results provide a theoretical basis for the Loewenstein's and Dempsey's *Al* sitting and pairing avoidance prirciples.

290. UPDATING AN UNDERGRADUATE PHUSICAL CHEMISTRY LAB. K.C. Morton, Department of Chemistry, Carson-Newman College, Jefferson City, TN 37725

An NSF (ILI) grant helped provide new equipment to improve a physical chemistry lab in a liberal arts college that graduates 5-6 BA chemistry majors/year and offers physical chemistry on alternating years. The items (total cost = $15,600) were; bomb and solution calorimeters, colloid osmometer, conductivity meter, and equation of state apparatus. The latter item is relatively new in the US and is of particular interest because it permits students to visually observe the behavior of a gas in the region of the critical point and to collect 70 or more P-V-T- data points in 3 hours. Students analyze their data in a subsequent exercise that has been developed using Wingz software on a Macintosh IIcx computer. Experimental isotherms are compared graphically and statistically to those predicted by the van der Waal's (vdW) equation; the effects of altering the vdW coefficients are readily displayed and investigated. The data are also analyzed using the Virial and the Clausius-Clapeyron equations.

291. PURE ROTATIONAL SCATTERING OF CO_2/CO DUE TO COLLSIONS WITH HOT HYDROGEN ATOMS. Chi-Kung Ni, Thomas Kreutz, and George Flynn, Chemistry Department, Columbia University, New York, N.Y. 10027.

Measurement of nascent Doppler line shapes after hot H atom/CO_2 or CO collisions provides information about molecules scattered *in* and *out* of rotational levels due to the collision process. Comparsions between IOS and ECS scaling approximations and the experimental data suggest that the line shapes from different rotational states correspond to different types of scattering, and that they are sensitive to the differential cross section.

292. STM INVESTIGATIONS OF THE SURFACE STRUCTURE OF $Bi_2Sr_2CuO_y$ and $Bi_2Sr_2Cu_{.9}Fe_{.1}O_y$
Chunming Niu and Charles M. Lieber,
Department of Chemistry, Columbia University, New York, N.Y. 10027

The surface structure and electronic properties of $Bi_2Sr_2CuO_y$(2201) and $Bi_2Sr_2Cu_{.9}Fe_{.1}O_y$ have been characterized by scanning tunneling microscopy.

Images with atomic resolution reveal nearly commensurate one-dimentional superstructure with a period of 5b. Moreover, high resolution images do not exhibit vacancies and disordered sites, that as expected based on the proposed models for the superlattice. The substitution of Cu with Fe decreases the modulation vector to 4.5b and causes disorder in the superstructure. A structure model that consistently explains these atomic resolution results are discussed.

293. CHANGE IN CRYSTAL HABIT IN THE VAPOR PHASE GROWTH OF ZnO. <u>Jaime Noriega</u>, D.E.Pg., Facultad de Química, UNAM, Ciudad Universitaria, 04510 México, D.F.

We report the final results for the growth of ZnO on the vapor phase obtained in a lab reactor after a long series of improvements in the equipment. We have increased the temperature of reaction and homogeneity in the reaction zone, under the new reaction conditions the crystals obtained show a distinct shape. This shows definite evidence of a different mechanism for crystal growth than in earlier samples; we attribute this change to have succesfully approached the roughening transaction of the crystal growth process, rendering the overall growth non dependent in the surface procesess and therefore reducing the presence of preferred orientation of the crystal habit. The new habit shows the emergence (by X-ray diffraction) of more facets and therefore overall shapes that appear less faceted.
The samples are characterized by X-ray diffraction, and photomography. Our present results confirm our previous work in the light of simplified models of crystal growth. We present the reaction temperature progress in the experimental program and the crystals obtained in each case.

294. PHOSPHORIC ROCK DISSOLUTIONS, AN EXPERIMENTAL REPORT. <u>Jaime Noriega</u>, D.E.Pg., Facultad de Química, UNAM, Ciudad Universitaria, 04510 México, D.F.

We present our experimental results on the dissolution reaction of phosphoric rock in mixtures of sulfuric and phosphoric acid. By measuring continously the electric conductivity of the solution during the batch reaction we obtained the progress of the reaction with time for different particle sizes of rock, temperature and acid concentrations.
With a much improved reaction system and instrumentation we carried on the reaction and compare the results with various models of heterogeneous solid fluid reactions. These models include the porosity of the rock itself and provide a better fit to the experimental data than the previously used shrinking-core model. We include data of conversion versus time, for different particle sizes, temperatures and concentrations of acids.
We discuss the results in the light of the models that include non homogeneities in the solid phase with its associated interfacial transport processes. The effect of the external surface area of the particles and the temperature, on the mechanism of reaction.

295. C-H$\cdots$O CONTACT INTERACTIONS IN MOLECULAR SOLIDS. J. J. Novoa and B. Tarron, Departmento de Quimica Fisica, Universidad de Barcelona, Barcelona, Spain; <u>M. -H. Whangbo</u>, Department of Chemistry, North Carolina State University, Raleigh, NC 27695; J. M. Williams, Chemistry and Materials Science Divisions, Argonne National Laboratory, Argonne, IL 60439.

To accurately evaluate the C-H$\cdots$O contact interaction energy we performed ab initio calculations on $CH_4\cdots OH_2$ at the SCF-MO and MP2 levels with various quality basis sets including a near Hartree-Fock limit (HFL) one. The interaction energies calculated with truncated basis sets have small basis set superposition errors and become close to the near HFL results only when they include diffuse functions. This is attributed to the necessity of properly representing the lone pair-electrons of oxygen that make short contact interaction with the C-H bond. Our MP2 calculations suggest that the conformation of $CH_4\cdots OH_2$ with one C-H$\cdots$O contact is the only minimum-energy structure. At the MP2 level with the near HFL basis set, the binding energy of $CH_4 \cdots OH_2$ is calculated to be 0.59 ± 0.05 kcal/mol. The potential energy curves of $CH_4 \cdots OH_2$ calculated as functions of the $H \cdots O$ contact angles are consistent with the observation that the most frequently found $H \cdots O$ contact angles of organic solids are in the range $|\phi| \leq \sim 60°$ and $|\theta| \leq \sim 30°$.

296. **A COMPARISON OF THREE CHEMICAL MECHANISMS USED TO MODEL MEXICO CITY AIR QUALITY.** Xóchitl Cruz-Núñez, María E. Ruiz, Domingo Salazar, Gustavo Sosa. Gerencia de Energéticos Alternos y Química Ambiental, Instituto Mexicano del Petróleo. Eje Central Lázaro Cárdenas 152, Apdo. Postal 14-805, 07730, México DF. MEXICO.

Good air quality models include photochemical mechanisms that take into account the particular species in the pollutants -primary and secondary. In this work, we compare three chemical mechanisms: the Carbon Bond IV (CBM-IV from the EKMA model), the mechanism proposed by Falls, Seinfeld and McRae (FSM), and the popular ERT mechanism (both within the CalTech algorithm). These have been tested under Mexico City conditions, as part of the development of a photochemical model for Mexico City air pollution. Results from all three mechanisms were obtained to simulate diurnal evolution of nitrogen oxides, hydrocarbons, ozone and other oxidants for a high ozone episode, and allowed to asses the advantages and disadvantages for each mechanism.

297. **OPTIMIZED BASIS SETS FOR AB INITIO CALCULATIONS.** Joseph W. Ochterski and G. A. Petersson, Hall-Atwater Laboratories of Chemistry, Wesleyan University, Middletown, Connecticut 06459

A major source of error in ab initio calculations is the inability of standard basis sets (STO-3G, 6-31G**, etc.) to deal with the changing sizes of orbitals in ions and molecules. In an attempt to address this problem, we have developed fully optimized 3-3q and 6-31q** contracted Gaussian basis sets for the atoms H through Ne and for their singly charged positive and negative ions. Approximate molecular optimized basis sets are obtained by interpolation of the atomic ion basis sets. Optimization of all exponents and contraction coefficients reduces the errors in ab initio energies by about a factor of two relative to the standard basis sets of the same size. Far more significant is the fact that the basis set truncation errors associated with the optimized basis sets follow a simple model very closely. The model uses the atomic numbers and orbital occupations to estimate the basis set truncation error. The accuracy of both SCF energies and correlation energies are dramatically improved with little increase in computational effort.

298. *AB INITIO* PROTON AFFINITIES COMPARED TO EXPERIMENTAL VALUES, INCLUDING PREDICTIONS FOR VARIOUS TRIAZENES. Judy L. Ozment, Department of Chemistry, and Ann Schmiedekamp, Department of Physics; Penn State University, Ogontz Campus; Abington, PA; 19001

We compile and correlate proton affinities for a series of molecules that have oxygen and/or nitrogen protonation sites. These are calculated by us and others at the *ab initio* basis levels of RHF/3-21G//3-21G, RHF/3-21+G//3-21G, RHF/6-31G*//6-31G*, RHF/6-31+G(d,p)//6-31G*, MPn//6-31G*//6-31G*, n=2,3, MPn/6-31+G(d,p)/6-31G*, n=2,3,4. The agreement of these calculations with experiment suggests accuracy limits for the proton affinities of small molecules containing nitrogen and oxygen. Scaling showing comparison trends suggests predictions.
Ab initio studies have been done at SCF levels for triazene, methyltriazene, dimethyltriazene, and acetyltriazene. With the exception of unsubstituted triazene, calculations have only been done with the 3-21G basis set. We now report *ab initio* proton affinities for triazene, 1- and 3-methyltriazene, and formyltriazene at RHF/3-21G//3-21G, RHF/3-21+G//3-21G, RHF/6-31G*, RHF/6-31+G(d,p)//6-31G*, MPn/6-31G*, n=2,3. Since experimental proton affinities are not available for any triazene, this study documents how well the smaller basis set calculations reproduce the trends in the higher level calculations, and then gives predictions for the experimental gas phase proton affinities.

299. TUNABLE DIODE LASER PROBE OF CHLORINE ATOMS PRODUCED FROM THE PHOTODISSOCIATION OF OF S_2Cl_2, J. Park, <u>Yongsik Lee</u>, and George W. Flynn, Department of Chemistry, Columbia University, New York, NY 10027.

Time resolved diode laser absorption spectroscopy is used to probe Cl atoms produced from the photolysis of S_2Cl_2 at excimer laser wavelengths 193 nm, 248 nm, and 308 nm. The diode laser was tuned to the Cl atom fine structure transition, $^2P_{1/2}(F^*=2) \leftarrow {}^2P_{3/2}(F=3)$. For 248 nm photolysis, the translational energy of nascent Cl atoms is measured to be 7±1 Kcal/mol from the Doppler lineshape of the diode laser absorption. Only 17 % of the available energy appears as translational recoil energy of photofragments. This result is consistent with a statistical model, suggesting a long-lived transition state complex. The relative yield for producing $Cl(^2P_{1/2})$ is 0.21±0.03 at 248 nm, 0.20±0.03 at 193 nm, and 0.48±0.06 at 308 nm. The rate constant for quenching of $Cl(^2P_{1/2})$ by argon is determined to be $(0.8\pm0.1)\times10^{-12}$ cm^3/molec sec.

300. EXPERIMENTS IN RAMAN SPECTROSCOPY FOR P. CHEM LAB. <u>Marsha I. Lester</u>, Robert J. Moore, Department of Chemistry, University of Pennsylvania, Philadelphia, Pennsylvania 19104-6323.

Raman spectroscopy is a useful technique for obtaining vibrational and, sometimes, rotational information about many types of molecular samples. Laser Raman spectroscopy can be used in the undergraduate physical chemistry laboratory to illustrate many important concepts of modern spectroscopy, including molecular symmetry and force fields, spectroscopic selection rules, Fermi resonance, polarization effects and nuclear spin statistics. We have developed a low resolution (3.5 1/cm) Raman spectrometer as well as a number of experiments for use in the physical chemistry teaching laboratory at the University of Pennsylvania. These include vibrational Raman spectroscopy of pure liquids and vapors. Experiments in resonance Raman and pure rotational Raman spectroscopy are under development. These experiments will be discussed, and a number of "myths" about Raman spectroscopy will be dispelled, among them: 1) Raman signals are weak (in fact, in neat liquids they can often be seen with the naked eye!); 2) Raman spectroscopy is difficult to set up and perform, and; 3) Raman spectrometers are extremely expensive.

301. AlH GAS-PHASE REACTION KINETICS. Louise R. Pasternack and Jane K. Rice, Chemistry Division, Naval Research Laboratory, Washington, DC 20375-5000.

The reactions of ground-state AlH ($^1\Sigma^+$) with O_2, H_2O, H_2, D_2, C_3H_8, C_2H_2, 2-butyne, C_2H_4 and 2-methyl-2-butene, are investigated at 298K. The AlH is produced by the photodissociation of triethylaluminum at 193nm and 248nm and detected by laser-induced fluorescence of the A $^1\Pi$-X $^1\Sigma^+$ transition. Absolute bimolecular rate constants are obtained from AlH concentration decay profiles as a function of reactant gas concentrations. The reactant list includes two oxygen containing species, H_2 and D_2 which result in pressure dependent association reactions, one saturated hydrocarbon with an undetectable reaction rate and finally, several saturated hydrocarbons which exhibit very rapid rates of reaction. Comparison of AlH reaction rates are made with those of the isovalent BH species, Al atoms, and the isoelectronic Si atom.

302. HIGH OVERTONE RESONANCE RAMAN SPECTRA OF PHOTODISSOCIATING NITROMETHANE IN SOLUTION. <u>David L. Phillips</u> and Anne B. Myers, Department of Chemistry, University of Rochester, Rochester, NY 14627

Resonance Raman spectra of nitromethane in cyclohexane, acetonitrile, and water solvents have been acquired using excitation frequencies of 218 and 200 nm. Fully deuterated nitromethane spectra were also obtained in both vapor and solution phases with 218 nm excitation. Resolvable Raman peaks are observed up to 15,000 cm^{-1}, which is approximately the lowest dissociation limit (to ground state CH_3O + NO). The solution and vapor spectra are similar in that the NO_2 symmetric stretch overtone progressions dominate; however, the higher signal-to-noise ratio of the solution phase spectra enables many weaker transitions to be observed as well. The Raman intensities, frequencies, and linewidths are interpreted to examine solvation effects on the potential energy surfaces and photodissociation dynamics.

303. A "HANDS-ON" TEACHING LASER. William F. Polik and Brad Mulder, Department of Chemistry, Hope College, Holland, MI 49423.

Lasers applications impact virtually every branch of science. In chemistry, the laser has revolutionized the fields of high resolution spectroscopy and molecular reaction dynamics. In order for laser experiments to be effectively introduced into the undergraduate laboratory curriculum, the underlying principles of laser operation must be taught and related to the useful properties of laser light. An open-cavity helium-neon laser is described which is used to teach the theory of laser operation through a "hands-on" approach. Students identify the components of a laser, align the laser, and study oscillator stability conditions. The emitted beam is characterized for polarization, transverse modes, and longitudinal modes. With a fundamental understanding and working knowledge of laser technology, students are well prepared for experiments illustrating the applications of lasers in chemistry.

304. ACCURATE MRCI STATE AND TRANSITION DIPOLE MOMENTS FOR THE 9 LOWEST-LYING ELECTRONIC STATES OF THE CuCl MOLECULE.
A. Ramírez Solís. SGIA del Instituto Mexicano del Petróleo. A.P. 14-805. 07730 México D.F., MEXICO. Also at Instituto de Física, Universidad Nacional Autónoma de México. México, D.F. MEXICO

The zeroth and first order perturbed (Möller-Plesset) wave functions for the 9 lowest-lying electronic states of CuCl were used to calculate the diagonal and transition dipole moments. For the diagonal dipole moments the MCSCF molecular orbitals (MO) are used, thus leading to very accurate results. In the case of the transition dipole moments, the MCSCF MOs of one state or the other were used in order to avoid the problems involved with non-orthoganal basis sets. The resulting figures for the zeroth and first order transition moments were averaged using both sets of optimized orbitals. The calculated dipole moments are meant to be used to obtain accurate transition probabilities between excited rovibronic levels of CuCl.

305. EXAMINATION OF THE C_3H_6 POTENTIAL ENERGY SURFACE USING QUADRATIC CONFIGURATION INTERACTION METHODS. Eric S. Replogle and John A. Pople*, Department of Chemistry, Carnegie Mellon University, Pittsburgh, PA 15213.

The potential energy surface of C_3H_6 is characterized with Quadratic Configuration Interaction (QCI) methods, with the major interest being the stereochemical and geometrical isomerization of cyclopropane. A key question is whether stereochemical isomerization proceeds via single or double rotation of the terminal methylene groups in the intermediate diradical, trimethylene. Both the conrotatory and disrotatory modes of the double rotation pathway are studied. In addition, the conversion of cyclopropane to propene via trimethylene is investigated.

306. THE EFFECT OF ELECTRONIC STATE ON THE REACTION CROSS SECTIONS OF TRANSITION METAL CATIONS: V^+ + ETHANE, ETHENE, AND PROPANE
Lary Sanders*, Scott D. Hanton, James C. Weisshaar
*Department of Chemistry, Wittenberg University, Springfield, OH. 45501.

We describe a crossed beam experiment which measures the cross section for reaction of state-specified V^+ with hydrocarbons. The desired electronic states are created by laser induced, resonance enhanced, two photon ionization of a neutral V beam. The resulting distribution of states is measured by photoelectron spectroscopy. The neutral V atoms are ionized within the intersection of the V and neutral reagent beams. Reactant and product ions are analyzed by pulsed time-of-flight mass spectrometry. We observe a dramatic dependence of cross section of the V⁺ electronic term. The second excited state is more reactive than either lower energy terms by a factor of ≥ 270, 80 and ≥ 6 for ethane, propane, and ethene, respectively. Comparison with earlier work shows that electronic energy is far more effective at promoting H_2 elimination than the same total amount of kinetic energy.

307. SIMULATION OF TRAPPING AND QUENCHING REACTION KINETICS IN LOW-DIMENSIONAL AND FRACTAL LATTICES. R. Schoonover and R. Kopelman. Department of Chemistry, University of Michigan, Ann Arbor, MI 48109.

The nearest-neighbor distribution of moving particles in the immediate vicinity of a trap or quencher, as well as density profiles and reaction kinetics, are studied for low-dimensional and fractal lattices. We find new subtle effects that depend on the lattice topology and on the mobility of both particles and traps/quenchers. Comparison with theory and experiments will be made.

308. SHORT RANGE SOLVATION BY CYCLIC DIETHERS. R.M. Scott, H. Zhu and M. Zhou, Dept. of Chemistry, Eastern Michigan University, Ypsilanti, Michigan, 48197.

Short range solvation involves effects on the reactivity of solutes by formation of hydrogen bonds between solute and solvent. We have reported the effects of numerous solvents on the formation of proton transfer complexes between diethyl amine and 2,4-dinitrophenol, and have devoted special attention to a variety of ethers. Equilibrium constants for formation of ether-amine hydrogen bonds display values for monoethers and noncyclic diethers of 60-100, while cyclic diethers have equilibrium constants up to more than 10 times those values. Various substituted cyclic diethers have been synthesized and studied in an attempt to explain this surprising result, and a hypothesis for the behavior is presented.

309. EXCITON KINETICS STUDIES ON ISOLATED POLYMER CHAINS. Zhong-You Shi and Raoul Kopelman, Department of Chemistry, University of Michigan, Ann Arbor, Michigan 48109-1055.

Triplet exciton homofusion and heterofusion on various time scales have been observed on extremely dilute samples of poly-1-vinylnaphthalene (P1VN) in poly(methylmethacrylate) (PMMA) at 77K. Depending on the laser pulse excitation scheme, classical or non-classical kinetic behavior is observed, which is in agreement with the theoretical expectations.

310.

REACTION PATH AND RATE CALCULATIONS OF DIFFERENT CHANNELS OF $H + HNO$ Maribel Soto and Michael Page, Naval Research Laboratory Washington, DC 20375, and Michael L. McKee, Department of Chemistry, Auburn University, Auburn, Al 36849

We have studied several reactive channels for $H + HNO$ using ab initio multi-configuration self-consistent field (MCSCF), single and doubles configuration interaction (CISD) and multi-reference configuration interaction (MRCI) methods with a correlation consistent valence double-zeta basis set. The reactions considered were:

$$H + HNO \longrightarrow H_2 + NO \qquad (1)$$

$$H + HNO \longrightarrow NH_2O(^2A') \qquad (2)$$

$$H + HNO \longrightarrow HNOH(cis, trans) \qquad (3)$$

Results obtained at the MRCI + ZPE level with structures optimized at the 5-in-5 MCSCF level predict that all three reactions have negligible barriers with exothermicities of -54.60 kcal/mol, -46.46 kcal/mol and -45.05 kcal/mol (for the *trans* product), for reactions 1-3 respectively. Other reactive channels which we explored were:

$$H + HNO \longrightarrow NH_2O \longrightarrow H_2 + NO$$

$$H + HNO \longrightarrow trans - HNOH \longrightarrow NH_2O$$

Furthermore, we have done rate calculations for reactions (1) and (2) using conventional transition state theory (TST) and improved canonical variational transition state theory (ICVTST).

311. CARBON CLUSTER ION REACTIONS: STRUCTURAL ISOMER EFFECTS ON REACTIVITY. Marianne B. Sowa, Paul A. Hintz, Scott L. Anderson. Department of Chemistry, State University of New York at Stony Brook, Stony Brook, NY 11794-3400.

Guided ion beam methods have been used to measure the product cross sections for reactions of mass-selected carbon cluster ions as a function of collision energy. Small carbon cluster ions are known to have structural isomers with significantly different properties. These differences are strongly manifested in the collision energy dependence of these reactions. For reactions with D_2, the cross section for the major product channel (C_nD^+ + D) depends strongly on both collision energy and cluster size. A two component cross section is observed for C_nD^+ formation as the collision energy is increased from 0.1 to 10 eV for C_3^+ through C_7^+. The two different components of these cross sections are attributed to two different isomers for these clusters, one reacting at low collision energy with the other being activated by increasing the collision energy. Product cross sections for the reaction of small carbon cluster ions with various neutral oxidizing molecules such as O_2, CO_2 and N_2O will be presented.

312. **INVESTIGATIONS OF Bi SUBSTITUTION IN $NdBa_2Cu_3O_y$.** Lawrence Suchow and Jian Tang, Chemistry Division, New Jersey Institute of Technology, Newark, New Jersey 07102.

Attempts to substitute Bi for Nd in orthorhombic $NdBa_2Cu_3O_y$ prepared in air or oxygen at about 950°C led instead to formation of Ba_2NdBiO_6, a new cubic compound with a = 0.8703 nm. The possibility was then explored of preparing superconducting $(Nd_{1-x}Bi_x)Ba_2Cu_3O_y$ by first forming the tetragonal phase at 880 - 950°C in nitrogen or argon followed by reheating in oxygen or air at 250 - 500°C in order to insert the additional oxygen required to yield the orthorhombic form while avoiding oxidation of Bi^{3+} to Bi^{5+}. X-ray diffraction studies, electrical conductivity measurements, and thermogravimetric analysis of products indicate that Bi does not enter the $NdBa_2Cu_3O_y$ lattice in either the tetragonal or the orthorhombic phase. Ba_2NdBiO_6 clearly forms on reheating in oxygen or air even at low temperatures, and evidence is presented that a poorly crystallized oxygen-deficient form of this compound is already present prior to the reheating.

313. AB-INITIO MO STUDIES ON Si_2H_6, SiH_3GeH_3, AND Ge_2H_6 RADICAL ANIONS AS PROTOTYPES OF POLYMER ANIONS WITH Si AND Ge BACKBONES
Tsukasa Tada and Reiko Yoshimura, Toshiba R & D Center, 1 Komukai Toshibacho, Saiwai-Ku, Kawasaki-City, Kanagawa 210, Japan

Relatively stable polysilane and polygermane radical anions were reported to be generated by ionizing radiation. As a first step in understanding these polymer anions, Si_2H_6, SiH_3GeH_3, and Ge_2H_6 radical anions have been investigated as prototypes of polymer anions with Si and Ge backbones by the ab-initio MO theory using the double-zeta quality basis set containing s, p-diffuse functions. The calculated results indicate that these anions have similar molecular and electronic structures (C_{2h}, ^2B_u for Si_2H_6 and Ge_2H_6 anions, C_s, $^2A'$ for SiH_3GeH_3 anions) with an unpaired electron occupying the Si-Si, Si-Ge, and Ge-Ge σ* antibonding orbitals, respectively. They are all found to be kinetically bound anions with positive vertical ionization potentials. Some of the differences among these radical anions are also studied.

314. A THEORETICAL STUDY OF RING-OPENING OF AZIRIDINONES BY NUCLEO-PHILES. Erach R. Talaty and Anthony E. Francis, Department of Chemistry, Wichita State University, Wichita, Kansas 67208.

The results of a semi-empirical study of the different modes of ring-opening of aziridinone by nucleophiles will be presented. Each potential energy surface was scanned by means of the MOPAC program, utilizing AM1 and PM3 Hamiltonians, to locate saddle points and energy minima. Experimental studies demonstrate dependence of the mode of ring-opening upon substituents on the ring as well as on the type of nucleophile. An attempt is made to correlate such experimental results with theoretical calculations and to discern the reasons for the high degree of selectivity often encountered in these ring-openings, which makes these heterocycles valuable intermediates in organic synthesis.

315. AB INITIO ENERGIES AND SHAPES OF $HSiS^+$, $HSiO^+$, HCP AND HSiP, AND THEIR ISOMERS, Erach R. Talaty and Melvin E. Zandler, Department of Chemistry, Wichita State University, Wichita, Kansas 67208.

Walsh's Rules predict that all of the compounds in the title should be linear by virtue of them containing ten valence electrons. Recent ab initio calculations by us on related molecules of the type HAB have indicated that several of them might be distinctly bent, but that the degree of bending as well as the barrier to isomerization of HAB to HBA varies considerably from one molecule to another. In an effort to discern any distinct patterns in these results, we have extended our calculations to include $HSiS^+$, $HSiO^+$, HCP and HSiP, and their isomers. Potential energy surfaces of each pair of isomers were scanned using the GAUSSIAN 86 program at several levels of theory, with and without electron correlation. Our best level of calculation, namely, MP2 (full core)/6-311G (2d, 2p), has enabled us to uncover more examples of bent molecules containing ten valence electrons.

316. ENERGY KINETICS IN NANOMETER CRYSTALS. Weihong Tan, Steve Smith and Raoul Kopelman, Department of Chemistry, University of Michigan, Ann Arbor, Michigan 48109-1055.

Aiming at nanometer light sources and exciton microscope, nanometer organic crystals have been produced in micropipettes and investigated. Nonlinear optical and spectroscopic effects are observed and will be reported for anthracene, tetracene, perylene, 9,10-diphenylanthracene, pyrene and some of these crystals doped in polymers.

317. **THE COMPLETE BASIS SET ENERGIES OF ONE AND TWO CARBON MOLECULES**. T. G. Tensfeldt and G. A. Petersson, Hall-Atwater Laboratories of Chemistry, Wesleyan University, Middletown, Connecticut 06459.

Accurate calculations of molecular energies require convergence of both the one-electron basis set expansion and the many-electron configuration interaction (CI) expansion. The complete basis set (CBS) method achieves the former by extrapolating each of the pair correlation energies to the complete basis set limit using the asymptotic convergence of pair natural orbital expansions (*J. Chem. Phys.* **1988**, *89*, 2193). To achieve the latter, we employ the quadratic CI approach of Pople (*J. Chem. Phys.* **1987**, 87, 5968). The combined CBS-QCI method using [6s6p3d2f] atomic pair natural orbital basis sets is the most accurate systematic method proposed to date having consistently achieved an accuracy of < 1 kcal/mol for chemical energy differences involving first-row atoms and hydrides, (*J. Chem. Phys.* **1991**, in press). We shall present a comprehensive study of all one and two carbon molecules including positive and negative ions, radicals, and transition state structures for a total of 67 species. We shall compare the energy changes obtained from the following computational methods: Dewar's AM1, UHF/STO-3G, MP2/6-31G**, CBS2/6-31G#, MP2/6-311+G**, CBS-QCI/6-311+G**, Pople's G1, QCI/[6s6p3d2f], and CBS-QCI/[6s6p3d2f], with the experimental values. This study includes predicted structures, ionization potentials, electron affinities, dissociation energies, activation energies, and excitation energies.

318. AN INTEGRATED WRITING PROGRAM IN THE PHYSICAL CHEMISTRY LABORATORY. T. M. Ticich, Department of Chemistry, Gustavus Adolphus College, St. Peter, Minnesota 56082.

Physical chemistry laboratory often provides students with their first exposure to scientific writing and critical thinking in most chemistry curricula. Getting students to think and communicate like mature scientists requires that instructors explicitly teach them how to formulate a good discussion and how to assemble a clear and concise presentation of their results in a journalistic style. The physical chemistry sequence at Gustavus Adolphus College uses writing workshops and a peer review process to teach and to reinforce these skills. The latter provides students with both good and bad models of scientific writing, and also teaches them about the review process in scientific journals. The scientific journal article, however, is not the only means of communication that professional chemists use. In the second semester course, students learn to present their results in formats appropriate for different audiences, such as a research proposal and a popular science article.

319. BEHAVIOR OF V^E AND Cp^E NEAR THE UCST OF POLYETHERS + ALKANE SYSTEMS. Luis Miguel Trejo & Alberto Allende. SQM. Depto. Física y Química Teórica, Facultad de Química, U.N.A.M. México, D.F., 04510, México

We have measured the coexistence curves, using opalescence points, for 2,5,8,11,14 Pentaoxapentadecane + n-Decane(I) (UCST = 19.53 ± 0.0.02°C, X ether ≠ 0.400 ±0.001) and for 2,5,8,11-tetraoxadodecane + n-pentadecane(II) (UCST= 21.51 ±0.02°C, X ether = 0.600± 0.001). For both systems we have determined the excess volume, V^E, and excess heat capacity, Cp^E, using respectively, a vibrating cell densimeter and a Picker flow microcalorimeters, at temperatures near the UCST. For Cp^E, the W-shape at 25°C, is enhanced enormously upon approaching UCST. The effect on V^E is less pronounced and is only modified 0.5°C near critical pont.

320. EFFECT OF BRANCHING AND CYCLING ON THE W-SHAPE Cp^E. Luis M. Trejo. SQM. Depto. de Física y Química Teórica, Facultad de Química, U.N.A.M. México, D. F., 04510 México.

Using a Picker flow microcalorimeter, we measured at 25°C the excess heat capacity, Cp^E, of polyethers + alkane systems modifing the lineartty with the branching or cycling of one of the components in order to see its effect on the W-shape Cp^E that exists with linear components. The results of Cp^E for the systems 2,5,8,11,14- Pentaoxapentadecane + 2,2-dimethylbutane, + 2,2,4-trimethylpentane, + cyclohexane (this at 17.5, 25 and 40°C); 2,5-dioxahexane + pristane and + squalane and p-dioxane + ristane and + squalane, compared with the literature ones for the linear, show the W-shape is reduced in going from linear to branched up to cycled components.

321. RESONANCE RAMAN SPECTROSCOPY OF SIMPLE AMIDES. Nancy E. Triggs and James J. Valentini, Department of Chemistry, Columbia University, New York, New York 10027

Resonance Raman spectroscopy has been used to probe the UV photoexitation of some simple amides. Spectra are reported for both the gas and the liquid phases of acetamide, N-methyl acetamide (NMA), N,N dimethyl acetamide (NNMA) and ε-caprolactam at excitation wavelengths of 266 nm and 212.8 nm. The UV absorption spectra of the amides studied show a weak $n \rightarrow \pi^*$ transition around 220 nm and a very strong $\pi \rightarrow \pi^*$ around 180 nm. In comparing the pre-resonant spectra at 266 nm with the resonant spectra at 212.8 nm one sees that it is the Amide II (C-N $_{st}$ and N-H $_{ipb}$) and the Amide III (C-N $_{st}$, N-H $_{ipb}$, C-CH$_3$ $_{st}$) that are most active. The effects of hydrogen bonding on the photoexcitation process are seen in comparing the gas phase spectra with the liquid phase spectra.

322. MOLECULAR-DYNAMICS SIMULATIONS OF RIGID-ROD AND HYDROXY-SUBSTITUTED RIGID-ROD POLYMERS. S. Trohalaki, AdTech Systems Research, Inc., 1342 N. Fairfield Rd., Dayton, OH 45432 and D.S. Dudis, Materials Laboratory, Wright Patterson Air Force Base, Dayton, OH 45433

The beneficial properties of the rigid-rod polymer poly(p-phenylene benzobisthiazole), PBT, such as thermo-oxidative and environmental stability, and excellent mechanical characteristics, are due to the inherent molecular modulus as well as the degree to which individual rod-like chains are aligned. Previous molecular-dynamics siulations (B.L. Farmer, unpublished results) show that PBT is surprisingly flexible despite its rod-like topology. Improved molecular orientation and, hence, physical properties may be effected by chemical substitutions that serve to stiffen the PBT chain; hydroxy substitutions on the two ortho positions of the phenyl ring adjacent to the nitrogens in the heterocycle would allow hydrogen bonding resulting in a "pseudo-ladder" chain, dihydroxy PBT (DPBT). Molecular-dynamics simulations performed on PBT and DPBT oligomers consisting of 8 repeat units enable comparison of conformational flexibility. Molecular-mechanics and molecular-dynamics calculations were performed using CHARMm through the QUANTA molecular modeling package offered by Polygen Corp., Waltham MA.

323. VARIATIONAL FORMULATION IN NETWORK THERMODYNAMICS.
Auster Valderrama-Cano
Depto. de Física y Química teórica, Div. Estudios de Posgrado, Facultad de Química, UNAM. Circuito Escolar, C.U., México, D.F. México.

The variational formulation of linear irreversible thermodynamics is used to obtain a Lagrangian representation. Under an adequate division of a complex system, fixed boundary conditions and a "coarse graining" definition of phenomenological coefficients, it is possible to derive pseudo-first order dynamic equations for such kind of systems. With that scheme and this model, the capacitive and constitutive equations of network thermodynamics are derived from an "action" through a variational principle. This variational formulation become useful to design an approximate method, of the kind of Ritz, to solve the dynamic equations of network thermodynamics. Some examples are discussed.

324. BARRIERS TO ROTATION IN OXIMES: A MOLECULAR ORBITAL STUDY. T.J. Venanzi, C. Ollendorf, A. Esslinger, Department of Chemistry, College of New Rochelle, New Rochelle, NY 10805. C.A. Venanzi, Chemistry Division, New Jersey Institute of Technology, University Heights, Newark, NJ 07102.

The barriers of rotation about the N-O bond in formaldoxime and (E)-propenaldoxime and about the C-C bond in (E)-propenaldoxime were calculated. 6-31G* and MP2/6-31G* optimized geometries were determined for the various structures. High level MP4/6-31G* and MP3/6-311++G** ab initio calculations at the optimized geometries as well as semi-empirical PM3 calculations indicate that the s-trans rotamer of formaldoxime is energetically preferred to the s-cis rotamer by 5-6.5 kcal/mol. The same result was found for rotation about the N-O bond in (E)-propenaldoxime. Similiar high level calculations for rotations around the C-C bond in (E)-propenaldoxime indicate the possibility of a gauche rotamer as the second stable rotamer of this system. This work was supported by a grant from the National Science Foundation.

325. PHOTOELIMINATION OF H_2 FROM 1,3-BUTADIENE. Bhawani Venkataraman, Jingyu Huang, James J. Valentini, Department of Chemistry, Columbia University, New York, New York 10027.

The product state distribution of H_2, formed by the photolysis of 1,3 - Butadiene at 212.8 nm., has been measured by 1 + 1 resonance enhanced multi-photon ionization (REMPI). Tunable vacuum ultraviolet radiation (~ 97 - 102 nm.), generated by frequency tripling the doubled output of a pulsed dye laser (~ 300 nm.), was used to excite the product H_2 from the ground $X\ ^1\Sigma_g^+$ to the $C\ ^1\Pi_u$ excited state. The 300 nm. radiation was used to ionize the H_2 from the $C\ ^1\Pi_u$ state to the ionization continuum. The H_2^+ was detected by time-of-flight mass spectroscopy. The rovibrational populations extracted from the REMPI spectra indicate a vibrationally excited but rotationally cold product. Prior calculations predict a colder vibrational distribution than the experimentally measured distributions. A vibrationally hot but rotationally cold product seems to indicate a "tight", symmetric transition state.

326. AB INITIO AND SEMIEMPIRICAL MOLECULAR ORBITAL STUDIES OF HYDROXAMIC ACIDS AND THEIR CONJUGATE BASES, Oscar N. Ventura, Laszlo Turi, and J.J. Dannenberg, Catedra de Quimica Cuantica, Facultad de Quimica, Universidad de la Republica, 11800 Montevideo, Uruguay, and Department of Chemistry, City University of New York, Hunter College and The Graduate School, 695 Park Avenue, New York, NY 10021.

The various tautomers of hydroxamic acid and its conjugate base have been studied in several conformations using both semiempirical (AM1 and PM3) and ab initio methods. Complete geometrical optimizations were carried out both at the Hartree-Fock and MP2 levels with a variety of basis sets. There was considerable discrepancies among the methods used. The HF calculations were generally found to be unreliab e as they predicted the Z-enolic form to be too unstable. MP2 optimization and increasing the basis set seemed to lower the relative energy of this form (which is predicted to be most stable using AM1). HF caluclations on the conjugate bases likewise proved to be unreliable.
The effects of substituents were studied primarily using the semiempirical methods. The effects of substituents on the relative energies of the various conformations and upon the acidities of the substituted acids will be discussed.

327. MOLECULAR ORBITALS OF METALLOPORPHYRINS: AN UNDERGRADUATE EXPERIMENT IN SPECTROSCOPY. Valerie A. Walters, Department of Chemistry, Lafayette College, 18042

The relative intensities of the α and β (visible) bands in the electronic absorption spectrum of metalloporphyrins is used as an indication of the energy difference between the a_{1u} and a_{2u} molecular orbitals. According to the Gouterman four-orbital model of porphyrins, these nearly degenerate orbitals and the higher energy e_g orbitals are important in forming the lower energy excited electronic states. In this laboratory experiment, the energy of the a_{2u} orbital relative to the a_{1u} orbital is altered by 1) substitution of metals of varying electronegativity, and 2) addition of electron-donating ligands. Students obtain absorption spectra of a series of metallated tetraphenylporphyrins in several ligating solvents to determine whether the a_{1u} or a_{2u} orbital of a given porphyrin is higher in energy, and to obtain an approximate quantitative measure of the energy difference. (A literature reference upon which the experiment is largely based is Spellane, P.J.; Gouterman, M.; Antipas, A.; Kim, S.; Liu, Y.C. Inorg. Chem. 1980, 19, 386-391.)

328. TWO-PHOTON ABSORPTION SPECTROSCOPY OF PHENANTHRENE. Valerie A. Walters, Kris Brubaker, and Mark Ledeboer, Department of Chemistry, Lafayette College, Easton, PA 18042

This physical chemistry laboratory experiment compares the two-photon fluorescence excitation spectrum of phenanthrene with the one-photon UV-Vis absorption spectrum. The two-photon spectrum is acquired with a nitrogen-pumped dye laser system. A series of vibrational bands between 28000-33000 cm^{-1} can be observed in both spectra, suggesting to the student that the two techniques can sometimes yield the same information, depending upon selection rules. A second series of bands appears in the two-photon spectrum between 37000-42000 cm^{-1}. In the UV-Vis spectrum, this region is completely dominated by other strong one-photon absorption bands. This illustrates that the two types of spectra can also provide complementary information. Students will also be asked to obtain the dispersed fluorescence spectrum. This is readily obtained with the nitrogen laser wavelength of 337 nm, which happens to coincide with one-photon absorption into the lower energy electronic state. The dispersed fluorescence spectrum yields the frequencies of the vibrational bands in the ground electronic state, which can be compared with excited state frequencies. This project was funded by the Pew Foundation and Lafayette College.

329. ENHANCEMENT OF CRITICAL CURRENT DENSITY OF $Bi_2Sr_2CaCu_2O_8$ SUPERCONDUCTOR BY METAL SUBSTITUTION. Yue Li Wang, Xian Liang Wu, Chia Chun Chen and Charles M. Lieber, Department of Chemistry, Columbia University, New York, N.Y. 10027

The effects of metal substitution on the superconducting transition temperature and critical current density in $Bi_2Sr_2CaCu_2O_8$ (2212) single crystal has been systematically studied. Pb-substitution enhances critical current density over an order of magnitude and increases the flux-lattice melting temperature. Structural studies further show that Pb-substitution causes disorder in the Bi-O layer of $Bi_2Sr_2CaCu_2O_8$. This disorder is proposed to increase the critical current density through enhanced pinning. In addition, studies of other metal substituted materials will be discussed.

330. PHOTOIONIZATION VIA TRANSIENT STATES: THE PHOTOPHYSICS OF AZULENE. Peter M. Weber and N. Thantu, Department of Chemistry, Brown University, Providence, RI 02912

The photophysics of electronic states with ultrashort lifetimes can be studied by the photoelectron spectrum arising from two photon ionization, if the one photon wavelength is tuned to the electronic transition frequency. Electronic relaxation processes on the timescale of the laser pulse duration are exposed by photoelectron peaks that are shifted from the origin transition. The potential of the technique is illustrated by studies of Azulene ionized via the S_2, S_3, and S_4 electronic states.

331. KINETICS OF $NF(a^1\Delta)$ AND $NF(X^3\Sigma)$
B. H. Weiller, R. F. Heidner, J. S. Holloway and J. B. Koffend
Chemistry and Spectroscopy Department
The Aerospace Corporation, P.O Box 92957, Los Angeles, CA 90009

The kinetics of NF(a) and NF(X) play a central role in any chemical laser system that exploits the long-lived NF(a) excited state. Of particular interest is the possibility of a chemical laser based on the NF(b→X) transition pumped by energy pooling of NF(a) with an excited partner. In this work we have conducted two sets of experiments, one to measure the rate constants for reaction of NF(X) with NF_2, NF_3 and NF(X) by laser induced fluorescence and the other to measure the rate constants for quenching of NF(a) by NF_3, O_2 and N_2F_4. Both NF(a) and NF(X) are formed by photolysis of NF_2 in a heated flow cell. Our experimental results will be discussed in light of recent literature and also in the context of the potential NF(b→X) system.

332. INFRARED LASER SPECTROSCOPY OF THE C_2H RADICAL

W-B. YAN AND T. AMANO

A $^2\Pi - {}^2\Sigma$ band of gas phase C_2H centered at 2928 cm^{-1} is observed using a high resolution difference-frequency laser spectrometer. The C_2H molecules were produced in a hollow cathode discharge in a mixture of C_2H_2 ($\sim$ 15 mTorr), H_2 ($\sim$ 0-150 mTorr), and He ($\sim$ 700 mTorr). Measurements were carried out using discharge modulation method at 3 kHz with a current of about 800 mA. Total laser pathlength was 36 m in a White cell arrangement. We have tentatively assigned this band to be the $3\nu_2 + \nu_3$ combination band of C_2H. In addition, two $^2\Pi - {}^2\Pi$ transitions of C_2H connecting known states were also recorded. The common lower level of the two bands is the $\tilde{X}$ (0 1 0) state, with the upper levels being the same as those of the known 3693 and 3600 cm^{-1} bands, respectively.

Address: Herzberg Institute of Astrophysics, National Research Council, Ottawa, Canada K1A 0R6

333. TWO-PHOTON DISSOCIATION OF FERROCENE AND NICKELOCENE AT 248 NM: STATISTICAL VS. NON-STATISTICAL BEHAVIOR, Rong Zhang and Robert L. Jackson, IBM Research Division, Almaden Research Center, 650 Harry Rd., San Jose, California 95120-6099

The gas-phase photodissociation of ferrocene and nickelocene has been studied at 248 nm (KrF* laser). A Nd:YAG-pumped dye laser was used to detect the metal atom and cyclopentadienyl radical (Cp) photoproducts by laser-induced fluorescence (LIF). For both ferrocene and nickelocene, dissociation occurs by a two-photon process. No single-photon chemistry was observed. In experiments conducted in the presence of He buffer gas, the relative yield of Cp formed from ferrocene was was found to be independent of He pressure, while the relative yield of Cp formed from nickelocene decays rapidly with increasing He pressure. The quenching of Cp formation observed upon photodissociation of nickelocene can be fit quantitatively using a time-dependent master equation formulation, where the dissociation of nickelocene is described by RRKM theory. This suggests that dissociation of nickelocene occurs via the ground state potential surface, following internal conversion from an electronically excited state. The lack of quenching observed upon photodissociation of ferrocene indicates that this species dissociates by a non-statistical process from an electronically excited state. The difference in the dissociation pathways for ferrocene and nickelocene is explained in terms of the symmetries of their respective excited states.

334. Quantum State Resolved N_2O Vibrational, Rotational and Translational Energy Transfer from Highly Vibrationally Excited NO_2 Donor. Liedong Zheng, James Z. Chou, George W. Flynn Department of Chemistry, Columbia University, New York, NY.

Quenching of highly vibrationally excited NO_2 by N_2O molecules has been studied. The ground state, first asymmetric stretch (n_3), and first symmetric stretch (n_1) vibrational states of N_2O were probed by CW IR diode laser rediation. The average amount of energy transferred per collision to the n_1 and n_3 states is 14±5 cm^{-1}, and 20±6 cm^{-1} respectively, with <1 cm^{-1}/collision of rotational and translational excitation. A relatively large amount of energy, 700±300 cm^{-1} per collision, was deposited into the translational degrees of freedom of the ground state.

335. ULTRAFAST MEASUREMENTS OF ENERGY FLOW AND STRUCTURAL DYNAMICS IN PHOTOEXCITED HEME PROTEINS, Huiping Zhu, Xiaobing Zhu, Robert Lingle Jr., and J. B. Hopkins, Department of Chemistry, Louisiana State University, Baton Rouge, Louisiana 70803.

A new method is demonstrated for distinguishing between energy relaxation and structural dynamics in photoexcited molecules in solution. The vibrational dynamics of photoexcited heme proteins is studied using the technique of two-color pump-probe ultrafast transient Raman spectroscopy. Analysis of Stokes and anti-Stokes Raman bands is used to simultaneously measure vibrational energy relaxation (vibrational cooling) and structural dynamics of the heme. The 1/e rate constant for cooling is 2-5ps in deoxyhemoglobin. Environmental effects on the vibrational cooling rate have been investigated. The results suggest a new interpretation for the mechanism of energy exchange between the heme and the protein. The ultrafast vibrational dynamics of photodissociation and geminate recombination of oxy and carboxyhemoglobin will also be presented.

336. A LOW COST MULTIPLE IMPACT PCHEM EXPERIMENT. Theresa Julia Zielinski, James Cooper, and Algis Ilgunas, Department of Chemistry, Niagara University, Niagara University, NY 14109

The rising cost of instrumentation makes it difficult to create new interesting and thought provoking experiments in PChem lab. To overcome this difficulty an experiment was designed around the departmental PC, a thermistor, an RS232 data device, a spreadsheet program, the data acquisition program provided with the RS232 device (student modified), a shareware curve fitting program and a standard polynomial fit program. Students learn the following concepts: the properties of a thermistor, computer data acquisition, Ascii file handling on a PC, transporting files into different types of software, multistep data analysis, nonlinear curve fitting, and calibration of a nonlinear device. The calibrated thermistor is later used to automatically record temperatures as a function of time for a two component phase diagram experiment where the RS232 device records both temperature and time. The consistent use of spreadsheet techniques allows rapid preparation of all graphs. A six pen color plotter is used for preparation of graphs for all lab reports.

337. AB INITIO STUDY OF K ADSORPTION ON THE Ag(100) SURFACE. Sanjukta Gayen and Walter C. Ermler, Department of Chemistry and Chemical Engineering, Stevens Institute of Technology, Hoboken, NJ 07030.

Ab initio Hartree-Fock calculations in the context of relativistic effective core potentials are reported for the chemisorption of potassium on silver. The Ag surface is modelled using a three-layer 14 atom cluster. Geometry optimization is carried out in four degrees of freedom. Mulliken population analyses indicate that there is no complete charge transfer from the adsorbate to the substrate, consistent with photoemission studies of Riffe et al.[1]

[1]D. M. Riffe, G. K. Wertheim and P. H. Citrin, Phys. Rev. Lett. __64__, 571 (1990).

338. AB INITIO FULL SOCVCI CALCULATIONS OF THE ELECTRONIC SPECTRA OF LiBe AND LiMg. M. M. Marino, Department of Chemistry, The Ohio The University, Columbus, OH 43210, W. C. Ermler, Department of Chemistry and Chemical Engineering, Stevens Institute of Technology, Hoboken, NJ 07030, C. W. Kern, Department of Chemistry, The Ohio State University, Columbus, OH 43210, and V. E. Bondybey, Department of Chemistry, The Ohio State University, Columbus, OH 43210 and Institut fuer Physikalische Chemie der T. U. Muenchen, 8046 Garching, Germany.

Ab initio full configuration interaction calculations including spin-orbit coupling and core-valence polarization effects (SOCVCI) are reported for the LiBe and LiMg metal dimers. Relativistic effective core potentials are used to incorporate SO and CV effects and the electronic states are defined in terms of their proper double group irreducible representations. Excitation energies, binding energies and spectroscopic constants for valence and Rydberg states are compared to results due to laser vaporization experiments.

339. COMPARISON OF EXPERIMENTAL AND THEORETICAL HYPERPOLARIZABILITIES, WITH EMPHASIS ON HF. John F. Ward, Randall Laboratory of Physics, University of Michigan, Ann Arbor, MI 48109-1120.

Nonlinear electric polarizabilities (hyperpolarizabilities) of atoms and small molecules have been measured over the last 30 years using the Kerr effect, optical third harmonic generation, and dc electric field induced second harmonic generation (dcSHG). Theoretical values have been computed by methods ranging from empirical to ab initio. HF constitutes an interesting test case where both second and third order polarizabilities have been measured to within 10% using dcSHG, and computed by Bartlett and Purvis using a sophisticated ab initio method (CHF SDQ-MBPT[4]). Experimental corrections and uncertainties will be critically discussed along with several other factors relevant to a comparison between experiment and theory. However, the theoretical results are about a factor of 2 smaller than the experimental data, and none of the factors considered seems to offer a resolution of this discrepancy.

340. CALCULATIONS OF STATIC HYPERPOLARIZABILITIES AND POLARIZABILITIES.
Ajit J. Thakkar, Department of Chemistry, University of New Brunswick, Fredericton, NB Canada E3B 6E2

It is still not easy to calculate reliable values of electric properties, such as static polarizabilities and hyperpolarizabilities, for molecules with more than a few electrons. One major problem is the difficulty of using large enough orbital basis sets, particularly in conjunction with a method that incorporates an adequate amount of electron correlation. Vibrational and rotational effects can also be important. Comparison with experiment is further clouded by the problems of removing the effects of intermolecular forces, and ensuring that the measured data have been reliably extrapolated to zero frequency. A discussion of these problems will be presented with particular reference to molecules which we have tackled in my group. The molecules discussed will include N_2, $I2$, CO_2 and linear polyenes.

341. FREQUENCY DEPENDENT AND CORRELATED HYPERPOLARIZABILITIES OF MOLECULES. H. Sekino and R.J. Bartlett, University of Florida, Quantum Theory Project, 362 Williamson Hall, Gainesville, FL 32611-2085

Frequency dependent hyperpolarizabilities of small molecules are investigated by ab-initio molecular orbital theory. The correlation effects are evaluated at the zero-frequency limit using the "relaxed density formalism" of Many-Body Perturbation Theory (MBPT) and Coupled Cluster (CC) Methods. The dispersion effects at the experimental frequency are estimated at the Hartree Fock level using general, high order Time Dependent Hartree Fock (TDHF) Theory. Results are reported for a series of small molecules using large basis sets and compared with experimental values.

342. FREQUENCY DEPENDENT HYPERPOLARIZABILITIES. Julia E. Rice, IBM Research Division, IBM Almaden Research Center, 650 Harry Road, San Jose, CA 95120; Nicholas C. Handy, University Chemical Laboratory, Lensfield Rd., Cambridge CB2 1EW, U.K.

This talk will concentrate on the evaluation of frequency dependent hyperpolarizabilities which allow for more direct comparison with experimental measurements. A method will be presented for the determination of frequency dependent hyperpolarizabilities using non-variational correlated wavefunctions, with details for second-order perturbation theory (MP2). Results will be presented for a variety of small molecules for the optical Kerr effect and for second harmonic generation, using a combination of static hyperpolarizabilities determined at high levels of theory in conjunction with an MP2 frequency dependent correction. The problems associated with the determination of accurate hyperpolarizabilities for the larger conjugated organic systems which are of relevance for non-linear optical devices will also be discussed.

343. NONLINEAR OPTICAL MEASUREMENTS OF MOLECULAR HYPERPOLARIZABILITIES. David P. Shelton, Physics Department, University of Nevada, Las Vegas, NV 89154.

The frequency dependence of the hyperpolarizability is an important consideration in the selection or design of molecular nonlinear optical materials for applications. The present state of gas phase measurements of the frequency dependence of the hyperpolarizabilities of atoms and small molecules will be summarized, and comparisons will be made with corresponding ab initio results. In condensed phases the interaction of a molecule with its neighbours may significantly alter the hyperpolarizability of the molecule. Measurements of the pair-hyperpolarizabilities of interacting atoms and molecules will be presented.

344. VIBRATIONAL ENERGY FLOW AT SOLID SURFACES. A. L. Harris, AT&T Bell Laboratories, Murray Hill, NJ 07974

Energy flow from the excited vibrational levels of molecules adsorbed at solid surfaces is followed with picosecond time resolution and submonolayer sensitivity using infrared-visible sum frequency generation (SFG) to monitor vibrational level populations. This technique can follow the pathway for vibrational energy relaxation at surfaces and distinguish among relaxation mechanisms. This talk will briefly review energy relaxation measurements on metal surfaces, where coupling to electron-hole pair excitations of the metal has received a great deal of attention as a relaxation mechanism. Recent vibrational energy relaxation measurements on stepped semiconductor surfaces will also be presented, showing that adsorbate-adsorbate energy transfer combined with energy relaxation at step edges can alter the relaxation dynamics of an entire adsorbate layer.

345. ELEMENTARY SURFACE TRANSPORT PROCESSES INVOLVING MOLECULAR CHEMISORBATES. J.E. Reutt-Robey, Department of Chemistry, University of Maryland, College Park, Maryland 20742.

Surface infrared spectroscopy is a sensitive probe of surface chemical species at the submonolayer level. In combination with fast pulsed molecular beam dosing, this macroscopic technique (MB-SIRS) is here used to provide microscopic information on molecular surface mobilities. The microscopic length scale of the measurement, variable from 10-100 A, is imparted from the periodically stepped crystal surfaces employed for the measurements. This talk will focus on recent applications of MB-SIRS methods to elementary transport processes involving CO and NO on stepped surfaces vicinal to Ni(100). Experiments which vary the initial dosing conditions and record the time evolution of the terrace and step site populations of chemisorbed species are described. Such investigations are used to quantify the contributions from the chemisorption event, diffusion along the terrace, and site-exchange processes involving chemisorbates at step sites in arriving at equilibrium surface distributions.

346. **STUDIES OF DYNAMICS AT CATALYST SURFACES BY SELECTIVE EXCITATION OF NUCLEAR MAGNETIC RESONANCES.**
T. M. Duncan, School of Chemical Engineering, Cornell University, Ithaca NY 14853, T. W. Root, Department of Chemical Engineering, University of Wisconsin, Madison, Wisconsin 53706, and A. M. Thayer, C&E News, Northeast News Bureau, Iselin NJ

Nuclear magnetic resonances with frequency shifts large relative to inhomogeneous broadening may be selectively excited and as such the targets of the 'hole burning' are labeled by their non-equilibrium polarization. Species at surfaces may be labeled on the basis of any property that causes a frequency shift in the magnetic resonance - chemical identity, spatial orientation, or proximity to heteronuclei. The labeling process requires 0.01 - 1 msec and thus is fast compared to conventional labeling with nuclear isotopes. Also, the labeling does not perturb the system chemically. An application of this method will be demonstrated with studies of H_2 and CO adsorbed on dispersed transition metals.

Work performed at AT&T Bell Laboratories, Murray Hill NJ

347. NMR STUDIES OF SIMPLE MOLECULES ON METAL SURFACES. C. A. Klug, C. P. Slichter Department of Physics, University of Illinois, 1110 W. Green Street, Urbana, IL., 61801, J. H. Sinfelt, Exxon Research & Engineering, Clinton Township, Route 22 East, Annandale, NJ 08801

Many researchers have applied a variety of experimental techniques to study molecules adsorbed on metal surfaces. Often this involves the use of single crystals under conditions of low coverage and gas pressure. Our group has used NMR to study the simple hydrocarbons acetylene and ethylene adsorbed on the surfaces of small supported Group VIII metal clusters typical of industrial catalysts. The use of highly disperse catalysts confined in a small volume allows one to study surface species under conditions of high coverage and gas pressure. This talk will introduce the NMR techniques employed and review some of our results, focussing on recent work studying surface reactions. The reactions studied include: the conversion of the low temperature species of ethylene to ethylidyne, hydrogen-deuterium exchange in ethylidyne, the reaction of vinylidene with hydrogen to form ethylidyne, and the dehydrogenation of vinylidene.

348. ONE- AND TWO-DIMENSIONAL HIGH RESOLUTION SOLID STATE NMR INVESTIGATIONS OF ZEOLITE LATTICE STRUCTURES. Colin A. Fyfe, Hiltrud Grondey, Hermann Gies, Yi Feng, George Kokotailo, Harald Strobl, Department of Chemistry, University of British Columbia, CANADA V6T 1Y6

The results of high-resolution ^{29}Si MAS NMR investigations of highly crystalline, highly siliceous zeolites will be described and compared to the data from diffraction studies of these systems. In general, sharp resonances are observed whose numbers and relative intensities reflect directly the numbers and occupancies of crystallographically inequivalent T-sites in the unit cell. These spectra are very sensitive to changes in local structure and environment. They may be used to determine space-group symmetry and to detect and characterize structural changes induced by temperature and sorbed organic molecules.
Combining the MAS experiment with 2D NMR techniques such as COSY and INADEQUATE makes it possible to deduce the three-dimensional connectivities in the lattice. Examples of the application of these experiments to typical zeolite frameworks will be presented and related to diffraction studies.

349. REACTIONS OF GROUND STATE AND ELECTRONICALLY EXCITED BARIUM ATOMS. H.F. Davis, A.G. Suits, H. Hou, and Y.T. Lee, Department of Chemistry, University of California, and Chemical Sciences Division, Lawrence Berkeley Laboratory Berkeley, CA 94720 USA.

By studying reactions of ground state and electronically excited Barium atoms using crossed supersonic beams, we can explore the role of Barium's second valence electron in chemical reactivity. The two general classes of reactions seen for Barium will be illustrated. As in the case of monovalent alkali atoms, reactions of Ba with NO_2, O_3, and ClO_2 are initiated by long range electron transfer. But due to the divalent character of Barium, the ensuing chemistry is considerably richer; two chemically distinct reaction channels are observed for each molecule. The angular and velocity distributions of these products will be discussed. Reactions with water or alcohols, on the other hand, can only result from a close collision since long range electron transfer is not possible. The effect of the initial Ba electronic state and the importance of H atom migration in these reactions will be discussed.

350. EFFECTS OF ELECTRONIC EXCITATION AND DIMERIZATION OF METALS ON PRODUCT STATE DISTRIBUTIONS. J.M. Parson, B.S. Cheong, R.P. Kampf, and M.D. Oberlander. Department of Chemistry, Ohio State University, Columbus, OH 43210.

Enhancement of electronic chemiluminescence by the introduction of dimers or metastable states of main-group metals in molecular beam reactions with halogen and oxygen containing molecules is surveyed. Examples are chosen from metals in groups 2, 14, and 15. The importance of symmetry and collisional lifetime in controlling energy disposal into product vibrational and electronic degrees of freedom is discussed. Recent experimental results are reported on alkaline earth atom reactions with hydrogen peroxide in which metastable atoms form exclusively metal hydroxide products, unlike ground state atoms which favor metal monoxide products. Chemiluminescent mechanisms involving reactions of arsenic and bismuth dimers with fluorine are also described.

351. HALF-COLLISION STUDIES OF THE INTERACTIONS OF ELECTRONICALLY EXCITED CADMIUM ATOMS WITH H_2 AND CH_4. I. Wallace, D. Funk, and W.H. Breckenridge, Department of Chemistry, University of Utah, Salt Lake City, Utah 84112.

Weakly bound van der Waals complexes of Cd atoms with H_2 and with CH_4 have been prepared by mixing cadmium vapor, carrier gas and either H_2 or CH_4 at the exit orifice of a supersonic nozzle. Information about the ground states as well as electronically excited states correlating with excited Cd(5s5p) atoms has been obtained by "action" spectroscopy, whereby atomic predissociation products are detected with one laser pulse while an excitation laser pulse is scanned in frequency through the absorption spectra of the complexes. While the Cd·CH_4 spectra are consistent with only van der Waals interactions in the excited states, the Cd-H_2 (Cd-D_2) spectra indicate chemically attractive interactions of Cd(5s5p$^1P_1$) atoms with the H-H bond.

352. STUDIES OF SOME GAS-PHASE METAL OXIDATION REACTIONS BY PHOTOELECTRON AND CHEMIELECTRON SPECTROSCOPY. J.M. Dyke, Department of Chemistry, The University, Southampton SO9 5NH, U.K.

In order to provide a reference against which the results of metal plus oxidant chemielectron spectra can be compared, chemielectron spectra have been recorded for some chemiionization reactions occurring in low pressure O atom/hydrocarbon gas-phase systems. Once these spectra have been briefly discussed and the method of analysis of the spectra outlined, a number of metal plus oxidant chemiionization reactions will be considered in some detail. The main oxidants used are $O_2(X^3\Sigma_g^-)$, $O_2(a^1\Delta g)$ and $O(^3P)$ and the metals under study include V, Ba, Sr and the lanthanides.

To complement these chemiionization studies, the study of a number of metal plus oxidant reactions with u.v. photoelectron spectroscopy will be described. The reactions M + N_2O and M + O_3 (where M=Li, Na or K) will be considered as examples to illustrate the information to be obtained from such studies.

353. GAS PHASE ALUMINUM ATOMS FROM PULSED LASER ABLATION: MECHANISMS AND APPLICATIONS, Brad R. Weiner, Hongxin Wang, and Aldo P. Salzberg, Department of Chemistry, University of Puerto Rico, Río Piedras, PR 00931.

Pulsed laser ablation of metal targets is a powerful and general method for the production of gas phase metal atoms. We report our results on the excimer laser ablation of aluminum targets at 193, 248 and 351 nm. Velocity distributions of the ablated Al atoms have been measured directly by monitoring their flight times from the target to the probe laser beam, in order to characterize the ablation mechanism. The velocity distributions resulting from the three wavelengths can be characterized as hyperthermal. Possible mechanisms for the production of these hyperthermal velocities will be discussed. The atoms have also been used as an atomic Al source for chemical studies, and these results will be presented.

354.
MOLECULAR DYNAMICS SIMULATIONS OF DIAMOND FILMS: STRUCTURE, GROWTH AND PROPERTIES. Donald W. Brenner, Code 6119, Naval Research Laboratory, Washington, DC 20375-5000.

Despite the tremendous technological advances made in the chemical vapor deposition (CVD) of diamond films, the underlying chemical and physical processes that allow diamond to be grown under apparently metastable conditions remain controversial. To explore the atomic-scale chemistry of the CVD of diamond films, we are using a many-body classical potential energy expression for reacting hydrocarbon molecules in molecular dynamics simulations of diamond films. This potential energy expression will be reviewed, and predictions of the energetics of surface reconstructions and molecular adsorption on diamond surfaces and the dynamics of atomic hydrogen and gas-phase molecules reacting with these surfaces will be presented. The relevance of these results to current models of CVD growth mechanisms will be discussed.

355.
CHEMISTRY WITH CONSTRAINTS: LOCAL DENSITY CALCULATIONS OF GROWTH REACTIONS ON THE DIAMOND 100 SURFACE. Koblar A. Jackson, Dept. of Physics, George Mason University, Fairfax, Virginia 22030

The chemistry of diamond film growth appears to be surprisingly similiar to that of gas phase hydrocarbon molecules, and empirical growth models based on gas phase data give a reasonable overall description of diamond film growth rates. Despite the similarity, however, there are important differences between the surface and the gas phase, mainly involving constraints to atomic relaxation and steric interference on the surface. Such effects can be expected to play a crucial role in reactions occurring at the surface, but are difficult to characterize experimentally. In this talk we discuss recent first-principles, quantum mechanical calculations based on the Local Density Approximation (LDA), which are being used to study diamond growth. The calculations are carried out on large cluster models of the diamond surface, and attempt to isolate specific reactions involved in current growth models to provide first-principles data for input into empirical modelling schemes. Topics to be covered include steric effects on the (100) surface and the role of absorbed hydrogen in the growth process.

356. **SEMIEMPIRICAL MOLECULAR ORBITAL AND TIGHT BINDING METHODS IN DIAMOND SURFACE REACTIVITY.** Steven M. Valone, Materials Science and Technology Division, MST-7, MS E549, Los Alamos National Laboratory, Los Alamos, NM 87545.

Bulk diamond, diamond surface morphology, and adsorbate chemistry have all been studied by a variety of semiempirical molecular orbital (MO) and tight binding (TB) methods with varying degrees of success. Based on some relative success and extensive work in related hydrocarbon systems, it was generally assumed that semiempirical MO and TB methods would convey a qualitatively correct picture of elementary reactions occurring at diamond surfaces. To assess their accuracy, several of these methods are benchmarked against the CH_4 + H to CH_3 + H_2 reaction. Generally, it has been found that, while equilibrium geometric properties of these systems appear reasonable enough, energetic and transition state properties lack chemical accuracy.

357. **MODELING DIAMOND GROWTH: FROM ELEMENTARY MECHANISMS TO REACTOR MODELS*.** M. E. Coltrin, Sandia National Laboratories, Albuquerque, NM, 87185; D. S. Dandy, E. Meeks, and R. J. Kee, Sandia National Laboratories, Livermore, CA 94551-0969

We describe a detailed numerical model of diamond growth. A formalism for treating elementary surface reaction mechanisms is developed and applied to diamond CVD. Our model includes adsorption of the growth species C, CH_3, and C_2H_2 and subsequent surface reactions. Development of such detailed reaction models is limited by experimental and theoretical information on reaction pathways, energetics, and thermochemistry. We discuss the range of data needed to develop a complete kinetic model for this system. The surface reaction mechanism and a gas-phase reaction mechanism taken from the combustion literature are combined into a complex model of the coupled chemical kinetics and convective and diffusive transport of mass and energy in a stagnation-point (pedestal) reactor.

*This work was performed at Sandia National Laboratories, supported by the U.S. Department of Energy under Contract No. DE-AC04-76DP00789.

358. **THEORY OF SCHOTTKY BARRIERS IN DIAMOND/METAL INTERFACES.** Steven C. Erwin, Dept. of Physics, University of Pennsylvania, Philadelphia PA 19104.

Although Schottky barriers were discovered nearly a century ago, many aspects of their behavior remain unexplained. Early models held that the Schottky-barrier height (SBH) depended linearly on the metal work function; this was contradicted by many experiments, which—for narrow gap semiconductors—showed almost no dependence on the metal species. Modern theories invoke a variety of Fermi-level pinning mechanisms, which fix the SBH at a value insensitive to both the metal species and the detailed interface geometry. This too is at odds with recent experiments, which show a 20-30% variation in SBH with interface orientation for Si/metal. We present recent results of theoretical studies on ideal diamond/metal interfaces which suggest that different interface orientations can, in principle, result in variations in the SBH of essentially 100% (barrier height changes of nearly 1 eV). The mechanism responsible for this is analogous to the mechanism that drives Group IV surface reconstructions, and may be observable in a wide variety of Group IV/metal interfaces.

359. **ABSORBED HYDROGEN AND FLUORINE NEAR THE DIAMOND/VAPOR INTERFACE.** Mark R. Pederson, Complex Systems Theory Branch, Naval Research Laboratory, Washington, DC 20375-5000: and Koblar A. Jackson, Dept. of Physics, George Mason University, Fairfax, VA 22030.

To identify quantitative and qualitative differences between diamond vapor deposition in hydrogenated and halogenated environments we have performed several similar local-density-based simulations on surfaces with extra adsorbed or absorbed passivators. Potential energy curves, reaction barriers and cohesive energies for extra atoms and molecules on the surface and in nearby voids will be presented. Differences between the halogenated and hydrogenated <111> results will be discussed. A comparison between the hydrogenated <111> and <100> surface will be presented also.

360. LOW TEMPERATURE STM IMAGING OF METAL SURFACES AND MANIPULATION OF ADSORBATES. P. Zeppenfeld*, D. M. Eigler, and C. P. Lutz, IBM Almaden Research Center, 650 Harry Road, San Jose, CA 95120-6099.

We report on recent progress in the imaging of close packed metal surfaces and and the manipulation of adsorbates using an STM operated at 4 Kelvin. We have accurately measured the atomic corrugation of the close-packed Pt(111) surface and have studied the dependence of the corrugation on the tunneling parameters and the state of the tip. We find that the corrugation amplitudes measured with different tips vary by as much as a factor of 100 under otherwise identical tunneling conditions. We discuss our findings in the light of recent theories of the STM imaging mechanism.
We have also used the STM to position single adsorbed atoms and molecules with atomic scale precision. The process for positioning adsorbates is characterized by a sharp threshold tip height below which the adsorbate will follow the movement of the tip and slide along the surface. We have successfully used this process to position xenon atoms on Ni(110) as well as CO molecules and single Pt atoms on Pt(111). Over a wide range of experimental conditions the threshold tip height does not depend on the sign or magnitude of the tunneling current, electric field, or tip bias voltage. We therefore conclude that the force responsible for sliding the adsorbate is due to the interatomic potential between the frontmost tip atoms and the adsorbate rather than an effect of the applied electric field or tunneling current.

*permanent address: Institut für Grenzflächenforschung und Vakuumphysik, KFA Forschungszentrum Jülich, Postfach 1913, 5170 Jülich, FRG

361. STM INDUCED ETCHING OF CONDUCTING OXIDES. Donna Saulys, Gregory Rudd, David Novak, and Eric Garfunkel, Department of Chemistry, and Laboratory for Surface Modification, Rutgers-the State University of New Jersey, P.O. Box 939, Piscataway, NJ 08855-0939.

Several quasi-low-dimensional conducting molybdenum oxide surfaces have been studied by scanning tunneling microscopy (STM). Periodic features imaged on the cleavage plane of these materials can be related to MoO_n octahedra or tetrahedra. An interesting superlattice with a repeat distance of 3-4 times larger than any periodicity in the bulk is observed on the $Na_{0.9}Mo_6O_{17}$ surface. Possible explanations include an ordering of surface alkali ions, a surface reconstruction, or surface charge density waves (CDW).

Although these oxides are relatively stable in ambient conditions, surface defects can be created by changing the voltage or current, or by pushing the tip into the surface. In $Rb_{0.3}MoO_3$ we have been able to perform nanolithography by poking holes and drawing lines which show no signs of change over many hours.[1] On the other hand, with $Na_{0.9}Mo_6O_{17}$, defects are created with a voltage jump method.[2] Changes in surface structure appear upon subsequent imaging and which are induced by the act of scanning the STM tip over the surface. Proposed mechanisms and kinetics of STM assisted etching will be discussed.

[1] E. Garfunkel, et al., *Science* **246**, 99 (1989).
[2] D. Saulys, et al., *J. Appl. Phys.*, **69** (3), 1707 (1991).

362. NANOFABRICATION WITH THE SCANNING TUNNELING MICROSCOPE. Alex de Lozanne, E.E. Ehrichs, and W.F. Smith, Dept. of Physics, Univ. of Texas, Austin, TX 78712.

There is increasing interest in the physical and chemical properties of structures that have more than a few atoms and yet are small enough to have properties that differ substantially from macroscopic bulk systems. Many techniques are being used to generate these nanostructures, which are typically only a few hundred angstroms in size. We have used the Scanning Tunneling Microscope (STM) to fabricate structures with linewidths as small as 10 nm. Our process is essentially a very localized chemical vapor deposition (CVD), where the organic precursor gas is broken with the electrons coming from the STM tip. We have also etched holes in silicon with diameters down to 20 nm. More recently we have built an STM inside an Scanning Electron Microscope (SEM). The latter allows us to accurately align the STM tip with contact pads on the surface of the substrate. This will make it possible to measure the electrical properties of 10 nm nickel wires written with the STM. This work is supported by the NSF (DMR-8553305), The Welch Foundation, and The Hertz Foundation.

363. PREPARATION OF PASSIVATED III-V SURFACES FOR STM NANOLITHOGRAPHY. J. A. Dagata, W. Tseng, J. Bennett, J. Schneir, and H. H. Harary, National Institute of Standards and Technology, Gaithersburg MD 20899.

Nanometer-scale pattern generation on semiconductor surfaces by a scanning tunneling microscope (STM) requires 1) substrates which possess a high degree of chemical and electrical homogeneity, and 2) a lithographic process which can be modulated by changes in the local current density or electric field beneath the scan tip. In this talk I will discuss the preparation and characterization of III-V semiconductor surfaces for STM nanolithography. Our current understanding of III-V surface passivation chemistry is obtained from STM, SEM, imaging time-of-flight SIMS, and XPS studies of sulfur-treated and MBE-grown substrates. STM direct patterning of ultrashallow oxide arrays, minimum feature width $\leq$ 50 nm, on these substrates is described as well as their anticipated use in the fabrication of low-dimensional heterostructures.

364. NANOMETER SCALE ETCHING OF MX$_2$ MATERIALS USING SCANNING TUNNELING MICROSCOPY. Jinlin Huang, Yun Kim, Charles M. Lieber, Department of Chemistry, Columbia University, Box 7, New York NY 10027

Scanning tunneling microscopy (STM) has been used to study the etching of transition metal dichalcogenide surfaces in air, vacuum and water vapour. In vacuum ($1 \times 10^{-7} - 10^{-9}$ Torr), the etching rate is much smaller than in air or in water vapour, and in addition, a strong bias dependance of etching rate is observed in water vapour. A mechanistic picture that explains these interesting results will be presented.

365. SCANNING TUNNELING MICROSCOPY OF DONOR GRAPHITE INTERCALATION COMPOUNDS. Stephen P. Kelty and Charles M. Lieber, Department of Chemistry, Columbia University, New York, NY 10027

Graphite intercalation compounds (GICs) have been the subject of theoretical and experimental interest for many years. Until recently, however, little was known about the surface region of these quasi-two dimensional materials. We have investigated these interesting materials using scanning tunneling microscopy (STM) to probe the surface electronic structure. Constant current and constant height mode images of stage-1 binary GICs (MC_8, M=K,Rb,Cs) exhibit a commensurate 2x2 superstructure while the stage-2 materials show an absence of any superstructure. Ternary GICs KM_xC_{4n} (M_x=Hg, $H_{0.8}$, $Tl_{1.5}$) show a variety of commensurate and incommensurate superstructures. The interpretation of these results is discussed in terms of the electronic and crystal structures of the materials.

366. **SURFACE SPECTROSCOPIC CHARACTERIZATION OF IODINE OVERLAYERS ON Pt(111).** Sam K. Jo* and J. M. White, Department of Chemistry, The University of Texas at Austin, Austin, Texas 78712.

The interaction of atomic iodine with Pt(111) was investigated. Coverage-dependent populations of two distinct states of adsorbed iodine, i.e., three-fold hollow and one-fold atop site adsorption states, were monitored and distinguished by temperature-programmed desorption (TPD), X-ray photoelectron spectroscopy (XPS), work function change ($\Delta\Phi$)and low-energy (<0.2 eV) photoelectron yield measurements. The results are in agreement with the adlattice structures suggested by LEED [Felter et al., *J. Electroanal. Chem.* 100 (1979) 473], XPS [DiCenzo et al., *Phys. Rev. B.* 30 (1984) 553] and STM [Schardt et al., *Science* 243 (1989) 1050] studies. The respective negative and positive $\Delta\Phi$'s upon the occupation of three-fold hollow and one-fold atop sites are consistent with a substrate polarization theory [E. Shustorovich, *Surf. Sci. Rep.* 6 (1986) 1].

*Current Address: Department of Electrical Engineering, Columbia University, New York, NY 10027.

367. HARPOONING REACTIONS OF EXCITED RARE GAS ATOMS, M. R. Bruce, W. B. Layne and J. W. Keto, Department of Physics, The University of Texas at Austin, Austin, Tx. 78712

Reactive quenching of the excited states of the rare gases has been studied for more than a decade. Early, Gundel et al [J. Chem. Phys. 64,4390(1976) proposed that reactions of the metastable states of argon ($3p^54s[3/2]_2$ and $3p^54s'[1/2]_0$) were analogous to reactions of alkali metal atoms with halogens and proceeded by a harpooning mechanism, which assumes the reaction proceeds through an ion intermediary, $Ar^* + Cl_2 \rightarrow Ar^+ + Cl_2^- \rightarrow ArCl^* + Cl$. Product branching fractions, vibrational energy disposal, velocity dependence, and vibrational excitation of reactants and propensity for ion-core conservation (N. Sadeghi, adjoining session) have now been studied for the reactions of metastable states of all the rare gases with many chloride compounds. This paper will present recent studies of the effect of excited state energy on the harpoon reaction. We present measurements of total binary and tertiary quench rates and branching fractions for the reaction, Xe ($5p^5np,np'$, n=6,7)+ Cl_2. Xenon atoms are excited by state-selective, two-photon absorption with a u.v. laser, and the time dependent fluorescence from the excited atom in the I.R., visible, and from XeCl* (B) products near 308 nm are then measured with subnanosecond time resolution. The energy dependence of the measured reaction rates are consistent with a multichannel harpoon model to be presented which is based on the formalism of Gislason and Sachs [J. Chem. Phys. 62, 2678 (1975)] .

368. STATE SELECTIVITY IN THE REACTIONS OF COPPER ATOMS WITH HALOGEN COMPOUNDS ; ION-CORE CONSERVATION. N. Sadeghi, Laboratoire de Spectrométrie Physique, Université Joseph-Fourier/Grenoble 1, B.P. 87, 38402 Saint-Martin d'Hères Cédex, France.

The chemiluminescence resulting from the gas phase reactions of state selected ground state Cu(^{2}S) and metastable Cu*(^{2}D) atoms with halogen compounds RX (X = F, Cℓ, Br) has been analyzed and branching ratios for the production of copper halides, CuX*, in different electronically excited states have been determined. CuX* molecules formed by the reaction of Cu*(^{2}D) atoms show strongly inverted vibrational distribution and a significant spin-orbit effect is also observed. For all of the halogen reactants used, the observed total reaction cross-sections for Cu(^{2}S) atoms were larger than those for metastable Cu*(^{2}D) atoms. A model based on direct harpooning with conservation of the copper atom's ionic core is proposed.

369. **FEMTOSECOND DYNAMICS OF Bi_2 DISSOCIATION.**
R.M. Bowman, J.J. Gerdy, G. Roberts and A.H. Zewail, A.A. Noyes Laboratory of Chemical Physics, California Institute of Technology, Pasadena, California 91125.

Abstract not received.

370. **SAMPLE REORIENTATION FOR HIGH-RESOLUTION NMR OF SOLIDS: DAS and DOR**
Karl T. Mueller, Chemical Sciences Division, Lawrence Berkeley Laboratory and Department of Chemistry, University of California, Berkeley, CA 94720.

Recent dynamic-angle spinning (DAS) and double rotation (DOR) technology has extended the capabilities of NMR to probe structure and fundamental processes in solid-state chemistry. New experimental techniques will be discussed including the relationship of the spectroscopic averaging to icosahedral symmetry. Results will be presented from a range of chemically relevant systems including aluminum-27, oxygen-17, and boron-11 in silicate minerals, molecular sieves, and inorganic glasses.

371. **CROSS POLARIZATION OF HALF-INTEGER SPINS IN ROTATING SAMPLES.** Alexander J. Vega, Du Pont Central Research and Development, P. O. Box 80356, Wilmington, Delaware 19880-0356.

The spin dynamics of cross polarization from a spin 1/2 to the central transition of a noninteger quadrupolar nucleus will be discussed. When the sample is rotating at the magic angle, the periodic sign changes of the first-order quadrupole splitting, Q, cause complications which sometimes prevent the accumulation of a cross-polarization signal. The severity of this effect depends on the rate of change of Q at the zero crossings. These passages are classified as adiabatic, intermediate, or sudden, depending on the relative magnitudes of the rf amplitude, the quadrupole coupling constant, and the rotor frequency. Irreversible decay of spin-locked magnetization of quadrupolar nuclei in the intermediate case competes with the cross-polarization process.

372. **MULTINUCLEAR NMR STUDY OF AN ALLOY SEMICONDUCTOR.** David B. Zax, Department of Chemistry, Cornell University, Ithaca, NY 14853; Shimon Vega, Department of Isotope Research, Weizmann Institute of Science, Rehovot, Israel 76100; David Zamir, Soreq Nuclear Research Center, Yavne, Israel.

We have observed NMR spectra and relaxation at each of the three lattice sites (Hg, Cd, and Te) in a variety of single crystal samples of nominal composition $Hg_{0.78}Cd_{0.22}Te$. Double resonance Hg-Te and Hg-Cd experiments were used to clarify the assignment of Te resonances. These results confirm earlier observations that in $Hg_xCd_{1-x}Te$ the cationic distribution is neither random nor explained by available theories of bonding in semiconductor alloys. Simple kinetic constraints also appear not to be responsible for the observed distributions. The amount of lattice Hg appears to be substantially less than what is suggested by the bulk formula. Relaxation measurements provide some clues as to the local conduction electron density and have been performed for all three of the nuclear sites as a function of sample properties (i.e. number and type of carriers) and temperature.

373. **SPECTROSCOPIC CHARACTERIZATION OF CO ON PALLADIUM COLLOIDS.** J.M. MILLAR, J.S. Bradley and E.W. Hill, Exxon Research and Engineering, Annandale NJ 08801.

Palladium colloids have been prepared with average diameters of 10, 18 and 75 Å using stabilizing agents isobutylaluminoxane, $Al(^iBu)_3$, and polyvinylpyrrolidone in polar and nonpolar solvents.

CO adsorption has been studied using infrared and ^{13}C NMR. Terminal carbonyls are observed by IR in the 10 Å case, both terminal and bridging are seen with the 18 Å and only bridging are seen at 75 Å.

Relatively broad lines produced by adsorbed CO are observed in the ^{13}C NMR of 10 and 18 Å Pd (15 and 20 ppm W_{HM}, respectively), indicating the heterogeneity of the bound CO environments. These lines are shown to be inhomogeneously broadened by meaurments of the transverse relaxation time, T_2, and holeburning experiments. The 75 Å sample shows adsorbed CO at 800 ppm using a novel saturation transfer experiment which takes advantage of the exchange between adsorbed and dissolved CO. Recent results of other Pt group metals will be shown.

374. MAGNETIC RESONANCE PROBES OF SHORT-RANGE ORDER IN NETWORK GLASSES.
Josef W. Zwanziger, Dept. of Chemistry, Indiana University, Bloomington, IN 47405

Network glasses appear to be ordered in some way on several length scales, and so it is necessary to use a variety of experimental approaches to fully elucidate their structure. NMR can yield the most direct and precise information on the short-range order in these systems; this information is necessary to successfully interpret experiments on medium-range order and the microscopic foundations of the bulk properties of these glasses. We discuss our recent efforts to discover the short-range chemical and configurational order in the IV-V-VI network glass $Ge_xAs_ySe_{1-x-y}$, as probed by selenium and germanium NMR.

375. NMR MEASUREMENTS OF ATOMIC CLUSTER DISTRIBUTIONS IN PSEUDOBINARY SEMICONDUCTOR ALLOYS. Robert Tycko and Gary Dabbagh, AT&T Bell Laboratories, Murray Hill, NJ 07974; Sarah R. Kurtz, Solar Energy Research Institute, Golden, CO 80401.

Semiconductor alloys such as $Ga_xIn_{1-x}P$ have the zinc blende crystal structure, in which anions (P) and cations (Ga, In) are placed separately on interpenetrating fcc lattices. Each anion has four cation nearest-neighbors, and vice versa. Electron diffraction and TEM measurements indicate that the two cations are not distributed randomly on their lattice in alloy thin films. Rapid magic angle spinning allows us to resolve ^{31}P NMR lines from the five possible $Ga_nIn_{4-n}P$ clusters in $Ga_xIn_{1-x}P$ and thereby obtain a quantitative measurement of the cation distribution statistics. Surprisingly, *our results on MOCVD-grown films are only consistent with very weak, if any, cation ordering*. In addition to these results, interesting observations on the effects of doping on the NMR spectra, the rationale for assigning the NMR lines, and the effects of cation substitution on the ^{31}P chemical shift will be presented.

POLYMER CHEMISTRY

Fourth Chemical Congress of North America

New York, NY
August 25 - 30, 1991

A. D. English, J. S. Riffle, A. Natansohn,
Program Chairman

SUNDAY MORNING
- **Tutorial on Polymeric Materials for Photonic and Optical Applications**
J. Torkelson, Presiding
Papers 1-5

SUNDAY AFTERNOON
- **Polymeric Materials for Photontonic and Optical Applications**
Nonlinear Optical Properties of Polymers: Theory and Experiment
G. C. Bjorklund, Presiding
Papers 6-11

SUNDAY EVENING
- **Poster Session: Polymer Synthesis**
Mendizabal-Mijares, Presiding
Papers 12-57

- **Poster Session: Polymer Characterization**
Jasso-Gastinel, Presiding
Papers 58-101

MONDAY MORNING AND AFTERNOON
- **Polymeric Materials for Photontonic and Optical Applications**
Synthesis and Chemistry of NLO Polymers I and II
G. Hadziioannou, M. Winnik, Presiding
Papers 102-108
126-132

- **Novel Trends in Thermosetting Systems**
S. J. Huang, C. S. P. Sung, Presiding
Papers 109-114
133-138

MONDAY MORNING
- **Diffusion Processes in Polymers**
Influence of Molecular Architecture on Diffusional Processes in Polymeric Systems
H. B. Hopfenberg, Presiding
Papers 115-120

- **ACS/APS Symposium: Thermal Reversible Gelation of Polymers**
M. S. Wolfe, G. B. McKenna, Presiding
Papers 121-125
145-149

- **Diffusion Processes in Polymers**
Molecular and Atomistic Modeling of Diffusion Processes in Polymer/Penetrant Systems
H. B. Hopfenberg, Presiding
Papers 139-144

TUESDAY MORNING AND AFTERNOON

**● Polymeric Materials for Phototonic and Optical Applications
Processing and Characterization of NLO Polymers**
G. Meredith, Presiding Papers 150-156

TUESDAY MORNING

● Novel Trends in Thermosetting Systems
H. L. Mei, Presiding Papers 157-162

**● Diffusion Processes in Polymers
Transport Properties in Molten and Concentrated Polymer Systems**
B. D. Freeman, Presiding Papers 163-168

● ACS/APS Symposium: Thermal Reversible Gelation of Polymers
G. Delmas, C. L. Jackson, Presiding Papers 169-173
197-201

TUESDAY AFTERNOON

**● Polymeric Materials for Phototonic and Optical Applications
Photonic Polymers for Device Application**
R. Lytel, Presiding Papers 174-180

**● Preceramic and Inorganic Hybrid Materials
New Polymers**
K. Wynne, Presiding Papers 181-190

**● Diffusion Processes in Polymers
Influence of Specific Interactions on Diffusional Processes in
Polymeric Systems**
B. D. Freeman, Presiding Papers 191-196

**● Panel Discussion: Interdisciplinary Science and Education for
the Future - Academic, Industrial, and Government Perspectives**
R. Pariser, Moderator Papers 202-204

TUESDAY EVENING

**● Poster Session: Polymeric Materials for Photonic and Optical
Applications**
 Papers 205-226

● Poster Session: Novel Trends in Thermosetting Systems
 Papers 227

● Poster Session: Preceramic and Inorganic Hybrid Materials
 Papers 228-232

**● Poster Session: Polymers of Geometrical Beauty - Combs and
Stars**
 Papers 233-239

**● Poster Session: ACS/APS Symposium: Thermoreversible
Gelation of Polymers**
 Papers 240-249

WEDNESDAY MORNING AND AFTERNOON
● Instrumental Methods in Polymer Characterization: Optical Methods
G. J. Vansco, J. Rabolt, Presiding

Papers 250-255
277-282

● Preceramic and Inorganic Hybrid Materials
Oxide Materials
K. Mauritz, Presiding

Papers 256-264

● Polymers of Geometrical Beauty - Combs and Stars
G. L. Baker, J. Roovers, Presiding

Papers 265-271
292-297

● Carl S. Marvel - Creative Polymer Chemistry Award Symposium: Polymers - Chemistry, Physics, and Biology
T. J. McCarthy, V. Percec, Presiding

Papers 272-276
298-302

WEDNESDAY AFTERNOON
● Preceramic and Inorganic Hybrid Materials
Hybrid Inorganic/Organic Materials
A. Brennan Presiding

Papers 283-291

THURSDAY MORNING AND AFTERNOON
● Instrumental Methods in Polymer Characterization: Optical Methods
B. Chase, Presiding

Papers 303-308
330-334

THURSDAY MORNING
● Preceramic and Inorganic Hybrid Materials
Silicon Nitride and Boron Nitride Preceramic Polymers
J. Harrod, Presiding

Papers 309-317

● Polymers of Geometrical Beauty - Combs and Stars
T. X. Neenan, B. Gordon, III, Presiding

Papers 318-324
344-349

● Conjugated Polymers Honoring C. B. Gorman - Winner of the Unilever Award for Outstanding Graduate Research in Polymer Chemistry
A. G. MacDiarmid, Presiding

Papers 325-329

● Preceramic and Inorganic Hybrid Materials
Silicon and Aluminum Polymeric Materials
R. Laine, Presiding

Papers 335-343

THURSDAY AFTERNOON
● **Atomistic Simulation of Polymer Materials**
G. P. Puglia, Presiding

Papers 350-356

FRIDAY MORNING
● **Preceramic and Inorganic Hybrid Materials Silicon Carbide and General Polymeric Precursors**
R. Paine, Presiding

Papers 357-364

● **Panel Discussion: Symposium on National and International Intersocietal Interactions in Polymer Science and Materials**
R. M. Ottenbrite, Moderator

● **Atomistic Simulation of Polymer Materials**
G. P. Puglia, Presiding

Papers 365-369

POLYMER CHEMISTRY

1. Introduction to Nonlinear Optics. **G. C. Bjorklund**.

   ## NO ABSTRACT

2. Rational Design of Organic NLO Materials. **T. J. Marks**.

   ## NO ABSTRACT

3. Characterization of NLO Materials. **G. R. Meredith**.

   ## NO ABSTRACT

4. Materials Requirements for Active and Passive Devices. **R. Lytel**.

   ## NO ABSTRACT

5. Introduction to Photorefractive Materials and Phenomena. **P. Gunter**.

   ## NO ABSTRACT

6. ORIENTATION OF ORGANIC MOLECULES IN SOL-GEL MATRICES FOR QUADRATIC NONLINEAR OPTICS
   Germain Pucetti, Eric Toussaere, Isabelle Ledoux and J. ZYSS
   Centre National d'Etudes des Télécommunications, 196 Av. Henri Ravera 92220 BAGNEUX - FRANCE
   Pascal Griesmar and Clément Sanchez, Laboratoire de Chimie de la Matière Condensée, 4 Place Jussieu 75252 PARIS - FRANCE

   Orientation of nonlinear organic molecules in ion-free sol-gel matrices upon application of an external D.C. electrical field is being evidenced. The quadratic nonlinear response of silicon oxide or transition metal oxide based gels containing organic molecules have been determined from Electric Field Induced Second Harmonic (EFISH) measurements. Large concentrations of optically Nonlinear Organic Molecules (NOM) have been either incorporated inside the macromolecular network or chemically bonded to the oxide backbone of the gels. Feasibility of permanently corona poled films has subsequently been demonstrated. Moreover EFISH measurements offer an original point of view on conformational changes occuring in the process of sol-gel transformations.

7. ELECTRON CORRELATED STATES AND NONLINEAR OPTICAL PROPERTIES OF LINEAR CHAINS. A.F. Garito and J.R. Heflin, Department of Physics, University of Pennsylvania, Philadelphia, PA 19104.

The electronic origin and mechanism of the comparatively large nonresonant second order $\chi^{(2)}(-\omega_3;\omega_1,\omega_2)$ and third order $\chi^{(3)}(-\omega_4;\omega_1,\omega_2,\omega_3)$ nonlinear, optical susceptibilities of conjugated organic and polymeric structures have been the center of intense scientific investigations for many years. In contrast to the considerable progress realized with second order optical processes, the first successful microscopic many-electron description has been achieved only recently for $\chi^{(3)}(-\omega_4;\omega_1,\omega_2,\omega_3)$ processes such as third harmonic generation, degenerate four wave mixing, and the optical Kerr effect in conjugated structures, and it is this recent advance which is the main subject of this paper.

8. LINEAR AND NONLINEAR OPTICAL PROPERTIES OF HIGHLY ORDERED CONJUGATED POLYMERS IN POLYETHYLENE: ORIENTATION BY MESOEPITAXY. C. Halvorson, D. Moses, T.W Hagler, K. Pakbaz and A.J. Heeger, Institute for Polymers and Organic Solids, University of California at Santa Barbara, Santa Barbara, CA 93103

Recent success in the synthesis of soluble PPV derivatives has enabled the fabrication of oriented films of blends of these semiconducting polymers by gel-processing in polyethylene. Since high degrees of orientation of the conjugated chains can be achieved, films of these blends can be prepared with controlled optical density and controlled dichroism. Such films are of interest for a wide range of optical properties. The nonlinear optical properties of MEH-PPV have been investigated through third harmonic generation (THG) spectroscopy with the pump frequency varied over a wide range of frequencies within the gap. We have characterized the THG response in thin spin-cast films of MEH-PPV, in nonoriented MEH-PPV/PE blends and in ordered and highly oriented MEH-PPV/PE blends. The ordered blends exhibit large values for $\chi^{(3)}$ in the infrared. The results will be presented and analyzed in the context of the effect of disorder on the THG response.

9. DETERMINATION OF THIRD-ORDER NONLINEAR SUSCEPTIBILITIES OF SEMICONDUCTING POLYMERS. COMPARISON WITH THEORETICAL CALCULATIONS.
J.MESSIER, F.CHARRA and C.SENTEIN
Laboratoire de Physique Electronique des Materiaux.CEA
GIF-SUR-YVETTE.91191 CEDEX.FRANCE

We report on the experimental determination of third-order nonlinear coefficients by third harmonic generation in conjugated polymers. We show the interest to measure the argument of the nonlinear coefficient γ. With the simultaneous use of the optical absorption data it is thus possible with a small number of parameters concerning the two-photon states to account for the absolute value of γ and for its variations with frequency. This is specially interesting for polymers exhibiting numerous conformations. We describe also quantum mechanical calculations of γ and show that it is possible to obtain a good estimate of it. As an axample we treat the case of polythiophene.

10. A SURVEY OF THE NLO–POLYMER PROGRAM AT THE MPI. W. Knoll, Max–Planck-Institut für Polymerforschung, Ackermannweg 10, D–W–6500 Mainz, Germany.

Linear and non–linear optical properties of novel polymeric materials developed over the past few years at the MPI for Polymer Research are described. Concentrating on conjugated π–electron systems, third harmonic generation and degenerate four wave mixing experiment, were performed in order to elucidate the third order non-linear optical susceptibilities $\chi^{(3)}$ and the relaxation times T_1 of some systems but also the structural parameters that determine the overall performance of the different chromophores in various supramolecular architectures.

Equally important in view of future device applications in optoelectronics or integrated optics is the availability of polymers that are highly transparent, adjustable in their index of refraction and their dispersion behavior and that can be processed, e.g. to build planar or channel waveguide structures. Particularly promising new systems in this respect are polyelectrolyte glasses that offer an extremely broad range of chemical manipulations to tailor materials with desired properties that satisfy special applicational needs. Other examples discussed include so-called "hairy-rod" polymers, e.g. helical backbones with flexible side-arms, that can be (self-)organized at the water surface in a nematic-analogue order and that can be transferred, layer by layer, onto a solid support to build easily-functionalized multilamellar films.

11. CUBIC NONLINEAR OPTICS OF POLYMER THIN FILMS. 1. MOLECULAR ENGINEERING AND PROCESSING OF POLYMERS WITH LARGE THIRD-ORDER OPTICAL PROPERTIES FOR PHOTONIC SWITCHING. Samson A. Jenekhe, Ashwini K. Agrawal, Chen-Jen Yang, John A. Osaheni, Wen-Chang Chen, and Michael F. Roberts, Center for Photoinduced Charge Transfer and Department of Chemical Engineering, University of Rochester, Rocheser, NY 14627-0166.

Nonlinear optical polymers with large third-order optical properties are currently of interest for photonic switching, optical interconnection, and all-optical signal processing. However, significant advances in nonlinear optical materials synthesis, optimization, thin film processing, and fabrication into device structures are required to realize this technological potential. My talk will describe aspects of our laboratory's approach to the molecular engineering and processing of cubic nonlinear optical polymers. Examples of conjugated polymer architecture under investigation include: homopolymers, random copolymers, block copolymers, rigid-rod and flexible chain structures, ladder and semi-ladder structures, polymer blends and molecular composites. Since conjugated polymers are generally difficult to process into thin films and device structures, one successful approach to the solution processing of different classes of conjugated polymers to optical quality thin films will be described.

12. COPOLYMERS OF NYLON 266 AND NYLON 66: SYNTHESIS AND CHARACTERIZATION. X. Chen and K. E. Gonsalves, Polymer Science Program, Institute of Materials Science and Department of Chemistry, University of Connecticut, Storrs, CT 06269.

Copolymers of nylon 266 and nylon 66 were synthesized via an interfacial polymerization reaction using two diamines (hexamethylene diamine and N-glyclhexanediamine) with adipoyl chloride. The polymers were characterized by NMR spectroscopy, thermal analysis and x-ray diffraction.

13. Polymerization of Mono-Substituted α,α'-Bis-(tetrahydrothiophenium)-p-xylene Dibromides. Bing R. Hsieh, Xerox Webster Research Center, 114-39D, Webster, NY 14580.

Sodium hydroxide treatment of methoxy α,α'-bis-(tetrahydrothiophenium)-p-xylene dibromide monomer yielded two xylylene intermediates which then copolymerized to give a polyelectrolyte precursor polymer. THe cast films of the polymer were thermally converted to methoxy Poly(phenylene vinylene) films. Xylylene intermediates were detected for the corresponding cyano monomer upon reacting with NaOH, but they did not polymerize due to their high stability. THe results from this were all in support of an anionic polymerization mechanism.

14. RING-OPENING POLYMERIZATION OF EPOXIDES CATALYZED BY DICOBALTOCTACARBONYL; James V. Crivello, Mingxin Fan; Department of Chemistry, Rensselaer Polytechnic Institute, Troy, NY 12180-3590.

$Co_2(CO)_8$ has been found to be a very efficient catalyst for the ring-opening polymerization of cyclic ethers, especially epoxides, as well as certain vinyl monomers in the presence of Si-H containing cocatalysts. The reaction conditions employed are very mild and similar to those used in $Co_2(CO)_8$ catalyzed hydrogenation and hydrosilylation reactions. Detailed investigations have been carried out to elucidate the nature of the active species for this catalytic system. Polymerization of epoxides and other monomers will be presented and efforts to elucidate the nature of the active species and the mechanism will be discussed.

15. THE SYNTHESIS OF NOVEL SILICON CONTAINING EPOXY MONOMERS AND OLIGOMERS; James V. Crivello, Daoshen Bi and Mingxin Fan; Department of Chemistry Rensselaer Polytechnic Institute, Troy, NY 12180-3590.

The discovery that the hydrosilylation of olefin with compounds containing two Si-H functional groups proceeds in a stepwise fashion has made possible the synthesis of a new series of interesting intermediates with both Si-H and epoxy groups in the same molecule. The use of rhodium catalysts has further provided a facile, high yield route to these novel compounds. Several examples are given for the synthesis of new and interesting monomers using the α-hydrogen- ω-epoxides as starting materials and the polymerization of these monomers using diaryliodonium and triarylsulfonium salt photoinitiators has been studied.

16. A NOVEL SYNTHESIS OF UV-CURABLE, MULTIFUNCTIONAL, AROMATIC ANALOGS OF VINYL ETHERS. James V. Crivello, Asha Ramdas, Department of Chemistry, Rensselaer Polytechnic Institute, Troy, NY 12180-3590

A novel and facile synthesis of multifunctional, aromatic vinyl ether analogs is reported. These materials, prepared by the condensation of 4-acetoxystyrene or 4-acetoxyisopropenylbenzene with halo compounds in the presence of base, can be cationically polymerized by using diaryliodonium salts as photoinitiators to produce crosslinked polymers. Relative reactivities of the monomers towards cationic polymerization and relative thermal stabilities were studied using photodifferential scanning calorimetry and thermo-gravimetric analysis respectively.

17. NOVEL PROPENYL PHENYL ETHER ANALOGS FOR UV-INDUCED CATIONIC POLYMERIZATION. James V. Crivello and Annina M. Carter, Department of Chemistry, Rensselaer Polytechnic Institute, Troy, NY 12180-3590.

Propenyl phenyl ether analogs are highly reactive monomers towards UV-induced cationic polymerization. Aromatic bispropenyl phenyl ether analogs are synthesized by the condensation of phenols and dibromoalkanes under phase transfer conditions. Difunctional monomers containing both a propenyl and a vinyl group are prepared by the reaction of phenols with 2-chloroethyl vinyl ether in the presence of base. Reactivity of the monomers was studied using photodifferential scanning calorimetry.

18. FLUORINATED POLYURETHANES: POLYMERIZATION, CHARACTERIZATION AND MECHANICAL PROPERTIES. Tai Ho, Dept. of Chemstry, George Mason University, Fairfax, VA 22030, and Kenneth J. Wynne, Chemistry Division, Office of Naval Research, Arlington, VA 22217

A fluorinated polyurethane based on an ethylene-fluoroalkyl-ethylene diol (1) and hexamethylene diisocyanate (2) was prepared. Kinetics of the polymerization reaction was studied using Fourier transform infrared

(FTIR) spectroscopy. The resulting polyurethane was characterized with viscometry, differential scanning calorimetry (DSC) and gel permeation chromatography (GPC). From intrinsic viscosity versus molecular weight plot, the exponential parameter in Mark-Houwink-Sakurada equation was found to be 0.89 suggesting a rigid structure for this material. Mechanical properties were assessed by stress-strain test coupled with DSC measurements. Significant effects of crystallinity on mechanical properties were found. Dynamic mechanical tests were also performed.
HO-$(CH_2)_2$-$(CF_2)_5$-$(CFCF_3)$-$(CH_2)_2$-OH   (1),   OCN-$(CH_2)_6$-NCO   (2)

19. FREE RADICAL GRAFTING OF 2-(DIMETHYLAMINO)ETHYL METHACRYLATE TO SQUALANE
     Jane B. Felmine, Warren E. Baker and Kenneth E. Russell, Department of Chemistry, Queen's University, Kingston, Ontario, Canada K7L 3N6.

The grafting of 2-(dimethylamino)ethyl methacrylate to squalane at 150°C using a peroxide initiator was investigated. The products of the reaction are partly soluble in squalane. ^{1}H nuclear magnetic resonance and Fourier transform infrared spectroscopy were used to follow the extent of grafting. Soluble products consist primarily of grafted material with average 1 to 7 monomer units per squalane residue. Homopolymer with degree of polymerization of 40 ± 10 and grafted squalane with higher monomer content were the main products of the insoluble fraction. Preliminary kinetic studies indicate a rate of disappearance of monomer of $7.1 \times 10^{-5} Ms^{-1}$ at 150°C for monomer and initiator concentrations of 0.26M and 0.03M respectively.

20. SYNTHESIS AND CHARACTERIZATION OF POLY(DIMETHYLSILOXANE-CO-METHYLPHENYLSILOXANE) RANDOM BLOCK COPOLYMERS AND ELASTOMERS, C.M. Kuo and S.J. Clarson, Department of Materials Science and Eng., University of Cincinnati, Cincinnati, OH 45221-0012

Random block copolymers are relatively rare due to the requirements for a specific reactivity of functional groups in the monomers or oligomers, as well as special reaction conditions. Depending on the block length, a random block copolymer may produce a homogeneous system with single phase morphology and also show physical and chemical properties that are a weighted average of each of the chemical subunits. Random block copolymers of poly(dimethylsiloxane-co-methylphenylsiloxane) (PDMS-PMPS) and their corresponding elastomers were prepared using a non-equilibrating tin(II) catalyst from well characterized oligomers with reactive hydroxyl end groups. The synthesis and characterization of both these functionally terminated PDMS-PMPS copolymers and their corresponding elastomers are discussed. The block composition and chain length of the polymers and hence network chain precursors were well defined. The Flory-Huggins interaction parameters for polydimethylsiloxane (PDMS) and polymethylphenylsiloxane (PMPS) were obtained from equilibrium swelling of the copolymer elastomers in oligomeric PDMS, and the data was compared with results obtained previously for PDMS-PMPS homopolymer blends.

21. NEW FLUORINATED COPOLYMERS: POLY(1,3-CYCLOHEXADIENE-*alt*-α FLUORO-ACRYLONITRILE). V. Panchalingam and John R. Reynolds, Center for Advanced Polymer Research, Department of Chemistry, The University of Texas at Arlington, Arlington, Texas 76019-0065.

1,3-Cyclohexadiene (1,3-CHD) and α-fluoroacrylonitrile (α-FAN) have been shown to form an alternating copolymer by a combination of NMR spectroscopic and elemental analyses when polymerized by AIBN. Structure-property correlations of poly(1,3-CHD/α-FAN) and the chlorinated analogue based on α-chloroacrylonitrile (α-CAN), [poly(1,3-CHD/α-CAN)], have been made. The fluorinated polymer is significantly more thermally stable than the chlorinated polymer due to the low propensity to HF elimination. NMR spectroscopic analyses show poly(1,3-CHD/α-FAN) to contain 1,2- and 1,4-linkages across the cyclohexene unit compared to the observation of only 1,4-linkages in poly(1,3-CHD/α-CAN).[1] ^{1}H 2D-COSY of poly(1,3-CHD/α-FAN) shows visible couplings between the adjacent olefinic protons suggesting 1,4-linkages. The results are compared with 3-methylcyclohexene as a model compound.

1. Panchalingam, V.; Reynolds, J. R. *Macromolecules*, **1988**, *21*, 960.

22. DEHYDROHALOGENATION STUDIES OF VINYLIDENE FLUORIDE/ TRIFLUOROETHYLENE COPOLYMER AND POLY (CHLOROFLUOROETHYLENE). V. Panchalingam and John R. Reynolds, Center for Advanced Polymer Research, Department of Chemistry; Linda E. Lopez and Howard J. Arnott, Center for Electron Microscopy, Department of Biology, The University of Texas at Arlington, Arlington, Texas 76019.

Dehydrohalogenation reactions were carried out on vinylidene fluoride/trifluoroethylene (VDF/TrFE) copolymer, and on poly(chlorofluoroethylene) (PCFE) in both solution and solid states in an attempt to prepare conductive poly(fluoroacetylene). The extents of reactions were determined by gravimetric analysis and the products were characterized by elemental analyses, FT-IR, and XPS. PCFE underwent quantitative elimination of only HCl under milder reaction conditions than the copolymer. XPS and elemental analyses of the elimination products formed indicate the incorporation of oxygen impurities in all cases. XPS and EDAX experiments show a higher degree of loss of fluorine on the surface of the copolymer film than in the bulk. In-situ vapor phase iodine doping of eliminated films obtained with PCFE shows three orders of magnitude increase in conductivity. The conductivity drops to the original value upon the removal of excess dopant. Electron micrographs of eliminated films show extensive surface cracking for the VDF/TrFE copolymer and smaller hole formation for the PCFE.

23. PREPARATION OF CARBON POWDERS BY PYROLYSIS OF GRAPHITIZABLE AND NON-GRAPHITIZABLE POLYMERS UNDER ARGON STREAM AND THEIR MANETIC PROPERTIES. K. Murata,* H. Ueda and K. Kawaguchi, National Chemical Laboratory for Industry, 1-1, Higashi, Tsukuba, Ibaraki 305 Japan.

The pyrolysis of three organic compounds (PVC, PAN, and tetramers) were carried out and the magnetic properties of the obtained carbon powders were investigated. A hysteresis loop was observed for some samples: Addition of activated carbon (AC) to PVC was required for an observation of the hysteresis, whereas that to tetramers was not effective. Good correlations between magnetic properties and XRD, ESR, conductivity analyses confirm that the hysteresis would come from a carbon structure but not metal impurities. However, the temperature dependence of Imax of PVC/C sample at the temperature range of 4.3-100K gave a possible linear relation, being illustrative of paramagnetism. Therefore, it seems likely that a major portion of the sample is still paramagnetic, although there might be an unusual high spin region in a small portion of the sample.

24. AQUEOUS SYNTHESIS OF SOLUBLE RIGID-CHAIN POLYMERS. AN IONIC POLY(P-PHENYLENE) ANALOG. Thomas I. Wallow and Bruce M. Novak, Department of Chemistry,University of California at Berkeley, Berkeley, CA, 94720

Poly(p-quaterphenylene 2,2'-dicarboxylic acid), (I) has been synthesized as a water-soluble rigid-chain polymer in water via a palladium(0)-mediated polycondensation. Various characterization methods indicate that (I) is a relatively high molecular-weight material and displays thermal characteristics typical of rigid-chain structures. This methodology represents the first example of a transition-metal catalyzed step-growth polymerization that proceeds homogeneously in water.

25. SYNTHESIS AND PROPERTIES OF POLY(ARYLENE ETHER BENZIMIDAZOLE)S. J.G. Smith Jr., J.W. Connell, and P.M. Hergenrother. NASA Langley Research Center, Mail Stop 226, Hampton, Virginia 23665-5225.

As a continuation of our work involving the incorporation of heterocyclic groups within the backbone of poly(arylene ether)s, a series of poly(arylene ether benzimidazole)s were synthesized by the aromatic nucleophilic displacement reaction of activated aromatic difluorides by the alkali metal bisphenates of bis(hydroxyphenylbenzimidazole)s in N,N-dimethylacetamide. The bis(benzimidazolephenol)s were obtained in 30-50 % yield from the reaction of aromatic bis(o-diamine)s and phenyl-4-hydroxybenzoate. The polymers had inherent viscosities as high as 1.99 dL/g with glass transition temperatures ranging from 264-352°C. No weight loss was observed below 300°C with 5% weight loss occurring at ~450°C in air and at ~475°C in nitrogen. Thin films at 23°C gave an average tensile strength, tensile modulus, and break elongation of 27.0 ksi, 800 ksi, and 10%, respectively. The chemistry, physical, and mechanical properties of these polymers will be discussed.

26. PREPARATION AND PROPERTIES OF POLY(ARYL SULFIDE PHENYLQUINOXALINES) D. R. McKean and J. W. Labadie, IBM Research Division, Almaden Research Center, San Jose, CA 95120

Poly(aryl sulfide phenylquinoxalines) have been prepared using the aromatic nucleophilic displacement of quinoxalinyl chlorides with mercaptans as the polymer forming reaction. Model reactions performed on 6-chloro and 6-nitroquinoxalines demonstrated that the displacement reactions proceed with high conversion to give the desired sulfide product. The displacement of the chloroquinoxaline goes rapidly at temperatures as low as 65°C. At higher temperatures complete conversion is observed. Displacement of the nitroquinoxaline is complicated by the oxidation of mercaptan with the nitrite ion which is produced during the reaction. Bis(aromatic mercaptans) react with bis(quinoxalinyl chlorides) to give high molecular weight polymer product in nearly quantitative yield. The polymer products have shown intrinsic viscosities as high as 4.0 dL/g. The poly(aryl sulfide phenylquinoxalines) are quite soluble in a variety of casting solvents and are readily solution processed. These materials have good thermal stability up to 500°C with glass transition temperatures in excess of 200°C.

27. CHEMICAL STABILIZATION OF POLY(VINYL CHLORIDE) BY PRETREATMENT WITH N-SUBSTITUTED MALEIMIDES. A. Velazquez and W. H. Starnes, Jr., Applied Science Program, Department of Chemistry, College of William and Mary, Williamsburg, VA 23185.

Poly(vinyl chloride) was allowed to react with various N-substituted maleimides (2.0 mol/mol of -CHCl-CH$_2$- units) in o-dichlorobenzene at 180 °C under nitrogen. The polymer samples were recovered by precipitation into methanol and purified by methanol extraction. Dehydrochlorination rates determined with solid specimens at 160 °C under nitrogen showed that the maleimide pretreatments had improved the thermal stability of the polymer by factors ranging up to ca. 6-7. As a function of the substituent on nitrogen, the order of pretreatment effectiveness was N-cyclohexyl ≃ N-o-bromophenyl > N-o-chlorophenyl > N-phenyl for the maleimides that were used. The stabilization effects resulting from the pretreatments are currently considered to be explicable in terms of a mechanism involving the deactivation, by maleimides, of thermally labile groups in the polymer.

28. VISIBLE-TRANSPARENT RIGID-ROD POLYMERS: SYNTHESIS AND CHARACTERIZATION OF DIAMANTANE-BASED POLYBENZAZOLES. Thuy D. Dang[a], Thomas G. Archibald[b], Aslam A. Malik[b], Francis O. Bonsu[b], Kurt Baum[b], Loon-Seng Tan[c], and Fred E. Arnold[c].
a. University of Dayton Research Institute, 300 College Park, Dayton, Ohio 45469-0001. b. Fluorochem, Inc., 680 S. Ayon Ave., Azusa, California 91702. c. Wright Laboratory, WL/MLBP, Wright Patterson AFB, Ohio 45433-6533.

The syntheses of diamantane-4,9-dicarboxylic acid a and its diacid chloride b and dimethylester c derivatives were described. Diamantane-based polybenzazoles, DMT-PBZT, DMT-PBO, DMT-PBI, were prepared from either a or c and 1,4-diamino-2,5-benzenedithiol dichloride, 2,4-diaminoresorcinol dihydrochloride and 1,2,3,4-tetraaminobenzene tetrahydrochloride, respectively, with intrinsic viscosities ranging from 2.7 dl/g to 10.2 dl/g. All the DMT-PBX polymers were white. The UV-VIS-NIR spectrum of a DMT-PBZT film, cast from methanesulfonic acid under reduced pressure, displayed no absorption bands in the visible and near-infrared regions. Thermal analyses results indicated that the diamantane cage is indeed very thermally stable.

29. COLORLESS RIGID-ROD POLYMERS CONTAINING BICYCLO[2.2.2] OCTANE MOIETIES. M. Dotrong[1], Minhhoa Dotrong[1], George J. Moore[2], and Robert C. Evers[2]. 1. University of Dayton Research Institute, 300 College Park, Dayton, OH 45469. 2. Wright Laboratory, Wright-Patterson AFB, OH 45433-6533.

A colorless rigid-rod benzobisthiazole polymer was synthesized through the polymerization in polyphosphoric acid of bicyclo[2.2.2]octane-1,4-dicarboxylic acid as well as the corresponding dimethyl ester or diacid chloride with 2,5-diamino-1,4-benzenedithiol dihydrochloride. Intrinsic viscosities as high as 30.6 dl/g (methane sulfonic acid, 30°C) were recorded. Polymer structure was substantiated by elemental analysis and by spectroscopic comparison of the polymer with model compounds. Due to interruption of the π-electron conjugation of the polymer chain with the non-aromatic bicyclo[2.2.2]octane moiety, water-white films of the rigid-rod polymer cast from methanesulfonic acid did not exhibit any absorption in the visible range of the electronic spectrum. Breakdown under thermogravimetric analysis in air took place in two steps with initial onset of degradation occurring at 423°C and onset of breakdown for the second steps being observed at 533°C. Thermogravimetric-mass spectral analysis in vacuo of the rigid-rod polymer substantiated early degradation of the bicyclo[2.2.2]octane moieties. No glass transition temperature was detected under differential scanning calorimetry.

30. STRUCTURE AND PROPERTIES OF DIACETYLENIC LIQUID CRYSTALLINE POLYMERS OBTAINED BY POLYMERIZATION IN THE MELT.   XJ. Zhang and A. Blumstein, Chemistry Department, University of Lowell, Lowell, MA 01854

A series of diacetylenic compounds were polymerized in the melt. General structures of the monomers are: R-OCO-(CH$_2$)$_8$-C≡C-C≡C-(CH$_2$)$_8$-COO-R, where R represents three different mesogenic groups. All polymers obtained display enantiotropic liquid crystalline phases. In contrast to many poly(diacetylenes), these polymers appear to display a low degree of crystallinity. Spectroscopic results indicate that polymerization proceeds through the opening of the triple bond and that the side groups are not affected by the elevated polymerization temperature. Backbone structure of these polymers was investigated by FTIR and solution NMR. The existence of the pendant acetylenic groups attached to the polymer backbone was observed. This observation suggests that unlike in the solid state polymerization (which proceeds through 1,4-addition), melt polymerization of these diacetylenic compounds proceeds via a 1,2-addition process.

31. FUNCTIONAL POLYESTERS AND COPOLYESTERS BASED ON THE 4,4'-DIHYDROXY-α-METHYLSTILBENE MESOGEN J.G. Jegal, C.H. Lin**, A. Blumstein* Polymer Science Program, Department of Chemistry, University of Lowell, Lowell, Massachusetts 01854

Liquid crystalline thermotropic main chain polyesters and copolyesters based on 4,4'-dihydroxy-α-methyl stilbene(DMS), 10,12-docosadiyndioic acid and azelaic acid were synthesized. High molecular weights of typical Mn = 57,000 have been obtained by interfacial polymerization. The mesomorphic properties of these polymers were studied and a phase diagram was established. The polymers studied display (depending on composition) nematic and smectic mesophases. Crosslinking of diacetylenic moieties and double bonds incorporated into the mainchain was achieved by means of UV-irradiation. X-ray diffraction of oriented crosslinked and uncrosslinked films and fibers was performed. Properties of crosslinked and uncrosslinked fibers and films were studied and compared.

32. POLYMERIZATION OF ETHYNYLPYRIDINE WITH BROMINE: A NEW ROUTE TO ELECTRICALLY CONDUCTING CONJUGATED POLYACETYLENE S. Subramanyam and A. Blumstein*, Department of Chemistry, Polymer Program, University of Lowell, Lowell, MA 01854.

A synthetic route to electrically conducting conjugated polyacetylenes is described. The polymerization reaction involves activation of the acetylenic triple bond in ethynylpyridine by formation of a donor-acceptor complex with bromine. The resulting polymer is highly conjugated and exhibits a characteristic emission band in its fluorescence spectrum. This is attributed to a possible formation of an intramolecular excimer. When doped with iodine, the polymer showed relatively high conductivity (10^{-2} S/cm) compared to the undoped state ($<10^{-9}$ S/cm).

33. SYNTHESIS AND CHARACTERIZATION OF A POLYIMIDE FROM 3,7-DIAMINOPHENOTHIAZINIUM CHLORIDE (THIONINE), Himansu M. Gajiwala and Robert Zand, Macromolecular Research Center, University of Michigan, Ann Arbor, MI 48109-2099

A new polyimide has been synthesized from the tricyclic heterocyclic molecule 3,7-diaminophenothiazinium chloride, thionine. In order to assess the reactivity of the thionine molecule, model compounds were first prepared by reacting thionine with pthalic anhydride. After characterizing the resulting diamic acid and its cyclized imide, the synthesis of the polymer from thionine and pyromellitic dianhydride was carried out in different solvents and under various experimental conditions. Optimum conditions for the synthesis of the polyamic acid were determined and the resulting polymer was characterized. The thermal stability of the resultant polyimide was also determined.

34. POLYAMIDES CONTAINING IMIDAZOLES. Kevin J. Bouck and Paul G. Rasmussen, Macromolecular Science and Engineering Program and Department of Chemistry, The University of Michigan, Ann Arbor, MI 48109.

Our interest in the diacid chloride, 1-methyl-4,5-imidazoledicarbonyl chloride (**1**), evolves from research in our laboratories on polyamides synthesized from AB type monomers derived from 2-amino-4,5-dicyanoimidazole (**2**).

The polymerization of the diacid chloride **1** with hexamethylenediamine (**3**) and neopentyldiamine (**4**), along with the characterization of these polymers, will be reported.

35. SYNTHESIS OF ARYL BIS(TRIALKYLTIN) MONOMERS. Jeffrey S. Moore and Gary A. Deeter, Department of Chemistry, University of Michigan, Ann Arbor, MI 48109.

We report a new synthesis of alkali trialkylstannates and demonstrate the use of these reagents in the synthesis of functionalized aryl bis(trialkyltin) monomers. This method is shown to have significant advantages over other approaches to aryl bis(trialkyltin) monomers (e.g. di-Grignard, di-Lithio, and using Palladium catalyzed stannylation). An example of the use of aryl bis(trialkyltin) monomers in a transition metal catalyzed step-growth polymerization is presented.

36. **TOWARDS STRUCTURALLY PERFECT POLY(2,5-PYRROLE)**
    Stefano Martina, Volker Enkelmann, Arnulf-Dieter Schlüter, Gerhard Wegner
    Max-Planck-Institut für Polymerforschung, Postfach 3148, D-6500, FRG

This paper describes attempts to apply the recently developed polycondensation reaction of bromo- and boronic acid (or trialkyltin)- substituted aromatic compounds to the pyrrole nucleus.

37. **A NOVEL ROUTE TO TWO HEAVILY FLUORINATED EPOXY RESINS FROM A FLUORODIIMIDEDIOL.** Henry S-W. Hu, Geo-Centers, Inc., Fort Washington, MD 20744 and James R. Griffith, Naval Research Laboratory, Washington, D.C. 20375

Two synthetic methodologies leading to heavily fluorinated epoxy resins are reported. The title compound $\underline{3}$, a fluorodiimidediol, was prepared through the condensation of the corresponding anhydride $\underline{1}$ and diamine $\underline{2}$. The conversion of $\underline{3}$ into diepoxy via dichlorohydrin $\underline{4}$ resulted in the formation of a network polymer $\underline{5}$ in which the imide structure has disappeared to yield amide components. On the other hand, the curing of a fluorodiepoxy $\underline{6}$ with equivalent $\underline{3}$ in the catalytic presence of tetramethylammonium bromide resulted in a network polymer $\underline{7}$ which has intact imide rings.

38. **THE SYNTHESIS AND PROPERTIES OF A SERIES OF SOLUBLE POLYIMIDES FOR USE IN MICROELECTRONICS.** Martin J. Wusik, Babita Jha and Maurice King, OCG Microelectronic Materials, Inc., Ardsley, NY 10502.

A series of fully imidized soluble fluorine containing polyimide copolymers were prepared. The polymers were synthesized from diaminophenylindane [DAPI] and two dianhydrides: biphenyl dianhydride [BPDA] and 4,4'-(hexafluoroisopropylidene) bis(phthalic anhydride) [6FDA]. An overall 1:1 molar ratio of dianhydride to diamine was employed with 6FDA making up 0, 10, 20, 30, 40, 50 or 100 mole % of the dianhydride component. All polymers were of comparable inherent viscosities. The effect of fluorine content on various polymer physical properties, most importantly dielectric constant (at low and elevated humidities) and coefficient of thermal expansion was examined.

39. **The Spontaneous Polymerization of Styrene In the Presence of Acid: Further Confirmation of the Mayo Mechanism.** W. C. Buzanowski, J. D. Graham, D. B. Priddy and Eric Shero, The Dow Chemical Company, Midland, MI 48640

Flory and Mayo have proposed alternate mechanisms to explain the spontaneous polymerization of styrene. Both of these mechanisms propose the formation of reactive styrene dimers capable of generating initiating radicals. However, the dimer intermediate proposed in the Mayo mechanism has defied all attempts of isolation. Since the Flory dimer would be expected to be stable to acid while the Mayo dimer would not, we studied spontaneous styrene polymerization in an acid environment to determine the effects on oligomer structure, polymerization rate, and polymer molecular weight. The results clearly show that the initiating Mayo dimer is deactivated by acid forming high levels of inactive dimer. The decreased initiation results in a lower concentration of growing polymer chains subsequently lowering the rate of termination (by chain coupling). The end result is a large shift of the spontaneous rate/molecular weight relationship for polystyrene.

40.
PREPARATION, CHARACTERIZATION AND TOUGHNESS OF ETHYNYL CONTAINING BLENDS
Brian J. Jensen, NASA Langley Research Center, Hampton, VA 23665-5225
As part of an effort to develop high performance structural resins with an attractive combination of properties for aerospace applications, a series of ethynyl-terminated polysulfones of different molecular weights were prepared and blended with a low molecular weight ethynyl-terminated coreactant. Upon heating above 200°C, these ethynyl containing materials react to form a chain extended and crosslinked network structure. This reaction renders the materials insoluble in common solvents but also reduces the toughness as compared to high molecular weight linear polysulfones. This work discusses the thermal characterization of these blends and the toughness of the resulting cured materials.

41. SYNTHESES OF NEW POLY(3-HYDROXYALKANOATE) CONJUGATES; PHA-CARBOHYDRATE AND PHA-SYNTHETIC POLYMERS.
Manssur Yalpani[1,2,3] Robert H. Marchessault[2], Frederick G. Morin[1,2], and Clevys J. Monasterios[2]
[1]Pulp and Paper Research Centre, and [2]Department of Chemistry, McGill University, 3420 University Street, Montreal, Quebec, Canada, H3A 2A3 [3]Present address: NutraSweet Co., 601 E. Kensington Road, Mount Prospect, IL 60056, U.S.A.

A new class of poly(3-hydroxyalkanoate) (PHA) conjugates has been prepared, based on modifications of the terminal PHA carboxyl function. Attachment of reduced molecular weight PHA to a series of carbohydrate and synthetic polymers, has yielded new types of branched conjugates. Thus, PHA has been condensed with chitosan, cellulose, poly(ethylene amine) and other polymers to afford conjugates with unusual properties. Thus, while neither PHA or chitosan are water soluble, the chitosan:PHA conjugate forms viscous, opaque aqueous solutions. The new PHA conjugates have been examined by ^{13}C NMR and X-ray spectroscopy.

42. SYNTHESIS AND CHARACTERIZATION OF ORDERED HETEROCYCLE-IMIDE COPOLYMERS CONTAINING ORTHO-PHENYLENE RINGS, J. W. Carson, Jr., J. Preston, Research Triangle Institute, Research Triangle Park, NC 27709; and S. S Sankar, Dept. Chem., NC State Univ., Raleigh, NC 27607.
Previous work reported data on very thermooxidatively stable fibers from ordered benzoxazole-imide copolymers derived from bis-benzoxazole diamines containing m- and p-phenylene rings bridging the heterocyclic rings. The focus here is on highly processable heat resistant resins containing o-phenylene rings. These polymers have been prepared in high molecular weight from the reaction of simple aromatic dianhydrides with aromatic diamines containing preformed bis-benzheterocyclic groups connected by an o-phenylene ring followed by cyclodehydration. For example, 2,2'-o-phenylene-bis(5-aminobenzoxazole), o-P5AB was polycondensed with benzophenone dianhydride, BPDA, to yield a polyamic-acid precursor which was converted to an ordered benzoxazole-imide copolymer via cyclodehydration.
Other polymers containing o-phenylene rings can be prepared from bis-benzheterocyclic diamines that are structural isomers, e.g., 2,2'-o-phenylene-bis(6-aminobenzoxazole) (o-P6AB). Additionally, diamines containing benzothiazole units have been prepared, e.g. 2,2'-o-bis(6-aminobenzothiazole).
The preparation of the various dinitro precursors to the bis-benzoheterocyclic diamines and their reduction to the diamines will be described along with NMR data for both. Characterization will be given for the ordered benzoheterocycle-imide copolymers, including DSC, TGA, T_g, etc., plus the inherent viscosities of the polyamic-acid precursors.

43. A HIGHLY CONDUCTING HYDROPOLYSILANE COPOLYMER; SYNTHESIS AND ELECTRICAL CONDUCTIVITY. Y.I. Lee[a], T.M. Hsu[a], F.G. Wakim[b] and S.P. Sawan[a*], Polymer Science Program, Department of Chemistry[a] and Electrical Engineering[b], University of Lowell, Lowell, MA 01854.

Recently, much attention has been paid to the semiconducting properties of polysilanes having a Si skeleton. Pristine organopolysilane solid films reveal high electric resistance ($\sigma = 10^{-12}$-10^{-14} ($\Omega^{-1}cm^{-1}$)). Polymethylphenylsilane (PMPS) and newly synthesized poly(hydromethyl-co-methylphenylsilylene) have been treated with strong electron acceptors such as $FeCl_3$ or SbF_5 to form conducting complexes with p-type electronic conductivities up to 0.3 ($\Omega^{-1}cm^{-1}$). The activation energy of undoped polysilanes was field independent and approximately 0.6 eV. In contrast to that of undoped polymers, the activation energy of poly(hydromethyl-co-methylphenyl-silylene) doped by $FeCl_3$ or SbF_5 was field dependent and decreased to 0.3 eV.

The evidence from these experiments indicates that both main chain and side chain, such as with phenyl groups, charge transport occurs with $FeCl_3$ or SbF_5 doping. However, the conductivity in doped polysilane dominantly depends on main chain electron transport rather than side group electron transport for these polymers.

44. SYNTHESIS OF A POLY(ESTER CROWN) BASED ON BIS(5-CARBOXY-1,3-PHENYLENE)-32-CROWN-10 AND 4,4'-ISOPROPYLINDENEDIPHENOL (BISPHENOL-A), Y. Delaviz and H. W. Gibson, Department of Chemistry, Virginia Polytechnic Institute & State University, Blacksburg, Virginia 24061.

A direct polycondensation reaction of bis((5-carboxy-1,3-phenylene)-32-crown-10 and 4,4'-isopropylidenediphenol (bisphenol-A) using tosyl chloride in pyridine in the presence of N,N-dimethylformamide formed a high molecular weight poly(ester crown). The polymer is soluble in common organic solvents such as chloroform and THF. It has a glass transition temperature (T_g) of 65°C; also it shows quite high thermal stability, 5% weight loss by TGA at 355°C in air.

45. SYNTHESIS OF N-SUBSTITUTED POLYBENZIMIDAZOLES BY CYCLODEHYDROGENATION OF PRECURSOR POLY(SCHIFF BASES). James J. Kane and Weimin Qian, Department of Chemistry, Wright State University, Dayton, OH 45387.

Three N-substituted polybenzimidazoles were synthesized by the one-pot reaction sequence between a tetraamine and the series of dialdehydes shown below. Product yields, polymer thermal properties and inherent viscosities are compared to those of the same series of PBI's synthesized by the previously reported two-step technique involving synthesis of the poly(o-anilino)amides and subsequent cyclodehydration to the PBI's.

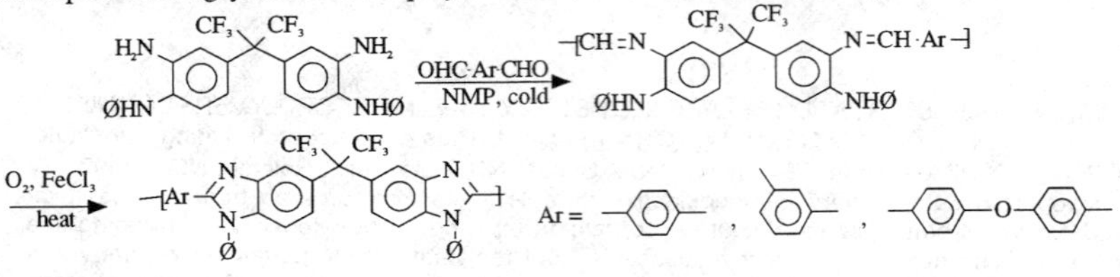

46. SYNTHESIS AND POLYMERIZATION OF HYDROXY POLYSTYRENE CARBOXYLIC ACID TELECHELOMERS II: A MODEL STUDY OF CONDENSATION POLYMERIZATION USING LOW TEMPERATURE TECHNIQUES Jeffrey G. Linert, Marc B. Goldfinger, and Kenneth B. Wagener, Department of Chemistry and Center for Macromolecular Science and Engineering, University of Florida, Gainesville, FL 32611-2046.

Polystyrene functionalized with regularly-spaced ester functionalities in the main chain has been synthesized using chain propagation/step propagation techniques and reported previously. This paper focuses on the condensation phase of the synthesis; low temperature condensation techniques were evaluated and optimized using tropic acid as a model for the hydroxy polystyrene carboxylic acid telechelomer. The investigation shows that polymerizations using 1,3-dicylclohexylcarbodiimide (DCC) or 1,3-diisopropylcarbodiimide (DIPC) is more effective than triphenylphosphine/hexachloroethane or triphenylphosphine dichloride. The molecular weights of the polymer produced in the dialkylcarbodiimide systems increased with decreasing temperature and solvent polarity. The molecular weights also increased with the addition of p-toluenesulfonic acid (PTSA) and N,N-dimethylaminopyridine (DMAP).

47. TRANSITION METAL CATALYZED POLYMERIZATION OF HETEROATOM SUBSTITUTED CYCLOHEXADIENES: PRECURSORS TO POLY(PARAPHENYLENE). D. L. Gin, V. P. Conticello, and R. H. Grubbs*, Department of Chemistry and Chemical Engineering, California Institute of Technology, Pasadena, California 91125.

Poly(paraphenylene) with greater than 95 % *para* linkages is synthesized by a two stage precursor process. *Cis*-5,6-bis(trimethylsiloxy)-1,3-cyclohexadiene (1) is polymerized regio- and stereoselectively by catalysts based on nickel to give the corresponding highly 1,4-linked polymer (2). This soluble precursor is then converted to the corresponding bis(acetoxy) polymer (3). Subsequent thermolysis gives linear poly(paraphenylene) (4) by the elimination of acetic acid. The synthesis and characterization of these materials, and their influence on the physical properties of the poly(paraphenylene) product will be discussed.

48. COPOLYMERIZATION OF TETRAHYDRO-1-(4-HYDROXY-1-NAPHTHYL)THIOPHENIUM HYDROXIDE INNER SALT AND β-BUTYROLACTONE. **D.L. Cangiano***, G. Odian City University of New York, College of Staten Island, S.I., N.Y. 10301. *Present address: Hoechst Celanese RLMTC, 86 Morris Avenue, Summit, N.J. 07901

This paper describes the copolymerization of tetrahydro-1-(4-hydroxy-1-naphthyl)thiophenium hydroxide inner salt, NZ zwitterion, with β-butyrolactone. The copolymer was determined to be random in nature, containing a higher percentage of NZ units. The four junction carbons of the copolymer were readily observed in the DEPT spectrum. The possibility of a block copolymer could be ruled out by the large signal intensities of the block junction linkages. Due to the low molecular weight of these materials (<4000 by GPC), both olefinic and naphthol end groups were visible in the ^{1}H and ^{13}C spectra of the copolymer. The presence of the proposed end groups were confirmed by preparation of model compounds, addition of bromine across the double bond, and acetylation of the naphthol group.

49. SYNTHESIS AND BIODEGRADABILITY STUDY OF FUNCTIONALIZED POLYSTYRENE DERIVATIVES Yan Zhang and Long Y. Chiang, Exxon Research and Engineering Company, Annandale, New Jersey 08801

Toward the study of potential biodegradable polymers consisting of a carbon-carbon backbone, we postulated model systems of functionalized molecular styrene and polystyrene derivatives for the evaluation of their bio-reactivity in the presence of culture mixtures of bacteria, fungi, and soil micro-organisms under controlled conditions. The model compounds were designed and synthesized through a sequence of pre-determined reactions. The functionalized polystyrene derivatives were prepared in different molecular weight ranges through ionic, free-radical, or thermal polymerization of the corresponding monomers. We will discuss the relationship between the biodegradability and the functionality on styrene based polymers as an approach to understand the reaction mechanism of enzyme induced biodegradation on polyolefin-like polymers.

50.   **COPOLYMERIZATION OF 1-ALLYL-2,5-DIMETHOXYBENZENE WITH ACRYLONITRIL** H.L. Lan, and Y. Okamoto; Department of Chemistry, Polytechnic University, 333 Jay Street, Brooklyn, NY 11201

Allyl monomers are generally known to be difficult to copolymerize with vinyl monomers by radical initiators. We have found that 1-allyl-2,5-dimethoxybenzene was copolymerized with acrylonitrile by a free radical initiator without a metal salt. The yield of the copolymer having the molar ratio of 1:2 of the allyl and acrylonitrile monomers, regardless of the initial monomer feeding ratio. Copolymerization with different feed ratios were carried out at 78°C without solvent in a sealed tube by using azobisisobutylonitrile (AIBN) as the initiator. The molecular weight was determined by gel permeation chromatography using tetrahydrofuran as solvent. Polystyrene was used as a standard and typical molecular weight values were found to be $Mn=8.80 \times 10^3$ and $Mw=2.60 \times 10^4$. Spectroscopic analyses (UV, NMR) of these copolymerization systems indicated that the charge transfer complex involves one molecule of allyl monomer with two molecules of acrylonitrile. This three molecule entity is then copolymerized in yield of 1:2 molar ratio copolymer of the allyl monomer and AN, respectively.

51.
MELAMINE-FORMALDEHYDE AEROGELS. C.T. Alviso and R.W. Pekala, Chemistry & Materials Science Department, Lawrence Livermore National Laboratory, Livermore, CA 94550

The ability to tailor the structure and properties of aerogels at the nanometer scale opens up exciting possibilities for these unique, low density materials. Traditional *inorganic* aerogels have been formed from the hydrolysis and condensation of metal alkoxides (e.g. tetramethoxysilane). Previously, we reported the synthesis of *organic* aerogels based upon the aqueous, polycondensation of resorcinol with formaldehyde. Although these aerogels exhibit minimal light scattering, their dark red color limits their use in certain optical applications. In this paper, we discuss the synthesis and characterization of melamine-formaldehyde aerogels --- a new type of organic aerogel that is both colorless and transparent.

52.   MICROWAVE INITIATED COPOLYMERIZATION MMA/HEMA KINETICS AND PROPERTIES. Joaquín Palacios, Jorge Sierra, Ma. Paz Rodríguez. División de Estudios de Posgrado, Facultad de Química, UNAM. Cd. Universitaria, 04510, México, D.F., México.

Copolymers with different compositions were prepared with AIBN, AMBN as initiators the activation energy was supplied by means of a microwave source. Several initiator concentrations and temperatures were tested for these reactions. The conversion data were analysed; and golbal specific rate constants calculated (k), with the values of (k) at temperatures in the range of 60-85°C the activation energy and the frequency factor were calcultaed for the 95% MMA copolymer. The influence of the initiator concentration and temperature on the average molecular weights and distributions is discussed. Density, viscosity and thermal properties were obtained for different copolymer compositions.
The use of microwave energy decreased by a factor of 10 the reaction time. High polymer yields were produced in less than 30 minutes, good mechanical properties, high and narrow distributions of molecular weights were obtained: the material has good filmability.

53.   CROSSLINKING OF POLYALLYLCARBONATES BY GAMMA IRRADIATION

Delia López V.* and Guillermina Burillo**

*Escuela de Ciencias Químicas, Universidad de Puebla, Apdo Post 1613 Puebla. México. Fax 91(22)468332. ** Instituto de Ciencias Nucleares UNAM, Ciudad Universitaria, México 04510 D.F.

The gamma ray induced crosslinking by quiescent polymerization of diallylcarbonate (DAC), monoethyleneglycol bisallylcarbonate (MEGBAC), propyleneglycol bisallylcarbonate (PGBAC), dipropyleneglycol bisallylcarbonate (DPGBAC), diethyleneglycol bisallylcarbonate (DEGBAC), and trallylcarbonate of glycerol (TACG).

Differences in crosslinking behavior of these monomers will be discussed and compared with those of multiacrylic monomers.

54.   NEW DIACETYLENE CONTAINING POLYESTERS.

Carmen Vasquez[1], Leticia Baños[1], Guillermina Burillo[2], and Takeshi Ogawa[1]*

Instituto de Investigaciones en Materiales[1], and Instituto de Ciencias Nucleares[2], Universidad Nacional Autonoma de Mexico, Apartado Postal 70-360, Coyoacan, Mexico DF 04510, MEXICO.

A diacetylenic glycol, m,m'-butadiynylenedibenzylalcohol, was synthesized and condensed with some dibasic acid chlorides to obtain new diacetylene containing polyesters with the following general structure;

$$----O-CH_2-\bigcirc-C\equiv C-C\equiv C-\bigcirc-CH_2-O-\overset{O}{\underset{\|}{C}}-R-\overset{O}{\underset{\|}{C}}----$$

where R is m- and p-phenylene, and $-(CH_2)_n-$ with n=4 and 6. Chemical and physical properties of these polyemrs are presented.

55.   GAMMA RAY POLYMERIZATION OF LONG CHAIN SUBSTITUTED CYCLOTRIPHOSPHAZENES.

Felipe Ramirez, Fernando Alonso, and Takeshi Ogawa*

Instituto de Investigaciones en Materiales, Universidad Nacional Autonoma de Mexico, Aptdo. Postal 70-360, Coyoacan, DF 04510, MEXICO.

Hexachlorocyclotriphosphazatriene was reacted with 10-undecen-1-ol, octylamine, dodecylamine, etc. to obtain cyclotriphosphazenes having long chain substituents. These octopus-like molecules were then irradiated by Gamma ray of Co-60, to obtain new polymers consisting of phosphazene rings connected with their long chain substituents. The polymerization was accelerated by the presence of chlorinated solvents such as chloroform, and insoluble polymers were mainly formed. Soluble polymers were also obtained depending on the irradiation conditions, and some of them were crystalline and showed an interesting morphological behavior, when observed under a polarized light microscope.

56. SYNTHESIS AND POLYMERIZATION OF DIACETYLENE CONTAINING BENZOIC ACID DERIVATIVES.
G. Burillo[1], M.E. Aguirre[1], P. Carreon[1], M.A. Lopez[2], and T. Ogawa[2]*
Instituto de Ciencias Nucleares[1], and Instituto de Investigaciones en Materiales[2], Universidad Nacional Autonoma de Mexico, Circuito Exterior, Ciudad Universitaria, Mexico DF, 04510, MEXICO.

In order to obtain new materials for nonlinear optical applications, various derivatives of diacetylene containing benzoic acids were prepared, and their polymerization under various conditions was investigated.

$X-\underset{\underset{O}{\|}}{C}-\underset{}{\bigcirc}-C\equiv C-C\equiv C-R$ R = alkyl, methoxyphenyl, alkylphenyl, etc.
    X = -OH, -O-alkyl, -NH-alkyl, -NH-$(CH_2)_n$-$N(CH_3)_2$

Some of them showed liquid crystalline phases, and some were light sensitive. Most of them polymerized at temperatures above their melting points.

57. SYNTHESIS OF SIDE-CHAIN LIQUID CRYSTALLINE POLYSILOXANES CONTAINING TRANS-CYCLO-HEXANE BASED MESOGENIC GROUPS. Chain-Shu Hsu and Yong-Hong Lu, Department of Applied Chemistry, National Chiao Tung University, Hsinchu, Taiwan 30050, R.O.C.

The synthesis and characterization of nine polysiloxanes containing 4-alkanyl-oxyphenyl trans-4-n-alkylcycloxane side groups are described. Six monomers which contain a pentenyloxy or a hexenyloxy flexible spacer display a nematic mesophase, while the other three monomers which contain an undecenyloxy flexible spacer display nematic, smectic A and smectic E mesophases. All synthesized polymers present two smectic mesophases except one containing 4-hexanyloxyphenyl trans-4-n-butylcyclohexanoate side groups presents one smectic mesophase and one containing 4-undecanyloxyphenyl trans-4-n-pentylcyclohexanoate side groups presents three smectic mesophases. Trans-cis isomerization of mesogens and side chain crystallization did not occur for any of the synthesized polymers . These results could be due to the cyclohexane based mesogenic units which exhibits conformational isomerism.

58. AMIDE SEGMENTAL DYNAMICS IN N-DEUTERATED POLY(p-PHENYLENE TEREPHTHALAMIDE). R. J. Schadt, E. J. Cain, K. H. Gardner, V. Gabara, S. R. Allen, and A. D. English, Du Pont Central Research and Development and Fibers, Experimental Station, Wilmington, Delaware, 19880-0356.

Deuterium NMR lineshape simulations of N-deuterated poly(p-phenylene terephthalamide), PPTA, are reported which illustrate both the geometry and time scale of amide segmental dynamics in PPTA. The experimental lineshape appears to be a composite of an essentially rigid powder pattern and a smaller fraction exhibiting anisotropic large angle reorientation. The fraction of the N-D bonds which are able to flip (reorient) increases only slightly for temperatures from -184°C to 228°C. The mobility of the amide linkage reflects the large degree of dynamic heterogeneity of PPTA, and the dynamic heterogeneity may correlate with the structural heterogeneity of the polymer crystals. The inter-relationship between the structural heterogeneity, the N-D mobility, and the previously reported terephthalate ring mobility will be examined.

59. STRUCTURE, CRYSTALLIZATION AND MELTING OF POLY(ARYL ETHER KETONE KETONE) PART I: STRUCTURE. KennCorwin H. Gardner, Benjamin S. Hsiao, Robert R. Matheson, Jr., Central Research and Development and Fibers, Du Pont, Experimental Station, Wilmington, DE 19880-0356.

The structure of melt and cold crystallized poly(aryl ether ketone ketones) prepared with different T/I ratios (100/0, 90/10, 80/20, 70/30, 60/40 and 50/50) has been investigated using X-ray diffraction techniques. Melt crystallized PEKKs (all T/I ratios) form a structure similar to that observed in other poly(aryl ether ketones) [Form 1; $a = 0.769$ nm, $b = 0.606$ nm and c (fiber axis) = 1.016 nm]. However, depending on composition, both 'TT' and 'TI' crystals were observed. In contrast to other poly(aryl ether ketones), a Form 2 crystalline modification ($a = 0.786$ nm, $b = 0.575$ and $c = 1.016$ nm) can be induced either by exposure to solvents or by cold crystallization. The two crystalline modifications differ from each other in the placement of the chain and, consequently, the interchain interactions.

60. STRUCTURE, CRYSTALLIZATION AND MELTING OF POLY(ARYL ETHER KETONE KETONE) PART II: CRYSTALLIZATION AND MELTING. Benjamin S. Hsiao, KennCorwin H. Gardner and Robert R. Matheson, Jr., Fibers and Central Research and Development, Du Pont, Experimental Station, Wilmington, Delaware 19880-0302.

The crystallization and melting behavior of poly(aryl ether ketones) with different T/I ratios (100/0, 90/10, 80/20, 70/30, 60/40 and 50/50) has been studied by thermal and optical analysis. The incorporation of I moiety decreases the rate of crystallization and the maximum crystallization rate temperature as determined from both DSC and spherulite growth rate studies. During the melting, the annealed PEKK specimens with 'TT' diads showed the double-melting behavior, whereas samples with 'TI' diads showed the triple-melting behavior. The highest melting transition of the 'TI' crystal is rate dependent and may be due to the recrystallization or reorganization process. The equilibrium melting temperatures of various PEKKs were estimated using the Hoffman-Weeks approach, and showed a linear correlation with the I content. However, this melting temperature depression cannot be explained by either the exclusion nor the inclusion model alone.

61. DEFORMATION AND FRACTURE OF SULFONATED POLYSTYRENE IONOMER/POLYSTYRENE BLENDS: EFFECT OF ION CONTENT. M. Bellinger, J.A. Sauer, and M. Hara, Dept. Mechanics and Materials Science, Rutgers University, P.O. Box 909, Piscataway, NJ 08855-0909.

The deformation and fracture behavior of sulfonated polystyrene ionomer/polystyrene blends have been investigated as a function of the ion content of the ionomer component. Tensile experiments show that the addition of 5 to 30 wt. % ionomer to polystyrene results in an increase in toughness ranging from 70% for low ion content blends to 100% for the higher ion content blends. This increase in fracture energy occurs without loss of modulus or strength common to rubber-modified polymers. Transmission electron microscopy (TEM) studies on thin films under simple tension show that the ionomer and polystyrene are phase separated and the dispersed ionomer particles adhere well to the matrix. The enhancement in toughness is attributed to the extensive matrix crazing generated from the ionomer-rich second phase particles. As the ion content of the of the ionomer-rich second phase increases, the average size of the particles increases and more matrix crazing develops.

62.
MORPHOLOGY OF POLYMER CHAINS BY TRANSMISSION ELECTRON MICROSCOPY. II. IMAGE CONTRAST ANALYSIS OF IODINATED POLYSTYRENE CHAINS. Y.P. Carignan, M. Bellinger, M. Hara, and F. Cosandey, 15 Parkview Ave., Basking Ridge, NJ 07920 and Dept. Mech. and Materials Sci., Rutgers University, Piscataway, NJ 08855-0909.

Thin films of tagged (iodinated) polystyrene chains dispersed in polystyrene were subjected to transmission electron microscopy (TEM) under operating voltages of 100kV and 60kV. Microdensitometry measurements were made using a graphics software package (LEXI) interfaced with the microdensitometer. The photomicrographs show a random distribution of darker spots of varying sizes on an otherwise relatively homogeneous background. The results provide new information on the value of the mass scattering cross section (S), the film thickness, the contrast difference achieved with the labeled chains, and the average dimensions of the labeled iodopolystyrene chains.

63. THE SOLUBILITY PARAMETERS OF AROMATIC POLYAMIDES. Shaul M. Aharoni, Polymer Science Laboratory, Allied-Signal Inc., Morristown, New Jersey 07962-1021.

The degree of swelling of several amorphous rigid aromatic polyamide networks was determined upon equilibration in many liquids, in several mixtures of solvents with non-solvent liquids and in mixtures of the three best solvents. The value of the solubility parameter, δ, was determined to be about 23.0 $(MPa)^{1/2}$, the dipolar contribution $\delta_p = 11.9$ $(MPa)^{1/2}$, and the hydrogen-bonding contribution $\delta_H = 7.9$ $(MPa)^{1/2}$. From these values, the contribution due to dispersion forces was calculated to be $\delta_d = 18.0$ $(MPa)^{1/2}$. These values are believed to closely approximate those of linear polyamides such as poly(p-benzamide), poly(p-phenylene terephthalamide) and poly(p-benzanilide terephthalamide). Due to their insolubility in practically all organic solvents, the solubility parameters of the linear aromatic polyamide was hitherto unknown exactly.

64. MODIFICATION OF INTERACTIONS BETWEEN AROMATIC POLYAMIDE CHAINS BY COMPLEXATION. A.G. Oertli, W.R. Meyer, P. Neuenschwander, U.W Suter, Institut für Polymere, ETH Zürich, CH-8092 Zürich, Switzerland.

In addition to the wide range of fully aromatic, rigid-rod polyamides, we have synthesized a new polymeric material containing pyrimidine moieties. The polymer can be spun into fibers by the dry-jet wet-spinning method and fiber X-ray studies show a packing of the chains reminiscent of that in modification II of PPTA. In addition, the chains can be "crosslinked" with transition metal cations such as nickel in freshly spun, still swollen fibers. Microanalysis for the determination of nickel content shows, that about every second pyrimidine ring is complexed. The "crosslinked" structure and its influence on fiber properties is being investigated. With the help of X-ray analysis on single crystals of low molecular weight compounds, we propose a model for the structure of the complexes in the fiber packing.

65. PREDICTIVE MODEL FOR THE VISCOSITIES OF HIGH MOLECULAR WEIGHT LIQUIDS: APPLICATION TO POLYMER PLASTICIZERS AND MODEL HYDROCARBONS. Deyan Wang, K.A. Mauritz and R.F. Storey, Dept. of Polymer Science, University of Southern Mississippi, Hattiesburg, MS 39406-0076.

A molecular shape-sensitive model of self-diffusion in high molecular weight liquids was constructed in which the molecular jump distance is weighted by a hole-creation probability that depends on free volume redistribution and an energetic barrier reflecting specific nearest-neighbor interactions. Perrin's equation relates the self-diffusion coefficient to the viscosity coefficient. An equivalent molecular radius/shape factor issues from theoretical conformational analysis and the maximum molecular cross-sectional areas in planes that are orthogonal to the three principal axes of inertia of complex, non-symmetric molecular shapes. A temperature-dependent model of steric volume, accounting for stretching and bending vibrations of covalently-linked atoms, was used in the calculate of free volume. The model was tested against experimental viscosity data for dialkyl phthalates, adipates, as well as model hydrocarbon systems.

66. GAS PERMEATION THROUGH NAFION MEMBRANES: THEORY AND EXPERIMENT. K.A. Mauritz, I.D. Stefanithis and Hong Chen, Department of Polymer Science, University of Southern Mississippi, Hattiesburg, MS 39406-0076.

A simple model for the permeation of simple gases through dehydrated perfluorosulfonate membranes containing fixed counterions of any valency (z), was rationalized. It is assumed that gas is mainly solubilized within the ionic clusters. While the rate-limiting factor for diffusion is tortuous intercluster hopping across largely crystalline TFE regions, permeability (P) differences between various ionic forms for the same *small* molecule are not attributed to interactive/size discrimination within the TFE regions, but to differences in solubility within clusters. Point-induced dipole and point-permanent dipole interactions between the cation and gas molecule dominate the energetics of solubility. Initial experimental studies of N_2 permeation in H^+, Na^+, Rb^+, Mg^{2+}, Ca^{2+}, and Al^{3+} indicate that P correlates with the ratio z^2/r^4, where r = N_2-cation center-center distance.

67. **DIFFUSION OF ALKYL ADIPATE PLASTICIZERS IN PVC.** Robson F. Storey, Brian B. Cole, P. Jeanene Broadway and Kenneth A. Mauritz, Department of Polymer Science, The University of Southern Mississippi, Southern Station Box 10076, Hattiesburg, MS 39406-0076.

Diffusion coefficients of several n-alkyl adipates and two commercial adipate plasticizers in PVC were measured over the temperature range $80 \leq T \leq 100$ (°C) by using a mass uptake technique. The values were observed to range from a low of 2.26 X 10^{-11} cm²/sec for a branched C_{10} adipate at 85 °C to a high of 2.14 X 10^{-8} cm²/sec for n-hexyl adipate at 88 °C. Diffusion coefficients were observed to increase with temperature, decrease with increase in alkyl group length, and decrease with increase in alkyl group branchiness. Diffusion temperature dependence was observed to follow an Arrhenius relationship over the temperature range studied, and the values of the apparent activation energies for diffusion ranged from 25 kcal/mole for n-hexyl and n-heptyl adipates to 42 kcal/mole for n-decyl adipate and 44 kcal/mole for a branched C_{10} adipate.

68. **SYNTHESIS AND DEGRADATION STUDIES OF POLY(ε-CAPROLACTONE) FUMARATE NETWORKS.** K.A. Mauritz, R.F. Storey, S.X. Liu, J.S. Wiggins, Department of Polymer Science, University of Southern Mississippi, Hattiesburg, MS 39406-0076.

Unsaturated poly(ε-caprolactone) fumarate oligomer prepromoted with cobalt naphthenate was crosslinked by different amounts (1, 3, 5, 7, 10 wt%) of 2-butanone peroxide. Hydrolytic degradation was carried out by immersion of the samples in an aqueous phosphate buffer solution (pH 7.0) at 37±1°C. The degradation rate decreases with increasing the amount of peroxide initiator. A surface erosion mechanism is proposed by the results of morphological studies with scanning electron microscopy and by the results of solid state NMR, indicating that gross degradation did not take place in the bulk of the materials.

69. **RELAXATION OF RUNS OF TRANS ROTATIONAL ISOMERS IN SINGLE RACEMIC POLY(VINYL CHLORIDE) CHAINS.** Ivet Bahar, Department of Chemical Engineering and Polymer Research Center, Boğaziçi University, 80815 Bebek, Istanbul, Turkey, and Kuei-Jen Lee and Wayne L. Mattice, Institute of Polymer Science, The University of Akron, Akron, Ohio 44325-3909.

A combination of rotational isomeric state (RIS) theory, molecular dynamics (MD) simulations, and dynamic rotational isomeric state (DRIS) theory has been used to investigate the dynamics of rotational isomerism in racemic poly(vinyl chloride) (PVC). RIS provides the information on the populations at equilibrium of the various rotational isomers. MD simulations of 16 ns duration, provide information on the activation energies for the transitions from one rotational isomeric state to another. DRIS combines the information from RIS and MD, and extends the time scale to 10^{-1} seconds. Relaxation occurs via a unimodal process (more exactly, by a family of modes with frequencies centered about one peak). In this respect, racemic PVC is qualitatively different from poly(oxymethylene), in which relaxation occurs via a bimodal process.

POLY

70. THE EFFECT OF THE LENGTH OF THE SIDE BRANCHES ON THE CHAIN STATISTICS OF SILOXANES.
N. A. Neuburger, W. L. Mattice, Institute of Polymer Science, The University of Akron, Akron, Ohio 44325-3909.

Molecular dynamic simulations were performed on a series of siloxane chains of the form poly-di(alkyl)siloxane. Analysis of the trajectories was performed by generating pair-wise probability maps. Using various minimization techniques a system of equations was evaluated to yield a set of statistical weights for each of the systems. Because of the complexity of these systems, the traditional rotational isomeric state treatment would be formidable.

71. HIGH-PRESSURE THERMAL ANALYSIS OF AN INCOMPATIBLE POLYMER BLEND OF POLYPROPYLENE-POLYAMIDE. Y. Maeda, Research Institute for Polymers and Textiles, Tsukuba, Ibaraki 305 JAPAN

Immiscible polymer blends of both crystalline polymers are practically important for the improvement of various properties or processabilities of existing plastics and high-performance polymers. The thermal properties of such polymer blend of polypropylene (PP) and polyamide(PA-6) have been studied under high pressures as wellas at atmospheric pressure. The PP/PA-6 blend, "Orgalloy R6000", supplied by Atochem, France, was used as test specimen. The sample composition has ca.60 vol% of polypropylene and 40 vol% of polyamide6. The thermal behaviors under pressures were studied by using a high-pressure DTA apparatus. Double melting peaks of PP and PA-6 were separately observed at elevated pressures up to 500 MPa in the first runs. The effect of pressure on the crystallization and melting behaviors of each component in the blend was dependent upon pressure. Although the thermal behaviors of PP component were almost independent of pressure, those in PA-6 changed with pressure. It was found that the crystallization of PA-6 in the blend is considerably supressed under pressures above 400 MPa and that the unstable PA-6 crystals are formed predominantly at such high pressures.

72. PRESSURE-INDUCED CRYSTAL MODIFICATION OF A LIQUID CRYSTALLINE POLYESTER.
Y. Maeda, Research Institute for Polymers and Textiles, Tsukuba, Ibaraki 305, JAPAN and A. Blumstein, Department of Chemistry, University of Lowell, Lowell, MA 01854.

It was found that a crystal polymorph of a thermotropic polyester, poly(4,4'-dioxy-2,2'-dimethylazoxybenzene dodecanedioyl)(labeled as DDA-9), is formed by cooling it from the nematic liquid crystal phase under hydrostatic pressures above 70 MPa. The crystal modification is stable under atmospheric pressure. The sample has a larger density and a higher melting point than those of the DDA-9 polyester formed under normal pressure.

73. MAIN CHAIN LIQUID CRYSTALLINE POLYELECTROLYTES DISPLAYING THERMOTROPIC AND LYOTROPIC BEHAVIOR
P. Cheng, S. Subramanyam, A. Blumstein*, Department of Chemistry, Polymer Program, University of Lowell, Lowell, MA 01854.

The synthesis and thermal properties of a series of main chain polyelectrolytes are described. The polyelectrolytes contain the trans-1,2-bis(4,4'-pyridyl)ethylene mesogen units connected by methylene spacers. All polymers in this homologous series possess a high degree of crystallinity at room temperature. They display thermotropic and lyotropic mesophases. The influence of the spacer length and of the counterion on the thermotropic and lyotropic liquid crystalline properties of these properties is described.

74. **ELECTRICALLY CONDUCTING IONIC POLYACETYLENES**
S.Subramanyam, M.S.Chetan, A.Blumstein*, *Department of Chemistry, Polymer Program* and S.Mil'shtein, *Department of Electrical Engineering University of Lowell, Lowell, MA 01854.*

A new family of electrically conducting polyelectrolytes were synthesized. These polymers are substituted and possess a polyacetylene backbone that is extensively conjugated. Incorporation of either donor or acceptor type dopants in these polymers result in substantial increase in conductivity (10^{-4} - 10^{-2} S.cm^{-1}) compared to the undoped state (<10^{-9} S.cm^{-1}). The influence of dopant type and dopant concentration on conductivity was also investigated for the present polymers.

75. **CRYSTALLIZATION KINETICS OF RANDOM ETHYLENE COPOLYMERS.** R.G. Alamo and L. Mandelkern, Institute of Molecular Biophysics, Department of Chemistry, Florida State University, Tallahassee, FL 32306.

The results of a study of the overall crystallization kinetics of a set of molecular weight and composition fractions of random ethylene copolymers will be reported. The influence on the crystallization process of molecular weight at a fixed co-unit content, as well as that of co-unit content at a fixed molecular weight is studied in a series of hydrogenated polybutadienes and one ethylene-hexene-1 copolymer having the most probable molecular weight distribution.
Compared to homopolymers, the crystallization isotherms do not superpose one with the other; deviations from the Avrami relation occur at low levels of crystallinity, and only relatively low levels of crystallinity can be obtained after long time crystallization. These phenomena can be explained by the changing composition of the melt during isothermal crystallization. Certain aspects of the crystallization process will be detailed which show that for these purposes a random copolymer behaves as if it were a homopolymer of much higher molecular weight.

76. CHARACTERIZATION OF THE HETEROGENEITY OF LINEAR LOW DENSITY POLYETHYLENE: TEMPERATURE RISING SUPER CRITICAL FLUID FRACTIONATION. Steven D. Smith, Mike M. Satkowski, Procter & Gamble Co., Miami Valley Labs, Cincinnati, Ohio 45239 Paul Ehrlich, Dept. Chem. Eng., State Univ. of New York at Buffalo, Buffalo, New York 14260 James J. Watkins, Val J. Krukonis, Phasex Corporation, 360 Merrimack St., Lawrence, MA. 01843

Super critical fluid processing of polymers is of growing importance. This paper will present one aspect of how super critical fluids may be utilized in the characterization of polymers. The chemical heterogeneity of polymers is of great importance to polymer chemists and engineers, but is not well characterized or appreciated. SCF fractionation of polymers has been demonstrated to fractionate polymers by molecular weight and by composition. Solubility differences due to molecular weight or composition have been taken advantage of to achieve the fractionations previously reported. We will present our experimental results from a study of the fractionation of linear low density polyethylene (LLDPE) as a function of SCF temperature and thus crystalline melting point.

77. THE PIVOT ALGORITHM WITH INTERACTION ENERGY. L.A.Johnson, R.A. Friesner, Department of Chemistry, Columbia University, New York, NY 10027

The pivot algorithm is a dynamic Monte Carlo method for generating self-avoiding polymer configurations. Configurations were generated on a cubic lattice with interaction energy. This algorithm scales a N, the length of the polymer chain, and represents an O(N) improvement over previous polymer generating algorithms. This allows statistical ensembles of polymer chains to be generated up to 14,000 chromophores in length. The results for polymer chains up to length 800 are presented.

78. NUMERICAL SIMULATION OF THE TIME DEPENDENCE OF ELECTRONIC ENERGY TRANSFER AND DEPOLARIZATION IN POLYMERS.
W. Samuel Parsons, S. E. Webber, Department of Chemistry and Biochemistry and Center for Polymer Studies, University of Texas at Austin, Austin, TX 78712.

Calculations of the time dependence of the depolarization of a donor excited state by down-chain electronic energy transfer (EET) and of the donor excited state survival probability in the presence of EET were carried out on polymer coils embedded in a cubic lattice having a low loading of chromophores and traps. EET between chromophores was assumed to occur via the Förster mechanism, with simultaneous trapping. It has been found that our calculations provide results for the root-mean-square radius of Gyration $<R_g^2>_a^{1/2}$ of poly(2-vinyl naphthalene-co-methyl methacrylate) (2VN/MMA) in a poly(methyl methacrylate) (PMMA) host that are consistent with excitation transport calculations of Fayer and co-workers.[12] The survival probability and depolarization are compared as tools for determining morphology and configuration of polymer coils.

79. COMPUTER SIMULATION OF DIRECT ENERGY TRANSFER IN POLYMERS
J. D. Byers, S.E. Webber, Department of Chemistry and Biochemistry and Center for Polymer Research, The University of Texas at Austin, Austin TX 78712, USA

The process of Direct Energy Transfer (DET) between a sensitized donor and an acceptor bound to the same polymer is a useful tool to study the morphology of the polymer system. Computer simulation of this process involves both the modelling of the polymer coil and the calculation of the probability of energy transfer with time in each coil. For this work the polymer coil was treated as a self-avoiding walk with a short range energy of interaction between non-bonded segments. Configurations were generated using the pivot algorithm employing the Metropolis importance sampling technique. The transfer probability was calculated based upon the Förster transfer rate and various donor-acceptor positions within the coil. These simulations show the dependance of the survival probability on the end-to-end distribution function and the overall morphology of the polymer coil.

80. PHENYL RING MOTIONS IN POLYCARBONATES. Huai Sun, Biosym Technologies, Inc., 10066 Barnes Canyon Road, San Diego, CA 92121

Phenyl ring motions in Bisphenol-A-Polycarbonates (BPAPC) have been investigated by performing molecular mechanics calculations and molecular dynamics simulations on 2,2-bis(4-hydroxyphenyl)propane (BHPP) and its clusters. The calculations are based on *ab initio* force fields developed recently. The calculated rotational minimum energy map shows that there are strong couplings between two phenyl groups and the minimum energy transition path is associated with coherent rotations of the phenyl rings. The transition barrier is found to be about 2.1 kcal/mol in the isolated molecule, but considerably higher in the condensed phases. Molecular dynamics simulations indicate that there are at least two motions of phenyl rings in the thermodynamic equilibrium states; a faster, small amplitude oscillation and a slower, large amplitude ring flip. Both motions are effectively influenced by temperature. The rings undergo 180° complete flips during 50 ps simulations at 500K but not at 300K. The results are compared with experimental results found in the literature.

81. SPECTROSCOPIC STUDIES OF ION IMPLANTED POLYCARBONATES. J.M. Sloan, Polymer Research Branch and M.F. Chen, Ceramics Research Branch, U.S. Army Materials Technology Laboratory, Watertown, MA 02172

The effect of high-energy ion implantation of boron, nitrogen and fluorine into a polycarbonate matrix was investigated by Fourier transform infrared spectroscopy and UV/VIS spectroscopy. A constant ion beam energy of 50 keV was used with dose rates varying from 1×10^{14} to 1×10^{16} ions/cm^2 to implant a Lexan polycarbonate. The FT-IR results showed initial degradation of the carbonate linkage at low dose rates. As the dose rates increased, a more violent and complete degradation of the Lexan material was observed such that carbonaceous deposits were formed below the surface. Solubility measurements of the surface material suggest a coating of this material is formed that protects the polymer from solvent attack. By observing the UV/VIS spectrum, one can follow the build-up of the latter material. Monitoring of the 350 nm band yeilds a ranking of activity of implanted ions where boron degrades the polycarbonate faster than nitrogen of fluorine.

82. UNIAXIALLY ORIENTED POLYAMIDE FILMS IN-PLANE PREPARED BY VACUUM DEPOSITION POLYMERIZATION METHODS. <u>Jiro Sakata</u>, Midori Mochizuki, Toyota Central Res. & Dev. Labs., Nagakute-cho, Aichi, 480-11, Japan

Polyamide films were deposited by a vacuum deposition polymerization (VDP) method from terephthaloyl chloride and two kinds of diamine on substrates which were rubbed with many kinds of polymers. Evident dichroism of absorption bands assigned to C=O groups was observed in IR spectra measured with polarized light parallel and perpendicular to the rubbing direction. From these results, it was found that uniaxially oriented polyamide films in-plane could be prepared by VDP. Orientation degrees of the films were affected by both critical surface tension of the rubbing materials and temperature of substrates, and its direction depended on the diamine monomers and the rubbing materials. The uniaxially oriented polyamide films were applied to alignment films for ferroelectric liquid crystals (FLC). The FLC devices showed bistability with longer memorization time than that using conventional polyimide films did. It is clear that utilization of the uniaxially oriented polyamide films makes it possible to realize the FLC devices possessing high memory capability.

83. The Facile Measurement of Phenolic End-Groups in Bisphenol A Polycarbonate Using GPC-UV Analysis. C. O. Mork and <u>D. B. Priddy</u>, The Dow Chemical Company, Midland, MI 48640

To maximize polycarbonate stability, it is extremely important that the polycarbonate (PC) manufacturing process be operated in such a way as to minimize uncapped polymer chain-ends. Insurance of complete end-capping during manufacture would require constant process monitoring using a fast and accurate phenolic end-group analysis technique for samples typically having <5% uncapped chain-ends. Published end-group analysis techniques are inaccurate if the polymer samples are contaminated with monomer or other small phenolic molecules. Accuracy can be improved by purifying the polymer before analysis, but this procedure is sometimes tedious. This paper describes the development of a facile GPC-UV technique for quantitative determination of low levels of phenolic uncapped chain-ends in PC. This technique improves accuracy because monomers and other small molecule impurities are separated and excluded from the PC during the analysis.

84. **POLYMER AND BIOPOLYMER MODIFICATIONS IN SUPERCRITICAL FLUIDS**
Manssur Yalpani, Domtar Research Centre, P.O. Box 300, Senneville, Quebec, Canada, H9X 3L7; Present address: NutraSweet Co., 601 E. Kensington Road, Mount Prospect, IL, 60056, U.S.A.

The use of supercritical fluid (SCF) technology for the extraction of low molecular weight substrates is well established. SCF applications have also been extended to high molecular weight species, e.g., for fractionations of synthetic polymers and polymerizations in SCFs. An attractive prospect for SCF methodology arises from its use as reaction medium. This paper illustrates the advantages SCF media offer for synthetic polymer and biopolymer modifications. Examples will be discussed, in which SCF-mediated reactions give rise to enhanced reactivities, higher degrees of selectivity, and homogeneous reactions for systems that are amenable only to heterogeneous modifications in conventional media.

85. RESIDUAL STRESS BEHAVIOR OF FUNCTIONALIZED POLY(AMIC ACID) PRECURSORS AND THEIR POLYIMIDES. M. Ree, S. Swanson, and W. Volksen, *IBM General Technology Division, Hopewell Junction, NY 12533 and IBM Almaden Research Center, San Jose, CA 95120*

Poly(amic acid) precursors are commonly functionalized with a crosslinkable monomer such as acrylate or methacrylate derivatives through covalent or ionic bond formation. Then, the crosslinkable functional group provides a photochemical patternability to a polyimide precursor, but may affect properties of resulting polyimide. In the present study, several poly(amic acid) precursors including fully rodlike, semi-rigid, and flexible polyimides were functionalized by attaching a methacrylate derivative, 2-(dimethylamino)ethyl methacrylate (DMAEM), through acid/amine complexation; PMDA-PDA, BPDA-PDA, PMDA-ODA, and BTDA-ODA poly(amic acid)s. For the poly(amic acid)s functionalized with DMAEM, their residual stress behavior on Si(100) wafers was dynamically measured during thermal imidizing and subsequent cooling over the range of 25°C ~ 400°C. Their stress behaviors were compared with those of the respective poly(amic acid)s and polyimides. The stress behavior of PMDA-PDA and BPDA-PDA polyimides was sensitivey changed with the history of DMAEM loading; namely, their stresses drastically increased with DMAEM loading. On the other hand, the stress of PMDA-ODA and BTDA-ODA polyimides was little varied. In addition, their stress relaxation was investigated at 25°C in air ambient with 50 %RH. The stress relaxation was also dependent upon the history of DMAEM loading. These stress behaviors are understood with consideration of structure/properties relationship and microcavities possibly formed by DMAEM degassing during thermal curing.

86. APPLICATION OF NMR TECHNIQUES FOR THE CHARACTERIZATION OF SILICON CONTAINING PHYNYLATED SOLUBLE ARAMIDS. Tamera S. Jahnke and David Walker, Department of Chemistry, Southwest Missouri State University, Springfield, MIssouri 65804, and Sanjiv S. Mohite, Center for Scientific Research, Southwest Missouri State University, Springfield, Missouri 65804.

The objective of this investigation is the complete characterization of a series of silicon containing phenylated soluble aramids by NMR spectroscopy. Seven new polymers were developed by Mohite, Maldar and Marvel in an attempt to improve the processibility of aramids [J. of Poly. Sci.: Part A: Poly. Chem. 26, 2777, (1988)]. The aramids were prepared by low temperature interfacial polycondensation of silicon containing diacid chlorides with diamines. These polymers were completely characterized by proton and carbon-13 NMR spectroscopy with the aid of two-dimensional NMR techniques. Three of the polymers had variable concentrations of two diamines but no characterization was done prior to this study to determine if these mole percentages were in fact part of the polymer. Using carbon-13 NMR we have been able to observe the ratios of different diamines within the polymer.

87. **CHARACTERIZATION OF INTERPHASES BETWEEN PMDA/4-BDAF POLYIMIDES AND SILVER SUBSTRATES USING SURFACE-ENHANCED RAMAN SCATTERING AND X-RAY PHOTOELECTRON SPECTROSCOPY,** W. H. Tsai and F. J. Boerio, Department of Materials Sci. and Eng., University of Cincinnati, Cincinnati, OH 45221-0012

The molecular structure of interphases formed by curing the polyamic acid of pyromellitic dianhydride (PMDA) and 2,2-bis[4-(4-aminophenoxy)-phenyl]-hexafluoropropane (4-BDAF) against silver substrates were determined by surface-enhanced Raman scattering (SERS) and x-ray photoelectron spectroscopy (XPS). SERS spectra of the cured polymer were more similar to the SERS spectra of the polyamic acid than to the normal Raman spectrum of the polyimide, indicating that curing of the polymer was inhibited by interaction of the polyamic acid with the substrate. These conclusions were substantiated by results obtained from XPS spectra of thin films of the polyamic acid of PMDA/4-BDAF cured against thick silver films. This was evidenced by observing a peak shifted upward by 3.2 eV from the main carbon peak near 284.6 eV which consisted of two overlapping peaks characteristic of the amide and carboxylate groups.

88. MEASUREMENT OF DIFFUSION COEFFICIENTS IN POLYMER MELTS USING SURFACED-ENHANCED RAMAN SCATTERING by P. P. Hong, F. J. Boerio, and S. J. Clarson Department of materials Science, University of Cincinnati Cincinnati, Ohio 45221

Interdiffusion of layers of polystyrene (PS) and deuterated polystyrene (DPS) was investigated using surface-enhanced Raman scattering (SERS). Silver island films having a thickness of about 45 A were evaporated onto glass slides. PS films were deposited on the silver films by spin-coating from solution. Films of DPS were deposited on clean glass slides by spin-coating from solution. The slides were immersed in water, resulting in delamination of the DPS films which were picked up on the PS-coated silver island films to form bilayers. When SERS spectra were obtained from the as-prepared bilayers, a band characteristic of the PS films adjacent to the surface was observed near 1014 cm^{-1} but no bands related to DPS were observed, indicating the short range nature of SERS. However, after the specimens were heated at 170°C for a few minutes, a band characteristic of DPS appeared near 976 cm^{-1}. After additional heating, equilibrium was reached and the relative intensities of the bands near 1014 and 976 cm^{-1} did not change with time. The time required to reach equilibrium and the thickness of the PS films were used to calculate the diffusion coefficient which were in good agreement with reported values.

89. **ON THE MISCIBILITY AND COCRYSTALLIZATION IN BLENDS OF POLY(PROPYLENE) AND POLY(BUTENE-1).** T.H. Lee, H. Marand, Chemistry Department, Virginia Polytechnic Institute & State University, Blacksburg, VA 24061.

The thermal behavior and morphology of poly(propylene)/poly(butene-1), (PP/PB-1), blends prepared by either freeze drying or melt mixing were investigated by differential scanning calorimetry, dynamic mechanical analysis, wide angle X-ray diffraction, scanning electron and polarized optical microscopy. A single, composition dependent glass transition temperature was observed for quenched blends and is indicative of at least some level of miscibility between PP and PB-1 in the melt state. The melting behavior of blend samples crystallized isothermally between the equilibrium melting temperatures of PB-1 and PP exhibits very strong dependences on both the composition and the crystallization temperature. Spherulitic isothermal growth rates of α phase PP were measured as a function of crystallization temperature and blend composition. At a given crystallization temperature, the growth rate of α phase PP crystals decreases with increasing concentration in PB-1. Such an observation is consistent with both a dilution effect at the crystal growth front and the observation of an equilibrium melting temperature depression for the PP crystals in the blends. Combined results of wide angle X-ray diffraction, DSC melting studies and temperature dependence of spherulitic growth rates lead us to suggest that partial cocrystallization of the PB-1 with PP is possible. Not only, can this explanation account for the observed melting temperature depression but also it brings further support to the idea that PP and PB-1 are at least partially miscible.

90. **FINITE THICKNESS EFFECTS IN THE THERMODYNAMICS OF DIBLOCK COPOLYMER THIN FILMS.** M. D. Foster, Dept. Polymer Science, University of Akron, Akron, OH 44325-3801; M. Sikka, N. Singh, F. S. Bates, Dept. Chemical Engineering. & Materials Science, University of Minnesota, Minneapolis, MN 55455; S. Satija, C. F. Majkrzak, RRD, National Institute of Standards and Technology, Gaithersburg, MD 20899.

The microstructure of thin films of a nearly symmetric (f=.55) model diblock copolymer, poly(ethylenepropylene) - poly(ethylethylene) has been studied in a temperature range about the bulk order-disorder transition temperature, T_{ODT}, (125°C) using neutron reflectometry. Microstructural change with temperature as well as the dynamics of disordering were found to depend strongly on film thickness, t, for thicknesses comparable to 3 - 12 times the bulk domain spacing, d_{bulk}. For a film thickness of approximately six times d_{bulk} the disordering process proceeded gradually over a broad range of temperature. This behavior contrasted markedly with that in the bulk where the transition has been shown by rheological and small angle neutron scattering measurements to be weakly first order. An unexpected, very slow, rearrangement of the film structure was also observed upon initially heating films of this thickness. The apparent domain spacing changed to a value about 50% larger than that in the bulk.

91. **IONIC CONDUCTIVITY OF POLY(ETHYLENE OXIDE) BOUND WITH A STERICALLY HINDERED PHENOLATE.** C. Foo and Y. Okamoto,† Department of Chemistry, Polytechnic University, 333 Jay Street, Brooklyn, New York 11201; T. Skotheim, Department of Applied Science, Brookhaven National Laboratory, Upton, Long Island, New York 11973.

The 2,6-di-t-butylphenol group bound covalently to the poly(ethylene oxide) are prepared and characterized. The Li and K salts of the phenolate are synthesized and the ionic conductivity of these ions are measured as a function of temperature. The conductivity for K is 10^{-6} s/cm at 50°C. This value is lower than the data of previously reported polymer systems in which the oligo(ethylene oxide) groups were bound as side chains to the polymer backbone. The ionic conductivity of these systems is associated with the local segmental movement of the polymer-ion complex. This segmental motion along the main chain is smaller than that along the side chain, as a result the Li and K ion conductivities of the presently reported polymer system are lower than those in which oligo(ethylene oxide) were covalently attached as side chains to the backbone.

92. ESR STUDIES ON THE POLYMERIZATION OF DIACETYLENE CONTAINING ACRYLATE AND METHACRYLATES.
Jimmy S. Hwang[1], Guillermina Burillo[1], and Takeshi Ogawa[2]*.
Instituto de Ciencias Nucleares[1], and Instituto de Investigaciones en Materiales[2], Universidad Nacional Autonoma de Mexico, Circuito Exterior, Ciudad Universitaria, Mexico DF 04510, MEXICO.

Diacetylene containing acrylate and methacrylates were prepared and their radicals formed during the polymerization at 70°C, and by Gamma ray irradiation at liquid nitrogen temperature, were analyzed by ESR spectroscopy. At 70°C, the methacrylate radicals were strongly interacted with the diacetylene groups. At low temperatures, the mathacrylate radicals do not interact with the diacetylenes, but for some monomers an intramolecular interaction was observed. These monomers showed blue color when irradiated by Gamma ray at liquid nitrogen temperature, but the color faded at above -100°C, and this was attributed to the free electrons delocalized in the diacetylene system, rather than the polymerization of diacetylene groups.

93. SOLUTION PROPERTIES OF MMA/HEMA COPOLYMERS.
Joaquín Palacios, Amparo Ortega, Jorge Sierra.
División de Estudios de Posgrado, Facultad de Química, UNAM. Cd. Universitaria, 04510, México, D.F. México.

Solution properties of methyl methacrylate/hydroxyethyl methacrylate (MMA/HEMA) copolymers, were studied in butanone, tetrahydrofurane and butanone-isopropanol solvents. Experiments were run at 25°C in an Ubbelholde viscometer. The GPC molecular weights and distributions data were combined with intrinsec viscosities, to calculate the hydrodynamic volumes of materials. Values of the a y K Sakurada constant are reported for these systems. The unperturbed dimensions were correlated with the composition of copolymers. Values for the 95, 90, 80% mole copolymers, are presented

94. POLYMERIC FLOCCULANTS OF THE ZWITTERIONIC TYPE.
Octavio Manero, Judith Cardoso and Teresa Orta*
Instituto de Investigaciones en Materiales, UNAM. AP.70-360,Coyoacán 04510 México, D.F. *Instituto de Ingeniería, UNAM.

It has been observed that polymeric flocculants present advantages with respect to more traditional inorganic flocculants. In this investigation, we present a comparative study between commercially-available water-treatment compounds and water-soluble polymers which contain ionic structures, such as N-oxide groups. These polymers present special characteristics, among them a high water solubility, high molecular expansion in solvents with high ionic strength and a suitable chemical stability. These properties can be useful in flocculation processes applied to water treatment for industrial waste effluents. Results show a better particle separation than that observed in commercial polyelectrolytes, with concentration around 1 ppm at low pH. Stable flocs are formed and their efficiency increases with convection. An extension to toxic metals separation is presently considered.

95. DYNAMIC MECHANICAL THERMAL ANALYSIS OF MOLD COMPOUNDS USED IN ELECTRONIC PACKAGES, Indira S. Adhihetty and R. Paddy Padmanabhan, Advanced Packaging Laboratory, B-136, Motorola, 5005 E. McDowell, Phoenix, Az. 85008.

Due to the increase in die sizes, the packaging of devices such as memory and microprocessors is becoming increasingly challenging. There has been a considerable effort to replace common mold compound materials used in several plastic packages with newer generations of mold compounds specifically designed for housing larger dice where the stress levels imposed during thermal cycling are high enough to cause one or all of the problems listed below, viz., cracks in the mold compound (mostly around the corners of the package), passivation cracking, metal movement and Al metallization corrosion. The objectives of this work were to evaluate the properties of some widely used mold compounds through Dynamic Mechanical Thermal Analysis and study the effects of test parameters upon these properties with the intent of establishing a baseline for future evaluations of newer mold compounds. Test variables like frame size, clamp type and clamping method did not have any effect on the mechanical properties of the samples within the test ranges in this study. However, frequency had a pronounced effect upon these properties. In particular, T_g increased by approximately 7° C for every ten fold increase in frequency, consistent with a theoretical prediction of about 7.5° C. The results also indicated that T_g was well outside the typical thermal cycle range and hence, mold compound materials such as these could be used without any risk of a drastic change in the properties during the course of the thermal cycle test.

96. MONITORING THE CURE OF PHOTOSENSTIVE POLYMERS USING A LUMINESCENT RHENIUM CARBONYL COMPLEX AS A SPECTROSCOPIC PROBE, Thomas G. Kotch, Alistair J. Lees, Department of Chemistry, S.U.N.Y.-University Center, Binghamton, NY 13902-6000, and Stephen J. Fuerniss, Kostas I. Papathomas and Randy Snyder, Systems Technology Division, IBM Corporation, Endicott, NY 13760

The complex fac-ClRe(CO)$_3$(Ph$_2$-phen) [Ph$_2$-phen = 4,7-diphenyl-1,10-phenanthroline] has been shown to be measurably luminescent in both acrylate and epoxy based photopolymers. The visible luminescence ($\lambda_{em.}$ = 565 - 585 nm) is attributed to a metal-to-ligand charge transfer state and is sensitive to changes in environmental rigidity. Upon curing of the model photoresist systems the emission band of the organometallic probe is seen to blue shift up to 1934 cm^{-1} and increase in intensity up to seven times. The unique luminescence characteristics of this organometallic complex make it useful as a non-reactive probe in cure monitoring of photopolymer systems.

97. STUDY OF MISCIBLE BLENDS OF POLY(VINYLIDENE FLUORIDE) AND POLY(3-HYDROXYBUTYRATE). S.L. Edie, H. Marand, Chemistry Department, Virginia Polytechnic Institute & State University, Blacksburg, VA 24061.

The thermal behavior and morphology of poly(vinylidene fluoride)/poly(3-hydroxybutyrate), (PVF$_2$/PHB), blends prepared by either solution or melt mixing were investigated by differential scanning calorimetry and polarized optical microscopy. A single, composition dependent glass transition temperature was observed for all blends regardless of the blend preparation conditions. Spherulitic isothermal growth rates of α phase PVF$_2$ were measured as a function of crystallization temperature and blend composition. At a given crystallization temperature, the growth rate of α phase PVF$_2$ crystals decreases with increasing concentration in PHB. Such an observation is consistent with both a dilution effect at the crystal growth front and the existence of an equilibrium melting temperature depression for the PVF$_2$ crystals in the blends. Study of the dynamic crystallization behavior of these blends under constant cooling rate indicates that the temperature at maximum crystallization rate for the PHB component is shifted to lower temperature when the blends are prepared by melt mixing instead of solution blending. These results can be explained by a decrease in the molecular weight of the poly(3-hydroxy butyrate) component during melt mixing. This explanation is confirmed by the observation of a lower intrinsic viscosity for the PHB exposed to melt blending conditions as compared to the PHB exposed to solution mixing conditions. The effect of the PVF$_2$ component on the degradation kinetics of PHB will be discussed.

98. CHARACTERIZATION AND FUNCTION OF HYDROPHOBIC POLYSACCHARIDE AGGREGATES. Kazunari Akiyoshi, Shigehiko Yamaguchi,[†] Shigeru Deguchi and Junzo Sunamoto, Department of Polymer Chemistry, Faculty of Engineering, Kyoto University, Kyoto 606, and [†]Tsukuba Research Laboratory, Nippon Oil & Fats Co., LTD., Tokodai 5-10, Tsukuba, Ibaraki 300-26

Naturally occurring polysaccharides are one of the most abundant and diverse families of biopolymers, and are widely applied to various fields. We have developed liposomes or oil-in-water droplets as coated by cell-specific polysaccharides which bear a hydrophobic anchor and utilized them in biotechnology and/or medicine. Recently we have found that palmytoyl- or cholesterol-substituted derivatives of naturally occurring polysaccharides such as pullulan, amylopectin, and dextran derivatives form self aggregates in an aqueous solution. In this work, the solution characteristics of pullulan derivatives bearing long alkyl chain or cholesterol moieties and the function of polysaccharide aggregate as a new host for hydrophobic guest molecules will be presented.

99. **MULTIPLE THERMAL TRANSITIONS IN MODEL COMPOUNDS FOR POLY(γ-ALKYL-α,L-GLUTAMATE)S.**
William H. Daly and Ioan I. Negulescu, Macromolecular Studies Group, Department of Chemistry, Louisiana State University, Baton Rouge, LA 70803-1804

A reinvestigation of multiple transitions occurring in poly(γ-hexadecyl-α,L-glutamate) and poly(γ-stearyl-α,L-glutamate) revealed that depending upon their thermal history there exists a large temperature domain in which the side chains can melt or crystallize. Hexadecyl alcohol, C16(OH), and N-phthaloyl-γ-stearyl-L-glutamate, NPSLG, were chosen to mime the melting of side chains in these polymers. A series of polymorphic transformations were evidenced in C16(OH) by FTIR, DSC and optical observations in polarized light while the transitions in NPSLG resembled closely that from poly(γ-stearyl-α,L-glutamate). Since in all cases the same enthalpic change was registered for all cooling cycles it has been concluded that 1) the same number of methylenic units is involved in the crystallization process and 2) the multiple transitions represent melting or formation of paraffinic crystallites of different degree of perfection.

100. A STUDY OF THIN BICOMPONENT POLYMER FILMS ON A METAL SURFACE BY ESCA. B. M. Ginzburg, A. O. Pozdnjakov, B. P. Redkov, D. G. Tochilnikov, Leningrad Branch of Mechanical Engineering Research Institute.

Thin films (3-5 μm) were casted on steel foils from diluted polymer compositions (two compatible polymers and a solvent). One composition contended polybenzimidazole-1 (PBI-I) plus polyorganosylaxane (POS), the other is POS+fluorine-epoxy polymer (FEP). After casting the phase separation of composite solutions occured during film drying: on the metal surface concentrates the polymer with polar groups (amine-groups in PBI-I and epoxy-groups in FEP) but on free surface dominates POS.
After the second casting of (FEP+POS) composition the new effect of gradient inversing in the second layer is observed. It can be explained by stronger interaction of FEP with itself on solid surface of the first layer. This effect can be of great importance for covers technology because the second covering changes the chemical composition and properties of surface.

101. MICRODEFORMATIONAL BEHAVIOUR OF ORIENTED BICOMPONENT POLYMER SYSTEMS. B. M. Ginzburg, Sh. Tuichiev, Tadjik State University, Dushanbe, USSR.

The study of high oriented bicomponent polymer systems (fibers and films) is carried out by SAXS method in combination with stretching of samples in diffractometer in situ. Two kinds of objects were used: 1. fibers from one component with second grafted one from the gas phase: nylon 6 with grafted polystyrene (N6/PS), isotactic polypropylene i-PP/PS, i-PP/polyacrylonitril; 2. films and fibers from polymer blends; i-PP+polyoxymethylene (POM), POM+high density PE, N6+low density PE, polyimide+polysulfonamide.
The analysis of experimental data was made in accordance with a new approach, which the competition of inhomogeneous deformation of long periods and mutual fibril slipping take into account. The more the part of grafted component the stronger the process of interfibrillar slipping. In case of oriented blends the technological prehistory of samples plays the dominant role. Some new possibilities of X-ray methods for study of uniaxial fiber-reinforced composites are discussed.

102. HIGHLY POLARIZABLE METALLIC COMPLEXES FOR NONLINEAR OPTICS. COBALTOUS COMPLEXES OF UNSYMMETRICAL HYDRAZONE IMINE GLYOXAL DERIVATIVES.
Thierry Thami, Pierre Bassoul, Michel A. Petit, Jacques Simon, Alain Fort, Marguerite Barzoukas, Albert Villaeys.

The synthesis of 1-p-phenylhydrazono-2-phenylimino-ethane derivatives unsymmetrically substituted in the para-position with an electron acceptor (-NO$_2$) and an electron donor (-OMe or -NMe$_2$) are described and the corresponding cobaltous complexes are prepared. X-ray diffraction on a single crystal of the dimethylamino-complex has been performed (space-group: $P\bar{1}$) showing that the coordinate site around the cobalt ion is nearly tetrahedral. The complex itself is approximately of C_2 symmetry. The inversion center of the space-group transforms one optical isomer into the other. The hyperpolarizability coefficients (β) of the ligands in their cis and trans-forms, and of the neutral cobaltous complexes, have been determined in solution by the Electric Field Induced Second Harmonic (EFISH) technique. The magnitude of the β-value obtained for the complex is larger than the value calculated from the tensorial addition of the molecular hyperpolarizability coefficients of the ligands. The importance of the cation on the nonlinear optical properties of metallo-organic complex is outlined.

103. INVESTIGATION OF NEW MOLECULES AND MATERIALS FOR QUADRATIC NONLINEAR OPTICS. J.F. Nicoud, Institut de Physique et Chimie des Matériaux de Strasbourg, Groupe des Matériaux Organiques, ICS -6, rue Boussingault 67083 Strasbourg cedex France.

Following our constant interest in the development of new materials for quadratic nonlinear optics, we have investigated the nonlinear optical properties of 4,4'-diphenylacetylene derivatives (tolans) 1 and zwiterrionic stilbazolium alkylsulfonate derivatives 2. Calculated and experimental hyperpolarizabilities ß of compounds 1 will be reported. A SHG screening by the powder test reveals a high proportion of non-centrosymmetrical crystal structure that can be related to the conformation of the tolan skeleton. Compounds of type 2 are zwiterrionic analogues of stilbazoliums salts that allow ß EFISH and dipole measurements contrary to ionic salts. We will describe two SHG highly efficient materials relative to their transparency range. The crystal struture of one of them points up the role of the highly polar pyridinium alkylsulfonate part of the molecule.

104. Preparation of Polymeric Films for NLO Applications. R. P. Foss, W. Tam, F.C. Zumsteg, Central Research & Development Du Pont Experimental Station, P.O. Box 80328, Wilmington, DE, 19880-0328.

The study of poled polymeric films containing attached organic molecules with delocalized π-electronic structures and large charge-transfer resonances is an active area of research. This intense interest is generated mainly from the potential for large nonlinear optical coefficients, ease of fabrication, and high optical quality. We will focus our talk on the preparation of side-chain and cross-linked polymers containing a variety of acceptor groups (-NO$_2$, SO$_2$C$_3$F$_7$, -CHO, CH=C(CN)$_2$, and C(CN)=C(CN)$_2$) , and a number of different delocalized π-electronic structures (phenyl, biphenyl, azoabenene and stilbene). We will briefly discuss the nonlinearity and temporal stability resulting from poling of these polymers..

105. SYNTHESIS OF SIDE-CHAIN AND MAIN-CHAIN NONLINEAR OPTICAL POLYMERS WITH SULFONYL ACCEPTOR GROUPS Douglas R. Robello, Jay S. Schildkraut, Nancy J. Armstrong, Thomas L. Penner, Werner Köhler, and Craig S. Willand, *Corporate Research Laboratories, Eastman Kodak Company, Rochester, New York 14650-2109*

Several novel polymers for second-order nonlinear optics have been prepared containing covalently-bound sulfonyl electron-acceptor groups. The synthetic flexibility provided by the sulfonyl group made possible the preparation of side-chain, main-chain, and amphiphilic polymers, each of which displayed substantial nonlinear optical activity after appropriate processing. We present details of the synthesis and characterization of these polymers, and discuss their nonlinear optical, electrical, and thermal properties.

106. **SYNTHETIC APPROACHES TO NLO POLYMERS.** R. Twieg, D. Burland, M. Lux, C. R. Moylan, C. Nguyen, P. Walsh, C. G. Willson, R. Zentel, IBM Research Division, Almaden Research Center, 650 Harry Road, San Jose, CA 95120-6099.

A gradual enhancement of basic understanding and a corresponding optimization of the functional properties of second-order nonlinear optical polymers has brought them closer to practical applications in a variety of integrated optoelectronic devices, such as modulators and harmonic generators. This progress is due to the simultaneous and synergistic optimization of chromophore, polymer and processing to deliver both a large and stable optical nonlinearity. Toward this end we have concentrated on the synthesis of polar tolan (diphenylacetylene) nonlinear chromophores and their implementation in a variety of linear and crosslinking epoxy resins. The amino donor is a common feature in these polar tolans and the acceptor groups examined include nitro, methylsulfonyl and perfluoroalkylsulfonyl. Changes in the acceptor groups have a profound influence on the physical properties of these polymers as well as their nonlinear optical properties. The synthetic chemistry, microscopic nonlinearities (as determined by both calculation and EFISH measurements) and incorporation of these tolan chromophores into epoxy polymers will be discussed.

107. NOVEL SIDECHAIN POLYMERS FOR ELECTRO-OPTIC APPLICATIONS. Diane E. Allen, Richard A. Keosian, Garo Khanarian, Jane Hudop, Stephen J. Meyer, Hoechst Celanese Corporation, 86 Morris Avenue, Summit, NJ 07901

A tremendous amount of interest surrounds the use of organic and polymeric materials for advanced electro-optic device and systems applications. The appeal of organic compounds lies in the flexibility in which one can tailor the molecular and macroscopic properties to achieve high activity, desirable mechanical characteristics, and processibility. To realize bulk second order activity the molecular dipole must be aligned non-centrosymmetrically. Since NLO active small molecules tend to crystallize with their dipoles paired, resulting in the loss of second order properties, it is desirable to incorporate the active units into a polymer and orient the dipoles via an external field. This paper will discuss the synthesis and evaluation of NLO active sidechain polymers.

108. SYNTHESIS AND NONLINEAR OPTICAL CHARACTERISTICS OF CHROMOPHORE-FUNCTIONALIZED POLYMERS HAVING CHROMOPHORE-CENTERED HYDROGEN-BONDING AND CROSSLINKING GROUPS. Y. Jin, and S. H. Carr, Department of Materials Science and Engineering and the Materials Research Center, T. J. Marks, Department of Chemistry and the Materials Research Center, W. Lin, and G. K. Wong, Department of Physics and the Materials Research Center, Northwestern University, Evanston, IL 60208-3113.

Major goals in the design and synthesis of poled, chromophore-functionalized polymeric second-order NLO materials include maximizing the chromophore number density and enhancing the temporal stability of poling-induced chromophore orientation. We report here a new class of NLO polymers in which the appended chromophore substituent is modified (hydroxylated) to allow direct chromophore involvement in both hydrogen-bonding and crosslinking processes. Thus, the glassy polymer poly(p-hydroxystyrene) has been functionalized to varying levels with the N-(3-hydroxy-4-nitrophenyl)-(S)-prolinoxy (HNPP) substituent. The time-dependent second harmonic generation characteristics of thin films of this material have been studied at λ = 1.064μm as a function of chromophore functionalization level, poling technique, film processing methodology, ambient atmosphere, and thermal crosslinking reagents. Detailed comparisons have also been drawn with both the O-benzylated and chromophore nonhydroxylated analogues.

109. **LIQUID CRYSTALLINE THERMOSETS: I. THERMAL BEHAVIOR OF MIXTURES** Andrea E. Hoyt and Samuel J. Huang, Institute of Materials Science, University of Connecticut, Storrs, CT 06269-3136

Binary mixtures of thermosetting liquid crystals based on chlorohydroquinone and methylhydroquinone were investigated in an attempt to lower transition temperatures by inducing eutectic behavior in the system. It was found that simply mixing the two monomers was insufficient to induce such behavior, but that this behavior could be induced by heating the monomers together for a brief time.

110. **TRIAZINE RIGID-ROD NETWORKS.** George G. Barclay[*], Christopher K. Ober[*], Kostas I. Papathomas [‡] and David W. Wang[‡], (*) Materials Science and Engineering, Bard Hall, Cornell University, Ithaca, NY 14853-1501, (‡) IBM Corporation, Systems Technology Division, Endicott, NY 13760-0553

Novel liquid crystalline (LC) triazine networks were prepared by the thermal cyclotrimerization of dicyanate compounds of ring substituted di(4-hydroxyphenyl)terephthalates. As the curing reaction proceeds a mesophase is formed. The resulting triazine networks have glass transition temperatures of ~190°C and show little decomposition until ~440°C. The LC phase can be aligned during the curing process under the influence of a magnetic field to obtain oriented triazine networks. The anisotropy of these networks was investigated by both wide angle x-ray diffraction and by measuring the coefficient of thermal expansion parallel and perpendicular to the direction of alignment.

111.

MOLECULAR COMPOSITES IN THERMOSETS. T.K. KWEI, K. LEVON, F. LIU, S. MUNUKUTLA, AND G. TESORO, DEPARTMENT OF CHEMISTRY, POLYTECHNIC UNIVERSITY, BROOKLYN, NY 11201.

A molecular composite comprising of a bis-maleimide thermoset matrix (Matrimid 5292-A,B) and amine terminated semiflexible polybezimidazole (SF-PBI) is described. The SF-PBI molecules were covalently linked to the matrix by Michael addition reaction of amine groups with the double bonds of the matrix. Composites were prepared by casting films from organic solvents, without the use of hostile solvents. Films with homogeneous morphology and improved mechanical properties (30% increase in tensile strength at SF-PBI/5292-A,B =10/90) were obtained. Phase separation was observed when bis-citraconimide(BCI) was used as the matrix instead of bis-maleimide.

112. RUBBER-MODIFIED BISMALEIMIDES. INFLUENCE OF A POLYSILOXANE RUBBER ON THE POLYMERIZATION PHASE SEPARATION PROCESS, AND GENERATED MORPHOLOGIES. A. Seris[1], J.P. Pascault[1] and Y. Camberlin[2], [1]Laboratoire Matériaux Macromoléculaires, INSA de Lyon, 20, Avenue A. Einstein, 69621 Villeurbanne, France, [2]Rhône Poulenc Recherches, BP 62 69192 Saint Fons, France

Bismaleimide monomers (BMI), were cured in the presence of a non reactive polymethyl phenyl siloxane. The presence of a non reactive polymethyl phenyl siloxane rubber has several effects on the formation of bismaleimide networks.
When BMI chain polymerization is thermally initiated, the polymerization rate is delayed and stopped, around 50 % monomer conversion even with small amounts of rubber. At the end of the reaction, no dispersed particles are visible. This behavior is explained by the fact that the rubber is a good radical transfer agent.
When BMI chain polymerization is initiated by imidazole, the behavior is quite different. The polymerization rate does not change at the beginning of the reaction, but after phase separation, the dispersed particles formed disturb and delay the gelation process. The generated morphologies depend on the initiator content and on the curing temperature.

113. MISCIBLE NOVOLAC/POLYMER BLENDS AND CURING STUDIES. H-I. Kim, J. R. Pennacchia, T. K. Kwei, and E. M. Pearce, Polymer Research Institute, Polytechnic University, Brooklyn, NY 11201.

Substituted ortho-methylene bridged novolacs, poly (1-hydroxyl-2,6-phenylene- methylenes) ((o,o-PHMP's), were previously synthesized and characterized. Miscible blends with a variety of methacrylates were studied in relation to glass transition temperature (T_g) phenomena and hydrogen bonding interactions. Formaldehyde-crosslinking studies were done at low temperature and high temperature, above and below the level critical solution temperatures, for PHMP blends with poly(methyl methacrylate co-styrene) copolymers. It was possible to demonstrate morphological control so that at certain degrees of H-bonding (from IR studies) single T_g (transparent) or two Tg (opaque) semi-interpenetrating polymer networks (semi-IPN's) could be obtained from the same compositions based on the cure temperatures used.

114. **MICRO-MORPHOLOGY CHARACTERIZATION OF UNSATURATED POLYESTER RESINS WITH LOW PROFILE ADDITIVES** Laurent Suspène and Yeong-Show Yang Cray Valley - TOTAL Group FRANCE

Unsaturated polyester (UP) resin is one of the most important thermoset resins for composite applications. One factor, limiting the development of those products, is the high polymerization shrinkage of the resin which may cause several molding problems. To date, in order to eliminate the shrinkage, low profile additives (LPA) are widely used in the curing of unsaturated polyester resins. Numerous investigations have been carried out to explain the mechanism of low profile behavior. The shrinkage compensation is mostly believed due to the formation of microvoids in the interfacial region between resin phase and additive phase.A Novel explanation of low-profile mechanism based on the concepts of phase diagram and fractionated phase separation will be proposed in this work. Based on this mechanism, a series of new acrylic modified low-profile additives for the unsaturated polyester resins have been developed. The results show that by adjusting the polarity of the additives, it is possible to adjust the phase diagrams and the corresponding shrinkage compensation.

115. THE INFLUENCE OF MOLECULAR ARCHITECTURE ON GAS TRANSPORT PROPERTIES OF LIQUID CRYSTALLINE POLYMERS. D. H. Weinkauf and D. R. Paul, Department of Chemical Engineering and Center for Polymer Research, The University of Texas at Austin, Austin, Texas 78712.

The supramolecular architecture of ordered liquid crystalline phases presents a unique opportunity to control the diffusional process in polymeric systems for use in barrier and membrane applications. A critical first step in developing liquid crystalline materials for such applications is to gain a fundamental understanding of the gas transport mechanism. In this investigation, several series of films based on copolyesters of hydroxybenzoic acid (HBA) and 2,6 hydroxynaphthoic acid (HNA) have been prepared in order to allow the examination of the effects of composition and processing. In general, the extraordinarily low permeabilities of these liquid crystalline materials are attributed to the packing efficiency of the rigid rod-like chains. The permeability and diffusivity of the simple gas penetrants decreases with increased orientation and annealing of extruded HBA/HNA films. The effect of the concentration of naphthyl structural units on dynamic mechanical behavior and transport coefficients of the HBA/HNA materials parallels the findings of an analogous study with amorphous, naphthyl based polymers. Film preparation and characterization are discussed.

116. CONDITIONING EFFECT OF CO_2 ON THE GAS TRANSPORT PROPERTIES OF POLYIMIDE ISOMERS. M.R. Coleman and W.J. Koros, Department of Chemical Engineering, University of Texas at Austin, Austin, Texas 78712

Past work has shown that glassy polymers exhibit long-lived hysteresis in gas transport properties when exposed to sufficient levels of highly sorbing penetrants. The hysteresis is generally referred to as "penetrant conditioning". The conditioning of a polycarbonate and a polyimide will be compared at pressures of CO_2 which results in a 7.8% volume dilation. Hysteresis in these two glassy polymers will also be compared for samples which have been conditioned with CO_2 at approximately 60 atmospheres. Penetrant induced hysteresis will be discussed for meta and para connected polyimide isomers conditioned with CO_2 at 60 atmospheres.

117.
Effects of Aryl-Substitution and Physical Blending on The Gas Permeability of Polymers, R. T. Chern, K. Ghosal, and S. Semar, Department of Chemical Engineering, North Carolina State University, Raleigh, NC 27695

Many new membrane materials for gas separation are polymers with aromatic backbone structures. Tailoring of the structure attached to the aromatic group thus offers not only a fruitful means for producing potentially desirable new materials but also a convenient route for studying the effects of molecular architecture on the permeability of these polymers. Physical blending, on the other hand, offers a convenient way of learning the effects of morphology on the transport properties. The effects of aryl halogenation, nitration, and amination on the permeabilities of polyesters and polyethers, together with the permeabilities of a series of blends of poly(1-trimethylsilylpropyne), a polymer with unusually large free volume, and poly(2,6-dimethyl phenylene oxide) will be presented and analyzed.

118. **DIFFUSION IN THE GAS SENSING OF PHTHALOCYANINE LANGMUIR - BLODGETT FILMS**

J. B. Lando, H. Y. Wang and W. H. Ko, Department of Macromolecular Science
Case Western Reserve University, Cleveland, OH 44106

Diffusion of chlorine gas in phthalocyanine Langmuir-Blodgett (L-B) films was studied and modeled. The phthalocyanine [$(C_6H_{13})_3$SiOSiPcOGePcOH] was specially designed for L-B film deposition and was used as sensing membrane for halogen gases. Si was used as the sensor substrate. It incorporated an interdigitated gold electrode in ohmic contact with the sensing film, a microheater and a diode to control the temperature of the film. The gas concentration was measured by the film conductivity change. The film temperature and thickness effects on gas sensitivity and response times were studied experimentally. The gas sensing process was modeled with various rate processes such as adsorption/desorption of the gas molecules on the film surface and then diffusion through the film. The experimental data can be fitted with the model and the results will be discussed.

119. **MECHANISMS OF GAS TRANSPORT IN POLY (1-TRIMETHYLSILYL-1 PROPYNE)**
S. R. Auvil, R. Srinivasan, and P. M. Burban, Air Products and Chemicals, Inc., 7201 Hamilton Boulevard, Allentown, PA 18195

The recent past witnessed the emergence of an extraordinary polymeric membrane material called poly trimethyl silyl propyne (PTMSP). Gas permeabilities in PTMSP are orders of magnitude larger than those in other glassy polymers. Reported in this paper are results from a systematic experimental study aimed at elucidating the mechanism(s) of gas transport in PTMSP. Based on gas solubility, permeability and other data, it is concluded that: (a) Fast diffusion (and not a large solubility) is responsible for the large permeabilities in PTMSP. (b) The polymer appears to have a "porous" structure, characterized by about 25 volume % of voids, and a chain-to-chain separation of at least 4 Å. Results are presented from mixture permeation experiments offering fairly definitive "proof" for the presence of continuous pores in PTMSP. For example, the permeability of a light gas such as He is drastically reduced by the co-permeation of a more strongly sorbing and/or more condensable species such as SF_6.

120. **THE EFFECT OF HIGHER ORDER MOLECULAR STRUCTURE UPON THE SORPTION AND TRANSPORT OF SMALL MOLECULES IN STIFF-CHAIN GLASSY POLYMERS**, Nathanael R. Miranda, Atsushi Morisato, Benny D. Freeman, Harold B. Hopfenberg, Department of Chemical Engineering, North Carolina State University, Raleigh, NC 27695-7905; Gianna Costa, and Saverio Russo, Centro Di Studi Chimico - Fisici Di Macromolecole Sintetiche E Naturali, Instituto Di Chimica Industriale, Universita' Di Genova, 16132 Genoa ITALY.

In PTBA, a high-permeability member of the family of substituted poly(acetylenes), changes in propane sorption vary systematically and dramatically with changes in *cis/trans* ratio as characterized by relative peak heights in the ^{13}C NMR solution spectrum of the polymers. As the *cis* content increases, the solubility and d-spacing, determined from wide angle x-ray diffraction studies, decreases. Based upon comparisons of anisotropic an isotropic arrangements of a single polymer, the secondary molecular structure of ordered liquid crystalline polymers seems to confer low solubility of small molecules. In both high permeability polymers such as the substituted polyacetylenes and low permeability polymers such as the liquid crystalline polymers, higher order (i.e. secondary, tertiary, and quaternary) molecular structure appears to influence markedly the sorption and transport of small molecules in stiff chain glassy polymers.

121. **THERMOREVERSIBLE GELATION: CRYSTALLIZATION AND OTHER MECHANISMS.** L. Mandelkern, and A. Prasad, Institute of Molecular Biophysics, Department of Chemistry, Florida State University, Tallahassee, FL 32306.

Several different mechanisms for thermoreversible gelation will be discussed, based on experimental results for polyethylenes and its copolymers in homogeneous and heterogeneous mixtures, as well as those for the polystyrenes. It is found that for the polystyrenes, thermoreversible gelation can occur over the complete range of chain stereo structures from syndiotactic to isotactic and epimerized samples between. In addition to crystallization being a primary mechanism, several others are shown to be operative for a given system depending on conditions. These include spinodal decomposition, segment-segment interaction and glass formation. Examples of each of these mechanisms will be presented in terms of the required structural and thermodynamic conditions.

122. **MICROPHASE SEPARATION AND REVERSIBLE GELATION.** L. Leibler Groupe de Physicochimie Théorique, E.S.P.C.I., 10, rue Vauquelin, 75231 Paris Cedex 05.

In many macromolecular systems such as block copolymers or weakly charged polyelectrolytes a microphase separation transition can be induced by change in temperature or polymer concentration. Below the transition these systems exhibit complex structures (spherical or wormlike micelles, vesicles etc.) more or less organized on mesoscopic scales. As a result of such an organization they behave in many respects as connected networks and have for example a finite shear modulus.
The aim of this talk is to discuss some structural aspects of block coplymer and weakly charged polyelectrolyte solutions and their relation to macroscopic gel-like behavior.

123. **Arrested melting due to strain. Effect of network formation in thermoreversible gels of polyolefins.** H. Phuong Nguyen and G. Delmas Case postale 8888, Succursale "A" Montréal (Québec) H3C 3P8

Gels of UHMWPE and of P4MP1 in various solvents have been investigated by their dissolution traces at low rate of temperature increase, v. The dissolution traces of the nascent or of the gel show a high temperature endotherm whose magnitude (10 to 30% of the total heat of fusion) varies with thermal history, concentration and nature of the solvent. Such endotherm is found in the melting trace of UHMWPE when v < 12 K/h be it nascent or recrystallized. It has gone unnoticed due to its very slow melting. Interpretation of the dissolution endotherms is made following that proposed for the melting trace. The elevated melting/dissolution (equilibrium) temperatures are associated with strained crystals extremely stable in decalin and alkanes. During cooling, the non-randomized melt recrystallizes as extended-chain crystals at high temperature. Those crystals constitute the network on which non-strained crystals grow during gel formation on quenching. A calorimetric characterization of a gel is proposed. The consequence of these results in other fields of polymer physics are discussed.

124. ASSOCIATION IN SOLUTIONS OF SYNDIOTACTIC POLY(METHYL METHACRYLATE) AS STUDIED BY SPECTROSCOPIC METHODS AND SMALL ANGLE X-RAY SCATTERING. Jiří Spěváček, Inés Fernandez-Pierola*, Josef Pleštil, Institute of Macromolecular Chemistry, Czechoslovak Academy of Sciences, 162 06 Prague 6, Czechoslovakia

According to ^{1}H NMR measurements, in the primary stage of self-association of syndiotactic (s) poly(methyl methacrylate) (PMMA) in solution there occurs double helix formation. In mixed solvents, where one component promotes association, while the other acts against it, there is a sharp transition between the associated and nonassociated states in a very narrow range of mixed solvent composition. The mechanism by which the solvent affects self-association of s-PMMA probably involves specific interactions of the solvent molecules with PMMA units. By small angle X-ray scattering it was confirmed that secondary aggregation of s-PMMA in toluene solution produces compact particles several hundred Å in diameter; about one third of the volume of these particles is occupied by solvent.

*Permanent address: Departamento de Química Física, Facultad de Ciencias Químicas, Universidad Complutense, 28040 Madrid, Spain

125. THERMOREVERSIBLE GELATION FROM POLY(ETHYLENE OXIDE)PARADICHLOROBENZENE AND THEIR INTERCALATE. J.J. Point.Faculté des Sciences.Université de Mons Hainaut,MONS,B7000,Belgique.

The phase diagram of the system poly(ethylene oxide) (PEO) paradichlorobenzene shows a bell shaped region which separates two eutectic points. In the solvent side,the eutectic composition is 0.25 % in PEO. In various conditions,supercooled dilute(from less than 1%)of peo (even of moderate molecular weight)in paradichlorobenzene come to gelation. Melting points of the gels(as determined by DSC,"ball drop method",microscopy) agree with these predicted from the phase diagram.No labile associations between the two species are observed prior to gelation;the gels are nucleated.In such gels(when freezed),we observe an elevation of the melting point of the solvent.It is thus suggested that these gels are lattice gels and that the intercalate crytals are fringed micelles.The properties of these gels and these gelation phenomena are very likely specific to our and related systems.

126. **SYNTHESIS OF CONTROLLED STRUCTURE POLYOLEFINS.** R.H. Grubbs, S. Marder, C. Gorman, E. Ginsburg, J. Moore, Division of Chemistry and Chemical Engineering, California Institute of Technology, Pasadena, CA 91125 USA.

Conjugated polymers show interesting electronic and optical properties. By their very nature, this class of polymers tend to be insoluble and difficult to process. Major advances have been made in the synthesis of soluble precursors and in the preparation of substituted analogs that allow for reasonable processing. Over the past few years, ring opening metathesis polymerization (ROMP) has been used in the preparation of polyacetylene. The most direct route involves the ring opening of cyclooctatetraene and its derivatives. This route results in the formation of a range of materials with variable electronic and optical properties. Transition metal complexes with tunable structure are the basis for the control of the structures of these polymers.

127. CHROMOPHORIC SELF-ASSEMBLED SUPERLATTICES. MULTILAYER CONSTRUCTION OF THIN FILM NONLINEAR OPTICAL MATERIALS. D. S. Allan, F. Kubota, Y. Orihashi, D. Li, T. J. Marks, Dept. of Chemistry and Materials Research Center, T. G. Zhang, W. P. Lin, G. K. Wong, Dept. of Physics and Materials Research Center, Northwestern University, Evanston IL 60208-3113.

An attractive and challenging approach to the construction of robust, thin film materials with large second-order optical nonlinearities would be the covalent self-assembly of aligned arrays of high-β molecular chromophores into multilayer superlattices. In this paper, we describe the design and synthesis of functionalized, high-β stilbene-based NLO chromophores, and the attendant chemistry employed to incorporate them into sequentially built-up multilayer structures on a variety of surfaces. Structural characterization techniques include optical and FTIR-ATR spectroscopy, advancing contact angle measurements, and polarized second harmonic generation (SHG) measurements. The details of film growth characteristics and microstructure are surveyed as a function of both the conditions as well as building blocks employed in the self-assembly process. The SHG data provide information on the degree of chromophore alignment and also indicate that nonresonant second-order macroscopic susceptibilities ($\chi^{2\omega}_{zzz}$) as large as ca. 10^{-7} esu can be achieved.

128. A NEW MOLECULAR DESIGN CONCEPT FOR POLED POLYMERS.
-UTILIZING THE OFF-DIAGONAL TENSOR COMPONENTS OF A MOLECULE-
T. Watanabe, M. Kagami, H. Yamamoto, A. Kidoguchi and S. Miyata Department of Material Systems Engineering, Tokyo University of Agriculture and Technology, koganei, Tokyo 184, Japan

Recently, a great deal of interest has arisen in poled polymers due to their large nonlinear optical effects. However, the poled polymers show a decay in their SHG activities. In order to overcome this problem, we have proposed a new molecular design concept where the nonlinear susceptibilities decays only a little even if dipolar orientation is relaxed when molecules possess not only diagonal tensors but also large off-diagonal tensors. In this presentation, the nonlinear susceptibilities of poled polymers possessing off-diagonal tensor components will be discussed by theoretical calculations, as well as some preliminary results.

129. **"Novel Polymers with Photorefractive and NLO Properties"**

Georges HADZIIOANNOU, Jurjen WILDEMAN, Jan HERREMA, Pieter SOETEMAN, Diny HISSINK, Jeannet BROUWER and Rien FLIPSE.

Laboratory of Polymer Chemistry, Department of Chemistry, University of Groningen, Nijenborgh 16, 9747 AG Groningen, The Netherlands

Recent effort on the synthesis of hybrid Organic-Inorganic small molecules and macromolecules towards materials with the desired Optical and Electronical properties will be described. More specifically the design of materials with high microscopic and macroscopic optical nonlinearities with the appropriate transmittance thus leading to various trade-offs between efficiency and transparency. The final goal within our research effort is the design of new high performance functional polymer materials for various optical and electronic components for the present macroscopic and the future Molecular Optoelectronic Devices.

130. THERMALLY STABLE ELECTRO-OPTIC POLYMERS. Susan Ermer, John T. Kenney, Jeong W. Wu, John F. Valley and Rick Lytel, Lockheed Missiles and Space Co., Research and Development Division, Palo Alto, CA 94304, and Anthony F. Garito, Department of Physics, University of Pennsylvania, Philadelphia, PA 19104-6396

Recently the development of higher speed electronic systems has led to proposals for the application of organic electro-optic polymers in integrated optical devices. In device applications, thermal stability is one of several major criteria in selecting appropriate polymers for optical applications because electronic assembly temperatures easily exceed 300°C. We describe an approach aimed at improving thermal stability in poled EO films. In this approach, the active EO moiety is treated as a solute in a host polymer solvent. By matching the Van der Waals volume of the guest to the free volume and morphology of the polyimide host, the thermal stability of guest-host systems is enhanced. These EO films are imidized and densified during high temperature curing under an applied dc poling field. Using this approach we have demonstrated thermally stable EO response in poled polyimide films at temperatures of 300°C for 2 hours and 200°C for over 80 hours.

131. DESIGN AND SYNTHESIS OF A NEW CLASS OF PHOTOCROSSLINKABLE NONLINEAR OPTICAL POLYMERS, Sukant Tripathy, Braja Mandal, Ru Jong Jeng, Jun Young Lee and Jayant Kumar*, Departments of Chemistry and *Physics, University of Lowell, Lowell, Massachusetts 01854

We have developed a new class of photocrosslinkable polymers with stable second and third order nonlinear optical (NLO) properties. The polymers have significant processing advantages over their inorganic counterparts. The NLO polymers have been designed as guest-host systems and as polymers with the NLO and photocrosslinkable units in the main chain and in the side chain. Cinnamoyl groups have been selected as photocrosslinking groups. Donor-acceptor substituted conjugated dyes functionalized with photocrosslinking units have been used as the NLO component. The stability of the poled and crosslinked polymers has been investigated by monitoring the decay of the second harmonic signal. No measurable decay in optical nonlinearity is observed for one of the photocrosslinked polymers up to temperatures of 160 °C for a period of several hours. These results are also confirmed by monitoring the stability of the UV-visible absorption spectrum at these temperatures. Channel waveguides and diffraction gratings have been fabricated using lithographic techniques.

132. **TOWARDS HIGH SUSCEPTIBILITIES IN SOLUBLE POLYACETYLENES FOR NON-LINEAR OPTICS.** J. Le Moigne, A. Hilberer, C. Strazielle, GMO, Institut de Physique et Chimie des Matériaux de Strasbourg and I.C.Sadron, 6, Rue Boussingault Strasbourg, 67083 Cedex France.

In order to increase the stiffness of a polyacetylenic chain in solution or in the melt, we have synthetized polyphenylacetylene derivatives; the stacking of the phenyl rings, in a plane orthogonal to the backbone, should preclude the torsional motion of the backbone. This paper describes the polymerization and the characterization results, in solid film or solution of a series of new polymer derivatives obtained from paraethynyl-benzoic-alkyl-esters. These polyacetylenes have been polymerized in solution by transition metal halides. All the polymers are highly soluble in common solvents. The UV-visible absorption spectra of the C_{16} derivative in THF exhibits a large absortion band in the visible range with a cut off at about 600nm. From the coupled on line GPC-light scattering measurements high molecular weigth polymers are characterized. The peak of the chromatogram is centrered on $M_w = 12.0 \times 10^4$ for the C_{16} derivative. The results are also discussed in term of stiffness of the backbone from the calculated values of the power term in the Mark-Houwink equation.

133. CHARACTERIZATION OF BIS-NADIMIDES OLIGOMERS AND HIGH Tg POLYIMIDE BASED SEMI IPN'S. V. Durand, T. Pascal, M. Senneron, B. Sillion
UMR 102/CEMOTA, B.P. 3 69390 VERNAISON France

A fully cyclized nadimide end-capped benzhydrolimide methylenedianiline oligomer, BBN, soluble in diglyme, was the crosslinkable resin for semi IPN'S prepared with three high Tg cyclized polyimides coming from 9,9'-bis-(4-aminophenyl)-fluorene, BAPF and benzophenone-tetracarboxylic dianhydride, BTDA, hexafluoro-isopropylidenediphthalic-anhydrid 6FDA and benzhydrol-tetracarboxylic-dimethylester, BHTDE. Three mixture containing 5, 10, 20 weight percent of linear polymer were studied for each polymer. TMA, viscoelastic behavior, K_{1C} and G_{1C} were determinated in connexion with electron scanning microscopy. BAPF/BHTDE carrying a benzhydrol group increases weaky Tg and K_{1C}, no phase segregation appears by E.S.M. For the mixture of BBN with BAPF/BTDA and BAPF/6FDA, a phase separation was observed. The Tg of the linear polymer appears when the concentration reaches 10%, K_{1C} and G_{1C} increase as a function of concentration. For example, BBN shows a Tg at 271°C, K_{1C} 0.89 MPa.m$^{0.5}$, $G1_C$ = 225 J m^{-2} ; BBN $\pm$ 20% BAPF-6FDA exhibits two Tg: 279 and 320°C, K_{1C} = 1.23 MPa m^{-1} and G_{1C} = 435 J m^{-2}.

134. ESTIMATION OF DENSITY AND GLASS TRANSITION OF "AMORPHOUS" POLYMETHYLENE. X. Qian and M.H. Litt, Department of Macromolecular Science, Case Western Reserve University, Cleveland, Ohio 44106.

A controversy exists in the literature as to what is the exact value for glass transition of the truly amorphous polymethylene (PM). This is caused by the fact that PM is highly crystalline and cannot be readily quenched to crystallinities below 50 %. The problem is further complicated by the occurrence of multiple secondary transitions in this polymer. We describe here an extrapolation method for obtaining both the density and glass transition of truly amorphous polymethylene. The measurements were made on a homologous series of polymers based on 4,4'-bis[(acryloyloxy)alkyloxy]biphenyl monomers. The monomers were photopolymerized in the isotropic state; the disordered structure was locked into the system due to crosslinking. Therefore, true amorphous methylenic structures could be obtained. The polymer can be treated as a block copolymer of polymethylene and biphenylbisacrylate. A density of 0.869 g/cm^3 and a Tg of -30 °C were obtained for polymethylene from the analysis, supporting the classification that the β transition of PM is its glass transition.

135. Crosslinking Reactions in Pigmented Olefinic Polymers
C. Houde, H.P. Schreiber and Alfred Rudin
Ecole Polytechnique, Universite de Montreal, c.p. 6079, Succ."A", Montreal Quebec, Canada H3C 3A7

Crosslinking reactions, initiated by dicumyl peroxide and a substituted alkyl peroxide, have been studied for a linear, low density polyethylene (LLDPE) and an ethylene-propylene rubber (EPR). Scanning calorimetric data were used to compare the reaction of pure matrix polymers with versions containing thermal stabilizers as well as titanium dioxide pigments. Pigments were selected on the basis of chromatographic analyses, showing one to have an acidic surface, another to be basic and third amphoteric. Crosslinking reactions with dicumyl peroxide were more rapid than with substituted peroxide, and terminated more rapidly in LLPDE than in EPR. Reaction enthapies for EPR were some 15% higher than for LLPDE systems. The presence of thermal stabilizers had no effect on crosslinking enthalpies or rates. Similarly, the acidic rutile pigment produced no significant effect on crosslinking parameters. In the presence of amphoteric, and particularly of basic rutiles, both kinetics and enthalpies of reaction were diminished. Acid/base coupling between these pigment surfaces and peroxides, or peroxide decomposition products, appears to control crosslinking reactions in the polyolefins.

136. POLYDICYCLOPENTADIENE, A NEW RIM THERMOSETTING RESIN, H. L. Mei, Hercules Incorporated, Research Center, c/o 1313 N. Market St., Wilmington, DE 19894-0001

Polydicyclopentadiene (PDCPD) can be made by metathesis polymerization through a reaction injection molding (RIM) process. The polymer is a highly crosslinked thermoset in which the olefinic unsaturation remains in the structure. The catalyst system is based on a tungsten VI compound activated by aluminum alkyls. Hercules has commercialized PDCPD product with the trademark METTON^R Liquid Molding Resin (LMR). It combines exceptional toughness, damage tolerance, and stiffness. The impact strength remains stable over the temperature range from -40 to 115°C. Large parts can be molded economically and conveniently. The material has found application in the recreation, automotive, and other markets.

137. A LOW-TEMPERATURE ROUTE TO POLYIMIDES. J.A. Moore and Ji-Heung Kim, Department of Chemistry, Rensselaer Polytechnic Institute, Troy, NY 12180-3590.

3-Dicyanomethylidene phthalide, a new phthalic anhydride analog, was synthesized by condensation of phthalic anhydride with malononitrile followed by treatment with phosphrous oxychloride. Several model reactions with aromatic amines and phenyl hydrazine disclosed unexpected chemistry producing the corresponding imide in tetrahydrofuran or N-methyl pyrrolidone solvent at room temperature. A difunctional pyromellitic dianhydride analog, 3,7-bis(2,2-dicyanomethylidene)pyromellitide, was also synthesized successfully using similar reaction conditions. The polymerization reaction of this new monomer with oxydianiline and the subsequent curing reaction to polyimide is discussed.

138. CURE CHARACTERISTICS OF TANNIN-BASED RESORCINOL-FORMALDEHYDE RESINS. Timothy G. Rials, USDA-Forest Service, Pineville, LA 71360.

Previous research in the utilization of condensed tannins as a replacement for phenol in wood adhesives has necessarily focused on the reaction with formaldehyde. As such, very little information is available on the curing behavior of this heterogeneous resin system, particularly in terms of the rheology of the process and its influence on network structure. This paper will discuss the application of dynamic mechanical analysis to study the curing characteristics of cold-setting, tannin-based resins; and the effect of formulation variables such as tannin content on the chemorheology of the curing reaction.

139. THERMODYNAMIC ANALYSIS OF GAS PERMEATION THROUGH GLASSY POLYMERIC MATERIALS. Mukesh Chhajer Tapan, Banerjee, and G. Glenn Lipscomb, Department of Chemical Engineering (ML 171), University of Cincinnati, Cincinnati, OH 45221-0171

A recently developed thermodynamic analysis of gas sorption in glassy polymeric materials is used to analyze gas permeation. The diffusive flux is calculated from the gradient in gas chemical potential from constraints imposed by the second law of thermodynamics. A constant diffusion coefficient is predicted if the penetrant partial molar volume and polymer bulk modulus are constants. Experimentally, however, the penetrant partial molar volume is found to increase with concentration. An *ad hoc* linear relationship between partial molar volume and concentration is assumed which yields diffusion coefficients that increase with gas concentration as observed experimentally. The final expressions for solubility and diffusivity are shown to be comparable to the dual-mode theory in ability to correlate and predict gas permeation rates in glassy polymeric materials.

140. STRESS EFFECTS ON DIFFUSION THROUGH SOLID POLYMERS

F. Doguieri, G.C. Sarti, Dipartimento di Ingegneria Chimica e di Processo Universita' di Bologna, Italy; R.G. Carbonell, Department of Chemical Eng., North Carolina State University, Raleigh, NC 27695.

A mathematical model is presented that describes the transport of a small penetrant in a solid polymeric matrix, that takes into account the effects of stress on the diffusive flux. Constitutive equations for the stress and for the chemical potential of the solute in the polymer are derived from the Helmholtz free energy for the system. The case of an elastic solid is considered. Numerical calculations are presented for the concentration, stress and deformation during diffusion in constrained and unconstrained rectangular films, cylinders, and constrained and unconstrained spheres. The effects of the stress field on the diffusion is illustrated. In addition, steady-state penetrant fluxes are calculated for the case of diffusion through the walls of a cylindrical tube. The extensions of the theory to linear viscoelastic solids is outlined.

141. A THEORY FOR SOLUBILITY IN STATIC SYSTEMS. A.A. Gusev, U.W.Suter, Eidgenössische Technische Hochschule Zürich, CH-8092 Zürich, Switzerland.

A theory for the solubility of small particles in static structures has been developed. The solubility at infinitesimal gas pressure (Henry's constant) as well as the pressure dependence of the solute concentration at elevated pressures has been found from the statistical equilibrium between the solute in the static matrix and the ideal gas phase. An application of the theory to the sorption of methane in the computed structures of glassy polycarbonate has resulted in a satisfactory agreement with experimental data.

142. COMPUTER ASSISTED ANALYSIS OF X-RAY SCATTERING FROM POLYMERIC GAS SEPARATION AND BARRIER MATERIALS. Solomon H. Jacobson, Hoechst Celanese, 86 Morris Ave., Summit NJ 07901.

Molecular modeling techniques have been employed to simulate WAXD behavior of polyimide materials. Simulation of X-ray scattering from single chain and dense glassy polymers are reported and compared to experimental results. These simulations indicated that the main peak observed in the WAXD spectra of Sixef®44 may be due to intramolecular chain backbone geometries. Moreover, the computed dense molecular model suggested that for this material the free space available for gas transport is in the from of large irregular voids, while neighboring chains can be spaced quite close to each other (3.4 Å). The observed shoulder in the WAXD spectra between 20 and 28 degrees is attributed to these close interchain interactions.

143. GAS SOLUBILITY IN POLYMERS AND BLENDS P. A. Rodgers and I. C. Sanchez, Chemical Engineering Dept. and Center for Polymer Research, The University of Texas at Austin, 78712

The Lattice Fluid model is used to predict and analyze the limiting solubility of gases in polymers and miscible polymer blends above their glass transition temperatures. Gas sorption in polymer blends and the respective pure polymers can be used to determine polymer-polymer interaction parameters. Analyzing gas solubility data for 5 different gases in polystyrene/poly(2,6 dimethyl phenylene oxide) blends yields a "bare" polymer-polymer interaction parameter of -20.1 ± 3.8 J/cm^3.

144. PREDICTIVE CAPABILITIES OF A FREE-VOLUME THEORY FOR POLYMER-SOLVENT DIFFUSION COEFFICIENTS. John M. Zielinski and J. L. Duda, Department of Chemical Engineering, Penn State University, University Park, PA 16802.

The Vrentas/Duda free-volume (FV) diffusion model accurately correlates polymer/solvent diffusion coefficients over wide ranges of concentration and temperature. Currently the model is self-predictive, i.e., limited diffusion data is required to estimate model parameters which can then be used to predict diffusion coefficient behavior over sundry conditions. In this work we present methods for estimating all of the model parameters without any diffusion data and examine the accuracy of the resulting predictions. This technique is the only one known which predicts polymer/solvent diffusion behavior without the use of any diffusion data.

145. **REVERSIBLE GELATION IN BINARY MIXTURES OF POLY(DI-N-ALKYLSILANES) AND AROMATIC SOLVENTS.** J. F. Rabolt, T. J. Lenk, V. M. Hallmark, R. L. Siemens, R. D. Miller, IBM Research Division, Almaden Research Center, 650 Harry Road, San Jose, California 95193-6099.

An investigation of binary mixtures of poly(di-n-butylsilane) and poly(di-n-pentylsilane) in aromatic solvents has shown that changes in backbone conformation result upon gelation. These conformational changes of the σ conjugated backbone result in dramatic alteration of the electronic structure as manifested by a shift in the UV absorption maxima from 317 nm to 360 nm. This spectroscopic shift has been attributed[1] to a change from a disordered helical conformation[2] prior to gelation to a planar zigzag structure in the gel state. Solvent factors which appear to affect the gel point will also be discussed.

[1] T. J. Lenk, R. L. Siemens, V. M. Hallmark, R. D. Miller and J. F. Rabolt, *MACROMOLECULES*, **23**, 3870 (1990).
[2] R. D. Miller, B. L. Farmer, W. Fleming, R. Sooriyakumaran and J. F. Rabolt, *JACS*, **109**, 2509 (1987).

146. **THERMOREVERSIBLE GELATION AND VITRIFICATION OF CONCENTRATED POLYMER SOLUTIONS UNDER POOR THERMODYNAMIC CONDITIONS**

B.A. Wolf and Th. Schneider
Institut für Physikalische Chemie der Universität and SFB 262, D-65 Mainz, Germany

With the system 2-propanol/poly(n-butyl methacrylate) (Θ = 24 °C, T_g of the pure polymer = 22 °C) one observes phase separation, thermoreversible gelation and glassy solidification within the same T-interval (− 30 to + 60 °C). Concentrated solutions gel thermoreversibly upon cooling some 60 to 40 °C before they become glassy. This partial freezing-in of motion can be monitored in dynamic experiments (rheology, dielectric relaxation spectroscopy and ^{2}H-NMR) as well as with calorimetric methods (DSC). It is interpreted in terms of non-crystalline physical crosslinks resulting from an uncommonly large heat of mixing. The T_g values are larger than calculated from additivity. The dynamic measurements also yield the characteristic relaxation times of different parts of the macromolecules. They range from 10^{-9} sec (NMR) to several days (viscometry); the corresponding activation enthalpies increase from 8 to 117 kJ/mol.

147. **THERMOREVERSIBLE GELATION OF SOLUTIONS OF STEREOREGULAR POLYSTYRENE**, A. Jacobs, F. Deberdt and H. Berghmans, Laboratory for Polymer Research, KULeuven, Celestijnenlaan 200 F, B-3001 Heverlee, Belgium

The structure formation and thermoreversible gelation of solutions of polystyrene is investigated. Much attention is paid to the phase behaviour. In cis-decalin, isotactic polystyrene forms crystalline, past-like gels by lamellar crystallization. At lower temperatures, transparent gels are formed by a two step mechanism. It is initiated by an intramolecular conformational change, followed by an intermolecular association. This was deduced from the combination of rheological and I.R. observations. The formation of a compound between polymer and solvent is suggested. At even lower temperatures, a vitrification takes place. From the T_g-concentration relation, one can conclude to the interference with a liquid-liquid demixing.
The analogy between the phase behaviour of this system and that of syndiotactic polystyrene in trans-decalin suggests the same complex mechanism for the last system.

148. **GELATION OF ATACTIC POLYSTYRENE IN CARBON DISULFIDE.**
Donald J. Plazek and In-Chul Chay, Materials Science Engineering,
University of Pittsburgh, Pittsburgh, Pennsylvania 15261.

There have been several reports of solutions of atactic polystyrene PS exhibiting gelation. However, high viscosities and finite storage moduli G′, dyne cm^{-2} are not definitive evidence for gelation. An equilibrium modulus or compliance is such evidence. The frequency independent G′ curves for a PS solution in carbon disulfide presented by Clark,

Wellinghoff, and Miller in Polymer Preprints (Am. Chem. Soc., Div. Polym. Chem.) 24(2), 86 (1983) appears convincing. Creep measurements made in a sealed frictionless apparatus on an 8.68% by wt. solution of PS in CS_2 substantiates the occurrence of this unexpected gelation. A steep decrease of a thousand-fold in compliance was observed to take place with a decrease of about 40°C in temperature. "Melting" of the gel took place at about -5°C, some 10°C above the temperature at which gelation was observed. Gelation of atactic polystyrene in other solvents has not been observed.

149. THERMOREVERSIBLE GELATION OF ISOTACTIC POLYSTYRENE. J.M. Guenet
Laboratoire d'Ultrasons et de Dynamiques des Fluides Complexes.
67070 STRASBOURG Cedex FRANCE.

The molecular structure of isotactic polystyrene (iPS) gels as revealed by neutron diffraction and small-angle neutron scattering will be presented. It will be shown that it bears close resemblance to nematic structures: one-dimensional order with the chains adopting worm-like statistics and enhanced rigidity (b≈ 8 nm instead of 2 nm in the amorphous state). Unlike crystalline polymers the chains do not retrieve their normal unperturbed conformation on melting but keep the conformation they possessed in the gel state. It is suggested that gelation arises from enhanced rigidification of the chains achieved by solvation of the helical form. The role of the solvent type is also discussed. Finally, the temperature-concentration phase diagram is given. The formation of spherulitic structures just above the gelation threshold that do not show crystalline order is discussed. From these results conclusions are drawn as to the gel formation mechanism.

150. SECOND HARMONIC GENERATION AND ELECTROCHROMISM STUDIES OF ELECTRICAL PROPERTIES OF NONLINEAR OPTICAL POLYMERS AND DEVICES. C. S. Willand, and M. Scozzafava, Eastman Kodak Company, Rochester, New York 14650-2021.

An understanding of electric-field poling behavior and device operation in single and multilayer nonlinear optical (NLO) devices cannot be accomplished without detailed knowledge of the electric fields producing these effects. Towards the end, we have developed a technique based on the electrochromic effect for probing the electric fields present within thin films subjected to an applied DC voltage. Coupling between the electric field and an applied AC voltage produces a modulation of the absorption coefficient whose magnitude is proportional to the electric field. Measurement of the resulting optical transmission provides information of the electric fields. The results of simultaneous electrochromism and second harmonic generation measurements on thin films of NLO polymers will be presented. These studies reveal the poling dynamics to be influenced by space charge present in the polymer during poling. Experiments on multilayer waveguides will also be discussed. We have performed work on two-layer films composed of NLO polymers and various buffer materials. Results indicate that the magnitude of the electric field within the polymer layer, and hence the induced optical nonlinearity, is strongly affected by the choice of buffer materials.

151.

ORIENTATION AND DECAY IN POLED POLYMER NONLINEAR OPTICAL MATERIALS. Kenneth D. Singer, Department of Physics, Case Western Reserve University, Cleveland, Ohio 44106; Lori A. King, AT&T Bell Laboratories, P.O. Box 900, Princeton, NJ 08540.

Poled polymer materials for second-order nonlinear optics have become the subject of intense interest due to their potential application as electro-optic and frequency conversion materials. The orientation of nonlinear optical moieties using electric field poling results in a noncentrosymmetric and polar structure which is necessary for second-order effects. The poling process results in a quasi-stable orientation described by simple thermodynamic considerations. The dynamics of the poling process and subsequent decay can be described by a phenomenological distribution of local relaxation energies. It is found from time and temperature dependent second harmonic generation measurements, that the distribution parameters describing the dynamical poling process are the precise ones governing their subsequent relaxation.

152. SECOND-ORDER NONLINEAR OPTICAL ONSET AND DECAY AND THEIR RELATIONSHIP TO POLYMER PHYSICS. Ali Dhinojwala, George K. Wong, and John M. Torkelson, Depts. of Chemical Engineering, Materials Science and Engineering and Physics and Astronomy, Northwestern University, Evanston, IL 60208

A model has been developed to explain the relationship between second-order nonlinear optical (NLO) properties associated with electric field poling onset and decay (after removal of the field) in contact-poled, doped polymer glasses. This model relies on a description of a dopant being in one of two kinds of sites, liquid-like, allowing rapid orientation and disorientation of dopants upon application and removal of the electric field, respectively, and locally "glassy", allowing no orientation at short poling times and only a slow, long-term decay of orientation upon removal of the field. For a 3 wt% DANS-doped polymethylmethacrylate sample at 30°C, it was determined that 27 to 29% of the dopants are in liquid-like regions, while at 55°C, 39 to 40% of the dopants are in liquid-like sites. Excellent agreement between the poling onset and decay results as well as the fact that decay data are fit well by the Williams-Watt model indicate that a direct understanding of the relationship between second-order NLO properties and polymer physics is possible.

153. SECOND-ORDER NLO AND RELAXATION PROPERTIES OF POLED POLYMERS. D. Y. Yoon, D. Jungbauer, I. Teraoka, B. Reck, R. Zentel, R. Twieg, J. D. Swalen, and C. G. Willson, IBM Almaden Research Center, San Jose, CA 95120-6099.

Two linear epoxy polymers based on bisphenol-A with 4-nitroaniline (BISA-NA) and 4-amino-4'-nitrotolane (BISA-ANT), respectively, both exhibit stretched exponential-type relaxation characteristics of polar orientation at temperatures below the glass transition temperatures, with decay times consistent with their dielectric α-dispersion characteristics. However, the dielectrically fitted Vogel-Fulcher temperature T_∞ of BISA-ANT is increased slightly due to poling, while T_∞ of BISA-NA is lowered substantially after the same poling procedure. This result suggests that the mobility of glassy polymers is affected by poling, possibly due to its effect on the Maier-Saupe thermotropic vs. electrostatic interactions among the neighboring chromophores. The BISA-ANT polymer exhibits a large doubling coefficient of $d_{33} \simeq 89$ pm/V for 1.06 μm fundamental and no decay for ca. two weeks at ambient temperature. Consistent with the strong Maier-Saupe interaction contributions in BISA-ANT, the oligomers with nearly twice the number density of nitrotolane moieties exhibit a smectic phase. Preventing the formation of this smectic phase by chemical crosslinks results in a very polar polymer exhibiting twice as large birefringence as that of BISA-ANT and little decay during ca. 4 weeks annealing at 80-90 °C.

154. **NONLINEAR OPTICAL PROPERTIES OF POLYMER/SILVER MICROSPHERE COMPOSITES.** Mark P. Andrews, McGill University, Montreal, Quebec, Canada and Mark G. Kuzyk, Department of Physics, Washington State University, Pullman, Washington.

Polymer/silver microsphere composites were prepared by vaporizing silver atoms into cold organic solvent solutions of polymethylmethacrylate (PMMA) in a rotatable cryostat. He-Ne laser quadratic electrooptic phase modulation spectroscopy was used to measure a third order nonlinear susceptibility of 1.6×10^{-14} esu from sub-10 nm silver particles hosted in rigid PMMA films. This value is enhanced by approximately 10^6 due to local field effects associated with excitation at the absorption maximum of the colloid dipolar surface plasmon mode. We discuss the origin of the nonlinearity in terms of enhanced local field effects and several mechanisms that can lead to an electric field dependence of the refractive index.

POLY

155. THE PHOTOREFRACTIVE EFFECT IN NON-LINEAR POLYMERS DOPED WITH CHARGE TRANSPORT AGENTS J. Campbell Scott, Stephen Ducharme, Robert J. Twieg and W.E. Moerner. IBM Research Division, Almaden Research Center, San Jose, CA 95120-6099.

The photorefractive effect has been observed in epoxy polymers containing non-linear chromophores, molecularly doped with charge transport agents. Photorefraction was established by a combination of hologram erasability, correlation with photoconductivity and electro-optic response, and enhancement in external fields in numerous samples. A useful property of these materials is that poling of the nonlinear chromophores is partially reversible and thus continuous realignment of the nonlinear molecules in the polymer permits external control of grating readout independent of the space-charge field formed.

156. SECOND-HARMONIC GENERATION FROM HYPERPOLARIZABLE AMPHIPHILES AT POLYMER-POLYMER INTERFACES. Jian P. Gao and Graham D. Darling[*], Department of Chemistry, McGill University, 801 Sherbrooke St. W., Montreal, QC, Canada H3A 2K6.

We report here the preparation, by a relatively simple and inexpensive spin-coating technique, of robust SHG-active films in which amphiphilic hyperpolarizable molecules (n-alkyl-NMe-4-Ph-CH=CH-4-Pyr "dyes") are permanently oriented between solid layers of hydrophobic (polystyrene) and hydrophilic (poly(ethylene-co-maleic acid)) polymers. Microscopic appearance, wettability (contact angle), optical absorbance and polarized-SHG properties suggest ionic interaction and partial penetration of the "pyridine" end of the dye into the acidic-hydrophilic polymer layer, with exclusion of the n-alkyl end, for a net orientation of the hyperpolarizable moiety to 28-30° from the vertical (for longest n-alkyl), comparing favourably with materials prepared by other techniques. This method shows promise for fabrication of SHG-efficient "organic waveguides" and multilayer laminates.

157. TOUGHENING OF HIGH PERFORMANCE EPOXIES USING DESIGNED CORE-SHELL RUBBER PARTICLES H.-J. Sue, E.I. Garcia-Meitin, Analytical & Eengineering Sciences Department, Texas Polymer Center, Dow Chemical U.S.A., Freeport, TX 77541; D.M. Pickelman and P.C. Yang, Advanced Polymeric Systems & Advanced Composites Laboratories Central Research, The Dow Chemical Company, Midland, MI 48674

The fracture behavior of high performance epoxies modified with seven types of designed rubber particles are examined using various microscopic techniques. It is found that the Mode-I plane strain fracture toughness values (K_{IC}) of the designed rubber-modified high crosslink density epoxy systems vary with the architecture of the interface between the matrix and the rubber particles. Accordingly, the toughening mechanisms observed among these systems are very different: rubber particle cavitation and matrix shear yielding are found in systems that exhibit higher K_{IC} values; only crack deflection is found in systems that possess lower K_{IC} values. Synergistic toughening can be obtained if both the matrix shear yielding and the crack deflection mechanisms are operative upon fracture. An approach for toughening highly crosslinked epoxies is addressed.

158. NETWORK STRUCTURE AND MECHANICAL PROPERTIES OF AMINE CURED EPOXY: AMINE EPOXIDE ADDITION REACTION VS. ETHERIFICATION, Dong W. Sohn and Nakho Sung[*], Laboratory for Materials and Interfaces, Department of Chemical Engineering, Tufts University, Medford, Massachusetts 02155-7021

Network structure of a stoichiometric mixture of diamine cured epoxy (DGEBA/DDS) was investigated by fluorescence dye labelling and IR spectroscopy techniques. The network structure of epoxy was varied via processing condition to produce a series of samples with different extent of epoxide cure. Fluorescence spectra from diaminoazobenzene (DAA) label provide the extent of amine epoxide addition reaction, and IR spectra give the overall extent of epoxide reaction. Comparison of both spectra indicates that the extent of amine addition is fairly close to the extent of epoxide reaction, but at later stage of cure, a significant degree of ether formation takes place making the extent of epoxide reaction to increase further even after the amine addition reaction stops. DSC study also confirms the ether formation at later stage of cure. Fracture toughness of cured epoxy was investigated in terms of network topology resulting from two different types of cure reactions. Preliminary results show that the fracture toughness initially increases to a maximum, then starts to decrease with cure time after amine epoxide addition reaction stops. The fracture toughness appears to decrease linearly with increasing extent of ether formation.

159. FLUORESCENCE OF AROMATIC DIAMINES FOR EPOXY CURE STUDIES. J.C. Song and C.S.P. Sung, Department of Chemistry, Institute of Materials Science, University of Connecticut, 97 North Eagleville Road, Storrs, CT 06269–3136.

Fluorescence behavior of two important aromatic diamine curing agents, 4,4'–diaminodiphenyl sulfone (DDS) and 4,4'–diaminodiphenyl methane (DDM) was studied during cure with diepoxide or tetraepoxide. As cure reaction proceeds, both emission and excitation spectra exhibit red shifts due to the conversion of primary amines to secondary and tertiary amines. The matrix effects such as polarity decrease and viscosity increase on the fluorescence spectra of tertiary amino–DDS are negligible. Excitation spectra provide sharper peaks than emission spectra. Excitation spectral shift is found to be linearly proportional to the extent of amine reaction. The spectral shift has also been correlated to the extent of epoxide reaction.

160. MEASUREMENT OF WATER UPTAKE IN CURED EPOXY BY EXTRINSIC AND INTRINSIC FLUOROPHORES. J. K. Yoo, X. D. Sun and C.S.P. Sung, Department of Chemistry, Institute of Materials Science, University of Connecticut, 97 N. Eagleville Rd., Storrs, CT 06269–3136

In order to monitor water–uptake in cured epoxy–diamine network, several polarity–sensitive extrinsic fluorophores have been evaluated. First, they were tested in n–butanol or N–methylpyrrolidone by adding successive amount of water. 1–Anilino–8–naphthalene sulfonic acid (ANS) and 9–anthroic acid (9AA) were chosen to be added in small amount during cure of diepoxide and aromatic diamine curing agent (diamino diphenyl sulfone). When immersed in water, fluorescence intensity decreases in a sensitive way. Surface water concentration has been estimated and correlated to the fluorescence intensity. In addition, intrinsic fluorescence was observed around 460 nm in epoxy–diamine during cure without any added extrinsic probe, probably due to some oxidation or degradation. This intrinsic fluorescence is found to be also sensitive to water–uptake.

161. A COMPARATIVE MECHANISTIC STUDY OF EPOXY/AMINE CURE KINETICS IN THERMAL AND MICROWAVE FIELDS. Jovan Mijović, Jony Wijaya and Arnon Fishbain, Chemical Engineering Department, Polytechnic University, Brooklyn, NY 11201.

A comprehensive mechanistic study of epoxy/amine cure kinetics in thermal and microwave fields was carried out. A model compound system and a multifunctional epoxy formulation were investigated. Model predictions and experimental results agreed famously. It was unambiguously shown that the reaction kinetics in thermal and microwave environments are identical.

162. MECHANISTIC MODELLING OF EPOXY/AMINE REACTIONS. Jovan Mijović, Arnon Fishbain and Jony Wijaya, Chemical Engineering Department, Polytechnic University, Brooklyn, NY 11201.

A fundamental kinetic study of epoxy/amine reactions was carried out on a model compound system consisting of aniline and phenyl glycidyl ether. A new mechanistic model was proposed and found to be in excellent agreement with experimental data from HPLC and FTIR. A negative substituiton effect was detected and all kinetic parameters evaluated. A parallel study of the model compound in the microwave field revealed no difference in reaction kinetics as a function of the type of energy source.

163. APPLICATION OF FLUORESCENCE TECHNIQUES TO THE STUDY OF DIFFUSIONAL PROCESSES IN COMPATIBLE POLYMER FILMS. Lora L. Spangler and John M. Torkelson, Depts. of Chemical Engineering and Materials Science and Engineering, Northwestern Univ., Evanston, IL 60208

Diffusion in compatible thin polymer films has been followed by monitoring changes in fluorescence intensity which occur as a result of a fluorescence quenching process. Specifically, we have investigated the interdiffusion of polyvinylmethylether and polystyrene, where the polystyrene chains have been terminally labeled with anthracene units. Experiments have been performed at temperatures 10°C to 40°C below the Tg of polystyrene (Tg=100°C) but 90°C to 120°C above the Tg of polyvinylmethylether (Tg=-31°C). At these conditions, these polymers form a compatible, miscible blend. A more rapid decrease in anthracene fluorescence as a function of temperature for a layered sample of the two polymers confirms the sensitivity of this technique to diffusion of polymer molecules over small distances. Preliminary modeling of data yield diffusion coefficients of polyvinylmethylether in polystyrene between 10^{-16} and $10^{-14} cm^2/sec$ at temperatures between 63°C and 92°C.

164. DIFFUSION IN POLYMERS BY FORCED RAYLEIGH SCATTERING. Hyuk Yu, Department of Chemistry, University of Wisconsin, Madison, WI 53706.

This paper deals with a review of what we have studied in the past decade by means of the Forced Rayleigh Scattering technique together with more recent results for a polyelectrolyte self diffusion study relative to the linear charge density dependence. It is divided into four sections. They are 1) the diffusion of small molecules in two polymer solutions, over the entire concentration range for one and a limited range for the other, 2) the probe chain diffusion in like chain matrix whereby the matrix molecular weight spans across a wide range including the asymptotic limit beyond which the probe chain diffusion no longer depends on the matrix molecular weight, with use of the system of polystyrene in toluene, 3) the self diffusion of polyisoprene in bulk state and its temperature dependence, and 4) a recent study on a polyelectrolyte system. With respect to the last case, we will present some novel way of demarcating the polyelectrolyte behavior in dilute solutions from the neutral polymer-like behavior in concentrated solutions, wherein the hydrophobic interactions of neutral chain segments appear to play a major role in determining the self diffusion processes.

165. MONOMER DIFFUSION AND THE KINETICS OF PROPAGATION IN FREE-RADICAL POLYMERIZATION Alessandro Faldi and Matthew Tirrell, Department of Chemical Engineering and Materials Science, University of Minnesota, Minneapolis, MN 55455; Timothy P. Lodge, Department of Chemistry, University of Minnesota, Minneapolis, MN 55455; Ernst D. von Meerwall, The University of Akron, Akron, OH 44325

We measured the diffusion coefficient of methyl methacrylate in solution with poly(methyl methacrylate) for different concentrations, temperatures, and polymer molecular weight using forced Rayleigh scattering and field-gradient nuclear magnetic resonance. The results are compared to two sets of kinetic data of the rate coefficient of propagation in the free-radical polymerization of methyl methacrylate in order to establish whether the kinetic step is diffusion-controlled. If the initiator efficiency and the reactant capture radius are considered constant, there is evidence that propagation is directly controlled by methyl methacrylate diffusion. The importance of diffusion measurements, as well as the need of more kinetic data, is demonstrated.

166. SMALL-ANGLE NEUTRON SCATTERING STUDY OF INTERDIFFUSION IN POLYSTYRENE-POLYXYLENYLETHER BLENDS. Ming-da Eu, George C. Summerfield, Robert Ullman, Macromolecular Research Center and Department of Nuclear Engineering, University of Michigan, Ann Arbor, MI 48109.

Blends of polystyrene and polyxylenyl ether prepared in a non-equilibrium state were allowed to interdiffuse at a temperature 20°C. above the glass temperature. Small-angle neutron scattering spectra of the original specimens before diffusion and after several intervals of interdiffusion were measured at room temperature. Fick's laws are not applicable to the system; however, diffusion constants and relaxation effects are obtainable using a linearized model proposed by Jäckle and Frisch. The rates of diffusion change as time proceeds owing to changes in the mean molecular environment as a function of time.

167. NON–LINEAR SWELLING OF PHANTOM NETWORKS. C.J. Durning, Department of Chemical Engineering & Applied Chemistry, Columbia University, New York NY 10027 and K.N. Morman Jr., Scientific Research Laboratories, Ford Motor Co., Dearborn MI 48121–2053.

We present a theory for the non–linear swelling of polymer gels. The approach is to solve the continuity equation for the liquid under the constraint that the instantaneous free energy be minimal. Constitutive equations are needed for the free energy density of the gel, W, and for the liquid diffusion current, J. In this preliminary calculation, we used the phantom network model for W and Fick's law for J. The theory is applied to the free swelling of a spherical gel to a semi–dilute equilibrium. We find a geometry dependent, non–local effect influencing the measurables (liquid mass uptake and gel diameter). This arises from the dependence of the network's deformation at the gel/liquid interface on the whole deformation field for the sample. The prediction is consistent with explanations of buckling instabilities during volume phase transitions in gels.

168. A MOLECULAR INTERPRETATION OF DIFFUSION IN POLYMERS. L.A. Errede and P.J. Henrich, 3M Corp. Res. Labs., 3M Center Bldg. 201-2N-22, St.Paul, Minn. 55133

We reported earlier that when liquid-saturated poly(styrene-co-divinybenzene) particles are allowed to evaporate at 23°C to the composition (α_g, in molecules per phenyl group) that marks completion of transition from the rubbery state to the glassy state, the residual volatile molecules are trapped in "n" (not more than six) different molecular environments, such that the kinetics of desorption with respect to the residual number (α_t) of these molecules is given thereafter by a linear combination of "n" exponential decay functions, and that the first-order rate constants (k_i) for decay of these populations are given by the relationship: Log k_i = Log k_o - mi, where "i" is the population identification number from 1 to 6, and k_o and m are constants characteristic of the molecular structures of the solvent and polymer respectively. We now report the results of experiments that were designed to show how α_g, the fractional distribution of the populations that comprise α_g, the set of k_i, and the constants k_o and m are affected by the temperature of evaporation from 20° to 80°C. The significance of these results, with respect to the mechanism of diffusion, in the glassy state will be discussed.

169. SURFACTANT ORGANOGELS STUDIED BY SMALL ANGLE NEUTRON SCATTERING. P. Terech, Institut Laue Langevin, 156X, 38042 Grenoble Cedex, France.

Aggregation of surfactant molecules in organic solvents can develop rod-like structures. Interactions of the chains is responsible of the onset of a three dimensional network. A large diversity of structural and rheological situations is found with the class of surfactant organogels. The small angle neutron scattering study of these systems provides structural information ranging from the internal molecular arrangement to the large scale structure of the overlapping chains. Two systems are presented : an hydroxy-substituted fatty acid and a binuclear copper(II) complex. The survey presents the large differences observed with these materials concerning the chain length and flexibility, the geometrical dispersity, the cross sectional shapes and the structures of the junction zones. These latter chain crosslinks can be structured from disordered chain entanglements to crystallites. This point distinguishes strong from weak gels with respect to the stability of the crosslinks for the time scales involved. Analogies with some polymeric systems can be drawn.

170. A REEXAMINATION OF THE FREEZING AND MELTING OF SOLVENT IN THERMOREVERSIBLE POLYMER GELS AND RELATED SYSTEMS. Catheryn L. Jackson and Gregory B. McKenna, National Institute of Standards and Technology, Polymers Division, Gaithersburg, MD 20899.

A presentation of the effect of small system size, or a confining geometry, on the melting temperature and enthalpy of fusion of pure organic liquids infused in model systems of porous glasses is made (J. Chem. Phys., 1990, 93, 9002). This work is fundamentally related to previous reports on solvent freezing and melting behavior in thermoreversible gels of isotactic-polystyrene (i-PS) in cis-decalin and in other polymer gels, and the use of solvent crystallization method (Macromolecules, 1989, 22, 3716) to characterize the gels. The organic liquids to be discussed include cis-decalin, trans-decalin and benzene, which are relevant to the gels of interest.

171. PHASE BEHAVIOR OF AQUEOUS POLY(VINYL-ALCOHOL) SOLUTION - A SMALL ANGLE NEUTRON SCATTERING STUDY. Wen-li Wu, National Institute of Standards and Technology, Gaithersburg, MD 20899; Hidenobu Kurokawa, Mitsuhiro Shibayama, Shunji Nomura, Kyoto Institute of Technology, Matsugasaki, Sakyoku, Kyoto, 606 Japan and Laurence D. Coyne, Food and Drug Administration, Rockville, MD 20852

Clear-opaque-clear transitions in aqueous poly(vinyl-alcohol) solutions were induced by adding different amounts of borate salts. PVA molecules are known to form ion complexes with anions such as borate and titanate. These complexes are responsible for the formation of both inter- and intrachain physical crosslinks which lead to the clear-opaque-clear transitions. The PVA single chain form factor and the second virial coefficient were determined from the small angle neutron scattering (SANS) results across these transitions. Measurements were conducted in both the semi-dilute and the dilute solution regions. The SANS results indicate that the collapse and expansion of individual chains coincide with the clear-opaque-clear transitions.

172. DYNAMICS OF REVERSIBLE NETWORKS. Michael Rubinstein[†], Ludwik Leibler[‡] and Ralph H. Colby[†], [†]Corporate Research Laboratories, Eastman Kodak Company, Rochester, NY 14650-2110, [‡]Groupe de Physico-Chimie Theorique, E.S.P.C.I., 10 rue Vauquelin, 75231 Paris, France.

We present a model for dynamics of entangled networks made up of linear chains with temporary cross-links. At times shorter than the lifetime of a cross-link such networks behave as elastic rubbers. At longer time scales they flow like polymer liquids. We discuss self-diffusion and stress relaxation in these systems and relate them to the microscopic time characterizing creation and destruction of cross-links. We show that when there are many cross-links per chain, the relaxation may be modeled by a "sticky reptation". The predictions of this model are in good agreement with the data of Stadler and de Lucca Freitas on model thermoplastic elastomers. If tie points are only at chain ends, we argue that relaxation by reptation may be inefficient when most of the time ends are linked and there are no free ends. Nevertheless chains can diffuse and stress can relax by an exchange of partners in cross-links.

173. GELATION AND PHASE SEPARATION OF PVC/PLASTICIZER SOLUTIONS. Hwee Khim Boo and Montgomery T. Shaw, Department of Chemical Engineering and Polymer Science Program, Institute of Materials Science, The University of Connecticut, Storrs, CT 06269-3136.

There are several details about the gelation of PVC which are of importance in understanding the long—term performance of plasticized formulations, as well as physical gelation in general. In this work we have investigated a number of aspects of the gelation step and the behavior of the gels using systems wherein the diluent/polymer interaction was varied over a wide range. A transition in the gels at around 70°C was found to be independent of plasticizer type and concentration but highly correlated with annealing temperature. The material responsible for this transition could not be detected with either WAXS or SAXS.

174. LOW-LOSS GRADED-INDEX AND SINGLE-MODE POLYMER OPTICAL FIBERS. Y. Koike and E. Nihei, Faculty of Science and Technology, Keio University, Yokohama 223, Japan

A low-loss and high-bandwidth graded-index (GI) polymer optical fiber (POF) was successfully obtained by the interfacial-gel copolymerization technique developed by us. The bandwidth of the GI POF is about 500 MHz·km which is one hundred times larger than that of the conventional step-index (SI) POF. The total attenuation of the light transmission is 150-200 dB/km at 652-nm wavelength. It is experimentally and theoretically confirmed that the GI POF is quite superior to the SI POF in the bandwidth, while the attenuation and mechanical properties of both fibers are comparable. We also proposed and prepared a low-loss single-mode POF in which the core diameter was 3-15 μ m and the attenuation was about 200 dB/km. As far as we know, this is the first single-mode POF.

175.
PLANAR POLYMER WAVEGUIDES FOR OPTICAL INTERCONNECTIONS
B. L. Booth, J. E. Marchegiano, C. T. Chang, T. K. Foreman, J. L. Hohman, S. L. Witman
E. I. du Pont de Nemours & Co. (Inc.), Wilmington, Delaware 19880-0357

Acrylate-based polymeric formulations combined with a unique assembly process under development at Du Pont offer broad capabilities as a generic passive integrated optic system. The technology referred to as Polyguide™ can create single and multimode waveguides and self-aligned interconnects which enable practical circuit board, connector, and component applications to be realizable.

176.   Guided-wave nonlinear optics in DCANP Langmuir-Blodgett films

P. Günter, Ch. Bosshard, M. Küpfer and M. Flörsheimer

Institute of Quantum Electronics, Nonlinear Optics Laboratory, Swiss Federal Institute of ETH Hönggerberg, CH-8093 Zürich, Switzerland

We will review the preparation of DCANP LB films and its physical properties, including the measurement of dispersion of refractive indices, nonlinear optical susceptibilities and waveguide attenuation losses. For the first time guide-wave second-harmonic generation of Nd:YAG lasers in a nonlinear optically active LB film will be reported. The first waveguiding experiments were carried out using a grating coupler. The grating was first etched into the pyrex substrate through ion etching. Subsequently the LB film was deposited with the dipping direction parallel to the grating grids. TM_0- and TE_0-modes were observed for layer thicknesses between 177 nm and 442 nm.

177. OPTICALLY NONLINEAR POLYMERS IN WAVEGUIDING PASSIVE AND ACTIVE DEVICES. G. R. Möhlmann, and W. H. G. Horsthuis, Akzo Research Laboratories Arnhem, Corporate Research, P. O. Box 9300, 6800 SB Arnhem, The Netherlands

Optically nonlinear side chain polymers possess attractive properties for application in passive and electro-optically active waveguiding devices for operation in the visible and near infrared wavelength ranges. To exhibit useful second order optically nonlinear effects, the polymers have to be poled with electric fields of the order of several 100's volts/μm. Nonresonant electro-optic coefficients of the order of 20 pm/V have been achieved at the wavelength of 1330 nm. Via successive spincoating of polymers with matched refractive indices, multilayer structures have been made, possessing slab waveguiding. By photo-bleaching through a mask, the refractive index can be changed, yielding monomode or multimode channel waveguides. Optically passive structures such as straight and bend waveguides and couplers, and active structures such as phase shifters, Mach-Zehnder interferometers and directional mode couplers, have thus been made. Switching voltages are of the order of 10 - 20 Volts, the corresponding modulation ratio's are of the order of 15 - 20 ds. The applied technology further permits the integration of light sources and detectors with waveguides and may open the way for mass production of cost effective optical components.

178.   NLO POLYMERS FOR FREQUENCY DOUBLING

Emiel G.J. Staring, G.L.J.A. Rikken, C.J.E. Seppen, S. Nijhuis, A.H.J. Venhuizen
          Philips Research Laboratories, P.O.Box 80000
              5600 JA Eindhoven, the Netherlands

An important area of application of second order NLO effects is frequency doubling or second harmonic generation (SHG) of the emission of a common GaAs/AlGaAs semiconductor solid state laser in order to achieve a 4-fold increase of information density in optical data storage systems. At our Labs the NLO research program is aimed at obtaining poled polymer materials for frequency doubling of these diode lasers. Our investigations showed that NLO molecules containing the sulphone electron acceptor group are especially suitable for our application. This contribution will discuss recent results on polymethylmethacrylate (PMMA) copolymer side-chain materials based on the 4-(dimethyl-amino)- 4'-(alkylsulphonyl)biphenyl chromophore, its synthesis, characterization and photochemical stability.

179.   RECENT DEVELOPMENTS IN EFFICIENT FREQUENCY DOUBLING IN POLED POLYMER FILMS.
        G.Khanarian, S.Meyer, R.A.Norwood, H.A.Goldberg, D.Holcomb and J.Stamatoff
        Hoechst Celanese, 86 Morris Avenue, Summit, NJ 07901

Frequency doubling in waveguides is becoming increasingly important for optical data storage, printers and display applications. Poled polymer frequency doublers have advantages in their relative ease of fabrication and integrability with other components. A polymer frequency doubler will require nonlinear optical dyes that have sufficient activity but which are also transparent at the second harmonic, polymers that can be poled and processed to make smooth waveguides and phase matching to give efficient doubling. A review will be given of the current status of polymer frequency doublers and compared with competing technologies.

180.   PHOTOCHEMICALLY DELINEATED REFRACTIVE INDEX PROFILES IN POLYMERIC SLAB WAVE-
        GUIDES. Keith A. Horn, David B. Schwind and James T. Yardley, Allied-Signal Inc.
        Research and Technology, 101 Columbia Road, Morristown, NJ 07962-1021.

   One promising new technique for the creation of waveguides and integrated optical devices in polymeric thin films uses photochemical reactions to spatially define refractive index patterns. In order to model the propagation of light through these devices, the graded index (GRIN) structures created by the photochemical reaction must be accurately determined. We have effectively modelled and experimentally tested the development of graded refractive index profiles in nitrone/PMMA slab waveguides. The rate equations developed by Simmons[1] for photochemical reactions in highly absorbing thin films were used with a "WKB model" to determine the effective refractive indices of these waveguide structures. The use of these methods to determine the quantum yields of solid phase photochemical reactions will also be discussed.

1. E.L. Simmons, J. Phys. Chem., 75, 588, 1971.

181. POLYPHOSPHAZENES MOLECULAR COMPOSITES. I. IN SITU POLYMERIZATION OF TETRAETHOXYSILANE. B. K. Coltrain, W. T. Ferrar, C. J. T. Landry, T. R. Molaire, Corporate Research Laboratories, Eastman Kodak Company, Rochester, NY 14650-2110

Composite materials have been prepared by combining polyphosphazene polymers with a silicon oxide precursor, tetraethoxysilane (TEOS), which is polymerized in situ. Poly[bis(methoxyethoxyethoxy)phosphazene] (MEEP), a polymer that flows at room temperature, forms a clear, flexible, free-standing film when TEOS is polymerized in situ. Improved mechanical properties of a fluoroalkoxyphosphazene, partially hydrolyzed poly[bis(trifluoroethoxy)phosphazene] (TFEP-OH), were evident in that a rubbery plateau region above T(1) is observed in the storage modulus.

$$[N=P(OCH_2CH_2OCH_2CH_2OCH_3)_2]_n \quad [P(OCH_2CF_3)_2=N]_{95x}[P(OCH_2CF_3)(OH)=N]_{5x}$$

182. REACTIONS OF POLYMERS WITH PENDANT CYCLOPHOSPHAZENES, C.W. Allen, M. Bahadur and D. E. Brown, Department of Chemistry, University of Vermont, Burlington, VT 05405-0125.

Chemical modification of carbon chain polymers with pendant cyclophosphazenes as substituents can be accomplished either by reactions of the monomer followed by polymerization or by reactions of the preformed polymer. The former route is limited by the electronic structure of the monomer or by chain transfer processes involving certain substituents on the phosphazene. Reactions of the preformed polymer, e.g. $[CH(ON_3P_3Cl_5)CH_2]_n$, with a variety of necleophiles lead to partial (with $HNMe_2$, $HNEt_2$, H_2NMe, OPh^-) or complete (with $OCH_2CF_3^-$) replacement of the chlorine atoms without significant degradation or cross linking. The thermal decomposition behavior of the new polymeric cyclophosphazene derivatives shows considerable variation depending on the ring substituents. Thus while the trifluoethoxy derivative undergoes random scission the phenoxy derivative leaves over 50% of a phosphazene containing char. The copolymerization of $2,4-N_3P_3F_4(C\equiv CPh)C_6H_4C(CH_3)=CH_2$ with styrene yields a copolymer with a pendant cyclophosphazene having an alkyne substituent. The alkyne can serve as a site to bind the dicolbalt hexacarbonyl fragment. The resulting organometallic derivative undergoes a reversible electrochemical reduction.

183. SYNTHESIS OF POLY(B-VINYLBORAZINE AND POLY(STRYENE-CO-B-VINYLBORAZINE) COPOLYMERS. Kai Su, Edward E. Remsen, Helen M. Thompson and Larry G. Sneddon, Department of Chemistry, University of Pennsylvania, Philadelphia, PA 19104-6323 and Physical Science Center, Monsanto Corporate Research, 700 Chesterfield Village Parkway, St. Louis, MO 63198

B-Vinylborazine has been found to readily polymerize in solution using free radical initiation with AIBN to yield soluble poly(B-vinylborazine) homopolymers. Molecular weight characterization using size exclusion chromatography/low angle laser light scattering (SEC/LALLS) suggests that poly(B-vinylborazine) has a heterogeneous chain structure consisting of linear and, possibly, branched polymer chains. AIBN initiation has also been employed to produce a range of new hybrid organic/inorganic poly(styrene-co-B-vinylborazine) (SVB) copolymers with controlled compositions and molecular weights. Random copolymerization of styrene and B-vinylborazine monomers was confirmed by relative reactivity ratios and excimer fluorescence measurements. SEC/LALLS in conjunction with ultraviolet absorption detection (SEC/LALLS/UV) suggests that the copolymerization conditions which promote copolymer compositional heterogeneity may also produce a fraction of branched copolymer molecules due to dehydropolymerization reactions of the B-vinylborazine monomer.

184. **NEW PREPARATIVE CHEMISTRY OF POLY(ALKYL/ARYLPHOSPHAZENES).** <u>Robert H. Neilson</u>, David L. Jinkerson, Sakthevil Karthikeyan, Remy Samuel, and Christopher E. Wood, Department of Chemistry, Texas Christian University, Fort Worth, TX 76129.

The title compounds, $[N=PR_2]_n$, are generally accessible via the condensation polymerization of suitable N-silylphosphoranimines, $Me_3SiN=PR_2X$. We are interested in investigating the overall scope and possible limitations of this condensation polymerization process as well as the physical and chemical properties of the resulting polymers. Some of our specific objectives, which are addressed in part by the results reported here, include the synthesis of: (1) a wide variety of new poly(alkyl/arylphosphazenes) for analysis of structure-property relationships, (2) a series of novel CF_3-substituted Si-N-P compounds that are possible precursors to new polyphosphazenes with enhanced thermal stability and surface properties, and (3) other new types of Si-N-P derivatives that will serve as polymer precursors, model compounds, probes of the polymerization mechanism, or potential end-blocking groups.

185. **POLYORGANOBORANES; SYNTHESIS and APPLICATIONS.**
<u>T. C. Chung</u>, Department of Materials Science, The Pennsylvania State University, University Park, PA 16802

Introduction of boron elements to hydrocarbon polymers represents a useful method to modify their physical and chemical properties. The borane groups in polymer are very versatile, which can serve as reactive reagents in chemical reactions and polymerization and also can be interconvertable to other functional groups. In this presentation, we would summarize our research results in the preparation of polyorganoboranes. A broad range of polymer structures, such as comb-like polymers, block copolymers and telechelic polymers have been prepared by transition metal polymerization and borane monomers. The unique features in this preparation will be pointed out. The stability of organoboranes in transition metal catalysts allows us to incorporate borane containing monomers to polyolefins. The high solubility of borane polymers in hydrocarbon solution, due to boron-carbon colvence bonding, maintains homogeneous reaction condition which is important for preparing high molecular weight polymers.

186. **POLY(THIONYLPHOSPHAZENES): A NEW CLASS OF INORGANIC POLYMERS WITH SKELETAL PHOSPHORUS, NITROGEN, AND SULPHUR(VI) ATOMS**
<u>Ian Manners</u>, Mong Liang, Chris Waddling, and Charles Honeyman, Department of Chemistry, University of Toronto, 80 St. George St., Toronto M5S 1A1, Ontario, Canada.

Inorganic polymers are attracting considerable current attention because of their unusual mechanical, electronic, and optical properties and their potential function as precursors to ceramic materials. In our programme at Toronto we are exploring the synthesis, material properties, and applications of new inorganic polymer systems. In this talk the synthesis of the first poly(thionylphosphazenes) (**I**), a new class of inorganic polymers which possess backbones of alternating phosphorus, nitrogen, and sulfur(VI) atoms, will be discussed. The chemistry of these macromolecules will then be described and compared to that of their recently prepared sulphur(IV) analogues, the poly(thiophosphazenes) **II**.

$$\left[\begin{array}{c} O \\ \| \\ -S=N-P=N-P=N- \\ | \quad | \quad | \\ R \quad R \quad R \end{array} \right]_n \qquad \left[\begin{array}{c} R \quad R \\ | \quad | \\ -S=N-P=N-P=N- \\ | \quad | \quad | \\ R \quad R \quad R \end{array} \right]_n$$
$$\qquad \mathbf{I} \qquad\qquad\qquad\qquad \mathbf{II}$$

187. **CHARACTERIZATION AND OXIDATION REACTIONS OF Si-B-C POLYMERS: EXPERIMENTAL AND THEORETICAL STUDIES.** <u>R. L. Jaffe</u> and S. Riccitiello, NASA Ames Research Center, Moffett Field, CA 94035, M.-T. Hsu and T. Chen, HC Chem Research and Services Corp., San Jose, CA and A. Komornicki, Polyatomics Research Institute, Mountain View, CA 94043

The IR spectra of Si-B-C polymers prepared from dichlorodimethylsilane and chloroboranes have been characterized by comparison with the spectra of model compounds and with theoretical spectra determined by *ab initio* quantum chemical methods. Assignments have been made based on both vibrational frequencies and intensities. The theoretical spectra provide alternative means of verifying the elemental composition and purity of the polymers. The thermal and oxidative stability of these materials has been studied and data presented. Comparisons are made to analogous properties of polycarbosilanes and boron nitride polymers.

188. PHOTOLABILE POLYSILANES: THE EFFECT OF ELECTRON TRANSFER ADDITIVES. R. D. Miller, M. Baier, N. Clecak, G. M. Wallraff, IBM Almaden Research Center, K95/801, 650 Harry Road, San Jose, CA 95120-6099

Substituted silane polymers are photochemically labile and the predominate photochemical pathways involve the extrusion of substituted silylenes and silicon-silicon bond homolysis. Recently, another process i.e. 1,1-photochemical reductive elimination has been identified as a chain cleaving pathway. Although chain scission is the predominate photochemical process for most polysilane derivatives, crosslinking becomes more important with pendant unsaturated substituents. Unfortunately, the efficiency of chain scission drops markedly in going from solution to the solid state; a feature which is detrimental for most imaging applications. We have approached this problem simultaneously from two directions: (i) the preparation of polysilanes derivatives which are intrinsically more photolabile, (ii) the use of small molecule additives to improve the photosensitivity, particular in the deep UV spectral region. Recent results in these areas and possible applications to deep UV microlithography will be discussed.

189. ANILINE SUBSTITUTED 2,2'-BIPYRIDINE LIGANDS FOR POLYMER MODIFIED ELECTRODES, Colin P. Horwitz and Zuo Qi, Department of Chemistry, Rensselaer Polytechnic Institute, Troy, NY 12180-3590

Electrode surfaces modified with transition metal complex polymer films prepared by electropolymerization techniques continue to be intensely active research area. We have synthesized the ligands shown below, which allow for the oxidative electropolymerization of some iron and ruthenium complexes. Ligand **2** is particularly interesting as only one of these ligands present on the metal complex is necessary to form a polymer film. The synthesis of the ligands and properties of the transition metal complex polymer films will be discussed.

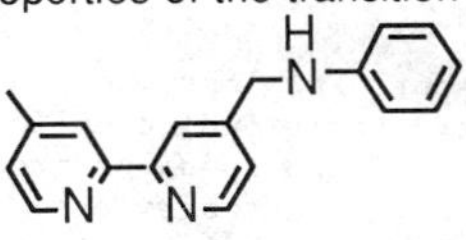

190. POLYSILYNE FUNCTIONALIZATION WITH TRIFLUOROMETHANESULFONIC ACID. David A. Smith, Patricia A. Bianconi, Corie A. Freed, David M. Goncalves; Department of Chemistry, The Pennsylvania State University, University Park, PA 16802

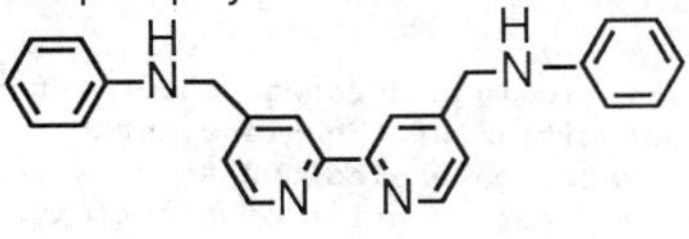

Recent discoveries in silicon polymer chemistry has opened a new ar . of research in three dimensional silicon backbone polymers known as the polysilynes. This new class of polymers has possible applications as ceramic precursors and photoresists. Current synthetic routes for polysilynes involves a reductive coupling (Wurtz) reaction. A sonochemically generated sodium/potassium emulsion in an appropriate solvent provides an efficient reductive environment to polymerize alkyl or aryl trichlorosilanes to the appropriate polymer. Unfortunately, this synthetic route does not allow the use of monomers with functional substituents, such as those with halogens, amines, alkoxides, esters, etc. Therefore, a route has been developed through which polyphenylsilyne can be selectively functionalized with triflic acid, producing a highly reactive triflated intermediate. This intermediate can be substituted with new functional groups (as sodium salts, lithium salt, Grignards, etc.) to give a new polymer with modified properties.

191. GAS SORPTION AND DIFFUSION IN HYDROGEN-BONDED POLYMERS: VINYL ALCOHOL/VINYL BUTYRAL COPOLYMERS. Mark J. Cibulsky, Michael J. Reimers, and Timothy A. Barbari, Department of Chemical Engineering, Johns Hopkins University, Baltimore, Maryland 21218

Gas sorption and diffusion in polymers depend on gas-polymer interactions, polymer-polymer interactions, and interchain packing (or free volume). Hydrogen bonding provides a level of molecular interaction between physical intermolecular forces and covalent bonding. We have examined a series of vinyl alcohol/vinyl butyral copolymers to assess the effect of internal hydrogen bonding on gas sorption and diffusion. FTIR spectroscopy can provide insight into the role that hydrogen bonding plays in determining gas sorption and diffusion. A much wider variety of transport behaviors is observed than with copolymers that do not hydrogen bond. Controlling this strong interaction may lead to the molecular design of new barrier or membrane structures.

192. DIFFUSION OF WATER VAPOR IN HYDROGEN BONDED POLYMER BLENDS. Thomas C. Gsell, Eli M. Pearce and T.K. Kwei, Department of Chemistry and Department of Chemical Engineering, Polytechnic University, Brooklyn, New York 11201

Water vaopr was used as diffusional probe to study interaction and segmental mobility in hydrogen bonded polymer blends. Modified polystyrenes containing p-(hexafluoro-2-hydroxy isoprooyl) groups as hydrogen bond donor were blended with poly (methyl methacrylate) and two styrene -acrylonitrile copolymers. In a number of blends, the diffusion coefficients were higher than the "average" values for the component polymers. The deviation in diffusion coefficient was opposite in sign to the deviation of the activation energy of diffusion from its "average" value. The residual activation energy was proportional to the excess volume.

193. INFLUENCE OF PLASTICIZERS ON THE DIFFUSION PROCESS OF WATER IN POLY(METHYL METHACRYLATE). S. Kalachandra, Dental Research Center, UNC, Chapel Hill, NC 27599.

Relatively little is understood about water transport in glassy polymers. Apart from general complexities of the glassy state, there are additional complications due to clustering of water molecules and to plasticization. The purpose of present work is to characterize water sorption of plasticized poly(methyl methacrylate) (PMMA) and also to find whether, there would be simple increases in the rate of water diffusion in the glassy state, i.e. up to Tg. Plasticized poly(methyl methacrylate) was made by r-irradiation of mixtures of monomer and various phthalates. Samples were immersed in water and uptake and diffusion coefficients (D) determined. More reliance was placed on determinations made in desorption because these did not involve complications due to loss of components by dissolution. Plasticizers were found to decrease the uptake of water. This decrease was attributed to the hydrophobic nature of the plasticizers and also to their ability to fill microvoids in PMMA which, in their absence would be available to water. The values of D increased monotonically with increasing volume fractions of plasticizer which is consistent with a "physical loosening of bonds". This investigation was supported by USPS Research Grant No. 06201.

194. DIFFUSION OF WATER INTO A PHOTOPOLYMER FILM Christopher R. Moylan, IBM Almaden Research Center, 650 Harry Road, San Jose, California, 95120- 6099.

Sorption of water vapor into thin films of a crosslinked dimethacrylate UV-cured polymer was measured by means of a quartz crystal microbalance apparatus. The diffusion coefficient and solubility were obtained. An upper limit to the water vapor diffusion coefficient, namely the diffusion coefficient for liquid water in intimate contact with the sample, was measured for reference purposes by both gravimetric and infrared spectral techniques. A water uptake value for thick films was obtained gravimetrically as a reference for the thin film value. Measurement of the diffusion coefficient allows the determination of the length of time that the photopolymer could act as an effective water vapor barrier.

POLY

195. DIFFUSION OF MULTI-COMPONENT SPECIES ACROSS POLYMERIC MEMBRANES. Steven M. Dinh, Bret Berner, and Yi-Ming Sun, *Basic Pharmaceutics Research, Ciba-Geigy Corporation, Ardsley, New York, 10502*

Understanding how to control the fluxes and the relative composition of multiple species permeating across a membrane would be invaluable to the design of controlled release systems. The diffusion of ethanol and water across polyethylene vinyl acetate (EVA) membranes was investigated as a case study. Ethanol and water fluxes across these membranes were measured by monitoring the changes in refractive indices of the donor and receiver solutions. Ethanol and water solubilities in these membranes were determined by a sorption and desorption method. Ethanol and water diffusivities across the membranes were calculated from the flux and solubility data. In a 37% vinyl acetate EVA membrane, the ratio of ethanol diffusivity to its activity coefficient was found to increase exponentially with ethanol activity in the membrane; whereas the ratio of water diffusivity to its activity coefficient decreased exponentially with water activity in the membrane. With these results, ethanol and water fluxes across the 37% vinyl acetate EVA membrane can be predicted for any set of compositions in the donor and receiver solutions. Surprisingly, ethanol and water fluxes across this membrane were found to be linearly dependent.

196. IMPROVING BARRIER PROPERTIES THROUGH SURFACE ENERGETICS. T. J. Fabish, T. L. Levendusky and T. W. Barefoot, *Alcoa Technical Center, Alcoa Center, PA 15069.*

Surface films that can enhance the barrier properties of low cost, low performance polymer films or coatings have considerable commercial potential for coated metal products. This progress report describes our initial work with gas/solid interactions to identify permeant/polymer systems that unambiguously manifest the energetics of the surface interaction independent of the mass diffusion process as a first step in evaluating the feasibility of very thin films for barrier control. A simple experiment shows a particular fluorocarbon surfactant to act to lower permeation of toluene in polycarbonate through surface solvation as opposed to a diffusion barrier.

197. DYNAMIC VISCOELASTICITY DURING THERMOREVERSIBLE GELATIN GELATION. Eric J. Amis, Donald F. Hodgson, and Qun Yu, *Department of Chemistry, University of Southern California, Los Angeles, CA 90089*

The frequency dependent dynamic storage and loss moduli, $G'(\omega)$ and $G''(\omega)$, respectively, have been measured for gelatin solutions which are in the course of thermoreversible gelation. Although very weak gels are formed under some conditions the macroscopic gelation time can be identified by the appearance of a frequency independent $\tan \delta \equiv G''/G'$. The frequency independent loss tangent corresponds to scaling of G' and G'' ($\sim \omega^\Delta$) with the same frequency exponent Δ. In contrast, the "extent of reaction" when $G' = G''$, which has been taken in some work to identify the gel point, is shown to be strongly frequency dependent. The dependence of the scaling exponent Δ on concentration, pH, ionic strength, and temperature of the gelation process has been investigated. These results will be compared with predictions based on mean field theories and percolation models. These results for the physical gel will also be compared to similar measurements on chemical gels of tetraethoxysilane.

198. PROBE DIFFUSION AND SELF-DIFFUSION IN THERMOREVERSIBLE GELS. Hyuk Yu, *Department of Chemistry, University of Wisconsin, Madison, WI 53706.*

A brief review of the diffusion studies by the technique of forced Rayleigh scattering (FRS) will be presented. It will be centered around two thermoreversible systems, gelatin gels and atactic polystyrene in carbon disulfide. In the first case the self diffusion is shown to be quenched below the gel point whereby we conclude that all the chains are involved in the gel structure formation. In the second case, the self diffusion is noted to be relatively free even below the gel point whereby we conclude that a substantial fraction of the chains is not involved in gel structure. The FRS signal profiles have shown a slow component tail which is separated from the fast one by three orders of magnitude and is independent of the fringe spacing. Hence it is not attributed to any diffusive process, although one can use its time constant to establish the upper limit of slowly diffusing component, i.e., not faster than 10^{-11} cm^2/s.

199. **DYNAMIC LIGHT SCATTERING STUDIES OF RESORCINOL FORMALDEHYDE GELS AS PRECURSORS OF ORGANIC AEROGELS.** Patricia M. Cotts, IBM Research Division, Almaden Research Center, 650 Harry Road, San Jose, California 95193-6099; Rick Pekala, Lawrence Livermore National Laboratory, Livermore, California.

Aerogels are exceptionally light foams formed by extraction of the solvent from a gel under supercritical conditions where surface tension forces are greatly reduced. This results in very little shrinkage of the gels when dried. The dried aerogels have extremely high surface area and have applications as catalyst supports and thermal and acoustic insulators. While most studies have used aerogels formed from silicon dioxide, we have focused on organic aerogels obtained by the reaction of resorcinol and formaldehyde (RF). We have studied the kinetics of the gel formation using dynamic light scattering. These studies encompass three areas: 1) initial reaction of RF into branched "clusters" as a function of resorcinol/catalyst (Na_2CO_3) ratio; 2) aggregation of the clusters into linear chains to form the gel network; and 3) the modulus of the completely gelled sample. The initial particle size increases linearly with R/C ratio as was observed by TEM on dried samples. The scattering intensity increases exponentially as the gel is formed as is observed for polymer gels formed by spinodal decomposition. The cooperative diffusion coefficient (which is directly related to the gel modulus) scales with concentration to a power of approximately unity, which is larger than the 3/4 usually observed for polymer networks. The larger exponent may be related to the greater rigidity of these gels relative to those formed from covalently bonded polymer chains.

200. **GEL-PROCESSING OF OPTO/ELECTROACTIVE POLYMERS,** Jeff Moulton and Paul Smith, University of California, Santa Barbara, Santa Barbara, CA 93106

Significant advances have been made in developing processable opto/electroactive polymers, either through the attachment of flexible side chains or preparing conjugated polymers via flexible precursor polymers. The increased solubility allows these materials to be readily co-processed with ultra-high molecular weight flexible chain polymers, i.e. ultra-high molecular weight polyethylene (UHMW-PE). The low UHMW-PE concentrations required to form coherent gels upon crystallization are advantageous, in that they allow for the formation of very high specific surface area substructures upon which the conjugated polymers may be absorbed. The low polyethylene entanglement densities also allow for very high composite draw ratios to be realized.

Discussion will focus on the enhancement of the optical, electrical and mechanical properties of polymer blends as understood in terms of the unique properties of ultra-high molecular weight thermoreversible gels.

201. THERMOREVERSIBLE GELS FROM DIBENZYLIDENE SORBITOL AND ORGANIC SOLVENTS. Gregory B. McKenna*, Francois Kern, Sauveur J. Candau, Laboratoire d'Ultrasons et de Dynamique des Fluides Complexes, Unité de Recherche Associée au C.N.R.S. n° 851, Université Louis Pasteur, Strasbourg, France; Jean-Claude Wittmann and Annette Thierry, Institut Charles Sadron, Strasbourg, France.

Rheological and optical measurements have been carried out on gel forming systems from 1,3: 2,4-Dibenzylidene Sorbitol (DBS) in the organic liquids dimethyl and dibutyl phthalate, dibutyl adipate and hexadecane. Visual observation of the gel formation indicates a different kinetics and gel structure depending on solvent and concentration. The rheological measurements verify the different kinetics and show a power law dependence of gel storage modulus $G'(\omega)$ and creep compliance $J(t)$ of the fully gelled systems show strong concentration dependences with $G' \sim \phi^4$ and $J \sim \phi^{-4}$.

*Permanent address: Polymers Division, NIST, Gaithersburg, MD 20899 USA

202. Academic Perspectives. R. Abbaschian.

NO ABSTRACT

203. Industrial Perspectives. **J. E. Nottke.**

NO ABSTRACT

204. NSF Perspectives. **N. M. Bikales.**

NO ABSTRACT

205. CUBIC NONLINEAR OPTICS OF POLYMER THIN FILMS. 2. STRUCTURE-$\chi^{(3)}$ RELATIONSHIPS IN RIGID-ROD POLYQUINOLINES. Ashwini K. Agrawal and Samson A. Jenekhe, Center for Photoinduced Charge Transfer and Department of Chemical Engineering, University of Rochester, Rochester, NY 14627-0166. Herman Vanherzeele and Jeffrey S. Meth, DuPont CR&D, P.O. Box 80356, Wilmington, DE 19880-0356.

Although third-order optical nonlinearities have been measured in many conjugated polymers, a fundamental understanding of the underlying structure-$\chi^{(3)}$ relationships is yet to be established. Conjugated rigid-rod polyquinolines are thermally stable (500 - 600°C), high mechanical strength, film- and fiber-forming materials that can be doped to high electronic conductivity. One of the most attractive features of the conjugated polyquinolines is the flexible synthetic scheme with which the basic molecular structure can be modified, thus making this class of polymers an ideal model system for investigating structure-$\chi^{(3)}$ relationships. We have prepared a series of homopolymers and random copolymers of polyquinolines with a systematically varied molecular structure, prepared their optical quality thin films, and measured their wavelength dependent $\chi^{(3)}$ by picosedond third harmonic generation in the range 0.9 - 2.4 µm.

206. **A STUDY OF THE COMPETITION BETWEEN ELECTRIC FIELD INDUCED MOLECULAR ORDERING AND CHEMICAL CROSS LINKING IN VITRIFIED NONLINEAR OPTICAL POLYMER FILMS.** Jon D. Bewsher, Geoffrey R. Mitchell, Polymer Science Centre, University of Reading, Whiteknights, Reading, RG6 2AF, U.K.

Recently, considerable effort has been put into synthesising polymeric thin films for use in nonlinear optical devices. The process of thermopoling guest/host systems is commonly used to generate the noncentrosymmetric distribution of nonlinear molecules, essential for the operation of all second order optical devices. However films manufactured from these materials ultimately suffer from a reduction in nonlinear optical performance due to a thermal relaxation of the molecular alignment. A multi-staged film preparation process of chemical cross linking under an electric field suggests one way in which this long term molecular disordering could be reduced. In this report we have set out to probe the interaction between the two, potentially competing, processes of electric field enhanced molecular ordering and chemical cross linking by experimentation and advanced molecular modelling.

207. **CONTROLLED SYNTHESIS OF METAL NANOCLUSTERS WITHIN BLOCK COPOLYMER MICRODOMAINS.** G. S. W. Craig, R. E. Cohen, Y. Ng Cheong Chan, R. R. Schrock, Departments of Chemical Engineering and Chemistry, Massachusetts Institute of Technology, Cambridge, MA 02139.

Sequential ring opening metathesis polymerization with a well-defined metal alkylidene catalyst yields low polydispersity diblock copolymers in which one of the blocks contains a metal complex (e.g. M = Pd, Pt) in each repeat unit. In static cast and spin cast films, these diblock copolymers undergo microphase separation at relatively low molecular weights, resulting in spherical, cylindrical, or lamellar microdomains of the metal containing blocks. Films were annealed to achieve a greater degree of orientation in the microdomains. Exposure to UV light ($\geq$ 300nm) or treatment with H_2 (60 psig, 100 °C) of the polymers containing Pd or Pt, respectively, reduces the metal atom and creates a metal cluster on the order of 20 - 80 Å within each microdomain. The Pd crystal lattice of the clusters is identical to bulk Pd, as determined by wide-angle x-ray scattering (WAXS). The microphase separation of these polymers was characterized with transmission electron microscopy (TEM) and small angle x-ray scattering (SAXS). The dielectric and nonlinear optical properties of the polymer films will be analyzed and compared to cluster size and concentration.

208. NONLINEAR OPTICAL PROPERTIES OF LINEAR EPOXY POLYMERS WITH PENDANT SULFONYL TOLAN GROUPS. M. Ebert, M. Lux, B. A. Smith, R. Twieg, C. G. Willson, D. Y. Yoon, IBM Almaden Research Center, San Jose, CA 95120-6099

The recent past has shown that the employment of poled amorphous polymers has been a very promising step towards the potential application of second harmonic generation and electro-optical devices. Several approaches have been taken to enhance the poling-induced nonlinear optical susceptibility and its long term stability for example by preparing crosslinked epoxy polymers and also by using linear epoxy polymers with specifically designed chromophores covalently attached to the backbone. Here we report on investigations of some nonlinear optical properties of a series of novel linear epoxy polymers. These polymers consist of tolan groups with attached sulfonyl groups, and are completely amorphous. The films were characterized by means of waveguiding experiments and second harmonic generation measurements. We will present experimental results on NLO properties, induced alignment and the relaxation characteristics as a function of temperature.

209. SYNTHESIS AND CHARACTERIZATION OF MONOMERS AND POLYMERS CONTAINING PHOTORESPONSIVE AZOBENZENE - BASED SIDEGROUPS
H. J. Haitjema, G. O. R. Alberda v. Ekenstein and Y. Y. Tan
P. J. Werkman, A. J. Koldijk, T. S. Boer
State University of Groningen, Nijenborgh 16, 9747 AG Groningen
The Netherlands

The effect of the incorporation of organic chromophores in a polymeric matrix has become a topic of considerable scientific interest in the last years. The environment of the chromophores, that is, the matrix, influences their photo-responsive behavior. These effects are rather well understood. We have devised a new kind of environment in the form of polymeric complexes, consisting of (poly)-acid/(poly)-base systems. The monomers that we aimed to synthesize and polymerize, are azobenzene - based compounds, acrylates and methacrylates with or without a spacer, with CO_2H or $N(CH_3)_2$ as end groups. The reason for the choice of these compounds is that we expect that the photoresponsive behavior (kinetics of thermal isomerization) of the azobenzene group might deviate from normal kinetics on interaction of the end groups with (poly)-acid/(poly)base systems.

210. POLING AND CHEMICAL-BINDING OF GLASS-EMBODIED CHROMOPHORES IN SUPPORTED SOL-GEL THIN-FILM GLASSES FOR SECOND HARMONIC GENERATION. Y. Haruvy[†], J. Byers, and S.E. Webber, Department of Chemistry and Biochemistry and Center for Polymer Research, The University of Texas at Austin, Austin, TX 78712, USA

By careful control of sol-gel reaction conditions and spin-casting speed it is possible to make films of poly(methylsiloxane) (PMSO) that are crack-free and of excellent optical clarity with a thickness in the 1-20 µm range. These films can also be used to embody laser dyes or molecules with a high hyperpolarizability, such as p-nitroaniline ($\beta = 35 \times 10^{-30}$ esu) or 4-dimethylamino-4'-nitrostilbene ($\beta = 450 \times 10^{-30}$ esu). We report our results on second harmonic generation (SHG) at 1064 nm (using a Nd:YAG mode-locked laser) in poled and unpoled PMSO glasses containing ca. 2-4 wt % of these molecules. While the overall efficiency of SHG seems quite good (ca. two to ten times that of powdered urea), the SHG efficiency of these films is a complicated function of sol-gel preparation method, poling protocol, and SHG-molecule content in the glass.

† On sabbatical leave from Soreq NRC, Yavne, 70600, Israel

211.

MODEL MOLECULES WITH DONOR-ACCEPTOR UNITS INTERMEDIATED BY SILICON CHAINS. Diny Hissink, Jeannet Brouwer, Rien Flipse and Georges Hadziioannou. Laboratory of Polymer Chemistry, Department of Chemistry, University of Groningen, Nijenborgh 16, 9747 AG Groningen, The Netherlands.

We aim to synthesize dye molecules with high non linear optical (NLO) properties and good transparency in the visible spectrum. By using a (σ–σ^*) conjugated system, i.e. a silicon backbone, we obtained transparent molecules (λ_{cutoff} < 400 nm.). This transparency is needed for applications such as frequency doubling of a 820 nm laser. We synthesized and characterized several donor-acceptor molecules intermediated by a silicon backbone with various donor and acceptor groups. We studied their spectroscopic properties (UV and fluorescence). With use of the absorption and emission maxima in different solvents we determined the difference in dipole moment from ground to excited state.

212. NOVEL THIRD ORDER NLO MATERIALS FROM 96% QUINOLINE. R.V. Honeychuck, Department of Chemistry, George Mason University, Fairfax, Virginia 22030, and Chemistry Division, Naval Research Laboratory, Washington, D.C. 20375.

Films from 96% quinoline and from 96% quinoline with poly(1,4-cyclohexylene carbonate), mesoporphyrin IX dimethyl ester, and Pd(II)mesoporphyrin IX dimethyl ester have been cast via heating and subsequent evaporation. The films are approximately 3 μm thick, with a maximum variation normal to the surface of 0.10 μm in 2 mm of surface travel. The optical quality of the samples formed from 96% quinoline alone is excellent, and variable from the quinoline/solute combinations. The third order nonlinear optical properties of these materials have been examined.

213. CUBIC NONLINEAR OPTICS OF POLYMER THIN FILMS. 5. WAVELENGTH DISPERSION OF THE $\chi^{(3)}$ OF POLY(P-PHENYLENE BENZOBISTHIAZOLE) BASED MOLECULAR COMPOSITES. Samson A. Jenekhe and Michael F. Roberts, Center for Photoinduced Charge Transfer and Department of Chemical Engineering, University of Rochester, Rochester, NY 14627-0166. Herman Vanherzeele and Jeffrey S. Meth, DuPont CR & D, P.O. Box 80356, Wilmington, DE 19880-0356.

Molecular composites of a conjugated polymer dispersed in the matrix of an optically transparent flexible chain polymer represent one approach to optimize the physical properties of cubic nonlinear optical materials, including the optical loss α and the device figure of merit $\chi^{(3)}/\alpha$. Composite materials are also excellent model systems for investigating the effects of composition and morphology on optical nonlinearities. We have prepared thin films of molecular composites of the rigid-rod conjugated polymer poly(p-phenylene benzobisthiazole) (PBZT) in the matrix of nylon 66 or poly(trimethyl hexamethylene terephthalamide). We will present the wavelength dependent $\chi^{(3)}$ of these composites as determined by picosecond third harmonic generation in the wavelength range 0.8 - 2.4 μm. Effect of the composite composition on the $\chi^{(3)}$ will also be discussed.

214. **MEASUREMENTS OF THE OPTO-ELECTRIC PROPERTIES OF MOLECULES OF POTENTIAL USE IN PHOTOACTIVE POLYMERS BY TIME RESOLVED MICROWAVE CONDUCTIVITY (TRMC).** S.A. Jonker and J.M. Warman, IRI, Delft University of Technology, Mekelweg 15, 2629 JB Delft, The Netherlands.

The application of the time-resolved microwave conductivity (TRMC) technique to the quantitative measurement of charge separation in flash-photolysed molecular systems is described. Marked changes in the photophysics (S_1 lifetimes and intersystem crossing efficiency) are found for a series of molecules consisting of an anilino donor and a nitrobenzene acceptor moiety as a function of the spacer which interconnects these moieties and the presence of alkyl substituents on the anilino nitrogen. Experiments on a hydroxy-cyanobiphenyl chromophore with and without a functional propylacrylate chain show the chain to have little influence on either the photophysics or the dipole relaxation time of the chromophoric unit.

215. **MACROMOLECULAR CHROMOPHORIC ASSEMBLIES WITH ENFORCED POLARITY: INORGANIC AND HETEROCYCLIC LINKAGES.** H.E. Katz, M.L. Schilling, C.E.D. Chidsey, T.M. Putvinski, W.L. Wilson, G.R. Scheller, W.T. Lavell. AT&T Bell Laboratories, Murray Hill, NJ 07974.

Materials with rigidly enforced dipolar orientation are of interest because of their unique combinations of electronic, optical, and mechanical properties. We have developed two new means of linking dipolar elements such that their relative orientations are almost the same, and their electronic and optical moments are additive. The first relies on the layer-by-layer deposition of chromophoric zirconium phosphonates/phosphates to form a polymeric multilayer in which the donor->acceptor direction is rigorously defined relative to the surface of a substrate. The second involves the chemical condensation of chromophores through the piperidine-4-imino linking group to give a multimeric, elongated oligomer. The polarity of these assemblies is confirmed by SHG linear in layer number squared for the zirconium layers and by measured dipole moment additivity for the piperidine-linked segments. Both examples are superior to analogous compositions in terms of moment additivity and relative orientational stability.

216. **SYNTHESIS, POLING AND OPTICAL CHARACTERIZATION OF POLYURETHANE FILMS BEARING NLO-ACTIVE CHROMOPHORES.** P. Kitipichai, R. Laperuta, Jr., G.M. Korenowski, and G. E. Wnek, Department of Chemistry, Rensselaer Polytechnic Institute, Troy, New York 12180-3590, I. Gorodisher, 3M Science Research Laboratory, St. Paul, MN 55144-1000.

Polymers doped with dipolar chromophores are currently of great interest for nonlinear optical applications. We have synthesized new polyurethanes in which chromophores having large quadratic hyperpolarizabilities are covalently attached or crosslinked to the backbone. Advantage is taken of the fast kinetics of isocyanate/alcohol reactions to electrically pole the polymers during synthesis. Monomers are deposited onto a transparent microscope slide coated with aluminium, and are then polymerized during the application of a dc electric field to orient the NLO moieties. The films require only about one minute to prepare. The d_{33} values of a poled polyurethane were obtained shortly after poling and ca. 6 months after poling at ambient temperature. While the relative second harmonic (SH) intensity gradually decreases with time, the SH intensity of a poled, crosslinked polyurethane is extremely stable. No relaxation of SH intensity is observed as a function of time at ambient temperature or after 1 hour at 100 °C.

217. SYNTHESIS, EXPERIMENTAL, AND THEORETICAL STUDIES OF NEW GROUP 6 PENTACARBONYL-BASED NLO CHROMOPHORES EXHIBITING LARGE AND UNUSUAL OPTICAL NONLINEARITIES. P. Lacroix, T. J. Marks, D. A. Kanis, M. A. Ratner, Department of Chemistry and the Materials Research Center, W. Lin and G. K. Wong, Department of Physics and the Materials Research Center, Northwestern University, Evanston, IL 60208-3113.

The design of efficient quadratic nonlinear optical organometallic chromophores containing transition metal substituents presents a great challenge, and to date it appears that few molecules are good candidates. We have approached this problem using perturbation theoretical calculations at the SCF-LCAO-ZINDO level to target and to understand the properties of candidate chromophores. We report here a combined synthetic, EFISH experimental, and theoretical study of a new class of group 6 pentacarbonyl-based stilbazole chromophores with large and unusual optical nonlinearities. In particular, we find that the $M(CO)_5$ fragment functions for the first time as an electron-withdrawing group and that the pattern of states contributing most to β_{zzz} does not follow patterns observed for typical organic chromophores. These observations, the good agreement between observed and calculated β_{zzz}, and a molecular orbital analysis suggest some new guidelines for the design of organometallic NLO chromophores.

218. OPTICAL FOUR-WAVE MIXING IN A SILVER COLLOID-POLYMER COMPOSITE. R. LaPeruta, P. Kitipichai, G. E. Wnek, and G. M. Korenowski, Department of Chemistry, Rensselaer Polytechnic Institute, Troy, New York 12180.

The first of a new group of composite nonlinear optical materials was synthesized. This material consists of 10 nm diameter silver colloids in a polyurethane containing a N,N-dihydroxyethyl-(p-tricyanoethylene)aniline moiety. With the addition of the silver colloid to the polymer host, the initially highly colored polymer host becomes optically transparent over the wavelength region of 480 nm through the near infrared. Degenerate optical four-wave mixing experiments ($\omega = \omega - \omega + \omega$) were performed in an optical waveguide configuration to measure the Kerr susceptibility $\chi^{(3)}$ of thin films of the silver colloid-polymer composite. At the wavelengths of 532 nm, 562 nm, and 570 nm, $\chi^{(3)}$ was measured as 3.65×10^{-10}, 7.91×10^{-11} and 5.86×10^{-11} (esu), respectively.

219. NOVEL PHOTOCHROMIC LIQUID CRYSTAL POLYMERS. C.H.Legge, M.J.Whitcombe, A.Gilbert and G.R.Mitchell, Polymer Science Centre, University of Reading, Whiteknights, Reading RG6 2AF, U.K.

The variation in optical properties of organic materials may be achieved directly at an electronic level or by reorganisation of the molecular structural arrangements. This contribution focuses on the latter option using novel liquid crystal copolymers containing a small fraction of photoactive units. There are two routes to reorganisation in a liquid crystal system, either variation in S, the long range orientational order parameter, or changes to the angular distribution of the director n. In this presentation we compare and contrast these alternatives through the use of copolymers containing non-mesogenic cinnamate based groups or mesogenic azobenzene derivatives as the photoactive element. The potential for use in optical processing, transmission and storage is highlighted.

220. Synthesis of Block Copolymers of Poly(styrene) and Poly(arylate) and their Optical Properties. H.Ohishi, S.Inaba, M.Kawabe and M.Kimura, R&D Laboratories-1, Nippon Steel Corporation, 1618 Ida, Kawasaki, 211, Japan.

Since the optical information storage systems such as Laser Vision , Compact Disk and Magnetic Optical Disk are developing quickly, interest in polymers suitable for optical applications has grown enormously. New synthesis methods of transparent block copolymers(PS-b-PAr) of poly(styrene)(PS) and poly(arylate) (PAr) were proposed, using functional macromers of PS at the room temperature and the ordinal pressure. The birefrigence of 50% drawn PS-b-PAr films whose PS-composition was 50% was about 1/100 lower than that of PAr, and the melt viscosity was 1/1000 lower at 513k. Because of their low birefrigences and melt viscosities, PS-b-PAr would be a suitable material for optical applications.

221. CUBIC NONLINEAR OPTICS OF POLYMER THIN FILMS. 4. STRUCTURE-$\chi^{(3)}$ RELATIONSHIPS IN POLYANILINES AND DERIVATIVES. John A. Osaheni and Samson A. Jenekhe, Center for Photoinduced Charge Transfer and Department of Chemical Engineering, University of Rochester, Rochester, NY 14627-0166. Herman Vanherzeele and Jeffrey S. Meth, DuPont CR&D, P.O. Box 80356, Wilmington, DE 19880-0356.

Polyanilines, a versatile family of polymers with variable electronic, optical and electrochemical properties, offer the potential for investigation of structure-$\chi^{(3)}$ relationships through the variation of the oxidation state and derivatization of the p-phenylene rings. We will report on the wavelength dependent $\chi^{(3)}$ of thin films of polyanilines and a series of alkyl, alkoxy, and sulfonated derivatives in their intermediate oxidation states (emeraldine base forms). Effects of oxidation state and three-photon resonance enhancement on $\chi^{(3)}$ will also be discussed.

222. **OPTO-ELECTRONIC PROPERTIES AND SECOND HARMONIC GENERATION BY σ-BOND SEPARATED DONOR-ACCEPTOR MOLECULES.** W. Schuddeboom[1], B. Krijnen[1], J.W. Verhoeven[1], E.G.J. Staring[2], G.L.J.A. Rikken[2], H. Oevering[3], S.A. Jonker[4] *(1)Laboratory of Organic Chemistry, University of Amsterdam, Nieuwe Achtergracht 129, 1018 WS Amsterdam, The Netherlands, (2) Philips Research Laboratories, PO Box 80000, 5600 JA, Eindhoven, The Netherlands, (3) DSM Research, PO Box 18, 6061 MD, Geleen, The Netherlands, (4) IRI, Delft University of Technology, Mekelweg 15, 2629 JB Delft, The Netherlands*

Whereas commonly studied organic molecules for nonlinear optics contain a π-conjugatively linked donor (D) / acceptor (A) pair we now report the behaviour of some molecules in which a rigid, saturated hydrocarbon bridge separates D and A groups by at least three sigma bonds. The electronic spectra of these systems and their hyperpolarizability values as determined by second-harmonic generation (SHG) measurements reveal that the chromophores cannot be regarded as electronically isolated, but that a non-negligible electronic interaction occurs, mediated via the interconnecting sigma bonds. The relative weakness of this through-bond interaction results in a smaller value for the oscillator strength of the intramolecular charge-transfer transition in such non-conjugated D/A systems as compared to fully π-conjugated ones. The lowering effect of the small transition dipole moment on the magnitude of the molecular hyperpolarizability, β, appears to be partly compensated, however, by the large difference between ground- and excited-state dipole moment achievable in these systems.

223. SIDE CHAIN COPOLYMERS FOR THIRD ORDER NONLINEAR OPTICAL APPLICATIONS. James R. Sounik, Robert A. Norwood, Jacquelyn Popolo, Douglas Holcomb, Hoechst Celanese Research Division, 86 Morris Avenue, Summit, NJ 07901.

Organic polymeric materials have received much attention for applications in nonlinear optical devices. Most recent work has focussed on the electro-optical applications of glassy polymeric materials in optical waveguide devices whereas the development of all optical waveguide devices has suffered from the lack of highly active materials which can be easily processed. There have been a number of reports in the literature on $\chi^{(3)}$ activity of porphyrin derivatives, namely tetrabenzporphyrins, phthalocyanines and naphthalocyanines, that show relatively high third order nonlinear optical responses. These molecules are planar pi-conjugated systems that have sharp absorption bands in the visible and near infrared and exhibit excellent thermal and chemical stabilities. It would be desirable to develop materials containing these types of porphyrin derivatives that show high activity as well as processability similar to those polymeric materials being developed for electro-optic applications. Here we present the chemistry and nonlinear optical properties of a series of silicon phthalocyanine methylmethacrylate copolymers prepared by the modification of the central silicon atom.

224. QUANTITATIVE STUDY OF MOLECULAR ORIENTATION OF HEMICYANINE LANGMUIR-BLODGETT FILMS BY FOURIER TRANSFORM INFRARED SPECTROSCOPY AND SECOND HARMONIC GENERATION. W-F. A. Su,* T. Kurata, H. Nobutoki, and H. Koezuka; Mitsubishi Electric Corporation, Japan 661; *Westinghouse Science & Technology Center, 1310 Beulah Road, Pittsburgh, Pennsylvania 15235.

The molecular orientation of amphiphilic hemicyanine Langmuir-Blodgett (LB) films has been studied quantitatively by Fourier Transform Infrared (FTIR) spectroscopy and Second Harmonic Generation (SHG) Maker fringe method. From the analysis of transmission and reflection absorption spectra of FTIR, the tilt angles of the hydrocarbon chain axis and the chromophore of amphiphilic hemicyanine LB film were estimated to be 24° and 43°, respectively. The SHG measurements yielded a chromophore tilt angle of 44°. Both techniques have shown consistent results for the molecular orientation of the chromophore of the hemicyanine LB film.

225. INTERNAL ELECTRIC FIELD AND DYE ORIENTATION IN PVDF/PMMA BLEND Naoto Tsutsumi, Yoshiaki Ueda and Tsuyoshi Kiyotsukuri, Department of Polymer Science & Engineering, Kyoto Institute of Technology, Kyoto 606, JAPAN

A melt-quenched compatible blend of 80 wt% of PVDF and 20 wt% of PMMA has the β-crystal form of PVDF, which grows larger by annealing at higher temperature. We measured the internal electric field E_i created between the β-crystallite dipoles of PVDF in this blend using the electrochromic peak shift of dye dissolved in the blend, and estimated the orientation of dyes by E_i from dichroic ratio of dye. We report the effects of E_i on thermal aging and annealing and the orientation of dye by E_i. The decreased E_i up to 60°C changes to increasing until it decreases again above 90°C, whereas pyroelectric coefficient C_{pyro} monotonically decreases with increasing annealing temperature. The decrease of E_i and C_{pyro} with increasing annealing temperature is ascribed to the relaxation of the aligned crystallite dipoles. The loss of F^- ion generated under high field poling is responsible to the significant increase of E_i.

226. CUBIC NONLINEAR OPTICS OF POLYMER THIN FILMS. 3. STRUCTURE-$\chi^{(3)}$ RELATIONSHIPS IN CONJUGATED AROMATIC POLYAZOMETHINES. Chen-Jen Yang and Samson A. Jenekhe, Center for Photoinduced Charge Transfer and Department of Chemical Engineering, University of Rochester, Rochester, NY 14627-0166. Herman Vanherzeele and Jeffrey S. Meth, DuPont CR & D, P.O. Box 80356, Wilmington, DE 19880-0356.

The third-order optical properties of conjugated aromatic p-phenylene polymers, such as poly(p-phenylene vinylene) (PPV) and derivatives, have been widely studied except the aromatic polyazomethines or Schiff base polymers with the basic structure $(-Ph-CH=N-Ph-N=CH-)_n$. We have synthesized a series of homopolymers and random copolymers of conjugated aromatic polyazomethines and successfully prepared their optical quality thin films from soluble complexes. We have investigated the cubic optical nonlinearities of these polymers by picosecond third harmonic generation and measured the wavelength dispersion of the $\chi^{(3)}$ in the range 0.9 - 2.4 μm. Trends in the structure-$\chi^{(3)}$ relationship as well as the effects of three-photon resonance enhancement of $\chi^{(3)}$ will be discussed.

227. CHROMATOGRAPHIC ANALYSIS OF OLIGOMER STRUCTURES DURING THE COPOLYMERIZATION OF EPOXY WITH BISPHENOL-A. Mark E. Smith, and Hatsuo Ishida, Department of Macromolecular Science, Case Western Reserve University, Cleveland, OH 44106.

The copolymerization of the diglycidyl ether of bisphenol-A and bisphenol-A results on the formation of linear, branched and crosslinked structures. Gel permeation chromatography allows the determination of the size distribution of these molecules as a function of reaction time. The analysis of molecular size distribution during the chemical reaction shows that the molecular growth follows linear copolymerization until a level of conversion near 60%. Further reaction leads to complex chromatographic results with an increase of polydispersity in molecular size. Intrinsic viscosity measurements on the equimolar samples reacted for 30 minutes and longer exhibit an increase is in agreement with the increase in molecular weight as determined by the chromatographic analysis. However, samples of excess bisphenol, and excess epoxide show that the direct analysis of chromatographic data is inaccurate for the type of molecular structures produced during this reaction. Direct analysis of chromatographic data for polydisperse samples containing branched structures results in significant error.

228. STRUCTURE-PROPERTY BEHAVIOR OF A SWOLLEN CERAMER, A. B. Brennan and F. Rabbani, Department of Materials Science and Engineering, University of Florida, Gainesville, FL 32611.

Sol-gel derived inorganic/organic hybrid materials have been prepared by numerous methods and materials. The structure/property behavior of CERAMERs, which represents one class of these materials, has been well characterized in terms of systems based upon various metal alkoxides reacted with silane end-capped oligomers of poly(tetramethene oxide). The nature of mixing between the PTMO chains and the inorganic structure is the objective of this research. It was postulated that one could better define the nature of interactions between phases by swelling the CERAMERs with reactive monomer such as tetraethylorthosilicate which was subsequently gelled using either an acid or base catalyst. The dynamic mechanical behavior, mechanical behavior and small angle scattering behavior will be reported for these materials.

229. CERAMIC COATINGS FOR C/C COMPOSITES VIA METALORGANIC/INORGANIC POLYMER BLENDS. K. E. Gonsalves*, R. Yazici+ and S. Han+. *Polymer Science Program, Institute of Materials Science & Department of Chemistry, University of Connecticut, Storrs, CT 06269. +Department of Materials Science & Engineering, Stevens Institute of Technology, Hoboken, NJ 07030.

Metalorganic and oligomeric precursors that yield Ti-C-N and Si-C-N systems have been synthesized and applied to PAN based graphite fabrics. Coated materials were pyrolyzed to produce protective ceramic coatings. Wetting characteristics of the precursors were determined by microstructural and elemental analyses. Stable amorphous coatings were obtained below 500°C. Crystalline phases required higher temperatures. Multilayered and composite coatings were prepared to improve adhesion and oxidation resistance. Weight loss experiments were carried out at elevated temperatures and under oxidizing atmospheres to assess the effectiveness of the coatings. X-ray diffraction and electron microscopy were used for structural characterization.

230. LASER INDUCED DECOMPOSITION OF PRECURSORS CONTAINING M-N BONDS FOR THE PREPARATION OF METAL-NITRIDE PRECERAMICS, POWDERS AND COATINGS, Chaitanya K. Narula and M. Matti Maricq, Chemistry Department, Ford Motor Company, Dearborn, MI 48121.

The decomposition of $Ti(NMe_2)_4$ takes place upon Nd-YAG laser irradiation (1.064 μm). The decomposition is accompanied by emission of visible light suggesting a multiphoton mechanism and yields a grey-black preceramic material which can be converted to TiN by heating at 1100°C in an ammonia atmosphere. The laser induced decomposition, carried out in the presence of MgO, Al_2O_3 or TiO_2, results in the formation of preceramic coatings on the oxide particles which can be converted to TiN coatings by pyrolysis under an ammonia atmosphere. $B[NMe_2]_3$ and $Si[NMe_2]_4$ do not undergo decomposition upon irradiation at 1.064 μm in the absence of metal-oxides or in the presence of MgO or Al_2O_3. In the presence of TiO_2, both amides undergo decomposition. The mechanistic aspects of decomposition process and characterization of various materials by NMR, XRD, TEM, EDS and HREM will be presented.

231.
SPECTROSCOPIC CHARACTERIZATION OF MIXED SILOXANE-OXIDE SYSTEMS.
Sandra Diré[1], Laurence Bois[2], Florence Babonneau[2], Giovanni Carturan[1]. [1]Dipartimento di Ingegneria dei Materiali, Università di Trento, 38050 Mesiano-Trento, Italia. [2]Chimie de la Matière Condensée, Université de Paris 6, 75005 Paris, France.

Mixed siloxane-oxide systems have been prepared from hydrolysis-condensation process of dimethyldiethoxysilane and various metallic alkoxides, $M(OR)_n$ with M=Si,Ti, Zr and Al. These preparations lead easily to transparent monolithic pieces. Structural investigations have been done on the Ti and Si containing systems essentially using MAS-NMR and X-ray absorption spectroscopies. According to experimental data, structural models can be proposed for both systems : the Si system seems to be a copolymer between the two kinds of precursor units, $(CH_3)_2SiO$ and SiO_2 while the Ti system appears as a composite materials with TiO_2-based particles and long siloxane chains. The rheological properties of the precursor sols, especially those containing transition metal ions, allow the easy formation of thick films. Such films have been successfully used as matrices for laser dyes.

232. SYNTHESIS AND CHARACTERIZATION OF POLYACRYLATE-INORGANIC HYBRID MATERIALS VIA THE SOL-GEL APPROACH, Yen Wei,* R. Bakthavatchalam, D.C. Yang, and C.K. Whitecar, *Department of Chemistry, Drexel University, Philadelphia, PA 19104*

New organic-inorganic hybrid materials have been synthesized by co-condensation of functionalized acrylate polymers with silicon or titanium alkoxides using the sol-gel approach. The acrylate polymers were prepared with controlled molecular weights by group-transfer polymerization of allyl methacrylate and by copolymerization of allyl methacrylate with methyl methacrylate. The allyl ester groups in these polymers were converted to triethoxysilyl groups by hydrosilylation in the presence of Speier's catalyst. Poly[3-(trimethoxysilyl)propyl methacrylate] and its copolymers with methyl methacrylate were also prepared via a free-radical polymerization. These polymers and copolymers with trialkoxysilyl groups were then co-condensed with silicon or titanium tetraalkoxide via an acid-catalyzed sol-gel route to yield the new monolithic hybrid materials. The materials were characterized by IR and Raman spectroscopy and by TGA and DSC. The thermal stability of the polymers in the glass matrices is higher than that of the pure polymers. No well-defined T_g was observed for the polymers in the matrices, suggesting that the polymer chains might be quite uniformly distributed in the materials. The hybrid materials showed good optical transparency and different mechanical properties from either the pure inorganic sol-gel glasses or the pure polyacrylates.

233. **CONTROL OF PROCESSABILITY & SOLID STATE PROPERTIES OF POLYMERS THROUGH ROTAXANE FORMATION.**
H. W. Gibson, C. Wu, Y. X. Shen, M. Bheda, J. Sze, P. Engen, A. Prasad, H. Marand, D. Loveday and G. Wilkes, Depts. of Chemistry & Chemical Engineering & NSF STC for High Performance Polymeric Adhesives & Composites, Virginia Polytechnic Institute and State University, Blacksburg, Virginia, 24061

Polyrotaxanes are two component systems comprised of macrocycles threaded by linear macromoleules. This novel architecture may be considered to be a physical analog of covalent types of copolymers. However, the lack of covalent bonding between the macrocycles and the linear backbone leads to novel behavior. Solubility enhancement of rigid rod types of linear polymers can be achieved without compromising backbone integrity. Motion of the cyclic species along the backbone can lead to phase separated morphologies. Thus these systems may present the polymer and materials scientist with a new strategy to design, optimize and control the processability of materials vis-a-vis the end use properties of the solid state. Several examples will be presented.

234.
EFFECT OF TOPOLOGY FOR CYCLIC AND LINEAR POLY(DIMETHYLSILOXANE) AND POLY(METHYLPHENYL-SILOXANE) BLENDS, C.M. Kuo and S.J. Clarson, Department of Materials Science and Eng., University of Cincinnati, Cincinnati, OH 45221-0012.

The equilibrium phase behavior of binary polymer blends of cyclic or linear polydimethyl-siloxanes (PDMS) with cyclic or linear polymethylphenylsiloxanes (PMPS) was investigated using a static wide angle light scattering technique. For each pair studied, the mixture contained at least one cyclic component. The cloud point measurements of all the blends showed an upper critical solution temperature (UCST), as has been observed previously for PDMS-PMPS blends in which both components were linear. The phase boundary (binodal curve), simulated by the measured cloud points, was analyzed to determine the interaction energy density parameters for pairs of polymers based on the Flory-Huggins theory for the free energy of mixing per unit volume of a polymer mixture. The effect of molecular topology is illustrated by the blend containing cyclic PDMS mixed with linear PMPS. This blend had slightly better compatibility than that of the corresponding blend where both components were linear. The phase boundary in the case of the cyclic PDMS with linear PMPS was at a lower temperature and thus had a lower value of the interaction energy density parameter.

235. RINGS VERSUS STRINGS: OFF-LATTICE, BEAD-STICK COMPUTER SIMULATIONS. Peter H. Verdier, Jack F. Douglas, and Gregory B. McKenna, Polymers Division, National Institute of Standards and Technology, Gaithersburg, Maryland 20899.

Mean-square radii of ring-shaped polymers with and without excluded volume interactions have been obtained by computer simulation of bead-stick models of from 8 to 98 sticks. The results are compared with mean-square dimensions obtained for linear, string-shaped polymers of the same chain lengths. The chain-length dependence of the expansion of the chains by excluded volume is found to be stronger than that predicted by field theory methods. For chains without excluded volume, the ratios of the mean-square radii of rings to those of strings of the same chain length are about 0.5, independent of chain length. In contrast, these ratios increase with increasing chain length for chains with excluded volume. Ratios of mean-square radius to mean-square end-to-end length for strings with excluded volume are found to be appreciably smaller than the known values for strings without excluded volume. Significant differences between on- and off-lattice simulations are noted and discussed.

236. THE SYNTHESIS AND THERMAL CHARACTERIZATION OF A SERIES OF MONODISPERSE, 1,3,5-PHENYLENE BASED, FLUORINATED DENDRIMERS. T. M. MILLER, T. X. NEENAN, R. ZAYAS, and H. E. BAIR. AT&T Bell Laboratories, Murray Hill, N. J. 07974.

Fluorination of organic materials gives rise to a unique set of properties including hydrophobicity, increased volatility, and lubricity caused by the weak interaction of fluorine atoms with its surroundings. We report here our initial results in the synthesis, characterization, and thermal behavior of the first examples of fluorinated starburst dendrimers. These materials consist of 4, 10, and 22 phenyl rings linked in their 1,3, and 5 positions by σ-bonds, and in which the outer 3, 6, and 12 respectively phenyl rings are perfluorinated. These dendrimers have high (D_{3h}) symmetry and no flexible linkages and should thus have well defined diameters. These diameters are 16, 22, and 28 Å for the dendrimers having 4, 10, and 22 phenyl rings.

237. CATIONIC CASCADE MOLECULES. Robert Engel, Kashturi Rengan, and Caroline Milne, Department of Chemistry and Biochemistry, Queens College, C.U.N.Y., Flushing, NY 11367

Our efforts have been directed toward the preparation of a series of cascade molecules which incorporate phosphorus or nitrogen as the core and branch sites. These efforts have resulted in the synthesis of dendrimers with multiple phosphonium or ammonium sites embedded in the fundamental skeleton. In addition to dendrimers bearing a phosphonium ion core surrounded by repeating branch phosphonium ion sites, a quinquedirectional core [P(V)] has been synthesized with associated phosphonium ion branch points. In this instance the core is neutral while the branch points are positively charged. Further, a phosphine oxide core has been used for a tridirectional cascade bearing phosphonium ion sites at the branch points. This material has been reduced at the core using silyl reducing agents to yield the phosphine core, which has been used in coordination with Au(I). Further, ammonium ion cascade structures have been prepared using hexamethylenetetramine as the core and for the branch sites. The syntheses and physical characteristics of these materials are to be discussed.

238. PHOTOCHEMICAL PROBES OF STARBURST DENDRIMERS AND THEIR UTILIZATION AS RESTRICTED REACTION SPACES FOR ELECTRON TRANSFER PROCESSES. G. Caminati, K.R. Gopidas, A.R. Leheny, N.J. Turro, Department of Chemistry, Columbia University, N.Y.,NY 10027 and Donald A. Tomalia, Michigan Molecular Institute, Midland, MI 48674.

Dynamics of the electron transfer quenching of photoexcited $Ru(phen)_3^{2+}$ by methyl viologen in solutions containing various anionic micelles and anionic starburst dendrimers were investigated by the single photon counting technique monitoring the luminescence decay of the excited complex. Analysis of the kinetics of luminescence quenching revealed that the quenching process in higher generation (3.5 G and higher) starburst dendrimer solutions obey a general kinetic model previously employed for micellar solutions. In these cases the intramicellar(or intrastarburst) quenching rate constants (k_q) were found to be unimolecular, and decrease with increasing size of the micelle (or starburst dendrimer). In the case of lower generation starburst dendrimers (2.5 G and lower) and smaller micelles (C-7 and C-8 alkyl sodium sulfates) the quenching reaction was found to be bimolecular in nature. This bimolecular quenching is attributed to the rapid exit of the probe from these macromolecules into the aqueous phase during its lifetime. Quenching studies using an anionic quencher such as $K_4[Fe(CN)_6]$ substantiates this conclusion.

239. SYNTHESIS AND CHARACTERIZATION OF MACROCYCLIC POLYSTYRENE. EFFECT OF MOLECULAR WEIGHT ON GLASS-TRANSITION TEMPERATURE. Janakiraman Sundararajan and Thieo E. Hogen-Esch. Loker Hydrocarbon Institute and Department of Chemistry, University of Southern California, University Park, Los Angeles, California, 90089-1661.

The synthesis is reported of highly pure monodisperse macrocyclic polystyrene (PS) by high dilution (~10^{-5} M) coupling of a two-ended living PS lithium capped with 2 - 3 equivalents of 2-vinylpyridine (2-VP) using 1,4-bis(bromomethyl) benzene (1,4 DBX) as the coupling agent. Analysis by SEC shows large differences in hydrodynamic radius for the two types of poly-mers. The ratio <G> defined as the ratio of the peak molecular weight for the cyclic and linear PS increased from a value of about .75 (70 ≤ DP ≤ 480) to .86 (DP = 40). Similar patterns were documented for macrocylic and linear P2VP. The glass transition temperatures of the macrocyclic PS samples (22 < DP < 480) were shown to increase with decreasing DP as shown previously for macrocyclic P2VP. Differences in Tg values of the cyclic and linear PS decrease from about 38º at a DP of 40 to about 7º at a DP of 310. The marked differences in the DP 100 - 300 range are much larger than found for macrocyclic P2VP. The results indicate that the Tg value is a valuable criterion to estimate the purity of PS macrocycles.

240. B+3/POLYSACCHARIDE GELS: INFLUENCE OF CROSSLINK CHEMISTRY ON RHEOLOGICAL PROPERTIES. Silvia Bucci, Germana Gallino, and Thomas P. Lockhart, ENIRICERCHE SpA, 20097 San Donato Milanese, ITALY

The rheological behavior of aqueous B+3 crosslinked hydroxypropyl guar (HPG) gels have been examined over a range of conditions. These gels undergo a fully reversible transition from the gel state to a viscoelastic solution over the temperature range from 25 - 75 C. From an analysis of the temperature dependence of the dynamic moduli an estimate of the enthalpy of the B+3/polymer crosslinking reaction and for the activation energy for reversible crosslink bond exchange has been derived. The value derived for the enthalpy of the crosslinking reaction from the macroscopic rheological measurement is virtually identical with that based on direct, B-11 NMR studies.

241. GELATION AND CRYSTALLIZATION IN SOLUTIONS OF PBLG. A. Dagan, M. Avichai, E. Gartstein, Y. Cohen, Department of Chemical Engineering, Technion, 32000 Haifa, Israel.

The state of aggregation of poly(γ-benzyl-L-glutamate) helices in benzyl alcohol solutions, undergoing gelation by cooling, were followed by small-angle x-ray scattering measurements. In a dilute solution, gelation is a consequence of aggregation of isolated helices, forming clusters which have an apparent fractal dimensionality of 2.5. In quenching a concentrated liquid crystalline (LC) solution, evidence is found for separation into two phases one of which is a very concentrated LC phase. In addition, from the same sample two different crystalline phases can be obtained, depending on the thermal history. Preliminary results of computer simulation of aggregation of rods are also presented.

242. A MOLECULAR INTERPRETATION OF THERMOREVERSIBLE GELATION IN POLYSTYRENE-LIQUID SYSTEMS. L.A. Errede, 3M Co., 3M Center, Bldg 201-2N-22, St.Paul MN 55133

Guenet reported that the average number ($\bar{\alpha}$) of adsorbed molecules per phenyl group of polystyrene in the polymer-rich phase of thermoreversible polystyrene-liquid (P-L) gels is characteristic of the liquid. We reported that the corresponding number (α_s) for polystyrene in "true solution" is also characteristic of the liquid, and that after all non-adsorbed molecules in crosslinked P-L systems have been eliminated, the residual number (α_s^c) varies with both α_s and the number of monomer units between crosslink junctions (or equivalent restraints). We now report that Guenet's $\bar{\alpha}$, for a given polymer tacticity and family of liquids, varies linearly with α_s; thus, the fraction of monomer units of polymer that undergoes some form of association during gelation can be established from the observed ratio $\bar{\alpha}/\alpha_s$. From this ratio one can also infer whether this association occurs by gain or loss of adsorbed molecules. The number of monomer units between quasi-crosslinkages, i.e. the microdomains of associated polymer, can be established using the equations that relate α_s to α_s^c and to $\bar{\alpha}$.

243.
PREPARATION AND PROPERTIES OF MACROPOROUS POLY(N-ISOPROPYLACRYLAMIDE) HYDROGELS
Xue Shen Wu, Allan S. Hoffman, and Paul Yager
Center for Bioengineering, FL-20, University of Washington, Seattle, WA 98195

Macroporous hydrogels are characterized by high pore volumes, high specific surface area, and large pore sizes. Besides these characteristics, macroporous hydrogels based on thermally reversible polymers respond to temperature changes much faster than hydrogels with normal porous structures.
Crosslinked poly(N-isopropylacrylamide) (polyNIPAAm) forms a thermally reversible hydrogel which shows a lower critical solution temperature (LCST) ca. 34° C in aqueous solutions; that is, the hydrogel swells below and deswells above the LCST. PolyNIPAAm hydrogels have recently been of increasing interest in the fields of controlled drug delivery, immobilization of enzymes and cells, and protein separations. However, the pore sizes of normal polyNIPAAm hydrogels are relatively small. PolyNIPAAm hydrogels with macroporous structures may be desirable for applications which involve protein permeation through the hydrogel pores. In addition, macroporous gels exhibit rapid deswelling and swelling which is also a useful property.
We have synthesized thermally reversible polyNIPAAm hydrogels with macroporous structures by a new method. These macroporous hydrogels have large pore volumes and large pore sizes. Compared with normal polyNIPAAm hydrogels, the macroporous polyNIPAAm hydrogels also have higher swelling ratios at a given temperature and exhibit much faster shrinking speeds. These thermally reversible macroporous hydrogels may be very useful in controlled drug delivery, immobilized enzyme or cell reactors, and protein separations. Peptides or proteins may behave as if they were in bulk solution within the large aqueous pores, and this may reduce their inactivation when such gels are used for their storage and later release or absorption and separation. The rapid volumeric response to temperature change may also be useful in microbotic applications.

244. REVERSIBLE GELATION OF MICROGELS DUE TO COLLOIDAL CRYSTALLIZATION. B.E. Rodriguez and M.S. Wolfe, Du Pont Co., Experimental Station, Wilmington, DE 19880-0356, and E.W. Kaler, University of Delaware, Dept. of Chemical Engineering, Newark, DE 19716.

Reversible gelation of swellable, submicron, internally-crosslinked polystyrene particles (microgels), has been shown to occur during creep. While in the good solvent, bromoform, liquid-like flow occurs at short times, at long times "solidification" occurs abruptly as evidenced by a nearly complete cessation of flow. From visual observations under shear, solidification was found to occur only after colloidal crystallization of the monodisperse microgels as evidenced by opalescence when illuminated by white light. This polycrystalline structure displays a solid-like recovery which is stable if unperturbed; it can be destroyed by exceeding a yield stress, but will reform at low stresses.

245. PTFE-MATRIX COMPOSITE HYDROGEL SHEETS WITH THERMOREVERSIBLE TRANSITIONS
S. Mitra, L.A.Errede and S. Ahmed, 235-3E-06, 3M Company, St. Paul, MN 55144 U.S.A.

Polytetrafluoroethylene based membranes are described which have incorporated in them fine particulates of crosslinked hydrogel materials that show thermoreversible critical swelling behavior. These mechanically strong sheets, with relatively high loadings of the hydrogels, have been used to study the dependence of the critical temperatures of the hydrogels on their compositions, e.g. comonomer ratios and crosslinker contents. The experiments show that in these systems, increased crosslinking and increased hydrophobicity of the gels result in the reduction of the critical transition temperatures.

246. **THERMOREVERSIBLE GELATION OF 60-CROWN-20 MACROCYCLES.** A. Prasad, H. Marand, M. Bheda and H.W. Gibson, Chemistry Department, NSF Science and Technology Center for High Performance Polymeric Adhesives and Composites, Virginia Polytechnic Institute & State University, Blacksburg, VA 24061.

A variety of polymers can form either thermoreversible gels or precipitated crystals in dilute or moderately dilute solution of suitable solvents. However, little is known about the thermoreversible gelation of small molecules. 60-crown-20 (60-C-20) which is a cyclic oligomer comprised of twenty ethylene oxide repeat units (Mn = Mw = 880 g/mol), forms thermoreversible gels and precipitated crystals in solvents such as carbon tetrachloride or acetone. We have studied the kinetic and thermodynamic characteristics of both the gelation and solution crystallization of 60-C-20 in CCl_4 by differential scanning calorimetry and the visual cloud point method. The resulting crystal structures were examined by wide angle X-ray diffraction.

The critical gelation concentration, c_o^* and the corresponding critical gelation temperature, T^*, for the 60-C-20/CCl_4 system are respectively, 0.035 v/v and -3 °C. Wet and dry gels are characterized by two melting endotherms differing in peak temperature by about 5 °C, indicating that the gels are crystalline in nature. On the contrary, 60-C-20 exhibits a single sharp melting endotherm when crystallized from the melt state. WAXD studies indicate that the crystal structure of 60-C-20 in dry gels and precipitates is identical to that of the bulk crystallized macrocycles. However, the diffraction peaks of gels and precipitated crystals are much sharper than those of the bulk crystallized material, suggesting that the presence of the solvent helps in producing a better chain packing.

247. CRYSTALLIZATION INDUCED GELATION IN POLY(ARYL ETHER KETONES).
Benjamin S. Hsiao, Du Pont Fibers Research, Experimental Station Wilmington, DE 19880-0302

Crystallization induced gelation is a commonly encountered problem in processing semi-crystalline polymers. In this work, we measured the complex moduli (G' and G") of poly(aryl ether ketones) during crystallization using the dynamic mechanical method. The samples were first equilibrated above the equilibrium temperature and were fast cooled into a supercooled state. As the supercooled liquid undergoes a liquid-crystal transition, both moduli increase several orders of magnitude. This behavior is closely associated with the crystallization kinetics characterized by DSC. Furthermore, we have demonstrated that blending of a liquid crystalline polymer suppresses the crystallization induced gelation in poly(aryl ether ketones).

248. Gelation/Crystallization of Polyethersulfone

K. Levon, S. Makhija and E.M. Pearce

Polytechnic University Brooklyn NY 11201

Polyethersulfone, an amorphous polymer, was showen to form a gel from methylene chloride solution. The gels became amorphous when heated above the T_g of polyethersulfone. Methylene chloride vapor treatment on thin films induced crystallization and the thermal behavior of the crystalline matrix was similar to the observed in the gels.

249. THERMOREVERSIBLE GELS WITH POLYETHYLENE BLOCK COPOLYMERS AND HYDROCARBON SOLVENTS
Boris D. Nahlovsky, Aerojet Corp. P.O. Box 13222, Sacramento, CA 95813-6000

Melting behavior of thermoreversible gels was studied. Hydrocarbon solvents, such as paraffinic oil, were gelled by polyethylene block copolymers (PEBCs). The PEBC used in this study was a triblock copolymer with quasi-polyethylene crystalline hard end blocks and amorphous center block of poly(ethylene-co-butene-1). PEBCs dissolve in hydrocarbon solvents above the melting transition (T_m) of the hard block. Thermoreversible gels are formed by cooling the solution below T_m. Endothermal melting transition was detected by DSC at the PEBC concentration in the 0.33 to 100% range. Mineral oil depressed T_m of the unplasticized polymer. T_m of the gel decreased with the increasing oil concentration from 104.4°C to approximately 81°C at 5% PEBC and did not decrease further at lower PEBC concentrations. Sharpness of T_m transition was increased by annealing several degrees Centigrade below T_m. Spherulites were observed by polarized light microscopy in annealed samples. These thermoreversible gels are apparently crosslinked by "physical crosslinks" in crystalline microdomains of quasi-polyethylene in a continuous phase of the hydrocarbon solvent.

250. STATIC AND DYNAMIC LIGHT SCATTERING FROM SOLUTIONS OF POLYMERS WITH DIFFERING ARCHITECTURE. Patricia M. Cotts,, IBM Research Division, Almaden Research Center, 650 Harry Road, San Jose, California 95193-6099.

Light scattering is a technique uniquely suited to the study of large molecules. While static (or intensity or classical) light scattering has been used to study polymers in solution for nearly 50 years, the more recent availability of the laser has led to dynamic light scattering, also known as quasielastic light scattering, photon correlation spectroscopy, or intensity fluctuation spectroscopy. With static light scattering, equilibrium molecular parameters such as the molecular weight, M_w, the radius of gyration, R_g, and the second virial coefficient, A_2 may be determined from the measured intensity of the scattered light at several scattering angles for a series of concentrations. Dynamic light scattering gives information about the dynamics of the solution (or melt) through the correlation function of the intensity fluctuations. In dilute solution, the decay of the correlation function is directly related to the diffusion coefficient, D_0, of the polymer in the solvent. The effective hydrodynamic radius, R_h, is calculated from the Stokes equation. Although obtained directly from a scattering technique, R_h is a hydrodynamic size, closely related to the radii obtained from older techniques such as viscometry and sedimentation equilibrium. The variation of R_g and R_h with M as a function of polymer architecture and solvent quality is discussed. Examples include flexible polymers in good and poor solvents, ring polymers, and semi-flexible polymers. The use of a multi-angle LS detector with size exclusion chromatography (SEC/LS) is also discussed.

251. MAGNET ENHANCED OPTICAL FALLING NEEDLE/SPHERE RHEOMETER. B. Chu and J. Wang, Department of Chemistry, State University of New York, Stony Brook, NY 11794-3400.

A falling needle/sphere rheometer with a precision optical monitor system under an artificial magnetic field has been developed. This instrument was equipped with a PC compatible computer for its operation, control, and data acquisition. The important advantages of this instrument over existing falling needle/sphere viscometers are: (1) an extremely wide viscosity range from the usual solvent viscosity (say, ~1.0 cP) to that of polymer melts (say, 10^{10} cP, in principle), could be achieved; (2) for very high viscosity samples, the measurement time could be reduced by about 4 orders of magnitude (e.g., from hours to seconds); (3) one single needle/sphere could cover a few orders of the viscosity range; (4) the measurement precision could be improved to the order of < 1%; (5) the instrument permits measurements of the shear rate dependence of viscosity using a single needle/sphere; (6) for non-Newtonian fluids, viscoelastic behavior could be measured.
A set of viscosity standards was used to calibrate the device. Linear calibration curves under both natural gravity field and fixed artificial magnetic field were obtained. The results, including factors which may affect the measurement precision, will be discussed.

252. REAL TIME SPECTROSCOPIC ELLIPSOMETRY: APPLICATIONS IN POLYMER AND THIN FILM GROWTH. R.W. Collins, Materials Research Laboratory and Department of Physics, The Pennsylvania State University, University Park PA 16802.

Real time ellipsometry has been applied extensively in both electrochemical and high pressure plasma environments to characterize growth kinetics and optical properties of thin films whose surfaces are inaccessible to modern electron and ion spectroscopies. Until recently, real time ellipsometry measurements have been confined to a single fixed wavelength, λ_0, owing to the excessive time required for spectral scanning. This has been a serious limitation, for example, in studies of electrochemically-deposited thin film conducting polymers and plasma-deposited semiconductors, for which the optical functions, $\{n(\lambda), k(\lambda)\}$ (namely the index of refraction and extinction coefficient vs. wavelength), provide important information on electronic characteristics. As a result, a novel rotating polarizer ellipsometer has been developed which employs a multichannel detection system to perform high speed real time spectroscopic ellipsometry (RTSE). The instrument has a potential acquisition time of 40 ms for a full spectrum (275-950 nm), and with it, thin film growth kinetics, optical functions, and thus electronic and microstructural information can be obtained from real time observations. In this presentation, applications of RTSE of relevance to polymer film growth will be described.

253. FOURIER TRANSFORM MID- AND FAR INFRARED IR SPECULAR REFLECTANCE STUDY OF OXIDIZED AND EXCHANGED POLYETHYLENE SURFACES. Hans Peter Brack and William M. Risen, Jr., Department of Chemistry, Brown University, Providence, Rhode Island 02912

The development of near normal incidence infrared and far infrared spectroscopy for the study of the polarizability and related physical characteristics of the surfaces of polymers is described. The optical measurement and data analysis issues will be discussed first. Then, the application of the method to study polyethlene, oxidized and oxidized-metal ion exchanged polyethylene surfaces will be described with emphasis on the far infrared behavior of the dielectric properties $\varepsilon'(\omega)$ and $\varepsilon''(\omega)$ and their relationship to those of metal ionomers and other disordered ionic systems. Finally, the potential for the use of low frequency dielectric data is calculations related to adhesion will be discussed and illustrated.

254. RECENT ADVANCES IN FOURIER TRANSFORM RAMAN SPECTROSCOPY, D. B. Chase, Central Research & Development, Du Pont Experimental Station, Wilmington, DE 19880-0328

The application of Raman spectroscopy to structural studies of polymers has been well documented over the last 25 years. The problems associated with attaining those Raman spectra have been equally defined. Background fluorescence caused by impurities in the polymers completely overwhelms the weak Raman scattering. The development of near infrared excited Raman scattering with detection by either interferometers or dispersive-CCD instruments has radically changed the situation. Spectra of many previously intractable polymers such as polyimides and polyamides can now be routinely acquired. The latest advances in these two areas of instrumentation will be discussed and applications to reading polymer systems will be shown.

255. MOBILITY OF ETHYL ACETATE IN POLY(VINYLIDENE FLUORIDE) MONITORED BY RHEO-PHOTOACOUSTIC FT-IR SPECTROSCOPY. B.W. Ludwig, M.W. Urban, Department of Polymers and Coatings, North Dakota State University, Fargo, ND 58105.

A newly developed rheo-photoacoustic FTIR spectroscopy was used to monitor the rate of diffusion of ethyl acetate (EAc) in poly(vinylidene fluoride) (PVF_2). The magnitude of diffusion depends on the degree of applied strain, the structural changes occurring in plastically deformed PVF_2, the concentration of EAc and morphology of the polymer. These changes are discussed in terms of the fractional free volume (FFV) theory and transport properties in polymer systems. It appears that the larger the FFV, the greater the rate of diffusion. This study also shows that the rate of diffusion of EAc may provide further information about morphological changes occurring in PVF_2.

256. SOL-GEL KINETICS BY NMR. *Roger A. Assink* and Bruce D. Kay, Sandia National Laboratories, Albuquerque, New Mexico 87185.

The chemical synthesis of advanced ceramic and glass materials by the sol-gel process has become an area of increasing activity in the field of material science. The sol-gel process provides a means to prepare homogeneous, high purity materials with tailored chemical and physical properties. Perhaps the single most powerful experimental technique for studying sol-gel chemistry is NMR spectroscopy. This paper presents recent NMR studies of silicon-based sol-gel kinetics. We discuss the results of the various models which are used to analyze the chemical kinetics of sol-gels. We then demonstrate the utility of NMR spectroscopy in investigating the influence that reaction conditions have on the pathways by which sol-gel derived materials are prepared.

257. SHEET AND TUBE ALKOXYSILOXANES. Li-Tain Yeh, Bruce A. Harrington, and Malcolm E. Kenney, Department of Chemistry, Case Western Reserve University, Cleveland, OH 44106 and Jesse Hefter, GTE Laboratories, Inc., Waltham, MA 02254.

The synthesis and characterization of a sheet ethoxysiloxane and a tube methoxysiloxane are reported. The sheet ethoxysiloxane is prepared by treating apophyllite, $KCa_4Si_8O_{20}(OH,F) \cdot 8H_2O$, with HCl and EtOH. It is postulated that the sheets in the ethoxide have a structure similar to the structure of the sheets in the parent silicate and that the sheets are associated in stacks. Electron microscope, X-ray powder pattern, and solid state ^{29}Si NMR data all support the proposed structure of the alkoxide. The tube methoxide is prepared by treating $K_2CuSi_4O_{10}$ with HCl and MeOH. It is postulated that the tubes in the methoxide have a structure similar to the structure of the tubes in the parent silicate and that the tubes are associated in bundles. Again electron microscope, X-ray powder pattern, and solid-state ^{29}Si NMR data support the proposed structure of the alkoxide. Other sheet and tube alkoxides are reported. Some sheet and tube alkoxides are of interest for the synthesis of ceramics.

258. SOL-GEL CONDENSATION OF TIN(IV) ALKOXIDE COMPOUNDS

M.J. Hampden-Smith[†] T.A. Wark[†], L.C Jones[†], C.J. Brinker[††] and A.L. Rheingold[†††]. Department of Chemistry[†], and Center for Micro-Engineered Ceramics, University of New Mexico, Albuquerque, New Mexico, 87131, Sandia National Laboratories, Albuquerque, New Mexico and Department of Chemistry, University of Delaware, Newark, Delaware.

In order to develop a better understanding of the evolution of structure on the hydolysis and condensation of metal alkoxide compounds, we have undertaken a study of the synthesis, characterization and reactivity of tin(IV) alkoxide compounds as a model system. Tin(VI) alkoxide compounds are prototypical examples of metal alkoxide compounds in general and have been chosen for study due to the presence of two NMR active nuclei which facilitate structure determination in solution. An overview of the solid state and solution structures of these compounds will be presented together with recent results concerning the hydrolysis and condensation of these species to form tin oxide films and powders.

259. **"Inverse" Organic-Inorganic Composite Materials. Free Radical Routes into Nonshrinking Sol-Gel Composites.** Bruce M. Novak* and Caroline Davies, Department of Chemistry, University of California at Berkeley, Berkeley, California 94720

In spite of its attractive features, the application of the sol-gel process in the synthesis of new composite materials is limited by insolubility of many important engineering polymers within the sol-gel solution. In addition, the extraordinary shrinkage which occurs upon drying the solvent-swollen gels (shrinkages of > 50 - 75% are common) precludes most molding processes and introduces considerable stress within these materials. The former problem can be circumvented by the *in situ* formation of both the inorganic and organic components. In order to address the latter problem, we have synthesized a series of tetraalkoxy orthosilicates possessing polymerizable groups in place of the standard ethoxide or methoxide groups commonly used in the sol-gel process. The hydrolysis and condensation of these siloxane derivatives to form the inorganic network liberates a polymerizable alcohol. In the presence of the appropriate catalyst (free radical or ROMP), and by using a stoichiometric amount of water and the corresponding alcohol as cosolvent, all components of these derivatives are polymerized. Since both the co-solvent and the liberated alcohol polymerize, gel drying is unnecessary and no gel shrinkage occurs.

260. ORGANIC-INORGANIC COMPOSITES PREPARED VIA THE SOL-GEL METHOD: SOME INSIGHTS INTO MORPHOLOGY CONTROL AND PHYSICAL PROPERTIES. Christine J.T. Landry and Bradley K. Coltrain, Corporate Research Laboratories, Eastman Kodak Company, Rochester, New York 14650-2110.

Composite materials possessing the desirable properties of both organic polymers and inorganic glasses are obtainable by in-situ hydrolysis and condensation of metal alkoxides within an organic polymeric media. The relative reaction rates of the alkoxides are governed by pH, solvent, water-to-alkoxide ratio, concentration, catalyst, and temperature. The kinetics will also be affected by the presence of an organic polymeric matrix. These factors play an important role in the gelation process and ultimately determine the morphology and physical properties of the composite. A model system consisting of tetraethoxysilane polymerized in the presence of poly(methyl methacrylate) is investigated with emphasis on pH and vitrification effects. Dynamic mechanical spectroscopy and transmission electron microscopy are utilized to investigate these composites. FTIR demonstrates the importance of hydrogen bonding between the silicate phase and PMMA under acidic conditions. ^{29}Si NMR provides information about the extent of reaction of the alkoxides within the composites.

261. CEMENT CHEMISTRY: A NEW APPROACH TO INORGANIC/ORGANIC HYBRID MATERIALS? M. C. Connaway-Wagner, E. D. Dimotakis, and W. G. Klemperer, Department of Chemistry, University of Illinois, Urbana, IL 61801.

Cement chemistry offers a potential route to new inorganic/organic hybrid materials, since the mild processing conditions involved are likely to minimize structural degradation of precursor molecules. Organophosphate cements are attractive systems to study since chemically analogous polyphosphate cement composites are well-established. To this end, MgO/phosphate and MgO/organophosphate cements have been prepared for the first time, and their properties and processibility compared.

262. **PHOTOCURABLE CEMENTS BASED ON POLYALKENOATES/ION-LEACHABLE GLASS HYBRID SYSTEMS.** S.B. Mitra, 3M Dental Products Div., 3M Center, St. Paul, MN. 55144-1000

This paper describes the chemistry and properties of a new type of water-based photocurable cement. The material is a hybrid composite; the matrix consists of a water-miscible polyalkenoic acid with pendent methacrylate groups, the filler is an ion-leachable fluoroaluminosilicate glass which decomposes in the presence of the polyacid to liberate simple and complex multivalent cations. These ions complex with the polycarboxylic acid moieties causing gelation and setting of the cement through the formation of ionically cross-linked salt hydrogel. The acid-base complexation reaction between the organic and inorganic portions of these composite cements results in the sustained release of fluoride ions over a prolonged period. The presence of the pendent methacrylates groups allows for a fast covalent cross-linking of the polymer chains by photopolymerization. Cements based on this system are now being used successfully in restorative dentistry. FTIR studies of the cement forming reaction as well as the physical properties of the photocured cements will be presented.

263. PMMA-IMPREGNATED SILICA GEL. L. C. Klein and B. Abramoff, Rutgers-The State University of New Jersey, Ceramics Department, P.O. Box 909, Piscataway, NJ 08855-0909.

Silica xerogels have been prepared from tetraethyl orthosilicate (TEOS) using the sol-gel process. Following drying by evaporation of the solvent, the result is a xerogel with 50 volume percent open porosity. Methyl methacrylate monomer has been introduced into xerogel monoliths and polymerized with ultraviolet radiation. The fully infiltrated silica gel - PMMA composite has a bulk density of about 1.6 g/cm^3. The resulting composite is lightweight and optically transparent. The PMMA lessens the sensitivity of the silica to moisture. The silica provides a high surface hardness to the PMMA. The composite has the flexure strength of bulk PMMA. The silica xerogel provides a rigid shape, and the infiltration to produce the composite is a net shape process. The application of this process so far has been to encapsulate organic dyes to make optical devices.

264. **NEW CERAMER MATERIALS WITH EMPHASIS ON HIGH REFRACTIVE INDEX AND ABRASION RESISTANCE COATINGS**
B. Wang, A. Gungor, A. B. Brennan, D. E. Rodrigues, J. E. McGrath, and G. L. Wilkes*
Polymer Materials and Interface Laboratory
Virginia Polytechnic Institute and State University, Blacksburg, VA 24061

ABSTRACT
Transparent high refractive index organic/inorganic hybrid ceramers have been prepared by incorporating functionalized low molecular weight organics with titanium or zirconium metal alkoxides by sol gel processing. These materials have been investigated as coating materials to improve the abrasion resistance of polycarbonate and related polymeric substrates. In order to improve the hardness of these coating materials, low molecular weight organics such as 4,4' diamino diphenylsulfone and bis(3-aminophenoxy-4,-phenyl) phosphine oxide that have been triethoxysilane capped are used in conjunction with additional metal alkoxides. The refractive index of these new ceramer systems based on titanium systems exhibit a linear relationship with titanium oxide content and can approach the value of 1.75 (sodium line). The curing temperature utilized and metal alkoxide content of the final ceramer coatings significantly enhance the final abrasion resistance behavior. Finally, these transparent coating materials display some level of very localized phase separation as noted by SAXS analysis.

265. **SOME STATIC AND DYNAMIC PROPERTIES OF STAR POLYMERS** Lewis J. Fetters, Exxon Research and Engineering Company, Annandale NJ 08801.

Various static and dynamical properties in solution and in the melt of model star shaped polymers are surveyed. Unlike small-molecule liquids, polymer melts can display complex and highly non-newtonian flow behavior which reflects the dynamics of the entangled chains. Star-branched polymers, which for f arms each containing Na monomeric units, joined covalently at a common branch point, provide good model systems for understanding the slow dynamics of branched polymers. The structure of dilute and semidilute solutions of 8, 12, and 18-arms stars in good solvent conditions was studied by small angle neutron scattering. Interparticle interactions influence the low angle scattering at concentrations below c*.

266. ANIONIC SYNTHESIS OF HETEROARM, STAR-BRANCHED STYRENE-BUTADIENE THERMOPLASTIC ELASTOMERS. Roderic P. Quirk and Bumjae Lee, Institute of Polymer Science, University of Akron, Akron, Ohio 44325

The living linking reaction of poly(styryl)lithium (PSLi) (M_n=15x10^3 g/mol) with 1,3-bis(1-phenylvinyl)benzene (MDDPE) proceeded in essentially quantitative yield to form the corresponding dianion. In the presence of one equivalent of lithium sec-butoxide per active center, the crossover reaction of the resulting diphenylalkyllithium species with butadiene produced heteroarm star-branched polymers with monomodal molecular weight distributions and high 1,4-microstructure. The resulting poly(butadienyl)lithium (PBDLi) arms (M_n=35x10^3 or 70x10^3 g/mol) were reacted with styrene to produce star-branched thermoplastic elastomers with the general structure (PS)$_2$MDPPE(PBD -b-PS)$_2$. The tensile and viscometric properties of these heteroarm star-branched polymers have been compared with the corresponding linear PS-b-(PBD)-b-PS (15-70-15) thermoplastic elastomer.

267. DIFFUSION OF STAR POLYMERS IN SOLUTIONS AND GELS, T. P. Lodge, J. Won, and N. A. Rotstein, Department of Chemistry, University of Minnesota, 207 Pleasant St. S.E., Minneapolis, MN 55455.

Dynamic light scattering has been used to determine the tracer diffusion of linear and star polystyrenes in polyvinylmethylether solutions and gels. Comparison of linear and star mobilities affords a direct test of the assumption, inherent in the reptation model, that linear chains prefer to move longitudinally. Comparison of diffusivities in solutions and gels provides a test of a concurrent assumption, that the mobility of the matrix is irrelevant to a reptating chain.

268. NEW COMB-SHAPED POLYMERS. M. A. H. Talukder, W. P. Norris, T. S. Stephens, R. Y. Yee, M. P. Nadler, R. A. Nissan, and G. A. Lindsay, Chemistry Division, Research Department, Naval Weapons Center, China Lake, CA 93555.

We have been developing new thermoplastic elastomers for use as binders for energetic composites. These comb-shaped polymers are based on the macromer concept and contain crystallizable teeth. Typically, they have acrylate or vinyl ether backbones, and are formed by free radical or cationic polymerization. Hence, the resulting polymer backbone is the elastomeric phase and the branch is the hard phase, giving a polymer with TPE properties without the usual (AB)n or ABA structure. An energetic macromer, such as linear nitramine acrylate (LNA) has been successfully synthesized and copolymerized with butyl acrylate and/or ethylhexyl acrylate. The polymers were analyzed by NMR, FTIR, DSC, DMA, and intrinsic viscosity. They have the following advantages: (1) permit high loadings, (2) possess good adhesion to oxidizer particles, (3) the Tg and Tm are easily controlled, (4) have low melt viscosity (due to internal plasticizaiton by side chains), and (5) strongly contribute to the desensitization of energetic compositions. Future studies will include varying the length of the spacer groups between the main chain and the crystallizable side chains and determining how this affects the thermo-mechanical properties of the polymers.

269. EQUILIBRIUM CONFIGURATION OF DENSE POLYMACROMONOMERS IN GOOD SOLVENT. Nily Dan- Brandon and Matthew Tirrell, Department of Chemical Engineering and Materials Science, University of Minnesota, Minneapolis, MN 55455 USA.

Polymacromonomers are synthesized by adding end functionalized branches onto a polymer backbone, forming a comblike structure. We are interested in polymacromonomers where the both the molecular weight and the branch density are high, so that the branches interact and are strongly stretched. Our aim is to evaluate the equilibrium configuration of such polymers in a good solvent by employing a scaling approach similar to that developed for star polymers and polymer micelles. We find that in some cases an there is an equilibrium backbone curvature, leading to helical configurations.

270. AGGREGATION OF HYDROPHOBICALLY MODIFIED AND COMB-TYPE POLYMERS
C. Maltesh, Qun Xu, K. Sivadasan and P. Somasundaran, Langmuir Center for Colloids and Interfaces, Columbia University, New York, NY 10027

Synthetic and natural polymers are being modified currently by the incorporation of hydrophobic groups into the backbone of the macromolecular chains. The molecular weight of such polymers is of the order of 10,000 but the rheological properties resulting from the use of such polymers are better than those obtained using high molecular weight polymers because the modified polymers form association complexes which are resistant to shear degradation. In this study, we have used spectrscopic and viscometric techniques to investigate the aggregation behavior of hydrophobically modified hydroxyethyl cellulose (HMHEC). Effect of the hydrophobic modification is also determined by comparing the behavior of HMHEC with that of unmodified hydroxyethyl cellulose. In addition, we present some preliminary investigations into the solution behavior of a commercially available comb-type polymer (DAPRAL GE 202).

271. AMPHIPHILIC STAR POLYMERS FROM TRI- OR TETRAISOCYANATES AND POLY(ETHYLENE GLYCOL) DERIVATIVES. Guangbin Zhou and Johannes Smid, Polymer Research Institute, Faculty of Chemistry, College of Environmental Science and Forestry, State University of New York, Syracuse, NY 13210.

Low molecular weight star polymers were made by reacting methoxypolyethylene glycols and nonylphenoxypolyethylene glycols (Igepals) with well defined tri- and tetraisocyanates made by the hydrosilylation of m-isopropenyl-α,α-dimethylbenzyl isocyanate (m-TMI) with a linear and a cyclic hydrogenmethylsiloxane. The polymers were characterized by ^{1}H and ^{13}C NMR, DSC and GPC, and shown to have exactly three and four arms, respectively. The large hydrophobic core causes the molecules to aggregate in water, and critical micelle concentrations were determined by surface tension measurements. Cloudpoints were obtained in the presence of salts. The Igepal-derived star polymers with hydrophobic end segments on each arm exhibit strong associative properties in water.

272. POLY(ETHYLENE OXIDE) STAR MOLECULES IN SURFACES FOR BIOLOGICAL STUDIES. P. Rempp, P. Lutz, Institut Charles Sadron CRM/EAHP 67083 Strasbourg, France. E. Merrill, A. Sagar, Dept. Chem. Eng., M.I.T., Cambridge, MA 02139.

PEO star molecules were formed by anionic polymerization by the core-first method (core = divinyl benzene DVB), with f PEO arms of molecular weight M_{arm} grown from the polycarbanion core; f varying from 10 to 200, M_{arm} from 3000 to >10,000. PEO star molecules are far more compact than linear counterparts of the same total molecular weight, resulting in chain segment densities within the volume of a single star up to 10-fold greater. Immobilization is readily accomplished by electron irradiation of thin layers of aqueous PEO star solutions in contact with a supporting material. The immobilized layer prevents access of biopolymers to, and non-specific binding on, the support. Tresylation of terminal hydroxyl groups permits attachment of specific biopolymers to the PEO arms which serve as tethers, allowing significant spatial mobility. Prior to attachment of biopolymers (peptide, enzyme, antibody), the PEO star surface displays the inertness (lack of binding of molecules or cells) expected from surfaces containing only PEO. The terminal hydroxyls of the stars after tresylation efficiently bind cysteine and argmine in the ratio expected in homogeneous systems.

273. **DESIGN AND SYNTHESIS OF NOVEL POLYMERS FOR MEDICINE.** Robert Langer, C. Laurencin, and L. Cima, Dept. of Chemical Engineering, MIT, Cambridge, MA 02139.

The design and synthesis of several novel polymer systems are described. First, we will review the polyanhydrides which are degradable polymers that can be made to last for one day to six years, and which are now being used in 53 hospitals for localized controlled release of anticancer agents for the potential treatment of malignant brain tumors. In particular, we will focus on new results obtained with localized controlled release of gentamycin for the potential treatment of osteomylitis, an application which is expected to enter clinical trials this year.

The development of polymer systems to aid in tissue replacement for liver disease and cartilage will also be presented. New data on how the nature of the polymeric substrate can be used to control the state of cell differentiation and growth will be discussed.

274. **THE THIRD-ORDER SUSCEPTIBILITY OF NEMATIC SOLUTIONS OF PBT.**
H. Mattoussi and G. C. Berry, Department of Chemistry, Carnegie Mellon University, Pittsburgh, PA 15213

The third-order optical susceptibility $\chi^{(3)}$ of nematic solutions of the rodlike poly(1,4-phenylene-2,6-benzobisthiazole), PBT, is studied by third harmonic generation (THG). The Maker Fringe Pattern (MFP) created as the slab sample was rotated in an incident beam was used to evaluate $\chi^{(3)}$. The nematic solution was prepared as a defect-free, fully aligned monodomain, and the THG data were obtained over a range of arrangements of the polarizations of the incident and third-harmonic beams and the orientation of the mean field nematic director **n**. These data are compared using refractive index measurements, independently measured on these materials, and necessary for MFP analysis. Coupling is observed between **n** and the incident fundamental polarization E_0, different geometries with respect to the relative orientation of E_0 and **n** providing different MFP's. In particular, with E_0 parallel to **n**, the THG signal is large, and depolarized. By contrast with E_0 perpendicular to **n**, the THG signal is small, and polarized. This behavior indicates that the electronic transitions involved in THG are not along the rodlike axis of the molecule.

275. DIFFUSION AND DYNAMICS IN MICROSTRUCTURED BLOCK COPOLYMER SOLUTIONS, N.P. Balsara, C.E. Eastman, M.D. Foster, P. Stepanek, T.P. Lodge, and M. Tirrell, Department of Chemical Engineering and Materials Science and Department of Chemistry, University of Minnesota, Minneapolis, Minnesota 55455

Abstract: This work explores the effect of heterogeneity of chemical composition on tracer diffusion, when the characteristic size of the heterogeneities approaches that of the diffusing molecule. A heterogeneous environment is created by the self-assembly of diblock copolymers in solution. The system chosen for this study is polystyrene-polyisoprene diblock copolymers in toluene, which is a common solvent for the two blocks. Above a certain critical concentration, these systems are known to microphase separate into swollen domains of polystyrene and polyisoprene. Diffusion of homopolystyrene through the microstructure is measured in this work. The characteristics of the microstructure are varied by studying block copolymers of different molecular weights and compositions. The tracer diffusion coefficients of the labelled polystyrenes are measured by forced Rayleigh scattering, while the microstructure of the matrix is inferred from small angle X-ray scattering measurements.

276. TWO-DIMENSIONAL POLYMERS. S.I. Stupp, S. Son, H.C. Lin, and L.S. Li, Department of Materials Science and Engineering, University of Illinois at Urbana-Champaign, Illinois 61801.

The synthesis of large molecular objects of "well defined shape" is an intriguing approach to materials chemistry for the next century. Macromolecules shaped as tubes, rods and sheets are examples and their potential in synthesis of materials is in part the inevitable three dimensional ordering we can expect dictated by their shape. It would be difficult to avoid, for example, assembly of tubes into an orientationally ordered phase or the stacking of sheets into layered structures. In this work homochiral recognition among strongly dipolar cyano groups appended to stereogenic centers has been used to transform an oligomer to a new type of high molar mass compound with structure and properties consistent with the formation of sheet-like macromolecules. These two-dimensional polymers can form through chemical reaction among the commonly handed nitriles and polymerization of acrylate groups resulting in the "double stitching" of ordered oligomer clusters. The supramolecular substance organizes into a smectic phase and exhibits remarkably stable second order nonlinear optical properties.

277. POLYSTYRENE-POLY(ETHYLENE OXIDE) BLOCK COPOLYMER MICELLES IN WATER. <u>Mitchell A. Winnik</u>, Renliang Xu, Department of Chemistry, University of Toronto, Toronto, Canada M5S 1A1

Polystyrene-poly(ethylene oxide) [PS-PEO] diblock copolymers and PEO-PS-PEO triblock copolymers form micelles in aqueous solution if the polymer composition is sufficiently rich in PEO. These micellar solutions have been examined by fluorescence probe techniques to obtain their CMC values, and by dynamic and static light scattering to characterize the micelles themselves. The micelles are spherical in shape and narrowly distributed in size. Their properties fit remarkably well to both the star model and to a mean field model of micellar structure. With these models χ parameters for the PEO-water and PEO-PS interactions can be obtained.

278. **A NEW SOLID-STATE SMALL-ANGLE LIGHT SCATTERING APPARATUS : DESCRIPTION AND APPLICATION TO PHASE SEPARATION KINETICS IN PC/PMMA BLENDS.** <u>F. Perreault</u> and R.E. Prud'homme, CERSIM, Chemistry Department, Laval University, Québec, Canada.

Solid-state small-angle light scattering (SALS) has been used for many years to study polymer morphology and orientation. Recent improvements in this field include the utilisation of bidimensional detectors coupled with a computer for data acquisition. In an effort to optimize such systems, a new apparatus has been constructed using a CCD (charged coupled device) camera as a detector. The high linearity, dynamic range, geometric stability and the low signal/noise ratio of the CCD detector justify our choice. Complete pictures can be taken every 2 s with a 400 x 600 pixel resolution; higher rates are possible with a lost of resolution. Special care has also been taken in the development of an original, IBM AT based, user-friendly software that controls both the acquisition and the treatment of the scattering data. As a typical application of the device, time-resolved light scattering studies were undertaken to elucidate the kinetics of phase separation through spinodal decomposition in bisphenol-A polycarbonate / poly(methyl methacrylate) (PC/PMMA) blends. The time evolution of the scattering halo was followed as a function of temperature jump. The early stage of the spinodal decomposition was found to be described adequately by the linearized theory of Cahn.

279. ANALYSIS OF NEAR SURFACE STRUCTURE IN POLY(ETHYLENETEREPHTHALATE) BY ATR-IR. <u>Dennis J. Walls</u>, Du Pont Company, Experimental Station, E326/135, Wilmington, DE 19880-0326.

Examinations of the relative amounts of surface crystallinity and amorphous structure and their respective surface orientations have been undertaken for poly (ethyleneterephthalate) (PET) films using polarized attenuated total reflectance infrared spectroscopy (ATR-IR). Depth information is obtained on a scale from ca. 0.5 μm to 5 μm. Uniaxially stretched films exhibit both increased surface extended trans structure and orientation in the MD direction relative to that of the bulk. The surface amorphous content only shows a slight decrease as a function of increasing uniaxial draw ratio, with no evidence for amorphous orientation being observed. Increased planarity of the terephthalate residues and uniplanarity of the benzene rings relative to the surface of the film is observed at the surface, with both characteristics increasing with increasing film draw ratio. All observations are consistent with increased surface crystallinity relative to the bulk and increased surface crystalline orientation at the surface.

280. KINETICS OF DNA CLEAVAGE BY CALICHEAMICIN FROM ELECTRIC BIREFRINGENCE

John S. Bowers, *Department of Polymer Science, University of Southern Mississippi, Hattiesburg, MS 39406-0076.*
George A. Ellestad and Wie-Dong Ding, *American Cyanamid Co. - Medical Research Division, Lederle Laboratories Pearl River NY 10965.*
Raymond Farinato, *American Cyanamid Co. - Chemical Research Division, Stamford CT 06904*
Robert K. Prud'homme, *Department of Chemical Engineering, Princeton University, Princeton NJ 08544-5263*

Calicheamicin γ_1^I is a highly potent naturally occurring antitumor drug whose activity is derived from its ability to cleave DNA. Gel electrophoresis studies of double stranded DNA demonstrate that, *in vitro*, it causes site specific single-stranded nicks and double-stranded cleavage at TCCT (primary sites) and similar (TCC?) base pair sequences. In transient electric birefringence (TEB), molecules are oriented by short (~0.1 ms) high (~1 kV/cm) electric field pulses. Information about size, flexibility, and polarizability of molecules can determined from the rise, steady state and decay of the birefringence. During digestion of DNA fragments by the drug, the changes in the steady state birefringence and the average decay time are determined by the extent of the cleavage of DNA fragments. Therefore with TEB measurements we followed the kinetics of cleavage *in situ*. Cleavage was studied with various fragment lengths (322 bp to 1857 bp) with and without EDTA; EDTA unexpectedly affected the rate of cleavage of DNA. Finally, a model was developed that allowed determination of the absolute number of double stranded cuts during digestion.

281. BIREFRINGENCE CHARACTERIZATION OF POLYMER SOLUTIONS UNDER TRANSIENT ELONGATIONAL FLOW. David J. Hunkeler, Tuan Q. Nguyen and Henning H. Kausch, Polymer Laboratory, Department of Materials Science, Swiss Federal Institute of Technology, MX-D, CH-1015 Lausanne, Switzerland

Flow birefringence has been observed for aqueous polyethyleneoxide solutions by forcing the liquid through a narrow contraction. The results indicate that the polymer coil is partially extended in a "transient" elongational flow and that a critical strain rate exists above which the chains begin to deform. This coil-to-stretch transition is dependent on the molecular weight of the polymer. Two independent birefringence domains have been observed for the dilute and semi-dilute regimes. These correspond to entangled network and isolated chain extension. The results are compared with data obtained under stagnant elongational flow conditions to evaluate the relative degree of chain extension under transient flow. Vertical and horizontal birefringence profiles in the vicinity of the orifice as well as optical photomicrographs of the shape of the birefringence zone will be discussed. Flow visualization of the streamlines will also be presented to infer the dominant flow characteristics (laminar, turbulent) in the vicinity of maximum chain extension. The results will be correlated with data on mechanochemical degradation which indicates a sharp propensity for midchain scission above a critical strain rate for chain fracture.

282. IN-SITU CURE MONITORING OF AROMATIC DIAMINE-EPOXY SYSTEM BY FIBER OPTIC FLUORESCENCE, Nakho Sung and H.J. Paik, Department of Chemical Engineering, Tufts University, Medford, MA 02155, C.S.P. Sung, Institute of Materials Science, University of Connecticut, Storrs, CT 06269-3136

In epoxy systems containing aromatic amine curing agent, such as diaminodiphenyl sulfone(DDS), fluorescence excitation spectra of DDS show a progressive red shift up to 24nm when the primary amine is reacted with epoxide to become secondary and tertiary amine. Relatively sharp excitation spectra allows us to easily determine the position and such peak position can be related to the overall extent of reaction for epoxide and amine. Using a custom-built fiber optic fluorimeter, in-situ cure reaction of epoxy ststem was monitored via a single distal end probe. Cure monitoring of neat epoxy matrix (DGEBA-DDS system) as well as glass fiber and graphite fiber prepreg is presented. Excitation spectral shift is also found to be temperature dependent making it necessary to do temperature corrections on the spectral data.

283. **REINFORCING ZIRCONIA PARTICLES PRECIPITATED INTO AN ELASTOMERIC MATERIAL.** S. B. Wang and J. E. Mark, Department of Chemistry and the Polymer Research Center, The University of Cincinnati, Cincinnati, OH 45221.

Particles of ZrO_2 were precipitated into a polysiloxane elastomer by the hydrolysis of zirconium n-propoxide, using NH_4OH or HCl as catalyst. Transmission electron micrographs of the resulting in-situ filled elastomers indicated that the particle diameters were similar to those obtained in silica precipitations, but with broader distributions of particle size. Stress-strain isotherms obtained on the elastomers in elongation showed that the particles obtained from either catalyst gave good reinforcement, but those prepared using HCl gave larger increases in ultimate strength and higher maximum extensibilities.

284. **STRUCTURAL FEATURES OF SOL GEL DERIVED HYBRID CERAMER MATERIALS BY SMALL ANGLE XRAY SCATTERING**

D. E. RODRIGUES, A. B. BRENNAN, C. BETRABET, B. WANG and G. L. WILKES
Department of Chemical Engineering
Virginia Polytechnic Institute & State University
Blacksburg, Va 24061.

ABSTRACT

The small angle xray scattering (SAXS) from several hybrid ceramer materials prepared in our laboratories is addressed. In the past, SAXS investigations have revealed that an interference peak exists at ca. 100 Å and this behavior has been used for establishing a morphological model. This paper tests several aspects of the validity of the model by varying the type of metal alkoxide, the oligomer or spacer length, the cure temperature and the metal alkoxide content in the ceramer. Finally the paper also provides some preliminary information on the development and growth of the inorganic structure in one specific ceramer system and considers the nature of the inorganic phase in view of mass and surface fractal analysis.

285. **MICROSTRUCTURAL EVOLUTION OF A SILICON OXIDE PHASE IN NAFION MEMBRANES BY AN *IN SITU* SOL-GEL REACTION.** K. A. Mauritz, I. D. Stefanithis, Polymer Science, Univ. of So. Mississippi, S S Box 10076, Hattiesburg, MS 39406-0076; G. L. Wilkes, Hao-Hsin Huang, Chemical Eng., Virginia Polytechnic Institute and State Univ, Blacksburg, VA 24061.

Microcomposites were produced via an *in situ* diffusion controlled, acid-catalyzed sol/gel reaction for tetraethoxysilane in pre-hydrated, methanol-swollen Nafion sulfonic acid films. Two strong thermal transitions identified with the polar cluster and microcrystalline domains were identified for microcomposite and unfilled membranes. Upon annealing at a temperature below the cluster transition temperature and quenching, this transition becomes suppressed for all membranes. SAXS and WAXD studies indicate persistence of the morphological template (phase-separated morphology) as well as eventual percolative intergrowth of silicon oxide clusters with increasing inorganic uptake. The permeability of simple gases through these microcomposites is sensitive to silicon oxide content. Large-scale morphology was investigated via environmental scanning electron microscopy (ESEM). The X-ray energy dispersive elemental microprobe attachment of the ESEM was used and indicated that there is significantly more silicon oxide in the near-surface regions then in the interior of the microcomposites.

286. **A MOLECULAR ORGANIC/INORGANIC SEMI-INTERPENETRATING NETWORK.** I. A. David and G. W. Scherer, Du Pont Central Research and Development, Experimental Station, P.O. Box 80328, Wilmington, DE 19880-0328

A 50/50 volume composite of SiO_2 gel with an organic polymer, poly(ethyl(oxazoline) (PEOX) was made by dissolving tetraethyl orthosilicate (TEOS) and the polymer in a common solvent, hydrolyzing the TEOS, then evaporating the solvent. The composite consisted of only one phase, as shown by electron microscopy, differential scanning calorimetry, and small angle X-ray scattering. On the basis of this evidence, we believe that it is a molecular organic/inorganic semi-interpenetrating network. The composite was a transparent glass. The degree of brittleness or flexibility of this polymer/inorganic combination depended on the relative amounts of PEOX and SiO_2, as did resistance to solvents for the polymer component. These characteristics were different from those of a similar composite of poly(vinyl alcohol) (PVOH) and SiO2 in which the components were covalently bound to each other. Further in-depth characterization and synthesis studies to extend the range of polymers and inorganics are underway.

287. DIELECTRIC AND MECHANICAL SPECTROSCOPY STUDIES OF POLY(VINYL ACETATE)/SILICON OXIDE MICROCOMPOSITES PREPARED VIA THE SOL-GEL METHOD. John J. Fitzgerald, Christine J. T. Landry, Robynn V. Schillace and John Pochan, Eastman Kodak Company, Rochester, NY 14650-2135

Blends of poly(vinyl acetate) (PVAc) and tetraethoxysilane (TEOS) were prepared by the acid catalyzed polymerization of TEOS. The local environment of the PVAc chains was probed using dielectric and mechanical spectroscopy. The results suggest that the PVAc and the silicate network interact strongly. While the glass transition temperature did not vary with increasing concentration of TEOS, there was a broadening of both the dielectric and mechanical loss peaks. Normalized loss factor curves were constructed and the empirical Kohlrausch-Williams-Watts function was fit to the data. The distribution parameter b was seen to decrease with increasing concentration of TEOS, indicating a broader distribution of relaxation times. The results suggest that the polymerized silicate network affects the molecular motion of the PVAc chains and the local environment is microheterogeneous.

288. METHACRYLATE PRECURSORS TO OXIDE GLASSES AND CERAMICS, M.H.E. Martin, C.K. Ober, MS&E, Cornell University, Ithaca, NY 14853-1501

A novel method for the preparation of silicate glasses and silicate ceramics materials composites from methacrylate polymer precursors is being investigated. Precursors to both glassy and crystalline materials have been prepared via the copolymerization of the K, Na, and Mg salts of methacrylic acid with various combinations of a siloxy-methacrylate and a methoxysilyl-methacrylate. Thermogravimetric/differential thermal, pyrolysis mass spectrometry, and high temperature x-ray diffraction analyses were used to monitor the conversion of the precursors during air pyrolysis. Some precursors demonstrated high (> 28 weight%) yields of glassy material; precursors to crystalline oxides showed yields as high as 27 weight%. The pyrolysis products were analyzed using x-ray diffraction, wavelength dispersive spectroscopy, and inductively coupled plasma mass spectrometry.

289. SYNTHESIS OF NEW MATERIALS: ORGANOCERAMICS. P. B. Messersmith and S. I. Stupp, Department of Materials Science and Engineering, University of Illinois, Urbana, Illinois 61801.

We have synthesized new materials in which polymer molecules reside within interlamellar cavities of inorganic crystals. These materials are called organoceramics and are synthesized by growing inorganic crystals from polymer solution, incorporating as much as 38 weight percent polymer into the product. Crystal morphology and structure of the organoceramic powders are affected by the type of polymer used. Precipitation of $Ca_2Al(OH)_6[X] \cdot nH_2O$ in the presence of poly(vinyl alcohol) yields an organoceramic in which the polymer chains reside in the spaces between crystal layers, forcing the layers apart by roughly 10 angstroms. The presence of poly(dimethyldiallylammonium chloride) results in formation of a different inorganic crystal, with the polymer possibly residing in channels that are a part of the crystal structure.

290. BIO-MIMETIC ROUTES TO COMPOSITES AND CERAMICS: INFLUENCE OF POLYMER MATRIX ON PRECIPITATION PROCESSES. Jeremy Burdon and **Paul Calvert**, Arizona Materials Laboratories, 4715 E. Ft. Lowell Road, Tucson AZ 85712.

We have been studying the precipitation of inorganic oxides by hydrolysis of metal alkoxides dissolved in various organic polymers. The polymer is expected to influence the particle size by limiting inward diffusion of the water and motion of the trapped reagents and growing particles. The polymer may also be catalytic for the precipitation reaction as we have found for the growth of oxide particles in a matrix of ethyleneaminoacrylate copolymer. The effect of polymer crystallinity will also be discussed.

291. RUBBER-CLAY HYBRID ——— SYNTHESIS AND PROPERTIES
A. Okada, K. Fukumori, A. Usuki, Y. Kojima, N. Sato, T. Kurauchi and O. Kamigaito
Toyota Central Research and Development Laboratories, Inc.,
Nagakute, Aichi, 480-11, Japan

We have reported that clay minerals give a molecular composite with nylon 6, termed as nylon 6-clay hybrid. This nylon 6-clay hybrid has excellent mechanical properties compared with nylon 6.
In this study a rubber-clay mineral hybrid was prepared from nitrile rubber and montmorillonite, a layered clay mineral. In this hybrid, negatively charged silicates and positively charged rubber end groups are directly bonded through ionic bonds and are uniformly mixed with each other. This hybrid exhibits superior mechanical properties and swelling resistance to the rubber systems reinforced by carbon black fillers.

292. **RING POLYMERS** Jacques Roovers, Institute for Environmental Chemistry, National Research Council, Ottawa CANADA

High molecular weight cyclic polymers have been a longstanding challenge for the polymer chemist. By making rings, chemists can prove their mastery of synthetic and analytcal techniques. Polymer physicists use the closed polymer chain in model calculations to contrast their work on linear chains and they expect unique properties of ring polymers. Ring polymers are prepared by interfacial condensation (polycarbonate), by equilibration reaction (dimethylsiloxane), by ring closure of two-ended living polymer (polystyrene, poly(2-vinylpyridine), polybutadiene) and by ring expansion (cationic, metathesis). Purification of ring polymers is based on chain expansion (living polymer), solubility difference, fractionation and preferential adsorption. Analysis of ring-linear mixtures relies on ultracentrifugation sedimentation, but most often on size exclusion chromatography. High resolution is required for meaningful analysis. Chemical proof for the ring structure is provided by a lack of end groups or by a specific ring opening reaction. The physical properties in dilute solution in conjunction with those of the linear polymer of identical molecular weight and molecular weight distribution constitute physical proof for the presence of rings. Good theoretical calculations are very helpful at this point.

293. CHARACTERIZATION OF BRANCHING ARCHITECTURE THROUGH "UNIVERSAL" RATIOS OF POLYMER SOLUTION PROPERTIES. J. F. Douglas, Polymers Division, National Institute of Standards and Technology, Gaithersburg, MD 20899; J. Roovers, Division of Chemistry, National Research Council of Canada, Ottawa, Ontario, Canada K1A OR6; and K. F. Freed, James Franck Institute and Dept. of Chemistry, University of Chicago, Chicago, Illinois 60637.

Experimental and Monte Carlo data for the dilute-solution properties of "lightly branched" polymers (stars, combs, rings,...) are compared with the renormalization group predictions of Douglas and Freed. The comparisons focus on "universal" dimensionless ratios of the mean dimensions of lightly branched polymers, relative to those of linear polymers having the same molecular weight. Complications associated with hydrodynamic solution properties and with the effect of ternary interactions are briefly discussed. Dimensionless ratios involving the polymer second virial coefficient A_2, are also tabulated and compared with theory.

294. LIGHT SCATTERING AND VISCOELASTICITY OF LINEAR AND RING POLYELECTROLYTES
Eric J. Amis and Donald F. Hodgson, *Department of Chemistry, University of Southern California, Los Angeles, CA 90089*

Matched pairs of linear and cyclic poly(2-vinyl pyridine) molecules of molecular weights ranging from 2.9×10^4 to 1.0×10^5 have been studied in dilute solution as charged polyelectrolytes. Acidified ethylene glycol was used as the solvent because it is a good solvent for the P2VP chain backbone and at the same time upon the addition of acid the pyridine groups are protonated to yield a well-defined polyelectrolyte. Simultaneous measurements of pH and knowledge of concentrations provides data necessary to calculate the polyelectrolyte charge density and the electrostatic screening lengths in the solution. Under high acid conditions (0.23 M HCl) normal polymer extrapolation of reduced viscosity yields intrinsic viscosity and extrapolation of dynamic light scattering yields D_0. In weakly acidic solvents (0.0012 M HCl) typical polyelectrolyte behavior is observed in both viscosity and light scattering. These results are interpreted in terms of an electrostatic chain expansion model.

295. Phase Diagrams and Spinodal Decomposition in Linear and Cyclic Polystyrene /Poly[Vinyl Methyl Ether] Blends, M. M. Santore, Depart. Chem. Eng., Lehigh U., Bethlehem, PA 18015 G. B. McKenna, and C. C. Han, National Institute of Standards and Technology, Gaithersburg, MD 20899

We investigated the role of cyclic backbone architecture in diffusion and the thermodynamic properties of polymer blends of chemically different components. This study of the PS/PVME (polystyrene /poly(vinylmethyl ether)) system at molecular weights exceeding entanglement compared a mixture of cyclic PS and linear PVME to that of the two linear components. Previous studies with linear chains indicated a lower critical solution temperature, i.e. phase separation on heating. In this investigation, cloud point studies revealed that blending cyclic PS instead of linear PS with PVME enhanced stability by raising the phase separation temperature by 7-8K. Negligible shifting of the critical composition (20% PS and 80% PVME) was observed. Time resolved light scattering near the critical composition was used to monitor the phase separation kinetics of spinodal decomposition following a rapid temperature jump into the two-phase region. Analyzing intensity as a function of time via Cahn-Hilliard-Cook theory for the early stage phase separation kinetics led to a mutual diffusion coefficient. The temperature dependence of the mutual diffusion coefficient confirmed the results of cloud point tests near the critical composition. Further, diffusion in exclusively linear and cycle-containing blends was similar when the two were compared at equal temperature increments above their respective spinodal temperatures. Though thermodynamic effects dominated the diffusion coefficient, the estimated mutual mobility was lower in the cycle-containing blend.

296. DYNAMICS OF RING DNA VERSUS LINEAR DNA FROM TRANSIENT ELECTRIC BIREFRINGENCE

John S. Bowers, *Department of Polymer Science, University of Southern Mississippi, Hattiesburg MS 39406-0076.*

Robert K. Prud'homme, *Department of Chemical Engineering, Princeton University, Princeton NJ 08544-5263.*

Absolutely monodisperse rings and linear DNA samples are obtainable by treating plasmid DNA with enzymes that control the form of DNA, such restriction enzymes, topoisomerases, and DNAase. Plasmids φX-174 (5386 bp, $M = 3.5 \times 10^6$), pMSG (7626 bp, $M = 4.96 \times 10^6$) and pGA482 (13.2 kbp, $M = 8.58 \times 10^6$) were cut with restriction enzymes to obtain monodisperse samples of linear DNA with the same or half the molecular weight as the original ring, and supercoiling was removed with Topoisomerase I or by nicking with DNAse. In Transient Electric Birefringence (TEB) molecules are oriented in electrical fields, and then allowed to return to a random configuration; the rise, steady state and field free decay of the birefringence contains information about the molecular weight, stiffness and polarizability of the molecules. By measuring the field free decay of the DNA samples, we tested polymer kinetic theory predictions for polymer dynamics from the Rouse/Zimm models for rings and linear polymers in dilute solution. We found excellent agreement with the theoretical prediction that rings behave as linear molecules of exactly half the ring length. We also investigated the effect of supercoiling of φX-174 DNA on its dynamics, and found that largest time constant, representing reorientation of the entire molecule, does not change with supercoiling.

297. **DYNAMIC MONTE CARLO SIMULATIONS OF RINGS AND BRANCHED MOLECULES.** Antonio Rey, Andrzej Sikorski, Andrjez Kolinski, **Jeffrey Skolnick,** Department of Biology, The Scripps Research Institute, 10666 North Torrey Pines Road, La Jolla, CA 92037.

In the context of a cubic lattice model of the polymer melt, dynamic Monte Carlo simulations of the dynamics of a melt of uncatenated rings with and without one self-knot, and a dilute solution of 3-arm branched molecules dissolved in a melt of linear chains have been simulated. For the case of a melt of rings, having a degree of polymerization, n, up to n=1536, we find that rings are far more mobile than the corresponding linear chains. In agreement with experiment, the entanglement spacing is a factor of about 4 times greater. As predicted by the theory of Cates and Deutsch, the equilibrium dimensions of a melt of unknotted rings are smaller than would be expected on the basis of Gaussian statistics. Unknotted rings and rings containing one self-knot exhibit virtually identical dynamic behavior. We have also simulated the dynamics of three arm stars of span size up to 400 beads dissolved in an n=800 linear chain matrix. Comparison with the dynamics of linear chains is made. An analysis of chain motion including a comparison with the conjectured arm retraction mechanism originally proposed by de Gennes will be presented.

298. SPONTANEOUS ADSORPTION OF POLYMERS FROM SOLUTION TO SOLUTION-POLYMER INTERFACES. Molly. S. Shoichet,[a] D.R. Iyengar,[a] Joan V. Brennan[b] and Thomas J. McCarthy,[a] Polymer Science and Engineering[a] and Chemistry Departments[b], University of Massachusetts, Amherst, Massachusetts 01003

Polystyrene, of molecular weight above a critical value of ~20 000, spontaneously and irreversibly adsorbs to poly(vinylidene fluoride) film from cyclohexane solution at 36 °C. A very thin (<10 Å - when dry) polystyrene overlayer forms as indicated by X-ray photoelectron spectroscopy. The tendency for adsorption is interpreted as analogous to standard polystyrene adsorptions from cyclohexane to inorganic substrates. Poly(L-lysine) (PLL) adsorbs to poly(tetrafluoroethylene-co-hexafluoropropylene) (FEP) film from pH 11 aqueous solution to form a thin (<10 Å) overlayer. Adsorption experiments using various molecular weight PLL samples, solutions of different pH and alcohol/water mixtures suggest that a decrease in (FEP-water) interfacial free energy and the unfolding of PLL from the α-helix structure drive the adsorption. FEP-PLL exhibits both increased water wettability and increased adhesion over FEP. The ε-NH_2 groups in FEP-PLL react as nucleophillic surface functionality.

299. MOLECULAR RECOGNITION DIRECTED SUPRAMOLECULAR POLYMER ARCHITECTURES. V. Percec and J. Heck, Department of Macromolecular Science, Case Western Reserve University, Cleveland, Ohio 44106.

Previous examples of self-assembling supramolecular structures have been formed by intermolecular noncovalent binding interactions. When polymeric molecules form supramolecular structures, intramolecular interactions must also be considered. This presentation will present examples and demonstrate the concept of self-assembling polymers where the structure is determined by intramolecular interactions in one case and a combination of intramolecular and intermolecular interactions in another case.

300. THE IMPACT OR MOLECULAR AND SUPERMOLECULAR STRUCTURE ON THE PROPERTIES OF COMPOSITES FOR ELECTRONICS. M.P. Zussman, Du Pont Electronics, Experimental Station, Wilmington, DE 19880

The interaction between composite structure and functionality as a printed wiring board substrate is illustrated by two examples: woven quartz reinforced modified polybutadiene and nonwoven aramid reinforced cyanate ester. Both systems yield laminates with desirable values for CTE (coefficient of thermal expansion), dielectric constant and glass transition temperature. Difficulties in processing the modified polybutadiene system are ascribed to the molecular structure of the resin. The nonwoven aramid reinforcement imparts low CTE and supports higher yields during printed wiring board fabrication by enabling better layer-to-layer registration than either woven glass or woven aramid. These macroscopic properties are attributed to the supermolecular structure of the nonwoven aramid.

301. PSEUDO-GRAFTING OF VINYL POLYMERS ONTO DNA. Mizuo Maeda, Akira Hirai, Chitoshi Nishimura and Makoto Takagi, Department of Chemical Science and Technology, Kyushu University, 6-10-1, Hakozaki, Higashi-ku, Fukuoka 812, Japan

A semi-synthetic macromolecular conjugate consisting of *Hind III* restriction fragments of λ phage DNA and polyacrylamide was prepared by copolymerization of acrylamide (AAm) with a DNA binding molecule having vinyl group (1) in the presence of the DNA fragments. Each fragment was grafted with PolyAAm chains with the aid of 1 moieties. The DNA fragments were dissociated from the conjugates with the addition of a strong intercalator, ethidium bromide under the conditions of high salt concentration. On the removal of ethidium, the DNA fragments were recombined with 1-containing copolymers which had been dissociated from the conjugates. The conjugation was thus confirmed to have a reversible nature based on non-covalent bonding. In contrast to this, 1-AAm copolymers prepared in the absence of DNA were found to have almost no ability to form a complex with the DNA. The difference in binding ability is discussed in relation to the structure of the copolymer.

302. NEW POLYMERS FROM ARTIFICIAL GENES: A PROGRESS REPORT. David A. Tirrell,[a,c] Maurille J. Fournier[b,c] and Thomas L. Mason,[b,c] Departments of Polymer Science and Engineering[a] and Biochemistry,[b] and Program in Molecular and Cellular Biology,[c] University of Massachusetts, Amherst, MA 01003 USA

Chemical methods of polymer synthesis are inherently limited by the statistical nature of polymerization processes. As a result, the polymers currently in use are not pure materials, but instead are mixtures characterized by distributions of the important structural variables (molecular weight, composition, sequence and stereochemistry). The preparation of pure polymeric materials, and the use of polymers in applications that require precise structural control, demand the introduction of new synthetic methods. With this in mind, we have begun to develop molecular biological approaches to the preparation of new polymeric materials. The fidelity of protein biosynthesis, coupled with recent advances in the synthesis, cloning and expression of genes, offers the prospect of a new synthetic technique of unprecedented precision and remarkable scope. This discussion will describe the design, synthesis and expression of several new families of genes that encode amino acid copolymers of some potential interest in materials science. In particular, we will consider the use of sequence-dependent secondary structures to control the folding of artificial protein chains in the solid state.

303. DYNAMIC INFRARED LINEAR DICHROISM OF POLYMERS. Isao Noda, A.E. Dowrey, and C. Marcott, The Procter & Gamble Company, PO Box 398707, Cincinnati, OH 45239-8707.

Dynamic infrared linear dichroism (DIRLD) is a rheo-optical phenomenon observed for many polymeric systems. When a small-amplitude oscillatory strain is applied to a polymer film, chemical constituents of the system undergo dynamic realignment. The reorientation of electric dipole-transition moments associated with the molecular vibrations of individual chemical functional groups results in the time-dependent variation of the absorption of a polarized IR beam. From the dynamic fluctuation of the absorbance difference between light polarized parallel and perpendicular to the applied strain, one can deduce the changes of the average orientation of dipole-transition moments. Interestingly, reorientations of submolecular structures under a given dynamic deformation do not necessarily proceed simultaneously; dynamic IR dichroism apparently measures the localized submolecular motions of individual functional groups. This result implies that the classical theory, which relates IR dichroism to the average orientation of a polymer chain by assuming a fixed angle between the chain axis and the dipole transition moment, breaks down under dynamic conditions. Applications of dynamic IR dichroism measurements in elucidating molecular motions of polymer segments will be discussed.

304. POLYMER CHARACTERIZATION BY STEP-SCAN FT-IR SPECTROSCOPY
Richard A. Palmer,[a] Rebecca M. Dittmar,[a] Vasilis G. Gregoriou,[a] and James L. Chao[b]

[a]Department of Chemistry, Duke University, Durham, NC 27706
[b]IBM Corporation, Research Triangle Park, NC 27709

The use of step-scan Fourier transform infrared (FT-IR) spectroscopy to obtain dynamic spectral information on polymeric systems is illustrated. Step-scan FT-IR simplifies the extraction of dynamic information due to the single modulation frequency that can be applied over the entire spectral range and the elimination of the time dependence of the spectral multiplexing. As a result, the only time dependence of the signal will be due to the sample. As a result, the application of FT-IR to both modulation/demodulation (phase-resolved) and impulse/response (time-resolved) dynamic measurements is simplified. The principle of step-scan FT-IR and its application to photoacoustic depth profiling of laminar polymers and to rheo-optical characterization of polymer films are discussed and illustrated. In both types of experiment it is the ease of extraction of the phase of the signal as its in phase and quadrature components that gives superiority to the step-scan technique.

305. DIFFUSION AVERAGE MOLAR MASS OF POLYDISPERSE POLYMERS: A PHOTON CORRELATION SPECTROSCOPY STUDY OF POLYSTYRENE IN GOOD AND THETA SOLVENTS. B. R. White and G. J. Vancso, Department of Chemistry, University of Toronto, Toronto, Ontario, Canada M5S 1A1.

Results from photon correlation spectroscopy measurements on dilute solutions of a polystyrene (PS) sample with a well characterized broad molar mass distribution function are reported. Dynamic Zimm plots were used to evaluate the center-of-mass translational diffusion coefficient ($D_{Z,0}$) of PS in tetrahydrofuran (THF) and cyclohexane. The magnitude of the diffusion average molar mass (M_D) was calculated based on the power law expression, which predicts the dependence of $D_{Z,0}$ on the molar mass. M_D values fall between the weight-average and the z-average molar masses. A novel procedure is proposed to determine the scaling parameters which relate the center-of-mass translational diffusion coefficient of macromolecules to the molar mass for heterogeneous polymer samples. A new light scattering polydispersity index $\Delta = M_D/M_W$ is also introduced.

306. CHARACTERIZATION OF PROTEIN SECONDARY STRUCTURE VIA FOURIER TRANSFORM RAMAN SPECTROSCOPY, T. M. Przybycien[1], H. M. Thompson[2], N. Summers[2], and R. Stegeman[2], [1]Howard P. Isermann Department of Chemical Engineering, Rensselaer Polytechnic Institute, Troy, NY 12180-3590, [2]Physical Sciences Center, Monsanto Corporate Research, Chesterfield, MO 63198

A method for the characterization of protein secondary structure content using Fourier transform Raman (FTR) spectroscopy and a deconvolution technique based on the maximum entropy method (MEM) has been developed. In this technique, a MEM estimate of the sample amide I band is constructed from the FTR spectrum. Estimates of the sample alpha-helix, beta-sheet, and random coil content are obtained from this construction via a constrained superposition of spectra from proteins with known secondary structures. Improvements over previous techniques include the use of FTR spectra with significantly reduced background fluorescence compared to conventional spectra, the elimination of spectral smoothing and baseline subtraction with the MEM technique, and a set of reference spectra which span a wider range of structure space. The method is demonstrated with a variety of protein spectra with varying signal-to-noise ratios.

307. DEPTH PROFILING USING VARIABLE ANGLE ATR, H. Ishida and R. Shick, Department of Macromolecular Science, Case Western Reserve University, Cleveland, Ohio 44106.

In polymer science it is quite often the case that it is desirable to have molecular information as a function of position within a sample. Two dimensional spatial information of the surface may be obtained using Hadamard transforms, but frequently it is interesting to have information in the depth direction. This is particularly true in anisotropic systems such as in composite interphase and polymer blend studies. It is the intent of this paper to illustrate the applicability of using variable angle ATR for obtaining spectral depth profile information. While a theoretical treatise has been proposed which would allow depth profiling using inverse Laplace transforms, it has not undergone rigorous experimental verification. A suitable model system is presented to validate the methodology. Applications to various polymeric systems are discussed.

308.
STRAIN-BIREFRINGENCE OF REAL NETWORKS: EFFECT OF SWELLING. Vassilios Galiatsatos, Institute of Polymer Science, University of Akron, Akron, OH 44325-3909.

The fact that a polymer network becomes uniaxially birefringent when it is subjected to uniaxial strain, forms the basis of an important technique for the optical characterization of elastomers. Any theory that attempts to explain the available experimental data on real networks must incorporate the effect of the topological constraints on the measured birefringence. This work will a) review the various theories of strain-birefringence and b) report new data on how the topological effects and the associated birefringence are dependent on swelling. The results, based on a recently developed theory of strain-birefringence, support the notion that analysis of experimental data is simplified if an optically isotropic solvent is employed.

309. POLYSILAZANES : A ROUTE TO CERAMICS. M. Schappacher, A. SOUM, Laboratoire de Chimie des Polymères Organiques, Université Bordeaux-1, 33405 Talence, France.

Polysilazanes are potential macromolecular precursors to silicon nitride and silicon carbonitride ceramics. However the usual synthesis by either ammonolysis or aminolysis of dihalosilanes leads to low molar mass oligomers.

In order to increase the chain length of the polysilazanes and, therefore, provides suitable ceramic synthons, the ring-opening polymerization of cyclosilazane monomers has been investigated. Various cyclosilazanes have been reacted with anionic initiators. These reactions give either polymers or linear and cyclic oligomers, depending on the size of the ring and on the structure of both the nitrogen and the silicon substituents.

310. CHARACTERIZATION OF PRECERAMIC PERHYDROPOLYSILAZANE. O. Funayama, T. Isoda, H. Kaya, T. Suzuki and Y. Tashiro, Tonen Corporation Corporate Research & Development Laboratory, 1-3-1, Nishi-Tsurugaoka, Ohi-Machi, Iruma-Gun, Saitama 354 Japan

Characterization of preceramic perhydropolysilazanes polymerized by heat treatment with or without ammonia of oligomer $(SiH_2NH)_n$ in the presence of pyridine has demonstrated that these polymers have great potential as precursors of silicon nitride. The perhydropolysilazanes are white solids that can be converted to ceramic material with high ceramic yields (>85wt%). The oligomer has been synthesized by the ammonolysis of dichlorosilane pyridine adduct. The polymers and the oligomer have been characterized by several techniques including total elemental analysis, molecular weight distribution measurement by GPC, ^{29}Si, ^{15}N and ^1H NMR spectroscopy, IR spectroscopy and thermogravimetric analysis. This paper will discuss the structures of these precursors and some properties of the pyrolysis products.

311. SYNTHESIS AND CERAMIC CONVERSION REACTIONS OF POLYBORAZYLENE. Paul J. Fazen, Edward E. Remsen and Larry G. Sneddon, Department of Chemistry, University of Pennsylvania, Philadelphia, PA 19104-6323 and Physical Science Center, Monsanto Corporate Research, 700 Chesterfield Village Parkway, St. Louis, MO 63198

Borazine has been found to readily dehydropolymerize in the liquid state at 70°C to produce the new inorganic polymer polyborazylene. The polymer appears to have a complex structure, having linear and branched chain segments, related to those of the organic polyphenylenes, in which the borazine rings are joined primarily by B-N linkages. The polymer is isolated as a white powder that is highly soluble in ethers and that according to SEC/LALLS analysis has Mw ranging from 2,100 g/mol to 7,600 g/mol and Mn between 980 g/mol and 3,400 g/mol. Polyborazylene has been found to convert to boron nitride with both high chemical and ceramic yields. TGA, DRIFT and X-ray diffraction studies of the ceramic conversion reaction suggest a simple process for the conversion of the polymer to boron nitride involving a 2-dimensional crosslinking-dehydrocoupling reaction between adjacent polymers.

312. SYNTHESIS AND PROCESSING OF PRECERAMIC BN POLYMERS R.T. Paine, J.F. Janik, T.T. Borek, D. A. Lindquist, D.M. Smith, T.T. Kodas, A.K. Datye and E. N. Duesler. Departments of Chemistry and Chemical and Nuclear Engineering and the Center for Microengineered Ceramics, University of New Mexico, Albuquerque, NM 87131.

Polyborazinyl amines have been shown to be good precursors for the formation of boron nitride powders. In addition, we have found that three point polymers are soluble in liquid NH_3 and the resulting solutions have been used to produce fibers, thin coatings and aerosol particles. In addition, two point polyborazinyl amines possessing a variety of functional groups on the borazine ring have solubility in organic solvents. These solutions have been processed into xerogels, aerogels, coatings and aerosols. The processing of the polymers and the characterizations of the final ceramic forms will be described in this paper. Furthermore, the utilization of these reactive polymers as reagents to form composites will be described.

313. THE RELATIONSHIPS BETWEEN THE STRUCTURE OF PRECERAMIC POLYSILAZANES, THE PYROLYSIS CONDITIONS, AND THEIR FINAL CERAMIC PRODUCTS. Yigal D. Blum, Robert B. Wilson, Gregory M. McDermott, and Albert S. Hirschon.

The final ceramic yields and compositions obtained after the pyrolysis of preceramic polysilazanes are dependent on the monomeric units and configuration of the precursors as well as the pyrolysis conditions. The structures of polymeric precursors are of most importance in controlling the multistep thermal reactivities that occur during the pyrolysis of the polymers and later on during the crystallization of the amorphous products (between 100° to 1600°C).

The role of the pyrolysis environment is critical to the development of the final product. Ammonia is necessary to eliminate any of the carbonaceous ceramic species. The ammonia eliminates the organic substituents before the polymer converts to ceramic products. However, different mechanisms are observed as a function of the organic substituents.

The molecular structures of the polymer-derived amorphous ceramic products were studied by FT-IR. The results reveal that even though silicon and nitrogen are the predominant elements in the final composition of the pyrolyzed material, the molecular structures of the amorphous ceramics as determined by FTIR are different and consequently different high-temperature behavior is expected.

General aspects of the relationship between polymer structures and pyrolytic characteristics and of structure-to-ceramic relationship will be discussed. Pyrolysis and crystallization effects will be presented.

314. SIO_2 AS A SOURCE OF SI CONTAINING COMPOUNDS/POLYMERS. Richard M. Laine[a], D. Jean Ray[b], Christopher Viney[b], Timothy R. Robinson[b], Martin L. Hoppe[a], Zhi-Fan Zhang[a]. Contribution from a) the Departments of Materials Science and Engineering and Chemistry; University of Michigan; Ann Arbor, MI 48109-2136; and b) the Department of Materials Science and Engineering; University of Washington; Seattle, WA 98195.

The search for new, polymeric materials that meet or exceed the physical properties of carbon-based polymers must eventually consider inorganic and organometallic materials. This presentation will describe methods of making hypervalent silicon compounds of use in the preparation of silicon preceramics and high temperature, liquid crystalline oligomers (and polymers?) directly from silica.

315. NITROGEN AND/OR BORON-CONTAINING POLYMERS AS PRECURSORS OF CERAMIC MATERIALS. Leon Maya, Chemistry Division, Oak Ridge National Laboratory, Oak Ridge, Tennessee 37831.

The use of polymeric precursors to produce ceramic materials affords considerable flexibility to control the morphology of the desired product. We have examined derivatization reactions of preformed polymers as a synthetic approach to produce precursors. Boron moieties have been bound to either polyethyleneimine or polybutadiene. In a different approach, polymeric cyanoborane, $(CNBH_2)n$, has been used as a single source for the chemical vapor deposition of boron nitride films on a variety of substrates. Attempts have been made to synthesize carbon nitride, an hypothetical compound predicted to have extreme hardness. These attempts have involved the pyrolysis of nitrogen containing precursors such as polymerized hydrogen cyanide.

Research sponsored by the Division of Materials Sciences, Office of Basic Energy Sciences, U.S. Department of Energy, under contract DE-AC05-84OR21400 with Martin Marietta Energy Systems, Inc.

316. BORON NITRIDE FIBERS FROM POLYBORATES.
Bruce Wade, David Mohr, N. Venkatasubramanian, Prashant Desai,
A. S. Abhiraman, *Polymer Education and Research Center,*
and Eugene C. Ashby, *School of Chemistry & Biochemistry,*
Georgia Institute of Technology, Atlanta, Georgia 30332.

The diffusion limitations which had existed in earlier studies (Economy, *et al.*) during nitriding of melt spun boron oxide fibers in ammonia may be avoided with sol-gel processed porous polyborate fibers. Polyborate precursors are formed with a low concentration of a high molecular weight polymer as a rheological aid which is also required to be fugitive during thermochemical conversion. Infrared and x-ray diffraction studies show that polyborate undergoes thermochemical conversion while nitriding in ammonia to turbostratic boron nitride. Small scale continuous melt spinning of 100μ diameter polyborate fibers has been performed.

317. APPLICATIONS OF POLYSILAZANES AS PRECURSORS TO SILICON NITRIDE
Stuart T. Schwab, Renee C. Graef, David L. Davidson, Yming Pan
Materials & Mechanics Department
6220 Culebra Road
San Antonio, Texas 78228-0510

Preceramic polymers provide a facile route to the production of a variety of ceramic articles. Because NICALON has proven to be a commercial success, much of the effort in preceramic polymer research has been focussed on the development of fiber precursors. There are, however, a number of other areas in which the utilization of preceramic polymers have the potential to yield viable commercial products.

Polysilazanes have been shown to be effective precursors to silicon nitride (Si_3N_4). Although polysilazane-derived fibers are currently being produced commercially, a number of potentially fruitful applications of these materials remain to be explored. At Southwest Research Institute, we have focussed on the use of polysilazanes as binders for Si_3N_4 powder processing, as precursors to the matrix in continuous fiber-reinforced composites, and as precursors to protective coatings on refractory substrates. Utilization of a low viscosity, thermosetting polysilazane has facilitated the development of the latter two applications.

318. NOVEL DENDRITIC MACROMOLECULES BY THE CONVERGENT GROWTH APPROACH: HYPERBRANCHED, BLOCK COPOLYMERS AND MONODISPERSE POLYESTERS. Craig J. Hawker[1,2], Karen L. Wooley[1] and Jean M.J. Fréchet[1]. (1) Dept. of Chemistry, Baker Laboratory, Cornell University, Ithaca, N.Y., 14853-1301. (2) Dept. of Chemistry, University of Queensland, St. Lucia, 4072, Queensland, Australia.

We have developed a convergent growth approach to novel families of dendritic or hyperbranched macromolecules. One of the unique characteristics of this approach is that it allows the control of number and placement of surface functional groups. By the coupling of different dendritic fragments to a central core, novel hyperbranched block copolymers were synthesized with specific regions of the periphery substituted by either bromo or cyano functionalities. This novel type of block copolymers is expected to possess unique properties due to the unsymmetrical placement of dipolar surface functionalities. The concept of the convergent growth has also been extended to the synthesis of monodisperse dendritic polyester macromolecules based on 3,5-dihydroxybenzoic acid. These macromolecules are the perfect analogs of the macromolecules obtained earlier from the one-step polymerization of 3,5-bis(trimethylsilyloxy)benzoyl chloride. These monodisperse dendritic polyesters will be useful in the comparisons of properties with analogous polymers obtained in the one-pot approach.

319
I THINK I SHALL NEVER SEE A POLYMER LOVELY AS A TREE. George R. Newkome, Gregory R. Baker, Charles N. Moorefield, and Mary Jane Saunders, Center for Molecular Design and Recognition, Departments of Chemistry and Biology, University of South Florida, Tampa, Florida 33620-5250.

The synthetic construction of cascade molecules and polymers has afforded novel entrance to unimolecular micelles and molecular aggregates, which possess well defined topologies and supramolecular size. Each has offered unique symmetry properties, which will be discussed.

320. THE SYNTHESIS AND CHARACTERIZATION OF A SERIES OF MONODISPERSE, 1,3,5-PHENYLENE BASED HYDROCARBON DENDRIMERS INCLUDING $C_{276}H_{186}$. T. M. Miller, T. X. Neenan, and H. E. Bair. AT&T Bell Laboratries, Murray Hill, N. J. 07974

The convergent synthesis of a series of monodisperse aromatic dendrimers having well defined diameters of 15-31 Å is described. These materials consist of 4, 10, 22, or 46 benzene rings linked symmetrically by σ-bonds. Increasingly large dendrimer arms are prepared stepwise via palladium-catalyzed coupling of arylboronic acids to 3,5-dibromotrimethylsilylbenzene. The aryltrimethylsilane is subsequently converted to a new arylboronic acid by reaction with boron tribromide followed by hydrolysis. Coupling of arylboronic acid dendrimer arms to 1,3,5-tribromobenzene or 1,3,5-tris(3,5-dibromophenyl)benzene is the final step in the synthesis. The properties of the dendrimers are discussed in terms of their solubilities and thermal stabilities.

321. SYNTHESIS OF RIGID DENDRIMERS THAT OVERCOME STERIC INHIBITION. Jeffrey S. Moore and Zhifu Xu, The Willard H. Dow Laboratories, Department of Chemistry, The University of Michigan, Ann Arbor, MI 48109

Rigid dendritic macromolecules of high-carbon content have been synthesized using a modified convergent synthetic approach. The repeat unit structure in the early generations is a 1-ethynyl-3,5-disubstituted benzene, and in later generations, the repeat unit has been enlarged in an effort to diminish steric crowding. The repetitive chemistry employed in the wedge synthesis involves palladium-catalyzed coupling of terminal acetylenes with aromatic dihalides that contain a terminal acetylene masked as its trimethylsilyl derivative. The protected acetylene is unmasked under mildly basic conditions to complete the repetitive cycle. Different peripheral groups have been tested to enhance the solubility of the dendritic molecules.

322. **SYNTHESIS AND SOLUTION PROPERTIES OF POLYSTYRENES WITH DENDRITIC END GROUPS.** Ivan Gitsov, K.L. Wooley, C.J. Hawker J.M.J. Fréchet, Department of Chemistry, Baker Laboratory, Cornell University, Ithaca, NY 14853-1301.

Polystyrenes with molecular weights in the range between 500 and 10^6 are synthesized via anionic polymerization initiated by naphthalene-potassium. The "living" polymers formed are end-capped with dendritic polyether groupings having molecular weight of 3275 by nucleophilic displacement with the corresponding dendritic bromide in vacuum. The materials obtained are characterized by ^{1}H-NMR and size exclusion chromatography coupled with differential viscometry and multi-angle laser light scattering. The results obtained show that the molecular weight characteristics of the polystyrene before and after end-capping with the dendrimer change significantly. The observed changes are much greater than would be expected for the simple addition of 2 dendritic groups with a total molecular weight of 6550. It is assumed that the incorporation of dendritic structures in the macromolecules strongly affects the size and conformation of the polymer chains in solution.

323. **TERMINAL MODIFICATIONS OF HYPERBRANCHED POLY(SILOXY SILANES).** Terrell W. Carothers, Lon J. Mathias, Department of Polymer Science, University of Southern Mississippi, Hattiesburg, MS 39406-0076.

A new synthetic approach to dendrimer polymers with silane-siloxane interiors is based on the hydrosilation of an AB_3 monomer, allyl tris(dimethylsiloxy)silane. Polyaddition with catalytic $H_2PtCl_6 \cdot 6H_2O$ (mixed anti- and Markovnikov addition) or Ashby catalyst (mostly anti-Markovnikov addition) gave moderate molecular weight polymers (ca 20,000 by ^{1}H NMR integration and size-exclusion chromatography). Subsequent addition of low molecular weight endcappers such as allyl phenyl ether and allyl glycidal ether to the reaction mixture gave quantitative endcapping of these novel dendrimers. Addition of an allyl-terminated oligomer of oxyethylene gave incomplete endcapping suggesting buried Si-H moieties in the silane-siloxane interior inaccessible to further reaction with high molecular weight molecules. Model compounds were also synthesized to help in NMR and IR elucidation of these novel dendritic polymers.

324. THE CONVERGENT SYNTHESIS OF FOUR GENERATIONS OF MONODISPERSE ARYLESTER DENDRIMERS. E. W. Kwock, T. X. Neenan, and T. M. Miller. AT&T Bell Laboratories, Murray Hill N. J. 07974

The convergent synthesis of a series of monodisperse dendrimers based upon symmetrically substituted benzene tricarboxylic acid esters is described. These materials consist of 4, 10, 22, and 46 benzene rings connected in a symmetric fashion and have molecular diameters of up to 45 Å. The synthesis proceeds in a stepwise convergent manner, building dendrimers arms, three of which are subsequently attached to a molecular core. The critical intermediate for the dendrimer arm synthesis is 5-(t-butyldimethylsiloxy)isophthaloyl dichloride, obtainable in three steps from 5-hydroxyisophthallic acid. Reaction of the dichloride with phenol, followed by removal of the silyl protecting group forms a new substituted phenol. Two equivalents of the latter may be further reacted with 5-(t-butyldimethylsiloxy)isophthaloyl dichloride. This process is repeated several times. The dendrimer arms formed by these reactions are coupled to 1,3,5-benzenetricarbonyl trichloride to form dendrimers. The dendrimers are soluble in a variety of organic solvents. Possible applications for these materials include particle size standards, novel rheological fluids or molecular inclusion hosts.

325. **CONJUGATED POLYMERS AS MATERIALS FOR NONLINEAR OPTICAL DEVICES.** Gregory L. Baker, Bellcore, 331 Newman Springs Road, Red Bank NJ 07701-7040.

Nonlinear optical devices prepared from conjugated polymers would in principle operate at the frequency of light, several orders of magnitude faster than electronic devices. Their potential as device materials depends on the magnitude of their nonresonant nonlinear response, their optical clarity, their optical damage threshold, and how amenable the polymers are to device fabrication.

We have tested many of these requirements by studying a model conjugated polymer, the polydiacetylene poly(4BCMU). Our results will be used to illustrate the positive features of conjugated polymers and where improvements in properties are needed.

326. APPLICATIONS OF CONDUCTING POLYMERS IN REDOX DEVICES AND INTELLIGENT MATERIALS SYSTEMS. R.H. Baughman, Research and Technology, Allied-Signal Inc., Morristown, NJ 07862 USA

Applications opportunities are described for conducting polymer devices, including (1) batteries and redox capacitors, (2) electromechanical actuators, (3) electrochromic windows and displays, and (4) indicators and sensors. Each of these potential application areas depends upon the dramatic property changes which occur during chemical or electrochemical doping. Since the properties profiles of conducting polymers can be reversibly changed either chemically or electrochemically, these polymers also provide exciting candidates for intelligent materials systems - where sensor, logic element, and actuator functions can be either partially or fully integrated. Results of device and properties investigations are used to evaluate future possibilities for conducting polymers in intelligent materials systems.

327. THE POLYANILINES: A NOVEL CLASS OF CONDUCTING POLYMERS. A.G. MacDiarmid, University of Pennsylvania, Philadelphia, PA 19104-6323; A.J. Epstein, Department of Physics, The Ohio State University, Columbus, OH 43210-1106.

The conductivity (doped form), degree of crystallinity, tensile strength and solubility of polyaniline (emeraldine oxidation state) is found to be greatly dependent on the past chemical and/or mechanical history of the polymer which significantly affects the ultra-structure of the material. The conductivity of the polymer increases monotonically with increase in molecular weight up to a value of ~ 160,000 (G.P.C.; M_p) after which it is essentially constant. A new form of polyaniline (polyaniline gel) which shows an anisotropic "shape-memory" effect under certain conditions when swelled in solvents in which it is insoluble, has been synthesized. Pernigraniline, the most highly oxidized form of polyaniline has been n-doped to give a material of conductivity ~0.5 S/cm. Doping (and undoping) occurs without degradation of the polymer backbone. A study of Donnan equilibrium effects on the protonic acid doping of polyaniline (emeraldine base) shows that the doping level depends not only on the pH of the aqueous dopant medium but also on the presence of dissolved neutral salts.

Work was supported in part by DARPA through a grant monitored by ONR and in part by NSF grant No. DMR-88-19885.

328. POLYMER SYNTHESIS WITH ORGANOMETALLIC REAGENTS. R.H. Grubbs, Division of Chemistry and Chemical Engineering, California Institute of Technology, Pasadena, CA 91125 USA.

Since the discovery that transition metal salts mixed with organoaluminum reagents would catalyze the polymerization of ethylene to crystalline polyethylene, organometallic complexes and reagents have played a major role in the polymer industry. Over the past 20 years a tremendous amount has been learned about the structures and mechanisms of reactions of complexes related to those proposed to be active in these systems. In the area of olefin metathesis and ring opening metathesis polymerization (ROMP), metal carbenes and metallacycles were proposed intermediates and over the past few years a number of complexes with these structures that will catalyze the olefin metathesis reaction have been prepared and studied. In contrast to the ill-defined classical catalysts based on Ziegler type catalysts, these are living polymerization systems. Examples from this work will be used to demonstrate the developments that are possible using preformed catalysts that can be used to synthesize polymers with predetermined structures. For example these catalysts can be used to prepare near monodispersed, non-branched polyethylene and the ROMP of cyclooctatetraene using these catalysts produces derivatives of polyacetylene with control of the structures and properties of the conducting materials.

329. SYNTHESIS AND PROPERTIES OF PARTIALLY SUBSTITUTED POLYACETYLENES DERIVED FROM THE RING-OPENING METATHESIS POLYMERIZATION OF MONOSUBSTITUTED CYCLOCTATETRAENES. Christopher B. Gorman, Eric J. Ginsburg, Jeffrey S. Moore, and Robert H. Grubbs, Division of Chemistry and Chemical Engineering, California Institute of Technology, 164-30, Pasadena, CA 91125 USA.

Ring-opening metathesis polymerization (ROMP) of substituted cyclooctatetraenes (COT) produces polyacetylenes that are substituted, on the average, every eight carbon atoms. These polymers are, in several cases, soluble and highly conjugated. The effect of the side group on the optical, electrical and materials properties of the polymer will be discussed, including synthesis and properties of chiral polyacetylenes.

330. FTIR AND FLUORESCENCE STUDIES OF SELF-ASSEMBLED SILOXANE MONO- AND MULTILAYERS, S. H. Chen, C. W. Frank, Department of Chemical Engineering, Stanford University, Stanford, CA 94305

A trichlorosilane monomer containing a long alkyl chain and a terminal methyl ester has been prepared according to the method described by Ulman. We have examined the kinetics of monolayer adsorption using ATR-FTIR and have modeled the process by an irreversible reaction with the order dependent upon the surface coverage. Fluorescence probe measurements and quenching studies allow the determination of the thermodynamics of the adsorption process. Numerical simulations are used to distinguish between random and patch-wise growth processes.

331. SPECTRAL BEHAVIOR OF DYE MOLECULES ORIENTED BY CHIRAL NEMATIC (ACETYL)(ETHYL)CELLULOSE SOLUTIONS. Jian-Xin Guo and Derek G. Gray, Pulp and Paper Research Institute of Canada, and Department of Chemistry, McGill University, 3420 University St., Montreal, PQ, CANADA H3A 2A7

Liquid crystal induced circular dichroism (LCICD) has been used to investigate the reversal of handedness with increasing acetyl content of lyotropic chiral nematic (acetyl)(ethyl)cellulose (AEC) liquid crystalline phases. An achiral dye, acridine orange, dissolved in these AEC mesophases showed detectable LCICD signals, presumably due to its orientation in the chiral nematic helicoidal structure. The sign of the LCICD dye peak was found to be opposite to that of the apparent CD peaks resulting from the reflection of circularly polarized light by the chiral nematic structure. Thus the LCICD peaks were positive for right-handed and negative for left-handed mesophases. The LCICD intensity is found to be dependent on dye concentration, temperature, sample thickness, and texture. The intensity also depends on the chiral nematic pitch which is related to the acetyl content, concentration of AEC and the solvent.

332. USING VIBRATIONAL SPECTROSCOPIES TO PROBE THE DIFFUSION OF SMALL MOLECULES IN POLYMER FILMS. N. E. Schlotter, Bellcore, Red Bank, New Jersey 07701.

Both infrared and Raman techniques can be used to monitor the diffusion of small molecules in polymer films. Both spectroscopies allow one to follow the diffusion process in approximately real time, without perturbing the diffusion process, and all of the chemical constituents can be monitored simultaneously. The Raman measurement is done using a waveguide geometry in which the guide is formed from a glassy polymer and a liquid/solid interface is formed on the guide surface. This minimizes background signal from the bulk liquid since the guided laser beam is predominately confined in the polymer film. The infrared measurement is done using a single bounce ATR geometry in which a semi-crystalline polymer film is formed on the surface of a ZnSe prism. The evanescent field is used to probe the film and the film is made thick enough to keep penetration of the light minimal in the solvent overlayer. Results of several studies will be presented.

333. MORPHOLOGY OF PEEK AT HIGH TEMPERATURES AND IN PEI BLENDS. S. D. Hudson, D. D. Davis, and A. J. Lovinger, AT&T Bell Laboratories, Murray Hill, NJ 07974.

PEEK (poly(aryl-ether-ether-ketone)) and PEI poly(ether-imide) are high temperature mutually miscible polymers. The semi-crystalline morphology of PEEK and PEEK/PEI blends was observed in thin films via bright field TEM and electron diffraction. PEEK has a unique double melting behavior associated with morphological inhomogeneity: high melting temperature crystals first grow in the open melt with lower melting crystals infilling the more restricted regions between these. At high crystallization temperatures (320°C), a transition in growth morphology is observed. At this temperature, some infilling lamellae grow together with the primary lamellae in the form of bundles 100-200nm wide. The resulting spherulite is more compact than observed at lower temperatures. In miscible blends with PEI, the identical morphology is observed at high temperatures, indicating that the amorphous PEI is rejected outside the spherulite. At lower temperatures, PEI segregates primarily to pockets between bundles of PEEK lamellae. The characteristic length scale for PEI phase separation decreases with decreasing temperature. At temperatures near the glass transition of the blend, a dendritic structure of small lamellae generally lying parallel to the plane of the thin film are observed in a single crystal texture. In spite of the various morphological transitions observed, electron diffraction shows no polymorphism of crystal structure.

334. FT-IR SPECTROSCOPIC STUDY OF THERMAL DEGRADATION IN PEEK (POLYETHERETHERKETONE) FILMS.

K. C. Cole, National Research Council Canada, Industrial Materials Institute, 75 boulevard de Mortagne, Boucherville, Québec, Canada J4B 6Y4, and I. G. Casella, Università degli Studi della Basilicata, Via N. Sauro 85, 85100 Potenza, Italy.

Polyetheretherketone (PEEK) is being increasingly used as a thermoplastic matrix in high-performance composites, but the high temperatures required to process this material ($\geq$ 380°C) can lead to thermal degradation, the nature of which is not well understood. In this work, Fourier transform infrared (FT-IR) spectroscopy has been used to study both qualitatively and quantitatively the changes which occur when PEEK film is exposed to temperatures in the range 400-500°C in air and nitrogen atmospheres. Spectra were measured in both transmission and attenuated total reflection. Two different degradation reactions have been detected. In a nitrogen atmosphere, only one is observed; it gives rise to a new peak at 1711 cm^{-1}, possibly corresponding to aldehyde groups. In an air atmosphere, a similar reaction is observed but it occurs about eight times faster. In addition there is a second reaction which produces another new peak at 1740 cm^{-1}, possibly corresponding to acid or ester groups. The rates of growth of these peaks were measured at different temperatures and used to determine activation energies for the different processes and conditions.

335. SYNTHESIS OF NANOMATERIALS VIA PRECERAMIC POLYMERS. Kenneth E. Gonsalves*, Peter R. Strutt+, Tongsan D. Xiao+, Paul G. Klemens#, *Department of Chemistry and Polymer Science Program, Institute of Materials Science; +Department of Metallurgy, Institute of Materials Science; #Department of Physics, University of Connecticut, Storrs, CT 06269 USA

A study has been made of basic mechanisms involved in the rapid synthesis of preceramic and ceramic nanoparticle powders. In this process an aerosol, formed from an ultrasonically atomized liquid organosilazane monomer, $[CH_3SiHNH]_n$, is injected into the beam of an industrial cw CO_2 laser. One critical feature examined was the rapid condensation of molecular species from the laser plume, in a process involving crosslinking/polycondensation reactions. In accompanying studies, a model has been formulated to determine the laser plume temperature, the cooling rate of condensing species and the particle diameter. These are obtained by analytical solution of heat conduction, momentum and mass conservation equations that consider heat loss by gas conduction, radiation, evaporation and convection.

336. Si-C-N-O CERAMIC FIBER COMPOSITIONS FROM POLYCARBOSILANE. J. Lipowitz and J. A. Rabe, Advanced Ceramics Research Department, Dow Corning Corporation, Midland, Michigan 48686-0995

Ceramic fibers are prepared by melt spinning, crosslinking, and pyrolysis of polycarbosilane with a nominal composition (MeHSiCH$_2$)$_x$. By variations in crosslinking (cure) and/or pyrolysis conditions, a broad range of ceramic fiber compositions, covering all of the Si, C, N, O compositional space (e.g. SiC, Si$_3$N$_4$, Si$_2$N$_2$O, SiO$_2$, and mixtures, as well as excess C) can be prepared. Pyrolysis in an inert atmosphere using non-oxidative cure methods provides Si, C compositions. Pyrolysis in an inert atmosphere after an oxidative cure (air cure) provides commercial NICALON™ Si-C-O fiber. Pyrolysis in an ammonia atmosphere after non-oxidative cure leads to Si_3N_4 fibers. Ammonia pyrolysis after air-cure leads to Si-N-O compositions. Mixtures of ammonia with argon lead to Si-C-N or Si-C-N-O compositions with partial carbon removal. The mechanism of carbon removal by ammonia will be discussed. This process occurs rapidly at 600 to 800°C.

337. SOME NOVEL REACTIONS FOR THE FORMATION OF Si-N BONDS. Hua Qin Liu and <u>John F. Harrod</u>, Department of Chemistry, McGill University, Montreal, Quebec, Canada H3A 2K6.

Amines in general react very sluggishly with silanes in the absence of catalysts. Hydrazine is an exception to this rule and we have found that some primary silanes react under mild conditions with hydrazine to yield oligomers which remain soluble, even after complete reaction of Si-H groups. The reaction of phenylsilane will be described in detail and evidence presented to support the conclusion that the predominant unit in the polymer produced in the uncatalyzed reaction is:

$$+(Ph)Si\underset{NH-NH}{\overset{NH-NH}{\diagup\diagdown}}Si(Ph)-NH-NH+_n$$

Evidence from GPC suggests that the average value of n is ca. 2.
Some results for catalyzed reactions, which give polymers with higher average molecular weights, will also be presented.

338. PRE-TREATMENT AND PYROLYSIS OF POLYSILAZANE PRE-CERAMIC BINDERS. <u>David L. Mohr</u>, Prashant Desai, Polymer Education and Research Center, Georgia Institute of Technology, and Thomas L. Starr, Materials Science and Technology Laboratory, Georgia Tech Research Institute, Atlanta, Georgia 30332.

Polysilazanes are of interest as polymeric precursors to ceramic materials. Pre-ceramic precursors used as binders in ceramic composites could improve processibility and reduce porosity in these systems, as well as change the physical morphology. Pre-treatment of these materials prior to pyrolysis can change the ceramic yield and the chemical and morphological structure of the derived product. The changes in polysilazane chemical structures that occurred during pre-treatment and pyrolysis were observed via diffuse reflectance FTIR and other techniques. Characteristics of the pyrolysis derived products were studied using x-ray diffraction, TEM, and other techniques. Ceramic samples incorporating these polysilazanes as binders exhibited significant increases in processed density compared to control samples.

339. ISOCYANATE-MODIFIED POLYSILAZANE CERAMIC PRECURSORS. <u>Joanne M. Schwark</u>, Hercules Incorporated, Hercules Advanced Materials and Systems Company, Research Center, Wilmington, DE 19894-0001

Liquid polysilazanes with a broad range of tailored viscosities have been prepared by the reaction of cyclic silazanes with mono- and multi-functional isocyanates. These isocyanate-modified polysilazanes may be thermoset by heating with dicumyl peroxide. Such thermoset polymers have high char yields and form mixtures of SiC and Si_3N_4 upon pyrolysis under nitrogen. The isocyanate-modified polysilazanes may be used in a variety of applications (e.g., binders, coatings, infiltrants) because their viscosity may be tailored to the desired end use.

340. CHARACTERIZATION OF HYDRIDOMETHYLPOLYSILAZANE - A PRECURSOR TO SiC/SiN CERAMICS. Regina M. Stewart, Neal R. Dando[1], Dietmar Seyferth and Anthony J. Perrotta[1], Department of Chemistry, Massachusetts Institute of Technology, Cambridge, MA 02139; [1]Alcoa Laboratories, Alcoa Technical Center, Alcoa Center, PA 15069

Preceramic polymer routes for the syntheses of nonoxide ceramics is an area of considerable research emphasis in modern materials science. These materials hold tremendous potential for use in fiber drawing, for the manufacture of complex shapes and as binders for ceramic powder processing. Concurrent with the development of divergent synthetic routes is the need for improved structural characterization techniques to investigate the chemical reaction mechanisms and phase development in these systems.

Nuclear magnetic resonance (NMR) has emerged as a powerful, non-destructive tool for characterizing oxide and nonoxide ceramics. The technique is particularly suited for following structure evolution through sol/gel/solid phase transitions.

The present investigation explores the use of 13C and 29Si NMR for characterizing the evolution of molecular structure in hydridomethylpolysilazane, a polymer precursor to SiC/SiN ceramics.

341. A NEW STRUCTURAL MODEL FOR ALUMOXANE MACROMOLECULES. Allen W. Apblett and Andrew R. Barron*. Department of Chemistry and Materials Research Laboratory, Harvard University, Cambridge, MA 02138.

Alumoxanes have been traditionally defined as oligomeric or polymeric materials consisting of an Al-O backbone with pendant organic substituents. We have determined that siloxy-substituted alumoxanes have the general formula $\{Al(O)_x(OH)_y(OSiR_3)_3\}_n$. Based on spectroscopic evidence, and confirmed by X-ray crystallography, we have proposed that these materials have a core structure analogous to that found in the minerals boehmite and diaspore, in which the aluminum centers are six-coordinate. Furthermore, we have proposed that this aluminum-oxygen core is encapsulated by terminal siloxy groups bound to four-coordinate aluminum atoms within fused six-membered rings. IR, NMR, XPS and X-ray crystallographic evidence will be presented and the relationship of the alumoxanes to both minerals and biological systems will be discussed.

342. POLYMERIC ROUTES TO ALUMINIUM OXIDES William S. Rees, Jr. and Werner Hesse, Department of Chemistry and Materials Research and Technology Center, Florida State University, Tallahassee, FL 32306-3006.

The preparation of a new class of organometallic polymers of the form $-[-Al(OR)-O(CH_2)_nO-]_{\overline{x}}$ is described. They pyrolytically convert to various phases of Al_2O_3 at mild (<500°C) temperatures, the morphology being controlled by method of polymer preparation.

343. METAL-POLYMER INTERACTION THROUGH METAL COORDINATION. R. D. Archer, Wenyan Tong, and Bing Wang* Department of Chemistry, University of Massachusetts, Amherst, MA 01003.

The formation of direct chemical bonds between dissimilar materials should provide maximum attraction between these materials. Whereas most adhesives depend on multiple weak attractive forces, direct bonding is thought to occur in many silicon coupling agents and in chromium, titanium, and zirconium coordination linkages between surface oxygen atoms on oxidized metal surfaces and the coordinated metal ions, which are in turn coordinated to groups compatible with organic polymers. Stress will be placed on our recent zirconium coupling systems, which provide inertness through the use of tetradentate chelating ligands. Compatibility with both polyolefins and polyesters have been obtained, with adhesion improvement by as much as a factor of 4.

*Current address: 3M Research Center, St. Paul, MN.

344. TOWARDS A MOLECULAR-SIZE "TINKERTOY" CONSTRUCTION SET: POLYMERIC STRUCTURES DERIVED FROM [1.1.1]PROPELLANE. Josef Michl, Richard W. Morrison, Gudipati S. Murthy, and Mohamed Ibrahim, Department of Chemistry and Biochemistry, University of Colorado, Boulder, CO 80309-0215.

[1.1.1] Propellane has been used as a building block for the production of oligomeric and polymeric linear rods, [n]staffanes, suitable for a controlled assembly of secondary regular supramolecular structures with repetitive structural motifs in two or three dimensions. An example is a regular square grid containing dirhodium tetracations as the nodes and terminal [n]staffanedicarboxylates as the rods.

Acknowledgement: This work was supported by the National Science Foundation, the Welch Foundation, and the Texas Advanced Research Program. Most of it was performed while the authors were at the University of Texas.

345. POLYROTAXANES: SYNTHESIS, CHARACTERIZATION & POTENTIAL USES. H. W. Gibson, C. Wu, Y. X. Shen, M. Bheda, J. Sze, P. Engen, A. Prasad, H. Marand, D. Loveday and G. Wilkes, Depts. of Chemistry & Chemical Engineering, & NSF STC for High Performance Polymeric Adhesives & Composites, Virginia Polytechnic Institute and State University, Blacksburg, Virginia, 24061

Polyrotaxanes are comprised of macrocycles that have been threaded by linear macromoleules, i.e., molecular analogs of strings of pearls. In this presentation synthetic methods for these unique polymers will be described along with the properties that this novel architecture conveys. Both step and chain growth systems will be included, e.g., polyesters, polyamides, polyurethanes, polyethers, polystyrene, polyacrylonitrile. Examples of enhanced solubility of linear polymers will be presented. The motion of the macrocycles along the linear backbone allows phase separation in some cases and some interesting examples will be outlined. A number of potential applications of these systems will be suggested, based on their unique structures and consequent behavior.

346. ON THE MORPHOLOGY AND CRYSTALLIZATION BEHAVIOR OF A POLYESTER ROTAXANE. H. Marand, A. Prasad, C. Wu, M. Bheda and H.W. Gibson, Chemistry Department and NSF Science & Technology Center for High Performance Polymeric Adhesives and Composites, Virginia Polytechnic Institute & State University, Blacksburg, VA 24061.

The thermal behavior and morphology of a polyester rotaxane obtained through statistical threading of 60-crown-20 macrocycles by linear poly(butylene sebacate) chains (50% wt/wt) were investigated by differential scanning calorimetry, wide angle X-ray diffraction and polarized optical microscopy. The behavior of the polyrotaxane was compared to that of the poly(butylene sebacate) (PBS) and crown ether (60-C-20) pure components and of their corresponding physical mixture (50/50) wt/wt. The physical mixture displays a macroscopically phase separated morphology and exhibits a crystallization and melting behavior identical to the pure components. On the other hand, both the crown ether and the poly(butylene sebacate) components of the rotaxane are characterized by slower crystallization rates and large melting point depressions when compared to the pure components. Wide angle X-ray diffraction studies indicate that the crystal structure of both the crown ether and the polyester are not affected by the formation of the rotaxane, but that the degree of crystallinity in either component is markedly decreased. Finally, dynamic crystallization studies indicate that the rate of crystallization of the polyrotaxane depends strongly on previous thermal history and especially on the residence time in the melt state. The latter observation is explained in terms of constrained brownian motions, along the linear polyester chain, of the macrocycles that were spatially segregated during the chain folded crystallization of the polyester component. A qualitative model for the crystallization of such a polyrotaxane will be presented.

347. **Stereoselection and Helicity in Living Polymerizations of Chiral Isocyanides.** Timothy J. Deming and Bruce M. Novak*, Department of Chemistry, University of California, Berkeley, California 94720.

Our efforts in the area of isocyanide polymerizations have focused on the aspects of understanding the mechanism of polymerization and applying this to the controlled synthesis of well-defined polymers. Recently, we have developed a highly active homogeneous catalyst system which promotes a true "living" polymerization of a variety of isocyanide monomers. This system, based on [(η^3-C_3H_5)Ni(OC(O)CF_3)]$_2$ (**I**) catalyst, is very robust and able to produce polyisocyanides of predetermined molecular weight and narrow molecular weight distribution without need of either inert atmosphere or high vacuum line techniques. Herein we plan to discuss the polymerizations of chiral isocyanide monomers and their relation to stereoselectivity in the polymerization and stability of helical conformations in solution.

348. **CHIRAL CLUES TO HELICAL SEQUENCE LENGTHS IN POLY(n-ALKYL ISOCYANATES)**
Christopher Andreola, Mark M. Green, Norman C. Peterson, Department of Chemistry and Polymer Research Institute, Polytechnic University, 333 Jay Street, Brooklyn, NY 11201; Shneior Lifson, Department of Chemical Physics, Weizman Institute of Science, Rehovat 76100, Israel.

Much evidence exists for a stable helical conformation for poly(n-alkyl isocyanates) in solution, the result of a high potential energy barrier for rotation away from the planar amide state, and the severe steric interference between the side chain and the carbonyl group present in that conformation. When the degree of polymerization increases beyond n=1000, the chains become more flexible, but the cause of the flexibility is controversial. We have found that the backbone conformation is highly sensitive to the stereospecific placement of deuterium α or β to the backbone nitrogen, inducing high optical activity and intense circular dichroic bands. We find that the temperature dependence of the optical rotations is consistent with a statistical thermodynamic model in which the average length between helix reversals increases as the temperature decreases.

349. NEW HYDROLYZABLE DIANIONIC INITIATOR FROM DIVINYL TETRAOXASPIROUNDECANE DERIVATIVE
<u>Barton J. Schober</u> and Bernard Gordon III, Polymer Science Program, Material Science & Engineering Department, The Pennsylvania State University, University Park, PA 16802

The use of multicharged initiators which contain reactive functionalities has been proposed to offer insight into ring and star polymers, as well as provide new routes to degradable and catenated polymers. In the pursuit of such an initiator a stable dianion, **7**, containing a tetraoxaspiroundecane moiety was prepared. Starting from 3,9-dimethyl-3,9-bis(3'-(1,1-diphenyletheny1))-2,4,8,10-tetraoxaspiro[5.5]undecane (**6**), the dianion was prepared by addition of two equivalents of n-butyllithium. The protonated dianion is shown, by NMR, to be hydrolyzable. Polymerization, work-up, and GPC analysis of polystryrene, initiated by **7**, revealed narrowly disperse polymer of half the molecular weight predicted by theory. It is argued that initiation proceeded equally from both charge centers on the initiator and that cleavage of the polymer occurred during work-up or analysis.

350. AB INITIO QUANTUM CHEMICAL POTENTIAL ENERGY FUNCTIONS FOR THE ATOMISTIC SIMULATION OF POLYMERIC MATERIALS. <u>R. L. Jaffe</u>, NASA Ames Research Center, Moffett Field, CA 94035.

Ab initio quantum chemical methods are used for the determination of potential energy functions for atomistic simulations of polymeric materials. Both the bonded and non-bonded terms in the polymer potential energy function can be provided from calculations using standard quantum chemical methods. In this presentation a procedure for determining potential energy functions is described using several aromatic polyimides as examples. The conformational and vibrational properties of these polymers are presented along with an estimate of selected mechanical properties. The advantages and disadvantages of several mathematical representations of the potential energy function will also be discussed.

351. **STRUCTURAL ANALYSIS OF POLYVINYLCHLORIDE USING AN AB INITIO DERIVED CLASSICAL FORCE FIELD.** P. J. Ludovice, Polygen Corporation, Waltham, MA 02254, R.L. Jaffe, NASA-Ames Research Center, Moffett Field, CA 94035, D.Y. Yoon, IBM-Almaden Research Center, San Jose, CA 95120.

Ab initio quantum chemical calculations have been used to describe the intra- and intermolecular portions of the potential energy surface for atactic polyvinylchloride (PVC). Intermolecular energies were calculated for complexes of 2-chloropropane and 1,3 dichloropropane, and intramolecular energies for 2,4-dichloropentane isomers. A classical potential energy function was then fitted to reproduce these ab initio energy calculations. This potential energy function is used in molecular mechanics calculations of atactic PVC in order elucidate the atomic level nature of PVC structure. Comparison of the simulated structure to that measured by wide angle x-ray scattering experiments is made. This new force field differs significantly from previous force-field descriptions for PVC, and its comparison to molecular mechanics results of a previously used PVC force field is also made.

352. **DETERMINATION OF THE CHEMICAL POTENTIALS OF POLYMERIC SYSTEMS FROM MONTE CARLO SIMULATIONS.** Sanat K. Kumar, Department of Materials Science and Engineering, Polymer Science Program, Pennsylvania State University, University Park, PA 16802; Igal Szleifer, Department of Chemistry, Cornell University, Ithaca, NY 14853; and Athanassios Z. Panagiotopoulos, School of Chemical Engineering, Cornell University, Ithaca, NY 14853.

We propose a new computer simulation technique, based on the Widom test particle method, to calculate the chemical potentials of components in a polymeric system. The technique is based on the insertion of test segments on to a polymer and is applicable for any chain length at gas and liquid-like densities. We perform sample calculations on homopolymers and show that the proposed technique allows for the enumeration of their thermodynamic behavior in the subcritical and supercritical temperature ranges.

353.

THEORETICAL STUDIES OF GAUCHE AND ECLIPSING INTERACTIONS FOR MODEL COMPOUNDS OF POLYMERS. P.N. Krishnan, L.A. Burke, G.R. Famini, J.O. Jensen, Department of Natural Sciences, Coppin State College, Baltimore, MD, 21216, Department of Chemistry, Rutgers University, Camden, NJ 08102, and the U.S. Army Chemical Research, Development and Engineering Center, Aberdeen Proving Ground, MD 21010.

Quantum chemical calculations on polymeric compounds are difficult to carry out because of the generally large amount of time required. One method of circumventing this is to examine small molecules in depth that contain many of the substituents and interaction types. To this end, conformational studies on a series of monomer units have been studied using several different molecular orbital methods. A series of methyl or chloro substituted propanes and butanes were studied in light of obtaining barrier heights to rotation and thermodynamic equilibria. The theoretical methods employed included semi-empirical (MNDO and AM1) and ab initio (3-21G* and 6-31G*). The barrier heights and the relative energies of the ani and gauche conformations will be discussed in terms of their contributions to polymer conformations.

354. Intra- and Inter-molecular Correlations involved in the Orthorhombic Pseudohexagonal "Rotator" Phase Transition in n-Heneicosane.

David C. Doherty and A.J, Hopfinger

Department of Chemistry
University of Illinois at Chicago
The University of Illinois at Chicago
P.O. Box 6998, m/c 111
Chicago, IL 60680

Current Address: Minnesota Supercomputer Ctr., Inc.
1200 Washington Ave. South
Minneapolis, MN 55415

A detailed molecular dynamics study of the B0 - Oh solid - solid phase transition in the n-heneicosane (C21H44) bilayer system has been performed. The transition of the 96 chain system is modeled as a series of five constant volume "microstates", each sampled for 50 ps, at a constant (experimental phase transition) temperature of 306K. Rotational jumps of the chains along their long axes are observed as the unit cell is expanded. Conformational properties in the Oh state correspond well with experiment.

Careful examination of the MD trajectories reveals details about the correlation of intra- and inter-molecular motions involved in the transition. Correlations which are manifest in the B0 state become amplified in the pseudohexagonal "rotator" phase and determine the characteristics of the motions of the chain rotations. Strong, coupled torsional wave-like motions appear to contribute to the overall chain rotation. The chain rotational motions thus involve an overall twisting motion, rather than a rigid-rod rotation.

355. STOCHASTIC DYNAMICS SIMULATIONS OF POLYMETHYLENE CHAINS IN THE BULK AND IN THIN FILMS. R.G. Winkler, D.Y. Yoon, P.J. Ludovice, T. Matsuda, and H. Morawitz, IBM Almaden Research Center, San Jose, CA 95120-6099.

Stochastic dynamics (SD) simulations of polymethylene (up to $C_{60}H_{122}$) melts have been performed in order to calculate primarily their static properties in the bulk and in thin films. The system consists of linear chains of united CH_2 mass points connected by rigid bonds, subjected to valence bond, torsional forces, and non-bonded interactions represented by a truncated Lennard-Jones potential. The mean square end-to-end distance and the distribution of conformations in the bulk were calculated and compared with results obtained by Monte Carlo simulations and by unperturbed rotational isomeric states model calculations. The comparison exhibits very good agreement between the results of the different approaches, and demonstrates the potential advantages of SD simulations for polymeric systems. For polymer melts confined by two walls, both the static and dynamic properties of the chains are little affected by the surface topography when the film thickness is relatively large (ca. 35 Å). Upon reducing the film thickness to 10 Å, chains confined by structureless surfaces undergo only minor changes. On the contrary, chains confined by corrugated surfaces exhibit nearly all-trans conformations, strong intermolecular orientational correlations, and a significant decrease in the apparent diffusivity from their bulk value.

356. DIRECT COMPUTATION OF STATISTICAL WEIGHTS FROM CONFORMATIONAL ANALYSIS FOR RIS MODEL: POLYDIMETHYLSILOXANE. S. Grigoras, Dow Corning Co., Central R&D, Midland, MI 48686

This paper reports the utilization of conformational analysis to obtain reliable statistical weights for the isomeric states. These statistical weights are used directly in the Rotational Isomeric States model to estimate the configurational properties of polymer chains. Polydimethylsiloxane is used as an example. By taking into consideration the flexibility of the Si-O-Si bond angle and the asymmetry of the deformational energy curve for this angle, the positive temperature coefficient is explained.

357. SILICON-ACETYLENE AND SILICON-OLEFIN POLYMERS AS PRECURSORS TO SiC. Sina Ijadi-Maghsoodi, Xianping Zhang, Yi Pang, Mitch Meyer, Mufit Akinc, and Thomas J. Barton. Ames Laboratory (Department of Energy), Iowa State University, Ames, Iowa 50011.

A series of silicon acetylene polymers with different pendant R groups were prepared. Each polymer was pyrolyzed to 1100°C in an inert atomsphere to give amorphous SiC/C ceramic chars. The major volatile products in the pyrolysis effluent were analyzed by GC-MS. Mechanistic aspects correlating the R group on the polymer structure, decomposition, and the ceramic char composition were studied. Increasing the mass and the size of the pendant R groups decreases the oxidative stability of the ceramic chars.

Catalytic polymerization of ethynyldiorganosilanes cleanly affords soluble poly(silylene)vinylenes which can be converted directly to SiC upon pyrolysis.

358. SOLID STATE MAS-NMR STUDY OF PRE-CERAMICS PRECURSORS AND PYROLYZED PRODUCTS. Florence Babonneau[a], Jacques Livage[a] and Richard M. Laine[b]. [a]Chimie de la Matière Condensée, Université Pierre et Marie Curie, 75005 Paris, France; [b]Dept of Materials Science and Engineering, University of Michigan, Ann Arbor, MI, 48109.

Many studies are currently devoted to the elaboration of non-oxide ceramics, carbides and nitrides, via polymeric route. With such approach, it should be possible to tailor a given material by an appropriate design of the polymeric precursor. To achieve this goal, it will be necessary to establish a fundamental understanding of the relations between the polymer molecular architecture and the selected ceramic products. Magic Angle Spinning Nuclear Magnetic Resonance (MAS-NMR) can provide very interesting information on the chemical changes that occur during the pyrolytic conversion of the polymeric precursors. Examples will be given in the Si-C and Si-C-N systems to show how variations in chemical composition can affect the selectivity to specific ceramics.

359.   CHEMICAL MODIFICATION OF PRECERAMIC POLYMERS: SOME EXAMPLES. Dietmar Seyferth, Heinrich Lang, Henry J. Tracy, Christine Sobon and Jutta Borm, Department of Chemistry, Massachusetts Institute of Technology, Cambridge, Massachusetts 02139.

Among the chemical modifications reported are the following: 1) Metalation of CH_2 groups in $[Me_2SiCH_2]_n$ and the Nicalon polycarbosilane (PCS), and reaction of the metalated polymers with $Me_2ViSiCl$ to introduce Me_2ViSi side chains which provided useful functionality for crosslinking. 2) Photochemical reaction of the liquid $[(MeSiH)_x-(MeSi)_y]_n$ polysilane with $(\eta^5-C_5H_5)_2MMe_2$ (M = Ti, Zr, Hf) to give crosslinked hybrid polymers containing these metals whose pyrolysis in argon gave SiC/MC composites. 3) Crosslinking of the Nicalon PCS by photochemical reaction with small amounts of polynuclear metal carbonyls to give SiC precursors whose pyrolysis in argon gave very high ceramic residue yields.

360.
HIGH MOLECULAR WEIGHT POLYCARBOSILANE AS A PRECURSOR TO OXYGEN-FREE SiC FIBERS. William Toreki, Christopher D. Batich, Guang J. Choi, University of Florida Department of Materials Science and Engineering Gainesville, FL 32611

There has been a lot of interest recently in the use of "Nicalon" and other types of SiC fiber as reinforcement in high temperature composites. Nicalon fiber is manufactured from a polycarbosilane precursor, and the finished fiber contains on the order of 10% oxygen by weight. It has been shown that this oxygen is partially responsible for the loss of mechanical properties observed when these fibers are heated above 1200°C. We have been able to utilize polycarbosilane of a higher molecular weight in order to produce SiC fibers with tensile strengths similar to Nicalon, but which contain little or no oxygen. Preliminary tests have shown these fibers to have better retention of strength at high temperatures. Details of the methodology used to prepare these fibers will be given, along with information concerning characterization by SEC, FTIR, XRD, SEM, TEM, Auger and elemental analysis. A discussion of the mechanical properties and prospects for use in composites will also be presented.

361.   THE PREPARATION, CHARACTERIZATION AND USE OF SILOXANES AS PRECURSORS
TO SILICON CARBIDE
Richard B. Taylor and Gregg A. Zank[*]
Advanced Ceramics Program, Dow Corning Corporation
Midland, Michigan 48640

The development of siloxane polymers that upon pyrolysis to 1800°C under inert atmospheres afford silicon carbide and carbon has been demonstrated. In this paper we will focus on the ability to control the levels of carbon in these ceramics and provide some insights into the pyrolysis chemistries at temperatures of 1200°C and above. We have found that within a "family" of siloxane resins facile control of the composition of the ceramic is accomplished by controlling the number of aryl groups in the polymer. Despite the fact that these polymers afford silicon carbide and excess carbon at 1800°C ^{29}Si MAS NMR has shown that the predominant silicon environment is that of $SiO_{4/2}$ at 1200 and 1400°C, similar to what has been documented for ethylsilicate, hydrogensilsesquioxane and methylsilsesquioxane sol-gels. The difference lies in the corresponding carbon phase produced in these materials and its ability to carbothermically reduce the SiO_2 and afford silicon carbide and carbon at temperatures above 1400°C.

362. **PREPARATION OF A LINEAR POLYCARBOSILANE VIA RING-OPENING POLYMERIZATION.** Hui-Jung Wu and Leonard V. Interrante, Department of Chemistry, Rensselaer Polytechnic Institute, Troy, NY 12180

A high molecular weight linear polycarbosilane, poly(dichlorosilaethylene), $(-SiCl_2CH_2-)_n$, has been prepared by ring-opening polymerization of 1,1,3,3-tetrachloro-1,3-disilacyclobutane. Reduction of this polymer with $LiAlH_4$ yields the corresponding poly(silaethylene), $(-SiH_2CH_2-)_n$. The structure of these polymers and the monomeric precursor have been investigated by IR, ^1H, ^{13}C, and ^{29}Si NMR spectroscopy and GPC. The result of these studies are consistent with expectations for high molecular weight linear polymers. The pyrolysis of the high molecular weight poly(silaethylene) was studied by TGA and was found to give a 76% ceramic yield, suggesting that this polymer has potential for use as a precursor to SiC ceramics.

363. **HIGH NITROGEN LOW HYDROGEN AMIDES & IMIDES FROM CYANOIMIDAZOLES.** Yang-Kook Kim, E. L. Thurber, Paul G. Rasmussen, Dept. of Chemistry and Macro. Science & Eng. Ctr., University of Michigan, Ann Arbor MI 48109.

Cyanoimidazoles are easily prepared from diaminomaleonitrile and appropriate electrophiles. We have developed the functionality and reactions necessary to prepare amides and imides based on these starting materials. The resulting polymers have unusually high nitrogen content and typically very low hydrogen content. Such organic/inorganic hybrid materials have high thermal stability and low flammability. The polyimide is reactive due to the ring strain implicit in fused five membered rings and can be solvolyized to polyamic acid and ester polymers. The sythesis and characterization of these materials will be described.

364. **INORGANIC PHASE BEHAVIORS MEDIATED BY THE ORGANIC POLYMER MATRICES.** Jun Lin, Patricia A. Bianconi, Department of Chemistry, The Pennsylvania State University, University Park, PA 16802.

Synthetic analogs of natural biominerilization processes have been further pursued by the investigation of inorganic phase behaviors in various synthetic polymer matrices: amorphous polymers but having strong binding sites to the metal ions, such as poly(2-vinyl pyrindine) and poly(2-vinyl pyridine-co-styrene); semicrystalline polymer and also forming crystalline complex with metal ions, such as poly(ethylene oxide). The detailed syntheses, characterization and discussion about the particle nucleation and growth mechanism in the polymer matrices will be described.

365. **SIZES OF THE SHORT-RANGE CORRELATIONS IN THE DYNAMICS OF ROTATIONAL ISOMERISM.** Wayne L. Mattice, Institute of Polymer Science, The University of Akron, Akron, Ohio 44325-3909.

The dynamics of rotational isomerism has been examined for several common chains. Molecular dynamics trajectories have been computed for the isolated chains, and also for the chains in special environments that provide a very large local viscosity. A full atomic description (including discrete hydrogen atoms) is used for the chain and the environment. For the isolated chains, there are examples in which there is no appreciable short-range correlation of dynamics of the rotational isomeric state transitions, and also examples where there is some degree of correlation between bonds that are either nearest neighbors or next-nearest neighbors. The qualitative nature of the correlations can be influenced strongly by a highly viscous medium. A chain that shows no correlations when it is isolated may show very strong next-nearest neighbor correlations in a highly viscous medium.

366.   **The Preparation of Dense Polymer Samples for Atomistic Modelling**
by

J.I. McKechnie, D. Brown and J.H.R. Clarke
Chemistry Department, UMIST, Manchester M60 1QD, U.K.

There is a general recognition of the need to take detailed account of the intermolecular forces and also the restricted flexibility of chains when modelling dense or glassy polymers. In previous work we have described the use of molecular dynamics to study an oriented polymer fiber, and most recently, to explore the mechanical properties of model polyethylene over a wide range of temperatures. Our current work summerizes the results of a detailed comparison of methods for both initial amorphous growth and also relaxation procedures in fluid and glassy samples. As a result of the non equilibrium conformational distribution in glassy samples no growth method could be found for the direct generation of realistic chain configurations.

367.   MOLECULAR DYNAMICS STUDIES OF AMPHOLYTIC IONOMERS AND THERE PENDANT CHAIN DERIVATIVES, A. C. Watterson, D. N. Chin, J. C. Salamone, Dept. of Chemistry/Polymer Science, University of Lowell, Lowell MA 01854.

Ampholytic ionomers are a class of polymers that have a charge density of 10 % or less of both cationic and anionic moieties. The small addition of ampholytic species gives rise to aggregates of the ion pairs into localized cluster regions. Small amounts of these polymers can have dramatic effects on aqueous systems in terms of flow behavior. It is the primary goal of this research to investigate the nature of the cluster in hopes of tailoring flow properties of aqueous systems. Here, the ion pair is incorporated into a neutral monomer, acrylamide. The ion pairs consist of the anionic monomer, 2-acrylamido-2-methacryl-amidopropanesulfonate (AMPS) and the cationic monomer, 2-methacrylamidopropyltrimethylammonium (MPTMA) and its carboxylate derivative where a pendant hydrophobic chain of various lengths was added.

The investigations of these aqueous systems was undertaken using an atomistic molecular dynamics simulation approach using the CHARMm force field potential. Because simulations are quite computational intensive for bulk waters, a stochastic simulation along with a distance dependant dielectric was used to mimic the viscous and solvation behavior of water. Studies include temperature dependance of NMR relaxation parameters, ion pair interaction energies and response under external stress (shear).

368.   DYNAMICS OF COUPLING AGENTS IN INTERFACIAL REGIONS. David S. Bohlen, and **Frank D. Blum**, Department of Chemistry and Materials Research Center, University of Missouri-Rolla, Rolla MO 65401.

The dynamics of silane coupling agents adsorbed on silica have been probed with deuterium NMR spectroscopy. Molecular modelling has been used to aid in the interpretation of the NMR data on deuterated γ-aminopropylsilane (DAPS). The DAPS used in the experimental studies is

$$(CH_3CH_2O)_3SiCH_2CH_2CD_2NH_2$$

In the simulation, DAPS has been placed in monolayer coverage on a silica surface and the dynamics followed using QUANTA and CHARMm. The dynamics of the individual C-H and N-H bond vectors were followed as a function of time and the results compared with the deuterium NMR data and the results from the slow motion theory of Freed in terms of the rate and mechanism of molecular motion. It is found that the difference in carbon chain length does not significantly affect the dynamics of the labelled species.

369. CHARACTERIZATION OF MATERIAL SURFACES VIA WETTABILITY ANALYSIS:
A Molecular Dynamics Simulation Method.
Gianni P. Puglia, Andrew J. Hoffman and Rudolph Potenzone Jr.
Polygen Corporation
200 Fifth ave. Waltham MA 02254 U.S.A.

...............Understanding the surface structure of polymeric materials is essential to improving the surface properties and applications of existing materials. The current techniques employed involve spectroscopy via contact angle measurements, Scanning Tunneling Microscopy [STM], and chemical analysis. The current methods although widely used are either costly[STM] or crude[contact angle] and do not provide the understanding required to modify existing materials so as to eliminate their deficiencies. The ability to study wettability using molecular dynamics simulation methodology provides tangible results of the chemistry at the surface of the material under investigation. The two materials that we have studied are polyethylene[unbranched] and polytetrafluoroethylene and we have simulated their relative interaction with water at 300K. The results correlate well with experimental data from contact angle measurements. Further work in this area will concern interactions with different solvents and an estimation of surface tension for the polymeric material.

POLYMERIC MATERIALS: SCIENCE & ENGINEERING

Fourth Chemical Congress of North America

New York, NY
August 25 - 30, 1991

G. R. Pilcher, R. A. Weiss,
Program Chairmen

MONDAY MORNING
- **Symposium in Interpenetrating Polymer Networks**
Session I. IPN Synthesis and Structure
G. Meyer, D. Klempner, Presiding   Papers 1-7

MONDAY MORNING
- **Symposium on Size Exclusion Chromatography, Field-Flow Fractionation, and Related Chromatographic Methods for Polymer Analysis**
Session I. Field-Flow Fractionation (Th-FFF, Flow-FFF)
M. Schure, Presiding   Papers 8-13

- **General Papers and New Concepts in Polymeric Materials I. New Polymer Synthesis**
G. R. Pilcher, Presiding   Papers 14-20

MONDAY AFTERNOON
- **Symposium on Interpenetrating Polymer Networks**
Session II. Miscibility and Phase Separation in IPN's
S. C. Kim, L. H. Sperling, Presiding   Papers 21-27

- **Symposium on Size Exclusion Chromatography, Field-Flow Fractionation, and Related Chromatographic Methods for Polymer Analysis**
Session II. SdFFFF, SEC
K. Ratanathanawongs, Presiding   Papers 28-32

- **Roy W. Tess Award Symposium Honoring K. L. Hoy**
J. W. Prane, Presiding   Papers 33-38

TUESDAY MORNING
- **Symposium on Interpenetrating Polymer Networks**
Session III. Structure-Property Relationships in IPN's
J. M. Liegeois, L. A. Utracki, Presiding   Papers 39-45

- **Section on Size Exclusion Chromatography, Field-Flow Fractionation, and Related Chromatographic Methods for Polymer Analysis**
  Session III. High-Temperature SEC, SEC Copolymer Analysis
  T. Harvard, Presiding .... Papers 46-51

- **General Papers and New Concepts in Polymeric Materials II: Polymers for High Performance**
  G. R. Pilcher, Presiding .... Papers 52-57

TUESDAY AFTERNOON

- **Symposium on Interpenetrating Polymer Networks**
  Session IV. IPN Transport and Permeability
  D. Hourston, H. L. Frisch, Presiding .... Papers 58-64

- **Symposium on Size Exclusion Chromatography, Field-Flow Fractionation and Related Chromatographic Methods for Polymer Analysis**
  Session IV: SEC/Viscometry/Data Treatment and Applications
  A. Rudin, Presiding .... Papers 65-70

- **Sherwin-Williams Student Award Symposium**
  M. J. Bowden, Presiding .... Papers 71-76

TUESDAY EVENING

- **Poster Session/Social Hour**
  G. R. Pilcher, Presiding .... Papers 77-86

WEDNESDAY MORNING

- **Symposium on Interpenetrating Polymer Networks**
  Session V. IPN's with Functionalized Triglyceride Oils
  S. Dirlikov, D. Klempner, Presiding .... Papers 87-93

- **Symposium on Size Exclusion Chromatography, Field-Flow Fractionation and Related Chromatographic Methods for Polymer Analysis**
  Session V: SEC/Viscometry/Light Scattering Data Treatment and Applications
  H. G. Barth, Presiding .... Papers 94-98

- **Symposium on Solid-State Characterization of Multicomponent Systems I and II**
  J. Runt, A. Natansohn, Presiding .... Papers 99-103

WEDNESDAY AFTERNOON
● **Symposium on Flow-Induced Structural Changes in Polymers Session I: Theory and Polymer Solutions**
R. A. Weiss, Presiding   Papers 104-108

● **General Papers and New Concepts in Polymeric Materials III. Composites**
G. R. Pilcher, Presiding   Papers 109-115

WEDNESDAY EVENING
● **SCI-MIX**
D. A. Cocuzzi, Presiding   Papers 121-131

THURSDAY MORNING
● **Symposium on Flow-Induced Structural Changes in Polymers Session II: Polymer Blends**
A. I. Nakatani, Presiding   Papers 132-137

THURSDAY MORNING AND AFTERNOON
● **Symposium on Container Coatings Systems I and II**
J. T. K. Woo, P. S. Sheih, Presiding   Papers 138-144
                                                                                       157-162

● **Symposium on Solid-State Characterization of Multicomponent Systems III and IV**
R. A. Register, E. Ueda, Presiding   Papers 145-149
                                                                                       163-168

● **General Papers and New Concepts in Polymeric Materials IV: Characterization**
G. R. Pilcher, Presiding   Papers 150-156

FRIDAY MORNING
● **General Papers and New Concepts in Polymeric Materials V: Physical and Kinetic Studies**
G. R. Pilcher, Presiding   Papers 169-174

● **Symposium on Reactive Processing**
R. C. Kowalski, Presiding   Papers 175-179

● **General Papers and New Concepts in Polymeric Materials VI**
D. A. Cocuzzi, Presiding   Papers 180-184

POLYMERIC MATERIALS: SCIENCE ENGINEERING

1. AN ENTIRELY RADICAL IN SITU SYNTHESIS OF INTERPENETRATING POLYMER NETWORKS
S.N. Derrough, C. Rouf, J.M. Widmaier, G.C. Meyer   Institut Charles Sadron (EAHP-CRM)   4, rue Boussingault   67000 Strasbourg, France.

Classically, the in situ synthesis of IPNs requires non-interfering polymerization modes to achieve distinct networks, held together by only physical entanglements. A system that has received much attention is the one based on polyurethanes, formed by a polycondensation reaction, and (meth)acrylic polymers, formed by a free-radical polymerization. By replacing the elastomeric polyurethane by poly(butyl acrylate), the difference in mechanisms vanishes, and a single randomly crosslinked network is formed. An alternative method is to consider monomers with quite different reactivity ratios. Studies were made on combinations of butyl acrylate and diallyl carbonate of bisphenol A, both radically polymerizable, using rather selective initiators. An in situ sequential process was elaborated: azobisisobutyronitrile (AIBN) was used to polymerize first butyl acrylate at 50°C, than temperature was raised up to 80°C to decompose benzoyle peroxide, allowing the second monomer to polymerize. It has been shown that AIBN is ineffective against allylic C=C double bonds and that decomposition of benzoyle peroxide at 50°C only forms a few radicals which preferently react with the acrylic monomer.

2. SYNTHESIS AND CHARACTERIZATION OF POLYPHOSPHAZENE INTERPENETRATING POLYMER NETWORKS. Karyn B. Visscher, Ian Manners, and Harry R. Allcock, Department of Chemistry, The Pennsylvania State University, University Park, PA 16802.

Polyphosphazenes are a broad, novel class of inorganic/organic macromolecules with the general formula $(NPR_2)_n$. The physical properties of polyphosphazenes can be understood in terms of a highly flexible backbone, with various physical or chemical characteristics tailored by the incorporation of specific substituent groups. As part of our program to synthesize new materials with hybrid macromolecular properties, we report the synthesis of several well characterized IPNs containing the phosphazene polymers poly(bis(2-(2-methoxyethoxy)ethoxy)phosphazene) and poly(bis(propyloxybenzoate)phosphazene) and several organic polymers including polystyrene, poly(methyl methacrylate), polyacrylonitrile and poly(acrylic acid).

3. STUDIES OF GRAFTED INTERPENETRATING POLYMER NETWORKS. Barry J. Bauer, Robert M. Briber, and Brian Dickens, Polymers Division, Materials Science and Engineering Laboratory, National Institute of Standards and Technology, Gaithersburg, MD 20899

A new class of interpenetrating polymer network (IPN) has been studied in which grafting reactions between the two components are varied. Four polymethylmethacrylates (PMMAs) with alkacrylate, methacrylate, acrylate, and -methyl styrene end groups were dissolved in styrene/divinyl benzene and polymerized. Small angle x-ray scattering was used to characterize the extent of phase separation. The uniformity of the IPNs is strongly dependent on the grafting efficiency. Grafted and nongrafted IPNs were also made from the PMMAs and polyethylene glycol diacrylates. Thermal studies showed one transition in the grafted samples and two distinct transitions in a nongrafted sample.

4. BIFUNCTIONAL BIOMIMETIC POLYMERS THROUGH INTERPENTRATING POLYMER NETWORK SYNTHESIS. Spiro D. Alexandratos and Corinne E. Grady, University of Tennessee, Department of Chemistry, Knoxville, TN 37996-1600.

Crosslinked bifunctional polymers are important reagents in ionic and molecular recognition processes when systems are defined wherein the ligands operate synergistically. Additionally, natural polymers such as enzymes owe their unique mechanisms of action to the ability of different groups on the polymer to cooperate. The synthesis of bifunctional polymers will be described within a polystyrene support to thus produce an interpenetrating polymer network. The choice of monomers is based on critical sites identified in a phytoagglutinin. Most recently, a toluene solution of vinylimidazole (VIm), methyl methacrylate (MMA), divinylbenzene, and AIBN was contacted with lightly crosslinked polystyrene beads and polymerized. The IPN was then hydrolyzed to give vinylimidazole/methacrylic acid (MMA) ligands. The unhydrolyzed VIm/MMA polymer complexes 59.8% of the Cu(II) in a dilute pH 5 aqueous solution while the hydrolyzed VIm/MMA IPN complexes 57.1%. Under identical conditions, the former complexes 12.5% of Co(II) in a cobalt nitrate solution while the VIm/MMA IPN complexes 75.7%.

5. HYDROPHOBIC-HYDROPHILIC IPN AND SIPN. S.J. Huang, F.O. Eschbach, Polymer Science Program, University of Connecticut, Box U-136, Storrs, CT 06269-3136.

Poly(2-hydroxyethyl methacrylate) hydrogels (PHEMA) in the water swollen state are very soft with high water content. Therefore, they are similar to natural tissue and are widely used as biomaterials. Unfortunately, PHEMA hydrogels have low mechanical strength and tear resistance. By incorporating polycaprolactone (PCL), the mechanical properties of the swollen networks are improved and their biocompatibility is preserved We investigate three ways of combining PHEMA and PCL. SIPN's of PHEMA and PCL diol were synthesized and compared with the corresponding copolymer PHEMA with unsaturated PCL grafted on the PHEMA backbone. Total extraction of the PCL diol and identification of the extract by FTIR confirmed the physical mixture morphology of the SIPN. Analogically, no sensitive weight loss after extraction confirmed the graft copolymeric structure. IPN's of PHEMA and PCL were synthesized within a two step reaction process. Mechanical and thermal characterization will be discussed.

6. SIMULTANEOUS INTERPENETRATING NETWORK PRODUCTS FROM LIGNIN, ORGANOSTANNANES AND HYDROXYL-CAPPED POLY(ETHYLENE OXIDE) AND ABA BLOCK COPOLYMERS CONTAINING HYDROXYL-CAPPED POLY(ETHYLENE OXIDE) WITH POLYDIMETHYLSILOXANE, C. E. Carraher, Jr., Dorothy Sterling, Thomas Ridgway, William Reiff, Bhoomin Pandya, Departments of Chemistry, Florida Atlantic University, Boca Raton, FL., University of Cincinnati, Cincinnati, Ohio and Mossbauer Spectroscopy Consultants, Burlington, MA. 01803

Simultaneous interpenetrating network materials were produced from the interfacial condensation technique employing organostannane halides, lignin and hydroxyl-capped poly(ethylene oxide) and related ABA block copolymers containing dimethylsiloxane. Physical and structural characterization includes Mossbauer, Infrared and Mass spectroscopies, elemental analysis, solubility, bulk electrical properties and thermal analysis. The products without the flexibilizing PEG are rigid whereas the SIN products can be pressed into flexible sheets.

7. PHENOLIC IPNS FOR LAMINATE APPLICATIONS. K. Yamamoto, Shimodate Research Laboratory, Hitachi Chemical Co., Ltd., Shimodate, Ibaraki-ken 308, JAPAN

Interpenetrating Polymer Networks (IPNs) composed of phenolics and vinyl compounds have properties, such as heat resistance and flexibility, of both components. Phenolic IPN varnishes prepared by dissolving phenolic prepolymers in monomeric acrylates have the property of rapid curing and are expected to offer the potential to improve productivity in commercial applications.
The author tried using these IPN varnishes for paper-based phenolic laminates, materials important in the electrical industry. The curing time of the varnishes was measured by differential scanning calorimeter (DSC), and it was found that the curing rate depended on the ratio of the components and the curing catalysts. Various formulations of varnish were proposed and the laminates were manufactured on a

laboratory scale. The results indicate that these laminates have advantages in productivity, processability and electrical properties compared with conventional commercial products.

8. ANALYSIS OF POLYMERS BY THERMAL FIELD-FLOW FRACTIONATION. Marcus N. Myers, Peter Chen, and J. Calvin Giddings, Field-Flow Fractionation Research Center, Department of Chemistry, University of Utah, Salt Lake City, Utah 84112.

Thermal field-flow fractionation is an FFF method in which the driving force is generated by a temperature gradient. The method is highly effective in the separation and analysis of lipophilic synthetic polymers. The thermal FFF technique is more flexible than GPC and, because the FFF channel has no packing, there is less risk of shear degradation, clogging, and system degradation.
In this paper examples will be given of polymer molecular weight analysis by thermal FFF. Preliminary studies on polyolefins will be reported. Factors affecting retention and separation, such as the channel cold wall temperature, will be discussed.

9. THERMAL FIELD-FLOW FRACTIONATION OF COPOLYMERS. Martin E. Schimpf, Department of Chemistry, Boise State University, Boise, Idaho 83725.

Thermal field-flow fractionation (ThFFF), like size-exclusion chromatography (SEC), separates macromolecules according to size. Unlike SEC, however, ThFFF retention is also influenced by chemical composition. This gives ThFFF an added dimension not present in SEC. Thus, two polymer components having similar sizes but differing in chemical composition will coelute with SEC but are resolved by ThFFF. Because retention in ThFFF is governed by both size and chemical composition, ThFFF may prove to be a powerful technique for characterizing copolymers. Unfortunately, thermal diffusion, the phenomenon that governs separation by chemical differences, is poorly understood and ill-characterized. Recent work has shown that thermal diffusion is more complex in copolymers than homopolymers. This paper will focus on studies of thermal diffusion in copolymers. Results for the ThFFF separation of block and random copolymers will be presented, and the outlook of ThFFF as a tool for the characterization of copolymers will be discussed.

10. GEL CONTENT DETERMINATION OF POLYMER USING FIELD-FLOW FRACTIONATION. Seungho Lee, Analytical Research, 3M Corporate Research Laboratories, St. Paul, MN 55144

The object of this study is to investigate the applicability of thermal field-flow fractionation (ThFFF) for characterization of gel-containing polymers. ThFFF channel is open and a sample can be injected without filtering. Gel content of a sample can be determined by subtracting the peak area of filtered sample from that of unfiltered one. ThFFF was used to monitor the effect of electron beam (EB) treatment on high molecular weight poly methyl methacrylate (PMMA). No sign of cross-linking was observed. A trend of polymer degradation was clearly observed; as EB dose increases, a low molecular weight shoulder grows and average molecular weight decreases. ThFFF was also used to determine the difference between two acrylate elastomers which are manufactured by the same procedure but show different mechanical properties. A study using SEC, SEC-Viscometry, and SEC-Light scattering photometry found no significant differences in molecular weight and in degree of branching between two acrylate elastomers. ThFFF result shows significant difference in gel content between two samples, 17 to 7 %.

11. SEPARATION OF WATER SOLUBLE SYNTHETIC AND BIOLOGICAL MACROMOLECULES BY FLOW
FIELD-FLOW FRACTIONATION. J. Calvin Giddings, Maria Anna Benincasa, Min-Kuang
Liu, and Ping Li, Field-Flow Fractionation Research Center, Department of
Chemistry, University of Utah, Salt Lake City, Utah 84112.

Because of the universal applicability of crossflow as a driving force in field-flow
fractionation, flow FFF can be utilized for the separation of most classes of macromolecules and particles ranging from less than 1000 molecular weight up to 50 μm particle
diameter. This paper focuses on water soluble macromolecules. Among the various types
of macromolecular materials separated by flow FFF at the Field-Flow Fractionation
Research Center are synthetic water soluble polymers (including both anionic and cationic
polyelectrolytes) and biological macromolecules, including low and high molecular weight
proteins along with various protein complexes and aggregates. Since most of these
separations take place based on differences in diffusion coefficient or, equivalently,
in Stokes diameter, it is a simple matter to measure the Stokes diameter or diameter
distribution of the sample. In this paper we will discuss both the specific applications
of flow FFF in this area and the general principles and theory that underlie them.

12. PARTICLE SIZE ANALYSIS USING FLOW FIELD-FLOW FRACTIONATION. S. Kim
Ratanathanawongs and J. Calvin Giddings, Field-Flow Fractionation Research Center,
Department of Chemistry, University of Utah, Salt Lake City, Utah 84112.

Field-flow fractionation (FFF) is a class of elution separation methods characterized by
the use of an externally applied field acting perpendicular to the direction of flow
through a thin channel. Different types of "fields" (e.g., gravitational, flow,
electrical, etc.) have been used, thus giving rise to different subtechniques of FFF.
  In the case of flow FFF, the force that acts perpendicular to the axis of separation
is provided by a second (crossflow) stream of carrier. The interaction between the
crossflow and the sample particles is determined by the effective diameter of the particles alone, thus producing a separation based on particle size but not on density. The
independence from density makes flow FFF, unlike sedimentation FFF, ideal for separating
and characterizing the size distribution of samples whose components span a broad density
spectrum. Analysis times are typically three minutes for particles of diameter >1 μm
and 30 minutes for diameter <1 μm. Flow FFF's versatility and breadth of applicability
will be demonstrated with a presentation of results obtained for a spectrum of materials
ranging from seed latex to different types of silicas and pollen grains.

13. SURFACE CONFORMATION OF COLLOID STABILIZING BLOCK COPOLYMERS
Karin D. Caldwell and Jenq-Thun Li, Center for Biopolymers at Interfaces, 2260 Merrill
Engineering Building, University of Utah, Salt Lake City, Utah 84112

Colloids aggregation is prevented by the adsorption of soluble polymers which form steric
repulsion layers on the particle surface. The Pluronic series of surfactants are PEO-PPO-PEO triblocks
whose adsorption to hydrophobic polymers is thought to occur via the central PPO block, leaving the PEO
chains in a mobile state on the colloid surface. The surface concentrations and ad-layer thicknesses which
result when such triblocks adsorb to polystyrene latex spheres of different diameters show a significant
variation with particle size. Thus, the Pluronic F108 is shown to adsorb with lower surface density and to
form thinner ad-layers the smaller the particle. The stability of these adsorption complexes is compared to
that formed between the latex and a PEO homopolymer.
  This investigation is based on the tandem use of two field-flow fractionation techniques,
namely the mass-sensitive sedimentation FFF and the size-sensitive flow FFF. The accuracy in the mass-
based surface density assignments is verified by isotope and fluorescent probe labelings of the soluble
polymer, while size assignments of the fractionated particles are verified by dynamic light scattering.

14. SYNTHESIS AND STRUCTURAL CHARACTERIZATION OF TITANOCENE-
CONTAINING POLYMERS DERIVED FROM ALPHA-AMINO ACIDS,
DIPEPTIDES AND TRIPEPTIDES

C. Carraher and L. Tisinger, Department of Chemistry
Florida Atlantic University, Boca Raton, FL 33431 and
GeoSyntec, Boynton Beach, FL 33426

Synthesis of titanocene-containing poly(amine esters) derived
from alpha-amino acids, dipeptides and tripeptides was accomplished
employing both the classical interfacial and aqueous solution

techniques. In all cases, higher yields and chain lengths were achieved using the interfacial technique. For the aqueous solution systems only oligomeric chains are formed but for the interfacial systems high polymers are formed. Such products are precursors to protein-containing "magic bullets" for use in drug delivery systems.

15. SYNTHESIS AND BIOLOGICAL CHARACTERIZATION OF POLYPHOSPHATE AND POLYPHOSPHONATE ESTERS FROM BIOLOGICALLY ACTIVE DIOLS. C. E. Carraher, D. Powers, B. Pandya, Florida Atlantic University, Boca Raton, FL 33431 and Wright State University, Dayton, OH 45435, Departments of Chemistry.

A wide variety of polymeric polyphosphate and polyphosphonate esters were synthesized using the interfacial and solution techniques. The products have degrees of polymerization of about one hundred. Diols were chosen that exhibit biological activity. The polyesters showed better inhibition of a number of cell lines including HeLa and L929 cancer cell lines in comparison with the most utilized cancer drug cis-DDP. The polymers also exhibit inhibition of a wide variety of bacteria with activity dependent on the nature of the bacteria and polyester.

16. **Liquid Crystalline Oligoesters Containing Single-Ring Aromatic Units Separated by Aliphatic Spacers.** Ganghui Teng, Cong Du, Dao-Zhang Wang, Adel F. Dimian, Polymers and Coatings Department, North Dakota State University, Fargo, ND 58105; Frank N. Jones*, NSF Coatings Research Center, Eastern Michigan University, Ypsilanti, MI 48197.

A series of oligomeric esters derived from terephthalic acid and a linear aliphatic diol (nGT, OH functional) and from hydroquinone and a linear aliphatic diacid (nAHQ, COOH functional) have been prepared and characterized by DSC, visual observation, crossed polarizing microscopy, and X-ray diffraction. All data indicate that both nGT and nAHQ series form liquid crystals at certain temperatures above T_m. nGT with n = 8-10, 12 appeared to be smectic C, while a cybotactic nematic mesophase was suggested for nGT with n = 4-7. nAHQ (n = 4, 8, 10) showed two mesophases, with the lower-temperature mesophase being smectic C. The mesogens are presumably formed from the aromatic units together with the semirigidly connected -COO- groups and the conformational isomerism of neighboring -CH$_2$- groups. Longer flexible spacer, lower molecular weight, and regular alternation of the aromatic unit and the flexible spacer are possibly other factors contributing to liquid crystallinity.

17. MODEL FILLED POLYMERS: SYNTHESIS OF UNIFORMLY CROSSLINKED POLYSTYRENE MICROBEADS. R. Salovey, Z.-Y. Ding, D. Z. Kriz and J. J. Aklonis*, Department of Chemical Engineering, *Department of Chemistry, University of Southern California, Los Angeles, California 90089-1211.

Monodisperse sized crosslinked PS beads prepared by reaction of styrene and DVB, in batch emulsion copolymerization in the absence of emulsifiers, are not uniformly crosslinked, because DVB is more reactive than styrene. For copolymerization of 1 to 10 mole% DVB and styrene, individual crosslinked PS microbeads vary in crosslink density by a factor exceeding two (decreasing with conversion). A semicontinuous copolymerization, involving incremental additions of DVB, produces uniformly crosslinked PS beads. For both copolymerization techniques, T_g correlates well with crosslink density and PS beads are spherical and monodisperse in size.

18. RIGID-ROD BENZOBISTHIAZOLE POLYMER WITH REACTIVE 2,6-DIMETHYLPHENOXY PENDENT GROUPS: SYNTHESIS AND CHARACTERIZATION. My Dotrong[1], Minhhoa Dotrong[1], and Robert C. Evers[2]. 1. University of Dayton Research Institute, 300 College Park, Dayton, OH 45469. 2. Wright Laboratory, Wright-Patterson AFB, OH 45433-6533.

Poly[benzo(1,2-d:4,5-d')bisthiazole-2,6-diyl]-1,4-phenylene with reactive 2,6-dimethylphenoxy pendent groups (PPBT) was synthesized through the polycondensation of 2,5-diamino-1,4-benzenedithiol dihydrochloride with 2-(2,6-dimethylphenoxy)terephthaloyl chloride in 83% polyphosphoric acid. Intrinsic viscosities as high as 15.5 dl/g in methanesulfonic acid were recorded, the maximum value being achieved when the polymerization was carried out at 165°C for 24 hours and subsequently at 185°C for 10 hours. The chemical structure of PPBT was determined by spectroscopic and elemental analysis. The thermal behavior of PPBT was studied by thermogravimetric analysis in air, thermogravimetric-mass spectral analysis, and differential scanning calorimetry. Upon treatment at 400°C for one hour in a nitrogen atmosphere, the resultant polymer was no longer soluble in methanesulfonic acid, possibly due to crosslinking through the reactive pendent groups.

19. RIGID-ROD BENZOBISTHIAZOLE POLYMER WITH REACTIVE 2,6-DIMETHYLPHENOXY PENDENT GROUPS: PROCESSING, MECHANICAL PROPERTIES AND MORPHOLOGY
Unnikrishnan Santhosh[1], Minhhoa Dotrong[2], Hyun H. Song[2], Charles Y-C Lee[3]. 1. AdTech Systems Research Inc., 1342 N. Fairfield Rd., Dayton, OH 45432. 2. University of Dayton Research Institute, 300 College Park, Dayton, OH 45469. 3. Wright Laboratory, Wright-Patterson AFB, OH 45433-6533.

Rigid-rod fibers, like poly p-(phenylenebenzobisthiazole) (PBZT), have low compressive strength which restricts their use in high-performance aerospace applications. 2,6-dimethyl phenoxy-PBZT was spun and heat-treated with the objective of studying the effect of crosslinking of the polymer on its compressive strength. Though they have the same backbone structure as PBZT, their tensile properties were only about half that of PBZT; their compressive properties were not very different. Scanning Electron Microscopy of the fiber revealed a skin-core structure, with a fibrillar morphology. Wide-angle X-ray measurements indicated existence of three-dimensional order in the fiber.

20. THE ULTRASONIC POLYMERIZATION OF ACRYLIC MONOMERS, J. O. STOFFER, O. C. SITTON AND HWEI-LING KAO, UNIVERSITY OF MISSOURI-ROLLA, ROLLA, MO 65401

Experiments with methyl methacrylate prove that polymerization can be initiated in solutions containing monomer and a special initiator with intense ultrasound. The number average molecular weight of the Poly(methyl methacrylate) that forms is about 300,000 gmol^{-1} as compared to polystyrene standards. The conversion of the polymerization is about 50% for poly(methyl methacrylate). Variations of the polymerization rate with times and the amount of the initiator are explained by a simple reaction mechanism. The rate constants for the polymerization mechanism obtained from the molecular weight data agree with the theoretical values.

21. KINETICS OF PHASE SEPARATION IN POLYURETHANE/POLYSTYRENE SEMI-1 IPNs. Xiongwei He, Jean-Michel Widmaier and Guy C. Meyer Institut Charles Sadron (EAHP-CRM), 4, rue Boussingault, F-67000 Strasbourg.

The aim is to correlate the phase separation observed in most IPNs to synthesis parameters like the degree of conversion of both systems. Thus, a series of semi-1 IPNs based on polyurethane (PUR) and polystyrene (PS) was prepared, in which PUR had different degrees of conversion at the onset of the polymerization of styrene. The conversion of the systems was followed by FTIR spectroscopy, and the appeerence of turbidity by light transmission measurements. For a given composition, phase separation occurs quite rapidly at as few as less than 2% styrene conversion, when the PUR network is not completely formed. Therefore, the presence of microgels of PUR is sufficent to induce incompatibility. The turbidity of the reaction medium still increases up to the gelation of PUR, and stays rather constant during the polymerization of styrene. On the contrary, transparent materials are obtained when the polymerization of styrene takes place in a completely crosslinked PUR network. Thus, the incompatibility of PUR and PS may be overcome by combining them through an adequate synthesis mode.

22. PHASE SEPARATION IN INTERPENETRATING POLYMER NETWORKS, R. S. Stein, O. Aoki, A. Hanyu, and Murali Sethumadhaven, University of Massachusetts, Amherst, MA 01002, and Russell Gaudiana, Polaroid Corporation, Cambridge, MA 02139

Three types of networks have been prepared: (1) bimodal network of poly(tetrahydrofuran) having two different molecular weights between crosslinks, (2) semi IPN's of linear polyvinylmethyl ether dissolved in crosslinked polystyrene, and (3) rigid rod polymers dissolved in a flexible chain network. The thermodynamics of miscibility and kinetics of phase separation were examined, primarily by scattering techniques. The miscibility of the network blend is less than that of the uncrosslinked one. However, the presence of crosslinks appreciably reduces the rate of phase separation. Thus, it is possible to obtain optically clear and stable IPN's with molecular level dispersion of components.

23. MISCIBILITY AND PROPERTIES OF CROSSLINKABLE POLYIMIDE BLENDS WITH RIGID AND SEMI-RIGID POLYIMIDES. M. Ree and D.Y. Yoon, IBM General Technology Division, Hopewell Junction, NY 12533 and IBM Almaden Research Center, San Jose, CA 95120

Crosslinkable polyimide blends with fully rigid and semi-rigid polyimides were prepared to meld all advantageous properties of rigid or semi-rigid and crosslinkable polyimides into one system through forming a semi-interpenetrated polymer network (semi-IPN). Four different blend systems were made by solution mixing of PMDA-PDA or PMDA-ODA polyimide precursor with a crosslinkable oligomeric isoimide or fluorinated imide, followed by solvent drying and thermal imidization. Their miscibility was studied in N-methylpyrrolidone (NMP) solvent as well as condensed state by optical microscopy. The miscibility of fully imidized samples was further examined by both optical microscopy and dynamic mechanical thermal analysis. The effect of polyimide precursor type on their miscibility was also investigated. For both rigid and semi-rigid polyimides, the poly(amic diethyl ester) precursor exhibited better miscibility in their ternary solution with NMP than the poly(amic acid) precursor, and finally gave fine phase separation in scale. The phase separation occurred primarily due to their phase demixing during solvent evaporation. The domain size, which was set during drying, was not changed significantly by thermal imidization. The domain size was depending upon blend system as well as composition, and was submicron or less for most cases in conventional drying/thermal imidization condition. For some compositions, optically clear blend films were obtained. In addition, properties of the fully imidized polyimide blends were studied by stress-strain analysis, residual stress analysis, and self-adhesion analysis.

24. PREDICTING THE PHASE INVERSION IN IMMISCIBLE LIQUID SYSTEMS. L.A. Utracki, National Research Council Canada, Industrial Materials Institute, 75 De Mortagne, Boucherville, Quebec, Canada, J4B 6Y4

Upon addition to the matrix of a small quantity of a second liquid the mixture shows a droplet/matrix structure. In macromolecular liquid systems, as the concentration of the dispersed phase increases above the percolation concentration, $\phi_{cr} \cong 0.156$, the initial morphology changes into co-continuous. With further increases, at the phase inversion concentration $\phi_{1I} = 1 - \phi_{2I}$, the maximum level of phase co-continuity is reached. It is postulated that ϕ_{2I} corresponds to the iso-viscous condition of systems where polymer-1 is dispersed in polymer-2 and the other where polymer-2 is dispersed in polymer-1. This assumption leads to the dependence: $\lambda \equiv \eta_1/\eta_2 = [(\phi_m - \phi_{2I})/(\phi_m - \phi_{1I})]$, where ϕ_m is the maximum packing volume fraction and $[\eta]$ is the intrinsic viscosity. In the new thermoplastic IPNs (TIPN) the co-continuity of phases most often is reached via the phase inversion mechanism, then the system is crosslinked with physical means (e.g., crystallization). Thus conditions for formation of TIPNs are well defined by the new derived relation. For most systems $\phi_m = 0.84$ and $[\eta] \cong 1.9$ was found. As a further consequence of the initial postulate, the viscosity-concentration dependence can be described assuming two mechanisms: the emulsion-like viscosity increase culminating at the inversion concentration, ϕ_{2I}, and the interlayer slip. The derived two-parameter equation was found to describe the observed η vs. ϕ dependencies very well.

25. EFFECT OF DIFFERENCES OF COMPONENT SOLUBILITY PARAMETERS AND GLASS TRANSITION TEMPERATURES ON THE PHASE MORPHOLOGY OF SIMULTANEOUS IPN's. H.L. Frisch and P. Zhou, Department of Chemistry, State University of New York, Albany, NY 12222.

Recently we prepared two classes of simultaneous interpenetrating polymer networks (IPN's): the members of the first class always containing as one network crosslinked poly(2,6-dimethyl-1,4-phenylene oxide) (PPO), a glassy material, and the second class members always containing as one network crosslinked aliphatic polycarbonate-urethane (PCU), a rubbery material. Except for polystyrene/PPO the linear polymers corresponding to these IPN's were wholly immiscible. On the other hand many but not all of the IPN's were fully miscible or showed large ranges of single phase morphology (as seen from thermal, mechanical or TEM studies). We approximately correlate the phase morphologies of these IPN's with differences in the solubility parameter and glass transition temperatures of the linear homologs of the IPN components. All of the IPN's show an intermediate composition (or an intermediate composition in the single phase region) in which the (Instron) tensile stress to break exhibits a maximum. All the IPN's show superior physical properties to the corresponding physical blends of the linear polymers.

26. MICROSTRUCTURAL ASPECTS OF IPN'S BASED ON BLOCK COPOLYMERS. R.P. Burford, J.J. Jones, Dept. of Polymer Science, Univ. of NSW, Kensington, NSW Australia 2033 and Y-W. Mai, Dept. of Mechanical Engineering, Sydney University, Sydney, NSW Australia 2006

The morphology of solvent-cast and cryosectioned bulk samples of two block copolymers, one a linear styrene butadiene diblock and the other a styrene butadiene styrene radial triblock, was determined by transmission electron microscopy. Changes in the styrene-rich domain dimensions and shape due to specimen preparation methods were recorded, and a measure of anistropy was found by sectioning in various directions. The morphology of associated semi and full-IPN's from crosslinked block copolymers and polystyrene was similarly measured, and further information regarding microphase dimensions obtained by SAXS. Differences between the structures of the IPN's and precursor block copolymers are shown, and a comparison made between the two types of block copolymer. The dynamic mechanical behaviour and fracture toughness of these IPN's are described. Semi-IPN's from highly crosslinked block copolymers have high fracture toughness combined with adequate transparency and stiffness.

27. RHEO-KINETICS AND MORPHOLOGY OF POLYURETHANE-UNSATURATED POLYESTER INTERPENETRATING POLYMER NETWORK. Y.C. Chou, L.J. Lee, Department of Chemical Engineering, The Ohio State University, Columbus, Ohio 43210.

Polyurethane-unsaturated polyester IPNs have been used in many industrial applications. The interactions between the two reaction systems and their influence on reaction kinetics, rheological changes and morphology of cured resins were studied. It was found that isocyanates had a strong catalytic effect on the reaction of polyester vinylene groups. Formation of urethane linkage between isocyanates and hydroxyl or carboxyl groups at the ends of polyester chains tended to increase the reactivity of polyester vinylene groups. Morphology of the cured IPNs showed that grafting played an important role in reducing the extent of phase separation between polyurethane and polyester networks.

28. OPTIMIZATION OF CHANNEL DIMENSIONS USED IN SEDIMENTATION FIELD-FLOW FRACTIONATION THROUGH COMPUTER SIMULATION. Mark R. Schure, Computer Applications Research, Rohm and Haas Company, 727 Norristown Road, Spring House, PA 19477.

In this talk we review recent advances in our understanding of the sedimentation field-flow fractionation experiment which we have obtained by using computer simulation techniques. Recent findings from our laboratory include the realizations that: 1) Peak shape is drastically affected by low retention conditions and under certain conditions will give two peaks; 2) zone purity is critically dependent on the direction the channel is spun relative to the flow direction; the channel should be spun in the direction opposite to that of channel flow for particles more dense than the carrier fluid (this has been verified by experiment); 3) thinner channels than those used in commercially available fractionators will give "cleaner" separations; the word "cleaner" will be explained in this talk; 4) on-column injection is better in terms of minimizing zone broadening than valve injection; 5) the choice of the endpiece angle is not critical in the design of the channel. Through these findings we will discuss some aspects of channel design including optimal breadth and width and the criticality of these dimensions.

29. CHARACTERIZATION OF LATEX AGGREGATES BY SEDIMENTATION FIELD-FLOW FRACTIONATION. Bhajendra N. Barman* and J. Calvin Giddings, Field-Flow Fractionation Research Center, Department of Chemistry, Univ. of Utah, Salt Lake City, UT 84112.
*Current address: Texaco, Inc., P.O. Box 1608, Port Arthur, TX 77641.

Sedimentation field-flow fractionation (SdFFF) is a technique capable of providing high resolution particle separation based on differences in effective particle mass. It is thus capable of providing narrow equal-mass fractions. In this paper SdFFF is exploited to separate successive aggregated clusters, each composed of some number n of uniform latex beads. Various aspects of aggregation phenomena can, through this separation, be tracked in considerable detail. In particular, results relating to the origin of colloidal aggregation as well as aggregate break-up kinetics are obtained and discussed. Steric perturbations in normal SdFFF are probed using latex samples having aggregated clusters. Examples are provided to show aggregate separation and also, as a second order effect, the differential elution of equal-mass aggregates having extended and compact conformations.

30. **Photon Scattering Correction in the Determination of Emulsion Particle Distribution** R.A. Arlauskas, D.R. Burtner, D.H. Klein, Alliance Pharmaceutical Corporation, 3040 Science Park Road, San Diego, CA 92121

Alliance Pharmaceutical Corporation's emulsion is a high density concentrated perfluorocarbon oil-in-water emulsion with a mean particle diameter of 0.25 μm and coefficent of variation of approximately 30%. Particle diameter distribution is measured by Sedimentation Field-Flow Fractionation (SdFFF)(FFFractionation Inc. S 101) and photosedimentation (Horiba CAPA-700). Both methods use optical detectors and because of emulsion particle size, a Mie correction is required. SdFFF allows collection of monosized fractions which can be analyzed by GC to measure the correction factor. The perfluorocarbon content of each eluant fraction was determined and the mass was ratioed to the detector signal. The ratio was plotted against the SdFFF calculated particle diameter and the resulting calibration curve used to correct for scattering effects in the photosedimentation detector (Horiba).

31. SIZE EXCLUSION CHROMATOGRAPHY OF CATIONIC, NONIONIC, AND ANIONIC COPOLYMERS OF VINYLPYRROLIDONE Chi-san Wu, Larry Senak, James F. Curry, GAF Chemicals Corporation, 1361 Alps Road, Wayne, NJ 07470.

Size exclusion chromatography, SEC, conditions for a cationic polymer, quaternized poly(vinylpyrrolidone-co-dimethylaminoethylmethacrylate), PVPDMAEMA, were optimized using 0.1M TRIS, pH 7 buffer containing 0.5M LiNO$_3$ with Waters Ultrahydrogel columns. Polymer recovery was 100%. SEC conditions for nonionic polymers poly(vinylpyrrolidone-co-vinylacetate), PVPVA, and poly(vinylpyrrolidone-co-DMAEMA-co-vinylcaprolactam), PVPDMAEMAVC, were optimized using Waters Ultrahydrogel columns with a 50:50 (v/v) MeOH/H$_2$O, 0.1M LiNO$_3$ mobile phase. Polymer recovery was 100%. Vinylpyrrolidone compositions of PVPVA ranging from 30 to 70 mole % were studied. SEC conditions for an anionic polymer, poly(vinylpyrrolidone-co-acetic acid), PVPAA, were optimized using Waters Ultrahydrogel columns with 0.1M TRIS, pH 9 buffer, 0.2M $LiNO_3$. Polymer recovery was 100%. Vinylpyrrolidone compositions of PVPAA ranging from 25 to 90 mole % were studied. The polymers studied were also investigated using nonaqueous SEC conditions.

32. APPLICATION OF SIZE EXCLUSION CHROMATOGRAPHY FOR CHARACTERIZATION OF COTTON FIBER. Judy D.Timpa, U. S. Department of Agriculture, Agriculture Research Service, Southern Regional Research Center, POB 19687, New Orleans, LA 70179.

Cotton fiber (~96% cellulose) has been technically difficult to characterize. In our laboratory, cotton fibers were dissolved in the solvent N,N-dimethylacetamide with lithium chloride (DMAC/LiCl). This procedure solubilizes fiber cell wall components directly without prior extraction or derivatization, processes that could lead to degradation of high molecular weight components. Molecular weight distributions were determined by size exclusion chromatography (SEC) with DMAC/LiCl as the mobile phase employing commercial SEC columns and instrumentation. A universal calibration was employed by incorporation of viscometer and refractive index detectors. Applications of this technique have focused on elucidation of the relationship between cotton fiber molecular composition and fiber quality. Primary and secondary wall compositions of cotton fiber have been monitored during fiber development. Fiber classification standards with a range of length and strengths have demonstrated correlation of weight average molecular weight with strength. Variations in molecular composition are being determined according to variety and influence of temperature and water deficit during development.

33. FOUR COMPONENT SOLUBILITY PARAMETERS: A NEW DIMENSION. John W. Van Dyk, Consultant, 106 Cambridge Drive, Wilmington, Delaware, 19803.

In spite of its many theoretical limitations, the three-component solubility parameter approach, of Hansen and Hoy, has been very useful in dealing with problems of polymer solubility. One of these limitations has been removed by replacing the single hydrogen bonding parameter with two - the proton donating and the proton accepting parameters proposed by Paul E. Rider (J.App.Poly.Sci.,25,2975(1980); Poly.Eng.&Sci.,23,810(1983)). The resulting four parameters have been satisfactorily correlated with both the solubility and solvency (as measured by inherent viscosity) of three methacrylate polymers. Data for 34 potential solvents were used in the correlation.

34. RHEOLOGICAL ADDITIVES IN COATING SYSTEMS: RHEOLOGICAL PROPERTIES AND THICKENING MECHANISMS, H.N. Nae, RHEOX, Inc., P.O. Box 700, Hightstown, NJ 08520

The interactive effect of the components of coating systems containing various rheological additives has been studied. The rheological properties of these systems are discussed, in particular, the interaction of the rheological additive with the solvent, the resin and the pigment. The interactions lead to various thickening mechanisms which determine the application properties of the coating system. The effect of each component may be presented as a power law relationship, and the overall viscosity is due to the combined effects of these interactions. Examples include conventional, high solids solvent-based and water-based coatings.

35. ASSOCIATIVE POLYMERS WITH NOVEL HYDROPHOBE STRUCTURES. Richard D. Jenkins, Brijnaresh R. Sinha, and David R. Bassett, Union Carbide Chemicals and Plastics Company, Inc. Technical Center, P.O. Box 8361, South Charleston WV 25303

We have prepared linear end-capped and comb nonionic urethane based associative polymers, and alkali-soluble hydrophobically modified emulsion polymers with novel hydrophobic structures. Polymers made with our new hydrophobes have enhanced efficiency in increasing the viscosity of a aqueous solution or latex dispersion, have enhanced adsorption to latex particles, and have enhanced interaction with surfactants as compared to thickeners made with traditional hydrophobes, such as nonylphenol or hexadecane. Solutions of associative polymers with the novel hydrophobes are linearly viscoelastic, and in the presence of nonionic surfactants, can show dramatic shear-thickening in the viscosity profile at moderate shear rates (i.e, at Weissenberg numbers on the order of 1 - 10). Nonylphenol and octylphenol based surfactants with five to six moles of ethoxylation increase the strength and number of network junctions in solution, as reflected in the increase in the viscoelastic properties of the solutions with added surfactant.

36. PHASE BEHAVIOR OF POLYMER - SUPERCRITICAL FLUID MIXTURES. Marc D. Donohue and Mark A. McHugh, The Johns Hopkins University, Department of Chemical Engineering, Baltimore, Maryland 21218.

The phase behavior of mixtures containing polymers, solvents, and supercritical fluids is unusual and not yet fully understood. Nonetheless, what is known about the thermodynamic and rheological behavior of these systems can be used to advantage in a number of chemical processes.

The phase behavior of these ternary mixtures is quite complicated. Polymer - solvent binary mixtures exhibit liquid - liquid immiscibilities with both a lower critical solution temperature and an upper critical solution temperature. By changing the pressure of the system, the LCST and UCST points become lines that can intersect the vapor-liquid curve and even the mixture critical curve. This gives rise to interesting phase diagrams with unexpected features.

This presentation will provide an overview of the competing physical phenomena that lead to the phase diagrams that have been observed experimentally.

37. MEASUREMENT OF PHASE PHENOMENA OF MIXTURES OF COATINGS MATERIALS AND SUPERCRITICAL CO_2. David C. Busby and Richard A. Engelman, Union Carbide Chemicals and Plastics Company Inc., P.O. Box 8361, South Charleston, WV 25303

An apparatus has been designed to facilitate observation of the phase behavior of mixtures of supercritical fluids with coatings materials. Incremental addition of supercritical fluid to the coating material and variation of the temperature and pressure of the mixture can be readily accomplished. Observation of the transition from a single phase to multiple phases defines the maximum supercritical fluid solubility for various conditions. Use of an RMF (rotating magnetic field) viscometer allows the *in situ* measurement of the viscosity of the mixture, so a viscosity reduction curve can be determined for the coating material, and density can be determined from a digital read out of the cell volume. Some basic trends in the interaction of supercritical CO_2 with coatings resins have been determined.

38. A PROCESS FOR THE POLYMERIZATION OF MONOMERS ON THE SURFACE OF HIDING PIGMENTS DISPERSED IN WATER. Kenneth L. Hoy and Oliver W. Smith, Union Carbide Technical Center, South Charleston, WV 25303

Since the development of water based latex paints in the 40's, paint chemist have had as a goals to emulate the flow, leveling, hiding and gloss properties of solvent base paints and yet retain the obvious advantages of latex paints. Paint technologists have often speculated these remaining properties could be attained by uniformly coating the pigment particles with a polymeric binder to form a "micro composite particles" which would exist as a stable dispersion until the time of application and film formation; in effect a "particle paint". Recently it has been found these micro composite systems (MCS) can be formed by the "in situ" polymerization on the surface of dispersed pigments upon which surface a amphophilic bilayer has been formed. It has been demonstrated that these MCS paint have much improve hiding and the potential to provide other sought paint properties.

39. INTERPENETRATING POLYMER NETWORKS: AN OVERVIEW. L. H. Sperling, Center for Polymer Science and Engineering, Materials Research Center, Whitaker Lab #5, Lehigh University, Bethlehem, Pennsylvania 18015.

There are many kinds of interpenetrating polymer networks, IPN's, based on different synthetic approaches. For example, there are sequential IPN's, simultaneous interpenetrating Networks, SIN's, semi-IPN's, latex IPN's, gradient IPN's, thermoplastic IPN's, and a host of related materials. Various groups are working on theory, synthesis, engineering, and processing. applications include ion exchange resins, false teeth, prosthetics, and plastics. New nomenclature for IPN's and related materials is being developed. Various terms will be defined. The State of the art, and probable future of IPN's will be described.

40. LOW FREQUENCY SOUND ABSORBING STUDIES ON FOAMS BASED ON POLYURETHANES AND INTERPENETRATING POLYMER NETWORKS (IPNs). D. Klempner, D. Sophiea, B. Suthar, K.C. Frisch, V. Sendijarevic, Polymer Technologies Inc., University of Detroit, Detroit, Michigan 48221.

Novel sound absorbing foam materials were prepared. Polyurethane (PU) and IPN foams were prepared from polyether polyols, utilizing two aromatic diisocyanates and a modified polyvinyl ester. Fillers were also utilized, including zinc powder as well as ground rubber of three different particle sizes. The sound and energy absorbing properties of these foam materials were evaluated and compared with commercial sound absorbing PU foam. The sound absorbing properties were found to depend on the chemical structure, type of surfactant, type of blowing agent (water and/or trichlorofluoromethane) and crosslinking agent. The dynamic mechanical analysis of the skin of the IPN foam materials showed a very broad tan & peak, suggesting microheterogeneity or dual phase continuity. The loss modulus behavior of the IPN foams also showed that the glass transition is broadened, indicative of a semi-miscible morphology. The sound absorption properties of these materials were superior to those of the commercial polyurethane foam designed for low frequency sound absorbing applications.

41. BIOCOMPATIBLE INTERPENETRATING POLYMER NETWORK OF POLYTETRAFLUOROETHYLENE AND POLYDIMETHYLSILOXANE. M.E. Dillon, Bio Med Sciences, Inc.,115 Research Drive Bethlehem, PA 18015

A thin film membrane comprised of a semi-interpenetrating polymer network of polytetrafluoroethylene and polydimethylsiloxane is produced by a solution casting method. The product is non-cytotoxic in both direct contact and liquid extract studies using L929 mouse fibroblast cells. Studies of hemolytic activity using rabbit blood have shown that the material is non-hemolytic, and it has been determined that the IPN does not cause compliment activation. The IPN films are semipermeable in that they allow moisture vapor transmission, but block the passage of liquid water and other high surface tension liquids. The permeability characteristics of PTFE/PDMS IPNs are a function of microstructure and composition, and can be tailored for specific applications because morphology and chemistry can be independently varied.

42. MORPHOLOGY OF P(nBAc)/PEP SIMULTANEOUS INTERPENETRATING POLYMER NETWORKS. Song Ma, Dept. Pure Appl. Chem., Strathclyde Univ., Glasgow G1 1XL, UK and Xinyi Tang, Dept. Chem.,Jilin Univ., Changchun, China

The poly(n-Butyl Acrylate)/polyEpoxide Resin Simultaneeous Interpenetrating Polymer Networks (P(nBAc)/PEp) SIN) synthesized with two chain reactions have been characterized by swelling, density, dynamic mechanical and TEM measurements. DMA spectra and TEM photographs show a clear two phase structure and indicate it's a partial compatible system. Due to the fact that there is a middle peaks between two peaks which reflect two components separately in DMA spectra it's believed that there is a third phase -- interfacial layer phase and the effects of composition and crosslink densities have been discussed. Based on the supposition that density increase is caused by interpenetration, interpenetration degrees were estimated from the theoretical and experimental density difference and the results are coordinate with DMA results.

43. KINETIC STUDY AND MORPHOLOGY BUILD UP IN METHACRYLATE URETHANE IPNs. J.M.Liégeois,D.Job,P.Landuyt, Univ.de Liège Belgium

The aim of this work is to rationalize the development of recipes for simultaneous IPNs by matching gel times and computing conditions for phase separation to occur after gel times.
The simultaneous IPN system based on methylmethacrylate, co-1,4 butanedioldimethacrylate and polycaprolactonetriol-hexamethylene diisocyanate has been studied under fast reaction conditions, an exothermic plateau being reached within about 10 minutes.
The kinetic of reaction of the individual systems has been established by monitoring the temperature under adiabatic reaction conditions, giving reaction orders, activation energies and preexponential factors. The dynamics of phase separation has been modelled based on Flory-Huggins equation, taking into account the temperature rise effect on free energy of mixing, reaction rates and molecular weight development.
In a trated example, conditions are shown where phase separation occurs after 240 seconds whereas gel times are within 120 seconds. Samples prepared under such conditions are transluscent to milky and are under further physical investigation.

44. COMPOSITES FROM POLYURETHANE AND EPOXY INTERPENETRATING POLYMER NETWORK. K.H.Hsieh, S.T.Lee, D.C.Liao, C.C.M. Ma, Department of Chemical Engineering, National Taiwan University, Taipei, Taiwan, R.O.C. 10764

Uni-directed glass-fiber reinforced composites of polyurethane (PU)-crosslinked epoxy and PU/epoxy interpenetrating polymer network with PU grafted on epoxy (Graft-IPN) were prepared. Their mechaincal properties were investigated and discussed.
The tensile strength of the PU-crosslinked epoxy or the PU/epoxy Graft-IPN composites was significantly increased in the longitudinal direction but decreased in the transverse direction. The effect of PU contents in the matrix composition on the tensile strength of the composites is not so remarkable with increasing PU content in the system.
The Izod impact property of these composites was greatly improved with the addition of glass-fiber in either transverse or longitudinal direction. The values of impact strength were strongly increased with increasing PU content in the matrix composition, especially, in the transverse direction.
The flexural strength of these composites had a similar change as the impact strength along with an increase of PU content in the system. In addition, the flexural strength of the composites in transverse direction was affected by various types of PU introduced into the system.

45. NEW SYNTHETIC POLYOLEFINE COMPOSITIONS. F.S.Dyachkovskii, Department of Polymer and Composite materials, Institute of Chemical Physics Academy of Sciences USSR, Chernogolovka, USSR

Polyolefines belong to the basic materials in plastic industry. From the point of view of polyolefines application it is very important to increase their thermostability, hardness and other physical-mechanical properties. This goal could be reached by using functionalized polyolefines, inorganic and organic fillers, chemically bonded with polymer matrix by the model of interpenetrating networks. These materials are characterised by higher thermostability and physical-mechanical properties. This talk will cover recent development and application of functionallized polyolefines, synthesis of inorganic or organic phase or network in polyolefine matrix. The attention will be paid to physical--mechanical and physical properties of such materials, their correlation with the nature of polymer matrix and composition content.

46. **MOLECULAR WEIGHT DETERMINATIONS OF POLY-4-METHYL-1-PENTENE**
Arja Lehtinen and **Hannele Jakosuo**, Neste OY, Technology Centre, P.O.Box 310, SF-06101 Porvoo and Technical University of Helsinki, Kemistintie 1, 02150 Espoo, Finland

A high temperature SEC method for the determination of molecular weight distribution of poly-4-methyl-1-pentene was developed. In this method, the samples are dissolved in a decalin-TCB mixture using microwave heating and run at 135 °C with TCB as solvent. The conditions chosen for the SEC analyses are based on studies of solubility of PMP in different solvents, stabilization of the solutions with different antioxidants and dissolution experiments in an air oven and a microwave oven. For the system calibration the samples were analyzed both by SEC and SEC/on-line viscometer combination. Also viscosity and light scattering measurements were made. Results for a set of isotactic poly-4-methyl-1-pentene samples with different melt flow rates are presented.

47. Long Chain Branching Analysis of LDPE by SEC/LALLS and SEC/CV Systems, Simon Pang and A. Rudin, University of Waterloo, Waterloo, Ontario, Canada

The quantitative long chain branching (LCB) characterization of several LDPE samples via size exclusion chromatography (SEC) coupled with on-line low angle laser light scattering (LALLS) and continuous viscometer (CV) detectors are reported in this paper. Recently, the use of SEC/LALLS and SEC/CV analysis have become popular in the industry due to their instrumental and theoretical development. This article describes and compares the methods for measuring the LCB parameter g' in the expression $g' = g^k$ through the two different combined SEC/LALLS and SEC/CV systems.

48. DEVELOPMENT OF GEL PERMEATION CHROMATOGRAPHY-FOURIER TRANSFORM INFRARED SPECTROSCOPY FOR CHARACTERIZATION OF ETHYLENE BASED POLYOLEFIN COPOLYMERS. Ronald P. Markovich, Dow Chemical, U.S.A., Freeport, Texas, 77541; Lonnie G. Hazlitt, Dow Chemical, U.S.A. Freeport, Texas, 77541; Linley Smith, Sam Houston University, Huntsville, Texas.

A Fourier transform infrared (FT-IR) spectrometer has been coupled with a high temperature gel permeation chromatography instrument to provide a powerful tool for the characterization of ethylene based polyolefin copolymers. Both GPC and FT-IR data are collected in real time in the presence of the eluting solvent. The combination of these devices provides the ability to simultaneously characterize the molecular weight/ molecular weight distribution and the chemical composition of each fraction. Systems analyzed to date include copolymers of ethylene with octene, hexene, butene, carbon monoxide, and acrylic acid. In every case, a strong indication of chemical structure (i.e. comonomer incorporation) as a function of molecular weight was evident from the analysis of the spectral and elution volume data. The relationship between comonomer incorporation and molecular weight for each class of copolymer can be of significant importance in explaining variations in the performance of different resins.

49. **CRITICAL CONDITIONS IN GEL PERMEATION CHROMATOGRAPHY** David J. Hunkeler, Tibor Macko and Dusan Berek, Polymer Institute, Slovak Academy of Sciences, 84 236 Bratislava, Czecho-Slovakia.

Binary eluents have been employed to investigate the retention of polystyrene and polymethylmethacrylate on a porous silica gel packing. It has been observed that the retention volume is sensitive to the mobil phase composition. At a highly specific "critical" condition the homopolymers eluted at an identical retention volume, independent of their molecular weight. This was confirmed for polymers from 10,000 to 1,000,000 daltons in toluene-methanol and THF-water systems. For the former, a critical composition of 68 vol% of toluene was observed for polystyrene, with more toluene rich mobil phases conforming to a classical exclusion regime and more methanol rich eluents characterized by polymer adsorption onto the column. For polymethylmethacrylate the corresponding critical composition was 27 vol% toluene. At the critical conditions for the homopolymers random and block copolymers were also analyzed to investigate the possibility of separating polystyrene-co-methylmethacrylate based on composition rather than molecular weight. A proposed mechanism for the molecular weight independent exclusion, and critical conditions, is also included.

50. **SIZE EXCLUSION CHROMATOGRAPHY AND END GROUP ANALYSIS FOR POLY METHYL METHACRYLATE.** Somnath Shetty and L.H. Garcia-Rubio. Chemical Engineering Department, University of South Florida Tampa, FL 33620

This paper reports on the combined size exclusion chromatography\end group analysis for poly methyl methacrylates produced at several temperatures and with different degrees of conversion. The spectroscopy analysis of the polymers, combined with information on the molecular weights, allow for the identification and quantification of the type and number of initiator fragments contained in the polymer. It is demonstrated that the residual initiator concentration may be obtained as an integral part of the analysis. The interpretation of the quantitative end group-molecular weight analysis results in the identification of the termination mechanisms and in the estimation of initiator efficiencies.

51. **COPOLYMER CHARACTERIZATION USING CONVENTIONAL SEC AND MOLAR MASS SENSITIVE DETECTORS**
F. Gores, C. Johann, P. Kilz, PSS Polymer Standards Service GmbH,
P. O. Box 3368, D-6500 Mainz, Germany

SEC characterization has made remarkable progress in recent years. Most applications, however, treat the analyte as a homo polymer. This paper describes a reliable and rapid method for the analysis of copolymers by size-exclusion chromatography (SEC). The properties of copolymers mainly depend on the choice of comonomers, molecular weight and composition.
In this presentation three on-line methods are used to characterize copolymers by SEC with respect to molar mass and composition distribution:
1. Conventional SEC utilizing multiple detection
2. On-line analysis of SEC fractions with a multi-angle laser light scattering detector
3. On-line viscometry
The validity and restrictions of each method will be discussed for different types of copolymers.

52. "Poly(Arylene Ether Phosphine Oxide)s I. Overview of Chemistry, Thermal Stability and Oxygen Plasma Resistance," C. D. Smith, H. Grubbs, H. F. Webster, J. P. Wightman, and J. E. McGrath, Virginia Polytechnic Institute & State University, NSF Science & Technology Center: High Performance Polymeric Adhesives and Composites and Dept. of Chemistry, Blacksburg, Virginia 24061-0212.

Poly(arylene ether phosphine oxide)s (PEPO) were explored in terms of their fire-resistance and aggressive oxygen plasma resistance. These thermoplastic materials with glass transition temperatures in the range of about 200-285°C showed substantial amounts of char yield at 800°C in air which was related to their excellent self-extinguishing characteristics relative to any other engineering thermoplastics. Additionally, the presence of phosphorus in the char after such high heating suggested that these materials might also be resistant to aggressive oxygen plasmas. Indeed, the PEPO systems showed extremely low amounts of etching in oxygen plasma when compared to other polymers in which the only structural change was the presence of the phosphine oxide unit. Only about 20% the height of thin films (ca. 1000Å) of PEPO was etched in the oxygen plasma after 15 minutes by ellipsometry, when compared to 100% etching for a number of other engineering thermoplastics (e.g., polysulfone, poly(ether ketone), polyimides). The presence of phosphorus residues after either burning or etching with oxygen plasma could play crucial roles in areas of fire resistance, electronic materials and aerospace applications.

53. "POLY(ARYLENE ETHER PHOSPHINE OXIDE)S II. PYROLYSIS STUDIES," H.J. Grubbs, C. D. Smith and J. E. McGrath, Virginia Polytechnic Institute & State University, NSF Science & Technology Center: High Performance Polymeric Adhesives and Composites and Dept. of Chemistry, Blacksburg, Virginia 24061-0212

Pyrolysis of poly(arylene ether phosphine oxide) (PEPO) thermoplastic materials was carried out in order to compare the degradation products with conventional poly(arylene ether)s (PAE), such as the polysulfones and polyketones. Both the volatile and char products were studied. Volatile products were identifies by gas chromatography/mass spectrometry techniques. Phosphorus content in both the pyrolysate (vapor) and the char residue was determined by neutron activation analysis (NAA).Compared to other PAE which were completely volatilized at 600 or 700°C, the PEPO afforded 60-70% char residue at the same temperatures. Very few phosphorus containing products could be found in the PEPO pyrograms, indicating most of the phosphorus was converted to non-volatile products. Further confirmation of phosphorus location after pyrolysis was obtained by NAA. Indeed, phosphorus content in the char was typically double that of the parent PEPO, while the phosphorus content in the volatile fraction was less than 2%.

54. "POLY(ARYLENE ETHER PHOSPHINE OXIDE)S III. INVESTIGATIONS OF SURFACE MODIFICATION BY OXYGEN PLASMA TREATMENT," H.F. Webster, C. D. Smith J. E. McGrath, and J. P. Wightman, Virginia Polytechnic Institute & State University, NSF Science & Technology Center: High Performance Polymeric Adhesives and Composites and Dept. of Chemistry, Blacksburg, Virginia 24061-0212

Surface analysis of the plasma treatment of poly(arlyene ether phosphine oxides) by XPS yielded results consistent with the loss of oxidized carbon species and the formation of a phosphate surface layer. The development of such a surface structure may be responsible for its resistance to degradation by oxygen plasma . The evolution of the surface stucture with time could be easily monitored using XPS and this technique showed that most of the surface conversion occurred within several minutes of oxygen plasma exposure. This response to oxygen plasma is quite different to that seen for a polysulfone analog and other engineering thermoplastics where the surface concentrations remain constant with various exposure times. The excellent structural properties of poly(arylene ether phosphine oxides) combined with resistance to polymer degradation by oxygen plasma may make it an ideal candidate material for space applications.

55. EFFECT OF MOISTURE CONTENT ON DIELECTRIC PROPERTIES IN MICROWAVE REGION FOR PRINTED CIRCUIT BOARD MATERIALS
A.Yamada, S.Yamaguchi, F.Baba, Y.Yamamoto,
Materials & Electronic Devices Laboratory, Mitsubishi Electric Corp., 8-1-1 Tsukaguchi-honmachi, Amagasaki, Hyogo, 661 Japan

Effect of moisture content on dielectric properties in microwave region of the printed circuit board materials based on PEEK(polyetheretherketone)/D-glass, BT(bismaleimidetriadine)/D-glass and conventional Epoxy/E-glass laminates were investigated. The measurements were performed by the cavity perturvation method at mainly 10.4 GHz. The result indicates that PEEK/D-glass, BT/D-glass laminates have lower relative dielectric constant and lower loss factor than those of conventional Epoxy/E-glass laminate. Moisture absorption were also performed on these laminates under four conditions. And the dry-wet cycle test was also carried out. Moisture content in PEEK/D-glass laminates were smaller than that of BT/D-glass or epoxy resin laminates. It was found that relative dielectric constant and loss factor for all laminates, regardless of the kind of laminates, are proportional only to the moisture content. It shoud be pointed out that moisture content showed a more significant effect on the loss factor than on the dielectric constant.

56. INTERPENETRATING POLYMER NETWORKS OF POLY(CARBONATE-URETHANE) AND POLYVINYL PYRIDINE. Peiguang Zhou, H.L. Frisch, Dept. of Chemistry, State University of New York at Albany, Albany, NY 12222.

Simultaneous interpenetrating polymer networks (IPN's), Pseudo IPN's and linear blends of alphatic poly(carbonate-urethane) (PCU) and polyvinyl pyridine (PVP) have been prepared and characterized by DSC, DMA and TEM. The full IPN's of PCU and PVP had a single phase morphology only above 50% PCU content by weight, as determined by both DSC and DMA and confirmed by transmission electron microscopy (TEM). However, in both pseudo IPN's of PCU and PVP and their linear blends there exist multiple glass transitions and melting points from DSC and DMA measurements indicating phases incompatability over the whole composition range. The full IPN's exhibited superior ultimate mechanical, properties and solvent resistance as compared to the pseudo IPN's, linear blends and the pure crosslinked PCU and PVP networks.

57. CHARACTERIZATION OF VIBRATION DAMPING OF LAMINATED STEEL SHEET WITH POLYURETHANE/ACRYLATE INTERPENETRATING POLYMER NETWORKS. C.J. Tung and T.J. Hsu, Institute of Materials Science & Engineering, National Sun Yat-Sen University, Kaohsiung 804, Taiwan; S.I.Chen, New Materials R&D Department, China Steel Corporation, Kaohsiung 802, Taiwan

A series of polyurethane/acrylate semi-II interpenetrating polymer networks (IPNs) have been prepared and each laminated between two steel sheets to form the laminate for vibration damping purpose. Comparisons have been made to demonstrate the effect of IPNs variables on the damping efficiency of the laminate. These variables include the composition of component in IPN, the crosslinking extent in polyurethane, the reaction sequence in IPN, and the curing temperature. We have found (1) polyurethane and acrylate homopolymers show a tangent delta of 1.0 over a narrow temperature range of ca. 20°C; (2) the IPNs exhibit a broader temperature range of 50°C but a lower damping capability between 0.4 and 0.7; (3) the laminate gives a loss factor even below 0.4, but is 2 to 3 order of magnitude far better than the pure steel sheet.

58. MORPHOLOGY/PROCESSING RELATIONSHIP IN INTERPENETRATING POLYMER BLENDS OF A SEBS BLOCKCOPOLYMER AND A POLYETHERESTER. H. Verhoogt, J. van Dam, A. Posthuma de Boer, Delft University of Technology, Department of Polymer Technology, Delft, The Netherlands.

Blending two polymers with (almost) equal viscosities over a broad range of shear rates results in blends with dual-phase continuity around the 50/50 composition. Such blends, also called Interpenetrating Polymer Blends (IPB), might be processable at different shear rates without a change of the morphology.
Blends of a SEBS blockcopolymer (Kraton G1657X) and a polyetherester (Arnitel EM400) were prepared on a two-roll mill at 200^{0}C and a shear rate of 750 (1/s). The morphologies of the formed blends show dual-phase continuity in a broad range of compositions (20 to 60% SEBS). During annealing at 200^{0}C a change in morphology has been observed, resulting in a decrease of dual-phase continuity. Processing of the blends at different shear rates results in a (partial) decrease of the dual-phase continuity.

59. IPN MEMBRANES FOR THE PERVAPORATION OF ETHANOL/WATER MIXTURE, Young Keun Lee, In Seung Sohn, Eun Jin Jeon and Sung Chul Kim, Department of Chemical Engineering, Korea Advanced Institute of Science & Technology, P.O. Box 150 Cheongryang, Seoul 130-650, Korea

Hydrophilic/hydrophobic, cationic/anionic heterogeneous membranes were prepared and the effect of the synthesis pressure and composition on the morphology were evaluated. The pervaporation characteristics and the sorption behavior of the membranes for the separation of ethanol/water mixture were measured and the effect of the IPN synthesis parameters including the composition, ionic concentration, hydrophilic/hydrophobic ratio, etc. were analyzed. The computer simulation of the actual continuous pervaporation process through rectangular channels were conducted and the optimum membrane characteristics to give the best performance was analyzed.

60. **GAS TRANSPORT IN IPN MEMBRANES ; EFFECT OF CROSSLINKED STATE AND ANNEALLING.** Doo Sung LEE[1], Won Kil KANG[1], Jeong Ho AHN[2] and Sung Chul KIM[3]
[1]Department of Textile Engineering, Sung Kyun Kwan University, Suwon, Kyungki 440-746. [2]Department of Chemical Engineering, Pohang Institute of Science & Technology, P.O. Box 125, Pohang, kyungbuk 790-600, [3]Department of Chemical Engineering, Korea Advanced Institute of Science & Technology, P.O. Box 131, Cheongryang, Seoul 130-650, (S. Korea)

A series of polyurethane(PU) –polystyrene(PS) IPN, semi–IPN (only PU component crosslinked) and linear blend membranes were prepared with varying synthesis temperature and composition. The permeability coefficient showed lower value and the separation factor showed higher value when it was prepared at low synthesis temperature due to the homogeneity increase as shown in the previous papers. The permeability coefficient showed minimum value and the separation factor showed maximum value at about 25 % PU composition. The permeability coefficient increased in the following sequence; IPN< <semi–IPN < linear blends. The thermal annealing was adopted to study the effect of homogeneity of two polymer components in the membrane on the gas permeation characteristics. The permeability coefficient increased and the synergistic value of separation factor at high PS composition decreased by annealing at high temperature due to the change of homogeneity of the membrane. The annealing effect was depended on the crosslinked nature of the membrane (IPN < semi–IPN < linear blends). The densities and the electron microscopys were also tested.

61. **GAS PERMEABILITY STUDIES OF LATEX INTERPENETRATING POLYMER NETWORK FILMS.** A.K. Holdsworth and D.J. Hourston, The Polymer Centre, Lancaster University, Lancaster, U.K.

The permeability of gases through polymer membranes is widely studied. However, much less work has been undertaken on the permeability of polymer blends towards different gases, and, to date, no work has been published with regard to any type of interpenetrating polymer network (IPN). In theory, the measurement of permeability coefficients offers a potentially subtle probe of blend morphology. In this paper, permeability measurements of carbon dioxide, nitrogen, oxygen, argon and helium through cast films of latex IPNs and latex blends will be reported. The latex IPNs involved are based on poly(ethyl acrylate)-poly(ethyl methacrylate), poly(ethyl acrylate)-poly(t-butyl acrylate) and poly(t-butyl acrylate)/poly(ethyl acrylate-co-isobornyl methacrylate). In addition to permeability data, dynamic mechanical analysis and electron microscopy results will be presented.

62. **FLOW IN LATEX INTERPENETRATING POLYMER NETWORKS.**
+M.S. Silverstein and ‡M. Narkis, +Materials Engineering and ‡Chemical Engineering, Technion, Haifa 32000, Israel

Elastomeric latex interpenetrating polymer networks (LIPN) can result from a two-stage emulsion polymerization process. Lightly crosslinked polyacrylates (PA) and lightly crosslinked polystyrene (PS) (75/25 PA/PS) can be combined in this manner with the styrene polymerized on a PA seed latex. The unique ability of these thermoset materials to flow and be processed using standard thermoplastic processing equipment lies in the formation of the IPN inside a 100 nm latex particle. The particles are held together by PS microdomains which disintegrate and reform above the PS glass transition. The flow in these materials can be described using a 'power-law' model, although the actual flow mechanism is not that of viscous flow on the molecular level but rather a stick-slip particle flow. These LIPN particles flow over a large temperature and shear rate range with a relatively temperature insensitive 'power-law' exponent and a relatively shear rate insensitive activation energy of flow.

63. LATEX PS/XPS SEMI-IPNs. N. Nemirovski and M. Narkis, Department of Chemical Engineering, Technion City, Haifa 32000, Israel.

Latex PS/XPS systems are structured by using a sequential emulsion polymerization method. The idea is to physically blend PS and XPS latices, or to proceed by a two-stage polymerization technique, where the seed is either PS or XPS. After coagulation and compression molding of the samples, PS merges into a continuous thermoplastic matrix in which inclusions of XPS domains are dispersed. The seeded semi-IPNs and the blend possess different structures. Hence three particular cases are actually under investigation. Different rheological and mechanical behavior in the three systems, along with a rather interesting property dependence on XPS content and crosslink density, have been noticed. This kind of behavior stems from the unique morphology of each system. Various techniques are instrumental to the understanding of structure and structure-property relations.

64. THE STUDY ON THE DAMPING PAINT OF PS/PA LATEX INTERPENETRATING POLYMER NETWORKS. Yuwei Li, Ruiying Liu, Jingyuan Wang, Xinyi Tang, Department of Chemistry, Jilin University, Changchun, 130023, People's Republic of China.

The damping paint of polystyrene/polyacrylate (PS/PA) has been synthesized by using two stage emulsion polymerization methods. Various parameters affecting damping value have been investigated systematically. The results show that Core/shell structure is an important factor affecting compatibility, and multicomponent copolymerization favors broadening the damping temperature range (-10-90°C). Moreover, changing the feeding sequence and adding inorganic fillers are the important ways to increase the damping value. The introduction of polar group into PA chain can increase effectively the adhesion of the paint.

65. IMPLEMENTATION OF A SYSTEMATIC APPROACH FOR QUANTITATIVE INTERPRETATION OF DATA FROM A SEC-VISCOMETER SYSTEM. Stephen T. Balke, Ruengsak Thitiratsakul, Raymond Lew, Paul Cheung, Department of Chemical Engineering and Applied Chemistry University of Toronto, Toronto, ON, M5S 1A4; Thomas H. Mourey, Research Laboratories, Eastman Kodak Company, Rochester, NY, 14650.

Mourey and Balke recently showed the development of a systematic approach for interpreting multi-detector SEC data[1]. This paper will summarize the approach and show our recent experience with attempting to apply it to the interpretation of data from a SEC equipped with a differential refractometer and a differential viscometer. The motivation for development of the approach is the conglomeration of numerical constants which even elementary analysis of the data demands. Constants include those for calibration, concentration, flow rate, transducer and resolution correction. Often the same results can be obtained by adjusting either of two constants. Quantitative interpretation of data from high temperature SEC analysis of polyolefin samples, including some samples of recycled waste plastic, is of particular interest. The results are required for polymer reaction engineering studies.

[1] T.H. Mourey and S.T. Balke, "A Strategy for Interpreting Multi-Detector SEC Data", First International GPC/Viscometry Symposium, Houston, TX, Apr. 24-26, 1991.

66. THE USE OF A SINGLE CAPILLARY VISCOMETER FOR ACCURATE DETERMINATION OF MOLECULAR WEIGHTS AND MARK-HOUWINK CONSTANTS. James LESEC*, Michele MILLEQUANT* and Trevor HAVARD**
*CNRS - ESPCI, 10 rue Vauquelin, 75231 PARIS 05 - FRANCE ** WATERS, 34 Maple street, Milford, MA 01757

The Waters 150 GPC/Viscometry system is equipped with an on-line single capillary viscometry detector. The detector's geometrical characteristics (18 μl internal volume, 2800 s^{-1} shear rate) allow measurements to be made near classical Ubbelohde viscometry conditions. When used in conjunction with a GPC/Viscometry software, it is possible to use universal calibration and to measure accurate molecular weights, Mark-Houwink K and alpha constants. Consequently, long-chain branching g' distribution can be determined for branched polymers. Initial trials of the system demonstrated that minor flow fluctuations occur as the samples elute through the detectors. The restriction between the columns and detectors and the change in specific viscosity of the eluting polymers are reponsible. The effect creates peak deformation which is observed as downstream peak shift for the viscometry detector. In order to determine the correct Mark-Houwink constants for broad polymers a detector offset is required that is different from the geometrical value. A prototype refractometer was designed to eliminate this phenomenon. The prototype was tested at room temperature (THF-30 C) using known narrow and broad distribution polystyrene samples. The modified system gave simular Mark-Houwink constants for all samples studied. This design allows the investigation of narrow distribution polymers with improved accuracy. Trials at high temperature (TCB-145 C) have also been carried out.

67. FLOW RATE EFFECTS IN MODERN SEC-VISCOMETRY DETECTORS.
Christian Jackson and Wallace W. Yau, E.I. DuPont de Nemours & Co., Central Research and Development, Experimental Station, P.O. Box 80228, Wilmington, Delaware 19880-0228, Max A. Haney, Viscotek Corporation, 1032 Russell Dr., Porter, Texas 77365.

An analysis of the effect of flow rate upsets on the single capillary SEC-Viscometry detector was recently presented by Lesec, Havard, et.al., at the First International GPC/Viscometry Symposium, Del Lago Resort, Tx, April 24-26, 1991. At high sample solution viscosity situations, local flow rate upsets may result from the passing of sample solution through the system flow resistances at the time of GPC Peak elution. This flow rate effect may cause a shift of the viscosity signal to a longer retention time with respect to the refractive index signal. This effective shift of the viscosity elution peak shape affects mainly the calculated Mark-Houwink exponent constant.

In this paper we seek a deeper understanding of the above phenomenom. We start with a review of the physics of flow in various viscometry detectors employing one capillary, two capillaries, and four capillaries. The results of a computer simulation study and relevant experimental results will be described to illustrate the effect of minor flow rate variations on the experimental determination of detector offsets and Mark-Houwink constants.

68. CHARACTERIZATION OF POLYMER MWD AND BRANCHING WITH GPC-VISCOMETRY IN THF AND DMF

C. Kuo, T. Provder, and M. E. Koehler, The Glidden Company, Member of ICI Paints, 16651 Sprague Road, Strongsville, Ohio, 44136

The characterization of molecular weight distribution, intrinsic viscosity-molecular weight behavior, and polymer chain branching for polymer standards and polymers of commercial interest are reported for THF, DMF, and DMF(LiBr) mobile phase. The instrument used in this study is the Millipore Waters single capillary GPC-Viscometer/Data Analysis System (GPCV). Results in THF are compared with previously reported results on the Glidden/GPC Viscometer/Data Analysis System (GPC/VIS).

In this paper, instrumentation, principles of operation, operational variable considerations, and data analysis methodology are discussed in the context of characterization information reported in THF, DMF, and DMF(LiBr).

69. The Characterization of Branched and Comb Polymers by SEC and Other Methods.
D. McIntyre, Institute of Polymer Science, The University of Akron, Akron, OH 44325-3909.

A variety of both randomly branched and grafted polymers have been studied by conventional solution techniques and by SEC to determine the limits of polymer size characterization using conventional methods of SEC with a variety of detectors.

70. MODERN SEC-VISCOMETRY AND NEW EXPERIMENTAL POSSIBILITIES FOR POLYMER CHARACTERIZATION, Wallace W. Yau, Howard G. Barth and Christian Jackson, E. I. du Pont de Nemours and Company, Central Research and Development, P. O. Box 80228, Wilmington, DE 19880-0228

With modern SEC-Viscometry, polymer intrinsic viscosity (IV) at every SEC retention volume can be measured to provide the IV distribution (IVD) for polymer characterization. Since IVD does not build on SEC MW-retention calibration, the IVD measurement is highly insensitive to adverse SEC conditions of flowrate and temperature fluctuations, column deteriorations, instrumental band broadening, sample overloading, etc. The high precision feature of IVD and the IVD-derived MWD results makes SEC-viscometry a prime candidate for polymer QC and process control. With some improvisation, modern SEC-viscometers can be easily made adaptable for studying rate or kinetic processes, such as polymerization reactions, polymer degradations, protein denaturing, biopolymer conformational transitions, etc. Furthermore, by coupling SEC-viscometry with a column of inert nonporous glass beads, one can intentionally create a sample concentration dispersion without actually causing any chromatographic separation. Under the circumstances, simultaneous viscosity and concentration detectors can be made on the unfractionated polymer sample at many different sample concentration levels dictated by the concentration elution curve profile. The basic Huggins' polymer-solvent interaction parameter can then be determined by simply plotting the concentration dependency of the viscosity data. This determination of Huggins' constant takes a minute or two in a single SEC-viscometry experiment (a measurement could take hours by conventional drop-time viscometers)!

71. "Change of Mechanism" Block Copolymerizations. The Formation of Block Copolymers Containing Helical and Amorphous Polyisocyanide and Elastomeric Polybutadiene Segments. Timothy J. Deming and Bruce M. Novak*, Department of Chemistry, University of California, Berkeley, California 94720.

The compound $[(\eta^3\text{-}C_3H_5)Ni(OC(O)CF_3)]_2$ (I), which has been used extensively as a butadiene polymerization catalyst and more recently as an isocyanide polymerization catalyst, has been sucessfully used in the preparation of polyisocyanide - polybutadiene block copolymers. Since both monomer polymerizations are living, this block copolymer synthesis is highly versatile with respect to polymer segment chain lengths and the types of monomers used. Block copolymers of [poly(α-methylbenzyl isocyanide) - polybutadiene] and [poly(*tert*-butyl isocyanide) - polybutadiene] have been prepared, and then characterized by using gel permeation chromatography, differential scanning calorimetry, ^{13}C NMR and scanning electron microscopy.

72. POLYMER COMPOSITES USING RECYCLED PET. K. S. Rebeiz, D. W. Fowler, D. R. Paul The University of Texas, Austin, TX 78712
Fundamental properties of polymer concrete (PC) and polymer mortar (PM) made from unsaturated polyester resins based on recycled poly(ethylene terephthalate), PET, are described. Mechanical properties investigated include strength, modulus of elasticity, ductility index, Poisson's ratio, coefficient of thermal expansion, glass transition temperature, shrinkage, and exotherm. Durability properties include chemical resistance, water absorption and sand-blast resistance. Resins based on recycled PET offer the possibility of a lower source cost for forming useful PC and PM based products. The recycling of PET in PC and PM would also help solve some of the solid waste problems posed by plastics and save energy.

73. ENERGY CONSUMING MICROMECHANISMS IN THE FRACTURE OF GLASSY POLYMERS I. CHAIN SCISSION IN POLYSTYRENE. N. Mohammadi, J. N. Yoo, A. Klein, and L. H. Sperling, Materials Research Center, Whitaker Lab #5, Lehigh University, Bethlehem, PA 18015.

The number of chain scissions per unit area that occur during the fracture of partially annealed latex films from $M_w \simeq$ 900,000 g/mol polystyrene particles of about 275 Å radius were measured and correlated to annealing times. A curve with four regimes was found. At short annealing times the curve is nearly flat, called the chain pull-out regime. In the second regime, the number of chains broken per unit area increases with a 0.8 power of annealing time as entanglement of the diffusing polymer chains increases in host neighboring particles. This is in good agreement with Wool's theory, which predicts a 0.75 power dependence. Then after reaching a peak, the number of scissions decreases, third regime, indicating a change in failure mechanism. The number of chain scissions increases again in the fourth regime, as final healing of the film interface takes place. Fracture surface analysis reveals a rough surface for short annealing times and a smooth surface for longer annealing times. The number of polymer chain scissions per unit area of fracture surface showed no dependence on initial molecular weights for $t >> \tau_r$ where t and τ_r are annealing and relaxation times respectively. The number of chain bridges crossing a unit area of interface was suggested as the basic molecular property.

74. THERMODYNAMIC INTERPRETATION OF PHASE BEHAVIOR IN A TRANSREACTING POLYESTER/POLYCARBONATE BLEND. Jeffrey S. Kollodge and Roger S. Porter, Polymer Science & Engineering Department, University of Massachusetts, Amherst, MA 01003.

A model polyester/polycarbonate blend system was used to quantitatively study the effect of transreaction on blend phase behavior. Poly(2-ethyl-2-methylpropylene terephthalate) (PEMPT) and Bisphenol-A-Polycarbonate (PC) were chosen for study due to their amorphous structure, solubility in polar solvents and phase behavior characteristics. When transreaction is prevented by addition of an inhibitor, the blend is a two phase system. In an unstabilized system, interchange reaction leads to the formation of a single phase blend. It was determined from proton NMR that only 2.5% of the terephthalate groups undergo transreaction during the phase transition. Predictions based on phase behavior studies lead to a defined upper limit of 5% reaction required for miscibility. Calculations based on block copolymer theory estimate the extent of reaction required for miscibility to be on the order of 1%.

75. SILICON-ASSISTED SYNTHESIS OF HIGH MOLECULAR WEIGHT POLYETHERS. K.E. Uhrich and J.M.J. Fréchet, Department of Chemistry, Baker Laboratory, Cornell University, Ithaca, New York 14853-1301.

The polycondensation of two AB monomers to form aromatic polyethers under mild conditions is described. Early work in the polycondensation of 3-bromomethylphenol under phase transfer conditions led to unsatisfactory results as the polymers obtained only had low molecular weights of up to 5000. It was postulated that silylation of the phenolic groups may decrease the possibility of phenolic side reactions, including C-alkylation, thus producing high molecular weight polymers. This approach was first investigated on a model system. The condensation of trimethylsilylphenol with benzyl bromide was performed using several nucleophiles under various conditions. Carbonates were found to be the most effective reagents for the condensation affording quantitative ether formation without any C-alkylation. This methodology applied to the polycondensation of 3-trimethylsilyloxybenzyl bromide afforded soluble, high molecular weight ($MW > 10^5$) polyethers free of C- alkylation. The mechanism of the carbonate initiated condensation requires a stoichiometric amount of carbonate, which is believed to attack the silicon of the trimethylsilyloxyphenol. The resulting pentacoordinate silicon may assist the phenolic oxygen in its nucleophilic attack of the benzyl bromide. The exact role of carbonate is still under investigation. This silicon-assisted condensation is being applied to similar A_2B monomers for the preparation of dendritic polyethers.

76. A NMR RELAXATION STUDY OF THE INFLUENCE OF CHARGE TRANSFER INTERACTIONS ON POLYMER BLEND PHASE STRUCTURE. Alexandra Simmons and Almeria Natansohn, Department of Chemistry, Queen's University, Kingston, Ontario, Canada K7L 3N6.

Miscible blends of poly((N-ethyl carbazol-3-yl) methyl methacrylate, an electron donor polymer and poly(2,-[(3,5-dinitrobenzoyl)oxy] ethyl methacrylate, an electron acceptor polymer were prepared. Two-phase decomplexed systems were obtained by heating above 185°C. Miscible blends have T_gs elevated with respect to the starting materials due to intermolecular charge transfer (CT) interactions, while decomplexed blends have two T_gs. $T_{1\rho}(H)$ and $T_{1\rho}(C)$ were determined by solid-state CP/MAS NMR. Miscible blends posess a short $T_{1\rho}(H)$ relative to the homopolymers, indicating intimate mixing and a reduced free volume and interproton distance due to pulling together of the chains by charge transfer interactions. Decomplexed blends exhibit two $T_{1\rho}(H)$s. The $T_{1\rho}(C)$s of the blends are longer than those of the components, reflecting reduced mobility due to interchain CT interactions.

77. ELABORACION DE PASTAS DE ALTO RENDIMIENTO: INFLUENCIA DE LA SULFONACION SOBRE LAS PROPIEDADES PAPELERAS. Carrasco, F., Garceau, J.J., Ahmed, A. y Kokta, B.V., Université du Québec à Trois-Rivières, C.P. 500, Trois-Rivières (Québec) G9A 5H7 Canada.

La fabricación de pastas químico-mecánicas en fase vapor comporta una serie de operaciones: impregnación de la madera con productos químicos, reacción con vapor de agua, descarga de las virutas (descompresión rápida o lenta), lavado, refinado y blanqueo.

Cada una de estas etapas tiene sus propias condiciones operatorias, lo cual dificulta sobremanera la comparación de distintos procesos de pasteado. Ahora bien, dichas condiciones operatorias afectan directamente a la velocidad de sulfonación de la madera, la cual influye notablemente sobre el rendimiento, las propiedades papeleras y la energía de refinado. Este estudio ha puesto de manifiesto que es posible, independientemente de las condiciones operatorias, relacionar las propiedades de las fibras (densidad, concentración de iones sulfonato y carboxilato, % de fibras largas) con las propiedades del papel (opacidad, longitud de ruptura, índice de desgarro y energía de refinado).

78. **ATOMIC FORCE MICROSCOPY ON COLD - EXTRUDED POLYETHYLENE**
S.N. Magonov[1,4], K. Qvarnstrom[2], D.M. Huong[1], V. Elings[3], H.-J. Cantow[1,2]
[1] Freiburg Materials Research Center (F·M·F) and [2] Institute of Macromolecular Chemistry; University of Freiburg, Stefan-Meier-Str. 31, D - 7800 Freiburg, FRG; [3] Digital Instruments Inc., Santa Barbara, CA 93110, U.S.A.; [4] Institute of Chemical Physics of the USSR Academy of Science, Kosygin ul.4, SU-117977 Moscow, USSR.

Results of Atomic Force Microscopy (AFM) studies on cold-extruded polyethylene (PE) are presented. The expected ordering of PE extended chains is favorable for AFM detection and interpretation of images. Pieces of extruded PE were embbedded in an epoxy resin and cut along the extrusion direction. AFM images from the range 700 x 700 nm down to the atomic scale were obtained on the flat surface prepared. In addition to the characterization of the supermolecular structure beyond the limits of conventional electron microscopy, the molecular organization can be directly depicted on the atomic scale. Large scale AFM micrographs of PE reveal the fibrillar morphology of the uniaxially oriented material. Microfibrils with diameters in the range of 20-90 nm are aligned parallel to the extrusion direction. In the smaller scale images the oriented patterns were assigned to the individual polymer chains. The surface texture pattern shows chain overlapping as a result of the extrusion process. Parts of extended chains were also found. AFM responses of individual methylene groups have been resolved.

79. **MECHANICAL PROPERTIES OF POLY(METHYL ACRYLATE-CO-DIMETHYL AMINOETHYL METHACRYLATED/EPOXY RESIN IPNS.** Jingyuan Wang, Yuwei Li, Peixiang Xing, Xinyi Tang. Department of Chemistry, Jilin University, Changchun, 130023, People's Republic of China.

Simultaneous interpenetrating polymer networks were perpared from poly (methyl acrylate-co-dimethyl aminoethyl methacrylate)/epoxy resin [P(MA-DMA)/EP]. Mechanical properties of IPNs were measured. A single transition temperature and a better compatibility was observed. This shows that the sample is the is the mixed polymer of cross-linking and interpenetrating networks. When P(MA-DMA)/EP = 20/80, there exists positive synergistic effect. (Tg) cal was calculated by FOX equation and it was found that (Tg)exp-(Tg)cal was proportional to dexp-dcal.(The difference in densities between the calculated and obtained experimentally). Moreover, the value of (Tg)exp-(Tg)cal can represent approximately the ratio of interpenetrating to cross-linking.

80. **CONJUGATE THREE-COMPONENT INTERPENETRATING POLYMER NETWORKS AND THEIR APPLICATIONS.** D. Jia, Y. Pang, Zh. Dai and X. Liang, Dept. of Polymer Materials Science and Engineering, South China University of Technology, Guangzhou 510641, P. R. China.

The conjugate three-component IPNs is a new concept developed by the authors recent year. They are a novel kind of special IPNs, where one component as a common network tightly weaves together other two independent networks by respect interpenetration with them at the same time, forming multicomponent polymeric materials with particular structures, morphologies, and then properties. This paper investigates the conjugate three-component IPNs, especially polybutadiene-based polyurethane/poly(styrene-co-divinylbenzene)/vulcanized rubber powder and polyether-polyurethane/poly(methyl methacrylate-co-ethylene glycol dimethacrylate)/vulcanized rubbers or poly(vinyl chloride), in regard to their syntheses, characterizations, properties, and applications in modifications of vulcanized rubber powders or in adhesion of polymeric materials, etc..

81. IPN - MODIFIED POROUS BEAD POLYMERS AS SUPPORTS FOR ENZYME IMMOBILIZATION
Anna B.Wójcik, Faculty of Chemistry
———————— and
Jerzy Łobarzewski, Department of Biochemistry
M.C.Skłodowska University, 20-031 Lublin, Poland

A set of porous bead materials differing in chemical nature and porous structure have been subjected to the IPN-modification by swelling the materials with aqueous solutions of acrylamide and bisacrylamide monomers and polymerizing in situ. The obtained sequential IPNs ,having free amine groups, have been used for immobilization of three enzymes: glucoamylase, peroxidase and urease. The yield of protein immobilization and specific activities of the supported enzymes seem to be affected by structure of the starting materials and, in less extent, by conditions of IPN preparation.

82. STUDY ON THE THREE-STAGE LATAX INTERPENETRATING POLYMER NETWORKS. Zhang Liucheng, Li Xiucuo, Liu Zhiqing, Department of Chemical Engineering, Hebei Institute of Technology, Tianjin, P.R. C.; 300130

Tree-stage LIPN have reported in many papent literatures, but the investigation of the particle diameter, morphology and dynamic mechanical properties has little reported yet in literature. In this paper the LIPN PBA/PS/PMMA was prepared by three-stage emulsion polymerization and its morphology was studied by RuO_4-staining technique The particle diameter and its distribution and the dynamic mechanical properties of LIPN PBA/PS/PMMA were determined. The influence of monomer feeding style and the dose of initiator and emulsifier on the morphology and particle diametr has been studied.

83. STRUCTURE AND PROPERTIES OF AB CROSSLINKED POLYMERS. Wenzhong Liu, Xiaozu Han, Jingjiang Liu, and Huarong Zhou, Changchun Institute of Applied Chemistry, Academia Sinica, Changchun, China 130022.

The effects of crosslink density, composition, structures and defect of networks and hydrogen bond on the T_g, morphology, mechanical properties of some AB crosslinked polymers are investigated. The crosslink density is one of the main factors effecting on the dynamic mechanical properties of the studied polymers. The sol fraction and free ends in crosslinked networks can improve the damping behavior in lower temperature range by decreasing the storage modulus and increasing the loss modulus. Adding small amount of acrylic acid in the polyurethane/PMMA crosslinked polymers, the phase compatibility and the damping factor could be increased due to the formation of hydrogen bonds. All of the AB crosslinked polymers studied here are multiphase ones with small polydisperse domains.

PMSE

84. COMPLEXITIES IN EVALUATING BEHAVIOR OF ASSOCIATIVE THICKENERS. J. Edward Glass; J. Philip Kaczmarski, Department of Polymers and Coatings, North Dakota State University, Fargo, North Dakota 58102.

Complexities in synthesizing, characterizing and evaluating aqueous solution behavior of hydrophobically ethoxylated urethanes (HEUR) are discussed. The effect of the extent of hydrophobe modification is shown to have a pronounced influence in aqueous surfactant solutions. Synthesis of HEUR thickeners by a step growth mechanism gives a broad molecular weight distribution which complicates modeling aqueous solution behavior.

85. COMPARISONS MADE ON STEADY AND DYNAMIC SHEAR INDUCED MORPHOLOGY IN BLENDS BY OPTICAL RHEOMETRY
K. Søndergaard, J. Lyngaae–Jørgensen, Department of Chemical Engineering Tech. Univ. of Denmark, Bldg. 229, DK–2800 Lyngby, Denmark

The experimental evaluation of strucutral changes in polymer blends is of fundamental importance in blend processing. In this paper comparisons are made on steady and dynamic shear induced morphologies in a polymer blend revealed by light scattering measurements during flow. A model blend system consisting of polystyrene and polymethylmethacrylate has been selected because of favorable features of the constituent components in relation to the purpose of the study. The measurements have been performed with a laser rheo–optical unit built on a Rheometrics Mechanical Spectrometer (RDII/800). The instrument allows rapid detection of the angular distribution of scattered light from thin samples of multiphase polymer melts during steady state and dynamic shear flow experiments. Comprehensive rheo–optical studies have previously been performed on the model blend in steady shear flow. The aim of this work is to establish an experimental technique to elucidate a samples structure and develop relationships between structure and the components of the stress tensor under transient shear flow conditions. Some preliminary experimental results obtained from the combined dynamic mechanical and optical measurements are presented with reference to steady state measurements.

86. SOLUTION PROPERTIES OF MMA/HEMA COPOLYMERS. Joaquín Palacios, Amparo Ortega, Jorge Sierra. División de Estudios de Posgrado, Facultad de Química, UNAM, Ciudad Universitaria, 04510 México, D.F. México.

Solution properties of methyl methacrylate/hydroxyethyl methacrylate (MMA/HEMA) copolymers, were studied in butanone, tetrahydrofurane and butanone-isopropanol solvents. Experiments were run at 25°C in an Ubbelholde viscometer. The GPC molecular weights and distributions data were combined with intrinsec viscosities, to calculate the hydrodynamic volumes of materials. Values of the a y K Sakurada constant are reported for these systems. The unperturbed dimensions were correlated with the composition of copolymers. Values for the 95, 90,80% mole copolymers, are presented.

87. TWO-PHASE EPOXY INTERPENETRATING THERMOSETS, Isabelle Frischinger and Stoil K. Dirlikov, Coatings Research Institute, Eastern Michigan University, Ypsilanti, MI 48197

Homogeneous mixtures of a liquid rubber based on prepolymers of epoxidated soybean oil with amines, diglycidyl ether of bisphenol A epoxy resins, and commercial diamines form, under certain conditions, two-phase thermosetting materials which consists of a rigid epoxy matrix and randomly distributed small soybean rubbery particles. Particle size varies in the range from 0.1 to 5 microns and depends on the soybean fraction, degree of rubber prepolymerization, nature of diamine, kinetics of curing, etc. Phase inversion phenomenon is observed at about 30% "soybean" content. Initial unoptimized results show that these two-phase thermosets have better toughness, similar to that of other rubber-modified epoxies, lower water absorption, and low sodium content. They exhibit slightly lower glass transition temperatures and Young's modili in comparison to the unmodified thermosets whereas their dielectric properties do not change. The epoxidized soybean oil is available at a price below that of commercial epoxy resins and appears very attractive for epoxy toughening in industrial scale.

88. POLY(ETHYLENE TEREPHTHALATE) SEMI-IPNS WITH FUNCTIONALIZED TRIGLYCERIDE OILS. L. W. Barrett and L. H. Sperling, Materials Research Center, Whitaker Laboratory #5, Lehigh University, Bethlehem, PA 18015.

Swelling and rubbery modulus studies show that castor oil ester/urethane networks have a solubility parameter of 9.8 $cal^{1/2}cm^{-3/2}$, very close to that of PET (9.7 $cal^{1/2}cm^{-3/2}$). A semi-IPN of PET and castor oil network has been produced which displays a single glass transition temperature by DSC, located near that predicted by the Fox equation, suggesting a miscible semi-IPN. Scanning DSC crystallization studies show that the semi-IPN enhances PET crystallization from the glass, due to the lowered glass transition temperature, but hinders crystallization from the melt.

89. INTERPENETRATING POLYMER NETWORK FORMED FROM CASTOR OIL POLYURETHAN AND VINYL COPOLYMERS
Peiwen Tan, Department of Chemical Engineering
Katholieke Universiteit Leuven, 3001 Heverlee, Belgium

The synthesis, morphology, physical and mechanical properties and formation kinetics of the graft Interpenetrating Polymer Network (IPN) formed from castor oil polyurethane (PU) and vinyl copolymer at room temperature are studied. It is shown that an initiator composed of benzoyl peroxide and dimethylanline is effective. When the weight percentage of PU is higher 50elastomeric behavior. The dynamic mechanical spectra of the graft IPNs show two Tg's for the IPN containing polystyrene or polymethyl methacrylate but only one Tg for the containing only polyacrylonitrile, indicating more molecular mixing in the latter IPN. The study of the formation kinetic of the graft IPNs indicated that castor oil not only reacts with diisocyanate (TDI) in the formation of polyurethane network, but also takes part in the formation of vinyl copolymer chain. The IPNs obtained is rather a grafted IPN than a semi-IPN.

90. STUDIES OF ROOM TEMPERATURE CURED CASTOR OIL POLYURETHANE/VINYL OR ACRYLIC POLYMER INTERPENETRATING POLYMER NETWORKS. H. Q. Xie, C. X. Zhang and G. G. Wang, Department of Chemistry, Huazhong University of Science and Technology, Wuhan, 430074, P. R. of China

The relative formation rates of castor oil polyurethane and acrylic or vinyl polymer during formation of the simultaneous grafted IPNs were compared by using IR spectrophotometry. The formation rate of polyurethane network is higher than that of the graft polymerization rate of methyl methacrylate. Time of gelation decreases with increasing polyurethane content. The rate of gel formation in the IPN containing poly(methyl acrylate) is higher than that in the IPN containing polystyrene. The relation between gel point and gel content of the IPNs with various compositions nearly agrees with the Flory-Stockmayer equation. There exists a maximum tensile strength of the IPNs, when the polyurethane content is about 50%. TEM microphotographs indicated that polyurethane existed as domains in the continuous phase of polyacrylate. A broad glass transition temperature of the IPN occurred in the dynamic mechanical spectrum of the IPN.

91. SIMULTANEOUS INTERPENETRATING POLYMER NETWORKS FROM CASTOR OIL POLYURETHANES AND POLY (STYRENT-METHYL METHACRYLATE-METHACRYLIC ACID). Peixiang Xing, Yuwei Li, Jingyuan Wang, Xinyi Tang. Department of Chemistry, Jilin University, Changchun, 130023, People's Republic of China.

A series of simultaneous interpenetrating networks from COPU/P(St-MMA-MAA) were prepared. The resulting interpenetrating polymer networks were characterized by using stress-strain analysis, transmission electron microscopy, dynamic mechanical spectroscopy and thermogravimetric analysis. The maximum ultimate tensile strength was observed at COPU/P(St-MMA-MAA) = 10/90 and it was shown that there exists some synergistic behavor. The dynamic mechanical behavior and morphologies of the samples show that the phase separation of the IPNs was extremely obvious. The addition of the internetworks grafting reagent can improve compatibility, and lead to a single peak of Tg transition, showing the occurrence of "forced compatibility".

92. SYNTHESES AND CHARACTERIZATION OF CTBN TOUGHENED EPOXY RESIN/CASTOR OIL POLYURETHANE IPN. <u>Xiaozu Han</u>, Ying Wang, Shixia Fan and Qingyu Zhang, Changchun Institute of Applied Chemistry, Academia Sinica, Changchun, China 130022.

CTBN toughened epoxy resin and castor oil polyurethane simultaneous joined IPN was synthesized and characterized by studying the stress-strain property, density, swelling behavior, dynamic mechanical property and morphology. The IPN has excellent mechanical properties and damping behavior, so it can be used as a good vibration- or sound-absorbing material. The tanδ-temperature curves show that there are two transition peaks, one of which is attributed to polyurethane network joining some epoxy resin chains, and another transition belongs to contribution of epoxy resin network. The morphology of the IPN shows that the domains (about 1μ) arised from CTBN rubber phase also exist in the IPN system.

93. EQUILIBRIUM AND NON-EQUILIBRIUM MICROPHASE STRUCTURE OF INTERPENETRATING POLYMER NETWORKS, <u>Yu. S. Lipatov</u>, Institute of Macromolecular Chemistry, Academy of Science of the Ukrainian SSR, Kiev 252 160, USSR.

The formation of both simultaneous and sequential IPNs proceeds under non-equilibrium conditions due to superposition of: (1) chemical reactions with their specific kinetics, and (2) physical processes of microphase separation due to cross-linking. Microphase separation is incomplete — the degree being dependent both on physical kinetics of spinodal decomposition and chemical kinetics for many systems. On the basis of structural and kinetic data, we established the mutual influence of IPN-forming components on the kinetics of formation of each network. We conclude that IPNs can exist in both quasi-equilibrium and non-equilibrium structures. For gradient IPNs the composition and ratio of separated phases was estimated.

94. STRUCTURAL ANALYSIS OF AGGREGATED POLYSACCHARIDES BY HPSEC/VISCOMETRY. <u>Marshall L. Fishman</u>, David T. Gillespie, and Branka Levaj, U.S. Department of Agriculture, ARS, ERRC, 600 East Mermaid Lane, Philadelphia, Pennsylvania 19118.

Three high performance size exclusion columns placed in series were calibrated in radii of gyration (R_g) and intrinsic viscosity (IV) x molecular weight (M) with a series of pullulan and dextran standards ranging in M from 853,000 to 10,000. Online detection was by differential refractive index (DRI) and viscometry (DV). By employing two forms of universal calibration, it was possible to obtain R_g, IV, and M for each injection of the unknown, in this case pectin, a group of complex plant polysaccharides. Curve fitting of chromatograms from DRI and DV detectors revealed that pectins were composed of 5 macromolecular sized species. For pectins extracted from a large number of sources, a Mark-Houwink plot gave a correlation coefficient of 0.2 whereas a plot of log IV against log R_g gave a correlation coefficient of 0.9. These results in addition to those from the analysis of several pectins from peach fruit indicated that pectins were aggregated and highly assymetric in shape. Thus it was concluded that scaling law exponents for pectins were effected by both shape and state of aggregation, rather than shape alone.

95. **ONE PARAMETER CORRELATION BETWEEN INTRINSIC VISCOSITY AND MOLECULAR WEIGHT**
R. Amin Sanayei and K.F. O'Driscoll
Institute for Polymer Research, University of Waterloo, Waterloo, Ont. Canada, N2L-3G1

Recent theoretical investigations of intrinsic viscosity of polymers lead to a single adjustable parameter model. $[\eta] = K_\theta \overline{M}_r^{1/2} + K' \overline{M}_w$ where $\overline{M}_w$ and $\overline{M}_r$ are weight average and radius average molecular weights of polymer respectively. This model indicates that $[\eta]$ is a function of the unperturbed average radius of the polymer coil ($\overline{M}_r^{1/2}$) and the expansion of the polymer coil which is proportional to $\overline{M}_w$. K_θ is constant and dependent only on the structure of the polymer coil and to some extent on temperature; whereas K' reflects the extent of the solvent-polymer interaction. The intrinsic viscosity of PSTY, PMMA, PE, and nitro-cellulose in different solvents with molecular weights ranging from 10^3 to 10^7 are well predicted by this model. In light of this new theory, a universal calibration curve for a GPC can be established without the need for K and a values.

96. Light Scattering Viscometry, Philip J. Wyatt and David W. Shortt, Wyatt Technology Corporation, Santa Barbara, CA 93103 USA

The applicability of universal calibration to a broad range of polymers permits, under special circumstances, the extraction of molecular weights from measurements of intrinsic viscometry. But absolute molecular weights may be obtained directly from light scattering measurements and do not require indirect measurements based on calibration standards. Nevertheless, recent interest in viscometric techniques suggests that these measurements per se may be of significance for macromolecular characterization. The hydrodynamic radius, for example, may be determined directly from such viscometric determinations. By combining differential light scattering (multiangle) with universal calibration, one may derive both intrinsic viscosity and relative viscosity values without the use of a viscometer. These deductions are particularly important for identifying the presence of microgels and similar structures since such structures generally correspond to molecular weights that cannot be derived from universal calibration. For such microgels, the viscometric parameters will confirm a hydrodynamic radius that decreases while the r.m.s. radius and weight averaged molecular weight increases.

97. HIGH-PRECISION CHARACTERIZATION OF POLYMER CONFORMATION BY SEC WITH AN ON-LINE VISCOMETRY AND LASER LIGHT SCATTERING
Christian Jackson, Howard G. Barth and Wallace W. Yau, E. I. du Pont de Nemours and Company, Central Research and Development, P. O. Box 80228, Wilmington, DE 19880-0228

The integration of on-line viscosity and light-scattering (LS) detectors with SEC enables the determination of polymer molecular weight, intrinsic viscosity and molecular size distributions in a single experiment. This information can then be used to measure polymer conformation and branching distributions. The coupling of a viscometer and a LS detector in one SEC instrument improves the precision and the dynamic range of SEC for polymer conformation studies. The results obtained with the integrated instrument can cover a wide molecular size range and are insensitive to adverse variations in SEC experimental conditions (such as flowrate inconsistency, band broadening, column deterioration, etc.). Additional precision can be obtained by column overloading to reduce the effect of detector baseline drift. The precision of the integrated instrument means that truly useful measurements of polymer conformation can be obtained. To demonstrate this precision, the Mark-Houwink exponent of a broad MWD polystyrene (NBS-706) in toluene was measured with 10 repeated sample injections. SEC-LS-VIS gave a value of 0.728 with a standard deviation of 0.005 or better, compared with the questionable value of "0.58" with a standard deviation of 0.43 from the same measurements made by SEC with an on-line multi-angle laser light scattering detector, SEC-MALLS. The high precision of the SEC-LS-VIS approach is especially valuable for the elucidation of architectural differences among star polymer samples.

98. SEC-VISCOMETRY-RIGHT ANGLE LASER LIGHT SCATTERING (SEC-VIS-RALLS).
Max A. Haney, Viscotek Corp., 1032 Russell Dr., Porter, Texas 77365.
Christian Jackson and Wallace W. Yau, E.I. DuPont de Nemours and Co., CR&D, Experimental Station, P.O. Box 80228, Wilmington, Del 19880-0228.

The two types of light scattering photometers currently being used for on-line SEC polymer detection are the low angle laser light scattering (LALLS), and the multi-angle laser light scattering (MALLS). In either of these two approaches, the attainment of the zero degree forward light scattering intensity is the targeted goal for the experimental determination of polymer molecular weight.

With the addition of an on-line viscometer detector, an attractive alternative to these "zero" angle techniques is now possible. We propose to use right angle (laser) light scattering (RALLS or RALS) in combination with the SEC-Viscometry technology. The key to the success of the SEC-VIS-RALS approach is the high s/n performance of the 90 degree scattering intensity, plus the unique opportunity of using the viscosity information for the correction of the light scattering angular asymmetry. The concept and the iterative computer algorithm for this light scattering asymmetry correction will be discussed and illustrated with the use of the SEC-VIS-RALLS results on polystyrene samples, obtained with a MALLS detector set at 90 degrees. Results were also obtained for aqueous dextran and PEO samples using the SEC-VIS-RALS method with a simple fluorescence detector as the RALS detector. Using the fluorescence detector in this way follows the work by R.M. Jones, B. Cunico, and M. Kunitani in a poster presentation at Pittsburgh Conference, Chicago, March 4-8, 1991.

99. SOLID STATE CHARACTERIZATION OF MULTICOMPONENT POLYMER SYSTEMS BY DIGITAL IMAGE ANALYSIS AND REAL TIME PULSED NMR. Toshio Nishi and Takafumi Hayashi, Department of Applied Physics, The University of Tokyo, 7-3-1 Hongo, Bunkyo-ku, Tokyo 113, JAPAN

There are many ways to characterize multicomponent polymer systems in solid state by scattering techniques, microscopy, relaxation phenomena and so on. They are mainly related to the high order structure or molecular motion in the system. We have recently developed digital image analysis(DIA) for the quantitative characterization of high order structures and real time pulsed NMR(RTPNMR) system for the characterization of molecular motion in multicomponent polymer systems.
Principles and application of DIA and RTPNMR to several systems will be given. They include DIA studies of phase separation mechanism in polymer blends and pattern formation of spherulite from crystalline-amorphous polymer blends. Application of RTPNMR for the studies on crosslinking process of epoxy resins and di-acetyenes will be also shown at the symposium.

100. NEOPENTYLDIAMINE POLYAMIDES AND BLENDS
E. E. Paschke, Amoco Chemical Company, PO Box 3011, Naperville, Illinois 60566

Polyamides based on neopentyldiamine have not been previously reported and they exhibit high glass transition temperatures and good thermal stability. The homopolymer of terephthalic acid and neopentyldiamine has a glass transition emperature of 188°C and a decomposition temperature of 378°C (1% weight loss). These amorphous polyamides are miscible with other amorphous and semi-crystalline polyamides and exhibit both high glass transition temperature and crystalline morphology. The 1/1 blend of nylon 66 has a high glass transition temperature and still is semi-crystalline since transamidolysis does not occur.

101. **Designing Polymer Blends Which Exhibit Synergistic Properties**
by *Eiji Ueda[1], Hugh Graham, Chuan Qin, Laurence A. Belfiore*,
Dept. of Chemical Engineering, Colorado State University, Fort Collins, Colorado, 80523, USA
[1]Asahi Chemical Industry, Okayama, JAPAN

This laboratory has been pursuing phenomenological correlations between macroscopic properties and molecular-level structure in strongly interacting polymer blends. Several polymer blend systems have been examined where the thermal and mechanical properties of some blends exceed those of either pure component. This is believed to be due to synergistic effects caused by the interaction of electron-donating and electron-accepting groups contained within the two components. Amorphous blends of poly(4-vinylpyridine-co-styrene) and poly(styrene-co-methacrylic acid) show a remarkable synergistic effect on Tg due to formation of hydrogen bonds near the stoichiometric equivalence of 4-vinyl pyridine and methacrylic acid groups within the mixtures. The Nylon-11/poly(N,N-dimethylacrylamide) system shows a strong synergistic increase in fracture strength. Both mechanical and thermal synergy are observed in the poly(2-vinylpyridine)/PEO blends.

102. Molecular and morphological characterization of 1-octene LLDPE (Tref) fractions and their blends.

F. Defoor, G. Groeninckx and H. Reynaers

Fractionation of ethylene/alpha-olefin copolymers (LLDPE) with respect to short chain branching using temperature rising-elution fractionation (TREF) is a very effective method for the elucidation of the complex chain microstructure of these copolymers. In the TREF-method, fractionation of short chain branched polyethylene molecules is obtained on the basis of their crystallizability which is directly related to the degree of short chain branching.

103. SYNCHROTRON SAXS STUDY OF MEAN-FIELD AND ISING CRITICAL BEHAVIOR OF POLY(2-CHLORO-STYRENE)/POLYSTYRENE BLENDS. Benjamin Chu, Qicong Ying, Kung Linliu, Ping Xie, Tong Gao, Yingjie Li, Dept. of Chem., State Univ. of New York at Stony Brook, Long Island,NY 11794-3400 and Takuhei Nose, Mamoru Okada, Dept. of Polym. Chem. of Tokyo Institute of Technology, Tokyo, Japan.

Critical fluctuations of a polymer blend poly(2-chloro-styrene)/polystyrene (P2ClS/PS) with 22.6 % weight fraction dibutyl phthalate (DBP) have been studied by synchrotron small angle X-ray scattering (SAXS). Critical SAXS patterns were carefully investigated in the neighborhood of its critical mixing point, both at and away from the critical solution concentration. The results clearly show a crossover in the static susceptibility from the mean-field to the Ising behavior at 2.8°K below the critical mixing temperature T_c of 407.65°K. The critical mixing temperature by extrapolation from SAXS patterns is consistent with independent phase separation measurements by using a modified centrifugal apparatus.

104. DOMAIN GROWTH DURING THE SPINODAL DECOMPOSITION UNDER SIMPLE SHEAR FLOW.
Mahesh A. Kotnis and M. Muthukumar
Polymer Science and Engineering
University of Massachusetts at Amherst, MA 01003 USA

We present a numerical study of the dynamics of spinodal decomposition of a symmetric polymer blend of protonated and deuterated polybutadienes under simple shear flow. From our numerical values we have generated computer images of the formation and growth of the phase separating domains.
We have also computed the structure factors along the directions parallel and perpendicular to the flow for various times during the phase separation. We see a strong anisotropy in the scattered intensity indicating an enhanced growth in the direction parallel to the flow.

105. CONFORMATIONAL REARRANGEMENTS OF FLEXIBLE POLYMER CHAINS IN FLOW.
M.J. Menasveta and D.A. Hoagland, Polymer Science and Engineering Department, University of Massachusetts, Amherst, Massachusetts 01003

During flow of dilute polymeric fluids, flexible molecules may be deformed from their equilibrium conformations, a process ultimately responsible for non-Newtonian rheology. With large elongational strain rates and limited vorticity, the level of deformation has often been assumed to approach the fully unraveled state. In this study, light scattering methods are employed to follow the deformation of dilute poly(styrene) chains in toluene under uniaxial elongational flows of varying strain rate; flow birefringence is measured simultaneously. The onset and saturation of birefringence does not correlate with complete unraveling of the poly(styrene) chains. Instead, much more limited stretching is found, even when the birefringence has saturated at high strain rate. The maximum stretching ratio (i.e., $R_{g,\parallel}/R_{g,eq}$) is of order 2 and varies linearly with molecular weight. It is suggested that the current measurements, performed at the stagnation point of an opposed jets flow device, reflect a limited residence time in the vicinity of the stagnation point, a time that may be sufficient for segmental orientation but not for chain unraveling.

106. RHEO-OPTICS OF FLOW INDUCED PHASE SEPARATION OF POLYMER SOLUTIONS. Jan van Egmond, Gerald G. Fuller, Department of Chemical Engineering, Stanford University, Stanford, CA 94305-5025

In recent years, flow-induced enhancement of concentration fluctuations in semi-dilute polymer solutions has received considerable attention in the literature. Of particular interest are the structure and dynamics of these concentration fluctuations, which may be investigated using non-intrusive optical techniques such as dichroism and small angle light scattering measurements. This paper investigates time dependent anisotropic growth and orientation of concentration fluctuations in a semi-dilute solution of polystyrene in dioctyl phthalate above the cloud point temperature in simple shear flow. For weak flows and small time it is found that, in the plane of shear, concentration fluctuations orient and grow perpendicular to the direction of extension of the flow. Our experimental findings of this anisotropic growth are the qualitative agreement with the predictions of the Helfand-Fredrickson model for the equations of motion for concentration fluctuations in polymer solutions.

107. PHASE BEHAVIOR OF A POLYMER SOLUTION STUDIED BY LIGHT SCATTERING DURING STEADY SHEAR AND FOLLOWING CESSATION OF SHEAR, Alan I. Nakatani, Dean A. Waldow, and Charles C. Han, National Institute of Standards and Technology, Polymers Division, Gaithersburg, MD 20899

An apparatus has been constructed for conducting light scattering experiments on polymeric samples under the influence of a well-defined simple shear field. A two-dimensional CCD array detector is used for quantitative measurement of anisotropic scattering patterns which commonly develop from sheared samples.
We describe results obtained from a 3% solution of polystyrene in dioctyl phthalate. This sample has been examined previously by a number of different researchers and has demonstrated shear induced phase separation behavior. The steady shear results obtained are consistent with the previous investigations. We also measure the kinetics of remixing following cessation of a steady shear. The results are analyzed using linearized Cahn-Hilliard-Cook theory to obtain values for the apparent diffusion coefficient and mobility.

108. **The Structure and Dynamics of a Semidilute Polymer Solution Under Shear Flow.**
Paul K. Dixon and David J. Pine, *Exxon Research and Engineering Co., Annandale NJ 08801,*
Xiao-Lun Wu, *Department of Physics, The University of Pittsburgh, Pittsburgh PA 15260.*

Elastic light scattering measurements of a semidilute polymer solution indicate that laminar shear flow enhances concentration fluctuations in both the weak ($\dot{\gamma}\tau < 1$) and strong ($\dot{\gamma}\tau > 1$) shear regimes, where $\dot{\gamma}$ is the shear rate and τ is the longest relaxation time of the solution. We observe anisotropic structure in the scattering that is qualitatively different from binary liquid mixtures under shear flow as a result of the strong coupling between the concentration fluctuations and the hydrodynamic flow (E. Helfand and G. H. Fredrickson, PRL **62**, 2468 (1989)., A. Onuki, PRL **62**, 2472 (1989)). Transient response measurements of this structure upon the termination of shear indicate that it relaxes in an exponential manner. The measured decay rate crosses over from diffusive behavior at small q to a rate limited by the stress relaxation time, τ, at large q.

109. PREPARATION OF TRANSPARENT GLASS FIBER REINFORCED PMMA MATRIX COMPOSITES.
J.O. Stoffer, D.E. Day, and K.D. Weaver, Materials Research Center,
University of Missouri-Rolla, Rolla, MO 65401

The Transparency Materials Improvement Program at the University of Missouri-Rolla Materials Research Center has been engaged in the development of the technology and proecssing necessary to produce optically transparent glass fiber reinforced PMMA matrix composites by adjusting the refractive index of the fiber to match that of PMMA. A pressure cured as-cast system for composite preparation is described. By curing at 65°C in '1000 psi N_2 for 18 hours, composites 4" x 6" and 0.25" thich containing up to 15 vol% glass fiber have been prepared. These composites give flexural bending strengths over four times greater than commercial PMMA of the same dimensions and maintain 89% optical transmission at 589.3 nm. at 19.5°C.

110. OPTICAL AND MECHANICAL PROPERTIES OF
POLY(METHYLMETHACRYLATE)/GLASS FIBER LAMINATES
R. K. Six, J. O. Stoffer, and D. E. Day, Univ. of Missouri-Rolla, Rolla, MO 65401

Transparent poly(methylmethacrylate)/glass fiber laminates have been constructed using a hot pressing technique. These composites exhibit tensile strengths which are significantly higher than commercial PMMA. When the laminates are pressed onto the surfaces of cast PMMA, three point bending strength is improved. The glass fiber in the composite decreases the optical clarity only slightly while producing a colored halo effect from observed objects and light sources.

111. LIMITATIONS OF CURRENT EPOXY POLYMERS AS MATRIX FOR STRUCTURAL COMPOSITES IN
OILFIELD APPLICATIONS. J. -F. Hwang, J. P. Dismukes, C. N. Marzinsky, and
R. R. Mueller, Exxon Research & Engineering Company, Annandale, NJ 08801

This paper describes the limitations of current epoxy polymers as the matrix in polymer composites for use as structural materials in corrosive oilfield applications. Typical needs for oilfield pipes are for long-term resistance to medium pressure CO_2 and H_2S gases (ca. 150 psi) and 5% salt water, at 150°F. This translates to a requirement for reversible tensile behavior, namely a composite tensile yield strength of about 55 ksi. Although many polymer composites have ultimate tensile strength in excess of 55 ksi, current materials exhibit weeping failure due to yielding below 20 ksi. This failure mode was determined to be preceded by matrix microcracking which is in turn attributed to insufficient matrix toughness. Current approaches to toughen epoxy polymers are reviewed in terms of property and processing requirements for oilfield applications. Of the systems thus far reported in the literature, advanced epoxies with single phase morphology show the most promise for further evaluation.

112. SURFACE MODIFICATION OF WOOD FIBERS AND THEIR PERFORMANCE IN
POLYSTYRENE COMPOSITES. D. Maldas and B. V. Kokta, Centre de
Recherche en Pâtes et Papiers, Université du Québec à Trois-Rivières, C.P. 500, Trois-Rivières, Québec, Canada, G9A 5H7

In the present study, wood fibers {e.g. chemithermomechanical pulp (CTMP) of hardwood aspen; sawdust aspen and sawdust of softwood spruce} were precoated with a maleic anhydride (MA) and/or poly{methylene (polyphenyl isocyanate)} [PMPPIC], polystyrene using benzoyl peroxide (BPO) as an initiator. The effect of the concentration of fiber, MA, PMPPIC and BPO, on the mechanical performance of modified fiber-filled polystyrene (PS201 and PS525) composites has been evaluated. As opposed to virgin polystyrene and unmodified fiber-filled composites, properties of the modified fiber-filled composites accelerate with the rise in the concentration of BPO, MA and/or PMPPIC up to a certain limit, and then either decrease or level off. Properties improve even more when mixtures of BPO, MA and PMPPIC are used compared to the use of only one of them. The optimum concentrations of BPO, MA, PMPPIC and fiber vary according to wood species, nature of fiber and grades of polystyrene.

113 COMPOSITE FILMS OF POLYANILINE AND POLY(VINLYCHLORIDE)/ POLY(VINYLIDENE CHLORIDE), <u>J. O. STOFFER,</u> O. C. SITTON, N.L.D. SOMASIRI AND L. C. HUNG, UNIV. OF MISSOURI-ROLLA, ROLLA, MO 65401

Polyaniline film was produced by a chemical deposition method. A continuous phase of polyaniline, 0.021 mm in thickness, was coated onto Saran Wrap^{TM} chemically. The mechanical properties of the composite film improved dramatically compared to bare polyaniline. Conductivity of the polyaniline coating was found to be 8.35 Scm^{-1} which was higher than the 5.5 Scm^{-1} reported by Focke, et al]. The conducting form of the polyaniline coating can be converted into insulating form by exposing the film to ammonia fumes. Digital Scanning Calorimeter test result shows polyaniline is amorphous. Diffusivity of aniline through Saran Wrap^{TM} in several solvent systems was investigated.

114. THE EFFECT OF PARTICLE SIZE DISTRIBUTION ON MECHNICAL PROPERTIES OF GRANULAR COMPOSITES. by <u>Alisa Buchman</u>, Arieh Sidess, Yizhak Holdengraber, RAFAEL - Armament Development Authority, P. O. Box 2250, Haifa, Israel.

Cast systems composed of high content of granular filler in polymer matrix are greatly influenced by the filler to binder ratio, the size and the size distribution of the filler particles. A physical model for prediction of the optimal binder content for a given distribution of particle size in order to achieve a system with optimal mechanical properties is presented. The model is based on the relation between the optimal binder volume and the grain volume by constants. The constants of the model were calculated from physical and mechanical tests produced on SiC/Epoxy composite. Different constants of the model were determined for each property tested. The model was verified by correlation between predicted and measured optimal binder content of the composite system. The results indicate the reliability of the model in predicting optimal mechanical properties without additional tests.

115. SENSOR PLATE FOR MANUFUCTURING THERMOPLASTIC COMPOSITES. By <u>Eli Buchman</u>, RAFAEL, P.O. Box 2250, Haifa 31021, Israel, Avraam, I. Isayev, Dr., The Akron University, Akron, Ohio 44325, USA

Manufacturing of thermoplastic composite materials is a costly process which demands an accurate and reliable control during production. Consolidation process of thermoplastic composite prepregs is usually performed in compression molding under high pressure and high temperature. Conventional control systems, such as ultrasonic sensors, could not be used because of the high melting temperatures of the thermoplastic matrix and a new control concept had to be developed. This concept is based on resistance measurments of the consolidating matrix exploiting the electrical conductivity of the carbon fibers. Various defects such as voids, porous material and cracks are identified by this resistance measurements. Other kinds of control devices such as pressure transducers, displacement measuring units and thermocouples are added to completed the quality control during the production process. C-scan ultrasonic measurments performed on the end product have proven the reliability of this control system.

116. ENHANCEMENT OF POLYMER MISCIBILITY THROUGH IONIC FUNCTIONAL GROUPS. Adi Eisenberg, and Attila Molnar, Department of Chemistry, McGill University, Montreal, Que. H3A 2K6.

Functionalization of polymers to a low level with ionic groups can be a useful method of compatibilizing a wide range of polymer pairs. Often commercial materials contain inherent functional groups so that miscibility can be enhanced by functionalizing only the other polymer pair. These principles are illustrated for a few interesting polymers, including Nafion®, urethanes and nylon 6. Using this technique, polymer blend morphology can be controlled from complete phase separation and crystallinity to complete miscibility and no crystallinity.

117. **MISCIBLE BLENDS OF SULFONATED POLYSTYRENE AND POLY(XYLENYL ETHER), Richard A. Register** and Theodore R. Bell, Department of Chemical Engineering, Princeton University, Princeton, NJ 08544.

Blends of poly(2,6-dimethyl 1,4-phenylene oxide (poly(xylenyl ether), PXE) and zinc-neutralized lightly sulfonated polystyrenes (ZnSPS) are miscible up to at least 7.8 mol% sulfonation, as evidenced by single glass transitions calorimetrically and mechanically. As in ZnSPS, the blends possess a microdomain morphology due to aggregation of ionic groups, which can be probed by small-angle x-ray scattering. The number of ionic groups per aggregate is reduced in the blend relative to the neat ZnSPS, and more ionic groups are dispersed in the matrix. The favorable PS/PXE mixing drives the morphological change in order to make more PS accessible to the PXE homopolymer.

118. CHARACTERIZATION OF BLOCK COPOLYMER IONOMERS BY SMALL ANGLE X-RAY SCATTERING AND DYNAMIC MECHANICAL THERMAL ANALYSIS, W.P. Steckle, Jr., X. Lu, and R.A. Weiss, Dept. of Chemical Engineering and Polymer Science Program, University of Connecticut, Storrs, CT 06269-3136

Microphase separation in block copolymer systems impart characteristics of a conventionally crosslinked system. The high temperature mechanical properties of a block copolymer may be enhanced by the incorporation of bonded ionic groups. As with phase separated block copolymers, these "ionic" networks are thermally reversible. Dynamic mechanical thermal analysis and small angle x-ray scattering were used to characterize the viscoelastic behavior and morphologies of lightly sulfonated poly(styrene-ethylene/butylene-styrene) triblock copolymer ionomers. The effect of sulfonation level, counterion, and film preparation (solution cast vs. compression molded) will be described.

119. INVESTIGATION OF A LIQUID CRYSTAL DISPERSED IN AN IONIC POLYMERIC MEMBRANE. J. A. Ratto[*], S. Ristori, M. Escoubes, M. Pineri, F. Volino, CEA-CENG, DRFMC/SESM/PCM, BP 85X 38041 Grenoble, France, R. B. Blumstein, Polymer Program, University of Lowell, Lowell, MA 01850 *(On leave from U. S. Army Natick RD&E Center, Natick, MA 01760)

Nuclear Magnetic Resonance (NMR) is used to study the novel polymer/ liquid crystal composite membrane consisting of a sulfonic perfluorinated ionomer, Nafion and 4-cyano'4-n-hexylbiphenyl (6CB) prepared by swelling methods. The weight fraction of 6CB in the membrane varied from 10 to 35%. Scanning electron microscopy (SEM) confirmed that this polymer dispersed liquid crystal (PDLC) system has phase separation and droplets in the range of .2 to 1 um radius. The temperature transition T_{I-N} is the same as the bulk 6CB for high concentrations of 6CB and slightly less for low concentrations. Also, the lineshape and distribution of the 6CB is related to the amount of water present in the membranes. The proton NMR lineshapes of the nematic phase in the dry membrane is reproduced by a simulated powder spectra. This indicates that the magnetic field (2.2T) does not orient the nematic phase inside the droplets. Moreover, the simulation shows that the order parameter of the PDLC system is dependent of the 6CB content. Fluorine NMR suggests that the polymer matrix is not affected by the presence of the liquid crystal (LC).

120. SUBSTITUENT INDUCED CROSSLINK AGGREGATION IN POLYDIMETHYLSILOXANE NETWORKS D. A. Hoffmann, A. D. Stein, J. E. Anderson, C. W. Frank, M. D. Fayer, Department of Chemical Engineering, Stanford University, Stanford, CA 94305

This paper describes a new mechanism of heterogeneous network formation and a novel approach for the study of network structure. A new class of polydimethylsiloxane networks formed from condensation of telechelic silanols with triethoxy silanes containing covalently bound fluorescence probes have been studied using photophysical and swelling measurements. We propose that self-condensation of the crosslinking agent can be controlled by the polarity of a fourth, unreactive subtituent on the silane unit.

121. SOLID STATE NMR STUDY OF THE MICROPHASE STRUCTURE OF PS–PMPS DIBLOCK COPOLYMERS. W.Z. Cai, K. Schmidt–Rohr, N. Egger, B. Gerharz, H.W. Spiess, Max–Planck–Institut für Polymerforschung, P.O.Box 3148, D–6500 Mainz, FRG

Solid state NMR–techniques have been used to study the micro–phase structure of polystyrene (PS) and polymethylphenylsiloxane (PMPS) diblock copolymers of different composition and different total molecular weight. The proton NMR–spectra display motionally narrowed lines of the PMPS component at ambient temperatures. The mobility of the segments in the copolymers is different, however, from that in the homopolymer. This is related to the shift of T_g for the siloxane rich phase. By means of proton spin diffusion with ^{13}C detection – combining a dipolar filter with various line–narrowing detecting techniques (1), the microdomain dimensions have been measured quantitatively. The results are compared with determination of the phase separation by SAXS. In addition solid state NMR is able to detect heterogeneities on a scale as small as 2 nm in systems that show only a single glass transition temperature T_g.

Reference
(1) Schmidt–Rohr, K., Clauss, J., Blümich, B., Spiess, H.W., Magn. Reson. Chem. 1990, 28, 53.

122. POLY (ARYL ETHER-KETONE) SYNTHESIS, VIA COMPETING S_NAR AND $S_{RN}1$ MECHANISMS, R.S. Mani and Dillip K. Mohanty, Department of Chemistry and Center for Applications in Polymer Science, Central Michigan University, Mt. Pleasant, MI 48859.

High molecular weight Poly (aryl ether-ketones) are prepared conveniently from bisfluoro monomers and bisphenols by nucleophilic aromatic substitution reactions. However, this is not the case with bischloro monomers and bisphenols. During the substitution of chlorine atom of bischloro monomer by certain phenoxide ion, a competition exists between two different mechanisms; S_NAR and $S_{RN}1$ resulting in the expected product and side product respectively. In this presentation we will provide evidence to show how to limit the side reaction occuring through $S_{RN}1$ mechanism thus increasing the yields of the expected product. Results from the model reactions conducted with bischloro momomers and phenols will be reported.

123. "Semicrystalline Poly(Arylene Sulfide)Ketones via Amorphous Ketimine Intermediates," K. R. Lyon and J. E. McGrath, Virginia Polytechnic Institute & State University, NSF Science & Technology Center: High Performance Polymeric Adhesives and Composites and Dept. of Chemistry, Blacksburg, Virginia 24061-0212 and J. F. Geibel, Phillips Petroleum, Bartlesville, Oklahoma 74004.

Semicrystalline poly(arylene sulfide)ketones were synthesized via acidic regeneration of polymeric ketimine intermediates. Ketimine derivation of the activated halide prior to polymerization was achieved via aniline reactions which were catalyzed with molecular sieves. The ketimine intermediate still activates the halide, in the same manner as the ketone, but can be selectively removed after the polymerization. Use of the ketimine modified monomers allows one to synthesize an amorphous polymer at moderate (e.g. 150°C) conditions and thus characterize the structure via classical means. It has been demonstrated that it is possible to essentially quantitatively hydrolyze the ketimine to regenerate the semicrystalline poly(arylene sulfide) ketone, even with aqueous hydrochloric acid. The resulting material can then be washed, dried and characterized. Properties of these materials were determined and compared to the poly(arylene sulfide)ketones generated via the conventional route.

124. MODIFICATION OF FLUOROPOLYMER SURFACES BY SODIUM NAPHTHALENIDE. Jenifer T. Marchesi, KiRyong Ha and Andrew Garton, Polymer Program and Chemistry Department, University of Connecticut, Storrs, CT 06269-3136

Adhesion of metals to fluoropolymers is necessary for electronics applications, and typically involves polymer pretreatment with sodium naphthalenide. We have previously quantified the detailed chemical functionality resulting from this treatment. In this investigation, three fluoropolymer types were studied: polytetrafluoroethylene (PTFE), a copolymer of TFE and perfluoroalkylvinyl ether (PFA), and a copolymer of TFE and fluorinated ethylene propylene (FEP). Samples of varying crystallinity were obtained by using films with different processing histories: extrusion, skiving, or lamination, with quenching or annealling. No significant effect of fluoropolymer type or crystallinity was found for the rate of etch, as determined gravimetrically, or for the chemical functionality, as determined by XPS and IR-IRS. Samples of PTFE with varying surface roughness were prepared by lamination to copper foils of different surface topography. Surface area played a major role in the amount of fluoropolymer etched. From 90° peel tests of copper deposited onto these surfaces, it was found that a combination of chemical modification and surface roughness provides the highest strength, with the exact value and locus of failure dependent upon depth of etch and surface topography.

125. **SURFACE MODIFICATION OF POLYCARBONATE BY FAR-UV RADIATION,** Mark R. Adams and Andrew Garton, Polymer Science and Dept. of Chemistry, University of Connecticut, Box U-136, Storrs, CT 06269

Far-ultraviolet (uv) radiation (<240nm) is particularly useful for surface modification of polymeric materials. Most polymers absorb 95% of incident far-uv radiation in the top 300nm of the surface, while for some like polycarbonate, the depth is of the order of 30nm. In the far-uv region, quantum yields of bond-scission are up to 50 times larger than mid and near-uv regions, resulting in a controlled, dry photoetching process. Polycarbonate undergoes a photo-Fries rearrangement followed by surface etching when exposed to a far-uv source such as a deuterium lamp. Fourier-Transform Infrared Spectroscopy, X-Ray Photoelectron Spectroscopy, and Scanning Electron Microscopy were employed to better understand this photo-ablative degradation pathway.

126. Aging of Primed Polyolefin Surfaces

Jiyue Yang and Andrew Garton, Polymer Program and Chemistry Department U-136, University of Connecticut, Storrs, CT 06269 - 3136

It is known that inorganic and organometallic primers, such as triphenylphosphine and cobalt (II) acetylacetonate, when applied as thin layers to polyolefin surfaces, can lead to strong durable bonds with cyanoacrylate adhesives. However, the strength of such bonds deteriorates if the primed surfaces are aged before application of the adhesive. This aging effect has been studied by various techniques with regard to the physical and chemical characteristics of the surface as well as the chemical stability of primer. The results show that the surface energy and the primer concentration in the surface region decrease during aging. We conclude that diffusion of primers into the bulk of the polymer is the most important factor responsible for the deterioration of bonding strength of the primed surfaces.

127. ANTIPLASTICIZATION OF POLYIMIDES BY A NONREACTIVE DILUENT
Edward Shockey and Andrew Garton, Polymer Program and Dep't. of Chemistry, Box U-136, University of Connecticut, Storrs, CT 06269-3136

We have described previously the modification of epoxy resin properties by stiff, polar diluents called "epoxy fortifiers". We have explored a range of linear polyimides blended with 4-(2-hydroxy-3-phenoxypropyl) acetanilide which proved effective in epoxy systems. Addition of 16% of this diluent increases the modulus 20% - 30%, increases the strength 5% - 10%, and decreases the elongation at break. Quantitative size exclusion chromatography analysis of the prepared films indicates that the diluent does not chemically polyimides, while processing temperature does effect molecular weight. Results show suppressed sub-Tg relaxations, deviation from volume additivity, and lower water sorption, all consistent with a loss of free volume. We are presently investigating a new series of diluents which provide structural variety for the mechanistic study and improved thermal stability as required in processing these materials.

128. NUMBERS OF POLYMER DEGREES GRANTED IN THE UNITED STATES. R. D. Deanin, Plastics Engineering Department, University of Lowell, Lowell, MA 01854; and William Sacks, Plastics Institute of America, Suite 100, 277 Fairfield Road, Fairfield, NJ 07004-1932.

In 1990, 34 colleges and universities granted the following degrees that named polymers or polymer applications: 32 Associates, 367 Bachelors, 116 Masters, and 87 Doctors, for a total of 602. In 1989, 42 schools reported the following degrees in polymers, regardless of the specific name of the degree: 216 Bachelors, 282 Masters, and 272 Doctors, for a total of 770. Combining the two surveys, minus overlaps, gave a total of 1161 polymer degrees per year. A rough estimate of 137 schools that give polymer courses suggested that 2740 graduates per year have had at least one course in polymers.

129. POLYMERIC FLOCCULANTS.
Octavio Manero, Judith Cardoso and Teresa Orta*.
Instituo de Investigaciones en Materiales, UNAM. AP 70-360, Coyoacán 04510 México, D. F. * Instituto de Ingeniería, UNAM.

It has been observed that polymeric flocculants present advantages with respect to more traditional inorganic flocculants. In this investigation, we present a comparative study between commercially-available water-treatment compounds and water-soluble polymers which contain ionic structures, such as N-oxide groups. These polymers present special characteristics, among them a high water solubility, high molecular expansion in solvents with high ionic strength and a suitable chemical stability. These properties can be useful in flocculation processes applied to water treatment for industrial waste effluents. Results show a better particle separation than that observed in commercial polyelectrolytes, with concentrations around 1 ppm at low pH. Stable flocs are formed and their efficiency increases with convection. An extension to toxic metals separation is presently considered.

130. GAMMA RAY POLYMERIZATION OF LONG CHAIN SUBSTITUTED CYCLOTRIPHOSPHA-
ZENES.

Felipe Ramirez, Fernando Alonso, and Takeshi Ogawa*
Instituto de Investigaciones en Materiales, Universidad Nacional Autónoma de México, Apartado Postal 70-360, Coyoacán, México DF 04510, MEXICO.

Hexachlorocyclotriphosphazatriene was reacted with 10-undecen-1-ol, octylamine, dodecylamine, etc, to obtain cyclotriphosphazenes having long chain substituents. These octopus-like molecules were then irradiated by Gamma ray of Co-60, to obtain new polymers consisting of phosphazene rings connected with their long chain substituents. The polymerization was accelerated by the presence of chlorinated solvents such as chloroform, and inoluble polymers were formed. Soluble polymers were also obtained depending on the irradiation conditions, and some of them were crystalline and showed an interesting morphological behavior, when observed under a polarized light microscope.

131. CROSSLINKING OF POLYALLYLCARBONATES BY GAMMA IRRADIATION

Delia López V* and Guillermina Burillo**
*Escuela de Ciencias Químicas,Universidad de Puebla,Apdo Post 1613 Pue. México.Fax 91(22)468332. ** Instituto de Ciencias Nucleares UNAM,Ciudad Universitaria,México 04510 D.F.

The gamma ray induced crosslinking by quiescent polymerization of diallylcarbonate (DAC), monoethyleneglicol bisallylcarbonate (MEGBAC),propyleneglicol bisallylcarbonate (PGBAC), dipropyleneglycol bisallylcarbonate (DPGBAC), diethyleneglycolbisallylcarbonate (DEGBAC), and triallylcarbonate of glycerol (TACG).

Differences in crosslinking behavior of these monomers will be discussed and compared with those of multiacrylic monomers.

132. FLOW-INDUCED PHASE CHANGES OF POLYMER BLENDS WITH UCST PHASE BEHAVIOR R. Wu, M.T. Shaw, R.A. Weiss, Department of Chemical Engineering & Polymer Science Program, The University of Connecticut, Storrs, CT 06269.

Phase transitions for blends of polystyrene (Mw=800, Mw/Mn=1.3)/polyisobutylene (Mw=500) and polystyrene (Mw=2000, Mw/Mn=1.06)/ poly methyl phenyl siloxane (Mw=2300, Mw/Mn=2.2) were studied in the quiescent state and under shear flow. The cloud points of blends were determined by turbidity measurements at a fixed cooling rate of 0.5 °C/min. The phase diagrams of both PS/PIB and PS/PMPS blends showed a UCST behavior with a bimodal miscibility gap at low temperature. The shear effects on the phase transition behavior of PS/PIB and PS/PMPS blends were investigated by a flow-turbidity apparatus which was constructed by modifying a Rheometrics Mechanical Spectrometer (RMS). No shear-induced phase separation was observed in the one phase region. For temperatures within the two-phase region, the transmitted light intensity increased abruptly when the shear was applied. The intensities decreased and approached to the initial values when the shear was removed. Several possible explanations will be discussed to explain these results. The one that appears to be most valid is that of shear-induced phase mixing.

133. FLUORESCENCE AND SHEAR-INDUCED MIXING IN A MISCIBLE POLYMER BLEND
Suresh Mani, M. F. Malone, H. H. Winter, J. L. Halary and L. Monnerie
Chemical Engineering Department, University of Massachusetts, Amherst, MA 01003 & E. S. P. C. I., 10 Rue Vauquelin, 75231 Paris, Cedex 05

The effects of shear flow on the phase behavior of a miscible blend of polystyrene with poly(vinyl methyl ether) have been investigated by *in-situ* fluorescence measurements under well-controlled temperature and shear conditions in a specially modified rheometer. For this purpose, approximately

1% of the polystyrene chains were labelled with the fluorescer anthracene and fluorescence quenching was used to detect the state of mixing during flow. Shear at a constant rate, during slow heating from the one-phase state increases the coexistence temperature above the quiescent spinodal temperature and this increase can be correlated with shear rate by $\Delta T/T \propto \dot{\gamma}^{0.59}$; within experimental error this is independent of the blend composition in the range 20 to 60% (w/w) polystyrene. Isothermal experiments in the two-phase region also show shear-induced mixing in good agreement with non-isothermal experiments done at slow heating rates. The *in-situ* fluorescence measurements provide a sensitive indicator of phase transitions that cannot be unambiguously identified by the measured shear stress or first normal stress difference.

134. PHASE BEHAVIOR OF A POLYMER BLEND SOLUTION DURING STEADY SHEAR AND CESSATION OF SHEAR STUDIED BY LIGHT SCATTERING, Alan I. Nakatani, Dean A. Waldow, and Charles C. Han, National Institute of Standards and Technology, Polymers Division, Gaithersburg, MD 20899

An apparatus has been constructed for conducting light scattering experiments on polymeric samples under the influence of a well-defined simple shear field. A two-dimensional CCD array detector is used for quantitative measurement of anisotropic scattering patterns which commonly develop from sheared samples.

We describe results obtained from an 8% solution of a polystyrene/polybutadiene blend (50:50) in dioctyl phthalate. This sample has been examined previously and has demonstrated shear induced mixing behavior. The steady shear results obtained are consistent with the previous investigations. We also measure the kinetics of phase separation following cessation of a steady shear. The kinetic results are much different in the directions parallel and normal to the original flow direction. We believe this is the first report of anisotropic behavior in the phase separation kinetics of a polymer blend.

135. EFFECT OF DEFORMATION HISTORY ON THE MORPHOLOGY OF BLENDS OF LIQUID CRYSTALLINE POLYMERS WITH THERMOPLASTICS, A.M. Sukhadia and D.G. Baird, Department of Chemical Engineering, Virginia Polytechnic Institute and State University, Blacksburg, VA 24061.

The concept of blending polymers to obtain materials with improved properties is well recognized. However, the properties of a blend depend to a large extent on the morphology achieved. In this paper we report studies aimed at investigating the effect of deformation history on the morphology of blends of thermotropic liquid crystalline polymers (LCPs) with thermoplastics. It was observed that blends extruded using a single-screw extruder had a skin-core fibril-droplet type morphology. However, blends extruded by a different processing method and using a static mixer consisted of infinitely long LCP fibrils in the matrix and no skin-core structure was observed. Also, uniaxial extensional deformation obtained by drawing of strands was seen to increase the LCP fibril number with a simultaneous decrease in fibril diameter. LCP fibrils once formed were also observed to breakup into droplets when processed through a capillary die with large L/D ratio. Reasons for the formation of these various morphologies are discussed.

136. SHEAR INDUCED STRUCTURAL TRANSITIONS IN TRIBLOCK COPOLYMER SBS
Diane B. Scott, Alan F. Waddon, Ye-Gang Lin, Frank F. Karasz, H. Henning Winter
Department of Chemical Engineering and Department of Polymer Science and Engineering
University of Massachusetts, Amherst, Massachusetts 01003

Shear induced re-alignment of phase separated domains in a triblock copolymer is studied by SAXS and TEM. The material studied is strongly phase separated triblock copolymer styrene-butadiene-styrene, with an equilibrium phase separated morphology of hexagonally packed cylinders of poly(styrene) in a poly(butadiene) matrix. We start with a well-defined initially ordered structure, where the cylinders are parallel to the shear plane and impose a well-defined shear flow to rotate the cylinders by 90^0, into the shear direction. To produce the initial orientation, a solution cast film is subjected to lubricated planar extension, producing a uniaxially oriented rectangular sheet. With novel sample preparation we are able to transform this rectangular geometry into axisymmetric geometry needed for large shear strain in a parallel plate rheometer. SAXS, with pinhole collimation, and electron microscopy are used to observe shear induced structural transitions as a function of applied strain on room temperature quenched samples. At all observed strains, the hexagonal faces are observed to have a preferred orientation with respect to the shear plane. We have also observed intermediate textures at small strains. By following the structural transitions from one aligned structure to another, we will discuss the processes that produce domain orientation under shear flow.

137. RHEOLOGICAL EVIDENCE OF FLOW STRUCTURING IN ENTRANCE FLOW OF POLYPROPYLENE MELTS. Yuan-Ze Xu. Institute of Chemistry, Academia Sinica, 100080 Beijing, China.

Rheological properties of PP melts with different M & MWD have been investigated in a capillary rheometer with length/diameter ratio of the capillary being 0, 6.4, 40, 100, and also a cone/plate and a falling sphere viscometers for comparison. Melt transition near 208 C was observed for all samples in entry flow. Below this temperature the flow drag increaces abruptly. The succeeding capillary flow may eliminate this exess stress. This phenomenon, which can be distinguished from the stress induced crystallization of PP melts observed above melting temperature, was not observed in pure shear flow in a rotational or a falling ball viscometers. An explanation was suggested by assuming the existence of weak mesomorphic state in the melts below 208 C which can be strengthened by molecular extention in converging flow while be reduced by rotational destructuring in shearing.

138. BARRIER POLYMERS FOR CONTAINER COATINGS. C. E. Rogers, Department of Macromolecular Science and the InterUniversity Center for Adhesives, Sealants and Coatings (ICASC), Case Western Reserve University, Cleveland, Ohio 44106

Polymeric materials are used extensively as protective coatings for a variety of packaging containers. We seek to describe relevant sorption and transport processes in those materials in terms of their dependence on the relative compositions and structures of the coatings materials and components of the substrates and packaged substances. The overall transport process is determined by three factors - the average mobility of penetrant molecules (diffusion coefficient), the number of penetrant molecules in the coating membrane (solubility coefficient) and the driving force causing flux (the concentration or chemical potential gradient). In the design of packaging coatings and films, it is desirable to minimize all three factors. There are a number of polymeric materials selection and modification procedures that can be utilized for that design purpose.

139. WATER COMPATIBLE PHENOLIC RESINS. John D. Fisher, Resin Division, Schenectady Chemicals, Inc., Schenectady, NY 12309

Recent pressures upon the coatings industry to reduce the volatile organic content of coatings for various applications has resulted in increasing demand for coatings components that are water compatible. Much of the work in the phenolic resin industry of the past ten years has been devoted to exploring the factors that determine the extent to which a phenolic resin is compatible with water. This talk will survey various routes that have been explored to render phenolic resins more water compatible. Included as part of the discussion will be the use of various phenolic monomers and their effect on water compatibility of the resulting polymer, in what way thermoplastic phenolic resins differ in water compatibility from thermosetting phenolic resins, formation of organic salts of phenolic resins to enhance water compatibility, use of water compatible solvents as bridging solvents, and the use of water compatible polymers as modifiers to enhance the water compatibility of phenolic resins.

140.
ORGANIC COATINGS FOR BEER, BEVERAGE AND FOOD METAL CONTAINERS, PETER J. PALACKDHARRY, DEXTER CORPORATION, PACKAGING PRODUCTS DIVISION, WAUKEGAN, ILLINOIS, 60085

Coating compositions for the inside of beer, beverage (B/B) and food cans must be FDA approved for direct food contact under 21CFR175.300. The fundamental purpose of the organic film is to protect the packed product from spoilage and to maintain its inherent flavor, color and texture over the designated shelf life. In the U.S., coatings for the inside of B/B cans, together with DWI (drawn and wall iron) food cans are based primarily on aqueous dispersions of acrylated epoxies which essentially are reaction products of acrylated copolymers with bisphenol-A. The inside coatings for can ends (lids) and some DRD (draw redraw) food cans are based on either solution types polyvinyl chloride (PVC) or dispersion type PVC. The precoated metal sheets are subjected to severe elongative and compressive stresses during the DRD can forming process or the fabrication of can ends. The integrity of the respective coating must be maintained during all the specific fabrication operations.

141. THE ANALYSIS OF PACKAGED FOOD SUBSTANCES BY GC/IR. Senja V. Compton, Jay R. Powell, and David A.C. Compton, Bio-Rad, Digilab Division, 237 Putnam Avenue, Cambridge, MA 02139

Packaging food substances inside polymeric materials (coated cans or flexible wrappers) always leads to concern that the packaging may impart some odor, flavor, color, or hazardous substance to the contents. The detection of any added substances is very difficult, since the body can detect changes to a food at extremely low levels, and the analyst is faced with a challenge equivalent to finding a needle in a haystack. Gas chromatography (GC) must be used to separate all the components of the mixture, and then an analytical technique such as mass or infrared spectroscopy is used in an attempt to identify all the components (GC/MS or GC/IR, respectively). GC/IR now has a sensitivity equivalent to GC/MS, and can often obtain a distinction between isomeric materials, very important when attempting to identify flavor agents. We have investigated a number of food substances by GC/IR, such as extracts of beer and coffee, and flavor oils. This talk will show the remarkable amount of information that can be achieved by such studies.

142 NOVEL WATERBORNE EPOXY RESINS FOR CAN AND COIL COATINGS WITH IMPROVED FLEXIBILITY, Duane S. Treybig and David P. S. Sheih, Coatings, Adhesives and Sealants Applications Lab, The Dow Chemical Company, Freeport, Texas 77541-3257

New two-piece can technologies are creating a market demand for more flexible can coatings. In response to the demand, several different water-borne epoxy resin systems have been prepared by systhesizing epoxy resins with new flexible backbones and converting them to onium forms.

More specifically, the coating epoxy resins were made by modifying the resin backbone with extended aliphatic spacers of the bisphenol structures. These flexible epoxy resins were then converted to onium forms by reacting the epoxide group with an aqueous mixture of Bronsted acid and nicotinamide. When the resulting aqueous dispersions are formulated and cured with melamine-formaldehyde resin, the resultant coatings exhibit improved flexibility, and pasteurization resistance as compared to coatings from D.E.R® 667 epoxy resin controls. The new epoxy resin backbone chemistry, water-borne conversion chemistry and coating properties are described in this paper.

143. PVC IN CONTAINER COATINGS AND APPROACHES TO ITS REPLACEMENT, M. Hickling, Manager, Packaging Coatings Development, Courtaulds Coatings, International Paint Ltd., North Woolwich Road, Silvertown, London E16 2AF.

PVC has established an important position as a component in container coatings being used particularly in highly flexible coatings offering low cost and relatively high application solids. In Europe, claims of potentially toxic by-products in its incineration have led to calls for PVC to be excluded from packaging compositions.

In many uses, alternative coatings can be used with few disadvantages other than cost. However, when extreme fabrication performance is demanded, more difficulties arise. Its wide applicability and well understood properties make it almost impossible to replace PVC directly and alternative formulating approaches have to be considered. This paper discusses the issues faced by formulators in attempts to develop high performance PVC-free container coatings. The types of approach adopted to date are considered together with their shortcomings and advantages.

144. MOBILITY ENHANCEMENT TECHNOLOGY FOR THE D&I ALUMINUM CAN MANUFACTURE, Sami B. Awad, Henkel Corp./Parker +Amchem, 32100 Stephenson HWY, Madison Heights, MI 48071

A novel clear coating technology to reduce the coefficient of static friction of cleaned aluminum can surfaces will be presented. The benefits and potential savings achieved on applying this new surface treatment in the D&I can plants will be discussed.

145. STRUCTURE AND PROPERTIES OF IONOMERS IN THE SOLID STATE. S.L. Cooper, S.A. Visser, R.A. Register[*], B.P. Grady. Department of Chemical Engineering, University of Wisconsin-Madison, Madison, WI 53706.

Ionomers are materials containing a small mole fraction of ionic groups covalently bonded to the polymer backbone. The ionic moieties cause the formation of a two phase material; the ions aggregate into a separate phase. The aggregates act as semipermanent crosslinks in these materials, resulting in a dramatic improvement in mechanical properties. A number of analytical techniques have been applied to these materials to understand the morphology in the solid state. An increase in the degree of local ordering around the cation as monitored by Extended X-ray Absorption Fine Spectroscopy (EXAFS) was found to correlate strongly with ultimate mechanical properties. A liquid-like hard sphere model of the ionic aggregates gives theoretical scattering patterns which quantitatively fit experimentally observed small angle x-ray scattering patterns. Small angle neutron scattering studies of model polyurethane ionomers show that no chain expansion occurs in the bulk as a result of ionic aggregation.

[*]Current address: Department of Chemical Engineering, Princeton University, Princeton, NJ 08544

146. PROPERTIES OF A NOVEL MODEL IONOMER SYSTEM: AMIC ACID IONOMERS. X. Yu[*], B.P. Grady, R.S. Reiner, S.L. Cooper. Department of Chemical Engineering, University of Wisconsin-Madison, Madison, WI 53706.

Model ionomer systems have been synthesized with ionic groups regularly spaced along the chain backbone. A novel model system with a hard segment derived from 3,3',4,4'benzophenonetetracarboxylic dianhydride has been synthesized. Differential scanning calorimetry (DSC) and dynamic mechanical thermal analysis have been used to analyze this material. The behavior of this system is very similar to model polyurethane ionomers. The phase separation between hard and soft segment increases as the neutralization level or soft segment molecular weight increases. Hard segment crystallinity is disrupted as the neutralization level increases. The plateau modulus is extended and increases in magnitude as the neutralization level increases due to the improved domain cohesion caused by the ionic groups.

[*]Chemistry Department, Nanjing University, People's Republic of China

147. Kinetics of Imidization of Poly (amic acid) in Miscible and Immiscible Blends - S. M. Makhija, E.M. Pearce and T.K.Kwei

Chemistry Department, Polytechnic University, Brooklyn, NY 11201

Blends of commercial poly (amic acid) LARC weren prepared with polyimide P84. The blends were shown to be miscible over certain composition range. Miscible composition of the blends wemained miscible after imid= ization. Immiscible composition of the blends showed two T_g after imidization. The kinetics of imidization in solid state remains unaffected while in the miscible state the kinetics becomes faster than that of the pure LARC. The kinetics was studied using FTIR.

148. COCRYSTALLIZATION IN BLENDS OF POLY(BUTYLENE TEREPHTHALATE) AND POLY(ESTER-ETHER) SEGMENTED BLOCK COPOLYMERS. Kevin P. Gallagher, Xuifang Zhang, James P. Runt, Department of Materials Science and Engineering, Penn State University, University Park, PA 16802 and Gia Huyuh-ba, E. I. DuPont de Nemours & Co., Wilmington, DE 19880 and J. S. Lin, Oak Ridge National Laboratory, Oak Ridge, TN 37831.

Blends of poly(butylene terephthalate) (PBT) with poly(ester-ether) block copolymers were investigated to provide insight into the state of mixing in these blends as well as the possibility for cocrystallization. An extensive range of blends were prepared by mixing PBT with copolymers possessing different hard segment concentrations (from 41-90 wt.%). Dielectric loss measurements show that amorphous phase miscibility increases as block copolymer hard segment concentration increases. Thermal analysis, small angle x-ray scattering and morphological evidence suggests that cocrystallization occurs in mixtures of PBT with segmented copolymers containing greater than 80 wt% hard segment under all crystallization conditions.

149. **Thermally Induced Morphological Changes in Segmented Polyurethanes.**
Yingjie Li[1], Jian Liu[1], Huichang Yang[2], Dezhu Ma[3] and Benjamin Chu[1,*]. 1. Chemistry Department, State University of New York at Stony Brook, Long Island, NY 11794; 2. Henan Institute of Chemistry, Zhengzhou, China; 3. Department of Materials Science, University of Science and Technology of China, Hefei, China.

By using small angle light scattering, it was found, for the first time, that spherulites could form from the *melt* by quenching the MDI-BD based segmented polyurethanes in the melt state to annealing temperatures above $120^\circ C$. The highest annealing temperature of spherulite formation (T_h) depended upon the hard-soft segment compatibility. The radius of the spherulite increased with increasing hard segment content. Synchrotron small angle x-ray scattering showed that the inter-domain spacing increased with increasing hard segment content and with increasing annealing temperature up to T_h. Above T_h, the inter-domain spacing decreased with increasing temperature.

150. CHARACTERIZATION OF CHEMICAL REACTIONS AT POLYMER/METAL INTERFACE BY SPUTTERING DEPTH PROFILING USING XPS. P. Zhang, J. Epp*, K.E. Newman, D.L. Allara, L. Cuddy Department of Materials Sci.& Eng., The Pennsylvania State University,University Park, PA 16802. *Alcoa Technical Center, ALCOA, Alcoa Center, PA 15069
Over the last 15 years, extensive efforts have been made to develop new polymeric materials possessing low density, high strength and modulus with long-term retention of those properties at high temperatures.Among the best materials discovered are two poly- mers, poly(p-phenylene benzobisthiazole)(PBT) and poly(p-phenylene benzobisoxazole)(PBO)

That annealing as-spun PBT and PBT fibers at temperatures from 630 to 700 C significantly enhances the properties of the fibers opens up a unique possibility of using such materials as reinforcements for metal matrix composites(MMCs). As part of our study on the feasibility of using PBT fibers for MMCs, we have employed XPS combined with ion sputtering to probe the interface between Al and PBT. It has been observed that aluminum oxide is the primary interface product in Al coated PBT film prior to high temperature heating. Al will react with PBT to form aluminum carbide and possibly nitide and sulfide when Al coated PBT film is heated at 600 C for 1 and 5 hours. Interdiffusion also occurs during the heating process. The chemical reaction and diffusion are responsible for the increased interface adhesion when Al coated PBT film is heated at high temperatures.

151. RAMAN STUDIES OF NEW MODEL COMPOUNDS OF POLY[N,N'-BIS(PHENOXYPHENYL)PYROMELLITIMIDE].
Ajay K. Saini, Howard H. Patterson, Chemistry Department, University of Maine, Orono, ME 04469 and Clifford M. Carlin, Department of Chemistry, University of North Carolina, Charlotte, NC 28223

Polyimides are a class of polymeric compounds formed by the condensation of an acid dianhydride and a primary diamine. For example 4,4'-oxydianiline (ODA) and pyromellitic dianhydride (PMDA) react to produce poly[N,N'-bis(phenoxyphenyl)pyromellitimide] (PI). In this paper we will report studies to make short chain model compounds of PI such as OPO which is composed of one PMDA and two ODA and POP which has one ODA and two PMDA moieties. We have subjected these model compounds to heating to determine their tendency to form imine type of bonds leading to crosslinkage. Raman studies show that the terminal amines for C=N bonds with imide carbonyls on curing. These results have substantiated our Raman results reported previously for vapor deposited polyimide films on a substrate.

152. CHARACTERIZATION OF BISPHENOL A - BASED CYANATE ESTER RESIN SYSTEMS
A. Osei-Owusu and G.C. Martin,
Department of Chemical Engineering and Materials Science
Syracuse University, Syracuse, NY 13244

J.T. Gotro
IBM Corporation, Endicott, NY 13760

The kinetics of cyclotrimerization and the effects of catalyst concentration and type on the network properties of bisphenol A - based cyanate ester resin systems were studied using Fourier Transform Infrared Spectroscopy, Differential Scanning Calorimetry, thermogravimetry, and extraction studies. The bisphenol A dicyanate was cured with 4 phr nonylphenol and metal concentrations ranging from 0 to 750 ppm. The overall polymerization reaction was described by second order kinetics. The final network properties were dependent on the catalyst concentration for the zinc catalysts while no effect of concentration on the network properties was observed for samples cured with cobalt and manganese. Gelation was found to occur at 60-65 percent conversion which is greater than the theoretical value of 50 percent predicted by the recursive modelling technique.

153. SURFACE PROFILING OF POLYMERS WITH FOURIER TRANSFORM INFRARED REFLECTANCE SPECTROSCOPY, G.C.Chen, L. J. Fina, Department of Materials Science and Engineering, College of Engineering, Rutgers University, Piscataway, NJ 08855.

Fourier transform infrared spectroscopy (FTIR) coupled with attenuated total reflectance spectroscopy (ATR) has been used to quantitatively depth profile polymer surfaces. Structural gradient information is contained in the absorption properties as a function of incident light angle. Calculations based on idealized polymeric systems are performed which support the use of the method under well defined conditions. The conformational composition of poly(ethylene terephthalate) (PET) as a function of distance from the surface is studied. The method is used to examine concentration gradients in PET that have been subjected to surface crystallization treatments. The method is also applied to uniaxially oriented atactic polystyrene where gradients in both absorption coefficient and refractive index become important.

154. **STRUCTURE AND ORIENTATION OF AQUEOUS SURFACTANT MONOLAYERS**
Yei-Shin Tung and Leslie J. Fina, Department of Materials Science and Engineering, College of Engineering, Rutgers University, Piscataway, NJ 08855, Tao Gao and Milton J. Rosen, Surfactant Research Institute, Brooklyn College, City University of New York, Brooklyn, New York 11210, Jose E. Valentini, E.I. du Pont de Nemours and Company, Imaging System Department, Brevard, North Carolina 28712

Reflection absorption infrared spectroscopy has been applied to the *in situ* measurement of soluble monomolecular films formed by surfactant molecules at the air-water interface. Relationships are established between the average chain orientation angle, anisotropic optical constants and the angle of incident light for s- and p- polarized radiation. Experimental infrared spectra of sodium dodecanesulfonate as a monolayer on pure water and on NaCl solutions of 0.1M and 0.5M total ionic strength are presented for both parallel and perpendicular polarization in the CH_2 and SO_2 stretching regions. Substantial changes in intensities for these two bands are related to the orientation change due to both surfactant concentrations and NaCl concentrations.

155. **RHEOLOGICAL STUDY OF POLYMER-POLYMER COMPLEXES OF AN IONOMER AND AN AMINO-CONTAINING POLYMER** Jian Wang[1], Dennis G. Peiffer[2] Wendel J. Shuely[3] and Benjamin Chu[1] 1. Department of Chemistry, State University of New York at Stony Brook, Stony Brook, NY 11794; 2. Exxon Research and Engineering Company, Corporate Research Science Laboratory, Clinton Township, Route 22 East, Annandale, NJ 08801; 3. Research Division, Chemical Research, Development, and Engineering Center, Aberdeen Proving Ground, MD 21010

The rheological properties of polymer-polymer complexes consisting of a random copolymer, poly(isobutyl methacrylate-tert--butyl aminoethyl methacrylate)(poly(iBMA-tBAEMA)), and an ionomer, a zinc salt of sulfonated polystyrene, in polar solvents have been studied. A phase separation region in term of the mixing ratio at a certain concentration was observed. In the miscible region, no phase separation was observed even at a high total polymer concentration of 0.1 g/ml. By using a magnetic needle rheometer the reduced viscosity of the mixtures in the miscible region was measured as a function of polymer composition, concentration, cosolvent quality, temperature and shear rate. The polymer-polymer complexes in solution showed a remarkably high viscosity as compared with that of the individual components.

156. **COMPATIBILITY OF TERNARY SEMI IPNs**
P.Rajalingam, N.Natchimuthu and Ganga Radhakrishnan
Central Leather Research Institute, Madras 600 020, INDIA
ABSTRACT

Multi phase polymers have been used extensively in the polymer industry to provide products with superior chemical and physical properties. Such systems, however, often experience phase stability problems due to lack of suitable bonding forces between them. The phase morphology of combined polymer systems is mainly due to the presence of specific interactions among the constituents and the study of such interactions is essential to understand the phase behaviour of a combined system. In this paper, we report the FTIR studies of semiinterpenetrating polymer networks based on nitrocellulose, polyurethanes and vinyl chloride-vinyl acetate copolymer. Compatibility of the ternary semi IPNs is interpreted in terms of specific interactions among the functional groups. Morphological features are studied with scanning electron microscope.

157. APPLICATION OF EPOXY RESINS IN CONTAINER COATINGS: AN OVERVIEW, Ronald S. Bauer, Shell Development Company, Westhollow Research Center, P.O. Box 1380, Houston, Texas 77251-1380

Almost since their commercial introduction in 1950, epoxy resin systems have been used as protective coatings for containers. It is estimated that in 1990 between about 38 to 43 million pounds of epoxy resin were consumed by the packaging industry. The unique properties of epoxy resin coatings have made them the predominate thermosetting resin for the interior of beer and beverage cans, hard to contain food products such as sauerkraut, tomato juice, and meat; and for chemical resistant linings of pails and drums. Epoxy resin container coatings are not only used to protect the metal of the container from corrosion; but they also protect the flavor of the contents, which can be affected by direct contact with metal. This paper reviews the chemistry and technology of epoxy resin systems used in container coating applications.

158. CONTAINER COATINGS: FDA CONSIDERATIONS. James L. Zawicki, ICI Americas Inc., Concord Pike and Murphy Road, Wilmington, DE 19897

Regulatory concerns have generated a need by coating manufacturers to take a closer look at their formulations and to reformulate in many cases to meet new environmental laws. This action often results in the use of new components, some of which may not be covered under existing Food and Drug Administration regulations. It is important to know where to find these regulations, how to verify the FDA status of components for food-contact coatings, and what to do to get regulatory approval if the components are not cleared for the intended use.

159. **THE ELECTRON BEAM-INDUCED CATIONIC POLYMERIZATION OF SILICONE-EPOXIDE RESINS**
J. V. Crivello, Department of Chemistry, Rensselaer Polytechnic Institute, Troy, New York 12180.

The rapid, efficient electron beam-induced cationic polymerization of multifunctional silicone-epoxy monomers has been carried out in the presence of diaryliodonium and triarylsulfonium salts. These polymerizations take place at low doses (1-3 Mrad) which make them attractive for commercial applications including coatings, inks and composites. The factors which contribute to the high reactivity of these monomers are discussed. A preliminary evaluation of the mechanical properties of the cured resins was carried out.

160.
MONITORING CURE IN INSIDE SPRAY COATINGS, PETER P. WINNER, DEXTER CORPORATION, PACKAGING PRODUCTS DIVISION, EAST WATER STREET, WAUKEGAN, ILLINOIS, 60085

A number of methods was evaluated to determine the degree of cure as a function of cure temperature and time conditions in the inside spray coatings for metal container for beer and beverage market. The following analytical methods were investigated in detail: FTIR, thermal analysis, GC for retained solvents, TGA, and spectrophotometric method. This paper will discuss the results of these evaluations and make recommendations for the preferred method.

161. THEORETICAL CALCULATION OF GRAFTING SITES IN THE EPOXY-G-ACRYLIC COPOLYMER, Jim T. K. Woo, and Alan Toman, The Glidden Company, Part of ICI Paints World Group, 16651 Sprague Road, Strongsville, OH, 44136.

A previous attempt to study the grafting sites in the epoxy-g-acrylic copolymer by ^{13}C NMR spectroscopy seemed to indicate that grafting occurs at the aliphatic backbone portion of the epoxy resins. However, no distinct graft peaks were identified. Model compound synthesis revealed that one of the graft peaks should be at 72 ppm down field from TMS. By preparing the graft copolymer in a hydrocarbon-butanol solvent mixture and by using a 400 MHz megacycle ^{13}NMR spectrometer at Case Western Reserve University, five graft peaks were identified. Grafting takes place at both aliphatic carbons in the epoxy backbone.

162. NOVEL EPOXY RESINS BASED ALKYLENEDIOXYDIPHENOLS - EFFECT OF BACKBONE FLEXIBILITY AND CROSSLINKING ON FLEXIBILITY OF CAN COATINGS. R. A. Dubois and P. S. Sheih, Coatings, Adhesives and Sealants Application Laboratory, Dow Chemical, Freeport, Texas 77541.

Due to an increasing trend in the can industry toward the use of precoated stock for drawing and forming operations, there is a clear and growing need for more flexible coatings. Conventional methods of improving the flexibility of epoxy coatings include blending and grafting of flexibilizers with bisphenol-based epoxy resins or using flexible aliphatic epoxies; however, such methods are often limited by the rigidity of the bisphenol core or they suffer from loss of other properties like adhesion, chemical resistance, or reactivity. This talk will describe a new class of epoxy resins based on alkylenedioxy diphenols which exhibit the flexibility of aliphatic epoxies while retaining other properties characteristic of aromatic epoxies. Correlations between structure of the alkylenedioxy bridging unit and backbone flexibility of the resin and ultimately flexibility of cured coatings will be presented. There will also be a discussion on the effect of crosslinking on this system.

163. HYDROGEN BONDED INTERPOLYMER COMPLEXES. Takuo Suzuki, E.M. Pearce, T.K. Kwei, Polytechnic University, Brooklyn, New York 11201.

Mutual precipitates of poly(N,N dimethylacrylamide) and poly(4-hydroxystyrene) from acetone and methanol were studied. As a result of hydrogen bonding between the component polymers, the complexes have high glass transition temperatures and short NMR spin-lattice relaxation times. The scale of homogeneity calculated from spin diffusion length is about 2-3 nm. The nature of the solvent from which the complexes are formed and the effect of thermal treatment on the properties of the precipitates are discussed.

164. THERMALLY INDUCED PHASE BEHAVIORS OF AROMATIC POLYBENZIMIDAZOLE/POLYIMIDE BLENDS

Soonja Choe and Tae-Kwang Ahn, Dept of Chemical Engineering, Inha University, Inchon 402-751, Korea

Mechanical properties and morphological behavior of partially miscible blends comprized of aromatic polybenzimidazole poly-2,2'(m-phenylene)-5,5'-bibenzimidazole (PBI) and the polyetherimide poly-2,2'-bis(3,4-dicarboxyphenoxy) phenyl propane-2-phenylene bisimide(PEI) have been studied using Instron and Scanning elctron microscopy. The miscibility/immiscibility phase boundary was determined from the morphological changes arised from the thermally induced phase separation. The differences of the mechanical properties between one and two phase regime have been compared. The phase boundary is in good agreement with the previous results which were observed from the DSC thermal and FTIR spectroscopic investigations.

165. CRYSTAL-AMORPHOUS INTERPHASES IN POLY(ETHYLENE OXIDE) BLENDS. Catherine A. Barron, Sanat K. Kumar, James P. Runt, The Pennsylvania State University, University Park, PA 16802, and John Fitzgerald, Eastman Kodak Company, Rochester, NY 14650.

A region of partial order between the crystalline lamellae and the interlamellar amorphous zone in semicrystalline polymers is necessary for the dissipation of order at the crystal surface. Such a region has been detected experimentally both in the case of homopolymers, as well as in miscible binary polymer blends with one crystallizable component. In a continuation of our study of the presence of the crystal-amorphous interphase in miscible binary blends we have investigated misible blends of PEO with poly(vinyl acetate), poly(methyl methacrylate) and poly(hydroxystyrene). PEO has relatively weak interactions with PVAc and PMMA, but much stronger hydrogen bonding interactions with PHS. Our results indicate the presence of a pure amorphous PEO interphase in the blends which have relatively weak interactions. However, blends which have stronger interactions show no evidence of an interphase suggesting that interactions have a strong effect on the structure of the interphase.

166. HIGH MELTING FRACTION FOUND IN THE SLOW MELTING OF ULTRA-HIGH MOLECULAR WEIGHT POLYETHYLENE.

H. Phuong-Nguyen and G. Delmas, Département de chimie, Université du Québec à Montréal, C.P.8888, Succursale "A", Montréal H3C 3P8 Canada.

Using the sensitive and stable Setaram microcalorimeter C80, the melting and dissolution traces of ultrahigh MW polyethylene GUR (UHMWPE) have been obtained at low rates of heating (v = 1 K/h). Slow heating permits to obtain three fractions associated respectively with low molecular weights (I), unconstrained cystals (II) and constrained chains (III). In this nascent sample, (III) is found highly superheatable not been able to melt below 200°C if v is higher than 12 K/h. Including H_3, the heat of fusion of (III), the overall crystallinity is high, respectively 0.84 and 0.94 for the as-received and annealed samples. The same fractions are found when dissolution is achieved with the same v in decalin. Interpretation of these results is made in terms of melting of strained crystals similar to melting under pressure or of crosslinked PE. A model for UHMWPE gels is presented.

167. DYNAMIC PROPERTIES OF A POLYSTYRENE/POLYMETHYLMETHACRYLATE/ GLASS BEADS SYSTEM N. Scherbakoff, H. Ishida, Case-Western Reserve University, Cleveland, OH 44106

It is proposed a new method of enhancement of the compatibilization of immiscible blends by combining the blending and reinforcing approaches. The major goal of this work is to understand the role of the filler surface on the compatibilization process. The polystyrene - polymethylmethacrylate - glass beads system was selected. A preliminary study on the Nylon 6/ polypropylene/glass beads system has showed that it is of vital importance to have the filler surface treated with a suitable coupling agent. The surface treatment gives the opportunity for both polymer components to interact with the filler surface and with each other. Dynamic mechanical analysis is used in this work. It allows to follow the glass transition, the modulus change and shifts in the glass transition. These results are correlated with the enhancements of the adhesion and polymer/polymer interactions.

168. CHEMICAL AND PHYSICAL CHARACTERIZATION OF AN INSULATOR CONTAINING TWO ELASTOMERIC COMPONENTS. Stephen N. Senzer, Anthony R. Cooper and Gerald B. McCauley. Lockheed Palo Alto Research Laboratories, 3251 Hanover St., Palo Alto, CA 94304.

Polymeric insulator materials are used in propulsion applications to protect the case from the heat and degradation products of the burning propellant. We have performed characterization experiments on a particular insulator material to investigate correlations between laboratory measurements and end use characteristics. The material studied was made from ethylene-propylene rubber (EPDM) and poly(chloroprene) (Neoprene) and contained Kevlar fibers. Chemical identification of the insulator components and insulator thermal degradation products were performed by mass spectrometry and Fourier transform infra-red spectroscopy. Degrees of cure of the rubber components at various stages of manufacture were determined by solvent extraction and swelling measurements. Thermal characterization of the cured insulators were performed by thermogravimetric analysis. Several unexpected results were obtained concerning the state of cure of the rubber components and correlations with end use performance are being studied.

169. **Kinetic Study of In-Situ Molecular Composite During Imidization of Poly(amic dialkylamide) Precursors,** Wansoo Huh, University of Dayton Research Institute, 300 College Park, Dayton OH 45469, Yih-Fang Chen, AdTech Systems Research, 1342 N. Fairfield Rd, Dayton OH 45432, and Loon-Seng Tan and Charles Y-C Lee, Materials Directorate, WL/MLBP, WPAFB OH 45433

The model compound for in-situ molecular composite, poly(amic dialkylamide)(PAA) with different backbone molecular weights and differnt leaving groups (pendent DNI-3A) were prepared and characterized in an attempt to understand the rod conversion process of imide for designing a optimum process scheme. The reaction kinetics and the chemorhelogy of these systems were studied using FTIR, TICA, DSC, and WAXD analysis. The two-stage first order reaction rates, k_1 and k_2, during imidization process were identified using FTIR analysis and these two reaction rate decreases with incresing molecular weight of PAA/DNI-3A backbone. Also, the length of aliphatic pendent side group affects the reaction rates of this system. These results suggest that the pendents with larger size, at least for aliphatic structures, have more thermal stability and act as a local plasticizer to reduce the viscosity for faster rod conversion. The reaction kinetics again is controlled by rheology.

170. The Kinetics of Sulfonation of Sulfite-Pretreated Aspen Treated with Vapor at High Temperature. Ahmed, A., Kokta, B.V. and Carrasco, F., Université du Québec à Trois-Rivières, C.P. 500, Trois-Rivières, Québec, G9A 5H7 CANADA.

The kinetic studies of sulfonation for ultra-high-yield pulp is important due to the close relation between pulp properties and its level of sulfonation. Aspen chips pretreated with Na_2SO_3 or $Na_2SO_3 + Na_2CO_3$ solution were treated with Vapor at temperature in the range of 180°C to 200°C for various times to study the kinetics of sulfonation. The concentration of Na_2SO_3 in pretreatment solution has significant effect on the rate of sulfonation. The rate of sulfonation was greatly affected by cooking temperature. The maximum level of sulfonation achieved at 195°C for 2.5 min is 78 m mol/kg which is substantially lower than 107 m mol/kg found in literature where low treatment temperature was used for long time. The results indicate that the high temperature process accelerates the sulfonation reaction in available sites and some probable sites disappear or change due to thermal degradation.

171. CRYSTALLIZATION KINETICS OF POLY(ETHYLENE TEREPHTHALATE) WITH FUNCTIONALIZED TRIGLYCERIDE OIL COMPOSITIONS. L. W. Barrett and L. H. Sperling, Materials Research Center, Whitaker Laboratory #5, Lehigh University, Bethlehem, PA 18015.

Castor oil and Vernonia oil are being used to increase the crystallization rate of poly(ethylene terephthalate) (PET). The Avrami equation has been applied to DSC data in order to calculate the isothermal crystalline growth rate continuously versus time, providing curves diagnostic of the crystallization process, which show that both oils significantly enhance PET crystallization. Bonding an aromatic sodium salt functionality onto castor oil results in further enhanced crystallization properties, including greater extent of crystallization and higher onset temperature in cooling from the melt. The aromatic sodium salt functionality is known to interchange with PET, thus probably forming PET/triglyceride graft copolymer *in-situ* during mixing.

172. **CURE/PROPERTY DIAGRAMS OF THERMOSETTING SYSTEMS.** X. Wang and J. K. Gillham, Polymer Materials Program, Department of Chemical Engineering, Princeton University, Princeton, New Jersey 08544.

The transformation of reactive thermosetting liquid to glassy solid generally involves two distinct macroscopic transitions: gelation and vitrification. Molecular gelation, which corresponds to the incipient formation of an infinite molecular network and occurs at a definite conversion for a simple system (according to Flory's theory of gelation), gives rise to long range elastic behavior in the macroscopic fluid. Vitrification occurs when the glass transition temperature (T_g) of the system rises to the cure temperature. The non-linear increase in the T_g vs. conversion relationship is made up of two effects: increase of molecular weight and increase of degree of crosslinking with increase of extent of cure. The occurrence of the two transitions during cure result in the Time-Temperature-Transformation (TTT) Cure Diagram and the Conversion-Temperature-Transformation Diagram. Furthermore, physical properties deep in the glassy state (e.g., density, modulus and physical aging rate) of an undercured material versus extent of cure are determined principally by the temperature interval T_g -T and are affected by the T_g and T_β- transitions (both of which increase with increase of extent of cure). These characteristics provide the basis for the T_g-Temperature Property (T_gTP) Diagram. These diagrams are intellectual frameworks for understanding relationships between properties, structure, transformations, states, cure path and reactants.

173. **ISOTHERMAL PHYSICAL AGING OF POLY(METHYLMETHACRYLATE): LOCALIZATION OF PERTURBATIONS IN THERMOMECHANICAL PROPERTIES.** Richard A. Venditti and John K. Gillham, Polymer Materials Program, Department of Chemical Engineering, Princeton University, Princeton, New Jersey 08544

The effect of isothermal physical aging on poly(methylmethacrylate) (PMMA) was investigated for isothermal aging temperatures (T_a) from T_g to T_g-120°C using a freely oscillating torsion pendulum (TBA). A single PMMA/glass fiber specimen, whose thermal history could be erased by heating above T_g(=116°C, 0.7 Hz), was used for all experiments; this facilitated comparison of the unaged and aged specimen. The modulus was observed to increase linearly with the logarithm of isothermal aging time. Thermomechanical properties of the aged vs. unaged specimen showed perturbations (e.g., increased modulus) principally in the vicinity of T_a. This suggests that different intermediate portions of the relaxation spectrum are specifically involved in the process of aging for different values of T_a.

174.
EFFECT OF MOLECULAR WEIGHT OF POLYDIMETHYLSILOXANE ON PHYSICAL PROPERTIES OF CURED EPOXY COMPOUNDS.
H.Ito, I.Takahashi and H.Kanegae, Manufacturing Development Laboratory, Mitsubishi Electric Corp., 8-1-1, Tsukaguchi-Honmachi, Amagasaki, Hyogo,661 Japan

The effect of the dispersed state of polydimethylsiloxane (PDMS) in the epoxy matrix on the physical properties of the epoxy compounds was investigated. The obtained conclusion was the following;
(A) With increasing the PDMS molecular weight , the change in the glass transition temperature, the coefficient of thermal expansion and the apparent crosslinking density was relatively remained in the following order; PDMS-MW640 << PDMS-MW1380 < PDMS-MW2040 < PDMS-MW4600. (B) The dispersed PDMS-MW1380 particle size was the smallest of all. (C) The PDMS-MW1380 was the most appropriate modifier capable of providing a significant heat shock resistance in this system.

PMSE

175. DISTRIBUTIVE MIXING AND ENERGY DISTRIBUTION IN TWIN SCREW EXTRUDERS.
David I. Bigio, University of Maryland, Mechanical Engineering Department, College Park, MD 20742.

As the area of condensed phase polymer reaction engineering continues to expand as a preferred approach for obtaining desired product properties and for eliminating the source of solvent emissions, twin screw extruders become an attractive apparatus for that process. This presentation begins to address the role distributive and dispersive mixing in twin screw extruders in obtaining successful polymer reactions and begins to model mixing which characterize the important machine and operating parameters. The nature of distributive mixing in the co-rotating, fully intermeshing and the counter-rotating, non-intermeshing twin screw extruders is experimentally examined. This presentation establishes a basis for evaluation of the extruder geometry and the relationship of distributive mixing and the chemical kinetics for an approach to modelling extruder reactions.

176. CHEMISTRY OF CONDENSATION REACTIONS CARRIED OUT IN POLYMER BLENDS. M. Lambla, C. Maier, R. Kowalski, B. Jones, Ecole d'Application, des Hauts Polymerès, 4, rue Boussingault, 67000 Strausbourg, France. Exxon Chemical Company, P.O. Box 45, Linden, NJ 07036. Polymers Group, Exxon Chemical International, Inc., Machelen Technology Center, Nieuwe Nijverheidslaan 2, 1830 Machelen, Belgium

Condensation reactions are used industrially for reactive blending, but there is still a need for new chemical systems. A common occurance is a lack of affinity between the components to be blended. The twin-screw extruder is an excellent tool for combining these components, due to its good control over mixing, temperature, residence time and its ability to operate without the use of solvents. This can be particularly so when the viscosities and weight fractions differ greatly. A model chemical system was chosen in order to investigate the interaction between kinetics, mixing and miscibility. A homologous series of small molecules gave a range of polarities which affected their miscibility. An insight into optimising the screw geometry and operating conditions was achieved as well.

177. **THE EFFECTS OF PROCESSING PARAMETERS AND ADHESION ON THE IMPACT PROPERTIES AND MORPHOLOGY OF RUBBER TOUGHENED POLYSTYRENE** N. C. Liu and W. E. Baker, Department of Chemistry, Queen's University, Kingston, ON., K7L 3N6, Canada

The effects of viscosity ratio of the components and interfacial chemical reaction on the rubber particle size and impact energy of polystyrene (PS)/acrylonitrile-butadiene (NBR) blends were investigated. Control of the viscosity ratio between the rubber phase and the PS matrix as measured by a processing torque ratio created during melt processing has a marked effect on the morphology of the blends. The rubber particle size decreased by a factor of 5 times as torque ratio dropped from 6.2 to 0.5. With the introduction of interfacial chemical reaction, the rubber particle size can be further reduced by a factor of up to 5 times. The unnotched impact energies of non-reactive blends were 1.5-3 times that of pure PS, those of reactive blends with the same particle size were 2-6 times that of pure PS. Interfacial adhesion not only greatly aids in reducing the rubber particle size during blending but also significantly improves the impact properties over blends without adhesion and the same particle size.

178. **CORRELATING CHEMORHEOLOGICAL CHANGES TO DIELECTRIC PROPERTIES DURING ISOTHERMAL EPOXY POLYMERIZATION.**
Joycelyn O. Simpson and Sue Ann Bidstrup, School of Chemical Engineering, Georgia Institute of Technology, Atlanta, Georgia 30332-0100

During the polymerization of a thermoset, valuable structural information can be obtained by following changes in dielectric properties. In order to implement dielectric analysis as a cure monitor, relations between dielectric properties and structure must be established. To develop fundamental structure-dielectric property relations, this research employs free volume theory which proposes that polymer chain segment mobility and ion mobility are dependent on the existence of sufficient unoccupied space to accomodate motion. The proposed free volume model relates dielectric and rheological properties by considering the fractional free volume required for polymer chain segment motion and for ion motion. This model has been successfully applied to a diglycidyl ether of bisphenol A epoxy resin isothermally cured with diamino-diphenyl sulfone at 177°C.

179. REACTIVE EXTRUSION IMIDIZATION OF ACRYLIC POLYMERS, Michael P. Hallden-Abberton, Rohm and Haas Co., Exploratory Plastics Research Department, P.O. Box 219, Bristol, Pennsylvania 19007, USA.

Polydimethylglutarimides can be prepared in an efficient manner by the use of a reactive extrusion process involving poly(methylmethacrylate) and primary amines such as methylamine in a plasticating extruder at high pressures and temperatures. Although dimethylglutarimide-containing polymers have been prepared previously, they, like this polymer, contain acid and anhydride functionality. It is proposed that the formation of acid and anhydride groups can be accounted for by an alkyl-oxygen cleavage reaction involving the primary amine and the alkyl-oxygen of the methyl methacrylate ester group. Analysis of the gaseous byproducts of this reaction, as well as experiments with dimethylamine and pMMA, serve to substantiate the proposed mechanism. The acid and anhydride-containing polydimethylglutarimide copolymers produced by this reaction can be further treated in a reactive extrusion process with specific alkylating agents such as orthoformate esters to prepare polydimethylglutarimides in which the acid and/or anhydride functionality is diminished or is absent (M. Hallden-Abberton, N. Bortnick, L. Cohen, W. Freed, and H. Fromuth, U.S. Pat. 4,727,117 (1988))..

180. **Dynamic Mechanical Analysis Studies of Reinforced Polyvinyl Chloride (PVC) Roofing Membranes.** R. M. Paroli and O. Dutt. *Institute for Research in Construction, National Research Council of Canada, Ottawa, Ontario, Canada K1A 0R6*

Dynamic mechanical thermal analysis (DMA) was used to study the effects of accelerated heat-aging at 100 °C and 130 °C on three PVC roofing membranes. The results obtained demonstrate that DMA, in conjunction with tensile testing, provides a more rigorous and accurate method of evaluating the quality of reinforced PVC roofing membranes.

181. STUDIES OF POLYMERS FOR SHOCK ABSORPTION APPLICATIONS. C.A. Byrne and W.X. Zukas, Polymer Research Branch, Army Materials Technology Laboratory, Watertown, MA02172

The purpose of this work was to evaluate polymers with potential shock damping properties. Many soft polyurethane elastomers were prepared in the laboratory and evaluated by thermal and mechanical analysis. These include mostly poly(propylene oxide) and polythioether based elastomers prepared with toluene diisocyanate. Tri- and tetrafunctional chain extenders were used in all cases, as well as some unusual secondary diamines, which disorganize the hard phase and soften the polymers. Comparisons were made to commercially available shock absorbing polymers. The synthesis of the polyurethanes is discussed in detail and the results of characterization of all of the polymers are described. The polymers exhibit T_g's ranging from -68 to -19 C, final softening temperatures between 115 and 246 C, Shore A hardness values ranging from 13-87, tensile strengths of 0.7 to 20 MPa and elongations of 175 to 2900 percent. Several display low Bashore rebound values in the temperature range of interest. Dynamic mechanical data were extrapolated to high frequency through the assembly of master curves and revealed that the polymers evaluated probably would exhibit good high frequency shock absorbing properties at room temperature or somewhat above room temperature, but that they would not display satisfactory behavior at temperatues significantly below room temperature.

182. SULFONIUM SALT DISTRIBUTION IN MODEL POLYMER FILMS BY ^{19}F MULTIPLE-QUANTUM NMR. B. E. SCRUGGS AND K. K. GLEASON, DEPT. OF CHEM. ENG., MASS. INST. TECH., CAMBRIDGE, MA. 02139.

Chemically amplified resists can provide high sensitivity to < 350 nm radiation, due to the catalytic nature of the resist chemistry. The efficiency of photoactive sulfonium salt conversion to an acid catalyst depends upon the dispersion of the salt within the resist's polymeric matrix. While SEM has been used to observe salt clusters > 500 Å in size, multiple-quantum (MQ) NMR has been used to quantify the number of individual fluorinated anions in sulfonium salt aggregates within polymeric films, providing information on a much smaller scale than SEM. The distributions observed with NMR ranged from individual anions to aggregates > 20 salt molecules. These distributions were studied as a function of salt loading and chemical identity of the polymeric matrix.

183. SOLVENT EFFECTS ON CONFIGURATIONS OF POLY(3-ALKYLTHIOPHENE)S. P. Love, Chem. Dept., Univ. of CT, Stamford, CT 06903; K. Yoshino, M. Onoda, Faculty of Eng., Osaka Univ., Osaka, Japan; and R. Sugimoto Osaka Research Center, Mitsui Toatsu Chem. Co., Osaka, Japan.

Solvatochromic shifts of λ_{max} in the spectra of poly(3-hexylthiophene), and poly(3-docosylthiophene) have been obtained for 20 solvents. For poly(3-hexylthiophene) λ_{max} is a function of Taft's π^* linear solvation energy relationship (LSER). For poly(3-docosylthiophene) there are marked deviations from the LSER equation. From an analysis of the LSER data and the spectral linewidths at half intensity it is shown that internal rotation changes of the polymer chains occur above a certain solvation energy. This results in greater ring coplanarity and shifts to longer wavelength λ_{max} values than expected in terms of the LSER correlations. The primary polymer solvation mechanism is solvent electron donor-polymer chain acceptor interation.

184. STRUCTURE OF POLYETHYLENE SHISH KEBABS IN SITU

D.M. Huong[1], M. Drechsler[2], M. Kunz[3], M. Moeller[4], H.-J. Cantow[1,3]
[1] Freiburg Materials Research Center (F·M·F), [2] Institute of Biophysics and [3] Institute of Macromolecular Chemistry; University of Freiburg, D - 7800 Freiburg, FRG; [4] Department of Chemical Technology, University of Twente, NL - 7500 AE Enschede, The Netherlands

Initial investigations of the morphology of polyethylene shish kebabs *in situ* demonstrated the presence of isolated micro shish kebabs in samples prepared from stirred solutions. Bright field micrographs by TEM showed parallel lamellae which apparently started to grow at the straight boundaries of the approximately 10 nm wide shish. The shish and the interior of the lamellae as well as the contact zone between them is completely crystalline, indicating the epitaxial crystallization of the overgrowth lamellae on the shish. By applying electron spectroscopic imaging (ESI) as a special technique in energy filtering electron microscopy, shishs perpendicular to the lamellae could be clearly visualized. In the ESI micrographs, enhanced contrast of the lamellae and shishs is caused by their higher carbon concentration compared to the amorphous matrix. The detected shish kebab morphology leads to the assumption that in vigorously stirred solutions, bundles of parallel micro shish kebabs build a polymer diffusion barrier which permits - by progressive crystallization - the lamellar growth only in external directions. In conventionally dried samples, the core of macro shish kebabs is probably formed by a bundle of micro shish kebabs.

PROFESSIONAL RELATIONS

Fourth Chemical Congress of North America

New York, NY
August 25 - 30, 1991

T. J. Kucera,
Program Chairman

TUESDAY MORNING AND AFTERNOON
● **Symposium on Whistle-Blowers, Advocates, and the Law**
G. B. Borowitz, Presiding   Papers 1-12

PROFESSIONAL RELATIONS

1.   GREED IN THE GROVES OF ACADEME. L. Minsky, National Coalition for Universities in the Public Interest, 1840 18th St NW, Washington, DC 20009

The contributions made by changes in patent and tax laws to the problem of fraud and misconduct is all but invisible and not taken into account when remedies are sought. The 1990 amendments to the Patenting Law altered the financial structure of science and added incentives to fraud and misconduct by giving ownership of patents to the universities. To competition for professional recognition and prestige is added a race for patentable products. Universities seeking to commercialize research are paying the price with more cases of fraud and misconduct. Implicated by their search for profits, their response to exposure by Congressional committees and whistleblowers becomes less than honest. Arguing that they are best able to monitor and punish abuses, universities are covering up the abuse and attacking the whistleblower, rather than exposing and punishing the perpetrator. Changes need to be made: we believe that only external agencies like the Congress, the NIH and the NSF can bring them about.

2.   TRUTH AND CONSEQUENCES: A PSYCHOLOGIST'S PERSPECTIVE ON WHISTLEBLOWING IN SCIENCE. Carolyn Phinney, Ph.D., Center for the Study of Youth Policy, 1015 E. Huron, University of Michigan, Ann Arbor, MI 48104-1689.

Although Congress and the press have focused a spotlight on the problem of scientific misconduct and how it is handled or mishandled by Universities, the scientific community has been slow to take this problem seriously and to take action to reduce misconduct. The author will identify and discuss one type of misconduct that she believes is relatively common -- the usurpation of credit for the creative ideas and research of scientists who are for some reason vulnerable by scientists who have power. The author will describe her own experiences as the alleged victim of such misconduct and as a whistleblower, focusing on the University's response to the allegations. The author will analyze some of the costs and benefits of different types of misconduct, to explain why some scientists and Universities may ignore or cover up allegations of misconduct. Finally, the author will propose methods to deter misconduct and cover ups.

3.   FROM DOUBTS TO CONVICTION OF FRAUD: TALES OF TRAUMA
Robert L. Sprague
Institute for Research on Human Development
University of Illinois
51 Gerty Drive
Champaign, IL 61820

This paper will present a brief summary of the first independent scientist in America to be indicted, found guilty, and sentenced for scientific fraud. Stephen E. Breuning, Ph.D., falsified data and journal articles regarding neuroleptic and stimulant medication for mentally retarded people who are some of the most vulnerable patients in our society because they typically do not have the ability to talk to their physicians about their reactions to medication. The National Institute of Mental Health investigated at glacier-like speed although thousands of impaired patients were at risk because of his scientific fraud. There was considerable trauma for me during the 6 years of the case: the University of Pittsburgh threatened to sue me for libel for my Congressional testimony, journals were reluctant to retract, and professional societies did not take action against him when the facts were reported to them.

4. THE FREEMAN CASE: PLAGIARISM AT THE ALBERT EINSTEIN COLLEGE OF MEDICINE OF YESHIVA UNIVERSITY (AECOM)/MONTEFIORE MEDICAL CENTER (MMC). <u>Heidi S. Weissmann, M.D.</u>, Saddle River, New Jersey

The author was an Associate Professor of Nuclear Medicine at AECOM and an attending physician at MMC. In 1987, she discovered her Division Chief had plagiarized a book chapter published by her and filed a copyright infringement lawsuit. The following morning she was met at her office by security guards who searched her and took her keys, and was advised to arrange to "pick up her things." In 1989, the U.S. Court of Appeals awarded the author judgment as a matter of law. Despite trial evidence clearly establishing the plagiarism, both AECOM and MMC have openly supported the plagiarist. A Congressional Report has strongly criticized both AECOM and MMC. The author will discuss the respective roles played in her case by the institutions, courts, media, public interest groups, and public officials, and outline proposals for the protection of whistleblowers.

5. THE BALTIMORE FIASCO: A CASE STUDY IN FRAUD. W. W. Stewart. National Institutes of Health, Building 8, Room BIIA-15, Bethesda, MD 20892.

An up-to-date recounting of a case of fraud and the notorious attempt to cover it up involving some of most prestigious universities in the country. The startling response of the scientific community as the scandal unfolded will be discussed.

6. THE FALSE CONFLICT: HELPING THE PUBLIC VS. HELPING THE CORPORATION. <u>Thomas Devine</u>, Legal Director, Government Accountability Project, 25 E. St., NW, Washington, D.C. 20001

"Whistleblowers" are those individuals whom publicly every politician likes, but privately few in positions of power will tolerate. They also are the point of the most dynamic expansion of free speech law in the country. Love them or hate them, increasingly it is certain that how management leaders react to whistleblowers will play a major role in the organization's success. This talk will concentrate on three themes. 1) Defining the whistleblower, legally and anthropologically: "Who are these guys?" 2) Why effective managers view whistleblowers as assets instead of threats: "It's better business to listen to the messenger than to kill him." 3) Techniques to turn dissent into a constructive resource. "The whistleblower: friend or foe? It's up to you."

7. TWO PERSONAL EXPERIENCES WITH INDUSTRIAL WHISTLE-BLOWING

<u>Robert T. Olsen</u>, 300 Country Lane RR-1, Eastham, MA 02642-3329

The author reports on two personal involvements in industrial whistle-blowing, the first concerning an internal operation and the second a public relationship; both were contended by letter and both resulted in his being dismissed. These and subsequent experiences are believed to have matured the author to be able to successfully negotiate difficult ethical situations face-to-face. In addition, he was active in the formation of the Division in '72-3 and he recommends that technologists faced with similar problems take advantage of confidential counselling with Local Section Professional Relation Committee members.

8. THE POLITICAL CONSEQUENCES FOR SCIENTISTS STUDYING AGENT ORANGE: A CASE STUDY. Jeanne Mager Stellman, School of Public Health, Columbia University, New York, N.Y. 10032

The possible effects of exposure to Agent Orange are both an important scientific question and a highly charged political issue. In 1983 we undertook an American Legion sponsored study of Vietnam veterans in their organization. The study was a comprehensive cross-sectional examination of the health and well-being of Vietnam veterans and only one aspect was designed to observe possible differences attributable to Agent Orange. The study was completed and resulted in a five paper series published in *Environmental Research*.

During the course of conducting the study, several tangentially related events occurred. The White House Agent Orange Working Group and the Centers for Disease Control declared it "impossible" to study Agent Orange effects in veterans and abandoned their efforts to do so, returning all unexpended funds. Also, under the auspices of the U.S. Eastern District Court we were asked to analyze extensive computerized files of troop movement data in Vietnam which the Court had obtained by Freedom of Information Act inquiry as part of the Agent Orange class action settlement. Access to these data uncovered a major source of troop tracking information, allowing calculation of probabilistic herbicide exposure measures. These were the same data abandoned by the government. The confluence of these events with the publication of the American Legion study, appears to be intimately related to intense political pressure on us, including a White House review, a Congressional hearing and personal professional attack, reaching far beyond the scope and implications of the comparatively modest cross-sectional study we completed. This paper documents the course of events and their implications for future scientists interested in independent research on Agent Orange.

9. FREEDOM OF SPEECH - THE DELICATE BALANCE: DEFENSE RESEARCH IN A DEMOCRATIC SOCIETY. R. D. Woodruff, Los Alamos National Laboratory, PO Box 1663, Los Alamos, New Mexico 87545

Perhaps one of the most difficult arenas in which to ensure that the public is fully informed is in my chosen field, defense research. Clearly a Nation needs the ability to preserve the secrecy of information vital to its own defense. Indeed, an entire structure of classification rules and regulations are in place to achieve this end. However, there is also a need for the citizenry to have sufficient information to make informed decisions with regard to the defense policy it will support. Yet there is no question that the freedom of speech of scientists involved in defense research is not, and cannot be, "unfettered."

My talk today, **Freedom of Speech - The Delicate Balance**, will discuss how an appropriate equilibrium can be maintained between freedom of speech and the secrecy requirements of national security.

10. ANATOMY OF A FRAUD CASE. Gene L. Trupin, 1000 Sunset Road, Piscataway, NJ, 08854-5510.

The author was an Assistant Professor of Anatomy at Rutgers Medical School, University of Medicine and Dentistry of NJ (UMDNJ). In 1984, the author filed a law suit in NJ Superior Court charging a colleague with fraud, plagiarism and theft of his research. UMDNJ administrators were charged with concealing the fraud and depriving the author of his livelihood. The case came to trial in 1987, and the author eventually received a financial settlement from UMDNJ. Trial documents and testimony produced an extensive public record of evidence. The author sent this material to the NIH Office of Scientific Integrity (OSI), requesting an investigation. The behavior of UMDNJ and OSI investigators will be examined in light of the trial record. The author will offer practical suggestions for whistleblower self-defense and make recommendations for promoting integrity in science.

11. LEGAL RIGHTS OF WHISTLEBLOWING PROFESSIONALS. A.G. Feliu, Paul, Hastings, Janofsky & Walker, 9 West 57th Street, New York, NY 10019.

Professionals make troublesome employees. Their allegiances extend beyond the employer's goals to include the mandates of his/her profession and its ethical code. Professional employees are, consequently, more likely than most employees to confront the issue of whistleblowing in the workplace.

The law in this area as, it applies to employees in general and to professional employees in particular will be reviewed.

12.   TO HELP WHISTLEBLOWERS, LEGALIZE PEG. <u>Alan C. Nixon</u>, 2140 Shattuck Ave., #511, Berkeley, CA, 94709.
   This paper will review efforts to legalize the Professional Employment Guidelines and suggest changes and strategies that might be adopted to get Congress to pass such legislation. One requirement is to get support from a large number of professional societies and organizations. This is necessary before members of Congress will spend their time introducing the bill. Not many business executives can be expected to favor the bill in spite of the fact that it could help to shield them from damaging publicity and lawsuits. Perhaps these factors might be used to convince the more forward thinking amongst them.
   Another way professional societies could strengthen the whole case in favor of whistleblowing would be to declare that they will support lawsuits against elements that seek to punish workers for their public spirited actions.

SMALL CHEMICAL BUSINESSES

Fourth Chemical Congress of North America

New York, NY
August 25 - 30, 1991

N. H. Giragosian,
Program Chairman

MONDAY MORNING AND AFTERNOON
● **Symposium on Starting & Operating Your Own Chemical Business**
N. H. Giragosian, Presiding

Papers 1-6
8-12

TUESDAY MORNING AND AFTERNOON
● **Symposium on True Stories of Small Chemical Businesses**
K. Greenlee, Presiding

Papers 13-20

WEDNESDAY MORNING AND AFTERNOON
● **Symposium on Opportunities for U. S. Chemical Businesses in Latin America**
M. Versteylen, Presiding

Papers 21-29

SMALL CHEMICAL BUSINESS

1. **SO YOU WANT TO START YOUR OWN CHEMICAL BUSINESS.** N.H. Giragosian

   NO ABSTRACT

2. **MARKETING RESEARCH.** N.H. Giragosian Delphi Marketing Services, Inc, 400 East 89th Street, New York, New York 10128.

   Market research is a vital function in determining what kind of products the would-be entrepreneur intends to manufacture. Included are:
   1. Estimated demand for the product(s)
   2. Who are your customers
   3. Pricing
   4. Forecast of future demand
   5. Determining channels of distribution
   6. Who is your competition

3. **SETTING YOUR STRATEGIES.** R. Polacek

   NO ABSTRACT

4. **CONSTRUCTING YOUR BUSINESS PLAN.** H. Avery

   NO ABSTRACT

5.
   FINDING THE MONEY
   Michael G. Mark, General Partner
   Unicorn Venture Funds
   6 Commerce Drive
   Cranford, New Jersey 07016

   The economic climate for raising investment capital has dramatically changed over the past several years. In particular, banks and other asset based lenders are increasingly reluctant to provide financing without substantial equity or subordinated debt in a deal. In addition, increasing environmental issues are making lenders and equity investors leary of chemical transactions.

   This talk will discuss the current economic environment and how to raise capital in it. The author will cite several recent examples of smaller transactions and discuss   (1) Deal Structure   (2) Pricing (3) Transaction costs and other related issues.

   Conclusions as to the future outlook of debt and equity markets will also be discussed.

6.
MANUFACTURING STRATEGIES. G.V. Austin. ChemDesign Corporation, 99 Development Road, Fitchburg, MA 01420.

This talk will examine the various strategies that need to be determined and implemented in order to manufacture your proposed products, such as:

1. Should one build a grass roots plant?
2. Should you lease an existing plant?
3. Should you have your products manufactured by a custom chemical manufacturer?

8. **CARRYING OUT R&D.** M. Strem

   NO ABSTRACT

9.
COMMERCIAL DEVELOPMENT. N.H. Giragosian. Delphi Marketing Services, Inc. 400 East 89th Street, Suite 2J, New York, New York 10128.

In order to take your products from the research stage to the marketing stage, it is necessary to determine the market for the new products, perform market development studies, and determine the business feasibility of the project. This talk will describe the various steps involved in the commercial development process and examine the business feasibility of each project.

10. **MARKETING YOUR PRODUCT(S).** S. Weber

    NO ABSTRACT

11.
HOW CHEMICAL ENTREPRENEURS CAN COPE WITH REGULATORY BURDENS
Dr. A. Wayne Tamarelli, Dock Resins Corporation, 1512 W. Elizabeth Ave., Linden, NJ 07036

Small chemical enterprises can cope with regulatory burdens. They can go beyond compliance and provide leadership in prevention of accidents and exposure hazards involving hazardous substances.

One company's experiences in combining both internal and external strengths to improve its safety and environmental programs will be discussed.

Many of the management control strengths prevalent in small organizations are helpful in plant safety and environmental programs. The closeness of top management to all employees can help transmit safety attitudes effectively if the hazards are properly recognized. These types of highly focused strengths are being sought by larger organizations as they move to improve their regulatory compliance programs.

On the other hand, smaller enterprises have resource constraints. The strongest impediments in small chemical businesses are staffing levels and regulatory complexity. One individual can be overwhelmed with multiple responsibilities normally borne by several committees or departments in larger organizations.

Nevertheless, most small organizations meet their safety obligations successfully by combining external specialized resources with their own results-oriented entrepreneurial personnel. The method will be summarized by the acrynom SAFE.

These outside resources include trade associations, consultants, suppliers, government agencies, and networking with other businesses.

Small chemical businesses can also benefit by volunteering to assist government agencies who have responsibility for safety and the environment. The agencies in turn tend to take a more positive and cooperative attitude toward businesses who are doing a responsible job.

The key to coping with regulatory burdens is persistent hard work: reading, attending meetings, sharing ideas and information. Only by staying ahead of emerging issues can one avoid falling behind later.

12. RAMBACH - THE HAPPY ALTERNATIVE. Dave Fishman. Rambach Corporation, Vesey Street, P. O. Box 5187, Newark, NJ 07105.

This author graduated from Miami University, then joined Rambach which is one of the largest dealers in "redundant" chemicals in the world. They specialize in helping dispose of chemicals no longer needed. By recycling they can save the owners tremendous expenses in sending the surpluses to landfills or to incinerators, and at the same time save money for others who happen to be shopping for those very same chemicals! As they are constantly discovering "gold" the Rambach people find their work not only rewarding but actually exciting.

13. FOUNDING FAIRFIELD CHEMICAL COMPANY. Hiram S. Allen, Fairfield Chemical Company, P. O. Box 20, Blythewood, SC 29016.

This is the story of Fairfield Chemical Company, from its beginning in 1965, as a part-time job; until its sale in 1988 - and its aftermath. Some of the main events and key people that contributed to the Company's success are discussed.

14. RAMBACH THE HAPPY ALTERNATIVE. Dave Fishman. Rambach Corporation, Vesey Street, P. O. Box 5187, Newark, NJ 07105.

This author graduated from Miami University, then joined Rambach which is one of the largest dealers in "redundant" chemicals in the world. They specialize in helping dispose of chemicals no longer needed. By recycling they can save the owners tremendous expenses in sending the surpluses to landfills or to incinerators, and at the same time save money for others who happen to be shopping for those very same chemicals! As they are constantly discovering "gold" the Rambach people find their work not only rewarding but actually exciting.

15. CRESCENT CHEMICAL AND ITS ORIGIN IN TRIDOM. Eric Rudnik, Crescent Chemical Company, Inc., 1324 Motor Parkway, Happauge, NY 11788.

This story could also be called "The Agony and the Ecstasy" or "The Nine Lives of a Chemical Importer." In the business of representing foreign companies you not only deal with all the tortures the EPA has become famous for, but also the politics of nations and big business.
It is a business of constant and unpredictable change. You develop new relationships constantly or you find yourself working for somebody else.

16. HAVE INTELLIGENCE, IMAGINATION & DILIGENCE? BE YOUR OWN BOSS! D. T. Meshri, Advance Research Chemicals, Inc., 1085 Fort Gibson Road, Catoosa, OK 74105.

Dr. Alfred Bader has been my mentor ever since I heard his start-up story of Aldrich Chemicals at one of the "true story" sessions of this Division. During his talk I realized that an intelligent, diligent, and caring chemist can make a difference, provided that he is the captain of his own ship. Hence, the seed of independence was sowed in my mind, and the consequent fruition of ARC will be discussed herein.

17. FAILING WITH A FLAIR. Elmer A. Fike, Fike Enterprises, Inc., P.O. Box 550, Nitro, WV 25143.

I have been in commercial chemistry for 51 years, and have learned a lot - especially how to avoid the problems of success - by failing. I failed in a big company by being outspoken, I failed in my own company by "going public," and failed in an attempt at public office - by being too honest. But, I succeeded in learning how to enjoy a life of troubles Let me tell you how.

18. RUNNING A SMALL BUSINESS WITHIN A LARGER COMPANY. Ric DuBoisson, PCR Incorporated, P. O. Box 1466, Gainesville, FL 32602.

A brief history of PCR Inc. will reveal the role of the Research Chemicals Division in helping to create a commercial chemical manufacturer. The PCR Research Chemicals Catalog is well known for its wide range of fluorine and silicon based reagents, and also highlights the technology practiced at PCR. The offspring is now larger than the parent, and the advantages and drawbacks of running a small company within a large company will be discussed.

19. ACQUISITION: THE PLEASANT(?) FATE OF MANY SMALL CHEMICAL BUSINESSES John Cochran, President, MTM Research Chemicals, Inc. 4131 Northwest 13th Street, Gainesville, FL 32609.

The acquisition of a small chemical business can bring benefits and problems for acquirer and the acquired. The manner in which both parties approach the acquisition and the expectations of both sides will determine if the experience will continue to be a pleasant one after the honey moon is over. The author explores this subject from both sides of the table, having been a stockholder in two businesses that were acquired, and now as a senior manager in a company that has made several acquisitions in recent years.

20. IODINE AND SODIUM NITRATE - THE STORY OF CHILEAN NITRATE. Fernando Encinas, President: SQM Iodine Corp., 150 Boush St., Suite 701, Norfolk, VA 23510.

Crude iodine was first produced in Chile in 1868, with world consumption reaching its highest level in 1989/1990. As recently as five years ago, irregularities in iodine production resulted in shortages and substantial price variations. Now, we are entering a new era of steady supply and stable prices, subject only to normal market conditions. New production facilities in the U.S. as well as in Chile have assured the chemical industry of steady and reliable supplies at market prices.

As a result of this assured supply, we are seeing increasing interest in the use of iodine in new applications, such as water treatment and biocides, as well as the traditional uses in X-ray contrast media, and iodophors for agricultural and other applications.

As President of SQM Iodine, the largest single producer of iodine in the world, I am here to assure the chemical industry that the outlook for iodine users has never been brighter.

21.
LATIN AMERICAN CHEMICALS APPROACH 2000 A.D.
Dr. Charles H. Kline
PanGraphion Inc.
389 Ski Trail, NJ 07405-2247

Experts forecast that chemical production in Central and South America will rise at 6.3% a year from 1989 to 1995 and at 8.4% a year from 1995 to 2000. The latter rate is the second highest in the world, exceeded only by that for the smaller countries of East Asia. European, North American, and Japanese megacompanies will control part of the new production. Locally owned and managed companies can control much of the rest if government policies and economic stability permit.

22.
TECHNOLOGY TRANSFER AND THE CHALENGE FOR SMALL CHEMICAL BUSINESS. U.S. AND
LATIN AMERICA APPROACH
Dr. Patrick Hoet
Bater World Trade Corporation
P.O. Box 784, Deerfield, Illinois 60015-0784

Technology transfer may be as fundamental as providing college-university or
technical education, but it may be as complex as starting a new high technology
manufacturing operation in an area where technical skilled personnel is difficult
to find. The chalenges vary depending on factors as the complexity of the technology, the level of technical or college education, ownership of the technology,
patented technology, propietary technical information, technical support, final
product sales country, local and international regulations and quality standards.
Concrete examples and options for US & Latin America will be discussed.

23.
INDUSTRIAL TECHNOLOGY FOR SMALL AND MEDIUM SIZE ENTERPRISES. THE CODETEC'S
EXPERIENCE
Eng. José Carlos Gerez
CODETEC, P.O. Box 6041, Campinas 13081, Brasil

The Brazilian Company for Technological Development (CODETEC) activities in the
area of fine chemistry originated in a Government decision to develop a program
of technological capabilities in the chemical pharmaceutical area. CODETEC under
contract with government and private companies has developed specific projects
ranging from bibliographic research to industrial implantation projects. Today
the program can point to impressive results in the pharmaceutical sector: sixty
products with an established industrial technology, twelve already in production
and more than forty in various stages of development. On the other hand, agrochemicals, dyes and organic intermediates are, as well as pharmaceutical products, part of CODETEC's work.

24.
THE INDUSTRIAL EXPLOITATION OF NATURAL PRODUCTS FROM THE TROPICAL RAIN FOREST. A
FIELD FOR SMALL CHEMICAL BUSINESS
Dr. Benjamin Gilbert
CODETEC, P.O. Box 6041, Campinas   13081, Brasil

The rapidly growing demand for natural materials by the food, cosmetic, agrochem-
ical and drug industries can absorb the product of the tropical rain forest, and
in fact, such natural materials command much higher prices that their synthetic
counterparts. Some thirty products have been identified for which immediate or
virtually immediate market exist. Based on previous experience in Latin America
small agile business investments are most likely to produce quick results.
Furthermore, the prices paid for natural products are usually less than one-tenth
of the sale prices of formulated products sold to the public. Therefore an excellent alternative to associate natural product exploitation with forest preserva-
tion is to have companies that formulate in the region a finished article of high
unit value.

25.
TECHNOLOGICAL SUPPORT FOR THE PRODUCTIVE SECTOR. THE CASE OF THE TECHNOLOGICAL
TRANSFER UNIT IN COSTA RICA
Eng. Jorge Monge Zeledón
Technological Transfer Unit. Universidad de Costa Rica
Ciudad Universitaria Rodrigo Facio, San José, Costa Rica

The Technological Transfer Unit (UTT) of the University of Costa Rica (UCR) promotes the linkage of this institution with the private sector in a sistematic
approach. The UCR holds eighty percent of the research performed in the country
and the UTT has been working on the development and transfer of technology focused on industrial needs, among other activities. The different aspects
covered include contract drafting, legal advice, standard regulations and the
development of technological innovation projects, among others. A case on a
Small Chemical Industry will be presented in detail.

26.
SMALL CHEMICAL BUSINESS IN COSTA RICA. THE CASE OF ARVI INC.
Dr. Eduardo Arguedas
ARVI Inc., P. O. Box 236
San Pedro, Costa Rica

ARVI Inc. is a small national chemical company specialized in natural products and fine chemicals. The management and financial system, the issues within the Government industrial policy, the relationship with the University of Costa Rica and other Science and Technology governmental institutions as well as with multinational companies and foreign investors and some examples of projects and products achieved are described.

27.
TECHNOLOGICAL SUPPORT FOR SMALL INDUSTRIES. THE CASE OF THE CENTER FOR TECHNOLOGICAL INNOVATION IN MEXICO
Dr. Jose Luis Solleiro
Center for Technological Innovation-UNAM
P.O. Box 20-103 Mexico D.F., Mexico

The Center for Technological Innovation (CIT) of the National Autonomous University of Mexico promotes the development and transfer of technology focused on industrial needs including: contract drafting and negotiation, patent drafting and follow-up, financial advice, feasibility studies, provision of technical assistance, and the development and transfer of specific technologies. The main areas of work have been biotechnology, applications of new materials and other chemical products. After eight years of activity, CIT has formalized 230 contracts with small and medium scale companies.

28.
THE ROLE OF RESEARCH CENTERS IN THE DEVELOPMENT OF CHEMICAL INDUSTRIES. THE CASE OF CELANESE MEXICANA INC.
Dr. Francisco Carrasco
CELANESE MEXICANA INC., Avenida Revolución 1425
Col. Tlacopac, Mexico D.F. 1040, Mexico

CELANESE MEXICANA Inc. has established some years ago its own local Center for Research and Development. The process leading to that accomplishment inside and outside the company, the relationship with the Mexican Government and other laboratories and R&D Centers, the management and financial system as well as some examples of the projects achieved are described.

29.   WITHDRAWN

COMMITTEE ON ENVIRONMENTAL IMPROVEMENT

Fourth Chemical Congress of North America

New York, NY
August 25 - 30, 1991

G. E. Bellen,
Program Chairman

TUESDAY MORNING
● **Symposium on Human Biomonitoring**
A. M. Ford, Presiding

Papers 1-14

WEDNESDAY MORNING
● **Symposium on Human Biomonitoring**
N. J. Karch, Presiding

Papers 15-21

WEDNESDAY AFTERNOON
● **Symposium on Approaches to Pollution Prevention**
G. E. Bellen, A. M. Ford, Presiding

Papers 22-28

ENVIRONMENTAL IMPROVEMENT

1. MEASURING TRACE LEVELS OF VOLATILE ORGANIC COMPOUNDS IN HUMAN BLOOD BY USING PURGE-AND-TRAP/GAS CHROMATOGRAPHY/MASS SPECTROMETRY. Michael A. Bonin, David L. Ashley, Fred L. Cardinali, Joan M. McCraw and Joe V. Wooten, Division of Environmental Health Laboratory Sciences, F17, Center for Environmental Health and Injury Control, Centers for Disease Control, Public Health Service, U.S. Department of Health and Human Services, Atlanta, GA 30333.

   Volatile organic compounds (VOCs) are a major public health concern because of the possible health effects associated with constant environmental exposure to them, especially from air in buildings. An analytic method was developed for measuring more than 30 VOCs in 10 mL of human blood at the low parts per trillion level. This method is sensitive enough to assess the normal body burden of VOCs among unexposed persons in the general population. This trace level sensitivity was achieved by coupling a purge-and-trap method to isolate the VOCs with the descriminating power of a high resolution magnetic sector mass spectrometer. To achieve the required reproducibility at trace levels of VOCs for a long-term study, special efforts were made to control contamination at each step in the analysis at the lowest possible level. A quality-control protocol was developed to ensure that the measurements were valid.

2. MEASUREMENT OF URINARY PHENOLIC METABOLITES OF ENVIRONMENTAL CONTAMINANTS IN THE U.S. POPULATION, Hill, R.H. Jr., Head, S.L., Shealy, D.B., Alley C.C., Bailey, S.L., Gregg, M., and Needham, L.L., Centers for Disease Control, Public Health Service, U.S. Department of Health and Human Services, Atlanta, Georgia 30333.

   Exposure to environmental contaminants is difficult to assess. We have developed a method to measure baseline or "normal" urinary concentrations of 12 analytes including: 2-isopropoxyphenol; 2,5-dichlorophenol; 2,4-dichlorophenol; 7-carbofuranphenol; 3,5,6-trichloro-2-pyridinol; 2,4,5-trichlorophenol ; 2,4,6-trichlorophenol ; 4-nitrophenol; 1-naphthol; 2-naphthol; 2,4-dichlorophenoxyacetic acid; and pentachlorophenol. This procedure is being used to determine these analytes in urine samples collected from 1,000 people who participated in the Third National Health and Nutrition Examination Survey (NHANES III). These measurements will be used as reference values for comparison with the values found in groups or communities that have the potential for exposure. We will report preliminary results of this study.

3. ANALYTICAL METHOD FOR BIOMONITORING OF ANILINE IN URINE AND COMPARISON WITH THE CLASSICAL COLORIMETRIC METHOD, Robert G. Orth and Jay M. Wendling, Monsanto Environmental Sciences Center, 800 N. Lindbergh Blvd., St. Louis, MO 63167

   For the development of valid exposure investigations many points should be understood. These include the metabolic pathways, the biological matrix to be evaluated, sampling and sample integrity, method of analysis and quantitation, and the use of the results. In the determination of exposure to aniline and the comparison to a biological exposure index, 50 mg/g creatinine, a semiquantitative analytical method is presently in use which measures the major metabolite p-aminophenol by a colorimetric method. In order to provide an analytical method which was more specific and provides faster analysis, an HPLC method was developed and compared to the colorimetric approach. In addition, because of observed pseudobackground of 5 mg/L for both methods, a gc/ms method was developed to determine whether the background was a coeluting interferent or from a background level of p-aminophenol from nonoccupational sources. The method detection limit for the HPLC method was 3 mg/g creatinine while the analytical detection limit was <0.5 mg/L.

4. BIOLOGICAL MONITORING OF WORKERS EXPOSED TO TRIAZINE HERBICIDES I. DETERMINATION OF SIMAZINE AND ITS METABOLITES IN URINE. J.G. Guillot, A. Leblanc, O. Samuel and J.P. Weber, CENTRE DE TOXICOLOGIE DU QUÉBEC, Le Centre Hospitalier de l'Université Laval, 2705, boulevard Laurier, Bureau 800, Québec, Qué. G1V 4G2.

Within the framework of biological monitoring of workers exposed to triazine herbicides, we have developed an analytical method for the determination of un-metabolised and metabolised triazines in urine capable of measuring amounts as low as 5 µg/L. The method consists in the extraction of the species with organic solvents after which the primary and secondary amino groups are methylated with iodomethane in the presence of sodium hydride. The derivatives are then concentrated and analysed by GC-MS in the selected ion monitoring mode. The advantage of this method is that it eliminates absorption problems often encountered at low concentrations. The reproducibility of the method on five urine samples spiked with 25 µg/L of each species (simazine and its metabolites I and III) varied from 10% within-day and 15% over the course of the project (45 days). The urinary levels of the un-metabolised simazine did not appear to be an appropriate marker of biological exposure. However, the levels of metabolites I and III were well correlated with the type of work and the protective gear used by the workers. Exposure was greatest for workers using a hand spray gun behind a tractor. We conclude that the biological monitoring of both metabolites (I and III) of simazine could be used to evaluate the workers' exposure and the effectiveness of safety practices. The data collected indicate that the workers absorbed amounts of simazine inferior to the acceptable daily intake value of 0.005mg/kg.

5. BIOLOGICAL MONITORING OF WORKERS EXPOSED TO TRIAZINE HERBICIDES. II: DETERMINATION OF HEXAZINONE AND ITS METABOLITES IN URINE OF EXPOSED FOREST PESTICIDE APPLICATORS. A. LeBlanc, O. Samuel, J.G. Guillot and J.P. Weber, CENTRE DE TOXICOLOGIE DU QUEBEC, Le Centre Hospitalier de l'Université Laval, 2705, boulevard Laurier, Bureau 800, Québec, Qc. G1V 4G2.

We propose an analytical method for the determination of un-metabolised and metabolised hexazinone in urine. The species are extracted with chloroform after which the secondary amino groups and hydroxyl functional groups of the metabolites are ethylated with iodoethane in the presence of sodium hydride. The derivatives are concentrated and analysed by GC-MS in the selected ion monitoring mode. The ethylation of the hydroxyl and amino groups allows us to differentiate un-metabolised hexazinone from all its principal metabolites and helps eliminating absorption problems on the GC column thus allowing detection limits inferior to 10 ug/l. Even though our new method allows us to detect small amounts of un-metabolised hexazinone, metabolite B appears to be a better marker of biological exposure since we have found amounts up to 15000 ug/l in the urine of exposed workers. Chemical and enzyme hydrolysis did not provide evidence that the hydroxylated metabolites are excreted as conjugates. Moreover, the levels of the different metabolites are correlated with the levels of hexazinone. The urine of workers involved in pesticide applications using spot guns was found to contain levels of metabolites of the order of 8 to 400 times greater than the workers involved in other types of applications. Many factors connected to the various working methods influence the degree of exposure, among which the wearing of individual protective equipments, smoking, eating and drinking while carrying out different types of applications, personal hygiene, etc. The biological monitoring of hexazinone and its metabolites proves to be an effective method for assessing the exposure of workers and in so doing helps in orienting preventive steps.

6. APPLICATIONS OF NEW HPLC/MS TECHNIQUES IN HUMAN BIOMONITORING. R. Brown and J. Johnston, California Public Health Foundation, Berkeley, CA 94704, W. Draper and R. Stephens, California Department of Health Services, Berkeley, CA 94704, and C. Becker, University of California, San Francisco, CA 94110.

The excretion products of most toxic organic chemicals are highly polar substances resulting from metabolic functionalization followed by conjugation to endogenous compounds including glucuronic acid, sulfuric acid and various amino acids. As such analytical techniques capable of separating and detecting polar metabolic products in urine and other body fluids offer the potential for recognition and quantification of exposure to a tremendous array of foreign compounds. Our laboratory has investigated applications of two such techniques, particle beam (PB) and thermospray (TSP) high pressure liquid chromatography coupled mass spectrometry (HPLC/MS), in human biomonitoring. This paper briefly reviews relevant features of PB and TSP mass spectrometry and describes LC/MS methods developed for determination of glucuronides, sulfates, and mercapturates, structures which encompass the major urinary metabolite classes. Also described is application of particle beam LC/MS to the measurement of human styrene exposures incurred under controlled conditions in a laboratory exposure chamber.

7. THE APPLICATION OF BIOLOGICAL MONITORING FOR ENVIRONMENTAL PUBLIC HEALTH Larry L Needham and James S. Holler, Centers for Disease Control, Atlanta, Georgia 30333

Over the last decade increasing concern has focused on human exposure to toxic chemicals. When confronted with the question of exposure and possible health effects, we have addressed the issue using various

biological monitoring techniques. Many of the compounds of interest--
pesticides, PCBs, dioxins--are typical of the compounds of concern
because of the publicity and widespread occurrence of these compounds.
These responses to environmental situations usually involve the
development of study protocols as the planning tool, the collection of
samples within the study populations, the application of laboratory
methods to obtain valid quantitative results and proper interpretation of
the data to determine if excessive exposure has occurred. Several case
studies are used to illustrate the importance of each of the phases of
the study and the critical nature of the biological monitoring activity.

8. HUMAN EXPOSURE TO HAZARDOUS WASTES: ASSESSMENT OF PREVALANCE, BIOAVAILABILITY AND BIOEFFECTIVE DOSE. M. W. Tabor and H. P. Kagen Inst. of Envtl. Hlth., Univ. of Cinc. Med. Ctr., Cincinnati, Ohio 45267-0056.

Exposure assessments require knowledge of three key elements: 1] prevalence and concentrations of the hazardous chemicals in the environment; 2] elucidation of routes and pathways of human exposure; and 3] determination of bioeffective [i.e. toxicologic significant] doses of the chemical and/or their bioactive metabolites in humans. At the Institute of Environmental Health of the University of Cincinnati Medical Center, the research program to assess human exposure to hazardous wastes is based on a strategy encompassing these key elements. These three elements are required to address the major goal of human biological monitoring, i.e. assessment of the risk to human health from environmental exposures. Based on this information, strategies can be employed to reduce human health risk to hazardous wastes. An important aspect in reducing these risks is the implementation of a prospective assessment of the effectiveness of these programs by a continuing human biological monitoring program. The strategy and requirements for this program will be presented. Supported in part by NIEHS 3P30 ES00159 and USEPA CR 814679.

9. Biomonitoring For Chromium Levels in the Urine of Individuals Living Adjacent to Chromate Production Waste Sites
A.H. Stern[1]; N.C.G. Freeman[2]; P. Pleben[3]; R. Boesch[4]; T. Wainman[2]; L.R. Korn[1]; T. Howell[5]; S. Shupack[5]; and P.J. Lioy[2]
1. NJ Dept. of Environmental Protection, Trenton, NJ; 2. Robert Wood Johnson Medical School, Piscataway, NJ; 3. Old Dominion University, Norfolk, VA; 4. NJ Dept. of Health, Trenton, NJ; 5. Villanova University, Villanova, PA.

The Chromium exposure of 50 individuals living adjacent to four chromate production waste sites in Hudson County, New Jersey was investigated through biomonitoring of urine chromium in conjunction with environmental sampling of residences and activity/lifestyle data collection. The potentially exposed population was compared to a control population of 24 individuals in two locations with no known chromium waste and no significant chromium use within 1/2 mile. Environmental data collection included surface wipe and vacuum samples and PM10 air samples. Chromium in spot urine samples was measured with precautions to eliminate background contamination using graphite furnace, Zeeman-effect atomic absorption spectrophotometry with creatinine correction of urine dilution. Geometric mean urine chormium concentrations in the total "exposed" and control populations (0.22 and 0.18 ug Cr/g creatinine respectively) were not significantly different. However, urine results frcm "exposed" subpopulations identified on the basis of Cr levels in wipe samples, outside play activites and work location were consistent with elevated environmental exposure to chromium.

10. ENVIRONMENTAL AND BIOLOGICAL MONITORING OF OCCUPATIONAL EXPOSURE TO CYCLOPHOSPHAMIDE IN THE PHARMACEUTICAL INDUSTRY. Christian Lévesque, DSC Sainte-Justine, 3175, Côte Ste-Catherine, Montréal, Qué. H3T 1C5 (CLSC Côte des Neiges, Santé au travail), Liliane Ferron, Alain LeBlanc, Jean-Philippe Weber, CENTRE DE TOXICOLOGIE DU QUÉBEC, CHUL, 2705, boul. Laurier, Québec, Qué. G1V 4G2.

The purpose of this study was to evaluate the efficiency of general and individual preventive measures implemented in a cyclophosphamide tablet manufacturing facility. Personal samplers were used to measure ambient levels at various production steps. Samples were collected on PVC filters (2 μm pore size, 37 mm diameter). Volumes of air sampled varied between 40 and 417 L. The total amount of particles on each filter was determined by differential weighing; cyclophosphamide was determined by gas chromatography. Exposure to cyclophosphamide was greatest at the granulation step. Biological monitoring was carried out by GC-MS determination of unchanged cyclophosphamide in the urine of workers. The majority of specimens analyzed contained measurable amounts of cyclophosphamide (1-45 µg/L). Since only a small proportion (ca. 10%) of cyclophosphamide is excreted unchanged, these results indicate that exposure is not negligible. Workers handling cyclophosphamide wear a full face mask with a supplied air respirator, a whole-body garment (including hood) made of non-porous material and two pairs of surgical gloves (changed after every 30 minutes). In the course of the granulation procedure, the worker is required to handle the paste to verify its consistency. Considering the type of protective gear used, exposure is possible through skin contamination by infiltration through the clothing, and by inhalation or skin exposure when workers remove their protective gear.

11. EXPOSURE OF FISHING PEOPLE TO CHLORINATED AROMATIC HYDROCARBONS IN LA BASSE CÔTE-NORD, QUÉBEC. John J. Ryan, Health and Welfare Canada, Ottawa, K1A 0L2 and Eric Dewailly, Cent. Hospit. Univ. Laval, Ste-Foy, Québec, G1V 2K8.

Marine animals such as finfish, shellfish, and seals from the lower north shore (la Basse Côte-Nord) in the gulf of St. Lawrence, Québec constitute a major source of both revenue and food to the people of this area. Data on the PCB and PCDD/PCDF content of marine life from this region is limited but indicates that its continued consumption may result in elevated exposure of these contaminants. Plasma blood samples collected from 185 individuals from 7 fishing villages showed mean total PCB values based on the 10 major congeners of 12.3 µg/kg wet weight i.e., about 7 times the normal for Québec of about 1.6 µg/kg. Plasma from 10 individuals showing the highest total PCB levels (mean 46.4 µg/kg wet weight or 7.9 mg/kg lipid weight) were also analysed for PCDDs/PCDFs and two co-planar PCBs, IUPAC #s 126 and 169. The co-planar PCBs 126 and 169 were found at mean levels of 770 and 500 ng/kg lipid (ppt) as compared to 38 and 15 ng/kg lipid from a pooled control sample. There was a significant correlation between 126 and 169 (r=0.90) and between the mono-ortho 118 and 126 (r=0.64). Using a plasma sample weight of 4 g, little or no 2,3,7,8-TCDD or 2,3,4,7,8-pentachlorodibenzofuran could be detected above 50 ng/kg lipid. These results suggest that fishermen with elevated total PCB levels also have elevated co-planar PCB exposure but probably not elevated PCDDs/PCDFs. Further work using more individuals and larger volumes of plasma is underway to define more clearly the degree of exposure.

12. HUMAN TISSUE SAMPLING FOLLOWED BY DIOXIN ANALYSIS TO ESTIMATE BODY BURDEN IN GENERAL POPULATION AND EXPOSED INDIVIDUALS. Arnold Schecter, Dept of Preventive Medicine, SUNY-Health Science Center at Syracuse, 88 Aldrich Ave., Binghamton, NY 13903

Beginning in the early 1980's in the US, fat, blood, and milk dioxin measurements from exposed workers and the general population have been used to estimate dioxin body burden. Data will be presented from exposed workers, where elevated chlorinated and brominated dioxins have been found up to 34 years after exposure, from Vietnamese living in Vietnam and U.S. Vietnam veterans who have been found to have elevated TCDD levels over 20 years after Agent Orange spraying and from tissue levels in a number of industrial and non-industrial countries. These include the USA, Germany, Canada, Japan, Vietnam, Thailand, Cambodia, the USSR, South Africa, India. In addition, dioxin levels in ancient frozen Eskimo tissue will be compared to modern tissue levels to address the "trace chemistry" hypothesis of dioxin formation.

13. Multivariate data analysis of PCDD/PCDF levels
in human blood samples from workers in a magnesium production plant

Hansson M, Rappe C. Institute of Environmental Ch~   University of Umeå. S-901 87 UMEÅ, Sweden.
Dybing E. National Institute for Public Health. D~   vironmental Medicin. N-0462 OSLO 4, Norway.
Grimstad T. Health Departmen~   N-3901 PORSGRUNN. Norway.
                                          WITHDRAWN

Significant amounts of polychlorinated dibenzo-p-dioxins and ..benzofurans (PCDD/PCDFs) are formed under various high-temperature processes, including the production of magnesium. They have been detected in effluents and exhausts, as well as in fish and sediment samples collected in the vicinity of the magnesium plant.

Ten workers, employed at the magnesium plant for 10 to 36 years, and a reference group consisting of ten unexposed persons with no declared exposure, were studied.

Isomer specific analyses of PCDD/PCDFs by means of HRGC/HRMS techniques were performed on coded blood plasma samples and 2,3,7,8-congeners of PCDD/PCDFs at ppt levels were detected. Multivariate data analysis was used to study 14 different congeners present in the samples and significant differences between the workers and the control group were found.

14.
CHLORINATED DIBENZODIOXINS AND DIBENZOFURANS IN THE BLOOD OF TWELVE TCDD EXPOSED PERSONS FROM JACKSONVILLE, ARKANSAS. S.J. Levin[1], P.C. Kahn[2], M. Hansson[3], and Christoffer Rappe[3]. [1]Mt. Sinai Medical Center, New York, New York; [2]Rutgers University, New Brunswick, New Jersey; [3]University of Umea, Umea, Sweden.

Blood plasma from eight men and four women associated with the Vertac site in Jacksonville, Arkansas, including two married couples, has been analyzed for all dioxins and furans having four or more chlorines. The 2378–TCDD levels ranged from 20 to 540 pg/g blood lipid in the women and from 220 to 1010 pg/g in the men. Other congeners are also elevated in some of the people, with OCDD, for example, ranging up to 2140 pg/p. The relationship of these data to other exposed cohorts in the United States and Europe will be discussed.

15.   BIOMONITORING: OPPORTUNITIES AND RESPONSIBILITIES. Nathan J. Karch, Karch & Associates, Inc., 1701 K Street, N.W., Washington, D.C. 20006.

The promise of biomonitoring is appealing: To identify markers of exposure, susceptibility or effect. With such information, uncertainties regarding exposure and dose can be resolved. By detecting subclinical signs of harm, the prospect for treatment may be enhanced. A range of prudent actions may be possible that otherwise are not justified or advisable. With such opportunities, however, we need to consider the impact of errors in attributing biomarkers to chemical exposure or to susceptibility. There are some notable markers, including immune parameters, urinary porphyrins, neuro-psychological tests, and others, that have been incorrectly attributed to chemical exposure, often in litigation. Some of the pitfalls and incorrect attributions will be discussed in an attempt to outline a scientifically responsible approach to identifying biomarkers.

16.   SISTER CHROMATID EXCHANGE TEST AS MONITOR OF HUMAN EXPOSURE. S.W. Pirages and A.G. Tohme, Karch & Associates, Inc., 1701 K Street, N.W., Suite 1000, Washington, D.C. 20006.

The Sister Chromatid Exchange (SCE) assay was developed as a simple laboratory technique for detecting genetic changes associated with chemicals. The SCE technique as a screening test for genotoxic activity is a useful laboratory procedure. However, use of the SCE assay as a monitor of human exposure has many limitations. The reliability and effectiveness of the SCE assay applied to human exposures is still uncertain. A review of the range of chemicals that have been tested will include evaluations of the sensitivity of the assay for determining environmental and occupational exposures and an emphasis on those factors, unrelated to chemical exposure, known to interfere with the predictive accuracy and precision of the assay.

17.   MEASUREMENT OF IMMUNE DISORDER IN PERSONS EXPOSED TO ORGANOCHLORINES. Peter R. McConnachie, Memorial Med Center, Springfield, Illinois and Arthur C. Zahalsky, Biological Sciences, Southern Illinois University, Edwardsville, Illinois, 62026.

Biomarkers of immune disorder and damage recommended by the Agency on Toxic Substances Disease Registry (ATSDR), U.S.P.H.S., consist of test tiers that evaluate lymphocyte phenotype frequencies and their functional activities. We have applied these tests to families and individuals exposed to polychlorinated aromatic hydrocarbons (PAH's), including technical chlordane (CHL), pentachlorophenol (PCP) and polychlorinated biphenyls (PCB's). Statistically significant immune alterations differing in the cluster of anomalies found were correlated with other clinical evidence of disorders in these persons. Evidence of phenotypic and functional immunodeficiency and potential autoimmune activation was detected up to ten years after exposure to CHL. Persons exposed to PCP showed depressed mitogenic responses, elevated activation markers, and deficient numbers of helper-inducer cells. Chronic exposure to PCB's resulted in the most anomalous profile amongst the PAH's including evidence of possible light chain immunopathy. The pattern of immune alteration found with each PAH supports the ATSDR's statements on the value of immune biomonitoring in detecting even subtle changes in immune cell ontogency and maturation.

18. THE USES AND ABUSES OF IMMUNOLOGICAL TESTING. D.K. Flaherty, Environmental Health Laboratory, Monsanto Agricultural Company, St. Louis, Missouri 63110.

Over the last ten years, there have been numerous allegations that select chemicals induce immune dysfunction in humans. There are now available sophisticated in vitro assays which probe the functional capability of individual immune effector cells. These assays can be used to: (1) determine the functional status of the immune system in relation to susceptibility to tumors and infection with xenobiotics; and (2) document immediate and delayed hypersensitivity reactions to chemicals. This talk will discuss the concept of immunological testing as a diagnostic tool. Included in the discussion will be examples of the accuracy, sensitivity and reproducibility of immunological assays. In addition, the concept of confounding factors in the interpretation of data will be explored.

19. PROBLEMS WITH THE USE OF URINARY CREATININE AS A STABLE METRIC FOR HUMAN BIOMONITORING STUDIES AND AN ALTERNATIVE STRATEGY. Jed M. Waldman and Timothy D. Buckley, Department of Env. Medicine, UMDNJ-RWJ Medical School, Piscataway, NJ 08854.

Because urinary output is highly variable among individuals and over time, elimination rates for biomarkers of exposure in urine are frequently normalized to creatinine excretion, i.e. ng xenobiotic per mg creatinine. This assumes that instantaneous creatinine elimination rates are more stable than the volume elimination rates of urine voids. However, there is a compelling body of literature that demonstrates the urinary creatinine is not eliminated at a constant rate. This variability, within and between individuals, confounds determinations of biomarker elimination rates. Data are presented herein that further substantiate earlier reports documenting the variability in creatinine elimination rates. We propose the use of a time-weighted approach to normalize urinary biomarker concentrations, using the time interval between successive voids, i.e. ng xenobiotic per hour. This approach provides a logistically feasible, scientifically sound and reliable alternative to the creatinine approach.

20. FUTURE ROLE OF ERYTHROCYTE PROTOPORPHYRIN (EP) AS A BIOCHEMICAL MARKER FOR DETECTING EXPOSURE TO LEAD. Patrick J. Parsons, Wadsworth Center for Laboratories and Research, New York State Department of Health, and Department of Environmental Health and Toxicology, School of Public Health, SUNY at Albany, Albany, NY 12201.

One of the effects of lead absorption on the hemopoietic system is an increased concentration of erythrocyte zinc protoporphyrin (ZPP). Accumulation of ZPP follows an approximately exponential relationship as a function of increased blood lead (BPb) concentrations. The ZPP or EP test has been widely used in the U.S. to screen young children for lead exposure. It is expected that the U.S. Centers for Disease Control will revise their action level for BPb in children from the current 25 µg/dL to perhaps 10 µg/dL in mid-1991; the current action level for EP is 35 µg/dL and is not expected to change. The change in the BPb action level will have important implications for pediatric lead screening programs many of which rely heavily on the EP test to detect elevated BPb levels. Data published recently by our laboratory (Parsons *et al* Clin. Chem. 1991:37 216-225) showed the sensitivity of the EP test drops significantly from 0.79 at current action levels to 0.43 at BPb levels ≥10 µg/dL. Thus EP would no longer be considered a reliable screening test for detecting BPb ≥10 µg/dL. Although the future of EP as a primary screening test for detecting elevated BPb levels looks bleak, it should not be overlooked that EP may still play a valuable role as a secondary test for detecting potential iron deficiency cases, and to prioritize further, asymptomatic BPb cases in the range 10-20 µg/dL.

21.   HUMAN CYTOCHROMES P450 AND THEIR ROLES IN ENVIRONMENTALLY-BASED DISEASE. Frank J. Gonzalez, National Cancer Institute, National Institutes of Health, Bethesda, Maryland 20892.

Human cytochromes P450, a superfamily of enzymes, are capable of oxidizing foreign compounds. These enzymes inactivate a variety of chemicals by two metabolic steps: transformation and conjugation. P450s are also responsible for metabolic activation of toxins, including promutagens and procarcinogens. The cellular composition of P450 enzymes can govern whether certain chemicals are inactivated and excreted or activated to metabolites which can bind to cellular macromolecules. To determine whether human P450s are involved in susceptibility or resistance to environmentally-based diseases, their activities toward the variety of chemicals humans are exposed to are being analyzed by cDNA expression. Methods are being developed to help predict the fate of a chemical after human exposure. Protocols are also being developed to measure expression of individual P450 forms in molecular epidemiological studies to determine if indeed levels of certain P450s govern human susceptibility or resistance to cancer and other chemically-induced diseases.

22.
Title: "Toward A Pollution Prevention Ethic: The EPA's Role and Program"

Abstract: Recently it was announced that the EPA's Pollution Prevention Office would become part of the Agency's Office of Toxic Substances. The Director of OTS (to become the Office of Pollution Prevention and Toxics) will discuss the Federal Government's effort to exert leadership in this important area. He will talk about specific pollution prevention initiatives related to the management of toxic chemicals, as well as discuss efforts to apply the "prevention ethic" to other EPA programs.

23.
USEPA's POLLUTIOIN PREVENTION RESEARCH AGENDA, Gregory Ondich,, USEPA Office of Research and Development, 401 M St. SW, (RD-681) Washington, DC 20460

In October 1990, Congress passed the Pollution Prevention Act and thereby established a national pollution prevention policy in the US. The Act declares that pollution prevention should be prevented or reduced by source reduction as the first priority in the pollution prevention hierarchy and that the USEPA should develop and implement a strategy to promote source reduction within the public and private sectors. Contained in the Agency strategy is the role of the Office of Research and Development (ORD) which is to identify research needs relating to pollution prevention, set priorities for research to target the most promising source reduction opportunities, and to establish a source reduction clearinghouse containing information on management, technical and operational approaches to source reduction. This talk will discuss ORD's most recent developments in the pollution Prevention information clearinghou, and the development of the ORD Pollution Prevention Research Strategy docu t.

24. EPA'S 33/50 PROJECT: POLLUTION PREVENTION THROUGH VOLUNTARY INITIATIVES. D.J. Sarokin, EPA Office of Toxic Substances, Washington DC 20460.

EPA has identified 17 high-priority toxic chemicals which annually account for 1.4 billion pounds of industrial wastes which are released to the environment or transferred to waste management facilities. The 33/50 Project sets national goals for reducing these wastes by 50% by 1995, with an interim goal of 33% reduction by 1992. Reductions will be achieved primarily through pollution prevention, and through voluntary commitments made by industry, rather than the conventional command-and-control approach. EPA has asked more than 6,000 companies to make numerical reduction commitments. The paper will review the status of these commitments, and the reasons companies elect to participate in the program.

25. INDUSTRIAL APPROACHES TO POLLUTION PREVENTION. A. M. Ford, R. A. Kimerle, A. F. Werner, Monsanto Company, 800 North Lindbergh Blvd., St. Louis, MO 63131

Many American companies have come to recognize that the use of end-of-pipe environmental control to effect major reductions in pollution is almost prohibitively expensive due to the need for both heavy research and development expenditures as well as heavy capital expenditures during the final phases of pollution reduction. Unfortunately, once a company begins a program of pollution reduction through end-of-pipe technology, resources are committed making major course changes difficult.

A more fruitful and no more expensive approach might be to spend research and development efforts on complete process redevelopment. Most chemical manufacturing processes in operation in the U.S. were designed before the public developed concern for major environmental issues.

In order to better understand the need and the justification for new process invention, the environmental impact of chemical products must be examined through their entire product life cycle.

26. THE ROLE OF THE CHEMIST IN PREVENTING POLLUTION. Robert B. Pojasek, Ph.D., Geraghty & Miller, Inc., One Corporate Drive, Andover, MA 01810

A large number of chemists have become involved in environmental studies and pollution control. Now it is important to understand where the chemist can assist in the developing field of pollution prevention. Each chemical utilized in an industrial operation or process has a function. The chemist and chemical engineer must help define this functionality before the process or operation can be changed to prevent losses of the chemical. Chemical losses to air, water, solid wastes, spills/leaks and accidents are the pollution which we are now seeking to control. The American Institute for Pollution Prevention is seeking to generate broad support in achieving widespread and expeditious adoption of pollution prevention concepts.

27. LIFE CYCLE ANALYSIS: A TOOL FOR POLLUTION PREVENTION, M.A. Curran, U.S. Environmental Protection Agency, Cincinnati, Ohio 45268

Life cycle analysis is an objective technical tool that can be used to identify and evaluate opportunities where environmental burdens associated with a specific product can be reduced. This tool can also be used to evaluate effects of resource management options which can be incorporated into the design of sustainable systems. Life cycle analyses take a holistic approach to studying a particular product by encompassing extraction and processing of raw materials; manufacturing, transportation, and distribution; use, reuse, and maintenance; recycling and composting; and final disposal. The results of life cycle studies can help manufacturers improve product designs as well as assist consumers in making purchasing decisions. The EPA's Pollution Prevention Research Branch in Cincinnati, Ohio, is supporting a research effort with Battelle to develop a standard method for inventorying product life cycle inputs and outputs. This talk will summarize the results of that effort.

28. MARKET DRIVEN POLLUTION PREVENTION? - GREEN MARKETING - A NATIONAL AND INTERNATIONAL STATUS REPORT, Gordon E. Bellen, Research and Development, NSF International, Ann Arbor, Michigan 48105.

Strong consumer attitudes favoring products perceived to reduce the toll on the environment have been reported in numerous surveys. The market response has been a proliferation of claims and product marks. The call for uniform national and international programs are most often based on product life-cycle analysis as a means for assessing a product's environmental impact from cradle to grave. This approach has been used in some national programs with success reported for market impacts. This presentation will focus on facts and fallacies of environmental labeling, as well as reporting on the status of national and international efforts his area.

ACS TASK FORCE

Fourth Chemical Congress of North America

New York, NY
August 25 - 30, 1991

R. W. Phifer,
Program Chairman

WEDNESDAY MORNING
- **Symposium on Laboratory Waste Management**

R. W. Phifer, Presiding

Paper 1

ACS TASK FORCE ON LABORATORY WASTE MANAGEMENT

1. PRACTICAL ASPECTS OF LABORATORY WASTE MANAGEMENT. R. W. Phifer, Environmental Assets Inc., 327 S. High Street, West Chester, Pennsylvania 19380; J. M. Harless, Techna Corporation, 44808 Helm Street, Plymouth, Michigan 48170.

The management of laboratory wastes is greatly influenced by regulations as well as logistics. Regulators have a difficult time determining how to regulate laboratory wastes, and laboratory personnel must both interpret the rules and make other decisions that affect safety, cost, and liability. This presentation will review practical aspects of laboratory waste management including quirks in the regulations, economics of waste management, and various management options for laboratory wastes.

COMMITTEE ON TECHNICIAN ACTIVITIES

Fourth Chemical Congress of North America

New York, NY
August 25 - 30, 1991

V. E. Burton,
Program Chairman

MONDAY MORNING AND AFTERNOON
- **National Technician Workshop**

V. E. Burton, Presiding

TUESDAY MORNING
- **45th National Technician Symposium**

V. E. Burton, Presiding   Papers 1-4

TUESDAY AFTERNOON
- **Poster Session**

V. E. Burton, Presiding   Papers 5-13

WEDNESDAY MORNING
- **45th National Technician Symposium**

V. E. Burton, Presiding   Papers 14-19

COMMITTEE ON TECHNICAL ACTIVITIES

1. THE MAKING OF A CONTINUING EDUCATION COURSE - START TO FINISH
   Ilene Bara, ICI Americas, Ruth Kliment, Hercules, and William Walker, E. I. du Pont de Nemours and Company, Wilmington, DE

   The need for affordable, pertinent courses on a technician level, prompted several people on the Delaware Technician Affiliate Group (TAG) to pursue a means to this end. A survey revealed a few of the areas where the need was greatest. Pharmacokinetics was one of the most requested courses. After soliciting Dr. DeBerardinis as instructor the remaining logistics, time, location, price, advertising, registration, etc. were worked out by a small committee. Dr. DeBerardinis added the assistance of Dr. Birmingham and between the two a very successful two day course was given to 34 people. Although the cost was kept low. a considerable profit was added to the scholarship program fulfilling our original goal.

2. A SUMMER INTERN PROGRAM FOR HIGH SCHOOL CHEMISTRY TEACHERS IN THE GREATER CINCINNATI AREA. Richard J. Sunberg, The Procter & Gamble Company, Miami Valley Laboratories, Cincinnati, Ohio 45239-8707

   The goal of this program is to improve the quality of high school chemistry education to meet future industrial needs for workforce-ready technical people in chemistry. The experience that a student has in a high school chemistry class can be a pivotal decision point that can direct that student toward a career in chemistry. The chemistry teacher plays a key role in making this introduction to chemistry a positive experience.

   This program was designed to better align the skills that industry needs in the work place and the skills that the students are being taught. The first chemistry teacher participant experienced first hand what skills are valued in a modern industrial research laboratory. Teachers who go through this program are encouraged and helped to share this summer experience with their students and fellow teachers. The learnings from designing this program and the first year of experience will be shared.

3. THE IMPORTANCE OF STATISTICS FOR THE TECHNICIAN.
   W. F. Stansbury and M. Aslam, Hoechst Celanese Corporation, Advanced Technology Group, 1901 Clarkwood Road, Corpus Christi, TX 78409

   The need for better communication between the technician and the chemist and between the technician and management is always a challenge at every level of industry. Experimental design and statistical process quality control can be used as tools by the technician to improve communication while at the same time improving quality and in turn productivity. When statistics are routinely used as part of the research method, rather than just used to confirm preconceived notions, the technician will be better able to identify the unusual variation in a given process and detect the causes of the variation. This would result in conclusions which are concise, measurable and can be presented to the chemist or management with confidence.

4. EMPLOYEE-SUPERVISOR PARTNERSHIP. E. L. Miller, Exxon Research and Development Laboratories, P. O. Box 2226, Baton Rouge, LA 70821

The Employee-Supervisor Partnership program is designed to provide the employee and supervisor with the opportunity to apply Quality principles to their working relationship. This is accomplished through clarification of their work expectations and goals, including customer requirements and work outputs, as well as understanding the impact of individual styles and preferences on their work. This paper highlights employee-supervisor partnership by explaining the concepts which comprise it: employee/work matching; preparation/guidance/direction; monitoring/facilitating/adding value; converting to achievement; feedback/assessment; reward/recognition.

5. VERSATILE ZYMARK ROBOT DILUTION SYSTEM, S. Duma and B. Maglia, E. I. du Pont de Nemours & Co., Imaging Systems Department, Parlin, New Jersey 08859

An automated system using a Zymate II laboratory robot, with PyTechnology and Accutrac*, has been used quite successfully for diluting gelatin solutions. The required heating and cooling stations, needed to handle gelatin solutions, are part of the system. All solution dispensing steps are gravimetrically monitored and the information is used to maintain dilution accuracy. The system is very versatile in that it has the capability of using multiple spiking and multiple diluents. Because of the system's hardware and software reliability, a high productivity rate can be achieved.

6. COMPUTER AUTOMATION IN THE LABORATORY. S. L. Felix, Separations Process Fundamentals Skill Center, Union Carbide Corporation, Technical Center, South Charleston, W. VA 25303

Within the past several years, the development of computer automated systems in the laboratory, involving interactive graphics, more extensive and higher quality data acquisition, and process control, has dramatically improved our understanding of experimental equipment. One such graphically oriented programming system is called LabVIEW. This talk will discuss the applications of LabVIEW incorporated with a separation of components using a wiped-film molecular still. Included will be examples of LabVIEW used as a controller-monitor along with data acquisition, process control and direct interfacing to instruments from hardware.

7.
Automation of Equipment for Manufacturing of Textile Structured Preforms, W. C. Walker, E. I. du Pont de Nemours & Co., Inc., Wilmington, DE 19880-0304

Taking an advanced material system from the laboratory to the marketplace reuires demonstrating the material sustem in a prototype structure or process. Prototype demonstration often means making the prototype by any available means. Yet automated prototype manufacture offers a number of advantages, as discussed in this poster. Two examples are presented of equipment automated to produce textile-structured preform prototypes for use in evaluation of material properties and process development.

8. Automation of Equipment for Manufacturing of Textile Structured Preforms, <u>W. C. Walker</u>, E. I. du Pont de Nemours & Co., Inc., Wilmington, DE 19880-0304

Taking an advanced material system from the laboratory to the marketplace reuires demonstrating the material sustem in a prototype structure or process. Prototype demonstration often means making the prototype by any available means. Yet automated prototype manufacture offers a number of advantages, as discussed in this poster. Two examples are presented of equipment automated to produce textile-structured preform prototypes for use in evaluation of material properties and process development.

9. WITHDRAWN

10. WITHDRAWN

11. WITHDRAWN

12. WITHDRAWN

13. WITHDRAWN

14. **DYNAMIC RHEOLOGICAL STUDIES ON THE LOW TEMPERATURE BEHAVIOR OF LUBRICANTS.** <u>G.M. Morton</u>, D.L. McGregor, and A.J. Stipanovic, Texaco Research and Development, P.O. Box 509, Beacon, New York, 12508, U.S.A.

Mineral oil based lubricants generally contain polymeric molecules, called pour point depressants (PPDs), which inhibit solidification at cold winter temperatures below 0°C. Using a Rheometrics RDS-2 Dynamic Fluids Spectrometer, the viscoelastic flow behavior of several lubricant model systems containing varying amounts and types of PPDs, have been characterized. In the absence of a PPD at low temperatures, paraffin wax containing oils experience a sharp increase in "elasticity" (G') over a narrow temperature range. When PPDs are added to these oils, a smaller increase in G' is observed over the same temperature range, and the fluid can be cooled an additional 10-20°C before the oil becomes completely elastic (solid). These results suggest that PPDs do not prevent wax crystallization, but rather modify crystal growth, inhibiting crystal interaction and network formation which causes solidification. This observation is consistent with other reports found in the technical literature.

15.
SAMPLE PREPARATION TECHNIQUES FOR THE ANALYSIS OF ORGANIC COMPOUNDS BY ION CHROMATOGRAPHY. T. L. Brown, Hercules, Inc., Research Center, Wilmington, DE 19894

Ion Chromatography (IC) is a separation technique used to analyze ions in solution. It has been applied to the analysis of anions, cations, transition metals, carbohydrates, carboxylic acids, and amino acids. However, in many cases, the sample analyte may not exist in a form that is compatible for direct IC detection and some form of sample pre-treatment is required. For example, some form of decomposition may be required to convert an organic sample into an ionic species. Solid phase extraction techniques and/or filtration might also be applied to isolate the ion of interest from the sample. This talk will focus on several pre-treatment methods that are used for the analysis of organic compounds in industrial applications. A Dionex Model 2000i/SP ion chromatograph equipped with a suppressed conductivity detector was used to characterize the samples described.

16. PHOTOSYNTHETIC WATER SPLITTING BY PLATINIZED CHLOROPLASTS. C. V. Tevault and E. Greenbaum, Oak Ridge National Laboratory, Oak Ridge, Tennessee 37831-6194.

Colloidal platinum precipitated on the thylakoid membrane of type C chloroplasts of spinach leaves is a proven catalyst for the photoevolution of molecular hydrogen via water splitting. Oxygen is a by-product of this light-induced reaction. The natural process of photosynthesis utilizes the Calvin-Benson cycle of green plants to reduce CO_2 to carbohydrates. In these experiments, that cycle is eliminated. Instead, metallic platinum is precipitated onto the surface of the photosynthetic membrane and becomes the catalyst for photoproduction of hydrogen and oxygen gases. Two approaches for the platinization of chloroplasts have been explored. With H_2O as the initial electron donor, photoprecipitation was achieved during the first light-on cycle. Precipitation was also accomplished by placing a sample of chloroplast membranes and hexachloroplatinate into a reactor cell with H_2 flowing in the cell headspace. Experimental procedures, parameters, equipment, and recent results will be presented for this photobiochemical study.

17. SYNTHESIS OF 4-ACETOXYSTYRENE MONOMER. W. F. Stansbury and M. Aslam, Hoechst Celanese Corporation, Advanced Technology Group, 1901 Clarkwood Road, Corpus Christi, TX 78409

The dehydration of 4-acetoxyphenylmethylcarbinol (APMC) for the commercial production of 4-acetoxystyrene monomer (ASM) has been investigated. ASM is an important intermediate for the synthesis of poly(4-hydroxystyrene) and other novel co-polymers and a potential intermediate for bronchodilators and beta-blockers. Numerous methods of dehydration of APMC to ASM have been studied at this site. In the first approach, the dehydration of APMC to ASM in a wiped film evaporator proceeded with 92% conversion, 67% selectivity and 62% yield. In the second approach, a continuous fed-batch (CFB) process was evaluated where dehydration of APMC to ASM proceeded with 94% conversion, 77% selectivity and 73% yield. In the third approach, CFB dehydration of APMC in the presence of acetic anhydride proceeded with 93% conversion, 90% selectivity and 84% yield.

18. MINIMIZING MASS DISCRIMINATION IN A SPLIT INJECTION SYSTEM FOR CAPILLARY GAS CHROMATOGRAPHY. M. H. Rood and Y. H. Liu Hoechst Celanese Corporation, Advanced Technology Group, 1901 Clarkwood Road, Corpus Christi, TX 78469

Although mass discrimination is observed during the split injection technique, the pattern of mass discrimination is unexpectedly consistent at four widely differing injection temperature when the sleeve is packed. The expected mass discrimation with varying injection block temperatures is observed when the same sleeve is not packed. The behavior of mass discrimination for cold on-column injection versus variable split injection has been investigated with straight chain alkanes from octane, C_8H_{18}, to nonadecane, $C_{19}H_{40}$, as the solutes in hexane and dodecane matrices. Results obtained from different injection temperature conditions, 50°C for on-column injection and 80, 120, 200, and 300°C for split injection are presented.

19. A NEW MORPHOLOGY OF PRECIPITATED CALCIUM CARBONATE, SYNTHESIS AND CHARACTERIZATION. M. J. Herman, J. D. Passaretti, Pfizer Specialty Minerals, Bethlehem, PA, 18017.

This presentation will discuss a new rhombohedral precipitated calcium carbonate morphology that has properties similar to TiO_2. This material may have a significant impact upon the paper industry which typically uses TiO_2 as a functional filler for opacity. The new rhombohedral morphology can be used in place of TiO_2 for many applications greatly reducing the cost of production.

The physical properties of this new material as well as application as a paper filling pigment will be discussed.

YOUNGER CHEMISTS COMMITTEE

Fourth Chemical Congress of North America

New York, NY
August 25 - 30, 1991

Program Chairman'

MONDAY AFTERNOON
● **Establishing an Academic Program**
C. Grissom, Presiding   Papers 1-4

TUESDAY MORNING AND AFTERNOON
● **Chemical Careers Insights Program (Roadshow) - Careers for Chemists in Industry**
S. Orlando, Presiding   Papers 5-13

YOUNGER CHEMIST COMMITTEE

1. **SUCCEEDING IN ACADEMIA -- THE GOOD, THE BAD, AND THE UGLY.** Paul G. Gassman, Department of Chemistry, University of Minnesota, Minneapolis, MN 55455.

   For those entering the academic world, it is the best of times and the worst of times. Opportunities abound. Starting salaries are escalating. Set-up packages have gone super-critical. It can all be yours! All you have to do is be perfect (or nearly so). If you're articulate, a potential Nobel laureate, and the quintessential entrepreneur, you have it made. Unfortunately, most of us are merely human and, as a result, fail to meet all of these requirements. For those who do pass through the entrance to academia, the view inside is not that of the ivory tower of legend. Instead, a great deal is expected from those to whom much has been given. The system can be wonderful and exciting. It can also be deadly. Who killed Professor Plum in the study with a letter of rejection? Maybe I can give you a Clue.

2. NON-FEDERAL FUNDING FOR NEW FACULTY. J. E. Rogers, Jr., Research Grants, American Chemical Society, 1155 16th St., NW, Washington, DC 20036.

   Research grants available from the ACS Petroleum Research Fund and other non-federal funding agencies will be discussed.

3. DO YOUNGER CHEMISTS GET A FAIR SHAKE? AN ANALYSIS OF NSF FUNDING OPPORTUNITIES FOR JUNIOR FACULTY. Kenneth G. Hancock, Director--Chemistry Division, National Science Foundation, Washington, DC 20550.

   The National Science Foundation and other agencies provide a range of grant programs responding to the needs of young faculty members who are initiating academic careers in research and education. These opportunities will be described and the success of junior faculty in the competition for research funds will be analyzed.

4. NIH FUNDING OPPORTUNITIES FOR THE ASSISTANT PROFESSOR. Michael Rogers, National Institutes of Health Westwood Building, Bethesda, MD 20892.

   NIH funding opportunities for new faculty in the chemical sciences will be discussed.

5. **FASCINATING FACTS ABOUT CHEMISTS.** R.Cyngier

   ABSTRACT NOT AVAILABLE

6. **OPPORTUNITIES FOR CHEMISTS IN MARKETING, SALES, AND TECHNICAL SERVICE.** W.H. Suits

   ABSTRACT NOT AVAILABLE

7. **TECHNICAL CAREER PATHS IN A SPECIALTY CHEMICAL COMPANY** K. Nordquist.

   ABSTRACT NOT AVAILABLE

8. **INFORMATION DISTRIBUTION: A MULTIFACETED CAREER OPPORTUNITY FOR CHEMISTS.** B. Lawlor.

ABSTRACT NOT AVAILABLE

9. **CAREERS IN SCIENTIFIC LIBRARIANSHIP.** B. Slutzky

ABSTRACT NOT AVAILABLE

10. **MANAGEMENT PATHS FOR TECHNICAL PROFESSIONALS.** B. Leonard.

ABSTRACT NOT AVAILABLE

11. **CHEMICAL CAREERS IN INDUSTRY: BETWEEN RESEARCH AND SALES.** D. B. Swanson

ABSTRACT NOT AVAILABLE

12. **STARTING A BUSINESS VS. WORKING FOR ONE.** J. Avolio.

ABSTRACT NOT AVAILABLE

13. **THE USE OF ACS EMPLOYMENT AND CAREER SERVICES.** J. Wheeland

NO ABSTRACT AVAILABLE.

KEYWORD INDEX

Reference by Division Code and Paper Number

horodithioate d(CGCTpS$_2$-TPS$_2$-	AAGCG): An Unusual Hairpin Loop.= +tide Phosp	BIOL 0123
apped Poly(Ethylene Oxide) and	ABA Block Copolymers Containing Hydroxyl-Cappe	PMSE 0006
ucing Anmimals. Ivermectin and	Abamectin Metabolism: Differences and Similarities.	AGRO 0128
d Experience with Catalytic Ozone	Abatement in Jet Aircraft.= +cial Development an	I&EC 0047
sual Hemiacetal from the Dorsal	Abdominal Glands of the Australian Spined Citrus B	AGRO 0035
lp Reduce Decay in Seal-Packa+	Abiotic Stresses of Heat or UV Illumination May He	AGFD 0076
emiluminescence from the Laser	Ablation of Yttrium Oxide and Chloride in Oxygen.=	PHYS 0255
uminum Atoms from Pulsed Laser	Ablation: Mechanisms and Applications.= +ase Al	PHYS 0353
Collisons.= Transparency and	Abrasion in High-Energy Nucleus-Nucleus	NUCL 0048
asis on High Refractive Index and	Abrasion-Resistance Coatings.= +erials with Emph	POLY 0264
Vapor Interface.=	Absorbed Hydrogen and Fluorine near the Diamond/	PHYS 0359
Efficacious, Well-	Absorbed Renin Inhibitors.= A New Class of	MEDI 0120
ch Graft Copolym+ Wicking and	Absorbency in Aluminum and Silicate-Modified Star	CARB 0042
Design for Packed Tower	Absorbers Using the PRSV Equation of State.=	CHED 0107
and I+ Low-Frequency Sound-	Absorbing Studies on Foams Based on Polyurethanes	PMSE 0040
Producing Animals.= Dermal	Absorption and Metabolism of Xenobiotics in Food-	AGRO 0095
Studies of Polymers for Shock	Absorption Applications.=	PMSE 0181
ect of Lipophilicity, Dosin+ Oral	Absorption of Di- and Tri-Peptidic Compounds: Eff	MEDI 0196
Factors Influencing the	Absorption of Small Peptides.= Renin Inhibitors:	MEDI 0197
quid Chromatography with UV-	Absorption Photometry and Particle-Beam Mass Spe	ENVR 0015
ed Combustion+ Thermal Swing	Absorption Process for Production of Oxygen-Enrich	I&EC 0023
nd Ketones of A+ The Near-UV	Absorption Spectra of Several Aliphatic Aldehydes a	ENVR 0024
raphy/Graphite Furnace Atomic	Absorption Spectrometry for the Determination of T	ANYL 0080
Two-Photon	Absorption Spectroscopy of Phenanthrene.=	PHYS 0328
Two-Photon	Absorption Spectroscopy of Phenanthrene.=	CHED 0085
-T$_c$ Cuprates as Probed by X-ray	Absorption.= +el for the Electronic States of High	PHYS 0053
	Abstract Scientific Concepts in the Courtroom.=	CHAL 0012
o Dimers Formed by Sulfur Atom	Abstraction from SPMe$_3$.= +ybdenum(III)-μ-Sulfid	INOR 0271
Modified Polymeric+ Selective	Abstraction of Mercaptan Compounds by Chemically	ANYL 0083
tabase and of a Printed Service of	Abstracts.= +e Joint Production of a Reaction Da	CINF 0010
Lunar Samples.= Initial	Abundances of Plutonium-244 in Meteorites and	NUCL 0083
ts of Nuclear Science—II. Initial	Abundances of Uranium-235 in Meteorites and Luna	NUCL 0082
Uses and	Abuses of Immunological Testing.=	CEI 0018
Law. Greed in the Groves of	Academe.= Whistle-Blowers, Advocates, and the	PROF 0001
Chemical Careers in	Academe.=	I&EC 0003
Chemical Careers in	Academe.=	CMEC 0025
Academic Program. Succeeding in	Academia: The Good, the Bad, and the Ugly.= +n	YCC 0001
Options for Search Services in an	Academic Library.= +d Access to Beilstein Online:	COMP 0012
Panel Discussion.	Academic Perspectives.=	POLY 0202
od, the Bad, an+ Establishing an	Academic Program. Succeeding in Academia: The Go	YCC 0001
POLYED—A Model for	Academic/Industrial Partnerships.=	CHED 0006
valuacion d+ Influencia del Nivel	Academico de los Alumnos en los Resultados de la E	CHED 0093
Force	Academy.= Twelve Days at the U.S. Air	CHED 0014
enoxyethyl)-1,3,4-Thiadiazoles as	Acaricides.= +d Potential Agrochemicals. 2-(1-Ph	AGRO 0071
Novel Substituted β-Ketoamide	ACAT Inhibitors: Chemistry and Biological Activit	MEDI 0108
of Novel Phosphorus-Containing	ACAT Inhibitors.= +Hypocholesterolemic Activity	MEDI 0110
Pyridylene Diurea Derivatives as	ACAT Inhibitors.= +mic Activity of Phenylene and	MEDI 0109
CoA: Cholesterol Acyltransferase (	ACAT) Inhibitors.= +eas: A Potent Series of Acyl	MEDI 0107
efrigerated Minced Channel Cat+	Acceleration of Lipid Oxidation during Cooking of R	AGFD 0124
ergy Transfer between Donor and	Acceptor Adlayers on Single-Crystal Surfaces.= +n	PHYS 0065
ar Optical Polymers with Sulfonyl	Acceptor Groups.= +ain and Main-Chain Nonline	POLY 0105
ion by σ-Bond-Separated Donor-	Acceptor Molecules.= +Second-Harmonic Generat	POLY 0222
The Natural Bond Orbital Donor-	Acceptor Perspective.= +ts in Hydrogen Bonding:	PHYS 0106
512FM Dextran+ Mechanism of	Acceptor Reactions of Leuconostoc mesenteroides B-	CARB 0028
Model Molecules with Donor-	Acceptor Units Intermediated by Silicon Chains.=	POLY 0211
d the Demonstration of a Separate	Acceptor.= +6-Deoxy- and 6-Fluoro-Analogues an	CARB 0028
Importance as Alternate Electron	Acceptors.= +g of Iron Oxides and Their Potential	GEOC 0007
es: Super-Homo-Michael-	Acceptors.= Cyclopropylthiocarbene-Metal Complex	ORGN 0216
sible Domain fo+ Unsymmetrical	Access to a Symmetrical Reaction Coordinate: A Pos	ORGN 0122
es in an Academic L+ Subsidized	Access to Beilstein Online: Options for Search Servic	COMP 0012
Ownership vs. Unrestricted	Access: The Library Dilemma.= Information	CINF 0045
Increasing Enzymatic	Accessibility on Lignocellulosic Materials.=	BIOT 0250
k: What Must We Yet Know To	Accomplish This Task? Relative Risk: A Blueprint fo	ENVR 0061
nto de+ Peliculas de Oxido sobre	Acero Inoxidable 304 y 316 Utilizadas en el Seguimie	ANYL 0055
cador una Pelicula de Oxido sobre	Acero Inoxidable 316.= +ando como Electrodo Indi	CHED 0163
Medir pH, por un Electrodo de	Acero Inoxidable 316/Pelicula de Oxido, en la Valida	ANYL 0066
co de Tecnicas Electroquimicas de	Aceros Microaleados en Acido Sulfurico.= +tadisti	PHYS 0281
Radical Reactivity of Cyclic	Acetals.=	ORGN 0156
paration and Spectral Chara+ 1-	Acetamido-2,x-Dimethylpyridinium N-Betaines: Pre	ORGN 0203
heo-Photoac+ Mobility of Ethyl	Acetate in Poly(vinylidene Fluoride) Monitored by R	POLY 0255

Combustion Byproducts of Ethyl	Acetate over Pt– Alumina Catalyst.=	COLL	0189
ectroscopy Studies of Poly(vinyl	Acetate)/Silicon Oxide Microcomposites Prepared vi	POLY	0287
2-O-Tetradecanoylphorbol-13-	Acetate–Induced H_2O_2 Production in Mouse Epiderm	AGFD	0069
Source Differentiation of Amyl	Acetate, Ethyl Acetate, Ethyl Butyrate, and Ethyl Ca	AGFD	0187
rentiation of Amyl Acetate, Ethyl	Acetate, Ethyl Butyrate, and Ethyl Caproate.= +ffe	AGFD	0187
Hydantoin Bioisosteres: Hydroxy	Acetic Acid Aldose Reductase Inhibitors.= +dress.	MEDI	0082
Radiolysis of Adsorbed	Acetic Acid on Na–Montmorillonite.=	NUCL	0072
Eastman Chemical Company	Acetic Anhydride Process.=	I&EC	0053
$CuSO_4$ en Agua/Agua–	Acetona.= Energia Libre de Transferencia de	PHYS	0236
Chemisorption of Cyanide and	Acetonitrile on Pt(III) Electrode Surfaces.=	COLL	0118
Synthesis of 4-	Acetoxystyrene Monomer.=	TECH	0017
Reactivity of an Iron Phosphide	Acetyl Complex $Cp(PPH_2)(CO)FeCOCH_3$-1 with Ele	INOR	0240
$Pr^i)(S_2COPr^i)CO(PMe$+ Agnostic	Acetyl Complexes: Crystal Structure of $Mo(\eta^2$-C(S)O	INOR	0036
and Ruthenium	Acetyl Complexes.= Catalytic Hydrosilation of Iron	INOR	0238
hesis of a Carbon Glycoside of N-	Acetyl Neuraminic Acid.= +emical Enzymatic Synt	CARB	0034
ules Oriented by Chiral Nematic (	Acetyl)(ethyl)cellulose Solutions.= +r of Dye Molec	POLY	0331
Production of	Acetyl–Xylan Esterase from Aspergillus niger.=	BIOT	0251
Convenient Synthesis of 2-	Acetyl-1-Methyl-5-Nitroimidazole.=	ORGN	0200
sotropic Reorientation of Vanadyl	Acetylacetonate in Toluene.= +th Study of the Ani	PHYS	0264
carcinogen, N-(Sulfonatooxy)-2-(	Acetylamino)Fluorene.= +s on the Ultimate Hepata	ORGN	0194
–Activity Relationships of Potent	Acetylcholinesterase Inhibitors: 8–Carba–physostigm	MEDI	0143
ischarge CVD in a Mixture Gas of	Acetylene and Hydrogen.= +hesized by DC Arc D	FUEL	0038
rganic Hybrid Materials. Silicon	Acetylene and Acetylene Olefin Polmers as Precursors t	POLY	0357
e Reaction of Oxygen Atoms with	Acetylene at 300 K.= +etylene Formation during th	FUEL	0065
An η^4-Benzene Species Mediates	Acetylene Cyclotrimerization.=	INOR	0210
of the Vinylidene–	Acetylene Isomerization Process.= Ab Initio Study	PHYS	0287
pe–Selective Zeol+ Conversion of	Acetylene to Higher Hydrocarbon over Modified Sha	CATL	0034
Cyclotetramerization of	Acetylene.= Mechanistic Studies of the Reppe	INOR	0331
mperature Diamond Growth from	Acetylene– Oxygen Flames.= +oepitaxial High–Te	FUEL	0037
e Analogue Inhibitors of Choline	Acetyltransferase and High–Affinity Choline Uptake.	MEDI	0142
y of Inhibitors of Spermidine N^8–	Acetyltransferase.= +Synthesis, and in Vitro Activit	MEDI	0020
tors of Spermidine/Spermine–N^1–	Acetyltransferase.= +Fumaramate Esters as Inhibi	MEDI	0021
aching L+ Critical Thinking and	Achievement in General Chemistry: Reaching and Te	CHED	0081
waps, Exchanges, and Toll Conv+	Achieving Purchasing Productivity through Global S	CMEC	0007
an: Program for the Great Lakes.	Achieving Zero Discharge of Persistent Toxic Substa	ENVR	0083
el Seguimiento de las Reacciones	Acido Base en Medio Acuoso.= +y 316 Utilizadas en	ANYL	0055
imicas de Aceros Microaleados en	Acido Sulfurico.= +tadistico de Tecnicas Electroqu	PHYS	0281
+dorreduccion con Zinc en Medio	Acido y Basico, Preparacion de Nitrado de Plata.=	CHED	0172
on Monoxide Fermentation by an	Acidogenic Anaerobe.= +nol Synthesis During Carb	BIOT	0015
Microtubules: Implications for +	Acousto–Conformational Transitions in Cytoskeletal	BIOT	0055
Spectral or Chromatographic Data	Acquisition and Analysis.= +Handling System for	COMP	0056
Analytical Courses.= Data	Acquisition and Computer Data Processing for	CHED	0119
f+ Macintosh Computers as Data	Acquisition Devices—A Tool for the Revitalization o	CHED	0160
Knowledge	Acquisition from Crystallographic Databases.=	COMP	0045
al Businesses.=	Acquisition—The Pleasant (?) Fate of Small Chemic	SCHB	0019
enic Dibenz[a,h]Acridine (DB[a,h]	ACR).= +l-8,9-Epoxide Diastereomers of Carcinog	ORGN	0208
phalosporin C in Cephalosporium	acremonium.= +ed Conversion of Penicillin N to Ce	BIOT	0085
omers of Carcinogenic Dibenz[a,h]	Acridine (DB[a,h]ACR).= +l-8,9-Epoxide Diastere	ORGN	0208
Interaction of a Ferrocene–Linked	Acridine Orange Compound with DNA.= +of the	ANYL	0109
,4-kl]-5a,6,7,8-Tetrahydrobenz[c]	Acridine 5N–Oxide, V.= +7,7-Dimethyl–Quino[2,3	ORGN	0198
n Strategy for C+ Evaluation of	Acridinium Chemiluminescence as an HPLC Detectio	ANYL	0084
omers Poly(styrene–Sodium–2-	Acrylamido–2–Methylpropane Sulfonate) in Nonaque	COLL	0181
ization of Diacetylene–Containing	Acrylate and Methacrylates.= +ies on the Polymer	POLY	0092
anical Properties of Poly(Methyl	Acrylate–co–Dimethyl Aminoethyl Methacrylated)/E	PMSE	0079
ynthetic Routes to Imidazole–5–	Acrylic Acids: Potent Small–Molecule Antiotensin II	ORGN	0119
Grafting Sites in the Epoxy–g–	Acrylic Copolymer.= Theoretical Calculation of	PMSE	0161
Ultrasonic Polymerization of	Acrylic Monomers.=	PMSE	0020
astor Oil Polyurethane/Vinyl or	Acrylic Polymer Interpenetrating Polymer Networks.	PMSE	0090
of	Acrylic Polymers.= Reactive Extrusion Imidization	PMSE	0179
ethoxybenzene with	Acrylonitrile.= Copolymerization of 1-Allyl-2,5-dim	POLY	0050
Programs, H+ Symposia on the	ACS Chemistry Olympiad and the NSF ChemSource	CHED	0011
se f+ Chemistry in Context: The	ACS College–Level Decision–Making Chemistry Cour	CHED	0051
ellence in Tea+ Award Address (	ACS Division of Analytical Chemistry Award for Exc	ANYL	0006
mical Instrum+ Award Address (	ACS Division of Analytical Chemistry Award in Che	ANYL	0011
trochemistry,+ Award Address (	ACS Division of Analytical Chemistry Award in Elec	ANYL	0096
trochemical+ Award Address (	ACS Division of Analytical Chemistry Award in Spec	ANYL	0001
The Use of	ACS Employment and Career Services.=	YCC	0013
mers. Dynamic Viscoelast+ APS–	ACS Symposium: Thermoreversible Gelation of Poly	POLY	0197
mers. Reversible Gelation+ APS–	ACS Symposium: Thermoreversible Gelation of Poly	POLY	0145
mers. Surfactant Organog+ APS–	ACS Symposium: Thermoreversible Gelation of Poly	POLY	0169
mers. Thermoreversible G+ APS–	ACS Symposium: Thermoreversible Gelation of Poly	POLY	0121
and Myosin–S1–NDP Binding to	Actin.= +echanism of Myosin Subfragment-1 (S1)	BIOL	0103
paration Studies of the Selected	Actinide Ions with 4–Benzoyl-5–Methyl-2–Phenyl–P	NUCL	0073
Production of	Actinide Isotopes via Transfer Reactions.=	NUCL	0016
to Heavy	Actinides.= Model for Transfer Reactions Leading	NUCL	0018
Synthetic Studies on	Actinobolin.=	ORGN	0004
on Ravidomycin Production by an	Actinomyces sp.= +Stress and Oxygen Availability	BIOT	0162
tional and Novel Applications of	Actinomycetes Fermentations from Shake Flask to P	BIOT	0083
	Actionable Information from Japan.=	CINF	0012
Metano.=	Activacion del Enlace C–H por Ni en la Molecula de	PHYS	0244

and NO.= QRRK Chemical	Activation Analysis on Reactions of CH with N_2, O_2,	PHYS	0229
ssessments of Electron–Transfer	Activation Barriers at Colloidal Titanium Dioxide Su	COLL	0132
y of Surface Potential Effect on	Activation Barriers of Unimolecular Transformations	COLL	0107
ecular Orbital Estimation of the	Activation Enthalpies for Intramolecular Hydrogen T	ORGN	0175
$IrCl_6^{2-}$ Ion.= Volumes of	Activation for the Oxidation of Benzenediols by the	INOR	0395
Photoinduced C–H	Activation in Metalorganic Systems.=	INOR	0167
lcohols: Carbon–Hydrogen Bond	Activation in Rhenium Alkoxide Complexes (η_5–C^5R^5	ORGN	0220
ase–Managed Organic Synthesis:	Activation of α–Chloro, α–Hydroxy, and N–Blocked	ORGN	0259
Nonmetals—Novel Techniques.	Activation of Ammonia on V_2O_5/TiO_2 Catalysts Used	COLL	0211
GTPase–Mediated	Activation of ATP Sulfurylase.=	BIOL	0062
tes on Rh(111).=	Activation of C–H, C–C, and C–O Bonds of Oxygena	CATL	0033
DNA	Activation of Nae I Endonuclease.=	BIOL	0125
Reactivity of (OEP)MX, and the	Activation of O_2 by (OEP)Y(μ–ME)$_2$AlMe$_2$.= +and	INOR	0294
	Activation of Wide–Pore Alumina Supports.=	PETR	0033
ersatile Glycosyl Donors. Selective	Activation over Thioglycosides.= +ides as Novel, V	CARB	0035
uclear Research+ Migration of	Activation Products in Discharge Pool in T.H.O.R. N	NUCL	0090
Coals by Chemical	Activation.= Production of Activated Carbons from	FUEL	0048
olloidal CdS: A+ Diffusion– and	Activation–Controlled Photoreactions Sensitized by C	COLL	0150
ilvC	Activator Protein: DNA Interactions.=	BIOL	0124
Benzoxazin+ Potassium Channel	Activators: Synthesis and SAR Studies of Novel 1,4–	MEDI	0099
Enzyme Mimics and	Activators.= Plenary Session. Multifunctional	ORGN	0265
munodeficientes.=	Actividad Inmunomoduladora del Zinc en Ratones In	AGFD	0144
al Searching Improve the Yield of	Activities?.= +in 3D Databases: Does Conformation	CINF	0026
Step Is Rate–Limiting for Skeletal	Actomyosin ATP Hydrolysis.= +e Bond–Splitting	BIOL	0141
Reacciones Acido Base en Medio	Acuoso.= +y 316 Utilizadas en el Seguimiento de las	ANYL	0055
y Nomenclature Teaching, Part I:	Acyclic Series.= +ctic Puzzles for Organic Chemistr	CHED	0165
trazole Ureas: A Potent Series of	Acyl CoA: Cholesterol Acyltransferase (ACAT) Inhib	MEDI	0107
Direct, One–Pot Synthesis of	Acyl Complexes Containing the Chiral Auxiliary (η^5–	INOR	0156
kyl)acyl, Diacyl, and (α–Ketoacyl)	acyl Complexes of Platinum.= +d Reactivity of (Al	INOR	0043
H_3+ Hydrosilation of Manganese	Acyl Complexes, $(CO)_5MnCOCH_2R$ (R = H,CH$_3$, OC	INOR	0337
lvents in Promoting the Alkyl to	Acyl Migratory Insertion Reaction in η–Cp(CO) (PPh	INOR	0109
diated Exchange Reactio+ Fatty	Acyl Substrate Preferences for Pancreatic Lipase–Me	AGFD	0036
Photochemistry of an N–	Acyl–1–Azadiene.= Solid– and Liquid–State	ORGN	0111
ctures, and Reactivity of (Alkyl)	acyl, Diacyl, and (α–Ketoacyl)acyl Complexes of Plat	INOR	0043
ificity of Penicillin G Amidase in	Acylating a Monocyclic β–Lactam Intermediate in th	BIOT	0161
Acylthiazolidi+ Highly Selective	Acylation of Di– and Polyhydroxyl Compounds by 3–	ORGN	0038
Termina+ Effect of N–Terminal	Acylation on Chemokinetic Activity of C5A Carboxyl	MEDI	0168
Enzymatic	Acylation.= Unusual Kinetic Properties of a Novel	BIOT	0160
ter En+ Asymmetric Alkylation,	Acylation, and Aldol Reaction of Chiral β–Lactam Es	ORGN	0046
d Polyhydroxyl Compounds by 3–	Acylthiazolidine–2–thiones.= +Acylation of Di– an	ORGN	0038
nt Series of Acyl CoA: Cholesterol	Acyltransferase (ACAT) Inhibitors.= +eas: A Pote	MEDI	0107
cid wi+ Solid–State NMR of N–	Acylureas Derived from the Reaction of Hyaluronic A	ORGN	0020
merican Chemicals Approach 2000	AD.= +ical Businesses in Latin America. Latin A	SCHB	0021
5–,7–Disubstituted	Adamantanones.= Face Selectivity in	ORGN	0238
nzyme–Activated C+ Derivatized	Adamantyl–Substituted 1,2–Dioxetanes: Improved E	BIOT	0167
Photodecomposition Pathways of	Adamantyldiazirine.= +nd Chemical Studies of the	ORGN	0163
Growth.= Rapid	Adaptation of a Murine Hybridoma to Serum–Free	BIOT	0163
Plant Allelochemical–	Adapted Glutathione Transferases in Lepidoptera.=	AGRO	0119
Asymmetric Michael	Addition Catalyzed by Chiral Azacrown Ethers.=	ORGN	0209
of Anions of+ Tandem Michael	Addition Ethoxycarbonylation–Carbonyl Deblocking	ORGN	0228
adicaloid Transit+ Nucleophilic	Addition of Alkyl Thiolate Ions to β–Nitrostyrenes: R	ORGN	0155
azoline Formed by Nucleophilic	Addition of Amine or Hydrazine to σ–Allenyl Comple	INOR	0041
everal Binucl+ Reactivity of the	Addition of Cu^{2+} to the Cation $Cu(tpen)^{2+}$ to Form S	INOR	0399
o–Substituted I+ Regioselective	Addition of Nitrile–Stabilized Carbanions to 4–Amin	ORGN	0010
anes: Synt+ Stereoselective SN2'	Addition of Organocuprate Reagents to Alkynyl Oxir	ORGN	0011
rotected Nitrone.=	Addition of Organolithium Compounds to N–THP–P	ORGN	0227
Copper(I) Complexe+ Oxidative	Addition of Small Molecules to mono– and Dinuclear	INOR	0044
stem Pt–Sn/ZnAl$_2$+ Effect of Sn	Addition on the Catalytic Properties of Bimetallic Sy	COLL	0127
m Heterobimetallics by Oxidative	Addition Pathways.= +ridged Iridium, Molybdenu	INOR	0385
[Ni(sal$_2$tim)].= Kinetics of the	Addition Reaction between N–N ligands and	INOR	0061
ine–Cured Epoxy: Amine Epoxide	Addition Reaction vs. +echanical Properties of Am	POLY	0158
η^3–Allyl Cations.= Nucleophilic	Addition to Cyclopentadienyl Manganese Dicarbonyl	INOR	0270
enes: Copper–Catalyzed Conjugate	Addition to Enones.= +Reactions of Alkyl Zirconoc	ORGN	0089
quilibrium Constants for Cyanide	Addition to Nicotinamide Arabinosides.= +on the E	BIOL	0013
e to Organocuprate Conjugate	Addition.= A Practical Reversed Polarity Alternativ	ORGN	0091
Ore Pellet+ Effect of Manganese	Additions in the Mechanical Properties of Magnetite	I&EC	0067
roduced during the D+ Standard	Additions Method for the Measurement of Sulfate P	FUEL	0055
yl to Al+ Lewis Acid Promoted	Additions of Allyl(cyclopentadienyl)Iron(II) Dicarbon	ORGN	0213
Synthesis.= Enzymatic Aldol	Additions: A Building Block Strategy for Asymmetric	BIOT	0090
e Using Wood Ash or Taconite as	Additive.= +minous Argonne Premium Coal Sampl	FUEL	0006
Analy+ Current Overview of Feed	Additives and Veterinary Drugs and Their Residual	AGFD	0102
and Thickening M+ Rheological	Additives in Coating Systems: Rheological Properties	PMSE	0034
Surface Layers from Lubricant	Additives in Friction/Wear Experiments.=	PETR	0140
n of Melted Sodium Silicate with	Additives on Quartz and Its Relation with Collapsibi	COLL	0079
Polyester Resins with Low–Profile	Additives.= +ogy Characterization of Unsaturated	POLY	0114
Electron Transfer	Additives.= Photolabile Polysilanes: The Effect of	POLY	0188
d for Excellence in Teac+ Award	Address (ACS Division of Analytical Chemistry Awar	ANYL	0006
d in Chemical Instrume+ Award	Address (ACS Division of Analytical Chemistry Awar	ANYL	0011
d in Electrochemistry, s+ Award	Address (ACS Division of Analytical Chemistry Awar	ANYL	0096
d in Spectrochemical A+ Award	Address (ACS Division of Analytical Chemistry Awar	ANYL	0001

Bristol–Myers Squibb).+ Award	Address (Edward E. Smissman Award, sponsored by	MEDI 0089
sponsored by Dexter C+ Award	Address. (Dexter Award in the History of Chemistry,	HIST 0011
sses Honoring J. F. Roth. Award	Address. (Industrial Chemistry Award). Some Adven	I&EC 0043
eptides for Insect Control. Award	Address. From Juvenile Hormones to Diuretic Horm	AGRO 0001
Farm Regulatory Issues. Award	Address. Global Change: A Catalyst for the Developm	AGRO 0051
an—Bristol–Myers Squibb Award	Address. Hydanto+ General and Edward E. Smissm	MEDI 0082
in Chemical Information. Award	Address. INVENTON: A System for Inventing Molec	CINF 0016
c/Bioorganic Synthesis. Keynote	Address. Synthesis of Carbohydrates Using Ezymes a	BIOT 0089
g Leads. Light–Directed, Spatially	Addressable Parallel Chemical Synthesis.= +f Dru	MEDI 0090
ard Winner—Award Presentation:	Addressing Polymers as Macromolecules.= +91 Aw	CHED 0027
aphic Studies of Organoaluminum	Adducts of (Dimethylaminomethyl)ferrocene.= +gr	INOR 0274
ique Dual Action Mechanism of	ADD17014, a Novel Triazoline Anticonvulsant, for Im	MEDI 0144
pyrophosphate+ Synthesis of P^1-(	adenosin-5'-yl) -P^2-[6-(ribosyl)-picolinamide-5'-yl]	CARB 0023
W-3902: A Potent and Selective	Adenosine A_1 Antagonist with Renal Protective and	MEDI 0112
pentyl- and N^6-(2-Thienyl)Ethyl-	Adenosine Agonists. +SAR in a Series of N^6-Cyclo	MEDI 0113
odification of the 5'-Position of	Adenosine Analogues Results in Dramatic Changes in	MEDI 0113
uted Thiazole and Benzothiazole	Adenosine Receptor Agonists: Potent Antihypertensi	MEDI 0111
cal Activity of Derivatives of Po+	Adenosine Receptor Prodrugs: Synthesis and Biologi	MEDI 0115
er Modeling of the Dog A_1 and A_2	Adenosine Receptors.= +nce Analysis and Comput	MEDI 0114
Catalysis. Mechanistic Studies of	Adenosine-5'-Phosphosulfate Kinase.= +nzymatic	BIOL 0008
ngement That Does Not Require	Adenosylcobalamin: The Lysine 2,3-Aminomutase R	BIOL 0023
Inactivators of S–	Adenosylhomocysteinase.= New Mechanism–Based	BIOL 0081
vated Irreversible Inhibitors of S–	Adenosylmethionine Decarboxylase.= +zyme–Acti	BIOL 0052
zymes Mandelate Racemase and	Adenylosuccinate Lyase: Odd Couples in the Evolutio	BIOL 0012
Proteins Designed for	Adherence to Cellulose.=	BIOT 0217
ity on the Wetting Properties and	Adhesion of Bacterial Surface Protein Layers.= +d	BIOT 0151
cts of Processing Parameters and	Adhesion on the Impact Properties and Morphology	PMSE 0177
Study of Polymer–Polymer	Adhesion.= Application of the Blister Test to the	COLL 0057
Complexes and	Adhesion.= Ion Beam Induced Interfacial Chemical	NUCL 0094
Diffusion of Alkyl	Adipate Plasticizers in PVC.=	POLY 0067
in the Urine of Individuals Living	Adjacent to Chromate Production Waste Sites.=	CEI 0009
nsfer between Donor and Acceptor	Adlayers on Single–Crystal Surfaces.= +nergy Tra	PHYS 0065
emistry Education: An Option for	Adoption.= +CPT–Chemical Education Option. Ch	CHED 0066
hesis an+ α2A-Subtype Selective	Adrenergic Agents. 1. Substituted Imidazolines, Synt	MEDI 0116
of Sodium Cations in Zeolite–Y+	Adsorbant Interactions and Spin–Lattice Relaxation	PHYS 0267
Picosecond Reorientation of	Adsorbates on Alkylsilane/Silica Surfaces.=	ANYL 0025
Energy Transfer from Organic	Adsorbates to Defect Sites in Silica Supports.=	PHYS 0135
etal Surfaces and Manipulation of	Adsorbates.= +–Temperature STM Imaging of M	PHYS 0360
Radiolysis of	Adsorbed Acetic Acid on Na–Montmorillonite.=	NUCL 0072
ques. Theoretical Studies of Single	Adsorbed Atoms in the STM.= +Related Techni	PHYS 0096
Surface and Bulk Diffusion of	Adsorbed Nickel on Ultrathin Thermally Grown Silic	COLL 0177
Dynamics of CH_3Br	Adsorbed on Pt(111).= Surface Photochemical	PHYS 0261
Spectroscopy of Species	Adsorbed on Zeolites.= UV Resonance Raman	COLL 0075
ergy–Loss Spectroscopy Studies of	Adsorbed Species on Metal Oxide Surfaces.= +En	COLL 0173
Vapor Interaction with Carbon	Adsorbents and Sucrose Crystals.= Rate of Water	CARB 0062
xchange Resins and Hydrophobic	Adsorbents.= +actions of Carbohydrates with Ion–E	CARB 0063
0).=	Adsorption and Chemistry of Cyclic CH_2N_2 on Pd(11	COLL 0114
ry(II) with Cr+ Investigation of	Adsorption and Complexation of Lead(II) and Mercu	COLL 0104
ilicon Surfaces.=	Adsorption and Decomposition of Diethylsilane on S	COLL 0008
	Adsorption and Growth on Si and Ge Surfaces.=	PHYS 0186
omplexes in Disperse Systems.=+	Adsorption and Luminescence of Charge–Transfer C	COLL 0110
ation Mechanisms. Calculation of	Adsorption Energies in Slitlike and Wedge–Shaped	FUEL 0008
r Interfacial Tension Measurem+	Adsorption Kinetics of Asphaltenes—A Liquid/Wate	PETR 0138
face.+ Equilibrium Constants for	Adsorption of Antifreeze Glycoprotein at the Ice Sur	BIOL 0041
NMR Spectrosc+ Homogeneous	Adsorption of Benzene on NaX and NaY Zeolites by	CATL 0025
oluminescence as a Probe of the	Adsorption of Carbonyl Compounds onto Cadmium S	COLL 0175
etter the Binding, the Better th+	Adsorption of Cellulases on Cellulose Surface: The B	BIOT 0186
olymer Interfaces.+ Spontaneous	Adsorption of Polymers from Solution to Solution-P	POLY 0298
upport and Particle Size for NO	Adsorption on Rh and Pt.= Study of the Effect of S	COLL 0186
Ab Initio Study of K	Adsorption on the Ag(100) Surface.=	PHYS 0337
of Guest Molecules w+ Surface	Adsorption Phenomena of Carbohydrates. Interaction	CARB 0061
from Layered Siloxene+ Oxygen	Adsorption Properties of Microporous Silica Derived	PETR 0144
Liquid–Solid	Adsorption Theory.= Verification of a Fundamental	COLL 0188
n.= Apparent Free Energy	Adsorption: Measurement of Surface–Bulk Associatio	COLL 0122
Dynamics of Block Copolymer	Adsorption.=	COLL 0073
udied by Optical Second Harm+	Adsorption, Orientation, and Order at Surfaces as St	ANYL 0024
Titanium Silicate.=	Adsorptive Properties of ETS-10: A New Large–Pore	PETR 0079
del Fenol con	Aductos Tonsil-Fe$(NO_3)_3$.= Estudio de la Nitracion	COLL 0120
Detection of	Adulterated Vanilla Extract.=	AGFD 0185
nt Products and Their+ Flavor	Adulteration. Isotope Distributions in Secondary Pla	AGFD 0194
n Raspberry Concentrate+ Flavor	Adulteration. Survey of Alpha–Ionone Enantiomers i	AGFD 0181
etection of Flavor and Fragrance	Adulteration.= Application of SNIF–NMR to the D	AGFD 0197
l Oils.= HRGC for Detection of	Adulterations of Italian Cold–Pressed Citrus Essentia	AGFD 0199
o-Year College Are Doing to Help	Advance Prehigh School Science Education.= +Tw	CHED 0076
sis in in Situ Steam+ Site Use of	Advective–Flux Probes for Enhanced Soil Gas Analy	ENVR 0052
ustrial Chemistry Award). Some	Adventures and Innovations in Industrial Catalysis.=	I&EC 0043
Chemistry and the Arts.	Adventures of a Chemical Collector.=	CONG 0001
Academe.= Whistle–Blowers,	Advocates, and the Law. Greed in the Groves of	PROF 0001
signed Sequencing Batch React+	Aerobic Biodegradation of Phenolics in Optimally De	BIOT 0138
a)(Phosp+ Ligand Effects on the	Aerobic Oxidation of Cyclohexene Catalyzed by (Aqu	INOR 0067

Ions.= Studies on Silica	Aerogels and Xerogels Doped with Transition-Metal	INOR 0081
yde Gels as Precursors of Organic	Aerogels.= +tering Studies of Resorcinol Formaldeh	POLY 0199
Melamine-Formaldehyde	Aerogels.=	POLY 0051
sters Formation on the Surface of	Aerosil.= +c Model of the Chromium(VI) Oxide Clu	COLL 0099
epared by the High-Temperature	Aerosol Decomposition (HTAD) Process.= +ysts Pr	CATL 0026
Particl+ Particle Size Effects in	Aerosol Generation, Transport, and Vaporization for	ANYL 0122
	Aerosols: Past, Present, Future?.=	CHED 0048
terization o+ Applications of GC-	AES/Boiling-Point Distribution Analysis for Charac	PETR 0119
Ammonia Fiber Explosion (	AFEX) Pretreatment of Municipal Solid Waste.=	BIOT 0016
de Exchange in HMG-CoA Lyase	Affects the Kinetics of Active Site Modification.=	BIOL 0064
g Predictions+ Ab Initio Proton	Affinites Compared to Experimental Values, Includin	PHYS 0298
Electron	Affinities.= Theoretical Study of Silicon Cluster	PHYS 0079
oline Acetyltransferase and High-	Affinity Choline Uptake.= +logue Inhibitors of Ch	MEDI 0142
Squalene Epoxidase by	Affinity Chromatography.= Purification of Pig Liver	BIOL 0070
Exchange and Immobilized Metal	Affinity FPLC.= +P-450 Patterns by Means of Ion	ANYL 0017
Fundamental Studies on	Affinity Partitioning.=	BIOT 0049
ied Phospholipids.=	Affinity Precipitation of Proteins with Ligand-Modif	BIOT 0020
paration—II. Bioseparation Using	Affinity Techniques.= +hic Methods of Protein Se	BIOT 0044
ontributions (SGCs) Increase with	Affinity.= +ts Confirms that Substituent Group C	MEDI 0088
.=	Affinity-Based Reversed Micellar Protein Extraction	BIOT 0045
Folding of Recombinant	aFGF.= Novel Scalable Process for Controlled	BIOT 0134
ationic Proteins Extracted from	Aflatoxin-Resistant/Susceptible Varieties of Corn.=	AGFD 0134
on Conditions in the Detection of	Aflatoxins.= +mization of Post-Column Derivatizati	AGFD 0111
STM and	AFM Extensions.=	PHYS 0209
Control by STM and	AFM.= Imaging DNA Molecules under Potential	PHYS 0157
in	Africa.= Pesticide Donations and the Disposal Crisis	AGRO 0044
Evolution of the Reaction ^{40}Ar +	Ag from E/A = 7 to 34 MeV.=	NUCL 0062
New	Ag(I) Compounds with a Hexaamine Ligand.=	INOR 0037
on the	Ag(100) Surface.= Ab Initio Study of K Adsorption	PHYS 0337
Chromatography on	Agarose Gels.= Optimization of Protein	ANYL 0030
of	Agarose Gels.= Transient Electric Birefringence	COLL 0182
Research	Agenda.= U.S. EPA's Pollution Prevention	CEI 0023
Method for Coal	Agglomerates' Strength Determination.=	FUEL 0042
eration of G+ Application of Oil	Agglomeration Technique to Soil Clean-Up and Gen	FUEL 0141
°From Large-Surfactant	Aggregate to Low Interfacial Tension.=	COLL 0078
or the Photosystem II Manganese	Aggregate.= +port for a Dimer-of-Dimers Model f	INOR 0408
Analysis. Structural Analysis of	Aggregated Polysaccharides by HPSEC/Viscometry.=	PMSE 0094
on.= Characterization of Latex	Aggregates by Sedimentation Field-Flow Fractionati	PMSE 0029
Characteristics of Protein	Aggregates in Escherichia coli.= Formation and	BIOT 0115
Energy Transfer in Molecular	Aggregates: Influence of Colored Noise.=	PHYS 0023
Hydrophobic Polysaccharide	Aggregates.= Characterization and Function of	POLY 0098
Redox-Switched Molecular	Aggregates.=	COLL 0020
Substituent-Induced Cross-Link	Aggregation in Polydimethylsiloxane Networks.=	PMSE 0120
-Type Polymers.=	Aggregation of Hydrophobically Modified and Comb	POLY 0270
	Aggregation of Magnetizable Colloids.=	COLL 0161
n Solubilization as Detected by J-	Aggregation.= +uirements of Premicellar Porphyri	ORGN 0161
e Content on the Physical	Aging of Gelatinized Cornstarch.= Effect of Moistur	CARB 0043
erturbatio+ Isothermal Physical	Aging of Poly(methylmethacrylate): Localization of P	PMSE 0173
	Aging of Primed Polyolefin Surfaces.=	PMSE 0126
ffect of High-Temperature Lean	Aging on the Performance of Pt, Rh/CeO$_2$ and Rare	PETR 0035
rmination and the Effects of Seed	Aging: Solid-State NMR and NMR Imaging.= +Ge	AGFD 0104
s with a Spacer between the	Aglycone and the Disaccharide.= Avermectin Analog	ORGN 0104
η^2-C(S)OPri)(S$_2$COPri)CO(PMe+	Agnostic Acetyl Complexes: Crystal Structure of Mo(	INOR 0036
Model for Histamine H$_3$	Agonist Activity.= Preliminary Conformational	MEDI 0070
n, a Novel 5-HT$_1$-Like	Agonist for the Treatment of Migraine.= Sumatripta	MEDI 0079
zamides: Potent Serotonin 5-HT$_4$	Agonists and/or 5-HT3 Antagonists.= +icyclic Ben	MEDI 0069
d N^6-(2-Thienyl)Ethyl-Adenosine	Agonists. +SAR in a Series of N^6-Cyclopentyl- an	MEDI 0113
Benzothiazole Adenosine Receptor	Agonists: Potent Antihypertensive Agents.= +and	MEDI 0111
Derivatives of Potent A$_1$-Selective	Agonists.= +: Synthesis and Biological Activity of	MEDI 0115
of Dopamine D1 Selective	Agonists.= Approaches Toward the Rational Design	MEDI 0124
stigations of the Magnitude of α-	Agostic and Steric Kinetic Isotope Effects in Carbon	INOR 0147
an	Agostic Complex.= Facile Reversible Metalation from	INOR 0384
,5-Cyclooctadiene Complexes into	Agostic η^4-1,3-Cyclooctadiene Complexes.= +η^4-1	INOR 0209
ion for Fertilizer and	Agrichemical Dealers.= Environmental Self-Evaluat	FERT 0018
p.). Herbert Hoover and Georgius	Agricola: The Distorting Mirrors of History.= +Cor	HIST 0011
r Sample Preparation in Food and	Agricultural Analyses.= +lid-Phase Extractions fo	AGFD 0175
al Capabilities and Applications in	Agricultural and Food Chemistry.= +phy: Analytic	AGFD 0201
ace/Groundwater and the Role of	Agricultural BMP in the 1990s.= +Presence in Surf	FERT 0019
.=	Agricultural Chemical Site Remediation/Regulations	AGRO 0143
Status of the Disposal of Waste	Agricultural Chemicals and Their Containers. Biolog	AGRO 0139
Status of the Disposal of Waste	Agricultural Chemicals and Their Containers. Curren	AGRO 0120
Status of the Disposal of Waste	Agricultural Chemicals and Their Containers. Introd	AGRO 0085
Status of the Disposal of Waste	Agricultural Chemicals and Their Containers. Pestici	AGRO 0102
Level.= Advances in Managing	Agricultural Chemicals in Groundwater at the Farm	AGRO 0052
Evaluation of PIC Formation in	Agricultural Product Bag Burns.= Laboratory	AGRO 0140
Recent Developments in Plastic	Agricultural-Chemical-Container Recycling.=	AGRO 0045
Product Formulating for	Agricultural, Professional, and Consumer Markets.=	FERT 0004
ts for Implementing Sustainable	Agriculture and the Role of Biodiversity in Reducing	AGRO 0082
nd Their Uses in Horticulture and	Agriculture.= +olled-Release Nitrogen Solutions a	FERT 0005
Sludge in	Agriculture.= Ontario's Program for Using Sewage	FERT 0021

Economic Implications for U.S.	Agriculture.= Potential Bans of Selected Pesticides:	AGRO	0083
ontaminated Soil Excavated from	Agrochemical Facilities.= +Remediate Pesticide-C	AGRO	0144
arch and Development Needs for	Agrochemical Retail Dealership Site Assessment and	AGRO	0142
Quantitative Determination of	Agrochemicals and Their Degradation Products in En	AGRO	0031
d Chemistry of New and Potential	Agrochemicals. 1,2,4-Triazolo[1,5a]P+ Synthesis an	AGRO	0046
d Chemistry of New and Potential	Agrochemicals. 2-(1-Phenoxyethyl)-+ Synthesis an	AGRO	0071
Transferencia de $CuSO_4$ en	Agua/Agua-Acetona.= Energia Libre de	PHYS	0236
Looking Back and Moving	Ahead in Chemistry Education Software Evaluation.=	CHED	0080
emical Engineering, 2nd Edition—	Aiba, Humphrey, and Millis.= +technology. Bioch	BIOT	0221
QSRR	Aided by Quantum Chemistry.=	ANYL	0100
Computer-	Aided Discovery of Unprecedented Reactions.=	CINF	0006
or-Based Approach to Computer-	Aided Molecule Design.= +ases. CAVEAT: A Vect	COMP	0044
roach to B+ Role of Computer-	Aided Process Design in Following an Integrated App	BIOT	0075
tabolites.= Etoposide: Studies	Aimed Toward the Mechanistic Role of Bioactive Me	MEDI	0012
sis Design.=	AIPHOS—An Integrated System for Organic Synthe	CINF	0005
Twelve Days at the U.S.	Air Force Academy.=	CHED	0014
for the Oxidation of Benzene–	Air Mixtures.= Tentative Detailed Chemical Scheme	FUEL	0084
Revised Hazard-Ranking System	Air Pathway.=	ENVR	0082
ogy+ Prioritization of Hazardous	Air Pollutants: The Need for Comparative Methodol	ENVR	0066
Sensitivity Analysis for an	Air Quality Model.=	ENVR	0026
nisms Used to Model Mexico City	Air Quality.= +mparison of Three Chemical Mecha	PHYS	0296
.=	Air Samplers and Complementary Analyses for PIXE	NUCL	0079
atalyzed Gasification of Carbon in	Air.= +Chemical Pretreatments on the Copper-C	FUEL	0028
of Oxygen-Enriched Combustion	Air.= +l Swing Absorption Process for Production	I&EC	0023
aces—III. Lateral Diffusion at the	Air-Water Interface.= +lutions, Blends, and Interf	COLL	0051
Cellulosic Polymers at the	Air-Water Interface.= Dynamic Properties of	COLL	0056
erene (C_{60}) in Helium, Argon,	Air, and Hydrogen at 25–1000°C.= Behavior of Full	FUEL	0021
Catalytic Ozone Abatement in Jet	Aircraft.= +cial Development and Experience with	I&EC	0047
The ULSI Window: A View of	Al and Cu Complexes for MOCVD.=	INOR	0125
Studies of Dealuminated+ ^1H-^{27}	Al Crosspolarization and Very Fast ^{27}Al MAS NMR	PETR	0006
Recent Developments in	Al in Conformational Analysis.= WIZARD–III	CINF	0034
Crosspolarization and Very Fast	^{27}Al MAS NMR Studies of Dealuminated Zeolite Y.=	PETR	0006
Solid-State	^{27}Al NMR Studies of Hydrated $AlPO_4$-8.=	PETR	0096
Theorectical Foundation of the	Al Sitting and Pairing Avoidance Principles in Zeolit	PHYS	0289
Ga_{13}, $GaAl_{12}$ and	Al_{13} Polyoxocations and Pillared Clays.=	PETR	0143
al Characterization of TiO_2 and	Al_2O_3 Prepared by Precipitation and Sol–Gel Method	PETR	0092
Trichloroethylene over a PdO on	Al_2O_3/Monolith.= Catalytic Oxidation of	COLL	0210
thylene Using 1.5 Pt on	Al_2O_3/Monolith.= Catalytic Oxidation of Trichloroe	COLL	0205
over H-[	Al]–ZSM-5 Catalysts.= Naphthene Transformation	PETR	0044
t.=	Al–Modified Sepiolite as Catalyst or Catalyst Suppor	PETR	0015
Mechanisms for the Formation of	Al–N Bonds on the Surface.= +Ammonia on Silica:	COLL	0045
RMER'; R,R' = Aryl, Alkyl; M =	Al, Ga, In; E = P, As.= +nts of Empirical Formula	INOR	0392
Reactive Collisions of	Al, Mg, Si, and C Atoms.=	PHYS	0089
ltrysosyl-N-(3-Phenylpropyl)-D-	Alaninamide.= +e Dipeptide Amide L-2,6-dimethy	MEDI	0152
cokinetics and Tissue Residues of	Albendazole in Cattle Administered via a Sustained	AGRO	0109
reeman Case: Plagiarism at the	Albert Einstein College of Medicine of Yeshiva Unive	PROF	0004
mocystis carinii and Anti-Candida	albicans Activities.= +ides with Potent Anti-Pneu	MEDI	0181
Tertiary Butyl	Alchohol on Silica Surfaces.= Chemisorption of a	COLL	0108
Oxidation and Reduction of	Alchol and 1-Propanol on $Cu_2O(100)$.=	COLL	0195
Behavior of Aqueous Poly(vinyl-	alcohol) Solution—A Small-Angle Neutron Scatterin	POLY	0171
Stereoselective Synthe+ Furfuryl	Alcohol-to-Pyranone Oxidation: An Approach to the	ORGN	0294
ydrogen-Bonded Polymers: Vinyl	Alcohol/Vinyl Butyral Copolymers.= +ffusion in H	POLY	0191
ater Swelling/Soluble Starches by	Alcoholic–Alkali Treatments.= +f Granular Cold W	CARB	0040
s an Inclusion Host: Resolution of	Alcohols by Inclusion Compound Formation.= +e a	ORGN	0041
f Ketones: A Facile Route to Ciral	Alcohols Using Discrete Rhodium Catalysts.=	INOR	0247
the Epimerization of Secondary	Alcohols: Carbon–Hydrogen Bond Activation in Rhen	ORGN	0220
Fluidization by Anesthetics and	Alcohols: Discrepancy between Core and Surface Res	MEDI	0032
Catalytic Conversion of Alkanes to	Alcohols: +l of Oxidation Pathways for Selective	CATL	0036
Preparation of Propargylic	Alcohols.= New Organoboron Reagent for the	ORGN	0109
s from Automobile Engines Using	Alcohols, Gasohol, and Other Oxygenates.= +ission	FUEL	0095
(p-$(NC)C_5H_4N$).=	Aldehyde Decarbonylation Catalyzed by cis-$Rh(CO)_2$	INOR	0255
ption Spectra of Several Aliphatic	Aldehydes and Ketones of Atmospheric Interest.=	ENVR	0024
pentadienyl)Iron(II) Dicarbonyl to	Aldehydes and Ketones.= +Additions of Allyl(cyclo	ORGN	0213
nic Bis(+ Synthesis of Branched	Aldehydes by Catalytic Hydroformylation with Catio	INOR	0154
Photoreduction of	Aldehydes with Semiconductor as Catalyst.=	ORGN	0247
n of Mixed Alkylmethylzincs with	Aldehydes.= +l as a Blocking Group in the Reactio	ORGN	0090
Enantiomerically Pure 1,2-Epoxy	Aldehydes, Thiiranes, and Aziridines.= +nthesis of	ORGN	0052
β-Elimination Reactions at	Aldehydic Sites in DNA.= Mechanistic Studies of	BIOL	0158
Search of the Biosynthetic Diels–	Alder Construction.= +ide Class of Mycotoxins: In	ORGN	0283
Conformation-Dependent Diels–	Alder Dienophile Face Selectivity.=	ORGN	0241
The Diels–	Alder Reaction of α-Cyano-1-Azadienes.=	ORGN	0234
ereofacial Selectivity in the Diels–	Alder Reaction of 2,3-Dialkoxy-1,3-Butadienes.=	ORGN	0233
Transannular Diels–	Alder Reaction: A Powerful Synthetic Strategy.=	ORGN	0288
Intramolecular Diels–	Alder Reactions of Vinylsulfonamides.=	ORGN	0231
ctivity in Electrophilic and Diels–	Alder Reactions.= +ctronic Control of Diastereosele	ORGN	0240
metric Synthesis.= Enzymatic	Aldol Additions: A Building Block Strategy for Asym	BIOT	0090
	Aldol Chemistry via Initial Double-Bond Migration.=	ORGN	0092
etric Alkylation, Acylation, and	Aldol Reaction of Chiral β-Lactam Ester Enolates.=	ORGN	0046
orinanes and the Synthesis and	Aldol Reactions of a Chiral 2-oxo-2-Propionyl-1,3,2-	ORGN	0050
arbene Complexes.+ Asymmetric	Aldol Reactions of Chelated Imidazolidone Fischer C	INOR	0112

Limits of	Aldolase Stereoselectivity.=	BIOL	0009
Bioisosteres: Hydroxy Acetic Acid	Aldose Reductase Inhibitors.= +dress. Hydantoin	MEDI	0082
Inhibition of	Aldosterone Biosynthesis.=	MEDI	0009
Carbon-11-Labeled	Alfentanil.= Synthetic Studies for	ORGN	0192
s Isolated from Marine	Algae.= Reactivity of the Vanadium Haloperoxidase	INOR	0097
abeled L-(+)-Lactice Acid Using	Algal Biomass.= Production of Stable Isotopically L	BIOT	0156
Polyhedra i+ Graph Theoretical	Algorithm To Label the Vertices of the Five Regular	COMP	0035
The Pivot	Algorithm with Interaction Energy.=	POLY	0077
Identification	Algorithm.= Enhanced Chemical Name	CINF	0041
	AlH Gas-Phase Reaction Kinetics.=	PHYS	0301
ver+ Automated Method for the	Alignment of Molecular Structures: Optimizing the O	CINF	0018
Hormigas:	Alimento Rico en Proteinas.=	AGFD	0205
Microbiologia de	Alimentos en el LNSP: Enfoque Prospectivo.=	AGFD	0206
O_4 (KMNP): A Centrosymmetric	Aliovalent Analogue of Potassium Titanyl Phosphate	INOR	0256
V Absorption Spectra of Several	Aliphatic Aldehydes and Ketones of Atmospheric Int	ENVR	0024
Ring Aromatic Units Separated by	Aliphatic Spacers.= +igoesters Containing Single-	PMSE	0016
een the Oxidation Reactions of	Alkali and Alkaline Earth Atoms.= Comparison betw	PHYS	0072
Flame Chemistry of	Alkali and Alkaline Earth Metals.=	PHYS	0073
rs.=	Alkali Earth Salts as Catalyst Supports and Stabilize	PETR	0034
$6-C_7H_8)_2$ and the Solubilization of	Alkali Metal Halides in Hydrocarbon Media.= +(η	INOR	0053
l η^6-Toluene Coordination to an	Alkali Metal: Strucutre of Lu{CH(SiMe$_3$)$_2$}$_3(\mu$-Cl)K(η	INOR	0053
nductors?.=	Alkali Metal-C$_{60}$ Films: The First 3-D Molecular Co	INOR	0354
anning Tunneling Microscopy of	Alkali Metals on III-V(110) Surfaces: From 1- and 2	PHYS	0183
Compounds of C$_{60}$ and	Alkali Metals.= Structural Studies of Binary	PHYS	0048
ng/Soluble Starches by Alcoholic-	Alkali Treatments.= +Granular Cold Water Swelli	CARB	0040
ne-Based Solids: C$_{60}$, C$_{70}$, and the	Alkali-Metal Fullerides.= +tural Studies of Fullere	PHYS	0049
Superconductivity in	Alkali-Metal-Doped Fullerenes.=	PHYS	0046
ate Ceramic Powders from Mixed	Alkali/Titanium Alkoxide Precursors.= +ous Titan	INOR	0278
idation Reactions of Alkali and	Alkaline Earth Atoms.= Comparison between the Ox	PHYS	0072
Flame Chemistry of Alkali and	Alkaline Earth Metals.=	PHYS	0073
ystallization of MFI Zeolites in an	Alkaline Fluoride Medium.= +re Hydrothermal Cr	PETR	0005
Cytochrome c Peroxidase(III) in	Alkaline pH.= Photo-Induced Reduction of Native	BIOL	0097
	Alkaline Phosphatase Kinetics.=	CHED	0183
New Inhibitor of	Alkaline Phosphatase.= Fumaric Acid Monoazide: A	BIOL	0057
on of Aryloxy- and Oxo-Aryloxy-	Alkaline-Earth Complexes.= +is and Characterizati	INOR	0223
f Ethanol to Hydrocarbons over	Alkaline-Free Synthesized ZSM-5 and Aluminated S	PETR	0065
ological Studies of Pumiliotoxin A	Alkaloids and Congeners.= +ynthesis and Pharmac	ORGN	0275
Mass Spectrometry of Erythrina	Alkaloids from the Foliage of Genetic Cloves of Thre	ANYL	0136
otal Synthesis of Amaryllidaceae	Alkaloids of the 5,11-Methanomorphanthridine Type	ORGN	0007
HPLC Separation of	Alkaloids.=	ANYL	0135
Pyrrolizidine	Alkaloids.= Synthesis of Macrolactone	ORGN	0285
dextrin Glucanotransferase by +	Alkalophilic Bacillus sp.= Production of Cyclomalto	BIOT	0157
tion Data for P+ Comparison of	Alkane Dehydrocyclization Activity and Characteriza	CATL	0007
iates: Estimates+ Mechanism of	Alkane Displacement from (Alkane)W(CO)$_5$ Intermed	INOR	0111
Properties of	Alkane Isomerization Products.=	PETR	0105
eneration of Pd/HM Catalysts for	Alkane Isomerization.= +ation by Coking and Reg	PETR	0130
des: The Effect of Surface Oxyg+	Alkane Rearrangement Reactions on Tungsten Carbi	CATL	0037
UCST of Polyethers +	Alkane Systems.= Behavior of V^E and Cp^E near the	PHYS	0319
LC+ Efficient Thermal Catalytic	Alkane Transfer-Dehydrogenation Using Rh(PMe$_3$)$_2$	INOR	0151
of Alkane Displacement from (	Alkane)W(CO)$_5$ Intermediates: Estimates of Alkane-	INOR	0111
(CO)$_5$ Intermediates: Estimates of	Alkane-W Bond Strengths.= +nt from (Alkane)W	INOR	0111
talysts for Aromatization of Light	Alkanes and Olefins.= +nt of High-Performance Ca	PETR	0063
hotochemical Functionalization of	Alkanes by Polyoxometalate Systems.= +Catlytic P	INOR	0286
and Thermal Dehydrogenation of	Alkanes Catalyzed by Rh(PMe$_3$)$_2$(CO)Cl.= +emical	INOR	0152
he Isomerization of C$_5$ and C$_6$ N-	Alkanes in Light Straight Run.= +te Catalysts for t	PETR	0113
Selective Catalytic Conversion of	Alkanes to Alcohols.= +l of Oxidation Pathways for	CATL	0036
ed for the Aromatization of Small	Alkanes.= +llium-Containing HZSM-5 Zeolites Us	PETR	0043
enation and Dehydrocyclization of	Alkanes.= +rol of Selectivity in Catalytic Dehydrog	CATL	0061
Enthalpic Properties of	Alkanes.= Optimal Graph Theoretical Models for	COMP	0034
Complexes.= Investigation of	Alkene Oxidation by Technetium(VII)-Oxo	INOR	0328
Chiral (z)-	Alkene.= Relative Conformational Stabilities of	ORGN	0172
rconocene Benzyne Complex with	Alkenes and Alkynes.= +molecular Coupling of a Zi	ORGN	0217
ymmetric Epoxidation of Simple	Alkenes Using a New Class of D$_4$-Symmetrical Chira	INOR	0300
zed Alkylations of Isobutane with	Alkenes.= +and Mechanism of Sulfuric Acid-Cataly	PETR	0085
cted Hydrocarbonylation of N-	Alkenylamides and α-Alkenyllactams.= Amide-Dire	INOR	0227
ion of N-Alkenylamides and α-	Alkenyllactams.= Amide-Directed Hydrocarbonylat	INOR	0227
lization (SiCAC) of Alkynes and	Alkenynes Catalyzed by Rh and Co-Rh Mixed Metal	ORGN	0218
Unsaturated Ruthenium Fluoro-	Alkoxide Complex.= +nthesis and Reactivity of an	INOR	0127
en Bond Activation in Rhenium	Alkoxide Complexes (η5-C^{5}R^5)Re(No)(PPh^3)(OCHRR	ORGN	0220
Sol-Gel Condensation of Tin(IV)	Alkoxide Compounds.=	POLY	0258
ders from Mixed Alkali/Titanium	Alkoxide Precursors.= +ous Titanate Ceramic Pow	INOR	0278
ethanol Synthesis Obtained by an	Alkoxide route.= +ivity of Cu/ZnO Catalysts for M	INOR	0071
Complexes and Mixed Phosphine-	Alkoxide/Thiolate Complexes.= +ium Phosphine	INOR	0129
n from Oxygen to Metal in Fe-Mn	Alkoxycarbyne Complexes.= +sm of Alkyl Migratio	INOR	0110
and Herbicidal Activity of 2-(2-(	Alkoxyimino)-Alkyl)-5-Pyrazolyl-1,3-Cyclohexadion	AGRO	0013
Sheet and Tube	Alkoxysiloxanes.=	POLY	0257
Synthesis of (R)- and (S)-3-	Alkybutyrolactones.=	ORGN	0210
Diffusion of	Alkyl Adipate Plasticizers in PVC.=	POLY	0067
e Catalytic Synthesis of Mixed	Alkyl Amines and Polyfunctional Amines.= Selectiv	I&EC	0055
stituted 2-Methylbiphenylmethyl	Alkyl and Dialkylamino Aryl Oxime Ethers.= +Sub	AGRO	0072

re–Activity Relationships of N–	Alkyl and N,N'–Dialkyl Substituted N,N'–Diarylguan	MEDI	0147
es in the Carbonylation of Metal	Alkyl Bonds Studied by Time–Resolved Infrared Spe	INOR	0169
Reactive (η^5–Indenyl) Ruthenium	Alkyl Complexes.= +ration Reactions of Unusually	INOR	0239
our–Coordinate Monomeric Zinc	Alkyl Derivatives Stabilized by Sterically Demanding	INOR	0085
ic Zinc Hydride and Magnesium	Alkyl Derivatives Stabilized by Sterically Demanding	INOR	0087
Hydrogen Insertion Reactions of	Alkyl Diazoacetates Catalyzed by Chiral Rhodium(II	ORGN	0042
istic Study on the Formation of	Alkyl Esters and Amidoesters of (Anilinomethylene)p	ORGN	0103
ynthesis and Characterization of	Alkyl Gallium Amide Anion: The Pathway to Pyrazo	INOR	0390
ts—II. Thermal Decomposition of	Alkyl Halides on Copper Surfaces.= +ns of Catalys	COLL	0204
Reactions of Aqueous	Alkyl Halides with Binary Metal Compounds.=	INOR	0080
s of Metal Powders with Aqueous	Alkyl Halides.= +ic Investigations into the Reaction	INOR	0248
etics Studies of the Reaction of	Alkyl Hydroperoxides with Vanadium(IV) in Acidic A	INOR	0276
Sequence Lengths in Poly(n–	alkyl Isocyanates).= Chiral Clues to Helical	POLY	0348
oxycarbyne Com+ Mechanism of	Alkyl Migration from Oxygen to Metal in Fe–Mn Alk	INOR	0110
otosubstitution of Fe(CO)$_5$ with	Alkyl Phosphines: Fe(CO)$_4$PR$_3$ and Fe(CO)$_3$(PR)$_2$ Ar	INOR	0253
ransit+ Nucleophilic Addition of	Alkyl Thiolate Ions to β–Nitrostyrenes: Radicaloid T	ORGN	0155
ating Solvents in Promoting the	Alkyl to Acyl Migratory Insertion Reaction in η–Cp(C	INOR	0109
di+ Transmetalation Reactions of	Alkyl Zirconocenes: Copper–Catalyzed Conjugate Ad	ORGN	0089
cal Formula RMER'; R,R' = Aryl,	Alkyl; M = Al, Ga, In; E = P, As.= +nts of Empiri	INOR	0392
son of Allyl, Formyl, and Vinyl to	Alkyl.= +Radical Reactions with Oxygen: Compari	PHYS	0228
al Activity of 2–(2–(Alkoxyimino)–	Alkyl)–5–Pyrazolyl–1,3–Cyclohexadiones.= +rbicid	AGRO	0013
, Structures, and Reactivity of (	Alkyl)acyl, Diacyl, and (α–Ketoacyl)acyl Complexes o	INOR	0043
in Model Compounds for Poly(γ–	alkyl–α,L–glutamate)s.= +iple Thermal Transitions	POLY	0099
lation and Phosphine–Promoted	Alkyl–CO Migration Reactions of Unusually Reactive	INOR	0239
and Conformational Analysis of	Alkyl–Substituted Quinazolinone CCK–B/Gastrin Re	MEDI	0066
Synthesis of 5–Perfluoroalkyl–3–	Alkyl–2–Aryl–1,3–Thiazolidines.=	ORGN	0040
General Synthesis of 4–	Alkyl/Aryl–Substituted Saccharins.=	ORGN	0036
Chemistry of Poly(	alkyl/arylphosphazenes).= New Preparative	POLY	0184
ivity of Electron–Rich Ruthenium	Alkyl, Hydride, and Aluminohydride Complexes.=	INOR	0244
Channels.= Interactions of N–	Alkylamides with Voltage–Sensitive Sodium	AGRO	0151
tinociceptive Agents: O– and N–	Alkylated Derivatives of the Dipeptide Amide L–2,6–	MEDI	0152
Mitomycin C.= Bioreductive	Alkylating Properties of BMY–25067, an Analogue of	MEDI	0014
Compounds.= Electrochemical	Alkylation and Methoxylation of Active Methylene	ORGN	0249
nsfer Catalyst.= Free–Radical	Alkylation in the Presence of H$_2$S as a Hydrogen Tra	PETR	0083
Propene	Alkylation of Biphenyl Catalyzed by Pillared Clays.=	PETR	0014
Solvolysis–Triggered	Alkylation of DNA Using a Silyl Phenol.=	ORGN	0191
Mitomycin C.= Bioreductive	Alkylation of Glutathione and Other Thiols by	BIOL	0088
s over Heterogeneous Catalysts.	Alkylation of Isobutane with Pentenes Using Sulfuric	PETR	0082
atalysts.=	Alkylation of Isoparaffins with Olefins over Zeolite C	PETR	0101
losilicate Zeolite+ Para–Selective	Alkylation of Mono Alkylbenzenes over Silylated Gal	PETR	0100
Sulfuric Acid–Catalyzed	Alkylation of Pentenes and Other Light Olefins.=	PETR	0084
Selective Agents for DNA	Alkylation.= Naphthoquinone Derivatives as	ORGN	0190
–Lactam Ester En+ Asymmetric	Alkylation, Acylation, and Aldol Reaction of Chiral β	ORGN	0046
erization of Short–Chain Hydr+	Alkylation, Aromatization, Oligomerization, and Isom	PETR	0082
erization of Short–Chain Hydr+	Alkylation, Aromatization, Oligomerization, and Isom	PETR	0130
erization of Short–Chain Hydr+	Alkylation, Aromatization, Oligomerization, and Isom	PETR	0043
erization of Short–Chain Hydr+	Alkylation, Aromatization, Oligomerization, and Isom	PETR	0112
erization of Short–Chain Hydr+	Alkylation, Aromatization, Oligomerization, and Isom	PETR	0007
erization of Short–Chain Hydr+	Alkylation, Aromatization, Oligomerization, and Isom	PETR	0100
erization of Short–Chain Hydr+	Alkylation, Aromatization, Oligomerization, and Isom	PETR	0025
erization of Short–Chain Hydr+	Alkylation, Aromatization, Oligomerization, and Isom	PETR	0061
hanism of Sulfuric Acid–Catalyzed	Alkylations of Isobutane with Alkenes.= +and Mec	PETR	0085
,10–Dihydroa+ Silicon–Mediated	Alkylations: A Convenient Synthesis of 9,9–Dialkyl–9	ORGN	0087
ra–Selective Alkylation of Mono	Alkylbenzenes over Silylated Gallosilicate Zeolites.=	PETR	0100
li+ Novel Epoxy Resins Based on	Alkylenedioxyphenols—Effect of Backbone Flexibi	PMSE	0162
rate Heteroa+ Mechanism of O6–	Alkylguanine–DNA Alkyltransferase: Effect of Subst	BIOL	0161
e of Photo–Induced Coupling of	Alkylidyne and Carbonyl Ligands by Nucleophiles an	INOR	0188
ec+ Proton–Induced Coupling of	Alkylidyne and Isocyanide Ligands: Synthetic and M	INOR	0150
g Group in the Reaction of Mixed	Alkylmethylzincs with Aldehydes.= +l as a Blockin	ORGN	0090
tive, One–Pot Synthesis of cis N–	Alkylperhydroquinolines from Cyclohexanones.= +c	ORGN	0031
o+ Chemisorption of Aluminum	Alkyls and Ammonia on Silica: Mechanisms for the F	COLL	0045
Si(100).= Reactions of Zinc	Alkyls on Semiconductor Substrates: GaAs(100) and	COLL	0048
Group IVB (Ti, Zr, Hf)	Alkyls with Aluminum Reagents.= Reactions of	INOR	0054
Formed by the Reaction of Metal	Alkyls with Deboronated Zeolites.= +atalytic Sites	PETR	0024
ry for the Determination of Trace	Alkylselenide Compounds.= +sorption Spectromet	ANYL	0080
Reorientation of Adsorbates on	Alkylsilane/Silica Surfaces.= Picosecond	ANYL	0025
in Binary Mixture of Poly(di–N–	alkylsilanes) and Aromatic Solvents.= +le Gelation	POLY	0145
Configurations of Poly(3–	Alkylthiophene)s.= Solvent Effects on	PMSE	0183
anism of O6–Alkylguanine–DNA	Alkyltransferase: Effect of Substrate Heteroatoms on	BIOL	0161
Tungsten–Iridium Tetranuclear+	Alkyne Coordination, Dimerization, and Scission on	CATL	0008
ilyl–Carbocyclization (SiCAC) of	Alkynes and Alkenynes Catalyzed by Rh and Co–Rh	ORGN	0218
ygen–Atom Transfer from N$_2$O to	Alkynes Coordinated to Zr.= +llic Compounds. Ox	INOR	0327
romoted Cyclization Reactions of	Alkynes: Asymmetric Synthesis and Pharmacological	ORGN	0275
f Fischer Carbene Complexes with	Alkynes.= +eoelectronic Effects in the Reactions o	INOR	0194
enzyne Complex with Alkenes and	Alkynes.= +molecular Coupling of a Zirconocene B	ORGN	0217
n of Organocuprate Reagents to	Alkynyl Oxiranes: Synthesis of a Precursor of (±)–Ve	ORGN	0011
of Symmetric Di(1–	Alkynyl)Tellurides.= Synthesis and characterization	INOR	0229
Tris(arsino)	allanes and the Crystal Structure of (t–Bu$_2$As)$_3$Ga.=	INOR	0231
sexta.= Identification of an	Allatostatin from the Tobacco Hornworm, Manduca	AGRO	0008
Lepidoptera.= Plant	Allelochemical–Adapted Glutathione Transferases in	AGRO	0119

evelo+ Effects of Cotton Plant	Allelochemicals and Nutrients on the Behavior and D	AGRO	0037
teractions of Endophyte-Induced	Allelochemicals with Resistance Mechanisms in Cate	AGRO	0114
Heliothis Resistance to Plant	Allelochemicals.=	AGRO	0118
Benzene Formation during	Allene Pyrolysis.=	FUEL	0069
and Diene-	Allenes.= Efficient Synthesis of ortho-Isotoluenes	ORGN	0098
Enyne-	Allenes.= Stereoselective Synthesis of Enynes and	ORGN	0099
ition of Amine or Hydrazine to σ-	Allenyl Complex of Platinum(II).= +cleophilic Add	INOR	0041
Unresponsiveness to Experimental	Allergic Encephalomyelitis.= +at Induce Immune	BIOL	0086
or—General Route to Novel Anti-	Allergy and Anti-Inflammatory Drugs.= +Inhibit	MEDI	0179
Precollege Education. Benefits of	Alliances to the Educational Process.= +erships in	CHED	0001
(ndash)α-Kainic Acid and (–)α-	Allokainic Acid.= Total Synthesis of	ORGN	0177
Synthesis of Altrose and	Allose.=	ORGN	0072
the Armyworm.= Search for	Allotostatins in Other Species: A Case History Using	AGRO	0007
: The New Generations of Carbon	Allotropes.= +into Rods, Rings, Nets, and Spheres	ORGN	0266
Subclasses.= Effects of Gm-	Allotropes on the Production of Human IgG	CARB	0018
s. Multinuclear NMR Study of an	Alloy Semiconductor.= +and Optical Spectroscopie	PHYS	0372
tructural Properties of Molecular	Alloys of Multiply Bonded Molybdenum and Tungst	INOR	0356
ns in Pseudobinary Semiconductor	Alloys.= +surements of Atomic Cluster Distributio	PHYS	0375
Memory	Alloys.= Reactivity of Nickel-Titanium Shape	INOR	0357
Selectivity in Catalysis: Clusters,	Alloys, and Poisoning. Comparison of Alkane Dehy	CATL	0007
Oxidation and Reduction of	Allyl Alcohol and 1-Propanol on $Cu_2O(100)$.=	COLL	0195
+ 4π Cycloaddition Reactions of	Allyl Cations and Cation Radicals Generated from 1,	ORGN	0123
adienyl Manganese Dicarbonyl η^3-	Allyl Cations.= +cleophilic Addition to Cyclopent	INOR	0270
ity Studies of Relevant Scandium	Allyl Species.= +esis, Characterization, and Reactiv	INOR	0280
ewis Acid Promoted Additions of	Allyl(cyclopentadienyl)Iron(II) Dicarbonyl to Aldehy	ORGN	0213
R_2 (M = Zn, Mg and R = Bz, Cp,	allyl).= +ntation-Recombination Reactions with M	INOR	0383
Copolymerization of 1-	Allyl-2,5-dimethoxybenzene with Acrylonitrile.=	POLY	0050
tions with Oxygen: Comparison of	Allyl, Formyl, and Vinyl to Alkyl.= +Radical Reac	PHYS	0228
Control in	Allylic Functionalization.= Absolute Stereochemical	ORGN	0292
tion-Induced Rearrangement of	Allylic Methoxyacetates: A Novel Type of Allylic Tra	ORGN	0211
Courses and Mechanisms of the	Allylic Rearrangement and Enoyl Reduction Catalyze	BIOL	0065
Methoxyacetates: A Novel Type of	Allylic Transposition Catalyzed by $Eu(fod)_3$.= +ic	ORGN	0211
of 4-	Allyloxyisoquinolines.= The Claisen Rearrangement	ORGN	0101
ctivation of O_2 by $(OEP)Y(\mu-ME)$	$2AlMe_2$.= +and Reactivity of (OEP)MX, and the A	INOR	0294
cal and Spectroscopic Studies of	Alpha-Hydroxytricarboxylic Acid Complexes of Iron(	INOR	0040
es, Essences, and Fla+ Survey of	Alpha-Ionone Enantiomers in Raspberry Concentrat	AGFD	0181
Base Hydrolysis of the Cation	$\alpha\beta$-anti-Chloro-[1,9-Bis(2-pyridyl)-2,5,8-Triazanona	INOR	0397
Characterization of	5αReductase in E. coli.= Cloning, Expression, and	BIOL	0071
Pha+ Clear Solution Synthesis of	$AlPO_4$ Molecular Sieves: Identification of Transient	PETR	0020
SSZ-24, the All-Silica Analog of	$AlPO_4$-5.= Aluminum- and Boron-Containing	PETR	0023
Cobalt in Molecular Sieve	$AlPO_4$-5.= Isomorphous Substitutions of Silicon and	PETR	0099
e of Transformation of VPI-5 to	$AlPO_4$-8 at Various Temperatures by Multinuclear S	PETR	0095
Hydrated	$AlPO_4$-8.= Solid-State ^{27}Al NMR Studies of	PETR	0096
mers: Poly(1,3-cyclohexadiene-	alt-α-fluoroacrylonitrile).= New Fluorinated Copoly	POLY	0021
Pathway.=	Alterations of the β-Lactam Antibiotic Biosynthetic	BIOT	0235
olites: The Effects of Systemically	Altering a Template Structure.= +Large-Pore Ze	PETR	0036
for the Graphs of	Alternant Hydrocarbons.= Eigenvalue Distributions	COMP	0029
nd Their Potential Importance as	Alternate Electron Acceptors.= +g of Iron Oxides a	GEOC	0007
Based Pest Control Strategies as	Alternatives to Chemical Pesticides: Prospects, Chall	AGRO	0080
ment of Chlorofluorocarbon	Alternatives.= Overview of the Commercial Develop	I&EC	0044
las P+ Elaboracion de Pastas de	Alto Rendimiento: Influencia de la Sulfonacion sobre	PMSE	0077
Synthesis of	Altrose and Allose.=	ORGN	0072
Characterization of Model	Alumina and Its Comparison with Other Aluminas.=	PETR	0072
Energy-Loss Spectrometry of	Alumina and Silica Catalyst Supports.= Electron	PETR	0073
Ethyl Acetate over Pt-	Alumina Catalyst.= Combustion Byproducts of	COLL	0189
talizadores NiW Soportados sobre	Alumina o Titania Modificadas con Fluor.= +de Ca	FUEL	0106
Activation of Wide-Pore	Alumina Supports.=	PETR	0033
etermination of Pillar Densities in	Alumina-Pillared Montmorillonite.= +Analytical D	PETR	0126
alysts—I. Methane Oxidation over	Alumina-Supported Noble Metal Catalysts.= +at	COLL	0185
odemetallization+ Selectivity of	Alumina-Supported Vanadium Catalysts in the Hydr	CATL	0044
olibdeno sobre Oxidos Mixtos de	Alumina-Titania.= +Tiofeno con Catalizadores de M	FUEL	0105
Its Comparison with Other	Aluminas.= Characterization of Model Alumina and	PETR	0072
Zirconium Phosphates,	Aluminas, and Mixed Oxides.= Open-Morphology	PETR	0051
th Porphyrins in Layered Lithium	Aluminate-Fatty Acid Systems.= +ron Transfer wi	INOR	0272
line-Free Synthesized ZSM-5 and	Aluminated Silicalite-1 Crystals.= +bons over Alka	PETR	0065
Supported on Mg, Cu, Zn, and Ni	Aluminates.= +tinum and Platinum-Tin Catalysts	COLL	0126
ular Sieves Based upon Silicate or	Alumino Phosphate Chemistry.= +tructural Molec	CATL	0028
Groups on a Surface of a Pure and	Alumino(Titano)-Containing Silicas.= +of Silanol	COLL	0112
h Ruthenium Alkyl, Hydride, and	Aluminohydride Complexes.= +vity of Electron-Ric	INOR	0244
MR Study of 18-Ring Large-Pore	Aluminophosphate Molecular Sieves.= +id-State N	PETR	0094
Paramagnetic Ions in Microporous	Aluminophosphate Molecular Sieves.= +itution of	PHYS	0231
ystallization of Very Large Pore	Aluminophosphate Phases and Their Physicochemica	PETR	0093
ms for the Fo+ Chemisorption of	Aluminum Alkyls and Ammonia on Silica: Mechanis	COLL	0045
ym+ Wicking and Absorbency in	Aluminum and Silicate-Modified Starch Graft Copol	CARB	0042
anisms and Applicat+ Gas-Phase	Aluminum Atoms from Pulsed Laser Ablation: Mech	PHYS	0353
Reactions of Boron and	Aluminum Atoms with Small Molecules.=	PHYS	0090
nt Technology for the D I	Aluminum Can Manufacture.= Mobility Enhanceme	PMSE	0144
Characterization Data for Pt-Sn-	Aluminum Catalysts.= +rocyclization Activity and	CATL	0007
to the Processes of Transparent	Aluminum Film Preparation through Aqueous Sol-G	COLL	0115
Polymeric Routes to	Aluminum Oxides.=	POLY	0342

Amorphous	Aluminum Phosphate Gels.=	PETR	0050
Zr, Hf) Alkyls with	Aluminum Reagents.= Reactions of Group IVB (Ti,	INOR	0054
II. Kinetic Studies of Boron and	Aluminum Species.= Gas–Phase Metal Reactions—	PHYS	0086
Mechanistic Studies of	Aluminum Thin Film Growth by MOCVD.=	INOR	0124
Charles Martin Hall—	Aluminum.=	CHAL	0027
by the Interaction of Morine with	Aluminum(III) in Aqueous Solution.= +ies Formed	INOR	0273
ilica Analog of AlPO₄-5.=	Aluminum– and Boron–Containing SSZ-24, the All–S	PETR	0023
[Et₂AlAs+ Preparation of Novel	Aluminum–Arsenic Compounds: Crystal Structure of	COLL	0137
arent Alumi+ New Approach of	Aluminum-27 MASNMR to the Processes of Transp	COLL	0115
encia del Nivel Academico de los	Alumnos en los Resultados de la Evaluacion de una	CHED	0093
New Structural Model for	Alumoxane Macromolecules.=	POLY	0341
hysostigmine for the Treatment of	Alzheimer's Disease.= +bstituted Derivatives of P	MEDI	0137
ase Isoenzymes from Penicillium	Amagasakiense by Ion–Exchange Chromatography.=	ANYL	0016
Inhibitor Activity in	Amaranthus Species.= Characterization of Trypsin	AGFD	0133
anthri+ First Total Synthesis of	Amaryllidaceae Alkaloids of the 5,11-Methanomorph	ORGN	0007
ubstances+ Activity of Selected	Amaryllidaceae Constituents and Related Synthetic S	AGFD	0165
Sediments in and near the	Amazon Delta.= Phosphorus Distribution in	GEOC	0011
hesion of Bacterial+ Effects of	Ambient Humidity on the Wetting Properties and Ad	BIOT	0151
se Substra+ Synthesis of Serine–	AMC–Carbamate: A Novel Fluorogenic Tryptophan	ORGN	0023
tional Flow of Information: North	America and Europe. Global Marketing of+ Interna	CINF	0011
U.S. Chemical Businesses in Latin	America. Latin American Chem+ Opportunities for	SCHB	0021
Chemical Business: U.S. and Latin	American Approach.= +d the Challenge for Small	SCHB	0022
Sheets from the Viewpoint of an	American Chemical Company.= WHMIS Data	CHAS	0035
School St+ Participation in the	American Chemical Society SEED Program for High	CHED	0143
sion of Chemical Education of the	American Chemical Society.= +itiative of the Divi	CHED	0110
York Section of the	American Chemical Society.= Centennial of the New	HIST	0029
al of the New York Section of the	American Chemical Society). +orating the Centenni	HIST	0025
usinesses in Latin America. Latin	American Chemicals Approach 2000 AD.= +ical B	SCHB	0021
Thomas to	American Chemistry.= Contributions of Arthur W.	HIST	0039
Science Literacy for All	Americans.=	ANYL	0010
Hens.= Metabolic Fate of	Ametryn in Rats, Lactating Goats, and Laying	AGRO	0111
Ionomer System:	Amic Acid Ionomers.= Properties of a Novel Model	PMSE	0146
Kinetics of Imidization of Poly(	amic Acid) in Miscible and Immiscible Blends.=	PMSE	0147
Behavior of Functionalized Poly(	amic Acid) Precursors and Their Polyimides.= +ess	POLY	0085
posite during Imidization of Poly(	amic dialkylamide)Precursors.= +tu Molecular Com	PMSE	0169
Advances in Penicillin–	Amidase Biocatalysts.=	BIOT	0130
strate Specificity of Penicillin G	Amidase in Acylating a Monocyclic β–Lactam Interm	BIOT	0161
New Biocatalyst of Penicillin–	Amidase.=	BIOT	0131
Characterization of Alkyl Gallium	Amide Anion: The Pathway to Pyrazol Gallates.=	INOR	0390
emistry and Stereocontrol During	Amide Iodolactonization.= +Protection on Stereoch	ORGN	0258
ted Derivatives of the Dipeptide	Amide L-2,6-dimethyltrysosyl-N-(3-Phenylpropyl)–	MEDI	0152
–phenylene Terephthalamide).=+	Amide Segmental Dynamics in N-Deuterated Poly(p	POLY	0058
dration-Induced Moti+ Nitroxyl	Amide Spin Labeling: Methyl Esterification– and Hy	AGFD	0012
ides and α–Alkenyllactams.=	Amide-Directed Hydrocarbonylation of N-Alkenylam	INOR	0227
um(V) Oxo Complexes of Amine–	Amide–Dithiol Ligands.= +chnetium(V) and Rheni	INOR	0181
High-Nitrogen Low-Hydrogen	Amides and Imides from Cyanoimidazoles.=	POLY	0363
α,β-Unsaturated	Amides.= New Synthesis of	ORGN	0255
Simple	Amides.= Resonance Raman Spectroscopy of	PHYS	0321
ha+ One-Pot Synthesis of Cyclic	Amidinium Tetrafluoroborates and Hexafluorophosp	ORGN	0204
Synthesis of 4–	Amido 2-Carboxytetrahydroquinolines.=	ORGN	0117
e Rhenium Hydrazine, Hydrazido,	Amido, and Ammonia Complexes.= +idation-Stat	INOR	0261
e Formation of Alkyl Esters and	Amidoesters of (Anilinomethylene)propanedioic Acid	ORGN	0103
of	Amiloride.= Molecular Dynamics Simulations	MEDI	0045
ide Reagents.= A Reductive	Amination-Lactamization Procedure Using Borohydr	ORGN	0242
active Chromium(III) Polydentate	Amine Complex Toward Base Hydrolysis.= +ry Re	INOR	0397
–Imine and Organotin–Platinum–	Amine Complexes.= +ly Active Organotin–Platinum	INOR	0234
roperties of Amine–Cured Epoxy:	Amine Epoxide Addition Reaction vs. +echanical P	POLY	0158
ed by Nucleophilic Addition of	Amine or Hydrazine to σ–Allenyl Complex of Platinu	INOR	0041
d Rhenium(V) Oxo Complexes +	Amine–Amide–Dithiol Ligands.= Technetium(V) an	INOR	0181
re and Mechanical Properties of	Amine–Cured Epoxy: Amine Epoxide Addition React	POLY	0158
lytic Synthesis of Mixed Alkyl	Amines and Polyfunctional Amines.= Selective Cata	I&EC	0055
ration of Primary and Secondary	Amines by Reversed–Phase Paired–Ion High–Perfor	ANYL	0089
econdary, and Tertiary Aromatic	Amines in Fossil Fuels Using Trifluoroacylation, I: A	FUEL	0060
ophilic Substitution by Aromatic	Amines on the Ultimate Hepatacarcinogen, N–(Sulfo	ORGN	0194
Films from the Reaction of	Amines with Metal Halides.= Synthesis of Nitride	INOR	0309
Chlorodeamination of Primary	Amines.= Convenient Hydrodeamination and	ORGN	0108
Alkyl Amines and Polyfunctional	Amines.= Selective Catalytic Synthesis of Mixed	I&EC	0055
Stereospecific Synthesis of	Amino Acid Conjugates of Ether Phospholipids.=	MEDI	0163
nvulsant, for Impairing Excitatory	Amino Acid Neurotransmission.= +iazoline Antico	MEDI	0144
Auxotrophic	Amino Acid Producers.= Culture Instability of	BIOT	0031
Fragments Bearing α–Hydroxy–β–	Amino Acid Residues.= +tric Synthesis of Peptide	ORGN	0048
oderma virde.=	Amino Acid Sequence of a 20K Xylanase from Trich	BIOT	0211
oy–Based Infant Fo+ Effect of	Amino Acid Supplementation on Protein Quality of S	AGFD	0209
I): A Kinetic Study.=	Amino Acid–Nucleotide Cross–Binding to Platinum(I	INOR	0182
is of Amphipathic Surfaces from	Amino Acids and Steroids for Construction of Hydro	ORGN	0066
–(Fluorenyl–9–Methoxycarbonyl)	Amino Acids and Their Derivatives as Anti–Inflamm	MEDI	0171
First Dissociation Constant of	Amino Acids by the Conductivity Method: I. MOPSO	BIOL	0033
f Ferrate(VI) and Ferrate(V) with	Amino Acids.= +tics and Mechanism. Reactivity o	INOR	0393
Mutagenesis with Unnatural	Amino Acids.=	BIOL	0029
ro, α–Hydroxy, and N–Blocked α–	Amino Acids.= +ic Synthesis: Activation of α–Chlo	ORGN	0259

taining Polymers Derived from α-	Amino Acids, Dipeptides, and Tripeptides.= +Con	PMSE	0014
anodeoxy, and Branched–Chain	Amino Sugars by Epoxy and Triflate Sugar Intermed	ORGN	0071
n of Sodium Chloride and Two	Amino Sulfonic Acids, HEPES and MOPSO, by EMF	BIOL	0034
Complexes of an Ionomer and an	Amino–Containing Polymer.= +f Polymer–Polymer	PMSE	0155
Nitric Oxide Synthase by Nε-	Amino–L–Homoarginine.= Inhibition of Inducible	MEDI	0106
rile–Stabilized Carbanions to 4-	Amino–Substituted Indole Chromium Complexes and	ORGN	0010
and Dopaminergic Activity of 2-	Amino-1,4-Epoxy-1,2,3,4-Tetrahydronaphthalenes.=	MEDI	0128
d Evaluation of a Series of 7-(3-	amino–3–aryl–1–pyrrolidinyl)–4–naphthyridones as A	MEDI	0061
CVD from a Copper(II)	Aminoalkoxide Precursor.= Chemistry of Thermal	COLL	0028
a+ Synthesis of Tritium–Labeled	Aminoalkoxychromones, NPC16377 and NPC13839,	MEDI	0150
thermodynamic Mixture of bis(4-	aminocyclohexyl) Methane Isomers.= +ine to a Non	I&EC	0056
no Sugars by Epo+ Syntheses of	Aminodeoxy, Cyanodeoxy, and Branched–Chain Ami	ORGN	0071
oly(Methyl Acrylate-co-Dimethyl	Aminoethyl Methacrylated)/Epoxy Resin IPNs.=	PMSE	0079
harmacological Evaluation of 1-	Aminoethyl-3,4-Dihydro-5-Hydroxy-1H-2-Benzopy	MEDI	0127
sis and Activity of Substituted (	Aminomethyl)–Piperidines: A Novel Class of Selectiv	MEDI	0025
Lysine 2,3-	Aminomutase Reaction.= An Organic Radical in the	BIOL	0077
enosylcobalamin: The Lysine 2,3-	Aminomutase Reaction.= +at Does Not Require Ad	BIOL	0023
Metal Cofactors of Lysine 2,3-	Aminomutase.=	BIOL	0078
Main Group Compounds.	Aminooxyalkylphosphonic Acids and Derivatives.=	INOR	0387
ecarboxylase:+ 2–Substituted 3-	Aminooxypropanamines as Inhibitors of Ornithine D	MEDI	0184
gues as Inhibitors for Enkephalin-	Aminopeptidase.= +–Containing Enkephalin Analo	MEDI	0151
urring in Acidic Solutions of a 21-	Aminosteroid (Tirilazad).= +adation Products Occ	ANYL	0085
	Aminosugar Attenuation of HIV Infection.=	AGFD	0193
Hydrazine, Hydrazido, Amido, and	Ammonia Complexes.= +idation–State Rhenium	INOR	0261
Municipal Solid Waste.=	Ammonia Fiber Explosion (AFEX) Pretreatment of	BIOT	0016
Effects of	Ammonia on CVD Diamond Film Growth.=	PHYS	0150
orption of Aluminum Alkyls and	Ammonia on Silica: Mechanisms for the Formation o	COLL	0045
tive Catalytic Re+ Activation of	Ammonia on V2O5/TiO2 Catalysts Used for the Selec	COLL	0211
Reduction of Nitric Oxide by	Ammonia over Fe–Y Zeolites.= Selective Catalytic	COLL	0212
Luftwaffe: Evolution of Metal-	Ammonia Reductions.= Synthetic Steroids and the	HIST	0013
Oxide with	Ammonia.= Selective Catalytic Reduction of Nitric	COLL	0207
tion Catalyzed by Brevibacterium	ammoniagenes Fatty Acid Synthase.= +noyl Reduc	BIOL	0065
Gold(I)	Ammonium Cations.= Classical and Nonclassical	INOR	0367
ole of Magnesium Phosphate in	Ammonium Control of Cephamycin C Biosynthesis.=	BIOT	0088
Steel Exposed to Inhibited Urea-	Ammonium Nitrate Solution.= +Corrosion of Mild	FERT	0001
taining bisQuaternary Piperidino-	Ammonium Salts.= +scle Relaxants of Ester–Con	MEDI	0047
Steel Exposed to Urea-	Ammonium Sulfate Suspensions.= Corrosion of Mild	FERT	0003
Parallels between Solid–State	Amorphization and Melting.= Thermodynamics	NUCL	0092
	Amorphous Aluminum Phosphate Gels.=	PETR	0050
Improvements on Both	Amorphous and Zeolitic Isomerization Catalyst.=	PETR	0115
Blends.= Crystal-	Amorphous Interphases in Poly(ethylene Oxide)	PMSE	0165
Poly(Arylene Sulfide) Ketones via	Amorphous Ketimine Intermediates.= +crystalline	PMSE	0123
olymers Containing Helical and	Amorphous Polyisocyanide and Elastomeric Polybuta	PMSE	0071
and Glass Transition of	Amorphous Polymethylene.= Estimation of Density	POLY	0134
Expression of Yeast	AMP Deaminase in E. coli.=	BIOL	0083
Pulsed	Amperometric Detection of Carbohydrates.=	CARB	0067
s for Construction+ Synthesis of	Amphipathic Surfaces from Amino Acids and Steroid	ORGN	0066
uorocarbon Media: A Strategy f+	Amphiphile–Mediated Polar Colloid Formation in Fl	COLL	0164
Generation from Hyperpolarizable	Amphiphiles at Polymer–Polymer Interfaces.= +ic	POLY	0156
nates and Poly(ethylene Glycol)+	Amphiphilic Star Polymers from Tri– or Tetraisocya	POLY	0271
Molecular Dynamics Studies of	Ampholytic Ionomers and Their Pendant–Chain Der	POLY	0367
himeric Antibodies in CHO Cells	Amplified with the Glutamine Synthetase System.=	BIOT	0165
on and Source Differentiation of	Amyl Acetate, Ethyl Acetate, Ethyl Butyrate, and Et	AGFD	0187
rosalicylate in Determining the α-	Amylase Activity.= +on the Reduction of 3,5-Dinit	BIOL	0111
m a C+ Phaseolus vulgaris α-	Amylase Inhibitors, Arcelins, and Lectins Evolved fro	AGFD	0041
Folding, Assembly, and	Amyloidosis of Transthyretin.= Progress on the	BIOL	0007
Nuclear Magnetic Relaxation of	Amylose and Dynamic Behavior of the Hydroxymeth	CARB	0049
of Environmental Organics Using	AM1 Simulated Sunlight Irradiation of Semiconduct	INOR	0318
nes.=	AM1 Studies of the Nucleation of Crystals of 1,3-Dio	PHYS	0247
with Positive Inject+ Residues of	Anabolic Steroids in Muscle Tissues from Carcasses	AGFD	0130
de Fermentation by an Acidogenic	Anaerobe.= +nol Synthesis During Carbon Monoxi	BIOT	0015
Polychlorinated Biphenyls in	Anaerobic Sediments.= Dechlorination of	BIOT	0120
yrido[3,4-c] [1,5+ Synthesis and	Analgesic Activity of Octahydro-1,2,3,4,4a,5,11,11a-P	MEDI	0148
is, Characterization, and Relative	Analgesic Potencies of the 6–Sulfate Conjugates of S	MEDI	0149
	Analisis Cuantitativo por Computadora.=	COMP	0051
la Facultad de Quimica, UNAM (	Analisis Estadistico de 10Anos).= +rera de QFB en	CHED	0092
Materias Biologicas.=	Analisis Longitudinal de Indices de Reprobacion en	CHED	0091
romatography of Biopolymers—I.	Analogies between Calcium Hydroxyapatite and Zirc	ANYL	0027
istry Enrollments: Toward a Path	Analytic Model of Science–Related Attitudes.= +m	CHED	0177
PIXE as an	Analytical/Educational Tool.=	NUCL	0078
Oxalate: A Student Experiment in	Analytical/Physical Chemistry.= +termination of	CHED	0171
The Challenges.=	Analyzing Exotic Organics in Mixed Nuclear Wastes:	ANYL	0060
pectrometry as a Technique for	Analyzing Functional Groups with Oxygen in Coal.=	FUEL	0061
Current Methodologies for	Analyzing Pesticides and Metabolites in Water.=	AGRO	0146
ial and Surface Charge of TiO2 (	Anatase) Suspensions as a Result of Exposure to Rad	COLL	0203
	Anatomy of a Fraud Case.=	PROF	0010
Lectins Evolved from a Common	Ancestor Gene.= +mylase Inhibitors, Arcelins, and	AGFD	0041
xcited State of Vanadium Oxides	Anchored onto SiO2 and Their Photoreactivity towar	INOR	0165
and+ Membrane Fluidization by	Anesthetics and Alcohols: Discrepancy between Core	MEDI	0032
-Trisubstituted-1,2,4-Triazoles as	Angiotensin II Antagonists.= +Evaluation of 3,4,5	MEDI	0105

Diacidic Nonpeptide	Angiotensin II Receptor Antagonists.=	MEDI 0103
Orally Active, 1,5-Naphhyridine	Angiotensin II Receptor Antagonists.= Potent,	MEDI 0102
yrimidines as Potent Orally Active	Angiotensin II Receptor Antagonists.= +lo[1,5-a]P	MEDI 0104
Depth Profiling Using Variable-	Angle ATR.=	POLY 0307
SEC-Viscometry-Right	Angle Laser Light Scattering (SEC-VIS-RALLS).=	PMSE 0098
ppl+ A New Solid-State Small-	Angle Light-Scattering Apparatus: Description and A	POLY 0278
Polystyrene-Polyxylenyl+ Small-	Angle Neutron Scattering Study of Interdiffusion in	POLY 0166
vinyl-alcohol) Solution—A Small-	Angle Neutron Scattering Study.= +Aqueous Poly(	POLY 0171
ant Organogels Studied by Small-	Angle Neutron Scattering.= +of Polymers. Surface	POLY 0169
Mediated Crystallization+ Small-	Angle Neutron-Scattering Studies of the Template-	PETR 0058
Medium: Relation with Small-	Angle Scattering and Direct Dipolar Energy Transfer	PHYS 0060
Copolymer Ionomers by Small-	Angle X-ray Scattering and Dynamic Mechanical Th	PMSE 0118
pectroscopic Methods and Small-	Angle X-ray Scattering.= +crylate) as Studied by S	POLY 0124
brid Ceramer Materials by Small-	Angle X-ray Scattering.= +of Sol-Gel-Derived Hy	POLY 0284
yclic Transition State and C-H-C	Angle.= +en Transfer as Functions of Size of the C	ORGN 0175
of Carbonic	Anhydrase.= Active-Site Ionicity and the Mechanism	BIOL 0063
of Carbonic	Anhydrase.= Cobalt Spectroscopy at the Active Site	PHYS 0256
Acetic	Anhydride Process.= Eastman Chemical Company	I&EC 0053
g Poly(vinyl Methyl Ether/Maleic	Anhydride) Copolymer.= +aporates Prepared Usin	COLL 0086
al Method for Biomonitoring of	Aniline in Urine and Comparison with the Classical C	CEI 0003
mer-Modified Electrodes.=	Aniline-Substituted 2,2'-Bipyridine Ligands for Poly	POLY 0189
lkyl Esters and Amidoesters of (	Anilinomethylene)propanedioic Acids by Direct Cond	ORGN 0103
otential Antipsychotic Drugs: II.	Animal Behavioral Testing of Novel Piperidine Deriv	MEDI 0084
Potential Cocai+ Synthesis and	Animal Biodistribution of Radioiodinated Cocaine: A	MEDI 0131
mmalian Cell Retention in a Sp+	Animal Cell Bioreactor Design and Optimization. Ma	BIOT 0007
istic Effects of Gas Sparging in	Animal Cell Bioreactors.= Quantitative and Mechan	BIOT 0009
Properties of Bubbles in	Animal Cell Culture Media.= Oxygen Transfer	BIOT 0008
Considerations.= Bacterial or	Animal Cell Fermentation?—Process and Design	BIOT 0076
Topbiology: Molecular Bases of	Animal Form. G. Edelman (Medicine and Physiology	CONG 0013
nds and Their Binding by C-Type	Animal Lectins.= +tructure of Carbohydrate Liga	BIOL 0148
rug Residues in Food Products of	Animal Origin. Application of Matrix+ Antibiotic/D	AGFD 0060
ibiotic/Drug Residues in Foods of	Animal Origin. Approaches to Detection and+ Ant	AGFD 0099
rug Residues in Food Products of	Animal Origin. Current Problems As+ Antibiotic/D	AGFD 0081
rug Residues in Food Products of	Animal Origin. Liquid Chromatograp+ Antibiotic/D	AGFD 0125
tibiotics and Drugs in Products of	Animal Origin.= +cation for Residue Analysis of An	AGFD 0083
Residues in Foods of	Animal Origin.= Incidence of Antibiotic/Drug	AGFD 0082
rug Residues in Foods of	Animal Origin.= Radioimmunoassay of Antibiotic/D	AGFD 0084
nfirmation of Sulfamethazine in	Animal Tissues by HPLC with Diode-Array Detectio	AGFD 0103
uent Analysis of Drug Residues in	Animal Tissues.= +to the Extraction and Subseq	AGFD 0060
Assessment.= Refinement of	Animal Toxicity Studies for Human Risk	CHAS 0019
of Xenobiotics in Food-Producing	Animals in the United States.= +pects of the Use	AGRO 0076
of Xenobiotics in Food-Producing	Animals. In Situ Techniques for Studying+ Fate	AGRO 0092
of Xenobiotics in Food-Producing	Animals. Introduction—The Uses of Veterin+ Fate	AGRO 0075
of Xenobiotics in Food-Producing	Animals. Metabolism and Residue Studies o+ Fate	AGRO 0108
ic Metabolism in Food-Producing	Animals.= +itu Techniques for Studying Xenobiot	AGRO 0092
in Plants and	Animals.= Comparative Metabolism of DPX-E9636	AGRO 0028
Xenobiotics in Food-Producing	Animals.= Dermal Absorption and Metabolism of	AGRO 0095
Studies in Farm	Animals.= In Vitro Models of Biotransformation	AGRO 0094
Computer	Animation.= Teaching Magnetic Resonance Using	CHED 0123
ate in RNase?+ Is the Vanadate	Anion an Analogue of the Phosphorane Transition St	BIOL 0061
	Anion Binding to Polar Molecules.=	PHYS 0224
rantium) Juice Using Neutral and	Anion Exchange Resins.= +Sour Orange (Citrus au	AGFD 0106
terization of Alkyl Gallium Amide	Anion: The Pathway to Pyrazol Gallates.= +Charac	INOR 0390
ole of Mobile-Phase+ High-pH	Anion-Exchange Separation of Carbohydrates: The R	CARB 0066
	Anion-Molecule and Electron-Molecule Complexes.=	PHYS 0056
Borate-Pillaring	Anionic Clay.= Preparation and Characterization of	PETR 0111
	Anionic Clays: Trends in Pillaring Chemistry.=	PETR 0127
graphy of Cationic, Nonionic, and	Anionic Copolymers of Vinylpyrrolidone.= +romato	PMSE 0031
lysis of Porphyrin Metalation at	Anionic Interfaces: Mode of Interfacial Binding and I	CATL 0060
ed Forms of Methylviologen from	Anionic Surfactant Solutions.= +eposition of Reduc	COLL 0077
ene-Butadiene Thermoplastic E+	Anionic Synthesis of Heteroarm, Star-Branched Styr	POLY 0266
oxylation of Cyanoacetic Acid by	Anionic Tungsten (0) Complexes.= Catalytic Decarb	INOR 0212
horus Chemistry. Phosphoranide	Anions (10-P-4) with Bidentate and Tridentate Liga	INOR 0344
6, SiH$_3$GeH$_3$, and Ge$_2$H$_6$ Radical	Anions as Prototypes of Polymer Anions with Si and	PHYS 0313
Molecular	Anions as Surface Probes.=	NUCL 0056
th+ Orienting Effect of Fluoride	Anions in the Synthesis of a New Gallophosphate of	PETR 0075
nylation-Carbonyl Deblocking of	Anions of Carbethoxy-Protected Cyanohydrins with	ORGN 0228
by [51]+ Interaction of Vanadate	Anions with Cu,Zn-Superoxide Dismutase as Probed	INOR 0184
al Anions as Prototypes of Polymer	Anions with Si and Ge Backbones.= +Ge$_2$H$_6$ Radic	PHYS 0313
e+ EPR Line-Width Study of the	Anisotropic Reorientation of Vanadyl Acetylacetonat	PHYS 0264
ultiple-Pulse+ NMR Imaging of	Anisotropic Solid-State Chemical Reactions Using M	CATL 0022
of Xenobiotics in Food-Producing	Anmimals. Ivermectin and Abamectin Meta+ Fate	AGRO 0128
Simulated	Annealing of Rings.=	ORGN 0169
M+ Effect of High-Temperature	Annealing on the Capillarity of Nylon Microporous	COLL 0037
Effect of Cross-Linked State and	Annealing.= Gas Transport in IPN Membranes:	PMSE 0060
nic Solid: Exciton Trapping and	Annihilation in Crystals of (CH$_3$)$_4$NMnCl$_3$(TMMC).=	PHYS 0131
staneum L. (Coleopt+ Toxicity of	Annona squamosa Seed Oil Extracts to Tribolium ca	AGRO 0036
Chemical Industry. To Be	Announced.= Environmental Successes in the	CMEC 0017
To Be	Announced.=	CMEC 0003
To Be	Announced.=	CMEC 0021

	To Be Announced.=	CMEC	0022
	To Be Announced.=	CMEC	0002
	To Be Announced.=	SCHB	0028
	To Be Announced.=	CMEC	0019
	To Be Announced.=	SCHB	0029
	To Be Announced.=	CMEC	0004
	To Be Announced.=	CMEC	0020
ith Two- and Three-Atom Intra-	Annular Tethers.= +stry of Bimetallic Complexes w	INOR	0149
Compounds.= Plenary Session.	Annulation Strategies for the Synthesis of Highly Su	ORGN	0286
Use of Cationic Surfactants in	Anodic Voltammetric and Coulometric Analysis of In	COLL	0002
for a Fast Electrochemical Met+	Anodic Voltammetry at Extreme Potentials: A Hope	ANYL	0095
a, UNAM (Analisis Estadistico de	10Anos).= +rrera de QFB en la Facultad de Quimic	CHED	0092
f in Situ Hum+ Fate of a PCB in	Anoxic Bottom Sediments: Characterizing the Role o	ENVR	0017
d T+ Biogeochemistry of Iron in	Anoxic Coastal Sediments: Cycling of Iron Oxides an	GEOC	0007
Preparation and Applications of	ansa-Bis(idenyl)metal Complexes Containing Chira	INOR	0214
Diabrotica.= Phytochemical	Antagonism of GABA-Based Resistances in	AGRO	0136
Fragment of norBNI in κ-Opioid	Antagonist Potency and Selectivity.= +rmacophore	MEDI	0157
New Dual PAF and Histamine	Antagonist Related to SCH 37370.= SCH 40338, a	MEDI	0161
tent and Selective Adenosine A_1	Antagonist with Renal Protective and Diuretic Activi	MEDI	0112
and Selective Competitive NMDA	Antagonist.= +Synthesis of NPC 12626: A Potent	MEDI	0083
thesis and Bin+ Sigma Receptor	Antagonists as Potential Antipsychotic Drugs: I. Syn	MEDI	0153
imal Behaviora+ Sigma Receptor	Antagonists as Potential Antipsychotic Drugs: II. An	MEDI	0084
D1 Selective	Antagonists.= Design and Synthesis of Dopamine	MEDI	0125
Receptor	Antagonists.= Diacidic Nonpeptide Angiotensin II	MEDI	0103
TXA_2	Antagonists.= Interphenylene 7-Oxabicycloheptane	MEDI	0160
ycine Site N-Methyl-D-Aspartate	Antagonists.= +renic Acid Analogues as Potent Gl	MEDI	0146
in 5-HT4 Agonists and/or 5-HT3	Antagonists.= +icyclic Benzamides: Potent Seroton	MEDI	0069
ine Angiotensin II Receptor	Antagonists.= Potent, Orally Active, 1,5-Naphthyrid	MEDI	0102
Muscarinic	Antagonists.= Solution Conformations of	MEDI	0138
ly Active Angiotensin II Receptor	Antagonists.= +lo[1,5-a]Pyrimidines as Potent Oral	MEDI	0104
-1,2,4-Triazoles as Angiotensin II	Antagonists.= +Evaluation of 3,4,5-Trisubstituted	MEDI	0105
nidines as Noncompetitive NMDA	Antagonists.= +ialkyl Substituted N,N'-Diarylgua	MEDI	0147
-Molecule Antiotensin II Receptor	Antagonists.= +zole-5-Acrylic Acids: Potent Small	ORGN	0119
Photosynthetic	Antennae.= Electronic Excitation Transport in	PHYS	0020
Chemical Compounds from	Anthers of Spathodea Campanulta Flowers.=	ORGN	0186
Meta-	Anthracite Coal.= Short-Range Structuring in a	FUEL	0040
Related 3'-Mercapto-	Anthracyclines.= Synthesis of 3-Thio-Sugars and	CARB	0032
al Activities of 1,2-Trimethylene-	Anthraquinones, Analogues of Mitomycin C.= +ogic	MEDI	0013
nhibitor—General Route to Novel	Anti-Allergy and Anti-Inflammatory Drugs.= +) I	MEDI	0179
Molecular Biology of	Anti-$\alpha(1\leftarrow6)$ Dextrans.=	CARB	0059
syn- and	anti-β-Phenylcysteines.= Enantioselective Routes to	ORGN	0115
nt Anti-Pneumocystis carinii and	Anti-Candida albicans Activities.= +ides with Pote	MEDI	0181
ase Hydrolysis of the Cation $\alpha\beta$-	anti-Chloro-[1,9-Bis(2-pyridyl)-2,5,8-Triazanonane]	INOR	0397
the Base Hydrolysis of the syn,	anti-cis-Dichloro (1,4,7,10-Tetraazacyclodecane) Cob	INOR	0398
nose as a Chiral Template and Its	Anti-HIV Activity in PBM Cells.= +Using D-Man	CARB	0022
s of Natural Products with	Anti-HIV Activity.= Structure-Activity Correlation	AGFD	0163
MAP 30: A New Plant-Derived	Anti-HIV Agent.=	AGFD	0190
Natural Products as	Anti-HIV Agents.=	AGFD	0164
rimidine Nucleosides as Potential	Anti-HIV Agents.= +omerically Pure Dioxolane-Py	MEDI	0051
Compounds as	Anti-HIV Agents.= Tannins and Related	AGFD	0166
Synthesis and Novel	Anti-Inflammatory Activity of Oxazinone, BF-389.=	MEDI	0175
no Acids and Their Derivatives as	Anti-Inflammatory Agents.= +thoxycarbonyl) Ami	MEDI	0171
Route to Novel Anti-Allergy and	Anti-Inflammatory Drugs.= +) Inhibitor—General	MEDI	0179
the Platelet Effects of the	Anti-Inflammatory DuP 697.= Characterization of	MEDI	0162
elated Lipopeptides with Potent	Anti-Pneumocystis carinii and Anti-Candida albican	MEDI	0181
yl)-3(2H)-Fura+ Synthesis and	Anti-Ulcer Activity of Novel 5-(2-Substituted Ethen	MEDI	0067
Derivatives with	Antiandrogenic Activity.= Flouorinated Pregnane	ORGN	0193
ounds Possessing Potent Class III	Antiarrhythmic Activity.= +rivatives. Novel Comp	MEDI	0095
mationally Restricted Class III	Antiarrhythmic Agent.= L-702,958: A Novel Confor	MEDI	0096
vel, Potent, and Selective Class III	Antiarrhythmic Agents.= +ne Spiropiperidines as No	MEDI	0191
vel, Potent, and Selective Class III	Antiarrhythmic Agents.= +Spiropiperidines as No	MEDI	0097
ines with Potential as Class II/III	Antiarrhythmic Agents.= +ity of Novel Arylpiperaz	MEDI	0094
squinazolin-4,4'-Diones and Their	Antibacterial Activity.= +cles: Synthesis of 3,3'-bi	MEDI	0057
Diazolo[3,2-b]Indazoles and Their	Antibacterial Activity.= +hesis of [1,3,4]Oxa(Thia)	MEDI	0056
Alterations of the β-Lactam	Antibiotic Biosynthetic Pathway.=	BIOT	0235
rugs by+ New Developments in	Antibiotic Production—I. Production of Antitumor D	BIOT	0108
ntion Ti+ New Developments in	Antibiotic Production—II. Prediction of HPLC Rete	BIOT	0127
Pharmaceutical Development of	Antibiotic Products.= Use of HPLC in	ANYL	0112
LC Tandem M+ Confirmation of	Antibiotic Residues in Edible Meats by On-Line HP	AGFD	0061
hes to Determination of β-Lactam	Antibiotic Residues in Milk and Tissues.= +proac	AGFD	0125
r Expression of a Ba+ Improving	Antibiotic Yields in Streptomyces by the Intracellula	BIOT	0087
he Synthesis of Loracarbef, a New	Antibiotic.= +nocyclic β-Lactam Intermediate in t	BIOT	0161
ortality in CD1 Immunodeficient	Antibiotic-Treated Mice.= +ve Bacteria and High M	AGFD	0146
l Origin. Application of Matrix+	Antibiotic/Drug Residues in Food Products of Anima	AGFD	0060
l Origin. Current Problems Ass+	Antibiotic/Drug Residues in Food Products of Anima	AGFD	0081
l Origin. Liquid Chromatograp+	Antibiotic/Drug Residues in Food Products of Anima	AGFD	0125
s Associated with the Detection of	Antibiotic/Drug Residues in Food.= +ent Problem	AGFD	0081
. Approaches to Detection and +	Antibiotic/Drug Residues in Foods of Animal Origin	AGFD	0099
Origin.= Incidence of	Antibiotic/Drug Residues in Foods of Animal	AGFD	0082
Origin.= Radioimmunoassay of	Antibiotic/Drug Residues in Foods of Animal	AGFD	0084

plication for Residue Analysis of	Antibiotics and Drugs in Products of Animal Origin.	AGFD 0083
ten+ Lipophilicity of Teicoplanin	Antibiotics as Revealed by RP-HPLC: Structure-Re	MEDI 0059
synthesis of Echinocandin-Type	Antibiotics: Distinct Origins of Proline Residues in L	MEDI 0058
ell+ Specificities of Monoclonal	Antibodies Binding to the polysaccharide Chains of C	CARB 0017
Can Be Limited to Chain-End +	Antibodies Can Either Read Epitopes Anywhere, or	CARB 0019
for the Production of Chimeric	Antibodies in CHO Cells Amplified with the Glutam	BIOT 0165
lex and t+ Library of Monoclonal	Antibodies to E. coli Pyruvate Dehydrogenase Comp	BIOL 0031
Direct Labeling of Monoclonal	Antibodies with Tc99m.= An Improved Method for	BIOT 0061
Catalytic	Antibodies.=	BIOL 0028
Random Coupling of Monoclonal	Antibodies.= +rbents Prepared by Site-Directed or	BIOT 0060
he Overproduction of Monoclonal	Antibodies.= +erogeneous Hybridoma Cultures for t	BIOT 0169
Single-Chain	Antibody Binding Sites.=	BIOT 0057
dvances in Antibody Technology.	Antibody Combining Sites: Prediction and Design.=	BIOT 0056
Cytotoxicity of ^{67}Cu-Porphyrin-	Antibody Conjugates to a Colon Carcinoma Cell Line	BIOT 0168
sion and Scaleup of Engineered	Antibody Fragments from Escherichia coli.= Expres	BIOT 0058
ediction and Desig+ Advances in	Antibody Technology. Antibody Combining Sites: Pr	BIOT 0056
	Antibody-Assisted Protein Folding.=	BIOT 0062
learance and Immunogenicity of	Antibody-Like Molecules by Immuno or Binding Ass	BIOT 0059
l+ Topoisomerase Inhibition and	Anticancer Activity of Novel 4'-Demethylepipodophy	MEDI 0011
ents Based+ QSAR Analysis of	Anticancer Benzothiopyranoindazole Intercalating Ag	MEDI 0018
C: A New Target for Antiviral and	Anticancer Therapy.= +nhibitors. Protein Kinase	MEDI 0072
d 3'-Methylidene-S+ Synthesis,	Anticancer, and Antiviral Activities of Various 2'- an	CARB 0038
as Inducers of	Anticarcinogenic Enzymes.= Phenolic Antioxidants	AGFD 0089
c Compounds in Food and Health.	Anticarcinogenicity of Flavonol Quercetin.= +noli	AGFD 0113
mple HPLC Assay for the Novel	Anticonvulsant D,L-3-Hydroxy-3-Ethyl-3-Phenylpr	ANYL 0082
f ADD17014, a Novel Triazoline	Anticonvulsant, for Impairing Excitatory Amino Acid	MEDI 0144
lCo$_2$S$_x$Se$_{2-x}$ (0 ≤ x ≤ 2.0) with +	Antiferro-to-Ferromagnetic Transition in Metallic T	INOR 0058
Pyrrolo[2,3-d]Pyrimidine	Antifolates.= Novel Synthetic Approach to	ORGN 0113
brium Constants for Adsorption of	Antifreeze Glycoprotein at the Ice Surface.= +uili	BIOL 0041
ifferences in Properties between	Antifreeze Glycoproteins and Antifreeze Proteins.=	AGFD 0035
ween Antifreeze Glycoproteins and	Antifreeze Proteins.= +fferences in Properties bet	AGFD 0035
s of Simplified Analogues of the	Antifungal Agent Cilofungin: A Total Synthesis App	MEDI 0183
tion of Echinocandin Analogues.	Antifungal and Antipneumocystis Activity of L-693,9	MEDI 0182
d to Chain-End Recognition of an	Antigen.= +Epitopes Anywhere, or Can Be Limite	CARB 0019
Le^a) as Human Tumor-Associated	Antigen.= +ed Type 1 Chain (Dimeric Le^a or Le^b/	CARB 0060
haride Fragments from the Forssman	Antigen.= +alysis of All Di-, Tri-, and Tetra-Sacc	CARB 0056
of the X(3-Fucosyl-Lactosamine)	Antigenic Determinant.= +ved in the Biosynthesis	CARB 0015
rides of the Inner+ Synthesis of	Antigenic Heptose and KDO Containing Oligosaccha	CARB 0058
uration of Gluco+ Carbohydrate	Antigens. Emil Fischer's Establishment of the Config	CARB 0011
Vaccines.= Carbohydrate	Antigens. Haemophilus Influenzae Type B Conjugate	CARB 0053
Biosynthesis of+ Carbohydrate	Antigens. Human Fucosyltransferases Involved in the	CARB 0015
enosine Receptor Agonists: Potent	Antihypertensive Agents.= +and Benzothiazole Ad	MEDI 0111
of Copper(II) Complexes with the	Antiinflammatory Drug Piroxicam.= +ty Constants	INOR 0407
	Antijuvenile Hormone Agents.=	AGRO 0003
pyrrolidinyl)-4-naphthyridones as	Antimicrobial Agents.= +of 7-(3-amino-3-aryl-1-	MEDI 0061
hemistry of Pectinolide: A Novel	Antimicrobial α-Pyrone from Hyptis pectinata (Lam	MEDI 0060
ceptor Assays for the Detection of	Antimicrobial Residues in Meat.= +arm Test II Re	AGFD 0085
. Synthesis and Properties of New	Antimony Precursors for OMCVD.= +c Precursors	COLL 0135
Phenolic Antioxidants as	Antimutagens in Plant Foods.=	AGFD 0045
mplexes.=	Antineoplastic Effect of Several Transition-Metal Co	INOR 0026
ecay of+ Limit on the Electron	Antineutrino Mass from Measurement of the Beta D	NUCL 0010
nalogues as Systemically Active	Antinociceptive Agents: O- and N-Alkylated Derivat	MEDI 0152
Biochemical Mechanism of the	Antinutritional Effects of Tannins.=	AGFD 0149
ylic Acids: Potent Small-Molecule	Antiotensin II Receptor Antagonists.= +zole-5-Acr	ORGN 0119
t Model Systems.=	Antioxidant Activity of Phenolic Compounds in Mea	AGFD 0019
s on Plant Colorant Carotenoids+	Antioxidant Activity of Tocopherols and Tocotrienol	AGRO 0039
Medicinal Plants and Their	Antioxidant Activity.= New Polyphenols from	AGFD 0017
Their	Antioxidant Activity.= Polyphenols from Teas and	AGFD 0048
of Copper+ Differences in the	Antioxidant Mechanism of Carnosine in the Presence	AGFD 0039
Identification and	Antioxidant Properties of Soybean Isoflavones.=	AGFD 0020
uation of Flavor in Untreated and	Antioxidant-Treated Meat.= +al and Sensory Eval	AGFD 0120
Phenolic	Antioxidants as Antimutagens in Plant Foods.=	AGFD 0045
Enzymes.= Phenolic	Antioxidants as Inducers of Anticarcinogenic	AGFD 0089
unds in Food and Health. Natural	Antioxidants from Plant Material.= +nolic Compo	AGFD 0015
Natural	Antioxidants from Spices.=	AGFD 0016
Phenothiazine	Antioxidants Structurally Related to Nafazatrom.=	MEDI 0174
Custom Design of Better in Vivo	Antioxidants Structurally Related to Vitamin E.=	AGFD 0155
Phenolic	Antioxidants.= Thermal Degradation of	AGFD 0049
Constituents.= Role of	Antioxidative Defense Systems by Phenolic Plants	AGFD 0018
f Selected Chemotherapeutics and	Antiparasitics.= +iaturized Multiresidue Analysis o	AGFD 0062
Chemical Aspects of	Antiperspirants and Deodorants.=	CHED 0049
luent.=	Antiplasticization of Polyimides by a Nonreactive Di	PMSE 0127
ater-Soluble A+ Antifungal and	Antipneumocystis Activity of L-693,989 and Other W	MEDI 0182
nia.= P-9236: A New Atypical	Antipsychotic Agent for the Treatment of Schizophre	MEDI 0085
yl)Heteroaryltropanes as Potential	Antipsychotic Agents.= +luation of N-(Aryloxyalk	MEDI 0130
Receptor Antagonists as Potential	Antipsychotic Drugs: I. Synthesis and Bin+ Sigma	MEDI 0153
Receptor Antagonists as Potential	Antipsychotic Drugs: II. Animal Behaviora+ Sigma	MEDI 0084
Mode of Action of Hypericin as an	Antiretroviral Agent.= +ents. Hypothesis for the	AGFD 0188
A: A Novel Target for Therapeutic	Antisense Chemistry.= +ructure of Messenger RN	BIOL 0120
Search for the Heparin	Antithrombin III Binding Site Precursor.=	CARB 0039

Thiazolylamino Acids as	Antitrypanosomal Compounds.=	MEDI	0062
Plant Phenolics as Cytotoxic	Antitumor Agents.=	AGFD	0092
otic Production—I. Production of	Antitumor Drugs by Recombinant Bacteria.= +tibi	BIOT	0108
e-S+ Synthesis, Anticancer, and	Antiviral Activities of Various 2'- and 3'-Methyliden	CARB	0038
	Antiviral Agents from Medicinal Plants.=	AGFD	0191
y.= Natural Products as	Antiviral Agents. Advances in Antiviral Chemotherap	AGFD	0162
of Hyperic+ Natural Products as	Antiviral Agents. Hypothesis for the Mode of Action	AGFD	0188
otein Kinase C: A New Target for	Antiviral and Anticancer Therapy.= +nhibitors. Pr	MEDI	0072
Antiviral Agents. Advances in	Antiviral Chemotherapy.= Natural Products as	AGFD	0162
AIDS-	Antiviral Natural Products.=	AGFD	0192
d Ribosome-Inactivating Protei+	Antiviral Studies with Trichosanthin, a Plant-Derive	AGFD	0189
Glass Transition in the Diamond	Anvil Cell.= High-Pressure NMR Studies of the	PHYS	0260
odies Can Either Read Epitopes	Anywhere, or Can Be Limited to Chain-End Recogni	CARB	0019
Resistance in	Aphids.= Esterase Genes Conferring Insecticide	AGRO	0137
Drug Discovery+ Applications of	API Liquid Chromatography/Mass Spectrometry to	ANYL	0139
e Contaminada con S. aureus y su	Aplicacion en Quesos.= +de Sodio a Partir de Lech	AGFD	0107
de Equipo de la Industria Qui+	Aplicaciones Recientes de CAD/CAM para el Diseno	COMP	0053
totoxic Halogenated Monoterpene	Aplysiapyranoid D.= +nt Total Synthesis of the Cy	ORGN	0001
mica a Nivel Med+ Desarrollo de	Apoyos Computarizados para la Ensenanza de la Qui	CHED	0124
ossil Fuels. High-Pressure TPR	Apparatus to Investigate Organic Sulfur Forms in Co	FUEL	0007
te Small-Angle Light-Scattering	Apparatus: Description and Application to Phase-Se	POLY	0278
and Polymerization in the Matrix	Apparently Caused by Arsenic Salts.= +Hydration	ENVR	0041
ons of Pectinic Polysaccharides in	Apple Tissue.= +tion-Induced Motional Perturbati	AGFD	0012
β-Damascenone Precursors from	Apples.=	AGFD	0142
Lepidopteran Pheromones in	Apples, Grapes, and Peaches.= Residue Analyses of	AGRO	0148
Urine of Exposed Forest Pesticide	Applicators.= +Hexazinone and Its Metabolites in	CEI	0005
Polymers. Dynamic Viscoelasti+	APS-ACS Symposium: Thermoreversible Gelation of	POLY	0197
Polymers. Reversible Gelation +	APS-ACS Symposium: Thermoreversible Gelation of	POLY	0145
Polymers. Surfactant Organoge+	APS-ACS Symposium: Thermoreversible Gelation of	POLY	0169
Polymers. Thermoreversible Ge+	APS-ACS Symposium: Thermoreversible Gelation of	POLY	0121
ion of Cyclohexene Catalyzed by (	Aqua)(Phosphine)Ruthenium(IV) Complexes.= +at	INOR	0067
Food-Producing	Aquatic Species.= Xenobiotic Metabolism in	AGRO	0112
Ecological Risk of	Aquatic-Habitat Degradation.=	ENVR	0062
Compounds.= Reactions of	Aqueous Alkyl Halides with Binary Metal	INOR	0080
Reactions of Metal Powders with	Aqueous Alkyl Halides.= +ic Investigations into the	INOR	0248
	Aqueous Coprecipitation of Zirconia Precursors.=	COLL	0123
nazo-III as Chromogenic Agent in	Aqueous Ethanolic Solution.= +thanum with Sulfo	ANYL	0074
Characterization of Organic and	Aqueous Guest/Host Solid-State Interactions by Mu	AGRO	0062
TiO_2 Semiconductor Particles in	Aqueous Media: A Look at Two Recent Controversie	COLL	0131
Temperatures.= Formation of	Aqueous Metal Chloride Complexes to High	ANYL	0008
Fertilizer/Ag-Chemical	Aqueous Mixtures.= Solar Evaporation of	FERT	0022
le Neutron S+ Phase Behavior of	Aqueous Poly(vinyl-alcohol) Solution—A Small-Ang	POLY	0171
uction by the Electrolysis of an	Aqueous Potassium Carbonate Electrolyte and the Im	PHYS	0286
minum Film Preparation through	Aqueous Sol-Gel Method.= +es of Transparent Alu	COLL	0115
termination by Gel Permeation in	Aqueous Solution.= +olymer Molecular Weight De	CHED	0210
10^{4-} with $S_2O_3^{2-}$ in Slightly Acidic	Aqueous Solution.= +dies of Reactions of $Fe_2(CN)$	INOR	0400
of Morine with Aluminum(III) in	Aqueous Solution.= +ies Formed by the Interaction	INOR	0273
Viscous Behavior of Xanthan	Aqueous Solutions from a Variant Strain of Xanthom	BIOT	0196
α-Ketoglutaric Acid in	Aqueous Solutions.= Gamma Radiolysis of	NUCL	0070
ides with Vanadium(IV) in Acidic	Aqueous Solutions.= +eaction of Alkyl Hydroperox	INOR	0276
Structure and Orientation of	Aqueous Surfactant Monolayers.=	PMSE	0154
oelectrochemical Measurements in	Aqueous Surfactant Solutions.= +mical and Spectr	COLL	0003
: An Ionic Poly(p-phenylene) A+	Aqueous Synthesis of Soluble Rigid-Chain Polymers	POLY	0024
fficultly Oxidizable Compounds in	Aqueous Systems.= +tric Analysis of Insoluble, Di	COLL	0002
Modeling and Application of an	Aqueous Two-Phased System to the A-B-E Fermen	BIOT	0159
Evolution of the Reaction	^{40}Ar + Ag from E/A = 7 to 34 MeV.=	NUCL	0062
Two-Photon Ionization of (NO)	nAr_m Clusters.=	PHYS	0251
of	Ara-C.= Synthesis of Chemoreversible Prodrugs	MEDI	0052
Cyanide Addition to Nicotinamide	Arabinosides.= +on the Equilibrium Constants for	BIOL	0013
n-Containing Phenylated Soluble	Aramids.= +iques for the Characterization of Silico	POLY	0086
mond Films Synthesized by DC	Arc Discharge CVD in a Mixture Gas of Acetylene an	FUEL	0038
s vulgaris α-Amylase Inhibitors,	Arcelins, and Lectins Evolved from a Common Ances	AGFD	0041
Products.= Synthesis of	Architecturally Complex Natural and Unnatural	ORGN	0140
terials.=	Architecture of Device and Sensor Made from Bioma	BIOT	0123
Crystalli+ Influence of Molecular	Architecture on Gas Transport Properties of Liquid	POLY	0115
l+ Characterization of Branching	Architecture through Universal Ratios of Polymer So	POLY	0293
utions of Polymers with Differing	Architecture.= +ynamic Light Scattering from Sol	POLY	0250
New Optical Memory	Architectures Based on Optical-Phased Arrays.=	BIOT	0025
Supramolecular Polymer	Architectures.= Molecular-Recognition-Directed	POLY	0299
NMR+ Characterization of (η_6-	Arene)Tricarbonylchromium Compounds by Proton	ORGN	0215
ective Synthesis of (erta$_6$-	Arene)Tricarbonylchromium Compounds.= Regiosel	ORGN	0214
dem Column+ Quantitation of	Argatroban, a Specific Thrombin Inhibitor, by a Tan	ANYL	0086
ciometrica de Clorhidrato de L-	Arginina en Producto Terminado Utilizando como El	CHED	0163
ara Cuantificar Clorhidrato de L-	Arginina en Tabletas en Medio.= +encio-Metrico p	ANYL	0066
Nongraphitizable Polymers under	Argon Stream and Their Magnetic Properties.=	POLY	0023
Complexes Isolated in Solid	Argon.= Infrared Spectra of Metal-Ethylene	PHYS	0074
of Fullerene (C_{60}) in Helium,	Argon, Air, and Hydrogen at 25-1000°C.= Behavior	FUEL	0021
the Gasification of a Bituminous	Argonne Premium Coal Sample Using Wood Ash or	FUEL	0006
Mass S+ Characterization of the	Argonne Premium Coal Samples by Field Ionization	FUEL	0062
from and Liquef+ Research with	Argonne Premium Coal Samples. Moisture Removal	FUEL	0001

ntents of Argonn+ Research with	Argonne Premium Coal Samples. Organic Oxygen Co	FUEL	0003
ples. Organic Oxygen Contents of	Argonne Premium Coal Samples.= +um Coal Sam	FUEL	0003
netics of Organic Sulfur Forms in	Argonne Premium Coals during Pyrolysis.= +ion Ki	FUEL	0059
e ^{13}C NMR Measurements on the	Argonne Premium Samples and Other Coals.= +tiv	FUEL	0004
xposed Persons from Jacksonville,	Arkansas.= +rans in the Blood of Twelve TCDD-E	CEI	0014
atabase Management at the U.S.	Army Chemical Research, Development, and Enginee	CINF	0037
ies: A Case History Using the	Armyworm.= Search for Allotostatins in Other Spec	AGRO	0007
	Arnold Beckman—Acidity Measurement.=	CHAL	0032
sphere Technology: Efficient	Aroma Enrichment in Fruit Cells.= Precursor Atmo	AGFD	0023
of Off-Flavors by	Aroma Extract Dilution Analysis.= Characterization	AGFD	0097
Inhibition of the P$_{450}$-Dependent	Aromatase by a Triazole Derivative.= +reoselective	MEDI	0007
Separation and Analysis of Some	Aromatase Inhibitors on Cellulose-Based Chiral Stat	ANYL	0133
Nonsteroidal	Aromatase Inhibitors.= Development of New	MEDI	0003
ew Binding Model of Azole–Type	Aromatase Inhibitors.= +ological Evaluation, and N	MEDI	0035
nd Inhibition of Cytochrome-P$_{450}$	Aromatase.= +ormone Biosynthesis. Mechanism a	MEDI	0006
Class of Nonsteroidal Inhibitors of	Aromatase.= +tructure–Activity Studies in a New	MEDI	0036
rimary, Secondary, and Tertiary	Aromatic Amines in Fossil Fuels Using Trifluoroacyl	FUEL	0060
N Nucleophilic Substitution by	Aromatic Amines on the Ultimate Hepatacarcinogen,	ORGN	0194
as Stable 3-Phosphate Mimics in	Aromatic Analogs of EPSP.= +of Various Moieties	ORGN	0028
s of UV-Curable, Multifunctional,	Aromatic Analogs of Vinyl Ethers.= +vel Synthesi	POLY	0016
omatography for the Separation of	Aromatic Compounds.= +scrimination Liquid Chr	ANYL	0098
ore+ In Situ Cure Monitoring of	Aromatic Diamine–Epoxy System by Fiber Optic Flu	POLY	0282
Fluorescence of	Aromatic Diamines for Epoxy Cure Studies.=	POLY	0159
ion of the Nature of Naphtheno–	Aromatic Groups in Heavy Petroleum Fractions by ^{1}	PETR	0147
arcinogenic Metal and Polycyclic	Aromatic Hydrocarbon–Induced Transformation of	CHAS	0007
ling the Growth of Polynuclear	Aromatic Hydrocarbons in Diffusion Flames.= Mode	FUEL	0109
of Fishing People to Chlorinated	Aromatic Hydrocarbons in La Basse Cote–Nord, Que	CEI	0011
x^{3}B+ Kinetics of the Reactions of	Aromatic Hydrocarbons with O(^{3}P) Atoms and CH$_2$(	FUEL	0082
ylene/Air Flames and the Role of	Aromatic Intermediates.= +ot Depostion from Eth	FUEL	0107
Salt.= Electrochemistry of	Aromatic Ketones in a Room–Temperature Molten	ANYL	0094
etal Carbonyls in the Presence of	Aromatic Ketones.= +hemically by d^8 Transition–M	INOR	0254
II: Synthesis and Evaluation of	Aromatic Mimics of the Tetrahedral Intermediate an	ORGN	0029
Oxidative Degradation of	Aromatic Pollutants by Ligninase Models.=	INOR	0187
dification of Interactions between	Aromatic Polyamide Chains by Complexation.=	POLY	0064
Solubility Parameters of	Aromatic Polyamides.=	POLY	0063
$-\chi^{(3)}$ Relationships in Conjugated	Aromatic Polyazomethines.= +n Films, 3: Structure	POLY	0226
mally Induced Phase Behaviors of	Aromatic Polybenzimidazole/Polyimide Blends.=	PMSE	0164
vating and Deactivating Groups in	Aromatic Rings.= +ctural Characterization of Acti	CHED	0096
ure of Poly(di-N-alkylsilanes) and	Aromatic Solvents.= +ble Gelation in Binary Mixt	POLY	0145
tial Benzannulation/Nucleophilic	Aromatic Substitution Mediated by Chromium(0).=	ORGN	0221
s: Nickel(O)–Cata+ Regiospecific	Aromatic Substitution Via Aryl 1,2-Dipolar Synthon	ORGN	0082
ligoesters Containing Single–Ring	Aromatic Units Separated by Aliphatic Spacers.=	PMSE	0016
Covalently Linked Polynuclear	Aromatic–Nucleoside and Oligonucleotide Complexes	PHYS	0021
uits.=	Aromaticity, Kekule Structures, and Conjugated Circ	COMP	0021
Flames.= Comparison of	Aromatics Formation in Decane and Kerosene	FUEL	0080
Influence of the Zinc Content on	Aromatics Formation over Zinc–Containing Ultra–	PETR	0048
cleation of Soot Particles.=	Aromatics Growth beyond the First Ring and the Nu	FUEL	0108
mbustion Chemistry: Single–Ring	Aromatics in Flames. Formation of Benzene in+ Co	FUEL	0078
and Large–Port Mordenites in C$_8$	Aromatics Isomerization.= +ic Properties of Small–	PETR	0132
N–Hexane in the Presence of	Aromatics over Pt/Zeolites.= Hydroisomerization of	PETR	0114
emicall+ Combustion Chemistry:	Aromatics Precursors and Formation. Analysis of Ch	FUEL	0063
Reacti+ Combustion Chemistry:	Aromatics Precursors and Formation. Studies on the	FUEL	0064
Industrial Catalysts for	Aromatics Reduction in Gas Oil.=	CATL	0048
ing Tr+ Synthesis of Substituted	Aromatics via Unsaturated Ketene Intermediates Us	ORGN	0274
r the Synthesis of Polysubstituted	Aromatics.= +Versatile New Metalation Director fo	ORGN	0083
n Flames.=	Aromatics, Fullerenes, and Soot as Charged Species i	FUEL	0111
Combustion Chemistry: Large	Aromatics, Fullerenes, and Soot. Soot Deposition from	FUEL	0107
by+ Simultaneous Detection of	Aromatics, Sulfur, and Hydrocarbons in Diesel Fuels	PETR	0118
en Zeolitas del Tipo Pentasil en la	Aromatizacion de Propano.= +Contenido de Galio	COLL	0125
f the Gallium Phase of GaHMFI	Aromatization Catalysts in the Course of the Prepara	PETR	0046
: Effect of Gallium.=	Aromatization of Ethane and of Propane on HZSM-5	PETR	0045
f High–Performance Catalysts for	Aromatization of Light Alkanes and Olefins.= +nt o	PETR	0063
nal Metallosilicate Ca+ Selective	Aromatization of Light Hydrocarbons on Polyfunctio	PETR	0062
ng HZSM-5 Zeolites Used for the	Aromatization of Small Alkanes.= +llium–Containi	PETR	0043
for Investigation of Light Olefin	Aromatization Reactions over HZSM-5 Zeolites.=	PETR	0061
uence of Support Preparation on	Aromatization Selectivity over Nonmicroporous Pt C	CATL	0030
–Containing Catalysts for C$_1$–C$_4$	Aromatization: Characterization and General Catalyst	PETR	0067
C$_4$ Paraffin Dehydrogenation and	Aromatization.= +f Reactions and Catalysts for C$_3$–	PETR	0066
d Ga–β as a Catalyst for Propane	Aromatization—Infrared Specroscopic Investigations	PETR	0047
Short–Chain Hydr+ Alkylation,	Aromatization, Oligomerization, and Isomerization of	PETR	0082
Short–Chain Hydr+ Alkylation,	Aromatization, Oligomerization, and Isomerization of	PETR	0130
Short–Chain Hydr+ Alkylation,	Aromatization, Oligomerization, and Isomerization of	PETR	0043
Short–Chain Hydr+ Alkylation,	Aromatization, Oligomerization, and Isomerization of	PETR	0112
Short–Chain Hydr+ Alkylation,	Aromatization, Oligomerization, and Isomerization of	PETR	0007
Short–Chain Hydr+ Alkylation,	Aromatization, Oligomerization, and Isomerization of	PETR	0100
Short–Chain Hydr+ Alkylation,	Aromatization, Oligomerization, and Isomerization of	PETR	0025
Short–Chain Hydr+ Alkylation,	Aromatization, Oligomerization, and Isomerization of	PETR	0061
al Tissues by HPLC with Diode–	Array Detection.= +tion of Sulfamethazine in Anim	AGFD	0103
plications for Neural–Like Protein	Array Devices.= +in Cytoskeletal Microtubules: Im	BIOT	0055
Bonded Phase HPLC with Diode	Array Multichannel Detection.= +Peaks in Normal	ANYL	0087

gradu+ Incorporation of a Diode	Array UV–Visible Spectrophotometer into the Under	CHED	0180
eling Spectroscopy of Metal Oxide	Arrays on Graphite.= +eling Microscopy and Tunn	COLL	0194
f Au and Pt Microfabricated Wire	Arrays.= +s for Selective Surface Derivatization o	INOR	0350
on Optical–Phased	Arrays.= New Optical Memory Architectures Based	BIOT	0025
ormation in Thermoreversible +	Arrested Melting Due to Strain: Effect of Network F	POLY	0123
Complex of Lanthanum with	Arsenazo–III.= Spectrophotometric Study of	ANYL	0075
reparation of Novel Aluminum–	Arsenic Compounds: Crystal Structure of [$Et_2AlAs(S$	COLL	0137
Marine Bioinorganic Chemistry.	Arsenic Cycling in the Sea.=	INOR	0136
the Matrix Apparently Caused by	Arsenic Salts.= +Hydration and Polymerization in	ENVR	0041
esent, and Future. The Search for	Arsenic.= +ime, III: Forensic Chemistry—Past, Pr	HIST	0001
Uptake and Biotransformation of	Arsenicals in the Marine Environment.=	INOR	0137
f the Hydrogen Chloride–Gallium	Arsenide Gas Surface Reaction.= +r–Beam Study o	COLL	0215
lexants sur la Fotta+ Separation	Arsenopyrite–Pyrite: Effet des Oxydants et des Comp	ANYL	0067
Trimethylgallium and	Arsine on GaAs(100).= HREELS Study of	COLL	0065
$(t-Bu_2As)_3Ga$.= Tris(	arsino)allanes and the Crystal Structure of	INOR	0231
zation of Monomeric and Dimeric	Arsino– and Phosphinogallanes.= +and Characteri	INOR	0308
haracterization, and Properties of	Arsino–Group 12 Organometallic Compounds.= +C	INOR	0219
Laboratory Examination of	Arson Evidence.= Pioneers in Forensic Science:	HIST	0007
Chemistry and Pre–Historic	Art (Videotape presentation).=	CONG	0002
The	Art and Science of String Instruments.=	CONG	0003
Moving Frontier.+ State–of–the–	Art Symposium: Solid–State Chemistry—A Rapidly	CHED	0020
Moving Frontier.+ State–of–the–	Art Symposium: Solid–State Chemistry—A Rapidly	CHED	0029
–Infected Mice with Transdermal	Artelinic Acid.= +reatment of Plasmodium Berghei	MEDI	0063
Phosphorus Chemistry.	Arthur Dock Fon Toy—Scientist, Colleague, Friend.=	INOR	0288
Contributions of	Arthur W. Thomas to American Chemistry.=	HIST	0039
an	Articial Enzyme.= Catalysis of Enolate Reactions by	ORGN	0053
tion Researc+ Pseudomorph and	Artifact Corrosion Studies as Tools for Biomineraliza	BIOT	0101
Chemistry and the	Arts. Adventures of a Chemical Collector.=	CONG	0001
in Costa Rica: The Case of	Arvi, Inc., Costa Rica.= Small Chemical Businesses	SCHB	0026
Synthesis of	Aryl bis(Trialkyltin) Monomers.=	POLY	0035
ystallization, and Melting of Poly(	aryl Ether Ketone Ketone), Part I: Structure.= +Cr	POLY	0059
stallization, and Melting of Poly(	aryl Ether Ketone Ketone), Part II: Crystallization a	POLY	0060
Gelation in Poly(	aryl Ether Ketones).= Crystallization–Induced	POLY	0247
and $S_{RN}1$ Mechanisms.= Poly(	aryl ether–ketone) Synthesis via Competing S_NAR	PMSE	0122
Role on Catalyst Stability in +	Aryl Group Interchange in Triarylphosphines and Its	INOR	0349
nylmethyl Alkyl and Dialkylamino	Aryl Oxime Ethers.= +Substituted 2–Methylbiphe	AGRO	0072
Properties of Poly(	aryl Sulfide Phenylquinoxalines).= Preparation and	POLY	0026
l(O)–Catalyzed Cross–Coupling of	Aryl Triflates with Grignard Reagents.= +ns: Nicke	ORGN	0082
ecific Aromatic Substitution Via	Aryl 1,2–Dipolar Synthons: Nickel(O)–Catalyzed Cro	ORGN	0082
Synthesis of Tri–	Aryl–Substituted Torands.=	ORGN	0064
Permeability of Pol+ Effects of	Aryl–Substitution and Physical Blending on the Gas	POLY	0117
ion of a Series of 7–(3–amino–3–	aryl–1–pyrrolidinyl)–4–naphthyridones as Antimicrob	MEDI	0061
5–Perfluoroalkyl–3–Alkyl–2–	Aryl–1,3–Thiazolidines.= Synthesis of	ORGN	0040
mpirical Formula RMER'; R,R' =	Aryl, Alkyl; M = Al, Ga, In; E = P, As.= +nts of E	INOR	0392
lymers of Poly(styrene) and Poly(	arylate) and Their Optical Properties.= +ock Copo	POLY	0220
y Various Arynes.= N– and 4–	Arylation of N–(4–Methoxyphenyl)–2–Butenamides b	ORGN	0224
Properties of Poly(	arylene Ether Benzimidazole)s.= Synthesis and	POLY	0025
emistry, Thermal Stabilit+ Poly(	Arylene Ether Phosphine Oxide)s—I. Overview of Ch	PMSE	0052
Studies.= Poly(	Arylene Ether Phosphine Oxide)s—II. Pyrolysis	PMSE	0053
s of Surface Modification+ Poly(	Arylene Ether Phosphine Oxide)s—III. Investigation	PMSE	0054
termedia+ Semicrystalline Poly(	Arylene Sulfide) Ketones via Amorphous Ketimine In	PMSE	0123
Four Generations of Monodisperse	Arylester Dendrimers.= +Convergent Synthesis of	POLY	0324
Chemistry of	Arylimido Complexes of Osmium and Ruthenium.=	INOR	0259
ynthesis and Characterization of	Aryloxy– and Oxo–Aryloxy–Alkaline–Earth Complex	INOR	0223
acterization of Aryloxy– and Oxo–	Aryloxy–Alkaline–Earth Complexes.= +is and Char	INOR	0223
N–Heteroaralkyl–Substituted–1–	Aryloxy–2–Propanolamine and Propylamine Derivat	MEDI	0095
Synthesis and Evaluation of N–(	Aryloxyalkyl)Heteroaryltropanes as Potential Anti	MEDI	0130
nd β–Blocking Activity of Novel	Arylpiperazines with Potential as Class II/III Antiarr	MEDI	0094
Selective Class III Antiarrhyt+	Arylpyranone Spiropiperidines as Novel, Potent, and	MEDI	0097
yl)–2–Butenamides by Various	Arynes.= N– and 4–Arylation of N–(4–Methoxyphen	ORGN	0224
$aAs(SiMe_3)$+ Reactions of (Me_3Si)	$3As$ with GaX_3 (X = Br, I): Crystal Structure of [I_2G	COLL	0140
.=	Asbestos en el Area Urbana de la Ciudad de Mexico?	ENVR	0025
loning and Sequence Analysis of	AscB and AscC, the Genes Coding for Yersinia pseud	BIOL	0110
Sequence Analysis of AscB and	AscC, the Genes Coding for Yersinia pseudotuberculo	BIOL	0110
Accumulation of Elements in the	Ascidiacea.= Marine Bioinorganic Chemistry.	INOR	0095
S Trace Metals Analysis of Fly	Ash by Inductively Coupled Plasma Atomic Emission	FUEL	0138
remium Coal Sample Using Wood	Ash or Taconite as Additive.= +minous Argonne P	FUEL	0006
Long Cure Times in Cement/Fly	Ash Solidification/Stabilization Can Result in Signif	ENVR	0041
	Ash Utilization and Disposal.=	FUEL	0135
an Energy from Waste and Coal:	Ash Utilization, Characterization, and Other Topics.	FUEL	0136
g Municipal Solid Waste Incinator	Ash.= +gth Portland Cement Concrete Containin	FUEL	0136
Regulation of the	Asialoglycoprotein Receptor by cGMP.=	BIOL	0154
of Polyphenolic Compounds of	Asian Origin.= Chemistry and Biological Activities	AGFD	0046
ymers Containing Photoresponsive	Asobenzene–Based Sidegroups.= +nomers and Pol	POLY	0209
Molecular Structure of	Asparagine by the Pseudospectral Gradient Method.=	PHYS	0275
Potent Glycine Site N–Methyl–D–	Aspartate Antagonists.= +renic Acid Analogues as	MEDI	0146
Regulatory Communication in	Aspartate Transcarbamolylase.=	BIOT	0242
Sulfonation of Sulfite–Pretreated	Aspen Treated with Vapor at High Temperature.=	PMSE	0170
Production of Bovine Chymosin in	Aspergillus niger. +i: Recent Improvements in	BIOT	0234
Esterase from	Aspergillus niger.= Production of Acetyl–Xylan	BIOT	0251

anol: A Novel Sestererpenoid from	Aspergillus variecolor.= +ereochemistry of Varicul	ORGN	0184
aste on the Thermal Properties of	Asphalt.= +Terephthalate (PET) from Plastics W	FUEL	0053
easure+ Adsorption Kinetics of	Asphaltenes—A Liquid/Water Interfacial Tension M	PETR	0138
3-Et+ Rapid and Simple HPLC	Assay for the Novel Anticonvulsant D,L-3-Hydroxy-	ANYL	0082
Improved Extraction and HPLC	Assay of Potato Glycoalkaloids.=	AGFD	0112
B-Protein	Assay.= Cancer Management: Versatility of the	BIOL	0069
n in Vitro Developmental Toxicity	Assay.= +ce on Metabolic Profile of Biphenyl in a	AGRO	0021
sting of Charm Test II Receptor	Assays for the Detection of Antimicrobial Residues in	AGFD	0085
Molecules by Immuno or Binding	Assays.= +and Immunogenicity of Antibody-Like	BIOT	0059
ms Based on Competitive Binding	Assays.= +C Postcolumn Reaction Detection Syste	ANYL	0090
and Fluorescence Studies of Self-	Assembled Siloxane Mono- and Multilayers.= +R	POLY	0330
Thin-Fil+ Chromophoric Self-	Assembled Superlattices: Multilayer Construction of	POLY	0127
Macromolecular Chromophoric	Assemblies with Enforced Polarity: Inorganic and He	POLY	0215
g Blocks of Future Biosen+ Self-	Assembling Lipid Bilayers on Solid Support: Buildin	BIOT	0102
nd Spheres: The New Generati+	Assembly of Carbon Atoms into Rods, Rings, Nets, a	ORGN	0266
Kinetics of	Assembly of Iron-Sulfur Clusters in Micellar Media.=	INOR	0076
cal Cofactor of+ Mechanism of	Assembly of the Dinuclear Iron Center: Tyrosyl Radi	BIOL	0022
-Size Tinkertoy Construction Set:	Assembly of the Rods and Spools.= +a Molecular	BIOT	0200
Self-	Assembly.= Pattern Recognition through	BIOT	0125
Progress on the Folding,	Assembly, and Amyloidosis of Transthyretin.=	BIOL	0007
graphical Information System To	Assess the Susceptibility of Surface Drinking Water	AGRO	0053
t Migration in Ground Water as	Assessed by EPA's Hazard-Ranking System: A Balan	ENVR	0081
al Competition among Polymers.	Assessing Cost and Performance in Intermaterials Co	CMEC	0009
ring New Ene+ Methodology for	Assessing the Feasibility of Developing and Transfer	FUEL	0142
of Cu+ Absence of Nucleophilic	Assistance by Solvent and Azide Ion to the Reaction	ORGN	0154
and Carbonyl Ligands by Nucl+	Assistance of Photo-Induced Coupling of Alkylidyne	INOR	0188
for the	Assistant Professor.= NIH Funding Opportunities	YCC	0004
J. J. Keller and	Associates: Publications and Products.=	CHAS	0029
ermination of Organic/Inorganic	Associations of Trace Elements in Oil Shale Kerogen	FUEL	0045
ures.=	Associative Polymers with Novel Hydrophobe Struct	PMSE	0035
Behavior of	Associative Thickeners.= Complexities in Evaluating	PMSE	0084
Quality	Assurance in Fertilizer and Pesticide Formulations.=	FERT	0009
Roots of Stevia salicifolia cav. (	Asteraceae).= New Ent-Atisene Glycosides from the	MEDI	0034
tructural Information I+ Recent	ASTM Standardization Developments for Chemical S	CINF	0038
Tannin and Relative	Astringency in Teas.= An Evaluation of Total	AGFD	0153
ne Fischer Carbene Complexes.+	Asymmetric Aldol Reactions of Chelated Imidazolido	INOR	0112
n of Chiral β-Lactam Ester Eno+	Asymmetric Alkylation, Acylation, and Aldol Reactio	ORGN	0046
Phosphines for	Asymmetric Catalysis.= New Electron-Rich Chiral	CATL	0004
Mechanistic Aspects of	Asymmetric Catalytic Hydrogenation.=	CATL	0003
New Class of D4-Sy+ Catalytic	Asymmetric Epoxidation of Simple Alkenes Using a	INOR	0300
Catalytic	Asymmetric Heck Reaction.=	CATL	0006
New Synthesis of	Asymmetric Hydrogenation Catalysts.=	INOR	0077
lexes and Their Roles in Catalytic	Asymmetric Hydrogenation of β-Keto Esters.= +p	ORGN	0044
ity in Catalysis: Stereoselectivity.	Asymmetric Hydrogenation Technology for the Prod	CATL	0001
ute to Ciral Alcohols Using Disc+	Asymmetric Hydrosilylation of Ketones: A Facile Ro	INOR	0247
hodium Phosphine Complexes for	Asymmetric Hydrosilylation of Ketones.= +hiral R	CATL	0005
zacrown Ethers.=	Asymmetric Michael Addition Catalyzed by Chiral A	ORGN	0209
Plenary Session. Studies in	Asymmetric Olefination and Ene Reactions.=	ORGN	0279
Mechanistic Study of New	Asymmetric Oxidation Catalysts.= Development and	ORGN	0132
cal Manganese-Tetraphenylpor+	Asymmetric Oxidations Catalyzed by a D4-Symmetri	ORGN	0045
	Asymmetric Reductions with Chiral NADH Models.=	ORGN	0264
of Practical	Asymmetric Syntheses.= Design and Development	ORGN	0297
yclization Reactions of Alkynes:	Asymmetric Synthesis and Pharmacological Studies o	ORGN	0275
oxaziridines.= Plenary Session.	Asymmetric Synthesis of α-Hydroxy Carbonyl Comp	ORGN	0293
	Asymmetric Synthesis of Calyculin A.=	ORGN	0289
oselectivity by Intramolecular C+	Asymmetric Synthesis of Lactones with High Enanti	ORGN	0042
elective Competitive NMDA A+	Asymmetric Synthesis of NPC 12626: A Potent and S	MEDI	0083
ositol (1,3,4,5)Tetrakisphospha+	Asymmetric Synthesis of P=1 Tethered Analogs of In	AGRO	0045
α-Hyd+ Novel Approach to the	Asymmetric Synthesis of Peptide Fragments Bearing	ORGN	0048
rahydropresecamine, and+ Total	Asymmetric Synthesis of 15,20-Dihydrosecodine, Tet	ORGN	0009
β-Lactams as Potential Precurs+	Asymmetric Synthesis of 3-Hydroxy-4-Substituted-	ORGN	0047
	Asymmetric Synthesis with 3-Fluoroazetidinone.=	ORGN	0081
: A Building Block Strategy for	Asymmetric Synthesis.= Enzymatic Aldol Additions	BIOT	0090
adrupole Coupling Constants and	Asymmetry Parameters in Bridging Metal Hydrides.	INOR	0382
Gelation of	Atactic Polystyrene in Carbon Disulfide.=	POLY	0148
ndanones+ Synthetic Studies of	Atipamezole Analogs: Fluorination of 5-Substituted I	ORGN	0078
a cav. (Asteraceae).= New Ent-	Atisene Glycosides from the Roots of Stevia salicifoli	MEDI	0034
r Slope Sediments of the Middle	Atlantic Bight.= Carbon and C-14 Budget for Uppe	GEOC	0016
the Middle	Atlantic Bight.= Seasonal Dissolved Gas Cycling in	GEOC	0003
nental Margin and the Open N.W.	Atlantic.= +Reactive Materials between the Conti	GEOC	0009
. New Developments in Modified	Atmosphere Packaging in Fruits and Vegetables.=	AGFD	0075
Modeling of a Modified-	Atmosphere Packaging System.=	AGFD	0077
t in Fruit Cells.= Precursor	Atmosphere Technology: Efficient Aroma Enrichmen	AGFD	0023
cules of Relevance in the Earth's	Atmosphere: Experimental Studies of the Spectrosco	PHYS	0017
ynamic Simulation Model of the	Atmospheric Deposition of Semivolatile Organic Com	GEOC	0004
iphatic Aldehydes and Ketones of	Atmospheric Interest.= +tion Spectra of Several Al	ENVR	0024
d.=	Atmospheric Nitrogen Oxides: A Bridesmaid Revisite	ENVR	0071
tile Organic Compounds to Model	Atmospheric Particulate Materials.= +of Semivola	ENVR	0059
n to Dru+ Electrospray/Ionspray	Atmospheric Pressure Ionization LC/MS: Applicatio	ANYL	0124
Developments in CE/MS Utilizing	Atmospheric Pressure Ionization.= +—II. Recent	ANYL	0138

Sulfido Dimers Formed by Sulfur	Atom Abstraction from SPMe$_3$.= +ybdenum(III)–μ–	INOR	0271
Desulfovi+ New Method for Iron	Atom Active–Site Replacement by a Nickel Atom in	INOR	0020
Ranges: Mea+ Kinetics of Metal	Atom and Radical Reactions over Wide Temperture	PHYS	0087
cations of Continuous–Flow Fast	Atom Bombardment Mass Spectroscopy as a Detecto	ANYL	0125
e–Site Replacement by a Nickel	Atom in Desulfovibrio Vulgaris (Hildenborough) Rub	INOR	0020
Complexes with Two– and Three–	Atom Intra–Annular Tethers.= +stry of Bimetallic	INOR	0149
of 3– and 4–	Atom Molecules.= Variational Rovibrational Studies	PHYS	0030
Deposition of Copper Films by H–	Atom Reaction with Cu(FOD)$_2$ and Cu(HFA)$_2$.=	PHYS	0259
Rhenium Complexes.= Oxygen	Atom Transfer between Molybdenum, Tungsten, and	INOR	0132
nometallic Compounds. Oxygen–	Atom Transfer from N$_2$O to Alkynes Coordinated to	INOR	0327
y Su+ Comparison of Atom–by–	Atom 3D Search Routines for Searching CAS Registr	CINF	0023
tures: Optimizing the Overlap of	Atom–Based Properties, in Particular the Steric and	CINF	0018
AS Registry Sub+ Comparison of	Atom–by–Atom 3D Search Routines for Searching C	CINF	0023
Implementation and Use of an	Atom–Mapping Procedure for Similarity Searching in	COMP	0043
with the	Atom–Probe FIM.= Atomic Scale Surface Analysis	PHYS	0101
romatography/Graphite Furnace	Atomic Absorption Spectrometry for the Determinat	ANYL	0080
ogenation of Diamond(100) Using	Atomic Beams.= +Diamond (and c–BN, SiC). Hal	PHYS	0176
nducto+ NMR Measurements of	Atomic Cluster Distributions in Pseudobinary Semico	PHYS	0375
Automatic Generation of 3D	Atomic Coordinates for Organic Molecules.=	CINF	0036
phy of Crudes and Residua Using	Atomic Emission Detection.= +re Gas Chromatogra	PETR	0121
sh by Inductively Coupled Plasma	Atomic Emission Spectroscopy.= +nalysis of Fly A	FUEL	0138
Related Materials.=	Atomic Force Microscope Studies of Biominerals and	INOR	0117
Deposition Processes.=	Atomic Force Microscope Studies of Underpotential	PHYS	0128
Scanning Tunneling and	Atomic Force Microscopy of Biological Surfaces.=	PHYS	0158
	Atomic Force Microscopy of Molecules on Surfaces.=	PHYS	0153
ylene.=	Atomic Force Microscopy on Cold–Extruded Polyeth	PMSE	0078
ar at the Nanometer Scale.=	Atomic Force Microscopy Studies of Friction and We	PHYS	0156
Forces of Materials Using	Atomic Force Microscopy.= Probing the Surface	PHYS	0154
Ion–Scanning Tunneling Micr+	Atomic Imaging: STM and Related Techniques. Field	PHYS	0182
ing of Individual Atoms and Cl+	Atomic Imaging: STM and Related Techniques. Imag	PHYS	0123
l Structure and Electronic Stat+	Atomic Imaging: STM and Related Techniques. Loca	PHYS	0206
–Temperature STM Imaging of+	Atomic Imaging: STM and Related Techniques. Low	PHYS	0360
ciples and Examples of Scannin+	Atomic Imaging: STM and Related Techniques. Prin	PHYS	0152
oretical Studies of Single Adsor+	Atomic Imaging: STM and Related Techniques. The	PHYS	0096
teractions—I. Structural Aspects.	Atomic Interactions between Proteins and Carbohyd	BIOL	0142
Surface Chemistry of GaAs	Atomic Layer Epitaxy.=	COLL	0069
ics of Diffusion and Corrosion at	Atomic Layer Resolution Studied by in Situ Electroc	PHYS	0127
canning Tunneling Microscopy.+	Atomic Scale Imaging of Molecules on Surfaces by S	PHYS	0159
FIM.=	Atomic Scale Surface Analysis with the Atom–Probe	PHYS	0101
Local	Atomic Structure of Defects on Oxide Surfaces.=	PHYS	0207
Processes at Surfaces Using	Atomic–Level Imaging.= Studies of Kinetic	PHYS	0098
on of Platinum from Pt (PF$_3$)$_4$ on	Atomically Clean Platinum Surfaces.= +he Depositi	COLL	0046
Polymer Samples for	Atomistic Modeling.= Preparation of Dense	POLY	0366
Quantum Chemical Potential +	Atomistic Simulation of Polymer Materials. Ab Initio	POLY	0350
the Short–Range Correlations in+	Atomistic Simulation of Polymer Materials. Sizes of	POLY	0365
otential Energy Functions for the	Atomistic Simulation of Polymeric Materials.= +P	POLY	0350
omatic Hydrocarbons with O(^{3}P)	Atoms and CH$_2$(x^{3}B$_1$) Radicals: Comparative Studies	FUEL	0082
Techniques. Imaging of Individual	Atoms and Clusters with the FIM.= +nd Related	PHYS	0123
c Substitutions at Nonstereogenic	Atoms by the Endocyclic Restriction Test.= +phili	ORGN	0120
Applicat+ Gas–Phase Aluminum	Atoms from Pulsed Laser Ablation: Mechanisms and	PHYS	0353
drocarbons of Transition–Metal	Atoms in Different Charge States: M, M+, and M^{2+}.=	PHYS	0199
Deposition.= Reactions of Metal	Atoms in Plasma Etching and Chemical Vapor	PHYS	0143
retical Studies of Single Adsorbed	Atoms in the STM.= +Related Techniques. Theo	PHYS	0096
	Atoms in Unusual Environments.=	INOR	0292
Phosphorus			
Generati+ Assembly of Carbon	Atoms into Rods, Rings, Nets, and Spheres: The New	ORGN	0266
Sulfur Formed by Coalescence of	Atoms on Re(0001) Surfaces.= +ensional Phases of	PHYS	0124
le Diode Laser Probe of Chlorine	Atoms Produced from the Photodissociation of S$_2$Cl$_2$	PHYS	0299
on during the Reaction of Oxygen	Atoms with Acetylene at 300 K.= +etylene Formati	FUEL	0065
f Electronically Excited Cadmium	Atoms with H$_2$ and CH$_4$.= +es of the Interactions o	PHYS	0351
tivity in the Reaction of Copper	Atoms with Halogen Compounds: Ion–Core Conserva	PHYS	0368
n Reactions of Transition–Metal	Atoms with Simple Molecules near Room Temperatu	PHYS	0198
and Aluminum	Atoms with Small Molecules.= Reactions of Boron	PHYS	0090
les Prepared from Solvated Metal	Atoms: Structure and Catalysis.= +imetallic Partic	CATL	0009
Sodium, Magnesium, and Copper	Atoms.= +of the Molecular Oxygen Reaction with	PHYS	0141
nd Electronically Excited Barium	Atoms.= +ctions—VI. Reactions of Ground State a	PHYS	0349
ng Reactions of Excited Rare Gas	Atoms.= +–Phase Metal Reactions—VII. Harpooni	PHYS	0367
ns of Alkali and Alkaline Earth	Atoms.= Comparison between the Oxidation Reactio	PHYS	0072
horus, Nitrogen, and Sulphur(VI)	Atoms.= +Inorganic Polymers with Skeletal Phosp	POLY	0186
e to Collisions with Hot Hydrogen	Atoms.= +re Rotational Scattering of CO$_2$/CO du	PHYS	0291
C	Atoms.= Reactive Collisions of Al, Mg, Si, and	PHYS	0089
of Small Molecules to Metal	Atoms.= Role of Cooperative Effects in the Binding	PHYS	0172
rs Containing One or More Boron	Atoms.= +Ruthenium and Osmium Carbonyl Cluste	INOR	0220
–Limiting for Skeletal Actomyosin	ATP Hydrolysis.= +e Bond–Splitting Step Is Rate	BIOL	0141
GTPase–Mediated Activation of	ATP Sulfurylase.=	BIOL	0062
by a Proton+ A Model for H+–	ATPase: Formation of Citraconyl Phosphate Driven	BIOL	0104
Variable–Angle	ATR.= Depth Profiling Using	POLY	0307
Poly(ethyleneterephthalate) by	ATR–IR.= Analysis of Near Surface Structure in	POLY	0279
.=	Atraniums: Novel Five–Coordinate Group 14 Cations	INOR	0023
Microcapsules of the Herbicides	Atrazine and Metribuzin: Preparation and Evaluation	AGRO	0063
Determination of	Atrazine Metabolites in Human Urine.=	AGRO	0032

aduate Experiment Introducing+	Attaining Optimal Conditions: An Advanced Undergr	CHED	0182
Aminosugar	Attenuation of HIV Infection.=	AGFD	0193
t in the 21s+ Public Policies and	Attitudes Will Affect Pesticide Use and Developmen	AGRO	0070
nalytic Model of Science–Related	Attitudes.= +mistry Enrollments: Toward a Path A	CHED	0177
Schizophrenia.= P-9236: A New	Atypical Antipsychotic Agent for the Treatment of	MEDI	0085
Selective Surface Derivatization of	Au and Pt Microfabricated Wire Arrays.= +s for	INOR	0350
+Polynuclear Selenide Clusters of	Au^+ and In^{3+}, Trapping Na^+ Ions in the Center.=	INOR	0203
Extremely Rapid	Au(I) Transfer Reactions between S–Donor Ligands.=	INOR	0235
tudies of the Chemical–Vapor +	Auger and Scanning Tunneling Microscopy (STM) S	COLL	0050
eacidifying Sour Orange (Citrus	aurantium) Juice Using Neutral and Anion Exchange	AGFD	0106
rtir de Leche Contaminada con S.	aureus y su Aplicacion en Quesos.= +de Sodio a Pa	AGFD	0107
e Dorsal Abdominal Glands of the	Australian Spined Citrus Bug.= +miacetal from th	AGRO	0035
pionate Metabolites of Siphonaria	Australis.= +nd Absolute Configuration of Polypro	ORGN	0003
cetate, Ethyl Acetate, Ethy+ The	Authentication and Source Differentiation of Amyl A	AGFD	0187
ttern–Recognition Techniques for	Authentication of Extracts and Essential Oils.= +a	AGFD	0198
logical Fluids Using Laboratory+	Automated Analysis of Drugs and Metabolites in Bio	ANYL	0119
	Automated Conformational Analysis.=	CINF	0035
	Automated Dynamic Interfacial Tension Technique.=	COLL	0091
IZARD III.=	Automated Learning in Conformational Analysis—W	CINF	0017
tructures: Optimizing the Over+	Automated Method for the Alignment of Molecular S	CINF	0018
	Automated Structure Design in 3D.=	COMP	0048
Experiences with	Automated 3D Design of Bioactive Molecules.=	COMP	0049
Organic Molecules.=	Automatic Generation of 3D Atomic Coordinates for	CINF	0036
erial Handling Equipment for the	Automatic Sortation of Postconsumer Plastics.= +t	I&EC	0041
Computer	Automation in the Laboratory.=	TECH	0006
tile–Structured Preforms—Part +	Automation of Equipment for Manufacturing of Tex	TECH	0007
tile–Structured Preforms—Part +	Automation of Equipment for Manufacturing of Tex	TECH	0008
ermination of Surface Tension by	Automation of the Drop–Weight Method.= +—Det	CHED	0118
xhaust Pollutant Emissions from	Automobile Engines Using Alcohols, Gasohol, and Ot	FUEL	0095
mical Properties of Pt–Containing	Automotive Emission Control Catalysts.= +sicoche	COLL	0208
ctions over a Typical Commercial	Automotive Exhaust Catalyst.= +and CO–NO Rea	COLL	0187
l Competition.= The Changing	Automotive Industry and Its Impact on Intermateria	CMEC	0015
e by Computer: Experiences with	AUTONOM 1.0.= +to a Systematic Chemical Nam	COMP	0019
dopteran Manduca sexta.=	Autoparalytic Peptides from Hemolymph of the Lepi	AGRO	0012
f Histidine/Carnosine on Fe(II)	Autoxidation and Enzymic or Nonenzymic Reduction	AGFD	0131
Food.= Factors Affecting Lipid	Autoxidation of a Spray–Dried Milkbase for Baby	AGFD	0158
Organophosphorus	Auxiliaries.= New Synthetic Methodology Based on	INOR	0377
Complexes Containing the Chiral	Auxiliary (η^5-C_5H_5)Fe(CO)(PPh$_3$): Synthesis and Ch	INOR	0156
Multivariable Controlled	Auxostat.= Experimental Verification of a	BIOT	0158
Culture Instability of	Auxotrophic Amino Acid Producers.=	BIOT	0031
ydrodynamic Stress and Oxygen	Availability on Ravidomycin Production by an Actin	BIOT	0162
one and the Disaccharide.=	Avermectin Analogs with a Spacer between the Aglyc	ORGN	0016
Beetle.= Mechanisms of	Avermectin Resistance in the Colorado Potato	AGRO	0154
HPLC Retention Times for Novel	Avermectins Using Calculated Lipophilicities.=	BIOT	0127
ation of the Al Sitting and Pairing	Avoidance Principles in Zeolites.= +rectical Found	PHYS	0289
Chymosin in Aspergillus niger var.	awamori.= +rovements in Production of Bovine	BIOT	0234
osium Honoring R. W. Ramette.	Award Address (ACS Division of Analytical Chemistr	ANYL	0006
w Face of Analytical Chemistry.	Award Address (ACS Division of Analytical Chemistr	ANYL	0011
y Award in Electrochemistry, s+	Award Address (ACS Division of Analytical Chemistr	ANYL	0096
w Face of Analytical Chemistry.	Award Address (ACS Division of Analytical Chemistr	ANYL	0001
red by Bristol–Myers Squibb). +	Award Address (Edward E. Smissman Award, sponso	MEDI	0089
mistry, sponsored by Dexter C+	Award Address. (Dexter Award in the History of Che	HIST	0011
tic Processes Honoring J. F. Roth.	Award Address. +stry Award New Industrial Cataly	I&EC	0043
s and Peptides for Insect Control.	Award Address. +rd Symposium: Juvenile Hormone	AGRO	0001
ernment/Farm Regulatory Issues.	Award Address. +—Advances in Research and in Gov	AGRO	0051
Smissman—Bristol–Myers Squibb	Award Address. Hydanto+ General and Edward E.	MEDI	0082
ensions in Chemical Information.	Award Address. +Honoring W. T. Wipke: New Dim	CINF	0016
Division of Analytical Chemistry	Award for Excellence in Teaching, cosponsored by E	ANYL	0006
Division of Analytical Chemistry	Award in Chemical Instrumentation, sponsored by th	ANYL	0011
Division of Analytical Chemistry	Award in Electrochemistry, sponsored by EG G Prin	ANYL	0096
ay. Personal Recollecti+ Dexter	Award in History of Chemistry, Honoring O. Hannaw	HIST	0009
ivision of Analytical Chemistry	Award in Spectrochemical Analysis, sponsored by the	ANYL	0001
xter Chemical Corp.). +s. (Dexter	Award in the History of Chemistry, sponsored by De	HIST	0011
J. F. Roth+ Industrial Chemistry	Award New Industrial Catalytic Processes Honoring	I&EC	0043
J. F. Roth+ Industrial Chemistry	Award New Industrial Catalytic Processes Honoring	I&EC	0052
J. F. Roth+ Industrial Chemistry	Award New Industrial Catalytic Processes Honoring	I&EC	0060
Shelf–Life+ Graduate Student	Award Presentation. Chemometric Approach to Milk	AGFD	0108
dying Seed+ Graduate Student	Award Presentation. Noninvasive Techniques for Stu	AGFD	0104
OLYED 1991 Award Winner—	Award Presentation: Addressing Polymers as Macrom	CHED	0027
into Middle, Vo+ POLYED 1990	Award Presentation: Integrating Polymer Education	CHED	0026
Precollege Chem+ POLYED 1990	Award Presentation: Introducing Macromolecules in	CHED	0025
nent Solubility P+ Roy W. Tess	Award Symposium Honoring K. L. Hoy. Four Compo	PMSE	0033
Address+ Excellence in Teaching	Award Symposium Honoring R. W. Ramette. Award	ANYL	0006
ensions in C+ Herman Skolnik	Award Symposium Honoring W. T. Wipke: New Dim	CINF	0016
vel Creative Polymer Chemistry	Award Symposium Polymers—Chemistry, Physics, an	POLY	0272
oly+ Sherwin–Williams Student	Award Symposium. Change of Mechanism Block Cop	PMSE	0071
Radical Stabilities.= Cope	Award Symposium. Effects of Structural Changes on	ORGN	0133
arvel Creative Polymer Chemistry	Award Symposium. Spontaneous Adsorpt+ C. S. M	POLY	0298
Expanded Porphyrins.= Cope	Award Symposium. Synthesis and Properties of	ORGN	0139
for+ Baxter, Burdick, Jackson	Award Symposium: Juvenile Hormones and Peptides	AGRO	0001

for+ Baxter, Burdick, Jackson	Award Symposium: Juvenile Hormones and Peptides	AGRO	0006
nteractions of+ Electrochemistry	Award Symposium—I, Honoring R. A. Osteryoung. I	ANYL	0093
eutral Specie+ Electrochemistry	Award Symposium—II. Reversible and Irreversible N	ANYL	0103
POLYED 1989 Polymer	Award Winner: Polymers Are Elementary!.=	CHED	0028
ymers as Macr+ POLYED 1991	Award Winner—Award Presentation: Addressing Pol	CHED	0027
d Address. (Industrial Chemistry	Award). Some Adventures and Innovations in Indust	I&EC	0043
SQUALOR—The Design of an	Award-Winning Laboratory Simulation.=	CHED	0079
rd Address (Edward E. Smissman	Award, sponsored by Bristol-Myers Squibb).+ Awa	MEDI	0089
Whole Plants.= Student	Awards Symposium. Fate of ^{14}C-Trichloroethylene in	ENVR	0056
d-Phase Sample Displacement +	Axial Compression Multicolumn Preparative Reverse	ANYL	0040
Human	Axillary Odors.= Precursors to the Characteristic	AGFD	0147
o[1,5-a]quinoxalinones, and Their	Aza Analogues.= +o[1,2-a]-quinoxalinones, Imidaz	MEDI	0118
'-Diones and Their Antibac+ Bis	Aza Heterocycles: Synthesis of 3,3'-bisquinazolin-4,4	MEDI	0057
Biological Evaluation of 8-	Aza-1-Deazapurine Nucleosides.= Synthesis and	MEDI	0029
Selection in the Reduction of 5-	Azaadamantan-2-one.= Electronic Control of Face	ORGN	0237
onists and/or 5-HT3 Ant+ New	Azabicyclic Benzamides: Potent Serotonin 5-HT4 Ag	MEDI	0069
Catalyzed by Chiral	Azacrown Ethers.= Asymmetric Michael Addition	ORGN	0209
ocyclic Nucleoside Analogue of 5-	Azacytidine.= +al Study of the Cyclopentenyl Carb	MEDI	0048
of an N-Acyl-1-	Azadiene.= Solid- and Liquid-State Photochemistry	ORGN	0111
α-Cyano-1-	Azadienes.= The Diels-Alder Reaction of	ORGN	0234
Activities of Novel	Azatricyclic Agents.= Serotonergic 5HT3 and 5HT4	MEDI	0187
es for the Construction of Novel	Azatricyclic Systems: Novel 5-HT3 and 5-HT4 Serot	ORGN	0104
Facile Synthesis of	Azetidine-2,4-Diones.=	ORGN	0034
for the Production of Chiral	Azetidinones.= Development of an Efficient Process	ORGN	0206
philic Assistance by Solvent and	Azide Ion to the Reaction of Cumyl Drivatives: Mech	ORGN	0154
ecomposition and Isomerization of	Azides and Isocyanates.= +nt. Dynamics of the D	PHYS	0188
Chiral	Azides.= The First Kinetic Resolution of	ORGN	0043
f Ribonucleotide Reductase by 2'-	Azido 2'-Deoxynucleotides.= +sm of Inactivation o	ORGN	0281
etal Salen Complexes B+ Olefin	Aziridination and Epoxidation Catalyzed by Chiral M	ORGN	0056
Thermal Transformations of	Aziridine on the Surface of Pyrogenous Silica.=	COLL	0113
-Epoxy Aldehydes, Thiiranes, and	Aziridines.= +nthesis of Enantiomerically Pure 1,2	ORGN	0052
of Ring-Opening of	Aziridinones by Nucleophiles.= Theoretical Study	PHYS	0314
ation, and New Binding Model of	Azole-Type Aromatase Inhibitors.= +ological Evalu	MEDI	0035
Substituted	Azomethine Ylides.= Hartree Fock Description of	ORGN	0171
es and C-Nucleosides, such as	AZT.= Synthetic Approaches to Modified Nucleosid	MEDI	0049
lar Organization: Naphthalene vs.	Azulene on Pt(111).= +al-Space Imaging of Molecu	PHYS	0125
Photophysics of	Azulene.= Photionization via Transient States: The	PHYS	0330
Stable Isotope Geochemistry of	Azusfres Systems.=	I&EC	0015
es en Geotermica de Fasico de los	Azusfres.= +rtamiento de los Gazas en Condensabl	I&EC	0013
d Inhibitors on+ Direct Solution	^{11}B NMR Observation of Tightly Bound Boronic Aci	BIOL	0082
Anomaly in	Ba$_2$MnCu$_3$O$_x$.= X-ray Diffraction and Magnetic	INOR	0257
of a Spray-Dried Milkbase for	Baby Food.= Factors Affecting Lipid Autoxidation	AGFD	0158
Possible To Escape from the Tar	Baby?.= Cyanomethylene and Benzyne: Is It	PHYS	0032
and Teaching.=	Bachelor's Degree Program at Berkeley in Chemistry	CHED	0069
anotransferase by an Alkalophil+	Bacillus sp.= Production of Cyclomaltodextrin Gluc	BIOT	0157
gents. Mechanism of Resistance to	Bacillus thuringiensis.= +Bioengineered Control A	AGRO	0153
Pendulum Swinging	Back?.= Testimony of Scientific Experts: Is the	CHAL	0016
l Moiety Directly Attached to the	Backbone at the P1 Site.= +ibitors with Cyclohexy	MEDI	0121
kylenedioxydiphenols—Effect of	Backbone Flexibility and Cross-Linking on Flexibilit	PMSE	0162
elaxation in σ-Delocalized Silicon	Backbone Network Polymers.= +on and Vibronic R	PHYS	0138
of Polymer Anions with Si and Ge	Backbones.= +Ge$_2$H$_6$ Radical Anions as Protoypes	PHYS	0313
Translocation of Gram-Negative	Bacteria and High Mortality in CD1 Immunodefici	AGFD	0146
Antitumor Drugs by Recombinant	Bacteria.= +tibiotic Production—I. Production of	BIOT	0108
Magnetotactic	Bacteria.= Inorganic Particles in	INOR	0118
Marine	Bacteria.= Novel Marine Siderophores from	INOR	0119
Biodegradable Plastics Made by	Bacteria.= Poly(β-hydroxyalkanoates):	CHED	0132
Hyperthermophilic	Bacteria.= Role of Tungsten in the Metabolism of	INOR	0099
harides of Salmonella and Shigella	Bacteria.= +Chains of Cell Envelope Lipopolysacc	CARB	0017
s as a Result of Sulfate-Reducing	Bacteria.= +s in Copper Sulfide Corrosion Product	BIOT	0100
y the Intracellular Expression of a	Bacterial Hemoglobin.= +Yields in Streptomyces b	BIOT	0087
Design Considerations.=	Bacterial or Animal Cell Fermentation?—Process and	BIOT	0076
Optical Spectroscopy of the	Bacterial Photosynthetic Reaction Center.=	PHYS	0170
etting Properties and Adhesion of	Bacterial Surface Protein Layers.= +dity on the W	BIOT	0151
Photosynthetic System in Nature,	Bacteriorhodopsin (bR).= +ic Energy by the Other	PHYS	0019
	Bacteriorhodopsin as an Intelligent Material.=	BIOT	0052
and Four-Wave Mixing in	Bacteriorhodopsin Materials.= Optical Bistability	BIOT	0081
rocessing.=	Bacteriorhodopsin Variants for Optical Information P	BIOT	0078
s and Optical Memories Based on	Bacteriorhodopsin.= +ics. Spatial Light Modulator	BIOT	0077
Memory Based on	Bacteriorhodopsin.= Semiconstant Holographic	BIOT	0079
Biochrom Films Based on 4-Keto-	Bacteriorhodopsin.= +ectrochromic Properties of	BIOT	0082
What Is the Future of	Baculovirus for Insect Control?.=	AGRO	0081
electi+ Improving the Efficacy of	Baculovirus Insecticides by Expressing with Insect S	AGRO	0009
eding in Academia: The Good, the	Bad, and the Ugly.= +n Academic Program. Succe	YCC	0001
on in Agricultural Product	Bag Burns.= Laboratory Evaluation of PIC Formati	AGRO	0140
Pesticide-	Bag Open Burning.= Emission Characterization of	AGRO	0141
PA's Hazard-Ranking System: A	Balance of Risk Assessment and Data Constraints.=	ENVR	0081
eedom of Speech—The Delicate	Balance: Defense Research in a Democratic Society.=	PROF	0009
Materia.=	BALANCE: Simulador de Procesos para Balances de	COMP	0052
Procesos para	Balances de Materia.= BALANCE: Simulador de	COMP	0052
The	Baltimore Fiasco: A Case Study in Fraud.=	PROF	0005

Left context	Title fragment	Section	Page
Synthesis and Characterization of	BaNb₇P₆O₃₃.= +obium Oxophosphate Compound:	INOR	0031
Charge Injection in Large–	Bandgap Semiconductor Colloids.= Sensitized	COLL	0149
or	Bane?.= Intermaterial Competition—Boon	CMEC	0012
The Protein Data	Bank—Present Status and Future Plans.=	CINF	0031
for U.S. Agriculture.= Potential	Bans of Selected Pesticides: Economic Implications	AGRO	0083
d State and Electronically Excited	Barium Atoms.= +ctions—VI. Reactions of Groun	PHYS	0349
Novel	Barium Compounds Useful for OMCVD.=	COLL	0158
River Lag+ Pesticide Residue in	Barrier Island Salt Marshes along the Florida Indian	ENVR	0019
om Polymeric Gas Separation and	Barrier Materials.= +alysis of X-ray Scattering fr	POLY	0142
ings Systems—I. Plenary Lecture.	Barrier Polymers for Container Coatings.= +Coat	PMSE	0138
Improving	Barrier Properties through Surface Energetics.=	POLY	0196
s of Electron–Transfer Activation	Barriers at Colloidal Titanium Dioxide Surfaces.=	COLL	0132
Theory of Schottky	Barriers in Diamond/Metal Interfaces.=	PHYS	0358
ce Potential Effect on Activation	Barriers of Unimolecular Transformations of Organic	COLL	0107
Study.=	Barriers to Rotation in Oximes: A Molecular Orbital	PHYS	0324
guimiento de las Reacciones Acido	Base en Medio Acuoso.= +y 316 Utilizadas en el Se	ANYL	0055
is(2–pyridyl)–2,5,8–Triazanona+	Base Hydrolysis of the Cation $\alpha\beta$–anti–Chloro–[1,9-B	INOR	0397
–Tetraaz+ Kinetic Study of the	Base Hydrolysis of the syn,anti–cis–Dichloro (1,4,7,10	INOR	0398
ydentate Amine Complex Toward	Base Hydrolysis.= +ry Reactive Chromium(III) Pol	INOR	0397
lication of Enantiopure Troger's	Base Methosulfate as an Inclusion Host: Resolution o	ORGN	0041
cture/Property Relationships in	Base Oil Processing: Monitoring Dewaxing Procedure	PETR	0151
eti+ Molecular Recognition via	Base Pairing: Noncovalent Approaches to Photosynth	ORGN	0059
Techniques for Fossil Fuels. Acid–	Base Properties of Coals and Coal Liquids.= +cal	FUEL	0039
Oligonucleotides by Mitomycin +	Base–Sequence Specificity of the Monoalkylation of	MEDI	0015
Properties of Ru(II) Schiff	Bases Complexes.= Synthesis and Electrochemical	INOR	0260
hysiology+ Topbiology: Molecular	Bases of Animal Form. G. Edelman (Medicine and P	CONG	0013
oxamic Acids and Their Conjugate	Bases.= +irical Molecular Orbital Studies of Hydr	PHYS	0326
ine and Related Long–Chain	Bases.= Inhibition of Protein Kinase C by Sphingos	MEDI	0075
ogenation of Precursor Poly(Schiff	Bases).= +ted Polybenzimidazoles by Cyclodehydr	POLY	0045
al(IV) Phosphates toward Organic	Bases, Metal Complexes, and Metal Oxides.= +Met	PETR	0107
	BASIC Biochemical Engineering.=	BIOT	0225
uccion con Zinc en Medio Acido y	Basico, Preparacion de Nitrado de Plata.= +dorred	CHED	0172
ted Aromatic Hydrocarbons in La	Basse Cote–Nord, Quebec.= +ng People to Chlorina	CEI	0011
Commingled	Batch Processing.=	I&EC	0026
n Optimally Designed Sequencing	Batch Reactors.= +ic Biodegradation of Phenolics i	BIOT	0138
le Hormones and Peptides for +	Baxter, Burdick, Jackson Award Symposium: Juveni	AGRO	0001
le Hormones and Peptides for +	Baxter, Burdick, Jackson Award Symposium: Juveni	AGRO	0006
nthesis of Enantiomerically Pure	Bay–Region 10,11–diol–8,9–Epoxide Diastereomers o	ORGN	0208
+	BC.= State–to–State Reaction Dynamics beyond A	PHYS	0162
y/Asymmetric Synthesis of (+)–	BCH–189 Using D–Mannose as a Chiral Template an	CARB	0022
Potential Energy Surfaces for	BCl Reactions.=	PHYS	0088
f Interphases between PMDA/4–	BDAF Polyimides and Silver Substrates Using Surfa	POLY	0087
on.= IPN–Modified Porous	Bead Polymers as Supports for Enzyme Immobilizati	PMSE	0081
Rings vs. Strings: Off–Lattice,	Bead–Stick Computer Simulations.=	POLY	0235
Poly(methylmethacrylate)/–Glass	Beads System.= +ical Properties of a Polystyrene/	PMSE	0167
Sensitivity Studies with Particle	Beam and Thermospray LC/MS.= Enhanced	ANYL	0123
Ion Reactions and RI–	Beam Experiments.= Intermediate–Energy Heavy	NUCL	0029
Adhesion.= Ion	Beam Induced Interfacial Chemical Complexes and	NUCL	0094
port, and Vaporization for Particle	Beam LC/MS.= +ts in Aerosol Generation, Trans	ANYL	0122
ombustion Studied by Molecular–	Beam Mass Spectrometry and Modeling.= +llant C	FUEL	0126
orption Photometry and Particle–	Beam Mass Spectrometry.= +graphy with UV–Abs	ENVR	0015
and Other Species by Molecular–	Beam Mass–Spectrometry during Hot–Filament–Ass	PHYS	0145
Molecular–	Beam Studies of Carbonaceous Clusters.=	FUEL	0110
Molecular	Beam Studies on Diamond Film Synthesis.=	PHYS	0203
nide Gas Surface R+ Molecular–	Beam Study of the Hydrogen Chloride–Gallium Arse	COLL	0215
poxide Resins.= Electron	Beam–Induced Cationic Polymerization of Silicone–E	PMSE	0159
: Composition Modif+ Use of Ion	Beams in the Analysis and Modification of Materials	NUCL	0091
: Elemental Analysis+ Use of Ion	Beams in the Analysis and Modification of Materials	NUCL	0077
: Elemental/Isotopi+ Use of Ion	Beams in the Analysis and Modification of Materials	NUCL	0064
: Laser Ionization M+ Use of Ion	Beams in the Analysis and Modification of Materials	NUCL	0041
s: Cluster Ion Impac+ Use of Ion	Beams in the Analysis and Modifications of Material	NUCL	0052
on of Diamond(100) Using Atomic	Beams.= +Diamond (and c–BN, SiC). Halogenati	PHYS	0176
aterials Using High–Energy Ion	Beams.= Composition and Structural Analysis of M	INOR	0008
c Synthesis of Peptide Fragments	Bearing α–Hydroxy–β–Amino Acid Residues.= +tri	ORGN	0048
by Chiral Metal Salen Complexes	Bearing Bulky Silyl Groups.= +xidation Catalyzed	ORGN	0056
acterization of Polyurethane Films	Bearing NLO–Active Chromophores.= +tical Char	POLY	0216
olecule+ Polymers of Geometrical	Beauty—Combs and Stars. Novel Dendritic Macrom	POLY	0318
Polymers of Geometrical	Beauty—Combs and Stars. Ring Polymers.=	POLY	0292
c Prop+ Polymers of Geometrical	Beauty—Combs and Stars. Some Static and Dynami	POLY	0265
Tink+ Polymers of Geometrical	Beauty—Combs and Stars. Toward a Molecular–Size	POLY	0344
Servicio Social y Programas de	Becas: Vinculo entre Universidades y Secotr Product	CHED	0114
	Beckman—Acidity Measurement.= Arnold	CHAL	0032
el–Fired Hybrid Fluidized–	Bed Boilers.= Commercial Application of Waste–Fu	FUEL	0122
U–Gas Fluidized–	Bed Gasification Process.= Fuel Evaluation for	FUEL	0120
Analytical Data.= Has a Release	Been Observed? HRS–Observed Releases and	ENVR	0080
Organic Coatings for	Beer, Beverage, and Food Metal Containers.=	PMSE	0140
on	Beet Sugar.= Large–Scale Production of PHB	CARB	0002
the Colorado Potato	Beetle.= Mechanisms of Avermectin Resistance in	AGRO	0154
Antipsychotic Drugs: II. Animal	Behavioral Testing of Novel Piperidine Derivatives.=	MEDI	0084
Matrices.= Inorganic Phase	Behaviors Mediated by the Organic Polymer	POLY	0364

Ble+ Thermally Induced Phase	Behaviors of Aromatic Polybenzimidazole/Polyimide	PMSE	0164
	Beilstein Current Facts on CD-ROM.=	COMP	0017
Beilstein Database. Status of the	Beilstein Database Activities.=	COMP	0009
Recent Developments of the	Beilstein Database on STN.=	COMP	0010
nts of Heterocyclic Organic Co+	Beilstein Database. Prediction of Normal Boiling Poi	COMP	0015
Activities.=	Beilstein Database. Status of the Beilstein Database	COMP	0009
Ring Index Searching of the	Beilstein Database.=	COMP	0018
Searches in the	Beilstein File.= Structure and Reaction Similarity	CINF	0039
cademic L+ Subsidized Access to	Beilstein Online: Options for Search Services in an A	COMP	0012
Using	Beilstein's Unique Problem-Solving Capabilities.=	COMP	0011
Data in the Formation of Rational	Belief.= +ting, Evidential Reason, and the Role of	ENVR	0075
os de Cupfer+ Hidroxilacion del	Benceno con H_2O_2 Catalizada por Complejos Derivad	COLL	0121
cesses—Or, How We Go from the	Bench Top to Commercial Reality.= +Forming Pro	PETR	0089
ine Dimer Formation on A-Tract	Bending and Its Biological Implications.= +of Thym	BIOL	0114
Project SEED	Benefits More Than High School Students.=	CHED	0145
+erships in Precollege Education.	Benefits of Alliances to the Educational Process.=	CHED	0001
Environmental	Benefits of Controlled-Release Nitrogen Sources.=	FERT	0006
Consumer View.= Risks and	Benefits of Pesticides: An Environmental and	AGRO	0059
Unanticipated	Benefits: Graphite Furnace AA.=	CHED	0184
Pesticides: Risks and	Benefits.=	AGRO	0058
enz-imidazol-2-YL] Carbamate (	Benomyl) in Hydrochloric Acid Solutions.= +-1H-B	AGRO	0041
riation of Diet in the Dominant	Benthic Community Component of a Tropical Coasta	BIOL	0084
datio+ Synthesis of Pyridines by	Bentonite-Supported Metallic Nitrates: Induced Oxi	ORGN	0245
ethyl [1-(Butylcarbamoyl)-1H-	Benz-imidazol-2-YL] Carbamate (Benomyl) in Hydr	AGRO	0041
) with Imines, Carbodiimides, and	Benzalazine.= +es $Cp(CO)_2M=C=CH_2$ (M=Mn, Re	INOR	0072
of Natural and Synthetic	Benzaldehyde by Deuterium NMR.= Differentiation	AGFD	0196
or 5-HT3 Ant+ New Azabicyclic	Benzamides: Potent Serotonin 5-HT4 Agonists and/	MEDI	0069
Hydrogen Bonding in Substituted	Benzamides.= +dies on the Role of Intramolecular	ANYL	0045
Complexation of	Benzamidinium Salts by a Synthetic Receptor.=	ORGN	0061
Mediated by Chrom+ Sequential	Benzannulation/Nucleophilic Aromatic Substitution	ORGN	0221
to a Receptor Binding Model for	Benzazepinone Calcium Entry Blockers: Application	MEDI	0188
Cyclohexene from	Benzene by Partial Hydrogenation.=	I&EC	0017
	Benzene Formation during Allene Pyrolysis.=	FUEL	0069
H-D Exchange during Selective	Benzene Hydrogenation to Cyclohexene.=	FUEL	0100
H-D Exchange during Selective	Benzene Hydrogenation to Cyclohexene.=	CATL	0041
romatics in Flames. Formation of	Benzene in Flames.= +n Chemistry: Single-Ring A	FUEL	0078
sc+ Homogeneous Adsorption of	Benzene on NaX and NaY Zeolites by NMR Spectro	CATL	0025
lopentadienyl Conversion during	Benzene Oxidation: Thermodynamic and Kinetic Me	FUEL	0085
Cyclotrimerization.= An η^{4-}	Benzene Species Mediates Acetylene	INOR	0210
Cyclohexene from	Benzene.= New Cyclohexanol Process via	PETR	0104
al Scheme for the Oxidation of	Benzene-Air Mixtures.= Tentative Detailed Chemic	FUEL	0084
n Flow-Reactor and Laminar-F+	Benzene/Toluene Oxidation Models: Studies Based o	FUEL	0083
Chemoe+ From Cyclopentadiene,	Benzene, and Cycloheptatriene to Carbohydrates: A	ORGN	0272
Activation for the Oxidation of	Benzenediols by the $IrCl_6^{2-}$ Ion.= Volumes of	INOR	0395
rinating Reagent.= N-Fluoro-o-	Benzenedisulfonimide (NFOBS): A Useful New Fluo	ORGN	0077
onmental Effects of Substituted	Benzenes Using Quantitative Structure-Retention Re	ANYL	0047
ings+ Kekule Structure Count in	Benzenoid Chains and the Number of Perfect Match	COMP	0033
vity against Liver Flukes of Some	Benzimidazole Derivatives.= +, Chemistry, and Acti	MEDI	0065
Poly(arylene Ether	Benzimidazole)s.= Synthesis and Properties of	POLY	0025
lphenoxy Pendent+ Rigid-Rod	Benzobisthiazole Polymer with Reactive 2,6-Dimethy	PMSE	0019
lphenoxy Pendent+ Rigid-Rod	Benzobisthiazole Polymer with Reactive 2,6-Dimethy	PMSE	0018
on of the $\chi^{(3)}$ of Poly(p-phenylene	Benzobisthiazole)-Based Molecular Composites.=	POLY	0213
ew Class of Nonste+ Substituted	Benzocycloalkenes: Structure-Activity Studies in a N	MEDI	0036
y Substitution on Biological Act+	Benzodiazepine Receptor Studies: Effect of Benzylox	MEDI	0019
Synthesis of 3,4-Disubstituted	Benzofurans via the Intramolecular Coupling of a Zir	ORGN	0217
,2-Dinitrosobenzene by Steady+	Benzofuroxan Phtochemistry: Direct Observation of 1	ORGN	0168
ation of Diacetylene-Containing	Benzoic Acid Derivatives.= Synthesis and Polymeriz	POLY	0056
istry Laboratory.=	Benzophenone Flash Photolysis in the Physical Chem	CHED	0086
1,1,4,5,6,7,8,8A-Octahydro-3H-2-	Benzopyran-3-one.= +actone Mosquito Repellent,	AGRO	0033
d Selective Class III Antiarrh+	Benzopyranone Spiropiperidines as Novel, Potent, an	MEDI	0191
1-3,4-Dihydro-5-Hydroxy-1H-2-	Benzopyrans as Dopamine D1 Selective Ligands.=	MEDI	0127
tivity of Tocotrienols and Related	Benzopyrans.= +hesis and Hypocholesterolemic Ac	MEDI	0190
Groups.= Stabilization of	Benzoquinodimethanes Using Organoruthenium	INOR	0051
ubstituted-3-Isoxazolyl)-2H-1,2-	Benzothiazine-3-Carboxamides, 1,1-Dioxide.= +5-S	MEDI	0178
vel N^6-Substituted Thiazole and	Benzothiazole Adenosine Receptor Agonists: Potent	MEDI	0111
o QSAR Analysis of Anticancer	Benzothiopyranoindazole Intercalating Agents Based	MEDI	0018
Novel Spiro[1,3,2-	Benzoxazaphosphole] Systems.=	INOR	0374
3,4,4a,5,11,11a-Pyrido[3,4-c] [1,5]	Benzoxazepine Fentanyl Derivatives.= +ahydro-1,2,	MEDI	0148
sis and SAR Studies of Novel 1,4-	Benzoxazine Derivatives.= +nel Activators: Synthe	MEDI	0099
e Selected Actinide Ions with 4-	Benzoyl-5-Methyl-2-Phenyl-Pyrazol-3-Thione and	NUCL	0073
tudy of N-(12-Dodecanoic Acid)	Benzoylformamide in Homogeneous and Microhetero	ORGN	0166
la+ Mechanisms of Resistance to	Benzoylphenyl Ureas and Other Insect Growth Regu	AGRO	0134
ack Moth to Pyrethroids and	Benzoylphenyl Ureas.= Resistance in the Diamondb	AGRO	0117
Photochemistry of	Benzyl Phenyl Sulfide.=	ORGN	0165
Reaction Rates of a Few	Benzyl-Type Radicals with O_2 and NO_x.=	FUEL	0081
Hydrolysis of Substituted	Benzylidenemalonodialdehydes.= Kinetics of the	ORGN	0152
pine Receptor Studies: Effect of	Benzyloxy Substitution on Biological Activity of Pyra	MEDI	0019
olecular Coupling of a Zirconocene	Benzyne Complex with Alkenes and Alkynes.= +m	ORGN	0217
Baby?.= Cyanomethylene and	Benzyne: Is It Possible To Escape from the Tar	PHYS	0032
ective Treatment of Plasmodium	Berghei-Infected Mice with Transdermal Artelinic A	MEDI	0063

c Enediyne Model Compounds for	Bergman Cyclization Studies.= +hesis of New Cycli	ORGN	0097
Bachelor's Degree Program at	Berkeley in Chemistry and Teaching.=	CHED	0069
Fed Erythrina poeppigiana and E.	berteroana Foliage.= +idines in the Milk of Goats	AGFD	0129
ructures.+ Zincophosphates and	Berylloarsenates—Zeolites and Other Microporous St	PETR	0055
for Propane Aromatiza+ Zeolite-	β and Isomorphously Substituted Ga–β as a Catalyst	PETR	0047
Isomorphously Substituted Ga-	β as a Catalyst for Propane Aromatization—Infrared	PETR	0047
Report of Some Recent Double-	Beta Decay Experiments.=	NUCL	0013
Double	Beta Decay of ^{100}Mo and ^{150}Nd.=	NUCL	0012
no Mass from Measurement of the	Beta Decay of Molecular Tritium.= +n Antineutri	NUCL	0010
Double	Beta Decay of U-238.=	NUCL	0061
Neutrino Mass.= The Tritium	Beta Decay Spectrum from End to End and the	NUCL	0011
Nuclear	Beta Decay.= Evidence for a Massive Neutrino in	NUCL	0003
no Science: Neutrino Masses from	Beta Decay—I. Neutri+ Recent Advances in Neutri	NUCL	0001
no Science: Neutrino Masses from	Beta Decay—II. Limit+ Recent Advances in Neutri	NUCL	0010
	Beta Solid Solution in the Bi_2O_3–SrO–CaO System.=	INOR	0201
tivity of Singlet Oxygen with α,	β Unsaturated Esters: Evidence for a Perepoxide Inte	ORGN	0164
ido-2,x-Dimethylpyridinium N-	Betaines: Preparation and Spectral Characteristics.=	ORGN	0203
Phenolic Compounds of Piper	betel as Flavoring and Nerve–Stimulating Agent.=	AGFD	0154
Removal from and Liquefaction of	Beulah–Zap Lignite.= +Coal Samples. Moisture	FUEL	0001
fragmentation Experiments at the	Bevalac.= +f a Time–Projection Chamber in Multi	NUCL	0034
ycling. Plastics Recycling: Beyond	Beverage Bottles.= +on Technology in Plastics Rec	I&EC	0037
nds in an Umbelliferous Vegetable	Beverage.= +Bound Phenolics and Other Compou	AGFD	0170
Organic Coatings for Beer,	Beverage, and Food Metal Containers.=	PMSE	0140
Activity of Oxazinone,	BF-389.= Synthesis and Novel Anti–Inflammatory	MEDI	0175
Femtosecond Dynamics of	Bi_2 Dissociation.=	PHYS	0369
Beta Solid Solution in the	Bi_2O_3–SrO–CaO System.=	INOR	0201
ration Depth in $YBa_2Cu_3O_x$ and	$Bi_2Sr_2CaCu_2O_x$ Superconductors by Electron Spin Re	PHYS	0284
ration Depth in $YBa_2Cu_3O_x$ and	$Bi_2Sr_2CaCu_2O_x$ Superconductors by Electron–Spin–R	PHYS	0050
troscopy Studies of Single–Crystal	$Bi_2Sr_2CaCu_2O_8$ High–T_c Superconductors.= +Spec	PHYS	0211
nt of Critical Current Density of	$Bi_2Sr_2CaCu_2O_8$ Superconductor by Metal Substitutio	PHYS	0329
face Structure of $Bi_2Sr_2CuO_y$ and	$Bi_2Sr_2Cu_{0.9}Fe_{0.1}O_y$.= +M Investigations of the Sur	PHYS	0292
ations of the Surface Structure of	$Bi_2Sr_2CuO_y$ and $Bi_2Sr_2Cu_{0.9}Fe_{0.1}O_y$.= +M Investig	PHYS	0292
	Bibliographic Databases for Scientists.=	CINF	0042
ESCA.= Study of Thin	Bicomponent Polymer Films on a Metal Surface by	POLY	0100
al Behavior of Oriented	Bicomponent Polymer Systems.= Microdeformation	POLY	0101
of the cis and trans Isomers of a	Bicyclic Lactone Mosquito Repellent, 1,1,4,5,6,7,8,8A	AGRO	0033
neous Catalytic Ca+ Synthesis of	Bicyclic Nitrogen Heterocycles by Means of Homoge	ORGN	0049
and –Isoquinolyl Compounds as	Bicyclic Ring–Constrained Analogues of Styryl–Based	MEDI	0023
Reactions of Carbenes and	Bicyclo[1.1.0]Butanes.= A New Mechanism for	ORGN	0147
Rigid–Rod Polymers Containing	Bicyclo[2.2.2] Octane Moieties.= Colorless	POLY	0029
osphoranide Anions (10–P–4) with	Bidentate and Tridentate Ligands.= +emistry. Ph	INOR	0344
Ring Strain of Manganese	Bidentate Phosphine Complexes.=	INOR	0193
Studies of Dolichos	Biflorus Lectins.= Structure–Function Relationship	BIOL	0149
trating Polymer Network Synt+	Bifunctional Biomimetic Polymers through Interpene	PMSE	0004
Synthesis of Macrocyclic	Bifunctional Chelating Agents.=	ORGN	0024
Ru(II) Complex with a	Bifunctional Ligand.= Cleavage of DNA by a Chiral	INOR	0232
e Construction: A Complementary	Bifurcate Approach to Taxusin and Taxol.= +axan	ORGN	0291
ediments of the Middle Atlantic	Bight.= Carbon and C–14 Budget for Upper Slope S	GEOC	0016
he Hudson River to the New York	Bight.= +c Carbon and Reactor–Derived ^{14}C from t	GEOC	0014
Middle Atlantic	Bight.= Seasonal Dissolved Gas Cycling in the	GEOC	0003
Bigness.=	+ialties for Niche Markets, or How to S	CMEC	0005
Design of a Molecular-	Bilayer Electrical Device.=	BIOT	0104
Charge Transport in Polymer	Bilayer Films.=	ANYL	0104
g of Metalloporphyrins in a Lipid	Bilayer.= +nsitive Ion–Conducting Device Consistin	BIOL	0098
es Ionic Currents across the Lipid	Bilayer.= +ice by Interfacial Electron–Transfer Gat	BIOL	0136
Mechanical Properties of Lipid	Bilayers Containing Light–Sensitive Compounds.=	BIOT	0106
Biosens+ Self–Assembling Lipid	Bilayers on Solid Support: Building Blocks of Future	BIOT	0102
Binding Studies of	Bile Acids by Various Types of Fiber.=	AGFD	0208
e Reac+ Selectivity of Supported	Bimetallic Catalysts Toward Catalytic Reforming: Th	CATL	0042
ses of Novel Rh–Co Mixed	Bimetallic Complexes and Their Catalysis.= Synthe	INOR	0226
tr+ Synthesis and Chemistry of	Bimetallic Complexes with Two– and Three–Atom In	INOR	0149
t+ Platinum–Tin and Gold–Tin	Bimetallic Particles Prepared from Solvated Metal A	CATL	0009
ion on the Catalytic Properties of	Bimetallic System Pt–Sn/$ZnAl_2O_4$.= +t of Sn Addit	COLL	0127
terization of Cationic Trinuclear	BINAP–Ruthenium(II) Complexes and Their Roles i	ORGN	0044
Structural Studies of	Binary Compounds of C_{60} and Alkali Metals.=	PHYS	0048
Alkyl Halides with	Binary Metal Compounds.= Reactions of Aqueous	INOR	0080
atic Sol+ Reversible Gelation in	Binary Mixture of Poly(di–N–alkylsilanes) and Arom	POLY	0145
How Proteins Recognize and	Bind Oligosaccharides.=	CARB	0012
hoderma reesei Cellobiohydrolases	Bind to and Degrade Cellulose?.= +. How Do Tric	BIOT	0206
m the Combustion of Quicklime	Binder–Enhanced Densified Refuse–Derived Fuel/Co	FUEL	0140
Polysilazane Preceramic	Binders.= Pretreatment and Pyrolysis of	POLY	0338
ation Cu(tpen)2+ to Form Several	Binuclear Complexes.= +Addition of Cu^{2+} to the C	INOR	0399
Functional Groups.= Cofacial	Binuclear Metal Complexes with Inwardly Directed	INOR	0242
lyanionic Chelating Ligands: C+	Binuclear Square Planar Cobalt(III) Complexes of Po	INOR	0243
Glass Formation in the	$BiO_{1.5}$–SrO–CaO–CuO System.=	INOR	0224
ntaining Chlorinate+ Integrated	Bio–Ozon Treatment of Pulp Bleaching Effluents Co	BIOT	0137
Toward the Mechanistic Role of	Bioactive Metabolites.= Etoposide: Studies Aimed	MEDI	0012
3D Design of	Bioactive Molecules.= Experiences with Automated	COMP	0049
Process D+ Integration of Rapid	Bioanalytical Techniques in Bioprocessing To Aid in	BIOT	0074
ionships between Structure and	Bioavailability in a Series of Metalloprotease Inhibito	MEDI	0195

mplex.=	Bioavailability of a Natural Chloroaniline Lignin Co	AGRO 0038
f Structural Modification on the	Bioavailability of Selected Peptide Analogues and Pe	MEDI 0192
Flavonoids on Mutagenicity and	Bioavailability of Xenobiotics in Foods.= Effect of	AGFD 0116
Peptides, Peptide Mimetics, and	Bioavailability. Influence of Structural Modificatio	MEDI 0192
: Chromatographic Retention and	Bioavailability.= +rtitioning Processes at Interfaces	ANYL 0102
Wastes: Assessment of Prevalence,	Bioavailability, and Bioeffective Dose.= +zardous	CEI 0008
Creosote Site–	Biobed Box Pilot–Scale Bioremediation Study.=	BIOT 0140
ion of a Novel Reaction Mediu+	Biocatalysis in Lyotropic Mesophases—Characterizat	BIOT 0248
ts of Mutations on the Thermo+	Biocatalyst Design for Stability and Specificity. Effec	BIOT 0188
Techniques for Probing the Me+	Biocatalyst Design for Stability and Specificity. New	BIOT 0112
view of Protein Chemical Cross+	Biocatalyst Design for Stability and Specificity. Over	BIOT 0216
of Insulin's C–Terminal B–Cha+	Biocatalyst Design for Stability and Specificity. Role	BIOT 0132
ctural Basis of Enzymatic Catal+	Biocatalyst Design for Stability and Specificity. Stru	BIOT 0239
New	Biocatalyst of Penicillin–Amidase.=	BIOT 0131
Flavor Compo+ Lipases: Useful	Biocatalysts for Enantioselective Reactions of Chiral	AGFD 0026
Food Industry Applications of	Biocatalysts Immobilized on Membranes.=	BIOT 0037
Application of Immobilized	Biocatalysts in Lipid Biotransformations.=	BIOT 0040
es. The Operation of Immobilized	Biocatalysts in Organic Solvents.= +d Opportuniti	BIOT 0035
Opportunities. T+ Immobilized	Biocatalysts: Preparation, Application, Problems, and	BIOT 0035
Advances in Penicillin–Amidase	Biocatalysts.=	BIOT 0130
Immobilized	Biocatalysts.= Bioreactor Design with	BIOT 0038
Compounds via	Biocatalytic Methods.= Enantiomerically Pure	BIOT 0092
lic Engineering and Production of	Biochemicals. Bioenergetics of+ Progress in Metabo	BIOT 0029
and Nucleotides: Chemistry and	Biochemistry/Asymmetric Synthesis of (+)–BCH–18	CARB 0022
nd Electrochromic Properties of	Biochrom Films Based on 4–Keto–Bacteriorhodopsin	BIOT 0082
Polytetrafluoroethylene and Po+	Biocompatible Interpenetrating Polymer Network of	PMSE 0041
s for Principles of Self–Organized	Biocomputers.= +Organisms as Possible Candidate	BIOT 0148
Enzyme/Protein Stabilization and	Bioconjugate Preparation.= +nking Techniques to	BIOT 0220
e Derivatives.= Synthesis and	Biodegradability Study of Functionalized Polystyren	POLY 0049
Poly(β–hydroxyalkanoates):	Biodegradable Plastics Made by Bacteria.=	CHED 0132
Contemporary Role of	Biodegradable Polymers in Health Care.=	CHED 0135
and Biodegradable Pol+ Truly	Biodegradable Polymers. Biodegradation of Polymers	CHED 0130
. Biodegradation of Polymers and	Biodegradable Polymers.= +odegradable Polymers	CHED 0130
Directions in R D of	Biodegradable Polymers.= Panel Discussion: Future	CHED 0136
tment: Past, Present, and Futur+	Biodegradation in Site Remediation and Waste Trea	AGRO 0121
ional Redesign of Cytochrome +	Biodegradation of Halogenated Hydrocarbons by Rat	BIOT 0118
ches and Field Applications—I.+	Biodegradation of Organic Wastes: Molecular Approa	BIOT 0117
ches and Field Applications—II+	Biodegradation of Organic Wastes: Molecular Approa	BIOT 0137
equencing Batch React+ Aerobic	Biodegradation of Phenolics in Optimally Designed S	BIOT 0138
Truly Biodegradable Polymers.	Biodegradation of Polymers and Biodegradable Poly	CHED 0130
Properties, Modifications, and	Biodegradation of Polysaccharides.= Production,	CHED 0131
ce Envir+ Expression of Cloned	Biodegradative Genes in Isolates from Deep Subsurfa	BIOT 0117
ial Cocai+ Synthesis and Animal	Biodistribution of Radioiodinated Cocaine: A Potent	MEDI 0131
nable Agriculture and the Role of	Biodiversity in Reducing Pesticide Use.= +ng Sustai	AGRO 0082
of Prevalence, Bioavailability, and	Bioeffective Dose.= +zardous Wastes: Assessment	CEI 0008
and Production of Biochemicals.	Bioenergetics of Clostridium thermosaccharolyticum	BIOT 0029
s Pests to Natural, Synthetic, and	Bioengineered Control Agents. +tance in Herbivorou	AGRO 0114
s Pests to Natural, Synthetic, and	Bioengineered Control Agents. +tance in Herbivorou	AGRO 0153
s Pests to Natural, Synthetic, and	Bioengineered Control Agents. +tance in Herbivorou	AGRO 0150
s Pests to Natural, Synthetic, and	Bioengineered Control Agents. +tance in Herbivorou	AGRO 0097
s Pests to Natural, Synthetic, and	Bioengineered Control Agents. +tance in Herbivorou	AGRO 0133
n. Hydrodynamics and Kinetics in	Biofilm Systems. Recent Advances.= +ed Corrosio	BIOT 0096
On–Line Monitoring of Microbial	Biofilms.=	BIOT 0099
: Cycling of Iron Oxides and T+	Biogeochemistry of Iron in Anoxic Coastal Sediments	GEOC 0007
	Bioinorganic Chemistry of Molybdenum.=	INOR 0005
in the Ascidiacea.= Marine	Bioinorganic Chemistry. Accumulation of Elements	INOR 0095
Marine	Bioinorganic Chemistry. Arsenic Cycling in the Sea.=	INOR 0136
s in the Oc+ Tutorial on Marine	Bioinorganic Chemistry. Distribution of Trace Metal	INOR 0001
Ocean.= Tutorial on Marine	Bioinorganic Chemistry. Manganese Chemistry in the	INOR 0009
eralization in Chitons.= Marine	Bioinorganic Chemistry. Mechanisms of Iron Biomin	INOR 0116
rization Studies of Platinum(II)+	Bioinorganic Chemistry. NMR and Kinetics of Isome	INOR 0403
ies at Sulfur–Bound Nickel: P_2S+	Bioinorganic Chemistry. Organometallic Functionalit	INOR 0177
	Bioinorganic Models of Tin Chelation.=	INOR 0052
uibb Award Address. Hydantoin	Bioisosteres: Hydroxy Acetic Acid Aldose Reductase	MEDI 0082
s de la Evaluacion de una Materia	Biologica.= +mico de los Alumnos en los Resultados	CHED 0093
n la Reprobacion en las Materias	Biologicas de la Carrera de QFB en la Facultad de Q	CHED 0092
Reprobacion en Materias	Biologicas.= Analisis Longitudinal de Indices de	CHED 0091
ive on the Role of Forensic	Biology in Crime Investigation.= Historical Perspect	HIST 0003
Molecular	Biology of Anti–α(1←6) Dextrans.=	CARB 0059
olymers—Chemistry, Physics, and	Biology. +Polymer Chemistry Award Symposium P	POLY 0272
tes of Cellulosic Food+ Edible	Biomass and Chemical Feedstocks from Prehydrolyza	BIOT 0153
Determination of Viable Cell	Biomass in Mammalian Cell Bioreactors.=	BIOT 0012
Economics of	Biomass Refining.= Impact of Byproducts on the	BIOT 0014
cient Hydrolysis of Lignocellulosic	Biomass.= +tion of Cellulases Composition for Effi	BIOT 0182
from	Biomass.= Production of Carbon Materials	FUEL 0052
L–(+)–Lactice Acid Using Algal	Biomass.= Production of Stable Isotopically Labeled	BIOT 0156
Pyrolysis and Gasification of	Biomass.= Transition Metals as Catalysts for	FUEL 0103
	Biomass–Fueled Gas Turbine Development.=	FUEL 0119
Coal: Energy from Sewage Sludge,	Biomass, and Municipal Waste. +from Waste and	FUEL 0115
Made from	Biomaterials.= Architecture of Device and Sensor	BIOT 0123

Degradable Polymers for	Biomedical Application.=	CHED 0134
adation of Erodible Polyesters for	Biomedical Applications.= +for the Controlled Degr	CHED 0133
mer Network Synt+ Bifunctional	Biomimetic Polymers through Interpenetrating Poly	PMSE 0004
uence of Polymer Matrix on Pr+	Biomimetic Routes to Composites and Ceramics: Infl	POLY 0290
ion. Hydrodynamics and Kineti+	Biomineralization and Microbially Influenced Corros	BIOT 0096
c Chemistry. Mechanisms of Iron	Biomineralization in Chitons.= Marine Bioinorgani	INOR 0116
act Corrosion Studies as Tools for	Biomineralization Research.= +domorph and Artif	BIOT 0101
Microscope Studies of	Biominerals and Related Materials.= Atomic Force	INOR 0117
l Materials.= Molecular and	Biomolecular Electronics. Design of Nonlinear Optica	BIOT 0173
Electronics.= Molecular and	Biomolecular Electronics. Photosynthetic Molecular	BIOT 0050
ization of Two-+ Molecular and	Biomolecular Electronics. Preparation and Character	BIOT 0121
yers on Solid S+ Molecular and	Biomolecular Electronics. Self-Assembling Lipid Bila	BIOT 0102
f Double-Well P+ Molecular and	Biomolecular Electronics. Some General Properties o	BIOT 0001
nd Optical Me+ Molecular and	Biomolecular Electronics. Spatial Light Modulators a	BIOT 0077
inkertoy Const+ Molecular and	Biomolecular Electronics. Toward a Molecular–Size T	BIOT 0200
tion Storage in+ Molecular and	Biomolecular Electronics. Very High Density Informa	BIOT 0023
	Biomolecular Materials and Design.=	BIOT 0051
hy: A New Tool for Large-Scale	Biomolecule Purification.= Perfusion Chromatograp	ANYL 0032
STM Studies of	Biomolecules and Metal Surfaces.=	ANYL 0023
ndividuals Living Adjacent to +	Biomonitoring for Chromium Levels in the Urine of I	CEI 0009
ith the+ Analytical Method for	Biomonitoring of Aniline in Urine and Comparison w	CEI 0003
e as a Stable Metric for Human	Biomonitoring Studies and an Alternative Strategy.=	CEI 0019
Responsibilities.= Human	Biomonitoring. Biomonitoring: Opportunities and	CEI 0015
rganic Compounds in+ Human	Biomonitoring. Measuring Trace Levels of Volatile O	CEI 0001
Human Biomonitoring.	Biomonitoring: Opportunities and Responsibilities.=	CEI 0015
Techniques in Human	Biomonitoring.= Applications of New HPLC/MS	CEI 0006
y for the Manufacture of	Biopharmaceuticals.= Displacement Chromatograph	ANYL 0041
Penetration.= Biochemical and	Biophysical Approaches to Improve Mucosal Peptide	MEDI 0194
rations by LC-CZE.= Rapid	Biopolymer Characterization. Two-Dimensional Sepa	ANYL 0126
phic Analysis in Rapid	Biopolymer Characterization.= Immunochromatogra	ANYL 0131
Polymer and	Biopolymer Modifications in Supercritical Fluids.=	POLY 0084
-Pressure Treatment and Use of	Biopolymers as Nonthermal Food Preservation Proce	AGFD 0029
LC/MS/MS.= Analysis of	Biopolymers by Electrospray Ionization LC/MS and	ANYL 0141
Phases.= Rapid HPLC of	Biopolymers with Micropellicular Stationary	ANYL 0128
ns in Process Chromatography of	Biopolymers: Operational and Instrumental Aspects.	ANYL 0042
d Characterization of Fluorinated	Biopolymers.= ic Resonance Imaging: Synthesis an	MEDI 0041
ative/Process Chromatography of	Biopolymers—I. Analogies between Calciu+ Prepar	ANYL 0027
ative/Process Chromatography of	Biopolymers—II. Modern High-Performan+ Prepar	ANYL 0037
lowing an Integrated Approach to	Bioprocess Development.= +Process Design in Fol	BIOT 0075
uconate by Immobil+ Integrated	Bioprocesses for the Production of Fructose and Ca-Gl	BIOT 0152
tegration: A Must for Tomorrow's	Bioprocesses.= +Upstream-Downstream Stages. In	BIOT 0069
apid Bioanalytical Techniques in	Bioprocessing To Aid in Process Design and Optimiz	BIOT 0074
Neat Organic Sol+ Nonaqueous	Bioprocessing: Protein Separation and Purification in	BIOT 0063
l Retention in a S+ Animal Cell	Bioreactor Design and Optimization. Mammalian Cel	BIOT 0007
	Bioreactor Design with Immobilized Biocatalysts.=	BIOT 0038
Retention in a Spinfilter Perfusion	Bioreactor: Analysis and Scaleup.= +malian Cell	BIOT 0007
Single-Pass Ceramic Matrix	Bioreactor.= Controlled Protein Secretion in a	BIOT 0011
in Mammalian Cell	Bioreactors.= Determination of Viable Cell Biomass	BIOT 0012
of Gas Sparging in Animal Cell	Bioreactors.= Quantitative and Mechanistic Effects	BIOT 0009
n Analogue of Mitomycin C.=	Bioreductive Alkylating Properties of BMY-25067, a	MEDI 0014
iols by Mitomycin C.=	Bioreductive Alkylation of Glutathione and Other Th	BIOL 0088
Pilot-Scale	Bioremediation Study.= Creosote Site-Biobed Box	BIOT 0140
upport: Building Blocks of Future	Biosensors and Molecular Devices.= +s on Solid S	BIOT 0102
issues, and Cells in the Context of	Biosensors for Use in Vitro and in Vivo.= +ators, T	BIOT 0039
Surface-Modified	Biosensors.=	BIOT 0107
ethods of Protein Separation—II.	Bioseparation Using Affinity Techniques.= +hic M	BIOT 0044
Novel Displacement Systems for	Bioseparations.=	ANYL 0038
	Bioseparations—Downstream Processing for Biotechn	BIOT 0224
H Structure Activity of Pheromone	Biosynthesis Activating Neuropeptide Analogs in	AGRO 0011
al Stereofacial Selectivity in the	Biosynthesis and Synthesis of the Brevianamide/Par	ORGN 0283
γ- and δ-Lacto+ Studies on the	Biosynthesis and the Biotechnological Production of	AGFD 0007
inct Origins of Proline Residues +	Biosynthesis of Echinocandin-Type Antibiotics: Dist	MEDI 0058
bolic Pathway Engineering: The	Biosynthesis of Tetracenomycins by Streptomyces.=	BIOT 0083
cosyltransferases Involved in the	Biosynthesis of the X(3-Fucosyl-Lactosamine) Antig	CARB 0015
dies of E3-Mediated Reducti+	Biosynthesis of 3,6-Dideoxyhexoses: Mechanistic Stu	BIOL 0100
Inhibitors of Steroid Hormone	Biosynthesis. Mechanism and Inhibition of Cytochro	MEDI 0006
Purine	Biosynthesis.= Carbocyclic Substrates for de Novo	BIOL 0076
Inhibition of Aldosterone	Biosynthesis.=	MEDI 0009
monium Control of Cephamycin C	Biosynthesis.= +e of Magnesium Phosphate in Am	BIOT 0088
Fungal Polyketide	Biosynthesis—From Enzymes to Genes.=	BIOT 0111
d Chemistry. The Sterol Molecule:	Biosynthesis, Structure, and Function.= +f Steroi	HIST 0012
ove.=	Biosynthetic Development of Flavor Precursors in Cl	AGFD 0211
ss of Mycotoxins: In Search of the	Biosynthetic Diels-Alder Construction.= +ide Cla	ORGN 0283
my+ Modeling the Formation of	Biosynthetic Enzymes and Cephamycin-C in Strepto	BIOT 0129
Antibiotic	Biosynthetic Pathway.= Alterations of the β-Lactam	BIOT 0235
ruction Using White Rot Fungi+	Biotechnological Approaches to Pesticide Waste Dest	AGRO 0126
dies on the Biosynthesis and the	Biotechnological Production of γ- and δ-Lactones.=	AGFD 0007
—Aiba, H+ Books for Teaching	Biotechnology. Biochemical Engineering, 2nd Edition	BIOT 0221
	Biotechnology: A Laboratory Course.=	BIOT 0227
Processing for	Biotechnology.= Bioseparations—Downstream	BIOT 0224

Pesticides in the 21st Century.	Biotechnology-Based Pest Control Strategies as Alte	AGRO	0080
Environment.= Uptake and	Biotransformation of Arsenicals in the Marine	INOR	0137
In Vitro Models of	Biotransformation Studies in Farm Animals.=	AGRO	0094
Biocatalysts in Lipid	Biotransformations.= Application of Immobilized	BIOT	0040
hord-Distribution Analysis of a	Biphasic Random Medium: Relation with Small-Ang	PHYS	0060
Propene Alkylation of	Biphenyl Catalyzed by Pillared Clays.=	PETR	0014
e Source on Metabolic Profile of	Biphenyl in an in Vitro Developmental Toxicity Assa	AGRO	0021
of Polychlorinated	Biphenyls in Anaerobic Sediments.= Dechlorination	BIOT	0120
Structural Organization of	Bipolar Lipids.=	BIOT	0122
.= Aniline-Substituted 2,2'-	Bipyridine Ligands for Polymer-Modified Electrodes	POLY	0189
Media.= Dynamic Magnetic	Birefringence as a Diagnostic Tool in Viscoelastic	COLL	0143
under Transient Elongational F+	Birefringence Characterization of Polymer Solutions	POLY	0281
Transient Electric	Birefringence of Agarose Gels.=	COLL	0182
crylamido-+ Transient Electric	Birefringence of Ionomers Poly(styrene-Sodium-2-A	COLL	0181
Strain-	Birefringence of Real Networks: Effect of Swelling.=	POLY	0308
DNA from Transient Electric	Birefringence.= Dynamics of Ring DNA vs. Linear	POLY	0296
Calicheamicin from Electric	Birefringence.= Kinetics of DNA Cleavage by	POLY	0280
cs from Transient Electric	Birefringence.= Polymers in Electric Fields: Dynami	COLL	0180
n.=	Bischler-Napieralski Reaction under Microwave Ove	ORGN	0195
e Copolymerization of Epoxy with	Bishpenol-A.= +s of Oligomer Structures during th	POLY	0227
ors of Pro+ Indolocarbazoles and	Bisindolylmaleimides as Potent and Selective Inhibit	MEDI	0074
the Polymer+ Rubber-Modified	bisMaleimides: Influence of a Polysiloxane Rubber on	POLY	0112
e Consisting of fused (μ-Hydroxo)	bisμ-carboxylato)diiron(III) Kernals.= +e Structur	INOR	0094
ystal and Catalytic Chemistry of	Bismuth Molybdate Hydrocarbon Oxidation Catalyst	CATL	0026
Coordination of	Bismuth(III) to Polyaminocarboxylate Ligands.=	INOR	0033
sed Semi-+ Characterization of	bisNadimides Oligomers and High-Tg Polyimide-Ba	POLY	0133
ent of Phenolic End-Groups in	bisPhenol A Polycarbonate Using GPC-UV Analysis.	POLY	0083
Characterization of	Bisphenol A-based Cyanate Ester Resin Systems.=	PMSE	0152
nd 4,4'-Isopropylindenediphenol (	bisPhenol-A).= +y-1,3-phenylene)-32-crown-10 a	POLY	0044
nt Mutant of R. rubrum Ribulose	Bisphosphate (RuBP) Carboxylase/Oxygenase.= +e	BIOL	0089
cle Relaxants of Ester-Containing	bisQuaternary Piperidino-Ammonium Salts.= +us	MEDI	0047
Heterocycles: Synthesis of 3,3'-	bisquinazolin-4,4'-Diones and Their Antibacterial Ac	MEDI	0057
sin Materials.= Optical	Bistability and Four-Wave Mixing in Bacteriorhodop	BIOT	0081
	Bisubstrate Reaction Templates.=	ORGN	0273
repeeled Potatoes.=	Bisulfite Replacement and Shelf-Life Extension of P	AGFD	0058
ic Effect on the Gasification of a	Bituminous Argonne Premium Coal Sample Using W	FUEL	0006
stillate Selectivity and Quality of	Bituminous Coals in the Catalytic Two-Stage Liquef	CATL	0053
ls: Engineering Materials. Carbon	Black from Coal by the Hydrocarb Process.= +Fue	FUEL	0047
o.). Teaching: A Witches' Brew of	Black Holes?.= +E. I. DuPont de Nemours and C	ANYL	0006
on Model for Marine Gas Turbine	Blades and Guide Vanes.= +orrosion: Life Predicti	FUEL	0099
he Cob Produced by an Improved	Blanching Technique.= +d Frozen Sweet Corn on t	AGFD	0136
getable Juices: Control of+ Cold	Blanching Treatments to Stabilize Raw Fruit and Ve	AGFD	0027
ed Bio-Ozon Treatment of Pulp	Bleaching Effluents Containing Chlorinated Phenolic	BIOT	0137
Applications in Pulp	Bleaching.= Practical Aspects of Xylanase	BIOT	0214
alysis of an Incompatible Polymer	Blend of Polypropylene-Polyamide.= +Thermal An	POLY	0071
Transfer Interactions on Polymer	Blend Phase Structure.= +the Influence of Charge	PMSE	0076
f+ Phase Behavior of a Polymer	Blend Solution during Steady Shear and Cessation o	PMSE	0134
a Miscible Polymer	Blend.= Fluorescence and Shear-Induced Mixing in	PMSE	0133
in PVDF/PMMA	Blend.= Internal Electric Field and Dye Orientation	POLY	0225
sreacting Polyester/Polycarbonate	Blend.= +erpretation of Phase Behavior in a Tran	PMSE	0074
of Aryl-Substitution and Physical	Blending on the Gas Permeability of Polymers.=	POLY	0117
Miscible Novolac/Polymer	Blends and Curing Studies.=	POLY	0113
mic Shear-Induced Morphology in	Blends by Optical Rheometry.= +Steady and Dyna	PMSE	0085
hip in Interpenetrating Polymer	Blends of a SEBS Block Copolymer and a Polyethere	PMSE	0058
n History on the Morphology of	Blends of Liquid Crystalline Polymers with Thermop	PMSE	0135
er-Ether)+ Cocrystallization in	Blends of Poly(butylene Terephthalate) and Poly(est	PMSE	0148
iscibility and Cocrystallization in	Blends of Poly(propylene) and Poly(butene-1).= +M	POLY	0089
oxybutyrate)+ Study of Miscible	Blends of Poly(vinylidene Fluoride) and Poly(3-hydr	POLY	0097
Ether).= Miscible	Blends of Sulfonated Polystyrene and Poly(xylenyl	PMSE	0117
Designing Polymer	Blends Which Exhibit Synergistic Properties.=	PMSE	0101
Hydrogen- and Deute+ Study of	Blends with Narrow Molecular-Weight Distribution:	COLL	0035
rties of Cross-Linkable Polyimide	Blends with Rigid and Semirigid Polyamides.= +pe	PMSE	0023
nduced Phase Changes of Polymer	Blends with UCST Phase Behavior.= +rs. Flow-I	PMSE	0132
Polystyrene Ionomer/Polystyrene	Blends: Effect of Ion Content.= +ure of Sulfonated	POLY	0061
—II. Quick Quenching of Polymer	Blends.= +lymer Solutions, Blends, and Interfaces	COLL	0032
eparation Kinetics in PC/PMMA	Blends.= +Description and Application to Phase-S	POLY	0278
Metalorganic/Inorganic Polymer	Blends.= Ceramic Coatings for C/C Composites via	POLY	0229
Polysaccharide Liquid-Crystal	Blends.= Characterization of Lyotropic	COLL	0017
Poly(ethylene Oxide)	Blends.= Crystal-Amorphous Interphases in	PMSE	0165
Hydrogen-Bonded Polymer	Blends.= Diffusion of Water Vapor in	POLY	0192
) and Poly(methylphenylsiloxane)	Blends.= +Cyclic and Linear Poly(dimethylsiloxane	POLY	0234
Gas Solubility in Polymers and	Blends.=	POLY	0143
Polymer	Blends.= Hydrogen-Bonding Interaction in	COLL	0034
in Miscible and Immiscible	Blends.= Kinetics of Imidization of Poly(amic Acid)	PMSE	0147
DPE (TREF) Fractions and Their	Blends.= +logical Characterization of 1-Octene LL	PMSE	0102
Temperatures and in PEI	Blends.= Morphology of PEEK at High	POLY	0333
Interface in Polyethylene and Its	Blends.= Nature of the Crystal/Amorphous	COLL	0036
and	Blends.= Neopentyldiamine Polyamides	PMSE	0100
styrene/Poly(vinyl Methyl Ether)	Blends.= +composition and Linear and Cyclic Poly	POLY	0295
ess of Ethynyl-Containing	Blends.= Preparation, Characterization, and Toughn	POLY	0040

n in Polystyrene–Polyxylenylether	Blends.= +utron Scattering Study of Interdiffusio	POLY	0166
ly(2–Chloro–Styrene)/Polystyrene	Blends.= +–Field and Ising Critical Behavior of Po	PMSE	0103
atic Polybenzimidazole/Polyimide	Blends.= +mally Induced Phase Behaviors of Arom	PMSE	0164
Gamble UERP Polymer Solutions,	Blends, and Interfaces—I. Dynamics at+ Proctor	COLL	0012
Gamble UERP Polymer Solutions,	Blends, and Interfaces—II. Quick Quenc+ Procter	COLL	0032
Gamble UERP Polymer Solutions,	Blends, and Interfaces—III. Lateral Diff+ Procter	COLL	0051
Gamble UERP Polymer Solutions,	Blends, and Interfaces—IV. Thin Polym+ Procter	COLL	0070
Adhesion.= Application of the	Blister Test to the Study of Polymer–Polymer	COLL	0057
Dynamics of	Block Copolymer Adsorption.=	COLL	0073
rating Polymer Blends of a SEBS	Block Copolymer and a Polyetherester.= +erpenet	PMSE	0058
attering an+ Characterization of	Block Copolymer Ionomers by Small-Angle X-ray Sc	PMSE	0118
Polystyrene–Poly(ethylene Oxide)	Block Copolymer Micelles in Water.= +Methods.	POLY	0277
hesis of Metal Nanoclusters within	Block Copolymer Microdomains.= +ntrolled Synt	POLY	0207
Dynamics in Microstructured	Block Copolymer Solutions.= Diffusion and	POLY	0275
polymer+ Change of Mechanism	Block Copolymerizations: The Formation of Block Co	PMSE	0071
o–methylphenylsiloxane) Random	Block Copolymers and Elastomers.= +hylsiloxane–c	POLY	0020
oreversible Gels with Polyethylene	Block Copolymers and Hydrocarbon Solvents.=	POLY	0249
+owth Approach: Hyperbranched	Block Copolymers and Monodisperse Polyesters.=	POLY	0318
lymerizations: The Formation of	Block Copolymers Containing Helical and Amorphou	PMSE	0071
Poly(Ethylene Oxide) and ABA	Block Copolymers Containing Hydroxyl–Capped Poly	PMSE	0006
and Their Optical+ Synthesis of	Block Copolymers of Poly(styrene) and Poly(arylate)	POLY	0220
and Poly(ester–Ether) Segmented	Block Copolymers.= +oly(butylene Terephthalate)	PMSE	0148
IPNs Based on	Block Copolymers.= Microstructural Aspects of	PMSE	0026
Prepared Using ROMP	Block Copolymers.= Semiconductor Clusters	INOR	0307
Colloid–Stabilizing	Block Copolymers.= Surface Conformation of	PMSE	0013
Phosphaalkynes: A New Building	Block in Organic Chemistry.= +nal Investigation of	COMP	0060
tic Aldol Additions: A Building	Block Strategy for Asymmetric Synthesis.= Enzyma	BIOT	0090
troscopies—A New Family on the	Block.= +er Corp.). Multiresonant Nonlinear Spec	ANYL	0001
of α–Chloro, α–Hydroxy, and N–	Blocked α–Amino Acids.= +ic Synthesis: Activation	ORGN	0259
Calcium–Entry	Blocker.= Synthesis of a Conformationally Rigid	ORGN	0114
r Benzazepinone Calcium Entry	Blockers: Application to the Design of Structurally N	MEDI	0188
diac Electrophysiological, and β–	Blocking Activity of Novel Arylpiperazines with Pote	MEDI	0094
ogen–Containing Neuromuscular	Blocking Agents.= Novel Quaternary–Enamino–Nitr	MEDI	0046
ylzincs with Aldeh+ Methyl as a	Blocking Group in the Reaction of Mixed Alkylmeth	ORGN	0090
Clusters as Building	Blocks for Extended Solids.= Reconstructed	INOR	0173
yers on Solid Support: Building	Blocks of Future Biosensors and Molecular Devices.=	BIOT	0102
Disaccharide–Derived Building	Blocks with Industrial Application Profiles.= Some	CARB	0003
ation of Hemicyanine Langmuir–	Blodgett Films by Fourier Transform Infrared Spect	POLY	0224
Phthalocyanine Langmuir–	Blodgett Films.= Diffusion in the Gas Sensing of	POLY	0118
in Langmuir–	Blodgett Films.= Fractals and Excitation–Transfer	PHYS	0110
DCANP Langmuir–	Blodgett Films.= Guided–Wave Nonlinear Optics in	POLY	0176
nt Interfaces Using the Langmuir–	Blodgett Technique.= +–Chain Polymer at Differe	COLL	0071
and Order in Langmuir–	Blodgett–Type Chromophore Systems.= Domains	PHYS	0111
e Organic Compounds in Human	Blood by Using Purge–and–Trap/Gas Chromatograp	CEI	0001
ioxins and Dibenzofurans in the	Blood of Twelve TCDD–Exposed Persons from Jacks	CEI	0014
ected by Use of Lecithin Suppl+	Blood Serum Lipid Concentrations of Humans as Aff	AGFD	0034
SP) in Mesoscale Coccolithophore	Blooms in the Gulf of Maine.= +iopropionate (DM	GEOC	0002
Groves of Academe.= Whistle–	Blowers, Advocates, and the Law. Greed in the	PROF	0001
To Help Whistle–	Blowers, Legalize PEG.=	PROF	0012
chologist's Perspective on Whistle	Blowing in Science.= +th and Consequences: A Psy	PROF	0002
Legal Rights of Whistle–	Blowing Professionals.=	PROF	0011
Industrial Whistle	Blowing.= Two Personal Experiences with	PROF	0007
and Determination of Methylene	Blue and Its Demethylated Metabolites from Milk.=	AGFD	0101
x Kinetics of Oxidation of Thymol	Blue with Bromate Ion in Acidic Solutions.= +ple	PHYS	0269
waters in Manufacture of Disperse	Blue 79.= +ction of Toxic Components and Waste	CMEC	0015
lish This Task? Relative Risk: A	Blueprint for Future Environmental Protection Effor	ENVR	0061
water and the Role of Agricultural	BMP in the 1990s.= +Presence in Surface/Ground	FERT	0019
uctive Alkylating Properties of	BMY–25067, an Analogue of Mitomycin C.= Biored	MEDI	0014
Preceramic	BN Polymers.= Synthesis and Processing of	POLY	0312
por Depositon of Diamond (and c–	BN, SiC). Characterization of Diam+ Chemical Va	PHYS	0118
or Deposition of Diamond (and c–	BN, SiC). Halogenation of Diamon+ Chemical Vap	PHYS	0176
or Deposition of Diamond (and c–	BN, SiC). Mole Fraction Measurem+ Chemical Vap	PHYS	0145
or Deposition of Diamond (and c–	BN, SiC). Molecular Dynamics Sim+ Chemical Vap	PHYS	0354
por Depostion of Diamond (and c–	BN, SiC). Nucleation and Growth o+ Chemical Va	PHYS	0091
or Deposition of Diamond (and c–	BN, SiC). Structure and Chemistry+ Chemical Vap	PHYS	0202
hemistry: An Electronic Bulletin	Board for High School Teachers of Chemistry in the	CHED	0111
Collection Systems? On–	Board Materials Densification.= What's New in	I&EC	0040
rowave Region for Printed Circuit	Board Materials.= +of Dielectric Properties in Mic	PMSE	0055
y Control o+ Application of the	BOC–MP Method for Evaluation of Reaction Pathwa	CATL	0032
ds of Washing sCD4-183 Inclusion	Bodies to Remove E. coli Proteins and En+ Metho	BIOT	0171
Fine–Dispersed Carbon	Bodies.= Interaction between Ozone and	COLL	0109
d by Dioxin Analysis to Estimate	Body Burden in General Population and Exposed In	CEI	0012
tational Suppression of Inclusion	Body Formation in the P22 Tailspike Protein.= Mu	BIOT	0113
Folding Properties of Inclusion	Body Mutants of 1L–1β.=	BIOT	0114
Volo	Bog.= Organic Geochemical Studies of Peat from	I&EC	0016
red Hybrid Fluidized-Bed	Boilers.= Commercial Application of Waste–Fuel–Fi	FUEL	0122
om Mole+ Prediction of Normal	Boiling Points of Heterocyclic Organic Compounds fr	COMP	0015
Sputtering Processes of Ion–	Bombarded Electronic Materials.=	NUCL	0043
ion Modification Produced by Ion–	Bombardment (Ion Mixing). +f Materials: Composit	NUCL	0091
of Continuous–Flow Fast Atom	Bombardment Mass Spectroscopy as a Detector for L	ANYL	0125

Silkworm,	Bombyx mori.= Prothoracicotropic Hormone of the	AGRO 0010
ary Alcohols: Carbon–Hydrogen	Bond Activation in Rhenium Alkoxide Complexes (η_5	ORGN 0220
Regioselective Carbon–Carbon	Bond Cleavage of Cubanes.=	ORGN 0256
thyl-1,3-penta+ Carbon–Carbon	Bond Formation during Thermolysis of Bis(2,4-Dime	INOR 0305
Graph Theory Applied to	Bond Length Variations in Organic Crystals.=	COMP 0031
No E+ Manipulation of Apparent	Bond Lengths as Determined by X-ray Diffraction:	INOR 0088
Double–	Bond Migration.= Aldol Chemistry via Initial	ORGN 0092
n Hydrogen Bonding: The Natural	Bond Orbital Donor–Acceptor Perspective.= +cts i	PHYS 0106
Sulfur–Sulfur	Bond Scission by a Copper Reagent.=	ORGN 0252
) and Si(100) Surfaces.= Role of	Bond Strain in the Chemistry of Hydrogen on Si(111	PHYS 0187
mediates: Estimates of Alkane–W	Bond Strengths.= +nt from (Alkane)W(CO)$_5$ Inter	INOR 0111
sotope Effects in Carbon–Carbon	Bond–Forming Reactions of Scandocene Derivatives.	INOR 0147
Tautomerization, and Novel C–C	Bond–Forming Reactions with η^2-Pyrrole Complexes	INOR 0114
cond–Harmonic Generation by σ–	Bond–Separated Donor–Acceptor Molecules.= +Se	POLY 0222
ctomyosin ATP Hydrolysis+ The	Bond–Splitting Step Is Rate–Limiting for Skeletal A	BIOL 0141
-ray Diffraction: No Evidence for	Bond–Stretch Isomerism.= +hs as Determined by X	INOR 0088
Hydrogen–	Bonded Complexes.= Ab Initio Calculation on	PHYS 0103
—IV. Plenary Lecture. Hydrogen–	Bonded Interpolymer Complexes.= +ent Systems	PMSE 0163
+of Molecular Alloys of Multiply	Bonded Molybdenum and Tungsten Compounds.=	INOR 0356
lysis of System Peaks in Normal	Bonded Phase HPLC with Diode Array Multichanne	ANYL 0087
ultichannel Detections in Normal	Bonded Phase HPLC.= +sional Separations and M	ANYL 0088
Vapor in Hydrogen–	Bonded Polymer Blends.= Diffusion of Water	POLY 0192
tion and Diffusion in Hydrogen–	Bonded Polymers: Vinyl Alcohol/Vinyl Butyral Copo	POLY 0191
hromophore–Centered Hydrogen–	Bonding and Crosslinking Groups.= +rs Having C	POLY 0108
abilit+ Contribution of Hydrogen	Bonding and Hydrophobic Interactions to Protein St	BIOT 0189
zones—Comparison with S+ H–	Bonding Control of Spin–State: Ferric Thiosemicarba	INOR 0268
ntations and Analysis of Hydrogen	Bonding from Molecular Dynamics Simulation.=	BIOL 0112
Steric Effects.= Structure and	Bonding in Cuprate Superconductors: Electronic and	CHED 0022
apidly Moving Frontier. Chemical	Bonding in Solids.= +Solid–State Chemistry—A R	CHED 0020
Role of Intramolecular Hydrogen	Bonding in Substituted Benzamides.= +dies on the	ANYL 0045
lectronic Structure and Chemical	Bonding in Ternary Transition–Metal Compounds.=	PHYS 0252
Hydrogen–	Bonding Interaction in Polymer Blends.=	COLL 0034
de–on (μ-N$_2$) vs. End–on (μ-N$_2$)	Bonding of Early Transition–Metal Dinitrogen Comp	INOR 0362
eric Effects in Metal–Phosphine	Bonding: Determination of the Enthalpy and Kinetic	INOR 0192
Cooperative Effects in Hydrogen	Bonding: The Natural Bond Orbital Donor–Acceptor	PHYS 0106
usters and Extended Metal–Metal	Bonding.= +ontier. New Structures with Metal Cl	CHED 0029
Relavtivistic Gold–Gold Covalent	Bonding.= +lts: Evidence for Energy Transfer and	INOR 0189
on of Precursors Containing M–N	Bonds for the Preparation of Metal–Nitride Precera	POLY 0230
C–H, C–C, and C–O	Bonds of Oxygenates on Rh(111).= Activation of	CATL 0033
anisms for the Formation of Al–N	Bonds on the Surface.= +Ammonia on Silica: Mech	COLL 0045
the Carbonylation of Metal Alkyl	Bonds Studied by Time–Resolved Infrared Spectral	INOR 0169
to Form Metal–Ligand Multiple	Bonds with Tungsten Chlorophosphine Complexes.=	INOR 0115
C–S and S–S	Bonds.= Grignard Reagent Promoted Scission of the	ORGN 0253
nactivated Carbon–Carbon Double	Bonds.= +hilic Mechanism for Isomerization of U	BIOL 0014
i(100): The Behavior of Dangling	Bonds.= Silicon Carbide Thin–Film Formation on S	PHYS 0204
Cleavage of the C–S and S–S	Bonds.= Single Electron Transfer Catalyzed	ORGN 0251
of Si–N	Bonds.= Some Novel Reactions for the Formation	POLY 0337
neering, 2nd Edition—Aiba, H+	Books for Teaching Biotechnology. Biochemical Engi	BIOT 0221
The Place of Reference	Books in a Computerized World.=	CHAS 0039
Intermaterial Competition—	Boon or Bane?.=	CMEC 0012
Di-t-Butylphosphino(diethyl)	borane.= Synthesis and Film–Growth Studies of	INOR 0230
Bis(3-tert-butylpyrazolyl–	borate) Nickel(II) Complexes.=	INOR 0035
Characterization of	Borate-Pillaring Anionic Clay.= Preparation and	PETR 0111
1-Substituted	Borepins and Their Mo(CO)$_3$ Complexes.=	ORGN 0212
=	Boretxino: Low–Energy Solar Neutrino Spectroscopy.	NUCL 0033
on the Stereochemistry of Nickel	Boride Reduction of 2,6-Dihalo-9-Oxabicyclononane	ORGN 0246
Precursors to Metal	Borides.= Metallaboranes as Low–Temperature	INOR 0101
Mo/Al$_2$O$_3$ Modificados con	Boro.= Estudio de los Catalizadores CoMo/Al$_2$O$_3$ y	FUEL 0104
ctamización Procedure Using	Borohydride Reagents.= A Reductive Amination–La	ORGN 0242
Reactions of	Boron and Aluminum Atoms with Small Molecules.=	PHYS 0090
eactions—II. Kinetic Studies of	Boron and Aluminum Species.= Gas–Phase Metal R	PHYS 0086
Clusters Containing One or More	Boron Atoms.= +Ruthenium and Osmium Carbonyl	INOR 0220
d+ Size–Dependent Properties of	Boron Cluster Ions: Reactivity, Structure, and Photo	PHYS 0262
, Hydrocarbon–Assisted	Boron Combustion.= Kinetics of High–Temperature	PHYS 0142
Intermediate–Sized	Boron Hydride Clusters.= Insertion of Nitrogen into	INOR 0388
Hot–Filament–Activated CVD of	Boron Nitride and Carbon.=	PHYS 0205
	Boron Nitride Fibers from Polyborates.=	POLY 0316
Characterizations of	Boron Nitride Films.= Depositions and	PHYS 0094
c Materials.= Nitrogen– and/or	Boron–Containing Polymers as Precursors of Cerami	POLY 0315
AlPO$_4$-5.= Aluminum– and	Boron–Containing SSZ-24, the All–Silica Analog of	PETR 0023
Solid–State NMR Investigation of	Boron–Substituted Zeolites and Silicates.= +clear	PETR 0136
ogeneous Gas–Phase Modeling of	Boron/Oxygen/Hydrogen/Carbon Combustion.=	FUEL 0097
R Observation of Tightly Bound	Boronic Acid Inhibitors on Serine Proteases: Evidenc	BIOL 0082
Diligence? Be Your Own	Boss.= Have Intelligence, Imagination, and	SCHB 0016
at	Boston University.= Teaching of Chemistry Program	CHED 0070
, and+ Phenolic Compounds in	Botanical Extracts Used in Foods, Flavors, Cosmetics	AGFD 0172
astics Recycling: Beyond Beverage	Bottles.= +on Technology in Plastics Recycling. Pl	I&EC 0037
Hum+ Fate of a PCB in Anoxic	Bottom Sediments: Characterizing the Role of in Situ	ENVR 0017
^{11}B NMR Observation of Tightly	Bound Boronic Acid Inhibitors on Serine Proteases:	BIOL 0082
of the Kinetic Fate of a Thiamin–	Bound Enamine Intermediate in Water.= +rvation	BIOL 0047

c Fate of Thiamin Diphosphate–	Bound Enamines on Brewers' Yeast Pyruvate Decarb	BIOL	0048
Nectarine.= Free and	Bound Flavor Constituents of White–Fleshed	AGFD	0024
Isolated and Fiber–	Bound Hemicelluloses.= Enzymatic Hydrolysis of	BIOT	0215
metallic Functionalities at Sulfur–	Bound Nickel: P_2S_2 vs. N_2S_2 Ligand Envi+ Organo	INOR	0177
liferous Vegetabl+ Glycosidically	Bound Phenolics and Other Compounds in an Umbe	AGFD	0170
General Electric College	Bound Program.=	CHED	0003
nductivity of Poly(ethylene Oxide)	Bound with a Sterically Hindered Phenolate.= +o	POLY	0091
.+ Electrified Immiscible Liquid	Boundaries: Conventional and Microscopic Interfaces	COLL	0004
R. Nuclear Research Reactor Site	Boundary.= +Products in Discharge Pool in T.H.O.	NUCL	0090
in Two– and Three–Dimensional	Bounded Crystal Lattices. +ate One–Electron States	BIOT	0028
roscopy at the Limit: Pushing the	Bounds of Sensitivity and Spatial Resolution.= +ct	ANYL	0012
nt Improvements in Production of	Bovine Chymosin in Aspergillus niger var. +i: Rece	BIOT	0234
of Recombinant	Bovine Somatotropin.= Oxidation and Naturation	BIOT	0135
Creosote Site–Biobed	Box Pilot–Scale Bioremediation Study.=	BIOT	0140
Subtilisin	BPN'.= Protein Stability Studies on	BIOT	0192
Excited–Stated Dynamics of Ru(	$BPY)_3^{2+}$ on Porous Vycor Glass.=	INOR	0160
pound, $[Cd(Se-2,4,6-i-Pr_3C_6H_2)_2($	bpy)].= +dies of a Novel Cadmium Selenolate Com	INOR	0078
le Properties of XCN (X = H, Cl,	Br) in Various Ionic States.= +metries and Molecu	PHYS	0238
em in Nature, Bacteriorhodopsin (	bR).= +c Energy by the Other Photosynthetic Syst	PHYS	0019
ns of $(Me_3Si)_3As$ with GaX_3 (X =	Br, I): Crystal Structure of $[I_2GaAs(SiMe_3)_2]_2$.= +o	COLL	0140
e cis-trans Isomerization of Gly^6–	Bradykinin and Analogs.= +MR Assessment of th	BIOL	0006
Sequential?.= Projectile	Brakup in 25–MeV/A Reactions: Prompt or	NUCL	0027
with Cationic Bis(+ Synthesis of	Branched Aldehydes by Catalytic Hydroformylation	INOR	0154
Methods.= Characterization of	Branched and Comb Polymers by SEC and Other	PMSE	0069
ific Control of Lepidopt+ Ethyl–	Branched Juvenile Hormones: Opportunities for Spec	AGRO	0004
Simulations of Rings and	Branched Molecules.= Dynamic Monte Carlo	POLY	0297
Synthesis of the First	Branched Quaterthienyls.=	ORGN	0202
ic Synthesis of Heteroarm, Star–	Branched Styrene–Butadiene Thermoplastic Elastom	POLY	0266
f Aminodeoxy, Cyanodeoxy, and	Branched–Chain Amino Sugars by Epoxy and Triflat	ORGN	0071
Synthesis of a	Branched–Chain Sugar Daunorubicin Analogue.=	CARB	0052
Effect of the Length of the Side	Branches on the Chain Statistics of Siloxanes.= The	POLY	0070
EC/CV Systems.= Long–Chain	Branching Analysis of LDPE by SEC/LALLS and S	PMSE	0047
Effect of	Branching and Cycling on the W–Shape Cp^E.=	PHYS	0320
Polymer Sol+ Characterization of	Branching Architecture through Universal Ratios of	POLY	0293
terization of Polymer MWD and	Branching with GPC–Viscometry in THF and DMF.	PMSE	0068
Phenolic Compounds of	Brassica Oilseeds.=	AGFD	0168
: The Codetec, Inc., Experience in	Brazil.= +for Small and Medium-Size Enterprises	SCHB	0023
ltireference Couple+ Symmetry	Breaking in Radicals Studied by the Fock–Space Mu	PHYS	0016
Interaction Studies of Symmetry	Breaking in Small Molecules.= Configuration	PHYS	0014
Symmetry	Breaking.= Coupled-Cluster Treatment of	PHYS	0015
Characterization of	Breck Structure Six (BSS).= Synthesis and	PETR	0002
iosynthesis and Synthesis of the	Brevianamide/Paraherquamide Class of Mycotoxins:	ORGN	0283
d Enoyl Reduction Catalyzed by	Brevibacterium ammoniagenes Fatty Acid Synthase.	BIOL	0065
rs and Co.). Teaching: A Witches'	Brew of Black Holes?.= +E. I. DuPont de Nemou	ANYL	0006
Diphosphate–Dependent Enzyme	Brewers' Yeast Pyruvate Decarboxylase.= +hiamin	BIOL	0032
Diphosphate–Bound Enamines on	Brewers' Yeast Pyruvate Decarboxylase.= +iamin	BIOL	0048
Atmospheric Nitrogen Oxides: A	Bridesmaid Revisited.=	ENVR	0071
on and Reactivity of Phosphido–	Bridged Iridium, Molybdenum Heterobimetallics by	INOR	0385
henylphosphino)ferrocene (dppf)–	Bridged Silver(I) Complexes.= +ome 1,1'-bis(Disp	INOR	0106
etal Complexes Containing Chiral	Bridging Groups.= +lications of ansa–Bis(idenyl)m	INOR	0214
nts and Asymmetry Parameters in	Bridging Metal Hydrides.= +pole Coupling Consta	INOR	0382
, Moderator.= Panel Discussion:	Bridging the Gap; Why More Is Better. D. L. Leister	I&EC	0058
ic DNA Typing Systems.=	Brief History of the Development and Use of Forens	HIST	0004
s—I. Neutrinos and Chemistry: A	Brief Tutorial.= +Neutrino Science: Solar Neutrino	NUCL	0019
mical Products from Geothermal	Brine Utilizing a Waste/Low-Grade Heat Source wit	I&EC	0010
Activities.=	Bringing Chemistry to Life through Decision-Making	CHED	0054
neral and Edward E. Smissman–	Bristol–Myers Squibb Award Address. Hydant+ Ge	MEDI	0082
tical Industry: Chemistry at the	Bristol–Myers Squibb Pharmaceutical Research Insti	CMEC	0028
E. Smissman Award, sponsored by	Bristol–Myers Squibb).+ Award Address (Edward	MEDI	0089
ory of Inhomogeneous Spectral	Broadening in Liquids and Glasses.= Molecular The	PHYS	0036
Studies of Electronic Line	Broadening in Liquids.= Statistical Mechanical	PHYS	0035
by Electron-Spin-Resonance Line	Broadening of Surface Paramagnetic Probes.= +s	PHYS	0050
by Electron Spin Resonance Line	Broadening of Surface Paramagnetic Probes.= +ors	PHYS	0284
of Oxidation of Thymol Blue with	Bromate Ion in Acidic Solutions.= +plex Kinetics	PHYS	0269
of a Mill Drier for Shrimp Head:	Bromatological Characteristics of Produced Meal.=	AGFD	0105
+a Mill-Drier for Shrimp Heads:	Bromatological Characteristics of Produced Meal.=	ENVR	0046
in the Presence of Cetyltrimethyl	Bromide (CTAB).= +rium(IV) with Sulfanilic Acid	ANYL	0073
f Tetradecyltrimethylammonium	Bromide Complexation with Polyacrylic Acid and Po	COLL	0074
Iridium	Bromide Complexes.= Fast Kinetics Investigation of	PHYS	0258
rization of Ethynylpyridine with	Bromine: A New Route to Electrically Conducting Co	POLY	0032
Extracting	Bromine.= Herbert H. Dow—Process of	CHAL	0019
Complexes of Vanadium	Bromoperoxidase.= Functional Vanadium Model	INOR	0016
Product Selectivity of Vanadium	Bromoperoxidase.=	INOR	0017
Structural Studies of Vanadium	Bromoperoxidases.=	INOR	0098
s of the Monophosphate Tungsten	Bronze $(PO_2)_4(WO_3)_{2m}$, m = 2.= +nsport Propertie	INOR	0352
Polytechnic University,	Brooklyn.= The Polymer Research Institute at the	HIST	0033
a New Continent of Chemistry. H.	Brown (Chemistry, 1979).= +y and Exploration of	CONG	0012
ated P+ Prevention of Enzymatic	Browning in Fruit Juice with Cyclodextrins and Sulf	AGFD	0030
able Juices: Control of Enzymatic	Browning.= +nts to Stabilize Raw Fruit and Veget	AGFD	0027
Structure Six (	BSS).= Synthesis and Characterization of Breck	PETR	0002

Left context	Keyword phrase	Ref
Production of Clean–Medium	Btu Gas from Gasification of Sludge Wastes.=	FUEL 0118
$2)_2$ and $Fe_2(CO)_6(\mu\text{-PPH}_2)(\mu\text{-P-t-}$	Bu_2).= +al Characterization of $Fe_2(CO)_5(\mu\text{-P-t-Bu}$	INOR 0050
aracterization of $Fe_2(CO)_5(\mu\text{-P-t-}$	$Bu_2)_2$ and $Fe_2(CO)_6(\mu\text{-PPH}_2)(\mu\text{-P-t-Bu}_2)$.= +al Ch	INOR 0050
Structure of (t-	$Bu_2As)_3Ga$.= Tris(arsino)allanes and the Crystal	INOR 0231
$GaAs(SiMe_3)_2]_2$ (R = t-$BuCH_2$,t-	Bu), and (t-$BuCH_2)_2Ga(Cl)\cdot As(SiMe_3)_3$.= +$_2Cl$, [$R_2$	COLL 0136
	Oxygen Transfer Properties of Bubbles in Animal Cell Culture Media.=	BIOT 0008
$)_2]_2$ (R = t-$BuCH_2$,t-Bu), and (t-	$BuCH_2)_2Ga(Cl)\cdot As(SiMe_3)_3$.= +$_2Cl$, [$R_2GaAs(SiMe_3$	COLL 0136
R_2Cl, [$R_2GaAs(SiMe_3)_2]_2$ (R = t-	$BuCH_2$,t-Bu), and (t-$BuCH_2)_2Ga(Cl)\cdot As(SiMe_3)_3$.=	COLL 0136
with a New Variable+ Ordered	Buckled–Dimer Configurations on Si(100) Observed	PHYS 0185
ocarbon Fuels: Fullerenes. Post-	Buckminsterfullerene View of Carbon Chemistry.=	FUEL 0019
Illustrate the Structure of	Buckminsterfullerene.= A Simple Model to	CHED 0102
Superconductivity.	Buckyball.=	PHYS 0045
lantic Bight.= Carbon and C-14	Budget for Upper Slope Sediments of the Middle At	GEOC 0016
and Development of the Tobacco	Budworm.= +micals and Nutrients on the Behavior	AGRO 0037
ds of the Australian Spined Citrus	Bug.= +miacetal from the Dorsal Abdominal Glan	AGRO 0035
gation of Phosphaalkynes: A New	Building Block in Organic Chemistry.= +nal Investi	COMP 0060
Enzymatic Aldol Additions: A	Building Block Strategy for Asymmetric Synthesis.=	BIOT 0090
Reconstructed Clusters as	Building Blocks for Extended Solids.=	INOR 0173
Lipid Bilayers on Solid Support:	Building Blocks of Future Biosensors and Molecular	BIOT 0102
= Some Disaccharide–Derived	Building Blocks with Industrial Application Profiles.	CARB 0003
Drugs.= Experiences	Building the Chapman and Hall Directionary of	CINF 0033
MDDR-3D and FCD-3D.=	Building 3D Structural Databases: Experiences with	COMP 0046
Kinetic Study and Morphology	Buildup in Methacrylate Urethane IPNs.=	PMSE 0043
ns of Polymethylene Chains in the	Bulk and in Thin Films.= +ic Dynamics Simulatio	POLY 0355
n: Measurement of Surface-	Bulk Association.= Apparent Free Energy Adsorptio	COLL 0122
rmally Grown Silic+ Surface and	Bulk Diffusion of Adsorbed Nickel on Ultrathin The	COLL 0177
n Polymers in Solution and in the	Bulk State.= +namics of Rotational Isomerization i	COLL 0018
ooligosaccharides—A Low-Calorie	Bulking Agent Prepared from Sucrose.= +is. Fruct	CARB 0001
al Metal Salen Complexes Bearing	Bulky Silyl Groups.= +xidation Catalyzed by Chir	ORGN 0056
hing of Chemistry: An Electronic	Bulletin Board for High School Teachers of Chemist	CHED 0111
ioxin Analysis to Estimate Body	Burden in General Population and Exposed Individu	CEI 0012
with Regulatory	Burdens.= How Chemical Entrepreneurs Can Cope	SCHB 0011
ones and Peptides for+ Baxter,	Burdick, Jackson Award Symposium: Juvenile Horm	AGRO 0001
ones and Peptides for+ Baxter,	Burdick, Jackson Award Symposium: Juvenile Horm	AGRO 0006
rgin Sediments: Implications for	Burial Fluxes and Phosphorus Residence Time in the	GEOC 0013
tosynthetic Units: Spectral–Hole-	Burning Studies.= +d Transport Dynamics of Pho	PHYS 0018
Pesticide–Bag Open	Burning.= Emission Characterization of	AGRO 0141
Agricultural Product Bag	Burns.= Laboratory Evaluation of PIC Formation in	AGRO 0140
Constructing Your	Business Plan.=	SCHB 0004
Starting a	Business vs. Working for One.=	YCC 0012
Running a Small	Business within a Larger Company.=	SCHB 0018
nd Operating Your Own Chemical	Business. So You Want to Start Your+ Starting a	SCHB 0001
the Challenge for Small Chemical	Business: U.S. and Lati+ Technology Transfer and	SCHB 0022
ant to Start Your Own Chemical	Business.= +ur Own Chemical Business. So You W	SCHB 0001
Operating the	Business.=	SCHB 0012
orest—A Field for Small Chemical	Business.= +al Products from the Tropical Rain F	SCHB 0024
Costa Rica.= Small Chemical	Businesses in Costa Rica: The Case of Arvi, Inc.,	SCHB 0026
Opportunities for U.S. Chemical	Businesses in Latin America. Latin American Chem	SCHB 0021
True Stories of Small Chemical	Businesses. Founding Fairfield Chemical Company.=	SCHB 0013
Small Chemical	Businesses.= Acquisition—The Pleasant (?) Fate of	SCHB 0019
MOs: A Nonmathematical	but Honest Approach.= Pictorialized Solid–State	CHED 0021
+in Differs from the Crystal State	but Spans the Range of Different Crystal Forms.=	BIOL 0002
oscandium S+ Polymerization of	Butadiene and Isoprene by Single–Component Organ	INOR 0280
n of the Low-Lying States of	Butadiene Radical Cation.= Theoretical Investigatio	PHYS 0237
teroarm, Star–Branched Styrene–	Butadiene Thermoplastic Elastomers.= +esis of He	POLY 0266
Photoelimination of H_2 form 1,3-	Butadiene.=	PHYS 0325
der Reaction of 2,3-Dialkoxy-1,3-	Butadienes.= +ereofacial Selectivity in the Diels–Al	ORGN 0233
r Cluster Cluster: Relationship to	Butane Hydrogenolysis Selectivities Exhibited by W-	CATL 0008
Carbenes and Bicyclo[1.1.0]	Butanes.= A New Mechanism for Reactions of	ORGN 0147
tation by an Acid+ Regulation of	Butanol Synthesis During Carbon Monoxide Fermen	BIOT 0015
on of N-(4-Methoxyphenyl)-2-	Butenamides by Various Arynes.= N- and 4-Arylati	ORGN 0224
ends of Poly(propylene) and Poly(	butene-1).= +iscibility and Cocrystallization in Bl	POLY 0089
opper(I) Oxide from Copper tert-	Butoxide.= +cal Vapor Deposition of Copper and C	INOR 0143
ediated Exchange Reactions with	Butteroil.= +e Preferences for Pancreatic Lipase-M	AGFD 0036
s of Cholesterol Removal from	Butteroil.= Polymer-Supported Tomatine as a Mean	AGFD 0135
Chemisorption of a Tertiary	Butyl Alchohol on Silica Surfaces.=	COLL 0108
ble for the Eclipsing in cis-2-tert-	Butyl-5-(tert-Butylsulfonyl)-1,3-Dioxane.= +onsi	ORGN 0150
lity and Stability of Methyl [1-(	Butylcarbamoyl)-1H-Benz-imidazol-2-YL] Carbama	AGRO 0041
crystallization in Blends of Poly(	butylene Terephthalate) and Poly(ester–Ether) Segm	PMSE 0148
Film–Growth Studies of Di-t-	Butylphosphino(diethyl)borane.= Synthesis and	INOR 0230
Bis(3-tert-	butylpyrazolyl–borate) Nickel(II) Complexes.=	INOR 0035
. Interactions of Sulfide Ion with	Butylpyridinium Chloride in Nonaqueous Media.=	ANYL 0093
psing in cis-2-tert-Butyl-5-(tert-	Butylsulfonyl)-1,3-Dioxane.= +onsible for the Ecli	ORGN 0150
ed Polymers: Vinyl Alcohol/Vinyl	Butyral Copolymers.= +ffusion in Hydrogen–Bond	POLY 0191
myl Acetate, Ethyl Acetate, Ethyl	Butyrate, and Ethyl Caproate.= +fferentiation of A	AGFD 0187
nium Hydroxide Inner Salt and β-	Butyrolactone.= +(4-hydroxy-1-naphthyl)thiophe	POLY 0048
anistic Studies on the Replicative	Bypass of DNA Photoproducts.= +Mutations: Mech	BIOL 0157
Vitro Studies of the Replicative	Bypass of the TpdU cis-syn Cyclobutane Photodime	BIOL 0113
Catalyst.= Combustion	Byproducts of Ethyl Acetate over Pt- Alumina	COLL 0189
Impact of	Byproducts on the Economics of Biomass Refining.=	BIOT 0014

Mussel Byssus.=	Metal-Binding Proteins of the	INOR 0138
with MR₂ (M = Zn, Mg and R = Bz, Cp, allyl).=	+ntation-Recombination Reactions	INOR 0383
Methanol to Olefins over Offret+	¹³C and ¹H MAS NMR Studies of the Conversion of	PETR 0042
Simulation of	¹³C and ¹H NMR Spectra.=	COMP 0002
Temperatures between 90 and+	¹³C CP/MAS NMR Studies of Catalytic Reactions at	COLL 0090
nic Carbon and Reactor-Derived	¹⁴C from the Hudson River to the New York Bight.=	GEOC 0014
in Heavy Petroleum Fractions by	¹³C NMR and Catalytic Dehydrogenation.= +oups	PETR 0147
amples and Other+ Quantitative	¹³C NMR Measurements on the Argonne Premium S	FUEL 0004
lites of N,N-Diethylbenzamide by	¹³C NMR Spectroscopy.= +ver Microsomal Metabo	AGRO 0026
osine)platinum(II) Chloride Using	¹³C NMR Spectroscopy.= +om cis-Diammine(guan	INOR 0251
tary Topological S+ Infrared and	¹³C NMR Spectrum Simulation: Role of Complemen	COMP 0006
	¹³C NMR Spectrum Simulation.=	COMP 0003
Retrieval Techniques for	¹³C NMR Spectrum Simulation.= Database	COMP 0004
Solution and Solid-State	¹³C NMR Studies of C_{60} and C_{70}.=	PHYS 0070
dstocks and Products by ¹H and	¹³C NMR.= Characterization of Delayed Coking Fee	FUEL 0076
namic Behavior of the Hydrox+	¹³C Nuclear Magnetic Relaxation of Amylose and Dy	CARB 0049
COOH)₂, n = 1-4 from ³¹P and	¹³C Solid-State NMR.= Structures of $Zr(O_3P(CH_2)_n$	INOR 0047
or the Production of Fructose and	Ca-Gluconate by Immobilized Cells.= +ioprocess f	BIOT 0152
Characterization of	Cabbage Proteases.= Purification and	AGRO 0019
CaO and	$CaCO_3$.= Catalysis of Char Gasification in O_2 by	FUEL 0030
Qui+ Aplicaciones Recientes de	CAD/CAM para el Diseno de Equipo de la Industria	COMP 0053
eractions of Electronically Excited	Cadmium Atoms with H_2 and CH_4.= +es of the Int	PHYS 0351
ounds onto Cadmium Sulfide and	Cadmium Selenide.= +sorption of Carbonyl Comp	COLL 0175
rystal NMR Studies of a Novel	Cadmium Selenolate Compound, [Cd(Se-2,4,6-i-Pr₃C	INOR 0078
tion of Carbonyl Compounds onto	Cadmium Sulfide and Cadmium Selenide.= +sorp	COLL 0175
Characterization and Control of	Cadmium, Lead, and Mercury from RDF Municipa	FUEL 0130
rations in Lung Tissue: A Com+	Cadmium, Lead, Nickel, Cobalt, and Copper Concent	ENVR 0040
lements in Silica-Saturated CaO-	CaF_2-SiO_2 Slags.= +lubility of Copper and Other E	I&EC 0066
u), Matte (Cu_2S), and Slag (CaO-	CaF_2-SiO_2) Under Controlled Pressures of SO_2.=	I&EC 0065
ation, and Theorectical Study of a	Caffeine Dimer.= +hesis, Spectroscopic Characteriz	INOR 0252
	Cage Effects in Organometallic Photochemistry.=	INOR 0158
ced Catalytic Reactions in Zeolite	Cages.= +talysis: Shape Selectivity—II. Photoindu	CATL 0020
Synthesis of Silicon-Containing	Cages.=	ORGN 0055
Templated Mazzite: A Stabilized	Calcined Structure.= Twelve-Ring Channel	PETR 0037
pillaries:+ In Situ Deposition of	Calcium Carbonate on the Internal Walls of Glass Ca	COLL 0163
ew Morphology of Precipitated	Calcium Carbonate: Synthesis and Characterization.	TECH 0019
ance Study of the System Water-	Calcium Chloride in a Heat Transformer.= +rform	I&EC 0009
Method for the Determination of	Calcium Cyanamide Using Precolumn Derivatization	ANYL 0113
Study+ Nature and Structure of	Calcium Dispersed on Carbon: XANES and EXAFS	FUEL 0032
inding Model for Benzazepinone	Calcium Entry Blockers: Application to the Design o	MEDI 0188
A Study on the Crystal Habit of	Calcium Hydrogen Phosphate Used to Phosphors.=	INOR 0361
e Separation+ Analogies between	Calcium Hydroxyapatite and Zirconium Oxide for th	ANYL 0027
ive Indicators for Measurement of	Calcium Ion Concentration.= +pment of NMR Act	ORGN 0149
Carbon Gasification Catalyzed by	Calcium.= +vidence on the Mechanism of the CO_2	FUEL 0029
and ENDOR Characterization of	Calcium-Binding Sites in Nephrocalcin with Vanady	BIOL 0045
Conformationally Rigid	Calcium-Entry Blocker.= Synthesis of a	ORGN 0114
ich Will Supplant the Hand-Held	Calculator?.= +uation Solvers or Spreadsheets: Wh	CHED 0117
tapas del Lavado Quimico de Una	Caldera por Tecnicas Electroquimicas.= +erentes E	PHYS 0257
Kinetics of DNA Cleavage by	Calicheamicin from Electric Birefringence.=	POLY 0280
standing Its Sequence-Selective+	Calicheamicin: Experimental progress Toward Under	ORGN 0298
balt, Copper, and Manganese from	California Continental Shelf Sediments.= +ron, Co	GEOC 0010
Management: A	California Perspective.= Regulatory Issues in Risk	CHAS 0021
anagement at the University of	California: What Was Revealed and How That Know	CHAS 0002
Influencia del Transporte de	Calor en Reacciones Solido-Gas No-Cataliticas.=	I&EC 0068
ommunity Component of a Trop+	Caloric Variation of Diet in the Dominant Benthic C	BIOL 0084
. Fructooligosaccharides—A Low-	Calorie Bulking Agent Prepared from Sucrose.= +s	CARB 0001
s and Properties of Novel, Zero-	Calorie Sugar Macronutrient Substitutes.= Synthesi	CARB 0026
ine-6-P Deaminase.=	Calorimetric Study of the Denaturation of Glucosam	BIOL 0106
lications of Differential Scanning	Calorimetry and Solvent Swelling for Studies of Coal	FUEL 0087
Time-Resolved Thermal Lens	Calorimetry with a Helium-Neon Laser.=	CHED 0089
nes as Examined by Photoacoustic	Calorimetry.= +of $CpMn(CO)_2$ with Tertiary Sila	INOR 0386
hange Reactions by Photoacoustic	Calorimetry.= +thalpy and Kinetics of Ligand Exc	INOR 0192
Asymmetric Synthesis of	Calyculin A.=	ORGN 0289
3D Search Facilities for the	Cambridge Structural Database (CSD).= Integrated	CINF 0021
	CAMEO: Recent Advances in Reaction Prediction.=	CINF 0008
ntrol Technologies.=	Camet Metal Monolith-Based Catalytic Emission Co	I&EC 0045
el Compounds Possessing Potent	cAMP and cGMP Phosphodiesterase Inhibitory Acti	MEDI 0118
Anthers of Spathodea	Campanulata Flowers.= Chemical Compounds from	ORGN 0186
a Variant Strain of Xanthomonas	campestris NRRL B-1459.= +eous Solutions from	BIOT 0196
It	Can't Happen Here!.=	CHAS 0015
H+ Role of Health and Welfare	Canada in Review of MSDS and Labels Required for	CHAS 0034
Offenses in	Canada.= Chemistry and Crime—Molecular	CHAL 0020
search and Casework in the Royal	Canadian Mounted Police.= +sic DNA Analysis Re	CHAL 0007
d Need+ Qualitative Ranking by	Cancer and Noncancer Risks Combined: Methods an	ENVR 0074
Oligosaccharide Processing in	Cancer Cells: Functional Studies.=	BIOL 0151
Sarcophytol A: A New	Cancer Chemopreventive Agent.=	AGFD 0087
ucuronide Analogues of 4-Hydr+	Cancer Chemopreventive Retinoid Metabolites: C-Gl	MEDI 0030
f Immunoglobulin G Possess S+	Cancer Detection: B-Protein and the Heavy Chains o	BIOL 0068
d Mutagens: Relevance to Human	Cancer Induction. +tally Significant Carcinogens an	CHAS 0007
say.=	Cancer Management: Versatility of the B-Protein As	BIOL 0069

Gallate (EGCG): A	Cancer Preventive Agent.=	(−)-Epigallocatechin	AGFD 0066
l Chloride Examples.= Issues in	Cancer Risk Assessment: Perchloroethylene and Viny		CHAS 0011
ti–Pneumocystis carinii and Anti–	Candida albicans Activities.=	+ides with Potent An	MEDI 0181
in Living Organisms as Possible	Candidates for Principles of Self–Organized Biocomp		BIOT 0148
e Regular Polyhedra in Order To	Canonically Name the Different Possible Isomers of		COMP 0035
O_2 by	CaO and $CaCO_3$.=	Catalysis of Char Gasification in	FUEL 0030
Bi_2O_3–SrO–	CaO System.=	Beta Solid Solution in the	INOR 0201
ther Elements in Silica–Saturated	$CaO–CaF_2–SiO_2$ Slags.=	+lubility of Copper and O	I&EC 0066
l (Cu), Matte (Cu_2S), and Slag (	$CaO–CaF_2–SiO_2$) Under Controlled Pressures of SO_2.		I&EC 0065
$BiO_{1.5}$–SrO–	CaO–CuO System.=	Glass Formation in the	INOR 0224
r Therapeutic Antisense+ The 5'	Cap Structure of Messenger RNA: A Novel Target fo		BIOL 0120
uid Chromatography: Analytical	Capabilities and Applications in Agricultural and Foo		AGFD 0201
oved Nuclear Magnetic Resonance	Capabilities for the Undergraduate Laboratory.=		CHED 0185
olvent Diffusion Coe+ Predictive	Capabilities of a Free–Volume Theory for Polymer–S		POLY 0144
Problem–Solving	Capabilities.=	Using Beilstein's Unique	COMP 0011
e on the Internal Walls of Glass	Capillaries: Influence of Temperature, Flow Rate, Ion		COLL 0163
gh–Temperature Annealing on the	Capillarity of Nylon Microporous Membranes.=		COLL 0037
Mixtures.=	Capillary Electrophoresis of Complex Carbohydrate		ANYL 0019
cules.=	Capillary Electrophoresis Separations for Small Mole		ANYL 0137
	Capillary Electrophoresis.=		AGFD 0179
nd Charge–Reversed Free Solution	Capillary Electrophoresis.=	+ns by Free Solution a	ANYL 0018
Scanning UV–VIS Detection for	Capillary Gas Chromatography Using a Remote Flow		ENVR 0060
on in a Split Injection System for	Capillary Gas Chromatography.=	+ss Discriminati	TECH 0018
fied Polymeric–Mercuric Resin in	Capillary Gas Chromatography.=	+Chemically Modi	ANYL 0083
tical Capabilities and Applicati+	Capillary Supercritical Fluid Chromatography: Analy		AGFD 0201
Molecular Wei+ Use of a Single–	Capillary Viscometer for Accurate Determination of		PMSE 0066
ontrolled Reaction Kinetics in a	Capillary: Reactant Segregation and Anomalous Tim		PHYS 0273
Enhancement in Fluorescence	Capillary/Fiber–Optic Sensors.=	Sensitivity	ANYL 0077
Organostannanes, and Hydroxyl–	Capped Poly(Ethylene Oxide) and ABA Block Copol		PMSE 0006
polymers Containing Hydroxyl–	Capped Poly(Ethylene Oxide) with Polydimethylsilox		PMSE 0006
cetate, Ethyl Butyrate, and Ethyl	Caproate.=	+fferentiation of Amyl Acetate, Ethyl A	AGFD 0187
Degradation Studies of Poly(ϵ–	caprolactone) Fumarate Networks.=	Synthesis and	POLY 0068
fication Using a Cation–Exchange	Capsule and Cation–Exchange HPLC.=	+lysin Puri	ANYL 0081
ministered via a Sustained Release	Captec Device.=	+ues of Albendazole in Cattle Ad	AGRO 0109
Reactions with	Captodative Olefins.=	1,3–Dipolar Cycloaddition	ORGN 0230
n de los Monocitos (FILM) Pro+	Caracterizacion del Factor Inhibidor de la Locomocio		BIOL 0102
de Carga por Espec+ Obtencion,	Caracterizacion, y Determinacion de la Transferencia		INOR 0190
Spectroscopic and Crystallographi	Caracterization of [M(tpen)]($ClO_4)_2$ Complexes.=		INOR 0057
Complexes.= Syntheses and	Caracterization of Transition–Metal Open Fulvalene		INOR 0113
Acetylcholinesterase Inhibitors: 8–	Carba–physostigmine Analogues.=	+hips of Potent	MEDI 0143
moyl)–1H–Benz–imidazol–2–YL]	Carbamate (Benomyl) in Hydrochloric Acid Solution		AGRO 0041
Determination of Methyl	Carbamate in Wines.=	Method for the	AGFD 0110
stra+ Synthesis of Serine–AMC–	Carbamate: A Novel Fluorogenic Tryptophanase Sub		ORGN 0023
Spirolactams (	Carbamates) as Phosphodiesterase Inhibitors.=		MEDI 0189
nthesis of Unsymmetric Ureas or	Carbamates, Sulfamides Using Triphosgene, CBMIT		ORGN 0106
ve Addition of Nitrile–Stabilized	Carbanions to 4–Amino–Substituted Indole Chromiu		ORGN 0010
Effects in the Reactions of Fischer	Carbene Complexes with Alkynes.=	+eoelectronic	INOR 0194
of Chelated Imidazolidone Fischer	Carbene Complexes.=	+ymmetric Aldol Reactions	INOR 0112
Phenoxydiazirines as Useful	Carbene Precursors for Photoaffinity Labeling.=		ORGN 0105
talysts for Enantioselective Metal	Carbene Transformations.=	+arkably Effective Ca	CATL 0002
Mechanism for Reactions of	Carbenes and Bicyclo[1.1.0]Butanes.= A New		ORGN 0147
Rhodium–Stabilized	Carbenoids.=	Heterocyclic Synthesis Using	ORGN 0276
arbonyl Deblocking of Anions of	Carbethoxy–Protected Cyanohydrins with Enones: Sy		ORGN 0228
Ultrafine Iron	Carbide as Liquefaction Catalyst Precursor.=		FUEL 0101
Ultrafine Iron	Carbide as Liquefaction Catalyst Precursor.=		CATL 0051
ior of Dangling Bonds.= Silicon	Carbide Thin–Film Formation on Si(100): The Behav		PHYS 0204
Precursors for LPCVD of Silicon	Carbide.=	Disilacyclobutanes as Single–Source	INOR 0105
Siloxanes as Precursors to Silicon	Carbide.=	+eparation, Characterization, and Use of	POLY 0361
ngement Reactions on Tungsten	Carbides: The Effect of Surface Oxygen on Reaction		CATL 0037
Studies of Six–Membered	Carbo– and Heterocycles.=	Recent Conformational	ORGN 0295
+Synthesis of Highly Substituted	Carboaromatic and Heteroaromatic Compounds.=		ORGN 0286
ed Pericyclic Elimination with a	Carbocation–Like Transition State for the Reaction o		ORGN 0153
the Synthesis of Medium–Sized	Carbocycles: Applications Toward the Total Synthes		ORGN 0282
gical Study of the Cyclopentenyl	Carbocyclic Nucleoside Analogue of 5–Azacytidine.=		MEDI 0048
zation of 5–, 6–, and 7–Membered	Carbocyclic Rings.=	+for Cyclization–O–Functionali	ORGN 0088
sis.=	Carbocyclic Substrates for de Novo Purine Biosynthe		BIOL 0076
Catalyzed by Rh+ Novel Silyl–	Carbocyclization (SiCAC) of Alkynes and Alkenynes		ORGN 0218
nic Acid with Isotopically Labeled	Carbodiimides.=	+ed from the Reaction of Hyaluro	ORGN 0020
$C=CH_2$ (M=Mn, Re) with Imines,	Carbodiimides, and Benzalazine.=	+es $Cp(CO)_2M=$	INOR 0072
roxyalkanoate) Conjugates: PHA–	Carbohydrate and PHA–Synthetic Polymers.=	+yd	POLY 0041
t of the Configuration of Glucos+	Carbohydrate Antigens. Emil Fischer's Establishmen		CARB 0011
e B Conjugate Vaccines.=	Carbohydrate Antigens. Haemophilus Influenzae Typ		CARB 0053
Involved in the Biosynthesis of +	Carbohydrate Antigens. Human Fucosyltransferases		CARB 0015
Structural Studies of Lectin–	Carbohydrate Interactions.=		BIOL 0146
omic Interactions bet+ Protein–	Carbohydrate Interactions—I. Structural Aspects. At		BIOL 0142
tructure of Carbohydr+ Protein–	Carbohydrate Interactions—II. Functional Aspects. S		BIOL 0148
Functional Aspects. Structure of	Carbohydrate Ligands and Their Binding by C–Type		BIOL 0148
of Complex	Carbohydrate Mixtures.=	Capillary Electrophoresis	ANYL 0019
us– and Sulfur–Modified	Carbohydrate Polymer Derivatives.=	New Phosphor	CARB 0030
talyzed Epoxid+ Stereoselective	Carbohydrate Synthesis via Palladium Hydroxide–Ca		ORGN 0074

ation of Microbial–Based Flavor+	Carbohydrate–Based Continuous High–Cell Ferment	CARB	0006
t Ripenin+ Naturally Occurring	Carbohydrate–Containing Regulators of Tomato Frui	AGFD	0031
mbinant Human+ Separation of	Carbohydrate–Mediated Microheterogeneity of Reco	ANYL	0018
for Decoding Information in Co+	Carbohydrate–Recognition Domains as Mechanisms	BIOL	0144
ccharides—A Low–Calorie Bulk+	Carbohydrates in Industrial Synthesis. Fructooligosa	CARB	0001
sis. Keynote Address. Synthesis of	Carbohydrates Using Ezymes as Catalysts.= +the	BIOT	0089
phobic Adsorbe+ Interactions of	Carbohydrates with Ion–Exchange Resins and Hydro	CARB	0063
Surface Adsorption Phenomena of	Carbohydrates. Interaction of Guest Molecules w+	CARB	0061
Benzene, and Cycloheptatriene to	Carbohydrates: A Chemoenzymatic Approach.= +,	ORGN	0272
Anion–Exchange Separation of	Carbohydrates: The Role of Mobile–Phase and Statio	CARB	0066
Interactions between Proteins and	Carbohydrates.= +–I. Structural Aspects. Atomic	BIOL	0142
of	Carbohydrates.= Pulsed Amperometric Detection	CARB	0067
or Chromatographic Separation of	Carbohydrates.= +ion of Cation–Exchange Resins f	CARB	0064
2,3,4–Tetrahydro–β–	Carboline.= Synthesis of 3–Carboxy–1–Pyridoxyl–1,	BIOL	0080
hyl+ Synthesis of Substituted 2–	Carbomethoxy–4'–Methylbiphenyls, 2–Cyano–4'–Met	ORGN	0112
tial Impact of Dissolved Organic	Carbon (DOC) Flux out of Margin Sediments on Oce	GEOC	0017
esorption.= Characterization of	Carbon Active Sites by Temperature–Programmed D	FUEL	0016
ite Behavior via+ Workshop on	Carbon Active Sites. Elucidation of Carbon–Active S	FUEL	0013
Water Vapor Interaction with	Carbon Adsorbents and Sucrose Crystals.= Rate of	CARB	0062
Spheres: The New Generations of	Carbon Allotropes.= +into Rods, Rings, Nets, and	ORGN	0266
mical Synthetic Fla+ The Use of	^{14}Carbon Analysis for the Determination of Petroche	AGFD	0186
of the Middle Atlantic Bight.=+	Carbon and C–14 Budget for Upper Slope Sediments	GEOC	0016
mposites: An Overview of Their+	Carbon and Graphite Matrices in Carbon–Carbon Co	FUEL	0051
Pentafluorosulfanyl (SF$_5$)	Carbon and Nitrogen Derivatives.= Trigeminal	INOR	0024
i+ Fluxes of Dissolved Inorganic	Carbon and Reactor–Derived ^{14}C from the Hudson R	GEOC	0014
The New Generati+ Assembly of	Carbon Atoms into Rods, Rings, Nets, and Spheres:	ORGN	0266
on Fuels: Engineering Materials.	Carbon Black from Coal by the Hydrocarb Process.=	FUEL	0047
Fine–Dispersed	Carbon Bodies.= Interaction between Ozone and	COLL	0109
Regioselective Carbon–	Carbon Bond Cleavage of Cubanes.=	ORGN	0256
,4–Dimethyl–1,3–pent+ Carbon–	Carbon Bond Formation during Thermolysis of Bis(2	INOR	0305
inetic Isotope Effects in Carbon–	Carbon Bond–Forming Reactions of Scandocene Der	INOR	0147
ost–Buckminsterfullerene View of	Carbon Chemistry.= +ocarbon Fuels: Fullerenes. P	FUEL	0019
Spectroscopy of Small	Carbon Clusters in Matrices.= Structure and	PHYS	0078
igh–Level Theoretical Studies of	Carbon Clusters with Coupled–Cluster Methods.= H	PHYS	0008
Supersonically Cooled	Carbon Clusters.= Infrared Laser Spectroscopy of	PHYS	0007
d Graphite Matrices in Carbon–	Carbon Composites: An Overview of Their Formation	FUEL	0051
in Supercritical	Carbon Dioxide Conditions.= Enzyme Inactivation	AGFD	0028
in	Carbon Disulfide.= Gelation of Atactic Polystyrene	POLY	0148
erization of Unactivated Carbon–	Carbon Double Bonds.= +hilic Mechanism for Isom	BIOL	0014
nce on the Mechanism of the CO$_2$	Carbon Gasification Catalyzed by Calcium.= +vide	FUEL	0029
hemical Enzymatic Synthesis of a	Carbon Glycoside of N–Acetyl Neuraminic Acid.=	CARB	0034
Copper–Catalyzed Gasification of	Carbon in Air.= +Chemical Pretreatments on the	FUEL	0028
Gasification of	Carbon in CO$_2$.= Transient Kinetic Study of the	FUEL	0009
Coal and	Carbon in the Postpetroleum World.=	FUEL	0054
Stable	Carbon Isotope Analysis of Coprocessing Materials.=	FUEL	0071
Residues.= TPOP–IRMS Stable	Carbon Isotope Analysis of Heavy Petroleum	PETR	0141
ity in Hydrocracking Reactions+	Carbon Materials as Catalysts—Activity and Selectiv	CATL	0046
hetic Diamond Films, and Related	Carbon Materials by Raman Spectroscopy.= +ynt	PHYS	0118
Production of	Carbon Materials from Biomass.=	FUEL	0052
s.= Design and Synthesis of	Carbon Molecular Sieves for Separation and Catalysi	PETR	0059
One– and Two–	Carbon Molecules.= Complete Basis-Set Energies fo	PHYS	0317
Catalysts for Copolymerization of	Carbon Monoxide and Ethylene.= +ine)palladium	INOR	0155
ion of Butanol Synthesis During	Carbon Monoxide Fermentation by an Acidogenic An	BIOT	0015
Ligation Dynamics of	Carbon Monoxide in Cytochrome c Oxidase.=	BIOL	0096
Methanol and	Carbon Monoxide Production from Natural Gas.=	FUEL	0096
rials. Development and Testing of	Carbon Pistons.= +arbon Fuels: Engineering Mate	FUEL	0050
ongraphitizable+ Preparation of	Carbon Powders by Pyrolysis of Graphitizable and N	POLY	0023
nization Behavior of Graphitizable	Carbon Precursors.= +itu Evaluation of the Carbo	FUEL	0049
a Route to	Carbon Subnitride.= Cyanoazacarbon Derivatives as	INOR	0310
Probing	Carbon Surfaces.= Use of Microcalorimetry for	FUEL	0018
Analysis of Porous	Carbon with ^{129}Xe NMR and N$_2$ Porosimetry.=	PETR	0070
uels: Diamond and Diamond–Like	Carbon. +w Materials Derived from Hydrocarbon F	FUEL	0034
uels: Diamond and Diamond–Like	Carbon. +w Materials Derived from Hydrocarbon F	FUEL	0035
uels: Diamond and Diamond–Like	Carbon. +w Materials Derived from Hydrocarbon F	FUEL	0037
uels: Diamond and Diamond–Like	Carbon. +w Materials Derived from Hydrocarbon F	FUEL	0036
uels: Diamond and Diamond–Like	Carbon. +w Materials Derived from Hydrocarbon F	FUEL	0038
m–Catalyzed CO$_2$ Gasification of	Carbon: Transient Kinetic Investigations.= Potassiu	FUEL	0031
Structure of Calcium Dispersed on	Carbon: XANES and EXAFS Study.= +ture and	FUEL	0032
w Family of Crystalline Phases of	Carbon.= +ond–Like Carbon. Carbophanes: A Ne	FUEL	0034
cleophilic Substitution at Tertiary	Carbon.= +of Cumyl Divratives: Mechanism of Nu	ORGN	0154
Nitride and	Carbon.= Hot–Filament–Activated CVD of Boron	PHYS	0205
roton Migration from Nitrogen to	Carbon.= +istic Investigation of Stereocontrolled P	ORGN	0296
ctivity of Platinum–Doped Glassy	Carbon.= +hesis, Characterization, and Catalytic A	INOR	0312
rbon Active Sites. Elucidation of	Carbon–Active Site Behavior via Desorption Techniq	FUEL	0013
Regioselective	Carbon–Carbon Bond Cleavage of Cubanes.=	ORGN	0256
of Bis(2,4–Dimethyl–1,3–penta+	Carbon–Carbon Bond Formation during Thermolysis	INOR	0305
Steric Kinetic Isotope Effects in	Carbon–Carbon Bond–Forming Reactions of Scandoc	INOR	0147
arbon and Graphite Matrices in	Carbon–Carbon Composites: An Overview of Their F	FUEL	0051
for Isomerization of Unactivated	Carbon–Carbon Double Bonds.= +hilic Mechanism	BIOL	0014
ects on Reactivity.=	Carbon–Cluster–Ion Reactions: Structural Isomer Eff	PHYS	0311

erization of Secondary Alcohols:	Carbon–Hydrogen Bond Activation in Rhenium Alko	ORGN	0220
ntioselectivity by Intramolecular	Carbon–Hydrogen Insertion Reactions of Alkyl Diazo	ORGN	0042
nal Groups on the Performance of	Carbon-Supported Catalysts.= +of Oxygen Functio	FUEL	0033
Synthetic Studies for	Carbon-11–Labeled Alfentanil.=	ORGN	0192
Organofluorides Using Fluorine-19	Carbon-13 Coupling.= +ion of Local Structure in	FUEL	0077
Synthesis of	Carbon-14– and Tritium–Labeled Panadiplon.=	MEDI	0133
Molecular–Beam Studies of	Carbonaceous Clusters.=	FUEL	0110
rolysis of an Aqueous Potassium	Carbonate Electrolyte and the Implications for Cold	PHYS	0286
I In Situ Deposition of Calcium	Carbonate on the Internal Walls of Glass Capillaries:	COLL	0163
orphology of Precipitated Calcium	Carbonate: Synthesis and Characterization.= +M	TECH	0019
rating Polymer Networks of Poly(	Carbonate–Urethane) and Polyvinyl Pyridine.= +et	PMSE	0056
Pellets as a Function of Time, pH,	Carbonate/Bicarbonate, and Oxygen Activities.=	INOR	0236
Mechanism of	Carbonic Anhydrase.= Active–Site Ionicity and the	BIOL	0063
Active Site of	Carbonic Anhydrase.= Cobalt Spectroscopy at the	PHYS	0256
cursor+ In Situ Evaluation of the	Carbonization Behavior of Graphitizable Carbon Pre	FUEL	0049
Production of Activated	Carbons from Coals by Chemical Activation.=	FUEL	0048
tion on Graphite and Microporous	Carbons: An Analysis of Experimental Data.= +orp	FUEL	0017
en Chemisorption on Microporous	Carbons.= +cation Mechanisms. Modeling of Oxyg	FUEL	0012
g the Reactive Surface Area of	Carbons.= Use of Transient Kinetics for Determinin	FUEL	0014
Gasification of Porous	Carbons/Chars.= Role of Activated Diffusion in the	FUEL	0010
ative Organic Chemistry of	Carbonyl and Organometallic Compounds.= Compar	CHED	0097
lusters: Ruthenium and Osmium	Carbonyl Clusters Containing One or More Boron At	INOR	0220
ers Using a Luminescent Rhenium	Carbonyl Complex as a Spectroscopic Probe.= +ym	POLY	0096
stitution Reactions for Group VI	Carbonyl Complexes of a New Ditertiary Phosphine	INOR	0333
Di– and Polynuclear Rhenium	Carbonyl Complexes.=	INOR	0107
e as a Probe of the Adsorption of	Carbonyl Compounds onto Cadmium Sulfide and Ca	COLL	0175
metric Synthesis of α–Hydroxy	Carbonyl Compounds Using Enantiopure N–Sulfonyl	ORGN	0293
l Addition Ethoxycarbonylation–	Carbonyl Deblocking of Anions of Carbethoxy–Prote	ORGN	0228
the	Carbonyl Group.= One-Pot (Gem-Dimethylation of	ORGN	0222
uced Coupling of Alkylidyne and	Carbonyl Ligands by Nucleophiles and by Electrophi	INOR	0188
igration Reactions of Unusuall+	Carbonylation and Phosphine–Promoted Alkyl–CO M	INOR	0239
cally by d^8 Transition–Metal C+	Carbonylation of Cyclohexane Catalyzed Photochemi	INOR	0254
Reactive Intermediates in the	Carbonylation of Metal Alkyl Bonds Studied by Tim	INOR	0169
ase–Transfer Catalysts+ Double	Carbonylation Reactions Catalyzed by Nickel and Ph	COLL	0117
ysts: New Mechanisms for Double	Carbonylation Reactions.= +Phase–Transfer Catal	COLL	0117
Means of Homogeneous Catalytic	Carbonylation.= +icyclic Nitrogen Heterocycles by	ORGN	0049
hemically by d^8 Transition–Metal	Carbonyls in the Presence of Aromatic Ketones.=	INOR	0254
Photocatalytic Behavior of Metal	Carbonyls on Porous Vycor Glass.=	INOR	0170
Metal	Carbonyls with NF_3.= Photochemical Reactions of	INOR	0195
emical–Vapor Deposition of Metal	Carbonyls.= +Microscopy (STM) Studies of the Ch	COLL	0050
ond and Diamond–Like Carbon.	Carbophanes: A New Family of Crystalline Phases of	FUEL	0034
nantioselect+ Chiral Rhodium(II)	Carboxamides: Remarkable Effective Catalysts for E	CATL	0002
Catalyzed by Chiral Rhodium(II)	Carboxamides.= +Reactions of Alkyl Diazoacetates	ORGN	0042
xazolyl)-2H-1,2-Benzothiazine-3-	Carboxamides, 1,1-Dioxide.= +5-Substituted-3-Iso	MEDI	0178
e.= Synthesis of 3-	Carboxy-1-Pyridoxyl-1,2,3,4-Tetrahydro-β-Carbolin	BIOL	0080
oly(ester crown) Based on bis(5-	Carboxy-1,3-phenylene)-32-crown-10 and 4,4'-Isopr	POLY	0044
on Chemokinetic Activity of C5A	Carboxyl Terminal Octapeptide Analogues.= +ation	MEDI	0168
m Ribulose Bisphosphate (RuBP)	Carboxylase/Oxygenase.= +ent Mutant of R. rubru	BIOL	0089
ydroxo)bisμ–carbo+ New Triiron	Carboxylate Core Structure Consisting of fused (μ–H	INOR	0094
cological Evaluation of Novel 6-	Carboxylate Esters of Prednisolone: Analogues of Me	MEDI	0170
-3,20-Dioxo-Pregna-1,4-Diene-6-	Carboxylate of Methyl 11β, 17α, 21-Trihydroxy	MEDI	0170
sisting of fused (μ–Hydroxo)bisμ–	carboxylato)diiron(III) Kernals.= +e Structure Con	INOR	0094
oute to Target Chiral Molecules+	Carboxylesterase–Mediated Reactions: A Versatile R	AGFD	0008
ylene.= Analysis of	Carboxylic Acid Metabolites of Polychlorotrifluoroeth	ANYL	0062
rization of Hydroxy Polystyrene	Carboxylic Acid Telechelomers, II: A Model Study of	POLY	0046
Facile Synthesis of a	Carboxylic Linker on Halofuginone.=	ORGN	0201
unds as Bicyclic Ring-Constrai+	Carboxynaphthyl, -Quinolyl and -Isoquinolyl Compo	MEDI	0023
f a True Reaction Intermediate of	Carboxypeptidase A.= +ructure at the Active Site o	BIOL	0011
Synthesis of 4-Amido 2-	Carboxytetrahydroquinolines.=	ORGN	0117
Photooxidation of Metal	Carbyne Complexes.=	INOR	0341
c Steroids in Muscle Tissues from	Carcasses with Positive Injection Sites.= +Anaboli	AGFD	0130
tive DNA Damage.=	Carcinogen–Mediated Oxidant Formation and Oxida	CHAS	0008
= Mechanisms of	Carcinogenesis and Mutagenesis by N-Nitrosamines.	CHAS	0010
Stomach, Lung, and Esophagus	Carcinogenesis by Green Tea.= Protection against	AGFD	0067
against Liver, Colon, and Tongue	Carcinogenesis by Plant Phenols.= +tective Effects	AGFD	0090
Orally Administered Green Tea on	Carcinogenesis by Ultraviolet B Light.= +fect of	AGFD	0065
+ol-8,9-Epoxide Diastereomers of	Carcinogenic Dibenz[a,h]Acridine (DB[a,h]ACR).=	ORGN	0208
lar and Cellular Mechanisms of	Carcinogenic Metal and Polycyclic Aromatic Hydroca	CHAS	0007
enicity and Modification of	Carcinogenic Response by Plant Phenols.= Carcinog	AGFD	0115
isms Underlying the Mutagenic,	Carcinogenic, and Chemopreventive Effects of Pheno	AGFD	0086
sponse by Plant Phenols.=	Carcinogenicity and Modification of Carcinogenic Re	AGFD	0115
of Environmentally Significant	Carcinogens and Mutagens: Relevance to Human Can	CHAS	0007
Exposures.= Case Studies of	Carcinogens: Environmental or Occupational	CHAS	0020
Mutagenesis and	Carcinogens.= Plant Phenolics as Inhibitors of	AGFD	0118
n-Antibody Conjugates to a Colon	Carcinoma Cell Line.= +toxicity of ^{67}Cu-Porphyri	BIOT	0168
of Novel Arylpipera+ Synthesis,	Cardiac Electrophysiological, and β–Blocking Activity	MEDI	0094
ses from the National Institute of	Cardiology (1984-1989).= +arative Study in 84 Ca	ENVR	0040
quinoxalinones,+ Synthesis and	Cardiovascular Effects of a Series of Imidazo[1,2-a]-	MEDI	0118
Personal	Care Industry.= Analytical Chemistry in the	CHED	0050
stry in the Cosmetic and Personal	Care Industry—I. Overview of Fragrance C+ Chemi	CHED	0046

stry in the Cosmetic and Personal	Care Industry—II. Silicone in Hair Conditi+ Chemi	CHED 0063
Polymers in Health	Care.= Contemporary Role of Biodegradable	CHED 0135
Distribution: A Multifaceted	Career Opportunity for Chemists.= Information	YCC 0008
	Career Options in Consulting.=	CMEC 0029
Technical	Career Paths in a Specialty Chemical Company.=	YCC 0007
The Life of a Chemist:	Career Possibilities in Mexico.=	CONG 0007
and	Career Services.= The Use of ACS Employment	YCC 0013
s Insights Program (Roadshow)—	Careers for Chemists in Industry.+ Chemical Career	YCC 0005
Chemical	Careers in Academe.=	I&EC 0003
Chemical	Careers in Academe.=	CMEC 0025
Nontraditional	Careers in Chemistry.=	CMEC 0027
Nontraditional	Careers in Chemistry.=	I&EC 0004
Chemical	Careers in Government.=	CMEC 0024
Chemical	Careers in Government.=	I&EC 0002
Chemical	Careers in Industry: Between Research and Sales.=	YCC 0011
Chemical	Careers in Industry.=	I&EC 0001
Chemical	Careers in Industry.=	CMEC 0023
	Careers in Scientific Librarianship.=	YCC 0009
t the Bristol-Myers Squibb Pha+	Careers in the Pharmaceutical Industry: Chemistry a	CMEC 0028
hemists in Industry.+ Chemical	Careers Insights Program (Roadshow)—Careers for C	YCC 0005
minacion de la Transferencia de	Carga por Espectroscopia IR de Complejos de Transf	INOR 0190
de Complejos de Transferencia de	Carga Pt(R-Nafin)·TCNQ.= +r Espectroscopia IR	INOR 0190
es with Potent Anti-Pneumocystis	carinii and Anti-Candida albicans Activities.= +id	MEDI 0181
Molecules.= Dynamic Monte	Carlo Simulations of Rings and Branched	POLY 0297
of Polymeric Systems from Monte	Carlo Simulations.= +of the Chemical Potentials	POLY 0352
Chester	Carlson—Electrophotography.=	CHAL 0028
in the Antioxidant Mechanism of	Carnosine in the Presence of Copper and Iron.= +s	AGFD 0039
nd Tocotrienols on Plant Colorant	Carotenoids.= +ioxidant Activity of Tocopherols a	AGRO 0039
en las Materias Biologicas de la	Carrera de QFB en la Facultad de Quimica, UNAM (	CHED 0092
e Go from the Bench Top to C+	Carrier and Catalyst-Forming Processes—Or, How W	PETR 0089
f Surface Properties on Charge-	Carrier Dynamics of Quantized Semiconductor Colloi	COLL 0151
Putting the Horse before the	Cart: Exploring the Melting Point of Mixtures.=	CHED 0159
the Past+ Collection of Favorite	Cartoons Appearing in CHEMTECH Magazine over	CHAL 0024
Favorite or Objectionable	Cartoons Related to Chemists or Chemicals.=	CHAL 0021
Favorite or Objectionable	Cartoons Related to Inventors or Patents.=	CHAL 0022
Favorite or Objectionable	Cartoons Related to Laws and Lawyers.=	CHAL 0023
Drying Fuels with the	Carver-Greenfield Process.=	FUEL 0116
ated Molecular Property Data for	CAS Registry Substances.= +enerated 3D and Rel	COMP 0041
3D Search Routines for Searching	CAS Registry Substances.= +on of Atom-by-Atom	CINF 0023
	Cascadas Reticulares.=	NUCL 0087
Cationic	Cascade Molecules.=	POLY 0237
er Simulation Studies of Collision	Cascades in Solids.= +olecular Dynamics Comput	NUCL 0091
on S. aureus y s+ Elaboracion de	Caseinato de Sodio a Partir de Leche Contaminada c	AGFD 0107
nsic DNA Analysis Research and	Casework in the Royal Canadian Mounted Police.=	CHAL 0007
he Application of PCR to Forensic	Casework Samples.= +New Genetic Markers for t	CHAL 0004
sa Seed Oil Extracts to Tribolium	castaneum L. (Coleopt+ Toxicity of Annona squamo	AGRO 0036
ng Polymer Network Formed from	Castor Oil Polyurethane and Vinyl Copolymers.=	PMSE 0089
ies of Room Temperature Cured	Castor Oil Polyurethane/Vinyl or Acrylic Polymer In	PMSE 0090
etrating Polymer Networks from	Castor Oil Polyurethanes and Poly(Styrene-Methyl	PMSE 0091
tadium L+ In Vivo and in Vitro	Catabolism Study of Juvenile Hormone in the Fifth S	AGRO 0022
Complexes.= Understanding the	Catalase-Like Reactivity of Dinuclear Manganese	INOR 0010
sil.=	Catalisis de Reacciones via Radicales Libres con Ton	COLL 0119
Reacciones Solido-Gas NO-	Cataliticas.= Influencia del Transporte de Calor en	I&EC 0068
roxilacion del Benceno con H2O2	Catalizada por Complejos Derivados de Cupferron.=	COLL 0121
con Boro.= Estudio de los	Catalizadores CoMo/Al2O3 y Mo/Al2O3 Modificados	FUEL 0104
Alumina-+ HDS de Tiofeno con	Catalizadores de Molibdeno sobre Oxidos Mixtos de	FUEL 0105
nia Modificadas con+ Estudio de	Catalizadores NiW Soportados sobre Alumina o Tita	FUEL 0106
d Plastic and Conserved Phases of	Catalysis by Glycosylases.= +Separately Controlle	BIOL 0152
Rate and Equilibri+ Interfacial	Catalysis by Phospholipase A2: Determination of the	BIOL 0019
c Characterization of+ Interfacial	Catalysis by Phospholipase A2: Structural and Kineti	BIOL 0020
c Reduction of NOx Using Zeolitic	Catalysis for High-Temperature Applications.= +ti	I&EC 0046
ase: The Importance of Dynamic	Catalysis for Successful Mechanism-Based Inhibitor	BIOL 0054
n, Characterization, and	Catalysis of a Modified ZSM-5 Zeolite.= Preparatio	CATL 0029
O3.=	Catalysis of Char Gasification in O2 by CaO and CaC	FUEL 0030
.=	Catalysis of Enolate Reactions by an Articial Enzyme	ORGN 0053
cking of Coal-Derived Distillate+	Catalysis of Metal-Modified Y-Zeolites for Hydrocra	CATL 0050
ces: Mode of Interfacial Binding+	Catalysis of Porphyrin Metalation at Anionic Interfa	CATL 0060
s: Product Selectivi+ Copper Ion	Catalysis of Propylene Oxidation on X and Y Zeolite	CATL 0017
Coordination Chemistry and	Catalysis of V, Mo, Co, Ni, and Cu Compounds: An E	CHED 0057
Photochemistry and Redox	Catalysis Using Re and Mo Complexes.=	INOR 0343
mical Model.= Oxygen Redox	Catalysis Using RuO2 Dispersions and the Electroche	COLL 0134
phosul+ Chemistry of Enzymatic	Catalysis. Mechanistic Studies of Adenosine-5'-Phos	BIOL 0008
Selectivity in Catalysis: Oxidation	Catalysis. Reaction-Transport Selectivity Models+	CATL 0056
n of Alkane Dehy+ Selectivity in	Catalysis: Clusters, Alloys, and Poisoning. Compariso	CATL 0007
Enzymes.= Targeted	Catalysis: Conferment of Substrate Selectivity on	BIOL 0090
I. Methane+ Surface Science of	Catalysis: Environmental Applications of Catalysts—	COLL 0185
II. Thermal+ Surface Science of	Catalysis: Environmental Applications of Catalysts—	COLL 0204
odel Compound+ Selectivity in	Catalysis: Fuel Chemistry—I. Characterization and M	CATL 0039
electivity Models+ Selectivity in	Catalysis: Oxidation Catalysis. Reaction-Transport S	CATL 0056
Catalytic Chemis+ Selectivity in	Catalysis: Reaction Pathway Control—I. Crystal and	CATL 0026

of the BOC-M+ Selectivity in	Catalysis: Reaction Pathway Control—II. Application	CATL 0032
Heterogeneous Ca+ Selectivity in	Catalysis: Shape Selectivity—I. Dynamic Sieving by	CATL 0013
ytic Reactions in+ Selectivity in	Catalysis: Shape Selectivity—II. Photoinduced Catal	CATL 0020
on Technology for+ Selectivity in	Catalysis: Stereoselectivity. Asymmetric Hydrogenati	CATL 0001
ity. Structural Basis of Enzymatic	Catalysis.= +lyst Design for Stability and Specific	BIOT 0239
ures and Innovations in Industrial	Catalysis.= +rial Chemistry Award). Some Advent	I&EC 0043
Better the Binding, the Better the	Catalysis.= +Cellulases on Cellulose Surface: The	BIOT 0186
lar Sieves for Separation and	Catalysis.= Design and Synthesis of Carbon Molecu	PETR 0059
for Asymmetric	Catalysis.= New Electron-Rich Chiral Phosphines	CATL 0004
Heterogeneous	Catalysis.= Oligomerization of Lower Olefins by	PETR 0009
vated Metal Atoms: Structure and	Catalysis.= +imetallic Particles Prepared from Sol	CATL 0009
Bimetallic Complexes and Their	Catalysis.= Syntheses of Novel Rh–Co Mixed	INOR 0226
g Intermediates in Heterogeneous	Catalysis.= +ns of n–Quantum Coherence in Probin	COLL 0089
e Steady States and Changes in	Catalyst Activity in Partial Oxidation Reactor Dynam	I&EC 0069
of Spray Drying in Production of	Catalyst and Catalyst Supports.= +terization. Role	PETR 0088
fication Method of Pt/KL Zeolite	Catalyst for Light Naphtha Reforming.= +w Modi	PETR 0064
rphously Substituted Ga–β as a	Catalyst for Propane Aromatization—Infrared Spectro	PETR 0047
ward Address. Global Change: A	Catalyst for the Development of Hydrologic Science.	AGRO 0051
: Carbon–Hydrogen Bond+ New	Catalyst for the Epimerization of Secondary Alcohols	ORGN 0220
Derivatives.= 18-Crown-6 as a	Catalyst in the Dialkylation of o–Nitrophenacyl	ORGN 0261
and Investigation of Hemin:	Catalyst of Oxidation Processes.= Immobilization	COLL 0102
Al–Modified Sepiolite as	Catalyst or Catalyst Support.=	PETR 0015
k Composition and Hydrocracking	Catalyst Performance.= +etween Zeolite Framewor	I&EC 0062
Liquefaction	Catalyst Precursor.= Ultrafine Iron Carbide as	FUEL 0101
Liquefaction	Catalyst Precursor.= Ultrafine Iron Carbide as	CATL 0051
phorous–Promoted Hydrotreating	Catalyst Precursors.= +S NMR, and XRD of Phos	COLL 0198
riarylphosphines and Its Role on	Catalyst Stability in Homogeneous Transition-Metal	INOR 0349
Catalysts.= Effects of	Catalyst Structure on the Performance of V_2O_5/TiO_2	COLL 0206
n and Characterization of Novel	Catalyst Support Materials: The Hydrous Metal Oxid	PETR 0018
Catalyst or	Catalyst Support.= Al–Modified Sepiolite as	PETR 0015
Alkali Earth Salts as	Catalyst Supports and Stabilizers.=	PETR 0034
Electrically Conductive	Catalyst Supports from Fibrous Precursors.=	PETR 0016
erization. In Situ Pore-Structur+	Catalyst Supports: Chemistry, Forming, and Charact	PETR 0068
erization. Microporous Oxides b+	Catalyst Supports: Chemistry, Forming, and Charact	PETR 0031
erization. Pillared Clays as Sup+	Catalyst Supports: Chemistry, Forming, and Charact	PETR 0013
erization. Role of Spray Drying +	Catalyst Supports: Chemistry, Forming, and Charact	PETR 0088
erization. Strong Solid Acid Cat+	Catalyst Supports: Chemistry, Forming, and Charact	PETR 0049
ing in Production of Catalyst and	Catalyst Supports.= +terization. Role of Spray Dry	PETR 0088
etry of Alumina and Silica	Catalyst Supports.= Electron Energy–Loss Spectrom	PETR 0073
TiO_2 to	Catalyst Supports.= Forming of High-Surface-Area	PETR 0091
Magnet+ Studies of Dynamics at	Catalyst Surfaces by Selective Excitation of Nuclear	PHYS 0346
f a Mixed Rhodium/Ruthenium	Catalyst System to Hydrogenate Crude Methylenedia	I&EC 0056
ays as Supports in Hydrocracking	Catalyst Systems.= +Characterization. Pillared Cl	PETR 0013
Curriculum.= Project	Catalyst: A New Technology-Enhanced Chemistry	ANYL 0009
Pentenes Using Sulfuric Acid as a	Catalyst: Chemistry and Reaction Mechanisms.=	PETR 0082
over Pt– Alumina	Catalyst.= Combustion Byproducts of Ethyl Acetate	COLL 0189
of H_2S as a Hydrogen Transfer	Catalyst.= Free-Radical Alkylation in the Presence	PETR 0083
Zeolitic Isomerization	Catalyst.= Improvements on Both Amorphous and	PETR 0115
Palladium Monolith	Catalyst.= Kinetics of Methanol Combustion on	COLL 0209
Conversion	Catalyst.= Magnetic Separation of Reduced Crude	I&EC 0063
Semiconductor as	Catalyst.= Photoreduction of Aldehydes with	ORGN 0247
Commercial Automotive Exhaust	Catalyst.= +and CO–NO Reactions over a Typical	COLL 0187
he Bench Top to+ Carrier and	Catalyst-Forming Processes—Or, How We Go from t	PETR 0089
Stable Heterogeneous Oligome+	Catalysts and Conditions for the Highly Efficient and	PETR 0008
m(III) RNA Transesterification	Catalysts and the Design of Artificial Restriction Enz	INOR 0406
and Vanadium/TiO_2(110) Model	Catalysts by Electron Spectroscopies.= +TiO_2(110)	COLL 0192
y Coke Formation of ZSM-5 Type	Catalysts during Oligomerization of Ethene.= +n b	PETR 0027
king and Regeneration of Pd/HM	Catalysts for Alkane Isomerization.= +ation by Co	PETR 0130
Industrial	Catalysts for Aromatics Reduction in Gas Oil.=	CATL 0048
velopment of High-Performance	Catalysts for Aromatization of Light Alkanes and Ole	PETR 0063
and General+ Zeolite-Containing	Catalysts for C_1–C_4 Aromatization: Characterization	PETR 0067
Investigation of Reactions and	Catalysts for C_3–C_4 Paraffin Dehydrogenation and A	PETR 0066
and E+ Bis(phosphine)palladium	Catalysts for Copolymerization of Carbon Monoxide	INOR 0155
oxamides: Remarkably Effective	Catalysts for Enantioselective Metal Carbene Transf	CATL 0002
ation and Reactivity of Cu/ZnO	Catalysts for Methanol Synthesis Obtained by an Alk	INOR 0071
Transition Metals as	Catalysts for Pyrolysis and Gasification of Biomass.=	FUEL 0103
es in+ Design of Pt–Mordenite	Catalysts for the Isomerization of C_5 and C_6 N–Alkan	PETR 0113
hase of GaHMFI Aromatization	Catalysts in the Course of the Preparation Procedure	PETR 0046
f Alumina-Supported Vanadium	Catalysts in the Hydrodemetallization (HDM) of NI–	CATL 0044
lybdate Hydrocarbon Oxidation	Catalysts Prepared by the High-Temperature Aeroso	CATL 0026
s of Platinum and Platinum–Tin	Catalysts Supported on Mg, Cu, Zn, and Ni Alumina	COLL 0126
Extruded Monolithic	Catalysts Supports.=	PETR 0090
Characterization of	Catalysts Supports.= Surface Area and Porosity	PETR 0069
lectivity of Supported Bimetallic	Catalysts Toward Catalytic Reforming: The Reaction	CATL 0042
tion of Ammonia on V_2O_5/TiO_2	Catalysts Used for the Selective Catalytic Reduction	COLL 0211
Hydrocarbons over Heterogeneous	Catalysts. +tion, and Isomerization of Short-Chain	PETR 0082
Hydrocarbons over Heterogeneous	Catalysts. +tion, and Isomerization of Short-Chain	PETR 0130
Hydrocarbons over Heterogeneous	Catalysts. +tion, and Isomerization of Short-Chain	PETR 0043
Hydrocarbons over Heterogeneous	Catalysts. +tion, and Isomerization of Short-Chain	PETR 0112
Hydrocarbons over Heterogeneous	Catalysts. +tion, and Isomerization of Short-Chain	PETR 0007

Hydrocarbons over Heterogeneous	Catalysts. +tion, and Isomerization of Short-Chain	PETR	0100
Hydrocarbons over Heterogeneous	Catalysts. +tion, and Isomerization of Short-Chain	PETR	0025
Hydrocarbons over Heterogeneous	Catalysts. +tion, and Isomerization of Short-Chain	PETR	0061
Using Syngas.= Selectivity in	Catalysts: Fuel Chemistry—II. Coal/Oil Coprocessing	CATL	0049
d by Nickel and Phase-Transfer	Catalysts: New Mechanisms for Double Carbonylatio	COLL	0117
di+ Sulfated Zirconia Superacid	Catalysts: Thermal Analysis and Characterization Stu	PETR	0028
ns by Homogeneous Nickel-Based	Catalysts.= +erization and Isomerization of Olefi	PETR	0007
haracterization. Strong Solid Acid	Catalysts.= +Supports: Chemistry, Forming, and C	PETR	0049
f Carbohydrates Using Ezymes as	Catalysts.= +thesis. Keynote Address. Synthesis o	BIOT	0089
Roth. Commercial Polypropylene	Catalysts.= +ial Catalytic Processes Honoring J. F.	I&EC	0052
zation Data for Pt-Sn-Aluminum	Catalysts.= +rocyclization Activity and Characteri	CATL	0007
uthenium-Based Fischer-Tropsch	Catalysts.= +lectivity Models and the Design of R	CATL	0056
ynamic Sieving by Heterogeneous	Catalysts.= +in Catalysis: Shape Selectivity—I. D	CATL	0013
Alumina-Supported Noble Metal	Catalysts.= +atalysts—I. Methane Oxidation over	COLL	0185
over Zeolite	Catalysts.= Alkylation of Isoparaffins with Olefins	PETR	0101
W-Ir and Mo-Ir Cluster-Derived	Catalysts.= +rogenolysis Selectivities Exhibited by	CATL	0008
Alcohols Using Discrete Rhodium	Catalysts.= +n of Ketones: A Facile Route to Ciral	INOR	0247
Modified Shape-Selective Zeolite	Catalysts.= +cetylene to Higher Hydrocarbon over	CATL	0034
New Asymmetric Oxidation	Catalysts.= Development and Mechanistic Study of	ORGN	0132
Diene Synthesis by Solid Acid	Catalysts.=	PETR	0029
ning Automotive Emission Control	Catalysts.= +sicochemical Properties of Pt-Contai	COLL	0208
kaline Earth-Doped Pt, Rh/CeO$_2$	Catalysts.= +ce of Pt, Rh/CeO$_2$ and Rare Earth/Al	PETR	0035
ure-Reactivity of Vanadium Oxide	Catalysts.= +Support and Promoters on the Struct	CATL	0057
Performance of V$_2$O$_5$/TiO$_2$	Catalysts.= Effects of Catalyst Structure on the	COLL	0206
Dispersed Metal	Catalysts.= Ethylene Dimerization over Highly	PETR	0010
Acid	Catalysts.= Evaluation of Acidity of Strong	ORGN	0148
olecular Organizates and Colloidal	Catalysts.= +-Controlled Reactive Processes on M	PHYS	0063
erformance of Carbon-Supported	Catalysts.= +of Oxygen Functional Groups on the P	FUEL	0033
electivity over Nonmicroporous Pt	Catalysts.= +port Preparation on Aromatization S	CATL	0030
ight Irradiation of Semiconductor	Catalysts.= +l Organics Using AM1 Simulated Sunl	INOR	0318
sivating Agents in Fluid-Cracking	Catalysts.= +ation of Metal Contaminants and Pas	CATL	0012
Vanadium Oxide	Catalysts.= Molecular Engineering of Supported	COLL	0196
H-[Al]-ZSM-5	Catalysts.= Naphthene Transformation over	PETR	0044
Hydrogenation	Catalysts.= New Synthesis of Asymmetric	INOR	0077
Oxide-Supported	Catalysts.= Reactions of Propane on Hydrous Metal	CATL	0040
Hydrocarbon Isomerization	Catalysts.= Scientific Foundations for Synthesis of	PETR	0135
on Polyfunctional Metallosilicate	Catalysts.= +Aromatization of Light Hydrocarbons	PETR	0062
Transition Metal Sulfide	Catalysts.= Structure-Function Relations in	CATL	0038
ity of HM Zeolite and Pd/HM	Catalysts.= Studies on the Activity and Surface Acid	PETR	0117
Acid	Catalysts.= Supported Fluorocarbonsulfonic	PETR	0053
Titania-Silica-Based DENO$_x$	Catalysts.= Surface Characterization of	PETR	0071
Triflic Acid-Modified Y-Zeolite	Catalysts.= Synthesis of MTBE and ETBE over	CATL	0016
Reactions+ Carbon Materials as	Catalysts—Activity and Selectivity in Hydrocracking	CATL	0046
is: Environmental Applications of	Catalysts—I. Methane+ Surface Science of Catalys	COLL	0185
is: Environmental Applications of	Catalysts—II. Thermal+ Surface Science of Catalys	COLL	0204
ynthesis, Molecular Structure, and	Catalytic Activity in the Oxidation of α-Olefins.=	INOR	0365
Synthesis, Characterization, and	Catalytic Activity of LaBo$_3$ (B = Ni, Mn, Co, Fe) P	PHYS	0225
Synthesis, Characterization, and	Catalytic Activity of Platinum-Doped Glassy Carbon	INOR	0312
Synthesis and	Catalytic Activity When Using a Type 2 Titanate.=	PETR	0017
h(PMe$_3$)$_2$LC+ Efficient Thermal	Catalytic Alkane Transfer-Dehydrogenation Using R	INOR	0151
s Shift Reaction over Ritheniu+	Catalytic and Spectroscopic Studies of the Water Ga	CATL	0027
	Catalytic Antibodies.=	BIOL	0028
Industrial	Catalytic Applications of Molecular Sieves.=	I&EC	0061
ometal Complexes—Synthetic and	Catalytic Applications.= +HCl from Chlorohydrid	INOR	0381
Using a New Class of D4-Sy+	Catalytic Asymmetric Epoxidation of Simple Alkenes	INOR	0300
	Catalytic Asymmetric Heck Reaction.=	CATL	0006
I) Complexes and Their Roles in	Catalytic Asymmetric Hydrogenation of β-Keto Este	ORGN	0044
ocycles by Means of Homogeneous	Catalytic Carbonylation.= +icyclic Nitrogen Heter	ORGN	0049
bon Oxidation Cat+ Crystal and	Catalytic Chemistry of Bismuth Molybdate Hydrocar	CATL	0026
Oxidation Pathways for Selective	Catalytic Conversion of Alkanes to Alcohols.= +l of	CATL	0036
de.=	Catalytic Conversion of Natural Rubber to Polyepoxi	ENVR	0045
ionic Tungsten (0) Complexes.=+	Catalytic Decarboxylation of Cyanoacetic Acid by An	INOR	0212
Inhib+ Mechanistic Proposal for	Catalytic Dehydrocyclization Rates on Pt/L-Zeolite:	CATL	0015
and the Control of Selectivity in	Catalytic Dehydrogenation and Dehydrocyclization o	CATL	0061
oleum Fractions by ^{13}C NMR and	Catalytic Dehydrogenation.= +oups in Heavy Petr	PETR	0147
s Designed to Interact with the	Catalytic Domain.= New Protein Kinase C Inhibitor	MEDI	0076
Argonne Premium Coal Sample+	Catalytic Effect on the Gasification of a Bituminous	FUEL	0006
Camet Metal Monolith-Based	Catalytic Emission Control Technologies.=	I&EC	0045
ion: Characterization and General	Catalytic Features.= +talysts for C$_1$-C$_4$ Aromatizat	PETR	0067
thesis of Branched Aldehydes by	Catalytic Hydroformylation with Cationic Bis(dioxap	INOR	0154
Asymmetric	Catalytic Hydrogenation.= Mechanistic Aspects of	CATL	0003
l Complexes.=	Catalytic Hydrosilation of Iron and Ruthenium Acety	INOR	0238
=	Catalytic Hydrotreatment of Coal-Derived Naphtha.	CATL	0052
Mechanism and Selectivity in	Catalytic Olefin Polymerization.=	CATL	0010
plexes.=	Catalytic Oligomerization of Ethylene by Ti(II) Com	INOR	0153
on Al$_2$O$_3$/Monolith.=	Catalytic Oxidation of Trichloroethylene over a PdO	COLL	0210
t on Al$_2$O$_3$/Monolith.=	Catalytic Oxidation of Trichloroethylene Using 1.5 P	COLL	0205
Development and Experience with	Catalytic Ozone Abatement in Jet Aircraft.= +cial	I&EC	0047
Model Hydrogenation on a	Catalytic Palladium Membrane.= Effects of a	CATL	0011
Molecular Sieve with High The+	Catalytic Performance of Pillared Interlayered Clay	PETR	0030

Chemistry Award New Industrial	Catalytic Processes Honoring J. F. Roth+ Industrial	I&EC	0043
Chemistry Award New Industrial	Catalytic Processes Honoring J. F. Roth+ Industrial	I&EC	0052
Chemistry Award New Industrial	Catalytic Processes Honoring J. F. Roth+ Industrial	I&EC	0060
in the Water-Restricted Enviro+	Catalytic Properties of a Combined Enzyme System	BIOL	0091
I_2+ Effect of Sn Addition on the	Catalytic Properties of Bimetallic System Pt-Sn/ZnA	COLL	0127
fluence of the Si/Al Ratio on the	Catalytic Properties of Small- and Large-Port Mord	PETR	0132
Superacid and	Catalytic Properties of Sulfated Zirconia.=	PETR	0133
^{13}C CP/MAS NMR Studies of	Catalytic Reactions at Temperatures between 90 and	COLL	0090
NMR Studies of Shape-Selective	Catalytic Reactions in Microporous Materials.= +te	CATL	0023
ape Selectivity—II. Photoinduced	Catalytic Reactions in Zeolite Cages.= +talysis: Sh	CATL	0020
ce of Energ+ Microwave-Induced	Catalytic Reactions of CO_2 and Water: A Clean Sour	CATL	0055
ce of Energ+ Microwave-Induced	Catalytic Reactions of CO_2 and Water: A Clean Sour	FUEL	0102
t Conditions.=	Catalytic Reactions of Hydrocarbons under Transien	PETR	0134
etallocene Complexes.=	Catalytic Reactions Using C2-Symmetrical Chiral M	INOR	0215
eve Frameworks and Products of	Catalytic Reactions Using Two-Dimensional NMR T	PETR	0060
	Catalytic Reduction in the Laboratory.=	CHED	0168
over Fe-Y Zeolites.= Selective	Catalytic Reduction of Nitric Oxide by Ammonia	COLL	0212
Selective	Catalytic Reduction of Nitric Oxide with Ammonia.=	COLL	0207
2 Catalysts Used for the Selective	Catalytic Reduction of NO.= +onia on V_2O_5/TiO	COLL	0211
or High-Temperatur+ Selective	Catalytic Reduction of NO_x Using Zeolitic Catalysis f	I&EC	0046
ted Bimetallic Catalysts Toward	Catalytic Reforming: The Reaction of Ethylene with	CATL	0042
ion of Furans.=	Catalytic Role of Metal Oxides in the Photodegradat	COLL	0190
New	Catalytic Route to Vinyl Esters.=	I&EC	0054
yls with Deboronated Z+ On the	Catalytic Sites Formed by the Reaction of Metal Alk	PETR	0024
Partial Oxidation of Methane by	Catalytic Supercritical Water Oxidation.= Selective	CATL	0043
functional Amines.= Selective	Catalytic Synthesis of Mixed Alkyl Amines and Poly	I&EC	0055
uality of Bituminous Coals in the	Catalytic Two-Stage Liquefaction Process.= +and Q	CATL	0053
e Bispho+ Chemical Rescue of a	Catalytically Deficient Mutant of R. rubrum Ribulos	BIOL	0089
Olefins.= Sulfuric Acid-	Catalyzed Alkylation of Pentenes and Other Light	PETR	0084
and Mechanism of Sulfuric Acid-	Catalyzed Alkylations of Isobutane with Alkenes.=	PETR	0085
erobic Oxidation of Cyclohexene	Catalyzed by (Aqua)(Phosphine)Ruthenium(IV) Com	INOR	0067
enylpo+ Asymmetric Oxidations	Catalyzed by a D_4-Symmetrical Manganese-Tetraph	ORGN	0045
rangement and Enoyl Reduction	Catalyzed by Brevibacterium ammoniagenes Fatty A	BIOL	0065
m of the CO_2 Carbon Gasification	Catalyzed by Calcium.= +vidence on the Mechanis	FUEL	0029
Asymmetric Michael Addition	Catalyzed by Chiral Azacrown Ethers.=	ORGN	0209
n Aziridination and Epoxidation	Catalyzed by Chiral Metal Salen Complexes Bearing	ORGN	0056
+Reactions of Alkyl Diazoacetates	Catalyzed by Chiral Rhodium(II) Carboxamides.=	ORGN	0042
Aldehyde Decarbonylation	Catalyzed by cis-Rh(CO)$_2$(p-(NC)C$_5$H$_4$N).=	INOR	0255
omerizations, and Polymerizations	Catalyzed by Co(III) and Rh(III) Complexes.= +ig	ORGN	0129
Polymerization of Epoxides	Catalyzed by Dicobaltoctacarbonyl.= Ring-Opening	POLY	0014
ovel Type of Allylic Transposition	Catalyzed by Eu(fod)$_3$.= +ic Methoxyacetates: A N	ORGN	0211
ouble Carbonylation Reactions	Catalyzed by Nickel and Phase-Transfer Catalysts: N	COLL	0117
Propene Alkylation of Biphenyl	Catalyzed by Pillared Clays.=	PETR	0014
iCAC) of Alkynes and Alkenynes	Catalyzed by Rh and Co-Rh Mixed Metal Complexe	ORGN	0218
rmal Dehydrogenation of Alkanes	Catalyzed by Rh(PMe$_3$)$_2$(CO)Cl.= +emical and The	INOR	0152
hloro+ Mechanism of the Acid-	Catalyzed Chlorination of 1-Methylpyrrole with N-C	ORGN	0158
Single Electron Transfer	Catalyzed Cleavage of the C-S and S-S Bonds.=	ORGN	0251
etic Investigations.+ Potassium-	Catalyzed CO_2 Gasification of Carbon: Transient Kin	FUEL	0031
ns of Alkyl Zirconocenes: Copper-	Catalyzed Conjugate Addition to Enones.= +eactio	ORGN	0089
,2-Dipolar Synthons: Nickel(O)-	Catalyzed Cross-Coupling of Aryl Triflates with Grig	ORGN	0082
nthesis via Palladium Hydroxide-	Catalyzed Epoxide Hydrogenolysis.= +ohydrate Sy	ORGN	0074
cal Pretreatments on the Copper-	Catalyzed Gasification of Carbon in Air.= +Chemi	FUEL	0028
taining Solid Superacids for Acid-	Catalyzed Hydrocarbon Conversions.= +lfate-Con	PETR	0102
ory—I. Support of Chem+ NSF-	Catalyzed Innovations in the Undergraduate Laborat	CHED	0146
ory—II. Undergraduate+ NSF-	Catalyzed Innovations in the Undergraduate Laborat	CHED	0179
Weight Determination b+ ILI-	Catalyzed Laboratory Innovation: Polymer Molecular	CHED	0210
Oxygen Spillover during Oxide-	Catalyzed Oxidation of Hydrocarbons.=	COLL	0174
Carbonylation of Cyclohexane	Catalyzed Photochemically by d^8 Transition-Metal C	INOR	0254
Homogeneous Metal-	Catalyzed Photochemistry in Organic Synthesis.=	INOR	0339
of Polyesters Formed in Enzyme-	Catalyzed Polycondensations.= +Molecular Weight	BIOT	0094
Cyclohexad+ Transition-Metal-	Catalyzed Polymerization of Heteroatom-Substituted	POLY	0047
l+ A Mild Osmium Tetraoxide-	Catalyzed Process for the Oxidation of Sulfides to Su	ORGN	0244
Homogeneous Transition-Metal-	Catalyzed Processes.= +ole on Catalyst Stability in	INOR	0349
hydridometal+ Phase-Transfer-	Catalyzed Reductive Elimination of HCl from Chloro	INOR	0381
o the Undergraduate Curr+ NSF	Catalyzes the Incorporation of NMR and GC/MS int	CHED	0149
preventive Effects of Phenols and	Catechols.= +Mutagenic, Carcinogenic, and Chemo	AGFD	0086
ls with Resistance Mechanisms in	Caterpillars.= +Endophyte-Induced Allelochemica	AGRO	0114
g of Refrigerated Minced Channel	Catfish Muscle.= +Lipid Oxidation during Cookin	AGFD	0124
Activity Relationships Studies of	Cathepsin B, a Proteolytic Enzyme Involved in Tum	MEDI	0017
azanona+ Base Hydrolysis of the	Cation $\alpha\beta$-anti-Chloro-[1,9-Bis(2-pyridyl)-2,5,8-Tri	INOR	0397
ty of the Addition of Cu^{2+} to the	Cation Cu(tpen)$^{2+}$ to Form Several Binuclear Compl	INOR	0399
edicacy of the Pentaphenylethyl	Cation in the Reaction of Perdeuterated Triphenylm	ORGN	0145
nation of 3-Hydrox+ Phthalide	Cation Is Not a Stable Species During Solution Proto	ORGN	0151
Photoformation of a Porphyrin	Cation Lattice by Interfacial Electron-Transfer Gate	BIOL	0136
on Reactions of Allyl Cations and	Cation Radicals Generated from 1,3-Dienes.= +diti	ORGN	0123
f Perdeuterated Triphenylmethyl	Cation with Diphenyldiazomethane or Diphenylketen	ORGN	0145
-Tetraazacyclodecane) Cobalt(III)	Cation.= +ysis of the syn,anti-cis-Dichloro (1,4,7,10	INOR	0398
States of Butadiene Radical	Cation.= Theoretical Investigation of the Low-Lying	PHYS	0237
y, and Cu Valence in the High+	Cation-Dopant Site Selectivity, Oxygen Stoichiometr	PHYS	0027
isteriolysin Purification Using a	Cation-Exchange Capsule and Cation-Exchange HPL	ANYL	0081

g a Cation-Exchange Capsule and	Cation-Exchange HPLC.= +lysin Purification Usin	ANYL	0081
tion of Carbohyd+ Utilization of	Cation-Exchange Resins for Chromatographic Separa	CARB	0064
Microporous Pillared Mica with	Cation-Incorporated Silicate Surfaces.=	PETR	0129
atalytic Hydroformylation with	Cationic Bis(dioxaphospholane)rhodium Complexes.=	INOR	0154
	Cationic Cascade Molecules.=	POLY	0237
Synthesis and Characterization+	Cationic Complexes of Tc-99m-Hydroxy-4-Pyrones:	MEDI	0042
Ru(II) Complex Containing the	Cationic Ligand N,N-Dimethylquaterpyridinium, Me	INOR	0233
trochemistry, and Spectroscopy of	Cationic Organometallic Compounds.= +ivity, Elec	INOR	0320
Electron Beam-Induced	Cationic Polymerization of Silicone-Epoxide Resins.=	PMSE	0159
Ether Analogs for UV-Induced	Cationic Polymerization.= Novel Propenyl Phenyl	POLY	0017
t/S+ Electrophoretic Analysis of	Cationic Proteins Extracted from Aflatoxin-Resistan	AGFD	0134
ulometric Analysis of In+ Use of	Cationic Surfactants in Anodic Voltammetric and Co	COLL	0002
ynthesis and Characterization of	Cationic Trinuclear BINAP-Ruthenium(II) Complex	ORGN	0044
ze Exclusion Chromatography of	Cationic, Nonionic, and Anionic Copolymers of Vinyl	PMSE	0031
Cycloaddition Reactions of Allyl	Cations and Cation Radicals Generated from 1,3-Die	ORGN	0123
ay Synthesis. Competitive Role of	Cations in Zeolite Crystallization.= +d Pillared Cl	PETR	0074
n-Lattice Relaxation of Sodium	Cations in Zeolite-Y Probed by Double-Rotation NM	PHYS	0267
ross-Sections of Transition-Metal	Cations: V+ + Ethane, Ethene, and Propane.= +C	PHYS	0306
14	Cations.= Atraniums: Novel Five-Coordinate Group	INOR	0023
Ammonium	Cations.= Classical and Nonclassical Gold(I)	INOR	0367
l Manganese Dicarbonyl η^3-Allyl	Cations.= Nucleophilic Addition to Cyclopentadieny	INOR	0270
by Polyoxometal+ Selective and	Catlytic Photochemical Functionalization of Alkanes	INOR	0286
issue Residues of Albendazole in	Cattle Administered via a Sustained Release Captec	AGRO	0109
on and Metabolism of Flunixin in	Cattle.= +uirements of the NADA: Residue Depleti	AGRO	0078
Tilmicosin in	Cattle.= Metabolism and Tissue Residues of	AGRO	0110
Biological Methods for Disposal of	Cattle-Dip Waste.= +cals and Their Containers.	AGRO	0139
m the Roots of Stevia salicifolia	cav. (Asteraceae).= New Ent-Atisene Glycosides fro	MEDI	0034
lecule Design.= 3D Databases.	CAVEAT: A Vector-Based Approach to Computer-A	COMP	0044
frared Transparent Chalcogenide:	CaYbInSe$_4$.= +cture and Spectroscopy of a New In	INOR	0030
es, Sulfamides Using Triphosgene,	CBMIT, and SBMIT.= +etric Ureas or Carbamat	ORGN	0106
Alkyl-Substituted Quinazolinone	CCK-B/Gastrin Receptor Ligands.= +l Analysis of	MEDI	0066
d Scanning Electron Microscopy (	CCSEM) in Coal Science.= +f Computer-Controlle	FUEL	0041
ium Selenolate Comp+ ^{77}Se and	^{113}Cd Single-Crystal NMR Studies of a Novel Cadm	INOR	0078
Cadmium Selenolate Compound, [	Cd(Se-2,4,6-i-Pr$_3$C$_6$H$_2$)$_2$(bpy)].= +dies of a Novel	INOR	0078
Beilstein Current Facts on	CD-ROM.=	COMP	0017
for Yersinia pseudotuberculosis	CDP-D-Glucose Oxidoreductase (E$_{od}$) and CDP-4-K	BIOL	0110
lucose Oxidoreductase (E$_{od}$) and	CDP-4-Keto-6-Deoxy-D-Glucose-3-Dehydrase (E$_1$)	BIOL	0110
dies of E$_3$-Mediated Reduction of	CDP-6-Deoxy-$\Delta^{3,4}$-Glucoseen.= +Mechanistic Stu	BIOL	0100
Photoelectrochemical Reactions of	CdS in Micelles and Reverse Micelles.= +esis and	COLL	0148
de Celdas Fotoelectroquimicas, I:	CdS Puro e Impurificado.= +Policristalinos dentro	PHYS	0221
oreactions Sensitized by Colloidal	CdS: A Flash Photolysis Study.= +Controlled Phot	COLL	0150
Implantation On-	CdTe.= Surface Modification by Metallic Ion	PHYS	0218
ve Bacteria and High Mortality in	CD1 Immunodeficient Antibiotic-Treated Mice.=	AGFD	0146
f Pt-Containing Aut+ Effect of	Ce on Performance and Physicochemical Properties o	COLL	0208
try—II. Recent Developments in	CE/MS Utilizing Atmospheric Pressure Ionization.=	ANYL	0138
MS Methods for Determination of	Ceftibuten in Biological Fluids.= +, and LC/LC/	ANYL	0118
uctores Policristalinos dentro de	Celdas Fotoelectroquimicas, I: CdS Puro e Impurifica	PHYS	0221
e and Fusarium Yellows-Resistant	Celery Cultivars.= +c Furocoumarins in Susceptibl	AGRO	0040
nd pH Stress in Denaturation of	Cellobiohydrolase I: Predenaturational Structural Ch	BIOT	0191
Site-Mutagenized	Cellobiohydrolase II.= Properties of Native and	BIOT	0240
–II. How Do Trichoderma reesei	Cellobiohydrolases Bind to and Degrade Cellulose?.=	BIOT	0206
Hemicellulases—I. Mechanism of	Cellulase Action: Current Understanding and Uncert	BIOT	0179
Molecular Regulation of	Cellulase Formation by Trichoderma reesei.=	BIOT	0183
Droplets.= Concentrating	Cellulase from a Water Solution onto Water Vapor	BIOT	0021
Novel Inducers for	Cellulase Production by Trichoderma reesei.=	BIOT	0184
ulase Action: Current Understa+	Cellulases and Hemicellulases—I. Mechanism of Cell	BIOT	0179
rma reesei Cellobiohydrolases B+	Cellulases and Hemicellulases—II. How Do Trichode	BIOT	0206
of Microbial	Cellulases and Xylanases.= Induction and Inducers	BIOT	0185
gnocellulosic B+ Optimization of	Cellulases Composition for Efficient Hydrolysis of Li	BIOT	0182
ing, the Better th+ Adsorption of	Cellulases on Cellulose Surface: The Better the Bind	BIOT	0186
Fungal	Cellulases.= Purification and Characterization of	BIOT	0187
of Cellulomonas fimi	Cellulases.= Structural and Functional Organization	BIOT	0207
New Intrigues of	Cellulolysis.=	BIOT	0209
Functional Organization of	Cellulomonas fimi Cellulases.= Structural and	BIOT	0207
by Chiral Nematic (Acetyl)(ethyl)	cellulose Solutions.= +r of Dye Molecules Oriented	POLY	0331
r th+ Adsorption of Cellulases on	Cellulose Surface: The Better the Binding, the Bette	BIOT	0186
to	Cellulose.= Proteins Designed for Adherence	BIOT	0217
iohydrolases Bind to and Degrade	Cellulose?.= +. How Do Trichoderma reesei Cellob	BIOT	0206
of Some Aromatase Inhibitors on	Cellulose-Based Chiral Stationary Phases.= +alysis	ANYL	0133
eedstocks from Prehydrolyzates of	Cellulosic Food Waste.= +Biomass and Chemical F	BIOT	0153
Dynamic Properties of	Cellulosic Polymers at the Air-Water Interface.=	COLL	0056
Semiflexible Polymers Such as the	Cellulosics.= +High-Performance Materials from	COLL	0015
Properties of the	Cellulosome of Clostridium thermocellum.=	BIOT	0180
rganic Hybrid Materials?.=	Cement Chemistry: A New Approach to Inorganic-O	POLY	0261
Incina+ High-Strength Portland	Cement Concrete Containing Municipal Solid Waste	FUEL	0136
tein Prenylation: Nature's Rubber	Cement for Membrane Attachment.= +stry of Pro	BIOL	0021
hemistry and Microstructure of	Cement Matrices for the Immobilization of Solidified	ENVR	0002
res and Pressures+ Reactions of	Cement Minerals with Water at Elevated Temperatu	INOR	0246
Physical Properties of Geothermal	Cement Systems.= +ng-Period Storage Effects on	I&EC	0014
Leaching Mechanism of Portland–	Cement-Based Waste Forms.= +r Diffusion Is the	ENVR	0011

Techniques Applied to	Cement-Solidified Hazardous Waste.= Petrographic	ENVR	0009
ctural Characterization of	Cement-Solidified Heavy Metal Wastes.= Microstru	ENVR	0008
y Characterization Techniques for	Cement-Solidified/Stabilized Metal Wastes.= +cop	ENVR	0005
idence That Long Cure Times in	Cement/Fly Ash Solidification/Stabilization Can Re	ENVR	0041
dified Waste Forms. Chemistry of	Cementitious Solidified Wastes.= +tructure of Soli	ENVR	0001
enn State Experi+ Chemistry of	Cementitious Systems for Waste Management: The P	ENVR	0006
e Characterization Techniques for	Cementitious Waste Forms. Use and Poten+ Surfac	ENVR	0007
Evaluacion Fisicoquimica de	Cementos Hidraulicos para Pozos Geotermicos.=	INOR	0063
ass Hybrid Syste+ Photocurable	Cements Based on Polyalkenoates/Ion-Leachable Gl	POLY	0262
X-ray Fluorescence Analysis of	Cements Used in Geothermal Wells.= Standards for	ANYL	0058
Chemical Society.=	Centennial of the New York Section of the American	HIST	0029
New York (Commemorating the	Centennial of the New York Section of the American	HIST	0025
f the Configuration of Glucose—A	Centennial Tribute.= +l Fischer's Establishment o	CARB	0011
-Scale Industries: The Case of the	Center for Technological Innovation in Mexico.=	SCHB	0027
lenes Modeling the Reaction	Center of Mo-co.= Reactions of Molybdenum Dithio	INOR	0404
Rhodobacter sphaeroides Reaction	Center: Polarization and Temperature Studies.=	BIOL	0094
Assembly of the Dinuclear Iron	Center: Tyrosyl Radical Cofactor of Ribonucleotide R	BIOL	0022
Learning	Center.= °From Computer Room to	CHED	0116
ch, Development, and Engineering	Center.= +ent at the U.S. Army Chemical Resear	CINF	0037
nd In^{3+}, Trapping Na$^+$ Ions in the	Center.= +Polynuclear Selenide Clusters of Au$^+$ a	INOR	0203
Photosynthetic Reaction	Center.= Optical Spectroscopy of the Bacterial	PHYS	0170
va University/Montefiore Medical	Center.= +rt Einstein College of Medicine of Yeshi	PROF	0004
n Attachment Reactions of Metal-	Centered Compounds.= +te Constants for Electro	PHYS	0234
olymers Having Chromophore-	Centered Hydrogen-Bonding and Crosslinking Group	POLY	0108
New Inorganic-	Centered Liquid Crystal Materials.=	INOR	0266
s Models for Nickel(III)-Cysteine	Centers in the Hudrogenase Enzymes.= +mplexes a	INOR	0179
ron Density at Transition-Metal	Centers via Poly 3'-(4-pyridyl)terthiophene-Based L	INOR	0027
Mutant Photosynthetic Reaction	Centers.= +mbination Dynamics in Wild-Type and	BIOL	0132
ol-Water Partition Coefficients by	Centrifugal Partition Chromatography.= +. Octan	ANYL	0068
Mg$_1$/3Nb$_2$/3)OPO$_4$ (KMNP): A	Centrosymmetric Aliovalent Analogue of Potassium T	INOR	0256
Splendor Solis: A Sixteenth-	Century Illuminated Chemical Manuscript.=	HIST	0036
the Centennial of the New Y+ a	Century of Chemistry in New York (Commemorating	HIST	0025
s as Alte+ Pesticides in the 21st	Century. Biotechnology-Based Pest Control Strategie	AGRO	0080
eveloped+ Pesticides in the 21st	Century. Outlook for Pesticides in Developing and D	AGRO	0056
tury.= Pesticides in the 21st	Century. Pesticide Regulatory Policy in the 21st Cen	AGRO	0066
Intracontinental Trade in the 21st	Century.= +inental Trade. Criteria for Successful	CMEC	0001
Regulatory Policy in the 21st	Century.= Pesticides in the 21st Century. Pesticide	AGRO	0066
Use and Development in the 21st	Century.= +cies and Attitudes Will Affect Pesticide	AGRO	0070
genase Constitutive Pseudomonas	cepacia.= +ion Products Using a Toluene Monooxy	BIOT	0119
zed Conversion of Penicillin N to	Cephalosporin C in Cephalosporium acremonium.=	BIOT	0085
athways Leading to Penicillin and	Cephalosporin.= +he Rate-Limiting Steps in the P	BIOT	0128
Penicillin N to Cephalosporin C in	Cephalosporium acremonium.= +ed Conversion of	BIOT	0085
osphate in Ammonium Control of	Cephamycin C Biosynthesis.= +e of Magnesium Ph	BIOT	0088
tion of Biosynthetic Enzymes and	Cephamycin-C in Streptomyces clavuligerus.= +ma	BIOT	0129
tures of Sol-Gel-Derived Hybrid	Ceramer Materials by Small-Angle X-ray Scattering	POLY	0284
e Index and Abrasion-Resi+ New	Ceramer Materials with Emphasis on High Refractiv	POLY	0264
Swollen	Ceramer.= Structure-Property Behavior of a	POLY	0228
	Ceramic Coatings for C/C Composites via Metalorga	POLY	0229
Synthesis and	Ceramic Conversion Reactions of Polyborazylene.=	POLY	0311
Si-C-N-O	Ceramic Fiber Compositions from Polycarbosilane.=	POLY	0336
ining Polymers as Precursors of	Ceramic Materials.= Nitrogen- and/or Boron-Conta	POLY	0315
Secretion in a Single-Pass	Ceramic Matrix Bioreactor.= Controlled Protein	BIOT	0011
ide Precurso+ Hydrous Titanate	Ceramic Powders from Mixed Alkali/Titanium Alkox	INOR	0278
to	Ceramic Powders.= Vapor-Phase Molecular Routes	INOR	0175
Polysilazane	Ceramic Precursors.= Isocyanate-Modified	POLY	0339
olysis Conditions, and Their Final	Ceramic Products.= +eramic Polysilazanes, the Pyr	POLY	0313
imetic Routes to Composites and	Ceramics: Influence of Polymer Matrix on Precipitat	POLY	0290
rials. Polysilazanes: A Route to	Ceramics.= Preceramic and Inorganic Hybrid Mate	POLY	0309
silazane—A Precursor to SiC/SiN	Ceramics.= +aracterization of Hydridomethylpoly	POLY	0340
Glasses and	Ceramics.= Methacrylate Precursors to Oxide	POLY	0288
Agents during Cooking-Extrusion	Cereals.= +ication of Phenolic Acids as Texturizing	AGFD	0150
nide Complex of Osmium(KV) by	Ceric Oxidation of Osmocene.= +hesis of an Isocya	INOR	0198
trophotometric Determination of	Cerium(IV) with Sulfanilic Acid in the Presence of C	ANYL	0073
s Complexants sur la Fottation de	ces Mineraux.= +-Pyrite: Effet des Oxydants et de	ANYL	0067
Solution during Steady Shear and	Cessation of Shear Studied by Light Scattering.=	PMSE	0134
uring Steady Shear and Following	Cessation of Shear.= +udied by Light Scattering d	PMSE	0107
Sulfanilic Acid in the Presence of	Cetyltrimethyl Bromide (CTAB).= +rium(IV) with	ANYL	0073
Cu$_4$(η-OCMe$_3$)$_6$[OC(	CF$_3$)$_3$]$_2$.= Synthesis and Structure of	INOR	0134
ucleotide Phosphorodithioate d(	CGCTpS$_2$-TPS$_2$-AAGCG): An Unusual Hairpin Loo	BIOL	0123
nds Possessing Potent cAMP and	cGMP Phosphodiesterase Inhibitory Activity. +pou	MEDI	0118
Receptor by	cGMP.= Regulation of the Asialoglycoprotein	BIOL	0154
ion Analysis on Reactions of	CH with N$_2$, O$_2$, and NO.= QRRK Chemical Activat	PHYS	0229
exes W(PMe$_3$)$_6$ and @(PMe$_3$)$_4$(η^{2-}	CH 2PMe$_2$)H with Phenols.= +ectron-rich Compl	INOR	0086
dene Complexes Cp(CO)$_2$M=C=	CH$_2$ (M=Mn, Re) with Imines, Carbodiimides, and B	INOR	0072
the H + CH$_4$ ← H$_2$ +	CH$_2$ Reaction.= Quasiclassical Trajectory Studies of	PHYS	0263
drocarbons with O(^{3}P) Atoms and	CH$_2$(x^{3}B$_1$) Radicals: Comparative Studies.= +ic Hy	FUEL	0082
NMR.= Structures of Zr(O$_3$P(	CH$_2$)$_n$COOH)$_2$, n = 1–4 from ^{31}P and ^{13}C Solid-State	INOR	0047
Cyclic	CH$_2$N$_2$ on Pd(110).= Adsorption and Chemistry of	COLL	0114
for the Combustion Intermediates	CH$_2$OH and CH$_3$O.= +Mass Spectrometric Studies	FUEL	0124
[cpRu(dppm)(	CH$_2$SO$_2$)]$^+$: A Strong Organometallic Electrophile.=	INOR	0249

Interactions (≤3 eV) of	CH_3^+ Ions on Platinum.= Ultra-Low-Energy	COLL	0216
XPS Studies of Interaction of	CH_3 with Mo(100) Surfaces.=	COLL	0217
$CO)(PPH_3)C(O)-C\equiv-R(R=CH_3,Si($	$CH_3)_3).=$ +nthesis and Chemistry of $(\eta^5-C_5H_5)FE($	INOR	0156
g and Annihilation in Crystals of (	$CH_3)_4NMnCl_3(TMMC).=$ +Solid: Exciton Trappin	PHYS	0131
ole Fraction Measurements of H,	CH_3, and Other Species by Molecular-Beam Mass-S	PHYS	0145
plexes, $(CO)_5MnCOCH_2R$ (R = H,	CH_3, OCH_3).= +rosilation of Manganese Acyl Com	INOR	0337
$H_5)FE(CO)(PPH_3)C(O)-C\equiv-R(R=$	$CH_3,Si(CH_3)_3).=$ +nthesis and Chemistry of $(\eta^5-C_5$	INOR	0156
Photochemical Dynamics of	CH_3Br Adsorbed on Pt(111).= Surface	PHYS	0261
tures of $\{[(CH_3CH)_3YbFe(CO)_4]\cdot 0\cdot$	$5CH_3CH]_n$ and $\{(CH_3CN)_3YbFe(CO)_4]_n.=$ +d Struc	INOR	0332
y: Syntheses and Structures of $\{[($	$CH_3CH)_3YbFe(CO)_4]\cdot 0.5CH_3CH]_n$ and $\{(CH_3CN)_3Yb$	INOR	0332
nd Solvation–Dynamics of	CH_3Cl^- and Cl^- at Multilayer Surfaces.= Femtoseco	PHYS	0084
$H)_3YbFe(CO)_4]\cdot 0.5CH_3CH]_n$ and $\{($	$CH_3CN)_3YbFe(CO)_4]_n.=$ +d Structures of $\{[(CH_3C$	INOR	0332
ustion Intermediates CH_2OH and	$CH_3O.=$ +Mass Spectrometric Studies for the Comb	FUEL	0124
Trajectory Studies of the H +	$CH_4 \leftarrow H_2 + CH_2$ Reaction.= Quasiclassical	PHYS	0263
Dehydrogenative Coupling of	CH_4 in a Hot-Wire Diffusion Column.=	PETR	0012
ted Cadmium Atoms with H_2 and	$CH_4.=$ +es of the Interactions of Electronically Exci	PHYS	0351
Synthesis of Na-	Chabazite and K-Mazzite.= Common Factors in the	PETR	0038
tate Compounds of $Na_3Cu_4Se_4$,+	Chalcogenide Synthesis in Molten Salts: New Solid-S	INOR	0351
he Three-Dimensi+ Molten Salt	Chalcogenide Synthesis of the Linear β-$CsBiS_2$ and t	INOR	0202
py of a New Infrared Transparent	Chalcogenide: $CaYbInSe_4.=$ +cture and Spectrosco	INOR	0030
lytical Chemistry Education.=	Challenges and Opportunities in Undergraduate Ana	ANYL	0007
Laboratory Safety.=	Challenges and Rewards of Developing Videotapes on	CHAS	0044
ts: Opportunities for Industry and	Challenges for Regulators.= +jor Chemical Acciden	CHAS	0016
Polymers: Opportunities and	Challenges in Photonics.= Nonlinear Optical	BIOT	0174
Nuclear Wastes: The	Challenges.= Analyzing Exotic Organics in Mixed	ANYL	0060
to Chemical Pesticides: Prospects,	Challenges, and Recent Developments.= +natives	AGRO	0080
Beval+ Use of a Time-Projection	Chamber in Multifragmentation Experiments at the	NUCL	0034
Total Synthesis of $(\pm)$-β-	Chamigrene.=	ORGN	0006
ovel 1,4-Benzoxazin+ Potassium	Channel Activators: Synthesis and SAR Studies of N	MEDI	0099
g Cooking of Refrigerated Minced	Channel Catfish Muscle.= +Lipid Oxidation durin	AGFD	0124
ow Fractionati+ Optimization of	Channel Dimensions Used in Sedimentation Field-Fl	PMSE	0028
Entrance	Channel Excitations in the ^{28}Si + ^{28}Si Reaction.=	NUCL	0039
in Ion	Channel Gating.= Simulation of the Role of Water	COLL	0084
Structure.= Twelve-Ring	Channel Templated Mazzite: A Stabilized Calcined	PETR	0037
Ion	Channeling.= Crystallographic Information from	NUCL	0066
Calculations of Different	Channels of H + $HNO_?$ Reaction Path and Rate	PHYS	0310
Voltage-Sensitive Sodium	Channels.= Interactions of N-Alkylamides with	AGRO	0151
s within One-Dimensional Zeolite	Channels.= +lite: Inhibited Deactivation of Pt Site	CATL	0015
Ions in Membrane	Channels.= Optical Spectroscopic Techniques for	BIOT	0141
Proteins.= Role of	Chaperones in Protein Folding—GroE Heat Shock	BIOT	0116
Experiences Building the	Chapman and Hall Directionary of Drugs.=	CINF	0033
Scientific Division of Routledge,	Chapman, and Hall.=	CHAS	0028
Catalysis of	Char Gasification in O_2 by CaO and $CaCO_3$.=	FUEL	0030
bustion Kinetics of Heterogeneous	Char Particle Populations.= +nisms. On the Com	FUEL	0026
anges during Oxidation of a Single	Char Particle.= +Mechanisms. Morphological Ch	FUEL	0025
Chemisorption of NO on	Char: Kinetics and Mechanism.=	FUEL	0011
Depositions and	Characterizations of Boron Nitride Films.=	PHYS	0094
sis and Related Techniques to the	Characterizations of Solids.= +Gravimetric Analy	INOR	0006
+B in Anoxic Bottom Sediments:	Characterizing the Role of in Situ Humic Matter.=	ENVR	0017
Laboratory Studies.= Activated	Charcoal Filtration of Pesticide Wastes: Field and	AGRO	0127
Colloids.= Sensitized	Charge Injection in Large-Bandgap Semiconductor	COLL	0149
ges in Zeta Potential and Surface	Charge of TiO_2 (Anatase) Suspensions as a Result of	COLL	0203
ansition-Metal Atoms in Different	Charge States: M, M+, and M^{2+}.= +ocarbons of Tr	PHYS	0199
ng Enzyme Stability in+ Surface	Charge Substitution: A General Strategy for Increasi	BIOT	0244
xation Study of the Influence of	Charge Transfer Interactions on Polymer Blend Phas	PMSE	0076
STM Studies of	Charge Transfer Salts.=	PHYS	0230
n Nonlinear Polymers Doped with	Charge Transport Agents.= +otorefractive Effect i	POLY	0155
	Charge Transport in Polymer Bilayer Films.=	ANYL	0104
Size and	Charge.= Electroacoustic Determination of Particle	COLL	0201
fluence of Surface Properties on	Charge-Carrier Dynamics of Quantized Semiconduct	COLL	0151
for Raman Spectroscopy Using	Charge-Coupled Device Detection.= Unique Optics	ANYL	0035
Mutant+ Electron Transfer and	Charge-Recombination Dynamics in Wild-Type and	BIOL	0132
roproteins by Free Solution and	Charge-Reversed Free Solution Capillary Electropho	ANYL	0018
o+ Kinetics and Mechanism of a	Charge-Sensitive Ion-Conducting Device Consisting	BIOL	0098
roup 1+ Multinuclear Inorganic	Charge-Transfer Complexes Based on Group 8 and G	INOR	0162
Adsorption and Luminescence of	Charge-Transfer Complexes in Disperse Systems.=	COLL	0110
ies of Double-Well Potential and	Charge-Transfer Molecules in Electric Fields—A Th	BIOT	0001
Rheology of Polydisperse,	Charged Colloidal Suspensions.= Structure and	COLL	0165
HPLC Analysis of Negatively	Charged Phospholipids.=	ANYL	0134
and Soot as	Charged Species in Flames.= Aromatics, Fullerenes,	FUEL	0111
lections of Frank C. Whitmore by	Charles D. Hurd.= +O. Hannaway. Personal Recol	HIST	0009
	Charles Martin Hall—Aluminum.=	CHAL	0027
Anti+ Evaluation and Testing of	Charm Test II Receptor Assays for the Detection of	AGFD	0085
Activity in Steam Gasification of	Chars.= +Composition of Parent Coal on Potassium	FUEL	0027
Preparation of a	Cheese Analog.=	AGFD	0138
Asymmetric Aldol Reactions of	Chelated Imidazolidone Fischer Carbene Complexes.	INOR	0147
Bifunctional	Chelating Agents.= Synthesis of Macrocyclic	ORGN	0024
alt(III) Complexes of Polyanionic	Chelating Ligands: Crystal Structures, NMR, and El	INOR	0243
nformation.= Engineered Metal	Chelation Sites as Probes of Protein Stability and Co	BIOT	0133
Bioinorganic Models of Tin	Chelation.=	INOR	0052

yacetates: A Novel Type of Ally+	Chelation-Induced Rearrangement of Allylic Methox	ORGN	0211
t Are New, a Revitalized Physical	Chem Lab Makes Its Debut.= +th Techniques Tha	CHED	0043
for P.	Chem Lab.= Experiments in Raman Spectroscopy	CHED	0088
n Reactions by Photoelectron and	Chemilelectron Spectroscopy.= +se Metal Oxidatio	PHYS	0352
for C+ Evaluation of Acridinium	Chemiluminescence as an HPLC Detection Strategy	ANYL	0084
ams by High-Resolution GC with	Chemiluminescence Detection.= +t Petroleum Stre	PETR	0124
um Oxide and Ch+ Origin of YO	Chemiluminescence from the Laser Ablation of Yttri	PHYS	0255
es: Improved Enzyme-Activated	Chemiluminescent Substrates in the Detection of Pro	BIOT	0167
Database and of a Printed Ser+	CHEMINFORM: The Joint Production of a Reaction	CINF	0010
Processes Involving Molecular	Chemisorbates.= Elementary Surface Transport	PHYS	0345
Surfaces.=	Chemisorption of a Tertiary Butyl Alchohol on Silica	COLL	0108
Silica: Mechanisms for the Fo+	Chemisorption of Aluminum Alkyls and Ammonia on	COLL	0045
) Electrode Surfaces.=	Chemisorption of Cyanide and Acetonitrile on Pt(III	COLL	0118
ism.=	Chemisorption of NO on Char: Kinetics and Mechan	FUEL	0011
s: An Analysis of Expe+	Oxygen Chemisorption on Graphite and Microporous Carbon	FUEL	0017
Mechanisms. Modeling of Oxygen	Chemisorption on Microporous Carbons.= +cation	FUEL	0012
Thiourea Derivatives	Chemisorption on Silica Surfaces.= Investigation of	COLL	0106
etal Surfaces Studied by STM.+	Chemisorption-Induced Phase Transformations of M	PHYS	0126
Interface between the Analytical	Chemist and the Chemical Engineer.=	I&EC	0051
Role of the	Chemist in Preventing Pollution.=	CEI	0026
terface, What Does the Practicing	Chemist Need? J. R.+ Panel Discussion: At the In	I&EC	0057
The Life of a	Chemist: Career Possibilities in Mexico.=	CONG	0007
Court Training for the Forensic	Chemist.=	CHAL	0019
Thomas Edison—	Chemist.=	CHAL	0026
Educating	Chemists about Nonlinear Regression.=	CHED	0193
	Chemists as Expert Witnesses in Patent Litigation.=	CHAL	0013
GC/MS: Doing What	Chemists Do!.=	CHED	0152
ding Opportunities+ Do Younger	Chemists Get a Fair Shake? An Analysis of NSF Fun	YCC	0003
Program (Roadshow)—Careers for	Chemists in Industry.+ Chemical Careers Insights	YCC	0005
Service.= Opportunities for	Chemists in Marketing, Sales, and Technical	YCC	0006
Cartoons Related to	Chemists or Chemicals.= Favorite or Objectionable	CHAL	0021
Chemistry as Practiced by	Chemists: The Project-Based Laboratory Course.=	CHED	0147
Industry. Fascinating Facts about	Chemists.= +Roadshow)—Careers for Chemists in	YCC	0005
d Career Opportunity for	Chemists.= Information Distribution: A Multifacete	YCC	0008
Scientists, including	Chemists.= Some Dichotomies of Creative	CONG	0009
Crossed Retorts: The	Chemists' Club of New York.= The Salamander and	HIST	0031
Dreams.= Future	Chemists, Future Chemistry. Profiles, Pathways, and	CONG	0004
Important to+ Incin: A Flexible	Chemkin Driver Program for Investigating Chemistry	I&EC	0024
d Compounds.= Regioselective	Chemo-Enzymatic Modifications of Polyhydroxylate	BIOT	0064
loheptatriene to Carbohydrates: A	Chemoenzymatic Approach.= +, Benzene, and Cyc	ORGN	0272
fect of N-Terminal Acylation on	Chemokinetic Activity of C5A Carboxyl Terminal Oc	MEDI	0168
ate Student Award Presentation.	Chemometric Approach to Milk Shelf-Life Predictio	AGFD	0108
Educating Undergraduates about	Chemometrics.=	CHED	0194
Sarcophytol A: A New Cancer	Chemopreventive Agent.=	AGFD	0087
he Mutagenic, Carcinogenic, and	Chemopreventive Effects of Phenols and Catechols.=	AGFD	0086
de Analogues of 4-Hyd+ Cancer	Chemopreventive Retinoid Metabolites: C-Glucuroni	MEDI	0030
Synthesis of	Chemoreversible Prodrugs of Ara-C.=	MEDI	0052
uring Isothermal E+ Correlating	Chemorheological Changes to Dielectric Properties d	PMSE	0178
-Terminal Modified Analogues of	Chemotactic Tripeptide F-Met-Leu-Phe-OH.= +C	MEDI	0166
Multiresidue Analysis of Selected	Chemotherapeutics and Antiparasitics.= +iaturized	AGFD	0062
Agents. Advances in Antiviral	Chemotherapy.= Natural Products as Antiviral	AGFD	0162
l Science.= Nontraditional	ChemSource Modules Show Chemistry Is the Centra	CHED	0018
Chemical Safety in the	ChemSource Program.=	CHED	0019
hemistry Olympiad and the NSF	ChemSource Programs, Honoring the Memory of Ma	CHED	0011
the 90s.= Minisymposium on	ChemSource. ChemSource: A Teaching Resource for	CHED	0017
Inservice Chemistry Teachers+	ChemSource: A Teaching Resource for Preservice and	CHED	0137
Minisymposium on ChemSource.	ChemSource: A Teaching Resource for the 90s.=	CHED	0017
of Favorite Cartoons Appearing in	CHEMTECH Magazine over the Past 20 Years.=	CHAL	0024
	Chester Carlson—Electrophotography.=	CHAL	0028
Acid to 3-Hydroxybenzoic Acid by	Chicken Kidney Microsomes.= +-Phenoxybenzoic	AGRO	0018
onomethoxine Residual Studies in	Chicken Tissues and Eggs.= +thoxine and Sulfam	AGFD	0127
Semduramicin in the	Chicken.= Tissue Residue Depletion Studies on	AGRO	0079
.). Pulse Voltammetry: Saturday's	Child.= +by EG G Princeton Applied Research Co	ANYL	0096
Story of	Chilean Nitrate.= Iodine and Sodium Nitrate—The	SCHB	0020
iation of Lipid Peroxidation and	Chilling Injury in Cucumber Fruit and Cell Cultures.	AGFD	0033
ofolate Synthase.= Domain	Chimeras in the Trifunctional Enzyme C_1-Tetrahydr	BIOT	0243
of Medium for the Production of	Chimeric Antibodies in CHO Cells Amplified with th	BIOT	0165
CD4(178)-PE40 Conformation: A	Chimeric Protein Having Potential as an AIDS Ther	ANYL	0005
e, a Constituent Isolated from	Chinese Herb.= Structural Modification of Edulinin	MEDI	0159
Stabilities of	Chiral (z)-Alkene.= Relative Conformational	ORGN	0172
cological Activit+ New Tocainide	Chiral Analogues: A Possible Dissociation of Pharma	MEDI	0098
	Chiral Analysis of Natural Flavor Compounds.=	AGFD	0184
Acyl Complexes Containing the	Chiral Auxiliary (η^5-C_5H_5)Fe(CO)(PPh$_3$): Synthesis a	INOR	0156
Addition Catalyzed by	Chiral Azacrown Ethers.= Asymmetric Michael	ORGN	0209
Process for the Production of	Chiral Azetidinones.= Development of an Efficient	ORGN	0206
The First Kinetic Resolution of	Chiral Azides.=	ORGN	0043
, Acylation, and Aldol Reaction of	Chiral β-Lactam Ester Enolates.= +tric Alkylation	ORGN	0046
enyl)metal Complexes Containing	Chiral Bridging Groups.= +lications of ansa-Bis(id	INOR	0214
alkyl Isocyanates).=	Chiral Clues to Helical Sequence Lengths in Poly(n–	POLY	0348
Molecular Coils That Form	Chiral Complexes.= Expanded Heterohelicenes:	ORGN	0065

oring Materials by the Analysis of	Chiral Components.=	+estigations of Natural Flav	AGFD 0182
for Enantioselective Reactions of	Chiral Flavor Compounds.=	+s: Useful Biocatalysts	AGFD 0026
n and Epoxidation Catalyzed by	Chiral Metal Salen Complexes Bearing Bulky Silyl G		ORGN 0056
Using C2-Symmetrical	Chiral Metallocene Complexes.=	Catalytic Reactions	INOR 0215
tions: A Versatile Route to Target	Chiral Molecules.=	+rboxylesterase–Mediated Reac	AGFD 0008
l Reinvestigation of	Chiral Muscarones.=	Synthesis and Pharmacologica	MEDI 0141
Asymmetric Reductions with	Chiral NADH Models.=		ORGN 0264
ior of Dye Molecules Oriented by	Chiral Nematic (Acetyl)(ethyl)cellulose Solutions.=		POLY 0331
New Electron-Rich	Chiral Phosphines for Asymmetric Catalysis.=		CATL 0004
ic Hydrosilylation of+ Discrete	Chiral Rhodium Phosphine Complexes for Asymmetr		CATL 0005
tive Catalysts for Enantioselect+	Chiral Rhodium(II) Carboxamides: Remarkably Effec		CATL 0002
Alkyl Diazoacetates Catalyzed by	Chiral Rhodium(II) Carboxamides.=	+Reactions of	ORGN 0042
Cleavage of DNA by a	Chiral Ru(II) Complex with a Bifunctional Ligand.=		INOR 0232
ase Inhibitors on Cellulose–Based	Chiral Stationary Phases.=	+alysis of Some Aromat	ANYL 0133
CH–189 Using D–Mannose as a	Chiral Template and Its Anti-HIV Activity in PBM		CARB 0022
g a New Class of D4-Symmetrical	Chiral Tetraphenylporphyrins.=	+le Alkenes Usin	INOR 0300
ich+ Synthesis and Chemistry of	Chiral Titanocenes: 1-Methyl, 2-Phenyltitanocene D		INOR 0281
nthesis and Aldol Reactions of a	Chiral 2-oxo-2-Propionyl-1,3,2-Oxazaphosphorinane		ORGN 0050
Using the	CHIRON Program.=	Heuristic Synthesis Planning	CINF 0002
vors and Fragrances.=	Chirospecific Analysis in the Origin Evaluation of Fla		AGFD 0183
isms of Iron Biomineralization in	Chitons.=	Marine Bioinorganic Chemistry. Mechan	INOR 0116
is Studies of Reinforced Polyvinyl	Chloride (PVC) Roofing Membranes.=	+cal Analys	PMSE 0180
rom 3,7-Diaminophenothiazinium	Chloride (Thionine).=	+cterization of a Polyimide f	POLY 0033
tion of the Interaction of Sodium	Chloride and Two Amino Sulfonic Acids, HEPES an		BIOL 0034
Waves.+ Reactions of Propargyl	Chloride and 1,5-Hexadiyne behind Reflected Shock		FUEL 0068
Formation of Aqueous Metal	Chloride Complexes to High Temperatures.=		ANYL 0008
ent: Perchloroethylene and Vinyl	Chloride Examples.=	Issues in Cancer Risk Assessm	CHAS 0011
y of the System Water–Calcium	Chloride in a Heat Transformer.=	Performance Stud	I&EC 0009
Sulfide Ion with Butylpyridinium	Chloride in Nonaqueous Media.=	+Interactions of	ANYL 0093
er Ablation of Yttrium Oxide and	Chloride in Oxygen.=	+iluminescence from the Las	PHYS 0255
udy of Galvanized Steel/Sodium	Chloride Interface by Electrochemical Impedance Sp		PHYS 0270
Mechanisms of Metal	Chloride Reactions with Dispersed Silica Surface.=		COLL 0219
Diammine(guanosine)platinum(II)	Chloride Using ^{13}C NMR Spectroscopy.=	+om cis-	INOR 0251
emical Stabilization of Poly(vinyl	Chloride) by Pretreatment with N-Substituted Male		POLY 0027
mers in Single Racemic Poly(vinyl	Chloride) Chains.=	+Runs of Trans Rotational Iso	POLY 0069
y(vinylchloride)/Poly(Vinylidene	Chloride).=	Composite Films of Polyaniline and Pol	PMSE 0113
ar-Beam Study of the Hydrogen	Chloride–Gallium Arsenide Gas Surface Reaction.=		COLL 0215
Tunichlorin, the Novel Nickel(II)	chlorin of Tunicates.=	+and Spectral Properties of	INOR 0139
Exposure of Fishing People to	Chlorinated Aromatic Hydrocarbons in La Basse Cot		CEI 0011
e Blood of Twelve TCDD-Expo+	Chlorinated Dibenzodioxins and Dibenzofurans in th		CEI 0014
methanes, and Haloacetonitriles in	Chlorinated Drinking Water.=	+acetic Acids, Halo	ENVR 0028
lp Bleaching Effluents Containing	Chlorinated Phenolic Compounds.=	+tment of Pu	BIOT 0137
tions for Di+ By-Products of the	Chlorination of Phenylalanine: Products and Implica		ENVR 0031
echanism of the Acid-Catalyzed	Chlorination of 1-Methylpyrrole with N-Chlorobenza		ORGN 0158
Tunable Diode Laser Probe of	Chlorine Atoms Produced from the Photodissociation		PHYS 0299
Phosphide by Molecular	Chlorine.=	Thermal and Laser Etching of Indium	COLL 0214
ydrolysis of the Cation αβ-anti-	Chloro–[1,9-Bis(2-pyridyl)-2,5,8-Triazanonane]-Chr		INOR 0397
Ising Critical Behavior of Poly(2-	Chloro–Styrene)/Polystyrene Blends.=	+Field and	PMSE 0103
Crystalline	Chloro–Trisodium Orthophosphate.=		INOR 0345
anic Synthesis: Activation of α-	Chloro, α-Hydroxy, and N-Blocked α-Amino Acids.=		ORGN 0259
Performed in Room-Temperature	Chloroaluminate Molten Salts.=	+tallic Chemistry	ANYL 0097
Bioavailability of a Natural	Chloroaniline Lignin Complex.=		AGRO 0038
ation of 1-Methylpyrrole with N-	Chlorobenzamides.=	+f the Acid-Catalyzed Chlorin	ORGN 0158
SNA on 2,4-Dinitro	Chlorobenzene, Part II.=		CHED 0170
PCDDS, PCDFS, Chlorophenols,	Chlorobenzenes, and Metals in Composts.=	+ces of	ENVR 0018
t Hydrodeamination and	Chlorodeamination of Primary Amines.=	Convenien	ORGN 0108
of Horseradish Peroxidase by 2-	Chloroethyl Sulfides and Thiodiglycol.=	Inhibition	BIOT 0249
Commercial Development of	Chlorofluorocarbon Alternatives.=	Overview of the	I&EC 0044
oroethylene Copolymer and Poly(	chlorofluoroethylene).=	+inylidene Fluoride/Triflu	POLY 0022
ductive Elimination of HCl from	Chlorohydridometal Complexes—Synthetic and Cata		INOR 0381
tems by R+ Characterization of	Chloromagnesium Species in Several Molten Salt Sys		INOR 0064
ements: Crystal Structures of Two	Chloromolybdates.=	+n Second-Row Transition El	INOR 0267
s an HPLC Detection Strategy for	Chlorophenol Determination.=	+miluminescence a	ANYL 0084
and Sources of PCDDS, PCDFS,	Chlorophenols, Chlorobenzenes, and Metals in Comp		ENVR 0018
nd Multiple Bonds with Tungsten	Chlorophosphine Complexes.=	+Form Metal-Liga	INOR 0115
Acti+ Phototactic Movements of	Chloroplasts and Their Relation with Photosynthetic		BIOL 0101
Platinized	Chloroplasts.=	Photosynthetic Water Splitting by	TECH 0016
uction of Chimeric Antibodies in	CHO Cells Amplified with the Glutamine Synthetase		BIOT 0165
Metabolic Responses of	CHO Cells to Different Cultivation Methods.=		BIOT 0164
Studies on Vibrio	cholerae Polysaccharides.=		CARB 0016
reas: A Potent Series of Acyl CoA:	Cholesterol Acyltransferase (ACAT) Inhibitors.=		MEDI 0107
of Inhibitors of	Cholesterol Esterase.=	Hypocholesterolemic Effects	BIOL 0018
ported Tomatine as a Means of	Cholesterol Removal from Butteroil.=	Polymer–Sup	AGFD 0135
ecognition Comp+ Synthesis of	Cholic Acid Derivatives for Formation of Molecular R		ORGN 0057
e Choline Analogue Inhibitors of	Choline Acetyltransferase and High-Affinity Choline		MEDI 0142
rase and High+ Redox Reactive	Choline Analogue Inhibitors of Choline Acetyltransfe		MEDI 0142
f Lecithin Supplements Varied in	Choline Content.=	+f Humans as Affected by Use o	AGFD 0034
etyltransferase and High-Affinity	Choline Uptake.=	+alogue Inhibitors of Choline Ac	MEDI 0142
Relaxation in Condensed Phases.	Chord–Distribution Analysis of a Biphasic Random		PHYS 0060
Potential Inhibitor of	Chorismate Synthase.=	Toward the Synthesis of a	ORGN 0030

Living Polymerizations of	Chrial Isocyanides.= Stereoselection and Helicity in	POLY	0347
of Individuals Living Adjacent to	Chromate Production Waste Sites.= +in the Urine	CEI	0009
Exposure.= Sister	Chromatid Exchange Test as Monitor of Human	CEI	0016
les in Foods.= Update on Gas	Chromatographic Analyses of Lipid Oxidation Volati	AGFD	0098
ring the Copolymerization of E+	Chromatographic Analysis of Oligomer Structures du	POLY	0227
erence Standards, a Key Factor in	Chromatographic Analysis of Pharmaceuticals.= +f	ANYL	0115
ectives in Protein Purifications—	Chromatographic and Nonchromatographic Methods	BIOT	0018
Lactam Antibiotic Resi+ Liquid	Chromatographic Approaches to Determination of β–	AGFD	0125
Gas	Chromatographic Columns in Food Analysis.=	AGFD	0176
–Handling System for Spectral or	Chromatographic Data Acquisition and Analysis.=	COMP	0056
tion+ Information Potential of	Chromatographic Data for Pharmacological Classifica	ANYL	0043
aluation Based on Headspace Gas	Chromatographic Data.= +and Flavor–Quality Ev	AGFD	0108
Mass Spectrometry: A Powerful	Chromatographic Detector.=	AGFD	0202
Gas	Chromatographic Detectors: New Developments.=	AGFD	0177
es i+ High–Performance Liquid	Chromatographic Determination of Eight Sulfonamid	AGFD	0128
alcium Cyanamide Usin+ Liquid	Chromatographic Method for the Determination of C	ANYL	0113
–Flow Fractionation, and Related	Chromatographic Methods for Polymer Analysis. +d	PMSE	0008
–Flow Fractionation, and Related	Chromatographic Methods for Polymer Analysis. +d	PMSE	0065
–Flow Fractionation, and Related	Chromatographic Methods for Polymer Analysis. +d	PMSE	0046
–Flow Fractionation, and Related	Chromatographic Methods for Polymer Analysis. +d	PMSE	0028
–Flow Fractionation, and Related	Chromatographic Methods for Polymer Analysis. +d	PMSE	0094
sing Monier–Williams and Liquid	Chromatographic Methods.= +ommercial Foods, U	AGFD	0109
Development of Fast Analytical	Chromatographic Methods.= Principles for the	ANYL	0117
Deconvolution of	Chromatographic Peaks Using a Neutral Network.=	ANYL	0048
roved High–Performance Liquid	Chromatographic Procedure for the Determination o	AGFD	0100
es of Penicillins+ Comparison of	Chromatographic Procedures for Determining Residu	AGFD	0063
elationships: Biological, Envir+	Chromatographic Quantitative Structure–Retention R	ANYL	0043
elationships: Biological, Envir+	Chromatographic Quantitative Structure–Retention R	ANYL	0068
elationships: Biological, Envir+	Chromatographic Quantitative Structure–Retention R	ANYL	0098
artitioning Processes at Interfaces:	Chromatographic Retention and Bioavailability.=	ANYL	0102
rom Molecular St+ Prediction of	Chromatographic Retention of Organic Compounds f	ANYL	0099
on of Cation–Exchange Resins for	Chromatographic Separation of Carbohydrates.= +i	CARB	0064
e Structural Isomers.=	Chromatographic Separation of Sucrose Monostearat	CARB	0050
lar Hydrogen Bonding i+ Liquid	Chromatographic Studies on the Role of Intramolecu	ANYL	0045
criptive Sensory Analysis and Gas	Chromatographic Techniques.= +termined by Des	AGFD	0161
tion and Analysis.= Perfusion	Chromatography®: A Novel Tool for Protein Purifica	BIOT	0073
Methyl Methacry+ Size Exclusion	Chromatography and End–Group Analysis for Poly	PMSE	0050
a Alkaloids from the Foliag+ Gas	Chromatography and Mass Spectrometry of Erythrin	ANYL	0136
ursors by Droplet Countercurrent	Chromatography and Mass Spectrometry.= +r Prec	AGFD	0203
d Pharmaceutical Analysis—III.	Chromatography and Pharmaceutical Analysis in the	ANYL	0132
armaceutical Analysis and New+	Chromatography and Pharmaceutical Analysis—I. Ph	ANYL	0110
ole of Chromatography for the +	Chromatography and Pharmaceutical Analysis—II. R	ANYL	0116
Chromatography and Pharmace+	Chromatography and Pharmaceutical Analysis—III.	ANYL	0132
Application of Size–Exclusion	Chromatography for the Characterization of Cotton F	PMSE	0032
rmaceutical Analysis—II. Role of	Chromatography for the Clinical Drug–Monitoring L	ANYL	0116
euticals.= Displacement	Chromatography for the Manufacture of Biopharmac	ANYL	0041
lar Shape Discrimination Liquid	Chromatography for the Separation of Aromatic Com	ANYL	0098
Thin–Layer	Chromatography in Food Analysis.=	AGFD	0180
/FTIR to Food Analysis: Persp+	Chromatography in Food Science. Applications of GC	AGFD	0200
ctions for Sample Preparation i+	Chromatography in Food Science. Solid–Phase Extra	AGFD	0175
Products.= Role of Process	Chromatography in Production of Recombinant	ANYL	0039
Enantioselective	Chromatography in the Real World.=	ANYL	0120
Ion	Chromatography in the 90s.=	AGFD	0204
n–Exchange Columns in Process	Chromatography of Biopolymers: Operational and In	ANYL	0042
en Calciu+ Preparative/Process	Chromatography of Biopolymers—I. Analogies betwe	ANYL	0027
erforman+ Preparative/Process	Chromatography of Biopolymers—II. Modern High–P	ANYL	0037
Copolymers of V+ Size Exclusion	Chromatography of Cationic, Nonionic, and Anionic	PMSE	0031
ic Emis+ High–Temperature Gas	Chromatography of Crudes and Residua Using Atom	PETR	0121
rsed–Phase Sample Displacement	Chromatography of Peptides.= +Preparative Reve	ANYL	0040
lecul+ Strategies for Preparative	Chromatography of Proteins and Pharmaceutical Mo	ANYL	0031
Reversed–Phase Elution	Chromatography of Proteins.= Preparative	ANYL	0029
Rapid Recycling Preparative	Chromatography of Proteins.=	ANYL	0028
Optimization of Protein	Chromatography on Agarose Gels.=	ANYL	0030
°From Paper	Chromatography to Drug Discovery.=	HIST	0019
Application of Micellar Liquid	Chromatography to QSAR Modeling of Organic Com	ANYL	0046
V–VIS Detection for Capillary Gas	Chromatography Using a Remote Flow Cell.= +U	ENVR	0060
Detection in Supercritical Fluid	Chromatography Using an Electrolytic Conductivity	ANYL	0091
vironmental Matrices by Liquid	Chromatography with UV–Absorption Photometry an	ENVR	0015
lecule Purification.= Perfusion	Chromatography: A New Tool for Large–Scale Biomo	ANYL	0032
ti+ Capillary Supercritical Fluid	Chromatography: Analytical Capabilities and Applica	AGFD	0201
efficients by Centrifugal Partition	Chromatography.= +. Octanol–Water Partition Co	ANYL	0068
drocarbons in Diesel Fuels by Gas	Chromatography.= +of Aromatics, Sulfur, and Hy	PETR	0118
–Performance Preparative Liquid	Chromatography.= +iopolymers—II. Modern High	ANYL	0037
Permeation	Chromatography.= Critical Conditions in Gel	PMSE	0049
High–Performance Size–Exclusion	Chromatography.= +eristics of Polysaccharides in	BIOT	0197
lity Using High–Temperature Gas	Chromatography.= +rmination of Motor Oil Volati	PETR	0122
termination Using Countercurrent	Chromatography.= +iquid Partition Coefficient De	ANYL	0070
njection System for Capillary Gas	Chromatography.= +ss Discrimination in a Split I	TECH	0018
Squalene Epoxidase by Affinity	Chromatography.= Purification of Pig Liver	BIOL	0070
onship Studies in Micellar Liquid	Chromatography.= +ture–Retention–Activity Relati	ANYL	0044

and Peptides by Perfusion	Chromatography.=	Rapid Separation of Proteins	ANYL 0127
ysis of Organic Compounds by Ion	Chromatography.=	+ation Techniques for the Anal	TECH 0015
c–Mercuric Resin in Capillary Gas	Chromatography.=	+Chemically Modified Polymeri	ANYL 0083
Amagasakiense by Ion–Exchange	Chromatography.=	+e Isoenzymes from Penicillium	ANYL 0016
ed–Ion High–Performance Liquid	Chromatography.=	+ines by Reversed–Phase Pair	ANYL 0089
Procedures via Spectroscopy and	Chromatography.=	+ocessing: Monitoring Dewaxing	PETR 0151
The Solute's View in	Chromatography.=		ANYL 0013
Development of Gel Permeation	Chromatography–Fourier Transform Infrared Spectr		PMSE 0048
on Spect+ Directly Coupled Gas	Chromatography/Graphite Furnace Atomic Absorpti		ANYL 0080
ery+ Applications of API Liquid	Chromatography/Mass Spectrometry to Drug Discov		ANYL 0139
od by Using Purge–and–Trap/Gas	Chromatography/Mass Spectrometry.=	+man Blo	CEI 0001
he Volatile Constituent+ Chrono–	Chromatography, a Three–Dimensional Analysis of t		AGFD 0022
ted Chromatog+ Size Exclusion	Chromatography, Field–Flow Fractionation, and Rela		PMSE 0008
ted Chromatog+ Size Exclusion	Chromatography, Field–Flow Fractionation, and Rela		PMSE 0065
ted Chromatog+ Size Exclusion	Chromatography, Field–Flow Fractionation, and Rela		PMSE 0046
ted Chromatog+ Size Exclusion	Chromatography, Field–Flow Fractionation, and Rela		PMSE 0028
ted Chromatog+ Size Exclusion	Chromatography, Field–Flow Fractionation, and Rela		PMSE 0024
s to 4–Amino–Substituted Indole	Chromium Complexes and the Total Synthesis of (-)		ORGN 0010
sis and Structure of Substituted	Chromium Complexes Containing the Thiocarbonyl L		INOR 0336
os+ Photochemical Conversion of	Chromium η^4–1,5–Cyclooctadiene Complexes into Ag		INOR 0209
nt Solutions at Low+ Removal of	Chromium from Soils by Soil Washing with Surfacta		ENVR 0013
Adjacent to+ Biomonitoring for	Chromium Levels in the Urine of Individuals Living		CEI 0009
: An Example of a Very Reactive	Chromium(III) Polydentate Amine Complex Toward		INOR 0397
2–pyridyl)–2,5,8–Triazanonane]–	Chromium(III): An Example of a Very Reactive Chro		INOR 0397
face of Ae+ Kinetic Model of the	Chromium(VI) Oxide Clusters Formation on the Sur		COLL 0099
romatic Substitution Mediated by	Chromium(0).=	+al Benzannulation/Nucleophilic A	ORGN 0221
uction Processes on the Surface of	Chromium–Containing Silica Gels.=	+idation–Red	COLL 0101
anthanum with Sulfonazo–III as	Chromogenic Agent in Aqueous Ethanolic Solution.=		ANYL 0074
Langmuir–Blodgett–Type	Chromophore Systems.=	Domains and Order in	PHYS 0111
Functionalized Polymers Having	Chromophore–Centered Hydrogen–Bonding and Cros		POLY 0108
nlinear Optical Characteristics of	Chromophore–Functionalized Polymers Having Chro		POLY 0108
up 6 Pentacarbonyl–Based NLO	Chromophores Exhibiting Large and Unusual Optica		POLY 0217
ical Binding of Glass–Embodied	Chromophores in Supported Sol–Gel Thin–Film Glas		POLY 0210
thane Films Bearing NLO–Active	Chromophores.=	+tical Characterization of Polyure	POLY 0216
organic and H+ Macromolecular	Chromophoric Assemblies with Enforced Polarity: In		POLY 0215
yer Construction of Thin–Film+	Chromophoric Self–Assembled Superlattices: Multila		POLY 0127
recursors to Flavor Compounds.	Chrono–Chromatography, a Three–Dimensional Anal		AGFD 0022
Site.=	Thorium–230 Chronology of Natural Waters at the Nevada Test		NUCL 0084
Mn Peroxidases in Phanerochaete	chrysosporium and in a Heterologous Host.=	+and	BIOT 0237
od Degradation by Phanerochaete	chrysosporium.=	+ydrolyzing Enzymes During Wo	BIOT 0208
ovements in Production of Bovine	Chymosin in Aspergillus niger var. +i: Recent Impr		BIOT 0234
The Test Case of α–	Chymotrypsin.=	Docking Ligands into Receptors:	COMP 0039
Zn^{2+} + H_2 Moeties.=	Ab Initio CI Study of the Interaction between the Zn, Zn^+ and		PHYS 0282
rores Detectados en Publicaciones	Cientificas.=	+enanza Experimental a Partir de Er	CHED 0178
Analogues of the Antifungal Agent	Cilofungin: A Total Synthesis Approach.=	+lified	MEDI 0183
Chemistry Teachers in the Greater	Cincinnati Area.=	+ern Program for High School	TECH 0002
of a Series of Substituted Phenyl–	Cinnolines.=	+is and Pollen–Suppressant Activity	AGFD 0042
on of Ketones: A Facile Route to	Ciral Alcohols Using Discrete Rhodium Catalysts.=		INOR 0247
in Microwave Region for Printed	Circuit Board Materials.=	+of Dielectric Properties	PMSE 0055
Conjugated	Circuits.=	Aromaticity, Kekule Structures, and	COMP 0021
Quantum Devices,	Circuits, and Applications.=	Semiconductor	BIOT 0003
lation and NMR Analyses of the	cis and trans Isomers of a Bicyclic Lactone Mosquito		AGRO 0033
Effective, One–Pot Synthesis of	cis N–Alkylperhydroquinolines from Cyclohexanones		ORGN 0031
uanosine by Nucleophiles from	cis–Diammine(guanosine)platinum(II) Chloride Using		INOR 0251
Base Hydrolysis of the syn,anti–	cis–Dichloro (1,4,7,10–Tetraazacyclodecane) Cobalt(I		INOR 0398
Easy Synthesis of	cis–Inositol.=		CARB 0029
Decarbonylation Catalyzed by	cis-Rh(CO)$_2$(p-(NCC_5H_4N).=	Aldehyde	INOR 0255
e Replicative Bypass of the TpdU	cis–syn Cyclobutane Photodimer of DNA.=	+of th	BIOL 0113
nd ^{13}C–NMR Assessment of the	cis–trans Isomerization of Gly^6–Bradykinin and Anal		BIOL 0006
Responsible for the Eclipsing in	cis–2–tert–Butyl–5–(tert–Butylsulfonyl)–1,3–Dioxane		ORGN 0150
	Cited Patent Searching: A Tutorial.=		CINF 0040
ork State: Oswego Harbor and +	Citizen Initiatives at Two Areas of Concern in New Y		ENVR 0084
el for H+–ATPase: Formation of	Citraconyl Phosphate Driven by a Proton Gradient.=		BIOL 0104
p Reduce Decay in Seal–Packaged	Citrus and Pepper Fruits.=	+Illumination May Hel	AGFD 0076
and Deacidifying Sour Orange (	Citrus aurantium) Juice Using Neutral and Anion Ex		AGFD 0106
l Glands of the Australian Spined	Citrus Bug.=	+miacetal from the Dorsal Abdomina	AGRO 0035
terations of Italian Cold–Pressed	Citrus Essential Oils.=	HRGC for Detection of Adul	AGFD 0199
d Health. Phenolic Compounds of	Citrus Fruits and Their Produc.=	+s in Foods an	AGFD 0167
echanisms Used to Model Mexico	City Air Quality.=	+mparison of Three Chemical M	PHYS 0296
Department at	City College of CUNY.=	History of the Chemistry	HIST 0028
in New York	City.=	One Hundred Years of Industrial Chemistry	HIST 0026
Asbestos en el Area Urbana de la	Ciudad de Mexico?.=		ENVR 0025
mbered M–As–M–As or M–As–M–	Cl Ring (M = Ga, In).=	+s Containing a Four–Me	COLL 0138
embered In–P–In–P or In–P–In–	Cl Ring.=	+vity of Compounds Containing a Four–M	COLL 0139
nes Catalyzed by Rh(PMe$_3$)$_2$(CO)	Cl.=	+mical and Thermal Dehydrogenation of Alka	INOR 0152
–Dynamics of CH$_3$Cl– and	Cl– at Multilayer Surfaces.=	Femtosecond Solvation	PHYS 0084
udies of the Spectroscopy of NO$_3$,	Cl$_2$O$_2$, and Cl$_2$O$_3$.=	+Atmosphere: Experimental St	PHYS 0017
e Spectroscopy of NO$_3$, Cl$_2$O$_2$, and	Cl$_2$O$_3$.=	+Atmosphere: Experimental Studies of th	PHYS 0017
rochemical Reduction of Ru(NH$_3$)	$_6$Cl$_3$.=	+in Naturally Occurring Fluids on the Elect	COLL 0005
BuCH$_2$,t-Bu), and (t-BuCH$_2$)$_2$Ga(	Cl)·As(SiMe$_3$)$_3$.=	+$_2$Cl, [R$_2$GaAs(SiMe$_3$)$_2$]$_2$ (R = t-	COLL 0136

for Si+ + SiX4 (X = F, Cl).=	An ab Initio Study of the Reaction Pathways	PHYS	0248
[Pt(terpy) Cl]+.=	Spectroscopic and Electrochemical Studies of	INOR	0216
trucutre of Lu{CH(SiMe3)2}3(μ− Cl)K(η^6−C7H8)2 and the Solubilization of Alkali Meta		INOR	0053
ecule Properties of XCN (X = H, Cl, Br) in Various Ionic States.=	+metries and Mol	PHYS	0238
Naturally Occurring Fertilizer Claims.=	Consumer Perspective of Organic or	FERT	0007
The	Claisen Rearrangement of 4-Allyloxyisoquinolines.=	ORGN	0101
Face Selection in the Thio−	Claisen Rearrangement.=	ORGN	0239
	Clandestine Laboratory Safety.=	CHAS	0047
s.=	Classical and Nonclassical Gold(I) Ammonium Cation	INOR	0367
n Urine and Comparison with the	Classical Colorimetric Method.= +oring of Aniline i	CEI	0003
hloride Using an ab Initio Derived	Classical Force Field.= +ral Analysis of Polyvinylc	POLY	0351
ographic Data for Pharmacological	Classification and Drug Design.= +ial of Chromat	ANYL	0043
A New	Classification of Organic Compounds.=	CHED	0167
the	Classroom.= Chemistry in Context: Report from	CHED	0053
stemically Altering a Tem+ From	Clathrasils to Large-Pore Zeolites: The Effects of Sy	PETR	0036
Formation of CO_2	Clathrate in the Deep Ocean.=	ENVR	0047
d Cephamycin-C in Streptomyces clavuligerus.=	+mation of Biosynthetic Enzymes an	BIOT	0129
Rubber−	Clay Hybrid—Synthesis and Properties.=	POLY	0291
son+ Interactions of Water with	Clay Minerals as Studied by ^{2}D Nuclear Magnetic Re	AGRO	0061
Organic Matter Associated with	Clay Minerals from Syncrude Sludge Pond Tailings.=	I&EC	0021
formance of Pillared Interlayered	Clay Molecular Sieve with High Thermal Stability.=	PETR	0030
Advances in Zeolites and Pillared	Clay Synthesis. Competitive Role of Cations in Zeol	PETR	0074
Advances in Zeolites and Pillared	Clay Synthesis. From Clathrasils to Large-Pore Zeo	PETR	0036
Advances in Zeolites and Pillared	Clay Synthesis. Investigation of the Crystallization	PETR	0093
Advances in Zeolites and Pillared	Clay Synthesis. Molecular Chemical Aspects of Sili	PETR	0054
Advances in Zeolites and Pillared	Clay Synthesis. Novel Approach to the Synthesis of	PETR	0019
Advances in Zeolites and Pillared	Clay Synthesis. Optimization of the Pillaring of a+	PETR	0142
Advances in Zeolites and Pillared	Clay Synthesis. Pillaring of Layered Inorganic Co	PETR	0106
Advances in Zeolites and Pillared	Clay Synthesis. Supergallery Pillared Clays.=	PETR	0125
Advances in Zeolites and Pillared	Clay Synthesis. Synthesis and Characterization of	PETR	0001
Borate-Pillaring Anionic	Clay.= Preparation and Characterization of	PETR	0111
g, and Characterization. Pillared	Clays as Supports in Hydrocracking Catalyst System	PETR	0013
Cracking Properties of Expanded	Clays Dried Using a Supercritical Fluid.= +ace and	PETR	0145
Anionic	Clays: Trends in Pillaring Chemistry.=	PETR	0127
Synthesis. Supergallery Pillared	Clays.= Advances in Zeolites and Pillared Clay	PETR	0125
Pillared	Clays.= Ga13, GaAl12 and Al13 Polyoxocations and	PETR	0143
by Pillared	Clays.= Propene Alkylation of Biphenyl Catalyzed	PETR	0014
	Clean Energy from Municipal Waste.=	FUEL	0093
Characterization, and Other To+	Clean Energy from Waste and Coal: Ash Utilization,	FUEL	0136
stics and Municipal Waste. Co+	Clean Energy from Waste and Coal: Energy from Pla	FUEL	0090
stics and Municipal Waste. Gas+	Clean Energy from Waste and Coal: Energy from Pla	FUEL	0091
stics and Municipal Waste. Rec+	Clean Energy from Waste and Coal: Energy from Pla	FUEL	0092
age Sludge, Biomass, and Mun+	Clean Energy from Waste and Coal: Energy from Sew	FUEL	0115
als. Metals Emissions Control +	Clean Energy from Waste and Coal: The Role of Met	FUEL	0129
udge: A Fascinating Feedstock for	Clean Energy.= +and Municipal Waste. Sewage Sl	FUEL	0115
num from Pt (PF3)4 on Atomically	Clean Platinum Surfaces.= +he Deposition of Plati	COLL	0046
ic Reactions of CO_2 and Water: A	Clean Source of Energy.= +rowave-Induced Catalyt	FUEL	0102
ic Reactions of CO_2 and Water: A	Clean Source of Energy.= +rowave-Induced Catalyt	CATL	0055
Wastes.= Production of	Clean-Medium Btu Gas from Gasification of Sludge	FUEL	0118
gglomeration Technique to Soil	Clean-Up and Generation of Good-Quality Solid and	FUEL	0141
Cells on Fibers for Waste	Clean-Up.=	FUEL	0133
lumns in Process Chromatograp+	Cleaning of Heavily Contaminated Ion-Exchange Co	ANYL	0042
Public Involvement in the	Cleanup of the Great Lakes.=	ENVR	0085
Pesticide Wastewater	Cleanup.=	AGRO	0123
Identification of Transient Pha+	Clear Solution Synthesis of AlPO4 Molecular Sieves:	PETR	0020
lecules by+ How to Monitor the	Clearance and Immunogenicity of Antibody-Like Mo	BIOT	0059
e.= Kinetics of DNA	Cleavage by Calicheamicin from Electric Birefringenc	POLY	0280
Carbon-Carbon Bond	Cleavage of Cubanes.= Regioselective	ORGN	0256
Bifunctional Ligand.=	Cleavage of DNA by a Chiral Ru(II) Complex with a	INOR	0232
Efficient Oxidative	Cleavage of DNA by Oxoruthenium(IV).=	INOR	0410
s Sequence-Selective Binding and	Cleavage of DNA.= +ess Toward Understanding It	ORGN	0298
the Recognition and	Cleavage of Phosphodiesters.= Polyaza Clefts for	ORGN	0054
Electron Transfer Catalyzed	Cleavage of the C-S and S-S Bonds.= Single	ORGN	0251
zoic Acid by Ch+ In Vitro Ether	Cleavage of 3-Phenoxybenzoic Acid to 3-Hydroxyben	AGRO	0018
ubstituted Nitro+ Study of Fast	Cleavage Reactions Following Reduction of Some α-S	ANYL	0107
Ether	Cleavage.= Easy Comparative Experiments for	CHED	0098
Phosphodiesters.=	Clefts for the Recognition and Cleavage of	ORGN	0054
with New Skeletons Related to a	Clerodanic Precursor from Mexican Salvia Species.=	ORGN	0280
Processes in Soil as a Function of	Climatic Parameter.= +radation and Translocation	AGRO	0025
+Role of Chromatography for the	Clinical Drug-Monitoring Laboratory in the 90s.=	ANYL	0116
Biochemical, Pharmacological, and	Clinical Properties.= +ptidase Inhibitors: Design,	MEDI	0002
aphi Caracterization of [M(tpen)](ClO4)2 Complexes.=	+pectroscopic and Crystallogr	INOR	0057
yl Me+ Improved Wettability of	Clofazimine in Coevaporates Prepared Using Poly(vin	COLL	0086
ubsurface Envir+ Expression of	Cloned Biodegradative Genes in Isolates from Deep S	BIOT	0117
e Genes Coding for Yersinia ps+	Cloning and Sequence Analysis of AscB and AscC, th	BIOL	0110
diene Resistant Drosophila: A+	Cloning of an Insensitive GABA Receptor from Cyclo	AGRO	0152
ctase in E. coli.=	Cloning, Expression, and Characterization of 5αRedu	BIOL	0071
: Valoracion Potenciometrica de	Clorhidrato de L-Arginina en Producto Terminado U	CHED	0163
otencio-Metrico para Cuantificar	Clorhidrato de L-Arginina en Tabletas en Medio.=	ANYL	0066
n.= Genes and Proteins of the	Clostridium thermocellum Subcellulosome Preparatio	BIOT	0181

Properties of the Cellulosome of	Clostridium thermocellum.=	BIOT	0180
rmation of Et+ Bioenergetics of	Clostridium thermosaccharolyticum in Relation to Fo	BIOT	0029
Production. Ethanol Tolerance of	Clostridium thermosaccharolyticum.= +ogical Fuel	BIOT	0013
Precursors in	Clove.= Biosynthetic Development of Flavor	AGFD	0211
aloids from the Foliage of Genetic	Cloves of Three Erythrina Species.= +rythrina Alk	ANYL	0136
Retorts: The Chemists'	Club of New York.= The Salamander and Crossed	HIST	0031
Isocyanates).= Chiral	Clues to Helical Sequence Lengths in Poly(n-alkyl	POLY	0348
Tungsten-Iridium Tetranuclear	Cluster Cores: Relationship to Butane Hydrogenolysi	CATL	0008
NMR Measurements of Atomic	Cluster Distributions in Pseudobinary Semiconducto	PHYS	0375
Theoretical Study of Silicon	Cluster Electron Affinities.=	PHYS	0079
	Cluster Energy Loss in Solids.=	NUCL	0053
Chemical Graph-Theoretic	Cluster Expansions: Diamagnetic Susceptibility.=	COMP	0030
	Cluster Impact on Solid Surfaces.=	NUCL	0055
Energy Effects on	Cluster Ion Chemistry.= Internal and Translational	PHYS	0242
Combinatorics of	Cluster Ion Enumeration.=	COMP	0028
s and Modifications of Materials:	Cluster Ion Impact Phenomena Used for Surface Mo	NUCL	0052
Dependent Properties of Boron	Cluster Ions: Reactivity, Structure, and Photodissocia	PHYS	0262
Semi-Metal	Cluster Ions.= Oxidation Reactions of Metal and	PHYS	0174
gand Transition-Metal Compo+	Cluster Mechanics: A Force-Field Study of Mixed Li	INOR	0075
k-Space Multireference Coupled-	Cluster Method.= +in Radicals Studied by the Foc	PHYS	0016
of Carbon Clusters with Coupled-	Cluster Methods.= +gh-Level Theoretical Studies	PHYS	0008
Luminescence Studies of Inorganic	Cluster Molecules.= +or Semiconductor Particles:	INOR	0317
Kinetic Processes with Metal	Cluster Participation in Shock Waves.=	PHYS	0175
tion and Analysis. Introduction to	Cluster Phenomena.= +Used for Surface Modifica	NUCL	0052
Coupled-	Cluster Treatment of Symmetry Breaking.=	PHYS	0015
ies Exhibited by W-Ir and Mo-Ir	Cluster-Derived Catalysts.= +rogenolysis Selectivit	CATL	0008
Reactivity.= Carbon-	Cluster-Ion Reactions: Structural Isomer Effects on	PHYS	0311
Chemical Analysis.= The	Cluster-Solid Interaction from the Perspective of	NUCL	0057
ontier. New Structures with Metal	Clusters and Extended Metal-Metal Bonding.=	CHED	0029
Properties of Nonmetallic Silver	Clusters and Photochemistry of Colloidal Silver Sols.	COLL	0130
Reconstructed	Clusters as Building Blocks for Extended Solids.=	INOR	0173
Ruthenium and Osmium Carbonyl	Clusters Containing One or More Boron Atoms	INOR	0220
odel of the Chromium(VI) Oxide	Clusters Formation on the Surface of Aerosil.= +c M	COLL	0099
of Small Carbon	Clusters in Matrices.= Structure and Spectroscopy	PHYS	0078
of Iron-Sulfur	Clusters in Micellar Media.= Kinetics of Assembly	INOR	0076
Cen+ Novel Polynuclear Selenide	Clusters of Au^+ and In^{3+}, Trapping Na^+ Ions in the	INOR	0203
Semiconductor	Clusters Prepared Using ROMP Block Copolymers.=	INOR	0307
el Theoretical Studies of Carbon	Clusters with Coupled-Cluster Methods.= High-Lev	PHYS	0008
nt Linkage of Manganese Oxo	Clusters with Nonzero Spin-Ground States.= Covale	INOR	0196
s of Fast (V No. V_0) Hydrogen	Clusters with Solid Materials.= Interaction Processe	NUCL	0054
Imaging of Individual Atoms and	Clusters with the FIM.= +nd Related Techniques.	PHYS	0123
Ruthenaborane and Osmaborane	Clusters: Ruthenium and Osmium Carbonyl Cluster	INOR	0220
Mixed	Clusters.= Electronic Structures and Stabilities of	PHYS	0214
Supersonically Cooled Carbon	Clusters.= Infrared Laser Spectroscopy of	PHYS	0007
ized Boron Hydride	Clusters.= Insertion of Nitrogen into Intermediate-S	INOR	0388
Carbonaceous	Clusters.= Molecular-Beam Studies of	FUEL	0110
III-V Semiconductor	Clusters.= Organometallic Precursors for Soluble	INOR	0313
ently Linked Porphyrin-Ru_{O_2}	Clusters.= Photoinduced Electron Transfer in Coval	COLL	0133
ies of Trimeric Vanadium Sulfide	Clusters.= +esis and Characterization of a New Ser	INOR	0372
$(NO)_nAr_m$	Clusters.= Two-Photon Ionization of	PHYS	0251
Metal-Based	Clusters.= Unique Complexation and Oxidation of	PHYS	0173
e Dehy+ Selectivity in Catalysis:	Clusters, Alloys, and Poisoning. Comparison of Alkan	CATL	0007
istic Studies of Reactions of $Fe_2($	$CN)_{10}^{4-}$ with $S_2O_3^{2-}$ in Slightly Acidic Aqueous Solu	INOR	0400
Fusion of	Co and Bi To Make Element 110.=	NUCL	0015
r(I) Complexes Derived from Cu-	Co Dimers.= +oxylation Reactions Involving Coppe	INOR	0213
nd Phosphine-Promoted Alkyl-	CO Migration Reactions of Unusually Reactive (η^5-In	INOR	0239
= Syntheses of Novel Rh-	Co Mixed Bimetallic Complexes and Their Catalysis.	INOR	0226
and Their Photoreactivity toward	CO Molecules.= +ium Oxides Anchored onto SiO_2	INOR	0165
Spectroscopic Characterization of	CO on Palladium Colloids.= +tical Spectroscopies.	PHYS	0373
n, sponsored by the Dow Chemical	Co. +Chemistry Award in Chemical Instrumentatio	ANYL	0011
EG G Princeton Applied Research	Co.). +ry Award in Electrochemistry, sponsored by	ANYL	0096
by E. I. DuPont de Nemours and	Co.). Teaching: A Witches' Brew of Black Holes?.=	ANYL	0006
y Steroid Research at G. D. Searle	Co.= +eroid Chemistry. Steroids and the Pill: Earl	HIST	0020
g the Reaction Center of Mo-	co.= Reactions of Molybdenum Dithiolenes Modelin	INOR	0404
and Polymerizations Catalyzed by	Co(III) and Rh(III) Complexes.= +igomerizations,	ORGN	0129
ure of $Mo(\eta^2-C(S)OPr^i)(S_2COPr^i)$	$CO(PMe_3)_2$.= +tic Acetyl Complexes: Crystal Struct	INOR	0036
ve-Induced Catalytic Reactions of	CO_2 and Water: A Clean Source of Energy.= +rowa	FUEL	0102
ve-Induced Catalytic Reactions of	CO_2 and Water: A Clean Source of Energy.= +rowa	CATL	0055
Evidence on the Mechanism of the	CO_2 Carbon Gasification Catalyzed by Calcium.=	FUEL	0029
Formation of	CO_2 Clathrate in the Deep Ocean.=	ENVR	0047
igations.= Potassium-Catalyzed	CO_2 Gasification of Carbon: Transient Kinetic Invest	FUEL	0031
omers.= Conditioning Effect of	CO_2 on the Gas Transport Properties of Polyimide in	POLY	0116
atings Materials and Supercritical	CO_2.= +nt of Phase Phenomena of Mixtures of Co	PMSE	0037
of Carbon in	CO_2.= Transient Kinetic Study of the Gasification	FUEL	0009
.+ Pure Rotational Scattering of	CO_2/CO due to Collisions with Hot Hydrogen Atoms	PHYS	0291
atory Insertion Reaction in η-Cp(	CO) $(PPh_3)FeMe^+$.= +oting the Alkyl to Acyl Migr	INOR	0109
onversion of a Cyclooxygenase (	CO) Inhibitor into a 5-Lipoxygenase (LO) Inhibitor-	MEDI	0179
on Using $Rh(PMe_3)_2LCl($	CO): Aspects of Selectivity and Mechanism.= +nati	INOR	0151
e Chiral Auxiliary $(\eta^5-C_5H_5)Fe($	$CO)(PPh_3)$: Synthesis and Chemistry of $(\eta^5-C_5H_5)FE$	INOR	0156
is and Chemistry of $(\eta^5-C_5H_5)FE($	$CO)(PPH_3)C(O)-C\equiv-R(R=CH_3,Si(CH_3)_3)$.= +nthes	INOR	0156

3).= Reactions of $H_2Os_3($	$CO)_{10}$ and $Ru_3(CO)_{12}$ with RC≡CPh (R=SiMe$_3$,SnBu	INOR	0241
of $H_2Os_3(CO)_{10}$ and $Ru_3($	$CO)_{12}$ with RC≡CPh (R=SiMe$_3$,SnBu$_3$).= Reactions	INOR	0241
Enthalpy of Reaction of CpMn(	$CO)_2$ with Tertiary Silanes as Examined by Photoaco	INOR	0386
Catalyzed by cis-Rh(	$CO)_2(p-(NC)C_5H_4N)$.= Aldehyde Decarbonylation	INOR	0255
of the Vinylidene Complexes Cp(	$CO)_2M=C=CH_2$ (M=Mn, Re) with Imines, Carbodii	INOR	0072
Their Mo(	$CO)_3$ Complexes.= 1-Substituted Borepins and	ORGN	0212
Phosphines: Fe(CO)$_4$PR$_3$ and Fe(	$CO)_3(PR)_2$ Are Single-Photon Products.= +h Alkyl	INOR	0253
d Structures of {[(CH$_3$CH)$_3$YbFe(	$CO)_4]$·0.5CH$_3$CH}$_n$ and {(CH$_3$CN)$_3$YbFe(CO)$_4$]$_n$.=	INOR	0332
0.5CH$_3$CH}$_n$ and {(CH$_3$CN)$_3$YbFe(	$CO)_4]_n$.= +d Structures of {[(CH$_3$CH)$_3$YbFe(CO)$_4$]·	INOR	0332
CO)$_5$ with Alkyl Phosphines: Fe(	$CO)_4$PR$_3$ and Fe(CO)$_3$(PR)$_2$ Are Single-Photon Prod	INOR	0253
Displacement from (Alkane)W(	$CO)_5$ Intermediates: Estimates of Alkane-W Bond St	INOR	0111
)$_3$(PR)+ Photosubstitution of Fe(	$CO)_5$ with Alkyl Phosphines: Fe(CO)$_4$PR$_3$ and Fe(CO	INOR	0253
ructural Characterization of Fe$_2$(	$CO)_5(\mu$-P-t-Bu$_2$)$_2$ and Fe$_2$(CO)$_6(\mu$-PPH$_2$)(μ-P-t-Bu	INOR	0050
n of Manganese Acyl Complexes, (	$CO)_5$MnCOCH$_2$R (R = H,CH$_3$, OCH$_3$).= +rosilatio	INOR	0337
scopy of UV-Photolyzed Cr(	$CO)_6$ in Solution.= Picosecond Transient IR Spectro	PHYS	0220
Thermal Decomposition of W(	$CO)_6$ on Pt (111).= TPD and XPS Study of the	COLL	0049
with the Reduction of Fe$_2$(PPh$_2$)$_2$(	$CO)_6$.= +udy of the Structural Changes Associated	ANYL	0106
of Fe$_2$(CO)$_5(\mu$-P-t-Bu$_2$)$_2$ and Fe$_2$(	$CO)_6(\mu$-PPH$_2$)(μ-P-t-Bu$_2$).= +al Characterization	INOR	0050
Alkanes Catalyzed by Rh(PMe$_3$)$_2$(	CO)Cl.= +mical and Thermal Dehydrogenation of	INOR	0152
phide Acetyl Complex Cp(PPH$_2$)(	CO)FeCOCH$_3$-1 with Electrophiles.= +an Iron Phos	INOR	0240
-vinylborazine) and Poly(styrene-	co-b-vinylborazine) Copolymers.= +hesis of Poly(b	POLY	0183
o-Phase Nitration of Toluene in a	Co-Current Packed Tubular Reactor.= +ng the Tw	I&EC	0006
perties of Poly(Methyl Acrylate-	co-Dimethyl Aminoethyl Methacrylated)/Epoxy Res	PMSE	0079
zation of Poly(dimethylsiloxane-	co-methylphenylsiloxane) Random Block Copolymer	POLY	0020
KA Investigation of CO-O$_2$ and	CO-NO Reactions over a Typical Commercial Autom	COLL	0187
rcial+ SSITKA Investigation of	CO-O$_2$ and CO-NO Reactions over a Typical Comme	COLL	0187
Pt, Ni-Fe-Pd, Fe-Co-Pt, and Fe-	Co-Pd.= +etism in Ternary Compounds of Ni-Fe-	PHYS	0253
unds of Ni-Fe-Pt, Ni-Fe-Pd, Fe-	Co-Pt, and Fe-Co-Pd.= +etism in Ternary Compo	PHYS	0253
d Alkenynes Catalyzed by Rh and	Co-Rh Mixed Metal Complexes.= +of Alkynes an	ORGN	0218
ic Activity of LaBo$_3$ (B = Ni, Mn,	Co, Fe) Perovskites.= +aracterization, and Catalyt	PHYS	0225
(B = Mn, Fe,	Co, Ni) Perovskites.= Electronic Structure of LaBo$_3$	PHYS	0288
hemistry and Catalysis of V, Mo,	Co, Ni, and Cu Compounds: An Example of Experim	CHED	0057
ol/Disulfide Exchange in HMG-	CoA Lyase Affects the Kinetics of Active Site Modifi	BIOL	0064
le Ureas: A Potent Series of Acyl	CoA: Cholesterol Acyltransferase (ACAT) Inhibitors.	MEDI	0107
Method for	Coal Agglomerates' Strength Determination.=	FUEL	0042
Microscope for	Coal Analysis.= NMR Imaging: A Chemical	FUEL	0075
	Coal and Carbon in the Postpetroleum World.=	FUEL	0054
ing Materials. Carbon Black from	Coal by the Hydrocarb Process.= +Fuels: Engineer	FUEL	0047
Oxidation.= Advances in	Coal Characterization by Programmed-Temperature	FUEL	0058
nance Techniques for the Study of	Coal Chemistry.= +d Electron Paramagnetic Reso	FUEL	0072
Naphthenes to	Coal during Coprocessing.= Hydrogen Transfer from	FUEL	0005
Utilization of	Coal Gasification Slag: An Overview.=	FUEL	0134
Unmodified Petroleum Coke and	Coal Gasification Slags.= +ase Characterization of	FUEL	0139
emistry of the Pore Structure of	Coal in the Presence of a Swelling Solvent Using a N	FUEL	0073
nts.=+ Hydrogen Transfer during	Coal Liquefaction Determined by ^{2}H/^{1}H Measureme	FUEL	0070
Analytical Techniques to Direct	Coal Liquefaction Oils.= Application of Advanced	FUEL	0044
Acid-Base Properties of Coals and	Coal Liquids.= +ical Techniques for Fossil Fuels.	FUEL	0039
pectrometry. Characterization of	Coal Macerals by Laser Desorption Mass Spectromet	FUEL	0088
Chemical Reaction on Low-Rank	Coal Mobile Component Generation.= Effect of	I&EC	0019
f Maceral Composition of Parent	Coal on Potassium Activity in Steam Gasification of	FUEL	0027
a Bituminous Argonne Premium	Coal Sample Using Wood Ash or Taconite as Additiv	FUEL	0006
ization of the Argonne Premium	Coal Samples by Field Ionization Mass Spectrometry	FUEL	0062
Research with Argonne Premium	Coal Samples. Moisture Removal from and Liquefa	FUEL	0001
Research with Argonne Premium	Coal Samples. Organic Oxygen Contents of Argonne	FUEL	0003
en Contents of Argonne Premium	Coal Samples.= +um Coal Samples. Organic Oxyg	FUEL	0003
Electron Microscopy (CCSEM) in	Coal Science.= +of Computer-Controlled Scanning	FUEL	0041
nd Solvent Swelling for Studies of	Coal Structure.= +ferential Scanning Calorimetry a	FUEL	0087
Determination of Sulfur Forms in	Coal Using Perchloric Acid.= +oduced during the	FUEL	0055
Clean Energy from Waste and	Coal: Ash Utilization, Characterization, and Other To	FUEL	0136
Clean Energy from Waste and	Coal: Energy from Plastics and Municipal Waste. Co	FUEL	0090
s+ Clean Energy from Waste and	Coal: Energy from Plastics and Municipal Waste. Ga	FUEL	0091
Clean Energy from Waste and	Coal: Energy from Plastics and Municipal Waste. Re	FUEL	0092
Clean Energy from Waste and	Coal: Energy from Sewage Sludge, Biomass, and Mun	FUEL	0115
T Clean Energy from Waste and	Coal: The Role of Metals. Metals Emissions Control	FUEL	0129
by	Coal.= Effect of Moisture on the Sorption of Gases	I&EC	0020
Extraction of	Coal.= Mechanism of Successive Solvolytic	I&EC	0018
Functional Groups with Oxygen in	Coal.= +ectrometry as a Technique for Analyzing	FUEL	0061
Meta-Anthracite	Coal.= Short-Range Structuring in a	FUEL	0040
NMR Imaging of	Coal.= Very High Frequency EPR Spectroscopy and	FUEL	0074
d Y-Zeolites for Hydrocracking of	Coal-Derived Distillates.= +lysis of Metal-Modifie	CATL	0050
Catalytic Hydrotreatment of	Coal-Derived Naphtha.=	CATL	0052
of Fluorescence Microscopy to	Coal-Derived Resid Characterization.= Application	FUEL	0043
2-Dimensional NMR Analyses of	Coal-Oil Hydroprocessing Products.= 1- and	PETR	0149
n Catalysts: Fuel Chemistry—II.	Coal/Oil Coprocessing Using Syngas.= Selectivity i	CATL	0049
ional Phases of Sulfur Formed by	Coalescence of Atoms on Re(0001) Surfaces.= +ens	PHYS	0124
sil Fuels. Acid-Base Properties of	Coals and Coal Liquids.= +ical Techniques for Fos	FUEL	0039
Activated Carbons from	Coals by Chemical Activation.= Production of	FUEL	0048
ulfur Forms in Argonne Premium	Coals during Pyrolysis.= +ion Kinetics of Organic S	FUEL	0059
Monitoring the Oxidation of	Coals in Storage.=	FUEL	0002
tivity and Quality of Bituminous	Coals in the Catalytic Two-Stage Liquefaction Proce	CATL	0053

vestigate Organic Sulfur Forms in	Coals.=	+els. High-Pressure TPR Apparatus to In	FUEL 0007
dilatometry to Solvent Swelling of	Coals.=	+Mass Spectrometry. Application of Micro	FUEL 0086
onne Premium Samples and Other	Coals.=	+tive ^{13}C NMR Measurements on the Arg	FUEL 0004
	Forms in Coals.=	Reactivity of Oxidized Organic Sulfur	FUEL 0057
munity Component of a Tropical	Coastal Lagoon.=	+et in the Dominant Benthic Com	BIOL 0084
Macrocrustacea in a Tropical	Coastal Lagoon.=	Energy Content of	BIOL 0105
from Freshwater and	Coastal Marine Sediments.=	N:P Release Ratios	GEOC 0012
iogeochemistry of Iron in Anoxic	Coastal Sediments:	Cycling of Iron Oxides and Their	GEOC 0007
Analysis of	Coated Fertilizers.=	Controlled-Release Nitrogen	FERT 0008
muni+ Electrochemical Studies of	Coated 4140 Steel Exposed to Mixed Microbial Com		BIOT 0097
ing M+ Rheological Additives in	Coating Systems: Rheological Properties and Thicken		PMSE 0034
f Epoxy Resins in Co+ Container	Coating Systems—II. Plenary Lecture. Application o		PMSE 0157
PVC in Container	Coatings and Approaches to Its Replacement.=		PMSE 0143
Containers.= Organic	Coatings for Beer, Beverage, and Food Metal		PMSE 0140
anic Polymer Blends.= Ceramic	Coatings for C/C Composites via Metalorganic/Inorg		POLY 0229
Phase Phenomena of Mixtures of	Coatings Materials and Supercritical CO_2.=	+nt of	PMSE 0037
ers for Container Co+ Container	Coatings Systems—I. Plenary Lecture. Barrier Polym		PMSE 0138
ne Epoxy Resins for Can and Coil	Coatings with Improved Flexibility.=	+el Waterbor	PMSE 0142
tion of Epoxy Resins in Container	Coatings: An Overview.=	+enary Lecture. Applica	PMSE 0157
Container	Coatings: FDA Considerations.=		PMSE 0158
e. Barrier Polymers for Container	Coatings.=	+Coatings Systems—I. Plenary Lectur	PMSE 0138
Nitride Preceramics, Powders, and	Coatings.=	+Bonds for the Preparation of Metal-	POLY 0230
Monitoring Cure in Inside Spray	Coatings.=		PMSE 0160
ve Index and Abrasion-Resistance	Coatings.=	+erials with Emphasis on High Refracti	POLY 0264
ross-Linking on Flexibility of Can	Coatings.=	+-Effect of Backbone Flexibility and C	PMSE 0162
tored Frozen Sweet Corn on the	Cob Produced by an Improved Blanching Technique		AGFD 0136
Substitutions of Silicon and	Cobalt in Molecular Sieve AlPO$_4$-5.=	Isomorphous	PETR 0099
nhydrase.=	Cobalt Spectroscopy at the Active Site of Carbonic A		PHYS 0256
oro (1,4,7,10-Tetraazacyclodecane)	Cobalt(III) Cation.=	+ysis of the syn,anti-cis-Dichl	INOR 0398
nds: C+ Binuclear Square Planar	Cobalt(III) Complexes of Polyanionic Chelating Liga		INOR 0243
Com+ Cadmium, Lead, Nickel,	Cobalt, and Copper Concentrations in Lung Tissue: A		ENVR 0040
tinental Shelf+ The Flux of Iron,	Cobalt, Copper, and Manganese from California Con		GEOC 0010
Complexes for Nonlinear Optics:	Cobaltous Complexes of Unsymmetrical Hydrazone I		POLY 0102
s Change the Lytic Activity of a	Cobra Venom Cytotoxin: Structure-Activity Implicat		BIOL 0085
sponsiven+ Regions of Modified	Cobra Venom Neurotoxin That Induce Immune Unre		BIOL 0086
dioiodinated Cocaine: A Potential	Cocaine Receptor Probe.=	+l Biodistribution of Ra	MEDI 0131
Biodistribution of Radioiodinated	Cocaine: A Potential Cocaine Receptor Probe.=	+l	MEDI 0131
X Zeolite: Reaction Control by the	Cocation.=	+eaction over Ruthenium-Exchanged	CATL 0027
opropionate (DMSP) in Mesoscale	Coccolithophore Blooms in the Gulf of Maine.=	+i	GEOC 0002
Stability of	Cochineal Pigment and Its Use in Food Products.=		AGFD 0139
German	Cockroach.=	Insecticide Resistance in the	AGRO 0138
halate) and Poly(ester-Ether) S+	Cocrystallization in Blends of Poly(butylene Terepht		PMSE 0148
oly(bute+ On the Miscibility and	Cocrystallization in Blends of Poly(propylene) and P		POLY 0089
ion, and Fluxional Behavior of [	cod]M{[Ph$_2$P(S)]$_n$[Ph$_2$P(O)]$_3$−nC}, where M = Rh and		INOR 0264
nd Medium-Size Enterprises: The	Codetec, Inc., Experience in Brazil.=	+for Small a	SCHB 0023
sis of AscB and AscC, the Genes	Coding for Yersinia pseudotuberculosis CDP-D-Gluc		BIOL 0110
ved 3-Hydroxy-3-Methylglutaryl-	Coenzyme A Reductase Inhibitors.=	+ophene-Deri	MEDI 0101
um+ Steric Course of the Methyl	Coenzyme M Reductase Reaction in Methanobacteri		ORGN 0025
N^5, N^{10}-Methenyltetrahydrofolate	Coenzyme.=	+hosphates: The Simplest Models of	ORGN 0204
ed Wettability of Clofazimine in	Coevaporates Prepared Using Poly(vinyl Methyl Eth		COLL 0086
irected Functional Groups.=	Cofacial Binuclear Metal Complexes with Inwardly D		INOR 0242
clear Iron Center: Tyrosyl Radical	Cofactor of Ribonucleotide Reductase.=	+the Dinu	BIOL 0022
Metal	Cofactors of Lysine 2,3-Aminomutase.=		BIOL 0078
xt o+ Immobilization of Proteins,	Cofactors, Mediators, Tissues, and Cells in the Conte		BIOT 0039
and Limitations of n-Quantum	Coherence in Probing Intermediates in Heterogeneou		COLL 0089
ted Geometries.= Exciton	Coherence-Size and Optical Nonlinearities in Restric		PHYS 0038
d Optical Spectroscopies. NMR,	Coherent Laser Spectroscopy, and Laser NMR Spect		PHYS 0040
New Applications of	Coherent Pulsed Electron Paramagnetic Resonance.=		PHYS 0116
Dense Molecular Fluids and L+	Coherent Raman Spectroscopy of High-Temperature		ANYL 0002
Photoionization with	Coherent Vacuum Ultraviolet Radiation.=		ANYL 0003
erborne Epoxy Resins for Can and	Coil Coatings with Improved Flexibility.=	+el Wat	PMSE 0142
Heterohelicenes: Molecular	Coils That Form Chiral Complexes.=	Expanded	ORGN 0065
rization of Unmodified Petroleum	Coke and Coal Gasification Slags.=	+ase Characte	FUEL 0139
igomerization o+ Deactivation by	Coke Formation of ZSM-5 Type Catalysts during Ol		PETR 0027
Zeolites: Effect of Si/Al Ratio and	Coke Formation.=	+proportionation Reaction over	PETR 0103
kane Isomeriza+ Deactivation by	Coking and Regeneration of Pd/HM Catalysts for Al		PETR 0130
R.= Characterization of Delayed	Coking Feedstocks and Products by ^{1}H and ^{13}C NM		FUEL 0076
Binding of Novel C-10	Colchicine Analogues.=	Synthesis and Tubulin	MEDI 0016
rocessed Fruits and Vegetables.	Cold Blanching Treatments to Stabilize Raw Fruit an		AGFD 0027
ectrolyte and the Implications for	Cold Fusion.=	+n Aqueous Potassium Carbonate El	PHYS 0286
Alkali+ Preparation of Granular	Cold Water Swelling/Soluble Starches by Alcoholic-		CARB 0040
Atomic Force Microscopy on	Cold-Extruded Polyethylene.=		PMSE 0078
ection of Adulterations of Italian	Cold-Pressed Citrus Essential Oils.=	HRGC for Det	AGFD 0199
racts to Tribolium castaneum L. (	Coleoptera: Tenebrionidae).=	+amosa Seed Oil Ext	AGRO 0036
rates.= ^{15}N NMR Studies of E.	coli β-Hydroxydecanoyl Thiol Ester Dehydrase: Dete		BIOL 0066
as a Probe of the Active Site of E.	coli β-Hydroxydecanoyl Thiol Ester Dehydrase.=		BIOL 0067
83 Inclusion Bodies to Remove E.	coli Proteins and Endotoxin.=	+of Washing sCD4-1	BIOT 0171
of Monoclonal Antibodies to E.	coli Pyruvate Dehydrogenase Complex and to the E1		BIOL 0031
5αReductase in E.	coli.=	Cloning, Expression, and Characterization of	BIOL 0071
-Specific Mutants Expressed in E.	coli.=	+B Complex and Binding Activities of Site	BIOL 0143

dy Fragments from Escherichia	coli.= Expression and Scaleup of Engineered Antibo	BIOT	0058
E.	coli.= Expression of Yeast AMP Deaminase in	BIOL	0083
Aggregates in Escherichia	coli.= Formation and Characteristics of Protein	BIOT	0115
ion of rRNA Synthesis in E.	coli.= Specialized Ribosomes and Feedback Regulat	BIOT	0245
Level in Recombinant Escherichia	coli.= +ngent Response as a Function of Induction	BIOT	0034
4-Epimerase from E.	coli.= Structure Determination of UDP-Galactose	BIOL	0147
Cytosolic PEPCK in E.	coli.= Subcloning and Expression of Rat Liver	BIOL	0107
s on Quartz and Its Relation with	Collapsibility.= +ted Sodium Silicate with Additive	COLL	0079
Dock Fon Toy—Scientist,	Colleague, Friend.= Phosphorus Chemistry. Arthur	INOR	0288
a Chemical	Collector.= Chemistry and the Arts. Adventures of	CONG	0001
mistry Faculty at the Two-Year	College Are Doing to Help Advance Prehigh School S	CHED	0076
General Electric	College Bound Program.=	CHED	0003
isciplinary Core Approach to	College Chemistry for Nonscience Students.= Interd	CHED	0176
ries and Inservic+ High School-	College Interface: Regional Instrumentation Laborato	CHED	0112
Department at City	College of CUNY.= History of the Chemistry	HIST	0028
lagiarism at the Albert Einstein	College of Medicine of Yeshiva University/Montefior	PROF	0004
s in Chemistry at Lebanon Valley	College.= +Chemical Education. Role of Computer	CHED	0115
Project SEED at Lehman	College.=	CHED	0140
Experiments at Roanoke	College.= Undergraduate Physical Chemistry Laser	CHED	0042
Chemistry in Context: The ACS	College-Level Decision-Making Chemistry Course fo	CHED	0051
r High School Students at Hunter	College, Chemistry Department.= +ED Program fo	CHED	0143
mber in Multi+ Nucleus-Nucleus	Collision Mechanisms. Use of a Time-Projection Cha	NUCL	0034
High-Energy Nucleus-Nucleus	Collisons.= Transparency and Abrasion in	NUCL	0048
s: Growth of Electronic Materia+	Colloid and Surface Chemistry of Advanced Material	COLL	0064
s: Growth of Electronic Materia+	Colloid and Surface Chemistry of Advanced Material	COLL	0153
s: Growth of Electronic Materia+	Colloid and Surface Chemistry of Advanced Material	COLL	0025
s: Growth of Electronic Materia+	Colloid and Surface Chemistry of Advanced Material	COLL	0006
s: Growth of Electronic Materia+	Colloid and Surface Chemistry of Advanced Material	COLL	0044
s: Growth of Electronic Materia+	Colloid and Surface Chemistry of Advanced Material	COLL	0135
y f+ Amphiphile-Mediated Polar	Colloid Formation in Fluorocarbon Media: A Strateg	COLL	0164
Mixing in a Silver	Colloid-Polymer Composite.= Optical Four-Wave	POLY	0218
Surface Conformation of	Colloid-Stabilizing Block Copolymers.=	PMSE	0013
ses on Molecular Organizates and	Colloidal Catalysts.= +-Controlled Reactive Proces	PHYS	0063
olled Photoreactions Sensitized by	Colloidal CdS: A Flash Photolysis Study.= +Contr	COLL	0150
Microgels due to	Colloidal Crystallization.= Reversible Gelation of	POLY	0244
Electrochemistry of	Colloidal Gold.=	COLL	0129
croheterogeneou+ Preparation of	Colloidal Metals by Reduction or Precipitation in Mi	COLL	0128
etic, and H+ Colloidal Particles:	Colloidal Particles in External Fields: Electric, Magn	COLL	0141
etic, and H+ Colloidal Particles:	Colloidal Particles in External Fields: Electric, Magn	COLL	0159
etic, and H+ Colloidal Particles:	Colloidal Particles in External Fields: Electric, Magn	COLL	0178
etic, and H+ Colloidal Particles:	Colloidal Particles in External Fields: Electric, Magn	COLL	0199
elds: Electric, Magnetic, and H+	Colloidal Particles: Colloidal Particles in External Fi	COLL	0141
elds: Electric, Magnetic, and H+	Colloidal Particles: Colloidal Particles in External Fi	COLL	0159
elds: Electric, Magnetic, and H+	Colloidal Particles: Colloidal Particles in External Fi	COLL	0178
elds: Electric, Magnetic, and H+	Colloidal Particles: Colloidal Particles in External Fi	COLL	0199
. Vectorial Electron Injection into	Colloidal Semiconductor Membranes.= +us Fluids	COLL	0147
er Clusters and Photochemistry of	Colloidal Silver Sols.= +perties of Nonmetallic Silv	COLL	0130
f a Weak Magnetic Field on the	Colloidal State of Several Materials in Stationary and	COLL	0160
lock Copolymers.=	Colloidal Structure of Microphase-Separated Multib	COLL	0053
c and Hy+ Response of Magnetic	Colloidal Suspension in Externally Imposed Magneti	COLL	0144
Polydisperse, Charged	Colloidal Suspensions.= Structure and Rheology of	COLL	0165
er and Transport Properties of	Colloidal Systems in External Fields.= Mass Transf	COLL	0142
n-Transfer Activation Barriers at	Colloidal Titanium Dioxide Surfaces.= +s of Electro	COLL	0132
ics and Transport Properties of	Colloids in Contact with Solid Surfaces in External F	COLL	0162
Magnetic, and Hydrodynamic—I.	Colloids in External Fields.= +nal Fields: Electric,	COLL	0141
ids+ Surfactants and Associated	Colloids: Electrochemistry in Microheterogeneous Flu	COLL	0166
ids+ Surfactants and Associated	Colloids: Electrochemistry in Microheterogeneous Flu	COLL	0019
ids+ Surfactants and Associated	Colloids: Electrochemistry in Microheterogeneous Flu	COLL	0040
ids+ Surfactants and Associated	Colloids: Electrochemistry in Microheterogeneous Flu	COLL	0001
ids+ Surfactants and Associated	Colloids: Electrochemistry in Microheterogeneous Flu	COLL	0128
ids+ Surfactants and Associated	Colloids: Electrochemistry in Microheterogeneous Flu	COLL	0147
aracterization of CO on Palladium	Colloids.= +tical Spectroscopies. Spectroscopic Ch	PHYS	0373
al Aspects of Conducting Polymer	Colloids.= +roheterogeneous Fluids. Electrochemic	COLL	0166
Aggregation of Magnetizable	Colloids.=	COLL	0161
mics of Quantized Semiconductor	Colloids.= +ace Properties on Charge-Carrier Dyna	COLL	0151
Large-Bandgap Semiconductor	Colloids.= Sensitized Charge Injection in	COLL	0149
rphyrin-Antibody Conjugates to a	Colon Carcinoma Cell Line.= +toxicity of ^{67}Cu-Po	BIOT	0168
Protective Effects against Liver,	Colon, and Tongue Carcinogenesis by Plant Phenol	AGFD	0090
RU) as Inhibitors of Experimental	Colonic Neoplasia.= +Quercetin (QU) and Rutin (	AGFD	0114
Avermectin Resistance in the	Colorado Potato Beetle.= Mechanisms of	AGRO	0154
pherols and Tocotrienols on Plant	Colorant Carotenoids.= +ioxidant Activity of Toco	AGRO	0039
on the Ultrafiltration of Polymeric	Colorants.= +mical and Engineering Perspectives	I&EC	0049
Aggregates: Influence of	Colored Noise.= Energy Transfer in Molecular	PHYS	0023
nd Comparison with the Classical	Colorimetric Method.= +oring of Aniline in Urine a	CEI	0003
Chemistry of Hair	Coloring.=	CHED	0065
.2] Octane Moieties.=	Colorless Rigid-Rod Polymers Containing Bicyclo[2.2	POLY	0029
at	Columbia.= One Hundred Years of Chemistry	HIST	0030
HPLC Analysis of Biological Mo+	Column and Instrumental Considerations for Rapid	ANYL	0129
Simplex Optimization of Post-	Column Derivatization Conditions in the Detection o	AGFD	0111
Program for Distillation	Column Design.= DESTIL—A User-Friendly	CHED	0122

Thrombin Inhibitor, by a Tandem	Column HPLC Procedure.= +gatroban, a Specific	ANYL	0086
Hot-Wire Diffusion	Column.= Dehydrogenative Coupling of CH_4 in a	PETR	0012
tration of DPX+ Use of a Novel	Column-Switching HPLC Method To Support Regis	AGRO	0030
Enzyme-Based HPLC	Columns as Probes of Enzyme Interactions.=	BIOT	0036
Gas Chromatographic	Columns in Food Analysis.=	AGFD	0176
vily Contaminated Ion-Exchange	Columns in Process Chromatography of Biopolymers	ANYL	0042
acterization of Branched and	Comb Polymers by SEC and Other Methods.= Char	PMSE	0069
New	Comb-Shaped Polymers.=	POLY	0268
Hydrophobically Modified and	Comb-Type Polymers.= Aggregation of	POLY	0270
	Combinatorics of Cluster Ion Enumeration.=	COMP	0028
ons.= Spectral Simulation.	Combinatorics of NMR and ESR Spectrum Simulati	COMP	0001
n Antibody Technology. Antibody	Combining Sites: Prediction and Design.= +nces i	BIOT	0056
Polymers of Geometrical Beauty—	Combs and Stars. Novel Dendritic Macromolecul+	POLY	0318
Geometrical Beauty—	Combs and Stars. Ring Polymers.= Polymers of	POLY	0292
Polymers of Geometrical Beauty—	Combs and Stars. Some Static and Dynamic Pro+	POLY	0265
Polymers of Geometrical Beauty—	Combs and Stars. Toward a Molecular-Size Tink+	POLY	0344
r Production of Oxygen-Enriched	Combustion Air.= +l Swing Absorption Process fo	I&EC	0023
lumina Catalyst.=	Combustion Byproducts of Ethyl Acetate over Pt- A	COLL	0189
rmation. Analysis of Chemically+	Combustion Chemistry: Aromatics Precursors and Fo	FUEL	0063
rmation. Studies on the Reacti+	Combustion Chemistry: Aromatics Precursors and Fo	FUEL	0064
deling. Thermal-Rate Constant+	Combustion Chemistry: Elementry Reactions and Mo	FUEL	0123
and Soot. Soot Deposition from+	Combustion Chemistry: Large Aromatics, Fullerenes,	FUEL	0107
mes. Formation of Benzene in +	Combustion Chemistry: Single-Ring Aromatics in Fla	FUEL	0078
omogeneous Gas-Phase Reactions:	Combustion Emission Modeling.= +n Oxides in H	FUEL	0127
ass Spectrometric Studies for the	Combustion Intermediates CH_2OH and $CH_3O.=$ +M	FUEL	0124
asification Mechanisms. On the	Combustion Kinetics of Heterogeneous Char Particle	FUEL	0026
s of PCBS and PCDDS from the	Combustion of Quicklime Binder-Enhanced Densifie	FUEL	0140
Kinetics of Methanol	Combustion on Palladium Monolith Catalyst.=	COLL	0209
Chemistry of Solid-Propellant	Combustion Studied by Molecular-Beam Mass Spect	FUEL	0126
Fuel-Rich	Combustion.= C_3H_3 Reaction Kinetics in	FUEL	0079
Organic Water during Stepped	Combustion.= Chemical and Isotopic Alteration of	ANYL	0061
Boron/Oxygen/Hydrogen/Carbon	Combustion.= +ogeneous Gas-Phase Modeling of	FUEL	0097
Hydrocarbon-Assisted Boron	Combustion.= Kinetics of High-Temperature,	PHYS	0142
rcury from RDF Municipal Waste	Combustors.= +Control of Cadmium, Lead, and Me	FUEL	0130
cesiva y un Simulador de Procesos	Comercial.= +lizando Programacion Cuadratica Su	I&EC	0072
ury of Chemistry in New York (	Commemorating the Centennial of the New York Sec	HIST	0025
min+ Technology of Recycling of	Commingled and Refined Commingled Plastics. Com	I&EC	0025
	Commingled Batch Processing.=	I&EC	0026
ucture and Properties of Recycled	Commingled Plastic Processed with Polystyrene.=	I&EC	0028
Economics of	Commingled Plastics Molding.=	I&EC	0035
d Refined Commingled Plastics.	Commingled Plastics Recycling—An Equipment Man	I&EC	0025
cling of Commingled and Refined	Commingled Plastics. Commin+ Technology of Recy	I&EC	0025
Continuously Extruded	Commingled Plastics.=	I&EC	0027
Research Results—Refined	Commingled Processing.=	I&EC	0029
d Training Used in a Universit+	Communicating Chemical Safety: Some Education an	CHAS	0048
Regulatory	Communication in Aspartate Transcarbamolylase.=	BIOT	0242
Services Provided by Hazard	Communication Resources, Inc.= Toxicological	CHAS	0024
Hazard	Communication Training.=	CHAS	0023
Steel Exposed to Mixed Microbial	Communities under Laboratory Conditions.= +40	BIOT	0097
f Diet in the Dominant Benthic	Community Component of a Tropical Coastal Lagoon	BIOL	0084
from the Residential	Community.= Collection of Plastics for Recycling	I&EC	0038
Innovatio+ Steroids, the Steroid	Community, and Upjohn in Perspective: A Profile of	HIST	0021
Producto Terminado Utilizando	como Electrodo Indicador una Pelicula de Oxido sob	CHED	0163
Estudio de los Catalizadores	$CoMo/Al_2O_3$ y Mo/Al_2O_3 Modificados con Boro.=	FUEL	0104
Eastman Chemical	Company Acetic Anhydride Process.=	I&EC	0053
sses. Founding Fairfield Chemical	Company.= +ue Stories of Small Chemical Busine	SCHB	0013
Larger	Company.= Running a Small Business within a	SCHB	0018
Chemical	Company.= Technical Career Paths in a Specialty	YCC	0007
t of an American Chemical	Company.= WHMIS Data Sheets from the Viewpoin	CHAS	0035
ar Forces: Do We Really Need to	Compare ab Initio Calculations with Experiment?.=	PHYS	0104
	Compatibility of Ternary Semi IPNs.=	PMSE	0156
ream via Reactive+ Feasibility of	Compatibilizing Components of the Plastic Waste St	I&EC	0033
Water-	Compatible Phenolic Resins.=	PMSE	0139
Study of Diffusional Processes in	Compatible Polymer Films.= +e Techniques to the	POLY	0163
Strain-	Compensated Heats of Hydrogenation.=	ORGN	0173
l ether-ketone) Synthesis via	Competing S_NAR and $S_{RN}1$ Mechanisms.= Poly(ary	PMSE	0122
ction Detection Systems Based on	Competitive Binding Assays.= +C Postcolumn Rea	ANYL	0090
g of Excited States by Ferrocen+	Competitive Energy and Electron Transfer Quenchin	INOR	0163
PC 12626: A Potent and Selective	Competitive NMDA Antagonist.= +Synthesis of N	MEDI	0083
tes and Pillared Clay Synthesis.	Competitive Role of Cations in Zeolite Crystallization	PETR	0074
Carga por Espectroscopia IR de	Complejos de Transferencia de Carga Pt(R-Nafin)·T	INOR	0190
Benceno con H_2O_2 Catalizada por	Complejos Derivados de Cupferron.= +xilacion del	COLL	0121
Air Samplers and	Complementary Analyses for PIXE.=	NUCL	0079
ntrolled Taxane Construction: A	Complementary Bifurcate Approach to Taxusin and	ORGN	0291
ng vs. de Novo Construction: Two	Complementary Methods with the Same Goal.= +i	COMP	0050
MR Spectrum Simulation: Role of	Complementary Topological Strategies.= +d ^{13}C N	COMP	0006
our-Year Spectroscopy Program	Complemented by the Microscale Approach in the G	CHED	0150
-Pyrite: Effet des Oxydants et des	Complexants sur la Fottation de ces Mineraux.=	ANYL	0067
Clusters.= Unique	Complexation and Oxidation of Metal-Based	PHYS	0173
pper(I).=	Complexation of a Mixed-Donor Ligand P_2S with Co	INOR	0029

c Receptor.=	Complexation of Benzamidinium Salts by a Syntheti	ORGN 0061
Investigation of Adsorption and	Complexation of Lead(II) and Mercury(II) with Cr	COLL 0104
orphyrin Analo+ Synthesis and	Complexation of Pyrrole–Containing Receptors and P	ORGN 0063
cyltrimethylammonium Bromide	Complexation with Polyacrylic Acid and Polymethac	COLL 0074
Health. Polyphenol	Complexation.= Phenolic Compounds in Foods and	AGFD 0001
Diacetylene LB Films by Polyion	Complexation.= Control of Photochromism of	BIOT 0146
en Aromatic Polyamide Chains by	Complexation.= +dification of Interactions betwe	POLY 0064
age Optical Memory with Polyion	Complexed Pyrene LB Films.= +otoamplified Stor	BIOT 0144
in the Presence of Oxalic Acid as	Complexing Agent.= +tion of Lanthanum by Oxine	ANYL 0054
hickeners.=	Complexities in Evaluating Behavior of Associative T	PMSE 0084
of the 90s: MSDS Right–to–Know	Compliance Made User Friendly.= +the Challenge	CHAS 0031
Future	Compliance.= Right–to–Know Support: Current and	CHAS 0026
eotermica de Fasico+ Estudio del	Comportamiento de los Gazas en Condensables en G	I&EC 0013
la Familia Scrophulariaceae de+	Composicion Quimica de las Plantas Pertenecientes a	ORGN 0185
etic Study of in Situ Molecular	Composite during Imidization of Poly(amic dialkylam	PMSE 0169
e)/Poly(Vinylidene Chloride).=+	Composite Films of Polyaniline and Poly(vinylchlorid	PMSE 0113
Transitions.= PTFE–Matrix	Composite Hydrogel Sheets with Thermoreversible	POLY 0245
hrin+ Inverse Organic–Inorganic	Composite Materials: Free Radical Routes into Nons	POLY 0259
r by Novel Ion–Selective Graphite	Composite Potentiometry.= +ection in Groundwater	ENVR 0044
Colloid–Polymer	Composite.= Optical Four–Wave Mixing in a Silver	POLY 0218
rix on Pr+ Biomimetic Routes to	Composites and Ceramics: Influence of Polymer Mat	POLY 0290
lar Structure on the Properties of	Composites for Electronics.= +lar and Supermolecu	POLY 0300
trating Polymer Network.=	Composites from Polyurethane and Epoxy Interpene	PMSE 0044
Polymers as Matrix for Structural	Composites in Oilfield Applications.= +rent Epoxy	PMSE 0111
Molecular	Composites in Thermosets.=	POLY 0111
Insights int+ Organic–Inorganic	Composites Prepared via the Sol–Gel Method: Some	POLY 0260
Polymer	Composites Using Recycled PET.=	PMSE 0072
ds.= Ceramic Coatings for C/C	Composites via Metalorganic/Inorganic Polymer Blen	POLY 0229
hite Matrices in Carbon–Carbon	Composites: An Overview of Their Formation, Struct	FUEL 0051
enzobisthiazole)–Based Molecular	Composites.= +n of the $\chi^{(3)}$ of Poly(p–phenylene B	POLY 0213
Mechanical Properties of Granular	Composites.= +ct of Particle Size Distribution on	PMSE 0114
Routes to Nonshrinking Sol–Gel	Composites.= +Composite Materials: Free Radical	POLY 0259
Polymer–Silver Microsphere	Composites.= Nonlinear Optical Properties of	POLY 0154
Fiber Reinforced PMMA Matrix	Composites.= Preparation of Transparent Glass	PMSE 0109
Thermoplastic	Composites.= Sensor Plate for Manufacturing	PMSE 0115
Their Performance in Polystyrene	Composites.= +e Modification of Wood Fibers and	PMSE 0112
Glass–Transition Temperature in	Composites.= +of Fiber/Matrix Interface on Local	COLL 0055
si+ Polyphosphazenes Molecular	Composites, I: In Situ Polymerization of Tetraethoxy	POLY 0181
ype B Conjuga+ Monosaccharide	Compositional Analysis of Haemophilus Influenzae T	ANYL 0021
ls, Chlorobenzenes, and Metals in	Composts.= +ces of PCDDS, PCDFS, Chloropheno	ENVR 0018
Containing	Compounds/Polymers.= SiO_2 as a Source of Silicon	POLY 0314
se Sample Displacement+ Axial	Compression Multicolumn Preparative Reversed–Pha	ANYL 0040
Analisis Cuantitativo por	Computadora.=	COMP 0051
Nivel Med+ Desarrollo de Apoyos	Computarizados para la Ensenanza de la Quimica a	CHED 0124
onal Analysis for RIS M+ Direct	Computation of Statistical Weights from Conformati	POLY 0356
Fruitful Interplay between	Computational and Experimental Chemistry.=	ORGN 0138
nergy Transport in Spatially Di+	Computational and Ultrafast Studies of Electronic E	PHYS 0133
ew Building Block in Organic +	Computational Investigation of Phosphaalkynes: A N	COMP 0060
	Computational Method for Polymer Simulations.=	PHYS 0112
Molecular Hyperpolarizabilities:	Computational Methods and Regularities.=	BIOT 0178
dergrad+ Integration of Modern	Computational Techniques and Graphics into the Un	CHED 0156
Resonance Using	Computer Animation.= Teaching Magnetic	CHED 0123
	Computer Automation in the Laboratory.=	TECH 0006
Data Acquisition and	Computer Data Processing for Analytical Courses.=	CHED 0119
drogen Bonding from Molecula+	Computer Graphics Presentations and Analysis of Hy	BIOL 0112
e Physical Chemistry Laborato+	Computer in Techniques–Oriented Experiments in th	CHED 0121
ical Education. Use of Personal	Computer Modeling in Undergraduate Organic Chem	CHED 0188
Recept+ Sequence Analysis and	Computer Modeling of the Dog A_1 and A_2 Adenosine	MEDI 0114
entication of Extracts an+ Use of	Computer Pattern–Recognition Techniques for Auth	AGFD 0198
A Simple	Computer Program to Fit Experimental Results.=	NUCL 0089
°From	Computer Room to Learning Center.=	CHED 0116
s.=	Computer Simulation Experiment: Chemical Kinetic	CHED 0157
olymers.=	Computer Simulation of Direct Energy Transfer in P	POLY 0079
in Polymers: Comparison of Th+	Computer Simulation of Electronic Energy Transfer	PHYS 0109
on Mixing). Molecular Dynamics	Computer Simulation Studies of Collision Cascades i	NUCL 0091
Field–Flow Fractionation through	Computer Simulation.= +Used in Sedimentation	PMSE 0028
	Computer Simulations for Instrumental Analysis.=	CHED 0192
tron Transfer and Proton Tran+	Computer Simulations of Quantum Dynamics in Elec	BIOL 0131
Off–Lattice, Bead–Stick	Computer Simulations.= Rings vs. Strings:	POLY 0235
o a Systematic Chemical Name fo	Computer: Experiences with AUTONOM 1.0.= +t	COMP 0019
ons.=	Computer–Aided Discovery of Unprecedented Reacti	CINF 0006
AT: A Vector–Based Approach to	Computer–Aided Molecule Design.= +ases. CAVE	COMP 0044
grated Approach to Bi+ Role of	Computer–Aided Process Design in Following an Inte	BIOT 0075
m Polymeric Gas Separation and+	Computer–Assisted Analysis of X–ray Scattering fro	POLY 0142
(CCSEM) in C+ Applications of	Computer–Controlled Scanning Electron Microscopy	FUEL 0041
perty Data fo+ Characteristics of	Computer–Generated 3D and Related Molecular Pro	COMP 0041
hysical Chemistry or Instrumen+	Computer–Interfaced Titration Experiment for the P	CHED 0161
uters in Chemical Education.	Computer–Managed Chemistry Instruction.= Comp	CHED 0077
	Computerization of Material and Chemical Data.=	PHYS 0160
Books in a	Computerized World.= The Place of Reference	CHAS 0039

the Revitalization of+	Macintosh Computers as Data Acquisition Devices—A Tool for	CHED	0160
ed Chemistry Instruction.=	Computers in Chemical Education. Computer-Manag	CHED	0077
Development and Symposium on	Computers in Chemical Education. +stry Laboratory	CHED	0155
rs in Chemistry at Lebanon Vall+	Computers in Chemical Education. Role of Compute	CHED	0115
omputer Modeling in Undergr+	Computers in Chemical Education. Use of Personal C	CHED	0188
ts in General Chemistry.=	Computers in Chemical Education. Using Spreadshee	CHED	0204
in Chemical Education. Role of	Computers in Chemistry at Lebanon Valley College.=	CHED	0115
mputers in Chemical Education.	Computers in the Physical Chemistry Laboratory: Le	CHED	0155
A Tool-Oriented Approach to	Computing.=	CHED	0078
Estudio de la Nitracion del Fenol	con Aductos Tonsil-Fe(NO$_3$)$_3$.=	COLL	0120
3 y Mo/Al$_2$O$_3$ Modificados	con Boro.= Estudio de los Catalizadores CoMo/Al$_2$O	FUEL	0104
de Alumina-+ HDS de Tiofeno	con Catalizadores de Molibdeno sobre Oxidos Mixtos	FUEL	0105
troquimica del 1-Nitro-Naftaleno	con Electrodos Metalicos y Semiconductores.= +ec	PHYS	0280
re Alumina o Titania Modificadas	con Fluor.= +de Catalizadores NiW Soportados sob	FUEL	0106
aras por Espectrometria de Masas	con Fuente de Impacto Electronico.= +e Tierras R	ANYL	0051
pfer+ Hidroxilacion del Benceno	con H$_2$O$_2$ Catalizada por Complejos Derivados de Cu	COLL	0121
de Secuencias de Destilacion	con Integracion de Energia.= Sintesis y Modificacion	I&EC	0074
io a Partir de Leche Contaminada	con S. aureus y s+ Elaboracion de Caseinato de Sod	AGFD	0107
Libres	con Tonsil.= Catalisis de Reacciones via Radicales	COLL	0119
eposito, Fusion, Oxidorreduccion	con Zinc en Medio Acido y Basico, Preparacion de N	CHED	0172
Iron, and Total Sulfur in Zinc Ore	Concentrates.= +rections in XRF Analysis of Zinc,	ANYL	0057
Ionone Enantiomers in Raspberry	Concentrates, Essences, and Flavors.= +of Alpha-	AGFD	0181
Water Vapor Droplets.=	Concentrating Cellulase from a Water Solution onto	BIOT	0021
hip Studies of Opi+ The Role of	Concepts and Models in Structure-Activity Relations	MEDI	0089
Abstract Scientific	Concepts in the Courtroom.=	CHAL	0012
	Concepts, Experiments, and Applications of Foams.=	CHED	0100
itizen Initiatives at Two Areas of	Concern in New York State: Oswego Harbor and the	ENVR	0084
ubstances in Great Lakes Areas of	Concern.= +Zero Discharge of Persistent Toxic S	ENVR	0083
Environmental	Concerns: A Future Perspective.=	AGRO	0067
-Like Transition State for the +	Concerted Pericyclic Elimination with a Carbocation	ORGN	0153
High-Strength Portland Cement	Concrete Containing Municipal Solid Waste Incina	FUEL	0136
lysis When Intense Lasers Miss+	Concrete Results through Virtual States: Surface Ana	NUCL	0044
omportamiento de los Gazas en	Condensables en Geotermica de Fasico de los Azusfre	I&EC	0013
hesis of Polycarbosilanes via the	Condensation of Cyclopentadienyl Sodium with Dialk	INOR	0199
Sol-Gel	Condensation of Tin(IV) Alkoxide Compounds.=	POLY	0258
chelomers, II: A Model Study of	Condensation Polymerization Using Low-Temperatu	POLY	0046
Melts.= Chemistry of	Condensation Reactions Carried Out in Polymer	PMSE	0176
lene)propanedioic Acids by Direct	Condensation.= +nd Amidoesters of (Anilinomethy	ORGN	0103
Ions in	Condensed Matter.= Stopping Power and Ranges of	NUCL	0037
nergy Transfer and Relaxation in	Condensed Phases. Chord-Distribution Analysis+ E	PHYS	0060
nergy Transfer and Relaxation in	Condensed Phases. Conformation and Dynamics+ E	PHYS	0107
nergy Transfer and Relaxation in	Condensed Phases. Energy Transfer and Relaxat+ E	PHYS	0080
nergy Transfer and Relaxation in	Condensed Phases. Excited-State Structure and+ E	PHYS	0018
nergy Transfer and Relaxation in	Condensed Phases. Time-Domain Analysis of th+ E	PHYS	0034
nergy Transfer and Relaxation in	Condensed Phases. Torsional Solitons in Organi+ E	PHYS	0129
e Edge: Structure and Function at	Condensed-Phase Interfaces.= +al Chemistry at th	ANYL	0026
re Industry—II. Silicone in Hair	Conditioners: Factors Controlling Hair Deposition.=	CHED	0063
Molten Urea.= Effect of	Conditioning Agents on the Corrosive Properties of	FERT	0011
perties of Polyimide Isomers.=	Conditioning Effect of CO$_2$ on the Gas Transport Pro	POLY	0116
e la Presion en las Propiedades de	Conduccion en Polimeros Zwitterionicos.= +ecto d	ORGN	0159
irements of the N+ Design and	Conduct of Studies to Meet Residue Chemistry Requ	AGRO	0078
mine: A New Route to Electrically	Conducting Conjugated Polyacetylene.= +with Bro	POLY	0032
nism of a Charge-Sensitive Ion-	Conducting Device Consisting of Metalloporphyrins i	BIOL	0098
d Electrical Conducti+ A Highly	Conducting Hydropolysilane Copolymer: Synthesis an	POLY	0043
Electrically	Conducting Ionic Polyacetylenes.=	POLY	0074
STM-Induced Etching of	Conducting Oxides.=	PHYS	0361
Fluids. Electrochemical Aspects of	Conducting Polymer Colloids.= +roheterogeneous	COLL	0166
nt Materials Sy+ Applications of	Conducting Polymers in Redox Devices and Intellige	POLY	0326
Perspectives for	Conducting Polymers.= Structure-Property	BIOT	0203
Class of	Conducting Polymers.= The Polyanilines: A Novel	POLY	0327
Precursors.= Electrically	Conductive Catalyst Supports from Fibrous	PETR	0016
s in the Oxidative Synthesis of	Conductive-Electroactive Polymers.= Use of Micelle	COLL	0168
ers by Time-Resolved Microwave	Conductivity (TRMC).= +se in Photoactive Polym	POLY	0214
matography Using an Electrolytic	Conductivity Detector.= +Supercritical Fluid Chro	ANYL	0091
todopi+ Transient Photoinduced	Conductivity in Single Crystals of YBa$_2$Cu$_3$O$_6$.3: Pho	PHYS	0011
n Constant of Amino Acids by the	Conductivity Method: I. MOPS+ First Dissociatio	BIOL	0033
terically Hindered Pheno+ Ionic	Conductivity of Poly(ethylene Oxide) Bound with a S	POLY	0091
polymer: Synthesis and Electrical	Conductivity.= +ly Conducting Hydropolysilane Co	POLY	0043
3-D Molecular	Conductors?.= Alkali Metal-C$_{60}$ Films: The First	INOR	0354
Targeted Catalysis:	Conferment of Substrate Selectivity on Enzymes.=	BIOL	0090
Esterase Genes	Conferring Insecticide Resistance in Aphids.=	AGRO	0137
l Energy Surface Using Quadratic	Configuration Interaction Methods.= +H$_6$ Potentia	PHYS	0305
king in Small Molecules.=	Configuration Interaction Studies of Symmetry Brea	PHYS	0014
Solvent.= Equilibrium	Configuration of Dense Polymacromonomers in Good	POLY	0269
il Fischer's Establishment of the	Configuration of Glucose—A Centennial Tribute.=	CARB	0011
onaria+ Synthesis and Absolute	Configuration of Polypropionate Metabolites of Siph	ORGN	0003
Solvent Effects on	Configurations of Poly(3-Alkylthiophene)s.=	PMSE	0183
able+ Ordered Buckled-Dimer	Configurations on Si(100) Observed with a New Vari	PHYS	0185
nary Survey of Various Data Sets	Confirms that Substituent Group Contributions (SG	MEDI	0088
Model for Land-Use Planning and	Conflict Resolution.= +Gaps in Development of a	ENVR	0043

Corporation.= The False Conflict: Helping the Public vs. Helping the	PROF	0006
Relaxation in Condensed Phases. Conformation and Dynamics of Partially Folded and	PHYS	0107
ion Characterization of Polymer Conformation by SEC with an On-Line Viscometry a	PMSE	0097
H on the Foaming Properties and Conformation of β-Lactoglobulin.= +erature and p	AGFD	0207
Copolymers.= Surface Conformation of Colloid-Stabilizing Block	PMSE	0013
Solution Phase Conformation of Rapamycin.=	MEDI	0169
Nucleic Acids Using Metal Com+ Conformation Specificity of Guanosine Oxidation in	BIOL	0122
eoselective Oxidation of Locked- Conformation Styrenes: Stereoselective Syntheses of	ORGN	0236
opic Studies of sCD4(178)-PE40 Conformation: A Chimeric Protein Having Potential	ANYL	0005
s Probes of Protein Stability and Conformation.= Engineered Metal Chelation Sites a	BIOT	0133
ce Selectivity.= Conformation-Dependent Diels-Alder Dienophile Fa	ORGN	0241
Tunable Nickel Complexes for Conformation-Specific Oxidation of Nucleic Acids.=	INOR	0183
tion of Statistical Weights from Conformational Analysis for RIS Model: Polydimethy	POLY	0356
zolinone CCK-+ Synthesis and Conformational Analysis of Alkyl-Substituted Quina	MEDI	0066
Saccharide Frag+ Synthesis and Conformational Analysis of All Di-, Tri-, and Tetra-	CARB	0056
namics Simulation Techniques for Conformational Analysis of β-Cyclodextrin.= +Dy	ORGN	0174
onding to the Cel+ Synthesis and Conformational Analysis of Oligosaccharides Corresp	CARB	0057
Automated Conformational Analysis.=	CINF	0035
Developments in AI in Conformational Analysis.= WIZARD-III Recent	CINF	0034
Automated Learning in Conformational Analysis—WIZARD III.=	CINF	0017
es from Papaya Latex.= Conformational Differences in the Cysteine Proteinas	BIOL	0074
in cis-2-tert-Bu+ Search for the Conformational Effect Responsible for the Eclipsing	ORGN	0150
s in the Synthesis of 2'+ Role of Conformational Factors and Sugar-Protecting Group	CARB	0024
Activity.= Preliminary Conformational Model for Histamine H$_3$ Agonist	MEDI	0070
Chains in Flow.= Conformational Rearrangements of Flexible Polymer	PMSE	0105
oid Receptors.= Studies on the Conformational Requirement of Naltrindole at δ-Opi	MEDI	0154
cophores in 3D Databases: Does Conformational Searching Improve the Yield of Activ	CINF	0026
Relative Conformational Stabilities of Chiral (z)-Alkene.=	ORGN	0172
Conformational Studies of Polyalkylsilanes.=	COMP	0059
and Heterocycles.= Recent Conformational Studies of Six-Membered Carbo-	ORGN	0295
Paraben Will Cause a T←R-Like Conformational Transition in Insulin Hexamers.=	BIOT	0150
ules: Implications for+ Acousto- Conformational Transitions in Cytoskeletal Microtub	BIOT	0055
Studies.= Conformational Versatility of Hyaluronan: ^{13}C-NMR	CARB	0047
try-Based Approach for Docking Conformationally Flexible Molecules from 2D or 3D	CINF	0028
cre+ Searching 3D Databases of Conformationally Flexible Molecules: A Flexible Pres	CINF	0027
ative Molecular Field Analysis of Conformationally Flexible Tyrosine Kinase Inhibitor	MEDI	0024
c Agent.= L-702,958: A Novel Conformationally Restricted Class III Antiarrhythmi	MEDI	0096
Synthesis of a Conformationally Rigid Calcium-Entry Blocker.=	ORGN	0114
Peptide Conformations and Protein Structures.=	BIOL	0004
Solution Conformations of Muscarinic Antagonists.=	MEDI	0138
'-yl]pyrophosphates. New NAD Congeners Containing C-Nucleoside Isosteres of Nico	CARB	0023
s of Pumiliotoxin A Alkaloids and Congeners.= +ynthesis and Pharmacological Studie	ORGN	0275
Individual Hexachlorobiphenyl Congeners.= Physicochemical Parameters of	ANYL	0101
$^1A_2$ and $^1B_1$ States of Ozone: A Conical Intersection and Its Effect on Photoabsorptio	PHYS	0191
l Zirconocenes: Copper-Catalyzed Conjugate Addition to Enones.= +Reactions of Alky	ORGN	0089
Alternative to Organocuprate Conjugate Addition.= A Practical Reversed Polarity	ORGN	0091
es of Hydroxamic Acids and Their Conjugate Bases.= +irical Molecular Orbital Studi	PHYS	0326
oxia Imagi+ 3-Iodobenzylamine Conjugate of 2-Nitroimidazole: A New Potential Hyp	NUCL	0076
er Networks and Their Applica+ Conjugate Three-Component Interpenetrating Polym	PMSE	0080
elivery of Methotrexate-Insulin Conjugate to H35 Hepatoma and IEC-6 Cells in Vitr	MEDI	0185
f Haemophilus Influenzae Type B Conjugate Vaccine.= +ide Compositional Analysis o	ANYL	0021
Haemophilus Influenzae Type B Conjugate Vaccines.= Carbohydrate Antigens.	CARB	0053
3: Structure-χ$^{(3)}$ Relationships in Conjugated Aromatic Polyazomethines.= +n Films,	POLY	0226
Structures, and Conjugated Circuits.= Aromaticity, Kekule	COMP	0021
ocarbonate in a Latent+ Cyclic Conjugated Enediynes via the Elimination of a Thion	ORGN	0096
tical S+ Pedagogical Approach to Conjugated Molecular Systems: Facile Graph Theore	COMP	0020
s.= Electron Localization in Conjugated π-Systems and Delocalization in π-Stack	BIOT	0201
Route to Electrically Conducting Conjugated Polyacetylene.= +with Bromine: A New	POLY	0032
evices.= Conjugated Polymers. Conjugated Polymers as Materials for Nonlinear Opt	POLY	0325
al Properties of Highly Ordered Conjugated Polymers in Polyethylene: Orientation by	POLY	0008
ials for Nonlinear Optical Devi+ Conjugated Polymers. Conjugated Polymers as Mater	POLY	0325
Electron Transfers within Conjugated Porphyrin-Flavin Ester.= Photoinduced	BIOL	0134
r Nonlinear Optical Properties for Conjugated, π-Electron Molecules.= +Third-Orde	BIOT	0177
Synthesis of Amino Acid Conjugates of Ether Phospholipids.= Stereospecific	MEDI	0163
lgesic Potencies of the 6-Sulfate Conjugates of Some 3-Substituted Morphine, Dihydr	MEDI	0149
icity of ^{67}Cu-Porphyrin-Antibody Conjugates to a Colon Carcinoma Cell Line.= +tox	BIOT	0168
f New Poly(3-hydroxyalkanoate) Conjugates: PHA-Carbohydrate and PHA-Synthetic	POLY	0041
onding: The Natural+ Chemical, Conjugative, and Cooperative Effects in Hydrogen B	PHYS	0106
Molecule with the Connectivity of a Cube.= Synthesis from DNA of a	BIOL	0162
th Halogen Compounds: Ion-Core Conservation.= +the Reaction of Copper Atoms wi	PHYS	0368
Separately Controlled Plastic and Conserved Phases of Catalysis by Glycosylases.=	BIOL	0152
zation: The Pew Physical C+ A Consortium-Based Approach to Laboratory Moderni	CHED	0125
lysis-Shock-Tub+ Thermal-Rate Constant Measurements by the Flash or Laser Photo	FUEL	0123
d: I. MOPS+ First Dissociation Constant of Amino Acids by the Conductivity Metho	BIOL	0033
Deuterim Quadrupole Coupling Constants and Asymmetry Parameters in Bridging+	INOR	0382
at the Ice Surface.+ Equilibrium Constants for Adsorption of Antifreeze Glycoprotein	BIOL	0041
tuent Effects on the Equilibrium Constants for Cyanide Addition to Nicotinamide Ara	BIOL	0013
Temperature-Dependent Rate Constants for Electron Attachment Reactions of Met	PHYS	0234
ammatory Drug Pirox+ Stability Constants of Copper(II) Complexes with the Antiinfl	INOR	0407
ation of the Rate and Equilibrium Constants.= +lysis by Phospholipase A$_2$: Determin	BIOL	0019

ular Weights and Mark-Houwink	Constants.= +for Accurate Determination of Molec	PMSE	0066
dentification Testing and Issues of	Constitutional Privacy.= +y: Update 1991. DNA I	CHAL	0001
Using a Toluene Monooxygenase	Constitutive Pseudomonas cepacia.= +ion Products	BIOT	0119
lyl Compounds as Bicyclic Ring-	Constrained Analogues of Styryl-Based Protein Tyro	MEDI	0023
l Structural Characterization of a	Constrained Ω-Loop Excised from Interleukin-1α.=	MEDI	0165
Swelling of	Constrained Polymer Gels.=	COLL	0184
arison of Acidity and Geometric	Constraint for Iso-Structural Molecular Sieves Based	CATL	0028
erconductivity.= Experimental	Constraints on the Theory of High-Temperature Sup	PHYS	0051
eactions on the+ Chemistry with	Constraints: Local Density Calculations of Growth R	PHYS	0355
ance of Risk Assessment and Data	Constraints.= +'s Hazard-Ranking System: A Bal	ENVR	0081
Career Options in	Consulting.=	CMEC	0029
Agricultural, Professional, and	Consumer Markets.= Product Formulating for	FERT	0004
ring Fertilizer Claims.=	Consumer Perspective of Organic or Naturally Occur	FERT	0007
An Environmental and	Consumer View.= Risks and Benefits of Pesticides:	AGRO	0059
sy Polymers—I. Chain+ Energy-	Consuming Micromechanisms in the Fracture of Glas	PMSE	0073
plication of Epoxy Resins in Co+	Container Coating Systems—II. Plenary Lecture. Ap	PMSE	0157
Replacement.= PVC in	Container Coatings and Approaches to Its	PMSE	0143
rrier Polymers for Container Co+	Container Coatings Systems—I. Plenary Lecture. Ba	PMSE	0138
re. Application of Epoxy Resins in	Container Coatings: An Overview.= +enary Lectu	PMSE	0157
	Container Coatings: FDA Considerations.=	PMSE	0158
nary Lecture. Barrier Polymers for	Container Coatings.= +Coatings Systems—I. Ple	PMSE	0138
Pesticide	Container Collection and Recycling.=	AGRO	0090
	Container Minimization and Reuse.=	AGRO	0087
Iowa's Plastic Pesticide	Container Recycling Program.=	AGRO	0091
Pesticide	Container Recycling.=	AGRO	0088
Plastic Agricultural-Chemical-	Container Recycling.= Recent Developments in	AGRO	0045
Pesticide	Container Regulations as Part of EPA Strategy.=	AGRO	0089
Agricultural Chemicals and Their	Containers. Biolog+ Status of the Disposal of Waste	AGRO	0139
Agricultural Chemicals and Their	Containers. Curre+ Status of the Disposal of Waste	AGRO	0120
Agricultural Chemicals and Their	Containers. Intro+ Status of the Disposal of Waste	AGRO	0085
Agricultural Chemicals and Their	Containers. Pestic+ Status of the Disposal of Waste	AGRO	0102
and Food Metal	Containers.= Organic Coatings for Beer, Beverage,	PMSE	0140
Pesticide and Liquid Fertilizer	Containment Facilities Handbook.= MWPS-37	AGRO	0042
oil Chemistry. Impact of Fertilizer	Containment on Stormwater Management.= +d S	FERT	0013
Retail Fertilizer Dealer	Containment.=	FERT	0020
einato de Sodio a Partir de Leche	Contaminada con S. aureus y s+ Elaboracion de Cas	AGFD	0107
by EPA's Hazard-Ranking Sy+	Contaminant Migration in Ground Water as Assessed	ENVR	0081
ing Catalys+ Migration of Metal	Contaminants and Passivating Agents in Fluid-Crack	CATL	0012
+brium Chemistry on Leaching of	Contaminants from Stabilized/Solidified Wastes.=	ENVR	0010
Assessments for Microbiological	Contaminants in Drinking Water.= Risk	ENVR	0070
olic Metabolites of Environmental	Contaminants in the U.S. +rement of Urinary Phen	CEI	0002
Soil Gas Analysis—Environmental	Contaminants. Review of the History of Applicat+	ENVR	0048
romatogra+ Cleaning of Heavily	Contaminated Ion-Exchange Columns in Process Ch	ANYL	0042
arming to Remediate Pesticide-	Contaminated Soil Excavated from Agrochemical Fac	AGRO	0144
lants in Quebec: Extensive Ana+	Contaminated Soil Samples from Wood-Preserving P	ENVR	0016
ate Petroleum- and Naphthalene-	Contaminated Soils.= +Soil Treatment to Remedi	BIOT	0139
tizing Agent for the Treatment of	Contaminated Waters.= +ntially Useful Photosensi	ENVR	0029
eothermal Water+ Sensitivity to	Contamination of Drilling Fluids Systems Based on G	I&EC	0011
duction of Rinsate and Nontarget	Contamination.= +icide Application Systems for Re	AGRO	0105
Gasoline and Diesel Fuel	Contamination.= Soil Gas Survey Methods for	ENVR	0053
nking Water Supplies to Pesticide	Contamination.= +the Susceptibility of Surface Dri	AGRO	0053
alth Care.=	Contemporary Role of Biodegradable Polymers in He	CHED	0135
a Aromatizacion+ Influencia del	Contenido de Galio en Zeolitas del Tipo Pentasil en l	COLL	0125
diators, Tissues, and Cells in the	Context of Biosensors for Use in Vitro and in Vivo.=	BIOT	0039
onscience Students. Chemistry in	Context: An Overview.= +Chemistry Course for N	CHED	0051
Chemistry in	Context: Report from the Classroom.=	CHED	0053
hemistry Course f+ Chemistry in	Context: The ACS College-Level Decision-Making C	CHED	0051
Chemistry in	Context: What Is in It and What's in It for You.=	CHED	0052
scovery and Exploration of a New	Continent of Chemistry. H. Brown (Chemi+ The Di	CONG	0012
of Reactive Materials between the	Continental Margin and the Open N.W+ Exchanges	GEOC	0009
of Sedimentary Organic Matter in	Continental Margin Ecosystems.= +mineralization	GEOC	0015
al Flu+ Forms of Phosphorus in	Continental Margin Sediments: Implications for Buri	GEOC	0013
r, and Manganese from California	Continental Shelf Sediments.= +ron, Cobalt, Coppe	GEOC	0010
cian Symposium. The Making of a	Continuing Education Course—Start to Finish.=	TECH	0001
Chemical Science and Engineering	Continuum.= +y and Chemical Engineering—the	I&EC	0048
n Electron Microscopy, II: Image	Contrast Analysis of Iodinated Polystyrene Chains.=	POLY	0062
Insect	Control?.= What Is the Future of Baculovirus for	AGRO	0081
one in Hair Conditioners: Factors	Controlling Hair Deposition.= +Industry—II. Silic	CHED	0063
stal Thin Films by Design: The+	Controlling the Structure of Ferroelectric Liquid Cry	BIOT	0202
= Chemistry of Ocean Margins.	Controls on Methane Fluxes from the Hudson River.	GEOC	0001
ous Media: A Look at Two Recent	Controversies.= +Semiconductor Particles in Aque	COLL	0131
Reaction-	Convection-Diffusion in Emulsion Voltammetry.=	COLL	0042
stitucion de Electrodo de Vidrio	Convencional para Medir pH, por un Electrodo de A	ANYL	0066
endritic Macromolecules by the	Convergent Growth Approach: Hyperbranched Block	POLY	0318
isperse Arylester Dendrime+ The	Convergent Synthesis of Four Generations of Monod	POLY	0324
ono-C+ Thermal and Enzymatic	Conversions of Precursors to Flavor Compounds. Chr	AGFD	0022
coside+ Thermal and Enzymatic	Conversions of Precursors to Flavor Compounds. Glu	AGFD	0050
cosidi+ Thermal and Enzymatic	Conversions of Precursors to Flavor Compounds. Gly	AGFD	0006
dies on+ Thermal and Enzymatic	Conversions of Precursors to Flavor Compounds. Stu	AGFD	0070
lobal Swaps, Exchanges, and Toll	Conversions.= +Purchasing Productivity through G	CMEC	0007

for Acid–Catalyzed Hydrocarbon	Conversions.=	PETR	0102
°From Doubts to	Sulfate–Containing Solid Superacids	PROF	0003
R.= Structures of Zr(O$_3$P(CH$_2$)	Conviction of Fraud: Tales of Trauma.=	INOR	0047
ration of Lipid Oxidation during	$_n$COOH)$_2$, n = 1–4 from ^{31}P and ^{13}C Solid-State NM	AGFD	0124
cids as Texturizing Agents during	Cooking of Refrigerated Minced Channel Catfish Mu	AGFD	0150
Spectroscopy of Supersonically	Cooking–Extrusion Cereals.= +ication of Phenolic A	PHYS	0007
g for Rotary Fertilizer Dryers and	Cooled Carbon Clusters.= Infrared Laser	FERT	0012
fer in Ultrahot Solids: Vibrational	Coolers and Their Effects on Performance.= +ghtin	PHYS	0250
ral+ Chemical, Conjugative, and	Cooling and Multiphonon Up–Pumping.= +Trans	PHYS	0106
s to Metal Atoms.= Role of	Cooperative Effects in Hydrogen Bonding: The Natu	PHYS	0172
NMR Measurement of	Cooperative Effects in the Binding of Small Molecule	PHYS	0069
New Five–	Cooperative Motions in Solids.=	INOR	0019
Atraniums: Novel Five–	Coordinate Complexes of Silicon(IV).=	INOR	0023
zed by Steri+ Three– and Four–	Coordinate Group 14 Cations.=	INOR	0085
of Two–	Coordinate Monomeric Zinc Alkyl Derivatives Stabili	INOR	0325
on Approaches to Low–	Coordinate Phosphines.= New Derivative Chemistry	INOR	0324
ccess to a Symmetrical Reaction	Coordinate Phosphorus Species.= Ring Fragmentati	ORGN	0122
n+ Rotation and Orientation of	Coordinate: A Possible Domain for Dynamic Effects.	INOR	0296
om Transfer from N$_2$O to Alkynes	Coordinated Imidazoles in Low-Spin Ferric Porphyri	INOR	0327
Generation of 3D Atomic	Coordinated to Zr.= +llic Compounds. Oxygen-At	CINF	0036
	Coordinates for Organic Molecules.= Automatic		
m in Indianapolis.=	Coordinating and Evaluating a Project SEED Progra	CHED	0141
l Migratory Insertion R+ Role of	Coordinating Solvents in Promoting the Alkyl to Acy	INOR	0109
lladium Hydride Dimers: Unusual	Coordination Behavior of LiBEt$_4$.= +Hydrides. Pa	INOR	0379
ompounds—The Practical Side.	Coordination between Theoretical and Practical Coor	CHED	0056
Ni, and Cu Compounds: An Ex+	Coordination Chemistry and Catalysis of V, Mo, Co,	CHED	0057
es as Models for Nickel(III)–Cy+	Coordination Chemistry of Ni(III)–Thiolate Complex	INOR	0179
between Theoretical and Practical	Coordination Chemistry.= +al Side. Coordination	CHED	0056
nity in the Area of Surface	Coordination Chemistry.= Applications and Opportu	CHED	0062
Polymers.= Metal	Coordination Complexes for Synthesis of New	CHED	0059
rder Nonlinear Optical Materia+	Coordination Compounds and Polymers as Second–O	INOR	0049
uinazolinone-2,3+ Mixed Metal	Coordination Compounds of the Heterocyclic 4(1H)Q	INOR	0079
Haloperoxidase.= Vanadium	Coordination Compounds: Modeling Vanadium	INOR	0018
by Light–Sensitive	Coordination Compounds.= Photocatalysis Induced	INOR	0340
ination between Theoret+ Metal	Coordination Compounds—The Practical Side. Coord	CHED	0056
orseradish P+ Comparison of the	Coordination Environment of Mn(III) and Mn(IV) H	BIOL	0095
te Ligands.=	Coordination of Bismuth(III) to Polyaminocarboxyla	INOR	0033
d Donors P$_2$S Toward Various+	Coordination Studies of a Tripodal Ligand with Mixe	INOR	0028
(SiMe$_3$)$_2$|$_3$+ Unusual η^6–Toluene	Coordination to an Alkali Metal: Strucutre of Lu{CH	INOR	0053
with the cupric Ion Intensity and	Coordination via Electron Spin Resonance.= +ucts	CATL	0017
Metal	Coordination.= Metal–Polymer Interaction through	POLY	0343
n–Iridium Tetranuclear+ Alkyne	Coordination, Dimerization, and Scission on Tungste	CATL	0008
fido Dimers Formed+ Labile and	Coordinatively Unsaturated Molybdenum(III)–μ–Sul	INOR	0271
es on Radical Stabilities.=	Cope Award Symposium. Effects of Structural Chang	ORGN	0133
f Expanded Porphyrins.=	Cope Award Symposium. Synthesis and Properties o	ORGN	0139
Entrepreneurs Can	Cope with Regulatory Burdens.= How Chemical	SCHB	0011
stilb+ Functional Polyesters and	Copolyesters Based on the 4,4'-Dihydroxy-α-methyl	POLY	0031
Dynamics of Block	Copolymer Adsorption.=	COLL	0073
Polymer Blends of a SEBS Block	Copolymer and a Polyetherester.= +erpenetrating	PMSE	0058
lidene Fluoride/Trifluoroethylene	Copolymer and Poly(chlorofluoroethylene).= +Viny	POLY	0022
and Molar Mass Sensitive Det+	Copolymer Characterization Using Conventional SEC	PMSE	0051
g an+ Characterization of Block	Copolymer Ionomers by Small-Angle X-ray Scatterin	PMSE	0118
yrene–Poly(ethylene Oxide) Block	Copolymer Micelles in Water.= +Methods. Polyst	POLY	0277
Metal Nanoclusters within Block	Copolymer Microdomains.= +ntrolled Synthesis of	POLY	0207
Transitions in Triblock	Copolymer SBS.= Shear–Induced Structural	PMSE	0136
Microstructured Block	Copolymer Solutions.= Diffusion and Dynamics in	POLY	0275
nd Silicate–Modified Starch Graft	Copolymer Superabsorbents.= +cy in Aluminum a	CARB	0042
n the Thermodynamics of Diblock	Copolymer Thin Films.= +nite Thickness Effects i	POLY	0090
ghly Conducting Hydropolysilane	Copolymer: Synthesis and Electrical Conductivity.=	POLY	0043
Methyl Ether/Maleic Anhydride)	Copolymer.= +aporates Prepared Using Poly(vinyl	COLL	0086
Sites in the Epoxy-g-Acrylic	Copolymer.= Theoretical Calculation of Grafting	PMSE	0161
ties.= Microwave–Initiated	Copolymerization MMA/HEMA Kinetics and Proper	POLY	0052
hosphine)palladium Catalysts for	Copolymerization of Carbon Monoxide and Ethylene	INOR	0155
of Oligomer Structures during the	Copolymerization of Epoxy with Bishpenol-A.= +s	POLY	0227
Cross–Linking, Modification, and	Copolymerization of Lignin with Low Molecular Wei	BIOT	0041
phthyl)thiophenium Hydroxide +	Copolymerization of Tetrahydro–1-(4–hydroxy-1–na	POLY	0048
ith Acrylonitrile.=	Copolymerization of 1–Allyl-2,5–dimethoxybenzene w	POLY	0050
er+ Change of Mechanism Block	Copolymerizations: The Formation of Block Copolym	PMSE	0071
hylphenylsiloxane) Random Block	Copolymers and Elastomers.= +hylsiloxane–co–met	POLY	0020
ible Gels with Polyethylene Block	Copolymers and Hydrocarbon Solvents.= +orevers	POLY	0249
Approach: Hyperbranched Block	Copolymers and Monodisperse Polyesters.= +owth	POLY	0318
izations: The Formation of Block	Copolymers Containing Helical and Amorphous Poly	PMSE	0071
Ethylene Oxide) and ABA Block	Copolymers Containing Hydroxyl–Capped Poly(Ethy	PMSE	0006
ion of Ordered Heterocycle–Imide	Copolymers Containing ortho–Phenylene Rings.=	POLY	0042
Applications.= Side–Chain	Copolymers for Third-Order Nonlinear Optical	POLY	0223
d Characterization.=	Copolymers of Nylon 266 and Nylon 66: Synthesis an	POLY	0012
heir Optical+ Synthesis of Block	Copolymers of Poly(styrene) and Poly(arylate) and T	POLY	0220
of Cationic, Nonionic, and Anionic	Copolymers of Vinylpyrrolidone.= +romatography	PMSE	0031
ylonitrile).= New Fluorinated	Copolymers: Poly(1,3-cyclohexadiene-alt-α-fluoroacr	POLY	0021
mers: Vinyl Alcohol/Vinyl Butyral	Copolymers.= +ffusion in Hydrogen-Bonded Poly	POLY	0191
oly(ester–Ether) Segmented Block	Copolymers.= +oly(butylene Terephthalate) and P	PMSE	0148

parated Multiblock	Copolymers.= Colloidal Structure of Microphase-Se	COLL 0053
Ethylene	Copolymers.= Crystallization Kinetics of Random	POLY 0075
tion of Ethylene-Based Polyolefin	Copolymers.= +ared Spectroscopy for Characteriza	PMSE 0048
Castor Oil Polyurethane and Vinyl	Copolymers.= +g Polymer Network Formed from	PMSE 0089
se via Synthesis of Well-Defin+	Copolymers.= Investigation of the Polymer Interpha	COLL 0033
Based on Block	Copolymers.= Microstructural Aspects of IPNs	PMSE 0026
Using ROMP Block	Copolymers.= Semiconductor Clusters Prepared	INOR 0307
e Structure of PS-PMPS Diblock	Copolymers.= +tate NMR Study of the Microphas	PMSE 0121
MMA/HEMA	Copolymers.= Solution Properties of	PMSE 0086
MMA/HEMA	Copolymers.= Solution Properties of	POLY 0093
Colloid-Stabilizing Block	Copolymers.= Surface Conformation of	PMSE 0013
Poly(styrene-co-b-vinylborazine)	Copolymers.= +hesis of Poly(b-vinylborazine) and	POLY 0183
of	Copolymers.= Thermal Field-Flow Fractionation	PMSE 0009
Chemical Vapor Deposition of	Copper and Copper(I) Oxide from Copper tert-Butox	INOR 0143
m of Carnosine in the Presence of	Copper and Iron.= +s in the Antioxidant Mechanis	AGFD 0039
-CaF$_2$-SiO$_2$ Slags.+ Solubility of	Copper and Other Elements in Silica-Saturated CaO	I&EC 0066
Dry Etching of	Copper at High Rates.=	COLL 0030
te Selectivity in the Reaction of	Copper Atoms with Halogen Compounds: Ion-Core C	PHYS 0368
on with Sodium, Magnesium, and	Copper Atoms.= +of the Molecular Oxygen Reacti	PHYS 0141
Target Fragmentation of	Copper by Intermediate-Energy Heavy Ions.=	NUCL 0025
Determination of Human Seric	Copper by Inverse Voltammetry at the Mercury Film	ANYL 0056
clooctadiene Cop+ Mechanism of	Copper Chemical-Vapor Deposition from the 1,5-Cy	COLL 0026
ubstr+ Photoredox Reactivity of	Copper Complexes and Photooxidations of Organic S	INOR 0342
mium, Lead, Nickel, Cobalt, and	Copper Concentrations in Lung Tissue: A Comparati	ENVR 0040
uoroacetylacetonate Complexes for	Copper CVD.= +and-Stabiilized Copper(I) Hexafl	COLL 0156
oom-Temperature Deposition of	Copper Films by H-Atom Reaction with Cu(FOD)$_2$ a	PHYS 0259
Chemical-Vapor Deposition of	Copper from Copper(I) Compounds.= Selective	COLL 0027
ure Chemical Vapor Deposition of	Copper from New Copper(I) Compounds.= +perat	INOR 0144
Chemical-Vapor Deposition of	Copper from 1,5-Cyclooctadiene Copper(I) Hexafluor	COLL 0029
d Y Zeolites: Product Selectivi+	Copper Ion Catalysis of Propylene Oxidation on X an	CATL 0017
als—Novel Techniques. Ultrathin	Copper Overlayers on TiO$_2$(110).= +try of Nonmet	COLL 0191
Sulfur-Sulfur Bond Scission by a	Copper Reagent.=	ORGN 0252
Tin Layer from	Copper Substrate.= Electrochemical Removal of a	PHYS 0285
lfate-Reduci+ Sulfur Isotopes in	Copper Sulfide Corrosion Products as a Result of Su	BIOT 0100
ecomposition of Alkyl Halides on	Copper Surfaces.= +ns of Catalysts—II. Thermal D	COLL 0204
Copper and Copper(I) Oxide from	Copper tert-Butoxide.= +cal Vapor Deposition of	INOR 0143
Precursors. Mechanistic Studies of	Copper Thin-Film Growth by MOCVD.= +tallic	COLL 0025
OMCVD of	Copper.= Decomposition Mechanisms Important in	COLL 0157
arboxylation Reactions Involving	Copper(I) Complexes Derived from Cu-Co Dimers.=	INOR 0213
Properties of Mononuclear	Copper(I) Complexes.= Dioxygen-Binding	INOR 0178
Molecules to mono- and Dinuclear	Copper(I) Complexes.= +dative Addition of Small	INOR 0044
n+ Deposition Characteristics of	Copper(I) Compounds for CVD by FT-IR and Rama	COLL 0031
Deposition of Copper from	Copper(I) Compounds.= Selective Chemical-Vapor	COLL 0027
r Deposition of Copper from New	Copper(I) Compounds.= +perature Chemical Vapo	INOR 0144
Cop+ Organic Ligand-Stabiilized	Copper(I) Hexafluoroacetylacetonate Complexes for	COLL 0156
ition from the 1,5-Cyclooctadiene	Copper(I) Hexafluoroacetylacetonate Complexes for	COLL 0026
f Copper from 1,5-Cyclooctadiene	Copper(I) Hexafluoroacetylacetonate.= +position o	COLL 0029
upon the Solid-State Emission of	Copper(I) Hilides.= +Crystallographic Symmetry	INOR 0157
ural and Luminescence Studies of	Copper(I) Intercalated Laser Dye Coumarins.= +ct	INOR 0245
l Vapor Deposition of Copper and	Copper(I) Oxide from Copper tert-Butoxide.= +ca	INOR 0143
P$_2$S with	Copper(I).= Complexation of a Mixed-Donor Ligand	INOR 0029
Thermal CVD from a	Copper(II) Aminoalkoxide Precursor.= Chemistry of	COLL 0028
ug Pirox+ Stability Constants of	Copper(II) Complexes with the Antiinflammatory Dr	INOR 0407
S Pseudothiolate C+ An N$_2$S$_2$-	Copper(II) Macrocycle and a Pentacoordinate CuN$_3$O	INOR 0262
Reactions of Alkyl Zirconocenes:	Copper-Catalyzed Conjugate Addition to Enones.=	ORGN 0089
+Chemical Pretreatments on the	Copper-Catalyzed Gasification of Carbon in Air.=	FUEL 0028
Stability.= New	Copper-Dioxygen Complexes with Thermal	INOR 0046
Ligands.= New	Copper-Dioxygen Species (Cu$_2$-O$_2$) Using Imidazole	INOR 0045
Shelf+ The Flux of Iron, Cobalt,	Copper, and Manganese from California Continental	GEOC 0010
Aqueous	Coprecipitation of Zirconia Precursors.=	COLL 0123
Analysis of	Coprocessing Materials.= Stable Carbon Isotope	FUEL 0071
ts: Fuel Chemistry—II. Coal/Oil	Coprocessing Using Syngas.= Selectivity in Catalys	CATL 0049
Naphthenes to Coal during	Coprocessing.= Hydrogen Transfer from	FUEL 0005
olicies Inviolate or in Violation?	Copyright and Technology: What Happens When Pa	CINF 0043
Journal Full Text:	Copyright Issues in Electronic Distribution.=	CINF 0046
Information Transfer Process.=	Copyright Protection and Risks in the International	CINF 0014
s on Information Ownership and	Copyright: Are Your Policies Inviolate or in Violation	CINF 0043
nd Alcohols: Discrepancy between	Core and Surface Responses.= +n by Anesthetics a	MEDI 0032
Students.= Interdisciplinary	Core Approach to College Chemistry for Nonscience	CHED 0176
s with Halogen Compounds: Ion-	Core Conservation.= +the Reaction of Copper Atom	PHYS 0368
ning Oligosaccharides of the Inner	Core Region of Lipopolysaccharides.= +O Contai	CARB 0058
carbo+ New Triiron Carboxylate	Core Structure Consisting of fused (μ-Hydroxo)bisμ-	INOR 0094
formance Epoxies Using Designed	Core-Shell Rubber Particles.= +ning of High-Per	POLY 0157
ten-Iridium Tetranuclear Cluster	Cores: Relationship to Butane Hydrogenolysis Select	CATL 0008
A Sulfonylurea Herbicide on	Corn (QSAR Analysis).=	AGRO 0050
Flavonolglyco+ HPLC Survey of	Corn (Zea mays L.) Populations and Inbreds for Silk	AGRO 0034
Quality of Stored Frozen Sweet	Corn on the Cob Produced by an Improved Blanchin	AGFD 0136
Conversion of	Corn Syrup Solids into γ-Cyclodextrin.=	CARB 0041
y of the Herbicide DPX-E9636 in	Corn: Uptake, Translocation, and Metabolism.= +t	AGRO 0027
Resistant/Susceptible Varieties of	Corn.= +tionic Proteins Extracted from Aflatoxin-	AGFD 0134

Herbicide in	Corn.= Synthesis of NC-319: A New Sulfonylurea	AGRO	0049
f DPX-E9636 Herbicide for Field	Corn.= +HPLC Method To Support Registration o	AGRO	0030
Pig	Corneal Proteases.=	BIOL	0043
Project SEED. The Sidney-	Cornell Project SEED Connection.=	CHED	0036
Physical Aging of Gelatinized	Cornstarch.= Effect of Moisture Content on the	CARB	0043
ry, sponsored by Dexter Chemical	Corp.).+ (Dexter Award in the History of Chemist	HIST	0011
s, sponsored by the Perkin-Elmer	Corp.). +emistry Award in Spectrochemical Analysi	ANYL	0001
Expert: Considerations of	Corporate In-House Counsel.= The Scientific	CHAL	0017
ation. Benefits of Alliances to t+	Corporate/Academic Partnerships in Precollege Educ	CHED	0001
Public vs. Helping the	Corporation.= The False Conflict: Helping the	PROF	0006
e Sorption of Orga+ Dispersion-	Corrected HPLC/FACP Technique for Measuring th	ANYL	0072
on of That of 9-Deazainosine.=	Corrected Structure of Pyrrolosine and Reconfirmati	CARB	0046
roperties during Isothermal Ep+	Correlating Chemorheological Changes to Dielectric P	PMSE	0178
ncy of+ Influence of Multipolar	Correlations and Surface Imperfections on the Efficie	PHYS	0063
Structure-Property	Correlations in Cuprate Superconductors.=	PHYS	0028
aterials. Sizes of the Short-Range	Correlations in the Dynamics of Rotational Isomeris	POLY	0365
Activity.= Structure-Activity	Correlations of Natural Products with Anti-HIV	AGFD	0163
emperture Ranges: Measurements,	Correlations, and Predictions.= +ions over Wide T	PHYS	0087
rface Dynamics of Diffusion and	Corrosion at Atomic Layer Resolution Studied by in	PHYS	0127
ect of Crystallization Induced on	Corrosion Behavior of Metglass 2826MB and MBF 1	PHYS	0219
ico de Una Ca+ Evaluacion de la	Corrosion de las Diferentes Etapas del Lavado Quim	PHYS	0257
sical Technologies of Fertilizers.	Corrosion of Mild Steel Exposed to Inhibited Urea-A	FERT	0001
Sulfate Suspensions.=	Corrosion of Mild Steel Exposed to Urea-Ammonium	FERT	0003
d Microbial Activity and Localized	Corrosion on Metal Surfaces.= +between Localize	BIOT	0098
ulfur Isotopes in Copper Sulfide	Corrosion Products as a Result of Sulfate-Reducing	BIOT	0100
earc+ Pseudomorph and Artifact	Corrosion Studies as Tools for Biomineralization Res	BIOT	0101
zation and Microbially Influenced	Corrosion. Hydrodynamics and Kineti+ Biominerali	BIOT	0096
urface Chemical Studies of Hot	Corrosion: Life Prediction Model for Marine Gas Tur	FUEL	0099
stems Based on Geothermal W+	Corrosive Effects on Drill Pipes by Drilling Fluids Sy	I&EC	0012
Conditioning Agents on the	Corrosive Properties of Molten Urea.= Effect of	FERT	0011
nt for an Economical Synthesis of	Cortisol: The Discovery of the Fluorosteroids.= +u	HIST	0022
Pill and	Cortisone.= Early Steroid Research in Mexico: The	HIST	0018
f Fragrance C+ Chemistry in the	Cosmetic and Personal Care Industry—I. Overview o	CHED	0046
Hair Conditi+ Chemistry in the	Cosmetic and Personal Care Industry—II. Silicone in	CHED	0063
l Extracts Used in Foods, Flavors,	Cosmetics, and Pharmaceuticals.= +ds in Botanica	AGFD	0172
ids from Methanol-Water Mixt+	Cosolvent and pH Effects on Sorption of Organic Ac	ENVR	0014
Water-Miscible Organic	Cosolvents on Kinetics of Plasmin.= Effects of	BIOL	0092
Award for Excellence in Teaching,	cosponsored by E. +vision of Analytical Chemistry	ANYL	0006
ition among Polymers. Assessing	Cost and Performance in Intermaterials Competition	CMEC	0009
Experiment.= Low-	Cost Multiple-Impact Physical Chemistry	PHYS	0336
Regulations Be Met in a	Cost-Effective Manner?.= Can Hazardous Material	CHAS	0033
Experiment.= Low-	Cost, Multiple-Impact Physical Chemistry	CHED	0162
Small Chemical Businesses in	Costa Rica: The Case of Arvi, Inc., Costa Rica.=	SCHB	0026
Rica: The Case of Arvi, Inc.,	Costa Rica.= Small Chemical Businesses in Costa	SCHB	0026
he Technological Transfer Unit in	Costa Rica.= +the Productive Sector: The Case of t	SCHB	0025
Correlation Spectroscopy (NQR-	COSY).= Two-Dimensional Pure Quadrupole	PHYS	0277
omatic Hydrocarbons in La Basse	Cote-Nord, Quebec.= +ng People to Chlorinated Ar	CEI	0011
graphy for the Characterization of	Cotton Fiber.= +ation of Size-Exclusion Chromato	PMSE	0032
ehavior and Develo+ Effects of	Cotton Plant Allelochemicals and Nutrients on the H	AGRO	0037
nts in Anodic Voltammetric and	Coulometric Analysis of Insoluble, Difficultly Oxidiza	COLL	0002
ion.=	Coulometric Initiation of Microemulsion Polymerizat	COLL	0169
Copper(I) Intercalated Laser Dye	Coumarins.= +ctural and Luminescence Studies of	INOR	0245
Corporate In-House	Counsel.= The Scientific Expert: Considerations of	CHAL	0017
dic Flavor Precursors by Droplet	Countercurrent Chromatography and Mass Spectrom	AGFD	0203
n Coefficient Determination Using	Countercurrent Chromatography.= +iquid Partitio	ANYL	0070
g.=	Countertrade—An Innovative Approach to Marketin	CMEC	0008
ides in Developing and Developed	Countries.= +the 21st Century. Outlook for Pestic	AGRO	0056
ACCP System in Underdeveloped	Countries.= +Problems in the Application of the H	AGFD	0079
an Spectroscopy Using Charge-	Coupled Device Detection.= Unique Optics for Ram	ANYL	0035
mic Absorption Spect+ Directly	Coupled Gas Chromatography/Graphite Furnace Ato	ANYL	0080
Analysis of Fly Ash by Inductively	Coupled Plasma Atomic Emission Spectroscopy.=	FUEL	0138
by the Fock-Space Multireference	Coupled-Cluster Method.= +in Radicals Studied	PHYS	0016
Studies of Carbon Clusters with	Coupled-Cluster Methods.= High-Level Theoretical	PHYS	0008
	Coupled-Cluster Treatment of Symmetry Breaking.=	PHYS	0015
and Adenylosuccinate Lyase: Odd	Couples in the Evolution of Mechanistic Motifs.=	BIOL	0012
Dynamics of	Coupling Agents in Interfacial Regions.=	POLY	0368
ridging+ Deuterim Quadrupole	Coupling Constants and Asymmetry Parameters in B	INOR	0382
Magnetic Relaxation	Coupling in Dynamically Heterogeneous Systems.=	PHYS	0232
nzofurans via the Intramolecular	Coupling of a Zirconocene Benzyne Complex with Al	ORGN	0217
cl+ Assistance of Photo-Induced	Coupling of Alkylidyne and Carbonyl Ligands by Nu	INOR	0188
hetic and Mec+ Proton-Induced	Coupling of Alkylidyne and Isocyanide Ligands: Synt	INOR	0150
ons: Nickel(O)-Catalyzed Cross-	Coupling of Aryl Triflates with Grignard Reagents.=	ORGN	0082
Dehydrogenative	Coupling of CH_4 in a Hot-Wire Diffusion Column.=	PETR	0012
ared by Site-Directed or Random	Coupling of Monoclonal Antibodies.= +rbents Prep	BIOT	0060
ted ortho Metalation and Cross-	Coupling Tactics: Application in Natural Product Sy	ORGN	0277
des Using Fluorine-19 Carbon-13	Coupling.= +ion of Local Structure in Organofluori	FUEL	0077
nt and Enoyl R+ Revised Steric	Courses and Mechanisms of the Allylic Rearrangeme	BIOL	0065
Processing for Analytical	Courses.= Data Acquisition and Computer Data	CHED	0119
in Analytical Chemistry	Courses.= Extensive Use of Spreadsheet Programs	CHED	0205
Physical Chemistry Survey	Courses.= Laboratory Experience for Students in	CHED	0215

	Court Training for the Forensic Chemist.=	CHAL	0019
the	Courtroom.= Abstract Scientific Concepts in	CHAL	0012
tudies on Mandelonitrile Lyase by	Covalent and Noncovalent Modification.= +istic S	BIOL	0075
nsfer and Relavtivistic Gold–Gold	Covalent Bonding.= +lts: Evidence for Energy Tra	INOR	0189
onzero Spin–Ground States.=	Covalent Linkage of Manganese Oxo Clusters with N	INOR	0196
c+ Photochemical Properties of	Covalently Attached Sensitizer/Mediator or Donor/A	INOR	0038
otoinduced Electron Transfer in	Covalently Linked Polynuclear Aromatic–Nucleoside	PHYS	0021
oinduced Electron Transfer in	Covalently Linked Porphyrin–RuO₂ Clusters.= Phot	COLL	0133
Studies of Pirlimycin in the Dairy	Cow.= +ducing Animals. Metabolism and Residue	AGRO	0108
Lactating Dairy	Cows.= Novel Sulfonamide Drug Metabolism in	AGRO	0113
Luprostiol in Dairy	Cows.= Residue Depletion and Metabolism of	AGRO	0130
Migratory Insertion Reaction in η–	Cp(CO) (PPh₃)FeMe⁺.= +oting the Alkyl to Acyl	INOR	0109
ons of the Vinylidene Complexes	Cp(CO)₂M=C=CH₂ (M=Mn, Re) with Imines, Carbo	INOR	0072
Systems.= Behavior of V^E and	Cp^E near the UCST of Polyethers + Alkane	PHYS	0319
W–Shape	Cp^E.= Effect of Branching and Cycling on the	PHYS	0320
n Iron Phosphide Acetyl Complex	Cp(PPH₂)(CO)FeCOCH₃-l with Electrophiles.= +a	INOR	0240
Complexes of HNSφ2 with	Cp*₂UCl₂.=	INOR	0366
Photochemistry of (	Cp)₂TiCl₂ on Porous Vycor Glass.=	INOR	0161
mperatures between 90 and+ ¹³C	CP/MAS NMR Studies of Catalytic Reactions at Te	COLL	0090
h MR₂ (M = Zn, Mg and R = Bz,	Cp, allyl).= +ntation–Recombination Reactions wit	INOR	0383
s in the Enthalpy of Reaction of	CpMn(CO)₂ with Tertiary Silanes as Examined by P	INOR	0386
Electrophile.= [	cpRu(dppm)(CH₂SO₂)]⁺: A Strong Organometallic	INOR	0249
ophiles.= Ruthenium Thiolates	cpRu(PR₃')₂SR: Synthesis and Reactions with Electr	INOR	0368
Educating Future Teachers: The	CPT–Chemical Education Option. Chemistry Educ	CHED	0066
Spectroscopy of UV–Photolyzed	Cr(CO)₆ in Solution.= Picosecond Transient IR	PHYS	0220
of the Excited–State Properties in	CR(III) Complexes.= +gnetic Field Perturbations	INOR	0159
and Passivating Agents in Fluid–	Cracking Catalysts.= +ation of Metal Contaminants	CATL	0012
a Supercritical Flui+ Surface and	Cracking Properties of Expanded Clays Dried Using	PETR	0145
	Cracking Properties of SAPO-37 and Faujasites.=	CATL	0045
roblems with the Use of Urinary	Creatinine as a Stable Metric for Human Biomonitor	CEI	0019
mers—Chemistry, Phys+ Marvel	Creative Polymer Chemistry Award Symposium Poly	POLY	0272
ntaneous Adsorpt+ C. S. Marvel	Creative Polymer Chemistry Award Symposium. Spo	POLY	0298
Some Dichotomies of	Creative Scientists, including Chemists.=	CONG	0009
n Study.=	Creosote Site–Biobed Box Pilot–Scale Bioremediatio	BIOT	0140
	Crescent Chemical and Its Origin in Tridom.=	SCHB	0015
Role of Forensic Biology in	Crime Investigation.= Historical Perspective on the	HIST	0003
Chemistry and	Crime—Molecular Offenses in Canada.=	CHAL	0020
uture. Lindberg+ Chemistry and	Crime, III: Forensic Chemistry—Past, Present, and F	HIST	0006
uture. The Sea+ Chemistry and	Crime, III: Forensic Chemistry—Past, Present, and F	HIST	0001
Disposal	Crisis in Africa.= Pesticide Donations and the	AGRO	0044
S Study of Mean–Field and Ising	Critical Behavior of Poly(2–Chloro–Styrene)/Polysty	PMSE	0103
hy.=	Critical Conditions in Gel Permeation Chromatograp	PMSE	0049
ductor by Meta+ Enhancement of	Critical Current Density of Bi₂Sr₂CaCu₂O₈ Supercon	PHYS	0329
of Periplasmic Proteins Using	Critical Fluid Disruption.= Downstream Processing	BIOT	0070
Surrogates for Natural Sorbents:	Critical Review of Experimental Evidence on Retenti	ANYL	0071
istry: Reaching and Teaching L+	Critical Thinking and Achievement in General Chem	CHED	0081
	Critique of DNA Evidence.=	CHAL	0002
Samples.=	Cross Polarization of Half–Integer Spins in Rotating	PHYS	0371
Amino Acid–Nucleotide	Cross–Binding to Platinum(II): A Kinetic Study.=	INOR	0182
r Synthons: Nickel(O)–Catalyzed	Cross–Coupling of Aryl Triflates with Grignard Reag	ORGN	0082
d Directed ortho Metalation and	Cross–Coupling Tactics: Application in Natural Prod	ORGN	0277
Cell Debris Removal by	Cross–Flow Microfiltration.=	BIOT	0048
works.= Substituent–Induced	Cross–Link Aggregation in Polydimethylsiloxane Net	PMSE	0120
mi+ Miscibility and Properties of	Cross–Linkable Polyimide Blends with Rigid and Se	PMSE	0023
Lec+ Formation of Homogeneous	Cross–Linked Lattices between Glycoconjugates and	BIOL	0145
Structure and Properties of AB	Cross–Linked Polymers.=	PMSE	0083
Polymers: Synthesis of Uniformly	Cross–Linked Polystyrene Microbeads.= +el Filled	PMSE	0017
in IPN Membranes: Effect of	Cross–Linked State and Annealing.= Gas Transport	PMSE	0060
e–Sp+ Comparative Study of the	Cross–Linking Activities of the 14 KDa β–Galactosid	BIOL	0109
olecular Ordering and Chemical	Cross–Linking in Vitrified Nonlinear Optical Polyme	POLY	0206
diation.=	Cross–Linking of Polyallylcarbonates by Gamma Irra	PMSE	0131
diation.=	Cross–Linking of Polyallylcarbonates by Gamma Irra	POLY	0053
Effect of Backbone Flexibility and	Cross–Linking on Flexibility of Can Coatings.= +–	PMSE	0162
ization and Bio+ Applications of	Cross–Linking Techniques to Enzyme/Protein Stabil	BIOT	0220
Glutaraldehyde	Cross–Linking: Fast and Slow Modes.=	BIOT	0218
.+ Overview of Protein Chemical	Cross–Linking: Implications for Protein Stabilization	BIOT	0216
f Lignin with Low Mo+ Enzyme	Cross–Linking, Modification, and Copolymerization o	BIOT	0041
Electronic State on the Reaction	Cross–Sections of Transition–Metal Cations: V+ + E	PHYS	0306
The Salamander and	Crossed Retorts: The Chemists' Club of New York.=	HIST	0031
/Polysaccharide Gels: Influence of	Crosslink Chemistry on Rheological Properties.=	POLY	0240
–Centered Hydrogen–Bonding and	Crosslinking Groups.= +ers Having Chromophore	POLY	0108
rs.=	Crosslinking Reactions in Pigmented Olefinic Polyme	POLY	0135
bazones—Comparison with Spin–	Crossover Analogs.= +in-State: Ferric Thiosemicar	INOR	0268
dies of Dealuminated+ ¹H–²⁷ Al	Crosspolarization and Very Fast ²⁷Al MAS NMR Stu	PETR	0006
Studies of Sterically	Crowded Porphyrins.= Crystallographic and EXAFS	INOR	0091
een Gystamine, Mexamine, and	Crown Ether on the Surface of Pyrogeneous Silica.=	COLL	0096
of Lead(II) and Mercury(II) with	Crown Ether on the Surface of Silochromium.= +on	COLL	0104
rown+ Synthesis of a Poly(ester	crown) Based on bis(5–Carboxy–1,3–phenylene)–32–c	POLY	0044
(5–Carboxy–1,3–phenylene)–32–	crown-10 and 4,4'–Isopropylindenediphenol (bisPhen	POLY	0044
of 60–	Crown–20 Macrocycles.= Thermoreversible Gelation	POLY	0246

phenacyl Derivatives.= 18-Crown-6 as a Catalyst in the Dialkylation of o-Nitro	ORGN	0261
Magnetic Separation of Reduced Crude Conversion Catalyst.=	I&EC	0063
atalyst System to Hydrogenate Crude Methylenedianiline to a Nonthermodynamic M	I&EC	0056
erature Gas Chromatography of Crudes and Residua Using Atomic Emission Detectio	PETR	0121
Characterization of Waxy Crudes and Waxes.=	PETR	0123
s: Reaction Pathway Control—I. Crystal and Catalytic Chemistry of Bismuth Molybda	CATL	0026
ved from 2-Methyl-thiosemicar+ Crystal and Molecular Structure of Compounds Deri	INOR	0032
Spectroscopy Studies of Single- Crystal $Bi_2Sr_2CaCu_2O_8$ High-T_c Superconductors.=	PHYS	0211
Polysaccharide Liquid- Crystal Blends.= Characterization of Lyotropic	COLL	0017
Investigation of a Liquid Crystal Dispersed in an Ionic Polymeric Membrane.=	PMSE	0119
but Spans the Range of Different Crystal Forms.= +in Differs from the Crystal State	BIOL	0002
Change in Crystal Habit in the Vapor-Phase Growth of ZnO.=	PHYS	0293
to Phosphors.= A Study on the Crystal Habit of Calcium Hydrogen Phosphate Used	INOR	0361
and Three-Dimensional Bounded Crystal Lattices. +ate One-Electron States in Two-	BIOT	0028
New Inorganic-Centered Liquid Crystal Materials.=	INOR	0266
Polyester.= Pressure-Induced Crystal Modification of a Liquid Crystalline	POLY	0072
e Comp+ ^{77}Se and ^{113}Cd Single- Crystal NMR Studies of a Novel Cadmium Selenolat	INOR	0078
Investigation of Liquid Crystal Photopolymerization.= Holographic	BIOT	0205
Novel Photochromic Liquid Crystal Polymers.=	POLY	0219
Solid State Structure, and Liquid Crystal Properties.= +etween Electronic Structure,	INOR	0269
cture of Insulin Differs from the Crystal State but Spans the Range of Different Cryst	BIOL	0002
l Aluminum-Arsenic Compounds: Crystal Structure of $[Et_2AlAs(SiMe_3)_2]_2$.= +of Nove	COLL	0137
$Me_3Si)_3As$ with GaX_3 (X = Br, I): Crystal Structure of $[I_2GaAs(SiMe_3)_2]_2$.= +ons of (	COLL	0140
Tris(arsino)allanes and the Crystal Structure of $(t-Bu_2As)_3Ga$.=	INOR	0231
x and Binding Activities of Sit+ Crystal Structure of an Oligosaccharide-FAB Comple	BIOL	0143
Me+ Agnostic Acetyl Complexes: Crystal Structure of $Mo(\eta^2-C(S)OPr^i)(S_2COPr^i)CO(P$	INOR	0036
ituted Nucleoside Analogues and Crystal Structure of 2'-Deoxy-2'-Methylidenecytidin	CARB	0038
Studies with a High-Resolution Crystal Structure.= Comparison of Ligand-Mapping	CARB	0013
sis of Naphthalene Containing + Crystal Structure-Based de Novo Design and Synthe	MEDI	0186
Second-Row Transition Elements: Crystal Structures of Two Chloromolybdates.= +n	INOR	0267
of Polyanionic Chelating Ligands: Crystal Structures, NMR, and Electrochemistry.=	INOR	0243
and Acceptor Adlayers on Single- Crystal Surfaces.= +ergy Transfer between Donor	PHYS	0065
Structure of Ferroelectric Liquid Crystal Thin Films by Design: The Molecules to Mat	BIOT	0202
ide) Blends.= Crystal-Amorphous Interphases in Poly(ethylene Ox	PMSE	0165
Its Blends.= Nature of the Crystal/Amorphous Interface in Polyethylene and	COLL	0036
Crystalline Chloro-Trisodium Orthophosphate.=	INOR	0345
atic Units Separated by+ Liquid Crystalline Oligoesters Containing Single-Ring Arom	PMSE	0016
n. Carbophanes: A New Family of Crystalline Phases of Carbon.= +ond-Like Carbo	FUEL	0034
and Lyotr+ Main-Chain Liquid Crystalline Polyelectrolytes Displaying Thermotropic	POLY	0073
Modification of a Liquid Crystalline Polyester.= Pressure-Induced Crystal	POLY	0072
roperties of Diacetylenic Liquid Crystalline Polymers Obtained by Polymerization in	POLY	0030
e Morphology of Blends of Liquid Crystalline Polymers with Thermoplastics.= +on th	PMSE	0135
as Transport Properties of Liquid Crystalline Polymers.= +lecular Architecture on G	POLY	0115
Potential New Ferroelect+ Liquid Crystalline Pyramidic Transition-Metal Complexes:	INOR	0135
l Signatures of Total Mercury in Crystalline Rock: Case Study—Waldoboro Pluton Co	ENVR	0055
Thermosetting Systems. Liquid Crystalline Thermosets, I: Thermal Behavior of Mixt	POLY	0109
yl Ether Ketone Ketone), Part II: Crystallization and Melting.= +Melting of Poly(ar	POLY	0060
ymers. Thermoreversible Gelation: Crystallization and Other Mechanisms.= +of Pol	POLY	0121
On the Morphology and Crystallization Behavior of a Polyester Rotaxane.=	POLY	0346
The Chemistry of NaY Crystallization from Sodium Silicate Solutions.=	PETR	0004
Gelation and Crystallization in Solutions of PBLG.=	POLY	0241
tglass 2826MB and+ Effect of Crystallization Induced on Corrosion Behavior of Me	PHYS	0219
ate) with Functionalized Trigly+ Crystallization Kinetics of Poly(Ethylene Terephthal	PMSE	0171
mers.= Crystallization Kinetics of Random Ethylene Copoly	POLY	0075
ow-Temperature Hydrothermal Crystallization of MFI Zeolites in an Alkaline Fluorid	PETR	0005
and Offre+ Kinetic Studies of the Crystallization of the Zeolites Mordenite, L, Omega,	PETR	0041
e Phases an+ Investigation of the Crystallization of Very Large Pore Aluminophosphat	PETR	0093
tudies of the Template-Mediated Crystallization of ZSM-5 Type Zeolite.= +ttering S	PETR	0058
petitive Role of Cations in Zeolite Crystallization.= +d Pillared Clay Synthesis. Com	PETR	0074
Influence on the Rate of ZSM-5 Crystallization.= Investigation of Seed Properties'	PETR	0039
due to Colloidal Crystallization.= Reversible Gelation of Microgels	POLY	0244
Ketones).= Crystallization-Induced Gelation in Poly(aryl Ether	POLY	0247
e Ketone), Part I: S+ Structure, Crystallization, and Melting of Poly(aryl Ether Keton	POLY	0059
e Ketone), Part II:+ Structure, Crystallization, and Melting of Poly(aryl Ether Keton	POLY	0060
omp+ Far IR Spectroscopic and Crystallographi Caracterization of $[M(tpen)](ClO_4)_2$ C	INOR	0057
owded Porphyrins.= Crystallographic and EXAFS Studies of Sterically Cr	INOR	0091
Knowledge Acquisition from Crystallographic Databases.=	COMP	0045
Crystallographic Information from Ion Channeling.=	NUCL	0066
s o+ Synthesis, NMR, and X-ray Crystallographic Studies of Organoaluminum Adduct	INOR	0274
ission of Copper(I+ Influence of Crystallographic Symmetry upon the Solid-State Em	INOR	0157
X-ray Crystallography: Instructional Materials.=	CHED	0158
X-ray Crystallography: Instructional Materials.=	PHYS	0271
d Phase Transition in Tetracene Crystals and Tetracene-Doped Polymeric Matrices.=	PHYS	0274
ton Trapping and Annihilation in Crystals of $(CH_3)_4NMnCl_3(TMMC)$.= +Solid: Exci	PHYS	0131
toinduced Conductivity in Single Crystals of $YBa_2Cu_3O_{6.3}$: Photodoping to the Metalli	PHYS	0011
Nucleation of Crystals of 1,3-Diones.= AM1 Studies of the	PHYS	0247
ases. Torsional Solitons in Organic Crystals.= +sfer and Relaxation in Condensed Ph	PHYS	0129
ation of Two-Dimensional Protein Crystals.= +ectronics. Preparation and Characteriz	BIOT	0121
Energy Kinetics in Nanometer Crystals.=	PHYS	0316
Variations in Organic Crystals.= Graph Theory Applied to Bond Length	COMP	0031

Cuprate	Crystals.= Growth and Characterization of High–T_c	PHYS	0026
Liquid	Crystals.= Phase Separation in Polymer–Dispersed	COLL	0014
Carbon Adsorbents and Sucrose	Crystals.= Rate of Water Vapor Interaction with	CARB	0062
SM–5 and Aluminated Silicalite–1	Crystals.= +bons over Alkaline–Free Synthesized Z	PETR	0065
genide Synthesis of the Linear β–	$CsBiS_2$ and the Three–Dimensional $K_2Bi_8Se_{13}$.= +o	INOR	0202
Cambridge Structural Database (	CSD).= Integrated 3D Search Facilities for the	CINF	0021
sence of Cetyltrimethyl Bromide (	CTAB).= +ium(IV) with Sulfanilic Acid in the Pre	ANYL	0073
yntheses and Characterization of	CTBN–Toughened Epoxy Resin/Castor Oil Polyuret	PMSE	0092
View of Al and	Cu Complexes for MOCVD.= The ULSI Window: A	INOR	0125
Catalysis of V, Mo, Co, Ni, and	Cu Compounds: An Example of Experimental Modul	CHED	0057
tivity, Oxygen Stoichiometry, and	Cu Valence in the High–T_c Superconductors.= +lec	PHYS	0027
r Films by H–Atom Reaction with	$Cu(FOD)_2$ and $Cu(HFA)_2$.= +Deposition of Coppe	PHYS	0259
tom Reaction with $Cu(FOD)_2$ and	$Cu(HFA)_2$.= +Deposition of Copper Films by H–A	PHYS	0259
+Addition of Cu^{2+} to the Cation	$Cu(tpen)^{2+}$ to Form Several Binuclear Complexes.=	INOR	0399
$Na_3Cu_4Se_4$, $Na_2Cu_2Se_2\cdot Cu_2O$ and	Cu_2-xTe_4S_{13}-y.= +ew Solid–State Compounds of	INOR	0351
cl+ Reactivity of the Addition of	Cu^{2+} to the Cation $Cu(tpen)^{2+}$ to Form Several Binu	INOR	0399
c6)$_4MnCl_4$][$TlCl_4$]$_2$ in Presence of	Cu^{2+}.= +e Characteristics of Mn^{2+} in Cubic [(Rb18	INOR	0039
New Copper–Dioxygen Species (	Cu_2-O_2) Using Imidazole Ligands.=	INOR	0045
Alcohol and 1–Propanol on	$Cu_2O(100)$.= Oxidation and Reduction of Allyl	COLL	0195
a–Saturated Metal (Cu), Matte (	Cu_2S), and Slag (CaO–CaF_2–SiO_2) Under Controlled	I&EC	0065
Synthesis and Structure of	$Cu_4(\eta$–$OCMe_3)_6[OC(CF_3)_3]_2$.=	INOR	0134
tions of Silica–Saturated Metal (	Cu), Matte (Cu_2S), and Slag (CaO–CaF_2–SiO_2) Unde	I&EC	0065
opper(I) Complexes Derived from	Cu–Co Dimers.= +boxylation Reactions Involving C	INOR	0213
Bi–Sr–	Cu–O System.= New Cubic Compound in the	PHYS	0227
cinoma Cell Line.+ Cytoxicity of	^{67}Cu–Porphyrin–Antibody Conjugates to a Colon Car	BIOT	0168
b+ Preparation and Reactivity of	Cu/ZnO Catalysts for Methanol Synthesis Obtained	INOR	0071
–Tin Catalysts Supported on Mg,	Cu, Zn, and Ni Aluminates.= +tinum and Platinum	COLL	0126
action of Vanadate Anions with	Cu,Zn–Superoxide Dismutase as Probed by ^{51}V NMR	INOR	0184
imicas Utilizando Programacion	Cuadratica Sucesiva y un Simulador de Procesos Com	I&EC	0072
l Metodo Potencio–Metrico para	Cuantificar Clorhidrato de L–Arginina en Tabletas e	ANYL	0066
Masas con Fue+ Determinacion	Cuantitativa de Tierras Raras por Espectrometria de	ANYL	0051
Analisis	Cuantitativo por Computadora.=	COMP	0051
Steroids in Mexico via	Cuba.= °From Ruzicka's Terpenes in Zurich to	HIST	0017
Cleavage of	Cubanes.= Regioselective Carbon–Carbon Bond	ORGN	0256
Connectivity of a	Cube.= Synthesis from DNA of a Molecule with the	BIOL	0162
cence Characteristics of Mn^{2+} in	Cubic [(Rb18c6)$_4MnCl_4$][$TlCl_4$]$_2$ in Presence of Cu^{2+}	INOR	0039
New	Cubic Compound in the Bi–Sr–Cu–O System.=	PHYS	0227
her Intergrowth of Hexagonal and	Cubic Faujasite.= +aracterization of VPI–6: Anot	PETR	0001
olecular Engineering and Proces+	Cubic Nonlinear Optics of Polymer Thin Films, I: M	POLY	0011
ructure–$\chi^{(3)}$ Relationships in Ri+	Cubic Nonlinear Optics of Polymer Thin Films, 2: St	POLY	0205
ructure–$\chi^{(3)}$ Relationships in C+	Cubic Nonlinear Optics of Polymer Thin Films, 3: St	POLY	0226
urcture–$\chi^{(3)}$ Relationships in Po+	Cubic Nonlinear Optics of Polymer Thin Films, 4: St	POLY	0221
avelength Dispersion of the $\chi^{(3)}$+	Cubic Nonlinear Optics of Polymer Thin Films, 5: W	POLY	0213
est Lying Electronic States of the	CuCl Molecule.= +pole Moments for the Nine Low	PHYS	0304
eroxidation and Chilling Injury in	Cucumber Fruit and Cell Cultures.= +on of Lipid P	AGFD	0033
usarium Yellows–Resistant Celery	Cultivars.= +c Furocoumarins in Susceptible and F	AGRO	0040
Cells to Different	Cultivation Methods.= Metabolic Responses of CHO	BIOT	0164
cers.=	Culture Instability of Auxotrophic Amino Acid Produ	BIOT	0031
Bubbles in Animal Cell	Culture Media.= Oxygen Transfer Properties of	BIOT	0008
Mammalian Cell	Culture.= Polystyrene Porous Microcarriers for	BIOT	0010
holipase A_2 from Mammalian Cell	Culture.= +Recombinant Human Secretory Phosp	BIOL	0072
Compounds by	Cultured Plant Cells.= Production of Phenolic	AGFD	0002
on of Heterogeneous Hybridoma	Cultures for the Overproduction of Monoclonal Antib	BIOT	0169
permum erythrorhizon Suspension	Cultures with Fungal Elicitors.= +uction in Lithos	BIOT	0229
condary Metabolism in Plant Cell	Cultures. Increasing Secondary Met+ Modifying Se	BIOT	0228
njury in Cucumber Fruit and Cell	Cultures.= +on of Lipid Peroxidation and Chilling I	AGFD	0033
ion of Chemicals from Plant Root	Cultures.= +on of Reactors for Large–Scale Product	BIOT	0233
Root	Cultures.= Sesquiterpene Induction from Plant	BIOT	0232
ary Metabolism in Hairy Root	Cultures.= Understanding and Manipulating Second	BIOT	0231
of Strained Cyclic	Cumulenes.= Recent Developments in the Synthesis	ORGN	0093
State for the Reaction of Tertiary	Cumyl Derivatives.= +arbocation–Like Transition	ORGN	0153
and Azide Ion to the Reaction of	Cumyl Drivatives: Mechanism of Nucleophilic Substi	ORGN	0154
acrocycle and a Pentacoordinate	CuN_3OS Pseudothiolate Complex Compared with Th	INOR	0262
City College of	CUNY.= History of the Chemistry Department at	HIST	0028
$BiO_{1.5}$–SrO–CaO–	CuO System.= Glass Formation in the	INOR	0224
izada por Complejos Derivados de	Cupferron.= +xilacion del Benceno con H_2O_2 Catal	COLL	0121
High–T_c	Cuprate Crystals.= Growth and Characterization of	PHYS	0026
lanes on Superconductivity.=	Cuprate Oxides: Effect of Nonmagnetic Chains and P	INOR	0358
cts.= Structure and Bonding in	Cuprate Superconductors: Electronic and Steric Effe	CHED	0022
Correlations in	Cuprate Superconductors: Structure–Property	PHYS	0028
r the Electronic States of High–T_c	Cuprates as Probed by X–ray Absorption.= +el fo	PHYS	0053
Correlation of Products with the	cupric Ion Intensity and Coordination via Electron S	CATL	0017
Ethers+ Novel Synthesis of UV–	Curable, Multifunctional, Aromatic Analogs of Vinyl	POLY	0016
n Foods and Health. Chemistry of	Curcumin and Curcuminoids.= +olic Compounds i	AGFD	0043
use Epid+ Inhibitory Effects of	Curcumin on Tumor Initiation and Promotion in Mo	AGFD	0088
ealth. Chemistry of Curcumin and	Curcuminoids.= +olic Compounds in Foods and H	AGFD	0043
rmaldehyde Resins.=	Cure Characteristics of Tannin–Based Resorcinol–Fo	POLY	0138
Monitoring	Cure in Inside Spray Coatings.=	PMSE	0160
+nistic Study of Epoxy/Amine =	Cure Kinetics in Thermal and Microwave Fields.=	POLY	0161
by Fiber Optic Fluore+ In Situ	Cure Monitoring of Aromatic Diamine–Epoxy System	POLY	0282

t Rhenium Car+ Monitoring the	Cure of Photosensitive Polymers Using a Luminescen	POLY	0096
for Epoxy	Cure Studies.= Fluorescence of Aromatic Diamines	POLY	0159
izati+ First Evidence That Long	Cure Times in Cement/Fly Ash Solidification/Stabil	ENVR	0041
=	Cure/Property Diagrams of Thermosetting Systems.	PMSE	0172
Studies of Room Temperature	Cured Castor Oil Polyurethane/Vinyl or Acrylic Poly	PMSE	0090
easurement of Water Uptake in	Cured Epoxy by Extrinsic and Intrinsic Fluorophores	POLY	0160
iloxane) on Physical Properties of	Cured Epoxy Compounds.= +ght of Poly(dimethyls	PMSE	0174
echanical Properties of Amine-	Cured Epoxy: Amine Epoxide Addition Reaction vs.	POLY	0158
bber Nettings Used for Packaging	Cured-Pork Products.= +trosamines in Elastic Ru	AGFD	0040
and	Curing Studies.= Miscible Novolac/Polymer Blends	POLY	0113
Right-to-Know Support:	Current and Future Compliance.=	CHAS	0026
y Meta+ Enhancement of Critical	Current Density of $Bi_2Sr_2CaCu_2O_8$ Superconductor b	PHYS	0329
Molecular Electronic Transition	Current Density.=	BIOT	0026
mposites in Oilfi+ Limitations of	Current Epoxy Polymers as Matrix for Structural Co	PMSE	0111
Beilstein	Current Facts on CD-ROM.=	COMP	0017
Metabolites in Water.=	Current Methodologies for Analyzing Pesticides and	AGRO	0146
Drugs and Their Residual Analy+	Current Overview of Feed Additives and Veterinary	AGFD	0102
ase Nitration of Toluene in a Co-	Current Packed Tubular Reactor.= +g the Two-Ph	I&EC	0006
ood Products of Animal Origin.	Current Problems Associated with the Detection of A	AGFD	0081
Molting Hormones fr+ Past and	Current Studies with Ponasterones, the First Insect-	HIST	0024
hemicals and Their Containers.	Current Technologies for Pesticide Waste Disposal: O	AGRO	0120
Background and	Current Trends.= Workplace Drug Testing:	HIST	0008
I. Mechanism of Cellulase Action:	Current Understanding and Uncertainty.= +ases—	BIOT	0179
on of Pesticide Wastes: Historical,	Current, and Future Technologies.= +cal Degradati	AGRO	0122
cial Electron-Transfer Gates Ionic	Currents across the Lipid Bilayer.= +ice by Interfa	BIOL	0136
d GC/MS into the Undergraduate	Curriculum: A Case Study.= +poration of NMR an	CHED	0149
Examples for the Undergraduate	Curriculum: A Progress Report.= +erials Chemistry	CHED	0023
n Interdisciplinary Middle School	Curriculum.= +ol Science Education. FACETS: A	CHED	0071
hotometer into the Undergraduate	Curriculum.= +a Diode Array UV-Visible Spectrop	CHED	0180
mentation into the Undergraduate	Curriculum.= +of Fourier Transform NMR Instru	CHED	0148
xperiments into Lower Division	Curriculum.= Incorporation of Multinuclear NMR E	CHED	0186
the	Curriculum.= Integrated Molecular Modeling in	CHED	0190
ndergraduate Physical Chemistry	Curriculum.= +Techniques and Graphics into the U	CHED	0156
Lasers in the Physical Chemistry	Curriculum.=	CHED	0126
ddle, Vocational, and High School	Curriculum.= +grating Polymer Education into Mi	CHED	0026
nhanced Chemistry	Curriculum.= Project Catalyst: A New Technology-E	ANYL	0009
Chemistry Education Emphasis	Curriculum—A Case Study.=	CHED	0068
Transferencia de	$CuSO_4$ en Agua/Agua-Acetona.= Energia Libre de	PHYS	0236
urally Related to Vitamin E.=	Custom Design of Better in Vivo Antioxidants Struct	AGFD	0155
uits and Vegetables. Physiology of	Cut Fruits and Vegetables.= +imally Processed Fr	AGFD	0010
etrabutylin: Application to the C_8	Cut Isomerization.= +of Mordenites Modified by T	PETR	0131
stics of Copper(I) Compounds for	CVD by FT-IR and Raman Spectroscopy.= +acteri	COLL	0031
ations in X-ray Lit+ Microwave	CVD Deposition of Diamond Thin Films with Applic	PHYS	0092
Effects of Ammonia on	CVD Diamond Film Growth.=	PHYS	0150
Simulation of	CVD Diamond Growth.=	PHYS	0181
Structural Characterization of	CVD Diamond.=	PHYS	0119
Chemistry of Thermal	CVD from a Copper(II) Aminoalkoxide Precursor.=	COLL	0028
On	CVD Growth of SiC.=	PHYS	0095
ynthesized by DC Arc Discharge	CVD in a Mixture Gas of Acetylene and Hydrogen.=	FUEL	0038
Hot-Filament-Activated	CVD of Boron Nitride and Carbon.=	PHYS	0205
	CVD of Precious Metals.=	INOR	0142
n of H_2 and H in a Hot-Filament	CVD Reactor of Diamond.= +d Density Distributio	PHYS	0146
Pedestal	CVD Reactors.= Modeling Diamond Growth in	PHYS	0179
ylacetonate Complexes for Copper	CVD.= +and-Stabiilized Copper(I) Hexafluoroacet	COLL	0156
for the Determination of Calcium	Cyanamide Using Precolumn Derivatization.= +od	ANYL	0113
Bisphenol A-based	Cyanate Ester Resin Systems.= Characterization of	PMSE	0152
on the Equilibrium Constants for	Cyanide Addition to Nicotinamide Arabinosides.=	BIOL	0013
Surfaces.= Chemisorption of	Cyanide and Acetonitrile on Pt(III) Electrode	COLL	0118
tion of Ternary Iron(II)-Diimine-	Cyanide Complexes.= +ics of Peroxodisulfate Oxida	INOR	0022
triles from Epoxides with Lithium	Cyanide in THF.= +s Preparation of β-Hydroxyni	ORGN	0107
es of Layered Rare Earth Gold(I)	Cyanide Salts: Evidence for Energy Transfer and Re	INOR	0189
of Organometallic	Cyanines.= Synthesis and Spectroscopic Properties	INOR	0335
The Diels-Alder Reaction of α-	Cyano-1-Azadienes.=	ORGN	0234
methoxy-4'-Methylbiphenyls, 2-	Cyano-4'-Methylbiphenyls and 2-(Tetrazol-5-yl-4'-	ORGN	0112
s.= Catalytic Decarboxylation of	Cyanoacetic Acid by Anionic Tungsten (0) Complexe	INOR	0212
bnitride.=	Cyanoazacarbon Derivatives as a Route to Carbon Su	INOR	0310
Epo+ Syntheses of Aminodeoxy,	Cyanodeoxy, and Branched-Chain Amino Sugars by	ORGN	0071
Anions of Carbethoxy-Protected	Cyanohydrins with Enones: Synthetic Applications.=	ORGN	0228
Amides and Imides from	Cyanoimidazoles.= High-Nitrogen Low-Hydrogen	POLY	0363
oup 8 and Group 10 Mixed-Metal	Cyanometalate Complexes.= +plexes Based on Gr	INOR	0162
pe from the Tar Baby?.=	Cyanomethylene and Benzyne: Is It Possible To Esca	PHYS	0032
Fu+ Inhibition of Oxidosqualene	Cyclase by Pseudosubstrates: Studies of 26- and 29-	ORGN	0021
Life	Cycle Analysis: A Tool for Pollution Prevention.=	CEI	0027
Nuclear Fuel	Cycle and Special Problems in Nuclear Energy.=	FUEL	0024
edure and Reaction-Regeneration	Cycles.= +ts in the Course of the Preparation Proc	PETR	0046
Radical Reactivity of	Cyclic Acetals.=	ORGN	0156
phospha+ One-Pot Synthesis of	Cyclic Amidinium Tetrafluoroborates and Hexafluoro	ORGN	0204
ethylph+ Effect of Topology for	Cyclic and Linear Poly(dimethylsiloxane) and Poly(m	POLY	0234
Adsorption and Chemistry of	Cyclic CH_2N_2 on Pd(110).=	COLL	0114
Thionocarbonate in a Latent +	Cyclic Conjugated Enediynes via the Elimination of a	ORGN	0096

Synthesis of Strained	Cyclic Cumulenes.= Recent Developments in the	ORGN 0093
Short-Range Solvation by	Cyclic Diethers.=	PHYS 0308
lization Stud+ Synthesis of New	Cyclic Enediyne Model Compounds for Bergman Cyc	ORGN 0097
Substituted	Cyclic Ethers.= Conversion of Lactones into	ORGN 0014
of Phosphorus Pentahalides with	Cyclic Ketones: A One-Step Preparation of 1,1,2-Tri	ORGN 0263
	Cyclic Oxyphosphoranes as Reactive Intermediates.=	INOR 0290
l Decomposition and Linear and	Cyclic Polystyrene/Poly(vinyl Methyl Ether) Blends.	POLY 0295
ransfer as Functions of Size of the	Cyclic Transition State and C-H-C Angle.= +en T	ORGN 0175
Synthesis of	Cyclic Vanadium(V) 1,3-Diol Complexes.=	INOR 0409
Characteristics of	Cyclic Vinyl Monomers.= Radical Ring-Opening	ORGN 0157
y Physical Chemistry Lab+ Two	Cyclic Voltammetry Experiments for the Introductor	CHED 0211
Seasonal Dissolved Gas	Cycling in the Middle Atlantic Bight.=	GEOC 0003
Chemistry. Arsenic	Cycling in the Sea.= Marine Bioinorganic	INOR 0136
on in Anoxic Coastal Sediments:	Cycling of Iron Oxides and Their Potential Importan	GEOC 0007
Effect of Branching and	Cycling on the W-Shape CpE.=	PHYS 0320
ropylamide: A Sim+ Dehydrative	Cyclization of Diphenyl Sulfoxide by Lithium Diisop	ORGN 0084
esis and+ Nucleophile-Promoted	Cyclization Reactions of Alkynes: Asymmetric Synth	ORGN 0275
e Model Compounds for Bergman	Cyclization Studies.= +hesis of New Cyclic Enediyn	ORGN 0097
mbered Carb+ New Strategy for	Cyclization-O-Functionalization of 5-, 6-, and 7-Me	ORGN 0088
A Novel General Method for	Cyclizing Peptides.=	BIOL 0039
idene Complexes Cp(CO)+ New	Cycloaddition and Metathesis Reactions of the Vinyl	INOR 0072
Radical+ Intramolecular $2\pi + 4\pi$	Cycloaddition Reactions of Allyl Cations and Cation	ORGN 0123
1,3-Dipolar	Cycloaddition Reactions with Captodative Olefins.=	ORGN 0230
ative Bypass of the TpdU cis-syn	Cyclobutane Photodimer of DNA.= +of the Replic	BIOL 0113
of Diazomethane with Platin(IV)	cyclobutanes.= +d Stereochemistry of the Reaction	INOR 0329
d by Nucleophilic+ Platinum-	Cyclobutanoneimine and Platinum-Pyrazoline Forme	INOR 0041
Methides from	Cyclobutenones.= Generation of ortho-Quinone	ORGN 0229
bstituted Polybenzimidazoles by	Cyclodehydrogenation of Precursor Poly(Schiff Bases	POLY 0045
for Conformational Analysis of β-	Cyclodextrin: +Dynamics Simulation Techniques	ORGN 0174
into γ-	Cyclodextrin.= Conversion of Corn Syrup Solids	CARB 0041
atic Browning in Fruit Juice with	Cyclodextrins and Sulfated Polysaccharides.= +zym	AGFD 0030
teraction of Guest Molecules with	Cyclodextrins.= +Phenomena of Carbohydrates. In	CARB 0061
Insensitive GABA Receptor from	Cyclodiene Resistant Drosophila: A Model System in	AGRO 0152
Cyclopentadiene, Benzene, and	Cycloheptatriene to Carbohydrates: A Chemoenzyma	ORGN 0272
Vinylcyclobutene to a	Cyclohexadiend.= Thermal Isomerization of a	ORGN 0144
orinated Copolymers: Poly(1,3-	cyclohexadiene-alt-α-fluoroacrylonitrile).= New Flu	POLY 0021
ation of Heteroatom-Substituted	Cyclohexadienes: Precursors to Poly(paraphenylene).	POLY 0047
xyimino)-Alkyl)-5-Pyrazolyl-1,3-	Cyclohexadiones.= +rbicidal Activity of 2-(2-(Alko	AGRO 0013
ition-Metal C+ Carbonylation of	Cyclohexane Catalyzed Photochemically by d^8 Trans	INOR 0254
d Polysiloxanes Containing trans-	Cyclohexane-Based Mesogenic Groups.= +in Liqui	POLY 0057
Benzene.= New	Cyclohexanol Process via Cyclohexene from	PETR 0104
N-Alkylperhydroquinolines from	Cyclohexanones.= +ective, One-Pot Synthesis of cis	ORGN 0031
ects on the Aerobic Oxidation of	Cyclohexene Catalyzed by (Aqua)(Phosphine)Ruthen	INOR 0067
.=	Cyclohexene from Benzene by Partial Hydrogenation	I&EC 0017
New Cyclohexanol Process via	Cyclohexene from Benzene.=	PETR 0104
Benzene Hydrogenation to	Cyclohexene.= H-D Exchange during Selective	FUEL 0100
Benzene Hydrogenation to	Cyclohexene.= H-D Exchange during Selective	CATL 0041
Oxygen Production by Triplet	Cyclohexenones.= Quantum Yields of Singlet	ORGN 0167
onships of Renin Inhibitors with	Cyclohexyl Moiety Attached to the Backbon	MEDI 0121
ies Using New F+ Preference for	Cyclohexylalanine at P1 Site by Renin: Kinetic Stud	BIOL 0035
hilic Bacillus sp.= Production +	Cyclomaltodextrin Glucanotransferase by an Alkalop	BIOT 0157
onversion of Chromium η^4-1,5-	Cyclooctadiene Complexes into Agostic η^4-1,3-Cycloo	INOR 0209
ne Complexes into Agostic η^4-1,3-	Cyclooctadiene Complexes.= +η^4-1,5-Cyclooctadie	INOR 0209
-Vapor Deposition from the 1,5-	Cyclooctadiene Copper(I) Hexafluoroacetylacetonate	COLL 0026
Deposition of Copper from 1,5-	Cyclooctadiene Copper(I) Hexafluoroacetylacetonate.	COLL 0029
olymerization of Monosubstituted	Cyclooctatetraenes.= +Ring-Opening Metathesis P	POLY 0329
(LO) Inhibito+ Conversion of a	Cyclooxygenase (CO) Inhibitor into a 5-Lipoxygenase	MEDI 0179
d Fenamates as Dual Inhibitors of	Cyclooxygenase and 5-Lipoxygenase.= +oxamic Aci	MEDI 0177
xy-2-Methyl-N-+ Synthesis and	Cyclooxygenase/5-Lipoxygenase Activity of 4-Hydro	MEDI 0178
f the putative Metabolites of the	Cyclopenta[a]-Phenanthrenes: Synthesis of the trans	ORGN 0118
uring Benzene Oxidation: Ther+	Cyclopentadiene and Cyclopentadienyl Conversion d	FUEL 0085
arbohydrates: A Chemoe+ From	Cyclopentadiene, Benzene, and Cycloheptatriene to C	ORGN 0272
ion: Ther+ Cyclopentadiene and	Cyclopentadienyl Conversion during Benzene Oxidat	FUEL 0085
tions.= Nucleophilic Addition to	Cyclopentadienyl Manganese Dicarbonyl η^3-Allyl Ca	INOR 0270
osilanes via the Condensation of	Cyclopentadienyl Sodium with Dialkyldichlorosilanes	INOR 0199
id Promoted Additions of Allyl(	cyclopentadienyl)Iron(II) Dicarbonyl to Aldehydes an	ORGN 0213
esis and Biological Study of the	Cyclopentenyl Carbocyclic Nucleoside Analogue of 5-	MEDI 0048
on of the SAR in a Series of N^6-	Cyclopentyl- and N^6-(2-Thienyl)Ethyl-Adenosine A	MEDI 0113
ing of Occupational Exposure to	Cyclophosphamide in the Pharmaceutical Industry.=	CEI 0010
Pendant	Cyclophosphazenes.= Reactions of Polymers with	POLY 0182
ity in Samarium Metal-Induced	Cyclopropanations: Synthesis of 1,25-Dihydroxychole	ORGN 0284
Tricyclic	Cyclopropenes.= Synthesis and Chemistry of Some	ORGN 0232
ith Low Fish To+ 1,4-Diaryl-1-	Cyclopropylbutanes: Highly Efficacious Insecticides w	AGRO 0073
mo-Michael-Acceptors.=	Cyclopropylthiocarbene-Metal Complexes: Super-Ho	ORGN 0216
tion of Released Cells to Overall	Cyclosporin A Productivity in an Immobilized Perfus	BIOT 0155
Studies of the Reppe	Cyclotetramerization of Acetylene.= Mechanistic	INOR 0331
Mediates Acetylene	Cyclotrimerization.= An η^4-Benzene Species	INOR 0210
of Long Chain Substituted	Cyclotriphosphazenes.= Gamma Ray Polymerization	PMSE 0130
of Long-Chain-Substituted	Cyclotriphosphazenes.= Gamma Ray Polymerization	POLY 0055
Generation from	Cysteine and Sugars.= Studies on Meat Flavor	AGFD 0074

mplexes as Models for Nickel(III)-	Cysteine Centers in the Hudrogenase Enzymes.=	INOR 0179
Reactivity Studies on	Cysteine Complexes of Iron(II) and Iron(III).=	INOR 0370
ational Differences in the	Cysteine Proteinases from Papaya Latex.= Conform	BIOL 0074
of a Subnanomolar, Orally Active	Cysteine–Derived Inhibitor.= +nhibitors: Synthesis	MEDI 0122
Carbon Monoxide in	Cytochrome c Oxidase.= Ligation Dynamics of	BIOL 0096
to-Induced Reduction of Native	Cytochrome c Peroxidase(III) in Alkaline pH.= Pho	BIOL 0097
gous Lepidoptera.=	Cytochrome P-450 Monoxygenase Genes in Oligopha	AGRO 0115
e and+ Analysis for Microsomal	Cytochrome P-450 Patters by Means of Ion Exchang	ANYL 0017
Regulation of a House Fly	Cytochrome P450 Monooxygenase.= Structure and	AGRO 0098
ocarbons by Rational Redesign of	Cytochrome P450cam.= +tion of Halogenated Hydr	BIOT 0118
esis. Mechanism and Inhibition of	Cytochrome–P$_{450}$ Aromatase.= +ormone Biosynth	MEDI 0006
lly Based Disease.= Human	Cytochromes P450 and Their Roles in Environmenta	CEI 0021
ng Cell-Sheet Morphogenesis and	Cytokinesis.= +quired for Cell-Shape Change duri	BIOL 0140
–Conformational Transitions in	Cytoskeletal Microtubules: Implications for Neural-L	BIOT 0055
Expression of Rat Liver	Cytosolic PEPCK in E. coli.= Subcloning and	BIOL 0107
Plant Phenolics as	Cytotoxic Antitumor Agents.=	AGFD 0092
Characterization of	Cytotoxic Ether Lipids.= Physicochemical	MEDI 0031
Efficient Total Synthesis of the	Cytotoxic Halogenated Monoterpene Aplysiapyranoid	ORGN 0001
Partial Structural Elucidation of	Cytotoxin D from Naja naja siamensis by Two–Dime	BIOL 0044
e Lytic Activity of a Cobra Venom	Cytotoxin: Structure–Activity Implications.= +e th	BIOL 0085
o a Colon Carcinoma Cell Line.+	Cytotoxicity of ^{67}Cu–Porphyrin–Antibody Conjugates t	BIOT 0168
imensional Separations by LC–	CZE.= Rapid Biopolymer Characterization. Two–D	ANYL 0126
e+ Application of Homonuclear	2D ^{51}V NMR to Determine the Effect of Temperatur	INOR 0228
s: Experiences with MDDR–	3D and FCD–3D.= Building 3D Structural Database	COMP 0046
teristics of Computer–Generated	3D and Related Molecular Property Data for CAS R	COMP 0041
Automatic Generation of	3D Atomic Coordinates for Organic Molecules.=	CINF 0036
lly Flexible Molecules from 2D or	3D Chemical Databases.= +Docking Conformationa	CINF 0028
milarity Searching in Databases of	3D Chemical Structures.= +pping Procedure for Si	COMP 0043
o Complementary Methods wit+	3D Database Searching vs. de Novo Construction: Tw	COMP 0050
s: A Flexible Prescre+ Searching	3D Databases of Conformationally Flexible Molecule	CINF 0027
to Computer–Aided Molecule D+	3D Databases. CAVEAT: A Vector–Based Approach	COMP 0044
g.=	3D Databases. Recent Advances in Molecular Dockin	COMP 0036
p+ Looking for Pharmacophores in	3D Databases: Does Conformational Searching Im	CINF 0026
Experiences with Automated	3D Design of Bioactive Molecules.=	COMP 0049
K$_2$PdSe$_{10}$: Two Interpenetrating	3D Frameworks of [Pd(Se$_4$)$_2$]$^{2-}$ and [Pd(Se$_6$)$_2$]$^{2-}$.=	INOR 0200
Two–Dimensional Infrared (	2D IR) Spectroscopy.=	PHYS 0117
	3D Metal Complexes with Uric Acid.=	INOR 0073
with Clay Minerals as Studied by	^{2}D Nuclear Magnetic Resonance Spectroscopy.= +r	AGRO 0061
ationally Flexible Molecules from	2D or 3D Chemical Databases.= +Docking Conform	CINF 0028
Searching: Techniques in	3D Query Formulation.= Flexible Queries in 3D	CINF 0025
Database (CSD).= Integrated	3D Search Facilities for the Cambridge Structural	CINF 0021
Comparison of Atom-by-Atom	3D Search Routines for Searching CAS Registry Sub	CINF 0023
Use of Most Restrictive Paths in	3D Search Strategy.=	CINF 0024
dling. Prioritizing the Hits from a	3D Search.= +imensional Chemical Structure Han	CINF 0020
= Flexible Queries in	3D Searching: Techniques in 3D Query Formulation.	CINF 0025
=	3D Shape Fitting and 3D–DB Searching by SPERM.	COMP 0040
Measures of	3D Shape Similarity.=	COMP 0042
D and FCD–3D.= Building	3D Structural Databases: Experiences with MDDR-3	COMP 0046
sidue Studies of Pirlimycin in the	Dairy Cow.= +ducing Animals. Metabolism and Re	AGRO 0108
in Lactating	Dairy Cows.= Novel Sulfonamide Drug Metabolism	AGRO 0113
Luprostiol in	Dairy Cows.= Residue Depletion and Metabolism of	AGRO 0130
and Oxidative DNA	Damage.= Carcinogen–Mediated Oxidant Formation	CHAS 0008
β–	Damascenone Precursors from Apples.=	AGFD 0142
Characterization of Vibration	Damping of Laminated Steel Sheet with Polyurethan	PMSE 0057
lymer Networks.= Study on the	Damping Paint of PS/PA Latex Interpenetrating Po	PMSE 0064
tion on Si(100): The Behavior of	Dangling Bonds.= Silicon Carbide Thin–Film Forma	PHYS 0204
ynthesis of Heavy Elements. The	Darmstadt Program To Investigate Heavy Elements.	NUCL 0014
High–Resolution NMR of Solids:	DAS and DOR.= +opies. Sample Reorientation for	PHYS 0370
Privacy Issues of DNA	Databanks and Databases.=	CHAL 0008
for the Cambridge Structural	Database (CSD).= Integrated 3D Search Facilities	CINF 0021
the Beilstein	Database Activities.= Beilstein Database. Status of	COMP 0009
NIST Structures and Properties	Database and Estimation System.=	COMP 0016
he Joint Production of a Reaction	Database and of a Printed Service of Abstracts.=	CINF 0010
esearch, Development+ Chemical	Database Management at the U.S. Army Chemical R	CINF 0037
The IDITIS Relational	Database of Protein Structure.=	CINF 0032
Beilstein	Database on STN.= Recent Developments of the	COMP 0010
m Simulation.=	Database Retrieval Techniques for ^{13}C NMR Spectru	COMP 0004
omplementary Methods wit+ 3D	Database Searching vs. de Novo Construction: Two C	COMP 0050
	Database Structure and Searching in MACCS-3D.=	CINF 0022
terocyclic Organic C+ Beilstein	Database. Prediction of Normal Boiling Points of He	COMP 0015
Activities.= Beilstein	Database. Status of the Beilstein Database	COMP 0009
hnologies from a Process Analysis	Database.= +in the Selection of Petrochemical Tec	I&EC 0064
Beilstein	Database.= Ring Index Searching of the	COMP 0018
Targets.= Using	Databases as an Aid in Developing Synthetic	COMP 0047
Bibliographic	Databases for Scientists.=	CINF 0042
Flexible Prescre+ Searching 3D	Databases of Conformationally Flexible Molecules: A	CINF 0027
cedure for Similarity Searching in	Databases of 3D Chemical Structures.= +pping Pro	COMP 0043
omputer–Aided Molecule+ 3D	Databases. CAVEAT: A Vector–Based Approach to C	COMP 0044
Synthesis Planning and Reaction	Databases. New Developments in the LHASA Prog	CINF 0001
	3D Databases. Recent Advances in Molecular Docking.=	COMP 0036

oking for Pharmacophores in 3D	Databases: Does Conformational Searching Improve	CINF	0026
.= Building 3D Structural	Databases: Experiences with MDDR-3D and FCD-3D	COMP	0046
olecules from 2D or 3D Chemical	Databases.= +Docking Conformationally Flexible M	CINF	0028
Crystallographic	Databases.= Knowledge Acquisition from	COMP	0045
Power of Selective Reaction	Databases.=	CINF	0009
and	Databases.= Privacy Issues of DNA Databanks	CHAL	0008
Reaction Retrieval from	Databases.=	CINF	0004
Branched-Chain Sugar	Daunorubicin Analogue.= Synthesis of a	CARB	0052
Sprague-	Dawley Rats.= Metabolism of Triallate in	AGRO	0023
Organizational Structure of	Dayton, Ohio, Project SEED.=	CHED	0142
3D Shape Fitting and 3D-	DB Searching by SPERM.=	COMP	0040
arcinogenic Dibenz[a,h]Acridine (	DB[a,h]ACR).= +1-8,9-Epoxide Diastereomers of C	ORGN	0208
Diamond Films Synthesized by	DC Arc Discharge CVD in a Mixture Gas of Acetylen	FUEL	0038
Nonlinear Optics in	DCANP Langmuir-Blodgett Films.= Guided-Wave	POLY	0176
esis of the Gene-Encoding T4	dCMP Hydroxymethylase.= Site-Directed Mutagen	BIOL	0160
ara Medir pH, por un Electrodo	de Acero Inoxidable 316/Pelicula de Oxido, en la Val	ANYL	0066
stico de Tecnicas Electroquimicas	de Aceros Microaleados en Acido Sulfurico.= +tadi	PHYS	0281
Microbiologia	de Alimentos en el LNSP: Enfoque Prospectivo.=	AGFD	0206
bre las P+ Elaboracion de Pastas	de Alto Rendimiento: Influencia de la Sulfonacion so	PMSE	0077
e Molibdeno sobre Oxidos Mixtos	de Alumina-Titania.= +Tiofeno con Catalizadores d	FUEL	0105
Quimica a Nivel Med+ Desarrollo	de Apoyos Computarizados para la Ensenanza de la	CHED	0124
uct+ Servicio Social y Programas	de Becas: Vinculo entre Universidades y Secotr Prod	CHED	0114
ria Qui+ Aplicaciones Recientes	de CAD/CAM para el Diseno de Equipo de la Indust	COMP	0053
Influencia del Transporte	de Calor en Reacciones Solido-Gas NO-Cataliticas.=	I&EC	0068
terminacion de la Transferencia	de Carga por Espectroscopia IR de Complejos de Tra	INOR	0190
IR de Complejos de Transferencia	de Carga Pt(R-Nafin)-TCNQ.= +r Espectroscopia	INOR	0190
a con S. aureus y s+ Elaboracion	de Caseinato de Sodio a Partir de Leche Contaminad	AGFD	0107
itania Modificadas con+ Estudio	de Catalizadores NiW Soportados sobre Alumina o T	FUEL	0106
onductores Policristalinos dentro	de Celdas Fotoelectroquimicas, I: CdS Puro e Impuri	PHYS	0221
Evaluacion Fisicoquimica	de Cementos Hidraulicos para Pozos Geotermicos.=	INOR	0063
des Complexants sur la Fottation	de ces Mineraux.= +-Pyrite: Effet des Oxydants et	ANYL	0067
tica: Valoracion Potenciometrica	de Clorhidrato de L-Arginina en Producto Terminad	CHED	0163
de Carga por Espectroscopia IR	de Complejos de Transferencia de Carga Pt(R-Nafin	INOR	0190
o de la Presion en las Propiedades	de Conduccion en Polimeros Zwitterionicos.= +ect	ORGN	0159
talizada por Complejos Derivados	de Cupferron.= +xilacion del Benceno con H_2O_2 Ca	COLL	0121
Energia Libre de Transferencia	de $CuSO_4$ en Agua/Agua-Acetona.=	PHYS	0236
rona.= Obtencion	de Derivados Hidrosolubles de la Dehidroepiandroste	MEDI	0038
y Modificacion de Secuencias	de Destilacion con Integracion de Energia.= Sintesis	I&EC	0074
Encuentro	de Dos Mundos a Traves de la Quimica.=	HIST	0035
por un Electrodo d+ Sustitucion	de Electrodo de Vidrio Convencional para Medir pH,	ANYL	0066
Determinacion del Mecanismo	de Electroreduccion de la Perezona.=	PHYS	0213
de Destilacion con Integracion	de Energia.= Sintesis y Modificacion de Secuencias	I&EC	0074
ntes de CAD/CAM para el Diseno	de Equipo de la Industria Quimica de Proceso.= +e	COMP	0053
nsenanza Experimental a Partir	de Errores Detectados en Publicaciones Cientificas.=	CHED	0178
Metales Pesados en Leche	de Establos del Sur del D.F.= Determinacion de	ENVR	0037
Estudio Sobre el Uso	de Eteres de Glicol en Mexico.=	ENVR	0038
as en Condensables en Geotermica	de Fasico de los Azusfres.= +rtamiento de los Gaz	I&EC	0013
soline Plant: An Engineering Tour	de Force.= +J. F. Roth. New Zealand Gas-to-Ga	I&EC	0060
Fundamentos de la Produccion	de Furfural.=	CARB	0051
acion+ Influencia del Contenido	de Galio en Zeolitas del Tipo Pentasil en la Aromatiz	COLL	0125
Laurylsulfate	de Gallium.= Flottation Ionique du	ANYL	0053
Estudio Sobre el Uso de Eteres	de Glicol en Mexico.=	ENVR	0038
pectrometria de Masas con Fuente	de Impacto Electronico.= +e Tierras Raras por Es	ANYL	0051
Analisis Longitudinal	de Indices de Reprobacion en Materias Biologicas.=	CHED	0091
Potenciometrica de Clorhidrato	de L-Arginina en Producto Terminado Utilizando co	CHED	0163
trico para Cuantificar Clorhidrato	de L-Arginina en Tabletas en Medio.= +tencio-Me	ANYL	0066
acion en las Materias Biologicas	de la Carrera de QFB en la Facultad de Quimica, UN	CHED	0092
Asbestos en el Area Urbana	de la Ciudad de Mexico?.=	ENVR	0025
Quimico de Una Ca+ Evaluacion	de la Corrosion de las Diferentes Etapas del Lavado	PHYS	0257
idad de Ratas Hembras.+ Efecto	de la Deficiencia o Exceso de Vitamina E en la Fertil	AGFD	0210
Derivados Hidrosolubles	de la Dehidroepiandrosterona.= Obtencion de	MEDI	0038
tectados en+ Enriquecimiento	de la Ensenanza Experimental a Partir de Errores De	CHED	0178
de los Alumnos en los Resultados	de la Evaluacion de una Materia Biologica.= +mico	CHED	0093
D/CAM para el Diseno de Equipo	de la Industria Quimica de Proceso.= +entes de CA	COMP	0053
acterizacion del Factor Inhibidor	de la Locomocion de los Monocitos (FILM) Producid	BIOL	0102
Tonsil-Fe$(NO_3)_3$.= Estudio	de la Nitracion del Fenol con Aductos	COLL	0120
Orden de Reaccion	de la Oxidacion de la Vitamina C.= Obtencion de	CHED	0164
Electroreduccion	de la Perezona.= Determinacion del Mecanismo de	PHYS	0213
olimeros Zwitterionicos.+ Efecto	de la Presion en las Propiedades de Conduccion en P	ORGN	0159
Fundamentos	de la Produccion de Furfural.=	CARB	0051
omputarizados para la Ensenanza	de la Quimica a Nivel Medio Superior.= +Apoyos C	CHED	0124
Traves	de la Quimica.= Encuentro de Dos Mundos a	HIST	0035
de Alto Rendimiento: Influencia	de la Sulfonacion sobre las Propiedades Papelaras.=	PMSE	0077
aracterizacion, y Determinacion	de la Transferencia de Carga por Espectroscopia IR d	INOR	0190
Reaccion de la Oxidacion	de la Vitamina C.= Obtencion del Orden de	CHED	0164
Ca+ Evaluacion de la Corrosion	de las Diferentes Etapas del Lavado Quimico de Una	PHYS	0257
iaceae d+ Composicion Quimica	de las Plantas Pertenecientes a la Familia Scrophular	ORGN	0185
y 316 Utilizadas en el Seguimiento	de las Reacciones Acido Base en Medio Acuoso.=	ANYL	0055
on de Caseinato de Sodio a Partir	de Leche Contaminada con S. aureus y s+ Elaboraci	AGFD	0107
Influencia del Nivel Academico	de los Alumnos en los Resultados de la Evaluacion d	CHED	0093

ensables en Geotermica de Fasico	de los Azusfres.= +rtamiento de los Gazas en Cond	I&EC	0013
Modificados con Boro.= Estudio	de los Catalizadores CoMo/Al$_2$O$_3$ y Mo/Al$_2$O$_3$	FUEL	0104
co+ Estudio del Comportamiento	de los Gazas en Condensables en Geotermica de Fasi	I&EC	0013
Factor Inhibidor de la Locomocion	de los Monocitos (FILM) Producido por E. +on del	BIOL	0102
Tierras Raras por Espectrometria	de Masas con Fuente de Impacto Electronico.= +e	ANYL	0051
para Balances	de Materia.= BALANCE: Simulador de Procesos	COMP	0052
D.F.= Determinacion	de Metales Pesados en Leche de Establos del Sur del	ENVR	0037
la Molecula	de Metano.= Activacion del Enlace C-H por Ni en	PHYS	0244
ntes a la Familia Scrophulariaceae	de Mexico.= +n Quimica de las Plantas Pertenecie	ORGN	0185
Ciudad	de Mexico?.= Asbestos en el Area Urbana de la	ENVR	0025
S de Tiofeno con Catalizadores	de Molibdeno sobre Oxidos Mixtos de Alumina-Titan	FUEL	0105
Sintesis de Tiofosfatos	de Naftilo, Posibles Garrapaticidas.=	MEDI	0064
ng, cosponsored by E. I. DuPont	de Nemours and Co.). Teaching: A Witches' Brew of	ANYL	0006
edio Acido y Basico, Preparacion	de Nitrado de Plata.= +dorreduccion con Zinc en M	CHED	0172
oal.= 3D Database Searching vs.	de Novo Construction: Two Complementary Method	COMP	0050
ining+ Crystal Structure-Based	de Novo Design and Synthesis of Naphthalene Conta	MEDI	0186
New Method for the	de Novo Design of Enzyme Inhibitors.=	COMP	0037
Carbocyclic Substrates for	de Novo Purine Biosynthesis.=	BIOL	0076
en el Seguimiento de+ Peliculas	de Oxido sobre Acero Inoxidable 304 y 316 Utilizadas	ANYL	0055
Electrodo Indicador una Pelicula	de Oxido sobre Acero Inoxidable 316.= +ando como	CHED	0163
e Acero Inoxidable 316/Pelicula	de Oxido, en la Validacion del Metodo Potencio-Met	ANYL	0066
onacion sobre las P+ Elaboracion	de Pastas de Alto Rendimiento: Influencia de la Sulf	PMSE	0077
atica Sucesiva y u+ Optimizacion	de Plantas Quimicas Utilizando Programacion Cuadr	I&EC	0072
n con Zinc en Medi+ Obtencion	de Plata por Electrodeposito, Fusion, Oxidorreduccio	CHED	0172
y Basico, Preparacion de Nitrado	de Plata.= +dorreduccion con Zinc en Medio Acido	CHED	0172
to de L-Arginina en+ Proyecto	de Practica: Valoracion Potenciometrica de Clorhidra	CHED	0163
de Equipo de la Industria Quimica	de Proceso.= +entes de CAD/CAM para el Diseno	COMP	0053
uadratica Sucesiva y un Simulador	de Procesos Comercial.= +lizando Programacion C	I&EC	0072
BALANCE: Simulador	de Procesos para Balances de Materia.=	COMP	0052
Tipo Pentasil en la Aromatizacion	de Propano.= +Contenido de Galio en Zeolitas del	COLL	0125
aterias Biologicas de la Carrera	de QFB en la Facultad de Quimica, UNAM (Analisis	CHED	0092
a Carrera de QFB en la Facultad	de Quimica, UNAM (Analisis Estadistico de 10Anos)	CHED	0092
so de Vitamina E en la Fertilidad	de Ratas Hembras.= +ecto de la Deficiencia o Exce	AGFD	0210
Obtencion del Orden	de Reaccion de la Oxidacion de la Vitamina C.=	CHED	0164
Catalisis	de Reacciones via Radicales Libres con Tonsil.=	COLL	0119
con Electrodos Metalic+ Estudio	de Reduccion Electroquimica del 1-Nitro-Naftaleno	PHYS	0280
Analisis Longitudinal de Indices	de Reprobacion en Materias Biologicas.=	CHED	0091
Deposito Electrolitico	de Samario y Plutonio.=	NUCL	0075
gia.= Sintesis y Modificacion	de Secuencias de Destilacion con Integracion de Ener	I&EC	0074
us y s+ Elaboracion de Caseinato	de Sodio a Partir de Leche Contaminada con S. aure	AGFD	0107
en Acido S+ Estudio Estadistico	de Tecnicas Electroquimicas de Aceros Microaleados	PHYS	0281
ue+ Determinacion Cuantitativa	de Tierras Raras por Espectrometria de Masas con F	ANYL	0051
idos Mixtos de Alumina-+ HDS	de Tiofeno con Catalizadores de Molibdeno sobre Ox	FUEL	0105
Sintesis	de Tiofosfatos de Naftilo, Posibles Garrapaticidas.=	MEDI	0064
r Espectroscopia IR de Complejos	de Transferencia de Carga Pt(R-Nafin)·TCNQ.=	INOR	0190
Energia Libre	de Transferencia de CuSO$_4$ en Agua/Agua-Acetona.=	PHYS	0236
ute+ Resultados Experimentales	de un Empaque Integral para Enriquecimiento en De	I&EC	0073
entes Etapas del Lavado Quimico	de Una Caldera por Tecnicas Electroquimicas.= +er	PHYS	0257
n los Resultados de la Evaluacion	de una Materia Biologica.= +mico de los Alumnos e	CHED	0093
rodo d+ Sustitucion de Electrodo	de Vidrio Convencional para Medir pH, por un Elect	ANYL	0066
Efecto de la Deficiencia o Exceso	de Vitamina E en la Fertilidad de Ratas Hembras.=	AGFD	0210
mica, UNAM (Analisis Estadistico	de 10Anos).= +rrera de QFB en la Facultad de Qui	CHED	0092
sing Neutral+ Debittering and	Deacidifying Sour Orange (Citrus aurantium) Juice U	AGFD	0106
Characterization of Activating and	Deactivating Groups in Aromatic Rings.= +ctural	CHED	0096
alysts during Oligomerization of+	Deactivation by Coke Formation of ZSM-5 Type Cat	PETR	0027
s over Heterogeneous Catalysts.	Deactivation by Coking and Regeneration of Pd/HM	PETR	0130
s Used fo+ Factors Affecting the	Deactivation of Gallium-Containing HZSM-5 Zeolite	PETR	0043
ates on Pt/L-Zeolite: Inhibited	Deactivation of Pt Sites within One-Dimensional Zeo	CATL	0015
Is Conventional Chemistry	Dead?.=	AGRO	0084
Assessments: Protecting the	Dealer and the Environment.= Environmental Site	FERT	0015
l Chemistry. The Retail Fertilizer	Dealer as Environmentalist.= +in Fertilizer and Soi	FERT	0017
Retail Fertilizer	Dealer Containment.=	FERT	0020
Fertilizer and Agrichemical	Dealers.= Environmental Self-Evaluation for	FERT	0018
Management for Retail Fertilizer	Dealers.= +l Solutions for Wastewater and Rinsate	FERT	0014
ent Needs for Agrochemical Retail	Dealership Site Assessment and Remediation.=	AGRO	0142
y Fast ^{27}Al MAS NMR Studies of	Dealuminated Zeolite Y.= +osspolarization and Ver	PETR	0006
Expression of Yeast AMP	Deaminase in E. coli.=	BIOL	0083
n of Glucosamine-6-P	Deaminase: Calorimetric Study of the Denaturatio	BIOL	0106
hesis and Biological Activity of 9-	Deaza-9-Substituted Guanines.= +Inhibitors: Synt	MEDI	0028
nd Reconfirmation of That of 9-	Deazainosine.= Corrected Structure of Pyrrolosine a	CARB	0046
Evaluation of 8-Aza-1-	Deazapurine Nucleosides.= Synthesis and Biological	MEDI	0029
rantium) Juice Using Neutral a+	Debittering and Deacidifying Sour Orange (Citrus au	AGFD	0106
n Ethoxycarbonylation-Carbonyl	Deblocking of Anions of Carbethoxy-Protected Cyan	ORGN	0228
the Reaction of Metal Alkyls with	Deboronated Zeolites.= +atalytic Sites Formed by	PETR	0024
Cell	Debris Removal by Cross-Flow Microfiltration.=	BIOT	0048
zed Physical Chem Lab Makes Its	Debut.= +th Techniques That Are New, a Revitali	CHED	0043
ere Did We Come From; Whe+ A	Decade of Chemical Safety Training, 1980-1991: Wh	CHAS	0043
Aromatics Formation in	Decane and Kerosene Flames.= Comparison of	FUEL	0080
$_5$H$_4$N).= Aldehyde	Decarbonylation Catalyzed by cis-Rh(CO)$_2$(p-(NC)C	INOR	0255
+mines as Inhibitors of Ornithine	Decarboxylase: Synthesis and Biological Activity.=	MEDI	0184

ibitors of S-Adenosylmethionine	Decarboxylase.= Enzyme-Activated Irreversible Inh	BIOL	0052
Enzyme Brewers' Yeast Pyruvate	Decarboxylase.= +hiamin Diphosphate-Dependent	BIOL	0032
mines on Brewers' Yeast Pyruvate	Decarboxylase.= +iamin Diphosphate-Bound Ena	BIOL	0048
rogenase Complex and to the E1	Decarboxylating Enzyme: A Monoclone That Is 98 In	BIOL	0031
gsten (0) Complexes.= Catalytic	Decarboxylation of Cyanoacetic Acid by Anionic Tun	INOR	0212
lexes Derived from Cu-Co Dim+	Decarboxylation Reactions Involving Copper(I) Comp	INOR	0213
der Nonlinear Optical Onset and	Decay and Their Relationship to Polymer Physics.=	POLY	0152
Double-Beta	Decay Experiments.= Report of Some Recent	NUCL	0013
Materials.= Orientation and	Decay in Poled Polymer Nonlinear Optical	POLY	0151
Illumination May Help Reduce	Decay in Seal-Packaged Citrus and Pepper Fruits.=	AGFD	0076
Double Beta	Decay of ^{100}Mo and ^{150}Nd.=	NUCL	0012
ss from Measurement of the Beta	Decay of Molecular Tritium.= +n Antineutrino Ma	NUCL	0010
Double Beta	Decay of U-238.=	NUCL	0061
Mass.= The Tritium Beta	Decay Spectrum from End to End and the Neutrino	NUCL	0011
Nuclear Beta	Decay.= Evidence for a Massive Neutrino in	NUCL	0003
ience: Neutrino Masses from Beta	Decay—I. Neutri+ Recent Advances in Neutrino Sc	NUCL	0001
ience: Neutrino Masses from Beta	Decay—II. Limit+ Recent Advances in Neutrino Sc	NUCL	0010
robic Sediments.=	Dechlorination of Polychlorinated Biphenyls in Anae	BIOT	0120
	Deciphering a Chemical Vapor Deposition Process.=	INOR	0122
rmulas: Enumeratio+ Studies in	Deciphering the Information Content of Chemical Fo	COMP	0027
A's Hazard-Ranking System and	Decision Methodology: What Is the Basis of EPA's L	ENVR	0077
Life through	Decision-Making Activities.= Bringing Chemistry to	CHED	0054
ontext: The ACS College-Level	Decision-Making Chemistry Course for Nonscience S	CHED	0051
xpert Witness. Overview of Legal	Decision-Making Process and the Role of Expert Op	CHAL	0011
Development and	Decline of Incandescent Gas Lighting.=	HIST	0038
alytic Model of Science-Relate+	Declining Chemistry Enrollments: Toward a Path An	CHED	0177
of RNA Minihelices in Relation to	Decoding Genetic Information.= +ry. Recognition	BIOL	0026
ion Domains as Mechanisms for	Decoding Information in Complex Sugar Structures.=	BIOL	0144
by the High-Temperature Aerosol	Decomposition (HTAD) Process.= +ysts Prepared	CATL	0026
th Experiment. Dynamics of the	Decomposition and Isomerization of Azides and Isocy	PHYS	0188
o+ Phase Diagrams and Spinodal	Decomposition and Linear and Cyclic Polystyrene/P	POLY	0295
f Copper.=	Decomposition Mechanisms Important in OMCVD o	COLL	0157
cations of Catalysts—II. Thermal	Decomposition of Alkyl Halides on Copper Surfaces.	COLL	0204
Adsorption and	Decomposition of Diethylsilane on Silicon Surfaces.=	COLL	0008
ilicon and Indium Phosphide.=	Decomposition of Mono- and Trimethylgallium on S	INOR	0123
for the Prepara+ Laser-Induced	Decomposition of Precursors Containing M-N Bonds	POLY	0230
Implicatio+ Kinetics of Thermal	Decomposition of Triethylgallium on GaAs(100) and	COLL	0068
XPS Study of the Thermal	Decomposition of W(CO)6 on Pt (111).= TPD and	COLL	0049
Propargyl Radicals.= Thermal	Decomposition of 1,7-Ocatdiyne as a Source of	FUEL	0067
main Growth during the Spinodal	Decomposition under Simple Shear Flow.= +s. Do	PMSE	0104
nds.=	Decontamination of Solutions of Hazardous Compou	CHAS	0006
utral Network.=	Deconvolution of Chromatographic Peaks Using a Ne	ANYL	0048
hosphorus Chemistry. A Life of	Dedication to Research in Industrial Chemistry.= P	INOR	0321
the	Deep Ocean.= Formation of CO_2 Clathrate in	ENVR	0047
degradative Genes in Isolates from	Deep Subsurface Environments.= +of Cloned Bio	BIOT	0117
from Organic Adsorbates to	Defect Sites in Silica Supports.= Energy Transfer	PHYS	0135
Local Atomic Structure of	Defects on Oxide Surfaces.=	PHYS	0207
of Speech—The Delicate Balance:	Defense Research in a Democratic Society.= +dom	PROF	0009
Role of Antioxidative	Defense Systems by Phenolic Plants Constituents.=	AGFD	0018
de Ratas Hembras.+ Efecto de la	Deficiencia o Exceso de Vitamina E en la Fertilidad	AGFD	0210
y Incidence and Acquired Immune	Deficiencies in Mice.= +termined Higher Mortalit	AGFD	0145
Chemical Rescue of a Catalytically	Deficient Mutant of R. rubrum Ribulose Bispho+	BIOL	0089
A-Bindin+ A Simple Method to	Define the Recognition Site of Sequence-Specific DN	BIOL	0129
Ionomer/Polystyrene Blends: E+	Deformation and Fracture of Sulfonated Polystyrene	POLY	0061
Organic Polycrystals under Shear	Deformation at High Pressure.= +ical Processes in	PHYS	0245
Liquid Crystalline P+ Effect of	Deformation History on the Morphology of Blends of	PMSE	0135
del for Segmental Orientation in	Deformed Polymer Networks: Characterization of Str	COLL	0016
tallic State in Polymers Lacking a	Degenerate Ground State.= +Disorder and the Me	PHYS	0039
	Degradable Polymers for Biomedical Application.=	CHED	0134
of Pesticides: Characterization of	Degradation and Translocation Processes in Soil as a	AGRO	0025
ydrolyzing Enzymes During Wood	Degradation by Phanerochaete chrysosporium.=	BIOT	0208
Spectroscopic Study of Thermal	Degradation in PEEK (Poly-ether-ether-ketone) Fil	POLY	0334
Models.= Oxidative	Degradation of Aromatic Pollutants by Ligninase	INOR	0187
ppl+ Methods for the Controlled	Degradation of Erodible Polyesters for Biomedical A	CHED	0133
between Selective Oxidation and	Degradation of Ethylene and Propylene over Silver.=	CATL	0058
and Future Technolo+ Chemical	Degradation of Pesticide Wastes: Historical, Current,	AGRO	0122
Thermal	Degradation of Phenolic Antioxidants.=	AGFD	0049
ation of Agrochemicals and Their	Degradation Products in Environmental Matrices by	AGRO	0031
ment of an Impurity Method for	Degradation Products Occurring in Acidic Solutions	ANYL	0085
se Cons+ Treatment of TCE and	Degradation Products Using a Toluene Monooxygena	BIOT	0119
te Networks.= Synthesis and	Degradation Studies of Poly(ϵ-caprolactone) Fumara	POLY	0068
on of Protected Gly+ Convenient	Degradation to Lower Furanose Sugars via Ozonizati	ORGN	0073
Aquatic-Habitat	Degradation.= Ecological Risk of	ENVR	0062
Compounds by Thermal Lignin	Degradation.= Formation of Smoke-Flavor	AGFD	0054
ei Cellobiohydrolases Bind to and	Degrade Cellulose?.= +. How Do Trichoderma rees	BIOT	0206
lavor Compounds.= Thermally	Degraded Thiamine: A Potent Source of Interesting F	AGFD	0071
Hidrosolubles de la	Dehidroepiandrosterona.= Obtencion de Derivados	MEDI	0038
P-4-Keto-6-Deoxy-D-Glucose-3-	Dehydrase (E_1).= +e Oxidoreductase (E_{od}) and CD	BIOL	0110
β-Hydroxydecanoyl Thiol Ester	Dehydrase: Determination of the Orientation of the A	BIOL	0066
li β-Hydroxydecanoyl Thiol Ester	Dehydrase.= +as a Probe of the Active Site of E. co	BIOL	0067

thium Diisopropylamide: A Sim+	Dehydrative Cyclization of Diphenyl Sulfoxide by Li	ORGN	0084
ta for P+ Comparison of Alkane	Dehydrocyclization Activity and Characterization Da	CATL	0007
in Catalytic Dehydrogenation and	Dehydrocyclization of Alkanes.= +rol of Selectivity	CATL	0061
echanistic Proposal for Catalytic	Dehydrocyclization Rates on Pt/L-Zeolite: Inhibited	CATL	0015
ing Enzyme: A M+ coli Pyruvate	Dehydrogenase Complex and to the E1 Decarboxylat	BIOL	0031
Inhibitors of Hydroxysteroid	Dehydrogenases.= Enzyme-Generated Irreversible	MEDI	0010
s and Catalysts for C_3-C_4 Paraffin	Dehydrogenation and Aromatization.= +f Reaction	PETR	0066
ontrol of Selectivity in Catalytic	Dehydrogenation and Dehydrocyclization of Alkanes.	CATL	0061
ient Photochemical and Thermal	Dehydrogenation of Alkanes Catalyzed by Rh(PMe$_3$)	INOR	0152
Reaction Kinetics of Propane	Dehydrogenation over Zn or Mo.=	I&EC	0075
rmal Catalytic Alkane Transfer-	Dehydrogenation Using Rh(PMe$_3$)$_2$LCl(L = ^{i}Pr$_3$,CO)	INOR	0151
ctions by ^{13}C NMR and Catalytic	Dehydrogenation.= +oups in Heavy Petroleum Fra	PETR	0147
fusion Column.=	Dehydrogenative Coupling of CH$_4$ in a Hot-Wire Dif	PETR	0012
/Trifluoroethylene Copolymer a+	Dehydrohalogenation Studies of Vinylidene Fluoride	POLY	0022
Oxides.=	Oxidative Deintercalation of Silver from Ternary Silver	INOR	0275
ivados de Cupfer+ Hidroxilacion	del Benceno con H$_2$O$_2$ Catalizada por Complejos Der	COLL	0121
n Geotermica de Fasico+ Estudio	del Comportamiento de los Gazas en Condensables e	I&EC	0013
en la Aromatizacion+ Influencia	del Contenido de Galio en Zeolitas del Tipo Pentasil	COLL	0125
Leche de Establos del Sur	del D.F.= Determinacion de Metales Pesados en	ENVR	0037
Activacion	del Enlace C-H por Ni en la Molecula de Metano.=	PHYS	0244
tos (FILM) Pr+ Caracterizacion	del Factor Inhibidor de la Locomocion de los Monoci	BIOL	0102
Estudio de la Nitracion	del Fenol con Aductos Tonsil-Fe(NO$_3$)$_3$.=	COLL	0120
orrosion de las Diferentes Etapas	del Lavado Quimico de Una Caldera por Tecnicas El	PHYS	0257
Determinacion	del Mecanismo de Electroreduccion de la Perezona.=	PHYS	0213
licula de Oxido, en la Validacion	del Metodo Potencio-Metrico para Cuantificar Clorh	ANYL	0066
s de la Evaluacion d+ Influencia	del Nivel Academico de los Alumnos en los Resultado	CHED	0093
Vitamina C.= Obtencion	del Orden de Reaccion de la Oxidacion de la	CHED	0164
Pesados en Leche de Establos	del Sur del D.F.= Determinacion de Metales	ENVR	0037
l Contenido de Galio en Zeolitas	del Tipo Pentasil en la Aromatizacion de Propano.=	COLL	0125
NO-Cataliticas.= Influencia	del Transporte de Calor en Reacciones Solido-Gas	I&EC	0068
Actividad Inmunomoduladora	del Zinc en Ratones Inmunodeficientes.=	AGFD	0144
dio de Reduccion Electroquimica	del 1-Nitro-Naftaleno con Electrodos Metalicos y Se	PHYS	0280
^{13}C NMR.= Characterization of	Delayed Coking Feedstocks and Products by ^{1}H and	FUEL	0076
halene: High-Pressure Effects and	Delayed Fluorescence ODMR.= +lly Mixed Napht	PHYS	0081
Socie+ Freedom of Speech—The	Delicate Balance: Defense Research in a Democratic	PROF	0009
b Waveguide+ Photochemically	Delineated Refractive Index Profiles in Polymeric Sla	POLY	0180
Hepatoma+ Receptor-Mediated	Delivery of Methotrexate-Insulin Conjugate to H35	MEDI	0185
Theory in the Study of Electron	Delocalization in Early Transition Metal Heteropoly-	COMP	0032
in Conjugated π-Systems and	Delocalization in π-Stacks.= Electron Localization	BIOT	0201
+n and Vibronic Relaxation in σ-	Delocalized Silicon Backbone Network Polymers.=	PHYS	0138
and near the Amazon	Delta.= Phosphorus Distribution in Sediments in	GEOC	0011
olatilization of Fenitrothion and	Deltamethrin from the Surface Microlayer of Water.	ENVR	0020
erivatives Stabilized by Sterically	Demanding poly(Pyrazolyl)hydroborato Ligands.=	INOR	0085
erivatives Stabilized by Sterically	Demanding tris(Pyrazolyl)hydroborato Ligands.=	INOR	0087
ination of Methylene Blue and Its	Demethylated Metabolites from Milk.= +d Determ	AGFD	0101
d Anticancer Activity of Novel 4'-	Demethylepipodophyllotoxin Analogues.= +tion an	MEDI	0011
te Balance: Defense Research in a	Democratic Society.= +dom of Speech—The Delica	PROF	0009
of Reducing Agents in Absence of	Denaturants.= +hydryl Content of Glycinin: Effect	AGFD	0141
al S+ Thermal and pH Stress in	Denaturation of Cellobiohydrolase I: Predenaturation	BIOT	0191
Calorimetric Study of the	Denaturation of Glucosamine-6-P Deaminase.=	BIOL	0106
the Reversibility of	Denaturation.= Facilitation of Protein Folding and	BIOT	0190
hotochemical Probes of Starbust	Dendrimers and Their Utilization as Restricted Reac	POLY	0238
Synthesis of Rigid	Dendrimers That Overcome Steric Inhibition.=	POLY	0321
3,5-Phenylene-Based, Fluorinated	Dendrimers.= +tion of a Series of Monodisperse 1,	POLY	0236
rations of Monodisperse Arylester	Dendrimers.= +Convergent Synthesis of Four Gene	POLY	0324
5-Phenylene-Based Hydrocarbon	Dendrimers, Including C$_{276}$H$_{186}$.= +nodisperse 1,3,	POLY	0320
Properties of Polystyrenes with	Dendritic End Groups.= Synthesis and Solution	POLY	0322
Approach: Hyperbranc+ Novel	Dendritic Macromolecules by the Convergent Growth	POLY	0318
Applications of PIXE in	Dendrochronology.=	NUCL	0081
Titania-Silica-Based	DENO$_x$ Catalysts.= Surface Characterization of	PETR	0071
ectroscopy of High-Temperature	Dense Molecular Fluids and Low-Temperature Mole	ANYL	0002
Equilibrium Configuration of	Dense Polymacromonomers in Good Solvent.=	POLY	0269
Preparation of	Dense Polymer Samples for Atomistic Modeling.=	POLY	0366
On-Board Materials	Densification.= What's New in Collection Systems?	I&EC	0040
on of Quicklime Binder-Enhanced	Densified Refuse-Derived Fuel/Coal Mixtures.= +i	FUEL	0140
Analytical Determination of Pillar	Densities in Alumina-Pillared Montmorillonite.=	PETR	0126
Polymethylene.= Estimation of	Density and Glass Transition of Amorphous	POLY	0134
yridyl)terth+ Tuning of Electron	Density at Transition-Metal Centers via Poly 3'-(4-p	INOR	0027
hemistry with Constraints: Local	Density Calculations of Growth Reactions on the Dia	PHYS	0355
CVD Reacto+ Temperature and	Density Distribution of H$_2$ and H in a Hot-Filament	PHYS	0146
Single-Particle	Density in Nuclear Dynamics.= Fluctuations of the	NUCL	0009
Using Scanning T+ Very High	Density Information Storage in Molecular Complexes	BIOT	0023
r.=	Density Matrix Theory of Ultrafast Electron Transfe	BIOT	0027
Enhancement of Critical Current	Density of Bi$_2$Sr$_2$CaCu$_2$O$_8$ Superconductor by Metal	PHYS	0329
he Heterogeneity of Linear Low-	Density Polyethylene: Temperature-Rising Supercrit	POLY	0076
Electronic	Density.= Equation for the Square Root of the	PHYS	0254
Current	Density.= Molecular Electronic Transition	BIOT	0026
Surface Science in	Dentifrice Formulations.= Role of Rheology and	CHED	0047
s Semiconductores Policristalinos	dentro de Celdas Fotoelectroquimicas, I: CdS Puro e	PHYS	0221
and	Deodorants.= Chemical Aspects of Antiperspirants	CHED	0049

l-D-Glucopyranoside and Its 6-	Deoxy- and 6-Fluoro-Analogues and the Demonstrat	CARB	0028
uctase (E$_{od}$) and CDP-4-Keto-6-	Deoxy-D-Glucose-3-Dehydrase (E$_1$).= +e Oxidored	BIOL	0110
3-Mediated Reduction of CDP-6-	Deoxy-$\Delta^{3,4}$-Glucoseen.= +Mechanistic Studies of E	BIOL	0100
ogues and Crystal Structure of 2'-	Deoxy-2'-Methylidenecytidine Hydrochloride.= +l	CARB	0038
leotide Reductase by 2'-Azido 2'-	Deoxynucleotides.= +sm of Inactivation of Ribonuc	ORGN	0281
clease DNA R+ Psoralen-Linked	Deoxyoligonucleotide Synthesis for UvrABC Endonu	BIOL	0126
ontrolled Synthesis of C-Linked	Deoxyribosides of 2-Hydroxypyridine and 2-Hydroxy	ORGN	0075
quirements of the NADA: Residue	Depletion and Metabolism of Flunixin in Cattle.=	AGRO	0078
Cows.= Residue	Depletion and Metabolism of Luprostiol in Dairy	AGRO	0130
Tissue Residue	Depletion Studies on Semduramicin in the Chicken.=	AGRO	0079
of Electronic Energy Transfer and	Depolarization in Polymers.= +Time Dependence	POLY	0078
NMR Relaxation, Fluorescence	Depolarization, and the Dynamics of Proteins and+	PHYS	0166
Plastics Recycling via PET	Depolymerization.=	I&EC	0032
of NH with the Surface of	Depositing Films.= Laser Studies of the Reactivity	COLL	0218
Metalorganic Chemical Vapor	Deposition Approaches to high-T$_c$ Superconducting F	INOR	0176
or CVD by FT-IR and Raman +	Deposition Characteristics of Copper(I) Compounds f	COLL	0031
nism of Copper Chemical-Vapor	Deposition from the 1,5-Cyclooctadiene Copper(I) H	COLL	0026
eering by Laser Chemical Vapor	Deposition of a Nonlinear Interface Optical Switch.=	BIOT	0024
lls of Glass Capillaries:+ In Situ	Deposition of Calcium Carbonate on the Internal Wa	COLL	0163
per tert-Butox+ Chemical Vapor	Deposition of Copper and Copper(I) Oxide from Cop	INOR	0143
h C+ Near-Room-Temperature	Deposition of Copper Films by H-Atom Reaction wit	PHYS	0259
Selective Chemical-Vapor	Deposition of Copper from Copper(I) Compounds.=	COLL	0027
w-Temperature Chemical Vapor	Deposition of Copper from New Copper(I) Compoun	INOR	0144
r(I) Hexafluo+ Chemical-Vapor	Deposition of Copper from 1,5-Cyclooctadiene Coppe	COLL	0029
n of Diamond+ Chemical Vapor	Deposition of Diamond (and c-BN, SiC). Halogenatio	PHYS	0176
ion Measurem+ Chemical Vapor	Deposition of Diamond (and c-BN, SiC). Mole Fract	PHYS	0145
Dynamics Simu+ Chemical Vapor	Deposition of Diamond (and c-BN, SiC). Molecular	PHYS	0354
nd Chemistry o+ Chemical Vapor	Deposition of Diamond (and c-BN, SiC). Structure a	PHYS	0202
Diagnostics for Chemical Vapor	Deposition of Diamond Films.= Laser Spectroscopic	PHYS	0151
in X-ray Lit+ Microwave CVD	Deposition of Diamond Thin Films with Applications	PHYS	0092
tion of Electro+ Chemical Vapor	Deposition of Low-Resistivity Metals for Interconnec	INOR	0126
M) Studies of the Chemical-Vapor	Deposition of Metal Carbonyls.= +Microscopy (ST	COLL	0050
echanism and Energetics in the	Deposition of Platinum from Pt (PF$_3$)$_4$ on Atomically	COLL	0046
ulation Model of the Atmospheric	Deposition of Semivolatile Organic Compounds.=	GEOC	0004
	Deposition of SiO$_2$ from Tetraethoxysilane.=	COLL	0009
r Routes to Inorganic Materials.	Deposition of Transition-Metal Thin Films from Org	INOR	0141
(η^3-C$_3$H$_5$)$_5$M+ Chemical Vapor	Deposition of Tungsten and Molybdenum Films from	INOR	0055
o Organometallic Chemical Vapor	Deposition of ZnSe: Effect of Precursor Chemistry	INOR	0171
ms in Plane Prepared by Vacuum	Deposition Polymerization Methods.= +lyamide Fil	POLY	0082
Deciphering a Chemical Vapor	Deposition Process.=	INOR	0122
Studies of Underpotential	Deposition Processes.= Atomic Force Microscope	PHYS	0128
ditioners: Factors Controlling Hair	Deposition.= +Industry—II. Silicone in Hair Con	CHED	0063
e Chemical Modeling of Diamond	Deposition.= +d Fluorescence Measurements for th	PHYS	0147
Mechanistic Aspects of Si/Ge	Deposition.=	COLL	0047
Etching and Chemical Vapor	Deposition.= Reactions of Metal Atoms in Plasma	PHYS	0143
ilms.=	Depositions and Characterizations of Boron Nitride F	PHYS	0094
	Deposito Electrolitico de Samario y Plutonio.=	NUCL	0075
tion of Diam+ Chemical Vapor	Depositon of Diamond (and c-BN, SiC). Characteriza	PHYS	0118
f Aromatic Intermediates.+ Soot	Depostion from Ethylene/Air Flames and the Role o	FUEL	0107
and Growth of+ Chemical Vapor	Depostion of Diamond (and c-BN, SiC). Nucleation	PHYS	0091
f the Magnetic-Field Penetration	Depth in YBa$_2$Cu$_3$O$_x$ and Bi$_2$Sr$_2$CaCu$_2$O$_x$ Supercond	PHYS	0284
f the Magnetic-Field Penetration	Depth in YBa$_2$Cu$_3$O$_x$ and Bi$_2$Sr$_2$CaCu$_2$O$_x$ Supercond	PHYS	0050
	Depth Profiling Using Variable-Angle ATR.=	POLY	0307
er/Metal Interface by Sputtering	Depth Profiling Using XPS.= +Reactions at Polym	PMSE	0150
n H$_2$O$_2$ Catalizada por Complejos	Derivados de Cupferron.= +xilacion del Benceno co	COLL	0121
a.= Obtencion de	Derivados Hidrosolubles de la Dehidroepiandrosteron	MEDI	0038
ng of Excited States by Ferrocene	Derivatives.= +ergy and Electron Transfer Quenchi	INOR	0163
ex Optimization of Post-Column	Derivatization Conditions in the Detection of Aflatox	AGFD	0111
s: Reagents for Selective Surface	Derivatization of Au and Pt Microfabricated Wire Ar	INOR	0350
ium Cyanamide Using Precolumn	Derivatization.= +od for the Determination of Calc	ANYL	0113
Light as a Reagent for Chemical	Derivatizations of Pesticides.=	AGRO	0147
Improved Enzyme-Activated C+	Derivatized Adamantyl-Substituted 1,2-Dioxetanes:	BIOT	0167
Food-Producing Animals.=	Dermal Absorption and Metabolism of Xenobiotics in	AGRO	0095
isk Assessments of Inhalation and	Dermal Exposures.= +Are the Keys to Scientific R	ENVR	0042
te-Pyrite: Effet des Oxydants et	des Complexants sur la Fottation de ces Mineraux.=	ANYL	0067
ation Arsenopyrite: Effet	des Oxydants et des Complexants sur la Fottation de	ANYL	0067
nza de la Quimica a Nivel Med+	Desarrollo de Apoyos Computarizados para la Ensena	CHED	0124
ratory Validations and Accurate	Descriptions of Analytical Procedures for Drug Resid	AGFD	0126
Peanut Flavor, as Determined by	Descriptive Sensory Analysis and Gas Chromatograp	AGFD	0161
Comparative Study of Molecular	Descriptors Derived from the Distance Matrix.=	COMP	0025
mistry Requirements of the N+	Design and Conduct of Studies to Meet Residue Che	AGRO	0078
ntheses.=	Design and Development of Practical Asymmetric Sy	ORGN	0297
oduction of Chemicals from Pl+	Design and Operation of Reactors for Large-Scale Pr	BIOT	0233
in a S+ Animal Cell Bioreactor	Design and Optimization. Mammalian Cell Retention	BIOT	0007
eriment Introducing Experimental	Design and Optimization.= +d Undergraduate Exp	CHED	0182
n Bioprocessing To Aid in Process	Design and Optimization.= +nalytical Techniques i	BIOT	0074
f HIV-1 Protease.=	Design and Preparation of C$_2$ Symmetric Inhibitors o	MEDI	0053
	Design and Status of GALLEX.=	NUCL	0031
kable Nonlinear Optical Polym+	Design and Synthesis of a New Class of Photocrosslin	POLY	0131
Separation and Catalysis.=	Design and Synthesis of Carbon Molecular Sieves for	PETR	0059

gonists.=	Design and Synthesis of Dopamine D1 Selective Anta	MEDI 0125
otein Kinase C.=	Design and Synthesis of Ether Lipid Inhibitors of Pr	MEDI 0073
able for Discovery of Peptide–+	Design and Synthesis of Large Peptide Libraries Suit	MEDI 0093
Opioid Receptor Selective Fluor+	Design and Synthesis of Naltrindole Analogues as δ–	MEDI 0155
rystal Structure–Based de Novo	Design and Synthesis of Naphthalene Containing Inh	MEDI 0186
e.=	Design and Synthesis of Novel Polymers for Medicin	POLY 0273
–Diagonal Tens+ New Molecular	Design Concept for Poled Polymers Utilizing the Off	POLY 0128
Fermentation?—Process and	Design Considerations.= Bacterial or Animal Cell	BIOT 0076
Equation of State.=	Design for Packed Tower Absorbers Using the PRSV	CHED 0107
ions on the Therm+ Biocatalyst	Design for Stability and Specificity. Effects of Mutat	BIOT 0188
for Probing the M+ Biocatalyst	Design for Stability and Specificity. New Techniques	BIOT 0112
tein Chemical Cros+ Biocatalyst	Design for Stability and Specificity. Overview of Pro	BIOT 0216
C–Terminal B–Cha+ Biocatalyst	Design for Stability and Specificity. Role of Insulin's	BIOT 0132
of Enzymatic Catal+ Biocatalyst	Design for Stability and Specificity. Structural Basis	BIOT 0239
3–Phosphate Mimics+ Inhibitor	Design I: Investigation of Various Moieties as Stable	ORGN 0028
ics o+ EPSP Synthase Inhibitor	Design II: Synthesis and Evaluation of Aromatic Mim	ORGN 0029
ole of Computer–Aided Process	Design in Following an Integrated Approach to Biopr	BIOT 0075
Automated Structure	Design in 3D.=	COMP 0048
cal Characteristics of Produced+	Design of a Mill Drier for Shrimp Head: Bromatologi	AGFD 0105
gical Characteristics of Produce+	Design of a Mill-Drier for Shrimp Heads: Bromatolo	ENVR 0046
	Design of a Molecular–Bilayer Electrical Device.=	BIOT 0104
Simulation.= SQUALOR—The	Design of an Award–Winning Laboratory	CHED 0079
ansesterification Catalysts and the	Design of Artificial Restriction Enzymes.= +A Tr	INOR 0406
Related to Vitamin E.= Custom	Design of Better in Vivo Antioxidants Structurally	AGFD 0155
Experiences with Automated 3D	Design of Bioactive Molecules.=	COMP 0049
Approaches Toward the Rational	Design of Dopamine D1 Selective Agonists.=	MEDI 0124
New Method for the de Novo	Design of Enzyme Inhibitors.=	COMP 0037
Orientation for the	Design of Molecular Electronics.= Study of Surface	BIOT 0053
and Biomolecular Electronics.	Design of Nonlinear Optical Materials.= Molecular	BIOT 0173
Rational	Design of Organic NLO Materials.=	POLY 0002
	Design of Photoactive Layered Phosphates.=	CATL 0021
es as a Driving Force in the	Design of Proteinase Inhibitors.= Chemical Principl	ORGN 0269
on of C5 and C6 N–Alkanes in L+	Design of Pt–Mordenite Catalysts for the Isomerizati	PETR 0113
sport Selectivity Models and the	Design of Ruthenium–Based Fischer–Tropsch Cataly	CATL 0056
Entry Blockers: Application to the	Design of Structurally Novel Agents.= +Calcium	MEDI 0188
The Use of Soil Gas Surveys to	Design Soil Vapor Extraction System.=	ENVR 0051
Bioreactor	Design with Immobilized Biocatalysts.=	BIOT 0038
Receptor Activity and Drug	Design. Pharmacology of Muscarinic Receptors.=	MEDI 0078
se (PNP+ Structure–Based Drug	Design: Inhibitors of Purine Nucleoside Phosphoryla	MEDI 0005
tric Liquid Crystal Thin Films by	Design: The Molecules to Materials Connection.=	BIOT 0202
y Combining Sites: Prediction and	Design.=	BIOT 0056
acological Classification and Drug	Design.= +nces in Antibody Technology. Antibod	ANYL 0043
ach to Computer–Aided Molecule	Design.= +ial of Chromatographic Data for Pharm	COMP 0044
Organic Synthesis	Design.= +ases. CAVEAT: A Vector–Based Appro	CINF 0005
Biomolecular Materials and	Design.= AIPHOS—An Integrated System for	BIOT 0051
Distillation Column	Design.= DESTIL—A User–Friendly Program for	CHED 0122
essful Mechanism–Based Inhibitor	Design.= +ortance of Dynamic Catalysis for Succ	BIOL 0054
Peptide Drug	Design.= Pharmacokinetic Considerations in	MEDI 0193
Plasma Reactor	Design.=	NUCL 0086
inc–Metallopeptidase Inhibitors:	Design, Biochemical, Pharmacological, and Clinical P	MEDI 0002
of Spermidine N8–Acetyltransf+	Design, Synthesis, and in Vitro Activity of Inhibitors	MEDI 0020
l Characterization of a Constrai+	Design, Synthesis, and Three–Dimensional Structura	MEDI 0165
High–Performance Epoxies Using	Designed Core–Shell Rubber Particles.= +ning of	POLY 0157
Proteins	Designed for Adherence to Cellulose.=	BIOT 0217
adation of Phenolics in Optimally	Designed Sequencing Batch Reactors.= +ic Biodegr	BIOT 0138
	Designed Spirocyclic Ionophores for Lithium.=	ORGN 0062
New Protein Kinase C Inhibitors	Designed to Interact with the Catalytic Domain.=	MEDI 0076
Properties.=	Designing Polymer Blends Which Exhibit Synergistic	PMSE 0101
Applications.=	Designing Polymeric Materials for All–Optical Device	BIOT 0175
ization of Coal Macerals by Laser	Desorption Mass Spectrometry.= +etry. Character	FUEL 0088
f Carbon–Active Site Behavior via	Desorption Techniques.= +ive Sites. Elucidation o	FUEL 0013
s by Temperature–Programmed	Desorption.= Characterization of Carbon Active Site	FUEL 0016
	Destabilized Ketenes.=	ORGN 0125
Column Design.=	DESTIL—A User–Friendly Program for Distillation	CHED 0122
Modificacion de Secuencias de	Destilacion con Integracion de Energia.= Sintesis y	I&EC 0074
cal Approaches to Pesticide Waste	Destruction Using White Rot Fungi.= +otechnologi	AGRO 0126
eplacement by a Nickel Atom in	Desulfovibrio Vulgaris (Hildenborough) Rubredoxin.	INOR 0020
ctroscopy as a Novel Method To	Detect Framework Substitution of Paramagnetic Ion	PHYS 0231
Experimental a Partir de Errores	Detectados en Publicaciones Cientificas.= +enanza	CHED 0178
(EP) as a Biochemical Marker for	Detecting Exposure to Lead.= +te Protoporphyrin	CEI 0020
Immunoassay for	Detecting PCB in Soil.= PCB–RISc: An On–Site	ENVR 0032
rom Oxidation of Probucol.=	Detection and Characterization of Radicals Derived f	BIOL 0099
of Animal Origin. Approaches to	Detection and Confirmation of Drug Residues in Mil	AGFD 0099
Automatic Sor+ Development of	Detection and Material Handling Equipment for the	I&EC 0041
Rem+ Rapid–Scanning UV–VIS	Detection for Capillary Gas Chromatography Using a	ENVR 0060
Square Wave	Detection for Flow–Injection Analysis.=	ANYL 0105
aphite Compos+ In Situ Mercury	Detection in Groundwater by Novel Ion–Selective Gr	ENVR 0044
ing an Electr+ Halogen–Selective	Detection in Supercritical Fluid Chromatography Us	ANYL 0091
o. Foundation). Thermo–Optical	Detection Methods in Analytical Laser Spectroscopy	ANYL 0011
	Detection of Adulterated Vanilla Extract.=	AGFD 0185

itrus Essential Oils.= HRGC for	Detection of Adulterations of Italian Cold-Pressed C	AGFD	0199
n Derivatization Conditions in the	Detection of Aflatoxins.= +mization of Post-Colum	AGFD	0111
ent Problems Associated with the	Detection of Antibiotic/Drug Residues in Food.=	AGFD	0081
m Test II Receptor Assays for the	Detection of Antimicrobial Residues in Meat.= +ar	AGFD	0085
Diesel Fuels by+ Simultaneous	Detection of Aromatics, Sulfur, and Hydrocarbons in	PETR	0118
Pulsed Amperometric	Detection of Carbohydrates.=	CARB	0067
pplication of SNIF-NMR to the	Detection of Flavor and Fragrance Adulteration.= A	AGFD	0197
Evanescent	Detection of Humidity.=	ANYL	0079
iethylbenzamide by ^{13}C NMR +	Detection of Liver Microsomal Metabolites of N,N-D	AGRO	0026
Paramagnetic Resonance	Detection of Phenolic Free Radicals.= Electron	AGFD	0004
hemiluminescent Substrates in the	Detection of Proteins and Nucleic Acids.= +ted C	BIOT	0167
hemiluminescence as an HPLC	Detection Strategy for Chlorophenol Determination.=	ANYL	0084
sa+ HPLC Postcolumn Reaction	Detection Systems Based on Competitive Binding As	ANYL	0090
Electrochemical and UV/Visible	Detection to the LC Separation and Determination o	AGFD	0101
noglobulin G Possess S+ Cancer	Detection: B-Protein and the Heavy Chains of Immu	BIOL	0068
C with Diode Array Multichannel	Detection.= +Peaks in Normal Bonded Phase HPL	ANYL	0087
ssues by HPLC with Diode-Array	Detection.= +tion of Sulfamethazine in Animal Ti	AGFD	0103
ution GC with Chemiluminescence	Detection.= +t Petroleum Streams by High-Resol	PETR	0124
rmined by HPSEC with Viscosity	Detection.= +of Dextran, Pullulan, and Pectin Dete	CARB	0045
d Residua Using Atomic Emission	Detection.= +re Gas Chromatography of Crudes an	PETR	0121
mploying HPLC Analysis with UV	Detection.= +Analysis in Pharmaceuticals When E	ANYL	0114
Nucleic Acids for Nonradioactive	Detection.= +of Photozyme and Its Use in Labeling	CARB	0037
Using Charge-Coupled Device	Detection.= Unique Optics for Raman Spectroscopy	ANYL	0035
nal Separations and Multichannel	Detections in Normal Bonded Phase HPLC.= +sio	ANYL	0088
bardment Mass Spectroscopy as a	Detector for Liquid-Phase Separation Systems.=	ANYL	0125
Using a Single	Detector.= A GC Method for Soil Gas Analysis	ENVR	0054
Using an Electrolytic Conductivity	Detector.= +Supercritical Fluid Chromatography	ANYL	0091
Indiana Silicon Sphere (ISiS) 4π	Detector.=	NUCL	0036
Chromatographic	Detector.= Mass Spectrometry: A Powerful	AGFD	0202
Kamiokande II	Detector.= Solar Neutrino Results from the	NUCL	0023
Gas Chromatographic	Detectors: New Developments.=	AGFD	0177
al SEC and Molar Mass Sensitive	Detectors.= +r Characterization Using Convention	PMSE	0051
SEC-Viscometry	Detectors.= Flow Rate Effects in Modern	PMSE	0067
pe+ Physical Characterization of	Detergents Using Fluorescent Probes: A Teaching Ex	CHED	0200
ectrometria de Masas con Fuen+	Determinacion Cuantitativa de Tierras Raras por Esp	ANYL	0051
ec+ Obtencion, Caracterizacion, y	Determinacion de la Transferencia de Carga por Esp	INOR	0190
blos del Sur del D.F.=	Determinacion de Metales Pesados en Leche de Esta	ENVR	0037
e la Perezona.=	Determinacion del Mecanismo de Electroreduccion d	PHYS	0213
3-Fucosyl-Lactosamine) Antigenic	Determinant.= +ved in the Biosynthesis of the X(	CARB	0015
roperties of Glycopeptide T-Cell	Determinants Derived from Hen Egg Lysozyme 81-9	CARB	0014
Permanganate.= Fundamental	Determinants in the Oxidation of DNA by	BIOL	0121
phosphate Insecticide Hydrolytic	Detoxification by Conventional Means and Reactive	AGRO	0124
Potential Technique for the Solar	Detoxification of Gas-Phase Wastes.= +er TiO$_2$: A	COLL	0082
Studies.= Gas-Phase Solar	Detoxification of Hazardous Wastes: Laboratory	ENVR	0035
+ide Segmental Dynamics in N-	Deuterated Poly(p-phenylene Terephthalamide).=	POLY	0058
metry Parameters in Bridging +	Deuterim Quadrupole Coupling Constants and Asym	INOR	0382
Integral para Enriquecimiento en	Deuterio.= +tados Experimentales de un Empaque	I&EC	0073
	Deuterium NMR Studies of Acid Sites in Zeolites.=	COLL	0197
Segmental Motion via	Deuterium NMR.= Characterization of Polymer	PHYS	0140
Synthetic Benzaldehyde by	Deuterium NMR.= Differentiation of Natural and	AGFD	0196
ght Distribution: Hydrogen- and	Deuterium-Labeled Poly(styrene) and Poly(vinyl me	COLL	0035
k for Pesticides in Developing and	Developed Countries.= +the 21st Century. Outloo	AGRO	0056
New Laboratory Experiments	Developed Using GC/MS Instrumentation.=	CHED	0187
Architecture of	Device and Sensor Made from Biomaterials.=	BIOT	0123
Materials for All-Optical	Device Applications.= Designing Polymeric	BIOT	0175
harge-Sensitive Ion-Conducting	Device Consisting of Metalloporphyrins in a Lipid Bi	BIOL	0098
roscopy Using Charge-Coupled	Device Detection.= Unique Optics for Raman Spect	ANYL	0035
roperties of Fluid+ A Microwave	Device for the On-Line Monitoring and Control of P	I&EC	0059
Electrical	Device.= Design of a Molecular-Bilayer	BIOT	0104
ed via a Sustained Release Captec	Device.= +ues of Albendazole in Cattle Administer	AGRO	0109
of Conducting Polymers in Redox	Devices and Intelligent Materials Systems.= +ions	POLY	0326
e III-V Compound Semiconductor	Devices by OMCVD.= +owth of High-Performanc	COLL	0010
Nonlinear Optical	Devices Using Organic Materials.=	BIOT	0176
as Materials for Nonlinear Optical	Devices.= +ated Polymers. Conjugated Polymers	POLY	0325
Future Biosensors and Molecular	Devices.= +s on Solid Support: Building Blocks of	BIOT	0102
f Nonlinear Optical Polymers and	Devices.= +mism Studies of Electrical Properties o	POLY	0150
ons for Neural-Like Protein Array	Devices.= +in Cytoskeletal Microtubules: Implicati	BIOT	0055
s for Interconnection of Electronic	Devices.= +r Deposition of Low-Resistivity Metal	INOR	0126
Passive	Devices.= Materials Requirements for Active and	POLY	0004
Semiconductor	Devices.= Meso-Scopic Effects in Small Metal and	BIOT	0002
Monoelectronic	Devices.= Molecular Engineering for	BIOT	0006
Waveguiding Passive and Active	Devices.= Optically Nonlinear Polymers in	POLY	0177
Computers as Data Acquisition	Devices—A Tool for the Revitalization of the Physica	CHED	0160
Phthalocyanines, Molecular	Devices, and Neural Systems.=	BIOT	0054
Semiconductor Quantum	Devices, Circuits, and Applications.=	BIOT	0003
Base Oil Processing: Monitoring	Dewaxing Procedures via Spectroscopy and Chromat	PETR	0151
Hannaway. Personal Recollecti+	Dexter Award in History of Chemistry, Honoring O.	HIST	0009
by Dexter Chemical Corp.). +. (	Dexter Award in the History of Chemistry, sponsored	HIST	0011
istory of Chemistry, sponsored by	Dexter Chemical Corp.).+ (Dexter Award in the H	HIST	0011
M Dextransucrase: Inhibition of	Dextran Synthesis by α-Methyl-D-Glucopyranoside	CARB	0028

C with Vi+ Fractal Properties of	Dextran, Pullulan, and Pectin Determined by HPSE	CARB	0045
Fluorescein–Labeled	Dextrans.= Fluorescence Polarization of	ANYL	0076
Anti-α(1←6)	Dextrans.= Molecular Biology of	CARB	0059
nostoc mesenteroides B–512FM	Dextransucrase: Inhibition of Dextran Synthesis by α	CARB	0028
Reaction Kinetics of Propane	Deydrogenation over Zinc or Molybdenum Supported	PETR	0087
GABA–Based Resistances in	Diabrotica.= Phytochemical Antagonism of	AGRO	0136
n Atoms with Acetylene at 300+	Diacetylene Formation during the Reaction of Oxyge	FUEL	0065
Control of Photochromism of	Diacetylene LB Films by Polyion Complexation.=	BIOT	0146
Studies on the Polymerization of	Diacetylene–Containing Acrylate and Methacrylates.	POLY	0092
Synthesis and Polymerization of	Diacetylene–Containing Benzoic Acid Derivatives.=	POLY	0056
New	Diacetylene–Containing Polyesters.=	POLY	0054
y+ Structures and Properties of	Diacetylenic Liquid Crystalline Polymers Obtained b	POLY	0030
nists.=	Diacidic Nonpeptide Angiotensin II Receptor Antago	MEDI	0103
es, and Reactivity of (Alkyl)acyl,	Diacyl, and (α–Ketoacyl)acyl Complexes of Platinum	INOR	0043
Magnetic Birefringence as a	Diagnostic Tool in Viscoelastic Media.= Dynamic	COLL	0143
en+ Laser–Induced Fluorescence	Diagnostics and Modeling of 10–Torr Methane/Oxyg	FUEL	0128
nd Films.= Laser Spectroscopic	Diagnostics for Chemical Vapor Deposition of Diamo	PHYS	0151
Filaments.= In Situ	Diagnostics of the Flame Growth of Diamond	PHYS	0149
Poled Polymers Utilizing the Off–	Diagonal Tensor Components of a Molecule.= +for	POLY	0128
n the Diels–Alder Reaction of 2,3–	Dialkoxy–1,3–Butadienes.= +ereofacial Selectivity i	ORGN	0233
ed Oxidation of Hantzsch Esters (	Dialkoxycarbonyldihidropyridines).= +trates: Induc	ORGN	0245
ationships of N–Alkyl and N,N'–	Dialkyl Substituted N,N'–Diarylguanidines as Nonco	MEDI	0147
ns: A Convenient Synthesis of 9,9–	Dialkyl-9,10–Dihydroanthracenes.= +ted Alkylatio	ORGN	0087
e during Imidization of Poly(amic	dialkylamide)Precursors.= +tu Molecular Composit	PMSE	0169
–Methylbiphenylmethyl Alkyl and	Dialkylamino Aryl Oxime Ethers.= +Substituted 2	AGRO	0072
Foliascens.= New	Dialkylated Scalarins from the Sponge Phyllospongia	ORGN	0182
Foliascens.= New	Dialkylated Scalarins from the Sponge Phyllospongia	ORGN	0188
18–Crown–6 as a Catalyst in the	Dialkylation of o–Nitrophenacyl Derivatives.=	ORGN	0261
of Cyclopentadienyl Sodium with	Dialkyldichlorosilanes.= +nes via the Condensation	INOR	0199
retic Cluster Expansions:	Diamagnetic Susceptibility.= Chemical Graph–Theo	COMP	0030
Synthesis and Characterization of	Diamantane–Based Polybenzazoles.= +d Polymers:	POLY	0028
he Solutions of the Iron–Ethylene	Diamine Tetraacetic Acid Complexes.= +otons in t	INOR	0090
itu Cure Monitoring of Aromatic	Diamine–Epoxy System by Fiber Optic Fluorescence	POLY	0282
Fluorescence of Aromatic	Diamines for Epoxy Cure Studies.=	POLY	0159
ical Evaluation of a Series of 2,4–	Diamino–6–[2–Phenylethenyl]Pyrido[2,3–d] Pyrimid	MEDI	0022
erization of a Polyimide from 3,7–	Diaminophenothiazinium Chloride (Thionine).= +t	POLY	0033
osine by Nucleophiles from cis-	Diammine(guanosine)platinum(II) Chloride Using ^{13}C	INOR	0251
Chemical Vapor Depositon of	Diamond (and c–BN, SiC). Characterization of Diam	PHYS	0118
Chemical Vapor Deposition of	Diamond (and c–BN, SiC). Halogenation of Diamond	PHYS	0176
Chemical Vapor Deposition of	Diamond (and c–BN, SiC). Mole Fraction Measurem	PHYS	0145
u+ Chemical Vapor Deposition of	Diamond (and c–BN, SiC). Molecular Dynamics Sim	PHYS	0354
f+ Chemical Vapor Deposition of	Diamond (and c–BN, SiC). Nucleation and Growth o	PHYS	0091
o+ Chemical Vapor Deposition of	Diamond (and c–BN, SiC). Structure and Chemistry	PHYS	0202
Derived from Hydrocarbon Fuels:	Diamond and Diamond–Like Carbon. +w Materials	FUEL	0034
Derived from Hydrocarbon Fuels:	Diamond and Diamond–Like Carbon. +w Materials	FUEL	0035
Derived from Hydrocarbon Fuels:	Diamond and Diamond–Like Carbon. +w Materials	FUEL	0037
Derived from Hydrocarbon Fuels:	Diamond and Diamond–Like Carbon. +w Materials	FUEL	0036
Derived from Hydrocarbon Fuels:	Diamond and Diamond–Like Carbon. +w Materials	FUEL	0038
of the Glass Transition in the	Diamond Anvil Cell.= High-Pressure NMR Studies	PHYS	0260
ents for the Chemical Modeling of	Diamond Deposition.= +d Fluorescence Measurem	PHYS	0147
Flame Growth of	Diamond Filaments.= In Situ Diagnostics of the	PHYS	0149
try during Hot–Filament–Assisted	Diamond Film Growth.= +Beam Mass-Spectrome	PHYS	0145
Effects of Ammonia on CVD	Diamond Film Growth.=	PHYS	0150
Kinetics of	Diamond Film Growth.=	PHYS	0177
Molecular Beam Studies on	Diamond Film Synthesis.=	PHYS	0203
Synthesis of Thick, Large–Area	Diamond Film.=	PHYS	0093
C OES and SEM Observations of	Diamond Films Synthesized by DC Arc Discharge	FUEL	0038
lecular Dynamics Simulations of	Diamond Films: Structure, Growth, and Properties.=	PHYS	0354
r Chemical Vapor Deposition of	Diamond Films.= Laser Spectroscopic Diagnostics fo	PHYS	0151
rization of Diamonds, Synthetic	Diamond Films, and Related Carbon Materials by Ra	PHYS	0118
, SiC). Nucleation and Growth of	Diamond from Hydrocarbon Gases.= +(and c–BN	PHYS	0091
omoepitaxial High–Temperature	Diamond Growth from Acetylene– Oxygen Flames.=	FUEL	0037
Modeling	Diamond Growth in Pedestal CVD Reactors.=	PHYS	0179
Filament–Assisted	Diamond Growth Kinetics.=	PHYS	0178
Reactor Models.= Modeling	Diamond Growth: From Elementary Mechanisms to	PHYS	0357
Simulation of CVD	Diamond Growth.=	PHYS	0181
Surface Chemistry of	Diamond Growth.=	PHYS	0180
What's on the Surface during	Diamond Growth?.=	PHYS	0122
tal and Tight Binding Methods in	Diamond Surface Reactivity.= +cal Molecular Orbi	PHYS	0356
ructure and Chemistry of Natural	Diamond Surfaces.= +amond (and c–BN, SiC). St	PHYS	0202
Microwave CVD Deposition of	Diamond Thin Films with Applications in X-ray Lit	PHYS	0092
n of the Nucleation and Growth of	Diamond Thin Films.= +tructural Characterizatio	PHYS	0121
ure Raman–Scattering Behavior in	Diamond.= +mond–Like Carbon. High–Temperat	FUEL	0035
ydrogen Binding and Diffusion in	Diamond.= +amond and Diamond–Like Carbon. H	FUEL	0036
Microscopy to the Study of	Diamond.= Applications of Transmission Electron	PHYS	0120
CVD	Diamond.= Structural Characterization of	PHYS	0119
n a Hot–Filament CVD Reactor of	Diamond.= +d Density Distribution of H$_2$ and H i	PHYS	0146
ations of Growth Reactions on the	Diamond(100) Surface.= +ts: Local Density Calcul	PHYS	0355
(and c–BN, SiC). Halogenation of	Diamond(100) Using Atomic Beams.= +Diamond	PHYS	0176
Hydrocarbon Fuels: Diamond and	Diamond–Like Carbon. +w Materials Derived from	FUEL	0034

Hydrocarbon Fuels: Diamond and	Diamond-Like Carbon. +w Materials Derived from	FUEL	0035
Hydrocarbon Fuels: Diamond and	Diamond-Like Carbon. +w Materials Derived from	FUEL	0037
Hydrocarbon Fuels: Diamond and	Diamond-Like Carbon. +w Materials Derived from	FUEL	0036
Hydrocarbon Fuels: Diamond and	Diamond-Like Carbon. +w Materials Derived from	FUEL	0038
Gas-Phase Analysis of	Diamond-Producing Plasmas.=	PHYS	0148
Theory of Schottky Barriers in	Diamond/Metal Interfaces.=	PHYS	0358
and Fluorine near the	Diamond/Vapor Interface.= Absorbed Hydrogen	PHYS	0359
yl Ureas.= Resistance in the	Diamondback Moth to Pyrethroids and Benzoylphen	AGRO	0117
arbon Mate+ Characterization of	Diamonds, Synthetic Diamond Films, and Related C	PHYS	0118
ne Derivativ+ New Hydrolyzable	Dianionic Initiator from Divinyl Tetraoxaspiroundeca	POLY	0349
ecticides with Low Fish To+ 1,4-	Diaryl-1-Cyclopropylbutanes: Highly Efficacious Ins	AGRO	0073
N,N'-Dialkyl Substituted N,N'-	Diarylguanidines as Noncompetitive NMDA Antagon	MEDI	0147
n of 2,3-Dialkoxy-1,3-Butadien+	Diastereofacial Selectivity in the Diels-Alder Reactio	ORGN	0233
luation of the Selective Protein+	Diastereomeric Preparation and Pharmacological Eva	MEDI	0026
y-Region 10,11-diol-8,9-Epoxide	Diastereomers of Carcinogenic Dibenz[a,h]Acridine (	ORGN	0208
Styrenes:+ Unexpectedly High	Diastereoselective Oxidation of Locked-Conformation	ORGN	0236
Rea+ Stereoelectronic Control of	Diastereoselectivity in Electrophilic and Diels-Alder	ORGN	0240
esis of 2-oxo-2-Propionyl-1,3,2-	Diaxaphosphorinanes and the Synthesis and Aldol R	ORGN	0050
Paradoxical Effect of	Diazinon on L-Tryptophan Metabolism in Mice.=	AGRO	0020
Mechanism-Based Inactivator 1-	Diazo-4-undecyl-2-one (Duo) as a Probe of the Act	BIOL	0067
gen Insertion Reactions of Alkyl	Diazoacetates Catalyzed by Chiral Rhodium(II) Carb	ORGN	0042
vi+ Synthesis of [1,3,4]Oxa(Thia)	Diazolo[3,2-b]Indazoles and Their Antibacterial Acti	MEDI	0056
Stereochemistry of the Reaction of	Diazomethane with Platin(IV)cyclobutanes.= +nd	INOR	0329
ide Diastereomers of Carcinogenic	Dibenz[a,h]Acridine (DB[a,h]ACR).= +ol-8,9-Epox	ORGN	0208
Monobrominated Polychlorinated	Dibenzo-p-Dioxins and Dibenzofurans.= +yash for	ANYL	0050
welve TCDD-Expo+ Chlorinated	Dibenzodioxins and Dibenzofurans in the Blood of T	CEI	0014
Chlorinated Dibenzodioxins and	Dibenzofurans in the Blood of Twelve TCDD-Expos	CEI	0014
hlorinated Dibenzo-p-Dioxins and	Dibenzofurans.= +yash for Monobrominated Polyc	ANYL	0050
opropylamide: A Simple Route to	Dibenzothiophene.= +yl Sulfoxide by Lithium Diis	ORGN	0084
Thermoreversible Gels from	Dibenzylidene Sorbitol and Organic Solvents.=	POLY	0201
Effects in the Thermodynamics of	Diblock Copolymer Thin Films.= +nite Thickness	POLY	0090
icrophase Structure of PS-PMPS	Diblock Copolymers.= +tate NMR Study of the M	PMSE	0121
Polymerization with Oligomeric	Diblock Macromonomer Stabilizers.= Emulsion	COLL	0039
Tetrahydrothiophenium)-p-xylene	Dibromides.= +ion of Mono-Substituted α,α'-bis(	POLY	0013
Reaction of 2,3-	Dibromocarbonyl Compounds with Nucleophiles.=	ORGN	0160
to Cyclopentadienyl Manganese	Dicarbonyl η^3-Allyl Cations.= Nucleophilic Addition	INOR	0270
of Allyl(cyclopentadienyl)Iron(II)	Dicarbonyl to Aldehydes and Ketones.= +Additions	ORGN	0213
es: 1-Methyl, 2-Phenyltitanocene	Dichloride and Its 1-,3-Analog.= +Chiral Titanocen	INOR	0281
Hydrolysis of the syn,anti-cis-	Dichloro (1,4,7,10-Tetraazacyclodecane) Cobalt(III) C	INOR	0398
Chemists.= Some	Dichotomies of Creative Scientists, including	CONG	0009
ethods. Dynamic Infrared Linear	Dichroism of Polymers.= +racterization: Optical M	POLY	0303
n of Epoxides Catalyzed by	Dicobaltoctacarbonyl.= Ring-Opening Polymerizatio	POLY	0014
afety Training, 1980-1991: Where	Did We Come From; Where Are We Going?.= +l S	CHAS	0043
Two Molecules).=	Didactic Method for Isomer Identification (between	CHED	0094
e Teaching, Part I: Acyclic Seri+	Didactic Puzzles for Organic Chemistry Nomenclatur	CHED	0165
Reductio+ Biosynthesis of 3,6-	Dideoxyhexoses: Mechanistic Studies of E_3-Mediated	BIOL	0100
oly(vinyl Acetate)/Silicon Oxide+	Dielectric and Mechanical Spectroscopy Studies of P	POLY	0287
ng Chemorheological Changes to	Dielectric Properties during Isothermal Epoxy Polym	PMSE	0178
Effect of Moisture Content of	Dielectric Properties in Microwave Region for Printe	PMSE	0055
ins: In Search of the Biosynthetic	Diels-Alder Construction.= +ide Class of Mycotox	ORGN	0283
Conformation-Dependent	Diels-Alder Dienophile Face Selectivity.=	ORGN	0241
The	Diels-Alder Reaction of α-Cyano-1-Azadienes.=	ORGN	0234
Diastereofacial Selectivity in the	Diels-Alder Reaction of 2,3-Dialkoxy-1,3-Butadie	ORGN	0233
Strategy.= Transannular	Diels-Alder Reaction: A Powerful Synthetic	ORGN	0288
Intramolecular	Diels-Alder Reactions of Vinylsulfonamides.=	ORGN	0231
reoselectivity in Electrophilic and	Diels-Alder Reactions.= +ectronic Control of Diaste	ORGN	0240
um Chemistry with Experiment.	Diene Excited-State Energetics, Structures, and Dyn	PHYS	0075
	Diene Synthesis by Solid Acid Catalysts.=	PETR	0029
ortho-Isotoluenes and	Diene-Allenes.= Efficient Synthesis of	ORGN	0098
ihydroxy-3,20-Dioxo-Pregna-1,4-	Diene-6-Carboxylate.= +of Methyl 11β, 17α, 21-Tr	MEDI	0170
tion Radicals Generated from 1,3-	Dienes.= +ition Reactions of Allyl Cations and Ca	ORGN	0123
ent Diels-Alder	Dienophile Face Selectivity.= Conformation-Depend	ORGN	0241
Comprehensive Analysis of	Diesel Fractions.=	PETR	0148
Methods for Gasoline and	Diesel Fuel Contamination.= Soil Gas Survey	ENVR	0053
atics, Sulfur, and Hydrocarbons in	Diesel Fuels by Gas Chromatography.= +of Arom	PETR	0118
Highly Potent, Orally Active	Diester Macrocyclic Human Renin Inhibitors.=	MEDI	0123
nt of a Tro+ Caloric Variation of	Diet in the Dominant Benthic Community Compone	BIOL	0084
istribution of Phytic Acid, Total	Dietary Fiber, Protein, and Starch in a Milled Hard	AGFD	0143
s of Experimental Colonic Neop+	Dietary Quercetin (QU) and Rutin (RU) as Inhibitor	AGFD	0114
Short-Range Solvation by Cyclic	Diethers.=	PHYS	0308
Studies of Di-t-Butylphosphino(	diethyl)borane.= Synthesis and Film-Growth	INOR	0230
Microsomal Metabolites of N,N-	Diethylbenzamide by ^{13}C NMR Spectroscopy.= +er	AGRO	0026
thiative Sulfonamidation of N,N-	Diethylbenzamides.= +Saccharins through the o-Li	ORGN	0035
s in the Presence of Sodium N,N-	Diethyldithiocarbamate.= +antalum Pentachloride	INOR	0092
Decomposition of	Diethylsilane on Silicon Surfaces.= Adsorption and	COLL	0008
valuacion de la Corrosion de las	Diferentes Etapas del Lavado Quimico de Una Calde	PHYS	0257
Factors 4F and (ISO)4F Interact	Differently with Oligoribonucleotide Analogs of Rabb	BIOL	0128
f+ Solution Structure of Insulin	Differs from the Crystal State but Spans the Range o	BIOL	0002
X-ray	Diffraction and Magnetic Anomaly in $Ba_2MnCu_3O_x$.=	INOR	0257
Fluorides.= X-ray Powder	Diffraction Studies of Polymeric Triorganotin(IV)	INOR	0065

engths as Determined by X-ray	Diffraction: No Evidence for Bond-Stretch Isomerism	INOR	0088
O6 by Neutron and X-ray Powder	Diffraction.= +ural Studies of Li4NpO5 and Li5Np	NUCL	0068
Studied by+ Surface Dynamics of	Diffusion and Corrosion at Atomic Layer Resolution	PHYS	0127
polymer Solutions.=	Diffusion and Dynamics in Microstructured Block Co	POLY	0275
ce Studied by Scannin+ Surface	Diffusion and Phase Transition on the Ge(111) Surfa	PHYS	0099
Gels.= Probe	Diffusion and Self-diffusion in Thermoreversible	POLY	0198
adical Polymerization+ Monomer	Diffusion and the Kinetics of Propagation in Free-R	POLY	0165
ends, and Interfaces—III. Lateral	Diffusion at the Air-Water Interface.= +lutions, Bl	COLL	0051
ers: A Photon Correlation Spec+	Diffusion Average Molar Mass of Polydisperse Polym	POLY	0305
ed-Enhanced+ Measurement of	Diffusion Coefficients in Polymer Melts Using Surfac	POLY	0088
Fluids. Micelle and Microemulsion	Diffusion Coefficients.= +in Microheterogeneous	COLL	0040
lume Theory for Polymer-Solvent	Diffusion Coefficients.= +Capabilities of a Free-Vo	POLY	0144
CH4 in a Hot-Wire	Diffusion Column.= Dehydrogenative Coupling of	PETR	0012
lear Aromatic Hydrocarbons in	Diffusion Flames.= Modeling the Growth of Polynuc	FUEL	0109
ke Carbon. Hydrogen Binding and	Diffusion in Diamond.= +amond and Diamond-Li	FUEL	0036
Reaction-Convection-	Diffusion in Emulsion Voltammetry.=	COLL	0042
hol/Vinyl Bu+ Gas Sorption and	Diffusion in Hydrogen-Bonded Polymers: Vinyl Alco	POLY	0191
.=	Diffusion in Polymers by Forced Rayleigh Scattering	POLY	0164
A Molecular Interpretation of	Diffusion in Polymers.=	POLY	0168
muir-Blodgett Films.=	Diffusion in the Gas Sensing of Phthalocyanine Lang	POLY	0118
rs.= Role of Activated	Diffusion in the Gasification of Porous Carbons/Cha	FUEL	0010
H2O	Diffusion in the H2O-D2O-Gelatin System.=	PHYS	0216
Probe Diffusion and Self-	diffusion in Thermoreversible Gels.=	POLY	0198
ment-B+ Determining Whether	Diffusion Is the Leaching Mechanism of Portland-Ce	ENVR	0011
Grown Silic+ Surface and Bulk	Diffusion of Adsorbed Nickel on Ultrathin Thermally	COLL	0177
	Diffusion of Alkyl Adipate Plasticizers in PVC.=	POLY	0067
	Diffusion of Multicomponent Species across Polymer	POLY	0195
tional Spectroscopies to Probe the	Diffusion of Small Molecules in Polymer Films.=	POLY	0332
	Diffusion of Star Polymers in Solutions and Gels.=	POLY	0267
	Diffusion of Water into a Photopolymer Film.=	POLY	0194
mer Blends.=	Diffusion of Water Vapor in Hydrogen-Bonded Poly	POLY	0192
Influence of Plasticizers on the	Diffusion Process of Water in Poly(methyl Methacry	POLY	0193
rescence Techniques to the Stu+	Diffusion Processes in Polymers. Application of Fluo	POLY	0163
iffusion in Hydrogen-Bonded +	Diffusion Processes in Polymers. Gas Sorption and D	POLY	0191
lar Architecture on Gas Transp+	Diffusion Processes in Polymers. Influence of Molecu	POLY	0115
alysis of Gas Permeation throu+	Diffusion Processes in Polymers. Thermodynamic An	POLY	0139
Stress Effects on	Diffusion through Solid Polymers.=	POLY	0140
End Distance Distributions and	Diffusion within Macromolecules Observed by Intram	PHYS	0061
Scaling Properties in Anomalous	Diffusion.=	PHYS	0130
Sensitized by Colloidal CdS: A+	Diffusion- and Activation-Controlled Photoreactions	COLL	0150
: Reactant Segregation and An+	Diffusion-Controlled Reaction Kinetics in a Capillary	PHYS	0273
perfections on the Efficiency of	Diffusion-Controlled Reactive Processes on Molecula	PHYS	0063
nce Techniques to the Study of	Diffusional Processes in Compatible Polymer Films.=	POLY	0163
Zeolites.=	Diffusivity of Hydrogen and Methane Molecules in A	CATL	0024
Activity of 7',7'-	Difluoroabscisic Acid.= Synthesis and Biological	AGRO	0017
olecular Graphics Te+ Study of	Difluoroacetylene Using a New Volume-Rendering M	COMP	0055
icomponent Polymer Systems by	Digital Image Analysis and Real-Time Pulsed NMR.	PMSE	0099
etition b+ Reduction of Vicinal	Dihalides by Photogenerated Ru(I) Complexes: Comp	INOR	0284
Nickel Boride Reduction of 2,6-	Dihalo-9-Oxabicyclononanes and Related Oxaadama	ORGN	0246
thrina poeppigiana+ Presence of	Dihydro-Erythroidines in the Milk of Goats Fed Ery	AGFD	0129
erocyclic 4(1H)Quinazolinone-2,3-	Dihydro-2-Thioxo.= +tion Compounds of the Het	INOR	0079
valuation of 1-Aminoethyl-3,4-	Dihydro-5-Hydroxy-1H-2-Benzopyrans as Dopamin	MEDI	0127
ent Synthesis of 9,9-Dialkyl-9,10-	Dihydroanthracenes.= +ted Alkylations: A Conveni	ORGN	0087
Microbial Hydroxylation of	Dihydroartemisinin Derivatives.=	BIOT	0170
imido)propyl]-1,6-hexanediamine	Dihydrochloride.= +imethyl-N,N'-bis[3-(2-Phthal	MEDI	0140
ihydrodiol of the Mutagen 15,16-	Dihydrocyclopenta[a]Phenanthren-17-one.= +,4-D	ORGN	0118
enes: Synthesis of the trans-3,4-	Dihydrodiol of the Mutagen 15,16-Dihydrocyclopent	ORGN	0118
of pKa Values of Isostructural η2-	Dihydrogen Complexes: Fe < Os < Ru.= +ve Order	INOR	0380
f Some 3-Substituted Morphine,	Dihydromorphine, and N-Methyl-Morphinium Anal	MEDI	0149
Stereoselective Recognition of 1,4-	Dihydropyridines by Their Receptor.= +ion of the	MEDI	0100
l Asymmetric Synthesis of 15,20-	Dihydrosecodine, Tetrahydropresecamine, and Tetra	ORGN	0009
d Copolyesters Based on the 4,4'-	Dihydroxy-α-methylstilbene Mesogen.= +esters an	POLY	0031
lopropanations: Synthesis of 1,25-	Dihydroxycholecalciferol.= +m Metal-Induced Cyc	ORGN	0284
cs and Photochemistry of Ru(II)-	Diimine Complexes in Mixed Solvents.= +otophysi	INOR	0279
ate Oxidation of Ternary Iron(II)-	Diimine-Cyanide Complexes.= +ics of Peroxodisulf	INOR	0022
sed (μ-Hydroxo)bisμ-carboxylato)	diiron(III) Kernals.= +e Structure Consisting of fu	INOR	0094
f Diphenyl Sulfoxide by Lithium	Diisopropylamide: A Simple Route to Dibenzothioph	ORGN	0084
Reactions of Meta-	Diisopropylbenzene on Acidic Molecular Sieves.=	CATL	0014
δ-	Diketones: Potential Synthons for Heterocycles.=	ORGN	0032
Access: The Library	Dilemma.= Information Ownership vs. Unrestricted	CINF	0045
Imagination, and	Diligence? Be Your Own Boss.= Have Intelligence,	SCHB	0016
V+ Preparation and Reaction of	Dilithio-2,4-Oxazolidinedione with α-Haloketones: A	ORGN	0223
Characterization of	Diltiazem Metabolites.= Synthesis and	MEDI	0129
Nonreactive	Diluent.= Antiplasticization of Polyimides by a	PMSE	0127
Local Polymer Dynamics in	Dilute Solution.= Optical and NMR Studies of	PHYS	0139
aste to Ethanol and Electricity via	Dilute Sulfuric Acid Hydrolysis.= +nversion of W	FUEL	0094
by Aroma Extract	Dilution Analysis.= Characterization of Off-Flavors	AGFD	0097
Versatile Zymark Robot	Dilution System.=	TECH	0005
ent Solubility Parameters: A New	Dimension.= +Honoring K. L. Hoy. Four Compon	PMSE	0033
rono-Chromatography, a Three-	Dimensional Analysis of the Volatile Constituents of	AGFD	0022

nching–Reaction Kinetics in Low–	Dimensional and Fractal Lattices.= +ing– and Que	PHYS	0307
parison and Evaluation of Two–	Dimensional and Three–Dimensional Measures of M	COMP	0023
ectron States in Two– and Three–	Dimensional Bounded Crystal Lattices. +ate One–El	BIOT	0028
Environment for Map+ Three–	Dimensional Chemical Structure Handling. Modeling	CINF	0029
ng the Hits from a 3D+ Three–	Dimensional Chemical Structure Handling. Prioritizi	CINF	0020
w Modifiaction of the Quasi–One–	Dimensional Compound KMo_4O_6.= +erties of a Ne	INOR	0359
Two–	Dimensional Electron Spin Resonance.=	PHYS	0114
stigations of Z+ One– and Two–	Dimensional High–Resolution Solid–State NMR Inve	PHYS	0348
Two–	Dimensional Infrared (2D IR) Spectroscopy.=	PHYS	0117
An+ Energy Transport in a One–	Dimensional Inorganic Solid: Exciton Trapping and	PHYS	0131
e Linear β–$CsBiS_2$ and the Three–	Dimensional $K_2Bi_8Se_{13}$.= +ogenide Synthesis of th	INOR	0202
ion between Structure and Low–	Dimensional Magnetic Interactions in Second–Row T	INOR	0267
n of Two–Dimensional and Three–	Dimensional Measures of Molecular Similarity.=	COMP	0023
Two–	Dimensional Microwave Spectroscopy.=	PHYS	0115
ssing Products.= 1– and 2–	Dimensional NMR Analyses of Coal–Oil Hydroproce	PETR	0149
tinum(II) Nucl+ One– and Two–	Dimensional NMR and Kinetics of Formation of Pla	INOR	0180
roteins D+ ^{31}P One– and Two–	Dimensional NMR of Duplex Oligonucleotides and P	INOR	0378
f Catalytic Reactions Using Two–	Dimensional NMR Techniques.= +and Products o	PETR	0060
e of Atoms on Re(0001) S+ Two–	Dimensional Phases of Sulfur Formed by Coalescenc	PHYS	0124
Two–	Dimensional Polymers.=	POLY	0276
ion and Characterization of Two–	Dimensional Protein Crystals.= +tronics. Preparat	BIOT	0121
rom Naja naja siamensis by Two–	Dimensional Proton NMR.= +tion of Cytotoxin D f	BIOL	0044
py (NQR–COSY).= Two–	Dimensional Pure Quadrupole Correlation Spectrosco	PHYS	0277
ound: Syn+ A Novel Quasi–Low–	Dimensional Reduced Niobium Oxophosphate Comp	INOR	0031
olymer Characterization. Two–	Dimensional Separations by LC–CZE.= Rapid Biop	ANYL	0126
Design, Synthesis, and Three–	Dimensional Structural Characterization of a Constra	MEDI	0165
tors of HIV-1 Protease by Three–	Dimensional Substructure Searching.= +vel Inhibi	COMP	0038
Low–	Dimensional Systems.= Relaxation and Reactions in	PHYS	0132
activation of Pt Sites within One–	Dimensional Zeolite Channels.= +lite: Inhibited De	CATL	0015
um Honoring W. T. Wipke: New	Dimensions in Chemical Information. Award Address	CINF	0016
ionati+ Optimization of Channel	Dimensions Used in Sedimentation Field–Flow Fract	PMSE	0028
Monomer and	Dimer Complexes with La and Nd Porphyrins.=	INOR	0221
w Variable+ Ordered Buckled–	Dimer Configurations on Si(100) Observed with a Ne	PHYS	0185
Site–Specific Effect of Thymine	Dimer Formation on A–Tract Bending and Its Biolog	BIOL	0114
h MR_2 (M =+ Rhodium Hydride	Dimer: Fragmentation–Recombination Reactions wit	INOR	0383
Valence [Ni(II)–NI(III)] Thiolate	Dimer.= +acterization, and Reactivity of a Mixed–	INOR	0060
d Theoretical Study of a Caffeine	Dimer.= +hesis, Spectroscopic Characterization, an	INOR	0252
ganese Aggregat+ Support for a	Dimer–of–Dimers Model for the Photosystem II Man	INOR	0408
haracterization of Monomeric and	Dimeric Arsino– and Phosphinogallanes.= +and C	INOR	0308
d An+ Extended Type 1 Chain (	Dimeric Le^a or Le^b/Le^a) as Human Tumor–Associate	CARB	0060
ects of Electronic Excitation and	Dimerization of Metals on Product–State Distributio	PHYS	0350
Catalysts.= Ethylene	Dimerization over Highly Dispersed Metal	PETR	0010
ranuclear+ Alkyne Coordination,	Dimerization, and Scission on Tungsten–Iridium Tet	CATL	0008
Catalyzed by Co(III) an+ Olefin	Dimerizations, Oligomerizations, and Polymerizations	ORGN	0129
ted Molybdenum(III)–μ–Sulfido	Dimers Formed by Sulfur Atom Abstraction from SP	INOR	0271
gregat+ Support for a Dimer–of–	Dimers Model for the Photosystem II Manganese Ag	INOR	0408
tal Hydrides. Palladium Hydride	Dimers: Unusual Coordination Behavior of $LiBEt_4$.=	INOR	0379
) Complexes Derived from Cu–Co	Dimers.= +boxylation Reactions Involving Copper(I	INOR	0213
Di–μ–oxo	Dimers.= Formation Pathway and Stability of	INOR	0185
Copolymerization of 1–Allyl–2,5–	dimethoxybenzene with Acrylonitrile.=	POLY	0050
ties of Poly(Methyl Acrylate–co–	Dimethyl Aminoethyl Methacrylated)/Epoxy Resin I	PMSE	0079
nate (DMSP)+ Measurements of	Dimethyl Sulfide (DMS) and Dimethylsulfoniopropio	GEOC	0002
M_2 Receptor Activity of N,N'–	Dimethyl-N,N'-bis[3-(2-Phthalimido)propyl]-1,6-he	MEDI	0140
]Acridine 5N+ Synthesis of 7,7–	Dimethyl–Quino[2,3,4–kl]–5a,6,7,8–Tetrahydrobenz[c	ORGN	0198
on during Thermolysis of Bis(2,4–	Dimethyl–1,3–pentadienyl)ruthenium.= +Formati	INOR	0305
Free Radical Grafting of 2–(	Dimethylamino)ethyl Methacrylate to Squalane.=	POLY	0019
of Organoaluminum Adducts of (	Dimethylaminomethyl)ferrocene.= +raphic Studies	INOR	0274
One–Pot (Gem–	Dimethylation of the Carbonyl Group.=	ORGN	0222
zole Polymer with Reactive 2,6–	Dimethylphenoxy Pendent Groups: Processing, Mech	PMSE	0019
zole Polymer with Reactive 2,6–	Dimethylphenoxy Pendent Groups: Synthesis and Ch	PMSE	0018
ectral Chara+ 1–Acetamido–2,x–	Dimethylpyridinium N–Betaines: Preparation and Sp	ORGN	0203
taining the Cationic Ligand N,N–	Dimethylquaterpyridinium, Me_2qpy^{2+}.= +plex Con	INOR	0233
ology for Cyclic and Linear Poly(	dimethylsiloxane) and Poly(methylphenylsiloxane) B	POLY	0234
ect of Molecular Weight of Poly(	dimethylsiloxane) on Physical Properties of Cured E	PMSE	0174
sis and Characterization of Poly(	dimethylsiloxane–co–methylphenylsiloxane) Random	POLY	0020
s of Dimethyl Sulfide (DMS) and	Dimethylsulfoniopropionate (DMSP) in Mesoscale C	GEOC	0002
of the Dipeptide Amide L–2,6–	dimethyltrysosyl–N–(3–Phenylpropyl)–D–Alaninamid	MEDI	0152
SNA on 2,4–	Dinitro Chlorobenzene, Part II.=	CHED	0170
of an ELISA+ Synthesis of 4,4'–	Dinitrocarbanilide–Hydrazone and the Development	BIOT	0166
onding of Early Transition–Metal	Dinitrogen Complexes.= +2) vs. End–on (μ–N_2) B	INOR	0362
t Based on the Reduction of 3,5–	Dinitrosalicylate in Determining the α–Amylase Acti	BIOL	0111
istry: Direct Observation of 1,2–	Dinitrosobenzene by Steady–State Spectroscopy—A	ORGN	0168
of Small Molecules to mono– and	Dinuclear Copper(I) Complexes.= +dative Addition	INOR	0044
Mechanism of Assembly of the	Dinuclear Iron Center: Tyrosyl Radical Cofactor of+	BIOL	0022
the Catalase–Like Reactivity of	Dinuclear Manganese Complexes.= Understanding	INOR	0010
Models.= Synthesis of	Dinucleating Ligands Directed toward Urease	INOR	0089
ated Medium.= Synthesis of	Dioctahedral 2:1–Layer Silicates in Acid and Fluorin	PETR	0128
Normal Bonded Phase HPLC with	Diode Array Multichannel Detection.= +Peaks in	ANYL	0087
Undergradu+ Incorporation of a	Diode Array UV–Visible Spectrophotometer into the	CHED	0180
the Photodissociation+ Tunable	Diode Laser Probe of Chlorine Atoms Produced from	PHYS	0299

h O(¹D).=	Diode Laser Probing of N₂O Following Collisions wit	PHYS	0278
in Animal Tissues by HPLC with	Diode–Array Detection.= +tion of Sulfamethazine	AGFD	0103
1,3–	Diol Complexes.= Synthesis of Cyclic Vanadium(V)	INOR	0409
erically Pure Bay–Region 10,11–	diol–8,9–Epoxide Diastereomers of Carcinogenic Dibe	ORGN	0208
Oxidation of 1,2–	Diols by Fe(VI): A Kinetic Study.=	INOR	0401
te Esters from Biologically Active	Diols.= +tion of Polyphosphate and Polyphosphona	PMSE	0015
nthesis of 3,3'–bisquinazolin–4,4'–	Diones and Their Antibacterial Activity.= +cles: Sy	MEDI	0057
of 1,3–	Diones.= AM1 Studies of the Nucleation of Crystals	PHYS	0247
Azetidine–2,4–	Diones.= Facile Synthesis of	ORGN	0034
Butyl–5–(tert–Butylsulfonyl)–1,3–	Dioxane.= +onsible for the Eclipsing in cis–2–tert–	ORGN	0150
silyl Reagents Using the 1,3,2λ–⁵–	Dioxaphospholane Methodology.= +with Trimethyl	ORGN	0051
droformylation with Cationic Bis(	dioxaphospholane)rhodium Complexes.= +alytic Hy	INOR	0154
ized Adamantyl–Substituted 1,2–	Dioxetanes: Improved Enzyme–Activated Chemilumi	BIOT	0167
Supercritical Carbon	Dioxide Conditions.= Enzyme Inactivation in	AGFD	0028
on Barriers at Colloidal Titanium	Dioxide Surfaces.= +s of Electron–Transfer Activati	COLL	0132
ltrathin Thermally Grown Silicon	Dioxide.= +Bulk Diffusion of Adsorbed Nickel on U	COLL	0177
nzothiazine–3–Carboxamides, 1,1–	Dioxide.= +5–Substituted–3–Isoxazolyl)–2H–1,2–Be	MEDI	0178
erization of Protonated trans Bis–	Dioxime Ruthenium Complexes.= +uctural Charact	INOR	0066
an Tissue Sampling Followed by	Dioxin Analysis to Estimate Body Burden in Genera	CEI	0012
nated Polychlorinated Dibenzo–p–	Dioxins and Dibenzofurans.= +ash for Monobromi	ANYL	0050
hyl 11β, 17α, 21–Trihydroxy–3,20–	Dioxo–Pregna–1,4–Diene–6–Carboxylate.= +of Met	MEDI	0170
nships of Enantiomerically Pure	Dioxolane–Pyrimidine Nucleosides as Potential Anti–	MEDI	0051
New Copper–	Dioxygen Complexes with Thermal Stability.=	INOR	0046
Ligands.= New Copper–	Dioxygen Species (Cu₂–O₂) Using Imidazole	INOR	0045
r(I) Complexes.=	Dioxygen–Binding Properties of Mononuclear Coppe	INOR	0178
al Methods for Disposal of Cattle–	Dip Waste.= +cals and Their Containers. Biologic	AGRO	0139
N–Alkylated Derivatives of the	Dipeptide Amide L–2,6–dimethyltrysosyl–N–(3–Phen	MEDI	0152
Interaction of Pt(II) and Pd(II)	Dipeptide Complexes with Nucleic Acid Constituents	INOR	0412
n to the Synthesis of Fluoroolefin	Dipeptide Isosteres.= +Peterson Olefination Reactio	ORGN	0079
mplexes of Pt(II) and Pd(II) with	Dipeptides and Substituted Pyrimidines.= +and Co	INOR	0411
ers Derived from α–Amino Acids,	Dipeptides, and Tripeptides.= +–Containing Polym	PMSE	0014
ated Triphenylmethyl Cation with	Diphenydiazomethane or Diphenylketene.= +euter	ORGN	0145
Sim+ Dehydrative Cyclization of	Diphenyl Sulfoxide by Lithium Diisopropylamide: A	ORGN	0084
ion with Diphenyldiazomethane or	Diphenylketene.= +euterated Triphenylmethyl Cat	ORGN	0145
hesis.+ New Applications of Bis(	Diphenylphosphino)methane in Organometallic Synt	INOR	0375
for Isomerization o+ Isopentenyl	Diphosphate Isomerase: An Electrophilic Mechanism	BIOL	0014
n of the Kinetic Fate of Thiamin	Diphosphate–Bound Enamines on Brewers' Yeast Py	BIOL	0048
e Determination of the Thiamin	Diphosphate–Dependent Enzyme Brewers' Yeast Pyr	BIOL	0032
Olefins.= 1,3–	Dipolar Cycloaddition Reactions with Captodative	ORGN	0230
Small–Angle Scattering and Direct	Dipolar Energy Transfer in Porous Solids.= +ith	PHYS	0060
Reaction Pathways Involving the	Dipolar Form.= +ter of Trimethylenemetanes: New	ORGN	0146
matic Substitution Via Aryl 1,2–	Dipolar Synthons: Nickel(O)–Catalyzed Cross–Coupl	ORGN	0082
rate MRCI State and Transition	Dipole Moments for the Nine Lowest Lying Electron	PHYS	0304
es: New Reaction Pathways Inv+	Dipole vs. Diradical Character of Trimethylenemetan	ORGN	0146
g the Dipolar Form.= Dipole vs.	Diradical Character of Trimethylenemetanes: New R	ORGN	0146
rypsin by N–Nitr+ Active–Site–	Directed Enzyme–Activated Substrate Inhibition of T	ORGN	0022
Metal Complexes with Inwardly	Directed Functional Groups.= Cofacial Binuclear	INOR	0242
and α–Alkenyllactams.= Amide–	Directed Hydrocarbonylation of N–Alkenylamides	INOR	0227
Redirected	Directed Lithiation, III.=	ORGN	0085
P Hydroxymethylase.= Site–	Directed Mutagenesis of the Gene–Encoding T4 dCM	BIOL	0160
unosorbents Prepared by Site–	Directed or Random Coupling of Monoclonal Antibod	BIOT	0060
cs: Application in N+ Combined	Directed ortho Metalation and Cross–Coupling Tacti	ORGN	0277
a Versatile New Metalation Dir+	Directed ortho Metalation: The O–Thiocarbamate as	ORGN	0083
Molecular–Recognition–	Directed Supramolecular Polymer Architectures.=	POLY	0299
Vaccines.= Study	Directed Toward the Development of Synthetic	CARB	0055
Isolaurepinnacin.= Studies	Directed Toward the Total Synthesis of	ORGN	0008
Dinucleating Ligands	Directed toward Urease Models.= Synthesis of	INOR	0089
Discovery of Drug Leads. Light–	Directed, Spatially Addressable Parallel Chemical Sy	MEDI	0090
Chapman and Hall	Directionary of Drugs.= Experiences Building the	CINF	0033
Panel Discussion: Future	Directions in R D of Biodegradable Polymers.=	CHED	0136
Perspective.= Future	Directions in the Use of Pesticides in IPM: Federal	AGRO	0068
te as a Versatile New Metalation	Director for the Synthesis of Polysubstituted Aromat	ORGN	0083
between the Aglycone and the	Disaccharide.= Avermectin Analogs with a Spacer	ORGN	0016
l Application Profiles.= Some	Disaccharide–Derived Building Blocks with Industria	CARB	0003
(ES MS).= Linkage Analysis in	Disaccharides by Electrospray Mass Spectrometry	CARB	0048
d Films Synthesized by DC Arc	Discharge CVD in a Mixture Gas of Acetylene and H	FUEL	0038
kes Areas of C+ Achieving Zero	Discharge of Persistent Toxic Substances in Great La	ENVR	0083
igration of Activation Products in	Discharge Pool in T.H.O.R. Nuclear Research+ M	NUCL	0090
ion by Anesthetics and Alcohols:	Discrepancy between Core and Surface Responses.=	MEDI	0032
symmetric Hydrosilylation of +	Discrete Chiral Rhodium Phosphine Complexes for A	CATL	0005
cile Route to Ciral Alcohols Using	Discrete Rhodium Catalysts.= +n of Ketones: A Fa	INOR	0247
Roles in Environmentally Based	Disease.= Human Cytochromes P450 and Their	CEI	0021
for the Treatment of Alzheimer's	Disease.= +bstituted Derivatives of Physostigmine	MEDI	0137
Recientes de CAD/CAM para el	Diseno de Equipo de la Industria Quimica de Proceso	COMP	0053
PCVD of Silicon Carbide.=	Disilacyclobutanes as Single–Source Precursors for L	INOR	0105
Surface	Disinfection by Chemical and Physical Treatments.=	AGFD	0032
ne: Products and Implications for	Disinfection of Wastewater.= +ation of Phenylalani	ENVR	0031
te Anions with Cu,Zn–Superoxide	Dismutase as Probed by ⁵¹V NMR.= +on of Vanada	INOR	0184
a Degenerate Ground S+ Paired	Disorder and the Metallic State in Polymers Lacking	PHYS	0039
Measurement of Immune	Disorder in Persons Exposed to Organochlorines.=	CEI	0017

Dynamics of Frenkel Excitons in	Disordered Systems: T = 0 K.=	+in Analysis of the	PHYS 0034
nic Energy Transport in Spatially	Disordered Systems.=	+Ultrafast Studies of Electro	PHYS 0133
nd Wastewaters in Manufacture of	Disperse Blue 79.=	+ction of Toxic Components a	CMEC 0018
s in a Surface Layer of	Disperse Silica.=	Reaction of Phosphorus Compound	COLL 0100
f Charge-Transfer Complexes in	Disperse Systems.=	Adsorption and Luminescence o	COLL 0110
sicochemical Structures of Liquid	Disperse Systems.=	+trochemical Methods and Phy	COLL 0022
Ozone and Fine-	Dispersed Carbon Bodies.=	Interaction between	COLL 0109
Investigation of a Liquid Crystal	Dispersed in an Ionic Polymeric Membrane.=		PMSE 0119
n the Surface of Hiding Pigments	Dispersed in Water.=	+lymerization of Monomers o	PMSE 0038
Phase Separation in Polymer-	Dispersed Liquid Crystals.=		COLL 0014
over Highly	Dispersed Metal Catalysts.=	Ethylene Dimerization	PETR 0010
ature and Structure of Calcium	Dispersed on Carbon: XANES and EXAFS Study.=		FUEL 0032
e Electrochemical R+ Effects of	Dispersed Phases in Naturally Occurring Fluids on th		COLL 0005
Chloride Reactions with	Dispersed Silica Surface.=	Mechanisms of Metal	COLL 0219
plication of Matrix Solid-Phase	Dispersion (MSPD) to the Extraction and Subsequen		AGFD 0060
lymer Thin Films, 5: Wavelength	Dispersion of the $\chi^{(3)}$ of Poly(p-phenylene Benzobist		POLY 0213
f Heavy-Ion Optical Potential.=+	Dispersion Relation and the Low-Energy Behavior o		NUCL 0040
easuring the Sorption of Organ+	Dispersion-Corrected HPLC/FACP Technique for M		ANYL 0072
n Redox Catalysis Using RuO_2	Dispersions and the Electrochemical Model.=	Oxyge	COLL 0134
Relaxation.=	Dispersive Energy Transport: Relation to Spin Glass		PHYS 0082
al Relationship of Some 1,1'-bis(	Disphenylphosphino)ferrocene (dppf)-Bridged Silver		INOR 0106
of Biopharmaceuticals.=	Displacement Chromatography for the Manufacture		ANYL 0041
eparative Reversed-Phase Sample	Displacement Chromatography of Peptides.=	+Pr	ANYL 0040
stimates+ Mechanism of Alkane	Displacement from (Alkane)W(CO)$_5$ Intermediates: E		INOR 0111
-Diammine(guanosine)+ Rates of	Displacement of Guanosine by Nucleophiles from cis		INOR 0251
Novel	Displacement Systems for Bioseparations.=		ANYL 0038
quid Crystalline Polyelectrolytes	Displaying Thermotropic and Lyotropic Behavior.=		POLY 0073
Iron Indu+ Feasible Solutions for	Disposal and Reuse of Pickle Liquors from Mexican		ENVR 0039
Pesticide Donations and the	Disposal Crisis in Africa.=		AGRO 0044
rward: Pesticide Management and	Disposal in the Mid-80s and the Mid-90s.=	+ng Fo	AGRO 0086
Containers. Biological Methods for	Disposal of Cattle-Dip Waste.=	+cals and Their	AGRO 0139
Containers. Biolog+ Status of the	Disposal of Waste Agricultural Chemicals and Their		AGRO 0139
Containers. Curre+ Status of the	Disposal of Waste Agricultural Chemicals and Their		AGRO 0120
Containers. Intro+ Status of the	Disposal of Waste Agricultural Chemicals and Their		AGRO 0085
Containers. Pestic+ Status of the	Disposal of Waste Agricultural Chemicals and Their		AGRO 0102
State Pesticide	Disposal Regulations and Programs.=		AGRO 0106
Technologies for Pesticide Waste	Disposal: Overview.=	+Their Containers. Current	AGRO 0120
Containers. Introduction to Waste	Disposal.=	+ste Agricultural Chemicals and Their	AGRO 0085
Ash Utilization and	Disposal.=		FUEL 0135
i/Al Ratio and Coke+ Toluene	Disproportionation Reaction over Zeolites: Effect of S		PETR 0103
f Organic-Pigment Thin Films by	Disruption of Micelles.=	+trochemical Formation o	COLL 0019
Proteins Using Critical Fluid	Disruption.=	Downstream Processing of Periplasmic	BIOT 0070
he Femtosecond Timesc+ Energy	Dissipation in Liquid-State Chemical Reactions on t		PHYS 0085
Domain.= Aspects of	Dissipative Reactions from Low Energy to the Fermi		NUCL 0007
tivity Method: I. MOPS+ First	Dissociation Constant of Amino Acids by the Conduc		BIOL 0033
: Statistical vs. N+ Two-Photon	Dissociation of Ferrocene and Nickelocene at 248 nm		PHYS 0333
nide Chiral Analogues: A Possible	Dissociation of Pharmacological Activities.=	+Tocai	MEDI 0098
Electromagnetic	Dissociation with Relativistic Heavy Ions.=		NUCL 0049
Femtosecond Dynamics of Bi$_2$	Dissociation.=		PHYS 0369
m Soils and Sediments: Reductive	Dissolution by Ti(III).=	+Iron Oxide Extraction fro	ENVR 0058
of Time, pH, Carbonate/Bicarb+	Dissolution Kinetics of UO$_{2.00}$ Pellets as a Function		INOR 0236
Phosphoric Rock	Dissolution: An Experimental Report.=		PHYS 0294
Seasonal	Dissolved Gas Cycling in the Middle Atlantic Bight.=		GEOC 0003
from the Hudson Ri+ Fluxes of	Dissolved Inorganic Carbon and Reactor-Derived ^{14}C		GEOC 0014
Sediments+ Potential Impact of	Dissolved Organic Carbon (DOC) Flux out of Margin		GEOC 0017
olecules Observed+ End-to-End	Distance Distributions and Diffusion within Macrom		PHYS 0061
ion Techniques+ Comparison of	Distance Geometry and Molecular Dynamics Simulat		ORGN 0174
formationally Flexible Molecul+	Distance Geometry-Based Approach for Docking Con		CINF 0028
Descriptors Derived from the	Distance Matrix.= Comparative Study of Molecular		COMP 0025
Rotational Resonance:	Distance Measurements in Large Proteins.=		PHYS 0196
Fixed	Distances.= Intramolecular Electron Transfer at		PHYS 0077
s in the Catalytic Two-Stage Li+	Distillate Selectivity and Quality of Bituminous Coal		CATL 0053
or Hydrocracking of Coal-Derived	Distillates.= +lysis of Metal-Modified Y-Zeolites f		CATL 0050
User-Friendly Program for	Distillation Column Design.= DESTIL—A		CHED 0122
Soil Gas Analysis in in Situ Steam	Distillation.= +vective-Flux Probes for Enhanced		ENVR 0052
rmal Brine Utiliz+ Production of	Distilled Water and Chemical Products from Geothe		I&EC 0010
ihydrogen Complexes: Fe < Os+	Distinctive Order of pK$_a$ Values of Isostructural η_2-D		INOR 0380
oover and Georgius Agricola:	Distorting Mirrors of History.= +Corp.). Herbert H		HIST 0011
bserved+ End-to-End Distance	Distributions and Diffusion within Macromolecules O		PHYS 0061
Hydrocarbons.= Eigenvalue	Distributions for the Graphs of Alternant		COMP 0029
Measurements of Atomic Cluster	Distributions in Pseudobinary Semiconductor Alloys.		PHYS 0375
Importance for Flavo+ Isotope	Distributions in Secondary Plant Products and Their		AGFD 0194
ous and Heterogen+ Recovery of	Distributions of Fluorescence Lifetimes in Homogene		PHYS 0134
issociation of+ Internal Energy	Distributions of SO (X$^3\Sigma^-$) from the 193-nm Photod		PHYS 0240
zation of Metals on Product-State	Distributions.= +Electronic Excitation and Dimeri		PHYS 0350
odal Electrophoretic Mobility	Distributions.= Occurrence and Importance of Polym		COLL 0202
ive Processing. Plenary Lecture.	Distributive Mixing and Energy Distribution in Twin		PMSE 0175
ynthesis of 3-Substituted and 1,3-	Disubstitued Pyrroles.= +Facile Method for the S		ORGN 0196
Face Selectivity in 5-,7-	Disubstituted Adamantanones.=		ORGN 0238
upling of a Zi+ Synthesis of 3,4-	Disubstituted Benzofurans via the Intramolecular Co		ORGN 0217

Carbon	Disulfide.= Gelation of Atactic Polystyrene in	POLY	0148
anic Precursor from Mexican S+	Diterpenoids with New Skeletons Related to a Clerod	ORGN	0280
VI Carbonyl Complexes of a New	Ditertiary Phosphine Ether.= +eactions for Group	INOR	0333
ire Arrays.= Solids and Films.	Dithiocarbamates: Reagents for Selective Surface De	INOR	0350
Oxo Complexes of Amine–Amide–	Dithiol Ligands.= +chnetium(V) and Rhenium(V)	INOR	0181
= Reactions of Molybdenum	Dithiolenes Modeling the Reaction Center of Mo–co.	INOR	0404
ivity of Phenylene and Pyridylene	Diurea Derivatives as ACAT Inhibitors.= +mic Act	MEDI	0109
agonist with Renal Protective and	Diuretic Activities.= +Selective Adenosine A_1 Ant	MEDI	0112
dress. From Juvenile Hormones to	Diuretic Activities.= +Insect Control. Award Ad	AGRO	0001
ene Derivatives in the Presence of	Divalent Metal Ions.= +ylboronate and Isopropylid	INOR	0263
Industry.=	Diversification for Financial Survival in the Fertilizer	FERT	0010
Source of Chemical	Diversity.= Peptide Expression Libraries as a	MEDI	0092
rolyzable Dianionic Initiator from	Divinyl Tetraoxaspiroundecane Derivative.= +Hyd	POLY	0349
r NMR Experiments into Lower	Division Curriculum.= Incorporation of Multinuclea	CHED	0186
e in Teac+ Award Address (ACS	Division of Analytical Chemistry Award for Excellenc	ANYL	0006
Instrume+ Award Address (ACS	Division of Analytical Chemistry Award in Chemical	ANYL	0011
emistry, s+ Award Address (ACS	Division of Analytical Chemistry Award in Electroch	ANYL	0096
emical A+ Award Address (ACS	Division of Analytical Chemistry Award in Spectroch	ANYL	0001
ROGRESS: An Initiative of the	Division of Chemical Education of the American Che	CHED	0110
Scientific	Division of Routledge, Chapman, and Hall.=	CHAS	0028
Models for the Ene-	Diyne Toxins with an Enol–Ether Trigger.=	ORGN	0102
ate in a Latent Z–Hex-3-ene-1,5-	diyne Unit.= +a the Elimination of a Thionucleara	ORGN	0096
ith GPC–Viscometry in THF and	DMF.= +zation of Polymer MWD and Branching w	PMSE	0068
surements of Dimethyl Sulfide (	DMS) and Dimethylsulfoniopropionate (DMSP) in M	GEOC	0002
nd Dimethylsulfoniopropionate (	DMSP) in Mesoscale Coccolithophore Blooms in the	GEOC	0002
	DNA Activation of Nae I Endonuclease.=	BIOL	0125
Selective Agents for	DNA Alkylation.= Naphthoquinone Derivatives as	ORGN	0190
Mechanism of O^6-Alkylguanine–	DNA Alkyltransferase: Effect of Substrate Heteroa	BIOL	0161
Laboratory.= Recombinant	DNA Analysis in the Undergraduate Biochemistry	CHED	0090
errelationship between Forensic	DNA Analysis Research and Casework in the Royal C	CHAL	0007
l Ligand.= Cleavage of	DNA by a Chiral Ru(II) Complex with a Bifunctiona	INOR	0232
Efficient Oxidative Cleavage of	DNA by Oxoruthenium(IV).=	INOR	0410
Determinants in the Oxidation of	DNA by Permanganate.= Fundamental	BIOL	0121
Birefringence.= Kinetics of	DNA Cleavage by Calicheamicin from Electric	POLY	0280
plex Oligonucleotides and Proteins	DNA Complexes.= +wo-Dimensional NMR of Du	INOR	0378
ructure and Dynamics of Ligand–	DNA Complexes.= +-State NMR Studies of the St	BIOL	0116
Formation and Oxidative	DNA Damage.= Carcinogen–Mediated Oxidant	CHAS	0008
Privacy Issues of	DNA Databanks and Databases.=	CHAL	0008
Critique of	DNA Evidence.=	CHAL	0002
ics of Ring DNA vs. Linear	DNA from Transient Electric Birefringence.= Dynam	POLY	0296
Mechanism of Action of the	DNA Helicase.=	BIOL	0117
DNA Technology: Update 1991.	DNA Identification Testing and Issues of Constitutio	CHAL	0001
	DNA in the Robert Golub Case.=	CHAL	0010
Sequence–Specific	DNA Interactions of a Tricationic Porphyrin.=	BIOL	0127
Dynamics by NMR.= Drug–	DNA Interactions: Probing Binding Modes and	PHYS	0197
ilvC Activator Protein:	DNA Interactions.=	BIOL	0124
and AFM.= Imaging	DNA Molecules under Potential Control by STM	PHYS	0157
Cube.= Synthesis from	DNA of a Molecule with the Connectivity of a	BIOL	0162
Enantioselective Recognition by	DNA of a Nonintercalating Ru(II) Complex Contai	INOR	0233
Mechanistic Studies on	DNA Photolyase.= Chemistry of Nucleic Acids.	BIOL	0155
udies on the Replicative Bypass of	DNA Photoproducts.= +Mutations: Mechanistic St	BIOL	0157
Inhibitors of	DNA Polymerases of Leishmania mexicana.=	BIOL	0118
nthesis for UvrABC Endonuclease	DNA Repair.= +n-Linked Deoxyoligonucleotide Sy	BIOL	0126
Studies of the T4	DNA Replication System.= Kinetic and Structural	BIOL	0159
nisms as Possible Can+ Specific	DNA Sequences and Their Complexes in Living Orga	BIOT	0148
Analytical Chemistry. High–Speed	DNA Sequencing in Ultrathin Gels.= +w Face of	ANYL	0033
Testing and Issues of+ Forensic	DNA Technology: Update 1991. DNA Identification	CHAL	0001
in	DNA Technology.= Patent Protection for Advances	CHAL	0006
pment and Use of Forensic	DNA Typing Systems.= Brief History of the Develo	HIST	0004
Legal Scrutiny of	DNA Typing Techniques.=	CHAL	0003
Alkylation of	DNA Using a Silyl Phenol.= Solvolysis–Triggered	ORGN	0191
ngence.= Dynamics of Ring	DNA vs. Linear DNA from Transient Electric Birefri	POLY	0296
Selective Binding and Cleavage of	DNA.= +ess Toward Understanding Its Sequence–	ORGN	0298
UV-Irradiated	DNA.= Determining Pyrimidine Photoproducts in	ANYL	0092
Structure and Association of	DNA.= Effects of High-Electric Fields on the	COLL	0179
Acridine Orange Compound with	DNA.= +of the Interaction of a Ferrocene–Linked	ANYL	0109
s-syn Cyclobutane Photodimer of	DNA.= +of the Replicative Bypass of the TpdU ci	BIOL	0113
Weight, Single–Stranded	DNA.= Laser Vaporization of High Molecular	ANYL	0130
Reactions at Aldehydic Sites in	DNA.= Mechanistic Studies of β–Elimination	BIOL	0158
onto	DNA.= Pseudografting of Vinyl Polymers	POLY	0301
ce.=	DNA-Based Paternity Testing: Five Years' Experien	CHAL	0009
Systems. Luminescence Probes of	DNA–Binding Interactions.= +sitive Metalorganic	INOR	0315
ognition Site of Sequence–Specific	DNA–Binding Proteins.= +thod to Define the Rec	BIOL	0129
GC/MS: Doing What Chemists	Do!.=	CHED	0152
t of Margin Sediments on Oceanic	DOC.= +Dissolved Organic Carbon (DOC) Flux ou	GEOC	0017
t of Dissolved Organic Carbon (	DOC) Flux out of Margin Sediments on Oceanic DOC	GEOC	0017
Phosphorus Chemistry. Arthur	Dock Fon Toy—Scientist, Colleague, Friend.=	INOR	0288
e Geometry–Based Approach for	Docking Conformationally Flexible Molecules from 2	CINF	0028
Chymotrypsin.=	Docking Ligands into Receptors: The Test Case of α–	COMP	0039
Molecular	Docking: 3D Databases. Recent Advances in	COMP	0036

Photochemical Study of N-(12- Dodecanoic Acid) Benzoylformamide in Homogeneou	ORGN	0166
sis and Computer Modeling of the Dog A_1 and A_2 Adenosine Receptors.= +nce Analy	MEDI	0114
r Electron and En+ How Are We Doing in Predicting Absolute Rates of Intramolecula	ORGN	0143
lty at the Two-Year College Are Doing to Help Advance Prehigh School Science Educ	CHED	0076
GC/MS: Doing What Chemists Do!.=	CHED	0152
Doing Your Market Research.=	SCHB	0002
Relationship Studies of Dolichos Biflorus Lectins.= Structure-Function	BIOL	0149
s in Disordered Systems+ Time- Domain Analysis of the Dynamics of Frenkel Exciton	PHYS	0034
etrahydrofolate Synthase.= Domain Chimeras in the Trifunctional Enzyme C_1-T	BIOT	0243
nergy Transfer Using Frequency- Domain Fluorometry.= +erved by Intramolecular E	PHYS	0061
l Reaction Coordinate: A Possible Domain for Dynamic Effects.= +ss to a Symmetrica	ORGN	0122
Structural Changes in Polymers. Domain Growth during the Spinodal Decomposition	PMSE	0104
Low Energy to the Fermi Domain.= Aspects of Dissipative Reactions from	NUCL	0007
Millisceond to Picosecond Time Domain.= Metalorganic Photochemistry in the	INOR	0287
d to Interact with the Catalytic Domain.= New Protein Kinase C Inhibitors Designe	MEDI	0076
omophore Systems.= Domains and Order in Langmuir-Blodgett-Type Chr	PHYS	0111
C+ Carbohydrate-Recognition Domains as Mechanisms for Decoding Information in	BIOL	0144
Caloric Variation of Diet in the Dominant Benthic Community Component of a Trop	BIOL	0084
ethods for Purification (WIMP)— Don't Be Fooled by the Name.= +munoaffinity M	CARB	0054
Pesticide Donations and the Disposal Crisis in Africa.=	AGRO	0044
Complex Compared with Their O- Donor Analogues.= +nate CuN_3OS Pseudothiolate	INOR	0262
ctronic Energy Transfer between Donor and Acceptor Adlayers on Single-Crystal Surf	PHYS	0065
g Tunneling Microscopy of Donor Graphite Intercalation Compounds.= Scannin	PHYS	0365
Complexation of a Mixed- Donor Ligand P_2S with Copper(I).=	INOR	0029
Reactions between S- Donor Ligands.= Extremely Rapid Au(I) Transfer	INOR	0235
N,S- Donor Ligands.= Reactions of Nickel(II) with	INOR	0405
Highly Vibrationally Excited NO_2 Donor.= +and Translational Energy Transfer from	PHYS	0334
Generation by σ-Bond-Separated Donor-Acceptor Molecules.= +Second-Harmonic	POLY	0222
onding: The Natural Bond Orbital Donor-Acceptor Perspective.= +cts in Hydrogen B	PHYS	0106
Chains.= Model Molecules with Donor-Acceptor Units Intermediated by Silicon	POLY	0211
y Attached Sensitizer/Mediator or Donor/Acceptor Complexes.= +erties of Covalentl	INOR	0038
of a Tripodal Ligand with Mixed Donors P_2S Toward Various Transition Metals.=	INOR	0028
osides as Novel, Versatile Glycosyl Donors. Selective Activation ove+ Phenylselenoglyc	CARB	0035
Versatile Glycosyl Donors.= 2-Oxo- and 2-Oximinoglycosyl Halides as	CARB	0025
Toward the Rational Design of Dopamine D1 Selective Agonists.= Approaches	MEDI	0124
Design and Synthesis of Dopamine D1 Selective Antagonists.=	MEDI	0125
5-Hydroxy-1H-2-Benzopyrans as Dopamine D1 Selective Ligands.= +l-3,4-Dihydro-	MEDI	0127
Acting Selectively at Dopamine Receptors.= Development of Agents	MEDI	0080
-Tetrahydronap+ Synthesis and Dopaminergic Activity of 2-Amino-1,4-Epoxy-1,2,3,4	MEDI	0128
quinolinones and Related Comp+ Dopaminergic and Serotonergic Activities of Imidazo	MEDI	0126
u Valence in the Hig+ Cation- Dopant Site Selectivity, Oxygen Stoichiometry, and C	PHYS	0027
Alkali-Metal- Doped Fullerenes.= Superconductivity in	PHYS	0046
nd Catalytic Activity of Platinum- Doped Glassy Carbon.= +hesis, Characterization, a	INOR	0312
Nonplanarity of Doped π Systems.=	PHYS	0233
etracene Crystals and Tetracene- Doped Polymeric Matrices.= +hase Transition in T	PHYS	0274
2 and Rare Earth/Alkaline Earth- Doped Pt, Rh/CeO_2 Catalysts.= +e of Pt, Rh/CeO	PETR	0035
tive Effect in Nonlinear Polymers Doped with Charge Transport Agents.= +otorefrac	POLY	0155
Silica Aerogels and Xerogels Doped with Transition-Metal Ions.= Studies on	INOR	0081
solution NMR of Solids: DAS and DOR.= +opies. Sample Reorientation for High-Re	PHYS	0370
i+ Unusual Hemiacetal from the Dorsal Abdominal Glands of the Australian Spined C	AGRO	0035
Encuentro de Dos Mundos a Traves de la Quimica.=	HIST	0035
High- Dose Dosimetry: Comparative Techniques.=	NUCL	0071
e, Bioavailability, and Bioeffective Dose.= +zardous Wastes: Assessment of Prevalenc	CEI	0008
High-Dose Dosimetry: Comparative Techniques.=	NUCL	0071
ompounds: Effect of Lipophilicity, Dosing Vehicle, and Prodrug Modification.= +ic C	MEDI	0196
ments in Efficient Frequency Doubling in Poled Polymer Films.= Recent Develop	POLY	0179
NLO Polymers for Frequency Doubling.=	POLY	0178
°From Doubts to Conviction of Fraud: Tales of Trauma.=	PROF	0003
nstrumentation, sponsored by the Dow Chemical Co. +Chemistry Award in Chemical I	ANYL	0011
Herbert H. Dow—Process of Extracting Bromine.=	CHAL	0029
Bioseparations— Downstream Processing for Biotechnology.=	BIOT	0224
g Critical Fluid Disruption.= Downstream Processing of Periplasmic Proteins Usin	BIOT	0070
tegrated Strategy for Improving Downstream Processing Yields for Production of Rec	BIOT	0071
Upstream Production with Downstream Purification Techniques.= Integrating	BIOT	0072
Separation Steps with Upstream- Downstream Stages. Integrat+ Integrating Primary	BIOT	0069
s(Disphenylphosphino)ferrocene (dppf)-Bridged Silver(I) Complexes.= +ome 1,1'-bi	INOR	0106
Electrophile.= [cpRu(dppm)(CH_2SO_2)]+: A Strong Organometallic	INOR	0249
ethod To Support Registration of DPX-E9636 Herbicide for Field Corn.= +HPLC M	AGRO	0030
m of Selectivity of the Herbicide DPX-E9636 in Corn: Uptake, Translocation, and Me	AGRO	0027
Comparative Metabolism of DPX-E9636 in Plants and Animals.=	AGRO	0028
Environmental Fate of DPX-E9636, a New Sulfonylurea Herbicide.=	AGRO	0029
of Adenosine Analogues Results in Dramatic Changes in Vasorelaxant Activity.= +on	MEDI	0113
Profiles, Pathways, and Dreams.= Future Chemists, Future Chemistry.	CONG	0004
Lipid Autoxidation of a Spray- Dried Milkbase for Baby Food.= Factors Affecting	AGFD	0158
ing Properties of Expanded Clays Dried Using a Supercritical Fluid.= +ace and Crack	PETR	0145
cs of Produce+ Design of a Mill Drier for Shrimp Head: Bromatological Characteristi	AGFD	0105
ics of Produc+ Design of a Mill- Drier for Shrimp Heads: Bromatological Characterist	ENVR	0046
thermal W+ Corrosive Effects on Drill Pipes by Drilling Fluids Systems Based on Geo	I&EC	0012
Sensitivity to Contamination of Drilling Fluids Systems Based on Geothermal Water	I&EC	0011
orrosive Effects on Drill Pipes by Drilling Fluids Systems Based on Geothermal Water	I&EC	0012

sess the Susceptibility of Surface	Drinking Water Supplies to Pesticide Contamination	AGRO	0053
Pesticides in	Drinking Water: A Farm Wife's Perspective.=	AGRO	0055
d Haloacetonitriles in Chlorinated	Drinking Water.= +acetic Acids, Halomethanes, an	ENVR	0028
Microbiological Contaminants in	Drinking Water.= Risk Assessments for	ENVR	0070
de Ion to the Reaction of Cumyl	Drivatives: Mechanism of Nucleophilic Substitution a	ORGN	0154
ormation of Citraconyl Phosphate	Driven by a Proton Gradient.= +for H+-ATPase: F	BIOL	0104
ational and Internatio+ Market-	Driven Pollution Prevention? Green Marketing: A N	CEI	0028
nt to+ Incin: A Flexible Chemkin	Driver Program for Investigating Chemistry Importa	I&EC	0024
= Chemical Principles as a	Driving Force in the Design of Proteinase Inhibitors.	ORGN	0269
by Automation of the+ Water	Drop Experiment—Determination of Surface Tension	CHED	0118
ace Tension by Automation of the	Drop-Weight Method.= +—Determination of Surf	CHED	0118
Glycosidic Flavor Precursors by	Droplet Countercurrent Chromatography and Mass S	AGFD	0203
Solution onto Water Vapor	Droplets.= Concentrating Cellulase from a Water	BIOT	0021
Juvenile Hormone of	Drosophila melanogaster.=	AGRO	0002
d Control Agents. Regulation of	Drosophila P450 Genes Involved in Insecticide Resist	AGRO	0097
eceptor from Cyclodiene Resistant	Drosophila: A Model System in a Model Insect.=	AGRO	0152
enoid Insect Growth Regulators in	Drosophila.= +ecular Genetics of Resistance to Juv	AGRO	0101
S-Transferases in	Drosophila.= Evolution of Glutathione	AGRO	0100
.= Receptor Activity and	Drug Design. Pharmacology of Muscarinic Receptors	MEDI	0078
orylase (PNP+ Structure-Based	Drug Design: Inhibitors of Purine Nucleoside Phosph	MEDI	0005
Pharmacological Classification and	Drug Design.= +ial of Chromatographic Data for	ANYL	0043
Peptide	Drug Design.= Pharmacokinetic Considerations in	MEDI	0193
Pharmaceutical Analysis and New	Drug Development.= +armaceutical Analysis—I.	ANYL	0110
atography/Mass Spectrometry to	Drug Discovery and Development.= +Liquid Chrom	ANYL	0139
°From Paper Chromatography to	Drug Discovery.=	HIST	0019
ptide Production for Discovery of	Drug Leads. Light-Direct+ New Technologies in Pe	MEDI	0090
Novel Sulfonamide	Drug Metabolism in Lactating Dairy Cows.=	AGRO	0113
nization LC/MS: Application to	Drug Monitoring in Biological Fluids and Tissues.=	ANYL	0124
plexes with the Antiinflammatory	Drug Piroxicam.= +ty Constants of Copper(II) Com	INOR	0407
action and Subsequent Analysis of	Drug Residues in Animal Tissues.= +to the Extr	AGFD	0060
tions of Analytical Procedures for	Drug Residues in Foods.= +s and Accurate Descrip	AGFD	0126
to Detection and Confirmation of	Drug Residues in Milk.= +mal Origin. Approaches	AGFD	0099
Analysis of Veterinary	Drug Residues.= Immunochemical Methods in the	AGFD	0064
Workplace	Drug Testing: Background and Current Trends.=	HIST	0008
Dynamics by NMR.=	Drug-DNA Interactions: Probing Binding Modes and	PHYS	0197
f Chromatography for the Clinical	Drug-Monitoring Laboratory in the 90s.= +Role o	ANYL	0116
borator+ Automated Analysis of	Drugs and Metabolites in Biological Fluids Using La	ANYL	0119
of Feed Additives and Veterinary	Drugs and Their Residual Analysis in Japan.= +ew	AGFD	0102
ction—I. Production of Antitumor	Drugs by Recombinant Bacteria.= +tibiotic Produ	BIOT	0108
esidue Analysis of Antibiotics and	Drugs in Products of Animal Origin.= +cation for R	AGFD	0083
gonists as Potential Antipsychotic	Drugs: I. Synthesis and Bin+ Sigma Receptor Anta	MEDI	0153
gonists as Potential Antipsychotic	Drugs: II. Animal Behaviora+ Sigma Receptor Anta	MEDI	0084
oduction—The Uses of Veterinary	Drugs.= +biotics in Food-Producing Animals. Intr	AGRO	0075
ti-Allergy and Anti-Inflammatory	Drugs.= +) Inhibitor—General Route to Novel An	MEDI	0179
Hall Directionary of	Drugs.= Experiences Building the Chapman and	CINF	0033
for Developing New	Drugs.= The Histamine H3-Receptor as a Target	MEDI	0081
	Dry Etching of Copper at High Rates.=	COLL	0030
Removal by	Dry Sorbents.= Fundamentals of Heavy Metal	FUEL	0132
Flighting for Rotary Fertilizer	Dryers and Coolers and Their Effects on Performanc	FERT	0012
	Drying Fuels with the Carver-Greenfield Process.=	FUEL	0116
Characterization. Role of Spray	Drying in Production of Catalyst and Catalyst Suppo	PETR	0088
ultinuclear Solid-State NMR and	DTA.= +o AlPO4-8 at Various Temperatures by M	PETR	0095
Flottation Ionique	du Laurylsulfate de Gallium.=	ANYL	0053
oline Anticonvulsant, fo+ Unique	Dual Action Mechanism of ADD17014, a Novel Triaz	MEDI	0144
Hydroxamic Acid Fenamates as	Dual Inhibitors of Cyclooxygenase and 5-Lipoxygena	MEDI	0177
37370.= SCH 40338, a New	Dual PAF and Histamine Antagonist Related to SCH	MEDI	0161
ivator 1-Diazo-4-undecyl-2-one (	Duo) as a Probe of the Active Site of E. +sed Inact	BIOL	0067
of the Anti-Inflammatory	DuP 697.= Characterization of the Platelet Effects	MEDI	0162
and Two-Dimensional NMR of	Duplex Oligonucleotides and Proteins DNA Complex	INOR	0378
in Teaching, cosponsored by E. I.	DuPont de Nemours and Co.). +ard for Excellence	ANYL	0006
Forensic	Dust Analysis.=	HIST	0005
es of Copper(I) Intercalated Laser	Dye Coumarins.= +ctural and Luminescence Studi	INOR	0245
pecies.=	Dye Laser Studies of Plasma and Flame-Generated S	CHED	0087
ethyl)cellu+ Spectral Behavior of	Dye Molecules Oriented by Chiral Nematic (Acetyl)(	POLY	0331
Internal Electric Field and	Dye Orientation in PVDF/PMMA Blend.=	POLY	0225
Effect on the Photoreduction of	Dyes by Photogenerated Ketyl Radicals in Photoelec	COLL	0081
	Dynamic Analysis of Regulation in Microorganisms.=	BIOT	0149
etradecyltrimethylammonium +	Dynamic and Steady-State Fluorescence Studies of T	COLL	0074
gnetic Relaxation of Amylose and	Dynamic Behavior of the Hydroxymethyl Group.=	CARB	0049
SP Synthase: The Importance of	Dynamic Catalysis for Successful Mechanism-Based	BIOL	0054
Coordinate: A Possible Domain for	Dynamic Effects.= +ss to a Symmetrical Reaction	ORGN	0122
of Pesticides in	Dynamic Flow Systems.= Modeling the Movement	AGRO	0064
aracterization: Optical Methods.	Dynamic Infrared Linear Dichroism of Polymers.=	POLY	0303
Automated	Dynamic Interfacial Tension Technique.=	COLL	0091
s with Differing Arch+ Static and	Dynamic Light Scattering from Solutions of Polymer	POLY	0250
maldehyde Gels as Precursors of+	Dynamic Light-Scattering Studies of Resorcinol For	POLY	0199
I in Viscoelastic Media.=	Dynamic Magnetic Birefringence as a Diagnostic Too	COLL	0143
Polyvinyl Chloride (PVC) Roof+	Dynamic Mechanical Analysis Studies of Reinforced	PMSE	0180
ly(methylmethacrylate)/-Glass +	Dynamic Mechanical Properties of a Polystyrene/Po	PMSE	0167
pounds Used in Electronic Pac+	Dynamic Mechanical Thermal Analysis of Mold Com	POLY	0095

Small-Angle X-ray Scattering and	Dynamic Mechanical Thermal Analysis.= +rs by	PMSE	0118
ched Molecules.=	Dynamic Monte Carlo Simulations of Rings and Bran	POLY	0297
-Water Interface.=	Dynamic Properties of Cellulosic Polymers at the Air	COLL	0056
Combs and Stars. Some Static and	Dynamic Properties of Star Polymers.= +eauty—	POLY	0265
45th Technician Symposium.	Dynamic Rheological Studies on the Low-Temperatu	TECH	0014
omparisons Made on Steady and	Dynamic Shear-Induced Morphology in Blends by O	PMSE	0085
in Catalysis: Shape Selectivity—I.	Dynamic Sieving by Heterogeneous Catalysts.=	CATL	0013
sition of Semivolatile Organic +	Dynamic Simulation Model of the Atmospheric Depo	GEOC	0004
	Dynamic Studies of Ternary Polymer Solutions.=	COLL	0054
oreversible Gelation of Polymers.	Dynamic Viscoelasticity during Thermoreversible Ge	POLY	0197
nder High Pressure: (6) Electro+	Dynamical Character of Physicochemical Processes u	PHYS	0222
Magnetic Relaxation Coupling in	Dynamically Heterogeneous Systems.=	PHYS	0232
n of Nuclear Magnet+ Studies of	Dynamics at Catalyst Surfaces by Selective Excitatio	PHYS	0346
utions, Blends, and Interfaces—I.	Dynamics at the Solid-Polymer Interface.= +r Sol	COLL	0012
State-to-State Reaction	Dynamics beyond A + BC.=	PHYS	0162
Probing Binding Modes and	Dynamics by NMR.= Drug-DNA Interactions:	PHYS	0197
ardment (Ion Mixing). Molecular	Dynamics Computer Simulation Studies of Collision	NUCL	0091
Polymers in Electric Fields:	Dynamics from Transient Electric Birefringence.=	COLL	0180
Studies of Local Polymer	Dynamics in Dilute Solution.= Optical and NMR	PHYS	0139
mputer Simulations of Quantum	Dynamics in Electron Transfer and Proton Transfer	BIOL	0131
Laureates Symposium. Reaction	Dynamics in Gases and at Surfaces. J. Polanyi (Chem	CONG	0011
Solutions.= Diffusion and	Dynamics in Microstructured Block Copolymer	POLY	0275
Statistics and	Dynamics in Multifragmentation.=	NUCL	0005
nd Optical Spectroscopies. Ligand	Dynamics in Myoglobin.= +Magnetic Resonance a	PHYS	0193
thalamide).= Amide Segmental	Dynamics in N-Deuterated Poly(p-phenylene Tereph	POLY	0058
nts of Energy Flow and Structural	Dynamics in Photoexcited Heme Proteins.= +reme	PHYS	0335
High-Frequency	Dynamics in Proteins.= Model Free Analysis of the	PHYS	0168
nsfer and Charge-Recombination	Dynamics in Wild-Type and Mutant Photosynthetic	BIOL	0132
Shear Flow.= Structure and	Dynamics of a Semidilute Polymer Solution under	PMSE	0108
Femtosecond	Dynamics of Bi₂ Dissociation.=	PHYS	0369
	Dynamics of Block Copolymer Adsorption.=	COLL	0073
Oxidase.= Ligation	Dynamics of Carbon Monoxide in Cytochrome c	BIOL	0096
Surface Photochemical	Dynamics of CH₃Br Adsorbed on Pt(111).=	PHYS	0261
Femtosecond Solvation-	Dynamics of CH₃Cl⁻ and Cl⁻ at Multilayer Surfaces.=	PHYS	0084
=	Dynamics of Coupling Agents in Interfacial Regions.	POLY	0368
r Resolution Studied by+ Surface	Dynamics of Diffusion and Corrosion at Atomic Laye	PHYS	0127
-Separation in w/o-Microemuls+	Dynamics of Electric-Field-Induced Transient Phase	COLL	0146
:+ Time-Domain Analysis of the	Dynamics of Frenkel Excitons in Disordered Systems	PHYS	0034
of Ethylene.=	Dynamics of H₂ Elimination in the Photodissociation	PHYS	0163
ational Relaxation and Structural	Dynamics of Heme Proteins.= +ical Studies of Vibr	PHYS	0169
cessful Intracontinental Trade i+	Dynamics of Intracontinental Trade. Criteria for Suc	CMEC	0001
MR Studies of the Structure and	Dynamics of Ligand-DNA Complexes.= +d-State N	BIOL	0116
ensed Phases. Conformation and	Dynamics of Partially Folded and Unfolded Protein	PHYS	0107
-State Structure and Transport	Dynamics of Photosynthetic Units: Spectral-Hole-Bu	PHYS	0018
rescence Depolarization, and the	Dynamics of Proteins and Membranes: Theory and E	PHYS	0166
face Properties on Charge-Carrier	Dynamics of Quantized Semiconductor Colloids.=	COLL	0151
	Dynamics of Reversible Networks.=	POLY	0172
ent Electric Birefringence.=	Dynamics of Ring DNA vs. Linear DNA from Transi	POLY	0296
e Short-Range Correlations in the	Dynamics of Rotational Isomerism.= +. Sizes of th	POLY	0365
Solution and in the Bulk State+	Dynamics of Rotational Isomerization in Polymers in	COLL	0018
Excited-Stated	Dynamics of Ru(BPY)₃²⁺ on Porous Vycor Glass.=	INOR	0160
NMR Studies of Structure and	Dynamics of Solid Polymers.= Multidimensional	PHYS	0137
um Chemistry with Experiment.	Dynamics of the Decomposition and Isomerization of	PHYS	0188
Quantum	Dynamics of the Hydrated Electron.=	PHYS	0190
ctions—I. Chemical Kinetics and	Dynamics of the Mesospheric Sodium Nightglow.=	PHYS	0071
	Dynamics of the O(³P) = NH₂ (ND₂) Reaction.=	PHYS	0189
Wettability Analysis: A Molecular	Dynamics Simulation Method.= +rial Surfaces via	POLY	0369
istance Geometry and Molecular	Dynamics Simulation Techniques for Conformationa	ORGN	0174
ydrogen Bonding from Molecular	Dynamics Simulation.= +ntations and Analysis of H	BIOL	0112
Molecular	Dynamics Simulations of Amiloride.=	MEDI	0045
Growth, and Propert+ Molecular	Dynamics Simulations of Diamond Films: Structure,	PHYS	0354
e Bulk and in Thin+ Stochastic	Dynamics Simulations of Polymethylene Chains in th	POLY	0355
bstituted Rigid-Ro+ Molecular-	Dynamics Simulations of Rigid-Rod and Hydroxy-Su	PHYS	0322
Pendant-Chain Der+ Molecular	Dynamics Studies of Ampholytic Ionomers and Their	POLY	0367
	Dynamics Studies of the Protein Echistatin.=	PHYS	0167
Probing Upconversion	Dynamics Using Site-Selective Spectroscopy.=	ANYL	0004
Ultrafast Vibrational	Dynamics Using Transient Infrared Spectroscopy.=	PHYS	0194
-State Energetics, Structures, and	Dynamics.= +stry with Experiment. Diene Excited	PHYS	0075
and	Dynamics.= Ab Initio Quantum Chemistry	PHYS	0165
Its Effect on Photoabsorption and	Dynamics.= +of Ozone: A Conical Intersection and	PHYS	0191
Density in Nuclear	Dynamics.= Fluctuations of the Single-Particle	NUCL	0009
Phase	Dynamics.= Fluorescence Probes of Solution	CHED	0201
Microtubule	Dynamics.= Interplay of Motors and	BIOL	0139
ivity in Partial Oxidation Reactor	Dynamics.= +y States and Changes in Catalyst Act	I&EC	0069
Macromolecular	Dynamics.= NMR Investigations of	PHYS	0195
lating Agents Based on Molecular	Dynamics.= +cer Benzothiopyranoindazole Interca	MEDI	0018
Unimolecular Reaction	Dynamics.=	PHYS	0164
P-D-Glucose Oxidoreductase (	E$_{od}$) and CDP-4-Keto-6-Deoxy-D-Glucose-3-Dehyd	BIOL	0110
Reactions of Alkali and Alkaline	Earth Atoms.= Comparison between the Oxidation	PHYS	0072
loxy- and Oxo-Aryloxy-Alkaline-	Earth Complexes.= +is and Characterization of Ary	INOR	0223

Left context	Keyword phrase	Code	Num
X-ray Studies of Layered Rare	Earth Gold(I) Cyanide Salts: Evidence for Energy Tr	INOR	0189
Alkaline	Earth Metals.= Flame Chemistry of Alkali and	PHYS	0073
Precursors to Rare	Earth Monochalcogenides.= Synthesis of Molecular	INOR	0311
Alkali	Earth Salts as Catalyst Supports and Stabilizers.=	PETR	0034
h/CeO2 and Rare Earth/Alkaline	Earth-Doped Pt, Rh/CeO2 Catalysts.= +ce of Pt, R	PETR	0035
mance of Pt, Rh/CeO2 and Rare	Earth/Alkaline Earth-Doped Pt, Rh/CeO2 Catalysts	PETR	0035
ed Molecules of Relevance in the	Earth's Atmosphere: Experimental Studies of the Sp	PHYS	0017
ntitative Determination of Rare	Earths in Rocks by Electron Impact Mass Spectrome	ANYL	0049
ss.=	Eastman Chemical Company Acetic Anhydride Proce	I&EC	0053
George	Eastman—Methods of Photography.=	CHAL	0030
=	EC '92: What Does It Mean to Chemical Literature?.	CINF	0013
hesis and Biological Evaluation of	Echinocandin Analogues. Antifungal and Ant+ Synt	MEDI	0182
roline Residues+ Biosynthesis of	Echinocandin-Type Antibiotics: Distinct Origins of P	MEDI	0058
Dynamics Studies of the Protein	Echistatin.=	PHYS	0167
o+ Utilization of Electron Spin	Echo-Modulation Spectroscopy as a Novel Method T	PHYS	0231
tional Effect Responsible for the	Eclipsing in cis-2-tert-Butyl-5-(tert-Butylsulfonyl)-	ORGN	0150
heoretical Studies of Gauche and	Eclipsing Interactions for Model Compounds of Poly	POLY	0353
olving.=	Ecological Knowledge and Environmental Problem S	ENVR	0063
	Ecological Risk of Aquatic-Habitat Degradation.=	ENVR	0062
tial Bans of Selected Pesticides:	Economic Implications for U.S. Agriculture.= Poten	AGRO	0083
he Fluorostero+ The Hunt for an	Economical Synthesis of Cortisol: The Discovery of t	HIST	0022
Impact of Byproducts on the	Economics of Biomass Refining.=	BIOT	0014
	Economics of Commingled Plastics Molding.=	I&EC	0035
Plastics Recycling Processing and	Economics.= +Technologies for Plastics Recycling.	I&EC	0031
ational Competition. R. Solow (	Economics, 1987).= Science, Technology, and Intern	CONG	0014
nic Matter in Continental Margin	Ecosystems.= +mineralization of Sedimentary Orga	GEOC	0015
Proteolytic Enzymes from	Ectothermic Organisms.=	BIOT	0194
olecular Bases of Animal Form. G.	Edelman (Medicine and Physiology, 1972).= +y: M	CONG	0013
nte+ Analytical Chemistry at the	Edge: Structure and Function at Condensed-Phase I	ANYL	0026
drolyzates of Cellulosic Food +	Edible Biomass and Chemical Feedstocks from Prehy	BIOT	0153
	Edible Films for Fruits and Vegetables.=	AGFD	0078
mation of Antibiotic Residues in	Edible Meats by On-Line HPLC Tandem Mass Spec	AGFD	0061
Thomas	Edison—Chemist.=	CHAL	0026
ogy. Biochemical Engineering, 2nd	Edition—Aiba, Humphrey, and Millis.= +otechnol	BIOT	0221
	Educating Chemists about Nonlinear Regression.=	CHED	0193
cation Option. Chemistry Educ+	Educating Future Teachers: The CPT-Chemical Edu	CHED	0066
ed for Reform.=	Educating High School Chemistry Teachers: The Ne	CHED	0067
	Educating Undergraduates about Chemometrics.=	CHED	0194
Safety	Education and Training at a University.=	CHAS	0049
nicating Chemical Safety: Some	Education and Training Used in a University Setting	CHAS	0048
ements for Lab+ Safety through	Education and Training. Regulatory Training Requir	CHAS	0042
atory—II. Undergraduate Science	Education at the National Science Foundation.=	CHED	0179
ium. The Making of a Continuing	Education Course—Start to Finish.= +cian Sympos	TECH	0001
Chemistry	Education Emphasis Curriculum—A Case Study.=	CHED	0068
	Education for the HAZMAT Worker.=	CHAS	0045
ms of the Institute for Chemical	Education in Revitalizing Science in the Nation's Sch	CHED	0138
resentation: Integrating Polymer	Education into Middle, Vocational, and High School	CHED	0026
iative of the Division of Chemical	Education of the American Chemical Society.= +it	CHED	0110
ure Teachers: The CPT-Chemistry	Education Option. Chemistry Educ+ Educating Fut	CHED	0066
Industry-	Education Partnership= Teachers in Industry: An	CHED	0004
Laboratory.= Science	Education Program at Oak Ridge National	CHAS	0050
Moving Ahead in Chemistry	Education Software Evaluation.= Looking Back and	CHED	0080
ademic Partnerships in Precollege	Education. Benefits of Alliances to+ Corporate/Ac	CHED	0001
n.= Computers in Chemical	Education. Computer-Managed Chemistry Instructio	CHED	0077
posium on Computers in Chemical	Education. +stry Laboratory Development and Sym	CHED	0155
hool Cu+ Prehigh School Science	Education. FACETS: An Interdisciplinary Middle Sc	CHED	0071
on Val+ Computers in Chemical	Education. Role of Computers in Chemistry at Leban	CHED	0115
ndergr+ Computers in Chemical	Education. Use of Personal Computer Modeling in U	CHED	0188
.= Computers in Chemical	Education. Using Spreadsheets in General Chemistry	CHED	0204
ical Education Option. Chemistry	Education: An Option for Adoption.= +CPT-Chem	CHED	0066
graduate Analytical Chemistry	Education.= Challenges and Opportunities in Under	ANYL	0007
Promotion of Science	Education.= Industry/Academic Interactions for the	CHED	0008
rams of the Institute for Chemical	Education.= +igh School Chemistry Outreach Prog	CHED	0073
p Advance Prehigh School Science	Education.= +Two-Year College Are Doing to Hel	CHED	0076
Handling of Waste Chemicals in	Educational Laboratories.=	ENVR	0036
ation. Benefits of Alliances to the	Educational Process.= +erships in Precollege Educ	CHED	0001
ry of Marjorie Gardner, Chemical	Educator. +mSource Programs, Honoring the Memo	CHED	0011
.= Structural Modification of	Edulinine, a Constituent Isolated from Chinese Herb	MEDI	0159
yers Squibb).+ Award Address (	Edward E. Smissman Award, sponsored by Bristol-M	MEDI	0089
Address. Hydanto+ General and	Edward E. Smissman—Bristol-Myers Squibb Award	MEDI	0082
a Fertilidad de Ratas Hembras.+	Efecto de la Deficiencia o Exceso de Vitamina E en l	AGFD	0210
on en Polimeros Zwitterionicos.+	Efecto de la Presion en las Propiedades de Conducci	ORGN	0159
Separation Arsenopyrite-Pyrite:	Effet des Oxydants et des Complexants sur la Fottat	ANYL	0067
ryl-1-Cyclopropylbutanes: Highly	Efficacious Insecticides with Low Fish Toxicity.=	AGRO	0073
A New Class of	Efficacious, Well-Absorbed Renin Inhibitors.=	MEDI	0120
on Treatment of Pulp Bleaching	Effluents Containing Chlorinated Phenolic Compoun	BIOT	0137
rbon Process.= Polysar's	Efforts in Exposing Precollege Students to a Hydroca	CHED	0010
Future Environmental Protection	Efforts.= +is Task? Relative Risk: A Blueprint for	ENVR	0061
Large-Scale Production of	Efrotomycin by Nocardia lactamdurans.=	BIOT	0109
in Electrochemistry, sponsored by	EG G Princeton Applied Research Co.). +ry Award	ANYL	0096

(-)-Epigallocatechin Gallate (	EGCG): A Cancer Preventive Agent.=	AGFD 0066
l Determinants Derived from Hen	Egg Lysozyme 81-96.= +ties of Glycopeptide T-Cel	CARB 0014
al Studies in Chicken Tissues and	Eggs.= +thoxine and Sulfamonomethoxine Residu	AGFD 0127
(sal$_2$tm).=	EHMO Calculations on the system Ni(sal$_2$en) and NI	INOR 0042
Hydrocarbons.=	Eigenvalue Distributions for the Graphs of Alternant	COMP 0029
an Case: Plagiarism at the Albert	Einstein College of Medicine of Yeshiva University/	PROF 0004
Asbestos en	el Area Urbana de la Ciudad de Mexico?.=	ENVR 0025
es Recientes de CAD/CAM para	el Diseno de Equipo de la Industria Quimica de Proc	COMP 0053
Microbiologia de Alimentos en	el LNSP: Enfoque Prospectivo.=	AGFD 0206
oxidable 304 y 316 Utilizadas en	el Seguimiento de las Reacciones Acido Base en Med	ANYL 0055
Estudio Sobre	el Uso de Eteres de Glicol en Mexico.=	ENVR 0038
Contaminada con S. aureus y s+	Elaboracion de Caseinato de Sodio a Partir de Leche	AGFD 0107
ia de la Sulfonacion sobre las P+	Elaboracion de Pastas de Alto Rendimiento: Influenc	PMSE 0077
Inhibition of Human PMN	Elastase by Monocyclic β-Lactams.= Mechanism of	BIOL 0050
Inhibitors of Human Leukocyte	Elastase: The Synthesis and Structure-Activity Relat	MEDI 0172
Leukocyte	Elastase.= Lazarus Inhibitors of Human	MEDI 0087
Human Leukocyte	Elastase.= Monocyclic β-Lactam Inhibitors of	MEDI 0173
actam Inhibitors of Human PMN	Elastase.= +e, Potent, and Specific Monocyclic β-L	BIOL 0049
ork Products.+ Nitrosamines in	Elastic Rubber Nettings Used for Packaging Cured-P	AGFD 0040
n of an Insulator Containing Two	Elastomeric Components.= +ysical Characterizatio	PMSE 0168
onia Particles Precipitated into an	Elastomeric Material.= +terials. Reinforcing Zirc	POLY 0283
nd Amorphous Polyisocyanide and	Elastomeric Polybutadiene Segments.= +Helical a	PMSE 0071
Styrene-Butadiene Thermoplastic	Elastomers.= +hesis of Heteroarm, Star-Branched	POLY 0266
Surface-Hydrophilic	Elastomers.=	COLL 0038
e) Random Block Copolymers and	Elastomers.= +hylsiloxane-co-methylphenylsiloxan	POLY 0020
Full SOCVCI Calculations of the	Elctronic Spectra of LiBe and LiMg.= Ab Initio	PHYS 0338
Transient	Electric Birefringence of Agarose Gels.=	COLL 0182
um-2-Acrylamido-+ Transient	Electric Birefringence of Ionomers Poly(styrene-Sodi	COLL 0181
Linear DNA from Transient	Electric Birefringence.= Dynamics of Ring DNA vs.	POLY 0296
by Calicheamicin from	Electric Birefringence.= Kinetics of DNA Cleavage	POLY 0280
Dynamics from Transient	Electric Birefringence.= Polymers in Electric Fields:	COLL 0180
General	Electric College Bound Program.=	CHED 0003
i+ On the Conversion of Solar to	Electric Energy by the Other Photosynthetic System	PHYS 0019
Blend.= Internal	Electric Field and Dye Orientation in PVDF/PMMA	POLY 0225
d Application of a Radiofrequency	Electric Field.= +Flow Rate, Ion Concentration, an	COLL 0163
orted, Nanometer-+ Influence of	Electric Fields on Organic Reactions at Zeolite-Supp	CATL 0018
DNA.= Effects of High-	Electric Fields on the Structure and Association of	COLL 0179
Birefringence.= Polymers in	Electric Fields: Dynamics from Transient Electric	COLL 0180
ult of Exposure to Radiofrequency	Electric Fields.= +(Anatase) Suspensions as a Res	COLL 0203
and Charge-Transfer Molecules in	Electric Fields—A Theoretical Study.= +otential	BIOT 0001
ho+ Measurements of the Opti-	Electric Properties of Molecules of Potential Use in P	POLY 0214
tudy of the Competition between	Electric-Field-Induced Molecular Ordering and Che	POLY 0206
ultiphase Polymer-Polymer-Sol+	Electric-Field-Induced Pearl-Chain Formation in M	COLL 0145
in w/o-Microemul+ Dynamics of	Electric-Field-Induced Transient Phase-Separation	COLL 0146
uorocarbon Media: A Strategy for	Electric-Field-Orientable Microparticles.= +n in Fl	COLL 0164
oidal Particles in External Fields:	Electric, Magnetic, and Hydrodynamic—I. +les: Coll	COLL 0141
oidal Particles in External Fields:	Electric, Magnetic, and Hydrodynamic—II. +es: Coll	COLL 0159
oidal Particles in External Fields:	Electric, Magnetic, and Hydrodynamic—III. +s: Coll	COLL 0178
oidal Particles in External Fields:	Electric, Magnetic, and Hydrodynamic—IV. +s: Coll	COLL 0199
ysilane Copolymer: Synthesis and	Electrical Conductivity.= +ly Conducting Hydropol	POLY 0043
Design of a Molecular-Bilayer	Electrical Device.=	BIOT 0104
δ.=	Electrical Properties and Structure of La$_{2-x}$Sr$_x$NiO$_4$+	INOR 0048
tic Fingerprinting of the Surface	Electrical Properties of a Single-Acid-Site Latex.=	COLL 0200
and Electrochromism Studies of	Electrical Properties of Nonlinear Optical Polymers a	POLY 0150
Size-Dependent	Electrical Properties of Organic Materials.=	BIOT 0252
e with Bromine: A New Route to	Electrically Conducting Conjugated Polyacetylene.=	POLY 0032
	Electrically Conducting Ionic Polyacetylenes.=	POLY 0074
us Precursors.=	Electrically Conductive Catalyst Supports from Fibro	PETR 0016
nversion of Waste to Ethanol and	Electricity via Dilute Sulfuric Acid Hydrolysis.=	FUEL 0094
nal and Microscopic Interfaces.+	Electrified Immiscible Liquid Boundaries: Conventio	COLL 0004
harge.=	Electroacoustic Determination of Particle Size and C	COLL 0201
rsible Neutral Species Transfer in	Electroactive Polymer Films.= +ersible and Irreve	ANYL 0103
dative Synthesis of Conductive-	Electroactive Polymers.= Use of Micelles in the Oxi	COLL 0168
tive Methylene Compounds.=	Electrochemical Alkylation and Methoxylation of Ac	ORGN 0249
o Films.=	Electrochemical and Langmuir Trough Studies of C$_6$	ANYL 0108
ments in Aqueous Surfactant S+	Electrochemical and Spectroelectrochemical Measure	COLL 0003
Separation and+ Application of	Electrochemical and UV/Visible Detection to the LC	AGFD 0101
ry in Microheterogeneous Fluids.	Electrochemical Aspects of Conducting Polymer Coll	COLL 0166
nditions.=	Electrochemical Behavior of PbS at the Flotation Co	PHYS 0217
y in Microheterogeneous Fluids.	Electrochemical Formation of Organic-Pigment Thin	COLL 0019
eel/Sodium Chloride Interface by	Electrochemical Impedance Spectroscopy.= +zed St	PHYS 0270
ructure and Its Effect on Electr+	Electrochemical Insights into Microemulsion Microst	COLL 0023
Ferrocene-Linked Acridine Ora+	Electrochemical Investigation of the Interaction of a	ANYL 0109
ials in Biological Membranes.=	Electrochemical Kinetics and Nonequilibrium Potent	COLL 0093
e Potentials: A Hope for a Fast	Electrochemical Method of Determination of Water.=	ANYL 0095
tures of Liquid Disperse System+	Electrochemical Methods and Physicochemical Struc	COLL 0022
ea+ Application of the Scanning	Electrochemical Microscope to the Measurement of R	COLL 0170
Using RuO$_2$ Dispersions and the	Electrochemical Model.= Oxygen Redox Catalysis	COLL 0134
ns of SDS/1-Pentanol/Cyclohe+	Electrochemical Processes in the Single-Phase Regio	COLL 0024
Complexes.= Synthesis and	Electrochemical Properties of Ru(II) Schiff Bases	INOR 0260

Naturally Occurring Fluids on the	Electrochemical Reduction of Ru(NH$_3$)$_6$Cl$_3$.= +in	COLL	0005
r Substrate.=	Electrochemical Removal of a Tin Layer from Coppe	PHYS	0285
yer Resolution Studied by in Situ	Electrochemical Scanning Tunneling Microscopy.=	PHYS	0127
Spectroscopic and	Electrochemical Studies of [Pt(terpy)Cl]$^+$.=	INOR	0216
d to Mixed Microbial Communi+	Electrochemical Studies of Coated 4140 Steel Expose	BIOT	0097
in B$_{12}$.=	Electrochemical Study of Electron Transfer to Vitam	BIOL	0135
ociated with the Reduction of +	Electrochemical Study of the Structural Changes Ass	ANYL	0106
. A. Osteryoung. Interactions of+	Electrochemistry Award Symposium—I, Honoring R	ANYL	0093
and Irreversible Neutral Species+	Electrochemistry Award Symposium—II. Reversible	ANYL	0103
rfactants and Associated Colloids:	Electrochemistry in Microheterogeneous Fluids+ Su	COLL	0166
rfactants and Associated Colloids:	Electrochemistry in Microheterogeneous Fluids+ Su	COLL	0019
rfactants and Associated Colloids:	Electrochemistry in Microheterogeneous Fluids+ Su	COLL	0040
rfactants and Associated Colloids:	Electrochemistry in Microheterogeneous Fluids+ Su	COLL	0001
rfactants and Associated Colloids:	Electrochemistry in Microheterogeneous Fluids+ Su	COLL	0128
rfactants and Associated Colloids:	Electrochemistry in Microheterogeneous Fluids+ Su	COLL	0147
mperature Molten Salt.=	Electrochemistry of Aromatic Ketones in a Room-Te	ANYL	0094
	Electrochemistry of Colloidal Gold.=	COLL	0129
ds: Crystal Structures, NMR, and	Electrochemistry.= +of Polyanionic Chelating Ligan	INOR	0243
nometal+ Photoinitiator Activity,	Electrochemistry, and Spectroscopy of Cationic Orga	INOR	0320
f Analytical Chemistry Award in	Electrochemistry, sponsored by EG G Princeton App	ANYL	0096
on 4-Keto-Ba+ The Photo- and	Electrochromic Properties of Biochrom Films Based	BIOT	0082
econd Harmonic Generation and	Electrochromism Studies of Electrical Properties of	POLY	0150
ocesses under High Pressure: (6)	Electrode Processes in Solution and Material Charac	PHYS	0222
Acetonitrile on Pt(III)	Electrode Surfaces.= Chemisorption of Cyanide and	COLL	0118
Voltammetry at the Mercury Film	Electrode.= +of Human Seric Copper by Inverse	ANYL	0056
Transfer at a Solid Semiconductor	Electrode-Liquid Junction.= +Interfacial Electron	COLL	0088
en from Anionic Surfactant Sol+	Electrodeposition of Reduced Forms of Methylviolog	COLL	0077
Medi+ Obtencion de Plata por	Electrodeposito, Fusion, Oxidorreduccion con Zinc en	CHED	0172
Ligands for Polymer-Modified	Electrodes.= Aniline-Substituted 2,2'-Bipyridine	POLY	0189
lite-Supported, Nanometer-Sized	Electrodes.= +ic Fields on Organic Reactions at Zeo	CATL	0018
ass Transport to Hydrodynamic	Electrodes.= Mechanism of Particulate-Enhanced M	COLL	0041
encional para Medir pH, por un	Electrodo de Acero Inoxidable 316/Pelicula de Oxido	ANYL	0066
r un Electrodo d+ Sustitucion de	Electrodo de Vidrio Convencional para Medir pH, po	ANYL	0066
ucto Terminado Utilizando como	Electrodo Indicador una Pelicula de Oxido sobre Ace	CHED	0163
uimica del 1-Nitro-Naftaleno con	Electrodos Metalicos y Semiconductores.= +ectroq	PHYS	0280
	Electroduction of o-Nitrobenzonitrile.=	ORGN	0250
Deposito	Electrolitico de Samario y Plutonio.=	NUCL	0075
Excess Heat Production by the	Electrolysis of an Aqueous Potassium Carbonate Elec	PHYS	0286
an Aqueous Potassium Carbonate	Electrolyte and the Implications for Cold Fusion.=	PHYS	0286
l Fluid Chromatography Using an	Electrolytic Conductivity Detector.= +Supercritica	ANYL	0091
Ions.=	Electromagnetic Dissociation with Relativistic Heavy	NUCL	0049
Potential Importance as Alternate	Electron Acceptors.= +g of Iron Oxides and Their	GEOC	0007
Cluster	Electron Affinities.= Theoretical Study of Silicon	PHYS	0079
Absolute Rates of Intramolecular	Electron and Energy Transfer?.= +ng in Predicting	ORGN	0143
e Beta Decay of+ Limit on the	Electron Antineutrino Mass from Measurement of th	NUCL	0010
e-Dependent Rate Constants for	Electron Attachment Reactions of Metal-Centered C	PHYS	0234
ilicone-Epoxide Resins.=	Electron Beam-Induced Cationic Polymerization of S	PMSE	0159
f Graph Theory in the Study of	Electron Delocalization in Early Transition Metal He	COMP	0032
ly 3'-(4-pyridyl)terth+ Tuning of	Electron Density at Transition-Metal Centers via Po	INOR	0027
Silica Catalyst Supports.=	Electron Energy-Loss Spectrometry of Alumina and	PETR	0073
nation of Rare Earths in Rocks by	Electron Impact Mass Spectrometry.= +ve Determi	ANYL	0049
heterogeneous Fluids. Vectorial	Electron Injection into Colloidal Semiconductor Mem	COLL	0147
Delocalization in π-Stacks.=	Electron Localization in Conjugated π-Systems and	BIOT	0201
Single-	Electron Logic: Molecular Implementation?.=	BIOT	0004
of Computer-Controlled Scanning	Electron Microscopy (CCSEM) in Coal Science.=	FUEL	0041
n of Municipal Waste by Scanning	Electron Microscopy and Optical Microscopy.= +io	FUEL	0137
Applications of Transmission	Electron Microscopy to the Study of Diamond.=	PHYS	0120
Scanning and Transmission	Electron Microscopy.= Analysis of Materials Using	INOR	0014
Polymer Chains by Transmission	Electron Microscopy, II: Image Contrast Analysis of	POLY	0062
ical Properties for Conjugated, π-	Electron Molecules.= +Third-Order Nonlinear Opt	BIOT	0177
olic Free Radicals.=	Electron Paramagnetic Resonance Detection of Phen	AGFD	0004
e Study of Coal Chem+ Advanced	Electron Paramagnetic Resonance Techniques for th	FUEL	0072
Applications of Coherent Pulsed	Electron Paramagnetic Resonance.= New	PHYS	0116
hrough Electron Propaga+ One-	Electron Pictures of Molecular Electronic Structure t	PHYS	0058
ular Electronic Structure through	Electron Propagator Theory.= +n Pictures of Molec	PHYS	0058
um/TiO$_2$(110) Model Catalysts by	Electron Spectroscopies.= +TiO$_2$(110) and Vanadi	COLL	0192
rganic Surfaces+ Applications of	Electron Spectroscopy in the Characterization of Ino	INOR	0013
ovel Method To+ Utilization of	Electron Spin Echo-Modulation Spectroscopy as a N	PHYS	0231
Sr$_2$CaCu$_2$O$_x$ Superconductors by	Electron Spin Resonance Line Broadening of Surface	PHYS	0284
on Intensity and Coordination via	Electron Spin Resonance.= +ucts with the cupric I	CATL	0017
Two-Dimensional	Electron Spin Resonance.=	PHYS	0114
e Models for Intermediate One-	Electron States in Two- and Three-Dimensional Bou	BIOT	0028
Polysilanes: The Effect of	Electron Transfer Additives.= Photolabile	POLY	0188
mics in Wild-Type and Mutant +	Electron Transfer and Charge-Recombination Dyna	BIOL	0132
Films.= Photoinduced	Electron Transfer and Energy Transfer in LB	BIOT	0103
lations of Quantum Dynamics in	Electron Transfer and Proton Transfer Reactions in	BIOL	0131
plexes: Competition between Back	Electron Transfer and Substrate Reduction.= +om	INOR	0284
icosecond Real-Time Interfacial	Electron Transfer at a Solid Semiconductor Electrod	COLL	0088
Intramolecular	Electron Transfer at Fixed Distances.=	PHYS	0077
S-S Bonds.= Single	Electron Transfer Catalyzed Cleavage of the C-S and	ORGN	0251

e Mechanisms and Photoinduced	Electron Transfer in Covalently Linked Polynuclear	PHYS 0021
uO_2 Clusters.= Photoinduced	Electron Transfer in Covalently Linked Porphyrin-R	COLL 0133
as Restricted Reaction Spaces for	Electron Transfer Processes.= +d Their Utilization	POLY 0238
rrocen+ Competitive Energy and	Electron Transfer Quenching of Excited States by Fe	INOR 0163
tal Complexes an+ Photoinduced	Electron Transfer Reactions between Transition-Me	INOR 0283
+of Thiyl Radical Formation and	Electron Transfer Reactions by Flash Photolysis.=	INOR 0396
m and Platinum Metal Films by	Electron Transfer to Photogenerated Organometallic	INOR 0319
Electrochemical Study of	Electron Transfer to Vitamin B_{12}.=	BIOL 0135
dies of Fast-Electro+ Biological	Electron Transfer. Experimental and Theoretical Stu	BIOL 0130
Ultrafast	Electron Transfer.= Density Matrix Theory of	BIOT 0027
n Microstructure and Its Effect on	Electron Transfer.= +l Insights into Microemulsio	COLL 0023
on Interfacial	Electron Transfer.= Kinetics Effects of ζ-Potential	COLL 0152
vin Ester.= Photoinduced	Electron Transfers within Conjugated Porphyrin-Fla	BIOL 0134
Films for Vectorial Photoinduced	Electron Transport.= +ally Polymerized Porphyrin	BIOT 0126
d Proteins.=	Electron Tunneling Pathways in Native and Modifie	BIOL 0133
Hydrated	Electron.= Quantum Dynamics of the	PHYS 0190
operties of Linear Chains.=	Electron-Correlated States and Nonlinear Optical Pr	POLY 0007
bed Species o+ High-Resolution	Electron-Energy-Loss Spectroscopy Studies of Adsor	COLL 0173
Temperature+ High-Resolution	Electron-Energy-Loss Spectroscopy Studies of High-	PHYS 0279
r Cement-Solidified/Stabilized+	Electron-Microscopy Characterization Techniques fo	ENVR 0005
Anion-Molecule and	Electron-Molecule Complexes.=	PHYS 0056
Superconductors.= Nonlinear	Electron-Phonon Interactions in High-Temperature	PHYS 0012
Catalysis.= New	Electron-Rich Chiral Phosphines for Asymmetric	CATL 0004
tallacycles in the Rections of the	Electron-rich Complexes $W(PMe_3)_6$ and @$(PMe_3)_4(\eta$	INOR 0086
ino+ Synthesis and Reactivity of	Electron-Rich Ruthenium Alkyl, Hydride, and Alum	INOR 0244
$Sr_2CaCu_2O_x$ Superconductors by	Electron-Spin-Resonance Line Broadening of Surfac	PHYS 0050
Mode-by-Mode Assessments of	Electron-Transfer Activation Barriers at Colloidal T	COLL 0132
rin Cation Lattice by Interfacial	Electron-Transfer Gates Ionic Currents across the Li	BIOL 0136
+and Theoretical Studies of Fast-	Electron-Transfer Reactions in Photosynthesis.=	BIOL 0130
ancement.= Ocean Feeding of	Electron-Treated Sludge for Ocean Productivity Enh	ENVR 0021
nd Metalloporphyrins.+ Effect of	Electron-Withdrawing Substituents on Porphyrins a	INOR 0301
ng in Cuprate Supercondcutors:	Electronic and Steric Effects.= Structure and Bondi	CHED 0022
Solids: C_{60}, C_{70}, and the Alkali+	Electronic and Structural Studies of Fullerene-Based	PHYS 0049
the Teaching of Chemistry: An	Electronic Bulletin Board for High School Teachers o	CHED 0111
ds for Infra-Red Reflow Tolerant	Electronic Components.= +astic Molding Compoun	CMEC 0014
of 5-Azaadamantan-2-one.=	Electronic Control of Face Selection in the Reduction	ORGN 0237
of the	Electronic Density.= Equation for the Square Root	PHYS 0254
vity Metals for Interconnection of	Electronic Devices.= +r Deposition of Low-Resisti	INOR 0126
Copyright Issues in	Electronic Distribution.= Journal Full Text:	CINF 0046
	Electronic Energy Relaxation at Interfaces.=	PHYS 0062
ation of the Time Dependence of	Electronic Energy Transfer and Depolarization in Po	POLY 0078
ptor Adlayers on Single-Crystal+	Electronic Energy Transfer between Donor and Acce	PHYS 0065
of Th+ Computer Simulation of	Electronic Energy Transfer in Polymers: Comparison	PHYS 0109
tational and Ultrafast Studies of	Electronic Energy Transport in Spatially Disordered	PHYS 0133
Product-State Distr+ Effects of	Electronic Excitation and Dimerization of Metals on	PHYS 0350
ntennae.=	Electronic Excitation Transport in Photosynthetic A	PHYS 0020
Statistical Mechanical Studies of	Electronic Line Broadening in Liquids.=	PHYS 0035
f Advanced Materials: Growth of	Electronic Materials from Organometallic Precursors	COLL 0064
f Advanced Materials: Growth of	Electronic Materials from Organometallic Precursors	COLL 0153
f Advanced Materials: Growth of	Electronic Materials from Organometallic Precursors	COLL 0025
f Advanced Materials: Growth of	Electronic Materials from Organometallic Precursors	COLL 0006
f Advanced Materials: Growth of	Electronic Materials from Organometallic Precursors	COLL 0044
f Advanced Materials: Growth of	Electronic Materials from Organometallic Precursors	COLL 0135
p 13 Compounds as Precursors for	Electronic Materials.= +w-Oxidation-State Grou	COLL 0153
Ion-Bombarded	Electronic Materials.= Sputtering Processes of	NUCL 0043
lysis of Mold Compounds Used in	Electronic Packages.= +c Mechanical Thermal Ana	POLY 0095
ion by σ-Bond-Separat+ Optic-	Electronic Properties and Second-Harmonic Generat	POLY 0222
uasi+ Synthesis, Structure, and	Electronic Properties of a New Modifiaction of the Q	INOR 0359
g Subsurface and Nonequilibrium	Electronic Properties of Semiconductors.= +Imagin	PHYS 0208
ransition-Metal Catio+ Effect of	Electronic State on the Reaction Cross-Sections of T	PHYS 0306
-Molecule R+ Vibrational Mode,	Electronic State, and Collision Energy Effects on Ion	PHYS 0241
X-ray Absorptio+ Model for the	Electronic States of High-T_c Cuprates as Probed by	PHYS 0053
nneling Mi+ Local Structure and	Electronic States of High-T_c Superconductors by Tu	PHYS 0206
oments for the Nine Lowest Lying	Electronic States of the CuCl Molecule.= +pole M	PHYS 0304
ry Transition-Metal Compound+	Electronic Structure and Chemical Bonding in Terna	PHYS 0252
pounds of Ni-Fe-Pt, Ni-Fe-Pd+	Electronic Structure and Magnetism in Ternary Com	PHYS 0253
Perovskites.=	Electronic Structure of $LaBO_3$ (B = Mn, Fe, Co, Ni)	PHYS 0288
-Electron Pictures of Molecular	Electronic Structure through Electron Propagator Th	PHYS 0058
tallomesogens: Relation between	Electronic Structure, Solid State Structure, and Liqu	INOR 0269
rs.=	Electronic Structures and Stabilities of Mixed Cluste	PHYS 0214
Molecular	Electronic Transition Current Density.=	BIOT 0026
te Tungsten Bro+ Magnetic and	Electronic Transport Properties of the Monophospha	INOR 0352
VI. Reactions of Ground State and	Electronically Excited Barium Atoms.= +ctions—	PHYS 0349
on Studies of the Interactions of	Electronically Excited Cadmium Atoms with H_2 and	PHYS 0351
de Masas con Fuente de Impacto	Electronico.= +e Tierras Raras por Espectrometria	ANYL 0051
Molecules for	Electronics Using STM.= Research on Organic	BIOT 0204
Molecular and Biomolecular	Electronics. Design of Nonlinear Optical Materials.=	BIOT 0173
Molecular and Biomolecular	Electronics. Photosynthetic Molecular Electronics.=	BIOT 0050
-+ Molecular and Biomolecular	Electronics. Preparation and Characterization of Two	BIOT 0121
S+ Molecular and Biomolecular	Electronics. Self-Assembling Lipid Bilayers on Solid	BIOT 0102

l P+ Molecular and Biomolecular	Electronics. Some General Properties of Double-Wel	BIOT	0001
e+ Molecular and Biomolecular	Electronics. Spatial Light Modulators and Optical M	BIOT	0077
st+ Molecular and Biomolecular	Electronics. Toward a Molecular-Size Tinkertoy Con	BIOT	0200
in+ Molecular and Biomolecular	Electronics. Very High Density Information Storage	BIOT	0023
ics. Photosynthetic Molecular	Electronics.= Molecular and Biomolecular Electron	BIOT	0050
n the Properties of Composites for	Electronics.= +lar and Supermolecular Structure o	POLY	0300
Design of Molecular	Electronics.= Study of Surface Orientation for the	BIOT	0053
Novel Side-Chain Polymers for	Electrooptic Applications.=	POLY	0107
Thermally Stable	Electrooptic Polymers.=	POLY	0130
A Comparison between Differen+	Electroorganic Synthesis in Micelles and Emulsions:	COLL	0043
Organometallic	Electrophile.= [cpRu(dppm)(CH$_2$SO$_2$)]$^+$: A Strong	INOR	0249
l Ligands by Nucleophiles and by	Electrophiles.= +upling of Alkylidyne and Carbony	INOR	0188
x Cp(PPH$_2$)(CO)FeCOCH$_3$-1 with	Electrophiles.= +an Iron Phosphide Acetyl Comple	INOR	0240
R: Synthesis and Reactions with	Electrophiles.= Ruthenium Thiolates cpRu(PR$_3$')$_2$S	INOR	0368
Nitrogen	Electrophiles.= Trapping Tetrazolinylidenes with	ORGN	0205
Control of Diastereoselectivity in	Electrophilic and Diels-Alder Reactions.= +ectronic	ORGN	0240
tenyl Diphosphate Isomerase: An	Electrophilic Mechanism for Isomerization of Unacti	BIOL	0014
Capillary	Electrophoresis of Complex Carbohydrate Mixtures.=	ANYL	0019
	Electrophoresis of Polyelectrolytes in Gels.=	COLL	0183
Capillary	Electrophoresis Separations for Small Molecules.=	ANYL	0137
Capillary	Electrophoresis.=	AGFD	0179
-Reversed Free Solution Capillary	Electrophoresis.= +ns by Free Solution and Charge	ANYL	0018
ed from Aflatoxin-Resistant/S+	Electrophoretic Analysis of Cationic Proteins Extract	AGFD	0134
cal Properties of a Single-Acid-+	Electrophoretic Fingerprinting of the Surface Electri	COLL	0200
and Importance of Polymodal	Electrophoretic Mobility Distributions.= Occurrence	COLL	0202
Chester Carlson—	Electrophotography.=	CHAL	0028
el Arylpipera+ Synthesis, Cardiac	Electrophysiological, and β-Blocking Activity of Nov	MEDI	0094
s Metalic+ Estudio de Reduccion	Electroquimica del 1-Nitro-Naftaleno con Electrodo	PHYS	0280
Estudio Estadistico de Tecnicas	Electroquimicas de Aceros Microaleados en Acido Su	PHYS	0281
mico de Una Caldera por Tecnicas	Electroquimicas.= +erentes Etapas del Lavado Qui	PHYS	0257
Determinacion del Mecanismo de	Electroreduccion de la Perezona.=	PHYS	0213
Peptide and Protein Analysis by	Electrospray and Continuous-Flow FAB LC/MS.=	ANYL	0140
Analysis of Biopolymers by	Electrospray Ionization LC/MS and LC/MS/MS.=	ANYL	0141
e Analysis in Disaccharides by	Electrospray Mass Spectrometry (ES MS).= Linkag	CARB	0048
ion LC/MS: Application to Dru+	Electrospray/Ionspray Atmospheric Pressure Ionizat	ANYL	0124
erties, in Particular the Steric and	Electrostatic Features.= +lap of Atom-Based Prop	CINF	0018
ect on Energetics of Surface Re+	Electrostatic Field of the Surface of SiO$_2$ and Its Eff	COLL	0116
orescence+ Investigation of the	Electrostatic Properties of Humic Substances by Flu	ENVR	0057
Extraction of Rf (	Element 104) and Its Homologues with TBP.=	NUCL	0074
Fusion of Co and Bi To Make	Element 110.=	NUCL	0015
Student Ideas about Chemical	Element, Acid, and Molecular Geometry.=	CHED	0174
is and Modification of Materials:	Elemental Analysis by Proton-Induced X-ray Emiss	NUCL	0077
is and Modification of Materials:	Elemental/Isotopic Analysis by Nuclear Reactions/S	NUCL	0064
Winner: Polymers Are	Elementary!.= POLYED 1989 Polymer Award	CHED	0028
onstan+ Combustion Chemistry:	Elementry Reactions and Modeling. Thermal-Rate C	FUEL	0123
tes.= Oatrim Products with	Elevated β-Glucan Contents as Natural Fat Substitu	AGFD	0037
Cement Minerals with Water at	Elevated Temperatures and Pressures in Geothermal	INOR	0246
Suspension Cultures with Fungal	Elicitors.= +uction in Lithospermum erythrorhizon	BIOT	0229
ntibiotics and Drugs in Product+	ELISA and Its Application for Residue Analysis of A	AGFD	0083
azone and the Development of an	ELISA to Nicarbazine.= +-Dinitrocarbanilide-Hydr	BIOT	0166
lm G+ Real-Time Spectroscopic	Ellipsometry: Applications in Polymer- and Thin-Fi	POLY	0252
nalysis, sponsored by the Perkin-	Elmer Corp.). +emistry Award in Spectrochemical A	ANYL	0001
olymer Solutions under Transient	Elongational Flow.= +ingence Characterization of P	POLY	0281
orkshop on Carbon Active Sites.	Elucidation of Carbon-Active Site Behavior via Deso	FUEL	0013
by Two-+ A Partial Structural	Elucidation of Cytotoxin D from Naja naja siamensis	BIOL	0044
roups in Heavy Petroleum Frac+	Elucidation of the Nature of Naphtheno-Aromatic G	PETR	0147
Structure	Elucidation.= Hypermedia Toolkit for	COMP	0008
Preparative Reversed-Phase	Elution Chromatography of Proteins.=	ANYL	0029
Fractal Theory of Radon	Emanation from Solids.=	NUCL	0088
and Chemical Binding of Glass-	Embodied Chromophores in Supported Sol-Gel Thin	POLY	0210
esidue-Method Evaluation Crit+	Emerging Residue-Analysis Techniques. Pesticide-R	AGRO	0145
c Acids, HEPES and MOPSO, by	EMF Measurements.= +de and Two Amino Sulfoni	BIOL	0034
bute.= Carbohydrate Antigens.	Emil Fischer's Establishment of the Configuration of	CARB	0011
nalysis by Proton-Induced X-ray	Emission (PIXE). +cation of Materials: Elemental A	NUCL	0077
rning.=	Emission Characterization of Pesticide-Bag Open Bu	AGRO	0141
ties of Pt-Containing Automotive	Emission Control Catalysts.= +sicochemical Proper	COLL	0208
Monolith-Based Catalytic	Emission Control Technologies.= Camet Metal	I&EC	0045
Crudes and Residua Using Atomic	Emission Detection.= +re Gas Chromatography of	PETR	0121
Multifragment	Emission in Light Ion Induced Reactions.=	NUCL	0026
Gas-Phase Reactions: Combustion	Emission Modeling.= +n Oxides in Homogeneous	FUEL	0127
c Symmetry upon the Solid-State	Emission of Copper(I) Hilides.= +Crystallographi	INOR	0157
ermination of Temperatures by C$_2$	Emission Spectroscopy.= +drocarbon Plasmas: Det	PHYS	0268
ductively Coupled Plasma Atomic	Emission Spectroscopy.= +nalysis of Fly Ash by In	FUEL	0138
ocarbon Composition of Oilfield	Emissions and Urban Nonmethane Organic Compoun	ENVR	0022
oal: The Role of Metals. Metals	Emissions Control Technologies for Waste Incinerato	FUEL	0129
Gasohol, an+ Exhaust Pollutant	Emissions from Automobile Engines Using Alcohols,	FUEL	0095
esultados Experimentales de un	Emission Integral para Enriquecimiento en Deuterio	I&EC	0073
Chemistry Education	Emphasis Curriculum—A Case Study.=	CHED	0068
oretical Hyperpolarizabilities, with	Emphasis on HF.= +son of Experimantal and The	PHYS	0339
esi+ New Ceramer Materials with	Emphasis on High Refractive Index and Abrasion-R	POLY	0264

ing Group 13 and 15 Elements of	Empirical Formula RMER'; R,R' = Aryl, Alkyl; M =	INOR	0392
	Employee–Supervisor Partnership.=	TECH	0004
orrection in the Determination of	Emulsion Particle Distribution.= +ton–Scattering C	PMSE	0030
acromonomer Stabilizers.=	Emulsion Polymerization with Oligomeric Diblock M	COLL	0039
usion in	Emulsion Voltammetry.= Reaction–Convection–Diff	COLL	0042
rganic Synthesis in Micelles and	Emulsions: A Comparison between Different Heterog	COLL	0043
Perfluorooctylbromide (PFOB)	Emulsions.=	COLL	0080
oquimicas de Aceros Microaleados	en Acido Sulfurico.= +tadistico de Tecnicas Electr	PHYS	0281
Transferencia de CuSO$_4$	en Agua/Agua–Acetona.= Energia Libre de	PHYS	0236
el Comportamiento de los Gazas	en Condensables en Geotermica de Fasico de los Azu	I&EC	0013
que Integral para Enriquecimiento	en Deuterio.= +tados Experimentales de un Empa	I&EC	0073
Asbestos	en el Area Urbana de la Ciudad de Mexico?.=	ENVR	0025
Microbiologia de Alimentos	en el LNSP: Enfoque Prospectivo.=	AGFD	0206
Inoxidable 304 y 316 Utilizadas	en el Seguimiento de las Reacciones Acido Base en M	ANYL	0055
nto de los Gazas en Condensables	en Geotermica de Fasico de los Azusfres.= +rtamie	I&EC	0013
Galio en Zeolitas del Tipo Pentasil	en la Aromatizacion de Propano.= +Contenido de	COLL	0125
Biologicas de la Carrera de QFB	en la Facultad de Quimica, UNAM (Analisis Estadist	CHED	0092
eficiencia o Exceso de Vitamina E	en la Fertilidad de Ratas Hembras.= +ecto de la D	AGFD	0210
C–H por Ni	en la Molecula de Metano.= Activacion del Enlace	PHYS	0244
oxidable 316/Pelicula de Oxido,	en la Validacion del Metodo Potencio–Metrico para C	ANYL	0066
Que Incrementan la Reprobacion	en las Materias Biologicas de la Carrera de QFB en l	CHED	0092
terionicos.+ Efecto de la Presion	en las Propiedades de Conduccion en Polimeros Zwit	ORGN	0159
on de Metales Pesados	en Leche de Establos del Sur del D.F.= Determinaci	ENVR	0037
Nivel Academico de los Alumnos	en los Resultados de la Evaluacion de una Materia B	CHED	0093
Indices de Reprobacion	en Materias Biologicas.= Analisis Longitudinal de	CHED	0091
usion, Oxidorreduccion con Zinc	en Medio Acido y Basico, Preparacion de Nitrado de	CHED	0172
ento de las Reacciones Acido Base	en Medio Acuoso.= +y 316 Utilizadas en el Seguimi	ANYL	0055
hidrato de L–Arginina en Tabletas	en Medio.= +tencio–Metrico para Cuantificar Clor	ANYL	0066
Glicol	en Mexico.= Estudio Sobre el Uso de Eteres de	ENVR	0038
en las Propiedades de Conduccion	en Polimeros Zwitterionicos.= +ecto de la Presion	ORGN	0159
ica de Clorhidrato de L–Arginina	en Producto Terminado Utilizando como Electrodo I	CHED	0163
Hormigas: Alimento Rico	en Proteinas.=	AGFD	0205
tal a Partir de Errores Detectados	en Publicaciones Cientificas.= +enanza Experimen	CHED	0178
ada con S. aureus y su Aplicacion	en Quesos.= +de Sodio a Partir de Leche Contamin	AGFD	0107
Inmunomoduladora del Zinc	en Ratones Inmunodeficientes.= Actividad	AGFD	0144
ia del Transporte de Calor	en Reacciones Solido–Gas NO–Cataliticas.= Influenc	I&EC	0068
ntificar Clorhidrato de L–Arginina	en Tabletas en Medio.= +tencio–Metrico para Cua	ANYL	0066
nfluencia del Contenido de Galio	en Zeolitas del Tipo Pentasil en la Aromatizacion de	COLL	0125
eaction of Fluoride with Human	Enamel: XPS, SEM, and Solution Characterization.=	COLL	0085
Kinetic Fate of a Thiamin–Bound	Enamine Intermediate in Water.= +ervation of the	BIOL	0047
of Thiamin Diphosphate–Bound	Enamines on Brewers' Yeast Pyruvate Decarboxylase	BIOL	0048
ing Agents.+ Novel Quaternary–	Enamino–Nitrogen–Containing Neuromuscular Block	MEDI	0046
stic Investigation+ Stereospecific	Enammonium–Iminium Rearrangements: A Mechani	ORGN	0296
entary Bifurcate Approach to T+	Enantiocontrolled Taxane Construction: A Complem	ORGN	0291
atase Inhibitors on Cell+ Direct	Enantiomeric Separation and Analysis of Some Arom	ANYL	0133
oxide Diastereom+ Synthesis of	Enantiomerically Pure Bay–Region 10,11–diol–8,9–Ep	ORGN	0208
Methods.=	Enantiomerically Pure Compounds via Biocatalytic	BIOT	0092
ucture–Activity Relationships of	Enantiomerically Pure Dioxolane–Pyrimidine Nucleo	MEDI	0051
Organic Synthesis: Synthesis of	Enantiomerically Pure 1,2–Epoxy Aldehydes, Thiiran	ORGN	0052
d Fla+ Survey of Alpha–Ionone	Enantiomers in Raspberry Concentrates, Essences, an	AGFD	0181
droxy Carbonyl Compounds Using	Enantiopure N–Sulfonyloxaziridines.= +s of α–Hy	ORGN	0293
ion Host: Resolu+ Application of	Enantiopure Troger's Base Methosulfate as an Inclus	ORGN	0041
=	Enantioselective Chromatography in the Real World.	ANYL	0120
markably Effective Catalysts for	Enantioselective Metal Carbene Transformations.=	CATL	0002
o+ Lipases: Useful Biocatalysts for	Enantioselective Reactions of Chiral Flavor Comp	AGFD	0026
alating Ru(II) Complex Contain+	Enantioselective Recognition by DNA of a Noninterc	INOR	0233
steines.=	Enantioselective Routes to syn– and anti–β–Phenylcy	ORGN	0115
Synthesis of Lactones with High	Enantioselectivity by Intramolecular Carbon–Hydrog	ORGN	0042
ying the Substra+ Increasing the	Enantioselectivity of Lipases and Esterases by Modif	BIOT	0091
nsiveness to Experimental Allergic	Encephalomyelitis.= +at Induce Immune Unrespo	BIOL	0086
ted Mutagenesis of the Gene–	Encoding T4 dCMP Hydroxymethylase.= Site–Direc	BIOL	0160
	Encuentro de Dos Mundos a Traves de la Quimica.=	HIST	0035
Decay Spectrum from End to	End and the Neutrino Mass.= The Tritium Beta	NUCL	0011
cromolecules Observed+ End–to–	End Distance Distributions and Diffusion within Ma	PHYS	0061
Polystyrenes with Dendritic	End Groups.= Synthesis and Solution Properties of	POLY	0322
ere, or Can Be Limited to Chain–	End Recognition of an Antigen.= +Epitopes Anywh	CARB	0019
Beta Decay Spectrum from	End to End and the Neutrino Mass.= The Tritium	NUCL	0011
Exclusion Chromatography and	End–Group Analysis for Poly Methyl Methacrylate.=	PMSE	0050
G Facile Measurement of Phenolic	End–Groups in bisPhenol A Polycarbonate Using	POLY	0083
Compounds. Side–on (μ–N$_2$) vs.	End–on (μ–N$_2$) Bonding of Early Transition–Metal D	INOR	0362
thin Macromolecules Observed +	End–to–End Distance Distributions and Diffusion wi	PHYS	0061
tereoselective Syntheses of Exo or	Endo Polycyclic Epoxides.= +ormation Styrenes: S	ORGN	0236
s at Nonstereogenic Atoms by the	Endocyclic Restriction Test.= +philic Substitution	ORGN	0120
onucleotide Synthesis for UvrABC	Endonuclease DNA Repair.= +n–Linked Deoxyolig	BIOL	0126
DNA Activation of Nae I	Endonuclease.=	BIOL	0125
Mechanisms in+ Interactions of	Endophyte–Induced Allelochemicals with Resistance	AGRO	0114
n Nephrocalcin with+ EPR and	ENDOR Characterization of Calcium–Binding Sites i	BIOL	0045
True Reaction Intermediate of +	Endor–Determined Structure at the Active Site of a	BIOL	0011
es to Remove E. coli Proteins and	Endotoxin.= +of Washing sCD4–183 Inclusion Bodi	BIOT	0171
Assessment.= Noncancer Health	Endpoints: Approaches to Quantitative Risk	ENVR	0068

Asymmetric Olefination and	Ene Reactions.= Plenary Session. Studies in	ORGN	0279
Models for the	Ene-Diyne Toxines with an Enol-Ether Trigger.=	ORGN	0102
nocarbonate in a Latent Z-Hex-3-	ene-1,5-diyne Unit.= +a the Elimination of a Thio	ORGN	0096
n Stud+ Synthesis of New Cyclic	Enediyne Model Compounds for Bergman Cyclizatio	ORGN	0097
Substituted	Enediyne Systems.= Synthesis and Reactivity of	ORGN	0095
in a Latent+ Cyclic Conjugated	Enediynes via the Elimination of a Thionocarbonate	ORGN	0096
$F_3)_4$ on Atomica+ Mechanism and	Energetics in the Deposition of Platinum from Pt (P	COLL	0046
Surface of SiO_2 and Its Effect on	Energetics of Surface Reactions.= +atic Field of the	COLL	0116
Surface	Energetics.= Improving Barrier Properties through	POLY	0196
Experiment. Diene Excited-State	Energetics, Structures, and Dynamics.= +stry with	PHYS	0075
gua-Acetona.=	Energia Libre de Transferencia de $CuSO_4$ en Agua/A	PHYS	0236
Destilacion con Integracion de	Energia.= Sintesis y Modificacion de Secuencias de	I&EC	0074
iP, and Their Isomer+ Ab Initio	Energies and Shapes of $HSiS^+$, $HSiO^+$, HCP, and HS	PHYS	0315
Complete Basis-Set	Energies fo One- and Two-Carbon Molecules.=	PHYS	0317
GeV/c+ Forward and Transverse	Energies in Relativistic Heavy Ion Collisions at 14.6	NUCL	0051
isms. Calculation of Adsorption	Energies in Slitlike and Wedge-Shaped Micropores.=	FUEL	0008
ollision Mechanisms: Intermediate	Energies. Heavy Residue Prop+ Nucleus-Nucleus C	NUCL	0024
Association.= Apparent Free	Energy Adsorption: Measurement of Surface-Bulk	COLL	0122
States by Ferrocen+ Competitive	Energy and Electron Transfer Quenching of Excited	INOR	0163
l Technologies from a Process +	Energy and Entropy in the Selection of Petrochemica	I&EC	0064
ispersion Relation and the Low-	Energy Behavior of Heavy-Ion Optical Potential.=	NUCL	0040
e Conversion of Solar to Electric	Energy by the Other Photosynthetic System in Natu	PHYS	0019
stal Lagoon.=	Energy Content of Macrocrustacea in a Tropical Coa	BIOL	0105
ns on the Femtosecond Timesc+	Energy Dissipation in Liquid-State Chemical Reactio	PHYS	0085
Lecture. Distributive Mixing and	Energy Distribution in Twin Screw Extruders.= +y	PMSE	0175
Photodissociation of+ Internal	Energy Distributions of SO ($X^3\Sigma^-$) from the 193-nm	PHYS	0240
Internal and Translational	Energy Effects on Cluster Ion Chemistry.=	PHYS	0242
de, Electronic State, and Collision	Energy Effects on Ion-Molecule Reactions.= +1 Mo	PHYS	0241
d+ Ultrafast Measurements of	Energy Flow and Structural Dynamics in Photoexcite	PHYS	0335
ptical Spectroscopies. Vibrational	Energy Flow at Solid Surfaces.= +esonance and O	PHYS	0344
Clean	Energy from Municipal Waste.=	FUEL	0093
ean Energy from Waste and Coal:	Energy from Plastics and Municipal Waste. Co+ Cl	FUEL	0090
ean Energy from Waste and Coal:	Energy from Plastics and Municipal Waste. Gas+ Cl	FUEL	0091
ean Energy from Waste and Coal:	Energy from Plastics and Municipal Waste. Re+ Cl	FUEL	0092
an Energy from Waste and Coal:	Energy from Sewage Sludge, Biomass, and Municipa	FUEL	0115
cterization, and Other To+ Clean	Energy from Waste and Coal: Ash Utilization, Chara	FUEL	0136
nd Municipal Waste. Co+ Clean	Energy from Waste and Coal: Energy from Plastics a	FUEL	0090
nd Municipal Waste. Gas+ Clean	Energy from Waste and Coal: Energy from Plastics a	FUEL	0091
nd Municipal Waste. Re+ Clean	Energy from Waste and Coal: Energy from Plastics a	FUEL	0092
ludge, Biomass, and Mun+ Clean	Energy from Waste and Coal: Energy from Sewage S	FUEL	0115
etals Emissions Control+ Clean	Energy from Waste and Coal: The Role of Metals. M	FUEL	0129
tio Quantum Chemical Potential	Energy Functions for the Atomistic Simulation of Po	POLY	0350
rcuture,+ Tuning the Multiplet	Energy Gap in Non-Kekule Molecules: Synthesis, St	ORGN	0128
Experiments.= Intermediate-	Energy Heavy Ion Reactions and RI-Beam	NUCL	0029
Copper by Intermediate-	Energy Heavy Ions.= Target Fragmentation of	NUCL	0025
Platinum.= Ultra-Low-	Energy Interactions (≤3 eV) of CH_3^+ Ions on	COLL	0216
alysis of Materials Using High-	Energy Ion Beams.= Composition and Structural An	INOR	0008
High-	Energy Ion Scattering for Materials Analysis.=	NUCL	0065
Surface Reactivity Using Low-	Energy Ion Scattering.= Studies of $TiO_2(110)$	COLL	0172
SiO_2 Layers.= Medium-	Energy Ion-Scattering Study of Nickel on Ultrathin	COLL	0176
	Energy Kinetics in Nanometer Crystals.=	PHYS	0316
Cluster	Energy Loss in Solids.=	NUCL	0053
sidue Properties in Intermediate-	Energy Nuclear Collisions with Gold.= +Heavy Re	NUCL	0024
and Abrasion in High-	Energy Nucleus-Nucleus Collisons.= Transparency	NUCL	0048
rational Spectrum, and Binding	Energy of the Lowest Singlet and Triplet States of M	PHYS	0276
and Municipal Waste. Recovery of	Energy or Chemicals from MSW.= +from Plastics	FUEL	0092
Conversion of Waste Polymers to	Energy Products.= +lastics and Municipal Waste.	FUEL	0090
c Phenomena. Synthesis of High-	Energy Projectile and Target Fragmentation.= +sti	NUCL	0046
Recoil Studies and High-	Energy Reaction Models.=	NUCL	0059
Electronic	Energy Relaxation at Interfaces.=	PHYS	0062
Borexino: Low-	Energy Solar Neutrino Spectroscopy.=	NUCL	0033
amination of the C_3H_6 Potential	Energy Surface Using Quadratic Configuration Intera	PHYS	0305
Potential	Energy Surfaces for BCl Reactions.=	PHYS	0088
Developing and Transferring New	Energy Technology to the Marketplace.= +bility of	FUEL	0142
Low-	Energy Tests of Neutrino Masses.=	NUCL	0020
Dissipative Reactions from Low	Energy to the Fermi Domain.= Aspects of	NUCL	0007
+Time Dependence of Electronic	Energy Transfer and Depolarization in Polymers.=	POLY	0078
ld(I) Cyanide Salts: Evidence for	Energy Transfer and Relavtivistic Gold-Gold Covale	INOR	0189
s. Chord-Distribution Analysis +	Energy Transfer and Relaxation in Condensed Phase	PHYS	0060
s. Conformation and Dynamics +	Energy Transfer and Relaxation in Condensed Phase	PHYS	0107
s. Energy Transfer and Relaxat+	Energy Transfer and Relaxation in Condensed Phase	PHYS	0080
s. Excited-State Structure and+	Energy Transfer and Relaxation in Condensed Phase	PHYS	0018
s. Time-Domain Analysis of the+	Energy Transfer and Relaxation in Condensed Phase	PHYS	0034
s. Torsional Solitons in Organic+	Energy Transfer and Relaxation in Condensed Phase	PHYS	0129
Relaxation in Condensed Phases.	Energy Transfer and Relaxation in Molecular Glasse	PHYS	0080
icro+ Toward Spatially Resolved	Energy Transfer and Scanning Molecular Exciton M	PHYS	0022
ers on Single-Crystal+ Electronic	Energy Transfer between Donor and Acceptor Adlay	PHYS	0065
l, Rotational, and Translational	Energy Transfer from Highly Vibrationally Excited N	PHYS	0334
Sites in Silica Supports.=	Energy Transfer from Organic Adsorbates to Defect	PHYS	0135
	Energy Transfer in Latex Microdomains.=	PHYS	0108

Electron Transfer and	Energy Transfer in LB Films.= Photoinduced	BIOT	0103
f Colored Noise.=	Energy Transfer in Molecular Aggregates: Influence o	PHYS	0023
mputer Simulation of Electronic	Energy Transfer in Polymers: Comparison of Theory	PHYS	0109
Computer Simulation of Direct	Energy Transfer in Polymers.=	POLY	0079
ngle Scattering and Direct Dipolar	Energy Transfer in Porous Solids.= +ith Small-A	PHYS	0060
ing and Multiphonon Up-Pump+	Energy Transfer in Ultrafast Solids: Vibrational Cool	PHYS	0250
ules Observed by Intramolecular	Energy Transfer Using Frequency-Domain Fluorome	PHYS	0061
ion and Its Implications for	Energy Transfer.= Photon Correlation and Localizat	PHYS	0064
es of Intramolecular Electron and	Energy Transfer?.= +ng in Predicting Absolute Rat	ORGN	0143
olid: Exciton Trapping and Ann+	Energy Transport in a One-Dimensional Inorganic S	PHYS	0131
d Ultrafast Studies of Electronic	Energy Transport in Spatially Disordered Systems.=	PHYS	0133
Relaxation.= Dispersive	Energy Transport: Relation to Spin Glass	PHYS	0082
A Fascinating Feedstock for Clean	Energy.= +and Municipal Waste. Sewage Sludge:	FUEL	0115
CO_2 and Water: A Clean Source of	Energy.= +rowave-Induced Catalytic Reactions of	FUEL	0102
CO_2 and Water: A Clean Source of	Energy.= +rowave-Induced Catalytic Reactions of	CATL	0055
in Nuclear	Energy.= Nuclear Fuel Cycle and Special Problems	FUEL	0024
Interaction	Energy.= The Pivot Algorithm with	POLY	0077
e of Glassy Polymers—I. Chain +	Energy-Consuming Micromechanisms in the Fractur	PMSE	0073
Catalyst Supports.= Electron	Energy-Loss Spectrometry of Alumina and Silica	PETR	0073
ies+ High-Resolution Electron-	Energy-Loss Spectroscopy Studies of Adsorbed Spec	COLL	0173
tur+ High-Resolution Electron-	Energy-Loss Spectroscopy Studies of High-Tempera	PHYS	0279
High-Pressure Effects+ Triplet	Energy-Transfer in Isotopically Mixed Naphthalene:	PHYS	0081
en el LNSP:	Enfoque Prospectivo.= Microbiologia de Alimentos	AGFD	0206
Chromophoric Assemblies with	Enforced Polarity: Inorganic and Heterocyclic Linkag	POLY	0215
Chemist and the Chemical	Engineer.= Interface between the Analytical	I&EC	0051
lysaccharides for Production—An	Engineer's Viewpoint.= +Properties of Microbial Po	BIOT	0199
li.= Expression and Scaleup of	Engineered Antibody Fragments from Escherichia co	BIOT	0058
in Stability and Conformation.=+	Engineered Metal Chelation Sites as Probes of Prote	BIOT	0133
Polymer Thin Films, I: Molecular	Engineering and Processing of Polymers with Large	POLY	0011
rgetics of+ Progress in Metabolic	Engineering and Production of Biochemicals. Bioene	BIOT	0029
a Nonlinear Interfac+ Molecular	Engineering by Laser Chemical Vapor Deposition of	BIOT	0024
mical Research, Development, and	Engineering Center.= +ent at the U.S. Army Che	CINF	0037
eering—the Chemical Science and	Engineering Continuum.= +y and Chemical Engin	I&EC	0048
Molecular	Engineering for Monoelectronic Devices.=	BIOT	0006
Biochemical	Engineering Fundamentals.=	BIOT	0223
Teaching of Chemical	Engineering in High School.=	CHED	0108
Derived from Hydrocarbon Fuels:	Engineering Materials. Carbon Bla+ New Materials	FUEL	0047
Derived from Hydrocarbon Fuels:	Engineering Materials. Developme+ New Materials	FUEL	0050
roduction of Novel Er+ Genetic	Engineering of Saccharopolyspora erythraea for the P	BIOT	0110
Catalysts.= Molecular	Engineering of Supported Vanadium Oxide	COLL	0196
lymeric Colorant+ Chemical and	Engineering Perspectives on the Ultrafiltration of Po	I&EC	0049
ealand Gas-to-Gasoline Plant: An	Engineering Tour de Force.= +J. F. Roth. New Z	I&EC	0060
terfacing Chemistry and Chemical	Engineering. Interfacing Chemistry and Chemic+ In	I&EC	0048
terfacing Chemistry and Chemical	Engineering. Panel Discussion: At the Interface+ In	I&EC	0057
y Streptom+ Metabolic Pathway	Engineering: The Biosynthesis of Tetracenomycins b	BIOT	0083
between Chemistry and Chemical	Engineering.= +mistry Perspective of the Interface	I&EC	0050
BASIC Biochemical	Engineering.=	BIOT	0225
Biochemical	Engineering.=	BIOT	0226
Phospholipase A_2	Engineering.=	BIOL	0016
erfacing Chemistry and Chemical	Engineering—the Chemical Science and Engineering	I&EC	0048
hing Biotechnology. Biochemical	Engineering, 2nd Edition—Aiba, Humphrey, and Mi	BIOT	0221
Analytical Chemistry for	Engineers.=	CHED	0154
tant Emissions from Automobile	Engines Using Alcohols, Gasohol, and Other Oxygena	FUEL	0095
Early	English Steroid History.=	HIST	0014
Random Mutagenesis to	Enhance Enzyme Activity in Organic Solvents.=	BIOT	0068
New Trifluoromethyl-Containing	Enkephalin Analogues as Inhibitors for Enkephalin	MEDI	0151
ciceptive Agents: O- and N-Alk+	Enkephalin Analogues as Systemically Active Antino	MEDI	0152
phalin Analogues as Inhibitors for	Enkephalin-Aminopeptidase.= +-Containing Enke	MEDI	0151
Activacion del	Enlace C-H por Ni en la Molecula de Metano.=	PHYS	0244
Toxines with an	Enol-Ether Trigger.= Models for the Ene-Diyne	ORGN	0102
Face Selection in	Enolate Chemistry.=	ORGN	0086
Catalysis of	Enolate Reactions by an Artical Enzyme.=	ORGN	0053
eaction of Chiral β-Lactam Ester	Enolates.= +tric Alkylation, Acylation, and Aldol R	ORGN	0046
oxy-Protected Cyanohydrins with	Enones: Synthetic Applications.= +ions of Carbeth	ORGN	0228
-Catalyzed Conjugate Addition to	Enones.= +Reactions of Alkyl Zirconocenes: Copper	ORGN	0089
f the Allylic Rearrangement and	Enoyl Reduction Catalyzed by Brevibacterium ammo	BIOL	0065
Process for Production of Oxygen-	Enriched Combustion Air.= +l Swing Absorption	I&EC	0023
idimensional NMR of Isotopically	Enriched Proteins.= +Optical Spectroscopies. Mult	PHYS	0113
Technology: Efficient Aroma	Enrichment in Fruit Cells.= Precursor Atmosphere	AGFD	0023
mistry.=	Enrichment of Precollege Science with Polymer Che	CHED	0009
rtir de Errores Detectados en P+	Enriquecimiento de la Ensenanza Experimental a Pa	CHED	0178
ales de un Empaque Integral para	Enriquecimiento en Deuterio.= +tados Experiment	I&EC	0073
ce-Relate+ Declining Chemistry	Enrollments: Toward a Path Analytic Model of Scien	CHED	0177
Apoyos Computarizados para la	Ensenanza de la Quimica a Nivel Medio Superior.=	CHED	0124
dos en+ Enriquecimiento de la	Ensenanza Experimental a Partir de Errores Detecta	CHED	0178
cifolia cav. (Asteraceae).= New	Ent-Atisene Glycosides from the Roots of Stevia sali	MEDI	0034
logy for Small and Medium-Size	Enterprises: The Codetec, Inc., Experience in Brazil.	SCHB	0023
Theoretical Models for	Enthalpic Properties of Alkanes.= Optimal Graph	COMP	0023
ital Estimation of the Activation	Enthalpies for Intramolecular Hydrogen Transfer as	ORGN	0175
e Bonding: Determination of the	Enthalpy and Kinetics of Ligand Exchange Reaction	INOR	0192

ilan+ Steric Contributions in the	Enthalpy of Reaction of CpMn(CO)2 with Tertiary S	INOR	0386
erpenetrating Polymer Networks.	Entirely Radical in Situ Synthesis of Interpenetratin	PMSE	0001
tion.=	Entrance Channel Excitations in the ^{28}Si + ^{28}Si Reac	NUCL	0039
al Evidence of Flow Structuring in	Entrance Flow of Polypropylene Melts.= +eologic	PMSE	0137
ial y Programas de Becas: Vinculo	entre Universidades y Secotr Productivo.= +cio Soc	CHED	0114
How Chemical	Entrepreneurs Can Cope with Regulatory Burdens.=	SCHB	0011
ies from a Process+ Energy and	Entropy in the Selection of Petrochemical Technolog	I&EC	0064
Rigid Calcium-	Entry Blocker.= Synthesis of a Conformationally	ORGN	0114
odel for Benzazepinone Calcium	Entry Blockers: Application to the Design of Structu	MEDI	0188
Polymer Market-	Entry Strategies.= Value Perception Analysis for	CMEC	0011
on Content of Chemical Formulas:	Enumeration of Fluorenoids/Fluoranthenoids.= +i	COMP	0027
Combinatorics of Cluster Ion	Enumeration.=	COMP	0028
he polysaccharide Chains of Cell	Envelope Lipopolysaccharides of Salmonella and Shi	CARB	0017
n of Strontium-90 in a High-Salt	Environment at the Nevada Test Site.= +erminatio	NUCL	0085
cal Structure Handling. Modeling	Environment for Map Interpretation.= +nal Chemi	CINF	0029
Comparison of the Coordination	Environment of Mn(III) and Mn(IV) Horseradish Pe	BIOL	0095
e System in the Water-Restricted	Environment of Nonaqueous Solvents.= +d Enzym	BIOL	0091
to Protect the	Environment.= Application Equipment Technology	AGRO	0125
Protecting the Dealer and the	Environment.= Environmental Site Assessments:	FERT	0015
s: New Complexes with NiS2O2	Environment.= Ni Chemistry of Oxothiolate Ligand	INOR	0204
erature- and Pressure-Controlled	Environment.= +terocyclic Compounds in a Temp	AGFD	0073
Arsenicals in the Marine	Environment.= Uptake and Biotransformation of	INOR	0137
ional Exposure to Cyclophosph+	Environmental and Biological Monitoring of Occupat	CEI	0010
Benefits of Pesticides: An	Environmental and Consumer View.= Risks and	AGRO	0059
of Soil Gas Techniques Used for	Environmental Application in Japan and Introductio	ENVR	0049
e+ Surface Science of Catalysis:	Environmental Applications of Catalysts—I. Methan	COLL	0185
al+ Surface Science of Catalysis:	Environmental Applications of Catalysts—II. Therm	COLL	0204
en Sources.=	Environmental Benefits of Controlled-Release Nitrog	FERT	0006
	Environmental Concerns: A Future Perspective.=	AGRO	0067
f Urinary Phenolic Metabolites of	Environmental Contaminants in the U.S. +rement o	CEI	0002
of Applicat+ Soil Gas Analysis—	Environmental Contaminants. Review of the History	ENVR	0048
g Quantitative Struc+ Predicting	Environmental Effects of Substituted Benzenes Usin	ANYL	0047
uences of Processing, Variety, and	Environmental Factors.= +spberry Phenolics: Infl	AGFD	0169
rea Herbicide.=	Environmental Fate of DPX-E9636, a New Sulfonylu	AGRO	0029
Soil Chemistry. Impact of Fertil+	Environmental Issues and Answers in Fertilizer and	FERT	0013
Soil Chemistry. The Retail Ferti+	Environmental Issues and Answers in Fertilizer and	FERT	0017
nd Their Degradation Products in	Environmental Matrices by LC/MS.= +hemicals a	AGRO	0031
f Polyaromatic Hydrocarbons in	Environmental Matrices by Liquid Chromatography	ENVR	0015
Case Studies of Carcinogens:	Environmental or Occupational Exposures.=	CHAS	0020
us Photocatalyzed Oxidations of	Environmental Organics Using AM1 Simulated Sunli	INOR	0318
Ecological Knowledge and	Environmental Problem Solving.=	ENVR	0063
ust W+ Quantitative Ranking of	Environmental Problems According to Risk: What M	ENVR	0061
ative Risk: A Blueprint for Future	Environmental Protection Efforts.= +is Task? Rel	ENVR	0061
evelopment, Quality Control, and	Environmental Protection.= +aboratory: Product D	FERT	0002
Biological Monitoring for	Environmental Public Health.= Application of	CEI	0007
on of Risk-Based Management of	Environmental Resources and Impacts.= +ementati	ENVR	0076
richemical Dealers.=	Environmental Self-Evaluation for Fertilizer and Ag	FERT	0018
ler and the Environment.=	Environmental Site Assessments: Protecting the Dea	FERT	0015
o Be Announced.=	Environmental Successes in the Chemical Industry. T	CMEC	0017
ention Relationships: Biological,	Environmental, and Physicochemical Applications—I	ANYL	0043
ention Relationships: Biological,	Environmental, and Physicochemical Applications—I	ANYL	0068
ention Relationships: Biological,	Environmental, and Physicochemical Applications—I	ANYL	0098
ry. The Retail Fertilizer Dealer as	Environmentalist.= +in Fertilizer and Soil Chemist	FERT	0015
es P450 and Their Roles in	Environmentally Based Disease.= Human Cytochrom	CEI	0021
Versatile Process for Producing	Environmentally Safe Shale Stabilizers.= Novel and	CARB	0009
lecular Mechanisms of Action of	Environmentally Significant Carcinogens and Mutage	CHAS	0007
d Feeds: Natural and Unnatural	Environments for Long-Chain Omega-3 Fatty Acids.	AGFD	0160
in Isolates from Deep Subsurface	Environments.= +of Cloned Biodegradative Genes	BIOT	0117
und Nickel: P2S2 vs. N2S2 Ligand	Environments.= +allic Functionalities at Sulfur-Bo	INOR	0177
Fiber-Optic Probes for Hostile	Environments.= Development of Raman	ANYL	0065
Phosphorus Atoms in Unusual	Environments.=	INOR	0292
ures and Pressures in Geothermal	Environments.= +with Water at Elevated Temperat	INOR	0246
Enynes and Enyne-Allenes.=	Stereoselective Synthesis of	ORGN	0099
Stereoselective Synthesis of	Enynes and Enyne-Allenes.=	ORGN	0099
Materials.= Increasing	Enzymatic Accessibility on Lignocellulosic	BIOT	0250
of a Novel	Enzymatic Acylation.= Unusual Kinetic Properties	BIOT	0160
y for Asymmetric Synthesis.=	Enzymatic Aldol Additions: A Building Block Strateg	BIOT	0090
s and Sulfated P+ Prevention of	Enzymatic Browning in Fruit Juice with Cyclodextrin	AGFD	0030
t and Vegetable Juices: Control of	Enzymatic Browning.= +nts to Stabilize Raw Frui	AGFD	0027
ne-5'-Phosphosul+ Chemistry of	Enzymatic Catalysis. Mechanistic Studies of Adenosi	BIOL	0008
nd Specificity. Structural Basis of	Enzymatic Catalysis.= +lyst Design for Stability a	BIOT	0239
ounds. Chrono-C+ Thermal and	Enzymatic Conversions of Precursors to Flavor Comp	AGFD	0022
ounds. Glucoside+ Thermal and	Enzymatic Conversions of Precursors to Flavor Comp	AGFD	0050
ounds. Glycosidi+ Thermal and	Enzymatic Conversions of Precursors to Flavor Comp	AGFD	0006
ounds. Studies on+ Thermal and	Enzymatic Conversions of Precursors to Flavor Comp	AGFD	0070
Probing	Enzymatic Glycoside Hydrolysis.=	ORGN	0267
emicelluloses.=	Enzymatic Hydrolysis of Isolated and Fiber-Bound H	BIOT	0215
ding and Implications Concerning	Enzymatic Metalation.= +: Mode of Interfacial Bin	CATL	0060
unds.= Regioselective Chemo-	Enzymatic Modifications of Polyhydroxylated Compo	BIOT	0064
its.=	Enzymatic Oxidation of Phenolic Compounds in Fru	AGFD	0151

e Functional Properties in Foo+	Enzymatic Phosphorylation of Soy Protein to Improv	AGFD	0140
Mechanism of	Enzymatic Phosphotriester Hydrolysis.=	BIOL	0010
Reactor: Experimental Verific+	Enzymatic Resolution Using a Multiphase Membrane	BIOT	0042
tyl Neuraminic Acid.+ Chemical	Enzymatic Synthesis of a Carbon Glycoside of N-Ace	CARB	0034
ounds.=	Enzymatic Synthesis of Monoterpenoid Flavor Comp	AGFD	0009
Chemical Modification on	Enzyme Activities and Stabilities.= Effect of	BIOT	0219
Mutagenesis to Enhance	Enzyme Activity in Organic Solvents.= Random	BIOT	0068
Thiamin Diphosphate-Dependent	Enzyme Brewers' Yeast Pyruvate Decarboxylase.=	BIOL	0032
Chimeras in the Trifunctional	Enzyme C_1-Tetrahydrofolate Synthase.= Domain	BIOT	0243
rization of Lignin with Low Mol+	Enzyme Cross-Linking, Modification, and Copolyme	BIOT	0041
Bead Polymers as Supports for	Enzyme Immobilization.= IPN-Modified Porous	PMSE	0081
Conditions.=	Enzyme Inactivation in Supercritical Carbon Dioxide	AGFD	0028
Design of	Enzyme Inhibitors.= New Method for the de Novo	COMP	0037
Columns as Probes of	Enzyme Interactions.= Enzyme-Based HPLC	BIOT	0036
dies of Cathepsin B, a Proteolytic	Enzyme Involved in Tumor Metastasis.= +hips Stu	MEDI	0017
Plenary Session. Multifunctional	Enzyme Mimics and Activators.=	ORGN	0265
	Enzyme Reactions on Insoluble Substrates.=	COLL	0052
Solvent Effects on	Enzyme Reactivity.=	BIOT	0247
A General Strategy for Increasing	Enzyme Stability in Polar Organic Solvents.= +on:	BIOT	0244
ers:+ Studies of Lysosomotropic	Enzyme Substrates with Human Fibroblast Monolay	BIOL	0079
alytic Properties of a Combined	Enzyme System in the Water-Restricted Environmen	BIOL	0091
Fermentation and	Enzyme Technology.=	BIOT	0222
ex and to the E1 Decarboxylating	Enzyme: A Monoclone That Is 98 Inhibitory.= +pl	BIOL	0031
Articial	Enzyme.= Catalysis of Enolate Reactions by an	ORGN	0053
ituted 1,2-Dioxetanes: Improved	Enzyme-Activated Chemiluminescent Substrates in t	BIOT	0167
sylmethionine Decarboxylase.=+	Enzyme-Activated Irreversible Inhibitors of S-Adeno	BIOL	0052
y N-Nitr+ Active-Site-Directed	Enzyme-Activated Substrate Inhibition of Trypsin b	ORGN	0022
Interactions.=	Enzyme-Based HPLC Columns as Probes of Enzyme	BIOT	0036
ar Weight of Polyesters Formed in	Enzyme-Catalyzed Polycondensations.= +Molecul	BIOT	0094
ysteroid Dehydrogenases.=	Enzyme-Generated Irreversible Inhibitors of Hydrox	MEDI	0010
of Cross-Linking Techniques to	Enzyme/Protein Stabilization and Bioconjugate Prep	BIOT	0220
g the Formation of Biosynthetic	Enzymes and Cephamycin-C in Streptomyces clavuli	BIOT	0129
se of Polysaccharide Hydrolyzing	Enzymes During Wood Degradation by Phanerochae	BIOT	0208
Proteolytic	Enzymes from Ectothermic Organisms.=	BIOT	0194
merically Pure 1,2-Epoxy Aldeh+	Enzymes in Organic Synthesis: Synthesis of Enantio	ORGN	0052
	Enzymes in Organic Synthesis.=	ORGN	0271
ddress. Synthesis of Carbohyd+	Enzymes in Organic/Bioorganic Synthesis. Keynote A	BIOT	0089
Ly+ The Muconate-Lactonizing	Enzymes Mandelate Racemase and Adenylosuccinate	BIOL	0012
Biosynthesis—From	Enzymes to Genes.= Fungal Polyketide	BIOT	0111
h Metalloporphyrins: From Model	Enzymes to Model Membranes.= +xygenations wit	ORGN	0136
teine Centers in the Hudrogenase	Enzymes.= +mplexes as Models for Nickel(III)-Cys	INOR	0179
oalanine with Tryptophan	Enzymes.= Effect of Indole on Reactions of Trifluor	BIOL	0060
he Design of Artificial Restriction	Enzymes.= +A Transesterification Catalysts and t	INOR	0406
Anticarcinogenic	Enzymes.= Phenolic Antioxidants as Inducers of	AGFD	0089
tal Complexes and Redox Sites in	Enzymes.= +sfer Reactions between Transition-Me	INOR	0283
Hyperthermophilic	Enzymes.= Structure-Function Relationships of	BIOT	0193
-Free Systems Using Immobilized	Enzymes.= +s in Nonaqueous Solvents and Solvent	BIOT	0065
Substrate Selectivity on	Enzymes.= Targeted Catalysis: Conferment of	BIOL	0090
nosine on Fe(II) Autoxidation and	Enzymic or Nonenzymic Reduction of Fe(III).= +r	AGFD	0131
sing: Protein Sepa+ Nonaqueous	Enzymology. Plenary Lecture. Nonaqueous Bioproces	BIOT	0063
Nonaqueous	Enzymology.= The How and Why of	BIOT	0067
of Erythrocyte Protoporphyrin (	EP) as a Biochemical Marker for Detecting Exposure	CEI	0020
Role in Chemical Proce+ OSHA-	EPA Interface in Health and Safety Issues. OSHA's	CHAS	0012
The OSHA-	EPA Interface: An Overview.=	CHAS	0014
OSHA and	EPA Programs.= Health and Safety Issues under	CHAS	0022
Part of	EPA Strategy.= Pesticide Container Regulations as	AGRO	0089
ology: What Is the Basis of E+	EPA's Hazard-Ranking System and Decision Method	ENVR	0077
in Ground Water as Assessed by	EPA's Hazard-Ranking System: A Balance of Risk A	ENVR	0081
ethodology: What Is the Basis of	EPA's Listing of Sites for Superfund? Overview of th	ENVR	0077
U.S.	EPA's Pollution Prevention Research Agenda.=	CEI	0023
a Pollution Prevention Ethic: The	EPA's Role and Program.= +Prevention. Toward	CEI	0022
oluntary Initiatives.=	EPA's 33/50 Project: Pollution Prevention through V	CEI	0024
duced H_2O_2 Production in Mouse	Epidermis by Green Tea Polyphenols.= +cetate-In	AGFD	0069
nitiation and Promotion in Mouse	Epidermis.= +tory Effects of Curcumin on Tumor I	AGFD	0088
Assignment of (±)-	Epiderstatin.= Total Synthesis and Stereochemical	ORGN	0183
Preventive Agent.= (-)-	Epigallocatechin Gallate (EGCG): A Cancer	AGFD	0066
of UDP-Galactose 4-	Epimerase from E. coli.= Structure Determination	BIOL	0147
gen Bond+ New Catalyst for the	Epimerization of Secondary Alcohols: Carbon-Hydro	ORGN	0220
s in Metalorganic Chemical Vapor	Epitaxy.= +tallic Precursors. Advances and Trend	COLL	0064
Layer	Epitaxy.= Surface Chemistry of GaAs Atomic	COLL	0069
d+ Antibodies Can Either Read	Epitopes Anywhere, or Can Be Limited to Chain-En	CARB	0019
n of Pig Liver Squalene	Epoxidase by Affinity Chromatography.= Purificatio	BIOL	0070
exes B+ Olefin Aziridination and	Epoxidation Catalyzed by Chiral Metal Salen Compl	ORGN	0056
D4-Sy+ Catalytic Asymmetric	Epoxidation of Simple Alkenes Using a New Class of	INOR	0300
es of Amine-Cured Epoxy: Amine	Epoxide Addition Reaction vs. +echanical Properti	POLY	0158
ure Bay-Region 10,11-diol-8,9-	Epoxide Diastereomers of Carcinogenic Dibenz[a,h]A	ORGN	0208
a Palladium Hydroxide-Catalyzed	Epoxide Hydrogenolysis.= +bohydrate Synthesis vi	ORGN	0074
Polymerization of Silicone-	Epoxide Resins.= Electron Beam-Induced Cationic	PMSE	0159
Ring-Opening Polymerization of	Epoxides Catalyzed by Dicobaltoctacarbonyl.=	POLY	0014
Inhibitors.= Reactivities of	Epoxides Toward Thiols: Applications to Papain	MEDI	0071

aration of β-Hydroxynitriles from	Epoxides with Lithium Cyanide in THF.= +s Prep	ORGN	0107
ntheses of Exo or Endo Polycyclic	Epoxides.= +ormation Styrenes: Stereoselective Sy	ORGN	0236
oughening of High-Performance	Epoxies Using Designed Core-Shell Rubber Particles	POLY	0157
esis of Enantiomerically Pure 1,2-	Epoxy Aldehydes, Thiiranes, and Aziridines.= +th	ORGN	0052
Branched-Chain Amino Sugars by	Epoxy and Triflate Sugar Intermediates.= +, and	ORGN	0071
rement of Water Uptake in Cured	Epoxy by Extrinsic and Intrinsic Fluorophores.=	POLY	0160
e) on Physical Properties of Cured	Epoxy Compounds.= +ght of Poly(dimethylsiloxan	PMSE	0174
Diamines for	Epoxy Cure Studies.= Fluorescence of Aromatic	POLY	0159
ites from Polyurethane and	Epoxy Interpenetrating Polymer Network.= Compos	PMSE	0044
ng Polymer Networks. Two-Phase	Epoxy Interpenetrating Themosets.= +erpenetrati	PMSE	0087
Novel Silicon-Containing	Epoxy Monomers and Oligomers.= Synthesis of	POLY	0015
ctric Properties during Isothermal	Epoxy Polymerization.= +ological Changes to Diele	PMSE	0178
in Oilfi+ Limitations of Current	Epoxy Polymers as Matrix for Structural Composites	PMSE	0111
ear Optical Properties of Linear	Epoxy Polymers with Pendant Sulfonyl Tolan Group	POLY	0208
acterization of CTBN-Toughened	Epoxy Resin/Castor Oil Polyurethane IPN.= +har	PMSE	0092
ect of Backbone Flexibili+ Novel	Epoxy Resins Based on Alkylenedioxydiphenols—Eff	PMSE	0162
ed Flexibilit+ Novel Waterborne	Epoxy Resins for Can and Coil Coatings with Improv	PMSE	0142
Route to Two Heavily Fluorinated	Epoxy Resins from a Fluorodiimidediol.= +Novel	POLY	0037
. Plenary Lecture. Application of	Epoxy Resins in Container Coatings: An Overview.=	PMSE	0157
Monitoring of Aromatic Diamine-	Epoxy System by Fiber Optic Fluorescence.= +re	POLY	0282
es during the Copolymerization of	Epoxy with Bishpenol-A.= +s of Oligomer Structur	POLY	0227
anical Properties of Amine-Cured	Epoxy: Amine Epoxide Addition Reaction vs. +ech	POLY	0158
on of Grafting Sites in the	Epoxy-g-Acrylic Copolymer.= Theoretical Calculati	PMSE	0161
minergic Activity of 2-Amino-1,4-	Epoxy-1,2,3,4-Tetrahydronaphthalenes.= +Dopa	MEDI	0128
omparative Mechanistic Study of	Epoxy/Amine = Cure Kinetics in Thermal and Micr	POLY	0161
Mechanistic Modeling of	Epoxy/Amine Reactions.=	POLY	0162
ing Sites in Nephrocalcin with +	EPR and ENDOR Characterization of Calcium-Bind	BIOL	0045
tion of Vanadyl Acetylacetonat+	EPR Line-Width Study of the Anisotropic Reorienta	PHYS	0264
Very High Frequency	EPR Spectroscopy and NMR Imaging of Coal.=	FUEL	0074
a Swelling Solvent Using a Novel	EPR Technique.= +ture of Coal in the Presence of	FUEL	0073
valuation of Aromatic Mimics of+	EPSP Synthase Inhibitor Design II: Synthesis and E	ORGN	0029
anism or Random, with a High+	EPSP Synthase: A Sequential-Ordered Kinetic Mech	BIOL	0053
is for Successful Mechanism-B+	EPSP Synthase: The Importance of Dynamic Catalys	BIOL	0054
te Mimics in Aromatic Analogs of	EPSP.= +of Various Moieties as Stable 3-Phospha	ORGN	0028
ity.=	Equation for the Square Root of the Electronic Dens	PHYS	0254
Nuclear	Equation of State and Finite-Range Interaction.=	NUCL	0008
Absorbers Using the PRSV	Equation of State.= Design for Packed Tower	CHED	0107
ant the Hand-Held Calculator?+	Equation Solvers or Spreadsheets: Which Will Suppl	CHED	0117
Synthesis.= Multiple	Equilibration Strategy for Polypropionate	ORGN	0100
(Cu2S), and Slag (CaO-CaF2-S+	Equilibrations of Silica-Saturated Metal (Cu), Matte	I&EC	0065
re of Interpenetrating Polymer+	Equilibrium and Nonequilibrium Microphase Structu	PMSE	0093
from Stabilized/Soli+ Effects of	Equilibrium Chemistry on Leaching of Contaminants	ENVR	0010
mers in Good Solvent.=	Equilibrium Configuration of Dense Polymacromono	POLY	0269
Glycoprotein at the Ice Surface.+	Equilibrium Constants for Adsorption of Antifreeze	BIOL	0041
ti+ 2'-Substituent Effects on the	Equilibrium Constants for Cyanide Addition to Nico	BIOL	0013
A2: Determination of the Rate and	Equilibrium Constants.= +lysis by Phospholipase	BIOL	0019
Preforms—Part+ Automation of	Equipment for Manufacturing of Textile-Structured	TECH	0007
Preforms—Part+ Automation of	Equipment for Manufacturing of Textile-Structured	TECH	0008
etection and Material Handling	Equipment for the Automatic Sortation of Postconsu	I&EC	0041
Protective	Equipment in the Laboratory.=	CHAS	0005
mmingled Plastics Recycling—An	Equipment Manufacturer's Perspective.= +tics. Co	I&EC	0025
Application	Equipment Technology to Protect the Environment.=	AGRO	0125
s de CAD/CAM para el Diseno de	Equipo de la Industria Quimica de Proceso.= +ente	COMP	0053
for the Controlled Degradation of	Erodible Polyesters for Biomedical Applications.=	CHED	0133
senanza Experimental a Partir de	Errores Detectados en Publicaciones Cientificas.=	CHED	0178
X and Y Have	Errors.= Treatment of Y = a + b X Data when Both	COMP	0057
Regioselective Synthesis of (	ertaa-Arene)Tricarbonylchromium Compounds.=	ORGN	0214
ngineering of Saccharopolyspora	erythraea for the Production of Novel Erythromycins	BIOT	0110
raphy and Mass Spectrometry of	Erythrina Alkaloids from the Foliage of Genetic Clov	ANYL	0136
roidines in the Milk of Goats Fed	Erythrina poeppigiana and E. +ce of Dihydro-Eryth	AGFD	0129
Foliage of Genetic Cloves of Three	Erythrina Species.= +rythrina Alkaloids from the	ANYL	0136
arker for Detec+ Future Role of	Erythrocyte Protoporphyrin (EP) as a Biochemical M	CEI	0020
eppigiana+ Presence of Dihydro-	Erythroidines in the Milk of Goats Fed Erythrina po	AGFD	0129
hraea for the Production of Novel	Erythromycins.= +ering of Saccharopolyspora eryt	BIOT	0110
n of a Soluble Form of the Murine	Erythropoietin Receptor.= +n and Characterizatio	BIOL	0108
geneity of Recombinant Human	Erythroproteins by Free Solution and Charge-Revers	ANYL	0018
lics Production in Lithospermum	erythrorhizon Suspension Cultures with Fungal Elici	BIOT	0229
ctrospray Mass Spectrometry (	ES MS).= Linkage Analysis in Disaccharides by Ele	CARB	0048
on a Metal Surface by	ESCA.= Study of Thin Bicomponent Polymer Films	POLY	0100
Benzyne: Is It Possible To	Escape from the Tar Baby?.= Cyanomethylene and	PHYS	0032
ered Antibody Fragments from	Escherichia coli.= Expression and Scaleup of Engine	BIOT	0058
Protein Aggregates in	Escherichia coli.= Formation and Characteristics of	BIOT	0115
f Induction Level in Recombinant	Escherichia coli.= +ngent Response as a Function o	BIOT	0034
n against Stomach, Lung, and	Esophagus Carcinogenesis by Green Tea.= Protectio	AGFD	0067
uantitativa de Tierras Raras con	Espectrometria de Masas con Fuente de Impacto Ele	ANYL	0051
de la Transferencia de Carga por	Espectroscopia IR de Complejos de Transferencia de	INOR	0190
olycrystals under Shear Deform+	ESR Investigation of Radical Processes in Organic P	PHYS	0245
ncies: Advantages and Prospects+	ESR Spectroscopy at Sub-9-GHz Microwave Freque	PHYS	0215
Combinatorics of NMR and	ESR Spectrum Simulations.= Spectral Simulation.	COMP	0001
ontaining Acrylate and Methac+	ESR Studies on the Polymerization of Diacetylene-C	POLY	0092

omers in Raspberry Concentrates,	Essences, and Flavors.= +of Alpha-Ionone Enanti	AGFD	0181
Metales Pesados en Leche de	Establos del Sur del D.F.= Determinacion de	ENVR	0037
icroaleados en Acido S+ Estudio	Estadistico de Tecnicas Electroquimicas de Aceros M	PHYS	0281
ltad de Quimica, UNAM (Analisis	Estadistico de 10Anos).= +rrera de QFB en la Facu	CHED	0092
-32-crown+ Synthesis of a Poly(	ester crown) Based on bis(5-Carboxy-1,3-phenylene)	POLY	0044
f+ coli β-Hydroxydecanoyl Thiol	Ester Dehydrase: Determination of the Orientation o	BIOL	0066
E. coli β-Hydroxydecanoyl Thiol	Ester Dehydrase.= +as a Probe of the Active Site of	BIOL	0067
ldol Reaction of Chiral β-Lactam	Ester Enolates.= +tric Alkylation, Acylation, and A	ORGN	0046
Bisphenol A-based Cyanate	Ester Resin Systems.= Characterization of	PMSE	0152
Conjugated Porphyrin-Flavin	Ester.= Photoinduced Electron Transfers within	BIOL	0134
um+ Novel Muscle Relaxants of	Ester-Containing bisQuaternary Piperidino-Ammoni	MEDI	0047
butylene Terephthalate) and Poly(	ester-Ether) Segmented Block Copolymers.= +ly(	PMSE	0148
Production of Acetyl-Xylan	Esterase from Aspergillus niger.=	BIOT	0251
Aphids.=	Esterase Genes Conferring Insecticide Resistance in	AGRO	0137
Inhibitors of Cholesterol	Esterase.= Hypocholesterolemic Effects of	BIOL	0018
nantioselectivity of Lipases and	Esterases by Modifying the Substrate and by Linking	BIOT	0091
and Materials.= Lipases and	Esterases in the Synthesis of New Pharmaceuticals	BIOT	0093
yl Amide Spin Labeling: Methyl	Esterification- and Hydration-Induced Motional Per	AGFD	0012
s: Induced Oxidation of Hantzsch	Esters (Dialkoxycarbonyldihidropyridines).= +trate	ORGN	0245
Study on the Formation of Alkyl	Esters and Amidoesters of (Anilinomethylene)propan	ORGN	0103
Imidazopyridine	Esters and Oxadiazoles.= Muscarinic Activity of	MEDI	0139
etyltra+ Polyamine Fumaramate	Esters as Inhibitors of Spermidine/Spermine-N[1]-Ac	MEDI	0021
yphosphate and Polyphosphonate	Esters from Biologically Active Diols.= +tion of Pol	PMSE	0015
tems Using Immo+ Syntheses of	Esters in Nonaqueous Solvents and Solvent-Free Sys	BIOT	0065
Meta+ Thermolysis of O-Ethyl	Esters of Phosphoramidic Acids: Generation of Ethyl	ORGN	0110
aluation of Novel 6-Carboxylate	Esters of Prednisolone: Analogues of Methyl 11β, 17α	MEDI	0170
Microemulsions with Fatty Acid	Esters or Vegetable Oils: Formulation, Optimization,	COLL	0083
glet Oxygen with α,β Unsaturated	Esters: Evidence for a Perepoxide Intermediate.=	ORGN	0164
New Catalytic Route to Vinyl	Esters.=	I&EC	0054
mmetric Hydrogenation of β-Keto	Esters.= +plexes and Their Roles in Catalytic Asy	ORGN	0044
Sucrose	Esters.= Synthesis and Purification of	CARB	0004
Followed by Dioxin Analysis to	Estimate Body Burden in General Population and Ex	CEI	0012
mina o Titania Modificadas con+	Estudio de Catalizadores NiW Soportados sobre Alu	FUEL	0106
l-Fe(NO$_3$)$_3$.=	Estudio de la Nitracion del Fenol con Aductos Tonsi	COLL	0120
3 Modificados con Boro.=	Estudio de los Catalizadores CoMo/Al$_2$O$_3$ y Mo/Al$_2$O	FUEL	0104
aftaleno con Electrodos Metalic+	Estudio de Reduccion Electroquimica del 1-Nitro-N	PHYS	0280
sables en Geotermica de Fasico+	Estudio del Comportamiento de los Gazas en Conden	I&EC	0013
ceros Microaleados en Acido S+	Estudio Estadistico de Tecnicas Electroquimicas de A	PHYS	0281
	Estudio Sobre el Uso de Eteres de Glicol en Mexico.=	ENVR	0038
yrite-Pyrite: Effet des Oxydants	et des Complexants sur la Fottation de ces Mineraux	ANYL	0067
Compounds: Crystal Structure of [	Et$_2$AlAs(SiMe$_3$)$_2$]$_2$.= +f Novel Aluminum-Arsenic	COLL	0137
e la Corrosion de las Diferentes	Etapas del Lavado Quimico de Una Caldera por Tecn	PHYS	0257
s.= Synthesis of MTBE and	ETBE over Triflic Acid-Modified Y-Zeolite Catalyst	CATL	0016
s of Metal Atoms in Plasma	Etching and Chemical Vapor Deposition.= Reaction	PHYS	0143
STM-Induced	Etching of Conducting Oxides.=	PHYS	0361
Dry	Etching of Copper at High Rates.=	COLL	0030
Chlorine.= Thermal and Laser	Etching of Indium Phosphide by Molecular	COLL	0214
Microscopy.+ Nanometer Scale	Etching of MX$_2$ Materials Using Scanning Tunneling	PHYS	0364
Estudio Sobre el Uso de	Eteres de Glicol en Mexico.=	ENVR	0038
Gallium.= Aromatization of	Ethane and of Propane on HZSM-5: Effect of	PETR	0045
f Transition-Metal Cations: V+ +	Ethane, Ethene, and Propane.= +Cross-Sections o	PHYS	0306
rolysis.+ Conversion of Waste to	Ethanol and Electricity via Dilute Sulfuric Acid Hyd	FUEL	0094
ticum in Relation to Formation of	Ethanol and Other Fermentation Products.= +roly	BIOT	0029
ape Effects in the Conversion of	Ethanol to Hydrocarbons over Alkaline-Free Synthes	PETR	0065
ts in Biological Fuel Production.	Ethanol Tolerance of Clostridium thermosaccharolyt	BIOT	0013
Pervaporation of	Ethanol/Water Mixture.= IPN Membranes for the	PMSE	0059
as Chromogenic Agent in Aqueous	Ethanolic Solution.= +thanum with Sulfonazo-III	ANYL	0074
atalysts during Oligomerization of	Ethene.= +n by Coke Formation of ZSM-5 Type C	PETR	0027
tion-Metal Cations: V+ + Ethane,	Ethene, and Propane.= +Cross-Sections of Transi	PHYS	0306
Activity of Novel 5-(2-Substituted	Ethenyl)-3(2H)-Furanones.= +sis and Anti-Ulcer	MEDI	0067
ion.= Novel Propenyl Phenyl	Ether Analogs for UV-Induced Cationic Polymerizat	POLY	0017
of Poly(arylene	Ether Benzimidazole)s.= Synthesis and Properties	POLY	0025
oxybenzoic Acid by Ch+ In Vitro	Ether Cleavage of 3-Phenoxybenzoic Acid to 3-Hydr	AGRO	0018
for	Ether Cleavage.= Easy Comparative Experiments	CHED	0098
llization, and Melting of Poly(aryl	Ether Ketone Ketone), Part I: Structure.= +Crysta	POLY	0059
zation, and Melting of Poly(aryl	Ether Ketone Ketone), Part II: Crystallization and M	POLY	0060
in Poly(aryl	Ether Ketones).= Crystallization-Induced Gelation	POLY	0247
Design and Synthesis of	Ether Lipid Inhibitors of Protein Kinase C.=	MEDI	0073
Cytotoxic	Ether Lipids.= Physicochemical Characterization of	MEDI	0031
Gystamine, Mexamine, and Crown	Ether on the Surface of Pyrogeneous Silica.= +en	COLL	0096
d(II) and Mercury(II) with Crown	Ether on the Surface of Silochromium.= +on of Lea	COLL	0104
Thermal Stability+ Poly(Arylene	Ether Phosphine Oxide)s—I. Overview of Chemistry,	PMSE	0052
Poly(Arylene	Ether Phosphine Oxide)s—II. Pyrolysis Studies.=	PMSE	0053
ace Modification+ Poly(Arylene	Ether Phosphine Oxide)s—III. Investigations of Surf	PMSE	0054
Amino Acid Conjugates of	Ether Phospholipids.= Stereospecific Synthesis of	MEDI	0163
with an Enol-	Ether Trigger.= Models for the Ene-Diyne Toxines	ORGN	0102
xes of a New Ditertiary Phosphine	Ether.= +eactions for Group VI Carbonyl Comple	INOR	0333
lic Polystyrene/Poly(vinyl Methyl	Ether) Blends.= +composition and Linear and Cyc	POLY	0295
ne Terephthalate) and Poly(ester-	Ether) Segmented Block Copolymers.= +ly(butyle	PMSE	0148
and Poly(xylenyl	Ether).= Miscible Blends of Sulfonated Polystyrene	PMSE	0117

mal Degradation in PEEK (Poly-	ether-ether-ketone) Films.= +scopic Study of Ther	POLY 0334
egradation in PEEK (Poly-ether-	ether-ketone) Films.= +scopic Study of Thermal D	POLY 0334
$S_{RN}1$ Mechanisms.= Poly(aryl	ether-ketone) Synthesis via Competing S_NAR and	PMSE 0122
Prepared Using Poly(vinyl Methyl	Ether/Maleic Anhydride) Copolymer.= +aporates	COLL 0086
ne Epoxide Addition Reaction vs.	Etherification.= +rties of Amine-Cured Epoxy: Ami	POLY 0158
by Chiral Azacrown	Ethers.= Asymmetric Michael Addition Catalyzed	ORGN 0209
Cyclic	Ethers.= Conversion of Lactones into Substituted	ORGN 0014
kyl and Dialkylamino Aryl Oxime	Ethers.= +Substituted 2-Methylbiphenylmethyl Al	AGRO 0072
ctional, Aromatic Analogs of Vinyl	Ethers.= +vel Synthesis of UV-Curable, Multifun	POLY 0016
n. Toward a Pollution Prevention	Ethic: The EPA's Role and Program.= +Preventio	CEI 0022
of+ Tandem Michael Addition	Ethoxycarbonylation-Carbonyl Deblocking of Anions	ORGN 0228
by Rheo-Photoac+ Mobility of	Ethyl Acetate in Poly(vinylidene Fluoride) Monitored	POLY 0255
Combustion Byproducts of	Ethyl Acetate over Pt- Alumina Catalyst.=	COLL 0189
Differentiation of Amyl Acetate,	Ethyl Acetate, Ethyl Butyrate, and Ethyl Caproate.=	AGFD 0187
n of Amyl Acetate, Ethyl Acetate,	Ethyl Butyrate, and Ethyl Caproate.= +fferentiatio	AGFD 0187
thyl Acetate, Ethyl Butyrate, and	Ethyl Caproate.= +fferentiation of Amyl Acetate, E	AGFD 0187
Ethyl Met+ Thermolysis of O-	Ethyl Esters of Phosphoramidic Acids: Generation of	ORGN 0110
osphoramidic Acids: Generation of	Ethyl Metaphosphate.= +of O-Ethyl Esters of Ph	ORGN 0110
Grafting of 2-(Dimethylamino)	ethyl Methacrylate to Squalane.= Free Radical	POLY 0019
ented by Chiral Nematic (Acetyl)(	ethyl)cellulose Solutions.= +r of Dye Molecules Ori	POLY 0331
Cyclopentyl- and N^6-(2-Thienyl)	Ethyl-Adenosine Agonists. +SAR in a Series of N^6-	MEDI 0113
or Specific Control of Lepidopte+	Ethyl-Branched Juvenile Hormones: Opportunities f	AGRO 0004
ticonvulsant D,L-3-Hydroxy-3-	Ethyl-3-Phenylpropionamide (HEPP) in Plasma and	ANYL 0082
-Ethyltoluene during the Toluene	Ethylation on Modified ZSM-5.= +sformation of p	PETR 0086
tive Oxidation and Degradation of	Ethylene and Propylene over Silver.= +ween Selec	CATL 0058
Catalytic Oligomerization of	Ethylene by Ti(II) Complexes.=	INOR 0153
Infrared Spectra of Metal-	Ethylene Complexes Isolated in Solid Argon.=	PHYS 0074
Random	Ethylene Copolymers.= Crystallization Kinetics of	POLY 0075
otons in the Solutions of the Iron-	Ethylene Diamine Tetraacetic Acid Complexes.=	INOR 0090
Catalysts.=	Ethylene Dimerization over Highly Dispersed Metal	PETR 0010
ri- or Tetraisocyanates and Poly(	ethylene Glycol) Derivatives.= +ar Polymers from T	POLY 0271
Reactions of	Ethylene over Ru-Y Zeolite.=	CATL 0031
es, and Hydroxyl-Capped Poly(	ethylene Oxide) and ABA Block Copolymers Contain	PMSE 0006
Interphases in Poly(	ethylene Oxide) Blends.= Crystal-Amorphous	PMSE 0165
tical Methods. Polystyrene-Poly(	ethylene Oxide) Block Copolymer Micelles in Water.	POLY 0277
en+ Ionic Conductivity of Poly(	ethylene Oxide) Bound with a Sterically Hindered Ph	POLY 0091
stry, Physics, and Biology. Poly(	ethylene Oxide) Star Molecules in Surfaces for biolog	POLY 0272
ontaining Hydroxyl-Capped Poly(	Ethylene Oxide) with Polydimethylsiloxane.= +s C	PMSE 0006
oreversible Gelation from Poly(	ethylene Oxide)Paradichlorobenzene and Their Inter	POLY 0125
ized Triglyceride Oils.= Poly(	Ethylene Terephthalate) Semi-IPNs with Functional	PMSE 0088
Crystallization Kinetics of Poly(	Ethylene Terephthalate) with Functionalized Trigl	PMSE 0171
alytic Reforming: The Reaction of	Ethylene with M-Ru/SiO$_2$.= +talysts Toward Cat	CATL 0042
erization of Carbon Monoxide and	Ethylene.= +ine)palladium Catalysts for Copolym	INOR 0155
Heterogeneous Oligomerization of	Ethylene.= +s for the Highly Efficient and Stable	PETR 0008
Photodissociation of	Ethylene.= Dynamics of H_2 Elimination in the	PHYS 0163
ectroscopy for Characterization of	Ethylene-Based Polyolefin Copolymers.= +ared Sp	PMSE 0048
mediates.+ Soot Depostion from	Ethylene/Air Flames and the Role of Aromatic Inter	FUEL 0107
Near Surface Structure in Poly(	ethyleneterephthalate) by ATR-IR.= Analysis of	POLY 0279
Mixed Metal Synthesis of 2-	Ethylhexyllithium.=	ORGN 0225
um Nitride Thin Films.=	Ethylimidotantalum Complex as Precursor to Tantal	INOR 0074
Secondary Transformation of p-	Ethyltoluene during the Toluene Ethylation on Modi	PETR 0086
rization, and Toughness of	Ethynyl-Containing Blends.= Preparation, Characte	POLY 0040
trically Con+ Polymerization of	Ethynylpyridine with Bromine: A New Route to Elec	POLY 0032
ole of Bioactive Metabolites.=	Etoposide: Studies Aimed Toward the Mechanistic R	MEDI 0012
Adsorptive Properties of	ETS-10: A New Large-Pore Titanium Silicate.=	PETR 0079
Molecular Sieve.= Synthesis of	ETS-10: A New Wide-Pore Titanium Silicate	PETR 0078
Allylic Transposition Catalyzed by	Eu(fod)$_3$.= +ic Methoxyacetates: A Novel Type of	ORGN 0211
ous Data Sets Confirms that S+	Eudismologic Analysis. 2. Preliminary Survey of Vari	MEDI 0088
Derivative of	EUO-Framework Zeolites.= Gallosilicate: A Novel	PETR 0076
rinary Medicines—Regulation in	Europe and the Importance of Pharmacokinetic Stud	AGRO 0077
f Information: North America and	Europe. Global Marketing of+ International Flow o	CINF 0011
Interactions (≤3	eV) of CH$_3^+$ Ions on Platinum.= Ultra-Low-Energy	COLL 0216
del Lavado Quimico de Una Cal+	Evaluacion de la Corrosion de las Diferentes Etapas	PHYS 0257
s Alumnos en los Resultados de la	Evaluacion de una Materia Biologica.= +mico de lo	CHED 0093
ara Pozos Geotermicos.=	Evaluacion Fisicoquimica de Cementos Hidraulicos p	INOR 0063
New Chemical	Evaluations under TSCA.=	CHAS 0013
	Evanescent Detection of Humidity.=	ANYL 0079
Mixtures.= Solar	Evaporation of Fertilizer/Ag-Chemical Aqueous	FERT 0022
Multipurpose	Event.= Chemistry Olympiad in Mexico: A	CHED 0109
ation of Ratio+ Theory Testing,	Evidential Reason, and the Role of Data in the Form	ENVR 0075
anes by an in S+ Microstructural	Evolution of a Silicon Oxide Phase in Nafion Membr	POLY 0285
la.=	Evolution of Glutathione S-Transferases in Drosophi	AGRO 0100
tors.=	Evolution of Heavy-Ion Recoil and Fragment Separa	NUCL 0038
ccinate Lyase: Odd Couples in the	Evolution of Mechanistic Motifs.= +and Adenylosu	BIOL 0012
ic Steroids and the Luftwaffe:	Evolution of Metal-Ammonia Reductions.= Synthet	HIST 0013
	Evolution of Regulations for Pesticide Rinsates.=	AGRO 0104
zation Catalysts in the Course o+	Evolution of the Gallium Phase of GaHMFI Aromati	PETR 0046
o 34 MeV.=	Evolution of the Reaction ^{40}Ar + Ag from $E/A = 7$ t	NUCL 0062
e Inhibitors, Arcelins, and Lectins	Evolved from a Common Ancestor Gene.= +mylas	AGFD 0041
es with the ZSM-5 Sructure.=	EXAFS Studies of Isomorphously Substituted Zeolit	PETR 0081

Crystallographic and	EXAFS Studies of Sterically Crowded Porphyrins.=	INOR	0091
Dispersed on Carbon: XANES and	EXAFS Study.= +ture and Structure of Calcium	FUEL	0032
Olympiad National	Exam.= How to Prepare Students for the Chemistry	CHED	0013
Mn(CO)₂ with Tertiary Silanes as	Examined by Photoacoustic Calorimetry.= +of Cp	INOR	0386
diate Pesticide–Contaminated Soil	Excavated from Agrochemical Facilities.= +Reme	AGRO	0144
R. W. Ramette. Award Address+	Excellence in Teaching Award Symposium Honoring	ANYL	0006
of Analytical Chemistry Award for	Excellence in Teaching, cosponsored by E. +vision	ANYL	0006
bras.+ Efecto de la Deficiencia o	Exceso de Vitamina E en la Fertilidad de Ratas Hem	AGFD	0210
Reflectance Study of Oxidized and	Exchanged Polyethylene Surfaces.= +IR Specular	POLY	0253
Shift Reaction over Ruthenium–	Exchanged X Zeolite: Reaction Control by the Cocat	CATL	0027
als: The Hydrous Metal Oxide Ion	Exchangers.= +of Novel Catalyst Support Materi	PETR	0018
erization of a Constrained Ω–Loop	Excised from Interleukin-1α.= +Structural Charact	MEDI	0165
tate Distr+ Effects of Electronic	Excitation and Dimerization of Metals on Product–S	PHYS	0350
ance and Optical Spectroscopies.	Excitation Methods Useful for Investigating Molecul	PHYS	0136
s at Catalyst Surfaces by Selective	Excitation of Nuclear Magnetic Resonances.= +ic	PHYS	0346
Tetracene Cry+ Flourescence and	Excitation Spectra Pattern and Phase Transition in	PHYS	0274
Electronic	Excitation Transport in Photosynthetic Antennae.=	PHYS	0020
Fractals and	Excitation–Transfer in Langmuir–Blodgett Films.=	PHYS	0110
Entrance Channel	Excitations in the ²⁸Si + ²⁸Si Reaction.=	NUCL	0039
line Anticonvulsant, for Impairing	Excitatory Amino Acid Neurotransmission.= +iazo	MEDI	0144
f Ground State and Electronically	Excited Barium Atoms.= +ctions—VI. Reactions o	PHYS	0349
the Interactions of Electronically	Excited Cadmium Atoms with H₂ and CH₄.= +es of	PHYS	0351
ces in Quantitative Two–Photon	Excited Fluorescence Spectroscopy.= Recent Advan	ANYL	0014
by Phosphorus Monoxide Laser–	Excited Molecular Fluorescence Spectrometry in a G	ANYL	0078
ransfer from Highly Vibrationally	Excited NO₂ Donor.= +and Translational Energy T	PHYS	0334
ns—VII. Harpooning Reactions of	Excited Rare Gas Atoms.= +–Phase Metal Reactio	PHYS	0367
bsurface and Nonequi+ Optically	Excited Scanning Tunneling Microscopy: Imaging Su	PHYS	0208
hip between the Geometry of the	Excited State of Vanadium Oxides Anchored onto Si	INOR	0165
d Electron Transfer Quenching of	Excited States by Ferrocene Derivatives.= +ergy an	INOR	0163
Ru + H₂ Reaction.= Role of	Excited States of Maximal d–Shell Occupancy in the	PHYS	0243
Ligand–Ligand Scharge–Transfer	Excited States of Zinc(II) Mixed–Ligand Complexes.	INOR	0164
emistry with Experiment. Diene	Excited–State Energetics, Structures, and Dynamics.	PHYS	0075
agnetic Field Perturbations of the	Excited–State Properties in CR(III) Complexes.=	INOR	0159
n Kinetics in Liquids+ Study of	Excited–State Relaxation and Photoinitiated Reactio	ANYL	0034
ptical Harmonic Measurements of	Excited–State Relaxation at Interfaces.= +econd O	COLL	0087
elaxation in Condensed Phases.	Excited–State Structure and Transport Dynamics of	PHYS	0018
Vycor Glass.=	Excited–Stated Dynamics of Ru(BPY)₃²⁺ on Porous	INOR	0160
Restricted Geometries.=	Exciton Coherence–Size and Optical Nonlinearities in	PHYS	0038
.=	Exciton Kinetics Studies on Isolated Polymer Chains	PHYS	0309
Transfer and Scanning Molecular	Exciton Microscopy.= +Spatially Resolved Energy	PHYS	0022
ne–Dimensional Inorganic Solid:	Exciton Trapping and Annihilation in Crystals of (C	PHYS	0131
alysis of the Dynamics of Frenkel	Excitons in Disordered Systems: T = 0 K.= +in An	PHYS	0034
for Poly Methyl Methacry+ Size	Exclusion Chromatography and End–Group Analysis	PMSE	0050
f Cotton+ Application of Size–	Exclusion Chromatography for the Characterization o	PMSE	0032
d Anionic Copolymers of V+ Size	Exclusion Chromatography of Cationic, Nonionic, an	PMSE	0031
arides by High–Performance Size–	Exclusion Chromatography.= +ristics of Polysacch	BIOT	0197
, and Related Chromatog+ Size	Exclusion Chromatography, Field–Flow Fractionation	PMSE	0008
, and Related Chromatog+ Size	Exclusion Chromatography, Field–Flow Fractionation	PMSE	0065
, and Related Chromatog+ Size	Exclusion Chromatography, Field–Flow Fractionation	PMSE	0046
, and Related Chromatog+ Size	Exclusion Chromatography, Field–Flow Fractionation	PMSE	0028
, and Related Chromatog+ Size	Exclusion Chromatography, Field–Flow Fractionation	PMSE	0094
Laboratory	Exercises for the Chemically Perplexed.=	CHED	0055
a Typical Commercial Automotive	Exhaust Catalyst.= +and CO–NO Reactions over	COLL	0187
es Using Alcohols, Gasohol, and+	Exhaust Pollutant Emissions from Automobile Engin	FUEL	0095
Blends Which	Exhibit Synergistic Properties.= Designing Polymer	PMSE	0101
enes: Stereoselective Syntheses of	Exo or Endo Polycyclic Epoxides.= +ormation Styr	ORGN	0236
ement of Sulfilimines from 3–	Exomethylenecephams.= [2,3] Sigmatropic Rearrang	ORGN	0033
Challenges.= Analyzing	Exotic Organics in Mixed Nuclear Wastes: The	ANYL	0060
ated Synthetic Substances against	Exotic RNA Viruses.= +ceae Constituents and Rel	AGFD	0165
rface and Cracking Properties of	Expanded Clays Dried Using a Supercritical Fluid.=	PETR	0145
m Chiral Complexes.=	Expanded Heterohelicenes: Molecular Coils That For	ORGN	0065
cterization of a New Hexapyrrolic	Expanded Porphyrin: Rubyrin.= +thesis and Chara	ORGN	0058
Synthesis and Properties of	Expanded Porphyrins.= Cope Award Symposium.	ORGN	0139
Graph–Theoretic Cluster	Expansions: Diamagnetic Susceptibility.= Chemical	COMP	0030
ry Laboratory Development. How	Expensive Are Lasers, Really?.= +hysical Chemist	CHED	0082
with Emphasis+ Comparison of	Experimantal and Theoretical Hyperpolarizabilities,	PHYS	0339
mpare ab Initio Calculations with	Experiment?.= +Forces: Do We Really Need to Co	PHYS	0104
ecimiento en Deute+ Resultados	Experimentales de un Empaque Integral para Enriqu	I&EC	0073
n–Making Process and the Role of	Expert Opinion.= +ess. Overview of Legal Decisio	CHAL	0011
	Expert Systems of Manufacture.=	I&EC	0070
Use of the	Expert Witness in Products Liability Litigation.=	CHAL	0015
Process and+ The Scientist as an	Expert Witness. Overview of Legal Decision–Making	CHAL	0011
Chemists as	Expert Witnesses in Patent Litigation.=	CHAL	0013
Counsel.= The Scientific	Expert: Considerations of Corporate In–House	CHAL	0017
Testimony of Scientific	Experts: Is the Pendulum Swinging Back?.=	CHAL	0016
ecursors.=	Expitaxial GaAs Film Growth from Single–Source Pr	COLL	0067
ain Forest—A+ The Industrial	Exploitation of Natural Products from the Tropical R	SCHB	0024
own (Chemi+ The Discovery and	Exploration of a New Continent of Chemistry. H. Br	CONG	0012
Pharmacophore	Exploration.=	CINF	0019
the Horse before the Cart:	Exploring the Melting Point of Mixtures.= Putting	CHED	0159

International Joint Ventures:	Exploring the Possibilities.=	CINF 0015
Waste.= Ammonia Fiber	Explosion (AFEX) Pretreatment of Municipal Solid	BIOT 0016
Segregation and Anomalous Time	Exponents.= +on Kinetics in a Capillary: Reactant	PHYS 0273
Process.= Polysar's Efforts in	Exposing Precollege Students to a Hydrocarbon	CHED 0010
	Exposure Assessment and Risk Assessment.=	ENVR 0072
vesters.= Developing Pesticide-	Exposure Mitigation Strategies for Handlers and Har	AGRO 0043
Hydrocarbons in La Basse Cot+	Exposure of Fishing People to Chlorinated Aromatic	C E I 0011
ical Monitoring of Occupational	Exposure to Cyclophosphamide in the Pharmaceutica	C E I 0010
lence, Bioavailability,+ Human	Exposure to Hazardous Wastes: Assessment of Preva	C E I 0008
Biochemical Marker for Detecting	Exposure to Lead.= +te Protoporphyrin (EP) as a	C E I 0020
natase) Suspensions as a Result of	Exposure to Radiofrequency Electric Fields.= +(A	COLL 0203
Monitor of Human	Exposure.= Sister Chromatid Exchange Test as	C E I 0016
ssments of Inhalation and Dermal	Exposures.= +Are the Keys to Scientific Risk Asse	ENVR 0042
on of Multiple-Pathway Chemical	Exposures.= +ted Approach to Risk Characterizati	ENVR 0073
Environmental or Occupational	Exposures.= Case Studies of Carcinogens:	CHAS 0020
Activities of Site-Specific Mutants	Expressed in E. +ride-FAB Complex and Binding	BIOL 0143
Recombinant Human Plasminogen	Expressed in Several Cell Lines.= +ycosylation of	ANYL 0020
acy of Baculovirus Insecticides by	Expressing with Insect Selective Proteins.= +Effic	AGRO 0009
STM and AFM	Extensions.=	PHYS 0209
od-Preserving Plants in Quebec:	Extensive Analysis and Search for Remediation Tech	ENVR 0016
l Chemistry Courses.=	Extensive Use of Spreadsheet Programs in Analytica	CHED 0205
l Particles: Colloidal Particles in	External Fields: Electric, Magnetic, and Hydrodynam	COLL 0141
l Particles: Colloidal Particles in	External Fields: Electric, Magnetic, and Hydrodynam	COLL 0159
l Particles: Colloidal Particles in	External Fields: Electric, Magnetic, and Hydrodynam	COLL 0178
l Particles: Colloidal Particles in	External Fields: Electric, Magnetic, and Hydrodynam	COLL 0199
and Hydrodynamic—I. Colloids in	External Fields.= +nal Fields: Electric, Magnetic,	COLL 0141
perties of Colloidal Systems in	External Fields.= Mass Transfer and Transport Pro	COLL 0142
in Contact with Solid Surfaces in	External Fields.= +Transport Properties of Colloids	COLL 0162
agnetic Colloidal Suspension in	Externally Imposed Magnetic and Hydrodynamic Fie	COLL 0144
us Protein Accumulation by St+	Extracellular Proteases and Their Role in Heterologo	BIOT 0084
Off-Flavors by Aroma	Extract Dilution Analysis.= Characterization of	AGFD 0097
Detection of Adulterated Vanilla	Extract.=	AGFD 0185
etic Analysis of Cationic Proteins	Extracted from Aflatoxin-Resistant/Susceptible Var	AGFD 0134
Herbert H. Dow—Process of	Extracting Bromine.=	CHAL 0029
Glycoalkaloids.= Improved	Extraction and HPLC Assay of Potato	AGFD 0112
ough Pressure-Induce+ Protein	Extraction and Recovery from Reversed Micelles thr	BIOT 0046
hase Dispersion (MSPD) to the	Extraction and Subsequent Analysis of Drug Residue	AGFD 0060
olution by+ Selective Iron Oxide	Extraction from Soils and Sediments: Reductive Diss	ENVR 0058
Solvolytic	Extraction of Coal.= Mechanism of Successive	I&EC 0018
f Oxalic Acid+ Selective Solvent	Extraction of Lanthanum by Oxine in the Presence o	ANYL 0054
with TBP.=	Extraction of Rf (Element 104) and Its Homologues	NUCL 0074
of Growth Substrate and Product	Extraction on Reactor Productivity.= +ey: Effects	BIOT 0030
Design Soil Vapor	Extraction System.= The Use of Soil Gas Surveys to	ENVR 0051
Protein	Extraction.= Affinity-Based Reversed Micellar	BIOT 0045
icultural Analyses.+ Solid-Phase	Extractions for Sample Preparation in Food and Agr	AGFD 0175
Techniques for Authentication of	Extracts and Essential Oils.= +attern-Recognition	AGFD 0198
city of Annona squamosa Seed Oil	Extracts to Tribolium castaneum L. (Coleopt+ Toxi	AGRO 0036
henolic Compounds in Botanical	Extracts Used in Foods, Flavors, Cosmetics, and Pha	AGFD 0172
Water Uptake in Cured Epoxy by	Extrinsic and Intrinsic Fluorophores.= +rement of	POLY 0160
Continuously	Extruded Commingled Plastics.=	I&EC 0027
	Extruded Monolithic Catalysts Supports.=	PETR 0090
on Cold-	Extruded Polyethylene.= Atomic Force Microscopy	PMSE 0078
nergy Distribution in Twin Screw	Extruders.= +y Lecture. Distributive Mixing and E	PMSE 0175
Shear.=	Extruding Waste Plastics with High Mixing and Low	I&EC 0030
xturizing Agents during Cooking-	Extrusion Cereals.= +cation of Phenolic Acids as Te	AGFD 0150
Reactive	Extrusion Imidization of Acrylic Polymers.=	PMSE 0179
Synthesis of Carbohydrates Using	Ezymes as Catalysts.= +thesis. Keynote Address.	BIOT 0089
Wheat Germ Initiation Factors	4F and (ISO)4F Interact Differently with Oligoribonu	BIOL 0128
Initiation Factors 4F and (ISO)	4F Interact Differently with Oligoribonucleotide Ana	BIOL 0128
bution in Model Polymer Films by	19F Multiple-Quantum NMR.= +onium Salt Distri	PMSE 0182
tructure of an Oligosaccharide-	FAB Complex and Binding Activities of Site-Specific	BIOL 0143
ctrospray and Continuous-Flow	FAB LC/MS.= Peptide and Protein Analysis by Ele	ANYL 0140
Di+ Analytical Science: The New	Face of Analytical Chemistry. Award Address (ACS	ANYL 0011
Di+ Analytical Science: The New	Face of Analytical Chemistry. Award Address (ACS	ANYL 0001
u+ Analytical Science: The New	Face of Analytical Chemistry. High-Speed DNA Seq	ANYL 0033
ic+ Analytical Science: The New	Face of Analytical Chemistry. Scanning Tunneling M	ANYL 0022
	Face Selection in Enolate Chemistry.=	ORGN 0086
2-one.= Electronic Control of	Face Selection in the Reduction of 5-Azaadamantan-	ORGN 0237
	Face Selection in the Thio-Claisen Rearrangement.=	ORGN 0239
	Face Selectivity in the Protonation of Glycals.=	ORGN 0069
es.=	Face Selectivity in 5-,7-Disubstituted Adamantanon	ORGN 0238
Diels-Alder Dienophile	Face Selectivity.= Conformation-Dependent	ORGN 0241
rehigh School Science Education.	FACETS: An Interdisciplinary Middle School Curric	CHED 0071
s.=	Facile and Selective Oxidation of Sulfides to Sulfone	CARB 0033
gsten(VI) Bis(sulfido) Complexes:	Facile Elimination of H2 from H2S.= +IV) and Tun	INOR 0369
Conjugated Molecular Systems:	Facile Graph Theoretical Solutions to Huckel MO Pa	COMP 0020
henol A Polycarbonate Using +	Facile Measurement of Phenolic End-Groups in bisP	POLY 0083
1,3-Disubstitued Py+ New and	Facile Method for the Synthesis of 3-Substituted and	ORGN 0196
lanar Iridium(I) Complexes.=	Facile Phosphine Ligand Exchange between Square P	INOR 0211
x.=	Facile Reversible Metalation from an Agostic Comple	INOR 0384

ic Hydrosilylation of Ketones: A	Facile Route to Ciral Alcohols Using Discrete Rhodiu	INOR 0247
Nucleophile Exchange: A Novel,	Facile Route to Highly Optically Pure Glycidol Deriv	ORGN 0207
one.=	Facile Synthesis of a Carboxylic Linker on Halofugin	ORGN 0201
	Facile Synthesis of Azetidine-2,4-Diones.=	ORGN 0034
of Denaturation.=	Facilitation of Protein Folding and the Reversibility	BIOT 0190
(CSD).= Integrated 3D Search	Facilities for the Cambridge Structural Database	CINF 0021
Liquid Fertilizer Containment	Facilities Handbook.= MWPS-37 Pesticide and	AGRO 0042
Soil Excavated from Agrochemical	Facilities.= +Remediate Pesticide-Contaminated	AGRO 0144
Recycling	Facility.= Operation of a Postconsumer Materials	I&EC 0042
SP Reference Standards, a Key	Factor in Chromatographic Analysis of Pharmaceutic	ANYL 0115
(FILM) Pr+ Caracterizacion del	Factor Inhibidor de la Locomocion de los Monocitos	BIOL 0102
terias Biologicas de la Carrera d+	Factores Que Incrementan la Reprobacion en las Ma	CHED 0092
Chemists in Industry. Fascinating	Facts about Chemists.= +Roadshow)—Careers for	YCC 0005
Beilstein Current	Facts on CD-ROM.=	COMP 0017
icas de la Carrera de QFB en la	Facultad de Quimica, UNAM (Analisis Estadistico de	CHED 0092
Advance Preh+ What Chemistry	Faculty at the Two-Year College Are Doing to Help	CHED 0076
Funding Opportunities for Junior	Faculty.= +Get a Fair Shake? An Analysis of NSF	YCC 0003
Nonfederal Funding for New	Faculty.=	YCC 0002
Modified	FAD.= Molecular Switches from	BIOT 0143
	Failing with a Flair.=	SCHB 0017
ties+ Do Younger Chemists Get a	Fair Shake? An Analysis of NSF Funding Opportuni	YCC 0003
all Chemical Businesses. Founding	Fairfield Chemical Company.= +ue Stories of Sm	SCHB 0013
Magnet-Enhanced Optical	Falling-Needle/Sphere Rheometer.=	POLY 0251
National Inventors Hall of	Fame.=	CHAL 0025
de las Plantas Pertenecientes a la	Familia Scrophulariaceae de Mexico.= +n Quimica	ORGN 0185
tion of a Seed from the Jatropha	Family as a Possible Source of Oil and Protein for H	AGFD 0038
Like Carbon. Carbophanes: A New	Family of Crystalline Phases of Carbon.= +ond-	FUEL 0034
Nonlinear Spectroscopies—A New	Family on the Block.= +er Corp.). Multiresonant	ANYL 0001
Biotransformation Studies in	Farm Animals.= In Vitro Models of	AGRO 0094
Chemicals in Groundwater at the	Farm Level.= Advances in Managing Agricultural	AGRO 0052
Pesticides in Drinking Water: A	Farm Wife's Perspective.=	AGRO 0055
Related to Pesticide Use on the	Farm: Methods to Monitor Small Watersheds for Pes	AGRO 0054
-Careers for Chemists in Industry.	Fascinating Facts about Chemists.= +Roadshow)-	YCC 0005
unicipal Waste. Sewage Sludge: A	Fascinating Feedstock for Clean Energy.= +and M	FUEL 0115
n Condensables en Geotermica de	Fasico de los Azuzfres.= +rtamiento de los Gazas e	I&EC 0013
(-)-	Fastiglinin-C.= Efficient Syntheses of (+/-)- and	ORGN 0179
β-Glucan Contents as Natural	Fat Substitutes.= Oatrim Products with Elevated	AGFD 0037
Student Awards Symposium.	Fate of ^{14}C-Trichloroethylene in Whole Plants.=	ENVR 0056
erizing the Role of in Situ Hum+	Fate of a PCB in Anoxic Bottom Sediments: Charact	ENVR 0017
scopic Observation of the Kinetic	Fate of a Thiamin-Bound Enamine Intermediate in	BIOL 0047
Laying Hens.= Metabolic	Fate of Ametryn in Rats, Lactating Goats, and	AGRO 0111
Environmental	Fate of DPX-E9636, a New Sulfonylurea Herbicide.=	AGRO 0029
ysimeter Studies on Long-Term	Fate of Pesticides: Characterization of Degradation a	AGRO 0025
Acquisition—The Pleasant (?)	Fate of Small Chemical Businesses.=	SCHB 0019
copic Observation of the Kinetic	Fate of Thiamin Diphosphate-Bound Enamines on B	BIOL 0048
itu Techniques for Studying X+	Fate of Xenobiotics in Food-Producing Animals. In S	AGRO 0092
oduction—The Uses of Veterin+	Fate of Xenobiotics in Food-Producing Animals. Intr	AGRO 0075
tabolism and Residue Studies of+	Fate of Xenobiotics in Food-Producing Animals. Me	AGRO 0108
ermectin and Abamectin Meta+	Fate of Xenobiotics in Food-Producing Anmimals. Iv	AGRO 0128
nteractions. Tertiary Structures of	Fatty Acid Binding Proteins.= +of Protein-Lipid I	BIOL 0015
ti+ Model Microemulsions with	Fatty Acid Esters or Vegetable Oils: Formulation, Op	COLL 0083
by Brevibacterium ammoniagenes	Fatty Acid Synthase.= +noyl Reduction Catalyzed	BIOL 0065
s in Layered Lithium Aluminate-	Fatty Acid Systems.= +on Transfer with Porphyrin	INOR 0272
onments for Long-Chain Omega-3	Fatty Acids.= +eds: Natural and Unnatural Envir	AGFD 0160
se-Mediated Exchange Reactio+	Fatty Acyl Substrate Preferences for Pancreatic Lipa	AGFD 0036
ergrowth of Hexagonal and Cubic	Faujasite.= +aracterization of VPI-6: Another Int	PETR 0001
and	Faujasites.= Cracking Properties of SAPO-37	CATL 0045
ine over the Past+ Collection of	Favorite Cartoons Appearing in CHEMTECH Magaz	CHAL 0024
ists or Chemicals.=	Favorite or Objectionable Cartoons Related to Chem	CHAL 0021
ors or Patents.=	Favorite or Objectionable Cartoons Related to Invent	CHAL 0022
and Lawyers.=	Favorite or Objectionable Cartoons Related to Laws	CHAL 0023
Experiences with MDDR-3D and	FCD-3D.= Building 3D Structural Databases:	COMP 0046
Container Coatings:	FDA Considerations.=	PMSE 0158
uctural η2-Dihydrogen Complexes:	Fe < Os < Ru.= +ve Order of pK_a Values of Isostr	INOR 0380
g the Chiral Auxiliary (η^5-C$_5$H$_5$)	Fe(CO)(PPh$_3$): Synthesis and Chemistry of (η^5-C$_5$H$_5$	INOR 0156
thesis and Chemistry of (η^5-C$_5$H$_5$)	FE(CO)(PPH$_3$)C(O)-C=-R(R=CH$_3$,Si(CH$_3$)$_3$).= +n	INOR 0156
lkyl Phosphines: Fe(CO)$_4$PR$_3$ and	Fe(CO)$_3$(PR)$_2$ Are Single-Photon Products.= +th A	INOR 0253
Fe(CO)$_5$ with Alkyl Phosphines:	Fe(CO)$_4$PR$_3$ and Fe(CO)$_3$(PR)$_2$ Are Single-Photon P	INOR 0253
CO)$_3$(PR)+ Photosubstitution of	Fe(CO)$_5$ with Alkyl Phosphines: Fe(CO)$_4$PR$_3$ and Fe(	INOR 0253
ffect of Histidine/Carnosine on	Fe(II) Autoxidation and Enzymic or Nonenzymic Red	AGFD 0131
ymic or Nonenzymic Reduction of	Fe(III).= +rnosine on Fe(II) Autoxidation and Enz	AGFD 0131
Aductos Tonsil-	Fe(NO$_3$)$_3$.= Estudio de la Nitracion del Fenol con	COLL 0120
Oxidation of 1,2-Diols by	Fe(VI): A Kinetic Study.=	INOR 0401
ure of Bi$_2$Sr$_2$CuO$_y$ and Bi$_2$Sr$_2$Cu$_{0.9}$Fe$_{0.1}$O$_y$.=	+M Investigations of the Surface Struct	PHYS 0292
chanistic Studies of Reactions of	Fe$_2$(CN)$_{10}^{4-}$ with S$_2$O$_3^{2-}$ in Slightly Acidic Aqueous	INOR 0400
d Structural Characterization of	Fe$_2$(CO)$_5$(μ-P-t-Bu$_2$)$_2$ and Fe$_2$(CO)$_6$(μ-PPH$_2$)(μ-P-t	INOR 0050
tion of Fe$_2$(CO)$_5$(μ-P-t-Bu$_2$)$_2$ and	Fe$_2$(CO)$_6$(μ-PPH$_2$)(μ-P-t-Bu$_2$).= +ral Characteriza	INOR 0050
Associated with the Reduction of	Fe$_2$(PPh$_2$)$_2$(CO)$_6$.= +udy of the Structural Changes	ANYL 0106
ctivity of LaBo$_3$ (B = Ni, Mn, Co,	Fe) Perovskites.= +aracterization, and Catalytic A	PHYS 0225
Fe-Pt, Ni-Fe-Pd, Fe-Co-Pt, and	Fe-Co-Pd.= +etism in Ternary Compounds of Ni-	PHYS 0253

mpounds of Ni-Fe-Pt, Ni-Fe-Pd,	Fe-Co-Pt, and Fe-Co-Pd.=	+etism in Ternary Co	PHYS 0253
igration from Oxygen to Metal in	Fe-Mn Alkoxycarbyne Complexes.=	+sm of Alkyl M	INOR 0110
ary Compounds of Ni-Fe-Pt, Ni-	Fe-Pd, Fe-Co-Pt, and Fe-Co-Pd.=	+etism in Tern	PHYS 0253
sm in Ternary Compounds of Ni-	Fe-Pt, Ni-Fe-Pd, Fe-Co-Pt, and Fe-Co-Pd.=	+eti	PHYS 0253
Nitric Oxide by Ammonia over	Fe-Y Zeolites.=	Selective Catalytic Reduction of	COLL 0212
$LaBO_3$ (B = Mn,	Fe, Co, Ni) Perovskites.=	Electronic Structure of	PHYS 0288
iquors from Mexican Iron Indu+	Feasible Solutions for Disposal and Reuse of Pickle L		ENVR 0039
de Acetyl Complex $Cp(PPH_2)(CO)$	$FeCOCH_3$-1 with Electrophiles.=	+an Iron Phosphi	INOR 0240
Chemistry in the	Federal Government.=		HIST 0032
of Pesticides in IPM:	Federal Perspective.=	Future Directions in the Use	AGRO 0068
dual Analy+ Current Overview of	Feed Additives and Veterinary Drugs and Their Resi		AGFD 0102
Specialized Ribosomes and	Feedback Regulation of rRNA Synthesis in E. coli.=		BIOT 0245
ctivity Enhancement.= Ocean	Feeding of Electron-Treated Sludge for Ocean Produ		ENVR 0021
g-Chain Omega-3 F+ Foods and	Feeds: Natural and Unnatural Environments for Lon		AGFD 0160
ste. Sewage Sludge: A Fascinating	Feedstock for Clean Energy.=	+and Municipal Wa	FUEL 0115
aracterization of Delayed Coking	Feedstocks and Products by ^1H and ^{13}C NMR.=	Ch	FUEL 0076
W Edible Biomass and Chemical	Feedstocks from Prehydrolyzates of Cellulosic Food		BIOT 0153
d Liquid Fuels from Poor-Quality	Feedstocks.=	+eneration of Good-Quality Solid an	FUEL 0141
Tunneling Microscopies: FEM,	FEES, FIM, A-P, STM, STS.=		PHYS 0097
Tunneling Microscopies: FEM,	FEES, FIM, A-P, STM, STS.=		PHYS 0097
tion Reaction in η-Cp(CO) (PPh_3)	$FeMe^+$.=	+oting the Alkyl to Acyl Migratory Inser	INOR 0109
	Femtosecond Dynamics of Bi_2 Dissociation.=		PHYS 0369
at Multilayer Surfaces.=	Femtosecond Solvation-Dynamics of CH_3Cl^- and Cl^-		PHYS 0084
al Spectroscopies. Multiple-Pulse	Femtosecond Spectroscopy.=	+esonance and Optic	PHYS 0066
-State Chemical Reactions on the	Femtosecond Timescale.=	+y Dissipation in Liquid	PHYS 0085
5-Lip+ Novel Hydroxamic Acid	Fenamates as Dual Inhibitors of Cyclooxygenase and		MEDI 0177
rol+ Kinetics of Volatilization of	Fenitrothion and Deltamethrin from the Surface Mic		ENVR 0020
Estudio de la Nitracion del	Fenol con Aductos $Tonsil-Fe(NO_3)_3$.=		COLL 0120
-Pyrido[3,4-c] [1,5]Benzoxazepine	Fentanyl Derivatives.=	+ahydro-1,2,3,4,4a,5,11,11a	MEDI 0148
	Fermentation and Enzyme Technology.=		BIOT 0222
ynthesis During Carbon Monoxide	Fermentation by an Acidogenic Anaerobe.=	+nol S	BIOT 0015
rate-Based Continuous High-Cell	Fermentation of Microbial-Based Flavorings.=	+yd	CARB 0006
Derivative b+ Development of a	Fermentation Process for the Production of an SCD4		BIOT 0086
o Formation of Ethanol and Other	Fermentation Products.=	+rolyticum in Relation t	BIOT 0029
in Grape Wine due to Malolactic	Fermentation.=	+e Odor-Active Compounds Found	AGFD 0137
wo-Phased System to the A-B-E	Fermentation.=	+and Application of an Aqueous T	BIOT 0159
Bacterial or Animal Cell	Fermentation?—Process and Design Considerations.=		BIOT 0076
l Applications of Actinomycetes	Fermentations from Shake Flask to Production-Scale		BIOT 0083
Shake Flask to Production-Scale	Fermenters. +of Actinomycetes Fermentations from		BIOT 0083
from Low Energy to the	Fermi Domain.=	Aspects of Dissipative Reactions	NUCL 0007
Survey of Important	Fermi Resonances.=		ANYL 0059
Collisions at	Fermilab.=	Search for Quark-Gluon Plasma in P-P	NUCL 0063
sm. Reactivity of Ferrate(VI) and	Ferrate(V) with Amino Acids.=	+tics and Mechani	INOR 0393
tics and Mechanism. Reactivity of	Ferrate(VI) and Ferrate(V) with Amino Acids.=		INOR 0393
ordinated Imidazoles in Low-Spin	Ferric Porphyrin Complexes.=	+Orientation of Co	INOR 0296
Bonding Control of Spin-State:	Ferric Thiosemicarbazones—Comparison with Spin-C		INOR 0268
kisphosphate from Glucose via the	Ferrier Rearrangement.=	+f Inositol (1,3,4,5)Tetra	AGRO 0045
	Ferritin Iron Storage and mRNA Regulation.=		INOR 0011
ome 1,1'-bis(Disphenylphosphino)	ferrocene (dppf)-Bridged Silver(I) Complexes.=		INOR 0106
N+ Two-Photon Dissociation of	Ferrocene and Nickelocene at 248 nm: Statistical vs.		PHYS 0333
er Quenching of Excited States by	Ferrocene Derivatives.=	+ergy and Electron Transf	INOR 0163
dducts of (Dimethylaminomethyl)	ferrocene.=	+raphic Studies of Organoaluminum A	INOR 0274
vestigation of the Interaction of a	Ferrocene-Linked Acridine Orange Compound with		ANYL 0109
h+ Controlling the Structure of	Ferroelectric Liquid Crystal Thin Films by Design: T		BIOT 0202
-Metal Complexes: Potential New	Ferroelectric Materials.=	+ne Pyramidal Transition	INOR 0135
aces Using the Lan+ Study of a	Ferroelectric Side-Chain Polymer at Different Interf		COLL 0071
$0 \le x \le 2.0$) with+ Antiferro-to-	Ferromagnetic Transition in Metallic $TlCo_2S_xSe_{2-x}$ (		INOR 0058
Characterization of	Ferrosilicates by Mossbauer Spectroscopy.=		NUCL 0069
ncia o Exceso de Vitamina E en la	Fertilidad de Ratas Hembras. +ecto de la Deficie		AGFD 0210
l Self-Evaluation for	Fertilizer and Agrichemical Dealers.=	Environmenta	FERT 0018
Quality Assurance in	Fertilizer and Pesticide Formulations.=		FERT 0009
ironmental Issues and Answers in	Fertilizer and Soil Chemistry. Impact of Fertil+ Env		FERT 0013
ironmental Issues and Answers in	Fertilizer and Soil Chemistry. The Retail Ferti+ Env		FERT 0017
Organic or Naturally Occurring	Fertilizer Claims.=	Consumer Perspective of	FERT 0007
MWPS-37 Pesticide and Liquid	Fertilizer Containment Facilities Handbook.=		AGRO 0042
er and Soil Chemistry. Impact of	Fertilizer Containment on Stormwater Management.		FERT 0013
er and Soil Chemistry. The Retail	Fertilizer Dealer as Environmentalist.=	+in Fertiliz	FERT 0017
Retail	Fertilizer Dealer Containment.=		FERT 0020
nd Rinsate Management for Retail	Fertilizer Dealers.=	+l Solutions for Wastewater a	FERT 0014
erformanc+ Flighting for Rotary	Fertilizer Dryers and Coolers and Their Effects on P		FERT 0012
Survival in the	Fertilizer Industry.=	Diversification for Financial	FERT 0010
he Impact of New Regulations on	Fertilizer Manufacture.=	+s Safety Management: T	FERT 0016
Solar Evaporation of	Fertilizer/Ag-Chemical Aqueous Mixtures.=		FERT 0022
d Their Uses in Hort+ Specialty	Fertilizers. Controlled-Release Nitrogen Solutions an		FERT 0005
ical and Physical Technologies of	Fertilizers. Corrosion of Mild Steel Exposed+ Chem		FERT 0001
of Coated	Fertilizers.=	Controlled-Release Nitrogen Analysis	FERT 0008
Scanning Tunneling Microscopy (	FI-STM).=	+and Related Techniques. Field Ion-	PHYS 0182
The Baltimore	Fiasco: A Case Study in Fraud.=		PROF 0005
Si-C-N-O Ceramic	Fiber Compositions from Polycarbosilane.=		POLY 0336
Solid Waste.= Ammonia	Fiber Explosion (AFEX) Pretreatment of Municipal		BIOT 0016

f Poly(methylmethacrylate)/Glass	Fiber Laminates.= +l and Mechanical Properties o	PMSE	0110
omatic Diamine-Epoxy System by	Fiber Optic Fluorescence.= +ure Monitoring of Ar	POLY	0282
aration of Transparent Glass	Fiber Reinforced PMMA Matrix Composites.= Prep	PMSE	0109
for the Characterization of Cotton	Fiber.= +ation of Size-Exclusion Chromatography	PMSE	0032
Types of	Fiber.= Binding Studies of Bile Acids by Various	AGFD	0208
Hydrolysis of Isolated and	Fiber-Bound Hemicelluloses.= Enzymatic	BIOT	0215
Development of Raman	Fiber-Optic Probes for Hostile Environments.=	ANYL	0065
mperature in+ The Influence of	Fiber/Matrix Interface on Local Glass-Transition Te	COLL	0055
on of Phytic Acid, Total Dietary	Fiber, Protein, and Starch in a Milled Hard Red Win	AGFD	0143
s+ Surface Modification of Wood	Fibers and Their Performance in Polystyrene Compo	PMSE	0112
Cells on	Fibers for Waste Clean-Up.=	FUEL	0133
Boron Nitride	Fibers from Polyborates.=	POLY	0316
and Single-Mode Polymer Optical	Fibers.= +Applications. Low-Loss Graded-Index	POLY	0174
Cells on Rotating	Fibers.=	BIOT	0154
a Precursor to Oxygen-Free SiC	Fibers.= High Molecular Weight Polycarbosilane as	POLY	0360
Enzyme Substrates with Human	Fibroblast Monolayers: Uptake, Hydrolysis, and Rele	BIOL	0079
Catalyst Supports from	Fibrous Precursors.= Electrically Conductive	PETR	0016
Kinase+ Comparative Molecular	Field Analysis of Conformationally Flexible Tyrosine	MEDI	0024
Internal Electric	Field and Dye Orientation in PVDF/PMMA Blend.=	POLY	0225
chrotron SAXS Study of Mean-	Field and Ising Critical Behavior of Poly(2-Chloro-S	PMSE	0103
Filtration of Pesticide Wastes:	Field and Laboratory Studies.= Activated Charcoal	AGRO	0127
astes: Molecular Approaches and	Field Applications—I+ Biodegradation of Organic W	BIOT	0117
astes: Molecular Approaches and	Field Applications—II+ Biodegradation of Organic W	BIOT	0137
Quantum Mechanics and Force	Field Calculations.= QSAR Parameters Derived from	MEDI	0044
tion of DPX-E9636 Herbicide for	Field Corn.= +HPLC Method To Support Registra	AGRO	0030
from the Tropical Rain Forest—A	Field for Small Chemical Business.= +al Products	SCHB	0024
: STM and Related Techniques.	Field Ion-Scanning Tunneling Microscopy (FI-STM)	PHYS	0182
rgonne Premium Coal Samples by	Field Ionization Mass Spectrometry.= +on of the A	FUEL	0062
tics of Surface Re+ Electrostatic	Field of the Surface of SiO_2 and Its Effect on Energe	COLL	0116
Influence of a Weak Magnetic	Field on the Colloidal State of Several Materials in S	COLL	0160
2+ Measurement of the Magnetic-	Field Penetration Depth in $YBa_2Cu_3O_x$ and Bi_2Sr	PHYS	0284
2+ Measurement of the Magnetic-	Field Penetration Depth in $YBa_2Cu_3O_x$ and Bi_2Sr	PHYS	0050
n CR(III) Complexes+ Magnetic	Field Perturbations of the Excited-State Properties i	INOR	0159
p+ Cluster Mechanics: A Force-	Field Study of Mixed Ligand Transition-Metal Com	INOR	0075
ation of a Radiofrequency Electric	Field.= +Flow Rate, Ion Concentration, and Applic	COLL	0163
n ab Initio Derived Classical Force	Field.= +ral Analysis of Polyvinylchloride Using a	POLY	0351
Thermal	Field-Flow Fractionation of Copolymers.=	PMSE	0009
mensions Used in Sedimentation	Field-Flow Fractionation through Computer Simulat	PMSE	0028
Analysis of Polymers by Thermal	Field-Flow Fractionation.= +for Polymer Analysis.	PMSE	0008
x Aggregates by Sedimentation	Field-Flow Fractionation.= Characterization of Late	PMSE	0029
Determination of Polymer Using	Field-Flow Fractionation.= Gel Content	PMSE	0010
Using Flow	Field-Flow Fractionation.= Particle Size Analysis	PMSE	0012
iological Macromolecules by Flow	Field-Flow Fractionation.= +luble Synthetic and B	PMSE	0011
ize Exclusion Chromatography,	Field-Flow Fractionation, and Related Chromatograp	PMSE	0008
ize Exclusion Chromatography,	Field-Flow Fractionation, and Related Chromatograp	PMSE	0065
ize Exclusion Chromatography,	Field-Flow Fractionation, and Related Chromatograp	PMSE	0046
ize Exclusion Chromatography,	Field-Flow Fractionation, and Related Chromatograp	PMSE	0028
ize Exclusion Chromatography,	Field-Flow Fractionation, and Related Chromatograp	PMSE	0094
e Competition between Electric-	Field-Induced Molecular Ordering and Chemical Cro	POLY	0206
Polymer-Polymer-So+ Electric-	Field-Induced Pearl-Chain Formation in Multiphase	COLL	0145
icroemul+ Dynamics of Electric-	Field-Induced Transient Phase-Separation in w/o-M	COLL	0146
on Media: A Strategy for Electric-	Field-Orientable Microparticles.= +in Fluorocarb	COLL	0164
anometer-+ Influence of Electric	Fields on Organic Reactions at Zeolite-Supported, N	CATL	0018
Effects of High-Electric	Fields on the Structure and Association of DNA.=	COLL	0179
ing Transition States with Force	Fields: Development of the Best Possible Parameters	ORGN	0170
ce.= Polymers in Electric	Fields: Dynamics from Transient Electric Birefringen	COLL	0180
es: Colloidal Particles in External	Fields: Electric, Magnetic, and Hydrodynamic—I. +l	COLL	0141
es: Colloidal Particles in External	Fields: Electric, Magnetic, and Hydrodynamic—II.	COLL	0159
es: Colloidal Particles in External	Fields: Electric, Magnetic, and Hydrodynamic—III.	COLL	0178
es: Colloidal Particles in External	Fields: Electric, Magnetic, and Hydrodynamic—IV.	COLL	0199
odynamic—I. Colloids in External	Fields.= +nal Fields: Electric, Magnetic, and Hydr	COLL	0141
netics in Thermal and Microwave	Fields.= +nistic Study of Epoxy/Amine — Cure Ki	POLY	0161
Colloidal Systems in External	Fields.= Mass Transfer and Transport Properties of	COLL	0142
posure to Radiofrequency Electric	Fields.= +(Anatase) Suspensions as a Result of Ex	COLL	0203
osed Magnetic and Hydrodynamic	Fields.= +c Colloidal Suspension in Externally Imp	COLL	0144
ct with Solid Surfaces in External	Fields.= +Transport Properties of Colloids in Conta	COLL	0162
ge-Transfer Molecules in Electric	Fields—A Theoretical Study.= +otential and Char	BIOT	0001
istribution of H_2 and H in a Hot-	Filament CVD Reactor of Diamond.= +Density D	PHYS	0146
Carbon.= Hot-	Filament-Activated CVD of Boron Nitride and	PHYS	0205
Mass-Spectrometry during Hot-	Filament-Assisted Diamond Film Growth.= +eam	PHYS	0145
	Filament-Assisted Diamond Growth Kinetics.=	PHYS	0178
Filamen+ Protein Production in	Filamentous Fungi. Heterologous Gene Expression in	BIOT	0234
eterologous Gene Expression in	Filamentous Fungi: Recent Improvements in Product	BIOT	0234
Expression of Restrictocin in	Filamentous Fungi.=	BIOT	0238
Growth of Diamond	Filaments.= In Situ Diagnostics of the Flame	PHYS	0149
the Beilstein	File.= Structure and Reaction Similarity Searches in	CINF	0039
erse Voltammetry at the Mercury	Film Electrode.= +of Human Seric Copper by Inv	ANYL	0056
g Bonds.+ Silicon Carbide Thin-	Film Formation on Si(100): The Behavior of Danglin	PHYS	0204
hores in Supported Sol-Gel Thin-	Film Glasses for Second-Harmonic Generation.=	POLY	0210
chanistic Studies of Copper Thin-	Film Growth by MOCVD.= +tallic Precursors. Me	COLL	0025

Aluminum Thin	Film Growth by MOCVD.= Mechanistic Studies of	INOR	0124
Expitaxial GaAs	Film Growth from Single-Source Precursors.=	COLL	0067
Hot-Filament-Assisted Diamond	Film Growth.= +Beam Mass-Spectrometry during	PHYS	0145
Diamond	Film Growth.= Effects of Ammonia on CVD	PHYS	0150
Kinetics of Diamond	Film Growth.=	PHYS	0177
As(100) and Implications for GaAs	Film Growth.= +osition of Triethylgallium on Ga	COLL	0068
plications in Polymer- and Thin-	Film Growth.= +me Spectroscopic Ellipsometry: Ap	POLY	0252
Multilayer Construction of Thin-	Film Nonlinear Optical Materials.= +Superlattices:	POLY	0127
cesses of Transparent Aluminum	Film Preparation through Aqueous Sol-Gel Method.	COLL	0115
Effect of Precursor Chemistry on	Film Properties.= +cal Vapor Deposition of ZnSe:	INOR	0171
Diamond	Film Synthesis.= Molecular Beam Studies on	PHYS	0203
Photopolymer	Film.= Diffusion of Water into a	POLY	0194
Diamond	Film.= Synthesis of Thick, Large-Area	PHYS	0093
la Locomocion de los Monocitos (	FILM) Producido por E. +n del Factor Inhibidor de	BIOL	0102
l)borane.= Synthesis and	Film-Growth Studies of Di-t-Butylphosphino(diethy	INOR	0230
rochromic Properties of Biochrom	Films Based on 4-Keto-Bacteriorhodopsin.= +lect	BIOT	0082
Characterization of Polyurethane	Films Bearing NLO-Active Chromophores.= +tical	POLY	0216
alt Distribution in Model Polymer	Films by ^{19}F Multiple-Quantum NMR.= +onium S	PMSE	0182
erroelectric Liquid Crystal Thin	Films by Design: The Molecules to Materials Connec	BIOT	0202
rmation of Organic-Pigment Thin	Films by Disruption of Micelles.= +trochemical Fo	COLL	0019
f Palladium and Platinum Metal	Films by Electron Transfer to Photogenerated Organ	INOR	0319
Hemicyanine Langmuir-Blodgett	Films by Fourier Transform Infrared Spectroscopy a	POLY	0224
emperature Deposition of Copper	Films by H-Atom Reaction with Cu(FOD)$_2$ and Cu(	PHYS	0259
hromism of Diacetylene LB	Films by Polyion Complexation.= Control of Photoc	BIOT	0146
Edible	Films for Fruits and Vegetables.=	AGFD	0078
Preparation of Polymeric	Films for NLO Applications.=	POLY	0104
rfacially Polymerized Porphyrin	Films for Vectorial Photoinduced Electron Transport	BIOT	0126
on of Tungsten and Molybdenum	Films from (η^3-C$_3$H$_5$)$_5$M (M = MO, W).= +Depositi	INOR	0055
Preparation of Silicon-Containing	Films from Molecular Precursors.= +ctions in the	INOR	0121
position of Transition-Metal Thin	Films from Organometallic Precursors.= +ials. De	INOR	0141
Halides.= Synthesis of Nitride	Films from the Reaction of Amines with Metal	INOR	0309
Uniaxially Oriented Polyamide	Films in Plane Prepared by Vacuum Deposition Poly	POLY	0082
nylidene Chloride).= Composite	Films of Polyaniline and Poly(vinylchloride)/Poly(Vi	PMSE	0113
Bicomponent Polymer	Films on a Metal Surface by ESCA.= Study of Thin	POLY	0100
cal Properties of Ultra-Thin MgO	Films on Mo(100).= +Characterization, and Chemi	COLL	0193
SEM Observations of Diamond	Films Synthesized by DC Arc Discharge CVD in a M	FUEL	0038
CVD Deposition of Diamond Thin	Films with Applications in X-ray Lithography.=	PHYS	0092
Voltammetry in Lipid	Films with Applications to Vesicles.=	COLL	0021
ace Derivatization o+ Solids and	Films. Dithiocarbamates: Reagents for Selective Surf	INOR	0350
nd Interfaces—IV. Thin Polymer	Films: From Chemistry to Materials Science in Integ	COLL	0070
Dynamics Simulations of Diamond	Films: Structure, Growth, and Properties.= +ular	PHYS	0354
Alkali Metal-C$_{60}$	Films: The First 3-D Molecular Conductors?.=	INOR	0354
Processes in Compatible Polymer	Films.= +e Techniques to the Study of Diffusional	POLY	0163
Transfer in Electroactive Polymer	Films.= +ersible and Irreversible Neutral Species	ANYL	0103
on of Inorganic Surfaces and Thin	Films.= +ctron Spectroscopy in the Characterizati	INOR	0013
trified Nonlinear Optical Polymer	Films.= +dering and Chemical Cross-Linking in Vi	POLY	0206
Bilayer	Films.= Charge Transport in Polymer	ANYL	0104
Nitride	Films.= Depositions and Characterizations of Boron	PHYS	0094
ine Langmuir-Blodgett	Films.= Diffusion in the Gas Sensing of Phthalocyan	POLY	0118
Studies of C$_{60}$	Films.= Electrochemical and Langmuir Trough	ANYL	0108
to Tantalum Nitride Thin	Films.= Ethylimidotantalum Complex as Precursor	INOR	0074
amics of Diblock Copolymer Thin	Films.= +nite Thickness Effects in the Thermodyn	POLY	0090
Langmuir-Blodgett	Films.= Fractals and Excitation-Transfer in	PHYS	0110
PEEK (Poly-ether-ether-ketone)	Films.= +oscopic Study of Thermal Degradation in	POLY	0334
etrating Polymer Network	Films.= Gas Permeability Studies of Latex Interpen	PMSE	0061
Langmuir-Blodgett	Films.= Guided-Wave Nonlinear Optics in DCANP	POLY	0176
Polypyrrole	Films.= Incorporation of Redox Compounds into	COLL	0013
al Vapor Deposition of Diamond	Films.= Laser Spectroscopic Diagnostics for Chemic	PHYS	0151
the Surface of Depositing	Films.= Laser Studies of the Reactivity of NH with	COLL	0218
nd Transition-Metal Nitride Thin	Films.= +Temperature Synthesis of Main-Group a	INOR	0104
aches to high-T$_c$ Superconducting	Films.= +rganic Chemical Vapor Deposition Appro	INOR	0176
th Polyion Complexed Pyrene LB	Films.= +otoamplified Storage Optical Memory wi	BIOT	0144
Energy Transfer in LB	Films.= Photoinduced Electron Transfer and	BIOT	0103
LB	Films.= Photoinduced Proton Transfer in	BIOT	0145
y Doubling in Poled Polymer	Films.= Recent Developments in Efficient Frequenc	POLY	0179
ne Chains in the Bulk and in Thin	Films.= +ic Dynamics Simulations of Polymethyle	POLY	0355
ion and Growth of Diamond Thin	Films.= +tructural Characterization of the Nucleat	PHYS	0121
Metal	Films.= Surface Chemistry of Monolayer	COLL	0063
on of Small Molecules in Polymer	Films.= +tional Spectroscopies to Probe the Diffusi	POLY	0332
f Diamonds, Synthetic Diamond	Films, and Related Carbon Materials by Raman Spec	PHYS	0118
onlinear Optics of Polymer Thin	Films, I: Molecular Engineering and Processing of Po	POLY	0011
nlinear Optics of Polymer Thin	Films, 2: Structure-$\chi^{(3)}$ Relationships in Rigid-Rod P	POLY	0205
onlinear Optics of Polymer Thin	Films, 3: Structure-$\chi^{(3)}$ Relationships in Conjugated	POLY	0226
onlinear Optics of Polymer Thin	Films, 4: Structure-$\chi^{(3)}$ Relationships in Polyanilines	POLY	0221
onlinear Optics of Polymer Thin	Films, 5: Wavelength Dispersion of the $\chi^{(3)}$ of Poly(p	POLY	0213
Studies.= Activated Charcoal	Filtration of Pesticide Wastes: Field and Laboratory	AGRO	0127
dual Atoms and Clusters with the	FIM.= +nd Related Techniques. Imaging of Indivi	PHYS	0123
Atom-Probe	FIM.= Atomic Scale Surface Analysis with the	PHYS	0101
FEM, FEES,	FIM, A-P, STM, STS.= Tunneling Microscopies:	PHYS	0097
Organization of Cellulomonas	fimi Cellulases.= Structural and Functional	BIOT	0207

Diversification for	Financial Survival in the Fertilizer Industry.=	FERT	0010
Obtaining	Financing (Alternative Methods).=	SCHB	0005
drodynamic—IV. Hydrodynamic	Fingerprinting as a Probe of Latex Characterization	COLL	0199
a Single-Acid-+ Electrophoretic	Fingerprinting of the Surface Electrical Properties of	COLL	0200
nuing Education Course—Start to	Finish.= +cian Symposium. The Making of a Conti	TECH	0001
Diblock Copolymer Thin Films.+	Finite Thickness Effects in the Thermodynamics of	POLY	0090
Nuclear Equation of State and	Finite-Range Interaction.=	NUCL	0008
Application of Waste-Fuel-	Fired Hybrid Fluidized-Bed Boilers.= Commercial	FUEL	0122
ctronic Effects in the Reactions of	Fischer Carbene Complexes with Alkynes.= +eoele	INOR	0194
actions of Chelated Imidazolidone	Fischer Carbene Complexes.= +ymmetric Aldol Re	INOR	0112
d the Design of Ruthenium-Based	Fischer-Tropsch Catalysts.= +lectivity Models an	CATL	0056
ose—A Centennial Tribut+ Emil	Fischer's Establishment of the Configuration of Gluc	CARB	0011
try: Schematic Method for	Fischer's Proof Teaching.= D-Glucose Stereochemis	CHED	0095
ic Metabolism in Food-Producing	Fish and Shellfish.= +thods for Studying Xenobiot	AGRO	0096
Lipid Oxidation of	Fish during Storage.=	AGFD	0123
Efficacious Insecticides with Low	Fish Toxicity.= +ryl-1-Cyclopropylbutanes: Highly	AGRO	0073
s in La Basse Cot+ Exposure of	Fishing People to Chlorinated Aromatic Hydrocarbon	CEI	0011
Geotermicos.= Evaluacion	Fisicoquimica de Cementos Hidraulicos para Pozos	INOR	0063
A Simple Computer Program to	Fit Experimental Results.=	NUCL	0089
3D Shape	Fitting and 3D-DB Searching by SPERM.=	COMP	0040
on, and Structure in the Chemical	Fixation Process.= +action Mechanisms, Distributi	ENVR	0012
Materials.= Construction of	FK-506 Analogs from Rapamycin-Derived	ORGN	0015
Immunosuppressant	FK-506.= Synthetic Studies Toward the	ORGN	0116
Failing with a	Flair.=	SCHB	0017
s.=	Flame Chemistry of Alkali and Alkaline Earth Metal	PHYS	0073
In Situ Diagnostics of the	Flame Growth of Diamond Filaments.=	PHYS	0149
Time-Dependent	Flame Model Utilizing the Method of Lines.=	FUEL	0125
Phosphorus	Flame Retardants for a Changing World.=	INOR	0346
Organoboron Compounds as	Flame Retardants.= Syntheses of Some	ORGN	0262
Net Reaction Rates of	Flame Species Pertinent to Fullerenes.= Fluxes and	FUEL	0112
Dye Laser Studies of Plasma and	Flame-Generated Species.=	CHED	0087
d on Flow-Reactor and Laminar-	Flame-Speed Data.= +dation Models: Studies Base	FUEL	0083
+ot Deposition from Ethylene/Air	Flames and the Role of Aromatic Intermediates.=	FUEL	0107
emistry: Single-Ring Aromatics in	Flames. Formation of Benzene in+ Combustion Ch	FUEL	0078
Flames. Formation of Benzene in	Flames.= +n Chemistry: Single-Ring Aromatics in	FUEL	0078
d Growth from Acetylene- Oxygen	Flames.= +oepitaxial High-Temperature Diamon	FUEL	0037
Charged Species in	Flames.= Aromatics, Fullerenes, and Soot as	FUEL	0111
Decane and Kerosene	Flames.= Comparison of Aromatics Formation in	FUEL	0080
eling of 10-Torr Methane/Oxygen	Flames.= +ced Fluorescence Diagnostics and Mod	FUEL	0128
anisms of Prompt NO in Methane	Flames.= +Measurements To Test Chemical Mech	FUEL	0098
Metal Ion Chemistry in	Flames.=	PHYS	0201
atic Hydrocarbons in Diffusion	Flames.= Modeling the Growth of Polynuclear Arom	FUEL	0109
Fullerenes in	Flames.= Production and Characterization of	FUEL	0020
Constant Measurements by the	Flash or Laser Photolysis-Shock-Tube Method: Resu	FUEL	0123
Laboratory.= Benzophenone	Flash Photolysis in the Physical Chemistry	CHED	0086
ns Sensitized by Colloidal CdS: A	Flash Photolysis Study.= +Controlled Photoreactio	COLL	0150
nd Electron Transfer Reactions by	Flash Photolysis.= +of Thiyl Radical Formation a	INOR	0396
ycetes Fermentations from Shake	Flask to Production-Scale Fermenters. +of Actinom	BIOT	0083
within Conjugated Porphyrin-	Flavin Ester.= Photoinduced Electron Transfers	BIOL	0134
Interactions of	Flavocytochrome c_{552}.=	BIOL	0093
enobiotics in Foods.= Effect of	Flavonoids on Mutagenicity and Bioavailability of X	AGFD	0116
and Health. Anticarcinogenicity of	Flavonol Quercetin.= +nolic Compounds in Food	AGFD	0113
Populations and Inbreds for Silk	Flavonolglycoside Contents.= +Corn (Zea mays L.)	AGRO	0034
ary Plant Products and Their I+	Flavor Adulteration. Isotope Distributions in Second	AGFD	0194
omers in Raspberry Concentrate+	Flavor Adulteration. Survey of Alpha-Ionone Enanti	AGFD	0181
SNIF-NMR to the Detection of	Flavor and Fragrance Adulteration.= Application of	AGFD	0197
Formation of Smoke-	Flavor Compounds by Thermal Lignin Degradation.=	AGFD	0054
atment.=	Flavor Compounds Formed from Lipids by Heat Tre	AGFD	0053
matic Conversions of Precursors to	Flavor Compounds. Chrono-C+ Thermal and Enzy	AGFD	0022
matic Conversions of Precursors to	Flavor Compounds. Glucoside+ Thermal and Enzy	AGFD	0050
matic Conversions of Precursors to	Flavor Compounds. Glycosidi+ Thermal and Enzy	AGFD	0006
matic Conversions of Precursors to	Flavor Compounds. Studies on+ Thermal and Enzy	AGFD	0070
es on the Formation of Meat-Like	Flavor Compounds.= +o Flavor Compounds. Studi	AGFD	0070
Chiral Analysis of Natural	Flavor Compounds.=	AGFD	0184
Monoterpenoid	Flavor Compounds.= Enzymatic Synthesis of	AGFD	0009
nantioselective Reactions of Chiral	Flavor Compounds.= +s: Useful Biocatalysts for E	AGFD	0026
: A Potent Source of Interesting	Flavor Compounds.= Thermally Degraded Thiamine	AGFD	0071
Free and Bound	Flavor Constituents of White-Fleshed Nectarine.=	AGFD	0024
ients to Minimize Warmed-Over-	Flavor Development in Meat Products.= +d Ingred	AGFD	0021
Studies on Meat	Flavor Generation from Cysteine and Sugars.=	AGFD	0074
emical and Sensory Evaluation of	Flavor in Untreated and Antioxidant-Treated Meat.	AGFD	0120
atograph+ Analysis of Glycosidic	Flavor Precursors by Droplet Countercurrent Chrom	AGFD	0203
Biosynthetic Development of	Flavor Precursors in Clove.=	AGFD	0211
Peptides as	Flavor Precursors in Model Maillard Reactions.=	AGFD	0072
Norisoprenoids: Important	Flavor Precursors.= Oxygenated C_{13}	AGFD	0025
Lipid Oxidation and	Flavor Stability.= Influence of Food Processes on	AGFD	0157
yrups.=	Flavor Volatiles and Related Compounds in Maple S	AGFD	0174
ursors of Varietal Grape and Wine	Flavor.= +to Flavor Compounds. Glycosidic Prec	AGFD	0006
Smoke	Flavor.= Contribution of Phenolic Compounds to	AGFD	0173
roducts and Their Importance for	Flavor-Origin Assignments.= +Secondary Plant P	AGFD	0194

o Milk Shelf-Life Prediction and	Flavor-Quality Evaluation Based on Headspace Gas	AGFD 0108
ct of Storage on Roasted Peanut	Flavor, as Determined by Descriptive Sensory Analys	AGFD 0161
Compounds of Piper betel as	Flavoring and Nerve-Stimulating Agent.= Phenolic	AGFD 0154
nents+ Investigations of Natural	Flavoring Materials by the Analysis of Chiral Compo	AGFD 0182
nation of Petrochemical Synthetic	Flavoring Materials.= +n Analysis for the Determi	AGFD 0186
sis for the Control of Natural	Flavoring Substances.= Use of Stable Isotope Analy	AGFD 0195
Fermentation of Microbial-Based	Flavorings.= +ydrate-Based Continuous High-Cell	CARB 0006
the Origin Evaluation of	Flavors and Fragrances.= Chirospecific Analysis in	AGFD 0183
Characterization of Off-	Flavors by Aroma Extract Dilution Analysis.=	AGFD 0097
berry Concentrates, Essences, and	Flavors.= +of Alpha-Ionone Enantiomers in Rasp	AGFD 0181
Constituents of Living Odors and	Flavors.= +e-Dimensional Analysis of the Volatile	AGFD 0022
Botanical Extracts Used in Foods,	Flavors, Cosmetics, and Pharmaceuticals.= +ds in	AGFD 0172
Constituents of White-	Fleshed Nectarine.= Free and Bound Flavor	AGFD 0024
ydiphenols—Effect of Backbone	Flexibility and Cross-Linking on Flexibility of Can C	PMSE 0162
Flexibility and Cross-Linking on	Flexibility of Can Coatings.= +-Effect of Backbone	PMSE 0162
and Coil Coatings with Improved	Flexibility.= +el Waterborne Epoxy Resins for Can	PMSE 0142
hemistry Important to+ Incin: A	Flexible Chemkin Driver Program for Investigating C	I&EC 0024
ch for Docking Conformationally	Flexible Molecules from 2D or 3D Chemical Databas	CINF 0028
Databases of Conformationally	Flexible Molecules: A Flexible Prescreening Techniqu	CINF 0027
Rearrangements of	Flexible Polymer Chains in Flow.= Conformational	PMSE 0105
rmationally Flexible Molecules: A	Flexible Prescreening Technique.= +bases of Confo	CINF 0027
Query Formulation.=	Flexible Queries in 3D Searching: Techniques in 3D	CINF 0025
Field Analysis of Conformationally	Flexible Tyrosine Kinase Inhibitors.= +Molecular	MEDI 0024
d Their Effects on Performance+	Flighting for Rotary Fertilizer Dryers and Coolers an	FERT 0012
Polymeric	Flocculants of the Zwitterionic Type.=	POLY 0094
Polymeric	Flocculants.=	PMSE 0129
Spectroscopic Studies of	Floppy Molecules: C3 and SiC2.=	PHYS 0235
rier Island Salt Marshes along the	Florida Indian River Lagoon.= +ide Residue in Bar	ENVR 0019
PbS at the	Flotation Conditions.= Electrochemical Behavior of	PHYS 0217
Use of Froth	Flotation for Plastics Recycling.=	I&EC 0036
	Flottation Ionique du Laurylsulfate de Gallium.=	ANYL 0053
nic Activity.=	Flouorinated Pregnane Derivatives with Antiandroge	ORGN 0193
ase Transition in Tetracene Cry+	Flourescence and Excitation Spectra Pattern and Ph	PHYS 0274
Spathodea Campanulta	Flowers.= Chemical Compounds from Anthers of	ORGN 0186
everal Materials in Stationary and	Flowing Systems.= +eld on the Colloidal State of S	COLL 0160
r Dynamics.=	Fluctuations of the Single-Particle Density in Nuclea	NUCL 0009
lective Detection in Supercritical	Fluid Chromatography Using an Electrolytic Conduc	ANYL 0091
pplicati+ Capillary Supercritical	Fluid Chromatography: Analytical Capabilities and A	AGFD 0201
lasmic Proteins Using Critical	Fluid Disruption.= Downstream Processing of Perip	BIOT 0070
Temperature-Rising Supercritical	Fluid Fractionation.= +Low-Density Polyethylene:	POLY 0076
Polymer-Supercritical	Fluid Mixtures.= Phase Behavior of	PMSE 0036
Supercritical	Fluid Processing Applied to Recycling.=	I&EC 0034
Clays Dried Using a Supercritical	Fluid.= +ace and Cracking Properties of Expanded	PETR 0145
minants and Passivating Agents in	Fluid-Cracking Catalysts.= +ation of Metal Conta	CATL 0012
y between Core and+ Membrane	Fluidization by Anesthetics and Alcohols: Discrepanc	MEDI 0032
Waste-Fuel-Fired Hybrid	Fluidized-Bed Boilers.= Commercial Application of	FUEL 0122
Fuel Evaluation for U-Gas	Fluidized-Bed Gasification Process.=	FUEL 0120
gh-Temperature Dense Molecular	Fluids and Low-Temperature Molecular Solids.=	ANYL 0002
n to Drug Monitoring in Biological	Fluids and Tissues.= +ization LC/MS: Applicatio	ANYL 0124
oring and Control of Properties of	Fluids during Processing.= +for the On-Line Monit	I&EC 0059
d Phases in Naturally Occurring	Fluids on the Electrochemical Reduction of Ru(NH3)	COLL 0005
ity to Contamination of Drilling	Fluids Systems Based on Geothermal Water as the C	I&EC 0011
Effects on Drill Pipes by Drilling	Fluids Systems Based on Geothermal Water.= +ive	I&EC 0012
rugs and Metabolites in Biological	Fluids Using Laboratory Robotics.= +Analysis of D	ANYL 0119
ochemistry in Microheterogeneous	Fluids+ Surfactants and Associated Colloids: Electr	COLL 0166
ochemistry in Microheterogeneous	Fluids+ Surfactants and Associated Colloids: Electr	COLL 0019
ochemistry in Microheterogeneous	Fluids+ Surfactants and Associated Colloids: Electr	COLL 0040
ochemistry in Microheterogeneous	Fluids+ Surfactants and Associated Colloids: Electr	COLL 0001
ochemistry in Microheterogeneous	Fluids+ Surfactants and Associated Colloids: Electr	COLL 0128
ochemistry in Microheterogeneous	Fluids+ Surfactants and Associated Colloids: Electr	COLL 0147
ination of Ceftibuten in Biological	Fluids.= +, and LC/LC/MS Methods for Determ	ANYL 0118
Photoreactivity in Supercritical	Fluids.=	ORGN 0131
Supercritical	Fluids.= Polymer and Biopolymer Modifications in	POLY 0084
Silahydrocarbon	Fluids.= Synthesis and Properties of Model	I&EC 0008
mistry, and Activity against Liver	Flukes of Some Benzimidazole Derivatives.= +, Che	MEDI 0065
due Depletion and Metabolism of	Flunixin in Cattle.= +uirements of the NADA: Resi	AGRO 0078
lumina o Titania Modificadas con	Fluor.= +de Catalizadores NiW Soportados sobre A	FUEL 0106
-(Sulfonatooxy)-2-(Acetylamino)	Fluorene.= +s on the Ultimate Hepatacarcinogen, N	ORGN 0194
hemical Formulas: Enumeration of	Fluorenoids/Fluoranthenoids.= +ion Content of C	COMP 0027
eir Derivatives as Anti-Inf+ N-(	Fluorenyl-9-Methoxycarbonyl) Amino Acids and Th	MEDI 0171
Fluorescence Polarization of	Fluorescein-Labeled Dextrans.=	ANYL 0076
al Wells.= Standards for X-ray	Fluorescence Analysis of Cements Used in Geotherm	ANYL 0058
e Polymer Blend.=	Fluorescence and Shear-Induced Mixing in a Miscibl	PMSE 0133
Sensitivity Enhancement in	Fluorescence Capillary/Fiber-Optic Sensors.=	ANYL 0077
oteins and+ NMR Relaxation,	Fluorescence Depolarization, and the Dynamics of Pr	PHYS 0166
ethane/Oxygen+ Laser-Induced	Fluorescence Diagnostics and Modeling of 10-Torr M	FUEL 0128
en+ Recovery of Distributions of	Fluorescence Lifetimes in Homogeneous and Heterog	PHYS 0134
ng of Diamond+ Laser-Induced	Fluorescence Measurements for the Chemical Modeli	PHYS 0147
anisms of Prom+ Laser-Induced	Fluorescence Measurements To Test Chemical Mech	FUEL 0098
n Transfer in Covalently Linked+	Fluorescence Mechanisms and Photoinduced Electro	PHYS 0021

racterization.= Application of	Fluorescence Microscopy to Coal–Derived Resid Cha	FUEL	0043
igh–Pressure Effects and Delayed	Fluorescence ODMR.= +lly Mixed Naphthalene: H	PHYS	0081
tudies.=	Fluorescence of Aromatic Diamines for Epoxy Cure S	POLY	0159
xtrans.=	Fluorescence Polarization of Fluorescein–Labeled De	ANYL	0076
ethylbenzenes in Solution, Using a	Fluorescence Probe.= +on Quenching of Triplet M	PHYS	0265
	Fluorescence Probes of Solution Phase Dynamics.=	CHED	0201
l Chemistry Lab.=	Fluorescence Quenching Experiments for the Physica	CHED	0202
roperties of Humic Substances by	Fluorescence Quenching.= +n of the Electrostatic P	ENVR	0057
noxide Laser–Excited Molecular	Fluorescence Spectrometry in a Graphite Furnace.=	ANYL	0078
antitative Two–Photon Excited	Fluorescence Spectroscopy.= Recent Advances in Qu	ANYL	0014
no– and Multilayer+ FT–IR and	Fluorescence Studies of Self–Assembled Siloxane Mo	POLY	0330
m+ Dynamic and Steady–State	Fluorescence Studies of Tetradecyltrimethylammoniu	COLL	0074
Processes in C+ Application of	Fluorescence Techniques to the Study of Diffusional	POLY	0163
ine–Epoxy System by Fiber Optic	Fluorescence.= +ure Monitoring of Aromatic Diam	POLY	0282
aracterization of Detergents Using	Fluorescent Probes: A Teaching Experiment.= +Ch	CHED	0200
ues as δ–Opioid Receptor Selective	Fluorescent Probes.= +hesis of Naltrindole Analog	MEDI	0155
phate of th+ Orienting Effect of	Fluoride Anions in the Synthesis of a New Gallophos	PETR	0075
ion of MFI Zeolites in an Alkaline	Fluoride Medium.= +re Hydrothermal Crystallizat	PETR	0005
ion Characterizatio+ Reaction of	Fluoride with Human Enamel: XPS, SEM, and Solut	COLL	0085
Miscible Blends of Poly(vinylidene	Fluoride) and Poly(3–hydroxybutyrate).= +udy of	POLY	0097
Ethyl Acetate in Poly(vinylidene	Fluoride) Monitored by Rheo–Photoacoustic FT–IR	POLY	0255
logenation Studies of Vinylidene	Fluoride/Trifluoroethylene Copolymer and Poly(chlo	POLY	0022
ing Agents fo+ Transition–Metal	Fluorides and Oxide Fluorides as Selective Fluorinat	INOR	0133
ition–Metal Fluorides and Oxide	Fluorides as Selective Fluorinating Agents for Hydro	INOR	0133
Polymeric Triorganotin(IV)	Fluorides.= X–ray Powder Diffraction Studies of	INOR	0065
Synthesis and Characterization of	Fluorinated Biopolymers.= +ic Resonance Imaging:	MEDI	0041
–α–fluoroacrylonitrile).= New	Fluorinated Copolymers: Poly(1,3–cyclohexadiene–alt	POLY	0021
odisperse 1,3,5–Phenylene–Based,	Fluorinated Dendrimers.= +tion of a Series of Mon	POLY	0236
A Novel Route to Two Heavily	Fluorinated Epoxy Resins from a Fluorodiimidediol.=	POLY	0037
2:1–Layer Silicates in Acid and	Fluorinated Medium.= Synthesis of Dioctahedral	PETR	0128
agnetic Resonance Imagin+ Novel	Fluorinated Polymeric Molecular Probes for F–19 M	MEDI	0041
rization, and Mechanical Prope+	Fluorinated Polyurethanes: Polymerization, Characte	POLY	0018
ing Groups in the Synthesis of 2'–	Fluorinated Purine Nucleosides.= +Sugar–Protect	CARB	0024
Heptafluorodiborate: A Useful	Fluorinating Agent.= N–Fluoropyridinium Pyridine	ORGN	0076
s and Oxide Fluorides as Selective	Fluorinating Agents for Hydrohalogenoalkanes.=	INOR	0133
nimide (NFOBS): A Useful New	Fluorinating Reagent.= N–Fluoro–o–Benzenedisulfo	ORGN	0077
Studies of Atipamezole Analogs:	Fluorination of 5–Substituted Indanones and Indand	ORGN	0078
Absorbed Hydrogen and	Fluorine near the Diamond/Vapor Interface.=	PHYS	0359
tructure in Organofluorides Using	Fluorine–19 Carbon–13 Coupling.= +ion of Local S	FUEL	0077
ity of an Unsaturated Ruthenium	Fluoro–Alkoxide Complex.= +ynthesis and Reactiv	INOR	0127
anoside and Its 6–Deoxy– and 6–	Fluoro–Analogues and the Demonstration of a Separ	CARB	0028
New Fluorinating Reagent.= N–	Fluoro–o–Benzenedisulfonimide (NFOBS): A Useful	ORGN	0077
Stereoselective Synthesis of 3–	Fluoro–3–Hydroxyalkylazetidinones.=	ORGN	0080
Poly(1,3–cyclohexadiene–alt–α–	fluoroacrylonitrile).= New Fluorinated Copolymers:	POLY	0021
Asymmetric Synthesis with 3–	Fluoroazetidinone.=	ORGN	0081
iated Polar Colloid Formation in	Fluorocarbon Media: A Strategy for Electric–Field–O	COLL	0164
Supported	Fluorocarbonsulfonic Acid Catalysts.=	PETR	0053
Fluorinated Epoxy Resins from a	Fluorodiimidediol.= +Novel Route to Two Heavily	POLY	0037
Renin: Kinetic Studies Using New	Fluorogenic Substrates.= +ylalanine at P1 Site by	BIOL	0035
Serine–AMC–Carbamate: A Novel	Fluorogenic Tryptophanase Substrate.= +thesis of	ORGN	0023
Protease Inhibition by Peptidyl	Fluoroketones.= NMR and Kinetic Studies of	BIOL	0005
ransfer Using Frequency–Domain	Fluorometry.= +served by Intramolecular Energy T	PHYS	0061
ation Reaction to the Synthesis of	Fluoroolefin Dipeptide Isosteres.= +Peterson Olefin	ORGN	0079
al Activity of N–(Substituted–2–	Fluorophenyl)–1,2,4–Triazolo–[1,5a]–Pyrimidine Sulf	AGRO	0048
Epoxy by Extrinsic and Intrinsic	Fluorophores.= +rement of Water Uptake in Cured	POLY	0160
Modification of	Fluoropolymer Surfaces by Sodium Naphthalenide.=	PMSE	0124
Useful Fluorinating Agent.= N–	Fluoropyridinium Pyridine Heptafluorodiborate: A	ORGN	0076
s of Cortisol: The Discovery of the	Fluorosteroids.= +unt for an Economical Synthesi	HIST	0022
alifornia Continental Shelf+ The	Flux of Iron, Cobalt, Copper, and Manganese from C	GEOC	0010
Dissolved Organic Carbon (DOC)	Flux out of Margin Sediments on Oceanic DOC.=	GEOC	0017
u Stea+ Site Use of Advective–	Flux Probes for Enhanced Soil Gas Analysis in in Sit	ENVR	0052
tinent to Fullerenes.=	Fluxes and Net Reaction Rates of Flame Species Per	FUEL	0112
diments: Implications for Burial	Fluxes and Phosphorus Residence Time in the Ocean	GEOC	0013
n Margins. Controls on Methane	Fluxes from the Hudson River.= Chemistry of Ocea	GEOC	0001
erived ^{14}C from the Hudson Ri+	Fluxes of Dissolved Inorganic Carbon and Reactor–D	GEOC	0014
Synthesis, Characterization, and	Fluxional Behavior of [cod]M{[Ph$_2$P(S)]$_n$[Ph$_2$P(O)]	INOR	0264
sion+ Trace Metals Analysis of	Fly Ash by Inductively Coupled Plasma Atomic Emis	FUEL	0138
and Regulation of a House	Fly Cytochrome P450 Monooxygenase.= Structure	AGRO	0098
rethroid Resistance in the House	Fly.= Characterization of a P450 Responsible for Py	AGRO	0099
ybrid MS/MS to the Analysis of	Flyash for Monobrominated Polychlorinated Dibenzo	ANYL	0050
f Leuconostoc mesenteroides B–	512FM Dextransucrase: Inhibition of Dextran Synthe	CARB	0028
Complete Basis–Set Energies	fo One– and Two–Carbon Molecules.=	PHYS	0317
s of Temperature and pH on the	Foaming Properties and Conformation of β–Lactoglo	AGFD	0207
ncy Sound–Absorbing Studies on	Foams Based on Polyurethanes and Interpenetrating	PMSE	0040
of	Foams.= Concepts, Experiments, and Applications	CHED	0100
Hartree	Fock Description of Substituted Azomethine Ylides.=	ORGN	0171
aking in Radicals Studied by the	Fock–Space Multireference Coupled–Cluster Method	PHYS	0016
ms by H–Atom Reaction with Cu(	FOD)$_2$ and Cu(HFA)$_2$.= +Deposition of Copper Fil	PHYS	0259
c Transposition Catalyzed by Eu(	fod)$_3$.= +ic Methoxyacetates: A Novel Type of Allyli	ORGN	0211
Taking the	Fog out of MSDSs.=	CHAS	0038

mation and Dynamics of Partially	Folded and Unfolded Protein Molecules.= +Confor	PHYS	0107
Peptide Models for Protein	Folding and Localization.=	BIOL	0003
Facilitation of Protein	Folding and the Reversibility of Denaturation.=	BIOT	0190
Process for Controlled	Folding of Recombinant aFGF.= Novel Scalable	BIOT	0134
	Folding of Trp Repressor Subdomains.=	BIOL	0115
1β.=	Folding Properties of Inclusion Body Mutants of 1L-	BIOT	0114
Probing the Mechanism of Protein	Folding.= +and Specificity. New Techniques for	BIOT	0112
Antibody-Assisted Protein	Folding.=	BIOT	0062
Role of Chaperones in Protein	Folding—GroE Heat Shock Proteins.=	BIOT	0116
Terminal B-Chain on Association,	Folding, and Stability.= +city. Role of Insulin's C-	BIOT	0132
Transthyretin.= Progress on the	Folding, Assembly, and Amyloidosis of	BIOL	0007
of Erythrina Alkaloids from the	Foliage of Genetic Cloves of Three Erythrina Species	ANYL	0136
na poeppigiana and E. berteroana	Foliage.= +idines in the Milk of Goats Fed Erythri	AGFD	0129
Sponge Phyllospongia	Foliascens.= New Dialkylated Scalarins from the	ORGN	0182
Sponge Phyllospongia	Folliascens.= New Dialkylated Scalarins from the	ORGN	0188
s Chemistry. Arthur Dock	Fon Toy—Scientist, Colleague, Friend.= Phosphoru	INOR	0288
ence. Applications of GC/FTIR to	Food Analysis: Perspectives and Limits.= +od Sci	AGFD	0200
in	Food Analysis.= Gas Chromatographic Columns	AGFD	0176
Thin-Layer Chromatography in	Food Analysis.=	AGFD	0180
actions for Sample Preparation in	Food and Agricultural Analyses.= +lid-Phase Extr	AGFD	0175
ercetin+ Phenolic Compounds in	Food and Health. Anticarcinogenicity of Flavonol Qu	AGFD	0113
Materia+ Phenolic Compounds in	Food and Health. Natural Antioxidants from Plant	AGFD	0015
Phenolic Compounds in	Food and Health. Tannin-Protein Interactions.=	AGFD	0148
Effect of Lipid Oxidation on	Food and Oil Quality.=	AGFD	0096
d Applications in Agricultural and	Food Chemistry.= +phy: Analytical Capabilities an	AGFD	0201
zed on Membranes.=	Food Industry Applications of Biocatalysts Immobili	BIOT	0037
Development i+ Uses of Oilseed	Food Ingredients to Minimize Warmed-Over-Flavor	AGFD	0021
Beer, Beverage, and	Food Metal Containers.= Organic Coatings for	PMSE	0140
HPLC Analysis of	Food Phenolic Compounds.=	AGFD	0005
se of Biopolymers as Nonthermal	Food Preservation Processes.= +e Treatment and U	AGFD	0029
Stability.= Influence of	Food Processes on Lipid Oxidation and Flavor	AGFD	0157
ix+ Antibiotic/Drug Residues in	Food Products of Animal Origin. Application of Matr	AGFD	0060
s+ Antibiotic/Drug Residues in	Food Products of Animal Origin. Current Problems A	AGFD	0081
ap+ Antibiotic/Drug Residues in	Food Products of Animal Origin. Liquid Chromatogr	AGFD	0125
Its Use in	Food Products.= Stability of Cochineal Pigment and	AGFD	0139
enobiotic Metabolism Studies in	Food Safety Evaluation: A Proposal for Regulatory U	AGRO	0132
lysis: Pers+ Chromatography in	Food Science. Applications of GC/FTIR to Food Ana	AGFD	0200
reparation i+ Chromatography in	Food Science. Solid-Phase Extractions for Sample P	AGFD	0175
HPLC in	Food Science.=	AGFD	0178
Analysis of Packaged	Food Substances by GC/IR.=	PMSE	0141
rom Prehydrolyzates of Cellulosic	Food Waste.= +Biomass and Chemical Feedstocks f	BIOT	0153
on of Antibiotic/Drug Residues in	Food.= +ent Problems Associated with the Detecti	AGFD	0081
Spray-Dried Milkbase for Baby	Food.= Factors Affecting Lipid Autoxidation of a	AGFD	0158
pects of the Use of Xenobiotics in	Food-Producing Animals in the United States.=	AGRO	0076
dying+ Fate of Xenobiotics in	Food-Producing Animals. In Situ Techniques for Stu	AGRO	0092
Veterin+ Fate of Xenobiotics in	Food-Producing Animals. Introduction—The Uses of	AGRO	0075
tudies o+ Fate of Xenobiotics in	Food-Producing Animals. Metabolism and Residue S	AGRO	0108
udying Xenobiotic Metabolism in	Food-Producing Animals.= +itu Techniques for St	AGRO	0092
Metabolism of Xenobiotics in	Food-Producing Animals.= Dermal Absorption and	AGRO	0095
in Meta+ Fate of Xenobiotics in	Food-Producing Ammimals. Ivermectin and Abamect	AGRO	0128
Xenobiotic Metabolism in	Food-Producing Aquatic Species.=	AGRO	0112
udying Xenobiotic Metabolism in	Food-Producing Fish and Shellfish.= +thods for St	AGRO	0096
nts for Long-Chain Omega-3 F+	Foods and Feeds: Natural and Unnatural Environme	AGFD	0160
minoid+ Phenolic Compounds in	Foods and Health. Chemistry of Curcumin and Curcu	AGFD	0043
istered+ Phenolic Compounds in	Foods and Health. Inhibitory Effect of Orally Admin	AGFD	0065
genic,+ Phenolic Compounds in	Foods and Health. Mechanisms Underlying the Muta	AGFD	0086
uits an+ Phenolic Compounds in	Foods and Health. Phenolic Compounds of Citrus Fr	AGFD	0167
Phenolic Compounds in	Foods and Health. Polyphenol Complexation.=	AGFD	0001
tion of Lipid Oxidation in Muscle	Foods by Nitrite and Nitrite-Free Compositions.=	AGFD	0121
C Antibiotic/Drug Residues in	Foods of Animal Origin. Approaches to Detection and	AGFD	0099
Antibiotic/Drug Residues in	Foods of Animal Origin.= Incidence of	AGFD	0082
Antibiotic/Drug Residues in	Foods of Animal Origin.= Radioimmunoassay of	AGFD	0084
Lipid Oxidation in	Foods via Redox Iron.=	AGFD	0094
Oxidation.= Lipid Oxidation in	Foods. Maillard Reaction Products and Lipid	AGFD	0119
Foods.= Lipid Oxidation in	Foods. Mechanisms of Lipid Oxidative Processes in	AGFD	0093
erent Foods+ Lipid Oxidation in	Foods. Sensory Evaluation of Lipid Oxidation in Diff	AGFD	0156
Peroxidation in Muscle	Foods: New Aspects.= Mechanism of Lipid	AGFD	0095
Lipid Oxidative Processes in	Foods.= Lipid Oxidation in Foods. Mechanisms of	AGFD	0093
on of Lipid Oxidation in Different	Foods.= +id Oxidation in Foods. Sensory Evaluati	AGFD	0156
Bioavailability of Xenobiotics in	Foods.= Effect of Flavonoids on Mutagenicity and	AGFD	0116
Improve Functional Properties in	Foods.= +matic Phosphorylation of Soy Protein to	AGFD	0140
l Procedures for Drug Residues in	Foods.= +s and Accurate Descriptions of Analytica	AGFD	0126
Plant	Foods.= Phenolic Antioxidants as Antimutagens in	AGFD	0045
in	Foods.= Role of Lipoxygenases in Lipid Oxidation	AGFD	0122
Heat-Processed	Foods.= Studies on the Formation of Furaneol in	AGFD	0052
of Lipid Oxidation Volatiles in	Foods.= Update on Gas Chromatographic Analyses	AGFD	0098
nds in Botanical Extracts Used in	Foods, Flavors, Cosmetics, and Pharmaceuticals.=	AGFD	0172
evels in a Variety of Commercial	Foods, Using Monier-Williams and Liquid Chromato	AGFD	0109
r Purification (WIMP)—Don't Be	Fooled by the Name.= +mmunoaffinity Methods fo	CARB	0054
Twelve Days at the U.S. Air	Force Academy.=	CHED	0014

d from Quantum Mechanics and	Force Field Calculations.= QSAR Parameters Derive	MEDI	0044
sing an ab Initio Derived Classical	Force Field.= +ral Analysis of Polyvinylchloride U	POLY	0351
Modeling Transition States with	Force Fields: Development of the Best Possible Pa	ORGN	0170
Chemical Principles as a Driving	Force in the Design of Proteinase Inhibitors.=	ORGN	0269
Materials.= Atomic	Force Microscope Studies of Biominerals and Related	INOR	0117
Deposition Processes.= Atomic	Force Microscope Studies of Underpotential	PHYS	0128
Scanning Tunneling and Atomic	Force Microscopy of Biological Surfaces.=	PHYS	0158
Atomic	Force Microscopy of Molecules on Surfaces.=	PHYS	0153
Atomic	Force Microscopy on Cold-Extruded Polyethylene.=	PMSE	0078
the Nanometer Scale.= Atomic	Force Microscopy Studies of Friction and Wear at	PHYS	0156
nciples and Examples of Scanning	Force Microscopy.= +and Related Techniques. Pri	PHYS	0152
Materials Using Atomic	Force Microscopy.= Probing the Surface Forces of	PHYS	0154
ne Plant: An Engineering Tour de	Force.= +J. F. Roth. New Zealand Gas-to-Gasoli	I&EC	0060
l Compo+ Cluster Mechanics: A	Force-Field Study of Mixed Ligand Transition-Meta	INOR	0075
Diffusion in Polymers by	Forced Rayleigh Scattering.=	POLY	0164
ment of Protein Interaction	Forces at a Solid-Liquid Interface.= Direct Measure	BIOT	0043
Probing the Surface	Forces of Materials Using Atomic Force Microscopy.=	PHYS	0154
lculations with+ Intermolecular	Forces: Do We Really Need to Compare ab Initio Ca	PHYS	0104
d Nicotiana taba+ Stability of a	Foreign Protein Production from Genetically Modifie	BIOT	0230
Introduction of	Foreign Substance into Cells Using Laser.=	BIOT	0172
Perspective on the Role of	Forensic Biology in Crime Investigation.= Historical	HIST	0003
kers for the Application of PCR to	Forensic Casework Samples.= +New Genetic Mar	CHAL	0004
Court Training for the	Forensic Chemist.=	CHAL	0019
berg+ Chemistry and Crime, III:	Forensic Chemistry—Past, Present, and Future. Lind	HIST	0006
Sea+ Chemistry and Crime, III:	Forensic Chemistry—Past, Present, and Future. The	HIST	0001
e Ro+ Interrelationship between	Forensic DNA Analysis Research and Casework in th	CHAL	0007
ification Testing and Issues of +	Forensic DNA Technology: Update 1991. DNA Ident	CHAL	0001
the Development and Use of	Forensic DNA Typing Systems.= Brief History of	HIST	0004
	Forensic Dust Analysis.=	HIST	0005
Laboratory.= Upgrading a	Forensic Science Program with a Microanalysis	CHED	0153
Evidence.= Pioneers in	Forensic Science: Laboratory Examination of Arson	HIST	0007
Milestones of	Forensic Toxicology.=	HIST	0002
s Metabolites in Urine of Exposed	Forest Pesticide Applicators.= +Hexazinone and It	CEI	0005
Products from the Tropical Rain	Forest—A Field for Small Chemical Business.= +al	SCHB	0024
s: Molecular Coils That	Form Chiral Complexes.= Expanded Heterohelicene	ORGN	0065
rangements of Organosilanes to	Form Metal-Ligand Multiple Bonds with Tungsten C	INOR	0115
and Characterization of a Soluble	Form of the Murine Erythropoietin Receptor.= +n	BIOL	0108
Cu2+ to the Cation Cu(tpen)2+ to	Form Several Binuclear Complexes.= +Addition of	INOR	0399
Photoelimination of H_2	form 1,3-Butadiene.=	PHYS	0325
iology: Molecular Bases of Animal	Form. G. Edelman (Medicine and Physiology+ Topb	CONG	0013
n Pathways Involving the Dipolar	Form.= +ter of Trimethylenemetanes: New Reactio	ORGN	0146
mination of the Steric Course of	Formal Nucleophilic Substitutions at Nonstereogenic	ORGN	0120
Melamine-	Formaldehyde Aerogels.=	POLY	0051
ion of Squalene.= Formation of	Formaldehyde and Malonaldehyde from Photooxidat	AGFD	0132
-Scattering Studies of Resorcinol	Formaldehyde Gels as Precursors of Organic Aerogel	POLY	0199
Tannin-Based Resorcinol-	Formaldehyde Resins.= Cure Characteristics of	POLY	0138
cycles in the Rections of the El+	Formations of Four- and five-Membered Oxametalla	INOR	0086
13 and 15 Elements of Empirical	Formula RMER'; R,R' = Aryl, Alkyl; M = Al, Ga, In	INOR	0392
otein Quality of Soy-Based Infant	Formulas Fed to Rats.= +d Supplementation on Pr	AGFD	0209
nformation Content of Chemical	Formulas: Enumeration of Fluorenoids/Fluorantheno	COMP	0027
Consumer Markets.= Product	Formulating for Agricultural, Professional, and	FERT	0004
Variational	Formulation in Network Thermodynamics.=	PHYS	0323
Techniques in 3D Query	Formulation.= Flexible Queries in 3D Searching:	CINF	0025
y Acid Esters or Vegetable Oils:	Formulation, Optimization, and Determination of the	COLL	0083
Pesticide	Formulations.= Quality Assurance in Fertilizer and	FERT	0009
Science in Dentifrice	Formulations.= Role of Rheology and Surface	CHED	0047
with Oxygen: Comparison of Allyl,	Formyl, and Vinyl to Alkyl.= +Radical Reactions	PHYS	0228
etra-Saccharide Fragments of the	Forssman Antigen.= +alysis of All Di-, Tri-, and T	CARB	0056
vy Ion Collisions at 14.6 GeV/c+	Forward and Transverse Energies in Relativistic Hea	NUCL	0051
Mid-80+ Looking Back, Looking	Forward: Pesticide Management and Disposal in the	AGRO	0086
nd Tertiary Aromatic Amines in	Fossil Fuels Using Trifluoroacylation, I: Analytical M	FUEL	0060
Novel Analytical Techniques for	Fossil Fuels. Acid-Base Properties of Coals and Coal	FUEL	0039
Novel Analytical Techniques for	Fossil Fuels. High-Pressure TPR Apparatus to Inv	FUEL	0007
Novel Analytical Techniques for	Fossil Fuels: Heteroatom Determination. Standard	FUEL	0055
Novel Analytical Techniques for	Fossil Fuels: Isotopes and Magnetic Resonance. Hy	FUEL	0070
Novel Analytical Techniques for	Fossil Fuels: Swelling Studes and Mass Spectromet	FUEL	0086
Novel Analytical Techniques for	Fossil Fuels: Swelling Studes and Mass Spectromet	FUEL	0088
+Policristalinos dentro de Celdas	Fotoelectroquimicas, I: CdS Puro e Impurificado.=	PHYS	0221
xydants et des Complexants sur la	Fottation de ces Mineraux.= +-Pyrite: Effet des O	ANYL	0067
Principles in Zeolit+ Theoretical	Foundation of the Al Sitting and Pairing Avoidance	PHYS	0289
Education at the National Science	Foundation.= +atory—II. Undergraduate Science	CHED	0179
onsored by the Dow Chemical Co.	Foundation). +ard in Chemical Instrumentation, sp	ANYL	0011
tion Catalysts.= Scientific	Foundations for Synthesis of Hydrocarbon Isomeriza	PETR	0135
ies of Small Chemical Businesses.	Founding Fairfield Chemical Company.= +ue Stor	SCHB	0013
rface Profiling of Polymers with	Fourier Transform Infrared Reflectance Spectroscopy	PMSE	0153
ique for Analyzing+ Pyrolysis-	Fourier Transform Infrared Spectrometry as a Techn	FUEL	0061
ne Langmuir-Blodgett Films by	Fourier Transform Infrared Spectroscopy and Second	POLY	0224
el Permeation Chromatography-	Fourier Transform Infrared Spectroscopy for Charac	PMSE	0048
ar Reflectance Study of Oxidiz+	Fourier Transform Mid- and Far-Infrared IR Specul	POLY	0253
ndergraduate+ Incorporation of	Fourier Transform NMR Instrumentation into the U	CHED	0148

f Protein Secondary Structure via	Fourier Transform Raman Spectroscopy.= +ation o	POLY	0306
Recent Advances in	Fourier Transform Raman Spectroscopy.=	POLY	0254
Molecules.= Laser and	Fourier Transform Spectroscopy of Transient	PHYS	0057
e and Immobilized Metal Affinity	FPLC.= +P-450 Patters by Means of Ion Exchang	ANYL	0017
Kinetics in Low-Dimensional and	Fractal Lattices.= +ing- and Quenching-Reaction	PHYS	0307
etermined by HPSEC with Vis+	Fractal Properties of Dextran, Pullulan, and Pectin D	CARB	0045
	Fractal Theory of Radon Emanation from Solids.=	NUCL	0088
ett Films.=	Fractals and Excitation-Transfer in Langmuir-Blodg	PHYS	0110
	Fractals.=	CHED	0034
Thermal Field-Flow	Fractionation of Copolymers.=	PMSE	0009
Used in Sedimentation Field-Flow	Fractionation through Computer Simulation.=	PMSE	0028
Polymers by Thermal Field-Flow	Fractionation.= +for Polymer Analysis. Analysis of	PMSE	0008
s by Sedimentation Field-Flow	Fractionation.= Characterization of Latex Aggregate	PMSE	0029
erature-Rising Supercritical Fluid	Fractionation.= +Low-Density Polyethylene: Temp	POLY	0076
Polymer Using Field-Flow	Fractionation.= Gel Content Determination of	PMSE	0010
Field-Flow	Fractionation.= Particle Size Analysis Using Flow	PMSE	0012
acromolecules by Flow Field-Flow	Fractionation.= +luble Synthetic and Biological M	PMSE	0011
on Chromatography, Field-Flow	Fractionation, and Related Chromatographic Method	PMSE	0008
on Chromatography, Field-Flow	Fractionation, and Related Chromatographic Method	PMSE	0065
on Chromatography, Field-Flow	Fractionation, and Related Chromatographic Method	PMSE	0046
on Chromatography, Field-Flow	Fractionation, and Related Chromatographic Method	PMSE	0028
on Chromatography, Field-Flow	Fractionation, and Related Chromatographic Method	PMSE	0094
nsuming Micromechanisms in the	Fracture of Glassy Polymers—I. Chain+ Energy-Co	PMSE	0073
rene Blends: E+ Deformation and	Fracture of Sulfonated Polystyrene Ionomer/Polysty	POLY	0061
is of the Second Pharmacophore	Fragment of norBNI in κ-Opioid Antagonist Potency	MEDI	0157
Recoil and	Fragment Separators.= Evolution of Heavy-Ion	NUCL	0038
How and Why Do Nuclei	Fragment?.=	NUCL	0006
Phosphorus Species.= Ring	Fragmentation Approaches to Low-Coordinate	INOR	0324
Heavy Ions.= Target	Fragmentation of Copper by Intermediate-Energy	NUCL	0025
on.=	Fragmentation of Gold Projectiles at 600 MeV/nucle	NUCL	0047
Ions.= Target	Fragmentation of Silver by 14.6 GeV/nucleon ^{16}O	NUCL	0050
igh-Energy Projectile and Target	Fragmentation.= +istic Phenomena. Synthesis of H	NUCL	0046
M =+ Rhodium Hydride Dimer:	Fragmentation-Recombination Reactions with MR$_2$ (	INOR	0383
symmetric Synthesis of Peptide	Fragments Bearing α-Hydroxy-β-Amino Acid Residu	ORGN	0048
Scaleup of Engineered Antibody	Fragments from Escherichia coli.= Expression and	BIOT	0058
l Di-, Tri-, and Tetra-Saccharide	Fragments of the Forssman Antigen.= +alysis of Al	CARB	0056
R to the Detection of Flavor and	Fragrance Adulteration.= Application of SNIF-NM	AGFD	0197
nal Care Industry—I. Overview of	Fragrance Chemistry and Application.= +nd Perso	CHED	0046
Evaluation of Flavors and	Fragrances.= Chirospecific Analysis in the Origin	AGFD	0183
P Relationship between Zeolite	Framework Composition and Hydrocracking Catalyst	I&EC	0062
y as a Novel Method To Detect	Framework Substitution of Paramagnetic Ions in Mic	PHYS	0231
Derivative of EUO-	Framework Zeolites.= Gallosilicate: A Novel	PETR	0076
aracterization of Molecular Sieve	Frameworks and Products of Catalytic Reactions Usi	PETR	0060
2PdSe$_{10}$: Two Interpenetrating 3D	Frameworks of [Pd(Se$_4$)$_2$]$^{2-}$ and [Pd(Se$_6$)$_2$]$^{2-}$.= +K	INOR	0200
num with Octahedral-Tetrahedral	Frameworks.= +Synthesis of Microporous Molybde	CATL	0019
nnaway. Personal Recollections of	Frank C. Whitmore by Charles D. Hurd.= +O. Ha	HIST	0009
Anatomy of a	Fraud Case.=	PROF	0010
°From Doubts to Conviction of	Fraud: Tales of Trauma.=	PROF	0003
in	Fraud.= The Baltimore Fiasco: A Case Study	PROF	0005
Constituents of	Fraxinus uhdei.= Secoiridoids and Other	ORGN	0187
Research in a Democratic Socie+	Freedom of Speech—The Delicate Balance: Defense	PROF	0009
lege of Medicine of Yeshiv+ The	Freeman Case: Plagiarism at the Albert Einstein Col	PROF	0004
Polyme+ Reexamination of the	Freezing and Melting of Solvent in Thermoreversible	POLY	0170
ain Analysis of the Dynamics of	Frenkel Excitons in Disordered Systems: T = 0 K.=	PHYS	0034
ix Infrared Spectra and ab Initio	Frequencies of Sulfur-Containing Reactive Species.=	PHYS	0031
roscopy at Sub-9-GHz Microwave	Frequencies: Advantages and Prospects.= +Spect	PHYS	0215
nt Developments in Efficient	Frequency Doubling in Poled Polymer Films.= Rece	POLY	0179
NLO Polymers for	Frequency Doubling.=	POLY	0178
Analysis of the High-	Frequency Dynamics in Proteins.= Model Free	PHYS	0168
Coal.= Very High	Frequency EPR Spectroscopy and NMR Imaging of	FUEL	0074
d on Polyurethanes and I+ Low-	Frequency Sound-Absorbing Studies on Foams Base	PMSE	0040
bilities of Molecules.=	Frequency-Dependent and Correlated Hyperpolariza	PHYS	0341
	Frequency-Dependent Hyperpolarizabilities.=	PHYS	0342
amolecular Energy Transfer Using	Frequency-Domain Fluorometry.= +served by Intr	PHYS	0061
N:P Release Ratios from	Freshwater and Coastal Marine Sediments.=	GEOC	0012
Force Microscopy Studies of	Friction and Wear at the Nanometer Scale.= Atomic	PHYS	0156
Lubricant Additives in	Friction/Wear Experiments.= Surface Layers from	PETR	0140
Toy—Scientist, Colleague,	Friend.= Phosphorus Chemistry. Arthur Dock Fon	INOR	0288
DESTIL—A User-	Friendly Program for Distillation Column Design.=	CHED	0122
-to-Know Compliance Made User	Friendly.= +the Challenge of the 90s: MSDS Right	CHAS	0031
te Chemistry—A Rapidly Moving	Frontier.+ State-of-the-Art Symposium: Solid-Sta	CHED	0020
te Chemistry—A Rapidly Moving	Frontier.+ State-of-the-Art Symposium: Solid-Sta	CHED	0029
l Competition in the 1990s+ New	Frontiers in Packaging: Polymer-Based Intermateria	CMEC	0016
Use of	Froth Flotation for Plastics Recycling.=	I&EC	0036
oved Blanchi+ Quality of Stored	Frozen Sweet Corn on the Cob Produced by an Impr	AGFD	0136
hydrates in Industrial Synthesis.	Fructooligosaccharides—A Low-Calorie Bulking Age	CARB	0001
Bioprocess for the Production of	Fructose and Ca-Gluconate by Immobilized Cells.=	BIOT	0152
n and Chilling Injury in Cucumber	Fruit and Cell Cultures.= +on of Lipid Peroxidatio	AGFD	0033
ing Treatments to Stabilize Raw	Fruit and Vegetable Juices: Control of Enzymatic Br	AGFD	0027
Changes on Minimally Processed	Fruit and Vegetable Surfaces.= Physicochemical	AGFD	0011

Kinetics of Thermal Softening of	Fruit and Vegetable Tissue.=	AGFD	0013
Efficient Aroma Enrichment in	Fruit Cells.= Precursor Atmosphere Technology:	AGFD	0023
ntion of Enzymatic Browning in	Fruit Juice with Cyclodextrins and Sulfated Polysacc	AGFD	0030
Processed Fruits and Vegetables.	Fruit Preservation by Combined Methods.= +mally	AGFD	0055
of Novel Minimally Processed	Fruit Products by Pectinase Infusion.= Preparation	AGFD	0059
-Containing Regulators of Tomato	Fruit Ripening.= +turally Occurring Carbohydrate	AGFD	0031
imental Chemistry.=	Fruitful Interplay between Computational and Exper	ORGN	0138
th. Phenolic Compounds of Citrus	Fruits and Their Produc.= +s in Foods and Heal	AGFD	0167
Stabilize+ Minimally Processed	Fruits and Vegetables. Cold Blanching Treatments to	AGFD	0027
ed Metho+ Minimally Processed	Fruits and Vegetables. Fruit Preservation by Combin	AGFD	0055
d Atmos+ Minimally Processed	Fruits and Vegetables. New Developments in Modifie	AGFD	0075
Vegetable+ Minimally Processed	Fruits and Vegetables. Physiology of Cut Fruits and	AGFD	0010
odified Atmosphere Packaging in	Fruits and Vegetables.= +New Developments in M	AGFD	0075
and Vegetables. Physiology of Cut	Fruits and Vegetables.= +imally Processed Fruits	AGFD	0010
Edible Films for	Fruits and Vegetables.=	AGFD	0078
Seal-Packaged Citrus and Pepper	Fruits.= +Illumination May Help Reduce Decay in	AGFD	0076
Compounds in	Fruits.= Enzymatic Oxidation of Phenolic	AGFD	0151
aracterization: Optical Methods.	FT-IR and Fluorescence Studies of Self-Assembled S	POLY	0330
Copper(I) Compounds for CVD by	FT-IR and Raman Spectroscopy.= +acteristics of	COLL	0031
in PEEK (Poly-ether-ether-ke+	FT-IR Spectroscopic Study of Thermal Degradation	POLY	0334
Monitored by Rheo-Photoacoustic	FT-IR Spectroscopy.= +Poly(vinylidene Fluoride)	POLY	0255
Step-Scan	FT-IR Spectroscopy.= Polymer Characterization by	POLY	0304
orous-Promoted Hydrotreating+	FT-IR, Solid-State MAS NMR, and XRD of Phosph	COLL	0198
Laboratory+ Use of a 300-MHz	FT-NMR in a Second-Semester Sophomore Unified	CHED	0181
ound Semiconductors.=	FTIR of Surface Reaction in OMCVD of III-V Comp	COLL	0011
ed in the Biosynthesis of the X(3-	Fucosyl-Lactosamine) Antigenic Determinant.=	CARB	0015
he X(3-Fucosyl-Lacto+ Human	Fucosyltransferases Involved in the Biosynthesis of t	CARB	0015
pound+ Selectivity in Catalysis:	Fuel Chemistry—I. Characterization and Model Com	CATL	0039
ngas.= Selectivity in Catalysts:	Fuel Chemistry—II. Coal/Oil Coprocessing Using Sy	CATL	0049
Gasoline and Diesel	Fuel Contamination.= Soil Gas Survey Methods for	ENVR	0053
Nuclear	Fuel Cycle and Special Problems in Nuclear Energy.=	FUEL	0024
n Process.=	Fuel Evaluation for U-Gas Fluidized-Bed Gasificatio	FUEL	0120
New Developments in Biological	Fuel Production. Ethanol Tolerance of Clostridium	BIOT	0013
rcial Application of Waste-	Fuel-Fired Hybrid Fluidized-Bed Boilers.= Comme	FUEL	0122
aste Plastics for the Production of	Fuel-Grade Gas.= +cipal Waste. Gasification of W	FUEL	0091
C3H3 Reaction Kinetics in	Fuel-Rich Combustion.=	FUEL	0079
hanced Densified Refuse-Derived	Fuel/Coal Mixtures.= +ion of Quicklime Binder-En	FUEL	0140
Biomass-	Fueled Gas Turbine Development.=	FUEL	0119
ulfur, and Hydrocarbons in Diesel	Fuels by Gas Chromatography.= +of Aromatics, S	PETR	0118
of Good-Quality Solid and Liquid	Fuels from Poor-Quality Feedstocks.= +eneration	FUEL	0141
rtiary Aromatic Amines in Fossil	Fuels Using Trifluoroacylation, I: Analytical Method	FUEL	0060
Drying	Fuels with the Carver-Greenfield Process.=	FUEL	0116
l Analytical Techniques for Fossil	Fuels. Acid-Base Properties of Coals and Coa+ Nove	FUEL	0039
l Analytical Techniques for Fossil	Fuels. High-Pressure TPR Apparatus to Inv+ Nove	FUEL	0007
terials Derived from Hydrocarbon	Fuels: Diamond and Diamond-Like Carbon. +w Ma	FUEL	0034
terials Derived from Hydrocarbon	Fuels: Diamond and Diamond-Like Carbon. +w Ma	FUEL	0035
terials Derived from Hydrocarbon	Fuels: Diamond and Diamond-Like Carbon. +w Ma	FUEL	0037
terials Derived from Hydrocarbon	Fuels: Diamond and Diamond-Like Carbon. +w Ma	FUEL	0036
terials Derived from Hydrocarbon	Fuels: Diamond and Diamond-Like Carbon. +w Ma	FUEL	0038
terials Derived from Hydrocarbon	Fuels: Engineering Materials. Carbon Bla+ New Ma	FUEL	0047
terials Derived from Hydrocarbon	Fuels: Engineering Materials. Developme+ New Ma	FUEL	0050
terials Derived from Hydrocarbon	Fuels: Fullerenes. Post-Buckminsterfulle+ New Ma	FUEL	0019
l Analytical Techniques for Fossil	Fuels: Heteroatom Determination. Standard+ Nove	FUEL	0055
l Analytical Techniques for Fossil	Fuels: Isotopes and Magnetic Resonance. Hy+ Nove	FUEL	0070
l Analytical Techniques for Fossil	Fuels: Swelling Studes and Mass Spectrometry. +ve	FUEL	0086
l Analytical Techniques for Fossil	Fuels: Swelling Studes and Mass Spectrometry. +ve	FUEL	0088
por Espectrometria de Masas con	Fuente de Impacto Electronico.= +e Tierras Raras	ANYL	0051
Dehydrogen+ Hydrogen Surface	Fugacities and the Control of Selectivity in Catalytic	CATL	0061
at 25-1000°C.= Behavior of	Fullerene (C60) in Helium, Argon, Air, and Hydrogen	FUEL	0021
ctronic and Structural Studies of	Fullerene-Based Solids: C60, C70, and the Alkali-Met	PHYS	0049
Characterization of	Fullerenes in Flames.= Production and	FUEL	0020
Derived from Hydrocarbon Fuels:	Fullerenes. Post-Buckminsterfulle+ New Materials	FUEL	0019
Flame Species Pertinent to	Fullerenes.= Fluxes and Net Reaction Rates of	FUEL	0112
Alkali-Metal-Doped	Fullerenes.= Superconductivity in	PHYS	0046
Vibrational Structure of	Fullerenes.=	PHYS	0006
Aromatics,	Fullerenes, and Soot as Charged Species in Flames.=	FUEL	0111
stion Chemistry: Large Aromatics,	Fullerenes, and Soot. Soot Deposition fro+ Combu	FUEL	0107
ds: C60, C70, and the Alkali-Metal	Fullerides.= +tural Studies of Fullerene-Based Soli	PHYS	0049
n of Transition-Metal Open	Fulvalene Complexes.= Syntheses and Caracterizatio	INOR	0113
rmine-N1-Acetyltra+ Polyamine	Fumaramate Esters as Inhibitors of Spermidine/Spe	MEDI	0021
Studies of Poly(ε-caprolactone)	Fumarate Networks.= Synthesis and Degradation	POLY	0068
e Phosphatase.=	Fumaric Acid Monoazide: A New Inhibitor of Alkalin	BIOL	0057
in-Carbohydrate Interactions—II.	Functional Aspects. +ype Animal Lectins.= Prote	BIOL	0148
Quantitative Analysis of Sulfur	Functional Forms and Reactions by XAFS Spectrosc	FUEL	0056
pported C+ Influence of Oxygen	Functional Groups on the Performance of Carbon-Su	FUEL	0033
etry as a Technique for Analyzing	Functional Groups with Oxygen in Coal.= +ectrom	FUEL	0061
Polymer Miscibility through Ionic	Functional Groups.= +y Lecture. Enhancement of	PMSE	0116
lexes with Inwardly Directed	Functional Groups.= Cofacial Binuclear Metal Comp	INOR	0242
	Functional Models for Tyrosinase.=	INOR	0208
Cellulases.= Structural and	Functional Organization of Cellulomonas fimi	BIOT	0207

Approaches to the Development of	Functional Organophosphorus Compounds.= +ent	INOR	0348
4,4'-Dihydroxy-α-methylstilb+	Functional Polyesters and Copolyesters Based on the	POLY	0031
ylation of Soy Protein to Improve	Functional Properties in Foods.= +matic Phosphor	AGFD	0140
Cancer Cells:	Functional Studies.= Oligosaccharide Processing in	BIOL	0151
Bromoperoxidase.=	Functional Vanadium Model Complexes of Vanadium	INOR	0016
ganic Chemistry. Organometallic	Functionalities at Sulfur–Bound Nickel: P_2S_2 vs. N_2S	INOR	0177
tive and Catlytic Photochemical	Functionalization of Alkanes by Polyoxometalate Sys	INOR	0286
New Strategy for Cyclization–O–	Functionalization of 5-, 6-, and 7-Membered Carbo	ORGN	0088
Acid.= Polysilyne	Functionalization with Trifluoromethanesulfonic	POLY	0190
Control in Allylic	Functionalization.= Absolute Stereochemical	ORGN	0292
P+ Residual Stress Behavior of	Functionalized Poly(amic Acid) Precursors and Their	POLY	0085
Characteristics of Chromophore–	Functionalized Polymers Having Chromophore–Cent	POLY	0108
and Biodegradability Study of	Functionalized Polystyrene Derivatives.= Synthesis	POLY	0049
oly(Ethylene Terephthalate) with	Functionalized Triglyceride Oil Compositions.= +P	PMSE	0171
Terephthalate) Semi-IPNs with	Functionalized Triglyceride Oils.= Poly(Ethylene	PMSE	0088
Synthesis and Reactivity of	Functionalized Zirconium Phosphonates.=	INOR	0084
substrates: Studies of 26- and 29-	Functionalized 2,3-Oxidosqualenes.= +e by Pseudo	ORGN	0021
aser Spectroscop+ Studies of the	Fundamental Aspects of Ion/Solid Collisions Using L	NUCL	0041
by Permanganate.=	Fundamental Determinants in the Oxidation of DNA	BIOL	0121
Verification of a	Fundamental Liquid–Solid Adsorption Theory.=	COLL	0188
	Fundamental Studies on Affinity Partitioning.=	BIOT	0049
ents.=	Fundamentals of Heavy Metal Removal by Dry Sorb	FUEL	0132
Biochemical Engineering	Fundamentals.=	BIOT	0223
	Fundamentos de la Produccion de Furfural.=	CARB	0051
Nonfederal	Funding for New Faculty.=	YCC	0002
Laboratory Development. NSF	Funding of Laboratory Innovations in Undergraduate	CHED	0196
a Fair Shake? An Analysis of NSF	Funding Opportunities for Junior Faculty.= +Get	YCC	0003
NIH	Funding Opportunities for the Assistant Professor.=	YCC	0004
Characterization of	Fungal Cellulases.= Purification and	BIOT	0187
orhizon Suspension Cultures with	Fungal Elicitors.= +uction in Lithospermum erythr	BIOT	0229
enes.=	Fungal Polyketide Biosynthesis—From Enzymes to G	BIOT	0111
rotein Production in Filamentous	Fungi. Heterologous Gene Expression in Filamen+ P	BIOT	0234
Gene Expression in Filamentous	Fungi: Recent Improvements in Production of Bovin	BIOT	0234
aste Destruction Using White Rot	Fungi.= +otechnological Approaches to Pesticide W	AGRO	0126
Filamentous	Fungi.= Expression of Restrictocin in	BIOT	0238
Studies on the Formation of	Furaneol in Heat-Processed Foods.=	AGFD	0052
e Synthesis of 3-Hydroxy-5(2H)-	Furanones.= +ione with α-Halketones: A Versatil	ORGN	0223
5-(2-Substituted Ethenyl)-3(2H)-	Furanones.= +is and Anti-Ulcer Activity of Novel	MEDI	0067
onvenient Degradation to Lower	Furanose Sugars via Ozonization of Protected Glycal	ORGN	0073
Photodegradation of	Furans.= Catalytic Role of Metal Oxides in the	COLL	0190
de	Furfural.= Fundamentos de la Produccion	CARB	0051
ch to the Stereoselective Synth+	Furfuryl Alcohol-to-Pyranone Oxidation: An Approa	ORGN	0294
Unanticipated Benefits: Graphite	Furnace AA.=	CHED	0184
Gas Chromatography/Graphite	Furnace Atomic Absorption Spectrometry for the De	ANYL	0080
cence Spectrometry in a Graphite	Furnace.= +oxide Laser-Excited Molecular Fluores	ANYL	0078
centrations of Photocarcinogenic	Furocoumarins in Susceptible and Fusarium Yellows	AGRO	0040
Effects of Spirostanic and	Furostanic Steroidal Glycosides.= Some Biological	MEDI	0037
Furocoumarins in Susceptible and	Fusarium Yellows-Resistant Celery Cultivars.= +c	AGRO	0040
Mechanisms: Overview. To	Fuse or Not To Fuse.= Nucleus-Nucleus Collision	NUCL	0004
Overview. To Fuse or Not To	Fuse.= Nucleus-Nucleus Collision Mechanisms:	NUCL	0004
ate Core Structure Consisting of	fused (μ-Hydroxo)bisμ-carboxylato)diiron(III) Kerna	INOR	0094
	Fusion of Co and Bi To Make Element 110.=	NUCL	0015
lyte and the Implications for Cold	Fusion.= +n Aqueous Potassium Carbonate Electro	PHYS	0286
ion de Plata por Electrodeposito,	Fusion, Oxidorreduccion con Zinc en Medio Acido y	CHED	0172
Solid Support: Building Blocks of	Future Biosensors and Molecular Devices.= +s on	BIOT	0102
Future Chemists,	Future Chemistry. Profiles, Pathways, and Dreams.=	CONG	0004
ys, and Dreams.=	Future Chemists, Future Chemistry. Profiles, Pathwa	CONG	0004
Current and	Future Compliance.= Right-to-Know Support:	CHAS	0026
Polymers.= Panel Discussion:	Future Directions in R D of Biodegradable	CHED	0136
ederal Perspective.=	Future Directions in the Use of Pesticides in IPM: F	AGRO	0068
sk? Relative Risk: A Blueprint for	Future Environmental Protection Efforts.= +is Ta	ENVR	0061
Power. Is There a	Future for Nuclear Power?.= Future of Nuclear	FUEL	0022
What Is the	Future of Baculovirus for Insect Control?.=	AGRO	0081
The	Future of Chemistry in a Developing Country.=	CONG	0005
ear Power?.=	Future of Nuclear Power. Is There a Future for Nucl	FUEL	0022
	Future of Pesticide R D.=	AGRO	0069
	Future Opportunities in Advanced Materials.=	INOR	0015
Environmental Concerns: A	Future Perspective.=	AGRO	0067
Status and	Future Plans.= The Protein Data Bank—Present	CINF	0031
Biochemical Marker for Detect+	Future Role of Erythrocyte Protoporphyrin (EP) as a	CEI	0020
ion. Chemistry Educ+ Educating	Future Teachers: The CPT–Chemical Education Opt	CHED	0066
e Wastes: Historical, Current, and	Future Technologies.= +cal Degradation of Pesticid	AGRO	0122
India.= Present Status and	Future Trends of Pesticide Usage and Research in	AGRO	059A
sic Chemistry—Past, Present, and	Future. Lindberg+ Chemistry and Crime, III: Foren	HIST	0006
sic Chemistry—Past, Present, and	Future. The Sea+ Chemistry and Crime, III: Foren	HIST	0001
ste Treatment: Past, Present, and	Future.= +egradation in Site Remediation and Wa	AGRO	0121
Aerosols: Past, Present,	Future?.=	CHED	0048
of (t-Bu$_2$As)	$_3$Ga.= Tris(arsino)allanes and the Crystal Structure	INOR	0231
= t-BuCH$_2$,t-Bu), and (t-BuCH$_2$)	$_2$Ga(Cl)·As(SiMe$_3$)$_3$.= +$_2$Cl, [R$_2$GaAs(SiMe$_3$)$_2$]$_2$ (R	COLL	0136
lays.=	Ga$_{13}$, GaAl$_{12}$ and Al$_{13}$ Polyoxocations and Pillared C	PETR	0143

and Isomorphously Substituted	Ga-β as a Catalyst for Propane Aromatization—Infra	PETR	0047
ER'; R,R' = Aryl, Alkyl; M = Al,	Ga, In; E = P, As.= +nts of Empirical Formula RM	INOR	0392
–M–As or M–As–M–Cl Ring (M =	Ga, In).= +ds Containing a Four-Membered M–As	COLL	0138
Ga_{13},	$GaAl_{12}$ and Al_{13} Polyoxocations and Pillared Clays.=	PETR	0143
Surface Chemistry of	GaAs Atomic Layer Epitaxy.=	COLL	0069
Growth on (001)	GaAs by OMCVD.= Real-Time Investigations on	COLL	0007
Expitaxial	GaAs Film Growth from Single-Source Precursors.=	COLL	0067
on GaAs(100) and Implications for	GaAs Film Growth.= +osition of Triethylgallium	COLL	0068
Numerical Modeling of	GaAs MOCVD.=	COLL	0066
mposition of Triethylgallium on	GaAs(100) and Implications for GaAs Film Growth.=	COLL	0068
Semiconductor Substrates:	GaAs(100) and Si(100).= Reactions of Zinc Alkyls on	COLL	0048
and Arsine on	GaAs(100).= HREELS Study of Trimethylgallium	COLL	0065
la: A+ Cloning of an Insensitive	GABA Receptor from Cyclodiene Resistant Drosophi	AGRO	0152
Phytochemical Antagonism of	GABA-Based Resistances in Diabrotica.=	AGRO	0136
olution of the Gallium Phase of	GaHMFI Aromatization Catalysts in the Course of th	PETR	0046
Determination of UDP–	Galactose 4-Epimerase from E. coli.= Structure	BIOL	0147
Vertebrate Lectin and Several β–	Galactoside-Specific Plant Lectins.= +ide-Specific	BIOL	0109
king Activities of the 14 KDa β–	Galactoside-Specific Vertebrate Lectin and Several β	BIOL	0109
on+ Influencia del Contenido de	Galio en Zeolitas del Tipo Pentasil en la Aromatizaci	COLL	0125
(–)-Epigallocatechin	Gallate (EGCG): A Cancer Preventive Agent.=	AGFD	0066
de Anion: The Pathway to Pyrazol	Gallates.= +Characterization of Alkyl Gallium Ami	INOR	0390
Design and Status of	GALLEX.=	NUCL	0031
is and Characterization of Alkyl	Gallium Amide Anion: The Pathway to Pyrazol Galla	INOR	0390
Study of the Hydrogen Chloride–	Gallium Arsenide Gas Surface Reaction.= +–Beam	COLL	0215
the Various Structural States of	Gallium in As-Synthesized and Modified MFI–Gallo	PETR	0077
Oxidation of Phosphines by	Gallium Peroxides: A Mechanistic Study.=	INOR	0389
in the Course o+ Evolution of the	Gallium Phase of GaHMFI Aromatization Catalysts	PETR	0046
on HZSM-5: Effect of	Gallium.= Aromatization of Ethane and of Propane	PETR	0045
de	Gallium.= Flottation Ionique du Laurylsulfate	ANYL	0053
rs Affecting the Deactivation of	Gallium-Containing HZSM-5 Zeolites Used for the A	PETR	0043
Anions in the Synthesis of a New	Gallophosphate of the LTA Type.= +ct of Fluoride	PETR	0075
ono Alkylbenzenes over Silylated	Gallosilicate Zeolites.= +–Selective Alkylation of M	PETR	0100
Zeolites.=	Gallosilicate: A Novel Derivative of EUO–Framework	PETR	0076
–Synthesized and Modified MFI–	Gallosilicates.= +Structural States of Gallium in As	PETR	0077
rochemical Impedanc+ Study of	Galvanized Steel/Sodium Chloride Interface by Elect	PHYS	0270
faces—I. Dynamics at t+ Proctor	Gamble UERP Polymer Solutions, Blends, and Inter	COLL	0012
faces—II. Quick Quenc+ Procter	Gamble UERP Polymer Solutions, Blends, and Inter	COLL	0032
faces—III. Lateral Diffu+ Procter	Gamble UERP Polymer Solutions, Blends, and Inter	COLL	0051
faces—IV. Thin Polyme+ Procter	Gamble UERP Polymer Solutions, Blends, and Inter	COLL	0070
Polyallylcarbonates by	Gamma Irradiation.= Cross-Linking of	POLY	0053
Polyallylcarbonates by	Gamma Irradiation.= Cross-Linking of	PMSE	0131
Inhibition of the α, β, and	γ Isozymes of Tyrosinase.=	BIOL	0058
s Solutions.=	Gamma Radiolysis of α-Ketoglutaric Acid in Aqueous	NUCL	0070
ed Cyclotriphosphazenes.=	Gamma Ray Polymerization of Long Chain Substitut	PMSE	0130
ted Cyclotriphosphazenes.=	Gamma Ray Polymerization of Long-Chain-Substitu	POLY	0055
R Tuning the Multiplet Energy	Gap in Non-Kekule Molecules: Synthesis, Structure,	ORGN	0128
Panel Discussion: Bridging the	Gap; Why More Is Better. D. L. Leister, Moderator.=	I&EC	0058
urce Precursors to the MOCVD of	GaP.= +rsors. Preparation of Potential Single-So	INOR	0304
ning an+ Data and Methodology	Gaps in Development of a Model for Land-Use Plan	ENVR	0043
nd Reactivity of R2GaAs(SiMe3)	2GaR2Cl, [R2GaAs(SiMe3)2]2 (R = t-BuCH2,t-Bu), a	COLL	0136
Honoring the Memory of Marjorie	Gardner, Chemical Educator. +mSource Programs,	CHED	0011
Posibles	Garrapaticidas.= Sintesis de Tiofosfatos de Naftilo,	MEDI	0064
ve-Flux Probes for Enhanced Soil	Gas Analysis in in Situ Steam Distillation.= +vecti	ENVR	0052
A GC Method for Soil	Gas Analysis Using a Single Detector.=	ENVR	0054
he History of Applications of Soil	Gas Analysis.= +ental Contaminants. Review of t	ENVR	0048
High-Resolution Soil	Gas Analysis.= A Sorbent Trapping Procedure for	ENVR	0050
of the History of Applicat+ Soil	Gas Analysis—Environmental Contaminants. Review	ENVR	0048
pooning Reactions of Excited Rare	Gas Atoms.= +–Phase Metal Reactions—VII. Har	PHYS	0367
Volatiles in Foods.= Update on	Gas Chromatographic Analyses of Lipid Oxidation	AGFD	0098
	Gas Chromatographic Columns in Food Analysis.=	AGFD	0176
y Evaluation Based on Headspace	Gas Chromatographic Data.= +and Flavor-Qualit	AGFD	0108
=	Gas Chromatographic Detectors: New Developments.	AGFD	0177
Descriptive Sensory Analysis and	Gas Chromatographic Techniques.= +termined by	AGFD	0161
thrina Alkaloids from the Foliag+	Gas Chromatography and Mass Spectrometry of Ery	ANYL	0136
tomic Emis+ High-Temperature	Gas Chromatography of Crudes and Residua Using A	PETR	0121
+UV-VIS Detection for Capillary	Gas Chromatography Using a Remote Flow Cell.=	ENVR	0060
d Hydrocarbons in Diesel Fuels by	Gas Chromatography.= +of Aromatics, Sulfur, an	PETR	0118
olatility Using High-Temperature	Gas Chromatography.= +rmination of Motor Oil V	PETR	0122
plit Injection System for Capillary	Gas Chromatography.= +ss Discrimination in a S	TECH	0018
meric-Mercuric Resin in Capillary	Gas Chromatography.= +Chemically Modified Poly	ANYL	0083
rption Spect+ Directly Coupled	Gas Chromatography/Graphite Furnace Atomic Abso	ANYL	0080
Seasonal Dissolved	Gas Cycling in the Middle Atlantic Bight.=	GEOC	0003
Fuel Evaluation for U–	Gas Fluidized-Bed Gasification Process.=	FUEL	0120
n of Clean-Medium Btu	Gas from Gasification of Sludge Wastes.= Productio	FUEL	0118
–Liquid, Liquid-Solid, and Solid–	Gas Interfaces.= +Spaces: Chemistry at the Liquid	ORGN	0126
Incandescent	Gas Lighting.= Development and Decline of	HIST	0038
Calor en Reacciones Solido–	Gas NO-Cataliticas.= Influencia del Transporte de	I&EC	0068
Arc Discharge CVD in a Mixture	Gas of Acetylene and Hydrogen.= +hesized by DC	FUEL	0038
Reduction in	Gas Oil.= Industrial Catalysts for Aromatics	CATL	0048
tion and Physical Blending on the	Gas Permeability of Polymers.= +of Aryl-Substitu	POLY	0117

olymer Network Films.=	Gas Permeability Studies of Latex Interpenetrating P	PMSE	0061
ers. Thermodynamic Analysis of	Gas Permeation through Glassy Polymeric Materials.	POLY	0139
and Experiment.=	Gas Permeation through Nafion Membranes: Theory	POLY	0066
r Routes to Inorganic Materials.	Gas Phase and Surface Chemical Reactions in the Pr	INOR	0121
Use of an MS or MSD to Study	Gas Phase Reactions.=	CHED	0198
Ionic Organic Chemistry in the	Gas Phase: what Can We Learn from Reactions with	ORGN	0121
Films.= Diffusion in the	Gas Sensing of Phthalocyanine Langmuir–Blodgett	POLY	0118
X-ray Scattering from Polymeric	Gas Separation and Barrier Materials.= +alysis of	POLY	0142
ectroscopic Studies of the Water	Gas Shift Reaction over Ruthenium–Exchanged X Ze	CATL	0027
	Gas Solubility in Polymers and Blends.=	POLY	0143
Diffusion Processes in Polymers.	Gas Sorption and Diffusion in Hydrogen–Bonded Po	POLY	0191
tive and Mechanistic Effects of	Gas Sparging in Animal Cell Bioreactors.= Quantita	BIOT	0009
drogen Chloride–Gallium Arsenide	Gas Surface Reaction.= +r-Beam Study of the Hy	COLL	0215
Contamination.= Soil	Gas Survey Methods for Gasoline and Diesel Fuel	ENVR	0053
System.= The Use of Soil	Gas Surveys to Design Soil Vapor Extraction	ENVR	0051
in Japan and In+ Review of Soil	Gas Techniques Used for Environmental Application	ENVR	0049
inked State and Annealing.=	Gas Transport in IPN Membranes: Effect of Cross–L	PMSE	0060
Mechanisms of	Gas Transport in Poly(1–Trimethylsilyl–1 Propyne).=	POLY	0119
nce of Molecular Architecture on	Gas Transport Properties of Liquid Crystalline Poly	POLY	0115
nditioning Effect of CO_2 on the	Gas Transport Properties of Polyimide Isomers.= Co	POLY	0116
Life Prediction Model for Marine	Gas Turbine Blades and Guide Vanes.= +orrosion:	FUEL	0099
Biomass–Fueled	Gas Turbine Development.=	FUEL	0119
for the Production of Fuel–Grade	Gas.= +cipal Waste. Gasification of Waste Plastics	FUEL	0091
Synthesis	Gas.= Biological Production of Hydrogen from	BIOT	0017
Petroleum	Gas.= Determination of Trace Metals in Liquefied	PETR	0139
from Natural	Gas.= Methanol and Carbon Monoxide Production	FUEL	0096
tion: Mechanisms and Applicat+	Gas–Phase Aluminum Atoms from Pulsed Laser Abla	PHYS	0353
=	Gas–Phase Analysis of Diamond–Producing Plasmas.	PHYS	0148
ganometallic Precursors. MOCVD	Gas–Phase Chemistry.= +tronic Materials from Or	COLL	0006
ron and Chemi+ Studies of Some	Gas–Phase Metal Oxidation Reactions by Photoelect	PHYS	0352
d Dynamics of the Mesospheric+	Gas–Phase Metal Reactions—I. Chemical Kinetics an	PHYS	0071
oron and Aluminum Species.= +	Gas–Phase Metal Reactions—II. Kinetic Studies of B	PHYS	0086
lecular Oxygen Reaction with S+	Gas–Phase Metal Reactions—III. Kinetics of the Mo	PHYS	0141
igand Interactions in ML_n^+ Sys+	Gas–Phase Metal Reactions—IV. Study of Ligand–L	PHYS	0171
s of Transition–Metal Atoms w+	Gas–Phase Metal Reactions—V. Association Reaction	PHYS	0198
d State and Electronically Exci+	Gas–Phase Metal Reactions—VI. Reactions of Groun	PHYS	0349
ions of Excited Rare Gas Atoms+	Gas–Phase Metal Reactions—VII. Harpooning React	PHYS	0367
arbon Combustio+ Homogeneous	Gas–Phase Modeling of Boron/Oxygen/Hydrogen/C	FUEL	0097
AlH	Gas–Phase Reaction Kinetics.=	PHYS	0301
itrogen Oxides in Homogeneous	Gas–Phase Reactions: Combustion Emission Modelin	FUEL	0127
: Laboratory Studies.=	Gas–Phase Solar Detoxification of Hazardous Wastes	ENVR	0035
que for the Solar Detoxification of	Gas–Phase Wastes.= +er TiO_2: A Potential Techni	COLL	0082
ailed Insight in the Mechanism of	Gas–Solid Reactions.= +chniques Will Provide Det	FUEL	0015
noring J. F. Roth. New Zealand	Gas-to-Gasoline Plant: An Engineering Tour de Forc	I&EC	0060
mpounds to Model Atmospheric+	Gas/Particle Partitioning of Semivolatile Organic Co	ENVR	0059
Water Vapor in a	Gaseous Mixture.= Chemical Modulator to Measure	ANYL	0052
posium. Reaction Dynamics in	Gases and at Surfaces. J. Polanyi (Chemistry, 1986).=	CONG	0011
of	Gases by Coal.= Effect of Moisture on the Sorption	I&EC	0020
th of Diamond from Hydrocarbon	Gases.= +(and c–BN, SiC). Nucleation and Grow	PHYS	0091
Laser–Polarized Rare	Gases.= Magnetic Resonance Experiments with	PHYS	0041
he Mechanism of the CO_2 Carbon	Gasification Catalyzed by Calcium.= +vidence on t	FUEL	0029
Catalysis of Char	Gasification in O_2 by CaO and $CaCO_3$.=	FUEL	0030
Energies in Slitlike and Wedge+	Gasification Mechanisms. Calculation of Adsorption	FUEL	0008
isorption on Microporous Carb+	Gasification Mechanisms. Modeling of Oxygen Chem	FUEL	0012
ring Oxidation of a Single Char+	Gasification Mechanisms. Morphological Changes du	FUEL	0025
cs of Heterogeneous Char Partic+	Gasification Mechanisms. On the Combustion Kineti	FUEL	0026
Sampl+ Catalytic Effect on the	Gasification of a Bituminous Argonne Premium Coal	FUEL	0006
Catalysts for Pyrolysis and	Gasification of Biomass.= Transition Metals as	FUEL	0103
atments on the Copper–Catalyzed	Gasification of Carbon in Air.= +Chemical Pretre	FUEL	0028
Transient Kinetic Study of the	Gasification of Carbon in CO_2.=	FUEL	0009
ons.= Potassium–Catalyzed CO_2	Gasification of Carbon: Transient Kinetic Investigati	FUEL	0031
al on Potassium Activity in Steam	Gasification of Chars.= +Composition of Parent Co	FUEL	0027
Activated Diffusion in the	Gasification of Porous Carbons/Chars.= Role of	FUEL	0010
Clean–Medium Btu Gas from	Gasification of Sludge Wastes.= Production of	FUEL	0118
Plastics and Municipal Waste.	Gasification of Waste Plastics for the Production of F	FUEL	0091
Fluidized–Bed	Gasification Process.= Fuel Evaluation for U–Gas	FUEL	0120
Utilization of Coal	Gasification Slag: An Overview.=	FUEL	0134
odified Petroleum Coke and Coal	Gasification Slags.= +ase Characterization of Unm	FUEL	0139
Two–Stage	Gasification: Effect of Hydropyrolysis Conditions.=	CATL	0054
Sludge for	Gasification.= Pretreatment of Municipal Sewage	FUEL	0117
utomobile Engines Using Alcohols,	Gasohol, and Other Oxygenates.= +issions from A	FUEL	0095
Soil Gas Survey Methods for	Gasoline and Diesel Fuel Contamination.=	ENVR	0053
J. F. Roth. New Zealand Gas-to-	Gasoline Plant: An Engineering Tour de Force.=	I&EC	0060
e by Interfacial Electron–Transfer	Gates Ionic Currents across the Lipid Bilayer.= +ic	BIOL	0136
Channel	Gating.= Simulation of the Role of Water in Ion	COLL	0084
unds of+ Theoretical Studies of	Gauche and Eclipsing Interactions for Model Compo	POLY	0353
$_3$)+ Reactions of $(Me_3Si)_3As$ with	GaX_3 (X = Br, I): Crystal Structure of $[I_2GaAs(SiMe$	COLL	0140
tudio del Comportamiento de los	Gazas en Condensables en Geotermica de Fasico de l	I&EC	0013
Detector.= A	GC Method for Soil Gas Analysis Using a Single	ENVR	0054
eum Streams by High–Resolution	GC with Chemiluminescence Detection.= +t Petrol	PETR	0124

aracterization o+ Applications of	GC–AES/Boiling–Point Distribution Analysis for Ch	PETR	0119
in Food Science. Applications of	GC/FTIR to Food Analysis: Perspectives and Limits	AGFD	0200
by	GC/IR.= Analysis of Packaged Food Substances	PMSE	0141
Experiments Developed Using	GC/MS Instrumentation.= New Laboratory	CHED	0187
s the Incorporation of NMR and	GC/MS into the Undergraduate Curriculum: A Case	CHED	0149
	GC/MS: Doing What Chemists Do!.=	CHED	0152
es of Polymer Anions with Si and	Ge Backbones.= +Ge_2H_6 Radical Anions as Protoyp	PHYS	0313
and	Ge Surfaces.= Adsorption and Growth on Si	PHYS	0186
ion and Phase Transition on the	Ge(111) Surface Studied by Scanning Tunneling Mic	PHYS	0099
Studies on Si_2H_6, SiH_3GeH_3, and	Ge_2H_6 Radical Anions as Prototypes of Polymer Anio	PHYS	0313
c Precursors. Surface Reactions of	Ge–Containing Organometallics on Si(100).= +talli	COLL	0044
nic Hybrid Materials via the Sol–	Gel Approach.= +cterization of Polyacrylate–Inorga	POLY	0232
cal Routes into Nonshrinking Sol–	Gel Composites.= +omposite Materials: Free Radi	POLY	0259
Sol–	Gel Condensation of Tin(IV) Alkoxide Compounds.=	POLY	0258
low Fractionation.=	Gel Content Determination of Polymer Using Field–F	PMSE	0010
lecular Chemical Aspects of Silica	Gel Formation.= +and Pillared Clay Synthesis. Mo	PETR	0054
Hybrid Materials. Sol–	Gel Kinetics by NMR.= Preceramic and Inorganic	POLY	0256
ation of Organic Molecules in Sol–	Gel Matrices for Quadratic Nonlinear Optics.= +t	POLY	0006
omposites Prepared via the Sol–	Gel Method: Some Insights into Morphology Control	POLY	0260
ocomposites Prepared via the Sol–	Gel Method.= +(vinyl Acetate)/Silicon Oxide Micr	POLY	0287
Preparation through Aqueous Sol–	Gel Method.= +s of Transparent Aluminum Film	COLL	0115
repared by Precipitation and Sol–	Gel Methods.= +racterization of TiO_2 and Al_2O_3 P	PETR	0092
Critical Conditions in	Gel Permeation Chromatography.=	PMSE	0049
Infrared Spect+ Development of	Gel Permeation Chromatography–Fourier Transform	PMSE	0048
olecular Weight Determination by	Gel Permeation in Aqueous Solution.= +olymer M	CHED	0210
e Intermediates Isolated from the	Gel Phase.= +ecular Sieves: Recrystallization of th	PETR	0019
n. Microporous Oxides by the Sol–	Gel Process: Synthesis and Applications.= +izatio	PETR	0031
Sol–	Gel Processing of Controlled–Pore–Size Oxides.=	PETR	0032
	Gel Processing of Opto/Electroactive Polymers.=	POLY	0200
on Membranes by an in Situ Sol–	Gel Reaction.= +n of a Silicon Oxide Phase in Nafi	POLY	0285
ture Characterization during Sol–	Gel Synthesis of Controlled–Porosity Materials.=	PETR	0068
hromophores in Supported Sol–	Gel Thin–Film Glasses for Second–Harmonic Genera	POLY	0210
PMMA–Impregnated Silica	Gel.=	POLY	0263
gle+ Structural Features of Sol–	Gel–Derived Hybrid Ceramer Materials by Small–An	POLY	0284
elasticity during Thermoreversible	Gelatin Gelation.= +of Polymers. Dynamic Visco	POLY	0197
Synthesis of Zeolite Y in the	Gelatin Solution.=	PETR	0003
H_2O Diffusion in the H_2O–D_2O–	Gelatin System.=	PHYS	0216
Content on the Physical Aging of	Gelatinized Cornstarch.= Effect of Moisture	CARB	0043
	Gelation and Crystallization in Solutions of PBLG.=	POLY	0241
olutions.=	Gelation and Phase Separation of PVC/Plasticizer S	POLY	0173
olutions und+ Thermoreversible	Gelation and Vitrification of Concentrated Polymer S	POLY	0146
ene and Thei+ Thermoreversible	Gelation from Poly(ethylene Oxide)Paradichlorobenz	POLY	0125
s) and Aromatic Sol+ Reversible	Gelation in Binary Mixture of Poly(di–N–alkylsilane	POLY	0145
Crystallization–Induced	Gelation in Poly(aryl Ether Ketones).=	POLY	0247
nterpretation of Thermoreversible	Gelation in Polystyrene–Liquid Systems.= +cular I	POLY	0242
	Gelation of Atactic Polystyrene in Carbon Disulfide.=	POLY	0148
Thermoreversible	Gelation of Isotactic Polystyrene.=	POLY	0149
Crystallization.= Reversible	Gelation of Microgels due to Colloidal	POLY	0244
CS Symposium: Thermoreversible	Gelation of Polymers. Dynamic Viscoelasti+ APS–A	POLY	0197
CS Symposium: Thermoreversible	Gelation of Polymers. Reversible Gelation+ APS–A	POLY	0145
CS Symposium: Thermoreversible	Gelation of Polymers. Surfactant Organoge+ APS–A	POLY	0169
CS Symposium: Thermoreversible	Gelation of Polymers. Thermoreversible G+ APS–A	POLY	0121
Thermoreversible	Gelation of Solutions of Stereoregular Polystyrene.=	POLY	0147
Thermoreversible	Gelation of 60–Crown–20 Macrocycles.=	POLY	0246
n of Polymers. Thermoreversible	Gelation: Crystallization and Other Mechanisms.=	POLY	0121
during Thermoreversible Gelatin	Gelation.= +of Polymers. Dynamic Viscoelasticity	POLY	0197
Reversible	Gelation.= Microphase Separation and	POLY	0122
	Gelation/Crystallization of Polyethersulfone.=	POLY	0248
vent in Thermoreversible Polymer	Gels and Related Systems.= +g and Melting of Sol	POLY	0170
udies of Resorcinol Formaldehyde	Gels as Precursors of Organic Aerogels.= +tering St	POLY	0199
Solvents.= Thermoreversible	Gels from Dibenzylidene Sorbitol and Organic	POLY	0201
rk Formation in Thermoreversible	Gels of Polyolefins.= +e to Strain: Effect of Netwo	POLY	0123
carbon Solve+ Thermoreversible	Gels with Polyethylene Block Copolymers and Hydro	POLY	0249
l Propertie+ B^{+3}/Polysaccharide	Gels: Influence of Crosslink Chemistry on Rheologica	POLY	0240
eed DNA Sequencing in Ultrathin	Gels.= +w Face of Analytical Chemistry. High–Sp	ANYL	0033
Phosphate	Gels.= Amorphous Aluminum	PETR	0050
and	Gels.= Diffusion of Star Polymers in Solutions	POLY	0267
in	Gels.= Electrophoresis of Polyelectrolytes	COLL	0183
Agarose	Gels.= Optimization of Protein Chromatography on	ANYL	0030
ce of Chromium–Containing Silica	Gels.= +idation–Reduction Processes on the Surfa	COLL	0101
Thermoreversible	Gels.= Probe Diffusion and Self–diffusion in	POLY	0198
Swelling of Constrained Polymer	Gels.=	COLL	0184
Agarose	Gels.= Transient Electric Birefringence of	COLL	0182
One–Pot (	Gem–Dimethylation of the Carbonyl Group.=	ORGN	0222
ovements in Prod+ Heterologous	Gene Expression in Filamentous Fungi: Recent Impr	BIOT	0234
Metal–Responsive	Gene Expression.=	INOR	0012
Evolved from a Common Ancestor	Gene.= +mylase Inhibitors, Arcelins, and Lectins	AGFD	0041
Site–Directed Mutagenesis of the	Gene–Encoding T4 dCMP Hydroxymethylase.=	BIOL	0160
l Properties+ Second Harmonic	Generation and Electrochromism Studies of Electrica	POLY	0150
roperties and Second–Harmonic	Generation by σ–Bond–Separated Donor–Acceptor M	POLY	0222

Left context	Title	Code	Number
Studies on Meat Flavor	Generation from Cysteine and Sugars.=	AGFD	0074
lymer-Pol+ Second-Harmonic	Generation from Hyperpolarizable Amphiphiles at Po	POLY	0156
f the Isomorphism Problem in the	Generation of Chemical Graphs.= +the Solution o	COMP	0024
l Esters of Phosphoramidic Acids:	Generation of Ethyl Metaphosphate.= +of O-Ethy	ORGN	0110
echnique to Soil Clean-Up and	Generation of Good-Quality Solid and Liquid Fuels f	FUEL	0141
Power Reactors.= New	Generation of Inherently Safe and Advanced Nuclear	FUEL	0023
tenones.=	Generation of ortho-Quinone Methides from Cyclobu	ORGN	0229
Molecules.= Automatic	Generation of 3D Atomic Coordinates for Organic	CINF	0036
died by Optical Second Harmonic	Generation.= +tation, and Order at Surfaces as Stu	ANYL	0024
ank Coal Mobile Component	Generation.= Effect of Chemical Reaction on Low-R	I&EC	0019
ilm Glasses for Second-Harmonic	Generation.= +hores in Supported Sol-Gel Thin-F	POLY	0210
ectroscopy and Second-Harmonic	Generation.= +by Fourier Transform Infrared Sp	POLY	0224
Particle Size Effects in Aerosol	Generation, Transport, and Vaporization for Particle	ANYL	0122
ings, Nets, and Spheres: The New	Generations of Carbon Allotropes.= +into Rods, R	ORGN	0266
e Convergent Synthesis of Four	Generations of Monodisperse Arylester Dendrimers.=	POLY	0324
Subcellulosome Preparation.=	Genes and Proteins of the Clostridium thermocellum	BIOT	0181
e Analysis of AscB and AscC, the	Genes Coding for Yersinia pseudotuberculosis CDP-	BIOL	0110
Esterase	Genes Conferring Insecticide Resistance in Aphids.=	AGRO	0137
ression of Cloned Biodegradative	Genes in Isolates from Deep Subsurface Environmen	BIOT	0117
P-450 Monoxygenase	Genes in Oligophagous Lepidoptera.= Cytochrome	AGRO	0115
ts. Regulation of Drosophila P450	Genes Involved in Insecticide Resistance.= +Agen	AGRO	0097
us Expression.= Xylanase	Genes of Streptomyces lividans and Their Homologo	BIOT	0213
New Polymers from Artificial	Genes: A Progress Report.=	POLY	0302
Enzymes to	Genes.= Fungal Polyketide Biosynthesis—From	BIOT	0111
rina Alkaloids from the Foliage of	Genetic Cloves of Three Erythrina Species.= +ryth	ANYL	0136
for the Production of Novel Ery+	Genetic Engineering of Saccharopolyspora erythraea	BIOT	0110
inihelices in Relation to Decoding	Genetic Information.= +ry. Recognition of RNA M	BIOL	0026
sic Case+ Development of New	Genetic Markers for the Application of PCR to Foren	CHAL	0004
rcetin and Quercetin Glycosides:	Genetic Regulation, Metabolic Processing, and in Vit	AGFD	0117
nd Acquired Immune Deficienc+	Genetically Determined Higher Mortality Incidence a	AGFD	0145
oreign Protein Production from	Genetically Modified Nicotiana tabacum Cell Lines.=	BIOT	0230
Lepidopteran.= Molecular	Genetics of Insecticide Resistance in a Leafrolling	AGRO	0116
gu+ Biochemistry and Molecular	Genetics of Resistance to Juvenoid Insect Growth Re	AGRO	0101
Organic	Geochemical Studies of Peat from Volo Bog.=	I&EC	0016
Stable Isotope	Geochemistry of Azusfres Systems.=	I&EC	0015
mission (PIXE). Staus of PIXE in	Geochemistry.= +ysis by Proton-Induced X-ray E	NUCL	0077
eptibility of Surface D+ Using a	Geographical Information System To Assess the Susc	AGRO	0053
ieve+ Comparison of Acidity and	Geometric Constraint for Iso-Structural Molecular S	CATL	0028
Optical Pulses.= Molecular	Geometric Phase Development and Phase-Controlled	PHYS	0067
itic Macromolecule+ Polymers of	Geometrical Beauty—Combs and Stars. Novel Dendr	POLY	0318
Polymers.= Polymers of	Geometrical Beauty—Combs and Stars. Ring	POLY	0292
and Dynamic Prop+ Polymers of	Geometrical Beauty—Combs and Stars. Some Static	POLY	0265
olecular-Size Tink+ Polymers of	Geometrical Beauty—Combs and Stars. Toward a M	POLY	0344
, Cl, Br) in Various Ionic States+	Geometries and Molecule Properties of XCN (X = H	PHYS	0238
Calculations of the	Geometries of Transition-Metal Complexes.=	PHYS	0192
Nonlinearities in Restricted	Geometries.= Exciton Coherence-Size and Optical	PHYS	0038
niques+ Comparison of Distance	Geometry and Molecular Dynamics Simulation Tech	ORGN	0174
Ancho+ Relationship between the	Geometry of the Excited State of Vanadium Oxides	INOR	0165
ubstituted Molecules Having That	Geometry.= +the Different Possible Isomers of S	COMP	0035
Element, Acid, and Molecular	Geometry.= Student Ideas about Chemical	CHED	0174
nally Flexible Molecul+ Distance	Geometry-Based Approach for Docking Conformatio	CINF	0028
	George Eastman—Methods of Photography.=	CHAL	0030
ical Corp.). Herbert Hoover and	Georgius Agricola: The Distorting Mirrors of History	HIST	0011
o de los Gazas en Condensables en	Geotermica de Fasico de los Azusfres.= +rtamient	I&EC	0013
os Hidraulicos para Pozos	Geotermicos.= Evaluacion Fisicoquimica de Cement	INOR	0063
ter and Chemical Products from	Geothermal Brine Utilizing a Waste/Low-Grade Hea	I&EC	0010
Effects on Physical Properties of	Geothermal Cement Systems.= +ng-Period Storage	I&EC	0014
ed Temperatures and Pressures in	Geothermal Environments.= +with Water at Elevat	INOR	0246
Drilling Fluids Systems Based on	Geothermal Water as the Continuous Phase.= +of	I&EC	0011
Drilling Fluids Systems Based on	Geothermal Water.= +ive Effects on Drill Pipes by	I&EC	0012
ce Analysis of Cements Used in	Geothermal Wells.= Standards for X-ray Fluorescen	ANYL	0058
Microbial	Geritol: From Iron to Infection.=	CHED	0061
ferently with Oligoribon+ Wheat	Germ Initiation Factors 4F and (ISO)4F Interact Dif	BIOL	0128
Insecticide Resistance in the	German Cockroach.=	AGRO	0138
ve Techniques for Studying Seed	Germination and the Effects of Seed Aging: Solid-St	AGFD	0104
Toward the Synthesis of	Germine.=	ORGN	0005
ortunities+ Do Younger Chemists	Get a Fair Shake? An Analysis of NSF Funding Opp	YCC	0003
ysical Chemistry Laboratory: Let's	Get It Right.= +Education. Computers in the Ph	CHED	0155
How To	Get Your Students Involved in Project SEED.=	CHED	0039
Hazards of Household Chemicals:	Getting the Chemical Kicks out of Your Kitchen.=	CHAS	0051
vistic Heavy Ion Collisions at 14.6	GeV/c per Nucleon.= +ansverse Energies in Relati	NUCL	0051
Silver by 14.6	GeV/nucleon ^{16}O Ions.= Target Fragmentation of	NUCL	0050
ct+ ESR Spectroscopy at Sub-9-	GHz Microwave Frequencies: Advantages and Prospe	PHYS	0215
of	Ginkgolide B.= Studies Toward the Total Synthesis	ORGN	0002
acetal from the Dorsal Abdominal	Glands of the Australian Spined Citrus Bug.= +mi	AGRO	0035
rene/Poly(methylmethacrylate)/-	Glass Beads System.= +ical Properties of a Polysty	PMSE	0167
rbonate on the Internal Walls of	Glass Capillaries: Influence of Temperature, Flow Ra	COLL	0163
Preparation of Transparent	Glass Fiber Reinforced PMMA Matrix Composites.=	PMSE	0109
m.=	Glass Formation in the $BiO_{1.5}$-SrO-CaO-CuO Syste	INOR	0224
on Polyalkenoates/Ion-Leachable	Glass Hybrid Systems.= +ocurable Cements Based	POLY	0262

Relation to Spin	Glass Relaxation.= Dispersive Energy Transport:	PHYS	0082
Pressure NMR Studies of the	Glass Transition in the Diamond Anvil Cell.= High-	PHYS	0260
Estimation of Density and	Glass Transition of Amorphous Polymethylene.=	POLY	0134
onent Solubility Parameters and	Glass Transition Temperatures on the Phase Morpho	PMSE	0025
Porous Vycor	Glass.= Excited-Stated Dynamics of Ru(BPY)$_3^{2+}$ on	INOR	0160
on Porous Vycor	Glass.= Photocatalytic Behavior of Metal Carbonyls	INOR	0170
Vycor	Glass.= Photochemistry of (Cp)$_2$TiCl$_2$ on Porous	INOR	0161
Poling and Chemical Binding of	Glass-Embodied Chromophores in Supported Sol-	POLY	0210
f Fiber/Matrix Interface on Local	Glass-Transition Temperature in Composites.= +o	COLL	0055
ne: Effect of Molecular Weight on	Glass-Transition Temperature.= +rocyclic Polystre	POLY	0239
Oxide	Glasses and Ceramics.= Methacrylate Precursors to	POLY	0288
nsfer and Relaxation in Molecular	Glasses and Polymer Solids.= +hases. Energy Tra	PHYS	0080
in Supported Sol-Gel Thin-Film	Glasses for Second-Harmonic Generation.= +hores	POLY	0210
Short-Range Order in Network	Glasses.= Magnetic Resonance Probes of	PHYS	0374
tral Broadening in Liquids and	Glasses.= Molecular Theory of Inhomogeneous Spec	PHYS	0036
alytic Activity of Platinum-Doped	Glassy Carbon.= +hesis, Characterization, and Cat	INOR	0312
nalysis of Gas Permeation through	Glassy Polymeric Materials.= +Thermodynamic A	POLY	0139
of Small Molecules in Stiff-Chain	Glassy Polymers.= +n the Sorption and Transport	POLY	0120
cromechanisms in the Fracture of	Glassy Polymers—I. Chain+ Energy-Consuming Mi	PMSE	0073
de	Glicol en Mexico.= Estudio Sobre el Uso de Eteres	ENVR	0038
egulatory Issues. Award Address.	Global Change: A Catalyst for the Development of H	AGRO	0051
ion: North America and Europe.	Global Marketing of Chemical Information: The STN	CINF	0011
	Global Sourcing within a Regulated Industry.=	CMEC	0006
rement. Chemical Specialties fo+	Global Sourcing—Strategic Issues in Chemical Procu	CMEC	0005
Purchasing Productivity through	Global Swaps, Exchanges, and Toll Conversions.=	CMEC	0007
bonucleotide Analogs of Rabbit α-	Globin mRNA.= +Interact Differently with Oligori	BIOL	0128
m Products with Elevated β-	Glucan Contents as Natural Fat Substitutes.= Oatri	AGFD	0037
Production of Cyclomaltodextr+	Glucanotransferase by an Alkalophilic Bacillus sp.=	BIOT	0157
e Production of Fructose and Ca-	Gluconate by Immobilized Cells.= +oprocess for th	BIOT	0152
xtran Synthesis by α-Methyl-D-	Glucopyranoside and Its 6-Deoxy- and 6-Fluoro-An	CARB	0028
of the Denaturation of	Glucosamine-6-P Deaminase.= Calorimetric Study	BIOL	0106
Glucose Sensing.=	Glucose Oxidase Immobilized with Polyurethane for	BIOT	0147
sakiense by+ Separation of the	Glucose Oxidase Isoenzymes from Penicillium Amaga	ANYL	0016
inia pseudotuberculosis CDP-D-	Glucose Oxidoreductase (E$_{od}$) and CDP-4-Keto-6-D	BIOL	0110
with Polyurethane for	Glucose Sensing.= Glucose Oxidase Immobilized	BIOT	0147
Fischer's Proof Teaching.= D-	Glucose Stereochemistry: Schematic Method for	CHED	0095
ol (1,3,4,5)Tetrakisphosphate from	Glucose via the Ferrier Rearrangement.= +f Inosit	AGRO	0045
	Glucose—A Centennial Tribute.= +l Fischer's Esta	CARB	0011
d) and CDP-4-Keto-6-Deoxy-D-	Glucose-3-Dehydrase (E$_1$).= +e Oxidoreductase (E$_o$	BIOL	0110
Reduction of CDP-6-Deoxy-$\Delta^{3,4}$-	Glucoseen.= +Mechanistic Studies of E$_3$-Mediated	BIOL	0100
in Recombinant β-	Glucosidase of Trichoderma reesei.= Developments	BIOT	0241
Precursors to Flavor Compounds.	Glucosides of Limonoids.= +ymatic Conversions of	AGFD	0050
Analysis and Reactivity of	Glucosones.=	AGFD	0051
Action of Sucrose-	Glucosyltransferases.= New Products from the	CARB	0007
entive Retinoid Metabolites: C-	Glucuronide Analogues of 4-Hydroxyphenylretinamid	MEDI	0030
of 4-Hydroxyphenylretinamide-	Glucuronide.= +bolites: C-Glucuronide Analogues	MEDI	0030
Search for Quark-	Gluon Plasma in P-P Collisions at Fermilab.=	NUCL	0063
Compounds for Poly(γ-alkyl-α,L-	glutamate)s.= +iple Thermal Transitions in Model	POLY	0099
in CHO Cells Amplified with the	Glutamine Synthetase System.= +meric Antibodies	BIOT	0165
.=	Glutaraldehyde Cross-Linking: Fast and Slow Modes	BIOT	0218
Bioreductive Alkylation of	Glutathione and Other Thiols by Mitomycin C.=	BIOL	0088
Evolution of	Glutathione S-Transferases in Drosophila.=	AGRO	0100
Plant Allelochemical-Adapted	Glutathione Transferases in Lepidoptera.=	AGRO	0119
t of the cis-trans Isomerization of	Gly^6-Bradykinin and Analogs.= +NMR Assessmen	BIOL	0006
ugars via Ozonization of Protected	Glycals.= +nient Degradation to Lower Furanose S	ORGN	0073
of	Glycals.= Face Selectivity in the Protonation	ORGN	0069
Analysis of Glycoprotein	Glycans.= HPLC of Glycoconjugates. HPLC	ANYL	0015
le Route to Highly Optically Pure	Glycidol Derivatives.= +le Exchange: A Novel, Faci	ORGN	0207
urenic Acid Analogues as Potent	Glycine Site N-Methyl-D-Aspartate Antagonists.=	MEDI	0146
enaturant+ Sulfhydryl Content of	Glycinin: Effect of Reducing Agents in Absence of D	AGFD	0141
Assay of Potato	Glycoalkaloids.= Improved Extraction and HPLC	AGFD	0112
us Cross-Linked Lattices between	Glycoconjugates and Lectins.= +ion of Homogeneo	BIOL	0145
Glycans.= HPLC of	Glycoconjugates. HPLC Analysis of Glycoprotein	ANYL	0015
etraisocyanates and Poly(ethylene	Glycol) Derivatives.= +ar Polymers from Tri- or T	POLY	0271
Applications.= Polyethylene	Glycol-Enhanced Protein Refolding: Model and New	BIOT	0136
nd Immunological Properties of	Glycopeptide T-Cell Determinants Derived from Hen	CARB	0014
tants for Adsorption of Antifreeze	Glycoprotein at the Ice Surface.= +uilibrium Cons	BIOL	0041
HPLC Analysis of	Glycoprotein Glycans.= HPLC of Glycoconjugates.	ANYL	0015
Role of Glycosylation in	Glycoprotein Hormone Function.=	BIOL	0150
s in Properties between Antifreeze	Glycoproteins and Antifreeze Proteins.= +fference	AGFD	0035
Papers.= The Mechanism of β-	Glycosidases: A Reassessment of Some Seminal	ORGN	0070
Probing Enzymatic	Glycoside Hydrolysis.=	ORGN	0267
Enzymatic Synthesis of a Carbon	Glycoside of N-Acetyl Neuraminic Acid.= +emical	CARB	0034
(Asteraceae).= New Ent-Atisene	Glycosides from the Roots of Stevia salicifolia cav.	MEDI	0034
, and+ Quercetin and Quercetin	Glycosides: Genetic Regulation, Metabolic Processing	AGFD	0117
and Furostanic Steroidal	Glycosides.= Some Biological Effects of Spirostanic	MEDI	0037
Interesting	Glycosides.= Synthesis of Biologically	ORGN	0287
ent Chromatograph+ Analysis of	Glycosidic Flavor Precursors by Droplet Countercurr	AGFD	0203
recursors to Flavor Compounds.	Glycosidic Precursors of Varietal Grape and Wine Fl	AGFD	0006
s in an Umbelliferous Vegetabl+	Glycosidically Bound Phenolics and Other Compound	AGFD	0170

elenoglycosides as Novel, Versatile	Glycosyl Donors. Selective Activation ove+	Phenyls	CARB 0035
Halides as Versatile	Glycosyl Donors.= 2-Oxo- and 2-Oximinoglycosyl		CARB 0025
Synthesis of Truncated	Glycosyl Phosphatidylinositides.=		CARB 0031
Conserved Phases of Catalysis by	Glycosylases.= +Separately Controlled Plastic and		BIOL 0152
Role of	Glycosylation in Glycoprotein Hormone Function.=		BIOL 0150
xpressed in Several Cell Lines+	Glycosylation of Recombinant Human Plasminogen E		ANYL 0020
Unsymmetrical Hydrazone Imine	Glyoxal Derivatives.= +cs: Cobaltous Complexes of		POLY 0102
Subclasses.= Effects of	Gm-Allotypes on the Production of Human IgG		CARB 0018
-Forming Processes—Or, How We	Go from the Bench Top to Commercial Reality.=		PETR 0089
ementary Methods with the Same	Goal.= +ing vs. de Novo Construction: Two Compl		COMP 0050
ydro-Erythroidines in the Milk of	Goats Fed Erythrina poeppigiana and E. +ce of Dih		AGFD 0129
Ametryn in Rats, Lactating	Goats, and Laying Hens.= Metabolic Fate of		AGRO 0111
d We Come From; Where Are We	Going?.= +l Safety Training, 1980-1991: Where Di		CHAS 0043
y Transfer and Relavtivistic Gold-	Gold Covalent Bonding.= +lts: Evidence for Energ		INOR 0189
Fragmentation of	Gold Projectiles at 600 MeV/nucleon.=		NUCL 0047
Chemistry with Experiment.	Gold 6.= Comparison of ab Initio Quantum		PHYS 0004
te-Energy Nuclear Collisions with	Gold.= +. Heavy Residue Properties in Intermedia		NUCL 0024
Electrochemistry of Colloidal	Gold.=		COLL 0129
Classical and Nonclassical	Gold(I) Ammonium Cations.=		INOR 0367
y Studies of Layered Rare Earth	Gold(I) Cyanide Salts: Evidence for Energy Transfer		INOR 0189
Energy Transfer and Relavtivistic	Gold-Gold Covalent Bonding.= +lts: Evidence for		INOR 0189
d Metal At+ Platinum-Tin and	Gold-Tin Bimetallic Particles Prepared from Solvate		CATL 0009
DNA in the Robert	Golub Case.=		CHAL 0010
ctroscopy Study of Polystyrene in	Good and Theta Solvents.= +oton Correlation Spe		POLY 0305
Polymacromonomers in	Good Solvent.= Equilibrium Configuration of Dense		POLY 0269
oil Clean-Up and Generation of	Good-Quality Solid and Liquid Fuels from Poor-Qua		FUEL 0141
am. Succeeding in Academia: The	Good, the Bad, and the Ugly.= +n Academic Progr		YCC 0001
Chemical Careers in	Government.=		I&EC 0002
Chemical Careers in	Government.=		CMEC 0024
Chemistry in the Federal	Government.=		HIST 0032
ter—Advances in Research and in	Government/Farm Regulatory Issues. +d Groundwa		AGRO 0051
ter—Advances in Research and in	Government/Farm Regulatory Issues. +d Groundwa		AGRO 0060
bisPhenol A Polycarbonate Using	GPC-UV Analysis.= +of Phenolic End-Groups in		POLY 0083
olymer MWD and Branching with	GPC-Viscometry in THF and DMF.= +zation of P		PMSE 0068
astics for the Production of Fuel-	Grade Gas.= +ipal Waste. Gasification of Waste Pl		FUEL 0091
mal Brine Utilizing a Waste/Low-	Grade Heat Source with a Heat Pump.= +Geother		I&EC 0010
Optical Applications. Low-Loss	Graded-Index and Single-Mode Polymer Optical Fib		POLY 0174
ne by the Pseudospectral	Gradient Method.= Molecular Structure of Asparagi		PHYS 0275
nyl Phosphate Driven by a Proton	Gradient.= +for H+-ATPase: Formation of Citraco		BIOL 0104
Approach to Milk Shelf-Life P+	Graduate Student Award Presentation. Chemometric		AGFD 0108
Techniques for Studying Seed +	Graduate Student Award Presentation. Noninvasive		AGFD 0104
num and Silicate-Modified Starch	Graft Copolymer Superabsorbents.= +cy in Alumi		CARB 0042
Studies of	Grafted Interpenetrating Polymer Networks.=		PMSE 0003
Squalane.= Free Radical	Grafting of 2-(Dimethylamino)ethyl Methacrylate to		POLY 0019
Theoretical Calculation of	Grafting Sites in the Epoxy-g-Acrylic Copolymer.=		PMSE 0161
Immunodefici+ Translocation of	Gram-Negative Bacteria and High Mortality in CD1		AGFD 0146
Numbers of Polymer Degrees	Granted in the United States.=		PMSE 0128
lcoholic-Alkali+ Preparation of	Granular Cold Water Swelling/Soluble Starches by A		CARB 0040
ution on Mechanical Properties of	Granular Composites.= +ct of Particle Size Distrib		PMSE 0114
Glycosidic Precursors of Varietal	Grape and Wine Flavor.= +to Flavor Compounds.		AGFD 0006
Odor-Active Compounds Found in	Grape Wine due to Malolactic Fermentation.= +e		AGFD 0137
teran Pheromones in Apples,	Grapes, and Peaches.= Residue Analyses of Lepidop		AGRO 0148
f the Five Regular Polyhedra in+	Graph Theoretical Algorithm To Label the Vertices o		COMP 0035
elationships.=	Graph Theoretical Approach to Structure-Property R		COMP 0014
cal Graph Theory in Chemistry.	Graph Theoretical Approach to Structure-Property R		COMP 0013
of Alkanes.= Optimal	Graph Theoretical Models for Enthalpic Properties		COMP 0034
gated Molecular Systems: Facile	Graph Theoretical Solutions to Huckel MO Paramet		COMP 0020
Organic Crystals.=	Graph Theory Applied to Bond Length Variations in		COMP 0031
oach to Structure+ Pedagogical	Graph Theory in Chemistry. Graph Theoretical Appr		COMP 0013
o Conjugated Mole+ Pedagogical	Graph Theory in Chemistry. Pedagogical Approach t		COMP 0020
n in Early Trans+ Applications of	Graph Theory in the Study of Electron Delocalizatio		COMP 0032
solut+ Applications of	Graph Theory. Molecular Similarity: Relative and Ab		COMP 0022
Susceptibility.= Chemical	Graph-Theoretic Cluster Expansions: Diamagnetic		COMP 0030
ame by Computer:+ °From the	Graphic Structure Input to a Systematic Chemical N		COMP 0019
Experience Using Molecular	Graphics in Teaching Organic Chemistry.=		CHED 0189
Computational Techniques and	Graphics into the Undergraduate Physical Chemistry		CHED 0156
nding from Molecula+ Computer	Graphics Presentations and Analysis of Hydrogen Bo		BIOL 0112
ew Volume-Rendering Molecular	Graphics Technique.= +Difluoroacetylene Using a N		COMP 0055
xpe+ Oxygen Chemisorption on	Graphite and Microporous Carbons: An Analysis of E		FUEL 0017
oundwater by Novel Ion-Selective	Graphite Composite Potentiometry.= +ection in Gr		ENVR 0044
Unanticipated Benefits:	Graphite Furnace AA.=		CHED 0184
ar Fluorescence Spectrometry in a	Graphite Furnace.= +oxide Laser-Excited Molecul		ANYL 0078
Tunneling Microscopy of Donor	Graphite Intercalation Compounds.= Scanning		PHYS 0365
Overview of Thei+ Carbon and	Graphite Matrices in Carbon-Carbon Composites: An		FUEL 0051
roscopy of Metal Oxide Arrays on	Graphite.= +eling Microscopy and Tunneling Spect		COLL 0194
Carbon Powders by Pyrolysis of	Graphitizable and Nongraphitizable Polymers under		POLY 0023
of the Carbonization Behavior of	Graphitizable Carbon Precursors.= +itu Evaluation		FUEL 0049
Eigenvalue Distributions for the	Graphs of Alternant Hydrocarbons.=		COMP 0029
em in the Generation of Chemical	Graphs.= +the Solution of the Isomorphism Probl		COMP 0024
etices in Liquids by Photothermal	Grating Spectroscopy.= +otoinitiated Reaction Kin		ANYL 0034

Charac+ Application of Thermal	Gravimetric Analysis and Related Techniques to the	INOR 0006
of Persistent Toxic Substances in	Great Lakes Areas of Concern.= +Zero Discharge	ENVR 0083
dial Action Plan: Program for the	Great Lakes. Achieving Zero Discharge of Pe+ Reme	ENVR 0083
of the	Great Lakes.= Public Involvement in the Cleanup	ENVR 0085
s, Advocates, and the Law.	Greed in the Groves of Academe.= Whistle–Blower	PROF 0001
t–Driven Pollution Prevention?	Green Marketing: A National and International Statu	CEI 0028
	Green Pigments with Perovskite Structure.=	INOR 0069
e–Induced Lung Tumorigenesis by	Green Tea and Its Components.= +ific Nitrosamin	AGFD 0068
ry Effect of Orally Administered	Green Tea on Carcinogenesis by Ultraviolet B Light.	AGFD 0065
roduction in Mouse Epidermis by	Green Tea Polyphenols.= +cetate–Induced H_2O_2 P	AGFD 0069
Esophagus Carcinogenesis by	Green Tea.= Protection against Stomach, Lung, and	AGFD 0067
Drying Fuels with the Carver–	Greenfield Process.=	FUEL 0116
Policy Implications of	Greenhouse Warming.=	ENVR 0064
S–S Bonds.=	Grignard Reagent Promoted Scission of the C–S and	ORGN 0253
ss–Coupling of Aryl Triflates with	Grignard Reagents.= +ns: Nickel(O)–Catalyzed Cro	ORGN 0082
Inhibitory Activity of	Griseolic Acid Derivatives.= Synthesis and PDE	MEDI 0117
Protein Folding—	GroE Heat Shock Proteins.= Role of Chaperones in	BIOT 0116
etal Reactions—VI. Reactions of	Ground State and Electronically Excited Barium Ato	PHYS 0349
n Polymers Lacking a Degenerate	Ground State.= +Disorder and the Metallic State i	PHYS 0039
Oxo Clusters with Nonzero Spin–	Ground States.= Covalent Linkage of Manganese	INOR 0196
g Sy+ Contaminant Migration in	Ground Water as Assessed by EPA's Hazard–Rankin	ENVR 0081
aging Agricultural Chemicals in	Groundwater at the Farm Level.= Advances in Man	AGRO 0052
os+ In Situ Mercury Detection in	Groundwater by Novel Ion–Selective Graphite Comp	ENVR 0044
Pesticides in Surface Water and	Groundwater—Advances in Research and in Gove	AGRO 0051
Pesticides in Surface Water and	Groundwater—Advances in Research and in Gove	AGRO 0060
Pesticides in Surface and	Groundwaters in the Midwest.=	AGRO 0065
and the Law. Greed in the	Groves of Academe.= Whistle–Blowers, Advocates,	PROF 0001
ystals.=	Growth and Characterization of High–T_c Cuprate Cr	PHYS 0026
cromolecules by the Convergent	Growth Approach: Hyperbranched Block Copolymers	POLY 0318
Soot Particles.– Aromatics	Growth beyond the First Ring and the Nucleation of	FUEL 0108
stic Studies of Copper Thin–Film	Growth by MOCVD.= +tallic Precursors. Mechani	COLL 0025
Aluminum Thin Film	Growth by MOCVD.= Mechanistic Studies of	INOR 0124
al Changes in Polymers. Domain	Growth during the Spinodal Decomposition under Si	PMSE 0104
axial High–Temperature Diamond	Growth from Acetylene– Oxygen Flames.= +oepit	FUEL 0037
Expitaxial GaAs Film	Growth from Single–Source Precursors.=	COLL 0067
Modeling Diamond	Growth in Pedestal CVD Reactors.=	PHYS 0179
mical Control of Nucleation and	Growth in the Conversion of Molecules to Solid–Stat	INOR 0145
Filament–Assisted Diamond	Growth Kinetics.=	PHYS 0178
In Situ Diagnostics of the Flame	Growth of Diamond Filaments.=	PHYS 0149
(and c–BN, SiC). Nucleation and	Growth of Diamond from Hydrocarbon Gases.=	PHYS 0091
acterization of the Nucleation and	Growth of Diamond Thin Films.= +tructural Char	PHYS 0121
hemistry of Advanced Materials:	Growth of Electronic Materials from Organometallic	COLL 0064
hemistry of Advanced Materials:	Growth of Electronic Materials from Organometallic	COLL 0153
hemistry of Advanced Materials:	Growth of Electronic Materials from Organometallic	COLL 0025
hemistry of Advanced Materials:	Growth of Electronic Materials from Organometallic	COLL 0006
hemistry of Advanced Materials:	Growth of Electronic Materials from Organometallic	COLL 0044
hemistry of Advanced Materials:	Growth of Electronic Materials from Organometallic	COLL 0135
conductor Devices by OMCVD.+	Growth of High–Performance III–V Compound Semi	COLL 0010
ffusion Flames.= Modeling the	Growth of Polynuclear Aromatic Hydrocarbons in Di	FUEL 0109
On CVD	Growth of SiC.=	PHYS 0095
	Growth of Silicide Layers on Si Surfaces.=	PHYS 0100
Vapor–Phase	Growth of ZnO.= Change in Crystal Habit in the	PHYS 0293
Real–Time Investigations on	Growth on (001) GaAs by OMCVD.=	COLL 0007
Adsorption and	Growth on Si and Ge Surfaces.=	PHYS 0186
nts: Local Density Calculations of	Growth Reactions on the Diamond(100) Surface.=	PHYS 0355
s of Resistance to Juvenoid Insect	Growth Regulators in Drosophila.= +ecular Genetic	AGRO 0101
oylphenyl Ureas and Other Insect	Growth Regulators.= +nisms of Resistance to Benz	AGRO 0134
ne.= Synthesis and Film–	Growth Studies of Di–t–Butylphosphino(diethyl)bora	INOR 0230
roduction from Whey: Effects of	Growth Substrate and Product Extraction on Reacto	BIOT 0030
Models.= Modeling Diamond	Growth: From Elementary Mechanisms to Reactor	PHYS 0357
Filament–Assisted Diamond Film	Growth.= +Beam Mass–Spectrometry during Hot–	PHYS 0145
d Pathways for Molecular–Weight	Growth.= +ation. Analysis of Chemically Activate	FUEL 0063
Film	Growth.= Effects of Ammonia on CVD Diamond	PHYS 0150
Kinetics of Diamond Film	Growth.=	PHYS 0177
0) and Implications for GaAs Film	Growth.= +osition of Triethylgallium on GaAs(10	COLL 0068
to Serum–Free	Growth.= Rapid Adaptation of a Murine Hybridoma	BIOT 0163
tions in Polymer– and Thin–Film	Growth.= +me Spectroscopic Ellipsometry: Applica	POLY 0252
Simulation of CVD Diamond	Growth.=	PHYS 0181
Surface Chemistry of Diamond	Growth.=	PHYS 0180
Diamond	Growth?.= What's on the Surface during	PHYS 0122
ions of Diamond Films: Structure,	Growth, and Properties.= +ular Dynamics Simulat	PHYS 0354
	GTPase–Mediated Activation of ATP Sulfurylase.=	BIOL 0062
Activity of 9–Deaza–9–Substituted	Guanines.= +Inhibitors: Synthesis and Biological	MEDI 0028
osine)+ Rates of Displacement of	Guanosine by Nucleophiles from cis–Diammine(guan	INOR 0251
om+ Conformation Specificity of	Guanosine Oxidation in Nucleic Acids Using Metal C	BIOL 0122
ucleophiles from cis–Diammine(	guanosine)platinum(II) Chloride Using ^{13}C NMR Spe	INOR 0251
from	Guayule.= Chemical Transformations of Triterpenes	ORGN 0019
a of Carbohydrates. Interaction of	Guest Molecules with Cyclodextrins.= +Phenomen	CARB 0061
rization of Organic and Aqueous	Guest/Host Solid–State Interactions by Multiple Qu	AGRO 0062
r–Blodgett Films.=	Guided–Wave Nonlinear Optics in DCANP Langmui	POLY 0176

ale Coccolithophore Blooms in the	Gulf of Maine.= +iopropionate (DMSP) in Mesosc	GEOC 0002
ruvate Level+ Improved Zooglan	Gum Properties by Modification of Succinate and Py	BIOT 0198
a+ Study of Interaction between	Gystamine, Mexamine, and Crown Ether on the Surf	COLL 0096
ing Feedstocks and Products by	^{1}H and ^{13}C NMR.= Characterization of Delayed Cok	FUEL 0076
l to Olefins over Offret+ ^{13}C and	^{1}H MAS NMR Studies of the Conversion of Methano	PETR 0042
Simulation of ^{13}C and	^{1}H NMR Spectra.=	COMP 0002
e Line-Narrowing Techniques and	^{1}H T$_1$ Weighting.= +eactions Using Multiple-Puls	CATL 0022
Change in Crystal	Habit in the Vapor-Phase Growth of ZnO.=	PHYS 0293
phors.= A Study on the Crystal	Habit of Calcium Hydrogen Phosphate Used to Phos	INOR 0361
Ecological Risk of Aquatic-	Habitat Degradation.=	ENVR 0062
Problems in the Application of the	HACCP System in Underdeveloped Countries.=	AGFD 0079
aride Compositional Analysis of	Haemophilus Influenzae Type B Conjugate Vaccine.=	ANYL 0021
.= Carbohydrate Antigens.	Haemophilus Influenzae Type B Conjugate Vaccines	CARB 0053
Chemistry of	Hair Coloring.=	CHED 0065
al Care Industry—II. Silicone in	Hair Conditioners: Factors Controlling Hair Depositi	CHED 0063
Conditioners: Factors Controlling	Hair Deposition.= +Industry—II. Silicone in Hair	CHED 0063
pS$_2$-TPS$_2$-AAGCG): An Unusual	Hairpin Loop.= +tide Phosphorodithioate d(CGCT	BIOL 0123
ting Secondary Metabolism in	Hairy Root Cultures.= Understanding and Manipula	BIOT 0231
ically Excited Cadmium Atoms+	Half-Collision Studies of the Interactions of Electron	PHYS 0351
Cross Polarization of	Half-Integer Spins in Rotating Samples.=	PHYS 0371
2-Oxo- and 2-Oximinoglycosyl	Halides as Versatile Glycosyl Donors.=	CARB 0025
the Solubilization of Alkali Metal	Halides in Hydrocarbon Media.= +(η^6-C$_7$H$_8$)$_2$ and	INOR 0053
. Thermal Decomposition of Alkyl	Halides on Copper Surfaces.= +ns of Catalysts—II	COLL 0204
Hydrogen	Halides Revisited.= Infrared Spectrum of the	CHED 0197
Reactions of Aqueous Alkyl	Halides with Binary Metal Compounds.=	INOR 0080
etal Powders with Aqueous Alkyl	Halides.= +ic Investigations into the Reactions of M	INOR 0248
Reaction of Amines with Metal	Halides.= Synthesis of Nitride Films from the	INOR 0309
the Chapman and	Hall Directionary of Drugs.= Experiences Building	CINF 0033
National Inventors	Hall of Fame.=	CHAL 0025
and	Hall.= Scientific Division of Routledge, Chapman,	CHAS 0028
Charles Martin	Hall—Aluminum.=	CHAL 0027
es in Chlorinated+ Formation of	Haloacetic Acids, Halomethanes, and Haloacetonitril	ENVR 0028
oacetic Acids, Halomethanes, and	Haloacetonitriles in Chlorinated Drinking Water.=	ENVR 0028
Linker on	Halofuginone.= Facile Synthesis of a Carboxylic	ORGN 0201
he Reaction of Copper Atoms with	Halogen Compounds: Ion-Core Conservation.= +t	PHYS 0368
Studies of	Halogen Substitution in High-T$_c$ Superconductors.=	PHYS 0239
hromatography Using an Electr+	Halogen-Selective Detection in Supercritical Fluid C	ANYL 0091
Cytochrome+ Biodegradation of	Halogenated Hydrocarbons by Rational Redesign of	BIOT 0118
t Total Synthesis of the Cytotoxic	Halogenated Monoterpene Aplysiapyranoid D.= +n	ORGN 0001
on of Diamond (and c-BN, SiC).	Halogenation of Diamond(100) Using Atomic Beams.	PHYS 0176
Free Radical	Halogenation of Hydridophosphoranes.=	ORGN 0248
io-2,4-Oxazolidinedione with α-	Haloketones: A Versatile Synthesis of 3-Hydroxy-5(2	ORGN 0223
Formation of Haloacetic Acids,	Halomethanes, and Haloacetonitriles in Chlorinated	ENVR 0028
Nonpeptide Derivatives of	Haloperidol.= Inhibition of HIV Protease with	MEDI 0054
Compounds: Modeling Vanadium	Haloperoxidase.= Vanadium Coordination	INOR 0018
Reactivity of the Vanadium	Haloperoxidases Isolated from Marine Algae.=	INOR 0097
dsheets: Which Will Supplant the	Hand-Held Calculator?.= +uation Solvers or Sprea	CHED 0117
Fertilizer Containment Facilities	Handbook.= MWPS-37 Pesticide and Liquid	AGRO 0042
How They Are	Handled.= A User's Perspective of the Issues and	CINF 0044
posure Mitigation Strategies for	Handlers and Harvesters.= Developing Pesticide-Ex	AGRO 0043
	Hands-On Teaching Laser.=	PHYS 0303
History of Chemistry, Honoring O.	Hannaway. +Charles D. Hurd.= Dexter Award in	HIST 0009
c Nitrates: Induced Oxidation of	Hantzsch Esters (Dialkoxycarbonyldihidropyridines).	ORGN 0245
	It Can't Happen Here!.=	CHAS 0015
opyright and Technology: What	Happens When Parallel Developments Intersect?.=	CINF 0043
Rambach—The	Happy Alternative.=	SCHB 0014
ncern in New York State: Oswego	Harbor and the St. +Initiatives at Two Areas of Co	ENVR 0084
r, Protein, and Starch in a Milled	Hard Red Winter Wheat.= +id, Total Dietary Fibe	AGFD 0143
f Electrical Properties+ Second	Harmonic Generation and Electrochromism Studies o	POLY 0150
ctronic Properties and Second-	Harmonic Generation by σ-Bond-Separated Donor-A	POLY 0222
hiles at Polymer-Pol+ Second-	Harmonic Generation from Hyperpolarizable Amphip	POLY 0156
aces as Studied by Optical Second	Harmonic Generation.= +tation, and Order at Surf	ANYL 0024
el Thin-Film Glasses for Second-	Harmonic Generation.= +hores in Supported Sol-G	POLY 0210
nfrared Spectroscopy and Second-	Harmonic Generation.= +by Fourier Transform I	POLY 0224
at Interfa+ Picosecond Optical	Harmonic Measurements of Excited-State Relaxation	COLL 0087
as-Phase Metal Reactions—VII.	Harpooning Reactions of Excited Rare Gas Atoms.=	PHYS 0367
Ylides.=	Hartree Fock Description of Substituted Azomethine	ORGN 0171
tion Strategies for Handlers and	Harvesters.= Developing Pesticide-Exposure Mitiga	AGRO 0043
ses and Analytical Data.=	Has a Release Been Observed? HRS-Observed Relea	ENVR 0080
cal Services Provided by	Hazard Communication Resources, Inc.= Toxicologi	CHAS 0024
	Hazard Communication Training.=	CHAS 0023
Revised	Hazard-Ranking System Air Pathway.=	ENVR 0082
What Is the Basis of EP+ EPA's	Hazard-Ranking System and Decision Methodology:	ENVR 0077
	Hazard-Ranking System and Relative Risk.=	ENVR 0079
und Water as Assessed by EPA's	Hazard-Ranking System: A Balance of Risk Assessm	ENVR 0081
? Overview of the Structure of the	Hazard-Ranking System.= +f Sites for Superfund	ENVR 0077
cess: Implications for the Revised	Hazard-Ranking System.= +'s Site Assessment Pro	ENVR 0078
e Methodology+ Prioritization of	Hazardous Air Pollutants: The Need for Comparativ	ENVR 0066
Decontamination of Solutions of	Hazardous Compounds.=	CHAS 0006
Cost-Effective Manner?.= Can	Hazardous Material Regulations Be Met in a	CHAS 0033

of California: What+	Studying Hazardous Materials Management at the University	CHAS	0002
	Hazardous Materials Support Programs.=	CHAS	0027
Chemical Inventory and	Hazardous Materials Tracking System.=	CHAS	0003
Applied to Cement–Solidified	Hazardous Waste.= Petrographic Techniques	ENVR	0009
ilability,+ Human Exposure to	Hazardous Wastes: Assessment of Prevalence, Bioava	CEI	0008
Solar Detoxification of	Hazardous Wastes: Laboratory Studies.= Gas–Phase	ENVR	0035
Immobilization of Solidified Toxic	Hazardous Wastes.= +of Cement Matrices for the	ENVR	0002
Characterization of Solidified	Hazardous Wastes.= Microscopic and Spectroscopic	ENVR	0004
Welfare Canada+ Regulation of	Hazardous Workplace Chemicals. Role of Health and	CHAS	0034
of MSDS and Labels Required for	Hazardous Workplace Chemicals.= +a in Review	CHAS	0034
Microbiological	Hazards and Opportunities.=	AGFD	0014
cal Kicks out of Your Kitchen.=+	Hazards of Household Chemicals: Getting the Chemi	CHAS	0051
Education for the	HAZMAT Worker.=	CHAS	0045
under	HAZWOPER.= Worker Training Requirements	CHAS	0018
pt of Hydrogen Back Spillover (	HBS): Toward the Development of High–Performanc	PETR	0063
Conductivity Method: I. MOPSO–	HCl at 25 °C.= +Constant of Amino Acids by the	BIOL	0033
alyzed Reductive Elimination of	HCl from Chlorohydridometal Complexes—Synthetic	INOR	0381
gies and Shapes of HSiS+, HSiO+,	HCP, and HSiP, and Their Isomers.= +Initio Ener	PHYS	0315
ysts in the Hydrodemetallization (	HDM) of NI–TPP and VO–TPP.= +nadium Catal	CATL	0044
re Oxidos Mixtos de Alumina–T+	HDS de Tiofeno con Catalizadores de Molibdeno sob	FUEL	0105
esign of a Mill Drier for Shrimp	Head: Bromatological Characteristics of Produced M	AGFD	0105
sign of a Mill–Drier for Shrimp	Heads: Bromatological Characteristics of Produced M	ENVR	0046
avor–Quality Evaluation Based on	Headspace Gas Chromatographic Data.= +and Fl	AGFD	0108
grams.=	Health and Safety Issues under OSHA and EPA Pro	CHAS	0022
Proces+ OSHA–EPA Interface in	Health and Safety Issues. OSHA's Role in Chemical	CHAS	0012
Labels Required for H+ Role of	Health and Welfare Canada in Review of MSDS and	CHAS	0034
Polymers in	Health Care.= Contemporary Role of Biodegradable	CHED	0135
Assessment.= Noncancer	Health Endpoints: Approaches to Quantitative Risk	ENVR	0068
Phenolic Compounds in Food and	Health. Anticarcinogenicity of Flavonol Quercetin+	AGFD	0113
henolic Compounds in Foods and	Health. Chemistry of Curcumin and Curcuminoid+ P	AGFD	0043
henolic Compounds in Foods and	Health. Inhibitory Effect of Orally Administered+ P	AGFD	0065
henolic Compounds in Foods and	Health. Mechanisms Underlying the Mutagenic,+ P	AGFD	0086
Phenolic Compounds in Food and	Health. Natural Antioxidants from Plant Materia+	AGFD	0015
henolic Compounds in Foods and	Health. Phenolic Compounds of Citrus Fruits an+ P	AGFD	0167
Compounds in Foods and	Health. Polyphenol Complexation.= Phenolic	AGFD	0001
Compounds in Food and	Health. Tannin–Protein Interactions.= Phenolic	AGFD	0148
Environmental Public	Health.= Application of Biological Monitoring for	CEI	0007
Pesticides in Public	Health.=	AGRO	0057
Seal–Packa+ Abiotic Stresses of	Heat or UV Illumination May Help Reduce Decay in	AGFD	0076
otassium Carbonate Ele+ Excess	Heat Production by the Electrolysis of an Aqueous P	PHYS	0286
e/Low–Grade Heat Source with a	Heat Pump.= +Geothermal Brine Utilizing a Wast	I&EC	0010
Protein Folding—GroE	Heat Shock Proteins.= Role of Chaperones in	BIOT	0116
ine Utilizing a Waste/Low–Grade	Heat Source with a Heat Pump.= +Geothermal Br	I&EC	0010
em Water–Calcium Chloride in a	Heat Transformer.= Performance Study of the Syst	I&EC	0009
Lipids by	Heat Treatment.= Flavor Compounds Formed from	AGFD	0053
of Furaneol in	Heat–Processed Foods.= Studies on the Formation	AGFD	0052
Microwave	Heating of Zeolite Synthesis Mixtures.=	PETR	0080
Strain–Compensated	Heats of Hydrogenation.=	ORGN	0173
cess Chromatogra+ Cleaning of	Heavily Contaminated Ion–Exchange Columns in Pro	ANYL	0042
idediol.+ A Novel Route to Two	Heavily Fluorinated Epoxy Resins from a Fluorodiim	POLY	0037
The Unbearable	Heaviness of Neutrinos.=	NUCL	0002
Catalytic Asymmetric	Heck Reaction.=	CATL	0006
: Which Will Supplant the Hand–	Held Calculator?.= +ation Solvers or Spreadsheets	CHED	0117
of Block Copolymers Containing	Helical and Amorphous Polyisocyanide and Elastome	PMSE	0071
Isocyanates).= Chiral Clues to	Helical Sequence Lengths in Poly(n–alkyl	POLY	0348
DNA	Helicase.= Mechanism of Action of the	BIOL	0117
des.= Stereoselection and	Helicity in Living Polymerizations of Chrial Isocyani	POLY	0347
ctivating Neuropeptide Analogs in	Helicoverpa zea.= +y of Pheromone Biosynthesis A	AGRO	0011
	Heliothis Resistance to Plant Allelochemicals.=	AGRO	0118
Lens Calorimetry with a	Helium–Neon Laser.= Time–Resolved Thermal	CHED	0089
Behavior of Fullerene (C60) in	Helium, Argon, Air, and Hydrogen at 25–1000°C.=	FUEL	0021
Dependence of α–	Helix Formation on Sidechain Structure.=	BIOL	0037
odel Peptide Approach to Triple–	Helix Perturbations.= +Structure and Function. M	BIOL	0001
Helping the Public vs.	Helping the Corporation.= The False Conflict:	PROF	0006
The False Conflict:	Helping the Public vs. Helping the Corporation.=	PROF	0006
amina E en la Fertilidad de Ratas	Hembras.= +ecto de la Deficiencia o Exceso de Vit	AGFD	0210
uctural Dynamics in Photoexcited	Heme Proteins.= +rements of Energy Flow and Str	PHYS	0335
ation and Structural Dynamics of	Heme Proteins.= +ical Studies of Vibrational Relax	PHYS	0169
Studies toward	Heme–Dependent Protein Mimics.=	INOR	0186
tebrate Oxygen–Carrying Proteins	Hemerythrin and Myohemerythrin.= +arine Inver	INOR	0140
e Australian Spined Ci+ Unusual	Hemiacetal from the Dorsal Abdominal Glands of th	AGRO	0035
urrent Understa+ Cellulases and	Hemicellulases—I. Mechanism of Cellulase Action: C	BIOT	0179
obiohydrolases B+ Cellulases and	Hemicellulases—II. How Do Trichoderma reesei Cell	BIOT	0206
and Fiber–Bound	Hemicelluloses.= Enzymatic Hydrolysis of Isolated	BIOT	0215
tudy of Molecular Orientation of	Hemicyanine Langmuir–Blodgett Films by Fourier T	POLY	0224
ation and Investigation of	Hemin: Catalyst of Oxidation Processes.= Immobiliz	COLL	0102
acellular Expression of a Bacterial	Hemoglobin.= +Yields in Streptomyces by the Intr	BIOT	0087
n and Binding–Site Mapping of	Hemolymph Juvenile Hormone–Binding Protein of L	AGRO	0014
Autoparalytic Peptides from	Hemolymph of the Lepidopteran Manduca sexta.=	AGRO	0012
ell–Wall Polysaccharide of the β–	Hemolytic Streptococci Group A.= +nding to the C	CARB	0057

Left fragment	Middle fragment	Code
–Cell Determinants Derived from	Hen Egg Lysozyme 81-96.= +ties of Glycopeptide T	CARB 0014
al Rotator-Phase Transition in n-	Heneicosane.= +the Orthorhombic Pseudohexagon	POLY 0354
Lactating Goats, and Laying	Hens.= Metabolic Fate of Ametryn in Rats,	AGRO 0111
Search for the	Heparin Antithrombin III Binding Site Precursor.=	CARB 0039
of	Heparinase.= Characterization of the Active Site	BIOL 0056
romatic Amines on the Ultimate	Hepatacarcinogen, N-(Sulfonatooxy)-2-(Acetylamino	ORGN 0194
in an in Vitro Devel+ Effect of	Hepatocyte Source on Metabolic Profile of Biphenyl	AGRO 0021
otrexate-Insulin Conjugate to H35	Hepatoma and IEC-6 Cells in Vitro.= +ry of Meth	MEDI 0185
e and Two Amino Sulfonic Acids,	HEPES and MOPSO, by EMF Measurements.= +d	BIOL 0034
-3-Ethyl-3-Phenylpropionamide (	HEPP) in Plasma and Urine.= +t D,L-3-Hydroxy	ANYL 0082
N-Fluoropyridinium Pyridine	Heptafluorodiborate: A Useful Fluorinating Agent.=	ORGN 0076
e Inner+ Synthesis of Antigenic	Heptose and KDO Containing Oligosaccharides of th	CARB 0058
stituent Isolated from Chinese	Herb.= Structural Modification of Edulinine, a Con	MEDI 0159
	Herbert H. Dow—Process of Extracting Bromine.=	CHAL 0029
ored by Dexter Chemical Corp.).	Herbert Hoover and Georgius Agricola: The Distortin	HIST 0011
yl)-1,2,4-Triazol+ Synthesis and	Herbicidal Activity of N-(Substituted-2-Fluorophen	AGRO 0048
Pyrazolyl-1,3-Cy+ Synthesis and	Herbicidal Activity of 2-(2-(Alkoxyimino)-Alkyl)-5-	AGRO 0013
t+ Mechanism of Selectivity of the	Herbicide DPX-E9636 in Corn: Uptake, Transloca	AGRO 0027
A Sulfonylurea	Herbicide for Corn (QSAR Analysis).=	AGRO 0050
pport Registration of DPX-E9636	Herbicide for Field Corn.= +HPLC Method To Su	AGRO 0030
Sulfonylurea	Herbicide in Corn.= Synthesis of NC-319: A New	AGRO 0049
Role of Agricultural BMP in th+	Herbicide Presence in Surface/Groundwater and the	FERT 0019
New Sulfonylurea	Herbicide.= Environmental Fate of DPX-E9636, a	AGRO 0029
Metabolism of S-53482, a New	Herbicide, in Rats.=	AGRO 0024
Polymeric Microcapsules of the	Herbicides Atrazine and Metribuzin: Preparation and	AGRO 0063
1,5a]Pyrimidine-2-Sulfonanilide	Herbicides: The Influence of Heterocyclic Substitutio	AGRO 0046
of Workers Exposed to Triazine	Herbicides, I: Determination of Simazine and Its Me	CEI 0004
of Workers Exposed to Triazine	Herbicides, II: Determination of Hexazinone and Its	CEI 0005
o[1,5a]Pyrimidine-2-Sulfonanilide	Herbides. The Influence of Heterocy+ 1,2,4-Triazol	AGRO 0047
lar Mechanisms of Resistance in	Herbivorous Pests to Natural, Synthetic, and Bioeng	AGRO 0114
lar Mechanisms of Resistance in	Herbivorous Pests to Natural, Synthetic, and Bioeng	AGRO 0153
lar Mechanisms of Resistance in	Herbivorous Pests to Natural, Synthetic, and Bioeng	AGRO 0150
lar Mechanisms of Resistance in	Herbivorous Pests to Natural, Synthetic, and Bioeng	AGRO 0097
lar Mechanisms of Resistance in	Herbivorous Pests to Natural, Synthetic, and Bioeng	AGRO 0133
	It Can't Happen Here!.=	CHAS 0015
Wipke: New Dimensions in Ch+	Herman Skolnik Award Symposium Honoring W. T.	CINF 0016
ine and Propylamine Deriva+ N-	Heteroaralkyl-Substituted-1-Aryloxy-2-Propanolam	MEDI 0095
moplastic+ Anionic Synthesis of	Heteroarm, Star-Branched Styrene-Butadiene Ther	POLY 0266
ly Substituted Carboaromatic and	Heteroaromatic Compounds.= +Synthesis of High	ORGN 0286
Pyrolysis of Nitrogen-Containing	Heteroaromatics.= +recursors of NO_x in the Rapid	FUEL 0114
Evaluation of N-(Aryloxyalkyl)	Heteroaryltropanes as Potential Antipsychotic Agent	MEDI 0130
ytical Techniques for Fossil Fuels:	Heteroatom Determination. Standard+ Novel Anal	FUEL 0055
Reduction of+ The Influence of	Heteroatom on the Stereochemistry of Nickel Boride	ORGN 0246
etal-Catalyzed Polymerization of	Heteroatom-Substituted Cyclohexadienes: Precursor	POLY 0047
kyltransferase: Effect of Substrate	Heteroatoms on Reactivity.= +kylguanine-DNA Al	BIOL 0161
Fo+ Mechanisms of Reactions at	Heteroatoms: Determination of the Steric Course of	ORGN 0120
o-Bridged Iridium, Molybdenum	Heterobimetallics by Oxidative Addition Pathways.=	INOR 0385
and Characterization of Ordered	Heterocycle-Imide Copolymers Containing ortho-Ph	POLY 0042
Synthesis of Bicyclic Nitrogen	Heterocycles by Means of Homogeneous Catalytic Ca	ORGN 0049
nes and Their Antibac+ Bis Aza	Heterocycles: Synthesis of 3,3'-bisquinazolin-4,4'-Dio	MEDI 0057
for	Heterocycles.= δ-Diketones: Potential Synthons	ORGN 0032
Six-Membered Carbo- and	Heterocycles.= Recent Conformational Studies of	ORGN 0295
- and 6-Ring Nitrogen and Sulfur	Heterocycles.= +el Approach to the Synthesis of 5	ORGN 0199
on Kinetics for the Formation of	Heterocyclic Compounds in a Temperature- and Pre	AGFD 0073
	Heterocyclic Homoconjugated Polyacetylenes.=	ORGN 0094
Enforced Polarity: Inorganic and	Heterocyclic Linkages.= +ophoric Assemblies with	POLY 0215
to the Stereoselective Synthesis of	Heterocyclic Natural Products.= +n: An Approach	ORGN 0294
tion of Normal Boiling Points of	Heterocyclic Organic Compounds from Molecular Str	COMP 0015
lide Herbicides: The Influence of	Heterocyclic Substitution on Biological Activity—I.=	AGRO 0046
ilide Herbides. The Influence of	Heterocyclic Substitution on Biological Activity—II.=	AGRO 0047
rbenoids.=	Heterocyclic Synthesis Using Rhodium-Stabilized Ca	ORGN 0276
Coordination Compounds of the	Heterocyclic 4(1H)Quinazolinone-2,3-Dihydro-2-Th	INOR 0079
Temper+ Characterization of the	Heterogeneity of Linear Low-Density Polyethylene:	POLY 0076
Protein Kinase C	Heterogeneity.= Natural Products as Probes of	MEDI 0077
idue Using Mass Spectrometry: A	Heterogeneous Analytical Model.= +of Vacuum Res	PETR 0150
Olefins by	Heterogeneous Catalysis.= Oligomerization of Lower	PETR 0009
rence in Probing Intermediates in	Heterogeneous Catalysis.= +ns of n-Quantum Cohe	COLL 0089
f Short-Chain Hydrocarbons over	Heterogeneous Catalysts. +tion, and Isomerization o	PETR 0082
f Short-Chain Hydrocarbons over	Heterogeneous Catalysts. +tion, and Isomerization o	PETR 0130
f Short-Chain Hydrocarbons over	Heterogeneous Catalysts. +tion, and Isomerization o	PETR 0043
f Short-Chain Hydrocarbons over	Heterogeneous Catalysts. +tion, and Isomerization o	PETR 0112
f Short-Chain Hydrocarbons over	Heterogeneous Catalysts. +tion, and Isomerization o	PETR 0007
f Short-Chain Hydrocarbons over	Heterogeneous Catalysts. +tion, and Isomerization o	PETR 0100
f Short-Chain Hydrocarbons over	Heterogeneous Catalysts. +tion, and Isomerization o	PETR 0025
f Short-Chain Hydrocarbons over	Heterogeneous Catalysts. +tion, and Isomerization o	PETR 0061
electivity.= I. Dynamic Sieving by	Heterogeneous Catalysts.= +in Catalysis: Shape S	CATL 0013
s. On the Combustion Kinetics of	Heterogeneous Char Particle Populations.= +nism	FUEL 0026
uction of Mon+ Manipulation of	Heterogeneous Hybridoma Cultures for the Overprod	BIOT 0169
or the Highly Efficient and Stable	Heterogeneous Oligomerization of Ethylene.= +s f	PETR 0008
ropoly Acids.=	Heterogeneous Oligomerization of Propene over Hete	PETR 0011

tor Particles in Aqueous Media:+	Heterogeneous Photocatalysis with TiO$_2$ Semiconduc	COLL	0131
mental Orga+ Kinetic Studies in	Heterogeneous Photocatalyzed Oxidations of Environ	INOR	0318
tion between Homogeneous and	Heterogeneous Reactions in Metal Vapor Oxidation.=	PHYS	0144
A Comparison between Different	Heterogeneous Systems.= +Micelles and Emulsions:	COLL	0043
Coupling in Dynamically	Heterogeneous Systems.= Magnetic Relaxation	PHYS	0232
ce Lifetimes in Homogeneous and	Heterogeneous Systems.= +tributions of Fluorescen	PHYS	0134
Complexes.= Expanded	Heterohelicenes: Molecular Coils That Form Chiral	ORGN	0065
roduction in Filamentous Fungi.	Heterologous Gene Expression in Filamentous Fungi	BIOT	0234
erochaete chrysosporium and in a	Heterologous Host.= +and Mn Peroxidases in Phan	BIOT	0237
Ligninase Expression in	Heterologous Hosts.=	BIOT	0236
ular Proteases and Their Role in	Heterologous Protein Accumulation by Streptomyces	BIOT	0084
of Propene over	Heteropoly Acids.= Heterogeneous Oligomerization	PETR	0011
lization in Early Transition Metal	Heteropoly- and Isopolyoxometalates.= +on Deloca	COMP	0032
or Binding Activity of a Series of	Heterosteroids.= Synthesis and Substance P Recept	MEDI	0158
ogram.=	Heuristic Synthesis Planning Using the CHIRON Pr	CINF	0002
a Thionocarbonate in a Latent Z-	Hex-3-ene-1,5-diyne Unit.= +a the Elimination of	ORGN	0096
New Ag(I) Compounds with a	Hexaamine Ligand.=	INOR	0037
Parameters of Individual	Hexachlorobiphenyl Congeners.= Physicochemical	ANYL	0101
ns of Propargyl Chloride and 1,5-	Hexadiyne behind Reflected Shock Waves.= +actio	FUEL	0068
nic Ligand-Stabiilized Copper(I)	Hexafluoroacetylacetonate Complexes for Copper CV	COLL	0156
the 1,5-Cyclooctadiene Copper(I)	Hexafluoroacetylacetonate Precursor.= +ition from	COLL	0026
rom 1,5-Cyclooctadiene Copper(I)	Hexafluoroacetylacetonate.= +position of Copper f	COLL	0029
midinium Tetrafluoroborates and	Hexafluorophosphates: The Simplest Models of N^5,	ORGN	0204
of VPI-6: Another Intergrowth of	Hexagonal and Cubic Faujasite.= +aracterization	PETR	0001
nformational Transition in Insulin	Hexamers.= +Paraben Will Cause a T←R-Like Co	BIOT	0150
s.= Hydroisomerization of N-	Hexane in the Presence of Aromatics over Pt/Zeolite	PETR	0114
Hydroisomerization of N-	Hexane on Nickel-Containing Zeolites.= Kinetics of	PETR	0116
bis[3-(2-Phthalimido)propyl]-1,6-	hexanediamine Dihydrochloride.= +methyl-N,N'-	MEDI	0140
Potent C5A-Related	Hexapeptides.=	MEDI	0167
sis and Characterization of a New	Hexapyrrolic Expanded Porphyrin: Rubyrin.= +the	ORGN	0058
Herbicides, II: Determination of	Hexazinone and Its Metabolites in Urine of Exposed	CEI	0005
polarizabilities, with Emphasis on	HF.= +son of Experimantal and Theoretical Hyper	PHYS	0339
Reactions of Group IVB (Ti, Zr,	Hf) Alkyls with Aluminum Reagents.=	INOR	0054
Reaction with Cu(FOD)$_2$ and Cu(	HFA)$_2$.= +Deposition of Copper Films by H-Atom	PHYS	0259
on of Monomers on the Surface of	Hiding Pigments Dispersed in Water.= +lymerizati	PMSE	0038
Fisicoquimica de Cementos	Hidraulicos para Pozos Geotermicos.= Evaluacion	INOR	0063
Obtencion de Derivados	Hidrosolubles de la Dehidroepiandrosterona.=	MEDI	0038
Complejos Derivados de Cupfer+	Hidroxilacion del Benceno con H$_2$O$_2$ Catalizada por	COLL	0121
l Atom in Desulfovibrio Vulgaris (	Hildenborough) Hubredoxin.= +cement by a Nicke	INOR	0020
Solid-State Emission of Copper(I)	Hilides.= +Crystallographic Symmetry upon the	INOR	0157
ne Oxide) Bound with a Sterically	Hindered Phenolate.= +onductivity of Poly(ethyle	POLY	0091
The History of	HIST, IV: Probationary Participants.=	HIST	0010
40338, a New Dual PAF and	Histamine Antagonist Related to SCH 37370.= SCH	MEDI	0161
Conformational Model for	Histamine H$_3$ Agonist Activity.= Preliminary	MEDI	0070
New Drugs.= The	Histamine H$_3$-Receptor as a Target for Developing	MEDI	0081
the Orientation of the Active-Site	Histidine Relative to Substrates.= +termination of	BIOL	0066
ymic or Nonenzymic+ Effect of	Histidine/Carnosine on Fe(II) Autoxidation and Enz	AGFD	0131
onocitos (FILM) Producido por E.	histolytica Mediante HPLC.= +comocion de los M	BIOL	0102
Chemistry and Pre-	Historic Art (Videotape presentation).=	CONG	0002
y in Crime Investigation.=	Historical Perspective on the Role of Forensic Biolog	HIST	0003
nal Aspects.=	Historical Reflections on Steroids: Lederle and Perso	HIST	0023
Degradation of Pesticide Wastes:	Historical, Current, and Future Technologies.= +cal	AGRO	0122
Industry: Early	History in New York.= The Pharmaceutical	HIST	0034
ntal Contaminants. Review of the	History of Applications of Soil Gas Analysis.= +e	ENVR	0048
	History of Chemistry Course—A New Approach.=	CHED	0175
nal Recollecti+ Dexter Award in	History of Chemistry, Honoring O. Hannaway. Perso	HIST	0009
Corp.). +s. (Dexter Award in the	History of Chemistry, sponsored by Dexter Chemical	HIST	0011
The	History of HIST, IV: Probationary Participants.=	HIST	0010
can Steroid Industry and Upda+	History of Steroid Chemistry. Prehistory of the Mexi	HIST	0016
arly Steroid Research at G. D.+	History of Steroid Chemistry. Steroids and the Pill: E	HIST	0020
iosynthesis, Structure, and Fun+	History of Steroid Chemistry. The Sterol Molecule: B	HIST	0012
of CUNY.=	History of the Chemistry Department at City College	HIST	0028
DNA Typing Systems.= Brief	History of the Development and Use of Forensic	HIST	0004
talline P+ Effect of Deformation	History on the Morphology of Blends of Liquid Crys	PMSE	0135
tins in Other Species: a Case	History Using the Armyworm.= Search for Allotosta	AGRO	0007
gricola: The Distorting Mirrors of	History.= +Corp.). Herbert Hoover and Georgius A	HIST	0011
Early English Steroid	History.=	HIST	0014
ructure Handling. Prioritizing the	Hits from a 3D Search.= +imensional Chemical St	CINF	0020
s a Chiral Template and Its Anti-	HIV Activity in PBM Cells.= +Using D-Mannose a	CARB	0022
Natural Products with Anti-	HIV Activity.= Structure-Activity Correlations of	AGFD	0163
Anti-	HIV Agent.= MAP 30: A New Plant-Derived	AGFD	0190
Natural Products as Anti-	HIV Agents.=	AGFD	0164
ne Nucleosides as Potential Anti-	HIV Agents.= +merically Pure Dioxolane-Pyrimidi	MEDI	0051
Anti-	HIV Agents.= Tannins and Related Compounds as	AGFD	0166
Aminosugar Attenuation of	HIV Infection.=	AGFD	0193
Haloperidol.= Inhibition of	HIV Protease with Nonpeptide Derivatives of	MEDI	0054
ds for Production of Recombinant	HIV Protease.= +ving Downstream Processing Yiel	BIOT	0071
Vitro Studies with	HIV RT Inhibitor U-9843.= Mode of Action in	MEDI	0055
Discovery of Novel Inhibitors of	HIV-1 Protease by Three-Dimensional Substructure	COMP	0038
Symmetric Inhibitors of	HIV-1 Protease.= Design and Preparation of C$_2$	MEDI	0053

Reverse Transcriptase Inhibitor of	HIV-1.= +of BI-RG-587: A Potent and Selective	ORGN 0027
of a Recombinant	HIV-2 Protease.= Purification and Characterization	BIOL 0042
Activity and Surface Acidity of	HM Zeolite and Pd/HM Catalysts.= Studies on the	PETR 0117
M Thiol/Disulfide Exchange in	HMG–CoA Lyase Affects the Kinetics of Active Site	BIOL 0064
Interconversion of $W_2(O_2C_8H_{16})_3$(	$HNMe_2)_2$ to $W_2(O_2C_8H_{16})_3$.= +ic Propellanes: The	INOR 0128
Different Channels of H +	HNO.= Reaction Path and Rate Calculations of	PHYS 0310
Complexes of	$HNS_{\varphi2}$ with $Cp*_2UCl_2$.=	INOR 0366
of Photosynthetic Units: Spectral–	Hole–Burning Studies.= +d Transport Dynamics	PHYS 0018
aching: A Witches' Brew of Black	Holes?.= +E. I. DuPont de Nemours and Co.). Te	ANYL 0006
lymerization.=	Holographic Investigation of Liquid Crystal Photopo	BIOT 0205
Semiconstant	Holographic Memory Based on Bacteriorhodopsin.=	BIOT 0079
Hubbard and	Holstein Models.= Quantum Simulations of	PHYS 0054
Results from the	Homestake Solar Neutrino Observatory.=	NUCL 0021
–Metal Complexes: Super–	Homo–Michael–Acceptors.= Cyclopropylthiocarbene	ORGN 0216
Synthase by Nε–Amino–L–	Homoarginine.= Inhibition of Inducible Nitric Oxide	MEDI 0106
Heterocyclic	Homoconjugated Polyacetylenes.=	ORGN 0094
ond and Diamond–Like Carbon.	Homoepitaxial High–Temperature Diamond Growth	FUEL 0037
aY Zeolites by NMR Spectrosc+	Homogeneous Adsorption of Benzene on NaX and N	CATL 0025
l Vapor O+ Competition between	Homogeneous and Heterogeneous Reactions in Meta	PHYS 0144
tions of Fluorescence Lifetimes in	Homogeneous and Heterogeneous Systems.= +tribu	PHYS 0134
canoic Acid) Benzoylformamide in	Homogeneous and Microheterogeneous Media.=	ORGN 0166
itrogen Heterocycles by Means of	Homogeneous Catalytic Carbonylation.= +icyclic N	ORGN 0049
onjugates and Lec+ Formation of	Homogeneous Cross–Linked Lattices between Glycoc	BIOL 0145
/Hydrogen/Carbon Combustio+	Homogeneous Gas–Phase Modeling of Boron/Oxygen	FUEL 0097
Oxidation of Nitrogen Oxides in	Homogeneous Gas–Phase Reactions: Combustion Em	FUEL 0127
rganic Synthesis.=	Homogeneous Metal–Catalyzed Photochemistry in O	INOR 0339
on and Isomerization of Olefins by	Homogeneous Nickel–Based Catalysts.= +erizati	PETR 0007
Route to Highly Optically Pur+	Homogeneous Nucleophile Exchange: A Novel, Facile	ORGN 0207
Its Role on Catalyst Stability in	Homogeneous Transition–Metal–Catalyzed Processes	INOR 0349
Streptomyces lividans and Their	Homologous Expression.= Xylanase Genes of	BIOT 0213
104) and Its	Homologues with TBP.= Extraction of Rf (Element	NUCL 0074
of Temperature+ Application of	Homonuclear 2D ^{51}V NMR to Determine the Effect	INOR 0228
Nonmathematical but	Honest Approach.= Pictorialized Solid–State MOs: A	CHED 0021
New Industrial Catalytic Processes	Honoring J. F. Roth+ Industrial Chemistry Award	I&EC 0043
New Industrial Catalytic Processes	Honoring J. F. Roth+ Industrial Chemistry Award	I&EC 0052
New Industrial Catalytic Processes	Honoring J. F. Roth+ Industrial Chemistry Award	I&EC 0060
Roy W. Tess Award Symposium	Honoring K. L. Hoy. Four Component Solubility P	PMSE 0033
er Award in History of Chemistry,	Honoring O. Hannaway. Personal Recollecti+ Dext	HIST 0009
ochemistry Award Symposium—I,	Honoring R. A. Osteryoung. Interactions of+ Electr	ANYL 0093
ce in Teaching Award Symposium	Honoring R. W. Ramette. Award Address+ Excellen	ANYL 0006
the NSF ChemSource Programs,	Honoring the Memory of Marjorie Gardner, Chemica	CHED 0011
erman Skolnik Award Symposium	Honoring W. T. Wipke: New Dimensions in C+ H	CINF 0016
exter Chemical Corp.). Herbert	Hoover and Georgius Agricola: The Distorting Mirror	HIST 0011
metry at Extreme Potentials: A	Hope for a Fast Electrochemical Method of Determin	ANYL 0095
Laboratory at Johns	Hopkins.= Physical Chemistry Instrumentation	CHED 0199
Laboratory at Johns	Hopkins.= Physical Chemistry Instrumentation	PHYS 0246
	Hormigas: Alimento Rico en Proteinas.=	AGFD 0205
Antijuvenile	Hormone Agents.=	AGRO 0003
al and Toxicological As+ Steroid	Hormone and Related Compound Residues: Analytic	AGRO 0129
Cytochr+ Inhibitors of Steroid	Hormone Biosynthesis. Mechanism and Inhibition of	MEDI 0006
Glycoprotein	Hormone Function.= Role of Glycosylation in	BIOL 0150
tro Catabolism Study of Juvenile	Hormone in the Fifth Stadium Larvae of Manduca s	AGRO 0022
Juvenile	Hormone of Drosophila melanogaster.=	AGRO 0002
Prothoracicotropic	Hormone of the Silkworm, Bombyx mori.=	AGRO 0010
the Molecular Action of Juvenile	Hormone Using Radioligands.= Hot JH: Studies of	AGRO 0005
apping of Hemolymph Juvenile	Hormone–Binding Protein of Larval Manduca sexta	AGRO 0014
kson Award Symposium: Juvenile	Hormones and Peptides for Insect Control. +ck, Jac	AGRO 0001
kson Award Symposium: Juvenile	Hormones and Peptides for Insect Control. +ck, Jac	AGRO 0006
sterones, the First Insect–Molting	Hormones from Plants.= +rrent Studies with Pona	HIST 0024
ol. Award Address. From Juvenile	Hormones to Diuretic Hormones.= +Insect Contr	AGRO 0001
idopt+ Ethyl–Branched Juvenile	Hormones: Opportunities for Specific Control of Lep	AGRO 0004
om Juvenile Hormones to Diuretic	Hormones.= +Insect Control. Award Address. Fr	AGRO 0001
Allatostatin from the Tobacco	Hornworm, Manduca sexta.= Identification of an	AGRO 0008
of Mixtures.= Putting the	Horse before the Cart: Exploring the Melting Point	CHED 0159
and Thiodiglycol.= Inhibition of	Horseradish Peroxidase by 2–Chloroethyl Sulfides	BIOT 0249
ironment of Mn(III) and Mn(IV)	Horseradish Peroxidase with Manganese–Porphyrin	BIOL 0095
rogen Solutions and Their Uses in	Horticulture and Agriculture.= +olled–Release Nit	FERT 0005
ase Methosulfate as an Inclusion	Host: Resolution of Alcohols by Inclusion Compound	ORGN 0041
ysosporium and in a Heterologous	Host.= +and Mn Peroxidases in Phanerochaete chr	BIOT 0237
Fiber–Optic Probes for	Hostile Environments.= Development of Raman	ANYL 0065
Heterologous	Hosts.= Ligninase Expression in	BIOT 0236
s+ Surface Chemical Studies of	Hot Corrosion: Life Prediction Model for Marine Ga	FUEL 0099
of CO_2/CO due to Collisions with	Hot Hydrogen Atoms.= +re Rotational Scattering	PHYS 0291
Hormone Using Radioligands.=+	Hot JH: Studies of the Molecular Action of Juvenile	AGRO 0005
sity Distribution of H_2 and H in a	Hot–Filament CVD Reactor of Diamond.= +d Den	PHYS 0146
Carbon.=	Hot–Filament–Activated CVD of Boron Nitride and	PHYS 0205
–Beam Mass–Spectrometry during	Hot–Filament–Assisted Diamond Film Growth.=	PHYS 0145
Coupling of CH_4 in a	Hot–Wire Diffusion Column.= Dehydrogenative	PETR 0012
Considerations of Corporate In–	House Counsel.= The Scientific Expert:	CHAL 0017
Structure and Regulation of a	House Fly Cytochrome P450 Monooxygenase.=	AGRO 0098

for Pyrethroid Resistance in the	House Fly.= Characterization of a P450 Responsible	AGRO	0099
ntrol, and Environmen+ The In-	House Laboratory: Product Development, Quality Co	FERT	0002
t of Your Kitchen.= Hazards of	Household Chemicals: Getting the Chemical Kicks ou	CHAS	0051
of Molecular Weights and Mark-	Houwink Constants.= +for Accurate Determination	PMSE	0066
Award Symposium Honoring K. L.	Hoy. +meters: A New Dimension.= Roy W. Tess	PMSE	0033
umental Considerations for Rapid	HPLC Analysis of Biological Molecules.= +nd Instr	ANYL	0129
	HPLC Analysis of Food Phenolic Compounds.=	AGFD	0005
HPLC of Glycoconjugates.	HPLC Analysis of Glycoprotein Glycans.=	ANYL	0015
.=	HPLC Analysis of Negatively Charged Phospholipids	ANYL	0134
Pharmaceuticals When Employing	HPLC Analysis with UV Detection.= +Analysis in	ANYL	0114
Relationships between Log P by	HPLC and Biological Activity.=	ANYL	0069
droxy-3-Et+ Rapid and Simple	HPLC Assay for the Novel Anticonvulsant D,L-3-Hy	ANYL	0082
Improved Extraction and	HPLC Assay of Potato Glycoalkaloids.=	AGFD	0112
Enzyme-Based	HPLC Columns as Probes of Enzyme Interactions.=	BIOT	0036
inium Chemiluminescence as an	HPLC Detection Strategy for Chlorophenol Determin	ANYL	0084
	HPLC in Food Science.=	AGFD	0178
Products.= Use of	HPLC in Pharmaceutical Development of Antibiotic	ANYL	0112
ans Isomers of+ Semipreparative	HPLC Isolation and NMR Analyses of the cis and tr	AGRO	0033
Aerosol Generation, Transport+	HPLC Mass Spectrometry—I. Particle Size Effects in	ANYL	0122
in CE/MS Utilizing Atmosphe+	HPLC Mass Spectrometry—II. Recent Developments	ANYL	0138
e of a Novel Column-Switching	HPLC Method To Support Registration of DPX-E96	AGRO	0030
Stationary Phases.= Rapid	HPLC of Biopolymers with Micropellicular	ANYL	0128
rotein Glycans.=	HPLC of Glycoconjugates. HPLC Analysis of Glycop	ANYL	0015
d on Competitive Binding Assa+	HPLC Postcolumn Reaction Detection Systems Base	ANYL	0090
n Inhibitor, by a Tandem Column	HPLC Procedure.= +gatroban, a Specific Thrombi	ANYL	0086
Calculated Lipo+ Prediction of	HPLC Retention Times for Novel Avermectins Using	BIOT	0127
	HPLC Separation of Alkaloids.=	ANYL	0135
Inbreds for Silk Flavonolglyco+	HPLC Survey of Corn (Zea mays L.) Populations and	AGRO	0034
dues in Edible Meats by On-Line	HPLC Tandem Mass Spectrometry.= +ibiotic Resi	AGFD	0061
Peaks in Normal Bonded Phase	HPLC with Diode Array Multichannel Detection.=	ANYL	0087
famethazine in Animal Tissues by	HPLC with Diode-Array Detection.= +tion of Sul	AGFD	0103
Antibiotics as Revealed by RP-	HPLC: Structure-Retention and Structure-Activity	MEDI	0059
ucido por E. histolytica Mediante	HPLC.= +comocion de los Monocitos (FILM) Prod	BIOL	0102
nge Capsule and Cation-Exchange	HPLC.= +lysin Purification Using a Cation-Excha	ANYL	0081
tections in Normal Bonded Phase	HPLC.= +sional Separations and Multichannel De	ANYL	0088
of Organ+ Dispersion-Corrected	HPLC/FACP Technique for Measuring the Sorption	ANYL	0072
Applications of New	HPLC/MS Techniques in Human Biomonitoring.=	CEI	0006
ullulan, and Pectin Determined by	HPSEC with Viscosity Detection.= +of Dextran, P	CARB	0045
of Aggregated Polysaccharides by	HPSEC/Viscometry.= +alysis. Structural Analysis	PMSE	0094
GaAs(100).=	HREELS Study of Trimethylgallium and Arsine on	COLL	0065
-Pressed Citrus Essential Oils.=+	HRGC for Detection of Adulterations of Italian Cold	AGFD	0199
Has a Release Been Observed?	HRS-Observed Releases and Analytical Data.=	ENVR	0080
io Energies and Shapes of HSiS+,	HSiO+, HCP, and HSiP, and Their Isomers.= +Init	PHYS	0315
hapes of HSiS+, HSiO+, HCP, and	HSiP, and Their Isomers.= +Initio Energies and S	PHYS	0315
b Initio Energies and Shapes of	HSiS+, HSiO+, HCP, and HSiP, and Their Isomers.=	PHYS	0315
Sumatriptan, a Novel 5-	HT$_1$-Like Agonist for the Treatment of Migraine.=	MEDI	0079
perature Aerosol Decomposition (	HTAD) Process.= +sts Prepared by the High-Tem	CATL	0026
Pairing in	HTC Materials.= Superconductivity. Theory of	PHYS	0001
vel Azatricyclic Systems: Novel 5-	HT3 and 5-HT4 Serotonergic Agents.= +ion of No	ORGN	0104
Agents.= Serotonergic	5HT3 and 5HT4 Activities of Novel Azatricyclic	MEDI	0187
otonin 5-HT4 Agonists and/or 5-	HT3 Antagonists.= +cyclic Benzamides: Potent Ser	MEDI	0069
Serotonergic 5HT3 and	5HT4 Activities of Novel Azatricyclic Agents.=	MEDI	0187
Benzamides: Potent Serotonin 5-	HT4 Agonists and/or 5-HT3 Antagonists.= +cyclic	MEDI	0069
clic Systems: Novel 5-HT3 and 5-	HT4 Serotonergic Agents.= +ion of Novel Azatricy	ORGN	0104
Quantum Simulations of	Hubbard and Holstein Models.=	PHYS	0054
le Graph Theoretical Solutions to	Huckel MO Parameters.= +olecular Systems: Faci	COMP	0020
ickel(III)-Cysteine Centers in the	Hudrogenase Enzymes.= +mplexes as Models for N	INOR	0179
and Reactor-Derived ^{14}C from the	Hudson River to the New York Bight.= +c Carbon	GEOC	0014
rols on Methane Fluxes from the	Hudson River.= Chemistry of Ocean Margins. Cont	GEOC	0001
Characterization of	Humic Acids from Sludge Pond Solids.=	I&EC	0022
Characterizing the Role of in Situ	Humic Matter.= +B in Anoxic Bottom Sediments:	ENVR	0017
of the Electrostatic Properties of	Humic Substances by Fluorescence Quenching.= +n	ENVR	0057
f Bacterial+ Effects of Ambient	Humidity on the Wetting Properties and Adhesion o	BIOT	0151
Evanescent Detection of	Humidity.=	ANYL	0079
l Engineering, 2nd Edition—Aiba,	Humphrey, and Millis.= +otechnology. Biochemica	BIOT	0221
One	Hundred Years of Chemistry at Columbia.=	HIST	0030
City.= One	Hundred Years of Industrial Chemistry in New York	HIST	0026
iscovery of the Fluorostero+ The	Hunt for an Economical Synthesis of Cortisol: The D	HIST	0022
gram for High School Students at	Hunter College, Chemistry Department.= +ED Pro	CHED	0143
Frank C. Whitmore by Charles D.	Hurd.= +O. Hannaway. Personal Recollections of	HIST	0009
Conformational Versatility of	Hyaluronan: ^{13}C-NMR Studies.=	CARB	0047
as Derived from the Reaction of	Hyaluronic Acid with Isotopically Labeled Carbodiim	ORGN	0020
ral Features of Sol-Gel-Derived	Hybrid Ceramer Materials by Small-Angle X-ray Sc	POLY	0284
Application of Waste-Fuel-Fired	Hybrid Fluidized-Bed Boilers = Commercial	FUEL	0122
rization of Polyacrylate-Inorganic	Hybrid Materials via the Sol-Gel Approach.= +acte	POLY	0232
posite+ Preceramic and Inorganic	Hybrid Materials. Polyphosphazenes Molecular Com	POLY	0181
s.= Preceramic and Inorganic	Hybrid Materials. Polysilazanes: A Route to Ceramic	POLY	0309
ipita+ Preceramic and Inorganic	Hybrid Materials. Reinforcing Zirconia Particles Prec	POLY	0283
n P+ Preceramic and Inorganic	Hybrid Materials. Silicon Acetylene and Silicon Olefi	POLY	0357
Preceramic and Inorganic	Hybrid Materials. Sol-Gel Kinetics by NMR.=	POLY	0256

ecer+ Preceramic and Inorganic	Hybrid Materials. Synthesis of Nanomaterials via Pr	POLY	0335
Approach to Inorganic–Organic	Hybrid Materials?.= Cement Chemistry: A New	POLY	0261
Macromolecules: The Utility of	Hybrid Molecules.= Molecular Recognition of	ORGN	0270
tion of High–Resolution MS and	Hybrid MS/MS to the Analysis of Flyash for Monob	ANYL	0050
lyalkenoates/Ion–Leachable Glass	Hybrid Systems.= +ocurable Cements Based on Po	POLY	0262
Rubber–Clay	Hybrid—Synthesis and Properties.=	POLY	0291
Manipulation of Heterogeneous	Hybridoma Cultures for the Overproduction of Mono	BIOT	0169
Rapid Adaptation of a Murine	Hybridoma to Serum–Free Growth.=	BIOT	0163
ol–Myers Squibb Award Address.	Hydantoin Bioisosteres: Hydroxy Acetic Acid Aldose	MEDI	0082
Solid–State ^{27}Al NMR Studies of	Hydrated AlPO$_4$–8.=	PETR	0096
Quantum Dynamics of the	Hydrated Electron.=	PHYS	0190
esult in Significant Reversal of	Hydration and Polymerization in the Matrix Apparen	ENVR	0041
lasma Samples Related to Their	Hydration Water Contents and Paramagnetic Metal	MEDI	0043
ling: Methyl Esterification– and	Hydration–Induced Motional Perturbations of Pectin	AGFD	0012
dation–State Rhenium Hydrazine,	Hydrazido, Amido, and Ammonia Complexes.= +i	INOR	0261
+cleophilic Addition of Amine or	Hydrazine to σ–Allenyl Complex of Platinum(II).=	INOR	0041
High–Oxidation–State Rhenium	Hydrazine, Hydrazido, Amido, and Ammonia Comple	INOR	0261
hesis of 4,4'–Dinitrocarbanilide–	Hydrazone and the Development of an ELISA to Nic	BIOT	0166
tous Complexes of Unsymmetrical	Hydrazone Imine Glyoxal Derivatives.= +cs: Cobal	POLY	0102
by Sterically+ Monomeric Zinc	Hydride and Magnesium Alkyl Derivatives Stabilized	INOR	0087
Intermediate–Sized Boron	Hydride Clusters.= Insertion of Nitrogen into	INOR	0388
tions with MR$_2$ (M =+ Rhodium	Hydride Dimer: Fragmentation–Recombination Reac	INOR	0383
tion Metal Hydrides. Palladium	Hydride Dimers: Unusual Coordination Behavior of L	INOR	0379
Synthesis of Magnesium	Hydride from Silanes.=	INOR	0206
mplexes.=	Hydride Thiolate Derivatives of Low–Valent Iron Co	INOR	0225
Electron–Rich Ruthenium Alkyl,	Hydride, and Aluminohydride Complexes.= +vity of	INOR	0244
ination Beha+ Transition Metal	Hydrides. Palladium Hydride Dimers: Unusual Coord	INOR	0379
try Parameters in Bridging Metal	Hydrides.= +pole Coupling Constants and Asymme	INOR	0382
Ceramics.+ Characterization of	Hydridomethylpolysilazane—A Precursor to SiC/SiN	POLY	0340
Free Radical Halogenation of	Hydridophosphoranes.=	ORGN	0248
erically Demanding tris(Pyrazolyl)	hydroborato Ligands.= +rivatives Stabilized by St	INOR	0087
rically Demanding poly(Pyrazolyl)	hydroborato Ligands.= +rivatives Stabilized by Ste	INOR	0085
s. Carbon Black from Coal by the	Hydrocarb Process.= +Fuels: Engineering Material	FUEL	0047
Treatment of MSW by	Hydrocarb Process.=	FUEL	0121
Urban Nonmethane Organic C+	Hydrocarbon Composition of Oilfield Emissions and	ENVR	0022
d Superacids for Acid–Catalyzed	Hydrocarbon Conversions.= Sulfate–Containing Soli	PETR	0102
nodisperse 1,3,5–Phenylene–Based	Hydrocarbon Dendrimers, Including C$_{276}$H$_{186}$.=	POLY	0320
Simultaneous Determination of	Hydrocarbon Distribution and Sulfur Content in Pet	PETR	0120
rb+ New Materials Derived from	Hydrocarbon Fuels: Diamond and Diamond–Like Ca	FUEL	0034
rb+ New Materials Derived from	Hydrocarbon Fuels: Diamond and Diamond–Like Ca	FUEL	0035
rb+ New Materials Derived from	Hydrocarbon Fuels: Diamond and Diamond–Like Ca	FUEL	0037
rb+ New Materials Derived from	Hydrocarbon Fuels: Diamond and Diamond–Like Ca	FUEL	0036
rb+ New Materials Derived from	Hydrocarbon Fuels: Diamond and Diamond–Like Ca	FUEL	0038
la+ New Materials Derived from	Hydrocarbon Fuels: Engineering Materials. Carbon B	FUEL	0047
e+ New Materials Derived from	Hydrocarbon Fuels: Engineering Materials. Developm	FUEL	0050
le+ New Materials Derived from	Hydrocarbon Fuels: Fullerenes. Post–Buckminsterful	FUEL	0019
ion and Growth of Diamond from	Hydrocarbon Gases.= +(and c–BN, SiC). Nucleat	PHYS	0091
py as an Analytical Tool for	Hydrocarbon Identification.= Derivative Spectrosco	PETR	0137
Foundations for Synthesis of	Hydrocarbon Isomerization Catalysts.= Scientific	PETR	0135
ization of Alkali Metal Halides in	Hydrocarbon Media.= +(η^6–C$_7$H$_8$)$_2$ and the Solubil	INOR	0053
onversion of Acetylene to Higher	Hydrocarbon over Modified Shape–Selective Zeolite	CATL	0034
hemistry of Bismuth Molybdate	Hydrocarbon Oxidation Catalysts Prepared by the H	CATL	0026
es by C$_2$ Emis+ Laser–Generated	Hydrocarbon Plasmas: Determination of Temperatur	PHYS	0268
g Precollege Students to a	Hydrocarbon Process.= Polysar's Efforts in Exposin	CHED	0010
rison of Allyl, Formyl, and Viny+	Hydrocarbon Radical Reactions with Oxygen: Compa	PHYS	0228
olyethylene Block Copolymers and	Hydrocarbon Solvents.= +oreversible Gels with P	POLY	0249
Kinetics of High–Temperature,	Hydrocarbon–Assisted Boron Combustion.=	PHYS	0142
ic Metal and Polycyclic Aromatic	Hydrocarbon–Induced Transformation of Mammalia	CHAS	0007
Biodegradation of Halogenated	Hydrocarbons by Rational Redesign of Cytochrome+	BIOT	0118
ection of Aromatics, Sulfur, and	Hydrocarbons in Diesel Fuels by Gas Chromatograph	PETR	0118
Growth of Polynuclear Aromatic	Hydrocarbons in Diffusion Flames.= Modeling the	FUEL	0109
Determination of Polyaromatic	Hydrocarbons in Environmental Matrices by Liquid	ENVR	0015
g People to Chlorinated Aromatic	Hydrocarbons in La Basse Cote–Nord, Quebec.= +n	CEI	0011
Comparison of Reactivity with	Hydrocarbons of Transition–Metal Atoms in Differen	PHYS	0199
Selective Aromatization of Light	Hydrocarbons on Polyfunctional Metallosilicate Ca	PETR	0062
in the Conversion of Ethanol to	Hydrocarbons over Alkaline–Free Synthesized ZSM–	PETR	0065
and Isomerization of Short–Chain	Hydrocarbons over Heterogeneous Catalysts. +tion,	PETR	0082
and Isomerization of Short–Chain	Hydrocarbons over Heterogeneous Catalysts. +tion,	PETR	0130
and Isomerization of Short–Chain	Hydrocarbons over Heterogeneous Catalysts. +tion,	PETR	0043
and Isomerization of Short–Chain	Hydrocarbons over Heterogeneous Catalysts. +tion,	PETR	0112
and Isomerization of Short–Chain	Hydrocarbons over Heterogeneous Catalysts. +tion,	PETR	0007
and Isomerization of Short–Chain	Hydrocarbons over Heterogeneous Catalysts. +tion,	PETR	0100
and Isomerization of Short–Chain	Hydrocarbons over Heterogeneous Catalysts. +tion,	PETR	0025
and Isomerization of Short–Chain	Hydrocarbons over Heterogeneous Catalysts. +tion,	PETR	0061
Catalytic Reactions of	Hydrocarbons under Transient Conditions.=	PETR	0134
tics of the Reactions of Aromatic	Hydrocarbons with O(^{3}P) Atoms and CH$_2$(x^{3}B$_1$) Rad	FUEL	0082
Graphs of Alternant	Hydrocarbons.= Eigenvalue Distributions for the	COMP	0029
Oxide–Catalyzed Oxidation of	Hydrocarbons.= Oxygen Spillover during	COLL	0174
o Polymer Plasticizers and Model	Hydrocarbons.= +lar Weight Liquids: Application t	POLY	0065
nyllactams.= Amide–Directed	Hydrocarbonylation of N–Alkenylamides and α–Alke	INOR	0227

ol-2-YL] Carbamate (Benomyl) in	Hydrochloric Acid Solutions.= +)-1H-Benz-imidaz	AGRO	0041
Tissue Residues of Ractopamine	Hydrochloride: A Phenethanolamine-Repartitioning	AGRO	0131
2'-Deoxy-2'-Methylidenecytidine	Hydrochloride.= +alogues and Crystal Structure of	CARB	0038
olite Framework Composition and	Hydrocracking Catalyst Performance.= +etween Ze	I&EC	0062
ion. Pillared Clays as Supports in	Hydrocracking Catalyst Systems.= +Characterizat	PETR	0013
of Metal-Modified Y-Zeolites for	Hydrocracking of Coal-Derived Distillates.= +lysis	CATL	0050
alysts—Activity and Selectivity in	Hydrocracking Reactions.= +rbon Materials as Cat	CATL	0046
Primary Amines.= Convenient	Hydrodeamination and Chlorodeamination of	ORGN	0108
orted Vanadium Catalysts in the	Hydrodemetallization (HDM) of NI-TPP and VO-T	CATL	0044
e Area γ-Mo$_2$N.=	Hydrodenitrogenation of Quinoline over High Surfac	CATL	0047
um Model Compounds Related to	Hydrodesulfurization.= +activity Studies of Vanadi	INOR	0197
te-Enhanced Mass Transport to	Hydrodynamic Electrodes.= Mechanism of Particula	COLL	0041
Externally Imposed Magnetic and	Hydrodynamic Fields.= +c Colloidal Suspension in	COLL	0144
gnetic, and Hydrodynamic—IV.	Hydrodynamic Fingerprinting as a Probe of Latex Ch	COLL	0199
vidomycin Produc+ Influences of	Hydrodynamic Stress and Oxygen Availability on Ra	BIOT	0162
nal Fields: Electric, Magnetic, and	Hydrodynamic—I. +les: Colloidal Particles in Exter	COLL	0141
nal Fields: Electric, Magnetic, and	Hydrodynamic—II. +es: Colloidal Particles in Exter	COLL	0159
nal Fields: Electric, Magnetic, and	Hydrodynamic—III. +s: Colloidal Particles in Exter	COLL	0178
nal Fields: Electric, Magnetic, and	Hydrodynamic—IV. +s: Colloidal Particles in Exter	COLL	0199
Microbially Influenced Corrosion.	Hydrodynamics and Kinetics in Biofilm Systems.	BIOT	0096
ranched Aldehydes by Catalytic	Hydroformylation with Cationic Bis(dioxaphospholan	INOR	0154
PTFE-Matrix Composite	Hydrogel Sheets with Thermoreversible Transitions.=	POLY	0245
rous Poly(N-isopropylacrylamide)	Hydrogels.= +eparation and Properties of Macropo	POLY	0243
= High-Nitrogen Low-	Hydrogen Amides and Imides from Cyanoimidazoles.	POLY	0363
Interface.= Absorbed	Hydrogen and Fluorine near the Diamond/Vapor	PHYS	0359
Diffusivity of	Hydrogen and Methane Molecules in A Zeolites.=	CATL	0024
(C$_{60}$) in Helium, Argon, Air, and	Hydrogen at 25-1000°C.= Behavior of Fullerene	FUEL	0021
O$_2$/CO due to Collisions with Hot	Hydrogen Atoms.= +re Rotational Scattering of C	PHYS	0291
ment of High+ The Concept of	Hydrogen Back Spillover (HBS): Toward the Develop	PETR	0063
mond and Diamond-Like Carbon.	Hydrogen Binding and Diffusion in Diamond.= +a	FUEL	0036
of Secondary Alcohols: Carbon-	Hydrogen Bond Activation in Rhenium Alkoxide Co	ORGN	0220
Protein Stabilit+ Contribution of	Hydrogen Bonding and Hydrophobic Interactions to	BIOT	0189
cs Presentations and Analysis of	Hydrogen Bonding from Molecular Dynamics Simula	BIOL	0112
dies on the Role of Intramolecular	Hydrogen Bonding in Substituted Benzamides.=	ANYL	0045
tive, and Cooperative Effects in	Hydrogen Bonding: The Natural Bond Orbital Donor	PHYS	0106
e+ Molecular-Beam Study of the	Hydrogen Chloride-Gallium Arsenide Gas Surface R	COLL	0215
n Processes of Fast (V No. V$_0$)	Hydrogen Clusters with Solid Materials.= Interactio	NUCL	0054
Biological Production of	Hydrogen from Synthesis Gas.=	BIOT	0017
Infrared Spectrum of the	Hydrogen Halides Revisited.=	CHED	0197
Nuclear Reaction Analysis of	Hydrogen in Materials: Principles and Applications.=	NUCL	0067
tivity by Intramolecular Carbon-	Hydrogen Insertion Reactions of Alkyl Diazoacetates	ORGN	0042
Evidence That	Hydrogen Ion Is (H$_3$O)$^+$.=	PHYS	0272
Bond Strain in the Chemistry of	Hydrogen on Si(111) and Si(100) Surfaces.= Role of	PHYS	0187
: Odor Control and P+ Effect of	Hydrogen Peroxide and Iron Salts on Sewage, Part 1	ENVR	0034
throline)Iron (II) by	Hydrogen Peroxide.= Oxidation of Tris(1,10-Phenan	INOR	0222
on the Crystal Habit of Calcium	Hydrogen Phosphate Used to Phosphors.= A Study	INOR	0361
ctivity in Catalytic Dehydrogen+	Hydrogen Surface Fugacities and the Control of Sele	CATL	0061
on Enthalpies for Intramolecular	Hydrogen Transfer as Functions of Size of the Cyclic	ORGN	0175
ion in the Presence of H$_2$S as a	Hydrogen Transfer Catalyst.= Free-Radical Alkylat	PETR	0083
otopes and Magnetic Resonance.	Hydrogen Transfer during Coal Liquefaction Determ	FUEL	0070
Coprocessing.=	Hydrogen Transfer from Naphthenes to Coal during	FUEL	0005
n.=	Hydrogen Vermiculite: Synthesis and Characterizatio	INOR	0056
n a Mixture Gas of Acetylene and	Hydrogen.= +hesized by DC Arc Discharge CVD i	FUEL	0038
Molecular-Weight Distribution:	Hydrogen- and Deuterium-Labeled Poly(styrene) an	COLL	0035
Ab Initio Calculation on	Hydrogen-Bonded Complexes.=	PHYS	0103
nt Systems—IV. Plenary Lecture.	Hydrogen-Bonded Interpolymer Complexes.= +ne	PMSE	0163
Diffusion of Water Vapor in	Hydrogen-Bonded Polymer Blends.=	POLY	0192
Gas Sorption and Diffusion in	Hydrogen-Bonded Polymers: Vinyl Alcohol/Vinyl Bu	POLY	0191
rs Having Chromophore-Centered	Hydrogen-Bonding and Crosslinking Groups.= +e	POLY	0108
	Hydrogen-Bonding Interaction in Polymer Blends.=	COLL	0034
/Ruthenium Catalyst System to	Hydrogenate Crude Methylenedianiline to a Nonther	I&EC	0056
New Synthesis of Asymmetric	Hydrogenation Catalysts.=	INOR	0077
eir Roles in Catalytic Asymmetric	Hydrogenation of β-Keto Esters.= +plexes and Th	ORGN	0044
Effects of a Model	Hydrogenation on a Catalytic Palladium Membrane.=	CATL	0011
e Chemicals and P+ Asymmetric	Hydrogenation Technology for the Production of Fin	CATL	0001
during Selective Benzene	Hydrogenation to Cyclohexene.= H-D Exchange	FUEL	0100
during Selective Benzene	Hydrogenation to Cyclohexene.= H-D Exchange	CATL	0041
Partial	Hydrogenation.= Cyclohexene from Benzene by	I&EC	0017
Asymmetric Catalytic	Hydrogenation.= Mechanistic Aspects of	CATL	0003
Strain-Compensated Heats of	Hydrogenation.=	ORGN	0173
er Cores: Relationship to Butane	Hydrogenolysis Selectivities Exhibited by W-Ir and	CATL	0008
um Hydroxide-Catalyzed Epoxide	Hydrogenolysis.= +bohydrate Synthesis via Palladi	ORGN	0074
Selective Fluorinating Agents for	Hydrohalogenoalkanes.= +s and Oxide Fluorides as	INOR	0133
Aromatics over Pt/Zeolites.=	Hydroisomerization of N-Hexane in the Presence of	PETR	0114
ing Zeolites.= Kinetics of	Hydroisomerization of N-Hexane on Nickel-Contain	PETR	0116
Transition State of Nucleoside	Hydrolase.=	BIOL	0073
A Catalyst for the Development of	Hydrologic Science.= +d Address. Global Change:	AGRO	0051
Hemicelluloses.= Enzymatic	Hydrolysis of Isolated and Fiber-Bound	BIOT	0215
ellulases Composition for Efficient	Hydrolysis of Lignocellulosic Biomass.= +tion of C	BIOT	0182
ydes.= Kinetics of the	Hydrolysis of Substituted Benzylidenemalonodialdeh	ORGN	0152

pyridyl)-2,5,8-Triazanona+ Base	Hydrolysis of the Cation $\alpha\beta$-anti-Chloro-[1,9-Bis(2-	INOR 0397
raaz+ Kinetic Study of the Base	Hydrolysis of the syn,anti-cis-Dichloro (1,4,7,10-Tet	INOR 0398
ate Amine Complex Toward Base	Hydrolysis.= +ry Reactive Chromium(III) Polydent	INOR 0397
Electricity via Dilute Sulfuric Acid	Hydrolysis.= +nversion of Waste to Ethanol and	FUEL 0094
Phosphotriester	Hydrolysis.= Mechanism of Enzymatic	BIOL 0010
Probing Enzymatic Glycoside	Hydrolysis.=	ORGN 0267
ting for Skeletal Actomyosin ATP	Hydrolysis.= +e Bond-Splitting Step Is Rate-Limi	BIOL 0141
n Fibroblast Monolayers: Uptake,	Hydrolysis, and Release.= +Substrates with Huma	BIOL 0079
of Organophosphate Insecticide	Hydrolytic Detoxification by Conventional Means an	AGRO 0124
Lipase from Pseudomonas sp. in	Hydrolytic Reactions.= Specificity of Lipoprotein	BIOT 0095
t Insects.=	Hydrolytic Resistance Mechanisms in Stored-Produc	AGRO 0135
xaspiroundecane Derivativ+ New	Hydrolyzable Dianionic Initiator from Divinyl Tetrao	POLY 0349
n Situ Release of Polysaccharide	Hydrolyzing Enzymes During Wood Degradation by	BIOT 0208
Studies of the Reaction of Alkyl	Hydroperoxides with Vanadium(IV) in Acidic Aqueo	INOR 0276
General Synthesis of	Hydroperoxides.=	ORGN 0260
Surface-	Hydrophilic Elastomers.=	COLL 0038
Hydrophobic-	Hydrophilic IPN and SIPN.=	PMSE 0005
Associative Polymers with Novel	Hydrophobe Structures.=	PMSE 0035
tes with Ion-Exchange Resins and	Hydrophobic Adsorbents.= +actions of Carbohydra	CARB 0063
and Steroids for Construction of	Hydrophobic Binding Sites and Self-Organizing Stru	ORGN 0066
ibution of Hydrogen Bonding and	Hydrophobic Interactions to Protein Stability.= +r	BIOT 0189
Prediction of the Solubility of	Hydrophobic Organic Compounds in Nonideal Solven	ENVR 0030
Characterization and Function of	Hydrophobic Polysaccharide Aggregates.=	POLY 0098
eaction Medium for Conversion of	Hydrophobic Substrates.= +terization of a Novel R	BIOT 0248
	Hydrophobic-Hydrophilic IPN and SIPN.=	PMSE 0005
Polymers.= Aggregation of	Hydrophobically Modified and Comb-Type	POLY 0270
Conducti+ A Highly Conducting	Hydropolysilane Copolymer: Synthesis and Electrical	POLY 0043
NMR Analyses of Coal-Oil	Hydroprocessing Products.= 1- and 2-Dimensional	PETR 0149
Gasification: Effect of	Hydropyrolysis Conditions.= Two-Stage	CATL 0054
Complexes.= Catalytic	Hydrosilation of Iron and Ruthenium Acetyl	INOR 0238
nCOCH$_2$R (R = H,CH$_3$, OCH$_3$)+	Hydrosilation of Manganese Acyl Complexes, (CO)$_5$M	INOR 0337
lcohols Using Disc+ Asymmetric	Hydrosillylation of Ketones: A Facile Route to Ciral A	INOR 0247
sphine Complexes for Asymmetric	Hydrosillylation of Ketones.= +hiral Rhodium Pho	CATL 0005
ed on Silica.=	Hydrosillylation Reactions Involving Si-H Groups Fix	COLL 0094
Nucleic Acids in the	Hydrosphere.= Identification and Quantification of	ENVR 0027
lkaline Fluo+ Low-Temperature	Hydrothermal Crystallization of MFI Zeolites in an A	PETR 0005
trating 3D Frameworks of [Pd(+	Hydrothermal Synthesis of K$_2$PdSe$_{10}$: Two Interpene	INOR 0200
unnel Structures.=	Hydrothermal Synthesis of Manganese Oxides with T	PETR 0056
m with Octahedral-Tetrahedral +	Hydrothermal Synthesis of Microporous Molybdenu	CATL 0019
lenide Compounds.=	Hydrothermal Synthesis of New Molybdenum Polyse	INOR 0355
d XRD of Phosphorous-Promoted	Hydrotreating Catalyst Precursors.= +S NMR, an	COLL 0198
Catalytic	Hydrotreatment of Coal-Derived Naphtha.=	CATL 0052
l Catalyst Support Materials: The	Hydrous Metal Oxide Ion Exchangers.= +of Nove	PETR 0018
Reactions of Propane on	Hydrous Metal Oxide-Supported Catalysts.=	CATL 0040
li/Titanium Alkoxide Precurso+	Hydrous Titanate Ceramic Powders from Mixed Alka	INOR 0278
pound Studies on Pd and Ni/Mo	Hydrous Titanium Oxides.= +tion and Model Com	CATL 0039
yclooxygenase and 5-Lip+ Novel	Hydroxamic Acid Fenamates as Dual Inhibitors of C	MEDI 0177
rical Molecular Orbital Studies of	Hydroxamic Acids and Their Conjugate Bases.= +i	PHYS 0326
hydroxy-1-naphthyl)thiophenium	Hydroxide Inner Salt and β-Butyrolactone.= +-(4-	POLY 0048
ohydrate Synthesis via Palladium	Hydroxide-Catalyzed Epoxide Hydrogenolysis.= +b	ORGN 0074
Structure Consisting of fused (μ-	Hydroxo)bisμ-carboxylato)diiron(III) Kernals.= +e	INOR 0094
During Solution Protonation of 3-	Hydroxphthalide.= +tion Is Not a Stable Species	ORGN 0151
Address. Hydantoin Bioisosteres:	Hydroxy Acetic Acid Aldose Reductase Inhibitors.=	MEDI 0082
f Oxochromate(V) Complex by α-	Hydroxy Acids.= +hange and Reduction Kinetics o	INOR 0402
-S+ Asymmetric Synthesis of α-	Hydroxy Carbonyl Compounds Using Enantiopure N	ORGN 0293
Synthesis and Polymerization of	Hydroxy Polystyrene Carboxylic Acid Telechelomers	POLY 0046
of Peptide Fragments Bearing α-	Hydroxy-β-Amino Acid Residues.= +tric Synthesis	ORGN 0048
ics Simulations of Rigid-Rod and	Hydroxy-Substituted Rigid-Rod Polymers.= +nam	PHYS 0322
merization of Tetrahydro-1-(4-	hydroxy-1-naphthyl)thiophenium Hydroxide Inner S	POLY 0048
f 1-Aminoethyl-3,4-Dihydro-5-	Hydroxy-1H-2-Benzopyrans as Dopamine D1 Select	MEDI 0127
e/5-Lipoxygenase Activity of 4-	Hydroxy-2-Methyl-N-(5-Substituted-3-Isoxazolyl)-	MEDI 0178
he Novel Anticonvulsant D,L-3-	Hydroxy-3-Ethyl-3-Phenylpropionamide (HEPP) in	ANYL 0082
Series of Thiophene-Derived 3-	Hydroxy-3-Methylglutaryl-Coenzyme A Reductase I	MEDI 0101
Cationic Complexes of Tc-99m-	Hydroxy-4-Pyrones: Synthesis and Characterizatio	MEDI 0042
ur+ Asymmetric Synthesis of 3-	Hydroxy-4-Substituted-β-Lactams as Potential Prec	ORGN 0047
tones: A Versatile Synthesis of 3-	Hydroxy-5(2H)-Furanones.= +ione with α-Haloke	ORGN 0223
thesis: Activation of α-Chloro, α-	Hydroxy, and N-Blocked α-Amino Acids.= +ic Syn	ORGN 0259
nd+ Syntheses of New Poly(3-	hydroxyalkanoate) Conjugates: PHA-Carbohydrate a	POLY 0041
Bacteria.= Poly(β-	hydroxyalkanoates): Biodegradable Plastics Made by	CHED 0132
Synthesis of 3-Fluoro-3-	Hydroxyalkylazetidinones.= Stereoselective	ORGN 0080
ion+ Analogies between Calcium	Hydroxyapatite and Zirconium Oxide for the Separat	ANYL 0027
of 3-Phenoxybenzoic Acid to 3-	Hydroxybenzoic Acid by Chicken Kidney Microsome	AGRO 0018
(vinylidene Fluoride) and Poly(3-	hydroxybutyrate).= +dy of Miscible Blends of Poly	POLY 0097
Evaluation of 3-	Hydroxycyprodime.= Synthesis and Biological	MEDI 0086
ion of the Orientation of+ coli β-	Hydroxydecanoyl Thiol Ester Dehydrase: Determinat	BIOL 0066
obe of the Active Site of E. coli β-	Hydroxydecanoyl Thiol Ester Dehydrase.= Is a Pr	BIOL 0067
ontrol During Amide+ Effect of	Hydroxyl Protection on Stereochemistry and Stereoc	ORGN 0258
m Lignin, Organostannanes, and	Hydroxyl-Capped Poly(Ethylene Oxide) and ABA B	PMSE 0006
A Block Copolymers Containing	Hydroxyl-Capped Poly(Ethylene Oxide) with Polydi	PMSE 0006
Microbial	Hydroxylation of Dihydroartemisinin Derivatives.=	BIOT 0170

lose and Dynamic Behavior of the	Hydroxymethyl Group.= +netic Relaxation of Amy	CARB	0049
the Gene-Encoding T4 dCMP	Hydroxymethylase.= Site-Directed Mutagenesis of	BIOL	0160
e+ Nonaqueous Preparation of β-	Hydroxynitriles from Epoxides with Lithium Cyanid	ORGN	0107
s: C-Glucuronide Analogues of 4-	Hydroxyphenylretinamide-o-Glucuronide.= +bolite	MEDI	0030
of C-Linked Deoxyribosides of 2-	Hydroxypyridine and 2-Hydroxyquinoline.= +esis	ORGN	0075
ides of 2-Hydroxypyridine and 2-	Hydroxyquinoline.= +esis of C-Linked Deoxyribos	ORGN	0075
ed Irreversible Inhibitors of	Hydroxysteroid Dehydrogenases.= Enzyme-Generat	MEDI	0010
Spectroscopic Studies of Alpha-	Hydroxytricarboxylic Acid Complexes of Iron(III).=	INOR	0040
heory and Practice. A Chemical	Hygiene Plan for R D Laboratories—A Case Study.=	CHAS	0001
e Convergent Growth Approach:	Hyperbranched Block Copolymers and Monodisperse	POLY	0318
Terminal Modifications of	Hyperbranched Poly(siloxy-silanes).=	POLY	0323
Novel Developments in	Hypercoordinate P(III) and P(V) Chemistry.=	INOR	0323
pothesis for the Mode of Action of	Hypericin as an Antiretroviral Agent.= +ents. Hy	AGFD	0188
	Hypermedia Toolkit for Structure Elucidation.=	COMP	0008
Calculations of Static	Hyperpolarizabilities and Polarizabilities.=	PHYS	0340
endent and Correlated	Hyperpolarizabilities of Molecules.= Frequency-Dep	PHYS	0341
Regularities.= Molecular	Hyperpolarizabilities: Computational Methods and	BIOT	0178
Frequency-Dependent	Hyperpolarizabilities.=	PHYS	0342
Measurements of Molecular	Hyperpolarizabilities.= Nonlinear Optical	PHYS	0343
n of Experimantal and Theoretical	Hyperpolarizabilities, with Emphasis on HF.= +so	PHYS	0339
ond-Harmonic Generation from	Hyperpolarizable Amphiphiles at Polymer-Polymer I	POLY	0156
the Metabolism of	Hyperthermophilic Bacteria.= Role of Tungsten in	INOR	0099
Relationships of	Hyperthermophilic Enzymes.= Structure-Function	BIOT	0193
ontaining ACAT+ Synthesis and	Hypocholesterolemic Activity of Novel Phosphorus-C	MEDI	0110
ylene Diurea De+ Synthesis and	Hypocholesterolemic Activity of Phenylene and Pyrid	MEDI	0109
lated Benzopyran+ Synthesis and	Hypocholesterolemic Activity of Tocotrienols and Re	MEDI	0190
rol Esterase.=	Hypocholesterolemic Effects of Inhibitors of Choleste	BIOL	0018
ral Products as Antiviral Agents.	Hypothesis for the Mode of Action of Hypericin as a	AGFD	0188
-Nitroimidazole: A New Potential	Hypoxia Imaging Agent.= +ylamine Conjugate of 2	NUCL	0076
ovel Antimicrobial α-Pyrone from	Hyptis pectinata (Lamiaceae).= +Pectinolide: A N	MEDI	0060
ctivation of Gallium-Containing	HZSM-5 Zeolites Used for the Aromatization of Sma	PETR	0043
efin Aromatization Reactions over	HZSM-5 Zeolites.= +for Investigation of Light Ol	PETR	0061
Ethane and of Propane on	HZSM-5: Effect of Gallium.= Aromatization of	PETR	0045
on of C_2-C_6 Olefins on	HZSM-5.= Comparative Study of the Oligomerizati	PETR	0026
of Antifreeze Glycoprotein at the	Ice Surface.= +ilibrium Constants for Adsorption	BIOL	0041
on and Applications of ansa-Bis(	idenyl)metal Complexes Containing Chiral Bridging	INOR	0214
The	IDITIS Relational Database of Protein Structure.=	CINF	0032
Conjugate to H35 Hepatoma and	IEC-6 Cells in Vitro.= +ry of Methotrexate-Insulin	MEDI	0185
Production of Human	IgG Subclasses.= Effects of Gm-Allotypes on the	CARB	0018
iperazines with Potential as Class	II/III Antiarrhythmic Agents.= +ity of Novel Arylp	MEDI	0094
ies of Mixed-Valance Titanium(	III/IV) Phosphate Compounds with Nasicon-Related	INOR	0353
Organometallic Routes to	III/V Compound Semiconductors.= Some New	INOR	0146
y of a Mixed-Valence [Ni(II)-NI(	IIII)] Thiolate Dimer.= +acterization, and Reactivit	INOR	0060
ratory Improvement by the NSF-	ILI Program.= +ry—I. Support of Chemistry Labo	CHED	0146
cular Weight Determination by+	ILI-Catalyzed Laboratory Innovation: Polymer Mole	CHED	0210
A Sixteenth-Century	Illuminated Chemical Manuscript.= Splendor Solis:	HIST	0036
Abiotic Stresses of Heat or UV	Illumination May Help Reduce Decay in Seal-Packa	AGFD	0076
	ilvC Activator Protein: DNA Interactions.=	BIOL	0124
onent Polymer Systems by Digital	Image Analysis and Real-Time Pulsed NMR.=	PMSE	0099
mission Electron Microscopy, II:	Image Contrast Analysis of Iodinated Polystyrene Ch	POLY	0062
+Scanning Tunneling Microscope	Images Showing Anomalous Periodic Structures.=	PHYS	0212
Have Intelligence,	Imagination, and Diligence? Be Your Own Boss.=	SCHB	0016
idazole: A New Potential Hypoxia	Imaging Agent.= +ylamine Conjugate of 2-Nitroim	NUCL	0076
STM and AFM.=	Imaging DNA Molecules under Potential Control by	PHYS	0157
ns Using Multiple-Pulse+ NMR	Imaging of Anisotropic Solid-State Chemical Reactio	CATL	0022
Spectroscopy and NMR	Imaging of Coal.= Very High Frequency EPR	FUEL	0074
: STM and Related Techniques.	Imaging of Individual Atoms and Clusters with the F	PHYS	0123
rbates+ Low-Temperature STM	Imaging of Metal Surfaces and Manipulation of Adso	PHYS	0360
Azulene on Pt(111)+ Real-Space	Imaging of Molecular Organization: Naphthalene vs.	PHYS	0125
eling Microscopy.+ Atomic Scale	Imaging of Molecules on Surfaces by Scanning Tunn	PHYS	0159
Electron Transfer to+ Patterned	Imaging of Palladium and Platinum Metal Films by	INOR	0319
canning Tunneling Microscopy:	Imaging Subsurface and Nonequilibrium Electronic P	PHYS	0208
NMR	Imaging: A Chemical Microscope for Coal Analysis.=	FUEL	0075
anning Tunneling Micr+ Atomic	Imaging: STM and Related Techniques. Field Ion-Sc	PHYS	0182
ndividual Atoms and C+ Atomic	Imaging: STM and Related Techniques. Imaging of I	PHYS	0123
ure and Electronic Stat+ Atomic	Imaging: STM and Related Techniques. Local Struct	PHYS	0206
rature STM Imaging o+ Atomic	Imaging: STM and Related Techniques. Low-Tempe	PHYS	0360
d Examples of Scannin+ Atomic	Imaging: STM and Related Techniques. Principles an	PHYS	0152
Studies of Single Adsor+ Atomic	Imaging: STM and Related Techniques. Theoretical	PHYS	0096
es for F-19 Magnetic Resonance	Imaging: Synthesis and Characterization of Fluorinat	MEDI	0041
ging: Solid-State NMR and NMR	Imaging.= +Germination and the Effects of Seed A	AGFD	0104
Using Atomic-Level	Imaging.= Studies of Kinetic Processes at Surfaces	PHYS	0098
iovascular Effects of a Series of	Imidazo[1,2-a]-quinoxalinones, Imidazo[1,5-a]quinox	MEDI	0118
f Imidazo[1,2-a]-quinoxalinones,	Imidazo[1,5-a]quinoxalinones, and Their Aza Analog	MEDI	0118
[1-(Butylcarbamoyl)-1H-Benz-	imidazol-2-YL] Carbamate (Benomyl) in Hydrochlor	AGRO	0041
(Cu_2-O_2) Using	Imidazole Ligands.= New Copper-Dioxygen Species	INOR	0045
ntiotensin+ Synthetic Routes to	Imidazole-5-Acrylic Acids: Potent Small-Molecule A	ORGN	0119
and Orientation of Coordinated	Imidazoles in Low-Spin Ferric Porphyrin Complexes	INOR	0296
Polyamides Containing	Imidazoles.=	POLY	0034
etric Aldol Reactions of Chelated	Imidazolidone Fischer Carbene Complexes.= +ymm	INOR	0112

drenergic Agents. 1. Substituted	Imidazolines, Synthesis and Structure–Activity Relat	MEDI	0116
Muscarinic Activity of	Imidazopyridine Esters and Oxadiazoles.=	MEDI	0139
rgic and Serotonergic Activities of	Imidazoquinolinones and Related Compounds.= +e	MEDI	0126
rization of Ordered Heterocycle–	Imide Copolymers Containing ortho–Phenylene Ring	POLY	0042
Low–Hydrogen Amides and	Imides from Cyanoimidazoles.= High–Nitrogen	POLY	0363
Reactive Extrusion	Imidization of Acrylic Polymers.=	PMSE	0179
Immiscible Blends.= Kinetics of	Imidization of Poly(amic Acid) in Miscible and	PMSE	0147
itu Molecular Composite during	Imidization of Poly(amic dialkylamide)Precursors.=	PMSE	0169
Studies of New Nitrido and	Imido Complexes of Osmium(VI).= Reactivity	ORGN	0219
Monomeric Osmium(II)	Imido Complexes.= Preparation and Reactivity of	INOR	0364
ally Active Organotin–Platinum–	Imine and Organotin–Platinum–Amine Complexes.=	INOR	0234
exes of Unsymmetrical Hydrazone	Imine Glyoxal Derivatives.= +cs: Cobaltous Compl	POLY	0102
$CO)_2M=C=CH_2$ (M=Mn, Re) with	Imines, Carbodiimides, and Benzalazine.= +es Cp(	INOR	0072
Stereospecific Enammonium–	Iminium Rearrangements: A Mechanistic Investigatio	ORGN	0296
ylcyclopentadienyl) Tantalum η^2–	Iminoacyl Chemistry.= +Mid–Valent Mono(peralk	INOR	0108
of Phenytoin, Thiophenytoin, and	Iminophenytoin.= +mparative Pharmacodynamics	MEDI	0134
Poly(amic Acid) in Miscible and	Immiscible Blends.= Kinetics of Imidization of	PMSE	0147
croscopic Interfaces.+ Electrified	Immiscible Liquid Boundaries: Conventional and Mi	COLL	0004
Inversion in	Immiscible Liquid Systems.= Predicting the Phase	PMSE	0024
of Oxidation Processes.=	Immobilization and Investigation of Hemin: Catalyst	COLL	0102
Defined Patterns.=	Immobilization of Proteins in Microlithographically	BIOT	0246
sues, and Cells in the Context o+	Immobilization of Proteins, Cofactors, Mediators, Tis	BIOT	0039
ture of Cement Matrices for the	Immobilization of Solidified Toxic Hazardous Wastes	ENVR	0002
n Silica Surface.=	Immobilization of Tetraazamacrocyclic Compounds o	COLL	0105
ers as Supports for Enzyme	Immobilization.= IPN–Modified Porous Bead Polym	PMSE	0081
s.= Application of	Immobilized Biocatalysts in Lipid Biotransformation	BIOT	0040
d Opportunities. The Operation of	Immobilized Biocatalysts in Organic Solvents.=	BIOT	0035
roblems, and Opportunities. T+	Immobilized Biocatalysts: Preparation, Application, P	BIOT	0035
Bioreactor Design with	Immobilized Biocatalysts.=	BIOT	0038
Inhibition on the Activity of	Immobilized Cells.= Effect of Acid Product	BIOT	172A
of Fructose and Ca–Gluconate by	Immobilized Cells.= +ioprocess for the Production	BIOT	0152
and Solvent–Free Systems Using	Immobilized Enzymes.= +s in Nonaqueous Solvents	BIOT	0065
rs by Means of Ion Exchange and	Immobilized Metal Affinity FPLC.= +P–450 Patte	ANYL	0017
Applications of Biocatalysts	Immobilized on Membranes.= Food Industry	BIOT	0037
Cyclosporin A Productivity in an	Immobilized Perfusion Reactor System.= +Overall	BIOT	0155
Glucose Oxidase	Immobilized with Polyurethane for Glucose Sensing.=	BIOT	0147
Mortality Incidence and Acquired	Immune Deficiencies in Mice.= +termined Higher	AGFD	0145
rines.= Measurement of	Immune Disorder in Persons Exposed to Organochlo	CEI	0017
Venom Neurotoxin That Induce	Immune Unresponsiveness to Experimental Allergic	BIOL	0086
ty of Antibody–Like Molecules by	Immuno or Binding Assays.= +and Immunogenici	BIOT	0059
Don't Be Fooled by the+ Weak	Immunoaffinity Methods for Purification (WIMP)—	CARB	0054
PCB–RISc: An On–Site	Immunoassay for Detecting PCB in Soil.=	ENVR	0032
ary Drug Residues.=	Immunochemical Methods in the Analysis of Veterin	AGFD	0064
er Characterization.=	Immunochromatographic Analysis in Rapid Biopolym	ANYL	0131
cteria and High Mortality in CD1	Immunodeficient Antibiotic–Treated Mice.= +ve Ba	AGFD	0146
w to Monitor the Clearance and	Immunogenicity of Antibody–Like Molecules by Imm	BIOT	0059
–Protein and the Heavy Chains of	Immunoglobulin G Possess Similar Properties.= +B	BIOL	0068
terminants Deri+ Synthesis and	Immunological Properties of Glycopeptide T–Cell De	CARB	0014
Uses and Abuses of	Immunological Testing.=	CEI	0018
omparison of the Performance of	Immunosorbents Prepared by Site–Directed or Rand	BIOT	0060
Synthetic Studies Toward the	Immunosuppressant FK–506.=	ORGN	0116
Rare Earths in Rocks by Electron	Impact Mass Spectrometry.= +ve Determination of	ANYL	0049
Refining.=	Impact of Byproducts on the Economics of Biomass	BIOT	0014
of Margin Sediments+ Potential	Impact of Dissolved Organic Carbon (DOC) Flux out	GEOC	0017
in Fertilizer and Soil Chemistry.	Impact of Fertilizer Containment on Stormwater Ma	FERT	0013
n the Properties of Composites +	Impact of Molecular and Supermolecular Structure o	POLY	0300
rocess Safety Management: The	Impact of New Regulations on Fertilizer Manufactur	FERT	0016
ip and Copyright: Are Your Pol+	Impact of New Technologies on Information Ownersh	CINF	0043
of the Merck Laboratories: The	Impact of Steroid Research.= Some Contributions	HIST	0015
ng Automotive Industry and Its	Impact on Intermaterial Competition.= The Changi	CMEC	0015
Cluster	Impact on Solid Surfaces.=	NUCL	0055
cations of Materials: Cluster Ion	Impact Phenomena Used for Surface Modification an	NUCL	0052
Low–Cost Multiple–	Impact Physical Chemistry Experiment.=	PHYS	0336
Low–Cost, Multiple–	Impact Physical Chemistry Experiment.=	CHED	0162
arameters and Adhesion on the	Impact Properties and Morphology of Rubber–Tough	PMSE	0177
trometria de Masas con Fuente de	Impacto Electronico.= +e Tierras Raras por Espec	ANYL	0051
t of Environmental Resources and	Impacts.= +tementation of Risk–Based Managemen	ENVR	0076
el Triazoline Anticonvulsant, for	Impairing Excitatory Amino Acid Neurotransmission	MEDI	0144
oride Interface by Electrochemical	Impedance Spectroscopy.= +zed Steel/Sodium Chl	PHYS	0270
ltipolar Correlations and Surface	Imperfections on the Efficiency of Diffusion–Control	PHYS	0063
Metallic Ion	Implantation On–CdTe.= Surface Modification by	PHYS	0218
Spectroscopic Studies of Ion–	Implanted Polycarbonates.=	POLY	0081
dure for Similarity Searching in+	Implementation and Use of an Atom–Mapping Proce	COMP	0043
c Methods for Polymer Analysis.	Implementation of a Systematic Approach for Quant	PMSE	0065
xt Steps To Allow for Immediate	Implementation of Risk–Based Management of Envi	ENVR	0076
Single–Electron Logic: Molecular	Implementation?.=	BIOT	0004
Considerations in Developing and	Implementing a Laboratory Safety Program.= +ful	CHAS	0004
f Biodiversit+ Key Elements for	Implementing Sustainable Agriculture and the Role o	AGRO	0082
during Cooking–Extrusion Cere+	Implication of Phenolic Acids as Texturizing Agents	AGFD	0150
Colloidal Suspension in Externally	Imposed Magnetic and Hydrodynamic Fields.= +c	COLL	0144

Phosphorylation of Soy Protein to	Improve Functional Properties in Foods.= +matic	AGFD	0140
That Knowledge Can Be Used To	Improve Laboratory Safety.= +Revealed and How	CHAS	0002
l and Biophysical Approaches to	Improve Mucosal Peptide Penetration.= Biochemica	MEDI	0194
s: Does Conformational Searching	Improve the Yield of Activities?.= +in 3D Database	CINF	0026
otoelectroquimicas, I: CdS Puro e	Impurificado.= +Policristalinos dentro de Celdas F	PHYS	0221
ing HPLC Analy+ Ruggedness of	Impurity Analysis in Pharmaceuticals When Employ	ANYL	0114
g in Acidic+ Development of an	Impurity Method for Degradation Products Occurrin	ANYL	0085
lear Selenide Clusters of Au+ and	In^{3+}, Trapping Na^+ Ions in the Center.= +Polynuc	INOR	0203
nthin, a Plant-Derived Ribosome-	Inactivating Protein.= +viral Studies with Trichosa	AGFD	0189
be of the Ac+ Mechanism-Based	Inactivator 1-Diazo-4-undecyl-2-one (Duo) as a Pro	BIOL	0067
New Mechanism-Based	Inactivators of S-Adenosylhomocysteinase.=	BIOL	0081
rn (Zea mays L.) Populations and	Inbreds for Silk Flavonolglycoside Contents.= +Co	AGRO	0034
Costa Rica: The Case of Arvi,	Inc., Costa Rica.= Small Chemical Businesses in	SCHB	0026
m-Size Enterprises: The Codetec,	Inc., Experience in Brazil.= +for Small and Mediu	SCHB	0023
Communication Resources,	Inc.= Toxicological Services Provided by Hazard	CHAS	0024
Development and Decline of	Incandescent Gas Lighting.=	HIST	0038
and	Incident Prevention.= Four Keys to Chemical Safety	CHAS	0017
igating Chemistry Important to+	Incin: A Flexible Chemkin Driver Program for Invest	I&EC	0024
Containing Municipal Solid Waste	Incinator Ash.= +gth Portland Cement Concrete	FUEL	0136
estigating Chemistry Important to	Incineration.= +e Chemkin Driver Program for Inv	I&EC	0024
Waste	Incineration.= Metals Behavior during Medical	FUEL	0131
ns Control Technologies for Waste	Incinerators.= +e Role of Metals. Metals Emissio	FUEL	0129
Methods of Washing sCD4-183	Inclusion Bodies to Remove E. coli Proteins and End	BIOT	0171
in.= Mutational Suppression of	Inclusion Body Formation in the P22 Tailspike Prote	BIOT	0113
Folding Properties of	Inclusion Body Mutants of $1L-1\beta$.=	BIOT	0114
n Host: Resolution of Alcohols by	Inclusion Compound Formation.= +e as an Inclusio	ORGN	0041
roger's Base Methosulfate as an	Inclusion Host: Resolution of Alcohols by Inclusion C	ORGN	0041
Pressure Thermal Analysis of an	Incompatible Polymer Blend of Polypropylene-Polya	POLY	0071
as de la Carrera d+ Factores Que	Incrementan la Reprobacion en las Materias Biologic	CHED	0092
n of 5-Substituted Indanones and	Indandiones.= +f Atipamezole Analogs: Fluorinatio	ORGN	0078
ogs: Fluorination of 5-Substituted	Indanones and Indandiones.= +f Atipamezole Anal	ORGN	0078
of [1,3,4]Oxa(Thia)Diazolo[3,2-b]	Indazoles and Their Antibacterial Activity.= +hesis	MEDI	0056
actions of Unusually Reactive (η^5-	Indenyl) Ruthenium Alkyl Complexes.= +ation Re	INOR	0239
with Emphasis on High Refractive	Index and Abrasion-Resistance Coatings.= +erials	POLY	0264
Applications. Low-Loss Graded-	Index and Single-Mode Polymer Optical Fibers.=	POLY	0174
ochemically Delineated Refractive	Index Profiles in Polymeric Slab Waveguides.= +ot	POLY	0180
Ring	Index Searching of the Beilstein Database.=	COMP	0018
Pesticide Usage and Research in	India.= Present Status and Future Trends of	AGRO	059A
nd Salt Marshes along the Florida	Indian River Lagoon.= +ide Residue in Barrier Isla	ENVR	0019
	Indiana Silicon Sphere (ISiS) 4π Detector.=	NUCL	0036
Project SEED Program in	Indianapolis.= Coordinating and Evaluating a	CHED	0141
inado Utilizando como Electrodo	Indicador una Pelicula de Oxido sobre Acero Inoxida	CHED	0163
Zeolite Based on a Solvatochromic	Indicator.= +nation of the Solvent-Like Nature of	COLL	0076
a+ Development of NMR Active	Indicators for Measurement of Calcium Ion Concentr	ORGN	0149
Time-Temperature	Indicators for Shelf-Life Monitoring.=	AGFD	0080
Analisis Longitudinal de	Indices de Reprobacion en Materias Biologicas.=	CHED	0091
for Topological	Indices.= Using Real Numbers as Vertex Invariants	COMP	0026
Microbial Production of	Indigo from Precursor Molecules.=	BIOT	0032
Thermal and Laser Etching of	Indium Phosphide by Molecular Chlorine.=	COLL	0214
rimethylgallium on Silicon and	Indium Phosphide.= Decomposition of Mono- and T	INOR	0123
es and the Total Synthesis of (-)-	Indolactam (V).= +uted Indole Chromium Complex	ORGN	0010
banions to 4-Amino-Substituted	Indole Chromium Complexes and the Total Synthesi	ORGN	0010
sis of Small, Biologically Active	Indole Derivatives.= Recent Advances in the Synthe	ORGN	0268
han Enzymes.= Effect of	Indole on Reactions of Trifluoroalaneine with Tryptop	BIOL	0060
and Selective Inhibitors of Pro+	Indolocarbazoles and Bisindolylmaleimides as Potent	MEDI	0074
	Indoor Radon Levels in New York and New Jersey.=	ENVR	0023
Cobra Venom Neurotoxin That	Induce Immune Unresponsiveness to Experimental A	BIOL	0086
reesei.= Novel	Inducers for Cellulase Production by Trichoderma	BIOT	0184
Phenolic Antioxidants as	Inducers of Anticarcinogenic Enzymes.=	AGFD	0089
Induction and	Inducers of Microbial Cellulases and Xylanases.=	BIOT	0185
moarginine.= Inhibition of	Inducible Nitric Oxide Synthase by N^g-Amino-L-Ho	MEDI	0106
Probing and	Inducing Silicon Surface Chemistry with the STM.=	PHYS	0184
ylanases.=	Induction and Inducers of Microbial Cellulases and X	BIOT	0185
Sesquiterpene	Induction from Plant Root Cultures.=	BIOT	0232
+ngent Response as a Function of	Induction Level in Recombinant Escherichia coli.=	BIOT	0034
gens: Relevance to Human Cancer	Induction. +tally Significant Carcinogens and Muta	CHAS	0007
ce Metals Analysis of Fly Ash by	Inductively Coupled Plasma Atomic Emission Spectr	FUEL	0138
M para el Diseno de Equipo de la	Industria Quimica de Proceso.= +entes de CAD/CA	COMP	0053
gical Support and Medium-Scale	Industries: The Case of the Center for Technological	SCHB	0027
f Science Education.=	Industry/Academic Interactions for the Promotion o	CHED	0008
on Protein Quality of Soy-Based	Infant Formulas Fed to Rats.= +d Supplementation	AGFD	0209
reatment of Plasmodium Berghei-	Infected Mice with Transdermal Artelinic Acid.=	MEDI	0063
Aminosugar Attenuation of HIV	Infection.=	AGFD	0193
Microbial Geritol: From Iron to	Infection.=	CHED	0061
Synthesis and Novel Anti-	Inflammatory Activity of Oxazinone, BF-389.=	MEDI	0175
ds and Their Derivatives as Anti-	Inflammatory Agents.= +thoxycarbonyl) Amino Aci	MEDI	0171
e to Novel Anti-Allergy and Anti-	Inflammatory Drugs.= +Inhibitor—General Rout	MEDI	0179
Platelet Effects of the Anti-	Inflammatory DuP 697.= Characterization of the	MEDI	0162
de Pastas de Alto Rendimiento:	Influencia de la Sulfonacion sobre las Propiedades P	PMSE	0077
o Pentasil en la Aromatizacion +	Influencia del Contenido de Galio en Zeolitas del Tip	COLL	0125

s Resultados de la Evaluacion d+	Influencia del Nivel Academico de los Alumnos en lo	CHED	0093
ido–Gas NO–Cataliticas.=	Influencia del Transporte de Calor en Reacciones Sol	I&EC	0068
ositional Analysis of Haemophilus	Influenzae Type B Conjugate Vaccine.= +ide Comp	ANYL	0021
rate Antigens. Haemophilus	Influenzae Type B Conjugate Vaccines.= Carbohyd	CARB	0053
atio+ Studies in Deciphering the	Information Content of Chemical Formulas: Enumer	COMP	0027
ortunity for Chemists.=	Information Distribution: A Multifaceted Career Opp	YCC	0008
Crystallographic	Information from Ion Channeling.=	NUCL	0066
Actionable	Information from Japan.=	CINF	0012
ains as Mechanisms for Decoding	Information in Complex Sugar Structures.= +Dom	BIOL	0144
lopments for Chemical Structural	Information Interchange.= +Standardization Deve	CINF	0038
Impact of New Technologies on	Information Ownership and Copyright: Are Your Pol	CINF	0043
Library Dilemma.=	Information Ownership vs. Unrestricted Access: The	CINF	0045
hysicochemical Applications—I.	Information Potential of Chromatographic Data for P	ANYL	0043
embranes Modeled on the Visual	Information Processing of the Retina.= +lecular M	BIOT	0124
Variants for Optical	Information Processing.= Bacteriorhodopsin	BIOT	0078
canning T+ Very High Density	Information Storage in Molecular Complexes Using S	BIOT	0023
urface D+ Using a Geographical	Information System To Assess the Susceptibility of S	AGRO	0053
n and Risks in the International	Information Transfer Process.= Copyright Protectio	CINF	0014
Legal Primer for Librarians and	Information Users.=	CINF	0047
pke: New Dimensions in Chemical	Information. Award Address. INVENTON: A+ Wi	CINF	0016
keting of+ International Flow of	Information: North America and Europe. Global Mar	CINF	0011
pe. Global Marketing of Chemical	Information: The STN Story.= +merica and Euro	CINF	0011
s in Relation to Decoding Genetic	Information.= +ry. Recognition of RNA Minihelice	BIOL	0026
Structural	Information.= Predicting Mass Spectra from	COMP	0005
oplastic Molding Compounds for	Infra–Red Reflow Tolerant Electronic Components.=	CMEC	0014
Two–Dimensional	Infrared (2D IR) Spectroscopy.=	PHYS	0117
f Complementary Topological S+	Infrared and ^{13}C NMR Spectrum Simulation: Role o	COMP	0006
urier Transform Mid– and Far–	Infrared IR Specular Reflectance Study of Oxidized a	POLY	0253
Carbon Clusters.=	Infrared Laser Spectroscopy of Supersonically Cooled	PHYS	0007
	Infrared Laser Spectroscopy of the C_2H Radical.=	PHYS	0332
zation: Optical Methods. Dynamic	Infrared Linear Dichroism of Polymers.= +racteri	POLY	0303
Polymers with Fourier Transform	Infrared Reflectance Spectroscopy.= +Profiling of	PMSE	0153
alyst for Propane Aromatization—	Infrared Specroscopic Investigations.= +β as a Cat	PETR	0047
Containing Reactive Sp+ Matrix	Infrared Spectra and ab Initio Frequencies of Sulfur–	PHYS	0031
ted in Solid Argon.=	Infrared Spectra of Metal–Ethylene Complexes Isola	PHYS	0074
Bonds Studied by Time–Resolved	Infrared Spectral Techniques.= +on of Metal Alkyl	INOR	0169
F+ Pyrolysis–Fourier Transform	Infrared Spectrometry as a Technique for Analyzing	FUEL	0061
gett Films by Fourier Transform	Infrared Spectroscopy and Second–Harmonic Genera	POLY	0224
omatography–Fourier Transform	Infrared Spectroscopy for Characterization of Ethyle	PMSE	0048
Dynamics Using Transient	Infrared Spectroscopy.= Ultrafast Vibrational	PHYS	0194
c Solvents Measured by Transient	Infrared Spectroscopy.= +ion Time of Ions in Proti	PHYS	0083
d.=	Infrared Spectrum of the Hydrogen Halides Revisite	CHED	0197
ucture and Spectroscopy of a New	Infrared Transparent Chalcogenide: $CaYbInSe_4$.=	INOR	0030
d Fruit Products by Pectinase	Infusion.= Preparation of Novel Minimally Processe	AGFD	0059
to Scientific Risk Assessments of	Inhalation and Dermal Exposures.= +Are the Keys	ENVR	0042
) Pr+ Caracterizacion del Factor	Inhibidor de la Locomocion de los Monocitos (FILM	BIOL	0102
clization Rates on Pt/L–Zeolite:	Inhibited Deactivation of Pt Sites within One–Dimen	CATL	0015
orrosion of Mild Steel Exposed to	Inhibited Urea–Ammonium Nitrate Solution.= +C	FERT	0001
thylepipodophyl+ Topoisomerase	Inhibition and Anticancer Activity of Novel 4'–Deme	MEDI	0011
al and Kinetic Characterization of	Inhibition and Substrate Specificity.= +$_2$: Structur	BIOL	0020
Kinetic Studies of Protease	Inhibition by Peptidyl Fluoroketones.= NMR and	BIOL	0005
	Inhibition of Aldosterone Biosynthesis.=	MEDI	0009
one Biosynthesis. Mechanism and	Inhibition of Cytochrome–P_{450} Aromatase.= +orm	MEDI	0006
roides B–512FM Dextransucrase:	Inhibition of Dextran Synthesis by α–Methyl–D–Glu	CARB	0028
ives of Haloperidol.=	Inhibition of HIV Protease with Nonpeptide Derivat	MEDI	0054
yl Sulfides and Thiodiglycol.=	Inhibition of Horseradish Peroxidase by 2–Chloroeth	BIOT	0249
β–Lactams.= Mechanism of	Inhibition of Human PMN Elastase by Monocyclic	BIOL	0050
Amino–L–Homoarginine.=	Inhibition of Inducible Nitric Oxide Synthase by N^g–	MEDI	0106
ates: Studies of 26– and 29–Fu+	Inhibition of Oxidosqualene Cyclase by Pseudosubstr	ORGN	0021
ic Compounds.=	Inhibition of Polyphenol Oxidase Activity by Phenol	AGFD	0152
elated Long–Chain Bases.=	Inhibition of Protein Kinase C by Sphingosine and R	MEDI	0075
=	Inhibition of the α, β, and γ Isozymes of Tyrosinase.	BIOL	0058
iazole Derivativ+ Stereoselective	Inhibition of the P_{450}–Dependent Aromatase by a Tr	MEDI	0007
ed Enzyme–Activated Substrate	Inhibition of Trypsin by N–Nitrosoamide Derivatives	ORGN	0022
–Induced H_2O_2 Production in+	Inhibition of 12–O–Tetradecanoylphorbol–13–Acetate	AGFD	0069
Effect of Acid Product	Inhibition on the Activity of Immobilized Cells.=	BIOT	172A
Overcome Steric	Inhibition.= Synthesis of Rigid Dendrimers That	POLY	0321
Characterization of Trypsin	Inhibitor Activity in Amaranthus Species.=	AGFD	0133
X–ray Studies of	Inhibitor Binding to Phospholipase A_2.=	MEDI	0004
as Stable 3–Phosphate Mimics +	Inhibitor Design I: Investigation of Various Moieties	ORGN	0028
matic Mimics o+ EPSP Synthase	Inhibitor Design II: Synthesis and Evaluation of Aro	ORGN	0029
for Successful Mechanism–Based	Inhibitor Design.= +ortance of Dynamic Catalysis	BIOL	0054
ersion of a Cyclooxygenase (CO)	Inhibitor into a 5–Lipoxygenase (LO) Inhibitor—Gen	MEDI	0179
of the Selective Protein Kinase C	Inhibitor NPC 15437.= +armacological Evaluation	MEDI	0026
Fumaric Acid Monoazide: A New	Inhibitor of Alkaline Phosphatase.=	BIOL	0057
Synthesis of a Potential	Inhibitor of Chorismate Synthase.= Toward the	ORGN	0030
d Selective Reverse Transcriptase	Inhibitor of HIV–1.= +of BI–RG–587: A Potent an	ORGN	0027
Phenolic	Inhibitor of Nitrosation.= Tocopherol as a Natural	AGFD	0091
630 as a Potent and Orally Active	Inhibitor of 5–Lipoxygenase.= +elopment of L–670,	MEDI	0180
with HIV RT	Inhibitor U–9843.= Mode of Action in Vitro Studies	MEDI	0055

and Selective 5-Lipoxygenase	Inhibitor.= A-69412, a Long-Lived, Orally Active,	MEDI 0176
r, Orally Active Cysteine-Derived	Inhibitor.= +nhibitors: Synthesis of a Subnanomola	MEDI 0122
bitor into a 5-Lipoxygenase (LO)	Inhibitor—General Route to Novel Anti-Allergy and	MEDI 0179
Argatroban, a Specific Thrombin	Inhibitor, by a Tandem Column HPLC Procedure.=	ANYL 0086
main.= New Protein Kinase C	Inhibitors Designed to Interact with the Catalytic Do	MEDI 0076
ntaining Enkephalin Analogues as	Inhibitors for Enkephalin-Aminopeptidase.= +-Co	MEDI 0151
	Inhibitors for Proline-Specific Proteases.=	BIOL 0030
es in a New Class of Nonsteroidal	Inhibitors of Aromatase.= +tructure-Activity Studi	MEDI 0036
Hypocholesterolemic Effects of	Inhibitors of Cholesterol Esterase.=	BIOL 0018
edox Reactive Choline Analogue	Inhibitors of Choline Acetyltransferase and High-Aff	MEDI 0142
roxamic Acid Fenamates as Dual	Inhibitors of Cyclooxygenase and 5-Lipoxygenase.=	MEDI 0177
ana.=	Inhibitors of DNA Polymerases of Leishmania mexic	BIOL 0118
Quercetin (QU) and Rutin (RU) as	Inhibitors of Experimental Colonic Neoplasia.=	AGFD 0114
Substructur+ Discovery of Novel	Inhibitors of HIV-1 Protease by Three-Dimensional	COMP 0038
Preparation of C_2 Symmetric	Inhibitors of HIV-1 Protease.= Design and	MEDI 0053
sis and Structure-Acti+ Lazarus	Inhibitors of Human Leukocyte Elastase: The Synthe	MEDI 0172
Lazarus	Inhibitors of Human Leukocyte Elastase.=	MEDI 0087
Monocyclic β-Lactam	Inhibitors of Human Leukocyte Elastase.=	MEDI 0173
nd Specific Monocyclic β-Lactam	Inhibitors of Human PMN Elastase.= +e, Potent, a	BIOL 0049
Enzyme-Generated Irreversible	Inhibitors of Hydroxysteroid Dehydrogenases.=	MEDI 0010
Multisubstrate Analogue	Inhibitors of Methyltransferases.= Potential Specific	MEDI 0027
Plant Phenolics as	Inhibitors of Mutagenesis and Carcinogens.=	AGFD 0118
Thiophosphate	Inhibitors of Myo-Inositol Monophosphatase.=	MEDI 0119
ted 3-Aminooxypropanamines as	Inhibitors of Ornithine Decarboxylase: Synthesis and	MEDI 0184
Synthesis of Ether Lipid	Inhibitors of Protein Kinase C.= Design and	MEDI 0073
aleimides as Potent and Selective	Inhibitors of Protein Kinase C.= +and Bisindolylm	MEDI 0074
Structure-Based Drug Design:	Inhibitors of Purine Nucleoside Phosphorylase (PNP	MEDI 0005
Enzyme-Activated Irreversible	Inhibitors of S-Adenosylmethionine Decarboxylase.=	BIOL 0052
Synthesis, and in Vitro Activity of	Inhibitors of Spermidine N^8-Acetyltransferase.=	MEDI 0020
olyamine Fumaramate Esters as	Inhibitors of Spermidine/Spermine-N^1-Acetyltransfe	MEDI 0021
ism and Inhibition of Cytochro+	Inhibitors of Steroid Hormone Biosynthesis. Mechan	MEDI 0006
mpetitive versus Testosterone.=	Inhibitors of Steroid 5α-Reductase Which Are Unco	MEDI 0008
thesis of Naphthalene Containing	Inhibitors of Thymidylate Synthase.= +ign and Syn	MEDI 0186
nd Analysis of Some Aromatase	Inhibitors on Cellulose-Based Chiral Stationary Phas	ANYL 0133
of Tightly Bound Boronic Acid	Inhibitors on Serine Proteases: Evidence for Multiple	BIOL 0082
-Activity Relationships of Renin	Inhibitors with Cyclohexyl Moiety Directly Attached	MEDI 0121
iral and Anti+ Protein Kinase C	Inhibitors. Protein Kinase C: A New Target for Antiv	MEDI 0072
l Substituted β-Ketoamide ACAT	Inhibitors: Chemistry and Biological Activity.= +ve	MEDI 0108
Clinica+ Zinc-Metallopeptidase	Inhibitors: Design, Biochemical, Pharmacological, and	MEDI 0002
Small Peptides.= Renin	Inhibitors: Factors Influencing the Absorption of	MEDI 0197
urine Nucleoside Phosphorylase	Inhibitors: Synthesis and Biological Activity of 9-Dea	MEDI 0028
e Cysteine+ Macrocyclic Renin	Inhibitors: Synthesis of a Subnanomolar, Orally Activ	MEDI 0122
ips of Potent Acetylcholinesterase	Inhibitors: 8-Carba-physostigmine Analogues.= +h	MEDI 0143
oxy Acetic Acid Aldose Reductase	Inhibitors.= +dress. Hydantoin Bioisosteres: Hydr	MEDI 0082
Well-Absorbed Renin	Inhibitors.= A New Class of Efficacious,	MEDI 0120
yl-Based Protein Tyrosine Kinase	Inhibitors.= +Ring-Constrained Analogues of Styr	MEDI 0023
in the Design of Proteinase	Inhibitors.= Chemical Principles as a Driving Force	ORGN 0269
ationally Flexible Tyrosine Kinase	Inhibitors.= +Molecular Field Analysis of Conform	MEDI 0024
Aromatase	Inhibitors.= Development of New Nonsteroidal	MEDI 0003
Macrocyclic Human Renin	Inhibitors.= Highly Potent, Orally Active Diester	MEDI 0123
Enzyme	Inhibitors.= New Method for the de Novo Design of	COMP 0037
Thiols: Applications to Papain	Inhibitors.= Reactivities of Epoxides Toward	MEDI 0071
lity in a Series of Metalloprotease	Inhibitors.= +ps between Structure and Bioavailabi	MEDI 0195
Phosphodiesterase	Inhibitors.= Spirolactams (Carbamates) as	MEDI 0189
lass of Selective Protein Kinase C	Inhibitors.= +minomethyl)-Piperidines: A Novel C	MEDI 0025
ylglutaryl-Coenzyme A Reductase	Inhibitors.= +iophene-Derived 3-Hydroxy-3-Meth	MEDI 0101
vel Phosphorus-Containing ACAT	Inhibitors.= +Hypocholesterolemic Activity of No	MEDI 0110
ylene Diurea Derivatives as ACAT	Inhibitors.= +mic Activity of Phenylene and Pyrid	MEDI 0109
g Model of Azole-Type Aromatase	Inhibitors.= +ological Evaluation, and New Bindin	MEDI 0035
holesterol Acyltransferase (ACAT)	Inhibitors.= +eas: A Potent Series of Acyl CoA: C	MEDI 0107
Phaseolus vulgaris α-Amylase	Inhibitors, Arcelins, and Lectins Evolved from a Com	AGFD 0041
Synthesis and PDE	Inhibitory Activity of Griseolic Acid Derivatives.=	MEDI 0117
Enhanced Monoamine Oxidase	Inhibitory Activity of Tranylcypromine.=	MEDI 0068
MP and cGMP Phosphodiesterase	Inhibitory Activity. +pounds Possessing Potent cA	MEDI 0118
ompounds in Foods and Health.	Inhibitory Effect of Orally Administered Green Tea o	AGFD 0065
nd Promotion in Mouse Epide+	Inhibitory Effects of Curcumin on Tumor Initiation a	AGFD 0088
Enzyme: A Monoclone That Is 98	Inhibitory.= +plex and to the E1 Decarboxylating	BIOL 0031
Glasses.= Molecular Theory of	Inhomogeneous Spectral Broadening in Liquids and	PHYS 0036
e+ Operation PROGRESS: An	Initiative of the Division of Chemical Education of th	CHED 0110
te: Oswego Harbor and+ Citizen	Initiatives at Two Areas of Concern in New York Sta	ENVR 0084
Prevention through Voluntary	Initiatives.= EPA's 33/50 Project: Pollution	CEI 0024
tiv+ New Hydrolyzable Dianionic	Initiator from Divinyl Tetraoxaspiroundecane Deriva	POLY 0349
f Hydroxamic Acids and Th+ Ab	Initio and Semiempirical Molecular Orbital Studies o	PHYS 0326
Ab	Initio Calculation on Hydrogen-Bonded Complexes.=	PHYS 0103
Vibrational Spectra.= Ab	Initio Calculations and the Interpretation of	PHYS 0033
o We Really Need to Compare ab	Initio Calculations with Experiment?.= +Forces: D	PHYS 0104
Optimized Basis Sets for ab	Initio Calculations.=	PHYS 0297
+ and Zn^{2+} + H_2 Moeties.= Ab	Initio CI Study of the Interaction between the Zn, Zn	PHYS 0282
s of Polyvinylchloride Using an ab	Initio Derived Classical Force Field.= +ral Analysi	POLY 0351
nd HSiP, and Their Isomer+ Ab	Initio Energies and Shapes of $HSiS^+$, $HSiO^+$, HCP, a	PHYS 0315

Matrix Infrared Spectra and ab	Initio Frequencies of Sulfur–Containing Reactive Spe	PHYS	0031
Spectra of LiBe and LiMg.= Ab	Initio Full SOCVCI Calculations of the Elctronic	PHYS	0338
Small Molecules by ab	Initio Methods.= Study of the Spectroscopy of	PHYS	0005
adical Anions as Protoypes+ Ab	Initio MO Studies on Si_2H_6, SiH_3GeH_3, and Ge_2H_6 R	PHYS	0313
alues, Including Predictions+ Ab	Initio Proton Affinites Compared to Experimental V	PHYS	0298
s for the Atomistic Simulati+ Ab	Initio Quantum Chemical Potential Energy Function	POLY	0350
Ab	Initio Quantum Chemistry and Dynamics.=	PHYS	0165
ison of Expe+ Comparison of ab	Initio Quantum Chemistry with Experiment. Compar	PHYS	0339
xcited–State+ Comparison of ab	Initio Quantum Chemistry with Experiment. Diene E	PHYS	0075
cs of the De+ Comparison of ab	Initio Quantum Chemistry with Experiment. Dynami	PHYS	0188
6.= Comparison of ab	Initio Quantum Chemistry with Experiment. Gold	PHYS	0004
vertone Spec+ Comparison of ab	Initio Quantum Chemistry with Experiment. High–O	PHYS	0029
lar Rearrang+ Comparison of ab	Initio Quantum Chemistry with Experiment. Molecu	PHYS	0013
e–Ion Photoe+ Comparison of ab	Initio Quantum Chemistry with Experiment. Negativ	PHYS	0055
m Theory of+ Comparison of ab	Initio Quantum Chemistry with Experiment. Quantu	PHYS	0161
scopy of Non+ Comparison of ab	Initio Quantum Chemistry with Experiment. Spectro	PHYS	0102
Conical Intersection and Its+ Ab	Initio Studies of the $^1A_2$ and $^1B_1$ States of Ozone: A	PHYS	0191
Surface.= Ab	Initio Study of K Adsorption on the Ag(100)	PHYS	0337
Spectrum, and Binding Ene+ Ab	Initio Study of the Molecular Structure, Vibrational	PHYS	0276
SiX_4 (X = F, Cl).= An ab	Initio Study of the Reaction Pathways for Si^+ +	PHYS	0248
Isomerization Process.= Ab	Initio Study of the Vinylidene–Acetylene	PHYS	0287
Flow–	Injection Analysis.= Square Wave Detection for	ANYL	0105
Colloids.= Sensitized Charge	Injection in Large–Bandgap Semiconductor	COLL	0149
neous Fluids. Vectorial Electron	Injection into Colloidal Semiconductor Membranes.=	COLL	0147
sues from Carcasses with Positive	Injection Sites.= +Anabolic Steroids in Muscle Tis	AGFD	0130
g Mass Discrimination in a Split	Injection System for Capillary Gas Chromatography.	TECH	0018
of Lipid Peroxidation and Chilling	Injury in Cucumber Fruit and Cell Cultures.= +on	AGFD	0033
del Zinc en Ratones	Inmunodeficientes.= Actividad Inmunomoduladora	AGFD	0144
Inmunodeficientes.= Actividad	Inmunomoduladora del Zinc en Ratones	AGFD	0144
se of the Center for Technological	Innovation in Mexico.= +Scale Industries: The Ca	SCHB	0027
n by+ ILI–Catalyzed Laboratory	Innovation: Polymer Molecular Weight Determinatio	CHED	0210
pjohn in Perspective: A Profile of	Innovation.= +oids, the Steroid Community, and U	HIST	0021
try Award). Some Adventures and	Innovations in Industrial Catalysis.= +rial Chemis	I&EC	0043
pport of Chemi+ NSF–Catalyzed	Innovations in the Undergraduate Laboratory—I. Su	CHED	0146
ndergraduate S+ NSF–Catalyzed	Innovations in the Undergraduate Laboratory—II. U	CHED	0179
ent. NSF Funding of Laboratory	Innovations in Undergraduate Physical Chemistry.=	CHED	0196
d Rinsate Management for Ret+	Innovative and Practical Solutions for Wastewater an	FERT	0014
Countertrade—An	Innovative Approach to Marketing.=	CMEC	0008
ssemblies with Enforced Polarity:	Inorganic and Heterocyclic Linkages.= +ophoric A	POLY	0215
Hudson Ri+ Fluxes of Dissolved	Inorganic Carbon and Reactor–Derived ^{14}C from the	GEOC	0014
up 8 and Group 1+ Multinuclear	Inorganic Charge–Transfer Complexes Based on Gro	INOR	0162
Particles: Luminescence Studies of	Inorganic Cluster Molecules.= +or Semiconductor	INOR	0317
into Nonshrin+ Inverse Organic–	Inorganic Composite Materials: Free Radical Routes	POLY	0259
od: Some Insights int+ Organic–	Inorganic Composites Prepared via the Sol–Gel Meth	POLY	0260
ay Synthesis. Pillaring of Layered	Inorganic Compounds.= +Zeolites and Pillared Cl	PETR	0106
haracterization of Polyacrylate–	Inorganic Hybrid Materials via the Sol–Gel Approach	POLY	0232
ular Composite+ Preceramic and	Inorganic Hybrid Materials. Polyphosphazenes Molec	POLY	0181
to Ceramics.= Preceramic and	Inorganic Hybrid Materials. Polysilazanes: A Route	POLY	0309
ticles Precipita+ Preceramic and	Inorganic Hybrid Materials. Reinforcing Zirconia Par	POLY	0283
licon Olefin P+ Preceramic and	Inorganic Hybrid Materials. Silicon Acetylene and Si	POLY	0357
NMR.= Preceramic and	Inorganic Hybrid Materials. Sol–Gel Kinetics by	POLY	0256
ials via Precer+ Preceramic and	Inorganic Hybrid Materials. Synthesis of Nanomater	POLY	0335
Intermediate	Inorganic Laboratory with a Difference.=	CHED	0151
Tutorial on Molecular Routes to	Inorganic Materials. Application of Thermal Gravi	INOR	0006
Tutorial on Molecular Routes to	Inorganic Materials. Applications of Electron Spect	INOR	0013
Thin Films+ Molecular Routes to	Inorganic Materials. Deposition of Transition–Metal	INOR	0141
l Reactions+ Molecular Routes to	Inorganic Materials. Gas Phase and Surface Chemica	INOR	0121
emistry.= Molecular Routes to	Inorganic Materials. Magnets Based on Molecular Ch	INOR	0100
Depositio+ Molecular Routes to	Inorganic Materials. Organometallic Chemical Vapor	INOR	0171
	Inorganic Particles in Magnetotactic Bacteria.=	INOR	0118
Polymer Matrices.=	Inorganic Phase Behaviors Mediated by the Organic	POLY	0364
gnetic+ Quantitative Analysis of	Inorganic Phosphates by Phosphorus–31 Nuclear Ma	INOR	0291
lphosphazenes): A New Class of	Inorganic Polymers with Skeletal Phosphorus, Nitrog	POLY	0186
Chain Length of	Inorganic Polyphosphates.= Factors Influencing the	INOR	0376
$C_8H_{16})_3(HNMe_2)_2$ to $W_2(O_2C_8H+$	Inorganic Propellanes: The Interconversion of $W_2(O_2$	INOR	0128
A Molecular Organic–	Inorganic Semiinterpenetrating Network.=	POLY	0286
Transport in a One–Dimensional	Inorganic Solid: Exciton Trapping and Annihilation	PHYS	0131
roscopy in the Characterization of	Inorganic Surfaces and Thin Films.= +ctron Spect	INOR	0013
New	Inorganic–Centered Liquid Crystal Materials.=	INOR	0266
Chemistry: A New Approach to	Inorganic–Organic Hybrid Materials?.= Cement	POLY	0261
roscale Approach in the General,	Inorganic, and Organic Instructional Laboratories.=	CHED	0150
esis of P=1 Tethered Analogs of	Inositol (1,3,4,5)Tetrakisphosphate from Glucose via	AGRO	0045
Inhibitors of Myo–	Inositol Monophosphatase.= Thiophosphate	MEDI	0119
Easy Synthesis of cis–	Inositol.=	CARB	0029
Peliculas de Oxido sobre Acero	Inoxidable 304 y 316 Utilizadas en el Seguimiento de	ANYL	0055
una Pelicula de Oxido sobre Acero	Inoxidable 316.= +ando como Electrodo Indicador	CHED	0163
ir pH, por un Electrodo de Acero	Inoxidable 316/Pelicula de Oxido, en la Validacion d	ANYL	0066
E °From the Graphic Structure	Input to a Systematic Chemical Name by Computer:	COMP	0019
venile Hormones and Peptides for	Insect Control. +ck, Jackson Award Symposium: Ju	AGRO	0001
venile Hormones and Peptides for	Insect Control. +ck, Jackson Award Symposium: Ju	AGRO	0006

Baculovirus for	Insect Control?.= What Is the Future of	AGRO	0081
trol Agents. Tribolium as a Model	Insect for Study of Resistance Mechanisms.= +on	AGRO	0133
Genetics of Resistance to Juvenoid	Insect Growth Regulators in Drosophila.= +ecular	AGRO	0101
o Benzoylphenyl Ureas and Other	Insect Growth Regulators.= +nisms of Resistance t	AGRO	0134
us Insecticides by Expressing with	Insect Selective Proteins.= +Efficacy of Baculovir	AGRO	0009
phila: A Model System in a Model	Insect.= +ceptor from Cyclodiene Resistant Droso	AGRO	0152
udies with Ponasterones, the First	Insect–Molting Hormones from Plants.= +rrent St	HIST	0024
yl and Dialkylamino Aryl Oxi+	Insecticidal Substituted 2-Methylbiphenylmethyl Alk	AGRO	0072
Evaluation of Organophosphate	Insecticide Hydrolytic Detoxification by Conventiona	AGRO	0124
= Molecular Genetics of	Insecticide Resistance in a Leafrolling Lepidopteran.	AGRO	0116
Esterase Genes Conferring	Insecticide Resistance in Aphids.=	AGRO	0137
	Insecticide Resistance in the German Cockroach.=	AGRO	0138
rosophila P450 Genes Involved in	Insecticide Resistance.= +Agents. Regulation of D	AGRO	0097
oving the Efficacy of Baculovirus	Insecticides by Expressing with Insect Selective Prot	AGRO	0009
propylbutanes: Highly Efficacious	Insecticides with Low Fish Toxicity.= +ryl-1-Cyclo	AGRO	0073
Stored-Product	Insects.= Hydrolytic Resistance Mechanisms in	AGRO	0135
Mycotoxin Resistance in	Insects.=	AGRO	0155
t Drosophila: A+ Cloning of an	Insensitive GABA Receptor from Cyclodiene Resistan	AGRO	0152
Hydride Clusters.=	Insertion of Nitrogen into Intermediate–Sized Boron	INOR	0388
oting the Alkyl to Acyl Migratory	Insertion Reaction in η–Cp(CO) (PPh$_3$)FeMe$^+$.=	INOR	0109
ntramolecular Carbon–Hydrogen	Insertion Reactions of Alkyl Diazoacetates Catalyzed	ORGN	0042
ching Resource for Preservice and	Inservice Chemistry Teachers.= +emSource: A Tea	CHED	0137
Instrumentation Laboratories and	Inservice Teacher Training.= +Interface: Regional	CHED	0112
Model for	Inservice Teaching.=	CHED	0074
echniques Will Provide Detailed	Insight in the Mechanism of Gas–Solid Reactions.=	FUEL	0015
ffect on Elect+ Electrochemical	Insights into Microemulsion Microstructure and Its E	COLL	0023
via the Sol–Gel Method: Some	Insights into Morphology Control and Physical Prope	POLY	0260
in Industry.+ Chemical Careers	Insights Program (Roadshow)—Careers for Chemists	YCC	0005
rals from S+ Characterization of	Insoluble Organic Matter Associated with Clay Mine	I&EC	0021
roteins.=	Insoluble Reagent for the Selective Precipitation of P	BIOL	0040
Enzyme Reactions on	Insoluble Substrates.=	COLL	0052
tric and Coulometric Analysis of	Insoluble, Difficultly Oxidizable Compounds in Aque	COLL	0002
Culture	Instability of Auxotrophic Amino Acid Producers.=	BIOT	0031
The Polymer Research	Institute at the Polytechnic University, Brooklyn.=	HIST	0033
nce in the Nat+ Programs of the	Institute for Chemical Education in Revitalizing Scie	CHED	0138
emistry Outreach Programs of the	Institute for Chemical Education.= +igh School Ch	CHED	0073
udy in 84 Cases from the National	Institute of Cardiology (1984-1989).= +arative St	ENVR	0040
Squibb Pharmaceutical Research	Institute.= +ustry: Chemistry at the Bristol-Myers	CMEC	0028
Computer–Managed Chemistry	Instruction.= Computers in Chemical Education.	CHED	0077
e General, Inorganic, and Organic	Instructional Laboratories.= +oscale Approach in th	CHED	0150
X-ray Crystallography:	Instructional Materials.=	CHED	0158
X-ray Crystallography:	Instructional Materials.=	PHYS	0271
pectroscopy as an Analytical Tool:	Instrument Requirements and Salient Results.= +S	NUCL	0042
ent for the Physical Chemistry or	Instrument Analysis Laboratory.= +tion Experim	CHED	0161
Computer Simulations for	Instrumental Analysis.=	CHED	0192
y of Biopolymers: Operational and	Instrumental Aspects.= +n Process Chromatograph	ANYL	0042
is of Biological Mo+ Column and	Instrumental Considerations for Rapid HPLC Analys	ANYL	0129
Optical Methods. Dynamic Infra+	Instrumental Methods in Polymer Characterization:	POLY	0303
Optical Methods. FT-IR and Fl+	Instrumental Methods in Polymer Characterization:	POLY	0330
Optical Methods. Polystyrene–+	Instrumental Methods in Polymer Characterization:	POLY	0277
Optical Methods. Static and Dy+	Instrumental Methods in Polymer Characterization:	POLY	0250
Laboratory at MIT.= Advanced	Instrumentation in the Undergraduate Chemistry	CHED	0128
tion of Fourier Transform NMR	Instrumentation into the Undergraduate Curriculum	CHED	0148
hool–College Interface: Regional	Instrumentation Laboratories and Inservice Teacher	CHED	0112
Physical Chemistry	Instrumentation Laboratory at Johns Hopkins.=	PHYS	0246
Physical Chemistry	Instrumentation Laboratory at Johns Hopkins.=	CHED	0199
Developed Using GC/MS	Instrumentation.= New Laboratory Experiments	CHED	0187
al Chemistry Award in Chemical	Instrumentation, sponsored by the Dow Chemical Co	ANYL	0011
The Art and Science of String	Instruments.=	CONG	0003
Physical Characterization of an	Insulator Containing Two Elastomeric Components.=	PMSE	0168
V(110) Surfaces: From 1– and 2-D	Insulators to 3-D Metals.= +Alkali Metals on III-	PHYS	0183
iated Delivery of Methotrexate–	Insulin Conjugate to H35 Hepatoma and IEC-6 Cells	MEDI	0185
Range of+ Solution Structure of	Insulin Differs from the Crystal State but Spans the	BIOL	0002
Like Conformational Transition in	Insulin Hexamers.= +Paraben Will Cause a T←R-	BIOT	0150
Stability and Specificity. Role of	Insulin's C–Terminal B–Chain on Association, Foldin	BIOT	0132
Cross Polarization of Half–	Integer Spins in Rotating Samples.=	PHYS	0371
Secuencias de Destilacion con	Integracion de Energia.= Sintesis y Modificacion de	I&EC	0074
s Experimentales de un Empaque	Integral para Enriquecimiento en Deuterio.= +tado	I&EC	0073
+Process Design in Following an	Integrated Approach to Bioprocess Development.=	BIOT	0075
rocessed Vegetable Snacks.=	Integrated Approach to Development of Minimally P	AGFD	0056
ltiple-Pathway Chemical E+ An	Integrated Approach to Risk Characterization of Mu	ENVR	0073
ning.=	Integrated Approach to Synthesis and Reaction Plan	CINF	0003
ches and Field Applications—II.	Integrated Bio–Ozon Treatment of Pulp Bleaching E	BIOT	0137
and Ca–Gluconate by Immobili+	Integrated Bioprocess for the Production of Fructose	BIOT	0152
hysical Chemistry Laboratory: An	Integrated Experimental Program.= +ers in the P	CHED	0045
mented by the Microscale+ An	Integrated Four-Year Spectroscopy Program Comple	CHED	0150
	Integrated Molecular Modeling in the Curriculum.=	CHED	0190
Chemistry to Materials Science in	Integrated Optics.= +Thin Polymer Films: From	COLL	0070
ssing Yields for Production of +	Integrated Strategy for Improving Downstream Proce	BIOT	0071
AIPHOS—An	Integrated System for Organic Synthesis Design.=	CINF	0005

ry Laboratory.=	Integrated Writing Program in the Physical Chemist	CHED 0213
ry Laboratory.=	Integrated Writing Program in the Physical Chemist	PHYS 0318
ructural Database (CSD).=	Integrated 3D Search Facilities for the Cambridge St	CINF 0021
YED 1990 Award Presentation:	Integrating Polymer Education into Middle, Vocation	CHED 0026
-Downstream Stages. Integrati+	Integrating Primary Separation Steps with Upstream	BIOT 0069
Prolog Approach to	Integrating Protein Sequence and Structure Data.=	CINF 0030
urification Techniques.=	Integrating Upstream Production with Downstream P	BIOT 0072
d Graphics into the Undergrad+	Integration of Modern Computational Techniques an	CHED 0156
processing To Aid in Process D+	Integration of Rapid Bioanalytical Techniques in Bio	BIOT 0074
h Upstream–Downstream Stages.	Integration: A Must for Tomorrow's Bioprocesses.=	BIOT 0069
Own Boss.= Have	Intelligence, Imagination, and Diligence? Be Your	SCHB 0016
Bacteriorhodopsin as an	Intelligent Material.=	BIOT 0052
ng Polymers in Redox Devices and	Intelligent Materials Systems.= +ions of Conducti	POLY 0326
ual States: Surface Analysis When	Intense Lasers Miss the Surface.= +ts through Virt	NUCL 0044
of Products with the cupric Ion	Intensity and Coordination via Electron Spin Resona	CATL 0017
itiation Factors 4F and (ISO)4F	Interact Differently with Oligoribonucleotide Analogs	BIOL 0128
Kinase C Inhibitors Designed to	Interact with the Catalytic Domain.= New Protein	MEDI 0076
Laboratory Practices: An	Interactive Student Evaluation.= New Approach for	CHED 0106
de)Paradichlorobenzene and Their	Intercalate.= +ble Gelation from Poly(ethylene Oxi	POLY 0125
uminescence Studies of Copper(I)	Intercalated Laser Dye Coumarins.= +ctural and L	INOR 0245
Direct Synthesis of Novel	Intercalated Layer Silicates.=	PETR 0108
cancer Benzothiopyranoindazole	Intercalating Agents Based on Molecular Dynamics.=	MEDI 0018
hates toward Organic Bases, Me+	Intercalating Properties of Layered Metal(IV) Phosp	PETR 0107
uperconducting Systems.=	Intercalation and Ion–Exchange Reactions in Oxide S	INOR 0360
rials, and Enhanced Physical P+	Intercalation Chemistry: Useful Reactions, New Mate	CHED 0030
Microscopy of Donor Graphite	Intercalation Compounds.= Scanning Tunneling	PHYS 0365
talyst Stability in+ Aryl Group	Interchange in Triarylphosphines and Its Role on Ca	INOR 0349
Chemical Structural Information	Interchange.= +Standardization Developments for	CINF 0038
tion of Low–Resistivity Metals for	Interconnection of Electronic Devices.= +r Deposi	INOR 0126
Optical	Interconnections.= Planar Polymer Waveguides for	POLY 0175
C_8H+ Inorganic Propellanes: The	Interconversion of $W_2(O_2C_8H_{16})_3(HNMe_2)_2$ to $W_2(O_2$	INOR 0128
gle Neutron Scattering Study of	Interdiffusion in Polystyrene–Polyxylenylether Blend	POLY 0166
y for Nonscience Students.=	Interdisciplinary Core Approach to College Chemistr	CHED 0176
Science Education. FACETS: An	Interdisciplinary Middle School Curriculum.= +ol	CHED 0071
d Thiamine: A Potent Source of	Interesting Flavor Compounds.= Thermally Degrade	AGFD 0071
Synthesis of Biologically	Interesting Glycosides.=	ORGN 0287
A Chemistry Perspective of the	Interface between Chemistry and Chemical Engineer	I&EC 0050
emical Engineer.=	Interface between the Analytical Chemist and the Ch	I&EC 0051
alvanized Steel/Sodium Chloride	Interface by Electrochemical Impedance Spectroscop	PHYS 0270
cal Reactions at Polymer/Metal	Interface by Sputtering Depth Profiling Using XPS.=	PMSE 0150
n Chemical Proces+ OSHA–EPA	Interface in Health and Safety Issues. OSHA's Role i	CHAS 0012
of the Crystal/Amorphous	Interface in Polyethylene and Its Blends.= Nature	COLL 0036
The Influence of Fiber/Matrix	Interface on Local Glass–Transition Temperature in	COLL 0055
l Vapor Deposition of a Nonlinear	Interface Optical Switch.= +ring by Laser Chemica	BIOT 0024
The OSHA–EPA	Interface: An Overview.=	CHAS 0014
Inservic+ High School–College	Interface: Regional Instrumentation Laboratories and	CHED 0112
rmodynamics at the Solid–Liquid	Interface.= +rement of Reaction Kinetics and The	COLL 0170
ateral Diffusion at the Air–Water	Interface.= +lutions, Blends, and Interfaces—III. L	COLL 0051
I. Dynamics at the Solid–Polymer	Interface.= +r Solutions, Blends, and Interfaces—	COLL 0012
the Diamond/Vapor	Interface.= Absorbed Hydrogen and Fluorine near	PHYS 0359
on Forces at a Solid–Liquid	Interface.= Direct Measurement of Protein Interacti	BIOT 0043
Polymers at the Air–Water	Interface.= Dynamic Properties of Cellulosic	COLL 0056
r Ion Transfer at a Liquid–Liquid	Interface.= +ential Studied by the Voltammetry fo	BIOT 0105
J. +ring. Panel Discussion: At the	Interface, What Does the Practicing Chemist Need?	I&EC 0057
mistry or Instrume+ Computer–	Interfaced Titration Experiment for the Physical Che	CHED 0161
Studies of Some Photosensitive	Interfaces by Surface Photovoltage Spectroscopy.=	BIOT 0142
ide–Chain Polymer at Different	Interfaces Using the Langmuir–Blodgett Technique.=	COLL 0071
bility+ Partitioning Processes at	Interfaces: Chromatographic Retention and Bioavaila	ANYL 0102
Porphyrin Metalation at Anionic	Interfaces: Mode of Interfacial Binding and Implicati	CATL 0060
rom Solution to Solution–Polymer	Interfaces.= +ontaneous Adsorption of Polymers f	POLY 0298
nd Function at Condensed–Phase	Interfaces.= +al Chemistry at the Edge: Structure a	ANYL 0026
ies: Conventional and Microscopic	Interfaces.= +ectrified Immiscible Liquid Boundar	COLL 0004
Electronic Energy Relaxation at	Interfaces.=	PHYS 0062
quid, Liquid–Solid, and Solid–Gas	Interfaces.= +Spaces: Chemistry at the Liquid–Li	ORGN 0126
of Semiconductor–Liquid	Interfaces.= Picosecond Nonlinear Optical Studies	PHYS 0283
nts of Excited–State Relaxation at	Interfaces.= +econd Optical Harmonic Measureme	COLL 0087
Amphiphiles at Polymer–Polymer	Interfaces.= +ic Generation from Hyperpolarizable	POLY 0156
Diamond/Metal	Interfaces.= Theory of Schottky Barriers in	PHYS 0358
P Polymer Solutions, Blends, and	Interfaces—I. Dynamics at+ Proctor Gamble UER	COLL 0012
P Polymer Solutions, Blends, and	Interfaces—II. Quick Quenc+ Procter Gamble UER	COLL 0032
P Polymer Solutions, Blends, and	Interfaces—III. Lateral Diff+ Procter Gamble UER	COLL 0051
P Polymer Solutions, Blends, and	Interfaces—IV. Thin Polym+ Procter Gamble UER	COLL 0070
n at Anionic Interfaces: Mode of	Interfacial Binding and Implications Concerning Enz	CATL 0060
tion of the Rate and Equilibri+	Interfacial Catalysis by Phospholipase A₂: Determina	BIOL 0019
and Kinetic Characterization of+	Interfacial Catalysis by Phospholipase A₂: Structural	BIOL 0020
Ion Beam Induced	Interfacial Chemical Complexes and Adhesion.=	NUCL 0094
or Elect+ Picosecond Real–Time	Interfacial Electron Transfer at a Solid Semiconduct	COLL 0088
ζ–Potential on	Interfacial Electron Transfer.= Kinetics Effects of	COLL 0152
f a Porphyrin Cation Lattice by	Interfacial Electron–Transfer Gates Ionic Currents ac	BIOL 0136
Dynamics of Coupling Agents in	Interfacial Regions.=	POLY 0368

of Asphaltenes—A Liquid/Water	Interfacial Tension Measurement.= +ption Kinetics	PETR	0138
Automated Dynamic	Interfacial Tension Technique.=	COLL	0091
Aggregate to Low	Interfacial Tension.= °From Large-Surfactant	COLL	0078
ial Photoinduced Electron Tran+	Interfacially Polymerized Porphyrin Films for Vector	BIOT	0126
erfacing Chemistry and Chemic+	Interfacing Chemistry and Chemical Engineering. Int	I&EC	0048
nel Discussion: At the Interface,+	Interfacing Chemistry and Chemical Engineering. Pa	I&EC	0057
istry and Chemical Engineering.	Interfacing Chemistry and Chemical Engineering—th	I&EC	0048
haracterization of VPI-6: Another	Intergrowth of Hexagonal and Cubic Faujasite.=	PETR	0001
to Olefins over Offretite/Erionite	Intergrowths.= +ies of the Conversion of Methanol	PETR	0042
atalytic Performance of Pillared	Interlayered Clay Molecular Sieve with High Therma	PETR	0030
Constrained Ω-Loop Excised from	Interleukin-1α.= +Structural Characterization of a	MEDI	0165
g Cost and Performance in Inte+	Intermaterial Competition among Polymers. Assessin	CMEC	0009
iers in Packaging: Polymer-Based	Intermaterial Competition in the 1990s.= +w Front	CMEC	0016
Effect of Recyclability on	Intermaterial Competition.=	CMEC	0010
tive Industry and Its Impact on	Intermaterial Competition.= The Changing Automo	CMEC	0015
	Intermaterial Competition—Boon or Bane?.=	CMEC	0012
New Way to Look at	Intermaterial Substitution.=	CMEC	0013
ssessing Cost and Performance in	Intermaterials Competition.= +among Polymers. A	CMEC	0009
with Donor-Acceptor Units	Intermediated by Silicon Chains.= Model Molecules	POLY	0211
metric Studies for the Combustion	Intermediates CH_2OH and CH_3O.= +Mass Spectro	FUEL	0124
n-Quantum Coherence in Probing	Intermediates in Heterogeneous Catalysis.= +ns of	COLL	0089
onds Studied by Tim+ Reactive	Intermediates in the Carbonylation of Metal Alkyl B	INOR	0169
lar Sieves: Recrystallization of the	Intermediates Isolated from the Gel Phase.= +ecu	PETR	0019
rface+ Sorption Complexes as the	Intermediates of Substitution Reactions on Silica Su	COLL	0095
romatics via Unsaturated Ketene	Intermediates Using Transition Metal-Mediated Pro	ORGN	0274
placement from $(Alkane)W(CO)_5$	Intermediates: Estimates of Alkane-W Bond Strengt	INOR	0111
Flames and the Role of Aromatic	Intermediates.= +ot Deposition from Ethylene/Air	FUEL	0107
Reactive	Intermediates.= Cyclic Oxyphosphoranes as	INOR	0290
Ketones via Amorphous Ketimine	Intermediates.= +crystalline Poly(Arylene Sulfide)	PMSE	0123
ugars by Epoxy and Triflate Sugar	Intermediates.= +, and Branched-Chain Amino S	ORGN	0071
Reaction of P+ Evidence for the	Intermedicacy of the Pentaphenylethyl Cation in the	ORGN	0145
mbic Pseudohexagon+ Intra- and	Intermolecular Correlation Involved in the Orthorho	POLY	0354
re ab Initio Calculations with +	Intermolecular Forces: Do We Really Need to Compa	PHYS	0104
in the Greater Cincin+ Summer	Intern Program for High School Chemistry Teachers	TECH	0002
Ion Chemistry.=	Internal and Translational Energy Effects on Cluster	PHYS	0242
/PMMA Blend.=	Internal Electric Field and Dye Orientation in PVDF	POLY	0225
193-nm Photodissociation of S+	Internal Energy Distributions of SO ($X^3\Sigma^-$) from the	PHYS	0240
ion of Calcium Carbonate on the	Internal Walls of Glass Capillaries: Influence of Tem	COLL	0163
duled for USA, July 11-22+ 1992	International Chemistry Olympiad Competition Sche	CHED	0139
U.S. Participation in the	International Chemistry Olympiad.=	CHED	0015
7).= Science, Technology, and	International Competition. R. Solow (Economics, 198	CONG	0014
d Europe. Global Marketing of+	International Flow of Information: North America an	CINF	0011
ight Protection and Risks in the	International Information Transfer Process.= Copyr	CINF	0014
ties.=	International Joint Ventures: Exploring the Possibili	CINF	0015
Green Marketing: A National and	International Status Report.= +lution Prevention?	CEI	0028
anostannanes, an+ Simultaneous	Interpenetrating Network Products from Lignin, Org	PMSE	0006
ology/Processing Relationship in	Interpenetrating Polymer Blends of a SEBS Block C	PMSE	0058
Permeability Studies of Latex	Interpenetrating Polymer Network Films.= Gas	PMSE	0061
tor Oil Polyurethane and Vinyl+	Interpenetrating Polymer Network Formed from Cas	PMSE	0089
ethylene and Po+ Biocompatible	Interpenetrating Polymer Network of Polytetrafluoro	PMSE	0041
nal Biomimetic Polymers through	Interpenetrating Polymer Network Synthesis.= +tio	PMSE	0004
from Polyurethane and Epoxy	Interpenetrating Polymer Network.= Composites	PMSE	0044
yurethane-Unsaturated Polyester	Interpenetrating Polymer Network.= +ology of Pol	PMSE	0027
oams Based on Polyurethanes and	Interpenetrating Polymer Networks (IPNs).= +n F	PMSE	0040
ca+ Conjugate Three-Component	Interpenetrating Polymer Networks and Their Appli	PMSE	0080
Polyurethanes an+ Simultaneous	Interpenetrating Polymer Networks from Castor Oil	PMSE	0091
te-Urethane) and Polyvinyl Pyr+	Interpenetrating Polymer Networks of Poly(Carbona	PMSE	0056
l in Situ Synthesis of Interpenet+	Interpenetrating Polymer Networks. Entirely Radica	PMSE	0001
Polymer Networks: An Overvi+	Interpenetrating Polymer Networks. Interpenetrating	PMSE	0039
e Separation in Polyurethane/+	Interpenetrating Polymer Networks. Kinetics of Phas	PMSE	0021
ocessing Relationship in Interpe+	Interpenetrating Polymer Networks. Morphology/Pr	PMSE	0058
xy Interpenetrating Themosets+	Interpenetrating Polymer Networks. Two-Phase Epo	PMSE	0087
erpenetrating Polymer Networks.	Interpenetrating Polymer Networks: An Overview.=	PMSE	0039
tirely Radical in Situ Synthesis of	Interpenetrating Polymer Networks.= +tworks. En	PMSE	0001
Sheet with Polyurethane/Acrylate	Interpenetrating Polymer Networks.= +ated Steel	PMSE	0057
uilibrium Microphase Structure of	Interpenetrating Polymer Networks.= +and Noneq	PMSE	0093
Flow in Latex	Interpenetrating Polymer Networks.=	PMSE	0062
of P (nBAc)/PEP Simultaneous	Interpenetrating Polymer Networks.= Morphology	PMSE	0042
Phase Separation in	Interpenetrating Polymer Networks.=	PMSE	0022
Studies of Grafted	Interpenetrating Polymer Networks.=	PMSE	0003
rethane/Vinyl or Acrylic Polymer	Interpenetrating Polymer Networks.= +r Oil Polyu	PMSE	0090
Damping Paint of PS/PA Latex	Interpenetrating Polymer Networks.= Study on the	PMSE	0064
Study on the Three-Stage Latex	Interpenetrating Polymer Networks.=	PMSE	0082
aracterization of Polyphosphazene	Interpenetrating Polymer Networks.= +sis and Ch	PMSE	0002
mer Networks. Two-Phase Epoxy	Interpenetrating Themosets.= +erpenetrating Poly	PMSE	0087
mal Synthesis of K_2PdSe_{10}: Two	Interpenetrating 3D Frameworks of $[Pd(Se_4)_2]^{2-}$ and	INOR	0200
s.= Investigation of the Polym+	Interphase via Synthesis of Well-Defined Copolymer	COLL	0033
d Silver Su+ Characterization of	Interphases between PMDA/4-BDAF Polyimides an	POLY	0087
Crystal-Amorphous	Interphases in Poly(ethylene Oxide) Blends.=	PMSE	0165
ists.=	Interphenylene 7-Oxabicycloheptane TXA_2 Antagon	MEDI	0160

Chemistry.= Fruitful	Interplay between Computational and Experimental	ORGN	0138
	Interplay of Motors and Microtubule Dynamics.=	BIOL	0139
enary Lecture. Hydrogen-Bonded	Interpolymer Complexes.= +nent Systems—IV. Pl	PMSE	0163
search and Casework in the Ro+	Interrelationship between Forensic DNA Analysis Re	CHAL	0007
pens When Parallel Developments	Intersect?.= +pyright and Technology: What Hap	CINF	0043
$^1B_1$ States of Ozone: A Conical	Intersection and Its Effect on Photoabsorption and D	PHYS	0191
e Orthorhombic Pseudohexagon+	Intra- and Intermolecular Correlation Involved in th	POLY	0354
lexes with Two- and Three-Atom	Intra-Annular Tethers.= +stry of Bimetallic Comp	INOR	0149
ic Yields in Streptomyces by the	Intracellular Expression of a Bacterial Hemoglobin.=	BIOT	0087
ntal Trade. Criteria for Successful	Intracontinental Trade in the 21st Century.= +ine	CMEC	0001
ontinental Trade i+ Dynamics of	Intracontinental Trade. Criteria for Successful Intrac	CMEC	0001
2-Pyridones.=	Intramolecular [4+4] Photocycloaddition of Tethered	ORGN	0235
with High Enantioselectivity by	Intramolecular Carbon-Hydrogen Insertion Reaction	ORGN	0042
isubstituted Benzofurans via the	Intramolecular Coupling of a Zirconocene Benzyne C	ORGN	0217
mides.=	Intramolecular Diels-Alder Reactions of Vinylsulfona	ORGN	0231
ng in Predicting Absolute Rates of	Intramolecular Electron and Energy Transfer?.=	ORGN	0143
=	Intramolecular Electron Transfer at Fixed Distances.	PHYS	0077
in Macromolecules Observed by	Intramolecular Energy Transfer Using Frequency-Do	PHYS	0061
tographic Studies on the Role of	Intramolecular Hydrogen Bonding in Substituted Be	ANYL	0045
of the Activation Enthalpies for	Intramolecular Hydrogen Transfer as Functions of S	ORGN	0175
Long-Range	Intramolecular Interactions.=	PHYS	0076
llyl Cations and Cation Radical+	Intramolecular $2\pi + 4\pi$ Cycloaddition Reactions of A	ORGN	0123
New	Intrigues of Cellulolysis.=	BIOT	0209
in Cured Epoxy by Extrinsic and	Intrinsic Fluorophores.= +rement of Water Uptake	POLY	0160
ameter Correlation between	Intrinsic Viscosity and Molecular Weight.= One-Par	PMSE	0095
Voltammetry Experiments for the	Introductory Physical Chemistry Laboratory.= +ic	CHED	0211
Using Real Numbers as Vertex	Invariants for Topological Indices.=	COMP	0026
dress. INVENTON: A System for	Inventing Molecular Structures.= +tion. Award Ad	CINF	0016
cal Information. Award Address.	INVENTON: A System for Inventing Molecular Stru	CINF	0016
National	Inventors Hall of Fame.=	CHAL	0025
Cartoons Related to	Inventors or Patents.= Favorite or Objectionable	CHAL	0022
System.= Chemical	Inventory and Hazardous Materials Tracking	CHAS	0003
Chemical	Inventory Management System.=	CHAS	0025
Radical Routes into Nonshrink+	Inverse Organic-Inorganic Composite Materials: Free	POLY	0259
ation of Human Seric Copper by	Inverse Voltammetry at the Mercury Film Electrode.	ANYL	0056
Predicting the Phase	Inversion in Immiscible Liquid Systems.=	PMSE	0024
al Incorporation into the Marine	Invertebrate Oxygen-Carrying Proteins Hemerythrin	INOR	0140
ents. The Darmstadt Program To	Investigate Heavy Elements.= +sis of Heavy Elem	NUCL	0014
High-Pressure TPR Apparatus to	Investigate Organic Sulfur Forms in Coals.= +els.	FUEL	0007
nd Copyright: Are Your Policies	Inviolate or in Violation? Copyright and Technology:	CINF	0043
Binuclear Metal Complexes with	Inwardly Directed Functional Groups.= Cofacial	INOR	0242
py, II: Image Contrast Analysis of	Iodinated Polystyrene Chains.= +Electron Microsco	POLY	0062
itrate.=	Iodine and Sodium Nitrate—The Story of Chilean N	SCHB	0020
Spectroscopic Characterization of	Iodine Overlayers on Pt(111).= Surface	PHYS	0366
ew Potential Hypoxia Imag+ 3-	Iodobenzylamine Conjugate of 2-Nitroimidazole: A N	NUCL	0076
y and Stereocontrol During Amide	Iodolactonization.= +Protection on Stereochemistr	ORGN	0258
n of 3,5-Dini+ Limitations of the	Iodometric Method and That Based on the Reductio	BIOL	0111
nd Adhesion.=	Ion Beam Induced Interfacial Chemical Complexes a	NUCL	0094
rials: Composition Modif+ Use of	Ion Beams in the Analysis and Modification of Mate	NUCL	0091
rials: Elemental Analysis+ Use of	Ion Beams in the Analysis and Modification of Mate	NUCL	0077
rials: Elemental/Isotopi+ Use of	Ion Beams in the Analysis and Modification of Mate	NUCL	0064
rials: Laser Ionization M+ Use of	Ion Beams in the Analysis and Modification of Mate	NUCL	0041
rials: Cluster Ion Impac+ Use of	Ion Beams in the Analysis and Modifications of Mate	NUCL	0052
of Materials Using High-Energy	Ion Beams.= Composition and Structural Analysis	INOR	0008
position Modification Produced by	Ion Bombardment (Ion Mixing). +f Materials: Com	NUCL	0091
lites: Product Selectivi+ Copper	Ion Catalysis of Propylene Oxidation on X and Y Zeo	CATL	0017
Water in	Ion Channel Gating.= Simulation of the Role of	COLL	0084
from	Ion Channeling.= Crystallographic Information	NUCL	0066
Metal	Ion Chemistry in Flames.=	PHYS	0201
Transition-Metal	Ion Chemistry.= Comparison of Main Group and	PHYS	0200
Effects on Cluster	Ion Chemistry.= Internal and Translational Energy	PHYS	0242
	Ion Chromatography in the 90s.=	AGFD	0204
nalysis of Organic Compounds by	Ion Chromatography.= +ation Techniques for the A	TECH	0015
rse Energies in Relativistic Heavy	Ion Collisions at 14.6 GeV/c per Nucleon.= +ansve	NUCL	0051
ators for Measurement of Calcium	Ion Concentration.= +pment of NMR Active Indic	ORGN	0149
ence of Temperature, Flow Rate,	Ion Concentration, and Application of a Radiofreque	COLL	0163
Contents and Paramagnetic Metal	Ion Concentrations.= +to Their Hydration Water	MEDI	0043
mer/Polystyrene Blends: Effect of	Ion Content.= +ure of Sulfonated Polystyrene Iono	POLY	0061
Combinatorics of Cluster	Ion Enumeration.=	COMP	0028
rome P-450 Patters by Means of	Ion Exchange and Immobilized Metal Affinity FPLC	ANYL	0017
Conventional Means and Reactive	Ion Exchange.= +ide Hydrolytic Detoxification by	AGRO	0124
terials: The Hydrous Metal Oxide	Ion Exchangers.= +of Novel Catalyst Support Ma	PETR	0018
mines by Reversed-Phase Paired-	Ion High-Performance Liquid Chromatography.=	ANYL	0089
odifications of Materials: Cluster	Ion Impact Phenomena Used for Surface Modificatio	NUCL	0052
Surface Modification by Metallic	Ion Implantation On-CdTe.=	PHYS	0218
ion of Thymol Blue with Bromate	Ion in Acidic Solutions.= +plex Kinetics of Oxidat	PHYS	0269
Multifragment Emission in Light	Ion Induced Reactions.=	NUCL	0026
Recoil Studies Revisited: Heavy	Ion Induced Reactions.=	NUCL	0060
ation of Products with the cupric	Ion Intensity and Coordination via Electron Spin Re	CATL	0017
Structures Formed by	Ion Irradiation.= Tribological Properties of	NUCL	0093

Evidence That Hydrogen	Ion Is (H$_3$O)$^+$.=	PHYS	0272
aterial Analysis Using Secondary	Ion Mass Spectrometry and Related Techniques.=	INOR	0007
Produced by Ion Bombardment (	Ion Mixing). +Materials: Composition Modification	NUCL	0091
Low-Energy Behavior of Heavy-	Ion Optical Potential.= +spersion Relation and the	NUCL	0040
ture of Free Radical+ Negative-	Ion Photoelectron Spectroscopic Studies of the Struc	PHYS	0055
Results with Theory+ Negative-	Ion Photoelectron Spectroscopy: Comparison of New	PHYS	0059
Intermediate-Energy Heavy	Ion Reactions and RI-Beam Experiments.=	NUCL	0029
Reactivity.= Carbon-Cluster-	Ion Reactions: Structural Isomer Effects on	PHYS	0311
Evolution of Heavy-	Ion Recoil and Fragment Separators.=	NUCL	0038
High-Energy	Ion Scattering for Materials Analysis.=	NUCL	0065
Reactivity Using Low-Energy	Ion Scattering.= Studies of TiO$_2$(110) Surface	COLL	0172
Assistance by Solvent and Azide	Ion to the Reaction of Cumyl Drivatives: Mechanism	ORGN	0154
l Studied by the Voltammetry for	Ion Transfer at a Liquid-Liquid Interface.= +entia	BIOT	0105
Pesticide Analysis in Water Using	Ion Trap Mass Spectrometry.= +gulatory Issues.	AGRO	0060
teryoung. Interactions of Sulfide	Ion with Butylpyridinium Chloride in Nonaqueous M	ANYL	0093
tes in Nephrocalcin with Vanadyl	Ion.= +OR Characterization of Calcium-Binding Si	BIOL	0045
Permanganate	Ion.= Reaction of Sulfides with	ORGN	0243
Benzenediols by the IrCl$_6^{2-}$	Ion.= Volumes of Activation for the Oxidation of	INOR	0395
Sputtering Processes of	Ion-Bombarded Electronic Materials.=	NUCL	0043
echanism of a Charge-Sensitive	Ion-Conducting Device Consisting of Metalloporphyr	BIOL	0098
Atoms with Halogen Compounds:	Ion-Core Conservation.= +the Reaction of Copper	PHYS	0368
om Penicillium Amagasakiense by	Ion-Exchange Chromatography.= +e Isoenzymes fr	ANYL	0016
eaning of Heavily Contaminated	Ion-Exchange Columns in Process Chromatography o	ANYL	0042
Systems.= Intercalation and	Ion-Exchange Reactions in Oxide Superconducting	INOR	0360
eractions of Carbohydrates with	Ion-Exchange Resins and Hydrophobic Adsorbents.=	CARB	0063
Powdered	Ion-Exchange Resins in Sugar Purification.=	CARB	0065
Spectroscopic Studies of	Ion-Implanted Polycarbonates.=	POLY	0081
e, and Collision Energy Effects on	Ion-Molecule Reactions.= +l Mode, Electronic Stat	PHYS	0241
+and Related Techniques. Field	Ion-Scanning Tunneling Microscopy (FI-STM).=	PHYS	0182
Layers.= Medium-Energy	Ion-Scattering Study of Nickel on Ultrathin SiO$_2$	COLL	0176
tection in Groundwater by Novel	Ion-Selective Graphite Composite Potentiometry.=	ENVR	0044
es of the Fundamental Aspects of	Ion/Solid Collisions Using Laser Spectroscopy.=	NUCL	0041
ith a Sterically Hindered Pheno+	Ionic Conductivity of Poly(ethylene Oxide) Bound w	POLY	0091
nterfacial Electron-Transfer Gates	Ionic Currents across the Lipid Bilayer.= +ice by I	BIOL	0136
nt of Polymer Miscibility through	Ionic Functional Groups.= +y Lecture. Enhanceme	PMSE	0116
We Learn from Reactions with+	Ionic Organic Chemistry in the Gas Phase: what Can	ORGN	0121
Soluble Rigid-Chain Polymers: An	Ionic Poly(p-phenylene) Analog.= +s Synthesis of	POLY	0024
Electrically Conducting	Ionic Polyacetylenes.=	POLY	0074
Liquid Crystal Dispersed in an	Ionic Polymeric Membrane.= Investigation of a	PMSE	0119
f XCN (X = H, Cl, Br) in Various	Ionic States.= +metries and Molecule Properties o	PHYS	0238
e the Effect of Temperature and	Ionic Strength on Chemical Exchange of Vanadate D	INOR	0228
Active-Site	Ionicity and the Mechanism of Carbonic Anhydrase.=	BIOL	0063
Flottation	Ionique du Laurylsulfate de Gallium.=	ANYL	0053
ysical Chemistry+ Multiphoton	Ionization and Scanning Tunneling Microscopy in Ph	CHED	0044
Biopolymers by Electrospray	Ionization LC/MS and LC/MS/MS.= Analysis of	ANYL	0141
/Ionspray Atmospheric Pressure	Ionization LC/MS: Application to Drug Monitoring i	ANYL	0124
d Modification of Materials: Laser	Ionization Mass Spectrometry. +in the Analysis an	NUCL	0041
e Premium Coal Samples by Field	Ionization Mass Spectrometry.= +on of the Argonn	FUEL	0062
Two-Photon	Ionization of (NO)$_n$Ar$_m$ Clusters.=	PHYS	0251
Sputter-Initiated Resonance	Ionization Spectroscopy as an Analytical Tool: Instru	NUCL	0042
S Utilizing Atmospheric Pressure	Ionization.= +—II. Recent Developments in CE/M	ANYL	0138
olymer-Polymer Complexes of an	Ionomer and an Amino-Containing Polymer.= +f P	PMSE	0155
Properties of a Novel Model	Ionomer System: Amic Acid Ionomers.=	PMSE	0146
cture of Sulfonated Polystyrene	Ionomer/Polystyrene Blends: Effect of Ion Content.=	POLY	0061
Dynamics Studies of Ampholytic	Ionomers and Their Pendant-Chain Derivatives.=	POLY	0367
racterization of Block Copolymer	Ionomers by Small-Angle X-ray Scattering and Dyn	PMSE	0118
cture. Structure and Properties of	Ionomers in the Solid State.= +s—III. Plenary Le	PMSE	0145
ransient Electric Birefringence of	Ionomers Poly(styrene-Sodium-2-Acrylamido-2-Me	COLL	0181
System: Amic Acid	Ionomers.= Properties of a Novel Model Ionomer	PMSE	0146
nces, and Fl+ Survey of Alpha-	Ionone Enantiomers in Raspberry Concentrates, Esse	AGFD	0181
Designed Spirocyclic	Ionophores for Lithium.=	ORGN	0062
Stopping Power and Ranges of	Ions in Condensed Matter.=	NUCL	0037
Techniques for	Ions in Membrane Channels.= Optical Spectroscopic	BIOT	0141
rk Substitution of Paramagnetic	Ions in Microporous Aluminophosphate Molecular Si	PHYS	0231
l Population Relaxation Time of	Ions in Protic Solvents Measured by Transient Infrar	PHYS	0083
rs of Au$^+$ and In^{3+}, Trapping Na$^+$	Ions in the Center.= +Polynuclear Selenide Cluste	INOR	0203
($\leq$3 eV) of CH$_3^+$	Ions on Platinum.= Ultra-Low-Energy Interactions	COLL	0216
hilic Addition of Alkyl Thiolate	Ions to β-Nitrostyrenes: Radicaloid Transition State	ORGN	0155
Studies of the Selected Actinide	Ions with 4-Benzoyl-5-Methyl-2-Phenyl-Pyrazol-3-	NUCL	0073
dent Properties of Boron Cluster	Ions: Reactivity, Structure, and Photodissociation St	PHYS	0262
Relativistic Heavy	Ions.= Electromagnetic Dissociation with	NUCL	0049
Semi-Metal Cluster	Ions.= Oxidation Reactions of Metal and	PHYS	0174
Doped with Transition-Metal	Ions.= Studies on Silica Aerogels and Xerogels	INOR	0081
Intermediate-Energy Heavy	Ions.= Target Fragmentation of Copper by	NUCL	0025
GeV/nucleon ^{16}O	Ions.= Target Fragmentation of Silver by 14.6	NUCL	0050
in the Presence of Divalent Metal	Ions.= +ylboronate and Isopropylidene Derivatives	INOR	0263
.=	Iowa's Plastic Pesticide Container Recycling Program	AGRO	0091
Use of Pesticides in	IPM: Federal Perspective.= Future Directions in the	AGRO	0068
Hydrophobic-Hydrophilic	IPN and SIPN.=	PMSE	0005
ater Mixture.=	IPN Membranes for the Pervaporation of Ethanol/W	PMSE	0059

Annealing.= Gas Transport in	IPN Membranes: Effect of Cross-Linked State and	PMSE 0060
xy Resin/Castor Oil Polyurethane	IPN.= +haracterization of CTBN-Toughened Epo	PMSE 0092
r Enzyme Immobilization.=	IPN-Modified Porous Bead Polymers as Supports fo	PMSE 0081
Microstructural Aspects of	IPNs Based on Block Copolymers.=	PMSE 0026
Phenolic	IPNs for Laminate Applications.=	PMSE 0007
thylene Terephthalate) Semi-	IPNs with Functionalized Triglyceride Oils.= Poly(E	PMSE 0088
Polyurethane/Polystyrene Semi-1	IPNs.= +tworks. Kinetics of Phase Separation in	PMSE 0021
High-Tg Polyimide-Based Semi-	IPNs.= +erization of bisNadimides Oligomers and	POLY 0133
Compatibility of Ternary Semi	IPNs.=	PMSE 0156
hase Morphology of Simultaneous	IPNs.= +Glass Transition Temperatures on the P	PMSE 0025
Methacrylate Urethane	IPNs.= Kinetic Study and Morphology Buildup in	PMSE 0043
Latex PS/XPS Semi-	IPNs.=	PMSE 0063
thyl Methacrylated)/Epoxy Resin	IPNs.= +ly(Methyl Acrylate-co-Dimethyl Aminoe	PMSE 0079
erpenetrating Polymer Networks (	IPNs).= +Foams Based on Polyurethanes and Int	PMSE 0040
ane Mono- and Multilayer+ FT-	IR and Fluorescence Studies of Self-Assembled Silox	POLY 0330
ysis Selectivities Exhibited by W-	Ir and Mo-Ir Cluster-Derived Catalysts.= +ogenol	CATL 0008
$h_2P(O)]_3$-nC}, where M = Rh and	Ir and n = 1, 2, and 3.= +or of [cod]M{[Ph$_2$P(S)]$_n$[P	INOR 0264
er(I) Compounds for CVD by FT-	IR and Raman Spectroscopy.= +acteristics of Copp	COLL 0031
vities Exhibited by W-Ir and Mo-	Ir Cluster-Derived Catalysts.= +ogenolysis Selecti	CATL 0008
cia de Carga por Espectroscopia	IR de Complejos de Transferencia de Carga Pt(R-Na	INOR 0190
of [M(tpen)](ClO$_4$)$_2$ Comp+ Far	IR Spectroscopic and Crystallographi Caracterization	INOR 0057
EEK (Poly-ether-ether-k+ FT-	IR Spectroscopic Study of Thermal Degradation in P	POLY 0334
Solution.= Picosecond Transient	IR Spectroscopy of UV-Photolyzed Cr(CO)$_6$ in	PHYS 0220
tored by Rheo-Photoacoustic FT-	IR Spectroscopy.= +oly(vinylidene Fluoride) Moni	POLY 0255
Step-Scan FT-	IR Spectroscopy.= Polymer Characterization by	POLY 0304
ransform Mid- and Far-Infrared	IR Specular Reflectance Study of Oxidized and Exch	POLY 0253
yleneterephthalate) by ATR-	IR.= Analysis of Near Surface Structure in Poly(eth	POLY 0279
Two-Dimensional Infrared (2D	IR) Spectroscopy.=	PHYS 0117
s-Promoted Hydrotreatin+ FT-	IR, Solid-State MAS NMR, and XRD of Phosphorou	COLL 0198
Oxidation of Benzenediols by the	IrCl$_6^{2-}$ Ion.= Volumes of Activation for the	INOR 0395
Fast Kinetics Investigation of	Iridium Bromide Complexes.=	PHYS 0258
ation, and Scission on Tungsten-	Iridium Tetranuclear Cluster Cores: Relationship to	CATL 0008
Exchange between Square Planar	Iridium(I) Complexes.= Facile Phosphine Ligand	INOR 0211
Reactivity of Phosphido-Bridged	Iridium, Molybdenum Heterobimetallics by Oxidativ	INOR 0385
Petroleum Residues.= TPOP-	IRMS Stable Carbon Isotope Analysis of Heavy	PETR 0141
Tris(1,10-Phenanthroline)	Iron (II) by Hydrogen Peroxide.= Oxidation of	INOR 0222
Catalytic Hydrosilation of	Iron and Ruthenium Acetyl Complexes.=	INOR 0238
derophore Analogues.=	Iron and Vanadium Complexes of Siderophores or Si	INOR 0120
m in Desulfovi+ New Method for	Iron Atom Active-Site Replacement by a Nickel Ato	INOR 0020
ganic Chemistry. Mechanisms of	Iron Biomineralization in Chitons.= Marine Bioinor	INOR 0116
Ultrafine	Iron Carbide as Liquefaction Catalyst Precursor.=	FUEL 0101
Ultrafine	Iron Carbide as Liquefaction Catalyst Precursor.=	CATL 0051
sm of Assembly of the Dinuclear	Iron Center: Tyrosyl Radical Cofactor of Ribonucleot	BIOL 0022
Low-Valent	Iron Complexes.= Hydride Thiolate Derivatives of	INOR 0225
ides and T+ Biogeochemistry of	Iron in Anoxic Coastal Sediments: Cycling of Iron Ox	GEOC 0007
e of Pickle Liquors from Mexican	Iron Industry.= +e Solutions for Disposal and Reus	ENVR 0039
=	Iron Limitation of Open-Ocean Primary Production.	INOR 0002
uctive Dissolution by+ Selective	Iron Oxide Extraction from Soils and Sediments: Red	ENVR 0058
ic Coastal Sediments: Cycling of	Iron Oxides and Their Potential Importance as Alter	GEOC 0007
CH$_3$-1 with El+ Reactivity of an	Iron Phosphide Acetyl Complex Cp(PPH$_2$)(CO)FeCO	INOR 0240
ffect of Hydrogen Peroxide and	Iron Salts on Sewage, Part 1: Odor Control and Phos	ENVR 0034
Ferritin	Iron Storage and mRNA Regulation.=	INOR 0011
Microbial Geritol: From	Iron to Infection.=	CHED 0061
ine in the Presence of Copper and	Iron.= +s in the Antioxidant Mechanism of Carnos	AGFD 0039
Redox	Iron.= Lipid Oxidation in Foods via	AGFD 0094
Cysteine Complexes of	Iron(II) and Iron(III).= Reactivity Studies on	INOR 0370
dditions of Allyl(cyclopentadienyl)	Iron(II) Dicarbonyl to Aldehydes and Ketones.=	ORGN 0213
oxodisulfate Oxidation of Ternary	Iron(II)-Diimine-Cyanide Complexes.= +ics of Per	INOR 0022
ntaining Proteins+ Tetranuclear	Iron(III) Complexes as Models for Nonheme-Iron-Co	INOR 0295
xytricarboxylic Acid Complexes of	Iron(III).= +Spectroscopic Studies of Alpha-Hydro	INOR 0040
Complexes of Iron(II) and	Iron(III).= Reactivity Studies on Cysteine	INOR 0370
mplexes as Models for Nonheme-	Iron-Containing Proteins.= +ranuclear Iron(III) Co	INOR 0295
r Protons in the Solutions of the	Iron-Ethylene Diamine Tetraacetic Acid Complexes.	INOR 0090
A Mechanism for the	Iron-Promoted Magnesium-Water Reaction.=	INOR 0394
Kinetics of Assembly of	Iron-Sulfur Clusters in Micellar Media.=	INOR 0076
rrections in XRF Analysis of Zinc,	Iron, and Total Sulfur in Zinc Ore Concentrates.=	ANYL 0057
Continental Shelf+ The Flux of	Iron, Cobalt, Copper, and Manganese from California	GEOC 0010
Photoproducts in UV-	Irradiated DNA.= Determining Pyrimidine	ANYL 0092
cs Using AM1 Simulated Sunlight	Irradiation of Semiconductor Catalysts.= +l Organi	INOR 0318
by Gamma	Irradiation.= Cross-Linking of Polyallylcarbonates	PMSE 0131
by Gamma	Irradiation.= Cross-Linking of Polyallylcarbonates	POLY 0053
Formed by Ion	Irradiation.= Tribological Properties of Structures	NUCL 0093
ases.= Enzyme-Generated	Irreversible Inhibitors of Hydroxysteroid Dehydrogen	MEDI 0010
rboxylase.= Enzyme-Activated	Irreversible Inhibitors of S-Adenosylmethionine Deca	BIOL 0052
Highly Selective Reversible and	Irreversible Ligands for the κ-Opioid Receptor.=	MEDI 0156
Polymer Films+ Reversible and	Irreversible Neutral Species Transfer in Electroactive	ANYL 0103
AXS Study of Mean-Field and	Ising Critical Behavior of Poly(2-Chloro-Styrene)/Po	PMSE 0103
Indiana Silicon Sphere (	ISiS) 4π Detector.=	NUCL 0036
ag+ Pesticide Residue in Barrier	Island Salt Marshes along the Florida Indian River L	ENVR 0019
Germ Initiation Factors 4F and (	ISO)4F Interact Differently with Oligoribonucleotide	BIOL 0128

ity and Geometric Constraint for	Iso-Structural Molecular Sieves Based upon Silicate	CATL	0028
uric Acid-Catalyzed Alkylations of	Isobutane with Alkenes.= +and Mechanism of Sulf	PETR	0085
talyst: Chemistry+ Alkylation of	Isobutane with Pentenes Using Sulfuric Acid as a Ca	PETR	0082
s.=	Isocyanate-Modified Polysilazane Ceramic Precursor	POLY	0339
n and Isomerization of Azides and	Isocyanates.= +nt. Dynamics of the Decompositio	PHYS	0188
Lengths in Poly(n-alkyl	Isocyanates).= Chiral Clues to Helical Sequence	POLY	0348
ion of Osmocen+ Synthesis of an	Isocyanide Complex of Osmium(KV) by Ceric Oxidat	INOR	0198
uced Coupling of Alkylidyne and	Isocyanide Ligands: Synthetic and Mechanistic Studi	INOR	0150
Polymerizations of Chrial	Isocyanides.= Stereoselection and Helicity in Living	POLY	0347
eparation of the Glucose Oxidase	Isoenzymes from Penicillium Amagasakiense by Ion-	ANYL	0016
Properties of Soybean	Isoflavones.= Identification and Antioxidant	AGFD	0020
f Cloned Biodegradative Genes in	Isolates from Deep Subsurface Environments.= +o	BIOT	0117
Total Synthesis of	Isolaurepinnacin.= Studies Directed Toward the	ORGN	0008
: Production, Properties, and U+	Isomalt (Palatinit®), a Versatile Alternative Sweetner	CARB	0005
Reactions: Structural	Isomer Effects on Reactivity.= Carbon-Cluster-Ion	PHYS	0311
Didactic Method for	Isomer Identification (between Two Molecules).=	CHED	0094
tion o+ Isopentenyl Diphosphate	Isomerase: An Electrophilic Mechanism for Isomeriza	BIOL	0014
ons in the Dynamics of Rotational	Isomerism.= +. Sizes of the Short-Range Correlati	POLY	0365
on: No Evidence for Bond-Stretch	Isomerism.= +hs as Determined by X-ray Diffracti	INOR	0088
Amorphous and Zeolitic	Isomerization Catalyst.= Improvements on Both	PETR	0115
Synthesis of Hydrocarbon	Isomerization Catalysts.= Scientific Foundations for	PETR	0135
k Stat+ Dynamics of Rotational	Isomerization in Polymers in Solution and in the Bul	COLL	0018
Cyclohexadiend.= Thermal	Isomerization of a Vinylcyclobutene to a	ORGN	0144
ynamics of the Decomposition and	Isomerization of Azides and Isocyanates.= +nt. D	PHYS	0188
f Pt-Mordenite Catalysts for the	Isomerization of C$_5$ and C$_6$ N-Alkanes in Light Strai	PETR	0113
NMR Assessment of the cis-trans	Isomerization of Gly^6-Bradykinin and Analogs.=	BIOL	0006
sed Catalys+ Oligomerization and	Isomerization of Olefins by Homogeneous Nickel-Ba	PETR	0007
matization, Oligomerization, and	Isomerization of Short-Chain Hydrocarbons over He	PETR	0082
matization, Oligomerization, and	Isomerization of Short-Chain Hydrocarbons over He	PETR	0130
matization, Oligomerization, and	Isomerization of Short-Chain Hydrocarbons over He	PETR	0043
matization, Oligomerization, and	Isomerization of Short-Chain Hydrocarbons over He	PETR	0112
matization, Oligomerization, and	Isomerization of Short-Chain Hydrocarbons over He	PETR	0007
matization, Oligomerization, and	Isomerization of Short-Chain Hydrocarbons over He	PETR	0100
matization, Oligomerization, and	Isomerization of Short-Chain Hydrocarbons over He	PETR	0025
matization, Oligomerization, and	Isomerization of Short-Chain Hydrocarbons over He	PETR	0061
An Electrophilic Mechanism for	Isomerization of Unactivated Carbon-Carbon Double	BIOL	0014
Vinylidene-Acetylene	Isomerization Process.= Ab Initio Study of the	PHYS	0287
Properties of Alkane	Isomerization Products.=	PETR	0105
Nuclesid+ NMR and Kinetics of	Isomerization Studies of Platinum(II) Complexes of	INOR	0403
n of Pd/HM Catalysts for Alkane	Isomerization.= +ation by Coking and Regeneratio	PETR	0130
w Developments in C$_5$-C$_7$ Paraffin	Isomerization.= +er Heterogeneous Catalysts. Ne	PETR	0112
-Port Mordenites in C$_8$ Aromatics	Isomerization.= +ic Properties of Small- and Large	PETR	0132
butylin: Application to the C$_8$ Cut	Isomerization.= +of Mordenites Modified by Tetra	PETR	0131
tion of Runs of Trans Rotational	Isomers in Single Racemic Poly(vinyl Chloride) Chai	POLY	0069
R Analyses of the cis and trans	Isomers of a Bicyclic Lactone Mosquito Repellent, 1,	AGRO	0033
lly Name the Different Possible	Isomers of Substituted Molecules Having That Geom	COMP	0035
SiO+, HCP, and HSiP, and Their	Isomers.= +Initio Energies and Shapes of HSiS+, H	PHYS	0315
Monostearate Structural	Isomers.= Chromatographic Separation of Sucrose	CARB	0050
ansport Properties of Polyimide	Isomers.= Conditioning Effect of CO$_2$ on the Gas Tr	POLY	0116
f bis(4-aminocyclohexyl) Methane	Isomers.= +line to a Nonthermodynamic Mixture o	I&EC	0056
ser Optical Tra+ Measuring the	Isometric Tension of Kinesin Molecules Using the La	BIOL	0137
Gr+ Toward the Solution of the	Isomorphism Problem in the Generation of Chemical	COMP	0024
Molecular Sieve AlPO$_4$-5.=	Isomorphous Substitutions of Silicon and Cobalt in	PETR	0099
opane Aromatiza+ Zeolite-β and	Isomorphously Substituted Ga-β as a Catalyst for Pr	PETR	0047
Srructure.= EXAFS Studies of	Isomorphously Substituted Zeolites with the ZSM-5	PETR	0081
Alkylation of	Isoparaffins with Olefins over Zeolite Catalysts.=	PETR	0101
Mechanism for Isomerization o+	Isopentenyl Diphosphate Isomerase: An Electrophilic	BIOL	0014
the Representation of	Isopolyanionic Structures.= Inexpensive Models for	CHED	0103
Transition Metal Heteropoly- and	Isopolyoxometalates.= +on Delocalization in Early	COMP	0032
Polymerization of Butadiene and	Isoprene by Single-Component Organoscandium Sy	INOR	0280
roperties of Macroporous Poly(N-	isopropylacrylamide) Hydrogels.= +paration and P	POLY	0243
cid and Its Phenylboronate and	Isopropylidene Derivatives in the Presence of Divalen	INOR	0263
phenylene)-32-crown-10 and 4,4'-	Isopropylindenediphenol (bisPhenol-A).= +y-1,3-	POLY	0044
rboxynaphthyl, -Quinolyl and -	Isoquinolyl Compounds as Bicyclic Ring-Constrained	MEDI	0023
ngeners Containing C-Nucleoside	Isosteres of Nicotinamide Riboside.= +ew NAD Co	CARB	0023
ynthesis of Fluoroolefin Dipeptide	Isosteres.= +Peterson Olefination Reaction to the S	ORGN	0079
stinctive Order of pK$_a$ Values of	Isostructural η^2-Dihydrogen Complexes: Fe < Os < R	INOR	0380
Thermoreversible Gelation of	Isotactic Polystyrene.=	POLY	0149
ges to Dielectric Properties during	Isothermal Epoxy Polymerization.= +ological Chan	PMSE	0178
te): Localization of Perturbatio+	Isothermal Physical Aging of Poly(methylmethacryla	PMSE	0173
Efficient Synthesis of ortho-	Isotoluenes and Diene-Allenes.=	ORGN	0098
Substances.= Use of Stable	Isotope Analysis for the Control of Natural Flavoring	AGFD	0195
Stable Carbon	Isotope Analysis of Coprocessing Materials.=	FUEL	0071
TPOP-IRMS Stable Carbon	Isotope Analysis of Heavy Petroleum Residues.=	PETR	0141
nments.= Flavor Adulteration.	Isotope Distributions in Secondary Plant Products an	AGFD	0194
A$_2$CuO$_4$-Based Superc+ Oxygen	Isotope Effect and Structural Phase Transitions in L	PHYS	0010
of α-Agostic and Steric Kinetic	Isotope Effects in Carbon-Carbon Bond-Forming Re	INOR	0147
Stable	Isotope Geochemistry of Azusfres Systems.=	I&EC	0015
ytical Techniques for Fossil Fuels:	Isotopes and Magnetic Resonance. Hy+ Novel Anal	FUEL	0070
esult of Sulfate-Reduci+ Sulfur	Isotopes in Copper Sulfide Corrosion Products as a R	BIOT	0100

Production of Actinide	Isotopes via Transfer Reactions.=	NUCL 0016
Combustion.= Chemical and	Isotopic Alteration of Organic Water during Stepped	ANYL 0061
copies. Multidimensional NMR of	Isotopically Enriched Proteins.= +Optical Spectros	PHYS 0113
Reaction of Hyaluronic Acid with	Isotopically Labeled Carbodiimides.= +ed from the	ORGN 0020
Biomass.= Production of Stable	Isotopically Labeled L-(+)-Lactice Acid Using Algal	BIOT 0156
cts+ Triplet Energy-Transfer in	Isotopically Mixed Naphthalene: High-Pressure Effe	PHYS 0081
-2-Methyl-N-(5-Substituted-3-	Isoxazolyl)-2H-1,2-Benzothiazine-3-Carboxamides,	MEDI 0178
Inhibition of the α, β, and γ	Isozymes of Tyrosinase.=	BIOL 0058
Impact of Fertil+ Environmental	Issues and Answers in Fertilizer and Soil Chemistry.	FERT 0013
The Retail Ferti+ Environmental	Issues and Answers in Fertilizer and Soil Chemistry.	FERT 0017
A User's Perspective of the	Issues and How They Are Handled.=	CINF 0044
and Vinyl Chloride Examples.=+	Issues in Cancer Risk Assessment: Perchloroethylene	CHAS 0011
s fo+ Global Sourcing—Strategic	Issues in Chemical Procurement. Chemical Specialtie	CMEC 0005
Journal Full Text: Copyright	Issues in Electronic Distribution.=	CINF 0046
Perspective.= Regulatory	Issues in Risk Management: A California	CHAS 0021
1. DNA Identification Testing and	Issues of Constitutional Privacy.= +y: Update 199	CHAL 0001
Privacy	Issues of DNA Databanks and Databases.=	CHAL 0008
Health and Safety	Issues under OSHA and EPA Programs.=	CHAS 0022
in Government/Farm Regulatory	Issues. +d Groundwater—Advances in Research and	AGRO 0051
PA Interface in Health and Safety	Issues. OSHA's Role in Chemical Proces+ OSHA-E	CHAS 0012
in Government/Farm Regulatory	Issues. +d Groundwater—Advances in Research and	AGRO 0060
ScienceGrasp!	It's Elementary.=	CHED 0075
for Detection of Adulterations of	Italian Cold-Pressed Citrus Essential Oils.= HRGC	AGFD 0199
Reactions of Group	IVB (Ti, Zr, Hf) Alkyls with Aluminum Reagents.=	INOR 0054
s in Food-Producing Ammimals.	Ivermectin and Abamectin Metabolism: Differences a	AGRO 0128
Peptides for I+ Baxter, Burdick,	Jackson Award Symposium: Juvenile Hormones and	AGRO 0001
Peptides for I+ Baxter, Burdick,	Jackson Award Symposium: Juvenile Hormones and	AGRO 0006
lve TCDD-Exposed Persons from	Jacksonville, Arkansas.= +rans in the Blood of Twe	CEI 0014
for Environmental Application in	Japan and Introduction of Novel Technique.= +ed	ENVR 0049
Actionable Information from	Japan.=	CINF 0012
gs and Their Residual Analysis in	Japan.= +ew of Feed Additives and Veterinary Dru	AGFD 0102
al Evaluation of a Seed from the	Jatropha Family as a Possible Source of Oil and Prot	AGFD 0038
New	Jersey Right-to-Know Program.=	CHAS 0037
New	Jersey.= Indoor Radon Levels in New York and	ENVR 0023
ith Catalytic Ozone Abatement in	Jet Aircraft.= +cial Development and Experience w	I&EC 0047
Reactions.= Using a Laminar	Jet To Determine the Overall Rate of Liquid–Liquid	I&EC 0007
he Use of Photoaffinity Analog of	JH II.= +ing Protein of Larval Manduca sexta by t	AGRO 0014
mone Using Radioligands.= Hot	JH: Studies of the Molecular Action of Juvenile Hor	AGRO 0005
°From Resume to	Job.=	CMEC 0026
°From Resume to	Job.=	I&EC 0005
Instrumentation Laboratory at	Johns Hopkins.= Physical Chemistry	CHED 0199
Instrumentation Laboratory at	Johns Hopkins.= Physical Chemistry	PHYS 0246
tribution.=	Journal Full Text: Copyright Issues in Electronic Dis	CINF 0046
Pyrethroid:	JS-88.= Synthesis and Activity of a New Type	AGRO 0015
+Sour Orange (Citrus aurantium)	Juice Using Neutral and Anion Exchange Resins.=	AGFD 0106
of Enzymatic Browning in Fruit	Juice with Cyclodextrins and Sulfated Polysaccharid	AGFD 0030
Stabilize Raw Fruit and Vegetable	Juices: Control of Enzymatic Browning.= +nts to	AGFD 0027
Competition Scheduled for USA,	July 11–22, 1992.= +rnational Chemistry Olympiad	CHED 0139
Semiconductor Electrode–Liquid	Junction.= +Interfacial Electron Transfer at a Solid	COLL 0088
of NSF Funding Opportunities for	Junior Faculty.= +Get a Fair Shake? An Analysis	YCC 0003
nd in Vitro Catabolism Study of	Juvenile Hormone in the Fifth Stadium Larvae of M	AGRO 0022
	Juvenile Hormone of Drosophila melanogaster.=	AGRO 0002
udies of the Molecular Action of	Juvenile Hormone Using Radioligands.= Hot JH: St	AGRO 0005
ng-Site Mapping of Hemolymph	Juvenile Hormone–Binding Protein of Larval Mandu	AGRO 0014
ick, Jackson Award Symposium:	Juvenile Hormones and Peptides for Insect Control.	AGRO 0001
ick, Jackson Award Symposium:	Juvenile Hormones and Peptides for Insect Control.	AGRO 0006
ect Control. Award Address. From	Juvenile Hormones to Diuretic Hormones.= +Ins	AGRO 0001
ol of Lepidopt+ Ethyl-Branched	Juvenile Hormones: Opportunities for Specific Contr	AGRO 0004
lecular Genetics of Resistance to	Juvenoid Insect Growth Regulators in Drosophila.=	AGRO 0101
Photoaffinity	Juvenoids.= Synthesis of Tritium-Labelled	AGRO 0016
Amino Acid Sequence of a	20K Xylanase from Trichoderma viride.=	BIOT 0211
Total Synthesis of (ndash)α-	Kainic Acid and (-)α-Allokainic Acid.=	ORGN 0177
Solar Neutrino Results from the	Kamiokande II Detector.=	NUCL 0023
Analysis of a 20-	Kd Xylanase of B. subtilis.= Structure-Function	BIOT 0212
ross-Linking Activities of the 14	KDa β-Galactoside-Specific Vertebrate Lectin and S	BIOL 0109
nthesis of Antigenic Heptose and	KDO Containing Oligosaccharides of the Inner Core	CARB 0058
ed Control Agents. Mechanisms of	Kdr Resistance.= +al, Synthetic, and Bioengineer	AGRO 0150
Structure of Polyethylene Shish	Kebabs in Situ.=	PMSE 0184
e Multiplet Energy Gap in Non-	Kekule Molecules: Synthesis, Structure, Reactivity, a	ORGN 0128
e Number of Perfect Matchings +	Kekule Structure Count in Benzenoid Chains and th	COMP 0033
Aromaticity,	Kekule Structures, and Conjugated Circuits.=	COMP 0021
J. J.	Keller and Associates: Publications and Products.=	CHAS 0029
droxo)bisμ-carboxylato)diiron(III)	Kernals.= +re Structure Consisting of fused (μ-Hy	INOR 0094
Catalysts. Shell Polygasoline and -	Kero Process.= +drocarbons over Heterogeneous	PETR 0025
ns of Trace Elements in Oil Shale	Kerogen.= +nation of Organic/Inorganic Associatio	FUEL 0045
Formation in Decane and	Kerosene Flames.= Comparison of Aromatics	FUEL 0080
uted Aromatics via Unsaturated	Ketene Intermediates Using Transition Metal–Media	ORGN 0274
Destabilized	Ketenes.=	ORGN 0125
e Sulfide) Ketones via Amorphous	Ketimine Intermediates.= +crystalline Poly(Arylen	PMSE 0123
Asymmetric Hydrogenation of β-	Keto Esters.= +lexes and Their Roles in Catalytic	ORGN 0044

Left fragment	Right fragment	Code	Num
es of Biochrom Films Based on 4-	Keto–Bacteriorhodopsin.= +ectrochromic Properti	BIOT	0082
Oxidoreductase (E_{od}) and CDP-4-	Keto-6-Deoxy-D-Glucose-3-Dehydrase (E_1).= +e	BIOL	0110
ity of (Alkyl)acyl, Diacyl, and (α-	Ketoacyl)acyl Complexes of Platinum.= +d Reactiv	INOR	0043
al Activit+ Novel Substituted β-	Ketoamide ACAT Inhibitors: Chemistry and Biologic	MEDI	0108
Gamma Radiolysis of α-	Ketoglutaric Acid in Aqueous Solutions.=	NUCL	0070
n, and Melting of Poly(aryl Ether	Ketone Ketone), Part I: Structure.= +Crystallizatio	POLY	0059
, and Melting of Poly(aryl Ether	Ketone Ketone), Part II: Crystallization and Melting	POLY	0060
tion in PEEK (Poly-ether-ether-	ketone) Films.= +scopic Study of Thermal Degrada	POLY	0334
Mechanisms.= Poly(aryl ether-	ketone) Synthesis via Competing S_NAR and $S_{RN}1$	PMSE	0122
Melting of Poly(aryl Ether Ketone	Ketone), Part I: Structure.= +Crystallization, and	POLY	0059
Melting of Poly(aryl Ether Ketone	Ketone), Part II: Crystallization and Melting.=	POLY	0060
Electrochemistry of Aromatic	Ketones in a Room–Temperature Molten Salt.=	ANYL	0094
f Several Aliphatic Aldehydes and	Ketones of Atmospheric Interest.= +tion Spectra o	ENVR	0024
icrystalline Poly(Arylene Sulfide)	Ketones via Amorphous Ketimine Intermediates.=	PMSE	0123
Asymmetric Hydrosilylation of	Ketones: A Facile Route to Ciral Alcohols Using Disc	INOR	0247
phorus Pentahalides with Cyclic	Ketones: A One–Step Preparation of 1,1,2–Trichloroc	ORGN	0263
onyls in the Presence of Aromatic	Ketones.= +hemically by d^8 Transition–Metal Carb	INOR	0254
or Asymmetric Hydrosilylation of	Ketones.= +hiral Rhodium Phosphine Complexes f	CATL	0005
n(II) Dicarbonyl to Aldehydes and	Ketones.= +Additions of Allyl(cyclopentadienyl)Iro	ORGN	0213
Poly(aryl Ether	Ketones).= Crystallization–Induced Gelation in	POLY	0247
ction of Dyes by Photogenerated	Ketyl Radicals in Photoelectrochemical (PEC) Cell.=	COLL	0081
ure and the Role of Biodiversit+	Key Elements for Implementing Sustainable Agricult	AGRO	0082
eu+ USP Reference Standards, a	Key Factor in Chromatographic Analysis of Pharmac	ANYL	0115
in Organic/Bioorganic Synthesis.	Keynote Address. +zymes as Catalysts.= Enzymes	BIOT	0089
Four	Keys to Chemical Safety and Incident Prevention.=	CHAS	0017
elopment and Validation Are the	Keys to Scientific Risk Assessments of Inhalation an	ENVR	0042
Behavior of	$KHSi_2O_5$.= Layered Silicates: The Protonation	PETR	0110
Chemicals: Getting the Chemical	Kicks out of Your Kitchen.= Hazards of Household	CHAS	0051
t, Present, and Future. Lindbergh	Kidnapping Case of 1932.= +ensic Chemistry—Pas	HIST	0006
–Hydroxybenzoic Acid by Chicken	Kidney Microsomes.= +–Phenoxybenzoic Acid to 3	AGRO	0019
he Determination of Neomycin in	Kidney Tissue.= +Chromatographic Procedure for t	AGFD	0100
Bases.= Inhibition of Protein	Kinase C by Sphingosine and Related Long–Chain	MEDI	0075
Probes of Protein	Kinase C Heterogeneity.= Natural Products as	MEDI	0077
valuation of the Selective Protein	Kinase C Inhibitor NPC 15437.= +armacological E	MEDI	0026
talytic Domain.= New Protein	Kinase C Inhibitors Designed to Interact with the Ca	MEDI	0076
t for Antiviral and Anti+ Protein	Kinase C Inhibitors. Protein Kinase C: A New Targe	MEDI	0072
A Novel Class of Selective Protein	Kinase C Inhibitors.= +minomethyl)–Piperidines:	MEDI	0025
tein Kinase C Inhibitors. Protein	Kinase C: A New Target for Antiviral and Anticance	MEDI	0072
Inhibitors of Protein	Kinase C.= Design and Synthesis of Ether Lipid	MEDI	0073
and Selective Inhibitors of Protein	Kinase C.= +and Bisindolylmaleimides as Potent	MEDI	0074
of Styryl–Based Protein Tyrosine	Kinase Inhibitors.= +Ring–Constrained Analogues	MEDI	0023
onformationally Flexible Tyrosine	Kinase Inhibitors.= +Molecular Field Analysis of C	MEDI	0024
s of Adenosine-5'-Phosphosulfate	Kinase.= +nzymatic Catalysis. Mechanistic Studie	BIOL	0008
asuring the Isometric Tension of	Kinesin Molecules Using the Laser Optical Trap.=	BIOL	0137
cation System.=	Kinetic and Structural Studies of the T4 DNA Repli	BIOL	0159
Phospholipase A_2: Structural and	Kinetic Characterization of Inhibition and Substrate	BIOL	0020
pectroscopic Observation of the	Kinetic Fate of a Thiamin–Bound Enamine Intermed	BIOL	0047
Spectroscopic Observation of the	Kinetic Fate of Thiamin Diphosphate–Bound Enam	BIOL	0048
asification of Carbon: Transient	Kinetic Investigations.= Potassium–Catalyzed CO_2 G	FUEL	0031
agnitude of α–Agostic and Steric	Kinetic Isotope Effects in Carbon–Carbon Bond–For	INOR	0147
e Oxidation: Thermodynamic and	Kinetic Mechanisitic Considerations.= +ing Benzen	FUEL	0085
nd Myosin–S1–NDP Binding to+	Kinetic Mechanism of Myosin Subfragment–1 (S1) a	BIOL	0103
Synthase: A Sequential–Ordered	Kinetic Mechanism or Random, with a High Degree	BIOL	0053
New	Kinetic Method Based on NMR Measurements.=	ANYL	0063
atization Reactions ov+ Use of a	Kinetic Model for Investigation of Light Olefin Arom	PETR	0061
ormation on the Surface of Ae+	Kinetic Model of the Chromium(VI) Oxide Clusters F	COLL	0099
Imaging.= Studies of	Kinetic Processes at Surfaces Using Atomic–Level	PHYS	0098
Shock Waves.=	Kinetic Processes with Metal Cluster Participation in	PHYS	0175
Unusual	Kinetic Properties of a Novel Enzymatic Acylation.=	BIOT	0160
The First	Kinetic Resolution of Chiral Azides.=	ORGN	0043
idations of Environmental Orga+	Kinetic Studies in Heterogeneous Photocatalyzed Ox	INOR	0318
Gas–Phase Metal Reactions—II.	Kinetic Studies of Boron and Aluminum Species.=	PHYS	0086
Fluoroketones.= NMR and	Kinetic Studies of Protease Inhibition by Peptidyl	BIOL	0005
Mordenite, L, Omega, and Offre+	Kinetic Studies of the Crystallization of the Zeolites	PETR	0041
exylalanine at P1 Site by Renin:	Kinetic Studies Using New Fluorogenic Substrates.=	BIOL	0035
late Urethane IPNs.=	Kinetic Study and Morphology Buildup in Methacry	PMSE	0043
Imidization of Poly(amic dialk+	Kinetic Study of in Situ Molecular Composite during	PMSE	0169
cis–Dichloro (1,4,7,10–Tetraaza+	Kinetic Study of the Base Hydrolysis of the syn,anti-	INOR	0398
Transient	Kinetic Study of the Gasification of Carbon in CO_2.=	FUEL	0009
Cross–Binding to Platinum(II): A	Kinetic Study.= Amino Acid–Nucleotide	INOR	0182
A	Kinetic Study.= Oxidation of 1,2–Diols by Fe(VI):	INOR	0401
the Mechanism+ Only Transient	Kinetic Techniques Will Provide Detailed Insight in	FUEL	0015
s of Metal Powders with Aqueo+	Kinetic/Mechanistic Investigations into the Reaction	INOR	0248
ion and Photoinitiated Reaction	Kinetices in Liquids by Photothermal Grating Spectr	ANYL	0034
e Metal Reactions—I. Chemical	Kinetics and Dynamics of the Mesospheric Sodium N	PHYS	0071
onducting Device Consisting of+	Kinetics and Mechanism of a Charge–Sensitive Ion–C	BIOL	0098
of Sulfuric Acid Composition on	Kinetics and Mechanism of Sulfuric Acid–Catalyzed	PETR	0085
nd Ferrate(V) with Amino Acids+	Kinetics and Mechanism. Reactivity of Ferrate(VI) a	INOR	0393
Chemisorption of NO on Char:	Kinetics and Mechanism.=	FUEL	0011
ed Polyester Interpenetr+ Rheo-	Kinetics and Morphology of Polyurethane–Unsaturat	PMSE	0027

Membranes.= Electrochemical	Kinetics and Nonequilibrium Potentials in Biological	COLL 0093
Copolymerization MMA/HEMA	Kinetics and Properties.= Microwave-Initiated	POLY 0052
nsaturated Radicals: n-C$_4$H$_5$ ++	Kinetics and Thermochemistry of the Oxidation of U	FUEL 0066
o the Measurement of Reaction	Kinetics and Thermodynamics at the Solid-Liquid In	COLL 0170
Hybrid Materials. Sol-Gel	Kinetics by NMR.= Preceramic and Inorganic	POLY 0256
n Transfer.=	Kinetics Effects of ζ-Potential on Interfacial Electro	COLL 0152
of Carbons.= Use of Transient	Kinetics for Determining the Reactive Surface Area	FUEL 0014
ds in a Temperature-+ Reaction	Kinetics for the Formation of Heterocyclic Compoun	AGFD 0073
Diffusion-Controlled Reaction	Kinetics in a Capillary: Reactant Segregation and An	PHYS 0273
ed Corrosion. Hydrodynamics and	Kinetics in Biofilm Systems. Recent Advances.=	BIOT 0096
C$_3$H$_3$ Reaction	Kinetics in Fuel-Rich Combustion.=	FUEL 0079
pping- and Quenching-Reaction	Kinetics in Low-Dimensional and Fractal Lattices.=	PHYS 0307
Energy	Kinetics in Nanometer Crystals.=	PHYS 0316
Application to Phase-Separation	Kinetics in PC/PMMA Blends.= +Description and	POLY 0278
ic Study of Epoxy/Amine = Cure	Kinetics in Thermal and Microwave Fields.= +nist	POLY 0161
Complexes.= Fast	Kinetics Investigation of Iridium Bromide	PHYS 0258
e in HMG-CoA Lyase Affects the	Kinetics of Active Site Modification.= +de Exchang	BIOL 0064
l Tension Measure+ Adsorption	Kinetics of Asphaltenes—A Liquid/Water Interfacia	PETR 0138
llar Media.=	Kinetics of Assembly of Iron-Sulfur Clusters in Mice	INOR 0076
	Kinetics of Diamond Film Growth.=	PHYS 0177
ctric Birefringence.=	Kinetics of DNA Cleavage by Calicheamicin from Ele	POLY 0280
nd Two-Dimensional NMR and	Kinetics of Formation of Platinum(II) Nucleotide.=	INOR 0180
echanisms. On the Combustion	Kinetics of Heterogeneous Char Particle Populations	FUEL 0026
d Boron Combustion.=	Kinetics of High-Temperature, Hydrocarbon-Assiste	PHYS 0142
el-Containing Zeolites.=	Kinetics of Hydroisomerization of N-Hexane on Nick	PETR 0116
e and Immiscible Blends.=	Kinetics of Imidization of Poly(amic Acid) in Miscibl	PMSE 0147
mplexes of Nuclesid+ NMR and	Kinetics of Isomerization Studies of Platinum(II) Co	INOR 0403
ermination of the Enthalpy and	Kinetics of Ligand Exchange Reactions by Photoacou	INOR 0192
Wide Temperture Ranges: Meas+	Kinetics of Metal Atom and Radical Reactions over	PHYS 0087
olith Catalyst.=	Kinetics of Methanol Combustion on Palladium Mon	COLL 0209
	Kinetics of NF(a$^1\Delta$) and NF(X$^3\Sigma$).=	PHYS 0331
m Coals durin+ Transformation	Kinetics of Organic Sulfur Forms in Argonne Premiu	FUEL 0059
Ion in Acidic Solution+ Complex	Kinetics of Oxidation of Thymol Blue with Bromate	PHYS 0269
Ligand Exchange and Reduction	Kinetics of Oxochromate(V) Complex by α-Hydrox	INOR 0402
Pressure, and Micellar Effects on	Kinetics of Peroxodisulfate Oxidation of Ternary Iro	INOR 0022
rpenetrating Polymer Networks.	Kinetics of Phase Separation in Polyurethane/Polyst	PMSE 0021
Organic Cosolvents on	Kinetics of Plasmin.= Effects of Water-Miscible	BIOL 0092
ionalized Trigly+ Crystallization	Kinetics of Poly(Ethylene Terephthalate) with Funct	PMSE 0171
ion+ Monomer Diffusion and the	Kinetics of Propagation in Free-Radical Polymerizat	POLY 0165
Mo.= Reaction	Kinetics of Propane Dehydrogenation over Zn or	I&EC 0075
lybdenum Supported+ Reaction	Kinetics of Propane Deydrogenation over Zinc or Mo	PETR 0087
Crystallization	Kinetics of Random Ethylene Copolymers.=	POLY 0075
Treated with Vapor at Hig+ The	Kinetics of Sulfonation of Sulfite-Pretreated Aspen	PMSE 0170
ds and [Ni(sal$_2$tim)].=	Kinetics of the Addition Reaction between N-N ligan	INOR 0061
emalonodialdehydes.=	Kinetics of the Hydrolysis of Substituted Benzyliden	ORGN 0152
Gas-Phase Metal Reactions—III.	Kinetics of the Molecular Oxygen Reaction with Sod	PHYS 0141
aterials.=	Kinetics of the Prehydrolysis of Xylan-Containing M	CARB 0044
with O(^{3}P) Atoms and CH$_2$(x^{3}B+	Kinetics of the Reactions of Aromatic Hydrocarbons	FUEL 0082
m on GaAs(100) and Implicatio+	Kinetics of Thermal Decomposition of Triethylgalliu	COLL 0068
e Tissue.=	Kinetics of Thermal Softening of Fruit and Vegetabl	AGFD 0013
ansfer Reactions by Flash Phot+	Kinetics of Thiyl Radical Formation and Electron Tr	INOR 0396
, Carbonate/Bicar+ Dissolution	Kinetics of UO$_2$.00 Pellets as a Function of Time, pH	INOR 0236
ethrin from the Surface Microl+	Kinetics of Volatilization of Fenitrothion and Deltam	ENVR 0020
ides with Vanadium(IV) in Aci+	Kinetics Studies of the Reaction of Alkyl Hydroperox	INOR 0276
Exciton	Kinetics Studies on Isolated Polymer Chains.=	PHYS 0309
AlH Gas-Phase Reaction	Kinetics.=	PHYS 0301
Alkaline Phosphatase	Kinetics.=	CHED 0183
Chemical	Kinetics.= Computer Simulation Experiment:	CHED 0157
Growth	Kinetics.= Filament-Assisted Diamond	PHYS 0178
the Chemical Kicks out of Your	Kitchen.= Hazards of Household Chemicals: Getting	CHAS 0051
sis of 7,7-Dimethyl-Quino[2,3,4-	kl]-5a,6,7,8-Tetrahydrobenz[c]Acridine 5N-Oxide, V	ORGN 0198
cture of K(Mg$_1$/3Nb$_2$/3)OPO$_4$ (	KMNP): +A Centrosymmetric Aliovalent Analogue of	INOR 0256
uasi-One-Dimensional Compound	KM$_o$O$_6$.= +erties of a New Modifiaction of the Q	INOR 0359
lenge of the 90s: MSDS Right-to-	Know Compliance Made User Friendly.= +the Chal	CHAS 0031
New Jersey Right-to-	Know Program.=	CHAS 0037
MSDS and Label Right-to-	Know Software.=	CHAS 0030
Right-to-	Know Support: Current and Future Compliance.=	CHAS 0026
ing to Risk: What Must We Yet	Know To Accomplish This Task? Relative Risk: A B	ENVR 0061
To	Know, To Do, and To Simulate.=	CHED 0113
es of Total Mercury in Crystalli+	Kriging Estimation of Differential Thermal Signatur	ENVR 0055
of Potassium Titanyl Phosphate (	KTP).= +A Centrosymmetric Aliovalent Analogue	INOR 0257
agonist with Renal Protective a+	KW-3902: A Potent and Selective Adenosine A$_1$ Ant	MEDI 0112
Methyl-D-Aspartate Antagonis+	Kynurenic Acid Analogues as Potent Glycine Site N-	MEDI 0146
Complexes with	La and Nd Porphyrins.= Monomer and Dimer	INOR 0221
o en Zeolitas del Tipo Pentasil en	la Aromatizacion de Propano.= +Contenido de Gali	COLL 0125
inated Aromatic Hydrocarbons in	La Basse Cote-Nord, Quebec.= +ng People to Chlor	CEI 0011
cion en las Materias Biologicas de	la Carrera de QFB en la Facultad de Quimica, UNA	CHED 0092
Asbestos en el Area Urbana de	la Ciudad de Mexico?.=	ENVR 0025
mico de Una Ca+ Evaluacion de	la Corrosion de las Diferentes Etapas del Lavado Qui	PHYS 0257
d de Ratas Hembras.+ Efecto de	la Deficiencia o Exceso de Vitamina E en la Fertilida	AGFD 0210

Derivados Hidrosolubles de	la Dehidroepiandrosterona.= Obtencion de	MEDI	0038
de Apoyos Computarizados para	la Ensenanza de la Quimica a Nivel Medio Superior.	CHED	0124
tados en+ Enriquecimiento de	la Ensenanza Experimental a Partir de Errores Detec	CHED	0178
los Alumnos en los Resultados de	la Evaluacion de una Materia Biologica.= +mico de	CHED	0093
ologicas de la Carrera de QFB en	la Facultad de Quimica, UNAM (Analisis Estadistico	CHED	0092
ca de las Plantas Pertenecientes a	la Familia Scrophulariaceae de Mexico.= +n Quimi	ORGN	0185
iencia o Exceso de Vitamina E en	la Fertilidad de Ratas Hembras.= +ecto de la Defic	AGFD	0210
Oxydants et des Complexants sur	la Fottation de ces Mineraux.= +-Pyrite: Effet des	ANYL	0067
CAM para el Diseno de Equipo de	la Industria Quimica de Proceso.= +entes de CAD/	COMP	0053
erizacion del Factor Inhibidor de	la Locomocion de los Monocitos (FILM) Producido p	BIOL	0102
por Ni en	la Molecula de Metano.= Activacion del Enlace C–H	PHYS	0244
Tonsil-Fe(NO$_3$)$_3$.= Estudio de	la Nitracion del Fenol con Aductos	COLL	0120
Orden de Reaccion de	la Oxidacion de la Vitamina C.= Obtencion del	CHED	0164
Electroreduccion de	la Perezona.= Determinacion del Mecanismo de	PHYS	0213
meros Zwitterionicos.+ Efecto de	la Presion en las Propiedades de Conduccion en Poli	ORGN	0159
Fundamentos de	la Produccion de Furfural.=	CARB	0051
putarizados para la Ensenanza de	la Quimica a Nivel Medio Superior.= +Apoyos Com	CHED	0124
de	la Quimica.= Encuentro de Dos Mundos a Traves	HIST	0035
ra d+ Factores Que Incrementan	la Reprobacion en las Materias Biologicas de la Carre	CHED	0092
e Alto Rendimiento: Influencia de	la Sulfonacion sobre las Propiedades Papelaras.=	PMSE	0077
cterizacion, y Determinacion de	la Transferencia de Carga por Espectroscopia IR de C	INOR	0190
idable 316/Pelicula de Oxido, en	la Validacion del Metodo Potencio-Metrico para Cua	ANYL	0066
de la Oxidacion de	la Vitamina C.= Obtencion del Orden de Reaccion	CHED	0164
Structure of	La$_2$-xSr$_x$NiO$_4$+δ.= Electrical Properties and	INOR	0048
d Structural Phase Transitions in	LA$_2$CuO$_4$-Based Superconductors.= +ope Effect an	PHYS	0010
New, a Revitalized Physical Chem	Lab Makes Its Debut.= +th Techniques That Are	CHED	0043
Practical	Lab Practical for General Chemistry.=	CHED	0058
e Case for the Advanced Teaching	Lab.= +l Chemistry Laboratory Development. Th	CHED	0041
Chem	Lab.= Experiments in Raman Spectroscopy for P.	CHED	0088
Physical Chemistry	Lab.= Experiments in Raman Spectroscopy for	PHYS	0300
Physical Chemistry	Lab.= Fluorescence Quenching Experiments for the	CHED	0202
Chemistry	Lab.= Updating an Undergraduate Physical	PHYS	0290
Chemistry	Lab.= Updating an Undergraduate Physical	CHED	0214
man. Nuclear Chemistry, the Met	Lab, and Nathan Sugarman—A Retrospective.=	NUCL	0058
MSDS and	Label Right–to–Know Software.=	CHAS	0030
Graph Theoretical Algorithm To	Label the Vertices of the Five Regular Polyhedra i	COMP	0035
Carbon–11–	Labeled Alfentanil.= Synthetic Studies for	ORGN	0192
C13839, a+ Synthesis of Tritium–	Labeled Aminoalkoxychromones, NPC16377 and NP	MEDI	0150
Hyaluronic Acid with Isotopically	Labeled Carbodiimides.= +ed from the Reaction of	ORGN	0020
Fluorescein–	Labeled Dextrans.= Fluorescence Polarization of	ANYL	0076
Production of Stable Isotopically	Labeled L–(+)–Lactice Acid Using Algal Biomass.=	BIOT	0156
Novel Spin–	Labeled Nucleoside.= Synthesis and Properties of a	ORGN	0026
Tritium–	Labeled Panadiplon.= Synthesis of Carbon–14– and	MEDI	0133
ion: Hydrogen– and Deuterium–	Labeled Poly(styrene) and Poly(vinyl methylether).=	COLL	0035
esis of Photozyme and Its Use in	Labeling Nucleic Acids for Nonradioactive Detection	CARB	0037
An Improved Method for Direct	Labeling of Monoclonal Antibodies with Tc99m.=	BIOT	0061
ced Moti+ Nitroxyl Amide Spin	Labeling: Methyl Esterification– and Hydration–Indu	AGFD	0012
Precursors for Photoaffinity	Labeling.= Phenoxydiazirines as Useful Carbene	ORGN	0105
Synthesis of Tritium–	Labelled Photoaffinity Juvenoids.=	AGRO	0016
Canada in Review of MSDS and	Labels Required for Hazardous Workplace Chemicals	CHAS	0034
(III)–μ–Sulfido Dimers Formed+	Labile and Coordinatively Unsaturated Molybdenum	INOR	0271
Electronic Structure of	LaBo$_3$ (B = Mn, Fe, Co, Ni) Perovskites.=	PHYS	0288
rization, and Catalytic Activity of	LaBo$_3$ (B = Ni, Mn, Co, Fe) Perovskites.= +aracte	PHYS	0225
terface: Regional Instrumentation	Laboratories and Inservice Teacher Training.= +In	CHED	0112
e Contributions of the Merck	Laboratories: The Impact of Steroid Research.= Som	HIST	0015
ulatory Training Requirements for	Laboratories.= +gh Education and Training. Reg	CHAS	0042
organic, and Organic Instructional	Laboratories.= +oscale Approach in the General, In	CHED	0150
Educational	Laboratories.= Handling of Waste Chemicals in	ENVR	0036
A Chemical Hygiene Plan for R D	Laboratories—A Case Study.= +ory and Practice.	CHAS	0001
Teachers.= Project	LABS—A Teacher–Scientist Interaction for K–12	CHED	0002
nd the Metallic State in Polymers	Lacking a Degenerate Ground State.= +Disorder a	PHYS	0039
Alterations of the β–	Lactam Antibiotic Biosynthetic Pathway.=	BIOT	0235
proaches to Determination of β–	Lactam Antibiotic Residues in Milk and Tissues.=	AGFD	0125
n, and Aldol Reaction of Chiral β–	Lactam Ester Enolates.= +tric Alkylation, Acylatio	ORGN	0046
Monocyclic β–	Lactam Inhibitors of Human Leukocyte Elastase.=	MEDI	0173
otent, and Specific Monocyclic β–	Lactam Inhibitors of Human PMN Elastase.= +e, P	BIOL	0049
ase in Acylating a Monocyclic β–	Lactam Intermediate in the Synthesis of Loracarbef,	BIOT	0161
Efrotomycin by Nocardia	lactamdurans.= Large-Scale Production of	BIOT	0109
ts.= A Reductive Amination–	Lactamization Procedure Using Borohydride Reagen	ORGN	0242
of 3–Hydroxy–4–Substituted–β–	Lactams as Potential Precursors of Norstatine, Statin	ORGN	0047
Elastase by Monocyclic β–	Lactams.= Mechanism of Inhibition of Human PMN	BIOL	0050
Metabolism in	Lactating Dairy Cows.= Novel Sulfonamide Drug	AGRO	0113
of Ametryn in Rats,	Lactating Goats, and Laying Hens.= Metabolic Fate	AGRO	0111
table Isotopically Labeled L–(+)–	Lactice Acid Using Algal Biomass.= +oduction of S	BIOT	0156
roperties and Conformation of β–	Lactoglobulin.= +erature and pH on the Foaming P	AGFD	0207
and trans Isomers of a Bicyclic	Lactone Mosquito Repellent, 1,1,4,5,6,7,8,8A–Octahy	AGRO	0033
Conversion of	Lactones into Substituted Cyclic Ethers.=	ORGN	0014
ular C+ Asymmetric Synthesis of	Lactones with High Enantioselectivity by Intramolec	ORGN	0042
nological Production of γ– and δ–	Lactones.= +s on the Biosynthesis and the Biotech	AGFD	0007
ylosuccinate L+ The Muconate–	Lactonizing Enzymes Mandelate Racemase and Aden	BIOL	0012

Biosynthesis of the X(3-Fucosyl-	Lactosamine) Antigenic Determinant.= +ed in the	CARB	0015
Production of	Lactosucrose and Its Properties.=	CARB	0008
Component of a Tropical Coastal	Lagoon.= +et in the Dominant Benthic Community	BIOL	0084
Tropical Coastal	Lagoon.= Energy Content of Macrocrustacea in a	BIOL	0105
hes along the Florida Indian River	Lagoon.= +ide Residue in Barrier Island Salt Mars	ENVR	0019
sistent Toxic Substances in Great	Lakes Areas of Concern.= +Zero Discharge of Per	ENVR	0083
ction Plan: Program for the Great	Lakes. Achieving Zero Discharge of Pe+ Remedial A	ENVR	0083
Great	Lakes.= Public Involvement in the Cleanup of the	ENVR	0085
α-Pyrone from Hyptis pectinata (	Lamiaceae).= +Pectinolide: A Novel Antimicrobial	MEDI	0060
d-Liquid Reactions.= Using a	Laminar Jet To Determine the Overall Rate of Liqui	I&EC	0007
udies Based on Flow-Reactor and	Laminar-Flame-Speed Data.= +idation Models: St	FUEL	0083
Phenolic IPNs for	Laminate Applications.=	PMSE	0007
rization of Vibration Damping of	Laminated Steel Sheet with Polyurethane/Acrylate I	PMSE	0057
(methylmethacrylate)/Glass Fiber	Laminates.= +l and Mechanical Properties of Poly	PMSE	0110
ps in Development of a Model for	Land-Use Planning and Conflict Resolution.= +Ga	ENVR	0043
Soil Excava+ Testing the Use of	Landfarming to Remediate Pesticide-Contaminated	AGRO	0144
ced Liquid Surfaces.=	Langmuir Lectures: Plenary Session. Surface-Introdu	COLL	0062
Electrochemical and	Langmuir Trough Studies of C_{60} Films.=	ANYL	0108
ular Orientation of Hemicyanine	Langmuir-Blodgett Films by Fourier Transform Infr	POLY	0224
Sensing of Phthalocyanine	Langmuir-Blodgett Films.= Diffusion in the Gas	POLY	0118
Excitation-Transfer in	Langmuir-Blodgett Films.= Fractals and	PHYS	0110
Nonlinear Optics in DCANP	Langmuir-Blodgett Films.= Guided-Wave	POLY	0176
r at Different Interfaces Using the	Langmuir-Blodgett Technique.= +-Chain Polyme	COLL	0071
Domains and Order in	Langmuir-Blodgett-Type Chromophore Systems.=	PHYS	0111
OEP)MX,+ Octaethylporphyrin	Lanthanide Chemistry: Synthesis and Reactivity of (	INOR	0294
nometallic Chemistry: S+ Direct	Lanthanide to Transition-Metal Interactions in Orga	INOR	0332
^{15}N NMR Study of	Lanthanide-Nitrate Complexes.=	INOR	0250
a Selective Solvent Extraction of	Lanthanum by Oxine in the Presence of Oxalic Acid	ANYL	0054
Study of Complex of	Lanthanum with Arsenazo-III.= Spectrophotometric	ANYL	0075
trophotometric Determination of	Lanthanum with Sulfonazo-III as Chromogenic Agen	ANYL	0074
nd the Design of+ Lead(II) and	Lanthanum(III) RNA Transesterification Catalysts a	INOR	0406
a	Larger Company.= Running a Small Business within	SCHB	0018
ile Hormone in the Fifth Stadium	Larvae of Manduca sexta.= +olism Study of Juven	AGRO	0022
ile Hormone-Binding Protein of	Larval Manduca sexta by the Use of Photoaffinity A	AGRO	0014
Evaluacion de la Corrosion de	las Diferentes Etapas del Lavado Quimico de Una Ca	PHYS	0257
Incrementan la Reprobacion en	las Materias Biologicas de la Carrera de QFB en la F	CHED	0092
eae d+ Composicion Quimica de	las Plantas Pertenecientes a la Familia Scrophulariac	ORGN	0185
ionicos.+ Efecto de la Presion en	las Propiedades de Conduccion en Polimeros Zwitter	ORGN	0159
Influencia de la Sulfonacion sobre	las Propiedades Papelaras.= +e Alto Rendimiento:	PMSE	0077
16 Utilizadas en el Seguimiento de	las Reacciones Acido Base en Medio Acuoso.= +y 3	ANYL	0055
YO Chemiluminescence from the	Laser Ablation of Yttrium Oxide and Chloride in Ox	PHYS	0255
ase Aluminum Atoms from Pulsed	Laser Ablation: Mechanisms and Applications.=	PHYS	0353
ent Molecules.=	Laser and Fourier Transform Spectroscopy of Transi	PHYS	0057
rfac+ Molecular Engineering by	Laser Chemical Vapor Deposition of a Nonlinear Inte	BIOT	0024
aracterization of Coal Macerals by	Laser Desorption Mass Spectrometry.= +etry. Ch	FUEL	0088
Studies of Copper(I) Intercalated	Laser Dye Coumarins.= +ctural and Luminescence	INOR	0245
Chlorine.= Thermal and	Laser Etching of Indium Phosphide by Molecular	COLL	0214
uate Physical Chemistry	Laser Experiments at Roanoke College.= Undergrad	CHED	0042
sis and Modification of Materials:	Laser Ionization Mass Spectrometry. +in the Analy	NUCL	0041
SEC-Viscometry-Right Angle	Laser Light Scattering (SEC-VIS-RALLS).=	PMSE	0098
with an On-Line Viscometry and	Laser Light Scattering.= +r Conformation by SEC	PMSE	0097
Coherent Laser Spectroscopy, and	Laser NMR Spectroscopy.= +ectroscopies. NMR,	PHYS	0040
on of Kinesin Molecules Using the	Laser Optical Trap.= +suring the Isometric Tensi	BIOL	0137
t Measurements by the Flash or	Laser Photolysis-Shock-Tube Method: Results for th	FUEL	0123
hotodissociation+ Tunable Diode	Laser Probe of Chlorine Atoms Produced from the P	PHYS	0299
$O(^{1}D)$.= Diode	Laser Probing of N_2O Following Collisions with	PHYS	0278
Deposition of Diamond Films.=+	Laser Spectroscopic Diagnostics for Chemical Vapor	PHYS	0151
Laboratory.= Time-Resolved	Laser Spectroscopy in the Undergraduate	CHED	0084
Laboratory.= Time-Resolved	Laser Spectroscopy in the Undergraduate	PHYS	0249
Clusters.= Infrared	Laser Spectroscopy of Supersonically Cooled Carbon	PHYS	0007
Infrared	Laser Spectroscopy of the C_2H Radical.=	PHYS	0332
l Detection Methods in Analytical	Laser Spectroscopy.= +undation). Thermo-Optica	ANYL	0011
ects of Ion/Solid Collisions Using	Laser Spectroscopy.= +es of the Fundamental Asp	NUCL	0041
Spectroscopies. NMR, Coherent	Laser Spectroscopy, and Laser NMR Spectroscopy.=	PHYS	0040
Species.= Dye	Laser Studies of Plasma and Flame-Generated	CHED	0087
ce of Depositing Films.=	Laser Studies of the Reactivity of NH with the Surfa	COLL	0218
-Stranded DNA.=	Laser Vaporization of High Molecular Weight, Single	ANYL	0130
Hands-On Teaching	Laser.=	PHYS	0303
Cells Using	Laser.= Introduction of Foreign Substance into	BIOT	0172
with a Helium-Neon	Laser.= Time-Resolved Thermal Lens Calorimetry	CHED	0089
mples by Phosphorus Monoxide	Laser-Excited Molecular Fluorescence Spectrometry	ANYL	0078
on of Temperatures by C_2 Emis+	Laser-Generated Hydrocarbon Plasmas: Determinati	PHYS	0268
ing M-N Bonds for the Prepara+	Laser-Induced Decomposition of Precursors Contain	POLY	0230
g of 10-Torr Methane/Oxygen+	Laser-Induced Fluorescence Diagnostics and Modelin	FUEL	0128
hemical Modeling of Diamond +	Laser-Induced Fluorescence Measurements for the C	PHYS	0147
Chemical Mechanisms of Prom+	Laser-Induced Fluorescence Measurements To Test	FUEL	0098
Experiments with	Laser-Polarized Rare Gases.= Magnetic Resonance	PHYS	0041
Applications of	Lasers in Chemistry.= Teaching the Principles and	CHED	0083
	Lasers in the Physical Chemistry Curriculum.=	CHED	0126
es: Surface Analysis When Intense	Lasers Miss the Surface.= +ts through Virtual Stat	NUCL	0044

Progress in Mode-Locked	Lasers.=	ANYL 0036
Development. How Expensive Are	Lasers, Really?.= +hysical Chemistry Laboratory	CHED 0082
ination of a Thionocarbonate in a	Latent Z-Hex-3-ene-1,5-diyne Unit.= +a the Elim	ORGN 0096
tions, Blends, and Interfaces—III.	Lateral Diffusion at the Air-Water Interface.= +lu	COLL 0051
ionation.= Characterization of	Latex Aggregates by Sedimentation Field-Flow Fract	PMSE 0029
amic Fingerprinting as a Probe of	Latex Characterization and Stability.= +Hydrodyn	COLL 0199
Gas Permeability Studies of	Latex Interpenetrating Polymer Network Films.=	PMSE 0061
Flow in	Latex Interpenetrating Polymer Networks.=	PMSE 0062
n the Damping Paint of PS/PA	Latex Interpenetrating Polymer Networks.= Study o	PMSE 0064
Study on the Three-Stage	Latex Interpenetrating Polymer Networks.=	PMSE 0082
Energy Transfer in	Latex Microdomains.=	PHYS 0108
	Latex PS/XPS Semi-IPNs.=	PMSE 0063
Proteinases from Papaya	Latex.= Conformational Differences in the Cysteine	BIOL 0074
l Properties of a Single-Acid-Site	Latex.= +ic Fingerprinting of the Surface Electrica	COLL 0200
es for U.S. Chemical Businesses in	Latin America. Latin American Chem+ Opportuniti	SCHB 0021
mall Chemical Business: U.S. and	Latin American Approach.= +d the Challenge for S	SCHB 0022
ical Businesses in Latin America.	Latin American Chemicals Approach 2000 AD.=	SCHB 0021
formation of a Porphyrin Cation	Lattice by Interfacial Electron-Transfer Gates Ionic	BIOL 0136
d Polymer Networks: Character+	Lattice Model for Segmental Orientation in Deforme	COLL 0016
dsorbant Interactions and Spin-	Lattice Relaxation of Sodium Cations in Zeolite-Y P	PHYS 0267
ate NMR Investigations of Zeolite	Lattice Structures.= +al High-Resolution Solid-St	PHYS 0348
Rings vs. Strings: Off-	Lattice, Bead-Stick Computer Simulations.=	POLY 0235
on of Homogeneous Cross-Linked	Lattices between Glycoconjugates and Lectins.= +i	BIOL 0145
ree-Dimensional Bounded Crystal	Lattices. +ate One-Electron States in Two- and Th	BIOT 0028
s in Low-Dimensional and Fractal	Lattices.= +ing- and Quenching-Reaction Kinetic	PHYS 0307
Transport in Ordered Polymer	Lattices.= Voltammetric Investigation of Proton	COLL 0167
nd at Surfaces. J. Polany+ Nobel	Laureates Symposium. Reaction Dynamics in Gases a	CONG 0011
Flottation Ionique du	Laurylsulfate de Gallium.=	ANYL 0053
sion de las Diferentes Etapas del	Lavado Quimico de Una Caldera por Tecnicas Electr	PHYS 0257
Marie	Lavoisier, the Other Parent of New Chemistry.=	HIST 0037
State: Oswego Harbor and the St.	Lawrence River at Massena.= +ncern in New York	ENVR 0084
Related to Laws and	Lawyers.= Favorite or Objectionable Cartoons	CHAL 0023
Thin-	Layer Chromatography in Food Analysis.=	AGFD 0180
Atomic	Layer Epitaxy.= Surface Chemistry of GaAs	COLL 0069
Removal of a Tin	Layer from Copper Substrate.= Electrochemical	PHYS 0285
Compounds in a Surface	Layer of Disperse Silica.= Reaction of Phosphorus	COLL 0100
iffusion and Corrosion at Atomic	Layer Resolution Studied by in Situ Electrochemical	PHYS 0127
Synthesis of Dioctahedral 2:1-	Layer Silicates in Acid and Fluorinated Medium.=	PETR 0128
Intercalated	Layer Silicates.= Direct Synthesis of Novel	PETR 0108
llared Clay Synthesis. Pillaring of	Layered Inorganic Compounds.= +Zeolites and Pi	PETR 0106
+ron Transfer with Porphyrins in	Layered Lithium Aluminate-Fatty Acid Systems.=	INOR 0272
, M+ Intercalating Properties of	Layered Metal(IV) Phosphates toward Organic Bases	PETR 0107
Design of Photoactive	Layered Phosphates.=	CATL 0021
ce, Raman, and X-ray Studies of	Layered Rare Earth Gold(I) Cyanide Salts: Evidence	INOR 0189
	i$_2$O$_5$.= Layered Silicates: The Protonation Behavior of KHS	PETR 0110
f Microporous Silica Derived from	Layered Siloxene by Oxidation.= +ion Properties o	PETR 0144
Experiments.= Surface	Layers from Lubricant Additives in Friction/Wear	PETR 0140
Growth of Silicide	Layers on Si Surfaces.=	PHYS 0100
esion of Bacterial Surface Protein	Layers.= +dity on the Wetting Properties and Adh	BIOT 0151
Nickel on Ultrathin SiO$_2$	Layers.= Medium-Energy Ion-Scattering Study of	COLL 0176
Lactating Goats, and	Laying Hens.= Metabolic Fate of Ametryn in Rats,	AGRO 0111
e Synthesis and Structure-Acti+	Lazarus Inhibitors of Human Leukocyte Elastase: Th	MEDI 0172
	Lazarus Inhibitors of Human Leukocyte Elastase.=	MEDI 0087
Photochromism of Diacetylene	LB Films by Polyion Complexation.= Control of	BIOT 0146
y with Polyion Complexed Pyrene	LB Films.= +otoamplified Storage Optical Memor	BIOT 0144
Energy Transfer in	LB Films.= Photoinduced Electron Transfer and	BIOT 0103
Photoinduced Proton Transfer in	LB Films.=	BIOT 0145
nd UV/Visible Detection to the	LC Separation and Determination of Methylene Blue	AGFD 0101
Two-Dimensional Separations by	LC-CZE.= Rapid Biopolymer Characterization.	ANYL 0126
n in+ LC/UV, LC/LC/UV, and	LC/LC/MS Methods for Determination of Ceftibute	ANYL 0118
ation of Ceftibuten in+ LC/UV,	LC/LC/UV, and LC/LC/MS Methods for Determin	ANYL 0118
by Electrospray Ionization	LC/MS and LC/MS/MS.= Analysis of Biopolymers	ANYL 0141
Atmospheric Pressure Ionization	LC/MS: Application to Drug Monitoring in Biologica	ANYL 0124
nd Vaporization for Particle Beam	LC/MS.= +ts in Aerosol Generation, Transport, a	ANYL 0122
Particle Beam and Thermospray	LC/MS.= Enhanced Sensitivity Studies with	ANYL 0123
pray and Continuous-Flow FAB	LC/MS.= Peptide and Protein Analysis by Electros	ANYL 0140
cts in Environmental Matrices by	LC/MS.= +hemicals and Their Degradation Produ	AGRO 0031
ray Ionization LC/MS and	LC/MS/MS.= Analysis of Biopolymers by Electrosp	ANYL 0141
Determination of Ceftibuten in +	LC/UV, LC/LC/UV, and LC/LC/MS Methods for	ANYL 0118
ehydrogenation Using Rh(PMe$_3$)	$_2$LCl(L = ^{i}Pr$_3$,CO): Aspects of Selectivity and Mecha	INOR 0151
TiO$_2$(001) Surfaces by Scanning+	Lcoal Structural Analysis of Reduced TiO$_2$(110) and	COLL 0171
ng-Chain Branching Analysis of	LDPE by SEC/LALLS and SEC/CV Systems.= Lo	PMSE 0047
xtended Type 1 Chain (Dimeric	Le^a or Le^b/Le^a) as Human Tumor-Associated Antige	CARB 0060
ded Type 1 Chain (Dimeric Le^a or	Le^b/Le^a) as Human Tumor-Associated Antigen.=	CARB 0060
nts Based on Polyalkenoates/Ion-	Leachable Glass Hybrid Systems.= +curable Ceme	POLY 0262
mining Whether Diffusion Is the	Leaching Mechanism of Portland-Cement-Based Wa	ENVR 0011
ion Sys+ Binding Chemistry and	Leaching Mechanisms of Stabilization and Solidificat	ENVR 0003
cts of Equilibrium Chemistry on	Leaching of Contaminants from Stabilized/Solidified	ENVR 0010
Marker for Detecting Exposure to	Lead.= +te Protoporphyrin (EP) as a Biochemical	CEI 0020
n Catalysts and the Design of +	Lead(II) and Lanthanum(III) RNA Transesterificatio	INOR 0406

Left Context	Title	Code	Number
Adsorption and Complexation of	Lead(II) and Mercury(II) with Crown Ether on the S	COLL	0104
zation and Control of Cadmium,	Lead, and Mercury from RDF Municipal Waste Com	FUEL	0130
Lung Tissue: A Com+ Cadmium,	Lead, Nickel, Cobalt, and Copper Concentrations in	ENVR	0040
one Calciu+ Structural Studies	Leading to a Receptor Binding Model for Benzazepin	MEDI	0188
Reactions	Leading to Formation of Heavy Elements.=	NUCL	0017
Model for Transfer Reactions	Leading to Heavy Actinides.=	NUCL	0018
e-Limiting Steps in the Pathways	Leading to Penicillin and Cephalosporin.= +he Rat	BIOT	0128
Production for Discovery of Drug	Leads. Light-Direct+ New Technologies in Peptide	MEDI	0090
Insecticide Resistance in a	Leafrolling Lepidopteran.= Molecular Genetics of	AGRO	0116
R+ Effect of High-Temperature	Lean Aging on the Performance of Pt, Rh/CeO$_2$ and	PETR	0035
y in the Gas Phase: what Can We	Learn from Reactions without Solvents?.= +emistr	ORGN	0121
°From Computer Room to	Learning Center.=	CHED	0116
Automated	Learning in Conformational Analysis—WIZARD III.=	CINF	0017
ole of Computers in Chemistry at	Lebanon Valley College.= +Chemical Education. R	CHED	0115
de Caseinato de Sodio a Partir de	Leche Contaminada con S. aureus y s+ Elaboracion	AGFD	0107
de Metales Pesados en	Leche de Establos del Sur del D.F.= Determinacion	ENVR	0037
of Humans as Affected by Use of	Lecithin Supplements Varied in Choline Content.=	AGFD	0034
-Galactoside-Specific Vertebrate	Lectin and Several β-Galactoside-Specific Plant Lec	BIOL	0109
Structural Studies of	Lectin-Carbohydrate Interactions.=	BIOL	0146
+mylase Inhibitors, Arcelins, and	Lectins Evolved from a Common Ancestor Gene.=	AGFD	0041
Their Binding by C-Type Animal	Lectins.= +tructure of Carbohydrate Ligands and	BIOL	0148
veral β-Galactoside-Specific Plant	Lectins.= +side-Specific Vertebrate Lectin and Se	BIOL	0109
ices between Glycoconjugates and	Lectins.= +ion of Homogeneous Cross-Linked Latt	BIOL	0145
of Dolichos Biflorus	Lectins.= Structure-Function Relationship Studies	BIOL	0149
ner Coating Systems—II. Plenary	Lecture. Application of Epoxy Resins in Co+ Contai	PMSE	0157
er Coatings Systems—I. Plenary	Lecture. Barrier Polymers for Container Coatings.=	PMSE	0138
i+ Reactive Processing. Plenary	Lecture. Distributive Mixing and Energy Distribution	PMSE	0175
icomponent Systems—II. Plenary	Lecture. Enh+ Solid-State Characterization of Mult	PMSE	0116
omponent Systems—IV. Plenary	Lecture. Hydrogen-Bonded Interpolymer Complexes	PMSE	0163
Nonaqueous Enzymology. Plenary	Lecture. Nonaqueous Bioprocessing: Protein Sepa+	BIOT	0063
ticomponent Systems—I. Plenary	Lecture. Soli+ Solid-State Characterization of Mul	PMSE	0099
omponent Systems—III. Plenary	Lecture. Structure and Properties of Ionomers in the	PMSE	0145
Surfaces.= Langmuir	Lectures: Plenary Session. Surface-Introduced Liquid	COLL	0062
Reflections on Steroids:	Lederle and Personal Aspects.= Historical	HIST	0023
an Expert Witness. Overview of	Legal Decision-Making Process and the Role of Expe	CHAL	0011
	Legal Primer for Librarians and Information Users.=	CINF	0047
	Legal Rights of Whistle-Blowing Professionals.=	PROF	0011
	Legal Scrutiny of DNA Typing Techniques.=	CHAL	0003
To Help Whistle-Blowers,	Legalize PEG.=	PROF	0012
Project SEED at	Lehman College.=	CHED	0140
Polymerases of	Leishmania mexicana.= Inhibitors of DNA	BIOL	0118
Gap; Why More Is Better. D. L.	Leister, Moderator.= Panel Discussion: Bridging the	I&EC	0058
Manipulation of Apparent Bond	Lengths as Determined by X-ray Diffraction: No Ev	INOR	0088
Chiral Clues to Helical Sequence	Lengths in Poly(n-alkyl Isocyanates).=	POLY	0348
Time-Resolved Thermal	Lens Calorimetry with a Helium-Neon Laser.=	CHED	0089
Genes in Oligophagous	Lepidoptera.= Cytochrome P-450 Monoxygenase	AGRO	0115
Glutathione Transferases in	Lepidoptera.= Plant Allelochemical-Adapted	AGRO	0119
des from Hemolymph of the	Lepidopteran Manduca sexta.= Autoparalytic Pepti	AGRO	0012
Peaches.= Residue Analyses of	Lepidopteran Pheromones in Apples, Grapes, and	AGRO	0148
Resistance in a Leafrolling	Lepidopteran.= Molecular Genetics of Insecticide	AGRO	0116
portunities for Specific Control of	Lepidopterans?.= +anched Juvenile Hormones: Op	AGRO	0004
e Physical Chemistry Laboratory:	Let's Get It Right.= +Education. Computers in th	CHED	0155
f Chemotactic Tripeptide F-Met-	Leu-Phe-OH.= +-Terminal Modified Analogues o	MEDI	0166
hanism of Acceptor Reactions of	Leuconostoc mesenteroides B-512FM Dextransucrase	CARB	0028
ti+ Lazarus Inhibitors of Human	Leukocyte Elastase: The Synthesis and Structure-Ac	MEDI	0172
Lazarus Inhibitors of Human	Leukocyte Elastase.=	MEDI	0087
Inhibitors of Human	Leukocyte Elastase.= Monocyclic β-Lactam	MEDI	0173
ienyl)Iron(II) Dicarbonyl to Ald+	Lewis Acid Promoted Additions of Allyl(cyclopentad	ORGN	0213
abases. New Developments in the	LHASA Program.= +is Planning and Reaction Dat	CINF	0001
Thermodynamic Properties of	Li$_3$.=	PHYS	0226
r Diffract+ Structural Studies of	Li$_4$NpO$_5$ and Li$_5$NpO$_6$ by Neutron and X-ray Powde	NUCL	0068
tructural Studies of Li$_4$NpO$_5$ and	Li$_5$NpO$_6$ by Neutron and X-ray Powder Diffraction.	NUCL	0068
Products	Liability Litigation.= Use of the Expert Witness in	CHAL	0015
ns of the Elctronic Spectra of	LiBe and LiMg.= Ab Initio Full SOCVCI Calculatio	PHYS	0338
Unusual Coordination Behavior of	LiBEt$_4$.= +Hydrides. Palladium Hydride Dimers:	INOR	0379
Legal Primer for	Librarians and Information Users.=	CINF	0047
Careers in Scientific	Librarianship.=	YCC	0009
Peptide Expression	Libraries as a Source of Chemical Diversity.=	MEDI	0092
and Synthesis of Large Peptide	Libraries Suitable for Discovery of Peptide-Macromo	MEDI	0093
Unrestricted Access: The	Library Dilemma.= Information Ownership vs.	CINF	0045
Dehydrogenase Complex and t+	Library of Monoclonal Antibodies to E. coli Pyruvate	BIOL	0031
or Search Services in an Academic	Library.= +d Access to Beilstein Online: Options f	COMP	0012
Agua/Agua-Acetona.= Energia	Libre de Transferencia de CuSO$_4$ en	PHYS	0236
Radicales	Libres con Tonsil.= Catalisis de Reacciones via	COLL	0119
=	Life Cycle Analysis: A Tool for Pollution Prevention.	CEI	0027
Replacement and Shelf-	Life Extension of Prepeeled Potatoes.= Bisulfite	AGFD	0058
Shelf-	Life Monitoring.= Time-Temperature Indicators for	AGFD	0080
The	Life of a Chemist: Career Possibilities in Mexico.=	CONG	0007
ry.= Phosphorus Chemistry. A	Life of Dedication to Research in Industrial Chemist	INOR	0321
metric Approach to Milk Shelf-	Life Prediction and Flavor-Quality Evaluation Based	AGFD	0108

emical Studies of Hot Corrosion:	Life Prediction Model for Marine Gas Turbine Blade	FUEL	0099
Bringing Chemistry to	Life through Decision–Making Activities.=	CHED	0054
thylbenzenes in Solution, Using+	Lifetime and Concentration Quenching of Triplet Me	PHYS	0265
of Distributions of Fluorescence	Lifetimes in Homogeneous and Heterogeneous Syste	PHYS	0134
ining Macrocycle: A New	Ligand Class.= Synthesis of a Tetraimidazole–Conta	ORGN	0060
es and Subst+ Studies on Mixed	Ligand Complexes of Pt(II) and Pd(II) with Dipeptid	INOR	0411
Excited States of Zinc(II) Mixed–	Ligand Complexes.= +d–Ligand Scharge–Transfer	INOR	0164
nance and Optical Spectroscopies.	Ligand Dynamics in Myoglobin.= +Magnetic Reso	PHYS	0193
ene Catalyzed by (Aqua)(Phosp+	Ligand Effects on the Aerobic Oxidation of Cyclohex	INOR	0067
(Oxo)(Phosphine)Ruthenium(I+	Ligand Effects on the redox Chemistry of a Series of	INOR	0068
ulfur–Bound Nickel: P2S2 vs. N2S2	Ligand Environments.= +allic Functionalities at S	INOR	0177
mate(V) Complex by α–Hydrox+	Ligand Exchange and Reduction Kinetics of Oxochro	INOR	0402
Complexes.= Facile Phosphine	Ligand Exchange between Square Planar Iridium(I)	INOR	0211
r of the Enthalpy and Kinetics of	Ligand Exchange Reactions by Photoacoustic Calori	INOR	0192
Reactions—IV. Study of Ligand–	Ligand Interactions in ML$_n$+ Systems.= +se Metal	PHYS	0171
cf Organosilanes to Form Metal–	Ligand Multiple Bonds with Tungsten Chlorophosph	INOR	0115
cmplex Containing the Cationic	Ligand N,N-Dimethylquaterpyridinium, Me$_2$qpy^{2+}.=	INOR	0233
Complexation of a Mixed–Donor	Ligand P2S with Copper(I).=	INOR	0029
cns of the Lowest π^* and Ligand–	Ligand Scharge–Transfer Excited States of Zinc(II)	INOR	0164
cs: A Force–Field Study of Mixed	Ligand Transition–Metal Compounds.= +Mechani	INOR	0075
rdination Studies of a Tripodal	Ligand with Mixed Donors P2S Toward Various Tran	INOR	0028
Complex with a Bifunctional	Ligand.= Cleavage of DNA by a Chiral Ru(II)	INOR	0232
Hexaamine	Ligand.= New Ag(I) Compounds with a	INOR	0037
exes Containing the Thiocarbonyl	Ligand.= +ructure of Substituted Chromium Compl	INOR	0336
cf the Structure and Dynamics of	Ligand–DNA Complexes.= +d–State NMR Studies	BIOL	0116
se Metal Reactions—IV. Study of	Ligand–Ligand Interactions in ML$_n$+ Systems.=	PHYS	0171
estigations of the Lowest π^* and	Ligand–Ligand Scharge–Transfer Excited States of Z	INOR	0164
ystal Structure.= Comparison of	Ligand–Mapping Studies with a High–Resolution Cr	CARB	0013
Precipitation of Proteins with	Ligand–Modified Phospholipids.= Affinity	BIOT	0020
ate Complexes for Cop+ Organic	Ligand–Stabiilized Copper(I) Hexafluoroacetylaceton	COLL	0156
Reaction between N–N	ligands and [Ni(sal$_2$tim)].= Kinetics of the Addition	INOR	0061
pects. Structure of Carbohydrate	Ligands and Their Binding by C–Type Animal Lecti	BIOL	0148
upling of Alkylidyne and Carbonyl	Ligands by Nucleophiles and by Electrophiles.=	INOR	0188
Synthesis of Dinucleating	Ligands Directed toward Urease Models.=	INOR	0089
–Substituted 2,2'–Bipyridine	Ligands for Polymer–Modified Electrodes.= Aniline	POLY	0189
e Reversible and Irreversible	Ligands for the κ–Opioid Receptor.= Highly Selectiv	MEDI	0156
α–Chymotrypsin.= Docking	Ligands into Receptors: The Test Case of	COMP	0039
plexes of Polyanionic Chelating	Ligands: Crystal Structures, NMR, and Electrochemi	INOR	0243
Ni Chemistry of Oxothiolate	Ligands: New Complexes with NiS$_2$O$_2$ Environment.=	INOR	0204
pling of Alkylidyne and Isocyanide	Ligands: Synthetic and Mechanistic Studies.= +u	INOR	0150
–4) with Bidentate and Tridentate	Ligands.= +emistry. Phosphoranide Anions (10–P	INOR	0344
Polyaminocarboxylate	Ligands.= Coordination of Bismuth(III) to	INOR	0033
Reactions between S–Donor	Ligands.= Extremely Rapid Au(I) Transfer	INOR	0235
anding tris(Pyrazolyl)hydroborato	Ligands.= +erivatives Stabilized by Sterically Dem	INOR	0087
Using Imidazole	Ligands.= New Copper–Dioxygen Species (Cu$_2$–O$_2$)	INOR	0045
N,S–Donor	Ligands.= Reactions of Nickel(II) with	INOR	0405
nding Profiles of Novel Piperidine	Ligands.= +tipsychotic Drugs: I. Synthesis and Bi	MEDI	0153
olinone CCK–B/Gastrin Receptor	Ligands.= +l Analysis of Alkyl–Substituted Quinaz	MEDI	0066
opyrans as Dopamine D1 Selective	Ligands.= +l-3,4–Dihydro–5–Hydroxy–1H–2–Benz	MEDI	0127
mplexes of Amine–Amide–Dithi+	Ligands.= Technetium(V) and Rhenium(V) Oxo Co	INOR	0181
nding poly(Pyrazolyl)hydroborato	Ligands.= +rivatives Stabilized by Sterically Dema	INOR	0085
3'–(4–pyridyl)terthiophene–Based	Ligands.= +y at Transition–Metal Centers via Poly	INOR	0027
Silyltellurolate and Selenolate	Ligands.= Volatile Metal Complexes with	INOR	0172
rogen–, Sulfur–, and Phosphorus–	Ligated Compounds. Side–on (μ–N$_2$) vs. End–+ Nit	INOR	0362
Unsaturated Ruthenium+ Oxy–	Ligated Compounds. Synthesis and Reactivity of an	INOR	0127
me c Oxidase.=	Ligation Dynamics of Carbon Monoxide in Cytochro	BIOL	0096
Incandescent Gas	Lighting: Development and Decline of	HIST	0038
sporium a+ Expression of Active	Lignin and Mn Peroxidases in Phanerochaete chryso	BIOT	0237
Chloroaniline	Lignin Degradation.= Bioavailability of a Natural	AGRO	0038
Compounds by Thermal	Lignin Degradation.= Formation of Smoke–Flavor	AGFD	0054
ification, and Copolymerization of	Lignin with Low Molecular Weight Components.=	BIOT	0041
etrating Network Products from	Lignin, Organostannanes, and Hydroxyl–Capped Pol	PMSE	0006
	Ligninase Expression in Heterologous Hosts.=	BIOT	0236
Aromatic Pollutants by	Ligninase Models.= Oxidative Degradation of	INOR	0187
and Liquefaction of Beulah–Zap	Lignite.= +Coal Samples. Moisture Removal from	FUEL	0001
osition for Efficient Hydrolysis of	Lignocellulosic Biomass.= +tion of Cellulases Comp	BIOT	0182
Accessibility on	Lignocellulosic Materials.= Increasing Enzymatic	BIOT	0250
Zirconia–	Lime Powders.= Stability of Suspensions of	COLL	0124
Elctronic Spectra of LiBe and	LiMg.= Ab Initio Full SOCVCI Calculations of the	PHYS	0338
ino Masses from Beta Decay—II.	Limit on the Electron Antineutrino Mass from Meas	NUCL	0010
Resolutio+ Spectroscopy at the	Limit: Pushing the Bounds of Sensitivity and Spatial	ANYL	0012
Iron	Limitation of Open–Ocean Primary Production.=	INOR	0002
Flavor Compounds. Glucosides of	Limonoids.= +ymatic Conversions of Precursors to	AGFD	0050
istry—Past, Present, and Future.	Lindbergh Kidnapping Case of 1932.= +ensic Chem	HIST	0006
Mechanical Studies of Electronic	Line Broadening in Liquids.= Statistical	PHYS	0035
ors by Electron–Spin–Resonance	Line Broadening of Surface Paramagnetic Probes.=	PHYS	0050
tors by Electron Spin Resonance	Line Broadening of Surface Paramagnetic Probes.=	PHYS	0284
Residues in Edible Meats by On–	Line HPLC Tandem Mass Spectrometry.= +ibiotic	AGFD	0061
A Microwave Device for the On–	Line Monitoring and Control of Properties of Fluid	I&EC	0059
On–	Line Monitoring of Microbial Biofilms.=	BIOT	0099

onformation by SEC with an On-	Line Viscometry and Laser Light Scattering.= +r C	PMSE 0097
jugates to a Colon Carcinoma Cell	Line.= +toxicity of ^{67}Cu-Porphyrin-Antibody Con	BIOT 0168
Reactions Using Multiple-Pulse	Line-Narrowing Techniques and ^{1}H T_1 Weighting.=	CATL 0022
f Vanadyl Acetylacetonate+ EPR	Line-Width Study of the Anisotropic Reorientation o	PHYS 0264
nd Spinodal Decomposition and	Linear and Cyclic Polystyrene/Poly(vinyl Methyl Eth	POLY 0295
rdered Conjugated Polymers in+	Linear and Nonlinear Optical Properties of Highly O	POLY 0008
and Viscoelasticity of	Linear and Ring Polyelectrolytes.= Light Scattering	POLY 0294
lt Chalcogenide Synthesis of the	Linear β-CsBiS$_2$ and the Three-Dimensional K$_2$Bi$_8$S	INOR 0202
Nonlinear Optical Properties of	Linear Chains.= Electron-Correlated States and	POLY 0007
ptical Methods. Dynamic Infrared	Linear Dichroism of Polymers.= +racterization: O	POLY 0303
Dynamics of Ring DNA vs.	Linear DNA from Transient Electric Birefringence.=	POLY 0296
Nonlinear Optical Properties of	Linear Epoxy Polymers with Pendant Sulfonyl Tolan	POLY 0208
erization of the Heterogeneity of	Linear Low-Density Polyethylene: Temperature-Risi	POLY 0076
ffect of Topology for Cyclic and	Linear Poly(dimethylsiloxane) and Poly(methylpheny	POLY 0234
tion.= Preparation of a	Linear Polycarbosilane via Ring-Opening Polymeriza	POLY 0362
minogen Expressed in Several Cell	Lines.= +ycosylation of Recombinant Human Plas	ANYL 0020
Modified Nicotiana tabacum Cell	Lines.= +eign Protein Production from Genetically	BIOT 0230
Method of	Lines.= Time-Dependent Flame Model Utilizing the	FUEL 0125
= Substituent-Induced Cross-	Link Aggregation in Polydimethylsiloxane Networks.	PMSE 0120
cibility and Properties of Cross-	Linkable Polyimide Blends with Rigid and Semirigid	PMSE 0023
ass Spectrometry (ES MS).=	Linkage Analysis in Disaccharides by Electrospray M	CARB 0048
Spin-Ground States.= Covalent	Linkage of Manganese Oxo Clusters with Nonzero	INOR 0196
olarity: Inorganic and Heterocyclic	Linkages.= +ophoric Assemblies with Enforced P	POLY 0215
of the Interaction of a Ferrocene-	Linked Acridine Orange Compound with DNA.=	ANYL 0109
Endonuclease DNA+ Psoralen-	Linked Deoxyoligonucleotide Synthesis for UvrABC	BIOL 0126
2 Stereocontrolled Synthesis of C-	Linked Deoxyribosides of 2-Hydroxypyridine and	ORGN 0075
rmation of Homogeneous Cross-	Linked Lattices between Glycoconjugates and Lectin	BIOL 0145
Cross-	Linked Polymers.= Structure and Properties of AB	PMSE 0083
Electron Transfer in Covalently	Linked Polynuclear Aromatic-Nucleoside and Oligon	PHYS 0021
rs: Synthesis of Uniformly Cross-	Linked Polystyrene Microbeads.= +l Filled Polyme	PMSE 0017
Electron Transfer in Covalently	Linked Porphyrin-RuO$_2$ Clusters.= Photoinduced	COLL 0133
Membranes: Effect of Cross-	Linked State and Annealing.= Gas Transport in IPN	PMSE 0060
Facile Synthesis of a Carboxylic	Linker on Halofuginone.=	ORGN 0201
omparative Study of the Cross-	Linking Activities of the 14 KDa β-Galactoside-Spec	BIOL 0109
ar Ordering and Chemical Cross-	Linking in Vitrified Nonlinear Optical Polymer Film	POLY 0206
Irradiation.= Cross-	Linking of Polyallylcarbonates by Gamma	PMSE 0131
Irradiation.= Cross-	Linking of Polyallylcarbonates by Gamma	POLY 0053
f Backbone Flexibility and Cross-	Linking on Flexibility of Can Coatings.= +Effect o	PMSE 0162
y Modifying the Substrate and by	Linking Resolutions in Series.= +s and Esterases b	BIOT 0091
and Bi+ Applications of Cross-	Linking Techniques to Enzyme/Protein Stabilization	BIOT 0220
Glutaraldehyde Cross-	Linking: Fast and Slow Modes.=	BIOT 0218
erview of Protein Chemical Cross-	Linking: Implications for Protein Stabilization.=	BIOT 0216
in with Low Mo+ Enzyme Cross-	Linking, Modification, and Copolymerization of Lign	BIOT 0041
s.= Specificity of Lipoprotein	Lipase from Pseudomonas sp. in Hydrolytic Reaction	BIOT 0095
strate Preferences for Pancreatic	Lipase-Mediated Exchange Reactions with Butteroil	AGFD 0036
reasing the Enantioselectivity of	Lipases and Esterases by Modifying the Substrate an	BIOT 0091
aceuticals and Materials.=	Lipases and Esterases in the Synthesis of New Pharm	BIOT 0093
ctions of Chiral Flavor Compou+	Lipases: Useful Biocatalysts for Enantioselective Rea	AGFD 0026
Baby Food.= Factors Affecting	Lipid Autoxidation of a Spray-Dried Milkbase for	AGFD 0158
nsisting of Metalloporphyrins in a	Lipid Bilayer.= +nsitive Ion-Conducting Device Co	BIOL 0098
er Gates Ionic Currents across the	Lipid Bilayer.= +ice by Interfacial Electron-Transf	BIOL 0136
s.= Mechanical Properties of	Lipid Bilayers Containing Light-Sensitive Compound	BIOT 0106
uture Biosens+ Self-Assembling	Lipid Bilayers on Solid Support: Building Blocks of F	BIOT 0102
Immobilized Biocatalysts in	Lipid Biotransformations.= Application of	BIOT 0040
of Lecithin Supp+ Blood Serum	Lipid Concentrations of Humans as Affected by Use	AGFD 0034
Voltammetry in	Lipid Films with Applications to Vesicles.=	COLL 0021
Design and Synthesis of Ether	Lipid Inhibitors of Protein Kinase C.=	MEDI 0073
Structural Aspects of Protein-	Lipid Interactions. Tertiary Structures of Fatty Acid	BIOL 0015
Influence of Food Processes on	Lipid Oxidation and Flavor Stability.=	AGFD 0157
ced Channel Cat+ Acceleration of	Lipid Oxidation during Cooking of Refrigerated Min	AGFD 0124
n in Foods. Sensory Evaluation of	Lipid Oxidation in Different Foods.= +id Oxidatio	AGFD 0156
	Lipid Oxidation in Foods via Redox Iron.=	AGFD 0094
s and Lipid Oxidation.=	Lipid Oxidation in Foods. Maillard Reaction Product	AGFD 0119
ative Processes in Foods.=	Lipid Oxidation in Foods. Mechanisms of Lipid Oxid	AGFD 0093
id Oxidation in Different Foods+	Lipid Oxidation in Foods. Sensory Evaluation of Lip	AGFD 0156
Role of Lipoxygenases in	Lipid Oxidation in Foods.=	AGFD 0122
ite-Free Composi+ Prevention of	Lipid Oxidation in Muscle Foods by Nitrite and Nitr	AGFD 0121
	Lipid Oxidation of Fish during Storage.=	AGFD 0123
Effect of	Lipid Oxidation on Food and Oil Quality.=	AGFD 0096
Chromatographic Analyses of	Lipid Oxidation Volatiles in Foods.= Update on Gas	AGFD 0098
	Lipid Oxidation: Effect on Meat Proteins.=	AGFD 0159
Maillard Reaction Products and	Lipid Oxidation.= Lipid Oxidation in Foods.	AGFD 0119
on in Foods. Mechanisms of	Lipid Oxidative Processes in Foods.= Lipid Oxidati	AGFD 0093
Fruit and Cell C+ Association of	Lipid Peroxidation and Chilling Injury in Cucumber	AGFD 0033
Mechanism of	Lipid Peroxidation in Muscle Foods: New Aspects.=	AGFD 0095
Flavor Compounds Formed from	Lipids by Heat Treatment.=	AGFD 0053
Cytotoxic Ether	Lipids.= Physicochemical Characterization of	MEDI 0031
Bipolar	Lipids.= Structural Organization of	BIOT 0122
-688,786 and a Series of Related	Lipopeptides with Potent Anti-Pneumocystis carinii	MEDI 0181
vel Avermectins Using Calculated	Lipophilicities.= +HPLC Retention Times for No	BIOT 0127

by RP-HPLC: Structure-Retent+	Lipophilicity of Teicoplanin Antibiotics as Revealed	MEDI	0059
i-Peptidic Compounds: Effect of	Lipophilicity, Dosing Vehicle, and Prodrug Modificat	MEDI	0196
ccharide Chains of Cell Envelope	Lipopolysaccharides of Salmonella and Shigella Bact	CARB	0017
arides of the Inner Core Region of	Lipopolysaccharides.= +O Containing Oligosacch	CARB	0058
ytic Reactions.= Specificity of	Lipoprotein Lipase from Pseudomonas sp. in Hydrol	BIOT	0095
ygenase (CO) Inhibitor into a 5-	Lipoxygenase (LO) Inhibitor—General Route to Nov	MEDI	0179
ynthesis and Cyclooxygenase/5-	Lipoxygenase Activity of 4-Hydroxy-2-Methyl-N-(5	MEDI	0178
Orally Active, and Selective 5-	Lipoxygenase Inhibitor.= A-69412, a Long-Lived,	MEDI	0176
hibitors of Cyclooxygenase and 5-	Lipoxygenase.= +xamic Acid Fenamates as Dual In	MEDI	0177
and Orally Active Inhibitor of 5-	Lipoxygenase.= +lopment of L-670,630 as a Potent	MEDI	0180
Role of	Lipoxygenases in Lipid Oxidation in Foods.=	AGFD	0122
Ultrafine Iron Carbide as	Liquefaction Catalyst Precursor.=	CATL	0051
Ultrafine Iron Carbide as	Liquefaction Catalyst Precursor.=	FUEL	0101
Hydrogen Transfer during Coal	Liquefaction Determined by $^2H/^1H$ Measurements.=	FUEL	0070
pes. Moisture Removal from and	Liquefaction of Beulah-Zap Lignite.= +Coal Sam	FUEL	0001
ical Techniques to Direct Coal	Liquefaction Oils.= Application of Advanced Analyt	FUEL	0044
Coals in the Catalytic Two-Stage	Liquefaction Process.= +and Quality of Bituminous	CATL	0053
Metals in	Liquefied Petroleum Gas.= Determination of Trace	PETR	0139
rption Kinetics of Asphaltenes—A	Liquid/Water Interfacial Tension Measurement.=	PETR	0138
s for Disposal and Reuse of Pickle	Liquors from Mexican Iron Industry.= +e Solution	ENVR	0039
apsule and Cation-Exchange +	Listeriolysin Purification Using a Cation-Exchange C	ANYL	0081
logy: What Is the Basis of EPA's	Listing of Sites for Superfund? Overview of the Stru	ENVR	0077
Chemical	Lists.= Chemical Accounting: Making Workplace	CHAS	0041
Science	Literacy for All Americans.=	ANYL	0010
Chemical	Literature?.= EC '92: What Does It Mean to	CINF	0013
Redirected Directed	Lithiation, III.=	ORGN	0085
ituted Saccharins through the o-	Lithiative Sulfonamidation of N,N-Diethylbenzamid	ORGN	0035
ansfer with Porphyrins in Layered	Lithium Aluminate-Fatty Acid Systems.= +ron Tr	INOR	0272
ydroxynitriles from Epoxides with	Lithium Cyanide in THF.= +s Preparation of β-H	ORGN	0107
ization of Diphenyl Sulfoxide by	Lithium Diisopropylamide: A Simple Route to Diben	ORGN	0084
for	Lithium.= Designed Spirocyclic Ionophores	ORGN	0062
Films with Applications in X-ray	Lithography.= +CVD Deposition of Diamond Thin	PHYS	0092
creased Phenolics Production in	Lithospermum erythrorhizon Suspension Cultures wi	BIOT	0229
Patent	Litigation.= Chemists as Expert Witnesses in	CHAL	0013
Liability	Litigation.= Use of the Expert Witness in Products	CHAL	0015
here: Experimental Stud+ Short-	Lived Molecules of Relevance in the Earth's Atmosp	PHYS	0017
Inhibitor.= A-69412, a Long-	Lived, Orally Active, and Selective 5-Lipoxygenase	MEDI	0176
Expression of Rat	Liver Cytosolic PEPCK in E. coli.= Subcloning and	BIOL	0107
s, Chemistry, and Activity against	Liver Flukes of Some Benzimidazole Derivatives.=	MEDI	0065
ide by ^{13}C NMR+ Detection of	Liver Microsomal Metabolites of N,N-Diethylbenzam	AGRO	0026
hy.= Purification of Pig	Liver Squalene Epoxidase by Affinity Chromatograp	BIOL	0070
enol+ Protective Effects against	Liver, Colon, and Tongue Carcinogenesis by Plant Ph	AGFD	0090
Xylanase Genes of Streptomyces	lividans and Their Homologous Expression.=	BIOT	0213
ve by Recombinant Streptomyces	lividans.= +for the Production of an SCD4 Derivati	BIOT	0086
in Accumulation by Streptomyces	lividans.= +s and Their Role in Heterologous Prote	BIOT	0084
evels in the Urine of Individuals	Living Adjacent to Chromate Production Waste Sites	CEI	0009
sis of the Volatile Constituents of	Living Odors and Flavors.= +e-Dimensional Analy	AGFD	0022
quences and Their Complexes in	Living Organisms as Possible Candidates for Princip	BIOT	0148
Stereoselection and Helicity in	Living Polymerizations of Chrial Isocyanides.=	POLY	0347
Chemical Modification of	LL-28249α.=	AGRO	0074
gical Characterization of 1-Octene	LLDPE (TREF) Fractions and Their Blends.= +lo	PMSE	0102
Microbiologia de Alimentos en el	LNSP: Enfoque Prospectivo.=	AGFD	0206
nhibitor into a 5-Lipoxygenase (	LO) Inhibitor—General Route to Novel Anti-Allergy	MEDI	0179
.=	Local Atomic Structure of Defects on Oxide Surfaces	PHYS	0207
he+ Chemistry with Constraints:	Local Density Calculations of Growth Reactions on t	PHYS	0355
ce of Fiber/Matrix Interface on	Local Glass-Transition Temperature in Composites.=	COLL	0055
Optical and NMR Studies of	Local Polymer Dynamics in Dilute Solution.=	PHYS	0139
The New York	Local Section Olympiad Selection Process.=	CHED	0012
g: STM and Related Techniques.	Local Structure and Electronic States of High-T_c Su	PHYS	0206
9 Carbon-13+ Determination of	Local Structure in Organofluorides Using Fluorine-1	FUEL	0077
r.= Photon Correlation and	Localization and Its Implications for Energy Transfe	PHYS	0064
tion in π-Stacks.= Electron	Localization in Conjugated π-Systems and Delocaliza	BIOT	0201
g of Poly(methylmethacrylate):	Localization of Perturbations in Thermomechanical P	PMSE	0173
and	Localization.= Peptide Models for Protein Folding	BIOL	0003
n Localized Microbial Activity and	Localized Corrosion on Metal Surfaces.= +betwee	BIOT	0098
on Meta+ Relationship between	Localized Microbial Activity and Localized Corrosion	BIOT	0098
Progress in Mode-	Locked Lasers.=	ANYL	0036
h Diastereoselective Oxidation of	Locked-Conformation Styrenes: Stereoselective Synt	ORGN	0236
zacion del Factor Inhibidor de la	Locomocion de los Monocitos (FILM) Producido por	BIOL	0102
Relationships with	Log P by HPLC and Biological Activity.=	ANYL	0069
Single-Electron	Logic: Molecular Implementation?.=	BIOT	0004
Biologicas.= Analisis	Longitudinal de Indices de Reprobacion en Materias	CHED	0091
New Way to	Look at Intermaterial Substitution.=	CMEC	0013
tor Particles in Aqueous Media: A	Look at Two Recent Controversies.= +Semiconduc	COLL	0131
ation Software Evaluation.=	Looking Back and Moving Ahead in Chemistry Educ	CHED	0080
ent and Disposal in the Mid-80+	Looking Back, Looking Forward: Pesticide Managem	AGRO	0086
Conformational Searching Impr+	Looking for Pharmacophores in 3D Databases: Does	CINF	0026
al in the Mid-80+ Looking Back,	Looking Forward: Pesticide Management and Dispos	AGRO	0086
aracterization of a Constrained Ω-	Loop Excised from Interleukin-1α.= +tructural Ch	MEDI	0165
S$_2$-AAGCG): An Unusual Hairpin	Loop.= +tide Phosphorodithioate d(CGCTpS$_2$-TP	BIOL	0123

Intermediate in the Synthesis of	Loracarbef, a New Antibiotic.= +nocyclic β–Lactam	BIOT	0161
nfluencia del Nivel Academico de	los Alumnos en los Resultados de la Evaluacion de u	CHED	0093
sables en Geotermica de Fasico de	los Azusfres.= +rtamiento de los Gazas en Conden	I&EC	0013
dos con Boro.= Estudio de	los Catalizadores CoMo/Al$_2$O$_3$ y Mo/Al$_2$O$_3$ Modifica	FUEL	0104
Estudio del Comportamiento de	los Gazas en Condensables en Geotermica de Fasico	I&EC	0013
tor Inhibidor de la Locomocion de	los Monocitos (FILM) Producido por E. +on del Fac	BIOL	0102
el Academico de los Alumnos en	los Resultados de la Evaluacion de una Materia Biolo	CHED	0093
Polymer	Lovely as a Tree.= I Think I Shall Never See a	POLY	0319
Single–Source Precursors for	LPCVD of Silicon Carbide.= Disilacyclobutanes as	INOR	0105
is of a New Gallophosphate of the	LTA Type.= +ct of Fluoride Anions in the Synthes	PETR	0075
to an Alkali Metal: Strucutre of	Lu[CH(SiMe$_3$)$_2$]$_3$(μ–Cl)K(η^6–C$_7$H$_8$)$_2$ and the Solubiliz	INOR	0053
Tm, Yb,	Lu) Structure Type.= Novel R$_2$Ba$_{1.25}$NiO$_{5.25}$ (R =	INOR	0034
Surface Layers from	Lubricant Additives in Friction/Wear Experiments.=	PETR	0140
he Low–Temperature Behavior of	Lubricants.= +. Dynamic Rheological Studies on t	TECH	0014
= Synthetic Steroids and the	Luftwaffe: Evolution of Metal–Ammonia Reductions.	HIST	0013
18c6)$_4$MnCl$_4$][TlCl$_4$]$_2$ in Presen+	Luminescence Characteristics of Mn^{2+} in Cubic [(Rb	INOR	0039
erse Systems.= Adsorption and	Luminescence of Charge–Transfer Complexes in Disp	COLL	0110
sensitive Metalorganic Systems.	Luminescence Probes of DNA–Binding Interactions.=	INOR	0315
	Luminescence Spectroscopy of F–Metal Complexes.=	INOR	0062
r Dye Coumarin+ Structural and	Luminescence Studies of Copper(I) Intercalated Lase	INOR	0245
els for Semiconductor Particles:	Luminescence Studies of Inorganic Cluster Molecules	INOR	0317
Rar+ Temperature–Dependent	Luminescence, Raman, and X–ray Studies of Layered	INOR	0189
Photosensitive Polymers Using a	Luminescent Rhenium Carbonyl Complex as a Spect	POLY	0096
of Uranium–235 in Meteorites and	Lunar Samples.= +ience—II. Initial Abundances	NUCL	0082
244 in Meteorites and	Lunar Samples.= Initial Abundances of Plutonium–	NUCL	0083
lt, and Copper Concentrations in	Lung Tissue: A Comparative Study in 84 Cases from	ENVR	0040
co–Specific Nitrosamine–Induced	Lung Tumorigenesis by Green Tea and Its Compone	AGFD	0068
Protection against Stomach,	Lung, and Esophagus Carcinogenesis by Green Tea.=	AGFD	0067
Metabolism of	Luprostiol in Dairy Cows.= Residue Depletion and	AGRO	0130
isulfide Exchange in HMG–CoA	Lyase Affects the Kinetics of Active Site Modificatio	BIOL	0064
nistic Studies on Mandelonitrile	Lyase by Covalent and Noncovalent Modification.=	BIOL	0075
the Inactivation of Mandelonitrile	Lyase by Phenylglyoxal.= +oducts Formed during	BIOL	0087
Racemase and Adenylosuccinate	Lyase: Odd Couples in the Evolution of Mechanistic	BIOL	0012
ole Moments for the Nine Lowest	Lying Electronic States of the CuCl Molecule.= +p	PHYS	0304
ical Investigation of the Low–	Lying States of Butadiene Radical Cation.= Theoret	PHYS	0237
ytes Displaying Thermotropic and	Lyotropic Behavior.= +uid Crystalline Polyelectrol	POLY	0073
Reaction Mediu+ Biocatalysis in	Lyotropic Mesophases—Characterization of a Novel	BIOT	0248
Characterization of	Lyotropic Polysaccharide Liquid–Crystal Blends.:	COLL	0017
Characterization of Degradatio+	Lysimeter Studies on Long–Term Fate of Pesticides:	AGRO	0025
An Organic Radical in the	Lysine 2,3–Aminomutase Reaction.=	BIOL	0077
Require Adenosylcobalamin: The	Lysine 2,3–Aminomutase Reaction.= +at Does Not	BIOL	0023
Metal Cofactors of	Lysine 2,3–Aminomutase.=	BIOL	0078
roblast Monolayers:+ Studies of	Lysosomotropic Enzyme Substrates with Human Fib	BIOL	0079
terminants Derived from Hen Egg	Lysozyme 81–96.= +ties of Glycopeptide T–Cell De	CARB	0014
pecific AA Residues Change the	Lytic Activity of a Cobra Venom Cytotoxin: Structur	BIOL	0085
olybdenum Films from (η^3–C$_3$H$_5$)	5M (M = MO, W).= +eposition of Tungsten and M	INOR	0055
of	M$_x$C$_{60}$.= Synthesis and Characterization	PHYS	0047
in	MACCS–3D.= Database Structure and Searching	CINF	0022
ctivity in Steam G+ Influence of	Maceral Composition of Parent Coal on Potassium A	FUEL	0027
ometry. Characterization of Coal	Macerals by Laser Desorption Mass Spectrometry.=	FUEL	0088
A Tool for the Revitalization of +	Macintosh Computers as Data Acquisition Devices—	CHED	0160
Energy Content of	Macrocrustacea in a Tropical Coastal Lagoon.=	BIOL	0105
iolate C+ An N$_2$S$_2$–Copper(II)	Macrocycle and a Pentacoordinate CuN$_3$OS Pseudoth	INOR	0262
Tetraimidazole–Containing	Macrocycle: A New Ligand Class.= Synthesis of a	ORGN	0060
60–Crown–20	Macrocycles.= Thermoreversible Gelation of	POLY	0246
Synthesis of	Macrocyclic Bifunctional Chelating Agents.=	ORGN	0024
ide Chemistry: Synthesis and R+	Macrocyclic Complexes. Octaethylporphyrin Lanthan	INOR	0294
Potent, Orally Active Diester	Macrocyclic Human Renin Inhibitors.= Highly	MEDI	0123
Triphosphazanes.= New	Macrocyclic Orthophenylene–Stabilized	INOR	0391
ynthesis and Characterization of	Macrocyclic Polystrene: Effect of Molecular Weight o	POLY	0239
omolar, Orally Active Cysteine+	Macrocyclic Renin Inhibitors: Synthesis of a Subnan	MEDI	0122
Synthesis of	Macrolactone Pyrrolizidine Alkaloids.=	ORGN	0285
rced Polarity: Inorganic and He+	Macromolecular Chromophoric Assemblies with Enfo	POLY	0215
NMR Investigations of	Macromolecular Dynamics.=	PHYS	0195
Suitable for Discovery of Peptide–	Macromolecular Interactions.= +Peptide Libraries	MEDI	0093
Soluble Synthetic and Biological	Macromolecules by Flow Field–Flow Fractionation.=	PMSE	0011
: Hyperbranc+ Novel Dendritic	Macromolecules by the Convergent Growth Approach	POLY	0318
Award Presentation: Introducing	Macromolecules in Precollege Chemistry.= +1990	CHED	0025
istributions and Diffusion within	Macromolecules Observed by Intramolecular Energy	PHYS	0061
Molecular Recognition of	Macromolecules: The Utility of Hybrid Molecules.=	ORGN	0270
Alumoxane	Macromolecules.= New Structural Model for	POLY	0341
sentation: Addressing Polymers as	Macromolecules.= +91 Award Winner—Award Pre	CHED	0027
on with Oligomeric Diblock	Macromonomer Stabilizers: Emulsion Polymerizati	COLL	0039
es of Novel, Zero–Calorie Sugar	Macronutrient Substitutes.= Synthesis and Properti	CARB	0026
s+ Preparation and Properties of	Macroporous Poly(N–isopropylacrylamide) Hydrogel	POLY	0243
toons Appearing in CHEMTECH	Magazine over the Past 20 Years.= +of Favorite Car	CHAL	0024
Monomeric Zinc Hydride and	Magnesium Alkyl Derivatives Stabilized by Sterically	INOR	0087
Synthesis of	Magnesium Hydride from Silanes.=	INOR	0206
hamycin C Biosynthesi+ Role of	Magnesium Phosphate in Ammonium Control of Cep	BIOT	0088
Iron–Promoted	Magnesium–Water Reaction.= A Mechanism for the	INOR	0394

ar Oxygen Reaction with Sodium,	Magnesium, and Copper Atoms.= +of the Molecul	PHYS	0141
heometer.=	Magnet-Enhanced Optical Falling-Needle/Sphere R	POLY	0251
Monophosphate Tungsten Bro+	Magnetic and Electronic Transport Properties of the	INOR	0352
Suspension in Externally Imposed	Magnetic and Hydrodynamic Fields.= +c Colloidal	COLL	0144
X-ray Diffraction and	Magnetic Anomaly in $Ba_2MnCu_3O_x$.=	INOR	0257
Viscoelastic Media.= Dynamic	Magnetic Birefringence as a Diagnostic Tool in	COLL	0143
Magnetic and Hy+ Response of	Magnetic Colloidal Suspension in Externally Imposed	COLL	0144
erals in S+ Influence of a Weak	Magnetic Field on the Colloidal State of Several Mat	COLL	0160
roperties in CR(III) Complexes+	Magnetic Field Perturbations of the Excited-State P	INOR	0159
Structure and Low-Dimensional	Magnetic Interactions in Second-Row Transition Ele	INOR	0267
rs under Argon Stream and Their	Magnetic Properties.= +Nongraphitizable Polyme	POLY	0023
ogeneous Systems.=	Magnetic Relaxation Coupling in Dynamically Heter	PHYS	0232
ion of the Hydrox+ ^{13}C Nuclear	Magnetic Relaxation of Amylose and Dynamic Behav	CARB	0049
Developments and Applications of	Magnetic Resonance and Optical Spectroscopies.	PHYS	0136
Developments and Applications of	Magnetic Resonance and Optical Spectroscopies.	PHYS	0193
Developments and Applications of	Magnetic Resonance and Optical Spectroscopies.	PHYS	0113
Developments and Applications of	Magnetic Resonance and Optical Spectroscopies.	PHYS	0372
Developments and Applications of	Magnetic Resonance and Optical Spectroscopies.	PHYS	0066
Developments and Applications of	Magnetic Resonance and Optical Spectroscopies.	PHYS	0166
Developments and Applications of	Magnetic Resonance and Optical Spectroscopies.	PHYS	0040
Developments and Applications of	Magnetic Resonance and Optical Spectroscopies.	PHYS	0370
Developments and Applications of	Magnetic Resonance and Optical Spectroscopies.	PHYS	0373
Developments and Applications of	Magnetic Resonance and Optical Spectroscopies.	PHYS	0344
ate Laborat+ Improved Nuclear	Magnetic Resonance Capabilities for the Undergradu	CHED	0185
ed Rare Gases.=	Magnetic Resonance Experiments with Laser-Polariz	PHYS	0041
meric Molecular Probes for F-19	Magnetic Resonance Imaging: Synthesis and Charact	MEDI	0041
Network Glasses.=	Magnetic Resonance Probes of Short-Range Order in	PHYS	0374
inerals as Studied by ^2D Nuclear	Magnetic Resonance Spectroscopy.= +r with Clay M	AGRO	0061
Teaching	Magnetic Resonance Using Computer Animation.=	CHED	0123
ques for Fossil Fuels: Isotopes and	Magnetic Resonance. Hy+ Novel Analytical Techni	FUEL	0070
Ultrasensitive	Magnetic Resonance.= New Approaches to	PHYS	0042
phates by Phosphorus-31 Nuclear	Magnetic Resonance.= +Analysis of Inorganic Phos	INOR	0291
by Selective Excitation of Nuclear	Magnetic Resonance.= +ics at Catalyst Surfaces	PHYS	0346
atalyst.=	Magnetic Separation of Reduced Crude Conversion C	I&EC	0063
d Bi_2Sr_2C+ Measurement of the	Magnetic-Field Penetration Depth in $YBa_2Cu_3O_x$ an	PHYS	0284
d Bi_2Sr_2C+ Measurement of the	Magnetic-Field Penetration Depth in $YBa_2Cu_3O_x$ an	PHYS	0050
rticles in External Fields: Electric,	Magnetic, and Hydrodynamic—I. +les: Colloidal Pa	COLL	0141
rticles in External Fields: Electric,	Magnetic, and Hydrodynamic—II. +es: Colloidal Pa	COLL	0159
rticles in External Fields: Electric,	Magnetic, and Hydrodynamic—III. +s: Colloidal Pa	COLL	0178
rticles in External Fields: Electric,	Magnetic, and Hydrodynamic—IV. +s: Colloidal Pa	COLL	0199
Fe-Pd+ Electronic Structure and	Magnetism in Ternary Compounds of Ni-Fe-Pt, Ni-	PHYS	0253
ne in the Mechanical Properties of	Magnetite Ore Pellets.= +ct of Manganese Additio	I&EC	0067
Aggregation of	Magnetizable Colloids.=	COLL	0161
Inorganic Particles in	Magnetotactic Bacteria.=	INOR	0118
r Routes to Inorganic Materials.	Magnets Based on Molecular Chemistry.= Molecula	INOR	0100
Lipid Oxidation in Foods.	Maillard Reaction Products and Lipid Oxidation.=	AGFD	0119
in Model	Maillard Reactions.= Peptides as Flavor Precursors	AGFD	0072
uly—Waldoboro Pluton Complex,	Maine.= +tal Mercury in Crystalline Rock: Case St	ENVR	0055
oithophore Blooms in the Gulf of	Maine.= +iopropionate (DMSP) in Mesoscale Cocc	GEOC	0002
a Revitalized Physical Chem Lab	Makes Its Debut.= +th Techniques That Are New,	CHED	0043
Pretreatment with N-Substituted	Maleimides.= +ilization of Poly(vinyl Chloride) by	POLY	0027
unds Found in Grape Wine due to	Malolactic Fermentation.= +e Odor-Active Compo	AGFD	0137
Formation of Formaldehyde and	Malonaldehyde from Photooxidation of Squalene.=	AGFD	0132
Viable Cell Biomass in	Mammalian Cell Bioreactors.= Determination of	BIOT	0012
Microcarriers for	Mammalian Cell Culture.= Polystyrene Porous	BIOT	0010
Secretory Phospholipase A_2 from	Mammalian Cell Culture.= +Recombinant Human	BIOL	0072
reactor Design and Optimization.	Mammalian Cell Retention in a Spinfilter Perfusion	BIOT	0007
a bon-Induced Transformation of	Mammalian Cells.= +Polycyclic Aromatic Hydroc	CHAS	0007
Chemical Education. Computer-	Managed Chemistry Instruction.= Computers in	CHED	0077
α-Hydroxy, and N-Blo+ Phase-	Managed Organic Synthesis: Activation of α-Chloro,	ORGN	0259
ack, Looking Forward: Pesticide	Management and Disposal in the Mid-80s and the M	AGRO	0086
evelopment+ Chemical Database	Management at the U.S. Army Chemical Research, D	CINF	0037
Studying Hazardous Materials	Management at the University of California: What+	CHAS	0002
utions for Wastewater and Rinsate	Management for Retail Fertilizer Dealers.= +l Sol	FERT	0014
e Implementation of Risk-Based	Management of Environmental Resources and Impac	ENVR	0076
	Management Paths for Technical Professionals.=	YCC	0010
Chemical Inventory	Management System.=	CHAS	0025
Managemen+ Laboratory Waste	Management. Practical Aspects of Laboratory Waste	LWM	0001
Regulatory Issues in Risk	Management: A California Perspective.=	CHAS	0021
rilizer Manufac+ Process Safety	Management: The Impact of New Regulations on Fe	FERT	0016
f Cementitious Systems for Waste	Management: The Penn State Experience.= +stry o	ENVR	0006
Cancer	Management: Versatility of the B-Protein Assay.=	BIOL	0069
ilizer Containment on Stormwater	Management.= +d Soil Chemistry. Impact of Fert	FERT	0013
tical Aspects of Laboratory Waste	Management.= +ratory Waste Management. Prac	LWM	0001
the Farm Level.= Advances in	Managing Agricultural Chemicals in Groundwater at	AGRO	0052
	Managing Our Nation's Nuclear Waste.=	CHAS	0046
rspective.=	Managing Pesticide Waste: A Developing Country Pe	AGRO	0107
Muconate-Lactonizing Enzymes	Mandelate Racemase and Adenylosuccinate Lyase: O	BIOL	0012
Modifica+ Mechanistic Studies on	Mandelonitrile Lyase by Covalent and Noncovalent	BIOL	0075
Formed during the Inactivation of	Mandelonitrile Lyase by Phenylglyoxal.= +oducts	BIOL	0087

	Mandelung Sums.=	CHED 0024
rmone-Binding Protein of Larval	Manduca sexta by the Use of Photoaffinity Analog o	AGRO 0014
Hemolymph of the Lepidopteran	Manduca sexta.= Autoparalytic Peptides from	AGRO 0012
from the Tobacco Hornworm,	Manduca sexta.= Identification of an Allatostatin	AGRO 0008
ne in the Fifth Stadium Larvae of	Manduca sexta.= +olism Study of Juvenile Hormo	AGRO 0022
H,CH_3, OCH_3+ Hydrosilation of	Manganese Acyl Complexes, $(CO)_5MnCOCH_2R$ (R =	INOR 0337
f Magnetite Ore Pellet+ Effect of	Manganese Additions in the Mechanical Properties o	I&EC 0067
ers Model for the Photosystem II	Manganese Aggregate.= +port for a Dimer-of-Dim	INOR 0408
Ring Strain of	Manganese Bidentate Phosphine Complexes.=	INOR 0193
Marine Bioinorganic Chemistry.	Manganese Chemistry in the Ocean.= Tutorial on	INOR 0009
e-Like Reactivity of Dinuclear	Manganese Complexes.= Understanding the Catalas	INOR 0010
ilic Addition to Cyclopentadienyl	Manganese Dicarbonyl η^3-Allyl Cations.= Nucleoph	INOR 0270
lux of Iron, Cobalt, Copper, and	Manganese from California Continental Shelf Sedime	GEOC 0010
Hydrothermal Synthesis of	Manganese Oxides with Tunnel Structures.=	PETR 0056
States.= Covalent Linkage of	Manganese Oxo Clusters with Nonzero Spin-Ground	INOR 0196
(IV) Horseradish Peroxidase with	Manganese-Porphyrin Model Compounds.= +d Mn	BIOL 0095
ns Catalyzed by a D_4-Symmetrical	Manganese-Tetraphenylporphyrin.= +ric Oxidatio	ORGN 0045
Cultures.= Understanding and	Manipulating Secondary Metabolism in Hairy Root	BIOT 0231
M Imaging of Metal Surfaces and	Manipulation of Adsorbates.= +-Temperature ST	PHYS 0360
ned by X-ray Diffraction: No E+	Manipulation of Apparent Bond Lengths as Determi	INOR 0088
for the Overproduction of Mon+	Manipulation of Heterogeneous Hybridoma Cultures	BIOT 0169
Met in a Cost-Effective	Manner?.= Can Hazardous Material Regulations Be	CHAS 0033
hesis of (+)-BCH-189 Using D-	Mannose as a Chiral Template and Its Anti-HIV Act	CARB 0022
Synthesis of	Mannostatin A.=	ORGN 0068
lastics Recycling—An Equipment	Manufacturer's Perspective.= +tics. Commingled P	I&EC 0025
Illuminated Chemical	Manuscript.= Splendor Solis: A Sixteenth-Century	HIST 0036
ndling. Modeling Environment for	Map Interpretation.= +nal Chemical Structure Ha	CINF 0029
	MAP 30: A New Plant-Derived Anti-HIV Agent.=	AGFD 0190
n of Microbial Polysaccharides on	Maple Syrup.= +robial Polysaccharides. Productio	BIOT 0195
Compounds in	Maple Syrups.= Flavor Volatiles and Related	AGFD 0174
P Purification and Binding-Site	Mapping of Hemolymph Juvenile Hormone-Binding	AGRO 0014
ementation and Use of an Atom-	Mapping Procedure for Similarity Searching in Data	COMP 0043
ucture.+ Comparison of Ligand-	Mapping Studies with a High-Resolution Crystal Str	CARB 0013
Materials between the Continental	Margin and the Open N.W+ Exchanges of Reactive	GEOC 0009
ary Organic Matter in Continental	Margin Ecosystems.= +mineralization of Sediment	GEOC 0015
rganic Carbon (DOC) Flux out of	Margin Sediments on Oceanic DOC.= +Dissolved O	GEOC 0017
ms of Phosphorus in Continental	Margin Sediments: Implications for Burial Fluxes an	GEOC 0013
on River.= Chemistry of Ocean	Margins. Controls on Methane Fluxes from the Huds	GEOC 0001
	.= Marie Lavoisier, the Other Parent of New Chemistry	HIST 0037
Haloperoxidases Isolated from	Marine Algae.= Reactivity of the Vanadium	INOR 0097
Novel Marine Siderophores from	Marine Bacteria.=	INOR 0119
ments in the Ascidiacea.=	Marine Bioinorganic Chemistry. Accumulation of Ele	INOR 0095
e Sea.=	Marine Bioinorganic Chemistry. Arsenic Cycling in th	INOR 0136
e Metals in the Oc+ Tutorial on	Marine Bioinorganic Chemistry. Distribution of Trac	INOR 0001
ry in the Ocean.= Tutorial on	Marine Bioinorganic Chemistry. Manganese Chemist	INOR 0009
Biomineralization in Chitons.= +	Marine Bioinorganic Chemistry. Mechanisms of Iron	INOR 0116
on of Arsenicals in the	Marine Environment.= Uptake and Biotransformati	INOR 0137
rrosion: Life Prediction Model for	Marine Gas Turbine Blades and Guide Vanes.= +o	FUEL 0099
er+ Metal Incorporation into the	Marine Invertebrate Oxygen-Carrying Proteins Hem	INOR 0140
Siderophore Production in	Marine Microorganisms.=	INOR 0004
Trace Metal Nutrition of	Marine Phytoplankton.=	INOR 0003
Freshwater and Coastal	Marine Sediments.= N:P Release Ratios from	GEOC 0012
Novel	Marine Siderophores from Marine Bacteria.=	INOR 0119
Metal Sulfide Complexes in	Marine Waters.=	GEOC 0006
rograms, Honoring the Memory of	Marjorie Gardner, Chemical Educator. +mSource P	CHED 0011
ination of Molecular Weights and	Mark-Houwink Constants.= +for Accurate Determ	PMSE 0066
oporphyrin (EP) as a Biochemical	Marker for Detecting Exposure to Lead.= +te Prot	CEI 0020
oid Industry and Update on R. E.	Marker—Illustrated Talk.= +of the Mexican Ster	HIST 0016
Development of New Genetic	Markers for the Application of PCR to Forensic Case	CHAL 0004
Doing Your	Market Research.=	SCHB 0002
ing: A National and Internation+	Market-Driven Pollution Prevention? Green Market	CEI 0028
Analysis for Polymer	Market-Entry Strategies.= Value Perception	CMEC 0011
orth America and Europe. Global	Marketing of Chemical Information: The STN Story	CINF 0011
	Marketing Your Product(s).=	SCHB 0010
ven Pollution Prevention? Green	Marketing: A National and International Status Rep	CEI 0028
Approach to	Marketing.= Countertrade—An Innovative	CMEC 0008
Opportunities for Chemists in	Marketing, Sales, and Technical Service.=	YCC 0006
ng New Energy Technology to the	Marketplace.= +bility of Developing and Transferri	FUEL 0142
Professional, and Consumer	Markets.= Product Formulating for Agricultural,	FERT 0004
Chemical Specialties for Niche	Markets, or How to Survive and Thrive in the Midst	CMEC 0005
ide Residue in Barrier Island Salt	Marshes along the Florida Indian River Lagoon.+	ENVR 0019
Charles	Martin Hall—Aluminum.=	CHAL 0027
s for biological Studies.= C. S.	Marvel Creative Polymer Chemistry Award Symposiu	POLY 0272
ion-Polymer Interfaces.= C. S.	Marvel Creative Polymer Chemistry Award Symposiu	POLY 0298
osspolarization and Very Fast ^{27}Al	MAS NMR Studies of Dealuminated Zeolite Y.=	PETR 0006
Olefins over Offret+ ^{13}C and ^{1}H	MAS NMR Studies of the Conversion of Methanol to	PETR 0042
drotreatin+ FT-IR, Solid-State	MAS NMR, and XRD of Phosphorous-Promoted Hy	COLL 0198
rolyzed Products.= Solid-State	MAS-NMR Study of Preceramics Precursors and Py	POLY 0358
erras Raras por Espectrometria de	Masas con Fuente de Impacto Electronico.= +e Ti	ANYL 0051
New Approach of Aluminum-27	MASNMR to the Processes of Transparent Alumi	COLL 0115

apillary Gas Chro+ Minimizing	Mass Discrimination in a Split Injection System for C	TECH	0018
it on the Electron Antineutrino	Mass from Measurement of the Beta Decay of Molec	NUCL	0010
Spec+ Diffusion Average Molar	Mass of Polydisperse Polymers: A Photon Correlation	POLY	0305
sing Conventional SEC and Molar	Mass Sensitive Detectors.= +r Characterization U	PMSE	0051
Predicting	Mass Spectra from Structural Information.=	COMP	0005
.=	Mass Spectrometric Analysis of Phenolic Compounds	AGFD	0003
rmediates CH+ Photoionization	Mass Spectrometric Studies for the Combustion Inte	FUEL	0124
Disaccharides by Electrospray	Mass Spectrometry (ES MS).= Linkage Analysis in	CARB	0048
stion Studied by Molecular-Beam	Mass Spectrometry and Modeling.= +llant Combu	FUEL	0126
rial Analysis Using Secondary Ion	Mass Spectrometry and Related Techniques.= +te	INOR	0007
oliag+ Gas Chromatography and	Mass Spectrometry of Erythrina Alkaloids from the F	ANYL	0136
Fossil Fuels: Swelling Studes and	Mass Spectrometry. +vel Analytical Techniques for	FUEL	0086
Fossil Fuels: Swelling Studes and	Mass Spectrometry. +vel Analytical Techniques for	FUEL	0088
tion of Materials: Laser Ionization	Mass Spectrometry. +in the Analysis and Modifica	NUCL	0041
ution of Vacuum Residue Using	Mass Spectrometry: A Heterogeneous Analytical Mod	PETR	0150
tector.=	Mass Spectrometry: A Powerful Chromatographic De	AGFD	0202
oa. Macerals by Laser Desorption	Mass Spectrometry. +etry. Characterization of C	FUEL	0088
Analysis in Water Using Ion Trap	Mass Spectrometry.= +gulatory Issues. Pesticide	AGRO	0060
untercurrent Chromatography and	Mass Spectrometry.= +r Precursors by Droplet Co	AGFD	0203
Coal Samples by Field Ionization	Mass Spectrometry.= +on of the Argonne Premium	FUEL	0062
Meats by On-Line HPLC Tandem	Mass Spectrometry.= +ibiotic Residues in Edible	AGFD	0061
n Photometry and Particle-Beam	Mass Spectrometry.= +ography with UV-Absorptio	ENVR	0015
arshs in Rocks by Electron Impact	Mass Spectrometry.= +ve Determination of Rare E	ANYL	0049
Sampling	Mass Spectrometry.= Rapid Analysis Using Direct	FUEL	0089
ol Generation, Transport+ HPLC	Mass Spectrometry—I. Particle Size Effects in Aeros	ANYL	0122
/MS Utilizing Atmosphe+ HPLC	Mass Spectrometry—II. Recent Developments in CE	ANYL	0138
s-Flow Fast Atom Bombardment	Mass Spectroscopy as a Detector for Liquid-Phase S	ANYL	0125
Systems in External Fields.=	Mass Transfer and Transport Properties of Colloidal	COLL	0142
hanism of Particulate-Enhanced	Mass Transport to Hydrodynamic Electrodes.= Mec	COLL	0041
End to End and the Neutrino	Mass.= The Tritium Beta Decay Spectrum from	NUCL	0011
her Species by Molecular-Beam	Mass-Spectrometry during Hot-Filament-Assisted D	PHYS	0145
bcr and the St. Lawrence River at	Massena.= +ncern in New York State: Oswego Har	ENVR	0084
ces in Neutrino Science: Neutrino	Masses from Beta Decay—I. Neutri+ Recent Advan	NUCL	0001
ces in Neutrino Science: Neutrino	Masses from Beta Decay—II. Limit+ Recent Advan	NUCL	0010
Low-Energy Tests of Neutrino	Masses.=	NUCL	0020
he Practicing Chemist Need? J. R.	Massingill, Moderator.= +Interface, What Does t	I&EC	0057
Evidence for a	Massive Neutrino in Nuclear Beta Decay.=	NUCL	0003
Chains and the Number of Perfect	Matchings in Square and Pentagonal Chains.= +d	COMP	0033
esultados de la Evaluacion de una	Materia Biologica.= +mico de los Alumnos en los R	CHED	0093
Balances de	Materia.= BALANCE: Simulador de Procesos para	COMP	0052
Celdas Fotoelectroquimicas, I:+	Materiales Semiconductores Policristalinos dentro de	PHYS	0221
to Inorganic-Organic Hybrid	Materials?.= Cement Chemistry: A New Approach	POLY	0261
crementan la Reprobacion en las	Materias Biologicas de la Carrera de QFB en la Facu	CHED	0092
Indices de Reprobacion en	Materias Biologicas.= Analisis Longitudinal de	CHED	0091
dation Products in Environmental	Matrices by LC/MS.= +hemicals and Their Degra	AGRO	0031
Hydrocarbons in Environmental	Matrices by Liquid Chromatography with UV-Absor	ENVR	0015
n of Organic Molecules in Sol-Gel	Matrices for Quadratic Nonlinear Optics.= +ntatio	POLY	0006
y and Microstructure of Cement	Matrices for the Immobilization of Solidified Toxic H	ENVR	0002
of Thei+ Carbon and Graphite	Matrices in Carbon-Carbon Composites: An Overview	FUEL	0051
s and Tetracene-Doped Polymeric	Matrices.= +Phase Transition in Tetracene Crystal	PHYS	0274
the Organic Polymer	Matrices.= Inorganic Phase Behaviors Mediated by	POLY	0364
Carbon Clusters in	Matrices.= Structure and Spectroscopy of Small	PHYS	0078
dration and Polymerization in the	Matrix Apparently Caused by Arsenic Salts.= +Hy	ENVR	0041
a Single-Pass Ceramic	Matrix Bioreactor.= Controlled Protein Secretion in	BIOT	0011
rsible Transitions.= PTFE-	Matrix Composite Hydrogel Sheets with Thermoreve	POLY	0245
Glass Fiber Reinforced PMMA	Matrix Composites.= Preparation of Transparent	PMSE	0109
Using Influence Coefficients for	Matrix Corrections in XRF Analysis of Zinc, Iron, an	ANYL	0057
s of Current Epoxy Polymers as	Matrix for Structural Composites in Oilfield Applicat	PMSE	0111
Sulfur-Containing Reactive Sp+	Matrix Infrared Spectra and ab Initio Frequencies of	PHYS	0031
nd Ceramics: Influence of Polymer	Matrix on Precipitation Processes.= +omposites a	POLY	0290
ction and Subse+ Application of	Matrix Solid-Phase Dispersion (MSPD) to the Extra	AGFD	0060
Density	Matrix Theory of Ultrafast Electron Transfer.=	BIOT	0027
rs Derived from the Distance	Matrix.= Comparative Study of Molecular Descripto	COMP	0025
of Silica-Saturated Metal (Cu),	Matte (Cu_2S), and Slag (CaO-CaF_2-SiO_2) Under Con	I&EC	0065
	Max Tishler—Medicinal Compounds.=	CHAL	0031
n.= Role of Excited States of	Maximal d-Shell Occupancy in the Ru + H_2 Reactio	PHYS	0243
Acid: Further Confirmation of the	Mayo Mechanism.= +of Styrene in the Presence of	POLY	0039
lyco+ HPLC Survey of Corn (Zea	mays L.) Populations and Inbreds for Silk Flavonolg	AGRO	0034
Ring Channel Templated	Mazzite: A Stabilized Calcined Structure.= Twelve-	PETR	0037
Na-Chabazite and K-	Mazzite.= Common Factors in the Synthesis of	PETR	0038
on Corrosion Behavior of Metglass	2826MB and MBF 15.= +Crystallization Induced	PHYS	0219
Behavior of Metglass 2826MB and	MBF 15.= +Crystallization Induced on Corrosion	PHYS	0219
sis and Characterization of (R_3P)	$MCl[(Ph_2P(S)]_n[Ph_2P(O)]_{3-n}C]$, Where M = Pd or P	INOR	0265
Databases: Experiences with	MDDR-3D and FCD-3D.= Building 3D Structural	COMP	0046
ynthesis and Characterization of	$Me_2GaC≡CSiMe_3$ and Its Reaction with Primary and	INOR	0217
N,N-Dimethylquaterpyridinium,	Me_2qpy^{2+}.= +plex Containing the Cationic Ligand	INOR	0233
of $[I_2GaAs(SiMe_3+$ Reactions of (	$Me_3Si)_3As$ with GaX_3 (X = Br, I): Crystal Structure	COLL	0140
he Activation of O_2 by (OEP)Y(μ-	$ME)_2AlMe_2$.= +and Reactivity of (OEP)MX, and t	INOR	0294
ogical Characteristics of Produced	Meal.= +f a Mill Drier for Shrimp Head: Bromatol	AGFD	0105
ogical Characteristics of Produced	Meal.= +a Mill-Drier for Shrimp Heads: Bromatol	ENVR	0046

Studies on	Meat Flavor Generation from Cysteine and Sugars.=	AGFD	0074
Phenolic Compounds in	Meat Model Systems.= Antioxidant Activity of	AGFD	0019
med–Over–Flavor Development in	Meat Products.= +od Ingredients to Minimize War	AGFD	0021
Lipid Oxidation: Effect on	Meat Proteins.=	AGFD	0159
ntreated and Antioxidant–Treated	Meat.= +al and Sensory Evaluation of Flavor in U	AGFD	0120
ction of Antimicrobial Residues in	Meat.= +arm Test II Receptor Assays for the Dete	AGFD	0085
unds. Studies on the Formation of	Meat–Like Flavor Compounds.= +o Flavor Compo	AGFD	0070
of Antibiotic Residues in Edible	Meats by On–Line HPLC Tandem Mass Spectromet	AGFD	0061
Determinacion del	Mecanismo de Electroreduccion de la Perezona.=	PHYS	0213
Chloride (PVC) Roof+ Dynamic	Mechanical Analysis Studies of Reinforced Polyvinyl	PMSE	0180
methacrylate)/–Glass+ Dynamic	Mechanical Properties of a Polystyrene/Poly(methyl	PMSE	0167
e Epox+ Network Structure and	Mechanical Properties of Amine–Cured Epoxy: Amin	POLY	0158
ct of Particle Size Distribution on	Mechanical Properties of Granular Composites.=	PMSE	0114
ight–Sensitive Compounds.=	Mechanical Properties of Lipid Bilayers Containing L	BIOT	0106
ect of Manganese Additions in the	Mechanical Properties of Magnetite Ore Pellets.=	I&EC	0067
imethyl Aminoethyl Methacryl+	Mechanical Properties of Poly(Methyl Acrylate–co–D	PMSE	0079
Glass Fiber Lamin+ Optical and	Mechanical Properties of Poly(methylmethacrylate)/	PMSE	0110
ymerization, Characterization, and	Mechanical Properties.= +ated Polyurethanes: Pol	POLY	0018
noxy Pendent Groups: Processing,	Mechanical Properties, and Morphology.= +hylphe	PMSE	0019
te)/Silicon Oxid+ Dielectric and	Mechanical Spectroscopy Studies of Poly(vinyl Aceta	POLY	0287
Liquids.= Statistical	Mechanical Studies of Electronic Line Broadening in	PHYS	0035
sed in Electronic Pac+ Dynamic	Mechanical Thermal Analysis of Mold Compounds U	POLY	0095
gle X–ray Scattering and Dynamic	Mechanical Thermal Analysis.= +rs by Small–An	PMSE	0118
rameters Derived from Quantum	Mechanics and Force Field Calculations.= QSAR Pa	MEDI	0044
ansition–Metal Compo+ Cluster	Mechanics: A Force–Field Study of Mixed Ligand Tr	INOR	0075
tion: Thermodynamic and Kinetic	Mechanisitic Considerations.= +ing Benzene Oxida	FUEL	0085
ILM) Producido por E. histolytica	Mediante HPLC.= +comocion de los Monocitos (F	BIOL	0102
GTPase–	Mediated Activation of ATP Sulfurylase.=	BIOL	0062
–Dialkyl–9,10–Dihydro+ Silicon–	Mediated Alkylations: A Convenient Synthesis of 9,9	ORGN	0087
ucleophilic Aromatic Substitution	Mediated by Chromium(0).= +al Benzannulation/N	ORGN	0221
Inorganic Phase Behaviors	Mediated by the Organic Polymer Matrices.=	POLY	0364
ormations of Organic Compounds	Mediated by Transition Metal Complexes.= +ansf	ORGN	0127
+tering Studies of the Template–	Mediated Crystallization of ZSM-5 Type Zeolite.=	PETR	0058
e to H35 Hepatoma+ Receptor–	Mediated Delivery of Methotrexate–Insulin Conjugat	MEDI	0185
Preferences for Pancreatic Lipase–	Mediated Exchange Reactions with Butteroil.= +e	AGFD	0036
n+ Separation of Carbohydrate–	Mediated Microheterogeneity of Recombinant Huma	ANYL	0018
Damage.= Carcinogen–	Mediated Oxidant Formation and Oxidative DNA	CHAS	0008
Media: A Strategy+ Amphiphile–	Mediated Polar Colloid Formation in Fluorocarbon	COLL	0164
mediates Using Transition Metal–	Mediated Processes.= +a Unsaturated Ketene Inter	ORGN	0274
iral Molecule+ Carboxylesterase–	Mediated Reactions: A Versatile Route to Target Ch	AGFD	0008
oses: Mechanistic Studies of E3–	Mediated Reduction of CDP–6–Deoxy–$\Delta^{3,4}$–Glucosee	BIOL	0100
An η^4–Benzene Species	Mediates Acetylene Cyclotrimerization.=	INOR	0210
Naphthols with Proton–	Mediating Side–Chains.=	ORGN	0162
bilization of Proteins, Cofactors,	Mediators, Tissues, and Cells in the Context of Biose	BIOT	0039
of Yeshiva University/Montefiore	Medical Center.= +rt Einstein College of Medicine	PROF	0004
Metals Behavior during	Medical Waste Incineration.=	FUEL	0131
Max Tishler–	Medicinal Compounds.=	CHAL	0031
Applications of the C–9 Rule of	Medicinal Effect (II).=	MEDI	0039
Piquancy to a New Rule of	Medicinal Effect.= °From Molecular Recognition of	MEDI	0040
New Polyphenols from	Medicinal Plants and Their Antioxidant Activity.=	AGFD	0017
Antiviral Agents from	Medicinal Plants.=	AGFD	0191
ses of Animal Form. G. Edelman (	Medicine and Physiology, 1972).= +: Molecular Ba	CONG	0013
at the Albert Einstein College of	Medicine of Yeshiva University/Montefiore Medical	PROF	0004
for	Medicine.= Design and Synthesis of Novel Polymers	POLY	0273
Metals in	Medicine.=	CHED	0060
e of Pharmacokineti+ Veterinary	Medicines—Regulation in Europe and the Importanc	AGRO	0077
ion, Oxidorreduccion con Zinc en	Medio Acido y Basico, Preparacion de Nitrado de Pl	CHED	0172
o de las Reacciones Acido Base en	Medio Acuoso.= +y 316 Utilizadas en el Seguimient	ANYL	0055
Ensenanza de la Quimica a Nivel	Medio Superior.= +Apoyos Computarizados para la	CHED	0124
ato de L–Arginina en Tabletas en	Medio.= +tencio–Metrico para Cuantificar Clorhidr	ANYL	0066
odo de Vidrio Convencional para	Medir pH, por un Electrodo de Acero Inoxidable 316	ANYL	0066
esign and Conduct of Studies to	Meet Residue Chemistry Requirements of the NADA	AGRO	0078
now Compliance Made User Fr+	Meeting the Challenge of the 90s: MSDS Right–to–K	CHAS	0031
	Melamine–Formaldehyde Aerogels.=	POLY	0051
d Rearrangements of the Natural	Melampolide Schkuhriolide: An Alternative Approac	ORGN	0018
Juvenile Hormone of Drosophila	melanogaster.=	AGRO	0002
Obtained by Polymerization in the	Melt.= +Diacetylenic Liquid Crystalline Polymers	POLY	0030
Study of the Wetting Action of	Melted Sodium Silicate with Additives on Quartz and	COLL	0079
in Thermoreversible+ Arrested	Melting Due to Strain: Effect of Network Formation	POLY	0123
ahigh Molecular Weight+ High–	Melting Fractions Found in the Slow Melting of Ultr	PMSE	0166
Structure, Crystallization, and	Melting of Poly(aryl Ether Ketone Ketone), Part I: S	POLY	0059
C Structure, Crystallization, and	Melting of Poly(aryl Ether Ketone Ketone), Part II:	POLY	0060
examination of the Freezing and	Melting of Solvent in Thermoreversible Polymer Gel	POLY	0170
ting Fractions Found in the Slow	Melting of Ultrahigh Molecular Weight Polyethylene	PMSE	0166
before the Cart: Exploring the	Melting Point of Mixtures.= Putting the Horse	CHED	0159
tone), Part II: Crystallization and	Melting.= +Melting of Poly(aryl Ether Ketone Ke	POLY	0060
Solid–State Amorphization and	Melting.= Thermodynamics Parallels between	NUCL	0092
iffusion Coefficients in Polymer	Melts Using Surfaced–Enhanced Raman Scattering.=	POLY	0088
Carried Out in Polymer	Melts.= Chemistry of Condensation Reactions	PMSE	0176
n Entrance Flow of Polypropylene	Melts.= +eological Evidence of Flow Structuring i	PMSE	0137

tion: Nature's Rubber Cement for	Membrane Attachment.= +stry of Protein Prenyla	BIOL	0021
Techniques for Ions in	Membrane Channels.= Optical Spectroscopic	BIOT	0141
Discrepancy between Core and+	Membrane Fluidization by Anesthetics and Alcohols:	MEDI	0032
r I+ Process of the Oscillation of	Membrane Potential Studied by the Voltammetry fo	BIOT	0105
Structure.= Dependence of	Membrane Protein Prediction Scales on Sidechain	BIOL	0038
tic Resolution Using a Multiphase	Membrane Reactor: Experimental Verification.=	BIOT	0042
Catalytic Palladium	Membrane.= Effects of a Model Hydrogenation on a	CATL	0011
Dispersed in an Ionic Polymeric	Membrane.= Investigation of a Liquid Crystal	PMSE	0119
f a Silicon Oxide Phase in Nafion	Membranes by an in Situ Sol–Gel Reaction.= +on o	POLY	0285
Polyphosphazene	Membranes for Chemical Separations.=	INOR	0070
Mixture.= IPN	Membranes for the Pervaporation of Ethanol/Water	PMSE	0059
essing of th+ Artificial Molecular	Membranes Modeled on the Visual Information Proc	BIOT	0124
ling.= Gas Transport in IPN	Membranes: Effect of Cross–Linked State and Annea	PMSE	0060
and the Dynamics of Proteins and	Membranes: Theory and Experiment.= +rization,	PHYS	0166
Gas Permeation through Nafion	Membranes: Theory and Experiment.=	POLY	0066
tion into Colloidal Semiconductor	Membranes.= +us Fluids. Vectorial Electron Injec	COLL	0147
across Polymeric	Membranes.= Diffusion of Multicomponent Species	POLY	0195
Polyvinyl Chloride (PVC) Roofing	Membranes.= +cal Analysis Studies of Reinforced	PMSE	0180
Capillarity of Nylon Microporous	Membranes.= +gh–Temperature Annealing on the	COLL	0037
iliorium Potentials in Biological	Membranes.= Electrochemical Kinetics and Nonequ	COLL	0093
Biocatalysts Immobilized on	Membranes.= Food Industry Applications of	BIOT	0037
s: From Model Enzymes to Model	Membranes.= +xygenations with Metalloporphyrin	ORGN	0136
tial Light Modulators and Optical	Memories Based on Bacteriorhodopsin.= +ics. Spa	BIOT	0077
Shape	Memory Alloys.= Reactivity of Nickel–Titanium	INOR	0357
Arrays.= New Optical	Memory Architectures Based on Optical–Phased	BIOT	0025
Semiconstant Holographic	Memory Based on Bacteriorhodopsin.=	BIOT	0079
emSource Programs, Honoring the	Memory of Marjorie Gardner, Chemical Educator.	CHED	0011
e Met Lab, and+ Symposium in	Memory of Nathan Sugarman. Nuclear Chemistry, th	NUCL	0058
Photoamplified Storage Optical	Memory with Polyion Complexed Pyrene LB Films.=	BIOT	0144
meric+ Selective Abstraction of	Mercaptan Compounds by Chemically Modified Poly	ANYL	0083
3-Thio-Sugars and Related 3'-	Mercapto–Anthracyclines.= Synthesis of	CARB	0032
.= Some Contributions of the	Merck Laboratories: The Impact of Steroid Research	HIST	0015
Chemically Modified Polymeric-	Mercuric Resin in Capillary Gas Chromatography.=	ANYL	0083
ective Graphite Compos+ In Situ	Mercury Detection in Groundwater by Novel Ion–Sel	ENVR	0044
er by Inverse Voltammetry at the	Mercury Film Electrode.= +of Human Seric Copp	ANYL	0056
Control of Cadmium, Lead, and	Mercury from RDF Municipal Waste Combustors.=	FUEL	0130
tial Thermal Signatures of Total	Mercury in Crystalline Rock: Case Study—Waldobor	ENVR	0055
terization of the First Monomeric	Mercury(II) Tetraalkylthiolate Complex.= +Charac	INOR	0059
d Complexation of Lead(II) and	Mercury(II) with Crown Ether on the Surface of Silo	COLL	0104
ceptor Reactions of Leuconostoc	mesenteroides B–512FM Dextransucrase: Inhibition o	CARB	0028
tor Devices.=	Meso–Scopic Effects in Small Metal and Semiconduc	BIOT	0002
rs in Polyethylene: Orientation by	Mesoepitaxy.= +ighly Ordered Conjugated Polyme	POLY	0008
4,4'-Dihydroxy-α-methylstilbene	Mesogen.= +yesters and Copolyesters Based on the	POLY	0031
taining trans–Cyclohexane–Based	Mesogenic Groups.= +ain Liquid Polysiloxanes Con	POLY	0057
Mediu+ Biocatalysis in Lyotropic	Mesophases—Characterization of a Novel Reaction	BIOT	0248
y:sulfoniopropionate (DMSP) in	Mesoscale Coccolithophore Blooms in the Gulf of Ma	GEOC	0002
ical Kinetics and Dynamics of the	Mesospheric Sodium Nightglow.= +tions—I. Chem	PHYS	0071
tisense+ The 5' Cap Structure of	Messenger RNA: A Novel Target for Therapeutic An	BIOL	0120
Material Regulations Be	Met in a Cost–Effective Manner?.= Can Hazardous	CHAS	0033
garman. Nuclear Chemistry, the	Met Lab, and Nathan Sugarman—A Retrospective.=	NUCL	0058
ues of Chemotactic Tripeptide F-	Met–Leu–Phe–OH.= +–Terminal Modified Analog	MEDI	0166
Short–Range Structuring in a	Meta–Anthracite Coal.=	FUEL	0040
Sieves.= Reactions of	Meta–Diisopropylbenzene on Acidic Molecular	CATL	0014
ls. Bioenergetics of+ Progress in	Metabolic Engineering and Production of Biochemica	BIOT	0029
and Laying Hens.=	Metabolic Fate of Ametryn in Rats, Lactating Goats,	AGRO	0111
to Production–Scale Fermenters.	Metabolic Pathway Engineering: The Biosynthesis of	BIOT	0083
ation: Nature+ Chemistry of New	Metabolic Pathways. Biochemistry of Protein Prenyl	BIOL	0021
Glycosides: Genetic Regulation,	Metabolic Processing, and in Vitro Effects on Oncoge	AGFD	0117
l+ Effect of Hepatocyte Source on	Metabolic Profile of Biphenyl in an in Vitro Deve	AGRO	0021
vation Methods.=	Metabolic Responses of CHO Cells to Different Culti	BIOT	0164
tics in Food–Producing Animals.	Metabolism and Residue Studies of Pirlimycin in the	AGRO	0108
drochloride: A Phenethanolami+	Metabolism and Tissue Residues of Ractopamine Hy	AGRO	0131
ttle.=	Metabolism and Tissue Residues of Tilmicosin in Ca	AGRO	0110
chniques for Studying Xenobiotic	Metabolism in Food–Producing Animals.= +itu Te	AGRO	0092
Xenobiotic	Metabolism in Food–Producing Aquatic Species.=	AGRO	0112
ethods for Studying Xenobiotic	Metabolism in Food–Producing Fish and Shellfish.=	AGRO	0096
g and Manipulating Secondary	Metabolism in Hairy Root Cultures.= Understandin	BIOT	0231
Novel Sulfonamide Drug	Metabolism in Lactating Dairy Cows.=	AGRO	0113
Diazinon on L–Tryptophan	Metabolism in Mice.= Paradoxical Effect of	AGRO	0020
ary Met+ Modifying Secondary	Metabolism in Plant Cell Cultures. Increasing Second	BIOT	0228
Xenobiotic	Metabolism in Vitro.= Methods for the Study of	AGRO	0093
Comparative	Metabolism of DPX–E9636 in Plants and Animals.=	AGRO	0028
he NADA: Residue Depletion and	Metabolism of Flunixin in Cattle.= +uirements of t	AGRO	0078
Role of Tungsten in the	Metabolism of Hyperthermophilic Bacteria.=	INOR	0099
Residue Depletion and	Metabolism of Luprostiol in Dairy Cows.=	AGRO	0130
	Metabolism of S–53482, a New Herbicide, in Rats.=	AGRO	0024
	Metabolism of Triallate in Sprague–Dawley Rats.=	AGRO	0023
als.= Dermal Absorption and	Metabolism of Xenobiotics in Food–Producing Anim	AGRO	0095
ical Models to Direct Xenobiotic	Metabolism Studies in Food Safety Evaluation: A Pr	AGRO	0132
In Vitro	Metabolism Studies of o–Tolunitrile.=	MEDI	0033

imals. Ivermectin and Abamectin	Metabolism: Differences and Similarities.= +Anm	AGRO	0128
Corn: Uptake, Translocation, and	Metabolism.= +ty of the Herbicide DPX-E9636 in	AGRO	0027
l Cultures. Increasing Secondary	Metabolite Production by Redirecting Transport.=	BIOT	0228
hylene Blue and Its Demethylated	Metabolites from Milk.= +d Determination of Met	AGFD	0101
utomated Analysis of Drugs and	Metabolites in Biological Fluids Using Laboratory Ro	ANYL	0119
Determination of Atrazine	Metabolites in Human Urine.=	AGRO	0032
rmination of Hexazinone and Its	Metabolites in Urine of Exposed Forest Pesticide Ap	CEI	0005
Determination of Simazine and Its	Metabolites in Urine.= +to Triazine Herbicides, I:	CEI	0004
Analyzing Pesticides and	Metabolites in Water.= Current Methodologies for	AGRO	0146
easurement of Urinary Phenolic	Metabolites of Environmental Contaminants in the U	CEI	0002
S Detection of Liver Microsomal	Metabolites of N,N-Diethylbenzamide by ^{13}C NMR	AGRO	0026
Analysis of Carboxylic Acid	Metabolites of Polychlorotrifluoroethylene.=	ANYL	0062
e Configuration of Polypropionate	Metabolites of Siphonaria Australis.= +nd Absolut	ORGN	0003
nthesi+ Synthesis of the putative	Metabolites of the Cyclopenta[a]-Phenanthrenes: Sy	ORGN	0118
ancer Chemopreventive Retinoid	Metabolites: C-Glucuronide Analogues of 4-Hydroxy	MEDI	0030
Mechanistic Role of Bioactive	Metabolites.= Etoposide: Studies Aimed Toward the	MEDI	0012
Diltiazem	Metabolites.= Synthesis and Characterization of	MEDI	0129
[12]	Metacyclophane.= Synthesis of (±)-Muscone from	ORGN	0181
quilibrations of Silica-Saturated	Metal (Cu), Matte (Cu_2S), and Slag ($CaO-CaF_2-SiO_2$	I&EC	0065
of Ion Exchange and Immobilized	Metal Affinity FPLC.= +P-450 Patters by Means	ANYL	0017
mediates in the Carbonylation of	Metal Alkyl Bonds Studied by Time-Resolved Infrar	INOR	0169
c Sites Formed by the Reaction of	Metal Alkyls with Deboronated Zeolites.= +atalyti	PETR	0024
lar Mechanisms of Carcinogenic	Metal and Polycyclic Aromatic Hydrocarbon-Induced	CHAS	0007
Oxidation Reactions of	Metal and Semi-Metal Cluster Ions.=	PHYS	0174
Meso-Scope Effects in Small	Metal and Semiconductor Devices.=	BIOT	0002
erture Ranges: Mea+ Kinetics of	Metal Atom and Radical Reactions over Wide Temp	PHYS	0087
ith Hydrocarbons of Transition-	Metal Atoms in Different Charge States: M, M+, and	PHYS	0199
Deposition.= Reactions of	Metal Atoms in Plasma Etching and Chemical Vapor	PHYS	0143
ciation Reactions of Transition-	Metal Atoms with Simple Molecules near Room Tem	PHYS	0198
Particles Prepared from Solvated	Metal Atoms: Structure and Catalysis.= +imetallic	CATL	0009
Binding of Small Molecules to	Metal Atoms.= Role of Cooperative Effects in the	PHYS	0172
tal Clusters and Extended Metal-	Metal Bonding.= +ontier. New Structures with Me	CHED	0029
Low-Temperature Precursors to	Metal Borides.= Metallaboranes as	INOR	0101
tive Catalysts for Enantioselective	Metal Carbene Transformations.= +arkably Effec	CATL	0002
otochemically by d^8 Transition-	Metal Carbonyls in the Presence of Aromatic Ketone	INOR	0254
Photocatalytic Behavior of	Metal Carbonyls on Porous Vycor Glass.=	INOR	0170
Photochemical Reactions of	Metal Carbonyls with NF_3.=	INOR	0195
he Chemical-Vapor Deposition of	Metal Carbonyls.= +Microscopy (STM) Studies of t	COLL	0050
Photooxidation of	Metal Carbyne Complexes.=	INOR	0341
n over Alumina-Supported Noble	Metal Catalysts.= +atalysts—I. Methane Oxidatio	COLL	0185
Highly Dispersed	Metal Catalysts.= Ethylene Dimerization over	PETR	0010
n Cross-Sections of Transition-	Metal Cations: V+ + Ethane, Ethene, and Propane.=	PHYS	0306
Electron Density at Transition-	Metal Centers via Poly 3'-(4-pyridyl)terthiophene-B	INOR	0027
and Conformation.= Engineered	Metal Chelation Sites as Probes of Protein Stability	BIOT	0133
Formation of Aqueous	Metal Chloride Complexes to High Temperatures.=	ANYL	0008
Surface.= Mechanisms of	Metal Chloride Reactions with Dispersed Silica	COLL	0219
and Semi-	Metal Cluster Ions.= Oxidation Reactions of Metal	PHYS	0174
Kinetic Processes with	Metal Cluster Participation in Shock Waves.=	PHYS	0175
ng Frontier. New Structures with	Metal Clusters and Extended Metal-Metal Bonding.	CHED	0029
	Metal Cofactors of Lysine 2,3-Aminomutase.=	BIOL	0078
er Reactions between Transition-	Metal Complexes and Redox Sites in Enzymes.= +f	INOR	0283
Applications of ansa-Bis(idenyl)	metal Complexes Containing Chiral Bridging Groups	INOR	0214
Optics.= Group IV Transition-	Metal Complexes for Third-Order Nonlinear	INOR	0334
Photoredox Chemistry of	Metal Complexes in Microheterogeneous Media.=	INOR	0316
Groups.= Cofacial Binuclear	Metal Complexes with Inwardly Directed Functional	INOR	0242
Ligands.= Volatile	Metal Complexes with Silyltellurolate and Selenolate	INOR	0172
3D	Metal Complexes with Uric Acid.=	INOR	0073
rystalline Pyramidic Transition-	Metal Complexes: Potential New Ferroelectric Mater	INOR	0135
Cyclopropylthiocarbene-	Metal Complexes: Super-Homo-Michael-Acceptors.=	ORGN	0216
mpounds Mediated by Transition	Metal Complexes.= +ansformations of Organic Co	ORGN	0127
Transition-	Metal Complexes.= Antineoplastic Effect of Several	INOR	0026
of Transition-	Metal Complexes.= Calculations of the Geometries	PHYS	0192
Oxidation in Nucleic Acids Using	Metal Complexes.= +tion Specificity of Guanosine	BIOL	0122
F-	Metal Complexes.= Luminescence Spectroscopy of	INOR	0062
atalyzed by Rh and Co-Rh Mixed	Metal Complexes.= +of Alkynes and Alkenynes C	ORGN	0218
4 Transition	Metal Complexes.= Organic Synthesis Using Group	ORGN	0134
Phosphates toward Organic Bases,	Metal Complexes, and Metal Oxides.= +Metal(IV)	PETR	0107
tudy of Mixed Ligand Transition-	Metal Compounds.= +Mechanics: A Force-Field S	INOR	0075
l Bonding in Ternary Transition-	Metal Compounds.= +ronic Structure and Chemica	PHYS	0252
Halides with Binary	Metal Compounds.= Reactions of Aqueous Alkyl	INOR	0080
Beverage, and Food	Metal Containers.= Organic Coatings for Beer,	PMSE	0140
-Cracking Catalys+ Migration of	Metal Contaminants and Passivating Agents in Fluid	CATL	0012
Polymers.=	Metal Coordination Complexes for Synthesis of New	CHED	0059
(1H)Quinazolinone-2,3+ Mixed	Metal Coordination Compounds of the Heterocyclic 4	INOR	0079
. Coordination between Theoret+	Metal Coordination Compounds—The Practical Side	CHED	0056
through	Metal Coordination.= Metal-Polymer Interaction	POLY	0343
on Group 8 and Group 10 Mixed-	Metal Cyanometalate Complexes.= +plexes Based	INOR	0162
Mixed-	Metal Derivatives of Tyrosinase.=	BIOL	0059
N_2) Bonding of Early Transition-	Metal Dinitrogen Complexes.= +2) vs. End-on (μ-	INOR	0362
ging of Palladium and Platinum	Metal Films by Electron Transfer to Photogenerated	INOR	0319

Surface Chemistry of Monolayer	Metal Films.=	COLL	0063
or mating Agents fo+ Transition-	Metal Fluorides and Oxide Fluorides as Selective Flu	INOR	0133
d Solids: C60, C70, and the Alkali-	Metal Fullerides.= +ural Studies of Fullerene-Base	PHYS	0049
)2 and the Solubilization of Alkali	Metal Halides in Hydrocarbon Media.= +(η^6-C7H8	INOR	0053
Reaction of Amines with	Metal Halides.= Synthesis of Nitride Films from the	INOR	0309
Delocalization in Early Transition	Metal Heteropoly- and Isopolyoxometalates.= +on	COMP	0032
l Coordination Beha+ Transition	Metal Hydrides. Palladium Hydride Dimers: Unusua	INOR	0379
symmetry Parameters in Bridging	Metal Hydrides.= +pole Coupling Constants and A	INOR	0382
f alkyl Migration from Oxygen to	Metal in Fe-Mn Alkoxycarbyne Complexes.= +sm o	INOR	0110
ygen-Carrying Proteins Hemer+	Metal Incorporation into the Marine Invertebrate Ox	INOR	0140
irect Lanthanide to Transition-	Metal Interactions in Organometallic Chemistry: Syn	INOR	0332
	Metal Ion Chemistry in Flames.=	PHYS	0201
and Transition-	Metal Ion Chemistry.= Comparison of Main Group	PHYS	0200
Water Contents and Paramagnetic	Metal Ion Concentrations.= +to Their Hydration	MEDI	0043
Xerogels Doped with Transition-	Metal Ions.= Studies on Silica Aerogels and	INOR	0081
atives in the Presence of Divalent	Metal Ions.= +ylboronate and Isopropylidene Deriv	INOR	0263
Technologies.= Camet	Metal Monolith-Based Catalytic Emission Control	I&EC	0045
mains.+ Controlled Synthesis of	Metal Nanoclusters within Block Copolymer Microdo	POLY	0207
s of Main-Group and Transition-	Metal Nitride Thin Films.= +emperature Synthesi	INOR	0104
Trace	Metal Nutrition of Marine Phytoplankton.=	INOR	0003
Caracterization of Transition-	Metal Open Fulvalene Complexes.= Syntheses and	INOR	0113
ni+ Studies of Some Gas-Phase	Metal Oxidation Reactions by Photoelectron and Che	PHYS	0352
py and Tunneling Spectroscopy of	Metal Oxide Arrays on Graphite.= +eling Microsco	COLL	0194
Support Materials: The Hydrous	Metal Oxide Ion Exchangers.= +of Novel Catalyst	PETR	0018
py Studies of Adsorbed Species on	Metal Oxide Surfaces.= +Energy-Loss Spectrosco	COLL	0173
Propane on Hydrous	Metal Oxide-Supported Catalysts.= Reactions of	CATL	0040
Catalytic Role of	Metal Oxides in the Photodegradation of Furans.=	COLL	0190
e mica+ Model Surface Studies of	Metal Oxides: Preparation, Characterization, and Ch	COLL	0193
anic Bases, Metal Complexes, and	Metal Oxides.= +Metal(IV) Phosphates toward Org	PETR	0107
New	Metal Oxides.= Synthesis and Characterization of	PETR	0057
Early-	Metal Phosphinidenes.=	INOR	0363
udies of Nonclassical Transition	Metal Polyhydrides: Rotational Tunnelling and Quan	PHYS	0068
t sensitive Metalorganic Systems.	Metal Polypyridines as Redox Photosensitizers.=	INOR	0282
nvestigations into the Reactions of	Metal Powders with Aqueous Alkyl Halides.= +ic I	INOR	0248
s of the Mesospheri+ Gas-Phase	Metal Reactions—I. Chemical Kinetics and Dynamic	PHYS	0071
l minum Species.= Gas-Phase	Metal Reactions—II. Kinetic Studies of Boron and A	PHYS	0086
gen Reaction with S+ Gas-Phase	Metal Reactions—III. Kinetics of the Molecular Oxy	PHYS	0141
cions in MLn+ Sys+ Gas-Phase	Metal Reactions—IV. Study of Ligand-Ligand Intera	PHYS	0171
t on-Metal Atoms w+ Gas-Phase	Metal Reactions—V. Association Reactions of Transi	PHYS	0198
Electronically Exci+ Gas-Phase	Metal Reactions—VI. Reactions of Ground State and	PHYS	0349
ted Rare Gas Atom+ Gas-Phase	Metal Reactions—VII. Harpooning Reactions of Exci	PHYS	0367
Fundamentals of Heavy	Metal Removal by Dry Sorbents.=	FUEL	0132
oxidation Catalyzed by Chiral	Metal Salen Complexes Bearing Bulky Silyl Groups.=	ORGN	0056
Bi2Sr2CaCu2O8 Superconductor by	Metal Substitution.= +Critical Current Density of	PHYS	0329
Relations in Transition	Metal Sulfide Catalysts.= Structure-Function	CATL	0038
	Metal Sulfide Complexes in Marine Waters.=	GEOC	0006
ent Polymer Films on a	Metal Surface by ESCA.= Study of Thin Bicompon	POLY	0100
+-Temperature STM Imaging of	Metal Surfaces and Manipulation of Adsorbates.=	PHYS	0360
duced Phase Transformations of	Metal Surfaces Studied by STM.= +emisorption-I	PHYS	0126
on	Metal Surfaces.= NMR Studies of Simple Molecules	PHYS	0347
tivity and Localized Corrosion on	Metal Surfaces.= +between Localized Microbial A	BIOT	0098
and	Metal Surfaces.= STM Studies of Biomolecules	ANYL	0023
Mixed	Metal Synthesis of 2-Ethylhexyllithium.=	ORGN	0225
Reactivity.= Early Transition-	Metal Tellurolates: Synthesis, Structure, and	INOR	0021
erials. Deposition of Transition-	Metal Thin Films from Organometallic Precursors.=	INOR	0141
= and Heterogeneous Reactions in	Metal Vapor Oxidation.= +n between Homogeneou	PHYS	0144
for Cement-Solidified/Stabilized	Metal Wastes.= +copy Characterization Techniques	ENVR	0005
Cement-Solidified Heavy	Metal Wastes.= Microstructural Characterization of	ENVR	0008
oluene Coordination to an Alkali	Metal: Strucutre of Lu[CH(SiMe3)2]3(μ-Cl)K(η^6-C7H	INOR	0053
ercalating Properties of Layered	Metal(IV) Phosphates toward Organic Bases, Metal C	PETR	0107
and the Luftwaffe: Evolution of	Metal-Ammonia Reductions.= Synthetic Steroids	HIST	0013
Oxidation of	Metal-Based Clusters.= Unique Complexation and	PHYS	0173
	Metal-Binding Proteins of the Mussel Byssus.=	INOR	0138
Conductors?.= Alkali	Metal-C60 Films: The First 3-D Molecular	INOR	0354
Synthesis.= Homogeneous	Metal-Catalyzed Photochemistry in Organic	INOR	0339
stituted Cyclohexad+ Transition-	Metal-Catalyzed Polymerization of Heteroatom-Sub	POLY	0047
lity in Homogeneous Transition-	Metal-Catalyzed Processes.= +ole on Catalyst Stab	INOR	0349
Electron Attachment Reactions of	Metal-Centered Compounds.= +ate Constants for	PHYS	0234
Superconductivity in Alkali-	Metal-Doped Fullerenes.=	PHYS	0046
Infrared Spectra of	Metal-Ethylene Complexes Isolated in Solid Argon.=	PHYS	0074
f Stereoselectivity in Samarium	Metal-Induced Cyclopropanations: Synthesis of 1,25-	ORGN	0284
ements of Organosilanes to Form	Metal-Ligand Multiple Bonds with Tungsten Chloro	INOR	0115
ne Intermediates Using Transition	Metal-Mediated Processes.= +a Unsaturated Kete	ORGN	0274
with Metal Clusters and Extended	Metal-Metal Bonding.= +ontier. New Structures	CHED	0029
l-Derived Distillate+ Catalysis of	Metal-Modified Y-Zeolites for Hydrocracking of Coa	CATL	0050
-N Bonds for the Preparation of	Metal-Nitride Preceramics, Powders, and Coatings.=	POLY	0230
halpy and Kin+ Steric Effects in	Metal-Phosphine Bonding: Determination of the Ent	INOR	0192
ion.=	Metal-Polymer Interaction through Metal Coordinat	POLY	0343
	Metal-Responsive Gene Expression.=	INOR	0012
in N+ Combined Directed ortho	Metalation and Cross-Coupling Tactics: Application	ORGN	0277

Binding+ Catalysis of Porphyrin	Metalation at Anionic Interfaces: Mode of Interfacial	CATL 0060
iocarbamate as a Versatile New	Metalation Director for the Synthesis of Polysubstitu	ORGN 0083
Facile Reversible	Metalation from an Agostic Complex.=	INOR 0384
w Metalation Di+ Directed ortho	Metalation: The O–Thiocarbamate as a Versatile Ne	ORGN 0083
mplications Concerning Enzymatic	Metalation.= +: Mode of Interfacial Binding and I	CATL 0060
D.F.= Determinacion de	Metales Pesados en Leche de Establos del Sur del	ENVR 0037
1–Nitro–Naftaleno con Electrodos	Metalicos y Semiconductores.= +ectroquimica del	PHYS 0280
etal Borides.=	Metallaboranes as Low–Temperature Precursors to M	INOR 0101
Trip+ Reaction of Platinum(IV)	Metallacyclopentanes with Trimethyl Phosphite and	INOR 0330
Complexes o+ Highly Polarizable	Metallic Complexes for Nonlinear Optics: Cobaltous	POLY 0102
Surface Modification by	Metallic Ion Implantation On–CdTe.=	PHYS 0218
yridines by Bentonite-Supported	Metallic Nitrates: Induced Oxidation of Hantzsch Es	ORGN 0245
und S+ Paired Disorder and the	Metallic State in Polymers Lacking a Degenerate Gro	PHYS 0039
YBa$_2$Cu$_3$O$_6$.$_3$: Photodoping to the	Metallic State.= +onductivity in Single Crystals of	PHYS 0011
–to–Ferromagnetic Transition in	Metallic TlCo$_2$S$_x$Se$_{2-x}$ (0 ≤ x ≤ 2.0) with the ThCr$_2$S	INOR 0058
esis and Reactivity Studies of 7-	Metallobenzonorbornadiene and 7-Metallobenzonorc	INOR 0258
etallobenzonorbornadiene and 7-	Metallobenzonorcaradiene Complexes.= +ies of 7-M	INOR 0258
C2-Symmetrical Chiral	Metallocene Complexes.= Catalytic Reactions Using	INOR 0215
Redox–Active Early	Metallocycles as Metalloligands.=	INOR 0025
as	Metalloligands.= Redox–Active Early Metallocycles	INOR 0025
ure, Solid State Structure, and+	Metallomesogens: Relation between Electronic Struct	INOR 0269
armacological, and Clinic+ Zinc-	Metallopeptidase Inhibitors: Design, Biochemical, Ph	MEDI 0002
Limiting Studies of	Metallophthalocyanines.= Synthetic and Optical	INOR 0303
haracterization of the First Ti(II)	Metalloporphyrin Complexes.= +, and Structural C	INOR 0298
–Conducting Device Consisting of	Metalloporphyrins in a Lipid Bilayer.= +nsitive Ion	BIOL 0098
Spectro+ Molecular Orbitals of	Metalloporphyrins: An Undergraduate Experiment in	PHYS 0327
M+ Selective Oxygenations with	Metalloporphyrins: From Model Enzymes to Model	ORGN 0136
g Substituents on Porphyrins and	Metalloporphyrins.= +fect of Electron–Withdrawin	INOR 0301
e and Bioavailability in a Series of	Metalloprotease Inhibitors.= +ps between Structur	MEDI 0195
t Hydrocarbons on Polyfunctional	Metallosilicate Catalysts.= +Aromatization of Ligh	PETR 0062
s to high-T$_c$ Superconducting Fi+	Metalorganic Chemical Vapor Deposition Approache	INOR 0176
ecursors. Advances and Trends in	Metalorganic Chemical Vapor Epitaxy.= +tallic Pr	COLL 0064
icosecond Time Domain.=	Metalorganic Photochemistry in the Millisceond to P	INOR 0287
Nanoclusters.=	Metalorganic Syntheses of Phosphide Semiconductor	INOR 0102
–Binding Intera+ Photosensitive	Metalorganic Systems. Luminescence Probes of DNA	INOR 0315
Photosensitizer+ Photosensitive	Metalorganic Systems. Metal Polypyridines as Redox	INOR 0282
s in Photopolym+ Photosensitive	Metalorganic Systems. Organometallic Photoinitiator	INOR 0338
ystems: An Ove+ Photosensitive	Metalorganic Systems. Photosensitive Metalorganic S	INOR 0166
alorganic Systems. Photosensitive	Metalorganic Systems: An Overview.= +sitive Met	INOR 0166
Photochemistry in	Metalorganic Systems.= Multielectron–Transfer	INOR 0285
Photoinduced C–H Activation in	Metalorganic Systems.=	INOR 0167
Coatings for C/C Composites via	Metalorganic/Inorganic Polymer Blends.= Ceramic	POLY 0229
lasma Atomic Emission+ Trace	Metals Analysis of Fly Ash by Inductively Coupled P	FUEL 0138
Biomass.= Transition	Metals as Catalysts for Pyrolysis and Gasification of	FUEL 0103
=	Metals Behavior during Medical Waste Incineration.	FUEL 0131
geneou+ Preparation of Colloidal	Metals by Reduction or Precipitation in Microhetero	COLL 0128
te and Coal: The Role of Metals.	Metals Emissions Control Technologies for Waste In	FUEL 0129
+r Deposition of Low–Resistivity	Metals for Interconnection of Electronic Devices.=	INOR 0126
hlorophenols, Chlorobenzenes, and	Metals in Composts.= +ces of PCDDS, PCDFS, C	ENVR 0018
Determination of Trace	Metals in Liquefied Petroleum Gas.=	PETR 0139
	Metals in Medicine.=	CHED 0060
Chemistry. Distribution of Trace	Metals in the Ocean.= +ial on Marine Bioinorganic	INOR 0001
Tunneling Microscopy of Alkali	Metals on III–V(110) Surfaces: From 1- and 2-D Ins	PHYS 0183
nic Excitation and Dimerization of	Metals on Product–State Distributions.= +Electro	PHYS 0350
rom Waste and Coal: The Role of	Metals. Metals Emissions Control+ Clean Energy f	FUEL 0129
n of Reaction Pathway Control on	Metals.= +of the BOC-MP Method for Evaluatio	CATL 0032
rs P$_2$S Toward Various Transition	Metals.= +of a Tripodal Ligand with Mixed Dono	INOR 0028
CVD of Precious	Metals.=	INOR 0142
Earth	Metals.= Flame Chemistry of Alkali and Alkaline	PHYS 0073
rom 1- and 2-D Insulators to 3-D	Metals.= +Alkali Metals on III–V(110) Surfaces: F	PHYS 0183
of C$_{60}$ and Alkali	Metals.= Structural Studies of Binary Compounds	PHYS 0048
Molecula de	Metano.= Activacion del Enlace C–H por Ni en la	PHYS 0244
phate Compoun+ Is Monomeric	Metaphosphate an Intermediate in Reactions of Phos	INOR 0293
amidic Acids: Generation of Ethyl	Metaphosphates.= +of O–Ethyl Esters of Phosphor	ORGN 0110
olytic Enzyme Involved in Tumor	Metastasis.= +hips Studies of Cathepsin B, a Prote	MEDI 0017
Derived from the Ring–Opening	Metathesis Polymerization of Monosubstituted Cyclo	POLY 0329
p(CO)+ New Cycloaddition and	Metathesis Reactions of the Vinylidene Complexes C	INOR 0072
al Abundances of Uranium–235 in	Meteorites and Lunar Samples.= +ience—II. Initi	NUCL 0082
of Plutonium–244 in	Meteorites and Lunar Samples.= Initial Abundances	NUCL 0083
Induced on Corrosion Behavior of	Metglass 2826MB and MBF 15.= +Crystallization	PHYS 0219
ics.=	Methacrylate Precursors to Oxide Glasses and Ceram	POLY 0288
of 2-(Dimethylamino)ethyl	Methacrylate to Squalane.= Free Radical Grafting	POLY 0019
Morphology Buildup in	Methacrylate Urethane IPNs.= Kinetic Study and	PMSE 0043
d–Group Analysis for Poly Methyl	Methacrylate.= +clusion Chromatography and En	PMSE 0050
ions of Syndiotactic Poly(methyl	Methacrylate) as Studied by Spectroscopic Methods	POLY 0124
n Process of Water in Poly(methyl	Methacrylate).= +ce of Plasticizers on the Diffusio	POLY 0193
ethanes and Poly(Styrene–Methyl	Methacrylate–Methacrylic Acid).= +stor Oil Polyur	PMSE 0091
Acrylate-co-Dimethyl Aminoethyl	Methacrylated)/Epoxy Resin IPNs.= +ly(Methyl	PMSE 0079
cetylene–Containing Acrylate and	Methacrylates.= +ies on the Polymerization of Dia	POLY 0092
ly(Styrene–Methyl Methacrylate–	Methacrylic Acid).= +tor Oil Polyurethanes and Po	PMSE 0091

= Selective Partial Oxidation of	Methane by Catalytic Supercritical Water Oxidation.	CATL	0043
cal Mechanisms of Prompt NO in	Methane Flames.= +Measurements To Test Chemi	FUEL	0098
ry of Ocean Margins. Controls on	Methane Fluxes from the Hudson River.= Chemist	GEOC	0001
ations of Bis(Diphenylphosphino)	methane in Organometallic Synthesis.= +w Applic	INOR	0375
Mixture of bis(4–aminocyclohexyl)	Methane Isomers.= +line to a Nonthermodynamic	I&EC	0056
Diffusivity of Hydrogen and	Methane Molecules in A Zeolites.=	CATL	0024
ncal Applications of Catalysts—I.	Methane Oxidation over Alumina–Supported Noble	COLL	0185
gnostics and Modeling of 10-Torr	Methane/Oxygen Flames.= +ced Fluorescence Dia	FUEL	0128
oenzyme M Reductase Reaction in	Methanobacterium thermoautotrophicum.= +hyl C	ORGN	0025
tural Gas.=	Methanol and Carbon Monoxide Production from Na	FUEL	0096
Catalyst.= Kinetics of	Methanol Combustion on Palladium Monolith	COLL	0209
ctivity of Cu/ZnO Catalysts for	Methanol Synthesis Obtained by an Alkoxide route.=	INOR	0071
R Studies of the Conversion of	Methanol to Olefins over Offretite/Erionite Intergrow	PETR	0042
on Sorption of Organic Acids from	Methanol–Water Mixtures.= +ent and pH Effects	ENVR	0014
yllidaceae Alkaloids of the 5,11–	Methanomorphanthridine Type: An Efficient Total S	ORGN	0007
The Simplest Models of N^5, N^{10}–	Methenyltetrahydrofolate Coenzyme.= +hosphates:	ORGN	0204
Generation of ortho–Quinone	Methides from Cyclobutenones.=	ORGN	0229
	Methionine Salvage Pathway.=	BIOL	0024
Metabolites in Water.= Current	Methodologies for Analyzing Pesticides and	AGRO	0146
Multiresidue	Methodologies.= Comparison Study of Regulatory	AGRO	0149
Auxiliaries.= New Synthetic	Methodology Based on Organophosphorus	INOR	0377
ing and Transferring New Ener+	Methodology for Assessing the Feasibility of Develop	FUEL	0142
nd–Use Planning an+ Data and	Methodology Gaps in Development of a Model for La	ENVR	0043
rd–Ranking System and Decision	Methodology: What Is the Basis of EPA's Listing of	ENVR	0077
ng Trifluoroacylation, I: Analytical	Methodology.= +matic Amines in Fossil Fuels Usi	FUEL	0060
utants: The Need for Comparative	Methodology.= +ioritization of Hazardous Air Poll	ENVR	0066
ng the 1,3,2λ⁵–Dioxaphospholane	Methodology.= +with Trimethylsilyl Reagents Usi	ORGN	0051
cm of Enantiopure Troger's Base	Methosulfate as an Inclusion Host: Resolution of Alc	ORGN	0041
Receptor–Mediated Delivery of	Methotrexate–Insulin Conjugate to H35 Hepatoma+	MEDI	0185
duced Rearrangement of Allylic	Methoxyacetates: A Novel Type of Allylic Transposit	ORGN	0211
s as Anti–Inf+ N–(Fluorenyl–9–	Methoxycarbonyl) Amino Acids and Their Derivative	MEDI	0171
Electrochemical Alkylation and	Methoxylation of Active Methylene Compounds.=	ORGN	0249
= N– and 4–Arylation of N–(4–	Methoxyphenyl)–2–Butenamides by Various Arynes.	ORGN	0224
L] C+ Solubility and Stability of	Methyl [1–(Butylcarbamoyl)–1H–Benz–imidazol–2–Y	AGRO	0041
Mechanical Properties of Poly(	Methyl Acrylate–co–Dimethyl Aminoethyl Methacryl	PMSE	0079
d Alkylmethylzincs with Aldehy+	Methyl as a Blocking Group in the Reaction of Mixe	ORGN	0090
Method for the Determination of	Methyl Carbamate in Wines.=	AGFD	0110
bacterium+ Steric Course of the	Methyl Coenzyme M Reductase Reaction in Methano	ORGN	0025
Nitroxyl Amide Spin Labeling:	Methyl Esterification– and Hydration–Induced Moti	AGFD	0012
and Cyclic Polystyrene/Poly(vinyl	Methyl Ether) Blends.= +composition and Linear	POLY	0295
porates Prepared Using Poly(vinyl	Methyl Ether/Maleic Anhydride) Copolymer.= +a	COLL	0086
and End–Group Analysis for Poly	Methyl Methacrylate.= +clusion Chromatography	PMSE	0050
Solutions of Syndiotactic Poly(	methyl Methacrylate) as Studied by Spectroscopic M	POLY	0124
iffusion Process of Water in Poly(	methyl Methacrylate).= +e of Plasticizers on the D	POLY	0193
Polyurethanes and Poly(Styrene–	Methyl Methacrylate–Methacrylic Acid).= +tor Oil	PMSE	0091
ional Transition in Insulin Hex+	Methyl Paraben Will Cause a T+–R–Like Conformat	BIOT	0150
rs of Prednisolone: Analogues of	Methyl 11β, 17α, 21–Trihydroxy–3,20–Dioxo–Pregna–	MEDI	0170
alogues as Potent Glycine Site N–	Methyl–D–Aspartate Antagonists.= +renic Acid An	MEDI	0146
ition of Dextran Synthesis by α–	Methyl–D–Glucopyranoside and Its 6–Deoxy– and 6–	CARB	0028
rphine, Dihydromorphine, and N–	Methyl–Morphinium Analogues.= +substituted Mo	MEDI	0149
enase Activity of 4–Hydroxy–2–	Methyl–N–(5–Substituted–3–Isoxazolyl)–2H–1,2–Ben	MEDI	0178
re of Compounds Derived from 2–	Methyl–thiosemicarbazide.= +d Molecular Structu	INOR	0032
Weight Determinations of Poly–4–	Methyl–1–Pentene.= +lymer Analysis. Molecular	PMSE	0046
Actinide Ions with 4–Benzoyl–5–	Methyl–2–Phenyl–Pyrazol–3–Thione and Tri–n–Octy	NUCL	0073
2–Acetyl–1–	Methyl–5–Nitroimidazole.= Convenient Synthesis of	ORGN	0200
emistry of Chiral Titanocenes: 1–	Methyl, 2–Phenyltitanocene Dichloride and Its 1–,3–	INOR	0281
centration Quenching of Triplet	Methylbenzenes in Solution, Using a Fluorescence Pr	PHYS	0265
Ox+ Insecticidal Substituted 2–	Methylbiphenylmethyl Alkyl and Dialkylamino Aryl	AGRO	0072
–Methylbiphenyls, 2–Cyano–4'–	Methylbiphenyls and 2–(Tetrazol–5–yl–4'–Methylbip	ORGN	0112
iphenyls and 2–(Tetrazol–5–yl–4'–	Methylbiphenyls.= +phenyls, 2–Cyano–4'–Methylb	ORGN	0112
ubstituted 2–Carbomethoxy–4'–	Methylbiphenyls, 2–Cyano–4'–Methylbiphenyls and 2	ORGN	0112
eparation and Determination of	Methylene Blue and Its Demethylated Metabolites fr	AGFD	0101
and Methoxylation of Active	Methylene Compounds.= Electrochemical Alkylation	ORGN	0249
st System to Hydrogenate Crude	Methylenedianiline to a Nonthermodynamic Mixture	I&EC	0056
beled Poly(styrene) and Poly(vinyl	methylether).= +n: Hydrogen– and Deuterium–La	COLL	0035
hiophene–Derived 3–Hydroxy–3–	Methylglutaryl–Coenzyme A Reductase Inhibitors.=	MEDI	0101
Activities of Various 2'– and 3'–	Methylidene–Substituted Nucleoside Analogues and	CARB	0038
Crystal Structure of 2'–Deoxy–2'–	Methylidenecytidine Hydrochloride.= +logues and	CARB	0038
thermal Physical Aging of Poly(	methylmethacrylate): Localization of Perturbations in	PMSE	0173
Properties of a Polystyrene/Poly(	methylmethacrylate)/–Glass Beads System.= +ical	PMSE	0167
nd Mechanical Properties of Poly(	methylmethacrylate)/Glass Fiber Laminates.= +l a	PMSE	0110
Poly(dimethylsiloxane) and Poly(	methylphenylsiloxane) Blends.= +yclic and Linear	POLY	0234
on of Poly(dimethylsiloxane–co–	methylphenylsiloxane) Random Block Copolymers an	POLY	0020
yrene–Sodium–2–Acrylamido–2–	Methylpropane Sulfonate) in Nonaqueous Solution.=	COLL	0181
Acid–Catalyzed Chlorination of 1–	Methylpyrrole with N–Chlorobenzamides.= +f the	ORGN	0158
s Based on 4,4'–Dihydroxy–α–	methylstilbene Mesogen.= +esters and Copolyester	POLY	0031
ate Analogue Inhibitors of	Methyltransferases.= Potential Specific Multisubstr	MEDI	0027
odeposition of Reduced Forms of	Methylviologen from Anionic Surfactant Solutions.=	COLL	0077
la de Oxido, en la Validacion del	Metodo Potencio–Metrico para Cuantificar Clorhidra	ANYL	0066
s of the Herbicides Atrazine and	Metribuzin: Preparation and Evaluation of Controlle	AGRO	0063

of Urinary Creatinine as a Stable	Metric for Human Biomonitoring Studies and an Alt	CEI	0019
Validacion del Metodo Potencio-	Metrico para Cuantificar Clorhidrato de L-Arginina	ANYL	0066
Projectile Brakup in 25-	MeV/A Reactions: Prompt or Sequential?.=	NUCL	0027
at 600	MeV/nucleon.= Fragmentation of Gold Projectiles	NUCL	0047
Interaction between Gystamine,	Mexamine, and Crown Ether on the Surface of Pyrog	COLL	0096
and Reuse of Pickle Liquors from	Mexican Iron Industry.= +e Solutions for Disposal	ENVR	0039
ed to a Clerodanic Precursor from	Mexican Salvia Species.= +th New Skeletons Relat	ORGN	0280
oid Chemistry. Prehistory of the	Mexican Steroid Industry and Update on R. E. Mark	HIST	0016
Leishmania	mexicana.= Inhibitors of DNA Polymerases of	BIOL	0118
mical Mechanisms Used to Model	Mexico City Air Quality.= +mparison of Three Che	PHYS	0296
Zurich to Steroids in	Mexico via Cuba.= °From Ruzicka's Terpenes in	HIST	0017
Chemistry Olympiad in	Mexico: A Multipurpose Event.=	CHED	0109
Early Steroid Research in	Mexico: The Pill and Cortisone.=	HIST	0018
s a la Familia Scrophulariaceae de	Mexico.= +n Quimica de las Plantas Perteneciente	ORGN	0185
en	Mexico.= Estudio Sobre el Uso de Eteres de Glicol	ENVR	0038
er for Technological Innovation in	Mexico.= +Scale Industries: The Case of the Cent	SCHB	0027
in	Mexico.= The Life of a Chemist: Career Possibilities	CONG	0007
de	Mexico?.= Asbestos en el Area Urbana de la Ciudad	ENVR	0025
e Hydrothermal Crystallization of	MFI Zeolites in an Alkaline Fluoride Medium.= +r	PETR	0005
in As-Synthesized and Modified	MFI-Gallosilicates.= +Structural States of Gallium	PETR	0077
ion Reactions with MR_2 (M = Zn,	Mg and R = Bz, Cp, allyl).= +ntation-Recombinat	INOR	0383
iovalent Analog+ Structure of K(	$Mg_1/3Nb_2/3)OPO_4$ (KMNP): A Centrosymmetric Al	INOR	0256
inum-Tin Catalysts Supported on	Mg, Cu, Zn, and Ni Aluminates.= +tinum and Plat	COLL	0126
Reactive Collisions of Al,	Mg, Si, and C Atoms.=	PHYS	0089
west Singlet and Triplet States of	MgC_2.= +Spectrum, and Binding Energy of the Lo	PHYS	0276
hemical Properties of Ultra-Thin	MgO Films on Mo(100).= +Characterization, and C	COLL	0193
nified Laboratory+ Use of a 300-	MHz FT-NMR in a Second-Semester Sophomore U	CHED	0181
Optical Spectroscopy at 1-	MHz Resolution.=	PHYS	0043
Microporous Pillared	Mica with Cation-Incorporated Silicate Surfaces.=	PETR	0129
t of Plasmodium Berghei-Infected	Mice with Transdermal Artelinic Acid.= +reatmen	MEDI	0063
Acquired Immune Deficiencies in	Mice.= +termined Higher Mortality Incidence and	AGFD	0145
L-Tryptophan Metabolism in	Mice.= Paradoxical Effect of Diazinon on	AGRO	0020
munodeficient Antibiotic-Treated	Mice.= +ve Bacteria and High Mortality in CD1 Im	AGFD	0146
otogenerated Ketyl Radicals in+	Micellar Effect on the Photoreduction of Dyes by Ph	COLL	0081
tion of+ Medium, Pressure, and	Micellar Effects on Kinetics of Peroxodisulfate Oxida	INOR	0022
of Organic Com+ Application of	Micellar Liquid Chromatography to QSAR Modeling	ANYL	0046
n-Activity Relationship Studies in	Micellar Liquid Chromatography.= +ture-Retentio	ANYL	0044
Iron-Sulfur Clusters in	Micellar Media.= Kinetics of Assembly of	INOR	0076
icelles through Pressure-Induced	Micellar Phase Transitions.= +ery from Reversed M	BIOT	0046
Affinity-Based Reversed	Micellar Protein Extraction.=	BIOT	0045
otentiometry and Polarography in	Micellar Systems.= +Microheterogeneous Fluids. P	COLL	0001
y in Microheterogeneous Fluids.	Micelle and Microemulsion Diffusion Coefficients.=	COLL	0040
eren+ Electroorganic Synthesis in	Micelles and Emulsions: A Comparison between Diff	COLL	0043
ectrochemical Reactions of CdS in	Micelles and Reverse Micelles.= +esis and Photoel	COLL	0148
lectroactive Polymers.= Use of	Micelles in the Oxidative Synthesis of Conductive-E	COLL	0168
ethylene Oxide) Block Copolymer	Micelles in Water.= +Methods. Polystyrene-Poly(	POLY	0277
ion and Recovery from Reversed	Micelles through Pressure-Induced Micellar Phase T	BIOT	0046
ment Thin Films by Disruption of	Micelles.= +trochemical Formation of Organic-Pig	COLL	0019
ns of CdS in Micelles and Reverse	Micelles.= +esis and Photoelectrochemical Reactio	COLL	0148
Ethers.= Asymmetric	Michael Addition Catalyzed by Chiral Azacrown	ORGN	0209
blocking of Anions of+ Tandem	Michael Addition Ethoxycarbonylation-Carbonyl De	ORGN	0228
Nitrogen and Su+ Tandem S_N2-	Michael Approach to the Synthesis of 5- and 6-Ring	ORGN	0199
Complexes: Super-Homo-	Michael-Acceptors.= Cyclopropylthiocarbene-Metal	ORGN	0216
ecnicas Electroquimicas de Aceros	Microaleados en Acido Sulfurico.= +tadistico de T	PHYS	0281
Science Program with a	Microanalysis Laboratory.= Upgrading a Forensic	CHED	0153
iformly Cross-Linked Polystyrene	Microbeads.= +el Filled Polymers: Synthesis of Un	PMSE	0017
Relationship between Localized	Microbial Activity and Localized Corrosion on Metal	BIOT	0098
On-Line Monitoring of	Microbial Biofilms.=	BIOT	0099
Induction and Inducers of	Microbial Cellulases and Xylanases.=	BIOT	0185
ed 4140 Steel Exposed to Mixed	Microbial Communities under Laboratory Conditions	BIOT	0097
	Microbial Geritol: From Iron to Infection.=	CHED	0061
atives.=	Microbial Hydroxylation of Dihydroartemisinin Deriv	BIOT	0170
eer's Vie+ Control Properties of	Microbial Polysaccharides for Production—An Engin	BIOT	0199
ial Polysaccharides. Production of	Microbial Polysaccharides on Maple Syrup.= +rob	BIOT	0195
olysac+ Control of Properties of	Microbial Polysaccharides. Production of Microbial P	BIOT	0195
ules.=	Microbial Production of Indigo from Precursor Molec	BIOT	0032
	Microbial Reduction of Uranium.=	GEOC	0008
inuous High-Cell Fermentation of	Microbial-Based Flavorings.= +ydrate-Based Cont	CARB	0006
d Kineti+ Biomineralization and	Microbially Influenced Corrosion. Hydrodynamics an	BIOT	0096
ospectivo.=	Microbiologia de Alimentos en el LNSP: Enfoque Pr	AGFD	0206
Risk Assessments for	Microbiological Contaminants in Drinking Water.=	ENVR	0070
	Microbiological Hazards and Opportunities.=	AGFD	0014
Use of	Microcalorimetry for Probing Carbon Surfaces.=	FUEL	0018
uzin: Preparation an+ Polymeric	Microcapsules of the Herbicides Atrazine and Metrib	AGRO	0063
Polystyrene Porous	Microcarriers for Mammalian Cell Culture.=	BIOT	0010
oly(vinyl Acetate)/Silicon Oxide	Microcomposites Prepared via the Sol-Gel Method.=	POLY	0287
nt Polymer Systems.=	Microdeformational Behavior of Oriented Bicompone	POLY	0101
Mass Spectrometry. Application of	Microdilatometry to Solvent Swelling of Coals.=	FUEL	0086
oclusters within Block Copolymer	Microdomains.= +ntrolled Synthesis of Metal Nan	POLY	0207
Energy Transfer in Latex	Microdomains.=	PHYS	0108

s of Soluble Polyimides for Use in	Microelectronics.= +hesis and Properties of a Serie	POLY	0038
heterogeneous Fluids. Micelle and	Microemulsion Diffusion Coefficients.= +in Micro	COLL	0040
t+ Electrochemical Insights into	Microemulsion Microstructure and Its Effect on Elec	COLL	0023
Coulometric Initiation of	Microemulsion Polymerization.=	COLL	0169
ansient Phase-Separation in w/o-	Microemulsion.= +cs of Electric-Field-Induced Tr	COLL	0146
S/1-Pentanol/Cyclohexane/Water	Microemulsion.= +the Single-Phase Regions of SD	COLL	0024
Oils: Formulation, Opti+ Model	Microemulsions with Fatty Acid Esters or Vegetable	COLL	0083
trface Derivatization of Au and Pt	Microfabricated Wire Arrays.= +s for Selective S	INOR	0350
Cross-Flow	Microfiltration.= Cell Debris Removal by	BIOT	0048
Reversible Gelation of	Microgels due to Colloidal Crystallization.=	POLY	0244
ετion of Carbohydrate-Mediated	Microheterogeneity of Recombinant Human Erythro	ANYL	0018
ετed Colloids: Electrochemistry in	Microheterogeneous Fluids+ Surfactants and Associ	COLL	0166
ετed Colloids: Electrochemistry in	Microheterogeneous Fluids+ Surfactants and Associ	COLL	0019
ετed Colloids: Electrochemistry in	Microheterogeneous Fluids+ Surfactants and Associ	COLL	0040
ετed Colloids: Electrochemistry in	Microheterogeneous Fluids+ Surfactants and Associ	COLL	0001
ated Colloids: Electrochemistry in	Microheterogeneous Fluids+ Surfactants and Associ	COLL	0128
ated Colloids: Electrochemistry in	Microheterogeneous Fluids+ Surfactants and Associ	COLL	0147
s by Reduction or Precipitation in	Microheterogeneous Liquids.= +of Colloidal Metal	COLL	0128
ylformamide in Homogeneous and	Microheterogeneous Media.= +canoic Acid) Benzo	ORGN	0166
of Metal Complexes in	Microheterogeneous Media.= Photoredox Chemistry	INOR	0316
nd Deltamethrin from the Surface	Microlayer of Water.= +tilization of Fenitrothion a	ENVR	0020
Immobilization of Proteins in	Microlithographically Defined Patterns.=	BIOT	0246
s—I. Chain+ Energy-Consuming	Micromechanisms in the Fracture of Glassy Polymer	PMSE	0073
olyester Resins with Low-Profil+	Micromorphology Characterization of Unsaturated P	POLY	0114
in	Microorganisms.= Dynamic Analysis of Regulation	BIOT	0149
Marine	Microorganisms.= Siderophore Production in	INOR	0004
τegy for Electric-Field-Orientable	Microparticles.= +n in Fluorocarbon Media: A Stra	COLL	0164
Biopolymers with	Micropellicular Stationary Phases.= Rapid HPLC of	ANYL	0128
	Microphase Separation and Reversible Gelation.=	POLY	0122
quilibrium and Nonequilibrium	Microphase Structure of Interpenetrating Polymer N	PMSE	0093
Solid-State NMR Study of the	Microphase Structure of PS-PMPS Diblock Copolym	PMSE	0121
Colloidal Structure of	Microphase-Separated Multiblock Copolymers.=	COLL	0053
ies in Slitlike and Wedge-Shaped	Micropores.= +s. Calculation of Adsorption Energ	FUEL	0008
titution of Paramagnetic Ions in	Microporous Aluminophosphate Molecular Sieves.=	PHYS	0231
Chemisorption on Graphite and	Microporous Carbons: An Analysis of Experimental D	FUEL	0017
eling of Oxygen Chemisorption on	Microporous Carbons.= +cation Mechanisms. Mod	FUEL	0012
e-Selective Catalytic Reactions in	Microporous Materials.= +te NMR Studies of Shap	CATL	0023
ealing on the Capillarity of Nylon	Microporous Membranes.= +gh-Temperature Ann	COLL	0037
ral+ Hydrothermal Synthesis of	Microporous Molybdenum with Octahedral-Tetrahed	CATL	0019
, Forming, and Characterization.	Microporous Oxides by the Sol-Gel Process: Synthes	PETR	0031
Silicate Surfaces.=	Microporous Pillared Mica with Cation-Incorporated	PETR	0129
xygen Adsorption Properties of	Microporous Silica Derived from Layered Siloxene by	PETR	0144
ylloarsenates—Zeolites and Other	Microporous Structures.= +ncophosphates and Ber	PETR	0055
y Program Complemented by the	Microscale Approach in the General, Inorganic, and	CHED	0150
NMR Imaging: A Chemical	Microscope for Coal Analysis.=	FUEL	0075
rpretation of Scanning Tunneling	Microscope Images Showing Anomalous Periodic Str	PHYS	0212
Materials.= Atomic Force	Microscope Studies of Biominerals and Related	INOR	0117
Processes.= Atomic Force	Microscope Studies of Underpotential Deposition	PHYS	0128
of the Scanning Electrochemical	Microscope to the Measurement of Reaction Kinetics	COLL	0170
Tunneling	Microscope.= Nanofabrication with the Scanning	PHYS	0362
lidified Hazardous Wastes.=	Microscopic and Spectroscopic Characterization of So	ENVR	0004
uid Boundaries: Conventional and	Microscopic Interfaces.= +ectrified Immiscible Liq	COLL	0004
Tunneling	Microscopies: FEM, FEES, FIM, A-P, STM, STS.=	PHYS	0097
uter-Controlled Scanning Electron	Microscopy (CCSEM) in Coal Science.= +of Comp	FUEL	0041
es. Field Ion-Scanning Tunneling	Microscopy (FI-STM).= +and Related Techniqu	PHYS	0182
Auger and Scanning Tunneling	Microscopy (STM) Studies of the Chemical-Vapor+	COLL	0050
cipal Waste by Scanning Electron	Microscopy and Optical Microscopy.= +ion of Muni	FUEL	0137
Tc Superconductors by Tunneling	Microscopy and Spectroscopy.= +c States of High-	PHYS	0206
ide Arrays+ Scanning Tunneling	Microscopy and Tunneling Spectroscopy of Metal Ox	COLL	0194
-Solidified/Stabilize+ Electron-	Microscopy Characterization Techniques for Cement	ENVR	0005
onization and Scanning Tunneling	Microscopy in Physical Chemistry Laboratory.= +I	CHED	0044
From 1- a+ Scanning Tunneling	Microscopy of Alkali Metals on III-V(110) Surfaces:	PHYS	0183
Tunneling and Atomic Force	Microscopy of Biological Surfaces.= Scanning	PHYS	0158
nds.= Scanning Tunneling	Microscopy of Donor Graphite Intercalation Compou	PHYS	0365
Atomic Force	Microscopy of Molecules on Surfaces.=	PHYS	0153
Atomic Force	Microscopy on Cold-Extruded Polyethylene.=	PMSE	0078
ometer Scale.= Atomic Force	Microscopy Studies of Friction and Wear at the Nan	PHYS	0156
Application of Fluorescence	Microscopy to Coal-Derived Resid Characterization.=	FUEL	0043
of Transmission Electron	Microscopy to the Study of Diamond.= Applications	PHYS	0120
ally Excited Scanning Tunneling	Microscopy: Imaging Subsurface and Nonequilibrium	PHYS	0208
l Chemistry. Scanning Tunneling	Microscopy: New Applications in Analytical Science.	ANYL	0022
and Examples of Scanning Force	Microscopy.= +and Related Techniques. Principles	PHYS	0152
plexes Using Scanning Tunneling	Microscopy.= +ormation Storage in Molecular Com	BIOT	0023
and Transmission Electron	Microscopy.= Analysis of Materials Using Scanning	INOR	0014
n Surfaces by Scanning Tunneling	Microscopy.= +omic Scale Imaging of Molecules o	PHYS	0159
Electron Microscopy and Optical	Microscopy.= +ion of Municipal Waste by Scanning	FUEL	0137
aterials Using Scanning Tunneling	Microscopy.= +nometer Scale Etching of MX$_2$ M	PHYS	0364
Photon Scanning Tunneling	Microscopy.=	PHYS	0210
Materials Using Atomic Force	Microscopy.= Probing the Surface Forces of	PHYS	0154
periments by Scanning Tunneling	Microscopy.= +Organic Unimolecular Rectifiers: Ex	BIOT	0005

ce Studied by Scanning Tunneling	Microscopy.= +se Transition on the Ge(111) Surfa	PHYS	0099
ectrochemical Scanning Tunneling	Microscopy.= +er Resolution Studied by in Situ El	PHYS	0127
r and Scanning Molecular Exciton	Microscopy.= +Spatially Resolved Energy Transfe	PHYS	0022
) Surfaces by Scanning Tunneling	Microscopy-Spectroscopy.= +O_2(110) and TiO_2(001	COLL	0171
i$_2$Sr$_2$CaC+ Scanning Tunneling	Microscopy/Spectroscopy Studies of Single-Crystal B	PHYS	0211
hains by Transmission Electron	Microscopy, II: Image Contrast Analysis of Iodinated	POLY	0062
Ion Exchange and+ Analysis for	Microsomal Cytochrome P-450 Patters by Means of	ANYL	0017
y ^{13}C NMR+ Detection of Liver	Microsomal Metabolites of N,N-Diethylbenzamide b	AGRO	0026
ybenzoic Acid by Chicken Kidney	Microsomes.= +-Phenoxybenzoic Acid to 3-Hydrox	AGRO	0018
Properties of Polymer-Silver	Microsphere Composites.= Nonlinear Optical	POLY	0154
olymers.=	Microstructural Aspects of IPNs Based on Block Cop	PMSE	0026
d Heavy Metal Wastes.=	Microstructural Characterization of Cement-Solidifie	ENVR	0008
nd Growth of Dia+ Surface and	Microstructural Characterization of the Nucleation a	PHYS	0121
Nafion Membranes by an in S+	Microstructural Evolution of a Silicon Oxide Phase in	POLY	0285
ical Insights into Microemulsion	Microstructure and Its Effect on Electron Transfer.=	COLL	0023
ation of Solidif+ Chemistry and	Microstructure of Cement Matrices for the Immobiliz	ENVR	0002
of Cementitiou+ Chemistry and	Microstructure of Solidified Waste Forms. Chemistry	ENVR	0001
Diffusion and Dynamics in	Microstructured Block Copolymer Solutions.=	POLY	0275
Interplay of Motors and	Microtubule Dynamics.=	BIOL	0139
	Microtubule-Based Motor Proteins.=	BIOL	0138
onal Transitions in Cytoskeletal	Microtubules: Implications for Neural-Like Protein A	BIOT	0055
ctive Polymers by Time-Resolved	Microwave Conductivity (TRMC).= +se in Photoa	POLY	0214
with Applications in X-ray Lit+	Microwave CVD Deposition of Diamond Thin Films	PHYS	0092
ontrol of Properties of Fluid+ A	Microwave Device for the On-Line Monitoring and C	I&EC	0059
e = Cure Kinetics in Thermal and	Microwave Fields.= +nistic Study of Epoxy/Amin	POLY	0161
SR Spectroscopy at Sub-9-GHz	Microwave Frequencies: Advantages and Prospects.=	PHYS	0215
	Microwave Heating of Zeolite Synthesis Mixtures.=	PETR	0080
under	Microwave Oven.= Bischler-Napieralski Reaction	ORGN	0195
ntent of Dielectric Properties in	Microwave Region for Printed Circuit Board Materia	PMSE	0055
Two-Dimensional	Microwave Spectroscopy.=	PHYS	0115
Water: A Clean Source of Energ+	Microwave-Induced Catalytic Reactions of CO_2 and	CATL	0055
Water: A Clean Source of Energ+	Microwave-Induced Catalytic Reactions of CO_2 and	FUEL	0102
Kinetics and Properties.=	Microwave-Initiated Copolymerization MMA/HEMA	POLY	0052
y of Oxidiz+ Fourier Transform	Mid- and Far-Infrared IR Specular Reflectance Stud	POLY	0253
m η^2-Iminoacyl Ch+ High- and	Mid-Valent Mono(peralkylcyclopentadienyl) Tantalu	INOR	0108
Management and Disposal in the	Mid-80s and the Mid-90s.= +ng Forward: Pesticide	AGRO	0086
d Disposal in the Mid-80s and the	Mid-90s.= +ng Forward: Pesticide Management an	AGRO	0086
How to Survive and Thrive in the	Midst of Bigness.= +ialties for Niche Markets, or	CMEC	0005
in the	Midwest.= Pesticides in Surface and Groundwaters	AGRO	0065
Agonist for the Treatment of	Migraine.= Sumatriptan, a Novel 5-HT$_1$-Like	MEDI	0079
igation of Stereocontrolled Proton	Migration from Nitrogen to Carbon.= +istic Invest	ORGN	0296
rbyne Com+ Mechanism of Alkyl	Migration from Oxygen to Metal in Fe-Mn Alkoxyca	INOR	0110
zard-Ranking Sy+ Contaminant	Migration in Ground Water as Assessed by EPA's Ha	ENVR	0081
n T.H.O.R. Nuclear Research R+	Migration of Activation Products in Discharge Pool i	NUCL	0090
ents in Fluid-Cracking Catalys+	Migration of Metal Contaminants and Passivating Ag	CATL	0012
Phosphine-Promoted Alkyl-CO	Migration Reactions of Unusually Reactive (η^5-Inden	INOR	0239
Double-Bond	Migration.= Aldol Chemistry via Initial	ORGN	0092
ts in Promoting the Alkyl to Acyl	Migratory Insertion Reaction in η-Cp(CO) (PPh$_3$)Fe	INOR	0109
Oxidation of Sulfides to Sulf+ A	Mild Osmium Tetraoxide-Catalyzed Process for the	ORGN	0244
logies of Fertilizers. Corrosion of	Mild Steel Exposed to Inhibited Urea-Ammonium N	FERT	0001
Suspensions.= Corrosion of	Mild Steel Exposed to Urea-Ammonium Sulfate	FERT	0003
	Milestones of Forensic Toxicology.=	HIST	0002
f β-Lactam Antibiotic Residues in	Milk and Tissues.= +proaches to Determination o	AGFD	0125
e of Dihydro-Erythroidines in the	Milk of Goats Fed Erythrina poeppigiana and E. +c	AGFD	0129
atio+ Chemometric Approach to	Milk Shelf-Life Prediction and Flavor-Quality Evalu	AGFD	0108
Confirmation of Drug Residues in	Milk.= +mal Origin. Approaches to Detection and	AGFD	0099
s Demethylated Metabolites from	Milk.= +d Determination of Methylene Blue and It	AGFD	0101
mination of Eight Sulfonamides in	Milk.= +rformance Liquid Chromatographic Deter	AGFD	0128
Autoxidation of a Spray-Dried	Milkbase for Baby Food.= Factors Affecting Lipid	AGFD	0158
eristics of Produce+ Design of a	Mill Drier for Shrimp Head: Bromatological Charact	AGFD	0105
teristics of Produce+ Design of a	Mill-Drier for Shrimp Heads: Bromatological Charac	ENVR	0046
ry Fiber, Protein, and Starch in a	Milled Hard Red Winter Wheat.= +id, Total Dieta	AGFD	0143
nd Edition—Aiba, Humphrey, and	Millis.= +otechnology. Biochemical Engineering, 2	BIOT	0221
anic Photochemistry in the	Millisceond to Picosecond Time Domain.= Metalorg	INOR	0287
l Modificatio+ Peptides, Peptide	Mimetics, and Bioavailability. Influence of Structura	MEDI	0192
and Evaluation of a 3-Phosphate	Mimic.= +Mimics of the Tetrahedral Intermediate	ORGN	0029
Multifunctional Enzyme	Mimics and Activators.= Plenary Session.	ORGN	0265
us Moieties as Stable 3-Phosphate	Mimics in Aromatic Analogs of EPSP.= +of Vario	ORGN	0029
esis and Evaluation of Aromatic	Mimics of the Tetrahedral Intermediate and Evaluat	ORGN	0029
Protein	Mimics.= Studies toward Heme-Dependent	INOR	0186
on during Cooking of Refrigerated	Minced Channel Catfish Muscle.= +Lipid Oxidati	AGFD	0124
Interactions of Water with Clay	Minerals as Studied by ^{2}D Nuclear Magnetic Resona	AGRO	0061
ganic Matter Associated with Clay	Minerals from Syncrude Sludge Pond Tailings.=	I&EC	0021
Pressures+ Reactions of Cement	Minerals with Water at Elevated Temperatures and	INOR	0246
omplexants sur la Fottation de ces	Mineraux.= +-Pyrite: Effet des Oxydants et des C	ANYL	0067
otherapeutics and Antiparasitic+	Miniaturized Multiresidue Analysis of Selected Chem	AGFD	0062
Multifragmentation.+ The MSU	Miniball: A Tool for the Characterization of Nuclear	NUCL	0035
Introducing	MINICUAD to Undergraduates.=	CHED	0120
Chemistry. Recognition of RNA	Minihelices in Relation to Decoding Genetic Informa	BIOL	0026
Physicochemical Changes on	Minimally Processed Fruit and Vegetable Surfaces.=	AGFD	0011

Infusion.= Preparation of Novel	Minimally Processed Fruit Products by Pectinase	AGFD 0059
nching Treatments to Stabilize+	Minimally Processed Fruits and Vegetables. Cold Bla	AGFD 0027
eservation by Combined Metho+	Minimally Processed Fruits and Vegetables. Fruit Pr	AGFD 0055
velopments in Modified Atmos+	Minimally Processed Fruits and Vegetables. New De	AGFD 0075
gy of Cut Fruits and Vegetables+	Minimally Processed Fruits and Vegetables. Physiolo	AGFD 0010
Approach to Development of	Minimally Processed Vegetable Snacks.= Integrated	AGFD 0056
heir Containers. Pesticide Rinsate	Minimization and Reuse.= +tural Chemicals and T	AGRO 0102
Container	Minimization and Reuse.=	AGRO 0087
s of Oilseed Food Ingredients to	Minimize Warmed-Over-Flavor Development in Mea	AGFD 0021
System for Capillary Gas Chro+	Minimizing Mass Discrimination in a Split Injection	TECH 0018
ching Resource for the 90s.=	Minisymposium on ChemSource. ChemSource: A Tea	CHED 0017
Georgius Agricola: The Distorting	Mirrors of History.= +Corp.). Herbert Hoover and	HIST 0011
ropylene) and Poly(bute+ On the	Miscibility and Cocrystallization in Blends of Poly(p	POLY 0089
de Blends with Rigid and Semir+	Miscibility and Properties of Cross-Linkable Polyimi	PMSE 0023
Lecture. Enhancement of Polymer	Miscibility through Ionic Functional Groups.= +y	PMSE 0116
zation of Poly(amic Acid) in	Miscible and Immiscible Blends.= Kinetics of Imidi	PMSE 0147
y(3-hydroxybutyrate)+ Study of	Miscible Blends of Poly(vinylidene Fluoride) and Pol	POLY 0097
xylenyl Ether).=	Miscible Blends of Sulfonated Polystyrene and Poly(	PMSE 0117
s.=	Miscible Novolac/Polymer Blends and Curing Studie	POLY 0113
Effects of Water-	Miscible Organic Cosolvents on Kinetics of Plasmin.=	BIOL 0092
Shear-Induced Mixing in a	Miscible Polymer Blend.= Fluorescence and	PMSE 0133
ace Analysis When Intense Lasers	Miss the Surface.= +ts through Virtual States: Surf	NUCL 0044
The	MIT Student's Outreach Program.=	CONG 0008
uate Chemistry Laboratory at	MIT.= Advanced Instrumentation in the Undergrad	CHED 0128
Developing Pesticide-Exposure	Mitigation Strategies for Handlers and Harvesters.=	AGRO 0043
oalkylation of Oligonucleotides by	Mitomycin C.= +-Sequence Specificity of the Mon	MEDI 0015
of BMY-25067, an Analogue of	Mitomycin C.= Bioreductive Alkylating Properties	MEDI 0014
Glutathione and Other Thiols by	Mitomycin C.= Bioreductive Alkylation of	BIOL 0088
ene-Anthraquinones, Analogues of	Mitomycin C.= +ogical Activities of 1,2-Trimethyl	MEDI 0013
g. Plenary Lecture. Distributive	Mixing and Energy Distribution in Twin Screw Extru	PMSE 0175
with High	Mixing and Low Shear.= Extruding Waste Plastics	I&EC 0030
Fluorescence and Shear-Induced	Mixing in a Miscible Polymer Blend.=	PMSE 0133
Optical Four-Wave	Mixing in a Silver Colloid-Polymer Composite.=	POLY 0218
Bistability and Four-Wave	Mixing in Bacteriorhodopsin Materials.= Optical	BIOT 0081
oduced by Ion Bombardment (Ion	Mixing). +f Materials: Composition Modification Pr	NUCL 0091
adores de Molibdeno sobre Oxidos	Mixtos de Alumina-Titania.= +Tiofeno con Cataliz	FUEL 0105
of Ligand-Ligand Interactions in	ML_n^+ Systems.= +se Metal Reactions—IV. Study	PHYS 0171
Solution Properties of	MMA/HEMA Copolymers.=	POLY 0093
Solution Properties of	MMA/HEMA Copolymers.=	PMSE 0086
-Initiated Copolymerization	MMA/HEMA Kinetics and Properties.= Microwave	POLY 0052
tion from Oxygen to Metal in Fe-	Mn Alkoxycarbyne Complexes.= +m of Alkyl Migra	INOR 0110
Expression of Active Lignin and	Mn Peroxidases in Phanerochaete chrysosporium +	BIOT 0237
he Coordination Environment of	Mn(III) and Mn(IV) Horseradish Peroxidase with M	BIOL 0095
tion Environment of Mn(III) and	Mn(IV) Horseradish Peroxidase with Manganese-Po	BIOL 0095
Luminescence Characteristics of	Mn^{2+} in Cubic [(Rb18c6)$_4$MnCl$_4$][TlCl$_4$]$_2$ in Presen	INOR 0039
talytic Activity of LaBo$_3$ (B = Ni,	Mn, Co, Fe) Perovskites.= +aracterization, and Ca	PHYS 0225
of LaBo$_3$ (B =	Mn, Fe, Co, Ni) Perovskites.= Electronic Structure	PHYS 0288
plexes Cp(CO)$_2$M=C=CH$_2$ (M=	Mn, Re) with Imines, Carbodiimides, and Benzalazin	INOR 0072
istics of Mn^{2+} in Cubic [(Rb18c6)	$_4$MnCl$_4$][TlCl$_4$]$_2$ in Presence of Cu^{2+}.= +Character	INOR 0039
Manganese Acyl Complexes, (CO)	$_5$MnCOCH$_2$R (R = H,CH$_3$, OCH$_3$).= +rosilation of	INOR 0337
Double Beta Decay of	^{100}Mo and ^{150}Nd.=	NUCL 0012
Catalysis Using Re and	Mo Complexes.= Photochemistry and Redox	INOR 0343
h Theoretical Solutions to Huckel	MO Parameters.= +olecular Systems: Facile Grap	COMP 0020
Anions as Protoypes+ Ab Initio	MO Studies on Si$_2$H$_6$, SiH$_3$GeH$_3$, and Ge$_2$H$_6$ Radical	PHYS 0313
Dehydrogenation over Zn or	Mo.= Reaction Kinetics of Propane	I&EC 0075
Their	Mo(CO)$_3$ Complexes.= 1-Substituted Borepins and	ORGN 0212
yl Complexes: Crystal Structure of	Mo(η^2-C(S)OPr^i)(S$_2$COPr^i)CO(PMe$_3$)$_2$.= +tic Acet	INOR 0036
Mo(VI) to	Mo(IV).= Photoreduction of Silica-Supported	COLL 0213
Silica-Supported	Mo(VI) to Mo(IV).= Photoreduction of	COLL 0213
CH$_3$ with	Mo(100) Surfaces.= XPS Studies of Interaction of	COLL 0217
rties of Ultra-Thin MgO Films on	Mo(100).= +Characterization, and Chemical Prope	COLL 0193
High Surface Area γ-	Mo$_2$N.= Hydrodenitrogenation of Quinoline over	CATL 0047
Characterization of	Mo$_4$S$_4$L$_6$ Complexes.= Synthesis and	INOR 0371
Modeling the Reaction Center of	Mo-co.= Reactions of Molybdenum Dithiolenes	INOR 0404
electivities Exhibited by W-Ir and	Mo-Ir Cluster-Derived Catalysts.= +rogenolysis S	CATL 0008
Catalizadores CoMo/Al$_2$O$_3$ y	Mo/Al$_2$O$_3$ Modificados con Boro.= Estudio de los	FUEL 0104
n Chemistry and Catalysis of V,	Mo, Co, Ni, and Cu Compounds: An Example of Exp	CHED 0057
um Films from (η^3-C$_3$H$_5$)$_5$M (M =	MO, W).= +Deposition of Tungsten and Molybden	INOR 0055
l Reaction on Low-Rank Coal	Mobile Component Generation.= Effect of Chemica	I&EC 0019
on of Carbohydrates: The Role of	Mobile-Phase and Stationary-Phase pH.= +eparati	CARB 0066
e of Polymodal Electrophoretic	Mobility Distributions.= Occurrence and Importanc	COLL 0202
inum Can Manufacture.=	Mobility Enhancement Technology for the D I Alum	PMSE 0144
e) Monitored by Rheo-Photoac+	Mobility of Ethyl Acetate in Poly(vinylidene Fluorid	POLY 0255
s from Organometallic Precursors.	MOCVD Gas-Phase Chemistry.= +tronic Material	COLL 0006
al Single-Source Precursors to the	MOCVD of GaP.= +rsors. Preparation of Potenti	INOR 0304
s of Copper Thin-Film Growth by	MOCVD.= +tallic Precursors. Mechanistic Studie	COLL 0025
Film Growth by	MOCVD.= Mechanistic Studies of Aluminum Thin	INOR 0124
Numerical Modeling of GaAs	MOCVD.=	COLL 0066
Complexes for	MOCVD.= The ULSI Window: A View of Al and Cu	INOR 0125
g Chemist Need? J. R. Massingill,	Moderator.= +Interface, What Does the Practicin	I&EC 0057

hy More Is Better. D. L. Leister,	Moderator.= Panel Discussion: Bridging the Gap; W	I&EC	0058
-Based Approach to Laboratory	Modernization: The Pew Physical Chemistry Project	CHED	0125
Interactions: Probing Binding	Modes and Dynamics by NMR.= Drug-DNA	PHYS	0197
Slow	Modes.= Glutaraldehyde Cross-Linking: Fast and	BIOT	0218
Electronic Properties of a New	Modifiaction of the Quasi-One-Dimensional Compou	INOR	0359
acion de Energia.= Sintesis y	Modificacion de Secuencias de Destilacion con Integr	I&EC	0074
portados sobre Alumina o Titania	Modificadas con Fluor.= +de Catalizadores NiW So	FUEL	0106
s CoMo/Al$_2$O$_3$ y Mo/Al$_2$O$_3$	Modificados con Boro.= Estudio de los Catalizadore	FUEL	0104
nds: An Example of Experimental	Modular Teaching.= +Mo, Co, Ni, and Cu Compou	CHED	0057
tilization of Electron Spin Echo-	Modulation Spectroscopy as a Novel Method To Det	PHYS	0231
Mixture.= Chemical	Modulator to Measure Water Vapor in a Gaseous	ANYL	0052
ecular Electronics. Spatial Light	Modulators and Optical Memories Based on Bacterio	BIOT	0077
Nontraditional ChemSource	Modules Show Chemistry Is the Central Science.=	CHED	0018
ween the Zn, Zn$^+$ and Zn^{2+} + H$_2$	Moeties.= Ab Initio CI Study of the Interaction bet	PHYS	0282
esign I: Investigation of Various	Moieties as Stable 3-Phosphate Mimics in Aromatic	ORGN	0028
Containing Bicyclo[2.2.2] Octane	Moieties.= Colorless Rigid-Rod Polymers	POLY	0029
Renin Inhibitors with Cyclohexyl	Moiety Directly Attached to the Backbone at the P1	MEDI	0121
ve Region for Printe+ Effect of	Moisture Content of Dielectric Properties in Microwa	PMSE	0055
ed Cornstarch.= Effect of	Moisture Content on the Physical Aging of Gelatiniz	CARB	0043
Effect of	Moisture on the Sorption of Gases by Coal.=	I&EC	0020
rgonne Premium Coal Samples.	Moisture Removal from and Liquefaction of Beulah-	FUEL	0001
relation Spec+ Diffusion Average	Molar Mass of Polydisperse Polymers: A Photon Cor	POLY	0305
tion Using Conventional SEC and	Molar Mass Sensitive Detectors.= +r Characteriza	PMSE	0051
c Mechanical Thermal Analysis of	Mold Compounds Used in Electronic Packages.=	POLY	0095
Elec+ Selection of Thermoplastic	Molding Compounds for Infra-Red Reflow Tolerant	CMEC	0014
Plastics	Molding.= Economics of Commingled	I&EC	0035
n of Diamond (and c-BN, SiC).	Mole Fraction Measurements of H, CH$_3$, and Other S	PHYS	0145
por Ni en la	Molecula de Metano.= Activacion del Enlace C-H	PHYS	0244
gands.= Hot JH: Studies of the	Molecular Action of Juvenile Hormone Using Radioli	AGRO	0005
Energy Transfer in	Molecular Aggregates: Influence of Colored Noise.=	PHYS	0023
Redox-Switched	Molecular Aggregates.=	COLL	0020
pic and Structural Properties of	Molecular Alloys of Multiply Bonded Molybdenum a	INOR	0356
onlinear Optical Materials.=	Molecular and Biomolecular Electronics. Design of N	BIOT	0173
etic Molecular Electronics.=	Molecular and Biomolecular Electronics. Photosynth	BIOT	0050
and Characterization of Two-D+	Molecular and Biomolecular Electronics. Preparation	BIOT	0121
bling Lipid Bilayers on Solid S+	Molecular and Biomolecular Electronics. Self-Assem	BIOT	0102
al Properties of Double-Well P+	Molecular and Biomolecular Electronics. Some Gener	BIOT	0001
t Modulators and Optical Mem+	Molecular and Biomolecular Electronics. Spatial Ligh	BIOT	0077
olecular-Size Tinkertoy Constr+	Molecular and Biomolecular Electronics. Toward a M	BIOT	0200
Density Information Storage in+	Molecular and Biomolecular Electronics. Very High	BIOT	0023
nce to Human Cancer Induction.	Molecular and Cellular Mechanisms of Carcinogenic	CHAS	0007
ctene LLDPE (TREF) Fractio+	Molecular and Morphological Characterization of 1-O	PMSE	0102
Syntheses of Related	Molecular and Solid-State Nitrides.=	INOR	0103
erties of Composites+ Impact of	Molecular and Supermolecular Structure on the Prop	POLY	0300
	Molecular Anions as Surface Probes.=	NUCL	0056
Biodegradation of Organic Wastes:	Molecular Approaches and Field Applications—I+	BIOT	0117
Biodegradation of Organic Wastes:	Molecular Approaches and Field Applications—II+	BIOT	0137
of Liquid Crystalli+ Influence of	Molecular Architecture on Gas Transport Properties	POLY	0115
cine and Physiology+ Topbiology:	Molecular Bases of Animal Form. G. Edelman (Medi	CONG	0013
hanistic Studie+ Unraveling the	Molecular Basis of Sunlight-Induced Mutations: Mec	BIOL	0157
.=	Molecular Beam Studies on Diamond Film Synthesis	PHYS	0203
	Molecular Biology of Anti-α(1→6) Dextrans.=	CARB	0059
lites and Pillared Clay Synthesis.	Molecular Chemical Aspects of Silica Gel Formation.	PETR	0054
Transport Processes Involving	Molecular Chemisorbates.= Elementary Surface	PHYS	0345
nic Materials. Magnets Based on	Molecular Chemistry.= Molecular Routes to Inorga	INOR	0100
Indium Phosphide by	Molecular Chlorine.= Thermal and Laser Etching of	COLL	0214
Expanded Heterohelicenes:	Molecular Coils That Form Chiral Complexes.=	ORGN	0065
Density Information Storage in	Molecular Complexes Using Scanning Tunneling Mic	BIOT	0023
eriment. Spectroscopy of Nonrigid	Molecular Complexes.= +um Chemistry with Exp	PHYS	0102
c dialk+ Kinetic Study of in Situ	Molecular Composite during Imidization of Poly(ami	PMSE	0169
	Molecular Composites in Thermosets.=	POLY	0111
henylene Benzobisthiazole)-Based	Molecular Composites.= +n of the $\chi^{(3)}$ of Poly(p-p	POLY	0213
etraethoxysi+ Polyphosphazenes	Molecular Composites, I: In Situ Polymerization of T	POLY	0181
The First 3-D	Molecular Conductors?.= Alkali Metal-C$_{60}$ Films:	INOR	0354
Matrix.= Comparative Study of	Molecular Descriptors Derived from the Distance	COMP	0025
ng the Off-Diagonal Tens+ New	Molecular Design Concept for Poled Polymers Utilizi	POLY	0128
g Blocks of Future Biosensors and	Molecular Devices.= +s on Solid Support: Buildin	BIOT	0102
Phthalocyanines,	Molecular Devices, and Neural Systems.=	BIOT	0054
Advances in	Molecular Docking.= 3D Databases. Recent	COMP	0036
Ion Bombardment (Ion Mixing).	Molecular Dynamics Computer Simulation Studies o	NUCL	0091
urfaces via Wettability Analysis: A	Molecular Dynamics Simulation Method.= +rial S	POLY	0369
rison of Distance Geometry and	Molecular Dynamics Simulation Techniques for Conf	ORGN	0174
nalysis of Hydrogen Bonding from	Molecular Dynamics Simulation.= +ntations and A	BIOL	0112
	Molecular Dynamics Simulations of Amiloride.=	MEDI	0045
on of Diamond (and c-BN, SiC).	Molecular Dynamics Simulations of Diamond Films:	PHYS	0354
and Their Pendant-Chain Der+	Molecular Dynamics Studies of Amphotylic Ionomers	POLY	0367
ole Intercalating Agents Based on	Molecular Dynamics.= +cer Benzothiopyranoindaz	MEDI	0018
pagat+ One-Electron Pictures of	Molecular Electronic Structure through Electron Pro	PHYS	0058
	Molecular Electronic Transition Current Density.=	BIOT	0026
r Electronics. Photosynthetic	Molecular Electronics.= Molecular and Biomolecula	BIOT	0050

Orientation for the Design of	Molecular Electronics.= Study of Surface	BIOT	0053
Optics of Polymer Thin Films, I:	Molecular Engineering and Processing of Polymers w	POLY	0011
csition of a Nonlinear Interface+	Molecular Engineering by Laser Chemical Vapor Dep	BIOT	0024
	Molecular Engineering for Monoelectronic Devices.=	BIOT	0006
e Catalysts.=	Molecular Engineering of Supported Vanadium Oxid	COLL	0196
ed Energy Transfer and Scanning	Molecular Exciton Microscopy.= +Spatially Resolv	PHYS	0022
ϵ Tyrosine Kinase+ Comparative	Molecular Field Analysis of Conformationally Flexibl	MEDI	0024
cpy of High-Temperature Dense	Molecular Fluids and Low-Temperature Molecular S	ANYL	0002
Lorus Monoxide Laser-Excited	Molecular Fluorescence Spectrometry in a Graphite F	ANYL	0078
frolling Lepidopteran.=	Molecular Genetics of Insecticide Resistance in a Lea	AGRO	0116
Growth Regu+ Biochemistry and	Molecular Genetics of Resistance to Juvenoid Insect	AGRO	0101
Controlled Optical Pulses.=	Molecular Geometric Phase Development and Phase-	PHYS	0067
Chemical Element, Acid, and	Molecular Geometry.= Student Ideas about	CHED	0174
ɪergy Transfer and Relaxation in	Molecular Glasses and Polymer Solids.= +hases. E	PHYS	0080
Experience Using	Molecular Graphics in Teaching Organic Chemistry.=	CHED	0189
ɛ Using a New Volume-Rendering	Molecular Graphics Technique.= +Difluoroacetylen	COMP	0055
ods and Regularities.=	Molecular Hyperpolarizabilities: Computational Meth	BIOT	0178
Measurements of	Molecular Hyperpolarizabilities.= Nonlinear Optical	PHYS	0343
	Molecular Identification in Surface Analysis.=	NUCL	0045
Single-Electron Logic:	Molecular Implementation?.=	BIOT	0004
Weak	Molecular Interactions.=	PHYS	0105
A	Molecular Interpretation of Diffusion in Polymers.=	POLY	0168
ɪ in Polystyrene-Liquid Sys+ A	Molecular Interpretation of Thermoreversible Gelatio	POLY	0242
Significant Carcinogens and M+	Molecular Mechanisms of Action of Environmentally	CHAS	0007
Pests to Natural, Synthetic, an+	Molecular Mechanisms of Resistance in Herbivorous	AGRO	0114
Pests to Natural, Synthetic, an+	Molecular Mechanisms of Resistance in Herbivorous	AGRO	0153
Pests to Natural, Synthetic, an+	Molecular Mechanisms of Resistance in Herbivorous	AGRO	0150
Pests to Natural, Synthetic, an+	Molecular Mechanisms of Resistance in Herbivorous	AGRO	0097
Pests to Natural, Synthetic, an+	Molecular Mechanisms of Resistance in Herbivorous	AGRO	0133
ation Processing of th+ Artificial	Molecular Membranes Modeled on the Visual Inform	BIOT	0124
Integrated	Molecular Modeling in the Curriculum.=	CHED	0190
nescence Studies of Inorganic C+	Molecular Models for Semiconductor Particles: Lumi	INOR	0317
Methods Useful for Investigating	Molecular Motion.= +al Spectroscopies. Excitation	PHYS	0136
of Kinesin Molecules Using the+	Molecular Motors. Measuring the Isometric Tension	BIOL	0137
Chemistry and Crime—	Molecular Offenses in Canada.=	CHAL	0020
mond Surface R+ Semiempirical	Molecular Orbital and Tight Binding Methods in Dia	PHYS	0356
alpies for Intramolecular Hydro+	Molecular Orbital Estimation of the Activation Enth	ORGN	0175
Th+ Ab Initio and Semiempirical	Molecular Orbital Studies of Hydroxamic Acids and	PHYS	0326
Oximes: A	Molecular Orbital Study.= Barriers to Rotation in	PHYS	0324
raduate Experiment in Spectros+	Molecular Orbitals of Metalloporphyrins: An Underg	PHYS	0327
between Electric-Field-Induced	Molecular Ordering and Chemical Cross-Linking in V	POLY	0206
Network.= A	Molecular Organic-Inorganic Semiinterpenetrating	POLY	0286
-Controlled Reactive Processes on	Molecular Organizates and Colloidal Catalysts.=	PHYS	0063
Pt(111)+ Real-Space Imaging of	Molecular Organization: Naphthalene vs. Azulene on	PHYS	0125
dgett Fi+ Quantitative Study of	Molecular Orientation of Hemicyanine Langmuir-Blo	POLY	0224
, and Copper A+ Kinetics of the	Molecular Oxygen Reaction with Sodium, Magnesium	PHYS	0141
	Molecular Particles of Solid-State Compounds.=	CHED	0033
es.= Synthesis of	Molecular Precursors to Rare Earth Monochalcogenid	INOR	0311
Polyoxoalkoxotitanates:	Molecular Precursors to Titanium Oxides.=	INOR	0174
e-Source Precursors to the MO+	Molecular Precursors. Preparation of Potential Singl	INOR	0304
of Silicon-Containing Films from	Molecular Precursors.= +ctions in the Preparation	INOR	0121
-Structure Tungsten Oxides from	Molecular Precursors.= +sis and Reactivity of Open	INOR	0314
in+ Novel Fluorinated Polymeric	Molecular Probes for F-19 Magnetic Resonance Imag	MEDI	0041
istry of Nonmetals—Novel Tec+	Molecular Processes at Solid Surfaces: Surface Chem	COLL	0211
istry of Nonmetals—Novel Tec+	Molecular Processes at Solid Surfaces: Surface Chem	COLL	0170
istry of Nonmetals—Novel Tec+	Molecular Processes at Solid Surfaces: Surface Chem	COLL	0191
puter-Generated 3D and Related	Molecular Property Data for CAS Registry Substanc	COMP	0041
ntum Chemistry with Experiment.	Molecular Rearrangements and Pseudorotation.=	PHYS	0013
Acid Derivatives for Formation of	Molecular Recognition Compounds.= +is of Cholic	ORGN	0057
y of Hybrid Molecules.=	Molecular Recognition of Macromolecules: The Utilit	ORGN	0270
Medicinal Effect.= °From	Molecular Recognition of Piquancy to a New Rule of	MEDI	0040
Approaches to Photosynthetic+	Molecular Recognition via Base Pairing: Noncovalent	ORGN	0059
hoderma reesei.=	Molecular Regulation of Cellulase Formation by Tric	BIOT	0183
Vapor-Phase	Molecular Routes to Ceramic Powders.=	INOR	0175
of Thermal Gravi+ Tutorial on	Molecular Routes to Inorganic Materials. Application	INOR	0006
s of Electron Spect+ Tutorial on	Molecular Routes to Inorganic Materials. Application	INOR	0013
of Transition-Metal Thin Films+	Molecular Routes to Inorganic Materials. Deposition	INOR	0141
and Surface Chemical Reactions+	Molecular Routes to Inorganic Materials. Gas Phase	INOR	0121
ased on Molecular Chemistry.= +	Molecular Routes to Inorganic Materials. Magnets B	INOR	0100
allic Chemical Vapor Deposition+	Molecular Routes to Inorganic Materials. Organomet	INOR	0171
tical Ultracentrifugation.=	Molecular Self-Association Studies by Modern Analy	BIOL	0036
ysicochemical Applications—III.	Molecular Shape Discrimination Liquid Chromatogra	ANYL	0098
ns of Silicon and Cobalt in	Molecular Sieve AlPO$_4$-5.= Isomorphous Substitutio	PETR	0099
tic Reaction+ Characterization of	Molecular Sieve Frameworks and Products of Cataly	PETR	0060
ance of Pillared Interlayered Clay	Molecular Sieve with High Thermal Stability.= +m	PETR	0030
Wide-Pore Titanium Silicate	Molecular Sieve.= Synthesis of ETS-10: A New	PETR	0078
3-Membered Ring Containing	Molecular Sieve.= Synthesis of VPI-7: A Novel	PETR	0098
ric Constraint for Iso-Structural	Molecular Sieves Based upon Silicate or Alumino Ph	CATL	0028
Design and Synthesis of Carbon	Molecular Sieves for Separation and Catalysis.=	PETR	0059
lear Solution Synthesis of AlPO$_4$	Molecular Sieves: Identification of Transient Phases.	PETR	0020

vel Approach to the Synthesis of	Molecular Sieves: Recrystallization of the Intermedia	PETR	0019
of	Molecular Sieves.= Industrial Catalytic Applications	I&EC	0061
enzene on Acidic	Molecular Sieves.= Reactions of Meta-Diisopropylb	CATL	0014
ng Large-Pore Aluminophosphate	Molecular Sieves.= +id-State NMR Study of 18-Ri	PETR	0094
n Microporous Aluminophosphate	Molecular Sieves.= +itution of Paramagnetic Ions i	PHYS	0231
cal Applications of Graph Theory.	Molecular Similarity: Relative and Absolute.= +mi	COMP	0022
d Three-Dimensional Measures of	Molecular Similarity.= +on of Two-Dimensional an	COMP	0023
C-H···O Contact Interactions in	Molecular Solids.	PHYS	0295
ular Fluids and Low-Temperature	Molecular Solids.= +gh-Temperature Dense Molec	ANYL	0002
ctral Gradient Method.=	Molecular Structure of Asparagine by the Pseudospe	PHYS	0275
Methyl-thiosemicar+ Crystal and	Molecular Structure of Compounds Derived from 2-	INOR	0032
t of Sm+ Effect of Higher Order	Molecular Structure upon the Sorption and Transpor	POLY	0120
rocyclic Organic Compounds from	Molecular Structure.= +al Boiling Points of Hete	COMP	0015
ntion of Organic Compounds from	Molecular Structure.= +of Chromatographic Rete	ANYL	0099
um Nitro Complexes, Synthesis,	Molecular Structure, and Catalytic Activity in the Ox	INOR	0365
ing Ene+ Ab Initio Study of the	Molecular Structure, Vibrational Spectrum, and Bind	PHYS	0276
ted Method for the Alignment of	Molecular Structures: Optimizing the Overlap of Ato	CINF	0018
VENTON: A System for Inventing	Molecular Structures.= +tion. Award Address. IN	CINF	0016
	Molecular Switches from Modified FAD.=	BIOT	0143
	Molecular Symmetry and Stereoisomer Number.=	CHED	0169
agogical Approach to Conjugated	Molecular Systems: Facile Graph Theoretical Solutio	COMP	0020
ning in Liquids and Glasses.=	Molecular Theory of Inhomogeneous Spectral Broade	PHYS	0036
easurement of the Beta Decay of	Molecular Tritium.= +n Antineutrino Mass from M	NUCL	0010
by High-Perf+ Determination of	Molecular Weight Characteristics of Polysaccharides	BIOT	0197
polymerization of Lignin with Low	Molecular Weight Components.= +fication, and Co	BIOT	0041
Laboratory Innovation: Polymer	Molecular Weight Determination by Gel Permeation	CHED	0210
Methods for Polymer Analysis.	Molecular Weight Determinations of Poly-4-Methyl-	PMSE	0046
sing Mass+ Measurement of the	Molecular Weight Distribution of Vacuum Residue U	PETR	0150
odel for the Viscosities of High	Molecular Weight Liquids: Application to Polymer P	POLY	0065
ical Properties of Cur+ Effect of	Molecular Weight of Poly(dimethylsiloxane) on Phys	PMSE	0174
at+ Some Factors Affecting the	Molecular Weight of Polyesters Formed in Enzyme-C	BIOT	0094
Macrocyclic Polystyrene: Effect of	Molecular Weight on Glass-Transition Temperature.	POLY	0239
Oxygen-Free SiC Fibers.= High	Molecular Weight Polycarbosilane as a Precursor to	POLY	0360
Synthesis of High	Molecular Weight Polyethers.= Silicon-Assisted	PMSE	0075
in the Slow Melting of Ultrahigh	Molecular Weight Polyethylene.= +Fractions Found	PMSE	0166
between Intrinsic Viscosity and	Molecular Weight.= One-Parameter Correlation	PMSE	0095
Laser Vaporization of High	Molecular Weight, Single-Stranded DNA.=	ANYL	0130
er for Accurate Determination of	Molecular Weights and Mark-Houwink Constants.=	PMSE	0066
opellant Combustion Studied by	Molecular-Beam Mass Spectrometry and Modeling.=	FUEL	0126
of H, CH3, and Other Species by	Molecular-Beam Mass-Spectrometry during Hot-Fil	PHYS	0145
	Molecular-Beam Studies of Carbonaceous Clusters.=	FUEL	0110
allium Arsenide Gas Surface Re+	Molecular-Beam Study of the Hydrogen Chloride-G	COLL	0215
Design of a	Molecular-Bilayer Electrical Device.=	BIOT	0104
Hydroxy-Substituted Rigid-Rod+	Molecular-Dynamics Simulations of Rigid-Rod and	PHYS	0322
ymer Architectures.=	Molecular-Recognition-Directed Supramolecular Pol	POLY	0299
y of the Rods and Sp+ Toward a	Molecular-Size Tinkertoy Construction Set: Assembl	BIOT	0200
ic Structures Derived+ Toward a	Molecular-Size Tinkertoy Construction Set: Polymer	POLY	0344
s and Characterizatio+ Toward a	Molecular-Size Tinkertoy Construction Set: Synthesi	ORGN	0124
e+ Study of Blends with Narrow	Molecular-Weight Distribution: Hydrogen- and Deut	COLL	0035
hemically Activated Pathways for	Molecular-Weight Growth.= +ation. Analysis of C	FUEL	0063
Anion-	Molecule and Electron-Molecule Complexes.=	PHYS	0056
le-5-Acrylic Acids: Potent Small-	Molecule Antiotensin II Receptor Antagonists.= +o	ORGN	0119
Anion-Molecule and Electron-	Molecule Complexes.=	PHYS	0056
sed Approach to Computer-Aided	Molecule Design.= +ases. CAVEAT: A Vector-Ba	COMP	0044
ous Ionic State+ Geometries and	Molecule Properties of XCN (X = H, Cl, Br) in Vari	PHYS	0238
d Collision Energy Effects on Ion-	Molecule Reactions.= +Mode, Electronic State, an	PHYS	0241
Synthesis from DNA of a	Molecule with the Connectivity of a Cube.=	BIOL	0162
of Steroid Chemistry. The Sterol	Molecule: Biosynthesis, Structure, and Function.=	HIST	0012
ying Electronic States of the CuCl	Molecule.= +pole Moments for the Nine Lowest L	PHYS	0304
Diagonal Tensor Components of a	Molecule.= +for Poled Polymers Utilizing the Off-	POLY	0128
Optics.= Investigation of New	Molecules and Materials for Quadratic Nonlinear	POLY	0103
Spectroscopy of Small	Molecules by ab Initio Methods.= Study of the	PHYS	0005
mmunogenicity of Antibody-Like	Molecules by Immuno or Binding Assays.= +and I	BIOT	0059
ons of WCl6, WOCl4, and MoOCl4	Molecules Chemically Fixed on Silica Surface.= +ti	COLL	0098
Research on Organic	Molecules for Electronics Using STM.=	BIOT	0204
ocking Conformationally Flexible	Molecules from 2D or 3D Chemical Databases.= +D	CINF	0028
nt Possible Isomers of Substituted	Molecules Having That Geometry.= +the Differe	COMP	0035
and Methane	Molecules in A Zeolites.= Diffusivity of Hydrogen	CATL	0024
l Potential and Charge-Transfer	Molecules in Electric Fields—A Theoretical Study.=	BIOT	0001
es to Probe the Diffusion of Small	Molecules in Polymer Films.= +tional Spectroscopi	POLY	0332
ar Optic+ Orientation of Organic	Molecules in Sol-Gel Matrices for Quadratic Nonline	POLY	0006
Sorption and Transport of Small	Molecules in Stiff-Chain Glassy Polymers.= +n the	POLY	0120
Biology. Poly(ethylene Oxide) Star	Molecules in Surfaces for biological Studies.= +d	POLY	0272
nsition-Metal Atoms with Simple	Molecules near Room Temperature.= +ions of Tra	PHYS	0198
f the Opti-Electric Properties of	Molecules of Potential Use in Photoactive Polymers	POLY	0214
xperimental Studi+ Short-Lived	Molecules of Relevance in the Earth's Atmosphere: E	PHYS	0017
NMR Studies of Simple	Molecules on Metal Surfaces.=	PHYS	0347
scopy.+ Atomic Scale Imaging of	Molecules on Surfaces by Scanning Tunneling Micro	PHYS	0159
Atomic Force Microscopy of	Molecules on Surfaces.=	PHYS	0153
of Organometallic	Molecules on Surfaces.= Photochemical Reactions	INOR	0168

)cellu+ Spectral Behavior of Dye	Molecules Oriented by Chiral Nematic (Acetyl)(ethyl	POLY 0331
rystal Thin Films by Design: The	Molecules to Materials Connection.= +tric Liquid C	BIOT 0202
Effects in the Binding of Small	Molecules to Metal Atoms.= Role of Cooperative	PHYS 0172
xe+ Oxidative Addition of Small	Molecules to mono- and Dinuclear Copper(I) Comple	INOR 0044
and Growth in the Conversion of	Molecules to Solid-State Compounds.= +ucleation	INOR 0145
AFM.= Imaging DNA	Molecules under Potential Control by STM and	PHYS 0157
hy of Proteins and Pharmaceutical	Molecules Using Porous Polymeric Packings.= +ap	ANYL 0031
the Isometric Tension of Kinesin	Molecules Using the Laser Optical Trap.= +suring	BIOL 0137
arbohydrates. Interaction of Guest	Molecules with Cyclodextrins.= +Phenomena of C	CARB 0061
by Silicon Chains.= Model	Molecules with Donor-Acceptor Units Intermediated	POLY 0211
bases of Conformationally Flexible	Molecules: A Flexible Prescreening Technique.=	CINF 0027
Spectroscopic Studies of Floppy	Molecules: C_3 and SiC_2.=	PHYS 0235
plet Energy Gap in Non-Kekule	Molecules: Synthesis, Structure, Reactivity, and Spin	ORGN 0128
ally Folded and Unfolded Protein	Molecules.= +Conformation and Dynamics of Parti	PHYS 0107
Anion Binding to Polar	Molecules.=	PHYS 0224
Coordinates for Organic	Molecules.= Automatic Generation of 3D Atomic	CINF 0036
for Small	Molecules.= Capillary Electrophoresis Separations	ANYL 0137
A Versatile Route to Target Chiral	Molecules.= +rboxylesterase-Mediated Reactions:	AGFD 0008
Cationic Cascade	Molecules.=	POLY 0237
apid HPLC Analysis of Biological	Molecules.= +nd Instrumental Considerations for R	ANYL 0129
and Two-Carbon	Molecules.= Complete Basis-Set Energies fo One-	PHYS 0317
Symmetry Breaking in Small	Molecules.= Configuration Interaction Studies of	PHYS 0014
Rings and Branched	Molecules.= Dynamic Monte Carlo Simulations of	POLY 0297
of Bioactive	Molecules.= Experiences with Automated 3D Design	COMP 0049
Hyperpolarizabilities of	Molecules.= Frequency-Dependent and Correlated	PHYS 0341
Spectroscopy of Transient	Molecules.= Laser and Fourier Transform	PHYS 0057
Precursor	Molecules.= Microbial Production of Indigo from	BIOT 0032
cence Studies of Inorganic Cluster	Molecules.= +or Semiconductor Particles: Lumines	INOR 0317
les: The Utility of Hybrid	Molecules.= Molecular Recognition of Macromolecu	ORGN 0270
-Bond-Separated Donor-Acceptor	Molecules.= +Second-Harmonic Generation by σ	POLY 0222
perties for Conjugated, π-Electron	Molecules.= +Third-Order Nonlinear Optical Pro	BIOT 0177
Atoms with Small	Molecules.= Reactions of Boron and Aluminum	PHYS 0090
Their Photoreactivity toward CO	Molecules.= +ium Oxides Anchored onto SiO_2 and	INOR 0165
and 4-Atom	Molecules.= Variational Rovibrational Studies of 3-	PHYS 0030
Identification (between Two	Molecules).= Didactic Method for Isomer	CHED 0094
de Tiofeno con Catalizadores de	Molibdeno sobre Oxidos Mixtos de Alumina-Titania.	FUEL 0105
$CsBiS_2$ and the Three-Dimensi+	Molten Salt Chalcogenide Synthesis of the Linear β-	INOR 0202
loromagnesium Species in Several	Molten Salt Systems by Raman Spectroscopy.= +h	INOR 0064
Ketones in a Room-Temperature	Molten Salt.= Electrochemistry of Aromatic	ANYL 0094
$_4Se_4$+ Chalcogenide Synthesis in	Molten Salts: New Solid-State Compounds of Na_3Cu	INOR 0351
om-Temperature Chloroaluminate	Molten Salts.= +tallic Chemistry Performed in Ro	ANYL 0097
Corrosive Properties of	Molten Urea.= Effect of Conditioning Agents on the	FERT 0011
th Ponasterones, the First Insect-	Molting Hormones from Plants.= +rent Studies wi	HIST 0024
Catalytic Chemistry of Bismuth	Molybdate Hydrocarbon Oxidation Catalysts Prepare	CATL 0026
lecular Alloys of Multiply Bonded	Molybdenum and Tungsten Compounds.= +of Mo	INOR 0356
Center of Mo-co.= Reactions of	Molybdenum Dithiolenes Modeling the Reaction	INOR 0404
apor Deposition of Tungsten and	Molybdenum Films from $(\eta^3-C_3H_5)_5M$ (M = MO, W	INOR 0055
of Phosphido-Bridged Iridium,	Molybdenum Heterobimetallics by Oxidative Additio	INOR 0385
thesis and Characterization of	Molybdenum Monothiocarbamate Complexes.= Syn	INOR 0093
Propylene Oxidation on	Molybdenum Oxide.= Effect of Support in	CATL 0059
Hydrothermal Synthesis of New	Molybdenum Polyselenide Compounds.=	INOR 0355
pane Deydrogenation over Zinc or	Molybdenum Supported on Silicalite.= +tics of Pro	PETR 0087
hermal Synthesis of Microporous	Molybdenum with Octahedral-Tetrahedral Framewo	CATL 0019
Bioinorganic Chemistry of	Molybdenum.=	INOR 0005
and Coordinatively Unsaturated	Molybdenum(III)-μ-Sulfido Dimers Formed by Sulfu	INOR 0271
Properties of a Reduced	Molybdenum-Pterin Complex.=	INOR 0277
t.=	Molybdenum-Technetium Solar Neutrino Experimen	NUCL 0022
Oxygen Atom Transfer between	Molybdenum, Tungsten, and Rhenium Complexes.=	INOR 0132
CI State and Transition Dipole	Moments for the Nine Lowest Lying Electronic State	PHYS 0304
iety of Commercial Foods, Using	Monier-Williams and Liquid Chromatographic Meth	AGFD 0109
Exchange Test as	Monitor of Human Exposure.= Sister Chromatid	CEI 0016
cide Use on the Farm: Methods to	Monitor Small Watersheds for Pesticides.= +Pesti	AGRO 0054
ody-Like Molecules by+ How to	Monitor the Clearance and Immunogenicity of Antib	BIOT 0059
ate in Poly(vinylidene Fluoride)	Monitored by Rheo-Photoacoustic FT-IR Spectrosco	POLY 0255
icrowave Device for the On-Line	Monitoring and Control of Properties of Fluids durin	I&EC 0059
	Monitoring Cure in Inside Spray Coatings.=	PMSE 0160
tionships in Base Oil Processing:	Monitoring Dewaxing Procedures via Spectroscopy a	PETR 0151
Application of Biological	Monitoring for Environmental Public Health.=	CEI 0007
tion LC/MS: Application to Drug	Monitoring in Biological Fluids and Tissues.= +iza	ANYL 0124
matography for the Clinical Drug-	Monitoring Laboratory in the 90s.= +Role of Chro	ANYL 0116
iber Optic Fluore+ In Situ Cure	Monitoring of Aromatic Diamine-Epoxy System by F	POLY 0282
On-Line	Monitoring of Microbial Biofilms.=	BIOT 0099
Environmental and Biological	Monitoring of Occupational Exposure to Cyclophosph	CEI 0010
5 to $AlPO_4$-8 at Variou+ In Situ	Monitoring of the Degree of Transformation of VPI-	PETR 0095
es, I: Determination+ Biological	Monitoring of Workers Exposed to Triazine Herbicid	CEI 0004
es, II: Determination+ Biological	Monitoring of Workers Exposed to Triazine Herbicid	CEI 0005
g a Luminescent Rhenium Car+	Monitoring the Cure of Photosensitive Polymers Usin	POLY 0096
	Monitoring the Oxidation of Coals in Storage.=	FUEL 0002
Shelf-Life	Monitoring.= Time-Temperature Indicators for	AGFD 0080
Base-Sequence Specificity of the	Monoalkylation of Oligonucleotides by Mitomycin+	MEDI 0015

Tranylcypromine.= Enhanced	Monoamine Oxidase Inhibitory Activity of	MEDI	0068
Phosphatase.= Fumaric Acid	Monoazide: A New Inhibitor of Alkaline	BIOL	0057
MS to the Analysis of Flyash for	Monobrominated Polychlorinated Dibenzo–p–Dioxin	ANYL	0050
Precursors to Rare Earth	Monochalcogenides.= Synthesis of Molecular	INOR	0311
Inhibidor de la Locomocion de los	Monocitos (FILM) Producido por E. +on del Factor	BIOL	0102
Chains of Cell+ Specificities of	Monoclonal Antibodies Binding to the polysaccharide	CARB	0017
enase Complex and t+ Library of	Monoclonal Antibodies to E. coli Pyruvate Dehydrog	BIOL	0031
Method for Direct Labeling of	Monoclonal Antibodies with Tc99m.= An Improved	BIOT	0061
–Directed or Random Coupling of	Monoclonal Antibodies.= +rbents Prepared by Site	BIOT	0060
ultures for the Overproduction of	Monoclonal Antibodies.= +erogeneous Hybridoma C	BIOT	0169
he E1 Decarboxylating Enzyme: A	Monoclone That Is 98 Inhibitory.= +plex and to t	BIOL	0031
e Elastase.=	Monocyclic β–Lactam Inhibitors of Human Leukocyt	MEDI	0173
st+ Stable, Potent, and Specific	Monocyclic β–Lactam Inhibitors of Human PMN Ela	BIOL	0049
icillin G Amidase in Acylating a	Monocyclic β–Lactam Intermediate in the Synthesis	BIOT	0161
of Human PMN Elastase by	Monocyclic β–Lactams.= Mechanism of Inhibition	BIOL	0050
Synthesis of Four Generations of	Monodisperse Arylester Dendrimers.= +Convergent	POLY	0324
erbranched Block Copolymers and	Monodisperse Polyesters.= +owth Approach: Hyp	POLY	0318
Characterization of a Series of	Monodisperse 1,3,5–Phenylene–Based Hydrocarbon D	POLY	0320
al Characterization of a Series of	Monodisperse 1,3,5–Phenylene–Based, Fluorinated D	POLY	0236
Molecular Engineering for	Monoelectronic Devices.=	BIOT	0006
Surface Chemistry of	Monolayer Metal Films.=	COLL	0063
Substrates with Human Fibroblast	Monolayers: Uptake, Hydrolysis, and Release.=	BIOL	0079
Surfactant	Monolayers.= Structure and Orientation of Aqueous	PMSE	0154
Combustion on Palladium	Monolith Catalyst.= Kinetics of Methanol	COLL	0209
Technologies.= Camet Metal	Monolith–Based Catalytic Emission Control	I&EC	0045
Extruded	Monolithic Catalysts Supports.=	PETR	0090
phyrins.=	Monomer and Dimer Complexes with La and Nd Por	INOR	0221
in Free–Radical Polymerization+	Monomer Diffusion and the Kinetics of Propagation	POLY	0165
Synthesis of 4–Acetoxystyrene	Monomer.=	TECH	0017
paration and Characterization of	Monomeric and Dimeric Arsino– and Phosphinogalla	INOR	0308
opic Characterization of the First	Monomeric Mercury(II) Tetraalkylthiolate Complex.	INOR	0059
ions of Phosphate Compoun+ Is	Monomeric Metaphosphate an Intermediate in React	INOR	0293
Preparation and Reactivity of	Monomeric Osmium(II) Imido Complexes.=	INOR	0364
ica+ Three– and Four–Coordinate	Monomeric Zinc Alkyl Derivatives Stabilized by Ster	INOR	0085
vatives Stabilized by Sterically+	Monomeric Zinc Hydride and Magnesium Alkyl Deri	INOR	0087
Silicon–Containing Epoxy	Monomers and Oligomers.= Synthesis of Novel	POLY	0015
ynthesis and Characterization of	Monomers and Polymers Containing Photoresponsive	POLY	0209
rocess for the Polymerization of	Monomers on the Surface of Hiding Pigments Disper	PMSE	0038
of Cyclic Vinyl	Monomers.= Radical Ring–Opening Characteristics	ORGN	0157
Synthesis of Aryl bis(Trialkyltin)	Monomers.=	POLY	0035
Acrylic	Monomers.= Ultrasonic Polymerization of	PMSE	0020
Dioxygen–Binding Properties of	Mononuclear Copper(I) Complexes.=	INOR	0178
rst Water–Soluble Porphyrinyl–	Mononucleoside and Other Nucleoside–Substituted P	ORGN	0189
ation Products Using a Toluene	Monooxygenase Constitutive Pseudomonas cepacia.=	BIOT	0119
House Fly Cytochrome P450	Monooxygenase.= Structure and Regulation of a	AGRO	0098
Myo–Inositol	Monophosphatase.= Thiophosphate Inhibitors of	MEDI	0119
onic Transport Properties of the	Monophosphate Tungsten Bronze (PO$_2$)$_4$(WO$_3$)$_{2m}$, m	INOR	0352
ilus Influenzae Type B Conjuga+	Monosaccharide Compositional Analysis of Haemoph	ANYL	0021
c Separation of Sucrose	Monostearate Structural Isomers.= Chromatographi	CARB	0050
ning Metathesis Polymerization of	Monosubstituted Cyclooctatetraenes.= +Ring–Ope	POLY	0329
esis of the Cytotoxic Halogenated	Monoterpene Aplysiapyranoid D.= +nt Total Synth	ORGN	0001
Enzymatic Synthesis of	Monoterpenoid Flavor Compounds.=	AGFD	0009
Characterization of Molybdenum	Monothiocarbamate Complexes.= Synthesis and	INOR	0093
ts for Copolymerization of Carbon	Monoxide and Ethylene.= +ine)palladium Catalys	INOR	0155
utanol Synthesis During Carbon	Monoxide Fermentation by an Acidogenic Anaerobe.	BIOT	0015
Ligation Dynamics of Carbon	Monoxide in Cytochrome c Oxidase.=	BIOL	0096
ological Samples by Phosphorus	Monoxide Laser–Excited Molecular Fluorescence Spe	ANYL	0078
Methanol and Carbon	Monoxide Production from Natural Gas.=	FUEL	0096
Cytochrome P-450	Monoxygenase Genes in Oligophagous Lepidoptera.=	AGRO	0115
Molecules.= Dynamic	Monte Carlo Simulations of Rings and Branched	POLY	0297
entials of Polymeric Systems from	Monte Carlo Simulations.= +of the Chemical Pot	POLY	0352
Acid on Na–	Montmorillonite.– Radiolysis of Adsorbed Acetic	NUCL	0072
llar Densities in Alumina–Pillared	Montmorillonite.= +Analytical Determination of Pi	PETR	0126
Reactions of WCl$_6$, WOCl$_4$, and	MoOCl$_4$ Molecules Chemically Fixed on Silica Surfac	COLL	0098
ds by the Conductivity Method: I.	MOPSO–HCl at 25 °C.= +Constant of Amino Aci	BIOL	0033
mino Sulfonic Acids, HEPES and	MOPSO, by EMF Measurements.= +de and Two A	BIOL	0034
C$_6$ N–Alkanes in+ Design of Pt–	Mordenite Catalysts for the Isomerization of C$_5$ and	PETR	0113
the Crystallization of the Zeolites	Mordenite, L, Omega, and Offretite.= +c Studies of	PETR	0041
operties of Small– and Large–Port	Mordenites in C$_8$ Aromatics Isomerization.= +ic Pr	PETR	0132
ynthesis and Characterization of	Mordenites Modified by Tetrabutylin: Application to	PETR	0131
Silkworm, Bombyx	mori.= Prothoracicotropic Hormone of the	AGRO	0010
cies Formed by the Interaction of	Morine with Aluminum(III) in Aqueous Solution.=	INOR	0273
onjugates of Some 3–Substituted	Morphine, Dihydromorphine, and N–Methyl–Morphi	MEDI	0149
Dihydromorphine, and N–Methyl–	Morphinium Analogues.= +ubstituted Morphine,	MEDI	0149
–Shape Change during Cell–Sheet	Morphogenesis and Cytokinesis.= +quired for Cell	BIOL	0140
cle.= Gasification Mechanisms.	Morphological Changes during Oxidation of a Single	FUEL	0025
= Thermally Induced	Morphological Changes in Segmented Polyurethanes.	PMSE	0149
(TREF) Fractio+ Molecular and	Morphological Characterization of 1-Octene LLDPE	PMSE	0102
eparation Process, and Generated	Morphologies.= +er on the Polymerization Phase–S	POLY	0112
Polyester Rotaxane.= On the	Morphology and Crystallization Behavior of a	POLY	0346

IPNs.=	Kinetic Study and Morphology Buildup in Methacrylate Urethane	PMSE	0043
l-Gel Method: Some Insights into	Morphology Control and Physical Properties.= +So	POLY	0260
eady and Dynamic Shear-Induced	Morphology in Blends by Optical Rheometry.= +St	PMSE	0085
ct of Deformation History on the	Morphology of Blends of Liquid Crystalline Polymer	PMSE	0135
netrating Polymer Networks.=	Morphology of P (nBAc)/PEP Simultaneous Interpe	PMSE	0042
PEI Blends.=	Morphology of PEEK at High Temperatures and in	POLY	0333
tron Microscopy, II: Image Con+	Morphology of Polymer Chains by Transmission Elec	POLY	0062
Interpenetra+ Rheo-Kinetics and	Morphology of Polyurethane-Unsaturated Polyester	PMSE	0027
hesis and Characterizatio+ New	Morphology of Precipitated Calcium Carbonate: Synt	TECH	0019
sion on the Impact Properties and	Morphology of Rubber-Toughened Polystyrene.=	PMSE	0177
sition Temperatures on the Phase	Morphology of Simultaneous IPNs.= +Glass Tran	PMSE	0025
Mixed Oxides.= Open-	Morphology Zirconium Phosphates, Aluminas, and	PETR	0051
ssing, Mechanical Properties, and	Morphology.= +hylphenoxy Pendent Groups: Proce	PMSE	0019
rpenetrating Polymer Networks.	Morphology/Processing Relationship in Interpenetra	PMSE	0058
am-Negative Bacteria and High	Mortality in CD1 Immunodeficient Antibiotic-Treate	AGFD	0146
Genetically Determined Higher	Mortality Incidence and Acquired Immune Deficienc	AGFD	0145
Pictorialized Solid-State	MOs: A Nonmathematical but Honest Approach.=	CHED	0021
ns Isomers of a Bicyclic Lactone	Mosquito Repellent, 1,1,4,5,6,7,8,8A-Octahydro-3H-2	AGRO	0033
Ferrosilicates by	Mossbauer Spectroscopy.= Characterization of	NUCL	0069
Use of	Most Restrictive Paths in 3D Search Strategy.=	CINF	0024
Resistance in the Diamondback	Moth to Pyrethroids and Benzoylphenyl Ureas.=	AGRO	0117
es in the Evolution of Mechanistic	Motifs.= +and Adenylosuccinate Lyase: Odd Coupl	BIOL	0012
Polymer Segmental	Motion via Deuterium NMR.= Characterization of	PHYS	0140
Useful for Investigating Molecular	Motion.= +al Spectroscopies. Excitation Methods	PHYS	0136
fication- and Hydration-Induced	Motional Perturbations of Pectinic Polysaccharides i	AGFD	0012
Phenyl Ring	Motions in Polycarbonates.=	POLY	0080
Cooperative	Motions in Solids.= NMR Measurement of	PHYS	0069
hromatograph+ Determination of	Motor Oil Volatility Using High-Temperature Gas C	PETR	0122
Microtubule-Based	Motor Proteins.=	BIOL	0138
Interplay of	Motors and Microtubule Dynamics.=	BIOL	0139
Molecules Using the+ Molecular	Motors. Measuring the Isometric Tension of Kinesin	BIOL	0137
d Casework in the Royal Canadian	Mounted Police.= +sic DNA Analysis Research an	CHAL	0007
tate-Induced H_2O_2 Production in	Mouse Epidermis by Green Tea Polyphenols.= +ce	AGFD	0069
umor Initiation and Promotion in	Mouse Epidermis.= +tory Effects of Curcumin on T	AGFD	0088
trol o+ Application of the BOC-	MP Method for Evaluation of Reaction Pathway Con	CATL	0032
the	MPI.= A Survey of the NLO-Polymer Program at	POLY	0010
on-Recombination Reactions with	MR_2 (M = Zn, Mg and R = Bz, Cp, allyl).= +ntati	INOR	0383
Nine Lowest Lying E+ Accurate	MRCI State and Transition Dipole Moments for the	PHYS	0304
Ferritin Iron Storage and	mRNA Regulation.=	INOR	0011
otide Analogs of Rabbit α-Globin	mRNA.= +Interact Differently with Oligoribonucle	BIOL	0128
Application of High-Resolution	MS and Hybrid MS/MS to the Analysis of Flyash fo	ANYL	0050
Use of an	MS or MSD to Study Gas Phase Reactions.=	CHED	0198
spray Mass Spectrometry (ES	MS).= Linkage Analysis in Disaccharides by Electro	CARB	0048
igh-Resolution MS and Hybrid	MS/MS to the Analysis of Flyash for Monobrominat	ANYL	0050
Use of an MS or	MSD to Study Gas Phase Reactions.=	CHED	0198
	MSDS and Label Right-to-Know Software.=	CHAS	0030
nd Welfare Canada in Review of	MSDS and Labels Required for Hazardous Workplac	CHAS	0034
eeting the Challenge of the 90s:	MSDS Right-to-Know Compliance Made User Frien	CHAS	0031
Taking the Fog out of	MSDSs.=	CHAS	0038
Matrix Solid-Phase Dispersion (	MSPD) to the Extraction and Subsequent Analysis o	AGFD	0060
clear Multifragmentation.+ The	MSU Miniball: A Tool for the Characterization of Nu	NUCL	0035
Treatment of	MSW by Hydrocarb Process.=	FUEL	0121
very of Energy or Chemicals from	MSW.= +from Plastics and Municipal Waste. Reco	FUEL	0092
lite Catalysts.= Synthesis of	MTBE and ETBE over Triflic Acid-Modified Y-Zeo	CATL	0016
e and Adenylosuccinate Ly+ The	Muconate-Lactonizing Enzymes Mandelate Racemas	BIOL	0012
physical Approaches to Improve	Mucosal Peptide Penetration.= Biochemical and Bio	MEDI	0194
Microphase-Separated	Multiblock Copolymers.= Colloidal Structure of	COLL	0053
ed Phase HPLC with Diode Array	Multichannel Detection.= +Peaks in Normal Bond	ANYL	0087
ultidimensional Separations and	Multichannel Detections in Normal Bonded Phase H	ANYL	0088
splacement+ Axial Compression	Multicolumn Preparative Reversed-Phase Sample Di	ANYL	0040
Solid-State Characterization of	Multicomponent Polymer Systems by Digital Image	PMSE	0099
Membranes.= Diffusion of	Multicomponent Species across Polymeric	POLY	0195
i+ Solid-State Characterization of	Multicomponent Systems—I. Plenary Lecture. Sol	PMSE	0099
Solid-State Characterization of	Multicomponent Systems—II. Plenary Lecture. Enha	PMSE	0116
Solid-State Characterization of	Multicomponent Systems—III. Plenary Lecture. Stru	PMSE	0145
Solid-State Characterization of	Multicomponent Systems—IV. Plenary Lecture. Hyd	PMSE	0163
ance and Optical Spectroscopies.	Multidimensional NMR of Isotopically Enriched Pro	PHYS	0113
namics of Solid Polymers.=	Multidimensional NMR Studies of Structure and Dy	PHYS	0137
ections in Normal Bonded Phas+	Multidimensional Separations and Multichannel Det	ANYL	0088
nic Systems.=	Multielectron-Transfer Photochemistry in Metalorga	INOR	0285
Information Distribution: A	Multifaceted Career Opportunity for Chemists.=	YCC	0008
ons.=	Multifragment Emission in Light Ion Induced Reacti	NUCL	0026
of a Time-Projection Chamber in	Multifragmentation Experiments at the Bevalac.=	NUCL	0034
	Multifragmentation of Nuclei with A=70.=	NUCL	0028
Statistics and Dynamics in	Multifragmentation.=	NUCL	0005
or the Characterization of Nuclear	Multifragmentation.= +e MSU Miniball: A Tool f	NUCL	0035
Plenary Session.	Multifunctional Enzyme Mimics and Activators.=	ORGN	0265
Novel Synthesis of UV-Curable,	Multifunctional, Aromatic Analogs of Vinyl Ethers.=	POLY	0016
ic Self-Assembled Superlattices:	Multilayer Construction of Thin-Film Nonlinear Opt	POLY	0127
mics of CH_3Cl^- and Cl^- at	Multilayer Surfaces.= Femtosecond Solvation-Dyna	PHYS	0084

f-Assembled Siloxane Mono- and	Multilayers.= +IR and Fluorescence Studies of Sel	POLY 0330
Based on Group 8 and Group 10+	Multinuclear Inorganic Charge-Transfer Complexes	INOR 0162
Curriculum.= Incorporation of	Multinuclear NMR Experiments into Lower Division	CHED 0186
ance and Optical Spectroscopies.	Multinuclear NMR Study of an Alloy Semiconductor	PHYS 0372
O$_4$-8 at Various Temperatures by	Multinuclear Solid-State NMR and DTA.= +o AlP	PETR 0095
iques for Analysis of Petroleum.	Multinuclear Solid-State NMR Investigation of Boro	PETR 0136
Enzymatic Resolution Using a	Multiphase Membrane Reactor: Experimental Verific	BIOT 0042
nduced Pearl-Chain Formation in	Multiphase Polymer-Polymer-Solvent Systems.=	COLL 0145
ot Solids: Vibrational Cooling and	Multiphonon Up-Pumping.= +Transfer in Ultrah	PHYS 0250
roscopy in Physical Chemistry +	Multiphoton Ionization and Scanning Tunneling Mic	CHED 0044
on Serine Proteases: Evidence for	Multiple Binding Mechanisms.= +c Acid Inhibitors	BIOL 0082
osilanes to Form Metal-Ligand	Multiple Bonds with Tungsten Chlorophosphine Com	INOR 0115
ynthesis.=	Multiple Equilibration Strategy for Polypropionate S	ORGN 0100
de Y to the Y2-Receptor by Using	Multiple Peptide Synthesis.= +Site of Neuropepti	MEDI 0091
/Host Solid-State Interactions by	Multiple Quantum NMR.= +ic and Aqueous Guest	AGRO 0062
vity in Partial Oxidation Reacto+	Multiple Steady States and Changes in Catalyst Acti	I&EC 0069
for Poly(γ-alkyl-α,L-glutamate+	Multiple Thermal Transitions in Model Compounds	POLY 0099
Low-Cost	Multiple-Impact Physical Chemistry Experiment.=	PHYS 0336
Low-Cost,	Multiple-Impact Physical Chemistry Experiment.=	CHED 0162
proach to Risk Characterization of	Multiple-Pathway Chemical Exposures.= +ted Ap	ENVR 0073
nance and Optical Spectroscopies.	Multiple-Pulse Femtosecond Spectroscopy.= +eso	PHYS 0066
-State Chemical Reactions Using	Multiple-Pulse Line-Narrowing Techniques and ^{1}H	CATL 0022
on in Model Polymer Films by ^{19}F	Multiple-Quantum NMR.= +onium Salt Distributi	PMSE 0182
thesis, Structure,+ Tuning the	Multiplet Energy Gap in Non-Kekule Molecules: Syn	ORGN 0128
the Efficiency of+ Influence of	Multipolar Correlations and Surface Imperfections on	PHYS 0063
Mexico: A	Multipurpose Event.= Chemistry Olympiad in	CHED 0109
dicals Studied by the Fock-Space	Multireference Coupled-Cluster Method.= +in Ra	PHYS 0016
and Antiparasitic+ Miniaturized	Multiresidue Analysis of Selected Chemotherapeutics	AGFD 0062
Comparison Study of Regulatory	Multiresidue Methodologies.=	AGRO 0149
ored by the Perkin-Elmer Corp.).	Multiresonant Nonlinear Spectroscopies—A New Fa	ANYL 0001
rases.= Potential Specific	Multisubstrate Analogue Inhibitors of Methyltransfe	MEDI 0027
Experimental Verification of a	Multivariable Controlled Auxostat.=	BIOT 0158
Encuentro de Dos	Mundos a Traves de la Quimica.=	HIST 0035
Pretreatment of	Municipal Sewage Sludge for Gasification.=	FUEL 0117
land Cement Concrete Containing	Municipal Solid Waste Incinerator Ash.= +gth Port	FUEL 0136
(AFEX) Pretreatment of	Municipal Solid Waste.= Ammonia Fiber Explosion	BIOT 0016
nd Optical+ Characterization of	Municipal Waste by Scanning Electron Microscopy a	FUEL 0137
um, Lead, and Mercury from RDF	Municipal Waste Combustors.= +Control of Cadmi	FUEL 0130
nd Coal: Energy from Plastics and	Municipal Waste. Co+ Clean Energy from Waste a	FUEL 0090
nd Coal: Energy from Plastics and	Municipal Waste. Gas+ Clean Energy from Waste a	FUEL 0091
nd Coal: Energy from Plastics and	Municipal Waste. Re+ Clean Energy from Waste a	FUEL 0092
rom Sewage Sludge, Biomass, and	Municipal Waste. +from Waste and Coal: Energy f	FUEL 0115
Clean Energy from	Municipal Waste.=	FUEL 0093
erization of a Soluble Form of the	Murine Erythropoietin Receptor.= +n and Charact	BIOL 0108
Rapid Adaptation of a	Murine Hybridoma to Serum-Free Growth.=	BIOT 0163
xadiazoles.=	Muscarinic Activity of Imidazopyridine Esters and O	MEDI 0139
Solution Conformations of	Muscarinic Antagonists.=	MEDI 0138
Design. Pharmacology of	Muscarinic Receptors.= Receptor Activity and Drug	MEDI 0078
Reinvestigation of Chiral	Muscarones.= Synthesis and Pharmacological	MEDI 0141
Synthesis of (±)-	Muscone from [12]Metacyclophane.=	ORGN 0181
Metal-Binding Proteins of the	Mussel Byssus.=	INOR 0138
ownstream Stages. Integration: A	Must for Tomorrow's Bioprocesses.= +Upstream-D	BIOT 0069
oblems According to Risk: What	Must We Yet Know To Accomplish This Task? Rela	ENVR 0061
he trans-3,4-Dihydrodiol of the	Mutagen 15,16-Dihydrocyclopenta[a]Phenanthren-17	ORGN 0118
Plant Phenolics as Inhibitors of	Mutagenesis and Carcinogens.=	AGFD 0118
Carcinogenesis and	Mutagenesis by N-Nitrosamines.= Mechanisms of	CHAS 0010
xymethylase.= Site-Directed	Mutagenesis of the Gene-Encoding T4 dCMP Hydro	BIOL 0160
Solvents.= Random	Mutagenesis to Enhance Enzyme Activity in Organic	BIOT 0068
	Mutagenesis with Unnatural Amino Acids.=	BIOL 0029
ts+ Mechanisms Underlying the	Mutagenic, Carcinogenic, and Chemopreventive Effec	AGFD 0086
Foods.= Effect of Flavonoids on	Mutagenicity and Bioavailability of Xenobiotics in	AGFD 0116
Properties of Native and Site-	Mutagenized Cellobiohydrolase II.=	BIOT 0240
ntally Significant Carcinogens and	Mutagens: Relevance to Human Cancer Induction.	CHAS 0007
Rescue of a Catalytically Deficient	Mutant of R. rubrum Ribulose Bispho+ Chemical	BIOL 0089
ation Dynamics in Wild-Type and	Mutant Photosynthetic Reaction Centers.= +mbin	BIOL 0132
Binding Activities of Site-Specific	Mutants Expressed in E. +ride-FAB Complex and	BIOL 0143
Body	Mutants of 1L-1β.= Folding Properties of Inclusion	BIOT 0114
in the P22 Tailspike Protein.=+	Mutational Suppression of Inclusion Body Formation	BIOT 0113
ability and Specificity. Effects of	Mutations on the Thermodynamics of Protein Reacti	BIOT 0188
ecular Basis of Sunlight-Induced	Mutations: Mechanistic Studies on the Replicative B	BIOL 0157
a+ Characterization of Polymer	MWD and Branching with GPC-Viscometry in THF	PMSE 0068
ent Facilities Handbook.=	MWPS-37 Pesticide and Liquid Fertilizer Containm	AGRO 0042
y.+ Nanometer Scale Etching of	MX$_2$ Materials Using Scanning Tunneling Microscop	PHYS 0364
ynthesis and Reactivity of (OEP)	MX, and the Activation of O$_2$ by (OEP)Y(μ-ME)$_2$Al	INOR 0294
	Mycotoxin Resistance in Insects.=	AGRO 0155
amide/Paraherquamide Class of	Mycotoxins: In Search of the Biosynthetic Diels-Alde	ORGN 0283
d Edward E. Smissman—Bristol-	Myers Squibb Award Address. Hydant+ General an	MEDI 0082
dustry: Chemistry at the Bristol-	Myers Squibb Pharmaceutical Research Institute.=	CMEC 0028
man Award, sponsored by Bristol-	Myers Squibb).+ Award Address (Edward E. Smiss	MEDI 0089
Thiophosphate Inhibitors of	Myo-Inositol Monophosphatase.=	MEDI 0119

pectroscopies. Ligand Dynamics in	Myoglobin.= +Magnetic Resonance and Optical S	PHYS	0193
arrying Proteins Hemerythrin and	Myohemerythrin.= +arine Invertebrate Oxygen–C	INOR	0140
ell–Sheet Morphoge+ Nonmuscle	Myosin Is Required for Cell–Shape Change during C	BIOL	0140
inding t+ Kinetic Mechanism of	Myosin Subfragment-1 (S1) and Myosin–S1–NDP B	BIOL	0103
f Myosin Subfragment-1 (S1) and	Myosin–S1–NDP Binding to Actin.= +echanism o	BIOL	0103
l Ester Dehydrase: Determinati+	^{15}N NMR Studies of E. coli β–Hydroxydecanoyl Thio	BIOL	0066
	^{15}N NMR Study of Lanthanide–Nitrate Complexes.=	INOR	0250
le Nitric Oxide Synthase by	N^g–Amino–L–Homoarginine.= Inhibition of Inducib	MEDI	0106
lusters of Au$^+$ and In^{3+}, Trapping	Na$^+$ Ions in the Center.= +Polynuclear Selenide C	INOR	0203
–State Compounds of Na$_3$Cu$_4$Se$_4$,	Na$_2$Cu$_2$Se$_2$·Cu$_2$O and Cu$_{12-x}$Te$_4$S$_{13-y}$.= +ew Solid	INOR	0351
New Solid–State Compounds of	Na$_3$Cu$_4$Se$_4$, Na$_2$Cu$_2$Se$_2$·Cu$_2$O and Cu$_{12-x}$Te$_4$S$_{13-y}$.=	INOR	0351
the Synthesis of	Na–Chabazite and K–Mazzite.= Common Factors in	PETR	0038
Acetic Acid on	Na–Montmorillonite.= Radiolysis of Adsorbed	NUCL	0072
Zeolite Synthesis in the	Na–THA System at Low Temperature.=	PETR	0022
Impregnated	Na–Y Zeolite.= Characterization of RuCl$_3$	CATL	0035
ide-5'-yl]pyrophosphates. New	NAD Congeners Containing C–Nucleoside Isosteres o	CARB	0023
Chemistry Requirements of the	NADA: Residue Depletion and Metabolism of Flunix	AGRO	0078
Chiral	NADH Models.= Asymmetric Reductions with	ORGN	0264
DNA Activation of	Nae I Endonuclease.=	BIOL	0125
Structurally Related to	Nafazatrom.= Phenothiazine Antioxidants	MEDI	0174
de Transferencia de Carga Pt(R–	Nafin)·TCNQ.= +Espectroscopia IR de Complejos	INOR	0190
tion of a Silicon Oxide Phase in	Nafion Membranes by an in Situ Sol–Gel Reaction.=	POLY	0285
Gas Permeation through	Nafion Membranes: Theory and Experiment.=	POLY	0066
ion Electroquimica del 1–Nitro–	Naftaleno con Electrodos Metalicos y Semiconductor	PHYS	0280
Sintesis de Tiofosfatos de	Naftilo, Posibles Garrapaticidas.=	MEDI	0064
lucidation of Cytotoxin D from	Naja naja siamensis by Two–Dimensional Proton NM	BIOL	0044
dation of Cytotoxin D from Naja	naja siamensis by Two–Dimensional Proton NMR.=	BIOL	0044
e Fluor+ Design and Synthesis of	Naltrindole Analogues as δ–Opioid Receptor Selectiv	MEDI	0155
Conformational Requirement of	Naltrindole at δ–Opioid Receptors.= Studies on the	MEDI	0154
Input to a Systematic Chemical	Name by Computer: Experiences with AUTONOM 1	COMP	0019
Enhanced Chemical	Name Identification Algorithm.=	CINF	0041
lyhedra in Order To Canonically	Name the Different Possible Isomers of Substituted	COMP	0035
(WIMP)—Don't Be Fooled by the	Name.= +mmunoaffinity Methods for Purification	CARB	0054
.+ Controlled Synthesis of Metal	Nanoclusters within Block Copolymer Microdomains	POLY	0207
Phosphide Semiconductor	Nanoclusters.= Metalorganic Syntheses of	INOR	0102
cope.=	Nanofabrication with the Scanning Tunneling Micros	PHYS	0362
Surfaces for STM	Nanolithography.= Preparation of Passivated III–V	PHYS	0363
nic Hybrid Materials. Synthesis of	Nanomaterials via Preceramic Polymers.= +Inorga	POLY	0335
Energy Kinetics in	Nanometer Crystals.=	PHYS	0316
anning Tunneling Microscopy.+	Nanometer Scale Etching of MX$_2$ Materials Using Sc	PHYS	0364
s of Friction and Wear at the	Nanometer Scale.= Atomic Force Microscopy Studie	PHYS	0156
c Reactions at Zeolite–Supported,	Nanometer–Sized Electrodes.= +ic Fields on Organi	CATL	0018
Potent, Orally Active, 1,5–	Naphhyridine Angiotensin II Receptor Antagonists.=	MEDI	0102
Pt/KL Zeolite Catalyst for Light	Naphtha Reforming.= +w Modification Method of	PETR	0064
Coal–Derived	Naphtha.= Catalytic Hydrotreatment of	CATL	0052
e Novo Design and Synthesis of	Naphthalene Containing Inhibitors of Thymidylate S	MEDI	0186
maging of Molecular Organization:	Naphthalene vs. Azulene on Pt(111)+ Real–Space I	PHYS	0125
–Transfer in Isotopically Mixed	Naphthalene: High–Pressure Effects and Delayed Flu	PHYS	0081
ent to Remediate Petroleum– and	Naphthalene–Contaminated Soils.= +l Soil Treatm	BIOT	0139
Surfaces by Sodium	Naphthalenide.= Modification of Fluoropolymer	PMSE	0124
lysts.=	Naphthene Transformation over H–[Al]–ZSM-5 Cata	PETR	0044
Hydrogen Transfer from	Naphthenes to Coal during Coprocessing.=	FUEL	0005
ac+ Elucidation of the Nature of	Naphtheno–Aromatic Groups in Heavy Petroleum Fr	PETR	0147
	Naphthols with Proton–Mediating Side–Chains.=	ORGN	0162
DNA Alkylation.=	Naphthoquinone Derivatives as Selective Agents for	ORGN	0190
of Tetrahydro-1-(4-hydroxy-1-	naphthyl)thiophenium Hydroxide Inner Salt and β–B	POLY	0048
–amino-3-aryl-1-pyrrolidinyl)-4–	naphthyridones as Antimicrobial Agents.= +of 7-(3	MEDI	0061
Bischler–	Napieralski Reaction under Microwave Oven.=	ORGN	0195
tions Using Multiple–Pulse Line	Narrowing Techniques and ^{1}H T$_1$ Weighting.= +ac	CATL	0022
I/IV) Phosphate Compounds with	Nasicon–Related Structures.= +lance Titanium(II	INOR	0353
and+ Symposium in Memory of	Nathan Sugarman. Nuclear Chemistry, the Met Lab,	NUCL	0058
lear Chemistry, the Met Lab, and	Nathan Sugarman—A Retrospective.= +man. Nuc	NUCL	0058
Managing Our	Nation's Nuclear Waste.=	CHAS	0046
tion in Revitalizing Science in the	Nation's Schools.= +Institute for Chemical Educa	CHED	0138
n Prevention? Green Marketing: A	National and International Status Report.= +lutio	CEI	0028
The U.S.	National Chemistry Olympiad: A Student's View.=	CHED	0016
Chemistry Olympiad	National Exam.= How to Prepare Students for the	CHED	0013
rative Study in 84 Cases from the	National Institute of Cardiology (1984–1989).= +a	ENVR	0040
	National Inventors Hall of Fame.=	CHAL	0025
at Oak Ridge	National Laboratory.= Science Education Program	CHAS	0050
graduate Science Education at the	National Science Foundation.= +atory—II. Under	CHED	0179
ontinuing Education Cour+ 45th	National Technician Symposium. The Making of a C	TECH	0001
Electron Tunneling Pathways in	Native and Modified Proteins.=	BIOL	0133
Properties	Native and Site–Mutagenized Cellobiohydrolase II.=	BIOT	0240
= Photo–Induced Reduction of	Native Cytochrome c Peroxidase(III) in Alkaline pH.	BIOL	0097
Oxidation and	Naturation of Recombinant Bovine Somatotropin.=	BIOT	0135
chemistry of Protein Prenylation:	Nature's Rubber Cement for Membrane Attachment	BIOL	0021
geneous Adsorption of Benzene on	NaX and NaY Zeolites by NMR Spectroscopy.=	CATL	0025
The Chemistry of	NaY Crystallization from Sodium Silicate Solutions.=	PETR	0004
dsorption of Benzene on NaX and	NaY Zeolites by NMR Spectroscopy.= +geneous A	CATL	0025

Networks.=	Morphology of P (nBAc)/PEP Simultaneous Interpenetrating Polymer	PMSE	0042
(R$_3$P)MCl{[Ph$_2$P(S)]$_n$[Ph$_2$P(O)]$_3$-	nC}, Where M = Pd or Pt and n = 1, 2, or 3.= +of	INOR	0265
of [cod]M{[Ph$_2$P(S)]$_n$[Ph$_2$P(O)]$_3$-	nC}, where M = Rh and Ir and n = 1, 2, and 3.=	INOR	0264
by cis-Rh(CO)$_2$(p-(	NC)C$_5$H$_4$N).= Aldehyde Decarbonylation Catalyzed	INOR	0255
Synthesis of	NC-319: A New Sulfonylurea Herbicide in Corn.=	AGRO	0049
hnology. Biochemical Engineering,	2nd Edition—Aiba, Humphrey, and Millis.= +otec	BIOT	0221
with La and	Nd Porphyrins.= Monomer and Dimer Complexes	INOR	0221
Double Beta Decay of ^{100}Mo and	^{150}Nd.=	NUCL	0012
Dynamics of the O(^{3}P) = NH$_2$ (	ND$_2$) Reaction.=	PHYS	0189
Total Synthesis of (	ndash)α-Kainic Acid and (-)α-Allokainic Acid.=	ORGN	0177
in	NdBa$_2$Cu$_3$O$_y$.= Investigations of Bi Substitution	PHYS	0312
bfragment-1 (S1) and Myosin-S1-	NDP Binding to Actin.= +chanism of Myosin Su	BIOL	0103
ein Separation and Purification in	Neat Organic Solvents.= +ous Bioprocessing: Prot	BIOT	0063
White-Fleshed	Nectarine.= Free and Bound Flavor Constituents of	AGFD	0024
What Does the Practicing Chemist	Need? J. R.+ Panel Discussion: At the Interface,	I&EC	0057
Optical Falling-	Needle/Sphere Rheometer.= Magnet-Enhanced	POLY	0251
HPLC Analysis of	Negatively Charged Phospholipids.=	ANYL	0134
Dye Molecules Oriented by Chiral	Nematic (Acetyl)(ethyl)cellulose Solutions.= +or of	POLY	0331
Susceptibility of	Nematic Solutions of PBT.= The Third-Order	POLY	0274
cosponsored by E. I. DuPont de	Nemours and Co.). Teaching: A Witches' Brew of Bla	ANYL	0006
rocedure for the Determination of	Neomycin in Kidney Tissue.= +Chromatographic P	AGFD	0100
Calorimetry with a Helium-	Neon Laser.= Time-Resolved Thermal Lens	CHED	0089
Characterization of	Neopentylantimony Compounds.= Synthesis and	INOR	0306
	Neopentyldiamine Polyamides and Blends.=	PMSE	0100
nhibitors of Experimental Colonic	Neoplasia.= +Quercetin (QU) and Rutin (RU) as I	AGFD	0114
ation of Calcium-Binding Sites in	Nephrocalcin with Vanadyl Ion.= +OR Characteriz	BIOL	0045
Piper betel as Flavoring and	Nerve-Stimulating Agent.= Phenolic Compounds of	AGFD	0154
Fullerenes.= Fluxes and	Net Reaction Rates of Flame Species Pertinent to	FUEL	0112
Carbon Atoms into Rods, Rings,	Nets, and Spheres: The New Generations of Carbon	ORGN	0266
Nitrosamines in Elastic Rubber	Nettings Used for Packaging Cured-Pork Products.=	AGFD	0040
Interpenetrating Polymer	Network Films.= Gas Permeability Studies of Latex	PMSE	0061
Melting Due to Strain: Effect of	Network Formation in Thermoreversible Gels of Poly	POLY	0123
Vinyl+ Interpenetrating Polymer	Network Formed from Castor Oil Polyurethane and	PMSE	0089
Short-Range Order in	Network Glasses.= Magnetic Resonance Probes of	PHYS	0374
atible Interpenetrating Polymer	Network of Polytetrafluoroethylene and Polydimethy	PMSE	0041
n σ-Delocalized Silicon Backbone	Network Polymers.= +on and Vibronic Relaxation i	PHYS	0138
Simultaneous Interpenetrating	Network Products from Lignin, Organostannanes, an	PMSE	0006
ine-Cured Epoxy: Amine Epoxi+	Network Structure and Mechanical Properties of Am	POLY	0158
through Interpenetrating Polymer	Network Synthesis.= +tional Biomimetic Polymers	PMSE	0004
Variational Formulation in	Network Thermodynamics.=	PHYS	0323
Semiinterpenetrating	Network.= A Molecular Organic-Inorganic	POLY	0286
Epoxy Interpenetrating Polymer	Network.= Composites from Polyurethane and	PMSE	0044
Peaks Using a Neutral	Network.= Deconvolution of Chromatographic	ANYL	0048
olyester Interpenetrating Polymer	Network.= +ology of Polyurethane-Unsaturated P	PMSE	0027
nes and Interpenetrating Polymer	Networks (IPNs).= +n Foams Based on Polyuretha	PMSE	0040
ponent Interpenetrating Polymer	Networks and Their Applications.= +te Three-Com	PMSE	0080
aneous Interpenetrating Polymer	Networks from Castor Oil Polyurethanes and Poly(S	PMSE	0091
l Pyr+ Interpenetrating Polymer	Networks of Poly(Carbonate-Urethane) and Polyviny	PMSE	0056
onships and Phase Transitions for	Networks of Semirigid Chains.= +ess-Strain Relati	COLL	0016
pene+ Interpenetrating Polymer	Networks. Entirely Radical in Situ Synthesis of Inter	PMSE	0001
vervi+ Interpenetrating Polymer	Networks. Interpenetrating Polymer Networks: An O	PMSE	0039
ane/+ Interpenetrating Polymer	Networks. Kinetics of Phase Separation in Polyureth	PMSE	0021
terp+ Interpenetrating Polymer	Networks. Morphology/Processing Relationship in In	PMSE	0058
osets+ Interpenetrating Polymer	Networks. Two-Phase Epoxy Interpenetrating Them	PMSE	0087
etworks. Interpenetrating Polymer	Networks: An Overview.= +enetrating Polymer N	PMSE	0039
rientation in Deformed Polymer	Networks: Characterization of Stress-Strain Relation	COLL	0016
Strain-Birefringence of Real	Networks: Effect of Swelling.=	POLY	0308
hesis of Interpenetrating Polymer	Networks.= +tworks. Entirely Radical in Situ Synt	PMSE	0001
Acrylate Interpenetrating Polymer	Networks.= +ated Steel Sheet with Polyurethane/	PMSE	0057
Dynamics of Reversible	Networks.=	POLY	0172
cture of Interpenetrating Polymer	Networks.= +and Nonequilibrium Microphase Stru	PMSE	0093
Polymer	Networks.= Flow in Latex Interpenetrating	PMSE	0062
eous Interpenetrating Polymer	Networks.= Morphology of P (nBAc)/PEP Simultan	PMSE	0042
Nonlinear Swelling of Phantom	Networks.=	POLY	0167
Polymer	Networks.= Phase Separation in Interpenetrating	PMSE	0022
Polymer	Networks.= Studies of Grafted Interpenetrating	PMSE	0003
Polymer Interpenetrating Polymer	Networks.= +r Oil Polyurethane/Vinyl or Acrylic	PMSE	0090
Latex Interpenetrating Polymer	Networks.= Study on the Damping Paint of PS/PA	PMSE	0064
Interpenetrating Polymer	Networks.= Study on the Three-Stage Latex	PMSE	0082
ation in Polydimethylsiloxane	Networks.= Substituent-Induced Cross-Link Aggreg	PMSE	0120
phazene Interpenetrating Polymer	Networks.= +sis and Characterization of Polyphos	PMSE	0002
Poly(ϵ-caprolactone) Fumarate	Networks.= Synthesis and Degradation Studies of	POLY	0068
Triazine Rigid-Rod	Networks.=	POLY	0110
Devices, and	Neural Systems.= Phthalocyanines, Molecular	BIOT	0054
tal Microtubules: Implications for	Neural-Like Protein Array Devices.= +in Cytoskele	BIOT	0055
f a Carbon Glycoside of N-Acetyl	Neuraminic Acid.= +emical Enzymatic Synthesis o	CARB	0034
-Enamino-Nitrogen-Containing	Neuromuscular Blocking Agents.= Novel Quaternary	MEDI	0046
heromone Biosynthesis Activating	Neuropeptide Analogs in Helicoverpa zea.= +y of P	AGRO	0011
tification of the Binding Site of	Neuropeptide Y to the Y2-Receptor by Using Multip	MEDI	0091
gions of Modified Cobra Venom	Neurotoxin That Induce Immune Unresponsiveness t	BIOL	0086

Impairing Excitatory Amino Acid	Neurotransmission.= +iazoline Anticonvulsant, for	MEDI	0144
ge (Citrus aurantium) Juice Using	Neutral and Anion Exchange Resins.= +Sour Oran	AGFD	0106
Chromatographic Peaks Using a	Neutral Network.= Deconvolution of	ANYL	0048
lms+ Reversible and Irreversible	Neutral Species Transfer in Electroactive Polymer F	ANYL	0103
Solar Neutrinos—II. SAGE Solar	Neutrino Experiment.= +ces in Neutrino Science:	NUCL	0030
Molybdenum–Technetium Solar	Neutrino Experiment.=	NUCL	0022
Evidence for a Massive	Neutrino in Nuclear Beta Decay.=	NUCL	0003
from End to End and the	Neutrino Mass.= The Tritium Beta Decay Spectrum	NUCL	0011
ent Advances in Neutrino Science:	Neutrino Masses from Beta Decay—I. Neutri+ Rec	NUCL	0001
ent Advances in Neutrino Science:	Neutrino Masses from Beta Decay—II. Limit+ Rec	NUCL	0010
Low-Energy Tests of	Neutrino Masses.=	NUCL	0020
Homestake Solar	Neutrino Observatory.= Results from the	NUCL	0021
Sudbury	Neutrino Observatory.=	NUCL	0032
Solar	Neutrino Results from the Kamiokande II Detector.=	NUCL	0023
—I. Neutri+ Recent Advances in	Neutrino Science: Neutrino Masses from Beta Decay	NUCL	0001
—II. Limit+ Recent Advances in	Neutrino Science: Neutrino Masses from Beta Decay	NUCL	0010
Chemistry+ Recent Advances in	Neutrino Science: Solar Neutrinos—I. Neutrinos and	NUCL	0019
Neutrino E+ Recent Advances in	Neutrino Science: Solar Neutrinos—II. SAGE Solar	NUCL	0030
trino Masses from Beta Decay—I.	Neutrino Science.= +ces in Neutrino Science: Neu	NUCL	0001
Borexino: Low-Energy Solar	Neutrino Spectroscopy.=	NUCL	0033
trino Science: Solar Neutrinos—I.	Neutrinos and Chemistry: A Brief Tutorial.= +Neu	NUCL	0019
The Unbearable Heaviness of	Neutrinos.=	NUCL	0002
vances in Neutrino Science: Solar	Neutrinos—I. Neutrinos and Chemistry+ Recent Ad	NUCL	0019
vances in Neutrino Science: Solar	Neutrinos—II. SAGE Solar Neutrino E+ Recent Ad	NUCL	0030
udies of Li_4NpO_5 and Li_5NpO_6 by	Neutron and X-ray Powder Diffraction.= +ural St	NUCL	0068
yrene–Polyxylenyl+ Small-Angle	Neutron Scattering Study of Interdiffusion in Polyst	POLY	0166
alcohol) Solution—A Small-Angle	Neutron Scattering Study.= +Aqueous Poly(vinyl-	POLY	0171
rganogels Studied by Small-Angle	Neutron Scattering.= +of Polymers. Surfactant O	POLY	0169
ed Crystallization+ Small-Angle	Neutron–Scattering Studies of the Template–Mediat	PETR	0058
n a High-Salt Environment at the	Nevada Test Site.= +ermination of Strontium-90 i	NUCL	0085
Natural Waters at the	Nevada Test Site.= Thorium-230 Chronology of	NUCL	0084
I Think I Shall	Never See a Polymer Lovely as a Tree.=	POLY	0319
Kinetics of	$NF(a^1\Delta)$ and $NF(X^3\Sigma)$.=	PHYS	0331
Kinetics of $NF(a^1\Delta)$ and	$NF(X^3\Sigma)$.=	PHYS	0331
with	NF_3.= Photochemical Reactions of Metal Carbonyls	INOR	0195
luoro-o-Benzenedisulfonimide (	NFOBS): A Useful New Fluorinating Reagent.= N-F	ORGN	0077
Quenching of	NH $A^3\Pi_i$ in High Rotational Levels.=	PHYS	0266
Studies of the Reactivity of	NH with the Surface of Depositing Films.= Laser	COLL	0218
Dynamics of the $O(^3P)$ =	NH_2 (ND_2) Reaction.=	PHYS	0189
Electrochemical Reduction of Ru(	$NH_3)_6Cl_3$.= +in Naturally Occurring Fluids on the	COLL	0005
sts Supported on Mg, Cu, Zn, and	Ni Aluminates.= +tinum and Platinum–Tin Cataly	COLL	0126
s with NiS_2O_2 Environment.=	Ni Chemistry of Oxothiolate Ligands: New Complexe	INOR	0204
Activacion del Enlace C-H por	Ni en la Molecula de Metano.=	PHYS	0244
d Reactivity of a Mixed-Valence [	Ni(II)–NI(IIII)] Thiolate Dimer.= +acterization, an	INOR	0060
–Cy+ Coordination Chemistry of	Ni(III)–Thiolate Complexes as Models for Nickel(III)	INOR	0179
ivity of a Mixed-Valence [Ni(II)–	NI(IIII)] Thiolate Dimer.= +acterization, and React	INOR	0060
the system	Ni(sal$_2$en) and NI(sal$_2$tm).= EHMO Calculations on	INOR	0042
between N–N ligands and [	Ni(sal$_2$tim)].= Kinetics of the Addition Reaction	INOR	0061
Ni(sal$_2$en) and	NI(sal$_2$tm).= EHMO Calculations on the system	INOR	0042
= Mn, Fe, Co,	Ni) Perovskites: Electronic Structure of LaBo$_3$ (B	PHYS	0288
Ternary Compounds of Ni–Fe–Pt,	Ni–Fe–Pd, Fe–Co–Pt, and Fe–Co–Pd.= +etism in	PHYS	0253
+etism in Ternary Compounds of	Ni–Fe–Pt, Ni–Fe–Pd, Fe–Co–Pt, and Fe–Co–Pd.=	PHYS	0253
e Hydrodemetallization (HDM) of	NI-TPP and VO–TPP.= +nadium Catalysts in th	CATL	0044
del Compound Studies on Pd and	Ni/Mo Hydrous Titanium Oxides.= +tion and Mo	CATL	0039
stry and Catalysis of V, Mo, Co,	Ni, and Cu Compounds: An Example of Experimenta	CHED	0057
d Catalytic Activity of LaBo$_3$ (B =	Ni, Mn, Co, Fe) Perovskites: +aracterization, an	PHYS	0225
the Development of an ELISA to	Nicarbazine.= +-Dinitrocarbanilide–Hydrazone and	BIOT	0166
Midst+ Chemical Specialties for	Niche Markets, or How to Survive and Thrive in the	CMEC	0005
nylation Reactions Catalyzed by	Nickel and Phase–Transfer Catalysts: New Mechanis	COLL	0117
Active-Site Replacement by a	Nickel Atom in Desulfovibrio Vulgaris (Hildenboroug	INOR	0020
oatom on the Stereochemistry of	Nickel Boride Reduction of 2,6–Dihalo-9–Oxabicyclo	ORGN	0246
on of Nucleic Acids.= Tunable	Nickel Complexes for Conformation-Specific Oxidati	INOR	0183
Zinc and	Nickel Octaethyloxophlorins.=	INOR	0302
Ion-Scattering Study of	Nickel on Ultrathin SiO_2 Layers.= Medium-Energy	COLL	0176
and Bulk Diffusion of Adsorbed	Nickel on Ultrathin Thermally Grown Silicon Dioxid	COLL	0177
Precursor-Based Synthesis of	Nickel Sulfides.=	INOR	0205
c Functionalities at Sulfur-Bound	Nickel: P_2S_2 vs. N_2S_2 Ligand Envir+ Organometalli	INOR	0177
orate)	Nickel(II) Complexes.= Bis(3-tert-butylpyrazolyl-b	INOR	0035
Reactions of	Nickel(II) with N,S-Donor Ligands.=	INOR	0405
operties of Tunichlorin, the Novel	Nickel(II)chlorin of Tunicates.= +l and Spectral Pr	INOR	0139
hiolate Complexes as Models for	Nickel(III)–Cysteine Centers in the Hudrogenase Enz	INOR	0179
Via Aryl 1,2–Dipolar Synthons:	Nickel(O)–Catalyzed Cross-Coupling of Aryl Triflate	ORGN	0082
zation of Olefins by Homogeneous	Nickel-Based Catalysts.= +erization and Isomeri	PETR	0007
erization of N-Hexane on	Nickel–Containing Zeolites.= Kinetics of Hydroisom	PETR	0116
Reactivity of	Nickel-Titanium Shape Memory Alloys.=	INOR	0357
issue: A Com+ Cadmium, Lead,	Nickel, Cobalt, and Copper Concentrations in Lung T	ENVR	0040
ton Dissociation of Ferrocene and	Nickelocene at 248 nm: Statistical vs. N+ Two-Pho	PHYS	0333
duction from Genetically Modified	Nicotiana tabacum Cell Lines.= +eign Protein Pro	BIOT	0230
Constants for Cyanide Addition to	Nicotinamide Arabinosides.= +on the Equilibrium	BIOL	0013
ntaining C–Nucleoside Isosteres of	Nicotinamide Riboside.= +ew NAD Congeners Co	CARB	0023

of Bovine Chymosin in Aspergillus	niger var. +i: Recent Improvements in Production	BIOT	0234
Aspergillus	niger.= Production of Acetyl-Xylan Esterase from	BIOT	0251
amics of the Mesospheric Sodium	Nightglow.= +tions—I. Chemical Kinetics and Dyn	PHYS	0071
or.=	NIH Funding Opportunities for the Assistant Profess	YCC	0004
	Ninety Years of Chemistry at Rockefeller.=	HIST	0027
Novel $R_2Ba_{1.25}NiO_{5.25}$	(R = Tm, Yb, Lu) Structure Type.=	INOR	0034
ce of Sodium N,+ Reduction of	Niobium and Tantalum Pentachlorides in the Presen	INOR	0092
uasi-Low-Dimensional Reduced	Niobium Oxophosphate Compound: Synthesis and C	INOR	0031
Ligands: New Complexes with	NiS_2O_2 Environment.= Ni Chemistry of Oxothiolate	INOR	0204
ation System.=	NIST Structures and Properties Database and Estim	COMP	0016
Estudio de la	Nitracion del Fenol con Aductos $Tonsil-Fe(NO_3)_3$.=	COLL	0120
io Acido y Basico, Preparacion de	Nitrado de Plata.= +dorreduccion con Zinc en Med	CHED	0172
^{15}N NMR Study of Lanthanide-	Nitrate Complexes.=	INOR	0250
sed to Inhibited Urea-Ammonium	Nitrate Solution.= +Corrosion of Mild Steel Expo	FERT	0001
Chilean	Nitrate.= Iodine and Sodium Nitrate—The Story of	SCHB	0020
Iodine and Sodium	Nitrate—The Story of Chilean Nitrate.=	SCHB	0020
y Bentonite-Supported Metallic	Nitrates: Induced Oxidation of Hantzsch Esters (Dia	ORGN	0245
r Re+ Modeling the Two-Phase	Nitration of Toluene in a Co-Current Packed Tubula	I&EC	0006
Selective Catalytic Reduction of	Nitric Oxide by Ammonia over Fe-Y Zeolites.=	COLL	0212
e.= Inhibition of Inducible	Nitric Oxide Synthase by N^g-Amino-L-Homoarginin	MEDI	0106
Selective Catalytic Reduction of	Nitric Oxide with Ammonia.=	COLL	0207
of Boron	Nitride and Carbon.= Hot-Filament-Activated CVD	PHYS	0205
Boron	Nitride Fibers from Polyborates.=	POLY	0316
Metal Halides.= Synthesis of	Nitride Films from the Reaction of Amines with	INOR	0309
of Boron	Nitride Films.= Depositions and Characterizations	PHYS	0094
nds for the Preparation of Metal-	Nitride Preceramics, Powders, and Coatings.= +Bo	POLY	0230
as Precursor to Tantalum	Nitride Thin Films.= Ethylimidotantalum Complex	INOR	0074
ain-Group and Transition-Metal	Nitride Thin Films.= +Temperature Synthesis of M	INOR	0104
Precursors to Silicon	Nitride.= Applications of Polysilazanes as	POLY	0317
Solid-State	Nitrides.= Syntheses of Related Molecular and	INOR	0103
Ternary	Nitrides.=	PHYS	0003
Reactivity Studies of New	Nitrido and Imido Complexes of Osmium(VI).=	ORGN	0219
d I+ Regioselective Addition of	Nitrile-Stabilized Carbanions to 4-Amino-Substitute	ORGN	0010
pid Oxidation in Muscle Foods by	Nitrite and Nitrite-Free Compositions.= +ion of Li	AGFD	0121
n in Muscle Foods by Nitrite and	Nitrite-Free Compositions.= +ion of Lipid Oxidatio	AGFD	0121
d Catalytic Ac+ Novel Palladium	Nitro Complexes, Synthesis, Molecular Structure, an	INOR	0365
Reduccion Electroquimica del 1-	Nitro-Naftaleno con Electrodos Metalicos y Semicon	PHYS	0280
Reduction of Some α-Substituted	Nitroalkanes.= +ast Cleavage Reactions Following	ANYL	0107
Electroduction of o-	Nitrobenzonitrile.=	ORGN	0250
Safe and Practical Synthesis of	Nitrocubanes.=	ORGN	0257
Controlled-Release	Nitrogen Analysis of Coated Fertilizers.=	FERT	0008
to the Synthesis of 5- and 6-Ring	Nitrogen and Sulfur Heterocycles.= +el Approach	ORGN	0199
yl (SF_5) Carbon and	Nitrogen Derivatives.= Trigeminal Pentafluorosulfan	INOR	0024
Trapping Tetrazolinylidenes with	Nitrogen Electrophiles.=	ORGN	0205
talytic Ca+ Synthesis of Bicyclic	Nitrogen Heterocycles by Means of Homogeneous Ca	ORGN	0049
Clusters.= Insertion of	Nitrogen into Intermediate-Sized Boron Hydride	INOR	0388
Cyanoimidazoles.= High-	Nitrogen Low-Hydrogen Amides and Imides from	POLY	0363
ns: Combustion E+ Oxidation of	Nitrogen Oxides in Homogeneous Gas-Phase Reactio	FUEL	0127
Atmospheric	Nitrogen Oxides: A Bridesmaid Revisited.=	ENVR	0071
d Agricult+ Controlled-Release	Nitrogen Solutions and Their Uses in Horticulture an	FERT	0005
Controlled-Release	Nitrogen Sources.= Environmental Benefits of	FERT	0006
controlled Proton Migration from	Nitrogen to Carbon.= +istic Investigation of Stereo	ORGN	0296
cursors of Ceramic Materials.=	Nitrogen- and/or Boron-Containing Polymers as Pre	POLY	0315
nds. Side-on (μ-N_2) vs. End-o+	Nitrogen-, Sulfur-, and Phosphorus-Ligated Compou	INOR	0362
s of NO_x in the Rapid Pyrolysis of	Nitrogen-Containing Heteroaromatics.= +recursor	FUEL	0114
s.+ Novel Quaternary-Enamino-	Nitrogen-Containing Neuromuscular Blocking Agent	MEDI	0046
olymers with Skeletal Phosphorus,	Nitrogen, and Sulphur(VI) Atoms.= +Inorganic P	POLY	0186
odobenzylamine Conjugate of 2-	Nitroimidazole: A New Potential Hypoxia Imaging A	NUCL	0076
2-Acetyl-1-Methyl-5-	Nitroimidazole.= Convenient Synthesis of	ORGN	0200
man Spectra of Photodissociating	Nitromethane in Solution.= +ertone Resonance Ra	PHYS	0302
N-THP-Protected	Nitrone.= Addition of Organolithium Compounds to	ORGN	0227
Catalyst in the Dialkylation of o-	Nitrophenacyl Derivatives.= 18-Crown-6 as a	ORGN	0261
tection against Tobacco-Specific	Nitrosamine-Induced Lung Tumorigenesis by Green	AGFD	0068
ckaging Cured-Pork Products.+	Nitrosamines in Elastic Rubber Nettings Used for Pa	AGFD	0040
Mutagenesis by N-	Nitrosamines.= Mechanisms of Carcinogenesis and	CHAS	0010
Inhibitor of	Nitrosation.= Tocopherol as a Natural Phenolic	AGFD	0091
trate Inhibition of Trypsin by N-	Nitrosoamide Derivatives.= +zyme-Activated Subs	ORGN	0022
tion of Alkyl Thiolate Ions to β-	Nitrostyrenes: Radicaloid Transition State or Princip	ORGN	0155
- and Hydration-Induced Moti+	Nitroxyl Amide Spin Labeling: Methyl Esterification	AGFD	0012
e la Evaluacion d+ Influencia del	Nivel Academico de los Alumnos en los Resultados d	CHED	0093
para la Ensenanza de la Quimica a	Nivel Medio Superior.= +Apoyos Computarizados	CHED	0124
s con+ Estudio de Catalizadores	NiW Soportados sobre Alumina o Titania Modificada	FUEL	0106
Second-Order	NLO and Relaxation Properties of Poled Polymers.=	POLY	0153
for	NLO Applications.= Preparation of Polymeric Films	POLY	0104
Group 6 Pentacarbonyl-Based	NLO Chromophores Exhibiting Large and Unusual O	POLY	0217
Novel Third-Order	NLO Materials from 96 Quinoline.=	POLY	0212
Characterization of	NLO Materials.=	POLY	0003
Rational Design of Organic	NLO Materials.=	POLY	0002
	NLO Polymers for Frequency Doubling.=	POLY	0178
Synthetic Approaches to	NLO Polymers.=	POLY	0106

Photorefractive and	NLO Properties.= Novel Polymers with	POLY	0129
on of Polyurethane Films Bearing	NLO-Active Chromophores.= +tical Characterizati	POLY	0216
A Survey of the	NLO-Polymer Program at the MPI.=	POLY	0010
utions of SO ($X^3\Sigma^-$) from the 193-	nm Photodissociation of $SOCl_2$.= +Energy Distrib	PHYS	0240
Ferrocene and Nickelocene at 248	nm: Statistical vs. N+ Two-Photon Dissociation of	PHYS	0333
Potent and Selective Competitive	NMDA Antagonist.= +Synthesis of NPC 12626: A	MEDI	0083
arylguanidines as Noncompetitive	NMDA Antagonists.= +ialkyl Substituted N,N'-Di	MEDI	0147
Annihilation in Crystals of (CH_3)	$_4NMnCl_3(TMMC)$.= +Solid: Exciton Trapping and	PHYS	0131
Ion Concentra+ Development of	NMR Active Indicators for Measurement of Calcium	ORGN	0149
s.= 1- and 2-Dimensional	NMR Analyses of Coal-Oil Hydroprocessing Product	PETR	0149
preparative HPLC Isolation and	NMR Analyses of the cis and trans Isomers of a Bicy	AGRO	0033
Heavy Petroleum Fractions by ^{13}C	NMR and Catalytic Dehydrogenation.= +oups in	PETR	0147
tures by Multinuclear Solid-State	NMR and DTA.= +o $AlPO_4$-8 at Various Tempera	PETR	0095
Simulation. Combinatorics of	NMR and ESR Spectrum Simulations.= Spectral	COMP	0001
F Catalyzes the Incorporation of	NMR and GC/MS into the Undergraduate Curriculu	CHED	0149
Peptidyl Fluoroketones.=	NMR and Kinetic Studies of Protease Inhibition by	BIOL	0005
cle+ One- and Two-Dimensional	NMR and Kinetics of Formation of Platinum(II) Nu	INOR	0180
ues.= Bioinorganic Chemistry.	NMR and Kinetics of Isomerization Studies of Platin	INOR	0403
Carbon with ^{129}Xe	NMR and N_2 Porosimetry.= Analysis of Porous	PETR	0070
Effects of Seed Aging: Solid-State	NMR and NMR Imaging.= +Germination and the	AGFD	0104
ly^6-Bradykinin a+ ^1H- and ^{13}C-	NMR Assessment of the cis-trans Isomerization of G	BIOL	0006
Incorporation of Multinuclear	NMR Experiments into Lower Division Curriculum.=	CHED	0186
eactions Using Multiple-Pulse+	NMR Imaging of Anisotropic Solid-State Chemical R	CATL	0022
Spectroscopy and	NMR Imaging of Coal.= Very High Frequency EPR	FUEL	0074
lysis.=	NMR Imaging: A Chemical Microscope for Coal Ana	FUEL	0075
Seed Aging: Solid-State NMR and	NMR Imaging.= +Germination and the Effects of	AGFD	0104
cratory+ Use of a 300-MHz FT-	NMR in a Second-Semester Sophomore Unified Lab	CHED	0181
crporation of Fourier Transform	NMR Instrumentation into the Undergraduate Curri	CHED	0148
d Sili+ Multinuclear Solid-State	NMR Investigation of Boron-Substituted Zeolites an	PETR	0136
	NMR Investigations of Macromolecular Dynamics.=	PHYS	0195
ral High-Resolution Solid-State	NMR Investigations of Zeolite Lattice Structures.=	PHYS	0348
.=	NMR Measurement of Cooperative Motions in Solids	PHYS	0069
in Pseudobinary Semiconducto+	NMR Measurements of Atomic Cluster Distributions	PHYS	0375
les and Other+ Quantitative ^{13}C	NMR Measurements on the Argonne Premium Samp	FUEL	0004
New Kinetic Method Based on	NMR Measurements.=	ANYL	0063
hibitors on+ Direct Solution ^{11}B	NMR Observation of Tightly Bound Boronic Acid In	BIOL	0082
^{31}P One- and Two-Dimensional	NMR of Duplex Oligonucleotides and Proteins DNA	INOR	0378
Spectroscopies. Multidimensional	NMR of Isotopically Enriched Proteins.= +Optical	PHYS	0113
Hyaluronic Acid wit+ Solid-State	NMR of N-Acylureas Derived from the Reaction of	ORGN	0020
Reorientation for High-Resolution	NMR of Solids: DAS and DOR.= +opies. Sample	PHYS	0370
ransfer Interactions on Polymer+	NMR Relaxation Study of the Influence of Charge T	PMSE	0076
ence and Optical Spectroscopies.	NMR Relaxation, Fluorescence Depolarization, and t	PHYS	0166
$d(CGCTpS_2$-TPS_2-AAGCG): +	NMR Spectra of Oligonucleotide Phosphorodithioate	BIOL	0123
Simulation of	NMR Spectra: Problems and Solutions.=	COMP	0007
Simulation of ^{13}C and ^1H	NMR Spectra.=	COMP	0002
leum Fractions+ Applications of	NMR Spectroscopy in Secondary Processing of Petro	PETR	0146
ent Laser Spectroscopy, and Laser	NMR Spectroscopy.= +ectroscopies. NMR, Coher	PHYS	0040
lchromium Compounds by Proton	NMR Spectroscopy.= +ion of (η_6-Arene)Tricarbony	ORGN	0215
of N,N-Diethylbenzamide by ^{13}C	NMR Spectroscopy.= +ver Microsomal Metabolites	AGRO	0026
zene on NaX and NaY Zeolites by	NMR Spectroscopy.= +geneous Adsorption of Ben	CATL	0025
e)platinum(II) Chloride Using ^{13}C	NMR Spectroscopy.= +om cis-Diammine(guanosin	INOR	0251
Topological S+ Infrared and ^{13}C	NMR Spectrum Simulation: Role of Complementary	COMP	0006
^{13}C	NMR Spectrum Simulation.=	COMP	0003
Techniques for ^{13}C	NMR Spectrum Simulation.= Database Retrieval	COMP	0004
^{77}Se and ^{113}Cd Single-Crystal	NMR Studies of a Novel Cadmium Selenolate Comp	INOR	0078
Deuterium	NMR Studies of Acid Sites in Zeolites.=	COLL	0197
Solution and Solid-State ^{13}C	NMR Studies of C_{60} and C_{70}.=	PHYS	0070
s between 90 and+ ^{13}C CP/MAS	NMR Studies of Catalytic Reactions at Temperature	COLL	0090
arization and Very Fast ^{27}Al MAS	NMR Studies of Dealuminated Zeolite Y.= +osspol	PETR	0006
ter Dehydrase: Determinat+ ^{15}N	NMR Studies of E. coli β-Hydroxydecanoyl Thiol Es	BIOL	0066
Solid-State ^{27}Al	NMR Studies of Hydrated $AlPO_4$-8.=	PETR	0096
Solution.= Optical and	NMR Studies of Local Polymer Dynamics in Dilute	PHYS	0139
ydrides: Rotational Tunnelling+	NMR Studies of Nonclassical Transition Metal Polyh	PHYS	0068
Solid-State	NMR Studies of Separation Media.=	ANYL	0064
in Microporous M+ Solid-State	NMR Studies of Shape-Selective Catalytic Reactions	CATL	0023
s.=	NMR Studies of Simple Molecules on Metal Surface	PHYS	0347
Polymers.= Multidimensional	NMR Studies of Structure and Dynamics of Solid	PHYS	0137
ins over Offret+ ^{13}C and ^1H MAS	NMR Studies of the Conversion of Methanol to Olef	PETR	0042
d Anvil Cell.= High-Pressure	NMR Studies of the Glass Transition in the Diamon	PHYS	0260
nd-DNA Complexe+ Solid-State	NMR Studies of the Structure and Dynamics of Liga	BIOL	0116
Hyaluronan: ^{13}C-	NMR Studies.= Conformational Versatility of	CARB	0047
tical Spectroscopies. Multinuclear	NMR Study of an Alloy Semiconductor.= +and Op	PHYS	0372
^{15}N	NMR Study of Lanthanide-Nitrate Complexes.=	INOR	0250
d Products.= Solid-State MAS-	NMR Study of Preceramics Precursors and Pyrolyze	POLY	0358
S Diblock Copoly+ Solid-State	NMR Study of the Microphase Structure of PS-PMP	PMSE	0121
te Molecular Sieve+ Solid-State	NMR Study of 18-Ring Large-Pore Aluminophospha	PETR	0094
-Containing Phe+ Application of	NMR Techniques for the Characterization of Silicon	POLY	0086
Reactions Using Two-Dimensional	NMR Techniques.= +s and Products of Catalytic	PETR	0060
plication of Homonuclear 2D ^{51}V	NMR to Determine the Effect of Temperature and I	INOR	0228
teration.= Application of SNI+	NMR to the Detection of Flavor and Fragrance Adul	AGFD	0197

Materials. Sol-Gel Kinetics by	NMR.=	Preceramic and Inorganic Hybrid	POLY 0256
e Analysis and Real-Time Pulsed	NMR.=	+onent Polymer Systems by Digital Imag	PMSE 0099
ensis by Two-Dimensional Proton	NMR.=	+tion of Cytotoxin D from Naja naja siam	BIOL 0044
ite-Y Probed by Double-Rotation	NMR.=	+tice Relaxation of Sodium Cations in Zeol	PHYS 0267
cks and Products by ^{1}H and ^{13}C	NMR.=	Characterization of Delayed Coking Feedsto	FUEL 0076
nteractions by Multiple Quantum	NMR.=	+ic and Aqueous Guest/Host Solid-State I	AGRO 0062
Motion via Deuterium	NMR.=	Characterization of Polymer Segmental	PHYS 0140
Benzaldehyde by Deuterium	NMR.=	Differentiation of Natural and Synthetic	AGFD 0196
Modes and Dynamics by	NMR.=	Drug-DNA Interactions: Probing Binding	PHYS 0197
oxide Dismutase as Probed by ^{51}V	NMR.=	+on of Vanadate Anions with Cu,Zn-Super	INOR 0184
1-4 from ^{31}P and ^{13}C Solid-State	NMR.=	Structures of $Zr(O_3P(CH_2)_nCOOH)_2$, n =	INOR 0047
r Films by ^{19}F Multiple-Quantum	NMR.=	+onium Salt Distribution in Model Polyme	PMSE 0182
ating Ligands: Crystal Structures,	NMR, and Electrochemistry.=	+of Polyanionic Chel	INOR 0243
aluminum Adducts o+ Synthesis,	NMR, and X-ray Crystallographic Studies of Organo		INOR 0274
atin+ FT-IR, Solid-State MAS	NMR, and XRD of Phosphorous-Promoted Hydrotre		COLL 0198
nce and Optical Spectroscopies.	NMR, Coherent Laser Spectroscopy, and Laser NMR		PHYS 0040
rmation of Volatile Precursors of	NO_x in the Rapid Pyrolysis of Nitrogen-Containing		FUEL 0114
Selective Catalytic Reduction of	NO_x Using Zeolitic Catalysis for High-Temperature		I&EC 0046
Radicals with O_2 and	NO_x.= Reaction Rates of a Few Benzyl-Type		FUEL 0081
rom Highly Vibrationally Excited	NO_2 Donor.= +and Translational Energy Transfer f		PHYS 0334
Aductos Tonsil-Fe(	$NO_3)_3$.= Estudio de la Nitracion del Fenol con		COLL 0120
tal Studies of the Spectroscopy of	NO_3, Cl_2O_2, and Cl_2O_3.= +Atmosphere: Experimen		PHYS 0017
Gases and at Surfaces. J. Polany+	Nobel Laureates Symposium. Reaction Dynamics in		CONG 0011
xidation over Alumina-Supported	Noble Metal Catalysts.= +atalysts—I. Methane O		COLL 0185
Efrotomycin by	Nocardia lactamdurans.= Large-Scale Production of		BIOT 0109
Influence of Colored	Noise.= Energy Transfer in Molecular Aggregates:		PHYS 0023
tic Puzzles for Organic Chemistry	Nomenclature Teaching, Part I: Acyclic Series.= +c		CHED 0165
ing the Multiplet Energy Gap in	Non-Kekule Molecules: Synthesis, Structure, Reactiv		ORGN 0128
s Enzymology. Plenary Lecture.	Nonaqueous Bioprocessing: Protein Separation and P		BIOT 0063
ous Bioprocessing: Protein Sepa+	Nonaqueous Enzymology. Plenary Lecture. Nonaque		BIOT 0063
The How and Why of	Nonaqueous Enzymology.=		BIOT 0067
with Butylpyridinium Chloride in	Nonaqueous Media.= +Interactions of Sulfide Ion		ANYL 0093
poxides with Lithium Cyanide +	Nonaqueous Preparation of β-Hydroxynitriles from E		ORGN 0107
o-2-Methylpropane Sulfonate) in	Nonaqueous Solution.= +ene-Sodium-2-Acrylamid		COLL 0181
g Immo+ Syntheses of Esters in	Nonaqueous Solvents and Solvent-Free Systems Usin		BIOT 0065
Water-Restricted Environment of	Nonaqueous Solvents.= +d Enzyme System in the		BIOL 0091
ative Risk Assessment.=	Noncancer Health Endpoints: Approaches to Quantit		ENVR 0068
alitative Ranking by Cancer and	Noncancer Risks Combined: Methods and Needs.=		ENVR 0074
—I. Perspectives in Protein Pur+	Nonchromatographic Methods of Protein Separation		BIOT 0018
—II. Bioseparation Using Affini+	Nonchromatographic Methods of Protein Separation		BIOT 0044
rifications—Chromatographic and	Nonchromatographic Methods.= +s in Protein Pu		BIOT 0018
Selectivity in	Nonchromatographic Protein Separations.=		BIOT 0019
Classical and	Nonclassical Gold(I) Ammonium Cations.=		INOR 0367
al Tunnelling+ NMR Studies of	Nonclassical Transition Metal Polyhydrides: Rotation		PHYS 0068
stituted N,N'-Diarylguanidines as	Noncompetitive NMDA Antagonists.= +ialkyl Sub		MEDI 0147
ar Recognition via Base Pairing:	Noncovalent Approaches to Photosynthetic Model Sy		ORGN 0059
delonitrile Lyase by Covalent and	Noncovalent Modification.= +istic Studies on Man		BIOL 0075
(II) Autoxidation and Enzymic or	Nonenzymic Reduction of Fe(III).= +rnosine on Fe		AGFD 0131
roscopy: Imaging Subsurface and	Nonequilibrium Electronic Properties of Semiconduc		PHYS 0208
ating Polyme+ Equilibrium and	Nonequilibrium Microphase Structure of Interpenetr		PMSE 0093
= Electrochemical Kinetics and	Nonequilibrium Potentials in Biological Membranes.		COLL 0093
	Nonfederal Funding for New Faculty.=		YCC 0002
y Pyrolysis of Graphitizable and	Nongraphitizable Polymers under Argon Stream and		POLY 0023
Iron(III) Complexes as Models for	Nonheme-Iron-Containing Proteins.= +tranuclear		INOR 0295
tory.=	Nonideal Solutions in the Physical Chemistry Labora		CHED 0101
drophobic Organic Compounds in	Nonideal Solvent Mixtures.= +the Solubility of Hy		ENVR 0030
lective Recognition by DNA of a	Nonintercalating Ru(II) Complex Containing the Cat		INOR 0233
te Student Award Presentation.	Noninvasive Techniques for Studying Seed Germinat		AGFD 0104
ion Chromatography of Cationic,	Nonionic, and Anionic Copolymers of Vinylpyrrolido		PMSE 0031
mperature Superconductors.=	Nonlinear Electron-Phonon Interactions in High-Te		PHYS 0012
r Chemical Vapor Deposition of a	Nonlinear Interface Optical Switch.= +ring by Lase		BIOT 0024
Copolymers for Third-Order	Nonlinear Optical Applications.= Side-Chain		POLY 0223
unctionalized Po+ Synthesis and	Nonlinear Optical Characteristics of Chromophore-F		POLY 0108
	Nonlinear Optical Devices Using Organic Materials.=		BIOT 0176
jugated Polymers as Materials for	Nonlinear Optical Devices.= +ated Polymers. Con		POLY 0325
olecular Electronics. Design of	Nonlinear Optical Materials.= Molecular and Biom		BIOT 0173
tilayer Construction of Thin-Film	Nonlinear Optical Materials.= +Superlattices: Mul		POLY 0127
ds and Polymers as Second-Order	Nonlinear Optical Materials.= +dination Compoun		INOR 0049
Decay in Poled Polymer	Nonlinear Optical Materials.= Orientation and		POLY 0151
polarizabilities.=	Nonlinear Optical Measurements of Molecular Hyper		PHYS 0343
onship to Polym+ Second-Order	Nonlinear Optical Onset and Decay and Their Relati		POLY 0152
hemical Cross-Linking in Vitrified	Nonlinear Optical Polymer Films.= +dering and C		POLY 0206
Studies of Electrical Properties of	Nonlinear Optical Polymers and Devices.= +mism		POLY 0150
of Side-Chain and Main-Chain	Nonlinear Optical Polymers with Sulfonyl Acceptor G		POLY 0105
lenges in Photonics.=	Nonlinear Optical Polymers: Opportunities and Chal		BIOT 0174
a New Class of Photocrosslinkable	Nonlinear Optical Polymers.= +and Synthesis of		POLY 0131
r+ Optimization of Third-Order	Nonlinear Optical Properties for Conjugated, π-Elect		BIOT 0177
ugated Polymers in+ Linear and	Nonlinear Optical Properties of Highly Ordered Conj		POLY 0008
Electron-Correlated States and	Nonlinear Optical Properties of Linear Chains.=		POLY 0007
ers with Pendant Sulfonyl Tola+	Nonlinear Optical Properties of Linear Epoxy Polym		POLY 0208

osphere Composites.=	Nonlinear Optical Properties of Polymer–Silver Micr	POLY	0154
Interfaces.= Picosecond	Nonlinear Optical Studies of Semiconductor–Liquid	PHYS	0283
Films.= Guided-Wave	Nonlinear Optics in DCANP Langmuir–Blodgett	POLY	0176
r Engineering and Proces+ Cubic	Nonlinear Optics of Polymer Thin Films, I: Molecula	POLY	0011
e–$\chi^{(3)}$ Relationships in Ri+ Cubic	Nonlinear Optics of Polymer Thin Films, 2: Structur	POLY	0205
e–$\chi^{(3)}$ Relationships in C+ Cubic	Nonlinear Optics of Polymer Thin Films, 3: Structur	POLY	0226
e–$\chi^{(3)}$ Relationships in P+ Cubic	Nonlinear Optics of Polymer Thin Films, 4: Sturctur	POLY	0221
gth Dispersion of the $\chi^{(3)}$+ Cubic	Nonlinear Optics of Polymer Thin Films, 5: Wavelen	POLY	0213
larizable Metallic Complexes for	Nonlinear Optics: Cobaltous Complexes of Unsymme	POLY	0102
in Sol–Gel Matrices for Quadratic	Nonlinear Optics.= +ntation of Organic Molecules	POLY	0006
tical Applications. Introduction to	Nonlinear Optics.= +terials for Photonic and Op	POLY	0001
Complexes for Third-Order	Nonlinear Optics.= Group IV Transition–Metal	INOR	0334
and Materials for Quadratic	Nonlinear Optics.= Investigation of New Molecules	POLY	0103
Soluble Polyacetylenes for	Nonlinear Optics.= Toward High Susceptibilities in	POLY	0132
gents.+ Photorefractive Effect in	Nonlinear Polymers Doped with Charge Transport A	POLY	0155
Active Devices.= Optically	Nonlinear Polymers in Waveguiding Passive and	POLY	0177
Educating Chemists about	Nonlinear Regression.=	CHED	0193
kin–Elmer Corp.). Multiresonant	Nonlinear Spectroscopies—A New Family on the Blo	ANYL	0001
Determination of Third-Order	Nonlinear Susceptibilities of Semiconducting Polyme	POLY	0009
	Nonlinear Swelling of Phantom Networks.=	POLY	0167
Coherence–Size and Optical	Nonlinearities in Restricted Geometries.= Exciton	PHYS	0038
b ting Large and Unusual Optical	Nonlinearities.= +–Based NLO Chromophores Exhi	POLY	0217
y.= Cuprate Oxides: Effect of	Nonmagnetic Chains and Planes on Superconductivit	INOR	0358
Pictorialized Solid-State MOs: A	Nonmathematical but Honest Approach.=	CHED	0021
oloidal Si+ Redox Properties of	Nonmetallic Silver Clusters and Photochemistry of C	COLL	0130
ic Surfaces: Surface Chemistry of	Nonmetals—Novel Techniques. +ar Processes at Sol	COLL	0211
id Surfaces: Surface Chemistry of	Nonmetals—Novel Techniques. +ar Processes at Sol	COLL	0170
id Surfaces: Surface Chemistry of	Nonmetals—Novel Techniques. +ar Processes at Sol	COLL	0191
of Oilfield Emissions and Urban	Nonmethane Organic Compounds in the Tulsa Area.	ENVR	0022
cn Aromatization Selectivity over	Nonmicroporous Pt Catalysts.= +port Preparation	CATL	0030
ge during Cell-Sheet Morphoge+	Nonmuscle Myosin Is Required for Cell-Shape Chan	BIOL	0140
Diacidic	Nonpeptide Angiotensin II Receptor Antagonists.=	MEDI	0103
Inhibition of HIV Protease with	Nonpeptide Derivatives of Haloperidol.=	MEDI	0054
d Transition State or Principle of	Nonperfect Synchronization?.= +tyrenes: Radicaloi	ORGN	0155
	Nonplanarity of Doped π Systems.=	PHYS	0233
A High-Purity	Nonpyrophoric Phenyllithium Solution.=	ORGN	0226
Use in Labeling Nucleic Acids for	Nonradioactive Detection.= +of Photozyme and Its	CARB	0037
Polyimides by a	Nonreactive Diluent.= Antiplasticization of	PMSE	0127
with Experiment. Spectroscopy of	Nonrigid Molecular Complexes.= +um Chemistry	PHYS	0102
ion–Making Chemistry Course for	Nonscience Students. +he ACS College-Level Decis	CHED	0051
ach to College Chemistry for	Nonscience Students.= Interdisciplinary Core Appro	CHED	0176
aterials: Free Radical Routes into	Nonshrinking Sol–Gel Composites.= +Composite M	POLY	0259
kelocene at 248 nm: Statistical vs.	Nonstatistical Behavior.= +on of Ferrocene and Nic	PHYS	0333
al Nucleophilic Substitutions on	Nonstereogenic Atoms by the Endocyclic Restriction	ORGN	0120
Development of New	Nonsteroidal Aromatase Inhibitors.=	MEDI	0003
Activity Studies in a New Class of	Nonsteroidal Inhibitors of Aromatase.= +tructure–	MEDI	0036
ems for Reduction of Rinsate and	Nontarget Contamination.= +icide Application Syst	AGRO	0105
atment and Use of Biopolymers as	Nonthermal Food Preservation Processes.= +e Tre	AGFD	0029
e Crude Methylenedianiline to a	Nonthermodynamic Mixture of bis(4-aminocyclohexy	I&EC	0056
	Nontraditional Careers in Chemistry.=	CMEC	0027
	Nontraditional Careers in Chemistry.=	I&EC	0004
y Is the Central Science.=	Nontraditional ChemSource Modules Show Chemistr	CHED	0018
Manganese Oxo Clusters with	Nonzero Spin–Ground States.= Covalent Linkage of	INOR	0196
nd Pharmacophore Fragment of	norBNI in κ-Opioid Antagonist Potency and Selectiv	MEDI	0157
Hydrocarbons in La Basse Cote+	Nord, Quebec.= +g People to Chlorinated Aromatic	CEI	0011
Oxygenated C_{13}	Norisoprenoids: Important Flavor Precursors.=	AGFD	0025
ectams as Potential Precursors of	Norstatine, Statine, and Their Analogs.= +ted-β-L	ORGN	0047
nternational Flow of Information:	North America and Europe. Global Marketing of+ I	CINF	0011
hool Teachers of Chemistry in the	Northwest.= +ectronic Bulletin Board for High Sc	CHED	0111
it+ 3D Database Searching vs. de	Novo Construction: Two Complementary Methods w	COMP	0050
ng+ Crystal Structure-Based de	Novo Design and Synthesis of Naphthalene Containi	MEDI	0186
New Method for the de	Novo Design of Enzyme Inhibitors.=	COMP	0037
Carbocyclic Substrates for de	Novo Purine Biosynthesis.=	BIOL	0076
Miscible	Novolac/Polymer Blends and Curing Studies.=	POLY	0113
DA A+ Asymmetric Synthesis of	NPC 12626: A Potent and Selective Competitive NM	MEDI	0083
lective Protein Kinase C Inhibitor	NPC 15437.= +armacological Evaluation of the Se	MEDI	0026
oalkoxychromones, NPC16377 and	NPC13839, as Probes of Sigma Receptors.= +Amin	MEDI	0150
abeled Aminoalkoxychromones,	NPC16377 and NPC13839, as Probes of Sigma Recep	MEDI	0150
Correlation Spectroscopy (	NQR–COSY).= Two-Dimensional Pure Quadrupole	PHYS	0277
Strain of Xanthomonas campestris	NRRL B–1459.= +eous Solutions from a Variant	BIOT	0196
S into the Undergraduate Curr+	NSF Catalyzes the Incorporation of NMR and GC/M	CHED	0149
S Chemistry Olympiad and the	NSF ChemSource Programs, Honoring the Memory o	CHED	0011
mistry Laboratory.=	NSF Funding of Laboratory Innovations in Undergra	CHED	0196
Get a Fair Shake? An Analysis of	NSF Funding Opportunities for Junior Faculty.=	YCC	0003
	NSF Perspectives.=	POLY	0204
boratory—I. Support of Chemi+	NSF-Catalyzed Innovations in the Undergraduate La	CHED	0146
boratory—II. Undergraduate S+	NSF-Catalyzed Innovations in the Undergraduate La	CHED	0179
y Laboratory Improvement by the	NSF-ILI Program.= +ory—I. Support of Chemistr	CHED	0146
ucture, and Properties+ Zeolite	NU-87: Aspects of Its Synthesis, Characterization, St	PETR	0021
Neutrino in	Nuclear Beta Decay.= Evidence for a Massive	NUCL	0003

in Memory of Nathan Sugarman.	Nuclear Chemistry, the Met Lab, and Nathan Sugar	NUCL	0058
Properties in Intermediate-Energy	Nuclear Collisions with Gold.= +. Heavy Residue	NUCL	0024
Single-Particle Density in	Nuclear Dynamics.= Fluctuations of the	NUCL	0009
Problems in	Nuclear Energy.= Nuclear Fuel Cycle and Special	FUEL	0024
ion.=	Nuclear Equation of State and Finite-Range Interact	NUCL	0008
Energy.=	Nuclear Fuel Cycle and Special Problems in Nuclear	FUEL	0024
ic Behavior of the Hydrox+ ^{13}C	Nuclear Magnetic Relaxation of Amylose and Dynam	CARB	0049
dergraduate Laborat+ Improved	Nuclear Magnetic Resonance Capabilities for the Un	CHED	0185
h Clay Minerals as Studied by ^{2}D	Nuclear Magnetic Resonance Spectroscopy.= +r wit	AGRO	0061
nic Phosphates by Phosphorus-31	Nuclear Magnetic Resonance.= +Analysis of Inorga	INOR	0291
Surfaces by Selective Excitation of	Nuclear Magnetic Resonances.= +ics at Catalyst	PHYS	0346
A Tool for the Characterization of	Nuclear Multifragmentation.= +e MSU Miniball:	NUCL	0035
Inherently Safe and Advanced	Nuclear Power Reactors.= New Generation of	FUEL	0023
Power?.= Future of	Nuclear Power. Is There a Future for Nuclear	FUEL	0022
There a Future for	Nuclear Power?.= Future of Nuclear Power. Is	FUEL	0022
Principles and Applications.=	Nuclear Reaction Analysis of Hydrogen in Materials:	NUCL	0067
s: Elemental/Isotopic Analysis by	Nuclear Reactions/Scattering. +ification of Material	NUCL	0064
cts in Discharge Pool in T.H.O.R.	Nuclear Research Reactor Site Boundary.= +Produ	NUCL	0090
d Li$_5$N+ Additional Aspects of	Nuclear Science—I. Structural Studies of Li$_4$NpO$_5$ an	NUCL	0068
235 in M+ Additional Aspects of	Nuclear Science—II. Initial Abundances of Uranium-	NUCL	0082
Managing Our Nation's	Nuclear Waste.=	CHAS	0046
Organics in Mixed	Nuclear Wastes: The Challenges.= Analyzing Exotic	ANYL	0060
Targeting Chemical	Nucleases.=	BIOL	0156
es to Soli+ Chemical Control of	Nucleation and Growth in the Conversion of Molecul	INOR	0145
on of Diamond (and c-BN, SiC).	Nucleation and Growth of Diamond from Hydrocarb	PHYS	0091
structural Characterization of the	Nucleation and Growth of Diamond Thin Films.=	PHYS	0121
AM1 Studies of the	Nucleation of Crystals of 1,3-Diones.=	PHYS	0247
beyond the First Ring and the	Nucleation of Soot Particles.= Aromatics Growth	FUEL	0108
Precursor Phases and	Nucleation of Zeolite TON.=	PETR	0040
How and Why Do	Nuclei Fragment?.=	NUCL	0006
Multifragmentation of	Nuclei with A=70.=	NUCL	0028
Pd(II) Dipeptide Complexes with	Nucleic Acid Constituents.= +raction of Pt(II) and	INOR	0412
hotozyme and Its Use in Labeling	Nucleic Acids for Nonradioactive Detection.= +of P	CARB	0037
and Quantification of	Nucleic Acids in the Hydrosphere.= Identification	ENVR	0027
cificity of Guanosine Oxidation in	Nucleic Acids Using Metal Complexes.= +tion Spe	BIOL	0122
Photolyase.= Chemistry of	Nucleic Acids. Mechanistic Studies on DNA	BIOL	0155
s in the Detection of Proteins and	Nucleic Acids.= +ted Chemiluminescent Substrate	BIOT	0167
ormation-Specific Oxidation of	Nucleic Acids.= Tunable Nickel Complexes for Conf	INOR	0183
y Ion Collisions at 14.6 GeV/c per	Nucleon.= +ansverse Energies in Relativistic Heav	NUCL	0051
hly Optically Pur+ Homogeneous	Nucleophile Exchange: A Novel, Facile Route to Hig	ORGN	0207
nes: Asymmetric Synthesis and+	Nucleophile-Promoted Cyclization Reactions of Alky	ORGN	0275
kylidyne and Carbonyl Ligands by	Nucleophiles and by Electrophiles.= +upling of Al	INOR	0188
f Displacement of Guanosine by	Nucleophiles from cis-Diammine(guanosine)platinum	INOR	0251
Compounds with	Nucleophiles.= Reaction of 2,3-Dibromocarbonyl	ORGN	0160
of Aziridinones by	Nucleophiles.= Theoretical Study of Ring-Opening	PHYS	0314
trostyrenes: Radicaloid Transiti+	Nucleophilic Addition of Alkyl Thiolate Ions to β-Ni	ORGN	0155
Platinum-Pyrazoline Formed by	Nucleophilic Addition of Amine or Hydrazine to σ-A	INOR	0041
e Dicarbonyl η^3-Allyl Cations.=+	Nucleophilic Addition to Cyclopentadienyl Manganes	INOR	0270
the Reaction of Cu+ Absence of	Nucleophilic Assistance by Solvent and Azide Ion to	ORGN	0154
f Cumyl Drivatives: Mechanism of	Nucleophilic Substitution at Tertiary Carbon.= +o	ORGN	0154
e Ultimate Hepatacarcinogen, +	Nucleophilic Substitution by Aromatic Amines on th	ORGN	0194
n of the Steric Course of Formal	Nucleophilic Substitutions at Nonstereogenic Atoms	ORGN	0120
Effect of Solvent on the	Nucleophilicity of Subtilisin.=	BIOT	0066
of the Cyclopentenyl Carbocyclic	Nucleoside Analogue of 5-Azacytidine.= +cal Study	MEDI	0048
and 3'-Methylidene-Substituted	Nucleoside Analogues and Crystal Structure of 2'-De	CARB	0038
tly Linked Polynuclear Aromatic-	Nucleoside and Oligonucleotide Complexes.= +alen	PHYS	0021
Transition State of	Nucleoside Hydrolase.=	BIOL	0073
w NAD Congeners Containing C-	Nucleoside Isosteres of Nicotinamide Riboside.=	CARB	0023
Drug Design: Inhibitors of Purine	Nucleoside Phosphorylase (PNP).= +ucture-Based	MEDI	0005
Biological Activity of 9-+ Purine	Nucleoside Phosphorylase Inhibitors: Synthesis and	MEDI	0028
n State Analysis.= Purine	Nucleoside Phosphorylase: Mechanism and Transitio	BIOL	0046
	Nucleoside Synthesis from Thioglycosides.=	ORGN	0067
Spin-Labeled	Nucleoside.= Synthesis and Properties of a Novel	ORGN	0026
yrinyl-Mononucleoside and Other	Nucleoside-Substituted Porphyrins.= +luble Porph	ORGN	0189
thetic Approaches to Modified	Nucleosides and C-Nucleosides, such as AZT.= Syn	MEDI	0049
mistry/Asymmetric Synthesis of+	Nucleosides and Nucleotides: Chemistry and Bioche	CARB	0022
ically Pure Dioxolane-Pyrimidine	Nucleosides as Potential Anti-HIV Agents.= +omer	MEDI	0051
Pentammineosmium(II) to	Nucleosides.= η^2-Binding of	INOR	0083
Synthesis of 2'-Fluorinated Purine	Nucleosides.= +d Sugar-Protecting Groups in the	CARB	0024
of 8-Aza-1-Deazapurine	Nucleosides.= Synthesis and Biological Evaluation	MEDI	0029
3-Substituted	Nucleosides.=	MEDI	0050
to Modified Nucleosides and C-	Nucleosides, such as AZT.= Synthetic Approaches	MEDI	0049
Study.= Amino Acid-	Nucleotide Cross-Binding to Platinum(II): A Kinetic	INOR	0182
etics of Formation of Platinum(II)	Nucleotide.= +d Two-Dimensional NMR and Kin	INOR	0180
ic Synthesis of+ Nucleosides and	Nucleotides: Chemistry and Biochemistry/Asymmetr	CARB	0022
ies of Platinum(II) Complexes of	Nucleoside-5'-Di- and Triphosphates and Their Thio	INOR	0403
s. Heavy Residue Pro+ Nucleus-	Nucleus Collision Mechanisms: Intermediate Energie	NUCL	0024
Not To Fuse.= Nucleus-	Nucleus Collision Mechanisms: Overview. To Fuse or	NUCL	0004
na. Synthesis of Hig+ Nucleus-	Nucleus Collision Mechanisms: Relativistic Phenome	NUCL	0046
lements. The Darmst+ Nucleus-	Nucleus Collision Mechanisms: Synthesis of Heavy E	NUCL	0014

tion Chamber in Mult+	Nucleus–Nucleus Collison Mechanisms. Use of a Time–Projec	NUCL	0034
High–Energy Nucleus–	Nucleus Collisons.= Transparency and Abrasion in	NUCL	0048
e Energies. Heavy Residue Prop+	Nucleus–Nucleus Collision Mechanisms: Intermediat	NUCL	0024
o Fuse or Not To Fuse.=	Nucleus–Nucleus Collision Mechanisms: Overview. T	NUCL	0004
Phenomena. Synthesis of High+	Nucleus–Nucleus Collision Mechanisms: Relativistic	NUCL	0046
Heavy Elements. The Darmsta+	Nucleus–Nucleus Collision Mechanisms: Synthesis of	NUCL	0014
e–Projection Chamber in Multi+	Nucleus–Nucleus Collison Mechanisms. Use of a Tim	NUCL	0034
Abrasion in High–Energy	Nucleus–Nucleus Collisons.= Transparency and	NUCL	0048
	Numerical Modeling of GaAs MOCVD.=	COLL	0066
ectronic Energy Transfer and D+	Numerical Simulation of the Time Dependence of El	POLY	0078
otton Plant Allelochemicals and	Nutrients on the Behavior and Development of the T	AGRO	0037
Trace Metal	Nutrition of Marine Phytoplankton.=	INOR	0003
re Annealing on the Capillarity of	Nylon Microporous Membranes.= +gh–Temperatu	COLL	0037
tion.= Copolymers of	Nylon 266 and Nylon 66: Synthesis and Characteriza	POLY	0012
Copolymers of Nylon 266 and	Nylon 66: Synthesis and Characterization.=	POLY	0012
GeV/nucleon	^{16}O Ions.= Target Fragmentation of Silver by 14.6	NUCL	0050
cf $Bi_2Sr_2CuO_y$ and $Bi_2Sr_2Cu_{0.9}Fe_{0.1}O_y$.=	+M Investigations of the Surface Structure	PHYS	0292
Science Education Program at	Oak Ridge National Laboratory.=	CHAS	0050
s Natural Fat Substitutes.=	Oatrim Products with Elevated β–Glucan Contents a	AGFD	0037
Chemicals.= Favorite or	Objectionable Cartoons Related to Chemists or	CHAL	0021
Patents.= Favorite or	Objectionable Cartoons Related to Inventors or	CHAL	0022
Lawyers.= Favorite or	Objectionable Cartoons Related to Laws and	CHAL	0023
Neutrino	Observatory.= Results from the Homestake Solar	NUCL	0021
Sudbury Neutrino	Observatory.=	NUCL	0032
Data.= Has a Release Been	Observed? HRS–Observed Releases and Analytical	ENVR	0080
epiandrosterona.=	Obtencion de Derivados Hidrosolubles de la Dehidro	MEDI	0038
erreduccion con Zinc en Medio+	Obtencion de Plata por Electrodeposito, Fusion, Oxid	CHED	0172
la Vitamina C.=	Obtencion del Orden de Reaccion de la Oxidacion de	CHED	0164
ansferencia de Carga por Espec+	Obtencion, Caracterizacion, y Determinacion de la Tr	INOR	0190
$Cu_4(\eta\text{–}OCMe_3)_6[$	$OC(CF_3)_3]_2$.= Synthesis and Structure of	INOR	0134
Thermal Decomposition of 1,7–	Ocatdiyne as a Source of Propargyl Radicals.=	FUEL	0067
ed States of Maximal d–Shell	Occupancy in the Ru + H_2 Reaction.= Role of Excit	PHYS	0243
ntal and Biological Monitoring of	Occupational Exposure to Cyclophosphamide in the	CEI	0010
Carcinogens: Environmental or	Occupational Exposures.= Case Studies of	CHAS	0020
Productivity Enhancement.=	Ocean Feeding of Electron–Treated Sludge for Ocean	ENVR	0021
e Hudson River.= Chemistry of	Ocean Margins. Controls on Methane Fluxes from th	GEOC	0001
Iron Limitation of Open–	Ocean Primary Production.=	INOR	0002
of Electron–Treated Sludge for	Ocean Productivity Enhancement.= Ocean Feeding	ENVR	0021
istribution of Trace Metals in the	Ocean.= +ial on Marine Bioinorganic Chemistry. D	INOR	0001
y. Manganese Chemistry in the	Ocean.= Tutorial on Marine Bioinorganic Chemistr	INOR	0009
Deep	Ocean.= Formation of CO_2 Clathrate in the	ENVR	0047
hosphorus Residence Time in the	Ocean.= +nts: Implications for Burial Fluxes and P	GEOC	0013
Flux out of Margin Sediments on	Oceanic DOC.= +Dissolved Organic Carbon (DOC)	GEOC	0017
s, $(CO)_5MnCOCH_2R$ (R = H,CH_3,	OCH_3).= +rosilation of Manganese Acyl Complexe	INOR	0337
omplexes $(\eta_5\text{–}C^5R^5)Re(No)(PPh^3)($	OCHRR').= +d Activation in Rhenium Alkoxide C	ORGN	0220
$Cu_4(\eta\text{–}$	$OCMe_3)_6[OC(CF_3)_3]_2$.= Synthesis and Structure of	INOR	0134
Zinc and Nickel	Octaethyloxophlorins.=	INOR	0302
IMe_2.= Macrocyclic Complexes.	Octaethylporphyrin Lanthanide Chemistry: Synthesi	INOR	0294
of Microporous Molybdenum with	Octahedral–Tetrahedral Frameworks.= +Synthesis	CATL	0019
nthesis and Analgesic Activity of	Octahydro-1,2,3,4,4a,5,11,11a–Pyrido[3,4–c] [1,5]Ben	MEDI	0148
squito Repellent, 1,1,4,5,6,7,8,8A–	Octahydro–3H–2–Benzopyran–3–one.= +actone Mo	AGRO	0033
Containing Bicyclo[2.2.2]	Octane Moieties.= Colorless Rigid–Rod Polymers	POLY	0029
hysicochemical Applications—II.	Octanol–Water Partition Coefficients by Centrifugal	ANYL	0068
ctivity of C5A Carboxyl Terminal	Octapeptide Analogues.= +ation on Chemokinetic A	MEDI	0168
phological Characterization of 1–	Octene LLDPE (TREF) Fractions and Their Blends.	PMSE	0102
nyl–Pyrazol–3–Thione and Tri–n–	Octylphosphine Oxide.= +enzoyl-5-Methyl-2-Phe	NUCL	0073
igand Schar+ Spectroscopic and	ODMR Investigations of the Lowest π^* and Ligand-L	INOR	0164
Effects and Delayed Fluorescence	ODMR.= +lly Mixed Naphthalene: High-Pressure	PHYS	0081
and Iron Salts on Sewage, Part 1:	Odor Control and Phosphate Removal.= +Peroxide	ENVR	0034
o Malolactic F+ Changes in the	Odor–Active Compounds Found in Grape Wine due t	AGFD	0137
he Volatile Constituents of Living	Odors and Flavors.= +e–Dimensional Analysis of t	AGFD	0022
Axillary	Odors.= Precursors to the Characteristic Human	AGFD	0147
try: Synthesis and Reactivity of (	OEP)MX, and the Activation of O_2 by (OEP)Y(μ–M	INOR	0294
MX, and the Activation of O_2 by (	OEP)Y(μ–ME)$_2$AlMe$_2$.= +and Reactivity of (OEP)	INOR	0294
ond and Diamond–Like Carbon.	OES and SEM Observations of Diamond Films Synt	FUEL	0038
Crime—Molecular	Offenses in Canada.= Chemistry and	CHAL	0020
Zeolites Mordenite, L, Omega, and	Offretite.= +c Studies of the Crystallization of the	PETR	0041
ersion of Methanol to Olefins over	Offretite/Erionite Intergrowths.= +ies of the Conv	PETR	0042
actic Tripeptide F-Met-Leu-Phe-	OH.= +–Terminal Modified Analogues of Chemot	MEDI	0166
Dayton,	Ohio, Project SEED.= Organizational Structure of	CHED	0142
Generation of G+ Application of	Oil Agglomeration Technique to Soil Clean–Up and	FUEL	0141
ha Family as a Possible Source of	Oil and Protein for Human Consumption.= +Jatrop	AGFD	0038
) with Functionalized Triglyceride	Oil Compositions.= +Poly(Ethylene Terephthalate	PMSE	0171
oxicity of Annona squamosa Seed	Oil Extracts to Tribolium castaneum L. (Coleopt+ T	AGRO	0036
nal NMR Analyses of Coal–	Oil Hydroprocessing Products.= 1- and 2-Dimensio	PETR	0149
mer Network Formed from Castor	Oil Polyurethane and Vinyl Copolymers.= +g Poly	PMSE	0089
–Toughened Epoxy Resin/Castor	Oil Polyurethane IPN.= +haracterization of CTBN	PMSE	0092
oom Temperature Cured Castor	Oil Polyurethane/Vinyl or Acrylic Polymer Interpene	PMSE	0090
Polymer Networks from Castor	Oil Polyurethanes and Poly(Styrene–Methyl Methacr	PMSE	0091
/Property Relationships in Base	Oil Processing: Monitoring Dewaxing Procedures via	PETR	0151

and	Oil Quality.= Effect of Lipid Oxidation on Food	AGFD	0096
Associations of Trace Elements in	Oil Shale Kerogen.= +nation of Organic/Inorganic	FUEL	0045
ograph+ Determination of Motor	Oil Volatility Using High-Temperature Gas Chromat	PETR	0122
in Gas	Oil.= Industrial Catalysts for Aromatics Reduction	CATL	0048
atrix for Structural Composites in	Oilfield Applications.= +rent Epoxy Polymers as M	PMSE	0111
Hydrocarbon Composition of	Oilfield Emissions and Urban Nonmethane Organic C	ENVR	0022
h Fatty Acid Esters or Vegetable	Oils: Formulation, Optimization, and Determination	COLL	0083
es to Direct Coal Liquefaction	Oils.= Application of Advanced Analytical Techniqu	FUEL	0044
n of Refinery Streams and Heavy	Oils.= +nt Distribution Analysis for Characterizatio	PETR	0119
an Cold-Pressed Citrus Essential	Oils.= HRGC for Detection of Adulterations of Itali	AGFD	0199
with Functionalized Triglyceride	Oils.= Poly(Ethylene Terephthalate) Semi-IPNs	PMSE	0088
tication of Extracts and Essential	Oils.= +attern-Recognition Techniques for Authen	AGFD	0198
-Flavor Development i+ Uses of	Oilseed Food Ingredients to Minimize Warmed-Over	AGFD	0021
Phenolic Compounds of Brassica	Oilseeds.=	AGFD	0168
Model for Investigation of Light	Olefin Aromatization Reactions over HZSM-5 Zeolite	PETR	0061
hiral Metal Salen Complexes B+	Olefin Aziridination and Epoxidation Catalyzed by C	ORGN	0056
izations Catalyzed by Co(III) an+	Olefin Dimerizations, Oligomerizations, and Polymer	ORGN	0129
ials. Silicon Acetylene and Silicon	Olefin Polmers as Precursors to SiC.= +rid Mater	POLY	0357
hedral Oligometallasilsesquioxa+	Olefin Polymerization by Vanadium-Containing Poly	INOR	0131
in Catalytic	Olefin Polymerization.= Mechanism and Selectivity	CATL	0010
Studies in Asymmetric	Olefination and Ene Reactions.= Plenary Session.	ORGN	0279
Di+ Application of the Peterson	Olefination Reaction to the Synthesis of Fluoroolefin	ORGN	0079
Pigmented	Olefinic Polymers.= Crosslinking Reactions in	POLY	0135
Oligomerization of Lower	Olefins by Heterogeneous Catalysis.=	PETR	0009
merization and Isomerization of	Olefins by Homogeneous Nickel-Based Catalysts.=	PETR	0007
Oligomerization of C$_2$-C$_6$	Olefins on HZSM-5.= Comparative Study of the	PETR	0026
of the Conversion of Methanol to	Olefins over Offretite/Erionite Intergrowths.= +ies	PETR	0042
Alkylation of Isoparaffins with	Olefins over Zeolite Catalysts.=	PETR	0101
tic Activity in the Oxidation of α-	Olefins.= +thesis, Molecular Structure, and Cataly	INOR	0365
Pentenes and Other Light	Olefins.= Sulfuric Acid-Catalyzed Alkylation of	PETR	0084
omatization of Light Alkanes and	Olefins.= +nt of High-Performance Catalysts for Ar	PETR	0063
Captodative	Olefins.= 1,3-Dipolar Cycloaddition Reactions with	ORGN	0230
eparated by+ Liquid Crystalline	Oligoesters Containing Single-Ring Aromatic Units S	PMSE	0016
E+ Chromatographic Analysis of	Oligomer Structures during the Copolymerization of	POLY	0227
Emulsion Polymerization with	Oligomeric Diblock Macromonomer Stabilizers.=	COLL	0039
s over Heterogeneous Catalysts.	Oligomerization and Isomerization of Olefins by Hom	PETR	0007
Comparative Study of the	Oligomerization of C$_2$-C$_6$ Olefins on HZSM-5.=	PETR	0026
n of ZSM-5 Type Catalysts during	Oligomerization of Ethene.= +n by Coke Formatio	PETR	0027
Catalytic	Oligomerization of Ethylene by Ti(II) Complexes.=	INOR	0153
fficient and Stable Heterogeneous	Oligomerization of Ethylene.= +s for the Highly E	PETR	0008
atalysis.=	Oligomerization of Lower Olefins by Heterogeneous C	PETR	0009
Heterogeneous	Oligomerization of Propene over Heteropoly Acids.=	PETR	0011
ydr+ Alkylation, Aromatization,	Oligomerization, and Isomerization of Short-Chain H	PETR	0082
ydr+ Alkylation, Aromatization,	Oligomerization, and Isomerization of Short-Chain H	PETR	0130
ydr+ Alkylation, Aromatization,	Oligomerization, and Isomerization of Short-Chain H	PETR	0043
ydr+ Alkylation, Aromatization,	Oligomerization, and Isomerization of Short-Chain H	PETR	0112
ydr+ Alkylation, Aromatization,	Oligomerization, and Isomerization of Short-Chain H	PETR	0007
ydr+ Alkylation, Aromatization,	Oligomerization, and Isomerization of Short-Chain H	PETR	0100
ydr+ Alkylation, Aromatization,	Oligomerization, and Isomerization of Short-Chain H	PETR	0025
ydr+ Alkylation, Aromatization,	Oligomerization, and Isomerization of Short-Chain H	PETR	0061
Co(III) an+ Olefin Dimerizations,	Oligomerizations, and Polymerizations Catalyzed by	ORGN	0129
haracterization of bisNadimides	Oligomers and High-Tg Polyimide-Based Semi-IPN	POLY	0133
Epoxy Monomers and	Oligomers.= Synthesis of Novel Silicon-Containing	POLY	0015
Vanadium-Containing Polyhedral	Oligometallasilsesquioxanes (POMSS).= +zation by	INOR	0131
nuclear Aromatic-Nucleoside and	Oligonucleotide Complexes.= +alently Linked Poly	PHYS	0021
S$_2$-AAGCG):+ NMR Spectra of	Oligonucleotide Phosphorodithioate d(CGCTpS$_2$-TP	BIOL	0123
Two-Dimensional NMR of Duplex	Oligonucleotides and Proteins DNA Complexes.=	INOR	0378
pecificity of the Monoalkylation of	Oligonucleotides by Mitomycin C.= +-Sequence S	MEDI	0015
Monoxygenase Genes in	Oligophagous Lepidoptera.= Cytochrome P-450	AGRO	0115
(ISO)4F Interact Differently with	Oligoribonucleotide Analogs of Rabbit α-Globin mR	BIOL	0128
al Studies.=	Oligosaccharide Processing in Cancer Cells: Function	BIOL	0151
s of Sit+ Crystal Structure of an	Oligosaccharide-FAB Complex and Binding Activitie	BIOL	0143
and Conformational Analysis of	Oligosaccharides Corresponding to the Cell-Wall Pol	CARB	0057
ic Heptose and KDO Containing	Oligosaccharides of the Inner Core Region of Lipopol	CARB	0058
Bind	Oligosaccharides.= How Proteins Recognize and	CARB	0012
Mechanistic Studies of	Oligosaccharyltransferase.=	BIOL	0055
ymposia on the ACS Chemistry	Olympiad and the NSF ChemSource Programs, Hono	CHED	0011
2+ 1992 International Chemistry	Olympiad Competition Scheduled for USA, July 11-2	CHED	0139
Chemistry	Olympiad in Mexico: A Multipurpose Event.=	CHED	0109
Students for the Chemistry	Olympiad National Exam.= How to Prepare	CHED	0013
The New York Local Section	Olympiad Selection Process.=	CHED	0012
The U.S. National Chemistry	Olympiad: A Student's View.=	CHED	0016
.S. Participation in the Chemistry	Olympiad.= +jorie Gardner, Chemical Educator. U	CHED	0011
Chemistry	Olympiad.= U.S. Participation in the International	CHED	0015
Important in	OMCVD of Copper.= Decomposition Mechanisms	COLL	0157
FTIR of Surface Reaction in	OMCVD of III-V Compound Semiconductors.=	COLL	0011
Precursors Useful for	OMCVD of ZnSe.=	COLL	0155
s of New Antimony Precursors for	OMCVD.= +c Precursors. Synthesis and Propertie	COLL	0135
pound Semiconductor Devices by	OMCVD.= +owth of High-Performance III-V Com	COLL	0010
for	OMCVD.= Novel Barium Compounds Useful	COLL	0158

(001) GaAs by	OMCVD.= Real-Time Investigations on Growth on	COLL	0007
ral Environments for Long-Chain	Omega-3 Fatty Acids.= +eds: Natural and Unnatu	AGFD	0160
ation of the Zeolites Mordenite, L,	Omega, and Offretite.= +c Studies of the Crystalliz	PETR	0041
es of Organoindium Precursors for	OMVPE.= +thesis, Characterization, and Properti	INOR	0218
rocessing, and in Vitro Effects on	Oncogenic Transformation.= +ulation, Metabolic P	AGFD	0117
Subsidized Access to Beilstein	Online: Options for Search Services in an Academic L	COMP	0012
Second-Order Nonlinear Optical	Onset and Decay and Their Relationship to Polym	POLY	0152
lture.=	Ontario's Program for Using Sewage Sludge in Agricu	FERT	0021
Unusual Ring-	Opened Rifamycin Derivatives.= Chemistry of Some	ORGN	0013
ng Process and the Role of Expert	Opinion.= +ess. Overview of Legal Decision-Maki	CHAL	0011
ophore Fragment of norBNI in κ-	Opioid Antagonist Potency and Selectivity.= +rmac	MEDI	0157
sis of Naltrindole Analogues as δ-	Opioid Receptor Selective Fluorescent Probes.= +e	MEDI	0155
Irreversible Ligands for the κ-	Opioid Receptor.= Highly Selective Reversible and	MEDI	0156
Requirement of Naltrindole at δ-	Opioid Receptors.= Studies on the Conformational	MEDI	0154
e-Activity Relationship Studies of	Opioids.= +le of Concepts and Models in Structur	MEDI	0089
e: An Alternative Approach to the	Oplopane Skeleton.= +l Melampolide Schkuhriolid	ORGN	0018
g+ Structure of K(Mg$_1$/3Nb$_2$/3)	OPO$_4$ (KMNP): A Centrosymmetric Aliovalent Analo	INOR	0256
Nonlinear Optical Polymers:	Opportunities and Challenges in Photonics.=	BIOT	0174
Biomonitoring. Biomonitoring:	Opportunities and Responsibilities.= Human	CEI	0015
Technical Service.=	Opportunities for Chemists in Marketing, Sales, and	YCC	0006
ting Major Chemical Accidents:	Opportunities for Industry and Challenges for Regula	CHAS	0016
ake? An Analysis of NSF Funding	Opportunities for Junior Faculty.= +Get a Fair Sh	YCC	0003
yl-Branched Juvenile Hormones:	Opportunities for Specific Control of Lepidopterans?	AGRO	0004
NIH Funding	Opportunities for the Assistant Professor.=	YCC	0004
America. Latin American Chem+	Opportunities for U.S. Chemical Businesses in Latin	SCHB	0021
Future	Opportunities in Advanced Materials.=	INOR	0015
f RNA Minihelices in Rela+ New	Opportunities in Biological Chemistry. Recognition o	BIOL	0026
Spreadsheets:	Opportunities in Physical Chemistry.=	CHED	0207
y Education.= Challenges and	Opportunities in Undergraduate Analytical Chemistr	ANYL	0007
ation, Application, Problems, and	Opportunities. T+ Immobilized Biocatalysts: Prepar	BIOT	0035
Microbiological Hazards and	Opportunities.=	AGFD	0014
n: A Multifaceted Career	Opportunity for Chemists.= Information Distributio	YCC	0008
Chemistry.= Applications and	Opportunity in the Area of Surface Coordination	CHED	0062
:Crystal Structure of Mo(η^2–C(S)	OPr^i)(S$_2$COPr^i)CO(PMe$_3$)$_2$.= +tic Acetyl Complexes	INOR	0036
e in Phot+ Measurements of	Opti-Electric Properties of Molecules of Potential Us	POLY	0214
Diamine-Epoxy System by Fiber	Optic Fluorescence.= +ure Monitoring of Aromatic	POLY	0282
Development of Raman Fiber–	Optic Probes for Hostile Environments.=	ANYL	0065
Fluorescence Capillary/Fiber-	Optic Sensors.= Sensitivity Enhancement in	ANYL	0077
eneration by σ-Bond-Separate+	Optic-Electronic Properties and Second-Harmonic G	POLY	0222
thacrylate)/Glass Fiber Lamina+	Optical and Mechanical Properties of Poly(methylme	PMSE	0110
cs in Dilute Solution.=	Optical and NMR Studies of Local Polymer Dynami	PHYS	0139
ymeric Materials for Photonic and	Optical Applications. Highly Polarizable Metal+ Pol	POLY	0102
ymeric Materials for Photonic and	Optical Applications. Introduction+ Tutorial on Pol	POLY	0001
ymeric Materials for Photonic and	Optical Applications. Low-Loss Graded-Index+ Pol	POLY	0174
ymeric Materials for Photonic and	Optical Applications. Orientation of Organic+ Pol	POLY	0006
ymeric Materials for Photonic and	Optical Applications. Second Harmonic Gener+ Pol	POLY	0150
ymeric Materials for Photonic and	Optical Applications. Synthesis of Controlled+ Pol	POLY	0126
Polypeptides for	Optical Applications: Construction of Repetitive	BIOT	0033
Third-Order Nonlinear	Optical Applications.= Side-Chain Copolymers for	POLY	0223
iorhodopsin Materials.=	Optical Bistability and Four-Wave Mixing in Bacter	BIOT	0081
ed Po+ Synthesis and Nonlinear	Optical Characteristics of Chromophore-Functionaliz	POLY	0108
ing NL+ Synthesis, Poling, and	Optical Characterization of Polyurethane Films Bear	POLY	0216
mical Co. Foundation). Thermo–	Optical Detection Methods in Analytical Laser Spect	ANYL	0011
Materials for All–	Optical Device Applications.= Designing Polymeric	BIOT	0175
Nonlinear	Optical Devices Using Organic Materials.=	BIOT	0176
ymers as Materials for Nonlinear	Optical Devices.= +ated Polymers. Conjugated Po	POLY	0325
Magnet-Enhanced	Optical Falling-Needle/Sphere Rheometer.=	POLY	0251
-Index and Single-Mode Polymer	Optical Fibers.= +Applications. Low-Loss Graded	POLY	0174
er Composite.=	Optical Four-Wave Mixing in a Silver Colloid-Polym	POLY	0218
elaxation at Interfa+ Picosecond	Optical Harmonic Measurements of Excited-State R	COLL	0087
Bacteriorhodopsin Variants for	Optical Information Processing.=	BIOT	0078
Planar Polymer Waveguides for	Optical Interconnections.=	POLY	0175
Synthetic and	Optical Limiting Studies of Metallophthalocyanines.=	INOR	0303
Electronics. Design of Nonlinear	Optical Materials.= Molecular and Biomolecular	BIOT	0173
struction of Thin-Film Nonlinear	Optical Materials.= +Superlattices: Multilayer Con	POLY	0127
ymers as Second-Order Nonlinear	Optical Materials.= +dination Compounds and In	INOR	0049
Polymer Nonlinear	Optical Materials.= Orientation and Decay in Poled	POLY	0151
ities.= Nonlinear	Optical Measurements of Molecular Hyperpolarizabil	PHYS	0343
ics. Spatial Light Modulators and	Optical Memories Based on Bacteriorhodopsin.=	BIOT	0077
Optical-Phased Arrays.= New	Optical Memory Architectures Based on	BIOT	0025
Films.+ Photoamplified Storage	Optical Memory with Polyion Complexed Pyrene LB	BIOT	0144
hods in Polymer Characterization:	Optical Methods. Dynamic Infr+ Instrumental Met	POLY	0303
hods in Polymer Characterization:	Optical Methods. FT-IR and F+ Instrumental Met	POLY	0330
hods in Polymer Characterization:	Optical Methods. Polystyrene-P+ Instrumental Met	POLY	0277
hods in Polymer Characterization:	Optical Methods. Static and D+ Instrumental Met	POLY	0250
Scanning Electron Microscopy and	Optical Microscopy.= +ion of Municipal Waste by	FUEL	0137
Exciton Coherence-Size and	Optical Nonlinearities in Restricted Geometries.=	PHYS	0038
res Exhibiting Large and Unusual	Optical Nonlinearities.= +-Based NLO Chromopho	POLY	0217
olym+ Second-Order Nonlinear	Optical Onset and Decay and Their Relationship to P	POLY	0152
oss-Linking in Vitrified Nonlinear	Optical Polymer Films.= +dering and Chemical Cr	POLY	0206

Electrical Properties of Nonlinear	Optical Polymers and Devices.= +mism Studies of	POLY	0150
+ain and Main-Chain Nonlinear	Optical Polymers with Sulfonyl Acceptor Groups.=	POLY	0105
Photonics.=	Nonlinear Optical Polymers: Opportunities and Challenges in	BIOT	0174
s of Photocrosslinkable Nonlinear	Optical Polymers.= +and Synthesis of a New Clas	POLY	0131
-Energy Behavior of Heavy-Ion	Optical Potential.= Dispersion Relation and the Low	NUCL	0040
zation of Third-Order Nonlinear	Optical Properties for Conjugated, π-Electron Molec	BIOT	0177
Polymers with Large Third-Order	Optical Properties for Photonic Switching.= +g of	POLY	0011
lymers in+ Linear and Nonlinear	Optical Properties of Highly Ordered Conjugated Po	POLY	0008
elated States and Nonlinear	Optical Properties of Linear Chains.= Electron-Corr	POLY	0007
ndant Sulfonyl Tola+ Nonlinear	Optical Properties of Linear Epoxy Polymers with Pe	POLY	0208
Composites.= Nonlinear	Optical Properties of Polymer-Silver Microsphere	POLY	0154
rene) and Poly(arylate) and Their	Optical Properties.= +lock Copolymers of Poly(sty	POLY	0220
pment and Phase-Controlled	Optical Pulses.= Molecular Geometric Phase Develo	PHYS	0067
Induced Morphology in Blends by	Optical Rheometry.= +Steady and Dynamic Shear-	PMSE	0085
d Order at Surfaces as Studied by	Optical Second Harmonic Generation.= +tation, an	ANYL	0024
ane Channels.=	Optical Spectroscopic Techniques for Ions in Membr	BIOT	0141
ations of Magnetic Resonance and	Optical Spectroscopies. +Developments and Applic	PHYS	0136
ations of Magnetic Resonance and	Optical Spectroscopies. +Developments and Applic	PHYS	0193
ations of Magnetic Resonance and	Optical Spectroscopies. +Developments and Applic	PHYS	0113
ations of Magnetic Resonance and	Optical Spectroscopies. +Developments and Applic	PHYS	0372
ations of Magnetic Resonance and	Optical Spectroscopies. +Developments and Applic	PHYS	0066
ations of Magnetic Resonance and	Optical Spectroscopies. +Developments and Applic	PHYS	0166
ations of Magnetic Resonance and	Optical Spectroscopies. +Developments and Applic	PHYS	0040
ations of Magnetic Resonance and	Optical Spectroscopies. +Developments and Applic	PHYS	0370
ations of Magnetic Resonance and	Optical Spectroscopies. +Developments and Applic	PHYS	0373
ations of Magnetic Resonance and	Optical Spectroscopies. +Developments and Applic	PHYS	0344
	Optical Spectroscopy at 1-MHz Resolution.=	PHYS	0043
Reaction Center.=	Optical Spectroscopy of the Bacterial Photosynthetic	PHYS	0170
Picosecond Nonlinear	Optical Studies of Semiconductor-Liquid Interfaces.=	PHYS	0283
ural Dynamics of He+ Ultrafast	Optical Studies of Vibrational Relaxation and Struct	PHYS	0169
eposition of a Nonlinear Interface	Optical Switch.= +ring by Laser Chemical Vapor D	BIOT	0024
inesin Molecules Using the Laser	Optical Trap.= +suring the Isometric Tension of K	BIOL	0137
Architectures Based on	Optical-Phased Arrays.= New Optical Memory	BIOT	0025
maging Subsurface and Nonequi+	Optically Excited Scanning Tunneling Microscopy: I	PHYS	0208
e and Active Devices.=	Optically Nonlinear Polymers in Waveguiding Passiv	POLY	0177
: A Novel, Facile Route to Highly	Optically Pure Glycidol Derivatives.= +le Exchange	ORGN	0207
ed Device Detection.= Unique	Optics for Raman Spectroscopy Using Charge-Coupl	ANYL	0035
Guided-Wave Nonlinear	Optics in DCANP Langmuir-Blodgett Films.=	POLY	0176
Rheo-	Optics of Flow-Induced Phase Separation.=	PMSE	0106
ing and Proces+ Cubic Nonlinear	Optics of Polymer Thin Films, I: Molecular Engineer	POLY	0011
tionships in Ri+ Cubic Nonlinear	Optics of Polymer Thin Films, 2: Structure-$\chi^{(3)}$ Rela	POLY	0205
tionships in C+ Cubic Nonlinear	Optics of Polymer Thin Films, 3: Structure-$\chi^{(3)}$ Rela	POLY	0226
tionships in P+ Cubic Nonlinear	Optics of Polymer Thin Films, 4: Sturcture-$\chi^{(3)}$ Rela	POLY	0221
sion of the $\chi^{(3)}$+ Cubic Nonlinear	Optics of Polymer Thin Films, 5: Wavelength Disper	POLY	0213
etallic Complexes for Nonlinear	Optics: Cobaltous Complexes of Unsymmetrical Hyd	POLY	0102
Matrices for Quadratic Nonlinear	Optics.= +ntation of Organic Molecules in Sol-Gel	POLY	0006
to Materials Science in Integrated	Optics.= +Thin Polymer Films: From Chemistry	COLL	0070
cations. Introduction to Nonlinear	Optics.= +terials for Photonic and Optical Appli	POLY	0001
Third-Order Nonlinear	Optics.= Group IV Transition-Metal Complexes for	INOR	0334
ials for Quadratic Nonlinear	Optics.= Investigation of New Molecules and Mater	POLY	0103
Polyacetylenes for Nonlinear	Optics.= Toward High Susceptibilities in Soluble	POLY	0132
periment Introducin+ Attaining	Optimal Conditions: An Advanced Undergraduate Ex	CHED	0182
perties of Alkanes.=	Optimal Graph Theoretical Models for Enthalpic Pro	COMP	0034
bic Biodegradation of Phenolics in	Optimally Designed Sequencing Batch Reactors.=	BIOT	0138
macion Cuadratica Sucesiva y u+	Optimizacion de Plantas Quimicas Utilizando Progra	I&EC	0072
Hydrolysis of Lignocellulosic Bi+	Optimization of Cellulases Composition for Efficient	BIOT	0182
c Methods for Polymer Analysis.	Optimization of Channel Dimensions Used in Sedime	PMSE	0028
eric Antibodies in CHO Cells +	Optimization of Medium for the Production of Chim	BIOT	0165
ons in the Detection o+ Simplex	Optimization of Post-Column Derivatization Conditi	AGFD	0111
Gels.=	Optimization of Protein Chromatography on Agarose	ANYL	0030
olites and Pillared Clay Synthesis.	Optimization of the Pillaring of a Saponite.= +Ze	PETR	0142
erties for Conjugated, π-Electr+	Optimization of Third-Order Nonlinear Optical Prop	BIOT	0177
Animal Cell Bioreactor Design and	Optimization. Mammalian Cell Retention in a S+	BIOT	0007
oducing Experimental Design and	Optimization.= +d Undergraduate Experiment Intr	CHED	0182
ing To Aid in Process Design and	Optimization.= +nalytical Techniques in Bioprocess	BIOT	0074
s or Vegetable Oils: Formulation,	Optimization, and Determination of the Structure.=	COLL	0083
	Optimized Basis Sets for ab Initio Calculations.=	PHYS	0297
in C in Cephalosporium acrem+	Optimized Conversion of Penicillin N to Cephalospor	BIOT	0085
nment of Molecular Structures:	Optimizing the Overlap of Atom-Based Properties, in	CINF	0018
Option. Chemistry Education: An	Option for Adoption.= +CPT-Chemical Education	CHED	0066
rs: The CPT-Chemical Education	Option. Chemistry Educ+ Educating Future Teache	CHED	0066
idized Access to Beilstein Online:	Options for Search Services in an Academic Library.	COMP	0012
Career	Options in Consulting.=	CMEC	0029
Gel Processing of	Opto/Electroactive Polymers.=	POLY	0200
: Effect of Lipophilicity, Dosing+	Oral Absorption of Di- and Tri-Peptidic Compounds	MEDI	0196
olo[1,5-a]Pyrimidines as Potent	Orally Active Angiotensin II Receptor Antagonists.=	MEDI	0104
ors: Synthesis of a Subnanomolar,	Orally Active Cysteine-Derived Inhibitor.= +nhibit	MEDI	0122
Inhibitors.= Highly Potent,	Orally Active Diester Macrocyclic Human Renin	MEDI	0123
ent of L-670,630 as a Potent and	Orally Active Inhibitor of 5-Lipoxygenase.= +elopm	MEDI	0180
r.= A-69412, a Long-Lived,	Orally Active, and Selective 5-Lipoxygenase Inhibito	MEDI	0176

Receptor Antagonists.= Potent,	Orally Active, 1,5-Naphhyridine Angiotensin II	MEDI	0102
Ultraviole+ Inhibitory Effect of	Orally Administered Green Tea on Carcinogenesis by	AGFD	0065
ebittering and Deacidifying Sour	Orange (Citrus aurantium) Juice Using Neutral and	AGFD	0106
n of a Ferrocene-Linked Acridine	Orange Compound with DNA.= +of the Interactio	ANYL	0109
Scientists Studying Agent	Orange: A Case Study.= Political Consequences for	PROF	0008
face R+ Semiempirical Molecular	Orbital and Tight Binding Methods in Diamond Sur	PHYS	0356
rogen Bonding: The Natural Bond	Orbital Donor-Acceptor Perspective.= +cts in Hyd	PHYS	0106
ntramolecular Hydro+ Molecular	Orbital Estimation of the Activation Enthalpies for I	ORGN	0175
tio and Semiempirical Molecular	Orbital Studies of Hydroxamic Acids and Their Conj	PHYS	0326
Molecular	Orbital Study.= Barriers to Rotation in Oximes: A	PHYS	0324
periment in Spectro+ Molecular	Orbitals of Metalloporphyrins: An Undergraduate Ex	PHYS	0327
C.= Obtencion del	Orden de Reaccion de la Oxidacion de la Vitamina	CHED	0164
nc Iron, and Total Sulfur in Zinc	Ore Concentrates.= +rections in XRF Analysis of Zi	ANYL	0057
echanical Properties of Magnetite	Ore Pellets.= +ct of Manganese Additions in the M	I&EC	0067
ent and pH Effects on Sorption of	Organic Acids from Methanol-Water Mixtures.=	ENVR	0014
s.= Energy Transfer from	Organic Adsorbates to Defect Sites in Silica Support	PHYS	0135
ormaldehyde Gels as Precursors of	Organic Aerogels.= +tering Studies of Resorcinol F	POLY	0199
ctions by M+ Characterization of	Organic and Aqueous Guest/Host Solid-State Intera	AGRO	0062
	Organic Approaches to Serotonin Analogs.=	ORGN	0254
d Metal(IV) Phosphates toward	Organic Bases, Metal Complexes, and Metal Oxides.=	PETR	0107
s+ Potential Impact of Dissolved	Organic Carbon (DOC) Flux out of Margin Sediment	GEOC	0017
earn from Reactions wit+ Ionic	Organic Chemistry in the Gas Phase: what Can We L	ORGN	0121
cyclic Seri+ Didactic Puzzles for	Organic Chemistry Nomenclature Teaching, Part I: A	CHED	0165
Compounds.= Comparative	Organic Chemistry of Carbonyl and Organometallic	CHED	0097
puter Modeling in Undergraduate	Organic Chemistry.= +ation. Use of Personal Com	CHED	0188
alkynes: A New Building Block in	Organic Chemistry.= +nal Investigation of Phospha	COMP	0060
Graphics in Teaching	Organic Chemistry.= Experience Using Molecular	CHED	0189
l Containers.=	Organic Coatings for Beer, Beverage, and Food Meta	PMSE	0140
Thermodynamic Properties of	Organic Compounds and Mixtures.= Estimating	COMP	0054
on Techniques for the Analysis of	Organic Compounds by Ion Chromatography.= +ati	TECH	0015
mal Boiling Points of Heterocyclic	Organic Compounds from Molecular Structure.=	COMP	0015
of Chromatographic Retention of	Organic Compounds from Molecular Structure.=	ANYL	0099
easuring Trace Levels of Volatile	Organic Compounds in Human Blood by Using Purg	CEI	0001
of the Solubility of Hydrophobic	Organic Compounds in Nonideal Solvent Mixtures.=	ENVR	0030
missions and Urban Nonmethane	Organic Compounds in the Tulsa Area.= +Oilfield E	ENVR	0022
omplexes.=+ Transformations of	Organic Compounds Mediated by Transition Metal C	ORGN	0127
ticle Partitioning of Semivolatile	Organic Compounds to Model Atmospheric Particula	ENVR	0059
A New Classification of	Organic Compounds.=	CHED	0167
matography to QSAR Modeling of	Organic Compounds.= +on of Micellar Liquid Chro	ANYL	0046
spheric Deposition of Semivolatile	Organic Compounds.= +lation Model of the Atmo	GEOC	0004
Unimolecular Transformations of	Organic Compounds.= +t on Activation Barriers of	COLL	0107
Effects of Water-Miscible	Organic Cosolvents on Kinetics of Plasmin.=	BIOL	0092
nsed Phases. Torsional Solitons in	Organic Crystals.= +sfer and Relaxation in Conde	PHYS	0129
Length Variations in	Organic Crystals.= Graph Theory Applied to Bond	COMP	0031
=	Organic Geochemical Studies of Peat from Volo Bog.	I&EC	0016
New Approach to Inorganic-	Organic Hybrid Materials?.= Cement Chemistry: A	POLY	0261
ach in the General, Inorganic, and	Organic Instructional Laboratories.= +oscale Appro	CHED	0150
on Relationships: A Physical	Organic Laboratory Experiment.= Structure-Functi	CHED	0212
ylacetonate Complexes for Cop+	Organic Ligand-Stabiilized Copper(I) Hexafluoroacet	COLL	0156
and Study of Superconductivity in	Organic Materials.= +of Pressure in the Creation	PHYS	0025
Nonlinear Optical Devices Using	Organic Materials.=	BIOT	0176
Properties of	Organic Materials.= Size-Dependent Electrical	BIOT	0252
S+ Characterization of Insoluble	Organic Matter Associated with Clay Minerals from	I&EC	0021
Remineralization of Sedimentary	Organic Matter in Continental Margin Ecosystems.=	GEOC	0015
f Organic Solutes on Sedimentary	Organic Matter.= +e for Measuring the Sorption o	ANYL	0072
Research on	Organic Molecules for Electronics Using STM.=	BIOT	0204
Nonlinear Optic+ Orientation of	Organic Molecules in Sol-Gel Matrices for Quadratic	POLY	0006
Atomic Coordinates for	Organic Molecules.= Automatic Generation of 3D	CINF	0036
Rational Design of	Organic NLO Materials.=	POLY	0002
Consumer Perspective of	Organic or Naturally Occurring Fertilizer Claims.=	FERT	0007
rgonne Premium Coal Samples.	Organic Oxygen Contents of Argonne Premium Coal	FUEL	0003
es: Chemistry at the Liquid-Li+	Organic Photochemistry in Restricted Reaction Spac	ORGN	0126
Advances in	Organic Photochemistry.=	ORGN	0142
Wavelength-Dependent	Organic Photochemistry.=	ORGN	0135
estigation of Radical Processes in	Organic Polycrystals under Shear Deformation at Hi	PHYS	0245
Behaviors Mediated by the	Organic Polymer Matrices.= Inorganic Phase	POLY	0364
Reaction.= An	Organic Radical in the Lysine 2,3-Aminomutase	BIOL	0077
Influence of Electric Fields on	Organic Reactions at Zeolite-Supported, Nanometer-	CATL	0018
+e for Measuring the Sorption of	Organic Solutes on Sedimentary Organic Matter.=	ANYL	0072
imental Evidence on Retention of	Organic Solutes.= +bents: Critical Review of Exper	ANYL	0071
th Silica Surface: Influence of the	Organic Solvents on the Reaction Rate.= +PCl5 wi	COLL	0097
on of Immobilized Biocatalysts in	Organic Solvents.= +d Opportunities. The Operati	BIOT	0035
paration and Purification in Neat	Organic Solvents.= +ous Bioprocessing: Protein Se	BIOT	0063
Enhance Enzyme Activity in	Organic Solvents.= Random Mutagenesis to	BIOT	0068
reasing Enzyme Stability in Polar	Organic Solvents.= +on: A General Strategy for Inc	BIOT	0244
Dibenzylidene Sorbitol and	Organic Solvents.= Thermoreversible Gels from	POLY	0201
Complexes and Photoxidations of	Organic Substrates.= +edox Reactivity of Copper	INOR	0342
rin+ Transformation Kinetics of	Organic Sulfur Forms in Argonne Premium Coals du	FUEL	0059
ure TPR Apparatus to Investigate	Organic Sulfur Forms in Coals.= +els. High-Press	FUEL	0007
Reactivity of Oxidized	Organic Sulfur Forms in Coals.=	FUEL	0057

Physical Propertie+ Overview of	Organic Superconductors: Syntheses, Structures, and	PHYS 0024
Physical Properties of the	Organic Superconductors.=	PHYS 0002
Integrated System for	Organic Synthesis Design.= AIPHOS—An	CINF 0005
omplexes.=	Organic Synthesis Using Group 4 Transition Metal C	ORGN 0134
y, and N–Bloc+ Phase–Managed	Organic Synthesis: Activation of α–Chloro, α–Hydrox	ORGN 0259
re 1,2-Epoxy Alde+ Enzymes in	Organic Synthesis: Synthesis of Enantiomerically Pu	ORGN 0052
Enzymes in	Organic Synthesis.=	ORGN 0271
Photochemistry in	Organic Synthesis.= Homogeneous Metal–Catalyzed	INOR 0339
echanisms and Synthesis.=	Organic Transformattions Involving Organocobalt: M	ORGN 0141
nning Tunnel+ Progress Toward	Organic Unimolecular Rectifiers: Experiments by Sca	BIOT 0005
plications—I+ Biodegradation of	Organic Wastes: Molecular Approaches and Field Ap	BIOT 0117
plications—II+ Biodegradation of	Organic Wastes: Molecular Approaches and Field Ap	BIOT 0137
ical and Isotopic Alteration of	Organic Water during Stepped Combustion.= Chem	ANYL 0061
l Routes into Nonshrin+ Inverse	Organic–Inorganic Composite Materials: Free Radica	POLY 0259
Gel Method: Some Insights into+	Organic–Inorganic Composites Prepared via the Sol–	POLY 0260
A Molecular	Organic–Inorganic Semiinterpenetrating Network.=	POLY 0286
le+ Electrochemical Formation of	Organic–Pigment Thin Films by Disruption of Micel	COLL 0019
thesis of Carbohyd+ Enzymes in	Organic/Bioorganic Synthesis. Keynote Address. Syn	BIOT 0089
Oil Shale Ke+ Determination of	Organic/Inorganic Associations of Trace Elements in	FUEL 0045
Analyzing Exotic	Organics in Mixed Nuclear Wastes: The Challenges.=	ANYL 0060
zed Oxidations of Environmental	Organics Using AM1 Simulated Sunlight Irradiation	INOR 0318
s and Their Complexes in Living	Organisms as Possible Candidates for Principles of S	BIOT 0148
Ectothermic	Organisms.= Proteolytic Enzymes from	BIOT 0194
d Reactive Processes on Molecular	Organizates and Colloidal Catalysts.= +–Controlle	PHYS 0063
Structural	Organization of Bipolar Lipids.=	BIOT 0122
Structural and Functional	Organization of Cellulomonas fimi Cellulases.=	BIOT 0207
Real–Space Imaging of Molecular	Organization: Naphthalene vs. Azulene on Pt(111)+	PHYS 0125
ED.=	Organizational Structure of Dayton, Ohio, Project SE	CHED 0142
Candidates for Principles of Self–	Organized Biocomputers.= +Organisms as Possible	BIOT 0148
drophobic Binding Sites and Self–	Organizing Structures.= +s for Construction of Hy	ORGN 0066
–ray Crystallographic Studies of	Organoaluminum Adducts of (Dimethylaminomethyl	INOR 0274
Syntheses of Some	Organoboron Compounds as Flame Retardants.=	ORGN 0262
Propargylic Alcohols.= New	Organoboron Reagent for the Preparation of	ORGN 0109
Synthesis of New Materials:	Organoceramics.=	POLY 0289
Disorder in Persons Exposed to	Organochlorines.= Measurement of Immune	CEI 0017
Transformattions Involving	Organocobalt: Mechanisms and Synthesis.= Organic	ORGN 0141
Reversed Polarity Alternative to	Organocuprate Conjugate Addition.= A Practical	ORGN 0091
Stereoselective S$_N$2' Addition of	Organocuprate Reagents to Alkynyl Oxiranes: Synt	ORGN 0011
ermination of Local Structure in	Organofluorides Using Fluorine–19 Carbon–13 Coupl	FUEL 0077
Gelation of Polymers. Surfactant	Organogels Studied by Small–Angle Neutron Scatter	POLY 0169
haracterization, and Properties of	Organoindium Precursors for OMVPE.= +thesis, C	INOR 0218
Nitrone.= Addition of	Organolithium Compounds to N–THP–Protected	ORGN 0227
r Routes to Inorganic Materials.	Organometallic Chemical Vapor Deposition of ZnSe:	INOR 0171
Characterization of+ Surface	Organometallic Chemistry on Zeolites—Synthesis and	PETR 0131
erature Chloroaluminate Molte+	Organometallic Chemistry Performed in Room–Temp	ANYL 0097
Transition–Metal Interactions in	Organometallic Chemistry: Syntheses and Structures	INOR 0332
gnitude of α–Agostic and Steric+	Organometallic Compounds. Investigations the Ma	INOR 0147
from N$_2$O to Alkynes Coordinat+	Organometallic Compounds. Oxygen–Atom Transfer	INOR 0327
of Some 1,1'–bis(Disphenylpho+	Organometallic Compounds. Structural Relationship	INOR 0106
Chemistry of Carbonyl and	Organometallic Compounds.= Comparative Organic	CHED 0097
stry, and Spectroscopy of Cationic	Organometallic Compounds.= +ivity, Electrochemi	INOR 0320
nd Properties of Arsino–Group 12	Organometallic Compounds.= +Characterization, a	INOR 0219
Spectroscopic Properties of	Organometallic Cyanines.= Synthesis and	INOR 0335
2)]+: A Strong	Organometallic Electrophile.= [cpRu(dppm)(CH$_2$SO	INOR 0249
nts.= Bioinorganic Chemistry.	Organometallic Functionalities at Sulfur–Bound Nick	INOR 0177
Photochemical Reactions of	Organometallic Molecules on Surfaces.=	INOR 0168
Cage Effects in	Organometallic Photochemistry.=	INOR 0158
sensitive Metalorganic Systems.	Organometallic Photoinitiators in Photopolymerizatio	INOR 0338
ductor Clusters.=	Organometallic Precursors for Soluble III–V Semicon	INOR 0313
owth of Electronic Materials from	Organometallic Precursors. +dvanced Materials: Gr	COLL 0064
owth of Electronic Materials from	Organometallic Precursors. +dvanced Materials: Gr	COLL 0153
owth of Electronic Materials from	Organometallic Precursors. +dvanced Materials: Gr	COLL 0025
owth of Electronic Materials from	Organometallic Precursors. +dvanced Materials: Gr	COLL 0006
owth of Electronic Materials from	Organometallic Precursors. +dvanced Materials: Gr	COLL 0044
owth of Electronic Materials from	Organometallic Precursors. +dvanced Materials: Gr	COLL 0135
Transition–Metal Thin Films from	Organometallic Precursors.= +ials. Deposition of	INOR 0141
ctron Transfer to Photogenerated	Organometallic Radicals.= +um Metal Films by Ele	INOR 0319
Polymer Synthesis with	Organometallic Reagents.=	POLY 0328
Semiconductors.= Some New	Organometallic Routes to III/V Compound	INOR 0146
(Diphenylphosphino)methane in	Organometallic Synthesis.= New Applications of Bis	INOR 0375
urface Reactions of Ge–Containing	Organometallics on Si(100).= +tallic Precursors. S	COLL 0044
on by Convention+ Evaluation of	Organophosphate Insecticide Hydrolytic Detoxificati	AGRO 0124
Methodology Based on	Organophosphorus Auxiliaries.= New Synthetic	INOR 0377
to the Development of Functional	Organophosphorus Compounds.= +ent Approaches	INOR 0348
Photochemical Reactions of	Organophosphorus Systems.= Free–Radical and	INOR 0347
New Reactions of	Organorhenium Compounds.=	ORGN 0130
Benzoquinodimethanes Using	Organoruthenium Groups.= Stabilization of	INOR 0051
d Isoprene by Single–Component	Organoscandium Systems: Synthesis, Characterizatio	INOR 0280
s with Tung+ Rearrangements of	Organosilanes to Form Metal–Ligand Multiple Bond	INOR 0115
Network Products from Lignin,	Organostannanes, and Hydroxyl–Capped Poly(Ethyle	PMSE 0006

e Organotin-Platinum-Imine and	Organotin-Platinum-Amine Complexes.= +ly Activ	INOR 0234
cterization of Biologically Active	Organotin-Platinum-Imine and Organotin-Platinum	INOR 0234
dia: A Strategy for Electric-Field-	Orientable Microparticles.= +in Fluorocarbon Me	COLL 0164
ptical Materials.=	Orientation and Decay in Poled Polymer Nonlinear O	POLY 0151
ugated Polymers in Polyethylene:	Orientation by Mesoepitaxy.= +ighly Ordered Conj	POLY 0008
Study of Surface	Orientation for the Design of Molecular Electronics.=	BIOT 0053
er- Lattice Model for Segmental	Orientation in Deformed Polymer Networks: Charact	COLL 0016
Internal Electric Field and Dye	Orientation in PVDF/PMMA Blend.=	POLY 0225
Structure and	Orientation of Aqueous Surfactant Monolayers.=	PMSE 0154
erric Porphyrin+ Rotation and	Orientation of Coordinated Imidazoles in Low-Spin F	INOR 0296
Quantitative Study of Molecular	Orientation of Hemicyanine Langmuir-Blodgett Fi	POLY 0224
otonic and Optical Applications.	Orientation of Organic Molecules in Sol-Gel Matrice	POLY 0006
ehydrase: Determination of the	Orientation of the Active-Site Histidine Relative to S	BIOL 0066
Performance Materials f+ Novel	Orientation Techniques for the Preparation of High-	COLL 0015
tical Second Harm+ Adsorption,	Orientation, and Order at Surfaces as Studied by Op	ANYL 0024
A Tool-	Oriented Approach to Computing.=	CHED 0078
Microdeformational Behavior of	Oriented Bicomponent Polymer Systems.=	POLY 0101
ctral Behavior of Dye Molecules	Oriented by Chiral Nematic (Acetyl)(ethyl)cellulose S	POLY 0331
matographic Data Acq+ Object-	Oriented Data-Handling System for Spectral or Chro	COMP 0056
orat+ Computer in Techniques-	Oriented Experiments in the Physical Chemistry Lab	CHED 0121
uum Deposition Pol+ Uniaxially	Oriented Polyamide Films in Plane Prepared by Vac	POLY 0082
of a New Gallophosphate of the+	Orienting Effect of Fluoride Anions in the Synthesis	PETR 0075
oxypropanamines as Inhibitors of	Ornithine Decarboxylase: Synthesis and Biological A	MEDI 0184
at.on in N+ Combined Directed	ortho Metalation and Cross-Coupling Tactics: Applic	ORGN 0277
le New Metalation Di+ Directed	ortho Metalation: The O-Thiocarbamate as a Versati	ORGN 0083
Efficient Synthesis of	ortho-Isotoluenes and Diene-Allenes.=	ORGN 0098
cle-Imide Copolymers Containing	ortho-Phenylene Rings.= +ion of Ordered Heterocy	POLY 0042
Generation of	ortho-Quinone Methides from Cyclobutenones.=	ORGN 0229
New Macrocyclic	Orthophenylene-Stabilized Triphosphazanes.=	INOR 0391
Crystalline Chloro-Trisodium	Orthophosphate.=	INOR 0345
cular Correlation Involved in the	Orthorhombic Pseudohexagonal Rotator-Phase Tran	POLY 0354
l π2-Dihydrogen Complexes: Fe <	Os < Ru.= +ve Order of pK_a Values of Isostructura	INOR 0380
ltammetry for I+ Process of the	Oscillation of Membrane Potential Studied by the Vo	BIOT 0105
Health and Safety Issues under	OSHA and EPA Programs.=	CHAS 0022
SHA's Role in Chemical Proces+	OSHA-EPA Interface in Health and Safety Issues. O	CHAS 0012
The	OSHA-EPA Interface: An Overview.=	CHAS 0014
rface in Health and Safety Issues.	OSHA's Role in Chemical Process Safety.= +Inte	CHAS 0012
onyl Cluste+ Ruthenaborane and	Osmaborane Clusters: Ruthenium and Osmium Carb	INOR 0220
Complexes of	Osmium and Ruthenium.= Chemistry of Arylimido	INOR 0259
borane Clusters: Ruthenium and	Osmium Carbonyl Clusters Containing One or More	INOR 0220
Synthesis and Reactivity of	Osmium Porphyrin Complexes.=	INOR 0299
tion of Sulfides to Sulf+ A Mild	Osmium Tetraoxide-Catalyzed Process for the Oxida	ORGN 0244
Reactivity of Monomeric	Osmium(II) Imido Complexes.= Preparation and	INOR 0364
hesis of an Isocyanide Complex of	Osmium(KV) by Ceric Oxidation of Osmocene.=	INOR 0198
and Imido Complexes of	Osmium(VI).= Reactivity Studies of New Nitrido	ORGN 0219
smium(KV) by Ceric Oxidation of	Osmocene.= +hesis of an Isocyanide Complex of O	INOR 0198
rd Symposium—I, Honoring R. A.	Osteryoung. +eous Media.= Electrochemistry Awa	ANYL 0093
as of Concern in New York State:	Oswego Harbor and the St. +Initiatives at Two Are	ENVR 0084
Pesticides in the 21st Century.	Outlook for Pesticides in Developing and Developed	AGRO 0056
The MIT Student's	Outreach Program.=	CONG 0008
cati+ Prehigh School Chemistry	Outreach Programs of the Institute for Chemical Edu	CHED 0073
Microwave	Oven.= Bischler-Napieralski Reaction under	ORGN 0195
Contribution of Released Cells to	Overall Cyclosporin A Productivity in an Immobiliz	BIOT 0155
Laminar Jet To Determine the	Overall Rate of Liquid-Liquid Reactions.= Using a	I&EC 0007
Dendrimers That	Overcome Steric Inhibition.= Synthesis of Rigid	POLY 0321
ecular Structures: Optimizing the	Overlap of Atom-Based Properties, in Particular the	CINF 0018
Characterization of Iodine	Overlayers on Pt(111).= Surface Spectroscopic	PHYS 0366
vel Techniques. Ultrathin Copper	Overlayers on $TiO_2(110)$.= +try of Nonmetals—No	COLL 0191
neous Hybridoma Cultures for the	Overproduction of Monoclonal Antibodies.= +eroge	BIOT 0169
ting Nitromethane in So+ High-	Overtone Resonance Raman Spectra of Photodissocia	PHYS 0302
emistry with Experiment. High-	Overtone Spectra: The Significance of Perturbations.	PHYS 0029
ene Ether Phosphine Oxide)s—I.	Overview of Chemistry, Thermal Stability, and Oxyg	PMSE 0052
d Their Residual Analy+ Current	Overview of Feed Additives and Veterinary Drugs an	AGFD 0102
and Personal Care Industry—I.	Overview of Fragrance Chemistry and Application.=	CHED 0046
Scientist as an Expert Witness.	Overview of Legal Decision-Making Process and the	CHAL 0011
roperties.= Superconductivity.	Overview of Organic Superconductors: Syntheses, Str	PHYS 0024
Project SEED.	Overview of Project SEED.=	CHED 0035
sign for Stability and Specificity.	Overview of Protein Chemical Cross-Linking: Implic	BIOT 0216
luorocarbon Alternatives.=	Overview of the Commercial Development of Chlorof	I&EC 0044
s Listing of Sites for Superfund?	Overview of the Structure of the Hazard-Ranking Sy	ENVR 0077
Carbon-Carbon Composites: An	Overview of Their Formation, Structure, and Propert	FUEL 0051
eus Collision Mechanisms.	Overview. To Fuse or Not To Fuse.= Nucleus-Nucl	NUCL 0004
udents. Chemistry in Context: An	Overview.= +Chemistry Course for Nonscience St	CHED 0051
Resins in Container Coatings: An	Overview.= +enary Lecture. Application of Epoxy	PMSE 0157
enetrating Polymer Networks: An	Overview.= +enetrating Polymer Networks. Interp	PMSE 0039
ensitive Metalorganic Systems: An	Overview.= +sitive Metalorganic Systems. Photos	INOR 0166
gies for Pesticide Waste Disposal:	Overview.= +Their Containers. Current Technolo	AGRO 0120
aste Forms. Use and Potential: An	Overview.= +ation Techniques for Cementitious W	ENVR 0007
The OSHA-EPA Interface: An	Overview.=	CHAS 0014
An	Overview.= Utilization of Coal Gasification Slag:	FUEL 0134

ew Technologies on Information	Ownership and Copyright: Are Your Policies Inviolat	CINF	0043
Dilemma.= Information	Ownership vs. Unrestricted Access: The Library	CINF	0045
terial Activi+ Synthesis of [1,3,4]	Oxa(Thia)Diazolo[3,2-b]Indazoles and Their Antibac	MEDI	0056
9-Oxabicyclononanes and Related	Oxaadamantanes.= +ride Reduction of 2,6-Dihalo-	ORGN	0246
Interphenylene 7-	Oxabicycloheptane TXA_2 Antagonists.=	MEDI	0160
ride Reduction of 2,6-Dihalo-9-	Oxabicyclononanes and Related Oxaadamantanes.=	ORGN	0246
Imidazopyridine Esters and	Oxadiazoles.= Muscarinic Activity of	MEDI	0139
trophotometric Determination of	Oxalate: A Student Experiment in Analytical/Physic	CHED	0171
anum by Oxine in the Presence of	Oxalic Acid as Complexing Agent.= +tion of Lanth	ANYL	0054
ns of Four- and five-Membered	Oxametallacycles in the Rections of the Electron-rich	INOR	0086
a Chiral 2-oxo-2-Propionyl-1,3,2-	Oxazaphosphorinane.= +is and Aldol Reactions of	ORGN	0050
Anti-Inflammatory Activity of	Oxazinone, BF-389.= Synthesis and Novel	MEDI	0175
on and Reaction of Dilithio-2,4-	Oxazolidinedione with α-Haloketones: A Versatile Sy	ORGN	0223
Steroidal	Oxazolines.= Synthesis and Reactions of	ORGN	0178
de Reaccion de la	Oxidacion de la Vitamina C.= Obtencion del Orden	CHED	0164
Carcinogen-Mediated	Oxidant Formation and Oxidative DNA Damage.=	CHAS	0008
Inhibition of Polyphenol	Oxidase Activity by Phenolic Compounds.=	AGFD	0152
Sensing.= Glucose	Oxidase Immobilized with Polyurethane for Glucose	BIOT	0147
Enhanced Monoamine	Oxidase Inhibitory Activity of Tranylcypromine.=	MEDI	0068
by+ Separation of the Glucose	Oxidase Isoenzymes from Penicillium Amagasakiense	ANYL	0016
in Cytochrome c	Oxidase.= Ligation Dynamics of Carbon Monoxide	BIOL	0096
Comparison between Selective	Oxidation and Degradation of Ethylene and Propylen	CATL	0058
Processes on Lipid	Oxidation and Flavor Stability.= Influence of Food	AGFD	0157
matotropin.=	Oxidation and Naturation of Recombinant Bovine So	BIOT	0135
anol on $Cu_2O(100)$.=	Oxidation and Reduction of Allyl Alcohol and 1-Prop	COLL	0195
Investigation of Alkene	Oxidation by Technetium(VII)-Oxo Complexes.=	INOR	0328
Models+ Selectivity in Catalysis:	Oxidation Catalysis. Reaction-Transport Selectivity	CATL	0056
ismuth Molybdate Hydrocarbon	Oxidation Catalysts Prepared by the High-Temperat	CATL	0026
c Study of New Asymmetric	Oxidation Catalysts.= Development and Mechanisti	ORGN	0132
annel Cat+ Acceleration of Lipid	Oxidation during Cooking of Refrigerated Minced Ch	AGFD	0124
oods. Sensory Evaluation of Lipid	Oxidation in Different Foods.= +id Oxidation in F	AGFD	0156
Lipid	Oxidation in Foods via Redox Iron.=	AGFD	0094
Lipid Oxidation.=	Oxidation in Foods. Maillard Reaction Products and	AGFD	0119
Processes in Foods.= Lipid	Oxidation in Foods. Mechanisms of Lipid Oxidative	AGFD	0093
dation in Different Foods+ Lipid	Oxidation in Foods. Sensory Evaluation of Lipid Oxi	AGFD	0156
Role of Lipoxygenases in Lipid	Oxidation in Foods.=	AGFD	0122
ee Composi+ Prevention of Lipid	Oxidation in Muscle Foods by Nitrite and Nitrite-Fr	AGFD	0121
mation Specificity of Guanosine	Oxidation in Nucleic Acids Using Metal Complexes.=	BIOL	0122
d Laminar-+ Benzene/Toluene	Oxidation Models: Studies Based on Flow-Reactor an	FUEL	0083
ms. Morphological Changes during	Oxidation of a Single Char Particle.= +Mechanis	FUEL	0025
ture, and Catalytic Activity in the	Oxidation of α-Olefins.= +nthesis, Molecular Struc	INOR	0365
ailed Chemical Scheme for the	Oxidation of Benzene-Air Mixtures.= Tentative Det	FUEL	0084
Volumes of Activation for the	Oxidation of Benzenediols by the $IrCl_6^{2-}$ Ion.=	INOR	0395
Monitoring the	Oxidation of Coals in Storage.=	FUEL	0002
Ligand Effects on the Aerobic	Oxidation of Cyclohexene Catalyzed by (Aqua)(Phosp	INOR	0067
l Determinants in the	Oxidation of DNA by Permanganate.= Fundamenta	BIOL	0121
Lipid	Oxidation of Fish during Storage.=	AGFD	0123
Intermediate Products from the	Oxidation of H_2S in Natural Waters.=	GEOC	0005
orted Metallic Nitrates: Induced	Oxidation of Hantzsch Esters (Dialkoxycarbonyldihid	ORGN	0245
during Oxide-Catalyzed	Oxidation of Hydrocarbons.= Oxygen Spillover	COLL	0174
pectedly High Diastereoselective	Oxidation of Locked-Conformation Styrenes: Stereos	ORGN	0236
Unique Complexation and	Oxidation of Metal-Based Clusters.=	PHYS	0173
er Oxidation.= Selective Partial	Oxidation of Methane by Catalytic Supercritical Wat	CATL	0043
hase Reactions: Combustion +	Oxidation of Nitrogen Oxides in Homogeneous Gas-P	FUEL	0127
lexes for Conformation-Specific	Oxidation of Nucleic Acids.= Tunable Nickel Comp	INOR	0183
Complex of Osmium(KV) by Ceric	Oxidation of Osmocene.= +hesis of an Isocyanide	INOR	0198
Enzymatic	Oxidation of Phenolic Compounds in Fruits.=	AGFD	0151
echanistic Study.=	Oxidation of Phosphines by Gallium Peroxides: A M	INOR	0389
tion of Radicals Derived from	Oxidation of Probucol.= Detection and Characteriza	BIOL	0099
raoxide-Catalyzed Process for the	Oxidation of Sulfides to Sulfones.= +d Osmium Tet	ORGN	0244
Facile and Selective	Oxidation of Sulfides to Sulfones.=	CARB	0033
ts on Kinetics of Peroxodisulfate	Oxidation of Ternary Iron(II)-Diimine-Cyanide Com	INOR	0022
ic Solution+ Complex Kinetics of	Oxidation of Thymol Blue with Bromate Ion in Acid	PHYS	0269
Al_2O_3/Monolith.= Catalytic	Oxidation of Trichloroethylene over a PdO on	COLL	0210
l Technique for+ Photoinitiated	Oxidation of Trichloroethylene over TiO_2: A Potentia	COLL	0082
Al_2O_3/Monolith.= Catalytic	Oxidation of Trichloroethylene Using 1.5 Pt on	COLL	0205
ydrogen Peroxide.=	Oxidation of Tris(1,10-Phenanthroline)Iron (II) by H	INOR	0222
tics and Thermochemistry of the	Oxidation of Unsaturated Radicals: n-C_4H_5 + O_2.=	FUEL	0066
	Oxidation of 1,2-Diols by Fe(VI): A Kinetic Study.=	INOR	0401
Effect of Lipid	Oxidation on Food and Oil Quality.=	AGFD	0096
Effect of Support in Propylene	Oxidation on Molybdenum Oxide.=	CATL	0059
opper Ion Catalysis of Propylene	Oxidation on X and Y Zeolites: Product Selectivity a	CATL	0017
cations of Catalysts-I. Methane	Oxidation over Alumina-Supported Noble Metal Cat	COLL	0185
n of Alkanes to Alc+ Control of	Oxidation Pathways for Selective Catalytic Conversio	CATL	0036
tion of Hemin: Catalyst of	Oxidation Processes.= Immobilization and Investiga	COLL	0102
tudies of Some Gas-Phase Metal	Oxidation Reactions by Photoelectron and Chemilele	PHYS	0352
ms.= Comparison between the	Oxidation Reactions of Alkali and Alkaline Earth Ato	PHYS	0072
r Ions.=	Oxidation Reactions of Metal and Semi-Metal Cluste	PHYS	0174
tal and T+ Characterization and	Oxidation Reactions of Si-B-C Polymers: Experimen	POLY	0187
ges in Catalyst Activity in Partial	Oxidation Reactor Dynamics.= +y States and Chan	I&EC	0069

matographic Analyses of Lipid	Oxidation Volatiles in Foods.= Update on Gas Chro	AGFD	0098
e+ Furfuryl Alcohol-to-Pyranone	Oxidation: An Approach to the Stereoselective Synth	ORGN	0294
Lipid	Oxidation: Effect on Meat Proteins.=	AGFD	0159
enyl Conversion during Benzene	Oxidation: Thermodynamic and Kinetic Mechanisitic	FUEL	0085
Reaction Products and Lipid	Oxidation.= Lipid Oxidation in Foods. Maillard	AGFD	0119
Programmed-Temperature	Oxidation.= Advances in Coal Characterization by	FUEL	0058
geneous Reactions in Metal Vapor	Oxidation.= +n between Homogeneous and Hetero	PHYS	0144
Derived from Layered Siloxene by	Oxidation.= +ion Properties of Microporous Silica	PETR	0144
by Catalytic Supercritical Water	Oxidation.= Selective Partial Oxidation of Methane	CATL	0043
romium-Containing Silica Gels+	Oxidation-Reduction Processes on the Surface of Ch	COLL	0101
for Electronic Materials.+ Low-	Oxidation-State Group 13 Compounds as Precursors	COLL	0153
ide, and+ Synthesis of High-	Oxidation-State Rhenium Hydrazine, Hydrazido, Am	INOR	0261
se-Tetraphenylpo+ Asymmetric	Oxidations Catalyzed by a D_4-Symmetrical Mangane	ORGN	0045
n Heterogeneous Photocatalyzed	Oxidations of Environmental Organics Using AM1 Si	INOR	0318
ock-Tube Method: Results for the	Oxidations of H_2 and D_2.= +Laser Photolysis-Sh	FUEL	0123
Dinuclear Copper(I) Complex e+	Oxidative Addition of Small Molecules to mono- and	INOR	0044
Molybdenum Heterobimetallics by	Oxidative Addition Pathways.= +ridged Iridium,	INOR	0385
Efficient	Oxidative Cleavage of DNA by Oxoruthenium(IV).=	INOR	0410
ninase Models.=	Oxidative Degradation of Aromatic Pollutants by Lig	INOR	0187
er Oxides.=	Oxidative Deintercalation of Silver from Ternary Silv	INOR	0275
Oxidant Formation and	Oxidative DNA Damage.= Carcinogen-Mediated	CHAS	0008
Foods. Mechanisms of Lipid	Oxidative Processes in Foods.= Lipid Oxidation in	AGFD	0093
eth+ Temperature-Programmed	Oxidative Pyroanalysis (TPOP): A New Analytical M	FUEL	0046
Repair Systems and	Oxidative Stress.=	CHAS	0009
ymers.= Use of Micelles in the	Oxidative Synthesis of Conductive-Electroactive Pol	COLL	0168
om the Laser Ablation of Yttrium	Oxide and Chloride in Oxygen.= +iluminescence fr	PHYS	0255
Tunneling Spectroscopy of Metal	Oxide Arrays on Graphite.= +eling Microscopy and	COLL	0194
Catalytic Reduction of Nitric	Oxide by Ammonia over Fe-Y Zeolites.= Selective	COLL	0212
Structure-Reactivity of Vanadium	Oxide Catalysts.= +Support and Promoters on the	CATL	0057
Supported Vanadium	Oxide Catalysts.= Molecular Engineering of	COLL	0196
tic Model of the Chromium(VI)	Oxide Clusters Formation on the Surface of Aerosil.=	COLL	0099
e Dissolution by+ Selective Iron	Oxide Extraction from Soils and Sediments: Reductiv	ENVR	0058
Transition-Metal Fluorides and	Oxide Fluorides as Selective Fluorinating Agents fo	INOR	0133
m Hydroxyapatite and Zirconium	Oxide for the Separation of Proteins.= +en Calciu	ANYL	0027
eposition of Copper and Copper(I)	Oxide from Copper tert-Butoxide.= +cal Vapor D	INOR	0143
Methacrylate Precursors to	Oxide Glasses and Ceramics.=	POLY	0288
ort Materials: The Hydrous Metal	Oxide Ion Exchangers.= +of Novel Catalyst Supp	PETR	0018
es of Poly(vinyl Acetate)/Silicon	Oxide Microcomposites Prepared via the Sol-Gel Me	POLY	0287
ostructural Evolution of a Silicon	Oxide Phase in Nafion Membranes by an in Situ Sol	POLY	0285
Ion-Exchange Reactions in	Oxide Superconducting Systems.= Intercalation and	INOR	0360
dies of Adsorbed Species on Metal	Oxide Surfaces.= +Energy-Loss Spectroscopy Stu	COLL	0173
on	Oxide Surfaces.= Local Atomic Structure of Defects	PHYS	0207
Inhibition of Inducible Nitric	Oxide Synthase by N^g-Amino-L-Homoarginine.=	MEDI	0106
T_c in Simple Superconducting	Oxide Systems.= Superconductivity. Synthesis and	PHYS	0009
Mixed Siloxane-	Oxide Systems.= Spectroscopic Characterization of	POLY	0231
Reduction of Nitric	Oxide with Ammonia.= Selective Catalytic	COLL	0207
on Molybdenum	Oxide.= Effect of Support in Propylene Oxidation	CATL	0059
Thione and Tri-n-Octylphosphine	Oxide.= +Benzoyl-5-Methyl-2-Phenyl-Pyrazol-3-	NUCL	0073
Hydroxyl-Capped Poly(Ethylene	Oxide) and ABA Block Copolymers Containing Hydr	PMSE	0006
Poly(ethylene	Oxide) Blends.= Crystal-Amorphous Interphases in	PMSE	0165
ethods. Polystyrene-Poly(ethylene	Oxide) Block Copolymer Micelles in Water.= +M	POLY	0277
nic Conductivity of Poly(ethylene	Oxide) Bound with a Sterically Hindered Phenolate.	POLY	0091
ysics, and Biology. Poly(ethylene	Oxide) Star Molecules in Surfaces for biological Stud	POLY	0272
Hydroxyl-Capped Poly(Ethylene	Oxide) with Polydimethylsiloxane.= +rs Containing	PMSE	0006
itle Gelation from Poly(ethylene	Oxide)/Paradichlorobenzene and Their Intercalate.=	POLY	0125
y+ Poly(Arylene Ether Phosphine	Oxide)s—I. Overview of Chemistry, Thermal Stabilit	PMSE	0052
Poly(Arylene Ether Phosphine	Oxide)s—II. Pyrolysis Studies.=	PMSE	0053
b Poly(Arylene Ether Phosphine	Oxide)s—III. Investigations of Surface Modification	PMSE	0054
Oxygen Spillover during	Oxide-Catalyzed Oxidation of Hydrocarbons.=	COLL	0174
on Hydrous Metal	Oxide-Supported Catalysts.= Reactions of Propane	CATL	0040
8-Tetrahydrobenz[c]Acridine 5N-	Oxide, V.= +7,7-Dimethyl-Quino[2,3,4-kl]-5a,6,7,	ORGN	0198
f the Excited State of Vanadium	Oxides Anchored onto SiO_2 and Their Photoreactivit	INOR	0165
astal Sediments: Cycling of Iron	Oxides and Their Potential Importance as Alternate	GEOC	0007
d Characterization. Microporous	Oxides by the Sol-Gel Process: Synthesis and Applic	PETR	0031
ivity of Open-Structure Tungsten	Oxides from Molecular Precursors.= +sis and React	INOR	0314
ustion E+ Oxidation of Nitrogen	Oxides in Homogeneous Gas-Phase Reactions: Comb	FUEL	0127
Catalytic Role of Metal	Oxides in the Photodegradation of Furans.=	COLL	0190
Synthesis of Manganese	Oxides with Tunnel Structures.= Hydrothermal	PETR	0056
Atmospheric Nitrogen	Oxides: A Bridesmaid Revisited.=	ENVR	0071
Superconductivity.= Cuprate	Oxides: Effect of Nonmagnetic Chains and Planes on	INOR	0358
Model Surface Studies of Metal	Oxides: Preparation, Characterization, and Chemica	COLL	0193
Fd and Ni/Mo Hydrous Titanium	Oxides.= +tion and Model Compound Studies on	CATL	0039
ases, Metal Complexes, and Metal	Oxides.= +Metal(IV) Phosphates toward Organic B	PETR	0107
Aluminas, and Mixed	Oxides.= Open-Morphology Zirconium Phosphates,	PETR	0051
Ternary Silver	Oxides.= Oxidative Deintercalation of Silver from	INOR	0275
Polymeric Routes to Aluminum	Oxides.=	POLY	0342
Precursors to Titanium	Oxides.= Polyoxoalkoxotitanates: Molecular	INOR	0174
Controlled-Pore-Size	Oxides.= Sol-Gel Processing of	PETR	0032
Conducting	Oxides.= STM-Induced Etching of	PHYS	0361
Metal	Oxides.= Synthesis and Characterization of New	PETR	0057

Left context	Title fragment	Ref	Num
c Analysis of Insoluble, Difficultly	Oxidizable Compounds in Aqueous Systems.= +tri	COLL	0002
IR Specular Reflectance Study of	Oxidized and Exchanged Polyethylene Surfaces.=	POLY	0253
Reactivity of	Oxidized Organic Sulfur Forms in Coals.=	FUEL	0057
el Seguimiento de+ Peliculas de	Oxido sobre Acero Inoxidable 304 y 316 Utilizadas en	ANYL	0055
ectrodo Indicador una Pelicula de	Oxido sobre Acero Inoxidable 316.= +ando como El	CHED	0163
cero Inoxidable 316/Pelicula de	Oxido, en la Validacion del Metodo Potencio-Metrico	ANYL	0066
udotuberculosis CDP-D-Glucose	Oxidoreductase (E_{od}) and CDP-4-Keto-6-Deoxy-D-	BIOL	0110
ata por Electrodeposito, Fusion,	Oxidorreduccion con Zinc en Medio Acido y Basico, P	CHED	0172
Catalizadores de Molibdeno sobre	Oxidos Mixtos de Alumina-Titania.= +Tiofeno con	FUEL	0105
of 26- and 29-Fu+ Inhibition of	Oxidosqualene Cyclase by Pseudosubstrates: Studies	ORGN	0021
of 26- and 29-Functionalized 2,3-	Oxidosqualenes.= +e by Pseudosubstrates: Studies	ORGN	0021
thyl Alkyl and Dialkylamino Aryl	Oxime Ethers.= +Substituted 2-Methylbiphenylme	AGRO	0072
Barriers to Rotation in	Oximes: A Molecular Orbital Study.=	PHYS	0324
Donors.= 2-Oxo- and 2-	Oximinoglycosyl Halides as Versatile Glycosyl	CARB	0025
ent Extraction of Lanthanum by	Oxine in the Presence of Oxalic Acid as Complexing	ANYL	0054
ganocuprate Reagents to Alkynyl	Oxiranes: Synthesis of a Precursor of (±)-Verrucosid	ORGN	0011
Covalent Linkage of Manganese	Oxo Clusters with Nonzero Spin-Ground States.=	INOR	0196
Technetium(V) and Rhenium(+	Oxo Complexes of Amine-Amide-Dithiol Ligands.=	INOR	0181
by Technetium(VII)-	Oxo Complexes.= Investigation of Alkene Oxidation	INOR	0328
Di-μ-	oxo Dimers.= Formation Pathway and Stability of	INOR	0185
e redox Chemistry of a Series of (	Oxo)(Phosphine)Ruthenium(IV) Complexes.= +th	INOR	0068
genation of Thioanisoles by (	Oxo)(phosphine)ruthenium(IV) Complexes.= S-Oxy	INOR	0130
Glycosyl Donors.= 2-	Oxo- and 2-Oximinoglycosyl Halides as Versatile	CARB	0025
Characterization of Aryloxy- and	Oxo-Aryloxy-Alkaline-Earth Complexes.= +is and	INOR	0223
Synthesis and+ Synthesis of 2-	oxo-2-Propionyl-1,3,2-Diaxaphosphorinanes and the	ORGN	0050
and Aldol Reactions of a Chiral 2-	oxo-2-Propionyl-1,3,2-Oxazaphosphorinane.= +is	ORGN	0050
change and Reduction Kinetics of	Oxochromate(V) Complex by α-Hydroxy Acids.=	INOR	0402
-Dimensional Reduced Niobium	Oxophosphate Compound: Synthesis and Characteriz	INOR	0031
DNA by	Oxoruthenium(IV).= Efficient Oxidative Cleavage of	INOR	0410
Environment.= Ni Chemistry of	Oxothiolate Ligands: New Complexes with NiS_2O_2	INOR	0204
f an Unsaturated Ruthenium F+	Oxy-Ligated Compounds. Synthesis and Reactivity o	INOR	0127
n Arsenopyrite-Pyrite: Effet des	Oxydants et des Complexants sur la Fottation de ces	ANYL	0067
pH, Carbonate/Bicarbonate, and	Oxygen Activities.= +Pellets as a Function of Time,	INOR	0236
Derived from Layered Siloxene+	Oxygen Adsorption Properties of Microporous Silica	PETR	0144
ten, and Rhenium Complexes.=+	Oxygen Atom Transfer between Molybdenum, Tungs	INOR	0132
Formation during the Reaction of	Oxygen Atoms with Acetylene at 300 K.= +etylene	FUEL	0065
ces of Hydrodynamic Stress and	Oxygen Availability on Ravidomycin Production by a	BIOT	0162
Carbons: An Analysis of Exper+	Oxygen Chemisorption on Graphite and Microporous	FUEL	0017
ication Mechanisms. Modeling of	Oxygen Chemisorption on Microporous Carbons.=	FUEL	0012
Premium Coal Samples. Organic	Oxygen Contents of Argonne Premium Coal Samples	FUEL	0003
Diamond Growth from Acetylene-	Oxygen Flames.= +oepitaxial High-Temperature	FUEL	0037
rbon-Supported C+ Influence of	Oxygen Functional Groups on the Performance of Ca	FUEL	0033
Analyzing Functional Groups with	Oxygen in Coal.= +ectrometry as a Technique for	FUEL	0061
ons in LA_2CuO_4-Based Superc+	Oxygen Isotope Effect and Structural Phase Transiti	PHYS	0010
n Carbides: The Effect of Surface	Oxygen on Reaction Pathways.= +tions on Tungste	CATL	0037
Chemistry, Thermal Stability, and	Oxygen Plasma Resistance.= +e)s—I. Overview of	PMSE	0052
gations of Surface Modification by	Oxygen Plasma Treatment.= +xide)s—III. Investi	PMSE	0054
Quantum Yields of Singlet	Oxygen Production by Triplet Cyclohexenones.=	ORGN	0167
per A+ Kinetics of the Molecular	Oxygen Reaction with Sodium, Magnesium, and Cop	PHYS	0141
the Electrochemical Model.=	Oxygen Redox Catalysis Using RuO_2 Dispersions and	COLL	0134
of Hydrocarbons.=	Oxygen Spillover during Oxide-Catalyzed Oxidation	COLL	0174
Cation-Dopant Site Selectivity,	Oxygen Stoichiometry, and Cu Valence in the High-	PHYS	0027
chanism of Alkyl Migration from	Oxygen to Metal in Fe-Mn Alkoxycarbyne Complexe	INOR	0110
ll Culture Media.=	Oxygen Transfer Properties of Bubbles in Animal Ce	BIOT	0008
on the Side Selectivity of Singlet	Oxygen with α,β Unsaturated Esters: Evidence for a	ORGN	0164
rocarbon Radical Reactions with	Oxygen: Comparison of Allyl, Formyl, and Vinyl to A	PHYS	0228
of Yttrium Oxide and Chloride in	Oxygen.= +iluminescence from the Laser Ablation	PHYS	0255
.= Organometallic Compounds.	Oxygen-Atom Transfer from N_2O to Alkynes Coordi	INOR	0327
on into the Marine Invertebrate	Oxygen-Carrying Proteins Hemerythrin and Myohem	INOR	0140
sorption Process for Production of	Oxygen-Enriched Combustion Air.= +l Swing Ab	I&EC	0023
olycarbosilane as a Precursor to	Oxygen-Free SiC Fibers.= High Molecular Weight P	POLY	0360
ecursors.=	Oxygenated C_{13} Norisoprenoids: Important Flavor Pr	AGFD	0025
and C-O Bonds of	Oxygenates on Rh(111).= Activation of C-H, C-C,	CATL	0033
sing Alcohols, Gasohol, and Other	Oxygenates.= +issions from Automobile Engines U	FUEL	0095
henium(IV) Complexes.= S-	Oxygenation of Thioanisoles by (Oxo)(phosphine)rut	INOR	0130
nzymes to Model M+ Selective	Oxygenations with Metalloporphyrins: From Model E	ORGN	0136
Cyclic	Oxyphosphoranes as Reactive Intermediates.=	INOR	0290
ing Chlorinate+ Integrated Bio-	Ozon Treatment of Pulp Bleaching Effluents Contain	BIOT	0137
ent and Experience with Catalytic	Ozone Abatement in Jet Aircraft.= +cial Developm	I&EC	0047
Interaction between	Ozone and Fine-Dispersed Carbon Bodies.=	COLL	0109
dies of the $^1A_2$ and $^1B_1$ States of	Ozone: A Conical Intersection and Its Effect on Phot	PHYS	0191
ion to Lower Furanose Sugars via	Ozonization of Protected Glycals.= +nient Degradat	ORGN	0073
$P(CH_2)_nCOOH)_2$, n = 1-4 from	^{31}P and ^{13}C Solid-State NMR.= Structures of ZrO_3	INOR	0047
gonucleotides and Proteins DN+	^{31}P One- and Two-Dimensional NMR of Duplex Oli	INOR	0378
May Help Reduce Decay in Seal-	Packaged Citrus and Pepper Fruits.= +llumination	AGFD	0076
Analysis of	Packaged Food Substances by GC/IR.=	PMSE	0141
ld Compounds Used in Electronic	Packages.= +c Mechanical Thermal Analysis of Mo	POLY	0095
Elastic Rubber Nettings Used for	Packaging Cured-Pork Products.= +trosamines in	AGFD	0040
opments in Modified Atmosphere	Packaging in Fruits and Vegetables.= +New Devel	AGFD	0075
Modified-Atmosphere	Packaging System.= Modeling of a	AGFD	0077

n in the 1990s+ New Frontiers	Packaging: Polymer-Based Intermaterial Competitio	CMEC	0016
of State.= Design for	Packed Tower Absorbers Using the PRSV Equation	CHED	0107
ration of Toluene in a Co-Current	Packed Tubular Reactor.= +ng the Two-Phase Nit	I&EC	0006
Molecules Using Porous Polymeric	Packings.= +aphy of Proteins and Pharmaceutical	ANYL	0031
0.= SCH 40338, a New Dual	PAF and Histamine Antagonist Related to SCH 3737	MEDI	0161
works.= Study on the Damping	Paint of PS/PA Latex Interpenetrating Polymer Net	PMSE	0064
acking a Degenerate Ground S+	Paired Disorder and the Metallic State in Polymers L	PHYS	0039
dary Amines by Reversed-Phase	Paired-Ion High-Performance Liquid Chromatograp	ANYL	0089
Foundation of the Al Sitting and	Pairing Avoidance Principles in Zeolites.= +rectical	PHYS	0289
Superconductivity. Theory of	Pairing in HTC Materials.=	PHYS	0001
i+ Molecular Recognition via Base	Pairing: Noncovalent Approaches to Photosynthet	ORGN	0059
tion, Properties, and+ Isomalt (	Palatinit®), a Versatile Alternative Sweetner: Produc	CARB	0005
ansfer to+ Patterned Imaging of	Palladium and Platinum Metal Films by Electron Tr	INOR	0319
Monoxide and+ Bis(phosphine)	palladium Catalysts for Copolymerization of Carbon	INOR	0155
oscopic Characterization of CO on	Palladium Colloids.= +tical Spectroscopies. Spectr	PHYS	0373
4.= Transition Metal Hydrides.	Palladium Hydride Dimers: Unusual Coordination B	INOR	0379
ctive Carbohydrate Synthesis via	Palladium Hydroxide-Catalyzed Epoxide Hydrogeno	ORGN	0074
Hydrogenation on a Catalytic	Palladium Membrane.= Effects of a Model	CATL	0011
Methanol Combustion on	Palladium Monolith Catalyst.= Kinetics of	COLL	0209
ructure, and Catalytic Ac+ Novel	Palladium Nitro Complexes, Synthesis, Molecular St	INOR	0365
Tritium-Labeled	Panadiplon.= Synthesis of Carbon-14- and	MEDI	0133
Efficient Total Synthesis of (±)-	Pancracine.= +Methanomorphanthridine Type: An	ORGN	0007
y Acyl Substrate Preferences for	Pancreatic Lipase-Mediated Exchange Reactions wit	AGFD	0036
	Panel Discussion. Academic Perspectives.=	POLY	0202
istry and Chemical Engineering.	Panel Discussion: At the Interface, What Does the P	I&EC	0057
tter. D. L. Leister, Moderator.=+	Panel Discussion: Bridging the Gap; Why More Is Be	I&EC	0058
radable Polymers.=	Panel Discussion: Future Directions in R D of Biodeg	CHED	0136
Thiols: Applications to	Papain Inhibitors.= Reactivities of Epoxides Toward	MEDI	0071
Cysteine Proteinases from	Papaya Latex.= Conformational Differences in the	BIOL	0074
Sulfonacion sobre las Propiedades	Papelaras.= +e Alto Rendimiento: Influencia de la	PMSE	0077
°From	Paper Chromatography to Drug Discovery.=	HIST	0019
Reassessment of Some Seminal	Papers.= The Mechanism of β-Glycosidases: A	ORGN	0070
de Procesos	para Balances de Materia.= BALANCE: Simulador	COMP	0052
on del Metodo Potencio-Metrico	para Cuantificar Clorhidrato de L-Arginina en Table	ANYL	0066
aciones Recientes de CAD/CAM	para el Diseno de Equipo de la Industria Quimica de	COMP	0053
mentales de un Empaque Integral	para Enriquecimiento en Deuterio.= +tados Experi	I&EC	0073
rollo de Apoyos Computarizados	para la Ensenanza de la Quimica a Nivel Medio Supe	CHED	0124
lectrodo de Vidrio Convencional	para Medir pH, por un Electrodo de Acero Inoxidabl	ANYL	0066
de Cementos Hidraulicos	para Pozos Geotermicos.= Evaluacion Fisicoquimica	INOR	0063
ns over Heterogeneous Catalysts.	Para-Selective Alkylation of Mono Alkylbenzenes ov	PETR	0100
ansition in Insulin He+ Methyl	Paraben Will Cause a T←R-Like Conformational Tr	BIOT	0150
elation from Poly(ethylene Oxide)	Paradichlorobenzene and Their Intercalate.= +le G	POLY	0125
Revising the Risk-Assessment	Paradigm.=	ENVR	0065
tabolism in Mice.=	Paradoxical Effect of Diazinon on L-Tryptophan Me	AGRO	0020
Reactions and Catalysts for C_3-C_4	Paraffin Dehydrogenation and Aromatization.= +f	PETR	0066
ysts. New Developments in C_5-C_7	Paraffin Isomerization.= +er Heterogeneous Catal	PETR	0112
h-Directed, Spatially Addressable	Parallel Chemical Synthesis.= +f Drug Leads. Lig	MEDI	0090
Technology: What Happens When	Parallel Developments Intersect?.= +pyright and	CINF	0043
Melting.= Thermodynamics	Parallels between Solid-State Amorphization and	NUCL	0092
etect Framework Substitution of	Paramagnetic Ions in Microporous Aluminophosphat	PHYS	0231
ir Hydration Water Contents and	Paramagnetic Metal Ion Concentrations.= +to The	MEDI	0043
nance Line Broadening of Surface	Paramagnetic Probes.= +s by Electron-Spin-Reso	PHYS	0050
nance Line Broadening of Surface	Paramagnetic Probes.= +ors by Electron Spin Reso	PHYS	0284
olutions of the Iron-Ethylene D+	Paramagnetic Relaxations of Water Protons in the S	INOR	0090
Radicals.= Electron	Paramagnetic Resonance Detection of Phenolic Free	AGFD	0004
f Coal Chem+ Advanced Electron	Paramagnetic Resonance Techniques for the Study o	FUEL	0072
Coherent Pulsed Electron	Paramagnetic Resonance.= New Applications of	PHYS	0116
and Molecular Weight.= One-	Parameter Correlation between Intrinsic Viscosity	PMSE	0095
s in Soil as a Function of Climatic	Parameter.= +radation and Translocation Processe	AGRO	0025
cohexadienes: Precursors to Poly(	paraphenylene).= +of Heteroatom-Substituted Cy	POLY	0047
uence of Maceral Composition of	Parent Coal on Potassium Activity in Steam Gasifica	FUEL	0027
Marie Lavoisier, the Other	Parent of New Chemistry.=	HIST	0037
Probationary	Participants.= The History of HIST, IV:	HIST	0010
d Sensitivity Studies with	Particle Beam and Thermospray LC/MS.= Enhance	ANYL	0123
n, Transport, and Vaporization for	Particle Beam LC/MS.= +ts in Aerosol Generatio	ANYL	0122
Fluctuations of the Single-	Particle Density in Nuclear Dynamics.=	NUCL	0009
in the Determination of Emulsion	Particle Distribution.= +ton-Scattering Correction	PMSE	0030
n Kinetics of Heterogeneous Char	Particle Populations.= +nisms. On the Combustio	FUEL	0026
by Nuclear Reactions/Scattering.	Particle Scattering for Measurement of Surface Com	NUCL	0064
nation.=	Particle Size Analysis Using Flow Field-Flow Fractio	PMSE	0012
Electroacoustic Determination of	Particle Size and Charge.=	COLL	0201
f Granular Composite+ Effect of	Particle Size Distribution on Mechanical Properties o	PMSE	0114
HPLC Mass Spectrometry—I.	Particle Size Effects in Aerosol Generation, Transpor	ANYL	0122
cy of the Effect of Support and	Particle Size for NO Adsorption on Rh and Pt.= Stu	COLL	0186
curing Oxidation of a Single Char	Particle.= +Mechanisms. Morphological Changes	FUEL	0025
t UV-Absorption Photometry and	Particle-Beam Mass Spectrometry.= +ography wit	ENVR	0015
alysis with TiO_2 Semiconductor	Particles in Aqueous Media: A Look at Two Recent C	COLL	0131
H+ Colloidal Particles: Colloidal	Particles in External Fields: Electric, Magnetic, and	COLL	0141
H+ Colloidal Particles: Colloidal	Particles in External Fields: Electric, Magnetic, and	COLL	0159
H+ Colloidal Particles: Colloidal	Particles in External Fields: Electric, Magnetic, and	COLL	0178

H+ Colloidal Particles: Colloidal	Particles in External Fields: Electric, Magnetic, and	COLL	0199
Inorganic	Particles in Magnetotactic Bacteria.=	INOR	0118
Molecular	Particles of Solid-State Compounds.=	CHED	0033
Materials. Reinforcing Zirconia	Particles Precipitated into an Elastomeric Material.=	POLY	0283
m-Tin and Gold-Tin Bimetallic	Particles Prepared from Solvated Metal Atoms: Stru	CATL	0009
ric, Magnetic, and H+ Colloidal	Particles: Colloidal Particles in External Fields: Elect	COLL	0141
ric, Magnetic, and H+ Colloidal	Particles: Colloidal Particles in External Fields: Elect	COLL	0159
ric, Magnetic, and H+ Colloidal	Particles: Colloidal Particles in External Fields: Elect	COLL	0178
ric, Magnetic, and H+ Colloidal	Particles: Colloidal Particles in External Fields: Elect	COLL	0199
ecular Models for Semiconductor	Particles: Luminescence Studies of Inorganic Cluster	INOR	0317
sing Designed Core-Shell Rubber	Particles.= +ning of High-Performance Epoxies U	POLY	0157
and the Nucleation of Soot	Particles.= Aromatics Growth beyond the First Ring	FUEL	0108
Compounds to Model Atmospheric	Particulate Materials.= +of Semivolatile Organic	ENVR	0059
mic Electrodes.= Mechanism of	Particulate-Enhanced Mass Transport to Hydrodyna	COLL	0041
de la Ensenanza Experimental a	Partir de Errores Detectados en Publicaciones Cienti	CHED	0178
aboracion de Caseinato de Sodio a	Partir de Leche Contaminada con S. aureus y s+ El	AGFD	0107
rtition Coefficients by Centrifugal	Partition Chromatography.= +. Octanol-Water Pa	ANYL	0068
rrent Chromato+ Liquid-Liquid	Partition Coefficient Determination Using Countercu	ANYL	0070
pplications—II. Octanol-Water	Partition Coefficients by Centrifugal Partition Chrom	ANYL	0068
Model Atmospheri+ Gas/Particle	Partitioning of Semivolatile Organic Compounds to	ENVR	0059
c Retention and Bioavailability+	Partitioning Processes at Interfaces: Chromatographi	ANYL	0102
Fundamental Studies on Affinity	Partitioning.=	BIOT	0049
Employee-Supervisor	Partnership.=	TECH	0004
Industry-Education	Partnership.= Teachers in Industry: An	CHED	0004
iances to+ Corporate/Academic	Partnerships in Precollege Education. Benefits of All	CHED	0001
Academic/Industrial	Partnerships.= POLYED—A Model for	CHED	0006
Total Synthesis of Racemic	Parvifolin.=	ORGN	0176
Protein Secretion in a Single-	Pass Ceramic Matrix Bioreactor.= Controlled	BIOT	0011
.= Preparation of	Passivated III-V Surfaces for STM Nanolithography	PHYS	0363
ration of Metal Contaminants and	Passivating Agents in Fluid-Cracking Catalysts.=	CATL	0012
Polymers in Waveguiding	Passive and Active Devices.= Optically Nonlinear	POLY	0177
Active and	Passive Devices.= Materials Requirements for	POLY	0004
st Insect-Molting Hormones fr+	Past and Current Studies with Ponasterones, the Fir	HIST	0024
CHEMTECH Magazine over the	Past 20 Years.= +of Favorite Cartoons Appearing in	CHAL	0024
Crime, III: Forensic Chemistry—	Past, Present, and Future. Lindber+ Chemistry and	HIST	0006
Crime, III: Forensic Chemistry—	Past, Present, and Future. The Sea+ Chemistry and	HIST	0001
emediation and Waste Treatment:	Past, Present, and Future.= +egradation in Site R	AGRO	0121
Aerosols:	Past, Present, Future?.=	CHED	0048
cion sobre las P+ Elaboracion de	Pastas de Alto Rendimiento: Influencia de la Sulfona	PMSE	0077
Chemists as Expert Witnesses in	Patent Litigation.=	CHAL	0013
Using Scientific Evidence in	Patent Proceedings.=	CHAL	0014
=	Patent Protection for Advances in DNA Technology.	CHAL	0006
Cited	Patent Searching: A Tutorial.=	CINF	0040
Related to Inventors or	Patents.= Favorite or Objectionable Cartoons	CHAL	0022
DNA-Based	Paternity Testing: Five Years' Experience.=	CHAL	0009
hemistry Enrollments: Toward a	Path Analytic Model of Science-Related Attitudes.=	CHED	0177
H + HNO.= Reaction	Path and Rate Calculations of Different Channels of	PHYS	0310
Management	Paths for Technical Professionals.=	YCC	0010
Technical Career	Paths in a Specialty Chemical Company.=	YCC	0007
Use of Most Restrictive	Paths in 3D Search Strategy.=	CINF	0024
Formation	Pathway and Stability of Di-μ-oxo Dimers.=	INOR	0185
Risk Characterization of Multiple-	Pathway Chemical Exposures.= +ted Approach to	ENVR	0073
ethod for Evaluation of Reaction	Pathway Control on Metals.= +of the BOC-MP M	CATL	0032
Selectivity in Catalysis: Reaction	Pathway Control—I. Crystal and Catalytic Chemist	CATL	0026
Selectivity in Catalysis: Reaction	Pathway Control—II. Application of the BOC-MP	CATL	0032
mycins by Streptom+ Metabolic	Pathway Engineering: The Biosynthesis of Tetraceno	BIOT	0083
Alkyl Gallium Amide Anion: The	Pathway to Pyrazol Gallates.= +Characterization of	INOR	0390
Biosynthetic	Pathway.= Alterations of the β-Lactam Antibiotic	BIOT	0235
Methionine Salvage	Pathway.=	BIOL	0024
Air	Pathway.= Revised Hazard-Ranking System	ENVR	0082
Analysis of Chemically Activated	Pathways for Molecular-Weight Growth.= +ation.	FUEL	0063
es to Alc+ Control of Oxidation	Pathways for Selective Catalytic Conversion of Alkan	CATL	0036
Study of the Reaction	Pathways for Si+ + SiX$_4$ (X = F, Cl).= An ab Initio	PHYS	0248
Electron Tunneling	Pathways in Native and Modified Proteins.=	BIOL	0133
methylenemetanes: New Reaction	Pathways Involving the Dipolar Form.= +ter of Tri	ORGN	0146
f the Rate-Limiting Steps in the	Pathways Leading to Penicillin and Cephalosporin.=	BIOT	0128
tudies of the Photodecomposition	Pathways of Adamantyldiazirine.= +nd Chemical S	ORGN	0163
re+ Chemistry of New Metabolic	Pathways. Biochemistry of Protein Prenylation: Natu	BIOL	0021
ct of Surface Oxygen on Reaction	Pathways.= +tions on Tungsten Carbides: The Effe	CATL	0037
bimetallics by Oxidative Addition	Pathways.= +ridged Iridium, Molybdenum Hetero	INOR	0385
Chemistry. Profiles,	Pathways, and Dreams: Future Chemists, Future	CONG	0004
Films by Electron Transfer to+	Patterned Imaging of Palladium and Platinum Metal	INOR	0319
r Microsomal Cytochrome P-450	Patters by Means of Ion Exchange and Immobilized	ANYL	0017
of	PBLG.= Gelation and Crystallization in Solutions	POLY	0241
plate and Its Anti-HIV Activity in	PBM Cells.= +Using D-Mannose as a Chiral Tem	CARB	0022
Electrochemical Behavior of	PbS at the Flotation Conditions.=	PHYS	0217
Solutions of	PBT.= The Third-Order Susceptibility of Nematic	POLY	0274
n to Phase-Separation Kinetics in	PC/PMMA Blends.= +Description and Applicatio	POLY	0278
e Role of in Situ Hum+ Fate of a	PCB in Anoxic Bottom Sediments: Characterizing th	ENVR	0017
for Detecting	PCB in Soil.= PCB-RISc: An On-Site Immunoassay	ENVR	0032

PCB in Soil.=	PCB-RISc: An On-Site Immunoassay for Detecting	ENVR	0032
e Binder-Enhance+ Analysis of	PCBS and PCDDS from the Combustion of Quicklim	FUEL	0140
nhance+ Analysis of PCBS and	PCDDS from the Combustion of Quicklime Binder-E	FUEL	0140
d Metals+ Levels and Sources of	PCDDS, PCDFS, Chlorophenols, Chlorobenzenes, an	ENVR	0018
Levels and Sources of PCDDS,	PCDFS, Chlorophenols, Chlorobenzenes, and Metals	ENVR	0018
Surface.= Peculiarities of	PCl$_5$ Reaction with Silanol Groups on Silica	INOR	0237
lvents on the React+ Reaction of	PCl$_5$ with Silica Surface: Influence of the Organic So	COLL	0097
tic Markers for the Application of	PCR to Forensic Casework Samples.= +New Gene	CHAL	0004
	PCR—Status of Validation Studies.=	CHAL	0005
and Model Compound Studies on	Pd and Ni/Mo Hydrous Titanium Oxides.= +tion	CATL	0039
(S)]$_n$[Ph$_2$P(O)]$_3$-nC}, Where M =	Pd or Pt and n = 1, 2, or 3.= +of (R$_3$P)MCl[{Ph$_2$P	INOR	0265
Ni-Fe-Pd, Fe-Co-Pt, and Fe-Co-	Pd.= +etism in Ternary Compounds of Ni-Fe-Pt,	PHYS	0253
on the Interaction of Pt(II) and	Pd(II) Dipeptide Complexes with Nucleic Acid Const	INOR	0412
Ligand Complexes of Pt(II) and	Pd(II) with Dipeptides and Substituted Pyrimidines.	INOR	0411
erpenetrating 3D Frameworks of [	Pd(Se$_4$)$_2$]$^{2-}$ and [Pd(Se$_6$)$_2$]$^{2-}$.= +2PdSe$_{10}$: Two Int	INOR	0200
Frameworks of [Pd(Se$_4$)$_2$]$^{2-}$ and [	Pd(Se$_6$)$_2$]$^{2-}$.= +2PdSe$_{10}$: Two Interpenetrating 3D	INOR	0200
CH$_2$N$_2$ on	Pd(110).= Adsorption and Chemistry of Cyclic	COLL	0114
on by Coking and Regeneration of	Pd/HM Catalysts for Alkane Isomerization.= +ati	PETR	0130
face Acidity of HM Zeolite and	Pd/HM Catalysts.= Studies on the Activity and Sur	PETR	0117
Compounds of Ni-Fe-Pt, Ni-Fe-	Pd, Fe-Co-Pt, and Fe-Co-Pd.= +etism in Ternary	PHYS	0253
Derivatives.= Synthesis and	PDE Inhibitory Activity of Griseolic Acid	MEDI	0117
Trichloroethylene over a	PdO on Al$_2$O$_3$/Monolith.= Catalytic Oxidation of	COLL	0210
ones in Apples, Grapes, and	Peaches.= Residue Analyses of Lepidopteran Pherom	AGRO	0148
rray Multic+ Analysis of System	Peaks in Normal Bonded Phase HPLC with Diode A	ANYL	0087
Chromatographic	Peaks Using a Neutral Network.= Deconvolution of	ANYL	0048
Effect of Storage on Roasted	Peanut Flavor, as Determined by Descriptive Sensory	AGFD	0161
mer-So+ Electric-Field-Induced	Pearl-Chain Formation in Multiphase Polymer-Poly	COLL	0145
Organic Geochemical Studies of	Peat from Volo Bog.=	I&EC	0016
Radicals in Photoelectrochemical (	PEC) Cell.= +n of Dyes by Photogenerated Ketyl	COLL	0081
perties of Dextran, Pullulan, and	Pectin Determined by HPSEC with Viscosity Detect	CARB	0045
y Processed Fruit Products by	Pectinase Infusion.= Preparation of Novel Minimall	AGFD	0059
timicrobial α-Pyrone from Hyptis	pectinata (Lamiaceae).= +Pectinolide: A Novel An	MEDI	0060
nduced Motional Perturbations of	Pectinic Polysaccharides in Apple Tissue.= +tion-I	AGFD	0012
tructure and Stereochemistry of	Pectinolide: A Novel Antimicrobial α-Pyrone from H	MEDI	0060
cal Graph Theory in Chemistry.	Pedagogical Approach to Conjugated Molecular Syste	COMP	0020
oretical Approach to Structure-+	Pedagogical Graph Theory in Chemistry. Graph The	COMP	0013
l Approach to Conjugated Mole+	Pedagogical Graph Theory in Chemistry. Pedagogica	COMP	0020
Modeling Diamond Growth in	Pedestal CVD Reactors.=	PHYS	0179
Study of Thermal Degradation in	PEEK (Poly-ether-ether-ketone) Films.= +oscopic	POLY	0334
Morphology of	PEEK at High Temperatures and in PEI Blends.=	POLY	0333
Legalize	PEG.= To Help Whistle-Blowers,	PROF	0012
Temperatures and in	PEI Blends.= Morphology of PEEK at High	POLY	0333
ndo como Electrodo Indicador una	Pelicula de Oxido sobre Acero Inoxidable 316.= +a	CHED	0163
Electrodo de Acero Inoxidable	316/Pelicula de Oxido, en la Validacion del Metodo P	ANYL	0066
Utilizadas en el Seguimiento de +	Peliculas de Oxido sobre Acero Inoxidable 304 y 316	ANYL	0055
Dissolution Kinetics of UO$_2$.00	Pellets as a Function of Time, pH, Carbonate/Bicarb	INOR	0236
nical Properties of Magnetite Ore	Pellets.= +ct of Manganese Additions in the Mecha	I&EC	0067
Reactions of Polymers with	Pendant Cyclophosphazenes.=	POLY	0182
es of Linear Epoxy Polymers with	Pendant Sulfonyl Tolan Groups.= +ptical Properti	POLY	0208
f Ampholytic Ionomers and Their	Pendant-Chain Derivatives.= +Dynamics Studies o	POLY	0367
h Reactive 2,6-Dimethylphenoxy	Pendent Groups: Processing, Mechanical Properties,	PMSE	0019
+Reactive 2,6-Dimethylphenoxy	Pendent Groups: Synthesis and Characterization.=	PMSE	0018
Scientific Experts: Is the	Pendulum Swinging Back?.= Testimony of	CHAL	0016
Steps in the Pathways Leading to	Penicillin and Cephalosporin.= +he Rate–Limiting	BIOT	0128
ctam I+ Substrate Specificity of	Penicillin G Amidase in Acylating a Monocyclic β-La	BIOT	0161
crem+ Optimized Conversion of	Penicillin N to Cephalosporin C in Cephalosporium a	BIOT	0085
Advances in	Penicillin-Amidase Biocatalysts.=	BIOT	0130
New Biocatalyst of	Penicillin-Amidase.=	BIOT	0131
dures for Determining Residues of	Penicillins.= +mparison of Chromatographic Proce	AGFD	0063
lucose Oxidase Isoenzymes from	Penicillium Amagasakiense by Ion–Exchange Chroma	ANYL	0016
tems for Waste Management: The	Penn State Experience.= +stry of Cementitious Sys	ENVR	0006
eoretical Studies of New Group 6	Pentacarbonyl-Based NLO Chromophores Exhibitin	POLY	0217
ction of Niobium and Tantalum	Pentachlorides in the Presence of Sodium N,N-Dieth	INOR	0092
id, On-Site Screening Test for	Pentachlorophenol in Soil.= Penta Soil RISc: A Rap	ENVR	0033
$_2$S$_2$-Copper(II) Macrocycle and a	Pentacoordinate CuN$_3$OS Pseudothiolate Complex C	INOR	0262
rmolysis of Bis(2,4-Dimethyl-1,3-	pentadienyl)ruthenium.= +Formation during The	INOR	0305
Derivatives.= Trigeminal	Pentafluorosulfanyl (SF$_5$) Carbon and Nitrogen	INOR	0024
Perfect Matchings in Square and	Pentagonal Chains.= +d Chains and the Number of	COMP	0033
ration+ Reaction of Phosphorus	Pentahalides with Cyclic Ketones: A One-Step Prepa	ORGN	0263
Complexes.= Reactions of	Pentamethylcyclopentadienylchromium(II)	INOR	0148
η^2-Binding of	Pentammineosmium(II) to Nucleosides.=	INOR	0083
Single-Phase Regions of SDS/1-	Pentanol/Cyclohexane/Water Microemulsion.= +he	COLL	0024
nce for the Intermedicacy of the	Pentaphenylethyl Cation in the Reaction of Perdeute	ORGN	0145
nido de Galio en Zeolitas del Tipo	Pentasil en la Aromatizacion de Propano.= +Conte	COLL	0125
erminations of Poly-4-Methyl-1-	Pentene.= +lymer Analysis. Molecular Weight Det	PMSE	0046
Acid–Catalyzed Alkylation of	Pentenes and Other Light Olefins.= Sulfuric	PETR	0084
y+ Alkylation of Isobutane with	Pentenes Using Sulfuric Acid as a Catalyst: Chemistr	PETR	0082
Rat Liver Cytosolic	PEPCK in E. coli.= Subcloning and Expression of	BIOL	0107
(±)-	Peperomin C.= Approaches to the Synthesis of	ORGN	0017
ecay in Seal-Packaged Citrus and	Pepper Fruits.= +Illumination May Help Reduce D	AGFD	0076

on the Bioavailability of Selected	Peptide Analogues and Peptidomimetics.= +cation	MEDI 0192
ntinuous-Flow FAB LC/MS.=	Peptide and Protein Analysis by Electrospray and Co	ANYL 0140
e Structure and Function. Model	Peptide Approach to Triple-Helix Perturbations.=	BIOL 0001
	Peptide Conformations and Protein Structures.=	BIOL 0004
Considerations in	Peptide Drug Design.= Pharmacokinetic	MEDI 0193
l Diversity.=	Peptide Expression Libraries as a Source of Chemica	MEDI 0092
to the Asymmetric Synthesis of	Peptide Fragments Bearing α-Hydroxy-β-Amino Ac	ORGN 0048
Design and Synthesis of Large	Peptide Libraries Suitable for Discovery of Peptide-	MEDI 0093
tructural Modificatio+ Peptides,	Peptide Mimetics, and Bioavailability. Influence of S	MEDI 0192
.=	Peptide Models for Protein Folding and Localization	BIOL 0003
Approaches to Improve Mucosal	Peptide Penetration.= Biochemical and Biophysical	MEDI 0194
ht-Direct+ New Technologies in	Peptide Production for Discovery of Drug Leads. Lig	MEDI 0090
roach to Triple-Helix Perturbat+	Peptide Structure and Function. Model Peptide App	BIOL 0001
e Y2-Receptor by Using Multiple	Peptide Synthesis.= +Site of Neuropeptide Y to th	MEDI 0091
Libraries Suitable for Discovery of	Peptide-Macromolecular Interactions.= +Peptide	MEDI 0093
ctions.=	Peptides as Flavor Precursors in Model Maillard Rea	AGFD 0072
Separation of Proteins and	Peptides by Perfusion Chromatography.= Rapid	ANYL 0127
mposium: Juvenile Hormones and	Peptides for Insect Control. +ck, Jackson Award Sy	AGRO 0001
mposium: Juvenile Hormones and	Peptides for Insect Control. +ck, Jackson Award Sy	AGRO 0006
Manduca sexta.= Autoparalytic	Peptides from Hemolymph of the Lepidopteran	AGRO 0012
Cyclizing	Peptides.= A Novel General Method for	BIOL 0039
Displacement Chromatography of	Peptides.= +Preparative Reversed-Phase Sample	ANYL 0040
Absorption of Small	Peptides.= Renin Inhibitors: Factors Influencing the	MEDI 0197
uence of Structural Modificatio+	Peptides, Peptide Mimetics, and Bioavailability. Infl	MEDI 0192
Oral Absorption of Di- and Tri-	Peptidic Compounds: Effect of Lipophilicity, Dosin	MEDI 0196
f Selected Peptide Analogues and	Peptidomimetics.= +cation on the Bioavailability o	MEDI 0192
of Protease Inhibition by	Peptidyl Fluoroketones.= NMR and Kinetic Studies	BIOL 0005
eavy Ion Collisions at 14.6 GeV/c	per Nucleon.= +ansverse Energies in Relativistic H	NUCL 0051
High- and Mid-Valent Mono(	peralkylcyclopentadienyl) Tantalum η^2-Iminoacyl Ch	INOR 0108
Synthesis Strategies by	Perceiving Structural Similarities.= Developing	CINF 0007
Strategies.= Value	Perception Analysis for Polymer Market-Entry	CMEC 0011
ion of Sulfur Forms in Coal Using	Perchloric Acid.= +oduced during the Determinat	FUEL 0055
ssues in Cancer Risk Assessment:	Perchloroethylene and Vinyl Chloride Examples.= I	CHAS 0011
lethyl Cation in the Reaction of	Perdeuterated Triphenylmethyl Cation with Dipheny	ORGN 0145
nsaturated Esters: Evidence for a	Perepoxide Intermediate.= +let Oxygen with α,β U	ORGN 0164
Electroreduccion de la	Perezona.= Determinacion del Mecanismo de	PHYS 0213
Synthesis of 5-	Perfluoroalkyl-3-Alkyl-2-Aryl-1,3-Thiazolidines.=	ORGN 0040
	Perfluorooctylbromide (PFOB) Emulsions.=	COLL 0080
lian Cell Retention in a Spinfilter	Perfusion Bioreactor: Analysis and Scaleup.= +ma	BIOT 0007
in Purification and Analysis.=	Perfusion Chromatography®: A Novel Tool for Prote	BIOT 0073
cale Biomolecule Purification.=	Perfusion Chromatography: A New Tool for Large-S	ANYL 0032
Proteins and Peptides by	Perfusion Chromatography.= Rapid Separation of	ANYL 0127
A Productivity in an Immobilized	Perfusion Reactor System.= +Overall Cyclosporin	BIOT 0155
Characterization of Preceramic	Perhydropolysilazane.=	POLY 0310
sition State for the+ Concerted	Pericyclic Elimination with a Carbocation-Like Tran	ORGN 0153
scope Images Showing Anomalous	Periodic Structures.= +Scanning Tunneling Micro	PHYS 0212
.= Downstream Processing of	Periplasmic Proteins Using Critical Fluid Disruption	BIOT 0070
emical Analysis, sponsored by the	Perkin-Elmer Corp.). +emistry Award in Spectroch	ANYL 0001
Reaction of Sulfides with	Permanganate Ion.=	ORGN 0243
Oxidation of DNA by	Permanganate.= Fundamental Determinants in the	BIOL 0121
and Physical Blending on the Gas	Permeability of Polymers.= +of Aryl-Substitution	POLY 0117
Polymer Network Films.= Gas	Permeability Studies of Latex Interpenetrating	PMSE 0061
Critical Conditions in Gel	Permeation Chromatography.=	PMSE 0049
ared Spect+ Development of Gel	Permeation Chromatography-Fourier Transform Infr	PMSE 0048
ular Weight Determination by Gel	Permeation in Aqueous Solution.= +olymer Molec	CHED 0210
Thermodynamic Analysis of Gas	Permeation through Glassy Polymeric Materials.=	POLY 0139
Experiment.= Gas	Permeation through Nafion Membranes: Theory and	POLY 0066
Green Pigments with	Perovskite Structure.=	INOR 0069
Mn, Fe, Co, Ni)	Perovskites.= Electronic Structure of LaBo$_3$ (B =	PHYS 0288
ty of LaBo$_3$ (B = Ni, Mn, Co, Fe)	Perovskites.= +aracterization, and Catalytic Activi	PHYS 0225
ol.= Inhibition of Horseradish	Peroxidase by 2-Chloroethyl Sulfides and Thiodiglyc	BIOT 0249
Mn(III) and Mn(IV) Horseradish	Peroxidase with Manganese-Porphyrin Model Comp	BIOL 0095
duction of Native Cytochrome c	Peroxidase(III) in Alkaline pH.= Photo-Induced Re	BIOL 0097
ression of Active Lignin and Mn	Peroxidases in Phanerochaete chrysosporium and in	BIOT 0237
and Cell C+ Association of Lipid	Peroxidation and Chilling Injury in Cucumber Fruit	AGFD 0033
Mechanism of Lipid	Peroxidation in Muscle Foods: New Aspects.=	AGFD 0095
trol and P+ Effect of Hydrogen	Peroxide and Iron Salts on Sewage, Part 1: Odor Con	ENVR 0034
ron (II) by Hydrogen	Peroxide.= Oxidation of Tris(1,10-Phenanthroline)I	INOR 0222
Phosphines by Gallium	Peroxides: A Mechanistic Study.= Oxidation of	INOR 0389
Micellar Effects on Kinetics of	Peroxodisulfate Oxidation of Ternary Iron(II)-Diimin	INOR 0022
Chemically	Perplexed.= Laboratory Exercises for the	CHED 0055
C+ Achieving Zero Discharge of	Persistent Toxic Substances in Great Lakes Areas of	ENVR 0083
Steroids: Lederle and	Personal Aspects.= Historical Reflections on	HIST 0023
Analytical Chemistry in the	Personal Care Industry.=	CHED 0050
Chemistry in the Cosmetic and	Personal Care Industry—I. Overview of Fragrance Ch	CHED 0046
Chemistry in the Cosmetic and	Personal Care Industry—II. Silicone in Hair Conditio	CHED 0063
s in Chemical Education. Use of	Personal Computer Modeling in Undergraduate Orga	CHED 0188
Blowing.= Two	Personal Experiences with Industrial Whistle	PROF 0007
hemistry, Honoring O. Hannaway.	Personal Recollections of Frank C. +in History of C	HIST 0009
t of Immune Disorder in	Persons Exposed to Organochlorines.= Measuremen	CEI 0017

Blood of Twelve TCDD-Exposed	Persons from Jacksonville, Arkansas.= +rans in the	CEI 0014
id Interaction from the	Perspective of Chemical Analysis.= The Cluster-Sol	NUCL 0057
Fertilizer Claims.= Consumer	Perspective of Organic or Naturally Occurring	FERT 0007
Chemical Enginee+ A Chemistry	Perspective of the Interface between Chemistry and	I&EC 0050
Handled.= A User's	Perspective of the Issues and How They Are	CINF 0044
Student	Perspective on the Project SEED Experience.=	CHED 0037
Investigation.= Historical	Perspective on the Role of Forensic Biology in Crime	HIST 0003
nd Consequences: A Psychologist's	Perspective on Whistle Blowing in Science.= +th a	PROF 0002
teroid Community, and Upjohn in	Perspective: A Profile of Innovation.= +oids, the S	HIST 0021
ethod Evaluation Criteria: State's	Perspective.= +s Techniques. Pesticide-Residue-M	AGRO 0145
g—An Equipment Manufacturer's	Perspective.= +tics. Commingled Plastics Recyclin	I&EC 0025
ural Bond Orbital Donor-Acceptor	Perspective.= +cts in Hydrogen Bonding: The Nat	PHYS 0106
Future	Perspective.= Environmental Concerns: A	AGRO 0067
Pesticides in IPM: Federal	Perspective.= Future Directions in the Use of	AGRO 0068
Developing Country	Perspective.= Managing Pesticide Waste: A	AGRO 0107
Wife's	Perspective.= Pesticides in Drinking Water: A Farm	AGRO 0055
Management: A California	Perspective.= Regulatory Issues in Risk	CHAS 0021
ns of GC/FTIR to Food Analysis:	Perspectives and Limits.= +od Science. Applicatio	AGFD 0200
Structure-Property	Perspectives for Conducting Polymers.=	BIOT 0203
ethods of Protein Separation—I.	Perspectives in Protein Purifications—Chromatograp	BIOT 0018
nant+ Chemical and Engineering	Perspectives on the Ultrafiltration of Polymeric Colo	I&EC 0049
Panel Discussion. Academic	Perspectives.=	POLY 0202
Industrial	Perspectives.=	POLY 0203
NSF	Perspectives.=	POLY 0204
posicion Quimica de las Plantas	Pertenecientes a la Familia Scrophulariaceae de Mex	ORGN 0185
Rates of Flame Species	Pertinent to Fullerenes.= Fluxes and Net Reaction	FUEL 0112
thylmethacrylate): Localization of	Perturbations in Thermomechanical Properties.=	PMSE 0173
nd Hydration-Induced Motional	Perturbations of Pectinic Polysaccharides in Apple T	AGFD 0012
(III) Complexes+ Magnetic Field	Perturbations of the Excited-State Properties in CR	INOR 0159
rtone Spectra: The Significance of	Perturbations.= +stry with Experiment. High-Ove	PHYS 0029
Peptide Approach to Triple-Helix	Perturbations.= +Structure and Function. Model	BIOL 0001
IPN Membranes for the	Pervaporation of Ethanol/Water Mixture.=	PMSE 0059
Determinacion de Metales	Pesados en Leche de Establos del Sur del D.F.=	ENVR 0037
esticides+ Biotechnology-Based	Pest Control Strategies as Alternatives to Chemical P	AGRO 0080
rnment/Farm Regulatory Issues.	Pesticide Analysis in Water Using Ion Trap Mass Sp	AGRO 0060
s Handbook.= MWPS-37	Pesticide and Liquid Fertilizer Containment Facilitie	AGRO 0042
ze and Nontarget Contaminati+	Pesticide Application Systems for Reduction of Rinsa	AGRO 0105
bolites in Urine of Exposed Forest	Pesticide Applicators.= +Hexazinone and Its Meta	CEI 0005
	Pesticide Container Collection and Recycling.=	AGRO 0090
Iowa's Plastic	Pesticide Container Recycling Program.=	AGRO 0091
	Pesticide Container Recycling.=	AGRO 0088
tegy.=	Pesticide Container Regulations as Part of EPA Stra	AGRO 0089
urface Drinking Water Supplies to	Pesticide Contamination.= +the Susceptibility of S	AGRO 0053
State	Pesticide Disposal Regulations and Programs.=	AGRO 0106
.=	Pesticide Donations and the Disposal Crisis in Africa	AGRO 0044
Fertilizer and	Pesticide Formulations.= Quality Assurance in	FERT 0009
Looking Back, Looking Forward:	Pesticide Management and Disposal in the Mid-80s	AGRO 0086
Future of	Pesticide R D.=	AGRO 0069
Pesticides in the 21st Century.	Pesticide Regulatory Policy in the 21st Century.=	AGRO 0066
g the Florida Indian River Lag+	Pesticide Residue in Barrier Island Salt Marshes alon	ENVR 0019
Chemicals and Their Containers.	Pesticide Rinsate Minimization and Reuse.= +tural	AGRO 0102
Evolution of Regulations for	Pesticide Rinsates.=	AGRO 0104
Status and Future Trends of	Pesticide Usage and Research in India.= Present	AGRO 059A
olicies and Attitudes Will Affect	Pesticide Use and Development in the 21st Century.	AGRO 0070
face-Water Research Related to	Pesticide Use on the Farm: Methods to Monitor Sma	AGRO 0054
e Role of Biodiversity in Reducing	Pesticide Use.= +ng Sustainable Agriculture and th	AGRO 0082
Biotechnological Approaches to	Pesticide Waste Destruction Using White Rot Fungi.	AGRO 0126
ntainers. Current Technologies for	Pesticide Waste Disposal: Overview.= +Their Co	AGRO 0120
Managing	Pesticide Waste: A Developing Country Perspective.=	AGRO 0107
Activated Charcoal Filtration of	Pesticide Wastes: Field and Laboratory Studies.=	AGRO 0127
chnolo+ Chemical Degradation of	Pesticide Wastes: Historical, Current, and Future Te	AGRO 0122
	Pesticide Wastewater Cleanup.=	NUCL 0123
Emission Characterization of	Pesticide-Bag Open Burning.=	AGRO 0141
se of Landfarming to Remediate	Pesticide-Contaminated Soil Excavated from Agroch	AGRO 0144
rs and Harvesters.= Developing	Pesticide-Exposure Mitigation Strategies for Handle	AGRO 0043
g Residue-Analysis Techniques.	Pesticide-Residue-Method Evaluation Criteria: State	AGRO 0145
Methodologies for Analyzing	Pesticides and Metabolites in Water.= Current	AGRO 0146
in the 21st Century. Outlook for	Pesticides in Developing and Developed Countries.=	AGRO 0056
	Pesticides in Drinking Water: A Farm Wife's Perspec	AGRO 0055
Modeling the Movement of	Pesticides in Dynamic Flow Systems.=	AGRO 0064
Future Directions in the Use of	Pesticides in IPM: Federal Perspective.=	AGRO 0068
	Pesticides in Public Health.=	AGRO 0057
est.=	Pesticides in Surface and Groundwaters in the Midw	AGRO 0065
nces in Research and in Gover+	Pesticides in Surface Water and Groundwater—Adva	AGRO 0051
nces in Research and in Gover+	Pesticides in Surface Water and Groundwater—Adva	AGRO 0060
Pest Control Strategies as Alte+	Pesticides in the 21st Century. Biotechnology-Based	AGRO 0080
s in Developing and Developed +	Pesticides in the 21st Century. Outlook for Pesticide	AGRO 0056
Policy in the 21st Century.=	Pesticides in the 21st Century. Pesticide Regulatory	AGRO 0066
Risks and Benefits of	Pesticides: An Environmental and Consumer View.=	AGRO 0059
r Studies on Long-Term Fate of	Pesticides: Characterization of Degradation and Tran	AGRO 0025

e.= Potential Bans of Selected	Pesticides: Economic Implications for U.S. Agricultur	AGRO	0083
egies as Alternatives to Chemical	Pesticides: Prospects, Challenges, and Recent Develo	AGRO	0080
	Pesticides: Risks and Benefits.=	AGRO	0058
to Monitor Small Watersheds for	Pesticides.= +Pesticide Use on the Farm: Methods	AGRO	0054
Derivatizations of	Pesticides.= Light as a Reagent for Chemical	AGRO	0147
sms of Resistance in Herbivorous	Pests to Natural, Synthetic, and Bioengineered Cont	AGRO	0114
sms of Resistance in Herbivorous	Pests to Natural, Synthetic, and Bioengineered Cont	AGRO	0153
sms of Resistance in Herbivorous	Pests to Natural, Synthetic, and Bioengineered Cont	AGRO	0150
sms of Resistance in Herbivorous	Pests to Natural, Synthetic, and Bioengineered Cont	AGRO	0097
sms of Resistance in Herbivorous	Pests to Natural, Synthetic, and Bioengineered Cont	AGRO	0133
Plastics Recycling via	PET Depolymerization.=	I&EC	0032
Recycled	PET.= Polymer Composites Using	PMSE	0072
of Polyethylene Terephthalate (	PET) from Plastics Waste on the Thermal Propertie	FUEL	0053
oroolefin Di+ Application of the	Peterson Olefination Reaction to the Synthesis of Flu	ORGN	0079
Analysis for the Determination of	Petrochemical Synthetic Flavoring Materials.= +n	AGFD	0186
and Entropy in the Selection of	Petrochemical Technologies from a Process Analysis	I&EC	0064
ed Hazardous Waste.=	Petrographic Techniques Applied to Cement–Solidifi	ENVR	0009
se Characterization of Unmodified	Petroleum Coke and Coal Gasification Slags.= +a	FUEL	0139
heno–Aromatic Groups in Heavy	Petroleum Fractions by ^{13}C NMR and Catalytic Deh	PETR	0147
oscopy in Secondary Processing of	Petroleum Fractions.= +plications of NMR Spectr	PETR	0146
Liquefied	Petroleum Gas.= Determination of Trace Metals in	PETR	0139
istribution and Sulfur Content in	Petroleum Residues.= +mination of Hydrocarbon D	PETR	0120
Isotope Analysis of Heavy	Petroleum Residues.= TPOP–IRMS Stable Carbon	PETR	0141
of Sulfur Components in Light	Petroleum Streams by High-Resolution GC with Che	PETR	0124
lytical Techniques for Analysis of	Petroleum. Applications of NMR Spe+ Modern Ana	PETR	0146
lytical Techniques for Analysis of	Petroleum. Multinuclear Solid–State+ Modern Ana	PETR	0136
lytical Techniques for Analysis of	Petroleum. Simultaneous Detection o+ Modern Ana	PETR	0118
cal Soil Treatment to Remediate	Petroleum– and Naphthalene–Contaminated Soils.=	BIOT	0139
to Laboratory Modernization: The	Pew Physical Chemistry Project.= +ed Approach	CHED	0125
ctroscopic Studies of sCD4(178)–	PE40 Conformation: A Chimeric Protein Having Pot	ANYL	0005
Deposition of Platinum from Pt (	PF$_3$)$_4$ on Atomically Clean Platinum Surfaces.= +e	COLL	0046
Perfluorooctylbromide (	PFOB) Emulsions.=	COLL	0080
he Role of Mobile–Phase+ High–	pH Anion–Exchange Separation of Carbohydrates: T	CARB	0066
nol–Water Mixt+ Cosolvent and	pH Effects on Sorption of Organic Acids from Metha	ENVR	0014
β–+ Effects of Temperature and	pH on the Foaming Properties and Conformation of	AGFD	0207
edenaturational S+ Thermal and	pH Stress in Denaturation of Cellobiohydrolase I: Pr	BIOT	0191
obile–Phase and Stationary–Phase	pH.= +eparation of Carbohydrates: The Role of M	CARB	0066
me c Peroxidase(III) in Alkaline	pH.= Photo–Induced Reduction of Native Cytochro	BIOL	0097
with Surfactant Solutions at Low	pH.= +al of Chromium from Soils by Soil Washing	ENVR	0013
rization of (R$_3$P)MCl{[Ph$_2$P(S)]$_n$[	Ph$_2$P(O)]$_3$–nC}, Where M = Pd or Pt and n = 1, 2, o	INOR	0265
Behavior of [cod]M{[Ph$_2$P(S)]$_n$[	Ph$_2$P(O)]$_3$–nC}, where M = Rh and Ir and n = 1, 2, a	INOR	0264
Characterization of (R$_3$P)MCl{[	Ph$_2$P(S)]$_n$[Ph$_2$P(O)]$_3$–nC}, Where M = Pd or Pt and	INOR	0265
d Fluxional Behavior of [cod]M{[	Ph$_2$P(S)]$_n$[Ph$_2$P(O)]$_3$–nC}, where M = Rh and Ir and	INOR	0264
0 Pellets as a Function of Time,	pH, Carbonate/Bicarbonate, and Oxygen Activities.=	INOR	0236
Vidrio Convencional para Medir	pH, por un Electrodo de Acero Inoxidable 316/Pelicu	ANYL	0066
3–hydroxyalkanoate) Conjugates:	PHA–Carbohydrate and PHA–Synthetic Polymers.=	POLY	0041
njugates: PHA–Carbohydrate and	PHA–Synthetic Polymers.= +hydroxyalkanoate) Co	POLY	0041
ve Lignin and Mn Peroxidases in	Phanerochaete chrysosporium and in a Heterologous	BIOT	0237
mes During Wood Degradation by	Phanerochaete chrysosporium.= +ydrolyzing Enzy	BIOT	0208
Nonlinear Swelling of	Phantom Networks.=	POLY	0167
and Pharmaceutical Analysis—I.	Pharmaceutical Analysis and New Drug Developmen	ANYL	0110
nalysis—III. Chromatography and	Pharmaceutical Analysis in the 1990s.= +eutical A	ANYL	0132
and Ne+ Chromatography and	Pharmaceutical Analysis—I. Pharmaceutical Analysis	ANYL	0110
y for the+ Chromatography and	Pharmaceutical Analysis—II. Role of Chromatograph	ANYL	0116
Pharmace+ Chromatography and	Pharmaceutical Analysis—III. Chromatography and	ANYL	0132
Products.= Use of HPLC in	Pharmaceutical Development of Antibiotic	ANYL	0112
yers Squibb Ph+ Careers in the	Pharmaceutical Industry: Chemistry at the Bristol–M	CMEC	0028
York.= The	Pharmaceutical Industry: Early History in New	HIST	0034
osure to Cyclophosphamide in the	Pharmaceutical Industry.= +g of Occupational Exp	CEI	0010
Technologies and the Need of	Pharmaceutical Industry.= Trends in Separation	ANYL	0111
Chromatography of Proteins and	Pharmaceutical Molecules Using Porous Polymeric P	ANYL	0031
istry at the Bristol–Myers Squibb	Pharmaceutical Research Institute.= +ustry: Chem	CMEC	0028
ses in the Synthesis of New	Pharmaceuticals and Materials.= Lipases and Estera	BIOT	0093
ggedness of Impurity Analysis in	Pharmaceuticals When Employing HPLC Analysis w	ANYL	0114
Production of Fine Chemicals and	Pharmaceuticals.= +ogenation Technology for the	CATL	0001
Analysis of	Pharmaceuticals.= Alternative Strategies to the	ANYL	0121
in Foods, Flavors, Cosmetics, and	Pharmaceuticals.= +ds in Botanical Extracts Used	AGFD	0172
or in Chromatographic Analysis of	Pharmaceuticals.= +ference Standards, a Key Fact	ANYL	0115
d Iminophenytoin+ Comparative	Pharmacodynamics of Phenytoin, Thiophenytoin, an	MEDI	0134
ign.=	Pharmacokinetic Considerations in Peptide Drug Des	MEDI	0193
in Europe and the Importance of	Pharmacokinetic Studies.= +edicines—Regulation	AGRO	0077
e in Cattle Administered via a +	Pharmacokinetics and Tissue Residues of Albendazol	AGRO	0109
alogues: A Possible Dissociation of	Pharmacological Activities.= +Tocainide Chiral An	MEDI	0098
+ial of Chromatographic Data for	Pharmacological Classification and Drug Design.=	ANYL	0043
Esters of Prednis+ Synthesis and	Pharmacological Evaluation of Novel 6–Carboxylate	MEDI	0170
Diastereomeric Preparation and	Pharmacological Evaluation of the Selective Protei	MEDI	0026
hydro–5–Hydrox+ Synthesis and	Pharmacological Evaluation of 1–Aminoethyl–3,4–Di	MEDI	0127
Muscarones.= Synthesis and	Pharmacological Reinvestigation of Chiral	MEDI	0141
ynes: Asymmetric Synthesis and	Pharmacological Studies of Pumiliotoxin A Alkaloids	ORGN	0275
e Inhibitors: Design, Biochemical,	Pharmacological, and Clinical Properties.= +ptidas	MEDI	0002

Activity and Drug Design.	Pharmacology of Muscarinic Receptors.= Receptor	MEDI 0078
	Pharmacophore Exploration.=	CINF 0019
tagonist+ Analysis of the Second	Pharmacophore Fragment of norBNI in κ–Opioid An	MEDI 0157
nal Searching Imp+ Looking for	Pharmacophores in 3D Databases: Does Conformatio	CINF 0026
: Mechanisms and Applica+ Gas–	Phase Aluminum Atoms from Pulsed Laser Ablation	PHYS 0353
Gas–	Phase Analysis of Diamond-Producing Plasmas.=	PHYS 0148
:bohydrates: The Role of Mobile–	Phase and Stationary-Phase pH.= +paration of Ca	CARB 0066
ation of Silicon-Containin+ Gas	Phase and Surface Chemical Reactions in the Prepar	INOR 0121
ıermodynamic Interpretation of	Phase Behavior in a Transreacting Polyester/Polycar	PMSE 0074
3teady Shear and Cessation of +	Phase Behavior of a Polymer Blend Solution during	PMSE 0134
ȝht Scattering during Steady Sh+	Phase Behavior of a Polymer Solution Studied by Li	PMSE 0107
·on—A Small-Angle Neutron S+	Phase Behavior of Aqueous Poly(vinyl–alcohol) Solut	POLY 0171
ures.=	Phase Behavior of Polymer-Supercritical Fluid Mixt	PMSE 0036
d State.=	Phase Behavior of Polyrotaxanes: Control of the Soli	COLL 0058
ɜes of Polymer Blends with UCST	Phase Behavior.= +rs. Flow-Induced Phase Chan	PMSE 0132
Matrices.= Inorganic	Phase Behaviors Mediated by the Organic Polymer	POLY 0364
ʲimide Ble+ Thermally Induced	Phase Behaviors of Aromatic Polybenzimidazole/Pol	PMSE 0164
ıges in Polymers. Flow-Induced	Phase Changes of Polymer Blends with UCST Phase	PMSE 0132
ke and Coal Gasification Slags.+	Phase Characterization of Unmodified Petroleum Co	FUEL 0139
metallic Precursors. MOCVD Gas–	Phase Chemistry.= +ronic Materials from Organo	COLL 0006
Solution	Phase Conformation of Rapamycin.=	MEDI 0169
Pulses.= Molecular Geometric	Phase Development and Phase-Controlled Optical	PHYS 0067
near and Cyclic Polystyrene/Po+	Phase Diagrams and Spinodal Decomposition and Li	POLY 0295
s+ Application of Matrix Solid–	Phase Dispersion (MSPD) to the Extraction and Sub	AGFD 0060
Fluorescence Probes of Solution	Phase Dynamics.=	CHED 0201
Preparative Reversed–	Phase Elution Chromatography of Proteins.=	ANYL 0029
etrating Polymer Networks. Two	Phase Epoxy Interpenetrating Themosets.= +rpen	PMSE 0087
d Agricultural Analyses+ Solid–	Phase Extractions for Sample Preparation in Food an	AGFD 0175
the Vapor–	Phase Growth of ZnO.= Change in Crystal Habit in	PHYS 0293
ystem Peaks in Normal Bonded	Phase HPLC with Diode Array Multichannel Detecti	ANYL 0087
ınel Detections in Normal Bonded	Phase HPLC.= +sional Separations and Multichan	ANYL 0088
ural Evolution of a Silicon Oxide	Phase in Nafion Membranes by an in Situ Sol-Gel R	POLY 0285
ture and Function at Condensed–	Phase Interfaces.= +l Chemistry at the Edge: Struc	ANYL 0026
Predicting the	Phase Inversion in Immiscible Liquid Systems.=	PMSE 0024
ınd Chem+ Studies of Some Gas–	Phase Metal Oxidation Reactions by Photoelectron a	PHYS 0352
namics of the Mesospheri+ Gas–	Phase Metal Reactions—I. Chemical Kinetics and Dy	PHYS 0071
and Aluminum Species.=+ Gas–	Phase Metal Reactions—II. Kinetic Studies of Boron	PHYS 0086
r Oxygen Reaction with+ Gas–	Phase Metal Reactions—III. Kinetics of the Molecula	PHYS 0141
Interactions in ML_n+ Sy+ Gas–	Phase Metal Reactions—IV. Study of Ligand–Ligand	PHYS 0171
Transition-Metal Atoms+ Gas–	Phase Metal Reactions—V. Association Reactions of	PHYS 0198
te and Electronically Exc+ Gas–	Phase Metal Reactions—VI. Reactions of Ground Sta	PHYS 0349
of Excited Rare Gas Atom+ Gas–	Phase Metal Reactions—VII. Harpooning Reactions	PHYS 0367
Combusti+ Homogeneous Gas–	Phase Modeling of Boron/Oxygen/Hydrogen/Carbon	FUEL 0097
Vapor–	Phase Molecular Routes to Ceramic Powders.=	INOR 0175
s Transition Temperatures on the	Phase Morphology of Simultaneous IPNs.= +Glas	PMSE 0025
Tubular R+ Modeling the Two–	Phase Nitration of Toluene in a Co–Current Packed	I&EC 0006
urse o+ Evolution of the Gallium	Phase of GaHMFI Aromatization Catalysts in the Co	PETR 0046
Secondary Amines by Reversed–	Phase Paired-Ion High-Performance Liquid Chroma	ANYL 0089
of Mobile-Phase and Stationary–	Phase pH.= +paration of Carbohydrates: The Role	CARB 0066
and Supercrit+ Measurement of	Phase Phenomena of Mixtures of Coatings Materials	PMSE 0037
AlH Gas–	Phase Reaction Kinetics.=	PHYS 0301
en Oxides in Homogeneous Gas–	Phase Reactions: Combustion Emission Modeling.=	FUEL 0127
Gas	Phase Reactions.= Use of an MS or MSD to Study	CHED 0198
hemical Processes in the Single–	Phase Regions of SDS/1-Pentanol/Cyclohexane/Wa	COLL 0024
lticolumn Preparative Reversed–	Phase Sample Displacement Chromatography of Pep	ANYL 0040
rks.=	Phase Separation in Interpenetrating Polymer Netwo	PMSE 0022
tals.=	Phase Separation in Polymer-Dispersed Liquid Crys	COLL 0014
g Polymer Networks. Kinetics of	Phase Separation in Polyurethane/Polystyrene Semi	PMSE 0021
Gelation and	Phase Separation of PVC/Plasticizer Solutions.=	POLY 0173
roscopy as a Detector for Liquid–	Phase Separation Systems.= +ardment Mass Spect	ANYL 0125
Rheo-Optics of Flow-Induced	Phase Separation.=	PMSE 0106
Laboratory Studies.= Gas–	Phase Solar Detoxification of Hazardous Wastes:	ENVR 0035
fer Interactions on Polymer Blend	Phase Structure.= +the Influence of Charge Trans	PMSE 0076
Critical Review of E+ Reversed–	Phase Supports as Surrogates for Natural Sorbents:	ANYL 0071
STM.+ Chemisorption-Induced	Phase Transformations of Metal Surfaces Studied by	PHYS 0126
hombic Pseudohexagonal Rotator–	Phase Transition in n-Heneicosane.= +the Orthor	POLY 0354
Excitation Spectra Pattern and	Phase Transition in Tetracene Crystals and Tetracen	PHYS 0274
Scannin+ Surface Diffusion and	Phase Transition on the Ge(111) Surface Studied by	PHYS 0099
Stress-Strain Relationships and	Phase Transitions for Networks of Semirigid Chains.	COLL 0016
en Isotope Effect and Structural	Phase Transitions in LA_2CuO_4-Based Superconducto	PHYS 0010
hrough Pressure-Induced Micellar	Phase Transitions.= +ery from Reversed Micelles t	BIOT 0046
r the Solar Detoxification of Gas–	Phase Wastes.= +r TiO_2: A Potential Technique fo	COLL 0082
ic Organic Chemistry in the Gas	Phase: what Can We Learn from Reactions without S	ORGN 0121
termediates Isolated from the Gel	Phase.= +ecular Sieves: Recrystallization of the In	PETR 0019
othermal Water as the Continuous	Phase.= +of Drilling Fluids Systems Based on Ge	I&EC 0011
etric Phase Development and	Phase-Controlled Optical Pulses.= Molecular Geom	PHYS 0067
hloro, α-Hydroxy, and N-Bloc+	Phase-Managed Organic Synthesis: Activation of α–C	ORGN 0259
Electric-Field-Induced Transient	Phase-Separation in w/o-Microemulsion.= +cs of	COLL 0146
: Description and Application to	Phase-Separation Kinetics in PC/PMMA Blends.=	POLY 0278
e Rubber on the Polymerization	Phase-Separation Process, and Generated Morpholog	POLY 0112

Left context	Title fragment	Journal	Page
eactions Catalyzed by Nickel and	Phase-Transfer Catalysts: New Mechanisms for Dou	COLL	0117
HCl from Chlorohydridometal +	Phase-Transfer-Catalyzed Reductive Elimination of	INOR	0381
Architectures Based on Optical-	Phased Arrays.= New Optical Memory	BIOT	0025
Application of an Aqueous Two-	Phased System to the A-B-E Fermentation.= +and	BIOT	0159
d Lectins Evolved from a Co+	Phaseolus vulgaris α-Amylase Inhibitors, Arcelins, an	AGFD	0041
Precursor	Phases and Nucleation of Zeolite TON.=	PETR	0040
ry Large Pore Aluminophosphate	Phases and Their Physicochemical Characterization.	PETR	0093
hemical R+ Effects of Dispersed	Phases in Naturally Occurring Fluids on the Electroc	COLL	0005
anes: A New Family of Crystalline	Phases of Carbon.= +ond-Like Carbon. Carboph	FUEL	0034
Controlled Plastic and Conserved	Phases of Catalysis by Glycosylases.= +Separately	BIOL	0152
Re(0001) Su+ Two-Dimensional	Phases of Sulfur Formed by Coalescence of Atoms on	PHYS	0124
sfer and Relaxation in Condensed	Phases. Chord-Distribution Analysis+ Energy Tran	PHYS	0060
sfer and Relaxation in Condensed	Phases. Conformation and Dynamics+ Energy Tran	PHYS	0107
sfer and Relaxation in Condensed	Phases. Energy Transfer and Relaxat+ Energy Tran	PHYS	0080
sfer and Relaxation in Condensed	Phases. Excited-State Structure and+ Energy Tran	PHYS	0018
sfer and Relaxation in Condensed	Phases. Time-Domain Analysis of th+ Energy Tran	PHYS	0034
sfer and Relaxation in Condensed	Phases. Torsional Solitons in Organi+ Energy Tran	PHYS	0129
Sieves: Identification of Transient	Phases.= +Solution Synthesis of AlPO4 Molecular	PETR	0020
Cellulose-Based Chiral Stationary	Phases.= +alysis of Some Aromatase Inhibitors on	ANYL	0133
Micropellicular Stationary	Phases.= Rapid HPLC of Biopolymers with	ANYL	0128
Large-Scale Production of	PHB on Beet Sugar.=	CARB	0002
emotactic Tripeptide F-Met-Leu-	Phe-OH.= +-Terminal Modified Analogues of Ch	MEDI	0166
tagen 15,16-Dihydrocyclopenta[a]	Phenanthren-17-one.= +,4-Dihydrodiol of the Mu	ORGN	0118
Synthesis and Structure of μ-	Phenanthrene Bis(TricarbonylChromium(0)).=	INOR	0207
Spectroscopy of	Phenanthrene.= Two-Photon Absorption	PHYS	0328
Spectroscopy of	Phenanthrene.= Two-Photon Absorption	CHED	0085
etabolites of the Cyclopenta[a]-	Phenanthrenes: Synthesis of the trans-3,4-Dihydrod	ORGN	0118
Oxidation of Tris(1,10-	Phenanthroline)Iron (II) by Hydrogen Peroxide.=	INOR	0222
of Ractopamine Hydrochloride: A	Phenethanolamine-Repartitioning Agent.= +sidues	AGRO	0131
Using a Silyl	Phenol.= Solvolysis-Triggered Alkylation of DNA	ORGN	0191
Bound with a Sterically Hindered	Phenolate.= +onductivity of Poly(ethylene Oxide)	POLY	0091
-Extrusion Cere+ Implication of	Phenolic Acids as Texturizing Agents during Cooking	AGFD	0150
ds.=	Phenolic Antioxidants as Antimutagens in Plant Foo	AGFD	0045
c Enzymes.=	Phenolic Antioxidants as Inducers of Anticarcinogeni	AGFD	0089
Thermal Degradation of	Phenolic Antioxidants.=	AGFD	0049
Production of	Phenolic Compounds by Cultured Plant Cells.=	AGFD	0002
Changes of	Phenolic Compounds during Plum Processing.=	AGFD	0171
oods, Flavors, Cosmetics, and +	Phenolic Compounds in Botanical Extracts Used in F	AGFD	0172
ogenicity of Flavonol Quercetin+	Phenolic Compounds in Food and Health. Anticarcin	AGFD	0113
ntioxidants from Plant Material+	Phenolic Compounds in Food and Health. Natural A	AGFD	0015
rotein Interactions.=	Phenolic Compounds in Food and Health. Tannin-P	AGFD	0148
y of Curcumin and Curcuminoi+	Phenolic Compounds in Foods and Health. Chemistr	AGFD	0043
y Effect of Orally Administered+	Phenolic Compounds in Foods and Health. Inhibitor	AGFD	0065
ms Underlying the Mutagenic, +	Phenolic Compounds in Foods and Health. Mechanis	AGFD	0086
Compounds of Citrus Fruits an+	Phenolic Compounds in Foods and Health. Phenolic	AGFD	0167
ol Complexation.=	Phenolic Compounds in Foods and Health. Polyphen	AGFD	0001
Enzymatic Oxidation of	Phenolic Compounds in Fruits.=	AGFD	0151
Antioxidant Activity of	Phenolic Compounds in Meat Model Systems.=	AGFD	0019
	Phenolic Compounds in Spices.=	AGFD	0047
	Phenolic Compounds of Brassica Oilseeds.=	AGFD	0168
ompounds in Foods and Health.	Phenolic Compounds of Citrus Fruits and Their Prod	AGFD	0167
Nerve-Stimulating Agent.=	Phenolic Compounds of Piper betel as Flavoring and	AGFD	0154
Contribution of	Phenolic Compounds to Smoke Flavor.=	AGFD	0173
Effluents Containing Chlorinated	Phenolic Compounds.= +tment of Pulp Bleaching	BIOT	0137
HPLC Analysis of Food	Phenolic Compounds.=	AGFD	0005
Oxidase Activity by	Phenolic Compounds.= Inhibition of Polyphenol	AGFD	0152
Mass Spectrometric Analysis of	Phenolic Compounds.=	AGFD	0003
Synthesis of	Phenolic Derivatives of Thioridazine.=	MEDI	0132
Using+ Facile Measurement of	Phenolic End-Groups in bisPhenol A Polycarbonate	POLY	0083
Resonance Detection of	Phenolic Free Radicals.= Electron Paramagnetic	AGFD	0004
Tocopherol as a Natural	Phenolic Inhibitor of Nitrosation.=	AGFD	0091
	Phenolic IPNs for Laminate Applications.=	PMSE	0007
s in th+ Measurement of Urinary	Phenolic Metabolites of Environmental Contaminant	CEI	0002
e Defense Systems by	Phenolic Plants Constituents.= Role of Antioxidativ	AGFD	0018
Water-Compatible	Phenolic Resins.=	PMSE	0139
Vegetabl+ Glycosidically Bound	Phenolics and Other Compounds in an Umbelliferous	AGFD	0170
Plant	Phenolics as Cytotoxic Antitumor Agents.=	AGFD	0092
Carcinogens.= Plant	Phenolics as Inhibitors of Mutagenesis and	AGFD	0118
eact+ Aerobic Biodegradation of	Phenolics in Optimally Designed Sequencing Batch R	BIOT	0138
Suspension Culture+ Increased	Phenolics Production in Lithospermum erythrorhizon	BIOT	0229
ironmental Fac+ Red Raspberry	Phenolics: Influences of Processing, Variety, and Env	AGFD	0169
c, and Chemopreventive Effects of	Phenols and Catechols.= +Mutagenic, Carcinogeni	AGFD	0086
Carcinogenic Response by Plant	Phenols.= Carcinogenicity and Modification of	AGFD	0115
@(PMe3)4(η2-CH 2PMe2)H with	Phenols.= +lectron-rich Complexes W(PMe3)6 and	INOR	0086
d Tongue Carcinogenesis by Plant	Phenols.= +tective Effects against Liver, Colon, an	AGFD	0090
Nafazatrom.=	Phenothiazine Antioxidants Structurally Related to	MEDI	0174
In Vitro Ether Cleavage of 3-	Phenoxybenzoic Acid to 3-Hydroxybenzoic Acid by C	AGRO	0018
Photoaffinity Labeling.=	Phenoxydiazirines as Useful Carbene Precursors for	ORGN	0105
nd Potential Agrochemicals. 2-(1-	Phenoxyethyl)-1,3,4-Thiadiazoles as Acaricides.=	AGRO	0071
del Compounds of Poly [N,N'-bis(	phenoxyphenyl)pyromellitimide].= +es of New Mo	PMSE	0151

merization.= Novel Propenyl	Phenyl Ether Analogs for UV-Induced Cationic Poly	POLY	0017
	Phenyl Ring Motions in Polycarbonates.=	POLY	0080
Photochemistry of Benzyl	Phenyl Sulfide.=	ORGN	0165
Activity of a Series of Substituted	Phenyl-Cinnolines.= +sis and Pollen-Suppressant	AGFD	0042
ns with 4-Benzoyl-5-Methyl-2-	Phenyl-Pyrazol-3-Thione and Tri-n-Octylphosphine	NUCL	0073
-Products of the Chlorination of	Phenylalanine: Products and Implications for Disinfe	ENVR	0031
acterization of Silicon-Containing	Phenylated Soluble Aramids.= +iques for the Char	POLY	0086
ropertiex of Quinic Acid and Its	Phenylboronate and Isopropylidene Derivatives in th	INOR	0263
and anti-β-	Phenylcysteines.= Enantioselective Routes to syn-	ORGN	0115
Hypocholesterolemic Activity of	Phenylene and Pyridylene Diurea Derivatives as ACA	MEDI	0109
Dispersion of the $\chi^{(3)}$ of Poly(p-	phenylene Benzobisthiazole)-Based Molecular Comp	POLY	0213
de Copolymers Containing ortho-	Phenylene Rings.= +on of Ordered Heterocycle-Imi	POLY	0042
ynamics in N-Deuterated Poly(p-	phenylene Terephthalamide).= +ide Segmental D	POLY	0058
Chain Polymers: An Ionic Poly(p-	phenylene) Analog.= +Synthesis of Soluble Rigid-	POLY	0024
n) Based on bis(5-Carboxy-1,3-	phenylene)-32-crown-10 and 4,4'-Isopropylindenedip	POLY	0044
a Series of Monodisperse 1,3,5-	Phenylene-Based Hydrocarbon Dendrimers, Includin	POLY	0320
of a Series of Monodisperse 1,3,5-	Phenylene-Based, Fluorinated Dendrimers.= +ion	POLY	0236
of a Series of 2,4-Diamino-6-[2-	Phenylethenyl]Pyrido[2,3-d] Pyrimidine Derivatives.	MEDI	0022
vation of Mandelonitrile Lyase by	Phenylglyoxal.= +oducts Formed during the Inacti	BIOL	0087
A High-Purity Nonpyrophoric	Phenyllithium Solution.=	ORGN	0226
ant D,L-3-Hydroxy-3-Ethyl-3-	Phenylpropionamide (HEPP) in Plasma and Urine.=	ANYL	0082
ide L-2,6-dimethyltrysosyl-N-(3-	Phenylpropyl)-D-Alaninamide.= +e Dipeptide Am	MEDI	0152
of Poly(aryl Sulfide	Phenylquinoxalines).= Preparation and Properties	POLY	0026
cnors. Selective Activation ove+	Phenylselenoglycosides as Novel, Versatile Glycosyl D	CARB	0035
+hiral Titanocenes: 1-Methyl, 2-	Phenyltitanocene Dichloride and Its 1-,3-Analog.=	INOR	0281
cmparative Pharmacodynamics of	Phenytoin, Thiophenytoin, and Iminophenytoin.=	MEDI	0134
alogs in+ Structure Activity of	Pheromone Biosynthesis Activating Neuropeptide An	AGRO	0011
ue Analyses of Lepidopteran	Pheromones in Apples, Grapes, and Peaches.= Resid	AGRO	0148
s and Peptides for Insect Control.	Philanthotoxin Analogs.= +ium: Juvenile Hormone	AGRO	0006
ductors.= Nonlinear Electron-	Phonon Interactions in High-Temperature Supercon	PHYS	0012
Computational Investigation of	Phosphaalkynes: A New Building Block in Organic+	COMP	0060
Alkaline	Phosphatase Kinetics.=	CHED	0183
Inhibitor of Alkaline	Phosphatase.= Fumaric Acid Monoazide: A New	BIOL	0057
nt Analogue of Potassium Titanyl	Phosphate (KTP).= +: A Centrosymmetric Aliovale	INOR	0256
Phosphoribulokinase's Sugar	Phosphate Binding Site.= Identification of	BIOL	0153
es Based upon Silicate or Alumino	Phosphate Chemistry.= +tructural Molecular Siev	CATL	0028
e an Intermediate in Reactions of	Phosphate Compounds in Water?.= +Metaphosphat	INOR	0293
ixed-Valance Titanium(III/IV)	Phosphate Compounds with Nasicon-Related Structu	INOR	0353
-ATPase: Formation of Citraconyl	Phosphate Driven by a Proton Gradient.= +for H+	BIOL	0104
Amorphous Aluminum	Phosphate Gels.=	PETR	0050
Biosynthesi+ Role of Magnesium	Phosphate in Ammonium Control of Cephamycin C	BIOT	0088
termediate and Evaluation of a 3-	Phosphate Mimic.= +Mimics of the Tetrahedral In	ORGN	0029
of Various Moieties as Stable 3-	Phosphate Mimics in Aromatic Analogs of EPSP.=	ORGN	0028
Sewage, Part 1: Odor Control and	Phosphate Removal.= +Peroxide and Iron Salts on	ENVR	0034
stal Habit of Calcium Hydrogen	Phosphate Used to Phosphors.= A Study on the Cry	INOR	0361
uantitative Analysis of Inorganic	Phosphates by Phosphorus-31 Nuclear Magnetic Res	INOR	0291
Properties of Layered Metal(IV)	Phosphates toward Organic Bases, Metal Complexes,	PETR	0107
Design of Photoactive Layered	Phosphates.=	CATL	0021
Open-Morphology Zirconium	Phosphates, Aluminas, and Mixed Oxides.=	PETR	0051
Synthesis of Truncated Glycosyl	Phosphatidylinositides.=	CARB	0031
-1 with El+ Reactivity of an Iron	Phosphide Acetyl Complex Cp(PPH2)(CO)FeCOCH3	INOR	0240
Laser Etching of Indium	Phosphide by Molecular Chlorine.= Thermal and	COLL	0214
Metalorganic Syntheses of	Phosphide Semiconductor Nanoclusters.=	INOR	0102
lgallium on Silicon and Indium	Phosphide.= Decomposition of Mono- and Trimethy	INOR	0123
et+ Formation and Reactivity of	Phosphido-Bridged Iridium, Molybdenum Heterobim	INOR	0385
nd Ki+ Steric Effects in Metal-	Phosphine Bonding: Determination of the Enthalpy a	INOR	0192
tivity of Low-Valent Vanadium	Phosphine Complexes and Mixed Phosphine-Alkoxid	INOR	0129
n of+ Discrete Chiral Rhodium	Phosphine Complexes for Asymmetric Hydrosilylatio	CATL	0005
Bidentate	Phosphine Complexes.= Ring Strain of Manganese	INOR	0193
yl Complexes of a New Diertiary	Phosphine Ether.= +eactions for Group VI Carbon	INOR	0333
Iridium(I) Complexes.= Facile	Phosphine Ligand Exchange between Square Planar	INOR	0211
mal Stability+ Poly(Arylene Ether	Phosphine Oxide)s—I. Overview of Chemistry, Ther	PMSE	0052
Poly(Arylene Ether	Phosphine Oxide)s—II. Pyrolysis Studies.=	PMSE	0053
dification+ Poly(Arylene Ether	Phosphine Oxide)s—III. Investigations of Surface Mo	PMSE	0054
rimethyl Phosphite and Triphenyl	Phosphine.= +m(IV) Metallacyclopentanes with T	INOR	0330
of Carbon Monoxide and+ Bis(	phosphine)palladium Catalysts for Copolymerization	INOR	0155
Cyclohexene Catalyzed by (Aqua)(	Phosphine)Ruthenium(IV) Complexes.= +ation of	INOR	0067
ox Chemistry of a Series of (Oxo)(	Phosphine)Ruthenium(IV) Complexes.= +the red	INOR	0068
ion of Thioanisoles by (Oxo)(	phosphine)ruthenium(IV) Complexes.= S-Oxygenat	INOR	0130
Phosphine Complexes and Mixed	Phosphine-Alkoxide/Thiolate Complexes.= +dium	INOR	0129
of Unusuall+ Carbonylation and	Phosphine-Promoted Alkyl-CO Migration Reactions	INOR	0239
Study.= Oxidation of	Phosphines by Gallium Peroxides: A Mechanistic	INOR	0389
New Electron-Rich Chiral	Phosphines for Asymmetric Catalysis.=	CATL	0004
bstitution of Fe(CO)5 with Alkyl	Phosphines: Fe(CO)4PR3 and Fe(CO)3(PR)2 Are Sing	INOR	0253
Two-Coordinate	Phosphines.= New Derivative Chemistry of	INOR	0325
ction with Primary and Secondary	Phosphines.= +on of Me2GaC≡CSiMe3 and Its Rea	INOR	0217
Early-Metal	Phosphinidenes.=	INOR	0363
nomeric and Dimeric Arsino- and	Phosphinogallanes.= +and Characterization of Mo	INOR	0308
tallacyclopentanes with Trimethyl	Phosphite and Triphenyl Phosphine.= +m(IV) Me	INOR	0330
Spirolactams (Carbamates) as	Phosphodiesterase Inhibitors.=	MEDI	0189

ssessing Potent cAMP and cGMP	Phosphodiesterase Inhibitory Activity. +pounds Po	MEDI	0118
Recognition and Cleavage of	Phosphodiesters.= Polyaza Clefts for the	ORGN	0054
and Properties of Analogs of	Phosphoinositides and Sphingolipids.= Synthesis	BIOL	0017
	Phospholipase A_2 Engineering.=	BIOL	0016
+Recombinant Human Secretory	Phospholipase A_2 from Mammalian Cell Culture.=	BIOL	0072
uilibri+ Interfacial Catalysis by	Phospholipase A_2: Determination of the Rate and Eq	BIOL	0019
ation of+ Interfacial Catalysis by	Phospholipase A_2: Structural and Kinetic Characteriz	BIOL	0020
Binding to	Phospholipase A_2.= X-ray Studies of Inhibitor	MEDI	0004
with Ligand-Modified	Phospholipids.= Affinity Precipitation of Proteins	BIOT	0020
Charged	Phospholipids.= HPLC Analysis of Negatively	ANYL	0134
Acid Conjugates of Ether	Phospholipids.= Stereospecific Synthesis of Amino	MEDI	0163
Functionalized Zirconium	Phosphonates.= Synthesis and Reactivity of	INOR	0084
hermolysis of O-Ethyl Esters of	Phosphoramidic Acids: Generation of Ethyl Metapho	ORGN	0110
anadate Anion an Analogue of the	Phosphorane Transition State in RNase?.= +the V	BIOL	0061
ands.= Phosphorus Chemistry.	Phosphoranide Anions (10-P-4) with Bidentate and	INOR	0344
Site.= Identification of	Phosphoribulokinase's Sugar Phosphate Binding	BIOL	0153
rt.=	Phosphoric Rock Dissolution: An Experimental Repo	PHYS	0294
NMR Spectra of Oligonucleotide	Phosphorodithioate d(CGCTpS$_2$-TPS$_2$-AAGCG):+	BIOL	0123
-State MAS NMR, and XRD of	Phosphorous-Promoted Hydrotreating Catalyst Prec	COLL	0198
m Hydrogen Phosphate Used to	Phosphors.= A Study on the Crystal Habit of Calciu	INOR	0361
	Phosphorus Atoms in Unusual Environments.=	INOR	0292
Polymers.= Application of	Phosphorus Chemistry to the Synthesis of High	INOR	0326
rch in Industrial Chemistry.=	Phosphorus Chemistry. A Life of Dedication to Resea	INOR	0321
tist, Colleague, Friend.=	Phosphorus Chemistry. Arthur Dock Fon Toy—Scien	INOR	0288
action.=	Phosphorus Chemistry. Mechanisms of the Wittig Re	INOR	0373
-4) with Bidentate and Trident+	Phosphorus Chemistry. Phosphoranide Anions (10-P	INOR	0344
Research in Industrial	Phosphorus Chemistry—A Reminiscence.=	INOR	0322
Disperse Silica.= Reaction of	Phosphorus Compounds in a Surface Layer of	COLL	0100
e Amazon Delta.=	Phosphorus Distribution in Sediments in and near th	GEOC	0011
.=	Phosphorus Flame Retardants for a Changing World	INOR	0346
noxide Laser+ Determination of	Phosphorus in Biological Samples by Phosphorus Mo	ANYL	0078
ations for Burial Flu+ Forms of	Phosphorus in Continental Margin Sediments: Implic	GEOC	0013
phorus in Biological Samples by	Phosphorus Monoxide Laser-Excited Molecular Fluo	ANYL	0078
-Step Preparation+ Reaction of	Phosphorus Pentahalides with Cyclic Ketones: A One	ORGN	0263
mplications for Burial Fluxes and	Phosphorus Residence Time in the Ocean.= +nts: I	GEOC	0013
Approaches to Low-Coordinate	Phosphorus Species.= Ring Fragmentation	INOR	0324
Chemistry and Biochemistry of	Phosphorus.=	INOR	0289
Polymer Derivatives.= New	Phosphorus- and Sulfur-Modified Carbohydrate	CARB	0030
ocholesterolemic Activity of Novel	Phosphorus-Containing ACAT Inhibitors.= +Hyp	MEDI	0110
End-o+ Nitrogen-, Sulfur-, and	Phosphorus-Ligated Compounds. Side-on (μ-N$_2$) vs.	INOR	0362
nalysis of Inorganic Phosphates by	Phosphorus-31 Nuclear Magnetic Resonance.= +A	INOR	0291
Inorganic Polymers with Skeletal	Phosphorus, Nitrogen, and Sulphur(VI) Atoms.=	POLY	0186
n: Inhibitors of Purine Nucleoside	Phosphorylase (PNP).= +ucture-Based Drug Desig	MEDI	0005
ctivity of 9-+ Purine Nucleoside	Phosphorylase Inhibitors: Synthesis and Biological A	MEDI	0028
Analysis.= Purine Nucleoside	Phosphorylase: Mechanism and Transition State	BIOL	0046
al Properties in Foo+ Enzymatic	Phosphorylation of Soy Protein to Improve Function	AGFD	0140
hanistic Studies of Adenosine-5'-	Phosphosulfate Kinase.= +zymatic Catalysis. Mec	BIOL	0008
Mechanism of Enzymatic	Phosphotriester Hydrolysis.=	BIOL	0010
cs of Azulene.=	Photionization via Transient States: The Photophysi	PHYS	0330
ilms Based on 4-Keto-Ba+ The	Photo- and Electrochromic Properties of Biochrom F	BIOT	0082
Ligands by Nucl+ Assistance of	Photo-Induced Coupling of Alkylidyne and Carbonyl	INOR	0188
eroxidase(III) in Alkaline pH.=	Photo-Induced Reduction of Native Cytochrome c P	BIOL	0097
ical Intersection and Its Effect on	Photoabsorption and Dynamics.= +of Ozone: A Con	PHYS	0191
h Tertiary Silanes as Examined by	Photoacoustic Calorimetry.= +of CpMn(CO)$_2$ wit	INOR	0386
of Ligand Exchange Reactions by	Photoacoustic Calorimetry.= +thalpy and Kinetics	INOR	0192
ne Fluoride) Monitored by Rheo-	Photoacoustic FT-IR Spectroscopy.= +oly(vinylide	POLY	0255
Design of	Photoactive Layered Phosphates.=	CATL	0021
of Molecules of Potential Use in	Photoactive Polymers by Time-Resolved Microwave	POLY	0214
rval Manduca sexta by the Use of	Photoaffinity Analog of JH II.= +ing Protein of La	AGRO	0014
Synthesis of Tritium-Labelled	Photoaffinity Juvenoids.=	AGRO	0016
Useful Carbene Precursors for	Photoaffinity Labeling.= Phenoxydiazirines as	ORGN	0105
n Complexed Pyrene LB Films.+	Photoamplified Storage Optical Memory with Polyio	BIOT	0144
Fusarium Ye+ Concentrations of	Photocarcinogenic Furocoumarins in Susceptible and	AGRO	0040
ion Compounds.=	Photocatalysis Induced by Light-Sensitive Coordinat	INOR	0340
Aqueous Media+ Heterogeneous	Photocatalysis with TiO$_2$ Semiconductor Particles in	COLL	0131
s Vycor Glass.=	Photocatalytic Behavior of Metal Carbonyls on Porou	INOR	0170
inetic Studies in Heterogeneous	Photocatalyzed Oxidations of Environmental Organic	INOR	0318
ydroxytricarboxylic Acid Comp+	Photochemical and Spectroscopic Studies of Alpha-H	INOR	0040
anes Catalyzed by Rh+ Efficient	Photochemical and Thermal Dehydrogenation of Alk	INOR	0152
	Photochemical Approach to Taxol.=	ORGN	0012
octadiene Complexes into Agos+	Photochemical Conversion of Chromium η^4-1,5-Cyclo	INOR	0209
Pt(111).= Surface	Photochemical Dynamics of CH$_3$Br Adsorbed on	PHYS	0261
xometal+ Selective and Catlytic	Photochemical Functionalization of Alkanes by Polyo	INOR	0286
heir Utilization as Restricted R+	Photochemical Probes of Starbust Dendrimers and T	POLY	0238
sitizer/Mediator or Donor/Acc+	Photochemical Properties of Covalently Attached Sen	INOR	0038
F$_3$.=	Photochemical Reactions of Metal Carbonyls with N	INOR	0195
s on Surfaces.=	Photochemical Reactions of Organometallic Molecule	INOR	0168
Systems.= Free-Radical and	Photochemical Reactions of Organophosphorus	INOR	0347
nzoylformamide in Homogen+ A	Photochemical Study of N-(12-Dodecanoic Acid) Be	ORGN	0166
ylation of Cyclohexane Catalyzed	Photochemically by d^8 Transition-Metal Carbonyls i	INOR	0254

s in Polymeric Slab Waveguides+	Photochemically Delineated Refractive Index Profile	POLY	0180
o Complexes.=	Photochemistry and Redox Catalysis Using Re and M	INOR	0343
Multielectron–Transfer	Photochemistry in Metalorganic Systems.=	INOR	0285
Homogeneous Metal–Catalyzed	Photochemistry in Organic Synthesis.=	INOR	0339
istry at the Liquid–Liq+ Organic	Photochemistry in Restricted Reaction Spaces: Chem	ORGN	0126
Time Domain.= Metalorganic	Photochemistry in the Millisecond to Picosecond	INOR	0287
.=	Photochemistry of $(Cp)_2TiCl_2$ on Porous Vycor Glass	INOR	0161
Solid– and Liquid–State	Photochemistry of an N–Acyl–1–Azadiene.=	ORGN	0111
	Photochemistry of Benzyl Phenyl Sulfide.=	ORGN	0165
of Nonmetallic Silver Clusters and	Photochemistry of Colloidal Silver Sols.= +perties	COLL	0130
ed Solvents.= Photophysics and	Photochemistry of Ru(II)–Diimine Complexes in Mix	INOR	0279
etry upon the Solid–State Emis+	Photochemistry. Influence of Crystallographic Symm	INOR	0157
Advances in Organic	Photochemistry.=	ORGN	0142
Cage Effects in Organometallic	Photochemistry.=	INOR	0158
Spectroscopy of Rhodopsin	Photochemistry.= Subpicosecond Transient	BIOT	0080
Wavelength–Dependent Organic	Photochemistry.=	ORGN	0135
Novel	Photochromic Liquid Crystal Polymers.=	POLY	0219
Complexation.= Control of	Photochromism of Diacetylene LB Films by Polyion	BIOT	0146
+and Synthesis of a New Class of	Photocrosslinkable Nonlinear Optical Polymers.=	POLY	0131
Leachable Glass Hybrid Syste+	Photocurable Cements Based on Polyalkenoates/Ion–	POLY	0262
Intramolecular [4+4]	Photocycloaddition of Tethered 2–Pyridones.=	ORGN	0235
pic and Chemical Studies of the	Photodecomposition Pathways of Adamantyldiazirine	ORGN	0163
Metal Oxides in the	Photodegradation of Furans.= Catalytic Role of	COLL	0190
of the TpdU cis–syn Cyclobutane	Photodimer of DNA.= +of the Replicative Bypass	BIOL	0113
tone Resonance Raman Spectra of	Photodissociating Nitromethane in Solution.= +er	PHYS	0302
Elimination in the	Photodissociation of Ethylene.= Dynamics of H_2	PHYS	0163
hlorine Atoms Produced from the	Photodissociation of S_2Cl_2.= +ode Laser Probe of C	PHYS	0299
ns of SO ($X^3\Sigma^-$) from the 193–nm	Photodissociation of $SOCl_2$.= +l Energy Distributio	PHYS	0240
er Ions: Reactivity, Structure, and	Photodissociation Studies.= +erties of Boron Clust	PHYS	0262
n Single Crystals of $YBa_2Cu_3O_6.3$:	Photodoping to the Metallic State.= +onductivity i	PHYS	0011
Development. Measurement of the	Photoelectric Effect.= +cal Chemistry Laboratory	CHED	0209
Photogenerated Ketyl Radicals in	Photoelectrochemical (PEC) Cell.= +on of Dyes by	COLL	0081
nc Reverse Micel+ Synthesis and	Photoelectrochemical Reactions of CdS in Micelles a	COLL	0148
ase Metal Oxidation Reactions by	Photoelectron and Chemilelectron Spectroscopy.=	PHYS	0352
of Free Radicals+ Negative–Ion	Photoelectron Spectroscopic Studies of the Structure	PHYS	0055
ults with Theory.+ Negative–Ion	Photoelectron Spectroscopy: Comparison of New Res	PHYS	0059
need Raman Scattering and X–ray	Photoelectron Spectroscopy.= +ing Surface–Enha	POLY	0087
ithium Aluminate–Fatty Acid +	Photoelectron Transfer with Porphyrins in Layered L	INOR	0272
	Photoelimination of H_2 form 1,3–Butadiene.=	PHYS	0325
Flow and Structural Dynamics in	Photoexcited Heme Proteins.= +rements of Energy	PHYS	0335
erfacial Electron–Transfer Gate+	Photoformation of a Porphyrin Cation Lattice by Int	BIOL	0136
n the Photoreduction of Dyes by	Photogenerated Ketyl Radicals in Photoelectrochemi	COLL	0081
etal Films by Electron Transfer to	Photogenerated Organometallic Radicals.= +um M	INOR	0319
ecuction of Vicinal Dihalides by	Photogenerated Ru(I) Complexes: Competition betwe	INOR	0284
George Eastman—Methods of	Photography.=	CHAL	0030
ms.=	Photoinduced C–H Activation in Metalorganic Syste	INOR	0167
Catalysis: Shape Selectivity—II.	Photoinduced Catalytic Reactions in Zeolite Cages.=	CATL	0020
$2Cu_3O_6.3$: Photodopi+ Transient	Photoinduced Conductivity in Single Crystals of YBa	PHYS	0011
r in LB Films.=	Photoinduced Electron Transfer and Energy Transfe	BIOT	0103
Fluorescence Mechanisms and	Photoinduced Electron Transfer in Covalently Linke	PHYS	0021
d Porphyrin–RuO_2 Clusters.=	Photoinduced Electron Transfer in Covalently Linke	COLL	0133
ransition–Metal Complexes an+	Photoinduced Electron Transfer Reactions between T	INOR	0283
Porphyrin–Flavin Ester.=	Photoinduced Electron Transfers within Conjugated	BIOL	0134
zed Porphyrin Films for Vectorial	Photoinduced Electron Transport.= +ally Polymeri	BIOT	0126
	Photoinduced Proton Transfer in LB Films.=	BIOT	0145
iO_2: A Potential Technique for +	Photoinitiated Oxidation of Trichloroethylene over T	COLL	0082
of Excited–State Relaxation and	Photoinitiated Reaction Kinetics in Liquids by Pho	ANYL	0034
oscopy of Cationic Organometal+	Photoinitiator Activity, Electrochemistry, and Spectr	INOR	0320
alorganic Systems. Organometallic	Photoinitiators in Photopolymerization.= +ve Met	INOR	0338
Combustion Intermediates CH_2+	Photoionization Mass Spectrometric Studies for the	FUEL	0124
adiation.=	Photoionization with Coherent Vacuum Ultraviolet R	ANYL	0003
sfer Additives.=	Photolabile Polysilanes: The Effect of Electron Tran	POLY	0188
tion.=	Photolithography, Chemistry, and Biological Recogni	BIOL	0027
arbonyl Compounds onto Cadm+	Photoluminescence as a Probe of the Adsorption of C	COLL	0175
Mechanistic Studies on DNA	Photolyase.= Chemistry of Nucleic Acids.	BIOL	0155
Benzophenone Flash	Photolysis in the Physical Chemistry Laboratory.=	CHED	0086
nsitized by Colloidal CdS: A Flash	Photolysis Study.= +Controlled Photoreactions Se	COLL	0150
ctron Transfer Reactions by Flash	Photolysis.= +of Thiyl Radical Formation and Ele	INOR	0396
asurements by the Flash or Laser	Photolysis–Shock–Tube Method: Results for the Oxi	FUEL	0123
ient IR Spectroscopy of UV–	Photolyzed $Cr(CO)_6$ in Solution.= Picosecond Trans	PHYS	0220
azography with UV–Absorption	Photometry and Particle–Beam Mass Spectrometry.=	ENVR	0015
Two–	Photon Absorption Spectroscopy of Phenanthrene.=	PHYS	0328
Two–	Photon Absorption Spectroscopy of Phenanthrene.=	CHED	0085
ions for Energy Transfer.=	Photon Correlation and Localization and Its Implicat	PHYS	0064
ass of Polydisperse Polymers: A	Photon Correlation Spectroscopy Study of Polystyren	POLY	0305
248 nm: Statistical vs.+	Photon Dissociation of Ferrocene and Nickelocene at	PHYS	0333
t Advances in Quantitative Two–	Photon Excited Fluorescence Spectroscopy. Recen	ANYL	0014
Two–	Photon Ionization of $(NO)_nAr_m$ Clusters.=	PHYS	0251
R_3 and $Fe(CO)_3(PR)_2$ Are Single–	Photon Products.= +h Alkyl Phosphines: $Fe(CO)_4P$	INOR	0253
	Photon Scanning Tunneling Microscopy.=	PHYS	0210

f Emulsion Particle Distributio+	Photon-Scattering Correction in the Determination o	PMSE	0030
e Metal+ Polymeric Materials for	Photonic and Optical Applications. Highly Polarizabl	POLY	0102
utorial on Polymeric Materials for	Photonic and Optical Applications. Introduction+ T	POLY	0001
−Index+ Polymeric Materials for	Photonic and Optical Applications. Low−Loss Graded	POLY	0174
ganic+ Polymeric Materials for	Photonic and Optical Applications. Orientation of Or	POLY	0006
Gener+ Polymeric Materials for	Photonic and Optical Applications. Second Harmonic	POLY	0150
trolled+ Polymeric Materials for	Photonic and Optical Applications. Synthesis of Con	POLY	0126
hird−Order Optical Properties for	Photonic Switching.= +g of Polymers with Large T	POLY	0011
Opportunities and Challenges in	Photonics.= Nonlinear Optical Polymers:	BIOT	0174
	Photooxidation of Metal Carbyne Complexes.=	INOR	0341
ehyde and Malonaldehyde from	Photooxidation of Squalene.= Formation of Formald	AGFD	0132
activity of Copper Complexes and	Photooxidations of Organic Substrates.= +edox Re	INOR	0342
Complexes in Mixed Solvents.=+	Photophysics and Photochemistry of Ru(II)−Diimine	INOR	0279
Transient States: The	Photophysics of Azulene.= Photoionization via	PHYS	0330
Diffusion of Water into a	Photopolymer Film.=	POLY	0194
Organometallic Photoinitiators in	Photopolymerization.= +ve Metalorganic Systems.	INOR	0338
Liquid Crystal	Photopolymerization.= Holographic Investigation of	BIOT	0205
Determining Pyrimidine	Photoproducts in UV−Irradiated DNA.=	ANYL	0092
on the Replicative Bypass of DNA	Photoproducts.= +Mutations: Mechanistic Studies	BIOL	0157
sion− and Activation−Controlled	Photoreactions Sensitized by Colloidal CdS: A Flash	COLL	0150
	Photoreactivity in Supercritical Fluids.=	ORGN	0131
des Anchored onto SiO$_2$ and Their	Photoreactivity toward CO Molecules.= +ium Oxi	INOR	0165
eterogeneous Media.=	Photoredox Chemistry of Metal Complexes in Microh	INOR	0316
tooxidations of Organic Substra+	Photoredox Reactivity of Copper Complexes and Pho	INOR	0342
Catalyst.=	Photoreduction of Aldehydes with Semiconductor as	ORGN	0247
dicals in+ Micellar Effect on the	Photoreduction of Dyes by Photogenerated Ketyl Ra	COLL	0081
V).=	Photoreduction of Silica−Supported Mo(VI) to Mo(I	COLL	0213
Novel Polymers with	Photorefractive and NLO Properties.=	POLY	0129
with Charge Transport Agents.+	Photorefractive Effect in Nonlinear Polymers Doped	POLY	0155
Introduction to	Photorefractive Materials and Phenomena.=	POLY	0005
onomers and Polymers Containing	Photoresponsive Asobenzene−Based Sidegroups.=	POLY	0209
Spectroscopy.= Studies of Some	Photosensitive Interfaces by Surface Photovoltage	BIOT	0142
Probes of DNA−Binding Intera+	Photosensitive Metalorganic Systems. Luminescence	INOR	0315
idines as Redox Photosensitizers+	Photosensitive Metalorganic Systems. Metal Polypyr	INOR	0282
Photoinitiators in Photopolym+	Photosensitive Metalorganic Systems. Organometallic	INOR	0338
Metalorganic Systems: An Ove+	Photosensitive Metalorganic Systems. Photosensitive	INOR	0166
sensitive Metalorganic Systems.	Photosensitive Metalorganic Systems: An Overview.=	INOR	0166
um Car+ Monitoring the Cure of	Photosensitive Polymers Using a Luminescent Rheni	POLY	0096
ms. Metal Polypyridines as Redox	Photosensitizers.= +osensitive Metalorganic Syste	INOR	0282
Riboflavin: A Potentially Useful	Photosensitizing Agent for the Treatment of Contam	ENVR	0029
: Fe(CO)$_4$PR$_3$ and Fe(CO)$_3$(PR)$_2$+	Photosubstitution of Fe(CO)$_5$ with Alkyl Phosphines	INOR	0253
st−Electron−Transfer Reactions in	Photosynthesis.= +l and Theoretical Studies of Fa	BIOL	0130
oroplasts and Their Relation with	Photosynthetic Activity in Vallisneria spiralis.= +l	BIOL	0101
Transport in	Photosynthetic Antennae.= Electronic Excitation	PHYS	0020
airing: Noncovalent Approaches to	Photosynthetic Model Systems.= +tion via Base P	ORGN	0059
and Biomolecular Electronics.	Photosynthetic Molecular Electronics.= Molecular	BIOT	0050
Spectroscopy of the Bacterial	Photosynthetic Reaction Center.= Optical	PHYS	0170
namics in Wild−Type and Mutant	Photosynthetic Reaction Centers.= +mbination Dy	BIOL	0132
to Electric Energy by the Other	Photosynthetic System in Nature, Bacteriorhodopsin	PHYS	0019
Solid−State Models of	Photosynthetic Systems.=	CHED	0031
ture and Transport Dynamics of	Photosynthetic Units: Spectral−Hole−Burning Studie	PHYS	0018
plasts.=	Photosynthetic Water Splitting by Platinized Chloro	TECH	0016
a Dimer−of−Dimers Model for the	Photosystem II Manganese Aggregate.= +port for	INOR	0408
lation with Photosynthetic Act+	Phototactic Movements of Chloroplasts and Their Re	BIOL	0101
Reaction Kinetics in Liquids by	Photothermal Grating Spectroscopy.= +otoinitiated	ANYL	0034
sensitive Interfaces by Surface	Photovoltage Spectroscopy.= Studies of Some Photo	BIOT	0142
Nonradioactive D+ Synthesis of	Photozyme and Its Use in Labeling Nucleic Acids for	CARB	0037
tion Protonation of 3−Hydrox+	Phthalide Cation Is Not a Stable Species During Solu	ORGN	0151
N,N'−Dimethyl−N,N'−bis[3−(2−	Phthalimido)propyl]−1,6−hexanediamine Dihydrochlo	MEDI	0140
Diffusion in the Gas Sensing of	Phthalocyanine Langmuir−Blodgett Films.=	POLY	0118
ems.	Phthalocyanines, Molecular Devices, and Neural Syst	BIOT	0054
enzene by Stead+ Benzofuroxan	Phtochemistry: Direct Observation of 1,2−Dinitrosob	ORGN	0168
eady−State Spectroscopy—A New	Phtochromic Reaction.= +2−Dinitrosobenzene by St	ORGN	0168
Scalarins from the Sponge	Phyllospongia Foliascens.= New Dialkylated	ORGN	0182
Scalarins from the Sponge	Phyllospongia Foliascens.= New Dialkylated	ORGN	0188
of Moisture Content on the	Physical Aging of Gelatinized Cornstarch.= Effect	CARB	0043
tion of Perturbatio+ Isothermal	Physical Aging of Poly(methylmethacrylate): Localiza	PMSE	0173
Effects of Aryl−Substitution and	Physical Blending on the Gas Permeability of Pol	POLY	0117
Two Elastomeri+ Chemical and	Physical Characterization of an Insulator Containing	PMSE	0168
scent Probes: A Teaching Exper+	Physical Characterization of Detergents Using Fluore	CHED	0200
d by Precipitation and Sol−Gel+	Physical Characterization of TiO$_2$ and Al$_2$O$_3$ Prepare	PETR	0092
ques That Are New, a Revitalized	Physical Chem Lab Makes Its Debut.= +th Techni	CHED	0043
Graphics into the Undergraduate	Physical Chemistry Curriculum.= +Techniques and	CHED	0156
Lasers in the	Physical Chemistry Curriculum.=	CHED	0126
Low−Cost Multiple−Impact	Physical Chemistry Experiment.=	PHYS	0336
Low−Cost, Multiple−Impact	Physical Chemistry Experiment.=	CHED	0162
ohns Hopkins.=	Physical Chemistry Instrumentation Laboratory at J	CHED	0199
ohns Hopkins.=	Physical Chemistry Instrumentation Laboratory at J	PHYS	0246
Spectroscopy for	Physical Chemistry Lab.= Experiments in Raman	PHYS	0300
Experiments for the	Physical Chemistry Lab.= Fluorescence Quenching	CHED	0202

Updating an Undergraduate	Physical Chemistry Lab.=	CHED 0214
Updating an Undergraduate	Physical Chemistry Lab.=	PHYS 0290
mposium on C+ Undergraduate	Physical Chemistry Laboratory Development and Sy	CHED 0155
sortium–Based+ Undergraduate	Physical Chemistry Laboratory Development. A Con	CHED 0125
xpensive Are L+ Undergraduate	Physical Chemistry Laboratory Development. How E	CHED 0082
rement of the+ Undergraduate	Physical Chemistry Laboratory Development. Measu	CHED 0209
unding of Labo+ Undergraduate	Physical Chemistry Laboratory Development. NSF F	CHED 0196
se for the Adv+ Undergraduate	Physical Chemistry Laboratory Development. The Ca	CHED 0041
imental Progr+ Polymers in the	Physical Chemistry Laboratory: An Integrated Exper	CHED 0045
ical Education. Computers in the	Physical Chemistry Laboratory: Let's Get It Right.=	CHED 0155
	Physical Chemistry Laboratory.=	CHED 0129
Flash Photolysis in the	Physical Chemistry Laboratory.= Benzophenone	CHED 0086
ques-Oriented Experiments in the	Physical Chemistry Laboratory.= +uter in Techni	CHED 0121
Program in the	Physical Chemistry Laboratory.= Integrated Writing	PHYS 0318
Program in the	Physical Chemistry Laboratory.= Integrated Writing	CHED 0213
Tool for the Revitalization of the	Physical Chemistry Laboratory.= +tion Devices—A	CHED 0160
Scanning Tunneling Microscopy in	Physical Chemistry Laboratory.= +Ionization and	CHED 0044
Nonideal Solutions in the	Physical Chemistry Laboratory.=	CHED 0101
Experiments for the Introductory	Physical Chemistry Laboratory.= +ic Voltammetry	CHED 0211
College.= Undergraduate	Physical Chemistry Laser Experiments at Roanoke	CHED 0042
ced Titration Experiment for the	Physical Chemistry or Instrumental Analysis Labora	CHED 0161
toratory Modernization: The Pew	Physical Chemistry Project.= +ed Approach to La	CHED 0125
Experience for Students in	Physical Chemistry Survey Courses.= Laboratory	CHED 0215
cry Innovations in Undergraduate	Physical Chemistry.= +. NSF Funding of Laborat	CHED 0196
Undergraduate	Physical Chemistry.= Different Approach to	CHED 0127
Calculations in a Course in	Physical Chemistry.= Quantum Chemical	CHED 0191
Spreadsheets in	Physical Chemistry.=	CHED 0208
Spreadsheets: Opportunities in	Physical Chemistry.=	CHED 0207
e-Function Relationships: A	Physical Organic Laboratory Experiment.= Structur	CHED 0212
+ght of Poly(dimethylsiloxane) on	Physical Properties of Cured Epoxy Compounds.=	PMSE 0174
Long-Period Storage Effects on	Physical Properties of Geothermal Cement Systems	I&EC 0014
	Physical Properties of the Organic Superconductors.=	PHYS 0002
uctors: Syntheses, Structures, and	Physical Properties.= +view of Organic Supercond	PHYS 0024
ns, New Materials, and Enhanced	Physical Properties.= +Chemistry: Useful Reactio	CHED 0030
ghts into Morphology Control and	Physical Properties.= +Sol-Gel Method: Some Insi	POLY 0260
d Steel Exposed+ Chemical and	Physical Technologies of Fertilizers. Corrosion of Mil	FERT 0001
Chemical and	Physical Treatments.= Surface Disinfection by	AGFD 0032
ps: Biological, Environmental, and	Physicochemical Applications—I. +tion Relationshi	ANYL 0043
ps: Biological, Environmental, and	Physicochemical Applications—II. +ion Relationshi	ANYL 0068
ps: Biological, Environmental, and	Physicochemical Applications—III. +on Relationshi	ANYL 0098
uit and Vegetable Surfaces.=	Physicochemical Changes on Minimally Processed Fr	AGFD 0011
Lipids.=	Physicochemical Characterization of Cytotoxic Ether	MEDI 0031
uminophosphate Phases and Their	Physicochemical Characterization.= +rge Pore Al	PETR 0093
obiphenyl Congeners.=	Physicochemical Parameters of Individual Hexachlor	ANYL 0101
Electro+ Dynamical Character of	Physicochemical Processes under High Pressure: (6)	PHYS 0222
Effect of Ce on Performance and	Physicochemical Properties of Pt-Containing Auto	COLL 0208
m+ Electrochemical Methods and	Physicochemical Structures of Liquid Disperse Syste	COLL 0022
nd Their Relationship to Polymer	Physics.= +r Nonlinear Optical Onset and Decay a	POLY 0152
Symposium Polymers—Chemistry,	Physics, and Biology. +Polymer Chemistry Award	POLY 0272
Processed Fruits and Vegetables.	Physiology of Cut Fruits and Vegetables.= +imally	AGFD 0010
Form. G. Edelman (Medicine and	Physiology, 1972).= +y: Molecular Bases of Animal	CONG 0013
holinesterase Inhibitors: 8-Carba+	physostigmine Analogues.= +hips of Potent Acetylc	MEDI 0143
of 7-Substituted Derivatives of	Physostigmine for the Treatment of Alzheimer's Dise	MEDI 0137
in a Milled Ha+ Corrosion of	Phytic Acid, Total Dietary Fiber, Protein, and Starch	AGFD 0143
ces in Diabrotica.=	Phytochemical Antagonism of GABA-Based Resistan	AGRO 0136
Trace Metal Nutrition of Marine	Phytoplankton.=	INOR 0003
Cation Radical+ Intramolecular	$2\pi + 4\pi$ Cycloaddition Reactions of Allyl Cations and	ORGN 0123
on Radical+ Intramolecular $2\pi +$	4π Cycloaddition Reactions of Allyl Cations and Cati	ORGN 0123
Indiana Silicon Sphere (ISiS)	4π Detector.=	NUCL 0036
Nonplanarity of Doped	π Systems.=	PHYS 0233
Laboratory Evaluation of	PIC Formation in Agricultural Product Bag Burns.=	AGRO 0140
olutions for Disposal and Reuse of	Pickle Liquors from Mexican Iron Industry.= +e S	ENVR 0039
(adenosin-5'-yl) $-P^2-[6-(ribosyl)-$	picolinamide-5'-yl]pyrophosphates. +nthesis of P^1-	CARB 0023
tor–Liquid Interfaces.=	Picosecond Nonlinear Optical Studies of Semiconduc	PHYS 0283
ed-State Relaxation at Interfac+	Picosecond Optical Harmonic Measurements of Excit	COLL 0087
at a Solid Semiconductor Electr+	Picosecond Real-Time Interfacial Electron Transfer	COLL 0088
e/Silica Surfaces.=	Picosecond Reorientation of Adsorbates on Alkylsilan	ANYL 0025
istry in the Milliseond to	Picosecond Time Domain.= Metalorganic Photochem	INOR 0287
yzed $Cr(CO)_6$ in Solution.=	Picosecond Transient IR Spectroscopy of UV-Photol	PHYS 0220
ut Honest Approach.=	Pictorialized Solid-State MOs: A Nonmathematical M	CHED 0021
lectron Propagat+ One-Electron	Pictures of Molecular Electronic Structure through E	PHYS 0058
	Pig Corneal Proteases.=	BIOL 0043
raphy.= Purification of	Pig Liver Squalene Epoxidase by Affinity Chromatog	BIOL 0070
Stability of Cochineal	Pigment and Its Use in Food Products.=	AGFD 0139
rochemical Formation of Organic–	Pigment Thin Films by Disruption of Micelles.=	COLL 0019
Crosslinking Reactions in	Pigmented Olefinic Polymers.=	POLY 0135
onomers on the Surface of Hiding	Pigments Dispersed in Water.= +lymerization of M	PMSE 0038
Green	Pigments with Perovskite Structure.=	INOR 0069
Mexico: The	Pill and Cortisone.= Early Steroid Research in	HIST 0018
eroid Chemistry. Steroids and the	Pill: Early Steroid Research at G. D. Searle Co.=	HIST 0020

and Analytical Determination of	Pillar Densities in Alumina-Pillared Montmorillonite	PETR	0126
in Zeo+ Advances in Zeolites and	Pillared Clay Synthesis. Competitive Role of Cations	PETR	0074
ore Ze+ Advances in Zeolites and	Pillared Clay Synthesis. From Clathrasils to Large-P	PETR	0036
ization+ Advances in Zeolites and	Pillared Clay Synthesis. Investigation of the Crystall	PETR	0093
of Sil+ Advances in Zeolites and	Pillared Clay Synthesis. Molecular Chemical Aspects	PETR	0054
hesis o+ Advances in Zeolites and	Pillared Clay Synthesis. Novel Approach to the Synt	PETR	0019
of a+ Advances in Zeolites and	Pillared Clay Synthesis. Optimization of the Pillaring	PETR	0142
ic C+ Advances in Zeolites and	Pillared Clay Synthesis. Pillaring of Layered Inorgan	PETR	0106
Advances in Zeolites and	Pillared Clay Synthesis. Supergallery Pillared Clays.=	PETR	0125
ion of+ Advances in Zeolites and	Pillared Clay Synthesis. Synthesis and Characterizat	PETR	0001
, Forming, and Characterization.	Pillared Clays as Supports in Hydrocracking Catalys	PETR	0013
Clay Synthesis. Supergallery	Pillared Clays.= Advances in Zeolites and Pillared	PETR	0125
Polyoxocations and	Pillared Clays.= Ga_{13}, $GaAl_{12}$ and Al_{13}	PETR	0143
Catalyzed by	Pillared Clays.= Propene Alkylation of Biphenyl	PETR	0014
The+ Catalytic Performance of	Pillared Interlayered Clay Molecular Sieve with High	PETR	0030
Surfaces.= Microporous	Pillared Mica with Cation-Incorporated Silicate	PETR	0129
on of Pillar Densities in Alumina-	Pillared Montmorillonite.= +nalytical Determinati	PETR	0126
TEM Analysis of	Pillared Rectorites.=	PETR	0109
Characterization of Borate-	Pillaring Anionic Clay.= Preparation and	PETR	0111
Anionic Clays: Trends in	Pillaring Chemistry.=	PETR	0127
lay Synthesis. Optimization of the	Pillaring of a Saponite.= +Zeolites and Pillared C	PETR	0142
olites and Pillared Clay Synthesis.	Pillaring of Layered Inorganic Compounds.= +Ze	PETR	0106
Creosote Site-Biobed Box	Pilot-Scale Bioremediation Study.=	BIOT	0140
n of Arson Evidence.=	Pioneers in Forensic Science: Laboratory Examinatio	HIST	0007
Agent.= Phenolic Compounds of	Piper betel as Flavoring and Nerve-Stimulating	AGFD	0154
nimal Behavioral Testing of Novel	Piperidine Derivatives.= +tipsychotic Drugs: II. A	MEDI	0084
esis and Binding Profiles of Novel	Piperidine Ligands.= +tipsychotic Drugs: I. Synth	MEDI	0153
of Substituted (Aminomethyl)-	Piperidines: A Novel Class of Selective Protein Kinas	MEDI	0025
f Ester-Containing bisQuaternary	Piperidino-Ammonium Salts.= +uscle Relaxants o	MEDI	0047
al W+ Corrosive Effects on Drill	Pipes by Drilling Fluids Systems Based on Geotherm	I&EC	0012
°From Molecular Recognition of	Piquancy to a New Rule of Medicinal Effect.=	MEDI	0040
etabolism and Residue Studies of	Pirlimycin in the Dairy Cow.= +ducing Animals. M	AGRO	0108
s with the Antiinflammatory Drug	Piroxicam.= +ty Constants of Copper(II) Complexe	INOR	0407
velopment and Testing of Carbon	Pistons.= +arbon Fuels: Engineering Materials. De	FUEL	0050
The	Pivot Algorithm with Interaction Energy.=	POLY	0077
	PIXE as an Analytical/Educational Tool.=	NUCL	0078
Applications of	PIXE in Dendrochronology.=	NUCL	0081
X-ray Emission (PIXE). Staus of	PIXE in Geochemistry.= +ysis by Proton-Induced	NUCL	0077
Statistical Summary of	PIXE Programs 1970-1980-1990.=	NUCL	0080
for	PIXE.= Air Samplers and Complementary Analyses	NUCL	0079
Proton-Induced X-ray Emission (	PIXE). +cation of Materials: Elemental Analysis by	NUCL	0077
s: Fe < Os+ Distinctive Order of	pK_a Values of Isostructural η_2-Dihydrogen Complexe	INOR	0380
e of Yeshiv+ The Freeman Case:	Plagiarism at the Albert Einstein College of Medicin	PROF	0004
and Practice. A Chemical Hygiene	Plan for R D Laboratories—A Case Study.= +ory	CHAS	0001
ischarge of Pe+ Remedial Action	Plan: Program for the Great Lakes. Achieving Zero D	ENVR	0083
Constructing Your Business	Plan.=	SCHB	0004
g Ligands: C+ Binuclear Square	Planar Cobalt(III) Complexes of Polyanionic Chelatin	INOR	0243
and Exchange between Square	Planar Iridium(I) Complexes.= Facile Phosphine Lig	INOR	0211
ions.=	Planar Polymer Waveguides for Optical Interconnect	POLY	0175
lly Oriented Polyamide Films in	Plane Prepared by Vacuum Deposition Polymerizatio	POLY	0082
ct of Nonmagnetic Chains and	Planes on Superconductivity.= Cuprate Oxides: Effe	INOR	0358
lopment of a Model for Land-Use	Planning and Conflict Resolution.= +Gaps in Deve	ENVR	0043
s in the LHASA Pro+ Synthesis	Planning and Reaction Databases. New Development	CINF	0001
volvement in the Remedial Action	Planning Process.= +-Table Discussion—Public In	ENVR	0086
Heuristic Synthesis	Planning Using the CHIRON Program.=	CINF	0002
Reaction	Planning.= Integrated Approach to Synthesis and	CINF	0003
and Future	Plans.= The Protein Data Bank—Present Status	CINF	0031
ses in Lepidoptera.=	Plant Allelochemical-Adapted Glutathione Transfera	AGRO	0119
and Develo+ Effects of Cotton	Plant Allelochemicals and Nutrients on the Behavior	AGRO	0037
Heliothis Resistance to	Plant Allelochemicals.=	AGRO	0118
odifying Secondary Metabolism in	Plant Cell Cultures. Increasing Secondary Met+ M	BIOT	0228
Cultured	Plant Cells.= Production of Phenolic Compounds by	AGFD	0002
Tocopherols and Tocotrienols on	Plant Colorant Carotenoids.= +ioxidant Activity of	AGRO	0039
Antimutagens in	Plant Foods.= Phenolic Antioxidants as	AGFD	0045
nd Several β-Galactoside-Specific	Plant Lectins.= +side-Specific Vertebrate Lectin a	BIOL	0109
Health. Natural Antioxidants from	Plant Material.= +nolic Compounds in Food and	AGFD	0015
	Plant Phenolics as Cytotoxic Antitumor Agents.=	AGFD	0092
rcinogens.=	Plant Phenolics as Inhibitors of Mutagenesis and Ca	AGFD	0118
of Carcinogenic Response by	Plant Phenols.= Carcinogenicity and Modification	AGFD	0115
on, and Tongue Carcinogenesis by	Plant Phenols.= +tective Effects against Liver, Col	AGFD	0090
otope Distributions in Secondary	Plant Products and Their Importance for Flavor-Ori	AGFD	0194
ale Production of Chemicals from	Plant Root Cultures.= +on of Reactors for Large-Sc	BIOT	0233
Sesquiterpene Induction from	Plant Root Cultures.=	BIOT	0232
th. New Zealand Gas-to-Gasoline	Plant: An Engineering Tour de Force.= +J. F. Ro	I&EC	0060
MAP 30: A New	Plant-Derived Anti-HIV Agent.=	AGFD	0190
iral Studies with Trichosanthin, a	Plant-Derived Ribosome-Inactivating Protein.= +v	AGFD	0189
d+ Composicion Quimica de las	Plantas Pertenecientes a la Familia Scrophulariaceae	ORGN	0185
a Sucesiva y u+ Optimizacion de	Plantas Quimicas Utilizando Programacion Cuadrat	I&EC	0072
DPX-E9636 in	Plants and Animals.= Comparative Metabolism of	AGRO	0028
New Polyphenols from Medicinal	Plants and Their Antioxidant Activity.=	AGFD	0017

Systems by Phenolic	Plants Constituents.= Role of Antioxidative Defense	AGFD	0018
Samples from Wood-Preserving	Plants in Quebec: Extensive Analysis and Search for	ENVR	0016
14C-Trichloroethylene in Whole	Plants.= Student Awards Symposium. Fate of	ENVR	0056
Antiviral Agents from Medicinal	Plants.=	AGFD	0191
st Insect-Molting Hormones from	Plants.= +rrent Studies with Ponasterones, the Fir	HIST	0024
Dye Laser Studies of	Plasma and Flame-Generated Species.=	CHED	0087
3-Phenylpropionamide (HEPP) in	Plasma and Urine.= +t D,L-3-Hydroxy-3-Ethyl-	ANYL	0082
of Fly Ash by Inductively Coupled	Plasma Atomic Emission Spectroscopy.= +nalysis	FUEL	0138
Reactions of Metal Atoms in	Plasma Etching and Chemical Vapor Deposition.=	PHYS	0143
Search for Quark-Gluon	Plasma in P-P Collisions at Fermilab.=	NUCL	0063
	Plasma Reactor Design.=	NUCL	0086
ry, Thermal Stability, and Oxygen	Plasma Resistance.= +e)s—I. Overview of Chemist	PMSE	0052
2's) of Water Protons in Human	Plasma Samples Related to Their Hydration Water C	MEDI	0043
f Surface Modification by Oxygen	Plasma Treatment.= +xide)s—III. Investigations o	PMSE	0054
Laser-Generated Hydrocarbon	Plasmas: Determination of Temperatures by C_2 Emis	PHYS	0268
Diamond-Producing	Plasmas.= Gas-Phase Analysis of	PHYS	0148
Cosolvents on Kinetics of	Plasmin.= Effects of Water-Miscible Organic	BIOL	0092
osylation of Recombinant Human	Plasminogen Expressed in Several Cell Lines.= +yc	ANYL	0020
l Arteli+ Effective Treatment of	Plasmodium Berghei-Infected Mice with Transderma	MEDI	0063
Recent Developments in	Plastic Agricultural-Chemical-Container Recycling.=	AGRO	0045
istence of Separately Controlled	Plastic and Conserved Phases of Catalysis by Glycosy	BIOL	0152
Iowa's	Plastic Pesticide Container Recycling Program.=	AGRO	0091
roperties of Recycled Commingled	Plastic Processed with Polystyrene.= +cture and P	I&EC	0028
ompatibilizing Components of the	Plastic Waste Stream via Reactive Processing.= +C	I&EC	0033
t Liquids: Application to Polymer	Plasticizers and Model Hydrocarbons.= +lar Weigh	POLY	0065
Diffusion of Alkyl Adipate	Plasticizers in PVC.=	POLY	0067
ymethyl Methacry+ Influence of	Plasticizers on the Diffusion Process of Water in Pol	POLY	0193
rom Waste and Coal: Energy from	Plastics and Municipal Waste. Co+ Clean Energy f	FUEL	0090
rom Waste and Coal: Energy from	Plastics and Municipal Waste. Gas+ Clean Energy f	FUEL	0091
rom Waste and Coal: Energy from	Plastics and Municipal Waste. Re+ Clean Energy f	FUEL	0092
Postconsumer	Plastics Collection.=	I&EC	0039
Community.= Collection of	Plastics for Recycling from the Residential	I&EC	0038
cipal Waste. Gasification of Waste	Plastics for the Production of Fuel-Grade Gas.=	FUEL	0091
tes): Biodegradable	Plastics Made by Bacteria.= Poly(β-hydroxyalkanoa	CHED	0132
Economics of Commingled	Plastics Molding.=	I&EC	0035
echnologies for Plastics Recycling.	Plastics Recycling Processing and Economics.= +T	I&EC	0031
	Plastics Recycling via PET Depolymerization.=	I&EC	0032
E Reclamation Technologies for	Plastics Recycling. Plastics Recycling Processing and	I&EC	0031
ction and Sortation Technology in	Plastics Recycling. Plastics Recycling: Beyo+ Colle	I&EC	0037
Technology in Plastics Recycling.	Plastics Recycling: Beyond Beverage Bottles.= +on	I&EC	0037
Use of Froth Flotation for	Plastics Recycling:=	I&EC	0036
mmingled Plastics. Commingled	Plastics Recycling—An Equipment Manufacturer's P	I&EC	0025
ylene Terephthalate (PET) in	Plastics Waste on the Thermal Properties of Asphalt	FUEL	0053
Extruding Waste	Plastics with High Mixing and Low Shear.=	I&EC	0030
ningled and Refined Commingled	Plastics. Commin+ Technology of Recycling of Com	I&EC	0025
Commingled	Plastics.= Continuously Extruded	I&EC	0027
cmatic Sortation of Postconsumer	Plastics.= +terial Handling Equipment for the Aut	I&EC	0041
cn Zinc en Medi+ Obtencion de	Plata por Electrodeposito, Fusion, Oxidorreduccion c	CHED	0172
Basico, Preparacion de Nitrado de	Plata.= +dorreduccion con Zinc en Medio Acido y	CHED	0172
Sensor	Plate for Manufacturing Thermoplastic Composites.=	PMSE	0115
Characterization of the	Platelet Effects of the Anti-Inflammatory DuP 697.=	MEDI	0162
he Reaction of Diazomethane with	Platin(IV)cyclobutanes.= +nd Stereochemistry of t	INOR	0329
Splitting by	Platinized Chloroplasts.= Photosynthetic Water	TECH	0016
Mg, Cu, Zn, and+ Properties of	Platinum and Platinum-Tin Catalysts Supported on	COLL	0126
d Energetics in the Deposition of	Platinum from Pt $(PF_3)_4$ on Atomically Clean Platin	COLL	0046
erned Imaging of Palladium and	Platinum Metal Films by Electron Transfer to Photo	INOR	0319
om Pt $(PF_3)_4$ on Atomically Clean	Platinum Surfaces.= +he Deposition of Platinum fr	COLL	0046
nd (α-Ketoacyl)acyl Complexes of	Platinum.= +nd Reactivity of (Alkyl)acyl, Diacyl, a	INOR	0043
of CH_3^+ Ions on	Platinum.= Ultra-Low-Energy Interactions (≤ 3 eV)	COLL	0216
s from cis-Diammine(guanosine)	platinum(II) Chloride Using ^{13}C NMR Spectroscopy.	INOR	0251
etics of Isomerization Studies of	Platinum(II) Complexes of Nucleside-5'-Di- and Tri	INOR	0403
MR and Kinetics of Formation of	Platinum(II) Nucleotide.= +d Two-Dimensional N	INOR	0180
otide Cross-Binding to	Platinum(II): A Kinetic Study.= Amino Acid-Nucle	INOR	0182
vdrazine to σ-Allenyl Complex of	Platinum(II).= +cleophilic Addition of Amine or H	INOR	0041
hosphite and Trip+ Reaction of	Platinum(IV) Metallacyclopentanes with Trimethyl P	INOR	0330
-Platinum-Imine and Organotin-	Platinum-Amine Complexes.= +ly Active Organotin	INOR	0234
line Formed by Nucleophilic A+	Platinum-Cyclobutanoneimine and Platinum-Pyrazo	INOR	0041
rization, and Catalytic Activity of	Platinum-Doped Glassy Carbon.= +hesis, Characte	INOR	0312
of Biologically Active Organotin-	Platinum-Imine and Organotin-Platinum-Amine Co	INOR	0234
atinum-Cyclobutanoneimine and	Platinum-Pyrazoline Formed by Nucleophilic Additi	INOR	0041
pared from Solvated Metal At+	Platinum-Tin and Gold-Tin Bimetallic Particles Pre	CATL	0009
nd+ Properties of Platinum and	Platinum-Tin Catalysts Supported on Mg, Cu, Zn, a	COLL	0126
Acquisition—The	Pleasant (?) Fate of Small Chemical Businesses.=	SCHB	0019
Container Coating Systems—II.	Plenary Lecture. +ntainer Coatings: An Overview.=	PMSE	0157
Container Coatings Systems—I.	Plenary Lecture. +lymers for Container Coatings.=	PMSE	0138
Extruders.= Reactive Processing.	Plenary Lecture. +rgy Distribution in Twin Screw	PMSE	0175
n of Multicomponent Systems—II.	Plenary Lecture. +s.= Solid-State Characterizatio	PMSE	0116
of Multicomponent Systems—IV.	Plenary Lecture. +.= Solid-State Characterization	PMSE	0163
ents.= Nonaqueous Enzymology.	Plenary Lecture. +Purification in Neat Organic Solv	BIOT	0063
n of Multicomponent Systems—I.	Plenary Lecture. +.= Solid-State Characterizatio	PMSE	0099

of Multicomponent Systems—III.	Plenary Lecture. +.= Solid-State Characterization	PMSE	0145
esis of Highly Substituted Carb+	Plenary Session. Annulation Strategies for the Synth	ORGN	0286
Carbonyl Compounds Using E+	Plenary Session. Asymmetric Synthesis of α-Hydroxy	ORGN	0293
d Cycloheptatriene to Carbohyd+	Plenary Session. From Cyclopentadiene, Benzene, an	ORGN	0272
atoms: Determination of the Ste+	Plenary Session. Mechanisms of Reactions at Hetero	ORGN	0120
Activators.=	Plenary Session. Multifunctional Enzyme Mimics and	ORGN	0265
nd Ene Reactions.=	Plenary Session. Studies in Asymmetric Olefination a	ORGN	0279
Surfaces.= Langmuir Lectures:	Plenary Session. Surface-Introduced Liquid	COLL	0062
nds Mediated by Transition M+	Plenary Session. Transformations of Organic Compou	ORGN	0127
during	Plum Processing.= Changes of Phenolic Compounds	AGFD	0171
ne Rock: Case Study—Waldoboro	Pluton Complex, Maine.= +tal Mercury in Crystalli	ENVR	0055
y	Plutonio.= Deposito Electrolitico de Samario	NUCL	0075
Initial Abundances of	Plutonium-244 in Meteorites and Lunar Samples.=	NUCL	0083
erization of Interphases between	PMDA/4-BDAF Polyimides and Silver Substrates U	POLY	0087
W(PMe3)6 and @(PMe3)4(η2-CH	2PMe2)H with Phenols.= +lectron-rich Complexes	INOR	0086
of Mo(η2-C(S)OPri)(S2COPri)CO(	PMe3)2.= +tic Acetyl Complexes: Crystal Structure	INOR	0036
ation of Alkanes Catalyzed by Rh(	PMe3)2(CO)Cl.= +mical and Thermal Dehydrogen	INOR	0152
sfer-Dehydrogenation Using Rh(	PMe3)2LCl(L = iPr3,CO): Aspects of Selectivity and	INOR	0151
ich Complexes W(PMe3)6 and @(	PMe3)4(η2-CH 2PMe2)H with Phenols.= +ectron-r	INOR	0086
the Electron-rich Complexes W(	PMe3)6 and @(PMe3)4(η2-CH 2PMe2)H with Phenol	INOR	0086
arent Glass Fiber Reinforced	PMMA Matrix Composites.= Preparation of Transp	PMSE	0109
	PMMA-Impregnated Silica Gel.=	POLY	0263
ism of Inhibition of Human	PMN Elastase by Monocyclic β-Lactams.= Mechan	BIOL	0050
ic β-Lactam Inhibitors of Human	PMN Elastase.= +e, Potent, and Specific Monocycl	BIOL	0049
the Microphase Structure of PS-	PMPS Diblock Copolymers.= +tate NMR Study of	PMSE	0121
Lipopeptides with Potent Anti-	Pneumocystis carinii and Anti-Candida albicans Act	MEDI	0181
urine Nucleoside Phosphorylase (	PNP).= +cture-Based Drug Design: Inhibitors of P	MEDI	0005
onophosphate Tungsten Bronze (	PO2)4(WO3)2m, m = 2.= +sport Properties of the M	INOR	0352
n the Milk of Goats Fed Erythrina	poeppigiana and E. +ce of Dihydro-Erythroidines i	AGFD	0129
in Catalysis: Clusters, Alloys, and	Poisoning. Comparison of Alkane Dehy+ Selectivity	CATL	0007
amics in Gases and at Surfaces. J.	Polanyi (Chemistry, 1986).= +sium. Reaction Dyn	CONG	0011
trategy f+ Amphiphile-Mediated	Polar Colloid Formation in Fluorocarbon Media: A S	COLL	0164
Anion Binding to	Polar Molecules.=	PHYS	0224
or Increasing Enzyme Stability in	Polar Organic Solvents.= +on: A General Strategy f	BIOT	0244
Fast Vibrational Relaxation for	Polar Solutes in Polar Solvents.=	PHYS	0037
Polar Solutes in	Polar Solvents.= Fast Vibrational Relaxation for	PHYS	0037
Addition.= A Practical Reversed	Polarity Alternative to Organocuprate Conjugate	ORGN	0091
ophoric Assemblies with Enforced	Polarity: Inorganic and Heterocyclic Linkages.=	POLY	0215
Hyperpolarizabilities and	Polarizabilities.= Calculations of Static	PHYS	0340
: Cobaltous Complexes o+ Highly	Polarizable Metallic Complexes for Nonlinear Optics	POLY	0102
cter sphaeroides Reaction Center:	Polarization and Temperature Studies.= +Rhodoba	BIOL	0094
Fluorescence	Polarization of Fluorescein-Labeled Dextrans.=	ANYL	0076
Samples.= Cross	Polarization of Half-Integer Spins in Rotating	PHYS	0371
Experiments with Laser-	Polarized Rare Gases.= Magnetic Resonance	PHYS	0041
eneous Fluids. Potentiometry and	Polarography in Micellar Systems.= +Microheterog	COLL	0001
Efficient Frequency Doubling in	Poled Polymer Films.= Recent Developments in	POLY	0179
Orientation and Decay in	Poled Polymer Nonlinear Optical Materials.=	POLY	0151
ew Molecular Design Concept for	Poled Polymers Utilizing the Off-Diagonal Tensor C	POLY	0128
Relaxation Properties of	Poled Polymers.= Second-Order NLO and	POLY	0153
k in the Royal Canadian Mounted	Police.= +sic DNA Analysis Research and Casework	CHAL	0007
I:+ Materiales Semiconductores	Policristalinos dentro de Celdas Fotoelectroquimicas,	PHYS	0221
	Policy Implications of Greenhouse Warming.=	ENVR	0064
Century. Pesticide Regulatory	Policy in the 21st Century.= Pesticides in the 21st	AGRO	0066
las Propiedades de Conduccion en	Polimeros Zwitterionicos.= +ecto de la Presion en	ORGN	0159
romophores in Supported Sol-+	Poling and Chemical Binding of Glass-Embodied Ch	POLY	0210
Films Bearing NL+ Synthesis,	Poling, and Optical Characterization of Polyurethane	POLY	0216
Orange: A Case Study.=	Political Consequences for Scientists Studying Agent	PROF	0008
d Phenyl-Cinnol+ Synthesis and	Pollen-Suppressant Activity of a Series of Substitute	AGFD	0042
Alcohols, Gasohol, an+ Exhaust	Pollutant Emissions from Automobile Engines Using	FUEL	0095
Degradation of Aromatic	Pollutants by Ligninase Models.= Oxidative	INOR	0187
Prioritization of Hazardous Air	Pollutants: The Need for Comparative Methodology.	ENVR	0066
Pollution Prevention. Toward a	Pollution Prevention Ethic: The EPA's Role and Pro	CEI	0022
U.S. EPA's	Pollution Prevention Research Agenda.=	CEI	0023
EPA's 33/50 Project:	Pollution Prevention through Voluntary Initiatives.=	CEI	0024
Ethic: The EPA+ Approaches to	Pollution Prevention. Toward a Pollution Prevention	CEI	0022
Industrial Approaches to	Pollution Prevention.=	CEI	0025
Life Cycle Analysis: A Tool for	Pollution Prevention.=	CEI	0027
and Internation+ Market-Driven	Pollution Prevention? Green Marketing: A National	CEI	0028
Preventing	Pollution.= Role of the Chemist in	CEI	0026
licon Acetylene and Silicon Olefin	Polmers as Precursors to SiC.= +rid Materials. Si	POLY	0357
dies of New Model Compounds of	Poly [N,N'-bis(phenoxyphenyl)pyromellitimide].=	PMSE	0151
aphy and End-Group Analysis for	Poly Methyl Methacrylate.= +clusion Chromatogr	PMSE	0050
y at Transition-Metal Centers via	Poly 3'-(4-pyridyl)terthiophene-Based Ligands.=	INOR	0027
New Preparative Chemistry of	Poly(alkyl/arylphosphazenes).=	POLY	0184
Kinetics of Imidization of	Poly(amic Acid) in Miscible and Immiscible Blends.=	PMSE	0147
tress Behavior of Functionalized	Poly(amic Acid) Precursors and Their Polyimides.=	POLY	0085
r Composite during Imidization of	Poly(amic dialkylamide)Precursors.= +tu Molecula	PMSE	0169
, Crystallization, and Melting of	Poly(aryl Ether Ketone Ketone), Part I: Structure.=	POLY	0059
e, Crystallization, and Melting of	Poly(aryl Ether Ketone Ketone), Part II: Crystallizat	POLY	0060
Gelation in	Poly(aryl Ether Ketones).= Crystallization-Induced	POLY	0247

AR and S$_{RN}$1 Mechanisms.=	Poly(aryl ether-ketone) Synthesis via Competing S$_N$	PMSE	0122
Preparation and Properties of	Poly(aryl Sulfide Phenylquinoxalines).=	POLY	0026
Copolymers of Poly(styrene) and	Poly(arylate) and Their Optical Properties.= +lock	POLY	0220
Synthesis and Properties of	Poly(arylene Ether Benzimidazole)s.=	POLY	0025
of Chemistry, Thermal Stability+	Poly(Arylene Ether Phosphine Oxide)s—I. Overview	PMSE	0052
Studies.=	Poly(Arylene Ether Phosphine Oxide)s—II. Pyrolysis	PMSE	0053
ations of Surface Modification +	Poly(Arylene Ether Phosphine Oxide)s—III. Investig	PMSE	0054
ine Intermediat+ Semicrystalline	Poly(Arylene Sulfide) Ketones via Amorphous Ketim	PMSE	0123
razine) Copolymer+ Synthesis of	Poly(b-vinylboratine) and Poly(styrene-co-b-vinylbo	POLY	0183
ade by Bacteria.=	Poly(β-hydroxyalkanoates): Biodegradable Plastics M	CHED	0132
in Blends of Poly(propylene) and	Poly(butene-1).= +Miscibility and Cocrystallization	POLY	0089
S Cocrystallization in Blends of	Poly(butylene Terephthalate) and Poly(ester-Ether)	PMSE	0148
enetrating Polymer Networks of	Poly(Carbonate-Urethane) and Polyvinyl Pyridine.=	PMSE	0056
/Trifluoroethylene Copolymer and	Poly(chlorofluoroethylene).= +Vinylidene Fluoride	POLY	0022
tle Gelation in Binary Mixture of	Poly(di-N-alkylsilanes) and Aromatic Solvents.=	POLY	0145
f Topology for Cyclic and Linear	Poly(dimethylsiloxane) and Poly(methylphenylsiloxa	POLY	0234
Effect of Molecular Weight of	Poly(dimethylsiloxane) on Physical Properties of Cur	PMSE	0174
rthesis and Characterization of	Poly(dimethylsiloxane-co-methylphenylsiloxane) Ran	POLY	0020
is and Degradation Studies of	Poly(ϵ-caprolactone) Fumarate Networks.= Synthes	POLY	0068
lene)-32-crown+ Synthesis of a	Poly(ester crown) Based on bis(5-Carboxy-1,3-pheny	POLY	0044
Poly(butylene Terephthalate) and	Poly(ester-Ether) Segmented Block Copolymers.=	PMSE	0148
rom Tri- or Tetraisocyanates and	Poly(ethylene Glycol) Derivatives.= +ar Polymers f	POLY	0271
tannanes, and Hydroxyl-Capped	Poly(Ethylene Oxide) and ABA Block Copolymers C	PMSE	0006
Interphases in	Poly(ethylene Oxide) Blends.= Crystal-Amorphous	PMSE	0165
: Optical Methods. Polystyrene-	Poly(ethylene Oxide) Block Copolymer Micelles in W	POLY	0277
ed Pheno+ Ionic Conductivity of	Poly(ethylene Oxide) Bound with a Sterically Hinder	POLY	0091
Chemistry, Physics, and Biology.	Poly(ethylene Oxide) Star Molecules in Surfaces for	POLY	0272
ers Containing Hydroxyl-Capped	Poly(Ethylene Oxide) with Polydimethylsiloxane.=	PMSE	0006
Thermoreversible Gelation from	Poly(ethylene Oxide)Paradichlorobenzene and Thei	POLY	0125
tionalized Triglyceride Oils.=	Poly(Ethylene Terephthalate) Semi-IPNs with Func	PMSE	0088
rigly+ Crystallization Kinetics of	Poly(Ethylene Terephthalate) with Functionalized T	PMSE	0171
of Near Surface Structure in	Poly(ethyleneterephthalate) by ATR-IR.= Analysis	POLY	0279
ansitions in Model Compounds for	Poly(γ-alkyl-α,L-glutamate)s.= +tiple Thermal Tr	POLY	0099
acryl+ Mechanical Properties of	Poly(Methyl Acrylate-co-Dimethyl Aminoethyl Meth	PMSE	0079
tion in Solutions of Syndiotactic	Poly(methyl Methacrylate) as Studied by Spectrosco	POLY	0124
the Diffusion Process of Water in	Poly(methyl Methacrylate).= +ce of Plasticizers on	POLY	0193
o+ Isothermal Physical Aging of	Poly(methylmethacrylate): Localization of Perturbati	PMSE	0173
cal and Mechanical Properties of	Poly(methylmethacrylate)/Glass Fiber Laminates.=	PMSE	0110
Linear Poly(dimethylsiloxane) and	Poly(methylphenylsiloxane) Blends.= +Cyclic and	POLY	0234
Sequence Lengths in	Poly(n-alkyl Isocyanates).= Chiral Clues to Helical	POLY	0348
on and Properties of Macroporous	Poly(N-isopropylacrylamide) Hydrogels.= +eparati	POLY	0243
velength Dispersion of the $\chi^{(3)}$ of	Poly(p-phenylene Benzobisthiazole)-Based Molecula	POLY	0213
ental Dynamics in N-Deuterated	Poly(p-phenylene Terephthalamide).= Amide Segm	POLY	0058
e Rigid-Chain Polymers: An Ionic	Poly(p-phenylene) Analog.= +s Synthesis of Solubl	POLY	0024
ed Cyclohexadienes: Precursors to	Poly(paraphenylene).= +of Heteroatom-Substitut	POLY	0047
and Cocrystallization in Blends of	Poly(propylene) and Poly(butene-1).= +Miscibility	POLY	0089
tabilized by Sterically Demanding	poly(Pyrazolyl)hydroborato Ligands.= +rivatives S	INOR	0085
yclodehydrogenation of Precursor	Poly(Schiff Bases).= +ted Polybenzimidazoles by C	POLY	0045
Hyperbranched	Poly(siloxy-silanes).= Terminal Modifications of	POLY	0323
ynthesis of Block Copolymers of	Poly(styrene) and Poly(arylate) and Their Optical Pr	POLY	0220
ydrogen- and Deuterium-Labeled	Poly(styrene) and Poly(vinyl methylether).= +n: H	COLL	0035
hesis of Poly(b-vinylborazine) and	Poly(styrene-co-b-vinylborazine) Copolymers.=	POLY	0183
m Castor Oil Polyurethanes and	Poly(Styrene-Methyl Methacrylate-Methacrylic Acid	PMSE	0091
ectric Birefringence of Ionomers	Poly(styrene-Sodium-2-Acrylamido-2-Methylpropan	COLL	0181
Polymers with Skeletal Phosp+	Poly(thionylphosphazenes): A New Class of Inorganic	POLY	0186
hanical Spectroscopy Studies of	Poly(vinyl Acetate)/Silicon Oxide Microcomposites P	POLY	0287
tuted+ Chemical Stabilization of	Poly(vinyl Chloride) by Pretreatment with N-Substi	POLY	0027
ational Isomers in Single Racemic	Poly(vinyl Chloride) Chains.= +Runs of Trans Rot	POLY	0069
n Coevaporates Prepared Using	Poly(vinyl Methyl Ether/Maleic Anhydride) Copolym	COLL	0086
terium-Labeled Poly(styrene) and	Poly(vinyl methylether).= +n: Hydrogen- and Deu	COLL	0035
n S+ Phase Behavior of Aqueous	Poly(vinyl-alcohol) Solution—A Small-Angle Neutro	POLY	0171
mposite Films of Polyanilina and	Poly(vinylchloride)/Poly(Vinylidene Chloride).= Co	PMSE	0113
te)+ Study of Miscible Blends of	Poly(vinylidene Fluoride) and Poly(3-hydroxybutyra	POLY	0097
ac+ Mobility of Ethyl Acetate in	Poly(vinylidene Fluoride) Monitored by Rheo-Photo	POLY	0255
Polystyrene and	Poly(xylenyl Ether).= Miscible Blends of Sulfonated	PMSE	0117
Mechanisms of Gas Transport in	Poly(1-Trimethylsilyl-1 Propyne).=	POLY	0119
New Fluorinated Copolymers:	Poly(1,3-cyclohexadiene-alt-α-fluoroacrylonitrile).=	POLY	0021
ield and Ising Critical Behavior of	Poly(2-Chloro-Styrene)/Polystyrene Blends.= +-F	PMSE	0103
Toward Structurally Perfect	Poly(2,5-pyrrole).=	POLY	0036
Configurations of	Poly(3-Alkylthiophene)s.= Solvent Effects on	PMSE	0183
drate and+ Syntheses of New	Poly(3-hydroxyalkanoate) Conjugates: PHA-Carbohy	POLY	0041
s of Poly(vinylidene Fluoride) and	Poly(3-hydroxybutyrate).= +udy of Miscible Blend	POLY	0097
Thermal Degradation in PEEK (	Poly-ether-ether-ketone) Films.= +scopic Study of	POLY	0334
olecular Weight Determinations of	Poly-4-Methyl-1-Pentene.= +lymer Analysis. M	PMSE	0046
lectrically Conducting Conjugated	Polyacetylene.= +with Bromine: A New Route to E	POLY	0032
operties of Partially Substituted	Polyacetylenes Derived from the Ring-Opening Meta	POLY	0329
Susceptibilities in Soluble	Polyacetylenes for Nonlinear Optics.= Toward High	POLY	0132
Electrically Conducting Ionic	Polyacetylenes.=	POLY	0074
Heterocyclic Homoconjugated	Polyacetylenes.=	ORGN	0094
ynthesis and Characterization of	Polyacrylate-Inorganic Hybrid Materials via the Sol-	POLY	0232

nium Bromide Complexation with	Polyacrylic Acid and Polymethacrylic Acid.= +mo	COLL	0074
Photocurable Cements Based on	Polyalkenoates/Ion–Leachable Glass Hybrid Syste	POLY	0262
Conformational Studies of	Polyalkylsilanes.=	COMP	0059
Cross–Linking of	Polyallylcarbonates by Gamma Irradiation.=	PMSE	0131
Cross–Linking of	Polyallylcarbonates by Gamma Irradiation.=	POLY	0053
of Interactions between Aromatic	Polyamide Chains by Complexation.= +dification	POLY	0064
sition Pol+ Uniaxially Oriented	Polyamide Films in Plane Prepared by Vacuum Depo	POLY	0082
Polymer Blend of Polypropylene–	Polyamide.= +hermal Analysis of an Incompatible	POLY	0071
Neopentyldiamine	Polyamides and Blends.=	PMSE	0100
	Polyamides Containing Imidazoles.=	POLY	0034
e Blends with Rigid and Semirigid	Polyamides.= +perties of Cross-Linkable Polyimid	PMSE	0023
Aromatic	Polyamides.= Solubility Parameters of	POLY	0063
midine/Spermine–N1–Acetyltran+	Polyamine Fumaramate Esters as Inhibitors of Sper	MEDI	0021
Coordination of Bismuth(III) to	Polyaminocarboxylate Ligands.=	INOR	0033
Chloride).= Composite Films of	Polyaniline and Poly(vinylchloride)/Poly(Vinylidene	PMSE	0113
4: Structure–χ(3) Relationships in	Polyanilines and Derivatives.= +lymer Thin Films,	POLY	0221
Polymers.= The	Polyanilines: A Novel Class of Conducting	POLY	0327
Planar Cobalt(III) Complexes of	Polyanionic Chelating Ligands: Crystal Structures, N	INOR	0243
+terization and Trace Analysis of	Polyanions with Direct Polyelectrolyte Titration.=	COLL	0072
es by Liquid+ Determination of	Polyaromatic Hydrocarbons in Environmental Matric	ENVR	0015
hosphodiesters.=	Polyaza Clefts for the Recognition and Cleavage of P	ORGN	0054
tionships in Conjugated Aromatic	Polyazomethines.= +n Films, 3: Structure–χ(3) Rela	POLY	0226
acterization of Diamantane–Based	Polybenzazoles.= +d Polymers: Synthesis and Char	POLY	0028
uced Phase Behaviors of Aromatic	Polybenzimidazole/Polyimide Blends.= +mally Ind	PMSE	0164
urs+ Synthesis of N–Substituted	Polybenzimidazoles by Cyclodehydrogenation of Prec	POLY	0045
Boron Nitride Fibers from	Polyborates.=	POLY	0316
Reactions of	Polyborazylene.= Synthesis and Ceramic Conversion	POLY	0311
us Polyisocyanide and Elastomeric	Polybutadiene Segments.= +Helical and Amorpho	PMSE	0071
Surface Modification of	Polycarbonate by Far UV Radiation.=	PMSE	0125
nolic End–Groups in bisPhenol A	Polycarbonate Using GPC–UV Analysis.= +of Phe	POLY	0083
Phenyl Ring Motions in	Polycarbonates.=	POLY	0080
Ion–Implanted	Polycarbonates.= Spectroscopic Studies of	POLY	0081
ibers.= High Molecular Weight	Polycarbosilane as a Precursor to Oxygen–Free SiC F	POLY	0360
Preparation of a Linear	Polycarbosilane via Ring–Opening Polymerization.=	POLY	0362
Compositions from	Polycarbosilane.= Si–C–N–O Ceramic Fiber	POLY	0336
dienyl Sodium wit+ Synthesis of	Polycarbosilanes via the Condensation of Cyclopenta	INOR	0199
Dechlorination of	Polychlorinated Biphenyls in Anaerobic Sediments.=	BIOT	0120
is of Flyash for Monobrominated	Polychlorinated Dibenzo–p–Dioxins and Dibenzofura	ANYL	0050
Acid Metabolites of	Polychlorotrifluoroethylene.= Analysis of Carboxylic	ANYL	0062
ers Formed in Enzyme–Catalyzed	Polycondensations.= +Molecular Weight of Polyest	BIOT	0094
n of Radical Processes in Organic	Polycrystals under Shear Deformation at High Press	PHYS	0245
nisms of Carcinogenic Metal and	Polycyclic Aromatic Hydrocarbon–Induced Transfor	CHAS	0007
elective Syntheses of Exo or Endo	Polycyclic Epoxides.= +ormation Styrenes: Stereos	ORGN	0236
f a Very Reactive Chromium(III)	Polydentate Amine Complex Toward Base Hydrolysi	INOR	0397
esin.=	Polydicyclopentadiene: A New RIM Thermosetting R	POLY	0136
ced Cross–Link Aggregation in	Polydimethylsiloxane Networks.= Substituent–Indu	PMSE	0120
ork of Polytetrafluoroethylene and	Polydimethylsiloxane.= +netrating Polymer Netw	PMSE	0041
mational Analysis for RIS Model:	Polydimethylsiloxane.= +tical Weights from Confor	POLY	0356
apped Poly(Ethylene Oxide) with	Polydimethylsiloxane.= +rs Containing Hydroxyl–C	PMSE	0006
iffusion Average Molar Mass of	Polydisperse Polymers: A Photon Correlation Spectro	POLY	0305
Structure and Rheology of	Polydisperse, Charged Colloidal Suspensions.=	COLL	0165
e Elementary!.=	POLYED 1989 Polymer Award Winner: Polymers Ar	CHED	0028
mer Education into Middle, Vo+	POLYED 1990 Award Presentation: Integrating Poly	CHED	0026
ol Chemistry Teachers Program.	POLYED 1990 Award Presentation: Introducing Mac	CHED	0025
Addressing Polymers as Macr+	POLYED 1991 Award Winner—Award Presentation:	CHED	0027
rships.=	POLYED—A Model for Academic/Industrial Partne	CHED	0006
Analysis of Polyanions with Direct	Polyelectrolyte Titration.= +terization and Trace	COLL	0072
Main–Chain Liquid Crystalline	Polyelectrolytes Displaying Thermotropic and Lyotro	POLY	0073
Electrophoresis of	Polyelectrolytes in Gels.=	COLL	0183
ty of Linear and Ring	Polyelectrolytes.= Light Scattering and Viscoelastici	POLY	0294
Rubber to	Polyepoxide.= Catalytic Conversion of Natural	ENVR	0045
ogy of Polyurethane–Unsaturated	Polyester Interpenetrating Polymer Network.= +ol	PMSE	0027
y Characterization of Unsaturated	Polyester Resins with Low–Profile Additives.= +og	POLY	0114
Crystallization Behavior of a	Polyester Rotaxane.= On the Morphology and	POLY	0346
of a Liquid Crystalline	Polyester.= Pressure–Induced Crystal Modification	POLY	0072
Phase Behavior in a Transacting	Polyester/Polycarbonate Blend.= +erpretation of	PMSE	0074
roxy–α–methylstilb+ Functional	Polyesters and Copolyesters Based on the 4,4'–Dihyd	POLY	0031
ontrolled Degradation of Erodible	Polyesters for Biomedical Applications.= +for the C	CHED	0133
ffecting the Molecular Weight of	Polyesters Formed in Enzyme–Catalyzed Polyconden	BIOT	0094
ck Copolymers and Monodisperse	Polyesters.= +owth Approach: Hyperbranched Blo	POLY	0318
New Diacetylene–Containing	Polyesters.=	POLY	0054
of a SEBS Block Copolymer and a	Polyetherester.= +erpenetrating Polymer Blends	PMSE	0058
CpE near the UCST of	Polyethers + Alkane Systems.= Behavior of VE and	PHYS	0319
Molecular Weight	Polyethers.= Silicon–Assisted Synthesis of High	PMSE	0075
Gelation/Crystallization of	Polyethersulfone.=	POLY	0248
Crystal/Amorphous Interface in	Polyethylene and Its Blends.= Nature of the	COLL	0036
lve+ Thermoreversible Gels with	Polyethylene Block Copolymers and Hydrocarbon So	POLY	0249
odel and New Applications.=	Polyethylene Glycol–Enhanced Protein Refolding: M	BIOT	0136
Structure of	Polyethylene Shish Kebabs in Situ.=	PMSE	0184
Study of Oxidized and Exchanged	Polyethylene Surfaces.= +IR Specular Reflectance	POLY	0253

ate on the Thermal P+ Effects of	Polyethylene Terephthalate (PET) from Plastics Wa	FUEL	0053
Ordered Conjugated Polymers in	Polyethylene: Orientation by Mesoepitaxy.= +ighly	POLY	0008
rogeneity of Linear Low-Density	Polyethylene: Temperature-Rising Supercritical Flui	POLY	0076
Cold-Extruded	Polyethylene.= Atomic Force Microscopy on	PMSE	0078
ng of Ultrahigh Molecular Weight	Polyethylene.= +Fractions Found in the Slow Melti	PMSE	0166
is of Mixed Alkyl Amines and	Polyfunctional Amines.= Selective Catalytic Synthes	I&EC	0055
tization of Light Hydrocarbons on	Polyfunctional Metallosilicate Catalysts.= +Aroma	PETR	0062
Tomatoes with Reduced	Polygalacturonase.= Evaluation of Transgenic	AGFD	0057
er Heterogeneous Catalysts. Shell	Polygasoline and -Kero Process.= +drocarbons ov	PETR	0025
the Vertices of the Five Regular	Polyhedra in Order To Canonically Name the Differe	COMP	0035
ization by Vanadium-Containing	Polyhedral Oligometallasilsesquioxanes (POMSS).=	INOR	0131
of Nonclassical Transition Metal	Polyhydrides: Rotational Tunnelling and Quantum E	PHYS	0068
y Selective Acylation of Di- and	Polyhydroxyl Compounds by 3-Acylthiazolidine-2-th	ORGN	0038
o-Enzymatic Modifications of	Polyhydroxylated Compounds.= Regioselective Chem	BIOT	0064
nd Properties of Cross-Linkable	Polyimide Blends with Rigid and Semirigid Polyamid	PMSE	0023
thesis and Characterization of a	Polyimide from 3,7-Diaminophenothiazinium Chlorid	POLY	0033
the Gas Transport Properties of	Polyimide Isomers.= Conditioning Effect of CO_2 on	POLY	0116
adimides Oligomers and High-Tg	Polyimide-Based Semi-IPNs.= +terization of bisN	POLY	0133
phases between PMDA/4-BDAF	Polyimides and Silver Substrates Using Surface-Enh	POLY	0087
Antiplasticization of	Polyimides by a Nonreactive Diluent.=	PMSE	0127
d Properties of a Series of Soluble	Polyimides for Use in Microelectronics.= +hesis an	POLY	0038
A Low-Temperature Route to	Polyimides.=	POLY	0137
(amic Acid) Precursors and Their	Polyimides.= +ress Behavior of Functionalized Poly	POLY	0085
of Diacetylene LB Films by	Polyion Complexation.= Control of Photochromism	BIOT	0146
fied Storage Optical Memory with	Polyion Complexed Pyrene LB Films.= +otoampli	BIOT	0144
ntaining Helical and Amorphous	Polyisocyanide and Elastomeric Polybutadiene Segm	PMSE	0071
Fungal	Polyketide Biosynthesis—From Enzymes to Genes.=	BIOT	0111
Configuration of Dense	Polymacromonomers in Good Solvent.= Equilibrium	POLY	0269
to the Study of Polymer-	Polymer Adhesion.= Application of the Blister Test	COLL	0057
ted Chromatographic Methods for	Polymer Analysis. +d-Flow Fractionation, and Rela	PMSE	0008
ted Chromatographic Methods for	Polymer Analysis. +d-Flow Fractionation, and Rela	PMSE	0065
ted Chromatographic Methods for	Polymer Analysis. +d-Flow Fractionation, and Rela	PMSE	0046
ted Chromatographic Methods for	Polymer Analysis. +d-Flow Fractionation, and Rela	PMSE	0028
ted Chromatographic Methods for	Polymer Analysis. +d-Flow Fractionation, and Rela	PMSE	0094
cal Fluids.=	Polymer and Biopolymer Modifications in Supercriti	POLY	0084
H_6 Radical Anions as Protoypes of	Polymer Anions with Si and Ge Backbones.= +Ge_2	PHYS	0313
ected Supramolecular	Polymer Architectures.= Molecular-Recognition-Dir	POLY	0299
dy of a Ferroelectric Side-Chain	Polymer at Different Interfaces Using the Langmuir-	COLL	0071
POLYED 1989	Polymer Award Winner: Polymers Are Elementary!.=	CHED	0028
Charge Transport in	Polymer Bilayer Films.=	ANYL	0104
ermal Analysis of an Incompatible	Polymer Blend of Polypropylene-Polyamide.= +Th	POLY	0071
f Charge Transfer Interactions on	Polymer Blend Phase Structure.= +the Influence o	PMSE	0076
ssation of+ Phase Behavior of a	Polymer Blend Solution during Steady Shear and Ce	PMSE	0134
Mixing in a Miscible	Polymer Blend.= Fluorescence and Shear-Induced	PMSE	0133
Relationship in Interpenetrating	Polymer Blends of a SEBS Block Copolymer and a P	PMSE	0058
Properties.= Designing	Polymer Blends Which Exhibit Synergistic	PMSE	0101
. Flow-Induced Phase Changes of	Polymer Blends with UCST Phase Behavior.= +rs	PMSE	0132
nterfaces—II. Quick Quenching of	Polymer Blends.= +lymer Solutions, Blends, and I	COLL	0032
sites via Metalorganic/Inorganic	Polymer Blends.= Ceramic Coatings for C/C Compo	POLY	0229
Hydrogen-Bonded	Polymer Blends.= Diffusion of Water Vapor in	POLY	0192
in	Polymer Blends.= Hydrogen-Bonding Interaction	COLL	0034
y, II: Image Con+ Morphology of	Polymer Chains by Transmission Electron Microscop	POLY	0062
Rearrangements of Flexible	Polymer Chains in Flow.= Conformational	PMSE	0105
Isolated	Polymer Chains.= Exciton Kinetics Studies on	PHYS	0309
troscopy.=	Polymer Characterization by Step-Scan FT-IR Spec	POLY	0304
c Infr+ Instrumental Methods in	Polymer Characterization: Optical Methods. Dynami	POLY	0303
nd F+ Instrumental Methods in	Polymer Characterization: Optical Methods. FT-IR a	POLY	0330
ene-P+ Instrumental Methods in	Polymer Characterization: Optical Methods. Polystyr	POLY	0277
d D+ Instrumental Methods in	Polymer Characterization: Optical Methods. Static an	POLY	0250
New Experimental Possibilities for	Polymer Characterization.= +EC-Viscometry and	PMSE	0070
hemistry, Phys+ Marvel Creative	Polymer Chemistry Award Symposium Polymers—C	POLY	0272
Adsorpt+ C. S. Marvel Creative	Polymer Chemistry Award Symposium. Spontaneous	POLY	0298
Science with	Polymer Chemistry.= Enrichment of Precollege	CHED	0009
rochemical Aspects of Conducting	Polymer Colloids.= +roheterogeneous Fluids. Elect	COLL	0166
Rheological Study of Polymer-	Polymer Complexes of an Ionomer and an Amino-Co	PMSE	0155
a Silver Colloid-	Polymer Composite.= Optical Four-Wave Mixing in	POLY	0218
	Polymer Composites Using Recycled PET.=	PMSE	0072
igh-Precision Characterization of	Polymer Conformation by SEC with an On-Line Vis	PMSE	0097
Numbers of	Polymer Degrees Granted in the United States.=	PMSE	0128
Sulfur-Modified Carbohydrate	Polymer Derivatives.= New Phosphorus- and	CARB	0030
NMR Studies of Local	Polymer Dynamics in Dilute Solution.= Optical and	PHYS	0139
Award Presentation: Integrating	Polymer Education into Middle, Vocational, and Hig	CHED	0026
onium Salt Distribution in Model	Polymer Films by ^{19}F Multiple-Quantum NMR.=	PMSE	0182
Study of Thin Bicomponent	Polymer Films on a Metal Surface by ESCA.=	POLY	0100
lends, and Interfaces—IV. Thin	Polymer Films: From Chemistry to Materials Science	COLL	0070
ffusional Processes in Compatible	Polymer Films.= +e Techniques to the Study of Di	POLY	0163
Species Transfer in Electroactive	Polymer Films.= +ersible and Irreversible Neutral	ANYL	0103
ing in Vitrified Nonlinear Optical	Polymer Films.= +dering and Chemical Cross-Link	POLY	0206
Frequency Doubling in Poled	Polymer Films.= Recent Developments in Efficient	POLY	0179
he Diffusion of Small Molecules in	Polymer Films.= +tional Spectroscopies to Probe t	POLY	0332

ng of Solvent in Thermoreversible Polymer Gels and Related Systems.= +g and Melti	POLY	0170
Swelling of Constrained Polymer Gels.=	COLL	0184
Metal- Polymer Interaction through Metal Coordination.=	POLY	0343
rfaces—I. Dynamics at the Solid- Polymer Interface.= +r Solutions, Blends, and Inte	COLL	0012
lymers from Solution to Solution- Polymer Interfaces.= +ntaneous Adsorption of Po	POLY	0298
arizable Amphiphiles at Polymer- Polymer Interfaces.= +c Generation from Hyperpol	POLY	0156
Oil Polyurethane/Vinyl or Acrylic Polymer Interpenetrating Polymer Networks.= +r	PMSE	0090
opolymers.= Investigation of t+ Polymer Interphase via Synthesis of Well-Defined C	COLL	0033
Proton Transport in Ordered Polymer Lattices.= Voltammetric Investigation of	COLL	0167
I Think I Shall Never See a Polymer Lovely as a Tree.=	POLY	0319
Value Perception Analysis for Polymer Market-Entry Strategies.=	CMEC	0011
ential+ Atomistic Simulation of Polymer Materials. Ab Initio Quantum Chemical Pot	POLY	0350
tions in+ Atomistic Simulation of Polymer Materials. Sizes of the Short-Range Correla	POLY	0365
Mediated by the Organic Polymer Matrices.= Inorganic Phase Behaviors	POLY	0364
posites and Ceramics: Influence of Polymer Matrix on Precipitation Processes.= +om	POLY	0290
ent of Diffusion Coefficients in Polymer Melts Using Surfaced-Enhanced Raman Sca	POLY	0088
Reactions Carried Out in Polymer Melts.= Chemistry of Condensation	PMSE	0176
lenary Lecture. Enhancement of Polymer Miscibility through Ionic Functional Groups	PMSE	0116
atalyzed Laboratory Innovation: Polymer Molecular Weight Determination by Gel Pe	CHED	0210
in THF a+ Characterization of Polymer MWD and Branching with GPC-Viscometry	PMSE	0068
of Latex Interpenetrating Polymer Network Films.= Gas Permeability Studies	PMSE	0061
ane and Vinyl+ Interpenetrating Polymer Network Formed from Castor Oil Polyureth	PMSE	0089
Biocompatible Interpenetrating Polymer Network of Polytetrafluoroethylene and Pol	PMSE	0041
olymers through Interpenetrating Polymer Network Synthesis.= +tional Biomimetic P	PMSE	0004
and Epoxy Interpenetrating Polymer Network.= Composites from Polyurethane	PMSE	0044
urated Polyester Interpenetrating Polymer Network.= +ology of Polyurethane-Unsat	PMSE	0027
olyurethanes and Interpenetrating Polymer Networks (IPNs).= +n Foams Based on P	PMSE	0040
hree-Component Interpenetrating Polymer Networks and Their Applications.= +te T	PMSE	0080
Simultaneous Interpenetrating Polymer Networks from Castor Oil Polyurethanes an	PMSE	0091
Polyvinyl Pyr+ Interpenetrating Polymer Networks of Poly(Carbonate-Urethane) and	PMSE	0056
s of Interpene+ Interpenetrating Polymer Networks. Entirely Radical in Situ Synthesi	PMSE	0001
ks: An Overvi+ Interpenetrating Polymer Networks. Interpenetrating Polymer Networ	PMSE	0039
olyurethane/+ Interpenetrating Polymer Networks. Kinetics of Phase Separation in P	PMSE	0021
ship in Interp+ Interpenetrating Polymer Networks. Morphology/Processing Relation	PMSE	0058
ng Themosets+ Interpenetrating Polymer Networks. Two-Phase Epoxy Interpenetrati	PMSE	0087
lymer Networks. Interpenetrating Polymer Networks: An Overview.= +enetrating Po	PMSE	0039
mental Orientation in Deformed Polymer Networks: Characterization of Stress-Strain	COLL	0016
Situ Synthesis of Interpenetrating Polymer Networks.= +tworks. Entirely Radical in	PMSE	0001
rethane/Acrylate Interpenetrating Polymer Networks.= +ated Steel Sheet with Polyu	PMSE	0057
hase Structure of Interpenetrating Polymer Networks.= +and Nonequilibrium Microp	PMSE	0093
Flow in Latex Interpenetrating Polymer Networks.=	PMSE	0062
Simultaneous Interpenetrating Polymer Networks.= Morphology of P (nBAc)/PEP	PMSE	0042
Interpenetrating Polymer Networks.= Phase Separation in	PMSE	0022
Interpenetrating Polymer Networks.= Studies of Grafted	PMSE	0003
Acrylic Polymer Interpenetrating Polymer Networks.= +r Oil Polyurethane/Vinyl or	PMSE	0090
f PS/PA Latex Interpenetrating Polymer Networks.= Study on the Damping Paint o	PMSE	0064
Latex Interpenetrating Polymer Networks.= Study on the Three-Stage	PMSE	0082
Polyphosphazene Interpenetrating Polymer Networks.= +sis and Characterization of	PMSE	0002
Orientation and Decay in Poled Polymer Nonlinear Optical Materials.=	POLY	0151
s Graded-Index and Single-Mode Polymer Optical Fibers.= +Applications. Low-Los	POLY	0174
d Decay and Their Relationship to Polymer Physics.= +r Nonlinear Optical Onset an	POLY	0152
ar Weight Liquids: Application to Polymer Plasticizers and Model Hydrocarbons.= +l	POLY	0065
A Survey of the NLO- Polymer Program at the MPI.=	POLY	0010
University, Brooklyn.= The Polymer Research Institute at the Polytechnic	HIST	0033
Preparation of Dense Polymer Samples for Atomistic Modeling.=	POLY	0366
ogram?.= Polymer Science: An Independent Undergraduate Pr	CHED	0099
Characterization of Polymer Segmental Motion via Deuterium NMR.=	PHYS	0140
Computational Method for Polymer Simulations.=	PHYS	0112
laxation in Molecular Glasses and Polymer Solids.= +hases. Energy Transfer and Re	PHYS	0080
cture through Universal Ratios of Polymer Solution Properties.= +Branching Archite	POLY	0293
Steady S+ Phase Behavior of a Polymer Solution Studied by Light Scattering during	PMSE	0107
Dynamics of a Semidilute Polymer Solution under Shear Flow.= Structure and	PMSE	0108
nd Vitrification of Concentrated Polymer Solutions under Poor Thermodynamic Cond	POLY	0146
Birefringence Characterization of Polymer Solutions under Transient Elongational Fl	POLY	0281
rodynamic—II. Shear Thinning of Polymer Solutions.= +Electric, Magnetic, and Hyd	COLL	0159
Dynamic Studies of Ternary Polymer Solutions.=	COLL	0054
ics at+ Proctor Gamble UERP Polymer Solutions, Blends, and Interfaces—I. Dynam	COLL	0012
Quenc+ Procter Gamble UERP Polymer Solutions, Blends, and Interfaces—II. Quick	COLL	0032
ral Diff+ Procter Gamble UERP Polymer Solutions, Blends, and Interfaces—III. Late	COLL	0051
Polym+ Procter Gamble UERP Polymer Solutions, Blends, and Interfaces—IV. Thin	COLL	0070
Polymer Synthesis with Organometallic Reagents.=	POLY	0328
racterization of Multicomponent Polymer Systems by Digital Image Analysis and Rea	PMSE	0099
Oriented Bicomponent Polymer Systems.= Microdeformational Behavior of	POLY	0101
ge Science.= Polymer Technology: An Effective Topic for Precolle	CHED	0007
roces+ Cubic Nonlinear Optics of Polymer Thin Films, I: Molecular Engineering and P	POLY	0011
in Ri+ Cubic Nonlinear Optics of Polymer Thin Films, 2: Structure-$\chi^{(3)}$ Relationships	POLY	0205
in C+ Cubic Nonlinear Optics of Polymer Thin Films, 3: Structure-$\chi^{(3)}$ Relationships	POLY	0226
in P+ Cubic Nonlinear Optics of Polymer Thin Films, 4: Structure-$\chi^{(3)}$ Relationships	POLY	0221
$\chi^{(3)}$+ Cubic Nonlinear Optics of Polymer Thin Films, 5: Wavelength Dispersion of the	POLY	0213
Gel Content Determination of Polymer Using Field-Flow Fractionation.=	PMSE	0010

Planar	Polymer Waveguides for Optical Interconnections.=	POLY 0175
t+ Rigid-Rod Benzobisthiazole	Polymer with Reactive 2,6-Dimethylphenoxy Penden	PMSE 0019
t+ Rigid-Rod Benzobisthiazole	Polymer with Reactive 2,6-Dimethylphenoxy Penden	PMSE 0018
oromer and an Amino-Containing	Polymer.= +f Polymer-Polymer Complexes of an I	PMSE 0155
opic Ellipsometry: Applications in	Polymer- and Thin-Film Growth.= +me Spectrosc	POLY 0252
0s+ New Frontiers in Packaging:	Polymer-Based Intermaterial Competition in the 199	CMEC 0016
Phase Separation in	Polymer-Dispersed Liquid Crystals.=	COLL 0014
Toward a Theory of	Polymer-Induced Protein Precipitation.=	BIOT 0022
2,2'-Bipyridine Ligands for	Polymer-Modified Electrodes.= Aniline-Substituted	POLY 0189
Blister Test to the Study of	Polymer-Polymer Adhesion.= Application of the	COLL 0057
mino-Co+ Rheological Study of	Polymer-Polymer Complexes of an Ionomer and an A	PMSE 0155
Hyperpolarizable Amphiphiles at	Polymer-Polymer Interfaces.= +ic Generation from	POLY 0156
l-Chain Formation in Multiphase	Polymer-Polymer-Solvent Systems.= +nduced Pear	COLL 0145
Nonlinear Optical Properties of	Polymer-Silver Microsphere Composites.=	POLY 0154
it.es of a Free-Volume Theory for	Polymer-Solvent Diffusion Coefficients.= +Capabil	POLY 0144
ormation in Multiphase Polymer-	Polymer-Solvent Systems.= +duced Pearl-Chain F	COLL 0145
Phase Behavior of	Polymer-Supercritical Fluid Mixtures.=	PMSE 0036
erol Removal from Butteroil.=	Polymer-Supported Tomatine as a Means of Cholest	AGFD 0135
ization of Chemical Reactions at	Polymer/Metal Interface by Sputtering Depth Profil	PMSE 0150
Inhibitors of DNA	Polymerases of Leishmania mexicana.=	BIOL 0118
spectives on the Ultrafiltration of	Polymeric Colorants.= +mical and Engineering Per	I&EC 0049
Preparation of	Polymeric Films for NLO Applications.=	POLY 0104
	Polymeric Flocculants of the Zwitterionic Type.=	POLY 0094
	Polymeric Flocculants.=	PMSE 0129
-alysis of X-ray Scattering from	Polymeric Gas Separation and Barrier Materials.=	POLY 0142
Applications.= Designing	Polymeric Materials for All-Optical Device	BIOT 0175
a:ions. Highly Polarizable Metal+	Polymeric Materials for Photonic and Optical Applic	POLY 0102
a:ions. Introduction+ Tutorial on	Polymeric Materials for Photonic and Optical Applic	POLY 0001
a:ions. Low-Loss Graded-Index+	Polymeric Materials for Photonic and Optical Applic	POLY 0174
a:ions. Orientation of Organic +	Polymeric Materials for Photonic and Optical Applic	POLY 0006
a:ions. Second Harmonic Gener+	Polymeric Materials for Photonic and Optical Applic	POLY 0150
a:ions. Synthesis of Controlled+	Polymeric Materials for Photonic and Optical Applic	POLY 0126
ns for the Atomistic Simulation of	Polymeric Materials.= +Potential Energy Functio	POLY 0350
of Gas Permeation through Glassy	Polymeric Materials.= +Thermodynamic Analysis	POLY 0139
ne Crystals and Tetracene-Doped	Polymeric Matrices.= +Phase Transition in Tetrace	PHYS 0274
Crystal Dispersed in an Ionic	Polymeric Membrane.= Investigation of a Liquid	PMSE 0119
Multicomponent Species across	Polymeric Membranes.= Diffusion of	POLY 0195
and Metribuzin: Preparation an+	Polymeric Microcapsules of the Herbicides Atrazine	AGRO 0063
nance Imagin+ Novel Fluorinated	Polymeric Molecular Probes for F-19 Magnetic Reso	MEDI 0041
aceutical Molecules Using Porous	Polymeric Packings.= +aphy of Proteins and Pharm	ANYL 0031
	Polymeric Routes to Aluminum Oxides.=	POLY 0342
eated Refractive Index Profiles in	Polymeric Slab Waveguides.= +otochemically Delin	POLY 0180
ize Tinkertoy Construction Set:	Polymeric Structures Derived from [1.1.1] Propellane	POLY 0344
cn of the Chemical Potentials of	Polymeric Systems from Monte Carlo Simulations.=	POLY 0352
Powder Diffraction Studies of	Polymeric Triorganotin(IV) Fluorides.= X-ray	INOR 0065
pounds by Chemically Modified	Polymeric-Mercuric Resin in Capillary Gas Chromat	ANYL 0083
Oligometallasilsesquiox+ Olefin	Polymerization by Vanadium-Containing Polyhedral	INOR 0131
ficant Reversal of Hydration and	Polymerization in the Matrix Apparently Caused by	ENVR 0041
Crystalline Polymers Obtained by	Polymerization in the Melt.= +Diacetylenic Liquid	POLY 0030
€ Prepared by Vacuum Deposition	Polymerization Methods.= +lyamide Films in Plan	POLY 0082
Ultrasonic	Polymerization of Acrylic Monomers.=	PMSE 0020
Component Organoscandium S+	Polymerization of Butadiene and Isoprene by Single-	INOR 0280
nd Methac+ ESR Studies on the	Polymerization of Diacetylene-Containing Acrylate a	POLY 0092
cid Derivatives.= Synthesis and	Polymerization of Diacetylene-Containing Benzoic A	POLY 0056
acarbonyl.= Ring-Opening	Polymerization of Epoxides Catalyzed by Dicobaltoct	POLY 0014
New Route to Electrically Cond+	Polymerization of Ethynylpyridine with Bromine: A	POLY 0032
ad+ Transition-Metal-Catalyzed	Polymerization of Heteroatom-Substituted Cyclohex	POLY 0047
cid Telechelomer+ Synthesis and	Polymerization of Hydroxy Polystyrene Carboxylic A	POLY 0046
osphazenes.= Gamma Ray	Polymerization of Long Chain Substituted Cyclotriph	PMSE 0130
hosphazenes.= Gamma Ray	Polymerization of Long-Chain-Substituted Cyclotrip	POLY 0055
ydrothiophenium)-p-xylene Di+	Polymerization of Mono-Substituted α,α'-bis(Tetrah	POLY 0013
g Pigments D+ A Process for the	Polymerization of Monomers on the Surface of Hidin	PMSE 0038
m the Ring-Opening Metathesis	Polymerization of Monosubstituted Cyclooctatetraen	POLY 0329
Electron Beam-Induced Cationic	Polymerization of Silicone-Epoxide Resins.=	PMSE 0159
urther Confirmati+ Spontaneous	Polymerization of Styrene in the Presence of Acid: F	POLY 0039
s Molecular Composites, I: In Situ	Polymerization of Tetraethoxysilane.= +sphazene	POLY 0181
of a Polysiloxane Rubber on the	Polymerization Phase-Separation Process, and Gener	POLY 0112
A Model Study of Condensation	Polymerization Using Low-Temperature Techniques	POLY 0046
er Stabilizers.= Emulsion	Polymerization with Oligomeric Diblock Macromonom	COLL 0039
operties during Isothermal Epoxy	Polymerization.= +ological Changes to Dielectric Pr	PMSE 0178
Microemulsion	Polymerization.= Coulometric Initiation of	COLL 0169
Catalytic Olefin	Polymerization.= Mechanism and Selectivity in	CATL 0010
cs of Propagation in Free-Radical	Polymerization.= +nomer Diffusion and the Kineti	POLY 0165
logs for UV-Induced Cationic	Polymerization.= Novel Propenyl Phenyl Ether Ana	POLY 0017
silane via Ring-Opening	Polymerization.= Preparation of a Linear Polycarbo	POLY 0362
ope+ Fluorinated Polyurethanes:	Polymerization, Characterization, and Mechanical Pr	POLY 0018
erizations, Oligomerizations, and	Polymerizations Catalyzed by Co(III) and Rh(III) Co	ORGN 0129
ion and Helicity in Living	Polymerizations of Chrial Isocyanides.= Stereoselect	POLY 0347
uced Electron Tra+ Interfacially	Polymerized Porphyrin Films for Vectorial Photoind	BIOT 0126
dable Polymers. Biodegradation of	Polymers and Biodegradable Polymers.= +odegra	CHED 0130

Gas Solubility in	Polymers and Blends.=	POLY	0143
al Properties of Nonlinear Optical	Polymers and Devices.= +mism Studies of Electric	POLY	0150
Polymer Award Winner:	Polymers Are Elementary!.= POLYED 1989	CHED	0028
—Award Presentation: Addressing	Polymers as Macromolecules.= +91 Award Winner	CHED	0027
onjugated Polymers. Conjugated	Polymers as Materials for Nonlinear Optical Devices	POLY	0325
i+ Limitations of Current Epoxy	Polymers as Matrix for Structural Composites in Oilf	PMSE	0111
ogen- and/or Boron–Containing	Polymers as Precursors of Ceramic Materials.= Nitr	POLY	0315
a+ Coordination Compounds and	Polymers as Second–Order Nonlinear Optical Materi	INOR	0049
IPN–Modified Porous Bead	Polymers as Supports for Enzyme Immobilization.=	PMSE	0081
Dynamic Properties of Cellulosic	Polymers at the Air–Water Interface.=	COLL	0056
Diffusion in	Polymers by Forced Rayleigh Scattering.=	POLY	0164
ation of Branched and Comb	Polymers by SEC and Other Methods.= Characteriz	PMSE	0069
for Polymer Analysis. Analysis of	Polymers by Thermal Field–Flow Fractionation.=	PMSE	0008
of Potential Use in Photoactive	Polymers by Time–Resolved Microwave Conductivity	POLY	0214
Colorless Rigid–Rod	Polymers Containing Bicyclo[2.2.2] Octane Moieties.=	POLY	0029
aracterization of Monomers and	Polymers Containing Photoresponsive Asobenzene–B	POLY	0209
ation of Titanocene–Containing	Polymers Derived from α–Amino Acids, Dipeptides, a	PMSE	0014
hotorefractive Effect in Nonlinear	Polymers Doped with Charge Transport Agents.=	POLY	0155
Degradable	Polymers for Biomedical Application.=	CHED	0134
tems—I. Plenary Lecture. Barrier	Polymers for Container Coatings.= +Coatings Sys	PMSE	0138
Novel Side–Chain	Polymers for Electrooptic Applications.=	POLY	0107
NLO	Polymers for Frequency Doubling.=	POLY	0178
Design and Synthesis of Novel	Polymers for Medicine.=	POLY	0273
Studies of	Polymers for Shock Absorption Applications.=	PMSE	0181
New	Polymers from Artificial Genes: A Progress Report.=	POLY	0302
ces.+ Spontaneous Adsorption of	Polymers from Solution to Solution–Polymer Interfa	POLY	0298
ylene Glycol+ Amphiphilic Star	Polymers from Tri– or Tetraisocyanates and Poly(eth	POLY	0271
of Chromophore–Functionalized	Polymers Having Chromophore–Centered Hydrogen–	POLY	0108
t Electric Birefringence.=	Polymers in Electric Fields: Dynamics from Transien	COLL	0180
Biodegradable	Polymers in Health Care.= Contemporary Role of	CHED	0135
s of Highly Ordered Conjugated	Polymers in Polyethylene: Orientation by Mesoepitax	POLY	0008
Sy+ Applications of Conducting	Polymers in Redox Devices and Intelligent Materials	POLY	0326
ics of Rotational Isomerization in	Polymers in Solution and in the Bulk State.= +nam	COLL	0018
Diffusion of Star	Polymers in Solutions and Gels.=	POLY	0267
Integrated Experimental Progr+	Polymers in the Physical Chemistry Laboratory: An	CHED	0045
Devices.= Optically Nonlinear	Polymers in Waveguiding Passive and Active	POLY	0177
Disorder and the Metallic State in	Polymers Lacking a Degenerate Ground State.=	PHYS	0039
f Diacetylenic Liquid Crystalline	Polymers Obtained by Polymerization in the Melt.=	POLY	0030
Novel Dendritic Macromolecule+	Polymers of Geometrical Beauty—Combs and Stars.=	POLY	0318
Ring Polymers.=	Polymers of Geometrical Beauty—Combs and Stars.	POLY	0292
Some Static and Dynamic Prop+	Polymers of Geometrical Beauty—Combs and Stars.	POLY	0265
Toward a Molecular–Size Tinke+	Polymers of Geometrical Beauty—Combs and Stars.	POLY	0344
Pseudografting of Vinyl	Polymers onto DNA.=	POLY	0301
ance Materials from Semiflexible	Polymers Such as the Cellulosics.= +High–Perform	COLL	0015
Synt+ Bifunctional Biomimetic	Polymers through Interpenetrating Polymer Network	PMSE	0004
lity and Solid–State Properties of	Polymers through Rotaxane Formation.= +ocessabi	POLY	0233
cipal Waste. Conversion of Waste	Polymers to Energy Products.= +lastics and Muni	FUEL	0090
phitizable and Nongraphitizable	Polymers under Argon Stream and Their Magnetic P	POLY	0023
oring the Cure of Photosensitive	Polymers Using a Luminescent Rhenium Carbonyl C	POLY	0096
lecular Design Concept for Poled	Polymers Utilizing the Off–Diagonal Tensor Compon	POLY	0128
Light Scattering from Solutions of	Polymers with Differing Architecture.= +ynamic	POLY	0250
ce Spectros+ Surface Profiling of	Polymers with Fourier Transform Infrared Reflectan	PMSE	0153
r Engineering and Processing of	Polymers with Large Third–Order Optical Properties	POLY	0011
Associative	Polymers with Novel Hydrophobe Structures.=	PMSE	0035
Reactions of	Polymers with Pendant Cyclophosphazenes.=	POLY	0182
ptical Properties of Linear Epoxy	Polymers with Pendant Sulfonyl Tolan Groups.=	POLY	0208
Novel	Polymers with Photorefractive and NLO Properties.=	POLY	0129
zenes): A New Class of Inorganic	Polymers with Skeletal Phosphorus, Nitrogen, and S	POLY	0186
nd Main–Chain Nonlinear Optical	Polymers with Sulfonyl Acceptor Groups.= +ain a	POLY	0105
gy of Blends of Liquid Crystalline	Polymers with Thermoplastics.= +on the Morpholo	PMSE	0135
the Stu+ Diffusion Processes in	Polymers. Application of Fluorescence Techniques to	POLY	0163
Intermaterial Competition among	Polymers. Assessing Cost and Performance in Int+	CMEC	0009
able Pol+ Truly Biodegradable	Polymers. Biodegradation of Polymers and Biodegrad	CHED	0130
linear Optical Devi+ Conjugated	Polymers. Conjugated Polymers as Materials for Non	POLY	0325
ow–Induced Structural Changes in	Polymers. Domain Growth during the Spinoda+ Fl	PMSE	0104
um: Thermoreversible Gelation of	Polymers. Dynamic Viscoelasti+ APS–ACS Symposi	POLY	0197
ow–Induced Structural Changes in	Polymers. Flow–Induced Phase Changes of Po+ Fl	PMSE	0132
onded+ Diffusion Processes in	Polymers. Gas Sorption and Diffusion in Hydrogen–B	POLY	0191
s Transp+ Diffusion Processes in	Polymers. Influence of Molecular Architecture on Ga	POLY	0115
um: Thermoreversible Gelation of	Polymers. Reversible Gelation+ APS–ACS Symposi	POLY	0145
um: Thermoreversible Gelation of	Polymers. Surfactant Organoge+ APS–ACS Symposi	POLY	0169
on throu+ Diffusion Processes in	Polymers. Thermodynamic Analysis of Gas Permeati	POLY	0139
um: Thermoreversible Gelation of	Polymers. Thermoreversible G+ APS–ACS Symposi	POLY	0121
rage Molar Mass of Polydisperse	Polymers: A Photon Correlation Spectroscopy Study	POLY	0305
Synthesis of Soluble Rigid–Chain	Polymers: An Ionic Poly(p–phenylene) Analog.= +s	POLY	0024
of Electronic Energy Transfer in	Polymers: Comparison of Theory and Experiment.=	PHYS	0109
sceptibilities of Semiconducting	Polymers: Comparison with Theoretical Calculations.	POLY	0009
+Oxidation Reactions of Si–B–C	Polymers: Experimental and Theoretical Studies.=	POLY	0187
Photonics.= Nonlinear Optical	Polymers: Opportunities and Challenges in	BIOT	0174
of Preceramic	Polymers: Some Examples.= Chemical Modification	POLY	0359

Visible–Transparent Rigid–Rod	Polymers: Synthesis and Characterization of Diaman	POLY	0028
styrene Microbead+ Model Filled	Polymers: Synthesis of Uniformly Cross–Linked Poly	PMSE	0017
Diffusion in Hydrogen–Bonded	Polymers: Vinyl Alcohol/Vinyl Butyral Copolymers.=	POLY	0191
rt Properties of Liquid Crystalline	Polymers.= +lecular Architecture on Gas Transpo	POLY	0115
amic Infrared Linear Dichroism of	Polymers.= +racterization: Optical Methods. Dyn	POLY	0303
Beauty—Combs and Stars. Ring	Polymers.= Polymers of Geometrical	POLY	0292
c and Dynamic Properties of Star	Polymers.= +eauty—Combs and Stars. Some Stati	POLY	0265
s of Nanomaterials via Preceramic	Polymers.= +Inorganic Hybrid Materials. Synthesi	POLY	0335
on of Polymers and Biodegradable	Polymers.= +odegradable Polymers. Biodegradati	CHED	0130
in	Polymers.= A Molecular Interpretation of Diffusion	POLY	0168
Modified and Comb–Type	Polymers.= Aggregation of Hydrophobically	POLY	0270
the Synthesis of High	Polymers.= Application of Phosphorus Chemistry to	INOR	0326
Transfer in	Polymers.= Computer Simulation of Direct Energy	POLY	0079
Olefinic	Polymers.= Crosslinking Reactions in Pigmented	POLY	0135
otocrosslinkable Nonlinear Optical	Polymers.= +and Synthesis of a New Class of Ph	POLY	0131
ll Molecules in Stiff–Chain Glassy	Polymers.= +n the Sorption and Transport of Sma	POLY	0120
ending on the Gas Permeability of	Polymers.= +of Aryl–Substitution and Physical Bl	POLY	0117
Opto/Electroactive	Polymers.= Gel Processing of	POLY	0200
Synthesis of New	Polymers.= Metal Coordination Complexes for	CHED	0059
d Hydroxy–Substituted Rigid–Rod	Polymers.= +namics Simulations of Rigid–Rod an	PHYS	0322
Structure and Dynamics of Solid	Polymers.= Multidimensional NMR Studies of	PHYS	0137
New Comb–Shaped	Polymers.=	POLY	0268
Crystal	Polymers.= Novel Photochromic Liquid	POLY	0219
gy Transfer and Depolarization in	Polymers.= +Time Dependence of Electronic Ener	POLY	0078
D of Biodegradable	Polymers.= Panel Discussion: Future Directions in R	CHED	0136
calized Silicon Backbone Network	Polymers.= +on and Vibronic Relaxation in σ–Delo	PHYS	0138
Acrylic	Polymers.= Reactive Extrusion Imidization of	PMSE	0179
Properties of Poled	Polymers.= Second–Order NLO and Relaxation	POLY	0153
Solid	Polymers.= Stress Effects on Diffusion through	POLY	0140
Cross–Linked	Polymers.= Structure and Properties of AB	PMSE	0083
Conducting	Polymers.= Structure–Property Perspectives for	BIOT	0203
Carbohydrate and PHA–Synthetic	Polymers.= +hydroxyalkanoate) Conjugates: PHA–	POLY	0041
BN	Polymers.= Synthesis and Processing of Preceramic	POLY	0312
Synthetic Approaches to NLO	Polymers.=	POLY	0106
Conducting	Polymers.= The Polyanilines: A Novel Class of	POLY	0327
ractions for Model Compounds of	Polymers.= +Studies of Gauche and Eclipsing Inte	POLY	0353
Thermally Stable Electrooptic	Polymers.=	POLY	0130
Two–Dimensional	Polymers.=	POLY	0276
sis of Conductive–Electroactive	Polymers.= Use of Micelles in the Oxidative Synthe	COLL	0168
mer Chemistry Award Symposium	Polymers—Chemistry, Physics, and Biology. +Poly	POLY	0272
hanisms in the Fracture of Glassy	Polymers—I. Chain+ Energy–Consuming Micromec	PMSE	0073
eration with Polyacrylic Acid and	Polymethacrylic Acid.= +monium Bromide Compl	COLL	0074
chastic Dynamics Simulations of	Polymethylene Chains in the Bulk and in Thin Films	POLY	0355
Transition of Amorphous	Polymethylene.= Estimation of Density and Glass	POLY	0134
Occurrence and Importance of	Polymodal Electrophoretic Mobility Distributions.=	COLL	0202
mes.= Modeling the Growth of	Polynuclear Aromatic Hydrocarbons in Diffusion Fla	FUEL	0109
n Transfer in Covalently Linked	Polynuclear Aromatic–Nucleoside and Oligonucleotid	PHYS	0021
Di– and	Polynuclear Rhenium Carbonyl Complexes.=	INOR	0107
ping Na+ Ions in the Cen+ Novel	Polynuclear Selenide Clusters of Au+ and In^{3+}, Trap	INOR	0203
New Synthetic	Polyolefin Compositions.=	PMSE	0045
haracterization of Ethylene–Based	Polyolefin Copolymers.= +ared Spectroscopy for C	PMSE	0048
Aging of Primed	Polyolefin Surfaces.=	PMSE	0126
Synthesis of Controlled–Structure	Polyolefins.= +Photonic and Optical Applications.	POLY	0126
ation in Thermoreversible Gels of	Polyolefins.= +e to Strain: Effect of Network Form	POLY	0123
	Polyorganoboranes: Synthesis and Applications.=	POLY	0185
nium Oxides.=	Polyoxoalkoxotitanates: Molecular Precursors to Tita	INOR	0174
Ga$_{13}$, GaAl$_{12}$ and Al$_{13}$	Polyoxocations and Pillared Clays.=	PETR	0143
a. Functionalization of Alkanes by	Polyoxometalate Systems.= +Catlytic Photochemic	INOR	0286
Construction of Repetitive	Polypeptides for Optical Applications.=	BIOT	0033
in Foods and Health.	Polyphenol Complexation.= Phenolic Compounds	AGFD	0001
Compounds.= Inhibition of	Polyphenol Oxidase Activity by Phenolic	AGFD	0152
ry and Biological Activities of	Polyphenolic Compounds of Asian Origin.= Chemist	AGFD	0046
Antioxidant Activity.= New	Polyphenols from Medicinal Plants and Their	AGFD	0017
ty.=	Polyphenols from Teas and Their Antioxidant Activi	AGFD	0048
n Mouse Epidermis by Green Tea	Polyphenols.= +cetate–Induced H$_2$O$_2$ Production i	AGFD	0069
d Biological Characterization of	Polyphosphate and Polyphosphonate Esters from Bio	PMSE	0015
Length of Inorganic	Polyphosphates.= Factors Influencing the Chain	INOR	0376
ynthesis and Characterization of	Polyphosphazene Interpenetrating Polymer Network	PMSE	0002
ns.=	Polyphosphazene Membranes for Chemical Separatio	INOR	0070
and Inorganic Hybrid Materials.	Polyphosphazenes Molecular Composites, I: In Situ P	POLY	0181
terization of Polyphosphate and	Polyphosphonate Esters from Biologically Active Dio	PMSE	0015
s and Absolute Configuration of	Polypropionate Metabolites of Siphonaria Australis.=	ORGN	0003
Strategy for	Polypropionate Synthesis.= Multiple Equilibration	ORGN	0100
Honoring J. F. Roth. Commercial	Polypropylene Catalysts.= +ial Catalytic Processes	I&EC	0052
w Structuring in Entrance Flow of	Polypropylene Melts.= +eological Evidence of Flo	PMSE	0137
an Incompatible Polymer Blend of	Polypropylene–Polyamide.= +Thermal Analysis of	POLY	0071
itive Metalorganic Systems. Metal	Polypyridines as Redox Photosensitizers.= +osens	INOR	0282
Compounds into	Polypyrrole Films.= Incorporation of Redox	COLL	0013
e–χ[(3)] Relationships in Rigid–Rod	Polyquinolines.= +Polymer Thin Films, 2: Structur	POLY	0205
Phase Behavior of	Polyrotaxanes: Control of the Solid State.=	COLL	0058

ntial Uses.=	Polyrotaxanes: Synthesis, Characterization, and Pote	POLY 0345
Function of Hydrophobic	Polysaccharide Aggregates.= Characterization and	POLY 0098
clonal Antibodies Binding to the	polysaccharide Chains of Cell Envelope Lipopolysacc	CARB 0017
egradation+ In Situ Release of	Polysaccharide Hydrolyzing Enzymes During Wood D	BIOT 0208
Characterization of Lyotropic	Polysaccharide Liquid-Crystal Blends.=	COLL 0017
Corresponding to the Cell-Wall	Polysaccharide of the β-Hemolytic Streptococci Grou	CARB 0057
lecular Weight Characteristics of	Polysaccharides by High-Performance Size-Exclusio	BIOT 0197
Structural Analysis of Aggregated	Polysaccharides by HPSEC/Viscometry.= +alysis.	PMSE 0094
Control Properties of Microbial	Polysaccharides for Production—An Engineer's View	BIOT 0199
otional Perturbations of Pectinic	Polysaccharides in Apple Tissue.= +tion-Induced M	AGFD 0012
charides. Production of Microbial	Polysaccharides on Maple Syrup.= +robial Polysac	BIOT 0195
Control of Properties of Microbial	Polysaccharides. Production of Microbial Polysac+	BIOT 0195
e with Cyclodextrins and Sulfated	Polysaccharides.= +zymatic Browning in Fruit Juic	AGFD 0030
ions, and Biodegradation of	Polysaccharides.= Production, Properties, Modificat	CHED 0131
Studies on Vibrio cholerae	Polysaccharides.=	CARB 0016
a Hydrocarbon Process.=	Polysar's Efforts in Exposing Precollege Students to	CHED 0010
of New Molybdenum	Polyselenide Compounds.= Hydrothermal Synthesis	INOR 0355
Additives.= Photolabile	Polysilanes: The Effect of Electron Transfer	POLY 0188
Isocyanate-Modified	Polysilazane Ceramic Precursors.=	POLY 0339
Pretreatment and Pyrolysis of	Polysilazane Preceramic Binders.=	POLY 0338
Applications of	Polysilazanes as Precursors to Silicon Nitride.=	POLY 0317
and Inorganic Hybrid Materials.	Polysilazanes: A Route to Ceramics.= Preceramic	POLY 0309
ong the Structure of Preceramic	Polysilazanes, the Pyrolysis Conditions, and Their Fi	POLY 0313
ed bisMaleimides: Influence of a	Polysiloxane Rubber on the Polymerization Phase-Se	POLY 0112
e+ Synthesis of Side-Chain Liquid	Polysiloxanes Containing trans-Cyclohexane-Bas	POLY 0057
ulfonic Acid.=	Polysilyne Functionalization with Trifluoromethanes	POLY 0190
Characterization of Macrocyclic	Polystrene: Effect of Molecular Weight on Glass-Tra	POLY 0239
Miscible Blends of Sulfonated	Polystyrene and Poly(xylenyl Ether).=	PMSE 0117
and Polymerization of Hydroxy	Polystyrene Carboxylic Acid Telechelomers, II: A Mo	POLY 0046
age Contrast Analysis of Iodinated	Polystyrene Chains.= +Electron Microscopy, II: Im	POLY 0062
Fibers and Their Performance in	Polystyrene Composites.= +e Modification of Wood	PMSE 0112
bility Study of Functionalized	Polystyrene Derivatives.= Synthesis and Biodegrada	POLY 0049
Gelation of Atactic	Polystyrene in Carbon Disulfide.=	POLY 0148
Correlation Spectroscopy Study of	Polystyrene in Good and Theta Solvents.= +oton	POLY 0305
ation and Fracture of Sulfonated	Polystyrene Ionomer/Polystyrene Blends: Effect of I	POLY 0061
thesis of Uniformly Cross-Linked	Polystyrene Microbeads.= +el Filled Polymers: Syn	PMSE 0017
ll Culture.=	Polystyrene Porous Microcarriers for Mammalian Ce	BIOT 0010
orphology of Rubber-Toughened	Polystyrene.= +ion on the Impact Properties and M	PMSE 0177
sy Polymers—I. Chain Scission in	Polystyrene.= +mechanisms in the Fracture of Glas	PMSE 0073
ommingled Plastic Processed with	Polystyrene.= +cture and Properties of Recycled C	I&EC 0028
Isotactic	Polystyrene.= Thermoreversible Gelation of	POLY 0149
Solutions of Stereoregular	Polystyrene.= Thermoreversible Gelation of	POLY 0147
n of Thermoreversible Gelation in	Polystyrene-Liquid Systems.= +cular Interpretatio	POLY 0242
haracterization: Optical Methods.	Polystyrene-Poly(ethylene Oxide) Block Copolymer	POLY 0277
attering Study of Interdiffusion in	Polystyrene-Polyxylenylether Blends.= +utron Sc	POLY 0166
amic Mechanical Properties of a	Polystyrene/Poly(methylmethacrylate)/-Glass Beads	PMSE 0167
omposition and Linear and Cyclic	Polystyrene/Poly(vinyl Methyl Ether) Blends.= +c	POLY 0295
s and Solution Properties of	Polystyrenes with Dendritic End Groups.= Synthesi	POLY 0322
ation Director for the Synthesis of	Polysubstituted Aromatics.= +Versatile New Metal	ORGN 0083
Research Institute at the	Polytechnic University, Brooklyn.= The Polymer	HIST 0033
penetrating Polymer Network of	Polytetrafluoroethylene and Polydimethylsiloxane.=	PMSE 0041
Network.= Composites from	Polyurethane and Epoxy Interpenetrating Polymer	PMSE 0044
Network Formed from Castor Oil	Polyurethane and Vinyl Copolymers.= +g Polymer	PMSE 0089
and Optical Characterization of	Polyurethane Films Bearing NLO-Active Chromopho	POLY 0216
Oxidase Immobilized with	Polyurethane for Glucose Sensing.= Glucose	BIOT 0147
oughened Epoxy Resin/Castor Oil	Polyurethane IPN.= +haracterization of CTBN-T	PMSE 0092
heo-Kinetics and Morphology of	Polyurethane-Unsaturated Polyester Interpenetratin	PMSE 0027
g of Laminated Steel Sheet with	Polyurethane/Acrylate Interpenetrating Polymer Ne	PMSE 0057
s. Kinetics of Phase Separation in	Polyurethane/Polystyrene Semi-1 IPNs.= +twork	PMSE 0021
m Temperature Cured Castor Oil	Polyurethane/Vinyl or Acrylic Polymer Interpenetra	PMSE 0090
bing Studies on Foams Based on	Polyurethanes and Interpenetrating Polymer Networ	PMSE 0040
olymer Networks from Castor Oil	Polyurethanes and Poly(Styrene-Methyl Methacryla	PMSE 0091
Mechanical Prope+ Fluorinated	Polyurethanes: Polymerization, Characterization, and	POLY 0018
Changes in Segmented	Polyurethanes.= Thermally Induced Morphological	PMSE 0149
cal Analysis Studies of Reinforced	Polyvinyl Chloride (PVC) Roofing Membranes.=	PMSE 0180
of Poly(Carbonate-Urethane) and	Polyvinyl Pyridine.= +etrating Polymer Networks	PMSE 0056
l Force+ Structural Analysis of	Polyvinylchloride Using an ab Initio Derived Classica	POLY 0351
y of Interdiffusion in Polystyrene-	Polyxylenylether Blends.= +utron Scattering Stud	POLY 0166
dral Oligometallasilsesquioxanes (	POMSS).= +tion by Vanadium-Containing Polyhe	INOR 0131
Past and Current Studies with	Ponasterones, the First Insect-Molting Hormones fro	HIST 0024
Sludge	Pond Solids.= Characterization of Humic Acids from	I&EC 0022
y Minerals from Syncrude Sludge	Pond Tailings.= +ganic Matter Associated with Cla	I&EC 0021
Activation Products in Discharge	Pool in T.H.O.R. Nuclear Research+ Migration of	NUCL 0090
C Survey of Corn (Zea mays L.)	Populations and Inbreds for Silk Flavonolglycoside C	AGRO 0034
cs of Heterogeneous Char Particle	Populations.= +nisms. On the Combustion Kineti	FUEL 0026
hing and Teaching Large Student	Populations.= +vement in General Chemistry: Reac	CHED 0081
del Benceno con H_2O_2 Catalizada	por Complejos Derivados de Cupferron.= +xilacion	COLL 0121
Analisis Cuantitativo	por Computadora.=	COMP 0051
los Monocitos (FILM) Producido	por E. +on del Factor Inhibidor de la Locomocion de	BIOL 0102
c en Medi+ Obtencion de Plata	por Electrodeposito, Fusion, Oxidorreduccion con Zin	CHED 0172

r Cuantitativa de Tierras Raras	por Espectrometria de Masas con Fuente de Impacto	ANYL	0051
ion de la Transferencia de Carga	por Espectroscopia IR de Complejos de Transferenci	INOR	0190
Activacion del Enlace C–H	por Ni en la Molecula de Metano.=	PHYS	0244
l Lavado Quimico de Una Caldera	por Tecnicas Electroquimicas.= +erentes Etapas de	PHYS	0257
rio Convencional para Medir pH,	por un Electrodo de Acero Inoxidable 316/Pelicula d	ANYL	0066
Activation of Wide–	Pore Alumina Supports.=	PETR	0033
te NMR Study of 18-Ring Large–	Pore Aluminophosphate Molecular Sieves.= +d–Sta	PETR	0094
the Crystallization of Very Large	Pore Aluminophosphate Phases and Their Physicoch	PETR	0093
stigation of the Chemistry of the	Pore Structure of Coal in the Presence of a Swelling	FUEL	0073
of ETS–10: A New Wide–	Pore Titanium Silicate Molecular Sieve.= Synthesis	PETR	0078
ETS–10: A New Large–	Pore Titanium Silicate.= Adsorptive Properties of	PETR	0079
T+ From Clathrasils to Large–	Pore Zeolites: The Effects of Systemically Altering a	PETR	0036
Controlled–	Pore–Size Oxides.= Sol–Gel Processing of	PETR	0032
hesis of Controlled–Por+ In Situ	Pore–Structure Characterization during Sol–Gel Synt	PETR	0068
ttings Used for Packaging Cured–	Pork Products.= +trosamines in Elastic Rubber Ne	AGFD	0040
NMR and N_2	Porosimetry.= Analysis of Porous Carbon with ^{129}Xe	PETR	0070
Surface Area and	Porosity Characterization of Catalysts Supports.=	PETR	0069
Sol–Gel Synthesis of Controlled–	Porosity Materials.= +ture Characterization during	PETR	0068
Immobilization.= IPN–Modified	Porous Bead Polymers as Supports for Enzyme	PMSE	0081
Porosimetry.= Analysis of	Porous Carbon with ^{129}Xe NMR and N_2	PETR	0070
in the Gasification of	Porous Carbons/Chars.= Role of Activated Diffusion	FUEL	0010
Polystyrene	Porous Microcarriers for Mammalian Cell Culture.=	BIOT	0010
d Pharmaceutical Molecules Using	Porous Polymeric Packings.= +aphy of Proteins an	ANYL	0031
Direct Dipolar Energy Transfer in	Porous Solids.= +ith Small–Angle Scattering and	PHYS	0060
$Ru(BPY)_3^{2+}$ on	Porous Vycor Glass.= Excited–Stated Dynamics of	INOR	0160
Metal Carbonyls on	Porous Vycor Glass.= Photocatalytic Behavior of	INOR	0170
Photochemistry of $(Cp)_2TiCl_2$ on	Porous Vycor Glass.	INOR	0161
Pyrrole–Containing Receptors and	Porphyrin Analogues.= +sis and Complexation of	ORGN	0063
nsfer Gate+ Photoformation of a	Porphyrin Cation Lattice by Interfacial Electron–Tra	BIOL	0136
ed Imidazoles in Low–Spin Ferric	Porphyrin Complexes.= +Orientation of Coordinat	INOR	0296
Osmium	Porphyrin Complexes.= Synthesis and Reactivity of	INOR	0299
Tra+ Interfacially Polymerized	Porphyrin Films for Vectorial Photoinduced Electron	BIOT	0126
Interfacial Binding+ Catalysis of	Porphyrin Metalation at Anionic Interfaces: Mode of	CATL	0060
dish Peroxidase with Manganese–	Porphyrin Model Compounds.= +Mn(IV) Horsera	BIOL	0095
cal Requirements of Premicellar	Porphyrin Solubilization as Detected by J–Aggregatio	ORGN	0161
of a New Hexapyrrolic Expanded	Porphyrin: Rubyrin.= +thesis and Characterization	ORGN	0058
a Tricationic	Porphyrin.= Sequence–Specific DNA Interactions of	BIOL	0127
a Cell Line+ Cytotoxicity of ^{67}Cu–	Porphyrin–Antibody Conjugates to a Colon Carcinom	BIOT	0168
Transfers within Conjugated	Porphyrin–Flavin Ester.= Photoinduced Electron	BIOL	0134
Transfer in Covalently Linked	Porphyrin–RuO_2 Clusters.= Photoinduced Electron	COLL	0133
ron–Withdrawing Substituents on	Porphyrins and Metalloporphyrins.= +fect of Elect	INOR	0301
d+ Photoelectron Transfer with	Porphyrins in Layered Lithium Aluminate–Fatty Aci	INOR	0272
and Properties of Expanded	Porphyrins.= Cope Award Symposium. Synthesis	ORGN	0139
of Sterically Crowded	Porphyrins.= Crystallographic and EXAFS Studies	INOR	0091
La and Nd	Porphyrins.= Monomer and Dimer Complexes with	INOR	0221
and Other Nucleoside–Substituted	Porphyrins.= +luble Porphyrinyl–Mononucleoside	ORGN	0189
Subst+ The First Water–Soluble	Porphyrinyl–Mononucleoside and Other Nucleoside–	ORGN	0189
c Properties of Small– and Large–	Port Mordenites in C_8 Aromatics Isomerization.=	PETR	0132
id Waste Incina+ High–Strength	Portland Cement Concrete Containing Municipal Sol	FUEL	0136
ion Is the Leaching Mechanism of	Portland–Cement–Based Waste Forms.= +r Diffus	ENVR	0011
Naftola,	Posibles Garrapaticidas.= Sintesis de Tiofosfatos de	MEDI	0064
eavy Chains of Immunoglobulin G	Possess Similar Properties.= +B–Protein and the H	BIOL	0068
Hydrocarbon Fuels: Fullerenes.	Post–Buckminsterfullerene View of Carbon Chemistr	FUEL	0019
tion o+ Simplex Optimization of	Post–Column Derivatization Conditions in the Detec	AGFD	0111
mpetitive Binding Assa+ HPLC	Postcolumn Reaction Detection Systems Based on Co	ANYL	0090
Operation of a	Postconsumer Materials Recycling Facility.=	I&EC	0042
	Postconsumer Plastics Collection.=	I&EC	0039
ent for the Automatic Sortation of	Postconsumer Plastics.= +terial Handling Equipm	I&EC	0041
Coal and Carbon in the	Postpetroleum World.=	FUEL	0054
One–	Pot (Gem–Dimethylation of the Carbonyl Group.=	ORGN	0222
ral Auxiliary (η^5–+ Direct, One–	Pot Synthesis of Acyl Complexes Containing the Chi	INOR	0156
Cyclohexanone+ Effective, One–	Pot Synthesis of cis N–Alkylperhydroquinolines from	ORGN	0031
s and Hexafluorophosph+ One–	Pot Synthesis of Cyclic Amidinium Tetrafluoroborate	ORGN	0204
al Composition of Parent Coal on	Potassium Activity in Steam Gasification of Chars.=	FUEL	0027
y the Electrolysis of an Aqueous	Potassium Carbonate Electrolyte and the Implication	PHYS	0286
tudies of Novel 1,4–Benzoxazine+	Potassium Channel Activators: Synthesis and SAR S	MEDI	0099
symmetric Aliovalent Analogue of	Potassium Titanyl Phosphate (KTP).= +: A Centro	INOR	0256
ansient Kinetic Investigations.=+	Potassium–Catalyzed CO_2 Gasification of Carbon: Tr	FUEL	0031
Resistance in the Colorado	Potato Beetle.= Mechanisms of Avermectin	AGRO	0154
HPLC Assay of	Potato Glycoalkaloids.= Improved Extraction and	AGFD	0112
Extension of Prepeeled	Potatoes.= Bisulfite Replacement and Shelf–Life	AGFD	0058
rization, and Relative Analgesic	Potencies of the 6–Sulfate Conjugates of Some 3–Sub	MEDI	0149
ido, en la Validacion del Metodo	Potencio–Metrico para Cuantificar Clorhidrato de L–	ANYL	0066
P Proyecto de Practica: Valoracion	Potenciometrica de Clorhidrato de L–Arginina en	CHED	0163
of norBNI in κ–Opioid Antagonist	Potency and Selectivity.= +rmacophore Fragment	MEDI	0157
ological Activity of Derivatives of	Potent A_1–Selective Agonists.= +: Synthesis and B	MEDI	0115
ructure–Activity Relationships of	Potent Acetylcholinesterase Inhibitors: 8–Carba–phy	MEDI	0143
d Development of L–670,630 as a	Potent and Orally Active Inhibitor of 5–Lipoxygenas	MEDI	0180
enal Protective a+ KW–3902: A	Potent and Selective Adenosine A_1 Antagonist with R	MEDI	0112
etric Synthesis of NPC 12626: A	Potent and Selective Competitive NMDA Antagonist	MEDI	0083

oles and Bisindolylmaleimides as	Potent and Selective Inhibitors of Protein Kinase C.	MEDI	0074
s Development of BI-RG-587: A	Potent and Selective Reverse Transcriptase Inhibitor	ORGN	0027
ries of Related Lipopeptides with	Potent Anti-Pneumocystis carinii and Anti-Candida	MEDI	0181
zole Adenosine Receptor Agonists:	Potent Antihypertensive Agents.= +and Benzothia	MEDI	0111
r+ Novel Compounds Possessing	Potent cAMP and cGMP Phosphodiesterase Inhibito	MEDI	0118
ves. Novel Compounds Possessing	Potent Class III Antiarrhythmic Activity.= +rivati	MEDI	0095
	Potent C5A-Related Hexapeptides.=	MEDI	0167
is+ Kynurenic Acid Analogues as	Potent Glycine Site N-Methyl-D-Aspartate Antagon	MEDI	0146
d Pyrazolo[1,5-a]Pyrimidines as	Potent Orally Active Angiotensin II Receptor Antago	MEDI	0104
se (ACAT)+ Tetrazole Ureas: A	Potent Series of Acyl CoA: Cholesterol Acyltransfera	MEDI	0107
t+ New Azabicyclic Benzamides:	Potent Serotonin 5-HT4 Agonists and/or 5-HT3 An	MEDI	0069
es to Imidazole-5-Acrylic Acids:	Potent Small-Molecule Antiotensin II Receptor Anta	ORGN	0119
hermally Degraded Thiamine: A	Potent Source of Interesting Flavor Compounds.= T	AGFD	0071
none Spiropiperidines as Novel,	Potent, and Selective Class III Antiarrhythmic Agent	MEDI	0191
none Spiropiperidines as Novel,	Potent, and Selective Class III Antiarrhythmic Agent	MEDI	0097
of Human PMN Elast+ Stable,	Potent, and Specific Monocyclic β-Lactam Inhibitors	BIOL	0049
Renin Inhibitors.= Highly	Potent, Orally Active Diester Macrocyclic Human	MEDI	0123
II Receptor Antagonists.=	Potent, Orally Active, 1,5-Naphhyridine Angiotensin	MEDI	0102
',4',5'-Tetraacetyl Riboflavin: A	Potentially Useful Photosensitizing Agent for the Tre	ENVR	0029
y in Microheterogenous Fluids.	Potentiometry and Polarography in Micellar Systems	COLL	0001
Ion-Selective Graphite Composite	Potentiometry.= +ection in Groundwater by Novel	ENVR	0044
n(IV) Fluorides.= X-ray	Powder Diffraction Studies of Polymeric Triorganoti	INOR	0065
d Li$_5$NpO$_6$ by Neutron and X-ray	Powder Diffraction.= +ural Studies of Li$_4$NpO$_5$ an	NUCL	0068
.=	Powdered Ion-Exchange Resins in Sugar Purification	CARB	0065
itizable+ Preparation of Carbon	Powders by Pyrolysis of Graphitizable and Nongraph	POLY	0023
urso+ Hydrous Titanate Ceramic	Powders from Mixed Alkali/Titanium Alkoxide Prec	INOR	0278
ations into the Reactions of Metal	Powders with Aqueous Alkyl Halides.= +ic Investig	INOR	0248
Zirconia-Lime	Powders.= Stability of Suspensions of	COLL	0124
Ceramic	Powders.= Vapor-Phase Molecular Routes to	INOR	0175
ion of Metal-Nitride Preceramics,	Powders, and Coatings.= +Bonds for the Preparat	POLY	0230
Stopping	Power and Ranges of Ions in Condensed Matter.=	NUCL	0037
	Power of Selective Reaction Databases.=	CINF	0009
Safe and Advanced Nuclear	Power Reactors.= New Generation of Inherently	FUEL	0023
Future of Nuclear	Power. Is There a Future for Nuclear Power?.=	FUEL	0022
Future for Nuclear	Power?.= Future of Nuclear Power. Is There a	FUEL	0022
Mass Spectrometry: A	Powerful Chromatographic Detector.=	AGFD	0202
Diels-Alder Reaction: A	Powerful Synthetic Strategy.= Transannular	ORGN	0288
Cementos Hidraulicos para	Pozos Geotermicos.= Evaluacion Fisicoquimica de	INOR	0063
on Phosphide Acetyl Complex Cp(	PPH$_2$)(CO)FeCOCH$_3$-1 with Electrophiles.= +an Ir	INOR	0240
O)$_5$(μ-P-t-Bu$_2$)$_2$ and Fe$_2$(CO)$_6$(μ-	PPH$_2$)(μ-P-t-Bu$_2$).= +al Characterization of Fe$_2$(C	INOR	0050
ociated with the Reduction of Fe$_2$(	PPh$_2$)$_2$(CO)$_6$.= +udy of the Structural Changes Ass	ANYL	0106
iral Auxiliary (η^5-C$_5$H$_5$)Fe(CO)(	PPh$_3$): Synthesis and Chemistry of (η^5-C$_5$H$_5$)FE(CO)	INOR	0156
xide Complexes (η_5-C^{5}R^5)Re(No)(	PPh^3)(OCHRR').= +d Activation in Rhenium Alko	ORGN	0220
d Chemistry of (η^5-C$_5$H$_5$)FE(CO)(	PPH$_3$)C(O)-C=-R(R=CH$_3$,Si(CH$_3$)$_3$).= -nthesis an	INOR	0156
Insertion Reaction in η-Cp(CO) (	PPh$_3$)FeMe$^+$.= +oting the Alkyl to Acyl Migratory	INOR	0109
5 with Alkyl Phosphines: Fe(CO)	4PR$_3$ and Fe(CO)$_3$(PR)$_2$ Are Single-Photon Products	INOR	0253
s.= Ruthenium Thiolates cpRu(	PR$_3$')$_2$SR: Synthesis and Reactions with Electrophile	INOR	0368
enation Using Rh(PMe$_3$)$_2$LCl(L =	^{i}Pr$_3$,CO): Aspects of Selectivity and Mechanism.=	INOR	0151
nolate Compound, [Cd(Se-2,4,6-i-	Pr$_3$C$_6$H$_2$)$_2$(bpy)].= +dies of a Novel Cadmium Sele	INOR	0078
phines: Fe(CO)$_4$PR$_3$ and Fe(CO)$_3$(	PR)$_2$ Are Single-Photon Products.= +h Alkyl Phos	INOR	0253
de L-Arginina en+ Proyecto de	Practica: Valoracion Potenciometrica de Clorhidrato	CHED	0163
y Course.= Chemistry as	Practiced by Chemists: The Project-Based Laborator	CHED	0147
New Approach for Laboratory	Practices: An Interactive Student Evaluation.=	CHED	0106
: At the Interface, What Does the	Practicing Chemist Need? J. R.+ Panel Discussion	I&EC	0057
Chemistry and	Pre-Historic Art (Videotape presentation).=	CONG	0002
Experiences of a SEED	Preceptor Team.=	CHED	0038
sphazenes Molecular Composite+	Preceramic and Inorganic Hybrid Materials. Polypho	POLY	0181
zanes: A Route to Ceramics.=	Preceramic and Inorganic Hybrid Materials. Polysila	POLY	0309
ing Zirconia Particles Precipita+	Preceramic and Inorganic Hybrid Materials. Reinforc	POLY	0283
cetylene and Silicon Olefin Po+	Preceramic and Inorganic Hybrid Materials. Silicon A	POLY	0357
Kinetics by NMR.=	Preceramic and Inorganic Hybrid Materials. Sol-Gel	POLY	0256
is of Nanomaterials via Precer+	Preceramic and Inorganic Hybrid Materials. Synthes	POLY	0335
of Polysilazane	Preceramic Binders.= Pretreatment and Pyrolysis	POLY	0338
Synthesis and Processing of	Preceramic BN Polymers.=	POLY	0312
Characterization of	Preceramic Perhydropolysilazane.=	POLY	0310
Chemical Modification of	Preceramic Polymers: Some Examples.=	POLY	0359
ls. Synthesis of Nanomaterials via	Preceramic Polymers.= +Inorganic Hybrid Materia	POLY	0335
ionships among the Structure of	Preceramic Polysilazanes, the Pyrolysis Conditions, a	POLY	0313
olid-State MAS-NMR Study of	Preceramics Precursors and Pyrolyzed Products.= S	POLY	0358
the Preparation of Metal-Nitride	Preceramics, Powders, and Coatings.= +Bonds for	POLY	0230
CVD of	Precious Metals.=	INOR	0142
acterizatio+ New Morphology of	Precipitated Calcium Carbonate: Synthesis and Char	TECH	0019
als. Reinforcing Zirconia Particles	Precipitated into an Elastomeric Material.= +teri	POLY	0283
on of TiO$_2$ and Al$_2$O$_3$ Prepared by	Precipitation and Sol-Gel Methods.= +aracterizati	PETR	0092
Colloidal Metals by Reduction or	Precipitation in Microheterogeneous Liquids.= +of	COLL	0128
Phospholipids.= Affinity	Precipitation of Proteins with Ligand-Modified	BIOT	0020
the Selective	Precipitation of Proteins.= Insoluble Reagent for	BIOL	0040
s: Influence of Polymer Matrix on	Precipitation Processes.= +omposites and Ceramic	POLY	0290
Polymer-Induced Protein	Precipitation.= Toward a Theory of	BIOT	0022
n: Introducing Macromolecules in	Precollege Chemistry.= +1990 Award Presentatio	CHED	0025

rporate/Academic Partnerships in	Precollege Education. Benefits of Alliances to+ Co	CHED	0001
Enrichment of	Precollege Science with Polymer Chemistry.=	CHED	0009
Effective Topic for	Precollege Science.= Polymer Technology: An	CHED	0007
Polysar's Efforts in Exposing	Precollege Students to a Hydrocarbon Process.=	CHED	0010
tion of Calcium Cyanamide Using	Precolumn Derivatization.= +od for the Determina	ANYL	0113
Enrichment in Fruit Cells.=	Precursor Atmosphere Technology: Efficient Aroma	AGFD	0023
apor Deposition of ZnSe: Effect of	Precursor Chemistry on Film Properties.= +cal V	INOR	0171
Skeletons Related to a Clerodanic	Precursor from Mexican Salvia Species.= +th New	ORGN	0280
Indigo from	Precursor Molecules.= Microbial Production of	BIOT	0032
o Alkynyl Oxiranes: Synthesis of a	Precursor of (±)-Verrucosidin.= +prate Reagents t	ORGN	0011
	Precursor Phases and Nucleation of Zeolite TON.=	PETR	0040
azoles by Cyclodehydrogenation of	Precursor Poly(Schiff Bases).= +ted Polybenzimid	POLY	0045
lar Weight Polycarbosilane as a	Precursor to Oxygen-Free SiC Fibers.= High Molecu	POLY	0360
of Hydridomethylpolysilazane—A	Precursor to SiC/SiN Ceramics.= +aracterization	POLY	0340
Ethylimidotantalum Complex as	Precursor to Tantalum Nitride Thin Films.=	INOR	0074
Copper(II) Aminoalkoxide	Precursor.= Chemistry of Thermal CVD from a	COLL	0028
pper(I) Hexafluoroacetylacetonate	Precursor.= +ition from the 1,5-Cyclooctadiene Co	COLL	0026
III Binding Site	Precursor.= Search for the Heparin Antithrombin	CARB	0039
Catalyst	Precursor.= Ultrafine Iron Carbide as Liquefaction	CATL	0051
Catalyst	Precursor.= Ultrafine Iron Carbide as Liquefaction	FUEL	0101
	Precursor-Based Synthesis of Nickel Sulfides.=	INOR	0205
naturation of Cellobiohydrolase I:	Predenaturational Structural Changes.= +ess in De	BIOT	0191
P): A New Analytical Method to	Predict the Performance of the Visbreaking Process.	FUEL	0046
Polymer-Solvent Diffusion Coe+	Predictive Capabilities of a Free-Volume Theory for	POLY	0144
ar Weight Liquids: Application +	Predictive Model for the Viscosities of High Molecul	POLY	0065
f Novel 6-Carboxylate Esters of	Prednisolone: Analogues of Methyl 11β, 17α, 21-Trih	MEDI	0170
: Kinetic Studies Using New Fl+	Preference for Cyclohexylalanine at P1 Site by Renin	BIOL	0035
nufacturing of Textile-Structured	Preforms—Part I.= +omation of Equipment for Ma	TECH	0007
nufacturing of Textile-Structured	Preforms—Part II.= +mation of Equipment for Ma	TECH	0008
17α, 21-Trihydroxy-3,20-Dioxo-	Pregna-1,4-Diene-6-Carboxylate.= +of Methyl 11β,	MEDI	0170
Flouorinated	Pregnane Derivatives with Antiandrogenic Activity.=	ORGN	0193
Institute for Chemical Educati+	Prehigh School Chemistry Outreach Programs of the	CHED	0073
rdisciplinary Middle School Cu+	Prehigh School Science Education. FACETS: An Inte	CHED	0071
ollege Are Doing to Help Advance	Prehigh School Science Education.= +Two-Year C	CHED	0076
History of Steroid Chemistry.	Prehistory of the Mexican Steroid Industry and Upd	HIST	0016
Kinetics of the	Prehydrolysis of Xylan-Containing Materials.=	CARB	0044
ass and Chemical Feedstocks from	Prehydrolyzates of Cellulosic Food Waste.= +Biom	BIOT	0153
J+ Topological Requirements of	Premicellar Porphyrin Solubilization as Detected by	ORGN	0161
ication of a Bituminous Argonne	Premium Coal Sample Using Wood Ash or Taconite	FUEL	0006
Characterization of the Argonne	Premium Coal Samples by Field Ionization Mass S	FUEL	0062
Liquef+ Research with Argonne	Premium Coal Samples. Moisture Removal from and	FUEL	0001
Argonn+ Research with Argonne	Premium Coal Samples. Organic Oxygen Contents of	FUEL	0003
anic Oxygen Contents of Argonne	Premium Coal Samples.= +um Coal Samples. Org	FUEL	0003
Organic Sulfur Forms in Argonne	Premium Coals during Pyrolysis.= +ion Kinetics of	FUEL	0059
R Measurements on the Argonne	Premium Samples and Other Coals.= +tive ^{13}C NM	FUEL	0004
Atta+ Biochemistry of Protein	Prenylation: Nature's Rubber Cement for Membrane	BIOL	0021
on Zinc en Medio Acido y Basico,	Preparacion de Nitrado de Plata.= +dorreduccion c	CHED	0172
s—I. Analogies between Calcium+	Preparative/Process Chromatography of Biopolymer	ANYL	0027
s—II. Modern High-Performan+	Preparative/Process Chromatography of Biopolymer	ANYL	0037
Shelf-Life Extension of	Prepeeled Potatoes.= Bisulfite Replacement and	AGFD	0058
ally Flexible Molecules: A Flexible	Prescreening Technique.= +bases of Conformation	CINF	0027
m Molecula+ Computer Graphics	Presentations and Analysis of Hydrogen Bonding fro	BIOL	0112
essed Fruits and Vegetables. Fruit	Preservation by Combined Methods.= +mally Proc	AGFD	0055
Biopolymers as Nonthermal Food	Preservation Processes.= +e Treatment and Use of	AGFD	0029
mSource: A Teaching Resource for	Preservice and Inservice Chemistry Teachers.= +e	CHED	0137
inated Soil Samples from Wood-	Preserving Plants in Quebec: Extensive Analysis and	ENVR	0016
ros Zwitterionicos.+ Efecto de la	Presion en las Propiedades de Conduccion en Polime	ORGN	0159
n of Adulterations of Italian Cold-	Pressed Citrus Essential Oils.= +GC for Detectio	AGFD	0199
cally Mixed Naphthalene: High-	Pressure Effects and Delayed Fluorescence ODMR.=	PHYS	0081
vity in Organic+ Importance of	Pressure in the Creation and Study of Superconducti	PHYS	0025
ctrospray/Ionspray Atmospheric	Pressure Ionization LC/MS: Application to Drug Mo	ANYL	0124
in CE/MS Utilizing Atmospheric	Pressure Ionization.= +—II. Recent Developments	ANYL	0138
Diamond Anvil Cell.= High-	Pressure NMR Studies of the Glass Transition in the	PHYS	0260
er Blend of Polypropyle+ High-	Pressure Thermal Analysis of an Incompatible Polym	POLY	0071
echniques for Fossil Fuels. High-	Pressure TPR Apparatus to Investigate Organic Sulf	FUEL	0007
hermal Food Preservatio+ High-	Pressure Treatment and Use of Biopolymers as Nont	AGFD	0029
ochemical Processes under High	Pressure: (6) Diffusion Processes in Solution and Ma	PHYS	0222
under Shear Deformation at High	Pressure.= +ical Processes in Organic Polycrystals	PHYS	0245
ompounds in a Temperature- and	Pressure-Controlled Environment.= +terocyclic C	AGFD	0073
rystalline Polyester.=	Pressure-Induced Crystal Modification of a Liquid C	POLY	0072
y from Reversed Micelles through	Pressure-Induced Micellar Phase Transitions.= +er	BIOT	0046
isulfate Oxidation of+ Medium,	Pressure, and Micellar Effects on Kinetics of Peroxod	INOR	0022
ter at Elevated Temperatures and	Pressures in Geothermal Environments.= +with Wa	INOR	0246
aO-CaF$_2$-SiO$_2$) Under Controlled	Pressures of SO$_2$.= +u), Matte (Cu$_2$S), and Slag (C	I&EC	0065
inetics of Sulfonation of Sulfite-	Pretreated Aspen Dried with Vapor at High Temp	PMSE	0170
ic Binders.=	Pretreatment and Pyrolysis of Polysilazane Preceram	POLY	0338
ation.=	Pretreatment of Municipal Sewage Sludge for Gasific	FUEL	0117
Fiber Explosion (AFEX)	Pretreatment of Municipal Solid Waste.= Ammonia	BIOT	0016
lization of Poly(vinyl Chloride) by	Pretreatment with N-Substituted Maleimides.= +i	POLY	0027
Effect of Thermal and Chemical	Pretreatments on the Copper-Catalyzed Gasification	FUEL	0028

azardous Wastes: Assessment of	Prevalence, Bioavailability, and Bioeffective Dose.=	CEI 0008
for Industry and Challenges for+	Preventing Major Chemical Accidents: Opportunities	CHAS 0016
Role of the Chemist in	Preventing Pollution.=	CEI 0026
Prevention. Toward a Pollution	Prevention Ethic: The EPA's Role and Program.=	CEI 0022
h Cyclodextrins and Sulfated P+	Prevention of Enzymatic Browning in Fruit Juice wit	AGFD 0030
itrite and Nitrite-Free Composi+	Prevention of Lipid Oxidation in Muscle Foods by N	AGFD 0121
U.S. EPA's Pollution	Prevention Research Agenda.=	CEI 0023
EPA's 33/50 Project: Pollution	Prevention through Voluntary Initiatives.=	CEI 0024
e EPA+ Approaches to Pollution	Prevention. Toward a Pollution Prevention Ethic: Th	CEI 0022
Incident	Prevention.= Four Keys to Chemical Safety and	CHAS 0017
Pollution	Prevention.= Industrial Approaches to	CEI 0025
Pollution	Prevention.= Life Cycle Analysis: A Tool for	CEI 0027
ation+ Market-Driven Pollution	Prevention? Green Marketing: A National and Intern	CEI 0028
(EGCG): A Cancer	Preventive Agent.= (-)-Epigallocatechin Gallate	AGFD 0066
Aging of	Primed Polyolefin Surfaces.=	PMSE 0126
Legal	Primer for Librarians and Information Users.=	CINF 0047
ctrochemistry, sponsored by EG G	Princeton Applied Research Co.). +ry Award in Ele	ANYL 0096
roperties in Microwave Region for	Printed Circuit Board Materials.= +of Dielectric P	PMSE 0055
n of a Reaction Database and of a	Printed Service of Abstracts.= +e Joint Productio	CINF 0010
for Comparative Methodology+	Prioritization of Hazardous Air Pollutants: The Need	ENVR 0066
nal Chemical Structure Handling.	Prioritizing the Hits from a 3D Search.= +imensio	CINF 0020
	Privacy Issues of DNA Databanks and Databases.=	CHAL 0008
sting and Issues of Constitutional	Privacy.= +y: Update 1991. DNA Identification Te	CHAL 0001
The History of HIST, IV:	Probationary Participants.=	HIST 0010
ble Gels.=	Probe Diffusion and Self-diffusion in Thermoreversi	POLY 0198
the Atom-	Probe FIM.= Atomic Scale Surface Analysis with	PHYS 0101
ssociation+ Tunable Diode Laser	Probe of Chlorine Atoms Produced from the Photodi	PHYS 0299
Hydrodynamic Fingerprinting as a	Probe of Latex Characterization and Stability.=	COLL 0199
Diazo-4-undecyl-2-one (Duo) as a	Probe of the Active Site of E. +ased Inactivator 1-	BIOL 0067
o Cad+ Photoluminescence as a	Probe of the Adsorption of Carbonyl Compounds ont	COLL 0175
ng Vibrational Spectroscopies to	Probe the Diffusion of Small Molecules in Polymer F	POLY 0332
in Solution, Using a Fluorescence	Probe.= +on Quenching of Triplet Methylbenzenes	PHYS 0265
bonyl Complex as a Spectroscopic	Probe.= +ymers Using a Luminescent Rhenium Car	POLY 0096
ine: A Potential Cocaine Receptor	Probe.= +l Biodistribution of Radioiodinated Coca	MEDI 0131
h Cu,Zn-Superoxide Dismutase as	Probed by ^{51}V NMR.= +on of Vanadate Anions wit	INOR 0184
on of Sodium Cations in Zeolite-Y	Probed by Double-Rotation NMR.= +tice Relaxati	PHYS 0267
nic States of High-T_c Cuprates as	Probed by X-ray Absorption.= +el for the Electro	PHYS 0053
am+ Site Use of Advective-Flux	Probes for Enhanced Soil Gas Analysis in in Situ Ste	ENVR 0052
luorinated Polymeric Molecular	Probes for F-19 Magnetic Resonance Imaging: Synth	MEDI 0041
Raman Fiber-Optic	Probes for Hostile Environments.= Development of	ANYL 0065
talorganic Systems. Luminescence	Probes of DNA-Binding Interactions.= +sitive Me	INOR 0315
HPLC Columns as	Probes of Enzyme Interactions.= Enzyme-Based	BIOT 0036
Natural Products as	Probes of Protein Kinase C Heterogeneity.=	MEDI 0077
ineered Metal Chelation Sites as	Probes of Protein Stability and Conformation.= Eng	BIOT 0133
Magnetic Resonance	Probes of Short-Range Order in Network Glasses.=	PHYS 0374
nes, NPC16377 and NPC13839, as	Probes of Sigma Receptors.= +Aminoalkoxychromo	MEDI 0150
Fluorescence	Probes of Solution Phase Dynamics.=	CHED 0201
as Restricted R+ Photochemical	Probes of Starburst Dendrimers and Their Utilization	POLY 0238
n of Detergents Using Fluorescent	Probes: A Teaching Experiment.= +Characterizatio	CHED 0200
oadening of Surface Paramagnetic	Probes.= +s by Electron-Spin-Resonance Line Br	PHYS 0050
oid Receptor Selective Fluorescent	Probes.= +hesis of Naltrindole Analogues as δ-Opi	MEDI 0155
oadening of Surface Paramagnetic	Probes.= +ors by Electron Spin Resonance Line Br	PHYS 0284
Molecular Anions as Surface	Probes.=	NUCL 0056
the STM.=	Probing and Inducing Silicon Surface Chemistry with	PHYS 0184
Drug-DNA Interactions:	Probing Binding Modes and Dynamics by NMR.=	PHYS 0197
Use of Microcalorimetry for	Probing Carbon Surfaces.=	FUEL 0018
	Probing Enzymatic Glycoside Hydrolysis.=	ORGN 0267
ions of n-Quantum Coherence in	Probing Intermediates in Heterogeneous Catalysis.=	COLL 0089
Diode Laser	Probing of N_2O Following Collisions with $O(^1D)$.=	PHYS 0278
d Specificity. New Techniques for	Probing the Mechanism of Protein Folding.= +an	BIOT 0112
ic Force Microscopy.=	Probing the Surface Forces of Materials Using Atom	PHYS 0154
e Spectroscopy.=	Probing Upconversion Dynamics Using Site-Selectiv	ANYL 0004
als Derived from Oxidation of	Probucol.= Detection and Characterization of Radic	BIOL 0099
Patent	Proceedings.= Using Scientific Evidence in	CHAL 0014
Equipo de la Industria Quimica de	Proceso.= +entes de CAD/CAM para el Diseno de	COMP 0053
ratica Sucesiva y un Simulador de	Procesos Comercial.= +lizando Programacion Cuad	I&EC 0072
BALANCE: Simulador de	Procesos para Balances de Materia.=	COMP 0052
s through Rotaxane+ Control of	Processability and Solid-State Properties of Polymer	POLY 0233
Furaneol in Heat-	Processed Foods.= Studies on the Formation of	AGFD 0052
emical Changes on Minimally	Processed Fruit and Vegetable Surfaces.= Physicoch	AGFD 0011
Preparation of Novel Minimally	Processed Fruit Products by Pectinase Infusion.=	AGFD 0059
atments to Stabilize+ Minimally	Processed Fruits and Vegetables. Cold Blanching Tre	AGFD 0027
by Combined Metho+ Minimally	Processed Fruits and Vegetables. Fruit Preservation	AGFD 0055
in Modified Atmos+ Minimally	Processed Fruits and Vegetables. New Developments	AGFD 0075
ruits and Vegetable+ Minimally	Processed Fruits and Vegetables. Physiology of Cut F	AGFD 0010
to Development of Minimally	Processed Vegetable Snacks.= Integrated Approach	AGFD 0056
s of Recycled Commingled Plastic	Processed with Polystyrene.= +cture and Propertie	I&EC 0028
stics Recycling. Plastics Recycling	Processing and Economics.= +Technologies for Pla	I&EC 0031
Supercritical Fluid	Processing Applied to Recycling.=	I&EC 0034
and Computer Data	Processing for Analytical Courses.= Data Acquisition	CHED 0119

Bioseparations—Downstream	Processing for Biotechnology.=	BIOT	0224
Oligosaccharide	Processing in Cancer Cells: Functional Studies.=	BIOL	0151
Sol–Gel	Processing of Controlled-Pore–Size Oxides.=	PETR	0032
Gel	Processing of Opto/Electroactive Polymers.=	POLY	0200
Fluid Disruption.= Downstream	Processing of Periplasmic Proteins Using Critical	BIOT	0070
NMR Spectroscopy in Secondary	Processing of Petroleum Fractions.= +plications of	PETR	0146
s, I: Molecular Engineering and	Processing of Polymers with Large Third–Order Opti	POLY	0011
Synthesis and	Processing of Preceramic BN Polymers.=	POLY	0312
odeled on the Visual Information	Processing of the Retina.= +lecular Membranes M	BIOT	0124
Properties and Morp+ Effects of	Processing Parameters and Adhesion on the Impact	PMSE	0177
etegy for Improving Downstream	Processing Yields for Production of Recombinant HI	BIOT	0071
Energy Distribution i+ Reactive	Processing. Plenary Lecture. Distributive Mixing and	PMSE	0175
cperty Relationships in Base Oil	Processing: Monitoring Dewaxing Procedures via Spe	PETR	0151
trol of Properties of Fluids during	Processing.= +for the On–Line Monitoring and Con	I&EC	0059
Information	Processing.= Bacteriorhodopsin Variants for Optical	BIOT	0078
during Plum	Processing.= Changes of Phenolic Compounds	AGFD	0171
Commingled Batch	Processing.=	I&EC	0026
Plastic Waste Stream via Reactive	Processing.= +Compatibilizing Components of the	I&EC	0033
Commingled	Processing.= Research Results—Refined	I&EC	0029
ε Genetic Regulation, Metabolic	Processing, and in Vitro Effects on Oncogenic Transf	AGFD	0117
imethylphenoxy Pendent Groups:	Processing, Mechanical Properties, and Morphology.	PMSE	0019
+spberry Phenolics: Influences of	Processing, Variety, and Environmental Factors.=	AGFD	0169
ad Interfaces—II. Quick Quenc+	Procter Gamble UERP Polymer Solutions, Blends, a	COLL	0032
ad Interfaces—III. Lateral Diff+	Procter Gamble UERP Polymer Solutions, Blends, a	COLL	0051
ad Interfaces—IV. Thin Polym+	Procter Gamble UERP Polymer Solutions, Blends, a	COLL	0070
ad Interfaces—I. Dynamics at +	Proctor Gamble UERP Polymer Solutions, Blends, a	COLL	0012
ing—Strategic Issues in Chemical	Procurement. Chemical Specialties fo+ Global Sourc	CMEC	0005
Lipophilicity, Dosing Vehicle, and	Prodrug Modification.= +ic Compounds: Effect of	MEDI	0196
Synthesis of Chemoreversible	Prodrugs of Ara–C.=	MEDI	0052
tives of Po+ Adenosine Receptor	Prodrugs: Synthesis and Biological Activity of Deriva	MEDI	0115
Fundamentos de la	Produccion de Furfural.=	CARB	0051
Amino Acid	Producers.= Culture Instability of Auxotrophic	BIOT	0031
●mocion de los Monocitos (FILM)	Producido por E. +on del Factor Inhibidor de la Loc	BIOL	0102
pounds of Citrus Fruits and Their	Producs.= +s in Foods and Health. Phenolic Com	AGFD	0167
Technological Support for the	Productive Sector: The Case of the Technological Tr	SCHB	0025
tron–Treated Sludge for Ocean	Productivity Enhancement.= Ocean Feeding of Elec	ENVR	0021
d Cells to Overall Cyclosporin A	Productivity in an Immobilized Perfusion Reactor Sy	BIOT	0155
Toll Con+ Achieving Purchasing	Productivity through Global Swaps, Exchanges, and	CMEC	0007
ad Product Extraction on Reactor	Productivity.= +ey: Effects of Growth Substrate a	BIOT	0030
culo entre Universidades y Secotr	Productivo.= +cio Social y Programas de Becas: Vin	CHED	0114
de Clorhidrato de L–Arginina en	Producto Terminado Utilizando como Electrodo Indi	CHED	0163
Formulating for Agricultural,	Professional, and Consumer Markets.= Product	FERT	0004
Legal Rights of Whistle–Blowing	Professionals.=	PROF	0011
Management Paths for Technical	Professionals.=	YCC	0010
Assistant Professor.=	NIH Funding Opportunities for the	YCC	0004
rated Polyester Resins with Low–	Profile Additives.= +ogy Characterization of Unsatu	POLY	0114
Hepatocyte Source on Metabolic	Profile of Biphenyl in an in Vitro Developmental To	AGRO	0021
ty, and Upjohn in Perspective: A	Profile of Ionization.= +oids, the Steroid Commun	HIST	0021
cally Delineated Refractive Index	Profiles in Polymeric Slab Waveguides.= +otochem	POLY	0180
c Drugs: I. Synthesis and Binding	Profiles of Novel Piperidine Ligands.= +tipsychoti	MEDI	0153
ks with Industrial Application	Profiles.= Some Disaccharide–Derived Building Bloc	CARB	0003
Chemists, Future Chemistry.	Profiles, Pathways, and Dreams.= Future	CONG	0004
d Reflectance Spectros+ Surface	Profiling of Polymers with Fourier Transform Infrare	PMSE	0153
Depth	Profiling Using Variable–Angle ATR.=	POLY	0307
-al Interface by Sputtering Depth	Profiling Using XPS.= +Reactions at Polymer/Me	PMSE	0150
Short Synthesis of the Synthetic	Progestin ST–1435.=	ORGN	0180
ry.+ Chemical Careers Insights	Program (Roadshow)—Careers for Chemists in Indus	YCC	0005
Bachelor's Degree	Program at Berkeley in Chemistry and Teaching.=	CHED	0069
Teaching of Chemistry	Program at Boston University.=	CHED	0070
Science Education	Program at Oak Ridge National Laboratory.=	CHAS	0050
A Survey of the NLO–Polymer	Program at the MPI.=	POLY	0010
●grated Four–Year Spectroscopy	Program Complemented by the Microscale Approach	CHED	0150
DESTIL—A User–Friendly	Program for Distillation Column Design.=	CHED	0122
Greater Cincin+ Summer Intern	Program for High School Chemistry Teachers in the	TECH	0002
merican Chemical Society SEED	Program for High School Students at Hunter College	CHED	0143
acin: A Flexible Chemkin Driver	Program for Investigating Chemistry Important to In	I&EC	0024
ge of Pe+ Remedial Action Plan:	Program for the Great Lakes. Achieving Zero Discha	ENVR	0083
Ontario's	Program for Using Sewage Sludge in Agriculture.=	FERT	0021
Evaluating a Project SEED	Program in Indianapolis.= Coordinating and	CHED	0141
Integrated Writing	Program in the Physical Chemistry Laboratory.=	PHYS	0318
Integrated Writing	Program in the Physical Chemistry Laboratory.=	CHED	0213
A Simple Computer	Program to Fit Experimental Results.=	NUCL	0089
Heavy Elements. The Darmstadt	Program To Investigate Heavy Elements.= +sis of	NUCL	0014
Upgrading a Forensic Science	Program with a Microanalysis Laboratory.=	CHED	0153
High School Chemistry Teachers	Program. POLYED 1990 Award Presentation: Intro	CHED	0025
d, an+ Establishing an Academic	Program. Succeeding in Academia: The Good, the Ba	YCC	0001
ention Ethic: The EPA's Role and	Program.= +Prevention. Toward a Pollution Prev	CEI	0022
●ry Improvement by the NSF–ILI	Program.= +ory—I. Support of Chemistry Laborat	CHED	0146
New Developments in the LHASA	Program.= +is Planning and Reaction Databases.	CINF	0001
ChemSource	Program.= Chemical Safety in the	CHED	0019

General Electric College Bound Program.=		CHED 0003
CHIRON Program.= Heuristic Synthesis Planning Using the		CINF 0002
Recycling Program.= Iowa's Plastic Pesticide Container		AGRO 0091
New Jersey Right-to-Know Program.=		CHAS 0037
tory: An Integrated Experimental Program.= +ers in the Physical Chemistry Labora		CHED 0045
The MIT Student's Outreach Program.=		CONG 0008
mplementing a Laboratory Safety Program.= +ful Considerations in Developing and I		CHAS 0004
Undergraduate Program?.= Polymer Science: An Independent		CHED 0099
de Plantas Quimicas Utilizando Programacion Cuadratica Sucesiva y un Simulador d		I&EC 0072
ecotr Product+ Servicio Social y Programas de Becas: Vinculo entre Universidades y S		CHED 0114
on Active Sites by Temperature- Programmed Desorption.= Characterization of Carb		FUEL 0016
Analytical Met+ Temperature- Programmed Oxidative Pyroanalysis (TPOP): A New		FUEL 0046
Coal Characterization by Programmed-Temperature Oxidation.= Advances in		FUEL 0058
Extensive Use of Spreadsheet Programs in Analytical Chemistry Courses.=		CHED 0205
Revitalizing Science in the Nat+ Programs of the Institute for Chemical Education in		CHED 0138
high School Chemistry Outreach Programs of the Institute for Chemical Education.=		CHED 0073
Statistical Summary of PIXE Programs 1970-1980-1990.=		NUCL 0080
Hazardous Materials Support Programs.=		CHAS 0027
and EPA Programs.= Health and Safety Issues under OSHA		CHAS 0022
and Programs.= State Pesticide Disposal Regulations		AGRO 0106
piad and the NSF ChemSource Programs, Honoring the Memory of Marjorie Gardne		CHED 0011
omena. Synthesis of High-Energy Projectile and Target Fragmentation.= +istic Phen		NUCL 0046
r Sequential?.= Projectile Brakup in 25-MeV/A Reactions: Prompt o		NUCL 0027
Fragmentation of Gold Projectiles at 600 MeV/nucleon.=		NUCL 0047
nts at the Beva+ Use of a Time- Projection Chamber in Multifragmentation Experime		NUCL 0034
pe Antibiotics: Distinct Origins of Proline Residues in L-671,329 and L-688,786.= +y		MEDI 0058
Inhibitors for Proline-Specific Proteases.=		BIOL 0030
d Structure Data.= Prolog Approach to Integrating Protein Sequence an		CINF 0011
) Dicarbonyl to Al+ Lewis Acid Promoted Additions of Allyl(cyclopentadienyl)Iron(II		ORGN 0213
Carbonylation and Phosphine- Promoted Alkyl-CO Migration Reactions of Unusual		INOR 0239
tric Synthesis an+ Nucleophile- Promoted Cyclization Reactions of Alkynes: Asymme		ORGN 0275
NMR, and XRD of Phosphorous- Promoted Hydrotreating Catalyst Precursors.=		COLL 0198
A Mechanism for the Iron- Promoted Magnesium-Water Reaction.=		INOR 0394
Grignard Reagent Promoted Scission of the C-S and S-S Bonds.=		ORGN 0253
Oxide C+ Effect of Support and Promoters on the Structure-Reactivity of Vanadium		CATL 0057
Role of Coordinating Solvents in Promoting the Alkyl to Acyl Migratory Insertion R		INOR 0109
urcumin on Tumor Initiation and Promotion in Mouse Epidermis.= +tory Effects of C		AGFD 0088
ic Interactions for the Promotion of Science Education.= Industry/Academ		CHED 0008
Schematic Method for Fischer's Proof Teaching.= D-Glucose Stereochemistry:		CHED 0095
mer Diffusion and the Kinetics of Propagation in Free-Radical Polymerization.= +no		POLY 0165
tronic Structure through Electron Propagator Theory.= +n Pictures of Molecular Elec		PHYS 0058
bstituted Ga-β as a Catalyst for Propane Aromatization—Infrared Spectroscopic Inves		PETR 0047
Reaction Kinetics of Propane Dehydrogenation over Zn or Mo.=		I&EC 0075
upported+ Reaction Kinetics of Propane Deydrogenation over Zinc or Molybdenum S		PETR 0087
Catalysts.= Reactions of Propane on Hydrous Metal Oxide-Supported		CATL 0040
Aromatization of Ethane and of Propane on HZSM-5: Effect of Gallium.=		PETR 0045
ations: V+ + Ethane, Ethene, and Propane.= +Cross-Sections of Transition-Metal C		PHYS 0306
Amidoesters of (Anilinomethylene) propanedioic Acids by Direct Condensation.= +d		ORGN 0103
egioselective Substitution of 1,2- Propanediol with Trimethylsilyl Reagents Using the		ORGN 0051
o Pentasil en la Aromatizacion de Propano.= +Contenido de Galio en Zeolitas del Tip		COLL 0125
of Allyl Alcohol and 1- Propanol on Cu2O(100).= Oxidation and Reduction		COLL 0195
aralkyl-Substituted-1-Aryloxy-2- Propanolamine and Propylamine Derivatives. +tero		MEDI 0095
ted Shock Waves.+ Reactions of Propargyl Chloride and 1,5-Hexadiyne behind Reflec		FUEL 0068
1,7-Ocatdiyne as a Source of Propargyl Radicals.= Thermal Decomposition of		FUEL 0067
for the Preparation of Propargylic Alcohols.= New Organoboron Reagent		ORGN 0109
ic Structures Derived from [1.1.1] Propellane.= +inkertoy Construction Set: Polymer		POLY 0344
NMe2)2 to W2(O2C8H+ Inorganic Propellanes: The Interconversion of W2(O2C8H16)3(H		INOR 0128
Mass Spec+ Chemistry of Solid- Propellant Combustion Studied by Molecular-Beam		FUEL 0126
d Clays.= Propene Alkylation of Biphenyl Catalyzed by Pillare		PETR 0014
Oligomerization of Propene over Heteropoly Acids.= Heterogeneous		PETR 0011
ionic Polymerization.= Novel Propenyl Phenyl Ether Analogs for UV-Induced Cat		POLY 0017
zation.= Investigation of Seed Properties' Influence on the Rate of ZSM-5 Crystalli		PETR 0039
d Isopropylidene D+ Uncoupling Propertiex of Quinic Acid and Its Phenylboronate an		INOR 0263
cos.+ Efecto de la Presion en las Propiedades de Conduccion en Polimeros Zwitterioni		ORGN 0159
uencia de la Sulfonacion sobre las Propiedades Papelaras.= +e Alto Rendimiento: Infl		PMSE 0077
owth Substrate and Product Ex+ Propionic Acid Production from Whey: Effects of Gr		BIOT 0030
esis and+ Synthesis of 2-oxo-2- Propionyl-1,3,2-Diaxaphosphorinanes and the Synth		ORGN 0050
dol Reactions of a Chiral 2-oxo-2- Propionyl-1,3,2-Oxazaphosphorinane.= +is and Al		ORGN 0050
ethyl-N,N'-bis[3-(2-Phthalimido) propyl]-1,6-hexanediamine Dihydrochloride.= +m		MEDI 0140
-1-Aryloxy-2-Propanolamine and Propylamine Derivatives. +teroaralkyl-Substituted		MEDI 0095
and Degradation of Ethylene and Propylene over Silver.= +ween Selective Oxidation		CATL 0058
Effect of Support in Propylene Oxidation on Molybdenum Oxide.=		CATL 0059
electivi+ Copper Ion Catalysis of Propylene Oxidation on X and Y Zeolites: Product S		CATL 0017
ocrystallization in Blends of Poly(propylene) and Poly(butene-1).= +iscibility and C		POLY 0089
Poly(1-Trimethylsilyl-1 Propyne).= Mechanisms of Gas Transport in		POLY 0119
LNSP: Enfoque Prospectivo.= Microbiologia de Alimentos en el		AGFD 0206
ery of Novel Inhibitors of HIV-1 Protease by Three-Dimensional Substructure Search		COMP 0038
NMR and Kinetic Studies of Protease Inhibition by Peptidyl Fluoroketones.=		BIOL 0005
Haloperidol.= Inhibition of HIV Protease with Nonpeptide Derivatives of		MEDI 0054
Inhibitors of HIV-1 Protease.= Design and Preparation of C2 Symmetric		MEDI 0053

217

Production of Recombinant HIV	Protease.= +ving Downstream Processing Yields for	BIOT 0071
Recombinant HIV-2	Protease.= Purification and Characterization of a	BIOL 0042
cumulation by St+ Extracellular	Proteases and Their Role in Heterologous Protein Ac	BIOT 0084
oronic Acid Inhibitors on Serine	Proteases: Evidence for Multiple Binding Mechanism	BIOL 0082
Inhibitors for Proline-Specific	Proteases.=	BIOL 0030
Pig Corneal	Proteases.=	BIOL 0043
Cabbage	Proteases.= Purification and Characterization of	AGRO 0019
Technology to	Protect the Environment.= Application Equipment	AGRO 0125
ocking of Anions of Carbethoxy-	Protected Cyanohydrins with Enones: Synthetic App	ORGN 0228
uranose Sugars via Ozonization of	Protected Glycals.= +nient Degradation to Lower F	ORGN 0073
Compounds to N-THP-	Protected Nitrone.= Addition of Organolithium	ORGN 0227
nformational Factors and Sugar-	Protecting Groups in the Synthesis of 2'-Fluorinated	CARB 0024
Environmental Site Assessments:	Protecting the Dealer and the Environment.=	FERT 0015
arcinogenesis by Green Tea.=	Protection against Stomach, Lung, and Esophagus C	AGFD 0067
uced Lung Tumorigenesis by G+	Protection against Tobacco-Specific Nitrosamine-Ind	AGFD 0068
n Transfer Process.= Copyright	Protection and Risks in the International Informatio	CINF 0014
ueprint for Future Environmental	Protection Efforts.= +is Task? Relative Risk: A Bl	ENVR 0061
Patent	Protection for Advances in DNA Technology.=	CHAL 0006
ing Amide+ Effect of Hydroxyl	Protection on Stereochemistry and Stereocontrol Dur	ORGN 0258
uality Control, and Environmental	Protection.= +aboratory: Product Development, Q	FERT 0002
enosine A$_1$ Antagonist with Renal	Protective and Diuretic Activities.= +Selective Ad	MEDI 0112
Carcinogenesis by Plant Phenols+	Protective Effects against Liver, Colon, and Tongue	AGFD 0090
	Protective Equipment in the Laboratory.=	CHAS 0005
es and Their Role in Heterologous	Protein Accumulation by Streptomyces lividans.=	BIOT 0084
Formation and Characteristics of	Protein Aggregates in Escherichia coli.=	BIOT 0115
ow FAB LC/MS.= Peptide and	Protein Analysis by Electrospray and Continuous-Fl	ANYL 0140
Possess+ Cancer Detection: B-	Protein and the Heavy Chains of Immunoglobulin G	BIOL 0068
ules: Implications for Neural-Like	Protein Array Devices.= +in Cytoskeletal Microtub	BIOT 0055
the B-	Protein Assay.= Cancer Management: Versatility of	BIOL 0069
otein Stabilization.+ Overview of	Protein Chemical Cross-Linking: Implications for Pr	BIOT 0216
Optimization of	Protein Chromatography on Agarose Gels.=	ANYL 0030
racterization of Two-Dimensional	Protein Crystals.= +ectronics. Preparation and Cha	BIOT 0121
Plans.= The	Protein Data Bank—Present Status and Future	CINF 0031
Dynamics Studies of the	Protein Echistatin.=	PHYS 0167
elles through Pressure-Induced+	Protein Extraction and Recovery from Reversed Mic	BIOT 0046
Micellar	Protein Extraction.= Affinity-Based Reversed	BIOT 0045
Peptide Models for	Protein Folding and Localization.=	BIOL 0003
Denaturation.= Facilitation of	Protein Folding and the Reversibility of	BIOT 0190
ues for Probing the Mechanism of	Protein Folding.= +and Specificity. New Techniq	BIOT 0112
Antibody-Assisted	Protein Folding.=	BIOT 0062
Role of Chaperones in	Protein Folding—GroE Heat Shock Proteins.=	BIOT 0116
ily as a Possible Source of Oil and	Protein for Human Consumption.= +Jatropha Fam	AGFD 0038
-PE40 Conformation: A Chimeric	Protein Having Potential as an AIDS Therapy.= +)	ANYL 0005
e.= Direct Measurement of	Protein Interaction Forces at a Solid-Liquid Interfac	BIOT 0043
Proteins.= Evaluating Starch-	Protein Interactions for Recovery of Recombinant	BIOT 0047
Food and Health. Tannin-	Protein Interactions.= Phenolic Compounds in	AGFD 0148
Chain Bases.= Inhibition of	Protein Kinase C by Sphingosine and Related Long-	MEDI 0075
Natural Products as Probes of	Protein Kinase C Heterogeneity.=	MEDI 0077
logical Evaluation of the	Protein Kinase C Inhibitor NPC 15437.= +armaco	MEDI 0026
h the Catalytic Domain.= New	Protein Kinase C Inhibitors Designed to Interact wit	MEDI 0076
w Target for Antiviral and Anti+	Protein Kinase C Inhibitors. Protein Kinase C: A Ne	MEDI 0072
ridines: A Novel Class of Selective	Protein Kinase C Inhibitors.= +minomethyl)-Pipe	MEDI 0025
= Protein Kinase C Inhibitors.	Protein Kinase C: A New Target for Antiviral and An	MEDI 0072
Lipid Inhibitors of	Protein Kinase C.= Design and Synthesis of Ether	MEDI 0073
Potent and Selective Inhibitors of	Protein Kinase C.= +and Bisindolylmaleimides as	MEDI 0074
and Adhesion of Bacterial Surface	Protein Layers.= +dity on the Wetting Properties	BIOT 0151
Heme-Dependent	Protein Mimics.= Studies toward	INOR 0186
of Partially Folded and Unfolded	Protein Molecules.= +Conformation and Dynamics	PHYS 0107
mph Juvenile Hormone-Binding	Protein of Larval Manduca sexta by the Use of Phot	AGRO 0014
Polymer-Induced	Protein Precipitation.= Toward a Theory of	BIOT 0022
Dependence of Membrane	Protein Prediction Scales on Sidechain Structure.=	BIOL 0038
mbrane Atta+ Biochemistry of	Protein Prenylation: Nature's Rubber Cement for Me	BIOL 0021
ana taba+ Stability of a Foreign	Protein Production from Genetically Modified Nicoti	BIOT 0230
us Gene Expression in Filamen+	Protein Production in Filamentous Fungi. Heterologo	BIOT 0234
atography®: A Novel Tool for	Protein Purification and Analysis.= Perfusion Chrom	BIOT 0073
omatographic+ Perspectives in	Protein Purifications—Chromatographic and Nonchr	BIOT 0018
Amino Acid Supplementation on	Protein Quality of Soy-Based Infant Formulas Fed t	AGFD 0209
ations on the Thermodynamics of	Protein Reactions.= +d Specificity. Effects of Mut	BIOT 0188
Polyethylene Glycol-Enhanced	Protein Refolding: Model and New Applications.=	BIOT 0136
Raman Spec+ Characterization of	Protein Secondary Structure via Fourier Transform	POLY 0306
Bioreactor.= Controlled	Protein Secretion in a Single-Pass Ceramic Matrix	BIOT 0011
Sol+ Nonaqueous Bioprocessing:	Protein Separation and Purification in Neat Organic	BIOT 0063
Nonchromatographic Methods of	Protein Separation—I. Perspectives in Protein Puri	BIOT 0018
Nonchromatographic Methods of	Protein Separation—II. Bioseparation Using Affinit	BIOT 0044
Nonchromatographic	Protein Separations.= Selectivity in	BIOT 0019
Prolog Approach to Integrating	Protein Sequence and Structure Data.=	CINF 0030
etal Chelation Sites as Probes of	Protein Stability and Conformation.= Engineered M	BIOT 0133
	Protein Stability Studies on Subtilisin BPN'.=	BIOT 0192
and Hydrophobic Interactions to	Protein Stability.= +ribution of Hydrogen Bonding	BIOT 0189
al Cross-Linking: Implications for	Protein Stabilization.= +erview of Protein Chemic	BIOT 0216

Database of	Protein Structure.= The IDITIS Relational	CINF	0032
Peptide Conformations and	Protein Structures.=	BIOL	0004
zymatic Phosphorylation of Soy	Protein to Improve Functional Properties in Foods.=	AGFD	0140
rained Analogues of Styryl-Based	Protein Tyrosine Kinase Inhibitors.= +Ring-Const	MEDI	0023
ilvC Activator	Protein: DNA Interactions.=	BIOL	0124
nt-Derived Ribosome-Inactivating	Protein.= +viral Studies with Trichosanthin, a Pla	AGFD	0189
Formation in the P22 Tailspike	Protein.= Mutational Suppression of Inclusion Body	BIOT	0113
pects. Atomic Interactions betw+	Protein-Carbohydrate Interactions—I. Structural As	BIOL	0142
spects. Structure of Carbohydra+	Protein-Carbohydrate Interactions—II. Functional A	BIOL	0148
tty Acid+ Structural Aspects of	Protein-Lipid Interactions. Tertiary Structures of Fa	BIOL	0015
Phytic Acid, Total Dietary Fiber,	Protein, and Starch in a Milled Hard Red Winter W	AGFD	0143
Hormigas: Alimento Rico en	Proteinas.=	AGFD	0205
Driving Force in the Design of	Proteinase Inhibitors.= Chemical Principles as a	ORGN	0269
Differences in the Cysteine	Proteinases from Papaya Latex.= Conformational	BIOL	0074
ects. Atomic Interactions between	Proteins and Carbohydrates.= +-I. Structural Asp	BIOL	0142
nclusion Bodies to Remove E. coli	Proteins and Endotoxin.= +of Washing sCD4-183 I	BIOT	0171
olarization, and the Dynamics of	Proteins and Membranes: Theory and Experiment.=	PHYS	0166
ent Substrates in the Detection of	Proteins and Nucleic Acids.= +ted Chemiluminesc	BIOT	0167
.= Rapid Separation of	Proteins and Peptides by Perfusion Chromatography	ANYL	0127
r Preparative Chromatography of	Proteins and Pharmaceutical Molecules Using Porou	ANYL	0031
	Proteins Designed for Adherence to Cellulose.=	BIOT	0217
R of Duplex Oligonucleotides and	Proteins DNA Complexes.= +wo-Dimensional NM	INOR	0378
ctrophoretic Analysis of Cationic	Proteins Extracted from Aflatoxin-Resistant/Suscep	AGFD	0134
ine Invertebrate Oxygen-Carrying	Proteins Hemerythrin and Myohemerythrin.= +ar	INOR	0140
Immobilization of	Proteins in Microlithographically Defined Patterns.=	BIOT	0246
some Preparation.= Genes and	Proteins of the Clostridium thermocellum Subcellulo	BIOT	0181
Metal-Binding	Proteins of the Mussel Byssus.=	INOR	0138
How	Proteins Recognize and Bind Oligosaccharides.=	CARB	0012
am Processing of Periplasmic	Proteins Using Critical Fluid Disruption.= Downstre	BIOT	0070
Affinity Precipitation of	Proteins with Ligand-Modified Phospholipids.=	BIOT	0020
nal NMR of Isotopically Enriched	Proteins.= +Optical Spectroscopies. Multidimensio	PHYS	0113
onium Oxide for the Separation of	Proteins.= +en Calcium Hydroxyapatite and Zirc	ANYL	0027
Structures of Fatty Acid Binding	Proteins.= +of Protein-Lipid Interactions. Tertiary	BIOL	0015
Sequence-Specific DNA-Binding	Proteins.= +thod to Define the Recognition Site of	BIOL	0129
and Proton Transfer Reactions in	Proteins.= +antum Dynamics in Electron Transfer	BIOL	0131
and Modified	Proteins.= Electron Tunneling Pathways in Native	BIOL	0133
for Recovery of Recombinant	Proteins.= Evaluating Starch-Protein Interactions	BIOT	0047
eeze Glycoproteins and Antifreeze	Proteins.= +fferences in Properties between Antifr	AGFD	0035
y Expressing with Insect Selective	Proteins.= +Efficacy of Baculovirus Insecticides b	AGRO	0009
Precipitation of	Proteins.= Insoluble Reagent for the Selective	BIOL	0040
Lipid Oxidation: Effect on Meat	Proteins.=	AGFD	0159
Microtubule-Based Motor	Proteins.=	BIOL	0138
High-Frequency Dynamics in	Proteins.= Model Free Analysis of the	PHYS	0168
Chromatography of	Proteins.= Preparative Reversed-Phase Elution	ANYL	0029
Chromatography of	Proteins.= Rapid Recycling Preparative	ANYL	0028
Folding—GroE Heat Shock	Proteins.= Role of Chaperones in Protein	BIOT	0116
Measurements in Large	Proteins.= Rotational Resonance: Distance	PHYS	0196
els for Nonheme-Iron-Containing	Proteins.= +tranuclear Iron(III) Complexes as Mod	INOR	0295
Dynamics in Photoexcited Heme	Proteins.= +rements of Energy Flow and Structural	PHYS	0335
and Structural Dynamics of Heme	Proteins.= +ical Studies of Vibrational Relaxation	PHYS	0169
the Context o+ Immobilization of	Proteins, Cofactors, Mediators, Tissues, and Cells in	BIOT	0039
nships Studies of Cathepsin B, a	Proteolytic Enzyme Involved in Tumor Metastasis.=	MEDI	0017
	Proteolytic Enzymes from Ectothermic Organisms.=	BIOT	0194
x mori.=	Prothoracicotropic Hormone of the Silkworm, Bomby	AGRO	0010
lation Relaxation Time of Ions in	Protic Solvents Measured by Transient Infrared Spe	PHYS	0083
Including Predictions+ Ab Initio	Proton Affinites Compared to Experimental Values,	PHYS	0298
Citraconyl Phosphate Driven by a	Proton Gradient.= +for H+-ATPase: Formation of	BIOL	0104
c Investigation of Stereocontrolled	Proton Migration from Nitrogen to Carbon.= +isti	ORGN	0296
carbonylchromium Compounds by	Proton NMR Spectroscopy.= +ion of (η6-Arene)Tri	ORGN	0215
ja siamensis by Two-Dimensional	Proton NMR.= +tion of Cytotoxin D from Naja na	BIOL	0044
	Photoinduced	Proton Transfer in LB Films.=	BIOT	0145
ynamics in Electron Transfer and	Proton Transfer Reactions in Proteins.= +antum D	BIOL	0131
Voltammetric Investigation of	Proton Transport in Ordered Polymer Lattices.=	COLL	0167
de Ligands: Synthetic and Mec+	Proton-Induced Coupling of Alkylidyne and Isocyani	INOR	0150
Materials: Elemental Analysis by	Proton-Induced X-ray Emission (PIXE). +cation of	NUCL	0077
Naphthols with	Proton-Mediating Side-Chains.=	ORGN	0162
s+ Structural Characterization of	Protonated trans Bis-Dioxime Ruthenium Complexe	INOR	0066
Layered Silicates: The	Protonation Behavior of $KHSi_2O_5$.=	PETR	0110
Face Selectivity in the	Protonation of Glycals.=	ORGN	0069
a Stable Species During Solution	Protonation of 3-Hydroxphthalide.= +ation Is Not	ORGN	0151
Forming Reactions with η^2-Pyr+	Protonation, Tautomerization, and Novel C-C Bond-	INOR	0114
elaxation Times ('T2's) of Water	Protons in Human Plasma Samples Related to Their	MEDI	0043
ramagnetic Relaxations of Water	Protons in the Solutions of the Iron-Ethylene Diami	INOR	0090
tec+ Future Role of Erythrocyte	Protoporphyrin (EP) as a Biochemical Marker for De	CEI	0020
3, and Ge_2H_6 Radical Anions as	Protypes of Polymer Anions with Si and Ge Backbo	PHYS	0313
ransient Kinetic Techniques Will	Provide Detailed Insight in the Mechanism of Gas-S	FUEL	0015
Toxicological Services	Provided by Hazard Communication Resources, Inc.=	CHAS	0024
Clorhidrato de L-Arginina en +	Proyecto de Practica: Valoracion Potenciometrica de	CHED	0163
Tower Absorbers Using the	PRSV Equation of State.= Design for Packed	CHED	0107
dy of the Microphase Structure of	PS-PMPS Diblock Copolymers.= +tate NMR Stu	PMSE	0121

Study on the Damping Paint of	PS/PA Latex Interpenetrating Polymer Networks.=	PMSE	0064
Latex	PS/XPS Semi-IPNs.=	PMSE	0063
of Atomic Cluster Distributions in	Pseudobinary Semiconductor Alloys.= +surements	PHYS	0375
	Pseudografting of Vinyl Polymers onto DNA.=	POLY	0301
n Involved in the Orthorhombic	Pseudohexagonal Rotator–Phase Transition in n–Hen	POLY	0354
uene Monooxygenase Constitutive	Pseudomonas cepacia.= +ion Products Using a Tol	BIOT	0119
ity of Lipoprotein Lipase from	Pseudomonas sp. in Hydrolytic Reactions.= Specific	BIOT	0095
s for Biomineralization Researc+	Pseudomorph and Artifact Corrosion Studies as Tool	BIOT	0101
rt. Molecular Rearrangements and	Pseudorotation.= +tum Chemistry with Experime	PHYS	0013
Structure of Asparagine by the	Pseudospectral Gradient Method.= Molecular	PHYS	0275
tion of Oxidosqualene Cyclase by	Pseudosubstrates: Studies of 26– and 29–Functionali	ORGN	0021
and a Pentacoordinate CuN$_3$OS	Pseudothiolate Complex Compared with Their O–Do	INOR	0262
C, the Genes Coding for Yersinia	pseudotuberculosis CDP-D-Glucose Oxidoreductase	BIOL	0110
UvrABC Endonuclease DNA R+	Psoralen–Linked Deoxyoligonucleotide Synthesis for	BIOL	0126
enc+ Truth and Consequences: A	Psychologist's Perspective on Whistle Blowing in Sci	PROF	0002
the Deposition of Platinum from	Pt (PF$_3$)$_4$ on Atomically Clean Platinum Surfaces.=	COLL	0046
Decomposition of W(CO)$_6$ on	Pt (111).= TPD and XPS Study of the Thermal	COLL	0049
Ph$_2$P(O)]$_3$–nC}, Where M = Pd or	Pt and n = 1, 2, or 3.= +of (R$_3$P)MCl}[Ph$_2$P(S)]$_n$[	INOR	0265
n Selectivity over Nonmicroporous	Pt Catalysts.= +port Preparation on Aromatizatio	CATL	0030
Surface Derivatization of Au and	Pt Microfabricated Wire Arrays.= +s for Selective	INOR	0350
Trichloroethylene Using 1.5	Pt on Al$_2$O$_3$/Monolith.= Catalytic Oxidation of	COLL	0205
Zeolite: Inhibited Deactivation of	Pt Sites within One–Dimensional Zeolite Channels.=	CATL	0015
e for NO Adsorption on Rh and	Pt.= Study of the Effect of Support and Particle Siz	COLL	0186
A+ Studies on the Interaction of	Pt(II) and Pd(II) Dipeptide Complexes with Nucleic	INOR	0412
s on Mixed Ligand Complexes of	Pt(II) and Pd(II) with Dipeptides and Substituted P	INOR	0411
Cyanide and Acetonitrile on	Pt(III) Electrode Surfaces.= Chemisorption of	COLL	0118
mplejos de Transferencia de Carga	Pt(R–Nafin)·TCNQ.= +r Espectroscopia IR de Co	INOR	0190
Studies of [	Pt(terpy)Cl]+.= Spectroscopic and Electrochemical	INOR	0216
ation: Naphthalene vs. Azulene on	Pt(111).= +al–Space Imaging of Molecular Organiz	PHYS	0125
CH$_3$Br Adsorbed on	Pt(111).= Surface Photochemical Dynamics of	PHYS	0261
Iodine Overlayers on	Pt(111).= Surface Spectroscopic Characterization of	PHYS	0366
Ethyl Acetate over	Pt– Alumina Catalyst.= Combustion Byproducts of	COLL	0189
d Physicochemical Properties of	Pt–Containing Automotive Emission Control Catalys	COLL	0208
nd C$_6$ N–Alkanes in+ Design of	Pt–Mordenite Catalysts for the Isomerization of C$_5$ a	PETR	0113
vity and Characterization Data for	Pt–Sn–Aluminum Catalysts.= +rocyclization Acti	CATL	0007
c Properties of Bimetallic System	Pt–Sn/ZnAl$_2$O$_4$.= +t of Sn Addition on the Catalyti	COLL	0127
g.+ New Modification Method of	Pt/KL Zeolite Catalyst for Light Naphtha Reformin	PETR	0064
tic Dehydrocyclization Rates on	Pt/L-Zeolite: Inhibited Deactivation of Pt Sites with	CATL	0015
the Presence of Aromatics over	Pt/Zeolites.= Hydroisomerization of N–Hexane in	PETR	0114
s of Ni-Fe-Pt, Ni-Fe-Pd, Fe-Co–	Pt, and Fe–Co–Pd.= +etism in Ternary Compound	PHYS	0253
in Ternary Compounds of Ni–Fe–	Pt, Ni-Fe-Pd, Fe-Co-Pt, and Fe-Co-Pd.= +etism	PHYS	0253
ean Aging on the Performance of	Pt, Rh/CeO$_2$ and Rare Earth/Alkaline Earth–Doped	PETR	0035
Rare Earth/Alkaline Earth–Doped	Pt, Rh/CeO$_2$ Catalysts.= +ce of Pt, Rh/CeO$_2$ and	PETR	0035
Molybdenum–	Pterin Complex.= Properties of a Reduced	INOR	0277
moreversible Transitions.=	PTFE–Matrix Composite Hydrogel Sheets with Ther	POLY	0245
Monitoring for Environmental	Public Health.= Application of Biological	CEI	0007
Pesticides in	Public Health.=	AGRO	0057
s.=	Public Involvement in the Cleanup of the Great Lake	ENVR	0085
Pr+ Round–Table Discussion–	Public Involvement in the Remedial Action Planning	ENVR	0086
se and Development in the 21st+	Public Policies and Attitudes Will Affect Pesticide U	AGRO	0070
The False Conflict: Helping the	Public vs. Helping the Corporation.=	PROF	0006
a Partir de Errores Detectados en	Publicaciones Cientificas.= +enanza Experimental	CHED	0178
J. J. Keller and Associates:	Publications and Products.=	CHAS	0029
Fractal Properties of Dextran,	Pullulan, and Pectin Determined by HPSEC with Vi	CARB	0045
tegrated Bio–Ozon Treatment of	Pulp Bleaching Effluents Containing Chlorinated Ph	BIOT	0137
Applications in	Pulp Bleaching.= Practical Aspects of Xylanase	BIOT	0214
Optical Spectroscopies. Multiple–	Pulse Femtosecond Spectroscopy.= +sonance and	PHYS	0066
mical Reactions Using Multiple–	Pulse Line–Narrowing Techniques and ^{1}H T$_1$ Weight	CATL	0022
Princeton Applied Research Co.).	Pulse Voltammetry: Saturday's Child.= +by EG G	ANYL	0096
	Pulsed Amperometric Detection of Carbohydrates.=	CARB	0067
New Applications of Coherent	Pulsed Electron Paramagnetic Resonance.=	PHYS	0116
s–Phase Aluminum Atoms from	Pulsed Laser Ablation: Mechanisms and Applications	PHYS	0353
al Image Analysis and Real–Time	Pulsed NMR.= +onent Polymer Systems by Digit	PMSE	0099
and Phase–Controlled Optical	Pulses.= Molecular Geometric Phase Development	PHYS	0067
is and Pharmacological Studies of	Pumiliotoxin A Alkaloids and Congeners.= +ynthes	ORGN	0275
w–Grade Heat Source with a Heat	Pump.= +Geothermal Brine Utilizing a Waste/Lo	I&EC	0010
nal Cooling and Multiphonon Up–	Pumping.= +Transfer in Ultrahot Solids: Vibratio	PHYS	0250
anges, and Toll Con+ Achieving	Purchasing Productivity through Global Swaps, Exch	CMEC	0007
ounds in Human Blood by Using	Purge–and–Trap/Gas Chromatography/Mass Spectr	CEI	0001
aphic+ Perspectives in Protein	Purifications—Chromatographic and Nonchromatogr	BIOT	0018
Novo	Purine Biosynthesis.= Carbocyclic Substrates for de	BIOL	0076
–Based Drug Design: Inhibitors of	Purine Nucleoside Phosphorylase (PNP).= +ucture	MEDI	0005
is and Biological Activity of 9–+	Purine Nucleoside Phosphorylase Inhibitors: Synthes	MEDI	0028
ransition State Analysis.=	Purine Nucleoside Phosphorylase: Mechanism and T	BIOL	0046
in the Synthesis of 2'–Fluorinated	Purine Nucleosides.= +d Sugar–Protecting Groups	CARB	0024
eldas Fotoelectroquimicas, I: CdS	Puro e Impurificado.= +Policristalinos dentro de C	PHYS	0221
tio+ Spectroscopy at the Limit:	Pushing the Bounds of Sensitivity and Spatial Resolu	ANYL	0012
renes: Synthesi+ Synthesis of the	putative Metabolites of the Cyclopenta[a]–Phenanth	ORGN	0118
lting Point of Mixtures.=	Putting the Horse before the Cart: Exploring the Me	CHED	0159
ng, Part I: Acyclic Seri+ Didactic	Puzzles for Organic Chemistry Nomenclature Teachi	CHED	0165

Left context	Middle (title)	Code	Num
placement.=	PVC in Container Coatings and Approaches to Its Re	PMSE	0143
in	PVC.= Diffusion of Alkyl Adipate Plasticizers	POLY	0067
of Reinforced Polyvinyl Chloride (	PVC) Roofing Membranes.= +cal Analysis Studies	PMSE	0180
Separation of	PVC/Plasticizer Solutions.= Gelation and Phase	POLY	0173
Dye Orientation in	PVDF/PMMA Blend.= Internal Electric Field and	POLY	0225
ew Ferroelect+ Liquid Crystalline	Pyramidic Transition–Metal Complexes: Potential N	INOR	0135
tive Synt+ Furfuryl Alcohol–to–	Pyranone Oxidation: An Approach to the Stereoselec	ORGN	0294
um Amide Anion: The Pathway to	Pyrazol Gallates.= +Characterization of Alkyl Galli	INOR	0390
4–Benzoyl–5–Methyl–2–Phenyl–	Pyrazol–3–Thione and Tri–n–Octylphosphine Oxide.	NUCL	0073
lobutanoneimine and Platinum–	Pyrazoline Formed by Nucleophilic Addition of Amin	INOR	0041
ngiotensin II Rece+ Substituted	Pyrazolo[1,5-a]Pyrimidines as Potent Orally Active A	MEDI	0104
bstitution on Biological Activity of	Pyrazoloquinolines.= +ies: Effect of Benzyloxy Su	MEDI	0019
zed by Sterically Demanding tris(	Pyrazolyl)hydroborato Ligands.= +rivatives Stabili	INOR	0087
ed by Sterically Demanding poly(	Pyrazolyl)hydroborato Ligands.= +rivatives Stabiliz	INOR	0085
of 2–(2–(Alkoxyimino)–Alkyl)–5–	Pyrazolyl–1,3–Cyclohexadiones.= +rbicidal Activity	AGRO	0013
Memory with Polyion Complexed	Pyrene LB Films.= +otoamplified Storage Optical	BIOT	0144
ization of a P450 Responsible for	Pyrethroid Resistance in the House Fly.= Character	AGRO	0099
Type	Pyrethroid: JS–88.= Synthesis and Activity of a New	AGRO	0015
in the Diamondback Moth to	Pyrethroids and Benzoylphenyl Ureas.= Resistance	AGRO	0117
Agent.= N–Fluoropyridinium	Pyridine Heptafluorodiborate: A Useful Fluorinating	ORGN	0076
arbonate–Urethane) and Polyvinyl	Pyridine.= +etrating Polymer Networks of Poly(C	PMSE	0056
Induced Oxidatio+ Synthesis of	Pyridines by Bentonite–Supported Metallic Nitrates:	ORGN	0245
2,4–Diamino–6–[2–Phenylethenyl]	Pyrido[2,3-d] Pyrimidine Derivatives.= +Series of	MEDI	0022
of Octahydro–1,2,3,4,4a,5,11,11a–	Pyrido[3,4-c] [1,5]Benzoxazepine Fentanyl Derivativ	MEDI	0148
n of Tethered 2–	Pyridones.= Intramolecular [4+4] Photocycloadditio	ORGN	0235
Synthesis of 3–Carboxy–1–	Pyridoxyl-1,2,3,4-Tetrahydro-β-Carboline.=	BIOL	0080
tion $\alpha\beta$–anti–Chloro–[1,9–Bis(2–	pyridyl)–2,5,8–Triazanonane]–Chromium(III): An Ex	INOR	0397
ion–Metal Centers via Poly 3'–(4–	pyridyl)terthiophene–Based Ligands.= +at Transit	INOR	0027
olemic Activity of Phenylene and	Pyridylene Diurea Derivatives as ACAT Inhibitors.=	MEDI	0109
to Pyrrolo[2,3-d]	Pyrimidine Antifolates.= Novel Synthetic Approach	ORGN	0113
6–[2–Phenylethenyl]]Pyrido[2,3-d]	Pyrimidine Derivatives.= +Series of 2,4–Diamino–	MEDI	0022
nantiomerically Pure Dioxolane–	Pyrimidine Nucleosides as Potential Anti–HIV Agen	MEDI	0051
Determining	Pyrimidine Photoproducts in UV–Irradiated DNA.=	ANYL	0092
orophenyl)–1,2,4–Triazolo–[1,5a]–	Pyrimidine Sulfonamides.= +N–(Substituted–2–Flu	AGRO	0048
e of Heter+ 1,2,4–Triazolo[1,5a]	Pyrimidine-2-Sulfonanilide Herbicides: The Influenc	AGRO	0046
of Heteroc+ 1,2,4–Triazolo[1,5a]	Pyrimidine-2-Sulfonanilide Herbides: The Influence	AGRO	0047
ec+ Substituted Pyrazolo[1,5-a]	Pyrimidines as Potent Orally Active Angiotensin II R	MEDI	0104
) with Dipeptides and Substituted	Pyrimidines.= +and Complexes of Pt(II) and Pd(II	INOR	0411
Fott+ Separation Arsenopyrite–	Pyrite: Effet des Oxydants et des Complexants sur la	ANYL	0067
erature–Programmed Oxidative	Pyroanalysis (TPOP): A New Analytical Method to P	FUEL	0046
nd Crown Ether on the Surface of	Pyrogeneous Silica.= +en Gystamine, Mexamine, a	COLL	0096
and Properties of	Pyrogenic Titanosilicas.= Peculiarities of Structure	COLL	0111
Aziridine on the Surface of	Pyrogenous Silica.= Thermal Transformations of	COLL	0113
Metals as Catalysts for	Pyrolysis and Gasification of Biomass.= Transition	FUEL	0103
of Preceramic Polysilazanes, the	Pyrolysis Conditions, and Their Final Ceramic Produ	POLY	0313
eparation of Carbon Powders by	Pyrolysis of Graphitizable and Nongraphitizable Poly	POLY	0023
e Precursors of NOx in the Rapid	Pyrolysis of Nitrogen–Containing Heteroaromatics.=	FUEL	0114
Pretreatment and	Pyrolysis of Polysilazane Preceramic Binders.=	POLY	0338
Oxide)s—II.	Pyrolysis Studies.= Poly(Arylene Ether Phosphine	PMSE	0053
Allene	Pyrolysis.= Benzene Formation during	FUEL	0069
n Argonne Premium Coals during	Pyrolysis.= +ion Kinetics of Organic Sulfur Forms i	FUEL	0059
as a Technique for Analyzing F+	Pyrolysis–Fourier Transform Infrared Spectrometry	FUEL	0061
y of Preceramics Precursors and	Pyrolyzed Products.= Solid-State MAS–NMR Stud	POLY	0358
of Poly [N,N'–bis(phenoxyphenyl)	pyromellitimide].= +es of New Model Compounds	PMSE	0151
inolide: A Novel Antimicrobial α–	Pyrone from Hyptis pectinata (Lamiaceae).= +Pect	MEDI	0060
omplexes of Tc–99m–Hydroxy–4–	Pyrones: Synthesis and Characterization.= +ionic C	MEDI	0042
2–[6–(ribosyl)–picolinamide–5'–yl]	pyrophosphates. +nthesis of P^1–(adenosin-5'-yl) –P	CARB	0023
Bond–Forming Reactions with η^2–	Pyrrole Complexes.= +merization, and Novel C–C	INOR	0114
Poly(2,5–	pyrrole).= Toward Structurally Perfect	POLY	0036
Synthesis and Complexation of	Pyrrole–Containing Receptors and Porphyrin Analog	ORGN	0063
Substituted and 1,3–Disubstituted	Pyrroles.= +Facile Method for the Synthesis of 3–	ORGN	0196
of 3–Substituted	Pyrroles.= New and Short Method for the Synthesis	ORGN	0197
Series of 7–(3–amino–3–aryl–1–	pyrrolidinyl)–4–naphthyridones as Antimicrobial Age	MEDI	0061
Synthesis of Macrolactone	Pyrrolizidine Alkaloids.=	ORGN	0285
Novel Synthetic Approach to	Pyrrolo[2,3-d]Pyrimidine Antifolates.=	ORGN	0113
osine.= Corrected Structure of	Pyrrolosine and Reconfirmation of That of 9–Deazain	CARB	0046
ependent Enzyme Brewers' Yeast	Pyruvate Decarboxylase.= +hiamin Diphosphate–D	BIOL	0032
ound Enamines on Brewers' Yeast	Pyruvate Decarboxylase.= +iamin Diphosphate–B	BIOL	0048
arboxylating Enzyme: A M+ coli	Pyruvate Dehydrogenase Complex and to the E1 Dec	BIOL	0031
by Modification of Succinate and	Pyruvate Levels.= +roved Zooglan Gum Properties	BIOT	0198
esonance Raman Studies of the	Q$_y$ Transitions in Rhodobacter sphaeroides Reaction	BIOL	0094
terias Biologicas de la Carrera de	QFB en la Facultad de Quimica, UNAM (Analisis Es	CHED	0092
CH with N$_2$, O$_2$, and NO.=	QRRK Chemical Activation Analysis on Reactions of	PHYS	0229
le Intercalating Agents Based +	QSAR Analysis of Anticancer Benzothiopyranoindazo	MEDI	0018
Corn (	QSAR Analysis).= A Sulfonylurea Herbicide for	AGRO	0050
icellar Liquid Chromatography to	QSAR Modeling of Organic Compounds.= +on of M	ANYL	0046
s and Force Field Calculations.=+	QSAR Parameters Derived from Quantum Mechanic	MEDI	0044
	QSRR Aided by Quantum Chemistry.=	ANYL	0100
olonic Neo+ Dietary Quercetin (	QU) and Rutin (RU) as Inhibitors of Experimental C	AGFD	0114
6 Potential Energy Surface Using	Quadratic Configuration Interaction Methods.= +H	PHYS	0305

Molecules in Sol-Gel Matrices for	Quadratic Nonlinear Optics.=	+ntation of Organic	POLY 0006
Molecules and Materials for	Quadratic Nonlinear Optics.=	Investigation of New	POLY 0103
= Two-Dimensional Pure	Quadrupole Correlation Spectroscopy (NQR-COSY).		PHYS 0277
ameters in Bridging+ Deuterim	Quadrupole Coupling Constants and Asymmetry Par		INOR 0382
Identification and	Quantification of Nucleic Acids in the Hydrosphere.=		ENVR 0027
,4-Dihydropyridines by Their +	Quantification of the Stereoselective Recognition of 1		MEDI 0100
ibitor, by a Tandem Column H+	Quantitation of Argatroban, a Specific Thrombin Inh		ANYL 0086
es on Charge-Carrier Dynamics of	Quantized Semiconductor Colloids.= +ace Properti		COLL 0151
ical Chemistry.=	Quantum Chemical Calculations in a Course in Phys		CHED 0191
he Atomistic Simulati+ Ab Initio	Quantum Chemical Potential Energy Functions for t		POLY 0350
Ab Initio	Quantum Chemistry and Dynamics.=		PHYS 0165
f Expe+ Comparison of ab Initio	Quantum Chemistry with Experiment. Comparison o		PHYS 0339
-State+ Comparison of ab Initio	Quantum Chemistry with Experiment. Diene Excited		PHYS 0075
he De+ Comparison of ab Initio	Quantum Chemistry with Experiment. Dynamics of t		PHYS 0188
Comparison of ab Initio	Quantum Chemistry with Experiment. Gold 6.=		PHYS 0004
e Spec+ Comparison of ab Initio	Quantum Chemistry with Experiment. High-Overton		PHYS 0029
rmng+ Comparison of ab Initio	Quantum Chemistry with Experiment. Molecular Rea		PHYS 0013
Photoe+ Comparison of ab Initio	Quantum Chemistry with Experiment. Negative-Ion		PHYS 0055
ory of+ Comparison of ab Initio	Quantum Chemistry with Experiment. Quantum The		PHYS 0161
of Non+ Comparison of ab Initio	Quantum Chemistry with Experiment. Spectroscopy		PHYS 0102
QSRR Aided by	Quantum Chemistry.=		ANYL 0100
eno+ Use and Limitations of n-	Quantum Coherence in Probing Intermediates in Het		COLL 0089
Semiconductor	Quantum Devices, Circuits, and Applications.=		BIOT 0003
Tran+ Computer Simulations of	Quantum Dynamics in Electron Transfer and Proton		BIOL 0131
	Quantum Dynamics of the Hydrated Electron.=		PHYS 0190
drides: Rotational Tunnelling and	Quantum Exchange.= +cal Transition Metal Polyhy		PHYS 0068
QSAR Parameters Derived from	Quantum Mechanics and Force Field Calculations.=		MEDI 0044
lid-State Interactions by Multiple	Quantum NMR.= +ic and Aqueous Guest/Host So		AGRO 0062
l Polymer Films by ^{19}F Multiple-	Quantum NMR.= +nium Salt Distribution in Mode		PMSE 0182
els.=	Quantum Simulations of Hubbard and Holstein Mod		PHYS 0054
um Chemistry with Experiment.	Quantum Theory of Chemical Reactions: Reactive Sc		PHYS 0161
iplet Cyclohexenones.=	Quantum Yields of Singlet Oxygen Production by Tr		ORGN 0167
t on Activation Barriers of Uni+	Quantum-Chemical Study of Surface Potential Effec		COLL 0107
a., and Translational Energy Tr+	Quantum-State-Resolved N_2O Vibrational, Rotation		PHYS 0334
Search for	Quark-Gluon Plasma in P-P Collisions at Fermilab.=		NUCL 0063
Sodium Silicate with Additives on	Quartz and Its Relation with Collapsibility.= +ted		COLL 0079
phate Compound: Syn+ A Novel	Quasi-Low-Dimensional Reduced Niobium Oxophos		INOR 0031
ries of a New Modifiaction of the	Quasi-One-Dimensional Compound KMo_4O_6.= +e		INOR 0359
$H_2 + CH_2$ Reaction.=	Quasiclassical Trajectory Studies of the $H + CH_4 \leftarrow$		PHYS 0263
scular Blocking Agents.= Novel	Quaternary-Enamino-Nitrogen-Containing Neuromu		MEDI 0046
Synthesis of the First Branched	Quaterthienyls.=		ORGN 0202
logicas de la Carrera d+ Factores	Que Incrementan la Reprobacion en las Materias Bio		CHED 0092
from Wood-Preserving Plants in	Quebec: Extensive Analysis and Search for Remediat		ENVR 0016
rocarbons in La Basse Cote-Nord,	Quebec.= +ng People to Chlorinated Aromatic Hyd		CEI 0011
Lab.= Fluorescence	Quenching Experiments for the Physical Chemistry		CHED 0202
ve Energy and Electron Transfer	Quenching of Excited States by Ferrocene Derivatives		INOR 0163
Blends, and Interfaces—II. Quick	Quenching of NH $A^3\Pi_i$ in High Rotational Levels.=		PHYS 0266
Blends, and Interfaces—II. Quick	Quenching of Polymer Blends.= +lymer Solutions,		COLL 0032
sin+ Lifetime and Concentration	Quenching of Triplet Methylbenzenes in Solution, U		PHYS 0265
umic Substances by Fluorescence	Quenching.= +n of the Electrostatic Properties of H		ENVR 0057
Simulation of Trapping- and	Quenching-Reaction Kinetics in Low-Dimensional an		PHYS 0307
erimental Colonic Neop+ Dietary	Quercetin (QU) and Rutin (RU) as Inhibitors of Exp		AGFD 0114
ion, Metabolic Processing, and +	Quercetin and Quercetin Glycosides: Genetic Regulat		AGFD 0117
Processing, and+ Quercetin and	Quercetin Glycosides: Genetic Regulation, Metabolic		AGFD 0117
t. Anticarcinogenicity of Flavonol	Quercetin.= +nolic Compounds in Food and Healt		AGFD 0113
Formulation.= Flexible	Queries in 3D Searching: Techniques in 3D Query		CINF 0025
Searching: Techniques in 3D	Query Formulation.= Flexible Queries in 3D		CINF 0025
e con S. aureus y su Aplicacion en	Quesos.= +de Sodio a Partir de Leche Contaminad		AGFD 0107
PCDDS from the Combustion of	Quicklime Binder-Enhanced Densified Refuse-Deriv		FUEL 0140
tarizados para la Ensenanza de la	Quimica a Nivel Medio Superior.= +Apoyos Compu		CHED 0124
crophulariaceae d+ Composicion	Quimica de las Plantas Pertenecientes a la Familia S		ORGN 0185
Diseno de Equipo de la Industria	Quimica de Proceso.= +entes de CAD/CAM para el		COMP 0053
la	Quimica.= Encuentro de Dos Mundos a Traves de		HIST 0035
errera de QFB en la Facultad de	Quimica, UNAM (Analisis Estadistico de 10Anos).=		CHED 0092
iva y u+ Optimizacion de Plantas	Quimicas Utilizando Programacion Cuadratica Suces		I&EC 0072
las Diferentes Etapas del Lavado	Quimico de Una Caldera por Tecnicas Electroquimic		PHYS 0257
el Analysis of Alkyl-Substituted	Quinazolinone CCK-B/Gastrin Receptor Ligands.=		MEDI 0066
pounds of the Heterocyclic 4(1H)	Quinazolinone-2,3-Dihydro-2-Thioxo.= +tion Com		INOR 0079
ne D+ Uncoupling Propertiex of	Quinic Acid and Its Phenylboronate and Isopropylide		INOR 0263
$^{15}N+$ Synthesis of 7,7-Dimethyl-	Quino[2,3,4-kl]-5a,6,7,8-Tetrahydrobenz[c]Acridine		ORGN 0198
Hydrodenitrogenation of	Quinoline over High Surface Area $\gamma-Mo_2N$.=		CATL 0047
96	Quinoline.= Novel Third-Order NLO Materials from		POLY 0212
r-Constra+ Carboxynaphthyl, -	Quinolyl and -Isoquinolyl Compounds as Bicyclic Rin		MEDI 0023
Generation of ortho-	Quinone Methides from Cyclobutenones.=		ORGN 0229
a]-quinoxalinones, Imidazo[1,5-a]	quinoxalinones, and Their Aza Analogues.= +o[1,2-		MEDI 0118
ts of a Series of Imidazo[1,2-a]-	quinoxalinones, Imidazo[1,5-a]quinoxalinones, and T		MEDI 0118
ith Oligoribonucleotide Analogs of	Rabbit α-Globin mRNA.= +Interact Differently w		BIOL 0128
Lactonizing Enzymes Mandelate	Racemase and Adenylosuccinate Lyase: Odd Couples		BIOL 0012
Total Synthesis of	Racemic Parvifolin.=		ORGN 0176
rans Rotational Isomers in Single	Racemic Poly(vinyl Chloride) Chains.= +Runs of T		POLY 0069

tabolism and Tissue Residues of	Ractopamine Hydrochloride: A Phenethanolamine-R	AGRO	0131
Ultraviolet	Radiation.= Photoionization with Coherent Vacuum	ANYL	0003
by Far UV	Radiation.= Surface Modification of Polycarbonate	PMSE	0125
σ-Delocalized Silicon Backbon+	Radiative Recombination and Vibronic Relaxation in	PHYS	0138
ogen Transfer Catalyst.= Free-	Radical Alkylation in the Presence of H_2S as a Hydr	PETR	0083
phorus Systems.= Free-	Radical and Photochemical Reactions of Organophos	INOR	0347
on Si_2H_6, SiH_3GeH_3, and Ge_2H_6	Radical Anions as Prototypes of Polymer Anions with	PHYS	0313
Low-Lying States of Butadiene	Radical Cation.= Theoretical Investigation of	PHYS	0237
he Dinuclear Iron Center: Tyrosyl	Radical Cofactor of Ribonucleotide Reductase.= +t	BIOL	0022
by Flash Phot+ Kinetics of Thiyl	Radical Formation and Electron Transfer Reactions	INOR	0396
ylate to Squalane.= Free	Radical Grafting of 2-(Dimethylamino)ethyl Methacr	POLY	0019
Free	Radical Halogenation of Hydridophosphoranes.=	ORGN	0248
ting Polymer Networks. Entirely	Radical in Situ Synthesis of Interpenetrating Polyme	PMSE	0001
An Organic	Radical in the Lysine 2,3-Aminomutase Reaction.=	BIOL	0077
e Kinetics of Propagation in Free-	Radical Polymerization.= +nomer Diffusion and th	POLY	0165
r Deform+ ESR Investigation of	Radical Processes in Organic Polycrystals under Shea	PHYS	0245
ea+ Kinetics of Metal Atom and	Radical Reactions over Wide Temperature Ranges: M	PHYS	0087
Formyl, and Vinyl+ Hydrocarbon	Radical Reactions with Oxygen: Comparison of Allyl,	PHYS	0228
	Radical Reactivity of Cyclic Acetals.=	ORGN	0156
l Monomers.=	Radical Ring-Opening Characteristics of Cyclic Viny	ORGN	0157
rganic Composite Materials: Free	Radical Routes into Nonshrinking Sol-Gel Composit	POLY	0259
Effects of Structural Changes on	Radical Stabilities.= Cope Award Symposium.	ORGN	0133
C_2H	Radical.= Infrared Laser Spectroscopy of the	PHYS	0332
Catalisis de Reacciones via	Radicales Libres con Tonsil.=	COLL	0119
hiolate Ions to β-Nitrostyrenes:	Radicaloid Transition State or Principle of Nonperfe	ORGN	0155
ction and Characterization of	Radicals Derived from Oxidation of Probucol.= Dete	BIOL	0099
ctions of Allyl Cations and Cation	Radicals Generated from 1,3-Dienes.= +dition Rea	ORGN	0123
of Dyes by Photogenerated Ketyl	Radicals in Photoelectrochemical (PEC) Cell.= +on	COLL	0081
Couple+ Symmetry Breaking in	Radicals Studied by the Fock-Space Multireference	PHYS	0016
Few Benzyl-Type	Radicals with O_2 and NO_x.= Reaction Rates of a	FUEL	0081
with $O(^3P)$ Atoms and $CH_2(x^3B_1)$	Radicals: Comparative Studies.= +ic Hydrocarbons	FUEL	0082
y of the Oxidation of Unsaturated	Radicals: $n-C_4H_5 + O_2$.= +ics and Thermochemistr	FUEL	0066
c Studies of the Structure of Free	Radicals.= +gative-Ion Photoelectron Spectroscopi	PHYS	0055
Detection of Phenolic Free	Radicals.= Electron Paramagnetic Resonance	AGFD	0004
to Photogenerated Organometallic	Radicals.= +um Metal Films by Electron Transfer	INOR	0319
as a Source of Propargyl	Radicals.= Thermal Decomposition of 1,7-Ocatdiyne	FUEL	0067
oncentration, and Application of a	Radiofrequency Electric Field.= +Flow Rate, Ion C	COLL	0163
ensions as a Result of Exposure to	Radiofrequency Electric Fields.= +(Anatase) Susp	COLL	0203
oods of Animal Origin.=	Radioimmunoassay of Antibiotic/Drug Residues in F	AGFD	0084
sis and Animal Biodistribution of	Radioiodinated Cocaine: A Potential Cocaine Recept	MEDI	0131
tion of Juvenile Hormone Using	Radioligands.= Hot JH: Studies of the Molecular Ac	AGRO	0005
llonite.=	Radiolysis of Adsorbed Acetic Acid on Na-Montmori	NUCL	0072
Solutions.= Gamma	Radiolysis of α-Ketoglutaric Acid in Aqueous	NUCL	0070
Fractal Theory of	Radon Emanation from Solids.=	NUCL	0088
Indoor	Radon Levels in New York and New Jersey.=	ENVR	0023
tural Products from the Tropical	Rain Forest—A Field for Small Chemical Business.=	SCHB	0024
Light Scattering (SEC-VIS-	RALLS).= SEC-Viscometry-Right Angle Laser	PMSE	0098
Environments.= Development of	Raman Fiber-Optic Probes for Hostile	ANYL	0065
strates Using Surface-Enhanced	Raman Scattering and X-ray Photoelectron Spectros	POLY	0087
r Melts Using Surfaced-Enhanced	Raman Scattering.= +fusion Coefficients in Polyme	POLY	0088
Surface Enhanced	Raman Spectra of C_{60} and C_{70}.=	COLL	0092
Sol+ High-Overtone Resonance	Raman Spectra of Photodissociating Nitromethane in	PHYS	0302
Experiments in	Raman Spectroscopy for P. Chem Lab.=	CHED	0088
Experiments in	Raman Spectroscopy for Physical Chemistry Lab.=	PHYS	0300
olecular Fluids and+ Coherent	Raman Spectroscopy of High-Temperature Dense M	ANYL	0002
Resonance	Raman Spectroscopy of Simple Amides.=	PHYS	0321
Zeolites.= UV Resonance	Raman Spectroscopy of Species Adsorbed on	COLL	0075
Detection.= Unique Optics for	Raman Spectroscopy Using Charge-Coupled Device	ANYL	0035
and Related Carbon Materials by	Raman Spectroscopy.= +ynthetic Diamond Films,	PHYS	0118
n Several Molten Salt Systems by	Raman Spectroscopy.= +hloromagnesium Species i	INOR	0064
y Structure via Fourier Transform	Raman Spectroscopy.= +ation of Protein Secondar	POLY	0306
ompounds for CVD by FT-IR and	Raman Spectroscopy.= +acteristics of Copper(I) C	COLL	0031
cting $YBa_2Cu_3O_7$ Using	Raman Spectroscopy.= Investigation of Supercondu	PHYS	0052
Transform	Raman Spectroscopy.= Recent Advances in Fourier	POLY	0254
,N'-bis(phenoxyphenyl)pyrome+	Raman Studies of New Model Compounds of Poly [N	PMSE	0151
sphaeroides Reacti+ Resonance	Raman Studies of the Q_y Transitions in Rhodobacter	BIOL	0094
-Like Carbon. High-Temperature	Raman-Scattering Behavior in Diamond.= +mond	FUEL	0035
ature-Dependent Luminescence,	Raman, and X-ray Studies of Layered Rare Earth G	INOR	0189
	Rambach—The Happy Alternative.=	SCHB	0014
ward Symposium Honoring R. W.	Ramette. +ack Holes?.= Excellence in Teaching A	ANYL	0006
iloxane-co-methylphenylsiloxane)	Random Block Copolymers and Elastomers.= +hyls	POLY	0020
ents Prepared by Site-Directed or	Random Coupling of Monoclonal Antibodies.= +rb	BIOT	0060
Crystallization Kinetics of	Random Ethylene Copolymers.=	POLY	0075
stribution Analysis of a Biphasic	Random Medium: Relation with Small-Angle Scatter	PHYS	0060
Organic Solvents.=	Random Mutagenesis to Enhance Enzyme Activity in	BIOT	0068
	Random, with a High Degree of Synergistic Binding	BIOL	0053
Stopping Power and	Ranges of Ions in Condensed Matter.=	NUCL	0037
eactions over Wide Temperature	Ranges: Measurements, Correlations, and Predictions	PHYS	0087
of Chemical Reaction on Low-	Rank Coal Mobile Component Generation.= Effect	I&EC	0019
Methods and Need+ Qualitative	Ranking by Cancer and Noncancer Risks Combined:	ENVR	0074

sk: What Must W+ Quantitative	Ranking of Environmental Problems According to Ri	ENVR	0061
Revised Hazard–	Ranking System Air Pathway.=	ENVR	0082
the Basis of+ EPA's Hazard–	Ranking System and Decision Methodology: What Is	ENVR	0077
Hazard–	Ranking System and Relative Risk.=	ENVR	0079
r as Assessed by EPA's Hazard–	Ranking System: A Balance of Risk Assessment and	ENVR	0081
w of the Structure of the Hazard–	Ranking System.= +Sites for Superfund? Overvie	ENVR	0077
plications for the Revised Hazard–	Ranking System.= +s Site Assessment Process: Im	ENVR	0078
Solution Phase Conformation of	Rapamycin.=	MEDI	0169
FK-506 Analogs from	Rapamycin–Derived Materials.= Construction of	ORGN	0015
minacion Cuantitativa de Tierras	Raras por Espectrometria de Masas con Fuente de I	ANYL	0051
n, and X-ray Studies of Layered	Rare Earth Gold(I) Cyanide Salts: Evidence for Ener	INOR	0189
Molecular Precursors to	Rare Earth Monochalcogenides.= Synthesis of	INOR	0311
Performance of Pt, Rh/CeO$_2$ and	Rare Earth/Alkaline Earth–Doped Pt, Rh/CeO$_2$ Cat	PETR	0035
Quantitative Determination of	Rare Earths in Rocks by Electron Impact Mass Spec	ANYL	0049
Harpooning Reactions of Excited	Rare Gas Atoms.= +–Phase Metal Reactions—VII.	PHYS	0367
Laser–Polarized	Rare Gases.= Magnetic Resonance Experiments with	PHYS	0041
+of Alpha-Ionone Enantiomers in	Raspberry Concentrates, Essences, and Flavors.=	AGFD	0181
y, and Environmental Fac+ Red	Raspberry Phenolics: Influences of Processing, Variet	AGFD	0169
Subcloning and Expression of	Rat Liver Cytosolic PEPCK in E. coli.=	BIOL	0107
de Vitamina E en la Fertilidad de	Ratas Hembras.= +ecto de la Deficiencia o Exceso	AGFD	0210
nolipase A$_2$: Determination of the	Rate and Equilibrium Constants.= +lysis by Phosp	BIOL	0019
HNO.= Reaction Path and	Rate Calculations of Different Channels of H +	PHYS	0310
Photolysis–Shock–Tu+ Thermal–	Rate Constant Measurements by the Flash or Laser	FUEL	0123
f Met+ Temperature-Dependent	Rate Constants for Electron Attachment Reactions o	PHYS	0234
Flow	Rate Effects in Modern SEC–Viscometry Detectors.=	PMSE	0067
Jet To Determine the Overall	Rate of Liquid–Liquid Reactions.= Using a Laminar	I&EC	0007
ents and Sucrose Crystals.=	Rate of Water Vapor Interaction with Carbon Adsorb	CARB	0062
Seed Properties' Influence on the	Rate of ZSM–5 Crystallization.= Investigation of	PETR	0039
Organic Solvents on the Reaction	Rate.= +PCl$_5$ with Silica Surface: Influence of the	COLL	0097
sis+ The Bond–Splitting Step Is	Rate-Limiting for Skeletal Actomyosin ATP Hydroly	BIOL	0141
icillin an+ Determination of the	Rate-Limiting Steps in the Pathways Leading to Pen	BIOT	0128
Influence of Temperature, Flow	Rate, Ion Concentration, and Application of a Radiof	COLL	0163
NO$_x$.= Reaction	Rates of a Few Benzyl–Type Radicals with O$_2$ and	FUEL	0081
from cis–Diammine(guanosine)+	Rates of Displacement of Guanosine by Nucleophiles	INOR	0251
Fluxes and Net Reaction	Rates of Flame Species Pertinent to Fullerenes.=	FUEL	0112
e Doing in Predicting Absolute	Rates of Intramolecular Electron and Energy Transfe	ORGN	0143
for Catalytic Dehydrocyclization	Rates on Pt/L-Zeolite: Inhibited Deactivation of Pt	CATL	0015
Dry Etching of Copper at High	Rates.=	COLL	0030
Role of Data in the Formation of	Rational Belief.= +ting, Evidential Reason, and the	ENVR	0075
Approaches Toward the	Rational Design of Dopamine D1 Selective Agonists.=	MEDI	0124
	Rational Design of Organic NLO Materials.=	POLY	0002
of Halogenated Hydrocarbons by	Rational Redesign of Cytochrome P450cam.= +tion	BIOT	0118
Inmunomoduladora del Zinc en	Ratones Inmunodeficientes.= Actividad	AGFD	0144
oy–Based Infant Formulas Fed to	Rats.= +d Supplementation on Protein Quality of S	AGFD	0209
in	Rats.= Metabolism of S-53482, a New Herbicide,	AGRO	0024
Sprague–Dawley	Rats.= Metabolism of Triallate in	AGRO	0023
Metabolic Fate of Ametryn in	Rats, Lactating Goats, and Laying Hens.=	AGRO	0111
Stress and Oxygen Availability on	Ravidomycin Production by an Actinomyces sp.=	BIOT	0162
Diffusion in Polymers by Forced	Rayleigh Scattering.=	POLY	0164
haracteristics of Mn^{2+} in Cubic [(	Rb18c6)$_4$MnCl$_4$][TlCl$_4$]$_2$ in Presence of Cu^{2+}.= +C	INOR	0039
H$_2$Os$_3$(CO)$_{10}$ and Ru$_3$(CO)$_{12}$ with	RC=CPh (R=SiMe$_3$,SnBu$_3$).= Reactions of	INOR	0241
admium, Lead, and Mercury from	RDF Municipal Waste Combustors.= +Control of C	FUEL	0130
Catalysis Using	Re and Mo Complexes.= Photochemistry and Redox	INOR	0343
um Alkoxide Complexes (η_5-C^{5}R^5)	Re(No)(PPh^3)(OCHRR').= +d Activation in Rheni	ORGN	0220
rmed by Coalescence of Atoms on	Re(0001) Surfaces.= +ensional Phases of Sulfur Fo	PHYS	0124
xes Cp(CO)$_2$M=C=CH$_2$ (M=Mn,	Re) with Imines, Carbodiimides, and Benzalazine.=	INOR	0072
Obtencion del Orden de	Reaccion de la Oxidacion de la Vitamina C.=	CHED	0164
tilizadas en el Seguimiento de las	Reacciones Acido Base en Medio Acuoso.= +y 316 U	ANYL	0055
del Transporte de Calor en	Reacciones Solido–Gas NO–Cataliticas.= Influencia	I&EC	0068
Catalisis de	Reacciones via Radicales Libres con Tonsil.=	COLL	0119
ievement in General Chemistry:	Reaching and Teaching Large Student Populations.=	CHED	0081
ntal/Isotopic Analysis by Nuclear	Reactions/Scattering. +ification of Materials: Eleme	NUCL	0064
n Models: Studies Based on Flow–	Reactor and Laminar–Flame–Speed Data.= +datio	FUEL	0083
Plasma	Reactor Design.=	NUCL	0086
alyst Activity in Partial Oxidation	Reactor Dynamics.= +y States and Changes in Cat	I&EC	0069
Elementary Mechanisms to	Reactor Models.= Modeling Diamond Growth: From	PHYS	0357
H$_2$ and H in a Hot–Filament CVD	Reactor of Diamond.= +d Density Distribution of	PHYS	0146
bstrate and Product Extraction on	Reactor Productivity.= +ey: Effects of Growth Su	BIOT	0030
ool in T.H.O.R. Nuclear Research	Reactor Site Boundary.= +Products in Discharge P	NUCL	0090
ivity in an Immobilized Perfusion	Reactor System.= +Overall Cyclosporin A Product	BIOT	0155
on Using a Multiphase Membrane	Reactor: Experimental Verification.= +tic Resoluti	BIOT	0042
in a Co–Current Packed Tubular	Reactor.= +ng the Two–Phase Nitration of Toluene	I&EC	0006
Dissolved Inorganic Carbon and	Reactor–Derived ^{14}C from the Hudson River to the N	GEOC	0014
om Pl+ Design and Operation of	Reactors for Large–Scale Production of Chemicals fr	BIOT	0233
mally Designed Sequencing Batch	Reactors.= +ic Biodegradation of Phenolics in Opti	BIOT	0138
CVD	Reactors.= Modeling Diamond Growth in Pedestal	PHYS	0179
Advanced Nuclear Power	Reactors.= New Generation of Inherently Safe and	FUEL	0023
in–End+ Antibodies Can Either	Read Epitopes Anywhere, or Can Be Limited to Cha	CARB	0019
Light as a	Reagent for Chemical Derivatizations of Pesticides.=	AGRO	0147
Alcohols.= New Organoboron	Reagent for the Preparation of Propargylic	ORGN	0109

Insoluble	Reagent for the Selective Precipitation of Proteins.=	BIOL 0040
Bonds.= Grignard	Reagent Promoted Scission of the C–S and S–S	ORGN 0253
BS): A Useful New Fluorinating	Reagent.= N-Fluoro-o-Benzenedisulfonimide (NFO	ORGN 0077
Copper	Reagent.= Sulfur–Sulfur Bond Scission by a	ORGN 0252
nd Pt Micro+ Dithiocarbamates:	Reagents for Selective Surface Derivatization of Au a	INOR 0350
S_N2' Addition of Organocuprate	Reagents to Alkynyl Oxiranes: Synthesis of a Precurs	ORGN 0011
Propanediol with Trimethylsilyl	Reagents Using the $1,3,2\lambda-5$-Dioxaphospholane Meth	ORGN 0051
Procedure Using Borohydride	Reagents.= A Reductive Amination–Lactamization	ORGN 0242
Organometallic	Reagents.= Polymer Synthesis with	POLY 0328
Alkyls with Aluminum	Reagents.= Reactions of Group IVB (Ti, Zr, Hf)	INOR 0054
Sorption of Redox	Reagents.= Redox Potential Alteration under the	COLL 0103
ng of Aryl Triflates with Grignard	Reagents.= +ns: Nickel(0)-Catalyzed Cross–Coupli	ORGN 0082
om the Bench Top to Commercial	Reality.= +Forming Processes—Or, How We Go fr	PETR 0089
E Intermolecular Forces: Do We	Really Need to Compare ab Initio Calculations with	PHYS 0104
It	Really Work?.= The Wittig Reaction: How Does	ORGN 0278
ment. How Expensive Are Lasers,	Really?.= +hysical Chemistry Laboratory Develop	CHED 0082
s and Mechanisms of the Allylic	Rearrangement and Enoyl Reduction Catalyzed by B	BIOL 0065
Type of Ally+ Chelation–Induced	Rearrangement of Allylic Methoxyacetates: A Novel	ORGN 0211
cephams.= [2,3] Sigmatropic	Rearrangement of Sulfilimines from 3-Exomethylene	ORGN 0033
The Claisen	Rearrangement of 4-Allyloxyisoquinolines.=	ORGN 0101
e Effect of Surface Oxy+ Alkane	Rearrangement Reactions on Tungsten Carbides: Th	CATL 0037
alamin: The Lysine 2+ B_{12}-Type	Rearrangement That Does Not Require Adenosylcob	BIOL 0023
phate from Glucose via the Ferrier	Rearrangement.= +f Inositol (1,3,4,5)Tetrakisphos	AGRO 0045
Thio–Claisen	Rearrangement.= Face Selection in the	ORGN 0239
istry with Experiment. Molecular	Rearrangements and Pseudorotation.= +tum Chem	PHYS 0013
Flow.= Conformational	Rearrangements of Flexible Polymer Chains in	PMSE 0105
and Multiple Bonds with Tungs+	Rearrangements of Organosilanes to Form Metal–Lig	INOR 0115
riolide: An Alter+ Acid–Induced	Rearrangements of the Natural Melampolide Schkuh	ORGN 0018
ospecific Enammonium–Iminium	Rearrangements: A Mechanistic Investigation of Ster	ORGN 0296
Mechanism of β-Glycosidases: A	Reassessment of Some Seminal Papers.= The	ORGN 0070
Muscarinic Receptors.=	Receptor Activity and Drug Design. Pharmacology of	MEDI 0078
hthalimido)propyl]-1,6-he+ M_2	Receptor Activity of N,N'-Dimethyl-N,N'-bis[3-(2-P	MEDI 0140
le and Benzothiazole Adenosine	Receptor Agonists: Potent Antihypertensive Agents.=	MEDI 0111
gs: I. Synthesis and Bin+ Sigma	Receptor Antagonists as Potential Antipsychotic Dru	MEDI 0153
gs: II. Animal Behaviora+ Sigma	Receptor Antagonists as Potential Antipsychotic Dru	MEDI 0084
Angiotensin II	Receptor Antagonists.= Diacidic Nonpeptide	MEDI 0103
1,5-Naphhyridine Angiotensin II	Receptor Antagonists.= Potent, Orally Active,	MEDI 0102
otent Orally Active Angiotensin II	Receptor Antagonists.= +lo[1,5-a]Pyrimidines as P	MEDI 0104
ent Small–Molecule Antiotensin II	Receptor Antagonists.= +zole–5-Acrylic Acids: Pot	ORGN 0119
The Histamine H_3-	Receptor as a Target for Developing New Drugs.=	MEDI 0081
n and Testing of Charm Test II	Receptor Assays for the Detection of Antimicrobial R	AGFD 0085
ds.= Synthesis and Substance P	Receptor Binding Activity of a Series of Heterosteroi	MEDI 0158
u+ Structural Studies Leading to a	Receptor Binding Model for Benzapepinone Calci	MEDI 0188
Asialoglycoprotein	Receptor by cGMP.= Regulation of the	BIOL 0154
Site of Neuropeptide Y to the Y2-	Receptor by Using Multiple Peptide Synthesis.=	MEDI 0091
Cloning of an Insensitive GABA	Receptor from Cyclodiene Resistant Drosophila: A	AGRO 0152
ed Quinazolinone CCK-B/Gastrin	Receptor Ligands.= +l Analysis of Alkyl-Substitut	MEDI 0066
ated Cocaine: A Potential Cocaine	Receptor Probe.= +l Biodistribution of Radioiodin	MEDI 0131
of Derivatives of Po+ Adenosine	Receptor Prodrugs: Synthesis and Biological Activity	MEDI 0115
Naltrindole Analogues as δ-Opioid	Receptor Selective Fluorescent Probes.= +hesis of	MEDI 0155
n Biological Ac+ Benzodiazepine	Receptor Studies: Effect of Benzyloxy Substitution o	MEDI 0019
by a Synthetic	Receptor: Complexation of Benzamidinium Salts	ORGN 0061
ible Ligands for the κ-Opioid	Receptor.= Highly Selective Reversible and Irrevers	MEDI 0156
orm of the Murine Erythropoietin	Receptor.= +n and Characterization of a Soluble F	BIOL 0108
of 1,4-Dihydropyridines by Their	Receptor.= +ion of the Stereoselective Recognition	MEDI 0100
Conjugate to H35 Hepatoma a+	Receptor-Mediated Delivery of Methotrexate–Insulin	MEDI 0185
mplexation of Pyrrole–Containing	Receptors and Porphyrin Analogues.= +sis and Co	ORGN 0063
Docking Ligands into	Receptors: The Test Case of α-Chymotrypsin.=	COMP 0039
Pharmacology of Muscarinic	Receptors.= Receptor Activity and Drug Design.	MEDI 0078
Selectively at Dopamine	Receptors.= Development of Agents Acting	MEDI 0080
g of the Dog A_1 and A_2 Adenosine	Receptors.= +nce Analysis and Computer Modelin	MEDI 0114
ment of Naltrindole at δ-Opioid	Receptors.= Studies on the Conformational Require	MEDI 0154
nd NPC13839, as Probes of Sigma	Receptors.= +Aminoalkoxychromones, NPC16377 a	MEDI 0150
e la Industria Qui+ Aplicaciones	Recientes de CAD/CAM para el Diseno de Equipo d	COMP 0053
stics Recycling Processing and +	Reclamation Technologies for Plastics Recycling. Pla	I&EC 0031
Polyaza Clefts for the	Recognition and Cleavage of Phosphodiesters.=	ORGN 0054
	Recognition and Replications in Model Systems.=	ORGN 0137
mplex Contai+ Enantioselective	Recognition by DNA of a Nonintercalating Ru(II) Co	INOR 0233
atives for Formation of Molecular	Recognition Compounds.= +is of Cholic Acid Deriv	ORGN 0057
formation in C+ Carbohydrate–	Recognition Domains as Mechanisms for Decoding In	BIOL 0144
or Can Be Limited to Chain-End	Recognition of an Antigen.= +Epitopes Anywhere,	CARB 0019
Hybrid Molecules.= Molecular	Recognition of Macromolecules: The Utility of	ORGN 0270
Effect.= °From Molecular	Recognition of Piquancy to a New Rule of Medicinal	MEDI 0040
rtunities in Biological Chemistry.	Recognition of RNA Minihelices in Relation to Deco	BIOL 0026
ntification of the Stereoselective	Recognition of 1,4-Dihydropyridines by Their Recep	MEDI 0100
A Simple Method to Define the	Recognition Site of Sequence-Specific DNA-Binding	BIOL 0129
ts a+ Use of Computer Pattern–	Recognition Techniques for Authentication of Extrac	AGFD 0198
Pattern	Recognition through Self-Assembly.=	BIOT 0125
es to Photosyntheti+ Molecular	Recognition via Base Pairing: Noncovalent Approach	ORGN 0059
Biological	Recognition.= Photolithography, Chemistry, and	BIOL 0027

Architectures.= Molecular-	Recognition-Directed Supramolecular Polymer	POLY	0299
How Proteins	Recognize and Bind Oligosaccharides.=	CARB	0012
Evolution of Heavy-Ion	Recoil and Fragment Separators.=	NUCL	0038
	Recoil Studies and High-Energy Reaction Models.=	NUCL	0059
ns.=	Recoil Studies Revisited: Heavy Ion Induced Reactio	NUCL	0060
onoring O. Hannaway. Personal	Recollections of Frank C. Whitmore by Charles D. H	HIST	0009
Controlled Folding of	Recombinant aFGF.= Novel Scalable Process for	BIOT	0134
Production of Antitumor Drugs by	Recombinant Bacteria.= +tibiotic Production—I.	BIOT	0108
Developments in	Recombinant β-Glucosidase of Trichoderma reesei.=	BIOT	0241
Oxidation and Naturation of	Recombinant Bovine Somatotropin.=	BIOT	0135
iochemistry Laboratory.=	Recombinant DNA Analysis in the Undergraduate B	CHED	0090
s a Function of Induction Level in	Recombinant Escherichia coli.= +ngent Response a	BIOT	0034
rocessing Yields for Production of	Recombinant HIV Protease.= +ving Downstream P	BIOT	0071
Characterization of a	Recombinant HIV-2 Protease.= Purification and	BIOL	0042
Mediated Microheterogeneity of	Recombinant Human Erythroproteins by Free Soluti	ANYL	0018
ral Cell Lines+ Glycosylation of	Recombinant Human Plasminogen Expressed in Seve	ANYL	0020
nd Structure Determination of a	Recombinant Human Secretory Phospholipase A_2 fro	BIOL	0072
raphy in Production of	Recombinant Products.= Role of Process Chromatog	ANYL	0039
Interactions for Recovery of	Recombinant Proteins.= Evaluating Starch-Protein	BIOT	0047
duction of an SCD4 Derivative by	Recombinant Streptomyces lividans.= +for the Pro	BIOT	0086
ized Silicon Backbon+ Radiative	Recombination and Vibronic Relaxation in σ-Delocal	PHYS	0138
Electron Transfer and Charge-	Recombination Dynamics in Wild-Type and Mutant	BIOL	0132
Hydride Dimer: Fragmentation-	Recombination Reactions with MR_2 (M = Zn, Mg an	INOR	0383
ed Structure of Pyrrolosine and	Reconfirmation of That of 9-Deazainosine.= Correct	CARB	0046
ded Solids.=	Reconstructed Clusters as Building Blocks for Exten	INOR	0173
e Synthesis of Molecular Sieves:	Recrystallization of the Intermediates Isolated from t	PETR	0019
ss Toward Organic Unimolecular	Rectifiers: Experiments by Scanning Tunneling Micr	BIOT	0005
mbered Oxametallacycles in the	Rections of the Electron-rich Complexes $W(PMe_3)_6$ a	INOR	0086
TEM Analysis of Pillared	Rectorites.=	PETR	0109
Effect of	Recyclability on Intermaterial Competition.=	CMEC	0010
ren+ Structure and Properties of	Recycled Commingled Plastic Processed with Polysty	I&EC	0028
Polymer Composites Using	Recycled PET.=	PMSE	0072
Materials	Recycling Facility.= Operation of a Postconsumer	I&EC	0042
Collection of Plastics for	Recycling from the Residential Community.=	I&EC	0038
lastics. Commin+ Technology of	Recycling of Commingled and Refined Commingled P	I&EC	0025
Rapid	Recycling Preparative Chromatography of Proteins.=	ANYL	0028
ies for Plastics Recycling. Plastics	Recycling Processing and Economics.= +Technolog	I&EC	0031
Container	Recycling Program.= Iowa's Plastic Pesticide	AGRO	0091
Plastics	Recycling via PET Depolymerization.=	I&EC	0032
amation Technologies for Plastics	Recycling. Plastics Recycling Processing and+ Recl	I&EC	0031
d Sortation Technology in Plastics	Recycling. Plastics Recycling: Beyo+ Collection an	I&EC	0037
ogy in Plastics Recycling. Plastics	Recycling: Beyond Beverage Bottles.= +on Technol	I&EC	0037
Pesticide Container	Recycling.=	AGRO	0088
and	Recycling.= Pesticide Container Collection	AGRO	0090
Agricultural-Chemical-Container	Recycling.= Recent Developments in Plastic	AGRO	0045
to	Recycling.= Supercritical Fluid Processing Applied	I&EC	0034
Plastics	Recycling.= Use of Froth Flotation for	I&EC	0036
d Plastics. Commingled Plastics	Recycling—An Equipment Manufacturer's Perspectiv	I&EC	0025
ariety, and Environmental Fac+	Red Raspberry Phenolics: Influences of Processing, V	AGFD	0169
ic Molding Compounds for Infra-	Red Reflow Tolerant Electronic Components.= +ast	CMEC	0014
tein, and Starch in a Milled Hard	Red Winter Wheat.= +id, Total Dietary Fiber, Pro	AGFD	0143
enated Hydrocarbons by Rational	Redesign of Cytochrome P450cam.= +tion of Halog	BIOT	0118
	Redirected Directed Lithiation, III.=	ORGN	0085
condary Metabolite Production by	Redirecting Transport.= +Cultures. Increasing Se	BIOT	0228
Photochemistry and	Redox Catalysis Using Re and Mo Complexes.=	INOR	0343
trochemical Model.= Oxygen	Redox Catalysis Using RuO_2 Dispersions and the Elec	COLL	0134
enium(I+ Ligand Effects on the	redox Chemistry of a Series of (Oxo)(Phosphine)Ruth	INOR	0068
Incorporation of	Redox Compounds into Polypyrrole Films.=	COLL	0013
tions of Conducting Polymers in	Redox Devices and Intelligent Materials Systems.=	POLY	0326
Lipid Oxidation in Foods via	Redox Iron.=	AGFD	0094
c Systems. Metal Polypyridines as	Redox Photosensitizers.= +osensitive Metalorgani	INOR	0282
dox Reagents.=	Redox Potential Alteration under the Sorption of Re	COLL	0103
Photochemistry of Colloidal Sil+	Redox Properties of Nonmetallic Silver Clusters and	COLL	0130
ne Acetyltransferase and High+	Redox Reactive Choline Analogue Inhibitors of Choli	MEDI	0142
the Sorption of	Redox Reagents.= Redox Potential Alteration under	COLL	0103
Transition-Metal Complexes and	Redox Sites in Enzymes.= +sfer Reactions between	INOR	0283
=	Redox-Active Early Metallocycles as Metalloligands.	INOR	0025
	Redox-Switched Molecular Aggregates.=	COLL	0020
Electrodos Metalic+ Estudio de	Reduccion Electroquimica del 1-Nitro-Naftaleno con	PHYS	0280
t or UV Illumination May Help	Reduce Decay in Seal-Packaged Citrus and Pepper F	AGFD	0076
Magnetic Separation of	Reduced Crude Conversion Catalyst.=	I&EC	0063
actant Sol+ Electrodeposition of	Reduced Forms of Methylviologen from Anionic Surf	COLL	0077
Properties of a	Reduced Molybdenum-Pterin Complex.=	INOR	0277
Novel Quasi-Low-Dimensional	Reduced Niobium Oxophosphate Compound: Synthe	INOR	0031
Transgenic Tomatoes with	Reduced Polygalacturonase.= Evaluation of	AGFD	0057
in+ Lcoal Structural Analysis of	Reduced $TiO_2(110)$ and $TiO_2(001)$ Surfaces by Scann	COLL	0171
Investigation of	Reduced V_2O_5 Xerogel.=	INOR	0082
dryl Content of Glycinin: Effect of	Reducing Agents in Absence of Denaturants.= +hy	AGFD	0141
n Products as a Result of Sulfate-	Reducing Bacteria.= +s in Copper Sulfide Corrosio	BIOT	0100
re and the Role of Biodiversity in	Reducing Pesticide Use.= +ng Sustainable Agricultu	AGRO	0082

of Inactivation of Ribonucleotide	Reductase by 2'-Azido 2'-Deoxynucleotides.= +ism	ORGN	0281
teres: Hydroxy Acetic Acid Aldose	Reductase Inhibitors.= +dress. Hydantoin Bioisos	MEDI	0082
xy-3-Methylglutaryl-Coenzyme A	Reductase Inhibitors.= +iophene-Derived 3-Hydro	MEDI	0101
urse of the Methyl Coenzyme M	Reductase Reaction in Methanobacterium thermoaut	ORGN	0025
rone.= Inhibitors of Steroid 5α-	Reductase Which Are Uncompetitive versus Testoste	MEDI	0008
adical Cofactor of Ribonucleotide	Reductase.= +the Dinuclear Iron Center: Tyrosyl R	BIOL	0022
llylic Rearrangement and Enoyl	Reduction Catalyzed by Brevibacterium ammoniagen	BIOL	0065
Aromatics	Reduction in Gas Oil.= Industrial Catalysts for	CATL	0048
Catalytic	Reduction in the Laboratory.=	CHED	0168
-Hydrox+ Ligand Exchange and	Reduction Kinetics of Oxochromate(V) Complex by α	INOR	0402
$Cu_2O(100)$.= Oxidation and	Reduction of Allyl Alcohol and 1-Propanol on	COLL	0195
echanistic Studies of E_3-Mediated	Reduction of CDP-6-Deoxy-$\Delta^{3,4}$-Glucoseen.= +M	BIOL	0100
ation and Enzymic or Nonenzymic	Reduction of Fe(III).= +rnosine on Fe(II) Autoxid	AGFD	0131
tural Changes Associated with the	Reduction of $Fe_2(PPh_2)_2(CO)_6$.= +udy of the Struc	ANYL	0106
Alkaline pH.= Photo-Induced	Reduction of Native Cytochrome c Peroxidase(III) in	BIOL	0097
in the Presence of Sodium N,N+	Reduction of Niobium and Tantalum Pentachlorides	INOR	0092
Zeolites.= Selective Catalytic	Reduction of Nitric Oxide by Ammonia over Fe-Y	COLL	0212
Selective Catalytic	Reduction of Nitric Oxide with Ammonia.=	COLL	0207
ts Used for the Selective Catalytic	Reduction of NO.= +onia on V_2O_5/TiO_2 Catalys	COLL	0211
Temperatur+ Selective Catalytic	Reduction of NO_x Using Zeolitic Catalysis for High-	I&EC	0046
esticide Application Systems for	Reduction of Rinsate and Nontarget Contamination.	AGRO	0105
ing Fluids on the Electrochemical	Reduction of $Ru(NH_3)_6Cl_3$.= +in Naturally Occurr	COLL	0005
Fast Cleavage Reactions Following	Reduction of Some α-Substituted Nitroalkanes.=	ANYL	0107
Manufacture of Disperse Blue +	Reduction of Toxic Components and Wastewaters in	CMEC	0018
Microbial	Reduction of Uranium.=	GEOC	0008
u(I) Complexes: Competition be+	Reduction of Vicinal Dihalides by Photogenerated R	INOR	0284
tereochemistry of Nickel Boride	Reduction of 2,6-Dihalo-9-Oxabicyclononanes and R	ORGN	0246
c Method and That Based on the	Reduction of 3,5-Dinitrosalicylate in Determining th	BIOL	0111
Control of Face Selection in the	Reduction of 5-Azaadamantan-2-one.= Electronic	ORGN	0237
eparation of Colloidal Metals by	Reduction or Precipitation in Microheterogeneous Li	COLL	0128
ntaining Silica Gel+ Oxidation-	Reduction Processes on the Surface of Chromium-Co	COLL	0101
k Electron Transfer and Substrate	Reduction.= +omplexes: Competition between Bac	INOR	0284
Asymmetric	Reductions with Chiral NADH Models.=	ORGN	0264
Evolution of Metal-Ammonia	Reductions.= Synthetic Steroids and the Luftwaffe:	HIST	0013
g Borohydride Reagents.= A	Reductive Amination-Lactamization Procedure Usin	ORGN	0242
raction from Soils and Sediments:	Reductive Dissolution by Ti(III).= +Iron Oxide Ext	ENVR	0058
tal+ Phase-Transfer-Catalyzed	Reductive Elimination of HCl from Chlorohydridome	INOR	0381
lases—II. How Do Trichoderma	reesei Cellobiohydrolases Bind to and Degrade Cellul	BIOT	0206
β-Glucosidase of Trichoderma	reesei.= Developments in Recombinant	BIOT	0241
Formation by Trichoderma	reesei.= Molecular Regulation of Cellulase	BIOT	0183
Trichoderma	reesei.= Novel Inducers for Cellulase Production by	BIOT	0184
t in Thermoreversible Polymer+	Reexamination of the Freezing and Melting of Solven	POLY	0170
y of Recycling of Commingled and	Refined Commingled Plastics. Commin+ Technolog	I&EC	0025
Research Results—	Refined Commingled Processing.=	I&EC	0029
isk Assessment.=	Refinement of Animal Toxicity Studies for Human R	CHAS	0019
n Analysis for Characterization of	Refinery Streams and Heavy Oils.= +nt Distributio	PETR	0011
of Biomass	Refining.= Impact of Byproducts on the Economics	BIOT	0014
s with Fourier Transform Infrared	Reflectance Spectroscopy.= +Profiling of Polymer	PMSE	0153
- and Far-Infrared IR Specular	Reflectance Study of Oxidized and Exchanged Polyet	POLY	0253
hloride and 1,5-Hexadiyne behind	Reflected Shock Waves.= +actions of Propargyl C	FUEL	0068
Aspects.= Historical	Reflections on Steroids: Lederle and Personal	HIST	0023
olding Compounds for Infra-Red	Reflow Tolerant Electronic Components.= +astic M	CMEC	0014
ene Glycol-Enhanced Protein	Refolding: Model and New Applications.= Polyethyl	BIOT	0136
Teachers: The Need for	Reform.= Educating High School Chemistry	CHED	0067
tallic Catalysts Toward Catalytic	Reforming: The Reaction of Ethylene with M-Ru/Si	CATL	0042
eolite Catalyst for Light Naphtha	Reforming.= +w Modification Method of Pt/KL Z	PETR	0064
aterials with Emphasis on High	Refractive Index and Abrasion-Resistance Coatings.=	POLY	0264
de+ Photochemically Delineated	Refractive Index Profiles in Polymeric Slab Wavegui	POLY	0180
Lipid Oxidation during Cooking of	Refrigerated Minced Channel Catfish Muscle.=	AGFD	0124
klime Binder-Enhanced Densified	Refuse-Derived Fuel/Coal Mixtures.= +ion of Quic	FUEL	0140
paration Procedure and Reaction-	Regeneration Cycles.= +s in the Course of the Pre	PETR	0046
iza+ Deactivation by Coking and	Regeneration of Pd/HM Catalysts for Alkane Isomer	PETR	0130
of Fischer Carbene Complexes +	Regio- and Stereoelectronic Effects in the Reactions	INOR	0194
High School-College Interface:	Regional Instrumentation Laboratories and Inservice	CHED	0112
Induce Immune Unresponsiven+	Regions of Modified Cobra Venom Neurotoxin That	BIOL	0086
cal Processes in the Single-Phase	Regions of SDS/1-Pentanol/Cyclohexane/Water Mi	COLL	0024
Interfacial	Regions.= Dynamics of Coupling Agents in	POLY	0368
ons to 4-Amino-Substituted Nitrile-Stabilized Carbani	Regioselective Addition of	ORGN	0010
anes.=	Regioselective Carbon-Carbon Bond Cleavage of Cub	ORGN	0256
olyhydroxylated Compounds.=	Regioselective Chemo-Enzymatic Modifications of P	BIOT	0064
rimethylsilyl Reagents Using t+	Regioselective Substitution of 1,2-Propanediol with T	ORGN	0051
chromium Compounds.=	Regioselective Synthesis of (erta$_6$-Arene)Tricarbonyl	ORGN	0214
olar Synthons: Nickel(O)-Catal+	Regiospecific Aromatic Substitution Via Aryl 1,2-Dip	ORGN	0082
hing HPLC Method To Support	Registration of DPX-E9636 Herbicide for Field Corn	AGRO	0030
Molecular Property Data for CAS	Registry Substances.= +enerated 3D and Related	COMP	0041
earch Routines for Searching CAS	Registry Substances.= +on of Atom-by-Atom 3D S	CINF	0023
Nonlinear	Regression.= Educating Chemists about	CHED	0193
State Pesticide Disposal	Regulations and Programs.=	AGRO	0106
Pesticide Container	Regulations as Part of EPA Strategy.=	AGRO	0089
Can Hazardous Material	Regulations Be Met in a Cost-Effective Manner?.=	CHAS	0033

Evolution of	Regulations for Pesticide Rinsates.=	AGRO	0104
Management: The Impact of New	Regulations on Fertilizer Manufacture.= +s Safety	FERT	0016
n of Transparent Glass Fiber	Reinforced PMMA Matrix Composites.= Preparatio	PMSE	0109
c Mechanical Analysis Studies of	Reinforced Polyvinyl Chloride (PVC) Roofing Memb	PMSE	0180
and Inorganic Hybrid Materials.	Reinforcing Zirconia Particles Precipitated into an E	POLY	0283
Synthesis and Pharmacological	Reinvestigation of Chiral Muscarones.=	MEDI	0141
The IDITIS	Relational Database of Protein Structure.=	CINF	0032
and Transverse Energies in	Relativistic Heavy Ion Collisions at 14.6 GeV/c per N	NUCL	0051
Dissociation with	Relativistic Heavy Ions.= Electromagnetic	NUCL	0049
us–Nucleus Collision Mechanisms:	Relativistic Phenomena. Synthesis of High+ Nucle	NUCL	0046
Evidence for Energy Transfer and	Relavtivistic Gold–Gold Covalent Bonding.= +lts:	INOR	0189
ino–Ammonium+ Novel Muscle	Relaxants of Ester–Containing bisQuaternary Piperid	MEDI	0047
Liquids+ Study of Excited–State	Relaxation and Photoinitiated Reaction Kinetics in	ANYL	0034
ms.=	Relaxation and Reactions in Low–Dimensional Syste	PHYS	0132
st Optical Studies of Vibrational	Relaxation and Structural Dynamics of Heme Protei	PHYS	0169
Electronic Energy	Relaxation at Interfaces.=	PHYS	0062
c Measurements of Excited–State	Relaxation at Interfaces.= +econd Optical Harmoni	COLL	0087
Systems.= Magnetic	Relaxation Coupling in Dynamically Heterogeneous	PHYS	0232
Fast Vibrational	Relaxation for Polar Solutes in Polar Solvents.=	PHYS	0037
Analysis+ Energy Transfer and	Relaxation in Condensed Phases. Chord–Distribution	PHYS	0060
Dynamics+ Energy Transfer and	Relaxation in Condensed Phases. Conformation and	PHYS	0107
d Relaxat+ Energy Transfer and	Relaxation in Condensed Phases. Energy Transfer an	PHYS	0080
cture and+ Energy Transfer and	Relaxation in Condensed Phases. Excited–State Stru	PHYS	0018
lysis of th+ Energy Transfer and	Relaxation in Condensed Phases. Time–Domain Ana	PHYS	0034
ir Organi+ Energy Transfer and	Relaxation in Condensed Phases. Torsional Solitons	PHYS	0129
sed Phases. Energy Transfer and	Relaxation in Molecular Glasses and Polymer Solids.	PHYS	0080
tive Recombination and Vibronic	Relaxation in σ–Delocalized Silicon Backbone Netwo	PHYS	0138
Hydrox+ ^{13}C Nuclear Magnetic	Relaxation of Amylose and Dynamic Behavior of the	CARB	0049
ingle Racemic Poly(vinyl Chlori+	Relaxation of Runs of Trans Rotational Isomers in S	POLY	0069
nt Interactions and Spin–Lattice	Relaxation of Sodium Cations in Zeolite–Y Probed b	PHYS	0267
Second–Order NLO and	Relaxation Properties of Poled Polymers.=	POLY	0153
r Interactions on Polyme+ NMR	Relaxation Study of the Influence of Charge Transfe	PMSE	0076
by Tra+ Vibrational Population	Relaxation Time of Ions in Protic Solvents Measured	PHYS	0083
Plasma Samples R+ Transverse	Relaxation Times ('T2's) of Water Protons in Human	MEDI	0043
Relation between T$_1$ and T$_2$	Relaxation Times.=	PHYS	0044
to Spin Glass	Relaxation.= Dispersive Energy Transport: Relation	PHYS	0082
amics of Proteins and+ NMR	Relaxation, Fluorescence Depolarization, and the Dyn	PHYS	0166
Iron–Ethylene D+ Paramagnetic	Relaxations of Water Protons in the Solutions of the	INOR	0090
and Analytical Data.= Has a	Release Been Observed? HRS–Observed Releases	ENVR	0080
tle Administered via a Sustained	Release Captec Device.= +ues of Albendazole in Ca	AGRO	0109
Controlled–	Release Nitrogen Analysis of Coated Fertilizers.=	FERT	0008
ulture and Agricult+ Controlled–	Release Nitrogen Solutions and Their Uses in Hortic	FERT	0005
of Controlled–	Release Nitrogen Sources.= Environmental Benefits	FERT	0006
ing Wood Degradation+ In Situ	Release of Polysaccharide Hydrolyzing Enzymes Dur	BIOT	0208
ion and Evaluation of Controlled–	Release Properties.= +ne and Metribuzin: Preparat	AGRO	0063
Sediments.= N:P	Release Ratios from Freshwater and Coastal Marine	GEOC	0012
nolayers: Uptake, Hydrolysis, and	Release.= +Substrates with Human Fibroblast Mo	BIOL	0079
n an Immobili+ Contribution of	Released Cells to Overall Cyclosporin A Productivity	BIOT	0155
Observed? HRS–Observed	Releases and Analytical Data.= Has a Release Been	ENVR	0080
Studi+ Short–Lived Molecules of	Relevance in the Earth's Atmosphere: Experimental	PHYS	0017
ficant Carcinogens and Mutagens:	Relevance to Human Cancer Induction. +tally Signi	CHAS	0007
	Reliability Theory Applied to the Process Industry.=	I&EC	0071
iral Rhodium(II) Carboxamides:	Remarkably Effective Catalysts for Enantioselective	CATL	0002
Achieving Zero Discharge of Pe+	Remedial Action Plan: Program for the Great Lakes.	ENVR	0083
ussion—Public Involvement in the	Remedial Action Planning Process.= +–Table Disc	ENVR	0086
esting the Use of Landfarming to	Remediate Pesticide–Contaminated Soil Excavated f	AGRO	0144
te+ Biological Soil Treatment to	Remediate Petroleum– and Naphthalene–Contamina	BIOT	0139
c Futur+ Biodegradation in Site	Remediation and Waste Treatment: Past, Present, an	AGRO	0121
Extensive Analysis and Search for	Remediation Technologies.= +g Plants in Quebec:	ENVR	0016
il Dealership Site Assessment and	Remediation.= +ent Needs for Agrochemical Reta	AGRO	0142
Agricultural Chemical Site	Remediation/Regulations.=	AGRO	0143
Cont+ Seasonal Variations in the	Remineralization of Sedimentary Organic Matter in	GEOC	0015
Chemistry—A	Reminiscence.= Research in Industrial Phosphorus	INOR	0322
lary Gas Chromatography Using a	Remote Flow Cell.= +UV–VIS Detection for Capil	ENVR	0060
ing sCD4–183 Inclusion Bodies to	Remove E. coli Proteins and En+ Methods of Wash	BIOT	0171
ive Adenosine A$_1$ Antagonist with	Renal Protective and Diuretic Activities.= +Select	MEDI	0112
roacetylene Using a New Volume–	Rendering Molecular Graphics Technique.= +ifluo	COMP	0055
Elaboracion de Pastas de Alto	Rendimiento: Influencia de la Sulfonacion sobre las P	PMSE	0077
ructure–Activity Relationships of	Renin Inhibitors with Cyclohexyl Moiety Directly At	MEDI	0121
of Small Peptides.–	Renin Inhibitors: Factors Influencing the Absorption	MEDI	0197
Σ Active Cysteine+ Macrocyclic	Renin Inhibitors: Synthesis of a Subnanomolar, Orall	MEDI	0122
Well–Absorbed	Renin Inhibitors.= A New Class of Efficacious,	MEDI	0120
Diester Macrocyclic Human	Renin Inhibitors.= Highly Potent, Orally Active	MEDI	0123
Cyclohexylalanine at P1 Site by	Renin: Kinetic Studies Using New Fluorogenic Subst	BIOL	0035
Optical Spectroscopies. Sample	Reorientation for High–Resolution NMR of Solids: D	PHYS	0370
Surfaces.= Picosecond	Reorientation of Adsorbates on Alkylsilane/Silica	ANYL	0025
Width Study of the Anisotropic	Reorientation of Vanadyl Acetylacetonate in Toluene	PHYS	0264
	Repair Systems and Oxidative Stress.=	CHAS	0009
s for UvrABC Endonuclease DNA	Repair.= +n–Linked Deoxyoligonucleotide Synthesi	BIOL	0126
drochloride: A Phenethanolamine–	Repartitioning Agent.= +idues of Ractopamine Hy	AGRO	0131

s of a Bicyclic Lactone Mosquito	Repellent, 1,1,4,5,6,7,8,8A–Octahydro–3H–2–Benzopy	AGRO 0033
Studies of the T4 DNA	Replication System.= Kinetic and Structural	BIOL 0159
Recognition and	Replications in Model Systems.=	ORGN 0137
ations: Mechanistic Studies on the	Replicative Bypass of DNA Photoproducts.= +Mut	BIOL 0157
Vivo and in Vitro Studies of the	Replicative Bypass of the TpdU cis–syn Cyclobutane	BIOL 0113
Mechanistic Studies of the	Reppe Cyclotetramerization of Acetylene.=	INOR 0331
Folding of Trp	Repressor Subdomains.=	BIOL 0115
d+ Factores Que Incrementan la	Reprobacion en las Materias Biologicas de la Carrera	CHED 0092
Longitudinal de Indices de	Reprobacion en Materias Biologicas.= Analisis	CHED 0091
um Ribulose Bispho+ Chemical	Rescue of a Catalytically Deficient Mutant of R. rubr	BIOL 0089
e Microscopy to Coal–Derived	Resid Characterization.= Application of Fluorescenc	FUEL 0043
for Burial Fluxes and Phosphorus	Residence Time in the Ocean.= +nts: Implications	GEOC 0013
Recycling from the	Residential Community.= Collection of Plastics for	I&EC 0038
as Chromatography of Crudes and	Residua Using Atomic Emission Detection.= +re G	PETR 0121
s and Veterinary Drugs and Their	Residual Analysis in Japan.= +ew of Feed Additive	AGFD 0102
c Acid) Precursors and Their Po+	Residual Stress Behavior of Functionalized Poly(ami	POLY 0085
thoxine and Sulfamonomethoxine	Residual Studies in Chicken Tissues and Eggs.=	AGFD 0127
ples, Grapes, and Peaches.=	Residue Analyses of Lepidopteran Pheromones in Ap	AGRO 0148
t+ ELISA and Its Application for	Residue Analysis of Antibiotics and Drugs in Produc	AGFD 0083
and Conduct of Studies to Meet	Residue Chemistry Requirements of the NADA: Resi	AGRO 0078
try Requirements of the NADA:	Residue Depletion and Metabolism of Flunixin in Ca	AGRO 0078
Dairy Cows.=	Residue Depletion and Metabolism of Luprostiol in	AGRO 0130
Chicken.= Tissue	Residue Depletion Studies on Semduramicin in the	AGRO 0079
rida Indian River Lag+ Pesticide	Residue in Barrier Island Salt Marshes along the Flo	ENVR 0019
s: Intermediate Energies. Heavy	Residue Properties in Intermediate–Energy Nuclear	NUCL 0024
oducing Animals. Metabolism and	Residue Studies of Pirlimycin in the Dairy Cow.=	AGRO 0108
r Weight Distribution of Vacuum	Residue Using Mass Spectrometry: A Heterogeneous	PETR 0150
thod Evaluation Crit+ Emerging	Residue–Analysis Techniques. Pesticide–Residue–Me	AGRO 0145
Analysis Techniques. Pesticide–	Residue–Method Evaluation Criteria: State's Perspec	AGRO 0145
ical Modifications of Specific AA	Residues Change the Lytic Activity of a Cobra Veno	BIOL 0085
and Subsequent Analysis of Drug	Residues in Animal Tissues.= +to the Extraction	AGFD 0060
m M+ Confirmation of Antibiotic	Residues in Edible Meats by On–Line HPLC Tande	AGFD 0061
tion of Matrix+ Antibiotic/Drug	Residues in Food Products of Animal Origin. Applica	AGFD 0060
t Problems As+ Antibiotic/Drug	Residues in Food Products of Animal Origin. Curren	AGFD 0081
Chromatograp+ Antibiotic/Drug	Residues in Food Products of Animal Origin. Liquid	AGFD 0125
the Detection of Antibiotic/Drug	Residues in Food.= +ent Problems Associated with	AGFD 0081
etection and+ Antibiotic/Drug	Residues in Foods of Animal Origin. Approaches to D	AGFD 0099
Incidence of Antibiotic/Drug	Residues in Foods of Animal Origin.=	AGFD 0082
assay of Antibiotic/Drug	Residues in Foods of Animal Origin.= Radioimmuno	AGFD 0084
of Analytical Procedures for Drug	Residues in Foods.= +s and Accurate Descriptions	AGFD 0126
biotics: Distinct Origins of Proline	Residues in L–671,329 and L–688,786.= +ype Anti	MEDI 0058
for the Detection of Antimicrobial	Residues in Meat.= +arm Test II Receptor Assays	AGFD 0085
rmination of β–Lactam Antibiotic	Residues in Milk and Tissues.= +proaches to Dete	AGFD 0125
tection and Confirmation of Drug	Residues in Milk.= +mal Origin. Approaches to De	AGFD 0099
Transformation of Solid Urban	Residues in Usable Products.=	CHED 0105
S Pharmacokinetics and Tissue	Residues of Albendazole in Cattle Administered via a	AGRO 0109
Carcasses with Positive Injecti+	Residues of Anabolic Steroids in Muscle Tissues from	AGFD 0130
aphic Procedures for Determining	Residues of Penicillins.= +mparison of Chromatogr	AGFD 0063
nolami+ Metabolism and Tissue	Residues of Ractopamine Hydrochloride: A Phenetha	AGRO 0131
Metabolism and Tissue	Residues of Tilmicosin in Cattle.=	AGRO 0110
Hormone and Related Compound	Residues: Analytical and Toxicological Aspects.=	AGRO 0129
Analysis of Veterinary Drug	Residues.= Immunochemical Methods in the	AGFD 0064
earing α–Hydroxy–β–Amino Acid	Residues.= +tric Synthesis of Peptide Fragments B	ORGN 0048
and Sulfur Content in Petroleum	Residues.= +mination of Hydrocarbon Distribution	PETR 0120
Analysis of Heavy Petroleum	Residues.= TPOP–IRMS Stable Carbon Isotope	PETR 0141
ally Modified Polymeric–Mercuric	Resin in Capillary Gas Chromatography.= +Chemic	ANYL 0083
minoethyl Methacrylated)/Epoxy	Resin IPNs.= +ly(Methyl Acrylate–co–Dimethyl A	PMSE 0079
A–based Cyanate Ester	Resin Systems.= Characterization of Bisphenol	PMSE 0152
Thermosetting	Resin.= Polydicyclopentadiene: A New RIM	POLY 0136
ation of CTBN–Toughened Epoxy	Resin/Castor Oil Polyurethane IPN.= +haracteriz	PMSE 0092
Carbohydrates with Ion–Exchange	Resins and Hydrophobic Adsorbents.= +actions of	CARB 0063
Backbone Flexibili+ Novel Epoxy	Resins Based on Alkylenedioxydiphenols—Effect of	PMSE 0162
ibilit+ Novel Waterborne Epoxy	Resins for Can and Coil Coatings with Improved Flex	PMSE 0142
Utilization of Cation–Exchange	Resins for Chromatographic Separation of Carbohydr	CARB 0064
o Two Heavily Fluorinated Epoxy	Resins from a Fluorodiimidediol.= +Novel Route t	POLY 0037
ary Lecture. Application of Epoxy	Resins in Container Coatings: An Overview.= +en	PMSE 0157
Powdered Ion–Exchange	Resins in Sugar Purification.=	CARB 0065
erization of Unsaturated Polyester	Resins with Low–Profile Additives.= +ogy Charact	POLY 0114
Resorcinol–Formaldehyde	Resins.= Cure Characteristics of Tannin–Based	POLY 0138
sing Neutral and Anion Exchange	Resins.= +Sour Orange (Citrus aurantium) Juice U	AGFD 0106
zation of Silicone–Epoxide	Resins.= Electron Beam–Induced Cationic Polymeri	PMSE 0159
Water–Compatible Phenolic	Resins.=	PMSE 0139
Antagonism of GABA–Based	Resistances in Diabrotica.= Phytochemical	AGRO 0136
usceptible and Fusarium Yellows–	Resistant Celery Cultivars.= +c Furocoumarins in S	AGRO 0040
ABA Receptor from Cyclodiene	Resistant Drosophila: A Model System in a Model In	AGRO 0152
roteins Extracted from Aflatoxin–	Resistant/Susceptible Varieties of Corn.= +ionic P	AGFD 0134
ical Vapor Deposition of Low–	Resistivity Metals for Interconnection of Electronic D	INOR 0126
nd–Mapping Studies with a High–	Resolution Crystal Structure.= +mparison of Liga	CARB 0013
ies of Adsorbed Species+ High–	Resolution Electron–Energy–Loss Spectroscopy Stud	COLL 0173
ies of High–Temperatur+ High–	Resolution Electron–Energy–Loss Spectroscopy Stud	PHYS 0279

ght Petroleum Streams by High-	Resolution GC with Chemiluminescence Detection.=	PETR	0124
f Flyash+ Application of High-	Resolution MS and Hybrid MS/MS to the Analysis o	ANYL	0050
s. Sample Reorientation for High-	Resolution NMR of Solids: DAS and DOR.= +pie	PHYS	0370
ethosulfate as an Inclusion Host:	Resolution of Alcohols by Inclusion Compound Form	ORGN	0041
The First Kinetic	Resolution of Chiral Azides.=	ORGN	0043
Procedure for High-	Resolution Soil Gas Analysis.= A Sorbent Trapping	ENVR	0050
e- and Two-Dimensional High-	Resolution Solid-State NMR Investigations of Zeolit	PHYS	0348
and Corrosion at Atomic Layer	Resolution Studied by in Situ Electrochemical Scann	PHYS	0127
xperimental Verific+ Enzymatic	Resolution Using a Multiphase Membrane Reactor: E	BIOT	0042
Land-Use Planning and Conflict	Resolution.= +Gaps in Development of a Model for	ENVR	0043
Optical Spectroscopy at 1-MHz	Resolution.=	PHYS	0043
Bounds of Sensitivity and Spatial	Resolution.=	ANYL	0012
otent Acetylcholine+ Syntheses,	Resolution, and Structure-Activity Relationships of P	MEDI	0143
ying the Substrate and by Linking	Resolutions in Series.= +s and Esterases by Modif	BIOT	0091
xciton Micro+ Toward Spatially	Resolved Energy Transfer and Scanning Molecular E	PHYS	0022
al Alkyl Bonds Studied by Time-	Resolved Infrared Spectral Techniques.= +n of Met	INOR	0169
Laboratory.= Time-	Resolved Laser Spectroscopy in the Undergraduate	PHYS	0249
Laboratory.= Time-	Resolved Laser Spectroscopy in the Undergraduate	CHED	0084
n Photoactive Polymers by Time-	Resolved Microwave Conductivity (TRMC).= +e i	POLY	0214
onal Energy T+ Quantum-State-	Resolved N_2O Vibrational, Rotational, and Translati	PHYS	0334
Helium-Neon Laser.= Time-	Resolved Thermal Lens Calorimetry with a	CHED	0089
ents and Applications of Magnetic	Resonance and Optical Spectroscopies. +Developm	PHYS	0136
ents and Applications of Magnetic	Resonance and Optical Spectroscopies. +Developm	PHYS	0193
ents and Applications of Magnetic	Resonance and Optical Spectroscopies. +Developm	PHYS	0113
ents and Applications of Magnetic	Resonance and Optical Spectroscopies. +Developm	PHYS	0372
ents and Applications of Magnetic	Resonance and Optical Spectroscopies. +Developm	PHYS	0066
ents and Applications of Magnetic	Resonance and Optical Spectroscopies. +Developm	PHYS	0166
ents and Applications of Magnetic	Resonance and Optical Spectroscopies. +Developm	PHYS	0040
ents and Applications of Magnetic	Resonance and Optical Spectroscopies. +Developm	PHYS	0370
ents and Applications of Magnetic	Resonance and Optical Spectroscopies. +Developm	PHYS	0373
ents and Applications of Magnetic	Resonance and Optical Spectroscopies. +Developm	PHYS	0344
at+ Improved Nuclear Magnetic	Resonance Capabilities for the Undergraduate Labor	CHED	0185
Electron Paramagnetic	Resonance Detection of Phenolic Free Radicals.=	AGFD	0004
Gases.= Magnetic	Resonance Experiments with Laser-Polarized Rare	PHYS	0041
cular Probes for F-19 Magnetic	Resonance Imaging: Synthesis and Characterization o	MEDI	0041
Tool: Instru+ Sputter-Initiated	Resonance Ionization Spectroscopy as an Analytical	NUCL	0042
erconductors by Electron-Spin-	Resonance Line Broadening of Surface Paramagnetic	PHYS	0050
perconductors by Electron Spin	Resonance Line Broadening of Surface Paramagnetic	PHYS	0284
Glasses.= Magnetic	Resonance Probes of Short-Range Order in Network	PHYS	0374
methane in Sol+ High-Overtone	Resonance Raman Spectra of Photodissociating Nitro	PHYS	0302
	Resonance Raman Spectroscopy of Simple Amides.=	PHYS	0321
on Zeolites.= UV	Resonance Raman Spectroscopy of Species Adsorbed	COLL	0075
hodobacter sphaeroides Reacti+	Resonance Raman Studies of the Q_y Transitions in R	BIOL	0094
s Studied by ^{2}D Nuclear Magnetic	Resonance Spectroscopy.= +r with Clay Minerals a	AGRO	0061
Advanced Electron Paramagnetic	Resonance Techniques for the Study of Coal Chemi	FUEL	0072
Teaching Magnetic	Resonance Using Computer Animation.=	CHED	0123
ossil Fuels: Isotopes and Magnetic	Resonance. Hy+ Novel Analytical Techniques for F	FUEL	0070
Proteins.= Rotational	Resonance: Distance Measurements in Large	PHYS	0196
nd Coordination via Electron Spin	Resonance.= +ucts with the cupric Ion Intensity a	CATL	0017
Electron Paramagnetic	Resonance.= New Applications of Coherent Pulsed	PHYS	0116
Magnetic	Resonance.= New Approaches to Ultrasensitive	PHYS	0042
Phosphorus-31 Nuclear Magnetic	Resonance.= +Analysis of Inorganic Phosphates by	INOR	0291
Two-Dimensional Electron Spin	Resonance.=	PHYS	0114
ve Excitation of Nuclear Magnetic	Resonances.= +ics at Catalyst Surfaces by Selecti	PHYS	0346
Survey of Important Fermi	Resonances.=	ANYL	0059
amic Light-Scattering Studies of	Resorcinol Formaldehyde Gels as Precursors of Orga	POLY	0199
Characteristics of Tannin-Based	Resorcinol-Formaldehyde Resins.= Cure	POLY	0138
chers+ ChemSource: A Teaching	Resource for Preservice and Inservice Chemistry Tea	CHED	0137
urce. ChemSource: A Teaching	Resource for the 90s.= Minisymposium on ChemSo	CHED	0017
ed Management of Environmental	Resources and Impacts.= +ementation of Risk-Bas	ENVR	0076
n.=	Resources for Material Safety Data Sheet Preparatio	CHAS	0040
Hazard Communication	Resources, Inc.= Toxicological Services Provided by	CHAS	0024
oring: Opportunities and	Responsibilities.= Human Biomonitoring. Biomonit	CEI	0015
Metal-	Responsive Gene Expression.=	INOR	0012
58: A Novel Conformationally	Restricted Class III Antiarrhythmic Agent.= L-702,9	MEDI	0096
ed Enzyme System in the Water-	Restricted Environment of Nonaqueous Solvents.=	BIOL	0091
and Optical Nonlinearities in	Restricted Geometries.= Exciton Coherence-Size	PHYS	0038
dimers and Their Utilization as	Restricted Reaction Spaces for Electron Transfer Pro	POLY	0238
-Liq+ Organic Photochemistry in	Restricted Reaction Spaces: Chemistry at the Liquid	ORGN	0126
alysts and the Design of Artificial	Restriction Enzymes.= +A Transesterification Cat	INOR	0406
reogenic Atoms by the Endocyclic	Restriction Test.= +philic Substitutions at Nonste	ORGN	0120
Use of Most	Restrictive Paths in 3D Search Strategy.=	CINF	0024
Expression of	Restrictocin in Filamentous Fungi.=	BIOT	0238
cademico de los Alumnos en los	Resultados de la Evaluacion de una Materia Biologic	CHED	0093
para Enriquecimiento en Deuter+	Resultados Experimentales de un Empaque Integral	I&EC	0073
°From	Resume to Job.=	I&EC	0005
°From	Resume to Job.=	CMEC	0026
lopment Needs for Agrochemical	Retail Dealership Site Assessment and Remediation.	AGRO	0142
Fertilizer and Soil Chemistry. The	Retail Fertilizer Dealer as Environmentalist.= +in	FERT	0017
	Retail Fertilizer Dealer Containment.=	FERT	0020

ater and Rinsate Management for	Retail Fertilizer Dealers.= +l Solutions for Wastew	FERT	0014
Phosphorus Flame	Retardants for a Changing World.=	INOR	0346
Compounds as Flame	Retardants.= Syntheses of Some Organoboron	ORGN	0262
Cascadas	Reticulares.=	NUCL	0087
ual Information Processing of the	Retina.= +lecular Membranes Modeled on the Vis	BIOT	0124
-Hyd+ Cancer Chemopreventive	Retinoid Metabolites: C-Glucuronide Analogues of 4	MEDI	0030
The Salamander and Crossed	Retorts: The Chemists' Club of New York.=	HIST	0031
Reaction	Retrieval from Databases.=	CINF	0004
Simulation.= Database	Retrieval Techniques for ^{13}C NMR Spectrum	COMP	0004
et Lab, and Nathan Sugarman—A	Retrospective.= +man. Nuclear Chemistry, the M	NUCL	0058
asible Solutions for Disposal and	Reuse of Pickle Liquors from Mexican Iron Industry	ENVR	0039
esticide Rinsate Minimization and	Reuse.= +tural Chemicals and Their Containers. P	AGRO	0102
Container Minimization and	Reuse.=	AGRO	0087
ization Can Result in Significant	Reversal of Hydration and Polymerization in the Ma	ENVR	0041
Reactions of CdS in Micelles and	Reverse Micelles.= +esis and Photoelectrochemical	COLL	0148
I-RG-587: A Potent and Selective	Reverse Transcriptase Inhibitor of HIV-1.= +of B	ORGN	0027
ns by Free Solution and Charge-	Reversed Free Solution Capillary Electrophoresis.=	ANYL	0018
Affinity-Based	Reversed Micellar Protein Extraction.=	BIOT	0045
in Extraction and Recovery from	Reversed Micelles through Pressure-Induced Micella	BIOT	0046
ugate Addition.= A Practical	Reversed Polarity Alternative to Organocuprate Conj	ORGN	0091
Proteins.= Preparative	Reversed-Phase Elution Chromatography of	ANYL	0029
rimary and Secondary Amines by	Reversed-Phase Paired-Ion High-Performance Liqu	ANYL	0089
ession Multicolumn Preparative	Reversed-Phase Sample Displacement Chromatograp	ANYL	0040
Sorbents: Critical Review of Ex+	Reversed-Phase Supports as Surrogates for Natural	ANYL	0071
Protein Folding and the	Reversibility of Denaturation.= Facilitation of	BIOT	0190
Receptor.= Highly Selective	Reversible and Irreversible Ligands for the κ-Opioid	MEDI	0156
hemistry Award Symposium—II.	Reversible and Irreversible Neutral Species Transfer	ANYL	0103
reversible Gelation of Polymers.	Reversible Gelation in Binary Mixture of Poly(di-N-	POLY	0145
stallization.=	Reversible Gelation of Microgels due to Colloidal Cry	POLY	0244
Microphase Separation and	Reversible Gelation.=	POLY	0122
Facile	Reversible Metalation from an Agostic Complex.=	INOR	0384
Dynamics of	Reversible Networks.=	POLY	0172
	Revising the Risk-Assessment Paradigm.=	ENVR	0065
Recoil Studies	Revisited: Heavy Ion Induced Reactions.=	NUCL	0060
Bridesmaid	Revisited.= Atmospheric Nitrogen Oxides: A	ENVR	0071
Halides	Revisited.= Infrared Spectrum of the Hydrogen	CHED	0197
uisition Devices—A Tool for the	Revitalization of the Physical Chemistry Laboratory.	CHED	0160
ith Techniques That Are New, a	Revitalized Physical Chem Lab Makes Its Debut.=	CHED	0043
stitute for Chemical Education in	Revitalizing Science in the Nation's Schools.= +In	CHED	0138
Safety.= Challenges and	Rewards of Developing Videotapes on Laboratory	CHAS	0044
Extraction of	Rf (Element 104) and Its Homologues with TBP.=	NUCL	0074
se+ Process Development of BI-	RG-587: A Potent and Selective Reverse Transcripta	ORGN	0027
kynes and Alkenynes Catalyzed by	Rh and Co-Rh Mixed Metal Complexes.= +of Al	ORGN	0218
$P(S)]_n[Ph_2P(O)]_3$-nC], where M =	Rh and Ir and n = 1, 2, and 3.= +or of [cod]M[(Ph$_2$	INOR	0264
rticle Size for NO Adsorption on	Rh and Pt.= Study of the Effect of Support and Pa	COLL	0186
kenynes Catalyzed by Rh and Co-	Rh Mixed Metal Complexes.= +of Alkynes and Al	ORGN	0218
n Catalyzed by cis-	Rh(CO)$_2$(p-(NC)C$_5$H$_4$N).= Aldehyde Decarbonylatio	INOR	0255
rizations Catalyzed by Co(III) and	Rh(III) Complexes.= +igomerizations, and Polyme	ORGN	0129
ogenation of Alkanes Catalyzed by	Rh(PMe$_3$)$_2$(CO)Cl.= +emical and Thermal Dehydr	INOR	0152
ransfer-Dehydrogenation Using	Rh(PMe$_3$)$_2$LCl(L = ^{i}Pr$_3$,CO): Aspects of Selectivity a	INOR	0151
of Oxygenates on	Rh(111).= Activation of C-H, C-C, and C-O Bonds	CATL	0033
Catalysis.= Syntheses of Novel	Rh-Co Mixed Bimetallic Complexes and Their	INOR	0226
Aging on the Performance of Pt,	Rh/CeO$_2$ and Rare Earth/Alkaline Earth-Doped Pt,	PETR	0035
Earth/Alkaline Earth-Doped Pt,	Rh/CeO$_2$ Catalysts.= +ce of Pt, Rh/CeO$_2$ and Rare	PETR	0035
on-Hydrogen Bond Activation in	Rhenium Alkoxide Complexes (η5-C^{5}R^5)Re(No)(PPh	ORGN	0220
e Polymers Using a Luminescent	Rhenium Carbonyl Complex as a Spectroscopic Prob	POLY	0096
Di- and Polynuclear	Rhenium Carbonyl Complexes.=	INOR	0107
een Molybdenum, Tungsten, and	Rhenium Complexes.= Oxygen Atom Transfer betw	INOR	0132
nthesis of High-Oxidation-State	Rhenium Hydrazine, Hydrazido, Amido, and Ammon	INOR	0261
l Ligands.= Technetium(V) a+	Rhenium(V) Oxo Complexes of Amine-Amide-Dithio	INOR	0181
aturated Polyester Interpenetra+	Rheo-Kinetics and Morphology of Polyurethane-Uns	PMSE	0027
	Rheo-Optics of Flow-Induced Phase Separation.=	PMSE	0106
vinylidene Fluoride) Monitored by	Rheo-Photoacoustic FT-IR Spectroscopy.= +Poly(	POLY	0255
al Properties and Thickening M+	Rheological Additives in Coating Systems: Rheologic	PMSE	0034
e Flow of Polypropylene Melts.+	Rheological Evidence of Flow Structuring in Entranc	PMSE	0137
cal Additives in Coating Systems:	Rheological Properties and Thickening Mechanisms.	PMSE	0034
fluence of Crosslink Chemistry on	Rheological Properties of +/Polysaccharide Gels: In	POLY	0240
echnician Symposium. Dynamic	Rheological Studies on the Low-Temperature Behav	TECH	0014
f an Ionomer and an Amino-Co+	Rheological Study of Polymer-Polymer Complexes o	PMSE	0155
Formulations.= Role of	Rheology and Surface Science in Dentifrice	CHED	0047
Suspensions.= Structure and	Rheology of Polydisperse, Charged Colloidal	COLL	0165
Falling-Needle/Sphere	Rheometer.= Magnet-Enhanced Optical	POLY	0251
Morphology in Blends by Optical	Rheometry.= +Steady and Dynamic Shear-Induced	PMSE	0085
e to Ciral Alcohols Using Discrete	Rhodium Catalysts.= +n of Ketones: A Facile Rout	INOR	0247
h Cationic Bis(dioxaphospholane)	rhodium Complexes.= +alytic Hydroformylation wit	INOR	0154
tion Reactions with MR$_2$ (M =+	Rhodium Hydride Dimer: Fragmentation-Recombina	INOR	0383
rosilylation of+ Discrete Chiral	Rhodium Phosphine Complexes for Asymmetric Hyd	CATL	0005
atalysts for Enantioselect+ Chiral	Rhodium(II) Carboxamides: Remarkably Effective C	CATL	0002
Diazoacetates Catalyzed by Chiral	Rhodium(II) Carboxamides.= +Reactions of Alkyl	ORGN	0042
Heterocyclic Synthesis Using	Rhodium-Stabilized Carbenoids.=	ORGN	0276

te Crude Methy+ Use of a Mixed	Rhodium/Ruthenium Catalyst System to Hydrogena	I&EC	0056
Studies of the Q_y Transitions in	Rhodobacter sphaeroides Reaction Center: Polarizati	BIOL	0094
Transient Spectroscopy of	Rhodopsin Photochemistry.= Subpicosecond	BIOT	0080
Heavy Ion Reactions and	RI–Beam Experiments.= Intermediate–Energy	NUCL	0029
nt for th+ 2',3',4',5'–Tetraacetyl	Riboflavin: A Potentially Useful Photosensitizing Age	ENVR	0029
o+ Mechanism of Inactivation of	Ribonucleotide Reductase by 2'–Azido 2'–Deoxynucle	ORGN	0281
enter: Tyrosyl Radical Cofactor of	Ribonucleotide Reductase.= +the Dinuclear Iron C	BIOL	0022
cleoside Isosteres of Nicotinamide	Riboside.= +ew NAD Congeners Containing C–Nu	CARB	0023
h Trichosanthin, a Plant–Derived	Ribosome–Inactivating Protein.= +viral Studies wit	AGFD	0189
esis in E. coli.= Specialized	Ribosomes and Feedback Regulation of rRNA Synth	BIOT	0245
is of P¹–(adenosin–5'–yl) –P²–[6–(	ribosyl)–picolinamide–5'–yl]pyrophosphates. +nthes	CARB	0023
Deficient Mutant of R. rubrum	Ribulose Bisphosphate (RuBP) Carboxylase/Oxygena	BIOL	0089
Chemical Businesses in Costa	Rica: The Case of Arvi, Inc., Costa Rica.= Small	SCHB	0026
The Case of Arvi, Inc., Costa	Rica.= Small Chemical Businesses in Costa Rica:	SCHB	0026
hnological Transfer Unit in Costa	Rica.= +the Productive Sector: The Case of the Tec	SCHB	0025
Hormigas: Alimento	Rico en Proteinas.=	AGFD	0205
Program at Oak	Ridge National Laboratory.= Science Education	CHAS	0050
Unusual Ring–Opened	Rifamycin Derivatives.= Chemistry of Some	ORGN	0013
S).= SEC–Viscometry–	Right Angle Laser Light Scattering (SEC–VIS–RALL	PMSE	0098
hemistry Laboratory: Let's Get It	Right.= +Education. Computers in the Physical C	CHED	0155
the Challenge of the 90s: MSDS	Right-to-Know Compliance Made User Friendly.=	CHAS	0031
New Jersey	Right-to-Know Program.=	CHAS	0037
MSDS and Label	Right-to-Know Software.=	CHAS	0030
iance.=	Right-to-Know Support: Current and Future Compl	CHAS	0026
Legal	Rights of Whistle–Blowing Professionals.=	PROF	0011
s–Linkable Polyimide Blends with	Rigid and Semirigid Polyamides.= +perties of Cros	PMSE	0023
Synthesis of a Conformationally	Rigid Calcium–Entry Blocker.=	ORGN	0114
Synthesis of	Rigid Dendrimers That Overcome Steric Inhibition.=	POLY	0321
Aqueous Synthesis of Soluble	Rigid–Chain Polymers: An Ionic Poly(p–phenylene) A	POLY	0024
ecular–Dynamics Simulations of	Rigid–Rod and Hydroxy–Substituted Rigid–Rod Poly	PHYS	0322
,6–Dimethylphenoxy Pendent +	Rigid–Rod Benzobisthiazole Polymer with Reactive 2	PMSE	0019
,6–Dimethylphenoxy Pendent +	Rigid–Rod Benzobisthiazole Polymer with Reactive 2	PMSE	0018
Triazine	Rigid–Rod Networks.=	POLY	0110
Octane Moieties.= Colorless	Rigid–Rod Polymers Containing Bicyclo[2.2.2]	POLY	0029
of Diama+ Visible–Transparent	Rigid–Rod Polymers: Synthesis and Characterization	POLY	0028
id–Rod and Hydroxy–Substituted	Rigid–Rod Polymers.= +namics Simulations of Rig	PHYS	0322
2: Structure–$\chi^{(3)}$ Relationships in	Rigid–Rod Polyquinolines.= +Polymer Thin Films,	POLY	0205
Structural Evidence for the	Rigidity of Torands.=	INOR	0297
Polydicyclopentadiene: A New	RIM Thermosetting Resin.=	POLY	0136
Carlo Simulations of	Rings and Branched Molecules.= Dynamic Monte	POLY	0297
Simulations.=	Rings vs. Strings: Off-Lattice, Bead-Stick Computer	POLY	0235
6–, and 7–Membered Carbocyclic	Rings.= +for Cyclization–O–Functionalization of 5–,	ORGN	0088
Simulated Annealing of	Rings.=	ORGN	0169
Deactivating Groups in Aromatic	Rings.= +ctural Characterization of Activating and	CHED	0096
mers Containing ortho–Phenylene	Rings.= +ion of Ordered Heterocycle–Imide Copoly	POLY	0042
bly of Carbon Atoms into Rods,	Rings, Nets, and Spheres: The New Generations of C	ORGN	0266
plication Systems for Reduction of	Rinsate and Nontarget Contamination.= +icide Ap	AGRO	0105
al Solutions for Wastewater and	Rinsate Management for Retail Fertilizer Dealers.=	FERT	0014
ls and Their Containers. Pesticide	Rinsate Minimization and Reuse.= +tural Chemica	AGRO	0102
Treatment of	Rinsate.=	AGRO	0103
Pesticide	Rinsates.= Evolution of Regulations for	AGRO	0104
aining Regulators of Tomato Fruit	Ripening.= +turally Occurring Carbohydrate–Cont	AGFD	0031
from Conformational Analysis for	RIS Model: Polydimethylsiloxane.= +tical Weights	POLY	0356
rophenol in Soil.= Penta Soil	RISc: A Rapid, On-Site Screening Test for Pentacblo	ENVR	0033
in Soil.= PCB–	RISc: An On-Site Immunoassay for Detecting PCB	ENVR	0032
nsity Polyethylene: Temperature–	Rising Supercritical Fluid Fractionation.= +ow–De	POLY	0076
rl–Ranking System: A Balance of	Risk Assessment and Data Constraints.= +'s Haza	ENVR	0081
ide Examples.= Issues in Cancer	Risk Assessment: Perchloroethylene and Vinyl Chlor	CHAS	0011
Exposure Assessment and	Risk Assessment.=	ENVR	0072
Approaches to Quantitative	Risk Assessment.= Noncancer Health Endpoints:	ENVR	0068
Studies for Human	Risk Assessment.= Refinement of Animal Toxicity	CHAS	0019
Use of Mechanistic Data in	Risk Assessment.=	ENVR	0069
n Drinking Water.=	Risk Assessments for Microbiological Contaminants i	ENVR	0070
cation Are the Keys to Scientific	Risk Assessments of Inhalation and Dermal Exposur	ENVR	0042
l E+ An Integrated Approach to	Risk Characterization of Multiple–Pathway Chemica	ENVR	0073
Regulatory Issues in	Risk Management: A California Perspective.=	CHAS	0021
Ecological	Risk of Aquatic–Habitat Degradation.=	ENVR	0062
Accomplish This Task? Relative	Risk: A Blueprint for Future Environmental Protecti	ENVR	0061
enmental Problems According to	Risk: What Must We Yet Know To Accomplish This	ENVR	0061
Relative	Risk.= Hazard–Ranking System and	ENVR	0079
Assessment of Relative	Risk.= It Is Possible To Do Quantitative	ENVR	0067
Revising the	Risk–Assessment Paradigm.=	ENVR	0065
or Immediate Implementation of	Risk–Based Management of Environmental Resource	ENVR	0076
nd Consumer View.=	Risks and Benefits of Pesticides: An Environmental a	AGRO	0059
Pesticides:	Risks and Benefits.=	AGRO	0058
anking by Cancer and Noncancer	Risks Combined: Methods and Needs.= +litative R	ENVR	0074
ess.= Copyright Protection and	Risks in the International Information Transfer Proc	CINF	0014
wego Harbor and the St. Lawrence	River at Massena.= +ncern in New York State: Os	ENVR	0084
Marshes along the Florida Indian	River Lagoon.= +ide Residue in Barrier Island Salt	ENVR	0019
tor–Derived ¹⁴C from the Hudson	River to the New York Bight.= +c Carbon and Reac	GEOC	0014

Methane Fluxes from the Hudson	River.= Chemistry of Ocean Margins. Controls on	GEOC 0001
5 Elements of Empirical Formula	RMER'; R,R' = Aryl, Alkyl; M = Al, Ga, In; E = P,	INOR 0392
logical Chemistry. Recognition of	RNA Minihelices in Relation to Decoding Genetic In	BIOL 0026
A Lead(II) and Lanthanum(III)	RNA Transesterification Catalysts and the Design of	INOR 0406
nthetic Substances against Exotic	RNA Viruses.= +ceae Constituents and Related Sy	AGFD 0165
e 5' Cap Structure of Messenger	RNA: A Novel Target for Therapeutic Antisense Che	BIOL 0120
e Phosphorane Transition State in	RNase?.= +the Vanadate Anion an Analogue of th	BIOL 0061
emical Careers Insights Program (	Roadshow)—Careers for Chemists in Industry.+ Ch	YCC 0005
Chemistry Laser Experiments at	Roanoke College.= Undergraduate Physical	CHED 0042
e Sensory+ Effect of Storage on	Roasted Peanut Flavor, as Determined by Descriptiv	AGFD 0161
DNA in the	Robert Golub Case.=	CHAL 0010
Versatile Zymark	Robot Dilution System.=	TECH 0005
iological Fluids Using Laboratory	Robotics.= +Analysis of Drugs and Metabolites in B	ANYL 0119
Phosphoric	Rock Dissolution: An Experimental Report.=	PHYS 0294
s of Total Mercury in Crystalline	Rock: Case Study—Waldoboro Pluton Complex, Mai	ENVR 0055
Ninety Years of Chemistry at	Rockefeller.=	HIST 0027
e Determination of Rare Earths in	Rocks by Electron Impact Mass Spectrometry.= +v	ANYL 0049
−Dynamics Simulations of Rigid-	Rod and Hydroxy–Substituted Rigid–Rod Polymers.	PHYS 0322
ethylphenoxy Pendent+ Rigid-	Rod Benzobisthiazole Polymer with Reactive 2,6–Dim	PMSE 0019
ethylphenoxy Pendent+ Rigid-	Rod Benzobisthiazole Polymer with Reactive 2,6–Dim	PMSE 0018
Triazine Rigid-	Rod Networks.=	POLY 0110
Moieties.= Colorless Rigid-	Rod Polymers Containing Bicyclo[2.2.2] Octane	POLY 0029
m+ Visible-Transparent Rigid-	Rod Polymers: Synthesis and Characterization of Dia	POLY 0028
and Hydroxy–Substituted Rigid-	Rod Polymers.= +namics Simulations of Rigid–Rod	PHYS 0322
cture–$\chi^{(3)}$ Relationships in Rigid-	Rod Polyquinolines.= +Polymer Thin Films, 2: Stru	POLY 0205
Construction Set: Assembly of the	Rods and Spools.= +a Molecular-Size Tinkertoy	BIOT 0200
thesis and Characterization of the	Rods.= +ular-Size Tinkertoy Construction Set: Syn	ORGN 0124
Assembly of Carbon Atoms into	Rods, Rings, Nets, and Spheres: The New Generati	ORGN 0266
Beilstein Current Facts on CD-	ROM.=	COMP 0017
Prepared Using	ROMP Block Copolymers.= Semiconductor Clusters	INOR 0307
nforced Polyvinyl Chloride (PVC)	Roofing Membranes.= +cal Analysis Studies of Rei	PMSE 0180
Vinyl or Acrylic Pol+ Studies of	Room Temperature Cured Castor Oil Polyurethane/	PMSE 0090
Atoms with Simple Molecules near	Room Temperature.= +ions of Transition–Metal	PHYS 0198
°From Computer	Room to Learning Center.=	CHED 0116
metallic Chemistry Performed in	Room-Temperature Chloroaluminate Molten Salts.=	ANYL 0097
H-Atom Reaction with C+ Near-	Room-Temperature Deposition of Copper Films by	PHYS 0259
of Aromatic Ketones in a	Room-Temperature Molten Salt.= Electrochemistry	ANYL 0094
oduction of Chemicals from Plant	Root Cultures.= +on of Reactors for Large-Scale Pr	BIOT 0233
Plant	Root Cultures.= Sesquiterpene Induction from	BIOT 0232
Secondary Metabolism in Hairy	Root Cultures.= Understanding and Manipulating	BIOT 0231
Equation for the Square	Root of the Electronic Density.=	PHYS 0254
Ent-Atisene Glycosides from the	Roots of Stevia salicifolia cav. (Asteraceae).= New	MEDI 0034
Xylanases from Streptomyces	roseiscleroticus: Purification and Characterization.=	BIOT 0210
e Waste Destruction Using White	Rot Fungi.= +otechnological Approaches to Pesticid	AGRO 0126
ts on Performanc+ Flighting for	Rotary Fertilizer Dryers and Coolers and Their Effec	FERT 0012
Cells on	Rotating Fibers.=	BIOT 0154
Half-Integer Spins in	Rotating Samples.= Cross Polarization of	PHYS 0371
in Low-Spin Ferric Porphyrin +	Rotation and Orientation of Coordinated Imidazoles	INOR 0296
Barriers to	Rotation in Oximes: A Molecular Orbital Study.=	PHYS 0324
s in Zeolite-Y Probed by Double-	Rotation NMR.= +ice Relaxation of Sodium Cation	PHYS 0267
e Correlations in the Dynamics of	Rotational Isomerism.= +. Sizes of the Short-Rang	POLY 0365
in the Bulk Stat+ Dynamics of	Rotational Isomerization in Polymers in Solution and	COLL 0018
or+ Relaxation of Runs of Trans	Rotational Isomers in Single Racemic Poly(vinyl Chl	POLY 0069
Quenching of NH $A^3\Pi_i$ in High	Rotational Levels.=	PHYS 0266
rge Proteins.=	Rotational Resonance: Distance Measurements in La	PHYS 0196
ith Hot Hydrogen Atoms.+ Pure	Rotational Scattering of CO_2/CO due to Collisions w	PHYS 0291
al Transition Metal Polyhydrides:	Rotational Tunnelling and Quantum Exchange.= +c	PHYS 0068
State-Resolved N_2O Vibrational,	Rotational, and Translational Energy Transfer from	PHYS 0334
e Orthorhombic Pseudohexagonal	Rotator-Phase Transition in n-Heneicosane.= +th	POLY 0354
te Properties of Polymers through	Rotaxane Formation.= +ocessability and Solid-Sta	POLY 0233
Behavior of a Polyester	Rotaxane.= On the Morphology and Crystallization	POLY 0346
Catalytic Processes Honoring J. F.	Roth. +Industrial Chemistry Award New Industrial	I&EC 0043
Catalytic Processes Honoring J. F.	Roth. +Industrial Chemistry Award New Industrial	I&EC 0052
Catalytic Processes Honoring J. F.	Roth. +Industrial Chemistry Award New Industrial	I&EC 0060
Remedial Action Planning Pro+	Round-Table Discussion—Public Involvement in the	ENVR 0086
Composite Materials: Free Radical	Routes into Nonshrinking Sol-Gel Composites.=	POLY 0259
Polymeric	Routes to Aluminum Oxides.=	POLY 0342
Vapor-Phase Molecular	Routes to Ceramic Powders.=	INOR 0175
lymer Matrix on Pr+ Biomimetic	Routes to Composites and Ceramics: Influence of Po	POLY 0290
Some New Organometallic	Routes to III/V Compound Semiconductors.=	INOR 0146
Molecule Antiotensin+ Synthetic	Routes to Imidazole–5–Acrylic Acids: Potent Small-	ORGN 0119
al Gravi+ Tutorial on Molecular	Routes to Inorganic Materials. Application of Therm	INOR 0006
on Spect+ Tutorial on Molecular	Routes to Inorganic Materials. Applications of Electr	INOR 0013
ion-Metal Thin Films+ Molecular	Routes to Inorganic Materials. Deposition of Transit	INOR 0141
e Chemical Reactions+ Molecular	Routes to Inorganic Materials. Gas Phase and Surfac	INOR 0121
lecular Chemistry.+ Molecular	Routes to Inorganic Materials. Magnets Based on Mo	INOR 0100
ical Vapor Depositio+ Molecular	Routes to Inorganic Materials. Organometallic Chem	INOR 0171
Enantioselective	Routes to syn- and anti-β-Phenylcysteines.=	ORGN 0115
son of Atom-by-Atom 3D Search	Routines for Searching CAS Registry Substances.=	CINF 0023
Scientific Division of	Routledge, Chapman, and Hall.=	CHAS 0028

Variational	Rovibrational Studies of 3- and 4-Atom Molecules.=	PHYS	0030
agnetic Interactions in Second-	Row Transition Elements: Crystal Structures of Two	INOR	0267
. Four Component Solubility Pa+	Roy W. Tess Award Symposium Honoring K. L. Hoy	PMSE	0033
ysis Research and Casework in the	Royal Canadian Mounted Police.= +sic DNA Anal	CHAL	0007
larin Antibiotics as Revealed by	RP-HPLC: Structure-Retention and Structure-Activ	MEDI	0059
and Feedback Regulation of	rRNA Synthesis in E. coli.= Specialized Ribosomes	BIOT	0245
Studies with HIV	RT Inhibitor U-9843.= Mode of Action in Vitro	MEDI	0055
mal d-Shell Occupancy in the	Ru + H$_2$ Reaction.= Role of Excited States of Maxi	PHYS	0243
Dihydrogen Complexes: Fe < Os <	Ru.= +ve Order of pK_a Values of Isostructural η_2-	INOR	0380
Excited-Stated Dynamics of	Ru(BPY)$_3^{2+}$ on Porous Vycor Glass.=	INOR	0160
nal Dihalides by Photogenerated	Ru(I) Complexes: Competition between Back Electro	INOR	0284
n by DNA of a Nonintercalating	Ru(II) Complex Containing the Cationic Ligand N,N	INOR	0233
Cleavage of DNA by a Chiral	Ru(II) Complex with a Bifunctional Ligand.=	INOR	0232
Electrochemical Properties of	Ru(II) Schiff Bases Complexes.= Synthesis and	INOR	0260
tophysics and Photochemistry of	Ru(II)-Diimine Complexes in Mixed Solvents.= Pho	INOR	0279
the Electrochemical Reduction of	Ru(NH$_3$)$_6$Cl$_3$.= +in Naturally Occurring Fluids on	COLL	0005
Reactions of H$_2$Os$_3$(CO)$_{10}$ and	Ru$_3$(CO)$_{12}$ with RC≡CPh (R=SiMe$_3$,SnBu$_3$).=	INOR	0241
ta y Quercetin (QU) and Rutin (	RU) as Inhibitors of Experimental Colonic Neoplasia	AGFD	0114
Reactions of Ethylene over	Ru-Y Zeolite.=	CATL	0031
he Reaction of Ethylene with M-	Ru/SiO$_2$.= +talysts Toward Catalytic Reforming: T	CATL	0042
y of Protein Prenylation: Nature's	Rubber Cement for Membrane Attachment.= +str	BIOL	0021
oducts.+ Nitrosamines in Elastic	Rubber Nettings Used for Packaging Cured-Pork Pr	AGFD	0040
ides: Influence of a Polysiloxane	Rubber on the Polymerization Phase-Separation Pro	POLY	0112
pcxies Using Designed Core-Shell	Rubber Particles.= +ning of High-Performance E	POLY	0157
Catalytic Conversion of Natural	Rubber to Polyepoxide.=	ENVR	0045
	Rubber-Clay Hybrid—Synthesis and Properties.=	POLY	0291
siloxane Rubber on the Polymer+	Rubber-Modified bisMaleimides: Influence of a Poly	POLY	0112
ac Properties and Morphology of	Rubber-Toughened Polystyrene.= +ion on the Imp	PMSE	0177
. Librum Ribulose Bisphosphate (	RuBP) Carboxylase/Oxygenase.= +ent Mutant of R	BIOL	0089
fovibrio Vulgaris (Hildenborough)	Rubredoxin.= +acement by a Nickel Atom in Desul	INOR	0020
lyically Deficient Mutant of R.	rubrum Ribulose Bisphosphate (RuBP) Carboxylase/	BIOL	0089
expyrrolic Expanded Porphyrin:	Rubyrin.= +thesis and Characterization of a New H	ORGN	0058
Characterization of	RuCl$_3$ Impregnated Na-Y Zeolite.=	CATL	0035
When Employing HPLC Analy+	Ruggedness of Impurity Analysis in Pharmaceuticals	ANYL	0114
d C$_6$ N-Alkanes in Light Straight	Run.= +ite Catalysts for the Isomerization of C$_5$ an	PETR	0113
=	Running a Small Business within a Larger Company.	SCHB	0018
Foly(vinyl Chlor+ Relaxation of	Runs of Trans Rotational Isomers in Single Racemic	POLY	0069
m and Osmium Carbonyl Cluste+	Ruthenaborane and Osmaborane Clusters: Rutheniu	INOR	0220
Hydrosilation of Iron and	Ruthenium Acetyl Complexes.= Catalytic	INOR	0238
f Unusually Reactive (η^5-Indenyl)	Ruthenium Alkyl Complexes.= +ration Reactions o	INOR	0239
ne Reactivity of Electron-Rich	Ruthenium Alkyl, Hydride, and Aluminohydride Com	INOR	0244
omne and Osmaborane Clusters:	Ruthenium and Osmium Carbonyl Clusters Containi	INOR	0220
of Protonated trans Bis-Dioxime	Ruthenium Complexes.= +ructural Characterization	INOR	0066
and Reactivity of an Unsaturated	Ruthenium Fluoro-Alkoxide Complex.= +ynthesis	INOR	0127
Reactions with Electrophiles.=	Ruthenium Thiolates cpRu(PR$_3$')$_2$SR: Synthesis and	INOR	0368
is 2,4-Dimethyl-1,3-pentadienyl)	ruthenium.= +Formation during Thermolysis of B	INOR	0305
Osmium and	Ruthenium.= Chemistry of Arylimido Complexes of	INOR	0259
of Cationic Trinuclear BINAP-	Ruthenium(II) Complexes and Their Roles in Cataly	ORGN	0044
e Catalyzed by (Aqua)(Phosphine)	Ruthenium(IV) Complexes.= +ation of Cyclohexen	INOR	0067
ry of a Series of (Oxo)(Phosphine)	Ruthenium(IV) Complexes.= +the redox Chemist	INOR	0068
anisoles by (Oxo)(phosphine)	ruthenium(IV) Complexes.= S-Oxygenation of Thio	INOR	0130
ectivity Models and the Design of	Ruthenium-Based Fischer-Tropsch Catalysts.= +l	CATL	0056
e Water Gas Shift Reaction over	Ruthenium-Exchanged X Zeolite: Reaction Control b	CATL	0027
op+ Dietary Quercetin (QU) and	Rutin (RU) as Inhibitors of Experimental Colonic Ne	AGFD	0114
via Cuba.= °From	Ruzicka's Terpenes in Zurich to Steroids in Mexico	HIST	0017
Covalently Linked Porphyrin-	RuO$_2$ Clusters.= Photoinduced Electron Transfer in	COLL	0133
Oxygen Redox Catalysis Using	RuO$_2$ Dispersions and the Electrochemical Model.=	COLL	0134
arement and Disposal in the Mid-	80s and the Mid-90s.= +g Forward: Pesticide Man	AGRO	0086
ne) Synthesis via Competing	S$_{N}$AR and S$_{RN}$1 Mechanisms.= Poly(aryl ether-keto	PMSE	0122
Ring Nitrogen and Su+ Tandem	S$_{N}$2-Michael Approach to the Synthesis of 5- and 6-	ORGN	0199
Oxiranes: Synt+ Stereoselective	S$_{N}$2' Addition of Organocuprate Reagents to Alkynyl	ORGN	0011
sis via Competing S$_{N}$AR and	S$_{RN}$1 Mechanisms.= Poly(aryl ether-ketone) Synthe	PMSE	0122
alysis of All Di-, Tri-, and Tetra-	Saccharide Fragments of the Forssman Antigen.=	CARB	0056
of N+ Synthesis of Substituted	Saccharins through the o-Lithiative Sulfonamidation	ORGN	0035
4-Alkyl/Aryl-Substituted	Saccharins.= General Synthesis of	ORGN	0036
ovel Er+ Genetic Engineering of	Saccharopolyspora erythraea for the Production of N	BIOT	0110
New Generation of Inherently	Safe and Advanced Nuclear Power Reactors.=	FUEL	0023
	Safe and Practical Synthesis of Nitrocubanes.=	ORGN	0257
or Producing Environmentally	Safe Shale Stabilizers.= Novel and Versatile Process	CARB	0009
Four Keys to Chemical	Safety and Incident Prevention.=	CHAS	0017
Resources for Material	Safety Data Sheet Preparation.=	CHAS	0040
Material	Safety Data Sheets—Effective in the Workplace?.=	CHAS	0036
Making	Safety Data Simple.=	CHAS	0032
	Safety Education and Training at a University.=	CHAS	0049
ictic Metabolism Studies in Food	Safety Evaluation: A Proposal for Regulatory Use.=	AGRO	0132
Chemical	Safety in the ChemSource Program.=	CHED	0019
Health and	Safety Issues under OSHA and EPA Programs.=	CHAS	0022
FA-EPA Interface in Health and	Safety Issues. OSHA's Role in Chemical Proces+ OS	CHAS	0012
in Fertilizer Manufac+ Process	Safety Management: The Impact of New Regulations	FERT	0016
g and Implementing a Laboratory	Safety Program.= +ful Considerations in Developin	CHAS	0004

Training Requirements for Labo+	Safety through Education and Training. Regulatory	CHAS	0042
om; Whe+ A Decade of Chemical	Safety Training, 1980–1991: Where Did We Come Fr	CHAS	0043
ersit+ Communicating Chemical	Safety: Some Education and Training Used in a Univ	CHAS	0048
OSHA's Role in Chemical Process	Safety.= +Interface in Health and Safety Issues.	CHAS	0012
Videotapes on Laboratory	Safety.= Challenges and Rewards of Developing	CHAS	0044
Clandestine Laboratory	Safety.=	CHAS	0047
n Be Used To Improve Laboratory	Safety.= +Revealed and How That Knowledge Ca	CHAS	0002
lan for R D Labora+ Laboratory	Safety—Theory and Practice. A Chemical Hygiene P	CHAS	0001
rino Science: Solar Neutrinos—II.	SAGE Solar Neutrino Experiment.= +ces in Neut	NUCL	0030
system Ni(	sal$_2$en) and NI(sal$_2$tm).= EHMO Calculations on the	INOR	0042
between N–N ligands and [Ni(	sal$_2$tim)].= Kinetics of the Addition Reaction	INOR	0061
Ni(sal$_2$en) and NI(	sal$_2$tm).= EHMO Calculations on the system	INOR	0042
Club of New York.= The	Salamander and Crossed Retorts: The Chemists'	HIST	0031
idation Catalyzed by Chiral Metal	Salen Complexes Bearing Bulky Silyl Groups.= +x	ORGN	0056
Research and	Sales.= Chemical Careers in Industry: Between	YCC	0011
Chemists in Marketing,	Sales, and Technical Service.= Opportunities for	YCC	0006
osides from the Roots of Stevia	salicifolia cav. (Asteraceae).= New Ent-Atisene Glyc	MEDI	0034
ool: Instrument Requirements and	Salient Results.= +Spectroscopy as an Analytical T	NUCL	0042
l Envelope Lipopolysaccharides of	Salmonella and Shigella Bacteria.= +Chains of Cel	CARB	0017
hyl)thiophenium Hydroxide Inner	Salt and β-Butyrolactone.= +-(4-hydroxy-1-napht	POLY	0048
and the Three-Dimensi+ Molten	Salt Chalcogenide Synthesis of the Linear β-CsBiS$_2$	INOR	0202
ltiple-Quantum NM+ Sulfonium	Salt Distribution in Model Polymer Films by ^{19}F Mu	PMSE	0182
nation of Strontium-90 in a High-	Salt Environment at the Nevada Test Site.= +rmi	NUCL	0085
sticide Residue in Barrier Island	Salt Marshes along the Florida Indian River Lagoon.	ENVR	0019
nesium Species in Several Molten	Salt Systems by Raman Spectroscopy.= +hloromag	INOR	0064
Room-Temperature Molten	Salt.= Electrochemistry of Aromatic Ketones in a	ANYL	0094
Alkali Earth	Salts as Catalyst Supports and Stabilizers.=	PETR	0034
Complexation of Benzamidinium	Salts by a Synthetic Receptor.=	ORGN	0061
t of Hydrogen Peroxide and Iron	Salts on Sewage, Part 1: Odor Control and Phosphat	ENVR	0034
red Rare Earth Gold(I) Cyanide	Salts: Evidence for Energy Transfer and Relavtivistic	INOR	0189
halcogenide Synthesis in Molten	Salts: New Solid-State Compounds of Na$_3$Cu$_4$Se$_4$, N	INOR	0351
rix Apparently Caused by Arsenic	Salts.= +Hydration and Polymerization in the Mat	ENVR	0041
uaternary Piperidino-Ammonium	Salts.= +uscle Relaxants of Ester-Containing bisQ	MEDI	0047
perature Chloroaluminate Molten	Salts.= +tallic Chemistry Performed in Room-Tem	ANYL	0097
STM Studies of Charge Transfer	Salts.=	PHYS	0230
Methionine	Salvage Pathway.=	BIOL	0024
lerodanic Precursor from Mexican	Salvia Species.= +th New Skeletons Related to a C	ORGN	0280
Deposito Electrolitico de	Samario y Plutonio.=	NUCL	0075
s+ Control of Stereoselectivity in	Samarium Metal-Induced Cyclopropanations: Synthe	ORGN	0284
	Air Samplers and Complementary Analyses for PIXE.=	NUCL	0079
ody Burden in+ Human Tissue	Sampling Followed by Dioxin Analysis to Estimate B	CEI	0012
Rapid Analysis Using Direct	Sampling Mass Spectrometry.=	FUEL	0089
Cracking Properties of	SAPO-37 and Faujasites.=	CATL	0045
Optimization of the Pillaring of a	Saponite.= +Zeolites and Pillared Clay Synthesis.	PETR	0142
nyl)Ethyl-+ Investigation of the	SAR in a Series of N^6-Cyclopentyl- and N^6-(2-Thie	MEDI	0113
hannel Activators: Synthesis and	SAR Studies of Novel 1,4-Benzoxazine Derivatives.=	MEDI	0099
t.=	Sarcophytol A: A New Cancer Chemopreventive Agen	AGFD	0087
per and Other Elements in Silica-	Saturated CaO–CaF$_2$–SiO$_2$ Slags.= +lubility of Cop	I&EC	0066
CaF$_2$-+ Equilibrations of Silica-	Saturated Metal (Cu), Matte (Cu$_2$S), and Slag (CaO-	I&EC	0065
esearch Co.). Pulse Voltammetry:	Saturday's Child.= +by EG G Princeton Applied R	ANYL	0096
ior of Poly(2-Chlo+ Synchrotron	SAXS Study of Mean-Field and Ising Critical Behav	PMSE	0103
Using Triphosgene, CBMIT, and	SBMIT.= +etric Ureas or Carbamates, Sulfamides	ORGN	0106
Triblock Copolymer	SBS.= Shear-Induced Structural Transitions in	PMSE	0136
Recombinant aFGF.= Novel	Scalable Process for Controlled Folding of	BIOT	0134
New Dialkylated	Scalarins from the Sponge Phyllospongia Foliascens.=	ORGN	0182
Folliascens.= New Dialkylated	Scalarins from the Sponge Phyllospongia	ORGN	0188
ography: A New Tool for Large-	Scale Biomolecule Purification.= Perfusion Chromat	ANYL	0032
Creosote Site–Biobed Box Pilot-	Scale Bioremediation Study.=	BIOT	0140
neling Microscopy.+ Nanometer	Scale Etching of MX$_2$ Materials Using Scanning Tun	PHYS	0364
from Shake Flask to Production-	Scale Fermenters.= +of Actinomycetes Fermentations	BIOT	0083
Tunneling Microscopy.+ Atomic	Scale Imaging of Molecules on Surfaces by Scanning	PHYS	0159
nological Support and Medium-	Scale Industries: The Case of the Center for Technol	SCHB	0027
Operation of Reactors for Large-	Scale Production of Chemicals from Plant Root Cult	BIOT	0233
lactamdurans.=	Scale Production of Efrotomycin by Nocardia	BIOT	0109
Large-	Scale Production of PHB on Beet Sugar.=	CARB	0002
Atomic	Scale Surface Analysis with the Atom–Probe FIM.=	PHYS	0101
and Wear at the Nanometer	Scale.= Atomic Force Microscopy Studies of Friction	PHYS	0156
Membrane Protein Prediction	Scales on Sidechain Structure.= Dependence of	BIOL	0038
herichia coli.= Expression and	Scaleup of Engineered Antibody Fragments from Esc	BIOT	0058
Perfusion Bioreactor: Analysis and	Scaleup.= +malian Cell Retention in a Spinfilter	BIOT	0007
	Scaling Properties in Anomalous Diffusion.=	PHYS	0130
Characterization by Step-	Scan FT-IR Spectroscopy.= Polymer	POLY	0304
nd Reactivity Studies of Relevant	Scandium Allyl Species.= +esis, Characterization, a	INOR	0280
arbon Bond-Forming Reactions of	Scandocene Derivatives.= +e Effects in Carbon–C	INOR	0147
Analysis of Materials Using	Scanning and Transmission Electron Microscopy.=	INOR	0014
New Applications of Differential	Scanning Calorimetry and Solvent Swelling for Stu	FUEL	0087
ment of Rea+ Application of the	Scanning Electrochemical Microscope to the Measure	COLL	0170
cations of Computer-Controlled	Scanning Electron Microscopy (CCSEM) in Coal Scie	FUEL	0041
erization of Municipal Waste by	Scanning Electron Microscopy and Optical Microscop	FUEL	0137
iques. Principles and Examples of	Scanning Force Microscopy.= +and Related Techn	PHYS	0152

lly Resolved Energy Transfer and	Scanning Molecular Exciton Microscopy.= +Spatia	PHYS 0022
Biological Surfaces.=	Scanning Tunneling and Atomic Force Microscopy of	PHYS 0158
malou+ On the Interpretation of	Scanning Tunneling Microscope Images Showing Ano	PHYS 0212
Nanofabrication with the	Scanning Tunneling Microscope.=	PHYS 0362
nd Related Techniques. Field Ion-	Scanning Tunneling Microscopy (FI-STM).= +a	PHYS 0182
e Chemical-Vapor+ Auger and	Scanning Tunneling Microscopy (STM) Studies of th	COLL 0050
troscopy of Metal Oxide Arrays+	Scanning Tunneling Microscopy and Tunneling Spec	COLL 0194
ry+ Multiphoton Ionization and	Scanning Tunneling Microscopy in Physical Chemist	CHED 0044
III-V(110) Surfaces: From 1- a+	Scanning Tunneling Microscopy of Alkali Metals on	PHYS 0183
ntercalation Compounds.=	Scanning Tunneling Microscopy of Donor Graphite I	PHYS 0365
and Nonequi+ Optically Excited	Scanning Tunneling Microscopy: Imaging Subsurface	PHYS 0208
w Face of Analytical Chemistry.	Scanning Tunneling Microscopy: New Applications in	ANYL 0022
age in Molecular Complexes Using	Scanning Tunneling Microscopy.= +ormation Stor	BIOT 0023
aging of Molecules on Surfaces by	Scanning Tunneling Microscopy.= +omic Scale Im	PHYS 0159
e Etching of MX_2 Materials Using	Scanning Tunneling Microscopy.= +nometer Scal	PHYS 0364
Photon	Scanning Tunneling Microscopy.=	PHYS 0210
lecular Rectifiers: Experiments by	Scanning Tunneling Microscopy.= +Organic Unimo	BIOT 0005
n the Ge(111) Surface Studied by	Scanning Tunneling Microscopy.= +se Transition o	PHYS 0099
tudied by in Situ Electrochemical	Scanning Tunneling Microscopy.= +er Resolution S	PHYS 0127
$O_2(110)$ and $TiO_2(001)$ Surfaces by	Scanning Tunneling Microscopy-Spectroscopy.=	COLL 0171
es of Single-Crystal Bi_2Sr_2CaCu+	Scanning Tunneling Microscopy/Spectroscopy Studi	PHYS 0211
atography Using a Re+ Rapid-	Scanning UV-VIS Detection for Capillary Gas Chrom	ENVR 0060
ight Angle Laser Light	Scattering (SEC-VIS-RALLS).= SEC-Viscometry-R	PMSE 0098
dium: Relation with Small-Angle	Scattering and Direct Dipolar Energy Transfer in Po	PHYS 0060
Ionomers by Small-Angle X-ray	Scattering and Dynamic Mechanical Thermal Analys	PMSE 0118
y of Chemical Reactions: Reactive	Scattering and Transition-State Theory.= +Theor	PHYS 0161
Polyelectrolytes.= Light	Scattering and Viscoelasticity of Linear and Ring	POLY 0294
Using Surface-Enhanced Raman	Scattering and X-ray Photoelectron Spectroscopy.=	POLY 0087
Solid-State Small-Angle Light-	Scattering Apparatus: Description and Application to	POLY 0278
rbon. High-Temperature Raman-	Scattering Behavior in Diamond.= +ond-Like Ca	FUEL 0035
ion Particle Distributi+ Photon-	Scattering Correction in the Determination of Emuls	PMSE 0030
lymer Solution Studied by Light	Scattering during Steady Shear and Following Cessat	PMSE 0107
High-Energy Ion	Scattering for Materials Analysis.=	NUCL 0065
ar Reactions/Scattering. Particle	Scattering for Measurement of Surface Composition	NUCL 0064
puter-Assisted Analysis of X-ray	Scattering from Polymeric Gas Separation and Barri	POLY 0142
Arch+ Static and Dynamic Light	Scattering from Solutions of Polymers with Differing	POLY 0250
drogen Atoms.+ Pure Rotational	Scattering of CO_2/CO due to Collisions with Hot Hy	PHYS 0291
as Precursors o+ Dynamic Light-	Scattering Studies of Resorcinol Formaldehyde Gels	POLY 0199
llization+ Small-Angle Neutron-	Scattering Studies of the Template-Mediated Crysta	PETR 0058
lyxylenyl+ Small-Angle Neutron	Scattering Study of Interdiffusion in Polystyrene-Po	POLY 0166
Layers.= Medium-Energy Ion-	Scattering Study of Nickel on Ultrathin SiO_2	COLL 0176
Solution—A Small-Angle Neutron	Scattering Study.= +Aqueous Poly(vinyl-alcohol)	POLY 0171
Light	Scattering Viscometry.=	PMSE 0096
Studied by Small-Angle Neutron	Scattering.= +of Polymers. Surfactant Organogels	POLY 0169
Methods and Small-Angle X-ray	Scattering.= +crylate) as Studied by Spectroscopic	POLY 0124
Rayleigh	Scattering.= Diffusion in Polymers by Forced	POLY 0164
-Line Viscometry and Laser Light	Scattering.= +r Conformation by SEC with an On	PMSE 0097
Using Surfaced-Enhanced Raman	Scattering.= +fusion Coefficients in Polymer Melts	POLY 0088
ssation of Shear Studied by Light	Scattering.= +Solution during Steady Shear and Ce	PMSE 0134
r Materials by Small-Angle X-ray	Scattering.= +of Sol-Gel-Derived Hybrid Cerame	POLY 0284
Using Low-Energy Ion	Scattering.= Studies of $TiO_2(110)$ Surface Reactivity	COLL 0172
Process for the Production of an	SCD4 Derivative by Recombinant Streptomyces livid	BIOT 0086
Havin+ Spectroscopic Studies of	sCD4(178)-PE40 Conformation: A Chimeric Protein	ANYL 0005
s and En+ Methods of Washing	sCD4-183 Inclusion Bodies to Remove E. coli Protein	BIOT 0171
Histamine Antagonist Related to	SCH 37370.= SCH 40338, a New Dual PAF and	MEDI 0161
nist Related to SCH 37370.=	SCH 40338, a New Dual PAF and Histamine Antago	MEDI 0161
e Lowest π^* and Ligand-Ligand	Scharge-Transfer Excited States of Zinc(II) Mixed-L	INOR 0164
Chemistry Olympiad Competition	Scheduled for USA, July 11-22, 1992.= +rnational	CHED 0139
D-Glucose Stereochemistry:	Schematic Method for Fischer's Proof Teaching.=	CHED 0095
ical Properties of Ru(II)	Schiff Bases Complexes.= Synthesis and Electrochem	INOR 0260
ehydrogenation of Precursor Poly(	Schiff Bases).= +ed Polybenzimidazoles by Cyclod	POLY 0045
otic Agent for the Treatment of	Schizophrenia.= P-9236: A New Atypical Antipsych	MEDI 0085
nts of the Natural Melampolide	Schkuhriolide: An Alternative Approach to the Oplop	ORGN 0018
for Chemical Educati+ Prehigh	School Chemistry Outreach Programs of the Institute	CHED 0073
ummer Intern Program for High	School Chemistry Teachers in the Greater Cincinnat	TECH 0002
Award Presentation: Intr+ High	School Chemistry Teachers Program. POLYED 1990	CHED 0025
Educating High	School Chemistry Teachers: The Need for Reform.=	CHED 0067
ETS: An Interdisciplinary Middle	School Curriculum.= +ol Science Education. FAC	CHED 0071
nto Middle, Vocational, and High	School Curriculum.= +grating Polymer Education i	CHED 0026
nary Middle School Cu+ Prehigh	School Science Education. FACETS: An Interdiscipli	CHED 0071
e Doing to Help Advance Prehigh	School Science Education.= +Two-Year College Ar	CHED 0076
Society SEED Program for High	School Students at Hunter College, Chemistry Depar	CHED 0143
Than High	School Students.= Project SEED Benefits More	CHED 0145
+ectronic Bulletin Board for High	School Teachers of Chemistry in the Northwest.=	CHED 0111
High	School.= Teaching of Chemical Engineering in	CHED 0108
aboratories and Inservic+ High	School-College Interface: Regional Instrumentation L	CHED 0112
evitalizing Science in the Nation's	Schools.= +Institute for Chemical Education in R	CHED 0138
Theory of	Schottky Barriers in Diamond/Metal Interfaces.=	PHYS 0358
	ScienceGrasp! It's Elementary.=	CHED 0075
Abstract	Scientific Concepts in the Courtroom.=	CHAL 0012

=	Scientific Division of Routledge, Chapman, and Hall.	CHAS 0028
Using	Scientific Evidence in Patent Proceedings.=	CHAL 0014
In-House Counsel.= The	Scientific Expert: Considerations of Corporate	CHAL 0017
Back?.= Testimony of	Scientific Experts: Is the Pendulum Swinging	CHAL 0016
Isomerization Catalysts.=	Scientific Foundations for Synthesis of Hydrocarbon	PETR 0135
Careers in	Scientific Librarianship.=	YCC 0009
t and Validation Are the Keys to	Scientific Risk Assessments of Inhalation and Derma	ENVR 0042
cision-Making Process and+ The	Scientist as an Expert Witness. Overview of Legal De	CHAL 0011
Project LABS—A Teacher—	Scientist Interaction for K-12 Teachers.=	CHED 0002
y. Arthur Dock Fon Toy—	Scientist, Colleague, Friend.= Phosphorus Chemistr	INOR 0288
Political Consequences for	Scientists Studying Agent Orange: A Case Study.=	PROF 0008
Bibliographic Databases for	Scientists.=	CINF 0042
Some Dichotomies of Creative	Scientists, including Chemists.=	CONG 0009
Sulfur-Sulfur Bond	Scission by a Copper Reagent.=	ORGN 0252
ture of Glassy Polymers—I. Chain	Scission in Polystyrene.= +mechanisms in the Frac	PMSE 0073
Grignard Reagent Promoted	Scission of the C-S and S-S Bonds.=	ORGN 0253
Coordination, Dimerization, and	Scission on Tungsten-Iridium Tetranuclear Cluster C	CATL 0008
Devices.= Meso-	Scopic Effects in Small Metal and Semiconductor	BIOT 0002
Synthesis of a New Type of	Scopolamine Analogues.=	MEDI 0136
a Soil RISc: A Rapid, On-Site	Screening Test for Pentachlorophenol in Soil.= Pent	ENVR 0033
and Energy Distribution in Twin	Screw Extruders.= +y Lecture. Distributive Mixing	PMSE 0175
lantas Pertenecientes a la Familia	Scrophulariaceae de Mexico.= +n Quimica de las P	ORGN 0185
Legal	Scrutiny of DNA Typing Techniques.=	CHAL 0003
s in the Single-Phase Regions of	SDS/1-Pentanol/Cyclohexane/Water Microemulsion	COLL 0024
el Cadmium Selenolate Compo+	^{77}Se and ^{113}Cd Single-Crystal NMR Studies of a Nov	INOR 0078
enetrating 3D Frameworks of [Pd(	Se$_4$)$_2$]$^{2-}$ and [Pd(Se$_6$)$_2$]$^{2-}$.= +$_2$PdSe$_{10}$: Two Interp	INOR 0200
meworks of [Pd(Se$_4$)$_2$]$^{2-}$ and [Pd(	Se$_6$)$_2$]$^{2-}$.= +$_2$PdSe$_{10}$: Two Interpenetrating 3D Fra	INOR 0200
mium Selenolate Compound, [Cd(	Se-2,4,6-i-Pr$_3$C$_6$H$_2$)$_2$(bpy)].= +dies of a Novel Cad	INOR 0078
Cycling in the	Sea.= Marine Bioinorganic Chemistry. Arsenic	INOR 0136
nation May Help Reduce Decay in	Seal-Packaged Citrus and Pepper Fruits.= +Illumi	AGFD 0076
Reaction Similarity	Searches in the Beilstein File.= Structure and	CINF 0039
3D Shape Fitting and 3D-DB	Searching by SPERM.=	COMP 0040
-by-Atom 3D Search Routines for	Searching CAS Registry Substances.= +on of Atom	CINF 0023
Databases: Does Conformational	Searching Improve the Yield of Activities?.= +in 3D	CINF 0026
apping Procedure for Similarity	Searching in Databases of 3D Chemical Structures.=	COMP 0043
Database Structure and	Searching in MACCS-3D.=	CINF 0022
Ring Index	Searching of the Beilstein Database.=	COMP 0018
ntary Methods wit+ 3D Database	Searching vs. de Novo Construction: Two Compleme	COMP 0050
Molecules: A Flexible Prescre+	Searching 3D Databases of Conformationally Flexible	CINF 0027
Cited Patent	Searching: A Tutorial.=	CINF 0040
Flexible Queries in 3D	Searching: Techniques in 3D Query Formulation.=	CINF 0025
Three-Dimensional Substructure	Searching.= +ovel Inhibitors of HIV-1 Protease by	COMP 0038
l: Early Steroid Research at G. D.	Searle Co.= +roid Chemistry. Steroids and the Pil	HIST 0020
ic Bight.=	Seasonal Dissolved Gas Cycling in the Middle Atlant	GEOC 0003
entary Organic Matter in Cont+	Seasonal Variations in the Remineralization of Sedim	GEOC 0015
erpenetrating Polymer Blends of a	SEBS Block Copolymer and a Polyetherester.=	PMSE 0058
s of LDPE by SEC/LALLS and	SEC/CV Systems.= Long-Chain Branching Analysi	PMSE 0047
Branching Analysis of LDPE by	SEC/LALLS and SEC/CV Systems.= Long-Chain	PMSE 0047
ei.=	Secoiridoids and Other Constituents of Fraxinus uhd	ORGN 0187
cas: Vinculo entre Universidades y	Secotr Productivo.= +cio Social y Programas de Be	CHED 0114
Bioreactor.= Controlled Protein	Secretion in a Single-Pass Ceramic Matrix	BIOT 0011
nation of a Recombinant Human	Secretory Phospholipase A$_2$ from Mammalian Cell C	BIOL 0072
gical Support for the Productive	Sector: The Case of the Technological Transfer Unit	SCHB 0025
.= Sintesis y Modificacion de	Secuencias de Destilacion con Integracion de Energia	I&EC 0074
ations in the Remineralization of	Sedimentary Organic Matter in Continental Margin	GEOC 0015
he Sorption of Organic Solutes on	Sedimentary Organic Matter.= +e for Measuring t	ANYL 0072
n of Channel Dimensions Used in	Sedimentation Field-Flow Fractionation through Co	PMSE 0028
rization of Latex Aggregates by	Sedimentation Field-Flow Fractionation.= Characte	PMSE 0029
Phosphorus Distribution in	Sediments in and near the Amazon Delta.=	GEOC 0011
d C-14 Budget for Upper Slope	Sediments of the Middle Atlantic Bight.= Carbon an	GEOC 0016
Carbon (DOC) Flux out of Margin	Sediments on Oceanic DOC.= +Dissolved Organic	GEOC 0017
Fate of a PCB in Anoxic Bottom	Sediments: Characterizing the Role of in Situ Humi	ENVR 0017
mistry of Iron in Anoxic Coastal	Sediments: Cycling of Iron Oxides and Their Potenti	GEOC 0007
osphorus in Continental Margin	Sediments: Implications for Burial Fluxes and Phosp	GEOC 0013
n Oxide Extraction from Soils and	Sediments: Reductive Dissolution by Ti(III).= +Iro	ENVR 0058
Biphenyls in Anaerobic	Sediments.= Dechlorination of Polychlorinated	BIOT 0120
and Coastal Marine	Sediments.= N:P Release Ratios from Freshwater	GEOC 0012
from California Continental Shelf	Sediments.= +ron, Cobalt, Copper, and Manganese	GEOC 0010
I Think I Shall Never	See a Polymer Lovely as a Tree.=	POLY 0319
Germination and the Effects of	Seed Aging: Solid-State NMR and NMR Imaging.=	AGFD 0104
Project	SEED at Lehman College.=	CHED 0140
Project	SEED at Stevens.=	CHED 0144
Project	SEED Benefits More Than High School Students.=	CHED 0145
Sidney-Cornell Project	SEED Connection.= Project SEED. The	CHED 0036
Project	SEED Experience.= Student Perspective on the	CHED 0037
of Oil an+ Total Evaluation of a	Seed from the Jatropha Family as a Possible Source	AGFD 0038
nvasive Techniques for Studying	Seed Germination and the Effects of Seed Aging: Sol	AGFD 0104
t+ Toxicity of Annona squamosa	Seed Oil Extracts to Tribolium castaneum L. (Coleop	AGRO 0036
Experiences of a	SEED Preceptor Team.=	CHED 0038
the American Chemical Society	SEED Program for High School Students at Hunter	CHED 0143

Evaluating a Project	SEED Program in Indianapolis.= Coordinating and	CHED 0141
stallization.= Investigation of	Seed Properties' Influence on the Rate of ZSM–5 Cry	PETR 0039
Project	SEED. Overview of Project SEED.=	CHED 0035
Connection.= Project	SEED. The Sidney–Cornell Project SEED	CHED 0036
Project	SEED.= Project SEED. Overview of	CHED 0035
Project	SEED.= How To Get Your Students Involved in	CHED 0039
Project	SEED.= Organizational Structure of Dayton, Ohio,	CHED 0142
lene Terephthalamide).= Amide	Segmental Dynamics in N–Deuterated Poly(p–pheny	POLY 0058
Characterization of Polymer	Segmental Motion via Deuterium NMR.=	PHYS 0140
ks: Character+ Lattice Model for	Segmental Orientation in Deformed Polymer Networ	COLL 0016
ephthalate) and Poly(ester–Ether)	Segmented Block Copolymers.= +oly(butylene Ter	PMSE 0148
Morphological Changes in	Segmented Polyurethanes.= Thermally Induced	PMSE 0149
de and Elastomeric Polybutadiene	Segments.= +Helical and Amorphous Polyisocyani	PMSE 0071
Kinetics in a Capillary: Reactant	Segregation and Anomalous Time Exponents.= +on	PHYS 0273
idable 304 y 316 Utilizadas en el	Seguimiento de las Reacciones Acido Base en Medio	ANYL 0055
Development of Agents Acting	Selectively at Dopamine Receptors.=	MEDI 0080
aship to Butane Hydrogenolysis	Selectivities Exhibited by W–Ir and Mo–Ir Cluster–D	CATL 0008
on on X and Y Zeolites: Product	Selectivity and Correlation of Products with the cupr	CATL 0017
Me$_3$)$_2$LCl(L = ^{i}Pr$_3$,CO): Aspects of	Selectivity and Mechanism.= +nation Using Rh(P	INOR 0151
atalytic Two–Stage L+ Distillate	Selectivity and Quality of Bituminous Coals in the C	CATL 0053
ng. Comparison of Alkane Dehy+	Selectivity in Catalysis: Clusters, Alloys, and Poisoni	CATL 0007
rization and Model Compound +	Selectivity in Catalysis: Fuel Chemistry—I. Characte	CATL 0039
–Transport Selectivity Models +	Selectivity in Catalysis: Oxidation Catalysis. Reaction	CATL 0056
‒. Crystal and Catalytic Chemist+	Selectivity in Catalysis: Reaction Pathway Control—	CATL 0026
‒I. Application of the BOC–MP+	Selectivity in Catalysis: Reaction Pathway Control—	CATL 0032
ic Sieving by Heterogeneous Ca+	Selectivity in Catalysis: Shape Selectivity—I. Dynam	CATL 0013
nduced Catalytic Reactions in +	Selectivity in Catalysis: Shape Selectivity—II. Photo	CATL 0020
e Hydrogenation Technology for+	Selectivity in Catalysis: Stereoselectivity. Asymmetri	CATL 0001
‒ Coprocessing Using Syngas.=	Selectivity in Catalysts: Fuel Chemistry—II. Coal/Oi	CATL 0049
ce Fugacities and the Control of	Selectivity in Catalytic Dehydrogenation and Dehydr	CATL 0061
Mechanism and	Selectivity in Catalytic Olefin Polymerization.=	CATL 0010
aterials as Catalysts—Activity and	Selectivity in Hydrocracking Reactions.= +rbon M	CATL 0046
ns.=	Selectivity in Nonchromatographic Protein Separatio	BIOT 0019
revianam+ Unusual Stereofacial	Selectivity in the Biosynthesis and Synthesis of the B	ORGN 0283
xy–1,3–Butadie+ Diastereofacial	Selectivity in the Diels–Alder Reaction of 2,3–Dialko	ORGN 0233
Face	Selectivity in the Protonation of Glycals.=	ORGN 0069
logen Compounds: Ion–C+ State	Selectivity in the Reaction of Copper Atoms with Ha	PHYS 0368
Face	Selectivity in 5-,7-Disubstituted Adamantanones.=	ORGN 0238
3ed Fische+ Reaction–Transport	Selectivity Models and the Design of Ruthenium–Ba	CATL 0056
ts in the Hydrodemetallization +	Selectivity of Alumina–Supported Vanadium Catalys	CATL 0044
3ter+ Solvent Effects on the Side	Selectivity of Singlet Oxygen with α,β Unsaturated E	ORGN 0164
Catalytic Reforming: The Reac+	Selectivity of Supported Bimetallic Catalysts Toward	CATL 0042
take, Translocat+ Mechanism of	Selectivity of the Herbicide DPX–E9636 in Corn: Up	AGRO 0027
Product	Selectivity of Vanadium Bromoperoxidase.=	INOR 0017
Conferment of Substrate	Selectivity on Enzymes.= Targeted Catalysis:	BIOL 0090
ort Preparation on Aromatization	Selectivity over Nonmicroporous Pt Catalysts.= +p	CATL 0030
κ–Opioid Antagonist Potency and	Selectivity.= +rmacophore Fragment of norBNI in	MEDI 0157
Dienophile Face	Selectivity.= Conformation–Dependent Diels–Alder	ORGN 0241
Selectivity in Catalysis: Shape	Selectivity—I. Dynamic Sieving by Heterogeneous Ca	CATL 0013
Z Selectivity in Catalysis: Shape	Selectivity—II. Photoinduced Catalytic Reactions in	CATL 0020
the High+ Cation–Dopant Site	Selectivity, Oxygen Stoichiometry, and Cu Valence in	PHYS 0027
s in the Cen+ Novel Polynuclear	Selenide Clusters of Au+ and In^{3+}, Trapping Na+ Ion	INOR 0203
o Cadmium Sulfide and Cadmium	Selenide.= +sorption of Carbonyl Compounds ont	COLL 0175
R Studies of a Novel Cadmium	Selenolate Compound, [Cd(Se–2,4,6–i–Pr$_3$C$_6$H$_2$)$_2$(bpy	INOR 0078
Silylellurolate and	Selenolate Ligands.= Volatile Metal Complexes with	INOR 0172
DC Arc Discharge+ OES and	SEM Observations of Diamond Films Synthesized by	FUEL 0038
uoride with Human Enamel: XPS,	SEM, and Solution Characterization.= +ction of Fl	COLL 0085
Depletion Studies on	Semduramicin in the Chicken.= Tissue Residue	AGRO 0079
300–MHz FT–NMR in a Second–	Semester Sophomore Unified Laboratory Course.=	CHED 0181
Compatibility of Ternary	Semi IPNs.=	PMSE 0156
Poly(Ethylene Terephthalate)	Semi–IPNs with Functionalized Triglyceride Oils.=	PMSE 0088
rs and High–Tg Polyimide–Based	Semi–IPNs.= +terization of bisNadimides Oligome	POLY 0133
Latex PS/XPS	Semi–IPNs.=	PMSE 0063
Metal and	Semi–Metal Cluster Ions.= Oxidation Reactions of	PHYS 0174
tion in Polyurethane/Polystyrene	Semi-1 IPNs.= +tworks. Kinetics of Phase Separa	PMSE 0021
rder Nonlinear Susceptibilities of	Semiconducting Polymers: Comparison with Theoret	POLY 0009
ter Distributions in Pseudobinary	Semiconductor Alloys.= +surements of Atomic Clus	PHYS 0375
Aldehydes with	Semiconductor as Catalyst.= Photoreduction of	ORGN 0247
Simulated Sunlight Irradiation of	Semiconductor Catalysts.= +l Organics Using AM1	INOR 0318
k Copolymers.=	Semiconductor Clusters Prepared Using ROMP Bloc	INOR 0307
Precursors for Soluble III–V	Semiconductor Clusters.= Organometallic	INOR 0313
ge–Carrier Dynamics of Quantized	Semiconductor Colloids.= +ace Properties on Char	COLL 0151
Injection in Large–Bandgap	Semiconductor Colloids.= Sensitized Charge	COLL 0149
gh–Performance III–V Compound	Semiconductor Devices by OMCVD.= +owth of Hi	COLL 0010
Small Metal and	Semiconductor Devices.= Meso–Scopic Effects in	BIOT 0002
facial Electron Transfer at a Solid	Semiconductor Electrode–Liquid Junction.= +Inter	COLL 0088
l Electron Injection into Colloidal	Semiconductor Membranes.= +us Fluids. Vectoria	COLL 0147
Syntheses of Phosphide	Semiconductor Nanoclusters.= Metalorganic	INOR 0102
eneous Photocatalysis with TiO$_2$	Semiconductor Particles in Aqueous Media: A Look a	COLL 0131
organic C+ Molecular Models for	Semiconductor Particles: Luminescence Studies of In	INOR 0317

ications.=	Semiconductor Quantum Devices, Circuits, and Appl	BIOT	0003
Reactions of Zinc Alkyls on	Semiconductor Substrates: GaAs(100) and Si(100).=	COLL	0048
tinuclear NMR Study of an Alloy	Semiconductor.= +and Optical Spectroscopies. Mul	PHYS	0372
Nonlinear Optical Studies of	Semiconductor–Liquid Interfaces.= Picosecond	PHYS	0283
oelectroquimicas, I:+ Materiales	Semiconductores Policristalinos dentro de Celdas Fot	PHYS	0221
taleno con Electrodos Metalicos y	Semiconductores.= +ectroquimica del 1–Nitro–Naf	PHYS	0280
OMCVD of III–V Compound	Semiconductors.= FTIR of Surface Reaction in	COLL	0011
uilibrium Electronic Properties of	Semiconductors.= +Imaging Subsurface and Noneq	PHYS	0208
to III/V Compound	Semiconductors.= Some New Organometallic Routes	INOR	0146
orhodopsin.=	Semiconstant Holographic Memory Based on Bacteri	BIOT	0079
morphous Ketimine Intermediat+	Semicrystalline Poly(Arylene Sulfide) Ketones via A	PMSE	0123
Structure and Dynamics of a	Semidilute Polymer Solution under Shear Flow.=	PMSE	0108
Methods in Diamond Surface R+	Semiempirical Molecular Orbital and Tight Binding	PHYS	0356
mic Acids and Th+ Ab Initio and	Semiempirical Molecular Orbital Studies of Hydroxa	PHYS	0326
High–Performance Materials from	Semiflexible Polymers Such as the Cellulosics.=	COLL	0015
A Molecular Organic–Inorganic	Semiinterpenetrating Network.=	POLY	0286
s: A Reassessment of Some	Seminal Papers.= The Mechanism of β–Glycosidase	ORGN	0070
of the cis and trans Isomers of+	Semipreparative HPLC Isolation and NMR Analyses	AGRO	0033
Phase Transitions for Networks of	Semirigid Chains.= +ess–Strain Relationships and	COLL	0016
Polyimide Blends with Rigid and	Semirigid Polyamides.= +perties of Cross–Linkable	PMSE	0023
eri+ Gas/Particle Partitioning of	Semivolatile Organic Compounds to Model Atmosph	ENVR	0059
of the Atmospheric Deposition of	Semivolatile Organic Compounds.= +lation Model	GEOC	0004
Films.= Diffusion in the Gas	Sensing of Phthalocyanine Langmuir–Blodgett	POLY	0118
Polyurethane for Glucose	Sensing.= Glucose Oxidase Immobilized with	BIOT	0147
Lipid Bilayers Containing Light–	Sensitive Compounds.= Mechanical Properties of	BIOT	0106
Induced by Light–	Sensitive Coordination Compounds.= Photocatalysis	INOR	0340
onventional SEC and Molar Mass	Sensitive Detectors.= +r Characterization Using C	PMSE	0051
ics and Mechanism of a Charge–	Sensitive Ion–Conducting Device Consisting of Metal	BIOL	0098
N–Alkylamides with Voltage–	Sensitive Sodium Channels.= Interactions of	AGRO	0151
	Sensitivity Analysis for an Air Quality Model.=	ENVR	0026
the Limit: Pushing the Bounds of	Sensitivity and Spatial Resolution.= +ctroscopy at	ANYL	0012
iber–Optic Sensors.=	Sensitivity Enhancement in Fluorescence Capillary/F	ANYL	0077
pray LC/MS.= Enhanced	Sensitivity Studies with Particle Beam and Thermos	ANYL	0123
ms Based on Geothermal Water+	Sensitivity to Contamination of Drilling Fluids Syste	I&EC	0011
ation–Controlled Photoreactions	Sensitized by Colloidal CdS: A Flash Photolysis Stud	COLL	0150
onductor Colloids.=	Sensitized Charge Injection in Large–Bandgap Semic	COLL	0149
roperties of Covalently Attached	Sensitizer/Mediator or Donor/Acceptor Complexes.=	INOR	0038
Architecture of Device and	Sensor Made from Biomaterials.=	BIOT	0123
osites.=	Sensor Plate for Manufacturing Thermoplastic Comp	PMSE	0115
Capillary/Fiber–Optic	Sensors.= Sensitivity Enhancement in Fluorescence	ANYL	0077
Solid–State Chemical	Sensors.=	CHED	0032
or, as Determined by Descriptive	Sensory Analysis and Gas Chromatographic Techniq	AGFD	0161
xidant–Treated+ Chemical and	Sensory Evaluation of Flavor in Untreated and Antio	AGFD	0120
ds.= Lipid Oxidation in Foods.	Sensory Evaluation of Lipid Oxidation in Different F	AGFD	0156
o+ Evidence for the Existence of	Separately Controlled Plastic and Conserved Phases	BIOL	0152
rs on Cell+ Direct Enantiomeric	Separation and Analysis of Some Aromatase Inhibito	ANYL	0133
ay Scattering from Polymeric Gas	Separation and Barrier Materials.= +alysis of X–r	POLY	0142
Carbon Molecular Sieves for	Separation and Catalysis.= Design and Synthesis of	PETR	0059
UV/Visible Detection to the LC	Separation and Determination of Methylene Blue an	AGFD	0101
naqueous Bioprocessing: Protein	Separation and Purification in Neat Organic Solvent	BIOT	0063
Microphase	Separation and Reversible Gelation.=	POLY	0122
et des Complexants sur la Fotta+	Separation Arsenopyrite–Pyrite: Effet des Oxydants	ANYL	0067
Phase	Separation in Interpenetrating Polymer Networks.=	PMSE	0022
Phase	Separation in Polymer–Dispersed Liquid Crystals.=	COLL	0014
er Networks. Kinetics of Phase	Separation in Polyurethane/Polystyrene Semi–1 IPN	PMSE	0021
–Field–Induced Transient Phase–	Separation in w/o–Microemulsion.= +cs of Electric	COLL	0146
ription and Application to Phase–	Separation Kinetics in PC/PMMA Blends.= +Desc	POLY	0278
Solid–State NMR Studies of	Separation Media.=	ANYL	0064
HPLC	Separation of Alkaloids.=	ANYL	0135
n Liquid Chromatography for the	Separation of Aromatic Compounds.= +scriminatio	ANYL	0098
neity of Recombinant Human +	Separation of Carbohydrate–Mediated Microheteroge	ANYL	0018
ase+ High–pH Anion–Exchange	Separation of Carbohydrates: The Role of Mobile–Ph	CARB	0066
hange Resins for Chromatographic	Separation of Carbohydrates.= +ion of Cation–Exc	CARB	0064
imultaneous Determination and	Separation of Primary and Secondary Amines by Rev	ANYL	0089
Chromatography.= Rapid	Separation of Proteins and Peptides by Perfusion	ANYL	0127
atite and Zirconium Oxide for the	Separation of Proteins.= +en Calcium Hydroxyap	ANYL	0027
Gelation and Phase	Separation of PVC/Plasticizer Solutions.=	POLY	0173
Magnetic	Separation of Reduced Crude Conversion Catalyst.=	I&EC	0063
Isomers.= Chromatographic	Separation of Sucrose Monostearate Structural	CARB	0050
Penicillium Amagasakiense by I+	Separation of the Glucose Oxidase Isoenzymes from	ANYL	0016
l Macromolecules by Flow Fiel+	Separation of Water–Soluble Synthetic and Biologica	PMSE	0011
er on the Polymerization Phase–	Separation Process, and Generated Morphologies.=	POLY	0112
. Integrati+ Integrating Primary	Separation Steps with Upstream–Downstream Stages	BIOT	0069
4–Benzoyl–5–Methyl–2–Pheny+	Separation Studies of the Selected Actinide Ions with	NUCL	0073
py as a Detector for Liquid–Phase	Separation Systems.= +ardment Mass Spectrosco	ANYL	0125
tical Industry.= Trends in	Separation Technologies and the Need of Pharmaceu	ANYL	0111
Phase	Separation.= Rheo–Optics of Flow–Induced	PMSE	0106
omatographic Methods of Protein	Separation—I. Perspectives in Protein Pur+ Nonchr	BIOT	0018
omatographic Methods of Protein	Separation—II. Bioseparation Using Affini+ Nonchr	BIOT	0044
Bonded Phas+ Multidimensional	Separations and Multichannel Detections in Normal	ANYL	0088

cterization. Two-Dimensional	Separations by LC–CZE.= Rapid Biopolymer Chara	ANYL	0126
Capillary Electrophoresis	Separations for Small Molecules.=	ANYL	0137
Chemical	Separations.= Polyphosphazene Membranes for	INOR	0070
Protein	Separations.= Selectivity in Nonchromatographic	BIOT	0019
Fragment	Separators.= Evolution of Heavy–Ion Recoil and	NUCL	0038
Al–Modified	Sepiolite as Catalyst or Catalyst Support.=	PETR	0015
g A_1 and A_2 Adenosine Recept+	Sequence Analysis and Computer Modeling of the Do	MEDI	0114
ing for Yersinia ps+ Cloning and	Sequence Analysis of AscB and AscC, the Genes Cod	BIOL	0110
Integrating Protein	Sequence and Structure Data.= Prolog Approach to	CINF	0030
Chiral Clues to Helical	Sequence Lengths in Poly(n–alkyl Isocyanates).=	POLY	0348
viride.= Amino Acid	Sequence of a 20K Xylanase from Trichoderma	BIOT	0211
ucleotides by Mitomycin+ Base–	Sequence Specificity of the Monoalkylation of Oligon	MEDI	0015
gress Toward Understanding Its	Sequence–Selective Binding and Cleavage of DNA.=	ORGN	0298
Porphyrin.=	Sequence–Specific DNA Interactions of a Tricationic	BIOL	0127
to Define the Recognition Site of	Sequence–Specific DNA–Binding Proteins.= +thod	BIOL	0129
as Possible Can+ Specific DNA	Sequences and Their Complexes in Living Organisms	BIOT	0148
f Phenolics in Optimally Designed	Sequencing Batch Reactors.= +ic Biodegradation o	BIOT	0138
ical Chemistry. High-Speed DNA	Sequencing in Ultrathin Gels.= +w Face of Analyt	ANYL	0033
ubstitution Mediated by Chrom+	Sequential Benzannulation/Nucleophilic Aromatic S	ORGN	0221
Reactions: Prompt or	Sequential?.= Projectile Brakup in 25–MeV/A	NUCL	0027
with a Hig+ EPSP Synthase: A	Sequential–Ordered Kinetic Mechanism or Random,	BIOL	0053
Fil+ Determination of Human	Seric Copper by Inverse Voltammetry at the Mercury	ANYL	0056
o and Boronic Acid Inhibitors on	Serine Proteases: Evidence for Multiple Binding Mec	BIOL	0082
oxhanase Substrat+ Synthesis of	Serine–AMC–Carbamate: A Novel Fluorogenic Trypt	ORGN	0023
eated Com+ Dopaminergic and	Serotonergic Activities of Imidazoquinolinones and R	MEDI	0126
ystems: Novel 5–HT3 and 5–HT4	Serotonergic Agents.= +tion of Novel Azatricyclic S	ORGN	0104
tricyclic Agents.=	Serotonergic 5HT3 and 5HT4 Activities of Novel Aza	MEDI	0187
Organic Approaches to	Serotonin Analogs.=	ORGN	0254
Azabicyclic Benzamides: Potent	Serotonin 5–HT4 Agonists and/or 5–HT3 Antagonist	MEDI	0069
by Use of Lecithin Supp+ Blood	Serum Lipid Concentrations of Humans as Affected	AGFD	0034
Murine Hybridoma to	Serum–Free Growth.= Rapid Adaptation of a	BIOT	0163
istein Online: Options for Search	Services in an Academic Library.= +d Access to Be	COMP	0012
Resources, Inc.= Toxicological	Services Provided by Hazard Communication	CHAS	0024
Career	Services.= The Use of ACS Employment and	YCC	0013
Universidades y Secotr Product+	Servicio Social y Programas de Becas: Vinculo entre	CHED	0114
	Sesquiterpene Induction from Plant Root Cultures.=	BIOT	0232
ighly Substituted Carb+ Plenary	Session. Annulation Strategies for the Synthesis of H	ORGN	0286
yl Compounds Using+ Plenary	Session. Asymmetric Synthesis of α–Hydroxy Carbon	ORGN	0293
eptatriene to Carbohy+ Plenary	Session. From Cyclopentadiene, Benzene, and Cycloh	ORGN	0272
etermination of the Ste+ Plenary	Session. Mechanisms of Reactions at Heteroatoms: D	ORGN	0120
Activators.= Plenary	Session. Multifunctional Enzyme Mimics and	ORGN	0265
Reactions.= Plenary	Session. Studies in Asymmetric Olefination and Ene	ORGN	0279
Langmuir Lectures: Plenary	Session. Surface–Introduced Liquid Surfaces.=	COLL	0062
ciated by Transition M+ Plenary	Session. Transformations of Organic Compounds Me	ORGN	0127
First Demonstration	Session.=	CONG	0006
chemistry of Variculanol: A Novel	Sestererpenoid from Aspergillus variecolor.= +ereo	ORGN	0184
	Setting Your Strategies.=	SCHB	0003
and Training Used in a University	Setting.= +ting Chemical Safety: Some Education	CHAS	0048
Pretreatment of Municipal	Sewage Sludge for Gasification.=	FUEL	0117
Ontario's Program for Using	Sewage Sludge in Agriculture.=	FERT	0021
, Biomass, and Municipal Waste.	Sewage Sludge: A Fascinating Feedstock for Clean E	FUEL	0115
rom Waste and Coal: Energy from	Sewage Sludge, Biomass, and Municipal Waste. +f	FUEL	0115
ogen Peroxide and Iron Salts on	Sewage, Part 1: Odor Control and Phosphate Remova	ENVR	0034
ding Protein of Larval Manduca	sexta by the Use of Photoaffinity Analog of JH II.=	AGRO	0014
the Lepidopteran Manduca	sexta.= Autoparalytic Peptides from Hemolymph of	AGRO	0012
Tobacco Hornworm, Manduca	sexta.= Identification of an Allatostatin from the	AGRO	0008
Fifth Stadium Larvae of Manduca	sexta.= +olism Study of Juvenile Hormone in the	AGRO	0022
Trigeminal Pentafluorosulfanyl (	SF_5) Carbon and Nitrogen Derivatives.=	INOR	0024
Substituent Group Contributions (	SGCs) Increase with Affinity.= +ts Confirms that	MEDI	0088
ctinomycetes Fermentations from	Shake Flask to Production–Scale Fermenters. +of A	BIOT	0083
Do Younger Chemists Get a Fair	Shake? An Analysis of NSF Funding Opportunities	YCC	0003
ociations of Trace Elements in Oil	Shale Kerogen.= +nation of Organic/Inorganic Ass	FUEL	0045
Producing Environmentally Safe	Shale Stabilizers.= Novel and Versatile Process for	CARB	0009
I Think I	Shall Never See a Polymer Lovely as a Tree.=	POLY	0319
n Energies in Slitlike and Wedge–	Shaped Micropores.= +. Calculation of Adsorptio	FUEL	0008
New Comb–	Shaped Polymers.=	POLY	0268
: Isomer+ Ab Initio Energies and	Shapes of HSiS+, HSiO+, HCP, and HSiP, and Thei	PHYS	0315
er Blend Solution during Steady	Shear and Cessation of Shear Studied by Light Scatt	PMSE	0134
by Light Scattering during Steady	Shear and Following Cessation of Shear.= +udied	PMSE	0107
sses in Organic Polycrystals under	Shear Deformation at High Pressure.= +ical Proce	PHYS	0245
odal Decomposition under Simple	Shear Flow.= +s. Domain Growth during the Spin	PMSE	0104
te Polymer Solution under	Shear Flow.= Structure and Dynamics of a Semidilu	PMSE	0108
ng Steady Shear and Cessation of	Shear Studied by Light Scattering.= +Solution duri	PMSE	0134
Magnetic, and Hydrodynamic—II.	Shear Thinning of Polymer Solutions.= +Electric,	COLL	0159
and Low	Shear.= Extruding Waste Plastics with High Mixing	I&EC	0030
Shear and Following Cessation of	Shear.= +udied by Light Scattering during Steady	PMSE	0107
Fluorescence and	Shear–Induced Mixing in a Miscible Polymer Blend.=	PMSE	0133
ns Made on Steady and Dynamic	Shear–Induced Morphology in Blends by Optical Rh	PMSE	0085
polymer SBS.=	Shear–Induced Structural Transitions in Triblock Co	PMSE	0136
	Sheet and Tube Alkoxysiloxanes.=	POLY	0257

r Cell-Shape Change during Cell-	Sheet Morphogenesis and Cytokinesis.= +quired fo	BIOL	0140
Data	Sheet Preparation.= Resources for Material Safety	CHAS	0040
on Damping of Laminated Steel	Sheet with Polyurethane/Acrylate Interpenetrating P	PMSE	0057
Company.= WHMIS Data	Sheets from the Viewpoint of an American Chemical	CHAS	0035
Matrix Composite Hydrogel	Sheets with Thermoreversible Transitions.= PTFE-	POLY	0245
Material Safety Data	Sheets—Effective in the Workplace?.=	CHAS	0036
anese from California Continental	Shelf Sediments.= +ron, Cobalt, Copper, and Mang	GEOC	0010
Bisulfite Replacement and	Shelf-Life Extension of Prepeeled Potatoes.=	AGFD	0058
Indicators for	Shelf-Life Monitoring.= Time-Temperature	AGFD	0080
Chemometric Approach to Milk	Shelf-Life Prediction and Flavor-Quality Evaluation	AGFD	0108
Excited States of Maximal d-	Shell Occupancy in the Ru + H_2 Reaction.= Role of	PHYS	0243
ons over Heterogeneous Catalysts.	Shell Polygasoline and -Kero Process.= +drocarb	PETR	0025
nce Epoxies Using Designed Core-	Shell Rubber Particles.= +ning of High-Performa	POLY	0157
olism in Food-Producing Fish and	Shellfish.= +thods for Studying Xenobiotic Metab	AGRO	0096
ge of Mechanism Block Copoly+	Sherwin-Williams Student Award Symposium. Chan	PMSE	0071
scopic Studies of the Water Gas	Shift Reaction over Ruthenium-Exchanged X Zeolite	CATL	0027
polysaccharides of Salmonella and	Shigella Bacteria.= +Chains of Cell Envelope Lipo	CARB	0017
Structure of Polyethylene	Shish Kebabs in Situ.=	PMSE	0184
Studies of Polymers for	Shock Absorption Applications.=	PMSE	0181
Folding—GroE Heat	Shock Proteins.= Role of Chaperones in Protein	BIOT	0116
luene/Methanol Mixtures.= A	Shock Tube Investigation of Soot Formation from To	FUEL	0113
Cluster Participation in	Shock Waves.= Kinetic Processes with Metal	PHYS	0175
d 1,5-Hexadiyne behind Reflected	Shock Waves.= +actions of Propargyl Chloride an	FUEL	0068
y the Flash or Laser Photolysis-	Shock-Tube Method: Results for the Oxidations of H	FUEL	0123
ing Tunneling Microscope Images	Showing Anomalous Periodic Structures.= +Scann	PHYS	0212
uce+ Design of a Mill Drier for	Shrimp Head: Bromatological Characteristics of Prod	AGFD	0105
duce+ Design of a Mill-Drier for	Shrimp Heads: Bromatological Characteristics of Pro	ENVR	0046
Excitations in the	^{28}Si + ^{28}Si Reaction.= Entrance Channel	NUCL	0039
Protoypes of Polymer Anions with	Si and Ge Backbones.= +Ge$_2$H$_6$ Radical Anions as	PHYS	0313
Adsorption and Growth on	Si and Ge Surfaces.=	PHYS	0186
^{28}Si +	^{28}Si Reaction.= Entrance Channel Excitations in the	NUCL	0039
Growth of Silicide Layers on	Si Surfaces.=	PHYS	0100
Reaction Pathways for	Si$^+$ + SiX$_4$ (X = F, Cl).= An ab Initio Study of the	PHYS	0248
E(CO)(PPH$_3$)C(O)-C=-R(R=CH$_3$,	Si(CH$_3$)$_3$).= +nthesis and Chemistry of (η^5-C$_5$H$_5$)F	INOR	0156
uckled-Dimer Configurations in	Si(100) Observed with a New Variable Temperature	PHYS	0185
istry of Hydrogen on Si(111) and	Si(100) Surfaces.= Role of Bond Strain in the Chem	PHYS	0187
arbide Thin-Film Formation on	Si(100): The Behavior of Dangling Bonds.= Silicon C	PHYS	0204
Ge-Containing Organometallics on	Si(100).= +tallic Precursors. Surface Reactions of	COLL	0044
r Substrates: GaAs(100) and	Si(100).= Reactions of Zinc Alkyls on Semiconducto	COLL	0048
in the Chemistry of Hydrogen on	Si(111) and Si(100) Surfaces.= Role of Bond Strain	PHYS	0187
oypes+ Ab Initio MO Studies on	Si$_2$H$_6$, SiH$_3$GeH$_3$, and Ge$_2$H$_6$ Radical Anions as Prot	PHYS	0313
ation and Oxidation Reactions of	Si-B-C Polymers: Experimental and Theoretical Stu	POLY	0187
bosilane.=	Si-C-N-O Ceramic Fiber Compositions from Polycar	POLY	0336
Reactions Involving	Si-H Groups Fixed on Silica.= Hydrosilylation	COLL	0094
Formation of	Si-N Bonds.= Some Novel Reactions for the	POLY	0337
n Reaction over Zeolites: Effect of	Si/Al Ratio and Coke Formation.= +proportionatio	PETR	0103
d Large-Port+ Influence of the	Si/Al Ratio on the Catalytic Properties of Small- an	PETR	0132
Mechanistic Aspects of	Si/Ge Deposition.=	COLL	0047
Reactive Collisions of Al, Mg,	Si, and C Atoms.=	PHYS	0089
on of Cytotoxin D from Naja naja	siamensis by Two-Dimensional Proton NMR.= +ti	BIOL	0044
e as a Precursor to Oxygen-Free	SiC Fibers.= High Molecular Weight Polycarbosilan	POLY	0360
n Olefin Polmers as Precursors to	SiC.= +rid Materials. Silicon Acetylene and Silico	POLY	0357
On CVD Growth of	SiC.=	PHYS	0095
C$_3$ and	SiC$_2$.= Spectroscopic Studies of Floppy Molecules:	PHYS	0235
Depositon of Diamond (and c-BN,	SiC). Characterization of Diam+ Chemical Vapor	PHYS	0118
eposition of Diamond (and c-BN,	SiC). Halogenation of Diamond+ Chemical Vapor D	PHYS	0176
eposition of Diamond (and c-BN,	SiC). Mole Fraction Measurem+ Chemical Vapor D	PHYS	0145
eposition of Diamond (and c-BN,	SiC). Molecular Dynamics Simu+ Chemical Vapor D	PHYS	0354
Depostion of Diamond (and c-BN,	SiC). Nucleation and Growth of+ Chemical Vapor	PHYS	0091
eposition of Diamond (and c-BN,	SiC). Structure and Chemistry o+ Chemical Vapor D	PHYS	0202
ethylpolysilazane—A Precursor to	SiC/SiN Ceramics.= +aracterization of Hydridom	POLY	0340
Novel Silyl-Carbocyclization (	SiCAC) of Alkynes and Alkenynes Catalyzed by Rh+	ORGN	0218
Formation on	Sidechain Structure.= Dependence of α-Helix	BIOL	0037
Protein Prediction Scales on	Sidechain Structure.= Dependence of Membrane	BIOL	0038
hotoresponsive Asobenzene-Based	Sidegroups.= +nomers and Polymers Containing P	POLY	0209
Complexes of Siderophores or	Siderophore Analogues.= Iron and Vanadium	INOR	0120
	Siderophore Production in Marine Microorganisms.=	INOR	0004
Novel Marine	Siderophores from Marine Bacteria.=	INOR	0119
Vanadium Complexes of	Siderophores or Siderophore Analogues.= Iron and	INOR	0120
Project SEED. The	Sidney-Cornell Project SEED Connection.=	CHED	0036
Silicon and Cobalt in Molecular	Sieve AlPO$_4$-5.= Isomorphous Substitutions of	PETR	0099
n+ Characterization of Molecular	Sieve Frameworks and Products of Catalytic Reactio	PETR	0060
lared Interlayered Clay Molecular	Sieve with High Thermal Stability.= +mance of Pil	PETR	0030
Titanium Silicate Molecular	Sieve.= Synthesis of ETS-10: A New Wide-Pore	PETR	0078
Ring Containing Molecular	Sieve.= Synthesis of VPI-7: A Novel 3-Membered	PETR	0098
int for Iso-Structural Molecular	Sieves Based upon Silicate or Alumino Phosphate Ch	CATL	0028
Synthesis of Carbon Molecular	Sieves for Separation and Catalysis.= Design and	PETR	0059
ion Synthesis of AlPO$_4$ Molecular	Sieves: Identification of Transient Phases.= +Solut	PETR	0020
h to the Synthesis of Molecular	Sieves: Recrystallization of the Intermediates Isolated	PETR	0019
	Molecular Sieves.= Industrial Catalytic Applications of	I&EC	0061

Acidic Molecular	Sieves.= Reactions of Meta-Diisopropylbenzene on	CATL	0014
Pore Aluminophosphate Molecular	Sieves.= +id-State NMR Study of 18-Ring Large-	PETR	0094
rous Aluminophosphate Molecular	Sieves.= +itution of Paramagnetic Ions in Micropo	PHYS	0231
sis: Shape Selectivity—I. Dynamic	Sieving by Heterogeneous Catalysts.= +in Cataly	CATL	0013
tic Drugs: I. Synthesis and Bind+	Sigma Receptor Antagonists as Potential Antipsycho	MEDI	0153
tic Drugs: II. Animal Behavioral+	Sigma Receptor Antagonists as Potential Antipsycho	MEDI	0084
6377 and NPC13839, as Probes of	Sigma Receptors.= +Aminoalkoxychromones, NPC1	MEDI	0150
3-Exomethylenecephams.= [2,3]	Sigmatropic Rearrangement of Sulfilimines from	ORGN	0033
timation of Differential Thermal	Signatures of Total Mercury in Crystalline Rock: Cas	ENVR	0055
ms of Action of Environmentally	Significant Carcinogens and Mutagens: Relevance to	CHAS	0007
tion/Stabilization Can Result in	Significant Reversal of Hydration and Polymerization	ENVR	0041
Ab Initio MO Studies on Si2H6,	SiH3GeH3, and Ge2H6 Radical Anions as Protoypes+	PHYS	0313
of Model	Silahydrocarbon Fluids.= Synthesis and Properties	I&EC	0008
ion of CpMn(CO)2 with Tertiary	Silanes as Examined by Photoacoustic Calorimetry.=	INOR	0386
from	Silanes.= Synthesis of Magnesium Hydride	INOR	0206
Poly(siloxy-	silanes).= Terminal Modifications of Hyperbranched	POLY	0323
Tit+ Comparison of Reactivity of	Silanol Groups on a Surface of a Pure and Alumino(	COLL	0112
PCl5 Reaction with	Silanol Groups on Silica Surface.= Peculiarities of	INOR	0237
Metal Ions.= Studies on	Silica Aerogels and Xerogels Doped with Transition-	INOR	0081
Containing SSZ-24, the All-	Silica Analog of AlPO4-5.= Aluminum- and Boron-	PETR	0023
Spectrometry of Alumina and	Silica Catalyst Supports.= Electron Energy-Loss	PETR	0073
ption Properties of Microporous	Silica Derived from Layered Siloxene by Oxidation.=	PETR	0144
is. Molecular Chemical Aspects of	Silica Gel Formation.= +and Pillared Clay Synthes	PETR	0054
PMMA-Impregnated	Silica Gel.=	POLY	0263
Surface of Chromium-Containing	Silica Gels.= +idation-Reduction Processes on the	COLL	0101
Adsorbates to Defect Sites in	Silica Supports.= Energy Transfer from Organic	PHYS	0135
he React+ Reaction of PCl5 with	Silica Surface: Influence of the Organic Solvents on t	COLL	0097
clic Compounds on	Silica Surface.= Immobilization of Tetraazamacrocy	COLL	0105
Reactions with Dispersed	Silica Surface.= Mechanisms of Metal Chloride	COLL	0219
Silanol Groups on	Silica Surface.= Peculiarities of PCl5 Reaction with	INOR	0237
l. Molecules Chemically Fixed on	Silica Surface.= +tions of WCl6, WOCl4, and MoOC	COLL	0098
Alchohol on	Silica Surfaces.= Chemisorption of a Tertiary Butyl	COLL	0108
Derivatives Chemisorption on	Silica Surfaces.= Investigation of Thiourea	COLL	0106
iates of Substitution Reactions on	Silica Surfaces.= +tion Complexes as the Intermed	COLL	0095
minum Alkyls and Ammonia on	Silica: Mechanisms for the Formation of Al-N Bonds	COLL	0045
Groups Fixed on	Silica.= Hydrosilylation Reactions Involving Si-H	COLL	0094
Surface Layer of Disperse	Silica.= Reaction of Phosphorus Compounds in a	COLL	0100
her on the Surface of Pyrogenous	Silica.= +en Gystamine, Mexamine, and Crown Et	COLL	0096
the Surface of Pyrogenous	Silica.= Thermal Transformations of Aziridine on	COLL	0113
Characterization of Titania-	Silica-Based DENOx Catalysts.= Surface	PETR	0071
of Copper and Other Elements in	Silica-Saturated CaO-CaF2-SiO2 Slags.= +lubility	I&EC	0066
(CaO-CaF2-+ Equilibrations of	Silica-Saturated Metal (Cu), Matte (Cu2S), and Slag	I&EC	0065
Photoreduction of	Silica-Supported Mo(VI) to Mo(IV).=	COLL	0213
inc or Molybdenum Supported on	Silicalite.= +tics of Propane Deydrogenation over Z	PETR	0087
nthesized ZSM-5 and Aluminated	Silicalite-1 Crystals.= +bons over Alkaline-Free Sy	PETR	0065
and Alumino(Titano)-Containing	Silicas.= +of Silanol Groups on a Surface of a Pure	COLL	0112
New Wide-Pore Titanium	Silicate Molecular Sieve.= Synthesis of ETS-10: A	PETR	0078
ural Molecular Sieves Based upon	Silicate or Alumino Phosphate Chemistry.= +truct	CATL	0028
Crystallization from Sodium	Silicate Solutions.= The Chemistry of NaY	PETR	0004
Cation-Incorporated	Silicate Surfaces.= Microporous Pillared Mica with	PETR	0129
etting Action of Melted Sodium	Silicate with Additives on Quartz and Its Relation wi	COLL	0079
Large-Pore Titanium	Silicate.= Adsorptive Properties of ETS-10: A New	PETR	0079
d Absorbency in Aluminum and	Silicate-Modified Starch Graft Copolymer Superabso	CARB	0042
sis of Dioctahedral 2:1-Layer	Silicates in Acid and Fluorinated Medium.= Synthe	PETR	0128
Layered	Silicates: The Protonation Behavior of KHSi2O5.=	PETR	0110
of Boron-Substituted Zeolites and	Silicates.= +clear Solid-State NMR Investigation	PETR	0136
Layer	Silicates.= Direct Synthesis of Novel Intercalated	PETR	0108
Growth of	Silicide Layers on Si Surfaces.=	PHYS	0100
and Inorganic Hybrid Materials.	Silicon Acetylene and Silicon Olefin Polmers as Prec	POLY	0357
Isomorphous Substitutions of	Silicon and Cobalt in Molecular Sieve AlPO4-5.=	PETR	0099
Mono- and Trimethylgallium on	Silicon and Indium Phosphide.= Decomposition of	INOR	0123
tronic Relaxation in σ-Delocalized	Silicon Backbone Network Polymers.= +on and Vi	PHYS	0138
e Behavior of Dangling Bonds.=+	Silicon Carbide Thin-Film Formation on Si(100): Th	PHYS	0204
ce Precursors for LPCVD of	Silicon Carbide.= Disilacyclobutanes as Single-Sour	INOR	0105
Use of Siloxanes as Precursors to	Silicon Carbide.= +eparation, Characterization, and	POLY	0361
ptor Units Intermediated by	Silicon Chains.= Model Molecules with Donor-Acce	POLY	0211
Theoretical Study of	Silicon Cluster Electron Affinities.=	PHYS	0079
SiO2 as a Source of	Silicon Containing Compounds/Polymers.=	POLY	0314
hel on Ultrathin Thermally Grown	Silicon Dioxide.= +Bulk Diffusion of Adsorbed Nic	COLL	0177
Precursors to	Silicon Nitride.= Applications of Polysilazanes as	POLY	0317
c Materials. Silicon Acetylene and	Silicon Olefin Polmers as Precursors to SiC.= +ri	POLY	0357
Microstructural Evolution of a	Silicon Oxide Phase in Nafion Membranes by an in S	POLY	0285
Indiana	Silicon Sphere (ISiS) 4π Detector.=	NUCL	0036
Probing and Inducing	Silicon Surface Chemistry with the STM.=	PHYS	0184
Diethylsilane on	Silicon Surfaces.= Adsorption and Decomposition of	COLL	0008
of	Silicon(IV).= New Five-Coordinate Complexes	INOR	0019
Polyethers.=	Silicon-Assisted Synthesis of High Molecular Weight	PMSE	0075
Synthesis of	Silicon-Containing Cages.=	ORGN	0055
Oligomers.= Synthesis of Novel	Silicon-Containing Epoxy Monomers and	POLY	0015
l Reactions in the Preparation of	Silicon-Containing Films from Molecular Precursors.	INOR	0121

niques for the Characterization of	Silicon–Containing Phenylated Soluble Aramids.=	POLY 0086
is of 9,9-Dialkyl-9,10-Dihydroa+	Silicon-Mediated Alkylations: A Convenient Synthes	ORGN 0087
and Personal Care Industry—II.	Silicone in Hair Conditioners: Factors Controlling Ha	CHED 0063
Cationic Polymerization of	Silicone–Epoxide Resins.= Electron Beam–Induced	PMSE 0159
ys L.) Populations and Inbreds for	Silk Flavonolglycoside Contents.= +Corn (Zea ma	AGRO 0034
Hormone of the	Silkworm, Bombyx mori.= Prothoracicotropic	AGRO 0010
th Crown Ether on the Surface of	Silochromium.= +on of Lead(II) and Mercury(II) wi	COLL 0104
escence Studies of Self–Assembled	Siloxane Mono– and Multilayers.= +IR and Fluor	POLY 0330
Characterization of Mixed	Siloxane–Oxide Systems.= Spectroscopic	POLY 0231
tion, Characterization, and Use of	Siloxanes as Precursors to Silicon Carbide.= +epara	POLY 0361
anches on the Chain Statistics of	Siloxanes.= The Effect of the Length of the Side Br	POLY 0070
orous Silica Derived from Layered	Siloxene by Oxidation.= +ion Properties of Microp	PETR 0144
Hyperbranched Poly(	siloxy–silanes).= Terminal Modifications of	POLY 0323
Target Fragmentation of	Silver by 14.6 GeV/nucleon ^{16}O Ions.=	NUCL 0050
Redox Properties of Nonmetallic	Silver Clusters and Photochemistry of Colloidal Silve	COLL 0130
Optical Four–Wave Mixing in a	Silver Colloid–Polymer Composite.=	POLY 0218
Oxidative Deintercalation of	Silver from Ternary Silver Oxides.=	INOR 0275
Properties of Polymer–	Silver Microsphere Composites.= Nonlinear Optical	POLY 0154
from Ternary	Silver Oxides.= Oxidative Deintercalation of Silver	INOR 0275
s and Photochemistry of Colloidal	Silver Sols.= +perties of Nonmetallic Silver Cluster	COLL 0130
MDA/4–BDAF Polyimides and	Silver Substrates Using Surface–Enhanced Raman Sc	POLY 0087
n of Ethylene and Propylene over	Silver.= +ween Selective Oxidation and Degradatio	CATL 0058
osphino)ferrocene (dppf)–Bridged	Silver(I) Complexes.= +Some 1,1'-bis(Disphenylph	INOR 0106
al Salen Complexes Bearing Bulky	Silyl Groups.= +xidation Catalyzed by Chiral Met	ORGN 0056
DNA Using a	Silyl Phenol.= Solvolysis–Triggered Alkylation of	ORGN 0191
ynes Catalyzed by Rh a+ Novel	Silyl–Carbocyclization (SiCAC) of Alkynes and Alken	ORGN 0218
ation of Mono Alkylbenzenes over	Silylated Gallosilicate Zeolites.= +–Selective Alkyl	PETR 0100
Volatile Metal Complexes with	Silyltellurolate and Selenolate Ligands.=	INOR 0172
ne Herbicides, I: Determination of	Simazine and Its Metabolites in Urine.= +to Triazi	C E I 0004
GaAs(SiMe$_3$)$_2$GaR$_2$Cl, [R$_2$GaAs(	SiMe$_3$)$_2$]$_2$ (R = t-BuCH$_2$,t-Bu), and (t-BuCH$_2$)$_2$Ga(C	COLL 0136
, I): Crystal Structure of [I$_2$GaAs(	SiMe$_3$)$_2$]$_2$.= +ns of (Me$_3$Si)$_3$As with GaX$_3$ (X = Br	COLL 0140
ds: Crystal Structure of [Et$_2$AlAs(	SiMe$_3$)$_2$]$_2$.= +f Novel Aluminum–Arsenic Compoun	COLL 0137
lkali Metal: Strucutre of Lu[CH(	SiMe$_3$)$_2$]$_3$(μ–Cl)K(η^6–C$_7$H$_8$)$_2$ and the Solubilization of	INOR 0053
tion, and Reactivity of R$_2$GaAs(	SiMe$_3$)$_2$GaR$_2$Cl, [R$_2$GaAs(SiMe$_3$)$_2$]$_2$ (R = t-BuCH$_2$,t-	COLL 0136
t-Bu), and (t-BuCH$_2$)$_2$Ga(Cl)·As(	SiMe$_3$)$_3$.= +$_2$Cl, [R$_2$GaAs(SiMe$_3$)$_2$]$_2$ (R = t-BuCH$_2$,	COLL 0136
Ru$_3$(CO)$_{12}$ with RC≡CPh (R=	SiMe$_3$,SnBu$_3$).= Reactions of H$_2$Os$_3$(CO)$_{10}$ and	INOR 0241
ectin Metabolism: Differences and	Similarities.= +Anmimals. Ivermectin and Abam	AGRO 0128
Perceiving Structural	Similarities.= Developing Synthesis Strategies by	CINF 0007
Conditions in the Detection of+	Simplex Optimization of Post–Column Derivatization	AGFD 0111
amacion Cuadratica Sucesiva y un	Simulador de Procesos Comercial.= +lizando Progr	I&EC 0072
BALANCE:	Simulador de Procesos para Balances de Materia.=	COMP 0052
To Know, To Do, and To	Simulate.=	CHED 0113
	Simulated Annealing of Rings.=	ORGN 0169
ironmental Organics Using AM1	Simulated Sunlight Irradiation of Semiconductor Cat	INOR 0318
Computer	Simulation Experiment: Chemical Kinetics.=	CHED 0157
y Analysis: A Molecular Dynamics	Simulation Method.= +rial Surfaces via Wettabilit	POLY 0369
Semivolatile Organic+ Dynamic	Simulation Model of the Atmospheric Deposition of	GEOC 0004
	Simulation of ^{13}C and ^{1}H NMR Spectra.=	COMP 0002
	Simulation of CVD Diamond Growth.=	PHYS 0181
Computer	Simulation of Direct Energy Transfer in Polymers.=	POLY 0079
s: Comparison of Th+ Computer	Simulation of Electronic Energy Transfer in Polymer	PHYS 0109
.=	Simulation of NMR Spectra: Problems and Solutions	COMP 0007
Chemical Potential+ Atomistic	Simulation of Polymer Materials. Ab Initio Quantum	POLY 0350
Range Correlations in+ Atomistic	Simulation of Polymer Materials. Sizes of the Short–	POLY 0365
nergy Functions for the Atomistic	Simulation of Polymeric Materials.= +Potential E	POLY 0350
ng.=	Simulation of the Role of Water in Ion Channel Gati	COLL 0084
ergy Transfer and+ Numerical	Simulation of the Time Dependence of Electronic En	POLY 0078
inetics in Low–Dimensional and+	Simulation of Trapping– and Quenching–Reaction K	PHYS 0307
. Molecular Dynamics Computer	Simulation Studies of Collision Cascades in Solids.=	NUCL 0091
ometry and Molecular Dynamics	Simulation Techniques for Conformational Analysis	ORGN 0174
m Simulations.= Spectral	Simulation. Combinatorics of NMR and ESR Spectru	COMP 0001
nfrared and ^{13}C NMR Spectrum	Simulation: Role of Complementary Topological Stra	COMP 0006
Fractionation through Computer	Simulation.= +Used in Sedimentation Field–Flow	PMSE 0028
^{13}C NMR Spectrum	Simulation.=	COMP 0003
onding from Molecular Dynamics	Simulation.= +ntations and Analysis of Hydrogen B	BIOL 0112
NMR Spectrum	Simulation.= Database Retrieval Techniques for ^{13}C	COMP 0004
Award–Winning Laboratory	Simulation.= SQUALOR—The Design of an	CHED 0079
Computer	Simulations for Instrumental Analysis.=	CHED 0192
Molecular Dynamics	Simulations of Amiloride.=	MEDI 0045
d Propert+ Molecular Dynamics	Simulations of Diamond Films: Structure, Growth, an	PHYS 0354
Quantum	Simulations of Hubbard and Holstein Models.=	PHYS 0054
d in Thin+ Stochastic Dynamics	Simulations of Polymethylene Chains in the Bulk an	POLY 0355
fer and Proton Tran+ Quantum	Simulations of Quantum Dynamics in Electron Trans	BIOL 0131
Rigid–Ro+ Molecular–Dynamics	Simulations of Rigid–Rod and Hydroxy–Substituted	PHYS 0322
Dynamic Monte Carlo	Simulations of Rings and Branched Molecules.=	POLY 0297
of NMR and ESR Spectrum	Simulations.= Spectral Simulation. Combinatorics	COMP 0001
Polymer	Simulations.= Computational Method for	PHYS 0112
ymeric Systems from Monte Carlo	Simulations.= +of the Chemical Potentials of Pol	POLY 0352
Bead–Stick Computer	Simulations.= Rings vs. Strings: Off–Lattice,	POLY 0235
and Binding Energy of the Lowest	Singlet and Triplet States of MgC$_2$.= +Spectrum,	PHYS 0276

s.= Quantum Yields of	Singlet Oxygen Production by Triplet Cyclohexenone	ORGN	0167
ffects on the Side Selectivity of	Singlet Oxygen with α,β Unsaturated Esters: Evidenc	ORGN	0164
cidas.=	Sintesis de Tiofosfatos de Naftilo, Posibles Garrapati	MEDI	0064
con Integracion de Energia.=	Sintesis y Modificacion de Secuencias de Destilacion	I&EC	0074
agnetic, and Hydrodynamic—III.	Sinusoidal Enhancement of Correlated Structures.=	COLL	0178
trostatic Field of the Surface of	SiO_2 and Its Effect on Energetics of Surface Reaction	COLL	0116
anadium Oxides Anchored onto	SiO_2 and Their Photoreactivity toward CO Molecules	INOR	0165
olymers.=	SiO_2 as a Source of Silicon Containing Compounds/P	POLY	0314
Deposition of	SiO_2 from Tetraethoxysilane.=	COLL	0009
Study of Nickel on Ultrathin	SiO_2 Layers.= Medium-Energy Ion-Scattering	COLL	0176
ts in Silica-Saturated $CaO-CaF_2-$	SiO_2 Slags.= +lubility of Copper and Other Elemen	I&EC	0066
atte (Cu_2S), and Slag ($CaO-CaF_2-$	SiO_2) Under Controlled Pressures of SO_2.= +u), M	I&EC	0065
of Polypropionate Metabolites of	Siphonaria Australis.= +nd Absolute Configuration	ORGN	0003
and	SIPN.= Hydrophobic-Hydrophilic IPN	PMSE	0005
an Exposure.=	Sister Chromatid Exchange Test as Monitor of Hum	CEI	0016
or Agrochemical Retail Dealership	Site Assessment and Remediation.= +ent Needs f	AGRO	0142
Hazard-Ranking+ Superfund's	Site Assessment Process: Implications for the Revised	ENVR	0078
Environment.= Environmental	Site Assessments: Protecting the Dealer and the	FERT	0015
tes. Elucidation of Carbon-Active	Site Behavior via Desorption Techniques.= +ive Si	FUEL	0013
H.O.R. Nuclear Research Reactor	Site Boundary.= +Products in Discharge Pool in T.	NUCL	0090
ence for Cyclohexylalanine at P1	Site by Renin: Kinetic Studies Using New Fluorogen	BIOL	0035
n of the Orientation of the Active-	Site Histidine Relative to Substrates.= +erminatio	BIOL	0066
PCB-RISc: An On-	Site Immunoassay for Detecting PCB in Soil.=	ENVR	0032
Anhydrase.= Active-	Site Ionicity and the Mechanism of Carbonic	BIOL	0063
trical Properties of a Single-Acid-	Site Latex.= +ic Fingerprinting of the Surface Elec	COLL	0200
ding+ Purification and Binding-	Site Mapping of Hemolymph Juvenile Hormone-Bin	AGRO	0014
yase Affects the Kinetics of Active	Site Modification.= +de Exchange in HMG-CoA L	BIOL	0064
Acid Analogues as Potent Glycine	Site N-Methyl-D-Aspartate Antagonists.= +renic	MEDI	0146
termined Structure at the Active	Site of a True Reaction Intermediate of Carboxypept	BIOL	0011
at the Active	Site of Carbonic Anhydrase.= Cobalt Spectroscopy	PHYS	0256
ne (Duo) as a Probe of the Active	Site of E. +ased Inactivator 1-Diazo-4-undecyl-2-o	BIOL	0067
Characterization of the Active	Site of Heparinase.=	BIOL	0056
M Identification of the Binding	Site of Neuropeptide Y to the Y2-Receptor by Using	MEDI	0091
ethod to Define the Recognition	Site of Sequence-Specific DNA-Binding Proteins.=	BIOL	0129
Antithrombin III Binding	Site Precursor.= Search for the Heparin	CARB	0039
t, and Futur+ Biodegradation in	Site Remediation and Waste Treatment: Past, Presen	AGRO	0121
Agricultural Chemical	Site Remediation/Regulations.=	AGRO	0143
w Method for Iron Atom Active-	Site Replacement by a Nickel Atom in Desulfovibrio	INOR	0020
Penta Soil RISc: A Rapid, On-	Site Screening Test for Pentachlorophenol in Soil.=	ENVR	0033
ce in the High+ Cation-Dopant	Site Selectivity, Oxygen Stoichiometry, and Cu Valen	PHYS	0027
l Gas Analysis in in Situ Steam+	Site Use of Advective-Flux Probes for Enhanced Soi	ENVR	0052
Phosphate Binding	Site.= Identification of Phosphoribulokinase's Sugar	BIOL	0153
Environment at the Nevada Test	Site.= +ermination of Strontium-90 in a High-Salt	NUCL	0085
tached to the Backbone at the P1	Site.= +ibitors with Cyclohexyl Moiety Directly At	MEDI	0121
at the Nevada Test	Site.= Thorium-230 Chronology of Natural Waters	NUCL	0084
Creosote	Site-Biobed Box Pilot-Scale Bioremediation Study.=	BIOT	0140
n of Trypsin by N-Nitr+ Active-	Site-Directed Enzyme-Activated Substrate Inhibitio	ORGN	0022
dCMP Hydroxymethylase.=	Site-Directed Mutagenesis of the Gene-Encoding T4	BIOL	0160
of Immunosorbents Prepared by	Site-Directed or Random Coupling of Monoclonal An	BIOT	0060
Properties of Native and	Site-Mutagenized Cellobiohydrolase II.=	BIOT	0240
Dynamics Using	Site-Selective Spectroscopy.= Probing Upconversion	ANYL	0004
A-Tract Bending and Its Biol+	Site-Specific Effect of Thymine Dimer Formation on	BIOL	0114
Complex and Binding Activities of	Site-Specific Mutants Expressed in E. +ride-FAB	BIOL	0143
struction of Hydrophobic Binding	Sites and Self-Organizing Structures.= +ds for Con	ORGN	0066
n.= Engineered Metal Chelation	Sites as Probes of Protein Stability and Conformatio	BIOT	0133
aracterization of Carbon Active	Sites by Temperature-Programmed Desorption.= Ch	FUEL	0016
t Is the Basis of EPA's Listing of	Sites for Superfund? Overview of the Structure of th	ENVR	0077
eboronated Z+ On the Catalytic	Sites Formed by the Reaction of Metal Alkyls with D	PETR	0024
n Reactions at Aldehydic	Sites in DNA.= Mechanistic Studies of β-Eliminatio	BIOL	0158
tion-Metal Complexes and Redox	Sites in Enzymes.= +sfer Reactions between Transi	INOR	0283
aracterization of Calcium-Binding	Sites in Nephrocalcin with Vanadyl Ion.= +OR Ch	BIOL	0045
Organic Adsorbates to Defect	Sites in Silica Supports.= Energy Transfer from	PHYS	0135
al Calculation of Grafting	Sites in the Epoxy-g-Acrylic Copolymer.= Theoretic	PMSE	0161
Deuterium NMR Studies of Acid	Sites in Zeolites.=	COLL	0197
olite: Inhibited Deactivation of Pt	Sites within One-Dimensional Zeolite Channels.=	CATL	0015
D Workshop on Carbon Active	Sites. Elucidation of Carbon-Active Site Behavior via	FUEL	0013
Technology. Antibody Combining	Sites: Prediction and Design.= +nces in Antibody	BIOT	0056
nt to Chromate Production Waste	Sites.= +in the Urine of Individuals Living Adjace	CEI	0009
Carcasses with Positive Injection	Sites.= +Anabolic Steroids in Muscle Tissues from	AGFD	0130
Single-Chain Antibody Binding	Sites.=	BIOT	0057
heoretical Foundation of the Al	Sitting and Pairing Avoidance Principles in Zeolites.	PHYS	0289
Reaction Pathways for Si^+	SiX_4 (X = F, Cl).= An ab Initio Study of the	PHYS	0248
Manuscript.= Splendor Solis: A	Sixteenth-Century Illuminated Chemical	HIST	0036
Nitrogen into Intermediate-	Sized Boron Hydride Clusters.= Insertion of	INOR	0388
ies in the Synthesis of Medium-	Sized Carbocycles: Applications Toward the Total Sy	ORGN	0282
at Zeolite-Supported, Nanometer-	Sized Electrodes.= +c Fields on Organic Reactions	CATL	0018
plitting Step Is Rate-Limiting for	Skeletal Actomyosin ATP Hydrolysis.= +e Bond-S	BIOL	0141
Class of Inorganic Polymers with	Skeletal Phosphorus, Nitrogen, and Sulphur(VI) Ato	POLY	0186
rnative Approach to the Oplopane	Skeleton.= +l Melampolide Schkuhriolide: An Alte	ORGN	0018
xican S+ Diterpenoids with New	Skeletons Related to a Clerodanic Precursor from Me	ORGN	0280

ew Dimensions in C+ Herman	Skolnik Award Symposium Honoring W. T. Wipke: N	CINF 0016
active Index Profiles in Polymeric	Slab Waveguides.= +otochemically Delineated Refr	POLY 0180
d Metal (Cu), Matte (Cu$_2$S), and	Slag (CaO–CaF$_2$–SiO$_2$) Under Controlled Pressures o	I&EC 0065
Utilization of Coal Gasification	Slag: An Overview.=	FUEL 0134
oleum Coke and Coal Gasification	Slags.= +ase Characterization of Unmodified Petr	FUEL 0139
Silica–Saturated CaO–CaF$_2$–SiO$_2$	Slags.= +lubility of Copper and Other Elements in	I&EC 0066
culation of Adsorption Energies in	Slitlike and Wedge–Shaped Micropores.= +s. Cal	FUEL 0008
bon and C–14 Budget for Upper	Slope Sediments of the Middle Atlantic Bight.= Car	GEOC 0016
Municipal Sewage	Sludge for Gasification.= Pretreatment of	FUEL 0117
n Feeding of Electron–Treated	Sludge for Ocean Productivity Enhancement.= Ocea	ENVR 0021
Sewage	Sludge in Agriculture.= Ontario's Program for Using	FERT 0021
Acids from	Sludge Pond Solids.= Characterization of Humic	I&EC 0022
with Clay Minerals from Syncrude	Sludge Pond Tailings.= +ganic Matter Associated	I&EC 0021
Gas from Gasification of	Sludge Wastes.= Production of Clean–Medium Btu	FUEL 0118
s, and Municipal Waste. Sewage	Sludge: A Fascinating Feedstock for Clean Energy.=	FUEL 0115
ste and Coal: Energy from Sewage	Sludge, Biomass, and Municipal Waste. +from Wa	FUEL 0115
s.= Award Address (Edward E.	Smissman Award, sponsored by Bristol–Myers Squib	MEDI 0089
bitors.= General and Edward E.	Smissman—Bristol–Myers Squibb Award Address.	MEDI 0082
Compounds to	Smoke Flavor.= Contribution of Phenolic	AGFD 0173
Degradation.= Formation of	Smoke–Flavor Compounds by Thermal Lignin	AGFD 0054
System Pt–Sn/ZnAl$_2$+ Effect of	Sn Addition on the Catalytic Properties of Bimetallic	COLL 0127
and Characterization Data for Pt–	Sn–Aluminum Catalysts.= +rocyclization Activity	CATL 0007
operties of Bimetallic System Pt–	Sn/ZnAl$_2$O$_4$.= +of Sn Addition on the Catalytic Pr	COLL 0127
	SNA on 2,4–Dinitro Chlorobenzene, Part II.=	CHED 0170
Minimally Processed Vegetable	Snacks.= Integrated Approach to Development of	AGFD 0056
with RC≡CPh (R=SiMe$_3$,	SnBu$_3$).= Reactions of H$_2$Os$_3$(CO)$_{10}$ and Ru$_3$(CO)$_{12}$	INOR 0241
e Adulteration.= Application of	SNIF–NMR to the Detection of Flavor and Fragranc	AGFD 0197
O$_2$) Under Controlled Pressures of	SO$_2$.= +u), Matte (Cu$_2$S), and Slag (CaO–CaF$_2$–Si	I&EC 0065
uimiento de+ Peliculas de Oxido	sobre Acero Inoxidable 304 y 316 Utilizadas en el Seg	ANYL 0055
o Indicador una Pelicula de Oxido	sobre Acero Inoxidable 316.= +ando como Electrod	CHED 0163
de Catalizadores NiW Soportados	sobre Alumina o Titania Modificadas con Fluor.=	FUEL 0106
Estudio	Sobre el Uso de Eteres de Glicol en Mexico.=	ENVR 0038
ento: Influencia de la Sulfonacion	sobre las Propiedades Papelaras.= +e Alto Rendimi	PMSE 0077
o con Catalizadores de Molibdeno	sobre Oxidos Mixtos de Alumina–Titania.= +Tiofen	FUEL 0105
ades y Secotr Product+ Servicio	Social y Programas de Becas: Vinculo entre Universid	CHED 0114
ation in the American Chemical	Society SEED Program for High School Students at	CHED 0143
ucation of the American Chemical	Society.= +itiative of the Division of Chemical Ed	CHED 0110
the American Chemical	Society.= Centennial of the New York Section of	HIST 0029
Defense Research in a Democratic	Society.= +dom of Speech—The Delicate Balance:	PROF 0009
Section of the American Chemical	Society). +orating the Centennial of the New York	HIST 0025
the 193–nm Photodissociation of	SOCl$_2$.= +l Energy Distributions of SO (X$^3\Sigma^-$) from	PHYS 0240
LiBe and LiMg.= Ab Initio Full	SOCVCI Calculations of the Elctronic Spectra of	PHYS 0338
y s+ Elaboracion de Caseinato de	Sodio a Partir de Leche Contaminada con S. aureus	AGFD 0107
and Spin–Lattice Relaxation of	Sodium Cations in Zeolite–Y Probed by Double–Rota	PHYS 0267
with Voltage–Sensitive	Sodium Channels.= Interactions of N–Alkylamides	AGRO 0151
vestigation of the Interaction of	Sodium Chloride and Two Amino Sulfonic Acids, HE	BIOL 0034
Pentachlorides in the Presence of	Sodium N,N–Diethyldithiocarbamate.= +Tantalum	INOR 0092
Fluoropolymer Surfaces by	Sodium Naphthalenide.= Modification of	PMSE 0124
and Dynamics of the Mesospheric	Sodium Nightglow.= +tions—I. Chemical Kinetics	PHYS 0071
Iodine and	Sodium Nitrate—The Story of Chilean Nitrate.=	SCHB 0020
Crystallization from	Sodium Silicate Solutions.= The Chemistry of NaY	PETR 0004
of the Wetting Action of Melted	Sodium Silicate with Additives on Quartz and Its Re	COLL 0079
Condensation of Cyclopentadienyl	Sodium with Dialkyldichlorosilanes.= +nes via the	INOR 0199
gence of Ionomers Poly(styrene–	Sodium–2–Acrylamido–2–Methylpropane Sulfonate)	COLL 0181
e Molecular Oxygen Reaction with	Sodium, Magnesium, and Copper Atoms.= +of th	PHYS 0141
Kinetics of Thermal	Softening of Fruit and Vegetable Tissue.=	AGFD 0013
Ahead in Chemistry Education	Software Evaluation.= Looking Back and Moving	CHED 0080
Right–to–Know	Software.= MSDS and Label	CHAS 0030
Spreadsheet	Software.= Teaching Biochemistry with	CHED 0206
on and Translocation Processes in	Soil as a Function of Climatic Parameter.= +radati	AGRO 0025
ues and Answers in Fertilizer and	Soil Chemistry. Impact of Fertil+ Environmental Iss	FERT 0013
ues and Answers in Fertilizer and	Soil Chemistry. The Retail Ferti+ Environmental Iss	FERT 0017
Oil Agglomeration Technique to	Soil Clean–Up and Generation of Good–Quality Solid	FUEL 0141
emediate Pesticide–Contaminated	Soil Excavated from Agrochemical Facilities.= +R	AGRO 0144
vective–Flux Probes for Enhanced	Soil Gas Analysis in in Situ Steam Distillation.=	ENVR 0052
A GC Method for	Soil Gas Analysis Using a Single Detector.=	ENVR 0054
w of the History of Applications of	Soil Gas Analysis.= +ental Contaminants. Revie	ENVR 0048
for High–Resolution	Soil Gas Analysis.= A Sorbent Trapping Procedure	ENVR 0050
view of the History of Applicati+	Soil Gas Analysis—Environmental Contaminants. Re	ENVR 0048
el Contamination.=	Soil Gas Survey Methods for Gasoline and Diesel Fu	ENVR 0053
System.= The Use of	Soil Gas Surveys to Design Soil Vapor Extraction	ENVR 0051
tion in Japan and In+ Review of	Soil Gas Techniques Used for Environmental Applica	ENVR 0049
achlorophenol in Soil.= Penta	Soil RISc: A Rapid, On–Site Screening Test for Pent	ENVR 0033
c: Extensive Ana+ Contaminated	Soil Samples from Wood–Preserving Plants in Quebe	ENVR 0016
halene–Contaminate+ Biological	Soil Treatment to Remediate Petroleum– and Napht	BIOT 0139
Gas Surveys to Design	Soil Vapor Extraction System.= The Use of Soil	ENVR 0051
oval of Chromium from Soils by	Soil Washing with Surfactant Solutions at Low pH.=	ENVR 0013
Detecting PCB in	Soil.= PCB–RISc: An On–Site Immunoassay for	ENVR 0032
Test for Pentachlorophenol in	Soil.= Penta Soil RISc: A Rapid, On–Site Screening	ENVR 0033
ctive Iron Oxide Extraction from	Soils and Sediments: Reductive Dissolution by Ti(III	ENVR 0058

ow+ Removal of Chromium from	Soils by Soil Washing with Surfactant Solutions at L	ENVR	0013
- and Naphthalene-Contaminated	Soils.= +l Soil Treatment to Remediate Petroleum	BIOT	0139
norganic Hybrid Materials via the	Sol-Gel Approach.= +acterization of Polyacrylate-I	POLY	0232
Radical Routes into Nonshrinking	Sol-Gel Composites.= +Composite Materials: Free	POLY	0259
ds.=	Sol-Gel Condensation of Tin(IV) Alkoxide Compoun	POLY	0258
	Sol-Gel Kinetics by NMR.= Preceramic and	POLY	0256
ientation of Organic Molecules in	Sol-Gel Matrices for Quadratic Nonlinear Optics.=	POLY	0006
ic Composites Prepared via the	Sol-Gel Method: Some Insights into Morphology Con	POLY	0260
Microcomposites Prepared via the	Sol-Gel Method.= +y(vinyl Acetate)/Silicon Oxide	POLY	0287
ilm Preparation through Aqueous	Sol-Gel Method.= +es of Transparent Aluminum F	COLL	0115
O_3 Prepared by Precipitation and	Sol-Gel Methods.= +aracterization of TiO_2 and Al_2	PETR	0092
zation. Microporous Oxides by the	Sol-Gel Process: Synthesis and Applications.= +ri	PETR	0031
	Sol-Gel Processing of Controlled-Pore-Size Oxides.=	PETR	0032
Nafion Membranes by an in Situ	Sol-Gel Reaction.= +on of a Silicon Oxide Phase in	POLY	0285
ructure Characterization during	Sol-Gel Synthesis of Controlled-Porosity Materials.=	PETR	0068
ied Chromophores in Supported	Sol-Gel Thin-Film Glasses for Second-Harmonic Ge	POLY	0210
l-Angle+ Structural Features of	Sol-Gel-Derived Hybrid Ceramer Materials by Smal	POLY	0284
O_2: A Potential Technique for the	Solar Detoxification of Gas-Phase Wastes.= +er Ti	COLL	0082
y Studies.= Gas-Phase	Solar Detoxification of Hazardous Wastes: Laborato	ENVR	0035
s Mixtures.=	Solar Evaporation of Fertilizer/Ag-Chemical Aqueou	FERT	0022
ience: Solar Neutrinos—II. SAGE	Solar Neutrino Experiment.= +ces in Neutrino Sc	NUCL	0030
Molybdenum-Technetium	Solar Neutrino Experiment.=	NUCL	0022
Results from the Homestake	Solar Neutrino Observatory.=	NUCL	0021
ctor.=	Solar Neutrino Results from the Kamiokande II Dete	NUCL	0023
Borexino: Low-Energy	Solar Neutrino Spectroscopy.=	NUCL	0033
ent Advances in Neutrino Science:	Solar Neutrinos—I. Neutrinos and Chemistry+ Rec	NUCL	0019
ent Advances in Neutrino Science:	Solar Neutrinos—II. SAGE Solar Neutrino E+ Rec	NUCL	0030
System i+ On the Conversion of	Solar to Electric Energy by the Other Photosynthetic	PHYS	0019
ing, and Characterization. Strong	Solid Acid Catalysts.= +Supports: Chemistry, Form	PETR	0049
Diene Synthesis by	Solid Acid Catalysts.=	PETR	0029
Fundamental Liquid-	Solid Adsorption Theory.= Verification of a	COLL	0188
and Generation of Good-Quality	Solid and Liquid Fuels from Poor-Quality Feedstock	FUEL	0141
Complexes Isolated in	Solid Argon.= Infrared Spectra of Metal-Ethylene	PHYS	0074
Analysis.= The Cluster-	Solid Interaction from the Perspective of Chemical	NUCL	0057
No. V_0) Hydrogen Clusters with	Solid Materials.= Interaction Processes of Fast (V	NUCL	0054
of Structure and Dynamics of	Solid Polymers.= Multidimensional NMR Studies	PHYS	0137
through	Solid Polymers.= Stress Effects on Diffusion	POLY	0140
Insight in the Mechanism of Gas-	Solid Reactions.= +hniques Will Provide Detailed	FUEL	0015
Interfacial Electron Transfer at a	Solid Semiconductor Electrode-Liquid Junction.=	COLL	0088
Beta	Solid Solution in the Bi_2O_3-SrO-CaO System.=	INOR	0201
on between Electronic Structure,	Solid State Structure, and Liquid Crystal Properties.	INOR	0269
and Properties of Ionomers in the	Solid State.= +s—III. Plenary Lecture. Structure	PMSE	0145
Control of the	Solid State.= Phase Behavior of Polyrotaxanes:	COLL	0058
nversions.= Sulfate-Containing	Solid Superacids for Acid-Catalyzed Hydrocarbon Co	PETR	0102
lf-Assembling Lipid Bilayers on	Solid Support: Building Blocks of Future Biosensors	BIOT	0102
perties of Colloids in Contact with	Solid Surfaces in External Fields.= +Transport Pro	COLL	0162
vel Tec+ Molecular Processes at	Solid Surfaces: Surface Chemistry of Nonmetals—No	COLL	0211
vel Tec+ Molecular Processes at	Solid Surfaces: Surface Chemistry of Nonmetals—No	COLL	0170
vel Tec+ Molecular Processes at	Solid Surfaces: Surface Chemistry of Nonmetals—No	COLL	0191
opies. Vibrational Energy Flow at	Solid Surfaces.= +esonance and Optical Spectrosc	PHYS	0344
Cluster Impact on	Solid Surfaces.=	NUCL	0055
Transformation of	Solid Urban Residues in Usable Products.=	CHED	0105
nt Concrete Containing Municipal	Solid Waste Incinator Ash.= +gth Portland Ceme	FUEL	0136
Pretreatment of Municipal	Solid Waste.= Ammonia Fiber Explosion (AFEX)	BIOT	0016
in a One-Dimensional Inorganic	Solid: Exciton Trapping and Annihilation in Crystals	PHYS	0131
yl-1-Azadiene.=	Solid- and Liquid-State Photochemistry of an N-Ac	ORGN	0111
Liquid-Liquid, Liquid-Solid, and	Solid-Gas Interfaces.= +Spaces: Chemistry at the	ORGN	0126
etics and Thermodynamics at the	Solid-Liquid Interface.= +rement of Reaction Kin	COLL	0170
Protein Interaction Forces at a	Solid-Liquid Interface.= Direct Measurement of	BIOT	0043
d Subse+ Application of Matrix	Solid-Phase Dispersion (MSPD) to the Extraction an	AGFD	0060
hromatography in Food Science.	Solid-Phase Extractions for Sample Preparation in F	AGFD	0175
nd Interfaces—I. Dynamics at the	Solid-Polymer Interface.= +r Solutions, Blends, a	COLL	0012
Beam Mass Spect+ Chemistry of	Solid-Propellant Combustion Studied by Molecular-	FUEL	0126
Solution and	Solid-State ^{13}C NMR Studies of C_{60} and C_{70}.=	PHYS	0070
.=	Solid-State ^{27}Al NMR Studies of Hydrated $AlPO_4$-8	PETR	0096
ynamics Parallels between	Solid-State Amorphization and Melting.= Thermod	NUCL	0092
ent Systems—I. Plenary Lecture.	Solid-State Characterization of Multicomponent Pol	PMSE	0099
tems—I. Plenary Lecture. Solid+	Solid-State Characterization of Multicomponent Sys	PMSE	0099
tems—II. Plenary Lecture. Enh+	Solid-State Characterization of Multicomponent Sys	PMSE	0116
tems—III. Plenary Lecture. Str+	Solid-State Characterization of Multicomponent Sys	PMSE	0145
tems—IV. Plenary Lecture. Hy+	Solid-State Characterization of Multicomponent Sys	PMSE	0163
e+ NMR Imaging of Anisotropic	Solid-State Chemical Reactions Using Multiple-Puls	CATL	0022
	Solid-State Chemical Sensors.=	CHED	0032
State-of-the-Art Symposium:	Solid-State Chemistry—A Rapidly Moving Frontier.	CHED	0020
State-of-the-Art Symposium:	Solid-State Chemistry—A Rapidly Moving Frontier.	CHED	0029
Synthesis in Molten Salts: New	Solid-State Compounds of $Na_3Cu_4Se_4$, $Na_2Cu_2Se_2$-Cu	INOR	0351
in the Conversion of Molecules to	Solid-State Compounds.= +ucleation and Growth	INOR	0145
Molecular Particles of	Solid-State Compounds.=	CHED	0033
ystallographic Symmetry upon the	Solid-State Emission of Copper(I) Hilides.= +Cr	INOR	0157
rganic and Aqueous Guest/Host	Solid-State Interactions by Multiple Quantum NMR	AGRO	0062

romoted Hydrotreatin+	FT-IR, Solid-State MAS NMR, and XRD of Phosphorous-P	COLL	0198
rsors and Pyrolyzed Products.=	Solid-State MAS-NMR Study of Preceramics Precu	POLY	0358
	Solid-State Models of Photosynthetic Systems.=	CHED	0031
Approach.= Pictorialized	Solid-State MOs: A Nonmathematical but Honest	CHED	0021
Molecular and	Solid-State Nitrides.= Syntheses of Related	INOR	0103
us Temperatures by Multinuclear	Solid-State NMR and DTA.= +o $AlPO_4$-8 at Vario	PETR	0095
on and the Effects of Seed Aging:	Solid-State NMR and NMR Imaging.= +Germinati	AGFD	0104
Zeolites and Sili+ Multinuclear	Solid-State NMR Investigation of Boron-Substituted	PETR	0136
o-Dimensional High-Resolution	Solid-State NMR Investigations of Zeolite Lattice St	PHYS	0348
Reaction of Hyaluronic Acid wit+	Solid-State NMR of N-Acylureas Derived from the	ORGN	0020
	Solid-State NMR Studies of Separation Media.=	ANYL	0064
ic Reactions in Microporous M+	Solid-State NMR Studies of Shape-Selective Catalyst	CATL	0023
mics of Ligand-DNA Complexe+	Solid-State NMR Studies of the Structure and Dyna	BIOL	0116
of PS-PMPS Diblock Copoly+	Solid-State NMR Study of the Microphase Structure	PMSE	0121
minophosphate Molecular Sieves+	Solid-State NMR Study of 18-Ring Large-Pore Alu	PETR	0094
H_2)$_2$, n = 1-4 from ^{31}P and ^{13}C	Solid-State NMR.= Structures of $Zr(O_3P(CH_2)_nCOO$	INOR	0047
e+ Control of Processability and	Solid-State Properties of Polymers through Rotaxan	POLY	0233
: Description and Appli+ A New	Solid-State Small-Angle Light-Scattering Apparatus	POLY	0278
try at the Liquid-Liquid, Liquid-	Solid, and Solid-Gas Interfaces.= +Spaces: Chemis	ORGN	0126
g Mechanisms of Stabilization and	Solidification Systems.= +Chemistry and Leachin	ENVR	0003
Cure Times in Cement/Fly Ash	Solidification/Stabilization Can Result in Significant	ENVR	0041
Techniques Applied to Cement-	Solidified Hazardous Waste.= Petrographic	ENVR	0009
Spectroscopic Characterization of	Solidified Hazardous Wastes.= Microscopic and	ENVR	0004
Characterization of Cement-	Solidified Heavy Metal Wastes.= Microstructural	ENVR	0008
Matrices for the Immobilization of	Solidified Toxic Hazardous Wastes.= +of Cement	ENVR	0002
Chemistry and Microstructure of	Solidified Waste Forms. Chemistry of Cementitious	ENVR	0001
Forms. Chemistry of Cementitious	Solidified Wastes.= +tructure of Solidified Waste	ENVR	0001
erization Techniques for Cement-	Solidified/Stabilized Metal Wastes.= +opy Charact	ENVR	0005
rte de Calor en Reacciones	Solido-Gas NO-Cataliticas.= Influencia del Transpo	I&EC	0068
lective Surface Derivatization of+	Solids and Films. Dithiocarbamates: Reagents for Se	INOR	0350
Conversion of Corn Syrup	Solids into γ-Cyclodextrin.=	CARB	0041
ctural Studies of Fullerene-Based	Solids: C_{60}, C_{70}, and the Alkali-Metal Fullerides.=	PHYS	0049
tion for High-Resolution NMR of	Solids: DAS and DOR.= +opies. Sample Reorienta	PHYS	0370
m+ Energy Transfer in Ultrahot	Solids: Vibrational Cooling and Multiphonon Pu-Pu	PHYS	0250
ipolar Energy Transfer in Porous	Solids.= +ith Small-Angle Scattering and Direct D	PHYS	0060
in Molecular Glasses and Polymer	Solids.= +hases. Energy Transfer and Relaxation	PHYS	0080
ng Frontier. Chemical Bonding in	Solids.= +Solid-State Chemistry—A Rapidly Movi	CHED	0020
niques to the Characterizations of	Solids.= +Gravimetric Analysis and Related Tech	INOR	0006
n Studies of Collision Cascades in	Solids.= +olecular Dynamics Computer Simulatio	NUCL	0091
Molecular	Solids.= C-H···O Contact Interactions in	PHYS	0295
Sludge Pond	Solids.= Characterization of Humic Acids from	I&EC	0022
Cluster Energy Loss in	Solids.=	NUCL	0053
and Low-Temperature Molecular	Solids.= +gh-Temperature Dense Molecular Fluids	ANYL	0002
from	Solids.= Fractal Theory of Radon Emanation	NUCL	0088
Motions in	Solids.= NMR Measurement of Cooperative	PHYS	0069
for Extended	Solids.= Reconstructed Clusters as Building Blocks	INOR	0173
Manuscript.= Splendor	Solis: A Sixteenth-Century Illuminated Chemical	HIST	0036
n in Condensed Phases. Torsional	Solitons in Organic Crystals.= +sfer and Relaxatio	PHYS	0129
International Competition. R.	Solow (Economics, 1987).= Science, Technology, and	CONG	0014
Photochemistry of Colloidal Silver	Sols.= +perties of Nonmetallic Silver Clusters and	COLL	0130
l)-1H-Benz-imidazol-2-YL] C+	Solubility and Stability of Methyl [1-(Butylcarbamoy	AGRO	0041
Gas	Solubility in Polymers and Blends.=	POLY	0143
A Theory for	Solubility in Static Systems.=	POLY	0141
turated $CaO-CaF_2-SiO_2$ Slags.+	Solubility of Copper and Other Elements in Silica-Sa	I&EC	0066
nideal Solven+ Prediction of the	Solubility of Hydrophobic Organic Compounds in No	ENVR	0030
ect of Differences of Component	Solubility Parameters and Glass Transition Tempera	PMSE	0025
	Solubility Parameters of Aromatic Polyamides.=	POLY	0063
oring K. L. Hoy. Four Component	Solubility Parameters: A New Dimension.= +Hon	PMSE	0033
rements of Premicellar Porphyrin	Solubilization as Detected by J-Aggregation.= +qui	ORGN	0161
e3)2|3(μ-Cl)K($η^6$-C_7H_8)$_2$ and the	Solubilization of Alkali Metal Halides in Hydrocarbo	INOR	0053
ty of L-693,989 and Other Water-	Soluble Analogues of L-688,786.= +ocystis Activi	MEDI	0182
of Silicon-Containing Phenylated	Soluble Aramids.= +iques for the Characterization	POLY	0086
ication and Characterization of a	Soluble Form of the Murine Erythropoietin Receptor	BIOL	0108
Organometallic Precursors for	Soluble III-V Semiconductor Clusters.=	INOR	0313
Toward High Susceptibilities in	Soluble Polyacetylenes for Nonlinear Optics.=	POLY	0132
hesis and Properties of a Series of	Soluble Polyimides for Use in Microelectronics.=	POLY	0038
leoside-Subst+ The First Water-	Soluble Porphyrinyl-Mononucleoside and Other Nuc	ORGN	0189
ylene) A+ Aqueous Synthesis of	Soluble Rigid-Chain Polymers: An Ionic Poly(p-phen	POLY	0024
Flow Fie+ Separation of Water-	Soluble Synthetic and Biological Macromolecules by	PMSE	0011
The	Solute's View in Chromatography.=	ANYL	0013
Relaxation for Polar	Solutes in Polar Solvents.= Fast Vibrational	PHYS	0037
Measuring the Sorption of Organic	Solutes on Sedimentary Organic Matter.= +e for	ANYL	0072
Evidence on Retention of Organic	Solutes.= +bents: Critical Review of Experimental	ANYL	0071
imetallic Particles Prepared from	Solvated Metal Atoms: Structure and Catalysis.=	CATL	0009
Short-Range	Solvation by Cyclic Diethers.=	PHYS	0308
Surfaces.= Femtosecond	Solvation-Dynamics of CH_3Cl^- and Cl^- at Multilayer	PHYS	0084
Like Nature of Zeolite Based on a	Solvatochromic Indicator.= +nation of the Solvent-	COLL	0076
ce of Nucleophilic Assistance by	Solvent and Azide Ion to the Reaction of Cumyl Driv	ORGN	0154
ree-Volume Theory for Polymer-	Solvent Diffusion Coefficients.= +apabilities of a F	POLY	0144
ophene)s.=	Solvent Effects on Configurations of Poly(3-Alkylthi	PMSE	0183

gen with α,β Unsaturated Ester+	Solvent Effects on Enzyme Reactivity.=	BIOT	0247
esence of Oxalic Acid+ Selective	Solvent Effects on the Side Selectivity of Singlet Oxy	ORGN	0164
n of the Freezing and Melting of	Solvent Extraction of Lanthanum by Oxine in the Pr	ANYL	0054
e Organic Compounds in Nonideal	Solvent in Thermoreversible Polymer Gels and Relat	POLY	0170
Effect of	Solvent Mixtures.= +the Solubility of Hydrophobi	ENVR	0030
erential Scanning Calorimetry and	Solvent on the Nucleophilicity of Subtilisin.=	BIOT	0066
pplication of Microdilatometry to	Solvent Swelling for Studies of Coal Structure.= +f	FUEL	0087
in Multiphase Polymer–Polymer–	Solvent Swelling of Coals.= +Mass Spectrometry. A	FUEL	0086
Coal in the Presence of a Swelling	Solvent Systems.= +duced Pearl–Chain Formation	COLL	0145
Polymacromonomers in Good	Solvent Using a Novel EPR Technique.= +ture of	FUEL	0073
ers in Nonaqueous Solvents and	Solvent.= Equilibrium Configuration of Dense	POLY	0269
romic Indic+ Examination of the	Solvent-Free Systems Using Immobilized Enzymes.=	BIOT	0065
ntheses of Esters in Nonaqueous	Solvent-Like Nature of Zeolite Based on a Solvatoch	COLL	0076
sertion R+ Role of Coordinating	Solvents and Solvent–Free Systems Using Immobiliz	BIOT	0065
elaxation Time of Ions in Protic	Solvents in Promoting the Alkyl to Acyl Migratory In	INOR	0109
Surface: Influence of the Organic	Solvents Measured by Transient Infrared Spectrosco	PHYS	0083
y(di-N-alkylsilanes) and Aromatic	Solvents on the Reaction Rate.= +PCl5 with Silica	COLL	0097
mobilized Biocatalysts in Organic	Solvents.= +ble Gelation in Binary Mixture of Pol	POLY	0145
and Purification in Neat Organic	Solvents.= +d Opportunities. The Operation of Im	BIOT	0035
	Solvents.= +ous Bioprocessing: Protein Separation	BIOT	0063
cted Environment of Nonaqueous	Solvents.= +d Enzyme System in the Water–Restri	BIOL	0091
of Polystyrene in Good and Theta	Solvents.= +oton Correlation Spectroscopy Study	POLY	0305
Solutes in Polar	Solvents.= Fast Vibrational Relaxation for Polar	PHYS	0037
II)–Diimine Complexes in Mixed	Solvents.= Photophysics and Photochemistry of Ru(	INOR	0279
Enzyme Activity in Organic	Solvents.= Random Mutagenesis to Enhance	BIOT	0068
Enzyme Stability in Polar Organic	Solvents.= +on: A General Strategy for Increasing	BIOT	0244
e Sorbitol and Organic	Solvents.= Thermoreversible Gels from Dibenzyliden	POLY	0201
ock Copolymers and Hydrocarbon	Solvents.= +oreversible Gels with Polyethylene Bl	POLY	0249
We Learn from Reactions without	Solvents?.= +emistry in the Gas Phase: what Can	ORGN	0121
and–Held Calculator?+ Equation	Solvers or Spreadsheets: Which Will Supplant the H	CHED	0117
l Phenol.=	Solvolysis–Triggered Alkylation of DNA Using a Sily	ORGN	0191
Mechanism of Successive	Solvolytic Extraction of Coal.=	I&EC	0018
Recombinant Bovine	Somatotropin.= Oxidation and Naturation of	BIOT	0135
Aromatics, Fullerenes, and	Soot as Charged Species in Flames.=	FUEL	0111
Aromatics, Fullerenes, and Soot.	Soot Deposition from Ethylene/Air Flames and the R	FUEL	0107
A Shock Tube Investigation of	Soot Formation from Toluene/Methanol Mixtures.=	FUEL	0113
Ring and the Nucleation of	Soot Particles.= Aromatics Growth beyond the First	FUEL	0108
Large Aromatics, Fullerenes, and	Soot. Soot Deposition fro+ Combustion Chemistry:	FUEL	0107
z FT–NMR in a Second–Semester	Sophomore Unified Laboratory Course.= +300–MH	CHED	0181
Estudio de Catalizadores NiW	Soportados sobre Alumina o Titania Modificadas con	FUEL	0106
Soil Gas Analysis.= A	Sorbent Trapping Procedure for High-Resolution	ENVR	0050
pports as Surrogates for Natural	Sorbents: Critical Review of Experimental Evidence	ANYL	0071
by Dry	Sorbents.= Fundamentals of Heavy Metal Removal	FUEL	0132
Gels from Dibenzylidene	Sorbitol and Organic Solvents.= Thermoreversible	POLY	0201
rs: Vinyl Alcohol/Vinyl Bu+ Gas	Sorption and Diffusion in Hydrogen–Bonded Polyme	POLY	0191
er Molecular Structure upon the	Sorption and Transport of Small Molecules in Stiff–	POLY	0120
tion Reactions on Silica Surface+	Sorption Complexes as the Intermediates of Substitu	COLL	0095
Effect of Moisture on the	Sorption of Gases by Coal.=	I&EC	0020
t+ Cosolvent and pH Effects on	Sorption of Organic Acids from Methanol–Water Mix	ENVR	0014
CP Technique for Measuring the	Sorption of Organic Solutes on Sedimentary Organic	ANYL	0072
Alteration under the	Sorption of Redox Reagents.= Redox Potential	COLL	0103
ing Equipment for the Automatic	Sortation of Postconsumer Plastics.= +terial Handl	I&EC	0041
Recycling: Beyo+ Collection and	Sortation Technology in Plastics Recycling. Plastics	I&EC	0037
ethanes and In+ Low–Frequency	Sound-Absorbing Studies on Foams Based on Polyur	PMSE	0040
a Debittering and Deacidifying	Sour Orange (Citrus aurantium) Juice Using Neutral	AGFD	0106
Global	Sourcing within a Regulated Industry.=	CMEC	0006
. Chemical Specialties fo+ Global	Sourcing—Strategic Issues in Chemical Procurement	CMEC	0005
Enzymatic Phosphorylation of	Soy Protein to Improve Functional Properties in Foo	AGFD	0140
lementation on Protein Quality of	Soy-Based Infant Formulas Fed to Rats.= +d Supp	AGFD	0209
Antioxidant Properties of	Soybean Isoflavones.= Identification and	AGFD	0020
ne vs. Azulene on Pt(111+ Real–	Space Imaging of Molecular Organization: Naphthale	PHYS	0125
in Radicals Studied by the Fock–	Space Multireference Coupled–Cluster Method.=	PHYS	0016
Avermectin Analogs with a	Spacer between the Aglycone and the Disaccharide.=	ORGN	0016
atic Units Separated by Aliphatic	Spacers.= +igoesters Containing Single–Ring Arom	PMSE	0016
Utilization as Restricted Reaction	Spaces for Electron Transfer Processes.= +d Their	POLY	0238
chemistry in Restricted Reaction	Spaces: Chemistry at the Liquid–Liquid, Liquid–Soli	ORGN	0126
Differs from the Crystal State but	Spans the Range of Different Crystal Forms.= +in	BIOL	0002
and Mechanistic Effects of Gas	Sparging in Animal Cell Bioreactors.= Quantitative	BIOT	0009
Compounds from Anthers of	Spathodea Campanulta Flowers.= Chemical	ORGN	0186
lar and Biomolecular Electronics.	Spatial Light Modulators and Optical Memories Bas	BIOT	0077
ing the Bounds of Sensitivity and	Spatial Resolution.= +ctroscopy at the Limit: Push	ANYL	0012
of Drug Leads. Light-Directed,	Spatially Addressable Parallel Chemical Synthesis.=	MEDI	0090
of Electronic Energy Transport in	Spatially Disordered Systems.= +Ultrafast Studies	PHYS	0133
olecular Exciton Micro+ Toward	Spatially Resolved Energy Transfer and Scanning M	PHYS	0022
RNA Synthesis in E. coli.=	Specialized Ribosomes and Feedback Regulation of r	BIOT	0245
Thrive in the Midst+ Chemical	Specialties for Niche Markets, or How to Survive and	CMEC	0005
	Specialties in Analytical Chemistry.=	CHED	0104
Technical Career Paths in a	Specialty Chemical Company.=	YCC	0007
lutions and Their Uses in Horti+	Specialty Fertilizers. Controlled-Release Nitrogen So	FERT	0005
polysaccharide Chains of Cell +	Specificities of Monoclonal Antibodies Binding to the	CARB	0017

Propane Aromatization—Infrared	Spectroscopic Investigations.= +-β as a Catalyst for	PETR	0047
ng Reactive Sp+ Matrix Infrared	Spectra and ab Initio Frequencies of Sulfur-Containi	PHYS	0031
Predicting Mass	Spectra from Structural Information.=	COMP	0005
Surface Enhanced Raman	Spectra of C_{60} and C_{70}.=	COLL	0092
Calculations of the Elctronic	Spectra of LiBe and LiMg.= Ab Initio Full SOCVCI	PHYS	0338
Solid Argon.= Infrared	Spectra of Metal-Ethylene Complexes Isolated in	PHYS	0074
$CTpS_2$-TPS_2-AAGCG):+ NMR	Spectra of Oligonucleotide Phosphorodithioate d(CG	BIOL	0123
igh-Overtone Resonance Raman	Spectra of Photodissociating Nitromethane in Solutio	PHYS	0302
of A+ The Near-UV Absorption	Spectra of Several Aliphatic Aldehydes and Ketones	ENVR	0024
ry+ Flourescence and Excitation	Spectra Pattern and Phase Transition in Tetracene C	PHYS	0274
Simulation of NMR	Spectra: Problems and Solutions.=	COMP	0007
with Experiment. High-Overtone	Spectra: The Significance of Perturbations.= +stry	PHYS	0029
Interpretation of Vibrational	Spectra.= Ab Initio Calculations and the	PHYS	0033
Simulation of ^{13}C and ^1H NMR	Spectra.=	COMP	0002
ral Nematic (Acetyl)(ethyl)cellu+	Spectral Behavior of Dye Molecules Oriented by Chi	POLY	0331
ular Theory of Inhomogeneous	Spectral Broadening in Liquids and Glasses.= Molec	PHYS	0036
ium N-Betaines: Preparation and	Spectral Characteristics.= +do-2,x-Dimethylpyridin	ORGN	0203
nted Data-Handling System for	Spectral or Chromatographic Data Acquisition and A	COMP	0056
(II)chlorin of Tu+ Structural and	Spectral Properties of Tunichlorin, the Novel Nickel	INOR	0139
R Spectrum Simulations.=	Spectral Simulation. Combinatorics of NMR and ES	COMP	0001
udied by Time-Resolved Infrared	Spectral Techniques.= +on of Metal Alkyl Bonds St	INOR	0169
ynamics of Photosynthetic Units:	Spectral-Hole-Burning Studies.= +nd Transport D	PHYS	0018
Analytical Chemistry Award in	Spectrochemical Analysis, sponsored by the Perkin-E	ANYL	0001
rfactant S+ Electrochemical and	Spectroelectrochemical Measurements in Aqueous Su	COLL	0003
Mass	Spectrometric Analysis of Phenolic Compounds.=	AGFD	0003
ates CH+ Photoionization Mass	Spectrometric Studies for the Combustion Intermedi	FUEL	0124
harides by Electrospray Mass	Spectrometry (ES MS).= Linkage Analysis in Disacc	CARB	0048
Studied by Molecular-Beam Mass	Spectrometry and Modeling.= +llant Combustion	FUEL	0126
nalysis Using Secondary Ion Mass	Spectrometry and Related Techniques.= +terial A	INOR	0007
ysis-Fourier Transform Infrared	Spectrometry as a Technique for Analyzing Function	FUEL	0061
ecies by Molecular-Beam Mass-	Spectrometry during Hot-Filament-Assisted Diamon	PHYS	0145
hite Furnace Atomic Absorption	Spectrometry for the Determination of Trace Alkylse	ANYL	0080
r-Excited Molecular Fluorescence	Spectrometry in a Graphite Furnace.= +oxide Lase	ANYL	0078
ts.= Electron Energy-Loss	Spectrometry of Alumina and Silica Catalyst Suppor	PETR	0073
Gas Chromatography and Mass	Spectrometry of Erythrina Alkaloids from the Foliag	ANYL	0136
I Liquid Chromatography/Mass	Spectrometry to Drug Discovery and Development.=	ANYL	0139
l Fuels: Swelling Studes and Mass	Spectrometry. +vel Analytical Techniques for Fossi	FUEL	0086
l Fuels: Swelling Studes and Mass	Spectrometry. +vel Analytical Techniques for Fossi	FUEL	0088
Materials: Laser Ionization Mass	Spectrometry. +in the Analysis and Modification of	NUCL	0041
of Vacuum Residue Using Mass	Spectrometry: A Heterogeneous Analytical Model.=	PETR	0150
Detector.= Mass	Spectrometry: A Powerful Chromatographic	AGFD	0202
Trap/Gas Chromatography/Mass	Spectrometry.= +man Blood by Using Purge-and-	CEI	0001
acerals by Laser Desorption Mass	Spectrometry.= +etry. Characterization of Coal M	FUEL	0088
sis in Water Using Ion Trap Mass	Spectrometry.= +gulatory Issues. Pesticide Analy	AGRO	0060
urrent Chromatography and Mass	Spectrometry.= +r Precursors by Droplet Counterc	AGFD	0203
Samples by Field Ionization Mass	Spectrometry.= +on of the Argonne Premium Coal	FUEL	0062
by On-Line HPLC Tandem Mass	Spectrometry.= +ibiotic Residues in Edible Meats	AGFD	0061
otometry and Particle-Beam Mass	Spectrometry.= +ography with UV-Absorption Ph	ENVR	0015
n Rocks by Electron Impact Mass	Spectrometry.= +ve Determination of Rare Earths i	ANYL	0049
Sampling Mass	Spectrometry.= Rapid Analysis Using Direct	FUEL	0089
eration, Transport+ HPLC Mass	Spectrometry—I. Particle Size Effects in Aerosol Gen	ANYL	0122
Utilizing Atmosphe+ HPLC Mass	Spectrometry—II. Recent Developments in CE/MS	ANYL	0138
on of a Diode Array UV-Visible	Spectrophotometer into the Undergraduate Curriculu	CHED	0180
th Sulfanilic Acid in the Presen+	Spectrophotometric Determination of Cerium(IV) wi	ANYL	0073
th Sulfonazo-III as Chromogeni+	Spectrophotometric Determination of Lanthanum wi	ANYL	0074
dent Experiment in Analytical+	Spectrophotometric Determination of Oxalate: A Stu	CHED	0171
with Arsenazo-III.=	Spectrophotometric Study of Complex of Lanthanum	ANYL	0075
omposition Pathways of Adama+	Spectroscopic and Chemical Studies of the Photodec	ORGN	0163
[M(tpen)]$(ClO_4)_2$ Comp+ Far IR	Spectroscopic and Crystallographi Caracterization of	INOR	0057
py)Cl]+.=	Spectroscopic and Electrochemical Studies of [Pt(ter	INOR	0216
st π^* and Ligand-Ligand Schar+	Spectroscopic and ODMR Investigations of the Lowe	INOR	0164
Alloys of Multiply Bonded Mo+	Spectroscopic and Structural Properties of Molecular	INOR	0356
nce and Optical Spectroscopies.	Spectroscopic Characterization of CO on Palladium C	PHYS	0373
on Pt(111).= Surface	Spectroscopic Characterization of Iodine Overlayers	PHYS	0366
ide Systems.=	Spectroscopic Characterization of Mixed Siloxane-Ox	POLY	0231
us Wastes.= Microscopic and	Spectroscopic Characterization of Solidified Hazardo	ENVR	0004
ic Me+ Synthesis, Structure, and	Spectroscopic Characterization of the First Monomer	INOR	0059
dy of a Caffeine Dim+ Synthesis,	Spectroscopic Characterization, and Theoretical Stu	INOR	0252
ition of Diamond Films.= Laser	Spectroscopic Diagnostics for Chemical Vapor Depos	PHYS	0151
and Thin-Film G+ Real-Time	Spectroscopic Ellipsometry: Applications in Polymer-	POLY	0252
thyl Methacrylate) as Studied by	Spectroscopic Methods and Small-Angle X-ray Scat	POLY	0124
hiamin-Bound Enamine Interm+	Spectroscopic Observation of the Kinetic Fate of a T	BIOL	0047
amin Diphosphate-Bound Ena+	Spectroscopic Observation of the Kinetic Fate of Thi	BIOL	0048
Rhenium Carbonyl Complex as a	Spectroscopic Probe.= +ymers Using a Luminescent	POLY	0096
Cyanines.= Synthesis and	Spectroscopic Properties of Organometallic	INOR	0335
Acid Comp+ Photochemical and	Spectroscopic Studies of Alpha-Hydroxytricarboxylic	INOR	0040
iC_2.=	Spectroscopic Studies of Floppy Molecules: C_3 and S	PHYS	0235
tes.=	Spectroscopic Studies of Ion-Implanted Polycarbona	POLY	0081
ion: A Chimeric Protein Having+	Spectroscopic Studies of sCD4(178)-PE40 Conformat	ANYL	0005
als+ Negative-Ion Photoelectron	Spectroscopic Studies of the Structure of Free Radic	PHYS	0055

n over Rutheni+ Catalytic and	Spectroscopic Studies of the Water Gas Shift Reactio	CATL	0027
K (Poly-ether-ether-ke+ FT-IR	Spectroscopic Study of Thermal Degradation in PEE	POLY	0334
Channels.= Optical	Spectroscopic Techniques for Ions in Membrane	BIOT	0141
ules in Poly+ Using Vibrational	Spectroscopies to Probe the Diffusion of Small Molec	POLY	0332
Magnetic Resonance and Optical	Spectroscopies. +Developments and Applications of	PHYS	0136
Magnetic Resonance and Optical	Spectroscopies. +Developments and Applications of	PHYS	0193
Magnetic Resonance and Optical	Spectroscopies. +Developments and Applications of	PHYS	0113
Magnetic Resonance and Optical	Spectroscopies. +Developments and Applications of	PHYS	0372
Magnetic Resonance and Optical	Spectroscopies. +Developments and Applications of	PHYS	0066
Magnetic Resonance and Optical	Spectroscopies. +Developments and Applications of	PHYS	0166
Magnetic Resonance and Optical	Spectroscopies. +Developments and Applications of	PHYS	0040
Magnetic Resonance and Optical	Spectroscopies. +Developments and Applications of	PHYS	0370
Magnetic Resonance and Optical	Spectroscopies. +Developments and Applications of	PHYS	0373
Magnetic Resonance and Optical	Spectroscopies and Applications of	PHYS	0344
(110) Model Catalysts by Electron	Spectroscopies.= +TiO_2(110) and Vanadium/TiO_2	COLL	0192
r Corp.). Multiresonant Nonlinear	Spectroscopies—A New Family on the Block.= +e	ANYL	0001
Pure Quadrupole Correlation	Spectroscopy (NQR–COSY).= Two-Dimensional	PHYS	0277
nitoring Dewaxing Procedures via	Spectroscopy and Chromatography.= +ocessing: Mo	PETR	0151
Very High Frequency EPR	Spectroscopy and NMR Imaging of Coal.=	FUEL	0074
+by Fourier Transform Infrared	Spectroscopy and Second-Harmonic Generation.=	POLY	0224
Fast Atom Bombardment Mass	Spectroscopy as a Detector for Liquid-Phase Separat	ANYL	0125
Electron Spin Echo-Modulation	Spectroscopy as a Novel Method To Detect Framewo	PHYS	0231
Identification.= Derivative	Spectroscopy as an Analytical Tool for Hydrocarbon	PETR	0137
-Initiated Resonance Ionization	Spectroscopy as an Analytical Tool: Instrument Requ	NUCL	0042
Advantages and Prospect+ ESR	Spectroscopy at Sub-9-GHz Microwave Frequencies:	PHYS	0215
Anhydrase.= Cobalt	Spectroscopy at the Active Site of Carbonic	PHYS	0256
nsitivity and Spatial Resolution+	Spectroscopy at the Limit: Pushing the Bounds of Se	ANYL	0012
Optical	Spectroscopy at 1-MHz Resolution.=	PHYS	0043
New Approaches in	Spectroscopy Experiments.=	CHED	0203
phy-Fourier Transform Infrared	Spectroscopy for Characterization of Ethylene-Based	PMSE	0048
Experiments in Raman	Spectroscopy for P. Chem Lab.=	CHED	0088
Experiments in Raman	Spectroscopy for Physical Chemistry Lab.=	PHYS	0300
ractions+ Applications of NMR	Spectroscopy in Secondary Processing of Petroleum F	PETR	0146
rfaces+ Applications of Electron	Spectroscopy in the Characterization of Inorganic Su	INOR	0013
Time-Resolved Laser	Spectroscopy in the Undergraduate Laboratory.=	CHED	0084
Time-Resolved Laser	Spectroscopy in the Undergraduate Laboratory.=	PHYS	0249
enide: CaYbInSe+ Structure and	Spectroscopy of a New Infrared Transparent Chalcog	INOR	0030
r Activity, Electrochemistry, and	Spectroscopy of Cationic Organometallic Compounds	INOR	0320
Luminescence	Spectroscopy of F-Metal Complexes.=	INOR	0062
Fluids and+ Coherent Raman	Spectroscopy of High-Temperature Dense Molecular	ANYL	0002
neling Microscopy and Tunneling	Spectroscopy of Metal Oxide Arrays on Graphite.=	COLL	0194
here: Experimental Studies of the	Spectroscopy of NO_3, Cl_2O_2, and Cl_2O_3.= +Atmosp	PHYS	0017
+um Chemistry with Experiment.	Spectroscopy of Nonrigid Molecular Complexes.=	PHYS	0102
Two-Photon Absorption	Spectroscopy of Phenanthrene.=	CHED	0085
Two-Photon Absorption	Spectroscopy of Phenanthrene.=	PHYS	0328
Subpicosecond Transient	Spectroscopy of Rhodopsin Photochemistry.=	BIOT	0080
Resonance Raman	Spectroscopy of Simple Amides.=	PHYS	0321
Structure and	Spectroscopy of Small Carbon Clusters in Matrices.=	PHYS	0078
Methods.= Study of the	Spectroscopy of Small Molecules by ab Initio	PHYS	0005
UV Resonance Raman	Spectroscopy of Species Adsorbed on Zeolites.=	COLL	0075
Clusters.= Infrared Laser	Spectroscopy of Supersonically Cooled Carbon	PHYS	0007
Reaction Center.= Optical	Spectroscopy of the Bacterial Photosynthetic	PHYS	0170
Infrared Laser	Spectroscopy of the C_2H Radical.=	PHYS	0332
Laser and Fourier Transform	Spectroscopy of Transient Molecules.=	PHYS	0057
.= Picosecond Transient IR	Spectroscopy of UV-Photolyzed $Cr(CO)_6$ in Solution	PHYS	0220
cale+ An Integrated Four-Year	Spectroscopy Program Complemented by the Micros	CHED	0150
esolution Electron-Energy-Loss	Spectroscopy Studies of Adsorbed Species on Metal	COLL	0173
esolution Electron-Energy-Loss	Spectroscopy Studies of High-Temperature Supercon	PHYS	0279
Oxid+ Dielectric and Mechanical	Spectroscopy Studies of Poly(vinyl Acetate)/Silicon	POLY	0287
Polymers: A Photon Correlation	Spectroscopy Study of Polystyrene in Good and The	POLY	0305
n.= Unique Optics for Raman	Spectroscopy Using Charge-Coupled Device Detectio	ANYL	0035
ry.+ Negative-Ion Photoelectron	Spectroscopy: Comparison of New Results with Theo	PHYS	0059
ction Methods in Analytical Laser	Spectroscopy.= +undation). Thermo-Optical Dete	ANYL	0011
ors by Tunneling Microscopy and	Spectroscopy.= +c States of High-T_c Superconduct	PHYS	0206
lated Carbon Materials by Raman	Spectroscopy.= +ynthetic Diamond Films, and Re	PHYS	0118
pies. Multiple-Pulse Femtosecond	Spectroscopy.= +esonance and Optical Spectrosco	PHYS	0066
ser Spectroscopy, and Laser NMR	Spectroscopy.= +ectroscopies. NMR, Coherent La	PHYS	0040
f Ion/Solid Collisions Using Laser	Spectroscopy.= +es of the Fundamental Aspects o	NUCL	0041
Neutrino	Spectroscopy.= Borexino: Low-Energy Solar	NUCL	0033
ium Compounds by Proton NMR	Spectroscopy.= +ion of (η_6-Arene)Tricarbonylchrom	ORGN	0215
al Molten Salt Systems by Raman	Spectroscopy.= +hloromagnesium Species in Sever	INOR	0064
Mossbauer	Spectroscopy.= Characterization of Ferrosilicates by	NUCL	0069
attering and X-ray Photoelectron	Spectroscopy.= +ing Surface-Enhanced Raman Sc	POLY	0087
ure via Fourier Transform Raman	Spectroscopy.= +ation of Protein Secondary Struct	POLY	0306
ds for CVD by FT-IR and Raman	Spectroscopy.= +acteristics of Copper(I) Compoun	COLL	0031
N-Diethylbenzamide by ^{13}C NMR	Spectroscopy.= +ver Microsomal Metabolites of N,	AGRO	0026
n NaX and NaY Zeolites by NMR	Spectroscopy.= +geneous Adsorption of Benzene o	CATL	0025
y ^{2}D Nuclear Magnetic Resonance	Spectroscopy.= +r with Clay Minerals as Studied b	AGRO	0061
$YBa_2Cu_3O_7$ Using Raman	Spectroscopy.= Investigation of Superconducting	PHYS	0052
n of Temperatures by C_2 Emission	Spectroscopy.= +drocarbon Plasmas: Determinatio	PHYS	0268

Scanning Tunneling Microscopy–	Spectroscopy.=	$+_2(110)$ and $TiO_2(001)$ Surfaces by	COLL 0171
ed by Rheo–Photoacoustic FT–IR	Spectroscopy.=	+Poly(vinylidene Fluoride) Monitor	POLY 0255
An Undergraduate Experiment in	Spectroscopy.=	+ar Orbitals of Metalloporphyrins:	PHYS 0327
Step–Scan FT–IR	Spectroscopy.=	Polymer Characterization by	POLY 0304
Using Site–Selective	Spectroscopy.=	Probing Upconversion Dynamics	ANYL 0004
al Forms and Reactions by XAFS	Spectroscopy.=	+itative Analysis of Sulfur Function	FUEL 0056
num(II) Chloride Using ^{13}C NMR	Spectroscopy.=	+om cis–Diammine(guanosine)plati	INOR 0251
Transform Raman	Spectroscopy.=	Recent Advances in Fourier	POLY 0254
o–Photon Excited Fluorescence	Spectroscopy.=	Recent Advances in Quantitative Tw	ANYL 0014
Photoelectron and Chemilelectron	Spectroscopy.=	+se Metal Oxidation Reactions by	PHYS 0352
rfaces by Surface Photovoltage	Spectroscopy.=	Studies of Some Photosensitive Inte	BIOT 0142
Liquids by Photothermal Grating	Spectroscopy.=	+otoinitiated Reaction Kinetics in	ANYL 0034
ace by Electrochemical Impedance	Spectroscopy.=	+zed Steel/Sodium Chloride Interf	PHYS 0270
er Transform Infrared Reflectance	Spectroscopy.=	+Profiling of Polymers with Fouri	PMSE 0153
Coupled Plasma Atomic Emission	Spectroscopy.=	+nalysis of Fly Ash by Inductively	FUEL 0138
IR)	Spectroscopy.=	Two–Dimensional Infrared (2D	PHYS 0117
Two–Dimensional Microwave	Spectroscopy.=		PHYS 0115
Using Transient Infrared	Spectroscopy.=	Ultrafast Vibrational Dynamics	PHYS 0194
s Measured by Transient Infrared	Spectroscopy.=	+ion Time of Ions in Protic Solvent	PHYS 0083
Dinitrosobenzene by Steady–State	Spectroscopy—A New Phtochromic Reaction.=	+2–	ORGN 0168
ctroscopies. NMR, Coherent Laser	Spectroscopy, and Laser NMR Spectroscopy.=	+e	PHYS 0040
The Tritium Beta Decay	Spectrum from End to End and the Neutrino Mass.=		NUCL 0011
Infrared	Spectrum of the Hydrogen Halides Revisited.=		CHED 0197
ogical S+ Infrared and ^{13}C NMR	Spectrum Simulation: Role of Complementary Topol		COMP 0006
^{13}C NMR	Spectrum Simulation.=		COMP 0003
Techniques for ^{13}C NMR	Spectrum Simulation.=	Database Retrieval	COMP 0004
Combinatorics of NMR and ESR	Spectrum Simulations.=	Spectral Simulation.	COMP 0001
Molecular Structure, Vibrational	Spectrum, and Binding Energy of the Lowest Singlet		PHYS 0276
form Mid– and Far–Infrared IR	Specular Reflectance Study of Oxidized and Exchang		POLY 0253
a Democratic Socie+ Freedom of	Speech—The Delicate Balance: Defense Research in		PROF 0009
ow–Reactor and Laminar–Flame–	Speed Data.=	+dation Models: Studies Based on Fl	FUEL 0083
ce of Analytical Chemistry. High–	Speed DNA Sequencing in Ultrathin Gels.=	+w Fa	ANYL 0033
by SPERM.=	3D Shape Fitting and 3D–DB Searching		COMP 0040
d in Vitro Activity of Inhibitors of	Spermidine N^8–Acetyltransferase.=	+Synthesis, an	MEDI 0020
umaramate Esters as Inhibitors of	Spermidine/Spermine–N^1–Acetyltransferase.=	+F	MEDI 0021
e Q_y Transitions in Rhodobacter	sphaeroides Reaction Center: Polarization and Temp		BIOL 0094
Indiana Silicon	Sphere (ISiS) 4π Detector.=		NUCL 0036
oms into Rods, Rings, Nets, and	Spheres: The New Generations of Carbon Allotropes.		ORGN 0266
of Phosphoinositides and	Sphingolipids.=	Synthesis and Properties of Analogs	BIOL 0017
ition of Protein Kinase C by	Sphingosine and Related Long–Chain Bases.=	Inhib	MEDI 0075
Natural Antioxidants from	Spices.=		AGFD 0016
Phenolic Compounds in	Spices.=		AGFD 0047
h+ The Concept of Hydrogen Back	Spillover (HBS): Toward the Development of Hig		PETR 0063
Hydrocarbons.=	Oxygen Spillover during Oxide–Catalyzed Oxidation of		COLL 0174
hod To+ Utilization of Electron	Spin Echo–Modulation Spectroscopy as a Novel Met		PHYS 0231
f Coordinated Imidazoles in Low–	Spin Ferric Porphyrin Complexes.=	+Orientation o	INOR 0296
Transport: Relation to	Spin Glass Relaxation.=	Dispersive Energy	PHYS 0082
–Induced Moti+ Nitroxyl Amide	Spin Labeling: Methyl Esterification– and Hydration		AGFD 0012
O_x Superconductors by Electron	Spin Resonance Line Broadening of Surface Paramag		PHYS 0284
ity and Coordination via Electron	Spin Resonance.=	+ucts with the cupric Ion Intens	CATL 0017
Two–Dimensional Electron	Spin Resonance.=		PHYS 0114
nthesis, Structure, Reactivity, and	Spin.=	+Energy Gap in Non–Kekule Molecules: Sy	ORGN 0128
emicarbazones—Comparison with	Spin–Crossover Analogs.=	+pin–State: Ferric Thios	INOR 0268
ese Oxo Clusters with Nonzero	Spin–Ground States.=	Covalent Linkage of Mangan	INOR 0196
Properties of a Novel	Spin–Labeled Nucleoside.=	Synthesis and	ORGN 0026
–Y+ Adsorbant Interactions and	Spin–Lattice Relaxation of Sodium Cations in Zeolite		PHYS 0267
x Superconductors by Electron–	Spin–Resonance Line Broadening of Surface Parama		PHYS 0050
with Sp+ H–Bonding Control of	Spin–State: Ferric Thiosemicarbazones—Comparison		INOR 0268
bdominal Glands of the Australian	Spined Citrus Bug.=	+miacetal from the Dorsal A	AGRO 0035
Mammalian Cell Retention in a	Spinfilter Perfusion Bioreactor: Analysis and Scaleup		BIOT 0007
tyrene/Po+ Phase Diagrams and	Spinodal Decomposition and Linear and Cyclic Polys		POLY 0295
mers. Domain Growth during the	Spinodal Decomposition under Simple Shear Flow.=		PMSE 0104
Half–Integer	Spins in Rotating Samples.=	Cross Polarization of	PHYS 0371
tosynthetic Activity in Vallisneria	spiralis.=	+loroplasts and Their Relation with Pho	BIOL 0101
Novel	Spiro[1,3,2–Benzoxazaphosphole] Systems.=		INOR 0374
Designed	Spirocyclic Ionophores for Lithium.=		ORGN 0062
ibitors.=	Spirolactams (Carbamates) as Phosphodiesterase Inh		MEDI 0189
s III Antiarrhy+ Benzopyranone	Spiropiperidines as Novel, Potent, and Selective Clas		MEDI 0191
s III Antiarrhyt+ Arylpyranone	Spiropiperidines as Novel, Potent, and Selective Clas		MEDI 0097
Some Biological Effects of	Spirostanic and Furostanic Steroidal Glycosides.=		MEDI 0037
emical Manuscript.=	Splendor Solis: A Sixteenth–Century Illuminated Ch		HIST 0036
imizing Mass Discrimination in a	Split Injection System for Capillary Gas Chromatogr		TECH 0018
Photosynthetic Water	Splitting by Platinized Chloroplasts.=		TECH 0016
sin ATP Hydrolysi+ The Bond–	Splitting Step Is Rate–Limiting for Skeletal Actomyo		BIOL 0141
by Sulfur Atom Abstraction from	$SPMe_3$.=	+ybdenum(III)–μ–Sulfido Dimers Formed	INOR 0271
Scalarins from the	Sponge Phyllospongia Foliascens.=	New Dialkylated	ORGN 0182
Scalarins from the	Sponge Phyllospongia Folliascens.=	New Dialkylated	ORGN 0188
ess (Edward E. Smissman Award,	sponsored by Bristol–Myers Squibb).+	Award Addr	MEDI 0089
ward in the History of Chemistry,	sponsored by Dexter Chemical Corp.).+	(Dexter A	HIST 0011
istry Award in Electrochemistry,	sponsored by EG G Princeton Applied Research Co.)		ANYL 0096

ard in Chemical Instrumentation,	sponsored by the Dow Chemical Co. +Chemistry Aw	ANYL	0011
ward in Spectrochemical Analysis,	sponsored by the Perkin-Elmer Corp.). +emistry A	ANYL	0001
on Set: Assembly of the Rods and	Spools.= +a Molecular-Size Tinkertoy Constructi	BIOT	0200
Metabolism of Triallate in	Sprague-Dawley Rats.=	AGRO	0023
Monitoring Cure in Inside	Spray Coatings.=	PMSE	0160
rg, and Characterization. Role of	Spray Drying in Production of Catalyst and Catalyst	PETR	0088
Affecting Lipid Autoxidation of a	Spray-Dried Milkbase for Baby Food.= Factors	AGFD	0158
Courses.= Extensive Use of	Spreadsheet Programs in Analytical Chemistry	CHED	0205
Teaching Biochemistry with	Spreadsheet Software.=	CHED	0206
Chemical Education. Using	Spreadsheets in General Chemistry.= Computers in	CHED	0204
	Spreadsheets in Physical Chemistry.=	CHED	0208
	Spreadsheets: Opportunities in Physical Chemistry.=	CHED	0207
Calculator?+ Equation Solvers or	Spreadsheets: Which Will Supplant the Hand-Held	CHED	0117
as an Analytical Tool: Instrum+	Sputter-Initiated Resonance Ionization Spectroscopy	NUCL	0042
ns at Polymer/Metal Interface by	Sputtering Depth Profiling Using XPS.= +Reactio	PMSE	0150
aterials.=	Sputtering Processes of Ion-Bombarded Electronic M	NUCL	0043
ino)ethyl Methacrylate to	Squalane.= Free Radical Grafting of 2-(Dimethylam	POLY	0019
Purification of Pig Liver	Squalene Epoxidase by Affinity Chromatography.=	BIOL	0070
ldehyde from Photooxidation of	Squalene.= Formation of Formaldehyde and Malona	AGFD	0132
ratory Simulation.=	SQUALOR—The Design of an Award-Winning Labo	CHED	0079
L. (Coleopt+ Toxicity of Annona	squamosa Seed Oil Extracts to Tribolium castaneum	AGRO	0036
e Number of Perfect Matchings in	Square and Pentagonal Chains.= +d Chains and th	COMP	0033
Chelating Ligands: C+ Binuclear	Square Planar Cobalt(III) Complexes of Polyanionic	INOR	0243
hine Ligand Exchange between	Square Planar Iridium(I) Complexes.= Facile Phosp	INOR	0211
Equation for the	Square Root of the Electronic Density.=	PHYS	0254
	Square Wave Detection for Flow-Injection Analysis.=	ANYL	0105
ard E. Smissman—Bristol-Myers	Squibb Award Address. Hydanto+ General and Edw	MEDI	0082
y: Chemistry at the Bristol-Myers	Squibb Pharmaceutical Research Institute.= +ustr	CMEC	0028
ward, sponsored by Bristol-Myers	Squibb).+ Award Address (Edward E. Smissman A	MEDI	0089
uthenium Thiolates cpRu(PR$_3$')	2SR: Synthesis and Reactions with Electrophiles.= R	INOR	0368
New Cubic Compound in the Bi-	Sr-Cu-O System.=	PHYS	0227
Bi$_2$O$_3$-	SrO-CaO System.= Beta Solid Solution in the	INOR	0201
Glass Formation in the BiO$_{1.5}$-	SrO-CaO-CuO System.=	INOR	0224
ituted Zeolites with the ZSM-5	Srructure.= EXAFS Studies of Isomorphously Subst	PETR	0081
ons over a Typical Commercial +	SSITKA Investigation of CO-O$_2$ and CO-NO Reacti	COLL	0187
um- and Boron-Containing	SSZ-24, the All-Silica Analog of AlPO$_4$-5.= Alumin	PETR	0023
tegies as Alte+ Pesticides in the	21st Century. Biotechnology-Based Pest Control Stra	AGRO	0080
nd Developed+ Pesticides in the	21st Century. Outlook for Pesticides in Developing a	AGRO	0056
Century.= Pesticides in the	21st Century. Pesticide Regulatory Policy in the 21st	AGRO	0066
sful Intracontinental Trade in the	21st Century.= +inental Trade. Criteria for Succes	CMEC	0001
icide Regulatory Policy in the	21st Century.= Pesticides in the 21st Century. Pest	AGRO	0066
icide Use and Development in the	21st Century.= +cies and Attitudes Will Affect Pest	AGRO	0070
ork State: Oswego Harbor and the	St. +Initiatives at Two Areas of Concern in New Y	ENVR	0084
Progestin	ST-1435.= Short Synthesis of the Synthetic	ORGN	0180
plexes for Co+ Organic Ligand-	Stabiilized Copper(I) Hexafluoroacetylacetonate Com	COLL	0156
Relative Conformational	Stabilities of Chiral (z)-Alkene.=	ORGN	0172
Electronic Structures and	Stabilities of Mixed Clusters.=	PHYS	0214
Structural Changes on Radical	Stabilities.= Cope Award Symposium. Effects of	ORGN	0133
Enzyme Activities and	Stabilities.= Effect of Chemical Modification on	BIOT	0219
lation Sites as Probes of Protein	Stability and Conformation.= Engineered Metal Che	BIOT	0133
Therm+ Biocatalyst Design for	Stability and Specificity. Effects of Mutations on the	BIOT	0188
g the M+ Biocatalyst Design for	Stability and Specificity. New Techniques for Probin	BIOT	0112
cal Cros+ Biocatalyst Design for	Stability and Specificity. Overview of Protein Chemi	BIOT	0216
al B-Cha+ Biocatalyst Design for	Stability and Specificity. Role of Insulin's C-Termin	BIOT	0132
tic Catal+ Biocatalyst Design for	Stability and Specificity. Structural Basis of Enzyma	BIOT	0239
e Antiinflammatory Drug Piroxi+	Stability Constants of Copper(II) Complexes with th	INOR	0407
phines and Its Role on Catalyst	Stability in Homogeneous Transition-Metal-Catalyze	INOR	0349
al Strategy for Increasing Enzyme	Stability in Polar Organic Solvents.= +on: A Gener	BIOT	0244
ically Modified Nicotiana taba+	Stability of a Foreign Protein Production from Genet	BIOT	0230
roducts.=	Stability of Cochineal Pigment and Its Use in Food P	AGFD	0139
Formation Pathway and	Stability of Di-μ-oxo Dimers.=	INOR	0185
midazol-2-YL] C+ Solubility and	Stability of Methyl [1-(Butylcarbamoyl)-1H-Benz-i	AGRO	0041
	Stability of Suspensions of Zirconia-Lime Powders.=	COLL	0124
Morine with Aluminum(III) in+	Stability of the Species Formed by the Interaction of	INOR	0273
Protein	Stability Studies on Subtilisin BPN'.=	BIOT	0192
hain on Association, Folding, and	Stability.= +city. Role of Insulin's C-Terminal B-C	BIOT	0132
obe of Latex Characterization and	Stability.= +Hydrodynamic Fingerprinting as a Pr	COLL	0199
olecular Sieve with High Thermal	Stability.= +mance of Pillared Interlayered Clay M	PETR	0030
drophobic Interactions to Protein	Stability.= +ribution of Hydrogen Bonding and Hy	BIOT	0189
Oxidation and Flavor	Stability.= Influence of Food Processes on Lipid	AGFD	0157
Thermal	Stability.= New Copper-Dioxygen Complexes with	INOR	0046
. Overview of Chemistry, Thermal	Stability, and Oxygen Plasma Resistance.= +e)s—I	PMSE	0052
ng Techniques to Enzyme/Protein	Stabilization and Bioconjugate Preparation.= +nki	BIOT	0220
stry and Leaching Mechanisms of	Stabilization and Solidification Systems.= +Chemi	ENVR	0003
ruthenium Groups.=	Stabilization of Benzoquinodimethanes Using Organo	INOR	0051
t with N-Substituted+ Chemical	Stabilization of Poly(vinyl Chloride) by Pretreatmen	POLY	0027
-Linking: Implications for Protein	Stabilization.= +erview of Protein Chemical Cross	BIOT	0216
Cold Blanching Treatments to	Stabilize Raw Fruit and Vegetable Juices: Control of	AGFD	0027
onomeric Zinc Alkyl Derivatives	Stabilized by Sterically Demanding poly(Pyrazolyl)hy	INOR	0085
nd Magnesium Alkyl Derivatives	Stabilized by Sterically Demanding tris(Pyrazolyl)hy	INOR	0087

Channel Templated Mazzite: A	Stabilized Calcined Structure.= Twelve-Ring	PETR	0037
egioselective Addition of Nitrile-	Stabilized Carbanions to 4-Amino-Substituted Indol	ORGN	0010
Using Rhodium-	Stabilized Carbenoids.= Heterocyclic Synthesis	ORGN	0276
Orthophenylene-	Stabilized Triphosphazanes.= New Macrocyclic	INOR	0391
n Leaching of Contaminants from	Stabilized/Solidified Wastes.= +brium Chemistry o	ENVR	0010
Supports and	Stabilizers.= Alkali Earth Salts as Catalyst	PETR	0034
ric Diblock Macromonomer	Stabilizers.= Emulsion Polymerization with Oligome	COLL	0039
ing Environmentally Safe Shale	Stabilizers.= Novel and Versatile Process for Produc	CARB	0009
Conformation of Colloid-	Stabilizing Block Copolymers.= Surface	PMSE	0013
rials.=	Stable Carbon Isotope Analysis of Coprocessing Mate	FUEL	0071
Residues.= TPOP-IRMS	Stable Carbon Isotope Analysis of Heavy Petroleum	PETR	0141
Thermally	Stable Electrooptic Polymers.=	POLY	0130
ions for the Highly Efficient and	Stable Heterogeneous Oligomerization of Ethylene.=	PETR	0008
Flavoring Substances.= Use of	Stable Isotope Analysis for the Control of Natural	AGFD	0195
	Stable Isotope Geochemistry of Azufres Systems.=	I&EC	0015
g Algal Biomass.= Production of	Stable Isotopically Labeled L-(+)-Lactice Acid Usin	BIOT	0156
e Use of Urinary Creatinine as a	Stable Metric for Human Biomonitoring Studies and	CEI	0019
rox+ Phthalide Cation Is Not a	Stable Species During Solution Protonation of 3-Hyd	ORGN	0151
ation over Zinc-Containing Ultra-	Stable Y Zeolites.= +c Content on Aromatics Form	PETR	0048
stigation of Various Moieties as	Stable 3-Phosphate Mimics in Aromatic Analogs of E	ORGN	0028
hibitors of Human PMN Elasta+	Stable, Potent, and Specific Monocyclic β-Lactam In	BIOL	0049
tems and Delocalization in π-	Stacks.= Electron Localization in Conjugated π-Sys	BIOT	0201
of Juvenile Hormone in the Fifth	Stadium Larvae of Manduca sexta.= +olism Study	AGRO	0022
Steps with Upstream-Downstream	Stages. Integrati+ Integrating Primary Separation	BIOT	0069
uels: Heteroatom Determination.	Standard Additions Method for the Measurement of	FUEL	0055
al Information I+ Recent ASTM	Standardization Developments for Chemical Structur	CINF	0038
ts Used in Geothermal Wells.=	Standards for X-ray Fluorescence Analysis of Cemen	ANYL	0058
s of Pharmaceu+ USP Reference	Standards, a Key Factor in Chromatographic Analysi	ANYL	0115
nd Biology. Poly(ethylene Oxide)	Star Molecules in Surfaces for biological Studies.=	POLY	0272
ly(ethylene Glycol+ Amphiphilic	Star Polymers from Tri- or Tetraisocyanates and Po	POLY	0271
Diffusion of	Star Polymers in Solutions and Gels.=	POLY	0267
Static and Dynamic Properties of	Star Polymers.= +eauty—Combs and Stars. Some	POLY	0265
E Anionic Synthesis of Heteroarm,	Star-Branched Styrene-Butadiene Thermoplastic	POLY	0266
ted R+ Photochemical Probes of	Starbust Dendrimers and Their Utilization as Restric	POLY	0238
Aluminum and Silicate-Modified	Starch Graft Copolymer Superabsorbents.= +cy in	CARB	0042
Total Dietary Fiber, Protein, and	Starch in a Milled Hard Red Winter Wheat.= +id,	AGFD	0143
ons.=	Starch-Based Thermoplastics in Industrial Applicati	CARB	0010
nant Proteins.= Evaluating	Starch-Protein Interactions for Recovery of Recombi	BIOT	0047
ular Cold Water Swelling/Soluble	Starches by Alcoholic-Alkali Treatments.= +f Gran	CARB	0040
Geometrical Beauty—Combs and	Stars. Novel Dendritic Macromolecule+ Polymers of	POLY	0318
Beauty—Combs and	Stars. Ring Polymers.= Polymers of Geometrical	POLY	0292
Geometrical Beauty—Combs and	Stars. Some Static and Dynamic Prop+ Polymers of	POLY	0265
Geometrical Beauty—Combs and	Stars. Toward a Molecular-Size Tink+ Polymers of	POLY	0344
a Continuing Education Course—	Start to Finish.= +an Symposium. The Making of	TECH	0001
hemical Business. So You Want to	Start Your Own Chemical Business.= +ur Own C	SCHB	0001
idue-Method Evaluation Criteria:	State's Perspective.= +s Techniques. Pesticide-Res	AGRO	0145
Glass.= Excited-	Stated Dynamics of Ru(BPY)$_3$$^{2+}$ on Porous Vycor	INOR	0160
aracterization: Optical Methods.	Static and Dynamic Light Scattering from Solutions	POLY	0250
Beauty—Combs and Stars. Some	Static and Dynamic Properties of Star Polymers.=	POLY	0265
Calculations of	Static Hyperpolarizabilities and Polarizabilities.=	PHYS	0340
A Theory for Solubility in	Static Systems.=	POLY	0141
otential Precursors of Norstatine,	Statine, and Their Analogs.= +ted-β-Lactams as P	ORGN	0047
oidal State of Several Materials in	Stationary and Flowing Systems.= +eld on the Coll	COLL	0160
hibitors on Cellulose-Based Chiral	Stationary Phases.= +alysis of Some Aromatase In	ANYL	0133
with Micropellicular	Stationary Phases.= Rapid HPLC of Biopolymers	ANYL	0128
es: The Role of Mobile-Phase and	Stationary-Phase pH.= +eparation of Carbohydrat	CARB	0066
adening in Liquids.=	Statistical Mechanical Studies of Electronic Line Bro	PHYS	0035
990.=	Statistical Summary of PIXE Programs 1970-1980-1	NUCL	0080
ocene and Nickelocene at 248 nm:	Statistical vs. N+ Two-Photon Dissociation of Ferr	PHYS	0333
RIS M+ Direct Computation of	Statistical Weights from Conformational Analysis for	POLY	0356
	Statistics and Dynamics in Multifragmentation.=	NUCL	0005
Importance of	Statistics for the Technician.=	TECH	0003
the Side Branches on the Chain	Statistics of Siloxanes.= The Effect of the Length of	POLY	0070
-Induced X-ray Emission (PIXE).	Staus of PIXE in Geochemistry.= +ysis by Proton	NUCL	0077
Blends b+ Comparisons Made on	Steady and Dynamic Shear-Induced Morphology in	PMSE	0085
a Polymer Blend Solution during	Steady Shear and Cessation of Shear Studied by Lig	PMSE	0134
tudied by Light Scattering during	Steady Shear and Following Cessation of Shear.=	PMSE	0107
rtial Oxidation Reacto+ Multiple	Steady States and Changes in Catalyst Activity in Pa	I&EC	0069
ethylammonium+ Dynamic and	Steady-State Fluorescence Studies of Tetradecyltrim	COLL	0074
ation of 1,2-Dinitrosobenzene by	Steady-State Spectroscopy—A New Phtochromic Re	ORGN	0168
anced Soil Gas Analysis in in Situ	Steam Distillation.= +vective-Flux Probes for Enh	ENVR	0052
ent Coal on Potassium Activity in	Steam Gasification of Chars.= +Composition of Par	FUEL	0027
of Fertilizers. Corrosion of Mild	Steel Exposed to Inhibited Urea-Ammonium Nitrate	FERT	0001
chemical Studies of Coated 4140	Steel Exposed to Mixed Microbial Communities und	BIOT	0097
ns.= Corrosion of Mild	Steel Exposed to Urea-Ammonium Sulfate Suspensio	FERT	0003
ibration Damping of Laminated	Steel Sheet with Polyurethane/Acrylate Interpenetra	PMSE	0057
Impedanc+ Study of Galvanized	Steel/Sodium Chloride Interface by Electrochemical	PHYS	0270
tion of Organic Water during	Stepped Combustion.= Chemical and Isotopic Altera	ANYL	0061
Total Synthesis and	Stereochemical Assignment of ($\pm$)-Epiderstatin.=	ORGN	0183
Absolute	Stereochemical Control in Allylic Functionalization.=	ORGN	0292

ffect of Hydroxyl Protection on	Stereochemistry and Stereocontrol During Amide Iod	ORGN	0258
Influence of Heteroatom on the	Stereochemistry of Nickel Boride Reduction of 2,6-D	ORGN	0246
al α-Pyrone fro+ Structure and	Stereochemistry of Pectinolide: A Novel Antimicrobi	MEDI	0060
th Plati+ On the Mechanism and	Stereochemistry of the Reaction of Diazomethane wi	INOR	0329
oil fro+ Structure and Absolute	Stereochemistry of Variculanol: A Novel Sestererpen	ORGN	0184
Proof Teaching.= D–Glucose	Stereochemistry: Schematic Method for Fischer's	CHED	0095
Protection on Stereochemistry and	Stereocontrol During Amide Iodolactonization.=	ORGN	0258
s: A Mechanistic Investigation of	Stereocontrolled Proton Migration from Nitrogen to	ORGN	0296
es of 2–Hydroxypyridine and 2+	Stereocontrolled Synthesis of C–Linked Deoxyribosid	ORGN	0075
ectrophilic and Diels-Alder Rea+	Stereoelectronic Control of Diastereoselectivity in El	ORGN	0240
arbene Complexes+ Regio- and	Stereoelectronic Effects in the Reactions of Fischer C	INOR	0194
esis of the Brevianam+ Unusual	Stereofacial Selectivity in the Biosynthesis and Synth	ORGN	0283
Molecular Symmetry and	Stereoisomer Number.=	CHED	0169
Gelation of Solutions of	Stereoregular Polystyrene.= Thermoreversible	POLY	0147
s of Chrial Isocyanides.=	Stereoselection and Helicity in Living Polymerization	POLY	0347
Hydroxide–Catalyzed Epoxide+	Stereoselective Carbohydrate Synthesis via Palladium	ORGN	0074
matase by a Triazole Derivative+	Stereoselective Inhibition of the P_{450}-Dependent Aro	MEDI	0007
by Their+ Quantification of the	Stereoselective Recognition of 1,4–Dihydropyridines	MEDI	0100
ents to Alkynyl Oxiranes: Synt+	Stereoselective S_N2' Addition of Organocuprate Reag	ORGN	0011
Locked–Conformation Styrenes:	Stereoselective Syntheses of Exo or Endo Polycyclic	ORGN	0236
es.=	Stereoselective Synthesis of Enynes and Enyne–Allen	ORGN	0099
Oxidation: An Approach to the	Stereoselective Synthesis of Heterocyclic Natural Pro	ORGN	0294
lazetidinones.=	Stereoselective Synthesis of 3–Fluoro–3–Hydroxyalky	ORGN	0080
ropanations: Synthes+ Control of	Stereoselectivity in Samarium Metal–Induced Cyclop	ORGN	0284
logy for+ Selectivity in Catalysis:	Stereoselectivity. Asymmetric Hydrogenation Techno	CATL	0001
Limits of Aldolase	Stereoselectivity.=	BIOL	0009
ts: A Mechanistic Investigation+	Stereospecific Enammonium–Iminium Rearrangemen	ORGN	0296
arbonyl Complexes of a New Di+	Stereospecific Substitution Reactions for Group VI C	INOR	0333
f Ether Phospholipids.=	Stereospecific Synthesis of Amino Acid Conjugates o	MEDI	0163
ased Properties, in Particular the	Steric and Electrostatic Features.= +lap of Atom-B	CINF	0018
$pMn(CO)_2$ with Tertiary Silan+	Steric Contributions in the Enthalpy of Reaction of C	INOR	0386
teroatoms: Determination of the	Steric Course of Formal Nucleophilic Substitutions a	ORGN	0120
Reaction in Methanobacterium+	Steric Course of the Methyl Coenzyme M Reductase	ORGN	0025
ngement and Enoyl R+ Revised	Steric Courses and Mechanisms of the Allylic Rearra	BIOL	0065
nation of the Enthalpy and Kin+	Steric Effects in Metal–Phosphine Bonding: Determi	INOR	0192
uperconductors: Electronic and	Steric Effects.= Structure and Bonding in Cuprate S	CHED	0022
That Overcome	Steric Inhibition.= Synthesis of Rigid Dendrimers	POLY	0321
the Magnitude of α–Agostic and	Steric Kinetic Isotope Effects in Carbon–Carbon Bon	INOR	0147
and EXAFS Studies of	Sterically Crowded Porphyrins.= Crystallographic	INOR	0091
Alkyl Derivatives Stabilized by	Sterically Demanding poly(Pyrazolyl)hydroborato Lig	INOR	0085
Alkyl Derivatives Stabilized by	Sterically Demanding tris(Pyrazolyl)hydroborato Lig	INOR	0087
cly(ethylene Oxide) Bound with a	Sterically Hindered Phenolate.= +onductivity of P	POLY	0091
Industry and Upda+ History of	Steroid Chemistry. Prehistory of the Mexican Steroid	HIST	0016
id Research at G. D.+ History of	Steroid Chemistry. Steroids and the Pill: Early Stero	HIST	0020
s, Structure, and Fu+ History of	Steroid Chemistry. The Sterol Molecule: Biosynthesi	HIST	0012
rofile of Innovatio+ Steroids, the	Steroid Community, and Upjohn in Perspective: A P	HIST	0021
Early English	Steroid History.=	HIST	0014
Analytical and Toxicological As+	Steroid Hormone and Related Compound Residues:	AGRO	0129
ition of Cytochr+ Inhibitors of	Steroid Hormone Biosynthesis. Mechanism and Inhib	MEDI	0006
istry. Prehistory of the Mexican	Steroid Industry and Update on R. E. Marker—Illus	HIST	0016
istry. Steroids and the Pill: Early	Steroid Research at G. D. Searle Co.= +roid Chem	HIST	0020
Early	Steroid Research in Mexico: The Pill and Cortisone.=	HIST	0018
k Laboratories: The Impact of	Steroid Research.= Some Contributions of the Merc	HIST	0015
us Testosterone.= Inhibitors of	Steroid 5α–Reductase Which Are Uncompetitive vers	MEDI	0008
Spirostanic and Furostanic	Steroidal Glycosides.= Some Biological Effects of	MEDI	0037
Synthesis and Reactions of	Steroidal Oxazolines.=	ORGN	0178
Synthesis of	Steroidal Thiazolidones.=	ORGN	0039
onia Reductions.= Synthetic	Steroids and the Luftwaffe: Evolution of Metal–Amm	HIST	0013
.= History of Steroid Chemistry.	Steroids and the Pill: Early Steroid Research at G.	HIST	0020
Surfaces from Amino Acids and	Steroids for Construction of Hydrophobic Binding Si	ORGN	0066
Terpenes in Zurich to	Steroids in Mexico via Cuba.= °From Ruzicka's	HIST	0017
ive Inject+ Residues of Anabolic	Steroids in Muscle Tissues from Carcasses with Posit	AGFD	0130
Historical Reflections on	Steroids: Lederle and Personal Aspects.=	HIST	0023
spective: A Profile of Innovatio+	Steroids, the Steroid Community, and Upjohn in Per	HIST	0021
story of Steroid Chemistry. The	Sterol Molecule: Biosynthesis, Structure, and Functio	HIST	0012
Project SEED at	Stevens.=	CHED	0144
e Glycosides from the Roots of	Stevia salicifolia cav. (Asteraceae).= New Ent–Atisen	MEDI	0034
Off–Lattice, Bead–	Stick Computer Simulations.= Rings vs. Strings:	POLY	0235
d Transport of Small Molecules in	Stiff–Chain Glassy Polymers.= +n the Sorption an	POLY	0120
betel as Flavoring and Nerve–	Stimulating Agent.= Phenolic Compounds of Piper	AGFD	0154
	STM and AFM Extensions.=	PHYS	0209
Potential Control by	STM and AFM.= Imaging DNA Molecules under	PHYS	0157
unneling Micr+ Atomic Imaging:	STM and Related Techniques. Field Ion–Scanning T	PHYS	0182
Atoms and C+ Atomic Imaging:	STM and Related Techniques. Imaging of Individual	PHYS	0123
lectronic Stat+ Atomic Imaging:	STM and Related Techniques. Local Structure and E	PHYS	0206
M Imaging o+ Atomic Imaging:	STM and Related Techniques. Low–Temperature ST	PHYS	0360
les of Scannin+ Atomic Imaging:	STM and Related Techniques. Principles and Examp	PHYS	0152
f Single Adsor+ Atomic Imaging:	STM and Related Techniques. Theoretical Studies o	PHYS	0096
f Adsorbates+ Low–Temperature	STM Imaging of Metal Surfaces and Manipulation o	PHYS	0360
$_2CuO_y$ and $Bi_2Sr_2Cu_{0.9}Fe_{0.1}O_y$.+	STM Investigations of the Surface Structure of Bi_2Sr	PHYS	0292

III-V Surfaces for	STM Nanolithography.= Preparation of Passivated	PHYS	0363
	STM Studies of Biomolecules and Metal Surfaces.=	ANYL	0023
	STM Studies of Charge Transfer Salts.=	PHYS	0230
s of Single Adsorbed Atoms in the	STM.= +Related Techniques. Theoretical Studie	PHYS	0096
ions of Metal Surfaces Studied by	STM.= +emisorption–Induced Phase Transformat	PHYS	0126
with a New Variable Temperature	STM.= +mer Configurations on Si(100) Observed	PHYS	0185
Chemistry with the	STM.= Probing and Inducing Silicon Surface	PHYS	0184
Electronics Using	STM.= Research on Organic Molecules for	BIOT	0204
canning Tunneling Microscopy (	STM) Studies of the Chemical–Vapor Deposition of	COLL	0050
nning Tunneling Microscopy (FI–	STM).= +and Related Techniques. Field Ion-Sca	PHYS	0182
	STM-Induced Etching of Conducting Oxides.=	PHYS	0361
FIM, A-P,	STM, STS.= Tunneling Microscopies: FEM, FEES,	PHYS	0097
ing of Chemical Information: The	STN Story.= +merica and Europe. Global Market	CINF	0011
Database on	STN.= Recent Developments of the Beilstein	COMP	0010
hains in the Bulk and in Thin +	Stochastic Dynamics Simulations of Polymethylene C	POLY	0355
Dopant Site Selectivity, Oxygen	Stoichiometry, and Cu Valence in the High-T_c Super	PHYS	0027
Green Tea.= Protection against	Stomach, Lung, and Esophagus Carcinogenesis by	AGFD	0067
atter.=	Stopping Power and Ranges of Ions in Condensed M	NUCL	0037
Ferritin Iron	Storage and mRNA Regulation.=	INOR	0011
l Cement Systems+ Long–Period	Storage Effects on Physical Properties of Geotherma	I&EC	0014
T+ Very High Density Information	Storage in Molecular Complexes Using Scanning	BIOT	0023
Descriptive Sensory+ Effect of	Storage on Roasted Peanut Flavor, as Determined by	AGFD	0161
rene LB Films.+ Photoamplified	Storage Optical Memory with Polyion Complexed Py	BIOT	0144
Lipid Oxidation of Fish during	Storage.=	AGFD	0123
in	Storage.= Monitoring the Oxidation of Coals	FUEL	0002
n Improved Blanchi+ Quality of	Stored Frozen Sweet Corn on the Cob Produced by a	AGFD	0136
Mechanisms in	Stored-Product Insects.= Hydrolytic Resistance	AGRO	0135
field Chemical Company.=+ True	Stories of Small Chemical Businesses. Founding Fair	SCHB	0013
pact of Fertilizer Containment on	Stormwater Management.= +d Soil Chemistry. Im	FERT	0013
Iodine and Sodium Nitrate—The	Story of Chilean Nitrate.=	SCHB	0020
f Chemical Information: The STN	Story.= +merica and Europe. Global Marketing o	CINF	0011
Si(100) Surfaces.= Role of Bond	Strain in the Chemistry of Hydrogen on Si(111) and	PHYS	0187
Complexes.= Ring	Strain of Manganese Bidentate Phosphine	PHYS	0193
queous Solutions from a Variant	Strain of Xanthomonas campestris NRRL B-1459.=	BIOT	0196
orks: Characterization of Stress–	Strain Relationships and Phase Transitions for Netw	COLL	0016
sible+ Arrested Melting Due to	Strain: Effect of Network Formation in Thermorever	POLY	0123
lling.=	Strain–Birefringence of Real Networks: Effect of Swe	POLY	0308
	Strain–Compensated Heats of Hydrogenation.=	ORGN	0173
in the Synthesis of	Strained Cyclic Cumulenes.= Recent Developments	ORGN	0093
Molecular Weight, Single–	Stranded DNA.= Laser Vaporization of High	ANYL	0130
Specialties f+ Global Sourcing—	Strategic Issues in Chemical Procurement. Chemical	CMEC	0005
otechnology-Based Pest Control	Strategies as Alternatives to Chemical Pesticides: Pro	AGRO	0080
Developing Synthesis	Strategies by Perceiving Structural Similarities.=	CINF	0007
g Pesticide–Exposure Mitigation	Strategies for Handlers and Harvesters.= Developin	AGRO	0043
ns and Pharmaceutical Molecul+	Strategies for Preparative Chromatography of Protei	ANYL	0031
Systems: Novel 5–H+ Synthetic	Strategies for the Construction of Novel Azatricyclic	ORGN	0104
rboaromatic and H+ Annulation	Strategies for the Synthesis of Highly Substituted Ca	ORGN	0286
cles: Applications Toward the +	Strategies in the Synthesis of Medium–Sized Carbocy	ORGN	0282
Alternative	Strategies to the Analysis of Pharmaceuticals.=	ANYL	0121
ole of Complementary Topological	Strategies.= +d ^{13}C NMR Spectrum Simulation: R	COMP	0006
Manufacturing	Strategies.=	SCHB	0006
Setting Your	Strategies.=	SCHB	0003
Market–Entry	Strategies.= Value Perception Analysis for Polymer	CMEC	0011
ol Additions: A Building Block	Strategy for Asymmetric Synthesis.= Enzymatic Ald	BIOT	0090
minescence as an HPLC Detection	Strategy for Chlorophenol Determination.= +milu	ANYL	0084
–, and 7–Membered Carb+ New	Strategy for Cyclization–O–Functionalization of 5-, 6	ORGN	0088
ation in Fluorocarbon Media: A	Strategy for Electric–Field–Orientable Microparticles	COLL	0164
s for Production of+ Integrated	Strategy for Improving Downstream Processing Yield	BIOT	0071
Charge Substitution: A General	Strategy for Increasing Enzyme Stability in Polar Or	BIOT	0244
Multiple Equilibration	Strategy for Polypropionate Synthesis.=	ORGN	0100
of EPA	Strategy.= Pesticide Container Regulations as Part	AGRO	0089
itoring Studies and an Alternative	Strategy.= +as a Stable Metric for Human Biomon	CEI	0019
Powerful Synthetic	Strategy.= Transannular Diels–Alder Reaction: A	ORGN	0288
Search	Strategy.= Use of Most Restrictive Paths in 3D	CINF	0024
aphitizable Polymers under Argon	Stream and Their Magnetic Properties.= +Nongr	POLY	0023
Components of the Plastic Waste	Stream via Reactive Processing.= +Compatibilizing	I&EC	0033
s for Characterization of Refinery	Streams and Heavy Oils.= +nt Distribution Analysi	PETR	0110
Components in Light Petroleum	Streams by High–Resolution GC with Chemilumines	PETR	0124
Method for Coal Agglomerates'	Strength Determination.=	FUEL	0042
Effect of Temperature and Ionic	Strength on Chemical Exchange of Vanadate Derivat	INOR	0228
icipal Solid Waste Incin+ High–	Strength Portland Cement Concrete Containing Mun	FUEL	0136
tes: Estimates of Alkane–W Bond	Strengths.= +nt from (Alkane)W(CO)$_5$ Intermedia	INOR	0111
Polysaccharide of the β–Hemolytic	Streptococci Group A.= +onding to the Cell–Wall	CARB	0057
Improving Antibiotic Yields in	Streptomyces by the Intracellular Expression of a Ba	BIOT	0087
ic Enzymes and Cephamycin–C in	Streptomyces clavuligerus.= +mation of Biosynthet	BIOT	0129
Expression.= Xylanase Genes of	Streptomyces lividans and Their Homologous	BIOT	0213
SCD4 Derivative by Recombinant	Streptomyces lividans.= +for the Production of an	BIOT	0086
rologous Protein Accumulation by	Streptomyces lividans.= +s and Their Role in Hete	BIOT	0084
cterization.= Xylanases from	Streptomyces roseiscleroticus: Purification and Chara	BIOT	0210
osynthesis of Tetracenomycins by	Streptomyces.= +lic Pathway Engineering: The Bi	BIOT	0083

uc+ Influences of Hydrodynamic	Stress and Oxygen Availability on Ravidomycin Prod	BIOT 0162
recursors and Their P+ Residual	Stress Behavior of Functionalized Poly(amic Acid) P	POLY 0085
	Stress Effects on Diffusion through Solid Polymers.=	POLY 0140
aturational S+ Thermal and pH	Stress in Denaturation of Cellobiohydrolase I: Preden	BIOT 0191
Repair Systems and Oxidative	Stress.=	CHAS 0009
er Networks: Characterization of	Stress–Strain Relationships and Phase Transitions fo	COLL 0016
ce Decay in Seal–Packa+ Abiotic	Stresses of Heat or UV Illumination May Help Redu	AGFD 0076
iffraction: No Evidence for Bond–	Stretch Isomerism.= +s as Determined by X-ray D	INOR 0088
The Art and Science of	String Instruments.=	CONG 0003
in Recombinant Escherichia col+	Stringent Response as a Function of Induction Level	BIOT 0034
Simulations.= Rings vs.	Strings: Off-Lattice, Bead-Stick Computer	POLY 0235
v+ Low–Level Determination of	Strontium-90 in a High–Salt Environment at the Ne	NUCL 0085
sing: Monitoring Dewaxing Proc+	Structure/Property Relationships in Base Oil Proces	PETR 0151
ent for Manufacturing of Textile–	Structured Preforms—Part I.= +mation of Equipm	TECH 0007
ent for Manufacturing of Textile–	Structured Preforms—Part II.= +ation of Equipm	TECH 0008
Short-Range	Structuring in a Meta-Anthracite Coal.=	FUEL 0040
+ Rheological Evidence of Flow	Structuring in Entrance Flow of Polypropylene Melts	PMSE 0137
Coordination to an Alkali Metal:	Strucutre of Lu[CH(SiMe$_3$)$_2$]$_3$(μ-Cl)K(η^6-C$_7$H$_8$)$_2$ and	INOR 0053
A–P, STM,	STS.= Tunneling Microscopies: FEM, FEES, FIM,	PHYS 0097
to Milk Shelf-Life+ Graduate	Student Award Presentation. Chemometric Approach	AGFD 0108
a for Studying Seed+ Graduate	Student Award Presentation. Noninvasive Technique	AGFD 0104
lock Copoly+ Sherwin–Williams	Student Award Symposium. Change of Mechanism B	PMSE 0071
hylene in Whole Plants.=	Student Awards Symposium. Fate of ^{14}C-Trichloroet	ENVR 0056
Practices: An Interactive	Student Evaluation.= New Approach for Laboratory	CHED 0106
tric Determination of Oxalate: A	Student Experiment in Analytical/Physical Chemistr	CHED 0171
olecular Geometry.=	Student Ideas about Chemical Element, Acid, and M	CHED 0174
e.=	Student Perspective on the Project SEED Experienc	CHED 0037
try: Reaching and Teaching Large	Student Populations.= +vement in General Chemis	CHED 0081
The MIT	Student's Outreach Program.=	CONG 0008
Olympiad: A	Student's View.= The U.S. National Chemistry	CHED 0016
Chemistry Teaching through the	Student's World.=	CHED 0005
SEED Program for High School	Students at Hunter College, Chemistry Department.=	CHED 0143
Exam.= How to Prepare	Students for the Chemistry Olympiad National	CHED 0013
Laboratory Experience for	Students in Physical Chemistry Survey Courses.=	CHED 0215
How To Get Your	Students Involved in Project SEED.=	CHED 0039
Efforts in Exposing Precollege	Students to a Hydrocarbon Process.= Polysar's	CHED 0010
Chemistry Course for Nonscience	Students. +he ACS College-Level Decision-Making	CHED 0051
ge Chemistry for Nonscience	Students.= Interdisciplinary Core Approach to Colle	CHED 0176
School	Students.= Project SEED Benefits More Than High	CHED 0145
chniques for Fossil Fuels: Swelling	Studes and Mass Spectrometry. +vel Analytical Te	FUEL 0086
chniques for Fossil Fuels: Swelling	Studes and Mass Spectrometry. +vel Analytical Te	FUEL 0088
ptics of Polymer Thin Films, 4:	Structure-$\chi^{(3)}$ Relationships in Polyanilines and Deri	POLY 0221
Spontaneous Copolymerization of	Styrene in the Presence of Acid: Further Confirmatio	POLY 0039
sis of Block Copolymers of Poly(	styrene) and Poly(arylate) and Their Optical Propert	POLY 0220
en- and Deuterium-Labeled Poly(	styrene) and Poly(vinyl methylether).= +: Hydrog	COLL 0035
ritical Behavior of Poly(2-Chloro-	Styrene)/Polystyrene Blends.= +Field and Ising C	PMSE 0103
esis of Heteroarm, Star–Branched	Styrene–Butadiene Thermoplastic Elastomers.= +h	POLY 0266
f Poly(b-vinylborazine) and Poly(	styrene-co-b-vinylborazine) Copolymers.= +hesis o	POLY 0183
stor Oil Polyurethanes and Poly(	Styrene-Methyl Methacrylate-Methacrylic Acid).=	PMSE 0091
Birefringence of Ionomers Poly(	styrene-Sodium-2-Acrylamido-2-Methylpropane Su	COLL 0181
idation of Locked–Conformation	Styrenes: Stereoselective Syntheses of Exo or Endo P	ORGN 0236
c Ring–Constrained Analogues of	Styryl-Based Protein Tyrosine Kinase Inhibitors.=	MEDI 0023
eche Contaminada con S. aureus y	su Aplicacion en Quesos.= +de Sodio a Partir de L	AGFD 0107
Prospect+ ESR Spectroscopy at	Sub-9-GHz Microwave Frequencies: Advantages and	PHYS 0215
of the Clostridium thermocellum	Subcellulosome Preparation.= Genes and Proteins	BIOT 0181
Production of Human IgG	Subclasses.= Effects of Gm-Allotypes on the	CARB 0018
PCK in E. coli.=	Subcloning and Expression of Rat Liver Cytosolic PE	BIOL 0107
Folding of Trp Repressor	Subdomains.=	BIOL 0115
Kinetic Mechanism of Myosin	Subfragment-1 (S1) and Myosin-S1-NDP Binding to	BIOL 0103
Renin Inhibitors: Synthesis of a	Subnanomolar, Orally Active Cysteine–Derived Inhib	MEDI 0122
Route to Carbon	Subnitride.= Cyanoazacarbon Derivatives as a	INOR 0310
Photochemistry.=	Subpicosecond Transient Spectroscopy of Rhodopsin	BIOT 0080
arch Services in an Academic L+	Subsidized Access to Beilstein Online: Options for Se	COMP 0012
eability of Po+ Effects of Aryl–	Substitution and Physical Blending on the Gas Perm	POLY 0117
atives: Mechanism of Nucleophilic	Substitution at Tertiary Carbon.= +of Cumyl Driv	ORGN 0154
epatacarcinogen,+ Nucleophilic	Substitution by Aromatic Amines on the Ultimate H	ORGN 0194
Studies of Halogen	Substitution in High-T$_c$ Superconductors.=	PHYS 0239
Investigations of Bi	Substitution in NdBa$_2$Cu$_3$O$_y$.=	PHYS 0312
annulation/Nucleophilic Aromatic	Substitution Mediated by Chromium(0).= +al Benz	ORGN 0221
l Method To Detect Framework	Substitution of Paramagnetic Ions in Microporous Al	PHYS 0231
Reagents Using t+ Regioselective	Substitution of 1,2-Propanediol with Trimethylsilyl	ORGN 0051
tor Studies: Effect of Benzyloxy	Substitution on Biological Activity of Pyrazoloquinol	MEDI 0019
des: The Influence of Heterocyclic	Substitution on Biological Activity—I.= +Herbici	AGRO 0046
des. The Influence of Heterocyclic	Substitution on Biological Activity—II.= +de Herbi	AGRO 0047
lexes of a New D+ Stereospecific	Substitution Reactions for Group VI Carbonyl Comp	INOR 0333
Complexes as the Intermediates of	Substitution Reactions on Silica Surfaces.= +tion	COLL 0095
)-Cata+ Regiospecific Aromatic	Substitution Via Aryl 1,2-Dipolar Synthons: Nickel(O	ORGN 0082
me Stability in+ Surface Charge	Substitution: A General Strategy for Increasing Enzy	BIOT 0244
aCu$_2$O$_8$ Superconductor by Metal	Substitution.= +Critical Current Density of Bi$_2$Sr$_2$C	PHYS 0329
Intermaterial	Substitution.= New Way to Look at	CMEC 0013

c Course of Formal Nucleophilic	Substitutions at Nonstereogenic Atoms by the Endoc	ORGN	0120
Sieve AlPO$_4$-5.= Isomorphous	Substitutions of Silicon and Cobalt in Molecular	PETR	0099
es and Esterases by Modifying the	Substrate and by Linking Resolutions in Series.=	BIOT	0091
n from Whey: Effects of Growth	Substrate and Product Extraction on Reactor Produc	BIOT	0030
e-DNA Alkyltransferase: Effect of	Substrate Heteroatoms on Reactivity.= +kylguanin	BIOL	0161
Site-Directed Enzyme-Activated	Substrate Inhibition of Trypsin by N-Nitrosoamide	ORGN	0022
d Exchange Reactio+ Fatty Acyl	Substrate Preferences for Pancreatic Lipase–Mediate	AGFD	0036
ween Back Electron Transfer and	Substrate Reduction.= +omplexes: Competition bet	INOR	0284
Catalysis: Conferment of	Substrate Selectivity on Enzymes.= Targeted	BIOL	0090
ating a Monocyclic β-Lactam I+	Substrate Specificity of Penicillin G Amidase in Acyl	BIOT	0161
Characterization of Inhibition and	Substrate Specificity.= +2: Structural and Kinetic	BIOL	0020
Layer from Copper	Substrate.= Electrochemical Removal of a Tin	PHYS	0285
Novel Fluorogenic Tryptophanase	Substrate.= +thesis of Serine–AMC–Carbamate: A	ORGN	0023
Carbocyclic	Substrates for de Novo Purine Biosynthesis.=	BIOL	0076
me-Activated Chemiluminescent	Substrates in the Detection of Proteins and Nucleic	BIOT	0167
4-BDAF Polyimides and Silver	Substrates Using Surface–Enhanced Raman Scatterin	POLY	0087
dies of Lysosomotropic Enzyme	Substrates with Human Fibroblast Monolayers: Upta	BIOL	0079
Zinc Alkyls on Semiconductor	Substrates: GaAs(100) and Si(100).= Reactions of	COLL	0048
Active-Site Histidine Relative to	Substrates.= +termination of the Orientation of the	BIOL	0066
m for Conversion of Hydrophobic	Substrates.= +terization of a Novel Reaction Mediu	BIOT	0248
Enzyme Reactions on Insoluble	Substrates.=	COLL	0052
es and Photooxidations of Organic	Substrates.= +edox Reactivity of Copper Complex	INOR	0342
ic Studies Using New Fluorogenic	Substrates.= +ylalanine at P1 Site by Renin: Kinet	BIOL	0035
ee of Synergistic Binding between	Substrates?.= +nism or Random, with a High Degr	BIOL	0053
1 Protease by Three-Dimensional	Substructure Searching.= +ovel Inhibitors of HIV-	COMP	0038
Tunneling Microscopy: Imaging	Subsurface and Nonequilibrium Electronic Propertie	PHYS	0208
ative Genes in Isolates from Deep	Subsurface Environments.= +of Cloned Biodegrad	BIOT	0117
Xylanase of B.	subtilis.= Structure–Function Analysis of a 20–Kd	BIOT	0212
Protein Stability Studies on	Subtilisin BPN'.=	BIOT	0192
of	Subtilisin.= Effect of Solvent on the Nucleophilicity	BIOT	0066
Imidazolines, Synthesis a+ α2A-	Subtype Selective Adrenergic Agents. 1. Substituted	MEDI	0116
ablishing an Academic Program.	Succeeding in Academia: The Good, the Bad, and the	YCC	0001
Announced.= Environmental	Successes in the Chemical Industry. To Be	CMEC	0017
um Properties by Modification of	Succinate and Pyruvate Levels.= +roved Zooglan G	BIOT	0198
ilizando Programacion Cuadratica	Sucesiva y un Simulador de Procesos Comercial.=	I&EC	0072
Nucleosides and C-Nucleosides,	such as AZT.= Synthetic Approaches to Modified	MEDI	0049
erials from Semiflexible Polymers	Such as the Cellulosics.= +High-Performance Mat	COLL	0015
with Carbon Adsorbents and	Sucrose Crystals.= Rate of Water Vapor Interaction	CARB	0062
Synthesis and Purification of	Sucrose Esters.=	CARB	0004
Chromatographic Separation of	Sucrose Monostearate Structural Isomers.=	CARB	0050
orie Bulking Agent Prepared from	Sucrose.= +is. Fructooligosaccharides—A Low-Cal	CARB	0001
New Products from the Action of	Sucrose–Glucosyltransferases.=	CARB	0007
	Sudbury Neutrino Observatory.=	NUCL	0032
Synthesis of a Branched-Chain	Sugar Daunorubicin Analogue.=	CARB	0052
ino Sugars by Epoxy and Triflate	Sugar Intermediates.= +, and Branched–Chain Am	ORGN	0071
Properties of Novel, Zero-Calorie	Sugar Macronutrient Substitutes.= Synthesis and	CARB	0026
Phosphoribulokinase's	Sugar Phosphate Binding Site.= Identification of	BIOL	0153
Resins in	Sugar Purification.= Powdered Ion-Exchange	CARB	0065
Decoding Information in Complex	Sugar Structures.= +Domains as Mechanisms for	BIOL	0144
Beet	Sugar.= Large–Scale Production of PHB on	CARB	0002
le of Conformational Factors and	Sugar-Protecting Groups in the Synthesis of 2'-Fluo	CARB	0024
Symposium in Memory of Nathan	Sugarman. Nuclear Chemistry, the Met Lab, and+	NUCL	0058
mistry, the Met Lab, and Nathan	Sugarman—A Retrospective.= +man. Nuclear Che	NUCL	0058
Synthesis of 3–Thio-	Sugars and Related 3'–Mercapto–Anthracyclines.=	CARB	0032
oxy, and Branched–Chain Amino	Sugars by Epoxy and Triflate Sugar Intermediates.=	ORGN	0071
nt Degradation to Lower Furanose	Sugars via Ozonization of Protected Glycals.= +nie	ORGN	0073
Cysteine and	Sugars.= Studies on Meat Flavor Generation from	AGFD	0074
Studies in Chicken Tissues and+	Sulfadimethoxine and Sulfamonomethoxine Residual	AGFD	0127
termination and Confirmation of	Sulfamethazine in Animal Tissues by HPLC with Di	AGFD	0103
symmetric Ureas or Carbamates,	Sulfamides Using Triphosgene, CBMIT, and SBMIT	ORGN	0106
ssues an+ Sulfadimethoxine and	Sulfamonomethoxine Residual Studies in Chicken Ti	AGFD	0127
etermination of Cerium(IV) with	Sulfanilic Acid in the Presence of Cetyltrimethyl Bro	ANYL	0073
ive Analgesic Potencies of the 6-	Sulfate Conjugates of Some 3–Substituted Morphine	MEDI	0149
Method for the Measurement of	Sulfate Produced during the Determination of Sulfur	FUEL	0055
Exposed to Urea–Ammonium	Sulfate Suspensions.= Corrosion of Mild Steel	FERT	0003
zed Hydrocarbon Conversions.=+	Sulfate–Containing Solid Superacids for Acid–Cataly	PETR	0102
Corrosion Products as a Result of	Sulfate–Reducing Bacteria.= +s in Copper Sulfide	BIOT	0100
Fruit Juice with Cyclodextrins and	Sulfated Polysaccharides.= +zymatic Browning in	AGFD	0030
ysis and Characterization Studi+	Sulfated Zirconia Superacid Catalysts: Thermal Anal	PETR	0028
Properties of	Sulfated Zirconia.= Superacid and Catalytic	PETR	0133
gents in Absence of Denaturant+	Sulfhydryl Content of Glycinin: Effect of Reducing A	AGFD	0141
SP)+ Measurements of Dimethyl	Sulfide (DMS) and Dimethylsulfoniopropionate (DM	GEOC	0002
bonyl Compounds onto Cadmium	Sulfide and Cadmium Selenide.= +sorption of Car	COLL	0175
Transition Metal	Sulfide Catalysts.= Structure–Function Relations in	CATL	0038
New Series of Trimeric Vanadium	Sulfide Clusters.= +esis and Characterization of a	INOR	0372
Metal	Sulfide Complexes in Marine Waters.=	GEOC	0006
duci+ Sulfur Isotopes in Copper	Sulfide Corrosion Products as a Result of Sulfate–Re	BIOT	0100
R. A. Osteryoung. Interactions of	Sulfide Ion with Butylpyridinium Chloride in Nonaq	ANYL	0093
Properties of Poly(aryl	Sulfide Phenylquinoxalines).= Preparation and	POLY	0026
Phenyl	Sulfide.= Photochemistry of Benzyl	ORGN	0165

at+ Semicrystalline Poly(Arylene	Sulfide) Ketones via Amorphous Ketimine Intermedi	PMSE	0123
h Peroxidase by 2-Chloroethyl	Sulfides and Thiodiglycol.= Inhibition of Horseradis	BIOT	0249
lyzed Process for the Oxidation of	Sulfides to Sulfones.= +d Osmium Tetraoxide-Cata	ORGN	0244
Facile and Selective Oxidation of	Sulfides to Sulfones.=	CARB	0033
Reaction of	Sulfides with Permanganate Ion.=	ORGN	0243
Nickel	Sulfides.= Precursor-Based Synthesis of	INOR	0205
nsaturated Molybdenum(III)-μ-	Sulfido Dimers Formed by Sulfur Atom Abstraction	INOR	0271
gsten(IV) and Tungsten(VI) Bis(	sulfido) Complexes: Facile Elimination of H_2 from H	INOR	0369
Sigmatropic Rearrangement of	Sulfilimines from 3-Exomethylenecephams.= [2,3]	ORGN	0033
ng Monier-Wi+ Determination of	Sulfite Levels in a Variety of Commercial Foods, Usi	AGFD	0109
The Kinetics of Sulfonation of	Sulfite-Pretreated Aspen Treated with Vapor at Hig	PMSE	0170
Alto Rendimiento: Influencia de la	Sulfonacion sobre las Propiedades Papelaras.= +e	PMSE	0077
accharins through the o-Lithiative	Sulfonamidation of N,N-Diethylbenzamides.= +S	ORGN	0035
Cows.= Novel	Sulfonamide Drug Metabolism in Lactating Dairy	AGRO	0113
atographic Determination of Eight	Sulfonamides in Milk.= +rformance Liquid Chrom	AGFD	0128
-1,2,4-Triazolo-[1,5a]-Pyrimidine	Sulfonamides.= +N-(Substituted-2-Fluorophenyl)	AGRO	0048
2,4-Triazolo[1,5a]Pyrimidine-2-	Sulfonanilide Herbicides: The Influence of Heterocyc	AGRO	0046
1,2,4-Triazolo[1,5a]Pyrimidine-2-	Sulfonanilide Herbides. The Influence of Heteroc+	AGRO	0047
-2-Acrylamido-2-Methylpropane	Sulfonate) in Nonaqueous Solution.= +ene-Sodium	COLL	0181
Miscible Blends of	Sulfonated Polystyrene and Poly(xylenyl Ether).=	PMSE	0117
E+ Deformation and Fracture of	Sulfonated Polystyrene Ionomer/Polystyrene Blends	POLY	0061
Vapor at Hig+ The Kinetics of	Sulfonation of Sulfite-Pretreated Aspen Treated with	PMSE	0170
e Ultimate Hepatacarcinogen, N-(	Sulfonatooxy)-2-(Acetylamino)Fluorene.= +s on th	ORGN	0194
etermination of Lanthanum with	Sulfonazo-III as Chromogenic Agent in Aqueous Eth	ANYL	0074
ss for the Oxidation of Sulfides to	Sulfones.= +d Osmium Tetraoxide-Catalyzed Proce	ORGN	0244
Sulfides to	Sulfones.= Facile and Selective Oxidation of	CARB	0033
odium Chloride and Two Amino	Sulfonic Acids, HEPES and MOPSO, by EMF Meas	BIOL	0034
by ^{19}F Multiple-Quantum N+	Sulfonium Salt Distribution in Model Polymer Films	PMSE	0182
Nonlinear Optical Polymers with	Sulfonyl Acceptor Groups.= +ain and Main-Chain	POLY	0105
ear Epoxy Polymers with Pendant	Sulfonyl Tolan Groups.= +ptical Properties of Lin	POLY	0208
ompounds Using Enantiopure N-	Sulfonyloxaziridines.= +of α-Hydroxy Carbonyl C	ORGN	0293
A	Sulfonylurea Herbicide for Corn (QSAR Analysis).=	AGRO	0050
Synthesis of NC-319: A New	Sulfonylurea Herbicide in Corn.=	AGRO	0049
DPX-E9636, a New	Sulfonylurea Herbicide.= Environmental Fate of	AGRO	0029
drative Cyclization of Diphenyl	Sulfoxide by Lithium Diisopropylamide: A Simple Ro	ORGN	0084
III)-μ-Sulfido Dimers Formed by	Sulfur Atom Abstraction from $SPMe_3$.= +ybdenum(	INOR	0271
Sulfur-	Sulfur Bond Scission by a Copper Reagent.=	ORGN	0252
Kinetics of Assembly of Iron-	Sulfur Clusters in Micellar Media.=	INOR	0076
igh-Resoluti+ Determination of	Sulfur Components in Light Petroleum Streams by H	PETR	0124
of Hydrocarbon Distribution and	Sulfur Content in Petroleum Residues.= +mination	PETR	0120
Su+ Two-Dimensional Phases of	Sulfur Formed by Coalescence of Atoms on Re(0001)	PHYS	0124
nsformation Kinetics of Organic	Sulfur Forms in Argonne Premium Coals during Pyro	FUEL	0059
uced during the Determination of	Sulfur Forms in Coal Using Perchloric Acid.= +od	FUEL	0055
Apparatus to Investigate Organic	Sulfur Forms in Coals.= +els. High-Pressure TPR	FUEL	0007
Reactivity of Oxidized Organic	Sulfur Forms in Coals.=	FUEL	0057
ectros+ Quantitative Analysis of	Sulfur Functional Forms and Reactions by XAFS Sp	FUEL	0056
sis of 5- and 6-Ring Nitrogen and	Sulfur Heterocycles.= +el Approach to the Synthe	ORGN	0199
F Analysis of Zinc, Iron, and Total	Sulfur in Zinc Ore Concentrates.= +rections in XR	ANYL	0057
s as a Result of Sulfate-Reduci+	Sulfur Isotopes in Copper Sulfide Corrosion Product	BIOT	0100
on (μ-N_2) vs. End-o+ Nitrogen-,	Sulfur-, and Phosphorus-Ligated Compounds. Side-	INOR	0362
Organometallic Functionalities at	Sulfur-Bound Nickel: P_2S_2 vs. N_2S_2 Ligand Environ	INOR	0177
ectra and ab Initio Frequencies of	Sulfur-Containing Reactive Species.= +Infrared Sp	PHYS	0031
New Phosphorus- and	Sulfur-Modified Carbohydrate Polymer Derivatives.=	CARB	0030
	Sulfur-Sulfur Bond Scission by a Copper Reagent.=	ORGN	0252
ltaneous Detection of Aromatics,	Sulfur, and Hydrocarbons in Diesel Fuels by Gas Ch	PETR	0118
f Isobutane with Pentenes Using	Sulfuric Acid as a Catalyst: Chemistry and Reaction	PETR	0082
m of Sulfuric Acid-C+ Effect of	Sulfuric Acid Composition on Kinetics and Mechanis	PETR	0085
Ethanol and Electricity via Dilute	Sulfuric Acid Hydrolysis.= +nversion of Waste to	FUEL	0094
Other Light Olefins.=	Sulfuric Acid-Catalyzed Alkylation of Pentenes and	PETR	0084
n on Kinetics and Mechanism of	Sulfuric Acid-Catalyzed Alkylations of Isobutane wit	PETR	0085
de Aceros Microaleados en Acido	Sulfurico.= +tadistico de Tecnicas Electroquimicas	PHYS	0281
ATP	Sulfurylase.= GTPase-Mediated Activation of	BIOL	0062
keletal Phosphorus, Nitrogen, and	Sulphur(VI) Atoms.= +Inorganic Polymers with S	POLY	0186
reatment of Migraine.=	Sumatriptan, a Novel 5-HT_1-Like Agonist for the T	MEDI	0079
Teachers in the Greater Cincin+	Summer Intern Program for High School Chemistry	TECH	0002
Mandelung	Sums.=	CHED	0024
al Organics Using AM1 Simulated	Sunlight Irradiation of Semiconductor Catalysts.=	INOR	0318
nraveling the Molecular Basis of	Sunlight-Induced Mutations: Mechanistic Studies on	BIOL	0157
arbene-Metal Complexes:	Super-Homo-Michael-Acceptors.= Cyclopropylthioc	ORGN	0216
Modified Starch Graft Copolymer	Superabsorbents.= +cy in Aluminum and Silicate-	CARB	0042
ia.=	Superacid and Catalytic Properties of Sulfated Zircon	PETR	0133
rization Studi+ Sulfated Zirconia	Superacid Catalysts: Thermal Analysis and Characte	PETR	0028
ions.= Sulfate-Containing Solid	Superacids for Acid-Catalyzed Hydrocarbon Convers	PETR	0102
Deposition Approaches to high-T_c	Superconducting Films.= +rganic Chemical Vapor	INOR	0176
ty. Synthesis and T_c in Simple	Superconducting Oxide Systems.= Superconductivi	PHYS	0009
xchange Reactions in Oxide	Superconducting Systems.= Intercalation and Ion-E	INOR	0360
Spectroscopy.= Investigation of	Superconducting $YBa_2Cu_3O_7$ Using Raman	PHYS	0052
.=	Superconductivity in Alkali-Metal-Doped Fullerenes	PHYS	0046
sure in the Creation and Study of	Superconductivity in Organic Materials.= +of Pres	PHYS	0025
	Superconductivity. Buckyball.=	PHYS	0045

eld Penetration Depth in YBa$_2$+	Superconductivity. Measurement of the Magnetic-Fi	PHYS	0050
tors: Syntheses, Structures, and+	Superconductivity. Overview of Organic Superconduc	PHYS	0024
conducting Oxide Systems.=	Superconductivity. Synthesis and T_c in Simple Super	PHYS	0009
als.=	Superconductivity. Theory of Pairing in HTC Materi	PHYS	0001
agnetic Chains and Planes on	Superconductivity.= Cuprate Oxides: Effect of Nonm	INOR	0358
e Theory of High-Temperature	Superconductivity.= Experimental Constraints on th	PHYS	0051
Current Density of Bi$_2$Sr$_2$CaCu$_2$O$_8$	Superconductor by Metal Substitution.= +Critical	PHYS	0329
YBa$_2$Cu$_3$O$_x$ and Bi$_2$Sr$_2$CaCu$_2$O$_x$	Superconductors by Electron Spin Resonance Line B	PHYS	0284
YBa$_2$Cu$_3$O$_x$ and Bi$_2$Sr$_2$CaCu$_2$O$_x$	Superconductors by Electron-Spin-Resonance Line B	PHYS	0050
and Electronic States of High-T_c	Superconductors by Tunneling Microscopy and Spec	PHYS	0206
ructure and Bonding in Cuprate	Superconductors: Electronic and Steric Effects.= St	CHED	0022
l Propertie+ Overview of Organic	Superconductors: Syntheses, Structures, and Physica	PHYS	0024
y, and Cu Valence in the High-T_c	Superconductors.= +lectivity, Oxygen Stoichiometr	PHYS	0027
opy Studies of High-Temperature	Superconductors.= +ectron-Energy-Loss Spectrosc	PHYS	0279
ractions in High-Temperature	Superconductors.= Nonlinear Electron-Phonon Inte	PHYS	0012
se Transitions in LA$_2$CuO$_4$-Based	Superconductors.= +ope Effect and Structural Pha	PHYS	0010
Organic	Superconductors.= Physical Properties of the	PHYS	0002
e-Crystal Bi$_2$Sr$_2$CaCu$_2$O$_8$ High-T_c	Superconductors.= +Spectroscopy Studies of Singl	PHYS	0211
in Cuprate	Superconductors.= Structure-Property Correlations	PHYS	0028
in High-T_c	Superconductors.= Studies of Halogen Substitution	PHYS	0239
Enzyme Inactivation in	Supercritical Carbon Dioxide Conditions.=	AGFD	0028
ixtures of Coatings Materials and	Supercritical CO$_2$.= +nt of Phase Phenomena of M	PMSE	0037
Halogen-Selective Detection in	Supercritical Fluid Chromatography Using an Electro	ANYL	0091
bilities and Applicati+ Capillary	Supercritical Fluid Chromatography: Analytical Capa	AGFD	0201
Polyethylene: Temperature-Rising	Supercritical Fluid Fractionation.= +Low-Density	POLY	0076
Phase Behavior of Polymer-	Supercritical Fluid Mixtures.=	PMSE	0036
	Supercritical Fluid Processing Applied to Recycling.=	I&EC	0034
of Expanded Clays Dried Using a	Supercritical Fluid.= +ace and Cracking Properties	PETR	0145
Photoreactivity in	Supercritical Fluids.=	ORGN	0131
Modifications in	Supercritical Fluids.= Polymer and Biopolymer	POLY	0084
idation of Methane by Catalytic	Supercritical Water Oxidation.= Selective Partial Ox	CATL	0043
asis of EPA's Listing of Sites for	Superfund? Overview of the Structure of the Hazard	ENVR	0077
r the Revised Hazard-Ranking+	Superfund's Site Assessment Process: Implications fo	ENVR	0078
and Pillared Clay Synthesis.	Supergallery Pillared Clays.= Advances in Zeolites	PETR	0125
anza de la Quimica a Nivel Medio	Superior.= +Apoyos Computarizados para la Ensen	CHED	0124
Chromophoric Self-Assembled	Superlattices: Multilayer Construction of Thin-Film	POLY	0127
osites+ Impact of Molecular and	Supermolecular Structure on the Properties of Comp	POLY	0300
n of Vanadate Anions with Cu,Zn-	Superoxide Dismutase as Probed by ^{51}V NMR.=	INOR	0184
Infrared Laser Spectroscopy of	Supersonically Cooled Carbon Clusters.=	PHYS	0007
Employee-	Supervisor Partnership.=	TECH	0004
lvers or Spreadsheets: Which Will	Supplant the Hand-Held Calculator?.= +uation So	CHED	0117
ans as Affected by Use of Lecithin	Supplements Varied in Choline Content.= +f Hum	AGFD	0034
the Center for T+ Technological	Support and Medium-Scale Industries: The Case of	SCHB	0027
and Pt.= Study of the Effect of	Support and Particle Size for NO Adsorption on Rh	COLL	0186
of Vanadium Oxide C+ Effect of	Support and Promoters on the Structure-Reactivity	CATL	0057
system II Manganese Aggregate+	Support for a Dimer-of-Dimers Model for the Photo	INOR	0408
echnological Tr+ Technological	Support for the Productive Sector: The Case of the T	SCHB	0025
Oxide.= Effect of	Support in Propylene Oxidation on Molybdenum	CATL	0059
aracterization of Novel Catalyst	Support Materials: The Hydrous Metal Oxide Ion Ex	PETR	0018
e Undergraduate Laboratory—I.	Support of Chemistry Laboratory Improvement by th	CHED	0146
er Nonmicroporou+ Influence of	Support Preparation on Aromatization Selectivity ov	CATL	0030
Hazardous Materials	Support Programs.=	CHAS	0027
n-Switching HPLC Method To	Support Registration of DPX-E9636 Herbicide for F	AGRO	0030
embling Lipid Bilayers on Solid	Support: Building Blocks of Future Biosensors and M	BIOT	0102
Right-to-Know	Support: Current and Future Compliance.=	CHAS	0026
Catalyst	Support.= Al-Modified Sepiolite as Catalyst or	PETR	0015
Alkali Earth Salts as Catalyst	Supports and Stabilizers.=	PETR	0034
l Review of Ex+ Reversed-Phase	Supports as Surrogates for Natural Sorbents: Critica	ANYL	0071
ed Porous Bead Polymers as	Supports for Enzyme Immobilization.= IPN-Modifi	PMSE	0081
Electrically Conductive Catalyst	Supports from Fibrous Precursors.=	PETR	0016
Characterization. Pillared Clays as	Supports in Hydrocracking Catalyst Systems.=	PETR	0013
In Situ Pore-Structur+ Catalyst	Supports: Chemistry, Forming, and Characterization.	PETR	0068
Microporous Oxides b+ Catalyst	Supports: Chemistry, Forming, and Characterization.	PETR	0031
Pillared Clays as Sup+ Catalyst	Supports: Chemistry, Forming, and Characterization.	PETR	0013
Role of Spray Drying+ Catalyst	Supports: Chemistry, Forming, and Characterization.	PETR	0088
Strong Solid Acid Cat+ Catalyst	Supports: Chemistry, Forming, and Characterization.	PETR	0049
oduction of Catalyst and Catalyst	Supports.= +terization. Role of Spray Drying in Pr	PETR	0088
Alumina	Supports.= Activation of Wide-Pore	PETR	0033
Alumina and Silica Catalyst	Supports.= Electron Energy-Loss Spectrometry of	PETR	0073
es to Defect Sites in Silica	Supports.= Energy Transfer from Organic Adsorbat	PHYS	0135
Extruded Monolithic Catalysts	Supports.=	PETR	0090
Catalyst	Supports.= Forming of High-Surface-Area TiO$_2$ to	PETR	0091
Characterization of Catalysts	Supports.= Surface Area and Porosity	PETR	0069
yl-Cinno+ Synthesis and Pollen-	Suppressant Activity of a Series of Substituted Phen	AGFD	0042
Tailspike Protein.= Mutational	Suppression of Inclusion Body Formation in the P22	BIOT	0113
Molecular-Recognition-Directed	Supramolecular Polymer Architectures.=	POLY	0299
Leche de Establos del	Sur del D.F.= Determinacion de Metales Pesados en	ENVR	0037
des Oxydants et des Complexants	sur la Fottation de ces Mineraux.= +-Pyrite: Effet	ANYL	0067
.= Studies on the Activity and	Surface Acidity of HM Zeolite and Pd/HM Catalysts	PETR	0117
teraction of Guest Molecules wi+	Surface Adsorption Phenomena of Carbohydrates. In	CARB	0061

e Results through Virtual States:	Surface Analysis When Intense Lasers Miss the Surf	NUCL	0044
Atomic Scale	Surface Analysis with the Atom–Probe FIM.=	PHYS	0101
Molecular Identification in	Surface Analysis.=	NUCL	0045
ltrathin Thermally Grown Silic+	Surface and Bulk Diffusion of Adsorbed Nickel on U	COLL	0177
Dried Using a Supercritical Flu+	Surface and Cracking Properties of Expanded Clays	PETR	0145
Pesticides in	Surface and Groundwaters in the Midwest.=	AGRO	0065
Nucleation and Growth of Dia+	Surface and Microstructural Characterization of the	PHYS	0121
sts Supports.=	Surface Area and Porosity Characterization of Cataly	PETR	0069
Quinoline over High	Surface Area γ–Mo_2N.= Hydrodenitrogenation of	CATL	0047
s for Determining the Reactive	Surface Area of Carbons.= Use of Transient Kinetic	FUEL	0014
Polymer Films on a Metal	Surface by ESCA.= Study of Thin Bicomponent	POLY	0100
NO_x Catalysts.=	Surface Characterization of Titania–Silica–Based DE	PETR	0071
s Waste Forms. Use and Potent+	Surface Characterization Techniques for Cementitiou	ENVR	0007
t Changes in Zeta Potential and	Surface Charge of TiO_2 (Anatase) Suspensions as a R	COLL	0203
Increasing Enzyme Stability in+	Surface Charge Substitution: A General Strategy for	BIOT	0244
icon–Containin+ Gas Phase and	Surface Chemical Reactions in the Preparation of Sil	INOR	0121
diction Model for Marine Gas +	Surface Chemical Studies of Hot Corrosion: Life Pre	FUEL	0099
Electronic Materia+ Colloid and	Surface Chemistry of Advanced Materials: Growth of	COLL	0064
Electronic Materia+ Colloid and	Surface Chemistry of Advanced Materials: Growth of	COLL	0153
Electronic Materia+ Colloid and	Surface Chemistry of Advanced Materials: Growth of	COLL	0025
Electronic Materia+ Colloid and	Surface Chemistry of Advanced Materials: Growth of	COLL	0006
Electronic Materia+ Colloid and	Surface Chemistry of Advanced Materials: Growth of	COLL	0044
Electronic Materia+ Colloid and	Surface Chemistry of Advanced Materials: Growth of	COLL	0135
	Surface Chemistry of Diamond Growth.=	PHYS	0180
	Surface Chemistry of GaAs Atomic Layer Epitaxy.=	COLL	0069
	Surface Chemistry of Monolayer Metal Films.=	COLL	0063
ular Processes at Solid Surfaces:	Surface Chemistry of Nonmetals—Novel Techniques	COLL	0211
ular Processes at Solid Surfaces:	Surface Chemistry of Nonmetals—Novel Techniques	COLL	0170
ular Processes at Solid Surfaces:	Surface Chemistry of Nonmetals—Novel Techniques	COLL	0191
Probing and Inducing Silicon	Surface Chemistry with the STM.=	PHYS	0184
cle Scattering for Measurement of	Surface Composition and Structure.= +ering. Parti	NUCL	0064
opolymers.=	Surface Conformation of Colloid-Stabilizing Block C	PMSE	0013
Opportunity in the Area of	Surface Coordination Chemistry.= Applications and	CHED	0062
rbamates: Reagents for Selective	Surface Derivatization of Au and Pt Microfabricated	INOR	0350
l) Surface Studied by Scanning+	Surface Diffusion and Phase Transition on the Ge(11	PHYS	0099
ments.=	Surface Disinfection by Chemical and Physical Treat	AGFD	0032
m To Assess the Susceptibility of	Surface Drinking Water Supplies to Pesticide Conta	AGRO	0053
What's on the	Surface during Diamond Growth?.=	PHYS	0122
mic Layer Resolution Studied b+	Surface Dynamics of Diffusion and Corrosion at Ato	PHYS	0127
trophoretic Fingerprinting of the	Surface Electrical Properties of a Single–Acid–Site L	COLL	0200
through	Surface Energetics.= Improving Barrier Properties	POLY	0196
	Surface Enhanced Raman Spectra of C_{60} and C_{70}.=	COLL	0092
Microscopy.= Probing the	Surface Forces of Materials Using Atomic Force	PHYS	0154
atalytic Dehydrogen+ Hydrogen	Surface Fugacities and the Control of Selectivity in C	CATL	0061
e of Multipolar Correlations and	Surface Imperfections on the Efficiency of Diffusion–	PHYS	0063
Phosphorus Compounds in a	Surface Layer of Disperse Silica.= Reaction of	COLL	0100
Wear Experiments.=	Surface Layers from Lubricant Additives in Friction/	PETR	0140
othion and Deltamethrin from the	Surface Microlayer of Water.= +tilization of Fenitr	ENVR	0020
r Ion Impact Phenomena Used for	Surface Modification and Analysis. +terials: Cluste	NUCL	0052
n–CdTe.=	Surface Modification by Metallic Ion Implantation O	PHYS	0218
e Oxide)s—III. Investigations of	Surface Modification by Oxygen Plasma Treatment.=	PMSE	0054
diation.=	Surface Modification of Polycarbonate by Far UV Ra	PMSE	0125
ormance in Polystyrene Compos+	Surface Modification of Wood Fibers and Their Perf	PMSE	0112
eactivity of Silanol Groups on a	Surface of a Pure and Alumino(Titano)–Containing S	COLL	0112
Oxide Clusters Formation on the	Surface of Aerosil.= +c Model of the Chromium(VI)	COLL	0099
ation–Reduction Processes on the	Surface of Chromium–Containing Silica Gels.= +id	COLL	0101
Reactivity of NH with the	Surface of Depositing Films.= Laser Studies of the	COLL	0218
olymerization of Monomers on the	Surface of Hiding Pigments Dispersed in Water.=	PMSE	0038
examine, and Crown Ether on the	Surface of Pyrogeneous Silica.= +en Gystamine, M	COLL	0096
tions of Aziridine on the	Surface of Pyrogenous Silica.= Thermal Transforma	COLL	0113
rcury(II) with Crown Ether on the	Surface of Silochromium.= +on of Lead(II) and Me	COLL	0104
ce Re+ Electrostatic Field of the	Surface of SiO_2 and Its Effect on Energetics of Surfa	COLL	0116
hesis and Characterization of M+	Surface Organometallic Chemistry on Zeolites—Synt	PETR	0131
Electronics.= Study of	Surface Orientation for the Design of Molecular	BIOT	0053
Tungsten Carbides: The Effect of	Surface Oxygen on Reaction Pathways.= +tions on	CATL	0037
in–Resonance Line Broadening of	Surface Paramagnetic Probes.= +s by Electron–Sp	PHYS	0050
pin Resonance Line Broadening of	Surface Paramagnetic Probes.= +ors by Electron S	PHYS	0284
on Pt(111).=	Surface Photochemical Dynamics of CH_3Br Adsorbed	PHYS	0261
e Photosensitive Interfaces by	Surface Photovoltage Spectroscopy.= Studies of Som	BIOT	0142
Quantum–Chemical Study of	Surface Potential Effect on Activation Barriers of Un	COLL	0107
Molecular Anions as	Surface Probes.=	NUCL	0056
m Infrared Reflectance Spectros+	Surface Profiling of Polymers with Fourier Transfor	PMSE	0153
uantized Semicon+ Influence of	Surface Properties on Charge–Carrier Dynamics of Q	COLL	0151
operties and Adhesion of Bacterial	Surface Protein Layers.= +dity on the Wetting Pr	BIOT	0151
Semiconductors.= FTIR of	Surface Reaction in OMCVD of III–V Compound	COLL	0011
n Chloride–Gallium Arsenide Gas	Surface Reaction.= +r–Beam Study of the Hydroge	COLL	0215
from Organometallic Precursors.	Surface Reactions of Ge–Containing Organometallics	COLL	0044
O_2 and Its Effect on Energetics of	Surface Reactions.= +atic Field of the Surface of Si	COLL	0116
.= Studies of $TiO_2(110)$	Surface Reactivity Using Low–Energy Ion Scattering	COLL	0172
ght Binding Methods in Diamond	Surface Reactivity.= +cal Molecular Orbital and Ti	PHYS	0356

ls: Discrepancy between Core and	Surface Responses.= +n by Anesthetics and Alcoho	MEDI	0032
Role of Rheology and	Surface Science in Dentifrice Formulations.=	CHED	0047
ions of Catalysts—I. Methane +	Surface Science of Catalysis: Environmental Applicat	COLL	0185
ions of Catalysts—II. Thermal+	Surface Science of Catalysis: Environmental Applicat	COLL	0204
rlayers on Pt(111).=	Surface Spectroscopic Characterization of Iodine Ove	PHYS	0366
ATR-IR.= Analysis of Near	Surface Structure in Poly(ethyleneterephthalate) by	POLY	0279
$_1O_y$.+ STM Investigations of the	Surface Structure of $Bi_2Sr_2CuO_y$ and $Bi_2Sr_2Cu_{0.9}Fe_0$.	PHYS	0292
Phase Transition on the Ge(111)	Surface Studied by Scanning Tunneling Microscopy.	PHYS	0099
cterization, and Chemica+ Model	Surface Studies of Metal Oxides: Preparation, Chara	COLL	0193
Experiment—Determination of	Surface Tension by Automation of the Drop-Weight	CHED	0118
Chemisorbates.= Elementary	Surface Transport Processes Involving Molecular	PHYS	0345
on of the C_3H_6 Potential Energy	Surface Using Quadratic Configuration Interaction M	PHYS	0305
rch and in Gove+ Pesticides in	Surface Water and Groundwater—Advances in Resea	AGRO	0051
rch and in Gove+ Pesticides in	Surface Water and Groundwater—Advances in Resea	AGRO	0060
act+ Reaction of PCl_5 with Silica	Surface: Influence of the Organic Solvents on the Re	COLL	0097
rption of Cellulases on Cellulose	Surface: The Better the Binding, the Better the Cata	BIOT	0186
Ag(100)	Surface.= Ab Initio Study of K Adsorption on the	PHYS	0337
Formation of Al-N Bonds on the	Surface.= +Ammonia on Silica: Mechanisms for the	COLL	0045
h Reactions on the Diamond(100)	Surface.= +ts: Local Density Calculations of Growt	PHYS	0355
sis When Intense Lasers Miss the	Surface.= +ts through Virtual States: Surface Analy	NUCL	0044
Antifreeze Glycoprotein at the Ice	Surface.= +uilibrium Constants for Adsorption of	BIOL	0041
Compounds on Silica	Surface.= Immobilization of Tetraazamacrocyclic	COLL	0105
with Dispersed Silica	Surface.= Mechanisms of Metal Chloride Reactions	COLL	0219
Silanol Groups on Silica	Surface.= Peculiarities of PCl_5 Reaction with	INOR	0237
lecules Chemically Fixed on Silica	Surface.= +tions of WCl_6, $WOCl_4$, and $MoOCl_4$ Mo	COLL	0098
Forming of High-	Surface-Area TiO_2 to Catalyst Supports.=	PETR	0091
Adsorption: Measurement of	Surface-Bulk Association.= Apparent Free Energy	COLL	0122
ides and Silver Substrates Using	Surface-Enhanced Raman Scattering and X-ray Pho	POLY	0087
	Surface-Hydrophilic Elastomers.=	COLL	0038
Lectures: Plenary Session.	Surface-Introduced Liquid Surfaces.= Langmuir	COLL	0062
	Surface-Modified Biosensors.=	BIOT	0107
the Farm: Method+ Advances in	Surface-Water Research Related to Pesticide Use on	AGRO	0054
MP in th+ Herbicide Presence in	Surface/Groundwater and the Role of Agricultural B	FERT	0019
efficients in Polymer Melts Using	Surfaced-Enhanced Raman Scattering.= +fusion Co	POLY	0088
mperature STM Imaging of Metal	Surfaces and Manipulation of Adsorbates.= +-Te	PHYS	0360
the Characterization of Inorganic	Surfaces and Thin Films.= +ctron Spectroscopy in	INOR	0013
rption, Orientation, and Order at	Surfaces as Studied by Optical Second Harmonic Ge	ANYL	0024
mic Scale Imaging of Molecules on	Surfaces by Scanning Tunneling Microscopy.= +o	PHYS	0159
educed $TiO_2(110)$ and $TiO_2(001)$	Surfaces by Scanning Tunneling Microscopy-Spectro	COLL	0171
Studies of Dynamics at Catalyst	Surfaces by Selective Excitation of Nuclear Magnet	PHYS	0346
Modification of Fluoropolymer	Surfaces by Sodium Naphthalenide.=	PMSE	0124
Potential Energy	Surfaces for BCl Reactions.=	PHYS	0088
ethylene Oxide) Star Molecules in	Surfaces for biological Studies.= +d Biology. Poly(	POLY	0272
Preparation of Passivated III-V	Surfaces for STM Nanolithography.=	PHYS	0363
ction+ Synthesis of Amphipathic	Surfaces from Amino Acids and Steroids for Constru	ORGN	0066
of Colloids in Contact with Solid	Surfaces in External Fields.= +Transport Properties	COLL	0162
d Phase Transformations of Metal	Surfaces Studied by STM.= +emisorption-Induce	PHYS	0126
Studies of Kinetic Processes at	Surfaces Using Atomic-Level Imaging.=	PHYS	0098
mi+ Characterization of Material	Surfaces via Wettability Analysis: A Molecular Dyna	POLY	0369
eaction Dynamics in Gases and at	Surfaces. J. Polanyi (Chemistry, 1986).= +sium. R	CONG	0011
y of Alkali Metals on III-V(110)	Surfaces: From 1- and 2-D Insulators to 3-D Metals	PHYS	0183
ec+ Molecular Processes at Solid	Surfaces: Surface Chemistry of Nonmetals—Novel T	COLL	0211
ec+ Molecular Processes at Solid	Surfaces: Surface Chemistry of Nonmetals—Novel T	COLL	0170
ec+ Molecular Processes at Solid	Surfaces: Surface Chemistry of Nonmetals—Novel T	COLL	0191
nd Chemistry of Natural Diamond	Surfaces.= +amond (and c-BN, SiC). Structure a	PHYS	0202
Surface-Introduced Liquid	Surfaces.= Langmuir Lectures: Plenary Session.	COLL	0062
Vibrational Energy Flow at Solid	Surfaces.= +esonance and Optical Spectroscopies.	PHYS	0344
sition of Alkyl Halides on Copper	Surfaces.= +ns of Catalysts—II. Thermal Decompo	COLL	0204
Diethylsilane on Silicon	Surfaces.= Adsorption and Decomposition of	COLL	0008
Ge	Surfaces.= Adsorption and Growth on Si and	PHYS	0186
Aging of Primed Polyolefin	Surfaces.=	PMSE	0126
on	Surfaces.= Atomic Force Microscopy of Molecules	PHYS	0153
Alchohol on Silica	Surfaces.= Chemisorption of a Tertiary Butyl	COLL	0108
Acetonitrile on Pt(III) Electrode	Surfaces.= Chemisorption of Cyanide and	COLL	0118
Cluster Impact on Solid	Surfaces.=	NUCL	0055
ers at Colloidal Titanium Dioxide	Surfaces.= +s of Electron-Transfer Activation Barri	COLL	0132
ceptor Adlayers on Single-Crystal	Surfaces.= +nergy Transfer between Donor and Ac	PHYS	0065
CH_3Cl^- and Cl^- at Multilayer	Surfaces.= Femtosecond Solvation-Dynamics of	PHYS	0084
ized and Exchanged Polyethylene	Surfaces.= +IR Specular Reflectance Study of Oxid	POLY	0253
Growth of Silicide Layers on Si	Surfaces.=	PHYS	0100
Adsorbed Species on Metal Oxide	Surfaces.= +Energy-Loss Spectroscopy Studies of	COLL	0173
Chemisorption on Silica	Surfaces.= Investigation of Thiourea Derivatives	COLL	0106
Oxide	Surfaces.= Local Atomic Structure of Defects on	PHYS	0207
$_3)_4$ on Atomically Clean Platinum	Surfaces.= +he Deposition of Platinum from Pt (PF	COLL	0046
Cation-Incorporated Silicate	Surfaces.= Microporous Pillared Mica with	PETR	0129
Metal	Surfaces.= NMR Studies of Simple Molecules on	PHYS	0347
Organometallic Molecules on	Surfaces.= Photochemical Reactions of	INOR	0168
Processed Fruit and Vegetable	Surfaces.= Physicochemical Changes on Minimally	AGFD	0011
on Alkylsilane/Silica	Surfaces.= Picosecond Reorientation of Adsorbates	ANYL	0025
and Localized Corrosion on Metal	Surfaces.= +between Localized Microbial Activity	BIOT	0098

Hydrogen on Si(111) and Si(100)	Surfaces.=	Role of Bond Strain in the Chemistry of	PHYS 0187
Microscopy of Biological	Surfaces.=	Scanning Tunneling and Atomic Force	PHYS 0158
f Substitution Reactions on Silica	Surfaces.=	+tion Complexes as the Intermediates o	COLL 0095
Metal	Surfaces.=	STM Studies of Biomolecules and	ANYL 0023
Coalescence of Atoms on Re(0001)	Surfaces.=	+ensional Phases of Sulfur Formed by	PHYS 0124
Carbon	Surfaces.=	Use of Microcalorimetry for Probing	FUEL 0018
Mo(100)	Surfaces.=	XPS Studies of Interaction of CH_3 with	COLL 0217
°From Large–	Surfactant Aggregate to Low Interfacial Tension.=		COLL 0078
of Aqueous	Surfactant Monolayers.= Structure and Orientation		PMSE 0154
oreversible Gelation of Polymers.	Surfactant Organogels Studied by Small-Angle Neut		POLY 0169
from Soils by Soil Washing with	Surfactant Solutions at Low pH.= +al of Chromium		ENVR 0013
emical Measurements in Aqueous	Surfactant Solutions.= +mical and Spectroelectroch		COLL 0003
s of Methylviologen from Anionic	Surfactant Solutions.= +eposition of Reduced Form		COLL 0077
y in Microheterogeneous Fluids.+	Surfactants and Associated Colloids: Electrochemistr		COLL 0166
y in Microheterogeneous Fluids.+	Surfactants and Associated Colloids: Electrochemistr		COLL 0019
y in Microheterogeneous Fluids.+	Surfactants and Associated Colloids: Electrochemistr		COLL 0040
y in Microheterogeneous Fluids.+	Surfactants and Associated Colloids: Electrochemistr		COLL 0001
y in Microheterogeneous Fluids.+	Surfactants and Associated Colloids: Electrochemistr		COLL 0128
y in Microheterogeneous Fluids.+	Surfactants and Associated Colloids: Electrochemistr		COLL 0147
Analysis of In+ Use of Cationic	Surfactants in Anodic Voltammetric and Coulometric		COLL 0002
x– Reversed–Phase Supports as	Surrogates for Natural Sorbents: Critical Review of E		ANYL 0071
Students in Physical Chemistry	Survey Courses.= Laboratory Experience for		CHED 0215
Contamination.= Soil Gas	Survey Methods for Gasoline and Diesel Fuel		ENVR 0053
Flavors.= Flavor Adulteration.	Survey of Alpha–Ionone Enantiomers in Raspberry C		AGFD 0181
ds for Silk Flavonolglyco+ HPLC	Survey of Corn (Zea mays L.) Populations and Inbre		AGRO 0034
	Survey of Important Fermi Resonances.=		ANYL 0059
A	Survey of the NLO–Polymer Program at the MPI.=		POLY 0010
nt Group Contribu+ Preliminary	Survey of Various Data Sets Confirms that Substitue		MEDI 0088
The Use of Soil Gas	Surveys to Design Soil Vapor Extraction System.=		ENVR 0051
Diversification for Financial	Survival in the Fertilizer Industry.=		FERT 0010
ties for Niche Markets, or How to	Survive and Thrive in the Midst of Bigness.= +ial		CMEC 0005
ar Optics.= Toward High	Susceptibilities in Soluble Polyacetylenes for Nonline		POLY 0132
nation of Third–Order Nonlinear	Susceptibilities of Semiconducting Polymers: Compa		POLY 0009
The Third–Order	Susceptibility of Nematic Solutions of PBT.=		POLY 0274
formation System To Assess the	Susceptibility of Surface Drinking Water Supplies to		AGRO 0053
Expansions: Diamagnetic	Susceptibility.= Chemical Graph–Theoretic Cluster		COMP 0030
tocarcinogenic Furocoumarins in	Susceptible and Fusarium Yellows–Resistant Celery		AGRO 0040
n in Lithospermum erythrorhizon	Suspension Cultures with Fungal Elicitors.= +uctio		BIOT 0229
Response of Magnetic Colloidal	Suspension in Externally Imposed Magnetic and Hyd		COLL 0144
urface Charge of TiO_2 (Anatase)	Suspensions as a Result of Exposure to Radiofrequen		COLL 0203
Stability of	Suspensions of Zirconia–Lime Powders.=		COLL 0124
Urea–Ammonium Sulfate	Suspensions.= Corrosion of Mild Steel Exposed to		FERT 0003
Polydisperse, Charged Colloidal	Suspensions.= Structure and Rheology of		COLL 0165
Key Elements for Implementing	Sustainable Agriculture and the Role of Biodiversit		AGRO 0082
a Medir pH, por un Electrodo d+	Sustitucion de Electrodo de Vidrio Convencional par		ANYL 0066
asing Productivity through Global	Swaps, Exchanges, and Toll Conversions.= +Purch		CMEC 0007
anchi+ Quality of Stored Frozen	Sweet Corn on the Cob Produced by an Improved Bl		AGFD 0136
Palatinit®), a Versatile Alternative	Sweetener: Production, Properties, and Uses.= +lt (		CARB 0005
Scanning Calorimetry and Solvent	Swelling for Studies of Coal Structure.= +ferential		FUEL 0087
on of Microdilatometry to Solvent	Swelling of Coals.= +Mass Spectrometry. Applicati		FUEL 0086
	Swelling of Constrained Polymer Gels.=		COLL 0184
Nonlinear	Swelling of Phantom Networks.=		POLY 0167
cture of Coal in the Presence of a	Swelling Solvent Using a Novel EPR Technique.=		FUEL 0073
ytical Techniques for Fossil Fuels:	Swelling Studes and Mass Spectrometry. +vel Anal		FUEL 0086
ytical Techniques for Fossil Fuels:	Swelling Studes and Mass Spectrometry. +vel Anal		FUEL 0088
Effect of	Swelling.= Strain–Birefringence of Real Networks:		POLY 0308
paration of Granular Cold Water	Swelling/Soluble Starches by Alcoholic–Alkali Treat		CARB 0040
Enriched Combustion+ Thermal	Swing Absorption Process for Production of Oxygen–		I&EC 0023
Is the Pendulum	Swinging Back?.= Testimony of Scientific Experts:		CHAL 0016
n of a Nonlinear Interface Optical	Switch.= +ring by Laser Chemical Vapor Depositio		BIOT 0024
Redox–	Switched Molecular Aggregates.=		COLL 0020
Molecular	Switches from Modified FAD.=		BIOT 0143
DP+ Use of a Novel Column–	Switching HPLC Method To Support Registration of		AGRO 0030
er Optical Properties for Photonic	Switching.= +g of Polymers with Large Third–Ord		POLY 0011
a	Swollen Ceramer.= Structure–Property Behavior of		POLY 0228
Synthesis and characterization of	Symmetric Di(1-Alkynyl)Tellurides.=		INOR 0229
Design and Preparation of C_2	Symmetric Inhibitors of HIV–1 Protease.=		MEDI 0053
Catalytic Reactions Using C2–	Symmetrical Chiral Metallocene Complexes.=		INOR 0215
Alkenes Using a New Class of D4–	Symmetrical Chiral Tetraphenylporphyrins.= +le		INOR 0300
ric Oxidations Catalyzed by a D4–	Symmetrical Manganese–Tetraphenylporphyrin.=		ORGN 0045
n fo+ Unsymmetrical Access to a	Symmetrical Reaction Coordinate: A Possible Domai		ORGN 0122
Molecular	Symmetry and Stereoisomer Number.=		CHED 0169
–Space Multireference Coupled+	Symmetry Breaking in Radicals Studied by the Fock		PHYS 0016
tion Interaction Studies of	Symmetry Breaking in Small Molecules.= Configura		PHYS 0014
Coupled–Cluster Treatment of	Symmetry Breaking.=		PHYS 0015
I+ Influence of Crystallographic	Symmetry upon the Solid-State Emission of Copper(		INOR 0157
NSF ChemSource Programs, Ho+	Symposia on the ACS Chemistry Olympiad and the		CHED 0011
olubility P+ Roy W. Tess Award	Symposium Honoring K. L. Hoy. Four Component S		PMSE 0033
s+ Excellence in Teaching Award	Symposium Honoring R. W. Ramette. Award Addres		ANYL 0006
in C+ Herman Skolnik Award	Symposium Honoring W. T. Wipke: New Dimensions		CINF 0016

r Chemistry, the Met Lab, and+	Symposium in Memory of Nathan Sugarman. Nuclea	NUCL	0058
stry Laboratory Development and	Symposium on Computers in Chemical Education.	CHED	0155
eative Polymer Chemistry Award	Symposium Polymers—Chemistry, Physics, and Biol	POLY	0272
Sherwin–Williams Student Award	Symposium. Change of Mechanism Block Copoly+	PMSE	0071
w–Temperatur+ 45th Technician	Symposium. Dynamic Rheological Studies on the Lo	TECH	0014
l Stabilities.= Cope Award	Symposium. Effects of Structural Changes on Radica	ORGN	0133
Plants.= Student Awards	Symposium. Fate of ^{14}C–Trichloroethylene in Whole	ENVR	0056
faces. J. Polany+ Nobel Laureates	Symposium. Reaction Dynamics in Gases and at Sur	CONG	0011
reative Polymer Chemistry Award	Symposium. Spontaneous Adsorpt+ C. S. Marvel C	POLY	0298
Porphyrins.= Cope Award	Symposium. Synthesis and Properties of Expanded	ORGN	0139
Cour+ 45th National Technician	Symposium. The Making of a Continuing Education	TECH	0001
axter, Burdick, Jackson Award	Symposium: Juvenile Hormones and Peptides for Ins	AGRO	0001
axter, Burdick, Jackson Award	Symposium: Juvenile Hormones and Peptides for Ins	AGRO	0006
ing Frontier.+ State-of-the-Art	Symposium: Solid–State Chemistry—A Rapidly Mov	CHED	0020
ing Frontier.+ State-of-the-Art	Symposium: Solid–State Chemistry—A Rapidly Mov	CHED	0029
Dynamic Viscoelasti+ APS-ACS	Symposium: Thermoreversible Gelation of Polymers.	POLY	0197
Reversible Gelation+ APS-ACS	Symposium: Thermoreversible Gelation of Polymers.	POLY	0145
Surfactant Organoge+ APS-ACS	Symposium: Thermoreversible Gelation of Polymers.	POLY	0169
Thermoreversible G+ APS-ACS	Symposium: Thermoreversible Gelation of Polymers.	POLY	0121
ons of+ Electrochemistry Award	Symposium—I, Honoring R. A. Osteryoung. Interacti	ANYL	0093
Specie+ Electrochemistry Award	Symposium—II. Reversible and Irreversible Neutral	ANYL	0103
plicative Bypass of the TpdU cis-	syn Cyclobutane Photodimer of DNA.= +of the Re	BIOL	0113
Enantioselective Routes to	syn- and anti-β-Phenylcysteines.=	ORGN	0115
dy of the Base Hydrolysis of the	syn,anti-cis-Dichloro (1,4,7,10-Tetraazacyclododecane)	INOR	0398
n State or Principle of Nonperfect	Synchronization?= +tyrenes: Radicaloid Transitio	ORGN	0155
ritical Behavior of Poly(2–Chlo+	Synchrotron SAXS Study of Mean–Field and Ising C	PMSE	0103
ssociated with Clay Minerals from	Syncrude Sludge Pond Tailings.= +ganic Matter A	I&EC	0021
y S+ Association in Solutions of	Syndiotactic Poly(methyl Methacrylate) as Studied b	POLY	0124
or Random, with a High Degree of	Synergistic Binding between Substrates?.= +nism	BIOL	0053
Which Exhibit	Synergistic Properties.= Designing Polymer Blends	PMSE	0101
II. Coal/Oil Coprocessing Using	Syngas.= Selectivity in Catalysts: Fuel Chemistry—	CATL	0049
n of Inducible Nitric Oxide	Synthase by N^g–Amino–L–Homoarginine.= Inhibitio	MEDI	0106
on of Aromatic Mimics o+ EPSP	Synthase Inhibitor Design II: Synthesis and Evaluati	ORGN	0029
or Random, with a Hig+ EPSP	Synthase: A Sequential–Ordered Kinetic Mechanism	BIOL	0053
Successful Mechanism–B+ EPSP	Synthase: The Importance of Dynamic Catalysis for	BIOL	0054
taining Inhibitors of Thymidylate	Synthase.= +ign and Synthesis of Naphthalene Con	MEDI	0186
Enzyme C$_1$–Tetrahydrofolate	Synthase.= Domain Chimeras in the Trifunctional	BIOT	0243
cterium ammoniagenes Fatty Acid	Synthase.= +noyl Reduction Catalyzed by Breviba	BIOL	0065
Inhibitor of Chorismate	Synthase.= Toward the Synthesis of a Potential	ORGN	0030
ells Amplified with the Glutamine	Synthetase System.= +meric Antibodies in CHO C	BIOT	0165
δ–Diketones: Potential	Synthons for Heterocycles.=	ORGN	0032
ubstitution Via Aryl 1,2–Dipolar	Synthons: Nickel(O)–Catalyzed Cross–Coupling of Ar	ORGN	0082
Conversion of Corn	Syrup Solids into γ–Cyclodextrin.=	CARB	0041
icrobial Polysaccharides on Maple	Syrup. = +robial Polysaccharides. Production of M	BIOT	0195
in Maple	Syrups.= Flavor Volatiles and Related Compounds	AGFD	0174
of Data f+ Implementation of a	Systematic Approach for Quantitative Interpretation	PMSE	0065
he Graphic Structure Input to a	Systematic Chemical Name by Computer: Experience	COMP	0019
N–Alk+ Enkephalin Analogues as	Systemically Active Antinociceptive Agents: O– and	MEDI	0152
arge–Pore Zeolites: The Effects of	Systemically Altering a Template Structure.= +L	PETR	0036
What's New in Collection	Systems? On–Board Materials Densification.=	I&EC	0040
of High-	T$_c$ Cuprate Crystals.= Growth and Characterization	PHYS	0026
for the Electronic States of High-	T$_c$ Cuprates as Probed by X–ray Absorption.= +el	PHYS	0053
Superconductivity. Synthesis and	T$_c$ in Simple Superconducting Oxide Systems.=	PHYS	0009
r Deposition Approaches to high-	T$_c$ Superconducting Films.= +ganic Chemical Vapo	INOR	0176
and Electronic States of High-	T$_c$ Superconductors by Tunneling Microscopy and Sp	PHYS	0206
etry, and Cu Valence in the High-	T$_c$ Superconductors.= +ectivity, Oxygen Stoichiom	PHYS	0027
gle–Crystal Bi$_2$Sr$_2$CaCu$_2$O$_8$ High-	T$_c$ Superconductors.= +pectroscopy Studies of Sin	PHYS	0211
Substitution in High-	T$_c$ Superconductors.= Studies of Halogen	PHYS	0239
m Genetically Modified Nicotiana	tabacum Cell Lines.= +eign Protein Production fro	BIOT	0230
ial Action Planning Pr+ Round-	Table Discussion—Public Involvement in the Remed	ENVR	0086
icar Clorhidrato de L–Arginina en	Tabletas en Medio.= +tencio–Metrico para Cuantif	ANYL	0066
Coal Sample Using Wood Ash or	Taconite as Additive.= +minous Argonne Premium	FUEL	0006
Metalation and Cross–Coupling	Tactics: Application in Natural Product Synthesis.=	ORGN	0277
erals from Syncrude Sludge Pond	Tailings.= +ganic Matter Associated with Clay Min	I&EC	0021
sion Body Formation in the P22	Tailspike Protein.= Mutational Suppression of Inclu	BIOT	0113
Fraud:	Tales of Trauma.= °From Doubts to Conviction of	PROF	0003
date on R. E. Marker—Illustrated	Talk.= +of the Mexican Steroid Industry and Up	HIST	0016
Specific Thrombin Inhibitor, by a	Tandem Column HPLC Procedure.= +gatroban, a	ANYL	0086
Edible Meats by On–Line HPLC	Tandem Mass Spectrometry.= +ibiotic Residues in	AGFD	0061
bonyl Deblocking of Anions of +	Tandem Michael Addition Ethoxycarbonylation–Car	ORGN	0228
– and 6–Ring Nitrogen and Sul+	Tandem S$_N$2–Michael Approach to the Synthesis of 5	ORGN	0199
An Evaluation of Total	Tannin and Relative Astringency in Teas.=	AGFD	0153
Cure Characteristics of	Tannin–Based Resorcinol–Formaldehyde Resins.=	POLY	0138
Compounds in Food and Health.	Tannin–Protein Interactions.= Phenolic	AGFD	0148
ts.=	Tannins and Related Compounds as Anti–HIV Agen	AGFD	0166
Antinutritional Effects of	Tannins.= Biochemical Mechanism of the	AGFD	0149
t Mono(peralkylcyclopentadienyl)	Tantalum η^2–Iminoacyl Chemistry.= +Mid–Valen	INOR	0108
Complex as Precursor to	Tantalum Nitride Thin Films.= Ethylimidotantalum	INOR	0074
N,+ Reduction of Niobium and	Tantalum Pentachlorides in the Presence of Sodium	INOR	0092
Possible To Escape from the	Tar Baby?.= Cyanomethylene and Benzyne: Is It	PHYS	0032

d Reactions: A Versatile Route to	Target Chiral Molecules.= +rboxylesterase–Mediate	AGFD	0008
hibitors. Protein Kinase C: A New	Target for Antiviral and Anticancer Therapy.= +n	MEDI	0072
The Histamine H$_3$-Receptor as a	Target for Developing New Drugs.=	MEDI	0081
cture of Messenger RNA: A Novel	Target for Therapeutic Antisense Chemistry.= +ru	BIOL	0120
ergy Heavy Ions.=	Target Fragmentation of Copper by Intermediate–En	NUCL	0025
^{16}O Ions.=	Target Fragmentation of Silver by 14.6 GeV/nucleon	NUCL	0050
sis of High-Energy Projectile and	Target Fragmentation.= +istic Phenomena. Synthe	NUCL	0046
ity on Enzymes.=	Targeted Catalysis: Conferment of Substrate Selectiv	BIOL	0090
	Targeting Chemical Nucleases.=	BIOL	0156
Synthetic	Targets.= Using Databases as an Aid in Developing	COMP	0047
e Yet Know To Accomplish This	Task? Relative Risk: A Blueprint for Future Environ	ENVR	0061
urine Transporting System.=	Taurine Derivatives as Potential Agents Active on Ta	MEDI	0135
as Potential Agents Active on	Taurine Transporting System.= Taurine Derivatives	MEDI	0135
ctions with η^2-Py+ Protonation,	Tautomerization, and Novel C–C Bond-Forming Rea	INOR	0114
pproach to T+ Enantiocontrolled	Taxane Construction: A Complementary Bifurcate A	ORGN	0291
ifurcate Approach to Taxusin and	Taxol.= +axane Construction: A Complementary B	ORGN	0291
Photochemical Approach to	Taxol.=	ORGN	0012
plementary Bifurcate Approach to	Taxusin and Taxol.= +axane Construction: A Com	ORGN	0291
Homologues with	TBP.= Extraction of Rf (Element 104) and Its	NUCL	0074
erizatio+ Cationic Complexes of	Tc-99m-Hydroxy-4-Pyrones: Synthesis and Charact	MEDI	0042
zofurans in the Blood of Twelve	TCDD-Exposed Persons from Jacksonville, Arkansas	CEI	0014
nooxygenase Cons+ Treatment of	TCE and Degradation Products Using a Toluene Mo	BIOT	0119
of Monoclonal Antibodies with	Tc99m.= An Improved Method for Direct Labeling	BIOT	0061
ced Lung Tumorigenesis by Green	Tea and Its Components.= +ific Nitrosamine–Indu	AGFD	0068
Chemistry of	Tea and Its Manufacturing.=	AGFD	0044
fect of Orally Administered Green	Tea on Carcinogenesis by Ultraviolet B Light.=	AGFD	0065
ion in Mouse Epidermis by Green	Tea Polyphenols.= +cetate-Induced H$_2$O$_2$ Product	AGFD	0069
agus Carcinogenesis by Green	Tea.= Protection against Stomach, Lung, and Esoph	AGFD	0067
Operation Chemistry:	Teacher Training in Chemistry.=	CHED	0072
ntation Laboratories and Inservice	Teacher Training.= +Interface: Regional Instrume	CHED	0112
Project LABS—A	Teacher-Scientist Interaction for K-12 Teachers.=	CHED	0002
rship.=	Teachers in Industry: An Industry–Education Partne	CHED	0004
ogram for High School Chemistry	Teachers in the Greater Cincinnati Area.= +ern Pr	TECH	0002
ic Bulletin Board for High School	Teachers of Chemistry in the Northwest.= +ectron	CHED	0111
on: Intr+ High School Chemistry	Teachers Program. POLYED 1990 Award Presentati	CHED	0025
emistry Educ+ Educating Future	Teachers: The CPT-Chemical Education Option. Ch	CHED	0066
School Chemistry	Teachers: The Need for Reform.= Educating High	CHED	0067
reservice and Inservice Chemistry	Teachers.= +emSource: A Teaching Resource for P	CHED	0137
Interaction for K-12	Teachers.= Project LABS—A Teacher-Scientist	CHED	0002
e. Award Address+ Excellence in	Teaching Award Symposium Honoring R. W. Ramett	ANYL	0006
	Teaching Biochemistry with Spreadsheet Software.=	CHED	0206
d Edition—Aiba, H+ Books for	Teaching Biotechnology. Biochemical Engineering, 2n	BIOT	0221
gents Using Fluorescent Probes: A	Teaching Experiment.= +Characterization of Deter	CHED	0200
Research and	Teaching in a Major University.=	CONG	0010
ment. The Case for the Advanced	Teaching Lab.= +l Chemistry Laboratory Develop	CHED	0041
General Chemistry: Reaching and	Teaching Large Student Populations.= +vement in	CHED	0081
Hands-On	Teaching Laser.=	PHYS	0303
ation.=	Teaching Magnetic Resonance Using Computer Anim	CHED	0123
	Teaching of Chemical Engineering in High School.=	CHED	0108
.=	Teaching of Chemistry Program at Boston University	CHED	0070
for High Schoo+ Improving the	Teaching of Chemistry: An Electronic Bulletin Board	CHED	0111
Molecular Graphics in	Teaching Organic Chemistry.= Experience Using	CHED	0189
istry Teachers+ ChemSource: A	Teaching Resource for Preservice and Inservice Chem	CHED	0137
ChemSource. ChemSource: A	Teaching Resource for the 90s.= Minisymposium on	CHED	0017
Chemistry.=	Teaching the Principles and Applications of Lasers in	CHED	0083
Chemistry	Teaching through the Student's World.=	CHED	0005
Uncommon Reactions: A Useful	Teaching Tool.=	CHED	0166
. I. DuPont de Nemours and Co.).	Teaching: A Witches' Brew of Black Holes?.= +E	ANYL	0006
in Chemistry and	Teaching.= Bachelor's Degree Program at Berkeley	CHED	0069
xample of Experimental Modular	Teaching.= +Mo, Co, Ni, and Cu Compounds: An E	CHED	0057
Method for Fischer's Proof	Teaching.= D-Glucose Stereochemistry: Schematic	CHED	0095
Model for Inservice	Teaching.=	CHED	0074
hemistry Award for Excellence in	Teaching, cosponsored by E. +vision of Analytical C	ANYL	0006
Organic Chemistry Nomenclature	Teaching, Part I: Acyclic Series.= +ctic Puzzles for	CHED	0165
Preceptor	Team.= Experiences of a SEED	CHED	0038
Polyphenols from	Teas and Their Antioxidant Activity.=	AGFD	0048
Astringency in	Teas.= An Evaluation of Total Tannin and Relative	AGFD	0153
	Technetium Solar Neutrino Experiment.=	NUCL	0022
Molybdenum—	Technetium(V) and Rhenium(V) Oxo Complexes of A	INOR	0181
mine-Amide-Dithiol Ligands.+	Technetium(VII)–Oxo Complexes.= Investigation of	INOR	0328
Alkene Oxidation by	Technician Symposium. Dynamic Rheological Studie	TECH	0014
s on the Low-Temperatur+ 45th	Technician Symposium. The Making of a Continuing	TECH	0001
Education Cour+ 45th National	Technician.=	TECH	0003
Importance of Statistics for the	Technologies and the Need of Pharmaceutical	ANYL	0111
Industry.= Trends in Separation	Technologies for Pesticide Waste Disposal: Overview	AGRO	0120
ls and Their Containers. Current	Technologies for Plastics Recycling. Plastics Recyclin	I&EC	0031
g Processing and+ Reclamation	Technologies for Waste Incinerators.= +e Role of	FUEL	0129
Metals. Metals Emissions Control	Technologies from a Process Analysis Database.=	I&EC	0064
in the Selection of Petrochemical	Technologies in Peptide Production for Discovery of	MEDI	0090
Drug Leads. Light-Direct+ New	Technologies of Fertilizers. Corrosion of Mild Steel E	FERT	0001
xposed+ Chemical and Physical			

ht: Are Your Po+ Impact of New	Technologies on Information Ownership and Copyrig	CINF	0043
Catalytic Emission Control	Technologies.= Camet Metal Monolith–Based	I&EC	0045
s: Historical, Current, and Future	Technologies.= +cal Degradation of Pesticide Waste	AGRO	0122
alysis and Search for Remediation	Technologies.= +g Plants in Quebec: Extensive An	ENVR	0016
Acido S+ Estudio Estadistico de	Tecnicas Electroquimicas de Aceros Microaleados en	PHYS	0281
vado Quimico de Una Caldera por	Tecnicas Electroquimicas.= +erentes Etapas del La	PHYS	0257
tructure–Reten+ Lipophilicity of	Teicoplanin Antibiotics as Revealed by RP–HPLC: S	MEDI	0059
oxy Polystyrene Carboxylic Acid	Telechelomers, II: A Model Study of Condensation P	POLY	0046
Symmetric Di(1–Alkynyl)	Tellurides.= Synthesis and characterization of	INOR	0229
Early Transition–Metal	Tellurolates: Synthesis, Structure, and Reactivity.=	INOR	0021
	TEM Analysis of Pillared Rectorites.=	PETR	0109
Catalysts Prepared by the High–	Temperature Aerosol Decomposition (HTAD) Proces	CATL	0026
n a Hot–Filament CVD Reactor+	Temperature and Density Distribution of H_2 and H i	PHYS	0146
MR to Determine the Effect of	Temperature and Ionic Strength on Chemical Exchan	INOR	0228
Conformation of β–+ Effects of	Temperature and pH on the Foaming Properties and	AGFD	0207
Microporous+ Effect of High–	Temperature Annealing on the Capillarity of Nylon	COLL	0037
Using Zeolitic Catalysis for High–	Temperature Applications.= +ic Reduction of NO_x	I&EC	0046
c Rheological Studies on the Low–	Temperature Behavior of Lubricants.= +Dynami	TECH	0014
rom New Copp+ Selective Low–	Temperature Chemical Vapor Deposition of Copper f	INOR	0144
c Chemistry Performed in Room–	Temperature Chloroaluminate Molten Salts.= +alli	ANYL	0097
r Acrylic Pol+ Studies of Room	Temperature Cured Castor Oil Polyurethane/Vinyl o	PMSE	0090
t Raman Spectroscopy of High–	Temperature Dense Molecular Fluids and Low–Temp	ANYL	0002
Reaction with C+ Near–Room–	Temperature Deposition of Copper Films by H–Atom	PHYS	0259
gen Flame+ Homoepitaxial High–	Temperature Diamond Growth from Acetylene– Oxy	FUEL	0037
sidua Using Atomic Emi+ High–	Temperature Gas Chromatography of Crudes and Re	PETR	0121
Motor Oil Volatility Using High–	Temperature Gas Chromatography.= +rmination of	PETR	0122
eolites in an Alkaline Flu+ Low–	Temperature Hydrothermal Crystallization of MFI Z	PETR	0005
terface on Local Glass–Transition	Temperature in Composites.= +of Fiber/Matrix In	COLL	0055
Time–	Temperature Indicators for Shelf–Life Monitoring.=	AGFD	0080
Rh/CeO_2 and+ Effect of High–	Temperature Lean Aging on the Performance of Pt,	PETR	0035
Dense Molecular Fluids and Low–	Temperature Molecular Solids.= +h–Temperature	ANYL	0002
Aromatic Ketones in a Room–	Temperature Molten Salt.= Electrochemistry of	ANYL	0094
erization by Programmed–	Temperature Oxidation.= Advances in Coal Charact	FUEL	0058
Metallaboranes as Low–	Temperature Precursors to Metal Borides.=	INOR	0101
d Diamond–Like Carbon.	Temperature Raman–Scattering Behavior in Diamon	FUEL	0035
A Low–	Temperature Route to Polyimides.=	POLY	0137
anipulation of Adsorbate+ Low–	Temperature STM Imaging of Metal Surfaces and M	PHYS	0360
00) Observed with a New Variable	Temperature STM.= +mer Configurations on Si(1	PHYS	0185
Reaction Center: Polarization and	Temperature Studies.= +Rhodobacter sphaeroides	BIOL	0094
straints on the Theory of High–	Temperature Superconductivity.= Experimental Con	PHYS	0051
oss Spectroscopy Studies of High–	Temperature Superconductors.= +ctron–Energy–L	PHYS	0279
–Phonon Interactions in High–	Temperature Superconductors.= Nonlinear Electron	PHYS	0012
n–Metal Nitride Thin F+ Low–	Temperature Synthesis of Main–Group and Transitio	INOR	0104
sation Polymerization Using Low–	Temperature Techniques.= +del Study of Conden	POLY	0046
with Simple Molecules near Room	Temperature.= +ions of Transition–Metal Atoms	PHYS	0198
cular Weight on Glass–Transition	Temperature.= +rocyclic Polystrene: Effect of Mole	POLY	0239
spen Treated with Vapor at High	Temperature.= +ulfonation of Sulfite–Pretreated A	PMSE	0170
System at Low	Temperature.= Zeolite Synthesis in the Na–THA	PETR	0022
of Heterocyclic Compounds in a	Temperature– and Pressure–Controlled Environment	AGFD	0073
X–ray Studies of Layered Rare+	Temperature–Dependent Luminescence, Raman, and	INOR	0189
Attachment Reactions of Meta+	Temperature–Dependent Rate Constants for Electron	PHYS	0234
zation of Carbon Active Sites by	Temperature–Programmed Desorption.= Characteri	FUEL	0016
POP): A New Analytical Meth+	Temperature–Programmed Oxidative Pyroanalysis (T	FUEL	0046
inear Low–Density Polyethylene:	Temperature–Rising Supercritical Fluid Fractionatio	POLY	0076
f Glass Capillaries: Influence of	Temperature, Flow Rate, Ion Concentration, and App	COLL	0163
Combustion.= Kinetics of High–	Temperature, Hydrocarbon–Assisted Boron	PHYS	0142
Morphology of PEEK at High	Temperatures and in PEI Blends.=	POLY	0333
Minerals with Water at Elevated	Temperatures and Pressures in Geothermal Environ	INOR	0246
Studies of Catalytic Reactions at	Temperatures between 90 and 450 K.= +AS NMR	COLL	0090
carbon Plasmas: Determination of	Temperatures by C_2 Emission Spectroscopy.= +dro	PHYS	0268
of VPI–5 to $AlPO_4$–8 at Various	Temperatures by Multinuclear Solid–State NMR and	PETR	0095
Parameters and Glass Transition	Temperatures on the Phase Morphology of Simultan	PMSE	0025
Chloride Complexes to High	Temperatures.= Formation of Aqueous Metal	ANYL	0008
and Radical Reactions over Wide	Temperture Ranges: Measurements, Correlations, an	PHYS	0087
89 Using D–Mannose as a Chiral	Template and Its Anti–HIV Activity in PBM Cells.=	CARB	0022
Effects of Systemically Altering a	Template Structure.= +Large–Pore Zeolites: The	PETR	0036
utron–Scattering Studies of the	Template–Mediated Crystallization of ZSM–5 Type Z	PETR	0058
= Twelve–Ring Channel	Templated Mazzite: A Stabilized Calcined Structure.	PETR	0037
Bisubstrate Reaction	Templates.=	ORGN	0273
bolium castaneum L. (Coleoptera:	Tenebrionidae).= +amosa Seed Oil Extracts to Tri	AGRO	0036
ment—Determination of Surface	Tension by Automation of the Drop–Weight Method	CHED	0118
enes—A Liquid/Water Interfacial	Tension Measurement.= +ption Kinetics of Asphalt	PETR	0138
l Tra+ Measuring the Isometric	Tension of Kinesin Molecules Using the Laser Optica	BIOL	0137
Automated Dynamic Interfacial	Tension Technique.=	COLL	0091
Low Interfacial	Tension.= °From Large–Surfactant Aggregate to	COLL	0078
lymers Utilizing the Off–Diagonal	Tensor Components of a Molecule.= +for Poled Po	POLY	0128
N–Deuterated Poly(p–phenylene	Terephthalamide).= Amide Segmental Dynamics in	POLY	0058
ermal P+ Effects of Polyethylene	Terephthalate (PET) from Plastics Waste on the Th	FUEL	0053
tion in Blends of Poly(butylene	Terephthalate) and Poly(ester–Ether) Segmented Blo	PMSE	0148
yceride Oils.= Poly(Ethylene	Terephthalate) Semi–IPNs with Functionalized Trigl	PMSE	0088

ization Kinetics of Poly(Ethylene	Terephthalate) with Functionalized Triglyceride Oil	PMSE 0171
ti+ Lysimeter Studies on Long-	Term Fate of Pesticides: Characterization of Degrada	AGRO 0025
rato de L-Arginina en Producto	Terminado Utilizando como Electrodo Indicador una	CHED 0163
Carboxyl Termina+ Effect of N-	Terminal Acylation on Chemokinetic Activity of C5A	MEDI 0168
Specificity. Role of Insulin's C-	Terminal B-Chain on Association, Folding, and Stab	BIOT 0132
y-silanes).=	Terminal Modifications of Hyperbranched Poly(silox	POLY 0323
Biological Evaluation of New C-	Terminal Modified Analogues of Chemotactic Tripep	MEDI 0166
okinetic Activity of C5A Carboxyl	Terminal Octapeptide Analogues.= +ation on Chem	MEDI 0168
nic Structure and Magnetism in	Ternary Compounds of Ni-Fe-Pt, Ni-Fe-Pd, Fe-Co-	PHYS 0253
cs of Peroxodisulfate Oxidation of	Ternary Iron(II)-Diimine-Cyanide Complexes.= +i	INOR 0022
	Ternary Nitrides.=	PHYS 0003
Dynamic Studies of	Ternary Polymer Solutions.=	COLL 0054
Compatibility of	Ternary Semi IPNs.=	PMSE 0156
of Silver from	Ternary Silver Oxides.= Oxidative Deintercalation	INOR 0275
tructure and Chemical Bonding in	Ternary Transition-Metal Compounds.= +tronic S	PHYS 0252
°From Ruzicka's	Terpenes in Zurich to Steroids in Mexico via Cuba.=	HIST 0017
ward the Total Synthesis of Novel	Terpenes.= +-Sized Carbocycles: Applications To	ORGN 0282
Studies of [Pt(	terpy)Cl]+.= Spectroscopic and Electrochemical	INOR 0216
al Centers via Poly 3'-(4-pyridyl)	terthiophene-Based Ligands.= +at Transition-Met	INOR 0027
ers: A New Dimension.= Roy W.	Tess Award Symposium Honoring K. +lity Paramet	PMSE 0033
inging Back?.=	Testimony of Scientific Experts: Is the Pendulum Sw	CHAL 0016
Which Are Uncompetitive versus	Testosterone.= Inhibitors of Steroid 5α-Reductase	MEDI 0008
a- Asymmetric Synthesis of P=1	Tethered Analogs of Inositol (1,3,4,5)Tetrakisphosph	AGRO 0045
Photocycloaddition of	Tethered 2-Pyridones.= Intramolecular [4+4]	ORGN 0235
- and Three-Atom Intra-Annular	Tethers.= +stry of Bimetallic Complexes with Two	INOR 0149
ons of the Iron-Ethylene Diamine	Tetraacetic Acid Complexes.= +otons in the Soluti	INOR 0090
nsitizing Agent for th+ 2',3',4',5'-	Tetraacetyl Riboflavin: A Potentially Useful Photose	ENVR 0029
the First Monomeric Mercury(II)	Tetraalkylthiolate Complex.= +Characterization of	INOR 0059
he syn,anti-cis-Dichloro (1,4,7,10-	Tetraazacyclodecane) Cobalt(III) Cation.= +is of t	INOR 0398
Immobilization of	Tetraazamacrocyclic Compounds on Silica Surface.=	COLL 0105
ation of Mordenites Modified by	Tetrabutylin: Application to the C8 Cut Isomerizatio	PETR 0131
Pattern and Phase Transition in	Tetracene Crystals and Tetracene-Doped Polymeric	PHYS 0274
ansition in Tetracene Crystals and	Tetracene-Doped Polymeric Matrices.= +Phase Tr	PHYS 0274
Engineering: The Biosynthesis of	Tetracenomycins by Streptomyces.= +lic Pathway	BIOT 0083
duction i+ Inhibition of 12-O-	Tetradecanoylphorbol-13-Acetate-Induced H2O2 Pro	AGFD 0069
dy-State Fluorescence Studies of	Tetradecyltrimethylammonium Bromide Complexati	COLL 0074
osites, I: In Situ Polymerization of	Tetraethoxysilane.= +sphazenes Molecular Comp	POLY 0181
Deposition of SiO2 from	Tetraethoxysilane.=	COLL 0009
t Synthesis of Cyclic Amidinium	Tetrafluoroborates and Hexafluorophosphates: The S	ORGN 0204
us Molybdenum with Octahedral-	Tetrahedral Frameworks.= +ynthesis of Microporo	CATL 0019
ation of Aromatic Mimics of the	Tetrahedral Intermediate and Evaluation of a 3-Pho	ORGN 0029
3-Carboxy-1-Pyridoxyl-1,2,3,4-	Tetrahydro-β-Carboline.= Synthesis of	BIOL 0080
Hydroxide+ Copolymerization of	Tetrahydro-1-(4-hydroxy-1-naphthyl)thiophenium	POLY 0048
imethyl-Quino[2,3,4-kl]-5a,6,7,8-	Tetrahydrobenz[c]Acridine 5N-Oxide, V.= +7,7-D	ORGN 0198
the Trifunctional Enzyme C1-	Tetrahydrofolate Synthase.= Domain Chimeras in	BIOT 0243
ty of 2-Amino-1,4-Epoxy-1,2,3,4-	Tetrahydronaphthalenes.= +Dopaminergic Activi	MEDI 0128
thesis of 15,20-Dihydrosecodine,	Tetrahydropresecamine, and Tetrahydrosecamine.=	ORGN 0009
cine, Tetrahydropresecamine, and	Tetrahydrosecamine.= +hesis of 15,20-Dihydroseco	ORGN 0009
cn of Mono-Substituted α,α'-bis(	Tetrahydrothiophenium)-p-xylene Dibromides.= +i	POLY 0013
Ligand Class.= Synthesis of a	Tetraimidazole-Containing Macrocycle: A New	ORGN 0060
hilic Star Polymers from Tri- or	Tetraisocyanates and Poly(ethylene Glycol) Derivativ	POLY 0271
ered Analogs of Inositol (1,3,4,5)	Tetrakisphosphate from Glucose via the Ferrier Rear	AGRO 0045
nd Scission on Tungsten-Iridium	Tetranuclear Cluster Cores: Relationship to Butane	CATL 0008
heme-Iron-Containing Proteins+	Tetranuclear Iron(III) Complexes as Models for Non	INOR 0295
e Dianionic Initiator from Divinyl	Tetraoxaspiroundecane Derivative.= +Hydrolyzabl	POLY 0349
ulfides to Sulf+ A Mild Osmium	Tetraoxide-Catalyzed Process for the Oxidation of S	ORGN 0244
by a D4-Symmetrical Manganese-	Tetraphenylporphyrin.= +ic Oxidations Catalyzed	ORGN 0045
w Class of D4-Symmetrical Chiral	Tetraphenylporphyrins.= +le Alkenes Using a Ne	INOR 0300
yano-4'-Methylbiphenyls and 2-(	Tetrazol-5-yl-4'-Methylbiphenyls.= +phenyls, 2-C	ORGN 0112
sterol Acyltransferase (ACAT) I+	Tetrazole Ureas: A Potent Series of Acyl CoA: Chole	MEDI 0107
Trapping	Tetrazolinylidenes with Nitrogen Electrophiles.=	ORGN 0205
Journal Full	Text: Copyright Issues in Electronic Distribution.=	CINF 0046
Equipment for Manufacturing of	Textile-Structured Preforms—Part I.= +omation of	TECH 0007
Equipment for Manufacturing of	Textile-Structured Preforms—Part II.= +mation of	TECH 0008
Implication of Phenolic Acids as	Texturizing Agents during Cooking-Extrusion Cere	AGFD 0150
sNadimides Oligomers and High-	Tg Polyimide-Based Semi-IPNs.= +erization of bi	POLY 0133
= a Continuing Education Cour+	45th National Technician Symposium. The Making o	TECH 0001
tudies on the Low-Temperature+	45th Technician Symposium. Dynamic Rheological S	TECH 0014
Zeolite Synthesis in the Na-	THA System at Low Temperature.=	PETR 0022
Co2SxSe2-x (0 ≤ x ≤ 2.0) with the	ThCr2Si2-Type Structure.= +nsition in Metallic Tl	INOR 0058
wo-Phase Epoxy Interpenetrating	Themosets.= +erpenetrating Polymer Networks. T	PMSE 0087
g Avoidance Principles in Zeolit+	Theoretical Foundation of the Al Sitting and Pairin	PHYS 0289
pectroscopic Characterization, and	Theoretical Study of a Caffeine Dimer.= +hesis, S	INOR 0252
ility.= Chemical Graph-	Theoretic Cluster Expansions: Diamagnetic Susceptib	COMP 0030
ive Regular Polyhedra i+ Graph	Theoretical Algorithm To Label the Vertices of the F	COMP 0035
tical Side. Coordination between	Theoretical and Practical Coordination Chemistry.=	CHED 0056
ph Theory in Chemistry. Graph	Theoretical Approach to Structure-Property Relation	COMP 0013
Relationships.= Graph	Theoretical Approach to Structure-Property	COMP 0014
y-g-Acrylic Copolymer.=	Theoretical Calculation of Grafting Sites in the Epox	PMSE 0161
ucting Polymers: Comparison with	Theoretical Calculations.= +ptibilities of Semicond	POLY 0009

omparison of Experimantal and	Theoretical Hyperpolarizabilities, with Emphasis on	PHYS	0339
Butadiene Radical Cation.=	Theoretical Investigation of the Low–Lying States of	PHYS	0237
Alkanes.= Optimal Graph	Theoretical Models for Enthalpic Properties of	COMP	0034
Molecular Systems: Facile Graph	Theoretical Solutions to Huckel MO Parameters.=	COMP	0020
–Cluster Methods.= High-Level	Theoretical Studies of Carbon Clusters with Coupled	PHYS	0008
ions in Phot+ Experimental and	Theoretical Studies of Fast–Electron–Transfer React	BIOL	0130
ions for Model Compounds of +	Theoretical Studies of Gauche and Eclipsing Interact	POLY	0353
B+ Synthesis, Experimental, and	Theoretical Studies of New Group 6 Pentacarbonyl–	POLY	0217
g: STM and Related Techniques.	Theoretical Studies of Single Adsorbed Atoms in the	PHYS	0096
B–C Polymers: Experimental and	Theoretical Studies.= +Oxidation Reactions of Si–	POLY	0187
y Nucleophiles.=	Theoretical Study of Ring–Opening of Aziridinones b	PHYS	0314
ies.=	Theoretical Study of Silicon Cluster Electron Affinit	PHYS	0079
er Molecules in Electric Fields—A	Theoretical Study.= +otential and Charge–Transf	BIOT	0001
mics of Proteins and Membranes:	Theory and Experiment.= +rization, and the Dyna	PHYS	0166
ansfer in Polymers: Comparison of	Theory and Experiment.= +f Electronic Energy Tr	PHYS	0109
Nafion Membranes:	Theory and Experiment.= Gas Permeation through	POLY	0066
D Labora+ Laboratory Safety—	Theory and Practice. A Chemical Hygiene Plan for R	CHAS	0001
Organic Crystals.= Graph	Theory Applied to Bond Length Variations in	COMP	0031
Reliability	Theory Applied to the Process Industry.=	I&EC	0071
e Capabilities of a Free–Volume	Theory for Polymer–Solvent Diffusion Coefficients.=	POLY	0144
A	Theory for Solubility in Static Systems.=	POLY	0141
Structure+ Pedagogical Graph	Theory in Chemistry. Graph Theoretical Approach to	COMP	0013
gated Mole+ Pedagogical Graph	Theory in Chemistry. Pedagogical Approach to Conju	COMP	0020
rly Trans+ Applications of Graph	Theory in the Study of Electron Delocalization in Ea	COMP	0032
nd Transition–State+ Quantum	Theory of Chemical Reactions: Reactive Scattering a	PHYS	0161
xperimental Constraints on the	Theory of High–Temperature Superconductivity.= E	PHYS	0051
quids and Glasses.= Molecular	Theory of Inhomogeneous Spectral Broadening in Li	PHYS	0036
Superconductivity.	Theory of Pairing in HTC Materials.=	PHYS	0001
Toward a	Theory of Polymer–Induced Protein Precipitation.=	BIOT	0022
Fractal	Theory of Radon Emanation from Solids.=	NUCL	0088
faces.=	Theory of Schottky Barriers in Diamond/Metal Inter	PHYS	0358
Density Matrix	Theory of Ultrafast Electron Transfer.=	BIOT	0027
Data in the Formation of Ratio+	Theory Testing, Evidential Reason, and the Role of	ENVR	0075
Chemical Applications of Graph	Theory. Molecular Similarity: Relative and Absolut	COMP	0022
e Scattering and Transition–State	Theory.= +Theory of Chemical Reactions: Reactiv	PHYS	0161
Comparison of New Results with	Theory.= +gative–Ion Photoelectron Spectroscopy:	PHYS	0059
ture through Electron Propagator	Theory.= +n Pictures of Molecular Electronic Struc	PHYS	0058
Liquid–Solid Adsorption	Theory.= Verification of a Fundamental	COLL	0188
essenger RNA: A Novel Target for	Therapeutic Antisense Chemistry.= +ructure of M	BIOL	0120
arget for Antiviral and Anticancer	Therapy.= +nhibitors. Protein Kinase C: A New T	MEDI	0072
tein Having Potential as an AIDS	Therapy.= +)–PE40 Conformation: A Chimeric Pro	ANYL	0005
Future of Nuclear Power. Is	There a Future for Nuclear Power?.=	FUEL	0022
ated Zirconia Superacid Catalysts:	Thermal Analysis and Characterization Studies.=	PETR	0028
of Polypropyle+ High–Pressure	Thermal Analysis of an Incompatible Polymer Blend	POLY	0071
ronic Pac+ Dynamic Mechanical	Thermal Analysis of Mold Compounds Used in Elect	POLY	0095
attering and Dynamic Mechanical	Thermal Analysis.= +rs by Small–Angle X–ray Sc	PMSE	0118
–Catalyzed Gasificatio+ Effect of	Thermal and Chemical Pretreatments on the Copper	FUEL	0028
Flavor Compounds. Chrono–C+	Thermal and Enzymatic Conversions of Precursors to	AGFD	0022
Flavor Compounds. Glucosides+	Thermal and Enzymatic Conversions of Precursors to	AGFD	0050
Flavor Compounds. Glycosidic+	Thermal and Enzymatic Conversions of Precursors to	AGFD	0006
Flavor Compounds. Studies on+	Thermal and Enzymatic Conversions of Precursors to	AGFD	0070
Molecular Chlorine.=	Thermal and Laser Etching of Indium Phosphide by	COLL	0214
Epoxy/Amine = Cure Kinetics in	Thermal and Microwave Fields.= +nistic Study of	POLY	0161
ydrolase I: Predenaturation St+	Thermal and pH Stress in Denaturation of Cellobioh	BIOT	0191
Liquid Crystalline Thermosets, I:	Thermal Behavior of Mixtures.= +etting Systems.	POLY	0109
Using Rh(PMe$_3$)$_2$LC+ Efficient	Thermal Catalytic Alkane Transfer–Dehydrogenation	INOR	0151
e 1,3,5–Phenylen+ Synthesis and	Thermal Characterization of a Series of Monodispers	POLY	0236
Precursor.= Chemistry of	Thermal CVD from a Copper(II) Aminoalkoxide	COLL	0028
tal Applications of Catalysts—II.	Thermal Decomposition of Alkyl Halides on Copper	COLL	0204
100) and Implicatio+ Kinetics of	Thermal Decomposition of Triethylgallium on GaAs(	COLL	0068
TPD and XPS Study of the	Thermal Decomposition of W(CO)$_6$ on Pt (111).=	COLL	0049
of Propargyl Radicals.=	Thermal Decomposition of 1,7–Ocatdiyne as a Source	FUEL	0067
FT–IR Spectroscopic Study of	Thermal Degradation in PEEK (Poly–ether–ether–ke	POLY	0334
	Thermal Degradation of Phenolic Antioxidants.=	AGFD	0049
h+ Efficient Photochemical and	Thermal Dehydrogenation of Alkanes Catalyzed by R	INOR	0152
	Thermal Field–Flow Fractionation of Copolymers.=	PMSE	0009
Analysis. Analysis of Polymers by	Thermal Field–Flow Fractionation.= +for Polymer	PMSE	0008
es to the Charac+ Application of	Thermal Gravimetric Analysis and Related Techniqu	INOR	0006
clohexadiend.=	Thermal Isomerization of a Vinylcyclobutene to a Cy	ORGN	0144
Laser.= Time–Resolved	Thermal Lens Calorimetry with a Helium–Neon	CHED	0089
Smoke–Flavor Compounds by	Thermal Lignin Degradation.= Formation of	AGFD	0054
(PET) from Plastics Waste on the	Thermal Properties of Asphalt.= +Terephthalate	FUEL	0053
riging Estimation of Differential	Thermal Signatures of Total Mercury in Crystalline	ENVR	0055
Kinetics of	Thermal Softening of Fruit and Vegetable Tissue.=	AGFD	0013
d Clay Molecular Sieve with High	Thermal Stability.= +mance of Pillared Interlayere	PETR	0030
Complexes with	Thermal Stability.= New Copper–Dioxygen	INOR	0046
ide)s—I. Overview of Chemistry,	Thermal Stability, and Oxygen Plasma Resistance.=	PMSE	0052
Oxygen–Enriched Combustion+	Thermal Swing Absorption Process for Production of	I&EC	0023
of Pyrogenous Silica.=	Thermal Transformations of Aziridine on the Surface	COLL	0113
–alkyl–α,L–glutamate+ Multiple	Thermal Transitions in Model Compounds for Poly(γ	POLY	0099

mentry Reactions and Modeling.	Thermal-Rate Constant Measurements by the Flash	FUEL	0123
nteresting Flavor Compounds.=+	Thermally Degraded Thiamine: A Potent Source of I	AGFD	0071
n of Adsorbed Nickel on Ultrathin	Thermally Grown Silicon Dioxide.= +Bulk Diffusio	COLL	0177
nted Polyurethanes.=	Thermally Induced Morphological Changes in Segme	PMSE	0149
ybenzimidazole/Polyimide Blen+	Thermally Induced Phase Behaviors of Aromatic Pol	PMSE	0164
	Thermally Stable Electrooptic Polymers.=	POLY	0130
Dow Chemical Co. Foundation).	Thermo-Optical Detection Methods in Analytical La	ANYL	0011
se Reaction in Methanobacterium	thermoautotrophicum.= +hyl Coenzyme M Reducta	ORGN	0025
and Proteins of the Clostridium	thermocellum Subcellulosome Preparation.= Genes	BIOT	0181
Clostridium	thermocellum.= Properties of the Cellulosome of	BIOT	0180
adicals: n-C_4H_5 ++ Kinetics and	Thermochemistry of the Oxidation of Unsaturated R	FUEL	0066
Diffusion Processes in Polymers.	Thermodynamic Analysis of Gas Permeation through	POLY	0139
rsion during Benzene Oxidation:	Thermodynamic and Kinetic Mechanisitic Considera	FUEL	0085
ted Polymer Solutions under Poor	Thermodynamic Conditions.= +cation of Concentra	POLY	0146
ε Transreacting Polyester/Poly+	Thermodynamic Interpretation of Phase Behavior in	PMSE	0074
	Thermodynamic Properties of Li_3.=	PHYS	0226
and Mixtures.= Estimating	Thermodynamic Properties of Organic Compounds	COMP	0054
ds in Contact with Solid Surfac+	Thermodynamics and Transport Properties of Colloi	COLL	0162
urement of Reaction Kinetics and	Thermodynamics at the Solid-Liquid Interface.=	COLL	0170
Finite Thickness Effects in the	Thermodynamics of Diblock Copolymer Thin Films.=	POLY	0090
f:city. Effects of Mutations on the	Thermodynamics of Protein Reactions.= +d Speci	BIOT	0188
rphization and Melting.=	Thermodynamics Parallels between Solid-State Amo	NUCL	0092
Network	Thermodynamics.= Variational Formulation in	PHYS	0323
-Carbon Bond Formation during	Thermolysis of Bis(2,4-Dimethyl-1,3-pentadienyl)ru	INOR	0305
cids: Generation of Ethyl Meta+	Thermolysis of O-Ethyl Esters of Phosphoramidic A	ORGN	0110
): Localization of Perturbations in	Thermomechanical Properties.= +thylmethacrylate	PMSE	0173
Sensor Plate for Manufacturing	Thermoplastic Composites.=	PMSE	0115
Star-Branched Styrene-Butadiene	Thermoplastic Elastomers.= +hesis of Heteroarm,	POLY	0266
eflow Tolerant Elec+ Selection of	Thermoplastic Molding Compounds for Infra-Red R	CMEC	0014
Starch-Based	Thermoplastics in Industrial Applications.=	CARB	0010
Liquid Crystalline Polymers with	Thermoplastics.= +on the Morphology of Blends of	PMSE	0135
rs. Dynamic Viscoelasticity during	Thermoreversible Gelation Gelation.= +of Polyme	POLY	0197
ntrated Polymer Solutions und+	Thermoreversible Gelation and Vitrification of Conce	POLY	0146
)Paradichlorobenzene and Their+	Thermoreversible Gelation from Poly(ethylene Oxide	POLY	0125
s+ A Molecular Interpretation of	Thermoreversible Gelation in Polystyrene-Liquid Sy	POLY	0242
	Thermoreversible Gelation of Isotactic Polystyrene.=	POLY	0149
coelasti+ APS-ACS Symposium:	Thermoreversible Gelation of Polymers. Dynamic Vis	POLY	0197
elation+ APS-ACS Symposium:	Thermoreversible Gelation of Polymers. Reversible G	POLY	0145
rganoge+ APS-ACS Symposium:	Thermoreversible Gelation of Polymers. Surfactant O	POLY	0169
rsible G+ APS-ACS Symposium:	Thermoreversible Gelation of Polymers. Thermoreve	POLY	0121
lar Polystyrene.=	Thermoreversible Gelation of Solutions of Stereoregu	POLY	0147
les.=	Thermoreversible Gelation of 60-Crown-20 Macrocyc	POLY	0246
reversible Gelation of Polymers.	Thermoreversible Gelation: Crystallization and Other	POLY	0121
nd Organic Solvents.=	Thermoreversible Gels from Dibenzylidene Sorbitol a	POLY	0201
n: Effect of Network Formation in	Thermoreversible Gels of Polyolefins.= +e to Strai	POLY	0123
lymers and Hydrocarbon Solve+	Thermoreversible Gels with Polyethylene Block Copo	POLY	0249
Self-diffusion in	Thermoreversible Gels.= Probe Diffusion and	POLY	0198
eezing and Melting of Solvent in	Thermoreversible Polymer Gels and Related Systems	POLY	0170
Composite Hydrogel Sheets with	Thermoreversible Transitions.= PTFE-Matrix	POLY	0245
t+ Bioenergetics of Clostridium	thermosaccharolyticum in Relation to Formation of E	BIOT	0029
Ethanol Tolerance of Clostridium	thermosaccharolyticum.= +ogical Fuel Production.	BIOT	0013
Molecular Composites in	Thermosets.=	POLY	0111
etting Systems. Liquid Crystalline	Thermosets, I: Thermal Behavior of Mixtures.=	POLY	0109
New RIM	Thermosetting Resin.= Polydicyclopentadiene: A	POLY	0136
mides Oligomer+ Novel Trends in	Thermosetting Systems. Characterization of bisNadi	POLY	0133
ts, I: Thermal+ Novel Trends in	Thermosetting Systems. Liquid Crystalline Thermose	POLY	0109
mance Epoxies+ Novel Trends in	Thermosetting Systems. Toughening of High-Perfor	POLY	0157
Cure/Property Diagrams of	Thermosetting Systems.=	PMSE	0172
Studies with Particle Beam and	Thermospray LC/MS.= Enhanced Sensitivity	ANYL	0123
talline Polyelectrolytes Displaying	Thermotropic and Lyotropic Behavior.= +uid Crys	POLY	0073
Study of Polystyrene in Good and	Theta Solvents.= +oton Correlation Spectroscopy	POLY	0305
ranching with GPC-Viscometry in	THF and DMF.= +zation of Polymer MWD and B	PMSE	0068
Epoxides with Lithium Cyanide in	THF.= +s Preparation of β-Hydroxynitriles from	ORGN	0107
l Activi+ Synthesis of [1,3,4]Oxa(	Thia)Diazolo[3,2-b]Indazoles and Their Antibacteria	MEDI	0056
micals. 2-(1-Phenoxyethyl)-1,3,4-	Thiadiazoles as Acaricides.= +d Potential Agroche	AGRO	0071
bservation of the Kinetic Fate of	Thiamin Diphosphate-Bound Enamines on Brewers'	BIOL	0048
e Structure Determination of the	Thiamin Diphosphate-Dependent Enzyme Brewers'	BIOL	0032
servation of the Kinetic Fate of a	Thiamin-Bound Enamine Intermediate in Water.=	BIOL	0047
pounds.= Thermally Degraded	Thiamine: A Potent Source of Interesting Flavor Com	AGFD	0071
nthesis of Novel N^6-Substituted	Thiazole and Benzothiazole Adenosine Receptor Ago	MEDI	0111
yl-2-Aryl-1,3-	Thiazolidines.= Synthesis of 5-Perfluoroalkyl-3-Alk	ORGN	0040
Synthesis of Steroidal	Thiazolidones.=	ORGN	0039
Synthesis and Reactions of Δ-	Thiazolines Substituted in the Position 2.=	ORGN	0037
ds.=	Thiazolylamino Acids as Antitrypanosomal Compoun	MEDI	0062
Synthesis of	Thick, Large-Area Diamond Film.=	PHYS	0093
of Associative	Thickeners.= Complexities in Evaluating Behavior	PMSE	0084
stems: Rheological Properties and	Thickening Mechanisms.= +dditives in Coating Sy	PMSE	0034
Copolymer Thin Films.+ Finite	Thickness Effects in the Thermodynamics of Diblock	POLY	0090
es of N^6-Cyclopentyl- and N^6-(2-	Thienyl)Ethyl-Adenosine Agonists. +SAR in a Seri	MEDI	0113
rically Pure 1,2-Epoxy Aldehydes,	Thiiranes, and Aziridines.= +nthesis of Enantiome	ORGN	0052

Surface by ESCA.= Study of	Thin Bicomponent Polymer Films on a Metal	POLY	0100
Studies of Aluminum	Thin Film Growth by MOCVD.= Mechanistic	INOR	0124
re of Ferroelectric Liquid Crystal	Thin Films by Design: The Molecules to Materials C	BIOT	0202
al Formation of Organic–Pigment	Thin Films by Disruption of Micelles.= +trochemic	COLL	0019
s. Deposition of Transition–Metal	Thin Films from Organometallic Precursors.= +ial	INOR	0141
ave CVD Deposition of Diamond	Thin Films with Applications in X-ray Lithography.	PHYS	0092
rization of Inorganic Surfaces and	Thin Films.= +ctron Spectroscopy in the Characte	INOR	0013
Precursor to Tantalum Nitride	Thin Films.= Ethylimidotantalum Complex as	INOR	0074
odynamics of Diblock Copolymer	Thin Films.= +nite Thickness Effects in the Therm	POLY	0090
oup and Transition–Metal Nitride	Thin Films.= +Temperature Synthesis of Main–Gr	INOR	0104
thylene Chains in the Bulk and in	Thin Films.= +ic Dynamics Simulations of Polyme	POLY	0355
ucleation and Growth of Diamond	Thin Films.= +tructural Characterization of the N	PHYS	0121
bic Nonlinear Optics of Polymer	Thin Films, I: Molecular Engineering and Processing	POLY	0011
bic Nonlinear Optics of Polymer	Thin Films, 2: Structure–$\chi^{(3)}$ Relationships in Rigid–	POLY	0205
bic Nonlinear Optics of Polymer	Thin Films, 3: Structure–$\chi^{(3)}$ Relationships in Conjug	POLY	0226
bic Nonlinear Optics of Polymer	Thin Films, 4: Stucture–$\chi^{(3)}$ Relationships in Polyan	POLY	0221
bic Nonlinear Optics of Polymer	Thin Films, 5: Wavelength Dispersion of the $\chi^{(3)}$ of P	POLY	0213
and Chemical Properties of Ultra–	Thin MgO Films on Mo(100).= +Characterization,	COLL	0193
ions, Blends, and Interfaces—IV.	Thin Polymer Films: From Chemistry to Materials S	COLL	0070
angling Bonds.= Silicon Carbide	Thin–Film Formation on Si(100): The Behavior of D	PHYS	0204
omophores in Supported Sol–Gel	Thin–Film Glasses for Second–Harmonic Generation	POLY	0210
rs. Mechanistic Studies of Copper	Thin–Film Growth by MOCVD.= +tallic Precurso	COLL	0025
try: Applications in Polymer– and	Thin–Film Growth.= +me Spectroscopic Ellipsome	POLY	0252
attices: Multilayer Construction of	Thin–Film Nonlinear Optical Materials.= +Superl	POLY	0127
	Thin–Layer Chromatography in Food Analysis.=	AGFD	0180
Tree.= I	Think I Shall Never See a Polymer Lovely as a	POLY	0319
aching and Teaching L+ Critical	Thinking and Achievement in General Chemistry: Re	CHED	0081
tic, and Hydrodynamic—II. Shear	Thinning of Polymer Solutions.= +Electric, Magne	COLL	0159
Face Selection in the	Thio–Claisen Rearrangement.=	ORGN	0239
s.= Synthesis of 3–	Thio–Sugars and Related 3'–Mercapto–Anthracycline	CARB	0032
Di– and Triphosphates and Their	Thioanalogues.= +II) Complexes of Nucleoside–5'–	INOR	0403
Complexes.= S–Oxygenation of	Thioanisoles by (Oxo)(phosphine)ruthenium(IV)	INOR	0130
rected ortho Metalation: The O–	Thiocarbamate as a Versatile New Metalation Direct	ORGN	0083
omium Complexes Containing the	Thiocarbonyl Ligand.= +ructure of Substituted Chr	INOR	0336
by 2–Chloroethyl Sulfides and	Thiodiglycol.= Inhibition of Horseradish Peroxidase	BIOT	0249
Nucleoside Synthesis from	Thioglycosides.=	ORGN	0067
Donors. Selective Activation over	Thioglycosides.= +ides as Novel, Versatile Glycosyl	CARB	0035
tion of+ coli β–Hydroxydecanoyl	Thiol Ester Dehyrase: Determination of the Orienta	BIOL	0066
Site of E. coli β–Hydroxydecanoyl	Thiol Ester Dehyrase.= +as a Probe of the Active	BIOL	0067
ts the Kinetics of Active Site M+	Thiol/Disulfide Exchange in HMG–CoA Lyase Affec	BIOL	0064
ordination Chemistry of Ni(III)–	Thiolate Complexes as Models for Nickel(III)–Cystei	INOR	0179
Complexes.= Hydride	Thiolate Derivatives of Low–Valent Iron	INOR	0225
a Mixed–Valence [Ni(II)–NI(IIII)]	Thiolate Dimer.= +acterization, and Reactivity of	INOR	0060
Nucleophilic Addition of Alkyl	Thiolate Ions to β–Nitrostyrenes: Radicaloid Transit	ORGN	0155
ith Electrophiles.= Ruthenium	Thiolates cpRu(PR$_3$')$_2$SR: Synthesis and Reactions w	INOR	0368
Glutathione and Other	Thiols by Mitomycin C.= Bioreductive Alkylation of	BIOL	0088
Reactivities of Epoxides Toward	Thiols: Applications to Papain Inhibitors.=	MEDI	0071
1-5-Methyl-2-Phenyl-Pyrazol-3-	Thione and Tri–n–Octylphosphine Oxide.= +enzoy	NUCL	0073
mpounds by 3–Acylthiazolidine–2–	thiones.= +Acylation of Di– and Polyhydroxyl Co	ORGN	0038
aminophenothiazinium Chloride (	Thionine).= +terization of a Polyimide from 3,7–Di	POLY	0033
ediynes via the Elimination of a	Thionocarbonate in a Latent Z–Hex–3–ene–1,5–diyne	ORGN	0096
mers with Skeletal Phos+ Poly(	thionylphosphazenes): A New Class of Inorganic Poly	POLY	0186
logical Evaluation of a Series of	Thiophene–Derived 3–Hydroxy–3–Methylglutaryl–Co	MEDI	0101
ydro–1–(4–hydroxy–1–naphthyl)	thiophenium Hydroxide Inner Salt and β–Butyrolact	POLY	0048
Pharmacodynamics of Phenytoin,	Thiophenytoin, and Iminophenytoin.= +mparative	MEDI	0134
phatase.=	Thiophosphate Inhibitors of Myo–Inositol Monophos	MEDI	0119
of	Thioridazine.= Synthesis of Phenolic Derivatives	MEDI	0132
mpounds Derived from 2–Methyl–	thiosemicarbazide.= +d Molecular Structure of Co	INOR	0032
ng Control of Spin–State: Ferric	Thiosemicarbazones—Comparison with Spin–Crossov	INOR	0268
Surfaces.= Investigation of	Thiourea Derivatives Chemisorption on Silica	COLL	0106
H)Quinazolinone–2,3–Dihydro–2–	Thioxo.= +tion Compounds of the Heterocyclic 4(1	INOR	0079
tions by Flash Phot+ Kinetics of	Thiyl Radical Formation and Electron Transfer Reac	INOR	0396
	Thomas Edison—Chemist.=	CHAL	0026
Contributions of Arthur W.	Thomas to American Chemistry.=	HIST	0039
evada Test Site.=	Thorium–230 Chronology of Natural Waters at the N	NUCL	0084
t Are New, a Revitalized Physi+	Though the Questions Are Old, with Techniques Tha	CHED	0043
Organolithium Compounds to N–	THP–Protected Nitrone.= Addition of	ORGN	0227
e Markets, or How to Survive and	Thrive in the Midst of Bigness.= +ialties for Nich	CMEC	0005
itation of Argatroban, a Specific	Thrombin Inhibitor, by a Tandem Column HPLC Pr	ANYL	0086
echanism of Resistance to Bacillus	thuringiensis.= +Bioengineered Control Agents. M	AGRO	0153
phthalene Containing Inhibitors of	Thymidylate Synthase.= +ign and Synthesis of Na	MEDI	0186
Its Biolo+ Site–Specific Effect of	Thymine Dimer Formation on A–Tract Bending and	BIOL	0114
omplex Kinetics of Oxidation of	Thymol Blue with Bromate Ion in Acidic Solutions.=	PHYS	0269
Ethylene by	Ti(II) Complexes.= Catalytic Oligomerization of	INOR	0153
tural Characterization of the First	Ti(II) Metalloporphyrin Complexes.= +, and Struc	INOR	0298
diments: Reductive Dissolution by	Ti(III).= +Iron Oxide Extraction from Soils and Se	ENVR	0058
Reactions of Group IVB (	Ti, Zr, Hf) Alkyls with Aluminum Reagents.=	INOR	0054
Photochemistry of (Cp)	2TiCl$_2$ on Porous Vycor Glass.=	INOR	0161
Determinacion Cuantitativa de	Tierras Raras por Espectrometria de Masas con Fuen	ANYL	0051
iempirical Molecular Orbital and	Tight Binding Methods in Diamond Surface Reactivi	PHYS	0356

lution ^{11}B NMR Observation of	Tightly Bound Boronic Acid Inhibitors on Serine Pro	BIOL	0082
Residues of	Tilmicosin in Cattle.= Metabolism and Tissue	AGRO	0110
cal Reactions on the Femtosecond	Timescale.= +y Dissipation in Liquid–State Chemi	PHYS	0085
Solvated Metal A+ Platinum–	Tin and Gold–Tin Bimetallic Particles Prepared from	CATL	0009
al A+ Platinum–Tin and Gold–	Tin Bimetallic Particles Prepared from Solvated Met	CATL	0009
erties of Platinum and Platinum–	Tin Catalysts Supported on Mg, Cu, Zn, and Ni Alu	COLL	0126
Bioinorganic Models of	Tin Chelation.=	INOR	0052
Electrochemical Removal of a	Tin Layer from Copper Substrate.=	PHYS	0285
Sol–Gel Condensation of	Tin(IV) Alkoxide Compounds.=	POLY	0258
d Sp+ Toward a Molecular–Size	Tinkertoy Construction Set: Assembly of the Rods an	BIOT	0200
rived+ Toward a Molecular–Size	Tinkertoy Construction Set: Polymeric Structures De	POLY	0344
izatio+ Toward a Molecular–Size	Tinkertoy Construction Set: Synthesis and Character	ORGN	0124
Potential and Surface Charge of	TiO$_2$ (Anatase) Suspensions as a Result of Exposure	COLL	0203
e+ Physical Characterization of	TiO$_2$ and Al$_2$O$_3$ Prepared by Precipitation and Sol–G	PETR	0092
terogeneous Photocatalysis with	TiO$_2$ Semiconductor Particles in Aqueous Media: A L	COLL	0131
Forming of High–Surface–Area	TiO$_2$ to Catalyst Supports.=	PETR	0091
idation of Trichloroethylene over	TiO$_2$: A Potential Technique for the Solar Detoxifica	COLL	0082
lysis of Reduced TiO$_2$(110) and	TiO$_2$(001) Surfaces by Scanning Tunneling Microscop	COLL	0171
l Structural Analysis of Reduced	TiO$_2$(110) and TiO$_2$(001) Surfaces by Scanning Tunn	COLL	0171
Scattering.= Studies of	TiO$_2$(110) Surface Reactivity Using Low–Energy Ion	COLL	0172
s. Ultrathin Copper Overlayers on	TiO$_2$(110).= +try of Nonmetals—Novel Technique	COLL	0191
s Mixtos de Alumina–+ HDS de	Tiofeno con Catalizadores de Molibdeno sobre Oxido	FUEL	0105
Sintesis de	Tiofosfatos de Naftilo, Posibles Garrapaticidas.=	MEDI	0064
Contenido de Galio en Zeolitas del	Tipo Pentasil en la Aromatizacion de Propano.=	COLL	0125
Solutions of a 21–Aminosteroid (	Tirilazad).= +adation Products Occurring in Acidic	ANYL	0085
Max	Tishler—Medicinal Compounds.=	CHAL	0031
am Alkoxide Precurso+ Hydrous	Titanate Ceramic Powders from Mixed Alkali/Titani	INOR	0278
Using a Type 2	Titanate.= Synthesis and Catalytic Activity When	PETR	0017
Reactivity of	Titanatranes.= Synthesis, Characterization, and	INOR	0191
NiW Soportados sobre Alumina o	Titania Modificadas con Fluor.= +de Catalizadores	FUEL	0106
sobre Oxidos Mixtos de Alumina–	Titania.= +iofeno con Catalizadores de Molibdeno	FUEL	0105
Surface Characterization of	Titania–Silica–Based DENO$_x$ Catalysts.=	PETR	0071
er Activation Barriers at Colloidal	Titanium Dioxide Surfaces.= +s of Electron–Transf	COLL	0132
tudies on Pd and Ni/Mo Hydrous	Titanium Oxides.= +tion and Model Compound S	CATL	0039
Molecular Precursors to	Titanium Oxides.= Polyoxoalkoxytitanates:	INOR	0174
Reactivity of Nickel–	Titanium Shape Memory Alloys.=	INOR	0357
ETS–10: A New Wide–Pore	Titanium Silicate Molecular Sieve.= Synthesis of	PETR	0078
ETS–10: A New Large–Pore	Titanium Silicate.= Adsorptive Properties of	PETR	0079
New Series of Mixed–Valance	Titanium(III/IV) Phosphate Compounds with Nasico	INOR	0353
a Surface of a Pure and Alumino(	Titano)–Containing Silicas.= +f Silanol Groups on	COLL	0112
d Structural Characterization of	Titanocene–Containing Polymers Derived from α–Am	PMSE	0014
nthesis and Chemistry of Chiral	Titanocenes: 1–Methyl, 2–Phenyltitanocene Dichlorid	INOR	0281
Properties of Pyrogenic	Titanosilicas.= Peculiarities of Structure and	COLL	0111
Aliovalent Analogue of Potassium	Titanyl Phosphate (KTP).= +: A Centrosymmetric	INOR	0256
nstrume+ Computer–Interfaced	Titration Experiment for the Physical Chemistry or I	CHED	0161
anions with Direct Polyelectrolyte	Titration.= +terization and Trace Analysis of Poly	COLL	0072
Mn^{2+} in Cubic [(Rb18c6)$_4$MnCl$_4$][	TlCl$_4$]$_2$ in Presence of Cu^{2+}.= +Characteristics of	INOR	0039
omagnetic Transition in Metallic	TlCo$_2$S$_x$Se$_{2-x}$ ($0 \le x \le 2.0$) with the ThCr$_2$Si$_2$–Type	INOR	0058
Novel R$_2$Ba$_{1.25}$NiO$_{5.25}$ (R =	Tm, Yb, Lu) Structure Type.=	INOR	0034
ion in Crystals of (CH$_3$)$_4$NMnCl$_3$(	TMMC).= +Solid: Exciton Trapping and Annihilat	PHYS	0131
Behavior and Development of the	Tobacco Budworm.= +micals and Nutrients on the	AGRO	0037
of an Allatostatin from the	Tobacco Hornworm, Manduca sexta.= Identification	AGRO	0008
igenesis by+ Protection against	Tobacco–Specific Nitrosamine–Induced Lung Tumor	AGFD	0068
of Pharmacological Activit+ New	Tocainide Chiral Analogues: A Possible Dissociation	MEDI	0098
ation.=	Tocopherol as a Natural Phenolic Inhibitor of Nitros	AGFD	0091
otenoid+ Antioxidant Activity of	Tocopherols and Tocotrienols on Plant Colorant Car	AGRO	0039
d Hypocholesterolemic Activity of	Tocotrienols and Related Benzopyrans.= +hesis an	MEDI	0190
idant Activity of Tocopherols and	Tocotrienols on Plant Colorant Carotenoids.= +iox	AGRO	0039
y Polymers with Pendant Sulfonyl	Tolan Groups.= +ptical Properties of Linear Epox	POLY	0208
logical Fuel Production. Ethanol	Tolerance of Clostridium thermosaccharolyticum.=	BIOT	0013
Compounds for Infra–Red Reflow	Tolerant Electronic Components.= +astic Molding	CMEC	0014
ugh Global Swaps, Exchanges, and	Toll Conversions.= +Purchasing Productivity thro	CMEC	0007
f Lu[CH(SiMe$_3$)$_2$]$_3$+ Unusual η^6–	Toluene Coordination to an Alkali Metal: Strucuture o	INOR	0053
ffect of Si/Al Ratio and Coke +	Toluene Disproportionation Reaction over Zeolites: E	PETR	0103
tion of p–Ethyltoluene during the	Toluene Ethylation on Modified ZSM–5.= +sforma	PETR	0086
ling the Two–Phase Nitration of	Toluene in a Co–Current Packed Tubular Reactor.=	I&EC	0006
d Degradation Products Using a	Toluene Monooxygenase Constitutive Pseudomonas c	BIOT	0119
ion of Vanadyl Acetylacetonate in	Toluene.= +th Study of the Anisotropic Reorientat	PHYS	0264
igation of Soot Formation from	Toluene/Methanol Mixtures.= A Shock Tube Invest	FUEL	0113
o–	Tolunitrile.= In Vitro Metabolism Studies of	MEDI	0033
Butteroil.= Polymer–Supported	Tomatine as a Means of Cholesterol Removal from	AGFD	0135
hydrate–Containing Regulators of	Tomato Fruit Ripening.= +turally Occurring Carbo	AGFD	0031
Evaluation of Transgenic	Tomatoes with Reduced Polygalacturonase.=	AGFD	0057
m Stages. Integration: A Must for	Tomorrow's Bioprocesses.= +Upstream–Downstrea	BIOT	0069
Zeolite	TON.= Precursor Phases and Nucleation of	PETR	0040
Effects against Liver, Colon, and	Tongue Carcinogenesis by Plant Phenols.= +tective	AGFD	0090
Libres con	Tonsil: Catalisis de Reacciones via Radicales	COLL	0119
Fenol con Aductos	Tonsil–Fe(NO$_3$)$_3$.= Estudio de la Nitracion del	COLL	0120
Spectroscopy as an Analytical	Tool for Hydrocarbon Identification.= Derivative	PETR	0137
fusion Chromatography: A New	Tool for Large–Scale Biomolecule Purification.= Per	ANYL	0032

Life Cycle Analysis: A	Tool for Pollution Prevention.=	CEI	0027
on Chromatography®: A Novel	Tool for Protein Purification and Analysis.= Perfusi	BIOT	0073
ntation.+ The MSU Miniball: A	Tool for the Characterization of Nuclear Multifragme	NUCL	0035
as Data Acquisition Devices—A	Tool for the Revitalization of the Physical Chemistry	CHED	0160
Birefringence as a Diagnostic	Tool in Viscoelastic Media.= Dynamic Magnetic	COLL	0143
ion Spectroscopy as an Analytical	Tool: Instrument Requirements and Salient Results.	NUCL	0042
Analytical/Educational	Tool.= PIXE as an	NUCL	0078
Teaching	Tool.= Uncommon Reactions: A Useful	CHED	0166
A	Tool–Oriented Approach to Computing.=	CHED	0078
Hypermedia	Toolkit for Structure Elucidation.=	COMP	0008
and Artifact Corrosion Studies as	Tools for Biomineralization Research.= +domorph	BIOT	0101
—Or, How We Go from the Bench	Top to Commercial Reality.= +Forming Processes	PETR	0089
elman (Medicine and Physiology+	Topbiology: Molecular Bases of Animal Form. G. Ed	CONG	0013
An Effective	Topic for Precollege Science.= Polymer Technology:	CHED	0007
ation, Characterization, and Other	Topics. +n Energy from Waste and Coal: Ash Utiliz	FUEL	0136
Novel 4'-Demethylepipodophyl+	Topoisomerase Inhibition and Anticancer Activity of	MEDI	0011
Vertex Invariants for	Topological Indices.= Using Real Numbers as	COMP	0026
olubilization as Detected by J+	Topological Requirements of Premicellar Porphyrin S	ORGN	0161
mulation: Role of Complementary	Topological Strategies.= +d ^{13}C NMR Spectrum Si	COMP	0006
e) and Poly(methylph+ Effect of	Topology for Cyclic and Linear Poly(dimethylsiloxan	POLY	0234
of	Torands.= Structural Evidence for the Rigidity	INOR	0297
Tri-Aryl-Substituted	Torands.= Synthesis of	ORGN	0064
e Diagnostics and Modeling of 10-	Torr Methane/Oxygen Flames.= +ced Fluorescenc	FUEL	0128
Relaxation in Condensed Phases.	Torsional Solitons in Organic Crystals.= +sfer and	PHYS	0129
and Characterization of CTBN-	Toughened Epoxy Resin/Castor Oil Polyurethane IP	PMSE	0092
erties and Morphology of Rubber-	Toughened Polystyrene.= +ion on the Impact Prop	PMSE	0177
ends in Thermosetting Systems.	Toughening of High-Performance Epoxies Using Des	POLY	0157
tion, Characterization, and	Toughness of Ethynyl-Containing Blends.= Prepara	POLY	0040
o-Gasoline Plant: An Engineering	Tour de Force.= +J. F. Roth. New Zealand Gas-t	I&EC	0060
State.= Design for Packed	Tower Absorbers Using the PRSV Equation of	CHED	0107
of Disperse Blue+ Reduction of	Toxic Components and Wastewaters in Manufacture	CMEC	0018
r the Immobilization of Solidified	Toxic Hazardous Wastes.= +of Cement Matrices fo	ENVR	0002
ing Zero Discharge of Persistent	Toxic Substances in Great Lakes Areas of Concern.=	ENVR	0083
enyl in an in Vitro Developmental	Toxicity Assay.= +ce on Metabolic Profile of Biph	AGRO	0021
ribolium castaneum L. (Coleopt+	Toxicity of Annona squamosa Seed Oil Extracts to T	AGRO	0036
Refinement of Animal	Toxicity Studies for Human Risk Assessment.=	CHAS	0019
cacious Insecticides with Low Fish	Toxicity.= +ryl-1-Cyclopropylbutanes: Highly Effi	AGRO	0073
mpound Residues: Analytical and	Toxicological Aspects.= +Hormone and Related Co	AGRO	0129
cation Resources, Inc.=	Toxicological Services Provided by Hazard Communi	CHAS	0024
Milestones of Forensic	Toxicology.=	HIST	0002
Models for the Ene-Diyne	Toxines with an Enol-Ether Trigger.=	ORGN	0102
Chemistry. Arthur Dock Fon	Toy—Scientist, Colleague, Friend.= Phosphorus	INOR	0288
of W(CO)$_6$ on Pt (111).=	TPD and XPS Study of the Thermal Decomposition	COLL	0049
+of the Replicative Bypass of the	TpdU cis-syn Cyclobutane Photodimer of DNA.=	BIOL	0113
ddition of Cu^{2+} to the Cation Cu(	tpen)$^{2+}$ to Form Several Binuclear Complexes.= +A	INOR	0399
tallographi Caracterization of [M(	tpen))(ClO$_4$)$_2$ Complexes.= +pectroscopic and Crys	INOR	0057
ammed Oxidative Pyroanalysis (	TPOP): A New Analytical Method to Predict the Per	FUEL	0046
vy Petroleum Residues.=	TPOP-IRMS Stable Carbon Isotope Analysis of Hea	PETR	0141
drodemetallization (HDM) of NI-	TPP and VO-TPP.= +nadium Catalysts in the Hy	CATL	0044
tion (HDM) of NI-TPP and VO-	TPP.= +nadium Catalysts in the Hydrodemetalliza	CATL	0044
s for Fossil Fuels. High-Pressure	TPR Apparatus to Investigate Organic Sulfur Forms	FUEL	0007
Phosphorodithioate d(CGCTpS$_2$-	TPS$_2$-AAGCG): An Unusual Hairpin Loop.= +tide	BIOL	0123
trometry for the Determination of	Trace Alkylselenide Compounds.= +sorption Spec	ANYL	0080
lyte Titrat+ Characterization and	Trace Analysis of Polyanions with Direct Polyelectro	COLL	0072
Organic/Inorganic Associations of	Trace Elements in Oil Shale Kerogen.= +nation of	FUEL	0045
an Blood by Using+ Measuring	Trace Levels of Volatile Organic Compounds in Hum	CEI	0001
	Trace Metal Nutrition of Marine Phytoplankton.=	INOR	0003
pled Plasma Atomic Emission +	Trace Metals Analysis of Fly Ash by Inductively Cou	FUEL	0138
Determination of	Trace Metals in Liquefied Petroleum Gas.=	PETR	0139
organic Chemistry. Distribution of	Trace Metals in the Ocean.= +ial on Marine Bioin	INOR	0001
Hazardous Materials	Tracking System.= Chemical Inventory and	CHAS	0003
Thymine Dimer Formation on A-	Tract Bending and Its Biological Implications.= +f	BIOL	0114
ria for Successful Intracontinental	Trade in the 21st Century.= +inental Trade. Crite	CMEC	0001
i+ Dynamics of Intracontinental	Trade. Criteria for Successful Intracontinental Trade	CMEC	0001
Fermentations from Shake Fla+	Traditional and Novel Applications of Actinomycetes	BIOT	0083
Reaction.= Quasiclassical	Trajectory Studies of the H + CH$_4$ ← H$_2$ + CH$_2$	PHYS	0263
al Characterization of Protonated	trans Bis-Dioxime Ruthenium Complexes.= +ructur	INOR	0066
^{3}C-NMR Assessment of the cis-	trans Isomerization of Gly^6-Bradykinin and Analogs.	BIOL	0006
d NMR Analyses of the cis and	trans Isomers of a Bicyclic Lactone Mosquito Repelle	AGRO	0033
yl Chlor+ Relaxation of Runs of	Trans Rotational Isomers in Single Racemic Poly(vin	POLY	0069
n Liquid Polysiloxanes Containing	trans-Cyclohexane-Based Mesogenic Groups.= +ai	POLY	0057
Phenanthrenes: Synthesis of the	trans-3,4-Dihydrodiol of the Mutagen 15,16-Dihydro	ORGN	0118
hetic Strategy.=	Transannular Diels-Alder Reaction: A Powerful Synt	ORGN	0288
Aspartate	Transcarbamoylase.= Regulatory Communication in	BIOT	0242
7: A Potent and Selective Reverse	Transcriptase Inhibitor of HIV-1.= +of BI-RG-58	ORGN	0027
dium Berghei-Infected Mice with	Transdermal Artelinic Acid.= +reatment of Plasmo	MEDI	0063
ad(II) and Lanthanum(III) RNA	Transesterification Catalysts and the Design of Artif	INOR	0406
tramolecular Electron and Energy	Transfer?.= +ng in Predicting Absolute Rates of In	ORGN	0143
Evolution of Glutathione S-	Transferases in Drosophila.=	AGRO	0100
Adapted Glutathione	Transferases in Lepidoptera.= Plant Allelochemical-	AGRO	0119

terizacion, y Determinacion de la	Transferencia de Carga por Espectroscopia IR de Co	INOR	0190
spectroscopia IR de Complejos de	Transferencia de Carga Pt(R-Nafin)·TCNQ.= +r E	INOR	0190
Energia Libre de	Transferencia de CuSO$_4$ en Agua/Agua–Acetona.=	PHYS	0236
he Feasibility of Developing and	Transferring New Energy Technology to the Marketp	FUEL	0142
Ester.= Photoinduced Electron	Transfers within Conjugated Porphyrin–Flavin	BIOL	0134
Profiling of Polymers with Fourier	Transform Infrared Reflectance Spectroscopy.=	PMSE	0153
Analyzing F+ Pyrolysis–Fourier	Transform Infrared Spectrometry as a Technique for	FUEL	0061
muir–Blodgett Films by Fourier	Transform Infrared Spectroscopy and Second–Harmo	POLY	0224
eation Chromatography–Fourier	Transform Infrared Spectroscopy for Characterizatio	PMSE	0048
ctance Study of Oxidiz+ Fourier	Transform Mid– and Far–Infrared IR Specular Refle	POLY	0253
duate+ Incorporation of Fourier	Transform NMR Instrumentation into the Undergra	CHED	0148
n Secondary Structure via Fourier	Transform Raman Spectroscopy.= +ation of Protei	POLY	0306
Recent Advances in Fourier	Transform Raman Spectroscopy.=	POLY	0254
Laser and Fourier	Transform Spectroscopy of Transient Molecules.=	PHYS	0057
Argonne Premium Coals during+	Transformation Kinetics of Organic Sulfur Forms in	FUEL	0059
c Aromatic Hydrocarbon–Induced	Transformation of Mammalian Cells.= +Polycycli	CHAS	0007
imary Formation and Secondary	Transformation of p–Ethyltoluene during the Toluen	PETR	0086
roducts.=	Transformation of Solid Urban Residues in Usable P	CHED	0105
itu Monitoring of the Degree of	Transformation of VPI-5 to AlPO$_4$-8 at Various Tem	PETR	0095
Naphthene	Transformation over H–[Al]–ZSM-5 Catalysts.=	PETR	0044
and in Vitro Effects on Oncogenic	Transformation.= +ulation, Metabolic Processing,	AGFD	0117
Pyrogenous Silica.= Thermal	Transformations of Aziridine on the Surface of	COLL	0113
Chemisorption–Induced Phase	Transformations of Metal Surfaces Studied by STM.	PHYS	0126
Complexes.= Plenary Session.	Transformations of Organic Compounds Mediated by	ORGN	0127
ctivation Barriers of Unimolecular	Transformations of Organic Compounds.= +t on A	COLL	0107
Chemical	Transformations of Triterpenes from Guayule.=	ORGN	0019
cr Enantioselective Metal Carbene	Transformations.= +arkably Effective Catalysts f	CATL	0002
s and Synthesis.= Organic	Transformattions Involving Organocobalt: Mechanism	ORGN	0141
εter–Calcium Chloride in a Heat	Transformer.= Performance Study of the System W	I&EC	0009
se.= Evaluation of	Transgenic Tomatoes with Reduced Polygalacturona	AGFD	0057
Hydrocarbons under	Transient Conditions.= Catalytic Reactions of	PETR	0134
	Transient Electric Birefringence of Agarose Gels.=	COLL	0182
rene–Sodium–2–Acrylamido–2+	Transient Electric Birefringence of Ionomers Poly(sty	COLL	0181
Ring DNA vs. Linear DNA from	Transient Electric Birefringence.= Dynamics of	POLY	0296
Electric Fields: Dynamics from	Transient Electric Birefringence.= Polymers in	COLL	0180
zation of Polymer Solutions under	Transient Elongational Flow.= +ingence Characteri	POLY	0281
Vibrational Dynamics Using	Transient Infrared Spectroscopy.= Ultrafast	PHYS	0194
ns in Protic Solvents Measured by	Transient Infrared Spectroscopy.= +ion Time of Io	PHYS	0083
)$_6$ in Solution.= Picosecond	Transient IR Spectroscopy of UV–Photolyzed Cr(CO	PHYS	0220
yzed CO$_2$ Gasification of Carbon:	Transient Kinetic Investigations.= Potassium–Catal	FUEL	0031
n in CO$_2$.=	Transient Kinetic Study of the Gasification of Carbo	FUEL	0009
Insight in the Mechanism+ Only	Transient Kinetic Techniques Will Provide Detailed	FUEL	0015
face Area of Carbons.= Use of	Transient Kinetics for Determining the Reactive Sur	FUEL	0014
Spectroscopy of	Transient Molecules.= Laser and Fourier Transform	PHYS	0057
amics of Electric–Field–Induced	Transient Phase–Separation in w/o–Microemulsion.=	COLL	0146
Molecular Sieves: Identification of	Transient Phases.= +Solution Synthesis of AlPO$_4$	PETR	0020
als of YBa$_2$Cu$_3$O$_6$.3: Photodopi+	Transient Photoinduced Conductivity in Single Cryst	PHYS	0011
y.= Subpicosecond	Transient Spectroscopy of Rhodopsin Photochemistr	BIOT	0080
Photionization via	Transient States: The Photophysics of Azulene.=	PHYS	0330
Molecular Electronic	Transition Current Density.=	BIOT	0026
g E+ Accurate MRCI State and	Transition Dipole Moments for the Nine Lowest Lyin	PHYS	0304
etic Interactions in Second–Row	Transition Elements: Crystal Structures of Two Chlo	INOR	0267
ause a T←R–Like Conformational	Transition in Insulin Hexamers.= +Paraben Will C	BIOT	0150
th+ Antiferro–to–Ferromagnetic	Transition in Metallic TlCo$_2$S$_x$Se$_2$-x (0 ≤ x ≤ 2.0) wi	INOR	0058
c Pseudohexagonal Rotator–Phase	Transition in n–Heneicosane.= +the Orthorhombi	POLY	0354
ation Spectra Pattern and Phase	Transition in Tetracene Crystals and Tetracene–Dop	PHYS	0274
ure NMR Studies of the Glass	Transition in the Diamond Anvil Cell.= High–Press	PHYS	0260
Organic Compounds Mediated by	Transition Metal Complexes.= +ansformations of	ORGN	0127
Organic Synthesis Using Group 4	Transition Metal Complexes.=	ORGN	0134
Electron Delocalization in Early	Transition Metal Heteropoly– and Isopolyoxometalat	COMP	0032
rs: Unusual Coordination Behav+	Transition Metal Hydrides. Palladium Hydride Dime	INOR	0379
g+ NMR Studies of Nonclassical	Transition Metal Polyhydrides: Rotational Tunnellin	PHYS	0068
Structure–Function Relations in	Transition Metal Sulfide Catalysts.=	CATL	0038
rated Ketene Intermediates Using	Transition Metal–Mediated Processes.= +a Unsatu	ORGN	0274
ification of Biomass.=	Transition Metals as Catalysts for Pyrolysis and Gas	FUEL	0103
ixed Donors P$_2$S Toward Various	Transition Metals.= +of a Tripodal Ligand with M	INOR	0028
Estimation of Density and Glass	Transition of Amorphous Polymethylene.=	POLY	0134
in+ Surface Diffusion and Phase	Transition on the Ge(111) Surface Studied by Scann	PHYS	0099
Phosphorylase: Mechanism and	Transition State Analysis.= Purine Nucleoside	BIOL	0046
as Functions of Size of the Cyclic	Transition State and C–H–C Angle.= +en Transfer	ORGN	0175
ination with a Carbocation–Like	Transition State for the Reaction of Tertiary Cumyl	ORGN	0153
n an Analogue of the Phosphorane	Transition State in RNase?.= +the Vanadate Anio	BIOL	0061
	Transition State of Nucleoside Hydrolase.=	BIOL	0073
s to β–Nitrostyrenes: Radicaloid	Transition State or Principle of Nonperfect Synchron	ORGN	0155
the Best Possible Pa+ Modeling	Transition States with Force Fields: Development of	ORGN	0170
/Matrix Interface on Local Glass–	Transition Temperature in Composites.= +of Fiber	COLL	0055
ect of Molecular Weight on Glass–	Transition Temperature.= +ocyclic Polystrene: Eff	POLY	0239
Solubility Parameters and Glass	Transition Temperatures on the Phase Morphology o	PMSE	0025
Reactivity with Hydrocarbons under	Transition–Metal Atoms in Different Charge States:	PHYS	0199
Roo+ Association Reactions of	Transition–Metal Atoms with Simple Molecules near	PHYS	0198

atalyzed Photochemically by d⁸	Transition-Metal Carbonyls in the Presence of Arom	INOR 0254
the Reaction Cross-Sections of	Transition-Metal Cations: V⁺ + Ethane, Ethene, and	PHYS 0306
h+ Tuning of Electron Density at	Transition-Metal Centers via Poly 3'-(4-pyridyl)tert	INOR 0027
tron Transfer Reactions between	Transition-Metal Complexes and Redox Sites in Enz	INOR 0283
Nonlinear Optics.= Group IV	Transition-Metal Complexes for Third-Order	INOR 0334
ct+ Liquid Crystalline Pyramidic	Transition-Metal Complexes: Potential New Ferroele	INOR 0135
Antineoplastic Effect of Several	Transition-Metal Complexes.=	INOR 0026
Calculations of the Geometries of	Transition-Metal Complexes.=	PHYS 0192
orce-Field Study of Mixed Ligand	Transition-Metal Compounds.= +Mechanics: A F	INOR 0075
and Chemical Bonding in Ternary	Transition-Metal Compounds.= +tronic Structure	PHYS 0252
End-on (μ-N₂) Bonding of Early	Transition-Metal Dinitrogen Complexes.= +₂) vs.	INOR 0362
elective Fluorinating Agents for+	Transition-Metal Fluorides and Oxide Fluorides as S	INOR 0133
mistry: S+ Direct Lanthanide to	Transition-Metal Interactions in Organometallic Che	INOR 0332
Comparison of Main Group and	Transition-Metal Ion Chemistry.=	PHYS 0200
and Xerogels Doped with	Transition-Metal Ions.= Studies on Silica Aerogels	INOR 0081
ure Synthesis of Main-Group and	Transition-Metal Nitride Thin Films.= +Temperat	INOR 0104
Syntheses and Caracterization of	Transition-Metal Open Fulvalene Complexes.=	INOR 0113
and Reactivity.= Early	Transition-Metal Tellurolates: Synthesis, Structure,	INOR 0021
organic Materials. Deposition of	Transition-Metal Thin Films from Organometallic P	INOR 0141
oatom-Substituted Cyclohexadi+	Transition-Metal-Catalyzed Polymerization of Heter	POLY 0047
Catalyst Stability in Homogeneous	Transition-Metal-Catalyzed Processes.= +Role on	INOR 0349
Reactions: Reactive Scattering and	Transition-State Theory.= +Theory of Chemical	PHYS 0161
s-Strain Relationships and Phase	Transitions for Networks of Semirigid Chains.= +es	COLL 0016
s for+ Acousto-Conformational	Transitions in Cytoskeletal Microtubules: Implication	BIOT 0055
+ope Effect and Structural Phase	Transitions in LA₂CuO₄-Based Superconductors.=	PHYS 0010
,L-glutamate+ Multiple Thermal	Transitions in Model Compounds for Poly(γ-alkyl-α	POLY 0099
onance Raman Studies of the Qy	Transitions in Rhodobacter sphaeroides Reaction Ce	BIOL 0094
Shear-Induced Structural	Transitions in Triblock Copolymer SBS.=	PMSE 0136
Pressure-Induced Micellar Phase	Transitions.= +ery from Reversed Micelles through	BIOT 0046
Sheets with Thermoreversible	Transitions.= PTFE-Matrix Composite Hydrogel	POLY 0245
Chemistry.= Internal and	Translational Energy Effects on Cluster Ion	PHYS 0242
₂O Vibrational, Rotational, and	Translational Energy Transfer from Highly Vibration	PHYS 0334
Mortality in CD1 Immunodefici+	Translocation of Gram-Negative Bacteria and High	AGFD 0146
racterization of Degradation and	Translocation Processes in Soil as a Function of Clim	AGRO 0025
ide DPX-E9636 in Corn: Uptake,	Translocation, and Metabolism.= +ty of the Herbic	AGRO 0027
pper-Catalyzed Conjugate Addi+	Transmetalation Reactions of Alkyl Zirconocenes: Co	ORGN 0089
Diamond.= Applications of	Transmission Electron Microscopy to the Study of	PHYS 0120
Materials Using Scanning and	Transmission Electron Microscopy.= Analysis of	INOR 0014
orphology of Polymer Chains by	Transmission Electron Microscopy, II: Image Contras	POLY 0062
-Nucleus Collisons.=	Transparency and Abrasion in High-Energy Nucleus	NUCL 0048
7 MASNMR to the Processes of	Transparent Aluminum Film Preparation through Aq	COLL 0115
d Spectroscopy of a New Infrared	Transparent Chalcogenide: CaYbInSe₄.= +cture an	INOR 0030
Composites.= Preparation of	Transparent Glass Fiber Reinforced PMMA Matrix	PMSE 0109
racterization of Diam+ Visible-	Transparent Rigid-Rod Polymers: Synthesis and Cha	POLY 0028
near Polymers Doped with Charge	Transport Agents.= +otorefractive Effect in Nonli	POLY 0155
a+ Excited-State Structure and	Transport Dynamics of Photosynthetic Units: Spectr	PHYS 0018
citon Trapping and Ann+ Energy	Transport in a One-Dimensional Inorganic Solid: Ex	PHYS 0131
d State and Annealing.= Gas	Transport in IPN Membranes: Effect of Cross-Linke	PMSE 0060
tric Investigation of Proton	Transport in Ordered Polymer Lattices.= Voltamme	COLL 0167
Electronic Excitation	Transport in Photosynthetic Antennae.=	PHYS 0020
Mechanisms of Gas	Transport in Poly(1-Trimethylsilyl-1 Propyne).=	POLY 0119
Charge	Transport in Polymer Bilayer Films.=	ANYL 0104
afast Studies of Electronic Energy	Transport in Spatially Disordered Systems.= +Ultr	PHYS 0133
Structure upon the Sorption and	Transport of Small Molecules in Stiff-Chain Glassy	POLY 0120
tes.= Elementary Surface	Transport Processes Involving Molecular Chemisorba	PHYS 0345
l Fields.= Mass Transfer and	Transport Properties of Colloidal Systems in Externa	COLL 0142
id Surfac+ Thermodynamics and	Transport Properties of Colloids in Contact with Sol	COLL 0162
f Molecular Architecture on Gas	Transport Properties of Liquid Crystalline Polymers.	POLY 0115
ioning Effect of CO₂ on the Gas	Transport Properties of Polyimide Isomers.= Condit	POLY 0116
n Bro+ Magnetic and Electronic	Transport Properties of the Monophosphate Tungste	INOR 0352
enium-Based Fische+ Reaction-	Transport Selectivity Models and the Design of Ruth	CATL 0056
m of Particulate-Enhanced Mass	Transport to Hydrodynamic Electrodes.= Mechanis	COLL 0041
Dispersive Energy	Transport: Relation to Spin Glass Relaxation.=	PHYS 0082
abolite Production by Redirecting	Transport.= +Cultures. Increasing Secondary Met	BIOT 0228
r Vectorial Photoinduced Electron	Transport.= +ally Polymerized Porphyrin Films fo	BIOT 0126
ize Effects in Aerosol Generation,	Transport, and Vaporization for Particle Beam LC/	ANYL 0122
NO-Cataliticas.= Influencia del	Transporte de Calor en Reacciones Solido-Gas	I&EC 0068
ntial Agents Active on Taurine	Transporting System.= Taurine Derivatives as Pote	MEDI 0135
yacetates: A Novel Type of Allylic	Transposition Catalyzed by Eu(fod)₃.= +ic Methox	ORGN 0211
erpretation of Phase Behavior in a	Transreacting Polyester/Polycarbonate Blend.=	PMSE 0074
and Amyloidosis of	Transthyretin.= Progress on the Folding, Assembly,	BIOL 0007
ons at 14.6 GeV/c+ Forward and	Transverse Energies in Relativistic Heavy Ion Collisi	NUCL 0051
s in Human Plasma Samples R+	Transverse Relaxation Times ('T2's) of Water Proton	MEDI 0043
Inhibitory Activity of	Tranylcypromine.= Enhanced Monoamine Oxidase	MEDI 0068
icide Analysis in Water Using Ion	Trap Mass Spectrometry.= +gulatory Issues. Pest	AGRO 0060
Molecules Using the Laser Optical	Trap.= +suring the Isometric Tension of Kinesin	BIOL 0137
man Blood by Using Purge-and-	Trap/Gas Chromatography/Mass Spectrometry.=	CEI 0001
nsional Inorganic Solid: Exciton	Trapping and Annihilation in Crystals of (CH₃)₄NMn	PHYS 0131
Selenide Clusters of Au⁺ and In³⁺,	Trapping Na⁺ Ions in the Center.= +Polynuclear	INOR 0203
Analysis.= A Sorbent	Trapping Procedure for High-Resolution Soil Gas	ENVR 0050

iles.=	Trapping Tetrazolinylidenes with Nitrogen Electroph	ORGN	0205
-Dimensional an+	Simulation of Trapping- and Quenching–Reaction Kinetics in Low	PHYS	0307
Tales of	Trauma.= °From Doubts to Conviction of Fraud:	PROF	0003
Encuentro de Dos Mundos a	Traves de la Quimica.=	HIST	0035
as a	Tree.= I Think I Shall Never See a Polymer Lovely	POLY	0319
acterization of 1–Octene LLDPE (	TREF) Fractions and Their Blends.= +ogical Char	PMSE	0102
Synthesis of Aryl bis(	Trialkyltin) Monomers.=	POLY	0035
Metabolism of	Triallate in Sprague–Dawley Rats.=	AGRO	0023
in+ Aryl Group Interchange in	Triarylphosphines and Its Role on Catalyst Stability	INOR	0349
hlero–[1,9–Bis(2–pyridyl)–2,5,8–	Triazanonane]–Chromium(III): An Example of a Ver	INOR	0397
Including Predictions for Various	Triazenes.= +es Compared to Experimental Values,	PHYS	0298
onitoring of Workers Exposed to	Triazine Herbicides, I: Determination of Simazine an	CEI	0004
onitoring of Workers Exposed to	Triazine Herbicides, II: Determination of Hexazinone	CEI	0005
	Triazine Rigid–Rod Networks.=	POLY	0110
e P$_{450}$–Dependent Aromatase by a	Triazole Derivative.= +reoselective Inhibition of th	MEDI	0007
tion of 3,4,5–Trisubstituted–1,2,4–	Triazoles as Angiotensin II Antagonists.= +Evalua	MEDI	0105
chanism of ADD17014, a Novel	Triazoline Anticonvulsant, for Impairing Excitatory A	MEDI	0144
: The Influence of Heter+	1,2,4–Triazolo[1,5a]Pyrimidine–2–Sulfonanilide Herbicides	AGRO	0046
The Influence of Heteroc+	1,2,4–Triazolo[1,5a]Pyrimidine–2–Sulfonanilide Herbides.	AGRO	0047
ubstituted–2–Fluorophenyl)–1,2,4–	Triazolo–[1,5a]–Pyrimidine Sulfonamides.= +N–(S	AGRO	0048
Structural Transitions in	Triblock Copolymer SBS.= Shear–Induced	PMSE	0136
d Bioengineered Control Agents.	Tribolium as a Model Insect for Study of Resistance	AGRO	0133
na squamosa Seed Oil Extracts to	Tribolium castaneum L. (Coleopt+ Toxicity of Anno	AGRO	0036
Irradiation.=	Tribological Properties of Structures Formed by Ion	NUCL	0093
uration of Glucose—A Centennial	Tribute.= +1 Fischer's Establishment of the Config	CARB	0011
Characterization of (η_6–Arene)	Tricarbonylchromium Compounds by Proton NMR+	ORGN	0215
Synthesis of (erta$_6$–Arene)	Tricarbonylchromium Compounds.= Regioselective	ORGN	0214
e of μ–Phenanthrene Bis(	TricarbonylChromium(0)).= Synthesis and Structur	INOR	0207
Interactions of a	Tricationic Porphyrin.= Sequence–Specific DNA	BIOL	0127
A One-Step Preparation of 1,1,2–	Trichlorocycloalkanes.= +ides with Cyclic Ketones:	ORGN	0263
Awards Symposium. Fate of ^{14}C–	Trichloroethylene in Whole Plants.= Student	ENVR	0056
Catalytic Oxidation of	Trichloroethylene over a PdO on Al$_2$O$_3$/Monolith.=	COLL	0210
for+ Photoinitiated Oxidation of	Trichloroethylene over TiO$_2$: A Potential Technique	COLL	0082
Catalytic Oxidation of	Trichloroethylene Using 1.5 Pt on Al$_2$O$_3$/Monolith.=	COLL	0205
nd Hemicellulases—II. How Do	Trichoderma reesei Cellobiohydrolases Bind to and D	BIOT	0206
Recombinant β–Glucosidase of	Trichoderma reesei.= Developments in	BIOT	0241
Cellulase Formation by	Trichoderma reesei.= Molecular Regulation of	BIOT	0183
Production by	Trichoderma reesei.= Novel Inducers for Cellulase	BIOT	0184
Xylanase from	Trichoderma viride.= Amino Acid Sequence of a 20K	BIOT	0211
ng Protei+ Antiviral Studies with	Trichosanthin, a Plant–Derived Ribosome–Inactivati	AGFD	0189
of Some	Tricyclic Cyclopropenes.= Synthesis and Chemistry	ORGN	0232
ions (10–P–4) with Bidentate and	Tridentate Ligands.= +emistry. Phosphoranide An	INOR	0344
in	Tridom.= Crescent Chemical and Its Origin	SCHB	0015
ics of Thermal Decomposition of	Triethylgallium on GaAs(100) and Implications for G	COLL	0068
hain Amino Sugars by Epoxy and	Triflate Sugar Intermediates.= +, and Branched–C	ORGN	0071
Catalyzed Cross-Coupling of Aryl	Triflates with Grignard Reagents.= +ns: Nickel(O)–	ORGN	0082
is of MTBE and ETBE over	Triflic Acid–Modified Y-Zeolite Catalysts.= Synthes	CATL	0016
atic Amines in Fossil Fuels Using	Trifluoroacylation, I: Analytical Methodology.= +m	FUEL	0060
Effect of Indole on Reactions of	Trifluoroalanine with Tryptophan Enzymes.=	BIOL	0060
Polysilyne Functionalization with	Trifluoromethanesulfonic Acid.=	POLY	0190
s Inhibitors for Enkephali+ New	Trifluoromethyl–Containing Enkephalin Analogues a	MEDI	0151
.= Domain Chimeras in the	Trifunctional Enzyme C$_1$-Tetrahydrofolate Synthase	BIOT	0243
trogen Derivatives.=	Trigeminal Pentafluorosulfanyl (SF$_5$) Carbon and Ni	INOR	0024
an Enol–Ether	Trigger.= Models for the Ene–Diyne Toxines with	ORGN	0102
Solvolysis–	Triggered Alkylation of DNA Using a Silyl Phenol.=	ORGN	0191
erephthalate) with Functionalized	Triglyceride Oil Compositions.= +Poly(Ethylene T	PMSE	0171
Semi-IPNs with Functionalized	Triglyceride Oils.= Poly(Ethylene Terephthalate)	PMSE	0088
nalogues of Methyl 11β, 17α, 21–	Trihydroxy–3,20–Dioxo–Pregna–1,4–Diene–6–Carbox	MEDI	0170
ed (μ–Hydroxo)bisμ–carbo+ New	Triiron Carboxylate Core Structure Consisting of fus	INOR	0094
haracterization of a New Series of	Trimeric Vanadium Sulfide Clusters.= +esis and C	INOR	0372
um(IV) Metallacyclopentanes with	Trimethyl Phosphite and Triphenyl Phosphine.=	INOR	0330
and Biological Activities of 1,2–	Trimethylene–Anthraquinones, Analogues of Mitomy	MEDI	0013
ving the+ Diradical Character of	Trimethylenemetanes: New Reaction Pathways Invol	ORGN	0146
HREELS Study of	Trimethylgallium and Arsine on GaAs(100).=	COLL	0065
Decomposition of Mono– and	Trimethylgallium on Silicon and Indium Phosphide.=	INOR	0123
titution of 1,2–Propanediol with	Trimethylsilyl Reagents Using the 1,3,2λ–5–Dioxapho	ORGN	0051
Transport in Poly(1–	Trimethylsilyl–1 Propyne).= Mechanisms of Gas	POLY	0119
and Characterization of Cationic	Trinuclear BINAP–Ruthenium(II) Complexes and T	ORGN	0044
Diffraction Studies of Polymeric	Triorganotin(IV) Fluorides.= X–ray Powder	INOR	0065
odified Analogues of Chemotactic	Tripeptide F–Met–Leu–Phe–OH.= +C–Terminal M	MEDI	0166
α–Amino Acids, Dipeptides, and	Tripeptides.= +–Containing Polymers Derived from	PMSE	0014
nes with Trimethyl Phosphite and	Triphenyl Phosphine.= +m(IV) Metallacyclopentane	INOR	0330
in the Reaction of Perdeuterated	Triphenylmethyl Cation with Diphenyldiazomethane	ORGN	0145
or Carbamates, Sulfamides Using	Triphosgene, CBMIT, and SBMIT.= +tric Ureas	ORGN	0106
omplexes of Nucleoside–5'–Di– and	Triphosphates and Their Thioanalogues.= +II) C	INOR	0403
Orthophenylene–Stabilized	Triphosphazanes.= New Macrocyclic	INOR	0391
ction. Model Peptide Approach to	Triple–Helix Perturbations.= +Structure and Fun	BIOL	0001
Singlet Oxygen Production by	Triplet Cyclohexenones.= Quantum Yields of	ORGN	0167
thalene: High–Pressure Effects +	Triplet Energy–Transfer in Isotopically Mixed Naph	PHYS	0081
and Concentration Quenching of	Triplet Methylbenzenes in Solution, Using a Fluoresc	PHYS	0265

Energy of the Lowest Singlet and	Triplet States of MgC_2.= +Spectrum, and Binding	PHYS	0276
rious+ Coordination Studies of a	Tripodal Ligand with Mixed Donors P_2S Toward Va	INOR	0028
$u_2As)_3Ga$.=	Tris(arsino)allanes and the Crystal Structure of (t-B	INOR	0231
tabilized by Sterically Demanding	tris(Pyrazolyl)hydroborato Ligands.= +erivatives S	INOR	0087
Peroxide.= Oxidation of	Tris(1,10-Phenanthroline)Iron (II) by Hydrogen	INOR	0222
Crystalline Chloro-	Trisodium Orthophosphate.=	INOR	0345
nthesis and Evaluation of 3,4,5-	Trisubstituted-1,2,4-Triazoles as Angiotensin II Anta	MEDI	0105
Chemical Transformations of	Triterpenes from Guayule.=	ORGN	0019
the Neutrino Mass.= The	Tritium Beta Decay Spectrum from End to End and	NUCL	0011
nt of the Beta Decay of Molecular	Tritium.= +n Antineutrino Mass from Measureme	NUCL	0010
and NPC13839, a+ Synthesis of	Tritium-Labeled Aminoalkoxychromones, NPC16377	MEDI	0150
Synthesis of Carbon-14- and	Tritium-Labeled Panadiplon.=	MEDI	0133
Synthesis of	Tritium-Labelled Photoaffinity Juvenoids.=	AGRO	0016
esolved Microwave Conductivity (	TRMC).= +e in Photoactive Polymers by Time-R	POLY	0214
solu+ Application of Enantiopure	Troger's Base Methosulfate as an Inclusion Host: Re	ORGN	0041
nthic Community Component of a	Tropical Coastal Lagoon.= +et in the Dominant Be	BIOL	0084
Macrocrustacea in a	Tropical Coastal Lagoon.= Energy Content of	BIOL	0105
ion of Natural Products from the	Tropical Rain Forest—A Field for Small Chemical B	SCHB	0024
ign of Ruthenium-Based Fischer-	Tropsch Catalysts.= +ectivity Models and the Des	CATL	0056
Electrochemical and Langmuir	Trough Studies of C_{60} Films.=	ANYL	0108
Folding of	Trp Repressor Subdomains.=	BIOL	0115
lymers and Biodegradable Pol+	Truly Biodegradable Polymers. Biodegradation of Po	CHED	0130
Synthesis of	Truncated Glycosyl Phosphatidylinositides.=	CARB	0031
e on Whistle Blowing in Scienc+	Truth and Consequences: A Psychologist's Perspectiv	PROF	0002
-Activated Substrate Inhibition of	Trypsin by N-Nitrosoamide Derivatives.= +nzyme	ORGN	0022
Characterization of	Trypsin Inhibitor Activity in Amaranthus Species.=	AGFD	0133
s of Trifluoroalanine with	Tryptophan Enzymes.= Effect of Indole on Reaction	BIOL	0060
Effect of Diazinon on L-	Tryptophan Metabolism in Mice.= Paradoxical	AGRO	0020
-Carbamate: A Novel Fluorogenic	Tryptophanase Substrate.= +thesis of Serine-AMC	ORGN	0023
New Chemical Evaluations under	TSCA.=	CHAS	0013
Sheet and	Tube Alkoxysiloxanes.=	POLY	0257
Methanol Mixtures.= A Shock	Tube Investigation of Soot Formation from Toluene/	FUEL	0113
lash or Laser Photolysis-Shock-	Tube Method: Results for the Oxidations of H_2 and	FUEL	0123
Toluene in a Co-Current Packed	Tubular Reactor.= +ng the Two-Phase Nitration of	I&EC	0006
Analogues.= Synthesis and	Tubulin Binding of Novel C-10 Colchicine	MEDI	0016
ethane Organic Compounds in the	Tulsa Area.= +Oilfield Emissions and Urban Nonm	ENVR	0022
nhibitory Effects of Curcumin on	Tumor Initiation and Promotion in Mouse Epidermi	AGFD	0088
a Proteolytic Enzyme Involved in	Tumor Metastasis.= +hips Studies of Cathepsin B,	MEDI	0017
imeric Le^a or Le^b/Le^a) as Human	Tumor-Associated Antigen.= +ed Type 1 Chain (D	CARB	0060
ecific Nitrosamine-Induced Lung	Tumorigenesis by Green Tea and Its Components.=	AGFD	0068
ced from the Photodissociation+	Tunable Diode Laser Probe of Chlorine Atoms Produ	PHYS	0299
c Oxidation of Nucleic Acids.=	Tunable Nickel Complexes for Conformation-Specifi	INOR	0183
n of Cyanoacetic Acid by Anionic	Tungsten (0) Complexes.= Catalytic Decarboxylatio	INOR	0212
Chemical Vapor Deposition of	Tungsten and Molybdenum Films from $(\eta^3-C_3H_5)_5M$	INOR	0055
Properties of the Monophosphate	Tungsten Bronze $(PO_2)_4(WO_3)_{2m}$, m = 2.= +nsport	INOR	0352
ne Rearrangement Reactions on	Tungsten Carbides: The Effect of Surface Oxygen on	CATL	0037
etal-Ligand Multiple Bonds with	Tungsten Chlorophosphine Complexes.= +Form M	INOR	0115
ultiply Bonded Molybdenum and	Tungsten Compounds.= +of Molecular Alloys of M	INOR	0356
Bacteria.= Role of	Tungsten in the Metabolism of Hyperthermophilic	INOR	0099
and Reactivity of Open-Structure	Tungsten Oxides from Molecular Precursors.= +sis	INOR	0314
ses, Structures, and Reactivity of	Tungsten(IV) and Tungsten(VI) Bis(sulfido) Comple	INOR	0369
Reactivity of Tungsten(IV) and	Tungsten(VI) Bis(sulfido) Complexes: Facile Elimina	INOR	0369
n, Dimerization, and Scission on	Tungsten-Iridium Tetranuclear Cluster Cores: Relati	CATL	0008
Transfer between Molybdenum,	Tungsten, and Rhenium Complexes.= Oxygen Atom	INOR	0132
rin, the Novel Nickel(II)chlorin of	Tunicates.= +l and Spectral Properties of Tunichlo	INOR	0139
tural and Spectral Properties of	Tunichlorin, the Novel Nickel(II)chlorin of Tunicates	INOR	0139
Chemistry of	Tunichromes.=	INOR	0096
ters via Poly 3'-(4-pyridyl)terth+	Tuning of Electron Density at Transition-Metal Cen	INOR	0027
olecules: Synthesis, Structure, R+	Tuning the Multiplet Energy Gap in Non-Kekule M	ORGN	0128
Manganese Oxides with	Tunnel Structures.= Hydrothermal Synthesis of	PETR	0056
Biological Surfaces.= Scanning	Tunneling and Atomic Force Microscopy of	PHYS	0158
n the Interpretation of Scanning	Tunneling Microscope Images Showing Anomalous P	PHYS	0212
Scanning	Tunneling Microscope.= Nanofabrication with the	PHYS	0362
M, STS.=	Tunneling Microscopies: FEM, FEES, FIM, A-P, ST	PHYS	0097
d Techniques. Field Ion-Scanning	Tunneling Microscopy (FI-STM).= +and Relate	PHYS	0182
al-Vapor+ Auger and Scanning	Tunneling Microscopy (STM) Studies of the Chemic	COLL	0050
es of High-T_c Superconductors by	Tunneling Microscopy and Spectroscopy.= +c Stat	PHYS	0206
f Metal Oxide Arrays+ Scanning	Tunneling Microscopy and Tunneling Spectroscopy o	COLL	0194
iphoton Ionization and Scanning	Tunneling Microscopy in Physical Chemistry Labora	CHED	0044
) Surfaces: From 1- a+ Scanning	Tunneling Microscopy of Alkali Metals on III-V(110	PHYS	0183
n Compounds.= Scanning	Tunneling Microscopy of Donor Graphite Intercalatio	PHYS	0365
equi+ Optically Excited Scanning	Tunneling Microscopy: Imaging Subsurface and Non	PHYS	0208
f Analytical Chemistry. Scanning	Tunneling Microscopy: New Applications in Analytic	ANYL	0022
ecular Complexes Using Scanning	Tunneling Microscopy.= +ormation Storage in Mol	BIOT	0023
olecules on Surfaces by Scanning	Tunneling Microscopy.= +omic Scale Imaging of M	PHYS	0159
of MX_2 Materials Using Scanning	Tunneling Microscopy.= +nometer Scale Etching	PHYS	0364
Photon Scanning	Tunneling Microscopy.=	PHYS	0210
ctifiers: Experiments by Scanning	Tunneling Microscopy.= +Organic Unimolecular Re	BIOT	0005
111) Surface Studied by Scanning	Tunneling Microscopy.= +se Transition on the Ge(	PHYS	0099
in Situ Electrochemical Scanning	Tunneling Microscopy.= +er Resolution Studied by	PHYS	0127

d TiO$_2$(001) Surfaces by Scanning	Tunneling Microscopy–Spectroscopy.= +O$_2$(110) an	COLL	0171
e–Crystal Bi$_2$Sr$_2$CaC+ Scanning	Tunneling Microscopy/Spectroscopy Studies of Singl	PHYS	0211
Proteins.= Electron	Tunneling Pathways in Native and Modified	BIOL	0133
nning Tunneling Microscopy and	Tunneling Spectroscopy of Metal Oxide Arrays on G	COLL	0194
or Metal Polyhydrides: Rotational	Tunnelling and Quantum Exchange.= +cal Transiti	PHYS	0068
Prediction Model for Marine Gas	Turbine Blades and Guide Vanes.= +orrosion: Life	FUEL	0099
Biomass–Fueled Gas	Turbine Development.=	FUEL	0119
tion of Trace Metals in the Oce+	Tutorial on Marine Bioinorganic Chemistry. Distribu	INOR	0001
ese Chemistry in the Ocean.=	Tutorial on Marine Bioinorganic Chemistry. Mangan	INOR	0009
. Application of Thermal Gravi+	Tutorial on Molecular Routes to Inorganic Materials	INOR	0006
. Applications of Electron Spect+	Tutorial on Molecular Routes to Inorganic Materials	INOR	0013
tical Applications. Introduction+	Tutorial on Polymeric Materials for Photonic and Op	POLY	0001
Neutrinos and Chemistry: A Brief	Tutorial.= +Neutrino Science: Solar Neutrinos—I.	NUCL	0019
Cited Patent Searching: A	Tutorial.=	CINF	0040
	Twelve Days at the U.S. Air Force Academy.=	CHED	0014
d Dibenzofurans in the Blood of	Twelve TCDD–Exposed Persons from Jacksontown, A	CEI	0014
ed Calcined Structure.=	Twelve–Ring Channel Templated Mazzite: A Stabiliz	PETR	0037
Mixing and Energy Distribution in	Twin Screw Extruders.= +y Lecture. Distributive	PMSE	0175
7–Oxabicycloheptane	TXA$_2$ Antagonists.= Interphenylene	MEDI	0160
and Use of Forensic DNA	Typing Systems.= Brief History of the Development	HIST	0004
Legal Scrutiny of DNA	Typing Techniques.=	CHAL	0003
Functional Models for	Tyrosinase.=	INOR	0208
of	Tyrosinase.= Inhibition of the α, β, and γ Isozymes	BIOL	0058
Mixed–Metal Derivatives of	Tyrosinase.=	BIOL	0059
Analogues of Styryl–Based Protein	Tyrosine Kinase Inhibitors.= +Ring–Constrained	MEDI	0023
lysis of Conformationally Flexible	Tyrosine Kinase Inhibitors.= +Molecular Field Ana	MEDI	0024
ly of the Dinuclear Iron Center:	Tyrosyl Radical Cofactor of Ribonucleotide Reductas	BIOL	0022
Behavior of V^E and C$_P^E$ near the	UCST of Polyethers + Alkane Systems.=	PHYS	0319
e Changes of Polymer Blends with	UCST Phase Behavior.= +rs. Flow–Induced Phas	PMSE	0132
Structure Determination of	UDP–Galactose 4–Epimerase from E. coli.=	BIOL	0147
Dynamics at+ Procter Gamble	UERP Polymer Solutions, Blends, and Interfaces—I.	COLL	0012
. Quick Quenc+ Procter Gamble	UERP Polymer Solutions, Blends, and Interfaces—II	COLL	0032
I. Lateral Diff+ Procter Gamble	UERP Polymer Solutions, Blends, and Interfaces—II	COLL	0051
V. Thin Polym+ Procter Gamble	UERP Polymer Solutions, Blends, and Interfaces—I	COLL	0070
emia: The Good, the Bad, and the	Ugly.= +n Academic Program. Succeeding in Acad	YCC	0001
Fraxinus	uhdei.= Secoiridoids and Other Constituents of	ORGN	0187
(2H)–Fura+ Synthesis and Anti–	Ulcer Activity of Novel 5–(2–Substituted Ethenyl)–3	MEDI	0067
MOCVD.= The	ULSI Window: A View of Al and Cu Complexes for	INOR	0125
s on Platinum.=	Ultra–Low–Energy Interactions ($\leq$3 eV) of CH$_3^+$ Ion	COLL	0216
s Formation over Zinc–Containing	Ultra–Stable Y Zeolites.= +nc Content on Aromatic	PETR	0048
ation, and Chemical Properties of	Ultra–Thin MgO Films on Mo(100).= +Characteriz	COLL	0193
Studies by Modern Analytical	Ultracentrifugation.= Molecular Self–Association	BIOL	0036
Density Matrix Theory of	Ultrafast Electron Transfer.=	BIOT	0027
ral Dynamics in Photoexcited H+	Ultrafast Measurements of Energy Flow and Structu	PHYS	0335
nd Structural Dynamics of He+	Ultrafast Optical Studies of Vibrational Relaxation a	PHYS	0169
patially Di+ Computational and	Ultrafast Studies of Electronic Energy Transport in S	PHYS	0133
ared Spectroscopy.=	Ultrafast Vibrational Dynamics Using Transient Infr	PHYS	0194
d Engineering Perspectives on the	Ultrafiltration of Polymeric Colorants.= +mical an	I&EC	0049
ursor.=	Ultrafine Iron Carbide as Liquefaction Catalyst Prec	FUEL	0101
ursor.=	Ultrafine Iron Carbide as Liquefaction Catalyst Prec	CATL	0051
ons Found in the Slow Melting of	Ultrahigh Molecular Weight Polyethylene.= +Fracti	PMSE	0166
n Up–Pum+ Energy Transfer in	Ultrahot Solids: Vibrational Cooling and Multiphono	PHYS	0250
New Approaches to	Ultrasensitive Magnetic Resonance.=	PHYS	0042
	Ultrasonic Polymerization of Acrylic Monomers.=	PMSE	0020
of Nonmetals—Novel Techniques.	Ultrathin Copper Overlayers on TiO$_2$(110).= +try	COLL	0191
. High–Speed DNA Sequencing in	Ultrathin Gels.= +w Face of Analytical Chemistry	ANYL	0033
ring Study of Nickel on	Ultrathin SiO$_2$ Layers.= Medium–Energy Ion–Scatte	COLL	0176
k Diffusion of Adsorbed Nickel on	Ultrathin Thermally Grown Silicon Dioxide.= +Bul	COLL	0177
d Green Tea on Carcinogenesis by	Ultraviolet B Light.= +fect of Orally Administere	AGFD	0065
Coherent Vacuum	Ultraviolet Radiation.= Photoionization with	ANYL	0003
olics and Other Compounds in an	Umbelliferous Vegetable Beverage.= +Bound Phen	AGFD	0170
onvencional para Medir pH, por	un Electrodo de Acero Inoxidable 316/Pelicula de Ox	ANYL	0066
Resultados Experimentales de	un Empaque Integral para Enriquecimiento en Deute	I&EC	0073
ogramacion Cuadratica Sucesiva y	un Simulador de Procesos Comercial.= +lizando Pr	I&EC	0072
tes Etapas del Lavado Quimico de	Una Caldera por Tecnicas Electroquimicas.= +eren	PHYS	0257
os Resultados de la Evaluacion de	una Materia Biologica.= +mico de los Alumnos en l	CHED	0093
izando como Electrodo Indicador	una Pelicula de Oxido sobre Acero Inoxidable 316.=	CHED	0163
c Mechanism for Isomerization of	Unactivated Carbon–Carbon Double Bonds.= +hili	BIOL	0014
e QFB en la Facultad de Quimica,	UNAM (Analisis Estadistico de 10Anos).= +rrera d	CHED	0092
The	Unanticipated Benefits: Graphite Furnace AA.=	CHED	0184
	Unbearable Heaviness of Neutrinos.=	NUCL	0002
ction: Current Understanding and	Uncertainty.= +ases—I. Mechanism of Cellulase A	BIOT	0179
	Uncommon Reactions: A Useful Teaching Tool.=	CHED	0166
teroid 5α–Reductase Which Are	Uncompetitive versus Testosterone.= Inhibitors of S	MEDI	0008
lboronate and Isopropylidene D+	Uncoupling Propertiex of Quinic Acid and Its Pheny	INOR	0263
m–Based Inactivator 1–Diazo–4–	undecyl–2–one (Duo) as a Probe of the Active Site of	BIOL	0067
plication of the HACCP System in	Underdeveloped Countries.= +Problems in the Ap	AGFD	0079
Challenges and Opportunities in	Undergraduate Analytical Chemistry Education.=	ANYL	0007
nant DNA Analysis in the	Undergraduate Biochemistry Laboratory.= Recombi	CHED	0090
Advanced Instrumentation in the	Undergraduate Chemistry Laboratory at MIT.=	CHED	0128

ion of NMR and GC/MS into the	Undergraduate Curriculum: A Case Study.= +porat	CHED	0149
erials Chemistry Examples for the	Undergraduate Curriculum: A Progress Report.=	CHED	0023
isible Spectrophotometer into the	Undergraduate Curriculum.= +a Diode Array UV-V	CHED	0180
m NMR Instrumentation into the	Undergraduate Curriculum.= +of Fourier Transfor	CHED	0148
Orbitals of Metalloporphyrins: An	Undergraduate Experiment in Spectroscopy.= +ar	PHYS	0327
ptimal Conditions: An Advanced	Undergraduate Experiment Introducing Experimenta	CHED	0182
tic Resonance Capabilities for the	Undergraduate Laboratory.= +oved Nuclear Magne	CHED	0185
Spectroscopy in the	Undergraduate Laboratory.= Time-Resolved Laser	CHED	0084
Spectroscopy in the	Undergraduate Laboratory.= Time-Resolved Laser	PHYS	0249
NSF-Catalyzed Innovations in the	Undergraduate Laboratory—I. Support of Chemi+	CHED	0146
NSF-Catalyzed Innovations in the	Undergraduate Laboratory—II. Undergraduate S+	CHED	0179
f Personal Computer Modeling in	Undergraduate Organic Chemistry.= +ation. Use o	CHED	0188
Techniques and Graphics into the	Undergraduate Physical Chemistry Curriculum.=	CHED	0156
Updating an	Undergraduate Physical Chemistry Lab.=	CHED	0214
Updating an	Undergraduate Physical Chemistry Lab.=	PHYS	0290
opment and Symposium on Co+	Undergraduate Physical Chemistry Laboratory Devel	CHED	0155
opment. A Consortium-Based +	Undergraduate Physical Chemistry Laboratory Devel	CHED	0125
opment. How Expensive Are La+	Undergraduate Physical Chemistry Laboratory Devel	CHED	0082
opment. Measurement of the P+	Undergraduate Physical Chemistry Laboratory Devel	CHED	0209
opment. NSF Funding of Labo+	Undergraduate Physical Chemistry Laboratory Devel	CHED	0196
opment. The Case for the Adva+	Undergraduate Physical Chemistry Laboratory Devel	CHED	0041
ts at Roanoke College.=	Undergraduate Physical Chemistry Laser Experimen	CHED	0042
ding of Laboratory Innovations in	Undergraduate Physical Chemistry.= +. NSF Fun	CHED	0196
Different Approach to	Undergraduate Physical Chemistry.=	CHED	0127
Independent	Undergraduate Program?.= Polymer Science: An	CHED	0099
Undergraduate Laboratory—II.	Undergraduate Science Education at the National Sc	CHED	0179
Educating	Undergraduates about Chemometrics.=	CHED	0194
Introducing MINICUAD to	Undergraduates.=	CHED	0120
Force Microscope Studies of	Underpotential Deposition Processes.= Atomic	PHYS	0128
ocked-Conformation Styrenes: +	Unexpectedly High Diastereoselective Oxidation of L	ORGN	0236
Dynamics of Partially Folded and	Unfolded Protein Molecules.= +Conformation and	PHYS	0107
red by Vacuum Deposition Pol+	Uniaxially Oriented Polyamide Films in Plane Prepa	POLY	0082
in a Second-Semester Sophomore	Unified Laboratory Course.= +300-MHz FT-NMR	CHED	0181
del Filled Polymers: Synthesis of	Uniformly Cross-Linked Polystyrene Microbeads.=	PMSE	0017
	Unimolecular Reaction Dynamics.=	PHYS	0164
unnel+ Progress Toward Organic	Unimolecular Rectifiers: Experiments by Scanning T	BIOT	0005
Effect on Activation Barriers of	Unimolecular Transformations of Organic Compound	COLL	0107
Granted in the	United States.= Numbers of Polymer Degrees	PMSE	0128
n Food-Producing Animals in the	United States.= +pects of the Use of Xenobiotics i	AGRO	0076
Molecules with Donor-Acceptor	Units Intermediated by Silicon Chains.= Model	POLY	0211
Containing Single-Ring Aromatic	Units Separated by Aliphatic Spacers.= +igoesters	PMSE	0016
sport Dynamics of Photosynthetic	Units: Spectral-Hole-Burning Studies.= +nd Tran	PHYS	0018
rogramas de Becas: Vinculo entre	Universidades y Secotr Productivo.= +cio Social y P	CHED	0114
can Chemical Society). New York	University and Chemistry in New York.= +Ameri	HIST	0025
us Materials Management at the	University of California: What Was Revealed and Ho	CHAS	0002
Education and Training Used in a	University Setting.= +ting Chemical Safety: Some	CHAS	0048
Major	University.= Research and Teaching in a	CONG	0010
a	University.= Safety Education and Training at	CHAS	0049
Boston	University.= Teaching of Chemistry Program at	CHED	0070
in College of Medicine of Yeshiva	University/Montefiore Medical Center.= +rt Einste	PROF	0004
Institute at the Polytechnic	University, Brooklyn.= The Polymer Research	HIST	0033
lags.+ Phase Characterization of	Unmodified Petroleum Coke and Coal Gasification S	FUEL	0139
Mutagenesis with	Unnatural Amino Acids.=	BIOL	0029
Foods and Feeds: Natural and	Unnatural Environments for Long-Chain Omega-3 F	AGFD	0160
Complex Natural and	Unnatural Products.= Synthesis of Architecturally	ORGN	0140
Computer-Aided Discovery of	Unprecedented Reactions.=	CINF	0006
Mutations: Mechanistic Studies+	Unraveling the Molecular Basis of Sunlight-Induced	BIOL	0157
eurotoxin That Induce Immune	Unresponsiveness to Experimental Allergic Encephal	BIOL	0086
Information Ownership vs.	Unrestricted Access: The Library Dilemma.=	CINF	0045
New Synthesis of α,β-	Unsaturated Amides.=	ORGN	0255
tivity of Singlet Oxygen with α,β	Unsaturated Esters: Evidence for a Perepoxide Inter	ORGN	0164
sis of Substituted Aromatics via	Unsaturated Ketene Intermediates Using Transition	ORGN	0274
med+ Labile and Coordinatively	Unsaturated Molybdenum(III)-μ-Sulfido Dimers For	INOR	0271
d Morphology of Polyurethane-	Unsaturated Polyester Interpenetrating Polymer Net	PMSE	0027
omorphology Characterization of	Unsaturated Polyester Resins with Low-Profile Addi	POLY	0114
ermochemistry of the Oxidation of	Unsaturated Radicals: n-C_4H_5 + O_2.= +ics and Th	FUEL	0066
s. Synthesis and Reactivity of an	Unsaturated Ruthenium Fluoro-Alkoxide Complex.=	INOR	0127
g Triphosgene, C+ Synthesis of	Unsymmetric Ureas or Carbamates, Sulfamides Usin	ORGN	0106
ordinate: A Possible Domain for+	Unsymmetrical Access to a Symmetrical Reaction Co	ORGN	0122
Optics: Cobaltous Complexes of	Unsymmetrical Hydrazone Imine Glyoxal Derivatives	POLY	0102
d Sensory Evaluation of Flavor in	Untreated and Antioxidant-Treated Meat.= +al an	AGFD	0120
/Bicar+ Dissolution Kinetics of	UO_2.00 Pellets as a Function of Time, pH, Carbonate	INOR	0236
Spectroscopy.= Probing	Upconversion Dynamics Using Site-Selective	ANYL	0004
xidation Volatiles in Foods.=	Update on Gas Chromatographic Analyses of Lipid O	AGFD	0098
the Mexican Steroid Industry and	Update on R. E. Marker—Illustrate+ Prehistory of	HIST	0016
of+ Forensic DNA Technology:	Update 1991. DNA Identification Testing and Issues	CHAL	0001
.=	Updating an Undergraduate Physical Chemistry Lab	CHED	0214
.=	Updating an Undergraduate Physical Chemistry Lab	PHYS	0290
nalysis Laboratory.=	Upgrading a Forensic Science Program with a Microa	CHED	0153
oids, the Steroid Community, and	Upjohn in Perspective: A Profile of Innovation.=	HIST	0021

Techniques.= Integrating	Upstream Production with Downstream Purification	BIOT 0072
ng Primary Separation Steps with	Upstream–Downstream Stages. Integrati+ Integrati	BIOT 0069
Microbial Reduction of	Uranium.=	GEOC 0008
–ience—II. Initial Abundances of	Uranium-235 in Meteorites and Lunar Samples.=	NUCL 0082
osition of Oilfield Emissions and	Urban Nonmethane Organic Compounds in the Tuls	ENVR 0022
Transformation of Solid	Urban Residues in Usable Products.=	CHED 0105
Asbestos en el Area	Urbana de la Ciudad de Mexico?.=	ENVR 0025
Corrosive Properties of Molten	Urea.= Effect of Conditioning Agents on the	FERT 0011
f Mild Steel Exposed to Inhibited	Urea–Ammonium Nitrate Solution.= +Corrosion o	FERT 0001
of Mild Steel Exposed to	Urea–Ammonium Sulfate Suspensions.= Corrosion	FERT 0003
ms of Resistance to Benzoylphenyl	Ureas and Other Insect Growth Regulators.= +nis	AGRO 0134
, C+ Synthesis of Unsymmetric	Ureas or Carbamates, Sulfamides Using Triphosgene	ORGN 0106
t-ansferase (ACAT)+ Tetrazole	Ureas: A Potent Series of Acyl CoA: Cholesterol Acyl	MEDI 0107
Pyrethroids and Benzoylphenyl	Ureas.= Resistance in the Diamondback Moth to	AGRO 0134
Directed toward	Urease Models.= Synthesis of Dinucleating Ligands	INOR 0089
Buildup in Methacrylate	Urethane IPNs.= Kinetic Study and Morphology	PMSE 0043
mer Networks of Poly(Carbonate–	Urethane) and Polyvinyl Pyridine.= +etrating Poly	PMSE 0056
3D Metal Complexes with	Uric Acid.=	INOR 0073
mon+ Problems with the Use of	Urinary Creatinine as a Stable Metric for Human Bio	CEI 0019
aminants in th+ Measurement of	Urinary Phenolic Metabolites of Environmental Cont	CEI 0002
for Biomonitoring of Aniline in	Urine and Comparison with the Classical Colorimetri	CEI 0003
Hexazinone and Its Metabolites in	Urine of Exposed Forest Pesticide Applicators.=	CEI 0005
ring for Chromium Levels in the	Urine of Individuals Living Adjacent to Chromate Pr	CEI 0009
of Simazine and Its Metabolites in	Urine.= +to Triazine Herbicides, I: Determination	CEI 0004
Human	Urine.= Determination of Atrazine Metabolites in	AGRO 0032
pionamide (HEPP) in Plasma and	Urine.= +t D,L-3-Hydroxy-3-Ethyl-3-Phenylpro	ANYL 0082
mpiad Competition Scheduled for	USA, July 11–22, 1992.= +rnational Chemistry Oly	CHED 0139
Residues in	Usable Products.= Transformation of Solid Urban	CHED 0105
Right-to-Know Compliance Made	User Friendly.= +the Challenge of the 90s: MSDS	CHAS 0031
Design.= DESTIL—A	User-Friendly Program for Distillation Column	CHED 0122
Handled.= A	User's Perspective of the Issues and How They Are	CINF 0044
Information	Users.= Legal Primer for Librarians and	CINF 0047
Estudio Sobre el	Uso de Eteres de Glicol en Mexico.=	ENVR 0038
graphic Analysis of Pharmaceu+	USP Reference Standards, a Key Factor in Chromato	ANYL 0115
Interactions of Vanadate with	Uteroferrin.=	BIOL 0051
•bre Acero Inoxidable 304 y 316	Utilizadas en el Seguimiento de las Reacciones Acido	ANYL 0055
gginina en Producto Terminado	Utilizando como Electrodo Indicador una Pelicula de	CHED 0163
ptimizacion de Plantas Quimicas	Utilizando Programacion Cuadratica Sucesiva y un S	I&EC 0072
es and Ketones of+ The Near–	UV Absorption Spectra of Several Aliphatic Aldehyd	ENVR 0024
enol A Polycarbonate Using GPC–	UV Analysis.= +of Phenolic End–Groups in bisPh	POLY 0083
n Employing HPLC Analysis with	UV Detection.= +Analysis in Pharmaceuticals Whe	ANYL 0114
•ka+ Abiotic Stresses of Heat or	UV Illumination May Help Reduce Decay in Seal–Pa	AGFD 0076
Polycarbonate by Far	UV Radiation.= Surface Modification of	PMSE 0125
rbed on Zeolites.=	UV Resonance Raman Spectroscopy of Species Adso	COLL 0075
by Liquid Chromatography with	UV-Absorption Photometry and Particle–Beam Mas	ENVR 0015
mnyl Ethers.+ Novel Synthesis of	UV-Curable, Multifunctional, Aromatic Analogs of V	POLY 0016
nyl Phenyl Ether Analogs for	UV-Induced Cationic Polymerization.= Novel Prope	POLY 0017
Photoproducts in	UV-Irradiated DNA.= Determining Pyrimidine	ANYL 0092
Transient IR Spectroscopy of	UV-Photolyzed $Cr(CO)_6$ in Solution.= Picosecond	PHYS 0220
z Using a Rem+ Rapid–Scanning	UV-VIS Detection for Capillary Gas Chromatograph	ENVR 0060
Incorporation of a Diode Array	UV–Visible Spectrophotometer into the Undergradua	CHED 0180
plication of Electrochemical and	UV/Visible Detection to the LC Separation and Dete	AGFD 0101
eoxyoligonucleotide Synthesis for	UvrABC Endonuclease DNA Repair.= +n-Linked D	BIOL 0126
a Application of Homonuclear 2D	^{51}V NMR to Determine the Effect of Temperature	INOR 0228
uperoxide Dismutase as Probed by	^{51}V NMR.= +on of Vanadate Anions with Cu,Zn-S	INOR 0184
Systems.= Behavior of	V^E and Cp^E near the UCST of Polyethers + Alkane	PHYS 0319
lus Influenzae Type B Conjugate	Vaccine.= +ide Compositional Analysis of Haemoph	ANYL 0021
Influenzae Type B Conjugate	Vaccines.= Carbohydrate Antigens. Haemophilus	CARB 0053
Development of Synthetic	Vaccines.= Study Directed Toward the	CARB 0055
amide Films in Plane Prepared by	Vacuum Deposition Polymerization Methods.= +ly	POLY 0082
Molecular Weight Distribution of	Vacuum Residue Using Mass Spectrometry: A Heter	PETR 0150
Photoionization with Coherent	Vacuum Ultraviolet Radiation.=	ANYL 0003
:h Nasic+ New Series of Mixed–	Valance Titanium(III/IV) Phosphate Compounds wi	INOR 0353
ation, and Reactivity of a Mixed–	Valence [Ni(II)–NI(III)] Thiolate Dimer.= +acteriz	INOR 0060
:y, Oxygen Stoichiometry, and Cu	Valence in the High-T_c Superconductors.= +lectivi	PHYS 0027
Derivatives of Low–	Valent Iron Complexes.= Hydride Thiolate	INOR 0225
–Iminoacyl C+ High– and Mid–	Valent Mono(peralkylcyclopentadienyl) Tantalum η^2	INOR 0108
ructure, and Reactivity of Low–	Valent Vanadium Phosphine Complexes and Mixed P	INOR 0129
able 316/Pelicula de Oxido, en la	Validacion del Metodo Potencio–Metrico para Cuant	ANYL 0066
s of+ Model Development and	Validation Are the Keys to Scientific Risk Assessmen	ENVR 0042
PCR—Status of	Validation Studies.=	CHAL 0005
Proc+ Importance of Laboratory	Validations and Accurate Descriptions of Analytical	AGFD 0126
mputers in Chemistry at Lebanon	Valley College.= +Chemical Education. Role of Co	CHED 0115
on with Photosynthetic Activity in	Vallisneria spiralis.= +loroplasts and Their Relati	BIOL 0101
ining en+ Proyecto de Practica:	Valoracion Potenciometrica de Clorhidrato de L-Arg	CHED 0163
ansition State in RNase?+ Is the	Vanadate Anion an Analogue of the Phosphorane Tr	BIOL 0061
as Probed by ^{51}+ Interaction of	Vanadate Anions with Cu,Zn-Superoxide Dismutase	INOR 0184
Strength on Chemical Exchange of	Vanadate Derivatives.= +Temperature and Ionic	INOR 0228
Interactions of	Vanadate with Uteroferrin.=	BIOL 0051
Catalysts b+ Characterization of	Vanadia/$TiO_2(110)$ and Vanadium/$TiO_2(110)$ Model	COLL 0192

Model Complexes of	Vanadium Bromoperoxidase.= Functional Vanadium	INOR 0016
Product Selectivity of	Vanadium Bromoperoxidase.=	INOR 0017
Structural Studies of	Vanadium Bromoperoxidases.=	INOR 0098
electivity of Alumina–Supported	Vanadium Catalysts in the Hydrodemetallization (H	CATL 0044
Analogues.= Iron and	Vanadium Complexes of Siderophores or Siderophore	INOR 0120
dium Haloperoxidase.=	Vanadium Coordination Compounds: Modeling Vana	INOR 0018
n Compounds: Modeling	Vanadium Haloperoxidase.= Vanadium Coordinatio	INOR 0018
Algae.= Reactivity of the	Vanadium Haloperoxidases Isolated from Marine	INOR 0097
Bromoperoxidase.= Functional	Vanadium Model Complexes of Vanadium	INOR 0016
zation, and Reactivity Studies of	Vanadium Model Compounds Related to Hydrodesul	INOR 0197
ers on the Structure–Reactivity of	Vanadium Oxide Catalysts.= +Support and Promot	CATL 0057
of Supported	Vanadium Oxide Catalysts.= Molecular Engineering	COLL 0196
Geometry of the Excited State of	Vanadium Oxides Anchored onto SiO_2 and Their Ph	INOR 0165
e, and Reactivity of Low–Valent	Vanadium Phosphine Complexes and Mixed Phosphi	INOR 0129
zation of a New Series of Trimeric	Vanadium Sulfide Clusters.= +esis and Characteri	INOR 0372
tion of Alkyl Hydroperoxides with	Vanadium(IV) in Acidic Aqueous Solutions.= +eac	INOR 0276
Synthesis of Cyclic	Vanadium(V) 1,3–Diol Complexes.=	INOR 0409
uiox+ Olefin Polymerization by	Vanadium–Containing Polyhedral Oligometallasilsesq	INOR 0131
zation of Vanadia/$TiO_2(110)$ and	Vanadium/$TiO_2(110)$ Model Catalysts by Electron S	COLL 0192
f the Anisotropic Reorientation of	Vanadyl Acetylacetonate in Toluene.= +th Study o	PHYS 0264
inding Sites in Nephrocalcin with	Vanadyl Ion.= +OR Characterization of Calcium–B	BIOL 0045
ne Gas Turbine Blades and Guide	Vanes.= +orrosion: Life Prediction Model for Mari	FUEL 0099
Detection of Adulterated	Vanilla Extract.=	AGFD 0185
te–Pretreated Aspen Treated with	Vapor at High Temperature.= +ulfonation of Sulfi	PMSE 0170
cting F+ Metalorganic Chemical	Vapor Deposition Approaches to high–T_c Supercondu	INOR 0176
echanism of Copper Chemical–	Vapor Deposition from the 1,5–Cyclooctadiene Coppe	COLL 0026
Engineering by Laser Chemical	Vapor Deposition of a Nonlinear Interface Optical Sw	BIOT 0024
m Copper tert–Butox+ Chemical	Vapor Deposition of Copper and Copper(I) Oxide fro	INOR 0143
nds.= Selective Chemical–	Vapor Deposition of Copper from Copper(I) Compou	COLL 0027
tive Low–Temperature Chemical	Vapor Deposition of Copper from New Copper(I) Co	INOR 0144
Copper(I) Hexafluo+ Chemical–	Vapor Deposition of Copper from 1,5–Cyclooctadiene	COLL 0029
genation of Diamond+ Chemical	Vapor Deposition of Diamond (and c–BN, SiC). Halo	PHYS 0176
Fraction Measurem+ Chemical	Vapor Deposition of Diamond (and c–BN, SiC). Mole	PHYS 0145
cular Dynamics Simu+ Chemical	Vapor Deposition of Diamond (and c–BN, SiC). Mole	PHYS 0354
cture and Chemistry o+ Chemical	Vapor Deposition of Diamond (and c–BN, SiC). Stru	PHYS 0202
oscopic Diagnostics for Chemical	Vapor Deposition of Diamond Films.= Laser Spectr	PHYS 0151
rconnection of Electro+ Chemical	Vapor Deposition of Low–Resistivity Metals for Inte	INOR 0126
y (STM) Studies of the Chemical–	Vapor Deposition of Metal Carbonyls.= +icroscop	COLL 0050
ms from (η^3–C_3H_5)$_5$M+ Chemical	Vapor Deposition of Tungsten and Molybdenum Fil	INOR 0055
istry+ Organometallic Chemical	Vapor Deposition of ZnSe: Effect of Precursor Chem	INOR 0171
Deciphering a Chemical	Vapor Deposition Process.=	INOR 0122
Plasma Etching and Chemical	Vapor Deposition.= Reactions of Metal Atoms in	PHYS 0143
acterization of Diam+ Chemical	Vapor Depositon of Diamond (and c–BN, SiC). Char	PHYS 0118
eation and Growth of+ Chemical	Vapor Depostion of Diamond (and c–BN, SiC). Nucl	PHYS 0091
Water Solution onto Water	Vapor Droplets.= Concentrating Cellulase from a	BIOT 0021
Trends in Metalorganic Chemical	Vapor Epitaxy.= +tallic Precursors. Advances and	COLL 0064
Surveys to Design Soil	Vapor Extraction System.= The Use of Soil Gas	ENVR 0051
to Measure Water	Vapor in a Gaseous Mixture.= Chemical Modulator	ANYL 0052
Diffusion of Water	Vapor in Hydrogen–Bonded Polymer Blends.=	POLY 0192
se Crystals.= Rate of Water	Vapor Interaction with Carbon Adsorbents and Sucro	CARB 0062
Heterogeneous Reactions in Metal	Vapor Oxidation.= +n between Homogeneous and	PHYS 0144
Change in Crystal Habit in the	Vapor–Phase Growth of ZnO.=	PHYS 0293
=	Vapor–Phase Molecular Routes to Ceramic Powders.=	INOR 0175
erosol Generation, Transport, and	Vaporization for Particle Beam LC/MS.= +ts in A	ANYL 0122
Single–Stranded DNA.= Laser	Vaporization of High Molecular Weight,	ANYL 0130
ine Chymosin in Aspergillus niger	var. +i: Recent Improvements in Production of Bov	BIOT 0234
s on Si(100) Observed with a New	Variable Temperature STM.= +mer Configuration	PHYS 0185
Depth Profiling Using	Variable–Angle ATR.=	POLY 0307
nthan Aqueous Solutions from a	Variant Strain of Xanthomonas campestris NRRL B	BIOT 0196
Bacteriorhodopsin	Variants for Optical Information Processing.=	BIOT 0078
s.=	Variational Formulation in Network Thermodynamic	PHYS 0323
Molecules.=	Variational Rovibrational Studies of 3– and 4–Atom	PHYS 0030
and Absolute Stereochemistry of	Variculanol: A Novel Sestererpenoid from Aspergillu	ORGN 0184
el Sestererpenoid from Aspergillu	variecolor.= +ereochemistry of Variculanol: A Nov	ORGN 0184
d by Use of Lecithin Supplements	Varied in Choline Content.= +f Humans as Affecte	AGFD 0034
es Results in Dramatic Changes in	Vasorelaxant Activity.= +on of Adenosine Analogu	MEDI 0113
e+ 3D Databases. CAVEAT: A	Vector–Based Approach to Computer–Aided Molecul	COMP 0044
y in Microheterogeneous Fluids.	Vectorial Electron Injection into Colloidal Semicondu	COLL 0147
Polymerized Porphyrin Films for	Vectorial Photoinduced Electron Transport.= +ally	BIOT 0126
r Compounds in an Umbelliferous	Vegetable Beverage.= +Bound Phenolics and Othe	AGFD 0170
ents to Stabilize Raw Fruit and	Vegetable Juices: Control of Enzymatic Browning.=	AGFD 0027
ulsions with Fatty Acid Esters or	Vegetable Oils: Formulation, Optimization, and Dete	COLL 0083
ment of Minimally Processed	Vegetable Snacks.= Integrated Approach to Develop	AGFD 0056
Minimally Processed Fruit and	Vegetable Surfaces.= Physicochemical Changes on	AGFD 0011
of Vegetable	Vegetable Tissue.= Kinetics of Thermal Softening	AGFD 0013
Minimally Processed Fruits and	Vegetables. Cold Blanching Treatments to Stabilize	AGFD 0027
Minimally Processed Fruits and	Vegetables. Fruit Preservation by Combined Method	AGFD 0055
s+ Minimally Processed Fruits and	Vegetables. New Developments in Modified Atmo	AGFD 0075
e+ Minimally Processed Fruits and	Vegetables. Physiology of Cut Fruits and Vegetabl	AGFD 0010
mosphere Packaging in Fruits and	Vegetables.= +New Developments in Modified At	AGFD 0075

b es. Physiology of Cut Fruits and	Vegetables.= +imally Processed Fruits and Vegeta	AGFD	0010
Edible Films for Fruits and	Vegetables.=	AGFD	0078
d;: Effect of Lipophilicity, Dosing	Vehicle, and Prodrug Modification.= +ic Compoun	MEDI	0196
nçe the Lytic Activity of a Cobra	Venom Cytotoxin: Structure–Activity Implications.=	BIOL	0085
v:n+ Regions of Modified Cobra	Venom Neurotoxin That Induce Immune Unresponsi	BIOL	0086
International Joint	Ventures: Exploring the Possibilities.=	CINF	0015
on Theory.=	Verification of a Fundamental Liquid–Solid Adsorpti	COLL	0188
Experimental	Verification of a Multivariable Controlled Auxostat.=	BIOT	0158
Membrane Reactor: Experimental	Verification.= +tic Resolution Using a Multiphase	BIOT	0042
Hydrogen	Vermiculite: Synthesis and Characterization.=	INOR	0056
s: Synthesis of a Precursor of (±)-	Verrucosidin.= +rate Reagents to Alkynyl Oxirane	ORGN	0011
s. and U+ Isomalt (Palatinit®), a	Versatile Alternative Sweetener: Production, Propertie	CARB	0005
Phenylselenoglycosides as Novel,	Versatile Glycosyl Donors. Selective Activation over	CARB	0035
2-Oximinoglycosyl Halides as	Versatile Glycosyl Donors.= 2-Oxo– and	CARB	0025
t:on: The O-Thiocarbamate as a	Versatile New Metalation Director for the Synthesis	ORGN	0083
e Shale Stabilizers.= Novel and	Versatile Process for Producing Environmentally Saf	CARB	0009
y esterase–Mediated Reactions: A	Versatile Route to Target Chiral Molecules.= +rbox	AGFD	0008
dinedione with α-Haloketones: A	Versatile Synthesis of 3-Hydroxy-5(2H)-Furanones.=	ORGN	0223
	Versatile Zymark Robot Dilution System.=	TECH	0005
Conformational	Versatility of Hyaluronan: 13C-NMR Studies.=	CARB	0047
Cancer Management:	Versatility of the B-Protein Assay.=	BIOL	0069
e 14 KDa β-Galactoside–Specific	Vertebrate Lectin and Several β-Galactoside–Specifi	BIOL	0109
Using Real Numbers as	Vertex Invariants for Topological Indices.=	COMP	0026
oretical Algorithm To Label the	Vertices of the Five Regular Polyhedra in Order To C	COMP	0035
Applications to	Vesicles.= Voltammetry in Lipid Films with	COLL	0021
Methods in the Analysis of	Veterinary Drug Residues.= Immunochemical	AGFD	0064
Overview of Feed Additives and	Veterinary Drugs and Their Residual Analysis in Jap	AGFD	0102
nimals. Introduction—The Uses of	Veterinary Drugs.= +biotics in Food-Producing A	AGRO	0075
Importance of Pharmacokineti+	Veterinary Medicines—Regulation in Europe and the	AGRO	0077
Determination of	Viable Cell Biomass in Mammalian Cell Bioreactors.=	BIOT	0012
dyurethan+ Characterization of	Vibration Damping of Laminated Steel Sheet with P	PMSE	0057
rgy Transfer in Ultrahot Solids:	Vibrational Cooling and Multiphonon Up-Pumping.=	PHYS	0250
Spectroscopy.= Ultrafast	Vibrational Dynamics Using Transient Infrared	PHYS	0194
rance and Optical Spectroscopies.	Vibrational Energy Flow at Solid Surfaces.= +eso	PHYS	0344
ergy Effects on Ion–Molecule R+	Vibrational Mode, Electronic State, and Collision En	PHYS	0241
rotic Solvents Measured by Tra+	Vibrational Population Relaxation Time of Ions in P	PHYS	0083
e+ Ultrafast Optical Studies of	Vibrational Relaxation and Structural Dynamics of H	PHYS	0169
Solvents.= Fast	Vibrational Relaxation for Polar Solutes in Polar	PHYS	0037
Interpretation of	Vibrational Spectra.= Ab Initio Calculations and the	PHYS	0033
Small Molecules in Poly+ Using	Vibrational Spectroscopies to Probe the Diffusion of	POLY	0332
tudy of the Molecular Structure,	Vibrational Spectrum, and Binding Energy of the Lo	PHYS	0276
	Vibrational Structure of Fullerenes.=	PHYS	0006
Quantum-State-Resolved N2O	Vibrational, Rotational, and Translational Energy Tr	PHYS	0334
onal Energy Transfer from Highly	Vibrationally Excited NO2 Donor.= +and Translati	PHYS	0334
Studies on	Vibrio cholerae Polysaccharides.=	CARB	0016
r+ Radiative Recombination and	Vibronic Relaxation in σ-Delocalized Silicon Backbo	PHYS	0138
s Competition b+ Reduction of	Vicinal Dihalides by Photogenerated Ru(I) Complexe	INOR	0284
Pre-Historic Art (	Videotape presentation).= Chemistry and	CONG	0002
Rewards of Developing	Videotapes on Laboratory Safety.= Challenges and	CHAS	0044
c d+ Sustitucion de Electrodo de	Vidrio Convencional para Medir pH, por un Electrod	ANYL	0066
+cio Social y Programas de Becas:	Vinculo entre Universidades y Secotr Productivo.=	CHED	0114
el Spectroscopy Studies of Poly(	vinyl Acetate)/Silicon Oxide Microcomposites Prepar	POLY	0287
l in Hydrogen-Bonded Polymers:	Vinyl Alcohol/Vinyl Butyral Copolymers.= +ffusio	POLY	0191
sessment: Perchloroethylene and	Vinyl Chloride Examples.= Issues in Cancer Risk As	CHAS	0011
Chemical Stabilization of Poly(	vinyl Chloride) by Pretreatment with N-Substituted	POLY	0027
l Isomers in Single Racemic Poly(	vinyl Chloride) Chains.= +Runs of Trans Rotationa	POLY	0069
from Castor Oil Polyurethane and	Vinyl Copolymers.= +g Polymer Network Formed	PMSE	0089
New Catalytic Route to	Vinyl Esters.=	I&EC	0054
bifunctional, Aromatic Analogs of	Vinyl Ethers.= +vel Synthesis of UV-Curable, Mu	POLY	0016
near and Cyclic Polystyrene/Poly(	vinyl Methyl Ether) Blends.= +composition and Li	POLY	0295
evaporates Prepared Using Poly(	vinyl Methyl Ether/Maleic Anhydride) Copolymer.=	COLL	0086
-Labeled Poly(styrene) and Poly(	vinyl methylether).= +: Hydrogen– and Deuterium	COLL	0035
Characteristics of Cyclic	Vinyl Monomers.= Radical Ring-Opening	ORGN	0157
Pseudografting of	Vinyl Polymers onto DNA.=	POLY	0301
Comparison of Allyl, Formyl, and	Vinyl to Alkyl.= +Radical Reactions with Oxygen:	PHYS	0228
hase Behavior of Aqueous Poly(	vinyl–alcohol) Solution—A Small-Angle Neutron Sca	POLY	0171
Copolyme+ Synthesis of Poly(b-	vinylborazine) and Poly(styrene-co-b-vinylborazine)	POLY	0183
borazine) and Poly(styrene-co-b-	vinylborazine) Copolymers.= +hesis of Poly(b-vinyl	POLY	0183
ite Films of Polyaniline and Poly(	vinylchloride)/Poly(Vinylidene Chloride).= +mpos	PMSE	0113
Thermal Isomerization of a	Vinylcyclobutene to a Cyclohexadiend.=	ORGN	0144
ine and Poly(vinylchloride)/Poly(	Vinylidene Chloride).= +mposite Films of Polyanil	PMSE	0113
and Metathesis Reactions of the	Vinylidene Complexes Cp(CO)2M=C=CH2 (M=Mn,	INOR	0072
rudy of Miscible Blends of Poly(	vinylidene Fluoride) and Poly(3-hydroxybutyrate).=	POLY	0097
ability of Ethyl Acetate in Poly(	vinylidene Fluoride) Monitored by Rheo–Photoacous	POLY	0255
a Dehydrohalogenation Studies of	Vinylidene Fluoride/Trifluoroethylene Copolymer	POLY	0022
Ab Initio Study of the	Vinylidene–Acetylene Isomerization Process.=	PHYS	0287
aionic, and Anionic Copolymers of	Vinylpyrrolidone.= +romatography of Cationic, No	PMSE	0031
Reactions of	Vinylsulfonamides.= Intramolecular Diels–Alder	ORGN	0231
Are Your Policies Inviolate or in	Violation? Copyright and Technology: What Happen	CINF	0043
from Trichoderma	viride.= Amino Acid Sequence of a 20K Xylanase	BIOT	0211

ic Substances against Exotic RNA	Viruses.= +ceae Constituents and Related Synthet	AGFD	0165
ing a Re+ Rapid-Scanning UV-	VIS Detection for Capillary Gas Chromatography Us	ENVR	0060
Light Scattering (SEC-	VIS-RALLS).= SEC-Viscometry-Right Angle Laser	PMSE	0098
to Predict the Performance of the	Visbreaking Process.= +A New Analytical Method	FUEL	0046
ce as a Diagnostic Tool in	Viscoelastic Media.= Dynamic Magnetic Birefringen	COLL	0143
e Gelation of Polymers. Dynamic	Viscoelasticity during Thermoreversible Gelatin Gela	POLY	0197
Light Scattering and	Viscoelasticity of Linear and Ring Polyelectrolytes.=	POLY	0294
Weig+ Use of a Single-Capillary	Viscometer for Accurate Determination of Molecular	PMSE	0066
erpretation of Data from an SEC-	Viscometer System.= +roach for Quantitative Int	PMSE	0065
rmation by SEC with an On-Line	Viscometry and Laser Light Scattering.= +r Confo	PMSE	0097
olymer Characte+ Modern SEC-	Viscometry and New Experimental Possibilities for P	PMSE	0070
Modern SEC-	Viscometry Detectors.= Flow Rate Effects in	PMSE	0067
MWD and Branching with GPC-	Viscometry in THF and DMF.= +ation of Polymer	PMSE	0068
Light Scattering	Viscometry.=	PMSE	0096
(SEC-VIS-RALLS).= SEC-	Viscometry-Right Angle Laser Light Scattering	PMSE	0098
ation+ Predictive Model for the	Viscosities of High Molecular Weight Liquids: Applic	POLY	0065
Correlation between Intrinsic	Viscosity and Molecular Weight.= One-Parameter	PMSE	0095
ectin Determined by HPSEC and	Viscosity Detection.= +of Dextran, Pullulan, and P	CARB	0045
m a Variant Strain of Xanthom+	Viscous Behavior of Xanthan Aqueous Solutions fro	BIOT	0196
rporation of a Diode Array UV-	Visible Spectrophotometer into the Undergraduate C	CHED	0180
and Characterization of Diama+	Visible-Transparent Rigid-Rod Polymers: Synthesis	POLY	0028
cular Membranes Modeled on the	Visual Information Processing of the Retina.= +le	BIOT	0124
Transfer to	Vitamin B_{12}.= Electrochemical Study of Electron	BIOL	0135
xidants Structurally Related to	Vitamin E.= Custom Design of Better in Vivo Antio	AGFD	0155
la Oxidacion de la	Vitamina C.= Obtencion del Orden de Reaccion de	CHED	0164
ecto de la Deficiencia o Exceso de	Vitamina E en la Fertilidad de Ratas Hembras.=	AGFD	0210
Thermoreversible Gelation and	Vitrification of Concentrated Polymer Solutions und	POLY	0146
ng and Chemical Cross-Linking in	Vitrified Nonlinear Optical Polymer Films.= +deri	POLY	0206
allization (HDM) of NI-TPP and	VO-TPP.= +nadium Catalysts in the Hydrodemet	CATL	0044
g Polymer Education into Middle,	Vocational, and High School Curriculum.= +gratin	CHED	0026
New	Volatile Compounds of Late Group 2 Elements.=	COLL	0154
ree-Dimensional Analysis of the	Volatile Constituents of Living Odors and Flavors.=	AGFD	0022
elenolate Ligands.=	Volatile Metal Complexes with Silyltellurolate and S	INOR	0172
ing+ Measuring Trace Levels of	Volatile Organic Compounds in Human Blood by Us	CEI	0001
Nitrogen-Contai+ Formation of	Volatile Precursors of NO_x in the Rapid Pyrolysis of	FUEL	0114
Flavor	Volatiles and Related Compounds in Maple Syrups.=	AGFD	0174
ic Analyses of Lipid Oxidation	Volatiles in Foods.= Update on Gas Chromatograph	AGFD	0098
aph+ Determination of Motor Oil	Volatility Using High-Temperature Gas Chromatogr	PETR	0122
the Surface Microl+ Kinetics of	Volatilization of Fenitrothion and Deltamethrin from	ENVR	0020
from	Volo Bog.= Organic Geochemical Studies of Peat	I&EC	0016
of N-Alkylamides with	Voltage-Sensitive Sodium Channels.= Interactions	AGRO	0151
f Cationic Surfactants in Anodic	Voltammetric and Coulometric Analysis of Insoluble,	COLL	0002
rdered Polymer Lattices.=	Voltammetric Investigation of Proton Transport in O	COLL	0167
st Electrochemical Met+ Anodic	Voltammetry at Extreme Potentials: A Hope for a Fa	ANYL	0095
f Human Seric Copper by Inverse	Voltammetry at the Mercury Film Electrode.= +o	ANYL	0056
ical Chemistry Lab+ Two Cyclic	Voltammetry Experiments for the Introductory Phys	CHED	0211
mbrane Potential Studied by the	Voltammetry for Ion Transfer at a Liquid-Liquid In	BIOT	0105
sicles.=	Voltammetry in Lipid Films with Applications to Ve	COLL	0021
eton Applied Research Co.). Pulse	Voltammetry: Saturday's Child.= +by EG G Princ	ANYL	0096
Emulsion	Voltammetry.= Reaction-Convection-Diffusion in	COLL	0042
redictive Capabilities of a Free-	Volume Theory for Polymer-Solvent Diffusion Coeff	POLY	0144
f Difluoroacetylene Using a New	Volume-Rendering Molecular Graphics Technique.=	COMP	0055
iols by the $IrCl_6^{2-}$ Ion.=	Volumes of Activation for the Oxidation of Benzened	INOR	0395
Pollution Prevention through	Voluntary Initiatives.= EPA's 33/50 Project:	CEI	0024
the Degree of Transformation of	VPI-5 to $AlPO_4$-8 at Various Temperatures by Mult	PETR	0095
ynthesis and Characterization of	VPI-6: Another Intergrowth of Hexagonal and Cubic	PETR	0001
Molecular Sieve.= Synthesis of	VPI-7: A Novel 3-Membered Ring Containing	PETR	0098
by a Nickel Atom in Desulfovibrio	Vulgaris (Hildenborough) Rubredoxin.= +acement	INOR	0020
Evolved from a Co+ Phaseolus	vulgaris α-Amylase Inhibitors, Arcelins, and Lectins	AGFD	0041
$Ru(BPY)_3^{2+}$ on Porous	Vycor Glass.= Excited-Stated Dynamics of	INOR	0160
Carbonyls on Porous	Vycor Glass.= Photocatalytic Behavior of Metal	INOR	0170
Porous	Vycor Glass.= Photochemistry of $(Cp)_2TiCl_2$ on	INOR	0161
in Crystalline Rock: Case Study—	Waldoboro Pluton Complex, Maine.= +al Mercury	ENVR	0055
rides Corresponding to the Cell-	Wall Polysaccharide of the β-Hemolytic Streptococci	CARB	0057
lcium Carbonate on the Internal	Walls of Glass Capillaries: Influence of Temperature,	COLL	0163
r Own Chemical Business. So You	Want to Start Your Own Chemical Business.= +u	SCHB	0017
ed Food Ingredients to Minimize	Warmed-Over-Flavor Development in Meat Product	AGFD	0021
Greenhouse	Warming.= Policy Implications of	ENVR	0064
li Proteins and En+ Methods of	Washing sCD4-183 Inclusion Bodies to Remove E. co	BIOT	0171
al of Chromium from Soils by Soil	Washing with Surfactant Solutions at Low pH.=	ENVR	0013
Biolog+ Status of the Disposal of	Waste Agricultural Chemicals and Their Containers.	AGRO	0139
Curre+ Status of the Disposal of	Waste Agricultural Chemicals and Their Containers.	AGRO	0120
Intro+ Status of the Disposal of	Waste Agricultural Chemicals and Their Containers.	AGRO	0085
Pestic+ Status of the Disposal of	Waste Agricultural Chemicals and Their Containers.	AGRO	0102
d Other To+ Clean Energy from	Waste and Coal: Ash Utilization, Characterization, an	FUEL	0136
Waste. Co+ Clean Energy from	Waste and Coal: Energy from Plastics and Municipal	FUEL	0090
Waste. Gas+ Clean Energy from	Waste and Coal: Energy from Plastics and Municipal	FUEL	0091
Waste. Re+ Clean Energy from	Waste and Coal: Energy from Plastics and Municipal	FUEL	0092
ss, and Mun+ Clean Energy from	Waste and Coal: Energy from Sewage Sludge, Bioma	FUEL	0115
ns Control+ Clean Energy from	Waste and Coal: The Role of Metals. Metals Emissio	FUEL	0129

Characterization of Municipal	Waste by Scanning Electron Microscopy and Optical	FUEL	0137
Handling of	Waste Chemicals in Educational Laboratories.=	ENVR	0036
Cells on Fibers for	Waste Clean–Up.=	FUEL	0133
nd Mercury from RDF Municipal	Waste Combustors.= +Control of Cadmium, Lead, a	FUEL	0130
rological Approaches to Pesticide	Waste Destruction Using White Rot Fungi.= +otech	AGRO	0126
Current Technologies for Pesticide	Waste Disposal: Overview.= +Their Containers.	AGRO	0120
Their Containers. Introduction to	Waste Disposal.= +ste Agricultural Chemicals and	AGRO	0085
y and Microstructure of Solidified	Waste Forms. Chemistry of Cementitiou+ Chemistr	ENVR	0001
tion Techniques for Cementitious	Waste Forms. Use and Poten+ Surface Characteriza	ENVR	0007
anism of Portland–Cement–Based	Waste Forms.= +r Diffusion Is the Leaching Mech	ENVR	0011
ncrete Containing Municipal Solid	Waste Incinator Ash.= +gth Portland Cement Co	FUEL	0136
Metals Behavior during Medical	Waste Incineration.=	FUEL	0131
missions Control Technologies for	Waste Incinerators.= +e Role of Metals. Metals E	FUEL	0129
Waste Managemen+ Laboratory	Waste Management. Practical Aspects of Laboratory	LWM	0001
istry of Cementitious Systems for	Waste Management: The Penn State Experience.=	ENVR	0006
t. Practical Aspects of Laboratory	Waste Management.= +ratory Waste Managemen	LWM	0001
erephthalate (PET) from Plastics	Waste on the Thermal Properties of Asphalt.= +T	FUEL	0053
unicipal Waste. Gasification of	Waste Plastics for the Production of Fuel–Grade Gas	FUEL	0091
Extruding	Waste Plastics with High Mixing and Low Shear.=	I&EC	0030
d Municipal Waste. Conversion of	Waste Polymers to Energy Products.= +lastics an	FUEL	0090
Adjacent to Chromate Production	Waste Sites.= +in the Urine of Individuals Living	CEI	0009
oilizing Components of the Plastic	Waste Stream via Reactive Processing.= +Compati	I&EC	0033
Acid Hydrolysis.+ Conversion of	Waste to Ethanol and Electricity via Dilute Sulfuric	FUEL	0094
gradation in Site Remediation and	Waste Treatment: Past, Present, and Future.= +e	AGRO	0121
nergy from Plastics and Municipal	Waste. Co+ Clean Energy from Waste and Coal: E	FUEL	0090
nergy from Plastics and Municipal	Waste. Gas+ Clean Energy from Waste and Coal: E	FUEL	0091
nergy from Plastics and Municipal	Waste. Re+ Clean Energy from Waste and Coal: E	FUEL	0092
e Sludge, Biomass, and Municipal	Waste. +from Waste and Coal: Energy from Sewag	FUEL	0115
Managing Pesticide	Waste: A Developing Country Perspective.=	AGRO	0107
ethods for Disposal of Cattle–Dip	Waste.= +cals and Their Containers. Biological M	AGRO	0139
Pretreatment of Municipal Solid	Waste.= Ammonia Fiber Explosion (AFEX)	BIOT	0016
Clean Energy from Municipal	Waste.=	FUEL	0093
Prehydrolyzates of Cellulosic Food	Waste.= +Biomass and Chemical Feedstocks from	BIOT	0153
Managing Our Nation's Nuclear	Waste.=	CHAS	0046
Cement–Solidified Hazardous	Waste.= Petrographic Techniques Applied to	ENVR	0009
Commercial Application of	Waste–Fuel–Fired Hybrid Fluidized–Bed Boilers.=	FUEL	0122
m Geothermal Brine Utilizing a	Waste/Low–Grade Heat Source with a Heat Pump.=	I&EC	0010
a Human Exposure to Hazardous	Wastes: Assessment of Prevalence, Bioavailability,	CEI	0008
Charcoal Filtration of Pesticide	Wastes: Field and Laboratory Studies.= Activated	AGRO	0127
emical Degradation of Pesticide	Wastes: Historical, Current, and Future Technologies	AGRO	0122
Detoxification of Hazardous	Wastes: Laboratory Studies.= Gas–Phase Solar	ENVR	0035
s––I+ Biodegradation of Organic	Wastes: Molecular Approaches and Field Application	BIOT	0117
s––II+ Biodegradation of Organic	Wastes: Molecular Approaches and Field Application	BIOT	0137
Organics in Mixed Nuclear	Wastes: The Challenges.= Analyzing Exotic	ANYL	0060
emistry of Cementitious Solidified	Wastes.= +tructure of Solidified Waste Forms. Ch	ENVR	0001
ion of Solidified Toxic Hazardous	Wastes.= +of Cement Matrices for the Immobilizat	ENVR	0002
minants from Stabilized/Solidified	Wastes.= +brium Chemistry on Leaching of Conta	ENVR	0010
ement–Solidified/Stabilized Metal	Wastes.= +copy Characterization Techniques for C	ENVR	0005
ation of Solidified Hazardous	Wastes.= Microscopic and Spectroscopic Characteriz	ENVR	0004
Cement–Solidified Heavy Metal	Wastes.= Microstructural Characterization of	ENVR	0008
Solar Detoxification of Gas–Phase	Wastes.= +er TiO$_2$: A Potential Technique for the	COLL	0082
from Gasification of Sludge	Wastes.= Production of Clean–Medium Btu Gas	FUEL	0118
ative and Practical Solutions for	Wastewater and Rinsate Management for Retail Fert	FERT	0014
Pesticide	Wastewater Cleanup.=	AGRO	0123
nd Implications for Disinfection of	Wastewater.= +ation of Phenylalanine: Products a	ENVR	0031
uction of Toxic Components and	Wastewaters in Manufacture of Disperse Blue 79.=	CMEC	0018
e Utiliz+ Production of Distilled	Water and Chemical Products from Geothermal Brin	I&EC	0010
in Gove+ Pesticides in Surface	Water and Groundwater—Advances in Research and	AGRO	0051
in Gove+ Pesticides in Surface	Water and Groundwater—Advances in Research and	AGRO	0060
ntaminant Migration in Ground	Water as Assessed by EPA's Hazard–Ranking System	ENVR	0081
ids Systems Based on Geothermal	Water as the Continuous Phase.= +of Drilling Flu	I&EC	0011
actions of Cement Minerals with	Water at Elevated Temperatures and Pressures in G	INOR	0246
ples Related to Their Hydration	Water Contents and Paramagnetic Metal Ion Concen	MEDI	0043
Tension by Automation of the +	Water Drop Experiment—Determination of Surface	CHED	0118
Isotopic Alteration of Organic	Water during Stepped Combustion.= Chemical and	ANYL	0061
nd Spectroscopic Studies of the	Water Gas Shift Reaction over Ruthenium–Exchange	CATL	0027
Simulation of the Role of	Water in Ion Channel Gating.=	COLL	0084
icizers on the Diffusion Process of	Water in Poly(methyl Methacrylate).= +ce of Plast	POLY	0193
–III. Lateral Diffusion at the Air–	Water Interface.= +utions, Blends, and Interfaces–	COLL	0051
Polymers at the Air–	Water Interface.= Dynamic Properties of Cellulosic	COLL	0056
Diffusion of	Water into a Photopolymer Film.=	POLY	0194
of Organic Acids from Methanol–	Water Mixtures.= +nt and pH Effects on Sorption	ENVR	0014
ethane by Catalytic Supercritical	Water Oxidation.= Selective Partial Oxidation of M	CATL	0043
mical Applications—II. Octanol–	Water Partition Coefficients by Centrifugal Partition	ANYL	0068
erse Relaxation Times ('T2's) of	Water Protons in Human Plasma Samples Related to	MEDI	0043
D+ Paramagnetic Relaxations of	Water Protons in the Solutions of the Iron–Ethylene	INOR	0090
Iron–Promoted Magnesium–	Water Reaction.: A Mechanism for the	INOR	0394
m: Metho+ Advances in Surface–	Water Research Related to Pesticide Use on the Far	AGRO	0054
Concentrating Cellulase from a	Water Solution onto Water Vapor Droplets.=	BIOT	0021
Photosynthetic	Water Splitting by Platinized Chloroplasts.=	TECH	0016

Susceptibility of Surface Drinking	Water Supplies to Pesticide Contamination.= +the	AGRO 0053
T Preparation of Granular Cold	Water Swelling/Soluble Starches by Alcoholic–Alkali	CARB 0040
insic Fluoroph+ Measurement of	Water Uptake in Cured Epoxy by Extrinsic and Intr	POLY 0160
atory Issues. Pesticide Analysis in	Water Using Ion Trap Mass Spectrometry.= +gul	AGRO 0060
from a Water Solution onto	Water Vapor Droplets.= Concentrating Cellulase	BIOT 0021
Chemical Modulator to Measure	Water Vapor in a Gaseous Mixture.=	ANYL 0052
Diffusion of	Water Vapor in Hydrogen–Bonded Polymer Blends.=	POLY 0192
and Sucrose Crystals.= Rate of	Water Vapor Interaction with Carbon Adsorbents	CARB 0062
Magnetic Reson+ Interactions of	Water with Clay Minerals as Studied by ^{2}D Nuclear	AGRO 0061
ed Catalytic Reactions of CO_2 and	Water: A Clean Source of Energy.= +rowave–Induc	CATL 0055
ed Catalytic Reactions of CO_2 and	Water: A Clean Source of Energy.= +rowave–Induc	FUEL 0102
Pesticides in Drinking	Water: A Farm Wife's Perspective.=	AGRO 0055
xide) Block Copolymer Micelles in	Water.= +Methods. Polystyrene–Poly(ethylene O	POLY 0277
e of Hiding Pigments Dispersed in	Water.= +lymerization of Monomers on the Surfac	PMSE 0038
mical Method of Generation of	Water.= +Potentials: A Hope for a Fast Electroche	ANYL 0095
ids Systems Based on Geothermal	Water.= +ive Effects on Drill Pipes by Drilling Flu	I&EC 0012
Pesticides and Metabolites in	Water.= Current Methodologies for Analyzing	AGRO 0146
tonitriles in Chlorinated Drinking	Water.= +acetic Acids, Halomethanes, and Haloace	ENVR 0028
in from the Surface Microlayer of	Water.= +tilization of Fenitrothion and Deltamethr	ENVR 0020
Contaminants in Drinking	Water.= Risk Assessments for Microbiological	ENVR 0070
–Bound Enamine Intermediate in	Water.= +ervation of the Kinetic Fate of a Thiamin	BIOL 0047
tions of Phosphate Compounds in	Water?.= +Metaphosphate an Intermediate in Reac	INOR 0293
erformance Study of the System	Water–Calcium Chloride in a Heat Transformer.= P	I&EC 0009
	Water–Compatible Phenolic Resins.=	PMSE 0139
Plasmin.= Effects of	Water–Miscible Organic Cosolvents on Kinetics of	BIOL 0092
ombined Enzyme System in the	Water–Restricted Environment of Nonaqueous Solve	BIOL 0091
s Activity of L–693,989 and Other	Water–Soluble Analogues of L–688,786.= +mocysti	MEDI 0182
er Nucleoside–Subst+ The First	Water–Soluble Porphyrinyl–Mononucleoside and Oth	ORGN 0189
ules by Flow Fiel+ Separation of	Water–Soluble Synthetic and Biological Macromolec	PMSE 0011
with Improved Flexibilit+ Novel	Waterborne Epoxy Resins for Can and Coil Coatings	PMSE 0142
Chronology of Natural	Waters at the Nevada Test Site.= Thorium–230	NUCL 0084
of H_2S in Natural	Waters.= Intermediate Products from the Oxidation	GEOC 0005
Marine	Waters.= Metal Sulfide Complexes in	GEOC 0006
r the Treatment of Contaminated	Waters.= +ntially Useful Photosensitizing Agent fo	ENVR 0029
Farm: Methods to Monitor Small	Watersheds for Pesticides.= +Pesticide Use on the	AGRO 0054
Square	Wave Detection for Flow–Injection Analysis.=	ANYL 0105
Composite.= Optical Four–	Wave Mixing in a Silver Colloid–Polymer	POLY 0218
Optical Bistability and Four–	Wave Mixing in Bacteriorhodopsin Materials.=	BIOT 0081
tt Films.= Guided–	Wave Nonlinear Optics in DCANP Langmuir–Blodge	POLY 0176
Permanent	Wave.=	CHED 0064
Planar Polymer	Waveguides for Optical Interconnections.=	POLY 0175
e Index Profiles in Polymeric Slab	Waveguides.= +otochemically Delineated Refractiv	POLY 0180
Optically Nonlinear Polymers in	Waveguiding Passive and Active Devices.=	POLY 0177
Optics of Polymer Thin Films, 5:	Wavelength Dispersion of the $\chi^{(3)}$ of Poly(p–phenyle	POLY 0213
	Wavelength–Dependent Organic Photochemistry.=	ORGN 0135
Participation in Shock	Waves.= Kinetic Processes with Metal Cluster	PHYS 0175
exadiyne behind Reflected Shock	Waves.= +actions of Propargyl Chloride and 1,5–H	FUEL 0068
and	Waxes.= Characterization of Waxy Crudes	PETR 0123
Characterization of	Waxy Crudes and Waxes.=	PETR 0123
ed+ Studies of the Reactions of	WCl_6, $WOCl_4$, and $MoOCl_4$ Molecules Chemically Fix	COLL 0098
y Training, 1980–1991: Where Did	We Come From; Where Are We Going?.= +l Safet	CHAS 0043
ular Electron and En+ How Are	We Doing in Predicting Absolute Rates of Intramolec	ORGN 0143
st–Forming Processes—Or, How	We Go from the Bench Top to Commercial Reality.=	PETR 0089
e Did We Come From; Where Are	We Going?.= +l Safety Training, 1980–1991: Wher	CHAS 0043
istry in the Gas Phase: what Can	We Learn from Reactions without Solvents?.= +em	ORGN 0121
ith+ Intermolecular Forces: Do	We Really Need to Compare ab Initio Calculations w	PHYS 0104
s According to Risk: What Must	We Yet Know To Accomplish This Task? Relative R	ENVR 0061
scopy Studies of Friction and	Wear at the Nanometer Scale.= Atomic Force Micro	PHYS 0156
dsorption Energies in Slitlike and	Wedge–Shaped Micropores.= +s. Calculation of A	FUEL 0008
erf+ Determination of Molecular	Weight Characteristics of Polysaccharides by High–P	BIOT 0197
ion of Lignin with Low Molecular	Weight Components.= +fication, and Copolymerizat	BIOT 0041
y Innovation: Polymer Molecular	Weight Determination by Gel Permeation in Aqueous	CHED 0210
for Polymer Analysis. Molecular	Weight Determinations of Poly–4–Methyl–1–Pentene	PMSE 0046
Measurement of the Molecular	Weight Distribution of Vacuum Residue Using Mass	PETR 0150
Blends with Narrow Molecular–	Weight Distribution: Hydrogen– and Deuterium–Lab	COLL 0035
ctivated Pathways for Molecular–	Weight Growth.= +tion. Analysis of Chemically A	FUEL 0063
the Viscosities of High Molecular	Weight Liquids: Application to Polymer Plasticizers	POLY 0065
nsion by Automation of the Drop–	Weight Method.= +—Determination of Surface Te	CHED 0118
ties of Cur+ Effect of Molecular	Weight of Poly(dimethylsiloxane) on Physical Proper	PMSE 0174
Factors Affecting the Molecular	Weight of Polyesters Formed in Enzyme–Catalyzed P	BIOT 0094
ic Polystyrene: Effect of Molecular	Weight on Glass–Transition Temperature.= +rocycl	POLY 0239
ee SiC Fibers.= High Molecular	Weight Polycarbosilane as a Precursor to Oxygen–Fr	POLY 0360
High Molecular	Weight Polyethers.= Silicon–Assisted Synthesis of	PMSE 0075
w Melting of Ultrahigh Molecular	Weight Polyethylene.= +Fractions Found in the Slo	PMSE 0166
Intrinsic Viscosity and Molecular	Weight.= One–Parameter Correlation between	PMSE 0095
of High Molecular	Weight, Single–Stranded DNA.= Laser Vaporization	ANYL 0130
Narrowing Techniques and ^1H T_1	Weighting.= +eactions Using Multiple–Pulse Line–	CATL 0022
urate Determination of Molecular	Weights and Mark–Houwink Constants.= +for Acc	PMSE 0066
irect Computation of Statistical	Weights from Conformational Analysis for RIS Mode	POLY 0356
uired for H+ Role of Health and	Welfare Canada in Review of MSDS and Labels Req	CHAS 0034

of Cements Used in Geothermal	Wells.= Standards for X-ray Fluorescence Analysis	ANYL	0058
rization of Material Surfaces via	Wettability Analysis: A Molecular Dynamics Simulat	POLY	0369
Using Poly(vinyl Me+ Improved	Wettability of Clofazimine in Coevaporates Prepared	COLL	0086
tives on Quartz an+ Study of the	Wetting Action of Melted Sodium Silicate with Addi	COLL	0079
cts of Ambient Humidity on the	Wetting Properties and Adhesion of Bacterial Surfac	BIOT	0151
Is in It and	What's in It for You.= Chemistry in Context: What	CHED	0052
ials Densification.=	What's New in Collection Systems? On-Board Mater	I&EC	0040
	What's on the Surface during Diamond Growth?.=	PHYS	0122
act Differently with Oligoribon+	Wheat Germ Initiation Factors 4F and (ISO)4F Inter	BIOL	0128
arch in a Milled Hard Red Winter	Wheat.= +id, Total Dietary Fiber, Protein, and St	AGFD	0143
rtland-Cement-B+ Determining	Whether Diffusion Is the Leaching Mechanism of Po	ENVR	0011
Propionic Acid Production from	Whey: Effects of Growth Substrate and Product Ex	BIOT	0030
s: A Psychologist's Perspective on	Whistle Blowing in Science.= +th and Consequence	PROF	0002
Industrial	Whistle Blowing.= Two Personal Experiences with	PROF	0007
he Groves of Academe.=	Whistle-Blowers, Advocates, and the Law. Greed in t	PROF	0001
To Help	Whistle-Blowers, Legalize PEG.=	PROF	0012
Legal Rights of	Whistle-Blowing Professionals.=	PROF	0011
Pesticide Waste Destruction Using	White Rot Fungi.= +otechnological Approaches to	AGRO	0126
Constituents of	White-Fleshed Nectarine.= Free and Bound Flavor	AGFD	0024
Personal Recollections of Frank C.	Whitmore by Charles D. +Honoring O. Hannaway.	HIST	0009
ican Chemical Company.=	WHMIS Data Sheets from the Viewpoint of an Amer	CHAS	0035
How and	Why Do Nuclei Fragment?.=	NUCL	0006
el Discussion: Bridging the Gap;	Why More Is Better. D. L. Leister, Moderator.= Pan	I&EC	0058
The How and	Why of Nonaqueous Enzymology.=	BIOT	0067
Modified Starch Graft Copolym+	Wicking and Absorbency in Aluminum and Silicate-	CARB	0042
adyl Acetylacetona+ EPR Line-	Width Study of the Anisotropic Reorientation of Van	PHYS	0264
Farm	Wife's Perspective.= Pesticides in Drinking Water: A	AGRO	0055
rge-Recombination Dynamics in	Wild-Type and Mutant Photosynthetic Reaction Cen	BIOL	0132
s+ Public Policies and Attitudes	Will Affect Pesticide Use and Development in the 21	AGRO	0070
in Insulin He+ Methyl Paraben	Will Cause a T←R-Like Conformational Transition	BIOT	0150
ly Transient Kinetic Techniques	Will Provide Detailed Insight in the Mechanism of G	FUEL	0015
n Solvers or Spreadsheets: Which	Will Supplant the Hand-Held Calculator?.= +uatio	CHED	0117
+mmercial Foods, Using Monier-	Williams and Liquid Chromatographic Methods.=	AGFD	0109
hanism Block Copol+ Sherwin-	Williams Student Award Symposium. Change of Mec	PMSE	0071
affinity Methods for Purification (	WIMP)—Don't Be Fooled by the Name.= +muno	CARB	0054
MOCVD.= The ULSI	Window: A View of Al and Cu Complexes for	INOR	0125
ctive Compounds Found in Grape	Wine due to Malolactic Fermentation.= +e Odor-A	AGFD	0137
Precursors of Varietal Grape and	Wine Flavor.= +to Flavor Compounds. Glycosidic	AGFD	0006
Carbamate in	Wines.= Method for the Determination of Methyl	AGFD	0110
POLYED 1989 Polymer Award	Winner: Polymers Are Elementary!.=	CHED	0028
s Macr+ POLYED 1991 Award	Winner—Award Presentation: Addressing Polymers a	CHED	0027
Design of an Award-	Winning Laboratory Simulation.= SQUALOR—The	CHED	0079
and Starch in a Milled Hard Red	Winter Wheat.= +id, Total Dietary Fiber, Protein,	AGFD	0143
ward Symposium Honoring W. T.	Wipke: New Dimensions in Chemical Information.	CINF	0016
tion of Au and Pt Microfabricated	Wire Arrays.= +s for Selective Surface Derivatiza	INOR	0350
of CH_4 in a Hot-	Wire Distribution Column.= Dehydrogenative Coupling	PETR	0012
de Nemours and Co.). Teaching: A	Witches' Brew of Black Holes?.= +E. I. DuPont	ANYL	0006
oporphyrins+ Effect of Electron-	Withdrawing Substituents on Porphyrins and Metall	INOR	0301
Use of the Expert	Witness in Products Liability Litigation.=	CHAL	0015
and+ The Scientist as an Expert	Witness. Overview of Legal Decision-Making Process	CHAL	0011
Chemists as Expert	Witnesses in Patent Litigation.=	CHAL	0013
The	Wittig Reaction: How Does It Really Work?.=	ORGN	0278
Mechanisms of the	Wittig Reaction.= Phosphorus Chemistry.	INOR	0373
Conformational Analysis—	WIZARD III.= Automated Learning in	CINF	0017
ational Analysis.=	WIZARD-III Recent Developments in Al in Conform	CINF	0034
osphate Tungsten Bronze $(PO_2)_4($	$WO_3)_{2m}$, m = 2.= +sport Properties of the Monoph	INOR	0352
tudies of the Reactions of WCl_6,	$WOCl_4$, and $MoOCl_4$ Molecules Chemically Fixed on	COLL	0098
onne Premium Coal Sample Using	Wood Ash or Taconite as Additive.= +minous Arg	FUEL	0006
de Hydrolyzing Enzymes During	Wood Degradation by Phanerochaete chrysosporium.	BIOT	0208
Compos+ Surface Modification of	Wood Fibers and Their Performance in Polystyrene	PMSE	0112
Contaminated Soil Samples from	Wood-Preserving Plants in Quebec: Extensive Anal	ENVR	0016
Really	Work?.= The Wittig Reaction: How Does It	ORGN	0278
=	Worker Training Requirements under HAZWOPER.	CHAS	0018
Education for the HAZMAT	Worker.=	CHAS	0045
ation+ Biological Monitoring of	Workers Exposed to Triazine Herbicides, I: Determin	CEI	0004
nation+ Biological Monitoring of	Workers Exposed to Triazine Herbicides, II: Determi	CEI	0005
Starting a Business vs.	Working for One.=	YCC	0012
Chemical Accounting: Making	Workplace Chemical Lists.=	CHAS	0041
anada+ Regulation of Hazardous	Workplace Chemicals. Role of Health and Welfare C	CHAS	0034
nd Labels Required for Hazardous	Workplace Chemicals.= +a in Review of MSDS a	CHAS	0034
	Workplace Drug Testing: Background and Current T	HIST	0008
Sheets—Effective in the	Workplace?.= Material Safety Data	CHAS	0036
rbon-Active Site Behavior via +	Workshop on Carbon Active Sites. Elucidation of Ca	FUEL	0013
Student's	World.= Chemistry Teaching through the	CHED	0005
Postpetroleum	World.= Coal and Carbon in the	FUEL	0054
Real	World.= Enantioselective Chromatography in the	ANYL	0120
Changing	World.= Phosphorus Flame Retardants for a	INOR	0346
Computerized	World.= The Place of Reference Books in a	CHAS	0039
Laboratory.= Integrated	Writing Program in the Physical Chemistry	CHED	0213
Laboratory.= Integrated	Writing Program in the Physical Chemistry	PHYS	0318

unctional Forms and Reactions by	XAFS Spectroscopy.= +itative Analysis of Sulfur F	FUEL	0056
of Calcium Dispersed on Carbon:	XANES and EXAFS Study.= +ture and Structure	FUEL	0032
Xanthom+ Viscous Behavior of	Xanthan Aqueous Solutions from a Variant Strain of	BIOT	0196
Solutions from a Variant Strain of	Xanthomonas campestris NRRL B-1459.= +eous	BIOT	0196
etries and Molecule Properties of	XCN (X = H, Cl, Br) in Various Ionic States.= +m	PHYS	0238
Analysis of Porous Carbon with	^{129}Xe NMR and N_2 Porosimetry.=	PETR	0070
In Situ Techniques for Studying	Xenobiotic Metabolism in Food-Producing Animal	AGRO	0092
pecies.=	Xenobiotic Metabolism in Food-Producing Aquatic S	AGRO	0112
Shellfish+ Methods for Studying	Xenobiotic Metabolism in Food-Producing Fish and	AGRO	0096
Methods for the Study of	Xenobiotic Metabolism in Vitro.=	AGRO	0093
se of Biological Models to Direct	Xenobiotic Metabolism Studies in Food Safety Evalu	AGRO	0132
egulatory Aspects of the Use of	Xenobiotics in Food-Producing Animals in the Unite	AGRO	0076
hniques for Studying+ Fate of	Xenobiotics in Food-Producing Animals. In Situ Tec	AGRO	0092
n—The Uses of Veterin+ Fate of	Xenobiotics in Food-Producing Animals. Introductio	AGRO	0075
and Residue Studies o+ Fate of	Xenobiotics in Food-Producing Animals. Metabolism	AGRO	0108
Absorption and Metabolism of	Xenobiotics in Food-Producing Animals.= Dermal	AGRO	0095
n and Abamectin Meta+ Fate of	Xenobiotics in Food-Producing Anmimals. Ivermecti	AGRO	0128
agenicity and Bioavailability of	Xenobiotics in Foods.= Effect of Flavonoids on Mut	AGFD	0116
ciple of Corresponding Sta+ The	Xenon Interaction Potential from the Extended Prin	COMP	0058
Investigation of Reduced V_2O_5	Xerogel.=	INOR	0082
Studies on Silica Aerogels and	Xerogels Doped with Transition-Metal Ions.=	INOR	0081
faces.=	XPS Studies of Interaction of CH_3 with Mo(100) Sur	COLL	0217
$W(CO)_6$ on Pt (111).= TPD and	XPS Study of the Thermal Decomposition of	COLL	0049
Sputtering Depth Profiling Using	XPS.= +Reactions at Polymer/Metal Interface by	PMSE	0150
of Fluoride with Human Enamel:	XPS, SEM, and Solution Characterization.= +ction	COLL	0085
IR, Solid-State MAS NMR, and	XRD of Phosphorous-Promoted Hydrotreating Cata	COLL	0198
ficients for Matrix Corrections in	XRF Analysis of Zinc, Iron, and Total Sulfur in Zinc	ANYL	0057
La_2-	$xSr_xNiO_4+\delta$.= Electrical Properties and Structure of	INOR	0048
u_4Se_4, $Na_2Cu_2Se_2 \cdot Cu_2O$ and Cu_{12}-	xTe_4S_{13}-y.= +w Solid-State Compounds of Na_3C	INOR	0351
Production of Acetyl-	Xylan Esterase from Aspergillus niger.=	BIOT	0251
Kinetics of the Prehydrolysis of	Xylan-Containing Materials.=	CARB	0044
Practical Aspects of	Xylanase Applications in Pulp Bleaching.=	BIOT	0214
Amino Acid Sequence of a 20K	Xylanase from Trichoderma viride.=	BIOT	0211
Homologous Expression.=	Xylanase Genes of Streptomyces lividans and Their	BIOT	0213
Analysis of a 20-Kd	Xylanase of B. subtilis.= Structure-Function	BIOT	0212
ation and Characterization.=	Xylanases from Streptomyces roseiscleroticus: Purific	BIOT	0210
Cellulases and	Xylanases.= Induction and Inducers of Microbial	BIOT	0185
'-bis(Tetrahydrothiophenium)-p-	xylene Dibromides.= +ion of Mono-Substituted α,α	POLY	0013
Polystyrene and Poly(	xylenyl Ether).= Miscible Blends of Sulfonated	PMSE	0117
Novel $R_2Ba_{1.25}NiO_{5.25}$ (R = Tm,	Yb, Lu) Structure Type.=	INOR	0034
etic-Field Penetration Depth in	$YBa_2Cu_3O_x$ and $Bi_2Sr_2CaCu_2O_x$ Superconductors by	PHYS	0284
etic-Field Penetration Depth in	$YBa_2Cu_3O_x$ and $Bi_2Sr_2CaCu_2O_x$ Superconductors by	PHYS	0050
Conductivity in Single Crystals of	$YBa_2Cu_3O_6.3$: Photodoping to the Metallic State.=	PHYS	0011
Investigation of Superconducting	$YBa_2Cu_3O_7$ Using Raman Spectroscopy.=	PHYS	0052
ses and Structures of {[(CH_3CH)	$_3YbFe(CO)_4$}$\cdot 0.5CH_3CH$}$_n$ and {$(CH_3CN)_3YbFe(CO)_4$}	INOR	0332
$(CO)_4$]$\cdot 0.5CH_3CH$}$_n$ and {(CH_3CN)	$_3YbFe(CO)_4$}$_n$.= +d Structures of {[$(CH_3CH)_3YbFe$	INOR	0332
Five	Years' Experience.= DNA-Based Paternity Testing:	CHAL	0009
Expression of	Yeast AMP Deaminase in E. coli.=	BIOL	0083
hate-Dependent Enzyme Brewers'	Yeast Pyruvate Decarboxylase.= +hiamin Diphosp	BIOL	0032
ate-Bound Enamines on Brewers'	Yeast Pyruvate Decarboxylase.= +iamin Diphosph	BIOL	0048
arins in Susceptible and Fusarium	Yellows-Resistant Celery Cultivars.= +c Furocoum	AGRO	0040
and AscC, the Genes Coding for	Yersinia pseudotuberculosis CDP-D-Glucose Oxidor	BIOL	0110
rt Einstein College of Medicine of	Yeshiva University/Montefiore Medical Center.=	PROF	0004
Azomethine	Ylides.= Hartree Fock Description of Substituted	ORGN	0171
ttrium Oxide and Ch+ Origin of	YO Chemiluminescence from the Laser Ablation of Y	PHYS	0255
a Century of Chemistry in New	York (Commemorating the Centennial of the New Yo	HIST	0025
Indoor Radon Levels in New	York and New Jersey.=	ENVR	0023
rom the Hudson River to the New	York Bight.= +c Carbon and Reactor-Derived ^{14}C f	GEOC	0014
Chemistry in New	York City.= One Hundred Years of Industrial	HIST	0026
The New	York Local Section Olympiad Selection Process.=	CHED	0012
Centennial of the New	York Section of the American Chemical Society.=	HIST	0029
orating the Centennial of the New	York Section of the American Chemical Society).	HIST	0025
at Two Areas of Concern in New	York State: Oswego Harbor and the St. +Initiatives	ENVR	0084
American Chemical Society). New	York University and Chemistry in New York.=	HIST	0025
University and Chemistry in New	York.= +American Chemical Society). New York	HIST	0025
in New	York.= The Pharmaceutical Industry: Early History	HIST	0034
Chemists' Club of New	York.= The Salamander and Crossed Retorts: The	HIST	0031
Your Own Chemical Business. So	You Want to Start Your Own Chemical Business.=	SCHB	0001
What's in It for	You.= Chemistry in Context: What Is in It and	CHED	0052
NSF Funding Opportunities+ Do	Younger Chemists Get a Fair Shake? An Analysis of	YCC	0003
Constructing	Your Business Plan.=	SCHB	0004
etting the Chemical Kicks out of	Your Kitchen.= Hazards of Household Chemicals: G	CHAS	0051
Doing	Your Market Research.=	SCHB	0002
and Diligence? Be	Your Own Boss.= Have Intelligence, Imagination,	SCHB	0016
Your+ Starting and Operating	Your Own Chemical Business. So You Want to Start	SCHB	0001
al Business. So You Want to Start	Your Own Chemical Business.= +ur Own Chemic	SCHB	0001
n Ownership and Copyright: Are	Your Policies Inviolate or in Violation? Copyright an	CINF	0043
Marketing	Your Product(s).=	SCHB	0010
Setting	Your Strategies.=	SCHB	0003
How To Get	Your Students Involved in Project SEED.=	CHED	0039

scence from the Laser Ablation of	Yttrium Oxide and Chloride in Oxygen.= +ilumine	PHYS	0255
from and Liquefaction of Beulah-	Zap Lignite.= +Coal Samples. Moisture Removal	FUEL	0001
nolglyc+ HPLC Survey of Corn (	Zea mays L.) Populations and Inbreds for Silk Flavo	AGRO	0034
uropeptide Analogs in Helicoverpa	zea.= +y of Pheromone Biosynthesis Activating Ne	AGRO	0011
cesses Honoring J. F. Roth. New	Zealand Gas-to-Gasoline Plant: An Engineering Tou	I&EC	0060
encia del Contenido de Galio en	Zeolitas del Tipo Pentasil en la Aromatizacion de Pro	COLL	0125
vity and Surface Acidity of HM	Zeolite and Pd/HM Catalysts.= Studies on the Acti	PETR	0117
ion of the Solvent-Like Nature of	Zeolite Based on a Solvatochromic Indicator.= +nat	COLL	0076
otoinduced Catalytic Reactions in	Zeolite Cages.= +talysis: Shape Selectivity—II. Ph	CATL	0020
w Modification Method of Pt/KL	Zeolite Catalyst for Light Naphtha Reforming.=	PETR	0064
Olefins over	Zeolite Catalysts.= Alkylation of Isoparaffins with	PETR	0101
on over Modified Shape-Selective	Zeolite Catalysts.= +cetylene to Higher Hydrocarb	CATL	0034
over Triflic Acid-Modified Y-	Zeolite Catalysts.= Synthesis of MTBE and ETBE	CATL	0016
Pt Sites within One-Dimensional	Zeolite Channels.= +lite: Inhibited Deactivation of	CATL	0015
is. Competitive Role of Cations in	Zeolite Crystallization.= +d Pillared Clay Synthes	PETR	0074
Catalyst+ Relationship between	Zeolite Framework Composition and Hydrocracking	I&EC	0062
olid-State NMR Investigations of	Zeolite Lattice Structures.= +al High-Resolution S	PHYS	0348
tion, Structure, and Properties+	Zeolite NU-87: Aspects of Its Synthesis, Characteriza	PETR	0021
mperature.=	Zeolite Synthesis in the Na-THA System at Low Te	PETR	0022
Microwave Heating of	Zeolite Synthesis Mixtures.=	PETR	0080
of	Zeolite TON.= Precursor Phases and Nucleation	PETR	0040
Synthesis of	Zeolite Y in the Gelatin Solution.=	PETR	0003
AS NMR Studies of Dealuminated	Zeolite Y.= +osspolarization and Very Fast ^{27}Al M	PETR	0006
hydrocyclization Rates on Pt/L-	Zeolite: Inhibited Deactivation of Pt Sites within On	CATL	0015
on over Ruthenium-Exchanged X	Zeolite: Reaction Control by the Cocation.= +eacti	CATL	0027
Na-Y	Zeolite.= Characterization of RuCl$_3$ Impregnated	CATL	0035
Catalysis of a Modified ZSM-5	Zeolite.= Preparation, Characterization, and	CATL	0029
Reactions of Ethylene over Ru-Y	Zeolite.=	CATL	0031
ed Crystallization of ZSM-5 Type	Zeolite.= +ttering Studies of the Template-Mediat	PETR	0058
atalyst for Propane Aromatizat+	Zeolite-β and Isomorphously Substituted Ga-β as a C	PETR	0047
: Characterization and General+	Zeolite-Containing Catalysts for C_1-C_4 Aromatization	PETR	0067
ric Fields on Organic Reactions at	Zeolite-Supported, Nanometer-Sized Electrodes.=	CATL	0018
e Relaxation of Sodium Cations in	Zeolite-Y Probed by Double-Rotation NMR.= +tic	PHYS	0267
hosphates and Berylloarsenates—	Zeolites and Other Microporous Structures.= +ncop	PETR	0055
le of Cations in Zeo+ Advances in	Zeolites and Pillared Clay Synthesis. Competitive Ro	PETR	0074
s to Large-Pore Ze+ Advances in	Zeolites and Pillared Clay Synthesis. From Clathrasil	PETR	0036
the Crystallization+ Advances in	Zeolites and Pillared Clay Synthesis. Investigation of	PETR	0093
ical Aspects of Sil+ Advances in	Zeolites and Pillared Clay Synthesis. Molecular Chem	PETR	0054
to the Synthesis o+ Advances in	Zeolites and Pillared Clay Synthesis. Novel Approach	PETR	0019
the Pillaring of a+ Advances in	Zeolites and Pillared Clay Synthesis. Optimization of	PETR	0142
ered Inorganic C+ Advances in	Zeolites and Pillared Clay Synthesis. Pillaring of Lay	PETR	0106
Pillared Clays.= Advances in	Zeolites and Pillared Clay Synthesis. Supergallery	PETR	0125
haracterization of+ Advances in	Zeolites and Pillared Clay Synthesis. Synthesis and C	PETR	0001
nvestigation of Boron-Substituted	Zeolites and Silicates.= +clear Solid-State NMR I	PETR	0136
tion of Benzene on NaX and NaY	Zeolites by NMR Spectroscopy.= +geneous Adsorp	CATL	0025
Catalysis of Metal-Modified Y-	Zeolites for Hydrocracking of Coal-Derived Distillate	CATL	0050
drothermal Crystallization of MFI	Zeolites in an Alkaline Fluoride Medium.= +re Hy	PETR	0005
udies of the Crystallization of the	Zeolites Mordenite, L, Omega, and Offretite.= +c St	PETR	0041
of Gallium-Containing HZSM-5	Zeolites Used for the Aromatization of Small Alkanes	PETR	0043
s of Isomorphously Substituted	Zeolites with the ZSM-5 Structure.= EXAFS Studie	PETR	0081
isproportionation Reaction over	Zeolites: Effect of Si/Al Ratio and Coke Formation.=	PETR	0103
Propylene Oxidation on X and Y	Zeolites: Product Selectivity and Correlation of Prod	CATL	0017
From Clathrasils to Large-Pore	Zeolites: The Effects of Systemically Altering a Temp	PETR	0036
atization Reactions over HZSM-5	Zeolites.= +for Investigation of Light Olefin Arom	PETR	0061
rzenes over Silylated Gallosilicate	Zeolites.= +-Selective Alkylation of Mono Alkylbe	PETR	0100
in	Zeolites.= Deuterium NMR Studies of Acid Sites	COLL	0197
Molecules in A	Zeolites.= Diffusivity of Hydrogen and Methane	CATL	0024
EUO-Framework	Zeolites.= Gallosilicate: A Novel Derivative of	PETR	0076
r Zinc-Containing Ultra-Stable Y	Zeolites.= +nc Content on Aromatics Formation ove	PETR	0048
N-Hexane on Nickel-Containing	Zeolites.= Kinetics of Hydroisomerization of	PETR	0116
of Metal Alkyls with Deboronated	Zeolites.= +atalytic Sites Formed by the Reaction	PETR	0024
Oxide by Ammonia over Fe-Y	Zeolites.= Selective Catalytic Reduction of Nitric	COLL	0212
d Pairing Avoidance Principles in	Zeolites.= +rectical Foundation of the Al Sitting an	PHYS	0289
Species Adsorbed on	Zeolites.= UV Resonance Raman Spectroscopy of	COLL	0075
ace Organometallic Chemistry on	Zeolites—Synthesis and Characterization of Mordeni	PETR	0131
atalytic Reduction of NO$_x$ Using	Zeolitic Catalysis for High-Temperature Application	I&EC	0046
Both Amorphous and	Zeolitic Isomerization Catalyst.= Improvements on	PETR	0115
at Lakes Areas of C+ Achieving	Zero Discharge of Persistent Toxic Substances in Gre	ENVR	0083
thesis and Properties of Novel,	Zero-Calorie Sugar Macronutrient Substitutes.= Syn	CARB	0026
the Time-Dependent Changes in	Zeta Potential and Surface Charge of TiO$_2$ (Anatase)	COLL	0203
and Four-Coordinate Monomeric	Zinc Alkyl Derivatives Stabilized by Sterically Dema	INOR	0085
and Si(100).= Reactions of	Zinc Alkyls on Semiconductor Substrates: GaAs(100)	COLL	0048
	Zinc and Nickel Octaethyloxophlorins.=	INOR	0302
taining Ultra-+ Influence of the	Zinc Content on Aromatics Formation over Zinc-Con	PETR	0048
ito, Fusion, Oxidorreduccion con	Zinc en Medio Acido y Basico, Preparacion de Nitrad	CHED	0172
Inmunomoduladora del	Zinc en Ratones Inmunodeficientes.= Actividad	AGFD	0144
ilized by Sterically+ Monomeric	Zinc Hydride and Magnesium Alkyl Derivatives Stab	INOR	0087
s of Propane Deydrogenation over	Zinc or Molybdenum Supported on Silicalite.= +tic	PETR	0087
of Zinc, Iron, and Total Sulfur in	Zinc Ore Concentrates.= +rections in XRF Analysis	ANYL	0057
charge-Transfer Excited States of	Zinc(II) Mixed-Ligand Complexes.= +nd-Ligand S	INOR	0164

tent on Aromatics Formation over	Zinc-Containing Ultra-Stable Y Zeolites.= +nc Con	PETR 0048
al, Pharmacological, and Clinica+	Zinc-Metallopeptidase Inhibitors: Design, Biochemic	MEDI 0002
x Corrections in XRF Analysis of	Zinc, Iron, and Total Sulfur in Zinc Ore Concentrate	ANYL 0057
Other Microporous Structures.+	Zincophosphates and Berylloarsenates—Zeolites and	PETR 0055
c Hybrid Materials. Reinforcing	Zirconia Particles Precipitated into an Elastomeric M	POLY 0283
Aqueous Coprecipitation of	Zirconia Precursors.=	COLL 0123
Characterization Studi+ Sulfated	Zirconia Superacid Catalysts: Thermal Analysis and	PETR 0028
Sulfated	Zirconia.= Superacid and Catalytic Properties of	PETR 0133
Stability of Suspensions of	Zirconia-Lime Powders.=	COLL 0124
Acidity of Substituted	Zirconias.=	PETR 0052
+en Calcium Hydroxyapatite and	Zirconium Oxide for the Separation of Proteins.=	ANYL 0027
Oxides.= Open-Morphology	Zirconium Phosphates, Aluminas, and Mixed	PETR 0051
of Functionalized	Zirconium Phosphonates.= Synthesis and Reactivity	INOR 0084
the Intramolecular Coupling of a	Zirconocene Benzyne Complex with Alkenes and Alk	ORGN 0217
nsmetalation Reactions of Alkyl	Zirconocenes: Copper-Catalyzed Conjugate Addition	ORGN 0089
Dehydrogenation over	Zn or Mo.= Reaction Kinetics of Propane	I&EC 0075
the Interaction between the Zn,	Zn$^+$ and Zn^{2+} + H$_2$ Moeties.= Ab Initio CI Study of	PHYS 0282
raction between the Zn, Zn$^+$ and	Zn^{2+} + H$_2$ Moeties.= Ab Initio CI Study of the Inte	PHYS 0282
ion of Vanadate Anions with Cu,	Zn-Superoxide Dismutase as Probed by ^{51}V NMR.=	INOR 0184
n Catalysts Supported on Mg, Cu,	Zn, and Ni Aluminates.= +tinum and Platinum-Ti	COLL 0126
ination Reactions with MR$_2$ (M =	Zn, Mg and R = Bz, Cp, allyl).= +ntation-Recomb	INOR 0383
y of the Interaction between the	Zn, Zn$^+$ and Zn^{2+} + H$_2$ Moeties.= Ab Initio CI Stud	PHYS 0282
Growth of	ZnO.= Change in Crystal Habit in the Vapor-Phase	PHYS 0293
ic Chemical Vapor Deposition of	ZnSe: Effect of Precursor Chemistry on Film Propert	INOR 0171
Precursors Useful for OMCVD of	ZnSe.=	COLL 0155
e and Pyruvate Level+ Improved	Zooglan Gum Properties by Modification of Succinat	BIOT 0198
m N$_2$O to Alkynes Coordinated to	Zr.= +llic Compounds. Oxygen-Atom Transfer fro	INOR 0327
id-State NMR.= Structures of	Zr(O$_3$P(CH$_2$)$_n$COOH)$_2$, n = 1-4 from ^{31}P and ^{13}C Sol	INOR 0047
Reactions of Group IVB (Ti,	Zr, Hf) Alkyls with Aluminum Reagents.=	INOR 0054
ns over Alkaline-Free Synthesized	ZSM-5 and Aluminated Silicalite-1 Crystals.= +bo	PETR 0065
H-[Al]-	ZSM-5 Catalysts.= Naphthene Transformation over	PETR 0044
erties' Influence on the Rate of	ZSM-5 Crystallization.= Investigation of Seed Prop	PETR 0039
ly Substituted Zeolites with the	ZSM-5 Sructure.= EXAFS Studies of Isomorphous	PETR 0081
ctivation by Coke Formation of	ZSM-5 Type Catalysts during Oligomerization of Eth	PETR 0027
plate-Mediated Crystallization of	ZSM-5 Type Zeolite.= +ttering Studies of the Tem	PETR 0058
Catalysis of a Modified	ZSM-5 Zeolite.= Preparation, Characterization, and	CATL 0029
e Toluene Ethylation on Modified	ZSM-5.= +sformation of p-Ethyltoluene during th	PETR 0086
°From Ruzicka's Terpenes in	Zurich to Steroids in Mexico via Cuba.=	HIST 0017
Polymeric Flocculants of the	Zwitterionic Type.=	POLY 0094
dades de Conduccion en Polimeros	Zwitterionicos.= +ecto de la Presion en las Propie	ORGN 0159
Versatile	Zymark Robot Dilution System.=	TECH 0005

AUTHOR INDEX

Reference by Division Code and Paper Number

Aaronson, A. M.	INOR 0346	Agarwal, V. K.	AGFD 0128	Alcala, P.	CATL 0041
		Agashe, K.	COLL 0188		FUEL 0100
Abatjoglou, A. G.	INOR 0349	Agathos, S. N.	BIOT 0155	Alder, A. C.	BIOT 0120
		Ager, A. L., Jr.	MEDI 0063	Al–Diab, S. S.	INOR 0234
Abbaschian, R.	POLY 0202	Agosta, W. C.	HIST 0027	Aldissi, M.	COLL 0166
Abdali, A.	CARB 0031	Agostinelli, R. M.	CHED 0123	Aldrich, J. R.	AGRO 0035
Abdalla, A. Y.	FUEL 0113			Aleklett, K.	NUCL 0024
Abdel–Magid, A. F.	ORGN 0242	Agoston, G. E.	ORGN 0213	Alemayehu, M.	ORGN 0145
		Agrawal, A. K.	POLY 0011	Alessi, D. A.	ORGN 0026
Abeles, R. H.	BIOL 0024		POLY 0205	Alex, A.	I&EC 0031
Abels, A. J.	CHAS 0030	Agreda, V. A.	I&EC 0053	Alex, R. F.	PETR 0123
Abhiraman, A. S.	POLY 0316	Aguilar C., A. E.	AGFD 0144	Alexander, J. M.	NUCL 0062
Abidi, S. L.	ANYL 0134		AGFD 0145	Alexander, M. H.	
Abola, E. E.	CINF 0031	Aguilar R., G.	COLL 0126		PHYS 0188
Aboul–Enein, H. Y.	ANYL 0133	Aguirre, M. E.	COLL 0127	Alexander, P.	BIOT 0192
			POLY 0056	Alexander, S.	INOR 0157
Abraha, A.	INOR 0184	Aguirre M., E.	AGFD 0107	Alexandratos, S. D.	
Abraham, A.	MEDI 0185	Aharoni, S. M.	POLY 0063		PMSE 0004
Abraham, I.	COLL 0086	Ahlafi, H.	CATL 0016	Alexiou, A.	FUEL 0113
Abraham, M. A.	CATL 0043	Ahmed, A.	PMSE 0077	Alfano, R. R.	BIOT 0080
	COLL 0189		PMSE 0170	Al–Laham, M. A.	
	COLL 0209	Ahmed, S.	POLY 0245		PHYS 0214
Abrahams, D. H.		Ahmed, Z.	AGRO 0074	Allan, D. S.	POLY 0127
	HIST 0026		ORGN 0071	Allan, P. W.	MEDI 0028
Abrahamson, H. B.		Ahn, C.	ORGN 0294	Allara, D. L.	PMSE 0150
	INOR 0040	Ahn, H–J.	BIOT 0013	Allcock, H. R.	INOR 0326
Abrajano, T. A., Jr.		Ahn, J. H.	PMSE 0060		PMSE 0002
	ANYL 0061	Ahn, S.	PETR 0016	Allen, C. L.	INOR 0281
Abramoff, B.	POLY 0263	Ahn, S. K.	MEDI 0051	Allen, C. R.	INOR 0209
Abrams, S. B.	CHAL 0006	Ahn, T–K.	PMSE 0164	Allen, C. W.	POLY 0182
Abrams, S. R.	AGRO 0017	Ahn, Y.	ORGN 0025	Allen, D. E.	POLY 0107
Abruzzo, G.	MEDI 0181	Ahsan, T.	FUEL 0039	Allen, E. E.	MEDI 0104
	MEDI 0182	Ahuja, S.	ANYL 0110	Allen, F. H.	CINF 0021
	MEDI 0183	Aichelin, J.	NUCL 0006		COMP 0045
Aceves H., J. M.		Aiello, R.	PETR 0020	Allen, H. S.	SCHB 0013
	PHYS 0213	Ailey–Trent, K. S.		Allen, S. G.	CHED 0027
Ackerman, M. S.			FUEL 0055	Allen, S. R.	POLY 0058
	MEDI 0124	Aishah, B.	BIOT 0152	Allende, A.	PHYS 0319
	MEDI 0125	Ajot, H.	PETR 0046	Alley, C. C.	CEI 0002
Ackerson, M. D.		Akagi, T.	AGRO 0050	Allison, S.	AGRO 0087
	BIOT 0017	Akers, K.	PHYS 0006	Allnutt, F.C.T.	BIOT 0156
	FUEL 0092	Akers, K. L.	COLL 0092	Alm, A.	ENVR 0061
Ackman, R. G.	AGFD 0160	Akester, J. D.	INOR 0270	Almlof, J.	INOR 0301
Acosta H., A.	ORGN 0180	Akhtar, M. H.	AGRO 0018	Almond, H. R., Jr.	
	ORGN 0197	Akhter, H.	ENVR 0041		ORGN 0296
Acquaye, J. H.	INOR 0067	Akimoto, H.	ORGN 0113	Al–Najjar, I. M.	
	INOR 0068	Akinc, M.	POLY 0357		INOR 0234
	INOR 0130	Akiyoshi, K.	POLY 0098	Al–Obeidi, F.	MEDI 0093
Acree, T. E.	AGFD 0137	Akiyoshi, M.	CATL 0049	Alon, R.	BIOL 0090
	AGFD 0142	Aklonis, J. J.	PMSE 0017	Alonso, F.	PMSE 0130
Adamovics, J. A.		Akrich, R.	FUEL 0080		POLY 0055
	ANYL 0121	Alam, T.	PHYS 0195	Alonso–Amigo, M. G.	
Adams, J.	ORGN 0006	Alamo, R. G.	POLY 0075		PHYS 0215
	ORGN 0027	Alario, F.	PETR 0046	Alper, H.	COLL 0117
Adams, J. P.	ORGN 0272	Alauddin, M. M.			INOR 0381
Adams, M. R.	INOR 0050		BIOT 0061	Alshehri, S.	INOR 0022
	PMSE 0125		NUCL 0076	Alsmeyer, D. C.	
Adams, M.W.W.		Albanese, J. A.	INOR 0328		INOR 0312
	INOR 0099	Albanese, M. M.		Altamirano, M. M.	
Adams, P.	MEDI 0127		MEDI 0161		BIOL 0106
Addison, A. W.	INOR 0262	Albarran, G.	NUCL 0070	Althaus, I. W.	MEDI 0055
Adebona, B.	AGRO 0140		NUCL 0072	Althaus, J. S.	MEDI 0126
Adefarati, A. A.		Albeniz, A. C.	INOR 0384	Altmann, E.	BIOL 0155
	MEDI 0058	Alberda v. Ekenstein, G.O.R.		Altomare, C.	MEDI 0059
Adhihetty, I. S.	POLY 0095		POLY 0209	Alva, S.	BIOT 0143
Adkins, M. D.	CHAL 0018	Albert, C.	FUEL 0117	Alvanipour, A.	COLL 0136
Afeyan, N.	ANYL 0127	Albizati, K. F.	ORGN 0003		COLL 0138
Afeyan, N. B.	ANYL 0032		ORGN 0100	Alvarez, D., Jr.	INOR 0303
	BIOT 0073	Albrecht, W.	AGFD 0007	Alves, A.	MEDI 0051
Affrossman, S.	CATL 0039		AGFD 0182	Alviso, C. T.	POLY 0051
Agarwal, M.	BIOL 0108	Albright, L. F.	PETR 0082		
			PETR 0085		

1

Alzamora, S. M.	AGFD 0055	Antalek, B.	PHYS 0216	Arnold, F. H.	BIOT 0068
		Anthony, R. G.	CATL 0040		BIOT 0133
Amador, C.	CHED 0024		FUEL 0061		BIOT 0244
Amano, T.	PHYS 0332		FUEL 0096	Arnold, J.	ENVR 0075
Ambrose, M. C., II			PETR 0017		INOR 0021
	FERT 0015	Antich, P. P.	MEDI 0041		INOR 0172
amEnde, D. J.	PETR 0085	Antonucci, J. M.		Arnold, R.	CHAL 0019
Amini, A.	PHYS 0131		ORGN 0156	Arnold, T. S.	AGRO 0108
Amiridis, M. D.			ORGN 0157	Arnott, H. J.	POLY 0022
	COLL 0212	Aoki, O.	PMSE 0022	Arreola, A. G.	AGFD 0028
Amis, E. J.	POLY 0197	Apblett, A. W.	POLY 0341	Arriola S., H.	NUCL 0069
	POLY 0294	Apen, P. G.	INOR 0310		NUCL 0089
Amyes, T. L.	ORGN 0153	Appelhans, A. D.		Arrivo, S. M.	PHYS 0220
	ORGN 0154		NUCL 0056	Arroyo, F.	ANYL 0054
An, G.	BIOT 0230	Appell, K. C.	MEDI 0158	Arroyo, P. A.	PETR 0048
Ananthapadmanabhan, K. P.		Appelman, E. H.		Arroyo-Reyna, A.	
	COLL 0074		NUCL 0068		BIOL 0074
Anantharaman, V.		Apple, T.	CATL 0031	Arteaga-Barajas, C.	
	MEDI 0095	Apple, T. M.	CATL 0035		COLL 0123
Anaya, A. L.	MEDI 0037	Applebaum, L. A.		Artman, L. D.	MEDI 0124
Andemichael, Y. W.			COLL 0175		MEDI 0125
	ORGN 0098	Applebaum, M.	I&EC 0030	Arvanitidou, E.	COLL 0183
	ORGN 0099	Appleby, S.	MEDI 0111	Arve, B.	ANYL 0030
Andersen, R. A.		Applegate, M. A.		Arzoumanidis, G. G.	
	INOR 0364		BIOT 0011		I&EC 0052
Anderson, A. B.		Appling, D. R.	BIOT 0243	Asali M., M.	CHED 0114
	FUEL 0036	Arafat, A.	PETR 0080	Asano, M.	MEDI 0099
Anderson, C. W.		Araki, K.	ORGN 0014	Asato, G.	AGRO 0074
	CHED 0147		ORGN 0080	Aschbacher, P. W.	
Anderson, G. D.			ORGN 0081		AGRO 0038
	CINF 0016	Araki, T.	MEDI 0109	Asgeirsson, B.	BIOT 0194
Anderson, J. E.	PMSE 0120	Aravind, S.	CARB 0031	Ash, D.	FERT 0022
Anderson, M.	BIOT 0246	Arce, E.	PHYS 0218	Ashby, E. C.	POLY 0316
Anderson, M. T.			PHYS 0219	Ashe, A. J., III	ORGN 0212
	INOR 0358	Arce Medina, E.		Ashe, B. M.	MEDI 0173
Anderson, M. W.			CHED 0122	Ashley, C. S.	PETR 0032
	CATL 0023		COMP 0052	Ashley, D. L.	CEI 0001
	PETR 0042	Archer, R. D.	CHED 0018	Ashley, K.	COLL 0093
Anderson, R. F.			CHED 0056	Ashraf, A.	COLL 0035
	GEOC 0016		POLY 0343		COLL 0053
Anderson, S. L.	PHYS 0174	Archibald, T. G.		Ashton, W. T.	MEDI 0105
	PHYS 0241		POLY 0028	Asiaie, R.	PETR 0074
	PHYS 0242	Arcuri, E. J.	BIOT 0084	Askari, C.	AGFD 0183
	PHYS 0262	Arechiga V., U.	I&EC 0068	Aslam, M.	TECH 0003
	PHYS 0311	Arellano, Z. M.	I&EC 0013		TECH 0017
Anderson, T. A.		Arendt, J. S.	CHAS 0016	Aspnes, D. E.	COLL 0007
	ENVR 0056	Arevalo M., X.	CHED 0164	Assawaweroonhakarn, P.	
Anderson, T. J.	COLL 0155		CHED 0178		FUEL 0090
Anderson, W.	MEDI 0068	Argaiz J., A.	AGFD 0055	Assefa, Z.	INOR 0189
Anderson, W. A.		Argentieri, T.	MEDI 0094	Assink, R. A.	PETR 0032
	BIOT 0038	Argentine, J. A.			POLY 0256
Andersson, L. A.			AGRO 0154	Astimar, A. A.	BIOT 0153
	INOR 0139	Argoudelis, C. J.		Astorga R., J.	COMP 0053
Andreola, C.	POLY 0348		BIOL 0080	Astroff, A. B.	BIOT 0166
Andrew, N. W.	AGRO 0128	Arguedas, E.	SCHB 0026		ORGN 0201
Andrews, D. R.	ORGN 0178	Arif, A. M.	INOR 0113	Atagi, L. M.	INOR 0115
Andrews, L.	PHYS 0031	Ariga, K.	ORGN 0054	Atkins, R. A.	ORGN 0031
	PHYS 0074	Arista, N. R.	NUCL 0053	Atkinson, D. E.	CHED 0206
Andrews, M. P.	POLY 0154	Arista-Reyes, R.		Attili, B. S.	FUEL 0138
Andrews, S. J.	PETR 0021		CHED 0094	Atwood, D. A.	INOR 0390
Andro, T. M.	ORGN 0093	Aristoff, P. A.	MEDI 0055		INOR 0392
Aneinikov, V. G.		Ariyaratne, K.	INOR 0366	Atwood, E. S.	ENVR 0054
	COLL 0109	Arlauskas, R. A.		Atwood, J. D.	INOR 0211
Anex, D.	PHYS 0203		PMSE 0030	Atwood, J. L.	INOR 0385
Angeles, E.	ORGN 0245	Armendariz H., H.			INOR 0390
Angelov, C. M.	INOR 0325		COLL 0126	Auble, R.	NUCL 0027
Angiola, A. J.	FUEL 0093		COLL 0127	Auerbach, R. A.	
Anglerot, D.	PETR 0075	Armenta L., M. T.			INOR 0284
Anguis T., C.	COLL 0119		ANYL 0055	Augelli-Szafran, C. E.	
Angus, J. C.	FUEL 0036	Armentrout, P. B.			MEDI 0108
	PHYS 0091		PHYS 0200	Auger, M.	PHYS 0196
Angyal, S. J.	CARB 0029	Armienta H., M. A.		Aunins, J. G.	BIOT 0071
Anicetti, V.	ANYL 0086		ANYL 0051	Austin, G.	SCHB 0006
Anker, L. S.	ANYL 0099	Armienta-Hernandez, M. A.			SCHB 0012
	COMP 0003		ANYL 0049	Auvil, S. R.	POLY 0119
Annand, R. R.	BIOL 0066	Armstrong, N. J.		Auza, L.	MEDI 0037
Annapoorna, K.			POLY 0105	Averin, D. V.	BIOT 0004
	INOR 0411	Armstrong, W. H.		Avery, H.	SCHB 0004
Annapragda, A.	COLL 0011		INOR 0129	Avichai, M.	POLY 0241
Annen, M. J.	PETR 0098		INOR 0372	Avignone, F. T., III	
Anpo, M.	INOR 0165		INOR 0408		NUCL 0013
Anslyn, E. V.	ORGN 0053	Arnett, E. M.	FUEL 0039	Aviram, A.	BIOT 0001
	ORGN 0054	Arnold, F. E.	POLY 0028	Avolio, J.	YCC 0012

Avouris, P.	PHYS 0184	Balasubramanian, V.		Barnes, C. J.	AGFD 0126
Awad, M.M.A.	MEDI 0189		MEDI 0150	Barnes de C., F. J.	
Awad, S. B.	PMSE 0144	Balazs, I.	CHAL 0009		CHED 0107
Awate, S. V.	PETR 0005	Balch, A. L.	INOR 0302	Barnes de Castro, F.	
Axenrod, T.	ORGN 0256	Baldeschwieler, J. D.			CONG 0005
Axon, S. A.	PETR 0081		PHYS 0159	Barnett, C.	BIOT 0240
Ayer, D. E.	MEDI 0133	Balding, S. D.	ORGN 0232	Baron, J.	PETR 0128
Azimi, S.	INOR 0250	Baldwin, J. J.	MEDI 0096	Baron, S.	FUEL 0022
Azumi, T.	INOR 0164		MEDI 0097	Barr, J.	AGRO 0093
Azzolina, B. A.	BIOL 0071		MEDI 0191	Barragan, D.	I&EC 0010
Baba, F.	PMSE 0055	Balentine, D. A.		Barragan, R. M.	
Babcock, E. E.	MEDI 0041		AGFD 0044		I&EC 0009
Babich, I. V.	COLL 0098	Baliga, B. S.	MEDI 0185	Barreto, J.	NUCL 0027
	INOR 0237	Balk, S. A.	INOR 0208	Barrett, L. W.	PMSE 0088
Babonneau, F.	POLY 0231	Balke, S. T.	PMSE 0065		PMSE 0171
	POLY 0358	Balko, B. A.	PHYS 0163	Barrier, J. W.	FUEL 0094
Babu, Y. S.	MEDI 0005	Balkovec, J. M.	MEDI 0182	Barrios, R.	ENVR 0040
Baca, A. A.	I&EC 0011	Ball, D.	CHED 0131	Barrish, J. C.	MEDI 0188
Baca A., A.	INOR 0063	Ball, J. W.	COMP 0003		ORGN 0114
	INOR 0246	Ballas, L. M.	MEDI 0075	Barron, A. R.	INOR 0389
Bachas, L. G.	ANYL 0090	Ballester, L.	ANYL 0059		POLY 0341
Bachmann, J. D.		Ballinger, M.	BIOL 0023	Barron, C. A.	PMSE 0165
	ENVR 0071	Ballinger, M. D.		Barteau, M. A.	CATL 0033
Baclaski, J.	BIOT 0033		BIOL 0077		COLL 0194
Bacquet, R. J.	MEDI 0186	Baloga, D. W.	AGFD 0177	Barth, A.	COMP 0010
Bacri, J-C.	COLL 0143	Balsara, N. P.	POLY 0275	Barth, H.	MEDI 0074
Baczynskyj, L.	AGRO 0108	Baltzis, B. C.	BIOT 0138	Barth, H. G.	PMSE 0070
Bader, A. R.	CONG 0001	Balu, K.	AGRO 0054		PMSE 0097
Bae, J. H.	CARB 0039	Banaszak Holl, M. N.		Barth, J. V.	PHYS 0126
Baenziger, J. U.			INOR 0103	Bartholmew, C. H.	
	BIOL 0150	Banerjee, T.	POLY 0139		FUEL 0030
Baer, C. S.	AGRO 0030	Bangov, I. P.	COMP 0024	Bartholomew, J.	
Baetz, A. L.	ANYL 0081	Banholzer, W. F.			INOR 0109
Baeza-Reyes, A.			PHYS 0146	Barthomeuf, D.	CATL 0045
	INOR 0397	Banin, A.	ANYL 0052		PETR 0142
Bagli, J.	MEDI 0095	Banker, A.	MEDI 0179	Bartizal, K.	MEDI 0181
Bahador, S. K.	PHYS 0222	Bankmann, M.	PETR 0091		MEDI 0182
Bahadur, M.	POLY 0182	Banks, B. J.	BIOT 0127		MEDI 0183
Bahar, I.	COLL 0015	Banks, H. D.	ORGN 0251	Bartlett, P. A.	COMP 0044
	COLL 0016	Banos, L.	INOR 0201	Bartlett, R. J.	PHYS 0008
	COLL 0018		POLY 0054		PHYS 0015
	POLY 0069	Banos L., L.	INOR 0069		PHYS 0341
Bahar, S.	BIOT 0157	Bansal, N.	PETR 0070	Bartlett, W. R.	CHED 0148
Bahn, P.	ANYL 0101	Banse, B. A.	COLL 0069	Bartolini, A.	MEDI 0098
Bai, M.	BIOL 0070	Banwenda, G. R.		Bartolo, R. G.	COLL 0017
Bai, Y-B.	BIOT 0142		CATL 0055	Barton, A.	PHYS 0041
Baier, M.	POLY 0188		FUEL 0102	Barton, G. J.	CINF 0030
Bailey, J. E.	BIOT 0087	Bara, I.	TECH 0001	Barton, T. J.	POLY 0357
	BIOT 0223	Barakat, K.	ORGN 0073	Bartram, M. E.	COLL 0045
Bailey, M. E.	AGFD 0119	Baraldi, M.	MEDI 0135	Bartus, R. T.	MEDI 0115
Bailey, S. L.	CEI 0002	Baran, E. F.	COLL 0029	Bartz, U.	MEDI 0036
Baiocchi, F. A.	INOR 0008	Barat, R. B.	I&EC 0024	Barzi, A.	MEDI 0029
Bair, H. E.	POLY 0236	Barba-Behrens, N.		Barzoukas, M.	POLY 0102
	POLY 0320		INOR 0079	Basch, H.	BIOL 0061
Baird, D. G.	PMSE 0135		INOR 0252		PHYS 0224
Bakac, A.	INOR 0276		INOR 0263	Base, K.	ORGN 0124
	INOR 0396	Barbar, D. C.	ORGN 0161	Basha, A.	ORGN 0227
Baker, A. D.	INOR 0232	Barbari, T. A.	POLY 0191	Basha, F. Z.	MEDI 0116
	INOR 0233	Barber, D. C.	CATL 0060	Bashir-Hashemi, A.	
Baker, B. F.	BIOL 0120	Barber, D. E.	INOR 0405		ORGN 0256
Baker, F. C.	AGRO 0004	Barbieri, A.	MEDI 0141		ORGN 0257
Baker, G. L.	POLY 0325	Barbour, J. F.	NUCL 0065	Basile, F.	ANYL 0014
Baker, H.	ORGN 0281	Barclay, G. G.	POLY 0110	Basmadjian, G. P.	
Baker, J. O.	BIOT 0191	Bard, A. J.	ANYL 0108		MEDI 0042
Baker, R.	MEDI 0119		ANYL 0109		MEDI 0131
	MEDI 0146		COLL 0170	Bass, J. D.	CHAL 0030
Baker, R. A.	AGFD 0059	Bardalaye, P. C.		Bass, M. B.	BIOT 0118
Baker, W. E.	PMSE 0177		AGFD 0103	Bassett, D. R.	PMSE 0035
	POLY 0019	Barden, T. C.	AGRO 0074	Bassett, J. M.	PETR 0131
Baker, W. R.	MEDI 0120	Barefoot, T. W.		Bassilakis, R.	FUEL 0052
	MEDI 0197		POLY 0196	Bassoul, P.	POLY 0102
Bakhtiar, S. N.	NUCL 0084	Baren, E. E.	COLL 0156	Bastos, C. M.	INOR 0150
Baktash, C.	NUCL 0027	Bari, S. S.	ORGN 0034		INOR 0188
Bakthavatchalam, R.			ORGN 0195	Basu, P.	PETR 0071
	POLY 0232	Barker, S. A.	AGFD 0060	Batchelor, B.	ENVR 0010
Balaban, A. T.	COMP 0026	Barkigia, K. M.	INOR 0091	Bates, F. S.	POLY 0090
Balaban, M. O.	AGFD 0028	Barksdale, J. M.		Batich, C. D.	POLY 0360
Balaji, V.	ORGN 0124		ANYL 0045	Batley, B. L.	MEDI 0121
Balani, S. K.	ORGN 0208	Barletta, G.	BIOL 0047	Bator, J. M.	MEDI 0171
Balasubramanian, K.		Barlow, J. L.	MEDI 0120	Batzl, C.	MEDI 0035
	COMP 0001		MEDI 0197		MEDI 0036
	COMP 0028	Barman, B. N.	PMSE 0029	Baucells, M.	ANYL 0057
		Barnes, A.	MEDI 0121	Bauder, A.	PHYS 0115

Bauer, A. J.	BIOL 0147	Bejar-Moscona, D.		Bergman, R. G.	ORGN 0095
Bauer, B. J.	PMSE 0003		ENVR 0039		ORGN 0127
Bauer, M.	COLL 0052	Belanger, P. C.	MEDI 0180	Bergo, C. H.	CHED 0101
Bauer, M. J.	ORGN 0232	BelBruno, J. J.	CHED 0126	Berka, R.	BIOT 0240
Bauer, R. S.	CHED 0009	Belderrain, T. R.		Berka, R. M.	BIOT 0234
	PMSE 0157		INOR 0035	Berkenkopf, J.	MEDI 0179
Baughman, L. L.		Beletski, I. P.	COLL 0101	Berkowitz, J.	FUEL 0124
	INOR 0373	Belfiore, L. A.	PMSE 0101	Berkowitz, W. F.	
Baughman, R. H.		Belford, R. L.	FUEL 0074		ORGN 0012
	BIOT 0203	Belfort, G.	BIOT 0042	Berks, A. H.	COMP 0017
	FUEL 0034		BIOT 0169	Berlin, K. D.	INOR 0374
	POLY 0326	Belknap, B.	BIOL 0141	Bernardo, M.	FUEL 0072
Baum, J.	BIOL 0001	Bell, A. T.	CATL 0032	Bernasconi, C. F.	
	PHYS 0167		CATL 0047		ORGN 0152
Baum, K.	POLY 0028		COLL 0206		ORGN 0155
Baum, T. H.	COLL 0029		COLL 0211	Bernath, P. F.	PHYS 0057
	COLL 0156		PETR 0004	Berner, B.	POLY 0195
Baumann, M.G.D.		Bell, M.	MEDI 0172	Berner, R. A.	GEOC 0011
	INOR 0356	Bell, M. R.	MEDI 0158	Bernier, P. P.	PHYS 0070
Baumgartner, J.			ORGN 0036	Bernier, R. L.	BIOT 0214
	CATL 0061	Bell, R. L.	MEDI 0176	Bernreuther, A.	
Baumgartner, J. E.		Bell, R. M.	MEDI 0075		AGFD 0184
	CATL 0015	Bell, T. R.	PMSE 0117	Bernstein, F. C.	
	CATL 0037	Bell, T. W.	INOR 0297		CINF 0031
Baumgold, J.	MEDI 0115		ORGN 0061	Bernstein, S.	HIST 0023
Bauschlicher, C. W., Jr.			ORGN 0062	Berry, A.	BIOT 0032
	PHYS 0005		ORGN 0063	Berry, A. D.	COLL 0135
	PHYS 0171		ORGN 0064		INOR 0306
Bautista Zuniga, F.			ORGN 0065	Berry, G. C.	POLY 0274
	ENVR 0039	Bellen, G. E.	CEI 0028	Berry, K. O.	CHED 0019
Bax, A.	PHYS 0113	Bellin, C. A.	ENVR 0014	Berson, J. A.	ORGN 0128
Baxter, S. M.	INOR 0282	Bellinger, M.	POLY 0061	Berthod, A.	ANYL 0070
Bayense, C. R.	PETR 0043		POLY 0062		COLL 0022
Bayer, E. A.	BIOL 0090	Belliotti, T. R.	MEDI 0177	Berti, M.	MEDI 0059
Bayer, H.	MEDI 0036	Belt, S. T.	INOR 0169	Bertrand, F.	NUCL 0027
Bayer, R. E.	CHED 0187	Belter, P. A.	BIOT 0224	Bertsch, B.	MEDI 0174
Bayes, K. D.	FUEL 0065	Belton, D. N.	PHYS 0122	Besse, J. P.	PETR 0127
Bays, D. E.	MEDI 0079	Beltran, M.	PHYS 0225	Bessiere, J.	ANYL 0053
Bazan, G. C.	INOR 0259	Belyakova, L. A.			ANYL 0067
Beach, D. B.	INOR 0126		COLL 0094	Betche, H. J.	MEDI 0074
Beach, J. W.	CARB 0022		COLL 0105	Bethune, D. S.	PHYS 0070
	MEDI 0051		COLL 0106	Betrabet, C.	POLY 0284
Beachley, O. T., Jr.			COLL 0108	Better, M.	BIOT 0058
	COLL 0153	Benbow, A. E.	CHED 0072	Beuchat, L. R.	AGFD 0032
Beak, P.	ORGN 0120	Bencsura, A. G.		Bewsher, P. J.	POLY 0206
Bean, M. F.	ANYL 0140		FUEL 0066	Beyer, E. M.	AGRO 0069
Beatty, C. L.	I&EC 0033	Bendele, A. M.	MEDI 0022	Beyer, T. A.	MEDI 0082
Beaty, J. A.	INOR 0251	Benincasa, M. A.		Bezman, R. D.	I&EC 0062
Beaulieu, F.	ORGN 0083		PMSE 0011	Bhadra, A. K.	ENVR 0034
Beaver, R. W.	AGFD 0111	Benitez A., C.	INOR 0069	Bhagwat, M.	BIOL 0158
Beazley, W. D.	CHAS 0047	Benkendorf, C. A.		Bhandari, G.	INOR 0214
Becerril, A.	NUCL 0075		CHAS 0024	Bharucha, K. N.	
Beck, M. J.	FUEL 0094	Benkovic, S. J.	BIOL 0159		ORGN 0095
Beck, W.	MEDI 0035	Bennett, J.	PHYS 0363	Bhat, Y. S.	PETR 0100
Becker, C.	CEI 0006	Bennett, L. L., Jr.		Bhatia, S. K.	BIOT 0246
Becker, C. H.	NUCL 0044		MEDI 0028	Bhatnagar, A.	MEDI 0035
Becker, D. P.	MEDI 0069	Benson, R. C.	AGRO 0027	Bhattacharyya, J.	
	MEDI 0187	Bent, B. E.	COLL 0204		ORGN 0203
	ORGN 0104		INOR 0124	Bhavani, A.K.D.	
Becker, J. M.	BIOT 0227	Ben Taarit, Y.	PETR 0027		MEDI 0057
Beckford, F.	INOR 0400	Bentley, H. W.	ENVR 0051	Bheda, M.	COLL 0058
Beckman, E. J.	I&EC 0034	Bentley, J.	COMP 0039		POLY 0233
Bednarek, J.	PHYS 0215	Bentley, W. E.	BIOT 0034		POLY 0246
Bednarski, M. D.		Bentrude, W. G.			POLY 0345
	CARB 0034		INOR 0347	Bhuvaneswaran, C.	
Bedolla, E.	I&EC 0065	Ben-Yehoshua, S.			MEDI 0017
	I&EC 0066		AGFD 0076	Bi, D.	POLY 0015
	I&EC 0067	Beratan, D. N.	BIOL 0133	Bi, X. X.	CATL 0051
Beebe, T. P., Jr.			BIOT 0173		FUEL 0101
	ANYL 0023	Bercaw, J. E.	INOR 0147	Bianchini, C.	INOR 0210
Beelen, T.P.M.	PETR 0054		INOR 0280	Bianconi, P. A.	INOR 0311
Beeman, R. W.	AGRO 0133	Berek, D.	PMSE 0049		POLY 0190
Beene, J. R.	NUCL 0027	Berelson, W. M.			POLY 0364
Begemann, M. H.			GEOC 0010	Biczo, G.	BIOT 0028
	CHED 0087	Berenbaum, M. R.		Bidstrup, S. A.	PMSE 0178
Begley, T. P.	BIOL 0155		AGRO 0115	Bidzilya, V. A.	COLL 0096
Begun, G.	ANYL 0065	Berg, H.	BIOT 0240		COLL 0104
	INOR 0064	Berger, D. M.	ORGN 0008	Biehl, E. R.	ORGN 0224
Behmke, F. D.	AGRO 0029	Berger, R.	AGFD 0122	Bielski, B.H.J.	INOR 0393
Behrens, M. A.	FUEL 0137	Berger, R. G.	AGFD 0023	Biely, P.	BIOT 0185
Beier, R. C.	BIOT 0166	Berghmans, H.	POLY 0147	Bierlein, J. D.	INOR 0256
	ORGN 0201	Bergman, R. G.	INOR 0167	Biersack, J.	NUCL 0037
			INOR 0364		

Name	Code	Num	Name	Code	Num	Name	Code	Num
Bierwagen, E. P.			Bocarsly, A. B.	INOR	0162	Bosch, P.	PETR	0092
	INOR	0147	Bock, K.	CARB	0014	Bosch G., P.	COLL	0126
Bieszk, N. C.	MEDI	0021	Bockimi, L.S.	PHYS	0227	Bose, A. K.	ORGN	0073
Bigelow, L. K.	PHYS	0093	Bock–Sickinger, A.				ORGN	0195
Bigham, W. S.	INOR	0370		MEDI	0091	Bose, R. N.	ANYL	0063
Bigio, D. I.	PMSE	0175	Boctor, A.	MEDI	0178		INOR	0182
Bigois, M.	FUEL	0046	Bodalbhai, L.	COLL	0023		INOR	0403
	PETR	0141	Boddeke, H.W.G.M.			Bosley, J. A.	BIOT	0040
Bihovsky, R.	MEDI	0071		MEDI	0078	Bosshard, C.	POLY	0176
Bikales, N. M.	POLY	0204	Bode, L. E.	AGRO	0125	Botch, B. H.	CHED	0116
Billah, M. M.	MEDI	0161	Bodepudi, V. R.			Botschwina, P.	PHYS	0103
Billig, E.	INOR	0349		ORGN	0086	Bott, S. G.	INOR	0390
Bilous, P. T.	CHAL	0007	Bodian, D.	COMP	0036	Botting, H. G.	AGFD	0209
Biolsi, L.	PHYS	0226	Bodor, N.	MEDI	0068	Botto, R. E.	FUEL	0075
Birch, A. J.	HIST	0013	Boer, T. S.	POLY	0209	Bottomley, L. A.		
Birge, R. R.	BIOT	0077	Boerio, F. J.	POLY	0087		CHED	0154
Birke, R. L.	BIOL	0135		POLY	0088	Boucher, C. R.	AGRO	0028
Birnbaum, G. I.			Boesch, R.	CEI	0009		AGRO	0029
	CARB	0038	Boese, W. T.	INOR	0254	Bouchet, F.	PETR	0026
Biscaye, P.	GEOC	0016	Bogatiriov, V. M.			Bouck, K. J.	POLY	0034
Bishop, A. R.	PHYS	0001		COLL	0100	Boucot, P.	FUEL	0046
Bishop, J.K.B.	GEOC	0009	Bogillo, V. I.	COLL	0107	Boudart, M.	CATL	0037
Bitterwolf, T. E.				COLL	0112	Bouhria, M.	FUEL	0080
	INOR	0149	Bogomaz, V. I.	COLL	0096	Bourke, J. B.	AGRO	0085
Bitterwolfe, T. E.				COLL	0104	Bourne, M. C.	AGFD	0013
	INOR	0281	Bogorad, P.	PHYS	0041	Boushehri, A.	COMP	0058
Bizzigotti, G. O.			Bohlen, D. S.	POLY	0368	Bouska, J.	MEDI	0176
	ENVR	0080	Bohler, R. J.	PETR	0118	Bousley, R.	MEDI	0108
Bjarnason, J. B.				PETR	0122		MEDI	0110
	BIOT	0194	Bohm, H. J.	COMP	0037	Boutonnet–Kizling, M.		
Bjork, C. K.	CHAL	0029	Bohn, M. A.	BIOT	0164		COLL	0128
Bjorklund, G. C.			Bohn, P. W.	ANYL	0026	Bowen, J. H.	CHAL	0007
	POLY	0001		COLL	0070	Bowers, J. S.	COLL	0180
Blachere, J. R.	FUEL	0136	Bohn, R.	PHYS	0031		POLY	0280
Black, P. D.	MEDI	0139	Bohnet, A.	NUCL	0006		POLY	0296
Blackburn, N. J.			Bohning, J. J.	HIST	0010	Bowers, L. D.	ANYL	0124
	INOR	0139		HIST	0031	Bowles, T. J.	NUCL	0010
Blackburn, R. L.			Bois, L.	POLY	0231	Bowman, M. K.		
	COLL	0132	Boland, J. J.	PHYS	0187		PHYS	0116
Blackman, G. S.			Boles, J. L.	FERT	0001	Bowman, R. M.		
	PHYS	0153		FERT	0003		PHYS	0369
Blackmond, D. G.			Bolin, H. R.	AGFD	0011	Box, P. F.	NUCL	0039
	COLL	0187	Bolivar, F.	BIOT	0130	Boyapati, V. L.	MEDI	0155
Blackwell, J.	ANYL	0027	Bollhardt, K. P. C.			Boyd, M. R.	AGFD	0192
Blake, A. B.	INOR	0266		ORGN	0141	Boyer, B. D.	ORGN	0206
	INOR	0269	Bollin, E., Jr.	AGRO	0030	Boyle, J.	PETR	0147
Blake, D.	COLL	0082	Bollinger, J. M.	ORGN	0281	Boyle, J. G.	FUEL	0069
Blamey, J.	INOR	0099	Bollinger, M.	BIOL	0022	Boylston, T. D.	AGFD	0161
Blanc, T.	ANYL	0018	Bolton, G. L.	MEDI	0110	Bozzelli, J. W.	ENVR	0013
Bland, J.	AGFD	0159	Bommarius, A. S.				FUEL	0063
Bland, R. G.	CARB	0009		BIOT	0248		FUEL	0085
Blaney, J.	CINF	0028	Bomo de Vivar R., A.				PHYS	0228
Blankenship, J.	MEDI	0020		ORGN	0187		PHYS	0229
Blankley, C. J.	MEDI	0110	Bonadies, L.	MEDI	0121	Braccolino, D. S.		
Blanski, R. L.	INOR	0131	Bonasia, P. J.	INOR	0172		ORGN	0028
Blas, J.	ORGN	0159	Bondybey, V. E.			Brack, H. P.	POLY	0253
Blauwhoff, P.M.M.				PHYS	0338	Bracken, D.	NUCL	0036
	PETR	0025	Bonilla–Marin, M.			Bradbury, R. H.		
	PETR	0112		PHYS	0289		MEDI	0102
Blizzard, T.	ORGN	0016	Bonin, M. A.	CEI	0001	Bradley, J. S.	PHYS	0373
Bloch, K.	HIST	0012	Bonjour, L.	MEDI	0127	Bradley, S. A.	COLL	0208
Blok, N.	MEDI	0171	Bonne, R.L.C.	CATL	0044	Bradley, S. M.	PETR	0143
Blough, N. V.	ENVR	0057	Bonnell, D. A.	COLL	0171	Bradshaw, J.	CINF	0024
Blout, E. R.	BIOL	0004		PHYS	0207	Brady, M. S.	AGFD	0082
Bluestone, S.	CHED	0157	Bonning, B.	AGRO	0009	Braezeale, W. H.		
Blum, F. D.	POLY	0368	Bonsu, F. O.	POLY	0028		CHED	0019
Blum, Y. D.	POLY	0313	Boo, H. K.	POLY	0173	Braga, R. A.	CHED	0154
Blumberg, P. M.			Booker, S.	ORGN	0281	Braghiroli, D.	MEDI	0135
	MEDI	0077	Booth, B. L.	POLY	0175	Brajter–Toth, A.		
Blumen, A.	PHYS	0130	Bordner, J.	MEDI	0082		COLL	0023
	PHYS	0132	Bordwell, F. G.	ORGN	0133	Bramble, F. Q.	AGRO	0031
Blumstein, A.	POLY	0030	Borek, T. T.	POLY	0312	Bramson, R. S.	CINF	0047
	POLY	0031	Borella, L.	MEDI	0067	Brand, R.	PETR	0091
	POLY	0032	Bores, G. M.	MEDI	0137	Brandes, S. D.	FUEL	0044
	POLY	0072	Borja, M. P.	INOR	0272	Brandhorst, T.	BIOT	0238
	POLY	0073	Borm, J.	POLY	0359	Brandsma, L.	INOR	0229
	POLY	0074	Bornemeier, D. A.			Brandt, K. M.	ORGN	0030
Blumstein, R. B.				MEDI	0177	Brandvold, T. A.		
	PMSE	0119	Bornhop, D. J.	ENVR	0060		INOR	0194
Boaz, N.	MEDI	0087	Bornmann, W. G.			Brashear, W. T.		
	MEDI	0172		ORGN	0009		ANYL	0062
Bobadilla, J.	I&EC	0064	Borutsky, P. N.			Braswell, E. H.	BIOL	0036
Bobbitt, J. L.	BIOL	0072		PETR	0135	Bratoeff, E. A.	ORGN	0193

Name	Code	Number	Name	Code	Number	Name	Code	Number
Brattsten, L. B.	AGRO	0114	Brookhart, M.	ORGN	0129	Buckley, A. M.	INOR	0081
Bratzler, R. L.	BIOT	0072	Brooks, D. W.	MEDI	0176	Buckley, T. D.	CEI	0019
Brauchle, C.	BIOT	0078		ORGN	0227	Buckwalter, B. L.		
Brauman, J. I.	ORGN	0121	Brouet, G.	PHYS	0231		AGRO	0074
Braunsdorf, R.	AGFD	0183	Brouwer, J.	POLY	0129	Bucovaz, E. T.	BIOL	0068
Braunstein, D. M.				POLY	0211		BIOL	0069
	CHED	0008	Brovko, V. N.	PETR	0135	Budowle, B.	CHAL	0005
Braunstein, M.	PHYS	0191	Brown, A. B.	ORGN	0263	Buettner, R.	ANYL	0109
Brause, K. A.	MEDI	0173	Brown, A. M.	AGRO	0028	Bugg, C. E.	MEDI	0005
Brawner, M.	BIOT	0084		AGRO	0029	Buhro, W. E.	INOR	0071
Brazwell, E. M.	BIOT	0094	Brown, B. R.	MEDI	0160		INOR	0102
Brechbiel, M. W.			Brown, C. A.	PHYS	0070	Buhrow, S. A.	MEDI	0096
	INOR	0033	Brown, D.	POLY	0366		MEDI	0191
Brechbiel, W. M.			Brown, D. E.	POLY	0182	Buitrago, A. A.	ENVR	0024
	ORGN	0024	Brown, D. W.	PHYS	0129	Bullin, J. A.	FUEL	0061
Breckenridge, W. H.			Brown, F. R.	ANYL	0091	Bunce, D. M.	CHED	0053
	PHYS	0351		CEI	0006		CHED	0054
Breen, J. J.	PHYS	0230	Brown, H. M.	AGRO	0069	Bunce, R. A.	ORGN	0199
Bregna, D. E.	MEDI	0085	Brown, H. S.	ENVR	0044	Bundle, D. R.	BIOL	0143
Brems, D.	BIOT	0150	Brown, L. M.	ORGN	0146		CARB	0013
Brems, D. N.	BIOT	0132	Brown, M. A.	ENVR	0015	Bungay, H. R.	BIOT	0014
Brendel, K.	AGRO	0093	Brown, N. F.	CATL	0033		BIOT	0149
Breneman, G. L.			Brown, P. R.	ANYL	0132		BIOT	0158
	CHED	0204	Brown, R.	BIOT	0240		BIOT	0225
Brenken, M.	CARB	0058	Brown, R. C.	PHYS	0142	Buono, R. A.	MEDI	0045
Brennan, A. B.	POLY	0228	Brown, S. D.	CHED	0194	Burbage, J. D.	ANYL	0025
	POLY	0264	Brown, S. J.	PETR	0007	Burban, P. M.	POLY	0119
	POLY	0284	Brown, T. L.	AGRO	0003	Burch, R. M.	MEDI	0025
Brennan, J.	ORGN	0206		TECH	0015		MEDI	0026
Brennan, J. V.	POLY	0298	Brown, T. M.	INOR	0092		MEDI	0171
Brenner, D. W.	PHYS	0154		INOR	0093	Burdi, D.	BIOL	0155
	PHYS	0354	Brown, W. M.	CHED	0010	Burdon, J.	POLY	0290
Brenner, H. C.	PHYS	0081	Brownawell, B. J.			Bures, M. G.	COMP	0038
	PHYS	0245		GEOC	0004	Burford, R. P.	PMSE	0026
Brescia, A.	ORGN	0204	Browne, D. S.	CHED	0123	Burgess, B. W.	BIOT	0071
Breslin, V. T.	ENVR	0018	Browner, R. F.	ANYL	0122	Burgess, J.	INOR	0022
Breslow, R.	ORGN	0265	Brubaker, K.	CHED	0085	Burgio, G. F.	NUCL	0009
Bretschneider, E.				PHYS	0328	Burgmayer, S.J.N.		
	COLL	0155	Bruce, M. R.	PHYS	0367		INOR	0277
Brewer, F.	BIOL	0109	Bruche, G.	AGFD	0183		INOR	0404
	BIOL	0145	Bruciaga, G.	CHED	0169	Burillo, G.	PMSE	0131
	BIOL	0146	Bruckenstein, S.				POLY	0053
Brewer, G.	INOR	0134		ANYL	0103		POLY	0054
Brewer, T. F.	COLL	0209	Bruehl, M.	PHYS	0037		POLY	0056
Brewster, M. E.			Bruening, R.	AGFD	0191		POLY	0092
	MEDI	0068	Bruice, T. W.	BIOL	0156	Burk, D. L.	CHAL	0001
Brezinsky, K.	FUEL	0083	Brumley, R. L.	ANYL	0033	Burk, M. J.	CATL	0004
Briber, R. M.	PMSE	0003	Brun, A. M.	BIOL	0134	Burkart, S. E.	AGRO	0073
Bribiesca Vazquez, S.			Brun, R.	MEDI	0062	Burke, F. P.	FUEL	0044
	COLL	0124	Brun, T. O.	PETR	0058		FUEL	0071
Bricout, J.	AGFD	0195	Brune, H.	PHYS	0126	Burke, H. M.	CHED	0002
Briend, M.	CATL	0045	Brunet, E.	ORGN	0209	Burke, J. A., Jr.		
Brigaud, T.	ORGN	0046	Bruning, J.	AGFD	0071		CHED	0151
Briggs, B. S.	BIOT	0161	Brunner, L. A.	ANYL	0119	Burke, L. A.	PHYS	0233
Bright, D. A.	INOR	0348	Bruno, J. W.	COLL	0198		POLY	0353
Briker, J.	FUEL	0042	Bruns, R. F.	MEDI	0066	Burke, L. D.	ORGN	0171
Brinckman, F. E.			Brunschwig, B. S.			Burke, T. R.	MEDI	0024
	INOR	0038		INOR	0038	Burke, T. R., Jr.		
Brindle, I. D.	INOR	0080	Bruschi, M.	INOR	0020		MEDI	0023
Brindle, P. A.	AGRO	0041	Brush, E. J.	BIOL	0075	Burkey, T. J.	INOR	0253
Brinker, C. J.	AGRO	0004		BIOL	0087	Burland, D.	POLY	0106
	PETR	0032	Brushmiller, J. G.			Burnett, D. S.	NUCL	0077
	PETR	0068		INOR	0040	Burnham, N. A.		
	POLY	0258	Bruzik, K. S.	BIOL	0017		PHYS	0154
Brinker, U. H.	ORGN	0147	Bryan, P.	BIOT	0192	Burns, J. L.	COLL	0017
Brito, E.	BIOT	0196	Bryant, D. R.	INOR	0349	Burns, M. R.	MEDI	0027
Britton, D. R.	MEDI	0127	Bryant, E. S.	BIOL	0043	Burns, S. J.	PHYS	0234
Broadway, P. J.			Bryant, R. G.	PHYS	0232	Burrell, A. K.	ORGN	0139
	POLY	0067	Brzychczyk, J.	NUCL	0036	Burrows, C. J.	BIOL	0122
Brock, P. J.	COLL	0029	Bucci, S.	POLY	0240		INOR	0089
Brockman, F. J.			Buchanan, M. V.				INOR	0183
	BIOT	0117	Buchert, J.	FUEL	0089		ORGN	0056
Brodberg, R. K.				BIOT	0215		ORGN	0057
	AGRO	0043	Buchman, A.	PMSE	0114	Bursian, N. R.	PETR	0066
Broder, J. D.	FUEL	0094	Buchman, E.	PMSE	0115		PETR	0135
Broder, M. F.	FERT	0020	Buchwald, S. L.			Burt, J. A.	INOR	0164
Brodsky, B.	BIOL	0001		ORGN	0134	Burtner, D. R.	PMSE	0030
Brody, A. L.	AGFD	0075		ORGN	0217	Burton, G. W.	AGFD	0155
Brodzinski, R. L.			Bucio-Vazquez, M. A.			Burton, K. A.	FUEL	0128
	NUCL	0013		ORGN	0268	Burushkina, T. N.		
Bronikowski, M.			Buckland, B. C.				COLL	0109
	NUCL	0050		BIOT	0071	Burwell, D. A.	INOR	0047
Bronstein, I.	BIOT	0167		BIOT	0109	Busby, D. C.	PMSE	0037

Busch, J. V.	CMEC 0009	Callis, C. F.	INOR 0291	Carpenter, B. K.		
Busch, K. W.	COLL 0160	Callstrom, M. R.			INOR	0331
	COLL 0163		INOR 0312		ORGN	0122
	COLL 0203	Calo, J. M.	FUEL 0010	Carpenter, J.	ORGN	0292
Busch, M. A.	COLL 0160		FUEL 0013	Carpenter, J. D.		
	COLL 0163	Calvert, J. M.	BIOT 0246		INOR	0030
	COLL 0203	Calvert, J. T.	AGFD 0039	Carpenter, W. D.		
Bush, D. M.	ORGN 0202	Calvert, P.	POLY 0290		AGRO	0058
Bushey, M. M.	ANYL 0126	Camacho, J.	BIOT 0140	Carr, P. W.	ANYL	0027
Buss, R. J.	COLL 0218	Camacho C., C.	ENVR 0038	Carr, S. A.	ANYL	0140
Bussmann–Holder, A.		Camberlin, Y.	POLY 0112	Carr, S. H.	POLY	0108
	PHYS 0012	Cambron, R. T.		Carr, S. W.	PETR	0006
Busso, M. E.	MEDI 0055		ANYL 0011		PETR	0081
Buszek, K. R.	ORGN 0210	Cameron, D. C.	BIOT 0198	Carraher, C. E.	PMSE	0014
Butcher, J. W.	MEDI 0096	Caminati, G.	POLY 0238		PMSE	0015
Butcher, R. J.	INOR 0262	Cammer, P. A.	ENVR 0066	Carraher, C. E., Jr.		
	INOR 0267	Campa, J.	INOR 0034		CHED	0006
Butenhoff, T. J.		Campbell, J. C.	BIOT 0032		PMSE	0006
	FUEL 0110	Campbell, N. F.		Carranco, I.	PHYS	0236
	PHYS 0235		AGFD 0140	Carrasco, F.	CARB	0044
Butera, J.	MEDI 0095	Campbell, S. J.	ENVR 0054		CARB	0051
Butera, R. A.	CHED 0127	Canard, G. M.	ORGN 0232		PMSE	0077
Butina, D.	MEDI 0079	Candau, S. J.	POLY 0201		PMSE	0170
Butler, A.	INOR 0016	Cangiano, D. L.		Carrasco S., M. A.		
	INOR 0017		POLY 0048		ANYL	0055
	INOR 0097	Canizal, G.	NUCL 0071	Carreon, P.	POLY	0056
	INOR 0119	Cann, J. R.	BIOL 0006	Carris, M. W.	COLL	0158
Butler, I. S.	INOR 0336	Cannon, D. W.	COLL 0201	Carroll, F. I.	MEDI	0075
Butler, L. G.	AGFD 0149	Cannon, P. L., Jr.		Carruthers, J. D.		
	CATL 0022		COLL 0024		PETR	0033
	ENVR 0004	Cantone, C. L.	MEDI 0105	Carruthers, N. I.		
	INOR 0382	Cantor, E. H.	MEDI 0118		ORGN	0178
Butron S., A.	I&EC 0070	Cantow, H–J.	PMSE 0078	Carson, J. W., Jr.		
	I&EC 0071		PMSE 0184		POLY	0042
Butruille, J. R.	PETR 0014	Cantrell, C. E.	CATL 0053	Carson, M. C.	AGFD	0099
Butterfield, D. A.		Cantwell, M.	AGFD 0010	Carta, G.	BIOT	0065
	MEDI 0149	Cao, J. G.	PETR 0105	Carter, A. M.	POLY	0017
Buttery, R. G.	AGFD 0024	Caperelli, C. A.	BIOL 0076	Carter, G. W.	MEDI	0167
Butzenlechner, M.		Capiris, T.	MEDI 0178		MEDI	0168
	AGFD 0194	Cappellacci, L.	MEDI 0029		MEDI	0176
Buxton, E.	ANYL 0033	Cappellani, E. P.		Carter, J. D.	INOR	0341
Buzanowski, W. C.			INOR 0380	Carter, J. P.	MEDI	0025
	POLY 0039	Cappelli, M. A.	FUEL 0035		MEDI	0026
Buzicky, G. D.	AGRO 0143	Capps, T. M.	AGRO 0111		MEDI	0171
Byers, J.	POLY 0210	Capson, T. L.	BIOL 0159	Carter, K. R.	POLY	0182
Byers, J. D.	POLY 0079	Caraco, N. F.	GEOC 0012	Carter, R. O., III		
Bykov, V. A.	BIOT 0006	Carbajal A., M. E.			PETR	0140
Bymaster, F.	MEDI 0139		ANYL 0055	Cartier, P. G.	ANYL	0031
Byrd, D.	CHAS 0040	Carberry, S. E.	BIOL 0128	Cartledge, F. K.		
Byrne, C. A.	PMSE 0181	Carbonell, R. G.			ENVR	0004
Byrne, J. W.	I&EC 0046		BIOT 0020		ENVR	0041
Cabal, M. P.	ORGN 0213		POLY 0140	Carturan, G.	POLY	0231
Caballero, C. R.		Cardellina, J. H., II		Cartwright, T.	BIOT	0076
	COLL 0157		AGFD 0192	Caruso, J.	INOR	0084
	COLL 0158	Cardenas P., J.	ORGN 0280	Casabianca, H.	PETR	0141
Cabell, L. A.	ORGN 0053	Cardin, J.	INOR 0208	Casanova, J.	CHED	0190
Cabrera, C. R.	INOR 0216	Cardinali, F. L.	CEI 0001	Casara, P.J.	BIOL	0052
Cadena, J. L.	ORGN 0230	Cardoso, J.	ORGN 0159	Casci, J. L.	PETR	0021
Cadisch, M.	COMP 0007		PMSE 0129	Casciano, C. N.	AGRO	0078
	COMP 0008		POLY 0094	Casella, I. G.	POLY	0334
Cadwell, L.	INOR 0123	Carey, J.	BIOL 0115	Casero, R. A., Jr.		
Cahero, I.	PHYS 0236	Carignan, Y. P.	POLY 0062		MEDI	0021
Cahill, D. G.	ANYL 0022	Carillo, E.	ORGN 0193	Casey, C. P.	ORGN	0130
Cahill, T. A.	NUCL 0080	Carino, F. A.	AGRO 0098	Casey, M.	BIOT	0024
Cai, W. Z.	PMSE 0121	Carlin, C. M.	PMSE 0151	Casey, T. S.	CHAS	0023
Cai, Y.	COLL 0207	Carlin, R. L.	INOR 0267	Cash, J. N.	AGFD	0058
Cain, E. J.	POLY 0058	Carlin, R. T.	ANYL 0097	Cashin, F. J.	FUEL	0093
Cain, G. A.	MEDI 0084	Carling, R. W.	MEDI 0146	Casteel, D. A.	ORGN	0260
	MEDI 0153		ORGN 0117	Castellanos R., M. A.		
Calatayud–Catano, M.		Carlini, E. J.	AGRO 0116		ENVR	0025
	AGFD 0138	Carlson, J. D.	BIOT 0062	Castellino, F. J.		
Calcagno, M.	BIOL 0106	Carlson, K. D.	PHYS 0024		ANYL	0020
Calcbrese, J. C.	INOR 0256	Carlson, M.	MEDI 0179	Casterline, J. L.		
Calderon V., H.		Carlson, R. H.	I&EC 0051		AGFD	0208
	AGFD 0105	Carmona, E.	INOR 0035	Castilla M., M. E.		
Calderon–Villagomez, H. E.			INOR 0036		ENVR	0040
	ENVR 0046	Carney, R. L.	AGRO 0003	Castilla V., P.	FUEL	0106
Caldwell, K. D.	PMSE 0013		AGRO 0008	Castillo, R.	MEDI	0064
Calemma, V.	FUEL 0027		AGRO 0012		MEDI	0065
Calkins, W. H.	FUEL 0003	Carothers, T. W.		Castillo A., S.	PHYS	0244
Callaway, J. A.	CINF 0031		POLY 0323	Castillo–Blum, S. E.		
Callender, R. H.		Carotti, A.	MEDI 0059		INOR	0263
	BIOT 0080				INOR	0273

Castillo-Blum, S. E.	INOR 0398	Chapman, G. L.	PETR 0148	Chen, X.	ANYL 0022
					BIOL 0122
Castillo-Rojas, S.		Chappell, E. L.	PHYS 0266		INOR 0183
	NUCL 0070	Chardon, C.	INOR 0210		PHYS 0231
Castillo V., P.	FUEL 0104	Charity, R.	NUCL 0027		PHYS 0240
	FUEL 0105	Charles, S. F.	AGRO 0078		POLY 0012
Castor, T. P.	BIOT 0070	Charleson, S.	MEDI 0180	Chen, X. Y.	PHYS 0167
Castro, C. M.	MEDI 0049	Charm, S. E.	AGFD 0081	Chen, Y.	ORGN 0022
Castro-Acuna, C. M.		Charra, F.	POLY 0009	Chen, Y-F.	PMSE 0169
	CHED 0109	Charton, B.	BIOL 0037	Chen, Y-J.	INOR 0141
Castro-Acuna, M.			BIOL 0038	Chen, Y-L.	BIOT 0151
	PHYS 0221		BIOT 0247	Chen, Y. L.	MEDI 0143
	PHYS 0280	Charton, M.	BIOL 0037	Chen, Z.	COLL 0020
Cate, C. D.	INOR 0111		BIOL 0038		INOR 0214
Cate, C. W.	INOR 0111		BIOT 0247		INOR 0215
Cates, G. D.	PHYS 0041	Chartrain, M. C.		Cheng, C-C.	BIOL 0122
Caufield, C. E.	MEDI 0169		BIOT 0109	Cheng, C-Y.	MEDI 0156
Caulton, K. G.	INOR 0127	Chase, D. B.	POLY 0254	Cheng, J.	INOR 0165
	INOR 0210	Chase, M. W.	PHYS 0160	Cheng, L. K.	INOR 0256
Cave, R. J.	PHYS 0237	Chatterjee, G.	ORGN 0216	Cheng, L-T.	BIOT 0173
Cayen, M. N.	AGRO 0078	Chatterjee, M.	ORGN 0190	Cheng, M-C.	INOR 0028
Cazers, A. R.	AGRO 0108	Chatterjee, S.	BIOT 0066		INOR 0029
Cazorla-Amoros, D.		Chau, F. T.	PHYS 0238		INOR 0041
	FUEL 0029	Chaudhary, A. G.			INOR 0043
	FUEL 0032		ORGN 0255	Cheng, P.	POLY 0073
Cea O., R.	INOR 0032	Chavez, M. L.	INOR 0201	Cheng, S.	PETR 0111
Cedeno C., L.	FUEL 0104		INOR 0224	Cheng, W.	BIOL 0146
	FUEL 0105	Chavez C., A.	I&EC 0074	Cheng, Y-C.	CARB 0038
	FUEL 0106	Chavez Nava, S.		Chenier, P. J.	ORGN 0232
Celamare, S.	MEDI 0059		ENVR 0026	Cheong, B. S.	PHYS 0350
Cerez, J. C.	SCHB 0023	Chavez T., R. H.		Chern, R. T.	POLY 0117
Cerf, D. C.	AGRO 0003		I&EC 0073	Chernov, A. A.	FUEL 0126
Cerkaukas, R. F.		Chavolla C., A.	I&EC 0070	Cheron, T. M.	INOR 0155
	AGRO 0040	Chawla, B.	PETR 0124	Chesta, C. A.	ORGN 0166
Cervantes, R.	ORGN 0245	Chawla, R. C.	AGRO 0140	Chester, T. L.	AGFD 0201
Cha, D. C.	INOR 0102	Chay, I-C.	POLY 0148	Chetan, M. S.	POLY 0074
Chabin, R.	BIOL 0049	Chbihi, A.	NUCL 0027	Cheung, P.	PMSE 0065
	BIOL 0050	Cheatham, S.	BIOL 0044	Chevalier, C.	FUEL 0084
Chai, Z. M.	PETR 0117	Chedekel, M. A.		Chevalier, S.	PETR 0142
Chaiken, J.	BIOT 0024		HIST 0008	Cheynis, B.	NUCL 0028
Chaikin, P.	PHYS 0002	Cheek, G. T.	ANYL 0094	Chhajer, M.	POLY 0139
Chalich, B. D.	CHAS 0003	Chen, B-C.	ORGN 0293	Chi, K. M	COLL 0027
Chamberlin, S. A.		Chen, C-C.	PHYS 0239	Chi, K. M.	COLL 0030
	ORGN 0221	Chen, C. C.	PHYS 0329		INOR 0144
Chambers, J. Q.		Chen, C-H.	PHYS 0128	Chianelli, R. R.	CATL 0038
	BIOT 0099	Chen, C-H.B.	BIOL 0156	Chianese, E. K.	PETR 0122
Chambon, B.	NUCL 0028	Chen, C-Y.	CATL 0014	Chiang, C-M.	COLL 0204
Chamizo, J. A.	CHED 0005	Chen, C. Z.	INOR 0361	Chiang, L. Y.	POLY 0049
Chamkasem, N.		Chen, D.	ANYL 0014	Chiang, M. Y.	INOR 0102
	AGRO 0149	Chen, E. N., Jr.		Chiang, S-J.D.	BIOT 0128
Chamulitrat, W.			PETR 0122	Chiang, T-C.	PHYS 0186
	AGFD 0012	Chen, F. R.	PETR 0133	Chiang, W.	INOR 0049
Chan, A. S.	INOR 0077	Chen, G. C.	PMSE 0153	Chiang, Y.	MEDI 0137
Chan, A.S.C.	CATL 0001	Chen, G. J.	I&EC 0008	Chiba, M.	AGRO 0040
Chan, C. C.	NUCL 0090	Chen, H.	BIOT 0156		AGRO 0041
Chan, H.S.O.	INOR 0107		FUEL 0068	Chibowski, E.	COLL 0163
Chan, L-M.	BIOL 0043		POLY 0066		COLL 0203
Chan, M. K.	INOR 0408	Chen, H-C.	AGFD 0190	Chidester, C. G.	
Chan, S.	BIOL 0096	Chen, J.	BIOL 0162		MEDI 0126
Chan, T-L.	ORGN 0181		INOR 0055	Chidsey, C.E.D.	
Chan, W-T.	PHYS 0090	Chen, J. K.	PETR 0114		PHYS 0127
Chandler, G. O.		Chen, J-L.	ORGN 0235		POLY 0215
	MEDI 0173	Chen, J. M.	I&EC 0046	Chien, J.C.W.	CHED 0059
Chang, C. D.	I&EC 0060	Chen, J. S.	AGFD 0028	Chien, W-J.	BIOL 0116
Chang, C. T.	POLY 0175	Chen, J-T.	INOR 0041	Chikate, R. C.	INOR 0268
Chang, H. N.	BIOT 0229		INOR 0043	Chin, C-K.	AGFD 0002
Chang, J.	MEDI 0179	Chen, K.	BIOT 0068	Chin, D. N.	POLY 0367
Chang, L. L.	MEDI 0105	Chen, K-H.	PHYS 0146	Chinake, R. C.	PHYS 0269
Ch'ang, L. Y.	AGRO 0097	Chen, M. F.	POLY 0081	Chiou, J. S.	MEDI 0032
Chang, N.	BIOT 0063	Chen, P.	PMSE 0008	Chipperfield, J. R.	
Chang, R.S.L.	MEDI 0103	Chen, Q.	AGFD 0048		INOR 0266
	MEDI 0104		ANYL 0135		INOR 0269
	MEDI 0105	Chen, S.	ANYL 0113	Chisholm, A.	CHED 0038
Chang, R. T.	INOR 0188		BIOT 0184	Chisholm, M. H.	
Chang, W-P.	INOR 0074		INOR 0382		INOR 0128
Chang, Y. J.	COLL 0087	Chen, S. H.	PETR 0099	Chiu, H-T.	INOR 0074
	PHYS 0062		POLY 0330	Chiu, Y. C.	COLL 0078
	PHYS 0283	Chen, S. I.	PMSE 0057	Chiu, Y-H.	PHYS 0241
Chao, H.	BIOT 0080	Chen, T.	ORGN 0243	Chiulli, R. J.	INOR 0239
Chao, J. L.	POLY 0304		PHYS 0152	Chlenov, M. A.	BIOT 0197
Chao, K. J.	PETR 0099		POLY 0187	Cho, H.	BIOT 0095
Chao, S. H.	PETR 0148	Chen, W.	PHYS 0088	Cho, H-C.	COLL 0009
		Chen, W-C.	POLY 0011	Cho, S-Y.	PHYS 0142

Cho, Y.	BIOL 0123	Churchill, M. R.		Cohen, Y.	POLY 0241
Cho, Y. H.	ORGN 0202		INOR 0066	Cohen de Lara, E.	
Choe, S.	PMSE 0164	Ciabatti, R.	MEDI 0059		CATL 0024
Choi, B. G.	MEDI 0051	Ciaccio, J. A.	ORGN 0107	Cohn, K. C.	CHED 0186
Choi, G. J.	POLY 0360	Cibulsky, M. J.	POLY 0191	Colby, R. H.	POLY 0172
Choi, H-J.	ORGN 0062	Cima, L.	POLY 0273	Cole, B. B.	POLY 0067
Choi, S-K.	BIOT 0180	Cina, J. A.	PHYS 0067	Cole, D. A.	BIOT 0168
Choi, Y. S.	PHYS 0164	Cisneros, G. R.	NUCL 0086	Cole, K. C.	POLY 0334
Chojnacki, J. A.		Ciszewski, L. A.		Coleman, M. R.	
	INOR 0212		CARB 0023		POLY 0116
Chojnacki, S. J.			CARB 0024	Colemenares-Landin, F.	
	INOR 0177	Ciszkowska, M.	ANYL 0095		PHYS 0243
Chomaz, P.	NUCL 0009	Civille, G. V.	AGFD 0156	Colin, H.	ANYL 0037
Chooback, L.	BIOL 0124	Claar, R. E.	CMEC 0015	Colizzo, F.	MEDI 0161
Choplin, A.	PETR 0131	Claeyssens, M.	BIOT 0206	Collar, J. I.	NUCL 0013
Choplin, L.	BIOT 0196	Clague, M. J.	INOR 0016	Collen, K. R.	INOR 0357
Chorny, V. Y.	COLL 0110	Clare, M.	MEDI 0152	Collette, J. W.	CHED 0007
Chott, R. C.	AGRO 0023	Claremon, D. A.		Colletti, S. L.	INOR 0214
Chottard, G.	INOR 0020		MEDI 0096		INOR 0215
Chou, J-C.	AGRO 0136		MEDI 0097	Collins, R. W.	POLY 0252
Chou, J-H.	INOR 0204		MEDI 0191	Collins, T. J.	INOR 0243
Chou, J. Z.	PHYS 0334	Clark, C. R.	ANYL 0045	Colombo, P.	ENVR 0011
Chou, Y. C.	PMSE 0027	Clark, J. E.	CINF 0012	Colton, F. B.	HIST 0020
Choudary, P.	AGRO 0009	Clark, J. M.	AGRO 0154	Colton, I.	BIOL 0104
Choudhry, V.	FUEL 0134	Clark, N. A.	BIOT 0202	Colton, K. H.	CHAL 0014
Choudhury, H.	ENVR 0073	Clark, R. F.	MEDI 0168	Colton, R. J.	PHYS 0154
Chowdhury, S.	BIOL 0107	Clark, S.	INOR 0266	Coltrain, B. K.	POLY 0181
Christ, G.	COLL 0091		INOR 0269		POLY 0260
Christiaens, P.	PHYS 0141	Clark, W. M.	BIOT 0049	Coltrin, M. E.	COLL 0066
Christian, H.	PHYS 0242	Clarke, J.H.R.	POLY 0366		PHYS 0179
Christian, J.	PHYS 0174	Clarke-Katzenburg, R. H.			PHYS 0357
Christian, J. J.	CINF 0031		CARB 0038	Colussi, D.	MEDI 0111
Christian, R. M.		Clarkson, R. B.	FUEL 0074	Colussi, D. J.	MEDI 0113
	MEDI 0182	Clarson, S. J.	POLY 0020	Combrink, K. D.	
Christiansen, J. A.			POLY 0088		ORGN 0291
	BIOT 0139		POLY 0234	Combs, J.	ORGN 0131
Christie, B. D.	CINF 0022	Class, Y.	ORGN 0294	Comer, F. I.	CARB 0022
	COMP 0046	Clausen, E. C.	BIOT 0017	Comita, P. B.	COLL 0046
Christopher, R. J.			FUEL 0092	Communal, F.	FUEL 0079
	AGRO 0078	Clausing, R. E.	PHYS 0119	Comor, M. V.	COLL 0151
Christos, T. E.	MEDI 0084	Clawson, D. K.	BIOL 0072	Compadre, C. M.	
	MEDI 0153	Clayton, T. W., Jr.			MEDI 0017
Christou, G.	INOR 0196		ORGN 0042	Compton, D.A.C.	
	INOR 0197	Clearfield, A.	PETR 0094		PMSE 0141
Christou, V.	INOR 0021		PETR 0106	Compton, S. V.	PMSE 0141
Chronister, C. W.		Cleary, D. G.	ORGN 0028	Conaway, W.	PHYS 0268
	ORGN 0263	Cecak, N.	POLY 0188	Confer, W. L.	MEDI 0026
Chrosniak, C. E.		Cleland, J. L.	BIOT 0136	Connaway-Wagner, M. C.	
	ENVR 0023	Clement, R. P.	AGRO 0078		POLY 0261
Chrunyk, B. A.	BIOT 0114	Clentsmith, G.K.B.		Connell, J. W.	POLY 0025
Chu, B.	COLL 0181		INOR 0379	Connelly, N. V.	BIOT 0007
	PMSE 0103	Clerc, J. T.	COMP 0008	Conner, J. R.	ENVR 0001
	PMSE 0149	Clesceri, L. S.	BIOT 0158	Conney, A. H.	AGFD 0065
	PMSE 0155	Cleveland, B.	NUCL 0021		AGFD 0067
	POLY 0251	Cleveland, T.	ENVR 0028		AGFD 0069
Chu, C. K.	CARB 0022	Clore, M.	PHYS 0113		AGFD 0088
	MEDI 0051	Clos, N.	INOR 0226	Connor, D. T.	MEDI 0177
Chu, G.	MEDI 0136	Closs, G. L.	ORGN 0143	Connor, J. R.	MEDI 0025
Chu, H-M.	I&EC 0075	Clyde, R.	FUEL 0133		MEDI 0026
	PETR 0087	Clyde, R. A.	BIOT 0154	Conrad, D.	ENVR 0005
Chu, J. S.	CINF 0044	Cnossen, J. E.	CATL 0026	Conrad, M.	BIOT 0125
Chu, P.	COLL 0057	Coale, K. H.	GEOC 0010	Conradi, M. S.	INOR 0102
Chu, P-J.	PETR 0094		INOR 0002	Constantinescu, A.	
Chuang, M-C.	PHYS 0146	Coates, I. H.	MEDI 0079		MEDI 0041
Chucholowski, A. W.		Cobb, J. T., Jr.	FUEL 0136	Conte, D.	MEDI 0098
	MEDI 0107	Cobb, S. H.	PHYS 0173	Conticello, V. P.	
Chui, W-K.	BIOT 0036	Cochran, J.	SCHB 0019		POLY 0047
Chuiko, A. A.	COLL 0096	Cochrane, B. J.	AGRO 0100	Contreras, L.	INOR 0036
	COLL 0100	Cocke, D. L.	ENVR 0003	Contreras, M. E.	
	COLL 0104	Cockroft, N. J.	ANYL 0004		COLL 0124
	COLL 0110	Cody, G. D.	FUEL 0086	Contreras, R.	INOR 0263
	COLL 0113	Coffield, J. E.	ANYL 0065	Contreras S., L.	
	COLL 0116		INOR 0064		AGFD 0038
Chun, G-T.	BIOT 0155	Coffman, H. R.	BIOL 0014		ORGN 0037
Chun, J-W.	FUEL 0096	Cogliano, J.	CHAS 0011	Contreras-Theurel, R.	
Chung, C.	NUCL 0090		CHAS 0020		INOR 0252
Chung, F-L.	AGFD 0068	Cohen, J.	MEDI 0120	Conway, J. D.	MEDI 0151
Chung, J. F.	ORGN 0214		MEDI 0197	Conway, P. G.	MEDI 0085
	ORGN 0215	Cohen, M. B.	AGRO 0115		MEDI 0130
Chung, T. C.	POLY 0185	Cohen, N.	ORGN 0284	Conyers, B.	ENVR 0031
Church, G. A., II		Cohen, R. E.	INOR 0307	Conzentino, P.	MEDI 0195
	AGRO 0042		POLY 0207	Cook, C. M.	MEDI 0165
Church, T. M.	GEOC 0007	Cohen, S. L.	COLL 0026	Cook, L.	MEDI 0084

Cook, L.	MEDI 0153	Cox, B.	MEDI 0111	Cruz-Nunez, X.	
Cook, S. L.	I&EC 0053	Cox, D. F.	COLL 0028		PHYS 0296
Coon, P.	CINF 0037		COLL 0195	Cruz-Sosa, F.	CHED 0094
Coon, P. A.	COLL 0008	Cox, D. I.	ORGN 0217		CHED 0095
Cooney, C.	BIOT 0074	Cox, G. B.	ANYL 0029		CHED 0096
Cooney, D. A.	MEDI 0048	Cox, J. A.	AGFD 0142		CHED 0097
Cooney, M. J.	ORGN 0094	Cox, J. C.	BIOT 0156	Cuan H., M. A.	PHYS 0244
Coons, D. E.	INOR 0388	Cox, P. A.	PETR 0021	Cucinotta, G.	MEDI 0188
Cooper, A. R.	PMSE 0168	Cox, T. D.	AGRO 0108	Cuddy, L. J.	PMSE 0150
Cooper, J.	CHED 0162	Coyle, C. L.	INOR 0371	Cuesta-Harvey, L.	
	PHYS 0336	Coyne, L. D.	POLY 0171		ORGN 0162
Cooper, S. L.	PMSE 0145	Crabtree, R. H.	INOR 0384	Cuevas G., R.	FUEL 0104
	PMSE 0146		INOR 0405		FUEL 0105
Cooper, T. A.	BIOT 0198	Cradock, J.	AGFD 0163		FUEL 0106
Cooper, W. T.	ANYL 0072	Cragg, P. J.	INOR 0297	Culleen, L. E.	CHAS 0013
	ANYL 0087	Craig, G.S.W.	POLY 0207	Cullen, D.	BIOT 0236
	ANYL 0088	Craig, J. R.	FUEL 0139	Cullen, T. G.	AGRO 0072
Coppock, R.	ENVR 0064	Craik, C. S.	MEDI 0054		AGRO 0073
Coppola, G. M.	MEDI 0101	Cramer, R. E.	INOR 0366	Cullen, W. R.	INOR 0137
Corcoran, E. W., Jr.		Cramer, S. M.	ANYL 0038	Culp, R. A.	AGFD 0187
	PETR 0057		ANYL 0048	Cumming, J. B.	
Corma, A.	PETR 0113		BIOT 0018		NUCL 0051
Cormerais, F. X.			BIOT 0042		NUCL 0059
	PETR 0115	Crampton, R. G.		Cumming, W. G.	
Corn, R. M.	ANYL 0024		CHED 0025		CHED 0018
Corneille, J. S.	COLL 0193	Crane, C. G.	INOR 0089	Cummings, D. G.	
Cornelius, R.	CHED 0115	Crans, D. C.	BIOL 0051		INOR 0070
Cornfeldt, M. L.			INOR 0184	Cummings, G.	INOR 0303
	MEDI 0137		INOR 0228	Cummings, L. H.	
Cornman, C. R.	INOR 0018		INOR 0409		CHAS 0035
Corrie, J.E.T.	ORGN 0026	Crawford, C. L.	PHYS 0258	Cummins, C. C.	
Corry, J. P.	BIOT 0233	Crawford, M. K.			INOR 0307
Corser, M. M.	CATL 0053		PHYS 0010	Cundy, K. C.	MEDI 0195
Cory, D. G.	CATL 0022	Crawford-Brown, D. J.		Cuomo, J. J.	PHYS 0092
Cory, M.	COMP 0039		ENVR 0075	Curley, R. W., Jr.	
Cosandey, F.	POLY 0062	Creech, D.	CHED 0039		MEDI 0030
Cosgrove, J. E.	PHYS 0149	Creech, D. L.	CHED 0139	Curran, M. A.	CEI 0027
Cosman, M.	PHYS 0021	Creed, R. K.	CHAS 0002	Currier, T. C.	BIOT 0162
Costa, G.	POLY 0120	Creegan, F. J.	CHED 0152	Curry, J. F.	PMSE 0031
Costales, M. J.	AGRO 0046	Creighton, J. R.		Curtis, C. W.	FUEL 0005
	AGRO 0047		COLL 0065	Curtis, W. R.	BIOT 0233
	AGRO 0048		COLL 0069	Curzon, P.	MEDI 0127
Costella, R. G.	AGRO 0064	Creuzet, F.	PHYS 0196	Cushing, T.	ORGN 0283
Costello, C. E.	AGRO 0034	Crimmins, M. T.		Cussler, E. L.	BIOT 0224
Costes, M.	PHYS 0089		ORGN 0002	Cusson, M.	AGRO 0007
Cotroneo, A.	AGFD 0199	Crippen, K. L.	AGFD 0098	Cutler, A. R.	INOR 0238
Cotter, D. J.	CHAL 0010		AGFD 0161		INOR 0239
Cotter, W. D.	INOR 0147	Critchlow, S. C.			INOR 0240
	INOR 0280		INOR 0115		INOR 0337
Cotts, P. M.	POLY 0199		INOR 0132	Cygler, M.	BIOL 0143
	POLY 0250		INOR 0271	Cyngier, R.	I&EC 0001
Coudurier, G.	PETR 0133	Crivello, J. V.	PMSE 0159		YCC 0005
Coughlan, M. P.			POLY 0014	Cyngler, R.	CMEC 0023
	BIOT 0209		POLY 0015	Cyr, M. J.	ORGN 0139
Coulomb, C.	AGFD 0197		POLY 0016	Czarniecki, M.	MEDI 0117
Court, J.	MEDI 0087		POLY 0017	Czerwinski, K. R.	
	MEDI 0172	Cromwell, E. F.			NUCL 0074
Court, J. J.	ORGN 0035		PHYS 0163	Czuchajowski, L.	
Cousins, B. L.	ANYL 0022	Crooks, P. A.	MEDI 0149		ORGN 0189
Cousins, L.	PHYS 0006	Cropek, D. M.	ANYL 0026	Daage, M.	CATL 0038
Cousins, L. M.	COLL 0092	Crosby, G. A.	CHED 0067	Dabbagh, G.	PHYS 0375
Coustal, S.	MEDI 0009		CHED 0110	Dabek, R. A.	CHAL 0022
Coutsolelos, A. G.			CHED 0129		CHAL 0027
	ENVR 0045		INOR 0164	Dabrowski, B.	PHYS 0009
	INOR 0221	Crosby, J. L.	CHED 0110	Dadali, A. A.	PHYS 0245
Couture, R.	AGFD 0106		CHED 0111	Daeseleire, E.	AGFD 0130
Covarrubias, A.	ORGN 0176	Crosley, D. R.	FUEL 0098	Dagan, A.	POLY 0241
Covert, K. J.	INOR 0158		PHYS 0266	Dagata, J. A.	PHYS 0363
Covey, D. F.	MEDI 0010	Cross, K. P.	CINF 0023	Dagdigian, P. J.	
Covick, L. A.	COMP 0055		COMP 0041		CHED 0199
Cowan, G. A.	NUCL 0061	Croteau, R.	AGFD 0009		PHYS 0189
Coward, J. K.	BIOL 0055	Crouse, D.	BIOT 0140		PHYS 0246
	MEDI 0027	Crowell, J. E.	COLL 0009	Dahr, R. K.	ORGN 0087
Cowdery-Corvan, R.			COLL 0047	Dai, S.	ANYL 0065
	INOR 0316	Crowley, T.	ENVR 0029		INOR 0064
Cowin, J. P.	PHYS 0084	Croze, E. M.	BIOT 0074	Dai, Z.	PMSE 0080
Cowley, A. H.	COLL 0067	Crum, A. D.	AGFD 0039	Dai, Z-G.	COLL 0215
	INOR 0146	Crump, P.	BIOT 0039	Dailey, G. C.	INOR 0185
	INOR 0292	Cruz, M.	INOR 0241	Dailey, O. D., Jr.	
	INOR 0390	Cruz-Almanza, R.			AGRO 0063
	INOR 0392		ORGN 0017	Daily, T.	NUCL 0021
Cowman, M. K.		Cruz Nunez, X.	ENVR 0026	Dais, P.	CARB 0049
	CARB 0047			Daivis, P. J.	COLL 0054

Name	Code	Num	Name	Code	Num	Name	Code	Num
Dale, B. E.	BIOT	0016	Davis, B. H.	CATL	0007	Delfau, J-L.	FUEL	0080
Dalidowicz, J. E.				CATL	0052	Delgado, G.	ORGN	0018
	AGRO	0131		PETR	0028	Della-Negra, S.	NUCL	0055
Dallanoce, C.	MEDI	0141	Davis, D. D.	POLY	0333	Dellwo, M. J.	PHYS	0168
Dallaria, J. F., Jr.			Davis, F. A.	ORGN	0077	Delmas, G.	PMSE	0166
	MEDI	0176		ORGN	0293		POLY	0123
Dalmia, B.	BIOT	0047	Davis, H.	BIOL	0111	Delmore, J. E.	NUCL	0056
Daly, W. H.	POLY	0099	Davis, H. F.	PHYS	0349	DeLoach, J. R.	BIOT	0166
Damavarapu, R.			Davis, J. D.	CHED	0150	Delorme, C.	MEDI	0009
	ORGN	0257	Davis, M. E.	CATL	0014	de los Santos, H.		
D'Amico, D. C.	ORGN	0030		COLL	0028		BIOL	0161
Damon, R. E.	MEDI	0101		PETR	0001	deLozanne, A.	PHYS	0362
D'Amore, M.	FUEL	0025		PETR	0098	Del Piero, G.	FUEL	0027
Damrauer, R.	ORGN	0055	Davis, P. J.	PETR	0068	DeLuke, D. A.	CHED	0063
Dan-Brandon, N.			Davis, P. M.	PETR	0032	Demain, A. L.	BIOT	0181
	POLY	0269	Davis, R.	NUCL	0021	Demerson, C.	MEDI	0179
Dando, N. R.	POLY	0340	Davis, R. J.	CATL	0030	Demeyer, A.	NUCL	0028
Dandy, D. S.	PHYS	0179	Davis, W. M.	INOR	0181		NUCL	0039
	PHYS	0357	Davison, A.	INOR	0181	De Micheli, C.	MEDI	0141
Dang, T. D.	POLY	0028	Day, C. L.	INOR	0298	DeMilo, A. B.	AGRO	0033
Danheiser, R. L.			Day, D. E.	PMSE	0109	Deming, T. J.	PMSE	0071
	ORGN	0286		PMSE	0110		POLY	0347
Danho, W.	BIOL	0042	Day, J.	FUEL	0045	Demodena, J. A.		
Daniel, L. W.	MEDI	0072	Day, V. W.	INOR	0174		BIOT	0087
Daniele, S.	COLL	0005	Dayalan, E.	COLL	0040	Demuth, H-U.	BIOL	0030
Daniels, D. H.	AGFD	0109	De, B.	MEDI	0124	Denekamp, C.	INOR	0052
Daniels, L. E.	FERT	0011	Deal, J. G.	MEDI	0186	Deng, H.	INOR	0220
Danielson, E.	INOR	0282	De Amici, M.	MEDI	0141		INOR	0332
Danishefsky, S. J.			Dean, A. M.	FUEL	0063	DeNinno, S. L.	MEDI	0176
	ORGN	0287		FUEL	0085	Denis, D.	MEDI	0180
Dankosh, H. E.	INOR	0331		PHYS	0228	Dennis, J. W.	BIOL	0151
Dannenberg, J. J.				PHYS	0229	Dennis, P. A.	AGRO	0012
	ORGN	0175	Dean, M. A.	CARB	0004	Dennison, P. R.		
	PHYS	0247		CARB	0050		PETR	0147
	PHYS	0326	Dean, T. R.	AGRO	0029	Denny, R. L.	AGRO	0086
Danzin, C.	BIOL	0052	de Angelis, M.	GEOC	0001	den Otter, G. J.		
Dao, L.	AGFD	0112	DeAngelis, R.	CATL	0035		PETR	0112
Daragan, V. K.	PETR	0066	Deanin, R. D.	PMSE	0128	Dentrone, L.	COLL	0013
Darby, W. L.	INOR	0267	Deason, M. E.	MEDI	0190	Deo, G.	CATL	0057
Dardon, E. H.	INOR	0245	de Belligny, P. C.				COLL	0196
Darensbourg, D. J.				AGRO	0100	de Pablo, L.	INOR	0201
	INOR	0212	Deberdt, F.	POLY	0147		INOR	0224
	INOR	0213	DeBernardis, J. F.			DePaula, J. C.	CHED	0084
Darensbourg, M. Y.				MEDI	0116	de Paula, J. C.	PHYS	0249
	INOR	0177		MEDI	0124	Depoy, R. E.	CHED	0215
	INOR	0225	DeCamp, D. L.	MEDI	0054	Derbyshire, F. J.		
Darke, P.	BIOT	0071	DeCanio, E. C.	COLL	0198		CATL	0051
Darling, C. L.	PHYS	0248	DeCanio, S.	FUEL	0115		FUEL	0043
Darling, G. D.	POLY	0156	Decker, E. A.	AGFD	0039		FUEL	0048
Das, J.	MEDI	0188	Deckman, B.	CMEC	0010		FUEL	0101
Das, K.	BIOT	0152	De Clercq, E.	MEDI	0048	DerMarderosian, A.		
	BIOT	0153	Decman, D. J.	NUCL	0011		AGFD	0153
Dasgupta, T.	INOR	0400	De Corte, B.	ORGN	0079	de Roos, K. B.	AGFD	0074
Datar, R.	BIOT	0076	Decowski, P.	NUCL	0039	DeRosa, C. T.	ENVR	0073
Date, T.	ORGN	0102	De Deken, J.	CATL	0041	De Rosa, M.	ORGN	0158
Datta, R.	BIOT	0015		FUEL	0100	DeRosier, T.	BIOL	0153
Datye, A. K.	PETR	0072	Deeds, D. A.	CHAS	0031	Deroski, B. R.	CINF	0031
	POLY	0312	Deeter, G. A.	POLY	0035	Derouane, E. G.		
Dauben, W. G.	ORGN	0135	Deeter, J. B.	MEDI	0066		PETR	0019
Daubenspeck, N.			Defoor, F.	PMSE	0102		PETR	0095
	INOR	0150	Degar, S.	AGFD	0188	de Roy, A.	PETR	0127
Daugherty, K. E.			Deguchi, S.	POLY	0098	Derrough, S. N.		
	FUEL	0138	De Guesquiere, A.				PMSE	0001
	FUEL	0140		AGFD	0130	de Ruiter, R.	PETR	0024
Dave, P. R.	ORGN	0256	de Gyves, J.	ANYL	0057	Desai, P.	POLY	0316
Davey, D. D.	MEDI	0118	DeHaven-Hudkins, D. L.				POLY	0338
David, B. V.	AGRO	059A		MEDI	0140	Desai, R.	MEDI	0172
David, D. E.	ORGN	0124	Deibel, M. R.	MEDI	0055	Desai, R. C.	ORGN	0035
David, I. A.	POLY	0286	Deisler, P. F., Jr.			Desai, S. R.	PHYS	0251
David, J.	ENVR	0019		ENVR	0074	Desai, V.	FUEL	0085
Davidson, A. D.			Deitz, V. R.	CARB	0062	Deschner, E. E.	AGFD	0114
	FERT	0008	de Jong, K. P.	PETR	0083	Des Courieres, T.		
Davidson, D. J.	ANYL	0020	deJong, R.L.P.	INOR	0229		PETR	0038
Davidson, D. L.			Delaglio, F.	PHYS	0113		PETR	0040
	POLY	0317	de la Mora, P.	PHYS	0288	Deshmukh, A. R.		
Davidson, E. R.			de la Motte, R.	CARB	0004		ORGN	0224
	PHYS	0013	de la Rosa, R. I.			DeShong, P.	ORGN	0294
Davies, C.	POLY	0259		INOR	0016	Deshpande, R.	PETR	0032
Davies, K.	CINF	0033	De la Torre A., N.				PETR	0068
Davies, K.J.A.	CHAS	0009		CHED	0114	Desikan, A.	CATL	0059
Davies, O. L.	CATL	0053		HIST	0035	Deslongchamps, P.		
Davis, A.	FUEL	0002	Delaviz, Y.	POLY	0044		ORGN	0288
	FUEL	0086	Del Bene, J. E.	PHYS	0033	de Souza, R. T.	NUCL	0035

d'Espinose, J-B.		Disbrow, G. L.	BIOT 0032	Dover, B. T.	ORGN 0225
	PETR 0015	Dismukes, J. P.	PMSE 0111	Dow, E. R.	BIOL 0072
Despres, S.	INOR 0165	Distel, J.	NUCL 0021	Dow, R. L.	ORGN 0183
De Tar, M. M.	PETR 0138	Dittmar, R. M.	POLY 0304	Dowd, P. F.	AGRO 0155
	PETR 0150	Dixon, C. N.	CATL 0043	Dower, W. J.	MEDI 0092
Deterding, L. J.		Dixon, D. D.	I&EC 0055	Dowle, M.	MEDI 0079
	ANYL 0125	Dixon, P. K.	PMSE 0108	Dowler, C. C.	AGRO 0063
de Teresa, C.	PHYS 0288	Dixon-Holland, D. E.		Dowling, K. C.	AGRO 0064
Detlefsen, D. J.	CHED 0081		AGFD 0083		AGRO 0124
Dettmann, H. D.		Djerassi, C.	HIST 0018	Downey, S. W.	NUCL 0043
	PETR 0149	Djukova, T. V.	BIOT 0079	Downie, R. H.	AGFD 0116
Dettweiler, G. R.		Dlott, D. D.	PHYS 0250	Dowrey, A.	COLL 0035
	AGFD 0023	Dobson, G. R.	INOR 0111	Dowrey, A. E.	COLL 0033
Deur, J.	FUEL 0127	Dobson, T.	INOR 0073		PHYS 0117
D'Eustachio, P.	CHAL 0002	Dobyns, K. A.	MEDI 0048		POLY 0303
DeVilbiss, E. D.		Dodge, T. C.	BIOT 0032	Doyemet, J. Y.	PETR 0045
	AGRO 0033	Doelwijt, A.	PETR 0065	Doyle, B. L.	NUCL 0065
Devine, C. D.	CHED 0176	Doetschman, D. C.		Doyle, J. J.	MEDI 0122
Devine, T.	PROF 0006		CHED 0121		MEDI 0123
Devol, A.	GEOC 0017	Dogra, P. V.	PETR 0146	Doyle, M. P.	CATL 0042
Devolder, P.	FUEL 0081	Dogueri, F.	POLY 0140		ORGN 0042
Devonshire, A. L.		Doherty, D. C.	POLY 0354	Doyle, T. W.	MEDI 0012
	AGRO 0137	Doherty, J.	BIOL 0049	Drago, R. S.	PETR 0049
Devore, T. C.	PHYS 0173		BIOL 0050	Drain, C. M.	BIOL 0098
de Vries, N.	INOR 0196	Doherty, J. B.	MEDI 0173		BIOL 0136
Dewailly, E.	CEI 0011	Dolan, G. J.	BIOT 0002	Drain, D.	NUCL 0028
Dhar, T.G.M.	ORGN 0068	Dolan, S.	BIOT 0163	Drake, J. M.	PHYS 0064
Dharmosetio, A. R.		Dolata, D. P.	CINF 0017		PHYS 0135
	COLL 0144		CINF 0034	Drake, S. R.	INOR 0019
Dhingra, S.	INOR 0203	Doll, G. L.	PHYS 0094		INOR 0052
Dhinojwala, A.	POLY 0152	Doll, R. J.	MEDI 0117	Draper, W. M.	CEI 0006
Dhurjati, P.	BIOT 0245	Dollinger, H. M.		Drawert, F.	AGFD 0023
Diachenko, G. W.			CHED 0133	Drechsler, M.	PMSE 0184
	AGFD 0109	Domagala, J. M.		Drew, M.G.B.	INOR 0297
Dias, J. R.	COMP 0020		MEDI 0061	Drickamer, K.	BIOL 0144
	COMP 0027	Dombrowski, A. W.		Driehuys, B.	PHYS 0041
Diaz, D.	CHED 0057		ORGN 0184	Driscoll, J. N.	ENVR 0054
	INOR 0020	Dominguez, J. M.		Drobny, G.	PHYS 0195
	INOR 0026		PETR 0109	Droske, J. P.	CHED 0006
	INOR 0076	Dominguez D., R.		Drossman, H.	ANYL 0033
Diaz, O.	BIOL 0102		CHED 0109	Drouin, S. D.	INOR 0380
Diaz de la Rubia, T.			CHED 0124	Drummond, S.	MEDI 0084
	NUCL 0091	Domokos, L.	COMP 0018		MEDI 0153
Diaz de la Torre, S.		Domser, C. A.	CHED 0076	Druzhko, A. B.	BIOT 0082
	PHYS 0219	Donadio, S.	BIOT 0110	Dryden, N.	INOR 0142
Di Bella, M.	MEDI 0135	Donaldson, G. K.		Dryer, F. L.	PHYS 0142
Dick, D. L.	MEDI 0031		BIOT 0116	Du, C.	PMSE 0016
Dickens, B.	PMSE 0003	Donato, M. G.	AGFD 0199	Dua, R. K.	BIOL 0060
DiDomenico, S.	MEDI 0124	Doney, J. J.	AGRO 0047	Dubb, H. E.	CHAL 0023
	MEDI 0125	Dong, M. W.	ANYL 0129		CHAL 0032
DiDonato, G. C.		Donnelly, J. R.	FUEL 0129	Dubis, E. N.	AGRO 0114
	ANYL 0139	Donnelly, V. M.		DuBois, G. E.	CARB 0026
Diebold, U.	COLL 0172		COLL 0068	Dubois, J-E.	COMP 0006
	COLL 0191		COLL 0214	Dubois, L. H.	COLL 0025
Dieckman, S. L.		Donoho, A. L.	AGRO 0110		INOR 0124
	FUEL 0075	Donohue, M. D.		Dubois, R.	PMSE 0162
Dieckmann, G. R.			PMSE 0036	DuBoisson, R.	SCHB 0018
	CHED 0081	Donovan, B. T.	ANYL 0109	Ducharme, S.	POLY 0155
Diederich, F.	ORGN 0266	Donovan, M. J.	BIOT 0084	Duclos, S.	INOR 0354
Diefenbacher, C. G.		Donovan, R. J.	INOR 0226	Duda, J. L.	POLY 0144
	ORGN 0013		ORGN 0218	Dudarev, S. V.	PETR 0039
Dijkstra, B. W.	MEDI 0004	Donster, B. B.	COLL 0109	Dudis, D. S.	PHYS 0322
Dikshitulu, S.	BIOT 0138	Dooley, K. M.	CATL 0022	Duesler, E. N.	POLY 0312
Dillon, A. C.	COLL 0008	Dordick, J. S.	BIOT 0067	Dufresne, C.	MEDI 0180
Dillon, M. E.	PMSE 0041	Dority, J. A.	MEDI 0158	Dufresne, L.	PETR 0063
Dilwith, R.	BIOT 0169	Dorn, C.	BIOL 0049	Dugo, G.	AGFD 0199
Dimian, A. F.	PMSE 0016		BIOL 0050	Dukat, W. W.	INOR 0133
Dimitrov, A. V.	INOR 0024	Dorn, H. C.	PHYS 0070	Dulcey, C. S.	BIOT 0246
Dimotakis, E. D.		Dorsey, J. G.	ANYL 0102	Dulski, R. E.	CHED 0177
	POLY 0261	Dorthe, G.	PHYS 0089	Duma, S.	TECH 0005
Ding, W-D.	POLY 0280	Dosch, R. G.	PETR 0017	Dumais, J.	PHYS 0095
Ding, Z-Y.	PMSE 0017	Dotrong, M.	PMSE 0018	du Manoir, J. R.	
Dinh, L.	ANYL 0052		POLY 0029		BIOT 0214
Dinh, S. M.	POLY 0195	Dotrong, M. H.	PMSE 0018	Dumesic, J. A.	COLL 0197
Dinkel, D.	ANYL 0014		PMSE 0019		COLL 0212
Dipple, K. A.	COLL 0154		POLY 0029	Dumont, F.	ENVR 0050
Dire, S.	POLY 0231	Dougherty, C.	CHED 0140	Dumont, J.	BIOT 0195
Di Renzo, F.	PETR 0038	Doughty, A.	FUEL 0114	Dunaiskis, A.	MEDI 0143
	PETR 0040	Douglas, J. F.	POLY 0235	Dunbar, P. G.	MEDI 0139
Dirlikov, S. K.	PMSE 0087		POLY 0293	Duncan, B.	PETR 0020
DiSalvo, F. J.	INOR 0275	Douglas, P.	COLL 0150	Duncan, M. A.	CHED 0089
	PHYS 0003	Douglas, T.	INOR 0313	Duncan, T. M.	PHYS 0346
Di Sanzo, F. P.	PETR 0124	Dourson, M.	ENVR 0068	Dunlap, R.	MEDI 0087

Dunlap, R.	MEDI	0172
Dunlop, D. S.	BIOT	0232
Dunn, D. A.	CINF	0026
Dunn, R. F.	MEDI	0046
Dunn, R. G.	CINF	0015
Dunn, R. W.	MEDI	0085
	MEDI	0130
Dupuis, M.	BIOT	0001
Dupuis, R. D.	COLL	0010
Durai-Swamy, K.		
	FUEL	0118
Duran, R. S.	COLL	0071
Duran D., C.	AGFD	0105
Durand, V.	POLY	0133
Duran-de-Bazua, C.		
	ENVR	0046
Durant, G. J.	MEDI	0070
	MEDI	0139
Durante, V. A.	CATL	0036
Durieux, D. O.	CATL	0041
	FUEL	0100
Durning, C. J.	COLL	0057
	POLY	0167
Dus, C. A.	AGFD	0156
Dutt, O.	PMSE	0180
Dutta, P. K.	CATL	0020
	COLL	0075
	COLL	0076
	INOR	0272
	PETR	0074
Dutton, C. J.	BIOT	0127
Dutton, P. L.	BIOL	0094
Duty, R. C.	COLL	0002
Dvoretzky, I.	CHED	0009
Dwinell, S.	AGRO	0103
Dwyer, D. W.	CHED	0121
Dyachkovskii, F. S.		
	PMSE	0045
Dybowski, C.	CATL	0029
	PETR	0070
Dyda, F.	BIOL	0032
Dyer, R. D.	MEDI	0177
Dyke, J. M.	PHYS	0352
Dypvik, T.	PETR	0027
Eads, D. D.	BIOL	0094
Eagleson, P. S.	AGRO	0051
Ealick, S. E.	CARB	0046
	MEDI	0005
Eang, S.	INOR	0073
Eapen, M. J.	PETR	0005
Earl, W. L.	PETR	0068
Earley, W.	MEDI	0195
Eastman, C. E.	POLY	0275
Eaton, H. C.	ENVR	0004
Ebata, T.	ORGN	0072
Eberhard, J.	ENVR	0043
Eberspacher, T. A.		
	INOR	0174
Ebert, K.	NUCL	0031
Ebert, M.	POLY	0208
Eby, R. J.	CARB	0053
Echegoyen, L.	COLL	0020
Eckrich, T. M.	ORGN	0206
Economou, T.	NUCL	0061
Economy, J.	INOR	0015
Edelstein, H.	CHED	0064
Edie, S. L.	POLY	0097
Ediger, M. D.	PHYS	0139
Edman, K.	ANYL	0122
Edmondson, D. E.		
	BIOL	0022
Edwards, B.	BIOT	0167
Edwards, J. C.	COLL	0198
	PETR	0136
Edwards, M. P.	MEDI	0102
Edwards, P. A.	ANYL	0099
Edye, L. A.	FUEL	0103
Egan, D. A.	MEDI	0120
	MEDI	0197
Egan, D. E.	ENVR	0081
Egger, N.	PMSE	0121
Egolf, L. M.	COMP	0015
Egolfopoulos, F. N.		
	FUEL	0083
Eguchi, M.	ORGN	0049
Ehart, O. R.	AGRO	0070
Ehrichs, E. E.	PHYS	0362
Ehrlich, G.	PHYS	0123
Ehrlich, K. C.	BIOL	0129
Ehrlich, P.	POLY	0076
Eichler, E.	CARB	0013
Eiff, S.	MEDI	0172
Eigler, D. M.	PHYS	0360
Eimouchnino, J.		
	PETR	0141
Eisenberg, A.	PMSE	0116
Eisenstein, O.	INOR	0210
Ej-Jennane, K.	PETR	0103
Ekberg, T.	CARB	0056
Ekerdt, J. G.	COLL	0067
	COLL	0213
	INOR	0146
Eklund, P. C.	CATL	0051
	FUEL	0101
Eksteen, R.	ANYL	0031
Elde, R.	MEDI	0155
Elias-Troy, E.	INOR	0273
Eliel, E. L.	ORGN	0150
Elings, V.	PMSE	0078
Elissalde, M. H.		
	BIOT	0166
	ORGN	0201
Elizalde Baez, F.		
	COLL	0122
Elizalde G., P.	AGFD	0038
	CHED	0106
	ORGN	0037
Elizalde T., J.	COMP	0051
Elk, S. B.	COMP	0035
El-Korchi, T.	ENVR	0008
Eller, R.	CMEC	0011
Ellerbrock, B. R.		
	MEDI	0139
Ellestad, G. A.	POLY	0280
Elliott, J. D.	ORGN	0115
Elliott, J. M.	AGFD	0036
	MEDI	0096
	MEDI	0097
	MEDI	0191
Elliott, S. R.	NUCL	0012
Elliott, W. L.	MEDI	0011
Ellis, A. B.	CHED	0023
	COLL	0175
	INOR	0356
	INOR	0357
Ellis, J.	MEDI	0155
Ellis, P. E.	CATL	0036
Ellsworth, P. E.		
	CHED	0123
Elmaimouni, L.	FUEL	0081
Elmore, S. W.	ORGN	0291
El-Nokaly, M.	COLL	0017
Elrod, V. A.	GEOC	0010
El-Sadany, S. K.		
	ORGN	0160
El-Sayed, M. A.		
	PHYS	0019
Elzey, T. K.	ORGN	0033
Emberger, R.	AGFD	0071
Emdee, J. L.	FUEL	0083
Emerson, A. B.	NUCL	0043
Emery, L. A.	MEDI	0061
Emler, W. C.	CHED	0191
Enas, J. D.	ORGN	0078
Encinas, F.	SCHB	0020
Engel, J. F.	AGRO	0073
Engel, K-H.	AGFD	0026
	AGFD	0182
Engel, R.	POLY	0237
Engelman, F. E.		
	COLL	0077
Engelman, R. A.		
	PMSE	0037
Engen, P.	COLL	0058
	POLY	0233
Engen, P.	POLY	0345
Englehardt, J.	PETR	0086
Engler, B. H.	PETR	0091
English, A. D.	PHYS	0140
	POLY	0058
Engstrom, R. G.		
	MEDI	0101
Enick, R. M.	I&EC	0034
Enkelmann, V.	POLY	0036
Epp, J.	PMSE	0150
Epperson, J. E.	PETR	0058
Epplin, J. L.	CINF	0027
Epstein, A. J.	INOR	0100
	POLY	0327
Epstein, A. L.	BIOT	0061
Eremenko, A. M.		
	COLL	0110
Erhardt, P. W.	MEDI	0118
Erickson, B. W.		
	INOR	0282
Erickson, J. W.	COMP	0038
Erickson, M. C.		
	AGFD	0124
	AGFD	0131
Erickson, R. H.	MEDI	0150
Ericson, J. F.	AGRO	0079
Eriks, K.	INOR	0109
Eriksen, K. A.	INOR	0371
Erion, M. D.	MEDI	0005
Erlich, H.	CHAL	0004
Erman, B.	COLL	0015
	COLL	0016
	COLL	0018
Ermer, S.	POLY	0130
Ermler, W. C.	PHYS	0212
	PHYS	0337
	PHYS	0338
Ermolin, N. E.	FUEL	0126
Erni, F.	ANYL	0111
Ernst, R. D.	INOR	0113
Errede, L. A.	POLY	0168
	POLY	0242
	POLY	0245
Ertl, G.	PHYS	0126
Erwin, M.L.G.	AGRO	0128
Erwin, S. C.	PHYS	0358
Escalante-Tovar, S.		
	INOR	0042
Escalona, S.	ORGN	0187
Escartin, E. F.	AGFD	0079
Eschbach, F. O.		
	PMSE	0005
Escobar, E.	BIOL	0084
	BIOL	0105
Escobar-Toledo, C. E.		
	I&EC	0064
Escobedo, J.	MEDI	0023
Escoubes, M.	PMSE	0119
Escudero, R.	INOR	0048
Eser, S.	FUEL	0086
Esesarte M., J.	CHED	0114
Eskew, N. A.	INOR	0377
Espejo, O.	MEDI	0037
Espenson, J. H.		
	INOR	0276
	INOR	0396
Esposito, P. A.	CINF	0031
Esquivel, J. R.	PHYS	0285
Esquivel R., B.	ORGN	0280
Essenberg, A. D.		
	MEDI	0107
Essenburg, A.	MEDI	0108
Essenburg, A. D.		
	MEDI	0110
Essenmacher, G. J.		
	ANYL	0009
Esslinger, A.	PHYS	0324
Estevez, V. A.	AGRO	045A
Estler, R. C.	CHED	0148
Estrada, A.	AGFD	0038
Estrada, C. A.	COLL	0193
Estrada, F.	PHYS	0252
	PHYS	0253

Estrada M., E.	ORGN 0250	Farnsworth, C. C.		ffrench-Constant, R. H.	
Estrella, C.	ORGN 0103		BIOL 0021		AGRO 0152
Estupinan, B.	BIOL 0073	Farone, A.	BIOT 0077	Field, L. M.	AGRO 0137
Etzler, M. E.	BIOL 0149	Farooq, O.	ORGN 0145	Fields, M.	INOR 0157
Eu, M-D.	POLY 0166	Farr, J. D.	INOR 0144	Fierro, J.L.G.	PETR 0010
Evans, B. M.	AGRO 0053	Farrauto, R. J.	COLL 0205		PETR 0102
Evans, D.	ENVR 0085		COLL 0210	Fife, W. K.	ORGN 0259
Evans, D. A.	ORGN 0289		I&EC 0047	Figg, D. J.	INOR 0062
Evans, D. F.	INOR 0052	Farrell, R.	BIOT 0236	Figueras, F.	PETR 0038
Evans, D. H.	ANYL 0106	Farrenkopf, B.	ANYL 0076		PETR 0040
	ANYL 0107		BIOL 0032	Figur, L. M.	MEDI 0126
	COLL 0077	Faruque, A.	CHED 0195	Fike, E.	SCHB 0017
Evans, D. L.	MEDI 0126	Faulkner, G. E.	ANYL 0004	Filby, R. H.	FUEL 0045
Evans, J.	BIOT 0114	Faunce, J. A.	ORGN 0092	Fildes, N.	CHAL 0004
Evans, J. C.	ORGN 0109	Fayer, M. D.	PMSE 0120	Filho, F. F.	AGFD 0041
Evans, J. F.	INOR 0013	Fazen, P. J.	POLY 0311	Filley, J.	BIOL 0022
Evans, L. A.	INOR 0116	Feaster, J. E.	CATL 0004	Filos, D.	BIOT 0094
Evans, S.	ORGN 0057	Feder-Davis, J.	CARB 0047	Fina, L. J.	PMSE 0153
Evans, S. A., Jr.		Fedunchak, W. B.			PMSE 0154
	INOR 0377		CHAL 0020	Findsen, L. A.	MEDI 0138
	ORGN 0050	Feeney, J.	ORGN 0026	Fine, L. W.	HIST 0030
	ORGN 0051	Feeney, R. E.	AGFD 0035	Fink, C.	MEDI 0111
Eveleth, D.	MEDI 0115		BIOL 0041	Fink, C. A.	MEDI 0113
Everett, K. M.	INOR 0277	Feenstra, R. M.		Fink, D. M.	MEDI 0130
Everett, R. R.	INOR 0097		PHYS 0099	Finke, P.	BIOL 0049
Evers, R. C.	PMSE 0018	Feher, F. J.	INOR 0131		BIOL 0050
	POLY 0029		INOR 0258	Finke, P. E.	MEDI 0173
Ewart, H. A.	CHED 0118	Fehlner, T. P.	INOR 0101	Finkelstein, J. A.	
Ewt, P. A.	BIOL 0033	Feigerle, C. S.	CHED 0044		ORGN 0119
Exner, H.	PETR 0116		PHYS 0251	Finley, F. N.	CHED 0074
Eyermann, C. J.		Feighery, W. G.		Finn, M. G.	INOR 0083
	COMP 0047		INOR 0054	Fireovid, R. L.	CMEC 0007
Eyrisch, O.	BIOT 0090	Feil, V. J.	AGRO 0113	Firouztale, E.	ANYL 0031
Ezell, E. F.	MEDI 0148	Feldman, J.	AGRO 0059	Fisanick, W.	CINF 0023
Fabish, T. J.	POLY 0196	Feliu, A. G.	PROF 0011		COMP 0041
Fabris, D.	NUCL 0028	Feliu, A. L.	ORGN 0192	Fischer, J. B.	MEDI 0147
Fagan, P.	PHYS 0197	Felix, S. L.	TECH 0006	Fischer, J. E.	PHYS 0048
Fagerson, I. S.	AGFD 0174	Fell, N. F.	COLL 0070	Fischer, R. A.	INOR 0141
Fahmy, M.	BIOT 0137	Felman, S. W.	MEDI 0067	Fischer, S. A.	INOR 0284
Fahrenholtz, S.	CHED 0039	Felmine, J. B.	POLY 0019	Fishbain, A.	POLY 0160
Failli, A.	MEDI 0179	Felsot, A. S.	AGRO 0144		POLY 0161
Fair, S. J.	FERT 0010	Feng, D.	BIOL 0034	Fisher, C.	AGFD 0047
Fair, W. E.	FERT 0002	Feng, Q. D.	FUEL 0118	Fisher, E. R.	COLL 0218
	FERT 0004	Feng, Y.	PHYS 0348	Fisher, J. D.	PMSE 0139
	FERT 0010	Feng, Z.	CATL 0040	Fisher, M.	ORGN 0016
	FERT 0014	Ferdelman, T. G.		Fisher, W.	ANYL 0014
Fairey, R.	GEOC 0010		GEOC 0006	Fishman, D.	SCHB 0014
Fairlamb, A. H.			GEOC 0007	Fishman, M.	CARB 0045
	MEDI 0062	Ferguson, C. P.	ORGN 0251	Fishman, M. L.	PMSE 0094
Fairman, R.	INOR 0039		ORGN 0252	Fisicaro, E.	COLL 0001
Fajer, J.	INOR 0091		ORGN 0253	Fitchett, A. W.	AGFD 0204
Fajula, F.	PETR 0038	Ferguson, D. T.		Fitz, D.	CATL 0055
	PETR 0040		AGRO 0141		FUEL 0102
Falck, J. R.	CARB 0031	Feringa, B. L.	INOR 0365	Fitz, N.	AGRO 0089
Faldi, A.	POLY 0165	Fermigier, M.	COLL 0161	Fitzgerald, J. J.	POLY 0287
Fallik, E.	AGFD 0076	Fernandez, H.	I&EC 0010	Fitzgerald, M.	AGRO 0127
Famini, G. R.	CINF 0037	Fernandez, J.	ANYL 0059	Fitzgerald, S.	FUEL 0045
	POLY 0353	Fernandez, M. L.		Fitzsimmons, B.	
Fan, J.	INOR 0160		BIOL 0084		MEDI 0180
	INOR 0161	Fernandez G., J. M.		Fix, R.	INOR 0104
Fan, M.	POLY 0014		INOR 0241	Fixari, B.	FUEL 0046
	POLY 0015	Fernandez-Pierola, I.			PETR 0141
Fan, S.	PMSE 0092		POLY 0124	Flagle, C. L.	CHAS 0025
Fan, Z.	FUEL 0090	Ferragina, C.	PETR 0107	Flaherty, D. K.	CEI 0018
Fang, L.	AGFD 0042	Ferrar, W. T.	POLY 0181	Flament, I. A.	AGFD 0022
Fang, T.	ORGN 0151	Ferraro, T.	AGFD 0088	Flanagan, K.	COMP 0048
Fanta, G. F.	CARB 0042	Ferrat T., G.	COLL 0125	Flath, R. A.	AGFD 0024
Farazdel, A.	BIOT 0001	Ferraudi, G.	INOR 0159	Flatt, L. S.	ORGN 0054
Farcasiu, D.	ORGN 0148		INOR 0287	Fleck, R.	BIOL 0058
Farcasiu, M.	CATL 0046	Ferrell, T. L.	PHYS 0210	Fleckenstein, L.	
Farina, S.	INOR 0371	Ferrer-Sueta, G.			MEDI 0063
Farinato, R.	POLY 0280		INOR 0057	Fleischer, J. M.	CINF 0008
Farinato, R. S.	COLL 0179		INOR 0399	Fleming, J. W.	FUEL 0125
Farkas, J.	COLL 0030	Ferrin, T.	COMP 0036		FUEL 0128
	COLL 0031	Ferron, L.	CEI 0010	Fleming, S. A.	ORGN 0165
	INOR 0144	Ferroni, E. L.	CHED 0183	Fletcher, J.C.Q.	
Farkas, M.	COMP 0007	Ferry, L. L.	COLL 0093		PETR 0011
	COMP 0008	Fesser, A.C.E.	AGFD 0085	Fletcher, M.	BIOT 0151
Farland, W. H.	ENVR 0068	Fessner, W-D.	BIOL 0009	Fletcher, S. R.	MEDI 0119
Farmer, P. J.	INOR 0177		BIOT 0090	Fleury, Y.	AGFD 0020
Farmer, P. S.	MEDI 0129	Fetters, L. J.	POLY 0265	Flick, G. J., Jr.	AGFD 0123
Farmer, S. G.	MEDI 0171	Fetting, F.	PETR 0116	Flipse, R.	POLY 0129
Farneth, W. E.	PHYS 0010	Feyereisen, R.	AGRO 0098		POLY 0211

Floess, J. K.	FUEL 0008	Fowler, D. W.	PMSE 0072	Friedli, A. C.	ORGN 0124		
	FUEL 0012	Fowler, F. W.	ORGN 0111	Friedman, M.	AGFD 0112		
	FUEL 0017		ORGN 0234	Friedman, S.	CMEC 0005		
Flook, T. F.	AGRO 0108	Fowler, T.	BIOT 0241	Friedman, S. B.			
Florent, T.	PETR 0019	Fox, J. W.	BIOT 0194		ENVR 0032		
Flores, F. X.	ORGN 0152	Fox, K. K.	PETR 0081		ENVR 0033		
Flores, H. E.	BIOT 0231	Fox, M. A.	COLL 0133	Frierson, M. R., III			
Flores, J-A.	INOR 0252		INOR 0317		MEDI 0175		
	PHYS 0254		ORGN 0131	Fries, D. S.	MEDI 0020		
Flores, J. A.	PHYS 0288	Fox, S.	INOR 0179	Fries, W. R.	CARB 0063		
Flores P., B.	ORGN 0196	Fragoso, L. O.	ANYL 0074	Friesner, R. A.	PHYS 0112		
Flores-Parra, A.		Fraleigh, S. P.	ANYL 0048		PHYS 0170		
	INOR 0263	Frame, G. M.	AGRO 0079		PHYS 0275		
Flores R., H.	AGFD 0038	Frame, K.	BIOT 0165		POLY 0077		
	CHED 0106	France, D.	AGRO 0074	Fripiat, J. J.	PETR 0015		
Flores R., L.	CHED 0106	Franchetti, P.	MEDI 0029	Frisch, H. L.	PMSE 0025		
Florsheimer, M.		Franchini, C.	MEDI 0098		PMSE 0056		
	POLY 0176	Francis, A. E.	PHYS 0314	Frisch, J.	INOR 0368		
Floss, H. G.	ORGN 0025	Francis, I. L.	AGFD 0006	Frisch, K. C.	PMSE 0040		
Floyd, D. M.	MEDI 0160	Francisco, H. L.		Frischinger, I.	PMSE 0087		
	MEDI 0188		MEDI 0038	Fritsch-Faules, I.			
Flynn, D. L.	MEDI 0069	Franck, R.	PETR 0142		INOR 0350		
	MEDI 0177	Franck, R. W.	ORGN 0069	Fritz, J. S.	ANYL 0083		
	MEDI 0187		ORGN 0070	Fronczek, F. R.	INOR 0242		
	ORGN 0104		ORGN 0101		INOR 0382		
Flynn, G.	PHYS 0291	Francl, M. M.	CHED 0156	Frontela, J.	PETR 0113		
Flynn, G. W.	PHYS 0230	Frank, B.	BIOT 0150	Frund, R.	BIOT 0041		
	PHYS 0278	Frank, C. W.	PMSE 0120	Fry, D. C.	MEDI 0165		
	PHYS 0299		POLY 0330	Fry, W. E.	PETR 0053		
	PHYS 0334	Frank, E. R.	ANYL 0022	Frye, G. C.	PETR 0032		
Fodor, S.P.A.	BIOL 0027	Franke, C.	MEDI 0087	Fryer, R. I.	MEDI 0019		
	MEDI 0090		MEDI 0172	Fryhle, C. B.	ORGN 0030		
Fogelsong, R.	MEDI 0075	Frankel, A.	I&EC 0041	Fryzuk, M. D.	INOR 0362		
Fogiel, A. J.	AGRO 0029	Frankel, H.	I&EC 0041		INOR 0379		
Foley, H. C.	CATL 0011	Frankel, R. B.	INOR 0118		INOR 0383		
	PETR 0059	Franklin, M. J.	BIOT 0099	Fu, H.	PHYS 0241		
	PETR 0070	Franklin, T.	COLL 0002	Fu, J. M.	INOR 0378		
Foley, J. P.	AGFD 0129	Franolic, J. D.	INOR 0060	Fu, J-M.	ORGN 0277		
	ANYL 0136		INOR 0179	Fu, P. P.	ORGN 0214		
Folting, K.	INOR 0128	Frantz, P.	COLL 0012		ORGN 0215		
	INOR 0210	Frati, W.	NUCL 0023	Fu, X.	ORGN 0182		
Fong, C. H.	AGFD 0050	Fratiello, A.	INOR 0250		ORGN 0188		
Fong, R. H.	INOR 0110	Frazier, J. W.	ORGN 0031	Fu, Y. C.	CATL 0049		
Font, J. L.	ORGN 0028	Frazier, R. B.	BIOL 0053	Fuchs, V. U.	ORGN 0027		
	ORGN 0029		BIOL 0054	Fuelster, R.	CHAL 0019		
Fontes Monteiro, J. L.		Frechet, J.M.J.	PMSE 0075	Fuentes, R. I.	PHYS 0095		
	PETR 0048		POLY 0322	Fuerniss, S. J.	POLY 0096		
Fontijn, A.	PHYS 0087	Freed, C. A.	POLY 0190	Fuesler, T. P.	AGRO 0029		
Foo, C.	POLY 0091	Freed, J. H.	PHYS 0114	Fuhr, B. J.	PETR 0123		
Foote, C. S.	ORGN 0167	Freed, K. F.	POLY 0293	Fuhr, F.	AGRO 0025		
Forano, C.	PETR 0127	Freedman, A.	PHYS 0176	Fuhrmann, M.	ENVR 0011		
Ford, A. M.	CEI 0025	Freeman, B. D.	POLY 0120	Fujihira, M.	BIOT 0103		
Ford, C.	BIOT 0047	Freeman, G.	ANYL 0014	Fujiki, H.	AGFD 0066		
Ford, D.	ORGN 0139	Freeman, N.C.G.			AGFD 0087		
Ford, M. E.	I&EC 0055		CEI 0009	Fujikura, T.	MEDI 0099		
Ford, P. C.	INOR 0169	Fregien, K.	ANYL 0124	Fujita, E.	INOR 0038		
Fordon, K. J.	INOR 0089	Frei, J.	MEDI 0184	Fujita, K.	CARB 0008		
Foreman, T. K.	POLY 0175	Freitag-Beeston, R. A.		Fujita, M.	AGFD 0102		
Forman, A.	CINF 0031		ORGN 0161	Fujita, S.	MEDI 0099		
Forney, C. E.	BIOT 0019	Freitas, D.	INOR 0184	Fujiya, K.	CATL 0049		
Forouzan, F.	ANYL 0079	Freitas, J. A.	FUEL 0037	Fukuchi, T.	AGRO 0071		
Fort, A.	POLY 0102	French, D. C.	FUEL 0075	Fukuda, K.	FUEL 0064		
Fort, R. C., Jr.	ORGN 0246	French, S. A.	BIOL 0079	Fukuda, M.	ENVR 0049		
Fortier, S.	COMP 0045	Frenette, R. L.	ORGN 0076	Fukumori, K.	POLY 0291		
Fortoul T., I.	ENVR 0040	Frenkel, K.	CHAS 0008	Fukunaga, T.	PETR 0064		
Foss, R. P.	POLY 0104	Frenklach, M.	FUEL 0108	Fukuoka, Y.	I&EC 0017		
Foster, A. C.	MEDI 0146		PHYS 0180		PETR 0104		
Foster, D. O.	AGFD 0155	Fret, J.M.J.	POLY 0318	Fukuyama, T.	ORGN 0290		
Foster, M. D.	POLY 0090	Freude, D.	PETR 0108	Full, G.	AGFD 0184		
	POLY 0275	Frey, P. A.	BIOL 0023	Fuller, G.	CHED 0145		
Foster, R.	INOR 0071		BIOL 0077	Fuller, G. G.	PMSE 0106		
Foubelo, F.	ORGN 0097		BIOL 0078	Fuller, R. C.	CHED 0132		
Fouda, H. G.	AGFD 0061		BIOL 0147	Fullwood, M.	FUEL 0030		
Foulkes, W.M.C.		Frezza, A.	PETR 0107	Fulton, S. P.	ANYL 0032		
	PHYS 0053	Friar, J. L.	NUCL 0010		BIOT 0073		
Fountain, M. G.		Frias, J. L.	I&EC 0010	Funatsu, K.	CINF 0005		
	ORGN 0263	Friberg, S. E.	COLL 0062	Funayama, O.	POLY 0310		
Fountoulakis, J. M.		Frie, S.	BIOT 0059	Fung, T. T.	AGFD 0133		
	MEDI 0181	Frieberg, D.	AGRO 0091	Funk, D.	PHYS 0351		
Fourney, R. M.	CHAL 0007	Fried, D.	PHYS 0255	Furenlid, L. R.	INOR 0091		
Fournier, M. J.	POLY 0302	Fried, J.	HIST 0022	Furet, P.	MEDI 0035		
Fournie-Zaluski, M. C.		Friedbacher, G.	INOR 0117	Furey, W.	BIOL 0032		
	MEDI 0002	Friedl, R. R.	PHYS 0017				

Furneaux, R. H.		Garcia, E.	COLL 0169	Gast, A. P.	COLL 0073
	ANYL 0064	Garcia, F.	ANYL 0138		COLL 0161
Furst, A.	COMP 0007	Garcia, J. G.	ORGN 0074	Gasteiger, J.	CINF 0007
Furth, P. S.	MEDI 0054		ORGN 0241		CINF 0010
Furuno, T.	BIOT 0121	Garcia, L.	BIOL 0102		CINF 0036
Futerko, P. M.	PHYS 0087		PETR 0044		COMP 0005
Fyfe, C. A.	PETR 0143	Garcia, M. C.	BIOL 0093	Gat, R.	PHYS 0091
	PHYS 0348	Garcia, N.	PHYS 0064	Gatenby, A. A.	BIOT 0116
Gabara, V.	POLY 0058	Garcia, R.	FUEL 0007	Gatzke, M.	PHYS 0041
Gabe, E. J.	CARB 0038	Garcia C., L.	I&EC 0070	Gau, C. S.	COLL 0056
Gabelica, Z.	PETR 0019		I&EC 0071	Gaucher, G. M.	BIOT 0111
	PETR 0077	Garcia C., M. A.		Gaudiana, R.	PMSE 0022
	PETR 0095		NUCL 0087	Gauss, J.	PHYS 0008
Gabrielsen, B.	AGFD 0165	Garcia-Casanova, Z.			PHYS 0015
Gaddamidi, V.	AGRO 0028		AGFD 0139	Gauthier, S.	MEDI 0166
Gaddy, J. L.	BIOT 0017	Garcia de la Mora, G. A.		Gavilan, G. I.	ENVR 0036
	FUEL 0092		ORGN 0196	Gayen, S.	PHYS 0337
Gadomski, J. E.			ORGN 0197	Geacintov, N. E.	
	ENVR 0027	Garcia G., F.	ANYL 0066		PHYS 0021
Gaffney, A. M.	CATL 0029		CHED 0163	Gebeyehu, G.	CARB 0037
Gaffney, B. J.	CHED 0199	Garcia-Jimenez, F.		Gebhard, L. A.	FUEL 0072
	PHYS 0246		BIOL 0101	Gebhard, R.	PHYS 0196
Gafney, H. D.	INOR 0160	Garcia-Meitin, E. I.		Gedridge, R. W., Jr.	
	INOR 0161		POLY 0157		INOR 0218
	INOR 0170	Garcia O., E. M.			INOR 0229
Gage, J. R.	ORGN 0289		PHYS 0257	Gegg, C.	BIOL 0149
Gailliot, F. P.	MEDI 0181	Garcia O., F.	CHED 0091	Geibel, J. F.	PMSE 0123
Gainer, J. L.	BIOT 0065		CHED 0092	Geier, G.	INOR 0235
Gaines, D. F.	INOR 0388		CHED 0093	Geierstanger, B.	
Gainor, J. A.	MEDI 0195	Garcia-Rubio, L. H.			PHYS 0197
Gaitanopoulos, D.			PMSE 0050	Geiger, F.	BIOT 0137
	ORGN 0119	Garcia-Tamayo, F.		Geise, R. J.	BIOT 0107
Gajewski, J. J.	ORGN 0144		AGFD 0144	Geiser, U.	PHYS 0024
Gajiwala, H. M.			AGFD 0145	Gelb, M. H.	BIOL 0020
	POLY 0033		AGFD 0146		BIOL 0021
Gajiwala, K.	BIOT 0189	Gard, D. R.	INOR 0291	Gelbard, G.	ORGN 0264
Gala, K. J.	MEDI 0046	Gard, J. K.	INOR 0291	Gembicki, S. A.	
Galazzo, J. L.	BIOT 0087	Gardiner, J. M.	MEDI 0049		I&EC 0061
Galbraith, W.	MEDI 0162	Gardner, K. H.	POLY 0058	Genack, A. Z.	PHYS 0064
Gale, P.	BIOL 0049		POLY 0059	Genesca L., J.	PHYS 0217
	BIOL 0050		POLY 0060		PHYS 0270
Gale, R. J.	ANYL 0093	Gardner, M.	CHED 0084	Geoffroy, G. L.	INOR 0072
Galiasso, R.	PETR 0044		PHYS 0249	George, C.	INOR 0306
	PETR 0077	Gardner, S. P.	CINF 0032	George, C. F.	INOR 0134
Galiatsatos, V.	POLY 0308	Garetz, W. V.	ENVR 0076	George, G. N.	FUEL 0057
Galindo, E.	BIOT 0196	Garfias V., F. J.			FUEL 0059
Galivan, J.	MEDI 0185		CHED 0107	George, S. M.	COLL 0008
Gallacher, J.	PETR 0147	Garfunkel, E.	COLL 0050		PHYS 0065
Gallagher, J. J.	ORGN 0096		COLL 0176	George, S. W.	AGRO 0030
Gallagher, K. P.			COLL 0177	Georger, J. H.	BIOT 0246
	PMSE 0148		PHYS 0361	Georgievsky, V. Y.	
Gallagher, M.	PHYS 0152	Garg, V. K.	BIOT 0072		PETR 0135
Gallagher, T.	BIOT 0192	Garigliano, N.	ANYL 0086	Georgina, E. P.	ORGN 0019
Gallaher, G.	COLL 0187	Garito, A. F.	BIOT 0174	Georgiou, G.	BIOT 0115
Gallant, S. R.	ANYL 0048		POLY 0007	Gerasimowicz, W. V.	
Gallegos, E. J.	CATL 0028		POLY 0130		AGRO 0061
Gallimore, W.	INOR 0039	Garland, D. A.	CHED 0173		AGRO 0062
Gallino, G.	POLY 0240	Garland, N. L.	PHYS 0086	Gerba, C. P.	ENVR 0070
Gallo, M. A.	ENVR 0069	Garlick, S. M.	COLL 0024	Gerber, R. G.	BIOT 0086
Gallucci, J.	INOR 0050	Garmer, D.	BIOL 0063	Gerdes, J. M.	ORGN 0078
Galuska, A. A.	NUCL 0094	Garmer, D. R.	PHYS 0256	Gerdy, J. J.	PHYS 0369
Galvan G., A.	NUCL 0071	Garner, C. M.	ORGN 0220	Gergens, D. D.	INOR 0258
Gamboa, G.	BIOL 0072	Garozzo, D.	CARB 0048	Gerhard, A.	INOR 0395
Gamerdinger, A. P.		Garren, K. W.	MEDI 0120	Gerharz, B.	PMSE 0121
	AGRO 0064		MEDI 0197	Gerlach, D.	AGRO 0019
Gammon, S. D.	ANYL 0009	Garritz, A.	CHED 0005	Gerlt, J. A.	BIOL 0158
	CHED 0080	Garrossian, M.	ORGN 0248	German, J. B.	AGFD 0122
Ganem, B.	ORGN 0267	Garroway, A. N.		Gerngross, U. T.	
Ganguly, A. K.	MEDI 0161		CATL 0022		BIOT 0181
Gans, B. I.	PHYS 0154	Garshasb, S.	ORGN 0178	Gerson, C. K.	CHAS 0044
Ganshaw, G. C.	BIOT 0032	Garton, A.	PMSE 0124	Gerstein, B. C.	CATL 0042
Gansow, O. A.	INOR 0033		PMSE 0125		COLL 0089
	ORGN 0024		PMSE 0126		COLL 0090
Gao, J.	BIOT 0230		PMSE 0127	Gerstner, J. M.	BIOT 0032
Gao, J. P.	POLY 0156	Gartstein, E.	POLY 0241	Gerth, D. L.	COMP 0056
Gao, P.	COLL 0118	Garvey, E. A.	GEOC 0014	Gerwick, B. C.	AGRO 0046
Gao, T.	PMSE 0103	Garzon S., L.	ENVR 0038		AGRO 0047
Gao, Z.	PETR 0030	Gasque, L.	INOR 0057		AGRO 0048
Garbelena, M.	COMP 0034	Gasque-Silva, L.		Gerz, R. R.	NUCL 0068
Garca de la Mora, G. A.			INOR 0026	Gerzeliev, I. M.	PETR 0101
	ORGN 0180	Gassman, P. G.	INOR 0301	Gesser, H. D.	PETR 0012
Garceau, J. J.	PMSE 0077		ORGN 0123	Gewirth, A. A.	PHYS 0128
Garces, J. M.	PETR 0110		YCC 0001		

Geyer, H. M., III		Giragosian, N. H.		Gomez–Gaytan, E.			
	MEDI 0085		SCHB 0001			CHED 0096	
Gharavi, A.	PHYS 0131		SCHB 0002	Gomez L., B.	ENVR 0025		
Ghasemian, F.	ORGN 0200		SCHB 0009	Gomez M., L. E.			
Ghazzi, M. A.	CATL 0035	Giral G., F.	ORGN 0186		ANYL 0082		
Ghelardini, C.	MEDI 0098	Girolami, G. S.	COLL 0025	Gomez R., H.	ANYL 0051		
Ghencui, A.	ORGN 0148		INOR 0143	Gomez–Ruiz, H.			
Ghiorso, A.	NUCL 0015		INOR 0153		ANYL 0049		
Ghodsi–Hovespian, S.			INOR 0244	Gomez V., P.	CHED 0114		
	MEDI 0139	Gitsov, I.	POLY 0322	Goncalves, D. M.			
Gholami, M. R.	PHYS 0258	Giusto, R. A.	ORGN 0178		POLY 0190		
Ghomashchi, F.		Giziewicz, J.	CARB 0038	Gonin, M.	NUCL 0028		
	BIOL 0020	Gladfelter, W. L.		Gonsalves, K. E.			
Ghomashehi, F.			INOR 0145		POLY 0012		
	BIOL 0021	Gladysz, J. A.	ORGN 0220		POLY 0229		
Ghosal, K.	POLY 0117	Glamkowski, E. J.			POLY 0335		
Ghosh, A.	INOR 0301		MEDI 0130	Gonthier, P.	NUCL 0028		
Ghosh, B. K.	INOR 0090		MEDI 0137	Gonzales, A. J.	MEDI 0055		
	MEDI 0043	Glasgow, J. I.	COMP 0045	Gonzales, F. J.	CEI 0021		
Ghosh, M.	ORGN 0195	Glashow, S. L.	NUCL 0001	Gonzales, L.	ORGN 0159		
Ghosh, S.	INOR 0339	Glass, J.	BIOT 0165	Gonzales H., A.	ORGN 0196		
Ghosh, S. K.	INOR 0182	Glass, J. E.	PMSE 0084	Gonzalez, E. F.	PETR 0017		
Ghoshal, M.	ANYL 0076	Glass, J. T.	PHYS 0121	Gonzalez, E. L.	HIST 0036		
Giammatteo, P. J.		Glasser, F. P.	ENVR 0002	Gonzalez, F.	PETR 0092		
	PETR 0136	Glassman, I.	FUEL 0083	Gonzalez, J. L.	ANYL 0053		
	PETR 0151	Glatz, A.	CHED 0014	Gonzalez, R. L.	BIOT 0131		
Giannetto, G.	PETR 0044	Glatz, A. C.	I&EC 0050	Gonzalez B., F. J.			
	PETR 0077	Glatz, C. E.	BIOT 0019		PHYS 0213		
Gibbons, D.	AGRO 0043	Glatzhofer, D. T.		Gonzalez C., M.			
Gibbons, P. C.	INOR 0102		INOR 0051		COMP 0051		
Gibbs, P.	BIOT 0074	Glaudemans, C.P.J.		Gonzalez Chavez, J. L.			
Gibson, H. W.	COLL 0058		CARB 0019		ANYL 0067		
	POLY 0044	Gleason, C.	MEDI 0181	Gonzalez P., G.	AGFD 0105		
	POLY 0233	Gleason, K. K.	PMSE 0182	Gonzalez–Pina, G.			
	POLY 0246	Gleason, W. S.	CHAS 0001		ENVR 0046		
	POLY 0345	Glemza, R.	PETR 0050	Gonzalez–Vergara, E.			
	POLY 0346	Glick, D. C.	FUEL 0002		CHED 0103		
Gibson, S. M.	AGFD 0150	Glidle, A.	ANYL 0104		CHED 0120		
Giddings, J. C.	PMSE 0008	Glinka, T.	ORGN 0283		INOR 0407		
	PMSE 0011	Gliozzi, A.	BIOT 0122	Goodings, J. M.			
	PMSE 0012	Gloffke, W.	INOR 0360		PHYS 0201		
	PMSE 0029	Glomset, J. A.	BIOL 0021	Goodman, D. W.			
Gier, T.	PETR 0055	Gluck, S. J.	ANYL 0068		COLL 0063		
Gier, T. E.	INOR 0256	Gluszak, T. J.	COLL 0197		COLL 0193		
Gierasch, L. M.	BIOL 0003	Gnagey, A.	AGRO 0101	Goodman, J. M.			
Giering, W. P.	INOR 0109	Gnep, N. S.	PETR 0026		ORGN 0170		
Gierlik, E.	NUCL 0039		PETR 0045	Goodnow, T.	COLL 0020		
Gies, H.	PHYS 0348	Goddard, J. D.	PHYS 0090	Goodson, T., Jr.			
Gilbert, A.	POLY 0219	Godinez M., L.	PHYS 0280		BIOL 0072		
Gilbert, B.	SCHB 0024	Goel, S. C.	INOR 0071	Goodson, T. L.	CHED 0141		
Gilbert, D. J.	INOR 0275		INOR 0102	Goodwin, D. G.	PHYS 0181		
Gilbert, K. E.	CHED 0188	Goetz, M. A.	ORGN 0184	Goodwin, G.	BIOL 0021		
Gilbert, L. I.	AGRO 0002	Goff, B.	COLL 0180	Goodwin, J. G., Jr.			
Giles, D. K.	AGRO 0105	Gokel, G. W.	COLL 0020		COLL 0187		
Gilje, J. W.	INOR 0366	Goldberg, H. A.		Gopalakrishman, R.			
Gilkes, N. R.	BIOT 0207		POLY 0179		FUEL 0030		
	BIOT 0217	Goldfarb, T. D.	ENVR 0018	Gopalsami, N.	FUEL 0075		
Gilles, M. K.	PHYS 0055	Goldfinger, M. B.		Gopidas, K. R.	POLY 0238		
Gillespie, D. T.	PMSE 0094		POLY 0046	Goralski, C. T.	ORGN 0109		
Gillespie, S.	BIOL 0018	Goldman, A. S.	INOR 0151	Gorbaty, M. L.	FUEL 0057		
Gillet, V. J.	COMP 0048		INOR 0152		FUEL 0059		
Gillham, J. K.	PMSE 0172		INOR 0254	Gorby, Y. A.	GEOC 0008		
	PMSE 0173		INOR 0255	Gordillo, B.	ORGN 0150		
Gilligan, P. J.	MEDI 0084	Goldman, B.	CHAS 0028	Gordon, B., III.	POLY 0349		
	MEDI 0153	Goldsmith, R. H.		Gordon, D. A.	MEDI 0190		
Gilliland, D.	ANYL 0014		HIST 0001	Gordon, J. I.	BIOL 0015		
Gilliland, G.	BIOT 0192	Goldstein, B.	MEDI 0059	Gordon, N. F.	ANYL 0032		
Gillum, A. M.	BIOT 0162	Goldstein, M. S.		Gordon, N. J.	ORGN 0050		
Gilmore, D. F.	CHED 0132		PETR 0033	Gordon, R. G.	INOR 0104		
Gilton, T. L.	PHYS 0084	Goldstin, B.	MEDI 0069	Gordon, T. D.	MEDI 0195		
Gimi, R. H.	ORGN 0079	Gole, J. L.	PHYS 0173	Gorenstein, D. G.			
Gin, D. L.	POLY 0047	Golebiowski, A.	ORGN 0272		BIOL 0123		
Ginsburg, E.	POLY 0126	Gollakota, S. G.			INOR 0378		
Ginsburg, E. J.	POLY 0329		CATL 0053	Gores, F.	PMSE 0051		
Ginter, D. M.	PETR 0004	Golovkova, L. P.		Gorlov, Y. I.	COLL 0113		
Ginther, C.	AGFD 0117		COLL 0096		COLL 0116		
Ginzburg, B. M.			COLL 0104	Gorman, C.	POLY 0126		
	POLY 0100	Gomenyuk, A. A.		Gorman, C. B.	POLY 0329		
	POLY 0101		COLL 0098	Gorodisher, I.	POLY 0216		
Giolando, S.	ENVR 0043	Gomez, B., Jr.	BIOT 0074	Gorte, R. J.	COLL 0186		
Giorni, A.	NUCL 0028	Gomez, E.	MEDI 0127	Gosalvez, D.	PHYS 0203		
	NUCL 0039	Gomez, R. P.	MEDI 0094	Gosser, D. K., Jr.			
		Gomez A., X.	PHYS 0270		BIOL 0135		

Gossett, L. S.	MEDI 0022	Greenlee, W. J.	MEDI 0104	Grubbs, R. H.	POLY 0328		
Gostomski, P. A.			MEDI 0122		POLY 0329		
	BIOT 0158		MEDI 0123	Gruen, D. M.	NUCL 0045		
Goswami, R.	MEDI 0158		ORGN 0112	Grushin, V.	INOR 0381		
Gothe, S. A.	CINF 0008	Greenlief, C. M.		Gruver, V. S.	PETR 0135		
Gotlieb, E.	ENVR 0013		COLL 0044	Gruys, K. J.	BIOL 0053		
Gotlieb, I.	ENVR 0013	Greenwood, J. M.		Gryczynski, I.	PHYS 0061		
Goto, S.	ORGN 0290		BIOT 0217	Grynkiewicz, G.			
Goto, Y.	PETR 0003	Greenwood, M.	CEI 0022		CARB 0032		
Gotro, J. T.	PMSE 0152	Gregg, B. T.	INOR 0337		CARB 0033		
Gottlieb, M. S.	INOR 0409	Gregg, M.	CEI 0002	Gryte, C. C.	COLL 0037		
Gottschall, D. W.		Gregorich, K.	NUCL 0074	Gschwend, P. M.			
	AGRO 0109	Gregoriou, V. G.			ENVR 0058		
Gougoutas, J.	MEDI 0188		POLY 0304	Gschwender, L. J.			
Gould, F.	AGRO 0118	Gregory, D.	BIOT 0056		I&EC 0008		
Goulet, M. T.	ORGN 0015	Greiner, S. H.	AGRO 0055	Gsell, T. C.	POLY 0192		
Goumri, A.	FUEL 0081	Gress, D. L.	ENVR 0008	Gu, K.	MEDI 0159		
Govil, G.	BIOT 0143	Grethlein, A. J.	BIOT 0015	Gu, Q–M.	BIOT 0092		
	BIOT 0147	Greve, J.	COLL 0179	Gu, X.	PHYS 0006		
Govindaswamy, K.		Gribble, N. R.	FUEL 0095	Gu, Y.	ORGN 0231		
	INOR 0059	Grierson, D.	ORGN 0234	Gu, Y. G.	ORGN 0099		
	INOR 0078	Grieser, F.	COLL 0152	Gu, Z–Q.	MEDI 0019		
Goyal, A.	FUEL 0120	Griesmar, P.	POLY 0006	Guadalupe, A. R.			
Grabiner, F. R.	AGFD 0080	Grifantini, M.	MEDI 0029		INOR 0216		
Grabski, A. C.	BIOT 0210	Griffin, R. G.	PHYS 0196	Guan, X. C.	AGRO 0007		
Grachek, S. J.	MEDI 0011	Griffioen, K. A.	NUCL 0039	Guarnieri, C. R.			
Gracia F., J.	COLL 0122	Griffith, A. L.	MEDI 0138		PHYS 0092		
Gracia–Mora, I.		Griffith, E. J.	INOR 0345	Guarnieri, F.	ORGN 0169		
	INOR 0026		INOR 0376	Guay, J.	MEDI 0180		
Gracia–Mora, J.		Griffith, J. R.	POLY 0037	Gueldner, R. C.			
	INOR 0020	Griffith, O. W.	MEDI 0106		AGRO 0034		
	INOR 0076	Griffiths, A.D.J.		Guenet, J. M.	POLY 0149		
Gradl, S.	AGFD 0194		ORGN 0090	Guerard, C. K.	NUCL 0013		
Grady, B. P.	PMSE 0145	Griggs, C. B.	NUCL 0081	Guerrero–Ruiz, A.			
	PMSE 0146	Grigoras, S.	POLY 0356		PETR 0010		
Grady, C. E.	PMSE 0004	Grillasca R., Y.	ORGN 0180		PETR 0102		
Graef, R. C.	POLY 0317		ORGN 0197	Guevara–Garcia, A.			
Gragg, C. E.	CINF 0038	Grim, M. D.	ORGN 0013		CHED 0102		
Graham, H.	PMSE 0101	Grim, S. O.	INOR 0264		CHED 0120		
Graham, J. D.	POLY 0039		INOR 0265	Guevara Garcia, M.			
Grampp, G. E.	BIOT 0011	Grimes, D.	MEDI 0179		ANYL 0058		
Grana, E.	MEDI 0141	Grindey, G. B.	MEDI 0022	Guevar–Garcia, A.			
Granger, J. N.	BIOL 0123	Grindstaff, W. K.			INOR 0407		
Granger, R. M.	INOR 0319		CHED 0175	Gui, M.	NUCL 0028		
Granick, S.	COLL 0012	Griskey, R. G.	I&EC 0048	Guida, W. C.	MEDI 0005		
Granite, E.	COLL 0174	Grissom, C. B.	CMEC 0024	Guigliarelli, B.	INOR 0020		
Grantier, D.	PHYS 0173		I&EC 0002	Guilford, W. J.	AGFD 0042		
Gratzel, M.	COLL 0147	Gritzali, M.	BIOT 0240	Guillot, J. G.	CEI 0004		
Graves, K. L.	BIOL 0160	Groen, J. C.	FUEL 0139		CEI 0005		
Graves, M.	BIOL 0042	Groeninckx, G.	PMSE 0102	Guimon, R. K.	NUCL 0085		
Gray, D. G.	POLY 0331	Groenke, D. A.	INOR 0358	Guinan, M. W.	NUCL 0091		
Gray, M. R.	BIOT 0129	Grohmann, A.	INOR 0367	Guinet, D.	NUCL 0028		
Grayeski, M. L.		Grohse, E. W.	FUEL 0047		NUCL 0039		
	ANYL 0084		FUEL 0121	Guisnet, M. R.	PETR 0026		
Grayson, R.	CHAS 0002	Grondey, H.	PHYS 0348		PETR 0045		
Greaney, M.	INOR 0058	Gronenborn, A.	PHYS 0113	Gulakowski, R. J.			
Greaney, M. A.	INOR 0371	Grootenhuis, P.D.J.			AGFD 0192		
Greasham, R. L.			COMP 0040	Gullikson, G. W.			
	BIOT 0071	Grootjans, J. J.	BIOL 0090		MEDI 0069		
	BIOT 0109	Grosch, W.	AGFD 0070		MEDI 0187		
Greeley, D. N.	MEDI 0165		AGFD 0097	Guner, O. F.	CINF 0025		
Green, A.	COLL 0150	Groshens, T.	INOR 0219	Gung, B. W.	ORGN 0172		
Green, B. G.	BIOL 0049	Groshens, T. J.	INOR 0230	Gungor, A.	POLY 0264		
	BIOL 0050	Gross, A.	AGFD 0152	Gun'ko, V. M.	COLL 0095		
Green, D. M.	COLL 0155	Gross, D. J.	BIOL 0128	Gunnlaugsdottir, H.			
Green, J. B.	FUEL 0060	Gross, K. C.	AGFD 0031		AGFD 0160		
Green, M.	CHED 0173	Gross, M. E.	INOR 0125	Gunter, P.	POLY 0005		
Green, M. E.	COLL 0084	Gross, P. M.	COLL 0138		POLY 0176		
Green, M. J.	MEDI 0161	Gross, R. B.	BIOT 0077	Guntert, M.	AGFD 0024		
Green, M. M.	POLY 0348	Gross, S. S.	MEDI 0106		AGFD 0071		
Green, S. A.	ENVR 0057	Grossman, M.	CHAL 0020	Guo, J–X.	POLY 0331		
Greenbaum, E.	BIOT 0050	Grothaus, J. T.	COLL 0035	Gupta, M. N.	BIOT 0220		
	TECH 0016	Grover, N.	INOR 0410	Gupta, V. K.	I&EC 0008		
Greenberg, S. S.		Groves, J. T.	BIOL 0127	Gurel, D.	HIST 0037		
	MEDI 0118		ORGN 0136	Gusev, A. A.	POLY 0141		
Greenblatt, M.	CHED 0032	Groy, T. L.	INOR 0092	Gusta, L. V.	AGRO 0017		
	INOR 0058		INOR 0093	Gustafson, K. R.			
	INOR 0081	Grozinger, K. G.			AGFD 0192		
	INOR 0352		ORGN 0027	Guth, H.	AGFD 0097		
	INOR 0359	Grubbs, H. J.	PMSE 0052	Gutierrez, C.	CHED 0169		
Greene, J. E.	INOR 0109		PMSE 0053		ORGN 0245		
Greene, R. J.	ANYL 0062	Grubbs, R. H.	POLY 0047	Gutierrez E., F.	CHED 0172		
Greenlee, W. J.	MEDI 0103		POLY 0126				

Name	Code	Number		Name	Code	Number		Name	Code	Number
Gutierrez L., R.	FUEL	0104		Halgeri, A. B.	PETR	0100		Han, O.	BIOL	0023
Gutierrez-Puebla, E.				Halgren, T. A.	MEDI	0122		Han, R.	INOR	0087
	INOR	0034		Hall, C. D.	AGRO	0041		Han, S.	POLY	0229
	INOR	0035		Hall, C. K.	BIOT	0022		Han, W.	ORGN	0077
	INOR	0036		Hall, G. G.	COMP	0029		Han, W-C.	MEDI	0160
	INOR	0065		Hall, K. A.	INOR	0271		Han, X.	PMSE	0083
Gutierrez-Ruiz, M.				Hall, K. W.	ANYL	0052			PMSE	0092
	ENVR	0039		Hall, M. B.	PHYS	0192		Hanai, T.	ANYL	0100
Gutman, D.	FUEL	0066		Hall, P. L.	MEDI	0158		Hancock, A. A.	MEDI	0116
Gutowski, M.	PHYS	0056		Hall, S. E.	MEDI	0160		Hancock, K. G.	YCC	0003
Guttenplan, J. B.				Hall, S. L.	COMP	0012		Handa-Corrigan, A.		
	CHAS	0010		Hall, T. W.	AGRO	0026			BIOT	0008
Guzi, P. J.	INOR	0371		Hallada, M. C.	CHED	0081		Handlon, A. L.	BIOL	0013
Guziec, F. S., Jr.				Hallberg, G. R.	AGRO	0065		Hands, R. M.	INOR	0391
	ORGN	0108		Hallden-Abberton, M. P.				Handy, N. C.	PHYS	0030
Guzman, S.	ORGN	0018			PMSE	0179		Hanebeck, W.	COMP	0005
Guzman C., L.	NUCL	0069		Halley, B. A.	AGRO	0128		Haneguelle, S.	AGFD	0197
Guzman C., M. L.				Halliday, W. R.	AGRO	0135		Hanessian, S.	CINF	0002
	COLL	0125		Hallmark, V. M.					ORGN	0279
Gyenes, A.	ORGN	0079			PHYS	0125		Haney, M. A.	PMSE	0067
Gyenes, F.	ORGN	0081			POLY	0145			PMSE	0098
Ha, K.	PMSE	0124		Hallock, R. B.	INOR	0055		Hankin, J.	ORGN	0055
Haagsma, N.	AGFD	0064		Halpern, J.	CATL	0003		Hanna, P. E.	ORGN	0261
Haas, E.	PHYS	0107		Halstead, J. A.	PHYS	0143		Hanna, P. K.	INOR	0155
Haase-Pettingell, C.					PHYS	0259		Hannak, P. G.	ENVR	0007
	BIOT	0113		Halterman, R. L.				Hannaway, O.	HIST	0011
Haasnoot, C.A.G.					INOR	0214		Hannink, N. J.	NUCL	0073
	COMP	0040			INOR	0215			NUCL	0074
Habdas, J.	ORGN	0189			INOR	0300		Hanrahan, R. J.		
Haberman, J. X.					ORGN	0045			PHYS	0258
	INOR	0188			ORGN	0240		Hansen, D. W., Jr.		
Habib, R. H.	CONG	0002		Haltiwanger, R. C.					MEDI	0152
Habif, S.	COLL	0080			INOR	0391		Hansen, R. J.	AGRO	0090
Hacker, N. P.	ORGN	0168		Halverson, K.	PHYS	0196		Hansler, U.	BIOL	0121
Hackett, M.	ENVR	0060		Halvorson, C.	POLY	0008		Hansma, P. K.	INOR	0117
Haddad, T. S.	INOR	0362		Halvorson, K. E.				Hanson, D. M.	PHYS	0080
Haddon, R. C.	INOR	0354			PHYS	0260		Hansson, M.	CEI	0014
Haddow, D. J.	BIOT	0010		Hamaguchi, H.	AGRO	0013		Hanton, S. D.	PHYS	0306
Hadley, S. R.	FUEL	0134		Hamed, E. A.	ORGN	0160		Hanusa, T. P.	INOR	0223
Hadziioannou, G.				Hamel, J-F.	BIOT	0074		Hanyu, A.	PMSE	0022
	POLY	0129		Hamelehle, K.	MEDI	0108		Hapeman-Somich, C. J.		
	POLY	0211		Hamelehle, K. L.					AGRO	0122
Haemers, A.	MEDI	0062			MEDI	0110			AGRO	0139
Haeusler, A.	MEDI	0035		Hamer, R.R.L.	MEDI	0137		Happer, W.	PHYS	0041
Haga, T.	AGRO	0050		Hameroff, S.	BIOT	0055		Hara, K.	CARB	0008
Hagaman, E. W.				Hamers, R. J.	ANYL	0022		Hara, M.	POLY	0061
	FUEL	0077			PHYS	0208			POLY	0062
Hagedorn, M.	AGFD	0196		Hamilton, D.	INOR	0157		Haraki, K. S.	CINF	0026
Hagel, K.	NUCL	0028		Hamilton, G.	MEDI	0083		Harary, H. H.	PHYS	0363
Hagen, K. S.	INOR	0094		Hamilton, J.	BIOT	0214		Harbison, G.	INOR	0059
Hager, G. T.	CATL	0051		Hamilton, P.	ORGN	0012		Harbison, G. S.	BIOL	0116
	FUEL	0101		Hamilton, R. W.					INOR	0078
Hagerman, A. E.					MEDI	0152			ORGN	0020
	AGFD	0148		Hamilton, T.	NUCL	0074			PHYS	0277
Haggblom, M. M.				Hammock, B. D.				Harcum, S. W.	BIOT	0034
	BIOT	0120			AGRO	0009		Hardcastle, F. D.		
Haggerty, B. S.	INOR	0148			AGRO	0032			COLL	0031
Hagler, T. W.	POLY	0008		Hammond, D. E.					INOR	0278
Hagmann, W.	BIOL	0049			GEOC	0010		Harder, S.	INOR	0229
	BIOL	0050		Hammond, D. L.				Hardgrove, G. L., Jr.		
Hagmann, W. K.					MEDI	0152			CHED	0045
	MEDI	0173		Hammond, M.	ORGN	0244		Hardy, L. W.	BIOL	0160
Hahn, F. M.	BIOL	0014		Hammond, M. L.				Hardysh, B. A.	MEDI	0189
Hahn, J. M.	ORGN	0237			MEDI	0182		Hargis, M.	AGRO	0100
Hahn, J-T.	MEDI	0075			MEDI	0183		Hargrave, K. D.		
Hahn, M.	CINF	0019		Hammond, W. B.					ORGN	0027
Hahn, R. C.	ORGN	0207			FUEL	0034		Hariyadi, P.	AGFD	0033
Hahn, R. L.	NUCL	0019		Hampden-Smith, M.				Harless, J. M.	LWM	0001
Hail, M.	ANYL	0141			COLL	0031		Harlow, R. L.	CATL	0004
Hain, M. A.	ORGN	0183		Hampden-Smith, M. J.					PHYS	0010
Haitjema, H. J.	POLY	0209			COLL	0027		Harman, W. D.	INOR	0114
Hajdu, J.	MEDI	0163			COLL	0030		Harmon, R. P.	CHAL	0003
Hajdu, R.	MEDI	0182			INOR	0144		Harnisch, H.	INOR	0321
Hajian, H.	TECH	0000			INOR	0278		Haro-Castellanos, J. A.		
Hakomori, S.	CARB	0060			POLY	0258			CHED	0094
Halarnkar, P. P.				Hampp, N.	BIOT	0078			CHED	0095
	AGRO	0022		Hamrick, J. T.	FUEL	0119			CHED	0096
Halary, J. L.	PMSE	0133		Hamrock, S.	ORGN	0124			CHED	0097
Halbert, M.	NUCL	0027		Han, C. C.	PMSE	0107			CHED	0165
Halbert, T. R.	INOR	0371			PMSE	0134			CHED	0166
Hale, J.	BIOL	0049			POLY	0295			CHED	0169
	BIOL	0050		Han, M.	COLL	0021		Harpel, M. R.	BIOL	0089
				Han, M-Y.	BIOT	0142		Harper, A.	INOR	0386

Harper, D. J.	CHED 0075	Hayes, D. M.	CHED 0202	Hendrickson, J. B.			
Harper, E. T.	CHED 0035	Hayes, L.	CATL 0039		CINF 0004		
Harriman, A.	ORGN 0059	Hayes, T. R.	COLL 0214	Hendrickson, W. H.			
Harriman, A. M.		Hayhurst, D. T.			INOR 0320		
	BIOL 0134		PETR 0078	Hendrickx, M.	PHYS 0141		
Harrington, B. A.			PETR 0079	Hendrix, J.	CONG 0008		
	POLY 0257	Haynes, D. R.	PHYS 0065	Hener, U.	AGFD 0183		
Harris, A. L.	PHYS 0344	Hays, J. A.	INOR 0298	Henick-Kling, T.			
Harris, B. D.	ORGN 0242	Hazlett, T.	COLL 0037		AGFD 0137		
Harris, C. B.	PHYS 0085	Hazlitt, L. G.	PMSE 0048	Henion, J. D.	ANYL 0138		
Harris, D. N.	MEDI 0160	He, J-W.	COLL 0193	Henk, L. L.	BIOT 0250		
Harris, G.	BIOL 0071	He, O-Y.	MEDI 0014	Hennes, P. O.	PETR 0139		
Harris, J. M.	ANYL 0011	He, X. W.	PMSE 0021	Hennig, H.	INOR 0340		
	ANYL 0034	He, Y.	CATL 0034	Henrich, E.	NUCL 0031		
Harris, J. N.	AGFD 0084	Head, S. L.	CEI 0002	Henrich, P. J.	POLY 0168		
Harris, K.	BIOT 0092	Heard, C.	I&EC 0009	Henrich, V. E.	COLL 0192		
Harris, S.	MEDI 0148	Heard, D. E.	FUEL 0098	Henrick, C. A.	AGRO 0003		
Harris, S. J.	PHYS 0178	Hearn, C. H.	ENVR 0024	Henriques, C.	CATL 0058		
Harris, T. D.	ANYL 0012	Heaslip, R. J.	MEDI 0189	Henry, D. R.	CINF 0022		
Harris, T. V.	CATL 0028	Heath, J. R.	PHYS 0007		CINF 0025		
Harris, W. E.	ANYL 0122	Heatherly, L.	PHYS 0119		COMP 0046		
Harrison, I.	PHYS 0261	Hebard, A. F.	INOR 0354	Hensens, O. D.	MEDI 0058		
Harrison, J. A.	PHYS 0154		PHYS 0046		MEDI 0181		
Harrison, W.T.A.		Heck, J.	POLY 0299	Hensley, D. C.	NUCL 0027		
	PETR 0055	Heck, J. V.	MEDI 0182	Hepel, M.	CHED 0119		
Harrod, J. F.	POLY 0337		MEDI 0183		COLL 0013		
Harshman, L.	AGRO 0009	Heck, R. M.	I&EC 0047	Hepel, T.	CHED 0119		
Hart, C.	CHAS 0049	Hedberg, A.	MEDI 0160	Heping, L.	COLL 0079		
Harte, D. C.	CHAS 0026		MEDI 0188	Herchen, H.	FUEL 0035		
Hartenstein, J.	MEDI 0074	Hedberg, K.	MEDI 0143	Hergenrother, P. M.			
Harter, W.	AGFD 0117	Hedges, A. R.	CARB 0061		POLY 0025		
Hartman, F. C.	BIOL 0089	Hediger, H. J.	COMP 0008	Herges, R.	CINF 0006		
Hartman, H. B.		Hedin, P. A.	AGRO 0037	Herman, M. F.	PHYS 0131		
	MEDI 0085	Heeger, A. J.	PHYS 0011	Herman, M. J.	TECH 0019		
Hartman, R. M.			POLY 0008	Herman, S. J.	CHED 0046		
	FUEL 0130	Heflin, J. R.	POLY 0007	Herman, Z.	AGFD 0050		
Hartman, T. G.	AGFD 0003	Hefter, J.	POLY 0257	Hermann, R. B.			
	AGFD 0170	Hegedus, M.	PETR 0134		BIOL 0072		
Hartmann, R. W.		Hehre, E. J.	BIOL 0152	Hernandez, A.	MEDI 0064		
	MEDI 0003	Hehre, W. J.	ORGN 0171		MEDI 0065		
	MEDI 0036	Heidlas, J.	AGFD 0007	Hernandez, B. E.			
Haruvy, Y	POLY 0210	Heidner, R. F.	PHYS 0331		MEDI 0037		
Hase, W. L.	PHYS 0088	Heier, R. F.	MEDI 0126	Hernandez, L.	MEDI 0060		
Hasegawa, S.	AGFD 0050	Heikkila, H.	CARB 0064	Hernandez-Arana, A.			
Hasha, D. L.	ORGN 0109	Heiman, A. S.	MEDI 0170		BIOL 0074		
Hashiguchi, T.	PETR 0013	Heimann, R. B.			BIOL 0106		
Hashimoto, H.	CARB 0008		ENVR 0005	Hernandez-Che, G.			
Hashimoto, M.	ORGN 0105	Heimbach, J.	BIOT 0071		INOR 0037		
Hashinger, B. M.		Hein, N. D.	AGRO 0084	Hernandez-Gallegos, Z.			
	AGRO 0028	Heinekey, D. M.			MEDI 0100		
Haskill, R.	ANYL 0005		PHYS 0068	Hernandez I., M.			
Haslam, E.	AGFD 0001	Heintz, R. A.	INOR 0148		AGFD 0210		
Hassanzadeh, P.		Heinzle, E.	BIOT 0137	Herndon, J. W.	ORGN 0216		
	PHYS 0031	Heiser, J. H.	ENVR 0011	Herndon, W. C.			
Hastie, S. B.	MEDI 0016	Held, G. A.	PHYS 0099		COMP 0023		
Hatanaka, Y.	ORGN 0105	Heller, D. N.	AGFD 0099		COMP 0034		
Hatano, H.	ANYL 0100	Helms, G. L.	BIOL 0065	Herrema, J.	POLY 0129		
Hatano, T.	AGFD 0017	Helsel, B. A.	INOR 0311	Herrera, P. S.	CATL 0048		
	AGFD 0046	Helson, H. E.	CINF 0008	Herron, N.	PHYS 0011		
Hatcher, P. G.	FUEL 0086	Helton, T. E.	FUEL 0096	Herschbach, D. R.			
	I&EC 0016	Hemken, H. G.	MEDI 0044		PHYS 0071		
Hatton, G. J.	I&EC 0059	Hemling, M. E.	ANYL 0140	Herschlag, D.	INOR 0293		
Hatton, T. A.	BIOT 0045	Hemmer, T.	PHYS 0188	Hersh, E. M.	MEDI 0093		
Hattori, M.	PETR 0144	Hemmi, G.	ORGN 0139	Hersh, W. H.	INOR 0110		
Hauserman, W. B.		Hemminger, J. C.		Herzfeld, J.	PHYS 0196		
	FUEL 0006		NUCL 0045	Hess, F. D.	AGRO 0081		
Haushalter, R.	CATL 0019	Hemmingson, J. A.		Hesse, R. H.	ORGN 0040		
Hausler, D. W.	PETR 0121		ANYL 0064	Hesse, W.	COLL 0158		
Havard, T.	PMSE 0066	Hen, S. J.	BIOT 0063		POLY 0342		
Hawari, J.	ENVR 0016	Hendayana, S.	ANYL 0046	Hester, D. M.	INOR 0192		
Hawbecker, D. E.		Hendershot, D. G.			INOR 0386		
	AGFD 0136		COLL 0135	Hettich, R. L.	FUEL 0089		
Hawk, E. L.	INOR 0075		INOR 0306	Hettinger, W. P., Jr.			
Hawker, C. J.	POLY 0318	Henderson, B.S.			I&EC 0063		
	POLY 0322		BIOL 0067	Heuer, J. L.	FUEL 0050		
Hawkins, C. J.	INOR 0095	Henderson, L. J., Jr.		Heveling, J.	PETR 0008		
Hawkins, N. C.	ENVR 0042		INOR 0343	Hianik, T.	BIOT 0106		
Hawks, S. E.	AGFD 0207	Henderson, R. A.		Hiatt, W. R.	AGFD 0057		
Haxton, W. C.	NUCL 0020		CHAL 0019	Hickle, L. A.	AGRO 0080		
Hay, P. J.	PHYS 0191	Henderson, S. J.		Hickling, M.	PMSE 0143		
Hayashi, T.	CATL 0006		PETR 0058	Hickman, J. J.	BIOT 0246		
	PMSE 0099	Henderson, W. W.		Hicks, K. B.	AGFD 0030		
Hayen, G. D.	CARB 0006		I&EC 0056	Hicks, M. G.	COMP 0015		

Hicks, R. F.	INOR 0141	Hodge, G.	CINF 0041	Honeyman, C.	POLY 0186
Hiel, G.	ORGN 0062	Hodges, L. M.	INOR 0114	Hong, C.	AGFD 0019
Hietala, S.	PETR 0032	Hodges, R. S.	ANYL 0040	Hong, D.	MEDI 0170
Higa, K. T.	INOR 0217	Hodgson, D. F.	POLY 0197	Hong, F-E.	INOR 0220
	INOR 0218		POLY 0294	Hong, F. T.	BIOT 0052
	INOR 0219	Hodgson, E.	AGRO 0118	Hong, G-P.	AGFD 0123
	INOR 0230	Hodkey, D. W.	ORGN 0015	Hong, J-Y.	AGFD 0067
	INOR 0231	Hoerle, S. L.	ANYL 0085	Hong, P. P.	POLY 0088
	INOR 0304	Hoet, P.	SCHB 0022	Honzawa, S.	AGRO 0050
	INOR 0308	Hoffgen, E. C.	CARB 0058	Hoogenboom, L.A.P.	
Higashino, Y.	AGRO 0071	Hoffman, A. E.	AGFD 0012		AGRO 0094
Highsmith, F. A.		Hoffman, A. J.	POLY 0369	Hope, E. G.	INOR 0133
	BIOT 0060	Hoffman, A. S.	POLY 0243	Hopfenberg, H. B.	
Higson, S.	BIOT 0059	Hoffman, D. C.	NUCL 0016		POLY 0120
Hihalic, Z.	COMP 0014		NUCL 0073	Hopfinger, A. J.	
Hilaireau, P.	ANYL 0037		NUCL 0074		MEDI 0018
Hilberer, A.	POLY 0132	Hoffman, D. J.	MEDI 0120		POLY 0354
Hill, C. L.	INOR 0286		MEDI 0197	Hopkins, J. B.	PHYS 0335
Hill, E. W.	PHYS 0373	Hoffman, D. M.		Hopkins, P. B.	ORGN 0075
Hill, J. C.	NUCL 0049		INOR 0104	Hopkins, P. D.	PETR 0022
Hill, P.	BIOT 0029	Hoffman, M. E.		Hopp, R.	AGFD 0071
	BIOT 0110		ENVR 0019	Hoppe, M. L.	POLY 0314
Hill, R.	FUEL 0140	Hoffman, M. M.		Hor, T.C.A.	INOR 0106
Hill, R. H., Jr.	C E I 0002		NUCL 0016	Hor, T.S.A.	INOR 0106
Hill, T. G.	INOR 0388	Hoffman, M. Z.	INOR 0279		INOR 0107
Hiller, C.	CINF 0036	Hoffman, W. E.		Horen, D.	NUCL 0027
Hillhouse, G. L.			MEDI 0126	Horenstein, B. A.	
	INOR 0327	Hoffmann, D. A.			BIOL 0073
Hillman, A. R.	ANYL 0103		PMSE 0120	Horie, M.	AGFD 0102
	ANYL 0104	Hoffmann, R.	CHED 0020	Horine, P. A.	INOR 0149
Hillman, D.	COMP 0046	Hofmann, S.	NUCL 0014	Horn, D.	COLL 0072
Hilton, L.	HIST 0026	Hogan, J. J.	NUCL 0060	Horn, K. A.	POLY 0180
Hilyard, K.	BIOT 0056	Hogberg, T.	MEDI 0080	Hornak, J. P.	CHED 0123
Himmel, M. E.	BIOT 0191	Hogen-Esch, T. E.			PHYS 0216
Hinks, D. G.	PHYS 0009		POLY 0239	Horne, D. M.	AGRO 0068
Hinman, C.	BIOL 0044	Hogg, J. A.	HIST 0021	Horne, P.	AGRO 0028
Hinman, C. L.	BIOL 0085	Hohman, J. L.	POLY 0175		AGRO 0029
Hinmann, C. L.		Hoke, R. A.	ORGN 0023	Horner, M. R.	CHED 0026
	BIOL 0086	Holcomb, D.	POLY 0179	Horner, R.	CHAS 0017
Hino, T.	ORGN 0044		POLY 0223	Hornish, R. E.	AGRO 0108
Hinton, C.	PETR 0007	Holcombe, T. C.		Horowitz, P. M.	
Hintz, P. A.	PHYS 0174		FUEL 0116		BIOT 0190
	PHYS 0262	Holden, H. M.	BIOL 0147	Horsthuis, W.H.G.	
	PHYS 0311	Holdengraber, Y.			POLY 0177
Hipps, J., Sr.	CHED 0142		PMSE 0114	Horton, D.	CARB 0052
Hirai, A.	POLY 0301	Holder, K. A.	PETR 0147	Horton, L. L.	PHYS 0119
Hirao, S.	PETR 0145	Holdgrun, X.	BIOT 0092	Horvath, C.	ANYL 0128
Hirose, M.	AGFD 0115	Holdsworth, A. K.		Horwitz, C. P.	INOR 0185
Hirschfeld, D.	FERT 0017		PMSE 0061		POLY 0189
Hirschmann, R.		Holland, D. C.	AGFD 0101	Hoss, W.	MEDI 0070
	HIST 0015	Holland, L. A.	BIOL 0033	Hoss, W. P.	MEDI 0139
Hirschon, A. S.	POLY 0313	Holland, P. M.	PHYS 0226	Hou, H.	PHYS 0349
Hissink, D.	POLY 0129	Holler, J. S.	C E I 0007	Houck, P. C.	FUEL 0118
	POLY 0211	Holley, C. A.	FUEL 0135	Houde, C.	POLY 0135
Hitaka, T.	ORGN 0113	Holloway, D.	MEDI 0179	Houdi, A. G.	MEDI 0149
Hittner, D. M.	CARB 0047	Holloway, J. H.	INOR 0133	Houk, C. C.	CHAS 0049
Hlasta, D.	MEDI 0087	Holloway, J. S.	PHYS 0331	Houpt, D. C.	AGRO 0022
	MEDI 0172	Holman, M.	INOR 0073	Hourston, D. J.	PMSE 0061
Hlasta, D. J.	ORGN 0035	Holmen, A.	PETR 0027	Howard, D. F.	AGRO 0104
Ho, C-T.	AGFD 0048	Holmes, D. L.	MEDI 0147	Howard, J. B.	FUEL 0020
	AGFD 0068	Holmes, J. L.	ANYL 0009		FUEL 0112
	AGFD 0069		CHED 0080	Howe, W. J.	COMP 0050
	AGFD 0072	Holmes, R. R.	INOR 0290	Howell, N. W.	ENVR 0024
	AGFD 0170	Holmes, W. D.	MEDI 0075	Howell, T.	C E I 0009
	AGFD 0171	Holms, J. H.	MEDI 0176		COMP 0056
Ho, F. M.	AGFD 0041	Holt, B. D.	ANYL 0061	Howells, S.	PHYS 0152
Ho, J.	INOR 0363	Holt, E. M.	INOR 0157	Hower, J. C.	FUEL 0043
	PHYS 0055	Holt, P. L.	CHED 0086	Hoy, K. L.	PMSE 0038
Ho, M.Y.K.	AGRO 0109	Holten, D.	BIOL 0132	Hoyt, A. E.	POLY 0109
Ho, P.	COLL 0218	Holtsberg, F.	AGRO 0100	Hrabak, O.	CARB 0002
	INOR 0122	Holtwick, J. B.	AGRO 0046	Hrib, N. J.	MEDI 0085
Ho, P.P.K.	MEDI 0174		AGRO 0047	Hrivnak, J. A.	BIOT 0233
Ho, T.	POLY 0018	Holtzapple, M. T.		Hruby, V. J.	MEDI 0093
Hoagland, D. A.			BIOT 0016	Hruz, P.	BIOL 0064
	COLL 0183	Hom, H.	PHYS 0238	Hsi, R.S.P.	MEDI 0133
	PMSE 0105	Homann, K. H.	FUEL 0111	Hsiao, B. S.	POLY 0059
Hoagland, P.	CARB 0045	Homeier, E. H.	PETR 0139		POLY 0060
Hoang, H.	ORGN 0222	Homer, R. W.	CINF 0027		POLY 0247
Hoatson, G. L.	PHYS 0136	Honda, J.	BIOT 0121	Hsiao, C-N.W.	MEDI 0124
Hochstrasser, R. M.		Honeychuck, R. V.		Hsiao, Y. M.	INOR 0177
	PHYS 0083		POLY 0212	Hsieh, B. R.	POLY 0013
	PHYS 0194	Honeycutt, R. C.		Hsieh, H-P.	ORGN 0057
Hodge, C. L.	ORGN 0232		AGRO 0054	Hsieh, K. H.	PMSE 0044

Hsieh, M-M.	ANYL 0102	Huggins, F. E.	FUEL 0056	Ibarra, P.	MEDI 0060		
Hsu, B-C.	INOR 0043	Huggins, J. W.	AGFD 0165	Ibrahim, A.	COLL 0117		
Hsu, C-S.	POLY 0057	Huggins, T.	ANYL 0138	Ibrahim, M.	BIOT 0200		
Hsu, C. W.	ANYL 0087	Hughes, A. N.	INOR 0375		POLY 0344		
	ANYL 0088	Hughes, D. E.	BIOT 0087	Ibrahim, P.	ORGN 0124		
Hsu, F-L.	ORGN 0252	Hughes, D. J.	ENVR 0084	Ichinose, R.	AGRO 0009		
Hsu, M-T.	POLY 0187	Hughes, D. W.	BIOT 0128	Idmoumaz, H.	INOR 0110		
Hsu, S.	ORGN 0084	Hughes, K. A.	INOR 0083	Iedema, M.	PHYS 0084		
Hsu, S. P.	PETR 0099	Hughes, K. D.	ANYL 0026	Iglesia, E.	CATL 0015		
Hsu, T. B.	ANYL 0101		COLL 0070		CATL 0037		
Hsu, T. J.	PMSE 0057	Hughes, L.	AGFD 0155		CATL 0056		
Hsu, T-L. C.	INOR 0285	Hughes, N.	ANYL 0103		CATL 0061		
Hsu, T. M.	POLY 0043	Huh, J-S.	INOR 0171	Ignacio, G.	ANYL 0056		
Hsu, W. L.	PHYS 0145	Huh, W.	PMSE 0169	Ignasiak, L.	FUEL 0042		
Hu, H.S-W.	POLY 0037	Huiming, H.	ANYL 0080		FUEL 0141		
Hu, L-Y.	MEDI 0147	Huizenga, D.	INOR 0208	Ignasiak, T.	FUEL 0141		
Hu, W-S.	BIOT 0224	Hultin, H. O.	AGFD 0094	Ihlenfeldt, W-D.			
Hu, X. C.	NUCL 0028	Humpheys, G. G.			CINF 0007		
Hu, Y.	BIOT 0170		PETR 0007	Iida, T.	AGFD 0127		
Hua, Q. X.	BIOL 0002	Humphrey, A. E.		Iizumi, Y.	MEDI 0109		
Huan, G.	INOR 0058		BIOT 0221	Ijadi-Maghsoodi, S.			
Huang, B-S.	MEDI 0047	Humphrey, G.	SCHB 0012		POLY 0357		
Huang, C.	COMP 0036	Humphrey, M. G.		Ijzerman, A. P.	MEDI 0114		
Huang, C-S.	CATL 0052		CATL 0008	Ikeda, S.	INOR 0164		
Huang, D.	CHED 0210	Humphries, K. A.		Ikeyama, T.	INOR 0164		
Huang, G-L.	INOR 0043		MEDI 0030	Ikura, M.	PHYS 0113		
Huang, G. X.	PETR 0117	Hunchak-Kariouk, K.		Ilgunas, A.	CHED 0162		
Huang, H-H.	POLY 0285		ENVR 0017		PHYS 0336		
Huang, H. I.	AGFD 0190	Hung, L. C.	PMSE 0113	Iliffe, F. J.	FERT 0021		
Huang, J.	PHYS 0263	Hunkeler, D. J.	PMSE 0049	Illyes, E. F.	AGRO 0079		
	PHYS 0325		POLY 0281	Impallomeni, G.			
	PHYS 0364	Hunsberger, L.	ANYL 0009		CARB 0048		
	ORGN 0214	Hunt, G. R.	BIOT 0109	Inaba, S.	POLY 0220		
	ORGN 0215	Hunt, J. E.	FUEL 0088	Inati, S.	PHYS 0068		
Huang, L. J.	ANYL 0050	Hunt, J. T.	MEDI 0188	Indig, G. L.	PHYS 0265		
Huang, L. Q.	AGFD 0065	Hunter, G. R.	AGRO 0149	Indivero, V. M.	CHED 0201		
Huang, M-T.	AGFD 0067	Hunter, K.	CINF 0046	Ingallina, P.	INOR 0226		
	AGFD 0069	Huong, D. M.	PMSE 0078	Ingebretsen, W. R.			
	AGFD 0088		PMSE 0184		MEDI 0118		
Huang, P.	AGFD 0190	Hupp, J. T.	COLL 0132	Inglett, G. E.	AGFD 0037		
Huang, P. L.	AGFD 0190	Hurlbut, J. A.	AGFD 0101	Ingold, K. U.	AGFD 0155		
Huang, Q.	BIOL 0135	Hurley, D.	I&EC 0027	Inskeep, P. B.	MEDI 0082		
Huang, S. J.	CHED 0130	Hurst, T.	CINF 0027	Interrante, L. V.			
	CHED 0134	Hurt, R. H.	FUEL 0026		INOR 0006		
	PMSE 0005	Hussain, W.	INOR 0266		INOR 0105		
	POLY 0109	Huston, J. S.	BIOT 0057		POLY 0362		
Huang, S-P.	INOR 0203	Huston, P.	INOR 0396	Inui, T.	PETR 0062		
Huang, S. X.	MEDI 0104	Hutchins, B.	BIOT 0059	Ione, K. G.	PETR 0039		
Huang, T-C.	AGFD 0171	Hutchins, C. W.		Irazoque P., G.	CHED 0034		
Huang, T. L.	MEDI 0020		MEDI 0116		CHED 0113		
Huang, T-M.	INOR 0041	Hutchinson, C. R.		Irwin, P. L.	AGFD 0012		
	INOR 0043		BIOT 0083	Isaacs, L. D.	INOR 0194		
Huang, X. L.	ORGN 0175		BIOT 0108	Isaacs, L. L.	FUEL 0053		
Huang, Z.	MEDI 0083	Hutchison, D. R.		Isarov, A. V.	COLL 0099		
Huanosta, A.	ORGN 0159		ORGN 0236	Isayev, A. I.	PMSE 0115		
Hubbard, A. T.	COLL 0118	Hutchison, J. R.		Isermann, H. P.			
Hubbard, C. D.	INOR 0022		CHED 0160		POLY 0306		
	INOR 0395	Hutte, R. S.	PETR 0120	Ishaaya, I.	AGRO 0134		
Hubbard, E.	BIOL 0058	Huttermann, A.		Ishida, H.	PMSE 0167		
Huber, D. L.	PHYS 0034		BIOT 0041		POLY 0227		
Hubler, T. L.	INOR 0149	Huve, L.	PETR 0128		POLY 0307		
Hucul, J. A.	BIOT 0032	Huynh, B. H.	BIOL 0022	Ishihara, M.	NUCL 0029		
Huddleston, M. J.		Huynh, H. B.	INOR 0208	Ishii, A.	MEDI 0112		
	ANYL 0140	Huynk, O. T.	ANYL 0018	Ishii, T.	BIOT 0123		
Hudkins, R. L.	MEDI 0140	Huyuh-ba, G.	PMSE 0148	Ismail, I.M.K.	FUEL 0021		
Hudop, J.	POLY 0107	Hwang, J-F.	PMSE 0111	Isobe, N.	AGRO 0024		
Hudson, B. P.	INOR 0315	Hwang, J. S.	PHYS 0264	Isoda, T.	POLY 0310		
Hudson, E.	ORGN 0292		POLY 0092	Israelachvili, J.	BIOT 0043		
Hudson, R. A.	BIOL 0086	Hwang, K. C.	COLL 0088		BIOT 0151		
	MEDI 0142	Hwang, L. S.	AGFD 0154	Issa, T. A.	INOR 0235		
	ORGN 0231	Hwang, S-J.	CATL 0042	Ito, H.	PMSE 0174		
Hudson, S. D.	POLY 0333		COLL 0089	Ito, N.	AGFD 0115		
Huerta C., A.	PHYS 0217	Hwu, S-J.	INOR 0030		MEDI 0109		
Huerta C., L.	INOR 0026		INOR 0031	Itoh, H.	PETR 0144		
Huestis, D. L.	FUEL 0062		INOR 0353	Iton, L. E.	PETR 0058		
Huffman, G. L.	FUEL 0131	Hybertsen, M. S.		Itterly, W.	AGRO 0111		
Huffman, G. P.	FUEL 0041		PHYS 0053	Ivahnenko, T.	AGRO 0053		
	FUEL 0056	Hyde, E. W.	BIOT 0077	Ives, J.	MEDI 0143		
Huffman, M. A.		Hyde, K. E.	CHED 0118	Ivey, D. G.	ENVR 0005		
	ORGN 0274	Hynes, J. T.	PHYS 0037	Ivie, G. W.	BIOT 0166		
Hugdahl, J. D.	INOR 0295	Iannone, M.	PHYS 0194	Iwaoka, K.	MEDI 0109		
Huger, F. P.	MEDI 0137	Ibanez, J.	PHYS 0218	Iwawaki, F.	PHYS 0097		
Huggins, F. E.	FUEL 0041	Ibarra, F.	MEDI 0065	Iyengar, D. R.	POLY 0298		

Iyengar, R.	AGFD 0152	Jenekhe, S. A.	POLY 0221	Johnson, R. C.	ENVR 0082		
Iyer, S.	ORGN 0257		POLY 0226	Johnson, R. D.	BIOT 0133		
Izgi, C. G.	BIOT 0077	Jeng, R. J.	POLY 0131		PHYS 0070		
Izumi, Y.	PETR 0129	Jenkins, J. N.	AGRO 0037	Johnson, R. P.	ORGN 0093		
Jackman, T. M.		Jenkins, R. D.	PMSE 0035	Johnson, R. W.	BIOL 0057		
	COMP 0059	Jenks, C. J.	COLL 0204	Johnson, T. A.	I&EC 0055		
Jackson, C.	PMSE 0067	Jennewein, H. K.		Johnson, T. J.	INOR 0127		
	PMSE 0070		MEDI 0086		INOR 0210		
	PMSE 0097	Jennings, P. W.		Johnson, T.J.A.			
	PMSE 0098		INOR 0329		BIOT 0218		
Jackson, C. L.	POLY 0170		INOR 0330	Johnson, W. P.	ANYL 0140		
Jackson, G. P.	AGRO 0022	Jennings, W.	AGFD 0176	Johnston, J.	CEI 0006		
Jackson, J.	ENVR 0083	Jensen, A. W.	ORGN 0165	Johnston, K.	ORGN 0131		
Jackson, J. D.	AGFD 0100	Jensen, B.	NUCL 0079	Johnston, M. V.			
Jackson, K. A.	PHYS 0355	Jensen, B. J.	POLY 0040		ANYL 0003		
	PHYS 0359	Jensen, J. K.	AGRO 0044	Jolad, S. D.	AGFD 0191		
Jackson, R. L.	PHYS 0333		AGRO 0107	Jolliffe, L. K.	BIOL 0108		
Jackson, T.	AGRO 0145	Jensen, J. O.	POLY 0353	Joly, J. F.	PETR 0046		
Jacobs, A.	POLY 0147	Jensen, K. F.	INOR 0171		PETR 0131		
Jacobsen, E. N.	ORGN 0132	Jeon, E. J.	PMSE 0059		PETR 0133		
Jacobson, J. M.	ANYL 0041	Jeong, L. S.	CARB 0022	Jones, A.	BIOT 0206		
Jacobson, K. A.		Jeong, N.	ORGN 0177	Jones, A. C.	PHYS 0268		
	MEDI 0114	Jerina, D. M.	ORGN 0208	Jones, A. G.	INOR 0181		
	MEDI 0115	Jerussi, T.	MEDI 0046	Jones, B.	PMSE 0176		
Jacobson, K. R.		Jha, B.	POLY 0038	Jones, C. A.	CATL 0029		
	CMEC 0011	Jho, C.	COLL 0091	Jones, C. R.	BIOL 0054		
Jacobson, R. A.	INOR 0298	Jhon, M. S.	COLL 0144		INOR 0378		
Jacobson, S. H.	POLY 0142	Ji, R.	MEDI 0159	Jones, E. B.	INOR 0359		
Jacoby, B.	COLL 0082	Ji, X-D.	MEDI 0115	Jones, E.R.H.	HIST 0014		
Jaenicke, J.	NUCL 0006	Jia, D.	PMSE 0080	Jones, E.T.T.	MEDI 0058		
Jaenicke, R.	BIOT 0193	Jiang, L.	BIOT 0142	Jones, F. N.	PMSE 0016		
Jaffe, F.	INOR 0348	Jiang, N.	BIOL 0113	Jones, J.	BIOT 0100		
Jaffe, R. L.	POLY 0187	Jiang, S.	ORGN 0213	Jones, J. D.	BIOL 0003		
	POLY 0350	Jiang, W.	BIOL 0103	Jones, J. J.	PMSE 0026		
	POLY 0351	Jiang, X-L.	FUEL 0038	Jones, J. M.	BIOT 0097		
Jagtoyen, M.	FUEL 0048	Jiang, Y. Y.	ORGN 0012	Jones, J. R.	PETR 0126		
Jahn, E.	PETR 0093	Jiao, J.	PETR 0030	Jones, K.	ORGN 0005		
Jahnke, T. S.	CHED 0149	Jimenez, R.	ORGN 0230	Jones, L. C.	POLY 0258		
	POLY 0086	Jimenez E., M.	ORGN 0185	Jones, L. J.	COLL 0138		
Jain, D.	BIOT 0109	Jimenez Otamendi, A.		Jones, M. M.	INOR 0251		
Jain, M. K.	BIOL 0019		CHED 0167	Jones, N. D.	BIOL 0072		
	BIOT 0015	Jin, G.	INOR 0380		MEDI 0066		
Jain, S. P.	MEDI 0131	Jin, Y.	POLY 0108	Jones, P. B.	ORGN 0199		
Jakosuo, H.	PMSE 0046	Jinkerson, D. L.		Jones, R. A.	COLL 0067		
Jakupca, M.	COLL 0075		POLY 0184		INOR 0146		
Jameison, F.	MEDI 0151	Jirkovsky, I. L.	MEDI 0067		INOR 0292		
James, D. G.	AGRO 0035	Jo, S. K.	PHYS 0366		INOR 0390		
James, M. O.	AGRO 0096	Job, D.	PMSE 0043		INOR 0392		
James M., G.	ENVR 0037	Jochum, C.	COMP 0009	Jones, R. L.	AGRO 0052		
	ENVR 0038	Joe, F. L., Jr.	AGFD 0109	Jones, S.	MEDI 0143		
Jan, S. T.	INOR 0300	Joens, J. A.	ENVR 0024	Jones, T. A.	CINF 0029		
Jane, J-L.	CARB 0040	Johann, C.	PMSE 0051	Jones, W. E., Jr.			
Janecki, T.	ORGN 0124	Johansen, J. D.	COLL 0140		INOR 0282		
Jang, W-L.	CATL 0034	Johansen, N. G.		Jong, L.	COLL 0034		
Janik, J. F.	POLY 0312		PETR 0120	Jonker, S. A.	POLY 0214		
Janik, T. S.	INOR 0066	John, V. T.	BIOT 0046		POLY 0222		
Jankowski, S.	INOR 0324		PETR 0114	Jonnalagadda, S. B.			
	ORGN 0110	Johnson, A. L.	MEDI 0084		PHYS 0269		
Jansen, J. C.	PETR 0080		MEDI 0153	Jordan, F.	BIOL 0031		
Jansen, M.	INOR 0275	Johnson, B.	CATL 0011		BIOL 0032		
Janson, C. A.	MEDI 0186	Johnson, C.	MEDI 0195		BIOL 0047		
Janson, J. C.	ANYL 0030	Johnson, C. E.	INOR 0230		BIOL 0048		
Jaramillo, D.	PHYS 0219	Johnson, C. R.	ORGN 0272		BIOL 0082		
Jarvie, T.	PHYS 0195	Johnson, C. S.	COLL 0132	Jordan, K. D.	PHYS 0076		
Jasinski, J. P.	INOR 0245	Johnson, D. C.	CARB 0067		PHYS 0172		
	INOR 0262	Johnson, D. L.	BIOL 0108	Jorgensen, J. D.			
Jasovsky, G.	CARB 0065	Johnson, H.	COLL 0012		PHYS 0009		
Jayasuriya, K.	COMP 0060	Johnson, J.	MEDI 0143	Jorgensen, W. L.			
Jayjock, M. A.	ENVR 0042	Johnson, J. A.	MEDI 0195		CINF 0008		
Jedju, T.	BIOT 0080	Johnson, J. R.	ORGN 0091	Jorgenson, J. W.			
Jedju, T. M.	PHYS 0193	Johnson, K.	ANYL 0141		ANYL 0125		
Jeffries, J. B.	FUEL 0098		BIOT 0066		ANYL 0126		
	PHYS 0147		CHED 0200	Joseleau, J. P.	BIOT 0208		
	PHYS 0266	Johnson, K. A.	BIOL 0117	Joseph, J.	BIOL 0099		
Jeffries, P. M.	INOR 0143	Johnson, K. S.	GEOC 0010	Joseph, J. T.	FUEL 0001		
Jeffries, T. W.	BIOT 0210	Johnson, L. A.	POLY 0077	Joseph, R. A.	CHED 0048		
Jegal, J. G.	POLY 0031	Johnson, M.	AGFD 0163	Joseph-Nathan, P.			
Jehoulet, C.	ANYL 0108		BIOT 0192		ORGN 0268		
Jelinek, R.	PHYS 0267	Johnson, M. E.	FUEL 0020	Joshi, B. D.	CHED 0208		
Jencks, W. P.	INOR 0293	Johnson, O.	CINF 0021	Joshi, P. N.	PETR 0005		
Jenekhe, S. A.	POLY 0011	Johnson, P. A.	COMP 0048	Joshua, H.	MEDI 0181		
	POLY 0205	Johnson, R. C.	ENVR 0077	Jousselin, H.	ORGN 0065		
	POLY 0213		ENVR 0079				

Joyce–Pruden, C.	ORGN 0247	Kameswaran, V.	AGRO 0074	Katiyar, S. S.	BIOL 0091	
Ju, S. C.	BIOT 0048	Kamienski, C. W.		Kato, K.	MEDI 0151	
Juang, C–L.	BIOL 0116		ORGN 0225	Kato, M.	ORGN 0249	
Juarez Calderon, J. M.		Kamigaito, O.	POLY 0291	Katsuno, H.	PETR 0064	
	CHED 0108	Kaminski, J. J.	MEDI 0161	Katz, H. E.	COLL 0164	
	CHED 0174	Kaminski, M. D.			POLY 0215	
Juarez F., M.	CHED 0124		CHAL 0011	Katz, L.	BIOT 0110	
Juarez O., C.	ENVR 0038	Kammel, D. W.	AGRO 0042	Katz, S. E.	AGFD 0082	
Juarez S., F.	ANYL 0051	Kammula, R. K.		Kaufman, R. A.		
Juarez-Sanchez, F.			MEDI 0020		ANYL 0076	
	ANYL 0049	Kamo, T.	FUEL 0070	Kausch, H. H.	POLY 0281	
Juaristi, E.	ORGN 0150	Kampf, J.	INOR 0018	Kavlock, R. J.	AGRO 0021	
	ORGN 0295	Kampf, J. W.	ORGN 0212	Kawabe, K.	PETR 0129	
Juillerat, M. A.	AGFD 0020	Kampf, R. P.	PHYS 0350	Kawabe, M.	POLY 0220	
Jung, B.	FUEL 0003	Kanaoka, Y.	ORGN 0105	Kawaguchi, K.	POLY 0023	
	INOR 0178	Kanatzidis, M. G.		Kawai, M.	MEDI 0167	
Jung, K–E.	ORGN 0260		INOR 0082		MEDI 0168	
Jung, M. E.	MEDI 0049		INOR 0200	Kawai, T.	INOR 0170	
	ORGN 0001		INOR 0202	Kawakami, H.	ORGN 0072	
Jungbauer, D.	POLY 0153		INOR 0203	Kawakami, Y.	MEDI 0018	
Juo, R. R.	BIOT 0167		INOR 0204	Kawakishi, S.	AGFD 0018	
Jurado, J. L.	ANYL 0054		INOR 0351	Kawamoto, T.	BIOT 0035	
Jurado Baizaval, J. L.			INOR 0355	Kawasaki W., S.		
	CHED 0104	Kane, J. J.	POLY 0045	Kawecki, R.	CHED 0164	
Jurcak, J. G.	MEDI 0085	Kaneda, N.	MEDI 0034	Kawooya, J. K.	ORGN 0080	
Jurs, P. C.	ANYL 0099	Kanegae, H.	PMSE 0174	Kay, B. D.	BIOT 0171	
	COMP 0003	Kaneko, H.	AGRO 0024	Kay, L. E.	POLY 0256	
	COMP 0015	Kaneko, T.	BIOT 0172	Kaya, H.	PHYS 0113	
Juszczak, A.	INOR 0099	Kaneski, C. R.	BIOL 0079	Kazi, A. B.	POLY 0310	
Kabacoff, C. M.		Kang, D.	INOR 0168	Kazlauskas, R. J.	MEDI 0163	
	BIOT 0156	Kang, W. K.	PMSE 0060		BIOL 0104	
Kabanyane, S.	ORGN 0088	Kanis, D.	BIOT 0178		BIOT 0091	
Kabat, E. A.	CARB 0059	Kanis, D. A.	POLY 0217	Kazmi, S.N.H.	ORGN 0071	
Kabat, M.	ORGN 0284	Kanner, J.	AGFD 0095	Kazmierski, W. M.		
Kacher, C. D.	NUCL 0074	Kantelinen, A.	BIOT 0215		MEDI 0093	
Kaczmarski, J. P.		Kanter–Cronin, M. R.		Keana, J.F.W.	MEDI 0147	
	PMSE 0084		CHAS 0005	Kearsley, S. K.	CINF 0018	
Kadaba, P. K.	MEDI 0144	Kanth, S. M.	CHAS 0012	Kebabian, J. W.		
Kaderli, S.	INOR 0178	Kanvinde, M. H.			MEDI 0124	
Kadkhodayan, B. A.			MEDI 0042		MEDI 0125	
	NUCL 0074		MEDI 0131		MEDI 0127	
Kadow, J. F.	MEDI 0012	Kao, H–L.	PMSE 0020	Keck, P.	BIOT 0057	
Kaesz, H. D.	INOR 0141	Kaplan, D.	CHED 0131	Kee, R. J.	COLL 0066	
Kagami, M.	POLY 0128	Kapteijn, F.	FUEL 0009		PHYS 0179	
Kagan, J.	ORGN 0202		FUEL 0015		PHYS 0357	
Kagen, H. P.	C E I 0008		FUEL 0031	Kee, T. P.	INOR 0259	
Kaghazchi, T.	I&EC 0007	Karas, J.	BIOL 0042	Keenan, R. M.	ORGN 0119	
Kahn, P. C.	C E I 0014	Karasawa, A.	MEDI 0112	Kehmani, K. C.		
Kahn, R.	CATL 0024	Karasz, F. F.	PMSE 0136		ANYL 0108	
Kahn, S. D.	ORGN 0171	Karayannis, N. M.		Keiter, E. A.	INOR 0382	
Kahwa, I. A.	INOR 0039		INOR 0073	Kelemen, S. R.	FUEL 0057	
Kaifer, A. E.	COLL 0020		I&EC 0052		FUEL 0059	
	COLL 0021	Karch, N. J.	C E I 0015	Keller, J.	PHYS 0254	
Kaila, N.	ORGN 0069	Karim, M. H.	PHYS 0229	Keller, J. A.	BIOT 0086	
Kalachandra, S.		Karl, V.	AGFD 0183	Keller, M. D.	GEOC 0002	
	POLY 0193	Karlin, K. D.	INOR 0044	Kelley, B. D.	BIOT 0045	
Kalantar, A. H.	COMP 0057		INOR 0045	Kelley, C.	INOR 0072	
Kalbus, G. E.	CHED 0171		INOR 0046	Kellis, J. T., Jr.		
Kalbus, L. H.	CHED 0171		INOR 0178		BIOT 0133	
Kaldor, S. W.	ORGN 0244	Karns, J. S.	AGRO 0121	Kellogg, C. H.	BIOT 0021	
Kaldor, U.	PHYS 0016		AGRO 0139	Kelly, J. W.	BIOL 0007	
Kale, N. J.	MEDI 0042	Karol, P. J.	NUCL 0048	Kelly, T. R.	ORGN 0273	
Kaler, E. W.	POLY 0244	Karpinski, J.	MEDI 0111	Kelly–Rowley, A. M.		
Kalghatgi, K.	ANYL 0128	Karslake, C.	INOR 0378		ORGN 0053	
Kalisz, H. M.	ANYL 0016	Karthikeyan, S.		Kelty, S. P.	PHYS 0365	
Kaliszan, R.	ANYL 0043		POLY 0184	Kelzenberg, J. C.		
Kaller, B. F.	ORGN 0004	Kasha, M.	CONG 0003		CATL 0042	
Kallo, D.	PETR 0086	Kashireninov, O. E.		Kemp, J. A.	MEDI 0146	
Kaloustian, M K.			PHYS 0144	Kemp, P.	GEOC 0003	
	ORGN 0204	Kashiwada, Y.	AGFD 0166	Kempf, A. J.	MEDI 0181	
Kalra, H.	PETR 0123	Kasi, S.	COLL 0026	Kempf, D. J.	ORGN 0106	
Kaluarachchi, K.		Kassini, E.	BIOT 0243	Kenealy, W. R.	BIOT 0238	
	INOR 0378	Kastner, M. E.	CHED 0158	Kennedy, J. A.	MEDI 0177	
Kaluza, Z.	ORGN 0073		PHYS 0271	Kennedy, J. T.	POLY 0130	
Kalyanaraman, B.		Kastra A., M. D.		Kenney, M. E.	POLY 0257	
	BIOL 0099		CHED 0092	Kenney, M. P.	CHAS 0029	
Kamat, P. V.	COLL 0149	Kaszynski, P.	ORGN 0124	Kenyon, W. H.	AGRO 0027	
	COLL 0190	Kataoka, H.	AGRO 0008	Keogh, R. A.	CATL 0052	
Kamaya, H.	MEDI 0032		AGRO 0010	Keosian, R. A.	POLY 0107	
Kamei, T.	BIOT 0144	Katayama, A.	AGRO 0126	Kerfoot, W. B.	ENVR 0052	
Kamermans, R.	NUCL 0039	Katerinopoulos, H. E.		Kergaye, A. A.	MEDI 0084	
			MEDI 0128			

Kergaye, A. A.	MEDI 0153	Kim, S. C.	PMSE 0059	Klemperer, W. G.		
Kerman, L. L.	MEDI 0130		PMSE 0060		POLY 0261	
Kern, C. W.	PHYS 0338	Kim, S-K.	PHYS 0164	Klempner, D.	PMSE 0040	
Kern, D. G.	FUEL 0058	Kim, Y.	BIOT 0150	Kleschick, W. A.		
Kern, F.	POLY 0201		INOR 0141		AGRO 0046	
Kern, R. D.	FUEL 0068		PHYS 0156		AGRO 0047	
Kernan, M. R.	AGFD 0191		PHYS 0364		AGRO 0048	
Kerr, J. S.	COMP 0047	Kim, Y-K.	POLY 0363	Klevan, L.	CARB 0037	
Kerrigan, J. E.	ORGN 0233	Kimball, S. D.	MEDI 0188	Klibanov, A. M.		
Kershishnik, E.	INOR 0108	Kimble, S. B.	INOR 0093		BIOT 0063	
Kershner, D. L.		Kimerle, R. A.	CEI 0025	Kliment, R.	TECH 0001	
	INOR 0155	Kimura, M.	POLY 0220	Kline, C. H.	SCHB 0021	
Kessler, H.	PETR 0075	Kimura, N.	BIOT 0172	Kline, P. C.	BIOL 0046	
Keston, A. S.	PHYS 0272	King, C. R.	I&EC 0049	Klinghofer, V.	MEDI 0120	
Keto, J. W.	PHYS 0367	King, G. P.	CHAS 0046		MEDI 0197	
Kettler, P. B.	INOR 0264	King, H. E., Jr.	CATL 0019	Klinowski, J.	PETR 0006	
	INOR 0265	King, J.	BIOT 0113		PETR 0042	
Kettner, C.	BIOL 0082	King, J.M.H.	BIOT 0099		PETR 0060	
Kevan, L.	CATL 0017	King, L. A.	POLY 0151		PETR 0081	
	CATL 0027	King, M.	POLY 0038		PETR 0096	
	PHYS 0050	King, M. L.	BIOT 0134	Kloczkowski, A.		
	PHYS 0231	King, R. B.	COMP 0032		COLL 0015	
	PHYS 0284	King, R. S.	BIOT 0074		COLL 0016	
Kevin, N. J.	ORGN 0112	King, S. R.	AGFD 0191	Kloosterman, D. A.		
Kexel, H.	AGFD 0194	King, T. S.	CATL 0042		AGRO 0108	
Kezdy, F. J.	MEDI 0055		COLL 0089	Kloss, J.	CINF 0042	
Khadzhiev, S. N.			COLL 0090	Kluepfel, D.	BIOT 0213	
	PETR 0101	King, W. P.	FUEL 0058	Klug, C. A.	PHYS 0347	
Khaledi, M. G.	ANYL 0044	Kinghorn, D.	MEDI 0034	Klusman, R. W.		
Khan, A.	BIOL 0111	Kingsbury, K. B.			ENVR 0048	
Khan, B. T.	INOR 0411		INOR 0341	Klyosov, A. A.	BIOT 0186	
	INOR 0412	Kingsley, W.	CHAS 0051	Knachel, H.	CHED 0142	
Khan, M. A.	INOR 0051	Kini, A. M.	PHYS 0024	Knaff, D. B.	BIOL 0093	
Khan, M. R.	FUEL 0115	Kinkel, J. N.	ANYL 0042	Knapp, D. A.	NUCL 0010	
	FUEL 0117	Kinoshita, T.	ANYL 0100	Knapp, J. A.	NUCL 0065	
	FUEL 0137	Kinsella, J. E.	AGFD 0207	Knapp, R.	MEDI 0093	
Khanarian, G.	POLY 0107	Kirchhoff, J. R.		Knapp, S.	ORGN 0067	
	POLY 0179		COLL 0003		ORGN 0068	
Khanin, V. V.	BIOT 0004	Kirchhoff, M. M.		Knight, D. S.	PHYS 0118	
Khau, V. V.	ORGN 0236		ORGN 0093	Knight, R. A.	COLL 0037	
Khawli, L. A.	BIOT 0061	Kirpekar, A. C.	BIOT 0088	Knight, W. B.	BIOL 0049	
Khelemshkaya, N.		Kirsch, P.	AGRO 0148		BIOL 0050	
	ANYL 0076	Kirschner, S.	CHED 0058		MEDI 0173	
Khelghatian, H. M.		Kirss, R. U.	INOR 0054	Knoll, W.	BIOT 0043	
	I&EC 0052		INOR 0055		POLY 0010	
Khurana, A. L.	AGFD 0005		INOR 0305	Knorr, D.	AGFD 0029	
Kidd, K. B.	INOR 0278	Kislin, B. S.	CINF 0016	Knowles, J.K.C.		
Kidoguchi, A.	POLY 0128	Kispert, L. D.	FUEL 0073		BIOT 0206	
Kiefer, J. J.	COLL 0074	Kiss, R. D.	BIOT 0031	Knox, L.	ENVR 0086	
Kiegiel, J.	ORGN 0284	Kitahata, S.	CARB 0008	Ko, C.	MEDI 0025	
Kiehart, D. P.	BIOL 0140	Kitipichai, P.	POLY 0216		MEDI 0026	
Kiers, N. H.	INOR 0365		POLY 0218	Ko, W. H.	POLY 0118	
Kies, C.	AGFD 0034	Kivlighn, S. D.	MEDI 0103	Kobayashi, K.	BIOT 0121	
Kietzmann, R.	PHYS 0111		MEDI 0105	Kobayashi, T.	BIOT 0181	
Kihara, S.	BIOT 0105	Kiyotsukuri, T.	POLY 0225	Kobayashi, Y.	COLL 0115	
Kijak, P. J.	AGFD 0099	Kizling, J.	COLL 0128	Kobilinsky, L.	CHED 0153	
Kilbourn, R. G.	MEDI 0106	Kjeldgaard, M.	CINF 0029	Koch, A.	ANYL 0108	
Kilburn, D. G.	BIOT 0010	Klabunde, K. J.		Koch, H. F.	CHED 0181	
	BIOT 0012		CATL 0009	Koch, S.	INOR 0059	
	BIOT 0207	Klafter, J.	PHYS 0130	Koch, S. A.	INOR 0078	
	BIOT 0217	Klapatch, T.	BIOT 0029	Koch, W. F.	BIOL 0033	
Killeen, K. P.	COLL 0006	Klares, U.	CARB 0025		BIOL 0034	
Kilpatrick, P. K.		Klassen, R.	AGFD 0116	Kochoyan, M.	BIOL 0002	
	BIOT 0020	Klasson, K. T.	BIOT 0017	Kodas, T.	COLL 0031	
Kilz, P.	PMSE 0051	Klayman, D. L.	MEDI 0063	Kodas, T. T.	COLL 0027	
Kim, A. J.	INOR 0382	Klazinga, A. H.	PETR 0025		COLL 0030	
Kim, G.	ORGN 0222	Klei, H.	BIOT 0033		INOR 0144	
	PHYS 0255	Klein, A.	PMSE 0073		INOR 0175	
Kim, H-I.	POLY 0113	Klein, D. H.	PMSE 0030		POLY 0312	
Kim, H. O.	MEDI 0051	Klein, D. J.	COMP 0021	Koehler, M. E.	PMSE 0068	
Kim, I. S.	ENVR 0015		COMP 0030	Koehler, R. T.	CINF 0016	
Kim, J.	COLL 0155	Klein, L. C.	PETR 0031	Koener, J. F.	AGRO 0098	
Kim, J-H.	POLY 0137		POLY 0263	Koenig, T. W.	INOR 0158	
Kim, J. J.	AGFD 0076	Klein, R. S.	CARB 0046	Koeppe, M. K.	AGRO 0027	
Kim, J-J.	BIOT 0159	Kleinert, H. D.	MEDI 0120	Koetzle, T. F.	CINF 0031	
Kim, K.	ORGN 0177		MEDI 0197	Koezuka, H.	POLY 0224	
Kim, K-W.	INOR 0200	Kleinow, K. M.	AGRO 0112	Koffend, J. B.	PHYS 0331	
Kim, L.	AGRO 0080	Kleinschroth, J.		Kogan, S. B.	PETR 0066	
Kim, M-H.	CATL 0014		MEDI 0074	Kohler, B.	MEDI 0013	
Kim, M-J.	BIOT 0095	Klemens, P. G.	POLY 0335	Kohler, W.	POLY 0105	
Kim, N. K.	ENVR 0072	Klemperer, W.	PHYS 0102	Kohno, E.	AGRO 0013	
Kim, P. S.	BIOT 0062	Klemperer, W. G.		Koike, Y.	POLY 0174	
Kim, S.	COLL 0051		INOR 0174	Koivula, A.	BIOT 0206	

Koizumi, M. PETR 0003	Kotasthane, A. N.	Kubota, F. POLY 0127
Kojima, Y. POLY 0291	PETR 0076	Kubrak, D. MEDI 0179
Kokotailo, G. PHYS 0348	Kotch, T. G. POLY 0096	Kucera, P. ANYL 0113
Kokta, B. V. PMSE 0077	Kotnis, M. A. PMSE 0104	Kucera, W. R. INOR 0325
PMSE 0112	Kotora, G. J. INOR 0199	Kuchenmeister, M.
PMSE 0170	Kottayil, S. G. MEDI 0149	PETR 0106
Kolb, C. E. PHYS 0071	Kouvarakis, A. MEDI 0128	Kudla, K. D. MEDI 0152
PHYS 0142	Kouzes, R. T. NUCL 0030	Kudzma, L. V. MEDI 0148
Kolchinskii, A. G.	Kovac, P. CARB 0019	Kuech, T. F. COLL 0064
COLL 0103	Kovar, P. MEDI 0120	Kuehne, M. E. ORGN 0009
Koldijk, A. J. POLY 0209	MEDI 0197	Kuester, J. L. FUEL 0090
Kole, P. L. ORGN 0208	Kowalski, R. PMSE 0176	Kugabalasooriar, S.
Kolinski, A. POLY 0297	Kozak, J. J. PHYS 0063	ORGN 0246
Kollitides, E. A.	Kozarich, J. W. BIOL 0012	Kuhn, M. W. ENVR 0051
FUEL 0116	Koziet, J. AGFD 0195	Kuhn, R. H. BIOT 172A
Kollmannsberger, H.	Kozlowski, J. F.	Kuibida, L. V. FUEL 0126
AGFD 0023	BIOL 0066	Kuiper, H. A. AGRO 0094
Kollodge, J. S. PMSE 0074	Kozub, G. M. COLL 0111	Kuipers, P. J. MEDI 0177
Kolodziejski, W.	Kradolfer, T. INOR 0235	Kuk, Y. PHYS 0100
PETR 0060	Kramer, J. FUEL 0141	Kulas, R. W. FUEL 0006
Kolotusha, T. P.	Kramer, M. AGFD 0057	Kulkarni, P. V. MEDI 0041
COLL 0105	Kramer, R. M. BIOL 0072	Kumar, A. BIOL 0091
Komernicki, A. POLY 0187	Kramer, S. J. AGRO 0008	ORGN 0293
Kominos, D. PHYS 0167	Krammer, G. AGFD 0184	Kumar, H.P.S. INOR 0024
Komisarcik, K. NUCL 0036	Krantz, A. ORGN 0269	Kumar, J. POLY 0131
Kommareddi, N.	Kranz, K. E. PETR 0082	Kumar, K. BIOL 0091
BIOT 0046	PETR 0084	ORGN 0024
Kondilenko, V. P.	Krasii, B. V. PETR 0135	Kumar, P. COLL 0188
COLL 0110	Kraus, V. PHYS 0023	Kumar, R. INOR 0242
Konja, G. CARB 0045	Krause, B. MEDI 0108	INOR 0274
Konopka, C-U. AGFD 0097	Krause, B. R. MEDI 0107	Kumar, S. MEDI 0015
Konuklar, G. AGFD 0033	MEDI 0110	ORGN 0208
Koo, K-M. INOR 0327	Krause, S. COLL 0145	Kumar, S. K. PMSE 0165
Koo, Y-E. PHYS 0273	Krauss, M. BIOL 0061	POLY 0352
Kook, S. K. PHYS 0080	BIOL 0063	Kumar, V. MEDI 0158
Kook, S-K. PHYS 0274	PHYS 0224	Kumari, P. ANYL 0118
Kool, L. B. INOR 0198	PHYS 0256	Kumazawa, N. BIOT 0123
INOR 0199	Kreek, S. A. NUCL 0074	Kumthekar, M. COLL 0207
Koontz, J. I. INOR 0114	Kreft, A. MEDI 0179	Kunchur, M. PHYS 0010
Kooser, R. G. CHED 0112	Kreis, P. AGFD 0183	Kundu, K. P. FUEL 0127
CHED 0155	Kretschmer, R. BIOL 0102	Kung, H. AGFD 0190
CHED 0210	Kreutner, W. MEDI 0161	Kuniholm, H. E.
Kopelman, R. PHYS 0022	Kreutz, T. PHYS 0291	NUCL 0081
PHYS 0273	Krieger, R. I. AGRO 0043	Kuniholm, P. I.
PHYS 0274	Krijnen, B. POLY 0222	NUCL 0081
PHYS 0307	Kripalani, K. J.	Kunin, R. CARB 0065
PHYS 0309	MEDI 0196	Kuntz, I. D. COMP 0036
PHYS 0316	Krishnagopalan, A.	Kunz, M. PMSE 0184
Kopke, T. AGFD 0183	PETR 0016	Kuo, C. PMSE 0068
Kopsel, M. AGFD 0071	Krishnamurthy, G.	Kuo, C. M. POLY 0020
Korczak, B. BIOL 0151	BIOL 0127	POLY 0234
Korda, A. ORGN 0049	Krishnan, P. N.	Kuo, J. AGFD 0191
Koreeda, M. ORGN 0118	POLY 0353	Kuo, S. C. BIOL 0137
ORGN 0211	Krishnan, T. R.	Kuo, S-J. AGFD 0036
Korendyke, N. P.	COLL 0086	Kupfer, M. POLY 0176
PHYS 0160	Kristof, P. V. PHYS 0090	Kurata, T. POLY 0224
Korenowski, G. M.	Kriven, W. M. INOR 0014	Kurauchi, T. POLY 0291
POLY 0216	Kriz, D. Z. PMSE 0017	Kurizki, G. PHYS 0064
POLY 0218	Krogh-Jespersen, K.	Kuroda, P. K. NUCL 0082
Korkowski, P. I.	PHYS 0233	NUCL 0083
ENVR 0036	Krogmann, C. CARB 0058	Kurokawa, H. POLY 0171
Korkowski, P.I. ENVR 0036	Kroo, E. FUEL 0052	Kurokawa, Y. COLL 0115
Korkowski P., I.	Kroto, H. W. FUEL 0019	Kurtz, D. M., Jr.
CHED 0106	Krueger, D. A. AGFD 0186	INOR 0094
Korn, L. R. CEI 0009	AGFD 0198	INOR 0140
Kornak, E. P. ORGN 0202	Krueger, R-G. AGFD 0186	Kurtz, S. R. PHYS 0375
Korobeinichev, O. P.	Krukonis, V. J. POLY 0076	Kuruc, M. P. BIOL 0040
FUEL 0126	Krupey, J. BIOL 0040	Kurys, B. E. MEDI 0130
Koros, W. J. POLY 0116	Krysan, D. ORGN 0274	Kusba, J. PHYS 0061
Korsrud, G. O. AGFD 0085	Krzycki, J. ORGN 0025	Kushida, T. PHYS 0255
Koseki, K. ORGN 0072	Kuan, D. P. ORGN 0235	Kusnadi, A. BIOT 0047
Koshi, M. FUEL 0064	Kube, D. BIOT 0150	Kustin, K. INOR 0394
Kosman, J. J. PETR 0119	Kubert, M. A. BIOL 0160	Kut, O. M. BIOT 0137
Kossoy, A. D. ANYL 0069	Kubiak, C. P. INOR 0319	Kutal, C. INOR 0166
Kosstrin, H. M.	Kubiak, G. D. PHYS 0202	Kuznicki, S. M.
FUEL 0142	Kubicek, C. P. BIOT 0183	PETR 0078
Kostichka, A. J.	Kubicek-Pranz, E. M.	PETR 0079
ANYL 0033	BIOT 0183	Kuzyk, M. G. BIOT 0175
Kostka, J. E. GEOC 0006	Kubo, A. CATL 0006	POLY 0154
GEOC 0007	Kubo, K. MEDI 0112	Kwast, E. ORGN 0283
Kostlan, C. R. MEDI 0177	Kubo, Y. ORGN 0059	Kwei, G. H. PHYS 0027
Kotasthane, A. N.	Kubo-Anderson, V.	Kwei, T. K. COLL 0034
PETR 0005	INOR 0250	PMSE 0147

Kwei, T. K.	PMSE 0163	Landgrave R., J.		Lawlor, B.	YCC 0008
	POLY 0111		COMP 0053	Lawrence, A. F.	
	POLY 0113	Lando, J. B.	POLY 0118		BIOT 0025
	POLY 0192	Landolph, J. R.	CHAS 0007	Lawrence, D. S.	
Kwiatek, P. J.	FUEL 0057	Landry, C.J.T.	POLY 0181		INOR 0186
Kwiatkowski, K.			POLY 0260		MEDI 0031
	NUCL 0036		POLY 0287	Lawrence, E. G.	
Kwini, M.	PHYS 0084	Landry, S. J.	BIOL 0003		BIOT 0139
Kwock, E. W.	POLY 0324	Landuyt, P.	PMSE 0043	Lawrence, G. D.	
Kwok, A.D–I.	INOR 0297	Lane, B.	MEDI 0167		INOR 0222
Kwok, T. J.	INOR 0154		MEDI 0168	Lawrence, K. B.	
	INOR 0209	Laneman, S. A.	CATL 0001		MEDI 0158
Kwon, C–H.	MEDI 0033		INOR 0077	Lawton, C.	BIOT 0033
	MEDI 0134	Lanford, W. A.	NUCL 0067	Layne, W. B.	PHYS 0367
Kwon, T. M.	COLL 0144	Lang, C. M.	CONG 0006	Lazarev, P. I.	BIOT 0006
Kydd, R. A.	PETR 0143	Lang, H.	POLY 0359	Lazaro, J.	PETR 0113
Kye, Y.	CATL 0031	Lang, M.	MEDI 0035	Leach, A. R.	CINF 0035
Kyle, K. R.	INOR 0167	Lang, N. D.	PHYS 0096	Leahy, M. M.	AGFD 0181
Kyotani, T.	FUEL 0032	Langer, R.	BIOL 0056	Lebedin, D. H.	INOR 0155
	I&EC 0020		POLY 0273	Le Beyec, Y.	NUCL 0055
Kyu, T.	COLL 0014	Langevine, C. M.		Leblanc, A.	CEI 0004
Labadie, J. W.	I&EC 0055		AGRO 0073		CEI 0005
	POLY 0026	Langford, R. E.	CHAS 0045		CEI 0010
Labat, G.	INOR 0187	Langhoff, C.	INOR 0334	LeBoulluec, K. L.	
Labgold, M.	BIOT 0239	Langhoff, S. R.	PHYS 0005		MEDI 0012
Labontz, J.	AGFD 0042		PHYS 0171	Lech, J.	AGFD 0170
Lachman, I. M.	PETR 0090	Langner, A.	PHYS 0216	Lech, J. J.	AGRO 0112
Lachowicz, E.	FUEL 0055	Langridge, R.	COMP 0036	Lechert, H.	PETR 0041
Lacort, G.	ANYL 0057	Lanni, C.	MEDI 0176	Lechowich, R. V.	
LaCount, R. B.	FUEL 0058	LaNoue, K. F.	MEDI 0115		AGFD 0014
LaCount, R. B., Jr.		Lansbury, P.	PHYS 0196	Leckband, D.	BIOL 0056
	FUEL 0058	Lanthier, P.	BIOT 0212		BIOT 0043
LaCourse, W. R.		Lantos, P. R.	CMEC 0012		BIOT 0151
	CARB 0067	LaPeruta, R.	POLY 0218	Ledeboer, M.	CHED 0085
Lacroix, P.	POLY 0217	Laperuta, R., Jr.			PHYS 0328
Ladduwahetty, T.			POLY 0216	Ledesma, R.	FUEL 0053
	MEDI 0119	Lapidot, N.	INOR 0283	Lednicer, D.	AGFD 0164
Ladinsky, H.	MEDI 0141	Lappe, R.	MEDI 0111	Ledoux, I.	POLY 0006
Ladner, E. P.	CATL 0046	Lara Lemus, A.	CHED 0094	Le Dred, R.	PETR 0128
Ladouceur, H. D.		Larkin, D. J.	INOR 0105	Lee, A. L.	BIOL 0043
	FUEL 0125	Larmann, J. P.	ANYL 0126		BIOT 0134
	FUEL 0128	Larochelle, J. H.		Lee, A.W.M.	ANYL 0089
Ladron de Guevara, O.			AGRO 0030	Lee, B.	POLY 0266
	BIOL 0102	Larsen, B. R.	ANYL 0117	Lee, C. C.	FUEL 0131
LaDuca, R. L.	INOR 0103	Larsen, R.	BIOL 0096	Lee, C. H.	PHYS 0011
Lafferty, C. J.	FUEL 0007	Larson, C. E.	COLL 0029	Lee, C. K.	NUCL 0021
Lafferty, N. L.	I&EC 0053		COLL 0156	Lee, C. Y.	AGFD 0136
Lafleur, A. L.	FUEL 0020	Larson, E.	INOR 0010		AGFD 0151
Lafyatis, D.	PETR 0070	Larson, R. A.	ENVR 0029	Lee, C. Y–C.	PMSE 0019
Lafyatis, D. S.	PETR 0059	Lartigue, J. G.	NUCL 0086		PMSE 0169
Lagally, M. G.	PHYS 0098	Laskin, J. D.	AGFD 0069	Lee, D.	NUCL 0074
La Ginestra, A.	PETR 0107	Lastovkin, G. A.		Lee, D. G.	ORGN 0243
Lago, M.	MEDI 0188		PETR 0135	Lee, D. S.	PMSE 0060
Lago, M. A.	ORGN 0115	Lastra A., M. D.		Lee, D. W.	BIOT 0010
Lah, M. S.	INOR 0018		AGFD 0144	Lee, E.	INOR 0163
Lahti, P. M.	INOR 0373		AGFD 0145	Lee, E. E.	CARB 0026
Lahti, R. A.	MEDI 0126		AGFD 0146	Lee, G–A.	BIOL 0005
Lah–Tusar, L.	COMP 0002		CHED 0091	Lee, H. C.	I&EC 0047
Laine, R. M.	POLY 0314		CHED 0093	Lee, H. J.	MEDI 0170
	POLY 0358	Laszlo, P.	CATL 0013	Lee, J.	BIOL 0055
Laird, E. R.	CINF 0008	Latli, B.	AGRO 0016		ORGN 0066
Lajmi, A.	INOR 0222	Lau, C. K.	MEDI 0180	Lee, J–G.	PHYS 0275
Lajolo, F. M.	AGFD 0041	Lauff, J. J.	BIOT 0032	Lee, J. M.	BIOT 0226
Lakis, R. E.	PETR 0073	Lauffer, D.	MEDI 0046		BIOT 0230
Lakowicz, J. R.	PHYS 0061	Lauher, J. W.	INOR 0075		CATL 0053
Lalazar, A.	AGFD 0076		ORGN 0111	Lee, J. Y.	POLY 0131
Lam, K. S.	MEDI 0093	Laurencin, C.	POLY 0273	Lee, K.	CATL 0047
Lam, P.Y–S.	COMP 0047	Laurent, M–H.	AGFD 0137	Lee, K. E.	INOR 0217
Lam, W. W.	ENVR 0005	Laurion, T. A.	FUEL 0137		INOR 0304
Lamanna, W. M.		Lavallee, D. K.	BIOT 0168	Lee, K–H.	AGFD 0092
	CATL 0010		CHED 0015		AGFD 0166
Lamb, S. I.	CHED 0206		CHED 0139		BIOL 0018
Lambert, S.	CATL 0052	Lavell, W. T.	POLY 0215	Lee, K–J.	POLY 0069
Lambla, M.	PMSE 0176	Lavie, D.	AGFD 0188	Lee, L. J.	PMSE 0027
Lamendola, J. F.		Lavie, G.	AGFD 0188	Lee, L. S.	ANYL 0071
	AGRO 0078	Lavin, E. H.	AGFD 0137		ENVR 0014
Lamy, A.	CATL 0045	Lavine, B. K.	ANYL 0046		ENVR 0030
Lan, H. L.	POLY 0050		CHED 0195	Lee, P. L.	PHYS 0069
Lan, J.	CARB 0037	Lavy, T. L.	AGRO 0127	Lee, P. W.	AGRO 0031
Lan, W.	BIOT 0192		AGRO 0146	Lee, R. T.	BIOL 0148
Lancet, M. S.	FUEL 0071	Law, C. K.	FUEL 0083	Lee, S.	PMSE 0010
Landa, E. R.	GEOC 0008	Lawhorne, S.	CINF 0037	Lee, S. H.	ORGN 0177
Lande, K.	NUCL 0021	Lawless, D.	COLL 0131	Lee, S. J.	MEDI 0175

Lee, S. M.	AGRO 0145	Leon y Leon, C. A.		Li, Y.	ORGN 0181
	AGRO 0149		FUEL 0028		ORGN 0247
Lee, S. S.	BIOT 0199		FUEL 0033		PHYS 0279
	I&EC 0052		I&EC 0020		PMSE 0064
Lee, S. T.	PMSE 0044	Leperchec, P.	FUEL 0046		PMSE 0079
Lee, T. H.	POLY 0089	Le Perchec, P.	PETR 0141		PMSE 0091
Lee, T. J.	PHYS 0276	Lepine-Frenette, C.			PMSE 0103
Lee, V.	MEDI 0188		ORGN 0006		PMSE 0149
Lee, V.H.L.	MEDI 0194	Lercari, F. A.	CHAS 0021	Li, Y. X.	CATL 0009
Lee, Y.	PHYS 0278	Lergenmuller, M.		Li, Z.	PHYS 0091
	PHYS 0299		CARB 0025	Lian, T.	PHYS 0194
Lee, Y. C.	BIOL 0148	Lerke, S. A.	ANYL 0106	Liang, M.	POLY 0186
Lee, Y. I.	POLY 0043	Lerner, R. A.	BIOL 0028	Liang, T. C.	BIOL 0005
Lee, Y. K.	PMSE 0059	Lesec, J.	PMSE 0066		BIOL 0035
Lee, Y. T.	PHYS 0163	Lester, M. I.	CHED 0088	Liang, X.	PMSE 0080
	PHYS 0203		PHYS 0300	Liang, Y.	AGFD 0042
	PHYS 0349	Leszczynski, J.	PHYS 0033		INOR 0051
Lee-Huang, S.	AGFD 0190	Leu, F.	BIOT 0134	Liang, Z.	ANYL 0078
Leenutaphong, D. L.		Leung, J.	BIOT 0110	Liao, D. C.	PMSE 0044
	BIOT 0087	Leung, K.W.P.	INOR 0106	Liao, J-H.	INOR 0355
Lees, A. J.	POLY 0096	Leung, L. K.	COLL 0175	Liao, M-Y.	PHYS 0277
Lees, J. F.	COLL 0153	Levaj, B.	PMSE 0094	Liao, R-F.	ORGN 0057
Leeson, P. D.	MEDI 0146	Le Van Mao, R.		Liberko, C. A.	BIOT 0201
	ORGN 0117		CATL 0016	Libutti, B. L.	CARB 0065
Lefebvre, F.	PETR 0131		PETR 0063	Licht, S.	ANYL 0079
Lefkowitz, L.	CHED 0200	Levendusky, T. L.			ENVR 0044
Legge, C. H.	POLY 0219		POLY 0196	Lichtenberger, D.	
Leheny, A. R.	PHYS 0135	Levery, S. B.	CARB 0060		PHYS 0152
	POLY 0238	Levesque, C.	CEI 0010	Lichtenthaler, F. W.	
Lehmann, D.	AGFD 0183	Levi, P. E.	AGRO 0118		CARB 0003
Lehmann F., P. A.		Levi, R.	MEDI 0106		CARB 0011
	ANYL 0082	Levie, B. E.	FUEL 0142		CARB 0025
	HIST 0016	Levin, B.	AGFD 0188	Liddell, M.	PETR 0131
	MEDI 0044	Levin, S. J.	CEI 0014	Lieber, C. M.	PHYS 0156
	MEDI 0088	Levis, R. J.	ANYL 0130		PHYS 0206
	MEDI 0100	Levitt, M. H.	PHYS 0196		PHYS 0211
Lehnert, J.	PHYS 0111	Levitz, P.	PHYS 0060		PHYS 0239
Lehrfeld, J.	AGFD 0143	Levon, K.	POLY 0111		PHYS 0279
Lehrman, R.	I&EC 0026		POLY 0248		PHYS 0292
Lehtinen, A.	PMSE 0046	Levy, J. N.	BIOT 0160		PHYS 0329
Lei, G.	CATL 0027	Levy, L. A.	ORGN 0149		PHYS 0364
Lei, H.	ORGN 0066	Levy, R.	PHYS 0167		PHYS 0365
Lei, W.	BIOT 0126	Lew, A. C.	AGRO 0072	Liebeskind, L. S.	
Leibler, L.	POLY 0122	Lew, R.	PMSE 0065		ORGN 0274
	POLY 0172	Lew, W.	ORGN 0001	Liegeois, J. M.	PMSE 0043
Leichtweis, H. C.		Lewandowski, G. A.		Liehr, M.	COLL 0026
	AGRO 0148		BIOT 0138	Liesch, J.	BIOL 0049
Leighton, J. L.	ORGN 0289	Lewandowski, Z.			BIOL 0050
Leighton, T.	AGFD 0117		BIOT 0096	Liesch, J. M.	MEDI 0181
Leipold, R. J.	BIOT 0245	Lewin, A. H.	MEDI 0075	Lieu, V. T.	CHED 0171
Leis, J. R.	INOR 0400	Lewin, M.	PHYS 0203	Lievens, J.	PETR 0099
Lemanski, J. R.		Lewin, S. Z.	HIST 0025	Lievense, J. C.	BIOT 0032
	I&EC 0049	Lewis, B. A.	AGFD 0133	Lifson, J. D.	AGFD 0189
LeMay, R. J.	MEDI 0055	Lewis, D.	BIOT 0168	Lifson, S.	POLY 0348
Lemieux, R. U.	CARB 0012	Lewis, M. A.	BIOT 0128	Ligler, F. S.	BIOT 0246
Lemley, A. T.	AGRO 0064	Lewis, R.	COMP 0036	Likharev, K. K.	
	AGRO 0124	Lewis, R. J.	CHAS 0039		BIOT 0004
Lemmo, A. V.	ANYL 0126	Lewis, V. P.	BIOT 0030	Liljenzin, J. O.	NUCL 0024
Le Moigne, J.	POLY 0132	Leyba, J. D.	NUCL 0016	Lillirud, K.	PETR 0020
Lemos, F.	PETR 0132	Leyden, J. J.	AGFD 0147	Lillquist, J.	BIOT 0114
Lengsfield, B. H.		Leyh, T. S.	BIOL 0062	Lim, B. B.	MEDI 0023
	PHYS 0014	Li, C. L.	PETR 0105		MEDI 0048
Lenk, T. J.	POLY 0145		PETR 0117	Lim, C. K.	ANYL 0089
le Noble, W. J.	ORGN 0086		PETR 0130	Lima G., R. M.	ENVR 0025
	ORGN 0237	Li, D.	ANYL 0063	Lin, A. J.	MEDI 0063
	ORGN 0238		INOR 0180	Lin, C.	ANYL 0118
	ORGN 0239		POLY 0127	Lin, C-E.	MEDI 0154
Lentini, G.	MEDI 0098	Li, G.	BIOT 0170		MEDI 0157
Lenz, R. W.	CHED 0132	Li, H.	PHYS 0259	Lin, C-H.	INOR 0110
Leo, G. C.	ORGN 0296	Li, H. B.	ORGN 0254	Lin, C. H.	POLY 0031
Leon, F.	CHED 0097	Li, J.	INOR 0056	Lin, C. T.	COMP 0049
Leonard, B.	NUCL 0079	Li, J. P.	AGRO 0008	Lin, G.	BIOL 0018
	YCC 0010		AGRO 0012	Lin, H. C.	POLY 0276
Leonard, J. C.	MEDI 0131	Li, J-T.	PMSE 0013	Lin, H-K.	BIOL 0020
Leonard, M. B.	INOR 0149	Li, L. D.	PETR 0130	Lin, J.	ORGN 0264
Leon C., F.	CHED 0106	Li, L. S.	POLY 0276		POLY 0364
	ORGN 0037	Li, M.	PHYS 0194	Lin, J. C.	PETR 0099
	ORGN 0196	Li, M-H.	BIOL 0001	Lin, J-L.	COLL 0204
Leone, C.	BIOL 0059	Li, P.	PMSE 0011	Lin, J. S.	PMSE 0148
Leong, Y-P.	INOR 0106	Li, R.	MEDI 0129	Lin, J-T.	PETR 0111
Leonhardt, B. A.		Li, T.	ORGN 0191	Lin, P.	ANYL 0072
	AGRO 0033	Li, T-J.	BIOT 0142	Lin, P. C.	AGRO 0141
		Li, X.	COMP 0005	Lin, R.	COLL 0176

Lin, R.	COLL 0177	Liu, L-K.	INOR 0107	Lopez-Villagomez, J. M.	
	INOR 0123	Liu, M-C.	CARB 0038		AGFD 0138
Lin, S.	PHYS 0020	Liu, M-K.	PMSE 0011	Lorenz, L. J.	ANYL 0112
Lin, S. Y.	AGRO 0073	Liu, R.	PMSE 0064	Lorenzo, J.	NUCL 0069
Lin, T-S.	CARB 0038	Liu, S-B.	CATL 0025	Lorimer, G. H.	BIOT 0116
Lin, W.	INOR 0244	Liu, S-T.	INOR 0028	Loring, R. F.	PHYS 0035
	POLY 0108		INOR 0029	Lotina-Hennsen, B.	
	POLY 0217	Liu, S. X.	POLY 0068		INOR 0263
Lin, W. P.	POLY 0127	Liu, W.	BIOT 0142	Lotman, M.	INOR 0215
Lin, Y.	PHYS 0080		PMSE 0083	Lotti, V. J.	MEDI 0103
Lin, Y-G.	PMSE 0136	Liu, Y.	I&EC 0019		MEDI 0104
Lin, Y-L.	BIOT 0058	Liu, Y. H.	TECH 0018		MEDI 0105
Lin, Z.	PHYS 0192	Liu, Y-J.	INOR 0082	Lou, B-S.	BIOL 0096
Linares-Solano, A.		Liucheng, Z.	PMSE 0082	Lou, Y.	NUCL 0028
	FUEL 0029	Livage, J.	POLY 0358	Loudon, M. H.	AGFD 0059
	FUEL 0032	Liverton, N. J.	MEDI 0119	Louis, J. B.	AGRO 0053
Lincoln, D. M.	I&EC 0054	Ljungdahl, L. G.		Lounsbury, B. B.	
Lind, J.	CHED 0084		BIOT 0180		AGRO 0106
	PHYS 0249	Llano, M.	INOR 0263	Love, G. D.	FUEL 0004
Lind, J. M.	MEDI 0094	Lleres, N.	NUCL 0028	Love, P.	PMSE 0183
Lindauer, R. F.	ANYL 0115	Lloyd, B. R.	INOR 0379	Loveday, D.	COLL 0058
Lindberg, A. A.	CARB 0017	Lloyd, P.	MEDI 0148		POLY 0345
Lindberg, B.	CARB 0016	Loar, J. M.	ENVR 0062	Lovejoy, D.	POLY 0233
Lindberg, P.	MEDI 0001	Lobanov, V. V.	COLL 0107	Lovejoy, N.	PHYS 0164
Linden, J. D.	BIOT 0251		COLL 0116	Loveland, W.	NUCL 0000
Lindenberg, K.	PHYS 0129	Lobarzewski, J.	PMSE 0081		NUCL 0024
Lindquist, D. A.		Lobato C., C.	AGFD 0107	Lovey, K.	MEDI 0165
	POLY 0312	Lobkovsky, E.	INOR 0128	Lovinger, A. J.	POLY 0333
Lindsay, A. E.	AGRO 0066		INOR 0197	Lovley, D. R.	GEOC 0008
Lindsay, G. A.	POLY 0268	Lochmuller, C. H.		Low, A. A.	PHYS 0192
Lindsay, N. E.	PETR 0140		ANYL 0013	Lowack, R. H.	ORGN 0205
Lindsay, S. M.	PHYS 0157	Locke, B.	PHYS 0194	Lowe, J. A.	PETR 0051
Lineberger, W. C.		Locke, P.	PHYS 0259	Lowe|, J. A., III	
	PHYS 0055	Locker, J. R.	CHED 0152		ORGN 0183
Linert, J. G.	POLY 0046	Lockhart, T. P.	POLY 0240	Lu, F.	FUEL 0056
Lingamurthy, S.		Lodge, T. P.	POLY 0165	Lu, G.	COLL 0047
	COLL 0081		POLY 0267	Lu, X.	BIOL 0014
Lingle, R., Jr.	PHYS 0335		POLY 0275		PMSE 0118
Lingnert, H.	AGFD 0157	Loeffler, R.	MEDI 0187	Lu, Y-H.	POLY 0057
Linhardt, R. J.	CARB 0039	Loewi, R.	MEDI 0071	Lu, Z.	MEDI 0150
Linliu, K.	PMSE 0103	Logan, M. E.	MEDI 0158	Luberoff, B. J.	CHAL 0024
Linn, B.	ORGN 0016	Loganathan, D.	CARB 0039		CHAL 0027
Linn, C. P.	ORGN 0023	Lohry, E. J.	FERT 0001		CHAL 0031
Linssen, J.P.H.	AGFD 0158		FERT 0016	Lucas, A. D.	AGRO 0032
Linz, F.	BIOT 0169	Lomeli F., C.	INOR 0026	Lucas, C.	BIOT 0059
Linzaga E., I.	ORGN 0198	Lomonosov, M. V.		Luckey, J. A.	ANYL 0033
Lioy, P. J.	CEI 0009		BIOT 0082	Ludemann, H-D.	
Lipatov, Y. S.	PMSE 0023	London, R. E.	ORGN 0149		BIOT 0041
Lipinski, C. A.	MEDI 0082	Long, A. K.	CINF 0001	Ludermann, C.	NUCL 0027
Lipkovskaya, N. A.		Long, A. R.	AGFD 0101	Ludovice, P. J.	POLY 0351
	COLL 0103	Long, C. G.	BIOL 0001		POLY 0355
Lipman, R.	MEDI 0015	Long, G. J.	INOR 0050	Ludwig, B. W.	POLY 0255
Lipowitz, J.	POLY 0336		INOR 0101	Ludwig, R. T.	ORGN 0064
Lippard, S. J.	CHED 0060	Long, T. J.	PHYS 0261	Luftman, H. S.	INOR 0007
Lipscomb, G. G.		Longmire, C. F.		Lugtenburg, J.	PHYS 0196
	POLY 0139		INOR 0377	Lui, A.S.T.	AGRO 0003
Lipscomb, R. D.		Longo, F. R.	COLL 0024	Luke, R.W.A.	MEDI 0102
	CHED 0007	Longridge, E. M.		Luk'yanov, D. B.	
Lira-Rocha, A.	MEDI 0037		INOR 0213		PETR 0061
	ORGN 0185	Lonyi, F.	PETR 0086	Luly, J. R.	MEDI 0167
Lis, R.	MEDI 0094	Loomis, A.	CHED 0209		MEDI 0168
Lisensky, G. C.	CHED 0023	Loomis, C. R.	MEDI 0075	Lumma, W. C., Jr.	
	COLL 0175	Looney, A.	INOR 0085		MEDI 0094
Lisi, P. J.	ANYL 0018	Looney, B. B.	ENVR 0053	Lundeen, J. E.	BIOT 0016
List, A. K.	INOR 0327	Lopez-Celis, I.	CHED 0095	Lunn, G.	CHAS 0006
Liston, D.	MEDI 0143	Lopez, E.	POLY 0022	Lunney, E.	MEDI 0121
Litt, M. H.	POLY 0134	Lopez, M. A.	POLY 0056	Luo, M-Z.	CARB 0038
Little, B.	BIOT 0100	Lopez-Celis, I.	CHED 0094	Luo, W.	COLL 0215
Little, R. D.	ORGN 0146		CHED 0096	Lupidi, G.	MEDI 0029
Liu, B.	PHYS 0105		CHED 0097	Lusby, W. R.	AGRO 0033
Liu, D.	BIOL 0076	Lopez de Compadre, R. L.			AGRO 0035
Liu, D.D.W.	AGRO 0111		MEDI 0017	Lussier, B. B.	MEDI 0158
Liu, F.	INOR 0315	Lopez I., F.	I&EC 0068	Luther, G. W., III	
	POLY 0111	Lopez-Izuna, F.			GEOC 0006
Liu, H.	INOR 0033		I&EC 0069		GEOC 0007
Liu, H. Q.	POLY 0337	Lopez-Lopez, J. M.		Luther, J. R.	BIOT 0019
Liu, H-W.	BIOL 0100		CHED 0098	Lutz, C. P.	PHYS 0360
	BIOL 0110	Lopez-Munguia, A.		Lutz, D.	AGFD 0008
Liu, J.	ANYL 0093		BIOT 0130	Lutz, M. A.	PHYS 0099
	PMSE 0083		BIOT 0131	Lutz, R.	POLY 0272
	PMSE 0149	Lopez V., D.	PMSE 0131	Lux, M.	POLY 0106
Liu, J. P.	ANYL 0019		POLY 0053		POLY 0208
Liu, J. Z.	PHYS 0052			Ly, D.	PETR 0063

Lykke, K. R.	FUEL 0088	MacCoss, M.	MEDI 0105	McKenna, G. B.	
	NUCL 0045	McCowan, J. R.			POLY 0170
Lyle, P. A.	PHYS 0020		MEDI 0066		POLY 0201
Lyman, C. E.	PETR 0073		MEDI 0174		POLY 0235
Lynch, J. J.	MEDI 0097	McCraw, J. M.	CEI 0001		POLY 0295
	MEDI 0191	McCreery, R. L.		McKenzie, K. D.	
Lynch, J. J., Jr.			INOR 0312		ENVR 0033
	MEDI 0096	McCulloch, D.	CINF 0026	Mackenzie, P. B.	
Lynch, M. J.	AGRO 0079	McCullough, B.			ORGN 0091
Lynch, V.	ORGN 0058		AGRO 0078		ORGN 0092
Lynd, L. R.	BIOT 0013	McCutchen, B. F.		MacKenzie, R.	BIOL 0143
	BIOT 0029		AGRO 0009		MEDI 0127
Lyngaae-Jorgensen, J.		McDermott, A.	PHYS 0196	MacKenzie, R. G.	
	PMSE 0085	McDermott, G. M.			MEDI 0124
Lyo, I-W.	PHYS 0184		POLY 0313		MEDI 0125
Lyon, K. R.	PMSE 0123	McDevitt, M. R.		McKeon, R.	FUEL 0117
Lyon, P. A.	AGFD 0202		INOR 0262	McKeon, R. J.	FUEL 0137
Lyons, B. A.	ORGN 0122	MacDiarmid, A. G.		Mackie, J. C.	FUEL 0114
Lyons, J. E.	CATL 0036		POLY 0327	McKinley, A. J.	
Lyons, L. J.	CHED 0180	McDiarmid, R.	PHYS 0075		ORGN 0124
Lysz, T.	AGFD 0088	MacDonald, A.	AGRO 0132	McKinney, K.	BIOT 0169
Lytel, R.	POLY 0004	McDougal, P. G.		McKinnon, J. T.	
	POLY 0130		ORGN 0233		FUEL 0020
Lytle, F. E.	ANYL 0014	Macedo, G.	CHED 0169	Macko, T.	PMSE 0049
Ma, C.	COLL 0160	McEllistrem, M. T.		McLarnon, C.	I&EC 0023
Ma, C.C.M.	PMSE 0044		ANYL 0022	McLaughlin, M. A.	
Ma, D.	PMSE 0149	McElroy, J. F.	MEDI 0084		AGFD 0208
Ma, L-J.	CATL 0025	McElwee-White, L.		McLaughlin, M. L.	
Ma, R.	INOR 0276		INOR 0341		ORGN 0241
Ma, S.	PMSE 0042	McEnaney, B.	FUEL 0016	McLean, A. D.	PHYS 0105
Mabuchi, T.	AGRO 0013		FUEL 0048	MacMahon, C. S.	
McAlpine, J. B.		McEvily, A. J.	AGFD 0152		PETR 0148
	BIOT 0110	McEvoy, M. A.	ORGN 0240	McMahon, J. B.	
McAteer, C. H.	CATL 0008	McEwen, W. E.			AGFD 0192
McBreen, P. H.			INOR 0373	McMahon, M.	FUEL 0115
	COLL 0114	Macey, D. J.	INOR 0116	McMahon, M. A.	
McBurney, R. N.		McFadden, D. C.			FUEL 0117
	MEDI 0147		MEDI 0182	McMarty, D.	INOR 0049
McCaffrey, R. R.		McFadden, D. L.		McMaster, J. H.	
	INOR 0070		PHYS 0234		MEDI 0052
McCann, J. M.	PETR 0118	McFarland, J. M.		McMillen, D. F.	
	PETR 0122		INOR 0385		FUEL 0062
McCarley, R. E.		McFarlane, C. S.		McMillin, D. R.	
	CHED 0029		MEDI 0180		INOR 0315
McCarroll, W. H.		McGaff, R. W.	INOR 0388	McMills, M. C.	ORGN 0179
	INOR 0359	McGinley, K.	AGFD 0147	McMordie, A.	BIOL 0155
McCarron, E. M.		McGorrin, R. J.		McMullen, D.	MEDI 0188
	PHYS 0010		AGFD 0200	McMurry, T. J.	INOR 0033
	PHYS 0011	McGowan, C. W.			ORGN 0024
McCarron, E. M., III			FUEL 0055	McNally, A.	BIOL 0031
	INOR 0256	McGown, S.	ANYL 0050	MacNeil, J. D.	AGFD 0085
McCarthy, B. A.		McGranaghan, M. B.		McNeil, J. N.	AGRO 0007
	ORGN 0240		CHED 0185	McNiel, M. B.	BIOT 0101
McCarthy, D. A.		McGrath, J. E.	PMSE 0052	McPhail, A. T.	COLL 0136
	INOR 0220		PMSE 0053		COLL 0137
McCarthy, J. E.			PMSE 0054		COLL 0138
	CINF 0031		PMSE 0123		COLL 0139
McCarthy, K. F.			POLY 0264		COLL 0140
	PHYS 0052	McGrath, P. S.	CHED 0003	McPheron, B. A.	
McCarthy, T. J.		McGraw, D.	ANYL 0036		AGRO 0116
	INOR 0202	McGregor, D. L.		McPherson, G. L.	
	POLY 0298		TECH 0014		PHYS 0131
McCartney, J. E.		McGuire, N. K.	PETR 0002	Macrae, A. R.	BIOT 0040
	BIOT 0057	Macguire, R. J.	ENVR 0020	McVey, G. R.	FERT 0006
McCauley, G. B.		McHale, P. J.	COMP 0046	McVicker, G. B.	
	PMSE 0168	McHugh, M. A.			PETR 0052
McCaulley, J. A.			PMSE 0036	Maczko, J.	I&EC 0025
	COLL 0068	Maciag, S. W.	AGRO 0027		I&EC 0035
McClelland, L. R.		Maciel, G. E.	AGFD 0104	Maddux, R. J.	ANYL 0114
	ENVR 0033	McIntyre, D.	PMSE 0069	Madey, T. E.	COLL 0172
McClure, D. B.	BIOL 0072	Mack, M.	ANYL 0042		COLL 0191
McClure, G.	CHAL 0004	MacKay, C. F.	CHED 0125	Madison, V. S.	MEDI 0165
McComsey, D. F.			CHED 0198	Madon, R. J.	CATL 0056
	ORGN 0296	Mackay, R. A.	COLL 0023	Madurawe, R. D.	
McConnachie, P. R.			COLL 0024		BIOT 0060
	CEI 0017	McKean, D. L.	ENVR 0073	Maeda, K.	BIOT 0105
McConville, D. H.		McKean, D. R.	POLY 0026	Maeda, M.	POLY 0301
	INOR 0362	McKechnie, J. I.		Maeda, S.	AGRO 0009
	INOR 0383		POLY 0366	Maeda, Y.	POLY 0071
McCormack, A. J.		McKelvey, S. A.			POLY 0072
	PETR 0118		BIOT 0233	Maga, J. A.	AGFD 0173
	PETR 0122	McKendrick, R. F.		Magda, D. J.	ORGN 0059
	PETR 0151		I&EC 0039	Magda, M. T.	NUCL 0018

MaGee, D.	ORGN	0088	Mancini, W. R.	CARB	0038	Mardones, M. A.	
Maggiore, C. J.	NUCL	0066	Mandal, B.	POLY	0131		COLL 0067
Maghari, A.	COMP	0058	Mandal, D.	BIOL	0109		INOR 0392
Maglia, B.	TECH	0005		BIOL	0145	Marek, F. L.	AGRO 0072
Magnera, T. F.	BIOT	0200	Mandava, N. B.				AGRO 0073
Magnolato, D.	AGFD	0020		AGRO	0056	Margalit, R.	BIOT 0053
Magnolo, S. K.	BIOT	0087		CHAS	0022	Margiatto, G.	ORGN 0016
Magnusson, G.	CARB	0056	Mandelkern, L.	POLY	0075	Margitfalvi, J. L.	
Magonov, S. N.	PMSE	0078		POLY	0121		PETR 0134
Maguire, J. A.	INOR	0151	Manero, O.	PMSE	0129	Margolin, A. L.	BIOT 0093
	INOR	0152		POLY	0094	Margulies, L.	PHYS 0021
Magyar, E. S.	CHED	0184	Maneval, D.	BIOT	0059	Mari, F.	INOR 0373
Magyar, J. G.	CHED	0184	Mangelsdorf, C. P.			Mariaca R., L.	PHYS 0281
Mahadevan, H.	BIOT	0022		FUEL	0136	Mariano, M. V.	ORGN 0019
Mahadevan, S.	AGRO	0018	Manhas, M. S.	CHED	0144	Maricq, M. M.	POLY 0230
Mahle, D. A.	ANYL	0062		ORGN	0034	Marin–Becerra, A.	
Mahler, L.	CMEC	0025		ORGN	0073		INOR 0057
	I&EC	0003		ORGN	0195		INOR 0260
Mahmood, A.	INOR	0181		ORGN	0255	Marinez, E.	INOR 0250
Mahoney, L. A.	ANYL	0105	Mani, R. S.	PMSE	0122	Marino, A. M.	MEDI 0196
Mahroof–Tahir, M.			Mani, S.	PMSE	0133	Marino, M. M.	PHYS 0338
	INOR	0046	Manjarrez, A.	CONG	0007	Marinos, V.	AGFD 0006
Mai, Y-W.	PMSE	0026	Mann, C. K.	ANYL	0087	Mariwala, R. K.	
Maictrinu, L.	PETR	0095		ANYL	0088		PETR 0059
Maier, C.	PMSE	0176	Mann, J. A., Jr.			Mark, J. E.	COLL 0015
Maier, L.	INOR	0387		PHYS	0091		COLL 0016
Maillard, M. C.	MEDI	0115	Mannaperuma, J.				POLY 0283
Maire, G.	COLL	0128		AGFD	0077	Mark, M.	SCHB 0005
Maistriau, L.	PETR	0019	Manners, I.	PMSE	0002	Markell, C.	AGFD 0175
Maiya, B. G.	ORGN	0139		POLY	0186	Markham, G. D.	
Majcherczyk, A.			Manning, W. B.				BIOL 0008
	BIOT	0041		ENVR	0032	Markham, J. R.	
Majeed, A. I.	PETR	0123	Manor, D.	BIOT	0080		PHYS 0149
Majid, A.	I&EC	0021	Mansfield, F. B.			Markovich, R. P.	
	I&EC	0022		BIOT	0097		PMSE 0048
Majkrzak, C. F.			Mansker, L.	INOR	0157	Marks, C. M.	FUEL 0037
	POLY	0090	Mansour, M.	PETR	0079	Marks, T. J.	BIOT 0178
Major, J. S.	MEDI	0102	Mansour, M. N.				INOR 0176
Major, L. J.	AGRO	0030		FUEL	0118		POLY 0002
Majors, P.	BIOT	0096	Mantlo, N. B.	MEDI	0103		POLY 0108
Majors, P. D.	PETR	0068	Manzer, L. E.	I&EC	0044		POLY 0127
Mak, T.C.W.	ORGN	0181	Mao, B.	PHYS	0021		POLY 0217
Makarovsky, Y.			Mao, Z.	INOR	0240	Markus, B.	MEDI 0033
	FUEL	0020	Mao, Z. P.	ORGN	0262	Markuszewski, R.	
Makela, O.	CARB	0018	Mapes, J. P.	ENVR	0032		FUEL 0055
Makhija, S.	POLY	0248		ENVR	0033	Marlow, B. J.	COLL 0200
Makhija, S. M.	PMSE	0147	Marambio–Dennett, E.			Marosan, C.	MEDI 0050
Makinen, M. W.				CHED	0098	Marquet, A.	MEDI 0009
	BIOL	0011		CHED	0165	Marquez, M.	ORGN 0158
Makowski, L.	BIOL	0146		CHED	0166	Marquez, V. E.	MEDI 0023
Makund, S.	INOR	0099		CHED	0168		MEDI 0048
Malabarba, A.	MEDI	0059		CHED	0170	Marquez V., M. S.	
Maldas, D.	PMSE	0112	Marand, H.	COLL	0058		ANYL 0066
Maldonado, L. A.				POLY	0089		CHED 0163
	ORGN	0176		POLY	0097		CHED 0172
	ORGN	0228		POLY	0233	Marr, T. G.	CHAL 0008
Maldonado M., A.				POLY	0246	Marrelli, J. D.	I&EC 0059
	ANYL	0055		POLY	0345	Marsh, G.	PETR 0151
Maldonado M., C.				POLY	0346	Marshall, C.	COMP 0048
	COLL	0125	Maraqah, H.	INOR	0056	Marshall, J. A.	ORGN 0011
Malek, M. A.	AGRO	0036	Marasco, C. J., Jr.			Marshall, L.	MEDI 0179
Malhotra, R.	FUEL	0052		MEDI	0073	Marshall, M. R.	
	FUEL	0062	Marchal, P.	BIOL	0052		AGFD 0028
Malhotra, V. M.			Marchegiano, J. E.			Marshall, P.	PHYS 0090
	FUEL	0001		POLY	0175	Marshall, W. E.	
Malik, A.	ORGN	0071	Marchesi, J. T.	PMSE	0124		AGFD 0140
Malik, A. A.	POLY	0028	Marchessault, R. H.			Marshman, R. W.	
Malisch, R.	AGFD	0062		CARB	0049		INOR 0409
Maliski, E. G.	CINF	0024		POLY	0041		ORGN 0219
Mallen, E. F.	ANYL	0104	Marcilly, C.	PETR	0103	Marsick, D. J.	CHAS 0040
Malley, M.	MEDI	0188		PETR	0132	Marth, C. F.	ORGN 0278
Mallouk, T. E.	CHED	0031		PETR	0142	Martin, A. M.	PETR 0114
Malloy, T. A.	ENVR	0018	Marcopulos, N.	MEDI	0095	Martin, A. R.	ORGN 0254
Malo, J. M.	PHYS	0257	Marcott, C.	COLL	0033	Martin, B. A.	BIOT 0067
Maloletneva, O. Y.				COLL	0035	Martin, C.	BIOT 0167
	BIOT	0197		PHYS	0117	Martin, D.	CARB 0003
Malone, M. F.	PMSE	0133		POLY	0303	Martin, D. L.	MEDI 0120
Maloof, S. R.	ENVR	0021	Marder, S.	POLY	0126		MEDI 0197
Maltby, P. A.	INOR	0380	Marder, S. R.	BIOT	0173	Martin, E. J.	AGRO 0140
Maltesh, C.	POLY	0270		INOR	0303		ANYL 0068
Mamantov, G.	ANYL	0065		INOR	0335	Martin, G. C.	PMSE 0152
	INOR	0064	Mardones, M.	INOR	0292	Martin, G. J.	AGFD 0197
Manceron, L.	PHYS	0074				Martin, J. C.	INOR 0344

Martin, J. G.	MEDI 0176	Mathieu, G.	GEOC 0014	Mehandru, S. P.			
Martin, L. R.	PHYS 0177	Mathieu-Pelta, I.			FUEL 0036		
Martin, M. B.	MEDI 0176		INOR 0377	Mehta, S.	CARB 0035		
Martin, M.H.E.			ORGN 0051	Mehta, V. D.	MEDI 0041		
	POLY 0288	Mathis, C. A.	ORGN 0078	Mei, H. L.	POLY 0136		
Martin, R. L.	PHYS 0191	Mathison, S.	CHED 0210	Meier, G. A.	AGRO 0073		
Martin, Y. C.	COMP 0049	Mathur, V. K.	I&EC 0023	Meijer, G.	PHYS 0070		
	MEDI 0124	Matrai, P. A.	GEOC 0002	Meijer, P.H.E.	BIOT 0252		
	MEDI 0125	Matsuda, K.	MEDI 0109	Meijer, R.	FUEL 0009		
Martina, S.	POLY 0036	Matsuda, T.	POLY 0355		FUEL 0031		
Martinelli, M. J.		Matsuhisa, A.	MEDI 0099	Meijer, R. J.	NUCL 0039		
	ORGN 0236	Matsui, H.	CATL 0050	Meikle, R. W.	AGRO 0046		
Martinez, A. P.	I&EC 0014		FUEL 0064	Meindl, G. L.	AGRO 0072		
Martinez, P.	BIOT 0244	Matsui, T.	MEDI 0050		AGRO 0073		
	ORGN 0193	Matsumoto, K.	ORGN 0072	Meisel, D.	COLL 0131		
Martinez, R.	ORGN 0019	Matsumoto, S.	AGRO 0071		COLL 0152		
	ORGN 0150	Matsumoto, Y.	MEDI 0099	Meislich, H.	HIST 0028		
	ORGN 0198	Matsumura, F.	AGRO 0126	Meldal, M.	CARB 0014		
	ORGN 0245	Matsunaga, H.	AGRO 0024	Meli, A.	INOR 0210		
Martinez, R. D.	ENVR 0024	Matsuoka, A.	PETR 0062	Melius, C. F.	FUEL 0078		
Martinez, T.	NUCL 0071	Matsushita, H.	ORGN 0072	Mellon, E. K.	CHED 0019		
Martinez-Magadan, J. M.		Mattern, G. C.	AGRO 0060	Melo, V.	AGFD 0205		
	PHYS 0243	Matthews, C. R.		Memoli, K.	MEDI 0067		
	PHYS 0282		BIOT 0112	Menasveta, M. J.			
Martinez V., A.	PHYS 0257	Matthews, D. A.			PMSE 0105		
Martin-Rovet, D.			MEDI 0186	Mendelsohn, L. G.			
	NUCL 0068	Mattice, J. D.	AGRO 0146		MEDI 0066		
Martorell, J.	COLL 0087	Mattice, W. L.	POLY 0069	Mendenhall, M. H.			
	PHYS 0062		POLY 0070		NUCL 0064		
	PHYS 0283		POLY 0365	Mendez, J. M.	ORGN 0228		
Martseniuk-Kucharuck, A. P.		Mattogno, G.	PETR 0107	Mendez S., J. M.			
	COLL 0101	Mattoussi, H.	POLY 0274		ORGN 0196		
Martynova, G. B.		Matula, T. I.	AGFD 0116		ORGN 0197		
	PETR 0135	Mauldin, S. C.	ORGN 0027	Menge, W.M.P.B.			
Martynski, T.	BIOT 0102	Mauritz, K. A.	POLY 0065		INOR 0191		
Marvin, C. H.	AGRO 0041		POLY 0066	Menon, A. N.	AGFD 0211		
Maryanoff, B. E.			POLY 0067	Menon, S.	BIOL 0058		
	ORGN 0296		POLY 0068	Menon, V. C.	I&EC 0020		
Maryanoff, C. A.			POLY 0285	Mercando, L. A.			
	ORGN 0242	Mauzerall, D.	BIOL 0136		INOR 0072		
	ORGN 0296		COLL 0088	Mercer, G. E.	FUEL 0045		
Marzec, A. J.	INOR 0243	Mauzerall, D. C.		Mercer-Smith, J. A.			
Marzinsky, C. N.			BIOL 0098		BIOT 0168		
	PMSE 0111	Maverick, A. W.		Meredith, G. R.			
Marzke, R. F.	PHYS 0260		INOR 0242		POLY 0003		
Marzoni, G. P.	MEDI 0186		INOR 0343	Mereen, E.	PETR 0046		
Mascarello, S. W.		May, R. F.	AGRO 0088	Mergens, W. J.	AGFD 0091		
	MEDI 0075	May, V.	BIOT 0027	Merkel, L.	MEDI 0111		
Maschhoff, B. L.		Maya, L.	POLY 0315	Merkel, L. A.	MEDI 0113		
	COLL 0172	Maycock, A.	BIOL 0049	Merkle, P. B.	GEOC 0004		
Mascho, J.	ANYL 0114		BIOL 0050	Merola, J. S.	INOR 0264		
Masel, R. I.	INOR 0123	Mayenez, C.	PETR 0077	Merriam, C. N.	I&EC 0038		
Mashima, K.	ORGN 0044	Mayer, J.	CHED 0131	Merrifield, J. H.			
Masiakowski, J. T.			COLL 0177		CHED 0063		
	PHYS 0284	Mayer, J. M.	INOR 0115	Merrill, E.	POLY 0272		
Masjedizadeh, M. R.			INOR 0132	Merrouche, A.	PETR 0075		
	ORGN 0146		INOR 0271	Meruelo, D.	AGFD 0188		
Mason, C. H.	AGRO 0136	Mayer, J. W.	NUCL 0081	Meshri, D.	SCHB 0016		
Mason, M. R.	INOR 0333	Mayer, R.	ENVR 0016	Messer, W. S., Jr.			
Mason, N. R.	MEDI 0066	Mayet C., L.	INOR 0026		MEDI 0070		
Mason, R. P.	MEDI 0041	Mayr, A.	INOR 0150		MEDI 0139		
Mason, R. S.	PETR 0126		INOR 0188	Messerle, L.	INOR 0108		
Mason, T. L.	POLY 0302	Mazumder, A.	BIOL 0158	Messersmith, P. B.			
Mass, J. H.	CMEC 0028	Mazur, Y.	AGFD 0188		POLY 0289		
Mass, J. L.	INOR 0103	Mazzaccaro, R. J.		Messier, J.	POLY 0009		
Massey, J. H.	AGRO 0127		ORGN 0263	Messner, R.	BIOT 0183		
Massiani, P.	PETR 0040	Mazzocchia, C.	CATL 0058	Metcalf, B. W.	MEDI 0008		
Massucci, M. A.		Mazzocchin, G. A.		Meth, J. S.	POLY 0205		
	PETR 0107		COLL 0005		POLY 0213		
Masters, A. F.	PETR 0007	Mead, D. A.	ANYL 0033		POLY 0221		
Masters, K. R.	PETR 0084	Meadows, K. A.			POLY 0226		
Masulaitis, A.	CHED 0038		INOR 0315	Mett, H.	MEDI 0184		
Mata, P.	COMP 0048	Meadows, R. P.		Metzger, R. M.	BIOT 0005		
Mata, R.	MEDI 0034	Meas, Y.	INOR 0378	Meunier, B.	INOR 0187		
Matchett, M. A.			NUCL 0075	Meyer, G. C.	PMSE 0001		
	INOR 0102		PHYS 0218		PMSE 0021		
Mate, C. M.	PHYS 0153		PHYS 0285	Meyer, M.	POLY 0357		
Matheson, R. R., Jr.		Mecklenburg, S. L.		Meyer, M. D.	MEDI 0124		
	POLY 0059		INOR 0282	Meyer, S.	POLY 0179		
	POLY 0060	Medforth, C. J.	INOR 0091	Meyer, S. J.	POLY 0107		
Mathew, R. M.	MEDI 0025	Medina, J. C.	COLL 0020	Meyer, T. J.	INOR 0282		
	MEDI 0026	Meeks, E.	PHYS 0357	Meyer, T. Y.	INOR 0108		
Mathias, L. J.	POLY 0323			Meyer, W. R.	POLY 0064		

Meyers, S. A.	COLL 0023		Mills, J. W.	CHED 0148		Mole, M. L.	AGRO 0021
Mezey, P. G.	COMP 0042		Mills, N. S.	CHED 0151		Mollison, K. W.	
Michael, J. V.	FUEL 0123		Mills, O. E.	AGFD 0098			MEDI 0167
Michaelides, M. R.			Mills, P. A.	CHED 0143			MEDI 0168
	MEDI 0125		Mills, R. L.	PHYS 0286		Molnar, A.	PMSE 0116
	MEDI 0127		Mills, S.	BIOL 0050		Monasterios, C. J.	
Michaels, G. S.	MEDI 0114		Mills, S. L.	MEDI 0131			POLY 0041
Michalczyk, M. J.			Milne, C.	POLY 0237		Monath, T. P.	AGFD 0165
	INOR 0206		Milne, G.W.A.	MEDI 0024		Monge, A.	INOR 0034
Michel, R. G.	ANYL 0078		Milne, T. A.	ENVR 0035			INOR 0035
Michelman, R. I.			Mil'shtein, S.	POLY 0074			INOR 0036
	INOR 0364		Milstein, O.	BIOT 0041			INOR 0065
Michl, J.	BIOT 0200		Miltz, D. J., Jr.	FUEL 0058		Monge, J.	SCHB 0025
	ORGN 0124		Minachev, K. M.			Monim, S. S.	COLL 0114
	POLY 0344			PETR 0067		Monnerie, L.	PMSE 0133
Micic, O.	COLL 0151		Minami, T.	ORGN 0102		Monque, R.	PETR 0044
Micich, T. J.	AGFD 0135		Minear, R. A.	ENVR 0027			PETR 0077
Middaugh, C. R.			Minero, C.	COLL 0001		Monsalud, L., Jr.	
	BIOT 0134			INOR 0318			BIOT 0149
Middlecamp, C. H.			Minsky, L.	PROF 0001		Montalvo, E. M.	
	ANYL 0009		Minton, L.	MEDI 0108			BIOT 0162
Midkiff, C. R., Jr.				MEDI 0110		Montalvo V., I.	AGFD 0210
	HIST 0007		Minton, M. A.	ORGN 0292		Monte, W. T.	AGRO 0046
Miebach, T.	ORGN 0147		Miranda, J.	NUCL 0080			AGRO 0047
Mielczarski, E.	CATL 0030		Miranda, N. R.	POLY 0120		Montgomery, J. A.	
Miessler, G. L.	CHED 0045		Miranda, R.	ORGN 0245			MEDI 0005
Mihalic, Z.	COMP 0025		Miranda R., R.	PHYS 0213			MEDI 0028
Mihelich, E. D.	BIOL 0072		Mirkin, C. A.	INOR 0027		Montgomery, J. A., Jr.	
Mijovic, J.	POLY 0160		Miskowski, T. A.				PHYS 0287
	POLY 0161			MEDI 0175		Montgomery, M. E.	
Mikita, M. A.	CHED 0186		Misra, R. N.	MEDI 0160			ANYL 0025
Mikula, R. J.	ENVR 0005		Mitchell, E. M.	CINF 0021		Montiel M., C.	I&EC 0072
Mikulski, C. M.			Mitchell, G. D.	FUEL 0002			I&EC 0074
	INOR 0073		Mitchell, G. R.	POLY 0206		Moock, T. E.	CINF 0022
Milder, S.	INOR 0038			POLY 0219			CINF 0025
Miley, H. S.	NUCL 0013		Mitchell, J. M.	AGFD 0069		Moog, R. S.	CHED 0203
Millar, D. M.	PETR 0110		Mitchell, P. J.	COLL 0185		Moon, J. B.	COMP 0050
Millar, J. M.	PHYS 0373		Mitchell, R. E.	FUEL 0026		Moon, M. W.	MEDI 0126
Millar, M.	INOR 0059		Mitchell, S. A.	PHYS 0198		Moon, R.	ORGN 0085
	INOR 0060		Mitchison, T. J.			Moore, A. W.	ANYL 0126
	INOR 0179			BIOL 0139		Moore, C.	ANYL 0050
Millequant, M.	PMSE 0066		Mitra, S.	POLY 0245		Moore, C. B.	INOR 0167
Miller, C. A.	AGRO 0008		Mitra, S. B.	POLY 0262			PHYS 0164
Miller, D. W.	ORGN 0214		Mitraki, A.	BIOT 0113		Moore, D. S.	ANYL 0002
	ORGN 0215		Mittelman, M. W.			Moore, E. A.	CHED 0080
Miller, E. L.	TECH 0004			BIOT 0099		Moore, G. J.	POLY 0029
Miller, G. C.	ENVR 0060		Mittelstaedt, W.			Moore, H. W.	ORGN 0229
Miller, J. A.	AGFD 0159			AGRO 0025		Moore, J.	POLY 0126
	FUEL 0078		Miwa, T.	ORGN 0113		Moore, J. A.	POLY 0137
Miller, J. B.	CATL 0022		Miyake, M.	AGFD 0050		Moore, J. L.	MEDI 0176
Miller, J. C.	PHYS 0251		Miyata, S.	POLY 0128		Moore, J. S.	POLY 0035
Miller, J. E.	COLL 0067		Mize, P. D.	ORGN 0023			POLY 0321
Miller, J. H.	FUEL 0109		Miziorko, H.	BIOL 0064			POLY 0329
Miller, J. R.	PHYS 0077			BIOL 0153		Moore, J. W.	ANYL 0009
Miller, J. S.	INOR 0100		Mizumoto, H.	MEDI 0112			CHED 0073
Miller, L.	BIOT 0237		Mlachowski, M. R.				CHED 0080
Miller, L. L.	BIOT 0201			INOR 0208			CHED 0138
Miller, M. J.	ORGN 0028		Moats, R. A.	INOR 0329		Moore, K. W.	MEDI 0146
	ORGN 0029		Moats, W. A.	AGFD 0125		Moore, P.	FUEL 0140
Miller, R. C., Jr.			Mochizuki, M.	POLY 0082		Moore, R.	CHED 0041
	BIOT 0207		Mody, T. D.	ORGN 0139		Moore, R. J.	CHED 0082
Miller, R. D.	POLY 0145		Moe, M. K.	NUCL 0012			CHED 0088
	POLY 0188		Moeller, M.	PMSE 0184			PHYS 0300
Miller, R. E.	CATL 0001		Moerner, W. E.	POLY C155		Mootoo, D. R.	CHED 0143
	INOR 0077		Moghaddas, S.	ANYL 0063		Moo-Young, M.	
Miller, R.J.D.	PHYS 0133			INOR 0182			BIOT 0038
	PHYS 0169		Mohammadi, N.			Mopper, K.	GEOC 0017
Miller, S.P.F.	BIOL 0079			PMSE 0073		Moral, E.	AGFD 0123
Miller, T. M.	CINF 0004		Mohammed, A. K.			Morales, A.	ANYL 0073
	POLY 0236			INOR 0343			ANYL 0075
	POLY 0320		Mohan, A.	ENVR 0028		Morales, J.	PHYS 0289
	POLY 0324		Mohanty, D. K.			Morales, M. R.	I&EC 0011
Miller, W. H.	CHED 0069			PMSE 0122			I&EC 0012
	PHYS 0161		Mohd Azemi, B.M.N.			Morales, R.	NUCL 0080
Miller, W. J.	ANYL 0021			BIOT 0152		Morales, U.	PHYS 0285
Miller Jr., R. C.			Mohite, S. S.	POLY 0086		Morales-Rios, M. S.	
	BIOT 0217		Mohlmann, G. R.				ORGN 0268
Millero, F. J.	GEOC 0005			POLY 0177		Morales Rosas, J. M.	
Millman, W. S.	COLL 0212		Mohr, D.	POLY 0316			ANYL 0058
Mills, A.	COLL 0134		Mohr, D. L.	POLY 0338		Morales-Rosas, J. M.	
	COLL 0150		Moinelo, S. R.	FUEL 0007			I&EC 0014
Mills, D. K.	INOR 0177		Mojica G., E.	COLL 0120		Morand, C.	NUCL 0039
Mills, I. M.	PHYS 0029		Molaire, T. R.	POLY 0181		Morawetz, R.	BIOT 0183

Morawitz, H.	POLY	0355	Moskovits, M.	COLL	0092	Muscate, A.	BIOL 0081
Morawski, J.	AGFD	0178		PHYS	0006	Mushrush, G. W.	
	AGRO	0147	Moskwa, A.	COLL	0085		ENVR 0023
Morcol, T.	BIOT	0060	Motsegood, K.	FUEL	0074	Musick, T. J.	AGRO 0038
Mordenti, J.	BIOT	0059	Moudder, A. H.			Musser, J.	MEDI 0179
More, K. L.	PHYS	0119		PHYS	0010	Mustafa, M.	COLL 0014
Morel, F.M.M.	ENVR	0057	Moulijn, J. A.	CATL	0044	Mustafi, D.	BIOL 0011
	INOR	0003		FUEL	0009		BIOL 0045
Moreland, S.	MEDI	0188		FUEL	0015	Mus–Veteau, I.	INOR 0020
	ORGN	0114		FUEL	0031	Muth, W. L.	BIOT 0085
Morell, A. G.	BIOL	0154	Moulton, J.	POLY	0200	Mutharasan, R.	BIOT 0162
Moreno, D. A.	COLL	0154	Moumi, C.	MEDI	0187	Muthukumar, M.	
Moreno, H.	CATL	0041	Mounts, T. L.	ANYL	0134		PMSE 0104
	FUEL	0100	Mourey, T. H.	PMSE	0065	Mutovkin, P. A.	
Moreno–Esparza, R.			Mouritsen, S.	CARB	0014		COLL 0097
	CHED	0057	Movassaghi, S.	ENVR	0033		INOR 0237
	INOR	0026	Mowlem, J. K.	COLL	0216	Mutter, M. S.	ORGN 0296
	INOR	0037	Moy, J.	INOR	0059	Myers, A. B.	PHYS 0302
	INOR	0042		INOR	0179	Myers, B. M.	AGRO 0146
	INOR	0057	Moye, H. A.	AGRO	0147	Myers, L. K.	INOR 0334
	INOR	0061	Moylan, C. R.	POLY	0106	Myers, M. N.	PMSE 0008
	INOR	0260		POLY	0194	Myers, M. R.	MEDI 0113
	INOR	0399	Mrozik, H.	ORGN	0016	Myers, R.	BIOL 0024
Moreno–Tellez, A.			Muckerman, J. T.			Myers, W. A.	NUCL 0082
	INOR	0076		PHYS	0263		NUCL 0083
Moresoli, C.	BIOT	0195	Mudd, R. L.	ENVR	0032	Myers, W. H.	INOR 0114
Moretto, L. G.	NUCL	0005	Muehlenbachs, K.			Mylchreest, I.	ANYL 0141
Morgado Cureno, D. M.				FUEL	0070	Mylvaganam, M.	
	CHED	0105	Mueller, D. D.	BIOL	0006		INOR 0362
Morgan, B. A.	MEDI	0195	Mueller, K. T.	PHYS	0370	Myrick, C.	AGRO 0142
Morgan, J. J.	INOR	0009	Mueller, P. H.	CHED	0147	Naciri, J.	COLL 0071
Morgan, K. R.	ANYL	0064	Mueller, R. R.	PMSE	0111	Nadasdi, T. T.	INOR 0025
Morgan, R. J.	INOR	0232	Mugrage, B. B.	ORGN	0013	Nadeau, R. G.	AGRO 0023
	INOR	0233	Mukamel, S.	PHYS	0038	Nadler, M. P.	POLY 0268
Morgan, R. W.	BIOT	0245	Mukasa, K.	BIOT	0123	Nadler, T.	ANYL 0028
Morgan, S. C.	ORGN	0163	Mukerjee, A.	ORGN	0239	Nae, H. N.	PMSE 0034
Morgan, T. K., Jr.			Mulcahy, L. J.	BIOL	0108	Nafie, L. A.	BIOT 0026
	MEDI	0094	Mulder, B.	PHYS	0303	Nagahara, H.	I&EC 0017
Mori, H.	AGFD	0090	Muller, D.	PETR	0093		PETR 0104
Morimoto, K.	AGRO	0049	Muller, H–J.	INOR	0141	Naganune, T.	BIOT 0121
Morin, A.	BIOT	0195	Muller, J. G.	INOR	0066	Nagase, H.	MEDI 0154
Morin, F. G.	POLY	0041		INOR	0067	Nagata, T.	AGFD 0127
Morisato, A.	POLY	0120		INOR	0068	Nagy, E.	PETR 0116
Morishima, T.	ORGN	0058	Muller, V.	CARB	0003	Nagy, J. O.	CARB 0034
	ORGN	0139	Muller, W.	ANYL	0042	Nagy, S.	AGFD 0167
Mork, C. O.	POLY	0083	Muller–Carrera, G.			Nahlovsky, B. D.	
Morley, K.	NUCL	0036		INOR	0263		POLY 0249
Morman, K. N., Jr.			Mullin, C. A.	AGRO	0136	Naidu, M. V.	AGRO 0029
	POLY	0167	Mullin, J.	CHED	0205	Naiini, A. A.	INOR 0191
Morosoli, R.	BIOT	0213	Mullins, D. E.	AGRO	0123	Nail, J. W.	COLL 0067
Morreal, C. E.	MEDI	0050	Mullis, J. O.	CHED	0154	Naim, M.	AGFD 0167
Morris, J. K.	MEDI	0126	Mulvaney, P.	COLL	0130	Nair, M. G.	MEDI 0185
Morris, R. H.	INOR	0380		COLL	0152	Naismith, R. W.	
Morris, S. E.	COLL	0167	Mumtaz, M. M.				MEDI 0175
Morrison, J. C.	BIOL	0069		ENVR	0073	Najafi, A.	NUCL 0076
Morrison, P. W., Jr.			Munavalli, S.	ORGN	0251	Najjar, M. S.	FUEL 0139
	PHYS	0149		ORGN	0252	Nakadaira, Y.	ORGN 0212
Morrison, R. A.	MEDI	0196		ORGN	0253	Nakagawa, Y.	BIOL 0045
Morrison, R. C.	ORGN	0225	Mundi, L.	CATL	0019		PETR 0036
	ORGN	0226	Munir, M.	CARB	0005	Nakai, S.	AGFD 0108
Morrison, R. W.			Munns, R. K.	AGFD	0101	Nakajima, S.	ORGN 0249
	ORGN	0124	Munoz, H.	ORGN	0103	Nakamura, A.	AGRO 0009
	POLY	0344	Munster, M.	BIOT	0164	Nakamura, M.	INOR 0296
Morrissey, D. J.				BIOT	0165	Nakamura, Y.	AGRO 0050
	NUCL	0046	Munukutla, S.	POLY	0111	Nakanishi, H.	BIOT 0124
Morrissey, J. J.	AGRO	0100	Mura, A.	MEDI	0087	Nakanishi, K.	AGRO 0006
Morrow, C. J.	BIOT	0094		MEDI	0172		HIST 0024
Morrow, J. R.	INOR	0406	Murai, A.	ORGN	0014		INOR 0096
Morse, J. G.	CHED	0068	Murai, S.	AGRO	0050	Nakano, R.	BIOT 0172
Morse, K. W.	CHED	0066	Murai, T.	ORGN	0139	Nakashima, N.	BIOT 0104
Morss, L. R.	NUCL	0068	Murata, K.	POLY	0023	Nakata, S.	COLL 0115
Morton, G. M.	TECH	0014	Murdoch, J.	BIOT	0173	Nakatani, A. I.	PMSE 0107
Morton, K. C.	CHED	0214	Murphy, C. K.	ORGN	0293		PMSE 0134
	PHYS	0290	Murphy, D. W.	CHED	0030	Nakatani, N.	AGFD 0016
Morton, L. A.	PETR	0012		INOR	0354	Nakatsuka, I.	AGRO 0024
Mosandl, A.	AGFD	0183	Murphy, E.	ORGN	0149	Nakazawa, H.	AGFD 0102
Mose, D. G.	ENVR	0023	Murphy, V. G.	BIOT	0250	Namdev, P. K.	BIOT 0129
Moseley, A. M.	MEDI	0146	Murray, H. H.	INOR	0372	Namiki, M.	AGFD 0018
Mosely, M. A.	ANYL	0125	Murray, R. E.	I&EC	0054	Nantermet, P. G.	
Moser, C. C.	BIOL	0094	Murray, W. T.	INOR	0377		ORGN 0002
Moser, P.	COLL	0179	Murthy, G. S.	ORGN	0124	Nappier, J. M.	AGRO 0108
Moser, W. R.	CATL	0026		POLY	0344	Nara, P. L.	AGFD 0190
Moses, D.	POLY	0008	Murthy, N. S.	BIOT	0203	Narang, S.	BIOL 0143

Narasimhan, N. I.	AGRO 0128	Newsam, J.	INOR 0058	Nonaka, G.	AGFD 0166
Narayanan, C. S.		Newton, M. G.	INOR 0094	Nonaka, H.	MEDI 0112
	AGFD 0211	Newton, M. W.	NUCL 0081	Nonokuchi, M.	ENVR 0049
Narayanan, V. L.		Newton, P.	ANYL 0114	Nordquist, K.	YCC 0007
	AGFD 0164	Newton, R.	ORGN 0005	Norell, J. R.	CARB 0006
Narkis, M.	PMSE 0062	Ng Cheong Chan, Y.		Noriega, J.	PHYS 0293
	PMSE 0063		POLY 0207		PHYS 0294
Narula, C. K.	POLY 0230	Nguyen, C.	POLY 0106	Norman, A. D.	INOR 0391
Nashed, E.	CARB 0019	Nguyen, D. C.	ANYL 0004	Norman, E. B.	NUCL 0003
Nasr, M.	AGFD 0163	Nguyen, D. T.	FERT 0001	Normandin, D.	MEDI 0188
Nastasi, M.	NUCL 0093		FERT 0003	Norris, W. P.	POLY 0268
Nastro, A.	PETR 0020		FERT 0011	Nortario, V.	AGFD 0117
Natansohn, A.	PMSE 0076	Nguyen, H. P.	POLY 0123	Norton, J.	BIOL 0022
Natchimuthu, N.		Nguyen, S. N.	INOR 0236	Norton, M. N.	FERT 0011
	PMSE 0156	Nguyen, T. Q.	POLY 0281	Norton, P. R.	INOR 0142
Natowitz, J. B.	NUCL 0028	Ni, C-K.	PHYS 0291	Norwood, R. A.	POLY 0179
Naulin, C.	PHYS 0089	Nicasio, C.	INOR 0035		POLY 0223
Nava E., N.	NUCL 0069	Nichols, D. E.	FERT 0001	Norwood, V.	FERT 0022
Nava–Lara, G.	CHED 0165		FERT 0003	Nosal, R.	MEDI 0069
Navaneeth Rao, T.			FERT 0011		MEDI 0187
	COLL 0081	Nichols, J.	PHYS 0056		ORGN 0104
Navarro O., A.	ORGN 0185	Nicklaus, M. C.		Nose, T.	PMSE 0103
Navia, M.	BIOL 0049		MEDI 0024	Nosker, T. J.	I&EC 0029
Nawamaki, T.	AGRO 0049	Nicolaides, C. P.		Nossal, N. G.	BIOL 0159
Nayak, S. K.	INOR 0253		PETR 0008	Nottke, J. E.	POLY 0203
Neamati, N.	CARB 0032	Nicolas, T. E.	ORGN 0101	Novak, B. M.	PMSE 0071
Nebbia, G.	NUCL 0028	Nicolis, N. G.	NUCL 0027		POLY 0024
Nedez, C.	PETR 0131	Nicolle, M. A.	PETR 0038		POLY 0259
Needham, L. L.		Nicoud, J. F.	POLY 0103		POLY 0347
	CEI 0002	Niedbala, H.	ORGN 0189	Novak, D.	PHYS 0361
	CEI 0007	Niederauer, M. Q.		Novak, M.	ORGN 0194
Needhan, M. D.			BIOT 0019	Novak, T. J.	ANYL 0084
	ANYL 0099	Niedome, Y.	BIOT 0145	Novaro, O.	PHYS 0243
Neels, K.	CMEC 0011	Nielsen, J.	BIOT 0109		PHYS 0282
Neenan, T. X.	INOR 0312		MEDI 0143	Novelli, A. C.	COLL 0024
	POLY 0236	Nierlich, F.	PETR 0009	Novelli, M. F.	PETR 0120
	POLY 0320	Nies, D. E.	MEDI 0177	Novick, S. G.	INOR 0372
	POLY 0324	Nieva, D.	I&EC 0013	Novoa, J. J.	PHYS 0295
Negron–Medoza, A.			I&EC 0015	Novotny, M.	ANYL 0019
	NUCL 0072	Nihei, E.	POLY 0174	Novotny, V. J.	PHYS 0153
Negron–Mendoza, A.		Nijhuis, S.	POLY 0178	Nowakowski, M. A.	
	NUCL 0070	Nikodijevic, O.	MEDI 0115		AGRO 0079
Negulescu, I. I.	POLY 0099	Nikolic, S.	COMP 0025	Noyes, R. T.	AGRO 0042
Neil, D. J.	CHAS 0027	Nikolov, Z. L.	BIOT 0047		AGRO 0102
Neilsen, W. D.	INOR 0329	Nilsson, U.	CARB 0056	Nozaki, K.	INOR 0164
Neilson, R. H.	INOR 0325	Nimlos, M. R.	COLL 0082	Nunan, J. G.	COLL 0208
	POLY 0184		ENVR 0035		PETR 0035
Nelson, D. L.	ORGN 0254	Nimmons, H. L.		Nunez, O.	ORGN 0187
Nelson, J.	MEDI 0179		ORGN 0045	Nurmia, M.	NUCL 0074
Nelson, J. W.	NUCL 0079	Nimura, N.	ANYL 0100	Nuzzo, R. G.	INOR 0124
Nelson, K. A.	CHED 0128	Ning, B.M.H.	COLL 0047	Oakley, E. O.	I&EC 0032
	PHYS 0066	Nishi, T.	PMSE 0099	Oballa, M. C.	CATL 0048
Nelson, M. A.	NUCL 0012	Nishikawa, O.	PHYS 0097	Obaya V., A.	CHED 0099
Nemeth, G.	BIOL 0044	Nishimichi, C.	PETR 0029	Obeng, Y. S.	ANYL 0108
Nemirovski, N.	PMSE 0063	Nishimura, C.	POLY 0301	Ober, C. K.	POLY 0110
Nenoff, T. M.	PETR 0055	Nishiyama, K.	BIOT 0146		POLY 0288
Neo, S-P.	INOR 0106	Nissan, R. A.	INOR 0229	Oberacker, D. A.	
Nero, V. P.	PETR 0148		POLY 0268		AGRO 0141
	PETR 0151	Niu, C.	PHYS 0292	Oberlander, M. D.	
Ness, R. O., Jr.	FUEL 0091	Nivens, D. E.	BIOT 0099		PHYS 0350
Nestor, J. J., Jr.		Niwas, S.	MEDI 0028	O'Brien, A. T.	HIST 0034
	MEDI 0193	Nixon, A. C.	PROF 0012	O'Brien, R. W.	COLL 0201
Neu, T.	ORGN 0097	Nnodimele, R.	COLL 0002	Ocampo, A.	ANYL 0113
Neuberger, N. A.		Noble, M. J.	COLL 0153	Occelli, M. L.	CATL 0012
	POLY 0070	Noble, R.	COLL 0082		PETR 0042
Neucere, J. N.	AGFD 0134	Nobutoki, H.	POLY 0224		PETR 0109
Neuenschwander, P.		Nocentini, G.	MEDI 0029		PETR 0145
	POLY 0064	Nocera, D. G.	CATL 0021	Ochoa I., A.	AGFD 0038
Neufeld, P.	CHAL 0002		INOR 0285	Ochterski, J. W.	
Neumark, D. M.		Noda, I.	COLL 0033		PHYS 0297
	PHYS 0059		COLL 0035	O'Conner, C. J.	INOR 0229
Neumeister, J.	PETR 0009		COLL 0036	O'Connor, C. T.	
Neuwirth, M.	ENVR 0005		COLL 0038		PETR 0011
Nevill, C. R.	ORGN 0282		PHYS 0117	O'Connor, K.	BIOT 0046
Newbury, N. R.			POLY 0303	O'Connor, K. J.	
	PHYS 0041	Nolan, L. L.	BIOL 0118		ORGN 0056
Newkome, G. R.		Noland, W. E.	ORGN 0261	O'Connor, K. J., Jr.	
	POLY 0319	Nolen, J.	NUCL 0038		CMEC 0003
Newman, J. P.	ENVR 0012	Noll, B. C.	INOR 0302	O'Connor, P. B.	
Newman, K. E.	PMSE 0150	Nollstadt, K.	MEDI 0182		PHYS 0090
Newman, R. C.	BIOT 0098	Nollstadt, K. M.		O'Dea, J. J.	ANYL 0105
Newmark, H.	AGFD 0118		MEDI 0183	Odian, G.	POLY 0048
		Nomura, S.	POLY 0171	Odier, L. M.	CARB 0029

O'Driscoll, K. F.	PMSE 0095	Orfanopoulos, M.	ORGN 0164	Paderes, G. D.	CINF 0008
Oertli, A. G.	POLY 0064	Orians, K. J.	INOR 0001	Padhye, S. B.	INOR 0268
Oesterhelt, D.	BIOT 0078	Orihashi, Y.	POLY 0127	Padia, J. K.	MEDI 0110
Oevering, H.	POLY 0222	Orlando, S.	CMEC 0026	Padilla, P.	BIOL 0102
Oganessian, Y. T.			I&EC 0005	Padilla-Higareda, F.	ORGN 0017
	NUCL 0017	Orlov, D. S.	PETR 0135	Padmanabhan, R. P.	
Ogawa, T.	PMSE 0130	Orna, M. V.	CHED 0137		POLY 0095
	POLY 0054	Ornstein, R. L.	BIOT 0118	Padwa, A.	ORGN 0276
	POLY 0055	Orpen, A. G.	INOR 0294	Paffett, M. F.	INOR 0144
	POLY 0056	Orta, T.	PMSE 0129	Page, M.	PHYS 0310
	POLY 0092		POLY 0094	Pahklov, E. M.	COLL 0112
Ogletree, M. L.	MEDI 0160	Ortega, A.	PMSE 0086	Pai, S. H.	BIOT 0048
Oh, J. R.	INOR 0220		POLY 0093	Paige, H. L.	I&EC 0008
Oh, K. H.	COLL 0083	Orth, A.	BIOT 0237	Paight, E. S.	INOR 0245
Oh, S. H.	COLL 0185	Orth, R. G.	CEI 0003	Paik, H. J.	POLY 0282
Oh, Y-C.	AGFD 0072	Orthner, C. L.	BIOT 0060	Paine, R. T.	POLY 0312
Oh, Y-I.	MEDI 0151	Ortiz, J. V.	PHYS 0058	Painter, P. C.	I&EC 0019
O'Halloran, T. V.		Ortiz Barrios, E.		Pakbaz, K.	POLY 0008
	INOR 0012		CHED 0103	Paklov, E. M.	COLL 0095
Ohashi, M.	INOR 0360	Ortiz de Montellano, P. R.		Palacios, J.	PMSE 0086
O'Haver, T. C.	CHED 0078		MEDI 0054		POLY 0052
Ohhata, I.	MEDI 0109	Ortiz O., V.	INOR 0026		POLY 0093
Ohishi, H.	POLY 0220	Orton, D. R.	BIOT 0009	Palacios A., J.	CHED 0099
Ohmer, J.	PETR 0091	Osaheni, J. A.	POLY 0011	Palackdharry, P. J.	
Ohshima, T.	AGRO 0013		POLY 0221		PMSE 0140
Ohta, H.	AGRO 0071	Osawa, T.	AGFD 0018	Palazzotto, M. C.	
Ohtsubo, K. O.	ENVR 0047		AGFD 0045		INOR 0320
Ojima, I.	INOR 0226	Osborne, M. P.	AGRO 0150	Palermo, R.	BIOL 0042
	INOR 0227	Osei-Owusu, A.		Paletsky, A. A.	FUEL 0126
	MEDI 0151		PMSE 0152	Palfreymann, N.	
	ORGN 0046	O'Shea, K.	ORGN 0131		COMP 0005
	ORGN 0047	Osman, J.	BIOT 0024	Palm, U.	AGFD 0183
	ORGN 0048	Ososkov, V.	ENVR 0013	Palma G., F. A.	NUCL 0087
	ORGN 0049	Ospina, S.	BIOT 0131	Palmer, C. L.	MEDI 0186
	ORGN 0218	Ossipov, M. H.	MEDI 0148	Palmer, D. A.	ANYL 0008
Okada, A.	POLY 0291	Osteryoung, J. G.		Palmer, J. G.	CHAS 0004
Okada, M.	PMSE 0103		ANYL 0105		CHAS 0048
Okamoto, H.	ENVR 0015		COLL 0167	Palmer, L. P.	AGRO 0090
Okamoto, P. R.	NUCL 0092	Osteryoung, R. A.		Palmer, R. A.	POLY 0304
Okamoto, Y.	POLY 0050		ANYL 0096	Palmer, R. T.	INOR 0207
	POLY 0091	Osuga, D. T.	AGFD 0035	Pamer, T.	CHED 0038
O'Konski, C. T.		O'Sullivan, M. C.		Pan, H.	ANYL 0118
	COLL 0179		BIOL 0065	Pan, J-M.	COLL 0172
Okuda, T.	AGFD 0017	Oth, D.	MEDI 0166		COLL 0191
	AGFD 0046	O'Toole, J. C.	ORGN 0081	Pan, Y.	POLY 0317
Olah, G. A.	ORGN 0145	Ottea, J. A.	AGRO 0151	Panagiotopoulos, A. Z.	
Oldham, A. A.	MEDI 0102	Otter, B. A.	CARB 0046		POLY 0352
Oleksiak, C.	COLL 0080	Ottinger, H. C.	COLL 0159	Panchalingam, V.	
Olinger, J. M.	ANYL 0026	Ottova, A.	BIOT 0102		POLY 0021
Oliva G., M.	CHED 0091	Ou, E.	INOR 0142		POLY 0022
	CHED 0092	Oukaci, R.	COLL 0187	Pandya, B.	PMSE 0006
	CHED 0093	Over, D. E.	INOR 0132		PMSE 0015
Oliver, J. E.	AGRO 0035	Overman, L. E.	ORGN 0007	Paneque, M.	INOR 0035
Oliver, J. P.	INOR 0274		ORGN 0008	Pang, S.	PMSE 0047
Ollendorf, W.	PHYS 0324		ORGN 0275	Pang, Y.	PMSE 0080
Ollis, D. F.	BIOT 172A	Owen, M.	MEDI 0079		POLY 0357
Olsen, R. T.	PROF 0007	Owens, F. H.	CHED 0002	Panigot, M. J.	MEDI 0030
Olson, G.	MEDI 0165	Owens, R. M.	FUEL 0005	Pankiewicz, K. W.	
Olvera L., E.	AGFD 0210	Owens, T. A.	ORGN 0161		CARB 0023
Omar, M. M.	AGFD 0172	Owens, W. H.	CARB 0026		CARB 0024
Omstead, D.	BIOT 0165	Own, Z. Y.	ORGN 0214	Pankow, J. F.	ENVR 0059
Omstead, D. R.	BIOT 0163		ORGN 0215	Pannuri, S.	BIOT 0232
	BIOT 0164	Owrutsky, J.	PHYS 0083	Pansegrau, P. D.	
Omstead, T. R.	COLL 0031		PHYS 0194		BIOL 0054
Ondeyka, D.	MEDI 0103	Oxford, A. W.	MEDI 0079		ORGN 0029
Ondich, G. G.	CEI 0023	Oyama, S. T.	CATL 0059	Papathakis, M. F.	
Ondrias, M.	BIOL 0096	O'Young, C. L.	PETR 0056		AGRO 0145
Ondrias, M. R.	BIOL 0097		PETR 0136		AGRO 0149
Ong, E.	BIOT 0217	Ozawa, F.	CATL 0006	Papathomas, K. I.	
Onoda, M.	PMSE 0183	Ozin, G. A.	PHYS 0267		POLY 0096
Onuchic, J. N.	BIOL 0133	Ozkan, U. S.	COLL 0207		POLY 0110
Oppenheimer, N. J.		Ozment, J. L.	CHED 0215	Papoulis, A.	ORGN 0063
	BIOL 0013		PHYS 0298	Paquette, L. A.	ORGN 0291
Oppenheimer, S.		Pace, C. N.	BIOT 0189	Parang, K.	ORGN 0200
	BIOT 0120	Pace, J. M.	ORGN 0212	Pardo V., G.	ENVR 0037
Opperman, K. A.		Pacheco M., G.	INOR 0257	Parent, Y. O.	PETR 0050
	INOR 0282	Pacheco-Sanchez, M. M.		Parida, S.	BIOT 0067
Oppermann, H.	BIOT 0057		CHED 0094	Park, C. W.	INOR 0174
Or, Y. S.	MEDI 0168	Pachuta, R. R.	AGFD 0202	Park, J.	PHYS 0299
Oranskaya, O. M.		Pack, R. T.	PHYS 0191	Park, J-B.	INOR 0099
	PETR 0066	Paddon-Row, M. N.		Park, S.	I&EC 0023
Orban, J.	PHYS 0195		PHYS 0076	Park, S. A.	ANYL 0018

Park, Y.	INOR 0351	Patterson, T.	AGFD 0042	Penner, T. L.	POLY 0105
Park, Y. C.	GEOC 0017	Paul, D. R.	PMSE 0072	Penney, C. M.	PHYS 0146
Park, Y-H.	ORGN 0047		POLY 0115	Pepper, D. R.	AGRO 0150
Parker, B. R.	FERT 0011	Paul, G. C.	ORGN 0144	Pepperrell, C. A.	
Parker, C.	COMP 0054	Paul, P. P.	INOR 0044		COMP 0043
Parker, D. H.	NUCL 0045	Paulsen, H.	CARB 0058	Percec, V.	POLY 0299
Parker, E.	AGRO 0074	Paulsen, M. D.	BIOT 0118	Pereda–Miranda, R.	
Parker, J. C.	CHAS 0011	Paulson, G. D.	AGRO 0113		MEDI 0034
Parker, O. J.	CHED 0204	Pautard–Cooper, A.			MEDI 0060
Parker, R. A.	MEDI 0190		INOR 0377	Pereira, C. J.	I&EC 0045
Parkin, G.	INOR 0085	Pauwels, J. F.	FUEL 0081	Peretti, S. W.	BIOT 172A
	INOR 0086	Pavia, D. L.	CHED 0079	Peretz, M.	BIOL 0083
	INOR 0087	Pavlath, A. E.	AGFD 0078	Perez, J.	ORGN 0043
	INOR 0088	Pavlik, P.	COMP 0049	Perez, J. A.	PETR 0044
	INOR 0369	Pavlik, R.	PETR 0031	Perez, J. J.	ORGN 0294
Parkin, I. P.	INOR 0128	Pawlak, W.	FUEL 0042	Perez, J. O.	PETR 0094
Parkin, K. L.	AGFD 0033		FUEL 0141	Perez, M.	PETR 0113
	AGFD 0036	Payne, G. T.	AGRO 0151	Perez, R.	PHYS 0227
Parks, D.	BIOT 0202	Payne, L. D.	AGFD 0129	Perez, S.	PHYS 0252
Parlow, A.	CINF 0010		ANYL 0136	Perez A., J. F.	ANYL 0055
Paroli, R. M.	PMSE 0180	Payne, S.	INOR 0094	Perez–Benitez, A.	
Parr, J.	INOR 0052		ORGN 0085		CHED 0102
Parravano, C.	CHED 0012	Pazik, J. C.	COLL 0135		CHED 0103
Parrott, W. L.	AGRO 0037		INOR 0306	Perez C., A. L.	ORGN 0187
Parry, R. J.	BIOL 0081	Peace, R. W.	AGFD 0209	Perez C., M. G.	CHED 0104
Parry, R. W.	INOR 0288	Pearce, B. C.	MEDI 0190	Perez–Caballero, G.	
Parson, J. M.	PHYS 0350	Pearce, E. M.	CHED 0006		ANYL 0073
Parson, W. W.	BIOL 0130		COLL 0034		ANYL 0074
Parsons, P. J.	C E I 0020		HIST 0033		ANYL 0075
Parsons, W. S.	POLY 0078		PMSE 0147	Perez–Cabellero, G.	
Partch, R.	CHED 0189		PMSE 0163		ANYL 0054
Parthasarathy, S.			POLY 0113	Perez G., C.	COLL 0119
	BIOL 0099		POLY 0192		COLL 0119
Partigianoni, C. M.			POLY 0248		COLL 0120
	INOR 0285	Pearce, R. J.	MEDI 0102		COLL 0121
Partridge, H.	PHYS 0005	Pearlman, R. S.		Perez G., R. M.	COLL 0119
	PHYS 0171		CINF 0020	Perez G., S.	COLL 0120
Pascal, T.	POLY 0133		CINF 0025		COLL 0121
Pascault, J. P.	POLY 0112	Pearson, N. R.	AGRO 0046		ENVR 0037
Paschal, J. W.	ORGN 0033		AGRO 0047	Perez M., V.	AGFD 0105
Paschke, E. E.	PMSE 0100	Pearson, W.	I&EC 0037	Perez–Mendoza, V. M.	
Passaretti, J. D.		Pease, E.	BIOT 0237		ENVR 0046
	TECH 0019	Peat, A. J.	ORGN 0172	Perez S., J. J.	ANYL 0055
Passoni, G.	FUEL 0027	Pechet, M. M.	ORGN 0040		ANYL 0066
Pastelin P., R.	AGFD 0145	Pecoraro, V. L.	INOR 0010		CHED 0163
	AGFD 0145		INOR 0018		CHED 0172
	AGFD 0146	Peden, C.H.F.	COLL 0031	Perez–Soler, R.	CARB 0032
	CHED 0091		INOR 0278	Peri, S. P.	ORGN 0260
	CHED 0092		PETR 0018	Periyasamy, S.	MEDI 0139
	CHED 0093	Pedersen, H.	AGFD 0002	Perkins, E. G.	AGFD 0096
Pasterczyk, J. W.			BIOT 0228	Perkins, M. T.	FUEL 0010
	COLL 0136	Pedersen, J.	BIOT 0056	Perkins, R.	I&EC 0040
Pasternack, L.	FUEL 0097	Pederson, M. R.		Perkovic, M.	INOR 0159
Pasternack, L. R.			PHYS 0359	Perna, P. J.	CARB 0001
	PHYS 0301	Pederson, R. L.	ORGN 0052	Perreault, F.	POLY 0278
Pasternak, A.	ORGN 0043	Pedraza, E.	I&EC 0067	Perrigan, R.	INOR 0250
Pastor, C.	NUCL 0028	Pedraza Acevedo, V.		Perrine, D. M.	ORGN 0202
Patarin, J.	PETR 0075		CHED 0122	Perrott, M. G.	PHYS 0237
Patch, R. J.	MEDI 0025	Peek, B. M.	INOR 0282	Perrotta, A. J.	POLY 0340
	MEDI 0083	Peeples, C. J.	ORGN 0199	Perry, J. W.	BIOT 0173
Patchett, A. A.	MEDI 0103	Peferoen, D.G.R.			BIOT 0176
	MEDI 0104		PETR 0083		INOR 0303
	MEDI 0122	Pei, S.	PHYS 0009	Perry, K. A.	GEOC 0006
	MEDI 0123	Peiffer, D. G.	COLL 0181	Perry, K. J.	INOR 0303
	ORGN 0112		PMSE 0155	Perry, T. A.	PHYS 0178
Patel, B. M.	AGRO 0147	Peilow, A. D.	BIOT 0040	Person, W. B.	PHYS 0033
Patel, D.	MEDI 0086	Peitz, D. J.	INOR 0207	Perumattam, J.	MEDI 0026
Patel, M.	INOR 0022	Pekala, R.	POLY 0199	Peruzzini, M.	INOR 0210
Patel, M. M.	MEDI 0160	Pekala, R. W.	POLY 0051	Perzynski, R.	COLL 0143
Patel, P. J.	MEDI 0142	Pekyardimci, S.		Peters, H. M.	CHAL 0021
Patel, S. S.	BIOL 0117		AGFD 0028		CHAL 0025
Patel–Misra, D.		Pelizzetti, E.	COLL 0001		CHAL 0026
	PHYS 0189		INOR 0318	Peters, S. B.	CHAL 0028
Patil, D. R.	BIOT 0067	Pellegrino, J. L.		Peterson, B. C.	ORGN 0236
Patil, S. A.	CARB 0046		BIOL 0108	Peterson, E. S.	INOR 0070
Patnaik, S.	INOR 0171	Pelletier, M. J.	ANYL 0035	Peterson, J. M.	BIOT 0014
Patrick, B.	COLL 0149	Pellin, M. J.	NUCL 0045	Peterson, M.	ORGN 0278
Patrick, K. S.	MEDI 0132	Pena, E.	INOR 0400	Peterson, N. C.	POLY 0348
Patrono, P.	PETR 0107	Peng, S–M.	INOR 0028	Peterson, T. H.	ORGN 0122
Patterson, H. H.			INOR 0029	Petersson, G. A.	
	INOR 0165	Pengsheng, H.	COLL 0079		PHYS 0287
	INOR 0189	Pennacchia, J. R.			PHYS 0297
	PMSE 0151		POLY 0113		PHYS 0317

Petit, M. A.	POLY 0102	Pirouz, P.	PHYS 0120	Portilla, M.	BIOL 0084
Petraco, N.	HIST 0005	Pirouzzadeh, B.			BIOL 0105
Petrides, D. P.	BIOT 0075		ORGN 0200		PETR 0092
Petrie, W. A.	COLL 0173	Pisanty, A.	CHED 0024	Portoghese, P. S.	
Petrillo, A.	INOR 0151		PHYS 0252		MEDI 0089
Petrosius, S. C.	PETR 0049		PHYS 0253		MEDI 0154
Petrovich, R.	BIOL 0023	Pitman, M. C.	CINF 0016		MEDI 0155
Petrovich, R. M.		Pitner, J. B.	ORGN 0023		MEDI 0157
	BIOL 0078	Pittarelli, L. A.	MEDI 0182	Portugal, E.	I&EC 0015
Petz, M.	AGFD 0063	Pitzele, B. S.	MEDI 0152	Poslusny, M.	FUEL 0140
Pewnim, T.	AGRO 0020	Pitzenberger, S.		Pospiech, E.	ANYL 0095
Pfefferle, L. D.	FUEL 0069		PHYS 0167	Poss, A. J.	ORGN 0076
Pfeifer, J.	ORGN 0122	Piwinski, J. J.	MEDI 0161	Posthuma de Boer, A.	
Pfennig, B. W.	INOR 0162	Pizarro, A.	INOR 0036		PMSE 0058
Phadke, R. S.	BIOT 0143	Planalp, R. P.	I&EC 0023	Poston, P. E.	ANYL 0011
	BIOT 0147	Plane, J.M.C.	PHYS 0072	Potapov, G. P.	COLL 0102
Phang, L-T.	INOR 0106	Plano, L. S.	PHYS 0148	Potember, R. S.	
Phebus, L.	MEDI 0174	Plass, W.	INOR 0023		BIOT 0023
Phifer, R. W.	CHAS 0042	Platz, M. S.	ORGN 0163	Potenzone, R., Jr.	
	LWM 0001	Plazek, D. J.	POLY 0148		POLY 0369
Philip, C. V.	FUEL 0061	Pleben, P.	CEI 0009	Potter, T. J.	PHYS 0150
Philipp, M.	CHED 0140	Plestil, J.	POLY 0124	Potter, T. L.	AGFD 0174
Philips, B. T.	MEDI 0097	Plowman, R. D.		Poulain, G. E.	PHYS 0244
Phillips, D. L.	PHYS 0302		AGRO 0000	Poulter, C. D.	BIOL 0014
Phillips, E.J.P.	GEOC 0008	Plumlee, K. W.	I&EC 0045	Pourquie, J.	BIOT 0182
Phillips, E. M.	I&EC 0031	Pluskey, S. T.	MEDI 0031	Pouyani, T.	ORGN 0020
Phillips, G. B.	MEDI 0094	Pluzhnikov, P. F.		Powell, J. R.	FUEL 0023
Phillips, J.	FUEL 0018		PHYS 0021		PMSE 0141
Phillips, J. B.	ANYL 0052	Plyuto, Y. V.	COLL 0097	Powell, K.	ANYL 0086
	BIOT 0046		COLL 0098	Powell, R. L.	INOR 0133
Phillips, L. G.	AGFD 0207		COLL 0099	Power, M. B.	INOR 0389
Phillips, P.	PHYS 0039		COLL 0219	Powers, D.	BIOT 0020
Phillips, R. S.	BIOL 0060		INOR 0237		PMSE 0015
Philpott, M. R.	PHYS 0153	Pocard, N. L.	INOR 0312	Powers, L.	BIOT 0051
Phinney, C.	PROF 0002	Pochan, J. M.	POLY 0287	Powers, T. S.	INOR 0112
Phippen, D. E.	FUEL 0065	Pochodylo, K. E.		Powles, M. A.	MEDI 0181
Phuong-Nguyen, H.			BIOT 0032		MEDI 0182
	PMSE 0166	Podkletnova, N. M.		Pozdniakov, R. U.	
Pickelman, D. M.			PETR 0066		INOR 0262
	POLY 0157	Poeppelmeier, K. R.		Pozdnjakov, A. O.	
Pickering, M.	CHED 0041		INOR 0358		POLY 0100
	CHED 0212	Pohl, C. A.	CARB 0066	Prakash, G.K.S.	
Piechuta, H.	MEDI 0180	Pohl, S.	CHED 0065		ORGN 0145
Pieles, U.	BIOL 0126	Point, J. J.	POLY 0125	Prasad, A.	COLL 0058
Pierce, B. M.	BIOT 0177	Pojasek, R. B.	CEI 0026		POLY 0121
Pierce, J. D.	ORGN 0124	Polacek, R.	SCHB 0002		POLY 0233
Pierce, S.	CHAL 0016		SCHB 0003		POLY 0246
Piercey, M F.	MEDI 0126	Polak, M. L.	PHYS 0055		POLY 0345
Pietrzak, R.	ENVR 0011	Polakowski, J.	MEDI 0120		POLY 0346
Pieza R., G.	COLL 0119		MEDI 0197	Prasad, G.	ORGN 0261
	COLL 0120	Polansky, B. F.	CINF 0014	Prasad, R.	MEDI 0147
	COLL 0121	Polce, B. A.	CHAS 0005	Prasada Rao, T.S.R.	
Piffeteau, A.	MEDI 0009	Policanski, D.	ENVR 0063		PETR 0100
Pike, R. D.	INOR 0205	Polichnowski, S. W.		Pratt, D. E.	AGFD 0015
Pike, R. M.	CHED 0150		I&EC 0053		AGFD 0049
Pileni, M-P.	COLL 0148	Polik, W. F.	CHED 0083	Pratt, D. W.	PHYS 0043
Pilgrim, J. S.	CHED 0089		PHYS 0303	Pratt, N. E.	ORGN 0100
Pimentel, G. C.	INOR 0167	Pollock, H. M.	PHYS 0154	Prencipe, M.	CHED 0047
Pimentel, P.	PHYS 0289	Polonskaya, I. N.			COLL 0085
Pinal, R.	ENVR 0014		COLL 0106	Prescott, J. H.	COLL 0199
	ENVR 0030	Pomerantz, M.	BIOT 0204	Preston, J.	POLY 0042
Pina P., C.	INOR 0069	Ponciano R., G.		Prestwich, G.	AGRO 0016
	INOR 0257		ENVR 0040	Prestwich, G. D.	
Pinder, D. N.	COLL 0054	Pond, D. M.	I&EC 0053		AGRO 0005
Pine, D. J.	PMSE 0108	Pond, W. G.	AGFD 0133		AGRO 0014
Pine, S. H.	ORGN 0222	Ponticello, G. S.			AGRO 0016
Pineri, M.	PMSE 0119		MEDI 0096		AGRO 045A
Pines, A.	PHYS 0267		MEDI 0097		BIOL 0070
Pinkes, J. R.	INOR 0238		MEDI 0191		ORGN 0020
Pinnavaia, T. J.		Poon, S. J.	PHYS 0010		ORGN 0021
	PETR 0014	Pop, E.	MEDI 0068	Prete, G.	NUCL 0028
	PETR 0125	Pope, C. J.	FUEL 0112	Preti, G.	AGFD 0147
Pinney, K. G.	ORGN 0011	Popiel, E. J.	PETR 0148	Pretsch, E.	COMP 0007
Pinto, B. M.	CARB 0035	Pople, J. A.	PHYS 0305		COMP 0008
	CARB 0057	Popolo, J.	POLY 0223	Price, G.	BIOT 0163
Piotrowski, A. M.		Popp, J. L.	BIOT 0067	Price, N. M.	INOR 0003
	INOR 0155	Porile, N. T.	NUCL 0025	Price, V.	ENVR 0053
Piotto, M. E.	BIOL 0123		NUCL 0034	Price, W. C.	HIST 0006
Pippin, C. G.	INOR 0033		NUCL 0050	Priddy, D. B.	POLY 0039
Pirages, S. W.	CEI 0016		NUCL 0063		POLY 0083
Piret, J. M.	BIOT 0010	Portela, M. F.	CATL 0058	Priebe, W.	CARB 0032
	BIOT 0012	Porter, J.	AGFD 0153		CARB 0033
Pirkle, R. J.	ENVR 0053	Porter, R. S.	PMSE 0074	Priem, B.	AGFD 0031

Printz, H.	AGRO 0025	Raatz, F.	PETR 0131	Ramm-Schmidt, L.	
Proal, J.	ANYL 0073		PETR 0132		CARB 0064
Proch, J.	MEDI 0175	Rabasco, D.	ORGN 0209	Ramos, S.	NUCL 0072
Prochaska, H. J.		Rabbani, F.	POLY 0228	Ramos M., A.	PHYS 0221
	AGFD 0089	Rabe, J. A.	POLY 0336		PHYS 0280
Prock, A.	INOR 0109	Rabideau, P. W.		Ramos-Nava, V.	
Proscia, J. W.	INOR 0309		ORGN 0087		INOR 0079
Pross, J.	ORGN 0247	Rabin, Y.	COLL 0159	Ramsey, J. M.	PHYS 0151
Prosser-McCartha, C. M.			COLL 0184	Ramsey, R. S.	ANYL 0092
	INOR 0286	Rabinovich, D.	INOR 0086	Ramsey, T. M.	INOR 0214
Protopopova, M. N.			INOR 0369		INOR 0215
	ORGN 0042	Rabitz, H.	PHYS 0142	Ramujachary, K. V.	
Prouty, W. F.	ANYL 0039	Rabolt, J. F.	POLY 0145		INOR 0359
Provder, T.	PMSE 0068	Radaelli, P.	PHYS 0009	Ranc, M. L.	COMP 0003
Prud'homme, R. E.		Radding, S. B.	CHAL 0022	Randic, M.	COMP 0013
	POLY 0278		CHAL 0028		COMP 0022
Prud'homme, R. K.		Radhakrishnan, G.		Randrup, J.	NUCL 0009
	COLL 0141		PMSE 0156	Rangappa, K. S.	
	COLL 0180	Radmer, R. J.	BIOT 0156		ORGN 0194
	POLY 0280	Radonovich, L. J.		Rantanen, T.	BIOT 0215
	POLY 0296		INOR 0207	Rao, G. N.	PETR 0076
Prusik, T.	AGFD 0080	Radousky, H. B.		Rao, J-L.	GEOC 0011
Pruski, M.	CATL 0042		PHYS 0052	Rao, K.L.V.	INOR 0401
	COLL 0090	Radovic, L. R.	FUEL 0014	Rao, M. P.	INOR 0401
Prusoff, W. H.	CARB 0038		FUEL 0028		INOR 0402
Przybycien, T. M.			FUEL 0033	Rao, P.S.C.	ANYL 0071
	BIOT 0135		I&EC 0020		ENVR 0014
	POLY 0306	Radzilowski, E. M.			ENVR 0030
Przyjazny, A.	ANYL 0090		MEDI 0097	Rao, T. N.	INOR 0262
Psaras, C.	MEDI 0148		MEDI 0191		INOR 0401
Pucetti, R.	POLY 0006	Rafael, C. B.	MEDI 0038		INOR 0402
Puckett, J.	MEDI 0182	Rafeiner, K.	COMP 0005	Rappe, C.	CEI 0014
Puddephatt, R. J.		Raffaelle, D. P.	PHYS 0260	Rasines, I.	INOR 0034
	INOR 0142	Ragan, I.	MEDI 0119	Rasmussen, G.	BIOT 0187
Pudzianowski, A.		Ragauskas, A.	CARB 0013	Rasmussen, P. G.	
	MEDI 0188	Raghavachari, K.			INOR 0310
	ORGN 0114		PHYS 0079		POLY 0034
Pugh, K. C.	BIOL 0006		PHYS 0214		POLY 0363
Puglia, G. P.	POLY 0369	Raghaven, R. S.		Ratajczyk, J. D.	
Puglisi, F. P.	COLL 0212		NUCL 0033		ORGN 0227
Pujado, P. R.	I&EC 0061	Rahbarnoohi, H.		Ratanathanawongs, S. K.	
Puls, J.	BIOT 0215		INOR 0274		PMSE 0012
Purdy, A. P.	INOR 0134	Rahman, M. H.	PHYS 0264	Rathbone, R. F.	
Puri, M.	PHYS 0284	Raina, A. K.	AGRO 0011		FUEL 0043
Purnell, J. H.	PETR 0126	Rajalingam, R.	PMSE 0156	Rathnamma, D. V.	
Putnam, J. E.	BIOL 0072	Rajczy, P.	BIOT 0028		FUEL 0099
Putvinski, T. M.		Rajh, T.	COLL 0151	Ratnam, C. V.	MEDI 0056
	POLY 0215	Rajski, S. R.	ORGN 0263	Ratnasamy, P.	PETR 0076
Putz, T.	AGRO 0025	Raju, B.	ORGN 0149	Ratner, M. A.	BIOT 0178
Pysnik, D.	CHED 0036	Raju, R. R.	BIOT 0219		POLY 0217
Qi, J. Z.	ORGN 0041	Raju, V. S.	ORGN 0195	Ratto, J. A.	PMSE 0119
Qi, Z.	POLY 0189	Rak, S. F.	BIOT 0201	Ratto, M.	BIOT 0215
Qian, L.	MEDI 0159	Ralat, M. D.	AGRO 0147	Rauch, F.	NUCL 0000
Qian, W.	POLY 0045	Ramanujachary, K. V.		Rauscher, D. J.	INOR 0210
Qian, X.	POLY 0134		INOR 0058	Raushel, F. M.	BIOL 0010
Qin, C.	PMSE 0101		INOR 0352	Rautonen, N.	CARB 0018
Qiubin, Z.	ANYL 0080	Ramao Ribeiro, F.		Rawlings, C. J.	CINF 0030
Quante, J. M.	ORGN 0023		PETR 0132	Rawluk, M.	PETR 0123
Quayle, W. H.	PETR 0051	Ramarathnam, N.		Ray, A.	CARB 0056
Queener, S. L.	BIOT 0085		AGFD 0018	Ray, D. J.	POLY 0314
Quesnelle, C. A.		Ramasamy, R.	INOR 0184	Ray, G. B.	BIOL 0095
	ORGN 0082	Ramaswamy, M.		Rayment, I.	BIOL 0147
Quevedo B., Q.	COMP 0053		BIOL 0126	Raymond, K. N.	
Quignard, A.	FUEL 0046	Ramdas, A.	POLY 0016		CHED 0061
Quilliam, M.	MEDI 0129	Ramette, R. W.	ANYL 0006		INOR 0120
Quin, G. S.	INOR 0324	Ramirez, F.	PMSE 0130	Reale, M. J.	CHAS 0038
Quin, L. D.	INOR 0324		POLY 0055	Ream, J. E.	BIOL 0053
	ORGN 0110	Ramirez C., J. M.			BIOL 0054
Quincy, D. A.	MEDI 0167		I&EC 0073		ORGN 0028
Quinn, D. M.	BIOL 0018	Ramirez L., A.	CHED 0124		ORGN 0029
Quintero, R.	BIOT 0130	Ramirez O., J.	INOR 0032	Reamer, R. A.	ORGN 0184
	BIOT 0131	Ramirez R., J. C.		Rebeiz, K. S.	PMSE 0072
Quiocho, F. A.	BIOL 0142		ORGN 0180	Rebek, J., Jr.	ORGN 0137
Quirasco B., M.		Ramirez S., J.	FUEL 0104	Reck, B.	POLY 0153
	CHED 0164		FUEL 0105	Reck, G.	INOR 0309
Quirk, R. P.	POLY 0266		FUEL 0106	Reck, G. P.	PHYS 0255
Quistad, G. B.	AGRO 0008	Ramirez-Solis, A.		Reck, R.	MEDI 0074
	AGRO 0012		PHYS 0244	Reddy, G. S.	ORGN 0040
Qureshi, A.	MEDI 0190		PHYS 0304	Reddy, K. V.	ORGN 0039
Qutubuddin, S.	COLL 0040	Ramli, E.	INOR 0117	Reddy, N. L.	MEDI 0147
Qvarnstrom, K.	PMSE 0078	Ramming, D. W.		Reddy, P. A.	ENVR 0032
Raatz, F.	PETR 0046		AGFD 0024	Reddy, P.S.N.	MEDI 0056
	PETR 0103				MEDI 0057

Reddy, R.	BIOT	0232	Rettig, W.	COLL	0071	Ripp, J. A.	ENVR	0050
Reddy, R. A.	ENVR	0033	Rettinger, K.	AGFD	0183	Risatti, J. R.	I&EC	0016
Reddy, R. T.	ORGN	0293	Reuillon, M.	FUEL	0080	Risby, T. H.	ANYL	0101
Reddy, V.R.K.	MEDI	0056	Reusser, F.	MEDI	0055	Risen, W. M., Jr.		
Redkov, B. P.	POLY	0100	Reuter, K. B.	PETR	0002		POLY	0253
Ree, M.	PMSE	0023	Reuther, I.	INOR	0368	Rishi, D. K.	ENVR	0015
	POLY	0085	Reutt-Robey, J. E.			Ristori, S.	PMSE	0119
Reed, B. B.	ORGN	0156		PHYS	0345	Rittenhouse, R. C.		
	ORGN	0157	Revesz, P.	NUCL	0081		ANYL	0009
Reed, C. P.	CATL	0035	Rey, A.	POLY	0297	Ritter, E. R.	FUEL	0085
Reed, G. H.	BIOL	0023	Reyes, M. V.	PETR	0137		I&EC	0024
	BIOL	0077	Reyes, S. C.	CATL	0056	Rius R., M.	HIST	0035
	BIOL	0078	Reyes Chumacero, A.			Riva, S.	BIOT	0064
Reed, M.	AGRO	0072		PHYS	0236	Rivera, J.	BIOL	0118
Reed, M. A.	BIOT	0003	Reyes Quiroz, R.			Rivera, L.	MEDI	0111
Rees, A. R.	BIOT	0056		AGFD	0139	Rivera, L. M.	MEDI	0113
Rees, W. S., Jr.	COLL	0154	Reynaers, H.	PMSE	0102	Rivera M., C.	ORGN	0186
	COLL	0155	Reynolds, E. R.	BIOT	0107	Rivera Santillan, R. E.		
	COLL	0157	Reynolds, J. R.	POLY	0021		ANYL	0053
	COLL	0158		POLY	0022	Rivera V., J. A.	CHED	0100
	POLY	0342	Reynolds, R.	CHAL	0004	Rivero, R. A.	ORGN	0112
Reese, J.	BIOT	0163		HIST	0004	Rivero S., O.	ENVR	0040
	FERT	0004	Reynolds, R. A.			Riviere, J. E.	AGRO	0095
Reeves, D. R.	AGRO	0108		INOR	0189	Rizo, J.	BIOL	0003
Reeves, J. H.	CHED	0159	Reynolds, S. K.	COLL	0029	Roat, R. M.	CHED	0152
	NUCL	0013		COLL	0156	Robb, E. W.	CHED	0144
Reeves, R. R.	PHYS	0259	Rezvani, A. B.	INOR	0040		ORGN	0255
Reeves, V. B.	AGFD	0099	Rhee, H.	ORGN	0010	Robbins, J.	INOR	0259
Regalbuto, J. R.			Rhee, K. S.	AGFD	0021	Robello, D. R.	POLY	0105
	COLL	0188	Rheingold, A. L.			Roberson, M. W.		
Regan, J.	ORGN	0254		INOR	0072		PHYS	0193
Regenass, U.	MEDI	0184		INOR	0148	Roberts, D. A.	MEDI	0102
Register, R. A.	PMSE	0117		INOR	0149	Roberts, G.	PHYS	0369
	PMSE	0145		INOR	0328	Roberts, G. D.	ANYL	0140
Regnier, F. E.	AGFD	0179		POLY	0258	Roberts, M. F.	POLY	0011
	ANYL	0028	Rhodes, C. A.	ORGN	0167		POLY	0213
	ANYL	0131	Rials, T. G.	POLY	0138	Roberts, S.	BIOT	0056
	BIOT	0073	Riaz, U.	INOR	0104	Roberts, W.	INOR	0394
Rego, J. A.	BIOT	0202	Ribeiro, F. H.	CATL	0037	Robertson, A.	CHAL	0031
Rehmat, A.	FUEL	0120	Ribeiro, M. F.	PETR	0132		CMEC	0027
Reibenspies, J. H.			Riccitiello, S.	POLY	0187		I&EC	0004
	INOR	0177	Rice, C.	CHAS	0018	Robertson, A. A.		
Reid, J. B.	ENVR	0043	Rice, D.	ANYL	0019		COLL	0068
Reid, M. D.	ORGN	0216	Rice, J. E.	PHYS	0342	Robertson, R.G.H.		
Reid, R. T.	INOR	0119	Rice, J. K.	PHYS	0301		NUCL	0010
Reife, A.	CMEC	0018	Rice, K. G.	BIOL	0148	Robins, M. J.	ORGN	0281
Reiff, W.	PMSE	0006	Richard, D. S.	AGRO	0002	Robinson, C. H.		
Reimer, J.	COLL	0211	Richard, J. P.	ORGN	0153		MEDI	0006
Reimer, J. A.	CATL	0047		ORGN	0154	Robinson, E. D.		
Reimer, K. J.	INOR	0136	Richard, M. A.	CATL	0041		ORGN	0179
Reimers, M. J.	POLY	0191		FUEL	0100	Robinson, G. N.		
Reineccius, G.	AGFD	0073	Richard, W. G.	ORGN	0166		PHYS	0071
Reineker, P.	PHYS	0493	Richards, D.	PHYS	0009	Robinson, L. A.		
Reiner, R. S.	PMSE	0146	Richards, G. N.	FUEL	0103		ORGN	0282
Reinikainen, T. R.			Richards, J. H.	BIOT	0239	Robinson, M. B.		
	BIOT	0206	Richards, P. B.	BIOT	0032		COLL	0008
Reis, K.	INOR	0314	Richmond, T.	BIOT	0239	Robinson, R. A.		
Rellick, G. F.	FUEL	0051	Richter-Mendau, J.				AGRO	0111
Remillieux, J.	NUCL	0052		PETR	0093	Robinson, R. E.		
Remoue, F.	PETR	0040	Rickard, C.E.F.	INOR	0243		FERT	0012
Rempp, P.	POLY	0272	Rico, A. G.	AGRO	0129	Robinson, T. R.		
Remsen, E. E.	POLY	0183	Rico, G.	BIOL	0102		POLY	0314
	POLY	0311	Ridenour, C. F.	AGFD	0104	Robles-Martinez, J. G.		
Remy, D. C.	MEDI	0096	Rider, P. J.	AGFD	0153		INOR	0190
	MEDI	0097	Ridgway, T.	PMSE	0006	Robson, B.	NUCL	0040
	MEDI	0191	Ridley, W. P.	AGRO	0023	Robyt, J. F.	CARB	0007
Rendleman, J. A., Jr.			Riede, J.	INOR	0367	Robyt, J. R.	CARB	0028
	CARB	0041	Rieger, P. T.	PHYS	0133	Rocha, J.	PETR	0006
Renfree, R.	I&EC	0028	Rieker, A.	ORGN	0250		PETR	0096
Renfro, D. H.	PETR	0121	Rieland, M.	INOR	0133	Rockefeller, C.	CHAS	0032
Rengan, K.	POLY	0237	Riendeau, D.	MEDI	0180	Roderick, R.	AGFD	0181
Renneke, R. F.	INOR	0286	Rigby, B. S.	MEDI	0175	Rodgers, P. A.	POLY	0143
Renner, M. W.	INOR	0091	Rikken, G.L.J.A.			Rodov, V.	AGFD	0076
Rennert, A. M.	CHED	0018		POLY	0178	Rodrigues, D. E.		
Renshaw, E.	NUCL	0036		POLY	0222		POLY	0264
Replogle, E. S.	PHYS	0305	Riley, D. P.	CATL	0005		POLY	0284
Resch, U.	COLL	0133		INOR	0247	Rodriguez, B. E.		
Resnick, L.	MEDI	0055	Riley, J. F.	CHAL	0021		POLY	0244
Rethwisch, D. G.				CHAL	0026	Rodriguez, C.	AGFD	0206
	BIOT	0067	Ripka, W. C.	COMP	0047	Rodriguez, J.	FUEL	0049
Rettig, S. J.	INOR	0362	Ripmeester, J. A.				FUEL	0076
	INOR	0379		I&EC	0021	Rodriguez, M. P.		
	INOR	0383		I&EC	0022		POLY	0052

Rodriguez, O.	PETR 0092	Rosario, A. G.	ORGN 0019	Ruiz-Loyola, B.			
Rodriguez, V.	MEDI 0034	Rose, D.	BIOL 0143		CHED 0168		
Rodriguez G., F. J.		Rose, J. B.	ENVR 0070		CHED 0169		
	PHYS 0270	Rose, J. D.	MEDI 0028		CHED 0170		
Rodriguez-Hahn, L.		Rose, P. A.	AGRO 0017	Ruiz-Ramirez, L.			
	ORGN 0280	Rose, R. L.	AGRO 0118		INOR 0026		
Rodriguez-Ramos, I.		Rose, R. N.	INOR 0180		INOR 0037		
	PETR 0010	Rosen, C-G.	BIOT 0069		INOR 0042		
	PETR 0102		BIOT 0076		INOR 0057		
Rodriguez-Vazquez, R.		Rosen, J. D.	AGRO 0060		INOR 0061		
	BIOT 0250	Rosen, R. T.	AGFD 0048		INOR 0260		
Rodriquez, J.	FUEL 0049		AGFD 0170		INOR 0399		
Rodriquez Robles, I.		Rosenberg, F. S.		Ruiz Santoyo, M. E.			
	INOR 0273		INOR 0174		ENVR 0026		
Roe, D.	COMP 0036	Rosenberg, S. H.		Ruiz-Valero, C.			
Roehl, R.	ANYL 0091		MEDI 0120		INOR 0034		
Roehr, J. E.	MEDI 0085		MEDI 0197		INOR 0035		
Rogers, C. E.	PMSE 0138	Rosenkranz, G.	HIST 0017	Rule, B. L.	MEDI 0162		
Rogers, D. W.	ORGN 0173	Rosenthal, G. L.		Rum, G.	COMP 0023		
Rogers, J. E., Jr.			INOR 0084	Runt, J. P.	PMSE 0148		
	YCC 0002	Roser, R. L.	FERT 0008		PMSE 0165		
Rogers, K. L.	MEDI 0085	Roshdy, T.	AGFD 0170	Runyan, M.	MEDI 0047		
	MEDI 0130	Ross, J. H.	AGRO 0043	Ruohonen, L.	BIOT 0206		
Rogers, M.	YCC 0004	Rosseinsky, M. J.		Rupert, D. R.	CHAS 0050		
Rogler, J. C.	AGFD 0149		INOR 0354	Ruscic, B.	FUEL 0124		
Roha, D.	COLL 0041	Rossi, A. R.	COMP 0059	Rush, D. K.	MEDI 0085		
Rohlfing, C. M.	PHYS 0079	Rossi, P. G.	AGRO 0030	Rush, J. E.	CINF 0045		
Rohlfing, E. A.	FUEL 0110	Rossi, R. A.	FUEL 0122	Rusinko, A.	CINF 0023		
	PHYS 0235	Rossky, P. J.	PHYS 0190		COMP 0041		
Rohmelt, J.	CINF 0010	Rossman, D. I.	ORGN 0251	Rusinko, A., III	CINF 0026		
Rohrbach, K. W.			ORGN 0252	Rusling, J. F.	CHED 0193		
	MEDI 0084		ORGN 0253	Russell, A.	BIOT 0066		
Rohrbaugh, D. K.		Rossmann, A.	AGFD 0194	Russell, J.	MEDI 0067		
	ORGN 0251	Rotella, D. P.	ORGN 0258	Russell, K. E.	POLY 0019		
	ORGN 0252	Roth, B. D.	MEDI 0108	Russell, T.	BIOL 0146		
	ORGN 0253	Roth, J. F.	I&EC 0043		COLL 0134		
Rohrer, G.	PHYS 0207	Rothberg, L.	BIOT 0080	Russo, L.	MEDI 0143		
Rohrer, G. S.	COLL 0171		PHYS 0193	Russo, S.	POLY 0120		
Roinestad, K. S.		Rothchild, R.	CHED 0153	Rutherford, J. S.			
	AGRO 0060	Rothe, E. W.	PHYS 0255		COMP 0031		
Rojas E., E.	AGFD 0038	Rotstein, N. A.	POLY 0267	Ruttenberg, K. C.			
Rojas H., A.	CHED 0104	Rouf, C.	PMSE 0001		GEOC 0013		
Rojas-Hernandez, A.		Rouhani, R.	MEDI 0106	Ruzicka, F.	BIOL 0023		
	ANYL 0054	Roura, M.	ANYL 0057	Ruzicka, F. J.	BIOL 0078		
	ANYL 0075	Rouseff, R.	AGFD 0106	Ryan, J. J.	CEI 0011		
Rojas M., N.	BIOL 0101	Rouseff, R. L.	AGFD 0167	Ryan, J. N.	ENVR 0058		
Rojo-Dominguez, A.		Rowe, G. T.	GEOC 0015	Ryba, D. W.	INOR 0169		
	BIOL 0106	Rowe, L. D.	BIOT 0166	Rye, R. R.	PHYS 0205		
Rokita, S.	BIOL 0121		ORGN 0201	Ryu, K.	BIOT 0067		
Rokita, S. E.	BIOL 0122	Rowell, R. L.	COLL 0199	Rzepecki, L. M.			
	INOR 0183		COLL 0200		INOR 0138		
	ORGN 0190	Rowland, R. S.	CINF 0021	Saab, N. H.	MEDI 0021		
	ORGN 0191		COMP 0045	Saam, B.	PHYS 0041		
Roknich, S.	MEDI 0070	Roy, C.	BIOT 0211	Saari, E. A.	CATL 0021		
Rokop, D. J.	NUCL 0022		CARB 0044	Saba, S.	ORGN 0204		
Rolison, D. R.	CATL 0018	Roy, D. M.	ENVR 0006	Sable, D. B.	INOR 0129		
Roloff, A.	INOR 0338	Roy, G.	CARB 0026	Sacchettini, J. C.			
Romaniec, M.P.M.		Roybal, J. E.	AGFD 0101		BIOL 0015		
	BIOT 0181	Rubingh, D. N.	COLL 0052	Sachinides, J. I.			
Romano, L. J.	ANYL 0130	Rubinstein, M.	POLY 0172		PETR 0007		
Romero M., A.	ENVR 0037	Rudd, G.	PHYS 0153	Sacks, W.	PMSE 0128		
	ENVR 0038		PHYS 0361	Sacristan-Fanjul, M.			
Romero O., R.	I&EC 0007	Rudello, D.	COLL 0005		INOR 0398		
Romig, W. R.	AGFD 0056	Rudin, A.	PMSE 0047	Sadana, A.	BIOT 0044		
Rommel, A.	AGFD 0169		POLY 0135		BIOT 0219		
Ronco, S.	INOR 0159	Rudnick, E.	SCHB 0015	Sadeghi, N.	PHYS 0368		
Rood, M. H.	TECH 0018	Rudolph, C.	CINF 0036	Sadowski, J.	CINF 0036		
Roongta, V.	INOR 0378		MEDI 0074	Sadron, I. C.	POLY 0132		
Roos, P. H.	ANYL 0017	Ruel, K.	BIOT 0208	Sady, W.	FUEL 0073		
Root, T. W.	COLL 0197	Rueter, M. A.	COLL 0048	Saehr, D.	PETR 0128		
	PHYS 0346	Ruger, J. R.	CHAL 0017	Saeki, M.	AGFD 0127		
Roots, W. R.	BIOT 0232	Ruhl, J. C.	ANYL 0107	Saezusa, H.	PETR 0003		
Roovers, J.	POLY 0292	Ruiz, C.	INOR 0036	Sagar, A.	POLY 0272		
	POLY 0293	Ruiz, J.	NUCL 0028	Sage, S. H.	ENVR 0084		
Roozen, J. P.	AGFD 0158	Ruiz, M. E.	PHYS 0296	Sagrero H., I.	CHED 0163		
Roques, B. P.	MEDI 0002	Ruiz-Loyola, B.			CHED 0172		
Rosales H., M. J.			CHED 0094	Saha, A. K.	ORGN 0203		
	INOR 0241		CHED 0095	Saiki, R.	CHAL 0004		
Rosales V., M. I.			CHED 0096	Saini, A. K.	PMSE 0151		
	INOR 0063		CHED 0097	St. Angelo, A. J.			
	INOR 0246		CHED 0098		AGFD 0120		
Rosan, A. M.	INOR 0270		CHED 0165	Saji, T.	COLL 0019		
Rosano, H. L.	COLL 0080		CHED 0166	Sakaba, H.	ORGN 0130		

Sakata, J.	POLY 0082		Sandoval G., C.	CHED 0106		Scallen, T. J.	MEDI 0101
Sakellaropoulos, G. P.			Sandoval M., R.			Scalzo, T. R.	COLL 0198
	CATL 0054			INOR 0032		Scapin, G.	BIOL 0015
Sakurai, H.	PETR 0129		Sanjurjo B., M.	ORGN 0186		Scattergood, E.	BIOT 0134
Sakurai, T.	PHYS 0182		Sankar, S. S.	POLY 0042		Scensy, P. M.	MEDI 0158
Salaita, G. N.	ENVR 0007		Sankaran, V.	INOR 0307		Schachtele, C.	MEDI 0074
Salamone, J. C.	POLY 0367		Sanogo, O.	FUEL 0080		Schadt, R. J.	POLY 0058
Salas C., P.	COLL 0126		Sanson, C.	INOR 0032		Schaefer, G. F.	CATL 0005
	COLL 0127		Sansone, E. B.	CHAS 0006			INOR 0247
Salazar, D.	PHYS 0296		Santerre, C. R.	AGFD 0058		Schaefer, H. F., III	
Salazar, F.	INOR 0263		Santhosh, U.	PMSE 0019			PHYS 0032
Salazar, S.	ENVR 0025		Santora, N. J.	CINF 0039		Schaefer, J.	PHYS 0069
Salcedo, R.	INOR 0032		Santora, V. J.	ORGN 0061		Schaefer, W. P.	INOR 0335
Salcido, J. E.	CHED 0089		Santore, M. M.	POLY 0295		Schaich, K. M.	AGFD 0004
Saldate, O.	AGFD 0206		Santos, R.	INOR 0059		Schamper, T.	CHED 0049
Saldivar O., L.	ENVR 0040		Santos, R. A.	INOR 0078		Schappacher, M.	
Saleh, M. H.	AGRO 0039		Santos, S. E.	ENVR 0036			POLY 0309
Salem, J. R.	PHYS 0070		Santos de Flores, E.			Scharf, S.	CHAL 0004
Salerno, T. A.	AGRO 0019			CHED 0097		Schaverien, C. J.	
Salgado-Zamora, H.			Santos S., E.	AGFD 0038			INOR 0053
	ORGN 0230		Santos S., E	ORGN 0037			INOR 0294
Salinas C. T., E.			Santoyo, E.	I&EC 0013		Schechtman, L. A.	
	INOR 0257		Santoyo, S. G.	I&EC 0014			COLL 0039
Salinas-Martinez de Lecea, C.			Santoyo G., S.	ANYL 0058		Scheck, B.	CHAL 0002
	FUEL 0029			INOR 0063		Schecter, A.	CEI 0012
	FUEL 0032			INOR 0246		Scheetz, B. E.	ENVR 0006
Salinas-Vazquez, M. R.			Santoyo-Guiterrez, S.			Scheigetz, J.	MEDI 0180
	CHED 0170			I&EC 0012		Schell, J.	BIOT 0163
Salina-Vazquez, M. R.			Santoyo-Guitterez, S.			Scheller, G. R.	POLY 0215
	CHED 0098			I&EC 0011		Schelly, Z. A.	COLL 0146
Salisbury, C.D.C.			Sanyal, I.	INOR 0045		Schenk, W. A.	INOR 0249
	AGFD 0085		Sapers, G. M.	AGFD 0027			INOR 0368
Salladay, D. G.	FERT 0013		Sarabu, R.	AGFD 0030		Schepartz, A.	CONG 0010
	FERT 0022		Sarantites, D. G.	MEDI 0165			ORGN 0270
Saller, H.	CINF 0010			NUCL 0027		Scherbakoff, N.	
Sallin, K. J.	MEDI 0176		Sargent, A. L.	PHYS 0192			PMSE 0167
Salmeron, M.	PHYS 0124		Sarid, D.	PHYS 0152		Scherer, G. W.	POLY 0286
Salmeron-Valverde, A.			Sarnoff, E.	HIST 0032		Schermerhorn, P. G.	
	INOR 0190		Sarofim, A. F.	FUEL 0025			AGFD 0099
Salmon, P. M.	BIOT 0109		Sarokin, D. J.	CEI 0024		Schevitz, R. W.	BIOL 0072
	MEDI 0181		Sarti, G. C.	POLY 0140		Schiavi, G. V.	MEDI 0141
Salmon, S. E.	MEDI 0093		Sarvas, H.	CARB 0018		Schieberle, P.	AGFD 0052
Salomon, R. G.	INOR 0339		Sarwar, G.	AGFD 0209		Schiksnis, R. A.	
Salovey, R.	PMSE 0017		Sasabe, H.	BIOT 0121			MEDI 0169
Salto, R.	MEDI 0054		Sasisekharan, R.			Schildkraut, J. S.	
Saltsburg, H.	COLL 0174			BIOL 0056			POLY 0105
Salzberg, A. P.	PHYS 0353		Satija, S.	POLY 0090		Schill, C. F.	CHAL 0013
Samanen, J.	ORGN 0115		Satishchandran, C.			Schillace, R. V.	POLY 0287
Samano, V.	ORGN 0281			BIOL 0008		Schilling, M. L.	COLL 0164
Samara, M. R.	BIOT 0251		Satkowski, M.	COLL 0036			POLY 0215
Sampson, J. C.	CHED 0071		Satkowski, M. M.			Schimmel, P.	BIOL 0026
Samson, R.	ENVR 0016			COLL 0033		Schimpf, M. E.	PMSE 0009
Samsonovich, A.				COLL 0035		Schinazi, R. F.	AGFD 0162
	BIOT 0055			COLL 0053			CARB 0022
Samuel, O.	CEI 0004			POLY 0076		Schirber, J. E.	MEDI 0051
	CEI 0005		Sato, A.	BIOT 0121			PHYS 0024
Samuel, P. L.	CHED 0070		Sato, N.	POLY 0291		Schirle-Keller, J-P.	PHYS 0025
Samuel, R.	POLY 0184		Sato, T.	AGRO 0049			AGFD 0073
Samulski, E. T.	COLL 0184		Satterlee, J. D.	BIOL 0093		Schiweck, H.	CARB 0005
Sanayei, R. A.	PMSE 0095		Sauder, D. G.	PHYS 0189		Schlegel, H. B.	PHYS 0088
Sanborn, W. B.	INOR 0092		Sauer, J. A.	POLY 0061			PHYS 0248
Sanchez, C.	POLY 0006		Saulnier, M. G.	MEDI 0012		Schlesinger, Z.	PHYS 0051
Sanchez, G.	AGFD 0129		Saulys, D.	PHYS 0361		Schleyer, P. v. R.	
Sanchez, I. C.	POLY 0143		Saulys, D. S.	COLL 0050			ORGN 0138
Sanchez, L.	INOR 0036		Saura-Llamas, I.			Schlick, S.	PHYS 0215
	INOR 0241			ORGN 0220		Schlotter, N. E.	
Sanchez, R. I.	MEDI 0017		Sauve, G.	MEDI 0166			POLY 0332
Sanchez E., O.	ORGN 0198		Savage, M. A.	MEDI 0152		Schluter, A-D.	POLY 0036
Sanchez-Mendoza, A. A.			Sawamura, M.	PHYS 0212		Schlutter, M.	PHYS 0053
	CHED 0168		Sawan, S. P.	CHED 0133		Schmaltz, T. G.	
Sandbaken, M.	BIOL 0153			POLY 0043			COMP 0030
Sander, L. C.	ANYL 0098		Sawrey, B. A.	CHED 0013		Schmarr, H-G.	AGFD 0183
Sandermann, H.				CHED 0017		Schmatz, D. M.	MEDI 0181
	AGRO 0038			CHED 0137			MEDI 0182
Sanders, D. K.	COLL 0090		Sawyer, J. E.	COLL 0189			MEDI 0183
Sanders, L.	PHYS 0306		Sax, M.	BIOL 0032			ORGN 0184
Sanders, N. D.	COLL 0202		Saxton, T.	CMEC 0010		Schmehl, R. H.	INOR 0284
Sanders, P. F.	AGRO 0053		Saykally, R. J.	PHYS 0007		Schmidbaur, H.	
Sanders, R. A.	AGFD 0057		Sayler, G. S.	BIOT 0099			INOR 0367
Sanderson, I.	ORGN 0117		Scahill, T.	BIOL 0044		Schmidhammer, H.	
Sandleback, B. L.			Scalamanna, R.	PETR 0044			MEDI 0086
	COMP 0030		Scalettar, R. T.	PHYS 0054		Schmidt, A.	PHYS 0069
Sando, K. M.	COMP 0055						

Schmidt, H-L.	AGFD	0194	Schuette, M. R.			Seiboth, B.	BIOT	0183
Schmidt, R.	COLL	0212		MEDI	0133	Seifert, J.	AGRO	0020
Schmidt, W. F.	AGRO	0033	Schulberg, M. T.			Sekerke, C.	MEDI	0108
Schmidt, W. K.	MEDI	0084		PHYS	0202	Sekhar, V.	MEDI	0052
Schmidt-Rohr, K.			Schulein, M.	BIOT	0187	Sekino, H.	PHYS	0341
	PMSE	0121	Schuler, M. A.	AGRO	0115	Self, M. F.	COLL	0139
Schmiedekamp, A.			Schultz, C. A.	GEOC	0017	Seligy, V.	BIOT	0212
	PHYS	0298	Schultz, P. G.	BIOL	0029	Selke, M.	INOR	0192
Schmitt, F-J.	BIOT	0043	Schulz, K. H.	COLL	0195	Selnick, H. G.	MEDI	0096
Schmitz, F. J.	ORGN	0182	Schulz, K. P.	COMP	0005		MEDI	0097
	ORGN	0188	Schumacher, W. A.				MEDI	0191
Schmitz, K. S.	COLL	0178		MEDI	0160	Semar, S.	POLY	0117
Schmolka, S. J.	MEDI	0175	Schure, M. R.	PMSE	0028	Semelhack, M. F.		
Schmuff, N. R.	COMP	0011	Schuster, D. I.	ORGN	0167		ORGN	0096
Schneemeyer, L. F.			Schuster, R.	PHYS	0126		ORGN	0097
	PHYS	0026	Schuth, F.	PETR	0047	Semkow, T. M.	NUCL	0088
Schneider, A.	BIOL	0009	Schwab, J. M.	BIOL	0065	Semmelhack, M. F.		
	BIOT	0090	Schwab, J. M.	BIOL	0066		ORGN	0010
Schneider, D. M.				BIOL	0067		ORGN	0102
	PHYS	0168	Schwab, S. T.	POLY	0317	Sen, N. P.	AGFD	0040
Schneider, E.	MEDI	0195	Schwabacher, A. W.				AGFD	0110
Schneider, H.	COLL	0012		ORGN	0066	Senak, L.	PMSE	0031
Schneider, P.	MEDI	0184	Schwark, J. M.	POLY	0339	Sendijarevic, V.		
Schneider, R. P.			Schwartz, A. T.	CHED	0051		PMSE	0040
	AGFD	0061		CHED	0053	Sendlinger, S. C.		
Schneider, T.	CARB	0025	Schwartz, M.	AGRO	0101		INOR	0197
	POLY	0146	Schwartz, R.	MEDI	0182	Sengupta, S.	ORGN	0082
Schneiders, G. E.			Schwartz, R. E.	MEDI	0181	Senior, D. J.	BIOT	0214
	AGRO	0029	Schwartz, W. E.			Senneron, M.	POLY	0133
Schneir, J.	PHYS	0363		ANYL	0031	Senseman, S. A.		
Schnorer, H.	PHYS	0132	Schweikert, A. W.				AGRO	0146
Schober, B. J.	POLY	0349		BIOL	0069	Sentein, C.	POLY	0009
Schobert, H. H.			Schweikert, E. A.			Senzer, S. N.	PMSE	0168
	CATL	0050		NUCL	0057	Seper, J. M.	CHED	0050
Schoen, K.	AGFD	0185	Schweitzer, C. T.			Seppen, C.J.E.	POLY	0178
Schoen, P. E.	BIOT	0246		INOR	0380	Sercel, A. D.	MEDI	0011
Schoenleber, R.			Schwenz, R. W.			Sergheraert, C.	MEDI	0076
	MEDI	0124		CHED	0197	Serio, M. A.	FUEL	0052
	MEDI	0125		CHED	0209	Seris, A.	POLY	0112
	MEDI	0127	Schwieger, W.	PETR	0108	Seris, J-L.	INOR	0187
Schoffers, E.	ORGN	0046	Schwind, D. B.	POLY	0180	Serpone, N.	COLL	0131
Schofield, K.	PHYS	0073	Scialdone, M. A.				INOR	0318
Schoichet, M. S.				ORGN	0272	Serra, D. L.	INOR	0031
	POLY	0298	Scoles, G.	PHYS	0104	Serrato-Rodriguez, J.		
Schooley, D. A.	AGRO	0001	Scott, A.	BIOT	0055		COLL	0123
	AGRO	0004	Scott, B. A.	INOR	0121	Serrato Rodriguez, J.		
	AGRO	0008	Scott, D. B.	PMSE	0136		COLL	0124
	AGRO	0022	Scott, J. C.	POLY	0155	Serrette, A.	INOR	0135
Schoonmaker, D.			Scott, J. G.	AGRO	0099	Sesin, D. F.	MEDI	0181
	INOR	0360	Scott, L.	ORGN	0046	Sesnie, J. C.	MEDI	0061
Schoonover, R.	PHYS	0307	Scott, L. T.	ORGN	0094	Sessler, J. L.	INOR	0295
Schorn, T. W.	MEDI	0122	Scott, R. M.	PHYS	0308		ORGN	0058
	MEDI	0123	Scott, T. W.	COLL	0087		ORGN	0059
Schou, C.	BIOT	0187		PHYS	0062		ORGN	0139
Schramm, V. L.				PHYS	0283	Seta, K.	ANYL	0141
	BIOL	0046	Scozzafava, M.	POLY	0150	Sethumadhaven, M.		
	BIOL	0073	Scranton, M. I.	GEOC	0001		PMSE	0022
	BIOL	0083	Scruggs, B. E.	PMSE	0182	Sethuram, B.	COLL	0081
Schramm, V.L.	BIOL	0107	Scully, F. E., Jr.				INOR	0401
Schreck, C. E.	AGRO	0033		ENVR	0031	Sethuraman, A.		
Schreder, K.	ORGN	0139	Scurrell, M. S.	PETR	0008		INOR	0402
Schreiber, H. P.			Seaborg, G. T.	NUCL	0024		FUEL	0048
	POLY	0135	Seals, J. V.	CINF	0011	Sethuraman, G.		
Schreier, P.	AGFD	0008	Sears, C. T.	CHED	0146		COLL	0160
	AGFD	0184	Sears, M. W.	BIOT	0171		COLL	0163
Schrier, D.	MEDI	0178	Secrist, J. A., III				COLL	0203
Schrier, D. J.	MEDI	0177		MEDI	0005	Setia, P.	AGRO	0083
Schrock, R. R.	INOR	0259		MEDI	0028	Setzer, N. N.	MEDI	0113
	INOR	0261	See, R. F.	INOR	0066	Severnak, S. A.	MEDI	0148
	INOR	0307	Seelye, J. A.	MEDI	0158	Sevilla, M. D.	AGFD	0012
	POLY	0207	Seeman, J.	CONG	0004	Sexton, J. D.	AGRO	0057
Schroder, U.	COLL	0072	Seeman, N. C.	BIOL	0162	Seyferth, D.	POLY	0340
Schroeder, H. A.			Sefton, M. A.	AGFD	0006		POLY	0359
	BIOT	0250	Segal, E.	CHAS	0051	Seymour, E.	COLL	0013
Schroeder, N. C.			Segal, E. B.	CHAS	0043	Shabtai, J.	PETR	0013
	NUCL	0022	Segal, L.	CHAS	0034	Shacklette, L. W.		
Schroeder, S. A.			Seger, N.	MEDI	0195		BIOT	0203
	INOR	0378	Sehnert, S. S.	ANYL	0101	Shadman, F.	FUEL	0132
Schubert, V.	AGFD	0183	Seib, K. E., Jr.	CHAL	0015	Shafagati, A.	AGRO	0140
Schuck, D. F.	ORGN	0155	Seib, P. A.	CARB	0040	Shafiee, A.	ORGN	0200
Schuddeboom, W.			Seiber, J. N.	AGRO	0032	Shah, A.	COLL	0188
	POLY	0222		AGRO	0120		FUEL	0041
			Seibl, J.	COMP	0008	Shah, A. S.	ORGN	0028

Shah, N.	FUEL 0041	Sher, P. M.	MEDI 0160	Siegfried, B. D.	AGRO 0138
	FUEL 0056	Sheridan, R. P.	CINF 0026	Siegl, P.K.S.	MEDI 0103
Shah, P. S.	CARB 0052	Sherman, M. C.			MEDI 0104
Shah, S.	BIOL 0049		CHED 0028		MEDI 0105
	BIOL 0050	Shero, E.	POLY 0039		MEDI 0122
Shah, S. K.	MEDI 0173	Shetty, S.	PMSE 0050		MEDI 0123
Shah, U.	MEDI 0179	Sheu, E. Y.	PETR 0138	Siemens, R. L.	POLY 0145
Shahidi, F.	AGFD 0019		PETR 0150	Siemiarczuk, A.	
	AGFD 0121	Sheu, J. J.	CARB 0009		PHYS 0134
	AGFD 0168	Shi, C-X.	ORGN 0174	Sierra, J.	PMSE 0086
Shaikh, B.	AGFD 0100	Shi, L.	PETR 0105		POLY 0052
Shakespeare, W. C.			PETR 0130		POLY 0093
	ORGN 0093	Shi, X.	PHYS 0071	Sierra, T.	BIOT 0202
Shakhashiri, B. Z.		Shi, Z-Y.	PHYS 0309	Siewert, R. M.	COLL 0185
	ANYL 0010	Shibamoto, T.	AGFD 0053	Sigman, D. S.	BIOL 0156
Shalaby, L. M.	AGRO 0031		AGFD 0132	Sigurskjold, B. W.	
Shalaby, S. W.	CHED 0135	Shibata, M.	BIOL 0112		CARB 0013
	CHED 0136	Shibayama, M.	POLY 0171	Sih, C. J.	BIOT 0092
Shaler, R. C.	HIST 0003	Shick, R.	POLY 0307	Sihver, L.	NUCL 0024
Shamsuddin, S.		Shieh, W-C.	ORGN 0067	Sii, D.	BIOT 0044
	INOR 0412	Shields, M. S.	BIOT 0119	Siirila, A. R.	ANYL 0026
Shaner, R. A.	CHED 0073	Shiels, J.	PHYS 0195	Sike, S.	COMP 0048
	CHED 0138	Shih, C.	MEDI 0022	Sikka, M.	POLY 0090
Shannon, C.	COLL 0118	Shih, F. F.	AGFD 0140	Sikorski, A.	POLY 0297
Shannon, M. D.		Shih, K-S.	INOR 0255	Sikorski, J. A.	BIOL 0053
	PETR 0021	Shim, J.	ORGN 0007		BIOL 0054
Shao, R.	BIOT 0202	Shimada, J.	MEDI 0112		ORGN 0028
Shapiro, B.	AGFD 0076	Shimshock, S. J.			ORGN 0029
Shapley, J. R.	CATL 0008		ORGN 0294	Silbergeld, E.	ENVR 0065
Shapley, P. A.	INOR 0370	Shin, H. K.	COLL 0027	Silberman, R. G.	
	ORGN 0219		INOR 0144		CHED 0055
Sharaf, S. M.	ORGN 0160	Shin, Y. K.	CATL 0021	Silbernagel, B. G.	
Shareck, F.	BIOT 0213	Shine, A. D.	COLL 0032		FUEL 0072
Sharkey, J. B.	HIST 0029	Shinkai, I.	ORGN 0297	Sillion, B.	POLY 0133
Sharma, A. K.	ORGN 0034	Shirai, M.	COLL 0179	Silver, R. G.	COLL 0208
Sharma, D. K.	I&EC 0018	Shirai, T.	AGFD 0115		COLL 0209
Sharma, M.	BIOL 0088	Shiralkar, V. P.	PETR 0005	Silverstein, M. S.	
Sharma, V. K.	INOR 0393		PETR 0076		PMSE 0062
Sharp, J. D.	BIOL 0072	Shirley, B. A.	BIOT 0189	Sim, S. J.	BIOT 0229
Sharp, L. L.	FUEL 0091	Shirota, O.	ANYL 0019	Simic, M. G.	AGFD 0093
Sharpe, L. R.	CHED 0180	Shively, E. R.	CINF 0011	Simic–Glavaski, B.	
Shashidhar, R.	COLL 0071	Shockey, E.	PMSE 0127		BIOT 0054
Shatkin, J. A.	ENVR 0044	Shoelson, S. E.	BIOL 0002	Simmonds, W.H.C.	
Shatlock, M. P.	PETR 0071	Shoemaker, S.	BIOT 0240		CMEC 0001
Shaul, G. M.	AGRO 0141	Shogren, R. L.	CARB 0043	Simmons, A.	PMSE 0076
Shaw, B. J.	AGRO 0097	Shoichet, B.	COMP 0036	Simmons, R. D.	
Shaw, F. V.	PETR 0088	Shore, S. G.	INOR 0220		AGRO 0078
Shaw, H.	COLL 0205		INOR 0332	Simms, S.	CARB 0037
	COLL 0210	Shorr, P.	ENVR 0028	Simon, J.	POLY 0102
Shaw, J. R.	INOR 0284	Shortt, D. W.	PMSE 0096	Simon, V.	COMP 0005
Shaw, M. T.	PMSE 0132	Showalter, D.	CONG 0006	Simone, C. M.	BIOL 0051
	POLY 0173	Showalter, H.D.H.		Simons, J.	PHYS 0056
Shaw, R.	BIOL 0096		MEDI 0011	Simpson, D. A.	ORGN 0294
Shaw, R. W.	PHYS 0151	Showler, A. T.	AGRO 0082	Simpson, H. J.	GEOC 0014
Shay, W. R.	ORGN 0218	Shpiro, E. S.	PETR 0067	Simpson, J. J.	NUCL 0002
Shea, G. T.	COMP 0044	Shriver–Lake, L. C.		Simpson, J. O.	PMSE 0178
Shealy, D. B.	CEI 0002		BIOT 0246	Simpson, K. M.	
Shearer, B. G.	MEDI 0025	Shtral, V. I.	PETR 0061		INOR 0218
	MEDI 0026	Shu, C–K.	AGFD 0072	Simpson, P. J.	MEDI 0174
	MEDI 0171	Shuely, W. J.	PMSE 0155	Simurov, A. V.	COLL 0094
Sheehan, E.	ANYL 0123	Shuh, D.	INOR 0141	Sinclair, S.	CINF 0008
Sheehy, R. E.	AGFD 0057	Shull, B. K.	ORGN 0211	Sinel'nik, A. P.	COLL 0113
Sheetz, M. P.	BIOL 0137	Shuman, S.	AGRO 0130	Sinerius, G.	BIOL 0009
Sheffer, G.	INOR 0071	Shupack, S.	CEI 0009		BIOT 0090
Sheih, D.P.S.	PMSE 0142	Shupack, S. I.	COMP 0056	Sinfelt, J. H.	PHYS 0347
Sheih, P. S.	PMSE 0162	Shupe, A.	CHED 0131	Singer, K. D.	POLY 0151
Sheldon, J.	MEDI 0111	Shusta, J. M.	ORGN 0219	Singh, I. D.	PETR 0146
Shelton, D. P.	PHYS 0343	Shustorovich, E.		Singh, J.	MEDI 0195
Shelton, D. R.	AGRO 0139		CATL 0032	Singh, M.	BIOT 0049
Shelton, V. M.	INOR 0406	Sibert, J. W.	INOR 0295	Singh, M. M.	CHED 0150
Shemetulskis, N.		Siddiqui, A. H.	ORGN 0039	Singh, N.	POLY 0090
	PHYS 0035	Siddiqui, F.	I&EC 0059	Singh, P.	BIOL 0057
Shemshedini, L.		Sidess, E.	PMSE 0114	Singh, R. P.	AGFD 0077
	AGRO 0101	Sidle, W. C.	ENVR 0055		AGRO 0041
Shen, C.	ANYL 0087	Sidorchouk, V. V.		Singh, S. B.	ORGN 0184
	ANYL 0088		COLL 0108	Singh, S. K.	I&EC 0018
Shen, C. Y.	INOR 0345	Sieber, W.	CINF 0003		MEDI 0132
Shen, G. S.	ORGN 0222	Sieburth, S. M.	AGRO 0073	Singh, V.	BIOT 0007
Shen, K.	PHYS 0173	Sieburth, S.McN.		Singhal, G.	INOR 0205
Shen, Y-F.	CATL 0012		ORGN 0235	Sinha, B. R.	PMSE 0035
Shen, Y. X.	COLL 0058	Siedle, A. R.	CATL 0010	Sinha, N. K.	AGFD 0058
	POLY 0233	Siegel, A.	ORGN 0085	Sinn, E.	INOR 0262
	POLY 0345	Siegel, M. E.	NUCL 0076		INOR 0266

Sinn, E.	INOR	0267	Smith, L. M.	ANYL	0033	Song, J. C.	POLY	0159
	INOR	0268	Smith, M. E.	POLY	0227	Sonowane, P. B.		
	INOR	0269	Smith, N. L.	AGFD	0136		INOR	0268
Sipes, I. G.	AGRO	0093	Smith, O. W.	PMSE	0038	Sophiea, D.	PMSE	0040
Siqueiros, J.	I&EC	0010	Smith, P.	POLY	0200	Sorensen, A. A.	INOR	0193
Sircar, I.	MEDI	0121	Smith, S.	CHED	0077	Sorgi, K. L.	ORGN	0296
Sircar, J. C.	MEDI	0177		PHYS	0316	Soriano–Correa, C.		
	MEDI	0178	Smith, S. D.	COLL	0033		INOR	0190
Sirkar, K. K.	CHED	0144		COLL	0035	Soricelli, C. L.	INOR	0404
Sisti, N. J.	ORGN	0234		COLL	0053	Sosa, A.	NUCL	0076
Sitrin, R. D.	BIOT	0134		POLY	0076	Sosa, G.	PHYS	0296
Sitton, O. C.	PMSE	0020	Smith, W. B.	ORGN	0147	Sosa–Torres, M. E.		
	PMSE	0113	Smith, W. F.	PHYS	0362		INOR	0397
Sivadasan, K.	POLY	0270	Smith, W. W.	MEDI	0186		INOR	0398
Six, R. K.	PMSE	0110	Smitrovich, J. H.			Sotelo L., A.	AGFD	0210
Sjiariel, B.	CATL	0016		ORGN	0089	Soto, M.	PHYS	0310
Skatrud, P. L.	BIOT	0235	Smoliar, L.	PHYS	0203	Sottos, N. R.	COLL	0055
Skeels, G. W.	PETR	0002	Smudde, G. H.	COLL	0217	Sou, J.	INOR	0315
Skinner, J. L.	PHYS	0036	Snail, K. A.	FUEL	0037	Soulard, M.	PETR	0075
	PHYS	0044	Snape, C. E.	CATL	0039	Soum, A.	POLY	0309
Skinner, P. N.	CHAS	0015		FUEL	0004	Sounik, J. R.	POLY	0223
Skinner, W. S.	AGRO	0012		FUEL	0007	Sousa–Aguiar, E. F.		
Skodras, G.	CATL	0054		PETR	0147		PETR	0048
Skolnick, J.	POLY	0297	Snarey, M.	BIOT	0127	Sousa S., V.	AGFD	0210
Skolnick, P.	MEDI	0019	Sneddon, L. G.	POLY	0183	Southard, D. A., Jr.		
Skoog, M. T.	INOR	0293		POLY	0311		ORGN	0232
Skora, J. G.	CINF	0031	Snider, B. G.	ANYL	0085	Sowa, M. B.	PHYS	0174
Skotheim, T.	POLY	0091	Snider, S. W.	AGRO	0046		PHYS	0262
Skotnicki, J. S.	MEDI	0189		AGRO	0047		PHYS	0311
Skouroumounis, G. K.			Snieckus, V.	ORGN	0082	Spada, A.	MEDI	0111
	AGFD	0006		ORGN	0083	Spada, A. P.	MEDI	0113
Skrzypek, W.	BIOL	0111		ORGN	0277	Spangler, L. L.	POLY	0163
Skulman, B. W.			Snook, M. E.	AGRO	0034	Spanier, A. M.	AGFD	0159
	AGRO	0146	Snyder, N. J.	ORGN	0033	Spanton, S.	BIOT	0110
Slagle, I. R.	FUEL	0066	Snyder, R.	POLY	0096	Sparks, B. D.	I&EC	0021
Slavin, A. J.	PHYS	0099	Sobon, C.	POLY	0359		I&EC	0022
Slavin, L. L.	INOR	0403	Sobotka, L. G.	NUCL	0027	Sparks, D. E.	CATL	0052
Slichter, C. P.	PHYS	0347	Sochet, L. R.	FUEL	0081	Spaulding, T.	MEDI	0046
Sloan, J. M.	POLY	0081	Soder, S. L.	CMEC	0029	Spears, D. R.	FUEL	0073
Slocum, D. W.	ORGN	0085	Soderlund, D. M.			Spears, K. G.	PHYS	0220
Slocum, M. G.	ORGN	0085		AGRO	0151	Specht, E. D.	PHYS	0119
Slutter, C.	BIOT	0239	Soedjak, H. S.	INOR	0097	Speer, T. M.	COLL	0137
Slutzky, B.	YCC	0009	Soeteman, P.	POLY	0129	Speights, R. M.	CARB	0001
Small, G. J.	PHYS	0018	Sofranko, J. A.	CATL	0029	Spence, K.V.N.	INOR	0039
Small, G. W.	COMP	0004	Softley, L. G.	FUEL	0094	Spencer, H. G.	BIOT	0037
Smalley, R. E.	PHYS	0004	Sohn, D. W.	POLY	0158	Spencer, M. D.	INOR	0153
	PHYS	0045	Sohn, I. S.	PMSE	0059	Spencer, R.G.S.		
Smallwood, J. K.			Sohrabi, M.	I&EC	0006		PHYS	0196
	MEDI	0174	Sokolov, I. M.	PHYS	0132	Sperber, D.	NUCL	0008
Smart, C. J.	COLL	0029	Solans, X.	INOR	0057	Spergel, S. H.	MEDI	0188
	COLL	0156	Solar, J. P.	PETR	0071		ORGN	0114
Smedley, J. M.	FUEL	0107	Soled, S. L.	PETR	0052	Sperling, L. H.	PMSE	0039
Smellie, A. S.	CINF	0035	Solis, A. R.	PHYS	0282		PMSE	0073
Smid, J.	POLY	0271	Solis–Mendiola, S.				PMSE	0088
Smirnov, V. N.	PHYS	0175		BIOL	0074		PMSE	0171
Smith, A. B., III			Solleiro, J. L.	SCHB	0027	Spero, D. M.	ORGN	0006
	ORGN	0140	Sollie, J. C.	PETR	0065	Speronello, B. K.		
Smith, A. L.	CHED	0117	Solomon, M. S.	ORGN	0075		I&EC	0046
Smith, A. R.	NUCL	0049	Solomon, P. R.	FUEL	0052	Spevacek, J.	POLY	0124
Smith, B. A.	POLY	0208		PHYS	0149	Spicer, R. L.	CATL	0052
Smith, B. F.	INOR	0062	Solorzano, Z. M.			Spichtinger, R.	PETR	0047
	NUCL	0073		ENVR	0036	Spielman, A. I.	AGFD	0147
Smith, C. D.	PMSE	0052	Soltanshahi, F.	CINF	0027	Spier, R. E.	BIOT	0008
	PMSE	0053	Soma, L. R.	AGFD	0103	Spiess, H. W.	PHYS	0137
	PMSE	0054	Somasiri, N.L.D.				PMSE	0121
Smith, C.F.C.	MEDI	0086		PMSE	0113	Spina, E.	CARB	0048
Smith, C. M.	CATL	0046	Somasundaram, L.			Spina, K. P.	MEDI	0120
Smith, D. A.	INOR	0299		AGRO	0067		MEDI	0197
	POLY	0190	Somasundaran, P.			Spinelli, W.	MEDI	0095
Smith, D. M.	PETR	0032		COLL	0074	Spiro, T. G.	BIOL	0094
	PETR	0068		POLY	0270		BIOL	0095
	POLY	0312	Sommese, A.	INOR	0324	Spittler, T. D.	AGRO	0148
Smith, G. B.	CINF	0027	Somogyvari, A. F.			Sponsler, M. B.	BIOT	0205
Smith, G. M.	CINF	0018		CATL	0048		INOR	0167
Smith, G. P.	FUEL	0098	Son, S.	POLY	0276	Spontak, R. J.	COLL	0017
Smith, J. D.	MEDI	0146	Sondergaard, K.				COLL	0053
Smith, J. G., Jr.				PMSE	0085	Spooner, S. P.	INOR	0316
	POLY	0025	Sonderhoff, S. A.			Spotts, J. M.	INOR	0335
Smith, J. L.	MEDI	0181		BIOT	0012	Sprague, J. R.	PHYS	0220
Smith, K.	MEDI	0062	Song, C.	CATL	0050	Sprague, R. L.	PROF	0003
Smith, K. M.	INOR	0091	Song, H. H.	PMSE	0019	Spratt, T. E.	BIOL	0161
	INOR	0139	Song, I. H.	ORGN	0238	Sproat, B. S.	BIOL	0126
Smith, L.	PMSE	0048	Song, J.	PHYS	0192	Sprock, M.	CATL	0042

Spry, D. O.	ORGN 0033	Steinberg, M.	FUEL 0024	Storm, D. A.	PETR 0150
Spyroylias, G.	INOR 0221		FUEL 0047	Storrs, S. B.	BIOT 0135
Squires, M. E.	INOR 0198		FUEL 0121	Story, J.M.E.	ENVR 0059
Srebnik, M.	ORGN 0090	Steiner, M. G.	MEDI 0123	Stote, R.	CHED 0131
Sreelatha, G.	INOR 0402	Steiner, S. J.	AGFD 0084	St Pierre, T. G.	INOR 0116
Srinivasan, R.	PETR 0028	Steinfeld, J. I.	CHED 0128	Stracener, D. W.	
	POLY 0119	Stellman, J. M.	PROF 0008		NUCL 0027
Srinivasan, S.	PETR 0072	Stellwagen, J.	COLL 0182	Stradley, S. J.	BIOL 0003
Srisodsuk, M.	BIOT 0206	Stellwagen, N. C.		Strang, C. J.	BIOT 0242
Srivastava, R. P.			COLL 0182	Strangl, H.	BIOT 0183
	MEDI 0163	Stemple, J. Z.	CATL 0018	Stranick, M.	COLL 0085
Staab, G. A.	I&EC 0033	Stencel, J. M.	CATL 0051	Stratakis, M.	ORGN 0164
Staal, G. B.	AGRO 0003		FUEL 0048	Stratton, W. J.	CHED 0055
Stachel, S. J.	FUEL 0003		FUEL 0101	Straub, K. M.	AGRO 0022
Stackhouse, P.	ENVR 0029	Stenger, H. G.	PETR 0073	Strausberg, S.	BIOT 0192
Stair, P.	COLL 0217	Stenius, P.	COLL 0128	Strauss, G.	AGFD 0150
Stajic, M.	HIST 0002	Stenzel, J. B.	INOR 0281	Strazielle, C.	POLY 0132
Stallman, B.	BIOT 0201	Stepanek, P.	POLY 0275	Street, I. P.	BIOL 0014
Stamatoff, J.	POLY 0179	Stephan, D. W.	INOR 0025	Street, J. P.	ORGN 0060
Stanek, J.	MEDI 0184		INOR 0363	Strekas, T. C.	INOR 0160
Stanescu, C.	ORGN 0107	Stephanopoulos, G. N.			INOR 0161
Stanfield, R.	MEDI 0108		BIOT 0011		INOR 0232
Stanfield, R. L.	MEDI 0107		BIOT 0031		INOR 0233
	MEDI 0110	Stephens, R. D.		Strelitz, R. A.	MEDI 0105
Stanitski, C. L.	CHED 0052		CEI 0006	Strem, M.	SCHB 0008
	CHED 0053		ENVR 0015		SCHB 0012
Stanley, R. J.	ANYL 0064	Stephens, T. S.	POLY 0268	Strickland, J. H.	
Stansberry, M.	MEDI 0117	Stephenson, G. J., Jr.			AGRO 0073
Stansbury, J. W.			NUCL 0010	Strickland, R. C.	
	ORGN 0156	Stephenson, T. A.			FUEL 0094
	ORGN 0157		CHED 0201	Stringer, D. A.	AGRO 0029
Stansbury, W. F.		Steplewski, Z.	BIOT 0168	Strobl, H.	PHYS 0348
	TECH 0003	Sterling, D.	PMSE 0006	Strohmaier, K. G.	
	TECH 0017	Stern, A. H.	CEI 0009		PETR 0037
Stanton, D. T.	COMP 0015	Stern, C. L.	INOR 0358	Strongin, D. R.	COLL 0216
Stanton, J. F.	PHYS 0008	Stevens, A. C.	CHAS 0024	Stroscio, J. A.	PHYS 0183
	PHYS 0015	Stevens, C.	PHYS 0268	Strothkamp, K. G.	
Stapelfeld, A.	MEDI 0152	Stevens, S. Y.	CARB 0026		BIOL 0058
Staretz, M. E.	MEDI 0016	Stevens, W. J.	PHYS 0224		BIOL 0059
Staring, E.G.J.	POLY 0178	Stevenson, E.	ENVR 0013	Stroud, M. R.	CARB 0060
	POLY 0222	Stevenson, G. I.		Strouse, E. E.	ORGN 0206
Stark, D. H.	BIOL 0072		ORGN 0117	Strouse, G. F.	INOR 0282
Starnes, W. H., Jr.		Stevens–Truss, R.		Struble, C. B.	AGRO 0092
	POLY 0027		BIOL 0085	Strupczewski, J. T.	
Starr, T. L.	POLY 0338		BIOL 0086		MEDI 0130
Starzak, M. E.	BIOT 0141	Steward, L. R.	BIOL 0044	Strutt, P. R.	POLY 0335
Stassi, P.	NUCL 0028	Stewart, A. L.	FUEL 0058	Struve, W. S.	PHYS 0020
Staunton, D.	BIOT 0056	Stewart, F. F.	INOR 0329	Strybuc, J. M.	MEDI 0189
Staver, M.	BIOT 0110		INOR 0330	Stryker, V.	CARB 0004
Staversky, R. J.		Stewart, J. M.	BIOL 0006	Strzelecki, A. R.	
	BIOT 0032	Stewart, K.	COMP 0039		INOR 0311
Stavric, B.	AGFD 0116	Stewart, R. M.	POLY 0340	Stuart, J. A.	BIOT 0077
Steadman, G.	INOR 0400	Stewart, T. N.	ENVR 0032	Stubbe, J.	BIOL 0022
Stebbings, A. L.		Stewart, W. W.	PROF 0005		ORGN 0281
	COMP 0048	Stieber, R. W.	BIOT 0088	Stubbins, J. F.	MEDI 0140
Stechel, E. B.	PHYS 0053	Stiefel, E. I.	INOR 0005	Stucky, G. D.	INOR 0117
Steckle, W. P., Jr.			INOR 0371		PETR 0055
	PMSE 0118		INOR 0372	Studabaker, W. B.	
Steehler, J. K.	CHED 0042	Stier, M. A.	MEDI 0061		ENVR 0032
Steele–Perkins, G.		Stiles, A. B.	PETR 0034	Stukus, M.	BIOT 0160
	BIOL 0020	Stiles, G. L.	MEDI 0114	Stulen, R. H.	PHYS 0202
Steensma, D. H.		Stipanovic, A. J.		Stupp, S. I.	POLY 0276
	ORGN 0272		TECH 0014		POLY 0289
Steer, J. G.	FUEL 0070	Stock, J. T.	HIST 0038	Stur, T.	INOR 0368
Stefanithis, I. D.		Stockert, R. J.	BIOL 0154	Sturm, R.	MEDI 0179
	POLY 0066	Stoeffl, W.	NUCL 0011	Sturtevant, J.	BIOT 0113
	POLY 0285	Stoffer, J. A.	PMSE 0020	Sturtevant, J. M.	
Steffan, R.	MEDI 0179	Stoffer, J. O.	PMSE 0109		BIOT 0188
Steffey, M.	MEDI 0127		PMSE 0110	Stutz, J.	ANYL 0086
Stegeman, R.	POLY 0306		PMSE 0113	Stutz, C.	COLL 0215
Steigerwald, M. L.		Stojek, Z.	ANYL 0095	Su, C-L.	AGRO 0115
	CHED 0033	Stojmenovic, I.	COMP 0033	Su, J.	ORGN 0182
	INOR 0173	Stolle, W. T.	MEDI 0133		ORGN 0188
Stein, A. D.	PMSE 0120	Stolow, A.	PHYS 0163	Su, K.	POLY 0183
Stein, H.	MEDI 0120	Stone, M. L.	INOR 0070	Su, T-L.	MEDI 0013
	MEDI 0197	Stoner, B. R.	PHYS 0121	Su, W-F.A.	POLY 0224
Stein, R.	MEDI 0191	Storace, L.	BIOL 0005	Su, Y. O.	BIOL 0095
Stein, R. B.	MEDI 0096	Storck, W.	PHYS 0111	Suarez, D. P.	AGRO 0072
Stein, R. S.	COLL 0036	Storey, R. F.	POLY 0065	Subramanian, M. V.	
	PMSE 0022		POLY 0067		AGRO 0046
Stein, S. E.	COMP 0016		POLY 0068		AGRO 0047
Steinberg, E. P.		Storm, D. A.	COLL 0198	Subramanian, R.	
	NUCL 0058		PETR 0138		INOR 0078

Subramanyam, C.	MEDI 0172	Swartz, S. G., Jr.	ORGN 0093	Tan, C–S.	INOR 0106	
	ORGN 0036	Syed, J. R.	CMEC 0013	Tan, L–S.	PMSE 0169	
Subramanyam, S.		Sykora, J.	INOR 0342		POLY 0028	
	POLY 0032	Sylvestre, P.	ENVR 0016	Tan, P.	PMSE 0089	
	POLY 0073	Sylwester, A.	CATL 0046	Tan, W.	PHYS 0316	
	POLY 0074	Synder, C. E., Jr.		Tan, Y. Y.	POLY 0209	
Suchow, L.	PHYS 0312		I&EC 0008	Tanaka, A.	BIOT 0035	
Sue, H–J.	POLY 0157	Szabo, A.	PHYS 0166	Tanaka, F.	CATL 0049	
Suffett, I. H.	ENVR 0017	Szabo, G.	PETR 0115	Tanaka, T.	AGFD 0090	
Sugie, S.	AGFD 0090	Szafran, Z.	CHED 0150	Tang, J.	PHYS 0312	
Sugimoto, M.	PETR 0064	Szafraniec, L. J.		Tang, J–S.	INOR 0324	
Sugimoto, R.	PMSE 0183		ORGN 0253	Tang, X.	PMSE 0042	
Suib, S. L.	CATL 0012	Szaniawski, A.	BIOT 0037		PMSE 0064	
Suits, A. G.	PHYS 0349	Szczepaniak, K.			PMSE 0079	
Suits, W. H.	YCC 0006		PHYS 0033	Tanis, D. O.	PMSE 0091	
Sukhadia, A. M.		Szczepanik, A. M.		Tanis, S. P.	CHED 0004	
	PMSE 0135		MEDI 0085	Tanner, R. D.	ORGN 0179	
Sullivan, J. P.	MEDI 0025	Szczepura, L. F.		Tanriseven, A.	BIOT 0021	
	MEDI 0026		INOR 0066	Taranenko, V. B.	CARB 0028	
Sullivan, M. E.	MEDI 0094	Szczesniak, M.	PHYS 0033		BIOT 0081	
Sullivan, T.	ANYL 0097	Sze, J.	COLL 0058	Tarazano, D. L.		
Sultana, M.	MEDI 0154		POLY 0233		INOR 0238	
Summerfelt, R. M.			POLY 0345		INOR 0239	
	AGRO 0012	Szentesi, I.	BIOT 0148	Tarr, D. A.	CHED 0045	
Summerfield, G. C.		Szewczak, M. R.		Tarron, B.	PHYS 0295	
	POLY 0166		MEDI 0085	Tartar, A.	MEDI 0076	
Summers, E. A.		Szleifer, I.	POLY 0352	Tarter, S. L.	NUCL 0081	
	BIOT 0168	Szmacinski, H.	PHYS 0061	Tasayco, M. L.	BIOL 0115	
Summers, J. B.	MEDI 0176	Szostak, R.	PETR 0020	Tashiro, Y.	POLY 0310	
Summers, N.	POLY 0306	Szymanski, T.	PETR 0089	Tassan–Got, L.	NUCL 0007	
Summers, R. G.		Szymocha, K.	FUEL 0042	Tasset, E. L.	PETR 0053	
	BIOT 0083		FUEL 0141	Tata, J. R.	MEDI 0123	
Sumner, L.	BIOT 0240	Tabares Rodriguez, L.		Tata, V. S.	INOR 0186	
Sun, C. N.	AGRO 0117		COMP 0052	Tatarchuk, B. J.		
Sun, H.	INOR 0279	Tabor, M. W.	CEI 0008		PETR 0016	
	POLY 0080	Tachibana, T.	PHYS 0121	Tate, L. R.	FERT 0015	
Sun, J. J.	ANYL 0083	Tada, T.	PHYS 0313		FERT 0018	
Sun, Y–M.	POLY 0195	Taghiei, M. M.	FUEL 0056	Taub, I. A.	INOR 0394	
Sunamoto, J.	POLY 0098	Tai, M–S.	BIOT 0057	Taub, R.	AGRO 0128	
Sunamoto, M.	INOR 0165	Taing, M.	ORGN 0229	Taube, D. J.	CATL 0041	
Sunberg, R. J.	TECH 0002	Takacs, I.	BIOT 0148		FUEL 0100	
Sundararajan, J.		Takagi, M.	POLY 0301	Tavera, F. J.	I&EC 0065	
	POLY 0239	Takahama, K.	PETR 0145		I&EC 0066	
Sundelof, J. G.	MEDI 0182	Takahashi, I.	PMSE 0174		I&EC 0067	
Sundram, U. N.		Takahashi, T.	ANYL 0078	Tavera, G.	AGFD 0206	
	ORGN 0003	Takai, M.	BIOT 0172	Tavizon, G.	INOR 0048	
Sung, C.S.P.	POLY 0159	Takaishi, H.	AGRO 0013		PHYS 0225	
Sung, N.	POLY 0158	Takaya, H.	ORGN 0044	Taylor, C. H.	CMEC 0006	
	POLY 0282	Takayama, K.	MEDI 0099	Taylor, J.	CMEC 0002	
Sung, W.	BIOT 0212	Takemori, A. E.		Taylor, J–S.	BIOL 0113	
Sunil, K. K.	PHYS 0172		MEDI 0154		BIOL 0114	
Sunkara, M.	PHYS 0091		MEDI 0157		BIOL 0157	
Suo, Y.	BIOL 0062	Takeoka, G. R.	AGFD 0024	Taylor, K. G.	CARB 0056	
Super, L.	BIOT 0163	Takeuchi, K. J.	INOR 0066	Taylor, P. R.	PHYS 0276	
Super, M. S.	I&EC 0034		INOR 0067	Taylor, R.	COMP 0043	
Suquet, H.	PETR 0142		INOR 0068	Taylor, R. B.	POLY 0361	
Surburg, H.	AGFD 0071		INOR 0130	Taylor, R. J.	PETR 0151	
Susnitzky, D. W.		Taladriz, L. M.	BIOT 0047	Taylor, W. G.	AGRO 0026	
	PETR 0110	Talalay, P.	AGFD 0089	(TBA),	CHED 0016	
Suspene, L.	POLY 0114	Talanquer A., V.			CHED 0037	
Suter, U. W.	POLY 0064		CHED 0034	Tchoubar, D.	PHYS 0060	
	POLY 0141		CHED 0113	Teater, C.	BIOL 0072	
Suthar, B.	PMSE 0040	Talaty, E. R.	PHYS 0314	Tedder, L. L.	COLL 0009	
Suto, M. J.	MEDI 0061		PHYS 0315	Teeri, T. T.	BIOT 0206	
Sutton, D. E.	ORGN 0226	Talbot, D.	BIOT 0033	Tehlewitz, B.	BIOT 0088	
Suuberg, E. M.	FUEL 0011	Taleb, D. A.	PETR 0045	Teixeira, J. L.	BIOT 0246	
	FUEL 0087	Tallent, J. R.	BIOT 0077	Tekle, E.	COLL 0146	
Suzuki, A.	AGRO 0010	Tallman, M.	CHED 0090	Telfer, S. J.	COMP 0044	
Suzuki, F.	MEDI 0112	Talukder, M.A.H.		Telford, J.	INOR 0120	
Suzuki, K.	AGRO 0049		POLY 0268	Tempesta, M. S.		
Suzuki, S.	AGRO 0071	Tam, S. W.	MEDI 0084		AGFD 0191	
	ORGN 0249		MEDI 0153	Temple, R. G.	FUEL 0095	
Suzuki, T.	PMSE 0163		MEDI 0155	Temps, F.	FUEL 0082	
	POLY 0310	Tam, W.	POLY 0104	Tener, G.	INOR 0073	
Swager, T. M.	INOR 0135	Tamai, N.	PHYS 0110	Teng, G.	PMSE 0016	
Swalen, J. D.	POLY 0153	Tamarelli, A. W.		Teng, H.	FUEL 0011	
Swaminathan, S.			SCHB 0011	Teng, M.	ORGN 0111	
	BIOL 0032	Tamariz, J.	ORGN 0103	Tenkanen, M.	BIOT 0215	
Swanson, D. B.	YCC 0011		ORGN 0230	Tenorio L. J., A.		
Swanson, S.	BIOT 0110	Tamor, M. A.	PHYS 0150		FUEL 0105	
	POLY 0085	Tamura, S.	AGRO 0074	Tensfeldt, T. G.		
		Tan, B.	AGRO 0039		PHYS 0317	

Teranishi, R.	AGFD 0024	Thuning-Roberson, C. A.		Torkelson, J. M.		
Teraoka, I.	POLY 0153		BIOT 0168		POLY 0152	
Terech, P.	POLY 0169	Thurber, E. L.	POLY 0363		POLY 0163	
Tertykh, V. A.	COLL 0094	Thut, P.	MEDI 0046	Tornaritis, M. J.		
	COLL 0108		MEDI 0047		ENVR 0045	
Terzian, R.	COLL 0131	Thwaites, M. W.		Torres, A.	MEDI 0037	
	INOR 0318		FUEL 0048	Torres, C.	CARB 0004	
Tesh, K. F.	INOR 0223	Ticich, T. M.	CHED 0213	Torres, H. M.	ORGN 0228	
Tesoro, G.	POLY 0111		PHYS 0318	Torres, L. G.	BIOT 0196	
Tevault, C. V.	BIOT 0050	Tidwell, T. T.	ORGN 0125	Torres, L. M., Jr.		
	TECH 0016	Tiemann, B. G.	BIOT 0173		FERT 0002	
Teweldemedhin, Z.		Tiemeier, D. C.	AGFD 0193	Torres, M. C.	CARB 0050	
	INOR 0058	Tien, H. T.	BIOT 0102	Tortorella, V.	MEDI 0098	
Texter, J.	COLL 0040		BIOT 0142	Toscano, R. A.	ORGN 0018	
	COLL 0042	Tien, M.	BIOT 0237		ORGN 0150	
	COLL 0169	Tierney, D. P.	FERT 0019	Toscano A., R.	INOR 0032	
Thaisrivongs, S.		Tierney, J. W.	FUEL 0049	Toschi, A.	AGRO 0008	
	MEDI 0053		FUEL 0076	Toseland, B. A.	I&EC 0056	
Thakkar, A. J.	PHYS 0340	Tietz, A. J.	BIOT 0085	Tosic, R.	COMP 0033	
Thami, T.	POLY 0102	Timinski, P. A.	INOR 0311	Toste, A. P.	ANYL 0060	
Thantu, N.	PHYS 0330	Timmerman, H.			CHED 0149	
Tharakan, J.	BIOT 0060		MEDI 0081	Tota, P. V.	PETR 0139	
Thayer, A. M.	PHYS 0346	Timmons, R. B.		Toth, K.	ORGN 0284	
Thayer, J. S.	INOR 0080		CATL 0034	Touchstone, J. C.		
	INOR 0248	Timpa, J. D.	PMSE 0032		AGFD 0180	
Theil, E. C.	INOR 0011	Timpe, R. C.	FUEL 0006	Touhara, K.	AGRO 0014	
Theisen, L. A.	INOR 0184	Tippetts, M. T.	CHED 0090	Touroude, R.	COLL 0128	
	INOR 0228	Tirrell, D. A.	POLY 0302	Toussaere, E.	POLY 0006	
Theolier, A.	PETR 0131	Tirrell, M.	POLY 0165	Tovar T., A.	ENVR 0040	
Theopold, K. H.			POLY 0269	Townsend, C. A.		
	INOR 0148		POLY 0275		ORGN 0298	
	INOR 0313	Tisdell, S. E.	ENVR 0018	Townsend, R. R.		
Therien, M.	MEDI 0180	Tisinger, L.	PMSE 0014		ANYL 0015	
Thermos, K.	MEDI 0128	Tittlebaum, M. E.		Townson, P. J.	INOR 0133	
Thiel, J. P.	INOR 0358		ENVR 0004	Toy, A.D.F.	INOR 0322	
Thien, M. P.	BIOT 0071	Tkachenko, O. P.		Toyoshima, C.	BIOT 0121	
Thierry, A.	POLY 0201		PETR 0067	Tracy, H. J.	POLY 0359	
Thitiratsakul, R.		Tkacz, J. S.	MEDI 0058	Trainor, C.	MEDI 0181	
	PMSE 0065	Tobe, M. L.	INOR 0397		MEDI 0182	
Thoden, J. B.	INOR 0264	Tobe, S. S.	AGRO 0007		MEDI 0183	
	INOR 0265	Tochilnikov, D. G.		Trammell, G. L.		
Thoennessen, M.			POLY 0100		CHED 0112	
	NUCL 0027	Todaro, A. B.	INOR 0240	Tran, A. D.	ANYL 0018	
Thomalla, M.	COLL 0043	Todd, R. J.	BIOT 0133	Trautmann, W.	NUCL 0006	
Thomann, H.	FUEL 0072	Todhunter, J. A.			NUCL 0047	
Thomas, A. L.	HIST 0039		CHAS 0014	Travers, C.	PETR 0103	
Thomas, C. G.	INOR 0156	Tognotti, L.	FUEL 0025		PETR 0131	
Thomas, G. A.	CATL 0052	Tohme, A. G.	CEI 0016		PETR 0132	
Thomas, J. B.	ORGN 0002	Tokmakoff, A.	PHYS 0065	Trayer, I. P.	ORGN 0026	
Thomas, J-P.	NUCL 0054	Toktarev, A. V.	PETR 0039	Traynham, J. G.		
Thomas, M. H.	AGFD 0099	Tolbert, L. M.	ORGN 0084		HIST 0009	
Thomas, R. L.	BIOT 0037		ORGN 0162	Traynor, J. R.	MEDI 0086	
Thomas, S.	MEDI 0127	Tomalia, D. A.	POLY 0238	Tregloan, P. A.	PETR 0007	
Thomas, S. D.	FUEL 0079	Toman, A.	PMSE 0161	Trehan, I. R.	ORGN 0034	
Thompson, H. M.		Tomas, H. P.	ANYL 0056	Trejo, L. M.	PHYS 0319	
	POLY 0183	Tomasz, M.	BIOL 0088		PHYS 0320	
	POLY 0306		MEDI 0014	Trejo C., L. M.	CHED 0100	
Thompson, J. L.			MEDI 0015	Tremblay, P.	MEDI 0166	
	CINF 0040	Tombacz, E.	COLL 0160	Trentham, D. R.		
Thompson, J. S.		Tomchak, L.	BIOL 0042		ORGN 0026	
	INOR 0211	Tomczuk, B.	MEDI 0158	Tressl, R.	AGFD 0007	
Thompson, K.	BIOT 0134	Tomellini, S. A.			AGFD 0182	
Thompson, L. K.			I&EC 0023	Trevor, D. J.	PHYS 0127	
	PHYS 0196	Tomer, K. B.	ANYL 0050	Treybig, D. S.	PMSE 0142	
Thompson, M. E.			ANYL 0125	Trick, C. G.	INOR 0004	
	INOR 0047	Tomigahara, Y.	AGRO 0024	Triegel, G.	AGRO 0054	
	INOR 0049	Tomita, A.	FUEL 0032	Trifunovich, I. D.		
	INOR 0334	Tomitori, M.	PHYS 0097		MEDI 0049	
Thompson, R.	MEDI 0182	Tomka, J.	CARB 0010	Triggs, N. E.	PHYS 0321	
Thompson, R. C.		Tonat, K.	ENVR 0065	Trimnell, D.	CARB 0042	
	ORGN 0291	Tong, C.B.S.	AGFD 0030	Trinajstic, N.	COMP 0014	
Thomson, J. S.	FUEL 0060	Tong, H. Y.	ANYL 0050		COMP 0025	
Thongsinthusak, T.		Tong, W.	POLY 0343	Trinidad, R.	PHYS 0225	
	AGRO 0043	Tonnesen, H. H.		Tripathy, S.	POLY 0131	
Thonnard, N.	NUCL 0042		AGFD 0043	Triplett, T. L.	BIOL 0003	
Thornton, J. M.		Topal, M. D.	BIOL 0125	Trivedi, B. K.	MEDI 0108	
	CINF 0032	Torardi, C. C.	CHED 0022		MEDI 0110	
Thornton, L. L.			PHYS 0028	Trohalaki, S.	PHYS 0322	
	AGRO 0053	Torchia, D.	PHYS 0113	Tromp, M.G.M.		
Thorp, H. H.	INOR 0410	Toreki, W.	POLY 0360		INOR 0098	
Thorson, J. S.	BIOL 0110	Torgerson, M. R.		Trouw, F.	PETR 0058	
Thrasher, J. S.	INOR 0024		CATL 0021	Truffy, J.	NUCL 0075	
				Truhlar, D. G.	PHYS 0165	

Name	Code	Num		Name	Code	Num		Name	Code	Num
Trujano S., A.	COMP	0053		Ujiie, M.	BIOT	0211		Vandenbosch, R.		
Trulli, T. K.	FUEL	0058		Ullman, J.	NUCL	0021			NUCL	0004
Trumbore, S. E.				Ullman, R.	POLY	0166		Vanden Bossche, H.		
	GEOC	0016		Umarani, R.	ORGN	0032			MEDI	0007
Trupin, G. L.	PROF	0010		Underiner, T. L.				van der Hoef, I.		
Tsai, C.	MEDI	0148			ORGN	0130			PHYS	0196
Tsai, F-Y.	INOR	0043		Unnikrishnan, S.				Vander Molen, R. H.		
Tsai, L. W.	AGRO	0004			PETR	0100			FUEL	0122
Tsai, M-D.	BIOL	0016		Unruh, B.	CINF	0011		VanDerveer, D.	ORGN	0233
Tsai, W. H.	POLY	0087			CINF	0043		van der Vies, S. M.		
Tsai, Y. S.	MEDI	0032		Unwin, P.	COLL	0170			BIOT	0116
Tsang, W.	FUEL	0067		Upton, R.	CINF	0033		Van de Water, C.		
Tsao, E. I.	BIOT	0164		Urabe, K.	PETR	0129			AGFD	0064
Tschirret-Guth, R. A.				Urban, M. W.	POLY	0255		Vandigrifft, V. L.		
	INOR	0017		Urban, P.	INOR	0249			COLL	0028
Tse, M. Y.	CATL	0055		Urbina-Duran, J. E.				VanDoremaele, G.		
	FUEL	0102			CHED	0096			BIOT	0173
Tseng, J.C.C.	BIOT	0173		Uruchurtu C., J.				Van Dyk, J. W.	PMSE	0033
Tseng, K-C.	MEDI	0039			PHYS	0281		Van Dyke, D. A.		
	MEDI	0040		Usher, J. J.	BIOT	0128			CHED	0161
Tseng, S-S.	AGRO	0074		Uskokovic, M. R.					CHED	0211
Tseng, W.	PHYS	0363			ORGN	0284		van Egmond, J.		
Tsiao, C.	CATL	0029		Usuki, A.	POLY	0291			PMSE	0106
Tsong, T. T.	PHYS	0101		Utley, D.	NUCL	0028		van Eldik, R.	INOR	0395
Tsushima, K.	ORGN	0014		Utracki, L. A.	PMSE	0024		van Galen, P.J.M.		
Tsutsumi, N.	POLY	0225		Utz, T.	INOR	0311			MEDI	0114
Tsuzuki, R.	MEDI	0099		Vaccari, G.	MEDI	0135		van Geerestein, V. J.		
Tsymbalov, S.	MEDI	0152		Vadakumcherry, S.					COMP	0040
Tudela, D.	INOR	0065			FUEL	0061		Van Hecke, G. R.		
Tuichiev, S.	POLY	0101		Vadgama, P.	BIOT	0039			CHED	0043
Tully, P. S.	ORGN	0091		Vala, M.	PHYS	0078			CHED	0182
Tulshian, D.	MEDI	0117		Valderrama-Cano, A.				VanHeertum, J. C.		
Tung, C. J.	PMSE	0057			PHYS	0323			AGRO	0046
Tung, Y-S.	PMSE	0154		Valdes de la Torre, T. B.					AGRO	0047
Tunstad, L.	INOR	0120			CHED	0165		Vanherzeele, H.		
Turbeville, W.	CATL	0020		Valdes M., J.	INOR	0032			POLY	0205
	COLL	0076		Valdes-Martinez, S. E.					POLY	0213
Turckes, M. K.	CHED	0139			AGFD	0138			POLY	0221
Turi, L.	PHYS	0247			AGFD	0139			POLY	0226
	PHYS	0326		Vale, M. G.	INOR	0261		van Hooff, J.H.C.		
Turk, T.	INOR	0317		Vale, R. D.	BIOL	0138			PETR	0043
Turkevich, A. L.				Valentin, J. R.	ANYL	0052		Vankai, V. A.	INOR	0094
	NUCL	0061		Valentin, L.	INOR	0216		VanLishout, Y.	FUEL	0008
Turler, A.	NUCL	0074		Valentine, K.	BIOL	0115		Van Meirvenne, N.		
Turnbull, A. R.	NUCL	0087		Valentine, K. G.					MEDI	0062
Turnbull, S. P., Jr.					INOR	0047		van Nieuwenhuizen, G. J.		
	MEDI	0148		Valentini, J. J.	PHYS	0162			NUCL	0039
Turner, C. M.	AGRO	0101			PHYS	0263		Van Nordstrand, R. A.		
Turner, S. R.	MEDI	0053			PHYS	0321			PETR	0023
Turnes, R.	BIOT	0157			PHYS	0325		van Oeveren, A.		
Turos, E.	ORGN	0213		Valenzuela Z., M.					ORGN	0042
Turro, N. J.	ORGN	0126			COLL	0126		Van Peteghem, C. H.		
	PHYS	0135			COLL	0127			AGFD	0130
	POLY	0238		Vali, V.	BIOT	0025		Van Rie, J.	AGRO	0153
Tusar, M.	COMP	0002		Valle, F.	BIOT	0130		Van Roey, P.	CARB	0022
Tway, C. L.	CATL	0035		Vallejo-Cordoba, B.					MEDI	0051
Tweldemedhin, Z. S.					AGFD	0108		van Roggen, A.	BIOT	0252
	INOR	0352		Valle-Vega, P.	AGFD	0139		Van Ryswyk, H.		
Twieg, R.	POLY	0106		Valle-Vega P., P.					ANYL	0007
	POLY	0153			AGFD	0138			CHED	0182
	POLY	0208		Valley, J. F.	POLY	0130		van Santen, R. A.		
Twieg, R. J.	POLY	0155		Valone, S. M.	PHYS	0356			PETR	0054
Tycko, R.	PHYS	0375		van Amerongen, H.				Vansco, G. J.	POLY	0305
Tyklar, Z.	INOR	0044			PHYS	0020		van Steenderen, P.		
Tyler, D. R.	INOR	0158		Van Antwerp, J.					CATL	0044
Tysoe, W.	INOR	0160			ANYL	0137		VanSweden, J. A.		
	INOR	0161		Vanasse, B.	MEDI	0166			CHED	0075
Tzeng, W-H.	INOR	0043		van Bekkum, H.				Vanysek, P.	COLL	0004
Uberoi, M.	FUEL	0132			PETR	0024		Van Zandt, M. C.		
Ubillas, R.	AGFD	0191			PETR	0065			ORGN	0272
Uboh, C.	AGFD	0103			PETR	0080		Vargas-Baca, I.	CHED	0057
Uchida, W.	MEDI	0099		Van Berkel, G. J.					INOR	0042
Ueda, E.	PMSE	0101			ANYL	0092			INOR	0061
Ueda, H.	POLY	0023		Van Bogaert, I.	MEDI	0062		Vargas Ch., V.	NUCL	0089
Ueda, I.	MEDI	0032		van Boom, J. H.				Vargas E., A.	COLL	0125
Ueda, M.	COLL	0146			CARB	0055		Vargha Butler, E. I.		
Ueda, Y.	POLY	0225		VanCamp, S. L.					COLL	0086
Uejima, A.	BIOT	0035			BIOL	0015		Varin, M.	MEDI	0166
Ugalde-Saldivar, V. M.				van Dam, J.	PMSE	0058		Varner, R.	NUCL	0027
	INOR	0397		van den Berg, J. P.				Varney, M. D.	MEDI	0186
Ugo, P.	COLL	0005			PETR	0025		Vartanian, M. A.		
Uhm, H. L.	INOR	0336			PETR	0112			MEDI	0177
Uhrich, K. E.	PMSE	0075								

Varvarin, A. M.		Viikari, L.	BIOT 0215	Waid, P.	MEDI 0025	
	COLL 0108	Viitanen, P. V.	BIOT 0116	Wainer, I. W.	ANYL 0120	
Vasquez, C.	POLY 0054	Vijh, A.	CONG 0009		BIOT 0036	
Vasquez, R.	BIOT 0053	Villaeys, A.	POLY 0102	Wainman, T.	CEI 0009	
Vaughan, D.E.W.		Villalobos-Penalosa, M.		Waite, J. H.	INOR 0138	
	PETR 0037		ENVR 0039	Wakarchuk, W.	BIOT 0212	
Vaughan, J. S.	PETR 0011	Villena, R.	INOR 0032	Wakeley, L. D.	ENVR 0009	
Vaughey, J. T.	INOR 0358	Vimalchand, P.	CATL 0053	Waki, M.	AGFD 0127	
Vaughn, S. N.	FUEL 0059	Vincente, L.	PHYS 0288	Wakim, F. G.	POLY 0043	
Vavrek, R. J.	BIOL 0006	Vinckier, C.	PHYS 0141	Walba, D. M.	BIOT 0202	
Vazquez M., A.	ORGN 0196	Viney, C.	POLY 0314	Walch, M.	BIOT 0097	
Vazquez-Martinez, A.		Vinje, K.	PETR 0020	Waldman, J. M.		
	CHED 0166	Vinodgopal, K.	COLL 0190		CEI 0019	
Vazquez-Olmos, A.		Vinogradoff, A. P.		Waldman, T. E.		
	INOR 0252		AGRO 0046		CATL 0005	
Vazquez-Ramos, C.			AGRO 0047		INOR 0247	
	INOR 0079	Viola, V. E.	NUCL 0026	Waldow, D. A.	PMSE 0107	
Veber, D. F.	MEDI 0192		NUCL 0036		PMSE 0134	
Vedage, G. A.	I&EC 0056	Visscher, K. B.	PMSE 0002	Wales, M. E.	BIOT 0242	
Vedejs, E.	ORGN 0278	Visser, S. A.	PMSE 0145	Walker, D.	POLY 0086	
Vederas, J. C.	BIOL 0065	Viswanath, R. S.		Walker, D. G.	FUEL 0054	
Vedres, D. D.	AGRO 0026		ENVR 0022	Walker, D. K.	FUEL 0058	
Vedrine, J. C.	PETR 0133	Vite, L.	CHED 0169	Walker, J.	INOR 0097	
Vega, A. J.	PHYS 0371	Vitous, J.	ORGN 0027	Walker, J. A.	FUEL 0067	
Vega, S.	PHYS 0372	Viveros, T.	PETR 0092	Walker, M. C.	BIOL 0054	
Velander, W. H.			PHYS 0225	Walker, W.	TECH 0001	
	BIOT 0060	Vizkelethy, G.	NUCL 0081	Walker, W. C.	TECH 0007	
Velazquez, A.	POLY 0027	Vizza, F.	INOR 0210		TECH 0008	
Velazquez, P.	MEDI 0064	Vogler, A.	INOR 0317	Wallace, D.W.R.		
Venanzi, C. A.	MEDI 0045	Vohs, J. M.	COLL 0048		GEOC 0003	
	ORGN 0057		COLL 0173	Wallace, G. G.	COLL 0129	
	ORGN 0174	Voigtman, E.	CHED 0192		COLL 0168	
Venanzi, T. J.	MEDI 0045	Vold, C. L.	FUEL 0037	Wallace, I.	PHYS 0351	
	PHYS 0324	Vold, R. L.	PHYS 0136	Wallace, J. S.	BIOT 0094	
Venditti, R. A.	PMSE 0173	Volgyi, A.	BIOT 0148	Wallow, T. I.	POLY 0024	
Venepalli, B. R.		Volino, F.	PMSE 0119	Wallraff, G. M.	COMP 0059	
	MEDI 0158	Volksen, W.	POLY 0085		POLY 0188	
Venero, A. F.	PETR 0069	Voll, R. J.	ORGN 0074	Walls, D. J.	POLY 0279	
Venhuizen, A.H.J.		von Borczyskowski, C.		Walmsley, F.	CHED 0151	
	POLY 0178		PHYS 0082	Walsh, A. M.	PHYS 0035	
Venkataraghavan, R.		von Meerwall, E. D.		Walsh, E. N.	INOR 0348	
	CINF 0026		POLY 0165	Walsh, P.	POLY 0106	
Venkataraman, B.		Vontor, T.	ORGN 0154	Walter, G. R.	ENVR 0051	
	PHYS 0325	VonVoigtlander, P. F.		Waltermire, R. E.		
Venkataraman, K.			MEDI 0126		ORGN 0294	
	AGRO 0128	Voogd, P.	PETR 0065	Walters, P. J.	CINF 0017	
Venkatasubramanian, N.		Vorinen, J.	COLL 0071	Walters, V. A.	CHED 0085	
	POLY 0316	Voronin, E. F.	COLL 0095		PHYS 0327	
Venkatraman, S.			COLL 0112		PHYS 0328	
	ORGN 0261	Vorres, K. S.	FUEL 0001	Walzer, J. F.	INOR 0131	
Ventura, O. N.	PHYS 0326	Voss, H.	BIOT 0248	Wamser, C. C.	BIOT 0126	
Venugopal, G.	COLL 0145	Vovelle, C.	FUEL 0080	Wan, B-Z.	I&EC 0075	
Venugopalan, M.		Voyta, J. C.	BIOT 0167		PETR 0087	
	CHED 0112	Vrakking, M.J.J.		Wan, J.K.S.	CATL 0055	
Venzani, C. A.	PHYS 0324		PHYS 0163		FUEL 0102	
Vera, E.	MEDI 0065	Vrana, R.	FUEL 0060	Wan, Z.	PHYS 0174	
Verburg, K. M.	MEDI 0197	Vrsanska, M.	BIOT 0185		PHYS 0242	
Vercellotti, J. R.		Vsevolodov, N. N.		Wanasundara, P.K.J.P.D.		
	AGFD 0098		BIOT 0079		AGFD 0019	
Verdier, P. H.	POLY 0235	Vyas, D. M.	MEDI 0012	Wand, A. J.	PHYS 0168	
Verhoeven, J. W.		Vyshenskii, S. V.		Wander, S. A.	INOR 0225	
	POLY 0222		BIOT 0004	Wandrekar, V.	ORGN 0189	
Verhoogt, H.	PMSE 0058	Wacholtz, W. F.		Wang, A.	CATL 0011	
Verkade, J. G.	CHED 0021		INOR 0284	Wang, B.	POLY 0264	
	INOR 0023	Wachs, I. E.	CATL 0057		POLY 0284	
	INOR 0191		COLL 0196		POLY 0343	
	INOR 0323	Wachter, G.	MEDI 0036	Wang, C. C.	I&EC 0075	
	INOR 0333	Wada, R.	NUCL 0028		PETR 0087	
Verkruijsse, H. D.		Waddell, E. L.	FERT 0013	Wang, C-G.	MEDI 0019	
	INOR 0229	Waddling, C.	POLY 0186	Wang, C-I.	BIOL 0114	
Verma, A. K.	AGFD 0113	Waddon, A. F.	PMSE 0136	Wang, C-K.	AGFD 0154	
Verma, M. P.	I&EC 0015	Wade, B.	POLY 0316	Wang, D.	POLY 0065	
Vernon, M.	COLL 0214	Wade, K.	PETR 0106	Wang, D.I.C.	BIOT 0000	
	COLL 0215	Wade, W. H.	COLL 0083		BIOT 0009	
Verzera, A.	AGFD 0199	Wagener, K. B.	POLY 0046		BIOT 0045	
Vesecky, S. M.	PHYS 0128	Wagner, B. D.	PHYS 0134		BIOT 0136	
Viades T., J.	CHED 0164	Wagner, F. W.	CARB 0004		BIOT 0222	
	CHED 0178		CARB 0050	Wang, D. W.	POLY 0110	
Viands, C. A.	BIOT 0023	Wagner, H. G.	FUEL 0082	Wang, D-Z.	PMSE 0016	
Viano, B.	NUCL 0028	Wagner, N. J.	COLL 0165	Wang, G. G.	PMSE 0090	
Vicente, L.	PHYS 0225	Wagner, W. J.	ORGN 0076	Wang, H.	ENVR 0060	
Vient, M. A.	NUCL 0012	Wahl, R. C.	MEDI 0195		FUEL 0108	
Viger, A.	MEDI 0009	Wahlstrom, D.	AGRO 0019		PHYS 0240	

Wang, H.	PHYS	0353	Warthen, J. D., Jr.			Weinhold, F. A.		
Wang, H-E.	INOR	0029		AGRO	0033		PHYS	0106
Wang, H. H.	PHYS	0024	Warzywoda, M.	BIOT	0182	Weinkauf, D. H.		
Wang, H. J.	BIOT	0037	Washington, G. E.				POLY	0115
Wang, H-L.	INOR	0028		COLL	0164	Weinstein, S. H.		
Wang, H. M.	CARB	0039	Wasserman, E. P.				MEDI	0196
Wang, H. Y.	POLY	0118		INOR	0167	Weis, K.	BIOL	0009
Wang, J.	BIOL	0097	Wassmer, K-H.			Weisburger, J. H.		
	COLL	0181		COLL	0072		AGFD	0086
	PMSE	0064	Waszczak, J. V.	PHYS	0026	Weiske, C.	CINF	0010
	PMSE	0079	Watanabe, K. A.			Weismiller, M. C.		
	PMSE	0091		CARB	0023		ORGN	0293
	PMSE	0155		CARB	0024	Weiss, C. A., Jr.		
	POLY	0251		MEDI	0013		AGRO	0061
Wang, J. D.	PETR	0106	Watanabe, S.	AGRO	0049	Weiss, D.	PHYS	0111
Wang, J-L.	MEDI	0155	Watanabe, T.	POLY	0128	Weiss, M. A.	BIOL	0002
Wang, K. K.	ORGN	0098	Waterman, S.	COMP	0044	Weiss, R.	ORGN	0205
	ORGN	0099	Waters, L. C.	AGRO	0097	Weiss, R. A.	PMSE	0118
Wang, K. W.	BIOT	0138	Waters, R. M.	AGRO	0033		PMSE	0132
Wang, R.	AGRO	0109		AGRO	0035	Weisshaar, J. C.		
	AGRO	0130	Watkins, D. M.	ORGN	0263		PHYS	0199
Wang, R-C.	PETR	0106	Watkins, J. J.	POLY	0076		PHYS	0306
Wang, R-J.	INOR	0106	Watkins, W. M.			Weissmann, H. S.		
Wang, R. Y.	PETR	0130		CARB	0015		PROF	0004
Wang, S.	COLL	0050	Watson, B. A.	COLL	0194	Weitekamp, D. P.		
	INOR	0353	Watson, D. G.	CINF	0021		PHYS	0042
Wang, S. B.	POLY	0283	Watson, P. L.	INOR	0328	Weitzberg, M.	MEDI	0171
Wang, S. J.	COLL	0078	Watson, R. F.	CHED	0179	Welch, J. T.	ORGN	0079
Wang, S. M.	ORGN	0214	Watterson, A. C.				ORGN	0080
	ORGN	0215		POLY	0367		ORGN	0081
Wang, T. C.	ENVR	0019	Watts, J. D.	PHYS	0008	Weller, M. R.	NUCL	0000
Wang, W.	BIOL	0035		PHYS	0015	Weller, P.	ENVR	0083
	FUEL	0074	Wayman, M.	BIOT	0184	Weller, R. A.	NUCL	0064
	ORGN	0277	Weaver, J. D.	PETR	0053	Wells, C.	MEDI	0067
Wang, W. G.	MEDI	0159	Weaver, J. H.	PHYS	0049	Wells, M.J.M.	ANYL	0047
Wang, W. T.	CARB	0054	Weaver, K. D.	PMSE	0109	Wells, R. L.	COLL	0136
Wang, W. Y.	INOR	0060	Weaver-Delamater, E. A.				COLL	0137
	INOR	0179		BIOT	0166		COLL	0138
Wang, X.	ORGN	0277		ORGN	0201		COLL	0139
	PMSE	0172	Webb, A. G.	FUEL	0074		COLL	0140
Wang, X. C.	ORGN	0106	Webb, C. F.	MEDI	0079	Wells, S. E.	ENVR	0078
Wang, X-Q.	BIOL	0141	Webb, J.	INOR	0116	Welsh, K. M.	MEDI	0186
Wang, Y.	CATL	0009	Webb, M. L.	MEDI	0160	Welsh, W.	INOR	0073
	COLL	0205	Webber, S.	MEDI	0186	Welsh, W. A.	PETR	0050
	INOR	0041	Webber, S. E.	PHYS	0109	Wemmer, D. E.	PHYS	0197
	INOR	0043		POLY	0078	Wen, Y-S.	INOR	0107
	PETR	0099		POLY	0079	Wender, I.	FUEL	0049
	PMSE	0092		POLY	0210		FUEL	0076
Wang, Y. L.	PHYS	0329	Weber, A. E.	MEDI	0122	Wendler, P.	BIOT	0236
Wang, Y-Z.	AGRO	0015		MEDI	0123	Wendling, J. M.		
Wang, Z.	ORGN	0116	Weber, E.	MEDI	0147		CEI	0003
Wang, Z. L.	COLL	0181	Weber, J-P.	CEI	0004	Weng, G.	BIOT	0080
	PHYS	0119		CEI	0005	Weng, W-Q.	INOR	0113
Wang, Z. Y.	AGFD	0065		CEI	0010	Went, G. T.	COLL	0206
	AGFD	0067	Weber, M.	BIOT	0110		COLL	0211
	AGFD	0088	Weber, N. E.	AGRO	0076	Went, M. S.	COLL	0211
Wangen, L. E.	INOR	0062	Weber, P. M.	PHYS	0330	Wentzlaff, T.	COLL	0204
Wann, M-H.	BIOL	0116	Weber, R. S.	COLL	0194	Werkhoff, P.	AGFD	0071
Ward, D. E.	ORGN	0004	Weber, S.	SCHB	0010	Werkman, P. J.		
Ward, J.	MEDI	0067	Webster, B. J.	BIOT	0098		POLY	0209
Ward, J. F.	PHYS	0339	Webster, H. F.	PMSE	0052	Werner, A. F.	CEI	0025
Ward, V.	AGRO	0009		PMSE	0054	Werner, H-J.	PHYS	0188
Ward, W. J., Jr.			Weed, H. C.	INOR	0236	Werner, T. C.	CHED	0202
	INOR	0373	Weenen, H.	AGFD	0051	Wertz, D.	FUEL	0001
Ward, Y. D.	ORGN	0092	Weerasinghe, C. A.			Wertz, D. A.	ORGN	0174
Ware, W. R.	PHYS	0134		AGRO	0130	Wertz, D. L.	FUEL	0040
Wark, D. L.	NUCL	0010	Wegner, G.	POLY	0036	Wery, J-P.	BIOL	0072
Wark, D. T.	CMEC	0016	Wei, D.	ORGN	0108	Wesley, I. V.	ANYL	0081
Wark, T. A.	POLY	0258	Wei, N.	INOR	0178	West, A. S.	CHAS	0036
Warman, J. M.	POLY	0214	Wei, Y.	POLY	0232	West, D.	CMEC	0008
Warnatz, J.	FUEL	0084	Weichman, B.	MEDI	0179	West, E. E.	MEDI	0021
Warner, C. R.	AGFD	0109	Weidman, T. W.			Westheimer, F. H.		
Warns, C.	PHYS	0023		PHYS	0138		INOR	0289
Warrack, B. M.	ANYL	0139	Weigand, W. A.			Westlake, D.W.S.		
Warrakah, E.	AGRO	0130		BIOT	0159		BIOT	0129
Warren, J.	AGRO	0023	Weigel, L. O.	ORGN	0031	Westmoreland, P. R.		
Warren, R.A.J.	BIOT	0207	Weigert, W. A.	CHED	0123		FUEL	0079
	BIOT	0217	Weiller, B. H.	INOR	0167	Weston, H.	BIOL	0049
Warren, W. S.	PHYS	0040		INOR	0195		BIOL	0050
Warrick, M. W.				PHYS	0331	Weston, H. D.	MEDI	0173
	BIOL	0072	Weiner, A. M.	PHYS	0178	Westrum, L. J.	ORGN	0042
Warshel, A.	BIOL	0131	Weiner, B. R.	PHYS	0240	Wetzel, R.	BIOL	0039
				PHYS	0353		BIOT	0114

Wever, R.	INOR 0098		Wildeman, J.	POLY 0129		Winstead, C. B.	
Wey, S-J.	ORGN 0056		Wildenhain, P.	NUCL 0021			PHYS 0173
Weyda, H.	PETR 0041		Wildt, T.	PETR 0009		Winter, H. H.	PMSE 0133
Whalen, M. J.	ENVR 0054		Wilen, S. H.	ORGN 0041			PMSE 0136
Whangbo, M-H.			Wilhelmy, J. B.	NUCL 0032		Winterhalter, P.	
	CHED 0022		Wilkerson, J. F.				AGFD 0024
	PHYS 0028			NUCL 0010			AGFD 0025
	PHYS 0295		Wilkes, G.	COLL 0058		Winters, R. T.	MEDI 0011
Wheeland, J.	YCC 0013			POLY 0233		Wintterlin, J.	PHYS 0126
Wheelock, G. D.				POLY 0345		Wipf, P.	MEDI 0052
	AGRO 0099		Wilkes, G. L.	POLY 0264			ORGN 0089
Whetten, R.	PHYS 0047			POLY 0284		Wipke, W. P.	CINF 0016
Whisnant, D. M.				POLY 0285		Wipke, W. T.	CINF 0009
	CHED 0207		Wilkins, R. M.	AGRO 0036		Wirick, C.	GEOC 0003
Whitaker, J. R.	AGFD 0041		Willand, C. S.	POLY 0105		Wirth, M. J.	ANYL 0025
Whitcombe, M. J.				POLY 0150		Wise, M. B.	FUEL 0089
	POLY 0219		Willermet, P. A.			Wise, M. L.	COLL 0008
White, A. D.	MEDI 0107			PETR 0140		Wise, S. A.	ANYL 0098
White, B. R.	POLY 0305		Willett, P.	COMP 0043		Wiseman, B. R.	
White, C.	FUEL 0040		Williams, A.	FUEL 0107			AGRO 0034
White, D. C.	BIOT 0099			FUEL 0113		Wislocki, P. G.	AGRO 0128
White, D. R.	FERT 0005		Williams, B. E.	PHYS 0121		Wisniewski, J. L.	
White, E. H.	ORGN 0022		Williams, B. W.				COMP 0019
White, H. D.	BIOL 0103			CHED 0200		Wisniewski, L.	PHYS 0068
	BIOL 0141		Williams, C. C.	COLL 0213		Witman, S. L.	POLY 0175
White, J. D.	ORGN 0285		Williams, C. E.	AGRO 0072		Witmer, M.	BIOL 0155
White, J. I.	MEDI 0097		Williams, D. J.	CMEC 0014		Wittenkeller, L.	
	MEDI 0191		Williams, D. R.	ORGN 0282			INOR 0184
White, J. M.	COLL 0065		Williams, E. T.	NUCL 0078		Wittkowski, R.	AGFD 0054
	PHYS 0366		Williams, G.	COLL 0150		Wittmann, J-C.	
White, J. R.	CHED 0001		Williams, J. L.	PETR 0090			POLY 0201
White, J. W.	PETR 0058		Williams, J. M.	PHYS 0024		Wnek, G. E.	POLY 0216
White, P.	ORGN 0150			PHYS 0295			POLY 0218
White, P. A.	ANYL 0115		Williams, K. B.	INOR 0309		Wohlfeil, E. R.	MEDI 0142
White, R. E.	MEDI 0196		Williams, K. E.	MEDI 0162		Wohn, F. K.	NUCL 0049
White, R. F.	MEDI 0181		Williams, L. T.	MEDI 0023		Wojcicki, A.	INOR 0050
Whitecar, C. K.			Williams, P. J.	AGFD 0006		Wojcik, A. B.	PMSE 0081
	POLY 0232			AGFD 0203		Wolczanski, P. T.	
Whitehair, S. J.			Williams, R. M.				INOR 0103
	PHYS 0092			ORGN 0283		Wold, A.	INOR 0205
Whitesell, J. K.	ORGN 0292		Williams, R. S.	INOR 0141		Wolf, B. A.	POLY 0146
Whitesides, G. M.			Williams, W. M.			Wolf, G. H.	PHYS 0260
	BIOT 0089			FERT 0019		Wolf, M. A.	ORGN 0172
Whitfield, J.	NUCL 0025		Williamson, R. L.			Wolf, M. O.	INOR 0027
Whitman, D. A.				PHYS 0192		Wolf, R. E.	AGRO 0125
	ANYL 0124		Willig, F.	PHYS 0111		Wolf, W. J.	AGFD 0141
Whitnell, R. M.			Willinger, R.	CHAS 0037		Wolfe, M. S.	POLY 0244
	PHYS 0037		Willis, R. A.	CHAS 0033		Wolfersberger, D. E.	
Whitney, C. C., III			Willner, I.	INOR 0283			FERT 0009
	MEDI 0162		Willoughby, R.	ANYL 0123		Wolff, J. A.	INOR 0181
Whitten, D. G.	CATL 0060		Willson, C. G.	POLY 0106		Wolfinger, T. F.	
	INOR 0316			POLY 0153			ENVR 0082
	ORGN 0161			POLY 0208		Wolfsberg, K.	NUCL 0022
	ORGN 0166		Wilmot, C. A.	MEDI 0085		Wolfson, G.	I&EC 0042
Whitten, W. B.	PHYS 0151		Wilschut, H. W.			Wolkow, R. A.	PHYS 0185
Whittern, D.	BIOT 0110			NUCL 0039		Wollman, E.	INOR 0350
Whittingham, M. S.			Wilson, D.	PHYS 0195		Wollman, S. T.	ANYL 0026
	INOR 0056		Wilson, E. B.	ANYL 0131		Woloszyn, T. F.	
	INOR 0314		Wilson, J. D.	ENVR 0067			ANYL 0099
	INOR 0360		Wilson, J. H.	CARB 0063		Wolpert, A. J.	CINF 0013
Whybrew, W. D.			Wilson, K. E.	MEDI 0181		Womelsdorf, J. F.	
	BIOL 0069		Wilson, K. R.	PHYS 0037			PHYS 0212
Wicker, R. K.	FUEL 0058		Wilson, M.	MEDI 0124		Won, J.	POLY 0267
Wickman, H.	BIOT 0151		Wilson, M. F.	PETR 0149		Wong, A.	INOR 0156
Wickramasinghe, H. K.			Wilson, P. J.	ORGN 0012		Wong, B. J.	CHAS 0003
	PHYS 0209		Wilson, R. B.	POLY 0313		Wong, C-H.	ORGN 0052
Wiczk, W.	PHYS 0061		Wilson, S. R.	CATL 0008			ORGN 0271
Widmaier, J-M.				ORGN 0043		Wong, C. Y.	INOR 0019
	PMSE 0001			ORGN 0169		Wong, D.W.S.	AGFD 0078
	PMSE 0021		Wilson, T.	PHYS 0265		Wong, G. K.	POLY 0108
Widstrom, N. W.			Wilson, T. G.	AGRO 0101			POLY 0127
	AGRO 0034		Wilson, W. L.	PHYS 0138			POLY 0152
Wiggins, J.	MEDI 0118			POLY 0215			POLY 0217
Wiggins, J. S.	POLY 0068		Winans, R. E.	FUEL 0088		Wong, G. S.	ENVR 0009
Wightman, J. P.			Wink, D. J.	INOR 0154		Wong, H.N.C.	ORGN 0181
	PMSE 0052			INOR 0209		Wong, J. K.	MEDI 0161
	PMSE 0054		Winkler, R. G.	POLY 0355		Wong, L.C.K.	CHAS 0019
Wijaya, J.	POLY 0160		Winner, P.	PMSE 0160		Wong, N. M.	COLL 0024
	POLY 0161		Winnik, M. A.	PHYS 0108		Wong, S.	MEDI 0175
Wijnen, P.W.J.G.				POLY 0277		Wong, S.H.Y.	ANYL 0116
	PETR 0054		Winograd, N.	NUCL 0041		Wong, S. J.	ENVR 0036
Wilchek, M.	BIOL 0090		Winschel, R. A.	FUEL 0044		Wong, S. S.	BIOT 0216
Wild, J. R.	BIOT 0242			FUEL 0071		Wong, S. T.	MEDI 0147

Wong-Leung, Y. L.		Wubbels, G. G.	CHED 0146	Yanagisawa, I.	MEDI 0099
	ANYL 0089		CHED 0196	Yanez, R.	NUCL 0024
Woo, J.T.K.	PMSE 0161	Wudl, F.	ANYL 0108	Yang, B.	PHYS 0241
Woo, L. K.	INOR 0298	Wulff, A.	CARB 0058	Yang, C. C.	BIOL 0125
	INOR 0299	Wulff, W. D.	INOR 0112	Yang, C. F.	BIOL 0075
Woo, S. L.	AGRO 0003		INOR 0194		BIOL 0087
Wood, C. D.	ENVR 0054		ORGN 0221	Yang, C-J.	POLY 0011
Wood, C. E.	POLY 0184	Wurpel, J.N.D.	MEDI 0134		POLY 0226
Wood, J. L.	CHAS 0041	Wurtzel, E.	CHED 0140	Yang, C-P.	MEDI 0053
Wood, S.	BIOL 0039	Wurz, P.	NUCL 0045	Yang, C. S.	AGFD 0065
Wood, T. M.	BIOT 0179	Wusik, M. J.	POLY 0038		AGFD 0067
Woodhouse, T. E.		Wyatt, P. J.	PMSE 0096	Yang, D.	MEDI 0069
	CATL 0060	Wynne, K. J.	POLY 0018		MEDI 0187
Woodin, R. L.	PHYS 0093	Xia, H.	ORGN 0229	Yang, D. C.	POLY 0232
Woodman, R.	PETR 0031	Xian, C. H.	COLL 0079	Yang, D-Y.	BIOL 0100
Woodruff, J.	MEDI 0070	Xiao, T. D.	POLY 0335	Yang, G. K.	INOR 0192
Woodruff, R. D.		Xiao, X.	ORGN 0021		INOR 0193
	PROF 0009	Xiaohua, B.	INOR 0157		INOR 0386
Woods, A. T.	MEDI 0130	Xie, H. Q.	PMSE 0090	Yang, H.	BIOL 0092
Woodward, J.	BIOT 0021	Xie, K.	FUEL 0068		PMSE 0149
Woodward, J. R.		Xie, M.	MEDI 0139	Yang, H. C.	BIOT 0200
	PHYS 0173	Xie, P.	PMSE 0103	Yang, J.	PMSE 0126
Woodward, K. N.		Xing, P.	PMSE 0079	Yang, L.	MEDI 0122
	AGRO 0075		PMSE 0091	Yang, M.	BIOT 0080
	AGRO 0077	Xing, S-B.	FUEL 0066	Yang, P. C.	POLY 0157
Wooley, K. L.	POLY 0318	Xiucuo, L.	PMSE 0082	Yang, R.	BIOT 0238
	POLY 0322	Xiuqi, L.	ANYL 0080	Yang, S. T.	BIOT 0030
Woolsey, N. F.	INOR 0207	Xu, C.	ANYL 0131	Yang, X.	ORGN 0013
Woolum, D. S.	NUCL 0077	Xu, D.	INOR 0141	Yang, Y.	COLL 0015
Wooten, J. V.	CEI 0001	Xu, L.	CATL 0052		ORGN 0254
Worden, R. M.	BIOT 0015		ORGN 0147	Yang, Y-S.	POLY 0114
Workman, J. M.			ORGN 0290	Yanni, J. M.	MEDI 0158
	INOR 0243	Xu, Q.	POLY 0270	Yannoni, C. S.	PHYS 0070
Worsnop, D. R.	PHYS 0071	Xu, R.	POLY 0277	Yao, B.	INOR 0250
Woska, D. C.	INOR 0109	Xu, S.	INOR 0170	Yao, J.	PETR 0063
Woski, S. A.	ORGN 0118	Xu, S. L.	ORGN 0229	Yao, Q.	INOR 0343
Woster, P. M.	MEDI 0021	Xu, Y.	AGFD 0068	Yao, Z.	MEDI 0120
Wovkulich, P. M.		Xu, Y-Z.	PMSE 0137		MEDI 0197
	ORGN 0284	Xu, Z.	POLY 0321	Yap, B. T.	BIOT 0153
Wozniak, G. J.	NUCL 0005	Xue, Z.	INOR 0141	Yappert, M. C.	ANYL 0077
Wright, C. D.	MEDI 0177	Ya, N-Q.	PHYS 0021	Yarar, B.	I&EC 0036
Wright, I. G.	BIOT 0161	Yabannavar, V. M.		Yardley, J. T.	POLY 0180
Wright, J.	CHED 0131		BIOT 0007	Yarmush, M. L.	
Wright, J. C.	ANYL 0001	Yacynych, A. M.			BIOT 0062
Wright, J.J.K.	MEDI 0190		BIOT 0107	Yarnold, C.	ORGN 0005
Wright, J. R.	CHED 0011	Yager, P.	POLY 0243	Yaron, D.	PHYS 0102
Wrighton, M. S.		Yaguchi, M.	BIOT 0211	Yaser, H. K.	ORGN 0292
	CHED 0062		BIOT 0212	Yasunaga, T.	MEDI 0109
	CHED 0128	Yakubovich, T. N.		Yates, J. T., Jr.	PHYS 0204
	INOR 0027		COLL 0102	Yau, W. W.	PMSE 0067
	INOR 0163	Yalpani, M.	CARB 0026		PMSE 0070
	INOR 0168		CARB 0030		PMSE 0097
	INOR 0350		POLY 0041		PMSE 0098
Wrolstad, R. E.	AGFD 0169		POLY 0084	Yazici, R.	POLY 0229
Wu, A-H.	ORGN 0145	Yamada, A.	PMSE 0055	Yee, D. K.	CATL 0041
Wu, C.	COLL 0058	Yamada, H.	AGRO 0024		FUEL 0100
	PHYS 0220	Yamada, M.	AGRO 0071	Yee, R. Y.	POLY 0268
	POLY 0233	Yamada, S.	ORGN 0038	Yegneswaran, P. K.	
	POLY 0345	Yamada, T.	ENVR 0049		BIOT 0088
	POLY 0346	Yamaguchi, E. S.		Yeh, H.J.C.	BIOT 0170
Wu, C-G.	INOR 0082		CHED 0145	Yeh, H. R.	BIOT 0249
Wu, C-H.	PHYS 0150	Yamaguchi, H.	BIOT 0124	Yeh, L-T.	POLY 0257
Wu, C-S.	PMSE 0031	Yamaguchi, S.	BIOT 0023	Yeh, Y.	AGFD 0035
Wu, D-R.	BIOT 0042		PMSE 0055		BIOL 0041
Wu, F.	BIOT 0071		POLY 0098	Yeh, Y-S.	INOR 0043
Wu, H-J.	POLY 0362	Yamaguchi, T.	PETR 0029	Yehoda, J. E.	PHYS 0092
Wu, J.	AGRO 0111	Yamamoto, D. M.		Yein, F. S.	AGRO 0108
Wu, J-F.	CATL 0025		MEDI 0127	Yeo, H.	AGFD 0132
Wu, J. W.	POLY 0130	Yamamoto, H.	POLY 0128	Yetter, R. A.	PHYS 0142
Wu, K.	ENVR 0010	Yamamoto, K.	PMSE 0007	Yeung, A. T.	BIOL 0126
Wu, M-C.	COLL 0193	Yamamoto, S.	AGRO 0049	Yeung, P.K.F.	MEDI 0129
Wu, R.	PMSE 0132		INOR 0164	Yi, C. S.	ORGN 0130
Wu, R. S.	ANYL 0076	Yamamoto, Y.	PMSE 0055	Yi, L.	PHYS 0152
Wu, S.	CATL 0031	Yamamura, H. I.		Yin, X-Y.	AGFD 0041
Wu, S-C.	MEDI 0156		MEDI 0140	Ying, Q.	PMSE 0103
Wu, T. C.	BIOT 0048	Yamanaka, S.	PETR 0144	Yoden, T.	MEDI 0099
Wu, V.	AGFD 0143	Yamashita, H.	FUEL 0032	Yokoyama, K.	BIOL 0021
Wu, W-L.	POLY 0171	Yamazaki, I.	PHYS 0110	Yokoyama, M.	PETR 0145
Wu, X. L.	PHYS 0329	Yamazaki, S.	BIOT 0134	Yong, G.	BIOL 0058
Wu, X-P.	INOR 0324	Yamazaki, T.	PHYS 0110		BIOL 0059
Wu, X. S.	POLY 0243	Yamdgani, R.	PETR 0143	Yoo, J. N.	PMSE 0073
Wu, Y. C.	BIOL 0033	Yan, W-B.	PHYS 0332	Yoo, S-E.	ORGN 0177
	BIOL 0034	Yan, Y. K.	INOR 0107	Yoon, D. Y.	PMSE 0023

Yoon, D. Y.	POLY 0153	Zaslonko, I. S.	PHYS 0175	Zhu, X. R.	ANYL 0011		
	POLY 0208	Zawicki, J. L.	PMSE 0158		ANYL 0034		
	POLY 0351	Zax, D. B.	PHYS 0372	Zhu, X-Y.	COLL 0065		
	POLY 0355	Zaya, M. J.	AGRO 0108	Zhu, Z.	PHYS 0308		
Yoon, K.	INOR 0088	Zayas, R.	POLY 0236	Zhu, Z. L.	PETR 0105		
Yoon, K. J.	MEDI 0170	Zegarski, B. R.	INOR 0124	Zhu, Z. Y.	ANYL 0077		
Yorke, A. F.	BIOT 0086	Zehani, S.	ORGN 0264	Ziegler, T.	CARB 0019		
Yoshida, T.	AGFD 0017	Zehavi, U.	AGFD 0167	Zielinska-Pfabe, M.			
	AGFD 0046	Zehe, A.	INOR 0190		NUCL 0040		
Yoshikawa, K.	BIOT 0123	Zeiler-Hilgart, G.		Zielinski, J. M.	COLL 0053		
Yoshimi, N.	AGFD 0090		AGFD 0070		POLY 0144		
Yoshimura, R.	PHYS 0313	Zelewski, L. M.	COLL 0212	Zielinski, T. J.	BIOL 0112		
Yoshino, H.	AGRO 0024	Zeliger, H. I.	CHAL 0012		CHED 0162		
Yoshino, K.	PMSE 0183	Zeng, L.	ORGN 0188		PHYS 0336		
Yoshitake, A.	AGRO 0024	Zeng, X.	BIOL 0048	Ziffer, H.	BIOT 0170		
Younathan, E. S.		Zeng, X-N.	AGFD 0147	Zilm, K. W.	PHYS 0068		
	ORGN 0074	Zentel, R.	POLY 0106	Zimmels, Y.	COLL 0142		
Young, G. A.	AGRO 0028		POLY 0153		COLL 0162		
	AGRO 0029	Zeppenfeld, P.	PHYS 0360	Zimmerman, H. E.			
Young, J. P.	ANYL 0065	Zewail, A. H.	PHYS 0369		ORGN 0142		
	INOR 0064	Zhan, Z-Y.	ORGN 0259	Zimmerman, R.			
Young, L. Y.	BIOT 0120	Zhang, C.	BIOL 0036		BIOT 0151		
Young, M.	BIOL 0143	Zhang, C. X.	PMSE 0090	Zinger, B.	BIOT 0201		
Young, M. B.	MEDI 0191	Zhang, F.	BIOL 0135	Zink, D. L.	ORGN 0184		
Young, M. S.	ENVR 0050	Zhang, F-Q.	FUEL 0038	Zink, J. I.	INOR 0141		
Young, P. A.	INOR 0142	Zhang, H.	AGFD 0122	Zinnen, M.	COMP 0054		
Young, P. R.	BIOT 0114		ORGN 0224	Zipp, A. P.	CHED 0054		
	MEDI 0176	Zhang, H. L.	ORGN 0222	Zitano, L.	MEDI 0181		
Young, S. M.	INOR 0391	Zhang, J.	BIOT 0205	Zmijewski, M. J., Jr.			
Younis, K.	COLL 0117		PHYS 0163		BIOT 0161		
Youssef, A-H.A.		Zhang, J-H.	INOR 0140	Zoeller, J. R.	I&EC 0053		
	ORGN 0160	Zhang, J-Z.	GEOC 0005	Zografi, G.	COLL 0056		
Yu, C.	PETR 0031	Zhang, M.	ORGN 0202	Zolotaryuk, A. V.			
Yu, C. G.	AGRO 0007	Zhang, P.	PMSE 0150		PHYS 0129		
Yu, G.	PHYS 0011	Zhang, Q.	PHYS 0207	Zones, S. I.	CATL 0028		
Yu, H.	COLL 0051		PMSE 0092		PETR 0023		
	COLL 0056	Zhang, R.	COLL 0046		PETR 0036		
	MEDI 0101		PHYS 0333	Zonta, F.	MEDI 0141		
	POLY 0164	Zhang, S.	BIOT 0008	Zopf, D. A.	CARB 0054		
	POLY 0198	Zhang, T. G.	POLY 0127	Zorner, P. S.	AGRO 0080		
Yu, J-S.	CATL 0017	Zhang, X.	PMSE 0148	Zou, J-Y.	CINF 0029		
Yu, M.	COLL 0217		POLY 0357	Zovinka, E. P.	INOR 0302		
Yu, M. J.	MEDI 0066	Zhang, X. J.	POLY 0030	Zub, Y. L.	COLL 0102		
	MEDI 0174	Zhang, X-Z.	BIOL 0103	ZuberBuhler, A. D.			
Yu, Q.	POLY 0197	Zhang, Y.	POLY 0049		INOR 0178		
Yu, S. J.	AGRO 0119	Zhang, Y-F.	FUEL 0038	Zukas, W. X.	PMSE 0181		
Yu, S. K-T.	FUEL 0060	Zhang, Z.	COLL 0192	Zumofen, G.	PHYS 0130		
Yu, T-C.	COLL 0210		INOR 0227	Zumsteg, F. C.	POLY 0104		
Yu, X.	PMSE 0146		ORGN 0049	Zuniga W., J.	COMP 0053		
Yuan, W.	BIOL 0020		ORGN 0218	Zupan, J.	COMP 0002		
Yuen, H. K.	BIOL 0053		PHYS 0211	Zussman, M. P.			
	BIOL 0054	Zhang, Z-F.	POLY 0314		POLY 0300		
Yuen, L. T.	CATL 0028	Zhao, B.	FUEL 0140	Zwanziger, J. W.			
Yu Ip, C. C.	ANYL 0021		ORGN 0277		PHYS 0374		
Yun, Y.	FUEL 0087	Zhao, J.	INOR 0194	Zyss, J.	POLY 0006		
Zabriskie, T. M.		Zhao, M.	ORGN 0047				
	BIOL 0065		ORGN 0048				
Zabrowski, D.	ORGN 0104	Zhao, R.	PHYS 0021				
Zabrowski, D. L.		Zhao, Y-B.	ORGN 0100				
	MEDI 0069	Zheng, G.	FUEL 0103				
Zaccagnino, A.	COLL 0085	Zheng, L.	ORGN 0182				
Zaera, F.	COLL 0049		PHYS 0334				
Zaffaroni, A.	HIST 0019	Zheng, L. M.	ORGN 0262				
Zafiris, G.	COLL 0186	Zhenzong, Y.	COLL 0079				
Zahalsky, A. C.	CEI 0017	Zhiqing, L.	PMSE 0082				
Zahirsky, K. E.	FERT 0007	Zhong, C.	BIOT 0201				
Zahniser, M. S.	PHYS 0071	Zhong, Q.	COLL 0171				
Zahos, S.	AGRO 0045	Zhong, S.	BIOL 0082				
Zale, S. E.	BIOT 0072	Zhou, G-B.	POLY 0271				
Zambias, R. A.	MEDI 0182	Zhou, H.	PMSE 0083				
	MEDI 0183	Zhou, J.	ANYL 0141				
Zamir, D.	PHYS 0372	Zhou, J-B.	COLL 0176				
Zand, R.	POLY 0033	Zhou, M.	PHYS 0308				
Zandler, M. E.	PHYS 0315	Zhou, P.	AGFD 0142				
Zank, G. A.	POLY 0361		PMSE 0025				
Zaragoza, L. J.	ENVR 0077		PMSE 0056				
	ENVR 0079	Zhou, Q.	MEDI 0136				
Zaragoza-Santamaria, L.		Zhu, D.	BIOL 0036				
	AGFD 0139	Zhu, H.	PHYS 0335				
Zarko, V. I.	COLL 0111	Zhu, K. J.	COLL 0034				
Zasadzinski, J.A.N.		Zhu, L.	PHYS 0278				
	PHYS 0158	Zhu, X.	PHYS 0220				
Zask, A.	ORGN 0223		PHYS 0335				

PRESIDING OFFICER INDEX

Reference by Division Code

Abul-Hajj, Y. J.	MEDI
Adams, V. D.	ENVR
Agarwal, V. K.	AGFD
Ahuja, S.	ANYL
Alexander, J. M.	NUCL
Allen, R. C.	MEDI
Andersson, L. A.	INOR
Anthony, R. G.	PETR
Antos, G. J.	PETR
Archer, R. D.	CHED
Arcuri, E. J.	BIOT
Armor, J. N.	I&EC
Auvil, S. R.	I&EC
Aviram, A.	BIOT
Avouris, P.	PHYS
Baetz, A. L.	ANYL
Bajpai, R. K.	BIOT
Baker, D. R.	AGRO
Baker, F. C.	AGRO
Baker, G.	POLY
Baker, G. L.	POLY
Baker, J. O.	BIOT
Barteau, M. A.	COLL
Barth, H. G.	PMSE
Bartlett, R. J.	PHYS
Baum, J.	PHYS
Begley, T.	BIOL
Bellen, G. E.	C E I
Berkowitz, W. F.	ORGN
Bitterwolf, T. E.	INOR
Bjorklund, G.	POLY
Bjorklund, G. C.	POLY
Blanchard, J. S.	BIOL
Bockstedt, R. J.	CMEC
Bohn, P. W.	COLL
Borowitz, G. B.	PROF
Bowers, L. D.	ANYL
Boyd, R.	PHYS
Bradley, S. A.	PETR
Brennan, A.	POLY
Breslow, R.	CONG
Brewer, C. F.	BIOL
Brezinsky, K.	FUEL
Brine, C. J.	AGFD
Brody, A. L.	AGFD
Buhro, W. E.	INOR
Bungay, H. R.	BIOT
Burgmayer, S.J.N.	INOR
Burnett, D. S.	NUCL
Burton, V. E.	TECH
Busch, M. A.	COLL
Butler, A.	INOR
Cahill, T. A.	NUCL
Carlin, R. T.	ANYL
Carr, P. W.	ANYL
Cashin, F. J.	FUEL
Charton, M.	BIOT
Chase, B.	POLY
Chastain, B. B.	HIST
Cheng, Y-T.	NUCL
Choplin, L.	BIOT
Chu, C. K.	AGFD
	CARB
Chung, C.	NUCL
Clark, C. R.	ANYL
Clarke, M. A.	CARB
Clearfield, A.	PETR
Coats, J. R.	AGRO
Cobb, J. T., Jr.	FUEL
Cocuzzi, D. A.	PMSE
Colton, R. J.	PHYS
Comita, P. B.	COLL
Conney, A. H.	AGFD
Converse, A. O.	BIOT
Cooper, W.	ENVR
Cothern, C. R.	ENVR
Coughlan, M. P.	BIOT
Coyle, R. T.	FUEL
Creighton, J. R.	COLL
Crowell, J. E.	COLL
Cullen, D.	BIOT
Cuomo, J. J.	PHYS
Curtis, C.	CATL
Cutler, H. G.	AGFD
Cyngier, R.	I&EC
Dailey, W. P., III	ORGN
Daniel, L. W.	MEDI
Datar, R.	BIOT
Davidovits, P.	PHYS
Davis, F. A.	ORGN
Davis, M.	CATL
Davis, R., Jr.	NUCL
Day, D. P.	FERT
Dean, A. M.	FUEL
Decanio, S.	FUEL
Deisler, P.	ENVR
Delfiner, J. D.	CHED
Delgado, G.	ORGN
Delmas, G.	POLY
Demuth, J. E.	PHYS
DePaula, J. C.	CHED
Dias, J. R.	COMP
Dirlikov, S.	PMSE
Di Sanzo, F. P.	PETR
Dordick, J. S.	BIOT
Dorsey, J. G.	ANYL
Dowling, N.	BIOT
Drake, J. M.	PHYS
Drobny, G. P.	PHYS
Duncan, M. A.	PHYS
Duncan, T. M.	PHYS
Durrell, W. S.	CMEC
Dutta, P.	CATL
Ermler, W.	PHYS
Ettre, L. S.	ANYL
Fair, W. E.	FERT
Farinato, R. S.	COLL
Farley, M. A.	CHAL
Farrow, P. E.	CMEC
Feenstra, R. M.	PHYS
Felsot, A. S.	AGRO
Fenyes, J.G.E.	AGRO
Flatt, J. H.	BIOT
Foley, H.	CATL
Fontijn, A.	PHYS
Ford, A. M.	C E I
Forster, D.	CATL
Freeman, B. D.	POLY
French, A.	CARB
Frisch, H. L.	PMSE
Fryzuk, M. D.	INOR
Fujiki, H.	AGFD
Fuller, G.	AGFD
Gabelica, Z.	PETR
Gaddy, J. L	FUEL
Gajewski, J. J.	ORGN
Galiasso, R.	PETR
Garegg, P. J.	CARB
Gast, A. P.	COLL
Gavalas, G. R.	FUEL
Gelb, M.	BIOL
Genack, A.	PHYS
Georgio, G.	BIOT
Georgiou, G.	BIOT
Gerber, S. M.	HIST
Gierasch, L.	BIOL
Gilding, T.	AGRO
Giragosian, N. H.	SCHB
Girolami, G.	INOR
Gladfelter, W.	INOR
Glass, J. T.	PHYS
Goddard, J. D.	PHYS
Gole, J. L.	PHYS
Goodman, J. L.	ORGN
Gordon, B., III.	POLY
Gortler, L.	HIST
Goswami, R.	MEDI
Granick, S.	COLL
Graves, D.	BIOT
Greasham, R. L.	BIOT
Greenbaum, E.	BIOT
Greenblatt, M.	INOR
Greenlee, K.	SCHB
Grethe, G.	CINF
Grey, P.	PETR
Griffith, E. J.	INOR
Griskey, R. G.	I&EC
Grissom, C.	Y C C
Gulari, E.	PETR
Guntert, M.	AGFD
Hadziioannou, G.	POLY
Hahn, R. L.	NUCL
Hamel, J-F.	BIOT
Hanesian, D.	I&EC
Harris, S. J.	PHYS
Harris, T. D.	ANYL
Harrod, J.	POLY
Harvard, T.	PMSE
Hawkins, D. R.	AGRO
Hayes, J. M.	CONG
Heaven, M. C.	PHYS
Heininger, S. A.	CONG
Heller, S.	COMP
Hendrick, C. A.	AGRO
Henion, J.	ANYL
Hersh, W. H.	INOR
Hicks, K. B.	AGFD
Himmel, M. E.	BIOT
Ho, C-T.	AGFD
Holley, C. A.	FUEL
Holt, E. M.	INOR
Holten, D.	BIOL
Honeycutt, R. C.	AGRO
Hopfenberg, H. B.	POLY
Hourston, D.	PMSE
Howard, J. B.	FUEL
Huang, M-T.	AGFD
Huang, S. J.	POLY
Hutson, D. H.	AGRO
Hwang, L. S.	AGFD
Iglesia, E.	CATL
Jackson, C. L.	POLY
Janson, J-C.	ANYL
Jasso-Gastinel, C. F.	POLY
Jennings, P. W.	INOR
Jennings, W. G.	AGFD
Jensen, J. K.	AGRO
Jensen, W. B.	HIST
Johnson, R. C.	ENVR
Johnson, R. E.	MEDI
Johnston, M. V.	ANYL
Jones, D.	CHED
Kablaoui, M. S.	FUEL
Kaifer, A.	COLL
Kaminski, M. D.	CHAL
Karch, N. J.	C E I

Name	Code	Name	Code	Name	Code
Katz, S. E.	AGFD	Morgan, B.	MEDI	Schulz, W. W.	I&EC
Kazlauskas, R. J.	BIOT	Morrison, P. W., Jr.		Schure, M.	PMSE
Kearney, P. C.	AGRO		FUEL	Schwartz, A. T.	CHED
Kenealy, W. R.	BIOT	Morss, L. R.	NUCL	Schwartz, J.	PETR
Kessler, D. G.	CMEC	Mullin, C. A.	AGRO	Schweikert, E. A.	NUCL
Kessler, H.	PETR	Munk, M.	COMP	Schwenz, R. W.	CHED
Khaledi, M. G.	ANYL	Murday, J. S.	PHYS	Scott, J. G.	AGRO
Khan, M. R.	FUEL	Murphy, R. M.	BIOT	Scranton, M. I.	GEOC
Kierstan, M.P.J.	BIOT	Nakatani, A. I.	PMSE	Seaborg, G. T.	NUCL
Kilburn, D. G.	BIOT	Natansohn, A.	PMSE	Secrist, J. A., III	AGFD
Kim, S. C.	PMSE	Neenan, T. X.	POLY	Seiber, J. N.	AGRO
King, E. T.	CINF	Newmark, H.	AGFD	Serianni, A. S.	CARB
Kirss, R. U.	INOR	Nicolaides, C. P.	PETR	Serpone, N.	INOR
Klabunde, K.	CATL	Nierlich, F.	PETR	Sessler, J. L.	INOR
Klein, M.	CATL	Noda, I.	COLL	Seydel, J. K.	MEDI
Klempner, D.	PMSE	Norman, A. D.	INOR	Shahidi, F.	AGFD
Koch, S.	INOR	Nosker, T. J.	I&EC	Shaikh, B.	AGFD
Kopelman, R.	PHYS	Novotny, J.	FERT	Shalaby, S. W.	CHED
Kowalski, R. C.	PMSE	Occelli, M. L.	PETR	Sharkey, J. B.	HIST
Kramer, S. J.	AGRO	Orlando, S.	YCC	Sheetz, M.	BIOL
Krueger, D.	AGFD	Osawa, T.	AGFD	Sheih, P. S.	PMSE
Kugler, E. L.	COLL	Osteryoung, J.	ANYL	Sheridan, R. P.	CINF
Kurtz, D.	INOR	Ottenbrite, R. M.	POLY	Sherman, M. C.	CHED
Kustin, K.	INOR	Paine, R.	POLY	Shields, M. S.	BIOT
Kutal, C.	INOR	Palmer, J. G.	CHAS	Shine, A. D.	COLL
Laine, R.	POLY	Pariser, R.	POLY	Shoemaker, S. P.	BIOT
Landolph, J. R.	CHAS	Parry, R. W.	INOR	Silver, R. G.	COLL
Lavine, B. K.	CHED	Paulson, G. D.	AGRO	Sinoliunas, S.	ENVR
Lawlor, B.	CINF	Pecoraro, V.	INOR	Skeels, G. W.	PETR
Lee, A.	BIOT	Pedersen, H.	BIOT	Skelly Frame, E. M.	
Lee, C. Y.	AGFD	Pederson, M. R.	PHYS		CHED
Lee, J. M.	BIOT	Pence, H. E.	CHED	Smith, V.	PHYS
Lee, S. M.	AGRO	Percec, V.	POLY	Sneddon, L. G.	POLY
Lee, T.	PHYS	Peters, H. M.	CHAL	Somasundaram, L.	
Lerner, R.	BIOL	Phifer, R. W.	LWM		AGRO
Levitz, P.	PHYS	Phillips, E. M.	I&EC		
Levon, K.	POLY	Pilcher, G. R.	PMSE	Sousa-Aguiar, E. F.	
Liao, J. C.	BIOT	Pinnavaia, T. J.	PETR		PETR
Liegeois, J. M.	PMSE	Piret, J. M.	BIOT	Spanier, A.	AGFD
Lindberg, A. A.	CARB	Pirrung, M. C.	MEDI	Spence, R. D.	ENVR
Lindenberg, K.	PHYS	Porile, N. T.	NUCL	Sperling, L. H.	PMSE
Lippmann, M.	ENVR	Potember, R. S.	BIOT	Squires, R. G.	I&EC
Loveland, W.	NUCL	Prane, J. W.	PMSE	Steinberg, M.	FUEL
Lovink, H. J.	PETR	Puglia, G. P.	POLY	Steliou, K.	ORGN
Lu, P. Y.	CHAS	Quin, L. D.	INOR	Stellwagen, N. C.	COLL
Lynd, L. R.	BIOT	Quistad, G. V.	AGRO	Stephan, D. W.	INOR
Lytel, R.	POLY	Qutubuddin, S.	COLL	Stephanopoulos, G.	
Lytle, F. E.	ANYL	Rabolt, J.	POLY		BIOT
McCarley, R. E.	CHED	Radovic, L. R.	FUEL	Stephenson, T. A.	CHED
McCarthy, T. J.	POLY	Ratanathanawongs, K.		Stone, D. B.	CHED
MacDiarmid, A. G.			PMSE	Stork, G.	HIST
	POLY	Rauch, F.	NUCL	Struble, C. B.	AGRO
Mackay, R.	COLL	Rees, W. S., Jr.	COLL	Stubbe, J.	BIOL
McKenna, G. B.	POLY	Register, R. A.	PMSE	Sturchio, J. L.	HIST
Mandava, N. B.	CHAS	Regnier, F. E.	ANYL	Sung, C.S.P.	POLY
Marder, S. R.	BIOT	Reimer, K. J.	INOR	Takeoka, G.	AGFD
Margolin, A.	BIOT	Reineccius, G. A.	AGFD	Takeuchi, K. J.	INOR
Mark, J. E.	COLL	Remillieux, R.	NUCL	Tanner, R. D.	BIOT
Marshall, J. A.	ORGN	Retcofsky, H. L.	FUEL	Tegeler, J.	MEDI
Martin, M. L.	PHYS	Rigby, J. H.	ORGN	Teranishi, R.	AGFD
Martin, R. L.	PHYS	Robson, H. E.	PETR	Tessler, M. M.	CARB
Martin, Y.	CINF	Rogers, J. W., Jr.	COLL	Texter, J.	COLL
Matkovich, M. W.		Roovers, J.	POLY	Theil, E. C.	INOR
	CHAS	Rosenthal, D.	CHED	Tien, H. T.	BIOT
Mauritz, K.	POLY	Rosenthal, G.	INOR	Tien, T.	BIOT
Mei, H. L.	POLY	Rowley, J. K.	NUCL	Timmerman, H.	MEDI
Mendizabal-Mijares, E.		Rudin, A.	PMSE	Torkelson, J.	POLY
	POLY	Runt, J.	PMSE	Toste, A. P.	ANYL
Meredith, G.	POLY	Rusling, J.	COLL	Tseng, C.	AGFD
Meyer, G.	PMSE	Russell, A.	BIOT	Turckes, M.	CHED
Michael, J. V.	PHYS	Saddler, J.	BIOT	Turckes, M. K.	CHED
Michl, J.	BIOT	Saferstein, R.	HIST	Tycko, R.	PHYS
Millar, M.	INOR	St. Angelo, A. J.	AGFD	Tyeklas, Z.	INOR
Miller, G. C.	ENVR	Sakurai, T.	PHYS	Uberoi, M.	FUEL
Miller, J. H.	FUEL	Salladay, D. G.	FERT	Ueda, E.	PMSE
Miller, R. S.	PHYS	Sapers, G. M.	AGFD	Utracki, L. A.	PMSE
Mills, J.	CMEC	Saris, F.	NUCL	Vancso, G. J.	POLY
Milne, G.W.A.	COMP	Sawrey, B. A.	CHED	Van den Berg, J. P.	
Mirau, P. A.	PHYS	Sawyer, J. E.	COLL		PETR
Mohrig, J.	ANYL	Scanlan, R. A.	AGFD	Vandenbosch, R.	NUCL
Moog, R. S.	CHED	Schaefer, H. F.	PHYS	Vaughan, D.E.W.	PETR
Moore, D. S.	ANYL	Schlegel, H. B.	PHYS	Vera, F.	BIOT
Moore, R.	CHED	Schmitz, K.	COLL	Vercellotti, J.	AGFD
				Versteylen, M.	SCHB

Vining, L. C.	BIOT
Viola, V. E.	NUCL
Vohs, J. M.	COLL
Voorhees, K. J.	ENVR
Vorres, K. S.	FUEL
Walsh, E. N.	INOR
Walters, V. A.	CHED
Warren, W. S.	PHYS
Weiss, R. A.	PMSE
Weller, M. R.	NUCL
Wells, R. L.	COLL
Wheeler, W. B.	AGRO
Whisnant, D. M.	CHED
White, D. C.	BIOT
Widlanski, T. S.	ORGN
Wilhelmy, J. B.	NUCL
Willett, P. S.	COMP
Willson, R.	BIOT
Winnik, M.	POLY
Winograd, N.	NUCL
Wipke, W. T.	COMP
Wolfe, M. S.	POLY
Wolfram, L.	PETR
Woo, J.T.K.	PMSE
Wood, J. L.	CHAS
Wubbels, G. G.	CHED
Wyman, C. E.	BIOT
Wynne, K.	POLY
Yamanaka, S.	PETR
Yates, J. T., Jr.	PHYS
Yu, H.	COLL
Zax, D. B.	PHYS
Zilm, K. W.	PHYS
Zografi, G.	COLL
Zumofen, G.	PHYS

CIC National Officers

C. Edward Capes, President
Donald B. Mutton, Vice-President
Ronald Gurak, Treasurer
Edward Piers, President, Canadian Society for Chemistry
H. Clarke Henry, President, Canadian Society for Chemical Engineering
John C. Cody, President, Canadian Society for Chemical Technology

1991-1992 Divisional Officers

Analytical Chemistry Division
Dr. Janis Gulens, MCIC
Chalk River Nuclear Laboratories
General Chemistry Branch
K0J 1J0

Biological Chemistry Division
Dr. Paul Harrison, MCIC
Department of Chemistry
McMaster University
Hamilton, ON
L8S 4M1

Biotechnology Division
Dr. Ron Neufeld, MCIC
McGill University
Dept. of Chem. Eng.
3480 University Street
Montreal, PQ
H3A 2A7

Catalysis Division
Dr. David E. Laycock, MCIC
Dow Chemical Canada Inc.
1086 Modeland Road
P.O. Box 1012
Sarnia, ON
N7T 7K7

Chemical Education Division
Dr. Michel Ringuet, MCIC
Départment de chimie-biologie
Université du Québec a
Trois-Rivieres
Trois-Rivieres, Québec
G9A 5H7

Economics and Business Management Division
Mr James E. Sigurdson, MCIC
Cheminfo Services Inc.
1706 Avenue Road, Suite 4
Toronto, ON
M5M 3Y6

Environment Division
Dr. Ronald R. Martin, MCIC
Department of Chemistry
University of Western Ontario
London, ON
N6A 5B7

Inorganic Chemistry Division
Dr. Peter Legzdins, FCIC
Department of Chemistry
University of British Columbia
Vancouver, BC
V6T 1Y6

Medicinal Chemistry Division
Dr. John Gillard, MCIC
Merck Frosst Centre
for Therapeutic Research
P.O. Box 1005
Dorval, PQ
H9R 4P8

Macromolecular Science and Engineering Division
Dr. Basil Favis, MCIC
École Polytechnique
Génie chimique
Case postale 6079, Succ. A
Montréal, PQ
H3C 3A7

Organic Chemistry Division
Dr. Robert McDonald, MCIC
Department of Chemistry
Mount Saint Vincent University
Halifax, NS
B3M 2J6

Protective Coatings Division
Mr. Bert Papenburg, MCIC
Peintures Chateau Inc.
6388 Avenue du Parc
Montreal, PQ
H3N 1W8

Physical and Theoretical Chemistry Division
Dr. Russell J. Boyd, FCIC
Department of Chemistry
Dalhousie University
Halifax, NS
B3H 4J3

Rubber Chemistry Division
Mr. J.C. David Baxter, MCIC
L.V. Lomas Chemical Co. Ltd
99 Summerlea Road
Brampton, ON
L6T 4V2

Systems and Control Division
Dr. Tom Harris, MCIC
Queen's University
Dept. of Chemical Engineering
Kingston, ON
K7L 3N6

Surface Science Division
Dr. Royston Paynter, MCIC
INRS-Energie
C.P. 1020
Varennes, Québec
J0L 2P0

Board of Directors

SOCIEDAD QUIMICA DE MEXICO
DRA. ELVIRA SANTOS DE FLORES–PRESIDENT
I. Q. GERMAN ESPINOSA CHAVARRIA
I. Q. JOSE LUIS PADILLA DE ALBA
Q. HECTOR C. BOLIVAR TERRAZAS
DR. JOAQUIN PALACIOS ALQUICIRA
M. EN C. MARIA DEL CORO ECHEVERRIA ORTEGA
DR. ANIBAL BASCUNAN BLASET
M. EN C. CLAUDIA MACERA VIVAR
M. EN C. EDUARDO MARAMBIO DENNETT
DR. HECTOR JAIME SALGADO ZAMORA
I. Q. I. JUAN MANUEL LOMELIN GALLARDO
DR. GUILLERMO MARROQUIN SUAREZ
DR. IGNACIO RODRIGUEZ ROBLES
DR. JAVIER GARFIAS AYALA
DRA. SARA ELVIRA MEZA GALINDO
DRA. MAGDALENA RIUS DE LAS POLA

INSTITUTO MEXICANO DE INGENIEROS QUIMICOS
RAUL MUNOZ LEOS–PRESIDENT
RICARDO TRUJILLO CABRERA
JUAN SANCHEZ NAVARRO SUAREZ
JOSE LUIS ZARAGOZA CUTIERREZ
CARLOS MIJARES LOPEZ
ENRIQUE BAZUA RUEDA

PUBLICATIONS AVAILABLE
A. Valiente, "Problemas de Balance de Materia y Energia en la Industria Alimentaria", Limusa, Mexico, 1986.

Diana Cruz, Jose Antonio Chamizo y A. Garritz, "Estructura de la Materia. Un Enfoque Quimico", Adison Wesley, Iberoamericana, 1985.

ACS National Officers

S. Allen Heininger, *president*
Ernest L. Eliel, *president-elect*
Paul G. Gassman, *immediate past pres.*
Joseph A. Dixon, *chairman, board of directors*
John Kistler Crum, *executive director*
Justin W. Collat, *deputy executive & secretary*
Brian A. Bernstein, *treasurer*

Divisional Officers

Division of Agricultural & Food Chemistry
C. Brine, Chairman; C.J. Mussinan,
Secretary-Treasurer
R&D, International Flavors & Fragrances
1515 Highway 36
Union Beach, NJ 07735.

Division of Agrochemicals
G. Paulson, Chairman; N.N. Ragsdale,
Secretary
Agri. Exp. Station, Symons Hall
University of Maryland
College Park, MD 20742

Division of Analytical Chemistry
J.G. Grasselli, Chairman; K.L. Busch, Secretary
School of Chemistry & Biochemistry
Georgia Institute of Technology
Atlanta, GA 30332

Division of Biochemical Technology
R.J. Huss, Chairman; A. Bose, Secretary-
Treasurer
Pfizer Central Research
470 Eastern Point Road
Groton, CT 06340

Division of Biological Chemistry
P.A. Frey, Chairman; R. Matthews, Secretary
Univ. of Michigan-Medical School
M5416 Medical Science I, Box 0606
Ann Arbor, MI 48109-0606

Division of Carbohydrate Chemistry
M.A. Clarke, Chairman; A.D. French Secretary
Southern Regional Research Center
PO Box 19687
New Orleans, LA 70179

Division of Cellulose, Paper & Textile
N.R. Bertoniere, Chairman; J.R. Obst
Secretary-Treasurer, Forest Products
Laboratory
USDA - Forest Service
1 Gifford Pinchot Drive
Madison, WI 53705-2398

Division of Chemical Education, Inc.
Dr. D. Kolb, Chairman; K.O. Berry Secretary
Univ. of Puget Sound, Dept of Chemistry
1500 N. Warner Street
Tacoma, WA 98416

Division of Chemical Health & Safety
H.H. Fawcett, Chairman; R.A. Hathaway,
Secretary,
Environmental Analysis
South 1810 Plaza Way East,
Cape Girandeau, MO 63702

Division of Chemical Information
B.B. Lide, Chairman; V. Veach, Secretary
P and TCS/3M
201 C2 12, 3M Center
St Paul, MN 55144

Division of Chemical Marketing and Economics
S.L. Sutliff, Chairman; J.H. Levy, Secretary
Sr. Market Research Analyst Allied-Signal Inc.
Engineered Plastics
P.O. Box 2332R
Morristown, NJ 07962-2332

Division of Chemistry & The Law, Inc.
J.F. Riley, Chairman; R.S. Berman, Secretary
Spensly, Horn, Jubas, and Lubitz
1880 Century Park East, Fifth Floor
Los Angeles, CA 90067

Division of Colloid & Surface Chemistry
E.L. Kugler, Chairman; A. Morfesis, Secretary
PPG Industries
Fiberglass Research Center
PO Box 2844,
Pittsburgh, PA 15320

Division of Computers in Chemistry
T. Pierce, III, Chairman; C.A. Shelley, Secretary
2754 Compass Drive, Ste. 375
Grand Junction, CO 81506